Understanding Biology

Fourth Edition

Kenneth A. Mason
University of Iowa

Tod Duncan

Jonathan B. Losos
William H. Danforth Distinguished University Professor and Director,
Living Earth Collaborative, Washington University

Mc Graw Hill

T0175603

UNDERSTANDING BIOLOGY

Published by McGraw Hill LLC, 1325 Avenue of the Americas, New York, NY 10019. Copyright ©2024 by McGraw Hill LLC. All rights reserved. Printed in the United States of America. No part of this publication may be reproduced or distributed in any form or by any means, or stored in a database or retrieval system, without the prior written consent of McGraw Hill LLC, including, but not limited to, in any network or other electronic storage or transmission, or broadcast for distance learning.

Some ancillaries, including electronic and print components, may not be available to customers outside the United States.

This book is printed on acid-free paper.

1 2 3 4 5 6 7 8 9 LWI 28 27 26 25 24 23

ISBN 978-1-266-10043-7
MHID 1-266-10043-1

Cover Image: *Anna Veselova/Shutterstock*

All credits appearing on page or at the end of the book are considered to be an extension of the copyright page.

The Internet addresses listed in the text were accurate at the time of publication. The inclusion of a website does not indicate an endorsement by the authors or McGraw Hill LLC, and McGraw Hill LLC does not guarantee the accuracy of the information presented at these sites.

mheducation.com/highered

Brief Contents

About the Authors iv

Changes to This Edition vi

Acknowledgments viii

A Learning Path to *Understanding Biology* ix

Contents xv

Part I The Molecular Basis of Life 1

1 The Science of Biology 1
2 The Nature of Molecules and the Properties of Water 21
3 The Chemical Building Blocks of Life 40

Part II Biology of the Cell 65

4 Cell Structure 65
5 Membranes 94
6 Energy and Metabolism 115
7 How Cells Harvest Energy 131
8 Photosynthesis 156
9 Cell Communication 179
10 How Cells Divide 198

Part III Genetics and Molecular Biology 221

11 Sexual Reproduction and Meiosis 221
12 Patterns of Inheritance 236
13 The Chromosomal Basis of Inheritance 256
14 DNA: The Genetic Material 276
15 Genes and How They Work 300
16 Control of Gene Expression 328
17 Biotechnology 354
18 Genomics 379

Part IV Evolution 404

19 Genes Within Populations 404
20 The Evidence for Evolution 429
21 The Origin of Species 452

Part V The Diversity of Life 474

22 Systematics and Phylogeny 474
23 Prokaryotes and Viruses 494
24 Protists 520
25 Fungi 547
26 Plants 568
27 Animal Diversity 593
28 Vertebrates 625

Part VI Plant Form and Function 655

29 Plant Form 655
30 Flowering Plant Reproduction 680
31 The Living Plant 705

Part VII Animal Form and Function 735

32 The Animal Body and How It Moves 735
33 The Nervous System 760
34 Fueling the Body's Metabolism 789
35 Maintaining Homeostasis 822
36 Reproduction and Development 861

Part VIII Ecology and Behavior 893

37 Behavioral Biology 893
38 Ecology of Individuals and Populations 919
39 Community Ecology and Ecosystem Dynamics 944
40 The Living World 979

Appendix: Answer Key A-1

Index I-1

About the Authors

Kenneth Mason

Kenneth Mason has held academic positions, as a teacher and researcher, at three different major universities. He began on the faculty of the University of Kansas, where he designed and established the genetics lab and taught and published on the genetics of pigmentation in amphibians. At Purdue University, he successfully developed and grew large introductory biology courses and collaborated with other faculty in an innovative biology, chemistry, and physics course supported by the National Science Foundation. At the University of Iowa, where his wife served as president of the university, he taught introductory biology and human genetics. His honor society memberships include Phi Sigma, Alpha Lambda Delta, and, by vote of Purdue pharmacy students, Phi Eta Sigma Freshman Honors Society. After eight years at the University of Iowa, Kenneth and his wife both retired; they maintain close ties to the institution as President Emerita and Lecturer Emeritus.

Lesley Howard

Tod Duncan was formerly Associate Clinical Associate Professor at the University of Colorado Denver, where he taught first-semester general biology and coordinated first- and second-semester general biology laboratories. He has also taught general microbiology, virology, the biology of cancer, medical microbiology, and cell biology. A bachelor's degree in cell biology with an emphasis on plant molecular and cellular biology from the University of East Anglia in England led to doctoral studies in cell-cycle control and then postdoctoral research on the molecular and biochemical mechanisms of DNA alkylation damage in vitro and in *Drosophila melanogaster.* Currently, he is interested in factors affecting retention and success of incoming first-year students from diverse backgrounds. He lives in Boulder, Colorado, with his Great Dane, Eddie.

Jonathan Losos

Jonathan Losos is the William H. Danforth Distinguished University Professor in the Department of Biology at Washington University and Director of the Living Earth Collaborative, a partnership between the university, the Saint Louis Zoo, and the Missouri Botanical Garden. Losos's research has focused on studying patterns of adaptive radiation and evolutionary diversification in lizards. He is a member of the National Academy of Sciences, a fellow of the American Academy of Arts and Science, and the recipient of several awards, including the Theodosius Dobzhanksy and David Starr Jordan Prizes, the Edward Osborne Wilson Naturalist Award, and the Daniel Giraud Elliot Medal, as well as receiving fellowships from the John Guggenheim and David and Lucile Packard Foundations. Losos has published more than 250 scientific articles and has written two books, *Lizards in an Evolutionary Tree: Ecology and Adaptive Radiation of Anoles* (University of California Press, 2009) and *Improbable Destinies: Fate, Chance, and the Future of Evolution* (Penguin-Random House, 2017). He is currently in the process of writing his next book, on scientific research on the ecology and evolution of domestic cats.

A Conceptual Approach

In the decade leading up to 2019, there was a gradual increase in the importance of a variety of forms of digital instruction. The use of online education was growing, and we lived through fads like the massive open online course. While much commentary was devoted to a "revolution" in instruction, in fact, it was a relatively slow evolutionary process. This was completely upended by the COVID pandemic. Institutions that had never considered online instruction were forced to move partly or entirely online. This was followed by a roller coaster ride of on again, off again, online and in-person instruction, and hybrid models using both modes of instruction. McGraw Hill was ideally positioned to respond to this crisis with the best online tools available from any publisher. This has allowed us to respond to the changing needs of faculty with a variety of tools, and of course, high-quality textbooks. One unexpected result of all of this turmoil has been the increasing importance of the actual textbook. As students were required to do more work on their own, the quality and accessibility of the textbook proved to be critical. While the pandemic may have revealed the importance of a high-quality textbook, the need will continue afterward. With the emergence of "active learning" in the majors biology classroom, the responsibility of learning basic information has moved back to the student, giving instructors the time and opportunity to increase critical thinking and higher-level learning in their courses.

Understanding Biology fits well into this ever-changing environment in two main ways. First, the book is the result of a thorough analysis using a variety of sources to determine what is actually being taught in most majors biology courses across the country. This allowed us to produce a book that goes against the trend of encyclopedic textbooks yet still has all of the material, in the appropriate depth, required for a majors biology course. This helps to reduce cost, and makes the text less intimidating for students.

The second way that *Understanding Biology* is ideal for the new landscape is that it was designed from the beginning using our knowledge of how people learn. There is a strong emphasis on concepts over disconnected facts, and an organization that provides the student with a clear path to success in learning difficult material. Each chapter begins with a Learning Path that introduces the major concepts for the chapter. Then within each section, larger concepts are broken down into more specific supporting concepts. Each of these secondary concepts comes with a learning objective that tells students what they should be able to do after completing the section. Each section has a brief review with a question to help students think about the concepts. This organization, and more important, the content, allows us to focus on promoting student understanding. The end-of-chapter material includes a review aligned with the organization of the chapter, and assessment based on a modified version of Bloom's Taxonomy of Understand, Apply, Synthesize questions.

The art program supports this approach with clear and accessible figures, and stepped-out figures where appropriate. This includes Scientific Thinking figures that walk a student through how a problem can be solved experimentally. We have also moved our "Connecting the Concepts" feature to become a tool for active learning online, which seems to be a better environment to explore the relationships of chapter material to core concepts. With this edition we have also added a Concept Overview feature. This opens the chapter with a simple figure containing the main higher-level concepts of the chapter organized as the top levels of a flow diagram. The diagram is then filled out at the end of the chapter with supporting conceptual statements. The Concept Overviews are expansions of the Learning Path in a graphic presentation. They provide students with a conceptual overview of the chapter, and then larger concepts are broken down into more specific supporting concepts. This is particularly helpful in the second half of the book, where we have combined some topics that might not appear to be connected, but in fact are related.

We're excited about the fourth edition of this innovative text, which provides a clear learning path for a new generation of students who have to deal with unprecedented challenges. The authors have extensive experience teaching undergraduate biology, which has provided a guide in producing a text that is up to date and beautifully illustrated and that incorporates a conceptual approach to learning. We have worked hard to provide explicit learning outcomes clearly tied to each section of the book, which are used as a basis for developing the assessments. We are also excited about the continually evolving digital environment that provides unique and engaging learning tools for modern students. We continue to work to closely integrate the text with its media support materials to provide instructors with an excellent complement to their teaching.

Changes to This Edition

A new pedagogical feature called "Concept Overview" was added to all chapters in this fourth edition of *Understanding Biology*. The chapter opening page contains a diagram laying out the main concepts in the chapter. In the online eBook in Connect, Progressive Concept Overviews are inserted in relevant places in the chapter and provide a visual presentation of concept statements that support the main concepts. These Progressive Concept Overviews are available in the Instructor Resources for students who are using a printed text. All of the Progressive Concept Overviews are combined in an end-of-chapter diagram that presents a conceptual hierarchy of the chapter, highlighting the important concepts.

Also, McGraw Hill is dedicated to creating products that foster a culture of belonging and are accessible to all the diverse global customers we serve. Within this edition, content has been reviewed to implement inclusive content guidelines around topics including generalizations and stereotypes, gender, abilities/disabilities, race/ethnicity, sexual orientation, diversity of names, and age. Similarly, the illustrations throughout were evaluated and revised to ensure meaningful text and images are distinguishable and perceivable by users with limited color vision and moderately low vision.

Part I The Molecular Basis of Life

Chapter 1 A number of edits were made for clarity. The section on the process of science was edited for clarity and content, including the social context of science.

Chapter 2 Some minor edits were made for clarity.

Chapter 3 Edits were made to clarify the structure of nucleic acids, and the process of protein folding. One figure was converted into two figures to better fit the text and to increase student understanding.

Part II Biology of the Cell

Chapter 4 The section on prokaryotic cell structure was rewritten to incorporate new information on organization, substructure, and compartments. This provides a different view of the complexity of prokaryotic cells. A new figure was added comparing bacterial flagella and the archaeal archaellum.

Chapter 5 A variety of edits were made for clarity and student comprehension.

Chapter 6 The material on the Second Law of Thermodynamics was rewritten for clarity and accessibility. This includes one new figure to clarify the exergonic and endergonic reactions. The nature of enzymes and how they function was also rewritten for clarity.

Chapter 7 Light edits were made for clarity and student understanding.

Chapter 8 The section on the light-independent reactions was rewritten for clarity and accuracy. This included edits to one figure for accuracy and understanding. This also makes this chapter more congruent with chapter 7 in terms of names of compounds that have multiple names.

Chapter 9 The chapter was edited for clarity, including content on the nature of steroid hormones, and how their structure affects their function.

Chapter 10 The section on eukaryotic chromosome structure was further updated to reflect current thinking on the organization of chromatin in the nucleus. This update includes a new figure on chromosome organization.

Part III Genetics and Molecular Biology

Chapter 11 Minor edits for clarity, including in several figures, should improve student understanding. The material on segregation without crossing over was deleted as outside the scope of a majors course.

Chapter 12 The historical material in the beginning of the chapter was streamlined, although not removed. The material on extensions to Mendel was reorganized and rewritten for both clarity and updated content. The material on quantitative traits was updated to provide a more modern view. A new figure was added to replace an old one on epistasis with a more approachable example for students.

Chapter 13 The chapter was reorganized to both provide a more logical flow of topics, and integrate new material that provides a more modern view of human genetics.

Chapter 14 The presentation of experiments underlying DNA structure was edited for clarity. The descriptions of DNA structure were edited to emphasize structure–function relationships.

Chapter 15 The chapter was edited for clarity and some content was updated. The discussion of the genetic code was rewritten for clarity, and information on the genome, transcriptome, and proteome was edited for currency.

Chapter 16 The introduction on transcription factors was edited for clarity and content. The influence of chromatin structure on gene regulation was updated. Material on the ubiquitin/proteasome regulation of protein turnover was added, including one new figure.

Chapter 17 The section on genetic tools and modern medicine was completely rewritten. New material on molecular tests for infectious diseases, including COVID-19, was added. This also clarifies the use of different kinds of tests.

Chapter 18 The chapter was edited extensively to update content on a very fast-moving field. The human genome section was rewritten to update all content. Discussion was added of reference genomes, and current knowledge of the human genome. The wheat genome project was also rewritten to include new material on sequencing complex genomes. The content on what makes us human was also updated.

Part IV Evolution

Chapter 19 The section on genetic variation in populations was revised reflecting new information based on widespread genomic investigation. Genomic variation in humans is now discussed in detail, quantifying the extent of variation that exists and how that variation is apportioned within and between populations.

Part V The Diversity of Life

Chapter 22 Information on roots of the eukaryotic tree was revised to reflect changes in the classification of the protists.

Chapter 23 The chapter was edited to reflect new information on the organization of prokaryotic cells, including compartmentalization. The material on prokaryotic genetics was compressed, and the figures were edited. New material on SARS-CoV-2 was added, including a figure showing the virus life cycle.

Chapter 24 Many changes were made to the chapter to improve clarity and readability.

Chapter 25 The chapter was edited for clarity and readability.

Chapter 26 The chapter was edited for clarity and readability.

Chapter 27 The discussion of relationships at the base of the phylogeny for all animals was revised to reflect new understanding and debate about relationships among sponges, ctenophores, and other animals. Additional changes were made to reflect other changes in understanding of phylogenetic relationships among animal taxa, such as the position of chaetognaths and relationships among protostome taxa. Aspects of taxonomy and natural history were updated in line with new findings.

Chapter 28 The discussion of human evolution was revised in light of new discoveries.

Part VI Plant Form and Function

Chapter 29 Minor edits were made throughout for readability, clarity, and accuracy.

Chapter 30 The chapter was edited for clarity and readability.

Chapter 31 The chapter was edited throughout for clarity, accuracy, and readability.

Part VII Animal Form and Function

Chapter 32 The chapter was edited for clarity. The material on bone structure and development was rewritten to improve student understanding.

Chapter 33 The description of sensory systems, including receptor types, was updated to make it more consistent and to improve student understanding. The material on the retina was edited extensively for clarity, and one figure was added.

Chapter 34 The section on lung function was edited for clarity. Information was added on how lung function is assessed. The cardiac cycle was edited for clarity and student understanding.

Chapter 35 The material on hormones was edited to provide a better focus on structure–function relationships. The material on fluid balance was edited to emphasize homeostatic mechanisms. The section on adaptive immunity was rewritten to include more history of vaccination and a more worldwide perspective. Material on vaccination and how the immune system responds was added. This includes a discussion of vaccination and the COVID pandemic.

Chapter 36 The material on nuclear reprogramming was edited to provide a better historical perspective. The material on cloning in mammals was edited to include more detail. The material on contraception was edited for clarity.

Part VIII Ecology and Behavior

Chapter 37 Several Review of Concept questions were replaced to better assess student understanding.

Chapter 38 Human population trends and other timely data were updated to stay current. Also, a new section (38.7) was added on pandemics and human health that covers the general topic, but extensively details the population biology of the COVID-19 pandemic.

Chapter 40 Data on biosphere impacts of humans were updated to stay current.

Acknowledgments

A revision of this scope relies on the talents and efforts of many people working behind the scenes, and we have benefited greatly from their assistance.

The copy editor, Beth Bulger, labored many hours and always improved the clarity and consistency of the text. She has made a tremendous contribution to the quality of the final product. We were fortunate to work with MPS Limited to update the art program and improve the layout of the pages. Our close collaboration resulted in a text that is pedagogically effective as well as more beautiful than any other biology text on the market.

We have the continued support of an excellent team at McGraw Hill. Lora Neyens, the Portfolio Manager for Biology, has been a steady leader during a time of change. Senior Product Developer, Liz Sievers, provided support in so many ways, it would be impossible to name them all. Kelly Hart, Lead Content Project Manager, and David Hash, Designer, ensured our text was elegantly designed. Kelly Brown, Senior Marketing Manager, is always a sounding board for more than just marketing, and many more people behind the scenes have all contributed to the success of our text. This includes the digital team, to whom we owe a great deal for their efforts to continue improving our Connect assessment tools.

Throughout this edition we have had the support of spouses who have seen less of us than they might have liked because of the pressures of getting this revision completed. They have adapted to the many hours this book draws us away from them, and, even more than us, looked forward to its completion.

In the end, the people we owe the most are the generations of students who have passed through our lecture halls. They have taught us at least as much as we have taught them, and their questions and suggestions continue to improve the text and supplementary materials.

Finally, we need to thank instructors from across the country who are continually sharing their knowledge and experience with us through market feedback, reviews, and symposia. Their feedback shaped this edition. All of these people took time to share their ideas and opinions to help us build a better edition of *Understanding Biology* for the next generation of introductory biology students, and they have our heartfelt thanks.

Reviewers for *Understanding Biology,* 4th edition

Christopher Allen *Lone Star College, University Park*
Lauri Carey *Illinois Valley Community College*
Christopher Chamberlain *St. Cloud Technical & Community College*
Mary Colon *Seminole State College*
Carolyn Danna *Stevenson University*
Jacqueline Dartley *Bergen Community College*
Sandra Fox-Moon *Anne Arundel Community College*
Robin Graham *Dallas College*
Linda Johnson *University of Maryland Eastern Shore*
Olga Ruiz Kopp *Utah Valley University*
Kimberly Maznicki *Seminole State College*
Rosa Moscarella *University of Massachusetts*
Terina Nusinov *Seminole State College*
Helene Peters *Brewton-Parker College*
Debra Rinne *Seminole State College*
Sherry Stewart *Navarro College*

A Learning Path to *Understanding Biology*

Understanding Biology and its online assets have been carefully thought out and crafted to help students and professors work efficiently and effectively through the material in the course, making the most of study time and furthering instructional goals.

The Learning Path

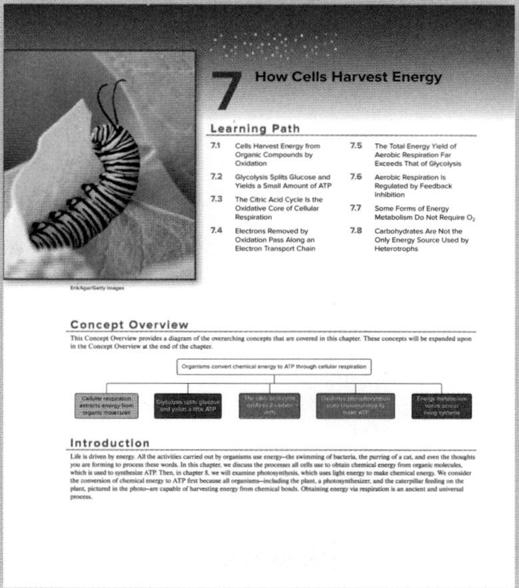
7 How Cells Harvest Energy

ErikAgar/Getty Images

Learning Path

7.1 Cells Harvest Energy from Organic Compounds by Oxidation
7.2 Glycolysis Splits Glucose and Yields a Small Amount of ATP
7.3 The Citric Acid Cycle Is the Oxidative Core of Cellular Respiration
7.4 Electrons Removed by Oxidation Pass Along an Electron Transport Chain
7.5 The Total Energy Yield of Aerobic Respiration Far Exceeds That of Glycolysis
7.6 Aerobic Respiration Is Regulated by Feedback Inhibition
7.7 Some Forms of Energy Metabolism Do Not Require O_2
7.8 Carbohydrates Are Not the Only Energy Source Used by Heterotrophs

Concept Overview

This Concept Overview provides a diagram of the overarching concepts that are covered in this chapter. These concepts will be expanded upon in the Concept Overview at the end of the chapter.

Organisms convert chemical energy to ATP through cellular respiration

Cellular respiration extracts energy from organic molecules

Glycolysis splits glucose and yields a little ATP

Introduction

Life is driven by energy. All the activities carried out by organisms use energy—the swimming of bacteria, the purring of a cat, and even the thoughts you are forming to process these words. In this chapter, we discuss the processes all cells use to obtain chemical energy from organic molecules, which is used to synthesize ATP. Then, in chapter 8, we will examine photosynthesis, which uses light energy to make chemical energy. We consider the conversion of chemical energy to ATP first because all organisms—including the plant, a photosynthesizer, and the caterpillar feeding on the plant, pictured in the photo—are capable of harvesting energy from chemical bonds. Obtaining energy via respiration is an ancient and universal process.

▲

Every chapter opens with a Learning Path that walks students through the main concepts in the chapter to enable their understanding of where the material fits in the context of other concepts in the chapter. A new Concept Overview provides a graphical presentation of the Learning Path, highlighting the main concepts in the chapter.

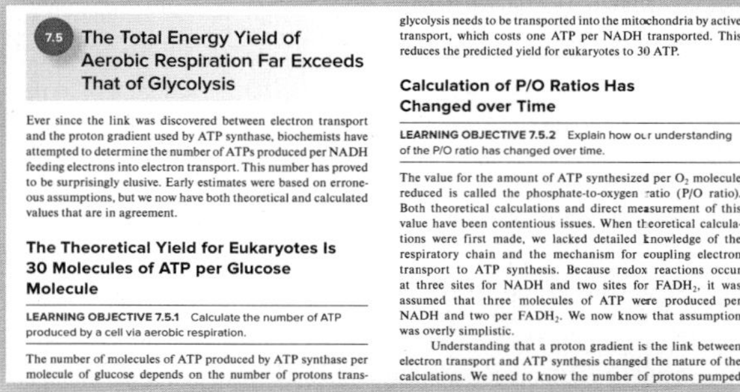
7.5 The Total Energy Yield of Aerobic Respiration Far Exceeds That of Glycolysis

Ever since the link was discovered between electron transport and the proton gradient used by ATP synthase, biochemists have attempted to determine the number of ATPs produced per NADH feeding electrons into electron transport. This number has proved to be surprisingly elusive. Early estimates were based on erroneous assumptions, but we now have both theoretical and calculated values that are in agreement.

The Theoretical Yield for Eukaryotes Is 30 Molecules of ATP per Glucose Molecule

LEARNING OBJECTIVE 7.5.1 Calculate the number of ATP produced by a cell via aerobic respiration.

The number of molecules of ATP produced by ATP synthase per molecule of glucose depends on the number of protons trans-

glycolysis needs to be transported into the mitochondria by active transport, which costs one ATP per NADH transported. This reduces the predicted yield for eukaryotes to 30 ATP.

Calculation of P/O Ratios Has Changed over Time

LEARNING OBJECTIVE 7.5.2 Explain how our understanding of the P/O ratio has changed over time.

The value for the amount of ATP synthesized per O_2 molecule reduced is called the phosphate-to-oxygen ratio (P/O ratio). Both theoretical calculations and direct measurement of this value have been contentious issues. When theoretical calculations were first made, we lacked detailed knowledge of the respiratory chain and the mechanism for coupling electron transport to ATP synthesis. Because redox reactions occur at three sites for NADH and two sites for $FADH_2$, it was assumed that three molecules of ATP were produced per NADH and two per $FADH_2$. We now know that assumption was overly simplistic.

Understanding that a proton gradient is the link between electron transport and ATP synthesis changed the nature of the calculations. We need to know the number of protons pumped

▲

Every concept is broken down into sections that cover skills or ideas students should master. Learning objectives at the beginning of each section help identify important concepts.

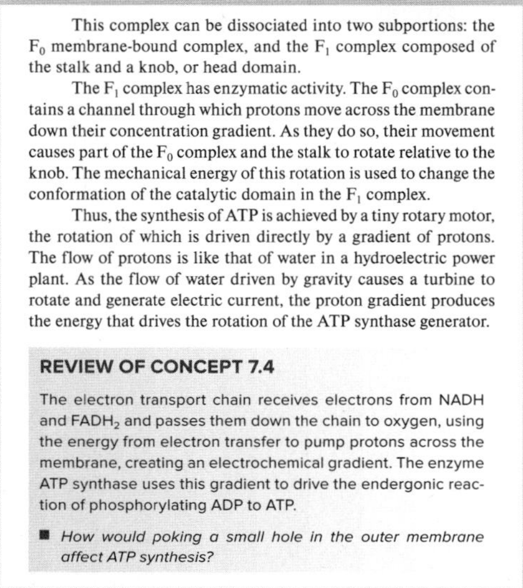
This complex can be dissociated into two subportions: the F_0 membrane-bound complex, and the F_1 complex composed of the stalk and a knob, or head domain.

The F_1 complex has enzymatic activity. The F_0 complex contains a channel through which protons move across the membrane down their concentration gradient. As they do so, their movement causes part of the F_0 complex and the stalk to rotate relative to the knob. The mechanical energy of this rotation is used to change the conformation of the catalytic domain in the F_1 complex.

Thus, the synthesis of ATP is achieved by a tiny rotary motor, the rotation of which is driven directly by a gradient of protons. The flow of protons is like that of water in a hydroelectric power plant. As the flow of water driven by gravity causes a turbine to rotate and generate electric current, the proton gradient produces the energy that drives the rotation of the ATP synthase generator.

REVIEW OF CONCEPT 7.4

The electron transport chain receives electrons from NADH and $FADH_2$ and passes them down the chain to oxygen, using the energy from electron transfer to pump protons across the membrane, creating an electrochemical gradient. The enzyme ATP synthase uses this gradient to drive the endergonic reaction of phosphorylating ADP to ATP.

- *How would poking a small hole in the outer membrane affect ATP synthesis?*

▲

At the end of each section, Review of Concept questions allow students to check their understanding before moving on to the next concept.

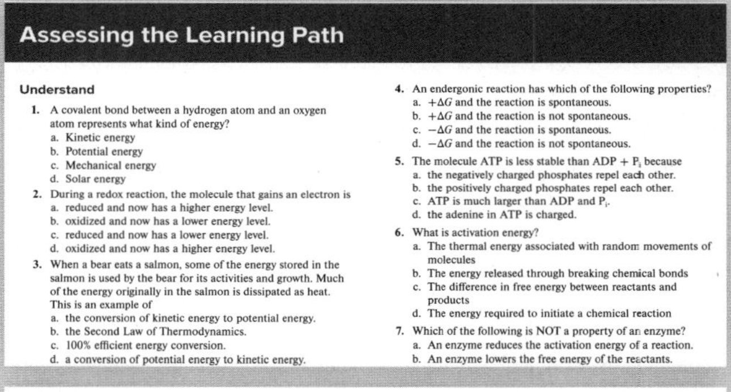
Assessing the Learning Path

Understand

1. A covalent bond between a hydrogen atom and an oxygen atom represents what kind of energy?
   a. Kinetic energy
   b. Potential energy
   c. Mechanical energy
   d. Solar energy
2. During a redox reaction, the molecule that gains an electron is
   a. reduced and now has a higher energy level.
   b. oxidized and now has a lower energy level.
   c. reduced and now has a lower energy level.
   d. oxidized and now has a higher energy level.
3. When a bear eats a salmon, some of the energy stored in the salmon is used by the bear for its activities and growth. Much of the energy originally in the salmon is dissipated as heat. This is an example of
   a. the conversion of kinetic energy to potential energy.
   b. the Second Law of Thermodynamics.
   c. 100% efficient energy conversion.
   d. a conversion of potential energy to kinetic energy.
4. An endergonic reaction has which of the following properties?
   a. $+\Delta G$ and the reaction is spontaneous.
   b. $+\Delta G$ and the reaction is not spontaneous.
   c. $-\Delta G$ and the reaction is spontaneous.
   d. $-\Delta G$ and the reaction is not spontaneous.
5. The molecule ATP is less stable than ADP + P_i because
   a. the negatively charged phosphates repel each other.
   b. the positively charged phosphates repel each other.
   c. ATP is much larger than ADP and P_i.
   d. the adenine in ATP is charged.
6. What is activation energy?
   a. The thermal energy associated with random movements of molecules
   b. The energy released through breaking chemical bonds
   c. The difference in free energy between reactants and products
   d. The energy required to initiate a chemical reaction
7. Which of the following is NOT a property of an enzyme?
   a. An enzyme reduces the activation energy of a reaction.
   b. An enzyme lowers the free energy of the reactants.

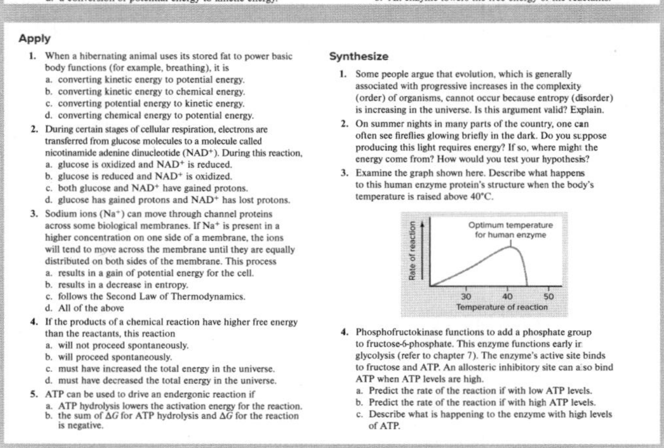
Apply

1. When a hibernating animal uses its stored fat to power basic body functions (for example, breathing), it is
   a. converting kinetic energy to potential energy.
   b. converting kinetic energy to chemical energy.
   c. converting potential energy to kinetic energy.
   d. converting chemical energy to potential energy.
2. During certain stages of cellular respiration, electrons are transferred from glucose molecules to a molecule called nicotinamide adenine dinucleotide (NAD^+). During this reaction,
   a. glucose is oxidized and NAD^+ is reduced.
   b. glucose is reduced and NAD^+ is oxidized.
   c. both glucose and NAD^+ have gained protons.
   d. glucose has gained protons and NAD^+ has lost protons.
3. Sodium ions (Na^+) can move through channel proteins across some biological membranes. If Na^+ is present in a higher concentration on one side of a membrane, the ions will tend to move across the membrane until they are equally distributed on both sides of the membrane. This process
   a. results in a gain of potential energy for the cell.
   b. results in a decrease in entropy.
   c. follows the Second Law of Thermodynamics.
   d. All of the above
4. If the products of a chemical reaction have higher free energy than the reactants, this reaction
   a. will not proceed spontaneously.
   b. will proceed spontaneously.
   c. must have increased the total energy in the universe.
   d. must have decreased the total energy in the universe.
5. ATP can be used to drive an endergonic reaction if
   a. ATP hydrolysis lowers the activation energy for the reaction.
   b. the sum of ΔG for ATP hydrolysis and ΔG for the reaction is negative.

Synthesize

1. Some people argue that evolution, which is generally associated with progressive increases in the complexity (order) of organisms, cannot occur because entropy (disorder) is increasing in the universe. Is this argument valid? Explain.
2. On summer nights in many parts of the country, one can often see fireflies glowing briefly in the dark. Do you suppose producing this light requires energy? If so, where might the energy come from? How would you test your hypothesis?
3. Examine the graph shown here. Describe what happens to this human enzyme protein's structure when the body's temperature is raised above 40°C.

Optimum temperature for human enzyme
Rate of reaction
30 40 50
Temperature of reaction

4. Phosphofructokinase functions to add a phosphate group to fructose-6-phosphate. This enzyme functions early in glycolysis (refer to chapter 7). The enzyme's active site binds to fructose and ATP. An allosteric inhibitory site can also bind ATP when ATP levels are high.
   a. Predict the rate of the reaction if with low ATP levels.
   b. Predict the rate of the reaction if with high ATP levels.
   c. Describe what is happening to the enzyme with high levels of ATP.

▲

Chapter concepts are assessed at three different levels at the end of the chapter. On a first pass through the chapter prior to class, students might focus on the "Understand" level. As they gain greater mastery of the material, they should challenge themselves with "Apply" and "Synthesize" questions that require higher cognitive skills.

Think Like a Scientist

Scientific Thinking figures throughout the text walk the student through a scientific experiment, laying out the Hypothesis, Predictions, Test Procedures, Results, and Conclusion. Some also challenge the students to devise further experiments.

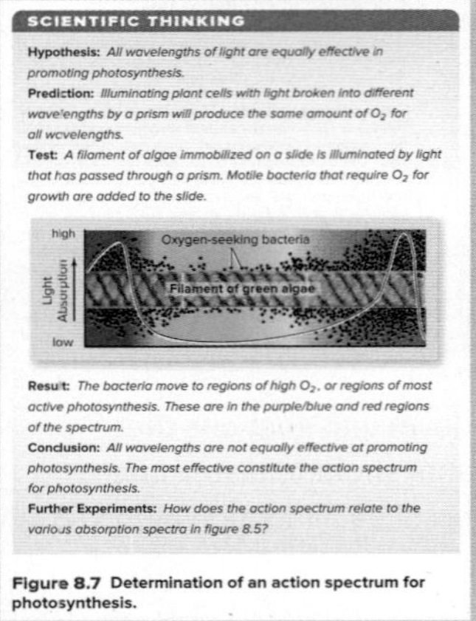

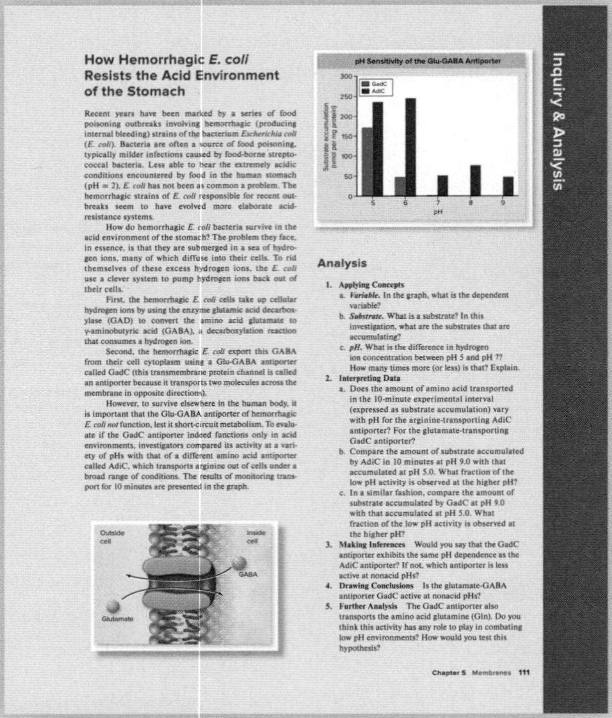

Inquiry & Analysis features at the ends of all chapters explore a scientific investigation in more detail, presenting experimental results and challenging students to interpret the data.

Concept Overviews

Each Concept Overview, a new feature in *Understanding Biology,* 4th edition, is a graphic representation of the Learning Path that guides students through the concepts in the chapter. Concept Overviews provide a conceptual framework of the chapter. When concept statements are placed in the context of a flow diagram, students see relationships and connections between concepts.

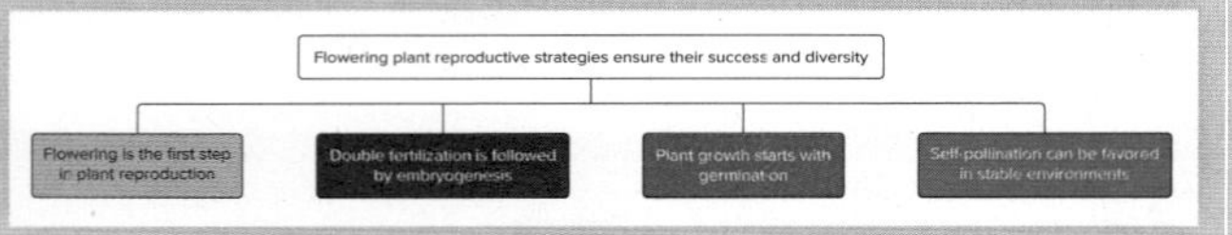

The Concept Overview that appears at the beginning of the chapter contains three to five broad concept statements that reveal the overarching structure of the chapter.

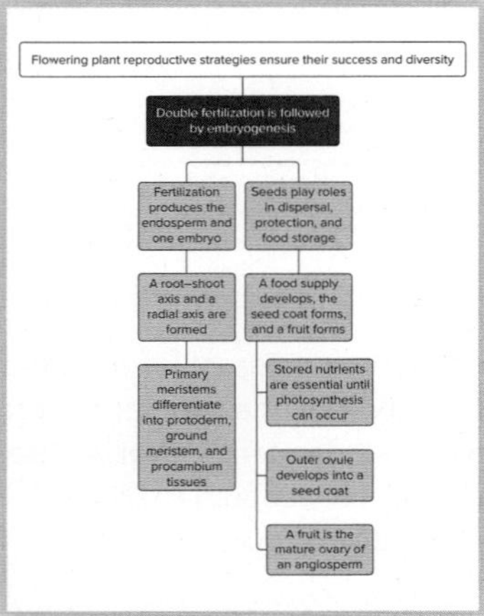

Progressive Concept Overviews are inserted throughout the chapter in the eBook, at the end of relevant sections. These diagrams identify concepts that support the overarching concept statements. The Progressive Concept Overviews are available in the Instructor Resources for students using a printed text.

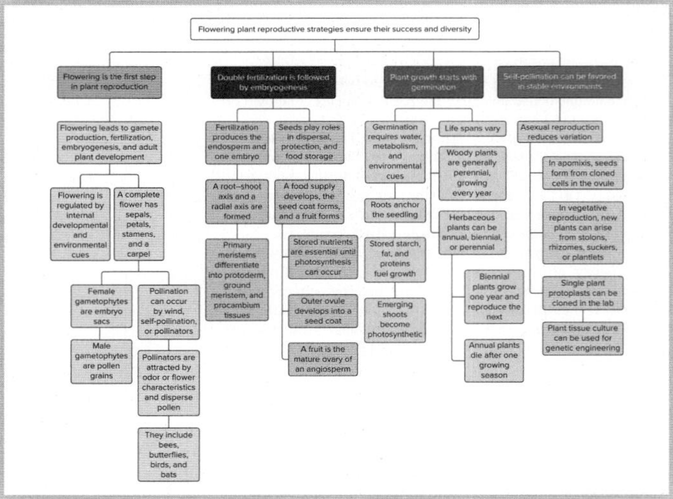

Each chapter's end-of-chapter Concept Overview brings together all of the concept statements in the Progressive Overview diagrams to reveal a conceptual overview of the chapter. These Concept Overviews differ from "concept maps" in that there is a hierarchical aspect important in developing a conceptual framework for the chapter.

Detailed Feedback in Connect®

Learning is a process of iterative development, of making mistakes, reflecting, and adjusting over time. The question bank and test bank in Connect® for *Understanding Biology,* 4th edition, are more than standard assessments; they are self-contained learning experiences that systematically build student learning over time.

For many students, choosing the right answer is not necessarily based on applying content correctly; it is more a matter of increasing their statistical odds of guessing. A major fault with this approach is students don't learn how to process the information correctly, mostly because they are repeating and reinforcing their mistakes rather than reflecting and learning from them. To help students develop problem-solving skills, all higher-level Bloom's questions in Connect are supported with hints, to help students focus on important information for answering the questions, as well as detailed feedback that walks students through the problem-solving process, using Socratic questions in a decision-tree-style framework to scaffold learning, where each step models and reinforces the learning process.

The feedback for each higher-level Bloom's question (Apply, Analyze, Evaluate) follows a similar process: Clarify Question, Gather Content, Choose Answer, Reflect on Process.

Unpacking the Concepts

We've taken problem solving a step further. In each chapter, two higher-level Bloom's questions in the question and test banks are broken out by the steps of the detailed feedback. Rather than leaving it up to the student to work through the Socratic questions, a second version of the question is presented in a stepwise format. Following the problem-solving steps, students need to answer questions about earlier steps, such as "What is the key concept addressed by the question?" before proceeding to answer the question. A professor can choose which version of the question to include in the assignment based on the problem-solving skills of the students. The Unpacking the Concepts questions are found under the Coursewide Content in Connect.

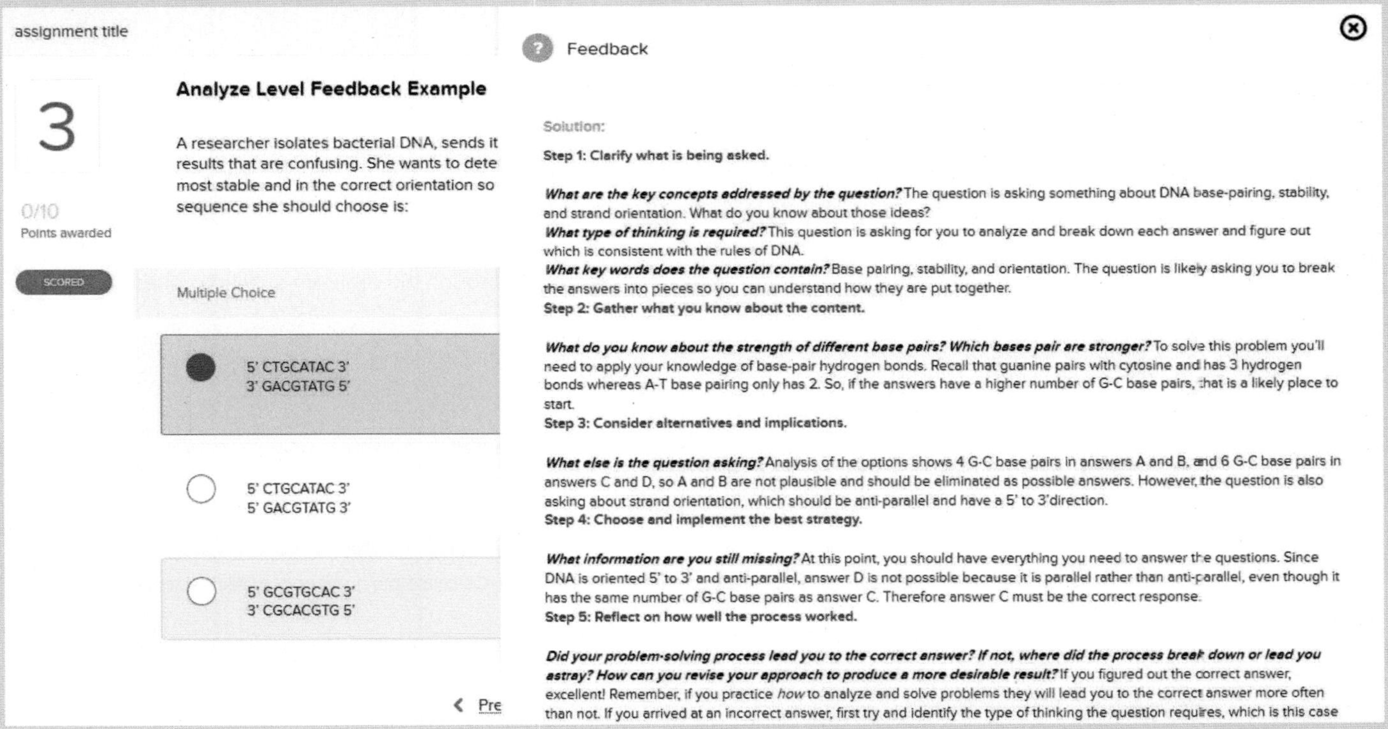

Data and Graphing Interactives

To help students develop analytical skills, Connect® for *Understanding Biology,* 4th edition, is enhanced with Data and Graphing Interactives found under the Coursewide Content. Students are presented with a scientific problem and the opportunity to manipulate variables in the interactive, analyze or evaluate data, or view different aspects of the problem. A series of questions follows the activity to assess whether the student understands and is able to interpret the data and results.

Quantitative Reasoning Question Bank

Under the Coursewide Content in Connect, the Quantitative Reasoning question bank provides more challenging algorithmic questions, intended to help students practice their quantitative reasoning skills. Hints and stepped-out solutions walk students through the problem-solving process.

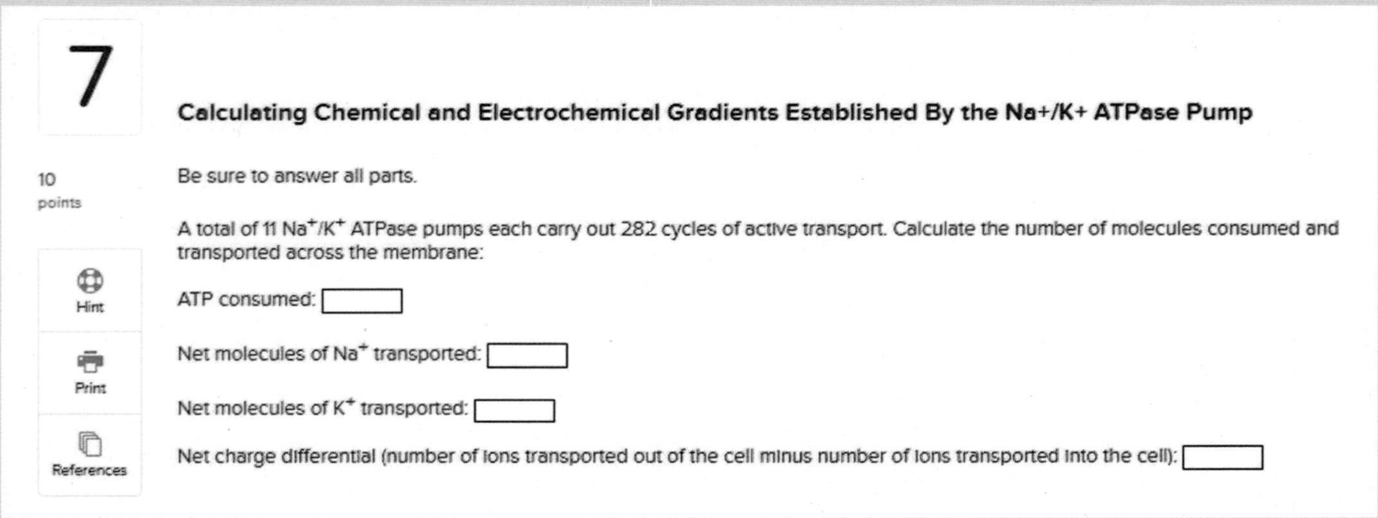

7

Calculating Chemical and Electrochemical Gradients Established By the Na+/K+ ATPase Pump

10 points

Hint

Print

References

Be sure to answer all parts.

A total of 11 Na^+/K^+ ATPase pumps each carry out 282 cycles of active transport. Calculate the number of molecules consumed and transported across the membrane:

ATP consumed: ____

Net molecules of Na^+ transported: ____

Net molecules of K^+ transported: ____

Net charge differential (number of ions transported out of the cell minus number of ions transported into the cell): ____

Virtual Labs and Lab Simulations

While the biological sciences are hands-on disciplines, instructors are now often being asked to deliver some of their lab components online, as full online replacements, supplements to prepare for in-person labs, or make-up labs.

These simulations help each student learn the practical and conceptual skills needed, then check for understanding and provide feedback. With an adaptive pre-lab assignment, found under Adaptive Learning Assignment, and post-lab assessment available under Coursewide Content, instructors can customize each assignment.

From the instructor's perspective, these simulations may be used in the lecture environment to help students visualize complex scientific processes, such as DNA technology or Gram staining, while at the same time providing a valuable connection between the lecture and lab environments.

McGraw Hill

Additional Assets in Connect®

Coursewide Content

There are book-specific question and test banks in Connect® for *Understanding Biology,* 4th edition, but there are also additional assets under the **Coursewide Content** section. In addition to the Unpacking the Concepts, Quantitative Reasoning Questions, and Data and Graphing Interactives mentioned earlier, this dropdown menu contains: Relevancy Modules, Virtual Labs Questions, BioNow Video Activities, and Biology NewsFlash Exercises.

SmartBook 2.0

Connect's SmartBook 2.0 provides an adaptive learning experience that combines eBook reading for comprehension and assessments that test understanding. Learning resources are also available at key points to further aid understanding. The reading experience and assessments adapt to individual student learning. This is an environment that develops self-awareness through meaningful, immediate feedback that improves student success.

Prep *for Majors Biology*

Connect's Prep is another adaptive learning experience. It is intended for use at the start of the majors biology course to get students up to speed on prerequisite material such as basic math skills, graphing, and statistics as well as introductory biology topics in chemistry and cell biology. An additional module, Fundamentals of Student Success, help students prepare for their college academic experience. Assessments determine a student's prerequisite knowledge and learning resources help to fill in gaps in knowledge.

Remote Proctoring & Browser-Locking Capabilities

Remote proctoring and browser-locking capabilities, hosted by Proctorio within Connect, provide control of the assessment environment by enabling security options and verifying the identity of the student.

Seamlessly integrated within Connect, these services allow instructors to control the assessment experience by verifying identification, restricting browser activity, and monitoring student actions.

Instant and detailed reporting gives instructors an at-a-glance view of potential academic integrity concerns, thereby avoiding personal bias and supporting evidence-based claims.

Writing Assignments

Available within McGraw Hill Connect®, the Writing Assignment tool delivers a learning experience to help students improve their written communication skills and conceptual understanding. As an instructor you can assign, monitor, grade, and provide feedback on writing more efficiently and effectively.

ReadAnywhere

Read or study when it's convenient for you with McGraw Hill's free ReadAnywhere app. Available for iOS or Android smartphones or tablets, ReadAnywhere gives users access to McGraw Hill tools, including the eBook and SmartBook 2.0 or Adaptive Learning Assignments in Connect. Take notes, highlight, and complete assignments offline–all of your work will sync when you open the app with WiFi access. Log in with your McGraw Hill Connect user-name and password to start learning–anytime, anywhere!

OLC-Aligned Courses

Implementing High-Quality Online Instruction and Assessment through Preconfigured Courseware

In consultation with the Online Learning Consortium (OLC) and our certified Faculty Consultants, McGraw Hill has created pre-configured courseware using OLC's quality scorecard to align with best practices in online course delivery. This turnkey courseware contains a combination of formative assessments, summative assessments, homework, and application activities, and can easily be customized to meet an individual's needs and course outcomes. For more information, visit https://www.mheducation.com/highered/olc.

Tegrity: Lectures 24/7

Mc Graw Hill | tegrity®

Tegrity in Connect is a tool that makes class time available 24/7 by automatically capturing every lecture. With a simple one-click start-and-stop process, you capture all computer screens and corresponding audio in a format that is easy to search, frame by frame. Students can replay any part of any class with easy-to-use, browser-based viewing on a PC, Mac, tablet, or other mobile device.

Educators know that the more students can see, hear, and experience class resources, the better they learn. In fact, studies prove it. Tegrity's unique search feature helps students efficiently find what they need, when they need it, across an entire semester of class recordings. Help turn your students' study time into learning moments immediately supported by your lecture. With Tegrity, you also increase intent listening and class participation by easing students' concerns about note-taking. Using Tegrity in Connect will make it more likely you will see students' faces, not the tops of their heads.

Test Builder in Connect

Available within Connect, Test Builder is a cloud-based tool that enables instructors to format tests that can be printed, administered within a Learning Management System, or exported as a Word document of the test bank. Test Builder offers a modern, streamlined interface for easy content configuration that matches course needs, without requiring a download.

Test Builder allows you to:

- access all test bank content from a particular title.
- easily pinpoint the most relevant content through robust filtering options.
- manipulate the order of questions or scramble questions and/or answers.
- pin questions to a specific location within a test.
- determine your preferred treatment of algorithmic questions.
- choose the layout and spacing.
- add instructions and configure default settings.

Test Builder provides a secure interface for better protection of content and allows for just-in-time updates to flow directly into assessments.

Create

Your Book, Your Way

McGraw Hill's Content Collections Powered by Create® is a self-service website that enables instructors to create custom course materials—print and eBooks—by drawing upon McGraw Hill's comprehensive, cross-disciplinary content. Choose what you want from our high-quality textbooks, articles, and cases. Combine it with your own content quickly and easily, and tap into other rights-secured, third-party content such as readings, cases, and articles. Content can be arranged in a way that makes the most sense for your course and you can include the course name and information as well. Choose the best format for your course: color print, black-and-white print, or eBook. The eBook can be included in your Connect course and is available on the free ReadAnywhere app for smartphone or tablet access as well. When you are finished customizing, you will receive a free digital copy to review in just minutes! Visit McGraw Hill Create®—www.mcgrawhillcreate.com—today and begin building!

Contents

About the Authors iv

Changes to This Edition vi

Acknowledgments viii

A Learning Path to *Understanding Biology* ix

Soames Summerhays/Natural Visions

PART I The Molecular Basis of Life 1

1 The Science of Biology 1

Concept 1.1 The Diversity of Life Is Overwhelming 2

Concept 1.2 Biology Is the Science of Life 2

Concept 1.3 Science Is Based on Both Observation and Reasoning 6

Concept 1.4 The Study of Evolution Is a Good Example of Scientific Inquiry 9

Concept 1.5 A Few Important Concepts Form the Core of Biology 14

Inquiry & Analysis 17

Retracing the Learning Path 18

Assessing the Learning Path 19

2 The Nature of Molecules and the Properties of Water 21

Concept 2.1 All Matter Is Composed of Atoms 22

Concept 2.2 The Elements in Living Systems Have Low Atomic Masses 25

Concept 2.3 Molecules Are Collections of Atoms Held Together by Chemical Bonds 27

Concept 2.4 The Properties of Water Result from Its Polar Nature 31

Concept 2.5 Water Molecules Can Dissociate into Ions 34

Inquiry & Analysis 36

Retracing the Learning Path 37

Assessing the Learning Path 38

3 The Chemical Building Blocks of Life 40

Concept 3.1 Carbon Provides the Framework of Biological Molecules 41

Concept 3.2 Carbohydrates Form Both Structural and Energy-Storing Molecules 43

Concept 3.3 Proteins Are the Tools of the Cell 47

Concept 3.4 Nucleic Acids Store and Express Genetic Information 55

Concept 3.5 Hydrophobic Lipids Form Fats and Membranes 58

Inquiry & Analysis 61

Retracing the Learning Path 62

Assessing the Learning Path 63

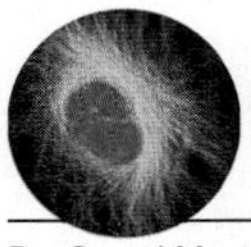

Dr. Gopal Murti/Science Source

PART II Biology of the Cell 65

4 Cell Structure 65

Concept 4.1 All Living Organisms Are Composed of Cells 66

Concept 4.2 Prokaryotic Cells Are Relatively Simple 69

Concept 4.3 Eukaryotic Cells Are Highly Compartmentalized 71

Concept 4.4 Membranes Organize the Cell Interior into Functional Compartments 75

Concept 4.5 Mitochondria and Chloroplasts Are Energy-Processing Organelles 79

Concept 4.6 An Internal Skeleton Supports the Shape of Cells 81

Concept 4.7 Extracellular Structures Protect Cells 84

Concept 4.8 Cell-to-Cell Connections Determine How Adjacent Cells Interact 86

Inquiry & Analysis 90

Retracing the Learning Path 91

Assessing the Learning Path 92

5 Membranes 94

Concept 5.1 Membranes Are Phospholipid Bilayers with Embedded Proteins 95

Concept 5.2 Phospholipids Provide a Membrane's Structural Foundation 98

Concept 5.3 Membrane Proteins Enable a Broad Range of Interactions with the Environment 99

Concept 5.4 Passive Transport Moves Molecules Across Membranes by Diffusion 101

Concept 5.5 Active Transport Across Membranes Requires Energy 105

Concept 5.6 Bulky Materials Cross Membranes Within Vesicles 108

Inquiry & Analysis 111

Retracing the Learning Path 112

Assessing the Learning Path 113

6 Energy and Metabolism 115

Concept 6.1 Energy Flows Through Living Systems 116

Concept 6.2 The Laws of Thermodynamics Govern All Energy Changes 117

Concept 6.3 ATP Is the Energy Currency of Cells 119

Concept 6.4 Enzymes Speed Up Reactions by Lowering Activation Energy 121

Concept 6.5 Metabolism Is the Sum of a Cell's Chemical Activities 125

Inquiry & Analysis 127

Retracing the Learning Path 128

Assessing the Learning Path 129

7 How Cells Harvest Energy 131

Concept 7.1 Cells Harvest Energy from Organic Compounds by Oxidation 132
Concept 7.2 Glycolysis Splits Glucose and Yields a Small Amount of ATP 136
Concept 7.3 The Citric Acid Cycle Is the Oxidative Core of Cellular Respiration 139
Concept 7.4 Electrons Removed by Oxidation Pass Along an Electron Transport Chain 142
Concept 7.5 The Total Energy Yield of Aerobic Respiration Far Exceeds That of Glycolysis 146
Concept 7.6 Aerobic Respiration Is Regulated by Feedback Inhibition 147
Concept 7.7 Some Forms of Energy Metabolism Do Not Require O_2 148
Concept 7.8 Carbohydrates Are Not the Only Energy Source Used by Heterotrophs 149
Inquiry & Analysis 152
Retracing the Learning Path 153
Assessing the Learning Path 154

8 Photosynthesis 156

Concept 8.1 Photosynthesis Uses Sunlight to Power the Synthesis of Organic Molecules 157
Concept 8.2 Experiments Revealed That Photosynthesis Is a Chemical Process 158
Concept 8.3 Pigments Capture Energy from Sunlight 160
Concept 8.4 Photosynthetic Pigments Are Organized into Photosystems 163
Concept 8.5 Energy from Sunlight Is Used to Produce a Proton Gradient 165
Concept 8.6 Carbon Fixation Incorporates CO_2 into Organic Molecules 170
Concept 8.7 Photorespiration Short-Circuits Photosynthesis 172
Inquiry & Analysis 175
Retracing the Learning Path 176
Assessing the Learning Path 177

9 Cell Communication 179

Concept 9.1 The Cells of Multicellular Organisms Communicate 180
Concept 9.2 Signal Transduction Begins with Cellular Receptors 182
Concept 9.3 Intracellular Receptors Respond to Signals by Regulating Gene Expression 184
Concept 9.4 Protein Kinase Receptors Respond to Signals by Phosphorylating Proteins 185
Concept 9.5 G Protein–Coupled Receptors Respond to Signals Through Effector Proteins 189
Inquiry & Analysis 194
Retracing the Learning Path 195
Assessing the Learning Path 196

10 How Cells Divide 198

Concept 10.1 Bacterial Cell Division Is Clonal 199
Concept 10.2 Eukaryotes Have Large, Linear Chromosomes 200
Concept 10.3 The Eukaryotic Cell Cycle Is Complex and Highly Organized 203
Concept 10.4 During Interphase, Cells Grow and Prepare for Mitosis 204
Concept 10.5 In Mitosis, Chromosomes Segregate 205
Concept 10.6 Events of the Cell Cycle Are Carefully Regulated 210
Concept 10.7 Cancer Is a Failure of Cell-Cycle Control 214
Inquiry & Analysis 217
Retracing the Learning Path 218
Assessing the Learning Path 219

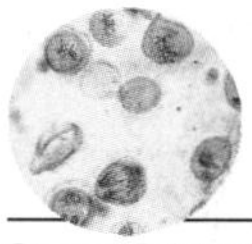

PART III Genetics and Molecular Biology 221

Steven P. Lynch

11 Sexual Reproduction and Meiosis 221

Concept 11.1 Sexual Reproduction Requires Meiosis 222
Concept 11.2 Meiosis Consists of Two Divisions with One Round of DNA Replication 223
Concept 11.3 The Process of Meiosis Involves Intimate Interactions Between Homologs 224
Concept 11.4 Meiosis Has Four Distinct Features 229
Concept 11.5 Genetic Variation Is the Evolutionary Consequence of Sex 231
Inquiry & Analysis 232
Retracing the Learning Path 233
Assessing the Learning Path 234

12 Patterns of Inheritance 236

Concept 12.1 Experiments Carried Out by Mendel Explain Inheritance 237
Concept 12.2 Mendel's Principle of Segregation Accounts for 3:1 Phenotypic Ratios 238
Concept 12.3 Mendel's Principle of Independent Assortment Asserts That Genes Segregate Independently 241
Concept 12.4 Probability Allows Us to Predict the Results of Crosses 243
Concept 12.5 Extending Mendel's Model Provides a Clearer View of Genetics in Action 245
Concept 12.6 Genotype Dictates Phenotype by Specifying Protein Sequences 250
Inquiry & Analysis 252
Retracing the Learning Path 253
Assessing the Learning Path 254

13 The Chromosomal Basis of Inheritance 256

Concept 13.1 Sex Linkage and the Chromosomal Theory of Inheritance 257
Concept 13.2 There Are Two Major Exceptions to Chromosomal Inheritance 259
Concept 13.3 Some Genes Do Not Assort Independently: Linkage 261

Concept 13.4 Genetic Crosses Provide Data for Genetic Maps 263
Concept 13.5 Changes in Chromosome Number Can Have Drastic Effects 265
Concept 13.6 Inheritance in Humans Can Be Studied by Analyzing Pedigrees 267
Inquiry & Analysis 272
Retracing the Learning Path 273
Assessing the Learning Path 274

14 DNA: The Genetic Material 276

Concept 14.1 DNA Is the Genetic Material 277
Concept 14.2 The DNA Molecule Is a Double Helix 279
Concept 14.3 Both Strands Are Copied During DNA Replication 283
Concept 14.4 Prokaryotes Organize the Enzymes Used to Duplicate DNA 286
Concept 14.5 Eukaryotic Chromosomes Are Large and Linear 291
Concept 14.6 Cells Repair Damaged DNA 293
Inquiry & Analysis 296
Retracing the Learning Path 297
Assessing the Learning Path 298

15 Genes and How They Work 300

Concept 15.1 Experiments Have Revealed the Nature of Genes 301
Concept 15.2 The Genetic Code Relates Information in DNA and Protein 303
Concept 15.3 Prokaryotes Exhibit All the Basic Features of Transcription 306
Concept 15.4 Eukaryotes Use Three Polymerases and Extensively Modify Transcripts 309
Concept 15.5 Eukaryotic Genes May Contain Noncoding Sequences 311
Concept 15.6 The Ribosome Is the Machine of Protein Synthesis 313
Concept 15.7 The Process of Translation Is Complex and Energy-Expensive 315
Concept 15.8 Mutations Are Heritable Changes in Genetic Material 321
Inquiry & Analysis 324
Retracing the Learning Path 325
Assessing the Learning Path 326

16 Control of Gene Expression 328

Concept 16.1 All Organisms Control Expression of Their Genes 329
Concept 16.2 Regulatory Proteins Control Genes by Interacting with Specific DNA Nucleotide Sequences 330
Concept 16.3 Prokaryotes Regulate Their Genes in Clusters 331
Concept 16.4 Transcription Factors Control Gene Transcription in Eukaryotes 336
Concept 16.5 Chromatin Structure Affects Gene Expression 339
Concept 16.6 Eukaryotic Genes Are Also Regulated After Transcription 341
Concept 16.7 Gene Regulation Determines How Cells Will Develop 346
Inquiry & Analysis 350
Retracing the Learning Path 351
Assessing the Learning Path 352

17 Biotechnology 354

Concept 17.1 Enzymes Allow the Creation of Recombinant Molecules In Vitro 355
Concept 17.2 The Polymerase Chain Reaction Is Used to Amplify Specific DNA Sequences 359
Concept 17.3 Molecular Tools Allow Us to Analyze and Modify Genetic Variation 362
Concept 17.4 Transgenic Organisms Are Used to Analyze Gene Function 364
Concept 17.5 Genetic Tools Are Changing Modern Medicine 366
Concept 17.6 Genetic Engineering Is Used in Industry and Agriculture 370
Inquiry & Analysis 375
Retracing the Learning Path 376
Assessing the Learning Path 377

18 Genomics 379

Concept 18.1 Mapping Identifies and Locates Functional Elements in Genomes 380
Concept 18.2 The Modernization of DNA Sequencing Has Accelerated Discovery 383
Concept 18.3 Genome Projects Reveal Insights into Medicine and Agriculture 386
Concept 18.4 Genome Annotation Assigns Functional Information to Genomes 388
Concept 18.5 Genome Comparisons Provide Information About Genomic Structure and Function 391
Concept 18.6 Comparative Genomics Informs Evolutionary Biology 396
Inquiry & Analysis 400
Retracing the Learning Path 401
Assessing the Learning Path 402

PART IV Evolution 404

Tetra Images/Getty Images

19 Genes Within Populations 404

Concept 19.1 Natural Populations Exhibit Genetic Variation 405
Concept 19.2 Frequencies of Alleles Can Change 407
Concept 19.3 Five Agents Are Responsible for Evolutionary Change 409
Concept 19.4 Selection Can Act on Traits Affected by Many Genes 414
Concept 19.5 Natural Selection Can Be Studied Experimentally 415
Concept 19.6 Fitness Is a Measure of Evolutionary Success 418
Concept 19.7 Evolutionary Processes Sometimes Maintain Variation 418
Concept 19.8 Sexual Selection Determines Reproductive Success 421

Inquiry & Analysis *425*
Retracing the Learning Path *426*
Assessing the Learning Path *427*

20 The Evidence for Evolution 429

Concept 20.1 The Beaks of Darwin's Finches Provide Evidence of Natural Selection 430
Concept 20.2 Peppered Moths and Industrial Melanism Illustrate Natural Selection in Action 433
Concept 20.3 Human-Initiated Artificial Selection Is Also a Powerful Agent of Change 434
Concept 20.4 Fossils Provide Direct Evidence of Evolution 436
Concept 20.5 Anatomical Evidence for Evolution Is Extensive and Persuasive 440
Concept 20.6 Genes Carry a Molecular Record of the Evolutionary Past 443
Concept 20.7 Natural Selection Favors Convergent Evolution in Similar Environments 444
Concept 20.8 Addressing Common Criticisms of Evolutionary Theory 446
Inquiry & Analysis *448*
Retracing the Learning Path *449*
Assessing the Learning Path *450*

21 The Origin of Species 452

Concept 21.1 The Biological Species Concept Highlights Reproductive Isolation 453
Concept 21.2 Natural Selection May Reinforce Reproductive Isolation 457
Concept 21.3 Natural Selection and Genetic Drift Play Key Roles in Speciation 459
Concept 21.4 Speciation Is Influenced by Geography 460
Concept 21.5 Adaptive Radiation Requires Both Speciation and Habitat Diversity 462
Concept 21.6 The Pace of Evolution Varies 467
Concept 21.7 Speciation and Extinction Have Molded Biodiversity Through Time 468
Inquiry & Analysis *470*
Retracing the Learning Path *471*
Assessing the Learning Path *472*

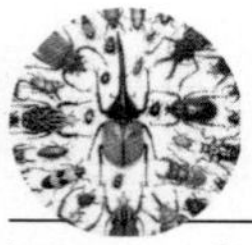

PART V The Diversity of Life 474

Imagemore Co, Ltd./Imagemore/Getty Images

22 Systematics and Phylogeny 474

Concept 22.1 Systematics Reconstructs Evolutionary Relationships 475
Concept 22.2 Cladistics Focuses on Traits Derived from a Common Ancestor 476
Concept 22.3 Classification Is a Labeling Process, Not an Evolutionary Reconstruction 480
Concept 22.4 Taxonomy Attempts to Classify Organisms in an Evolutionary Context 482
Concept 22.5 The Largest Taxa Are Domains 485
Inquiry & Analysis *490*
Retracing the Learning Path *491*
Assessing the Learning Path *492*

23 Prokaryotes and Viruses 494

Concept 23.1 Prokaryotes Are the Most Ancient Organisms 495
Concept 23.2 Prokaryotes Have an Organized but Simple Structure 497
Concept 23.3 The Genetics of Prokaryotes Focuses on DNA Transfer 501
Concept 23.4 Prokaryotic Metabolism Is Diverse 504
Concept 23.5 Bacteria Cause Important Human Diseases 505
Concept 23.6 Viruses Are Not Organisms 506
Concept 23.7 Bacterial Viruses Infect by DNA Injection 510
Concept 23.8 Animal Viruses Infect by Endocytosis 511
Inquiry & Analysis *516*
Retracing the Learning Path *517*
Assessing the Learning Path *518*

24 Protists 520

Concept 24.1 Protists, the First Eukaryotes, Arose by Endosymbiosis 521
Concept 24.2 Protists Are a Very Diverse Group 523
Concept 24.3 The Rough Outlines of Protist Phylogeny Are Becoming Clearer 525
Concept 24.4 Excavata Are Flagellated Protists Lacking Mitochondria 526
Concept 24.5 SAR: Stramenopiles and Alveolates Exhibit Secondary Endosymbiosis 529
Concept 24.6 SAR: Rhizaria Have Silicon Exoskeletons or Limestone Shells 535
Concept 24.7 Archaeplastida Are Descended from a Single Endosymbiosis Event 536
Concept 24.8 Amoebozoa and Opisthokonta Are Closely Related 539
Inquiry & Analysis *542*
Retracing the Learning Path *543*
Assessing the Learning Path *544*

25 Fungi 547

Concept 25.1 Fungi Have Unique Reproductive and Nutritional Strategies 548
Concept 25.2 Fungi Have an Enormous Ecological Impact 550
Concept 25.3 Fungi Are Important Plant and Animal Pathogens 553
Concept 25.4 Fungi Are Taxonomically Diverse 554
Concept 25.5 Microsporidia Are Unicellular Parasites 556
Concept 25.6 Chytridiomycota and Relatives: Fungi with Zoospores 556
Concept 25.7 Zygomycota Produce Zygotes 558
Concept 25.8 Glomeromycota Are Asexual Plant Symbionts 559

Concept 25.9 Basidiomycota Are the Mushroom Fungi 560
Concept 25.10 Ascomycota Are the Most Diverse Phylum of Fungi 560
Inquiry & Analysis 564
Retracing the Learning Path 565
Assessing the Learning Path 566

26 Plants 568

Concept 26.1 Land Plants Are Multicellular Autotrophs Adapted to Terrestrial Life 569
Concept 26.2 Bryophytes Have a Dominant Gametophyte Generation 571
Concept 26.3 Seedless Vascular Plants Evolved Roots, Stems, and Leaves 573
Concept 26.4 Lycophytes Have a Dominant Sporophyte Generation 575
Concept 26.5 Pterophytes Are Ferns and Their Relatives 576
Concept 26.6 Seed Plants Were a Key Step in Plant Evolution 579
Concept 26.7 Gymnosperms Are Plants with "Naked Seeds" 580
Concept 26.8 Angiosperms Are Flowering Plants 584
Inquiry & Analysis 588
Retracing the Learning Path 589
Assessing the Learning Path 590

27 Animal Diversity 593

Concept 27.1 The Diversity of Animal Body Plans Arose by a Series of Evolutionary Innovations 594
Concept 27.2 Molecular Data Are Clarifying the Animal Phylogenetic Tree 601
Concept 27.3 True Tissue Evolved in Simple Animals 603
Concept 27.4 Flatworms and Rotifers Are Very Simple Bilaterians 605
Concept 27.5 Mollusks and Annelids Are the Largest Groups of Lophotrochozoans 607
Concept 27.6 Lophophorates Are Very Simple Marine Organisms 611
Concept 27.7 Nematodes and Arthropods Are Both Large Groups of Ecdysozoans 612
Concept 27.8 Deuterostomes Are Composed of Echinoderms and Chordates 617
Inquiry & Analysis 620
Retracing the Learning Path 621
Assessing the Learning Path 623

28 Vertebrates 625

Concept 28.1 Nonvertebrate Chordates Do Not Form Bone 626
Concept 28.2 Almost All Chordates Are Vertebrates 627
Concept 28.3 Fishes Are the Earliest and Most Diverse Vertebrates 627
Concept 28.4 Amphibians Are Moist-Skinned Descendants of the Early Tetrapods 632
Concept 28.5 Reptiles Are Fully Adapted to Terrestrial Living 633
Concept 28.6 Birds Are Essentially Flying Reptiles 637
Concept 28.7 Mammals Are the Least Diverse of Vertebrates 640
Concept 28.8 Primates Include Lemurs, Monkeys, Apes, and Humans 643
Inquiry & Analysis 650
Retracing the Learning Path 651
Assessing the Learning Path 652

PART VI Plant Form and Function 655

Susan Singer

29 Plant Form 655

Concept 29.1 Meristems Articulate the Body Plan 656
Concept 29.2 Plants Contain Three Main Tissues 659
Concept 29.3 Roots Have Four Growth Zones 664
Concept 29.4 Stems Provide Support for Aboveground Organs 669
Concept 29.5 Leaves Are a Plant's Photosynthetic Organs 673
Inquiry & Analysis 676
Retracing the Learning Path 677
Assessing the Learning Path 678

30 Flowering Plant Reproduction 680

Concept 30.1 Reproduction Starts with Flowering 681
Concept 30.2 Flowers Attract Pollinators 683
Concept 30.3 Fertilization Leads to Embryogenesis 688
Concept 30.4 Seeds Protect Angiosperm Embryos 690
Concept 30.5 Fruits Promote Seed Dispersal 692
Concept 30.6 Germination Begins Seedling Growth 693
Concept 30.7 Plant Life Spans Vary Widely 696
Concept 30.8 Asexual Reproduction Is Common Among Flowering Plants 697
Inquiry & Analysis 700
Retracing the Learning Path 701
Assessing the Learning Path 702

31 The Living Plant 705

Concept 31.1 Water Moves Through Plants Based on Potential Differences 706
Concept 31.2 Roots Absorb Minerals and Water 709
Concept 31.3 Xylem Transports Water from Root to Shoot 711
Concept 31.4 Transpiration Rate Reflects Environmental Conditions 713
Concept 31.5 Plants Are Adapted to Water Stress 715
Concept 31.6 Phloem Transports Organic Molecules 716
Concept 31.7 Plants Require a Variety of Nutrients 717
Concept 31.8 Plants Use Hormones to Regulate Growth 719
Concept 31.9 Plant Growth Is Responsive to Light 724
Concept 31.10 Plant Growth Is Sensitive to Gravity 727
Inquiry & Analysis 730
Retracing the Learning Path 731
Assessing the Learning Path 733

Stockbyte/Getty Images

PART VII Animal Form and Function 735

32 The Animal Body and How It Moves 735

Concept 32.1 The Vertebrate Body Has a Hierarchical Organization 736
Concept 32.2 Epithelial Tissue Covers Body Surfaces 738
Concept 32.3 Nerve Tissue Conducts Signals Rapidly 740
Concept 32.4 Connective Tissue Supports the Body 740
Concept 32.5 Muscle Tissue Powers the Body's Movements 742
Concept 32.6 Skeletal Systems Anchor the Body's Muscles 744
Concept 32.7 Vertebrate Endoskeletons Are Made of Bone 745
Concept 32.8 Muscles Contract Because Their Myofilaments Slide 749
Concept 32.9 Animal Locomotion Takes Many Forms 753
Inquiry & Analysis 756
Retracing the Learning Path 757
Assessing the Learning Path 758

33 The Nervous System 760

Concept 33.1 The Nervous System Directs the Body's Actions 761
Concept 33.2 Neurons Maintain a Resting Potential Across the Plasma Membrane 762
Concept 33.3 Action Potentials Propagate Nerve Impulses 764
Concept 33.4 Synapses Are Where Neurons Communicate with Other Cells 766
Concept 33.5 The Central Nervous System Includes the Brain and Spinal Cord 769
Concept 33.6 The Peripheral Nervous System Consists of Both Sensory and Motor Neurons 771
Concept 33.7 Sensory Receptors Provide Information About the Body's Environment 773
Concept 33.8 Mechanoreceptors Sense Touch and Pressure 774
Concept 33.9 Sounds and Body Position Are Sensed by Vibration Detectors 775
Concept 33.10 Taste, Smell, and pH Senses Utilize Chemoreceptors 778
Concept 33.11 Vision Employs Photoreceptors to Perceive Objects at a Distance 779
Inquiry & Analysis 785
Retracing the Learning Path 786
Assessing the Learning Path 787

34 Fueling the Body's Metabolism 789

Concept 34.1 Vertebrate Digestive Systems Are Tubular Tracts 790
Concept 34.2 Food Is Processed as It Passes Through the Digestive Tract 791
Concept 34.3 The Digestive Tract Is Regulated by the Nervous System and Hormones 796
Concept 34.4 Respiratory Systems Promote Efficient Exchange of Gases 797
Concept 34.5 Gills Provide for Efficient Gas Exchange in Water 798
Concept 34.6 Lungs Are the Respiratory Organs of Terrestrial Vertebrates 800
Concept 34.7 Oxygen and Carbon Dioxide Are Transported by Fundamentally Different Mechanisms 803
Concept 34.8 Circulating Blood Carries Metabolites and Gases to the Tissues 806
Concept 34.9 Vertebrate Circulatory Systems Put a Premium on Efficient Circulation 808
Concept 34.10 The Four Chambers of the Heart Contract in a Cycle 811
Concept 34.11 The Circulatory Highway Is Composed of Arteries, Capillaries, and Veins 814
Inquiry & Analysis 817
Retracing the Learning Path 818
Assessing the Learning Path 820

35 Maintaining Homeostasis 822

Concept 35.1 Homeostasis Maintains a Constant Internal Environment 823
Concept 35.2 Hormones Are Chemical Messages That Direct Body Processes 827
Concept 35.3 The Pituitary and the Hypothalamus Are the Body's Control Centers 831
Concept 35.4 Peripheral Endocrine Glands Play Major Roles in Homeostasis 834
Concept 35.5 Animals Are Osmoconformers or Osmoregulators 838
Concept 35.6 The Kidney Maintains Osmotic Homeostasis in Mammals 839
Concept 35.7 Hormones Control Osmoregulation 844
Concept 35.8 The Immune System Defends the Body 845
Concept 35.9 Cell-Mediated Immunity Involves Helper and Killer T Cells 850
Concept 35.10 In Humoral Immunity, B Cells Produce Protective Antibodies 852
Inquiry & Analysis 856
Retracing the Learning Path 857
Assessing the Learning Path 859

36 Reproduction and Development 861

Concept 36.1 Mammals Are Viviparous 862
Concept 36.2 The Human Male Reproductive System Is Typical of Mammals 863
Concept 36.3 The Human Female Reproductive System Undergoes Cyclic Gamete Development 866
Concept 36.4 The First Step in Development Is Fertilization 871
Concept 36.5 Cells of the Early Embryo Are Totipotent 874
Concept 36.6 Cleavage Leads to the Blastula Stage 878
Concept 36.7 Gastrulation Forms the Basic Body Plan of the Embryo 880
Concept 36.8 The Body's Organs Form in Organogenesis 882
Concept 36.9 Human Development Takes Nine Months 885
Inquiry & Analysis 889
Retracing the Learning Path 890
Assessing the Learning Path 891

Mike Powles/Getty Images

PART VIII Ecology and Behavior 893

37 Behavioral Biology 893

Concept 37.1 An Animal's Genome Influences Its Behavior 894
Concept 37.2 Learning Also Influences Behavior 896
Concept 37.3 Thinking Directs the Behavior of Many Animals 897
Concept 37.4 Migratory Behavior Is Both Innate and Learned 898
Concept 37.5 Animal Communication Plays a Key Role in Ecological and Social Behavior 900
Concept 37.6 Behavior Evolves Adaptively 903
Concept 37.7 Behavioral Ecology Examines the Adaptive Significance of Behaviors 904
Concept 37.8 Behavioral Strategies Have Evolved to Maximize Reproductive Success 907
Concept 37.9 Some Behaviors Decrease Fitness to Benefit Other Individuals 909
Concept 37.10 Group Living Has Evolved in Both Insects and Vertebrates 912
Inquiry & Analysis 914
Retracing the Learning Path 915
Assessing the Learning Path 916

38 Ecology of Individuals and Populations 919

Concept 38.1 Populations Are Groups of a Single Species in One Place 920
Concept 38.2 Population Growth Depends upon Members' Age and Sex 923
Concept 38.3 Evolution Favors Life Histories That Maximize Lifetime Reproductive Success 926
Concept 38.4 Environment Limits Population Growth 928
Concept 38.5 Resource Availability Regulates Population Growth 930
Concept 38.6 Earth's Human Population Is Growing Explosively 933
Concept 38.7 Pandemics and Human Health 936
Inquiry & Analysis 939
Retracing the Learning Path 940
Assessing the Learning Path 942

39 Community Ecology and Ecosystem Dynamics 944

Concept 39.1 Competition Shapes How Species Live Together in Communities 945
Concept 39.2 Predator-Prey Relationships Foster Coevolution 949
Concept 39.3 Cooperation Among Species Can Lead to Coevolution 953
Concept 39.4 Ecological Succession Is a Consequence of Habitat Alteration 958
Concept 39.5 Chemical Elements Move Through Ecosystems in Biogeochemical Cycles 959
Concept 39.6 Energy Flows Through Ecosystems in One Direction 965
Concept 39.7 Biodiversity May Increase Ecosystem Stability 969
Inquiry & Analysis 974
Retracing the Learning Path 975
Assessing the Learning Path 977

40 The Living World 979

Concept 40.1 Ecosystems Are Shaped by Sun, Wind, and Water 980
Concept 40.2 Earth Has 14 Major Terrestrial Ecosystems, Called Biomes 984
Concept 40.3 Freshwater Habitats Occupy Less Than 2% of Earth's Surface 986
Concept 40.4 Marine Habitats Dominate the Earth 987
Concept 40.5 Humanity's Pollution Is Severely Impacting the Biosphere 989
Concept 40.6 Human Activity Is Altering Earth's Climate 991
Inquiry & Analysis 996
Retracing the Learning Path 997
Assessing the Learning Path 999

Appendix: Answer Key A-1

Index I-1

McGraw Hill connect®

Instructors

Student Success Starts with You

Tools to enhance your unique voice

Want to build your own course? No problem. Prefer to use an OLC-aligned, prebuilt course? Easy. Want to make changes throughout the semester? Sure. And you'll save time with Connect's auto-grading, too.

65%
Less Time Grading

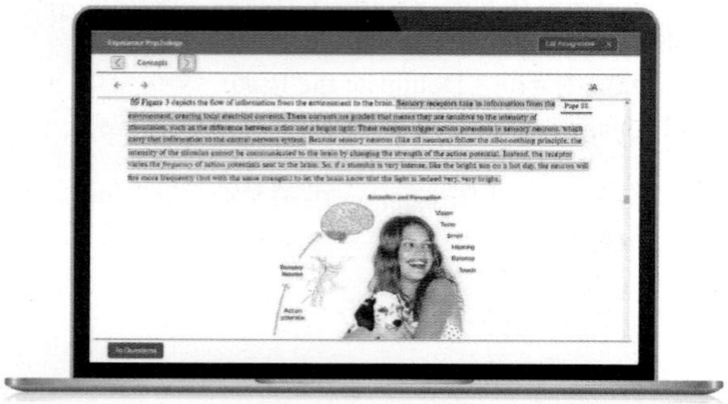

Laptop: Getty Images; Woman/dog: George Doyle/Getty Images

A unique path for each student

In Connect, instructors can assign an adaptive reading experience with SmartBook® 2.0. Rooted in advanced learning science principles, SmartBook 2.0 delivers each student a personalized experience, focusing students on their learning gaps, ensuring that the time they spend studying is time well-spent.
mheducation.com/highered/connect/smartbook

Affordable solutions, added value

Make technology work for you with LMS integration for single sign-on access, mobile access to the digital textbook, and reports to quickly show you how each of your students is doing. And with our Inclusive Access program, you can provide all these tools at the lowest available market price to your students. Ask your McGraw Hill representative for more information.

Solutions for your challenges

A product isn't a solution. Real solutions are affordable, reliable, and come with training and ongoing support when you need it and how you want it. Visit **supportateverystep.com** for videos and resources both you and your students can use throughout the term.

SUPPORT AT every step

Students
Get Learning that Fits You

Effective tools for efficient studying

Connect is designed to help you be more productive with simple, flexible, intuitive tools that maximize your study time and meet your individual learning needs. Get learning that works for you with Connect.

Study anytime, anywhere

Download the free ReadAnywhere® app and access your online eBook, SmartBook® 2.0, or Adaptive Learning Assignments when it's convenient, even if you're offline. And since the app automatically syncs with your Connect account, all of your work is available every time you open it. Find out more at **mheducation.com/readanywhere**

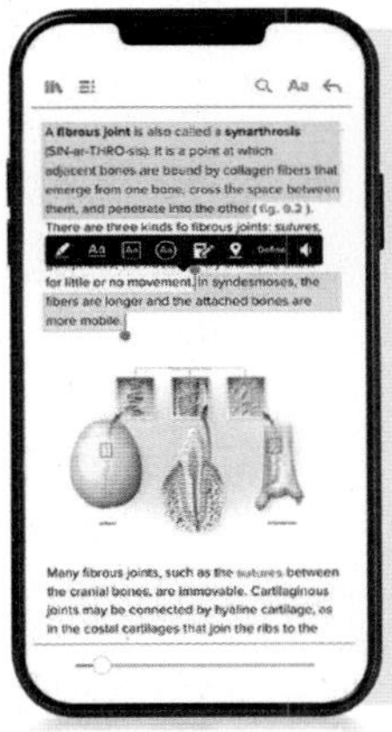

"I really liked this app—it made it easy to study when you don't have your text-book in front of you."

- Jordan Cunningham,
Eastern Washington University

iPhone: Getty Images

Everything you need in one place

Your Connect course has everything you need—whether reading your digital eBook or completing assignments for class, Connect makes it easy to get your work done.

Learning for everyone

McGraw Hill works directly with Accessibility Services Departments and faculty to meet the learning needs of all students. Please contact your Accessibility Services Office and ask them to email accessibility@mheducation.com, or visit **mheducation.com/about/accessibility** for more information.

Soames Summerhays/Natural Visions

1 The Science of Biology

Learning Path

1.1 The Diversity of Life Is Overwhelming

1.2 Biology Is the Science of Life

1.3 Science Is Based on Both Observation and Reasoning

1.4 The Study of Evolution Is a Good Example of Scientific Inquiry

1.5 A Few Important Concepts Form the Core of Biology

Concept Overview

This Concept Overview provides a diagram of the overarching concepts that are covered in this chapter. These concepts will be expanded upon in the Concept Overview at the end of the chapter.

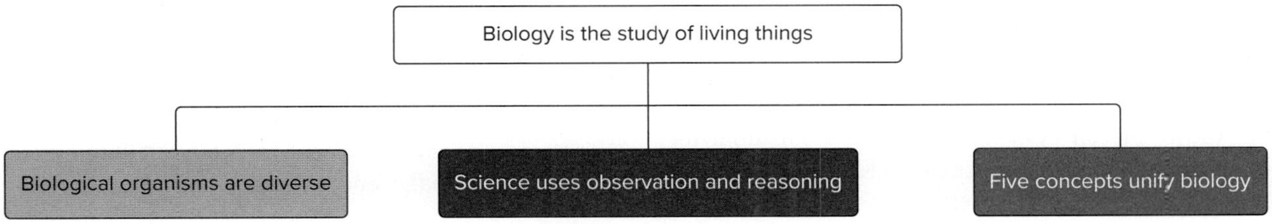

Biology is the study of living things
Biological organisms are diverse
Science uses observation and reasoning
Five concepts unify biology

Introduction

You are about to embark on a journey of discovery about the nature of life. Almost two centuries ago, a young English naturalist named Charles Darwin set sail on a similar journey on board H.M.S. *Beagle*; the photo on this page shows a replica of this ship. Darwin's observations during the voyage of the *Beagle* influenced his development of the theory of evolution by natural selection, which has become the core of the science of biology. Before we begin, however, let's take a moment to think about what biology is and why it's important.

1.1 The Diversity of Life Is Overwhelming

Biology is the study of living things—literally the science of life. The living world is teeming with a breathtaking variety of creatures—whales, butterflies, mushrooms, plants, bacteria—which despite their obvious differences share features common to all living organisms. We will begin with a brief history of how we classify this diversity.

Biological Diversity Can Be Organized by Evolutionary History

LEARNING OBJECTIVE 1.1.1 Describe the three domains of life.

The amazing diversity of life inspired a long history of classification systems to bring order to this diversity. This culminated in the mid-1700s with the Linnaean classification system, which used observable similarities to group organisms into hierarchical groups. This system used seven levels (kingdom, phylum, class, order, family, genus, species) to classify all organisms. One change from previous systems was the introduction of two-part, or binomial, names. For instance, humans are *Homo sapiens*, which is the genus and species description from the longer, complete classification. Classification based on shared evolutionary history often, but not always, produces results similar to that based on shared characteristics. Although the Linnaean system does not always reflect evolutionary relationships, it remains useful today because it is the only universal system, and its long history means that an extraordinary amount of biodiversity has been classified this way. These issues are considered in detail in chapter 22.

The highest level of organization in the Linnaean system is the kingdom. With the advent of microscopes, we also defined two basic cell types: those that have genetic material in a membrane-bounded nucleus (eukaryotes), and those that lack this membrane-bounded nucleus (prokaryotes). This was accommodated within the Linnaean system by creating a kingdom for prokaryotes, specifically Bacteria (once called Kingdom Monera). This was later upended by the discovery that there were actually two kinds of prokaryotes, called Bacteria (or Eubacteria) and Archaea (figure 1.1; discussed in detail in chapter 23).

This change led to the addition of a new taxonomic rank above kingdom called a domain. All living organisms can be divided into three domains: Eubacteria, Archaea, and Eukarya. Even this system has been challenged as there is evidence that eukaryotes arose from within Archaea, leading some to advocate for a two-domain system. This remains an area of active research, with many new prokaryotes being identified based on the ability to analyze genetic material without actually visualizing or culturing organisms.

The kingdoms you are probably most familiar with contain plants, animals, and fungi (figure 1.2), and these represent single lines of evolutionary descent. The group that has been problematic is the protists, which were grouped based on the shared similarity of being eukaryotic and unicellular. This was formerly a kingdom, but clearly does not represent a single line of evolutionary descent. This has led to attempts to find clear groups with shared evolutionary history. We will discuss this in detail in chapters 22 and 24.

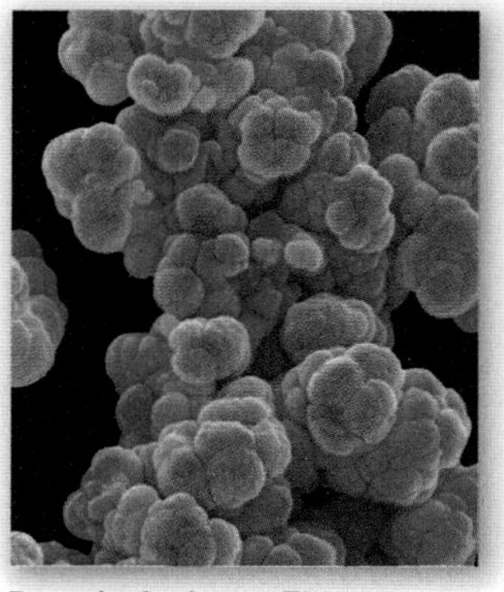

Domain Archaea. This prokaryotic domain includes this methanogen, which manufactures methane as a result of its metabolic activity.

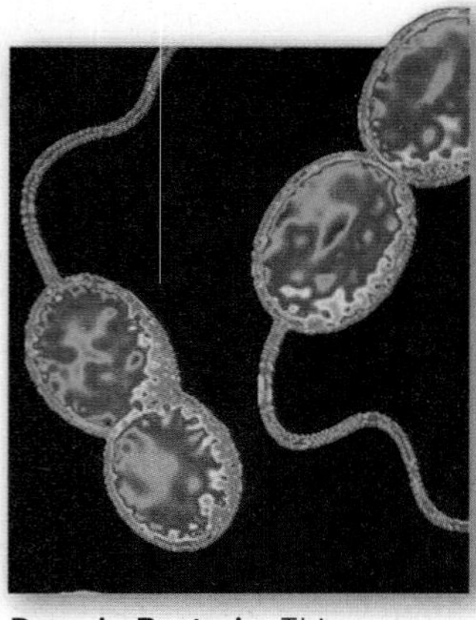

Domain Bacteria. This prokaryotic domain includes this purple sulfur bacteria, which can use light energy to drive the synthesis of organic compounds (false color).

Figure 1.1 The two prokaryotic domains. Bacteria and archaea share the feature of lacking a membrane-bounded nucleus. Organisms from both of these domains are single-celled.

(Archaea): Power and Syred/Science Source; (Bacteria): Alfred Pasieka/Science Source

The work of biologists affects your everyday life: what you eat, what happens to you when you go to the hospital, and how our society will handle environmental issues such as climate change. Unifying the diverse systems studied by biologists are the shared characteristics of all living things that have been shaped by the process of evolution by natural selection. Keeping this theme in mind will help you to manage the complexity and diversity of biology.

REVIEW OF CONCEPT 1.1

The living world is incredibly diverse. Various systems of classification have been proposed, which remains an area of active research. The oldest branching of the tree of life is into Bacteria, Archaea, and Eukarya.

- *What are some shared features of living systems?*

1.2 Biology Is the Science of Life

In its broadest sense, biology is the study of living things. So it would seem that biologists would have no problem defining life. In fact, it is quite difficult to provide a simple definition of life.

Life Defies Simple Definition

LEARNING OBJECTIVE 1.2.1 Describe five fundamental properties of life.

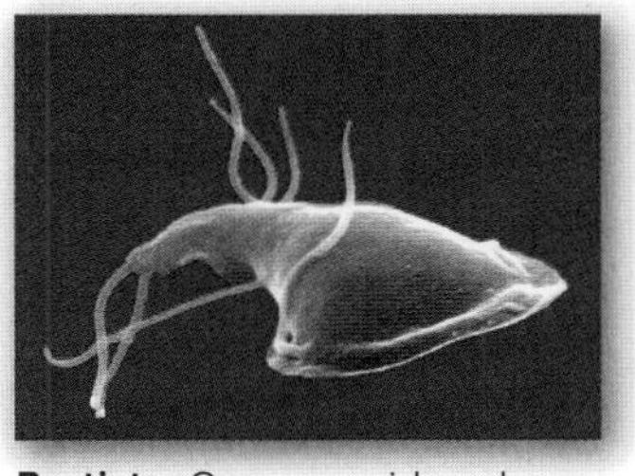

Protists. Once considered a kingdom, this term is still used informally to refer to unicellular eukaryotes.

Fungi. This kingdom contains mostly multicellular nonphotosynthetic organisms that digest their food externally, such as mushrooms.

Plantae. This kingdom contains photosynthetic multicellular organisms that are terrestrial, such as the flowering plant pictured here.

Animalia. Organisms in this kingdom are nonphotosynthetic multicellular organisms that digest their food internally, such as this ram.

Figure 1.2 The eukaryotic domain. Eukaryotes consist of most familiar organisms and many that are not. The eukaryotes can be divided into three kingdoms: Fungi, Plantae, and Animalia. Protists are a diverse group of unicellular eukaryotic organisms with no exclusive common ancestor.

(Protista): Dr. Stan Erlandsen and Dr. Dennis Feely/CDC; (Fungi): Russell Illig/Getty Images; (Plantae): Iconotec/Glow Images; (Animalia): Alan and Sandy Carey/Getty Images

What does it mean to be alive? What properties define a living organism? These questions are not as simple as they appear, because some of the most obvious properties of living organisms are also properties of many nonliving things—for example, *complexity* (a computer is complex), *movement* (clouds move in the sky), and *response to stimulation* (a soap bubble pops if you touch it). To appreciate why these three properties, so common among living things, do not help us to define life, imagine a mushroom standing next to a television: the television seems more complex than the mushroom, the picture on the television screen is moving but the mushroom just stands there, and the television responds to a remote-control device but the mushroom continues to just stand there—yet it is the mushroom that is alive.

All living things also share five more fundamental properties, passed down over millions of years from the first organisms to evolve on Earth: *cellular organization; energy utilization; homeostasis; growth, development, and reproduction;* and *heredity.*

1. **Cellular organization.** All living things are composed of one or more cells. Often too tiny to see, cells carry out the basic activities of living. Some cells have simple interiors, whereas others have complex organizations, but all are able to grow and reproduce. Many organisms possess only a single cell, like the paramecium in figure 1.3; your body contains about 10 trillion to 100 trillion cells (depending on your size).
2. **Energy utilization.** All living things use energy. Moving, growing, thinking—everything you do requires energy. Where does all this energy come from? It is captured from sunlight by plants and algae through photosynthesis. To get the energy that powers our lives, we extract it from plants or from plant-eating animals. That's what the kingfisher is doing in figure 1.4, eating a fish that ate algae.
3. **Homeostasis.** All living things maintain relatively constant internal conditions so that their complex processes can be better coordinated. Although the environment often varies considerably, organisms act to keep their interior conditions relatively constant, a process called *homeostasis*. Your body acts to maintain an internal temperature of 37°C (98.6°F), however hot or cold the weather might be.
4. **Growth, development, and reproduction.** All living things can grow and reproduce, although all members of a species may not reproduce. Bacteria increase in size and simply split into two, as often as every 15 minutes. Multicellular organisms grow by increasing the number of cells, and most produce different kinds of cells during development.
5. **Heredity.** All organisms possess a genetic system that is based on the replication and duplication of a long molecule called *DNA (deoxyribonucleic acid)*. The information that determines

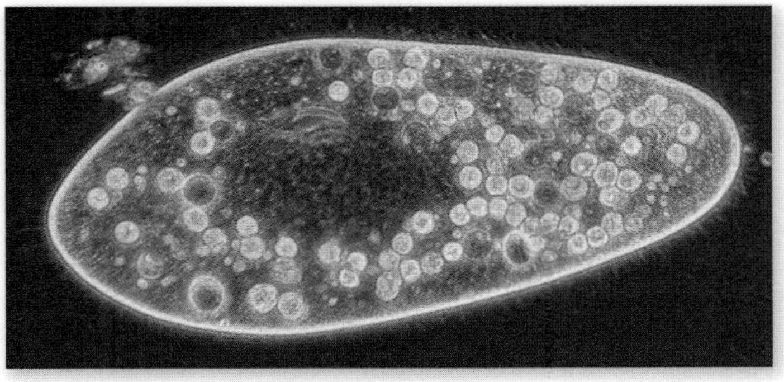

Figure 1.3 Cellular organization. This paramecium is a complex single-celled protist that has just ingested several yeast cells. Like this paramecium, many organisms consist of just a single cell, while others are composed of trillions of cells.

Melba Photo Agency/Alamy Stock Photo

Figure 1.4 Energy utilization. This kingfisher obtains the energy it needs to move, grow, and carry out its body processes by eating fish. It harvests the energy from food using chemical processes that occur within cells.
Armin Floreth/imagebroker/Alamy Stock Photo

what an individual organism will be like is contained in a code dictated by the order of the subunits making up the DNA molecule. Because DNA is copied from one generation to the next, any change in a gene is also preserved and passed on to future generations. The transmission of characteristics from parent to offspring is a process called *heredity*. All organisms interact with other organisms and the nonliving environment in ways that influence their survival, and as a consequence, organisms evolve adaptations to their environments.

Living Systems Show Hierarchical Organization

LEARNING OBJECTIVE 1.2.2 Describe the hierarchical nature of living systems.

Life's organisms interact with each other at many levels, in ways simple and complex. A key factor organizing these interactions is their degree of complexity. The organization of the biological world is hierarchical—that is, each level builds on the level below it, from the very simplest level of individual

Figure 1.5 Hierarchical organization of living systems. Life is highly organized, from the simplest atoms to complex, multicellular organisms. Along this hierarchy of structure, atoms form molecules, which are used to form organelles, which in turn form the functional subsystems within cells. Cells are organized into tissues, then into organs and organ systems such as the nervous system. This organization extends beyond individual organisms to populations, communities, ecosystems, and finally the entire biosphere.
(organelle): Keith R. Porter/Science Source; (cell): STEVE GSCHMEISSNER/SCIENCE PHOTO LIBRARY/Alamy Stock Photo; (tissue): Ed Reschke/Getty Images; (organism): Russell Illig/Getty Images; (population): George Ostertaga/gefotostock/Alamy Stock Photo; (species top): USDA Natural Resources Conservation Service; (species bottom): U.S. Department of Agriculture (USDA); (community): Ryan McGinnis/Alamy Stock Photo; (ecosystem): Steven P. Lynch/McGraw Hill; (biosphere): Goddard Space Flight Center/NASA

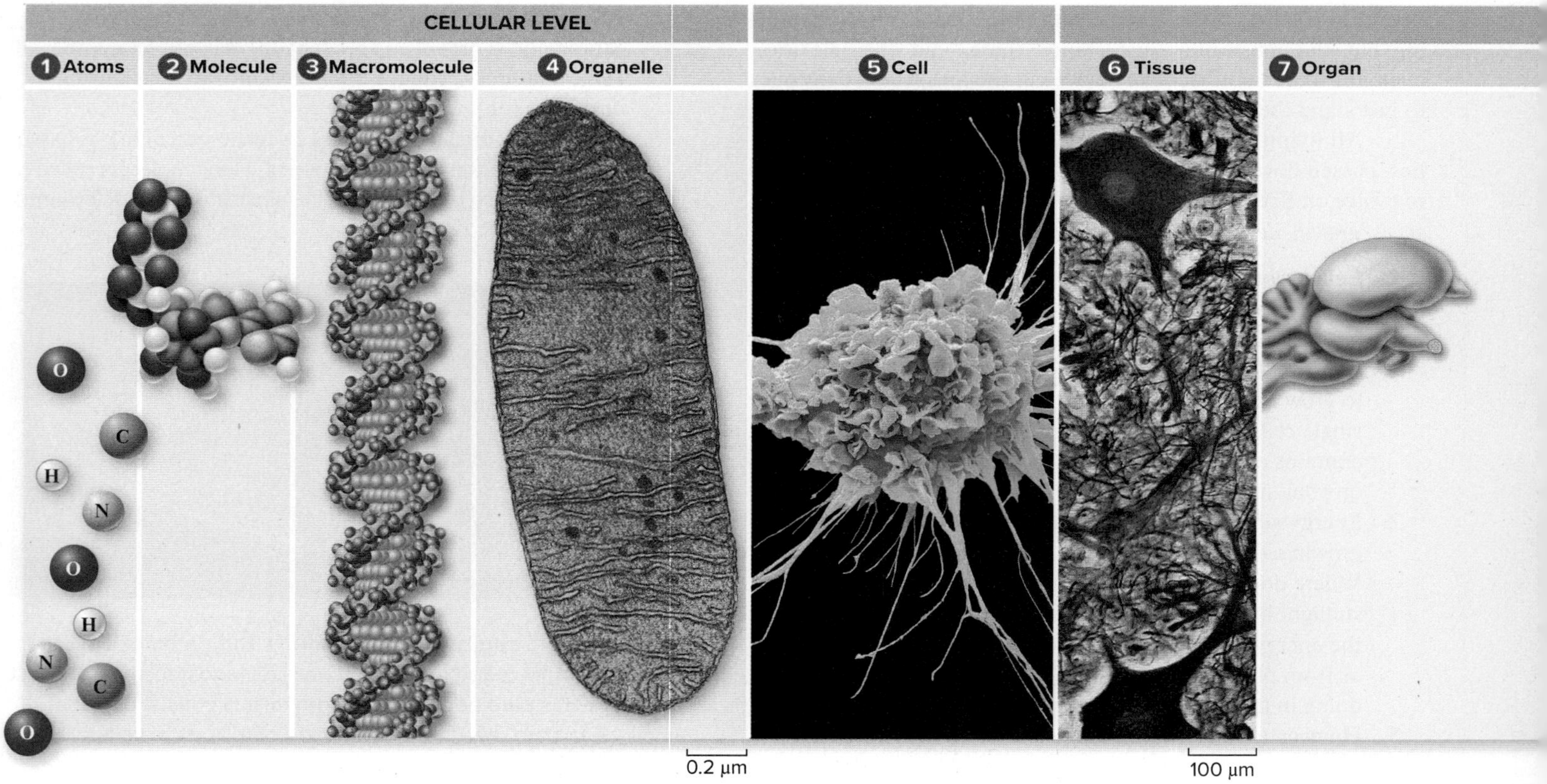

atoms to the vastly complex level of interacting ecosystems (figure 1.5):

The Cellular Level. At the cellular level, **atoms** 1, the fundamental elements of matter, are joined together by chemical bonds into stable assemblies called **molecules** 2. Large, complex molecules are called **macromolecules** 3. DNA, which stores the hereditary information, is a macromolecule. Complex biological molecules are assembled into tiny structures called **organelles** 4, within which cellular activities are organized. A mitochondrion is an organelle within which the cell extracts energy from food molecules. Membrane-bounded units called **cells** 5 are the basic units of life. Bacteria are composed of single cells. Animals, plants, and many other organisms are multicellular—composed of many cells.

The Organismal Level. Cells of multicellular organisms exhibit three levels of organization. The most basic level is that of **tissues** 6, which are groups of similar cells that act as a functional unit. Nerve tissue is one kind of tissue, specialized to carry electrical signals. Tissues, in turn, are grouped into **organs** 7 —body structures composed of several different tissues that act as a structural and functional unit. Your brain is an organ composed of nerve cells and cells that nourish and support them, as well as a variety of associated connective tissues that form both protective coverings and a network of blood vessels to bring oxygen and nutrients to the brain. At the third level of organization, organs are grouped into **organ systems** 8. The nervous system, for example, consists of sensory organs, the brain and spinal cord, and a network of neurons that convey signals between the brain and the other organs and tissues of the body.

The Populational Level. Individual **organisms** 9 occupy several hierarchical levels within the living world. The most basic of these is the **population** 10—a group of organisms of the same species living in the same place. All populations of a particular kind of organism together form a **species** 11, its members are similar in appearance and able to interbreed. At a higher level of biological organization, a **biological community** 12 consists of all the populations of different species living together in one place.

The Ecosystem Level. At the highest tier of biological organization, a biological community and the physical habitat (soil composition, available water, temperature range, wind, and a host of other environmental influences) within which it lives together constitute an ecological system, or **ecosystem** 13. The entire planet can be thought of as a global ecosystem we call the **biosphere** 14.

Novel Properties Emerge from More Complex Organization

LEARNING OBJECTIVE 1.2.3 Discuss how living systems display emergent properties.

At each higher level in the living hierarchy, novel properties emerge—properties that were not present at the simpler level of organization. These **emergent properties** result from the way in

which components interact and often cannot be guessed just by looking at the parts themselves. You have the same array of cell types as a giraffe, for example—so examining a collection of its individual cells gives little clue about what your body is like.

The emergent properties of life are not magical or supernatural. They are the natural consequences of the hierarchy or structural organization that is the hallmark of life. Both water (which makes up 50 to 75% of your body's weight) and ice are made of H_2O molecules, but one is liquid and the other is solid, because the H_2O molecules in ice are more organized. Two proteins—long chains of amino acids—may contain the same number of each amino acid, yet one might act as an enzyme to promote a chemical reaction whereas the other might not; the enzymatic activity is an emergent property, reflecting the information contained in the *sequence* of the amino acids.

Functional properties emerge from more complex organization. Metabolism is an emergent property of life. The chemical reactions within a cell arise from interactions between molecules that are orchestrated by the orderly environment of the cell's interior. Consciousness is an emergent property of the brain that results from the interactions of many neurons in different parts of the brain.

This description of the common features and organization of living systems begins to get at the nature of what it is to be alive. The rest of this book illustrates and expands on these basic ideas to provide you with a more complete account of living systems.

REVIEW OF CONCEPT 1.2

Biology is a unifying science that brings together all branches of science to study living systems. Life does not have a simple definition, but living systems share a number of properties that together describe life. Biologists organize living systems hierarchically, from the subcellular level to the entire biosphere, with emergent properties arising at each stage that cannot be guessed from studying its parts.

- *Can you name an emergent property at the population level?*

1.3 Science Is Based on Both Observation and Reasoning

Much like life itself, the nature of science defies simple description. For many years scientists have written about the "scientific method" as though there were a single way of doing science. This oversimplification has contributed to nonscientists' confusion about the nature of science. Although there is no single scientific method, we could say that there is a scientific mindset that involves skepticism and the importance of objective, verifiable facts.

At its core, science is concerned with developing an increasingly accurate understanding of the world by using observation and reasoning. To begin with, we assume that natural forces acting now have always acted, that the fundamental nature of the universe has not changed since its inception, and that it is not changing now.

There is a corollary to these assumptions about the action of natural forces: science does not address the so-called supernatural. This would seem to set science and religion in opposition, and there are some who hold this view. Others take the position that science and religion address different questions or shed light on different aspects of the human condition. Regardless of the position you take, questions of religion, the existence of ghosts, astrology, and any other supernatural phenomena are outside the scope of this book.

The Scientific Process Involves Observation and Both Deductive and Inductive Reasoning

LEARNING OBJECTIVE 1.3.1 Distinguish between deductive and inductive reasoning.

Virtually all science begins with observation of the natural world. This might also involve measurement or use technology that extends our ability to make observations. The use of technology has become increasingly important in biology, as we will learn throughout our explorations.

Descriptive science

The common version of the scientific process is that observations lead to hypotheses, which in turn make experimentally testable predictions. In this way, scientists develop an increasingly accurate understanding of nature. We discuss this way of doing science later in this section, but it is important to understand that much of science is purely descriptive: in order to understand anything, the first step is to describe it completely. Much of biology is concerned with arriving at an increasingly accurate description of nature.

The study of biodiversity is an example of descriptive science that has implications for other aspects of biology, in addition to societal implications. Efforts are currently under way to classify all life on Earth. This ambitious project is purely descriptive, but it will lead to a greater understanding of biodiversity and how human activity affects biodiversity.

A new approach to biodiversity uses massive DNA sequencing of environmental samples, followed by using computer software to identify organisms in the sample. This has revealed an even greater than anticipated diversity of both prokaryotic and eukaryotic organisms than was suspected based on culturing organisms in the laboratory. At this point our ability to generate these data is outrunning our ability to directly analyze any of these organisms.

One of the most important accomplishments of molecular biology at the dawn of the 21st century was completing the sequencing of the human genome. Since the initial project was completed, this descriptive work has been continuing as we refine and extend our knowledge of the human genome. All of these data are helping biologists to explore new areas of human biology and to generate hypotheses about both basic human biology and the understanding of disease states.

Logical reasoning

Scientists are often portrayed in popular culture as unemotional, purely logical individuals. Popular television has a long history of such characters, most recently Dr. Temperance Brennan in the drama *Bones* and Dr. Sheldon Cooper in the sitcom *The Big Bang Theory.* While entertaining, this is not an accurate portrayal of most scientists. However, the use of logical reasoning is critical to the scientific process.

The form of reasoning that applies general principles to predict specific results is called **deductive reasoning.** This form of reasoning arose from mathematics and philosophy and is used to test the validity of general ideas in all branches of knowledge. For example, if all mammals by definition have hair, and you find an animal that does not have hair, then you may conclude that this animal is not a mammal. A biologist uses deductive reasoning to infer the species of an unknown specimen from its characteristics. More than 2200 years ago, the Greek scientist Eratosthenes used Euclidean geometry and deductive reasoning to accurately estimate the circumference of the Earth. This is a powerful example of how the use of observation and logic can help us understand the natural world.

When logic flows in the opposite direction, from specific cases to general principles, we call it **inductive reasoning.** Inductive reasoning uses specific observations to construct general scientific principles. For example, if poodles have hair and terriers have hair, and every dog that you observe has hair, then you may conclude that all dogs have hair. Inductive reasoning leads to generalizations that can then be tested. Inductive reasoning first became important to science in the 17th century in Europe, when Francis Bacon, Isaac Newton, and others, including Margaret Cavendish and Jeanne Dumée, began to use the results of particular experiments to infer general principles about how the world operates.

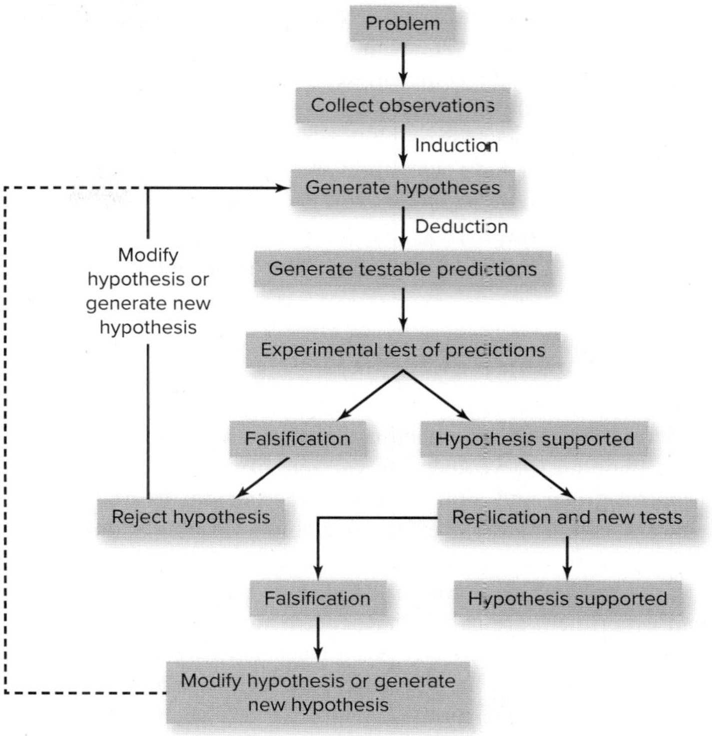

Figure 1.6 How experimental science is done. This provides a general flowchart for testing hypotheses by experimentation. A problem of interest is identified and observations are collected. Inductive reasoning leads to development of one or more potential explanations (hypotheses). Experimental results will either support or falsify a hypothesis. A hypothesis that is supported is still subject to further replication and testing, leading to either eventual rejection or further support. Falsified hypotheses can be modified or rejected in favor of a new hypothesis.

Hypothesis-Driven Science Makes and Tests Predictions

LEARNING OBJECTIVE 1.3.2 Illustrate how experimentation can be used to test hypotheses.

Experimental scientists use a combination of inductive and deductive reasoning. We use induction to go from a variety of specific observations to form a general proposal to explain the observations. We call the proposal that explains our initial observations a **hypothesis.** This hypothesis is more than an explanation; to be useful, it must generate testable predictions. We use the process of deduction to generate predictions that usually take an "if X, then Y" form. These predictions from one or more alternative hypotheses can be tested by experiments. This leads to a hypothesis being either supported or rejected. A hypothesis that is supported has not been "proven" and is subject to replication and additional testing. If it continues to be supported, our confidence in this as an explanation increases. If a hypothesis must be rejected, we need to either formulate a new hypothesis or modify the existing one (figure 1.6).

Because experiments do not actively "prove" a hypothesis, this process is ongoing as new data are generated. Even foundational ideas in a field will evolve with new information. For example, geneticists George Beadle and Edward Tatum studied the nature of genetic information to arrive at the "one-gene/one-enzyme" hypothesis (refer to chapter 15). This states that a gene represents the genetic information necessary to make a single enzyme. As investigators learned more about the molecular nature of genetic information, the hypothesis was refined to "one-gene/one-polypeptide," because enzymes can be made up of more than one polypeptide. With still more discoveries about the nature of genetic information, other investigators found that a single gene can specify more than one polypeptide, and the hypothesis was refined again.

Testing hypotheses

We call the test of a hypothesis an experiment. Suppose you enter a dark room. To understand why it is dark, you propose several hypotheses. The first might be "The room is dark because the light switch is turned off." Other hypotheses could involve whether the switch is functional, whether the light bulb is burned out, or even loss of power to the entire room. To evaluate these hypotheses, you would conduct experiments designed to eliminate one or more of the hypotheses.

For example, you might test your hypotheses by flipping the light switch. If you do so and the room is still dark, you have disproved the first hypothesis: something other than the setting of the light switch must be the reason for the darkness. Note that a test such as this does not prove that any of the other hypotheses are true; it merely demonstrates that the one being tested is not. A successful experiment is one in which one or more of the alternative hypotheses are demonstrated to be inconsistent with the results and are thus rejected.

As you proceed through this text, you will encounter many hypotheses that have withstood the test of experiment. Many will continue to do so; others will be revised as biologists make new observations. Biology, like all science, is in a constant state of change, with new ideas appearing and replacing or refining old ones.

The importance of controls

Often scientists are interested in learning about processes that are influenced by many factors, or **variables**. To evaluate alternative hypotheses, we keep the variable of interest constant in a **control group** and alter it in the **experimental treatment**. This allows us to isolate the effects of a single variable in our experiment, so any difference in the outcome is due to the influence of this variable. The factors or conditions that you manipulate in an experiment are called **independent variables,** and the factor observed or measured is called the **dependent variable**.

For example, in a trial to test the efficacy of a vaccine against a virus, the hypothesis being tested is whether the vaccine prevents infection. The independent variable is whether a participant received a placebo (control) or the vaccine (experimental), and the dependent variable is the incidence of disease in the two groups. Notice in this case there are a variety of possible "end points" that could be analyzed, including any infection, a reduction in severity of infection, hospitalizations, or deaths.

Summing up: An example

A hypothesis that does not make testable predictions will not do much to advance our understanding. It is also true that a hypothesis that explains everything you have observed is quite possibly dead wrong. Or, as Thomas Huxley, a contemporary and supporter of Charles Darwin, put it: "The great tragedy of science, the slaying of a beautiful theory by an ugly fact."

Let's put this together with an important example from the early history of microbiology. It was known that nutrient broth left sitting exposed to air becomes contaminated. Two hypotheses were proposed to explain this observation: spontaneous generation and the germ hypothesis. Spontaneous generation was an old idea that had been used to explain a variety of observations, including why maggots appear in rotting meat. This theory held that there was an inherent property in organic molecules that could lead to the spontaneous generation of life. The more recent, at the time, germ hypothesis proposed that preexisting microorganisms that were present in the air could contaminate the nutrient broth.

These competing hypotheses were tested by a number of experiments that involved filtering air and boiling the broth to kill any contaminating germs. The definitive experiment was performed by Louis Pasteur, who constructed flasks with curved necks that could be exposed to air, but that would trap any contaminating germs. When such flasks were boiled to sterilize them, they remained sterile, but if the curved neck was broken off, they became contaminated (figure 1.7).

SCIENTIFIC THINKING

Question: *What is the source of contamination that occurs in a flask of nutrient broth left exposed to the air?*

Germ Hypothesis: *Preexisting microorganisms present in the air contaminate nutrient broth.*

Prediction: *Sterilized broth will remain sterile if microorganisms are prevented from entering flask.*

Spontaneous Generation Hypothesis: *Living organisms will spontaneously generate from nonliving organic molecules in broth.*

Prediction: *Organisms will spontaneously generate from organic molecules in broth after sterilization.*

Test: *Use swan-necked flasks to prevent entry of microorganisms. To ensure that broth can still support life, break swan-neck after sterilization.*

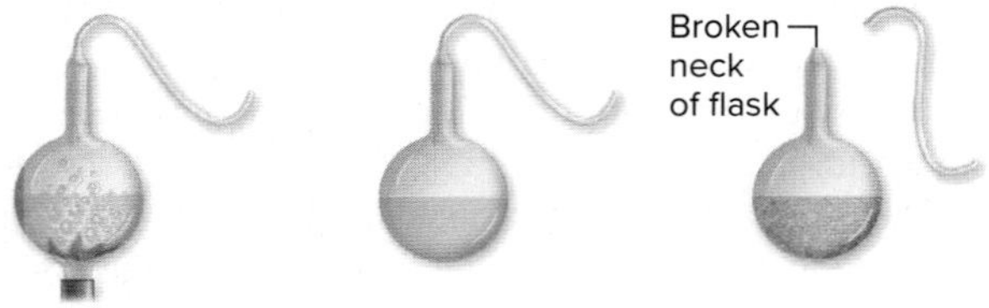
Broken neck of flask
Flask is sterilized by boiling the broth.
Unbroken flask remains sterile.
Broken flask becomes contaminated after exposure to germ-laden air.

Result: *No growth occurs in sterile swan-necked flasks. When the neck is broken off and the broth is exposed to air, growth occurs.*

Conclusion: *Growth in broth is of preexisting microorganisms.*

Figure 1.7 Experiment to test spontaneous generation versus germ hypothesis.

This result was predicted by the germ hypothesis—that when the sterile flask is exposed to air, airborne germs are deposited in the broth and grow. The spontaneous generation hypothesis predicted no difference in results with exposure to air. This experiment disproved the hypothesis of spontaneous generation and supported the hypothesis of airborne germs under the conditions tested.

Theories Are the Solid Conclusions of Science

LEARNING OBJECTIVE 1.3.3 Discuss how scientists use models to describe, explain, and test theories.

As we have discussed, a successful hypothesis should generate predictions that allow for experimental tests. If an experiment produces results inconsistent with predictions, the hypothesis must be rejected or modified. In contrast, if the predictions are supported by experimental testing, the hypothesis is supported. The more experimentally supported predictions a hypothesis makes, the stronger our confidence in its validity.

The nature of scientific theories

Scientists use the word **theory** in two main ways. The first meaning of *theory* is essentially deductive, a proposed explanation for some natural phenomenon, based on general principles. Thus, we speak of the principle first proposed by Newton as the "theory of gravity." Such theories often bring together concepts that were previously thought to be unrelated.

The second meaning of *theory* is essentially inductive: a body of interconnected concepts, supported by inductive scientific reasoning and experimental evidence, that explains the facts in some area of study. For example, quantum theory in physics brings together a set of ideas about the nature of the universe derived from diverse experimental observations and serves as a guide to further questions and experiments.

To scientists, theories are the solid ground of science, expressing ideas about which they are most certain. By contrast, to the general public the word *theory* usually implies the opposite—a lack of knowledge or a guess ("it's only a *theory* . . ."). Not surprisingly, this difference often results in confusion. In this text, theory will always be used in its scientific sense, in reference to an accepted general principle or body of knowledge.

Reductionism

Scientists often use the philosophical approach known as **reductionism** to understand a complex system by reducing it to its working parts. Reductionism has limits when applied to living systems, however—the complex interworking of many interconnected functions leads to emergent properties that cannot be predicted based on the workings of the parts. For example, ribosomes—complex cellular machines that make proteins—can be disassembled into their constituent parts. However, examination of the parts in isolation would not lead to predictions about the nature of protein synthesis. On a higher level, understanding the physiology of a single Canada goose would not lead to predictions about flocking behavior. Biologists are just beginning to come to grips with this problem and to think about ways of dealing with the whole as well as the workings of the parts. The emerging field of systems biology focuses on this different approach.

Biological models

Biologists construct models in many different ways for a variety of uses. Geneticists construct models of interacting networks of proteins that control gene expression, often even drawing cartoon figures to represent that which we cannot see. Population biologists build models of how evolutionary change occurs. Cell biologists build models to explain cell communication and the events leading from an external signal to internal events. Structural biologists build models of the structure of proteins and macromolecular complexes in cells.

Models provide a way to organize how we think about a problem. Models can also get us closer to the larger picture and away from the extreme reductionist approach. The working parts are provided by the reductionist analysis, but the model shows how they fit together. Often these models suggest other experiments that can be performed to refine or test the model.

Science as a social construct

Research results are written up and submitted for publication in scientific journals, where other scientists review the experiments and conclusions. This process of evaluation of others' work lies at the heart of modern science. It reduces the incidence of faulty research or false claims being given the authority of scientific fact, and provides a starting point for examining the reproducibility of experimental results. Results that cannot be reproduced are not taken seriously for long.

Even with evaluation by other scientists, there are issues with the reproducibility of experiments in multiple branches of science. Some of this has to do with misusing statistics, and some has to do with other issues, including the pressure to publish and the unwillingness of journals to publish negative results. There have also been a number of well-publicized examples of outright scientific fraud. While these are serious issues, that they are being debated indicates that science can be self-correcting.

There are also examples of theories whose acceptance led to resistance to alternative explanations. Thomas Kuhn described the progress of science as accepted dogma that is occasionally overturned by revolutionary new ideas. This is a great oversimplification of his ideas, but in some instances this appears to have been the case. The theory of relativity in physics is a good example. Our understanding of gravity that goes back to Newton was eventually replaced by Einstein's theory of general relativity.

This positive view of science and scientists should not obscure the fact that, like all other areas of human endeavor, the history of science includes examples of misuse of its perceived authority by scientists and politicians. The eugenics movement in the 20th century was used to justify forced sterilization of so-called "mentally unfit" minorities, and the history of medical science includes examples of terrible experiments on minorities without consent. These abuses have led to more robust systems of consent for experiments involving humans, but provide a warning to be vigilant about abuses.

REVIEW OF CONCEPT 1.3

Much of science is descriptive, amassing observations to gain an accurate view. Both deductive and inductive reasoning are used in science. Scientific hypotheses are suggested explanations for observed phenomena. When a hypothesis has been extensively tested and no contradictory information has been found, it becomes an accepted theory. Theories are coherent explanations of observed data, and they may be modified by new information.

- *How does a scientific theory differ from a hypothesis?*

1.4 The Study of Evolution Is a Good Example of Scientific Inquiry

Darwin's theory of evolution explains how organisms on Earth have changed over time and acquired a diversity of new forms. We devote three chapters to a detailed examination of evolution, but it is worth considering this theory now as an example of how a scientist develops a hypothesis and how a scientific theory is tested and gains acceptance.

The Idea of Evolution Existed Prior to Darwin

LEARNING OBJECTIVE 1.4.1 Describe ideas about evolution proposed before Darwin.

Charles Robert Darwin (1809–1882; figure 1.8) was an English naturalist who, after 30 years of study and observation, wrote one of the most famous and influential books of all time. This book, *On the Origin of Species by Means of Natural Selection,* created a sensation when it was published, and the ideas Darwin expressed in it have played a central role in the development of human thought.

Birth of the idea of evolution

In Darwin's time, most people believed that the different kinds of organisms and their individual structures resulted from direct actions of a Creator (many people still believe this). Species were thought to have been specially created and to be unchangeable over the course of time. This was the view of Carolus Linnaeus (1708–1778), the Swedish biologist who established the system of naming organisms that is still in use.

By the first part of the 18th century, many more kinds of organisms were being discovered than previously, as well as many fossil animals and plants. These discoveries gradually began to trigger discussions of evolution—the possibility that living things have changed during the history of life on Earth. The great French biologist Georges-Louis Leclerc, Comte de Buffon (1707–1788), spoke explicitly, a century before Darwin, of natural affinities between kinds of organisms, writing of "the universal kinship of all generations born from a common mother." He could see no explanation for the common features of all mammals except their evolution from a common ancestor.

Within 50 years these ideas led Jean Baptiste de Lamarck (1744–1829) to explicitly propose evolution as a theory to account for the patterns observed in nature. In 1801 he suggested that all species, including human beings, were descended from other species. Lamarck thought of life as having evolved progressively from simple to more complex forms, and he was the first to propose a coherent theory of evolution.

Lamarck's theory was based on the incorrect idea that organs and structures became stronger through use, and that the strengthened character was then passed on to offspring—the **theory of inheritance of acquired characteristics.** Although incorrect, Lamarck's theory called wide attention to the possibility of evolution and, by doing so, set the stage for the acceptance of the correct, and much simpler, explanation proposed by Charles Darwin half a century later.

Darwin attributed evolution to what he called *natural selection,* which he proposed as a coherent, logical explanation. His book *On the Origin of Species* was a best seller in its day and brought his ideas to wide public attention.

Figure 1.8 Charles Darwin. This newly rediscovered photograph, taken in 1881, the year before Darwin died, appears to be the last ever taken of the great biologist.

Huntington Library/Superstock

Darwin Gathered Information During the Voyage of the *Beagle*

LEARNING OBJECTIVE 1.4.2 Identify important observations made by Darwin on the *Beagle*.

The story of Darwin and his theory of evolution begins in 1831, when Darwin was 22 years old. He was part of a five-year navigational mapping expedition around the coasts of South America (figure 1.9), aboard H.M.S. *Beagle.* During this long voyage, Darwin had the chance to study a wide variety of plants and animals on continents, islands, and distant seas. Repeatedly, Darwin saw that the characteristics of similar species varied somewhat from place to place. These geographical patterns suggested to him that lineages change gradually as species migrate from one area to another. On the Galápagos Islands, 960 km (600 miles) off the coast of Ecuador, Darwin encountered a variety of finches on the various islands. The 14 species, although related, differed slightly in appearance, particularly in their beaks (figure 1.10).

Darwin thought it was reasonable to assume that all these birds had descended from a common ancestor that had arrived from the South American mainland several million years ago. Eating different foods on these islands, the finches' beaks had changed during their descent—"descent with modification," or evolution. (These finches are discussed in more detail in chapters 20 and 21.)

In a more general sense, Darwin was struck by the fact that the plants and animals on these relatively young volcanic islands resembled those on the nearby coast of South America. If each one of these plants and animals had been created independently and simply placed on the Galápagos Islands, why didn't they resemble the plants and animals of islands with similar climates—such as those off the coast of Africa? Why did they resemble those of the adjacent South American coast instead?

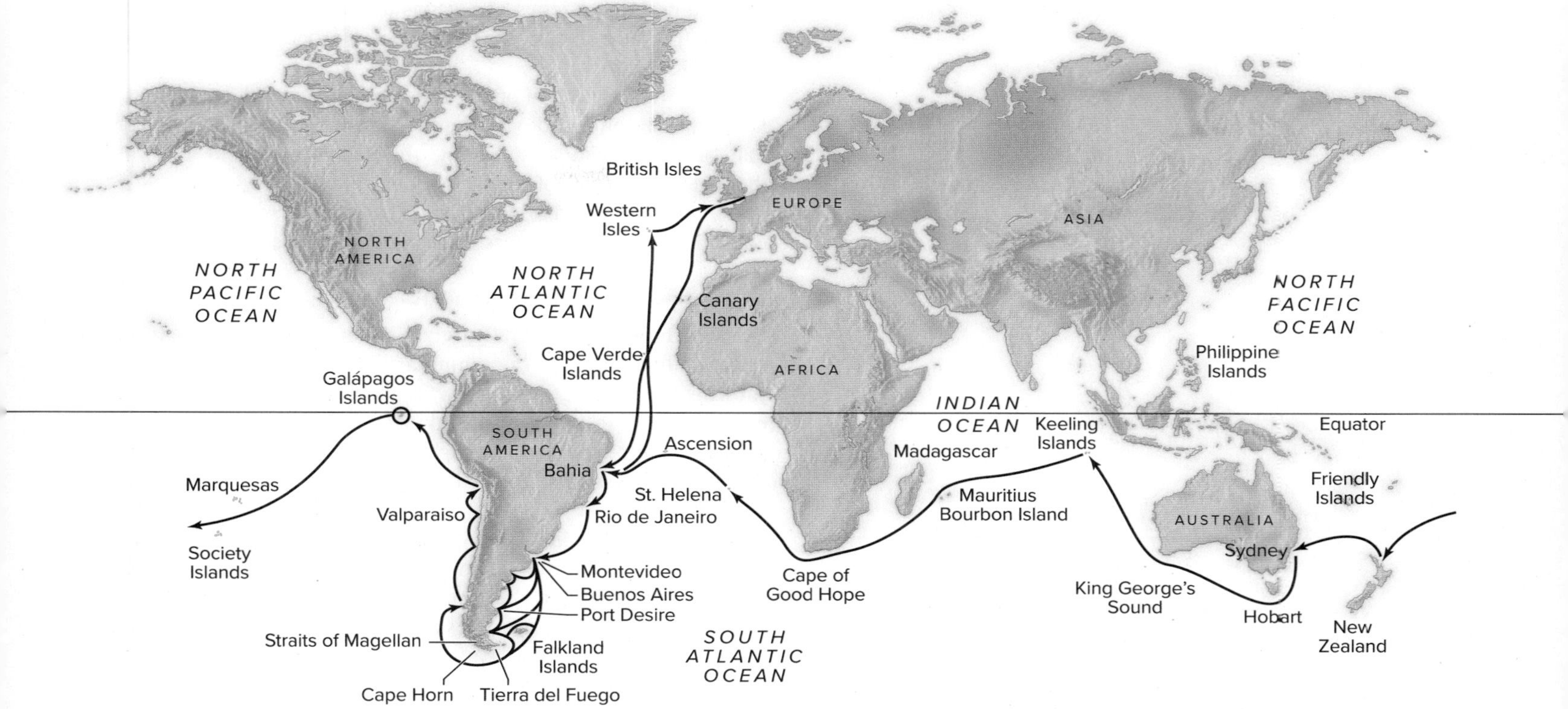

Figure 1.9 The five-year voyage of H.M.S. *Beagle*. Most of the time was spent exploring the coasts and coastal islands of South America, such as the Galápagos Islands. Darwin's studies of the animals of the Galápagos Islands played a key role in his eventual development of the concept of evolution by means of natural selection.

Figure 1.10 Three Galápagos finches and what they eat. On the Galápagos Islands, Darwin observed 14 species of finches that differ mainly in their beaks and feeding habits. These three finches eat very different food items, and Darwin surmised that the shapes of their bills are evolutionary adaptations that improve their ability to eat the foods available in their specific habitats.

Darwin Proposed Natural Selection as a Mechanism for Evolution

LEARNING OBJECTIVE 1.4.3 Describe Darwin's theory of evolution by natural selection.

Darwin's two great achievements were to observe evidence that evolution has occurred and to formulate a hypothesis that explains this evolution as the consequence of natural selection.

Darwin and Malthus

Of key importance to the development of Darwin's insight into natural selection was his study of Thomas Malthus's *An Essay on the Principle of Population* (1798). In this book Malthus stated that populations of plants and animals (including human beings) tend to increase geometrically, but humans are able to increase their food supply only arithmetically. Put another way, Malthus argued that population increases by a multiplying factor; for example, in

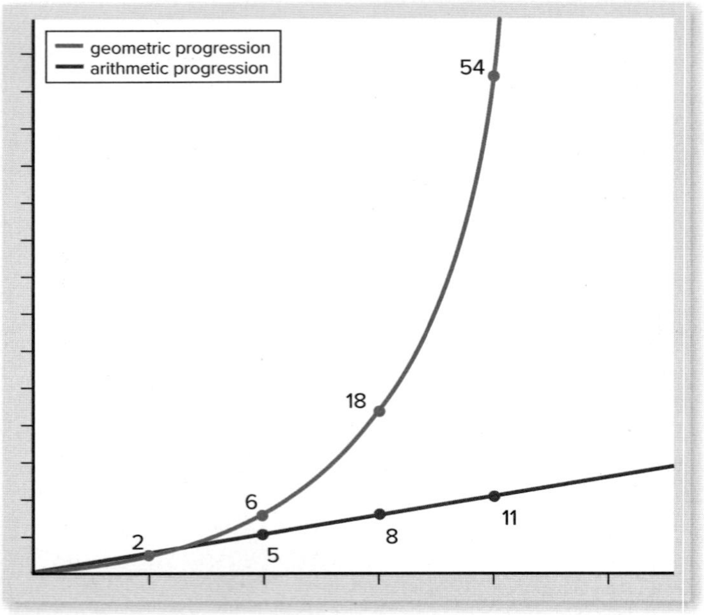

Figure 1.11 Geometric and arithmetic progressions. A geometric progression increases by a constant factor (for example, ×3 for each step), whereas an arithmetic progression increases by a constant difference (for example, +3 for each step). Malthus contended that the human growth curve is geometric, but the human food production curve is only arithmetic.

the series 2, 6, 18, 54, the multiplying factor is 3. By contrast, Malthus believed that a food supply increases by an additive factor; for example, the series 2, 5, 8, 11 adds 3 to each starting number. Figure 1.11 shows the difference that these two types of relationships produce.

Malthus pointed out that because populations increase geometrically, virtually any kind of animal or plant, if it reproduced unchecked, would cover the entire surface of the world surprisingly quickly. This does not happen, Malthus argued, because death limits population numbers so that populations of species remain fairly constant year after year.

Sparked by Malthus's ideas, Darwin saw that although every organism has the potential to produce more offspring than can survive, only a limited number actually do survive and produce further offspring. Combining this observation with what he had seen on the voyage of the *Beagle,* as well as with his own experiences in breeding domestic animals, Darwin made an important association: individuals possessing physical, behavioral, or other attributes that give them an advantage in their environment are more likely to survive and reproduce than are those with less advantageous traits. By surviving, these individuals gain the opportunity to pass on their favorable characteristics to their offspring. As the frequency of these characteristics increases in the population, the nature of the population as a whole gradually changes. Darwin called this process *selection.*

Natural selection

Darwin was thoroughly familiar with variation in domesticated animals, and he began *On the Origin of Species* with a detailed discussion of pigeon breeding. He knew that animal breeders selected certain varieties of pigeons and other animals, such as dogs, to produce certain characteristics, a process Darwin called **artificial selection.**

Artificial selection often produces a great variation in traits. Domestic pigeon breeds, for example, show much greater variety than all of the wild species found throughout the world. Darwin thought that this type of change could occur in nature, too. Surely if pigeon breeders could foster variation by artificial selection, nature could do the same—a process Darwin called **natural selection.**

Darwin drafts his argument

Darwin drafted the overall argument for evolution by natural selection in a book-length preliminary manuscript in 1842. After showing the manuscript to a few of his closest scientific friends, however, Darwin put it in a drawer and for 16 years turned to other research. No one knows for sure why Darwin did not publish his initial manuscript—it is very thorough and outlines his ideas in detail.

The stimulus that finally brought Darwin's hypothesis into print was an essay he received in 1858 from Alfred Russel Wallace (1823–1913), a young naturalist who was in Indonesia. It concisely set forth the hypothesis of evolution by means of natural selection, a hypothesis Wallace had developed independently of Darwin. After receiving Wallace's essay, friends of Darwin arranged for a joint presentation of the two men's ideas at a seminar in London. Darwin then completed his own book, expanding the 1842 manuscript he had written so long ago and submitting it for publication the following year.

The Predictions of Darwin's Theory Have Been Well Tested

LEARNING OBJECTIVE 1.4.4 Identify how evolution has been tested over time.

More than 130 years have elapsed since Darwin's death in 1882. During this period, the evidence supporting his theory has grown progressively stronger. We briefly explore some of this evidence here; in chapter 20, we will return to the theory of evolution by natural selection and examine the evidence in more detail.

The fossil record

Darwin predicted that the fossil record would yield intermediate links between the great groups of organisms—for example, between fishes and the amphibians thought to have arisen from them, and between reptiles and birds. Furthermore, natural selection predicts the relative positions in time of such transitional forms. We now know the fossil record to a degree unthinkable in the 19th century, and paleontologists have found what appear to be transitional forms at the predicted positions in time.

Recent discoveries of microscopic fossils have extended the known history of life on Earth back to about 3.5 billion years ago (BYA). The discovery of other fossils has supported Darwin's predictions and has shed light on how organisms have, over this enormous time span, evolved from the simple to the complex. For vertebrate animals especially, the fossil record is rich and exhibits a graded series of changes in form, where the evolutionary sequence can be discerned (figure 1.12).

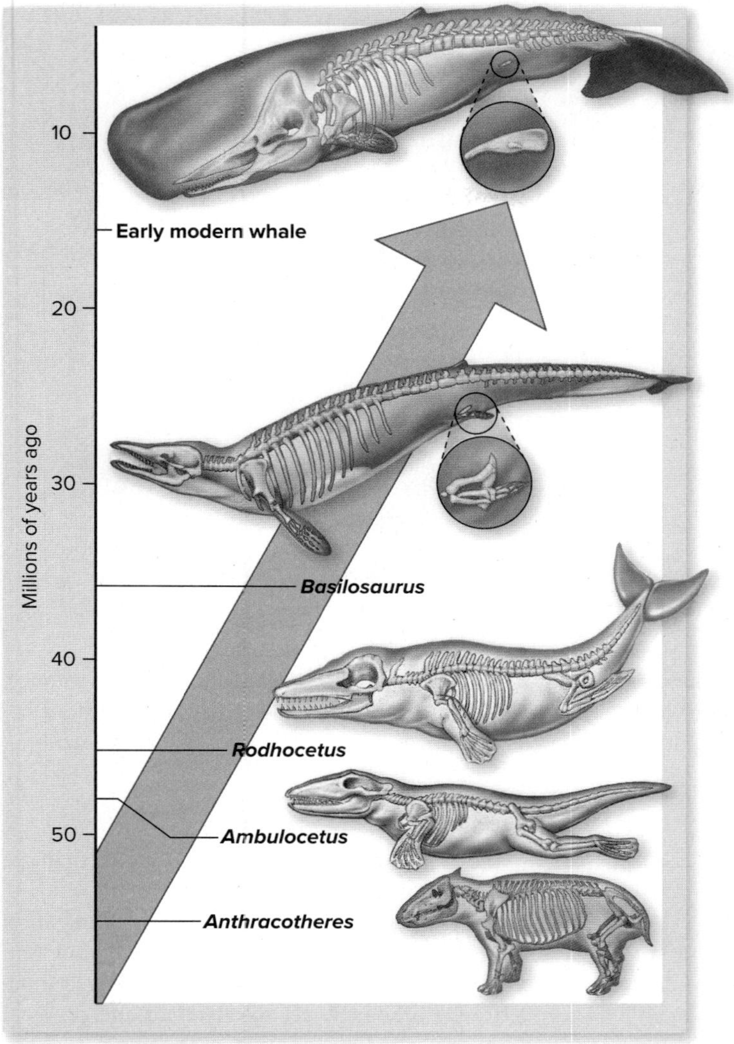

Figure 1.12 The evolutionary history of whales. Over a period of 35 million years, modern whales have evolved from the piglike ancestors of a hippopotamus, with intermediate steps preserved in the fossil record.

Comparative anatomy

Comparative studies of animals have provided strong evidence for Darwin's theory. In many different types of vertebrates, for example, the same bones are present, indicating their shared evolutionary past. Thus, the forelimbs shown in figure 1.13 are all constructed from the same basic array of bones, modified for different purposes.

These bones are said to be **homologous** in the different vertebrates; that is, they have the same evolutionary origin, but they now differ in structure and function. They are contrasted with **analogous** structures, such as the wings of birds and butterflies, which have similar function but different evolutionary origins.

Molecular evidence

Evolutionary patterns are also revealed at the molecular level. By comparing the genomes (that is, the sequences of all the genes) of different groups of animals or plants, we can more precisely specify the degree of relationship among the groups. A series of evolutionary changes over time should involve a continual accumulation of genetic changes in the DNA.

With the recent advances in genome sequencing, this prediction is now subject to direct test (figure 1.14). The result is clear: for a broad array of vertebrates, the more distantly related two organisms are, the greater their genomic distance.

This pattern of divergence over time is also apparent at the protein level. Comparing the hemoglobin amino acid sequences of different species, the pattern is again clear. Rhesus monkeys, which (like humans) are primates, have fewer differences from humans in the 146-amino-acid hemoglobin β-chain than do more distantly related mammals, such as dogs. Nonmammalian vertebrates, such as birds and frogs, differ even more.

The sequences of some genes, such as the ones specifying the hemoglobin proteins, have been determined in many organisms, and the entire time course of their evolution can be laid out with confidence by tracing the origins of particular nucleotide changes in the gene sequence. The pattern of descent obtained is called a **phylogenetic tree.** It represents the evolutionary history of the gene, its "family tree." Molecular phylogenetic

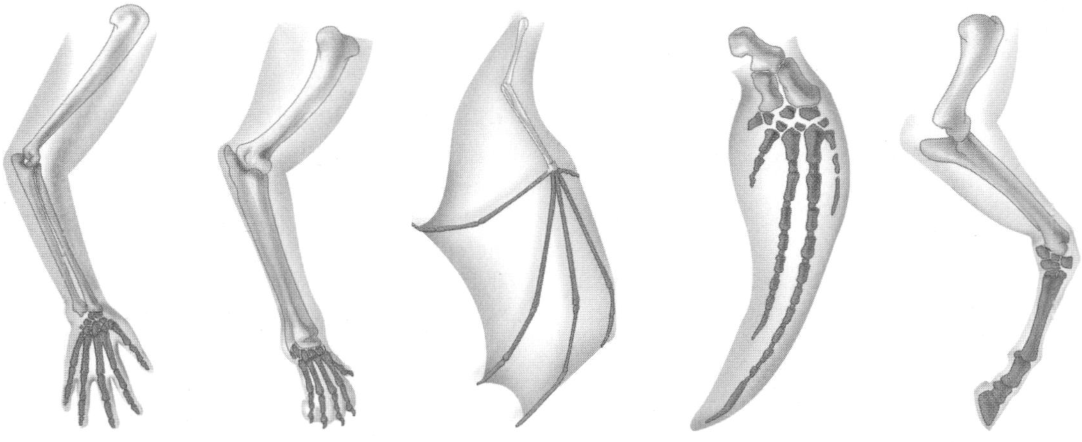

Figure 1.13 Homology among vertebrate limbs. The forelimbs of these five vertebrates show the ways in which the relative proportions of the forelimb bones have changed in relation to the particular way of life of each organism.

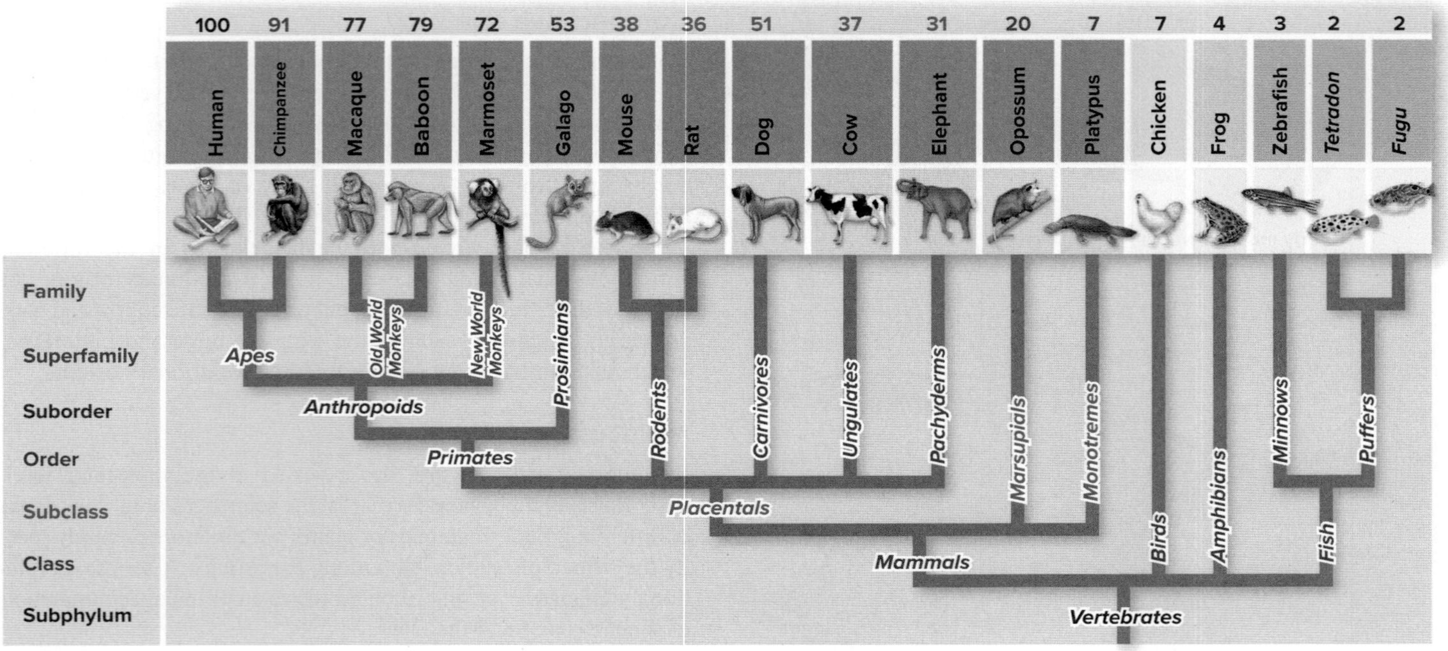

Figure 1.14 Evolution of the vertebrate genome. Genomic scientists have recently investigated the similarity of 44 representative regions scattered around the vertebrate genome (1% of the total genome). Moving from right to left, genomic similarity (expressed along the top as % sequence similarity to humans) increases as taxonomic distance from humans decreases—just as Darwin's theory predicts.

trees agree well with those derived from the fossil record, which is strong, direct evidence of evolution. The pattern of accumulating DNA changes represents, in a real sense, the footprints of evolutionary history.

REVIEW OF CONCEPT 1.4

Darwin observed differences in related organisms and proposed the hypothesis of evolution by natural selection to explain these differences. The predictions generated by natural selection have been tested and continue to be tested by analysis of the fossil record, comparative anatomy, and even the DNA of living organisms.

- *Does Darwin's theory of evolution by natural selection explain the origin of life?*

1.5 A Few Important Concepts Form the Core of Biology

The depth and breadth of material in introductory biology create challenges for students new to college. Consequently, in recent years there have been initiatives to emphasize conceptual understanding over factual recall, and to find ways to actively engage students in the learning process. In this book five major concepts unify our discussion of diverse biological topics. We will introduce these concepts here, but they form the underlying logic behind the organization of the entire book. At the beginning of each chapter, in the Learning Path, we use concepts that are derived from these five most fundamental concepts.

The five concepts are consistent with the past organization of this text and with national attempts–such as Vision and Change–to elaborate critical biological concepts. The five fundamental concepts are (1) life is subject to chemical and physical laws; (2) structure determines function; (3) living systems transform energy and matter; (4) living systems depend on information transactions; and (5) evolution explains the unity and diversity of life.

Five Concepts Unify Biology

LEARNING OBJECTIVE 1.5.1 Describe five unifying concepts of biology as a science.

1. Life is subject to chemical and physical laws.

It may seem obvious, but it is important to emphasize that living systems operate according to known chemical and physical principles. For this reason, almost all introductory textbooks, including this one, begin with several sections on chemistry. This is because biological systems are the ultimate application of some very complex chemistry. However, no new chemical or physical laws are found in biology, just the consistent application of familiar chemical principles and laws. For the student, this means that some knowledge of atomic structure, chemical bonding, thermodynamics, kinetics, and many other topics from basic chemistry and physics is crucial for understanding biological systems.

It may seem that some physics and chemistry would only be relevant in the "cell and molecular" sections of the book, but in fact, those principles are applied throughout the book. The movement of water in plants depends on the basic chemistry of water, the kidney is an osmotic machine, energy flow and nutrient cycling in ecosystems are driven by thermodynamic laws, and the cycling of many important elements involves biogeochemical cycles.

2. Structure determines function.

A major unifying theme of biology is the relationship between structure and function. We can think about this relationship in several ways. The proper functioning of molecules, of cells, and of tissues and organs arises from their three-dimensional structure. We can learn about the function of molecules by studying their structure, and we will encounter many examples of this throughout this book. We can also investigate the function of a molecule by experimentally altering its structure, then analyzing the effect on function.

We can also work in the opposite direction: if we know the function of a protein and find a similar protein in a different organism, we can infer that the function will be similar. For example, suppose that we know the structure of a human cell's surface receptor for insulin, the hormone that controls the uptake of glucose. We then find a similar molecule in the membrane of a cell from a very different species, such as a worm. We might conclude that this membrane molecule acts as a receptor for an insulin-like molecule produced by the worm. This would also imply an evolutionary relationship between glucose uptake in worms and in humans.

3. Living systems transform energy and matter.

From single cells to the highest level of biological organization, the biosphere, living systems have a constant need for energy. If we trace this all the way back, the original energy source for the biosphere is the Sun. Without this energy, living systems would not exhibit their characteristic highly organized state. This sounds simple, but it means that the basic nature of life is to constantly transform both energy and matter. We break down "food" molecules for energy, then use this energy to build up other complex molecules.

The energy from the Sun is trapped by photosynthetic organisms, which use this energy to reduce CO_2 and produce organic compounds. The rest of us, who need a constant source of energy and carbon, can oxidize these organic compounds back to CO_2, releasing energy to drive the processes of life. As all of these energy transactions are inefficient, a certain amount of energy is also randomized as heat.

This constant input of energy allows living systems to function far from thermodynamic equilibrium. At equilibrium, you are a pool of amino acids, nucleotides, and other small molecules, and not the complex dynamic system reading this sentence. Nonequilibrium systems also can exhibit the property of self-organization not present in equilibrium systems. Macromolecular complexes, such as the spindle necessary for chromosome separation, can self-organize (figure 1.15). A flock of birds, a school of fish, and the bacteria in a biofilm all also display self-organization, exhibiting properties not seen in the individual parts alone.

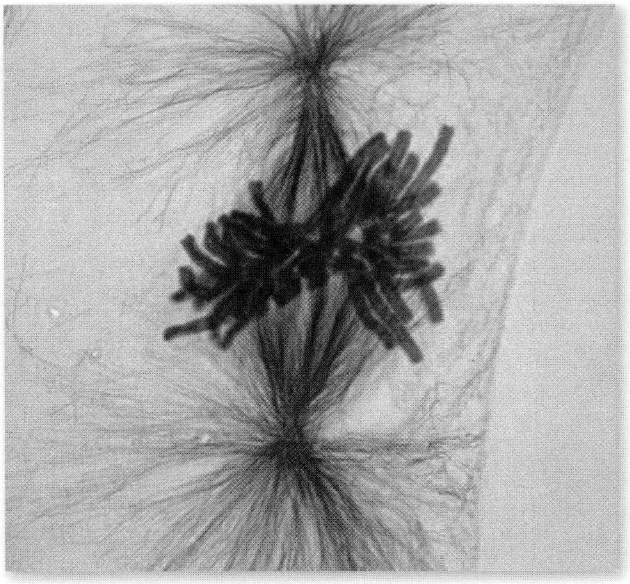

Figure 1.15 The spindle. In this dividing cell, microtubules have organized themselves into a spindle (stained red), pulling each chromosome (stained blue) to the central plane of the dividing cell.

Waheeb Heneen, the Swedish University of Agricultural Sciences

4. Living systems depend on information transactions.

The most obvious form of information in living systems is the genetic information carried in every cell in the form of **deoxyribonucleic acid (DNA).** Each DNA molecule is formed from two long chains of building blocks, called nucleotides, wound around each other (figure 1.16). Four different nucleotides are found in DNA, and the sequence in which they occur encodes the information to make and maintain a cell.

The continuity of life from one generation to the next—heredity—depends on the faithful copying of a cell's DNA into daughter cells. The entire set of DNA instructions that specifies a cell is called its genome. The sequence of the human genome, 3 billion nucleotides long, was decoded in rough-draft form in 2001.

The importance of information goes beyond genomes and their inheritance. Cells are highly complex nanomachines that receive, process, and respond to information. The information stored in DNA is used to direct the synthesis of cellular components, and the particular set of components can differ from cell to cell. The way proteins fold in space is a form of information that is three-dimensional, and interesting properties emerge from the interaction of these shapes in macromolecular complexes. The control of gene expression allows the differentiation of cell types in time and space, leading to changes over developmental time into different tissue types—even though all cells in an organism carry the same genetic information.

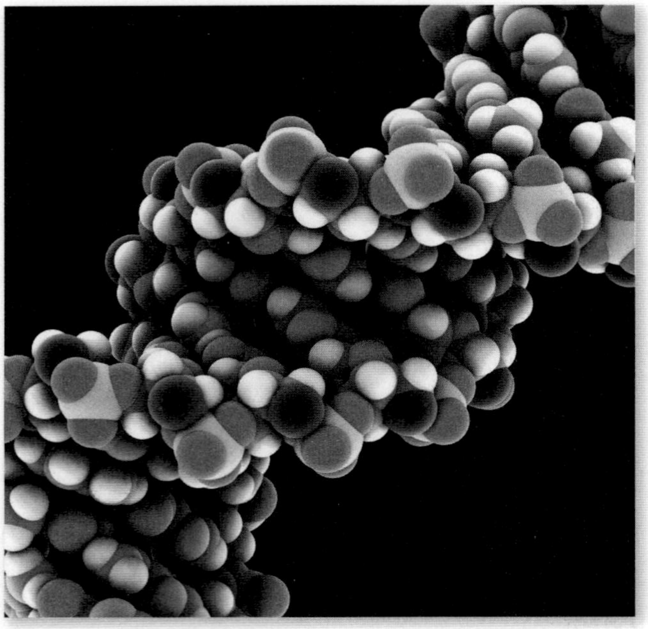

Figure 1.16 A model of DNA. All organisms store their hereditary information as sequences of DNA subunits, much as this textbook stores information as sequences of alphabet letters.
LAGUNA DESIGN/Science Photo Library/Alamy Stock Photo

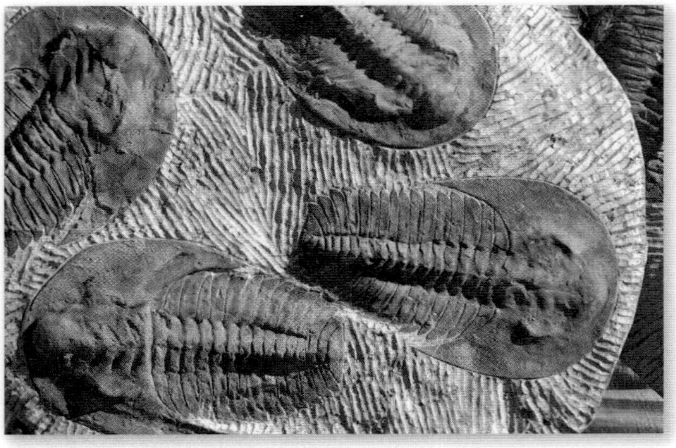

Figure 1.17 Fossil trilobites. Fossil organisms preserved in the Earth's rocks provide a vivid record of evolution.
Dave Sangster/Getty Images

Living systems are able to collect information about the environment, both internal and external, and then respond to this information. As you are reading this page, you are familiar with this process, but it also occurs at the level of cells, in terms of both single-celled organisms and the cells of multicellular organisms. Cells acquire information about their environment, send and receive signals, and respond to all of this information with signal transduction systems that can change cell morphology, behavior, or physiology (the subject of chapter 9).

5. Evolution explains the unity and diversity of life.

Biologists agree that all organisms alive today on Earth descended from a simple cellular organism that arose about 3.5 BYA. Some of the characteristics of that organism have been preserved through evolutionary history into the present. The storage of hereditary information in DNA, for example, is common to all living things.

The retention of conserved characteristics in a long line of descent implies that they are important to the success of the organism. We see this at the visible, morphological level, and also at the level of molecules. A quick glance at figure 1.14 shows us that all mammals have four limbs, and overall anterior to posterior polarity. Regarding molecules, in the same figure it is clear that more related species have more similar genomes. This is also true of many important proteins, such as the oxygen carrier hemoglobin, which is very similar in all organisms that use it to carry oxygen.

The unity of life that we see in certain key characteristics shared by many related life-forms contrasts with the incredible diversity of living things in the varied environments of Earth. The underlying unity of biochemistry and genetics argues that all life evolved from the same origin event. The incredible diversity of life apparent today arose by evolutionary change, much of it visible in the fossil record (figure 1.17).

REVIEW OF CONCEPT 1.5

Understanding biology requires higher-level concepts. We are using five unifying concepts throughout the book: life is subject to chemical and physical laws, structure determines function, living systems transform energy and matter, living systems depend on information transactions, and evolution explains the unity and diversity of life.

- *Why are there no unique laws of chemistry found in biology?*

Does the Presence of One Species Limit the Population Size of Others?

Implicit in Darwin's theory of evolution is the idea that species in nature compete for limited resources. Does this really happen? Some of the best evidence of competition between species comes from experimental field studies, studies conducted not in the laboratory but out in natural populations. By setting up experiments in which two species occur either alone or together, scientists can determine whether the presence of one species has a negative impact on the size of the population of the other species. The experiment discussed here concerns a variety of seed-eating rodents that live in North American deserts. In 1988, researchers set up a series of 50-m × 50-m enclosures to investigate the effect of kangaroo rats on smaller seed-eating rodents. Kangaroo rats were removed from half of the enclosures but not from the other enclosures. The walls of all the enclosures had holes that allowed rodents to come and go, but in plots without kangaroo rats the holes were too small to allow the kangaroo rats to enter.

The graph displays data collected over the course of the next three years as researchers monitored the number of the smaller rodents present in the enclosures. To estimate the population sizes, researchers determined how many small rodents could be captured in a fixed interval. Data were collected for each enclosure immediately after the kangaroo rats were removed in 1988, as well as at three-month intervals thereafter. The graph presents the relative population size—that is, the total number of captures averaged over the number of enclosures. (For example, if a total of 30 rats were captured from three enclosures, the average would be 10 rats.) The data show the number of small rodents for several years after removal of the kangaroo rats.

Rick & Nora Bowers/Alamy Stock Photo

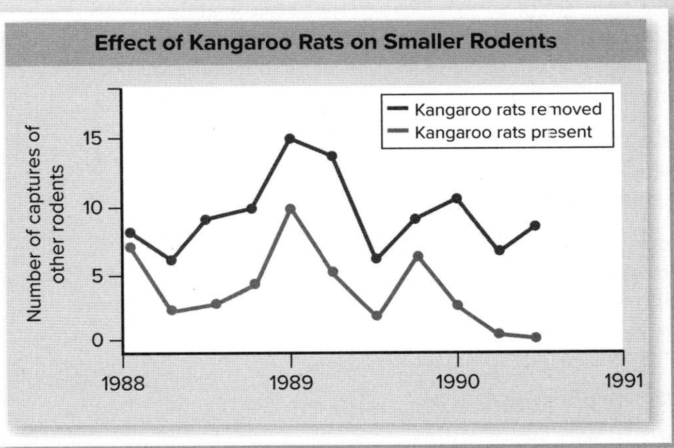

Analysis

1. **Applying Concepts**
   a. ***Variable.*** In the graph, what is the dependent variable?
   b. ***Relative magnitude.*** Which of the two kinds of enclosures maintains the higher population of small rodents? Does it have kangaroo rats, or have they been removed?
2. **Interpreting Data**
   a. What is the average number of small rodents in each of the two plots immediately after kangaroo rats were removed? After one year? After two?
   b. At what point is the difference between the two kinds of enclosures the greatest?
3. **Making Inferences**
   a. What precisely is the observed impact of kangaroo rats on the population size of small rodents?
   b. Examine the magnitude of the difference between the number of small rodents in the two plots. Is there a trend?
4. **Drawing Conclusions** Do these results support the hypothesis that kangaroo rats compete with other small rodents to limit their population sizes?
5. **Further Analysis**
   a. Can you think of any cause other than competition that would explain these results? Suggest an experiment that could eliminate or confirm this alternative.
   b. Do the populations of the two kinds of enclosures change in synchrony (that is, grow and shrink at the same times) over the course of a year? If so, why might this happen? How would you test this hypothesis?

Retracing the Learning Path

CONCEPT 1.1 The Diversity of Life Is Overwhelming

1.1.1 Biological Diversity Can Be Organized by Evolutionary History Life on Earth is very diverse but has traditionally been grouped into six kingdoms based on shared characteristics.

CONCEPT 1.2 Biology Is the Science of Life

1.2.1 Life Defies Simple Definition Although life is difficult to define, living systems have eight characteristics in common: they are capable of movement; are complex and highly ordered; can respond to stimuli; are composed of one or more cells; use energy to accomplish work; can maintain relatively constant internal conditions (homeostasis); can grow, develop, and reproduce; and can transmit genetic information to their offspring, making them capable of evolutionary adaptation to the environment.

1.2.2 Living Systems Show Hierarchical Organization The hierarchical organization of living systems progresses from atoms to complex organisms to the entire biosphere.

1.2.3 Novel Properties Emerge from More Complex Organization As biological systems become more complex, emergent properties arise that could not be predicted from the sum of the parts.

CONCEPT 1.3 Science Is Based on Both Observation and Reasoning

1.3.1 The Scientific Process Involves Observation and Both Deductive and Inductive Reasoning Science is concerned with developing an increasingly accurate description of nature through observation and experimentation. Science uses deductive reasoning, applying general principles to predict specific results, and inductive reasoning, using specific observations to construct general scientific principles.

1.3.2 Hypothesis-Driven Science Makes and Tests Predictions A hypothesis is constructed based on observations, and it must generate experimentally testable predictions. Experiments involve a manipulated variable and a control. Hypotheses are rejected if their predictions cannot be verified by observation or experiment.

1.3.3 Theories Are the Solid Conclusions of Science Reductionism attempts to understand a complex system by breaking it down into its component parts, but parts may act differently when isolated from the larger system. Biologists construct models to explain living systems. A model provides a different way to study a problem, and it may suggest experimental approaches. Scientists use the word *theory* in deductive and inductive ways: as a proposed explanation for a natural phenomenon and as a body of concepts that explains facts in an area of study.

CONCEPT 1.4 The Study of Evolution Is a Good Example of Scientific Inquiry

1.4.1 The Idea of Evolution Existed Prior to Darwin A century before Darwin, naturalists suggested that living things had evolved over the course of Earth's history.

1.4.2 Darwin Gathered Information During the Voyage of the *Beagle* During the voyage of H.M.S. *Beagle,* Darwin had an opportunity to observe worldwide patterns of diversity.

1.4.3 Darwin Proposed Natural Selection as a Mechanism for Evolution Darwin noted that species produce many offspring, of which only a limited number survive and reproduce. He proposed that individuals possessing traits that increase survival and reproductive success become more numerous in populations over time. This is the essence of descent with modification (natural selection). Alfred Russel Wallace independently came to the same conclusions from his own studies.

1.4.4 The Predictions of Darwin's Theory Have Been Well Tested Natural selection has been tested using data from many fields. Among these are the fossil record; comparative anatomy and the study of homologous structures; and molecular data that provide evidence for changes in DNA and proteins over time. Taken together, these findings strongly support evolution by natural selection.

CONCEPT 1.5 A Few Important Concepts Form the Core of Biology

1.5.1 Five Concepts Unify Biology

1. **Life is subject to chemical and physical laws.** All living systems function based on the laws of chemistry and physics.
2. **Structure determines function.** The function of macromolecules is dictated by and dependent on their structure. Similarity of structure and function may indicate an evolutionary relationship.
3. **Living systems transform energy and matter.** Living systems have a constant need for energy, which is ultimately provided by the Sun. The nature of life is to constantly transform energy. We break down food molecules to provide energy to build up complex structures.
4. **Living systems depend on information transactions.** Hereditary information found in the DNA molecule is passed on from one generation to the next. This information is read out to produce proteins, which themselves have information in their structures. Living systems can also acquire information about their environment.
5. **Evolution explains the unity and diversity of life.** The underlying similarities in biochemistry and genetics support the contention that all life evolved from a single source. The diversity found in living systems arises by evolutionary change.

Concept Overview

This Concept Overview diagrams the key concepts that were discussed in this chapter.

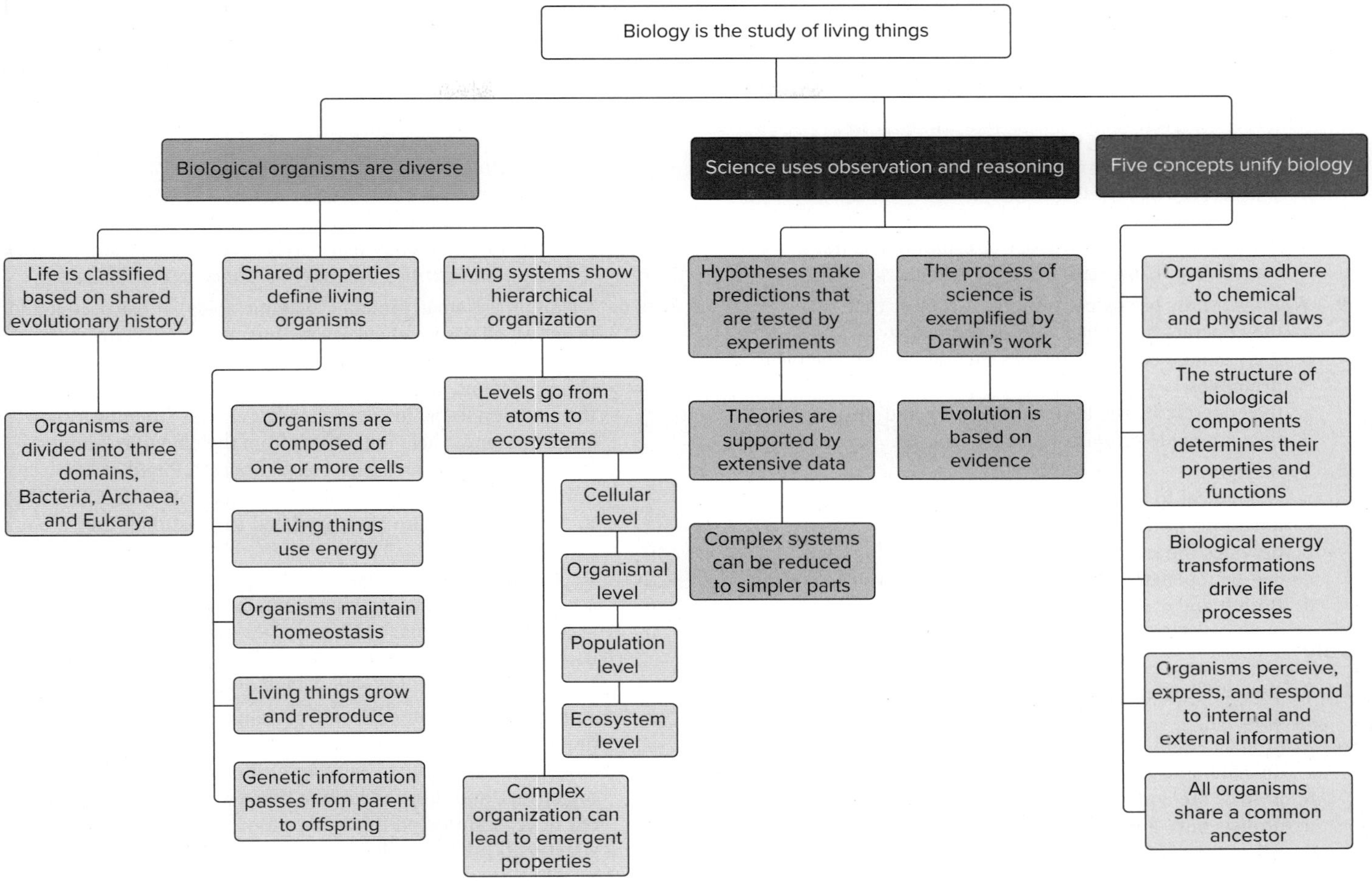

Assessing the Learning Path

Understand

1. Humans and yeast, single-celled organisms with a nucleus, are both members of
   a. Domain Archaea.
   b. Kingdom Animalia.
   c. Domain Eukarya.
   d. None of the above
2. Which of the following is NOT a property of life?
   a. The use of energy that ultimately comes from the Sun
   b. Movement
   c. The maintenance of relatively constant internal conditions in a variable environment
   d. The ability to grow, develop, and reproduce using instructions found in DNA
3. Emergent properties
   a. result from the way components within a hierarchical level interact.
   b. explain why our cells are similar to those of a mountain goat, despite very different body plans.
   c. arise at each level of hierarchical organization.
   d. All of the above
4. The process of inductive reasoning involves
   a. the use of general principles to predict a specific result.
   b. the generation of specific predictions based on a belief system.
   c. the use of specific observations to develop general principles.
   d. the use of general principles to support a hypothesis.
5. A hypothesis in biology is best described as
   a. a possible explanation of an observation.
   b. an observation that supports a theory.
   c. a general principle that explains some aspect of life.
   d. an unchanging statement that correctly predicts some aspect of life.
6. The idea of evolution by natural selection
   a. is an example of how a scientist develops a hypothesis based on previous knowledge and observations.
   b. was developed independently by Charles Darwin and Alfred Russel Wallace.
   c. is supported by modern-day molecular evidence.
   d. All of the above

7. How is the process of natural selection different from that of artificial selection?
   a. Natural selection produces more variation.
   b. Natural selection makes an individual better adapted.
   c. Artificial selection is a result of human intervention.
   d. Artificial selection results in better adaptations.
8. How does the fossil record help support the theory of evolution by natural selection?
   a. It demonstrates that complex organisms have become simplified and more efficient over time.
   b. It provides evidence that organisms have changed over time.
   c. It shows that diversity existed millions of years ago.
   d. It doesn't support the theory of evolution, as no transitional forms have been identified.
9. When we study biological systems, it is clear that
   a. they operate by unique laws of chemistry.
   b. their unique chemistry allows them to violate chemical principles.
   c. they operate by the laws of chemistry and physics.
   d. they are made of unique forms of matter.
10. Your DNA
   a. is identical to the DNA from organisms classified in other kingdoms of life.
   b. provides the instructions for your cells to function.
   c. can be replicated faithfully and passed to your offspring.
   d. Both b and c

Apply

1. Protists and bacteria are grouped into different domains because
   a. protists are multicellular organisms and bacteria are unicellular organisms.
   b. protists are composed of cells and bacteria are not.
   c. protists have a nucleus and bacteria do not.
   d. All of the above
2. A group of giant pandas inhabit a bamboo forest in China. The male members of the group have testes; testes are organs that are part of the reproductive system. The testes are composed of several cell types, including Leydig cells, which secrete testosterone molecules. Which of the following is a correct representation of the hierarchy of biological organization from least complex to most complex?
   a. Testosterone, Leydig cells, testes, reproductive system, panda, panda population
   b. Forest community, panda population, panda, reproductive system, testes, Leydig cells, testosterone
   c. Testosterone, testes, Leydig cells, panda, reproductive system, panda community
   d. Reproductive system, testosterone, Leydig cells, testes, panda, panda population
3. Birds are vertebrate animals that have feathers and reproduce by laying eggs. While on a nature walk, you come across a vertebrate animal that lays eggs but does not have feathers. You conclude that this animal is not a bird. This is an example of
   a. inductive reasoning.
   b. deductive reasoning.
   c. descriptive science.
   d. reductionism.
4. You are conducting an experiment to test the hypothesis that apple trees will produce larger fruit when exposed to lullabies at night. You have 10 experimental trees that are exposed to music at night. Which of the following would make the best control group?
   a. The control group should be exposed to lullabies during the day.
   b. The control group should not be exposed to lullabies at all.
   c. The control group should be exposed to a different type of music, such as rap.
   d. The control group should be exposed to music all the time.
5. In which of the following situations could evolution by natural selection occur?
   a. A population of island finches possesses almost no genetic variability.
   b. Over several generations a population of snails has access to unlimited resources.
   c. Over generations almost every member of a population of howler monkeys is able to produce offspring.
   d. Over generations a population of mountain gorillas with genes for thick body hair survives cold and cloudy weather better than gorillas that have genes for less body hair.
6. Ubiquitin is a small protein (76 amino acids); it has been found in almost all tissues of eukaryotic organisms and is involved in protein degradation in the cell. The amino acid sequence of ubiquitin found in humans has one amino acid difference from the ubiquitin found in *Caenorhabditis elegans* (roundworm) and three amino acid differences from the ubiquitin found in *Saccharomyces cerevisiae* (brewer's yeast). Based on this information, which of the following statements is ACCURATE?
   a. Humans are more closely related to yeast than they are to roundworms.
   b. Yeast, roundworms, and humans all share a common ancestor.
   c. These data tell us nothing about evolutionary history, because molecular evidence relies only on DNA.
   d. Molecular data can tell us little about evolution, because it is usually incompatible with data derived from the fossil record.
7. Aquaporins are water channel proteins that have been found in membranes of prokaryotes, protists, fungi, plants, and animals. These proteins have very similar structure in all these organisms, and they function to allow water to move into and out of cells. Based on this information, which of the following statements are ACCURATE? (Select all that apply.)
   a. All of these organisms are related.
   b. Any alterations in the structure of aquaporins would not affect their function in transporting water.
   c. Water transport is important to life.
   d. The DNA that contains the hereditary information for aquaporins must be conserved.

Synthesize

1. Scientists have recently estimated that there are 10^{21} stars in our universe. If 1 in 10 of the stars that are like our Sun has planets, if 1 in 10,000 of these planets is capable of supporting life, and if 1 of each million life-supporting planets evolves an intelligent life-form, how many planets in the universe support intelligent life? Can you think of a reasonable hypothesis to explain why we haven't heard from anyone out there?
2. Based on Apply question 2, which level of hierarchical organization is the lowest level that carries out all of the activities we associate with life? Explain your answer.
3. The Cape Verde Islands are as far from the coast of Africa as the Galápagos Islands are from the coast of South America, and both island groups have similar climates. Do you think Darwin would have found animals on the Cape Verde Islands to be similar to those on the Galápagos? Explain.
4. Exobiology is the study of life on other planets. In recent years scientists have sent various spacecraft into the galaxy in search of extraterrestrial life. Assuming that all life shares common properties, what should exobiologists be looking for as they explore other worlds?

Fuse/Getty Images

2 The Nature of Molecules and the Properties of Water

Learning Path

2.1 All Matter Is Composed of Atoms

2.2 The Elements in Living Systems Have Low Atomic Masses

2.3 Molecules Are Collections of Atoms Held Together by Chemical Bonds

2.4 The Properties of Water Result from Its Polar Nature

2.5 Water Molecules Can Dissociate into Ions

Concept Overview

This Concept Overview provides a diagram of the overarching concepts that are covered in this chapter. These concepts will be expanded upon in the Concept Overview at the end of the chapter.

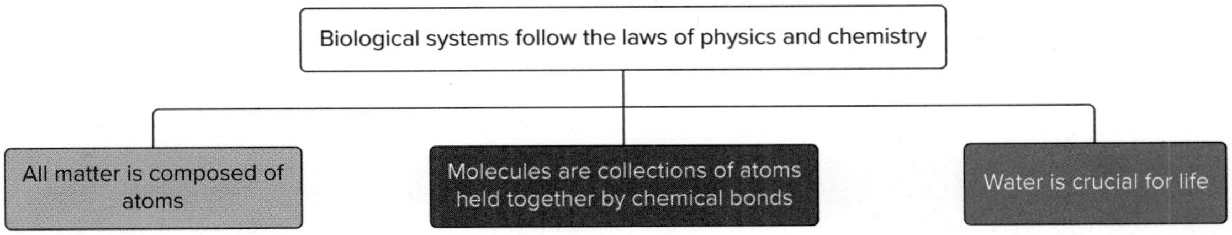

Introduction

Every atom in your body was formed in a star. About 12.5 billion years ago (BYA), an enormous explosion signaled the beginning of the universe. This explosion started a process of chemical evolution, star building, and planetary formation that eventually led to the formation of Earth about 4.5 BYA. Around a billion years later, life began on Earth and started to diversify. To understand the nature of life on Earth, we first need to understand the nature of the star-born substances that form the building blocks of all life. Chemistry is the study of the properties of these substances. To understand how living systems are assembled, we must first understand a little about atomic structure, about how atoms can be linked together by chemical bonds to make molecules, about the ways in which these small molecules are joined together to make larger molecules, and how they interact with their surroundings, until finally we arrive at the structures of cells and then of organisms. Our study of life on Earth, therefore, begins with atoms and how they form molecules. Organisms are chemical machines, and to understand them we must begin with atomic structure.

2.1 All Matter Is Composed of Atoms

Any substance in the universe that has mass and occupies space is defined as *matter.* All matter is composed of extremely small particles called **atoms.** Because of their size, atoms are difficult to study. Objects as small as atoms can be "seen" only indirectly, by using complex technology such as tunneling microscopy (figure 2.1).

Atoms Are Composed of Three Kinds of Subatomic Particles

LEARNING OBJECTIVE 2.1.1 Describe the structure of the Bohr atom.

The Bohr atom has electrons orbiting a central nucleus

We now know a great deal about the complexities of atomic structure, but the simple view put forth in 1913 by the physicist Niels Bohr provides a good starting point for understanding atomic theory. Bohr proposed that every atom possesses an orbiting cloud of tiny subatomic particles called *electrons* whizzing around a core, like the planets of a miniature solar system. At the center of each atom is a small, very dense nucleus formed of two other kinds of subatomic particles: *protons* and *neutrons* (figure 2.2).

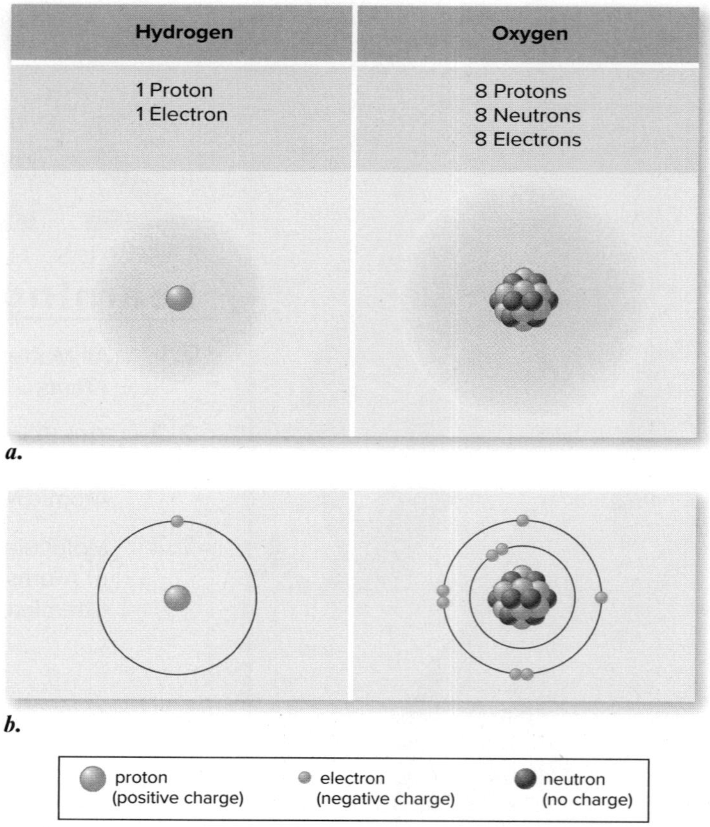

Hydrogen
Oxygen
1 Proton
1 Electron
8 Protons
8 Neutrons
8 Electrons
a.
b.
proton
(positive charge)
electron
(negative charge)
neutron
(no charge)

Figure 2.2 Basic structure of atoms. All atoms have a nucleus consisting of protons and neutrons, except hydrogen, the smallest atom, which usually has only one proton and no neutrons in its nucleus. Oxygen typically has eight protons and eight neutrons in its nucleus. In the simple "Bohr model" of atoms pictured here, electrons spin around the nucleus at a relatively great distance. ***a.*** Atoms are depicted as a nucleus with a cloud of electrons (not shown to scale). ***b.*** The electrons are shown in discrete energy levels. These are described in greater detail in the text.

Atomic number defines elements

Within the nucleus, the cluster of protons and neutrons is held together by a force that works only over short, subatomic distances. Each proton carries a positive (+) charge, and each neutron has no charge. Each electron carries a negative (−) charge. Typically, an atom has one electron for each proton and is, thus, electrically neutral.

Figure 2.1 Scanning tunneling microscope image. The scanning tunneling microscope is a nonoptical way of imaging that allows atoms to be visualized. This image shows a lattice of oxygen atoms (dark blue) on a rhodium crystal (light blue).
Bruker Corporation

Different atoms are defined by their *atomic number,* which is the number of protons. Atoms with the same atomic number (that is, the same number of protons) have the same chemical properties and are said to belong to the same element. Formally speaking, an element is any substance that cannot be broken down to any other substance by ordinary chemical means. The chemical behavior of an atom is due to the number and configuration of electrons, as we will discuss later in this section.

Atomic mass

The terms *mass* and *weight* are often used interchangeably, but they have slightly different meanings. *Mass* refers to the amount of a substance, but *weight* refers to the force gravity exerts on a substance. An object has the same mass whether it is on the Earth or the Moon, but its weight is greater on the Earth, because the

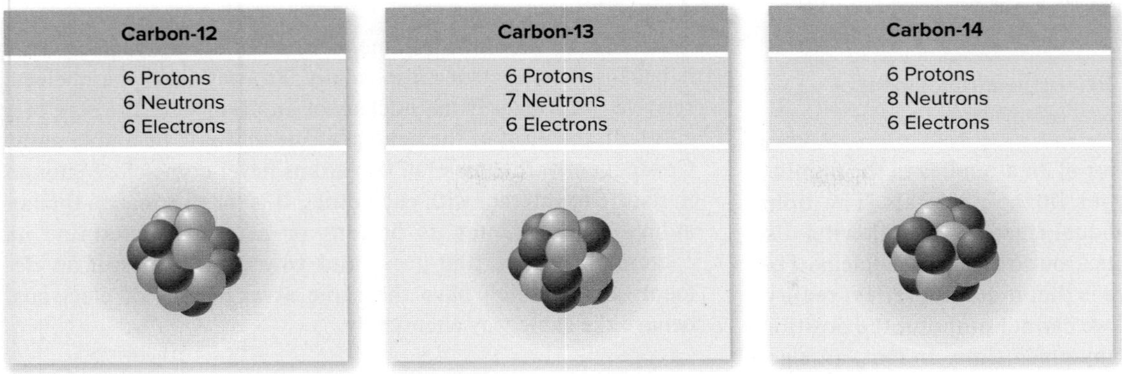
Carbon-12
6 Protons
6 Neutrons
6 Electrons
Carbon-13
6 Protons
7 Neutrons
6 Electrons
Carbon-14
6 Protons
8 Neutrons
6 Electrons

Figure 2.3 The three most abundant isotopes of carbon. Isotopes of a particular element have different numbers of neutrons.

Earth's gravitational force is greater than the Moon's. The *atomic mass* of an atom is equal to the sum of the masses of its protons and neutrons. Atoms that occur naturally on Earth contain from 1 to 94 protons and up to 146 neutrons.

The mass of atoms and subatomic particles is measured in units called *daltons.* To give you an idea of just how small these units are, note that it takes 602 million million billion (6.02×10^{23}) daltons to make 1 gram (g). A proton weighs approximately 1 dalton (actually 1.007 daltons), as does a neutron (1.009 daltons). In contrast, electrons weigh only 1/1840 of a dalton, so they contribute almost nothing to the overall mass of an atom. The total weight of all the electrons in your body is less than that of an eyelash.

Ions

The positive charges in the nucleus of an atom are neutralized, or counterbalanced, by negatively charged electrons, which are located in regions called **orbitals,** which lie at varying distances around the nucleus. Atoms with the same number of protons and electrons are electrically neutral; that is, they have no net charge and are therefore called *neutral atoms.*

Electrons are maintained in their orbitals by their attraction to the positively charged nucleus. Sometimes other forces overcome this attraction, and an atom loses one or more electrons. In other cases, atoms gain additional electrons. Atoms in which the number of electrons does not equal the number of protons are known as *ions,* and they are charged particles. An atom having more protons than electrons has a net positive charge and is called a **cation.** For example, an atom of sodium (Na) that has lost one electron becomes a sodium ion (Na^+), with a charge of +1. An atom having fewer protons than electrons carries a net negative charge and is called an **anion.** A chlorine atom (Cl) that has gained one electron becomes a chloride ion (Cl^-), with a charge of −1.

Isotopes

Although all atoms of an element have the same number of protons, they may not all have the same number of neutrons. Atoms of a single element that possess different numbers of neutrons are called **isotopes** of that element.

Most elements in nature exist as mixtures of different isotopes. Carbon (C), for example, has three isotopes, all containing six protons (figure 2.3). Over 99% of the carbon found in nature exists as an isotope that also contains six neutrons. Because the total mass of this isotope is 12 daltons (6 from protons plus 6 from neutrons), it is referred to as carbon-12 and is symbolized ^{12}C. Most of the rest of the naturally occurring carbon is carbon-13, an isotope with seven neutrons. The rarest carbon isotope is carbon-14, with eight neutrons. Unlike the other two isotopes, carbon-14 is unstable; this means that its nucleus tends to break up into elements with lower atomic numbers. This nuclear breakup, which emits a significant amount of energy, is called *radioactive decay,* and isotopes that decay in this fashion are **radioactive isotopes.** A number of researchers contributed to the early work on radioactive elements, including Marie Curie, the first woman awarded a Nobel prize, and one of only four people to win two awards.

Some radioactive isotopes are more unstable than others, and therefore they decay more readily. For any given isotope, however, the rate of decay is constant. The decay time is usually expressed as the *half-life,* the time it takes for one-half of the atoms in a sample to decay. Carbon-14, for example, often used in the carbon dating of fossils and other organic materials, has a half-life of 5730 years. A sample of carbon containing 1 g of carbon-14 today would contain 0.5 g of carbon-14 after 5730 years, 0.25 g 11,460 years from now, 0.125 g 17,190 years from now, and so on. By determining the ratios of the different isotopes of carbon and other elements in biological samples and in rocks, scientists are able to accurately determine when these materials formed.

Radioactivity has many useful applications in modern biology. Radioactive isotopes are one way to label, or "tag," a specific molecule and then follow its progress, either in a chemical reaction or in living cells and tissue. The downside, however, is that the energetic subatomic particles emitted by radioactive substances have the potential to severely damage living cells, producing genetic mutations and, at high doses, cell death. Consequently, exposure to radiation is carefully controlled and regulated. Scientists who work with radioactivity follow strict handling protocols and wear radiation-sensitive badges to monitor their exposure over time to help ensure a safe level of exposure.

Electrons Determine the Chemical Behavior of Atoms

LEARNING OBJECTIVE 2.1.2 Relate the arrangement of electrons in an atom to its chemical behavior.

The key to the chemical behavior of an atom lies in the number and arrangement of its electrons in their orbitals. The Bohr model of the atom shows individual electrons as following discrete, or distinct, circular orbits around a central nucleus. The trouble with this simple picture is that it doesn't reflect reality. Modern physics indicates that we cannot pinpoint the position of any individual electron at any given time. In fact, an electron could be anywhere, from close to the nucleus to infinitely far away from it.

A particular electron, however, is more likely to be in some areas than in others. An orbital is defined as the area around a nucleus where an electron is most likely to be found. These orbitals represent probability distributions for electrons, that is, regions more likely to contain an electron. Some electron orbitals near the nucleus are spherical (*s*-orbitals); others are dumbbell-shaped (*p*-orbitals; figure 2.4). Still other orbitals, farther away from the nucleus, may have different shapes. Regardless of its shape, no orbital can contain more than two electrons.

Almost all of the volume of an atom is empty space. This is because the electrons are usually far away from the nucleus, relative to its size. If the nucleus of an atom were the size of a golf ball, the orbit of the nearest electron would be a mile away. Consequently, the nuclei of two atoms never come close enough in nature to interact with each other. It is for this reason that an atom's electrons, not its protons or neutrons, determine its chemical behavior, and it explains why the isotopes of an element, all of which have the same arrangement of electrons, behave the same way chemically.

Atoms Contain Discrete Energy Levels

LEARNING OBJECTIVE 2.1.3 Explain how energy is quantized in atoms.

Because electrons are attracted to the positively charged nucleus, it takes work to keep them in their orbitals, just as it takes work to hold a grapefruit in your hand against the pull of gravity. The formal definition of energy is the ability to do work.

A grapefruit held above the ground is said to possess *potential energy* because of its position. If you release it, the grapefruit falls, and its potential energy is reduced. If you carried the grapefruit to the top of a building, you would increase its potential energy. Electrons also have a potential energy that is related to their position. To oppose the attraction of the nucleus and move the electron to a more distant orbital requires an input of energy, which results in an electron with greater potential energy. The chlorophyll that makes plants green captures energy in this way from light during photosynthesis. As you'll learn in chapter 8, light energy excites electrons in the chlorophyll molecule. Moving an electron closer to the nucleus has the opposite effect: energy is released,

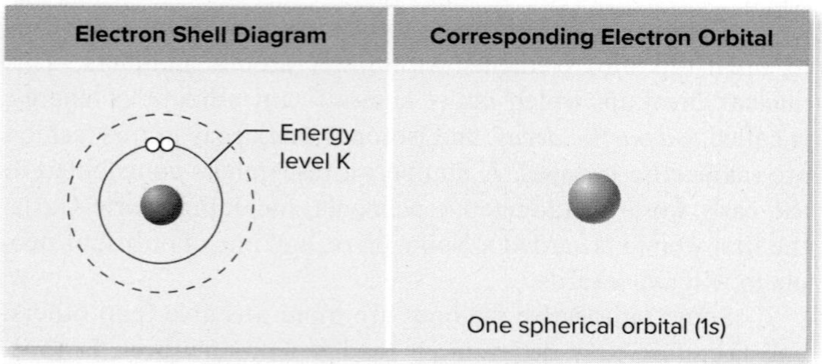

a.

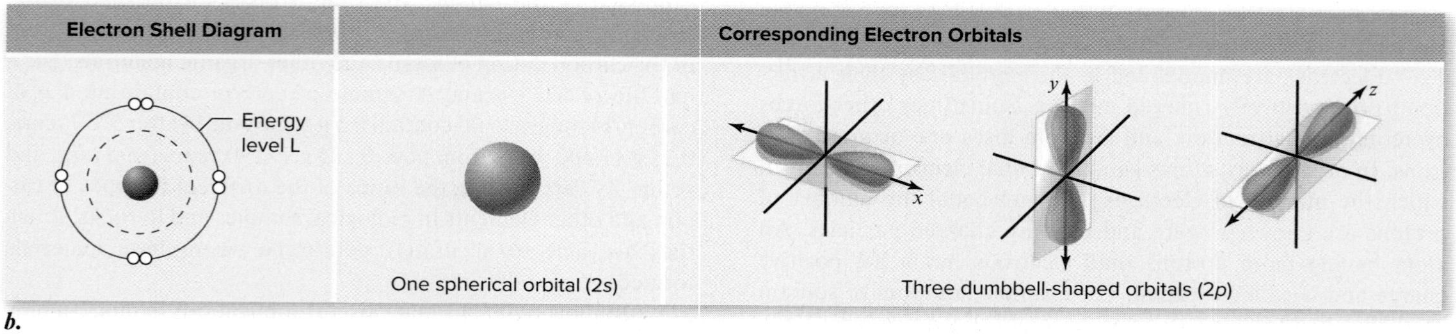

b.

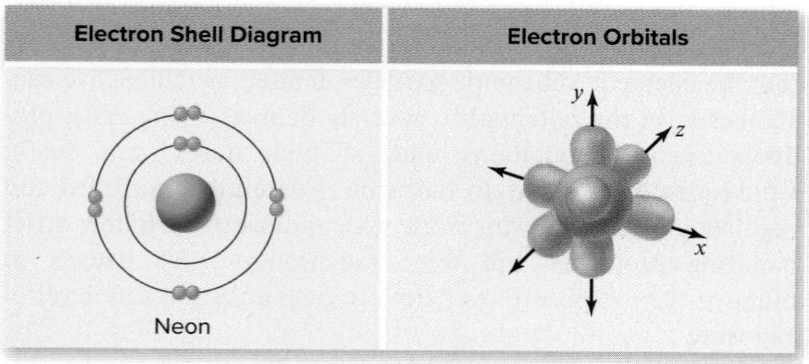

c.

Figure 2.4 Electron orbitals. ***a.*** The lowest energy level, or electron shell—the one nearest the nucleus—is level K. It is occupied by a single *s*-orbital, referred to as 1*s*. ***b.*** The next highest energy level, L, is occupied by four orbitals: one *s*-orbital (referred to as the 2*s* orbital) and three *p*-orbitals (each referred to as a 2*p* orbital). Each orbital holds two paired electrons with opposite spin. Thus, the K level is populated by two electrons, and the L level is populated by a total of eight electrons. ***c.*** The neon atom shown has the L and K energy levels completely filled with electrons and is thus unreactive.

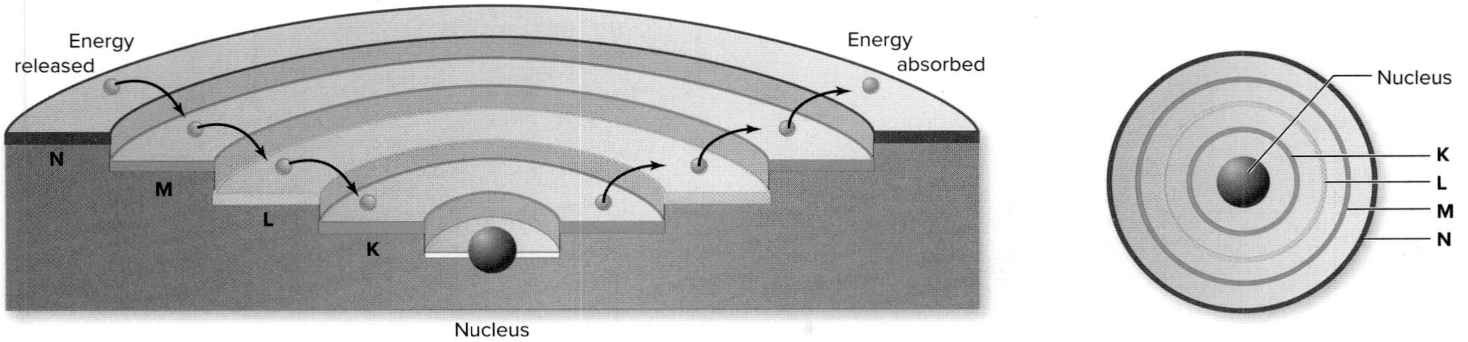

Figure 2.5 Atomic energy levels. Electrons have energy of position. When an atom absorbs energy, an electron moves to a higher energy level, farther from the nucleus. When an electron falls to lower energy levels, closer to the nucleus, energy is released. The first two energy levels are the same as shown in figure 2.4.

usually as radiant energy (heat or light), and the electron ends up with less potential energy (figure 2.5).

One of the initially surprising aspects of atomic structure is that electrons within the atom have discrete **energy levels.** These discrete levels correspond to quanta (singular, *quantum*), which means specific amounts of energy. To use the grapefruit analogy, it is as though a grapefruit could be raised only to particular floors of a building. Every atom exhibits a ladder of potential energy values, a discrete set of orbitals at particular energetic "distances" from the nucleus.

Because the amount of energy an electron possesses is related to its distance from the nucleus, electrons that are the same distance from the nucleus have the same energy, even if they occupy different orbitals. Such electrons are said to occupy the same energy level. The energy levels are denoted with letters K, L, M, and so on (figure 2.5). Be careful not to confuse energy levels, which are drawn as rings to indicate an electron's *energy,* with orbitals, which have a variety of three-dimensional shapes and indicate an electron's most likely *location.* Electron orbitals are arranged so that as they are filled, this fills each energy level in successive order. This filling of orbitals and energy levels is what is responsible for the chemical reactivity of elements.

During some chemical reactions, electrons are transferred from one atom to another. In such reactions, the loss of an electron is called **oxidation,** and the gain of an electron is called **reduction.**

Notice that when an electron is transferred in this way, it keeps its energy of position. In organisms, chemical energy is stored in high-energy electrons that are transferred from one atom to another in reactions involving oxidation and reduction (described in chapter 7). When the processes of oxidation and reduction are coupled, which often happens, one atom or molecule is oxidized while another is reduced in the same reaction. We call these combinations *redox reactions.*

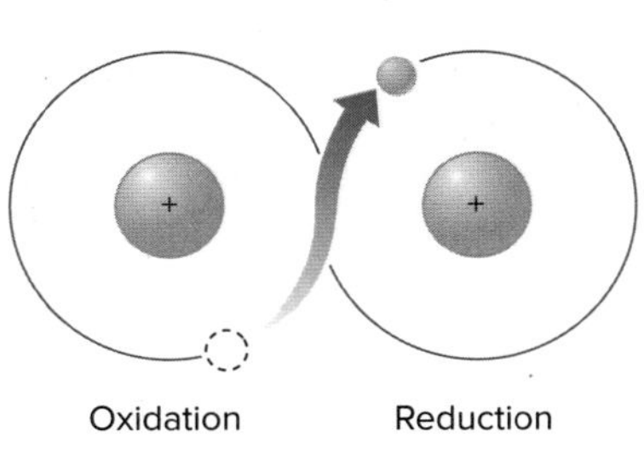

REVIEW OF CONCEPT 2.1

An atom consists of a nucleus of protons and neutrons surrounded by a cloud of electrons. Elements are characterized by their atomic number, which is the number of protons in an atom. Atoms of an element with different numbers of neutrons are called isotopes. The chemical behavior of an element is determined by electrons, which are located in orbitals. As electron orbitals are filled, successive energy levels are filled. No orbital can contain more than two electrons, but multiple orbitals may have the same energy level, and thus contain electrons with the same energy.

- *If the number of protons exceeds the number of electrons, is the charge on the atom positive or negative?*

2.2 The Elements in Living Systems Have Low Atomic Masses

Ninety-two elements occur naturally, each with a different number of protons and a different arrangement of electrons. Understanding the relationship between atomic number and chemical behavior was one of the great discoveries of science, as fundamental to chemistry as evolution is to biology.

The Periodic Table Reflects the Electronic Structure of Atoms

LEARNING OBJECTIVE 2.2.1 Relate the periodic table to the chemical reactivity of different elements.

In the mid-19th century there was much discussion of the chemical properties of the elements. In 1863 there were 56 known elements, with a new one being discovered practically every year. Patterns linking their chemical properties to atomic number were proposed by several investigators, but none succeeded in predicting any of the new elements. Success came in 1869, when chemist Dmitri Mendeleev arranged the known elements in an atomic number table in a new way and discovered one of the great generalizations of science: the elements exhibit a pattern

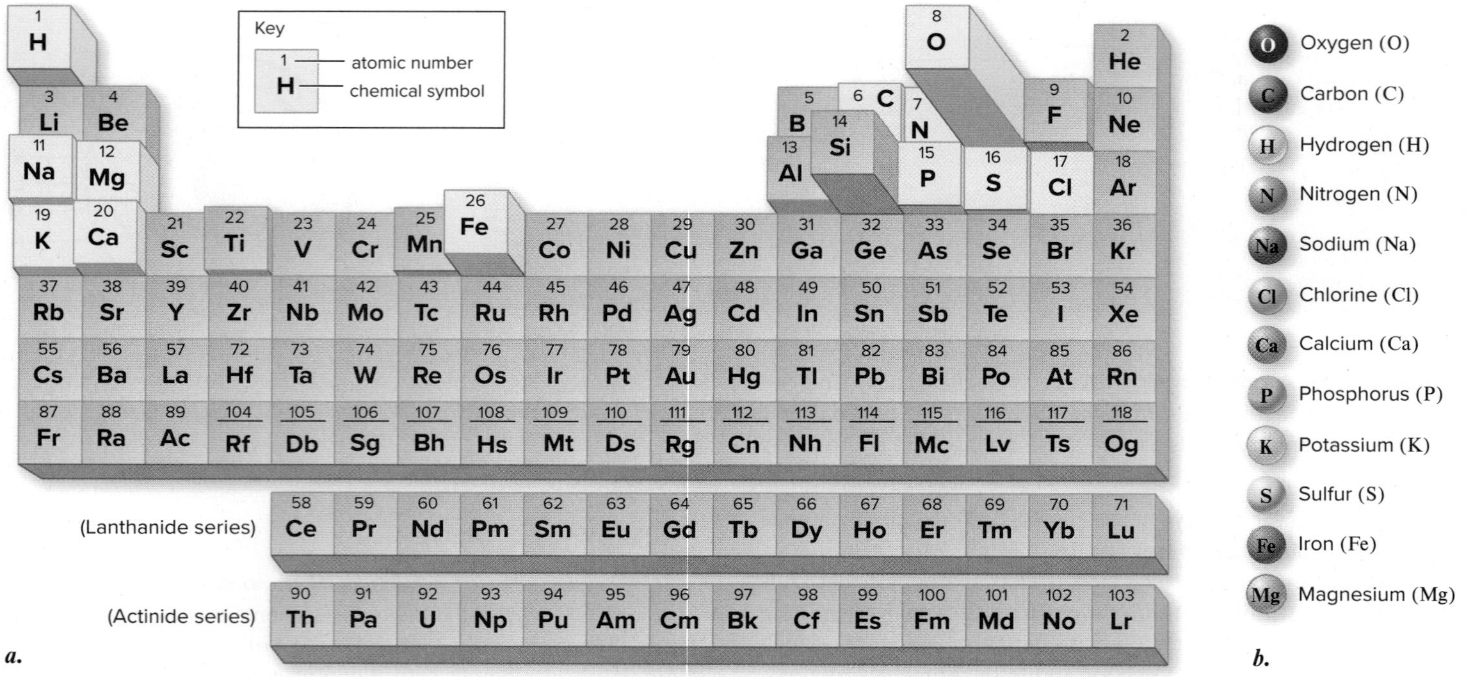

Figure 2.6 Periodic table of elements. ***a.*** The height of each element indicates its frequency in the Earth's crust. Elements shaded in green are found in living systems in more than trace amounts. ***b.*** The 12 common elements found in living systems are listed in descending order of abundance, shown with the colors used in figures throughout the text.

of chemical properties that repeats itself in groups of eight. This periodically repeating pattern lent the table its name: the **periodic table of elements** (figure 2.6). Using this table, Mendeleev successfully predicted the discovery of the elements germanium, gallium, and scandium and accurately described what their chemical properties would be.

The eight-element periodicity that Mendeleev found is based on the interactions of the electrons in the outermost energy level of the different elements. These electrons are called **valence electrons,** and their interactions are the basis for the elements' differing chemical properties. For most of the atoms important to life, the outermost energy level can contain no more than eight electrons; the chemical behavior of an element reflects how many of the eight positions are filled. Elements possessing all eight electrons in their outer energy level (two for helium) are *inert,* or nonreactive. These elements, which include helium (He), neon (Ne), and argon (Ar), are termed the *noble gases.* In sharp contrast, elements with seven electrons (one fewer than the maximum number of eight) in their outer energy level, such as fluorine (F), chlorine (Cl), and bromine (Br), are highly reactive. They tend to gain the extra electron needed to fill the energy level. Elements with only one electron in their outer energy level, such as lithium (Li), sodium (Na), and potassium (K), are also very reactive. They tend to lose the single electron in their outer level.

The octet rule predicts chemical behavior

Mendeleev's periodic table leads to a useful generalization, the **octet rule,** or *rule of eight* (Latin *octo,* "eight"): Atoms tend to establish completely full outer energy levels. For the main group elements of the periodic table, the rule of eight is accomplished by one filled *s*-orbital and three filled *p*-orbitals (figure 2.7). The exception to this is He, in the first row, which needs only two

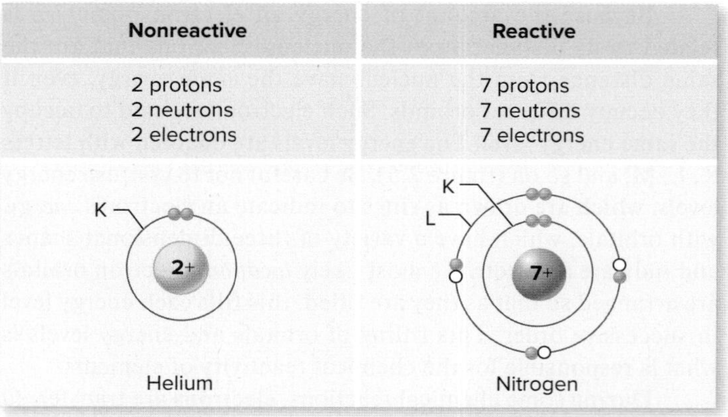

Figure 2.7 Electron energy levels for helium and nitrogen. Green balls represent electrons; the pink or blue ball represents the nucleus with number of protons indicated by number of + charges. Note that the helium atom has a filled K shell and is thus unreactive, whereas the nitrogen atom has five electrons in the L shell, three of which are unpaired, making it reactive.

electrons to fill the 1*s* orbital. Most chemical behavior of biological interest can be predicted quite accurately from this simple rule, combined with the tendency of atoms to balance positive and negative charges. For instance, you read earlier that sodium ion (Na^+) has lost an electron in its outer energy level and chloride ion (Cl^-) has gained an extra electron in its outer energy level.

Only 12 elements are common in organisms

Of the 94 naturally occurring elements on Earth, only 12 (C, H, O, N, P, S, Na, K, Ca, Mg, Fe, and Cl) are found in living systems in more than trace amounts (0.01% or higher). These elements all have low atomic masses, with atomic numbers less than 27. It is also worth noting that Na, K, Ca, Fe, and Cl are all found as ions in cells.

The distribution of elements in living systems is by no means accidental. The most common elements inside organisms are not the elements that are the most abundant in the Earth's crust. For example, silicon, aluminum, and iron constitute 39.2% of the Earth's crust, but they exist in only trace amounts in the human body.

Of the 12 elements found in significant amounts in living systems, the first four elements (oxygen, carbon, hydrogen, and nitrogen) constitute 96.3% of the weight of your body. The majority of molecules that make up your cells are compounds of carbon, or organic compounds. These organic compounds contain primarily these four elements (CHON), plus P and S. But the most common molecule in your body is water, which explains why oxygen, not carbon, is the most abundant element in your body. Water is so important that we devote two sections of this chapter (sections 2.4 and 2.5) to it.

The 6 elements C, H, O, N, P, and S are the building blocks for the important organic molecules that make up living systems, as we will discuss in chapter 3. The remaining elements are all found as ions in living systems and have varied roles, including in animal nervous systems, in regulating water balance, and as functional parts of enzymes. We will encounter examples of all of these roles and more over the course of our explorations of biology.

Some trace elements, such as copper (Cu), zinc (Zn), molybdenum (Mo), and iodine (I), are critical, even though they are present in only tiny amounts. Copper, molybdenum, and zinc are key components of many enzymes, and iodine is an essential component of thyroid hormone. Iodine is a good example of how deficiency in a trace element can have significant effects on human biology (figure 2.8).

SCIENTIFIC THINKING

Observation: *In the early 20th century in the U.S., many areas had a high incidence of enlarged thyroid, or goiter, in both adults and children. The incidence in young girls was twice that in young boys.*

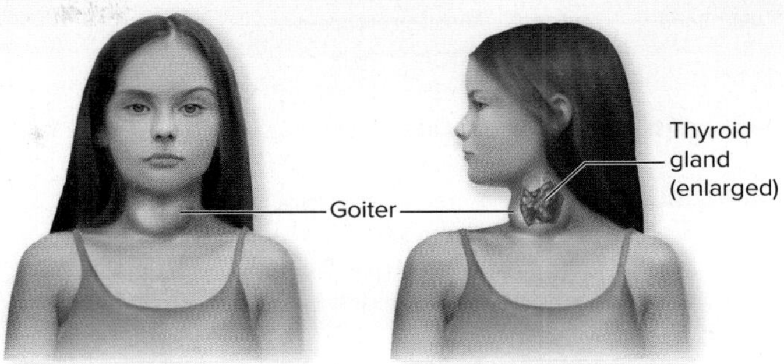

Hypothesis: *Goiter is caused by an iodine deficiency.*
Test: *Schoolchildren were enrolled in a trial using NaI to supplement their diets. Girls whose parents gave consent were put into the treatment group, and those whose parents did not give consent formed the control group.*
Result: *The table shows the results for treatment with 200 mg of NaI for 10 days.*

	Treatment	*Control*
Initially normal (n)	*(908)*	*(1257)*
unchanged %	*99.8*	*72.4*
increased %	*0.2*	*27.6*
Initially enlarged (n)	*(1282)*	*(1048)*
unchanged %	*39.5*	*72*
increased %	*0.2*	*14*
decreased %	*60.3*	*13.8*

Conclusion: *The data show that iodine supplementation can both prevent goiters and reduce the size of preexisting goiters.*
Further Experiments: *This was a very early clinical trial. What changes would make this a better experiment?*

Figure 2.8 Iodine deficiency can lead to enlarged thyroid.

REVIEW OF CONCEPT 2.2

The periodic table shows the elements in terms of atomic number and repeating chemical properties. Only 12 elements are found in significant amounts in living organisms: C, H, O, N, P, S, Na, K, Ca, Mg, Fe, and Cl.

- *Why are the noble gases more stable than other elements in the periodic table?*

2.3 Molecules Are Collections of Atoms Held Together by Chemical Bonds

A **molecule** is a group of atoms held together in a stable association by energy. When a molecule contains atoms of more than one element, it is called a compound. There are two main ways to form compounds: by transferring electrons to form ionic compounds and by sharing electrons to form covalent compounds. In an ionic compound, the ions are held together by electrostatic forces, also called **ionic bonds.** The term used for the chemical bonds formed by atoms sharing electrons is **covalent bonds,** as the electrons are spending time in the valence shell of each atom. In general, when we say chemical bond, we really mean covalent bond. Another interaction between molecules important in biological systems is the hydrogen bond, which is also based on electrostatic forces and discussed in section 2.3.3. Lastly, an even weaker kind of interaction called a *van der Waals* attraction is important in biological systems (table 2.1).

Ionic Bonds Form Crystals

LEARNING OBJECTIVE 2.3.1 Explain how ionic bonds promote crystal formation.

We saw in section 2.1 that atoms can form ions by gaining or losing electrons. Ions with opposite charge are attracted to each other by electrostatic force, forming ionic bonds. If you have ever

TABLE 2.1	Bonds and Interactions	
Name	**Basis of Interaction**	**Strength**
Covalent bond	Sharing of electron pairs	Strong
Ionic bond	Attraction of opposite charges	↑
Hydrogen bond	Sharing of H atom	↕
Hydrophobic interaction	Forcing of hydrophobic portions of molecules together in presence of polar substances	↓
Van der Waals attraction	Weak attractions between atoms due to oppositely polarized electron clouds	Weak

played with two magnets, you have observed that the positive pole of one magnet is attracted to the negative pole, and repelled by the positive pole, of the other magnet.

A familiar compound formed by ionic bonds is everyday table salt (NaCl). The sodium and chlorine atoms of table salt are ions. In figure 2.9, you see that a sodium atom gives up the sole electron in its outermost shell (the shell underneath has eight), while a chlorine atom gains an electron to complete its outermost shell. Recall from section 2.2 that an atom is more stable when its outermost electron shell is filled (with two electrons in the innermost shell or eight electrons in shells that are farther out from the nucleus). This reaction, in which sodium is oxidized and chlorine is reduced, produces stable ions with opposite charge.

When metallic sodium and gaseous chlorine are put in contact, they react swiftly and explosively, as the sodium atoms donate electrons to chlorine to form Na^+ and Cl^- ions. These ions are then attracted to each other, and can form NaCl. The electrical attractive force holding NaCl together, however, is not directed specifically between individual Na^+ and Cl^- ions, and no individual sodium chloride molecules form. Instead, the force exists between any one ion and all neighboring ions of the opposite charge. The ions aggregate in a crystal matrix with a precise geometry, which we know as a salt crystal. Because of the lack of directionality, ionic bonds do not hold together most atoms in biological molecules, but many ions are important in biological systems.

Although the ionic bonds holding the crystal together are relatively strong, when a salt such as NaCl is placed in water, the electrical attraction of the water molecules, for reasons we will discuss in section 2.4, disrupts the forces holding the ions in their crystal matrix, causing the salt to dissolve into a roughly equal mixture of free Na^+ and Cl^- ions. There are a number of important ions in biological systems, such as Ca^{2+}, which is involved in cell signaling, and K^+ and Na^+, which are involved in the conduction of nerve impulses.

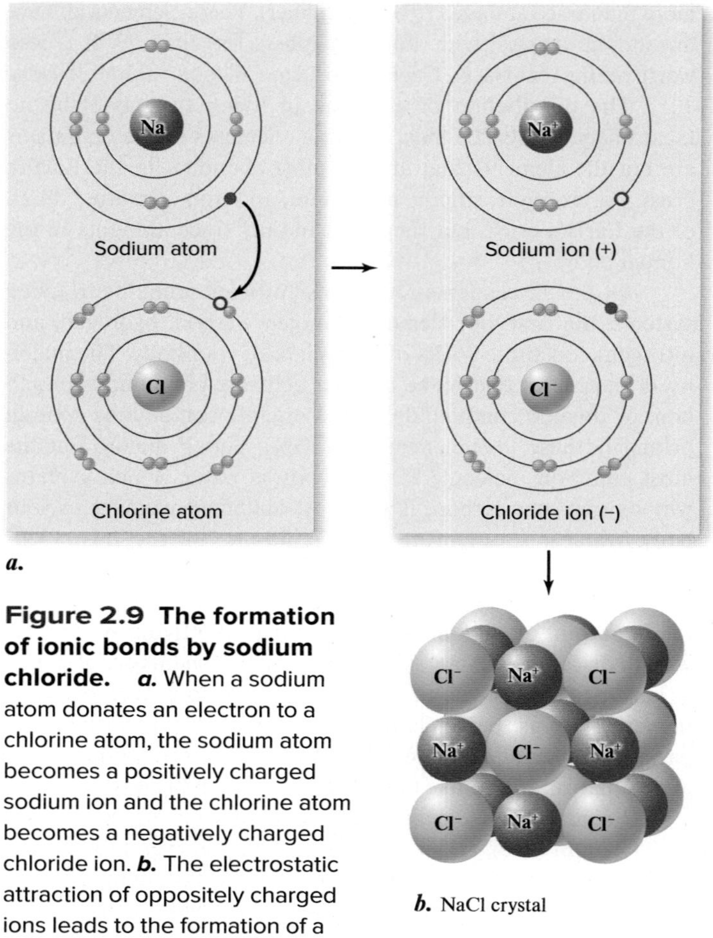

Figure 2.9 The formation of ionic bonds by sodium chloride. ***a.*** When a sodium atom donates an electron to a chlorine atom, the sodium atom becomes a positively charged sodium ion and the chlorine atom becomes a negatively charged chloride ion. ***b.*** The electrostatic attraction of oppositely charged ions leads to the formation of a lattice of Na^+ and Cl^-.

Covalent Bonds Build Stable Molecules

LEARNING OBJECTIVE 2.3.2 Explain how covalent bonds hold atoms together.

Strong, stable chemical bonds called **covalent bonds** form when two atoms share one or more pairs of valence electrons. Consider gaseous hydrogen (H_2) as an example. Each hydrogen atom has an unpaired electron and an unfilled outer energy level; for these reasons, the hydrogen atom is unstable. However, when two hydrogen atoms are in close association, each atom's electron is attracted to both nuclei. In effect, the nuclei are able to share their electrons. The result is a diatomic (two-atom) molecule of hydrogen gas.

The molecule formed by the two hydrogen atoms is stable for three reasons:

1. **It has no net charge.** The diatomic molecule formed as a result of this sharing of electrons is not charged, because it still contains two protons and two electrons.
2. **The octet rule is satisfied.** Each of the two hydrogen atoms can be considered to have two orbiting electrons in its outer energy level. This state satisfies the octet rule, because each shared electron is included in the outer energy level of both atoms.

3. **It has no unpaired electrons.** The bond between the two atoms also pairs the two free electrons.

Unlike ionic bonds, covalent bonds are highly directional, forming between two individual atoms, and so give rise to true, discrete molecules.

The strength of covalent bonds

The strength of a covalent bond depends on the number of shared electrons. Thus, *double bonds,* which satisfy the octet rule by allowing two atoms to share two pairs of electrons, are stronger than *single bonds,* in which only one electron pair is shared. In practical terms, more energy is required to break a double bond than a single bond. The strongest covalent bonds are *triple bonds,* such as those that link the two nitrogen atoms of nitrogen gas molecules (N_2).

Molecules with several covalent bonds

A vast number of biological compounds are composed of more than two atoms. An atom that requires two, three, or four additional electrons to fill its outer energy level completely may acquire them by sharing its electrons with two or more other atoms.

For example, the carbon atom (C) contains six electrons, four of which are in its outer energy level and are unpaired. To satisfy the octet rule, a carbon atom must form four covalent bonds. As four covalent bonds may form in many ways, carbon atoms are found in many different kinds of molecules, such as CO_2 (carbon dioxide), CH_4 (methane), and C_2H_5OH (ethanol).

Polar and nonpolar covalent bonds

Atoms differ in their affinity for electrons, a property called **electronegativity.** In general, electronegativity increases left to right across a row of the periodic table and decreases down a column. Thus, the elements in the upper-right corner have the highest electronegativity.

For bonds between identical atoms—for example, between two hydrogen or two oxygen atoms—the affinity for electrons is obviously the same and the electrons are equally shared. Such bonds are termed **nonpolar.**

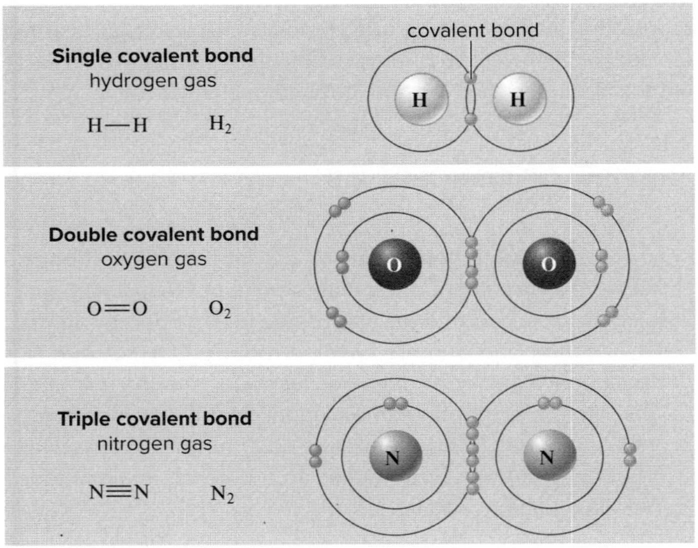

covalent bond
Single covalent bond
hydrogen gas
H—H
H_2
H
H
Double covalent bond
oxygen gas
O═O
O_2
O
O
Triple covalent bond
nitrogen gas
N≡N
N_2
N
N

TABLE 2.2 Relative Electronegativities of Some Important Atoms

Atom	Electronegativity
O	3.5
N	3.0
C	2.5
H	2.1

For atoms that differ greatly in electronegativity, electrons are not shared equally. The shared electrons are more likely to be closer to the atom with greater electronegativity. Thus, although the molecule is still electrically neutral (same number of protons as electrons), the distribution of charge is not uniform, resulting in regions of partial negative charge near the more electronegative atom and regions of partial positive charge near the less electronegative atom. Such bonds are termed **polar covalent bonds,** and the molecules are called polar molecules. The partial charge observed in a polar covalent bond is relatively small—far less than the unit charge of an ion. We can predict polarity of bonds by knowing the relative electronegativity of their atoms (table 2.2). Notice that although C and H differ slightly in electronegativity, this small difference is negligible, and C—H bonds are considered nonpolar.

Hydrogen Bonds Can Form Between Polar Molecules

LEARNING OBJECTIVE 2.3.3 Predict which kinds of molecules will form hydrogen bonds with each other.

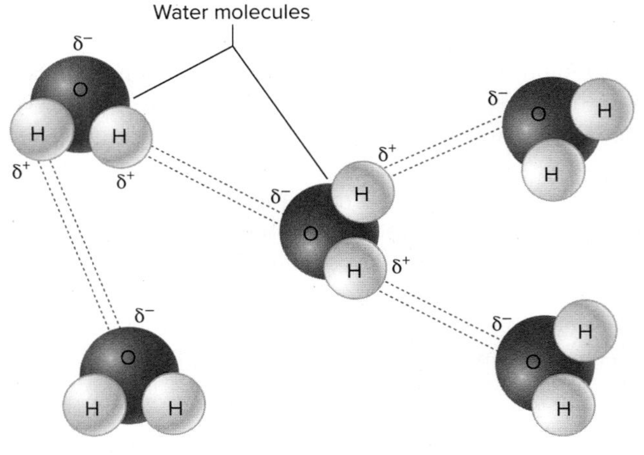

Water molecules
δ^-
δ^+
O
H

Polar molecules have an unequal distribution of electric charge—that is, a partially positive "end" and a partially negative "end." This allows polar molecules to form weak chemical associations with each other. In water, two hydrogen atoms are bound to the

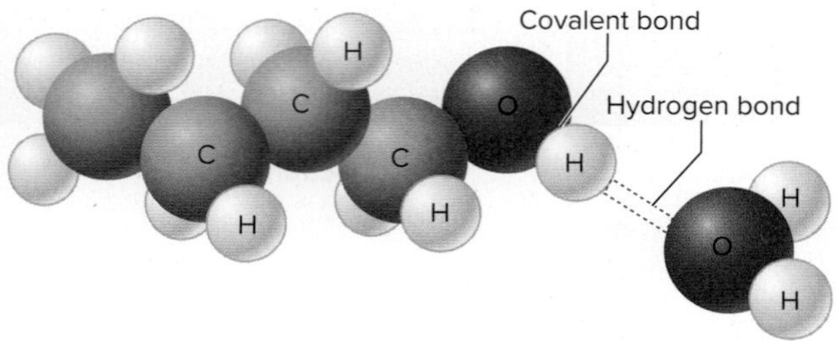

much more electronegative oxygen. The hydrogen atoms will have a partial positive charge (δ^+) and the oxygen atoms a partial negative charge (δ^-), leading to weak interactions we call **hydrogen bonds** (shown as dotted lines). Thus each water molecule is interacting via hydrogen bonding with other adjacent water molecules. These are not true chemical bonds, but the cumulative effect of hydrogen bonding is the major determinant of many of the properties of water, which we will explore in section 2.4.

It is important to note that this is not limited to water, but can occur whenever hydrogen is bound to a more electronegative atom, usually oxygen (O) or nitrogen (N). Hydrogen bonds can occur between different molecules, as we saw for water, but also between different parts of a large molecule.

Key properties of hydrogen bonds

Two key properties of hydrogen bonds cause them to play an important role in the molecules found in organisms. First, they are weak and so are not effective over long distances like more powerful covalent and ionic bonds. Hydrogen bonds are too weak to form stable molecules by themselves. Instead, they act like Velcro, forming a tight bond by the additive effects of many weak interactions. Second, hydrogen bonds are highly directional. In chapter 3, we will discuss the role of hydrogen bonding in maintaining the structures of large biological molecules such as proteins and DNA.

Van der Waals Attractions Draw Surfaces Together

LEARNING OBJECTIVE 2.3.4 Distinguish between a chemical bond and van der Waals attractions.

Another important kind of weak chemical attraction is a nondirectional attractive force called **van der Waals forces** (or van der Waals attractions). These chemical forces come into play only when two atoms are very close to one another. The attraction is very weak, and it disappears if the atoms move even a little apart. It becomes significant when numerous atoms in one molecule simultaneously come close to numerous atoms of another molecule—that is, when the shapes of the molecules match precisely. For example, this interaction is important when antibodies in your blood recognize the shape of an invading virus as foreign.

Chemical Reactions Alter Bonds

LEARNING OBJECTIVE 2.3.5 Identify three factors that influence which chemical reactions occur within cells.

The formation and breaking of chemical bonds are termed *chemical reactions.* All chemical reactions involve the shifting of atoms from one molecule or ionic compound to another, without any change in the number or identity of the atoms. For convenience, we refer to the original molecules before the reaction starts as *reactants,* and the molecules resulting from the chemical reaction as *products.* For example:

$$6H_2O + 6CO_2 \rightarrow C_2H_{12}O_6 + 6O_2$$
$$\textit{reactants} \quad \rightarrow \quad \textit{products}$$

You may recognize this reaction as a simplified form of the photosynthesis reaction, in which water and carbon dioxide are combined to produce glucose and oxygen. Most animal life ultimately depends on this reaction, discussed in detail in chapter 8.

The extent to which chemical reactions occur is influenced by three important factors:

1. **Temperature.** Heating the reactants increases the rate of a reaction, because the reactants collide with one another more often. (Care must be taken that the temperature is not so high that it destroys the molecules.)
2. **Concentration of reactants and products.** Reactions proceed more quickly when more reactants are available, allowing more frequent collisions. An accumulation of products typically slows the reaction and in reversible reactions may speed the reaction in the reverse direction.
3. **Catalysts.** A catalyst is a substance that increases the rate of a reaction. It doesn't alter the reaction's equilibrium between reactants and products, but it does shorten the time needed to reach equilibrium, often dramatically. In living systems these catalysts, called **enzymes,** catalyze almost every chemical reaction.

Many reactions in nature are reversible. This means that the products may themselves be reactants, allowing the reaction to proceed in reverse. We can write the preceding reaction in the reverse order:

$$C_6H_{12}O_6 + 6O_2 \rightarrow 6H_2O + 6CO_2$$
$$\textit{reactants} \quad \rightarrow \quad \textit{products}$$

This reaction is a simplified version of oxidation by cellular respiration, in which glucose is broken down into water and carbon dioxide in the presence of oxygen. Virtually all organisms carry out forms of glucose oxidation, covered in chapter 7.

REVIEW OF CONCEPT 2.3

An ionic bond is an attraction between ions of opposite charge in an ionic compound. A covalent bond is formed when two atoms share one or more pairs of electrons. Complex biological compounds are formed in large part by atoms that can form one or more covalent bonds: C, H, O, and N. A polar covalent bond involves unequal sharing of electrons. Nonpolar bonds exhibit equal sharing of electrons. Hydrogen bonds form when H in a polar bond interacts with an O or N in a polar bond either on a different molecule or within the same molecule.

- *How is a polar covalent bond different from an ionic bond?*

2.4 The Properties of Water Result from Its Polar Nature

Of all the common molecules, only water exists as a liquid at the relatively low temperatures that prevail on the Earth's surface. Three-fourths of the Earth is covered by liquid water. When life was beginning, water provided a medium in which other molecules could move around and interact, without being held in place by strong covalent or ionic bonds. Life evolved in water for 2 billion years before spreading to land. And even today, life is inextricably tied to water. About two-thirds of any organism's body is composed of water, and all organisms require a water-rich environment, either inside or outside them, for growth and reproduction. It is no accident that tropical rainforests are bursting with life, while dry deserts appear almost lifeless except when water becomes temporarily plentiful, such as after a rainstorm.

Water's Structure Facilitates Hydrogen Bonding

LEARNING OBJECTIVE 2.4.1 Explain how the structure of water leads to hydrogen bond formation.

Water has a simple molecular structure, consisting of an oxygen atom bound to two hydrogen atoms by two single covalent bonds (figure 2.10). The resulting molecule is stable: it satisfies the octet rule, has no unpaired electrons, and carries no net electric charge.

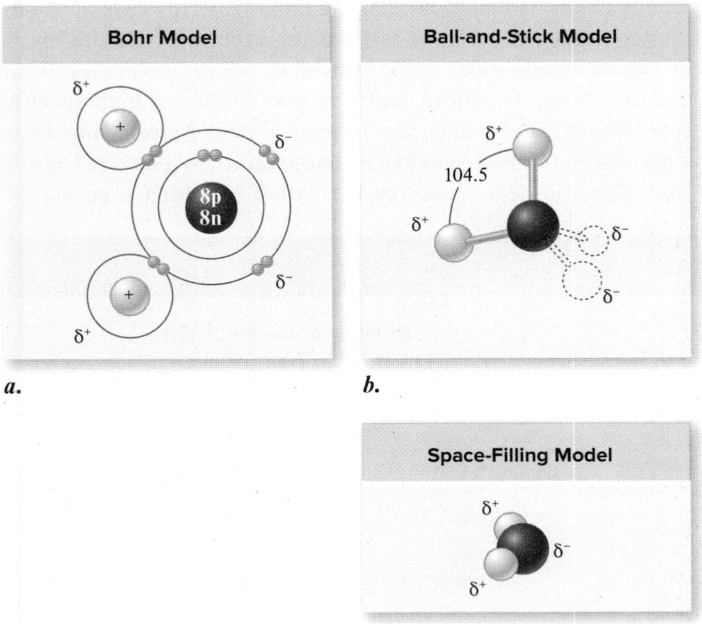
Bohr Model
δ+
δ−
8p
8n
δ−
δ+
a.
Ball-and-Stick Model
δ+
104.5
δ+
δ−
δ−
b.
Space-Filling Model
δ+
δ−
δ+
c.

Figure 2.10 Water has a simple molecular structure. ***a.*** Each water molecule is composed of one oxygen atom and two hydrogen atoms. The oxygen atom shares one electron with each hydrogen atom. ***b.*** The greater electronegativity of the oxygen atom makes the water molecule polar: water carries two partial negative charges (δ^-) near the oxygen atom and two partial positive charges (δ^+), one on each hydrogen atom. ***c.*** Space-filling model shows what the molecule would look like if it were visible.

As we saw in section 2.3.3, the key chemical property of water is its ability to form weak chemical associations, called *hydrogen bonds*. These bonds form between the partially negative O atoms and the partially positive H atoms of two water molecules. Although these bonds have only 5 to 10% of the strength of covalent bonds, they are important to DNA and protein structure and thus are responsible for much of the chemical organization of living systems.

The electronegativity of O is much greater than that of H (refer to table 2.2), and so the bonds between these atoms are highly polar. *The polarity of water underlies water's chemistry and the chemistry of life.*

If we consider the shape of a water molecule, the two polar covalent bonds have a partial charge at each end: δ^- at the oxygen end and δ^+ at the hydrogen end. The most stable arrangement of these charges is a *tetrahedron* (*a pyramid with a triangle as its base*), in which the two negative and two positive charges are approximately equidistant from one another. The oxygen atom lies at the center of the tetrahedron, the hydrogen atoms occupy two of the apexes (corners), and the partial negative charges occupy the other two apexes (figure 2.10*b*). In water, the partial negative charges occupy more space than the partial positive regions, so the oxygen–hydrogen bond angle is slightly compressed.

Water Molecules Are Both Cohesive and Adhesive

LEARNING OBJECTIVE 2.4.2 Distinguish adhesion from cohesion.

The polarity of water allows water molecules to be attracted to one another; that is, water is *cohesive*. The oxygen end of each water molecule, which is δ^-, is attracted to the hydrogen end, which is δ^+, of other molecules. The attraction produces hydrogen bonds among water molecules (figure 2.11). Each hydrogen bond is individually very weak and transient, lasting on average only a hundred-billionth (10^{-11}) of a second. The cumulative effects of large numbers of these bonds, however, can be enormous. Water forms an abundance of hydrogen bonds, which are responsible for many of water's important physical properties (table 2.3).

Water's cohesion is responsible for its being a liquid, not a gas, at moderate temperatures. The cohesion of liquid water is also responsible for its **surface tension.** Some small insects can walk on water (figure 2.12) because at the air–water interface all the surface water molecules are hydrogen-bonded to molecules below them.

The polarity of water causes it to be attracted to other polar or charged molecules as well. This attraction for other polar substances is called *adhesion.* Water adheres to any substance with which it can form hydrogen bonds. This property explains why substances containing polar molecules get "wet" when they are immersed in water but those that are composed of nonpolar molecules (such as oils) do not.

These properties of cohesion and adhesion together are responsible for the phenomenon of capillary action. This is the movement of water up a narrow tube against the force of gravity, or water being drawn into very small openings in a porous medium. In the case of the glass tube, the water molecules are attracted to the polar surface of the glass and pulled upward until this upward force is balanced by gravity. The narrower the tube, the greater the surface area of water in contact with the glass relative to the weight

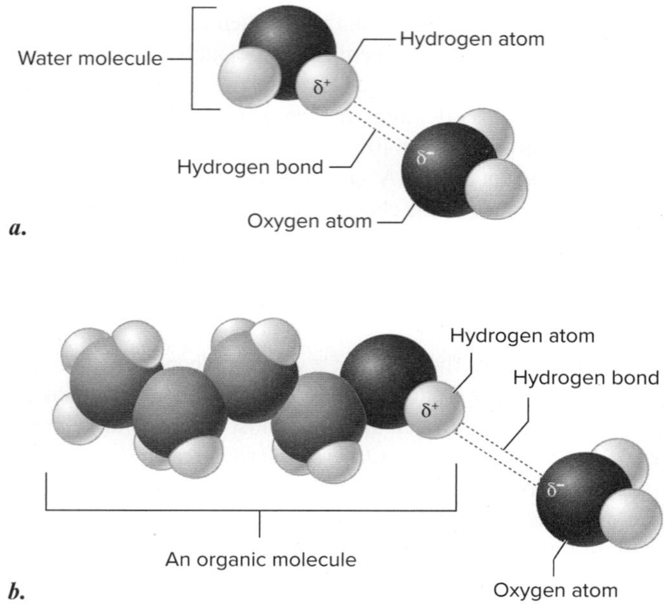

Figure 2.11 Structure of a hydrogen bond. ***a.*** Hydrogen bond between two water molecules. ***b.*** Hydrogen bond between an organic molecule (*n*-butanol) and water. H in *n*-butanol forms a hydrogen bond with oxygen in water. This kind of hydrogen bond is possible anytime H is bound to a more electronegative atom (table 2.2).

Figure 2.12 Cohesion. Some insects, such as this water strider, literally walk on water. Because the surface tension of the water is greater than the force of one foot, the strider glides atop the surface of the water rather than sinking. The high surface tension of water is due to hydrogen bonding between water molecules.

Hermann Eisenbeiss/Science Source

of the column. So water will move higher in a narrower tube (figure 2.13). Another familiar example of capillary action is the absorption of water by a paper towel, and we will encounter this phenomenon in many contexts throughout the book.

Water Has High Specific Heat

LEARNING OBJECTIVE 2.4.3 Explain why water heats up so slowly.

Water moderates temperature through two properties: its high specific heat and its high heat of vaporization. As we will see, water also has the unusual property of being less dense in its solid form, ice, than as a liquid. Water also acts as a solvent for polar molecules and exerts an organizing effect on nonpolar molecules. All these properties result from its polar nature.

The temperature of any substance is a measure of how rapidly its individual molecules are moving. In the case of water, a large input of thermal energy is required to break the many hydrogen bonds that keep individual water molecules from moving about. Therefore, water is said to have a high **specific heat,** which is defined as the amount of heat 1 g of a substance must absorb or lose to change its temperature by 1 degree Celsius (°C). Specific heat measures the extent to which a substance

TABLE 2.3 The Properties of Water

Property	Explanation	Example of Benefit to Life
Cohesion/Adhesion	Hydrogen bonds cause water molecules to be attracted to other polar or charged species.	Leaves pull water upward from the roots; seeds swell and germinate.
High specific heat	Hydrogen bonds absorb heat when they break and release heat when they form, minimizing temperature changes.	Water stabilizes the temperature of organisms and the environment.
High heat of vaporization	Many hydrogen bonds must be broken for water to evaporate.	Evaporation of water cools body surfaces.
Lower density of ice	Water molecules in an ice crystal are spaced relatively far apart because of hydrogen bonding.	Because ice is less dense than water, lakes do not freeze solid, allowing fish and other life in lakes to survive the winter.
Solubility	Polar water molecules are attracted to ions and polar compounds, making these compounds soluble.	Many kinds of molecules can move freely in cells, permitting a diverse array of chemical reactions.
Hydrophobic exclusion	Water repels hydrophobic compounds, forcing them to associate together.	Biological membranes have a bilayer structure with hydrophobic interior.

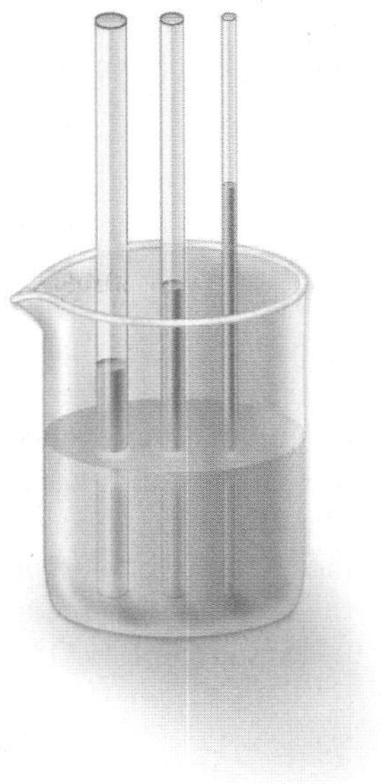

Figure 2.13 Adhesion. Capillary action causes the water within a narrow tube to rise above the surrounding water level; the adhesion of the water to the glass surface, which draws water upward, is stronger than the force of gravity, which tends to pull it down. The narrower the tube, the greater the surface area available for adhesion for a given volume of water, and the higher the water rises in the tube.

resists changing its temperature when it absorbs or loses heat. Because a polar substance tends to form hydrogen bonds, the more polar it is, the higher its specific heat. The specific heat of water (1 cal/g/°C) is twice that of most carbon compounds and nine times that of iron. Only ammonia, which is more polar than water and forms very strong hydrogen bonds, has a higher specific heat than water (1.23 cal/g/°C). Still, only 20% of the hydrogen bonds are broken as water heats from 0° to 100°C.

Because of its high specific heat, water heats up more slowly than almost any other compound and holds its temperature longer. Because organisms have a high water content, water's high specific heat allows them to maintain a relatively constant internal temperature. The heat generated by the chemical reactions inside cells would destroy the cells if not for the absorption of this heat by the water within them.

Water Has a High Heat of Vaporization

LEARNING OBJECTIVE 2.4.4 Explain why sweating cools.

The **heat of vaporization** is defined as the amount of energy required to change 1 g of a substance from a liquid to a gas. A considerable amount of heat energy (586 cal) is required to accomplish this change in water. As water changes from a liquid to a gas, it requires energy (in the form of heat) to break its many hydrogen bonds. Absorbing this energy lowers the temperature. It is for this reason that the evaporation of water from a surface cools that surface. Many organisms dispose of excess body heat by evaporative cooling—for example, through sweating in humans and many other vertebrates.

Solid Water Is Less Dense Than Liquid Water

LEARNING OBJECTIVE 2.4.5 Explain why ice floats.

At low temperatures, water molecules are locked into a crystal-like lattice of hydrogen bonds, forming solid ice. Interestingly, ice is less dense than liquid water, because the hydrogen bonds in ice space the water molecules relatively far apart. This unusual feature enables icebergs to float. If water did not have this property, nearly all bodies of water would be ice, with only the shallow surface melting every year. The buoyancy of ice is important ecologically, because it means bodies of water freeze from the top down and not the bottom up. Because ice floats on the surface of lakes in the winter and the water beneath the ice remains liquid, fish and other animals keep from freezing, allowing them to survive winter in cold climates.

Water Is a Good Solvent for Polar Molecules

LEARNING OBJECTIVE 2.4.6 Explain why salt dissolves in water.

Water molecules gather closely around any substance that bears an electric charge, whether that substance carries a full charge (ion) or a charge separation (polar molecule). For example, sucrose (table sugar) is composed of molecules that contain polar hydroxyl (OH) groups. A sugar crystal dissolves rapidly in water, because water molecules can form hydrogen bonds with individual hydroxyl groups of the sucrose molecules. Therefore, sucrose is said to be *soluble* in water. Water is termed the *solvent,* and sugar is called the *solute.* Every time a sucrose molecule dissociates, or breaks away, from a solid sugar crystal, water molecules surround it in a cloud, forming a *hydration shell* that prevents it from associating with other sucrose molecules. Hydration shells also form around ions such as Na^+ and Cl^- (figure 2.14).

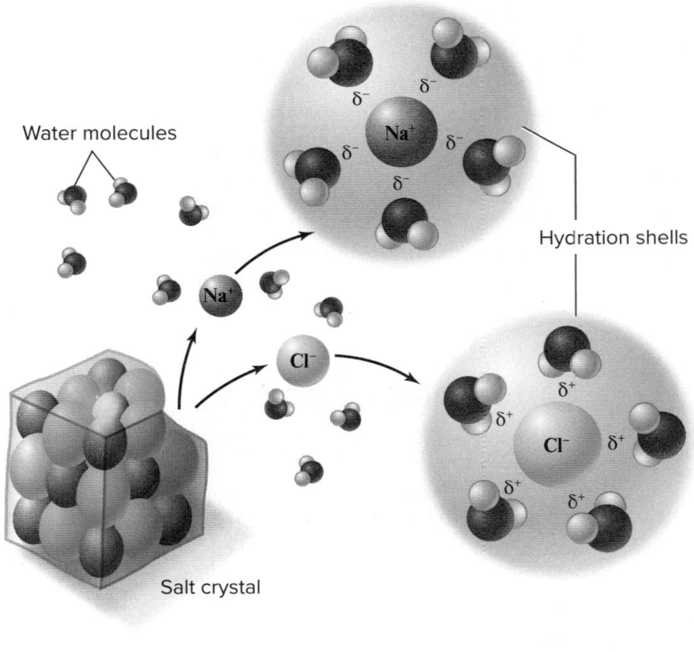

Figure 2.14 Why salt dissolves in water. When a crystal of table salt dissolves in water, individual Na^+ and Cl^- ions break away from the salt lattice and become surrounded by polar water molecules whose partial charges are attracted to the charges of the ions. Surrounded by hydration shells, Na^+ and Cl^- never re-enter the salt lattice.

Water Organizes Nonpolar Molecules

LEARNING OBJECTIVE 2.4.7 Explain why oil will not dissolve in water.

Water molecules always tend to form the maximum possible number of hydrogen bonds. When nonpolar molecules such as oils, which do not form hydrogen bonds, are placed in water, the water molecules act to exclude them. The nonpolar molecules aggregate, or clump together, thus minimizing their disruption of the hydrogen bonding of water. In effect, they shrink from contact with water, and for this reason they are referred to as **hydrophobic** (Greek *hydros,* "water," and *phobos,* "fearing"). In contrast, polar molecules, which readily form hydrogen bonds with water, are said to be **hydrophilic** ("water-loving").

The tendency of nonpolar molecules to aggregate in water is known as **hydrophobic exclusion.** By forcing the hydrophobic portions of molecules together, water causes these molecules to assume particular shapes. This property has a major impact on the structure of proteins, DNA, and biological membranes. In fact, the interaction of nonpolar molecules with water acts as a critical organizing influence in all living systems.

REVIEW OF CONCEPT 2.4

Because of its polar covalent bonds, water can form hydrogen bonds with itself and with other polar molecules. Hydrogen bonding is responsible for water's cohesion, the force that holds water molecules together, and its adhesion, which is its ability to "stick" to other polar molecules. Capillary action results from both of these properties. Water has a high specific heat, so it does not change temperature rapidly, which helps living systems maintain a near-constant temperature. Water's high heat of vaporization allows cooling by evaporation. Solid water is less dense than liquid water, because the hydrogen bonds space the molecules farther apart. Polar molecules are soluble in a water solution, but water tends to exclude nonpolar molecules.

- *If water were made of C and H instead of H and O, would it still be cohesive and adhesive? Explain why or why not.*

2.5 Water Molecules Can Dissociate into Ions

The covalent bond holding one hydrogen atom to the oxygen atom in a water molecule can spontaneously be broken. In pure water at 25°C, only 1 out of every 550 million water molecules undergoes this process. When it happens, a proton (hydrogen atom nucleus) dissociates from the molecule. Because the dissociated proton lacks the negatively charged electron it was sharing, its positive charge is no longer counterbalanced, and it becomes a hydrogen ion, H^+. The rest of the dissociated water molecule, which has retained the shared electron from the covalent bond, is negatively charged and forms a hydroxide ion, OH^-. This process of spontaneous ion formation is called *ionization:*

$$\underset{\textit{water}}{H_2O} \rightarrow \underset{\textit{hydroxide ion}}{OH^-} + \underset{\textit{hydrogen ion (proton)}}{H^+}$$

At 25°C, 1 liter (L) of water contains one ten-millionth (or 10^{-7}) mole of H^+ ions. A **mole** (mol) is defined as the weight of a substance in grams that corresponds to the atomic masses of all of the atoms in a molecule of that substance. In the case of H^+, the atomic mass is 1, and a mole of H^+ ions would weigh 1 g. One mole of any substance always contains 6.02×10^{23} molecules of the substance. Therefore, the **molar concentration** of hydrogen ions in pure water, represented as $[H^+]$, is 10^{-7} mol/L. (In reality, the H^+ usually associates with another water molecule to form a hydronium ion, H_3O^+.)

The pH Scale Measures Hydrogen Ion Concentration

LEARNING OBJECTIVE 2.5.1 Calculate the pH of a solution based on the molar concentration of H^+.

The concentration of hydrogen ions, and concurrently of hydroxide ions, in a solution is described by the terms *acidity* and *basicity,* respectively. Pure water, having an $[H^+]$ of 10^{-7} mol/L, is considered to be neutral, that is, neither acidic nor basic. Recall that for every H^+ ion formed when water dissociates, an OH^- ion is also formed, meaning that the dissociation of water produces H^+ and OH^- in equal amounts.

The *pH scale* (figure 2.15) is a more convenient way to express the hydrogen ion concentration of a solution. This scale

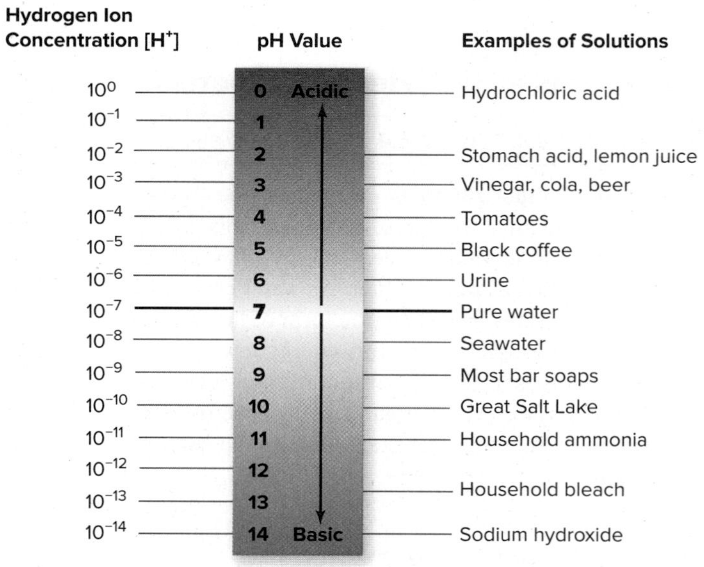

Figure 2.15 The pH scale. The pH value of a solution indicates its concentration of hydrogen ions. Solutions with a pH less than 7 are acidic, whereas those with a pH greater than 7 are basic. The scale is logarithmic, which means that a pH change of 1 represents a 10-fold change in the concentration of hydrogen ions.

defines *pH,* which stands for "power of hydrogen," as the negative logarithm of the hydrogen ion concentration in the solution:

$$pH = -\log [H^+]$$

Because the logarithm of the hydrogen ion concentration is simply the exponent of the molar concentration of H^+, the pH equals the exponent times −1. For water, therefore, an $[H^+]$ of 10^{-7} mol/L corresponds to a pH value of 7. This is the neutral point—a balance between H^+ and OH^-—on the pH scale. This balance occurs because the dissociation of water produces equal amounts of H^+ and OH^-.

Note that because the pH scale is *logarithmic,* a difference of 1 on the scale represents a 10-fold change in $[H^+]$. A solution with a pH of 4 therefore has 10 times the $[H^+]$ of a solution with a pH of 5 and 100 times the $[H^+]$ of a solution with a pH of 6.

Acids

Any substance that dissociates in water to increase the $[H^+]$ (and lower the pH) is called an **acid.** The stronger an acid is, the more hydrogen ions it produces and the lower its pH. For example, hydrochloric acid (HCl), which is abundant in your stomach, ionizes completely in water. A 1 molar solution of HCl dissociates to form 10^0 mol/L of H^+, giving the solution a pH of 0. The pH of champagne, which bubbles because of the carbonic acid dissolved in it, is about 2.

Bases

A substance that combines with H^+ when dissolved in water, and thus lowers the $[H^+]$, is called a **base.** Therefore, basic (or alkaline) solutions have pH values above 7. Very strong bases, such as sodium hydroxide (NaOH), have pH values of 12 or more. Many common cleaning substances, such as ammonia and bleach, accomplish their action because of their high pH.

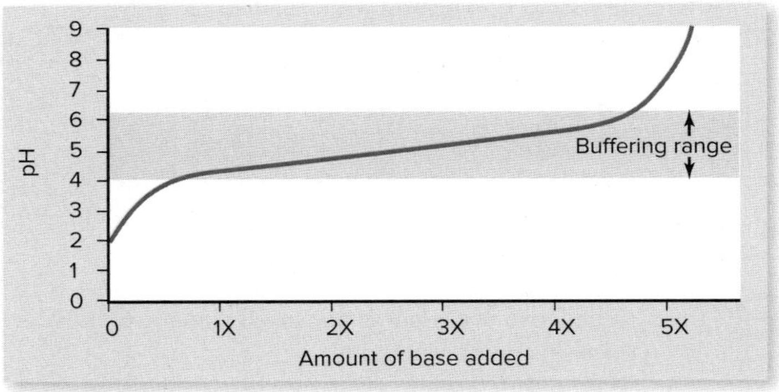

Figure 2.16 Buffers minimize changes in pH. Adding a base to a solution neutralizes some of the acid present and so raises the pH. Thus, as the curve moves to the right, reflecting more and more base, it also rises to higher pH values. A buffer makes the curve rise or fall very slowly over a portion of the pH scale, called the "buffering range" of that buffer.

Buffers help stabilize pH

The pH inside almost all living cells, and in the fluid surrounding cells in multicellular organisms, is fairly close to neutral, 7. Most of the enzymes in living systems are extremely sensitive to pH. Often even a small change in pH will alter their shape, thereby disrupting their activities. For this reason, it is important that a cell maintain a constant pH level.

But the chemical reactions of life constantly produce acids and bases within cells. Furthermore, many animals eat substances that are acidic or basic. Cola drinks, for example, are moderately strong (although dilute) acidic solutions. Despite such variations in the concentrations of H^+ and OH^-, the pH of an organism is kept at a relatively constant level by buffers (figure 2.16).

A **buffer** is a substance that resists changes in pH. Buffers act by releasing hydrogen ions when a base is added and absorbing hydrogen ions when acid is added, with the overall effect of keeping $[H^+]$ relatively constant.

Within organisms, most buffers consist of pairs of substances, one an acid and the other a base. The key buffer in human blood is an acid–base pair consisting of carbonic acid (acid) and bicarbonate (base). These two substances interact in a pair of reversible reactions. First, carbon dioxide (CO_2) and H_2O join to form carbonic acid (H_2CO_3), which in a second reaction dissociates to yield bicarbonate ion (HCO_3^-) and H^+.

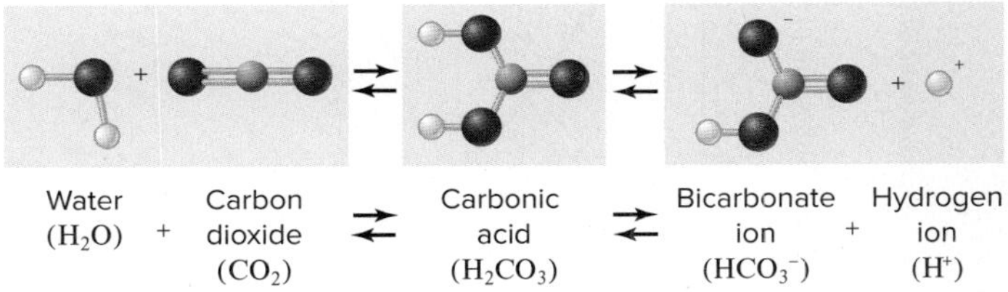

If some acid or other substance adds H^+ to the blood, the HCO_3^- acts as a base and removes the excess H^+ by forming H_2CO_3. Similarly, if a basic substance removes H^+ from the blood, H_2CO_3 dissociates, releasing more H^+ into the blood. The forward and reverse reactions that interconvert H_2CO_3 and HCO_3^- thus stabilize the blood's pH.

The reaction of carbon dioxide and water to form carbonic acid is a crucial one, because it permits carbon, essential to life, to enter water from the air. The Earth's oceans are rich in carbon because of the reaction of carbon dioxide with water.

REVIEW OF CONCEPT 2.5

Water dissociates to form H^+ and OH^-. Acidic solutions have a high $[H^+]$, and basic solutions have a low $[H^+]$ (and therefore a high $[OH^-]$). The pH of a solution is the negative logarithm of its $[H^+]$. Low pH values indicate acids, and high pH values indicate bases. Even small changes in pH can be harmful to life. Buffer systems in organisms help to maintain pH within a narrow range.

- *A change of 2 pH units indicates what change in $[H^+]$?*

Using Radioactive Decay to Date the Iceman

In the fall of 1991, two Austrian hikers found the corpse of an adult male who appeared to have been naturally mummified. It was clear the body was very old, frozen in an icy trench long ago and only now released as the ice melted. In the years since this unusual find, scientists have learned a great deal about the dead man, whom they named Ötzi, because he was found in the Ötztal Alps. They were able to determine how old he was when he died, his health at that time, the clothing he wore, what he ate, and that he died from an arrow that ripped through his back. Its tip is still embedded in the back of his left shoulder. From the distribution of chemicals in his teeth and bones, we know he lived his life within 60 km of where he died.

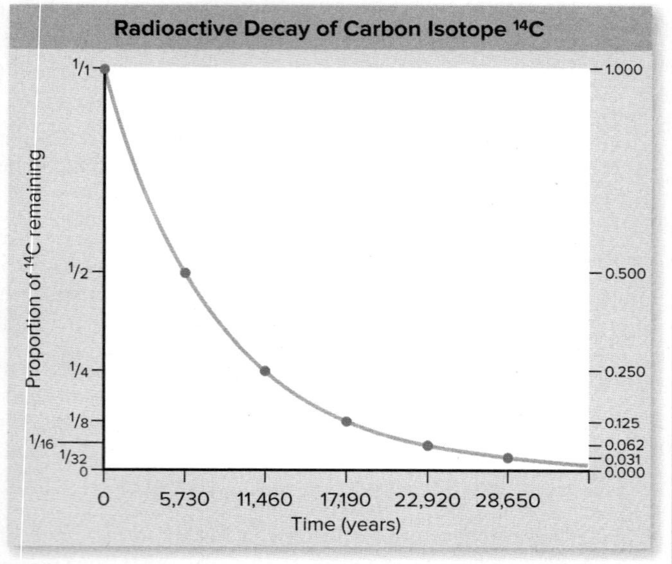

Photo Archives South Tyrol Museum of Archaeology-www.iceman.it

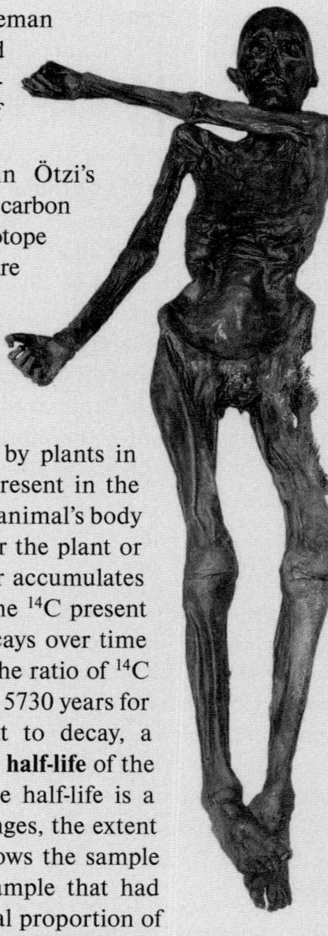

Photo Archives South Tyrol Museum of Archaeology-www.iceman.it

When did this Iceman die? Scientists answered this key question by measuring the degree of decay of the short-lived carbon isotope ^{14}C in Ötzi's body. Although most carbon atoms are the stable isotope ^{12}C, a tiny proportion are the unstable radioactive isotope ^{14}C, created by cosmic rays bombarding nitrogen-14 (^{14}N) atoms. This proportion of ^{14}C is captured by plants in photosynthesis and is present in the carbon molecules of the animal's body that eats the plant. After the plant or animal dies, it no longer accumulates any more carbon, and the ^{14}C present at the time of death decays over time back to ^{14}N. Gradually the ratio of ^{14}C to ^{12}C decreases. It takes 5730 years for half of the ^{14}C present to decay, a length of time called the **half-life** of the ^{14}C isotope. Because the half-life is a constant that never changes, the extent of radioactive decay allows the sample to be dated. Thus, a sample that had one-quarter of its original proportion of ^{14}C remaining would be approximately 11,460 years old (two half-lives).

The graph displays the radioactive decay curve of the carbon isotope ^{14}C. Scientists know it takes 5730 years for half of the ^{14}C present in a sample to decay to nitrogen-14 (^{14}N). When Ötzi's carbon isotopes were analyzed, researchers determined that the ratio of ^{14}C to ^{12}C, also written as the fraction $^{14}C/^{12}C$, in Ötzi's body was 0.435 of the fraction found in tissues of a person who has died recently.

Analysis

1. **Applying Concepts** What proportion of the ^{14}C present in Ötzi's body when he died is still there today? When he died, it would have been 1.0.
2. **Interpreting Data** Plot this proportion on the ^{14}C radioactive decay curve above. How many half-lives does this point represent?
3. **Making Inferences** If Ötzi were indeed a recent corpse, made to look old by the harsh weather conditions found on the high mountain pass, what would you expect the ratio of ^{14}C to ^{12}C to be, relative to that in your own body?
4. **Drawing Conclusions** When did Ötzi the Iceman die?

Retracing the Learning Path

CONCEPT 2.1 All Matter Is Composed of Atoms

2.1.1 Atoms Are Composed of Three Kinds of Subatomic Particles Each element is defined by its atomic number, the number of protons in the nucleus. Electrically neutral atoms have the same number of protons as electrons. Atoms that gain or lose electrons are called ions. Atomic mass is the sum of the mass of protons and neutrons. Isotopes are forms of an element with different numbers of neutrons, and thus different atomic mass. Radioactive isotopes are unstable.

2.1.2 Electrons Determine the Chemical Behavior of Atoms Electrons are found in orbitals around the nucleus. *s*-orbitals are spherical; other orbitals have different shapes, such as the dumbbell-shaped *p*-orbitals.

2.1.3 Atoms Contain Discrete Energy Levels The potential energy of electrons increases as distance from the nucleus increases. Energy levels correspond to quanta (singular, *quantum*) of energy. The loss of electrons from an atom is called oxidation. The gain of electrons is called reduction. Electrons can be transferred from one atom to another in coupled redox reactions.

CONCEPT 2.2 The Elements in Living Systems Have Low Atomic Masses

2.2.1 The Periodic Table Reflects the Electronic Structure of Atoms Atoms tend to establish completely full outer energy levels (the octet rule). Elements with filled outermost orbitals are inert. Twelve of these elements are found in living organisms in greater than trace amounts: C, H, O, N, P, S, Na, K, Ca, Mg, Fe, and Cl. Compounds of carbon are called organic compounds. The majority of molecules in living systems are composed of C bound to H, O, and N.

CONCEPT 2.3 Molecules Are Collections of Atoms Held Together by Chemical Bonds

2.3.1 Ionic Bonds Form Crystals Chemical bonds form due to the attraction between atoms. Ions with opposite electric charges form ionic bonds, as in NaCl.

2.3.2 Covalent Bonds Build Stable Molecules Covalent bonds are formed by atoms sharing electrons. This forms stable molecules by filling outer electron shells with no net charge. Covalent bonds may be single, double, or triple, depending on the number of electrons shared. Nonpolar covalent bonds involve equal sharing of electrons between atoms. Polar covalent bonds involve unequal sharing of electrons.

2.3.3 Hydrogen Bonds Can Form Between Polar Molecules Hydrogen bonds occur when the positive end of one polar molecule is attracted to the negative end of another.

2.3.4 Van der Waals Attractions Draw Surfaces Together Van der Waals attractions draw two atoms toward each other weakly when the two atoms are very close to each other.

2.3.5 Chemical Reactions Alter Bonds Temperature, reactant concentration, and the presence of catalysts affect reaction rates. Most biological reactions are reversible.

CONCEPT 2.4 The Properties of Water Result from Its Polar Nature

2.4.1 Water's Structure Facilitates Hydrogen Bonding Hydrogen bonds are weak interactions between a partially positive H in one molecule and a partially negative O in another molecule.

2.4.2 Water Molecules Are Both Cohesive and Adhesive Cohesion is the tendency of water molecules to adhere to one another due to hydrogen bonding. The cohesion of water is responsible for its surface tension. Adhesion occurs when water molecules adhere to other polar molecules. Capillary action results from water's adhesion to the sides of narrow tubes, combined with its cohesion.

2.4.3 Water Has High Specific Heat The specific heat of water is high because it takes a considerable amount of energy to disrupt hydrogen bonds.

2.4.4 Water Has a High Heat of Vaporization Breaking hydrogen bonds to turn liquid water into vapor takes a lot of energy. Many organisms lose excess heat through evaporative cooling, such as by sweating.

2.4.5 Solid Water Is Less Dense Than Liquid Water The hydrogen bonds in ice result in the molecules being spaced farther apart in the solid phase of water than in the liquid phase. As a result, ice floats.

2.4.6 Water Is a Good Solvent for Polar Molecules Water's polarity makes it a good solvent for polar substances and ions. Polar molecules or portions of molecules are attracted to water (hydrophilic). Molecules that are nonpolar are repelled by water (hydrophobic).

2.4.7 Water Organizes Nonpolar Molecules Nonpolar molecules will aggregate to avoid water. This maximizes the hydrogen bonds that water can make. This hydrophobic exclusion can affect the structure of DNA, proteins, and biological membranes.

CONCEPT 2.5 Water Molecules Can Dissociate into Ions

2.5.1 The pH Scale Measures Hydrogen Ion Concentration Water can dissociate into H^+ and OH^-. The concentration of H^+, shown as $[H^+]$, in pure water is 10^{-7} mol/L. The pH scale is a quantitative measure of $[H^+]$. pH is defined as the negative logarithm of $[H^+]$. Pure water has a pH of 7. A difference of 1 pH unit means a 10-fold change in $[H^+]$. Acids have a greater $[H^+]$ and therefore a lower pH; bases have a lower $[H^+]$ and therefore a higher pH. Organisms need to maintain a constant pH in cells and extracellular fluid, using buffers that resist changes in pH.

Concept Overview

This Concept Overview diagrams the key concepts that were discussed in this chapter.

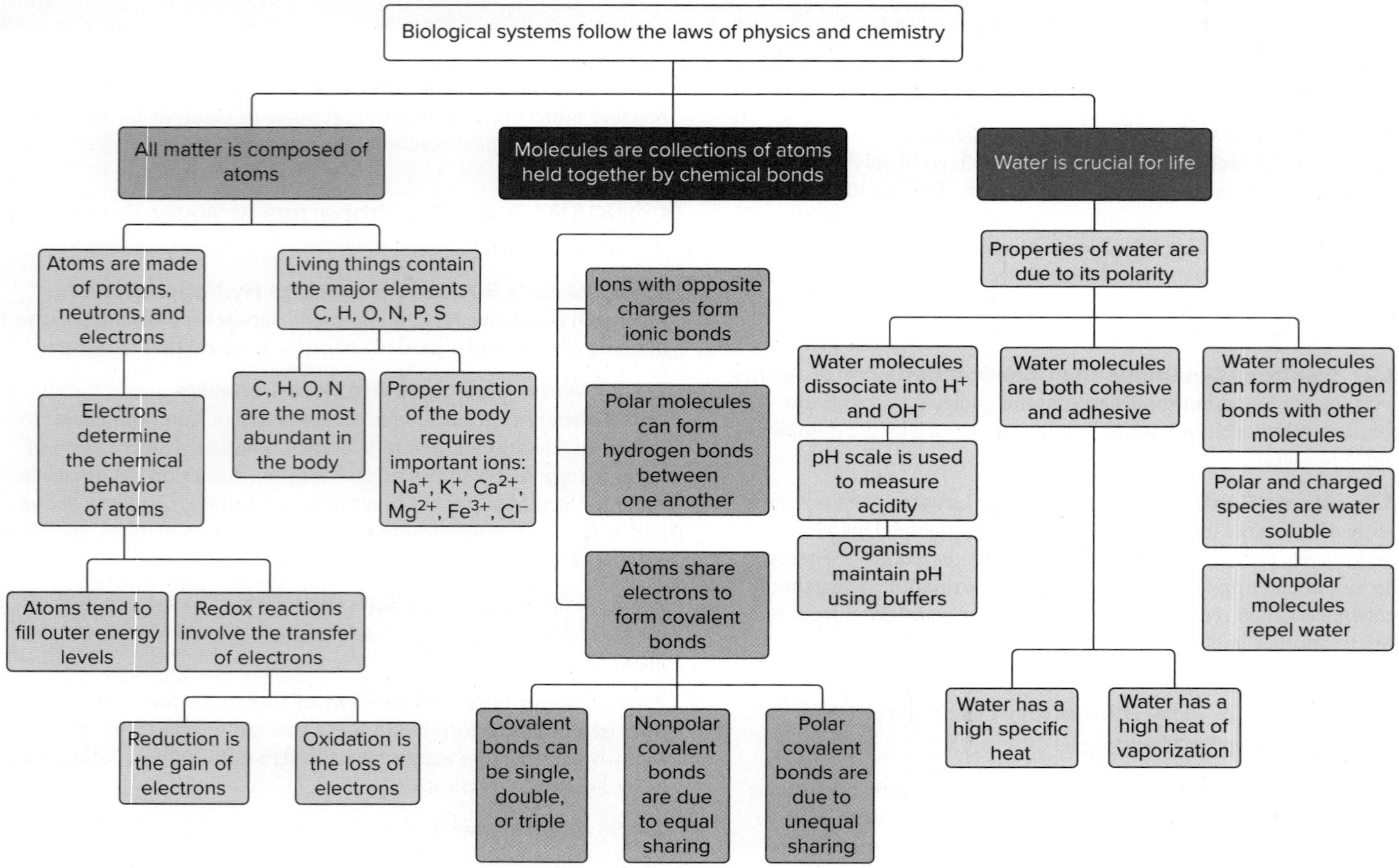

Assessing the Learning Path

Understand

1. The isotopes carbon-12 and carbon-14 differ in
   a. the number of neutrons.
   b. the number of protons.
   c. the number of electrons.
   d. Both b and c
2. An atom with a net positive charge must have more
   a. protons than neutrons.
   b. protons than electrons.
   c. electrons than neutrons.
   d. electrons than protons.
3. A completely filled L energy level contains how many more electrons than a completely filled K energy level?
   a. two
   b. four
   c. six
   d. eight
4. Ca^{2+}
   a. is an anion.
   b. is a neutral isotope.
   c. is an atom that has lost two electrons.
   d. is reduced.
5. The property that distinguishes an atom of one element (carbon, for example) from an atom of another element (oxygen, for example) is
   a. the number of electrons.
   b. the number of protons.
   c. the number of neutrons.
   d. the combined number of protons and neutrons.
6. An atom with one valence electron
   a. would tend to easily lose an electron.
   b. would be inert.
   c. would be found in the right-most column of the periodic table.
   d. would easily gain an extra electron.
7. Which of the following is NOT accurate about the elements commonly found in living organisms?
   a. They have a low atomic mass.
   b. They have atomic numbers less than 27.
   c. They have a completely filled outer electron shell.
   d. They are incorporated into organic molecules.

8. Ionic bonds arise from
   a. shared valence electrons.
   b. attractions between valence electrons.
   c. charge attractions between valence electrons.
   d. attractions between ions of opposite charge.
9. A hydrogen bond is
   a. weaker than a covalent bond.
   b. stronger than an ionic bond.
   c. the same as a van der Waals attraction.
   d. stronger than a covalent bond.
10. Van der Waals attractions
   a. involve electrical attraction, as do ionic and hydrogen bonds.
   b. are stronger than hydrogen bonds.
   c. are not directional.
   d. involve only two atoms.
11. A chemical reaction will go faster if
   a. the temperature is lowered.
   b. there are more products present.
   c. there are fewer reactants present.
   d. a catalyst is present.
12. Two adjacent water molecules are able to form hydrogen bonds with each other because
   a. oxygen is more electronegative than hydrogen.
   b. hydrogen is more electronegative than oxygen.
   c. hydrogen and oxygen have equal affinity for electrons.
   d. the hydrogen atoms have a partial negative charge and the oxygen atoms have a partial positive charge.
13. The difference between cohesion and adhesion is that
   a. cohesion involves the formation of hydrogen bonds, but adhesion does not.
   b. in cohesion water molecules are attracted to each other, but in adhesion they are attracted to other substances.
   c. cohesion promotes the capillary action of water, but adhesion opposes it.
   d. adhesion promotes surface tension, but cohesion opposes it.
14. Why are buffers important in living systems?
   a. Many chemical reactions are affected by the acidity of the solution in which they occur.
   b. Chemical reactions of life produce acids and bases within cells.
   c. Enzyme activity is affected by pH.
   d. All of the above
15. Buffer systems
   a. will always remove H^+ from solution.
   b. will always add H^+ to solution.
   c. will help keep pH relatively constant.
   d. will stop water from ionizing.

Apply

1. The maximum number of electrons found in a dumbbell-shaped *p*-orbital
   a. is greater than the number found in a spherical *s*-orbital.
   b. is less than the number found in a spherical *s*-orbital.
   c. is the same as the number found in a spherical *s*-orbital.
   d. has no relation to the number found in a spherical *s*-orbital.
2. An atom of iron (Fe) has 26 protons, 30 neutrons, and 26 electrons. An iron atom
   a. has an atomic mass of 82.
   b. has an atomic number of 26.
   c. is an ion.
   d. has an atomic number of 56.
3. Iodine (I)
   a. is present in living systems in more than trace amounts.
   b. has reactivity similar to that of chlorine.
   c. is an isotope of chlorine.
   d. All of the above
4. Using the periodic table (figure 2.6), which of the following atoms would you predict could form a positively charged ion (cation)?
   a. Fluorine (F)   c. Potassium (K)
   b. Neon (Ne)   d. Sulfur (S)
5. Refer to the element pictured. How many covalent bonds could this atom form?

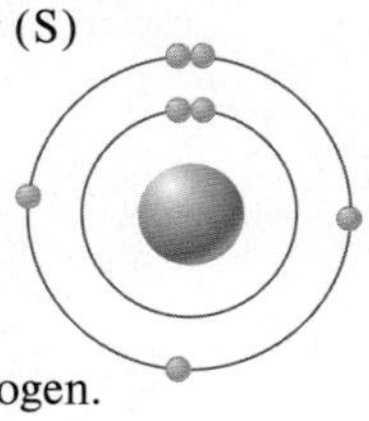

   a. Two   c. Four
   b. Three   d. None
6. Hydrogen bonds are formed
   a. between any molecules that contain hydrogen.
   b. only between water molecules.
   c. when hydrogen is part of a polar covalent bond.
   d. when two atoms of hydrogen share an electron.
7. A molecule with polar covalent bonds would
   a. be soluble in water.
   b. not be soluble in water.
   c. contain atoms with very similar electronegativity.
   d. contain atoms that have gained or lost electrons.
8. Which of the following would NOT be soluble in water?
   a. Polar molecules
   b. Ions
   c. Molecules that contained many hydroxyl groups
   d. Nonpolar molecules
9. If the pH of a solution changes from pH 5 to pH 3, this means the H^+ concentration is now
   a. 20-fold higher.   c. 20-fold lower.
   b. 100-fold higher.   d. 100-fold lower.

Synthesize

1. The half-life of radium-226 is 1620 years. If a sample of material contains 16 milligrams of radium-226, how long will it take for the sample to contain 1 milligram of radium-226?
2. Over half of your body weight is oxygen atoms. Why do you suppose that is so? In what molecule does most of this oxygen reside?
3. Why are the atoms of a stable molecule linked together by covalent bonds, but not by ionic bonds?
4. A popular theme in science fiction literature has been the idea of silicon-based life-forms in contrast to our carbon-based life. Evaluate the possibility of silicon-based life, based on the chemical structure and potential for chemical bonding of a silicon atom.
5. Recent efforts by NASA to search for signs of life on Mars have focused on searching for evidence of liquid water rather than looking directly for biological organisms (living or fossilized). Use your knowledge of the influence of water on life on Earth to construct an argument justifying this approach.
6. Champagne, a carbonic acid buffer, has a pH of about 2. How can we drink such a strong acid?

Deco/Alamy Stock Photo

3 The Chemical Building Blocks of Life

Learning Path

3.1 Carbon Provides the Framework of Biological Molecules

3.2 Carbohydrates Form Both Structural and Energy-Storing Molecules

3.3 Proteins Are the Tools of the Cell

3.4 Nucleic Acids Store and Express Genetic Information

3.5 Hydrophobic Lipids Form Fats and Membranes

Concept Overview

This Concept Overview provides a diagram of the overarching concepts that are covered in this chapter. These concepts will be expanded upon in the Concept Overview at the end of the chapter.

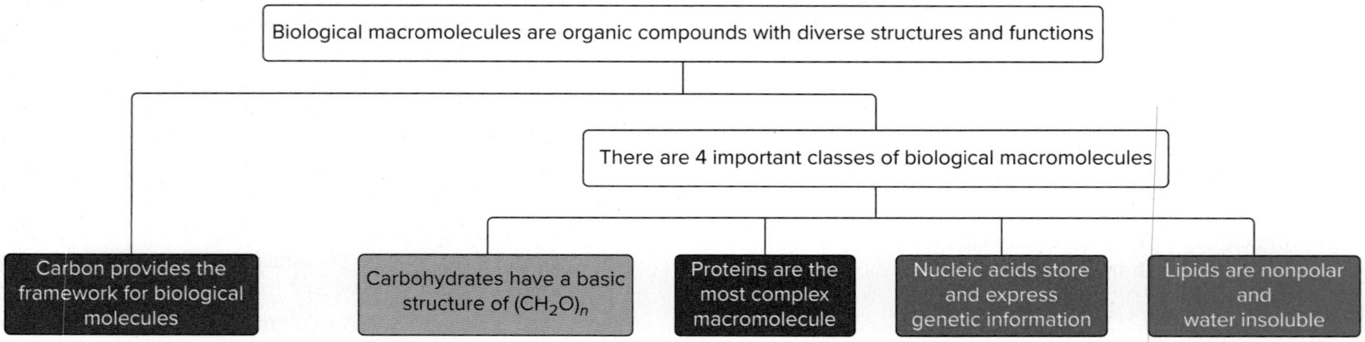

Introduction

Despite the complexity and diversity of living systems, all organisms share the same basic chemistry. In the last chapter, we saw how this involves a limited number of elements; now we will see that organisms consist of a few basic kinds of molecules. These molecules are all built on a carbon framework, making them organic molecules. Because of their large size, they are called macromolecules, and they are usually synthesized within cells as long chains of subunit molecules that are linked together like the cars of a railway train. We recognize four classes of biological macromolecules: carbohydrates, proteins, nucleic acids, and lipids.

3.1 Carbon Provides the Framework of Biological Molecules

The cells of living organisms contain an incredible diversity of molecules, but these share many common features. They are primarily compounds of carbon, or **organic molecules**. These organic compounds are made up primarily of the common elements: C, H, O, N, P, and S. Complex molecules are built up using smaller, simpler subunits by similar chemical reactions. Because carbon atoms can form up to four covalent bonds, organic molecules can form long chains of repeating carbon atoms, branched chains, or even rings, which allow for a nearly infinite variety of shapes.

Functional Groups Provide Chemical Flexibility

LEARNING OBJECTIVE 3.1.1 Define different functional groups based on their chemical properties.

Hydrocarbons

Molecules consisting only of carbon and hydrogen are called *hydrocarbons.* Because the oxidation of hydrocarbon compounds results in a net release of energy, they make good fuels. Gasoline, for example, is rich in hydrocarbons. Propane gas, another hydrocarbon, consists of a chain of three carbon atoms, with eight hydrogen atoms bound to it. The empirical formula (which lists the number of atoms in a molecule as subscripts) for propane is C_3H_8. Its structural formula is

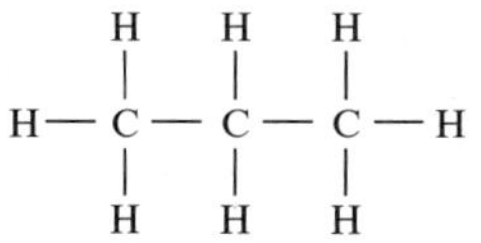

propane structural formula

Theoretically speaking, the length of a chain of carbon atoms is unlimited. As described in the rest of this chapter, the four main types of biological molecules often consist of very long chains of carbon-containing compounds.

Functional groups

Carbon and hydrogen atoms both have very similar electronegativities. Electrons in C—C and C—H bonds are therefore evenly distributed, with no significant differences in charge over the molecular surface. For this reason, hydrocarbons are nonpolar. Most biological molecules produced by cells, however, also contain other atoms. Because these other atoms frequently have different electronegativities, molecules containing them exhibit regions of partial positive or negative charge that are polar. These molecules can be thought of as possessing a C—H core to which specific molecular groups, called **functional groups,** are attached (figure 3.1). One such common functional group is —OH, called a *hydroxyl group.*

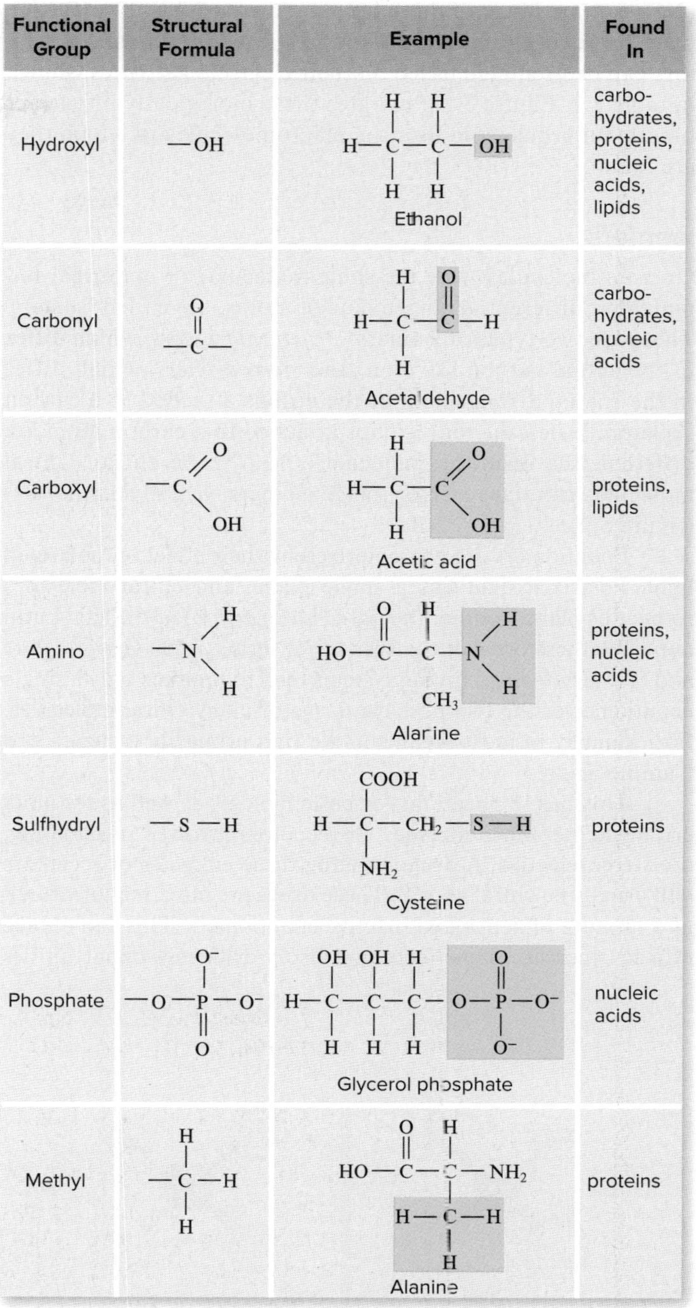
Functional Group
Structural Formula
Example
Found In
Hydroxyl
Ethanol
carbohydrates, proteins, nucleic acids, lipids
Carbonyl
Acetaldehyde
carbohydrates, nucleic acids
Carboxyl
Acetic acid
proteins, lipids
Amino
Alanine
proteins, nucleic acids
Sulfhydryl
Cysteine
proteins
Phosphate
Glycerol phosphate
nucleic acids
Methyl
Alanine
proteins

Figure 3.1 The primary functional chemical groups. These groups tend to act as units during chemical reactions and give specific chemical properties to the molecules that possess them. Carboxyl groups, for example, make a molecule more acidic.

Functional groups have definite chemical properties that they retain no matter where they occur. Both the hydroxyl and carbonyl (C═O) groups, for example, are polar because of the electronegativity of the oxygen atoms (refer to chapter 2). Other common functional groups are the acidic carboxyl (COOH) and phosphate (PO_4) groups, and the basic amino

(NH_2) group. Many of these functional groups can also participate in hydrogen bonding. Hydrogen bond donors and acceptors can be predicted based on their electronegativities, shown in table 2.2. Figure 3.1 illustrates these biologically important functional groups and lists the macromolecules in which they are found.

Isomers

Organic molecules with the same molecular or empirical formula, but different arrangements of atoms, are called **isomers**. There are two types of isomers: *structural isomers*, which differ in the actual carbon skeleton, and *stereoisomers*, which differ in the spatial arrangement of the groups attached to a carbon skeleton. When the four groups attached to a carbon atom are different, the resulting molecule is said to be chiral. Chiral molecules can have mirror image isomers we call *enantiomers* (figure 3.2).

Enantiomers are characterized by their effect on polarized light. Polarized light has a single plane, and chiral molecules rotate this plane to either the right (Latin *dextro*) or the left (Latin *levo*). We therefore call the two chiral forms *d* for *dextrorotatory* and *l* for *levorotatory*. Living systems tend to produce only a single enantiomer of the two possible forms of many chiral molecules. For example, in most organisms we find primarily D-sugars and L-amino acids.

This has implications for basic biology, as well as for modern medicine. Most enzymes, or receptors involved in signaling, are stereoselective. A drug targeting these enzymes or receptors will usually be chiral as well. These drugs are often manufactured as a mixture of active and inactive enantiomers. This can reduce efficacy, increase side-effects, and also provide new opportunities for patenting. For instance, a common drug used to treat acid reflux is Prilosec, which is a mixture of active and inactive forms. The competing drug Nexium consists of only the active enantiomer.

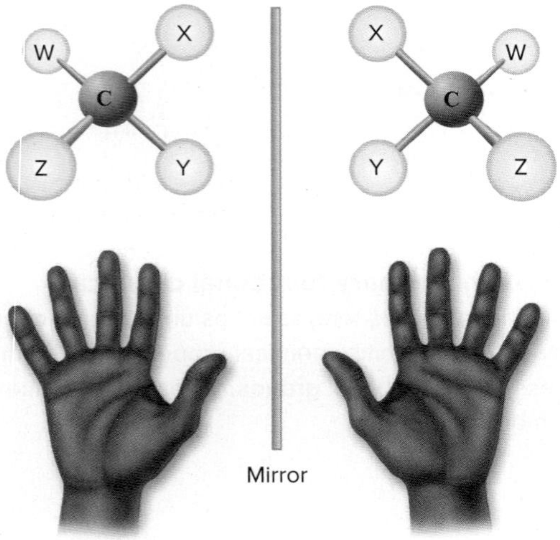

Figure 3.2 Chiral molecules. When carbon is bound to four different groups, the resulting molecule is said to be chiral (from Greek *cheir,* meaning "hand"). A chiral molecule has stereoisomers that are mirror images. The two molecules shown have the same four groups but cannot be superimposed, much as your two hands cannot be superimposed. These types of stereoisomers are called *enantiomers.*

Macromolecules

Biological macromolecules are commonly grouped into four general classes (table 3.1): carbohydrates, proteins, nucleic acids, and lipids. The first three of these are all **polymers** constructed by joining together many small, similar chemical subunits called **monomers**. A long polymer of linked monomers forms a molecule analogous to a long train consisting of many railroad cars linked together.

Different polymers are determined by their constituent monomers. Complex carbohydrates such as starch are polymers composed of simple, ring-shaped sugars. Nucleic acids (DNA and RNA) are polymers of nucleotides, and proteins are polymers of amino acids.

The class of macromolecules that do not fit this simple monomer/polymer framework are the lipids. This diverse group of molecules shares the chemical feature of being hydrophobic and includes fats, oils, and steroids. As we will discover, these molecules also have diverse functions.

Biological Macromolecules Are Polymers

LEARNING OBJECTIVE 3.1.2 Contrast hydrolysis and dehydration reactions.

The chains of carbohydrates, proteins, and nucleic acids are all built in the same way, assembled via chemical reactions termed *dehydration reactions*, and broken down by *hydrolysis reactions*. The reaction that joins fatty acids to glycerol in triglycerides is also a dehydration reaction.

The dehydration reaction

Despite the differences between monomers of these major polymers, the basic chemistry of their synthesis is similar: to form a covalent bond between two monomers, an —OH group is removed from one monomer, and a hydrogen atom (H) is removed from the other (figure 3.3*a*). For example, this simple chemistry is the same for linking amino acids together to make a protein or assembling glucose units together to make starch. This reaction is also used to link fatty acids to glycerol in lipids. This chemical reaction is called condensation, or a **dehydration reaction,** because the removal of —OH and —H is the same as the removal of a molecule of water (H_2O). For every subunit added to a macromolecule, one water molecule is removed. These and other biochemical reactions require that the reacting substances are held close together and that the correct chemical bonds are stressed and broken. This process of positioning and stressing, termed *catalysis,* is carried out within cells by enzymes.

The hydrolysis reaction

Cells disassemble macromolecules into their constituent subunits through reactions that are the reverse of dehydration—a molecule

TABLE 3.1 Macromolecules

Macromolecule	Subunit	Function	Example
CARBOHYDRATES			
Starch, glycogen	Glucose	Energy storage	Potatoes
Cellulose	Glucose	Structural support in plant cell walls	Paper; strings of celery
Chitin	Modified glucose	Structural support	Crab shells
PROTEINS			
Functional	Amino acids	Catalysis; transport	Hemoglobin
Structural	Amino acids	Support	Hair; silk
NUCLEIC ACIDS			
DNA	Nucleotides	Encodes genes	Chromosomes
RNA	Nucleotides	Needed for gene expression	Messenger RNA
LIPIDS			
Fats	Glycerol and three fatty acids	Energy storage	Butter; corn oil; soap
Phospholipids	Glycerol, two fatty acids, phosphate, and polar R groups	Cell membranes	Phosphatidylcholine
Prostaglandins	Five-carbon rings with two nonpolar tails	Chemical messengers	Prostaglandin E (PGE)
Steroids	Four fused carbon rings	Membranes; hormones	Cholesterol; estrogen
Terpenes	Long carbon chains	Pigments; structural support	Carotene; rubber

of water is added instead of removed (figure 3.3*b*). In this reaction, called **hydrolysis,** a hydrogen atom is attached to one subunit and a hydroxyl group to the other, breaking a specific covalent bond in the macromolecule.

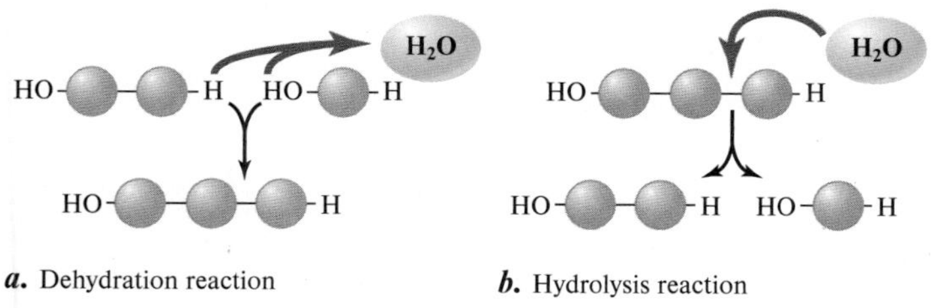

Figure 3.3 Making and breaking macromolecules. ***a.*** Biological macromolecules are polymers formed by linking monomers together through dehydration reactions. This process releases a water molecule for every bond formed. ***b.*** Breaking the bond between subunits involves hydrolysis, which reverses the loss of a water molecule by dehydration.

REVIEW OF CONCEPT 3.1

Functional groups account for differences in chemical properties in organic molecules. Isomers are compounds with the same empirical formula but different structures. This difference may affect biological function. Macromolecules are polymers consisting of long chains of similar subunits that are joined by dehydration reactions and are broken down by hydrolysis reactions.

- *What is the relationship between dehydration and hydrolysis?*

3.2 Carbohydrates Form Both Structural and Energy-Storing Molecules

Carbohydrates are a loosely defined group of molecules that all contain carbon, hydrogen, and oxygen in the molar ratio 1:2:1. Their empirical formula is $(CH_2O)_n$, where n is the number of

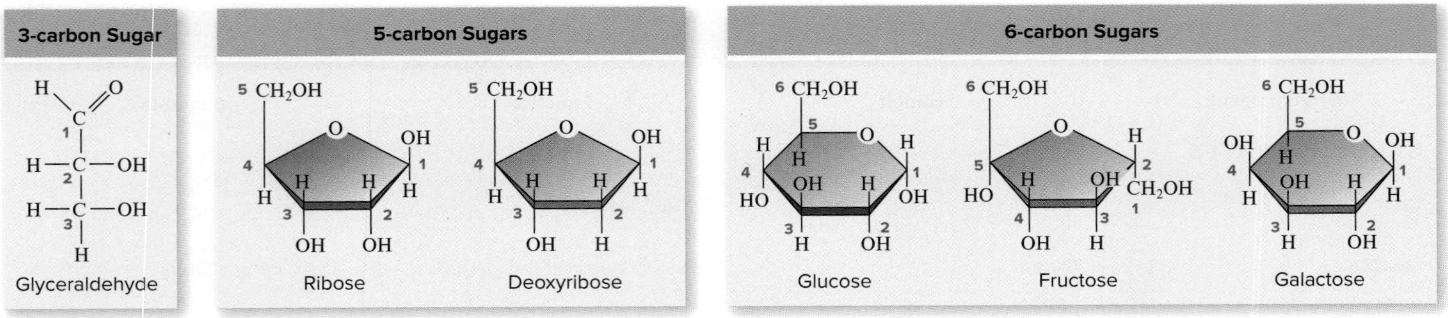

Figure 3.4 Monosaccharides. Monosaccharides, or simple sugars, can contain as few as 3 carbon atoms and are often used as building blocks to form larger molecules. The 5-carbon sugars ribose and deoxyribose are components of nucleic acids. The carbons are conventionally numbered from the more oxidized end.

carbon atoms. Because they contain many carbon–hydrogen (C—H) bonds, which release energy when oxidation occurs, carbohydrates are well suited for energy storage. Sugars are among the most important energy-storage molecules, and they exist in several different forms.

Monosaccharides Are Simple Sugars

LEARNING OBJECTIVE 3.2.1 Distinguish between structural isomers and stereoisomers.

The simplest of the carbohydrate sugars are the **monosaccharides** (Greek *mono,* "single," and Latin *saccharum,* "sugar"). Simple sugars contain as few as three carbon atoms, but those that play the central role in energy storage have six (figure 3.4). The empirical formula of 6-carbon sugars is

$$C_6H_{12}O_6 \quad \text{or} \quad (CH_2O)_6$$

Six-carbon sugars can exist as a straight chain, but dissolved in water (an aqueous environment), they almost always form rings.

The most important of the 6-carbon monosaccharides for energy storage is glucose. Glucose has seven energy-storing C—H bonds (figure 3.5). Depending on the orientation of the carbonyl group (C═O) when the ring is closed, glucose can exist in two different forms: alpha (α) or beta (β).

Sugar isomers have structural differences

Glucose is not the only sugar with the formula $C_6H_{12}O_6$. Both structural isomers and stereoisomers of this simple 6-carbon skeleton exist in nature. Fructose is a structural isomer that differs in the position of the carbonyl carbon (C═O); galactose is a stereoisomer that differs in the position of —OH and —H groups relative to the ring (figure 3.6). These differences often account for substantial functional differences between the isomers. Your taste buds can discern them: fructose tastes much sweeter than glucose, despite the fact that both sugars have identical chemical composition. Enzymes that act on different sugars can distinguish both the structural isomers and the stereoisomers of this basic 6-carbon skeleton. The different stereoisomers of glucose are also important in the polymers that can be made using glucose as a monomer, as you will see later in this section.

Disaccharides Are Transport Sugars

LEARNING OBJECTIVE 3.2.2 Distinguish between monosaccharides and disaccharides.

Both animals and plants transport sugars within their bodies. In humans, the glucose that circulates in the blood does so as a simple monosaccharide. In plants and many other animals, however, glucose is converted into a transport form before it is moved from place to place within the body. In such a form, it is less readily metabolized during transport.

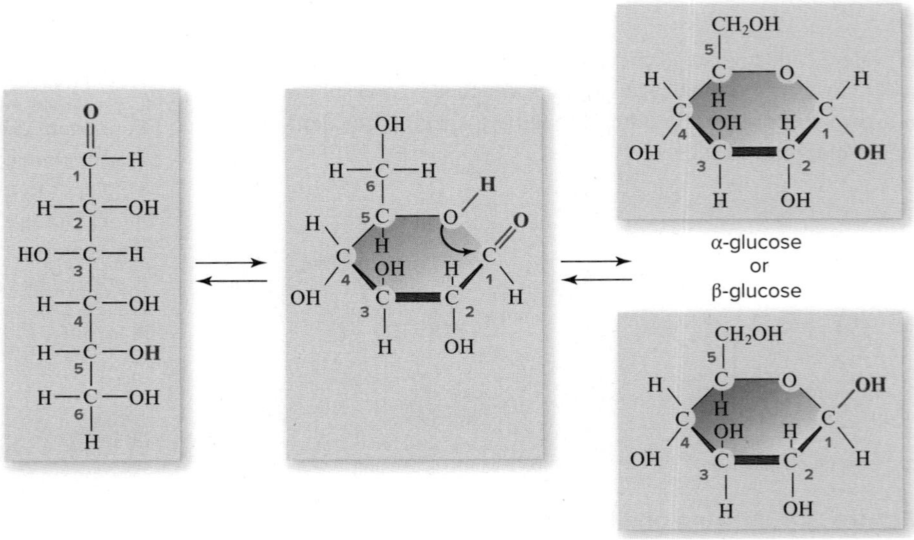

Figure 3.5 Structure of the glucose molecule. Glucose is a linear 6-carbon molecule that forms a six-membered ring in solution. Ring closure occurs such that two forms can result: α-glucose and β-glucose. These structures differ only in the position of the —OH bound to carbon 1.

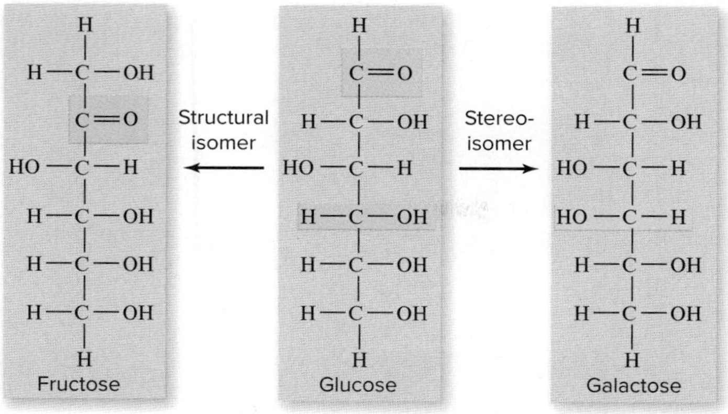

Figure 3.6 Structural isomers and stereoisomers. The sugars glucose, fructose, and galactose are isomers with the empirical formula $C_6H_{12}O_6$. A structural isomer of glucose, such as fructose, has identical chemical groups bonded to different carbon atoms. A stereoisomer of glucose, such as galactose, has identical chemical groups bonded to the same carbon atoms but in different orientations (the —OH at carbon 4).

Transport forms of sugars are commonly made by linking two monosaccharides together to form a **disaccharide** (Greek *di,* "two"). Disaccharides serve as effective reservoirs of glucose because the enzymes that normally use glucose in the organism cannot break the bond linking the two monosaccharide subunits. Enzymes that can do so are typically present only in the tissue destined to use the glucose.

Glucose forms a variety of transport disaccharides. In plants, glucose forms a disaccharide with its structural isomer fructose. The resulting disaccharide is *sucrose,* or table sugar (figure 3.7*a*). Sucrose is the form most plants use to transport glucose, and it is the sugar that most humans eat.

In mammals, glucose is linked to its stereoisomer galactose, forming the disaccharide *lactose,* or milk sugar. Many mammals supply energy to their young in the form of lactose. Adults often have greatly reduced levels of lactase, the enzyme required to cleave lactose into its two monosaccharide components, and thus they cannot metabolize lactose efficiently. This effectively reserves the energy stored in lactose for the offspring.

Polysaccharides Are Building Materials and Energy-Storage Compounds

LEARNING OBJECTIVE 3.2.3 Explain why humans can digest starch but not cellulose, but a cow can digest both.

Polysaccharides are longer sugar polymers made up of monosaccharides that have been joined through dehydration reactions.

Starches and glycogen

Organisms store the metabolic energy contained in monosaccharides by first converting them into disaccharides, such as *maltose* (figure 3.7*b*). These are then linked together into insoluble storage polysaccharides called **starches** (figure 3.8).

Starches differ mainly in how the long-chain polymers branch. The starch with the simplest structure is *amylose.* It is composed of many hundreds of α-glucose molecules linked together in long, unbranched chains. Each linkage occurs between the carbon 1 (C-1) of one glucose molecule and the C-4 of another, making them α-1→4 linkages (figure 3.8*a*). The long chains tend to coil up in water, a property that renders amylose insoluble. Potato starch is about 20% amylose (figure 3.8*b*).

Most plant starch, including the remaining 80% of potato starch, is a somewhat more complicated variant of amylose called *amylopectin* (figure 3.8*b*). Pectins are branched polysaccharides with the branches occurring at bonds between the C-1 of one molecule and the C-6 of another (α-1→6 linkages). These short branches consist of 20 to 30 glucose subunits.

The molecule comparable to starch in animals is **glycogen.** Like amylopectin, glycogen is an insoluble polysaccharide containing branched amylose chains. Glycogen has a much longer average chain length and more branches than plant starch (figure 3.8*c*).

Cellulose

Although some chains of sugars store energy, others serve as structural material for cells. For glucose molecules to link together in a chain, the glucose subunits must be of the same form. Starches are α-glucose chains. *Cellulose* is a β-glucose chain (figure 3.9*a*). The bonds between adjacent glucose molecules in cellulose still extend between the C-1 of the first glucose and the C-4 of the next glucose, but in cellulose these are both β-1→4 linkages.

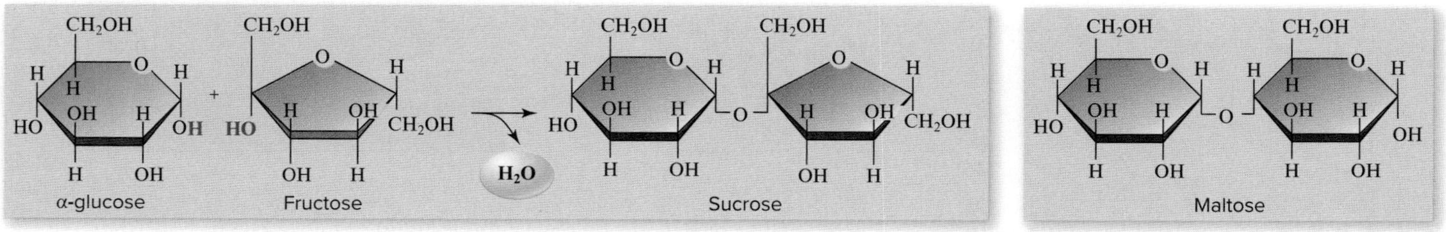

a. *b.*

Figure 3.7 How disaccharides form. Some disaccharides are used to transport glucose from one part of an organism's body to another; one example is sucrose (***a***), which is found in sugarcane. Other disaccharides, such as maltose (***b***), are used in grain for storage.

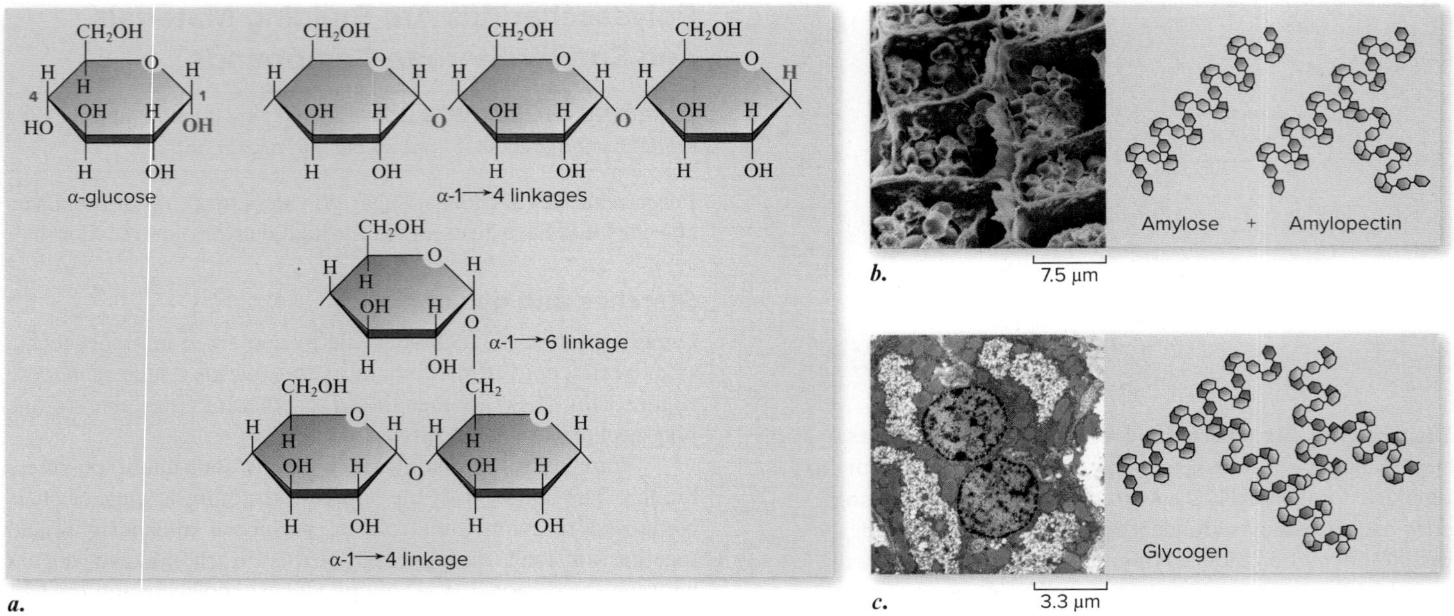

Figure 3.8 Polymers of glucose: Starch and glycogen. ***a.*** Starch chains consist of polymers of α-glucose subunits joined by α-1→4 glycosidic linkages. These chains can be branched by forming similar α-1→6 glycosidic bonds. These storage polymers then differ primarily in their degree of branching. ***b.*** Starch is found in plants and is composed of amylose and amylopectin, which are unbranched and branched, respectively. The branched form is insoluble and forms starch granules in plant cells. ***c.*** Glycogen is found in animal cells and is highly branched and insoluble, forming glycogen granules.

(b): Asa Thoresen/Science Source; (c): J.L. Carson/CMSP Biology/Newscom

The properties of a β-glucose chain are very different from those of starch. Long, unbranched β-linked chains make tough fibers. Cellulose is the chief component of plant cell walls (figure 3.9*b*). It is chemically similar to amylose, with one important difference: the starch-hydrolyzing enzymes that occur in most organisms cannot break the bond between two β-glucose units, because they recognize only α linkages.

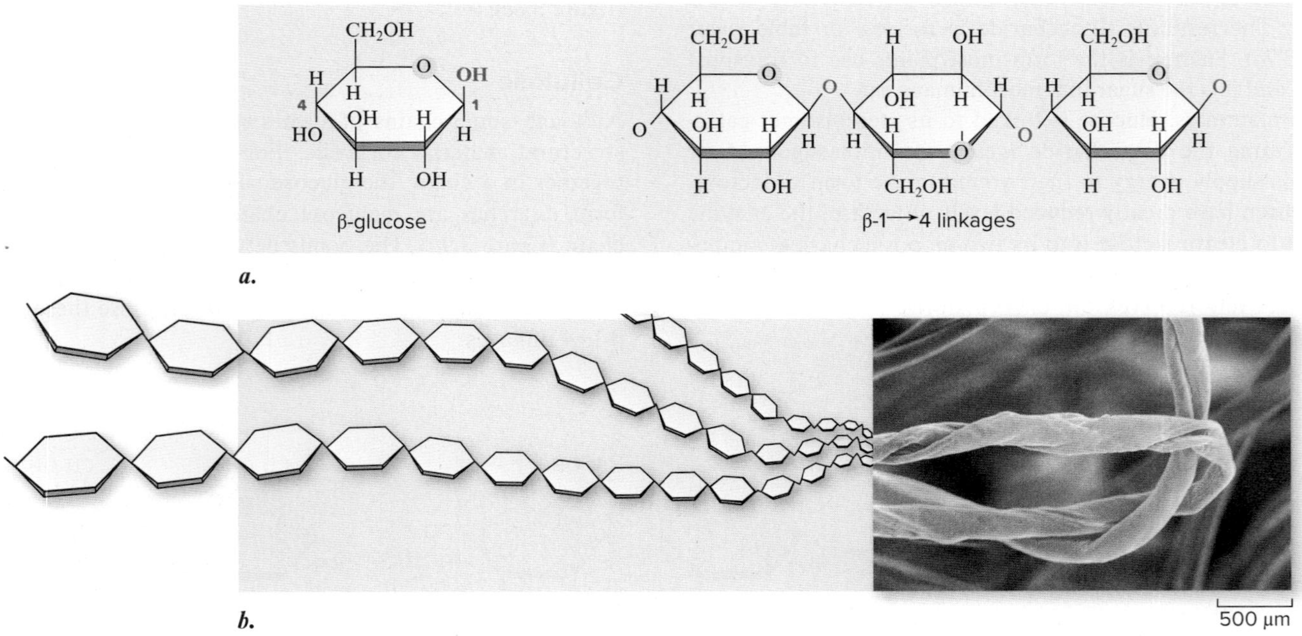

Figure 3.9 Polymers of glucose: Cellulose. Starch chains consist of α-glucose subunits, and cellulose chains consist of β-glucose subunits. ***a.*** Thus, the bonds between adjacent glucose molecules in cellulose are β-1→4 glycosidic linkages. ***b.*** Cellulose is unbranched and forms long fibers. Cellulose fibers can be very strong and are quite resistant to metabolic breakdown, which is one reason wood is such a good building material.

(b): Jim Zuckerman/age fotostock

Figure 3.10 Chitin. Chitin is the principal structural element in the external skeletons of many invertebrates, such as this lobster.
OAR/National Undersea Research Program (NURP)

Because its unbranched structure forms long, tough fibers, and most animals cannot degrade it, cellulose makes an excellent structural material. Some animals, such as cows, are able to break down cellulose helped by symbiotic bacteria and protists in their digestive tracts. These organisms produce an enzyme that can cleave the β-1→4 linkages, converting cellulose to glucose for further metabolism.

Chitin

Chitin, the structural material found in arthropods and many fungi, is a polymer of *N*-acetylglucosamine, a derivative of glucose. When cross-linked by proteins, it forms a tough, resistant surface material that serves as the hard exoskeleton of insects and crustaceans (figure 3.10). Few animals are able to digest chitin in their stomachs, although most possess a chitinase enzyme, probably to protect against fungi.

REVIEW OF CONCEPT 3.2

Monosaccharides have carbon atoms typically arranged in a ring form. Disaccharides consist of two linked monosaccharides; polysaccharides are long chains of monosaccharides. Starches are branched polymers of α-glucose used for energy storage. Cellulose in plants consists of unbranched chains of β-glucose that are not easily digested.

- *How do the structures of starch, glycogen, and cellulose affect their function?*

3.3 Proteins Are the Tools of the Cell

When you think of the important cellular functions, you are almost always thinking about functions performed by proteins. The most diverse group of macromolecules, proteins are also critical to most aspects of cell physiology, and they form the building blocks for larger-scale structures important to your physiology. In every chapter of this book we will encounter proteins in different contexts.

Proteins Have Diverse Functions

LEARNING OBJECTIVE 3.3.1 List seven essential functions of proteins in the life of a cell.

Proteins are the most diverse group of biological macromolecules, both chemically and functionally. Because proteins have so many different functions in cells, we could not begin to list them all. We can, however, group these functions into the following seven categories. The descriptions provided here are a summary only; for each, details are covered in later chapters.

1. **Enzyme catalysis.** Enzymes are biological catalysts that facilitate specific chemical reactions. Most enzymes are three-dimensional, globular proteins that fit snugly around the molecules they act on. This fit facilitates chemical reactions by stressing particular chemical bonds.
2. **Defense.** Other globular proteins use their shapes to "recognize" foreign microbes and cancer cells. White blood cells destroy foreign cells, and others make antibody proteins that attach to invading cells.
3. **Transport.** A variety of globular proteins transport small molecules and ions. Red blood cells contain the transport protein hemoglobin, which transports oxygen in the blood. On the surfaces of individual cells, membrane transport proteins help move ions and molecules across the membrane.
4. **Support.** Protein fibers play many important structural roles. In humans, these fibers include keratin, the fibrin in blood clots, and collagen. The last one, collagen, forms the matrix of skin, ligaments, tendons, cartilage, and bones.
5. **Motion.** Muscles contract through the sliding motion of two kinds of protein filaments: actin and myosin. Proteins also play key roles in the cell's cytoskeleton and in moving materials within cells.
6. **Regulation.** Small proteins called hormones serve as intercellular messengers in animals. Proteins also play many regulatory roles within the cell—turning on and shutting off genes, for example. Proteins also act as receptors for extracellular signals.
7. **Storage.** Calcium and iron are stored in the body by binding as ions to storage proteins found in cells.

Proteins Are Polymers of Amino Acids

LEARNING OBJECTIVE 3.3.2 Illustrate how peptide bonds are formed.

Proteins are long, linear polymers of amino acids. The sequence of the amino acids in a protein is referred to as its **primary structure.** Proteins have incredibly varied primary structures, as there are 20 different commonly occurring amino acids, any one of which can occupy any position in the protein chain.

Amino acids, as their name suggests, contain an amino group (—NH_2) and an acidic carboxyl group (—COOH). The specific order of amino acids determines the protein's structure and function. Many scientists believe amino acids were among the first biologically important molecules formed on the early Earth. It seems highly likely that the oceans that existed early in the history of the Earth contained a wide variety of amino acids.

Amino acid structure

The generalized structure of an amino acid is shown here as amino and carboxyl groups bonded to a central carbon atom, with an additional hydrogen and a functional side group indicated by R. These four components completely occupy the four valence bonds of the central carbon:

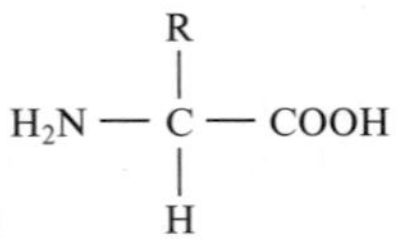

The unique character of each amino acid is determined by the nature of the R group. Notice that unless the R group is an H atom, as in glycine, amino acids are chiral and can exist as two enantiomeric forms: D or L. In living systems, only the L-amino acids are found in proteins.

The R group also determines the chemistry of amino acids. Thus, serine, in which the R group is —CH_2OH, is a polar molecule. By contrast, alanine, which has —CH_3 as its R group, is nonpolar. The 20 common amino acids are grouped into five chemical classes, based on their R group:

1. *Nonpolar amino acids,* such as leucine, often have R groups that contain —CH_2 or —CH_3.
2. *Polar uncharged amino acids,* such as threonine, have R groups that contain oxygen (or —OH).
3. *Charged amino acids,* such as glutamic acid, have R groups that contain acids or bases that can ionize.
4. *Aromatic amino acids,* such as phenylalanine, have R groups that contain an organic (carbon) ring with alternating single and double bonds. These are quite nonpolar.
5. Some amino acids have *unique properties*. Some examples are methionine, which is often the first amino acid in a chain of amino acids; proline, which causes kinks in chains; and cysteine, which links chains together.

Each amino acid affects the shape of a protein differently, depending on the chemical nature of its side group. For example, portions of a protein chain with numerous nonpolar amino acids tend to fold into the interior of the protein by hydrophobic exclusion (more on this later in this section).

Forming peptide bonds

In addition to its R group, each amino acid is ionized at physiological pH, with a positive amino (NH_3^+) group at one end and a negative carboxyl (COO^-) group at the other. The amino groups on one amino acid can undergo a dehydration reaction with the carboxyl group on another amino acid to form a covalent bond. The covalent bond that links two amino acids is called a **peptide bond** (figure 3.11). The two amino acids linked by such a bond are

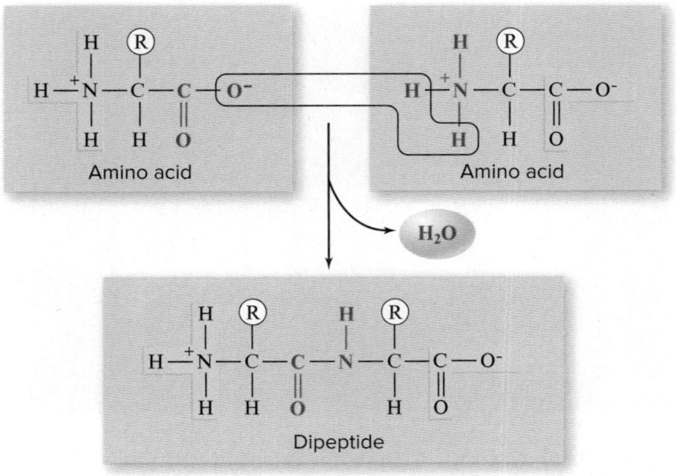

Figure 3.11 The peptide bond. A peptide bond is formed when the amino end of one amino acid is joined to the carboxyl end of another. Reacting amino and carboxyl groups are shown in red, and nonreacting groups are highlighted in green. Notice that the resulting dipeptide still has an amino end and a carboxyl end. Because of the partial double-bond nature of peptide bonds, the resulting peptide chain cannot rotate freely around these bonds.

not free to rotate around the N—C linkage because the peptide bond has a partial double-bond character. This is different from the N—C and C—C bonds to the central carbon of the amino acid. This lack of rotation about the peptide bond is one factor that determines the structural character of the coils and other regular shapes formed by chains of amino acids.

An unbranched chain of amino acids linked by peptide bonds is called a **polypeptide**. If a polypeptide can fold into a structure with some function, such as the antimicrobial enzyme lysozyme in your saliva, we also call it a protein. However, some proteins, such as the oxygen-carrying protein hemoglobin in your blood, consist of more than one polypeptide.

The pioneering work of Frederick Sanger in the early 1950s provided the evidence that each kind of protein has a specific amino acid sequence. Using chemical methods to remove successive amino acids and then identify them, Sanger succeeded in determining the amino acid sequence of insulin. In so doing he demonstrated clearly that this protein has a defined sequence, which is the same for all insulin molecules. Although many different amino acids occur in nature, only 20 commonly occur in proteins. Figure 3.12 illustrates these 20 amino acids and their side groups.

There Are Four Levels of Protein Structure

LEARNING OBJECTIVE 3.3.3 Describe the four levels of protein structure and how each is stabilized.

The structure of proteins is usually discussed in terms of a hierarchy of four levels: *primary, secondary, tertiary,* and *quaternary* (figure 3.13). We will examine this view and then integrate it with a more modern approach arising from our increasing knowledge of protein structure.

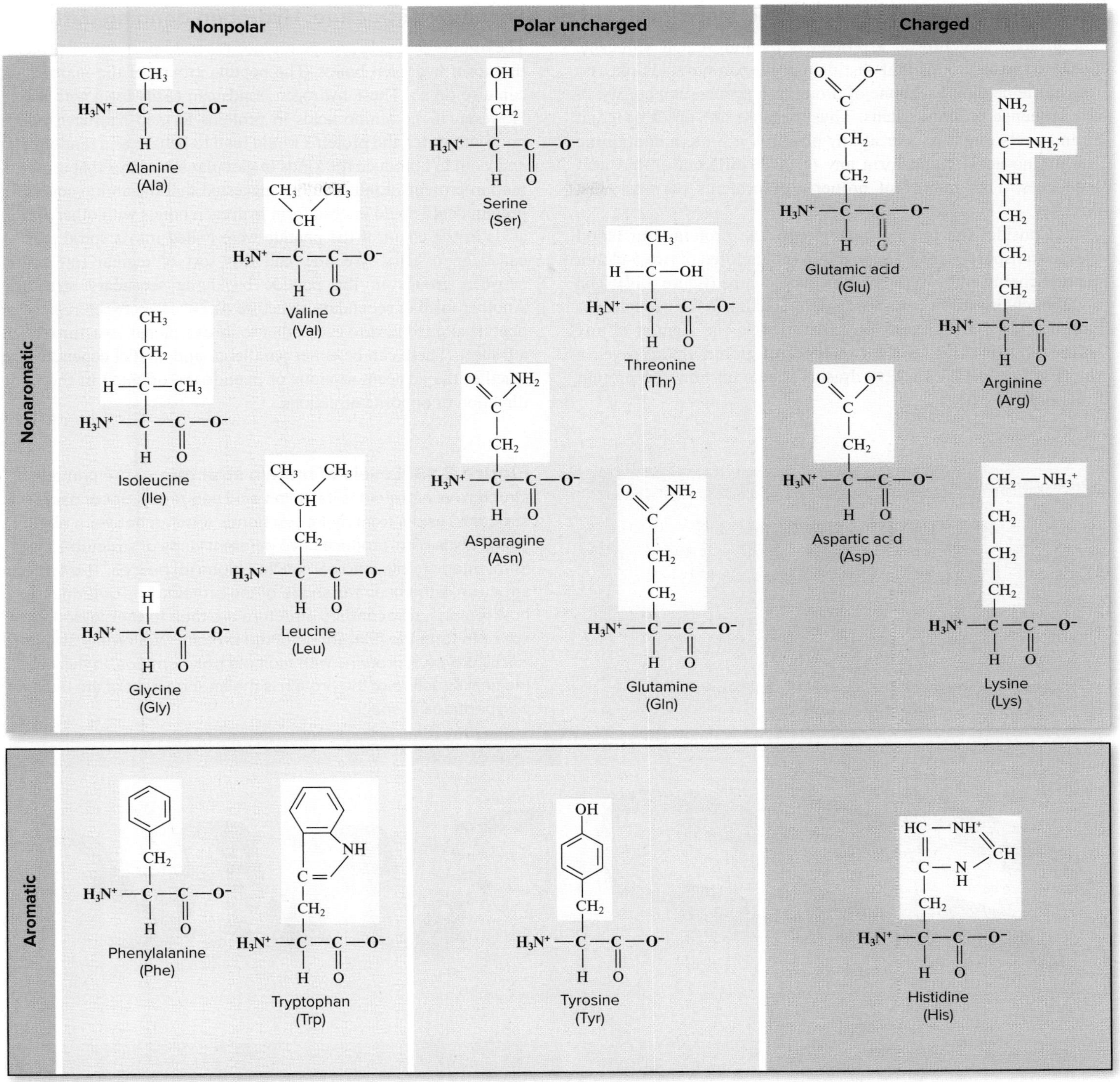

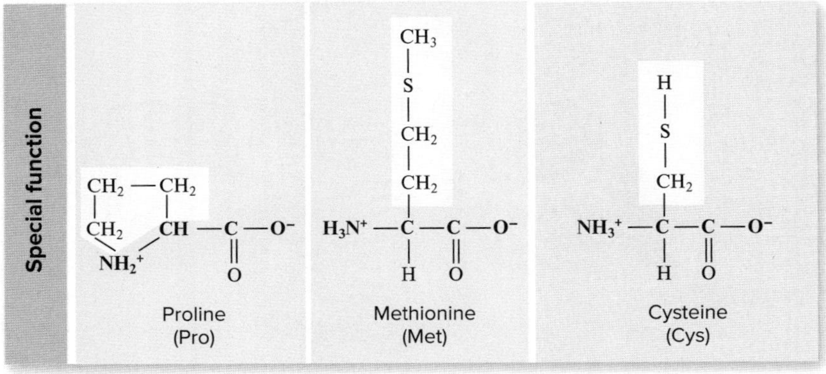

Figure 3.12 The 20 common amino acids. All amino acids have the same chemical backbone but differ in the side, or R, group. Seven of the amino acids are nonpolar, because they have $—CH_2$ or $—CH_3$ in their R groups. Two of the seven contain ring structures with alternating double and single bonds, which classifies them also as aromatic. Another five are polar, because they have oxygen or a hydroxyl group in their R groups. Five others are capable of ionizing to a charged form. The remaining three special-function amino acids have chemical properties that allow them to help form links between protein chains or kinks in proteins.

Primary structure: Amino acid sequence

The **primary structure** of a protein is its amino acid sequence. Because the R groups that distinguish the amino acids play no role in the peptide backbone of proteins, a protein can consist of any sequence of amino acids. Thus, because any of 20 different amino acids might appear at any position, a protein containing 100 amino acids could form any of 20^{100} different amino acid sequences. This important property of proteins permits great diversity.

Consider the protein hemoglobin, the protein your blood uses to transport oxygen. Hemoglobin is composed of two α-globin peptide chains and two β-globin peptide chains (figure 3.14). The α-globin chains differ from the β-globin chains in the sequence of amino acids. Furthermore, any alteration in the identity of any amino acid in either of the two types of globin chains—even a single amino acid—can have drastic effects on how the protein functions.

Secondary structure: Hydrogen bonding patterns

The amino acid side groups are not the only portions of proteins that form hydrogen bonds. The peptide groups of the main chain can also do so. These hydrogen bonds can easily form with water. However, if the amino acids in proteins formed hydrogen bonds only with water, the proteins would tend to behave as a random coil and wouldn't produce the kinds of globular structures that are common in proteins. Linus Pauling suggested that the amino acids of a protein chain could instead form hydrogen bonds with other amino acids in the chain, if the peptide were coiled into a spiral that he called an ***α* helix.** We now call this sort of regular interaction between groups in the peptide backbone **secondary structure.** Another form of secondary structure can occur between regions of peptide aligned next to each other to form a planar structure called a **β sheet.** These can be either parallel or antiparallel depending on whether the adjacent sections of peptide are oriented in the same direction or opposite directions.

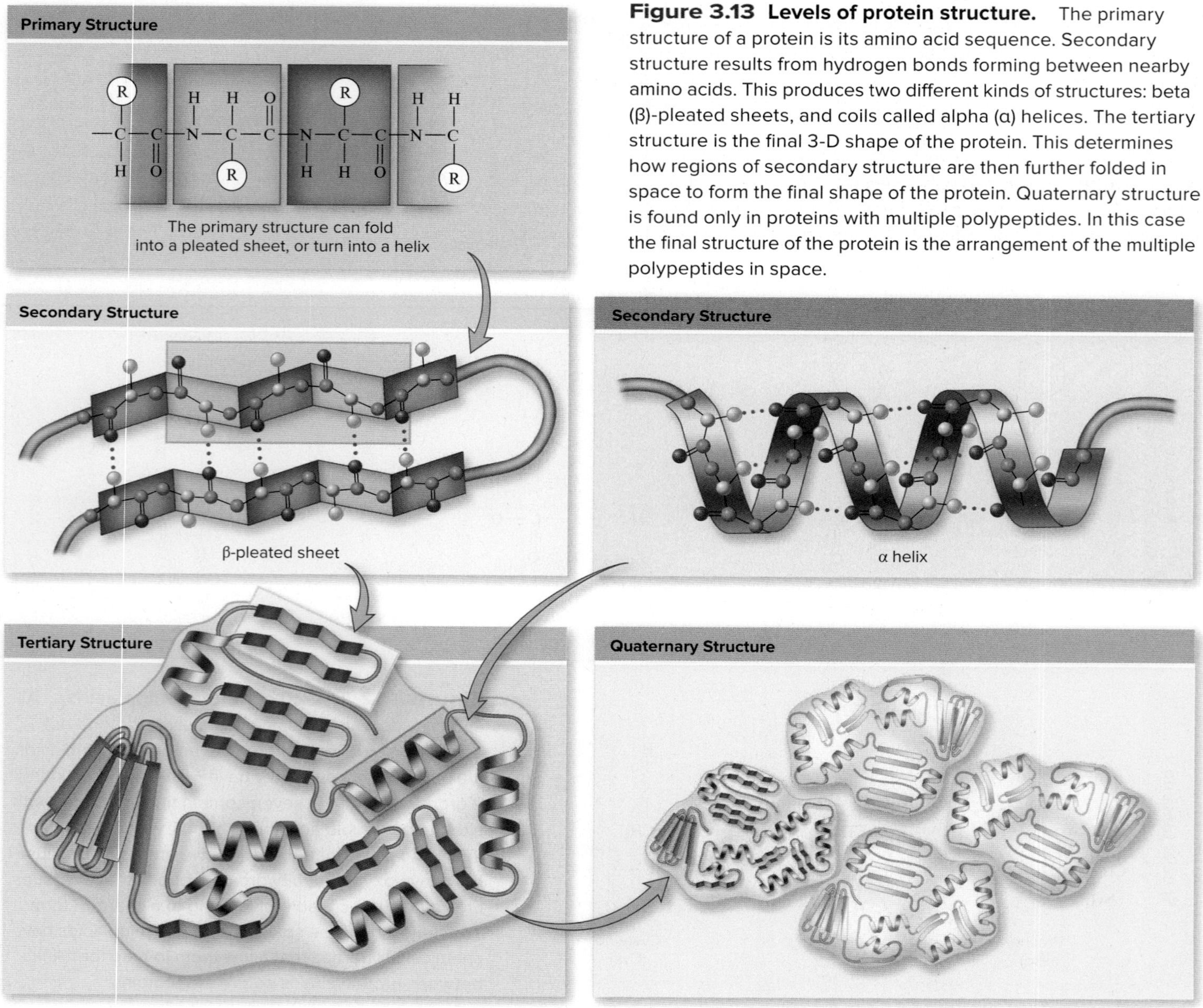

Primary Structure
The primary structure can fold into a pleated sheet, or turn into a helix
Secondary Structure
β-pleated sheet
Secondary Structure
α helix
Tertiary Structure
Quaternary Structure

Figure 3.13 Levels of protein structure. The primary structure of a protein is its amino acid sequence. Secondary structure results from hydrogen bonds forming between nearby amino acids. This produces two different kinds of structures: beta (β)-pleated sheets, and coils called alpha (α) helices. The tertiary structure is the final 3-D shape of the protein. This determines how regions of secondary structure are then further folded in space to form the final shape of the protein. Quaternary structure is found only in proteins with multiple polypeptides. In this case the final structure of the protein is the arrangement of the multiple polypeptides in space.

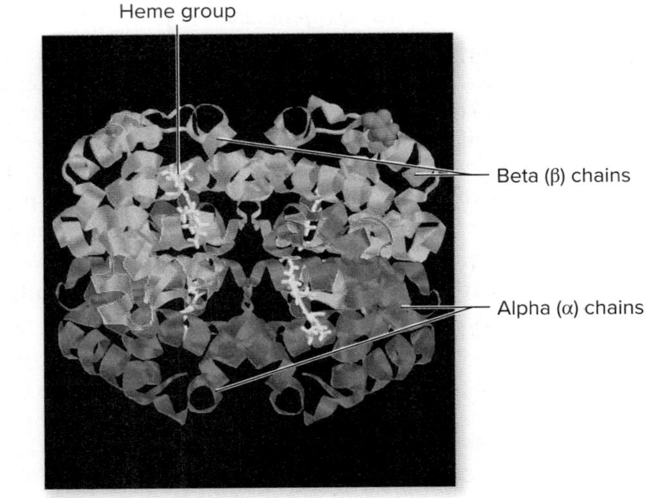

Figure 3.14 The hemoglobin molecule. The hemoglobin molecule is composed of four protein chain subunits: two copies of the "alpha chain" and two copies of the "beta chain." Each chain is associated with a heme group, and each heme group has a central iron atom, shown here in red, which can bind to a molecule of oxygen.

Kenneth Eward/BioGrafx/Science Source

These two kinds of secondary structure create regions of the protein that are cylindrical (α helices) and planar (β sheets). A protein's final structure can include regions of each type of secondary structure. For example, DNA-binding proteins usually have regions of α helix that can lie across DNA and interact directly with the bases of DNA. Porin proteins that form holes in membranes are composed of β sheets arranged to form a pore in the membrane. Finally, in hemoglobin the α- and β-globin peptide chains that make up the final molecule each have characteristic regions of helical and β-sheet secondary structure.

Tertiary structure: Folding into shape

The final folded shape of a globular protein is called its **tertiary structure.** A protein is initially driven into its tertiary structure by hydrophobic exclusion from water. Ionic bonds between oppositely charged R groups bring regions into close proximity, and disulfide bonds (covalent links between two cysteine R groups) lock particular regions together. The final folding of a protein is determined by its primary structure—the chemical nature of its side groups.

The tertiary structure of a protein is stabilized by a number of forces, including hydrogen bonding between R groups of different amino acids, electrostatic attraction between R groups with opposite charge (also called salt bridges), hydrophobic exclusion of nonpolar R groups, and covalent bonds in the form of disulfides (figure 3.15). The stability of a protein, once it has folded into its tertiary shape, is strongly influenced by how well its interior fits together. When two nonpolar chains in the interior are very close together, they experience a form of molecular attraction called van der Waals forces. Individually quite weak, these forces can add up to a strong attraction when many of them come into play, like the combined strength of the hundreds of hooks and loops of a strip of Velcro. These forces are

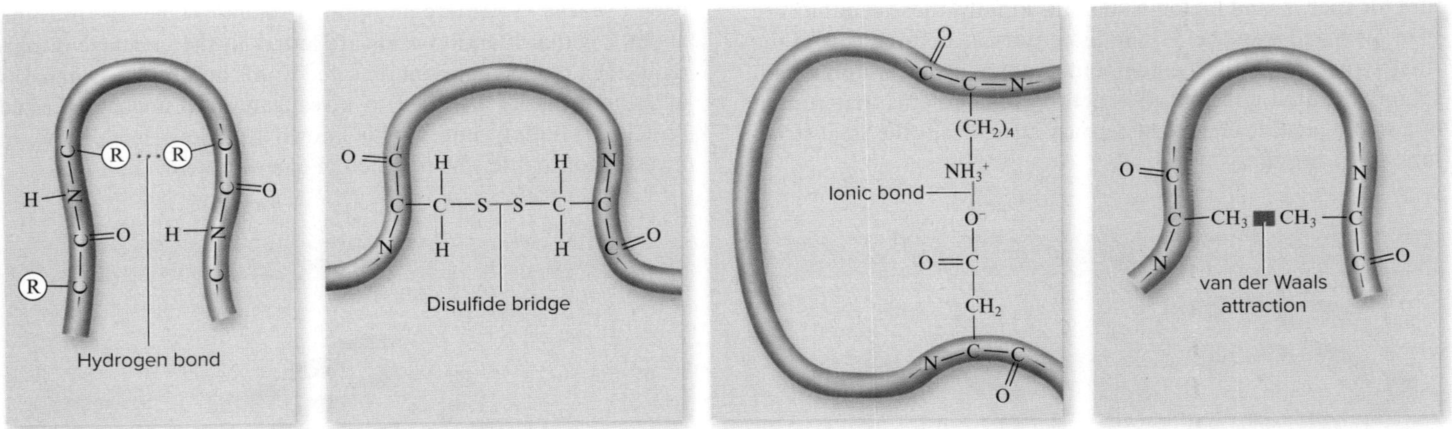

a. *b.* *c.* *d.*

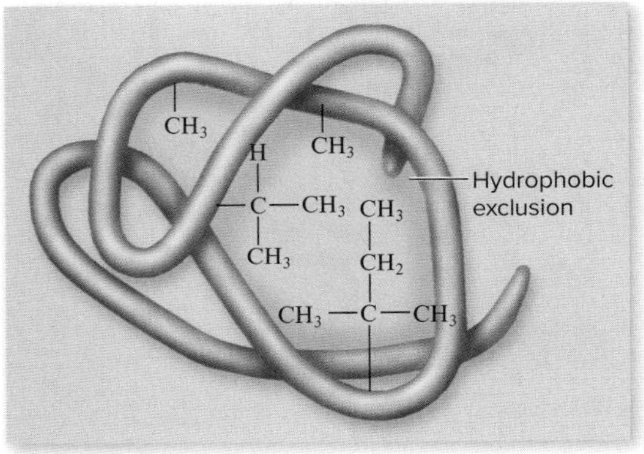

e.

Figure 3.15 Interactions that contribute to a protein's shape. Aside from the bonds that link together the amino acids in a protein, several other weaker forces and interactions determine how a protein will fold. ***a.*** Hydrogen bonds can form between the different amino acids. ***b.*** Covalent disulfide bridges can form between two cysteine side chains. ***c.*** Ionic bonds can form between groups with opposite charges. ***d.*** Van der Waals forces, which are weak attractions between atoms due to oppositely polarized electron clouds, can occur. ***e.*** Polar portions of the protein tend to gather on the outside of the protein and interact with water, whereas the hydrophobic portions of the protein, including nonpolar amino acid chains, are shoved toward the interior of the protein.

effective only over short distances, however. No "holes" or cavities exist in the interior of proteins. The variety of nonpolar amino acids, each with a different-size R group with its own distinctive shape, allows nonpolar chains to fit very precisely within the protein interior.

It is therefore not surprising that changing a single amino acid can drastically alter the structure, and thus the function, of a protein. The sickle-cell version of hemoglobin (HbS), for example, is due to a change at position B6 from glutamic acid (very polar) to valine (nonpolar) in the β-globin chain. This change substitutes a nonpolar amino acid for a polar one on the surface of the protein, leading the protein to become sticky and form clumps. More than 700 structural variants of hemoglobin are known, with up to 7% of the world's population being carriers of forms that are medically important.

Quaternary structure: Subunit arrangements

When two or more polypeptide chains associate to form a functional protein, the individual chains are referred to as subunits of the protein. The arrangement of these subunits is termed its **quaternary structure.** In proteins composed of subunits, the interfaces where the subunits touch one another are often nonpolar, and they play a key role in transmitting information between the subunits about individual subunit activities.

The protein hemoglobin is composed of two α-chain subunits and two β-chain subunits (figure 3.14). Each α- and β-globin chain has a primary structure consisting of a specific sequence of amino acids. This then assumes a characteristic secondary structure consisting of α helices and β sheets that are then pushed by hydrophobic exclusion into a specific tertiary structure for each α- and β-globin subunit. Finally, these subunits are then arranged into their final quaternary structure. This is the final structure of the hemoglobin protein. Only proteins with subunits exhibit quaternary structure. For proteins that consist of a single peptide chain, the tertiary structure is the final structure of the protein.

Motifs and Domains Organize Secondary Structure

LEARNING OBJECTIVE 3.3.4 Explain the role of motifs and domains in determining protein structure.

As biologists discovered the three-dimensional structure of proteins (an even more laborious task than determining the amino acid sequence), they noticed two classes of similarities between otherwise dissimilar proteins, which have come to be called motifs and domains.

Motifs

The smaller of these similar structures are called **motifs,** or "supersecondary structure." The term *motif* is borrowed from the arts and refers to a recurring thematic element in music or design.

One very common protein motif is the β-α-β motif, which creates a fold or crease; this is the so-called Rossmann fold at the core of nucleotide-binding sites in a wide variety of proteins. A second motif that occurs in many proteins is the β barrel, which is a β sheet folded around to form a tube. A third type of motif, the helix-turn-helix, consists of two α helices separated by a bend. This motif is important because many proteins use it to bind to the DNA double helix (figure 3.16).

Motifs indicate a logic to structure that investigators still do not fully understand. In large measure motifs seem to represent reuse by evolution of everyday workable solutions to recurrent problems, but in individual instances the design may have been tweaked to achieve an optimal solution. One way to think about it is that if amino acids are letters in the language of proteins, then motifs represent repeated words or phrases. Databases of protein motifs are used to investigate new, uncharacterized proteins. Finding motifs with known functions may allow an investigator to infer the function of a new protein.

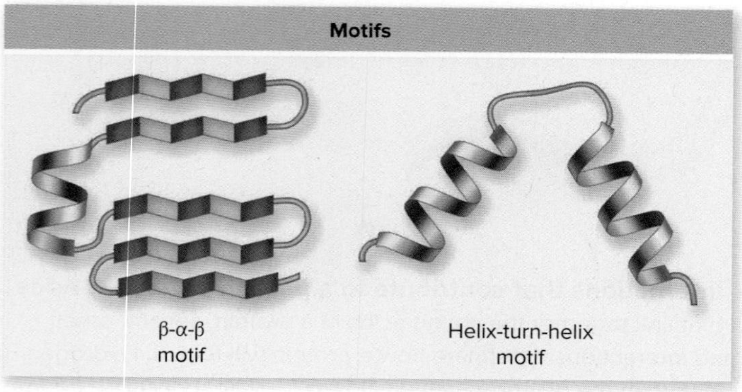

Motifs
β-α-β
motif
Helix-turn-helix
motif

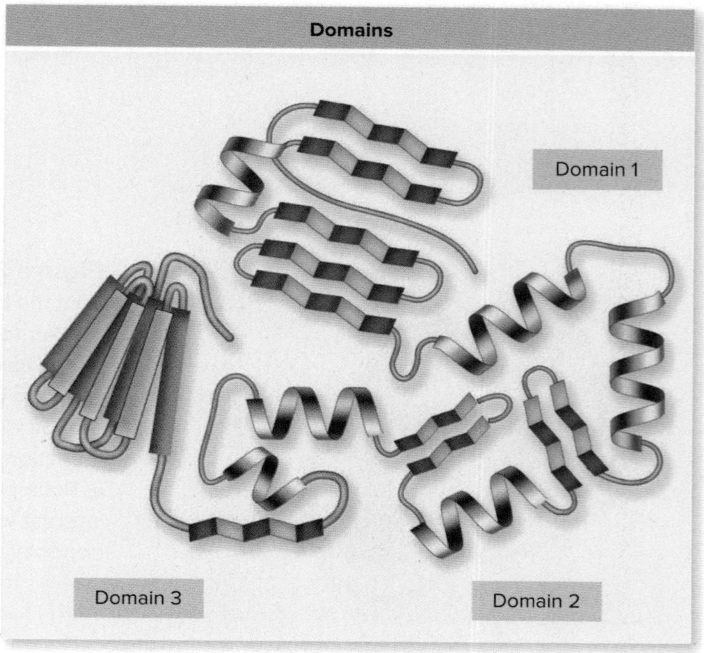

Domains
Domain 1
Domain 3
Domain 2

Figure 3.16 Motifs and domains. The elements of secondary structure can combine, fold, or crease to form motifs. These motifs are found in different proteins and can be used to predict function. Proteins also are made of larger domains, which are functionally distinct parts of a protein. The arrangement of these domains in space is the tertiary structure of a protein.

Domains

Domains of proteins are functional units within a larger structure. They can be thought of as substructure within the tertiary structure of a protein (figure 3.16). To continue the metaphor, amino acids are letters in the protein language, motifs are words or phrases, and domains are paragraphs, or even chapters in a book.

Most proteins are made up of multiple domains that perform different parts of the protein's function. In many cases these domains can be physically separated. For example, transcription factors (discussed in chapter 16) are proteins that bind to DNA and initiate its transcription. If the DNA-binding region of a particular factor is exchanged with that of a different transcription factor, then the specificity of the factor for DNA can be changed without changing its ability to stimulate transcription. Such "domain-swapping" experiments have been performed with many transcription factors, and they demonstrate very clearly that the DNA-binding and activation domains are functionally separate.

The functional domains of a protein may also help the protein fold into its proper shape. As a polypeptide chain folds, its domains take their proper shape, each more or less independently of the others. This action can be demonstrated experimentally by artificially producing the fragment of a polypeptide that forms the domain in the intact protein and showing that the fragment by itself folds to form the same structure it exhibits in the intact protein. A single polypeptide chain connects the domains of a protein, like a rope tied into several adjacent knots.

The Process of Folding Relies on Chaperone Proteins

LEARNING OBJECTIVE 3.3.5 Describe the role of chaperone proteins in protein folding.

The traditional view of protein folding was that it was a spontaneous process that was not affected by other cellular proteins. A protein randomly goes through different configurations as hydrophobic interactions with water shove nonpolar amino acids into the interior until the protein arrives at a final structure. It is now clear that this view is too simple. Protein chains can fold in so many different ways that a random process would simply take too long. In addition, as the open chain folds its way toward its final form, nonpolar, "sticky" interior portions are exposed during intermediate stages. If these intermediate forms are placed in a test tube in an environment identical to that inside a cell, they stick to other, unwanted protein partners, forming a gluey mess.

How do cells avoid having their proteins clump into a mass? An early clue came with the discovery of mutations in bacterial cells that prevent the assembly of replicating viruses. The mutations affected a newly identified class of proteins named **chaperone proteins,** for their ability to accompany a protein on its path to a properly folded state.

Molecular biologists have now identified many proteins that act as molecular chaperones. This class of proteins has multiple subclasses that aid in protein folding and assembly. They seem to be essential for viability, illustrating their fundamental importance. Many are *heat shock proteins,* produced in greatly increased amounts when cells are exposed to elevated temperature. High temperatures cause proteins to unfold, and heat shock chaperone proteins help the cell's proteins refold properly.

One class of these proteins, called **chaperonins,** has been extensively studied. In the bacterium *Escherichia coli (E. coli),* one example is the essential protein GroE chaperonin. In mutants in which the GroE chaperonin is inactivated, fully 30% of the bacterial proteins fail to fold properly. Chaperonins associate to form a large, macromolecular complex that resembles a cylindrical container. Proteins can move into the container, and the container itself can change its shape considerably (figure 3.17). Experiments have shown that an improperly folded protein can enter the chaperonin and be refolded. Although we don't know exactly how this happens, it seems to involve changes in the hydrophobicity of the interior of the chamber.

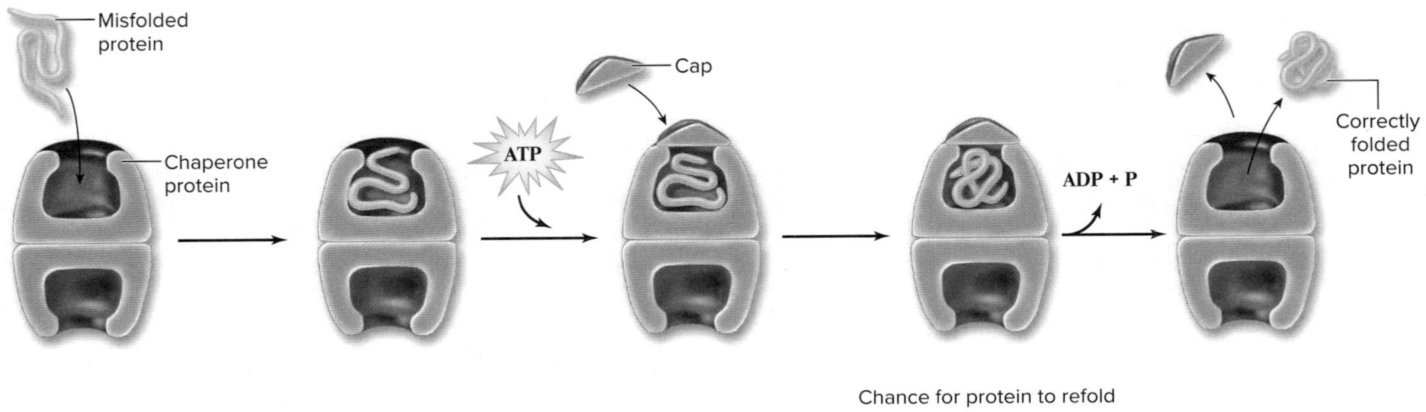

Figure 3.17 How one type of chaperone protein works. This barrel-shaped chaperonin is from the GroE family of chaperone proteins. It is composed of two identical rings, each with seven identical subunits and each subunit with three distinct domains. An incorrectly folded protein enters one chamber of the barrel, and a cap seals the chamber. Energy from the hydrolysis of ATP fuels structural alterations to the chamber, changing it from hydrophobic to hydrophilic. This change allows the protein to refold. After a short time, the protein is ejected, either folded or unfolded, and the cycle can repeat itself.

Some diseases result from improper folding

Cystic fibrosis (CF) is a hereditary disorder caused by a mutation in the gene for a protein that moves ions across cell membranes. As a result, people with cystic fibrosis have thicker than normal mucus, resulting in serious lung and organ problems. The most common variant, found in around 90% of CF patients, is the deletion of a single amino acid, which affects the protein's ability to fold properly. This results in a large number of misfolded proteins, which are destroyed—an example of protein homeostasis in which cells eliminate proteins that have not folded properly.

Environmental Changes Can Cause a Loss of Protein Structure

LEARNING OBJECTIVE 3.3.6 Explain how altered environmental conditions lead to denaturation of proteins.

If a protein's environment is altered, the protein may change its shape or even unfold completely. This process is called **denaturation.** Proteins can be denatured when the pH, temperature, or ionic concentration of the surrounding solution changes. Note that these environmental changes affect the interactions detailed in figure 3.15, leading to the loss of structure.

Denatured proteins are usually biologically inactive. This action is particularly significant in the case of enzymes. Because practically every chemical reaction in a living organism is catalyzed by a specific enzyme, it is vital that a cell's enzymes work properly. The traditional methods of food preservation—salt curing and pickling—involve denaturation of proteins. Prior to the general availability of refrigerators and freezers, the only practical way to keep microorganisms from growing in food was to keep the food in a solution containing a high concentration of salt or vinegar, which denatured the enzymes of most microorganisms and prevented them from growing.

When a protein's normal environment is re-established after denaturation, a small protein may spontaneously refold into its natural shape, driven by the interactions between its nonpolar amino acids and water (figure 3.18). This process is termed *renaturation,* and it was first established for the enzyme ribonuclease (RNase). The renaturation of RNase led to the doctrine that primary structure determines tertiary structure. Larger proteins can rarely refold spontaneously, however, because of the complex nature of their final shape, so this simple idea needs to be qualified.

The fact that some proteins can spontaneously renature implies that tertiary structure is strongly influenced by primary structure. In an extreme example, the *E. coli* ribosome can be taken apart and put back together experimentally. Although this process requires temperature and ion concentration shifts, it indicates an amazing degree of self-assembly. That complex structures can arise by self-assembly is a key idea in the study of modern biology.

It is important to distinguish denaturation from **dissociation.** For proteins with quaternary structure, the subunits may be dissociated without losing their individual tertiary

SCIENTIFIC THINKING

Hypothesis: *The 3-D structure of a protein arises from the primary structure of the protein and the solution conditions.*

Prediction: *If a protein is denatured and allowed to renature under native conditions, it will refold into the native structure.*

Test: *Ribonuclease is treated with a reducing agent to break disulfide bonds and is then treated with urea to completely unfold the protein. The disulfide bonds are reformed under nondenaturing conditions to assess whether or not the protein refolds properly.*

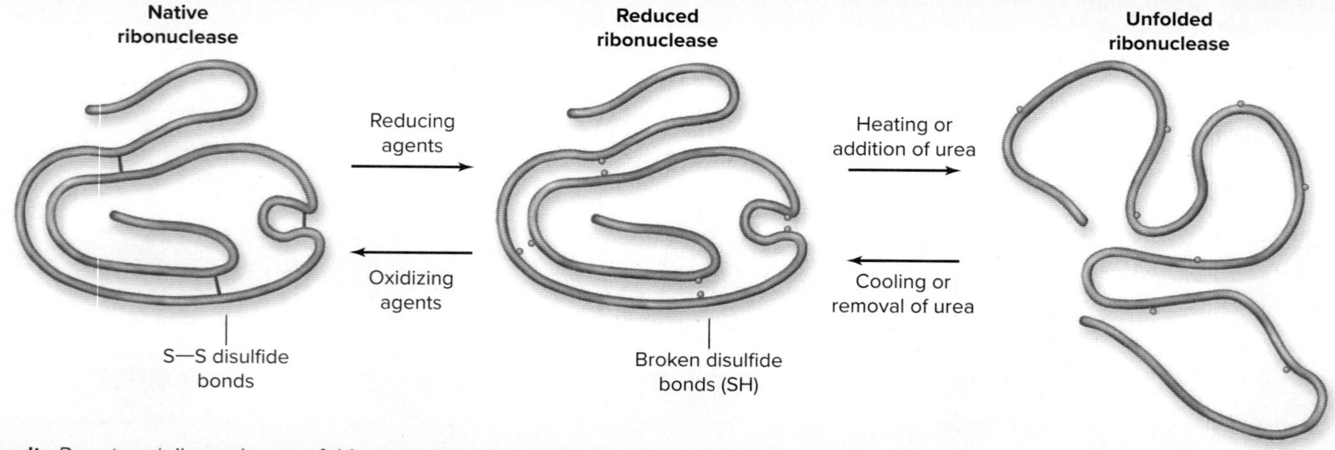

Result: *Denatured ribonuclease refolds properly under nondenaturing conditions.*

Conclusion: *The hypothesis is supported. The information in the primary structure (amino acid sequence) is sufficient for refolding to occur. This implies that protein folding results in the thermodynamically stable structure.*

Further Experiments: *If the disulfide bonds were allowed to reform under denaturing conditions, would we get the same result? How can we rule out that the protein had not been completely denatured and therefore retained some structure?*

Figure 3.18 Primary structure determines tertiary structure.

structure. For example, the four subunits of hemoglobin may dissociate into four individual molecules (two α-globins and two β-globins) without denaturation of the folded globin proteins. They readily reassume their four-subunit quaternary structure.

REVIEW OF CONCEPT 3.3

Proteins are molecules with diverse functions. They are constructed from 20 different kinds of amino acids. Protein structure can be viewed at four levels: (1) the amino acid sequence, or primary structure; (2) coils and sheets, called secondary structure; (3) the three-dimensional shape, called tertiary structure; and (4) individual polypeptide subunits associated in a quaternary structure.

- *How does our knowledge of protein structure help us to predict the function of unknown proteins?*

3.4 Nucleic Acids Store and Express Genetic Information

The life of a cell depends on its ability to produce a large number of proteins, each with a specific sequence. The information necessary to produce the correct proteins at the correct time is encoded by the cell within nucleic acids. Two main varieties of nucleic acids are **deoxyribonucleic acid (DNA)** (figure 3.19) and **ribonucleic acid (RNA)** (figure 3.20). DNA encodes the genetic information used to assemble proteins, and RNA has a variety of roles in gene expression (as discussed in detail in chapter 15).

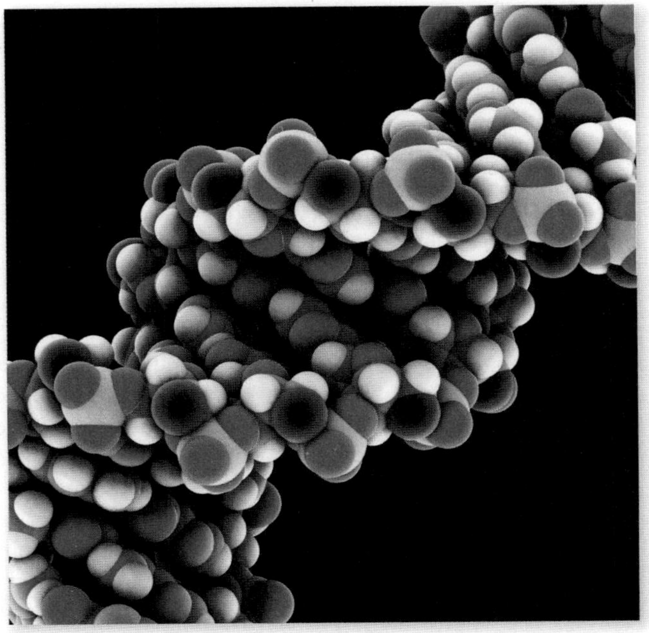

Figure 3.19 DNA. A space-filling model of DNA.

LAGUNA DESIGN/Science Photo Library/Alamy Stock Photo

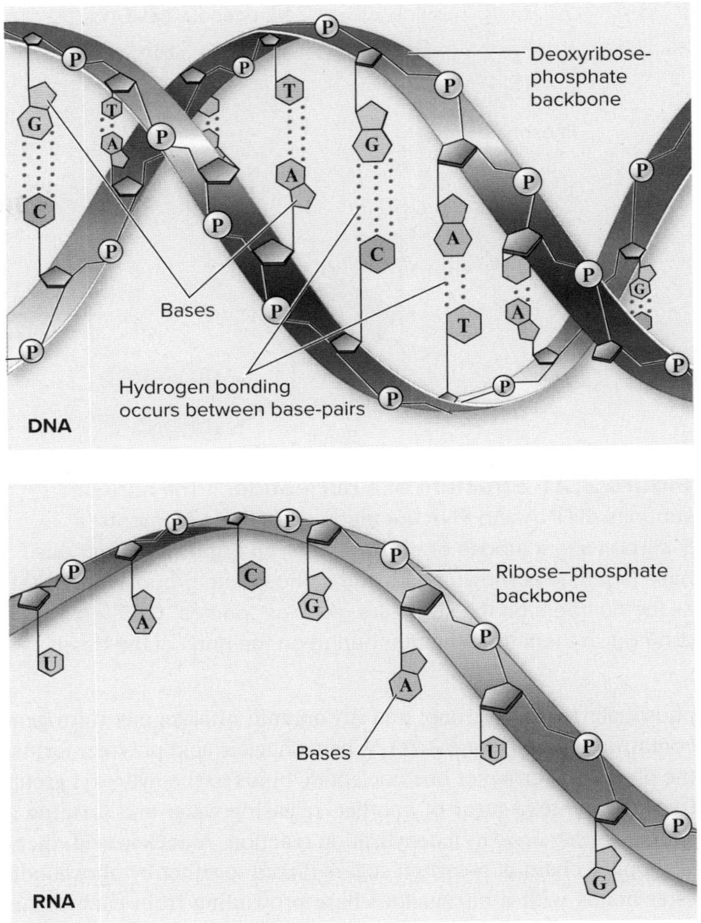

Figure 3.20 DNA versus RNA. DNA forms a double helix, uses deoxyribose as the sugar in its sugar–phosphate backbone, and uses thymine among its nitrogenous bases. RNA is usually single-stranded, uses ribose as the sugar in its sugar–phosphate backbone, and uses uracil in place of thymine.

The Information Molecules of the Cell Are Long Chains of Nucleotides

LEARNING OBJECTIVE 3.4.1 Compare and contrast the structures of DNA and RNA.

Unique among macromolecules, nucleic acids are able to serve as templates to produce precise copies of themselves. This characteristic allows genetic information to be preserved during cell division and copied during the manufacture of proteins. DNA, found primarily in the nuclear region of cells, contains the genetic information necessary to specify all of the proteins in a cell. RNA is also copied from DNA, and has a variety of roles in gene expression. We will explore the structure and function of DNA and RNA in later chapters, but introduce the basics of nucleic acid structure here.

Nucleic acids are nucleotide polymers

Nucleic acids are long polymers of repeating subunits called **nucleotides.** Each nucleotide consists of three components: a pentose, or 5-carbon sugar (ribose in RNA and deoxyribose in DNA); a

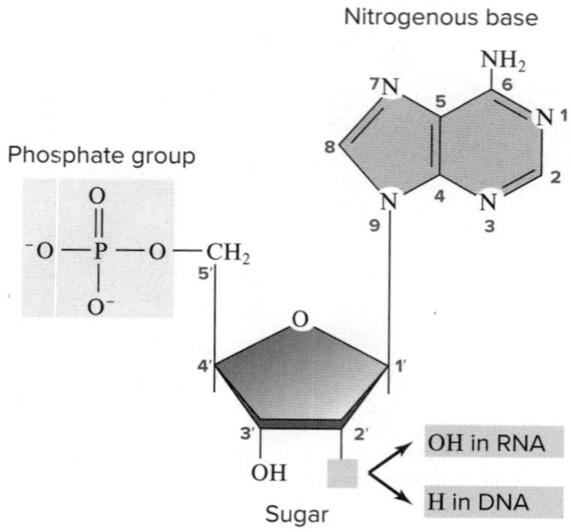

Figure 3.21 Structure of a nucleotide. The nucleotide subunits of DNA and RNA are made up of three elements: a 5-carbon sugar (ribose or deoxyribose), an organic nitrogenous base (adenine is shown here), and a phosphate group. Notice that all the numbers on the sugar are given as "primes" (1′, 2′, etc.) to distinguish them from the numbering on the rings of the bases.

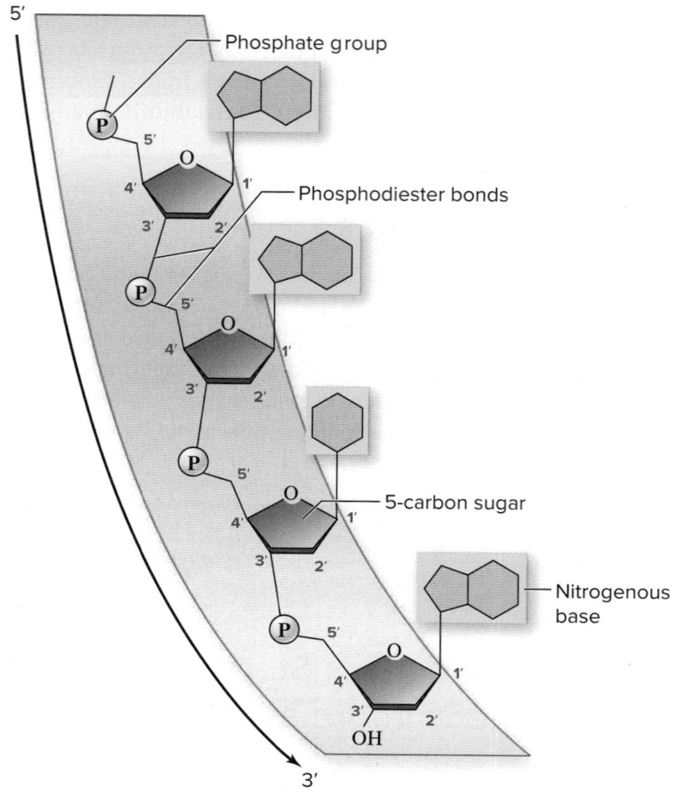

Figure 3.23 The structure of a polynucleotide chain. In a nucleic acid, nucleotides are linked to one another via phosphodiester bonds formed between the phosphate of one nucleotide and the sugar of the next nucleotide. The organic bases protrude from this sugar–phosphate backbone. The backbone also has different ends: a 5′ phosphate end and a 3′ hydroxyl end (the numbers come from the numbers in the sugars).

phosphate (—PO_4) group; and an organic nitrogenous (nitrogen-containing) base (figure 3.21). When a nucleic acid polymer forms, the phosphate group of one nucleotide binds to the hydroxyl group from the pentose sugar of another, releasing water and forming a *phosphodiester bond* by a dehydration reaction. A **nucleic acid,** then, is simply a chain of 5-carbon sugars linked together by phosphodiester bonds with a nitrogenous base protruding from each sugar. These chains of nucleotides, *polynucleotides,* have different ends: a phosphate on one end and an —OH from a sugar on the other end. We conventionally refer to these ends as 5′ ("five-prime," —PO_4) and 3′ ("three-prime," —OH), taken from the carbon numbering of the sugar, as illustrated in figure 3.21.

Two types of nitrogenous bases occur in nucleotides (figure 3.22). The first type, *purines,* are large, double-ring molecules found in both DNA and RNA; the two types of purines are adenine (A) and guanine (G). The second type, *pyrimidines,* are smaller, single-ring molecules; they include cytosine (C, in both DNA and RNA), thymine (T, in DNA only), and uracil (U, in RNA only).

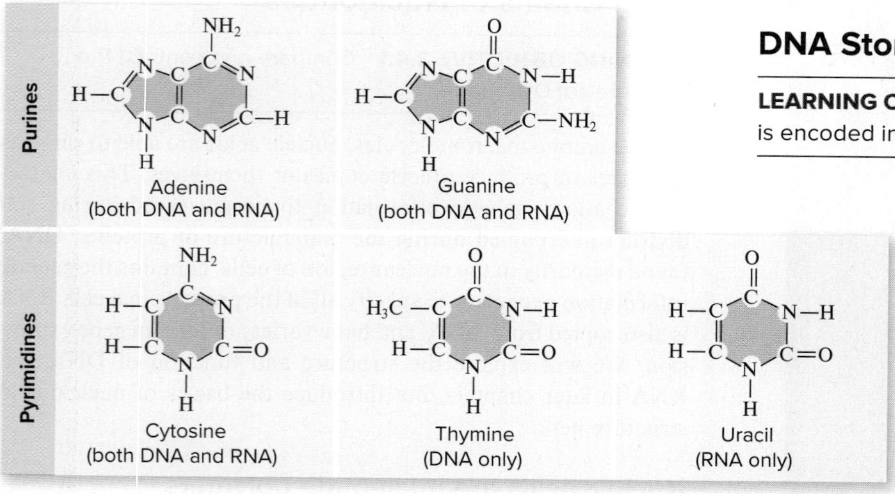

Figure 3.22 The structure of the nitrogenous bases. The nitrogenous bases are either purines (2 rings) or pyrimidines (1 ring). The base thymine is round in DNA, and replaced by uracil in RNA.

DNA Stores the Genetic Information

LEARNING OBJECTIVE 3.4.2 Explain how genetic information is encoded in the structure of DNA.

The function of DNA as an information-storage molecule is intimately tied to its structure (figure 3.23). Organisms use sequences of nucleotides in DNA to encode the information specifying the amino acid sequences of their proteins. This method of encoding information is very similar to the way in which sequences of letters encode information in a sentence. A sentence written in English consists of a combination of the 26 different letters of the alphabet in a certain order; the code of a DNA molecule consists of different combinations of the four types of nucleotides in a certain order, such as the sequence CGCTTACG.

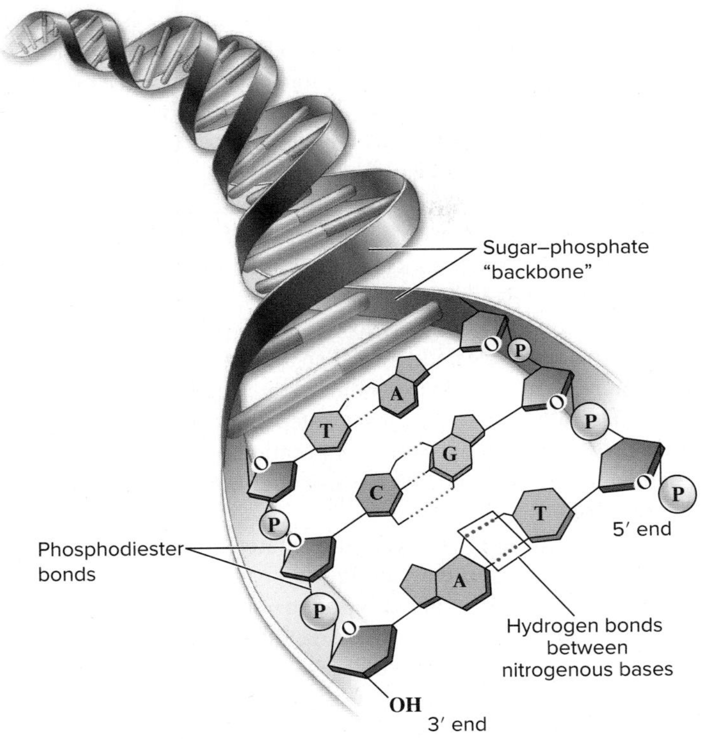

Figure 3.24 The structure of DNA. DNA consists of two polynucleotide chains running in opposite directions wrapped about a single helical axis. Hydrogen bond formation (dotted lines) between the nitrogenous bases, called base-pairing, causes the two chains of DNA to bind to each other and form a double helix.

DNA molecules in organisms exist not as single chains folded into complex shapes, like proteins, but rather as two chains of nucleotides wrapped about each other—a long, linear molecule in eukaryotes and a circular molecule in most prokaryotes. The two nucleotide chains of a DNA polymer wind around each other like the outside and inside rails of a spiral staircase. Such a spiral shape is called a helix, and a helix composed of two chains is called a **double helix.** Each step of DNA's helical staircase is composed of a **base-pair**. The pair consists of a base in one chain attracted by hydrogen bonds to a base opposite it on the other chain (figure 3.24).

The base-pairing rules are rigid: adenine can pair only with thymine (in DNA) or with uracil (in RNA), and cytosine can pair only with guanine. The bases that participate in base-pairing are said to be **complementary** to each other. Additional details of the structure of DNA and how it interacts with RNA in the production of proteins are presented in chapters 14 and 15.

RNA Has Many Roles in a Cell

LEARNING OBJECTIVE 3.4.3 Describe four significant roles RNA plays in the cell's utilization of information stored in DNA.

RNA is similar to DNA, but with two major chemical differences. First, RNA molecules contain ribose sugars, in which the C-2 carbon is bonded to a hydroxyl group. (In DNA, a hydrogen atom replaces this hydroxyl group.) Second, RNA molecules use uracil in place of thymine. Uracil has a similar structure to thymine, except that one of its carbons lacks a methyl (—CH_3) group.

RNA is produced by transcription (copying) from DNA and is usually single-stranded (figure 3.20). The role of RNA in cells is quite varied: it carries information in the form of messenger RNA (mRNA), it is part of the ribosome in the form of ribosomal RNA (rRNA), and it carries amino acids in the form of transfer RNA (tRNA).

There has been a revolution of late in how we view RNA. Enzymes have been found in which RNA, not protein, has catalytic activity. New roles for RNA are being discovered as we refine our view of cells at the molecular level. We now know that newly discovered forms of RNA, micro-RNA and small interfering RNAs, are involved in regulating gene expression and may even help protect the genome from invading viruses (explored in more detail in chapter 16). All of this is changing how we view the role of RNA in cellular metabolism. We are even finding that much more of our own genome is being copied into RNA than is being used to make proteins.

Other Nucleotides Are Vital Components of Energy Reactions

LEARNING OBJECTIVE 3.4.4 Describe how adenosine triphosphate stores and releases chemical energy.

Nucleotides play other critical roles in the life of a cell. For example, **adenosine triphosphate (ATP)** is called the energy currency of the cell. The hydrolysis of ATP releases energy, which can be used to drive energetically unfavorable processes. Cells use the energy from ATP hydrolysis in a variety of transactions, the way we use money in society. ATP hydrolysis can provide energy for energetically unfavorable chemical reactions, for transport across membranes, or for the movement of cells.

ATP is a ribonucleotide triphosphate (figure 3.25), meaning it contains ribose, with three phosphate groups attached to the 5′ carbon of the ribose. The hydrolysis of ATP produces adenosine diphosphate (ADP) plus phosphate and releases energy. This reaction releases energy because the negatively charged phosphates are repelled by each other, making the overall molecule unstable. Like a coiled spring, they are poised to push apart. Thus, when a phosphate is removed, energy is released. This also means that cells must obtain energy from food molecules to synthesize ATP, as we will discuss in chapter 6.

Two other important nucleotide-containing molecules are **nicotinamide adenine dinucleotide (NAD^+)** and **flavin adenine**

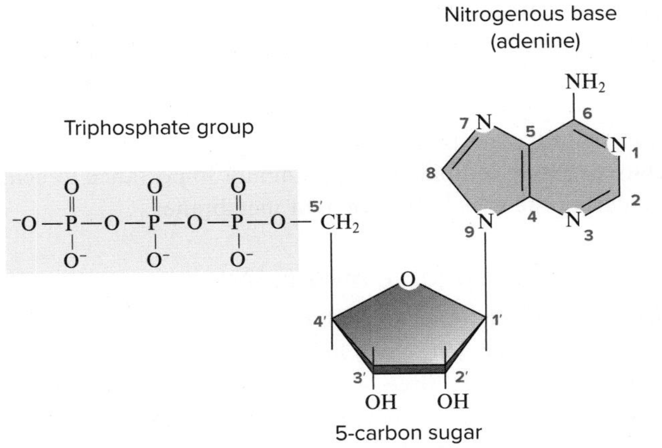

Figure 3.25 ATP. Adenosine triphosphate (ATP) contains adenine, a 5-carbon sugar, and three phosphate groups.

dinucleotide (FAD). These molecules function as electron carriers in a variety of cellular processes. You will see the action of these molecules in detail when we discuss respiration and photosynthesis in chapters 7 and 8.

REVIEW OF CONCEPT 3.4

A nucleic acid is a polymer composed of alternating phosphate and 5-carbon sugar groups with a nitrogenous base protruding from each sugar. In DNA, this sugar is deoxyribose. In RNA, the sugar is ribose. RNA also contains the base uracil instead of thymine. DNA is a double-stranded helix that stores hereditary information as a specific sequence of nucleotide bases. RNA is a single-stranded molecule consisting of a transcript of a DNA sequence that directs protein synthesis.

- *If a DNA molecule can form a double strand, what prevents an RNA transcript of that DNA molecule from forming a double strand?*

3.5 Hydrophobic Lipids Form Fats and Membranes

Fats Consist of Fatty Acids Attached to Glycerol

LEARNING OBJECTIVE 3.5.1 Distinguish between triglycerides that form solid fats and those that form liquid oils.

Lipids are a somewhat loosely defined group of molecules with one main chemical characteristic: they are insoluble in water. Storage fats such as animal fat are one kind of lipid. Oils such as those from olives, corn, and coconut are also lipids, as are waxes such as beeswax and earwax. Even some vitamins are lipids! Lipids have a very high proportion of nonpolar carbon–hydrogen (C—H) bonds, and so long-chain lipids cannot fold up like a protein to confine their nonpolar portions away from the surrounding aqueous environment. Instead, when they are placed in water, many lipid molecules spontaneously cluster together and expose any polar (hydrophilic) groups they have to the surrounding water, while confining the nonpolar (hydrophobic) parts of the molecules together within the cluster. You may have noticed this effect when you add oil to a pan containing water, and the oil beads up into cohesive drops on the water's surface. This spontaneous assembly of lipids is of paramount importance to cells, as it underlies the structure of cellular membranes.

Fats are hydrophobic molecules

Fats are lipids built from two kinds of molecules: fatty acids and glycerol. Fatty acids are long-chain hydrocarbons with a carboxyl group (COOH) at one end. Glycerol is a three-carbon polyalcohol (three —OH groups). A fat molecule consists of a glycerol molecule with three fatty acids attached by dehydration synthesis, one to each carbon of the glycerol backbone. Because it contains three fatty acids, a fat molecule is commonly called a **triglyceride** (the more accurate chemical name is *triacylglycerol*). This basic structure is depicted in figure 3.26. The three fatty acids of a triglyceride need not be identical, and often they are very different from one another. The hydrocarbon chains of fatty acids vary in length. The most common are even-numbered chains of 14 to 20 carbons. The many C—H bonds of fats serve as a form of long-term energy storage.

If all of the internal carbon atoms in a fatty acid chain are bonded to at least two hydrogen atoms, the fatty acid is said to be **saturated,** which refers to its having all the hydrogen atoms possible. A fatty acid that has double bonds between one or more pairs of successive carbon atoms is said to be **unsaturated.** Fatty acids with one double bond are called *monounsaturated,* and those with more than one double bond are termed **polyunsaturated.** Most naturally occurring unsaturated fatty acids have double bonds with a *cis* (same-side) configuration, in which the carbon chain is on the same side before and after the double bond (as in figure 3.26*b*).

When fats are partially hydrogenated industrially, this can produce double bonds with a *trans* (opposite-side) configuration, in which the carbon chain is on opposite sides before and after the double bond. These are the so-called ***trans* fats.** *Trans* fats have been linked to elevated levels of low-density lipoprotein (LDL; "bad cholesterol") and lowered levels of high-density lipoprotein (HDL; "good cholesterol"), a condition associated with an increased risk for coronary heart disease.

Having double bonds in its fatty acid chains changes the behavior of the fat molecule, because free rotation cannot occur about a C═C double bond as it can with a C—C single bond. This characteristic affects the melting point of the fat–that is, whether the fatty acid is a solid fat or a liquid oil at room temperature. Fats containing polyunsaturated fatty acids have low melting points, because their fatty acid chains bend at the double bonds, preventing the fat molecules from aligning closely with one another. Most unsaturated fats, such as plant oils, are liquid at room temperature.

Organisms contain many other kinds of lipids besides fats (figure 3.27). *Terpenes* are long-chain lipids that are components of many biologically important pigments, such as chlorophyll and the visual pigment retinal. Rubber is also a terpene. *Steroids,* another class of lipid, are composed of four carbon rings. Most animal cell membranes contain the steroid cholesterol. Other steroids, such as testosterone and estrogen, function as hormones in multicellular animals. *Prostaglandins* are a group of about 20 lipids that are modified fatty acids, with two nonpolar "tails" attached to a 5-carbon ring. Prostaglandins act as local chemical messengers in many vertebrate tissues. Chapters 35 and 36 explore the effects of some of these complex fatty acids.

Fats are excellent energy-storage molecules

Most fats contain over 40 carbon atoms. The ratio of energy-storing C—H bonds in fats is more than twice that of carbohydrates (refer to section 3.2), making fats much more efficient molecules for storing chemical energy. On average, fats yield about 9 kilocalories (kcal) of chemical energy per gram, as compared with about 4 kcal/g for carbohydrates.

Most fats produced by animals (except some fish oils) are saturated, whereas most plant fats are unsaturated. The exceptions are the tropical plant oils (palm oil and coconut oil), which are saturated even though they are liquid at room temperature.

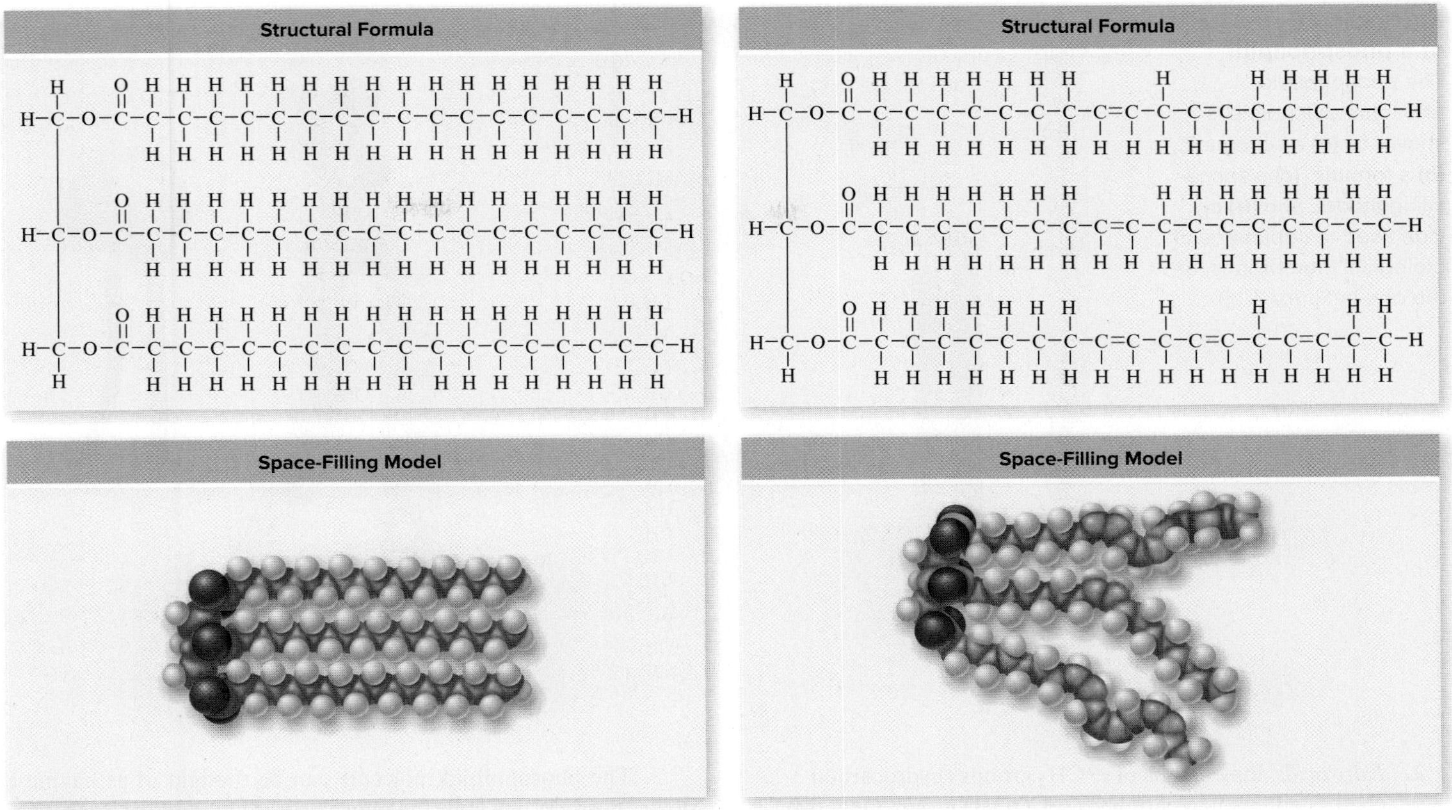
Structural Formula
Space-Filling Model
Structural Formula
Space-Filling Model

a. *b.*

Figure 3.26 Saturated and unsaturated fats. ***a.*** A saturated fat is composed of triglycerides that contain three saturated fatty acids (the kind that have no double bonds). A saturated fat therefore has the maximum number of hydrogen atoms bonded to its carbon chain. Most animal fats are saturated. ***b.*** Unsaturated fat is composed of triglycerides that contain one or more unsaturated fatty acids (the kind that have one or more double bonds). These have fewer than the maximum number of hydrogen atoms bonded to the carbon chain. This example includes both a monounsaturated and two polyunsaturated fatty acids. Plant fats are typically unsaturated. The many kinks of the double bonds prevent the triglycerides from closely aligning, which makes plant fats liquid at room temperature.

An oil may be converted into a solid fat by chemically adding hydrogen. Most peanut butter is artificially hydrogenated to make the peanut fats solidify, preventing them from separating out as oils while the jar sits on the store shelf. However, artificially hydrogenating unsaturated fats produces the *trans*-fatty acids described previously.

When an animal consumes carbohydrate, any excess is converted into glycogen or fat and is reserved for future use. This can cause problems in both humans and domestic animals as they age. If decreasing energy expenditures are not balanced by decreasing consumption, they will gain weight. While this is an oversimplification, there is some truth to it.

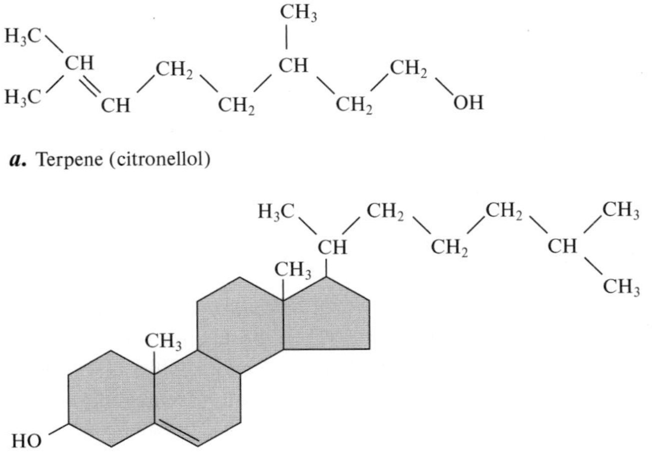

a. Terpene (citronellol)

b. Steroid (cholesterol)

Figure 3.27 Other kinds of lipids. ***a.*** Terpenes are found in biological pigments, such as chlorophyll and retinal, and ***b.*** steroids play important roles in membranes and as the basis for a class of hormones involved in chemical signaling.

Phospholipids Form Membranes

LEARNING OBJECTIVE 3.5.2 Explain why the lipid bilayer of a biological membrane forms spontaneously.

Complex lipid molecules called **phospholipids** are among the most important molecules of the cell, because they form the core of all biological membranes. An individual phospholipid can be thought of as a substituted triglyceride, that is, a triglyceride with a phosphate replacing one of the fatty acids. The basic structure of a phospholipid includes three kinds of subunits:

1. *Glycerol,* a 3-carbon alcohol, in which each carbon bears a hydroxyl group. Glycerol forms the backbone of the phospholipid molecule, just as it does for a fat molecule.

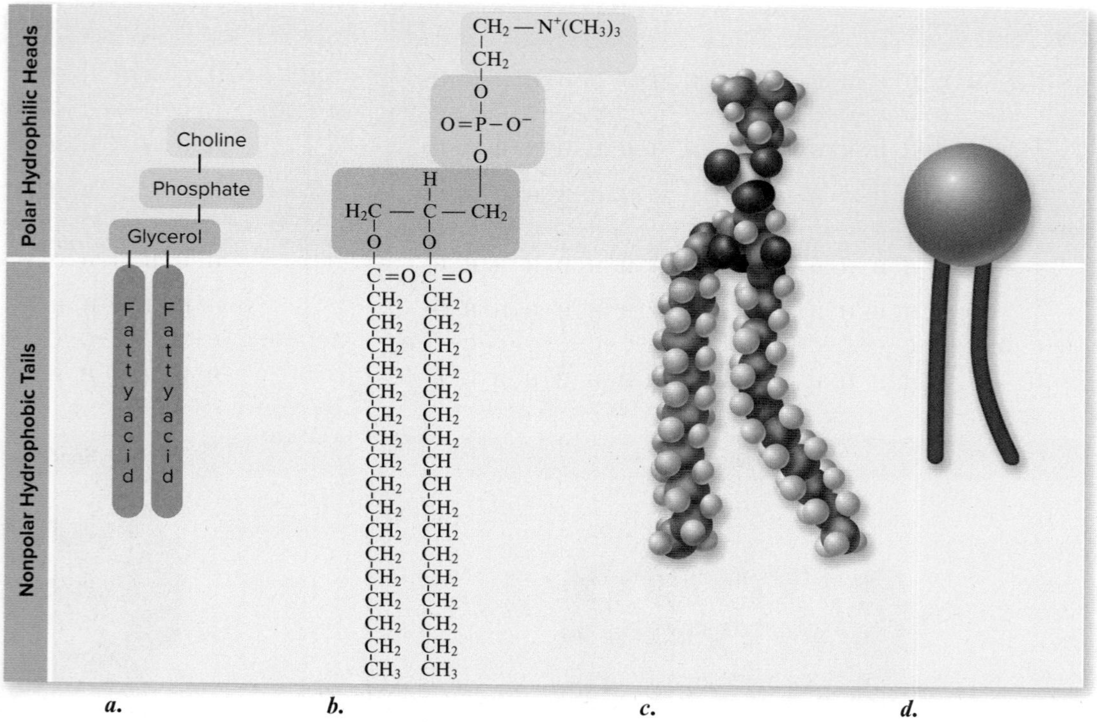

Figure 3.28 Portraits of a phospholipid. The phospholipid phosphatidylcholine is shown as **(*a*)** a schematic, **(*b*)** a formula, **(*c*)** a space-filling model, and **(*d*)** an icon used in depictions of biological membranes, as the one in figure 3.29.

2. *Fatty acids,* long chains of $—CH_2$ groups (hydrocarbon chains) ending in a carboxyl (—COOH) group. Only two fatty acids are attached to the glycerol backbone in a phospholipid molecule, rather than the three in a fat.
3. *A phosphate group* ($—PO_4^{2-}$) attached to one end of the glycerol. The charged phosphate group usually has a charged organic molecule linked to it, such as choline, ethanolamine, or the amino acid serine.

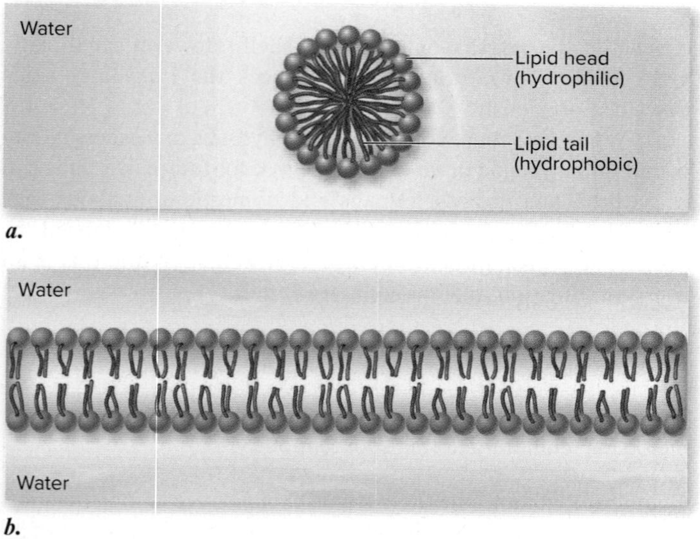

Figure 3.29 Lipids spontaneously form micelles or lipid bilayers in water. In an aqueous environment, amphipathic lipid molecules orient so that their polar (hydrophilic) heads are in the polar medium, water, and their nonpolar (hydrophobic) tails are held away from the water. ***a.*** Detergents form droplets called micelles, or ***b.*** phospholipid molecules can arrange themselves into two layers, called a bilayer.

The phospholipid molecule can be thought of as having a polar "head" at one end (the phosphate group) and two long, very nonpolar "tails" at the other (figure 3.28). We call molecules with both polar and nonpolar regions amphipathic. This structure is essential for how these molecules function, although it first appears paradoxical. Why would a molecule need to be soluble in water but also not soluble in water? The formation of a membrane shows the unique properties of such a structure.

In water, amphipathic lipids will aggregate away from water to form different structures. One such structure is the spherical micelle, with the hydrophobic portion facing inward (figure 3.29*a*). This is how detergent molecules work to make grease soluble in water, with the grease within the nonpolar interior of the micelle, and the polar surface of the micelle soluble in water. Because of their long fatty acid tails, phospholipids form a more complex bilayer structure in water. Two layers of phospholipid line up, with the hydrophobic tails of each layer pointing toward one another, or inward, and the hydrophilic heads oriented outward (figure 3.29*b*). Lipid bilayers are the basic framework of biological membranes, the subject of chapter 5.

REVIEW OF CONCEPT 3.5

Triglycerides are made of fatty acids linked to glycerol. Fats can contain twice as many C—H bonds as carbohydrates. Oxidation of fats releases energy. Because the C—H bonds in lipids are nonpolar, they are not water-soluble and aggregate together in water. Phospholipids replace one fatty acid with a hydrophilic phosphate group. This allows them to spontaneously form bilayers, which are the basis of biological membranes.

- *Why do phospholipids form membranes, whereas triglycerides form insoluble droplets?*

How Does pH Affect a Protein's Function?

Red blood cells carry oxygen to all parts of your body. These cells are red because they are chock-full of a large, iron-rich protein called *hemoglobin.* The iron atoms in each hemoglobin molecule provide a place for oxygen gas molecules to stick to the protein. When oxygen levels are highest (in the lungs), oxygen atoms bind to hemoglobin tightly, and a large percentage of the hemoglobin molecules in a cell possess bound oxygen atoms. When oxygen levels are lower (in the tissues of the body), hemoglobin doesn't bind oxygen atoms as tightly, and consequently hemoglobin releases its oxygen to the tissues. What causes this difference between lungs and tissues in how hemoglobin loads and unloads oxygen? Oxygen concentration is not the only factor that might be responsible. A protein's function can be affected by pH, and blood pH differs between lungs and body tissues. Tissues are slightly more acidic (that is, they have more H^+ ions and a lower pH). Their metabolic activities release CO_2 into the blood, which quickly becomes converted to carbonic acid, lowering the pH.

The graph displays the so-called dissociation curve for oxygen and hemoglobin. This general type of curve relates the binding of two species to the concentration of one. This can be used to analyze the binding of substrate to enzyme, or a signaling molecule to its receptor. In this case, oxygen binding to hemoglobin is related to oxygen concentration, so the *y*-axis shows percent O_2 saturation of hemoglobin at increasing O_2 concentration (*x*-axis). As the concentration of oxygen rises, more is bound to hemoglobin until the hemoglobin is saturated with oxygen (100% bound). The graph shows the curves for oxygen binding to hemoglobin at three different pH values (7.6, 7.4, 7.2), corresponding to the blood pH that might be expected in resting, exercising, and very active muscle tissue, respectively.

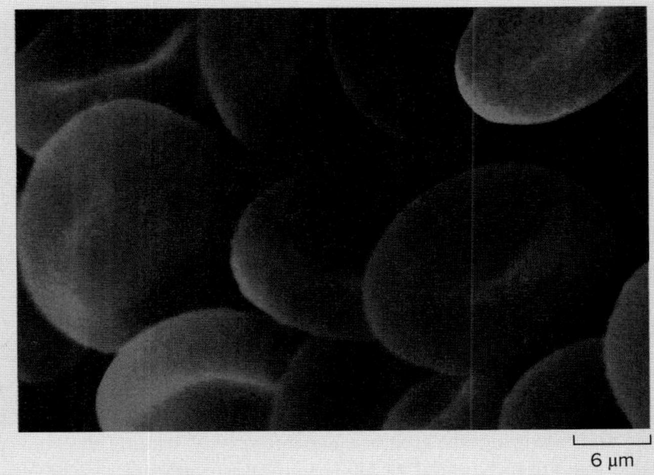

Kenneth Eward/Science Source

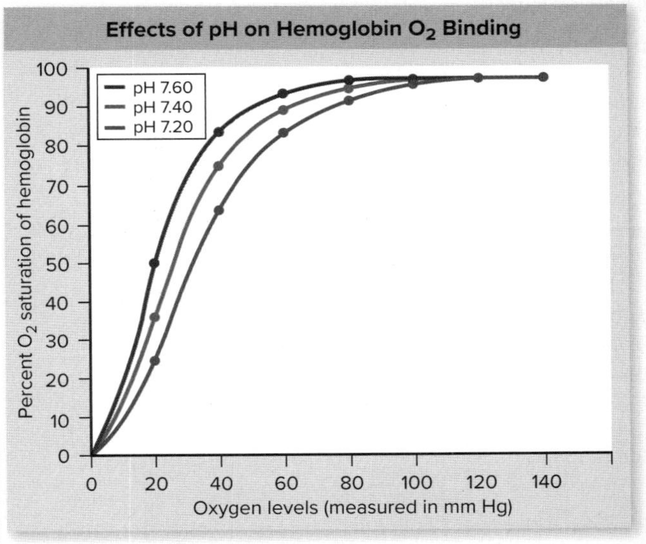

Analysis

1. **Applying Concepts** Which of the three pH values represents the highest concentration of hydrogen ions? Is this value more acidic or more basic than the other two? Is the relationship between hemoglobin saturation and oxygen concentration linear?
2. **Interpreting Data** What is the percent O_2 saturation of hemoglobin for each of the three pH concentrations at saturation? At an oxygen level of 20 mm Hg? At 40 mm Hg? At 60 mm Hg?
3. **Making Inferences** At an oxygen level of 40 mm Hg, is more O_2 bound to hemoglobin than at a pH of 7.6 or 7.0?
4. **Drawing Conclusions** How does pH affect the release of oxygen from hemoglobin?

Retracing the Learning Path

CONCEPT 3.1 Carbon Provides the Framework of Biological Molecules

3.1.1 Functional Groups Provide Chemical Flexibility Carbon can form four bonds. Hydrocarbons consist of carbon and hydrogen, and their bonds store considerable energy. Functional groups are small, molecular entities that confer specific chemical characteristics when attached to a hydrocarbon. Carbon and hydrogen have similar electronegativity, so C—H bonds are not polar. Oxygen and nitrogen have greater electronegativity, leading to polar bonds between O or N and C or H. Structural isomers are molecules with the same formula but different structures; stereoisomers differ in how groups are attached. Enantiomers are mirror-image stereoisomers.

3.1.2 Biological Macromolecules Are Polymers Most important biological macromolecules are polymers–long chains of monomer units. Biological polymers are formed by the elimination of water (H and OH) from two monomers (dehydration reaction). They are broken down by adding water (hydrolysis).

CONCEPT 3.2 Carbohydrates Form Both Structural and Energy-Storing Molecules

3.2.1 Monosaccharides Are Simple Sugars The empirical formula of a carbohydrate is $(CH_2O)_n$. Carbohydrates are used for energy storage and as structural molecules. Simple sugars contain three to six or more carbon atoms. The general formula for 6-carbon sugars is $C_6H_{12}O_6$, and many isomeric forms are possible. Living systems often have enzymes for converting isomers from one to the other.

3.2.2 Disaccharides Are Transport Sugars Plants convert glucose into the disaccharide sucrose for transport within their bodies. Lactating female mammals produce the disaccharide lactose to nourish their young.

3.2.3 Polysaccharides Are Building Materials and Energy-Storage Compounds Glucose is used to make three important polymers: glycogen (in animals) and starch and cellulose (in plants). Chitin is a related structural material found in arthropods and many fungi.

CONCEPT 3.3 Proteins Are the Tools of the Cell

3.3.1 Proteins Have Diverse Functions Most enzymes are proteins. Proteins also provide defense, transport, motion, and regulation, among many other roles.

3.3.2 Proteins Are Polymers of Amino Acids Twenty common amino acids each contain amino and carboxyl groups that are joined by peptide bonds to make polypeptides. Unique R groups determine their properties.

3.3.3 There Are Four Levels of Protein Structure Protein structure is defined by the following hierarchy: primary (amino acid sequence), secondary (hydrogen bonding patterns), tertiary (three-dimensional folding), and quaternary (associations between two or more polypeptides).

3.3.4 Motifs and Domains Organize Secondary Structure Motifs are similar structural elements found in dissimilar proteins. Domains are functional subunits or sites within a tertiary structure.

3.3.5 The Process of Folding Relies on Chaperone Proteins Chaperone proteins assist in the folding of proteins. Heat shock proteins are an example of chaperone proteins.

3.3.6 Environmental Changes Can Cause a Loss of Protein Structure Changes to the environment, such as temperature, pH, and salt concentration, can denature proteins by interfering with the weak forces holding proteins together. Some proteins will regain their structure when returned to normal conditions, showing that amino acid sequence strongly influences structure.

CONCEPT 3.4 Nucleic Acids Store and Express Genetic Information

3.4.1 The Information Molecules of the Cell Are Long Chains of Nucleotides Deoxyribonucleic acid (DNA) and ribonucleic acid (RNA) are polymers composed of nucleotide monomers. Cells use nucleic acids for information storage and transfer. Nucleic acids contain four different nucleotide bases. In DNA these are adenine, guanine, cytosine, and thymine. In RNA, thymine is replaced by uracil.

3.4.2 DNA Stores the Genetic Information DNA exists as a double helix held together by specific base-pairs: adenine with thymine and guanine with cytosine. The nucleic acid sequence constitutes the genetic code.

3.4.3 RNA Has Many Roles in a Cell RNA is made by copying DNA. mRNA is used to assemble polypeptides, tRNA interacts with amino acids and mRNA, and rRNA is part of the ribosome. Other forms of RNA are involved in gene expression.

3.4.4 Other Nucleotides Are Vital Components of Energy Reactions The hydrolysis of ATP provides energy in cells; NAD^+ and FAD transport electrons in cellular processes.

CONCEPT 3.5 Hydrophobic Lipids Form Fats and Membranes

3.5.1 Fats Consist of Fatty Acids Attached to Glycerol Lipids are insoluble in water, because they have a high proportion of nonpolar C—H bonds. Fats are excellent energy-storage molecules.

3.5.2 Phospholipids Form Membranes Phospholipids contain two fatty acids and one phosphate attached to glycerol. In phospholipid-bilayer membranes, the phosphate heads are hydrophilic and cluster on the two faces of the membrane, and the hydrophobic tails are in the center.

Concept Overview

This Concept Overview diagrams the key concepts that were discussed in this chapter.

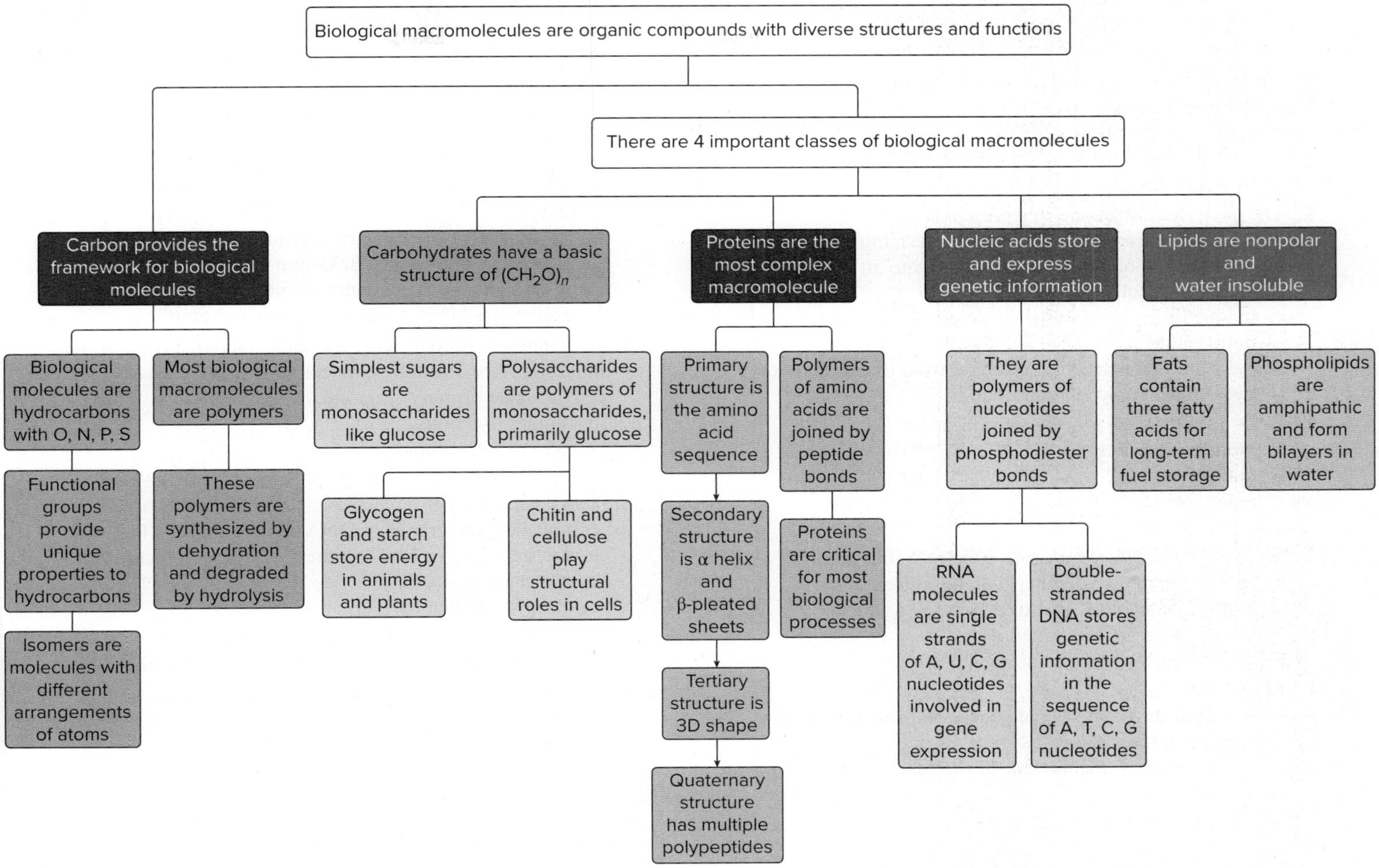

Assessing the Learning Path

Understand

1. The four kinds of organic molecules are
   a. hydroxyls, carboxyls, aminos, and phosphates.
   b. proteins, carbohydrates, lipids, and nucleic acids.
   c. DNA, RNA, sugars, and amino acids.
   d. carbon, hydrogen, oxygen, and nitrogen.
2. Which of the following is a hydroxyl group?
   a. —OH
   b. $—CH_3$
   c. —C=O
   d. $—NH_2$
3. Both animals and plants utilize disaccharides to
   a. store energy.
   b. transport sugars.
   c. provide structural support.
   d. break down polysaccharides.
4. Plant cells store energy in the form of ________, and animal cells store energy in the form of ________.
   a. fructose; glucose
   b. disaccharides; monosaccharides
   c. cellulose; chitin
   d. starch; glycogen
5. Why can cows use cellulose as a food source whereas you can't?
   a. Cows produce the enzymes that can break β-1→4 linkages.
   b. Cows have symbiotic microbes in their digestive system that can break β-1→4 linkages.
   c. Cows have excess starch-digesting enzymes in their saliva.
   d. Cows' broad teeth allow them to mechanically break apart the cellulose.

6. Which of the following atoms or groups would be found bound to the central carbon of an amino acid?
   a. An ($—NH_2$) amino group
   b. A unique R group
   c. A hydrogen atom
   d. All of the above
7. Amino acids are linked together to form a protein by
   a. disulfide bridges.
   b. hydrophobic exclusion.
   c. peptide bonds.
   d. hydrogen bonds.
8. The quaternary structure of a protein is the
   a. first four amino acids at the amino terminus.
   b. organization of a polypeptide chain into an α helix or a β-pleated sheet.
   c. unique three-dimensional shape of the fully folded polypeptide.
   d. overall protein structure resulting from the aggregation of two or more polypeptide subunits.
9. DNA and RNA are polymers composed of
   a. monosaccharides.
   b. nucleotides.
   c. amino acids.
   d. fatty acids.
10. Which of the following are found in DNA, but not in RNA?
    a. Ribose, thymine, adenine, guanine, cytosine
    b. 5′ phosphate and 3′ hydroxyl ends
    c. Deoxyribose and thymine
    d. A 5-carbon sugar and a nitrogenous base
11. RNA
    a. functions in the storage of genetic information in the nucleus of the cell.
    b. is found only in the nucleus of the cell.
    c. is usually a double-stranded helix.
    d. can function as an enzyme.
12. A triglyceride is a _______, which is composed of ________.
    a. lipid; fatty acids and glucose
    b. lipid; fatty acids and glycerol
    c. carbohydrate; fatty acids
    d. lipid; cholesterol
13. Triglycerides, sterols, and terpene are all lipids because
    a. they are all hydrophobic.
    b. they are all amphipathic.
    c. they are all energy-storage molecules.
    d. they are all produced by dehydration synthesis.

Apply

1. The addition of which of the following functional groups to a hydrocarbon would NOT make it water-soluble?
   a. Methyl group
   b. Amino group
   c. Sulfhydryl group
   d. All of the above would make it soluble.
2. While characterizing a newly discovered plant species, you find a previously unknown carbohydrate polymer. From what you have learned in this chapter, you predict that
   a. the monomers are most likely ribose.
   b. the monomers must be glucose.
   c. it is synthesized by a hydrolysis reaction.
   d. it is synthesized by a dehydration reaction.
3. Which of the following amino acids would NOT be expected to occur in the interior of a protein like hemoglobin?
   a. Alanine
   b. Leucine
   c. Valine
   d. Serine
4. In a protein, a glycine is replaced with a histidine. This would
   a. always change the primary structure of the protein, never change tertiary structure or function.
   b. sometimes change the primary structure of a protein, always affect tertiary structure, and sometimes affect function.
   c. never change the primary structure and always affect secondary and tertiary structure and function.
   d. always change the primary structure of a protein and sometimes affect tertiary structure and function.
5. Two different proteins have the same domain as part of their final structure. From this, we can infer that they have
   a. the same primary structure.
   b. similar functions.
   c. very different functions.
   d. the same primary structure but different functions.
6. Two adjacent water molecules, an α helix, and two complementary strands of DNA are all similar in that they
   a. are stabilized or held together by hydrogen bonds.
   b. are macromolecules.
   c. are hydrophilic.
   d. are put together by dehydration synthesis.
7. What chemical property of lipids accounts for their insolubility in water?
   a. The COOH group of fatty acids
   b. The large number of nonpolar C—H bonds
   c. The branching of saturated fatty acids
   d. The C—C bonds found in unsaturated fatty acids
8. Partial hydrogenation of vegetable oil would result in which of the following?
   a. The oil becoming more solid at room temperature
   b. Adding double bonds to the oil
   c. Producing *trans* double bonds in the fatty acid
   d. Both a and c

Synthesize

1. Why do you suppose organisms contain primarily D-sugars but L-amino acids? Why not the same chiral form?
2. Plants make both starch and cellulose. Would you predict that the enzymes involved in starch synthesis could also be used by the plant for cellulose synthesis? Construct an argument to explain this based on the structure and function of the enzymes and the polymers synthesized.
3. If a protein's primary structure dictated its tertiary structure and thus its function, why would it need a chaperonin to fold correctly?
4. Of all possible DNA nucleotide sequences, what sequence of base-pairs would most easily dissociate into single strands upon gentle heating of the DNA double helix? Why did you choose this sequence?
5. The membranes of arctic fish must remain fluid at low temperatures. Would these fish be more likely to have a high percentage of saturated or of unsaturated fatty acids in their phospholipids? Long or short fatty acids? Explain.

Part II Biology of the Cell

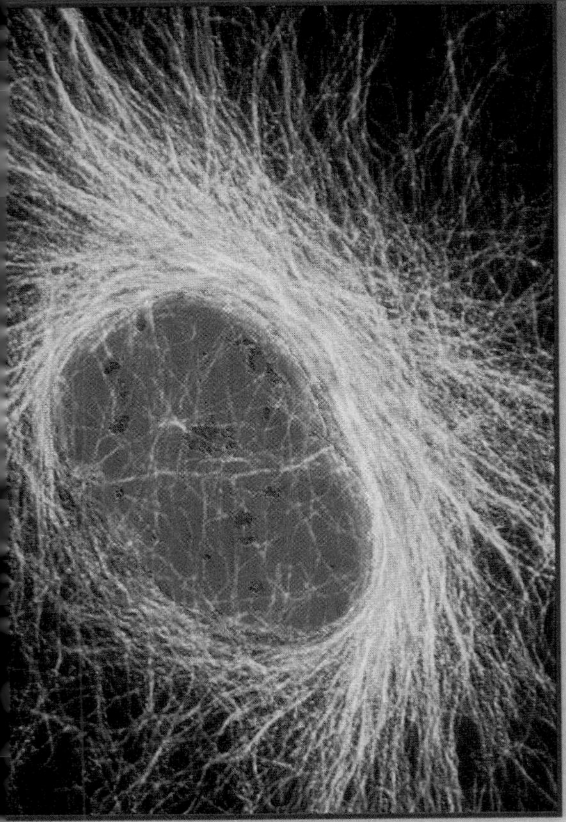

Dr. Gopal Murti/Science Source

4 Cell Structure

Learning Path

4.1 All Living Organisms Are Composed of Cells

4.2 Prokaryotic Cells Are Relatively Simple

4.3 Eukaryotic Cells Are Highly Compartmentalized

4.4 Membranes Organize the Cell Interior into Functional Compartments

4.5 Mitochondria and Chloroplasts Are Energy-Processing Organelles

4.6 An Internal Skeleton Supports the Shape of Cells

4.7 Extracellular Structures Protect Cells

4.8 Cell-to-Cell Connections Determine How Adjacent Cells Interact

Concept Overview

This Concept Overview provides a diagram of the overarching concepts that are covered in this chapter. These concepts will be expanded upon in the Concept Overview at the end of the chapter.

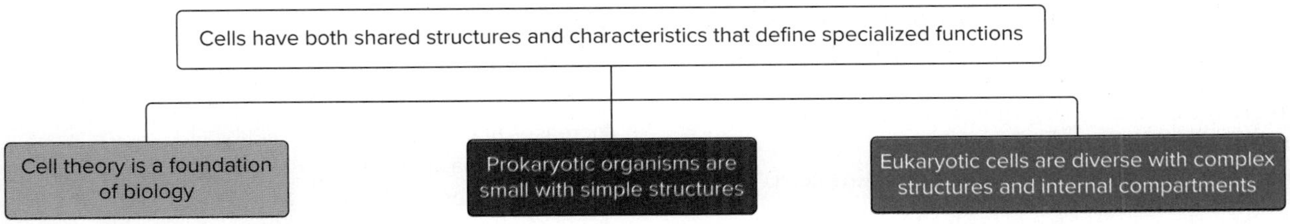
Cells have both shared structures and characteristics that define specialized functions
Cell theory is a foundation of biology
Prokaryotic organisms are small with simple structures
Eukaryotic cells are diverse with complex structures and internal compartments

Introduction

All organisms are composed of cells. The gossamer wing of a butterfly is a thin sheet of cells and so is the glistening outer layer of your eyes. The hamburger or tomato you eat is composed of cells, and its contents soon become part of your cells. Some organisms consist of a single cell too small to see with the unaided eye. Other organisms, such as humans, are composed of many specialized cells, such as the fibroblast cell shown in the striking fluorescence micrograph on the previous page. Cells are so much a part of life that we cannot imagine an organism that is not cellular in nature. In this chapter we take a close look at the internal structure of cells. In the next six chapters, we will focus on cells in action—how they communicate with their environment, grow, and reproduce.

4.1 All Living Organisms Are Composed of Cells

All living things are composed of cells, almost all of them too small to see with the naked eye. Although there are exceptions, a typical eukaryotic cell is 10 to 100 micrometers (μm) (10 to 100 millionths of a meter) in diameter, and most prokaryotic cells are only 1 to 10 μm in diameter.

The Cell Theory Is the Unifying Foundation of Biology

LEARNING OBJECTIVE 4.1.1 Discuss the three principles of cell theory.

Because cells are so small, they were not discovered until the invention of the microscope in the 17th century. Robert Hooke was the first to observe cells in 1665, naming the shapes he saw in cork *cellulae* (Latin, "small rooms"). This has come down to us as *cells.* Another early microscopist, Anton van Leeuwenhoek, first observed living cells, which he termed "animalcules," or little animals. After these early efforts, a century and a half passed before biologists fully recognized the importance of cells. In 1838, botanist Matthias Schleiden stated that all plants "are aggregates of fully individualized, independent, separate beings, namely the cells themselves." In 1839, Theodor Schwann reported that all animal tissues also consist of individual cells. Thus, the cell theory was born.

The cell theory developed from the observation that all organisms are composed of cells. Although it sounds simple, it is a far-reaching statement about the organization of life. In its modern form, the **cell theory** includes the following three principles:

1. All organisms are composed of one or more cells, and the life processes of metabolism and heredity occur within these cells.
2. Cells are the smallest living things, the basic units of organization of all organisms.
3. New cells arise only by division of preexisting cells.

Cells are thought to have evolved spontaneously over 3.5 BYA. Biologists have concluded that no cells are originating spontaneously at present. Rather, life on Earth represents a continuous line of descent from those early cells.

Cell Size Is Limited

LEARNING OBJECTIVE 4.1.2 Illustrate how the surface-area-to-volume ratio limits cell size.

Most cells are relatively small. Why? The reason relates to the diffusion of substances into and out of cells. The rate of this diffusion is affected by a number of variables, including (1) the surface area available for diffusion, (2) temperature, (3) concentration gradient of the diffusing substance, and (4) distance over which diffusion must occur. These are related by an equation known as Fick's Law of Diffusion, described in chapter 34. As the size of a cell increases, the length of time for diffusion from the outside membrane to the interior of the cell increases as well. This soon becomes a problem, as larger cells need to synthesize more macromolecules, have correspondingly higher energy requirements, and produce a greater quantity of waste. Molecules used for energy and biosynthesis must be transported through the membrane. Any metabolic waste produced must be removed, also passing through the membrane. The rate at which this transport occurs depends on both the distance to the membrane and the area of membrane available.

The advantage of small cell size is readily apparent in terms of the **surface-area-to-volume ratio.** As a cell's size increases, its volume increases much more rapidly than its surface area. For a spherical cell, the surface area is proportional to the square of the radius, whereas the volume is proportional to the cube of the radius. Thus, if the radii of two cells differ by a factor of 10, the larger cell will have 10^2, or 100, times the surface area but 10^3, or 1000, times the volume of the smaller cell (figure 4.1).

The membrane surrounding the cell plays a key role in controlling cell function, because the cell surface provides the only opportunity for interaction with the environment, as all substances enter and exit a cell via this surface. Because small cells have more surface area per unit of volume than large ones, their control over cell contents is more effective.

Not all cells are small. Some larger cells function quite efficiently because they have structural features that increase surface area. For example, some cells, such as skeletal muscle cells, have more than one nucleus, allowing genetic information to be spread around a large cell. Cells in the nervous system called neurons are long, slender cells, some extending more than a meter in length.

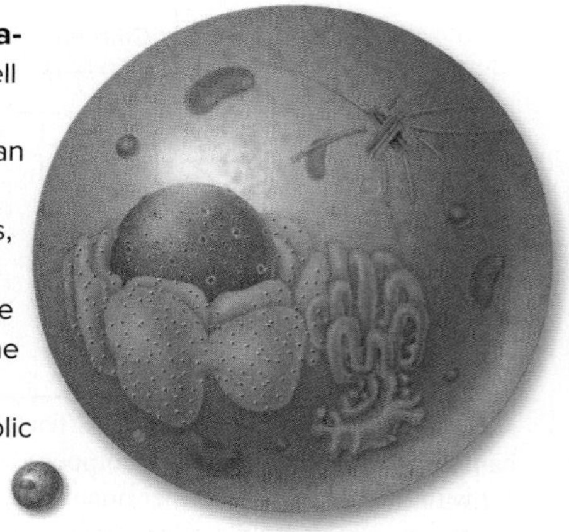

Figure 4.1 Surface-area-to-volume ratio. As a cell gets larger, its volume increases at a faster rate than its surface area. If the cell radius increases by 10 times, the surface area increases by 100 times, but the volume increases by 1000 times. The surface area must be large enough to meet the metabolic needs of the volume.

Cell radius (r)	1 unit	10 units
Surface area ($4\pi r^2$)	12.57 units	1257 units
Volume ($\frac{4}{3}\pi r^3$)	4.189 units	4189 units
Surface area / Volume	3	0.3

Although they are long, they are thin, so that any given point within the cell is close to the plasma membrane. This permits rapid diffusion between the inside and the outside of the cell. For the same reason, many cells in your body are shaped flat, like a thin plate.

Structural features that can dramatically increase a cell's surface area are finger-like projections called *microvilli*. The cells that line the small intestine are covered with microvilli.

Microscopes Allow Us to Visualize Cells

LEARNING OBJECTIVE 4.1.3 Describe the tools biologists use to visualize cells.

Other than egg cells, not many cells are visible to the naked eye (figure 4.2). Most are less than 50 µm in diameter, far smaller than the period at the end of this sentence. How do we study cells if they are too small to see? To visualize cells, we need the aid of technology.

Light microscopes

One way to overcome the limitations of our eyes is to increase magnification so that small objects appear larger. Modern *light microscopes,* which operate with visible light, use two magnifying lenses (and a variety of correcting lenses) to achieve very high magnification and clarity (table 4.1). The first lens focuses the image of the object on the second lens, which magnifies it again and focuses it on the back of the eye. Microscopes that magnify in stages using several lenses are called *compound microscopes.*

Electron microscopes

Light microscopes, even compound ones, are not powerful enough to resolve many of the structures within cells. Why not just add another magnifying stage to the microscope to increase its resolving power? The reason we can't is the limited resolution of the human eye. *Resolution* is the minimum distance two points can be apart and still be distinguished as two separate points. When two objects are closer together than about 100 µm, the light reflected from each strikes the same photoreceptor cell at the rear of the eye. Only when the objects are farther than 100 µm apart can the light from each strike different cells, allowing your eye to resolve them as two distinct objects rather than one.

Making matters worse, when light beams reflecting from the two images are closer than a few hundred micrometers, they start to overlap each other. The only way two light beams can get closer together and still be resolved is if their wavelengths are shorter. One way to avoid overlap is to use a beam of electrons rather than a beam of light. Electrons have a much shorter wavelength, and an *electron microscope,* employing electron beams, has 1000 times the resolving power of a light microscope. In *transmission electron microscopes,* the electrons used to visualize the specimens are transmitted through the material and are capable of resolving objects only 0.2 nm apart—which is only twice the diameter of a hydrogen atom!

A second kind of electron microscope, the *scanning electron microscope,* bounces beams of electrons off the surface of the specimen. The electrons reflected back, and others that the specimen itself emits, are amplified and transmitted to a screen, where the image can be viewed and photographed as a striking three-dimensional image.

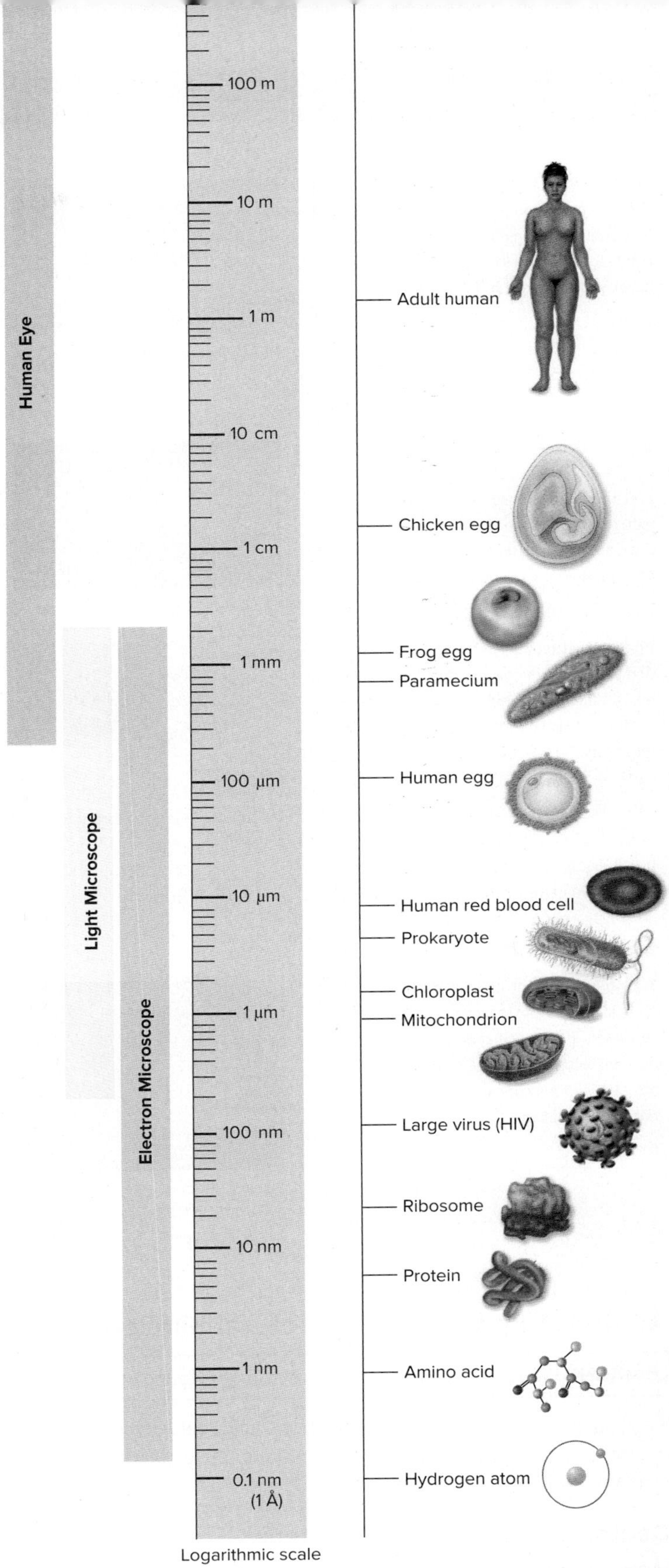

Figure 4.2 The size of cells and their contents. Except for vertebrate eggs, which can typically be seen with the unaided eye, most cells are microscopic in size. Prokaryotic cells are generally 1 to 10 µm across.
$1\ m = 10^2\ cm = 10^3\ mm = 10^6\ \mu m = 10^9\ nm$

TABLE 4.1 Microscopes

LIGHT MICROSCOPES

Bright-field microscope:
Light is transmitted through a specimen, giving little contrast. Staining specimens improves contrast but requires that cells be fixed (not alive), which can distort or alter components.

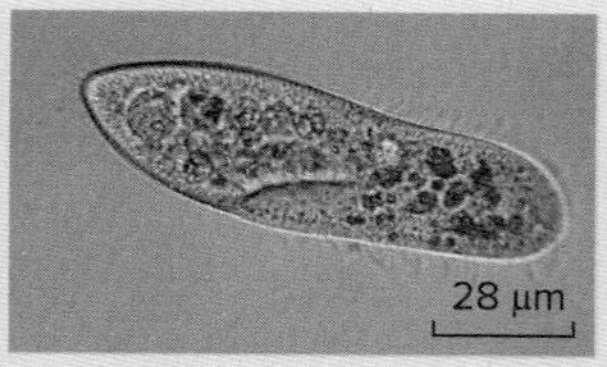

Nancy Nehring/E+/Getty Images

Differential-interference–contrast microscope:
Polarized light is split into two beams that have slightly different paths through the sample. Combining these two beams produces greater contrast, especially at the edges of structures.

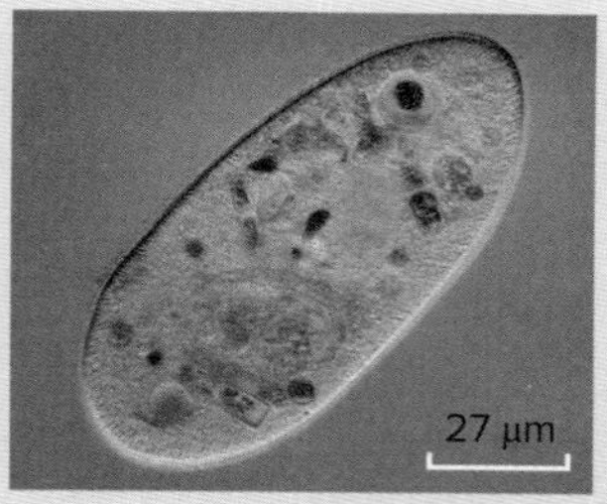

Blickwinkel/Fox/Alamy Stock Photo

Dark-field microscope:
Light is directed at an angle toward the specimen. A condenser lens transmits only light reflected off the specimen. The field is dark, and the specimen is light against this dark background.

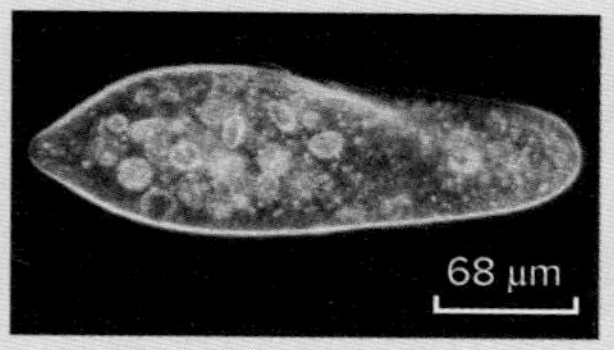

Laguna Design/Science Source

Fluorescence microscope:
Fluorescent stains absorb light at one wavelength, then emit it at another. Filters transmit only the emitted light.

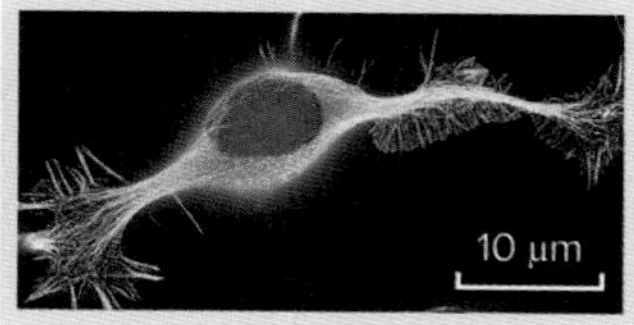

Dr. Torsten Wittmann/Science Source

Phase-contrast microscope:
Components of the microscope bring light waves out of phase, which produces differences in contrast and brightness when the light waves recombine.

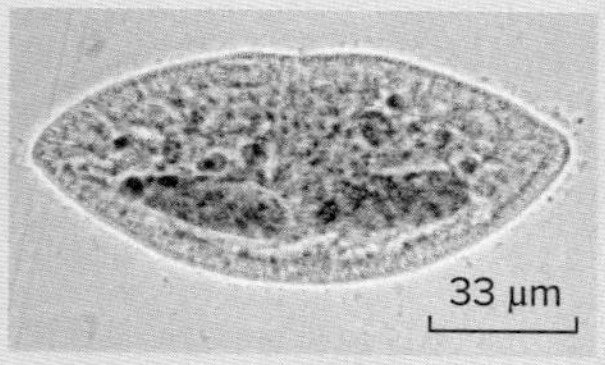

De Agostini Picture Library/age fotostock

Confocal microscope:
Light from a laser is focused to a point and scanned across the fluorescently stained specimen in two directions. This produces clear images of one plane of the specimen. Other planes of the specimen are excluded to prevent the blurring of the image. Multiple planes can be used to reconstruct a 3-D image.

Med. Mic. Sciences, Cardiff Uni./Wellcome Images

ELECTRON MICROSCOPES

Transmission electron microscope:
A beam of electrons is passed through the specimen. Electrons that pass through are used to expose film. Areas of the specimen that scatter electrons appear dark. False coloring enhances the image.

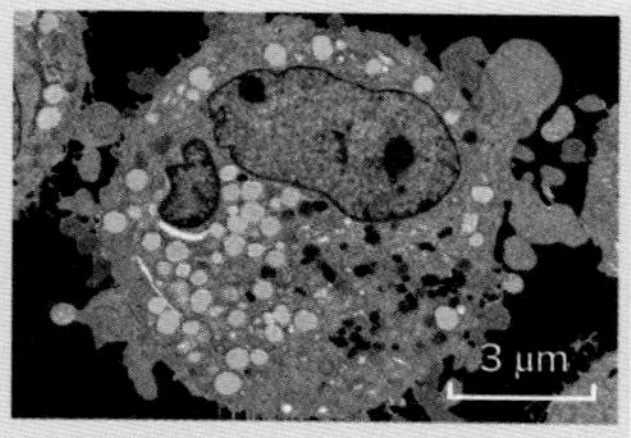

Science Photo Library – STEVE GSCHMEISSNER./Brand X Pictures/Getty Images

Scanning electron microscope:
An electron beam is scanned across the surface of the specimen, and electrons are knocked off the surface. Thus, the topography of the specimen determines the contrast and the content of the image. False coloring enhances the image.

Steve Gschmeissner/Science Source

All Cells Exhibit Basic Structural Similarities

LEARNING OBJECTIVE 4.1.4 Identify similarities found in all cells.

The general plan of cellular organization varies between different organisms, but despite these modifications, all cells share four fundamental features.

Centrally located genetic material

Every cell contains DNA for genetic material. In **prokaryotes,** the simplest organisms, this consists primarily of a single circular molecule of DNA. It is typically found near the center of the cell in an area called the **nucleoid.** This region is not surrounded by a membrane.

By contrast, eukaryotic cells are more complex, with DNA organized into linear chromosomes segregated into a nucleus, which is surrounded by a double-membrane structure called the nuclear envelope (described in section 4.3). In all cells, the DNA contains the genes that code for the proteins synthesized by the cell.

The cytoplasm

A semifluid matrix called the **cytoplasm** fills the interior of the cell. The cytoplasm contains all of the sugars, amino acids, and proteins needed for cellular activities. Although it is an aqueous medium, cytoplasm is more like Jell-O than water due to the high concentration of proteins and other macromolecules. The fluid part of the cytoplasm, which contains organic molecules and ions in solution, is called the **cytosol.** Membrane-bound **organelles,** usually specialized for a particular function and containing specific proteins, create compartments in the cytoplasm.

Ribosomes

Ribosomes are large, macromolecular machines composed of RNA and protein that synthesize all cellular proteins. Ribosomes use information from the genome and other accessory molecules to direct the synthesis of proteins by sequential addition of amino acids (discussed in chapter 15).

The plasma membrane

A **plasma membrane** encloses every cell, separating its contents from the surroundings. The plasma membrane is a phospholipid bilayer about 5 to 10 nm (5 to 10 billionths of a meter) thick, with proteins embedded in it. Viewed in cross section with the electron microscope, such membranes appear as two dark lines separated by a lighter area. Viewed in cross section with the electron microscope, membranes appear as two dark lines separated by a lighter area. This arises from the tail-to-tail organization of phospholipid molecules that make up the membrane.

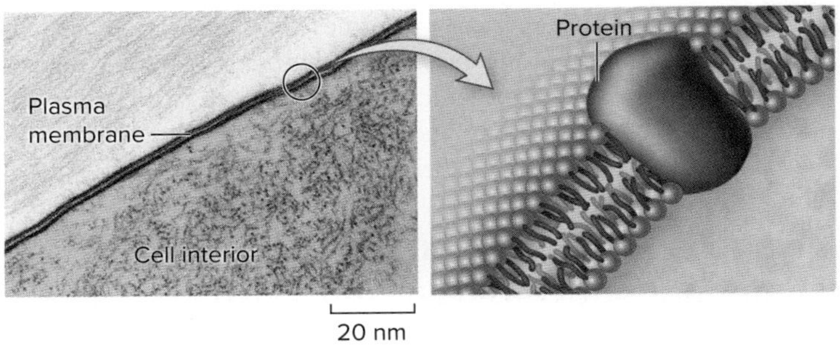

Left: Don W. Fawcett/Science Source

The proteins of the plasma membrane are generally responsible for a cell's ability to interact with the environment. Membrane proteins give cells identity, and provide for a variety of functions, including transport and communication with other cells and the environment. The role of the plasma membrane is so important that we will study it in detail in chapter 5.

REVIEW OF CONCEPT 4.1

All organisms are single cells or aggregates of cells that arise from preexisting cells. Cell size is limited primarily by the efficiency of diffusion across the plasma membrane. All cells are bounded by a plasma membrane and filled with cytoplasm. The genetic material is found in the central portion of the cell; in eukaryotic cells, it is contained in a membrane-bounded nucleus.

- *Would finding life on Mars change our view of cell theory?*

4.2 Prokaryotic Cells Are Relatively Simple

When cells were visualized with microscopes, two basic cellular architectures were recognized: eukaryotic and prokaryotic. These terms refer to the presence or absence, respectively, of a membrane-bounded nucleus that contains genetic material. In addition to lacking a nucleus, prokaryotic cells do not have an internal membrane system or numerous membrane-bounded organelles. Prokaryotes have two main domains: archaea and bacteria, introduced here and discussed in detail in chapter 23.

Prokaryotic Cells Have Organized Substructure

LEARNING OBJECTIVE 4.2.1 Distinguish between bacteria and archaea.

Prokaryotic cells were once considered to be "bags of enzymes," but our view of their structure has changed significantly over the years. They are often described as simple cells consisting of a plasma membrane containing cytoplasm with genetic material and ribosomes, surrounded by a rigid **cell wall** (figure 4.3). While accurate as far as it goes, this is an oversimplification.

The shape of prokaryotic cells is determined primarily by a cell wall, not a cytoskeleton like that in eukaryotes. However, prokaryotes do have molecules related to the important cytoskeletal

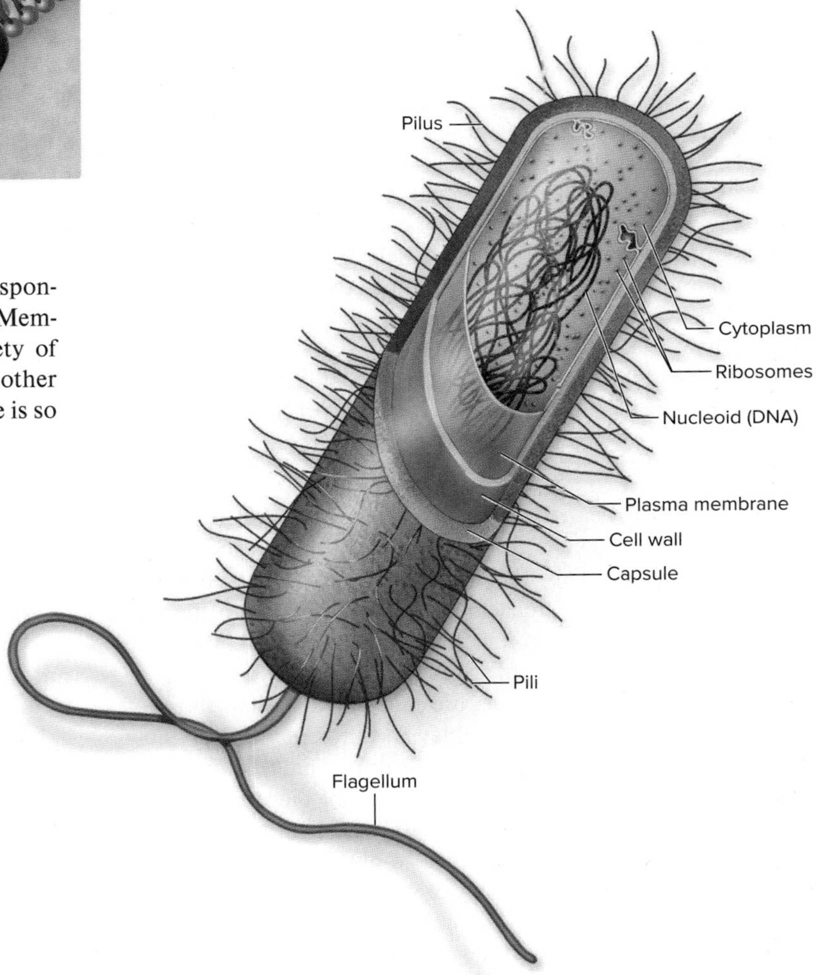

Figure 4.3 Structure of a prokaryotic cell. Generalized cell organization of a prokaryote. The nucleoid is visible as a dense central region segregated from the cytoplasm. Some prokaryotes have hairlike growths, called pili (singular, *pilus*), on the outside of the cell.

proteins actin and tubulin, and these proteins can affect cell-wall structure. For instance, in rod-shaped cells, actin-like MreB fibers running the length of the cell can lead to cell-wall fibers with the same orientation. When MreB protein is removed, these cells become spherical. During cell division, cell-wall deposition is influenced by the tubulin-like FtsZ protein (refer to chapter 10).

The plasma membrane of a prokaryotic cell carries out some of the functions organelles perform in eukaryotic cells. For example, some photosynthetic bacteria, such as the cyanobacterium *Prochloron*, have an extensively folded plasma membrane, with the folds extending into the cell's interior. These membrane folds contain the bacterial pigments connected with photosynthesis. In eukaryotic plant cells, photosynthetic pigments are found in the thylakoid membrane of the chloroplast.

Bacteria

Most bacterial cells are encased by a strong cell wall composed of peptidoglycan, which consists of a carbohydrate matrix cross-linked with short polypeptide units (refer to chapter 23 for details). Cell walls protect the cell, maintain its shape, and prevent excessive uptake or loss of water. The exception is the class *Mollicutes,* which includes the common genus *Mycoplasma,* which lack a cell wall. Plants, fungi, and most protists also have cell walls, but peptidoglycan is unique to bacteria. For this reason, antibiotic drugs used to treat bacterial infection often target their cell walls. For example, penicillin and vancomycin interfere with the ability of bacteria to cross-link the peptides in their peptidoglycan cell wall. Like removing all the nails from a wooden house, this destroys the integrity of the structural matrix, which can no longer prevent water from rushing in and swelling the cell to bursting.

Some bacteria also secrete a jelly-like protective capsule of polysaccharide around the cell. Many disease-causing bacteria have such a capsule, which enables them to adhere to teeth, skin, food—or practically any surface that can support their growth.

Archaea

We have much to learn about the physiology and structure of archaea. Many archaeans are difficult to culture in the laboratory, as shown recently by the decade-long effort required to culture a single deep-sea archaean. It is much easier to identify archaeal genomes in environmental samples, so we know more about archaeal genomics than about archaeal physiology.

The cell walls of archaea are more diverse than those of bacteria, containing polysaccharides and proteins, and possibly even inorganic components. One feature that distinguishes archaea from bacteria is the structure of their membrane lipids. Archaeal membrane lipids include saturated hydrocarbons that are covalently at both ends attached to glycerol, which forms a monolayer membrane (refer to chapter 23 for details). These features seem to confer greater thermal stability to archaeal membranes, although the trade-off seems to be an inability to alter the degree of saturation of the hydrocarbons—limiting their response to changing environmental temperatures.

The cellular machinery that replicates DNA and synthesizes proteins in archaea is more closely related to eukaryotic systems than to bacterial systems. Even though they share a similar overall cellular architecture with bacteria, archaea appear to be more closely related on a molecular basis to eukaryotes.

Internal compartments

While it is still accurate to say that there are no organelles common to all prokaryotes, particular species do have internal structures that form compartments. These compartments can be formed by membranes, or can be protein-based.

One example of a compartment formed from membrane is the *magnetosome,* found in bacteria that can move along a magnetic field. Magnetosomes consist of spherical membranes containing iron oxide crystals. These are arrayed in a chain by attachment to an external fiber, which allows the cell to align with a magnetic field (figure 4.4). Both bacterial and archaeal species may have internal membrane-bound storage structures. Prokaryotes can also have infoldings of the plasma membrane that serve to segregate metabolic reactions.

Some bacteria also contain cellular compartments bounded by a semipermeable protein shell. These **bacterial microcompartments (BMCs)** range in size from 40 to 400nm, and act to isolate a specific metabolic process, or to store a particular substance. Most of these BMCs segregate enzymes used to degrade particular substances, but in one case, photosynthetic bacteria, the isolated reactions build up organic compounds from CO_2. While the structure is quite different, these compartments are functional analogs of eukaryotic organelles.

Flagella in Bacteria and in Archaea Have Independent Origins

LEARNING OBJECTIVE 4.2.2 Compare and contrast the bacterial flagella and the archaellum.

Bacterial, archaeal, and eukaryotic cells all have external structures for motility—long filaments made of proteins—that superficially appear similar, but actually are three evolutionarily distinct

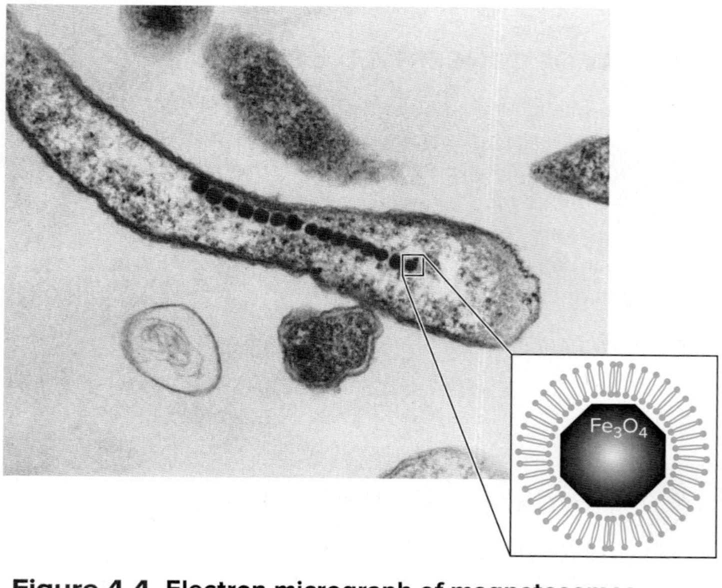

Figure 4.4 Electron micrograph of magnetosomes. Magnetosomes are membrane-enclosed spheres containing iron oxide crystals. These allow magnetotactic bacteria to align with a magnetic field.

Dennis Kunkel Microscopy/Science Source

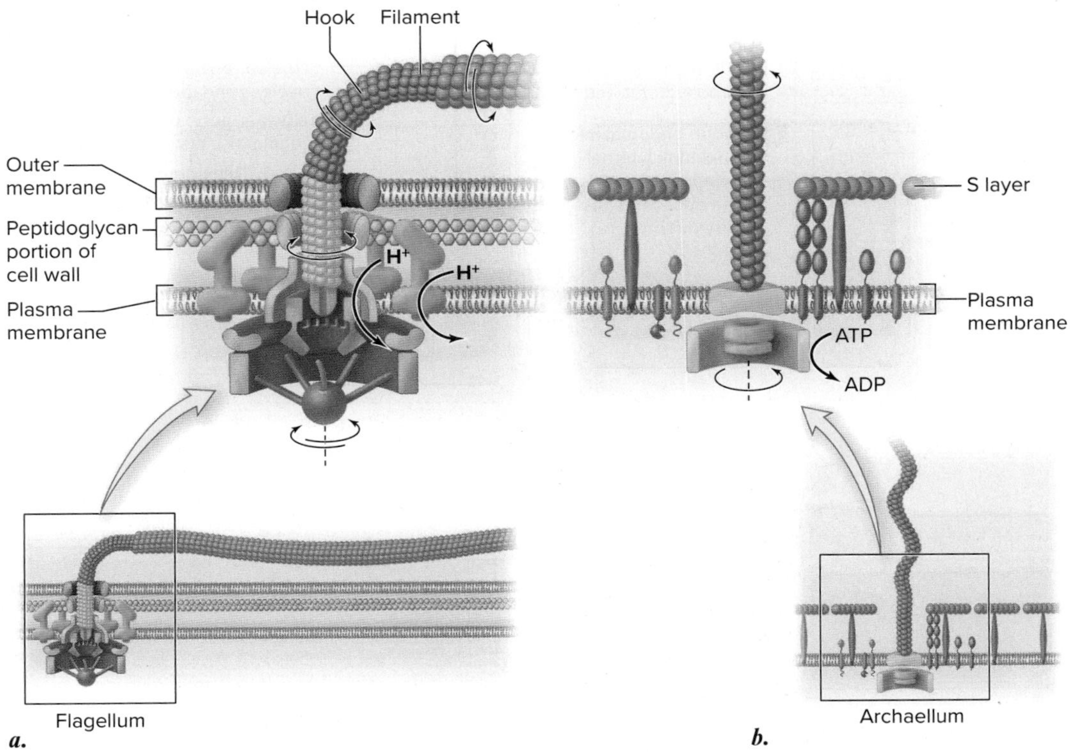

Figure 4.5 Comparison of bacterial flagellum and archaellum. ***a.*** The bacterial flagellum consists of concentric discs embedded in the membrane and cell wall. These are connected to a protein fiber composed of subunits of flagellin. The structure rotates powered by the flow of protons into the cell. ***b.*** The archaellum, found in all motile archaea, is built on components related to those in a type of bacterial pilus. In this case, rotation of the protein fiber is powered by the hydrolysis of ATP.

structures. Bacterial **flagella** (singular *flagellum*) consists of protein rings embedded in the plasma membrane, and a cell wall with long protein fibers extending from this (figure 4.5*a*). The analogous archaeal structure, now called an **archaellum,** consists of a structure that appears related to a different bacterial external structure known as a *pilus* (refer to chapter 23). This is formed by a disk of membrane proteins with protein filaments extending from the cell (figure 4.5*b*). Both bacterial and archaeal structures rotate like a propellor, while eukaryotes have whiplike flagella. However, the motive force for rotation is different: bacteria use a proton gradient similar to that used in eukaryotic mitochondria (refer to chapter 7), while archaea use the hydrolysis of ATP. The independent evolution of different structures with similar function is called convergent evolution, discussed in more detail in chapter 20.

REVIEW OF CONCEPT 4.2

The two domains of prokaryotes are archaea and bacteria. Bacteria cell walls are composed of peptidoglycan, while archaea have cell walls made from a variety of polysaccharides and peptides, as well as membranes containing unusual lipids. Some prokaryotes have membrane-bounded organelles, and semipermeable protein compartments. Bacteria and archaea have rotating motility structures that evolved independently.

- *What features do bacteria and archaea share?*

4.3 Eukaryotic Cells Are Highly Compartmentalized

For the first 1 billion years of life on Earth, all organisms were prokaryotes, cells with very simple interiors. About 1.5 BYA, a new kind of cell appeared for the first time, the eukaryotic cell. Eukaryotic cells are much larger than and profoundly different from prokaryotic cells, with a complex interior organization. All cells alive today, except bacteria and archaea, are of this new kind.

Organelles and Internal Membranes Organize the Interior of Eukaryotic Cells

LEARNING OBJECTIVE 4.3.1 List the structural elements unique to eukaryotic cells.

Figures 4.6 and 4.7 present cross-sectional diagrams of idealized animal and plant cells. It is clear that both of these types of eukaryotic cell are more complex than the prokaryotic cell shown in figure 4.3. The **plasma membrane** ❶ encases a semifluid matrix called the **cytoplasm** ❷, which contains the nucleus and various cell structures called organelles. An **organelle** is a specialized structure within which particular cell processes occur. Each organelle, such as a **mitochondrion** ❸, has a specific function in the eukaryotic cell. The organelles are anchored at specific

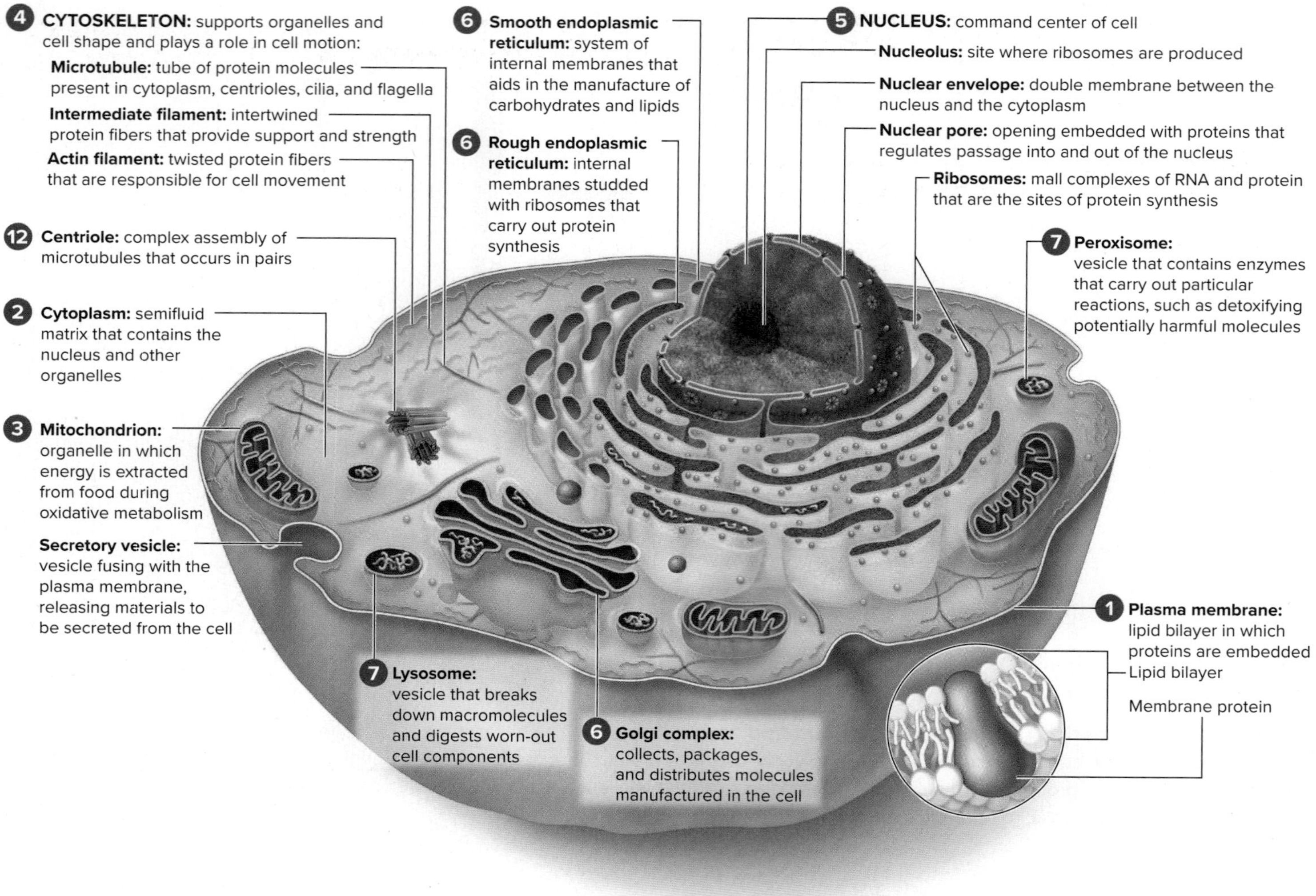
4 CYTOSKELETON: supports organelles and cell shape and plays a role in cell motion:
Microtubule: tube of protein molecules present in cytoplasm, centrioles, cilia, and flagella
Intermediate filament: intertwined protein fibers that provide support and strength
Actin filament: twisted protein fibers that are responsible for cell movement
12 Centriole: complex assembly of microtubules that occurs in pairs
2 Cytoplasm: semifluid matrix that contains the nucleus and other organelles
3 Mitochondrion: organelle in which energy is extracted from food during oxidative metabolism
Secretory vesicle: vesicle fusing with the plasma membrane, releasing materials to be secreted from the cell
6 Smooth endoplasmic reticulum: system of internal membranes that aids in the manufacture of carbohydrates and lipids
6 Rough endoplasmic reticulum: internal membranes studded with ribosomes that carry out protein synthesis
5 NUCLEUS: command center of cell
Nucleolus: site where ribosomes are produced
Nuclear envelope: double membrane between the nucleus and the cytoplasm
Nuclear pore: opening embedded with proteins that regulates passage into and out of the nucleus
Ribosomes: mall complexes of RNA and protein that are the sites of protein synthesis
7 Peroxisome: vesicle that contains enzymes that carry out particular reactions, such as detoxifying potentially harmful molecules
1 Plasma membrane: lipid bilayer in which proteins are embedded
Lipid bilayer
Membrane protein
7 Lysosome: vesicle that breaks down macromolecules and digests worn-out cell components
6 Golgi complex: collects, packages, and distributes molecules manufactured in the cell

Figure 4.6 Structure of an animal cell. The plasma membrane encases the animal cell, which contains the cytoskeleton and various cell organelles and interior structures suspended in a semifluid matrix called the cytoplasm.

locations in the cytoplasm by an interior scaffold of protein fibers, the **cytoskeleton** 4.

One of the organelles is very visible when these cells are examined with a microscope, filling the center of the cell like the pit of a peach. Seeing it, the English botanist Robert Brown in 1831 called it the **nucleus** 5 (plural, *nuclei*), from the Latin word for "kernel." Inside the nucleus, the DNA is wound tightly around proteins and packaged into compact units called chromosomes. It is the nucleus that gives **eukaryotes** their name, from the Greek words *eu,* "true," and *karyon,* "nut"; by way of contrast, the earlier-evolving bacteria and archaea are called prokaryotes ("before the nut").

If you examine the organelles in figures 4.6 and 4.7, you will discover that most of them form separate compartments within the cytoplasm, bounded by their own membranes. *The hallmark of the eukaryotic cell is this compartmentalization.* This internal compartmentalization is achieved by an extensive **endomembrane system** 6 that weaves through the cell interior, providing extensive surface area for many membrane-associated cell processes to occur.

Vesicles 7 (small, membrane-bounded sacs that store and transport materials) form in the cell either by budding off of the endomembrane system or by incorporating lipids and protein in the cytoplasm. These many closed-off compartments allow different processes to proceed simultaneously without interfering with one another, just as rooms do in a house. Thus, the organelles called *lysosomes* are recycling centers. Their very acid interiors break down old organelles, and the component molecules are recycled. This acid would be very destructive if released into the cytoplasm. Similarly, chemical isolation is essential to the function of the organelles called *peroxisomes.* Toxic chemicals are degraded and food molecules are processed within peroxisomes by enzymes that remove electrons and associated hydrogen atoms. If not isolated within the peroxisomes, these enzymes would tend to short-circuit chemical reactions occurring in the cytoplasm, which often involve adding hydrogen atoms to molecules.

Comparing figure 4.6 with figure 4.7 reveals a set of common features, and a few that are specific to each type of cell. For example, the cells of plants, fungi, and many protists have strong, thick exterior **cell walls** 8 composed of cellulose or chitin fibers, but the cells of animals lack cell walls. All plants and many kinds of protists have **chloroplasts** 9, within which photosynthesis occurs. No animal or fungal cells contain chloroplasts. Plant cells also contain a large **central vacuole** 10 that stores water, as well as cytoplasmic connections through openings in the cell wall called **plasmodesmata** 11. **Centrioles** 12 are present in animal cells but absent in plant and fungal cells. Some kinds of animal cells possess finger-like projections called *microvilli.* Many animal and

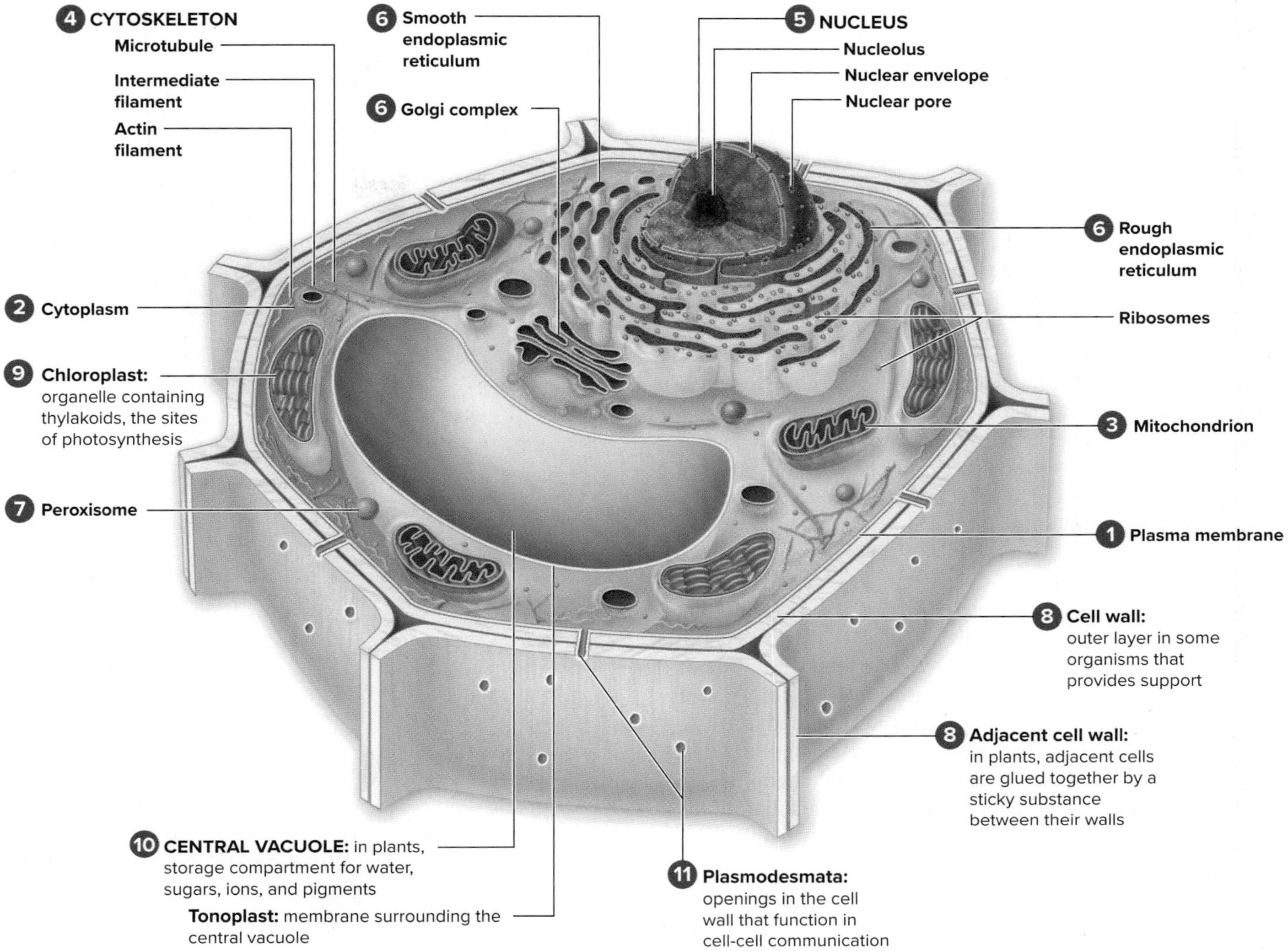

4 CYTOSKELETON
Microtubule
Intermediate filament
Actin filament
6 Smooth endoplasmic reticulum
6 Golgi complex
5 NUCLEUS
Nucleolus
Nuclear envelope
Nuclear pore
6 Rough endoplasmic reticulum
Ribosomes
2 Cytoplasm
9 Chloroplast: organelle containing thylakoids, the sites of photosynthesis
3 Mitochondrion
7 Peroxisome
1 Plasma membrane
8 Cell wall: outer layer in some organisms that provides support
8 Adjacent cell wall: in plants, adjacent cells are glued together by a sticky substance between their walls
10 CENTRAL VACUOLE: in plants, storage compartment for water, sugars, ions, and pigments
Tonoplast: membrane surrounding the central vacuole
11 Plasmodesmata: openings in the cell wall that function in cell-cell communication

Figure 4.7 Structure of a plant cell. Most mature plant cells contain large central vacuoles that occupy a major portion of the internal volume of the cell, as well as organelles called chloroplasts, within which photosynthesis takes place. The cells of plants, fungi, and some protists have cell walls. Flagella occur in sperm of a few plant species but are otherwise absent in plant and fungal cells. Centrioles are also absent in plant and fungal cells.

protist cells possess *flagella,* which aid in movement, or *cilia,* which have many different functions. Flagella occur in sperm of a few plant species but are otherwise absent in plant and fungal cells.

We will now journey into the interior of a typical eukaryotic cell and explore it in more detail, using diagrams that highlight the particular organelle we are examining. Although the various organelles are color-coded for easier identification, remember that most are actually colorless.

The Nucleus Acts as the Cell's Information Center

LEARNING OBJECTIVE 4.3.2 Relate the structure of the nucleus to its function.

The largest and most easily seen organelle within a eukaryotic cell is the **nucleus,** first described by the botanist Robert Brown in 1831. Nuclei are roughly spherical in shape, and in animal cells they are typically located in the central region of the cell (figure 4.8). In some cells, a network of fine cytoplasmic filaments seems to cradle the nucleus in this position.

The nucleus is the repository of the genetic information that enables the synthesis of nearly all proteins of a living eukaryotic cell. Most eukaryotic cells possess a single nucleus, although the cells of fungi and some other groups may have several to many nuclei. Mammalian erythrocytes (red blood cells) lose their nuclei when they mature. Many nuclei exhibit a dark-staining zone called a **nucleolus,** which is a region where intensive synthesis of ribosomal RNA (rRNA) is taking place.

The nuclear envelope

The surface of the nucleus is bounded by *two* phospholipid bilayer membranes, which together make up the **nuclear envelope** (figure 4.8*a*). The outer membrane of the nuclear envelope is continuous with the cytoplasm's interior membrane system, called the *endoplasmic reticulum* (described in section 4.4).

Scattered over the surface of the nuclear envelope are what appear as shallow depressions in the electron micrograph but are, in fact, structures called **nuclear pores** (figure 4.8*b, c*). These pores form 50 to 80 nm apart at locations where the two membrane layers of the nuclear envelope pinch together. The structure consists of a central framework with eightfold symmetry that is embedded in the nuclear envelope. This is bounded by a cytoplasmic face with eight fibers and a nuclear face with a complex ring that forms a basket beneath the central ring. The pore allows ions and small molecules to diffuse freely between nucleoplasm and cytoplasm while controlling the passage of proteins and RNA–protein complexes. Transport across the pore is controlled and consists mainly of the import of proteins that function in the nucleus and the export to the cytoplasm of RNA and RNA–protein complexes formed in the nucleus.

The inner surface of the nuclear envelope is covered with a network of fibers that make up the nuclear lamina (figure 4.8*d*). This is composed of intermediate filament fibers called *nuclear lamins.* This structure gives the nucleus its shape and is involved in the deconstruction and reconstruction of the nuclear envelope that accompanies cell division.

DNA packaging

In both prokaryotes and eukaryotes, DNA is the molecule that stores genetic information. In eukaryotes, the DNA is divided into multiple linear chromosomes, which are organized with proteins into a complex structure called **chromatin.** It is becoming clear that the very structure of chromatin affects the function of DNA. Changes in gene expression that do not involve changes in DNA sequence, so-called epigenetic changes, involve alterations in chromatin structure (refer to chapter 16). Although still not fully understood, this offers an exciting new view of many old ideas.

Chromatin is also more organized in the nucleus than was once thought. The state of the DNA in chromatin must also change over the course of cell division, as we will see in chapter 10. Chromosomes become compacted into a more highly condensed state that forms the X-shaped chromosomes visible in the light microscope.

The nucleolus

Before cells can synthesize proteins in large quantity, they must first construct a large number of ribosomes to carry out this synthesis. Hundreds of copies of the genes encoding the rRNAs are clustered together on the chromosome, facilitating ribosome construction. By transcribing RNA molecules from this cluster, the cell rapidly generates large numbers of the molecules needed to assemble ribosomes.

The clusters of rRNA genes, the RNAs they produce, and the ribosomal proteins all come together within the nucleus during ribosome production. These ribosomal assembly areas are easily visible within the nucleus as one or more dark-staining regions called **nucleoli** (singular, *nucleolus*). Nucleoli can be seen under the light microscope even when the chromosomes are uncoiled.

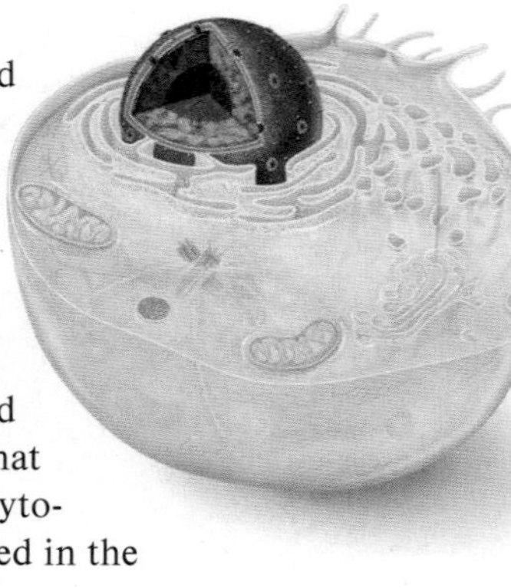
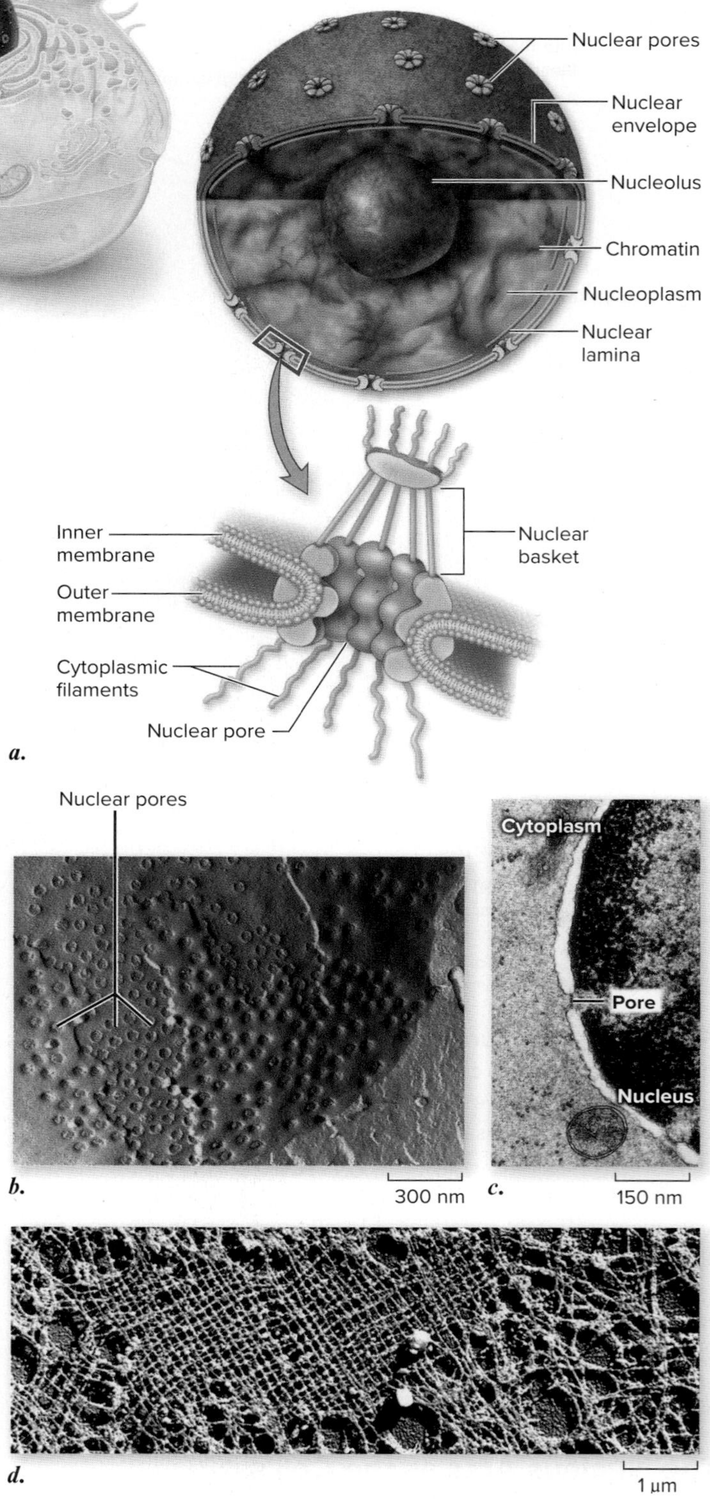

Figure 4.8 The nucleus. ***a.*** The nucleus consists of a double membrane called the nuclear envelope, enclosing a fluid matrix containing chromatin. Nuclear pores extend through the two membrane layers of the envelope. ***b.*** A freeze-fracture electron micrograph showing many nuclear pores. ***c.*** A transmission electron micrograph of the envelope showing a single nuclear pore. ***d.*** The nuclear lamina is visible as a dense network of fibers made of intermediate filaments. The nucleus has been colored purple in the micrographs.

(b): Don W. Fawcett/Science Source; (c): J. R. Factor/Science Source; (d): Dr. Ueli Aebi

Ribosomes Are the Cell's Protein Synthesis Machinery

LEARNING OBJECTIVE 4.3.3 Describe the structure of a ribosome.

Although the DNA in a cell's nucleus encodes the amino acid sequence of each protein in the cell, the proteins are not assembled there. A simple experiment demonstrates this: if a brief

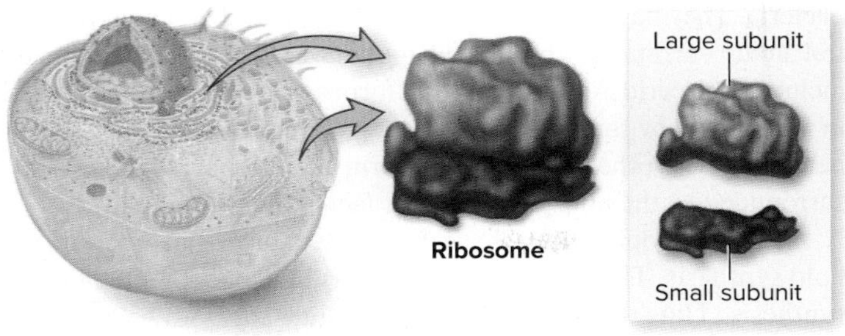

Figure 4.9 A ribosome. Ribosomes consist of a large subunit and a small subunit composed of rRNA and protein. The individual subunits are synthesized in the nucleus and then move through the nuclear pores to the cytoplasm, where they assemble to translate mRNA. Ribosomes serve as sites of protein synthesis.

pulse of radioactive amino acid is administered to a cell, the radioactivity shows up associated with newly made protein in the cytoplasm, not in the nucleus. When investigators first carried out these experiments, they found that protein synthesis is associated with large RNA–protein complexes (the organelles we now call **ribosomes**) outside the nucleus.

Ribosomes are among the most complex molecular assemblies found in cells. Each ribosome is composed of two subunits (figure 4.9), and each subunit is composed of a combination of RNA, called **ribosomal RNA (rRNA),** and several dozen different proteins. The subunits join to form a functional ribosome only when they are actively synthesizing proteins. This complicated process requires the two other main forms of RNA: **messenger RNA (mRNA),** which carries coding information from DNA, and **transfer RNA (tRNA),** which carries amino acids. Ribosomes use the information in mRNA to direct the synthesis of a protein. This process is often regulated by other small RNA molecules, the subject of much recent research. Protein synthesis is examined in detail in chapter 15.

Ribosomes are found either free in the cytoplasm or associated with internal membranes called endoplasmic reticulum, as described in section 4.4. Free ribosomes synthesize proteins that are found in the cytoplasm, nuclear proteins, mitochondrial proteins, and proteins in other organelles not derived from the endomembrane system. Membrane-associated ribosomes synthesize membrane proteins, proteins found in the endomembrane system, and proteins destined for export from the cell.

Ribosomes can be thought of as "universal organelles," because they are found in all cell types from all three domains of life. As we build a picture of the minimal essential functions for cellular life, ribosomes will be on the short list. Life is protein-based, and ribosomes are the factories that make proteins.

REVIEW OF CONCEPT 4.3

Eukaryotic cells exhibit compartmentalization with an endomembrane system and organelles that carry out specialized functions. The nucleus contains the cell's genetic material and consists of a double membrane connected to the endomembrane system. Material moves between the nucleus and cytoplasm through nuclear pores. Ribosomes synthesize proteins using information stored in DNA. Ribosomes are a universal organelle found in all known cells.

- *Would you expect cells in different organs in complex animals to have the same structure?*

4.4 Membranes Organize the Cell Interior into Functional Compartments

The interior of a eukaryotic cell is packed with membranes so thin that they are invisible under the low resolving power of light microscopes. This endomembrane system fills the cell, dividing it into compartments, channeling the passage of molecules through the interior of the cell, and providing surfaces for the synthesis of lipids and some proteins. The presence of these membranes in eukaryotic cells marks one of the fundamental distinctions between eukaryotes and prokaryotes.

The Endoplasmic Reticulum Is a Highway That Weaves Throughout the Cell

LEARNING OBJECTIVE 4.4.1 Distinguish between rough ER and smooth ER.

The largest of the internal membranes is called the **endoplasmic reticulum (ER).** *Endoplasmic* means "within the cytoplasm," and *reticulum* is Latin for "a little net." Like the plasma membrane, the ER is composed of a phospholipid bilayer with associated proteins. It weaves in sheets and tubules through the interior of the cell, creating a series of channels between its folds (figure 4.10). The two largest compartments in eukaryotic cells are the region within the ER, called the **cisternal space** or **lumen,** and the region exterior to it, the cytosol. The cytosol is the fluid component of the cytoplasm containing dissolved organic molecules such as proteins and ions.

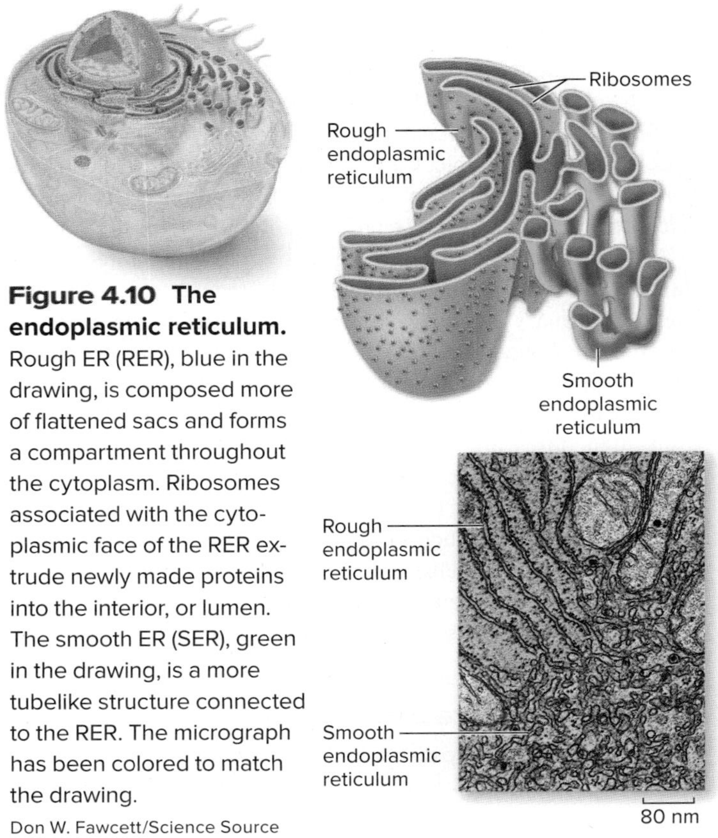

Figure 4.10 The endoplasmic reticulum. Rough ER (RER), blue in the drawing, is composed more of flattened sacs and forms a compartment throughout the cytoplasm. Ribosomes associated with the cytoplasmic face of the RER extrude newly made proteins into the interior, or lumen. The smooth ER (SER), green in the drawing, is a more tubelike structure connected to the RER. The micrograph has been colored to match the drawing.

Don W. Fawcett/Science Source

Rough ER is a site of protein synthesis

The **rough ER (RER)** gets its name from its surface appearance, dotted with small bumps, which are ribosomes. In the electron microscope, RER appears to be composed of flattened sacs, the surfaces of which are bumpy with ribosomes (figure 4.10).

The proteins synthesized on the surface of the RER are destined to be exported from the cell, sent to lysosomes or vacuoles (described later in this section), or embedded in the plasma membrane. These proteins enter the cisternal space as a first step in the pathway that will sort proteins to their eventual destinations. This pathway also involves vesicles and the Golgi apparatus, described later in this section. The sequence of the protein being synthesized determines whether the ribosome assembling it will become associated with the ER or remain a cytoplasmic ribosome.

In the ER, newly synthesized proteins can be modified by the addition of short-chain carbohydrates to form **glycoproteins.** Those proteins destined for secretion are separated from other products and later packaged into vesicles. The ER also manufactures membranes by producing membrane proteins and phospholipid molecules. The membrane proteins are inserted into the ER's own membrane, which can then expand and pinch off in the form of vesicles to be transferred to other locations.

Smooth ER has multiple roles

Regions of the ER with relatively few bound ribosomes are referred to as **smooth ER (SER).** The SER usually forms a network of tubules that appears distinct from, but still connected to, the flattened sacs of the RER. The membranes of the SER contain many embedded enzymes. Enzymes anchored within the SER, for example, catalyze the synthesis of a variety of carbohydrates and lipids. Steroid hormones are synthesized in the SER as well. The majority of membrane lipids are assembled in the SER and then sent to the parts of the cell that need membrane components.

The SER also has an important role storing intracellular Ca^{2+}. By sequestering intracellular Ca^{2+} in the SER, cells keep the cytosolic level of Ca^{2+} low. This allows a release of Ca^{2+} from the SER, or an influx of external Ca^{2+}, to be used as a signal. In muscle cells, for example, Ca^{2+} is used to trigger muscle contraction. In other cells, Ca^{2+} released from SER stores is involved in diverse signaling pathways (refer to chapter 9).

The SER plays a role in the detoxification of foreign substances by chemically modifying them to reduce toxicity. In the liver, the primary organ involved in detoxification, the enzymes involved are concentrated in SER. This function evolved as a form of protection, but it can neutralize substances that are ingested for therapeutic reasons, such as penicillin. Thus, relatively high doses are prescribed for some drugs to offset our body's efforts to remove them.

The ratio of SER to RER depends on a cell's function. In multicellular animals such as ourselves, this ratio varies greatly. Cells that carry out extensive lipid synthesis, such as those in the testes, intestine, and brain, have abundant SER. Cells that synthesize proteins that are secreted, such as antibodies, have much more extensive RER.

Lipid droplets store neutral lipids

Lipid droplets are widely distributed organelles involved in energy storage. They are found in plants, animals, fungi, and even some bacteria. They have a distinct structure, consisting of a core of neutral lipids surrounded by a lipid monolayer. These neutral lipids include triglycerides and sterols that form an oil phase made soluble in the cytosol by the phospholipid layer. Lipid droplets form when neutral lipids synthesized in the ER form an oil droplet that buds off surrounded by the phospholipid layer facing the cytoplasm.

Lipid droplets also have associated proteins involved in lipid synthesis. This allows them to act as a secondary site of lipid synthesis. Thus these organelles form a hub for both energy and membrane metabolism. Their size will vary with the needs of the cell, growing and shrinking as lipids are added or used.

The Golgi Apparatus Sorts and Packages Proteins

LEARNING OBJECTIVE 4.4.2 Explain the role of the Golgi body in the endomembrane system.

Flattened stacks of membranes, often interconnected with one another, form a complex called the **Golgi body.** These structures are named for Camillo Golgi, the 19th-century physician who first identified them. The number of stacked membranes within the Golgi body ranges from 1 or a few, in protists, to 20 or more in animal cells and to several hundred in plant cells. They are especially abundant in glandular cells, which manufacture and secrete substances. The Golgi body is often referred to as the **Golgi apparatus** (figure 4.11).

The Golgi apparatus is the post office of the cell. It functions in the collection, packaging, and distribution of molecules

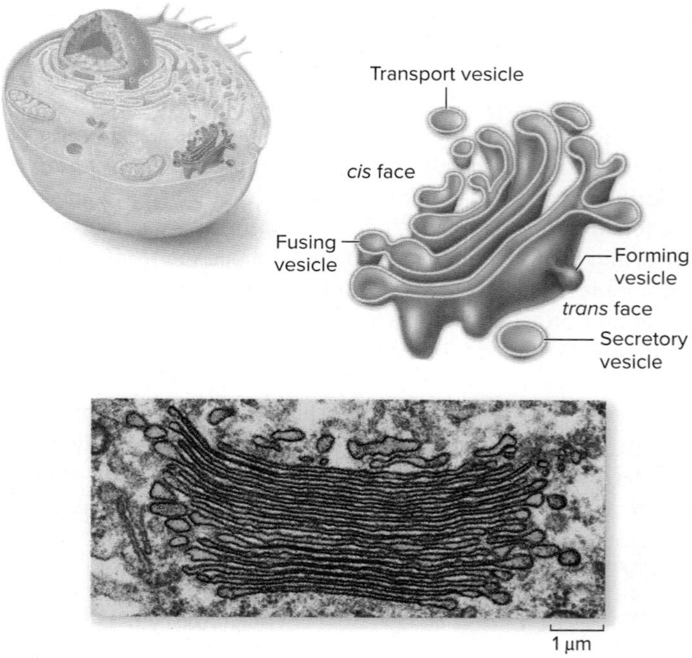

Figure 4.11 The Golgi apparatus. The Golgi apparatus is a smooth, concave, membranous structure. It receives material for processing in transport vesicles on the *cis* face and sends the material packaged in transport or secretory vesicles off the *trans* face. The substance in a vesicle could be for export out of the cell or for distribution to another region within the same cell.
Dennis Kunkel Microscopy/Science Source

synthesized at one location and used at another within the cell or even outside of it. A Golgi apparatus has a front and a back, with distinctly different membrane compositions at these opposite ends. The front, or receiving, end is called the *cis* face and is usually located near ER. Materials arrive at the *cis* face in transport vesicles that bud off the ER and exit at the *trans* face, where they are discharged in **secretory vesicles** (figure 4.12). How material transits through the Golgi apparatus has been a source of much contention. Models include maturation of the individual cisternae from *cis* to *trans,* transport between cisternae by vesicles, and direct tubular connections. Although there is probably transport of material by all of these, it now appears that the primary mechanism is cisternal maturation.

Proteins and lipids manufactured on the rough and smooth ER membranes are transported into the Golgi apparatus and modified as they pass through it. The most common alteration is the addition or modification of short sugar chains, forming glycoproteins and glycolipids. In many instances, enzymes in the Golgi apparatus modify existing glycoproteins and glycolipids made in the ER by cleaving a sugar from a chain or by modifying one or more of the sugars. These are then packaged into small, membrane-bounded vesicles that pinch off from the *trans* face of the Golgi apparatus. These vesicles then diffuse to other locations in the cell, distributing the newly synthesized molecules to their appropriate destinations.

Another function of the Golgi apparatus is the synthesis of cell-wall components. Noncellulose polysaccharides that form part of the cell wall of plants are synthesized in the Golgi apparatus and sent to the plasma membrane, where they can be added to the cellulose that is assembled on the exterior of the cell. Other polysaccharides secreted by plants are also synthesized in the Golgi apparatus.

Other Organelles Carry Out Degradation and Recycling

LEARNING OBJECTIVE 4.4.3 Explain how cells compartmentalize destructive enzymes.

Membrane-bounded digestive vesicles, called **lysosomes,** are also components of the endomembrane system (figure 4.13). Arising from the Golgi apparatus, they contain high levels of degrading enzymes, which catalyze the rapid breakdown of proteins, nucleic acids, lipids, and carbohydrates. Throughout the lives of eukaryotic cells, lysosomal enzymes break down old organelles and recycle their component molecules. This makes room for newly formed organelles. For example, mitochondria are replaced in some tissues every 10 days.

The digestive enzymes in the lysosome are optimally active at acid pH. Lysosomes are activated by fusing with a food vesicle produced by *phagocytosis* (a specific type of endocytosis; refer to chapter 5) or by fusing with an old or worn-out organelle. The fusion event activates proton pumps in the lysosomal membrane, resulting in a lower internal pH. As the interior pH falls, the arsenal of digestive enzymes contained in the lysosome is activated. This leads to the degradation of macromolecules in the food vesicle or the destruction of the old organelle.

A number of human genetic disorders, collectively called lysosomal storage disorders, affect lysosomes. For example, the

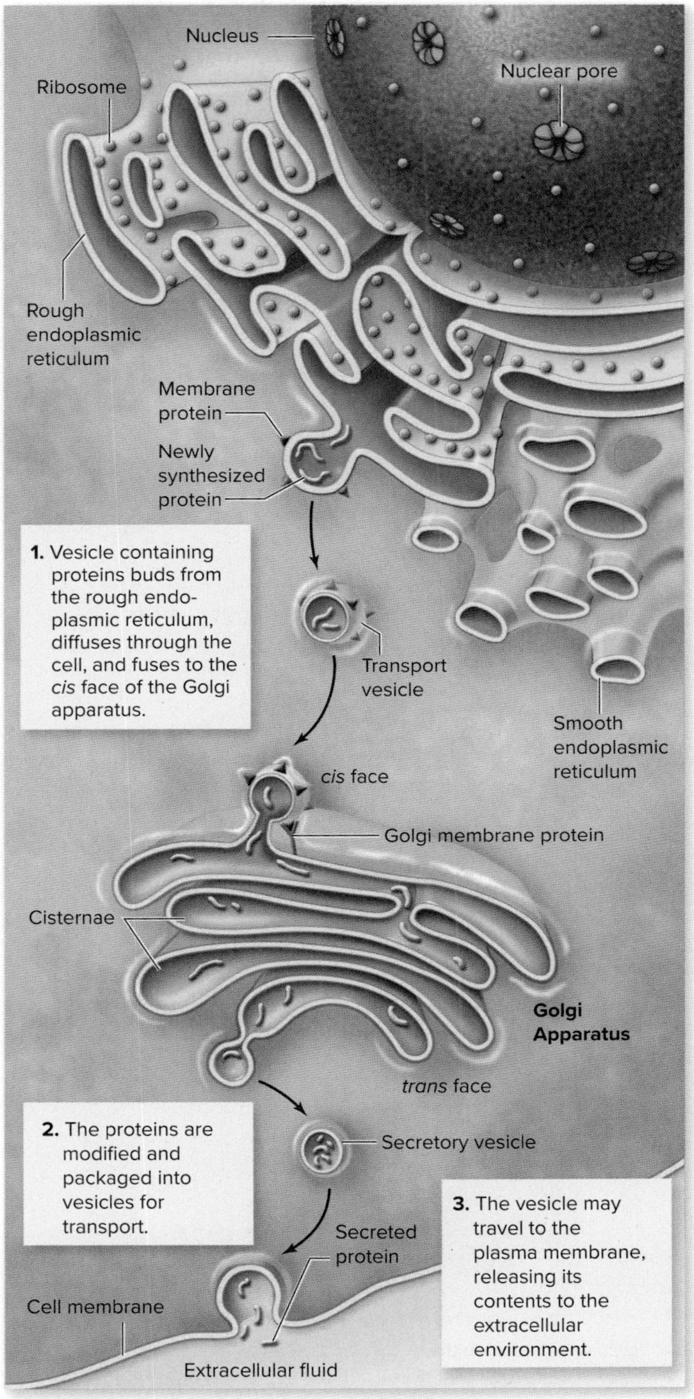

Figure 4.12 Protein transport through the endomembrane system. Proteins synthesized by ribosomes on the RER are translocated into the internal compartment of the ER. These proteins may be used at a distant location within the cell or secreted from the cell. They are transported within vesicles that bud off the RER. These transport vesicles travel to the *cis* face of the Golgi apparatus. There they can be modified and packaged into vesicles that bud off the *trans* face of the Golgi apparatus. Vesicles leaving the *trans* face transport proteins to other locations in the cell, or they fuse with the plasma membrane, releasing their contents to the extracellular environment.

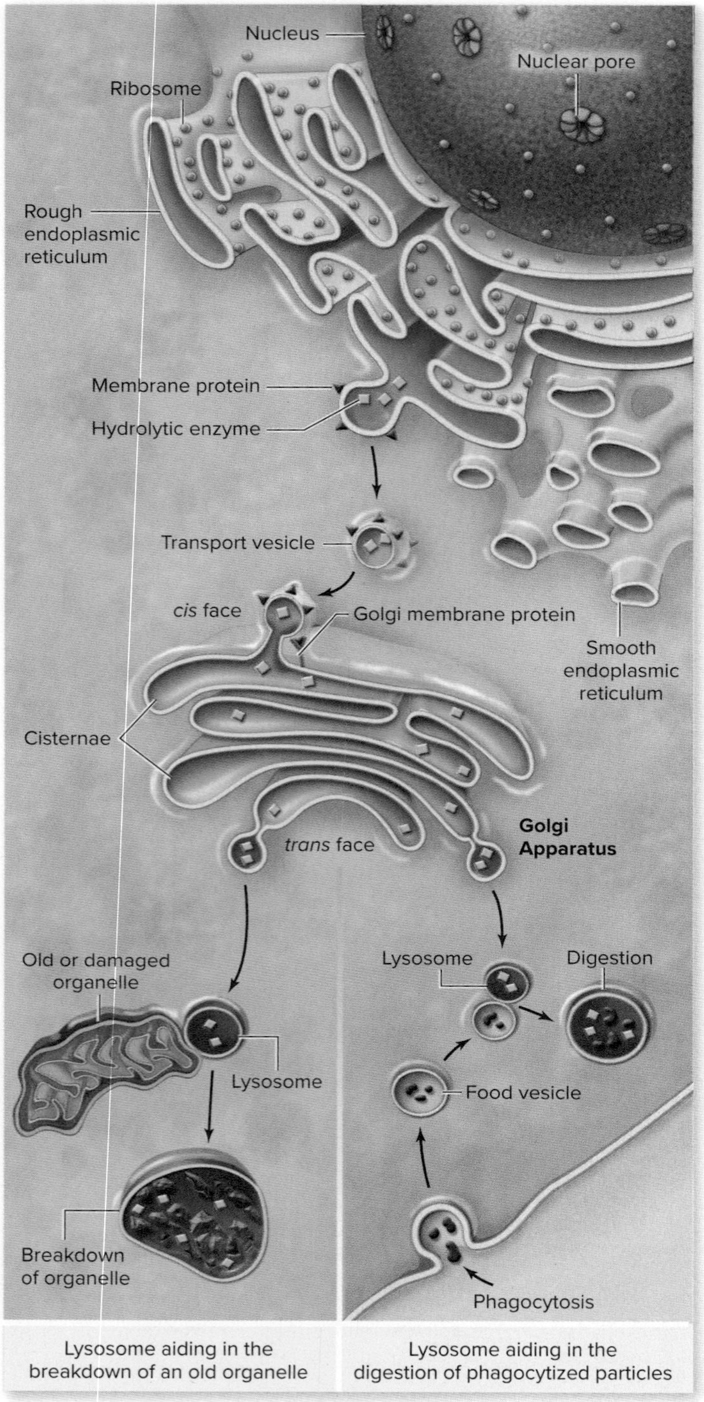
Nucleus
Nuclear pore
Ribosome
Rough endoplasmic reticulum
Membrane protein
Hydrolytic enzyme
Transport vesicle
cis face
Golgi membrane protein
Smooth endoplasmic reticulum
Cisternae
trans face
Golgi Apparatus
Old or damaged organelle
Lysosome
Digestion
Lysosome
Food vesicle
Breakdown of organelle
Phagocytosis
Lysosome aiding in the breakdown of an old organelle
Lysosome aiding in the digestion of phagocytized particles

Figure 4.13 Lysosomes. Lysosomes are formed from vesicles budding off the Golgi apparatus. They contain hydrolytic enzymes that digest particles or cells taken into the cell by phagocytosis and break down old organelles.

genetic abnormality called Tay-Sachs disease is caused by the loss of function of a single lysosomal enzyme (hexosaminidase). This enzyme is necessary to break down a membrane glycolipid found in nerve cells. The accumulation of glycolipid in lysosomes degrades nerve cell function, leading to a variety of clinical symptoms such as seizures and muscle rigidity.

In addition to breaking down organelles within cells, lysosomes eliminate other cells that the cell has engulfed by phagocytosis. When a white blood cell, for example, phagocytizes a passing pathogen, lysosomes fuse with the resulting "food vesicle," releasing their enzymes into the vesicle and degrading the material within.

Microbodies

Eukaryotic cells contain a variety of enzyme-bearing vesicles called **microbodies.** These are found in the cells of plants, animals, fungi, and protists. The distribution of enzymes into microbodies is one of the principal ways eukaryotic cells organize their metabolism.

An important type of microbody is the **peroxisome,** a small, spherical organelle that contains crystalline arrays of enzymes involved in the oxidation of fatty acids and the detoxification of harmful chemicals. Peroxisomes get their name from the hydrogen peroxide produced as a by-product of the activities of the peroxisome's digestive and detoxifying oxidative enzymes. Hydrogen peroxide is dangerous to cells because of its violent chemical reactivity. Peroxisomes contain the enzyme catalase, which breaks down hydrogen peroxide into its harmless constituents—water and oxygen. If the cell's oxidative enzymes were not isolated within microbodies, they would tend to short-circuit the metabolism of the cytoplasm, which often involves adding hydrogen atoms to oxygen.

Peroxisomes can form either by simple division of mature peroxisomes or from the fusion of ER-derived vesicles, which then import peroxisomal proteins to form a mature peroxisome.

Proteasomes

Not all cell compartmentalization is within membrane-bounded organelles. Compartmentalization also occurs on a much smaller scale. Cells recycle their proteins in large, cylindrical complexes called **proteasomes** (figure 4.14). Cells mark misfolded, damaged, or no longer needed proteins for destruction by attaching a tag to them—a 76-amino-acid protein called ubiquitin (the Inquiry &

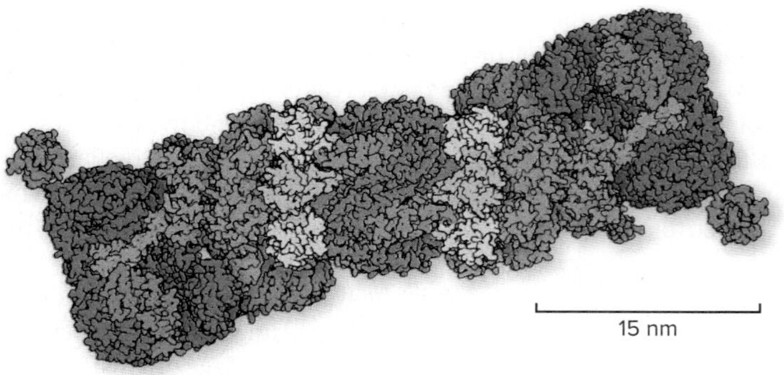
15 nm

Figure 4.14 The *Drosophila* proteasome. The central complex contains the proteolytic activity, and the flanking regions act as regulators. Proteins enter one end of the cylinder and are cleaved to peptide fragments that exit the other end.

Image of PDB ID 4B4T (Unverdorben, P., et al., "Deep classification of a large cryo-EM dataset defines the conformational landscape of the 26S proteasome," PNAS (2014) 111: 5444-5549). ©2015 David Goodsell & RCSB Protein Data Bank (www.rcsb.org). Molecule of the Month DOI 10.22/rcsb_pdp/mom_2013_10

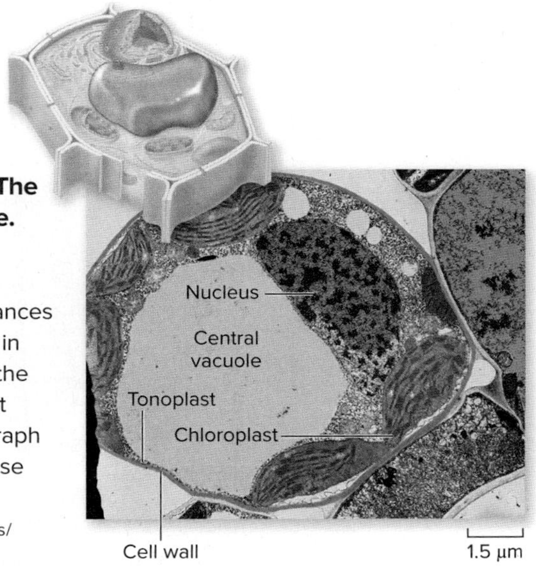

Nucleus
Central vacuole
Tonoplast
Chloroplast
Cell wall
1.5 µm

Figure 4.15 The central vacuole. A plant's central vacuole stores dissolved substances and can expand in size to increase the tonicity of a plant cell. The micrograph is shown with false color.
Biophoto Associates/Science Source

Analysis section at the end of this chapter describes this process). Proteins enter one end of the proteasome tube, are digested in the central region, and exit the other end as amino acids or peptide fragments. Prokaryotes do not have ubiquitin and thus were thought to lack proteasomes. Later research revealed that archaea do have the core degradative particle, although it is regulated differently. The only bacteria that appear to have a similar core particle are the actinomycetes.

Plant vacuoles

Plant cells have specialized membrane-bounded structures called **vacuoles.** The most conspicuous example is the large central vacuole found in most plant cells (figure 4.15). In fact, *vacuole* means "blank space," referring to its appearance in the light microscope. The membrane surrounding this vacuole is called the **tonoplast,** because it contains channels for water that are used to help the cell maintain its tonicity, or osmotic balance (see osmosis in chapter 5). This allows the cell to expand and contract depending on conditions. For many years biologists assumed that only one type of vacuole existed and that it served multiple functions. Studies of tonoplast transporters and the isolation of vacuoles from a variety of cell types have led to a more complex view of vacuoles. These studies have made it clear that different cells have different vacuolar types. These vacuoles are specialized, depending on the function of the cell.

REVIEW OF CONCEPT 4.4

The endoplasmic reticulum (ER) is an extensive internal membrane system that organizes the cell's biosynthetic activities. Proteins from the RER are transported by vesicles to the Golgi apparatus, where they are modified, packaged, and distributed to their final location. Lysosomes, peroxisomes, and proteasomes are vesicles that compartmentalize destructive enzymes. Vacuoles are membrane-bounded structures with roles ranging from storage to cell growth in plants.

- *How do ribosomes on the RER differ from cytoplasmic ribosomes?*

4.5 Mitochondria and Chloroplasts Are Energy-Processing Organelles

Mitochondria Metabolize Organic Compounds to Generate ATP

LEARNING OBJECTIVE 4.5.1 Describe the structure of a mitochondrion.

Mitochondria and chloroplasts are the ATP-generating organelles of the cell. They share both structural and functional similarities. Structurally, they are both surrounded by a double membrane, and both contain their own DNA and protein synthesis machinery. Functionally, they are both intimately involved in energy metabolism, as we will explore in detail in later chapters on energy metabolism and photosynthesis.

Mitochondria (singular, *mitochondrion*) vary in both size and shape, often existing as part of a dynamic network, and are found in all types of eukaryotic cells (figure 4.16). Within eukaryotic cells, mitochondria metabolize organic compounds, including sugars to generate ATP. As we will learn in chapter 7, mitochondria are bounded by two membranes: a smooth outer membrane and an inner folded membrane with numerous contiguous layers called **cristae** (singular, *crista*) that play a key role in ATP generation.

The cristae partition the mitochondrion into two compartments: a **matrix,** lying inside the inner membrane, and an outer compartment, or **intermembrane space,** lying between the two

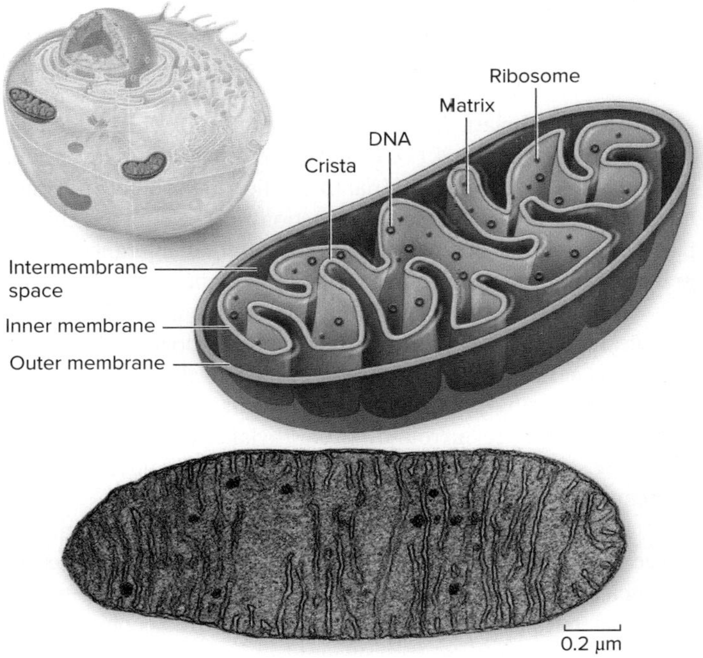

Ribosome
Matrix
DNA
Crista
Intermembrane space
Inner membrane
Outer membrane
0.2 µm

Figure 4.16 Mitochondria. The inner membrane of a mitochondrion is shaped into folds called cristae that greatly increase the surface area for oxidative metabolism. A mitochondrion in cross section and cut lengthwise is shown colored red in the micrograph.
Keith R. Porter/Science Source

mitochondrial membranes. On the surface of the inner membrane, and also embedded within it, are proteins that carry out oxidative metabolism, the oxygen-requiring process by which energy in macromolecules is used to produce ATP (chapter 7).

Mitochondria have their own DNA. This circular DNA molecule contains several genes that produce proteins essential to the mitochondrion's role in oxidative metabolism. Thus, the mitochondrion, in many respects, acts as a cell within a cell, maintaining its own genetic information specifying proteins for its unique functions. The mitochondria are not fully autonomous, however, because most of the genes that encode the enzymes used in oxidative metabolism are located in the cell nucleus.

A eukaryotic cell does not produce brand-new mitochondria each time the cell divides. Instead, the mitochondria themselves divide in two, doubling in number, and these are partitioned between the new cells. Most of the components required for mitochondrial division are encoded by genes in the nucleus and are translated into proteins by cytoplasmic ribosomes. Mitochondrial replication therefore is impossible without nuclear participation, and mitochondria thus cannot be grown in a cell-free culture.

Chloroplasts Use Light to Generate ATP and Sugars

LEARNING OBJECTIVE 4.5.2 Differentiate between mitochondria and chloroplasts.

Plant cells and cells of other eukaryotic organisms that carry out photosynthesis typically contain from one to several hundred **chloroplasts.** Chloroplasts use light to generate ATP and sugars. This bestows an obvious advantage on the organisms that possess them: they can manufacture their own food. Chloroplasts contain the photosynthetic pigment chlorophyll, which gives most plants their green color.

The chloroplast, like the mitochondrion, is surrounded by two membranes (figure 4.17). However, chloroplasts are larger and more complex than mitochondria. In addition to the outer and inner membranes, which lie in close association with each other, chloroplasts have closed compartments of stacked membranes called **grana** (singular, *granum*), which lie inside the inner membrane.

A chloroplast may contain a hundred or more grana, and each granum may contain from a few to several dozen disk-shaped structures called **thylakoids.** On the surface of the thylakoids are the light-capturing photosynthetic pigments, to be discussed in depth in chapter 8. Surrounding the thylakoid is a fluid matrix called the *stroma.* The enzymes used to synthesize glucose during photosynthesis are found in the stroma.

Like mitochondria, chloroplasts contain DNA, but many of the genes that specify chloroplast components are also located in the nucleus. Some of the elements used in photosynthesis, including the specific protein components necessary to accomplish the reaction, are synthesized entirely within the chloroplast.

Other DNA-containing organelles in plants, called *leucoplasts,* lack pigment and a complex internal structure. In root cells and some other plant cells, leucoplasts may serve as starch-storage sites. A leucoplast that stores starch (amylose) is sometimes termed an **amyloplast.** These organelles—chloroplasts, leucoplasts, and amyloplasts—are collectively called **plastids.** All plastids are produced by the division of existing plastids.

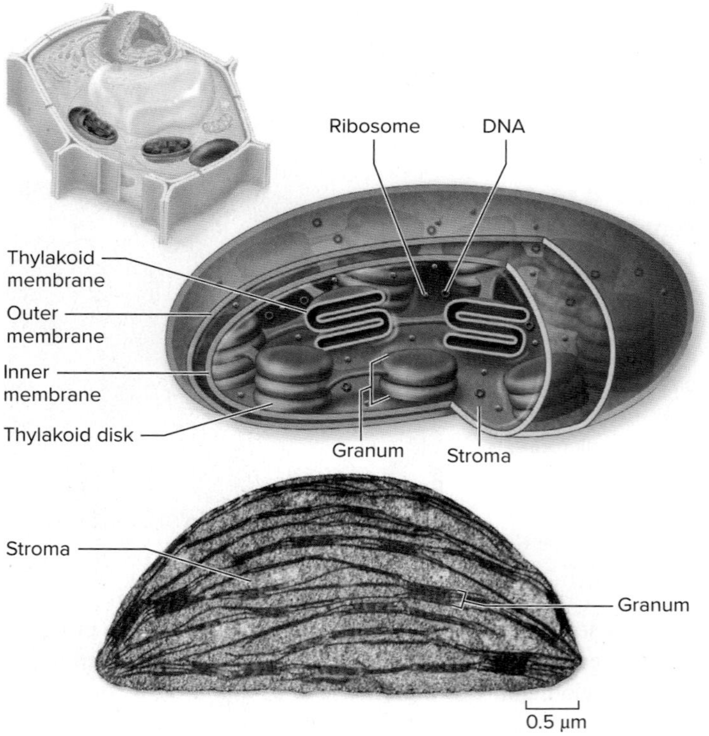

Figure 4.17 Chloroplast structure. The inner membrane of a chloroplast surrounds a membrane system of stacks of closed chlorophyll-containing vesicles called thylakoids, within which photosynthesis occurs. Thylakoids are typically stacked one on top of the other in columns called grana. The chloroplast has been colored green in the micrograph.
Jeremy Burgess/Science Source

Mitochondria and Chloroplasts Arose by Endosymbiosis

LEARNING OBJECTIVE 4.5.3 Describe how mitochondria might have evolved from ancient bacteria.

Symbiosis is a close relationship between organisms of different species that live together. The theory of **endosymbiosis** proposes that some of today's eukaryotic organelles evolved as a consequence of a symbiosis arising between two cells that were originally each free-living. One cell, a prokaryote, was engulfed by and became part of another cell, which was the precursor of modern eukaryotes (figure 4.18).

According to this endosymbiont theory, now widely accepted by biologists, the engulfed prokaryotes provided their hosts with certain advantages associated with their special metabolic abilities. Eventually many of the genes of the prokaryote transferred to the host eukaryotic chromosome.

Two key eukaryotic organelles are believed to be the descendants of these endosymbiotic prokaryotes: mitochondria, which are thought to have originated as bacteria capable of carrying out oxidative metabolism, and chloroplasts, which apparently arose from photosynthetic bacteria. The extensive evidence supporting this theory is discussed in chapter 24.

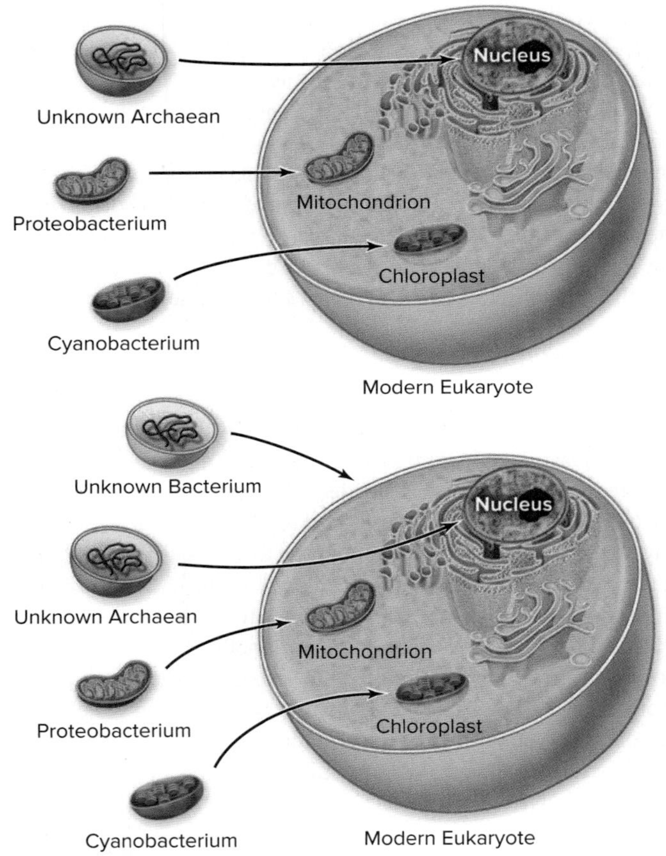

Figure 4.18 Possible origins of eukaryotic cells. Both mitochondria and chloroplasts are thought to have arisen by endosymbiosis, whereby a free-living cell is taken up but not digested. The nature of the engulfing cell is unknown. Two possibilities are shown. The engulfing cell (top) is an archaean that gave rise to the nuclear genome and cytoplasmic contents. The engulfing cell (bottom) consists of a nucleus derived from an archaean in a bacterial cell. This could arise by a fusion event or by engulfment of the archaean by the bacterium.

REVIEW OF CONCEPT 4.5

Mitochondria and chloroplasts both have an outer membrane and an extensive inner membrane compartment. Both also have their own DNA but also have nuclear-encoded proteins. Mitochondria metabolize sugar to produce ATP. Chloroplasts harness light energy to produce ATP and synthesize sugars. Both mitochondria and chloroplasts arose by endosymbiosis, whereby a prokaryotic cell was engulfed by a eukaryotic precursor.

- *Many proteins in mitochondria and chloroplasts are encoded by nuclear genes. How might this have come about?*

4.6 An Internal Skeleton Supports the Shape of Cells

Eukaryotic cells can take on an amazing array of shapes, from neurons with projections running from your spinal cord to your toes to skin cells that appear as compact, elongated cubes. This flexibility of shape is made possible by a complex internal skeleton, a structure made even more impressive by the cell's ability to reorganize portions of it, depending on the cell's needs.

The Cytoskeleton Is Composed of Three Types of Protein Fibers

LEARNING OBJECTIVE 4.6.1 Contrast the structure and function of the three protein fibers of the cytoskeleton.

The cytoplasm of all eukaryotic cells is crisscrossed by a network of three kinds of protein fibers that support the shape of the cell and anchor organelles to fixed locations. This network, called the **cytoskeleton,** is a dynamic system, constantly assembling and disassembling. Individual fibers consist of polymers of identical protein subunits that attract one another and spontaneously assemble into long chains. Fibers disassemble in the same way, as one subunit after another breaks away from one end of the chain.

Actin filaments (microfilaments)

Actin filaments are long fibers about 7 nm in diameter. Each filament is composed of two protein chains loosely twined together like two strands of pearls. Each "pearl," or subunit, on the chain is the globular protein **actin.** Actin filaments exhibit polarity; that is, they have plus (+) and minus (−) ends. These designate the direction of growth of the filaments. Actin molecules spontaneously form these filaments, even in a test tube. Cells regulate the rate of actin polymerization through other proteins that act as switches, turning on polymerization when appropriate.

Microtubules

Microtubules, the largest of the cytoskeletal elements, are hollow tubes about 25 nm in diameter, each composed of a ring of 13 protein protofilaments. Globular proteins consisting of dimers of α- and β-*tubulin* subunits polymerize to form the 13 protofilaments. The protofilaments are arrayed side by side around a central core, giving the microtubule its characteristic tube shape.

Microtubules are in a constant state of flux, continually polymerizing and depolymerizing. The average half-life of a microtubule ranges from as long as 10 minutes in a nondividing animal cell to as short as 20 seconds in a dividing animal cell. The ends of the microtubule are designated as plus (+; away from the nucleation center) or minus (−; toward the nucleation center).

Intermediate filaments

The most durable element of the cytoskeleton is a system of tough, fibrous protein molecules twined together in an overlapping arrangement. These *intermediate filaments* are characteristically 8 to 10 nm in diameter–between the sizes of actin filaments and microtubules. Once formed, intermediate filaments are stable and only rarely break down.

Intermediate filaments constitute a mixed group of cytoskeletal fibers. The most common type, composed of protein subunits called *vimentin,* provides structural stability for many kinds of cells. *Keratin,* another class of intermediate filament, is found in epithelial cells (cells that line organs and body cavities) and associated structures such as hair and fingernails.

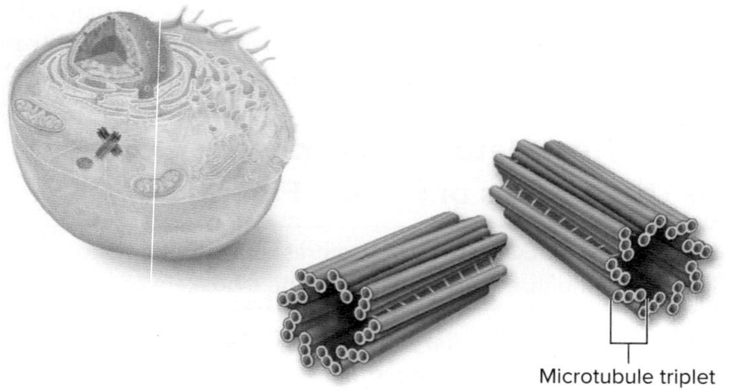
Microtubule triplet

Figure 4.19 Centrioles. Each centriole is composed of nine triplets of microtubules. Centrioles are usually not found in plant cells. In animal cells they help to organize microtubules.

Centrosomes are microtubule-organizing centers

Centrioles are barrel-shaped organelles found in the cells of animals and most protists. They occur in pairs, usually located at right angles to each other near the nuclear membranes (figure 4.19). The region surrounding the pair in almost all animal cells is referred to as a *centrosome.* Surrounding the centrioles in the centrosome is the **pericentriolar material,** which contains ring-shaped structures composed of tubulin. The pericentriolar material can nucleate the assembly of microtubules in animal cells. Structures with this function are called *microtubule-organizing centers.* The centrosome is also responsible for the reorganization of microtubules that occurs during cell division. The centrosomes of plants and fungi lack centrioles but still contain microtubule-organizing centers. You will learn more about the actions of the centrosomes when examining the process of cell division in chapter 10.

The Cytoskeleton Helps Move Materials Within Cells

LEARNING OBJECTIVE 4.6.2 Explain how animal cells use cytoskeletal elements to move materials within the cell.

Actin filaments and microtubules often orchestrate their activities to affect cellular processes. For example, during cell reproduction (refer to chapter 10), newly replicated chromosomes move to opposite sides of a dividing cell because they are attached to shortening microtubules. Then, in animal cells, a belt of actin pinches the cell in two by contracting like a purse string.

Muscle cells also use actin filaments, which slide along filaments of the motor protein myosin when a muscle contracts. The fluttering of an eyelash, the flight of an eagle, and the awkward crawling of a baby all depend on these cytoskeletal movements within muscle cells.

Not only is the cytoskeleton responsible for the cell's shape and movement, but it also provides a scaffold that holds certain enzymes and other macromolecules in defined areas of the cytoplasm. For example, many of the enzymes involved in cell metabolism bind to actin filaments, as do ribosomes. By moving and anchoring particular enzymes near one another, the cytoskeleton, like the endoplasmic reticulum, helps organize the cell's activities.

Molecular motors

All eukaryotic cells must move materials from one place to another in the cytoplasm. One way cells do this is by using the channels of the endoplasmic reticulum as an intracellular highway. Material can also be moved using vesicles loaded with cargo that can move along the cytoskeleton like it was a railroad track. For example, nerve cells have long projections that extend away from the cell body. Vesicles can move along tracks from the cell body to the end of the cell.

Four components are required to move material along microtubules: (1) a vesicle or an organelle that is to be transported, (2) a motor protein that provides the energy-driven motion, (3) a connector molecule that connects the vesicle to the motor molecule, and (4) microtubules on which the vesicle will ride like a train on a rail (figure 4.20).

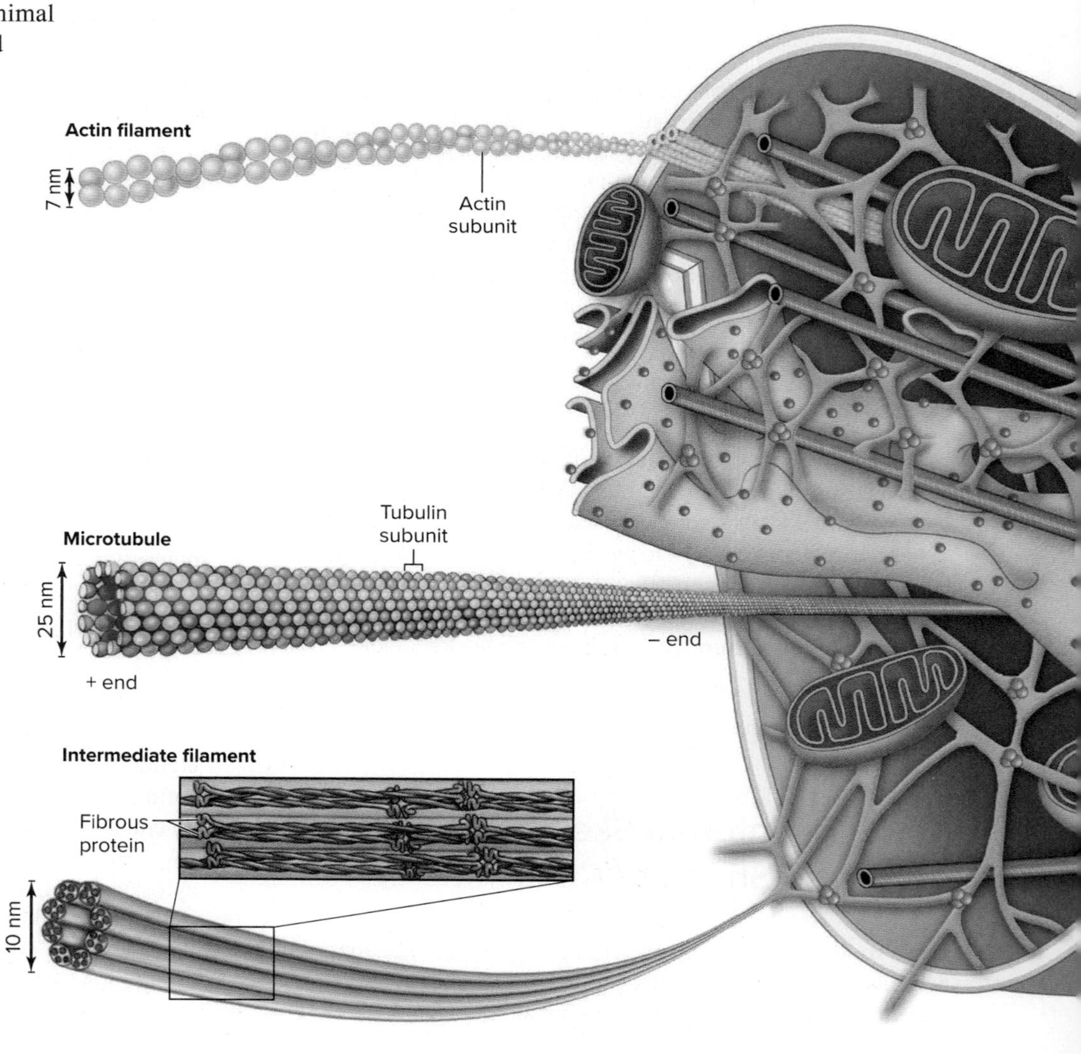
Actin filament
7 nm
Actin subunit
Microtubule
Tubulin subunit
25 nm
+ end
– end
Intermediate filament
Fibrous protein
10 nm

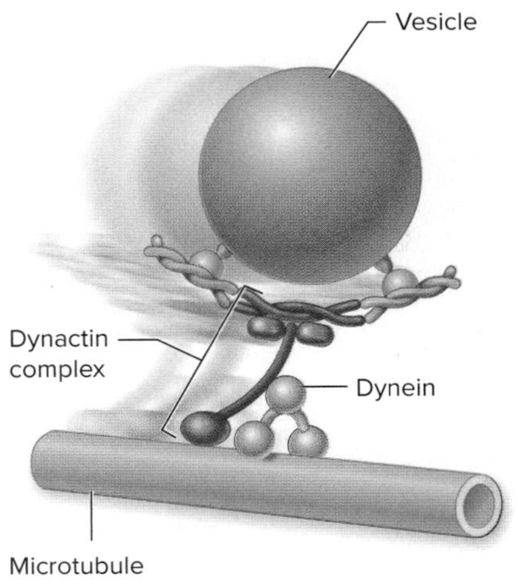

Figure 4.20 Molecular motors. Vesicles can be transported along microtubules using motor proteins that use ATP to generate force. The vesicles are attached to motor proteins by connector molecules, such as the dynactin complex shown here. The motor protein dynein moves the connected vesicle along microtubules.

The direction a vesicle is moved depends on the type of motor protein involved and the fact that microtubules are organized with their plus ends toward the periphery of the cell. In one case, a protein called kinectin binds vesicles to the motor protein *kinesin* (figure 4.21). Kinesin uses ATP to power its movement toward the cell periphery, dragging the vesicle with it as it travels along the microtubule toward the plus end. As nature's tiniest motors, these proteins pull the transport vesicles along the microtubular tracks. Another set of vesicle proteins, called the dynactin complex, binds vesicles to the motor protein *dynein* (illustrated in figure 4.20), which directs movement in the opposite direction along microtubules toward the minus end, inward toward the cell's center. (Dynein is also involved in the movement of eukaryotic flagella, as discussed later in this section.) The destination of a particular transport vesicle and its content is thus determined by the nature of the linking protein embedded within the vesicle's membrane.

SCIENTIFIC THINKING

Hypothesis: *Kinesin molecules can act as molecular motors and move along microtubules using energy from ATP.*

Test: *A microscope slide is covered with purified kinesin. Purified microtubules are added in a buffer containing ATP. The microtubules are monitored under a microscope using a video recorder to capture any movement.*

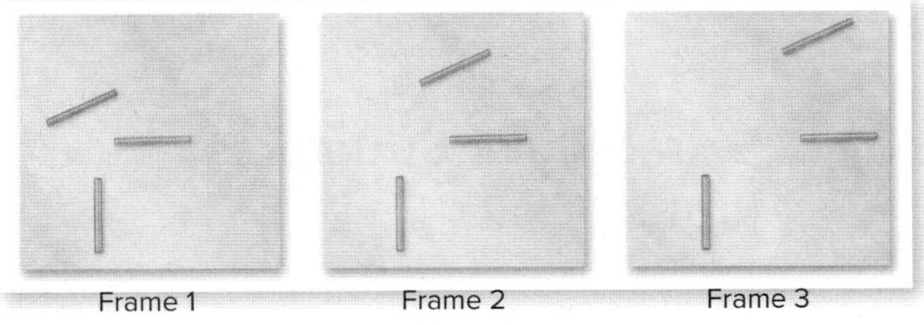

Result: *Over time, the movement of individual microtubules can be observed in the microscope. This is shown schematically in the figure by the movement of specific microtubules shown in color.*

Conclusion: *Kinesin acts as a molecular motor moving along (in this case actually moving) microtubules.*

Further Experiments: *Are there any further controls that are not shown in this experiment? What additional conclusions could be drawn by varying the amount of kinesin sticking to the slide?*

Figure 4.21 Demonstration of kinesin as molecular motor. Microtubules can be observed moving over a slide coated with kinesin.

Eukaryotic Cells Use Cytoskeletal Elements to Crawl or Swim

LEARNING OBJECTIVE 4.6.3 Contrast how an animal cell crawls with how a protist uses flagella to swim.

Essentially all cell motion is tied to the movement of actin filaments, microtubules, or both. Intermediate filaments act as intracellular tendons, preventing excessive stretching of cells. Actin filaments play a major role in determining the shape of cells. Because actin filaments can form and dissolve so readily, they enable some cells to change shape quickly.

The arrangement of actin filaments in the cell cytoplasm allows cells to literally crawl along a substrate! This kind of motility is important during animal development, and in adult animals for such diverse processes as inflammation, blood clotting, and wound healing. White blood cells, in particular, exhibit this ability. Produced in the bone marrow, these cells are released into the circulatory system and then eventually crawl out of blood vessels and into the tissues to destroy potential pathogens.

At the leading edge of a crawling cell, actin filaments rapidly polymerize, and their extension forces the edge of the cell forward. This extended region is stabilized when microtubules polymerize into the newly formed region. Overall forward movement of the cell is then achieved through the action of the protein **myosin,** which is best known for its role in muscle contraction. Myosin motors along the actin filaments contract, pulling the contents of the cell toward the newly extended front edge.

Cells crawl when these steps occur continuously, with a leading edge extending and stabilizing, and then motors contracting to pull the remaining cell contents along. Receptors on the cell surface can detect molecules outside the cell and stimulate extension in specific directions, allowing cells like white blood cells to move toward particular targets.

Flagella and cilia aid movement

In section 4.2, we saw that bacterial and archaeal flagella evolved independently, and now we see a third kind of flagellum evolved in eukaryotes: the eukaryotic flagellum, consisting of a circle of nine microtubule pairs surrounding two central microtubules. This arrangement is referred to as the *9 + 2 structure* (figure 4.22).

As pairs of microtubules move past each other using arms composed of the motor protein dynein, the eukaryotic flagellum *undulates,* rather than rotates. When examined carefully, each flagellum proves to be an outward projection of the cell's interior, containing cytoplasm and enclosed by the plasma membrane.

Figure 4.22 Flagella and cilia. A eukaryotic flagellum originates directly from a basal body. The flagellum has two microtubules in its core connected by radial spokes to an outer ring of nine paired microtubules with dynein arms (9 + 2 structure). The basal body consists of nine microtubule triplets connected by short protein segments. The structure of cilia is similar to that of flagella, but cilia are usually shorter.
William Dentler, University of Kansas

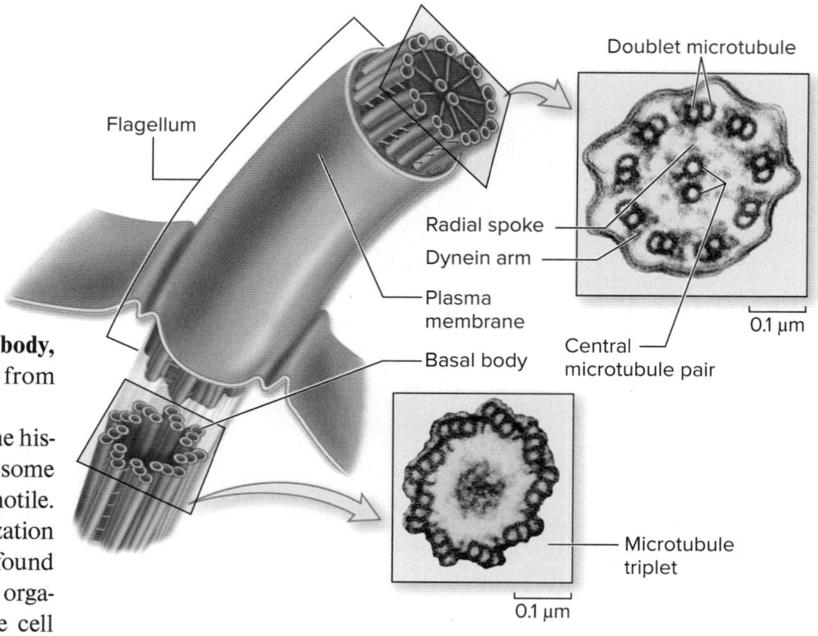

The microtubules of the flagellum are derived from a **basal body,** situated just below the point where the flagellum protrudes from the surface of the cell.

The flagellum's microtubular structure evolved early in the history of eukaryotes. Today the cells of many multicellular and some unicellular eukaryotes no longer possess flagella and are nonmotile. Other structures, called **cilia** (singular, *cilium*), with an organization similar to the 9 + 2 arrangement of microtubules, can still be found within them. Cilia are short cellular projections that are often organized in rows. They are more numerous than flagella on the cell surface but have the same internal structure. Often the beating of rows of cilia moves water over the tissue surface (figure 4.23).

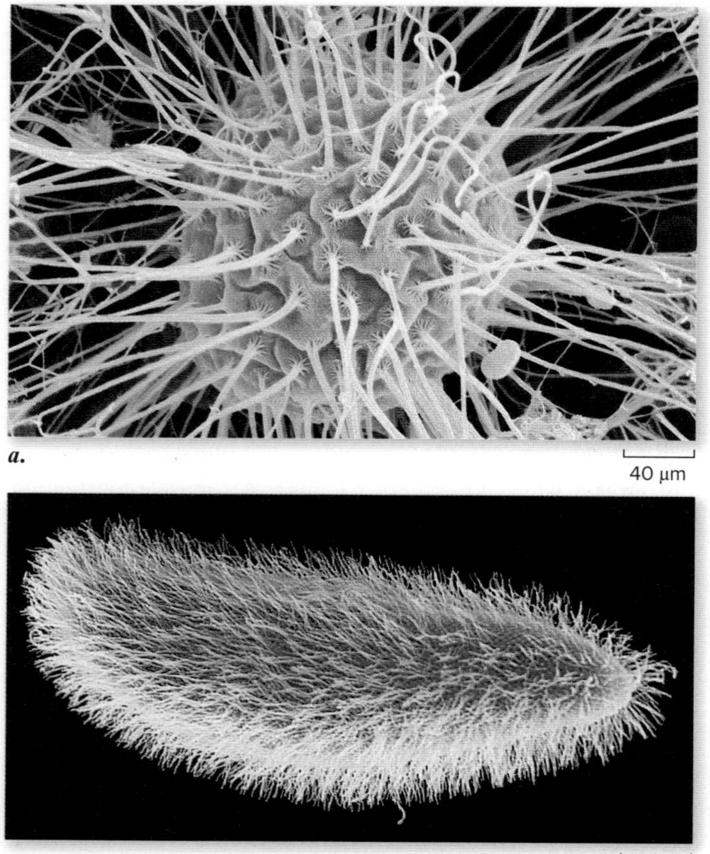

Figure 4.23 Flagella and cilia. ***a.*** A green alga with numerous flagella that allow it to move through the water. ***b.*** Paramecia are covered with many cilia, which beat in unison to move the cell. The cilia can also be used to move fluid into the paramecium's mouth to ingest material.
SPL/Science Source

REVIEW OF CONCEPT 4.6

The three cytoskeleton elements are microfilaments (actin), microtubules, and intermediate filaments. These fibers interact to modulate cell shape, permit cell movement, and move materials within the cytoplasm. Vesicles can move along microtubules via molecular motors. Cell movement involves actin and myosin for crawling, or cilia and flagella, made of microtubules, for swimming. Eukaryotic cilia and flagella are composed of bundles of microtubules, which undulate, in a 9 + 2 array.

- *How many different ways are cytoskeletal elements involved in movement?*

4.7 Extracellular Structures Protect Cells

Plant Cell Walls Provide Protection and Support

LEARNING OBJECTIVE 4.7.1 Contrast primary and secondary plant cell walls.

The cells of plants, fungi, and many types of protists have cell walls, which protect and support the cells. The cell walls of these eukaryotes are chemically and structurally different from prokaryotic cell walls. In plants and protists, the cell walls are composed of fibers of the polysaccharide cellulose, whereas in fungi, the cell walls are composed of chitin.

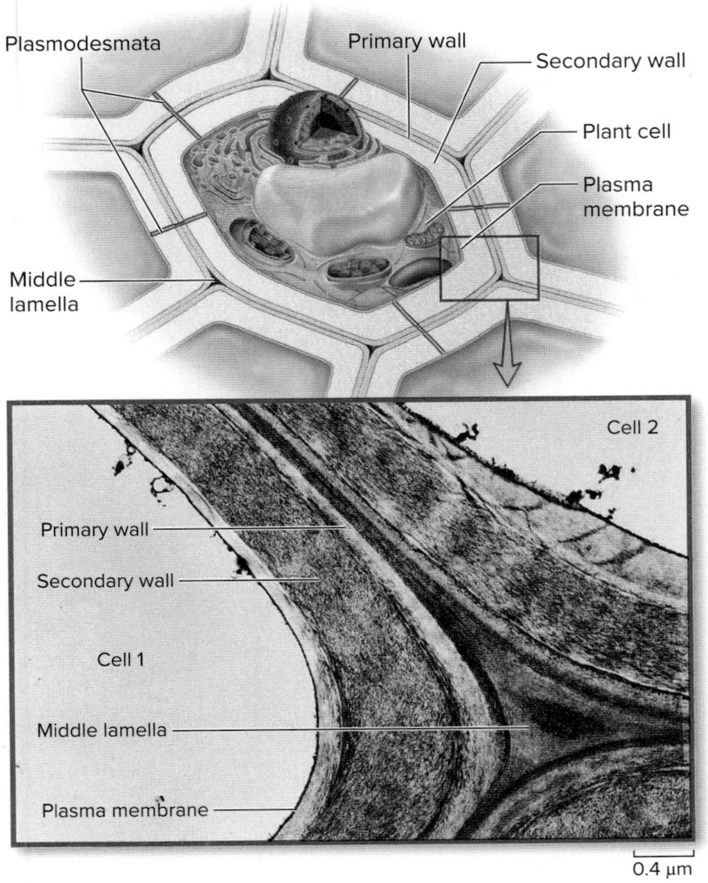

Figure 4.24 Cell walls in plants. Plant cell walls are thick, strong, and rigid. Primary cell walls are laid down when the cell is young. Thicker secondary cell walls may be added later when the cell is fully grown.

Biophoto Associates/Science Source

In plants, **primary walls** are laid down when the cell is still growing. Between the walls of adjacent cells a sticky substance, called the **middle lamella,** glues the cells together (figure 4.24). Some plant cells produce strong **secondary walls,** which are deposited inside the primary walls of fully expanded cells.

Animal Cells Secrete an Extracellular Matrix

LEARNING OBJECTIVE 4.7.2 Explain how integrins link the cytoskeleton to the extracellular matrix of animal cells.

Animal cells lack the cell walls that encase the cells of plants, fungi, and most protists. Instead, animal cells secrete an elaborate mixture of glycoproteins into the space around them, forming the *extracellular matrix* (*ECM;* figure 4.25). The fibrous protein collagen, the same protein found in cartilage, tendons, and ligaments, may be abundant in the ECM. Strong fibers of collagen and another fibrous protein, elastin, are embedded within a complex web of other glycoproteins, called proteoglycans, to form a protective layer over the cell surface.

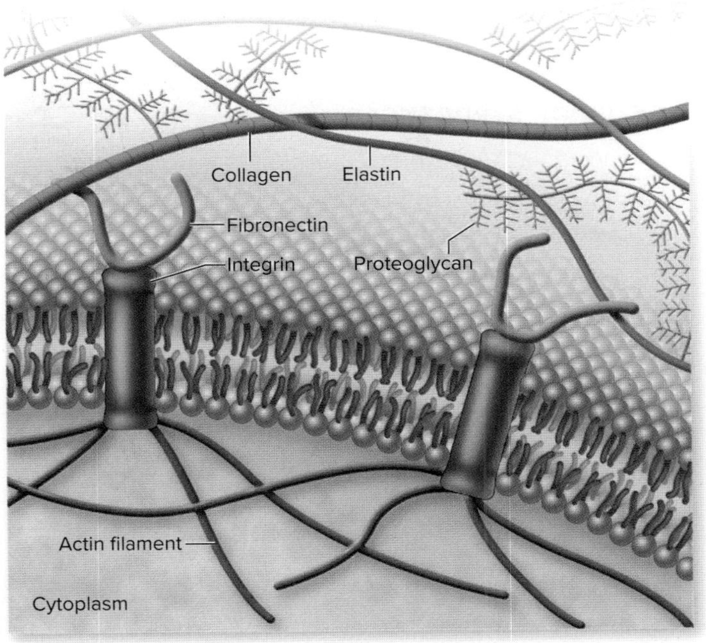

Figure 4.25 The extracellular matrix. Animal cells are surrounded by an extracellular matrix composed of various glycoproteins that give the cells support, strength, and resilience.

The ECM of some cells is attached to the plasma membrane by a third kind of glycoprotein, *fibronectin.* Fibronectin molecules bind not only to ECM glycoproteins but also to proteins called **integrins.** Integrins are an integral part of the plasma membrane, extending into the cytoplasm, where they are attached to the microfilaments and intermediate filaments of the cytoskeleton. Linking ECM and cytoskeleton, integrins allow the ECM to influence cell behavior in important ways. They can alter gene expression and cell migration patterns by a combination of mechanical and chemical signaling pathways. In this way, the ECM can help coordinate the behavior of all the cells in a particular tissue.

Table 4.2 compares and reviews three types of cells.

REVIEW OF CONCEPT 4.7

Plant cells have a cellulose-based cell wall. Animal cells lack a cell wall. In animal cells, the cytoskeleton is linked to a web of glycoproteins called the extracellular matrix.

- *What is the relationship between extracellular structures and the cytoskeleton?*

TABLE 4.2	A Comparison of Prokaryotic, Animal, and Plant Cells		
	Prokaryote	**Animal**	**Plant**
EXTERIOR STRUCTURES			
Cell wall	Present (protein-polysaccharide)	Absent	Present (cellulose)
Cell membrane	Present	Present	Present
Flagella/cilia	Flagella may be present	May be present (9 + 2 structure)	Absent except in sperm of a few species (9 + 2 structure)
INTERIOR STRUCTURES			
Endoplasmic reticulum	Absent	Usually present	Usually present
Ribosomes	Present	Present	Present
Microtubules	Absent	Present	Present
Centrioles	Absent	Present	Absent
Golgi apparatus	Absent	Present	Present
Nucleus	Absent	Present	Present
Mitochondria	Absent	Present	Present
Chloroplasts	Absent	Absent	Present
Chromosomes	Single; circle of DNA	Multiple; DNA–protein complex	Multiple; DNA–protein complex
Lysosomes	Absent	Usually present	Present
Vacuoles	Absent	Absent or small	Usually a large single vacuole

4.8 Cell-to-Cell Connections Determine How Adjacent Cells Interact

A basic feature of multicellular animals is the formation of diverse kinds of *tissue,* such as skin, blood, or muscle, where cells are organized in specific ways. Cells must also be able to communicate with each other and have markers of individual identity. All of these functions—connections between cells, markers of cellular identity, and cell communication—involve membrane proteins and proteins secreted by cells. As an organism develops, the cells acquire their identities by carefully controlling the *expression* of those genes, turning on the specific set of genes that encodes the functions of each cell type. Table 4.3 provides a summary of the kinds of connections found between cells that are explored in this section.

Surface Proteins Give Cells Their Identity

LEARNING OBJECTIVE 4.8.1 Explain how multicellular organisms are able to differentiate between the cells of different tissues.

One key set of genes functions to mark the surfaces of cells, identifying them as being of a particular type. When cells make contact, they "read" each other's cell-surface markers and react accordingly. Cells that are part of the same tissue type recognize each other, and they frequently respond by forming connections between their surfaces to better coordinate their functions.

Glycolipids

Most tissue-specific cell-surface markers are glycolipids, that is, lipids with carbohydrate heads. The glycolipids on the surface of red blood cells are responsible for the A, B, and O blood types.

TABLE 4.3	Cell-to-Cell Connections and Cell Identity		
Type of Connection	**Structure**	**Function**	**Example**
Surface markers	Variable, integral proteins or glycolipids in plasma membrane	Identify the cell	MHC complexes, blood groups, antibodies
Septate junctions **Tight junctions**	Tightly bound, leakproof, fibrous claudin protein seal that surrounds cell	Hold cells together such that materials pass through but not between the cells	Junctions between epithelial cells in the gut
Adhesive junctions (desmosomes)	Variant cadherins, desmocollins, bind to intermediate filaments	Create strong, flexible connections between cells; found in vertebrates	Epithelium
Adhesive junctions (adherens junctions)	Classical cadherins, bind to microfilaments of cytoskeleton	Connect cells together; oldest form of cell junction, found in all multicellular organisms	Tissues with high mechanical stress, such as the skin
Adhesive junctions (hemidesmosomes, focal adhesions)	Integrin proteins bind cell to extracellular matrix	Provide attachment to substrate	Involved in cell movement and important during development
Communicating junctions (gap junctions)	Six transmembrane connexon/ pannexin proteins create a pore	Allow passage of small molecules from cell to cell in a tissue	Excitable tissues such as heart muscle
Communicating junctions (plasmodesmata)	Cytoplasmic connections between gaps in adjoining plant cell walls	Communicating junctions between plant cells	Plant tissues

MHC proteins

One example of the function of cell-surface markers is the recognition of "self" and "nonself" cells by the immune system. This function is vital for multicellular organisms, which need to defend themselves against invading or malignant cells. The immune system of vertebrates uses a particular set of markers to distinguish self from nonself cells, encoded by genes of the *major histocompatibility complex (MHC).* Cell recognition in the immune system is covered in chapter 35.

Cell Junctions Mediate Cell-to-Cell Adhesion

LEARNING OBJECTIVE 4.8.2 Relate the structure of different types of junctions to their function.

The evolution of multicellularity required the acquisition of molecules that can connect cells to each other. It appears that multicellularity arose independently in different lineages, but the types of connections between cells are remarkably conserved, and many of the proteins have ancient origins.

The nature of the physical connections between the cells of a tissue largely determines what the tissue is like. Indeed, a tissue's proper functioning often depends critically on how the individual cells are arranged within it. Cell junctions can be characterized by both their visible structure in the microscope and the proteins involved in the junction.

Adhesive junctions

Adhesive junctions appear to have been the first to evolve. Primitive forms can even be found in sponges, and they are found in all animal species. They mechanically attach the cytoskeleton of a cell to the cytoskeletons of other cells or to the extracellular matrix. These junctions are found in tissues subject to mechanical stress, such as muscle and skin epithelium.

Adherens junctions are found in animals ranging from jellyfish to vertebrates. They are based on the protein **cadherin,** which is a Ca^{2+}-dependent adhesion molecule with very wide phylogenetic distribution. Cadherin has a single transmembrane domain, as well as an extracellular domain that interacts with other cadherins on adjacent cells to join them together (figure 4.26).

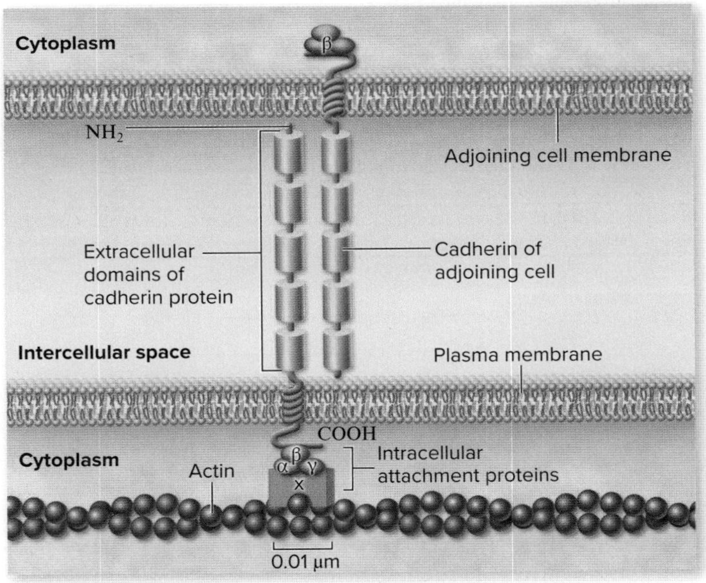

Figure 4.26 A cadherin-mediated junction. The cadherin molecule is anchored to actin in the cytoskeleton and passes through the membrane to interact with the cadherin of an adjoining cell.

Cadherins are divided into types I and II, and when cells bearing only type I are mixed with cells bearing only type II, they sort into populations joined by I-to-I or by II-to-II interactions. On the cytoplasmic side, the cadherins interact indirectly through other proteins with actin to form flexible connections (figure 4.26).

Desmosomes are a cadherin-based junction unique to vertebrates. They contain the cadherins desmocolin and desmoglein, which interact with intermediate filaments of cytoskeletons instead of actin. Desmosomes join adjacent cells (figure 4.27*b*). These connections support tissues against mechanical stress.

Hemidesmosomes and **focal adhesions** connect cells to the basal lamina or other ECM. In this case, the proteins that interact with the ECM are called integrins. The integrins are members of a large superfamily of cell-surface receptors that bind to a protein component of the extracellular matrix. These junctions also connect to the cytoskeleton of cells: actin filaments at focal adhesions and intermediate filaments at hemidesmosomes.

Septate and tight junctions

Septate junctions are found in both invertebrates and vertebrates and form a barrier that can seal off a sheet of cells. The proteins found at these junctions have been given different names in different systems; in *Drosophila,* the proteins include Discs large and Neurexin. Their wide distribution indicates that they probably evolved soon after or with adherens junctions.

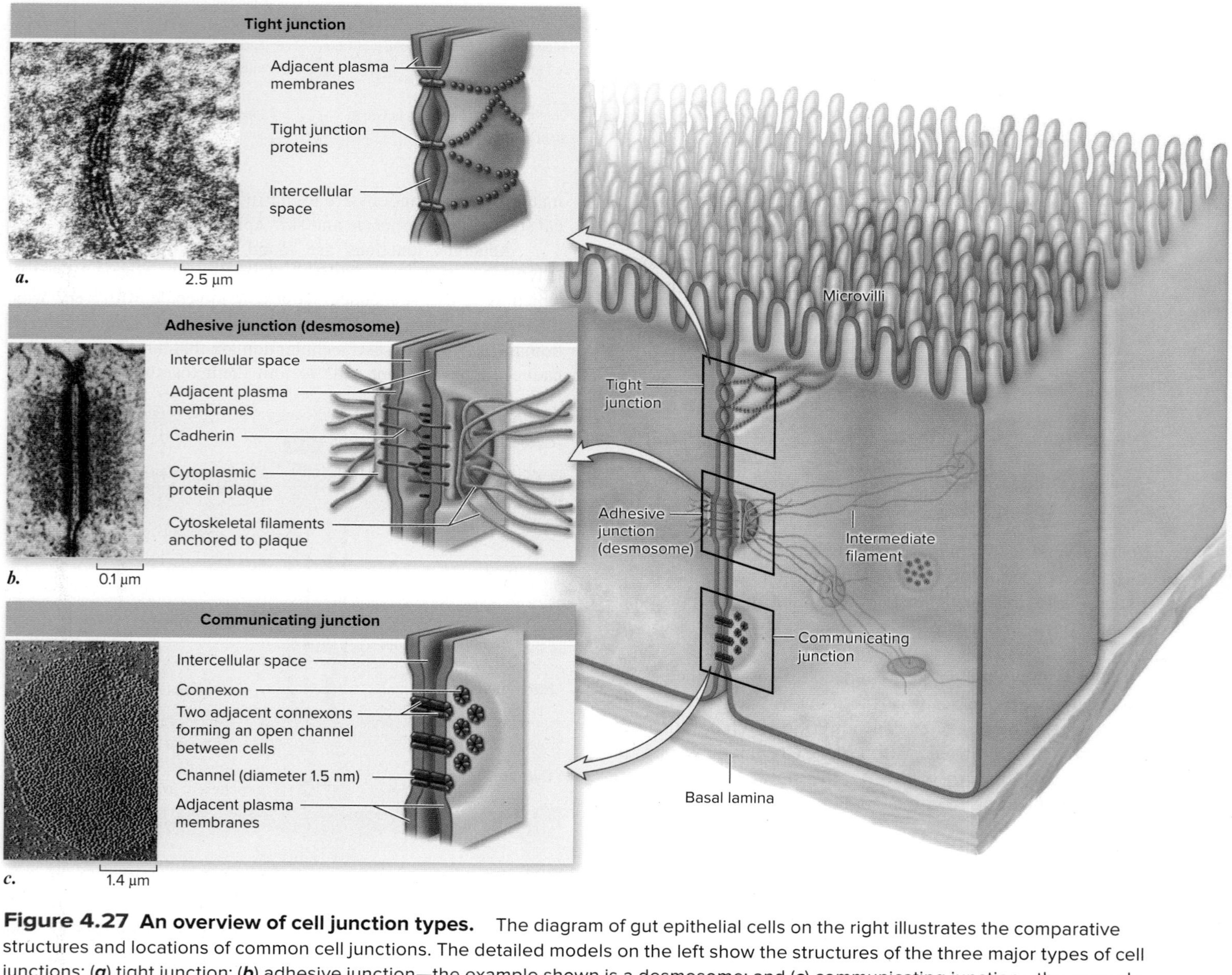

Figure 4.27 An overview of cell junction types. The diagram of gut epithelial cells on the right illustrates the comparative structures and locations of common cell junctions. The detailed models on the left show the structures of the three major types of cell junctions: (***a***) tight junction; (***b***) adhesive junction—the example shown is a desmosome; and (***c***) communicating junction—the example shown is a gap junction.

(a): Daniel Goodenough; (b, c): Don W. Fawcett/Science Source

Tight junctions are unique to vertebrates and contain proteins called claudins because of their ability to occlude, or block, substances from passing between cells. This form of junction between cells acts as a wall within the tissue, keeping molecules on one side (figure 4.27*a*).

Creating sheets of cells. The cells that line an animal's digestive tract are organized in a sheet only one cell thick. One surface of the sheet faces the inside of the tract, and the other faces the extracellular space, where blood vessels are located. Tight junctions encircle each cell in the sheet, like a belt cinched around a person's waist. The junctions between neighboring cells are so securely attached that there is no space between them for leakage. This forces nutrients from food in the digestive tract to pass directly through the sheet of cells to enter the bloodstream. This also partitions the plasma membranes of this sheet of cells into separate compartments. Transport proteins in the membrane facing inside carry nutrients you consume into the cytoplasm of the cells. Other proteins, located in the membrane on the opposite side, transport those nutrients from the cytoplasm to the extracellular fluid, where they can enter the bloodstream.

Communicating junctions

The proteins involved in the junctions previously described can be found in some single-celled organisms as well. The evolution of multicellularity also led to a new form of cellular connection: the *communicating junctions.* These junctions allow communication between cells by diffusion through small openings. Communicating junctions permit small molecules, such as glucose or amino acids, and ions to pass from one cell to the other. In animals, these direct communication channels between cells are called *gap junctions,* and in plants, *plasmodesmata.*

Gap junctions in animals. **Gap junctions** are found in both invertebrates and vertebrates. In invertebrates they are formed by proteins known as pannexins. Vertebrates have both pannexin-based gap junctions and an additional type based on similar proteins called connexons. In each case, a structure is formed by complexes of six identical transmembrane proteins (figure 4.27*c*). The proteins are arranged in a circle to create a channel through the plasma membrane that protrudes several nanometers from the cell surface. A gap junction forms when the connexons/pannexins of two cells align perfectly, creating an open channel that spans the plasma membranes of both cells.

Gap junction channels are dynamic structures that can open or close in response to a variety of factors, including Ca^{2+} and H^+ ions. This gating serves at least one important function. When a cell is damaged, its plasma membrane often becomes leaky. Ions in high concentrations outside the cell, such as Ca^{2+}, flow into the damaged cell and close its gap junction channels. This isolates the cell and prevents the damage from spreading.

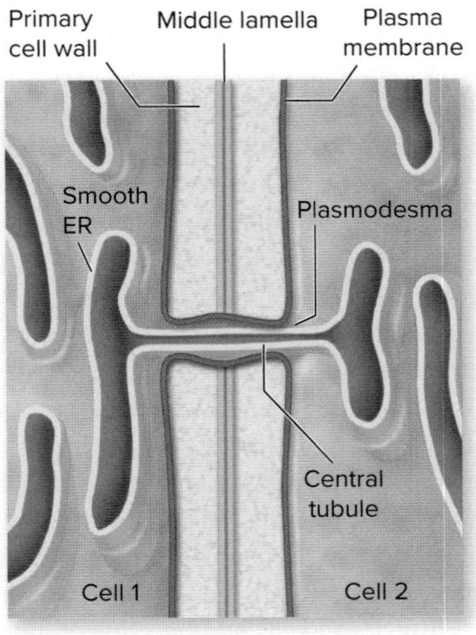

Figure 4.28 Plasmodesmata. Plant cells can communicate through specialized openings in their cell walls, called plasmodesmata, where the cytoplasm of adjoining cells is connected.

Plasmodesmata in plants. In plants, cell walls separate every cell from all others. Cell–cell junctions occur only at holes or gaps in the walls, where the plasma membranes of adjacent cells can come into contact with one another. Cytoplasmic connections that form across the touching plasma membranes are called **plasmodesmata** (singular, *plasmodesma;* figure 4.28). The majority of living cells within a higher plant are connected to their neighbors by these junctions.

Plasmodesmata function much as gap junctions do in animal cells, although their structure is more complex. Unlike gap junctions, plasmodesmata are lined with plasma membrane and contain a central tubule that connects the ER of the two cells.

REVIEW OF CONCEPT 4.8

The three types of cell junction are (1) tight junctions, which make watertight sheets of cells; (2) adhesive junctions, which provide strength and flexibility; and (3) communicating junctions, including gap junctions in animals and plasmodesmata in plants, which allow the passage of some materials between cells. Cell identity is conferred by surface glycoproteins, including the MHC proteins of the immune system.

- *How do cell junctions help to form tissues?*

Control of Endosome Formation

Among the many organelles found in eukaryotic cells, the most biochemically diverse are the small endosomes that recycle damaged cell components, detoxify potentially dangerous chemicals, and remove low-density lipoprotein (LDL) from the bloodstream. A complex protein network governs the fusing of early endosomes (EEs) to form particular classes of functional "late" endosomes such as lysosomes and peroxisomes, as well as the tethering of endosomes to particular cellular locations, and the mobility of endosomes that carry materials to the cell's plasma membrane. The two endosomes you see in the micrograph contain recycling cell fragments (red), banks of enzymes (light green), and LDL particles (dark green); an LDL particle awaiting endocytosis can be seen below, between the endosomes.

How does a cell regulate this complex system? Recent evidence has implicated a small GTPase called Rab5. As you will discover in chapter 9, a variety of GTPases play important roles in a cell's interior lines of communication. In this instance, Rab5 appears to be necessary for EE formation and the regulation of both endosome number and function.

To investigate this issue, researchers used the technique of RNA interference (RNAi) to turn off the *Rab5* gene. Discussed in chapter 16, this approach uses synthetic RNAs complementary to specific genes to shut off expression of these genes. These are called siRNA for "small interfering RNA." Treatment with siRNA leads to the destruction of the mRNA—when examined five days after RNAi treatment, the cells had an 80% reduction in *Rab5* mRNA levels.

In the RNAi-treated cells, the researchers observed a loss of endosomes of all types: EE, late endosomes, and lysosomes. This led to the question "Does this also affect uptake by endocytosis?" To address this, the researchers used a dye that stained LDL particles to monitor their uptake in liver cells. The results are shown in the graph (Rab5 KD are RNAi-treated cells).

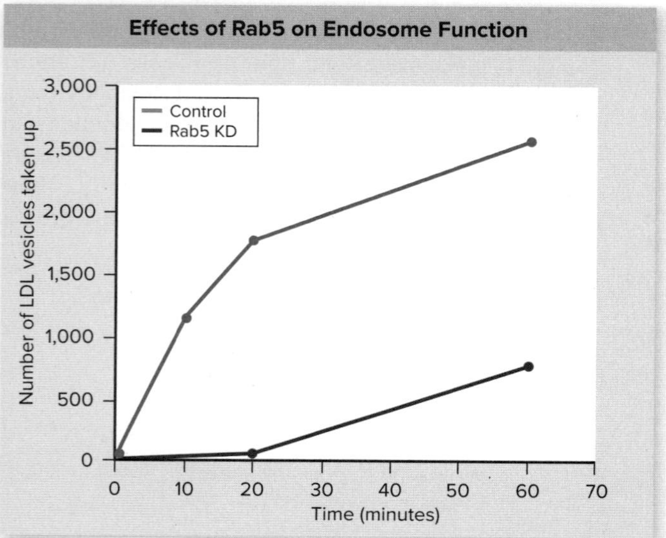
Effects of Rab5 on Endosome Function
Control
Rab5 KD
Number of LDL vesicles taken up
3,000
2,500
2,000
1,500
1,000
500
0
0
10
20
30
40
50
60
70
Time (minutes)

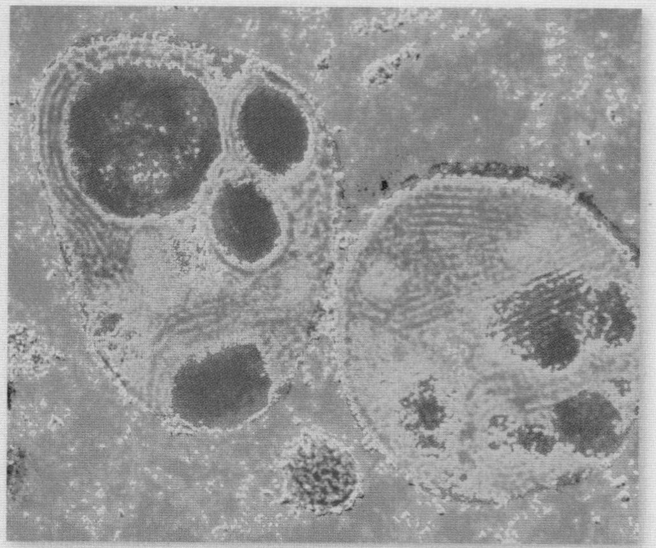
12,500×

Phototake Inc./Alamy Stock Photo

Analysis

1. **Applying Concepts**
   a. ***Variable.*** In the graph, what is the dependent variable?
   b. Is it necessary for investigators to sample equal numbers of controls and Rab5 KD cells? Explain.
2. **Interpreting Data**
   a. How does the number of LDL particles taken up by Rab5 KD cells (which lack Rab5) change over the hour of the experiment?
   b. How does the number of LDL particles taken up by control cells change with time?
   c. At 20 minutes into the experiment, how many LDL particles had been taken up by the control cells? By the Rab5 KD cells?
3. **Making Inferences** What would you say is the principal difference in LDL uptake behavior between the Rab5 KD cells and the control cells?
4. **Drawing Conclusions**
   a. Does the silencing of the *Rab5* gene by siRNA result in disruption of the ability of liver endosomes to carry out endocytosis of LDL particles?
   b. If the cells treated with the siRNA are allowed to recover, they show no permanent effects from the treatment. What does this say about the endosome system?
5. **Further Analysis** In the mouse system where this was analyzed, the researchers were able to administer the RNAi treatment directly to liver cells in adult animals. What would you predict for levels of serum LDL in liver cells with RNAi treatment versus serum LDL levels in liver cells of control animals?

Retracing the Learning Path

CONCEPT 4.1 All Living Organisms Are Composed of Cells

4.1.1 The Cell Theory Is the Unifying Foundation of Biology Cells arise only by division of preexisting cells.

4.1.2 Cell Size Is Limited As cell size increases, diffusion becomes inefficient.

4.1.3 Microscopes Allow Us to Visualize Cells Magnification gives better resolution than is possible with the naked eye, and it is improved by employing light of shorter wavelengths.

4.1.4 All Cells Exhibit Basic Structural Similarities All cells have centrally located DNA, a semifluid cytoplasm, and an enclosing plasma membrane.

CONCEPT 4.2 Prokaryotic Cells Are Relatively Simple

4.2.1 Prokaryotic Cells Have Organized Substructure Some species have membrane-bound organelles, but none are universal. Many bacteria have compartments formed by semipermeable protein shells, and by membrane infoldings. Prokaryotes lack a cytoskeleton, but have proteins related to eukaryotic actin and tubulin.

4.2.2 Flagella in Bacteria and Archaea Have Independent Origins Bacterial flagella and the archaeal archaellum arose by convergent evolution. They both rotate, but the motive force is different.

CONCEPT 4.3 Eukaryotic Cells Are Highly Compartmentalized

4.3.1 Organelles and Internal Membranes Organize the Interior of Eukaryotic Cells

4.3.2 The Nucleus Acts as the Cell's Information Center The nucleus contains DNA and is surrounded by an envelope of two phospholipid bilayers; its pores allow exchange of small molecules.

4.3.3 Ribosomes Are the Cell's Protein Synthesis Machinery Ribosomes translate mRNA to produce polypeptides. They are found in the cytoplasm of all cells.

CONCEPT 4.4 Membranes Organize the Cell Interior into Functional Compartments

4.4.1 The Endoplasmic Reticulum Is a Highway That Weaves Throughout the Cell The rough ER (RER), studded with ribosomes, synthesizes and modifies proteins and manufactures membranes. The smooth endoplasmic reticulum (SER) is involved in carbohydrate and lipid synthesis and in detoxification.

4.4.2 The Golgi Apparatus Sorts and Packages Proteins Golgi bodies receive vesicles from the ER, modify and package macromolecules, and export them via transport vesicles.

4.4.3 Other Organelles Carry Out Degradation and Recycling Lysosomes contain digestive enzymes that break down macromolecules and recycle old organelles. The proteasome destroys proteins that are damaged or no longer needed. Plants use vacuoles for both storage and water balance. Microbodies segregate a variety of metabolic processes.

CONCEPT 4.5 Mitochondria and Chloroplasts Are Energy-Processing Organelles

4.5.1 Mitochondria Metabolize Organic Compounds to Generate ATP Mitochondria have a double-membrane structure, contain their own DNA, and can divide independently. The inner membrane is extensively folded. Proteins on the surface and in the inner membrane carry out metabolism to produce ATP.

4.5.2 Chloroplasts Use Light to Generate ATP and Sugars Chloroplasts have a double membrane, contain DNA, and divide independently. Chloroplasts capture light energy via thylakoid membranes arranged in stacks and use it to synthesize glucose.

4.5.3 Mitochondria and Chloroplasts Arose by Endosymbiosis The endosymbiont theory proposes that both mitochondria and chloroplasts were once prokaryotes engulfed by other cells.

CONCEPT 4.6 An Internal Skeleton Supports the Shape of Cells

4.6.1 The Cytoskeleton Is Composed of Three Types of Protein Fibers The cytoskeleton consists of protein fibers that support the shape of the cell and anchor organelles: actin filaments, or microfilaments, are involved in cellular movement; microtubules move materials within a cell; intermediate filaments serve a wide variety of functions.

4.6.2 The Cytoskeleton Helps Move Materials Within Cells Molecular motors, such as kinesin and dynein, move vesicles along microtubules.

4.6.3 Eukaryotic Cells Use Cytoskeletal Elements to Crawl or Swim Cell crawling occurs as actin forces the cell membrane forward, whereas myosin pulls the cell body forward. Flagella and cilia allow cells to swim. Eukaryotic flagella have a 9 + 2 structure and arise from a basal body. Cilia are shorter and more numerous than flagella.

CONCEPT 4.7 Extracellular Structures Protect Cells

4.7.1 Plant Cell Walls Provide Protection and Support Plants have cell walls composed of cellulose fibers. The middle lamella, between cell walls, holds adjacent cells together.

4.7.2 Animal Cells Secrete an Extracellular Matrix Glycoproteins are the main component of the extracellular matrix.

CONCEPT 4.8 Cell-to-Cell Connections Determine How Adjacent Cells Interact

4.8.1 Surface Proteins Give Cells Their Identity
Glycolipids and MHC proteins help distinguish self from nonself.

4.8.2 Cell Junctions Mediate Cell-to-Cell Adhesion
Cell junctions include tight junctions, anchoring junctions, and communicating junctions. In animals, gap junctions allow the passage of small molecules between cells. In plants, plasmodesmata penetrate the cell wall and connect cells.

Concept Overview

This Concept Overview diagrams the key concepts that were discussed in this chapter.

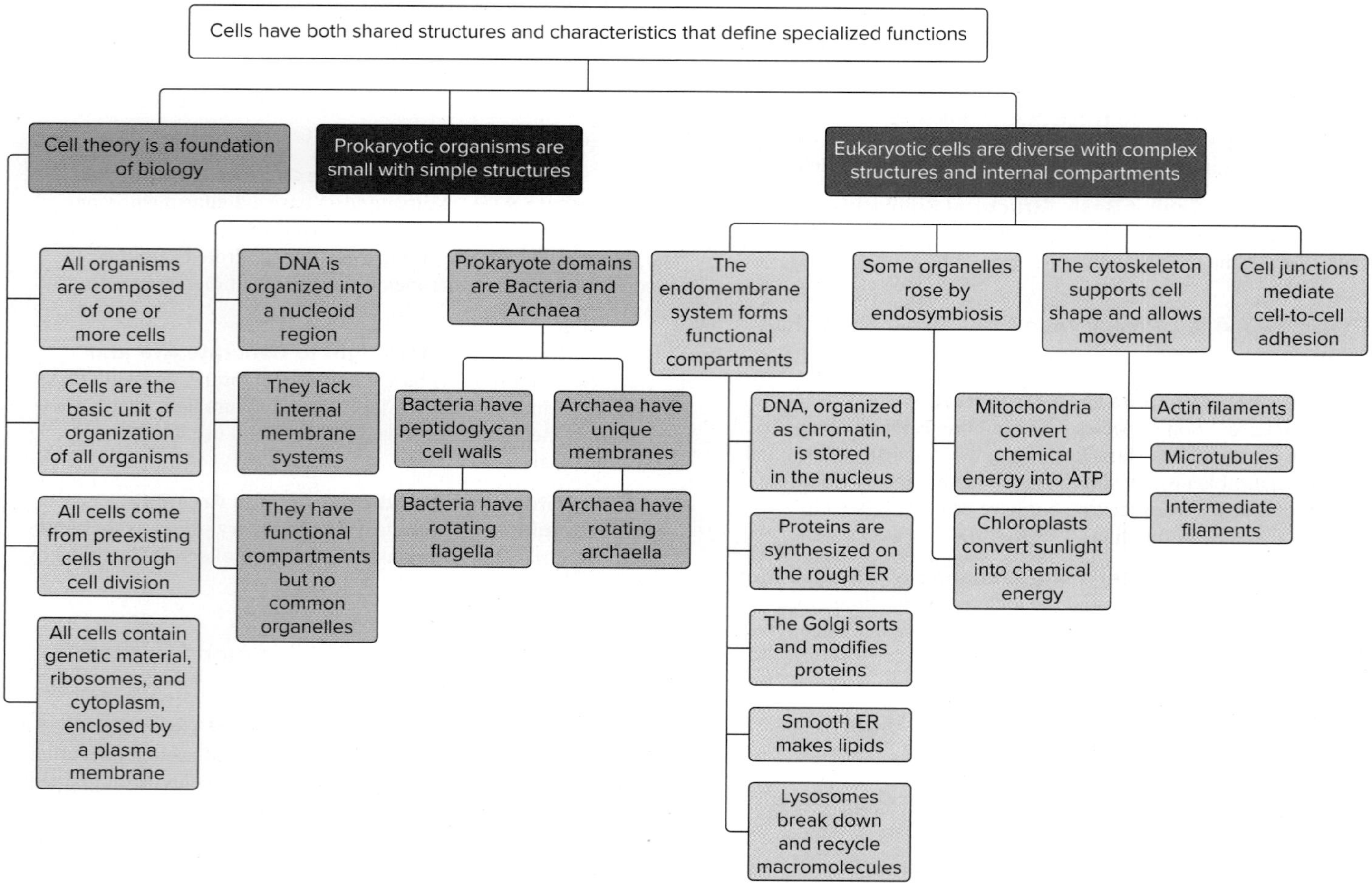

Assessing the Learning Path

Understand

1. Which of the following statements is NOT part of the cell theory?
   a. All organisms are composed of one or more cells.
   b. All cells come from other cells by division.
   c. Cells are the smallest living things.
   d. All cells are composed of molecules and organelles.
2. The most important factor that limits the size of a cell is the
   a. quantity of proteins and organelles a cell can make.
   b. concentration of water in the cytoplasm.
   c. surface-area-to-volume ratio of the cell.
   d. amount of DNA in the cell.
3. Which of the following is NOT a characteristic of all prokaryotic cells?
   a. Ribosomes
   b. Cell wall
   c. DNA
   d. Pili
4. Which of the following is found in eukaryotic cells but not in prokaryotic cells?
   a. Cell wall
   b. Plasma membrane
   c. Endoplasmic reticulum
   d. Ribosomes

5. Which of the following nuclear structures is NOT correctly matched with its function?
   a. The nucleolus–site of rRNA synthesis
   b. Nuclear pores–allow passage of molecules into and out of the nucleus
   c. Nuclear envelope–separates the contents of the nucleus from the cytoplasm
   d. Nuclear lamina–produces ribosomes
6. A protein that is secreted from the cell takes which pathway?
   a. Rough ER → transport vesicle → Golgi → secretory vesicle → plasma membrane
   b. Cytosol → plasma membrane
   c. Smooth ER → rough ER → transport vesicle → Golgi → lysosome
   d. Golgi → transport vesicle → rough ER → transport vesicle → plasma membrane
7. Mitochondria
   a. are fully autonomous.
   b. have a highly folded inner membrane.
   c. produce ATP and sugars.
   d. All of the above
8. The cytoskeleton includes
   a. microtubules made of actin filaments.
   b. microfilaments made of tubulin.
   c. intermediate filaments made of twisted fibers of vimentin and keratin.
   d. smooth endoplasmic reticulum.
9. Microfilaments
   a. do not exhibit polarity as observed in microtubules.
   b. grow from the centrosome in animal cells.
   c. are essential to cell division and muscle contraction.
   d. are used to transport vesicles through the endomembrane system.
10. Which of the following statements about the plant cell wall is NOT true?
    a. It functions in support and protection.
    b. It lies just inside the plasma membrane.
    c. The primary cell wall is formed while a cell is growing.
    d. It is composed primarily of polysaccharides.
11. Which of the following types of cell connection is correctly matched with its function?
    a. Adherens junction–blocks substances from passing between cells
    b. Plasmodesmata–communication between cells
    c. Septate junction–attaches the cytoskeleton of one cell to the cytoskeleton of another cell
    d. All are correctly matched.

Apply

1. You are examining a cell of unknown origin in the transmission electron microscope. You observed a variety of internal membrane structures. From this you conclude that the cell
   a. cannot be eukaryotic.
   b. is most likely bacterial.
   c. is most likely eukaryotic.
   d. is most likely archaeal.
2. You find a single-celled organism that has cell walls, has no nucleus, and is not susceptible to the antibiotic penicillin. What can you reasonably conclude about this organism?
   a. It lacks DNA.
   b. It has a cell wall composed of peptidoglycan.
   c. It is more likely archaeal and not bacterial.
   d. It does not make proteins.
3. Human plasma cells secrete large amounts of antibody proteins. You would expect plasma cells to have many
   a. cytosolic ribosomes.
   b. ribosomes attached to the rough ER.
   c. mitochondria.
   d. nucleoli.
4. The size and number of organelles in a cell correlate with cell function. Leydig cells of the testes produce the steroid hormone testosterone. Leydig cells are likely to have abundant
   a. smooth endoplasmic reticulum.
   b. rough endoplasmic reticulum.
   c. lysosomes.
   d. ribosomes.
5. Tubers are modified plant structures that are used to store nutrients for growth. If you make a section through a potato tuber and view it microscopically, you should see abundant
   a. mitochondria.
   b. chloroplasts.
   c. amyloplasts.
   d. cristae.
6. Different motor proteins such as kinesin and myosin are similar in that they can
   a. interact with microtubules.
   b. use energy from ATP to produce movement.
   c. interact with actin.
   d. Both a and b
7. You isolate a mutant plant that has abnormally weak adhesion between cells. This plant most likely has a problem
   a. forming plasmodesmata.
   b. producing the middle lamella.
   c. laying down secondary cell wall.
   d. producing cellulose fibers.
8. Some people affected with kidney failure have a mutation in a gene encoding a protein found in tight junctions. Which of the following statements about these individuals is accurate?
   a. The cells in their kidneys are not communicating with each other correctly.
   b. Molecules are moving through cells instead of between adjacent cells in their kidneys.
   c. The cells in their kidneys are not forming an adequate barrier to molecules.
   d. Both b and c

Synthesize

1. Suppose you are using a computer program to design a new single-celled organism. Discuss why a flat, platelike cell will be more efficient in transporting materials than a spherical, ball-like cell of the same volume.
2. You have been given a culture of a prokaryotic organism. How would you determine if it was a bacterium or an archaean?
3. Imagine that a mutation occurs that inhibits the deconstruction of nuclear lamin intermediate filaments. What would the effects of this mutation be on the affected cell?
4. Explain why a chloroplast cannot live independently of a eukaryotic cell.
5. What evolutionary mechanism could account for the appearance of the 9 + 2 and other similar structures in three distinct cellular components: centrioles, flagella, and cilia?

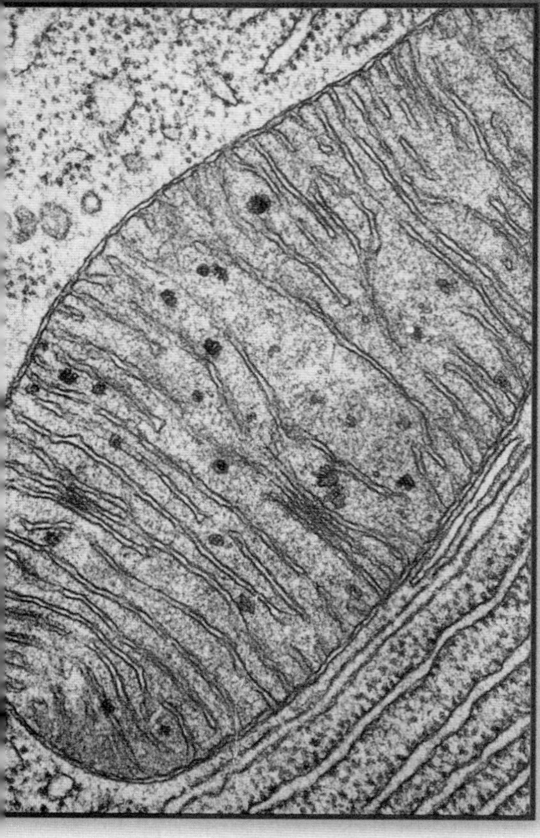

Keith R. Porter/Science Source

5 Membranes

Learning Path

5.1 Membranes Are Phospholipid Bilayers with Embedded Proteins

5.2 Phospholipids Provide a Membrane's Structural Foundation

5.3 Membrane Proteins Enable a Broad Range of Interactions with the Environment

5.4 Passive Transport Moves Molecules Across Membranes by Diffusion

5.5 Active Transport Across Membranes Requires Energy

5.6 Bulky Materials Cross Membranes Within Vesicles

Concept Overview

This Concept Overview provides a diagram of the overarching concepts that are covered in this chapter. These concepts will be expanded upon in the Concept Overview at the end of the chapter.

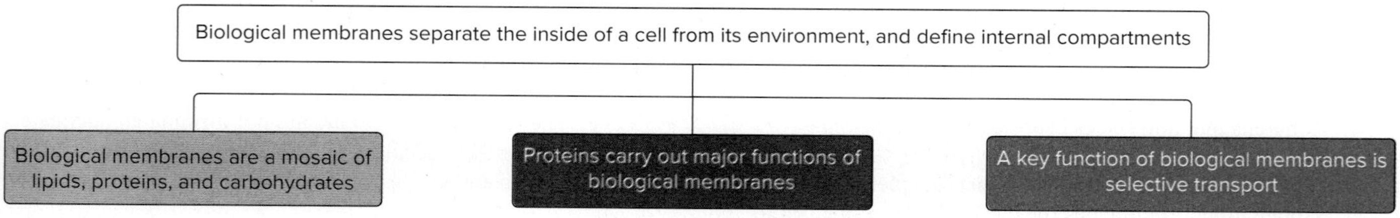

Biological membranes separate the inside of a cell from its environment, and define internal compartments
Biological membranes are a mosaic of lipids, proteins, and carbohydrates
Proteins carry out major functions of biological membranes
A key function of biological membranes is selective transport

Introduction

A cell's interactions with its environment are critical to its survival– a give-and-take that never ceases. Without these interactions, life could not exist. Living cells are encased within a lipid membrane through which few water-soluble substances can freely pass. The membrane permits their passage into and out of the cell through protein doorways that admit only specific substances. Biologists call the delicate skin of lipids with embedded protein molecules that encases the cell a **plasma membrane**. Eukaryotic cells also contain internal membranes, like the endoplasmic reticulum pictured on the chapter image and the mitochondrion you see embedded within it. This chapter examines the structure and function of these remarkable membranes.

5.1 Membranes Are Phospholipid Bilayers with Embedded Proteins

A plasma membrane made of phospholipid sheets only 5 to 10 nm thick encases all living cells. More than 10,000 of these sheets piled on one another would just equal the thickness of one page of this book. Biologists have long known the molecular components of membranes—lipids, proteins, and other molecules—but for many years the organization of these membrane components remained elusive.

Biological Membranes Are Fluid Mosaics

LEARNING OBJECTIVE 5.1.1 Explain the fluid mosaic model of membrane structure.

The lipid layer that forms the foundation of a cell's membranes is actually a bilayer formed of two **phospholipid sheets** (figure 5.1). For many years biologists thought that the protein components of the plasma membrane covered the inner and outer surfaces of the phospholipid bilayer like a coat of paint. An early model portrayed the plasma membrane as a sandwich, with a phospholipid bilayer between two layers of globular protein.

In 1972, S. Jonathan Singer and Garth J. Nicolson revised the model in a simple but profound way: they proposed that the globular proteins are *inserted* into the lipid bilayer, with their nonpolar segments in contact with the nonpolar interior of the bilayer and their polar portions protruding out from the membrane surface. In this model, called the *fluid mosaic model,* a mosaic of proteins floats in or on the fluid lipid bilayer like boats on a pond (figure 5.2).

Cellular Membranes Are Assembled from Four Major Components

LEARNING OBJECTIVE 5.1.2 Describe the four major components of biological membranes.

A eukaryotic cell contains many membranes. Although they are not all identical, they share the same fundamental architecture. Cell membranes are assembled from four components (table 5.1):

1. **Phospholipid bilayer.** Every cell membrane is composed of phospholipids in a bilayer. The other components of the membrane are embedded within the bilayer, which provides a flexible matrix and, at the same time, imposes a barrier to permeability.

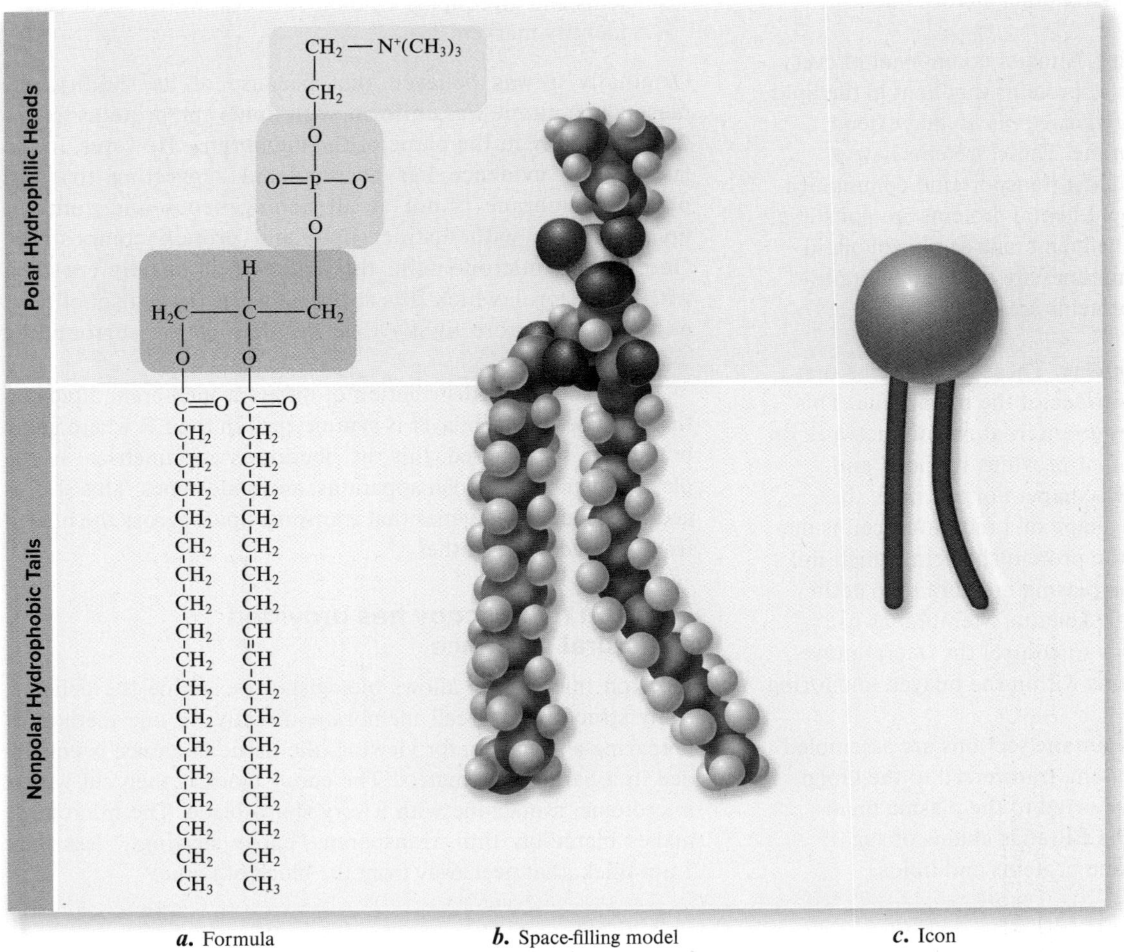

Polar Hydrophilic Heads
Nonpolar Hydrophobic Tails
a. Formula
b. Space-filling model
c. Icon

Figure 5.1 Different views of phospholipid structure. Phospholipids are composed of glycerol (pink) linked to two fatty acids and a phosphate group. The phosphate group (yellow) can have additional molecules attached, such as the positively charged choline (green) shown. Phosphatidylcholine is a common component of membranes; it is shown in **(a)** with its chemical formula, **(b)** as a space-filling model, and **(c)** as the icon that is used in most of the figures in this chapter.

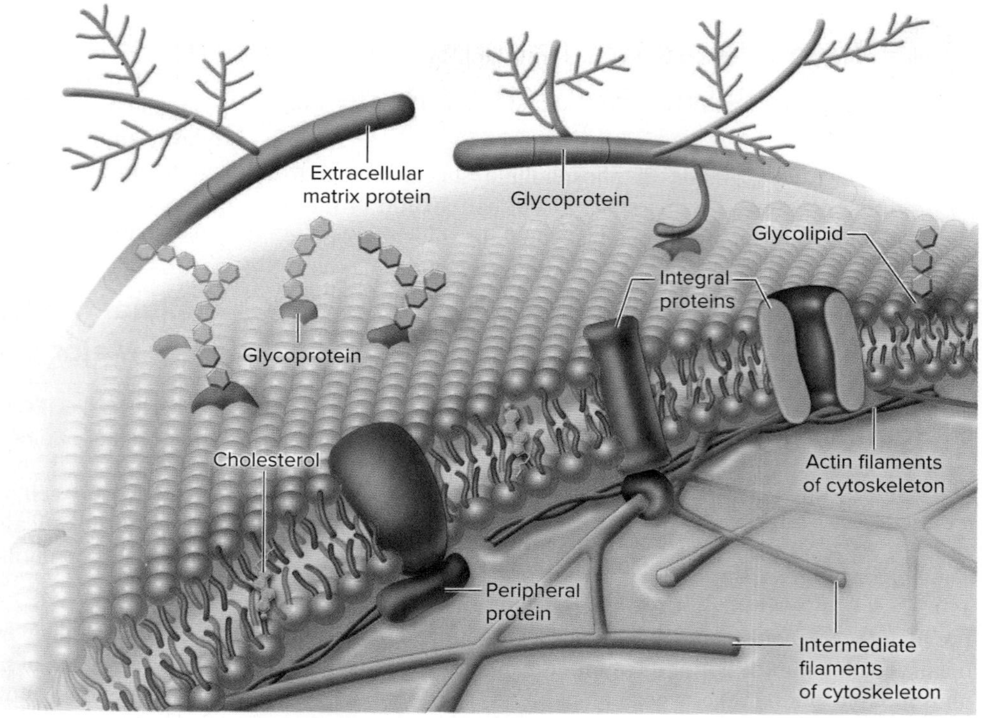

Figure 5.2 The fluid mosaic model of cell membranes. The cell membrane is composed of a phospholipid bilayer (animal cells also include cholesterol) and proteins. Integral proteins are embedded in this phospholipid bilayer, and peripheral proteins are associated with the surface of the membrane. Transmembrane proteins protrude through the membrane anchored by nonpolar regions in the hydrophobic interior. Surface markers include integral proteins with carbohydrate chains linked to the extracellular portion of these proteins (glycoproteins). Other surface markers include phospholipids modified by the addition of carbohydrates (glycolipids). Inside the cell, actin filaments and intermediate filaments interact with membrane proteins. Outside the cell, many animal cells have an elaborate extracellular matrix composed primarily of glycoproteins.

Animal cell membranes also contain a significant amount of cholesterol, a steroid with a polar hydroxyl group (—OH). Plant cells have other sterols, but little or no cholesterol.

2. **Integral membrane proteins.** A major component of every membrane is a collection of proteins that float in the lipid bilayer, including *transmembrane proteins* that extend through the entire membrane. These proteins have a variety of functions, including transport and communication across the membrane. Integral proteins are not static; they can move about in the membrane as phospholipid molecules do. Some membranes are crowded with proteins, but in others, the proteins are more sparsely distributed.
3. **Peripheral membrane proteins.** These are proteins that are associated with the surface of the membrane. This class of proteins includes an internal protein network on the cytoplasmic surface that provides support and reinforces the membrane's shape. For example, the characteristic biconcave shape of a red blood cell is due to a scaffold formed by the protein spectrin, which links proteins embedded in the plasma membrane to actin filaments in the cell's cytoskeleton. Membranes use networks of other proteins to control the lateral movements of some key proteins within the bilayer, anchoring them to specific sites.
4. **Cell-surface markers.** Membrane sections are assembled in the endoplasmic reticulum, transferred to the Golgi apparatus, and then transported to the plasma membrane. During passage, the ER adds chains of sugar molecules to the membrane proteins and lipids, converting them into **glycoproteins** and **glycolipids.** Different cell types exhibit different varieties of glycoproteins and glycolipids on their surfaces, which act as cell identity markers.

Originally it was believed that because of its fluidity, the plasma membrane was uniform, with lipids and proteins free to diffuse rapidly in the plane of the membrane. However, in the last decade evidence has accumulated suggesting that the plasma membrane is not at all homogeneous and contains microdomains with distinct lipid and protein composition. One type of microdomain, the *lipid raft,* is heavily enriched with cholesterol, which fills space between the phospholipids, packing them more tightly together than in the surrounding membrane.

Although the distribution of different membrane lipids on the two sides of the bilayer is symmetrical in the ER where membranes are synthesized, this distribution is asymmetrical in the plasma membrane, Golgi apparatus, and endosomes. This shift is accomplished by enzymes that transport lipids across the bilayer from one face to the other.

Electron microscopy has provided structural evidence

Electron microscopy allows biologists to examine the delicate, filmy structure of a cell membrane directly. In one method of preparing a specimen for viewing, the tissue of choice is embedded in a hard epoxy matrix. The epoxy block is then cut with a microtome, a machine with a very sharp blade. The microtome makes incredibly thin, transparent "epoxy shavings," less than 1 μm thick, that peel away from the block of tissue.

TABLE 5.1	Components of the Cell Membrane			
Component	**Examples**	**Function**	**How It Works**	**Example**
Phospholipid bilayer	Phospholipid molecules, plus cholesterol in animal cells	Provides permeability barrier, matrix for proteins	Excludes water-soluble molecules from nonpolar interior of bilayer and cell	Bilayer of cell is impermeable to large, water-soluble molecules, such as glucose
Integral membrane proteins (transmembrane) proteins	Carrier proteins	Actively or passively transport molecules across membrane	Move specific molecules across membrane by conformational changes	Glycophorin carrier for sugar transport; Na^+/K^+ pump
	Channel proteins	Passively transport molecules across membrane	Create a selective tunnel that acts as a passage through membrane	Sodium and potassium channels in nerve, heart, and muscle cells
	Receptor proteins	Transmit information into cell	Signal molecules bind outside cell; initiates signal transduction pathway inside cell	Specific receptors bind peptide hormones and neurotransmitters
Peripheral membrane proteins	Spectrin proteins	Determine shape of cell	Form supporting scaffold beneath membrane, anchored to both membrane and cytoskeleton	Red blood cell
	Clathrin proteins	Anchor proteins to specific sites, important for receptor-mediated endocytosis	Proteins line coated pits and facilitate binding to specific molecules	Localization of low-density lipoprotein receptor within coated pits
Cell-surface markers	Glycoproteins	"Self" recognition	A specific protein/carbohydrate chain that confers cell identity	Major histocompatibility complex protein recognized by immune system
	Glycolipid	Tissue recognition	A specific lipid/carbohydrate chain that confers tissue identity	A, B, O blood group markers

These shavings are placed on a grid, and a beam of electrons is directed through the grid. At the high magnification an electron microscope provides, resolution is good enough to reveal the double layers of a membrane. False color can be added to the micrograph to enhance detail.

Freeze-fracturing a specimen provides a way to dramatically visualize the inside surface of the membrane (figure 5.3). The tissue is embedded in a medium and quick-frozen with liquid nitrogen. The frozen tissue is then "tapped" with a knife, causing a crack between the phospholipid layers of membranes. Protein and carbohydrate chains, pits, pores, channels, or any other structure affiliated with the membrane will pull apart (whole, usually) and stick with one side of the split membrane.

Next, a very thin coating of platinum is evaporated onto the fractured surface, forming a replica, or "cast," of the surface. After the topography of the membrane has been preserved in the cast, the tissue is dissolved away, and the cast is examined with electron microscopy, creating a textured and three-dimensional picture of the membrane.

REVIEW OF CONCEPT 5.1

Cellular membranes contain (1) a phospholipid bilayer, (2) transmembrane proteins, (3) a supporting network of internal proteins, and (4) cell-surface markers composed of glycoproteins and glycolipids. The fluid mosaic model of membrane structure includes both the fluid nature of the membrane and the mosaic composition of proteins floating in the phospholipid bilayer.

- *If the plasma membrane were just a phospholipid bilayer, how would this affect its function?*

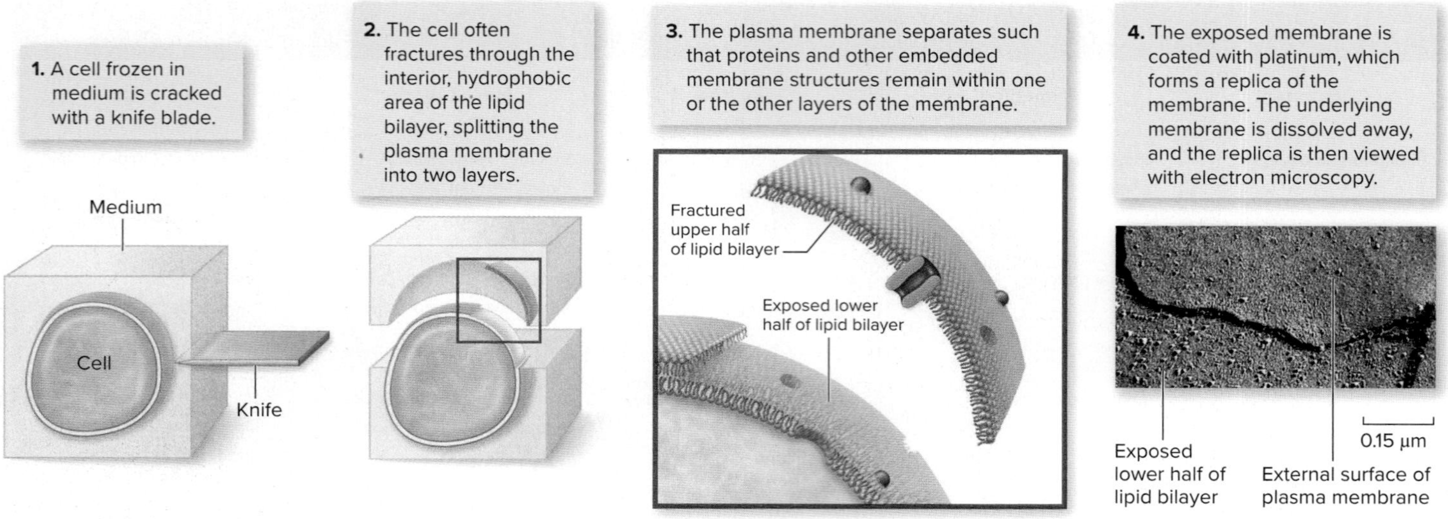
1. A cell frozen in medium is cracked with a knife blade.
Medium
Cell
Knife
2. The cell often fractures through the interior, hydrophobic area of the lipid bilayer, splitting the plasma membrane into two layers.
3. The plasma membrane separates such that proteins and other embedded membrane structures remain within one or the other layers of the membrane.
Fractured upper half of lipid bilayer
Exposed lower half of lipid bilayer
4. The exposed membrane is coated with platinum, which forms a replica of the membrane. The underlying membrane is dissolved away, and the replica is then viewed with electron microscopy.
0.15 μm
Exposed lower half of lipid bilayer
External surface of plasma membrane

Figure 5.3 Viewing a plasma membrane with freeze-fracture microscopy.
Don W. Fawcett/Science Source

5.2 Phospholipids Provide a Membrane's Structural Foundation

Triglycerides are formed by joining three fatty acid chains to the polyalcohol glycerol, and these fatty acid chains make them hydrophobic. Phospholipids are built from a carbon skeleton similar to that in triglycerides, except with phosphate replacing one of the fatty acid chains. This change makes a critical alteration to the chemical behavior of phospholipids: they are amphipathic, that is, both hydrophobic and hydrophilic.

The phosphate in a phospholipid can also be modified with polar organic molecules. By varying the polar organic group and the fatty acid chains, a large variety of lipids can be constructed on this simple molecular framework. Mammalian membranes, for example, contain hundreds of chemically distinct phospholipids.

The Lipid Bilayer Forms Spontaneously

LEARNING OBJECTIVE 5.2.1 Explain how lipid bilayers form spontaneously.

The phosphate groups of these lipids are charged, and other molecules attached to them are also charged, or polar. This creates a huge change in the molecule's physical properties compared with those of a triglyceride: the strongly polar phosphate end is hydrophilic, or "water-loving," but the fatty acid end is strongly nonpolar and hydrophobic, or "water-fearing." The two nonpolar fatty acids extend in one direction, roughly parallel to each other, and the polar phosphate group points in the other direction. To represent this, phospholipids are often diagrammed as a polar head with two dangling nonpolar tails (figure 5.1*c*).

What happens when a collection of phospholipid molecules is placed in water? The polar water molecules repel the long, nonpolar tails of the phospholipids while seeking partners for hydrogen bonding. Because of the polar nature of the water molecules, the nonpolar tails of the phospholipids end up packed closely together, sequestered as far as possible from the water. When *two* layers form with the tails facing each other, no tails ever come in contact with water. The resulting structure is the phospholipid bilayer (figure 5.4). Phospholipid bilayers form spontaneously, driven by the tendency of water molecules to form the maximum number of hydrogen bonds.

The nonpolar interior of a lipid bilayer impedes the passage of any water-soluble polar or charged substances through the bilayer, just as a layer of oil impedes the passage of a drop of water. This barrier to water-soluble substances is the key biological property of the lipid bilayer.

The phospholipid bilayer is fluid

A lipid bilayer is stable, because water's affinity for hydrogen bonding never stops. Just as surface tension holds a soap bubble together, even though it is made of a liquid, so the hydrogen bonding of water holds a membrane together. Although water continually drives phospholipid molecules into the bilayer configuration, it does not have any effect on the mobility of phospholipids relative to their lipid and nonlipid neighbors in the bilayer. Because phospholipids

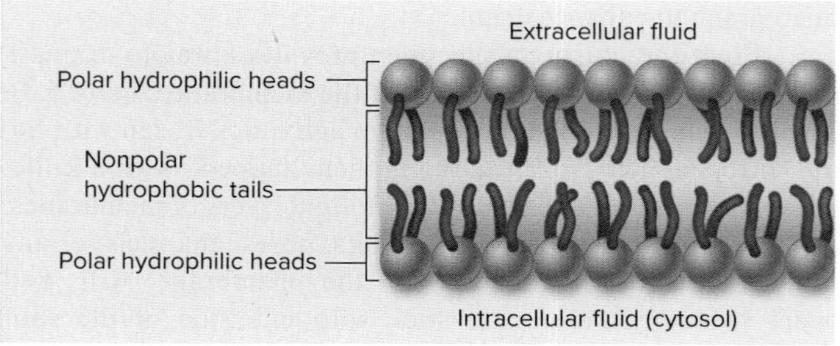
Extracellular fluid
Polar hydrophilic heads
Nonpolar hydrophobic tails
Polar hydrophilic heads
Intracellular fluid (cytosol)

Figure 5.4 The lipid bilayer. The basic structure of every plasma membrane is a double layer of phospholipids. When placed in a watery environment (the blue screened areas), the phospholipid molecules spontaneously aggregate to form a bilayer with a nonpolar interior (the tan screened area).

interact relatively weakly with one another, individual phospholipids and unanchored proteins are comparatively free to move about within the membrane, like ships floating on a lake. This can be demonstrated vividly by fusing cells and watching their proteins intermix with time, as done in the experiment described in figure 5.5.

The degree of fluidity of the plasma membrane can be altered by changing the fatty acid composition. Unsaturated fats make the membrane more fluid—the "kinks" introduced by the double bonds keep them from packing tightly.

In animal cells, cholesterol may make up as much as 50% of membrane lipids in the outer layer. The cholesterol can fill gaps left by unsaturated fatty acids. This has the effect of decreasing membrane fluidity, but it increases the strength of the membrane. Overall this leads to a plasma membrane with intermediate fluidity that is more durable and less permeable.

Changes in the environment can have drastic effects on the membranes of single-celled organisms such as bacteria. Increasing temperature makes a membrane more fluid, and decreasing temperature makes it less fluid. Bacteria have evolved mechanisms to maintain a constant membrane fluidity despite fluctuating temperatures. Some bacteria contain enzymes called *fatty acid desaturases,* which can introduce double bonds into membrane fatty acids. Genetic studies, involving either the inactivation of these enzymes or the introduction of them into cells that normally lack them, indicate that the action of these enzymes confers cold tolerance. At colder temperatures, the double bonds introduced by fatty acid desaturase make the membrane more fluid, counteracting the environmental effect of reduced temperature.

SCIENTIFIC THINKING

Hypothesis: *The plasma membrane is fluid, not rigid.*

Prediction: *If the membrane is fluid, membrane proteins may diffuse laterally.*

Test: *Fuse mouse and human cells, then observe the distribution of membrane proteins over time by labeling specific mouse and human proteins.*

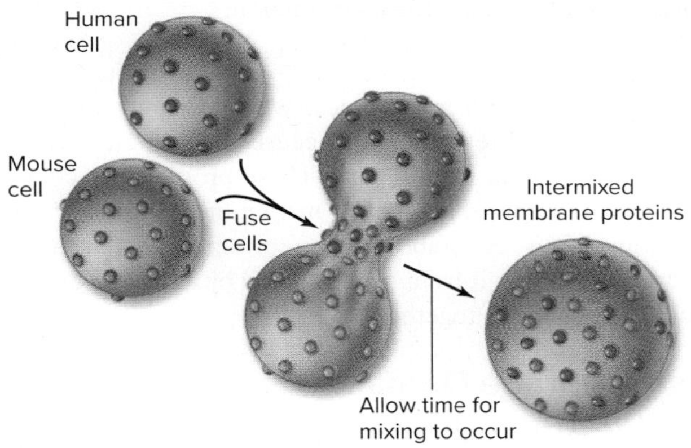

Result: *Over time, hybrid cells show increasingly intermixed proteins.*

Conclusion: *At least some membrane proteins can diffuse laterally in the membrane.*

Further Experiments: *Can you think of any other explanation for these observations? What if newly synthesized proteins were inserted into the membrane during the experiment? How could you use this basic experimental design to rule out this or other possible explanations?*

Figure 5.5 Proteins move about in membranes. Protein movement within membranes can be demonstrated easily by labeling the plasma membrane proteins of a mouse cell with fluorescent antibodies and then fusing that cell with a human cell. At first, all of the mouse proteins are located on the mouse side of the fused cell. However, within an hour the labeled and unlabeled proteins are intermixed throughout the hybrid cell's plasma membrane.

REVIEW OF CONCEPT 5.2

Biological membranes consist of a phospholipid bilayer. Each phospholipid has a hydrophilic (phosphate) head and a hydrophobic (lipid) tail. In water, phospholipids spontaneously form bilayers, with phosphate groups facing out and lipid tails facing in, where they are sequestered from water.

- *Would a phospholipid bilayer form in nonpolar solvent?*

5.3 Membrane Proteins Enable a Broad Range of Interactions with the Environment

The phospholipid bilayer is a fluid structure that forms the basis for all biological membranes. Different cellular membranes may contain different phospholipids, but the main functional distinctions in membranes arise due to the proteins present in the membrane. Proteins provide for the great diversity in membrane functions in different cellular contexts.

Membrane Proteins Have Many Functions

LEARNING OBJECTIVE 5.3.1 List six key functional classes of membrane proteins.

Membrane proteins are either integral proteins embedded in the membrane, or peripheral proteins associated with one surface of the membrane. Integral proteins are also often transmembrane proteins, which have an extracellular domain, a transmembrane domain, and an intracellular domain. Membrane proteins provide the distinct character of the plasma membrane of specific cell types, as well as the internal membranes in eukaryotic cells. We will see the function of membrane proteins in many contexts, but we will focus here on on six key functional classes of membrane protein (figure 5.6):

1. **Transporters.** Membranes are very selective, allowing only certain solutes to enter or leave the cell through channels or carriers composed of proteins. These proteins are found in most cells, but we will encounter specific examples critical to the functions of the nervous system and the kidney.
2. **Enzymes.** Cells carry out many chemical reactions on the interior surface of the plasma membrane, using enzymes attached to the membrane. Important examples of these proteins are the electron transport chains found in both

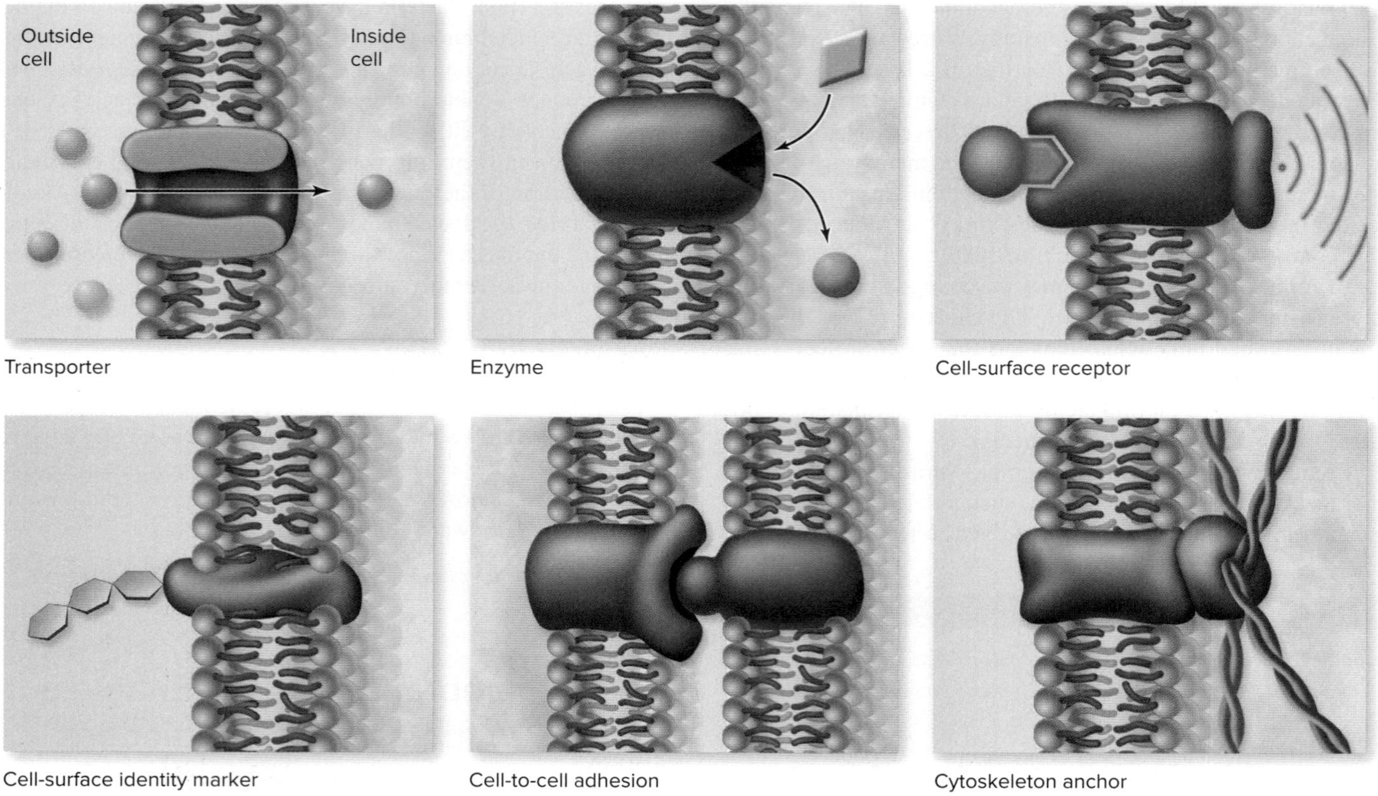
Outside cell
Inside cell
Transporter
Enzyme
Cell-surface receptor
Cell-surface identity marker
Cell-to-cell adhesion
Cytoskeleton anchor

Figure 5.6 Functions of plasma membrane proteins. Membrane proteins act as transporters, enzymes, cell-surface receptors, and cell-surface identity markers, as well as aiding in cell-to-cell adhesion and securing the cytoskeleton.

mitochondria and chloroplasts that are critical to cellular respiration and photosynthesis (refer to chapters 7 and 8).

3. **Cell-surface receptors.** Membranes are exquisitely sensitive to chemical messages, which are detected by receptor proteins anchored to their surfaces. The nature of cell-surface receptors found on a cell will determine what signals the cell can detect and respond to (refer to chapter 9).
4. **Cell-surface identity markers.** Membranes carry cell-surface markers that identify them to other cells. Most cell types carry their own ID tags, combinations of cell-surface proteins and glycoproteins. In vertebrates, the function of the immune system requires the ability to distinguish self from nonself (refer to chapter 35).
5. **Cell-to-cell adhesion proteins.** Cells use specific proteins to glue themselves to one another. Some adhere by forming temporary interactions, but others form a more permanent bond. Most cells secrete an extracellular matrix, which is a complex mix of proteins and carbohydrates that varies with cell and tissue type.
6. **Cytoskeleton anchors.** Surface proteins that interact with other cells are often firmly anchored to the cytoskeleton of the cell interior by linking proteins. These proteins and the adhesion proteins mentioned in item 5 were both important in the evolution of multicellularity. The presence of cellular junctions (refer to chapter 4) was critical for multicellularity.

Bending membranes

In chapter 4 we saw that intracellular membranes take on elaborate structures, like the folds and tubular networks of the endoplasmic reticulum. These structures can be affected by proteins that associate with membranes. Wedge-shaped transmembrane proteins called reticulons will cause a membrane to bend, or if there are enough, form a tube. These proteins are concentrated in tubular regions, and where membrane folds occur. Flattened regions can be stabilized by proteins that coat the membrane, preventing it from bending, and by adhesion proteins that hold adjacent folds together.

Transmembrane Domains Contain Nonpolar Amino Acids

LEARNING OBJECTIVE 5.3.2 Explain how proteins associate with fluid biological membranes.

Transmembrane domains

Integral membrane proteins can actually span the lipid bilayer. These proteins consist of a cytoplasmic domain inside the cell, **transmembrane domains** that cross the membrane, and an extracellular domain outside the cell. Because transmembrane domains are in contact with the nonpolar interior of the membrane, they consist primarily of nonpolar amino acids.

The polar water molecules exclude nonpolar amino acids, keeping transmembrane domains within the interior of the lipid bilayer. The extracellular and cytoplasmic domains tend to have more charged and polar amino acids that can interact with water. Any movement of the protein out of the membrane, in either direction, brings the nonpolar regions of the protein into contact with water, which "shoves" the protein back into the interior.

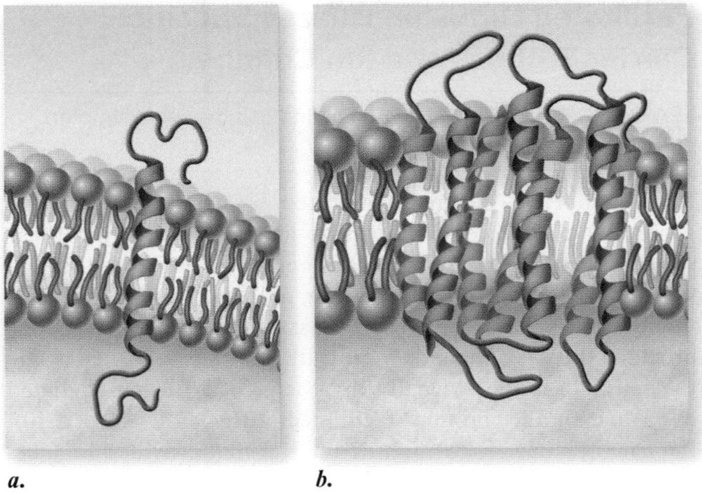
a.
b.

Figure 5.7 Transmembrane domains. Integral membrane proteins have at least one hydrophobic transmembrane domain (shown in blue) to anchor them in the membrane. ***a.*** Protein with a single transmembrane domain. ***b.*** Receptor protein with seven transmembrane domains.

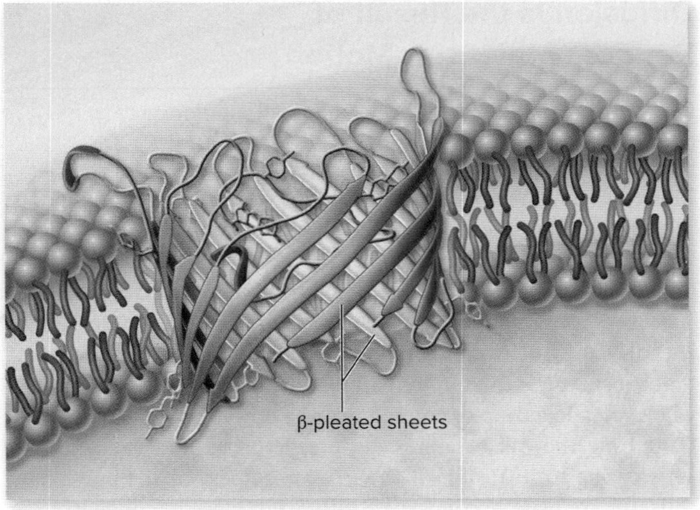
β-pleated sheets

Figure 5.8 A pore protein. The transmembrane protein porin creates large, open tunnels (pores) in a membrane. Sixteen strands of β-pleated sheets run antiparallel to one another, creating a β barrel in the membrane that allows water molecules to pass through.

These forces prevent the transmembrane proteins from simply popping out of the membrane and floating away. Transmembrane domains often form α helices (refer to chapter 3) (figure 5.7*a*).

Proteins need only a single transmembrane domain to be anchored in the membrane. For example, cadherin proteins, which are involved in cell adhesion (refer to figure 4.26), contain a single transmembrane domain. A single transmembrane domain will anchor a protein in a membrane, but many proteins actually contain more than one transmembrane domain. This can be used as a way to characterize these proteins. A specific class of protein will have a consistent number of transmembrane domains. For example, a type of receptor protein you will encounter in chapter 9, G protein–coupled receptors, have a characteristic structure with seven transmembrane domains (figure 5.7*b*).

Pores

Some transmembrane proteins have extensive nonpolar regions with secondary configurations of β-pleated sheets (refer to chapter 3) instead of α helices. The β sheets form a characteristic motif, folding back and forth in a cylinder so that the sheets arrange themselves like a pipe through the membrane, with the polar environment in the interior of the β sheets spanning the membrane. This so-called *β barrel,* open on both ends, is a common feature of the porin class of proteins that are found within the outer membrane of some bacteria. The openings allow polar water molecules to pass through the membrane (figure 5.8).

Proteins can be anchored to lipids

Some membrane proteins are attached to the surface of the membrane by special molecules that associate strongly with phospholipids. Like a ship tied to a floating dock, these anchored proteins are free to move about on the surface of the membrane tethered to a phospholipid. The anchoring molecules are modified lipids that have (1) nonpolar regions that insert into the internal portion of the lipid bilayer and (2) chemical bonding domains that link directly to proteins.

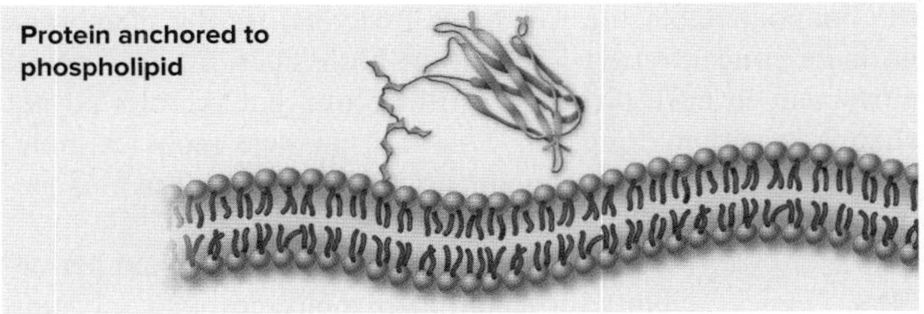
Protein anchored to phospholipid

REVIEW OF CONCEPT 5.3

Membrane proteins confer differences between membranes of different cells. Their functions include transport, enzymatic action, signal reception, cell-to-cell interactions, cell identity, and anchoring the cytoskeleton. Peripheral proteins can be anchored in the membrane by modified lipids. Integral membrane proteins span the membrane anchored by one or more hydrophobic transmembrane domains.

- *How could you program a computer to find transmembrane domains in protein sequence data?*

5.4 Passive Transport Moves Molecules Across Membranes by Diffusion

One of the most important functions of the plasma membrane is to control the contents of the cell. This involves controlling the transport of materials across the membrane. We will begin by considering the movement of a substance that is permeable in the phospholipid bilayer, then consider how proteins can extend the range of material transported.

Diffusion Is the Result of Random Molecular Motion

LEARNING OBJECTIVE 5.4.1 Explain the importance of a concentration to simple diffusion.

Molecules and ions dissolved in water are in constant random motion, called *Brownian movement.* This random motion results in a net movement of these substances from regions of high concentration to regions of lower concentration, a process called **diffusion.**

Net movement of a substance down its concentration gradient, driven by diffusion, will continue until the concentration of that substance is the same in all regions. Consider what happens when you add a drop of colored ink to a bowl of water (figure 5.9). Over time the ink becomes dispersed throughout the solution. This is due to diffusion of the ink molecules. In the context of cells, we are usually concerned with differences in concentration of molecules across the plasma membrane.

Many, but not all, substances can move into and out of cells. The major barrier to crossing a cell's plasma membrane is the hydrophobic interior of the bilayer that repels polar molecules but not nonpolar molecules. For a nonpolar molecule, if a concentration difference exists on the two sides of a membrane, the nonpolar molecule will move freely across the membrane until its concentration is equal on both sides. At this point, movement in both directions still occurs, but there is no net change in either direction. This free passage includes molecules such as O_2 and nonpolar organic molecules such as steroid hormones.

The plasma membrane has only limited permeability for small polar molecules, and very limited permeability for larger polar molecules and ions. The movement of water, one of the most important polar molecules, is discussed later in this section.

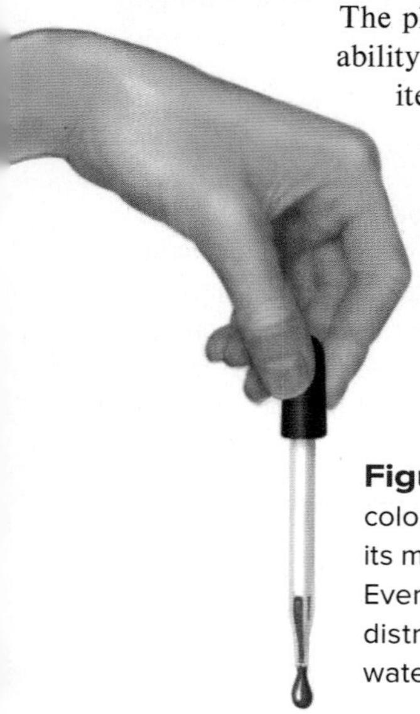

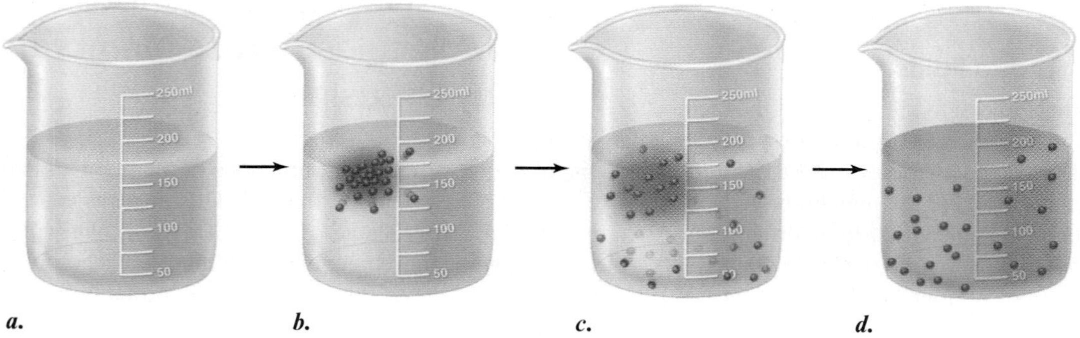

Figure 5.9 Diffusion. ***a.*** If a drop of colored ink is dropped into a beaker of water, its molecules dissolve **(*b*)** and diffuse **(*c*)**. Eventually diffusion results in an even distribution of ink molecules throughout the water **(*d*)**.

Facilitated Diffusion Utilizes Specific Carrier Proteins and Ion Channels

LEARNING OBJECTIVE 5.4.2 Distinguish between simple diffusion and facilitated diffusion.

Many important molecules required by cells are polar and so cannot easily cross the plasma membrane. How do these molecules enter the cell? They gain entry by diffusing through specific protein channels or carrier proteins embedded within the plasma membrane. Their passage requires no energy, provided there is a higher concentration of the molecule outside the cell than inside. We call this process of diffusion mediated by a membrane protein **facilitated diffusion** (figure 5.10). **Ion channels** have a hydrophilic interior that provides an aqueous channel through which ions can pass when the channel is open. **Carrier proteins,** in contrast to channels, bind specifically to the molecule they assist, much as an enzyme binds to its substrate. These channels and carriers are usually selective for one type of molecule, and thus the cell membrane is said to be **selectively permeable.**

Facilitated diffusion of ions through channels

Because of their charge, ions are repelled by nonpolar molecules such as those that make up the interior of the plasma membrane's lipid bilayer. Therefore, ions cannot move between the cytoplasm of a cell and the extracellular fluid without the assistance of membrane transport proteins.

Ion channels possess a hydrated interior that spans the membrane. Ions can diffuse through the channel in either direction, depending on their relative concentration across the membrane. Some channel proteins can be opened or closed in response to a stimulus. These channels are called *gated channels*; depending on the nature of the channel, the stimulus can be either chemical or electrical.

Three conditions determine the direction of net movement of the ions: (1) their relative concentrations on either side of the membrane, (2) the voltage difference across the membrane and for the gated channels, (3) the state of the gate (open or closed). A voltage difference is an electric potential difference across the membrane called a *membrane potential.* Changes in membrane potential form the basis for transmission of signals in the nervous system and some other tissues. (We discuss this topic in detail

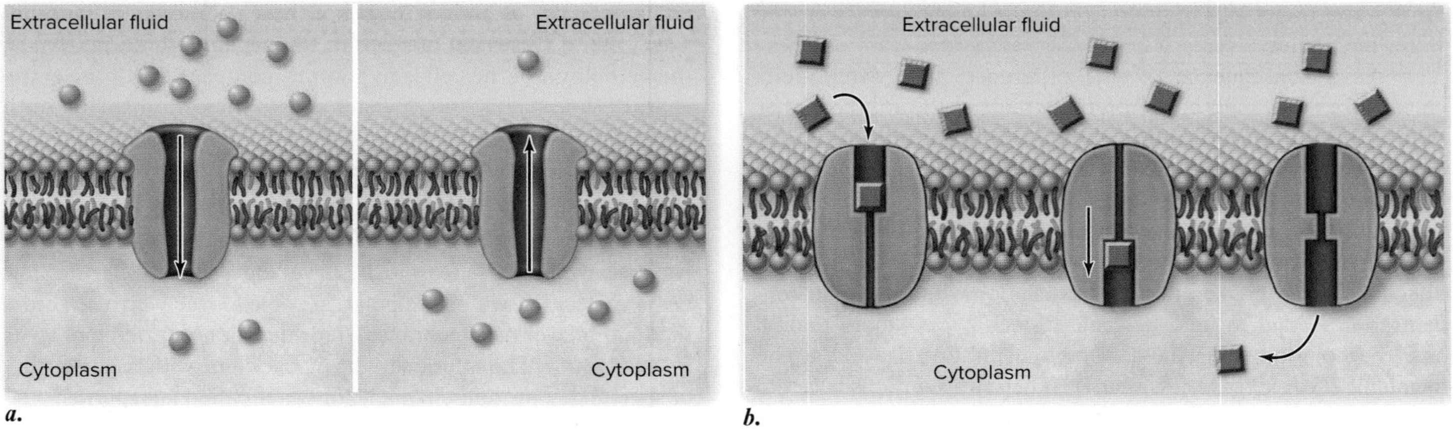

Figure 5.10 Facilitated diffusion. Diffusion can be facilitated by membrane proteins. ***a.*** The movement of ions through a channel is shown. On the left the concentration is higher outside the cell, so the ions move into the cell. On the right the situation is reversed. In both cases, transport continues until the concentration is equal on both sides of the membrane. At this point, ions continue to cross the membrane in both directions, but there is no net movement in either direction. ***b.*** Carrier proteins bind specifically to the molecules they transport. In this case the concentration is higher outside the cell, so molecules bind to the carrier on the outside. The carrier's shape changes, allowing the molecule to cross the membrane. This is reversible, so net movement continues until the concentration is equal on both sides of the membrane.

in chapter 33.) Each type of channel is specific for a particular ion, such as calcium (Ca^{2+}), sodium (Na^+), potassium (K^+), or chloride (Cl^-), or in some cases, for more than one cation or anion. Ion channels play an essential role in signaling by the nervous system.

Facilitated diffusion by carrier proteins

Channels are not the only way into cells. Carrier proteins can also transport both ions and other solutes, such as some sugars and amino acids, across the plasma membrane.

Transport by a carrier protein is still a form of diffusion and therefore requires a concentration difference across the membrane. However, diffusion across a membrane using a carrier protein differs from simple diffusion in one key respect: as a concentration gradient increases, transport by simple diffusion shows a linear increase in rate of transport. For transported molecules bound to carrier proteins, on the other hand, as the concentration gradient increases, a point is reached where all carriers are occupied and the rate of transport can increase no further, having reached *saturation.*

This situation is somewhat like that of a stadium (the cell) where a crowd must pass through turnstiles (the carrier protein) to enter. When ticket holders (transported molecules) are passing through all gates at maximum speed, the rate at which they enter cannot increase, no matter how many are waiting outside.

Facilitated diffusion in red blood cells

The function of carrier proteins is illustrated by the glucose transporter in vertebrate red blood cells (RBCs). The glucose transporter is a transmembrane protein that binds to a glucose molecule outside the cell and then changes its conformation, which moves the glucose through the bilayer and releases it on the inside of the plasma membrane. After the glucose is released, the transporter reverts to its original shape, and it is then available to bind the next glucose molecule that comes along outside the cell. This does not require energy, but to maintain a concentration gradient to favor inward transport, the cell adds a phosphate group to the glucose on entry. This modified glucose does not bind to the carrier. This maintains a steep concentration gradient for unphosphorylated glucose, favoring its entry into the cell.

Osmosis Is the Movement of Water Across Membranes

LEARNING OBJECTIVE 5.4.3 Predict the direction of osmotic movement of water.

The cytoplasm of a cell contains ions and molecules, such as sugars and amino acids, dissolved in water. The mixture of these substances and water is called an *aqueous solution.* Water is termed the **solvent,** and the substances dissolved in the water are **solutes.** Both water and solutes tend to diffuse from regions of high concentration to ones of low concentration; that is, they diffuse down their concentration gradients.

When two regions are separated by a membrane, what happens depends on whether the solutes can pass freely through that membrane. Most solutes, including ions and sugars, are not lipid-soluble and therefore are unable to cross the lipid bilayer. Importantly, a concentration gradient of these solutes can lead to the net movement of water across the membrane.

Why does this happen? Water molecules interact with dissolved solutes by forming hydration shells around the charged or polar solute molecules. As a direct result of this, when a membrane separates two solutions with different concentrations of charged or polar solutes, the concentrations of *free* water molecules on the two sides of the membrane also differ—the side with higher solute concentration has tied up more water molecules in hydration shells and so has fewer free water molecules.

As a consequence of this difference, free water molecules move down their concentration gradient, toward the higher solute concentration. This net diffusion of water across a membrane toward a higher solute concentration is called **osmosis** (figure 5.11).

The concentration of *all* solutes in a solution determines the **osmotic concentration** of the solution. If two solutions have unequal osmotic concentrations, the solution with the higher concentration is said to be **hypertonic** (Greek *hyper,* "more than"), and the solution with the lower concentration is said to be **hypotonic** (Greek *hypo,* "less than"). When two solutions have the same osmotic concentration, the solutions are **isotonic** (Greek *iso,* "equal"). The terms *hyperosmotic, hypoosmotic,* and *isosmotic* are also used to describe these conditions.

A cell in any environment can be thought of as a plasma membrane separating two solutions: the cytoplasm and the extracellular fluid. The direction and extent of any diffusion of water across the plasma membrane are determined by comparing the osmotic strength of these solutions. Put another way, water diffuses out of a cell in a hypertonic solution (that is, the cytoplasm of the cell is hypotonic, compared with the extracellular fluid). This loss of water causes the cell to shrink until the osmotic concentrations of the cytoplasm and the extracellular fluid become equal (figure 5.12).

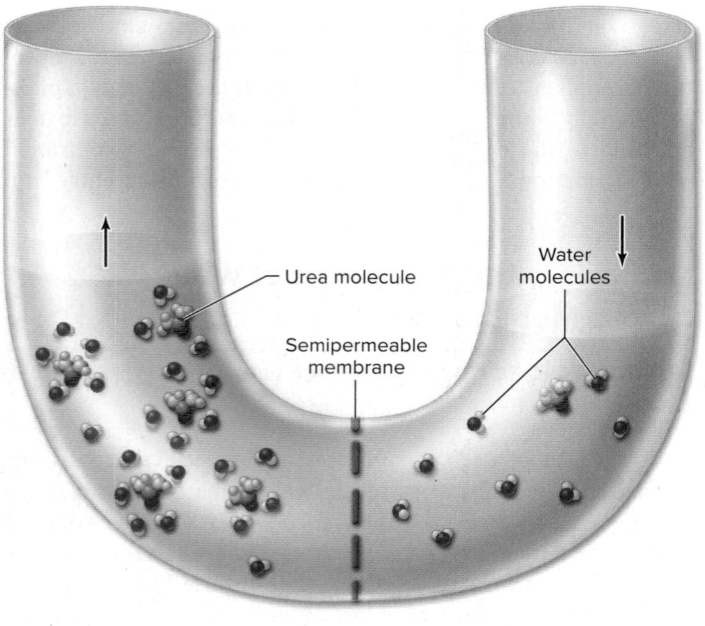

Figure 5.11 Osmosis. Concentration differences in charged or polar molecules that cannot cross a selectively permeable membrane result in the movement of water, which can cross the membrane. Water molecules form hydrogen bonds with charged or polar molecules, creating a hydration shell around them in solution. A higher concentration of polar molecules (urea), shown on the left side of the membrane, leads to water molecules gathering around each urea molecule. These water molecules are no longer free to diffuse across the membrane. The polar solute has reduced the concentration of free water molecules, creating a gradient. This causes a net movement of water by diffusion from right to left in the U-tube, raising the level on the left and lowering it on the right.

Aquaporins: Water channels

As we have discussed osmosis, have you been wondering how water molecules, which are polar, are able to freely diffuse across the lipid bilayer of membranes? This question puzzled biologists for a long time. The solution to the puzzle came with the discovery of specialized protein channels for water called **aquaporins.**

A simple experiment demonstrates the key role of aquaporins in admitting water into cells. If an amphibian egg is placed in hypotonic spring water (the solute concentration in the cell is higher than that of the surrounding water), the egg does not swell. Within an as-yet-undeveloped egg, the genes encoding aquaporins have not yet been expressed. If aquaporin mRNA is then injected into the egg, the amphibian channel proteins are expressed and appear in the egg's plasma membrane. Water can now diffuse into the egg, causing it to swell.

More than 11 kinds of aquaporins have been found in mammals. These fall into two general classes: those that are specific for only water and those that allow other small hydrophilic molecules, such as glycerol or urea, to cross the membrane as well. This latter class explains how some membranes allow the easy passage of small hydrophilic substances.

The hereditary disease nephrogenic diabetes insipidus (NDI) results in the excretion of large volumes of dilute urine. There are two forms, one that affects the insertion of aquaporins in the membrane of kidney tubules, and another that inactivates the actual aquaporin protein. This clearly illustrates that these channels are critical to kidney function.

Osmosis Can Generate Significant Pressure

LEARNING OBJECTIVE 5.4.4 Discuss three ways organisms maintain osmotic balance.

What happens to a cell in a hypotonic solution (that is, where the cell's cytoplasm is hypertonic relative to the extracellular fluid)? In this situation, water diffuses into the cell from the extracellular fluid, causing the cell to swell. The pressure of the cytoplasm pushing out against the cell membrane, or *hydrostatic pressure,* increases. The amount of water that enters the cell depends on the difference in solute concentration between the cell and the extracellular fluid. This is measured as **osmotic pressure,** defined as the force needed to stop osmotic flow.

If the membrane is strong enough, the cell reaches an equilibrium, where the osmotic pressure, which tends to drive water into the cell, is exactly counterbalanced by the hydrostatic pressure, which tends to drive water back out of the cell. However, a plasma membrane by itself cannot withstand large internal pressures, and an isolated cell under such conditions would burst like an overinflated balloon (figure 5.12).

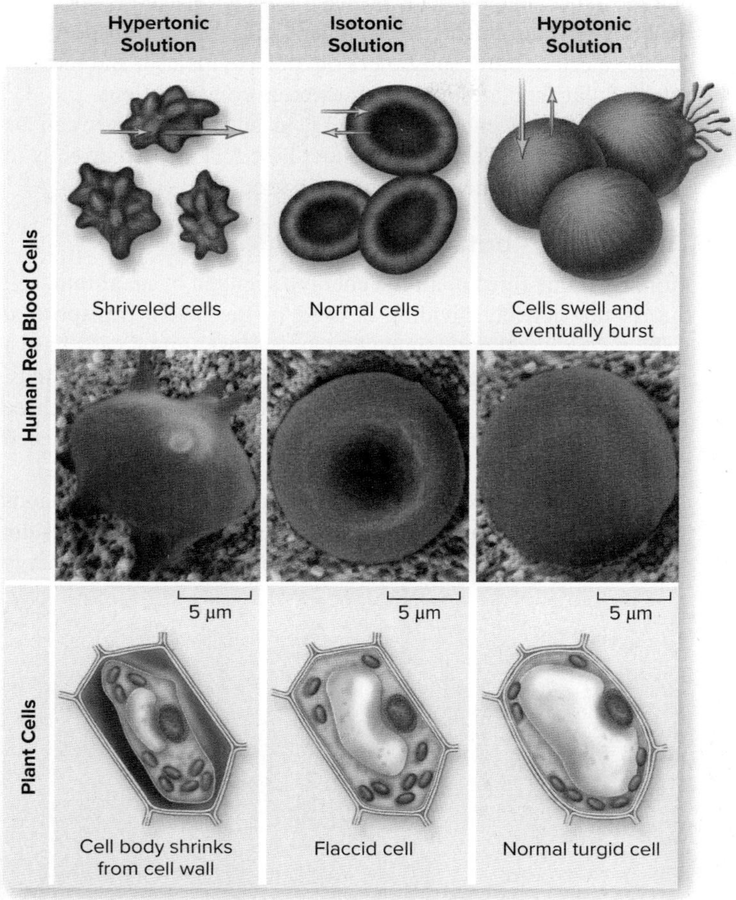
Hypertonic Solution
Isotonic Solution
Hypotonic Solution
Human Red Blood Cells
Shriveled cells
Normal cells
Cells swell and eventually burst
5 μm
5 μm
5 μm
Plant Cells
Cell body shrinks from cell wall
Flaccid cell
Normal turgid cell

Figure 5.12 How solutes create osmotic pressure. In a hypertonic solution, water moves out of the cell, causing the cell to shrivel. In an isotonic solution, water diffuses into and out of the cell at the same rate, with no change in cell size. In a hypotonic solution, water moves into the cell. Blue arrows (top) show the direction and amount of water movement. As water enters the cell from a hypotonic solution, pressure is applied to the plasma membrane until the cell ruptures. Water enters the cell due to osmotic pressure from the higher solute concentration in the cell. Osmotic pressure is measured as the force needed to stop osmosis. The strong cell wall of plant cells can withstand the hydrostatic pressure to keep the cell from rupturing. This is not the case with animal cells.
David M. Phillips/Science Source

Accordingly, it is important for animal cells, which are encased only within plasma membranes, to maintain osmotic balance. In contrast, the cells of prokaryotes, fungi, plants, and many protists are surrounded by strong cell walls that can withstand high internal pressures without bursting.

Maintaining osmotic balance

Organisms have developed many strategies for solving the dilemma posed by being hypertonic to their environment and therefore exposed to a steady influx of water by osmosis.

Extrusion. Some single-celled eukaryotes, such as the protist *Paramecium,* use organelles called contractile vacuoles to remove water. Each vacuole collects water from various parts of the cytoplasm and transports it to the central part of the vacuole, near the cell surface. The vacuole possesses a small pore that opens to the outside of the cell. By contracting rhythmically, the vacuole pumps out (extrudes) through this pore the water that is continuously drawn into the cell by osmotic forces.

Isosmotic regulation. Some organisms that live in the ocean adjust their internal concentration of solutes to match that of the surrounding seawater. Because they are isosmotic with respect to their environment, no net flow of water occurs into or out of these cells.

Many terrestrial animals solve the problem in a similar way, by circulating a fluid through their bodies that bathes cells in an isotonic solution. The blood in your body, for example, contains a high concentration of the protein albumin, which elevates the solute concentration of the blood to match that of your cells' cytoplasm.

Turgor. Most plant cells are hypertonic to their immediate environment, containing a high concentration of solutes in their central vacuoles. The resulting internal hydrostatic pressure, known as **turgor pressure,** presses the plasma membrane firmly against the interior of the cell wall, making the cell rigid.

REVIEW OF CONCEPT 5.4

Passive transport involves diffusion, which requires a concentration gradient. Hydrophobic molecules move by simple diffusion directly through the membrane. Polar molecules and ions move by facilitated diffusion through channel or carrier proteins. Channel proteins form a hydrophilic pore through the membrane for ions, whereas carrier proteins bind to the transported molecule. Water passes by osmosis through the membrane via aquaporins in response to solute concentration differences inside and outside the cell.

- *If you require intravenous (IV) medication in the hospital, what should the concentration of solutes in the IV solution be, relative to your blood cells?*

5.5 Active Transport Across Membranes Requires Energy

Diffusion, facilitated diffusion, and osmosis are all passive transport processes that move materials down their concentration gradients. However, to gather food molecules and other substances, cells must also be able to move substances across the plasma membrane *up* their concentration gradients. This process requires the expenditure of energy, typically from ATP, and is therefore called **active transport.**

Active Transport Utilizes Specific Carrier Proteins

LEARNING OBJECTIVE 5.5.1 Distinguish between active and passive transport.

Active transport uses energy to power the movement of materials across a membrane against a concentration gradient. Like facilitated diffusion, active transport involves highly selective protein carriers within the membrane that bind to the transported substance, typically an ion, a sugar, an amino acid, or a nucleotide. These carrier proteins are called **uniporters** if they transport a single type of molecule and symporters or antiporters if they transport two different molecules together. **Symporters** transport two molecules in the same direction, and **antiporters** transport two molecules in opposite directions. These terms are also sometimes used to describe facilitated diffusion carriers.

Active transport is one of the most important activities carried out by a cell. It enables the cell to take up additional molecules of a substance that is already present in its cytoplasm in concentrations higher than in the extracellular fluid. Active transport also enables a cell to move substances out of its cytoplasm and into the extracellular fluid, despite higher external concentrations.

The use of energy from ATP in active transport can be direct or indirect. Let's first consider how ATP is used directly to move ions against their concentration gradients.

The sodium–potassium pump

More than one-third of all the energy expended by an animal cell that is not actively dividing is used in the active transport of sodium (Na^+) and potassium (K^+) ions. Most animal cells have a low internal concentration of Na^+, relative to their surroundings, and a high internal concentration of K^+. They maintain these concentration differences by actively pumping Na^+ out of the cell and pumping K^+ in.

The remarkable protein that transports these two ions across the cell membrane is known as the **sodium-potassium pump** (figure 5.13). This antiporter carrier protein uses the energy

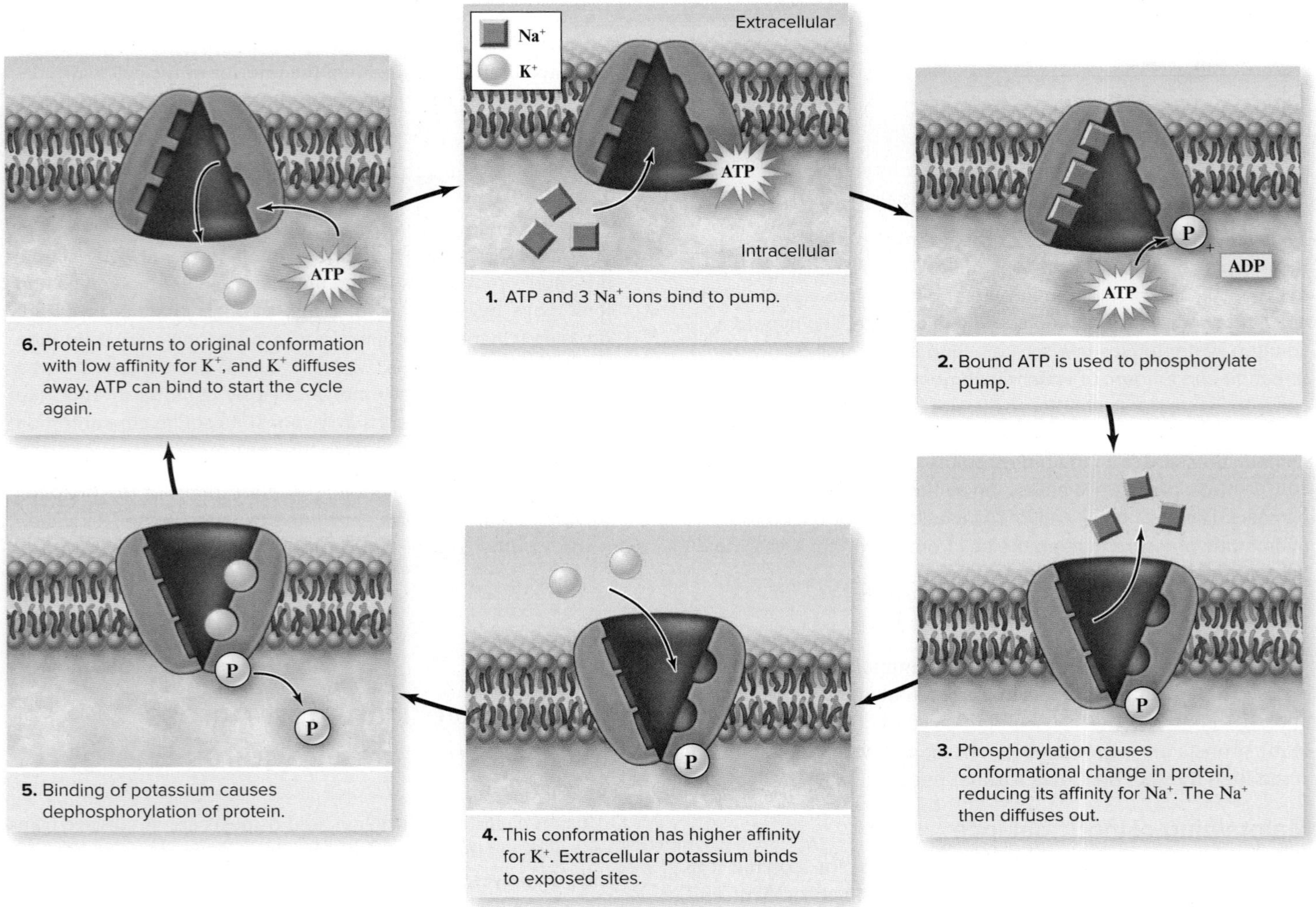

Figure 5.13 The sodium–potassium pump. The sodium–potassium pump is a protein carrier that transports sodium (Na^+) and potassium (K^+) across the plasma membrane. For every three Na^+ transported out of the cell, two K^+ are transported into it. Energy for the process is provided by ATP hydrolysis. The pump changes conformation between high affinity for Na^+ and high affinity for K^+ based on phosphorylation state.

stored in ATP to power the simultaneous movement of these two ions by changing the conformation of the carrier protein, which in turn changes its affinity first for Na^+ ions and then for K^+ ions. This is an excellent illustration of how subtle changes in the shape of a protein affect its function.

The most important characteristic of the sodium-potassium pump is that it is an active transport mechanism, transporting Na^+ and K^+ from areas of low concentration to areas of high concentration. This transport is the opposite of passive transport by diffusion, and it can be achieved only by the constant expenditure of metabolic energy. The sodium-potassium pump works through the following series of conformational changes in the antiporter transmembrane protein, summarized in figure 5.13:

Step 1. With ATP bound to the pump, three Na^+ bind to the cytoplasmic side of the protein.

Step 2. The bound ATP is hydrolyzed to ADP, transferring a phosphate to the pump. The protein is now phosphorylated.

Step 3. The phosphorylated protein undergoes a conformational change, exposing the three Na^+ to the outside. In this conformation, the pump has a low affinity for Na^+, allowing the three bound Na^+ ions to diffuse into the extracellular fluid.

Step 4. The phosphorylated protein has a high affinity for K^+, two of which bind to the extracellular side of the pump after Na^+ diffuses away.

Step 5. The binding of the K^+ causes hydrolysis of the bound phosphate and a conformational change back to the binding pocket facing the cytoplasm.

Step 6. The pump binds ATP, returning to a state with a low affinity for K^+ and high affinity for Na^+. The bound K^+ diffuse away, and three new Na^+ can bind to initiate another cycle.

In every cycle, three Na^+ leave the cell and two K^+ enter. The changes in protein conformation that occur during the cycle are rapid, enabling each carrier to transport as many as 300 Na^+ per second. The sodium-potassium pump appears to exist in all animal cells, although cells vary widely in the number of pump proteins they contain.

Coupled Transport Uses Ion Gradients to Move Molecules Against Their Concentration Gradients

LEARNING OBJECTIVE 5.5.2 Explain the energetics of coupled transport.

Some molecules are moved against their concentration gradient by using the energy stored in ATP indirectly, via a gradient of a different molecule. In this process, called *coupled transport,* the energy released as one molecule moves down its concentration gradient is used to move a different molecule against its gradient. As you just saw, the energy stored in ATP molecules can be used to create a gradient of Na^+ and K^+ across the membrane. These gradients can then be used to power the transport of other molecules across the membrane.

As one example, let's consider the active transport of glucose across the membrane in animal cells. Glucose is such an important molecule that it has a variety of transporters, one of which we discussed earlier in this section. In a multicellular organism, intestinal epithelial cells can have a far higher concentration of glucose inside the cell than outside, so these cells need to be able to transport glucose against its concentration gradient in order to absorb the glucose from metabolized food. This requires a different transporter than the passive uniporter involved in the facilitated diffusion of glucose.

Coupled transport

The active transport of glucose is carried out by a symporter that uses the Na^+ gradient produced by the sodium-potassium pump as a source of energy to power the movement of glucose into the cell. In this system, both glucose and Na^+ simultaneously bind to the symporter transport protein. Na^+ then passes into the cell *down* its concentration gradient, carrying glucose along with it into the cell (figure 5.14).

Countertransport

In the cotransport of Na^+ and glucose, both molecules move in the same direction across the membrane. In a related active transport process called *countertransport,* the inward movement of Na^+ is coupled with the outward movement of another substance, such as Ca^{2+} or H^+. As in cotransport, both Na^+ and the other substance bind to the same transport protein,

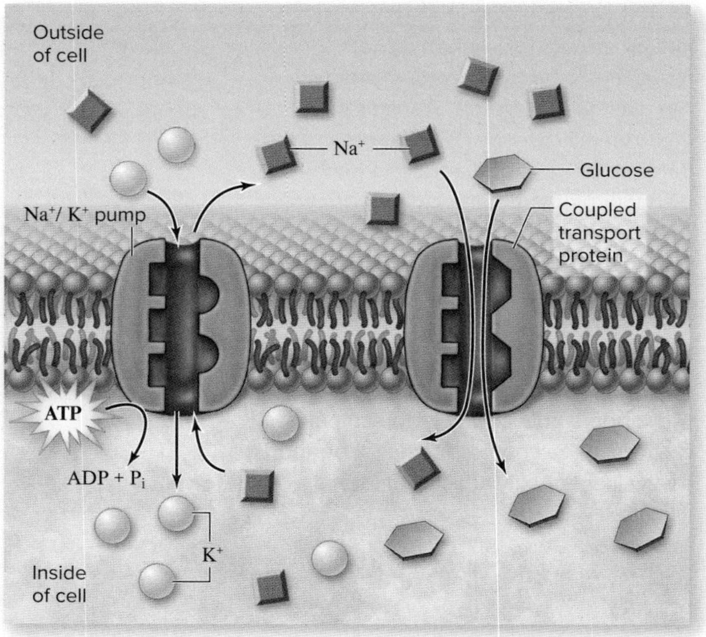

Figure 5.14 Coupled transport. A membrane protein transports Na^+ into the cell down a concentration gradient maintained by the Na^+/K^+ pump, at the same time transporting a glucose molecule. The gradient driving the Na^+ entry allows sugar molecules to be transported against their concentration gradient.

which in this case is an antiporter—the substances bind on opposite sides of the membrane and are moved in opposite directions. In countertransport, the cell uses the energy released as Na^+ moves down its concentration gradient into the cell to eject a substance against its concentration gradient. In both cotransport and countertransport, the potential energy in the concentration gradient of one molecule is used to transport another molecule against its concentration gradient. They differ only in the direction that the second molecule moves relative to the first.

REVIEW OF CONCEPT 5.5

Active transport requires both a carrier protein and energy to move molecules against a concentration gradient. The Na^+/K^+ pump uses ATP to move Na^+ in one direction and K^+ in the other to create and maintain the concentration differences of these ions. In coupled transport, a concentration gradient of one molecule is used to move a different molecule against a gradient.

- *How are active transport and facilitated diffusion similar? How are they different?*

5.6 Bulky Materials Cross Membranes Within Vesicles

The lipid nature of a cell's plasma membranes creates an interesting problem for the cell. The substances that cells require for growth are mostly large polar molecules that cannot cross the hydrophobic barrier a lipid bilayer creates. How do these substances get into cells? Two processes are involved in this **bulk transport:** *endocytosis* and *exocytosis.*

Endocytosis and Exocytosis Are Inverse Processes

LEARNING OBJECTIVE 5.6.1 Explain how endocytosis can be molecule-specific.

In **endocytosis,** the plasma membrane envelops food particles and fluids. Cells use three major types of endocytosis: phagocytosis, pinocytosis, and receptor-mediated endocytosis (figure 5.15). Like active transport, these processes also require energy expenditure.

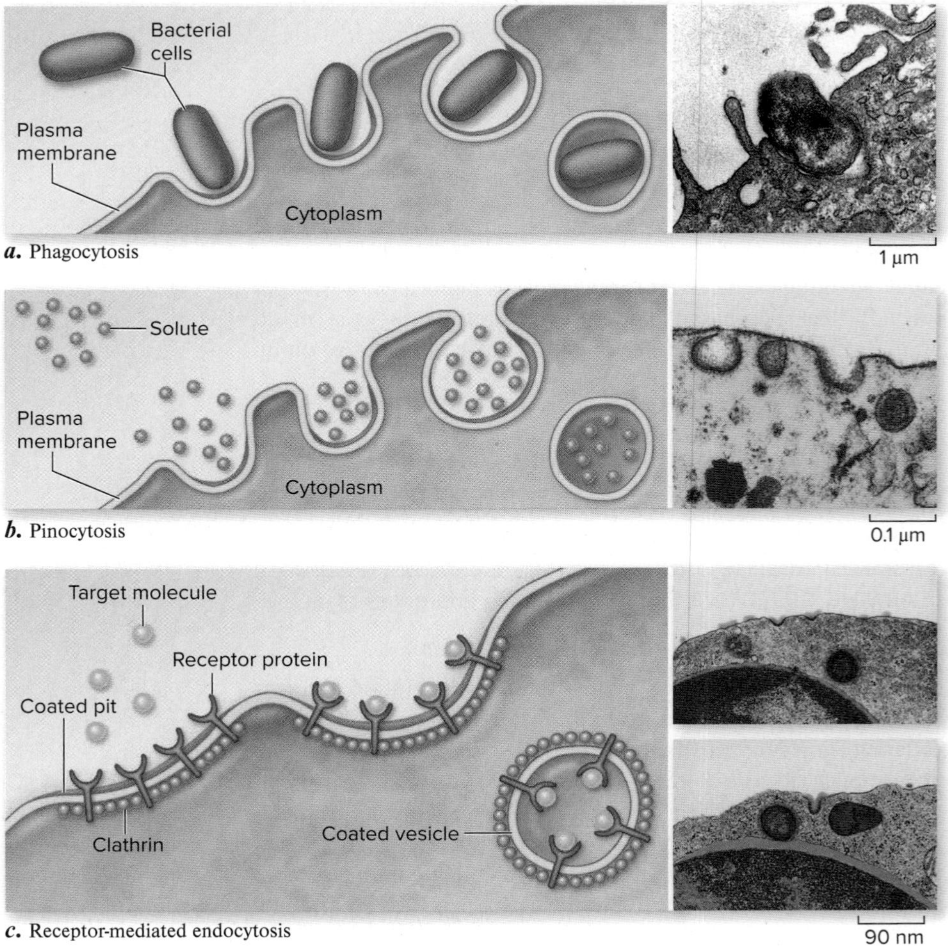

Figure 5.15 Endocytosis. Both ***a.*** phagocytosis and ***b.*** pinocytosis are forms of endocytosis. ***c.*** In receptor-mediated endocytosis, cells have pits coated with the protein clathrin that initiate endocytosis when target molecules bind to receptor proteins in the plasma membrane. Transmission electron microscopy (TEM) photo inserts (false color has been added to enhance distinction of structures): ***a.*** Phagocytosis of a bacterium, *Rickettsia tsutsugamushi,* by a mouse peritoneal mesothelial cell. ***b.*** Pinocytosis in a smooth muscle cell. ***c.*** A coated pit appears in the plasma membrane of a developing egg cell, covered with a layer of proteins. When an appropriate collection of molecules gathers in the coated pit, the pit deepens and will eventually seal off to form a vesicle.

(a): Dr. Edwin P. Ewing, Jr./CDC
(b-d): Don W. Fawcett/Science Source

If the material the cell takes in is particulate (made up of discrete particles), such as an organism or some other fragment of organic matter (figure 5.15*a*), the process is called **phagocytosis** (Greek *phagein,* "to eat," and *cytos,* "cell"). If the material the cell takes in is liquid (figure 5.15*b*), the process is called **pinocytosis** (Greek *pinein,* "to drink"). Pinocytosis is common among animal cells. Mammalian egg cells, for example, "nurse" from surrounding cells; the nearby cells secrete nutrients, which the maturing egg cell takes up by pinocytosis.

Virtually all eukaryotic cells constantly carry out these kinds of endocytotic processes, trapping particles and extracellular fluid in vesicles and ingesting them. Endocytosis rates vary from one cell type to another. They can be surprisingly high; some types of white blood cells ingest up to 25% of their cell volume each hour.

Receptor-mediated endocytosis

Sometimes endocytosis is targeted at specific molecules. In these instances the targeted molecules are transported into cells by **receptor-mediated endocytosis.** These molecules first bind to specific receptors in the plasma membrane—the binding is quite specific, with the target molecule shape fitting snugly into its receptor. Different cell types contain a characteristic battery of receptor types, each targeted at a different kind of molecule.

The portion of the receptor molecule that protrudes into the membrane is locked in place within an indented pit coated on the cytoplasmic side with the protein *clathrin.* Each pit acts as a molecular mousetrap, closing over to form an internal vesicle when the right molecule enters the pit (figure 5.15*c*). The trigger that releases the trap is the binding of the properly fitted target molecule to the embedded receptor. When binding occurs, the cell reacts by initiating endocytosis; the process is highly specific and very fast. The vesicle is now inside the cell carrying its cargo.

One important type of molecule that is taken up by receptor-mediated endocytosis is low-density lipoprotein (LDL). LDL molecules bring cholesterol into the cell, where it can be incorporated into membranes. This transport is important, as cholesterol plays a key role in determining the stiffness of the body's membranes. In the human genetic disease familial hypercholesterolemia, the LDL receptors lack tails, so they are never fastened in the clathrin-coated pits and as a result do not trigger vesicle formation. The cholesterol stays in the bloodstream of affected individuals, accumulating as plaques inside arteries and leading to heart attacks.

It is important to understand that endocytosis in itself does not bring substances directly into the cytoplasm of a cell. The material brought is still separated from the cytoplasm by the membrane of the vesicle.

Exocytosis

The reverse of endocytosis is **exocytosis,** the discharge of material from vesicles at the cell surface (figure 5.16). In plant cells, exocytosis is an important means of exporting the materials needed to construct the cell wall through the plasma membrane. Among protists, contractile vacuole discharge is considered a form of exocytosis. In animal cells, exocytosis provides a mechanism for secreting many hormones, neurotransmitters, digestive enzymes, and other substances.

The mechanisms for transport across cell membranes are summarized in table 5.2.

REVIEW OF CONCEPT 5.6

Large molecules and other bulky materials can enter a cell by endocytosis and leave the cell by exocytosis. These processes require energy. Endocytosis may be mediated by specific receptor proteins in the membrane that trigger the formation of vesicles.

- *What feature unites transport by receptor-mediated endocytosis, transport by a carrier, and catalysis by an enzyme?*

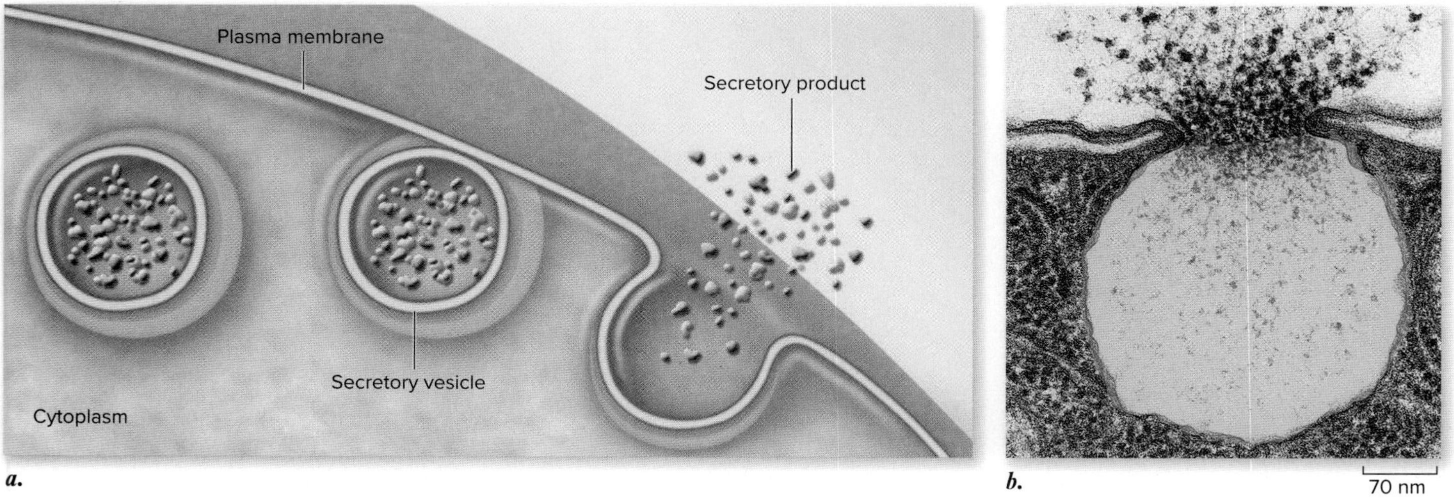

Figure 5.16 Exocytosis. ***a.*** Proteins and other molecules are secreted from cells in small packets called vesicles, whose membranes fuse with the plasma membrane, releasing their contents outside the cell. ***b.*** A false color transmission electron micrograph showing exocytosis.

b): Dr. Birgit Satir

TABLE 5.2 Mechanisms for Transport Across Cell Membranes

Process		How It Works	Example
PASSIVE PROCESSES			
Diffusion			
Direct	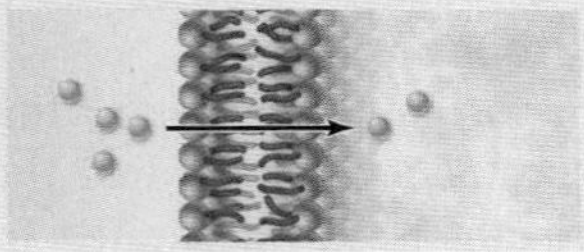	Random molecular motion produces net migration of nonpolar molecules toward region of lower concentration	Movement of oxygen into cells
Facilitated Diffusion			
Protein channel	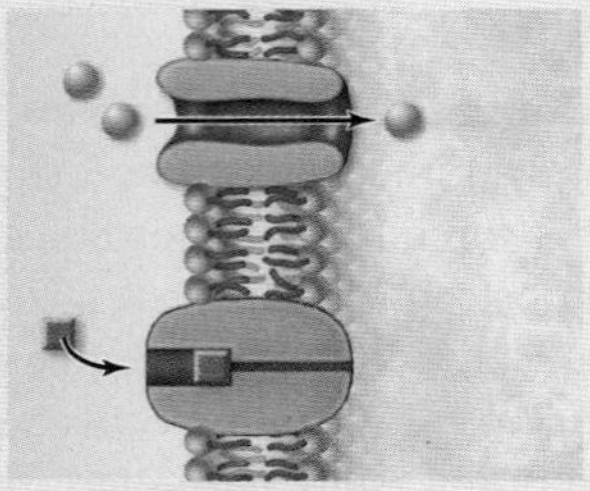	Polar molecules or ions move through a protein channel; net movement is toward region of lower concentration	Movement of ions into or out of cell
Protein carrier		Molecule binds to carrier protein in membrane and is transported across; net movement is toward region of lower concentration	Movement of glucose into cells
Osmosis			
Aquaporins	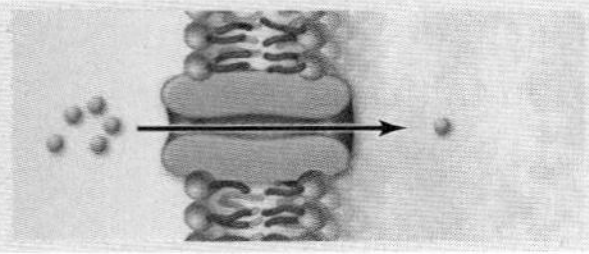	Diffusion of water across the membrane via osmosis; requires osmotic gradient	Movement of water into cells placed in a hypotonic solution
ACTIVE PROCESSES			
Active Transport			
Protein carrier			
Na^+/K^+ pump	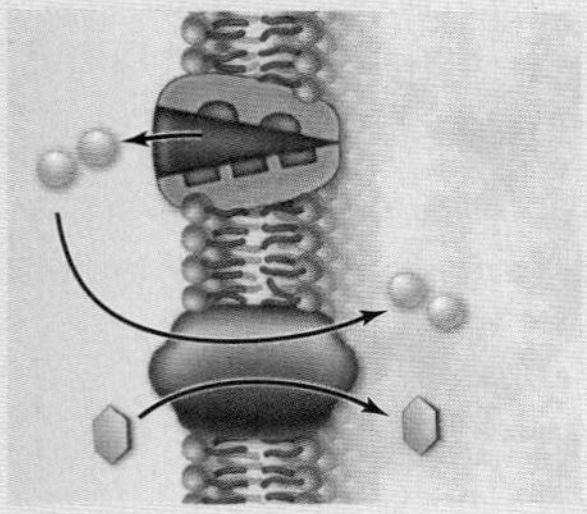	Carrier uses energy to move a substance across a membrane against its concentration gradient	Movement of Na^+ and K^+ against their concentration gradients
Coupled transport		Molecules are transported across a membrane against their concentration gradients by the cotransport of sodium ions or protons down their concentration gradients	Coupled uptake of glucose into cells against its concentration gradient using a Na^+ gradient
Endocytosis			
Membrane vesicle			
Phagocytosis	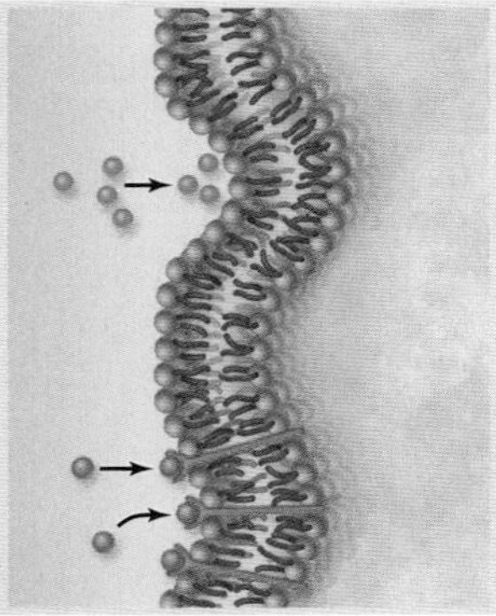	Particle is engulfed by membrane, which folds around it and forms a vesicle	Ingestion of bacteria by white blood cells
Pinocytosis		Fluid droplets are engulfed by membrane, which forms vesicles around them	"Nursing" of human egg cells
Receptor-mediated endocytosis		Endocytosis triggered by a specific receptor, forming clathrin-coated vesicles	Cholesterol uptake
Exocytosis			
Membrane vesicle	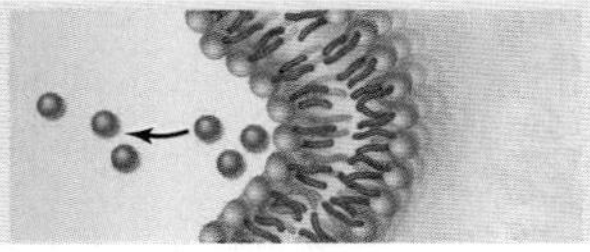	Vesicles fuse with plasma membrane and eject contents	Secretion of mucus; release of neurotransmitters

How Hemorrhagic *E. coli* Resists the Acid Environment of the Stomach

Recent years have been marked by a series of food poisoning outbreaks involving hemorrhagic (producing internal bleeding) strains of the bacterium *Escherichia coli* (*E. coli*). Bacteria are often a source of food poisoning, typically milder infections caused by food-borne streptococcal bacteria. Less able to bear the extremely acidic conditions encountered by food in the human stomach (pH = 2), *E. coli* has not been as common a problem. The hemorrhagic strains of *E. coli* responsible for recent outbreaks seem to have evolved more elaborate acid-resistance systems.

How do hemorrhagic *E. coli* bacteria survive in the acid environment of the stomach? The problem they face, in essence, is that they are submerged in a sea of hydrogen ions, many of which diffuse into their cells. To rid themselves of these excess hydrogen ions, the *E. coli* use a clever system to pump hydrogen ions back out of their cells.

First, the hemorrhagic *E. coli* cells take up cellular hydrogen ions by using the enzyme glutamic acid decarboxylase (GAD) to convert the amino acid glutamate to γ-aminobutyric acid (GABA), a decarboxylation reaction that consumes a hydrogen ion.

Second, the hemorrhagic *E. coli* export this GABA from their cell cytoplasm using a Glu-GABA antiporter called GadC (this transmembrane protein channel is called an antiporter because it transports two molecules across the membrane in opposite directions).

However, to survive elsewhere in the human body, it is important that the Glu-GABA antiporter of hemorrhagic *E. coli not* function, lest it short-circuit metabolism. To evaluate if the GadC antiporter indeed functions only in acid environments, investigators compared its activity at a variety of pHs with that of a different amino acid antiporter called AdiC, which transports arginine out of cells under a broad range of conditions. The results of monitoring transport for 10 minutes are presented in the graph.

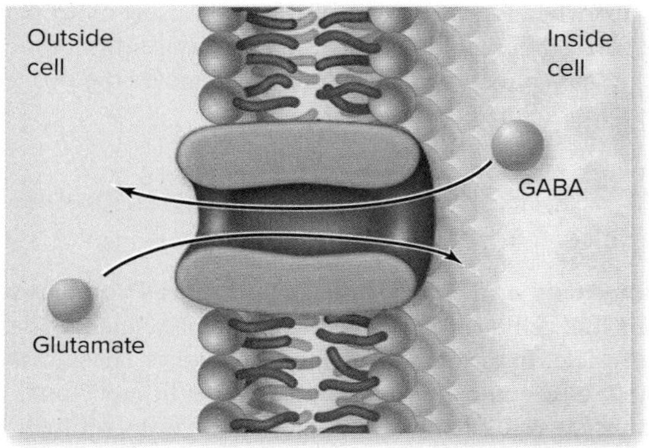

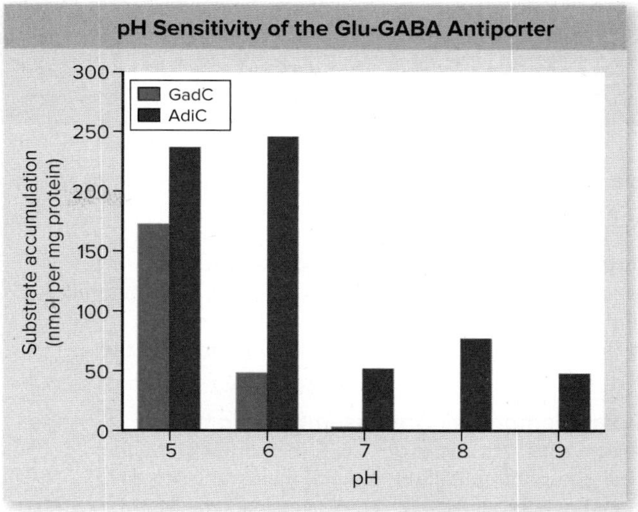

Analysis

1. **Applying Concepts**
   a. ***Variable.*** In the graph, what is the dependent variable?
   b. ***Substrate.*** What is a substrate? In this investigation, what are the substrates that are accumulating?
   c. ***pH.*** What is the difference in hydrogen ion concentration between pH 5 and pH 7? How many times more (or less) is that? Explain.
2. **Interpreting Data**
   a. Does the amount of amino acid transported in the 10-minute experimental interval (expressed as substrate accumulation) vary with pH for the arginine-transporting AdiC antiporter? For the glutamate-transporting GadC antiporter?
   b. Compare the amount of substrate accumulated by AdiC in 10 minutes at pH 9.0 with that accumulated at pH 5.0. What fraction of the low pH activity is observed at the higher pH?
   c. In a similar fashion, compare the amount of substrate accumulated by GadC at pH 9.0 with that accumulated at pH 5.0. What fraction of the low pH activity is observed at the higher pH?
3. **Making Inferences** Would you say that the GadC antiporter exhibits the same pH dependence as the AdiC antiporter? If not, which antiporter is less active at nonacid pHs?
4. **Drawing Conclusions** Is the glutamate-GABA antiporter GadC active at nonacid pHs?
5. **Further Analysis** The GadC antiporter also transports the amino acid glutamine (Gln). Do you think this activity has any role to play in combating low pH environments? How would you test this hypothesis?

Retracing the Learning Path

CONCEPT 5.1 Membranes Are Phospholipid Bilayers with Embedded Proteins

5.1.1 Biological Membranes Are Fluid Mosaics
Membranes are sheets of phospholipid bilayers with hydrophobic regions oriented inward and hydrophilic regions oriented outward. In the fluid mosaic model, proteins float on or in the lipid bilayer.

5.1.2 Cellular Membranes Are Assembled from Four Major Components In eukaryotic cells, membranes have four components: a phospholipid bilayer, integral membrane proteins, peripheral membrane proteins, and cell-surface markers. Many integral proteins are also transmembrane proteins. Peripheral membrane proteins form an interior protein network connected to cytoskeletal filaments. Membranes contain glycoproteins and glycolipids on the surface that act as cell identity markers.

CONCEPT 5.2 Phospholipids Provide a Membrane's Structural Foundation

5.2.1 The Lipid Bilayer Forms Spontaneously
Phospholipids are composed of two fatty acids and a phosphate group linked to a 3-carbon glycerol molecule. The phosphate group is polar and hydrophilic; the fatty acids are nonpolar and hydrophobic, and they orient away from the hydrophilic environment. The nonpolar interior of the lipid bilayer impedes the passage of water and water-soluble substances. The phospholipid bilayer is fluid. Hydrogen bonding of water keeps the membrane in its bilayer configuration; however, phospholipids and unanchored proteins can diffuse laterally. Membrane fluidity can change and depends on the fatty acid composition of the membrane. Unsaturated fats tend to make the membrane more fluid. Temperature also affects fluidity.

CONCEPT 5.3 Membrane Proteins Enable a Broad Range of Interactions with the Environment

5.3.1 Membrane Proteins Have Many Functions
Transporters are integral membrane proteins that carry specific substances through the membrane. Enzymes often occur on the interior surface of the membrane. Cell-surface receptors respond to external chemical messages and change conditions inside the cell; cell identity markers on the surface allow recognition of the body's cells as "self." Cell-to-cell adhesion proteins glue cells together; surface proteins that interact with other cells anchor to the cytoskeleton.

5.3.2 Transmembrane Domains Contain Nonpolar Amino Acids Surface proteins are attached to the surface by nonpolar regions that associate with nonpolar regions of phospholipids. Transmembrane proteins may cross the bilayer a number of times, and each membrane-spanning region is called a transmembrane domain. Such a domain is composed of hydrophobic amino acids usually arranged in α helices. In porins and certain other proteins, β-pleated sheets in the nonpolar region form a pipelike passageway having a polar environment.

CONCEPT 5.4 Passive Transport Moves Molecules Across Membranes by Diffusion

5.4.1 Diffusion Is the Result of Random Molecular Motion Simple diffusion is the passive movement of a substance along a concentration gradient. Biological membranes pose a barrier to hydrophilic polar molecules, while they allow hydrophobic substances to diffuse freely.

5.4.2 Facilitated Diffusion Utilizes Specific Carrier Proteins and Ion Channels Ions and large hydrophilic molecules cross the phospholipid bilayer, with the help of proteins, in facilitated diffusion. These proteins can be channels or carriers. Channels are specific for different ions and allow diffusion based on concentration or electrical gradients across the membrane. Carrier proteins bind to the molecules they transport, much as enzymes bind substrates. The rate of transport by a carrier is limited by the number of carriers in the membrane.

5.4.3 Osmosis Is the Movement of Water Across Membranes The direction of movement due to osmosis depends on the solute concentration on either side of the membrane.

5.4.4 Osmosis Can Generate Significant Pressure
Solutions can be isotonic, hypotonic, or hypertonic. Cells in an isotonic solution are in osmotic balance; cells in a hypotonic solution will gain water; and cells in a hypertonic solution will lose water. Aquaporins are water channels that facilitate the diffusion of water.

CONCEPT 5.5 Active Transport Across Membranes Requires Energy

5.5.1 Active Transport Utilizes Specific Carrier Proteins
Active transport uses specialized protein carriers that couple a source of energy to transport. Uniporters transport a specific molecule in one direction; symporters transport two molecules in the same direction; and antiporters transport two molecules in opposite directions. The sodium–potassium pump moves three Na^+ out of the cell and two K^+ into the cell against their concentration gradients using ATP. This pump appears to be almost universal in animal cells.

5.5.2 Coupled Transport Uses Ion Gradients to Move Molecules Against Their Concentration Gradients Coupled transport occurs when the energy released by a diffusing molecule is used to transport a different molecule against its concentration gradient in the same direction. Countertransport moves the two molecules in opposite directions.

CONCEPT 5.6 Bulky Materials Cross Membranes Within Vesicles

5.6.1 Endocytosis and Exocytosis Are Inverse Processes
In endocytosis, the cell membrane surrounds material and pinches off to form a vesicle. In receptor-mediated endocytosis, specific molecules bind to receptors on the cell membrane. In exocytosis, material in a vesicle is discharged when the vesicle fuses with the membrane.

Concept Overview

This Concept Overview diagrams the key concepts that were discussed in this chapter.

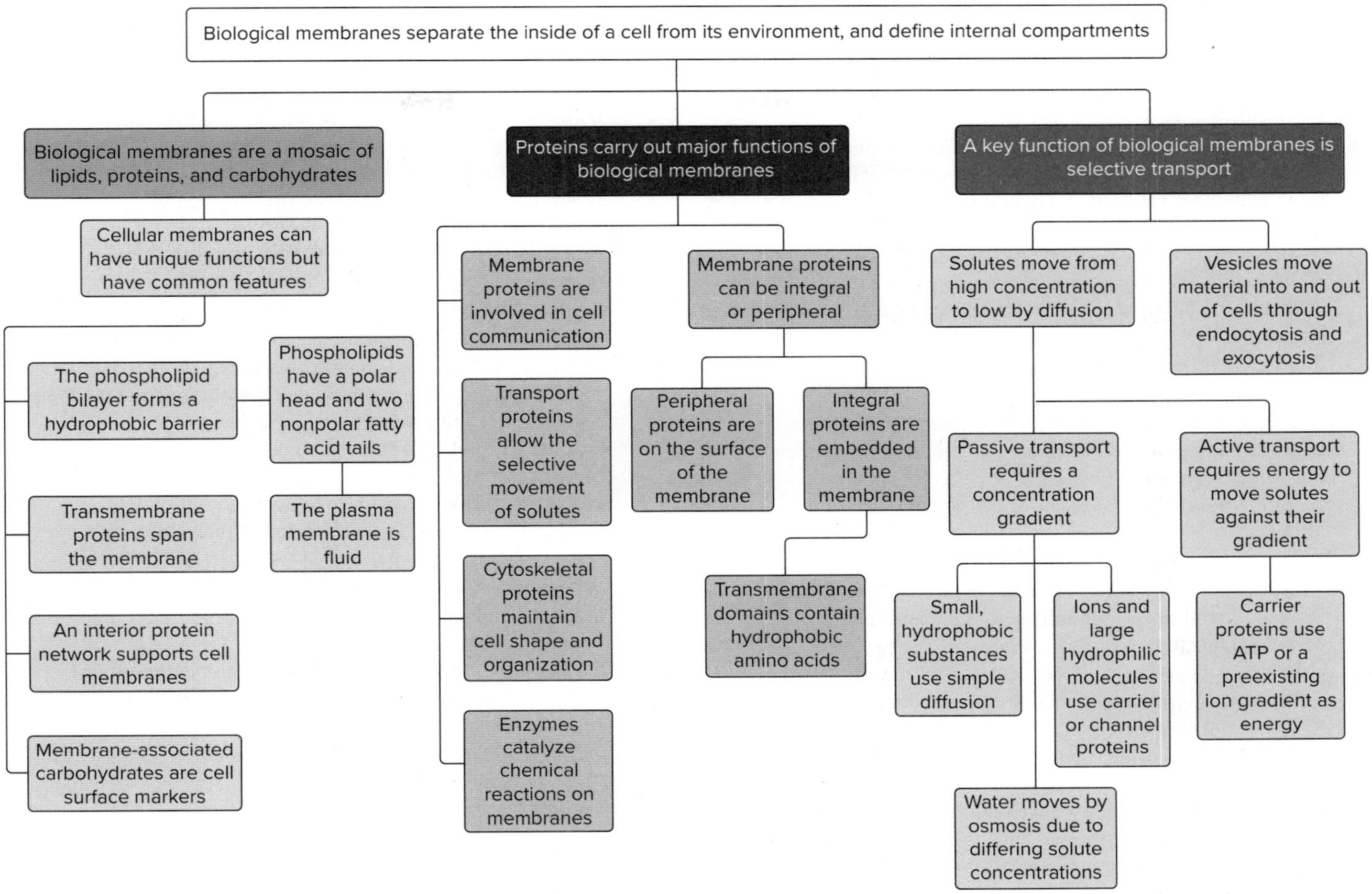

Assessing the Learning Path

Understand

1. Which of the following components is NOT typically part of a plasma membrane?
   a. Phospholipids
   b. Cholesterol
   c. Glycoproteins
   d. Cellulose
2. Which of the following statements about biological membranes are true? (Select all that apply.)
   a. Hydrophobic tails of phospholipids face toward each other; hydrophilic heads face out.
   b. Phospholipids can move laterally in their half of the bilayer.
   c. The carbohydrate group attached to membrane glycoproteins faces the cytosol.
   d. They are selectively permeable.
3. Which of the following statements about membrane phospholipids are true? (Select all that apply.)
   a. They make the membrane fluid.
   b. They spontaneously associate in water to form bilayers.
   c. They flip readily from one face of the bilayer to the other.
   d. They have hydrophilic heads.
4. Which of the following is NOT a key functional class of membrane protein?
   a. Membrane-anchored enzymes
   b. Storage proteins
   c. Cell identity markers
   d. Cytoskeleton anchors
5. What is unique about porins, compared with most other integral membrane proteins?
   a. They have a β-barrel motif, which allows them to make water tunnels in membranes.
   b. Their transmembrane domains are composed largely of α helices.
   c. Their transmembrane domains contain a high percentage of hydrophilic amino acids.
   d. They have a single transmembrane domain, which loosely anchors them into the membrane.
6. The hydrophobic interior of the phospholipid bilayer
   a. results in a membrane that is selectively permeable.
   b. allows nonpolar molecules free entrance into and exit from the cell.

c. rejects the movement of gases across the membrane.
d. ensures that nothing can get into or out of the cell, allowing it to maintain homeostasis.

7. Facilitated diffusion
a. moves molecules against their concentration gradients.
b. requires transmembrane proteins.
c. requires ATP.
d. All of the above

8. If you place an animal cell into a hypotonic solution, it will swell until it pops. However, if you put a plant cell into the same hypotonic solution, it will almost never pop. Why not?
a. The plant cell wall can withstand considerable hydrostatic pressure.
b. The cell wall blocks osmosis.
c. Plant and animal cells respond oppositely to hypotonic solutions.
d. Plant cells produce molecules that make their cytoplasm isotonic to the solution.

9. In coupled transport, molecules are moved against their concentration gradient. What is the direct energy source used in coupled transport?
a. Enzymes
b. ATP
c. A chemical gradient
d. Heat

10. The sodium–potassium pump
a. works through a series of conformational changes to move sodium and potassium ions across a membrane.
b. is a symporter.
c. moves sodium down its concentration gradient and potassium against its concentration gradient.
d. All of the above

11. Which of the following statements about bulk transport is ACCURATE?
a. Pinocytosis selectively brings dissolved ions into cells that need them.
b. Endocytosis is common in prokaryotic cells but is rarely observed in eukaryotic cells.
c. Proteins made in the endomembrane system are released from the cell by exocytosis.
d. Bulk transport is a passive process, as it does not require energy.

Apply

1. An animal cell is missing an enzyme that attaches sugar molecules to membrane proteins. Which of the following will be a result?
a. The cell may not be recognized by other cells with which it normally interacts.
b. The cell will be unable to form correct membrane-to-cytoskeleton connections.
c. There will likely be no effect, as sugar groups are purely decorative.
d. A normal cell wall will not be produced.

2. A bacterial cell that can alter the composition of saturated and unsaturated fatty acids in its membrane lipids is adapted to a cold environment. If this cell is moved to a warmer environment, it will react by
a. increasing the amount of cholesterol in its membrane.
b. altering the amount of protein present in the membrane.
c. increasing the degree of saturated fatty acids in its membrane.
d. increasing the degree of unsaturated fatty acids in its membrane.

3. EGFR is an integral membrane protein found in some human cell types. When an EGF molecule binds to the extracellular EGFR, it transmits a signal to the interior of the cell, telling it to divide. EGFR is
a. a cell-surface receptor.
b. a cell–cell adhesion protein.
c. a transporter.
d. a cell-surface identity marker.

4. Glut5 transports the disaccharide fructose down its concentration gradient across the plasma membranes of cells lining the small intestine. Glut5 undergoes a conformation change during the transport process. Which of the following describes the movement of fructose?
a. Simple diffusion
b. Facilitated diffusion via a carrier protein
c. Facilitated diffusion via a channel protein
d. Response to a membrane potential

5. Which of the following is an example of active transport?
a. The cystic fibrosis transmembrane regulator (CFTR) moves chloride ions across cell membranes against their concentration gradient.
b. Folate moves into cells through a carrier protein, down its concentration gradient.
c. Nitrous oxide (a gas) moves into and out of cells.
d. Water is lost from a cell in a hypertonic solution.

6. A genetically engineered mouse has a defective LDL receptor. Which of the following statements is ACCURATE?
a. The mouse will not be able to transport LDL cholesterol into its cells.
b. The mouse will have high levels of blood cholesterol.
c. The mouse will have an abnormal lipid composition in its membranes.
d. All of the above

Synthesize

1. Early models of membrane structure viewed the plasma membrane as a sandwich of a phospholipid bilayer between two layers of globular protein. Although initially accepted as a working idea, this model is fundamentally at odds with what we know about proteins. Why doesn't this model work?

2. The distribution of lipids in the ER membrane is symmetrical; that is, it is the same in both layers of the membrane. The Golgi apparatus and plasma membrane do not have symmetrical distribution of membrane lipids. What kinds of processes could achieve this outcome?

3. Membrane proteins that interact with molecules in the external environment are often anchored to the cytoskeleton. If the nonpolar segments of the membrane protein are firmly anchored within the lipid bilayer, how are interactions with external molecules communicated to the cytoskeleton?

4. Why is the lipid bilayer of a cell freely permeable to water, which is quite polar, but not freely permeable to ammonia, which is also polar and about the same molecular size?

5. Active transport allows cells to maintain higher concentrations of many different molecules than found in the cell's surroundings. You might then expect to find many different transmembrane channels carrying out active transport. However, almost all active transport is carried out by only two such channels, the sodium–potassium pump and the proton pump. Why do you suppose cells couple so many transport processes to these two channels, rather than using a variety of uniporters?

6. Exocytosis is an important process in plants. How do you explain this importance, given that plant cells are encased in thick and rigid cell walls?

Robert Caputo/Aurora Photos/Cavan images

6 Energy and Metabolism

Learning Path

6.1 Energy Flows Through Living Systems

6.2 The Laws of Thermodynamics Govern All Energy Changes

6.3 ATP Is the Energy Currency of Cells

6.4 Enzymes Speed Up Reactions by Lowering Activation Energy

6.5 Metabolism Is the Sum of a Cell's Chemical Activities

Concept Overview

This Concept Overview provides a diagram of the overarching concepts that are covered in this chapter. These concepts will be expanded upon in the Concept Overview at the end of the chapter.

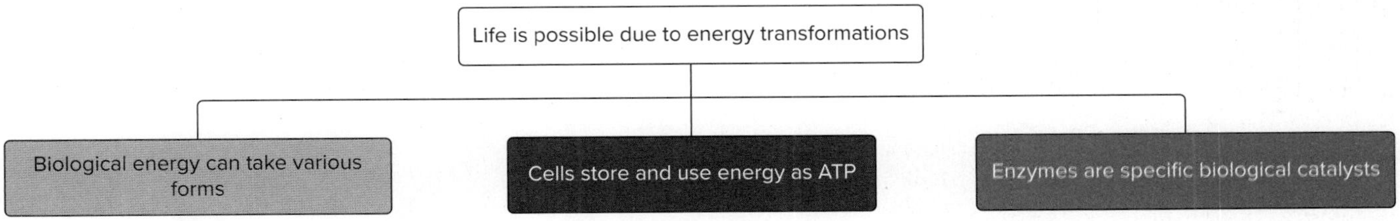
Life is possible due to energy transformations
Biological energy can take various forms
Cells store and use energy as ATP
Enzymes are specific biological catalysts

Introduction

Life requires energy. This sounds simple, but it hides an enormous amount of chemical detail. Life is made possible by the continual transformation of energy during all of the activities that make up living. Each of the significant properties by which we define life—order, growth, reproduction, responsiveness, and internal regulation—requires a constant supply of energy. The energy that the lion extracts from its meal of a giraffe will be used to run its cells, power its roar, fuel its running, and build a bigger lion. Deprived of a source of energy, life stops. Therefore, a comprehensive study of life would be impossible without discussing *bioenergetics,* the analysis of how energy powers the activities of living systems. In this chapter, we focus on energy—what it is and how it changes during chemical reactions. In chapters 7 and 8, we will examine how organisms capture, store, and use energy.

6.1 Energy Flows Through Living Systems

Energy is a word that we commonly use without thinking about its meaning. The statements "I don't have much energy today" and "The world is facing an energy crisis" use the word in related but different ways. In fact, **energy** has a precise definition: the capacity to do work.

Energy May Be Stored or Used to Do Work

LEARNING OBJECTIVE 6.1.1 Differentiate between kinetic and potential energy.

We think of energy as existing in two states: kinetic energy and potential energy (figure 6.1). **Kinetic energy** is the energy of motion. Moving objects perform work by causing other matter to move. **Potential energy** is stored energy. Objects that are not actively moving but have the capacity to do so possess potential energy. A boulder perched on a hilltop has gravitational potential energy. As it begins to roll downhill, some of its potential energy is converted into kinetic energy. Much of the work that living organisms carry out involves transforming potential energy into kinetic energy.

Energy can take many forms: mechanical energy, heat, sound, electric current, light, or radioactivity. Because it can exist in so many forms, energy can be measured in many ways. Heat is the most convenient way of measuring energy, because all other forms of energy can be converted into heat. In fact, the term *thermodynamics* means "heat changes."

The unit of heat most commonly employed in biology is the kilocalorie (kcal). One kilocalorie is equal to 1000 calories (cal). One calorie is the heat required to raise the temperature of 1 gram of water 1 degree Celsius (°C). (You are probably more used to seeing the term *Calorie* with a capital C. This is used on food labels and is actually the same as a kilocalorie.) Another energy unit, often used in physics, is the *joule;* 1 joule equals 0.239 cal.

Energy flows into most ecosystems from the Sun. It is estimated that sunlight provides the Earth with more than 13×10^{23} calories per year, or 40 million billion calories per second! Plants, algae, and certain kinds of bacteria capture a fraction of this energy through photosynthesis. There are also ecosystems where bacteria extract energy from the oxidation of inorganic compounds.

In chemical reactions, breaking chemical bonds requires energy, and forming covalent bonds releases energy. If a reaction forms a more ordered, higher energy state, as photosynthesis does, an input of energy is required. In photosynthesis, energy absorbed from sunlight is used to combine small molecules (water and carbon dioxide) into more complex ones (sugars). In the process, energy from sunlight is stored as potential energy in the structure of the sugar molecules.

Oxidation–Reduction Reactions Transfer Energy

LEARNING OBJECTIVE 6.1.2 Differentiate between oxidation and reduction reactions.

During a chemical reaction, the energy stored in chemical bonds may be used to make new bonds. In some of these reactions, electrons pass from one atom or molecule to another. An atom or a molecule that loses an electron is said to be oxidized, and

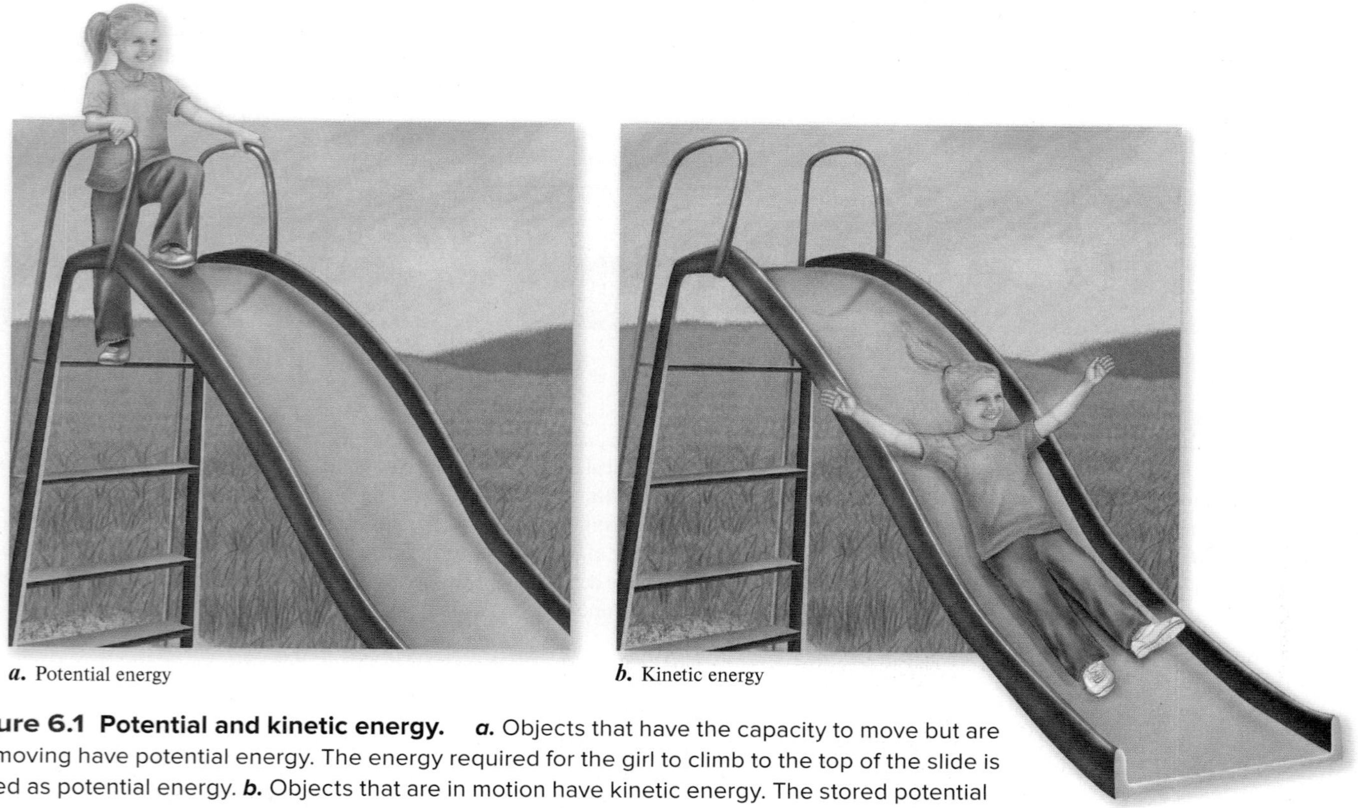

a. Potential energy
b. Kinetic energy

Figure 6.1 Potential and kinetic energy. ***a.*** Objects that have the capacity to move but are not moving have potential energy. The energy required for the girl to climb to the top of the slide is stored as potential energy. ***b.*** Objects that are in motion have kinetic energy. The stored potential energy is released as kinetic energy as the girl slides down.

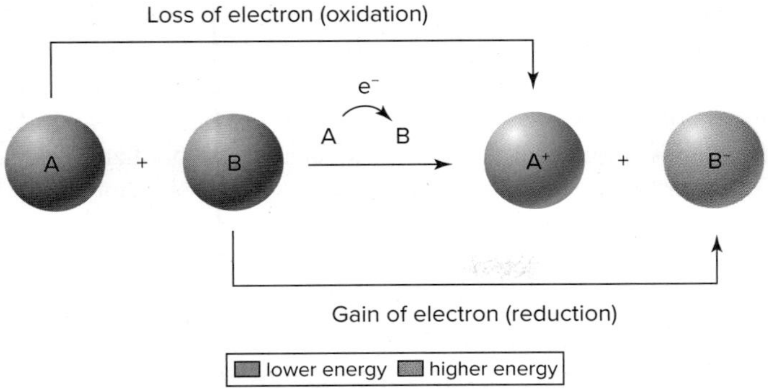
Loss of electron (oxidation)
e⁻
A
B
A
B
A⁺
B⁻
Gain of electron (reduction)
lower energy
higher energy

Figure 6.2 Oxidation–reduction. Oxidation is the loss of an electron; reduction is the gain of an electron. In this example, the charges of molecules A and B appear as superscripts in each molecule. Molecule A loses energy as it loses an electron, and molecule B gains that energy as it gains an electron.

the process by which this occurs is called **oxidation.** Conversely, an atom or a molecule that gains an electron is said to be reduced, and the process is called **reduction.** The reduced form of a molecule has a higher level of potential energy than the oxidized form (figure 6.2).

Oxidation and reduction always take place together, because every electron that is lost by one atom through oxidation is gained by another atom through reduction. Therefore, chemical reactions of this sort are called **oxidation–reduction,** or **redox, reactions.** Redox reactions transfer electrons when bonds are made or broken, and with these electrons the potential energy that the electrons hold. It is for this reason that the reduced form of the molecule has a higher level of energy than the oxidized form.

Oxidation–reduction reactions play a key role in the flow of energy through biological systems. In chapters 7 and 8, you will learn the details of how organisms derive energy from the oxidation of organic compounds via respiration, as well as from the energy in sunlight via photosynthesis.

REVIEW OF CONCEPT 6.1

Energy is defined as the capacity to do work. Energy is either stored (potential) or energy of motion (kinetic). The Sun is the ultimate source of energy for living systems. Organisms derive energy from oxidation–reduction reactions. Oxidation is the loss of electrons; reduction is the gain of electrons.

- *What energy source might ecosystems at the bottom of the ocean use?*

6.2 The Laws of Thermodynamics Govern All Energy Changes

Thermodynamics is the branch of chemistry concerned with energy changes. Cells are governed by the laws of physics and chemistry, so we must understand these laws in order to understand how cells function. All activities of living organisms—growing, running, thinking, singing, reading these words—involve changes in energy. Two universal laws, which we call the laws of thermodynamics, govern all energy changes in the universe, from a nuclear reaction to a bird flying through the air.

The First Law States That Energy Cannot Be Created or Destroyed

LEARNING OBJECTIVE 6.2.1 Define thermodynamics, and state the First Law of Thermodynamics.

The **First Law of Thermodynamics** can be stated simply: The total energy of the universe is constant. In any system, energy lost by the system is gained by the surroundings, and energy gained by the system is lost by the surroundings. Thus the energy of a system plus that of its surroundings is constant. Energy can be changed in form (from potential to kinetic, for example), but cannot be created or destroyed.

This means that living systems cannot create the energy needed for life, but must acquire it in some way. The lion eating a giraffe in this chapter's opening photo is acquiring energy. The lion is transferring some of the potential energy stored in the giraffe's tissues to its own body, just as the giraffe obtained the potential energy stored in the plants it ate while it was alive, and those plants captured energy from sunlight.

Within any living organism, chemical potential energy stored in some molecules can be shifted to other molecules and stored in different chemical bonds. It can also be converted into other forms, such as kinetic energy, light, or electricity. During each conversion, some of the energy dissipates into the environment as **heat,** which is a measure of the random motion of molecules (and therefore a measure of one form of kinetic energy). Energy continuously flows through the biological world in one direction, with new energy from the Sun constantly entering the system to replace the energy dissipated as heat.

Heat can be harnessed to do work only when there is a heat gradient—that is, a temperature difference between two areas. Cells are too small to maintain significant internal temperature differences, so heat energy is incapable of doing the work of cells. Instead, cells must rely on chemical reactions for energy.

Although the total amount of energy in the universe remains constant, the energy available to do work decreases as more of it is progressively lost as heat.

The Second Law States That Energy Transactions Are Not 100% Efficient

LEARNING OBJECTIVE 6.2.2 Define entropy, and state the Second Law of Thermodynamics.

The **Second Law of Thermodynamics** addresses the efficiency of energy transformations. Energy cannot be transformed from one form to another with 100% efficiency; some energy is always unavailable. This unavailable energy manifests as an increase in the random motion of molecules, an increase in the number of energy states available to atoms, or increased dispersal of energy in the system. This is often characterized as an increase in the randomness or disorder of the system. We measure this as an increase in **entropy** (S in section 6.2.3).

Energy transformations proceed spontaneously to convert matter from a more ordered, less stable form to a less ordered, but more stable form. Consider a glass of ice water in a room at 25°C. The water molecules in the ice form an ordered, regular array, but the water molecules in the liquid are more disordered. There are far fewer ways that the molecules can be arranged in the crystal than in the liquid water. Thus, as the ice absorbs heat from the surroundings, the hydrogen bonds holding the crystal together are broken, which increases the number of molecular states each water molecule can occupy. Put simply, the water is less ordered (higher entropy). When all of the ice has melted, the entropy of the water will be maximal for this system (figure 6.3). You can order the images in figure 6.3 based on experience, but the underlying reason is the second law. For this reason, it is sometimes called time's arrow.

Chemical Reactions Can Be Predicted Based on Changes in Free Energy

LEARNING OBJECTIVE 6.2.3 Use the definition of free energy to differentiate between endergonic and exergonic reactions.

It takes energy to break the chemical bonds that hold the atoms in a molecule together. Heat energy, because it increases atomic motion, makes it easier for the atoms to pull apart. For this reason, both chemical bonding and heat have a significant thermodynamic influence on a molecule, the former reducing disorder and the latter increasing it. The net effect, the amount of energy actually available to break and subsequently form other chemical bonds, is called the *free energy* of that molecule.

In a more general sense, **free energy** is defined as the maximum energy available to do work in any system. In a molecule within a cell, where pressure and volume usually do not change, the free energy is denoted by the symbol G (for "Gibbs free energy"), which limits the system being considered to the cell. G is equal to the total energy contained in a molecule's chemical bonds (called **enthalpy** and designated H) reduced by the term (TS), which measures the degree of disorder in the system, where S is the symbol for *entropy* and T is the absolute temperature expressed in the Kelvin scale ($K = °C + 273$):

$$G = H - TS$$

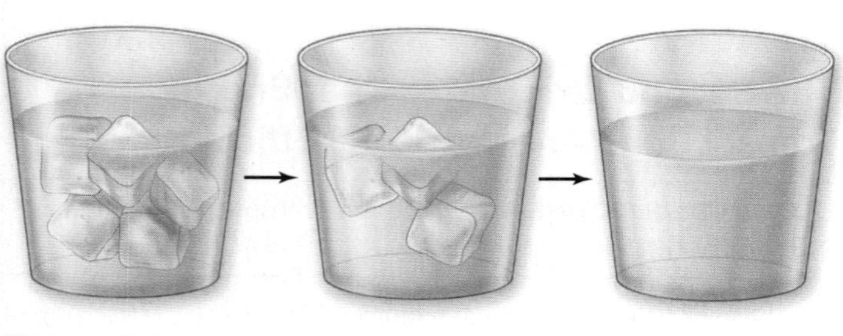

Figure 6.3 Entropy increases as ice melts. The entropy of the water molecules increases as the ice melts. As the ice absorbs heat from the surroundings, the hydrogen bonds holding the crystal together are broken and the water molecules become less ordered (higher entropy). The molecules in the liquid water are continually breaking and re-forming hydrogen bonds.

Chemical reactions may break some bonds in the reactants and form new bonds in the products. Consequently, reactions can produce changes in free energy. When a chemical reaction occurs under conditions of constant temperature, pressure, and volume—as do most biological reactions—the change (symbolized by the Greek capital letter delta, Δ) in free energy (ΔG) is simply

$$\Delta G = \Delta H - T\Delta S$$

We can use the change in free energy, or ΔG, to predict whether a chemical reaction is spontaneous or not. For some reactions, the ΔG is positive, which means that the products of the reaction contain *more* free energy than the reactants; the bond energy (H) is greater, or the disorder (S) in the system is lower. Such reactions do not proceed spontaneously, because they require an input of energy. Any reaction that requires an input of energy is said to be **endergonic** (figure 6.4*a*).

For other reactions, the ΔG is negative. In this case, the products of the reaction contain less free energy than the reactants; either the bond energy is lower or the disorder is greater, or both. Such reactions tend to proceed spontaneously. These reactions release the excess free energy as heat and are thus said to be **exergonic** (figure 6.4*b*). Any chemical reaction tends to proceed spontaneously if

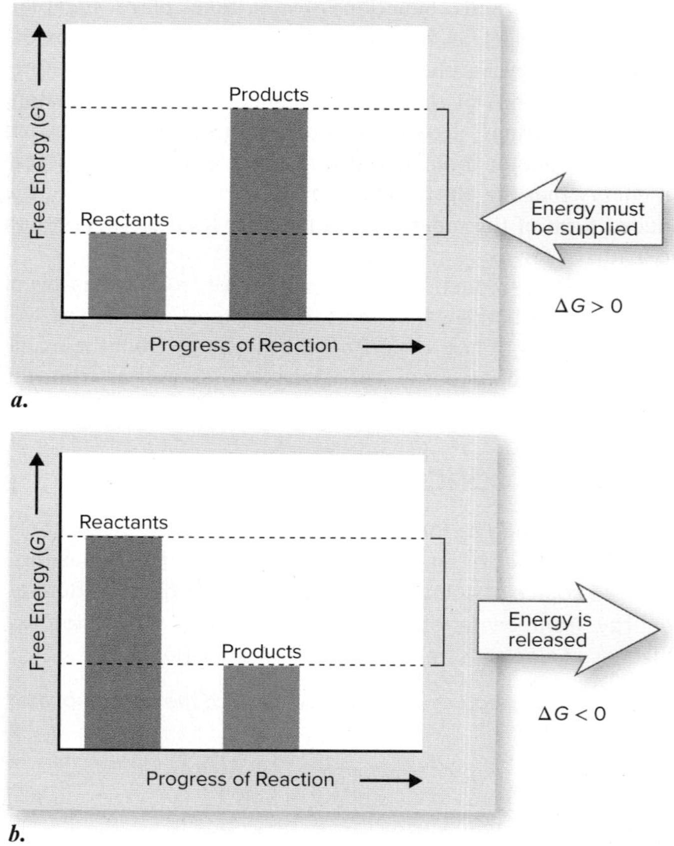

Figure 6.4 Energy in chemical reactions. ***a.*** In an endergonic reaction, the products of the reaction contain more energy than the reactants, and the extra energy must be supplied for the reaction to proceed. ***b.*** In an exergonic reaction, the products contain less energy than the reactants, and the excess energy is released.

the difference in disorder ($T\Delta S$) is *greater* than the difference in bond energies between reactants and products (ΔH).

Because chemical reactions are reversible, a reaction that is exergonic in the forward direction will be endergonic in the reverse direction. For each reaction, an equilibrium exists between the relative amounts of reactants and products. This is expressed quantitatively by the equilibrium constant (K_{eq}), a ratio of the concentrations of products divided by the concentrations of reactants. For exergonic reactions, ΔG will be negative, and K_{eq} will be >1, meaning that equilibrium favors the products. For endergonic reactions, ΔG will be positive, and K_{eq} will be <1, meaning that equilibrium favors the reactants.

The concept of equilibrium is very important in biological systems, but not because they are at equilibrium. In fact, the opposite is true: biological systems exist far from equilibrium, maintained by a constant input of energy.

Spontaneous chemical reactions require activation energy

It is important to recognize that while changes in free energy allow us to predict the direction of a chemical reaction, they do not tell us how fast the reaction will occur. In fact, a spontaneous reaction might not occur for thousands of years. This is the case because for a chemical reaction to occur, there must be an initial input of energy that we call **activation energy** (figure 6.5). Before new chemical bonds can form, even bonds that contain less energy, existing bonds must first be broken, which requires an input of energy. If the activation energy is high, even a spontaneous reaction will occur very slowly. This is why, although the oxidation of carbohydrates is highly exergonic, the sugar in your pantry does not spontaneously explode.

Note that the amount of activation energy needed for a reaction is independent of the change in free energy (figure 6.5). So changes in free energy allow us to predict the direction of a reaction, but not the rate of the reaction. This also means that increasing the rate of a reaction requires either providing the activation energy or lowering the activation energy. We will return to this topic in section 6.4.

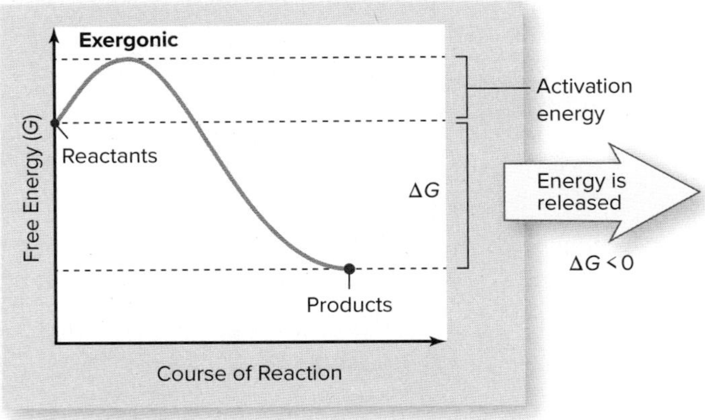

Figure 6.5 Activation energy. Even spontaneous (exergonic) reactions will not occur without an initial input of energy. This is initial energy barrier is called activation energy. This affects how fast a reaction will occur, but not whether or not the reaction is spontaneous.

REVIEW OF CONCEPT 6.2

The First Law of Thermodynamics states that energy cannot be created or destroyed. The Second Law states that the loss of energy results in greater disorder, or entropy. Free-energy changes (ΔG) can predict whether chemical reactions take place. Reactions with a negative ΔG occur spontaneously, and those with a positive ΔG do not. Even spontaneous reactions require activation energy to occur.

- *If entropy in the world always increases, what has prevented the world from becoming totally disordered?*

6.3 ATP Is the Energy Currency of Cells

In chapter 3 we introduced adenosine triphosphate (ATP) as both a precursor for RNA and also the energy currency for the cell. We call ATP the energy currency, because the energy released by ATP hydrolysis can power most endergonic transactions in cells.

Cells Store and Release Energy in the Bonds of ATP

LEARNING OBJECTIVE 6.3.1 Explain how ATP hydrolysis can provide energy for cellular processes.

The process of energy metabolism in cells is complex, as we shall discuss in chapters 7 and 8. Cells have a constant need for energy, so the output of energy metabolism is an intermediate storage molecule that can be made and used cyclically. The molecule selected by evolution to fulfill this role in all cells is ATP. Each ATP molecule is composed of three parts (figure 6.6). A 5-carbon sugar, ribose, serves as the framework, with the other two subunits attached to opposite sides of the ribose ring. These other components are the nitrogenous base adenine and a chain of three phosphates.

How ATP stores energy

The bonds joining the two terminal phosphates to ATP are often called high-energy bonds, but this does not reflect the energy of the actual bonds. This refers to the fact that hydrolysis of each of these terminal phosphates is exergonic, releasing a significant amount of energy that can be used to drive other endergonic cellular processes.

The key to how ATP stores energy lies in the triphosphate group. Each individual phosphate is highly negatively charged, and thus they strongly repel one another. This electrostatic repulsion makes the covalent bonds joining the phosphates unstable. The molecule can be thought of as a "coiled spring," with the phosphates straining away from one another.

The activation energy for ATP hydrolysis is actually quite high, but as reactions using ATP involve enzymes, this is not an issue (refer to section 6.4). It does mean that ATP is relatively stable in cells, but can be hydrolyzed when needed to release energy that can be used by cells. In most reactions involving ATP, only the terminal high-energy phosphate bond is hydrolyzed, cleaving off a single phosphate. This hydrolysis reaction produces the products *adenosine*

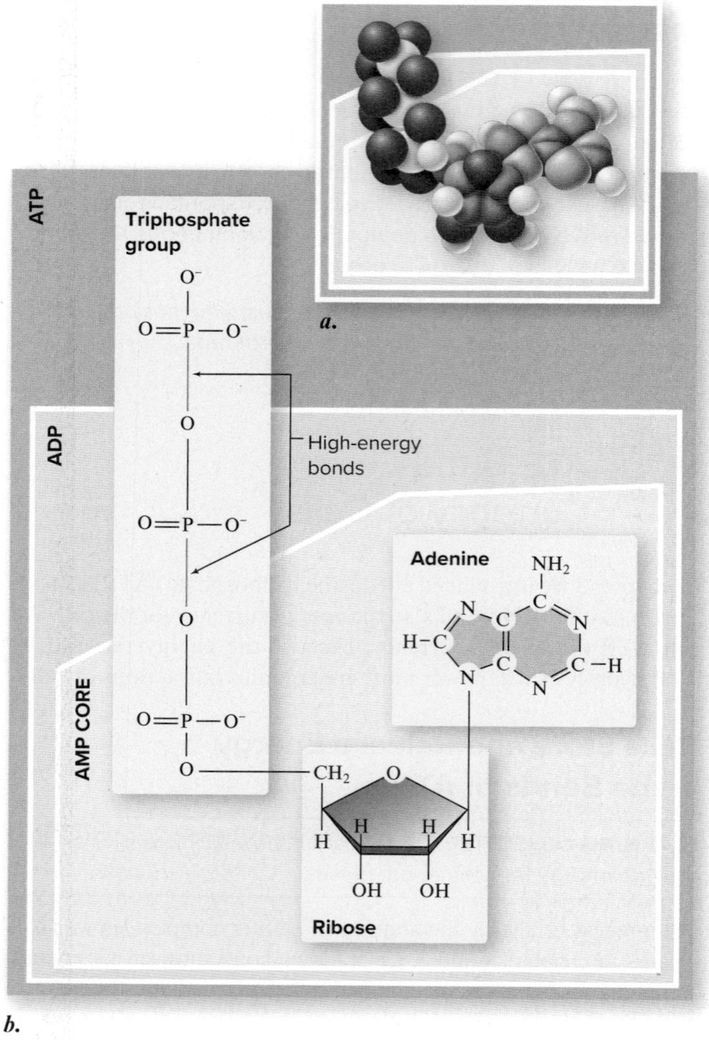

Figure 6.6 The ATP molecule.

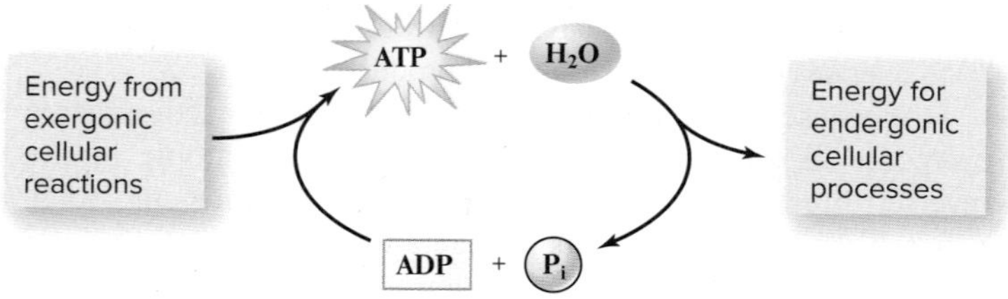

Figure 6.7 The ATP cycle. ATP is synthesized and hydrolyzed in a cyclic fashion.

diphosphate (*ADP*) plus an **inorganic phosphate (P_i),** and under cellular conditions releases 7.3 kcal/mol of energy. The liberated phosphate group is usually transferred to some intermediate molecule, which when dephosphorylated will release the phosphate as P_i.

Both of the two terminal phosphates can be hydrolyzed to release energy, leaving *adenosine monophosphate* (*AMP*); hydrolysis of the remaining phosphate in AMP does not yield significant energy.

ATP hydrolysis can drive endergonic reactions

Endergonic reactions do not proceed spontaneously, because their products possess more free energy than their reactants. However, if ATP hydrolysis releases more energy than a reaction consumes, the coupled reactions will proceed with a net release of energy ($-\Delta G$). Because almost all the endergonic reactions in cells require less energy than is released by ATP hydrolysis, this can provide sufficient energy for most cellular needs.

The use of ATP by cells is a continual cycle (figure 6.7). The synthesis of ATP is endergonic, and cells use energy from exergonic reactions to drive the synthesis of ATP from ADP + P_i (refer to chapter 7). The exergonic hydrolysis of ATP provides energy when and where a cell needs it for any endergonic process (figure 6.8).

Because of this cyclic use, cells do not need to maintain large stores of ATP. Instead, they typically have only a few seconds' supply of ATP at any given time, continually producing more from

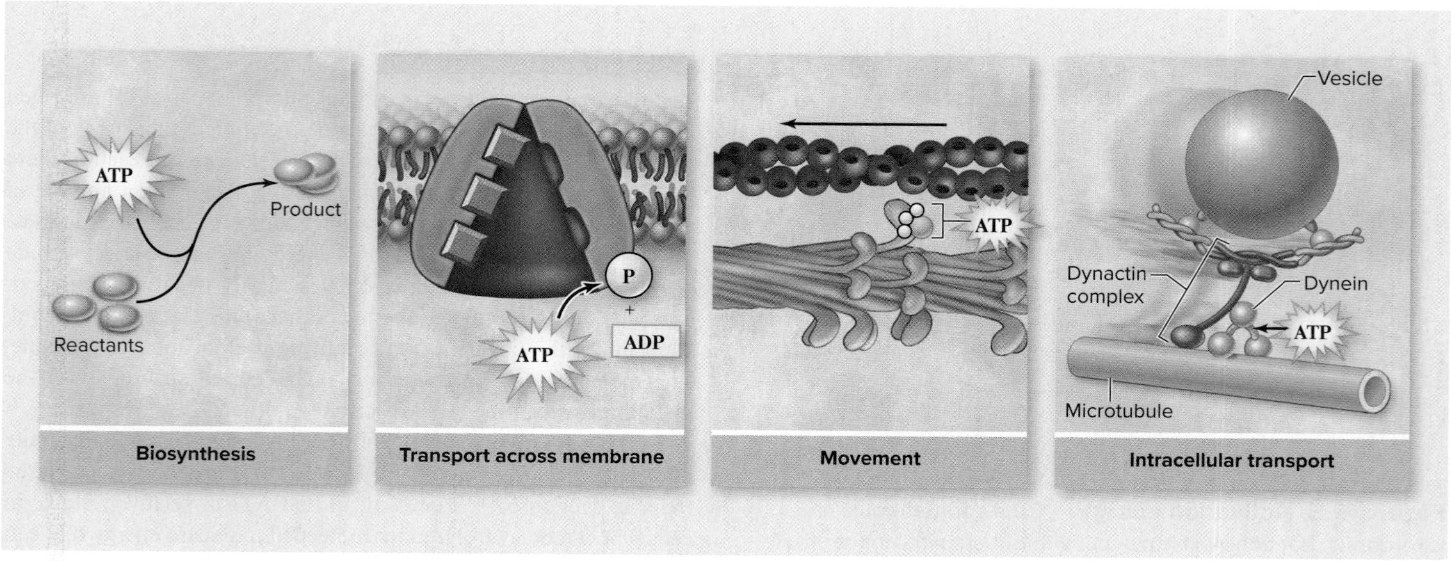

Figure 6.8 How cells use ATP. These examples show the versatility of ATP in powering diverse cellular processes. We will see ATP's role throughout this text.

ADP and P_i. If you obtained new ATP solely from your diet, you would need to consume roughly your body weight in ATP every day!

REVIEW OF CONCEPT 6.3

ATP is a nucleotide with three phosphate groups. Endergonic cellular processes can be driven by coupling to the exergonic hydrolysis of the two terminal phosphates, which act like a coiled spring. Hydrolysis of the terminal phosphate groups releases energy a cell can use.

- *If the molecular weight of ATP is 507.18 g/mol, and the ΔG for hydrolysis is −7.3 kcal/mol, how much energy is released over the course of the day by a 100-kg man?*

6.4 Enzymes Speed Up Reactions by Lowering Activation Energy

We have learned that free energy allows us to predict the direction of chemical reactions, but not the rate of the reaction. Biological systems need a way to increase the rate of reactions that occur very slowly.

Catalysts Act By Lowering Activation Energy

LEARNING OBJECTIVE 6.4.1 Explain how catalysts increase the rate of chemical reactions.

The rate of an exergonic reaction depends on the activation energy required for the reaction to begin. Reactions with larger activation energies tend to proceed more slowly, because fewer molecules succeed in getting over the initial energy hurdle. The rate of reactions can be increased in two ways: (1) by increasing the energy of reacting molecules or (2) by lowering activation energy. A **catalyst** is a substance that increases the rate of a reaction without being altered by the reaction. Catalysts achieve this by lowering the activation energy for a reaction. Chemists often drive important industrial reactions by increasing the energy of the reacting molecules by simply heating up the reactants. The other strategy is to use a catalyst to lower the activation energy (figure 6.9).

How catalysts work

Activation energies are not constant. Stressing particular chemical bonds can make them easier to break. The process of influencing chemical bonds in a way that lowers the activation energy needed to initiate a reaction is called **catalysis,** and substances that accomplish this are known as *catalysts* (figure 6.9).

Catalysts cannot violate the basic laws of thermodynamics; they cannot, for example, make an endergonic reaction proceed spontaneously. By reducing the activation energy, a catalyst accelerates both the forward and the reverse reactions by exactly the same amount. Therefore, a catalyst does not alter the proportion of reactant that is ultimately converted into product.

To understand this, imagine a bowling ball resting in a shallow depression on the side of a hill. Only a narrow rim of dirt below the ball prevents it from rolling down the hill. Now imagine digging away that rim of dirt. If you remove enough dirt from below the ball, it will start to roll down the hill—but removing dirt from below the ball will *never* cause the ball to roll up the hill. Removing the lip of dirt simply allows the ball to move freely; gravity determines the direction it then travels.

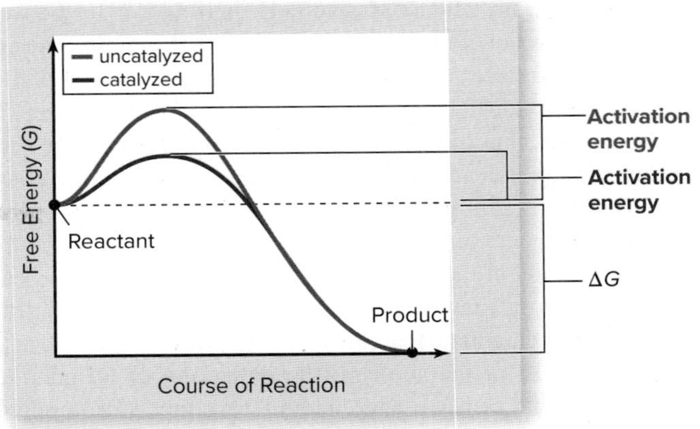

Figure 6.9 Activation energy and catalysis. Exergonic reactions do not necessarily proceed rapidly, because activation energy must be supplied to destabilize existing chemical bonds. Catalysts accelerate particular reactions by lowering the amount of activation energy required to initiate the reaction. Catalysts do not alter the free-energy change produced by the reaction.

Similarly, the direction in which a chemical reaction proceeds is determined solely by the difference in free energy between reactants and products. Like digging away the soil below the bowling ball on the hill, catalysts reduce the energy barrier that is preventing the reaction from proceeding. Only exergonic reactions can proceed spontaneously, and catalysts cannot change that. What catalysts *can* do is make a reaction proceed much faster.

Enzymes Are Specific Catalysts

LEARNING OBJECTIVE 6.4.2 Explain how enzymes lower activation energies, as well as the consequences of doing so.

The catalysts active in biological systems are **enzymes.** Like all catalysts, they accelerate chemical reactions, but are not consumed by the reaction. The most important features of enzymes are their specificity and their efficiency. Every enzyme can recognize a molecule, which we call the substrate, and catalyze a specific reaction. The great efficiency is highlighted by the fact that some enzymes can allow a reaction to occur a million times faster than it would with no enzyme. Most enzymes are proteins, although increasing evidence indicates that some enzymes are actually RNA molecules, as discussed later in this section.

Enzymes have great specificity because they have a unique three-dimensional shape. This enables an enzyme to stabilize a temporary association between **substrates**—the molecules that will undergo the reaction. By bringing two substrates together in the correct orientation or by stressing particular chemical bonds of a substrate, an enzyme lowers the activation energy required for new bonds to form. This allows the reaction to proceed much more quickly than it would without the enzyme.

Consider any chemical reaction; there is a state between reactants and products that is called the transition state. The activation energy is necessary to reach the transition state, and allow a reaction to proceed to completion. Enzymes can stabilize this transition state, thus lowering the activation energy and accelerating the reaction. Because an enzyme is not consumed or changed by a reaction, it can act repeatedly; therefore, enzymes are not required in large amounts.

Thousands of different kinds of enzymes are known, each catalyzing one or a few specific chemical reactions. By facilitating particular chemical reactions, the enzymes in a cell determine the course of metabolism—the collection of all chemical reactions—in that cell. We will encounter examples of important enzymes and their roles in cells throughout this book.

Different types of cells contain different sets of enzymes, and this difference contributes to structural and functional variations among cell types. For example, the chemical reactions taking place within a red blood cell differ from those that occur within a nerve cell, in part because different cell types contain different arrays of enzymes.

The Active Sites of Enzymes Conform to Fit the Shapes of Substrates

LEARNING OBJECTIVE 6.4.3 Describe how enzymes bind to their substrates.

Most enzymes are globular proteins with one or more pockets or clefts, called **active sites,** on their surface (figure 6.10).

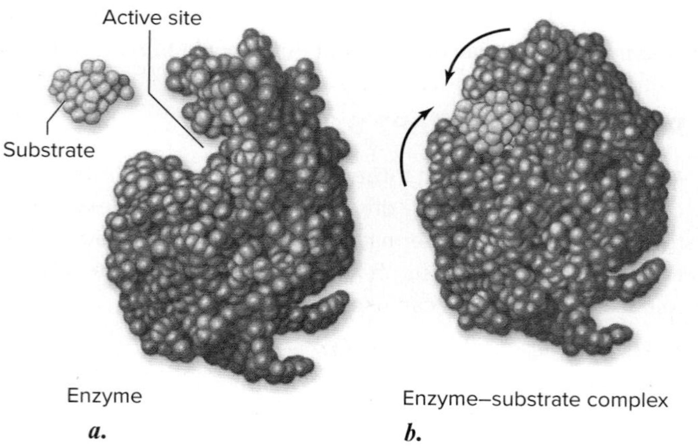
Active site
Substrate
Enzyme
a.
Enzyme–substrate complex
b.

Figure 6.11 Enzyme binding its substrate. ***a.*** The active site of the enzyme lysozyme fits the shape of its substrate, a peptidoglycan that makes up bacterial cell walls. ***b.*** When the substrate, indicated in yellow, slides into the groove of the active site, the protein is induced to alter its shape slightly and bind the substrate more tightly. This alteration of the shape of the enzyme to better fit the substrate is called induced fit.

Substrates bind to the enzyme at these active sites, forming an **enzyme-substrate complex** (figure 6.11). For catalysis to occur within the complex, a substrate molecule must fit precisely into an active site. When that happens, amino acid side groups of the enzyme end up very close to certain bonds of the substrate. These side groups interact chemically with the substrate, usually

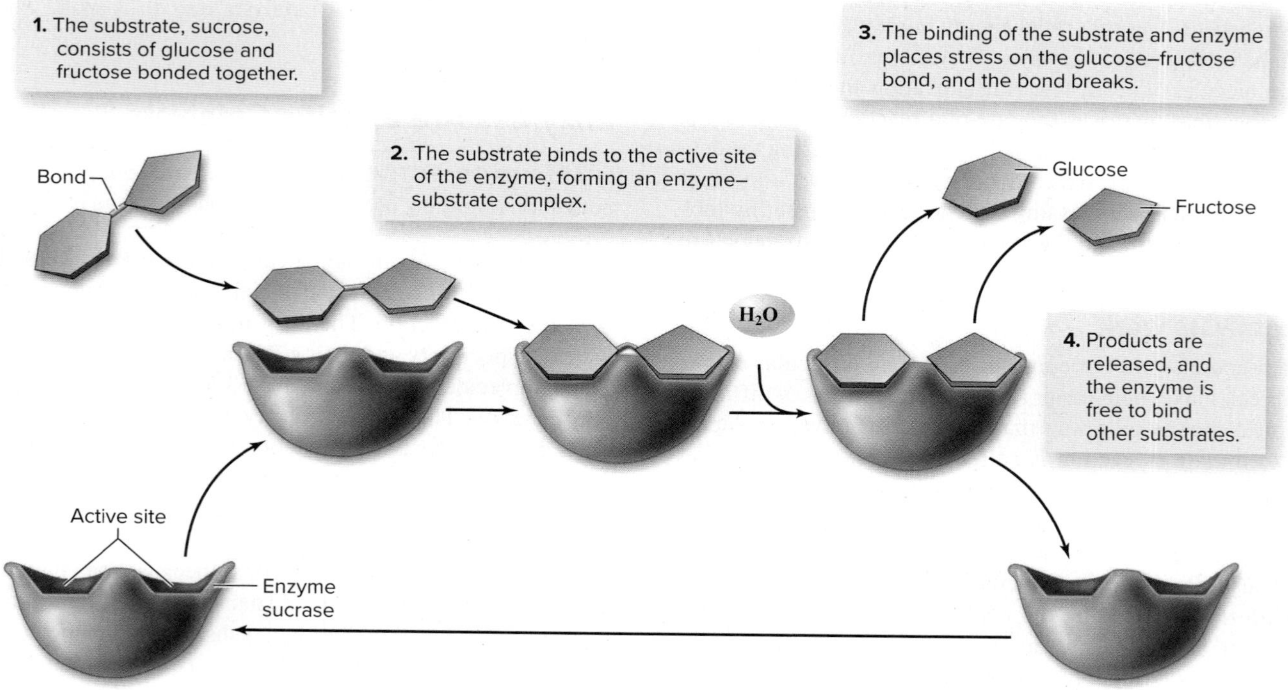
1. The substrate, sucrose, consists of glucose and fructose bonded together.
Bond
2. The substrate binds to the active site of the enzyme, forming an enzyme–substrate complex.
H_2O
3. The binding of the substrate and enzyme places stress on the glucose–fructose bond, and the bond breaks.
Glucose
Fructose
4. Products are released, and the enzyme is free to bind other substrates.
Active site
Enzyme sucrase

Figure 6.10 The catalytic cycle of an enzyme. Enzymes increase the speed at which chemical reactions occur, but they are not altered permanently themselves as they do so. In the reaction illustrated here, the enzyme sucrase is splitting the sugar sucrose into two simpler sugars: glucose and fructose.

stressing or distorting a particular bond and consequently lowering the activation energy needed to break the bond. After the bonds of the substrates are broken, or new bonds are formed, the substrates have been converted to products. These products then dissociate from the enzyme, leaving the enzyme ready to bind its next substrate and begin the cycle again.

Proteins are not rigid. The binding of a substrate induces the enzyme to adjust its shape slightly, leading to a better *induced fit* between enzyme and substrate (figure 6.11). This interaction may also facilitate the binding of other substrates; in such cases, one substrate "activates" the enzyme to receive other substrates.

Enzymes Occur in Many Forms

LEARNING OBJECTIVE 6.4.4 Describe the different types of molecules that may act as enzymes.

Although many enzymes are suspended in the cytoplasm of cells, not attached to any structure, other enzymes function as integral parts of cell membranes and organelles. Enzymes may also form associations called *multienzyme complexes* to carry out reaction sequences. And, as mentioned earlier in this section, evidence exists that some enzymes may consist of RNA rather than being only protein.

Multienzyme complexes

Often, several enzymes catalyzing different steps of a sequence of reactions are associated with one another in noncovalently bonded assemblies called **multienzyme complexes.** The bacterial pyruvate dehydrogenase multienzyme complex, shown in figure 6.12, contains enzymes that carry out three sequential reactions in oxidative metabolism. Each complex has multiple copies of each of the three enzymes—60 protein subunits in all. The many subunits work together to form a molecular machine.

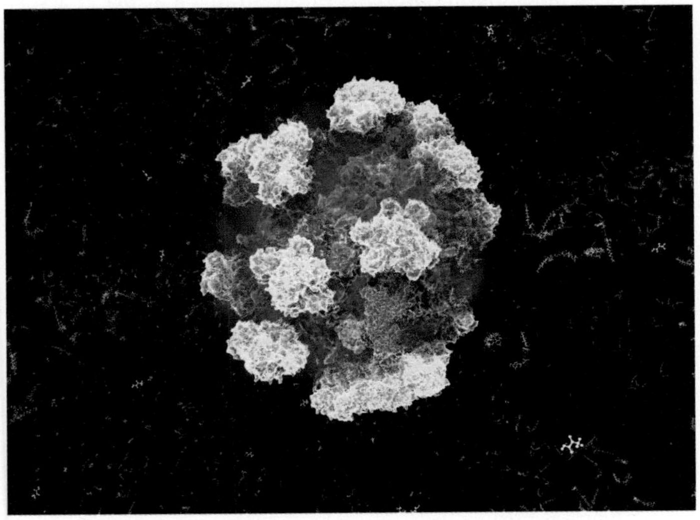

Figure 6.12 The pyruvate dehydrogenase enzyme. Pyruvate dehydrogenase, which catalyzes the oxidation of pyruvate, is a complex multisubunit enzyme. This image is an artistic computer rendering based on available data.
Science Photo Library/Alamy Stock Photo

Multienzyme complexes offer the following significant advantages in catalytic efficiency:

1. The rate of any enzyme reaction is limited by how often the enzyme collides with its substrate. If a series of sequential reactions occurs within a multienzyme complex, the product of one reaction can be delivered to the next enzyme without being released so that it diffuses away.
2. Because the reacting substrate doesn't leave the complex while it goes through the series of reactions, unwanted side reactions are prevented.
3. All of the reactions that take place within the multienzyme complex can be controlled as a unit.

In addition to pyruvate dehydrogenase, which controls entry to the citric acid cycle during aerobic respiration (refer to chapter 7), several other multienzyme complexes catalyze key processes in the cell. One well-studied system is the fatty acid synthetase complex that catalyzes the synthesis of fatty acids from two-carbon precursors. Seven different enzymes make up this multienzyme complex, and the intermediate reaction products remain associated with the complex for the entire series of reactions.

Nonprotein enzymes

Until a few years ago, most biology textbooks contained statements such as "Proteins called enzymes are the catalysts of biological systems." We now know that this statement is incomplete.

Thomas J. Cech and colleagues at the University of Colorado reported in 1981 that certain reactions involving RNA molecules appear to be catalyzed in cells by RNA itself, rather than by enzymes. This initial observation has been corroborated by additional examples of RNA catalysis and was the basis for a shared Nobel Prize in Chemistry in 1989. Like enzymes, these RNA catalysts, which are called "ribozymes," greatly accelerate the rate of particular biochemical reactions and show extraordinary substrate specificity.

Research has revealed at least two sorts of ribozymes. Some ribozymes have folded structures and catalyze reactions on themselves, a process called *intra*molecular catalysis. Other ribozymes act on other molecules without being changed themselves, a process called *inter*molecular catalysis.

The most striking example of the role of RNA as enzyme comes from recent work on the structure and function of the ribosome. For many years, it was thought that the RNA molecules of ribosomes provided a structural framework for proteins that carried out the catalysis of protein synthesis within the ribosome. It is now clear that ribosomal RNA molecules themselves play the key catalytic role, with proteins providing the framework that correctly orients the RNA subunits with respect to each other. The ribosome itself is a ribozyme.

The ability of RNA, an informational molecule, to act as a catalyst has stirred great excitement, because it seems to answer the question "Which came first, the protein or the nucleic acid?" It now seems at least possible that RNA evolved first and may have catalyzed the formation of the first proteins.

Enzyme cofactors

Enzyme function is often assisted by additional chemical components known as **cofactors.** These can be metal ions that are often found in the active site participating directly in catalysis. For example, the metallic ion zinc is used by some enzymes, such as protein-digesting carboxypeptidase, to draw electrons away from their position in covalent bonds, making the bonds less stable and easier to break. Iron, molybdenum, and manganese ion are also used as cofactors.

When the cofactor is a nonprotein organic molecule, it is called a **coenzyme.** Many of the small organic molecules essential in our diets that we call vitamins function as coenzymes. For example, the B vitamins B_6 and B_{12} both function as coenzymes for a number of different enzymes.

In numerous oxidation–reduction reactions that are catalyzed by enzymes, the electrons pass in pairs from the active site of the enzyme to a coenzyme that serves as the electron acceptor. The coenzyme then transfers the electrons to a different enzyme, which releases them (and the energy they bear) to the substrates in another reaction. Often, the electrons combine with protons (H^+) to form hydrogen atoms. In this way, coenzymes shuttle energy in the form of hydrogen atoms from one enzyme to another in a cell. The role of coenzymes and the specifics of their action will be explored in detail in chapters 7 and 8.

Cell Conditions Determine Enzyme Activity

LEARNING OBJECTIVE 6.4.5 Explain the effects of temperature and pH on an enzyme-catalyzed reaction.

Any chemical or physical factor that alters an enzyme's three-dimensional shape—such as temperature, pH, and the binding of regulatory molecules—can affect the enzyme's ability to catalyze a chemical reaction.

Temperature

Increasing the temperature of an uncatalyzed reaction increases its rate because the additional heat increases random molecular movement, which increases the fraction of molecules with kinetic energy above the activation energy. The rate of an enzyme-catalyzed reaction increases with temperature, too, but only up to a point called the *optimum temperature* (figure 6.13*a*). Below this temperature, the hydrogen bonds and hydrophobic interactions that determine the enzyme's shape are not flexible enough to permit the induced fit that is optimum for catalysis. Above the optimum temperature, these forces are too weak to maintain the enzyme's shape, and the protein will become denatured.

Most human enzymes have an optimum temperature between 35°C and 40°C—a range that includes normal body temperature. In contrast, prokaryotes found in hotsprings have enzymes with temperature optimums of 70°C or even higher. The molecular biology technique of PCR (refer to chapter 17), used in many diagnostic tests for infectious agents, takes advantage of just such a temperature-stable DNA polymerase. For both humans and the hotspring prokaryotes, the enzymes in their cells have evolved to have temperature optima corresponding to their environment.

pH

Ionic interactions between oppositely charged amino acid residues, such as glutamic acid (−) and lysine (+), also hold enzymes together. These interactions are sensitive to the

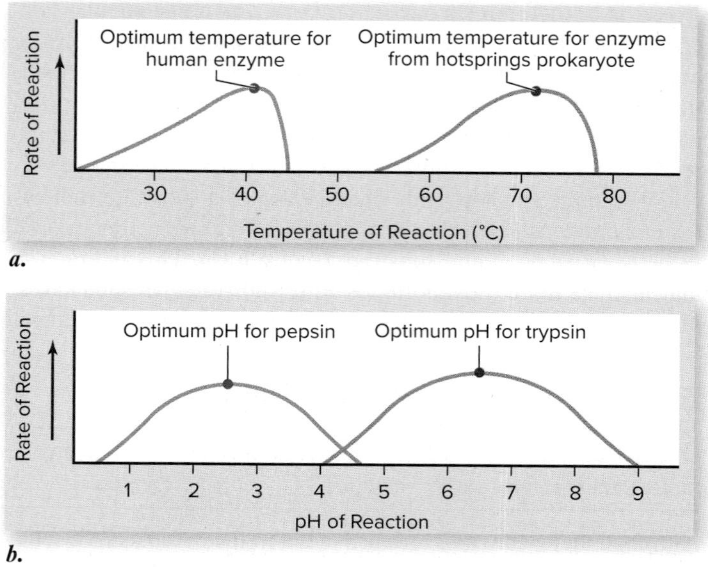

Figure 6.13 Enzyme sensitivity to the environment. The activity of an enzyme is influenced by both **(*a*)** temperature and **(*b*)** pH. Most human enzymes, such as the protein-degrading enzyme trypsin, work best at temperatures of about 40°C and within a pH range of 6 to 8. Pepsin works in the acidic environment of the stomach and has a lower optimum pH.

hydrogen ion concentration of the fluid in which the enzyme is dissolved, because changing that concentration shifts the balance between positively and negatively charged amino acid residues. For this reason, most enzymes have an *optimum pH* that usually ranges from pH 6 to 8.

Enzymes able to function in very acidic environments are proteins that maintain their three-dimensional shape even in the presence of high hydrogen ion concentrations. The enzyme pepsin, for example, digests proteins in the stomach at pH 2, a very acidic level (figure 6.13*b*).

Inhibitors and activators

Enzyme activity can also be affected by molecules other than the substrate that bind specifically to the enzyme. This allows a cell to regulate which of its enzymes are active and which are inactive over time. A substance that binds to an enzyme and decreases its activity is called an **inhibitor.** Enzyme inhibition occurs in two ways:

1. **Competitive inhibitors** compete with the substrate for the same active site, occupying the active site and thus preventing substrates from binding (figure 6.14*a*).
2. **Noncompetitive inhibitors** bind to a site on the enzyme other than the active site, changing the shape of the enzyme and making it unable to bind to the substrate (figure 6.14*b*).

We call the binding site for noncompetitive inhibitors an **allosteric site,** and such inhibitors are called **allosteric inhibitors.** The allosteric site acts as a switch that can shut off the activity of an enzyme based on the binding of an inhibitor.

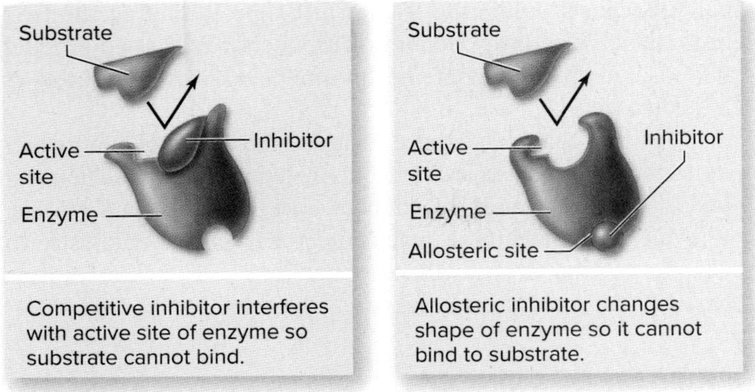

a. Competitive inhibition ***b.*** Noncompetitive inhibition

Figure 6.14 How enzymes can be inhibited. ***a.*** In competitive inhibition, the inhibitor has a shape similar to the substrate and competes for the active site of the enzyme. ***b.*** In noncompetitive inhibition, the inhibitor binds to the enzyme at its allosteric site, a place away from the active site, effecting a conformational change in the enzyme, making it less able to bind to the substrate.

The activity of an allosteric enzyme depends on its conformation: with no inhibitor, it is in an active conformation; bound to an inhibitor, it is in an inactive conformation. Enzymes of this sort provide metabolic flexibility and allow cells to increase metabolic efficiency and to control changes in their characteristics during development. Very often, the end product of a biochemical pathway acts as an inhibitor of an early reaction in the pathway, a process called feedback inhibition (discussed later, in section 6.5).

This kind of control can also be used to "turn on" the activity of an enzyme. **Allosteric activators** bind to allosteric sites and keep an enzyme in an active conformation, which *increases* the activity of the enzyme.

REVIEW OF CONCEPT 6.4

Enzymes are specific catalysts that accelerate chemical reactions in cells. Enzymes bind their substrates based on molecular shape, providing specificity. Enzyme activity is affected by temperature, pH, and the presence of inhibitors or activators. Some enzymes require an inorganic cofactor or an organic coenzyme.

- *Can an enzyme make an endergonic reaction exergonic?*

6.5 Metabolism Is the Sum of a Cell's Chemical Activities

Organisms contain thousands of different kinds of enzymes that catalyze a bewildering variety of reactions. The total of all chemical reactions carried out by an organism is called **metabolism.** Those chemical reactions that expend energy to build up molecules are called *anabolic* reactions, or **anabolism.** Reactions that harvest energy by breaking down molecules are called *catabolic* reactions, or **catabolism.** We have already seen examples of this in chapter 3, where we saw that biopolymers are built (anabolism) by dehydration reactions, and degraded (catabolism) by hydrolysis reactions.

Biochemical Pathways Organize Chemical Reactions in Cells

LEARNING OBJECTIVE 6.5.1 Describe how chemical reactions can be organized into pathways.

Many of these reactions in a cell are organized into sequences called **biochemical pathways,** where the product of one enzyme is the substrate for the next enzyme (figure 6.15). Each step requires a different enzyme specific for each substrate. Biochemical pathways are the organizational units of metabolism, which allow cells to integrate and regulate metabolic activity.

Many sequential enzyme steps in biochemical pathways take place in specific compartments of the cell; for example, the steps of the citric acid cycle (refer to chapter 7) occur in the matrix inside mitochondria in eukaryotes. By determining where many of the enzymes that catalyze these steps are located, we can "map out" a model of metabolic processes in the cell.

In the earliest cells, the first biochemical processes probably involved energy-rich molecules scavenged from the environment.

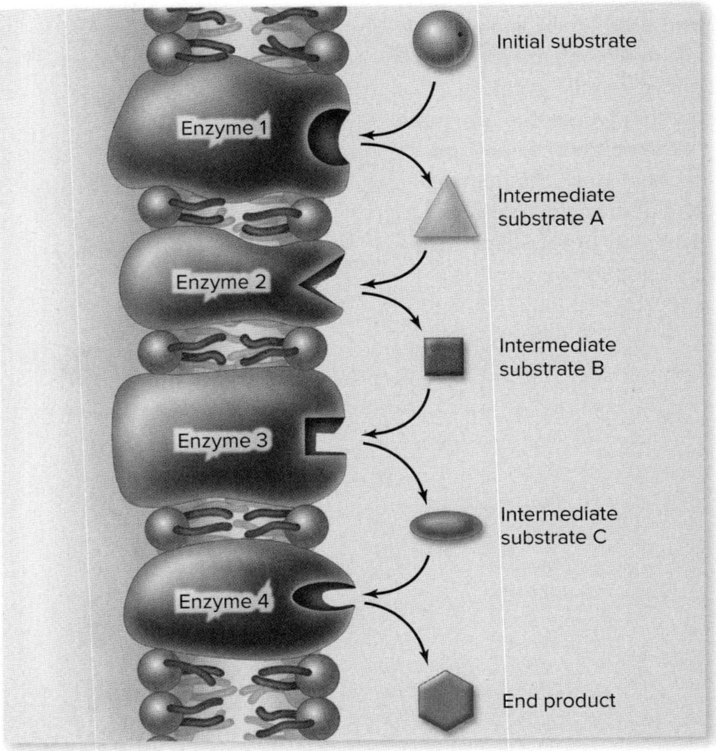

Figure 6.15 A biochemical pathway. Enzyme 1 converts the initial substrate into an intermediate that is a substrate for enzyme 2. Enzyme 2 converts this into a new intermediate that is a substrate for enzyme 3. Each enzyme in the pathway acts on the product of the previous stage. The enzymes may be either soluble or arranged in a membrane as shown.

Most of the molecules necessary for these processes are thought to have existed independently in the "organic soup" of the early oceans.

Feedback Inhibition Can Regulate Pathway Output

LEARNING OBJECTIVE 6.5.2 Explain the function of allosteric proteins.

For a biochemical pathway to operate efficiently, its activity must be coordinated and regulated by the cell. Not only is it unnecessary to synthesize a compound when plenty is already present, but doing so would waste energy and raw materials that could be put to use elsewhere in the cell. It is to the cell's advantage, therefore, to temporarily shut down biochemical pathways when their products are not needed.

The regulation of simple biochemical pathways often depends on an elegant feedback mechanism: the end-product of the pathway binds to an allosteric site on the enzyme that catalyzes the first reaction in the pathway. This mode of regulation is called **feedback inhibition** (figure 6.16). By inhibiting the first enzyme in the pathway, not only is production of the end-product halted, but production of the intermediates in the pathway is halted as well.

In this chapter, we have reviewed the basics of energy and its transformations as carried out in living systems. Chemical bonds are the primary location of energy storage and release. Enzymes facilitate these reactions by serving as catalysts. In Chapters 7 and 8, you will learn the details of how organisms harvest, store, and utilize energy.

REVIEW OF CONCEPT 6.5

Metabolism is the sum of all chemical reactions in a cell. Anabolic reactions use energy to build molecules. Catabolic reactions release energy by breaking down molecules. In a metabolic pathway, the product of one reaction is the substrate for the next.

- *Is a catabolic pathway likely to be subject to feedback inhibition?*

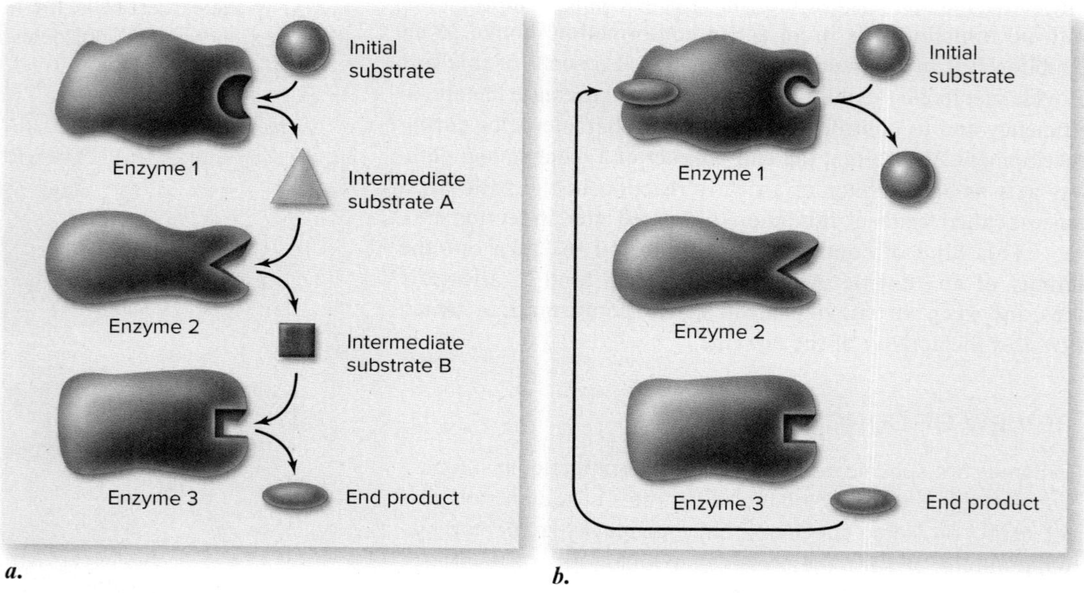

Figure 6.16 Feedback inhibition. ***a.*** A biochemical pathway with no feedback inhibition. ***b.*** A biochemical pathway in which the final end-product becomes the allosteric inhibitor for the first enzyme in the pathway. In other words, the formation of the pathway's final end-product stops the pathway. The pathway could be the synthesis of an amino acid, a nucleotide, or another important cellular molecule.

Do Enzymes Physically Attach to Their Substrates?

When scientists first began to examine the chemical activities of organisms, no one knew that biochemical reactions were catalyzed by enzymes. The first enzyme was discovered in 1833 by French chemist Anselme Payen. He was studying how beer is made from barley: first, barley is pressed and gently heated so its starches break down into simple two-sugar units; then, yeasts convert these units into ethanol. Payen found that the initial breakdown requires a chemical factor that is not alive and that does not seem to be used up during the process—a catalyst. He called this first enzyme *diastase* (we call it amylase today).

Did this catalyst operate at a distance, increasing reaction rate all around it, much as raising the temperature of nearby molecules might do? Or did it operate in physical contact, actually attaching to the molecules whose reaction it catalyzed (its "substrate")?

How Substrate Level Affects Reaction Rate

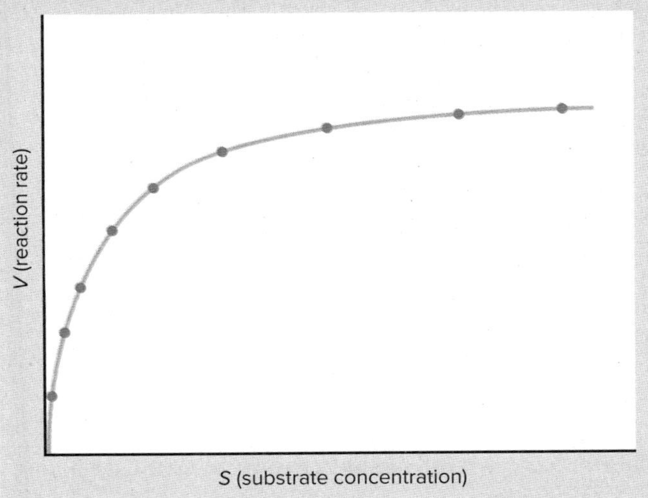

Reciprocal Plot

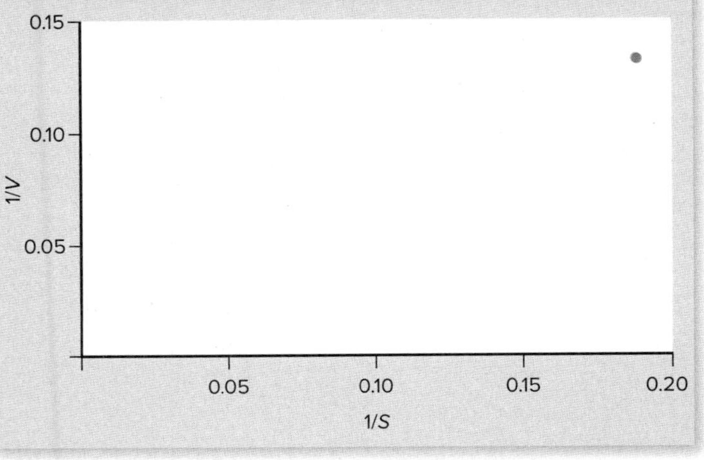

Trial	S	1/S	V	1/V
1	5	0.200	7.7	0.130
2	10	—	15.4	—
3	25	—	23.1	—
4	50	—	30.8	—
5	75	—	38.5	—
6	125	—	40.7	—
7	200	—	46.2	—
8	275	—	47.7	—
9	350	—	48.5	—

The answer was discovered in 1903 by French chemist Victor Henri. He saw that the hypothesis that an enzyme physically binds to its substrate makes a clear and testable prediction: in a solution of substrate and enzyme, there must be a maximum reaction rate, one faster than which the reaction cannot proceed. When all the enzyme molecules are working full tilt, the reaction simply cannot go any faster, no matter how much more substrate you add to the solution. Henri tested this prediction by measuring the rate of reaction (V) at different substrate concentrations (S). The data are shown in the graph titled "How Substrate Level Affects Reaction Rate."

Analysis

1. **Interpreting Data and Making Inferences** As S increases, does V increase? Is this increase linear? Is there a maximum reaction rate?
2. **Drawing Conclusions** Does this result provide support for the hypothesis that an enzyme binds physically to its substrate? Explain. If the hypothesis were incorrect, what would you expect the graph to look like?
3. **Further Analysis** If the smaller amounts by which V increases are strictly the result of fewer unoccupied enzymes being available at higher values of S, then the curve in Henri's experiment should show a pure exponential decline in V—this means, in mathematical terms, that a reciprocal plot ($1/V$ versus $1/S$) should be a straight line. If some other factor were also at work that reacts differently to substrate concentration, then the reciprocal plot would curve upward or downward. Fill in the reciprocal values in the table, and then plot the values on the lower graph ($1/S$ on the x-axis and $1/V$ on the y-axis). Is a reciprocal plot of Henri's data a straight line?

Retracing the Learning Path

CONCEPT 6.1 Energy Flows Through Living Systems

6.1.1 Energy May Be Stored or Used to Do Work Energy is the capacity to do work. Potential energy is stored energy, and kinetic energy is the energy of motion. Energy takes many forms: mechanical, heat, sound, electric current, or light. Energy is measured in units of heat known as kilocalories. Photosynthesis stores light energy from the Sun as potential energy in the structure of sugar molecules. Breaking bonds requires energy, and making bonds releases energy.

6.1.2 Oxidation–Reduction Reactions Transfer Energy Oxidation is a reaction involving the loss of electrons, and reduction is the gain of electrons. These two reactions take place together and are therefore termed redox reactions.

CONCEPT 6.2 The Laws of Thermodynamics Govern All Energy Changes

6.2.1 The First Law States That Energy Cannot Be Created or Destroyed Virtually all activities of living organisms require energy. Thermodynamics is the study of energy changes. Energy changes form as it moves through organisms and their biochemical systems, but it is not created or destroyed.

6.2.2 The Second Law States That Energy Transactions Are Not 100% Efficient The disorder, or entropy, of the universe is continuously increasing. In an open system like the Earth, which is receiving energy from the Sun, this may not be the case. To increase order, however, energy must be expended. In energy conversions, some energy is always lost as heat.

6.2.3 Chemical Reactions Can Be Predicted Based on Changes in Free Energy Free energy (G) is the energy available to do work in any system. Changes in free energy (ΔG) predict the direction of reactions. Reactions with a negative ΔG are spontaneous (exergonic) reactions, and reactions with a positive ΔG are not spontaneous (endergonic). Endergonic chemical reactions absorb energy from the surroundings, whereas exergonic reactions release energy to the surroundings. Activation energy is the energy required to destabilize chemical bonds and initiate chemical reactions. Even exergonic reactions require this activation energy.

CONCEPT 6.3 ATP Is the Energy Currency of Cells

6.3.1 Cells Store and Release Energy in the Bonds of ATP The energy of ATP is stored in the bonds between its terminal phosphate groups. The negative phosphate groups repel each other, making the covalent bonds joining them unstable. Enzymes catalyze the hydrolysis of ATP to produce ADP and phosphate, which have less potential energy than ATP. ATP is continuously cycled. It is synthesized with energy from exergonic reactions, and ATP hydrolysis releases energy to drive endergonic processes.

CONCEPT 6.4 Enzymes Speed Up Reactions by Lowering Activation Energy

6.4.1 Catalysts Act By Lowering Activation Energy Activation energy controls the rate of a chemical reaction. Reactions with high activation energy proceed slowly. A catalyst can lower the activation energy, allowing a reaction to proceed faster.

6.4.2 Enzymes Are Specific Catalysts An enzyme is a catalyst. Enzymes lower the activation energy needed to initiate specific chemical reactions in the cell.

6.4.3 The Active Sites of Enzymes Conform to Fit the Shapes of Substrates The binding causes the enzyme to adjust its shape to the substrate so that there is a better fit.

6.4.4 Enzymes Occur in Many Forms Enzymes can be free in the cytosol or exist as multienzyme complexes bound to membranes and organelles in a biochemical pathway. Although most enzymes are proteins, some are actually RNA molecules, called ribozymes. Cofactors are nonorganic metals necessary for enzyme function. Coenzymes are nonprotein organic molecules, such as certain vitamins, needed for enzyme function. Often, coenzymes serve as electron acceptors or donors in different processes.

6.4.5 Cell Conditions Determine Enzyme Activity An enzyme's functionality depends on its ability to maintain its three-dimensional shape, which can be disrupted by temperature and pH. The activity of enzymes can be decreased by inhibitors. Competitive inhibitors compete for the enzyme's active site. Noncompetitive inhibitors bind to the allosteric site, changing the structure of the enzyme to inhibit it. Allosteric binding can also activate an enzyme.

CONCEPT 6.5 Metabolism Is the Sum of a Cell's Chemical Activities

6.5.1 Biochemical Pathways Organize Chemical Reactions in Cells Metabolism is the sum of all biochemical reactions in a cell. Anabolic reactions require energy to build up molecules, and catabolic reactions break down molecules and release energy. Chemical reactions in biochemical pathways use the product of one reaction as the substrate for the next.

6.5.2 Feedback Inhibition Can Regulate Pathway Output Biosynthetic pathways are often regulated by the end-product of the pathway. Feedback inhibition occurs when the end-product of a reaction combines with the first enzyme's allosteric site to shut down the enzyme's activity.

Concept Overview

This Concept Overview diagrams the key concepts that were discussed in this chapter.

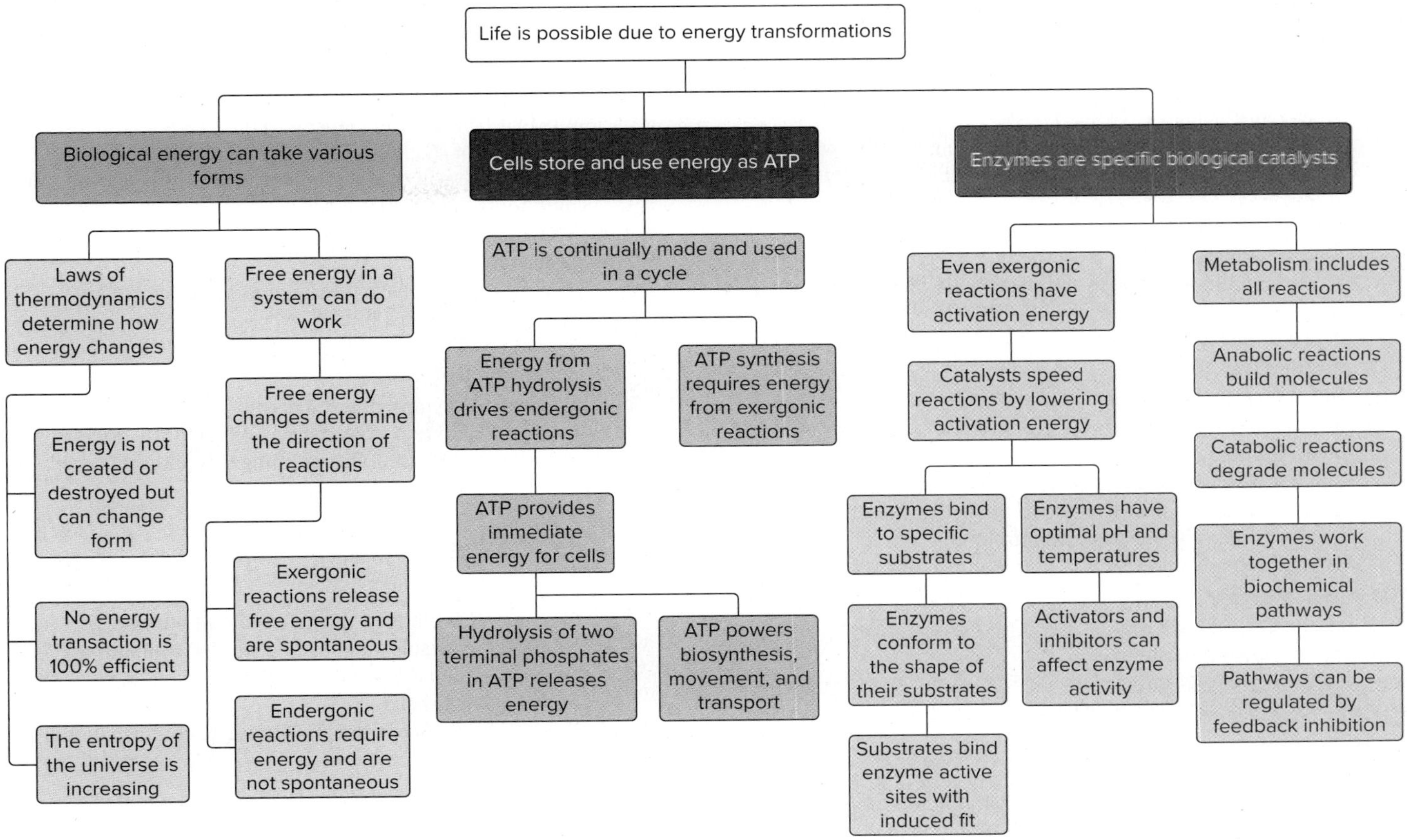

Assessing the Learning Path

Understand

1. A covalent bond between a hydrogen atom and an oxygen atom represents what kind of energy?
   a. Kinetic energy
   b. Potential energy
   c. Mechanical energy
   d. Solar energy
2. During a redox reaction, the molecule that gains an electron is
   a. reduced and now has a higher energy level.
   b. oxidized and now has a lower energy level.
   c. reduced and now has a lower energy level.
   d. oxidized and now has a higher energy level.
3. When a bear eats a salmon, some of the energy stored in the salmon is used by the bear for its activities and growth. Much of the energy originally in the salmon is dissipated as heat. This is an example of
   a. the conversion of kinetic energy to potential energy.
   b. the Second Law of Thermodynamics.
   c. 100% efficient energy conversion.
   d. a conversion of potential energy to kinetic energy.
4. An endergonic reaction has which of the following properties?
   a. $+\Delta G$ and the reaction is spontaneous.
   b. $+\Delta G$ and the reaction is not spontaneous.
   c. $-\Delta G$ and the reaction is spontaneous.
   d. $-\Delta G$ and the reaction is not spontaneous.
5. The molecule ATP is less stable than ADP + P_i because
   a. the negatively charged phosphates repel each other.
   b. the positively charged phosphates repel each other.
   c. ATP is much larger than ADP and P_i.
   d. the adenine in ATP is charged.
6. What is activation energy?
   a. The thermal energy associated with random movements of molecules
   b. The energy released through breaking chemical bonds
   c. The difference in free energy between reactants and products
   d. The energy required to initiate a chemical reaction
7. Which of the following is NOT a property of an enzyme?
   a. An enzyme reduces the activation energy of a reaction.
   b. An enzyme lowers the free energy of the reactants.

c. An enzyme does not change as a result of the reaction.
d. An enzyme works in both the forward and reverse directions of a reaction.

8. Which statement about the influence of temperature on enzymes is NOT true?
a. All enzymes have the same intrinsic optimal temperature.
b. Raising the temperature may increase the activity of an enzyme.
c. Some enzymes are stable at the boiling point of water.
d. Raising the temperature may decrease the activity of an enzyme.

9. Coenzymes
a. can be metal ions.
b. can bind in active sites and participate directly in a catalytic reaction.
c. are sometimes vitamins.
d. All of the above

10. In competitive inhibition,
a. two enzymes compete with each other for a substrate.
b. an inhibitor molecule binds to an allosteric site on an enzyme, causing a change in the active site.
c. an inhibitor molecule binds to the active site of an enzyme, so the substrate cannot bind.
d. the products of a reaction both compete for the active site.

11. Anabolism is
a. the gain of a proton.
b. metabolism in animals.
c. the buildup of molecules.
d. the breakdown of molecules.

Apply

1. When a hibernating animal uses its stored fat to power basic body functions (for example, breathing), it is
a. converting kinetic energy to potential energy.
b. converting kinetic energy to chemical energy.
c. converting potential energy to kinetic energy.
d. converting chemical energy to potential energy.

2. During certain stages of cellular respiration, electrons are transferred from glucose molecules to a molecule called nicotinamide adenine dinucleotide (NAD^+). During this reaction,
a. glucose is oxidized and NAD^+ is reduced.
b. glucose is reduced and NAD^+ is oxidized.
c. both glucose and NAD^+ have gained protons.
d. glucose has gained protons and NAD^+ has lost protons.

3. Sodium ions (Na^+) can move through channel proteins across some biological membranes. If Na^+ is present in a higher concentration on one side of a membrane, the ions will tend to move across the membrane until they are equally distributed on both sides of the membrane. This process
a. results in a gain of potential energy for the cell.
b. results in a decrease in entropy.
c. follows the Second Law of Thermodynamics.
d. All of the above

4. If the products of a chemical reaction have higher free energy than the reactants, this reaction
a. will not proceed spontaneously.
b. will proceed spontaneously.
c. must have increased the total energy in the universe.
d. must have decreased the total energy in the universe.

5. ATP can be used to drive an endergonic reaction if
a. ATP hydrolysis lowers the activation energy for the reaction.
b. the sum of ΔG for ATP hydrolysis and ΔG for the reaction is negative.
c. the sum of ΔG for ATP hydrolysis and ΔG for the reaction is positive.
d. ATP hydrolysis raises the activation energy for the reaction.

6. The enzyme aromatase is found in the cytosol of some cells and converts testosterone to estrogen. You decide to test aromatase from a particular cell type, and your lab partner admits that he drastically increased the pH in all the test tubes. Which of the following is a likely result?
a. The enzyme will be denatured and the substrate will not bind to its active site.
b. The enzyme will convert testosterone to estrogen at a faster rate.
c. The mistake will have no effect, as enzymes are not sensitive to changes in pH.
d. The ΔG will be lowered and the reaction will proceed spontaneously.

7. Consider the enzyme-catalyzed pathway below. It is regulated by feedback inhibition of enzyme 1. Which of the following statements is NOT accurate regarding this pathway?

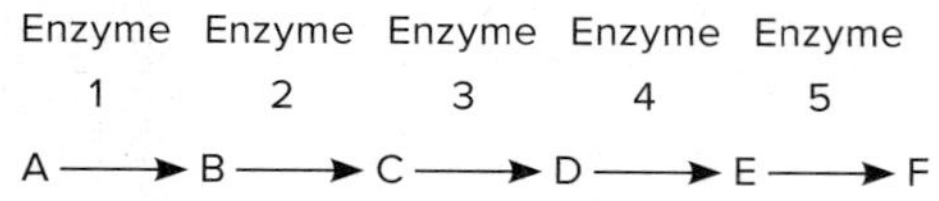

a. Compound F binds to the active site of enzyme 1.
b. Enzyme 1 contains an allosteric site for compound F.
c. When compound F is made in high enough levels, it shuts off its own synthesis.
d. When F binds to enzyme 1, it causes a conformational change.

Synthesize

1. Some people argue that evolution, which is generally associated with progressive increases in the complexity (order) of organisms, cannot occur because entropy (disorder) is increasing in the universe. Is this argument valid? Explain.

2. On summer nights in many parts of the country, one can often see fireflies glowing briefly in the dark. Do you suppose producing this light requires energy? If so, where might the energy come from? How would you test your hypothesis?

3. Examine the graph shown here. Describe what happens to this human enzyme protein's structure when the body's temperature is raised above 40°C.

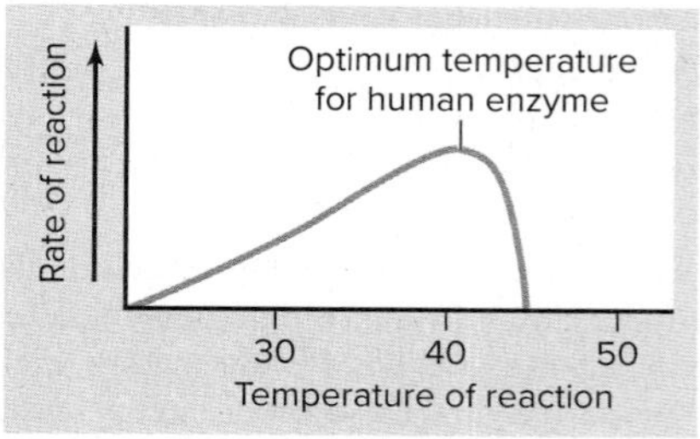

4. Phosphofructokinase functions to add a phosphate group to fructose-6-phosphate. This enzyme functions early in glycolysis (refer to chapter 7). The enzyme's active site binds to fructose and ATP. An allosteric inhibitory site can also bind ATP when ATP levels are high.
a. Predict the rate of the reaction if with low ATP levels.
b. Predict the rate of the reaction if with high ATP levels.
c. Describe what is happening to the enzyme with high levels of ATP.

ErikAgar/Getty Images

7 How Cells Harvest Energy

Learning Path

7.1 Cells Harvest Energy from Organic Compounds by Oxidation

7.2 Glycolysis Splits Glucose and Yields a Small Amount of ATP

7.3 The Citric Acid Cycle Is the Oxidative Core of Cellular Respiration

7.4 Electrons Removed by Oxidation Pass Along an Electron Transport Chain

7.5 The Total Energy Yield of Aerobic Respiration Far Exceeds That of Glycolysis

7.6 Aerobic Respiration Is Regulated by Feedback Inhibition

7.7 Some Forms of Energy Metabolism Do Not Require O_2

7.8 Carbohydrates Are Not the Only Energy Source Used by Heterotrophs

Concept Overview

This Concept Overview provides a diagram of the overarching concepts that are covered in this chapter. These concepts will be expanded upon in the Concept Overview at the end of the chapter.

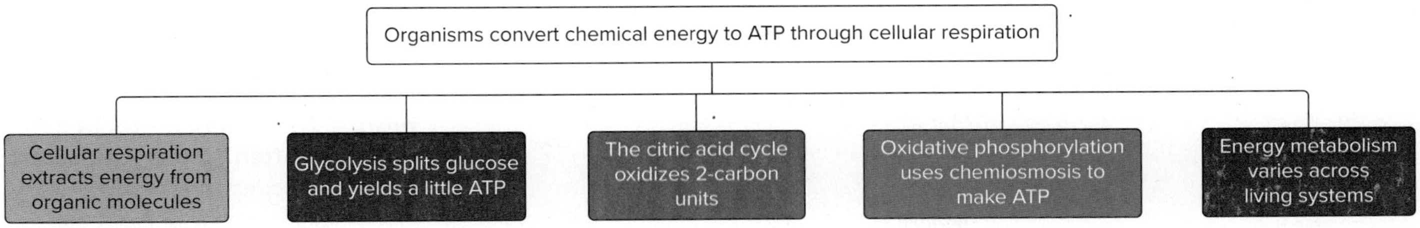
Organisms convert chemical energy to ATP through cellular respiration
Cellular respiration extracts energy from organic molecules
Glycolysis splits glucose and yields a little ATP
The citric acid cycle oxidizes 2-carbon units
Oxidative phosphorylation uses chemiosmosis to make ATP
Energy metabolism varies across living systems

Introduction

Life is driven by energy. All the activities carried out by organisms use energy—the swimming of bacteria, the purring of a cat, and even the thoughts you are forming to process these words. In this chapter, we discuss the processes all cells use to obtain chemical energy from organic molecules, which is used to synthesize ATP. Then, in chapter 8, we will examine photosynthesis, which uses light energy to make chemical energy. We consider the conversion of chemical energy to ATP first because all organisms—including the plant, a photosynthesizer, and the caterpillar feeding on the plant, pictured in the photo—are capable of harvesting energy from chemical bonds. Obtaining energy via respiration is an ancient and universal process.

7.1 Cells Harvest Energy from Organic Compounds by Oxidation

Plants, algae, and some bacteria can use the energy of sunlight to form organic compounds through photosynthesis, converting radiant energy into chemical energy. These organisms, along with a few others that use chemical energy in a similar way, are called **autotrophs** ("self-feeders"). All other organisms are **heterotrophs** ("fed by others") that obtain organic compounds by eating either autotrophs or other heterotrophs. Most of the organisms on Earth—all animals and fungi, and most protists and prokaryotes—are heterotrophs. Autotrophs also extract energy from organic compounds—they just have the additional capacity to use the radiant energy to synthesize these compounds. The central process in this energy metabolism is **cellular respiration,** which involves the oxidation of organic compounds and using the energy released to form ATP.

A reminder before we begin about naming organic acids: an acid dissociates into its conjugate base and H^+. We name the conjugate with the suffix "-ate." For example, pyruvic acid dissociates into pyruvate and H^+. Under physiological conditions most organic acids exist as their conjugate bases.

The Potential Energy Stored in Organic Molecules Resides Largely in C—H Bonds

LEARNING OBJECTIVE 7.1.1 Distinguish between oxidation and reduction reactions.

Most foods contain a variety of carbohydrates, proteins, and fats, all molecules with considerable energy in their complex structures. The rearrangement of these complex molecules to the simple molecules CO_2 and H_2O makes this chemical energy available to the cell.

The job of extracting energy from the complex organic mixture in most foods occurs in stages. First, enzymes break down the large molecules into smaller ones, a process called digestion (refer to chapter 34). Then, other enzymes dismantle these fragments a bit at a time, with the energy released by these reactions stored in molecules of ATP.

The reactions that break down these molecules share a common feature: they are all oxidations. Energy metabolism is therefore concerned with oxidation reactions, so to understand metabolism we must follow the fate of the electrons removed from oxidized food molecules. These reactions are also **dehydrogenations.** That is, the electrons obtained by oxidizing food molecules are accompanied by protons so that what is really transferred is a hydrogen atom, not just an electron.

Redox reactions

In chapter 6, you learned that an atom that loses electrons is said to be *oxidized,* and an atom accepting electrons is said to be *reduced.* Oxidation reactions are always coupled with reduction reactions, and these paired reactions are called *redox reactions.* Cells utilize enzyme-facilitated redox reactions to take energy from food sources and convert it to ATP. This is possible because the oxidation of food molecules is exergonic, and some of the released energy can be used to synthesize ATP.

As we discussed in chapter 6, oxidation–reduction reactions play a key role in the flow of energy through biological systems, because the electrons that pass from one atom to another carry potential energy with them. The amount of energy an electron possesses depends on its orbital position, or energy level, around the atom's nucleus. When this electron departs from one atom and moves to another in a redox reaction, the electron's potential energy is transferred with it.

Figure 7.1 shows how an enzyme catalyzes a redox reaction involving an energy-rich substrate molecule, with the help of a coenzyme, **nicotinamide adenine dinucleotide (NAD^+).** In this reaction, NAD^+ accepts a pair of electrons from the substrate,

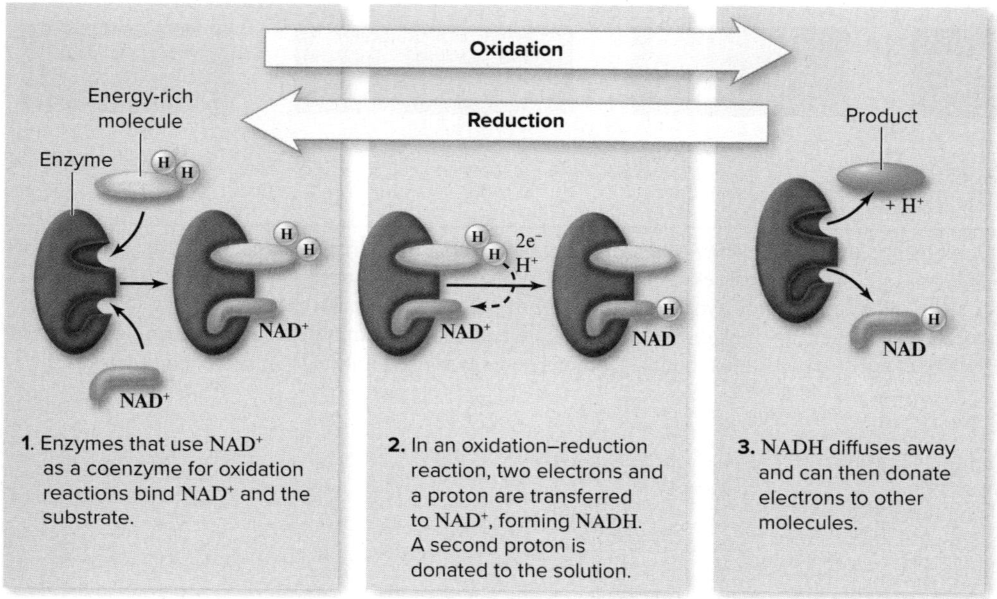

Oxidation
Reduction
Energy-rich molecule
Enzyme
H
H
NAD+
2e−
H+
NAD
Product
+ H+
1. Enzymes that use NAD+ as a coenzyme for oxidation reactions bind NAD+ and the substrate.
2. In an oxidation–reduction reaction, two electrons and a proton are transferred to NAD+, forming NADH. A second proton is donated to the solution.
3. NADH diffuses away and can then donate electrons to other molecules.

Figure 7.1 Oxidation–reduction reactions often employ cofactors. Cells use a coenzyme called nicotinamide adenine dinucleotide (NAD^+) to carry out many oxidation–reduction reactions. Two electrons and a proton are transferred to NAD^+ with another proton donated to the solution. Molecules that gain electrons are said to be reduced, and ones that lose energetic electrons are said to be oxidized. NAD^+ oxidizes energy-rich molecules by acquiring their electrons (in the figure, this proceeds 1 → 2 → 3) and then reduces other molecules by giving the electrons to them (in the figure, this proceeds 3 → 2 → 1). NADH is the reduced form of NAD^+.

along with a proton, to form NADH (this process will be described in more detail shortly). The oxidized product is now released from the enzyme's active site, as is NADH.

In the overall process of cellular energy-harvesting, dozens of redox reactions take place, and a number of molecules, including NAD^+, act as electron acceptors. During each transfer of electrons, energy is released. A substantial portion of this energy is captured and used to make ATP or to form other chemical bonds; the rest is lost as heat.

At the end of this process, electrons at a high energy level have lost much of their energy, and they are transferred to a final electron acceptor (figure 7.2). When this acceptor is oxygen, the process is called **aerobic respiration.** When the final electron acceptor is an inorganic molecule other than oxygen, the process is called **anaerobic respiration,** and when it is an organic molecule, the process is called **fermentation.**

"Burning" carbohydrates

Chemically, there is little difference between the catabolism of carbohydrates in a cell and the burning of wood in a fireplace. In both instances, the reactants are carbohydrates and oxygen, and the products are carbon dioxide, water, and energy:

$$\underset{\text{glucose}}{C_6H_{12}O_6} + \underset{\text{oxygen}}{6\ O_2} \rightarrow \underset{\text{carbon dioxide}}{6\ CO_2} + \underset{\text{water}}{6\ H_2O} + \text{energy (heat and ATP)}$$

The change in free energy (ΔG) for this reaction is -686 kcal/mol (or -2870 kJ/mol) under standard conditions. In the conditions that exist inside a cell, the energy released can be as high as -720 kcal/mol of glucose.

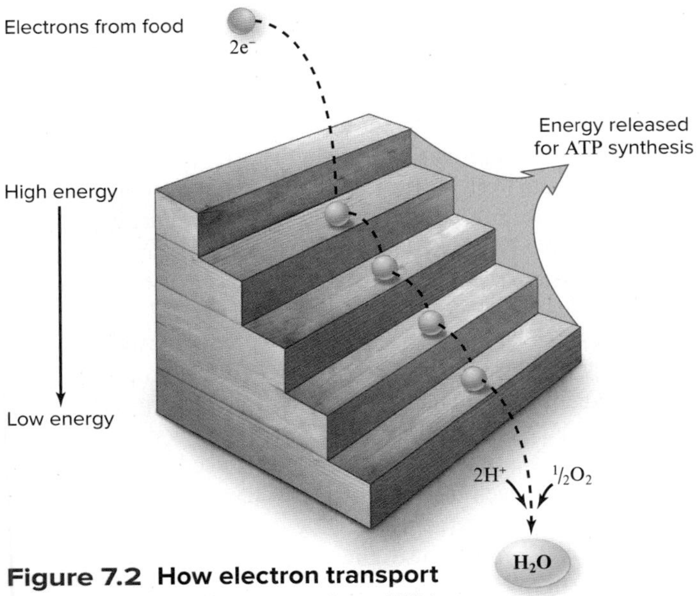
Electrons from food
2e⁻
Energy released
for ATP synthesis
High energy
Low energy
2H⁺
½O₂
H₂O

Figure 7.2 How electron transport works. This diagram shows how ATP is generated when electrons transfer from one energy level to another. Rather than releasing a single explosive burst of energy, electrons "fall" to lower and lower energy levels in steps, releasing stored energy with each fall as they tumble to the lowest (most electronegative) electron acceptor, O_2.

The same amount of energy is released whether glucose is catabolized or burned, but when it is burned, most of the energy is released as heat. Cells harvest useful energy from the catabolism of glucose by using a portion of the energy to drive the production of ATP.

Electron Carriers Transfer Electrons from One Molecule to Another

LEARNING OBJECTIVE 7.1.2 Describe the structure of NAD^+ and explain its role in energy metabolism.

During aerobic respiration, glucose is oxidized to CO_2. If the electrons were given directly to O_2, the reaction would be combustion, and cells would burst into flames. Instead, as you have just learned, the cell transfers the electrons to intermediate electron carriers, then eventually to O_2.

Many forms of electron carriers are used in this process: (1) soluble carriers that move electrons from one molecule to another and act as coenzymes, (2) membrane-bound carriers that form a redox chain, and (3) carriers that move within the membrane. The common feature of all of these carriers is that they can be reversibly oxidized and reduced. Some of these carriers, such as the iron-containing cytochromes, can carry just electrons, and some carry both electrons and protons.

NAD^+ is one of the most important carriers of electrons (and protons). As shown on the left in figure 7.3, the NAD^+ molecule is composed of two nucleotides bound together. The two nucleotides that make up NAD^+, nicotinamide monophosphate (NMP) and adenosine monophosphate (AMP), are joined head-to-head by their phosphate groups. The two nucleotides serve different functions in the NAD^+ molecule: AMP acts as the core, providing a shape recognized by many enzymes; NMP is the active part of the molecule because it is readily reduced—that is, it easily accepts electrons.

When NAD^+ acquires two electrons and a proton from the active site of an enzyme, it is reduced to NADH, shown on the right in figure 7.3. The NADH molecule now carries the two energetic electrons and can supply them to other molecules and reduce them.

This ability to supply high-energy electrons is critical both to energy metabolism and to the biosynthesis of many organic molecules, including fats and sugars. In animals, when ATP is plentiful, the reducing power of the accumulated NADH is diverted to supplying fatty acid precursors with high-energy electrons, reducing them to form fats, and so, storing the energy of the electrons.

Metabolism harvests energy in stages

It is generally true that the larger the release of energy in any single step, the more of that energy is released as heat and the less is available to be channeled into more useful paths. In the combustion of gasoline in a car, the same total amount of energy is released whether all of the gasoline in the gas tank explodes at once, or it is oxidized in a series of very small explosions inside the cylinders. But by releasing the energy in gasoline a little at a time, the energy is used more efficiently and will push the pistons and ultimately move the car.

Figure 7.3 NAD$^+$ and NADH. This dinucleotide serves as an "electron shuttle" during cellular respiration. NAD$^+$ accepts a pair of electrons and a proton from catabolized macromolecules and is reduced to NADH.

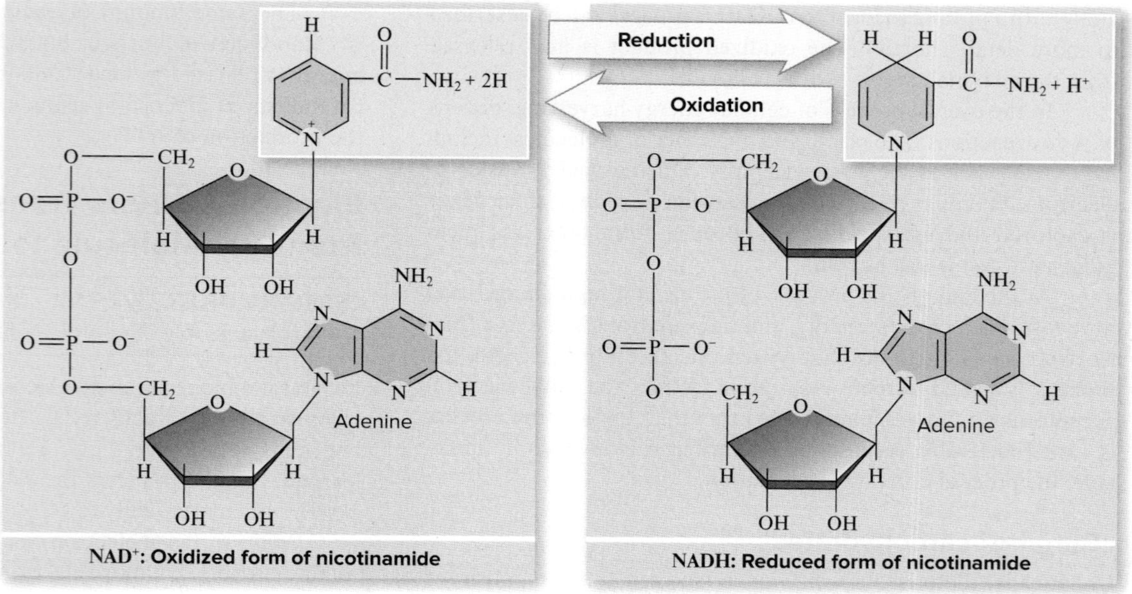

Reduction
Oxidation
Adenine
Adenine
NAD+: Oxidized form of nicotinamide
NADH: Reduced form of nicotinamide

The same principle applies to the oxidation of glucose inside a cell. If all of the electrons were transferred to oxygen in one explosive step, releasing all of the free energy at once, the cell would recover very little of that energy in a useful form. Instead, cells burn their fuel much as a car does, a little at a time.

The electrons in the C–H bonds of glucose are oxidized in stages in the series of enzyme-catalyzed reactions collectively referred to as glycolysis and the citric acid cycle. The electrons are removed by transferring them to NAD$^+$, as described earlier in this section, or to other electron carriers.

Rather than releasing the energy from all of these oxidation reactions at once, the electrons are each passed to another set of electron carriers called the electron transport chain, located in the mitochondrial inner membrane. The redox reactions of the transport chain release energy that is converted to potential energy in the form of an electrochemical gradient. We will examine this process in section 7.4.

The Central Role of Respiration Is to Produce ATP

LEARNING OBJECTIVE 7.1.3 Contrast substrate-level phosphorylation with oxidative phosphorylation.

The many steps of cellular respiration have as their ultimate goal the production of ATP. The previous chapter introduced ATP as the energy currency of the cell. Cells use ATP to power most of those activities that require work–one of the most obvious of which is movement. Tiny fibers within muscle cells pull against one another when muscles contract. Mitochondria can move a meter or more along the narrow nerve cells that extend from your spine to your feet. Chromosomes are pulled apart by microtubules during cell division. All of these movements require the expenditure of energy by ATP hydrolysis. Cells also use ATP to drive endergonic reactions that would otherwise not occur spontaneously (see figure 6.4).

Cells make ATP by two fundamentally different mechanisms

The synthesis of ATP can be accomplished by two distinct mechanisms: one that involves chemical coupling with an intermediate bound to phosphate and another that relies on an electrochemical gradient of protons for the potential energy to phosphorylate ADP.

1. In *substrate-level phosphorylation,* ATP is formed by transferring a phosphate group directly to ADP from a phosphate-bearing intermediate, or substrate (figure 7.4). If the ΔG for hydrolysis of phosphate for a molecule is more negative than the ΔG for hydrolysis of phosphate from ATP, then phosphate can be transferred to ADP to form ATP. During **glycolysis,** the initial breakdown of glucose (discussed in section 7.2), the chemical bonds of glucose are rearranged to produce intermediates that can transfer phosphate to ADP.
2. In **oxidative phosphorylation,** ATP is synthesized by the enzyme **ATP synthase,** using energy from a proton (H$^+$) gradient. This gradient is formed by high-energy electrons removed by the oxidation of glucose passing down an electron transport chain (described in section 7.4). These electrons reach their lowest energy level when they are transferred to oxygen; hence the term *oxidative phosphorylation.* ATP synthase uses the energy from the proton gradient to catalyze the reaction:

$$ADP + P_i \rightarrow ATP$$

Eukaryotes and aerobic prokaryotes produce the vast majority of their ATP this way.

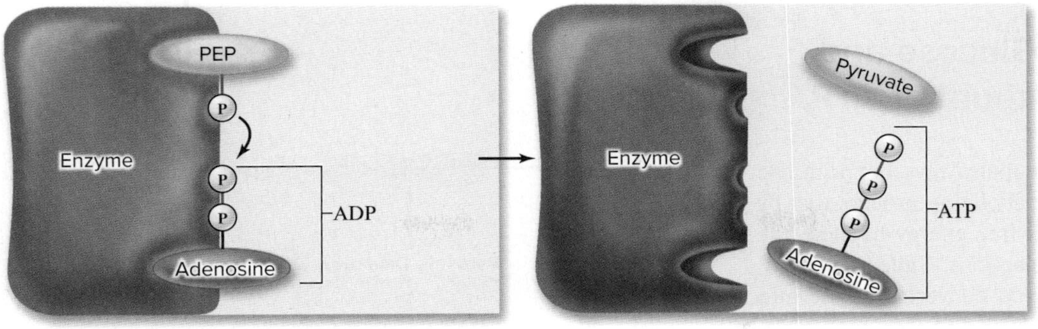

Figure 7.4 Substrate-level phosphorylation. Some molecules, such as phosphoenolpyruvate (PEP), possess a high-energy phosphate (P) bond similar to the bonds in ATP. When PEP's phosphate group is transferred enzymatically to ADP, the energy in the bond is conserved, and ATP is created.

In most organisms, these two processes are combined. To harvest energy to make ATP from glucose in the presence of oxygen, the cell carries out a complex series of enzyme-catalyzed reactions that remove energetic electrons via oxidation reactions. These electrons are then used in an electron transport chain that passes the electrons down a series of carriers while translocating protons into the intermembrane space. The final electron acceptor in aerobic respiration is oxygen, and the resulting proton gradient provides energy for the enzyme ATP synthase to phosphorylate ADP to ATP. The details of this complex process (figure 7.5) will be covered in the remainder of this chapter.

REVIEW OF CONCEPT 7.1

Cells acquire energy by the complete oxidation of glucose. In these redox reactions, protons as well as electrons are transferred, in dehydrogenation reactions. Electron carriers aid in the gradual, stepwise release of the energy from oxidation. Energy from this process is captured as ATP. Synthesis of ATP occurs by two mechanisms: substrate-level phosphorylation and oxidative phosphorylation.

- *Why don't cells just transfer the electrons from glucose directly to oxygen, releasing a significant amount of energy?*

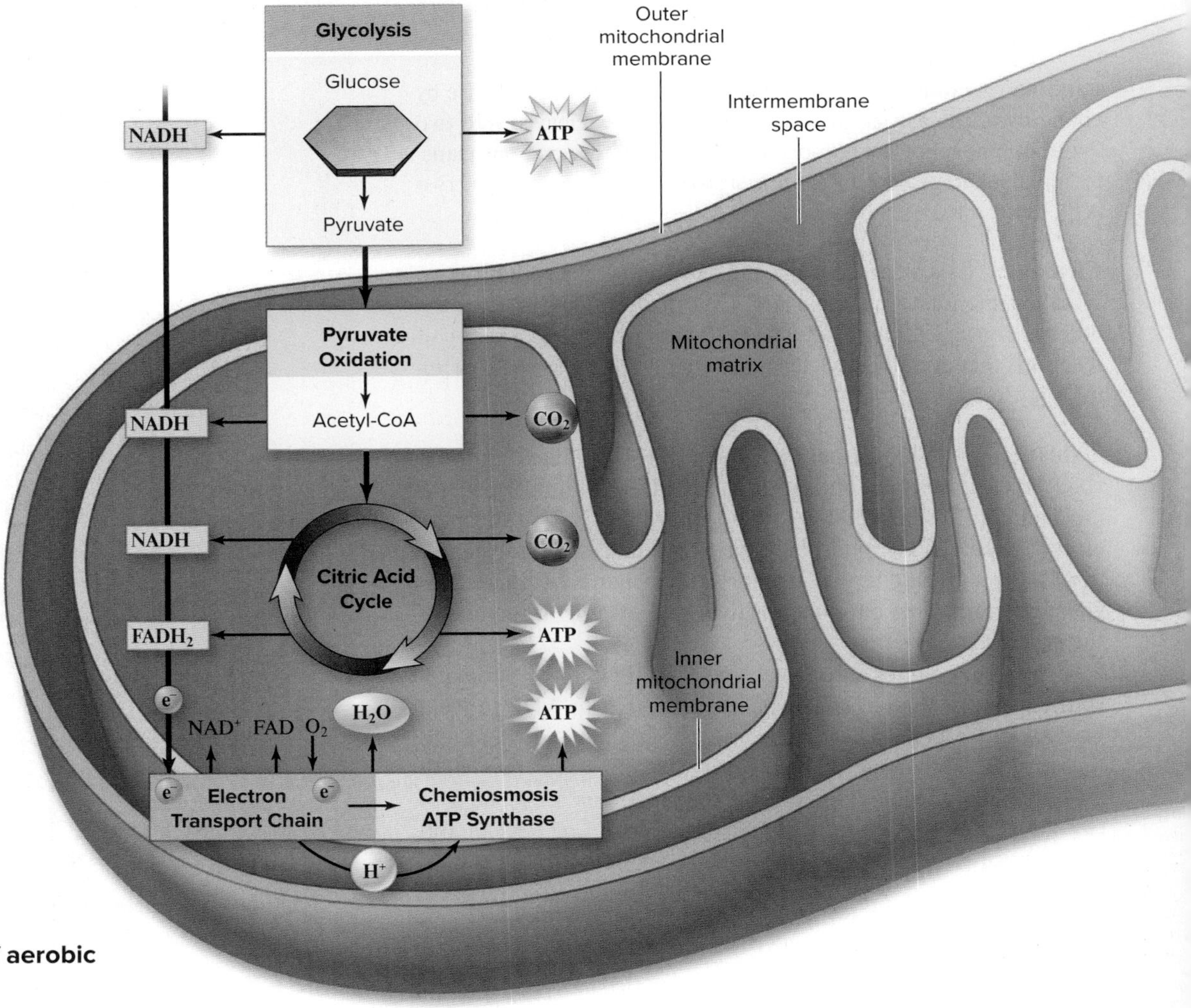

Figure 7.5 An overview of aerobic respiration.

7.2 Glycolysis Splits Glucose and Yields a Small Amount of ATP

Glucose molecules can be dismantled in many ways, but primitive organisms evolved a simple glucose-catabolizing process that by shuffling a few bonds releases enough free energy to drive the synthesis of ATP in enzyme-coupled reactions. Called *glycolysis,* this process occurs in the cytoplasm and converts glucose into two 3-carbon molecules of pyruvate (figure 7.6). For each molecule of glucose that passes through this transformation, the cell nets two ATP molecules.

Glycolysis Converts Glucose into Two Pyruvates and Yields Two ATP and Two NADH

LEARNING OBJECTIVE 7.2.1 Explain the process of glycolysis, including its energy yield.

The first half of glycolysis consists of five sequential reactions that convert one molecule of glucose into two molecules of the 3-carbon compound **glyceraldehyde-3-phosphate (G3P).** These reactions require the expenditure of ATP, so they constitute an endergonic process. In the second half of glycolysis, five more reactions convert G3P into pyruvates in an energy-yielding process that generates ATP.

Priming reactions The first three reactions "prime" glucose by changing it into a compound that can be readily cleaved into two 3-carbon phosphorylated molecules. Two of these reactions transfer a phosphate from ATP, so this step requires the cell to use two ATP molecules.

Cleavage This 6-carbon diphosphate sugar is then split into two 3-carbon monophosphate sugars. One of these is G3P, and the other is converted into G3P. The G3P then undergoes a series of reactions that eventually yields more energy than was spent priming (figure 7.7).

Oxidation and ATP formation Each G3P is oxidized, transferring two electrons (and one proton) to NAD^+, forming NADH. A molecule of P_i is also added to G3P to produce 1,3-bisphosphoglycerate (BPG). The phosphate incorporated can be transferred to ADP by substrate-level phosphorylation (figure 7.4) in an exergonic reaction. This allows a positive yield of ATP at the end of the process.

Another four reactions convert BPG into pyruvate. In the process, the phosphates are transferred to ADP to yield two ATP per G3P. The entire process is shown in detail in figure 7.7. Each glucose molecule is split into two G3P molecules, so the overall reaction sequence has a net yield of two molecules of ATP, as well as two molecules of NADH and two of pyruvates:

4 ATP (2 ATP for each of the 2 G3P molecules)
−2 ATP (used in the two reactions in the first step)

2 ATP (net yield for the entire process)

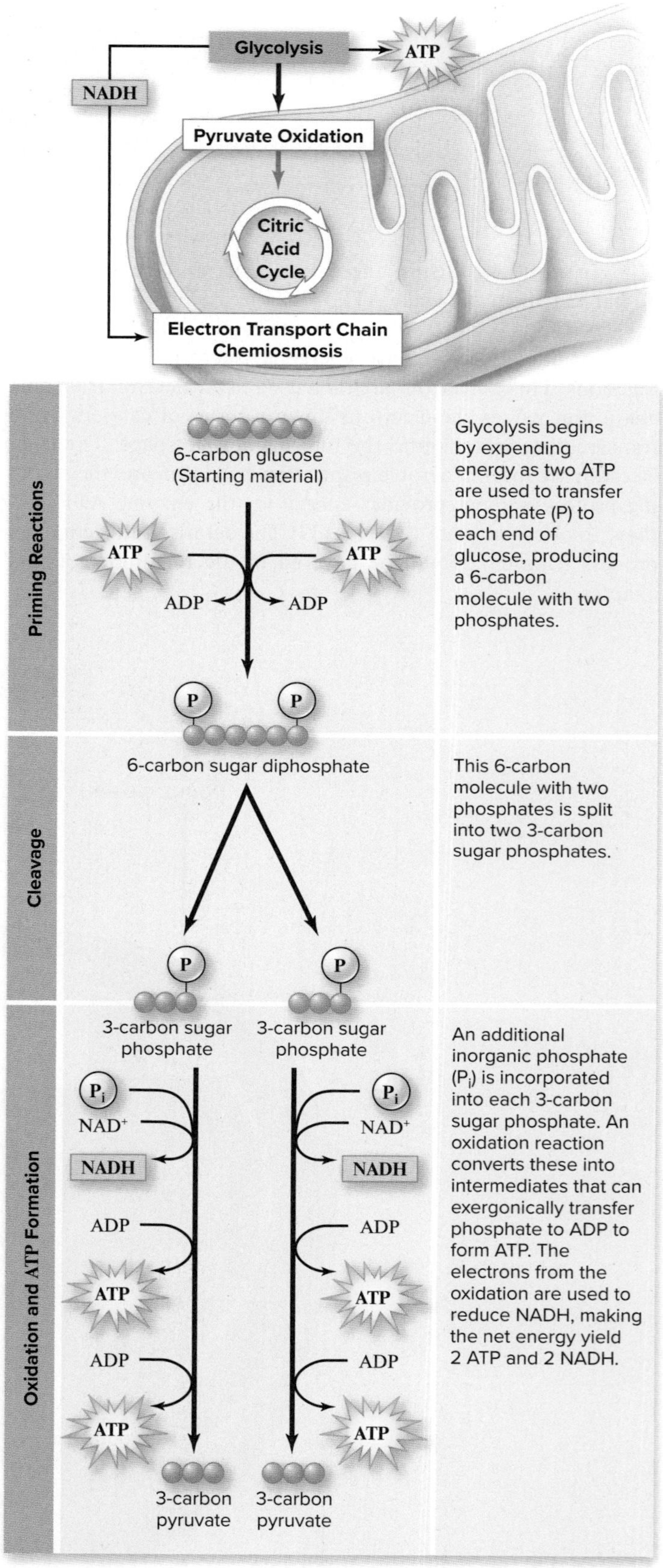

Figure 7.6 How glycolysis works.

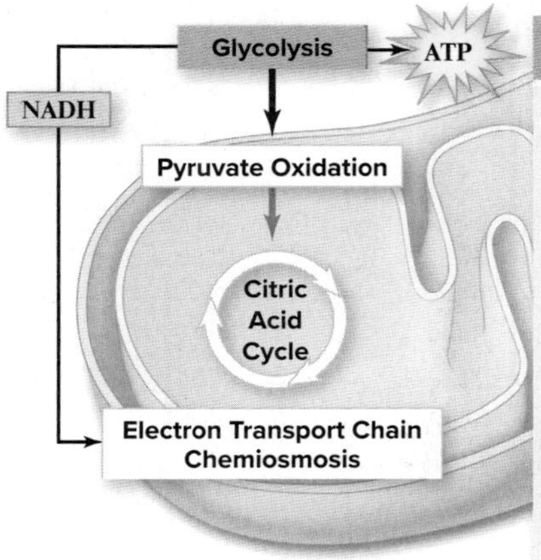

Glycolysis
ATP
NADH
Pyruvate Oxidation
Citric Acid Cycle
Electron Transport Chain
Chemiosmosis

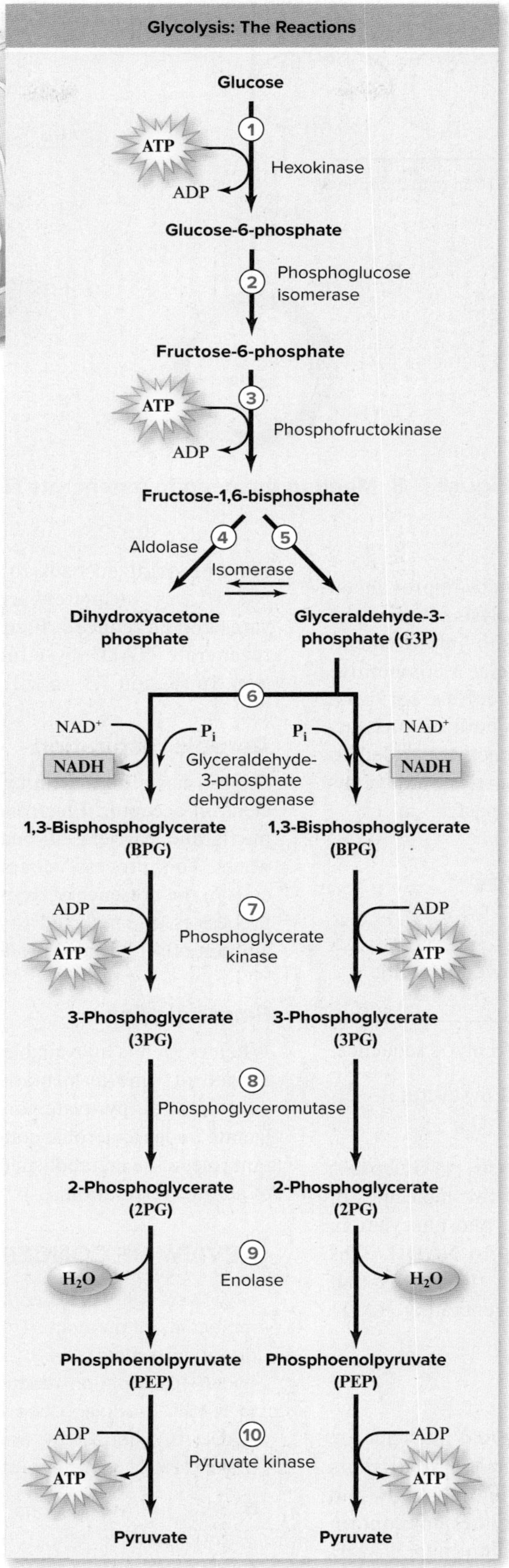

Glycolysis: The Reactions
Glucose
ATP
ADP
1
Hexokinase
Glucose-6-phosphate
2
Phosphoglucose isomerase
Fructose-6-phosphate
ATP
ADP
3
Phosphofructokinase
Fructose-1,6-bisphosphate
Aldolase
4
5
Isomerase
Dihydroxyacetone phosphate
Glyceraldehyde-3-phosphate (G3P)
6
NAD+
NADH
Pi
Glyceraldehyde-3-phosphate dehydrogenase
Pi
NAD+
NADH
1,3-Bisphosphoglycerate (BPG)
1,3-Bisphosphoglycerate (BPG)
ADP
ATP
7
Phosphoglycerate kinase
ADP
ATP
3-Phosphoglycerate (3PG)
3-Phosphoglycerate (3PG)
8
Phosphoglyceromutase
2-Phosphoglycerate (2PG)
2-Phosphoglycerate (2PG)
H2O
9
Enolase
H2O
Phosphoenolpyruvate (PEP)
Phosphoenolpyruvate (PEP)
ADP
ATP
10
Pyruvate kinase
ADP
ATP
Pyruvate
Pyruvate

1. Phosphorylation of glucose by ATP.

2–3. Rearrangement, followed by a second ATP phosphorylation.

4–5. The 6-carbon molecule is split into two 3-carbon molecules—one G3P, another that is converted into G3P in another reaction.

6. Oxidation followed by phosphorylation produces two NADH molecules and two molecules of BPG, each with a phosphate that can be transferred to ADP.

7. Transfer of two phosphate groups to two ADP molecules produces two ATP and two 3PG molecules.

8–9. Removal of water yields two PEP molecules, each with a phosphate group that can be transferred to ADP.

10. Transfer of two phosphate groups to two ADP molecules produces two ATP and two pyruvate molecules.

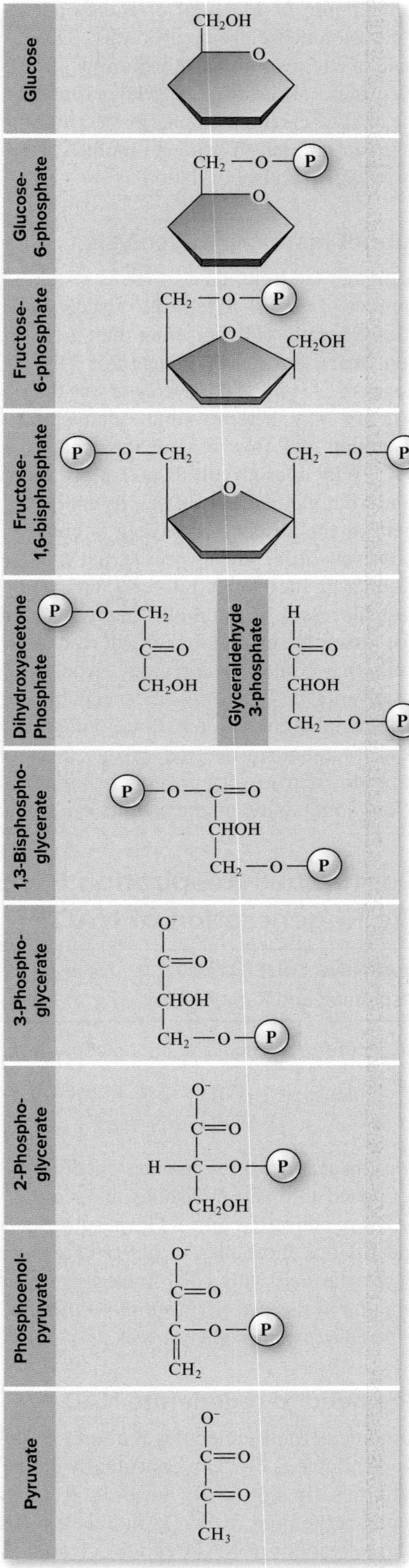

Glucose
Glucose-6-phosphate
Fructose-6-phosphate
Fructose-1,6-bisphosphate
Dihydroxyacetone Phosphate
Glyceraldehyde 3-phosphate
1,3-Bisphosphoglycerate
3-Phosphoglycerate
2-Phosphoglycerate
Phosphoenolpyruvate
Pyruvate

Figure 7.7 The glycolytic pathway. The first five reactions convert a molecule of glucose into two molecules of G3P. The second five reactions convert G3P into pyruvate.

The hydrolysis of one molecule of ATP yields a ΔG of -7.3 kcal/mol under standard conditions. Thus, cells harvest a maximum of 14.6 kcal of energy per mole of glucose from glycolysis. This yield of energy is only the beginning: the pyruvates can be further oxidized, and the NADH carries high-energy electrons that can be used to generate more ATP by oxidative phosphorylation.

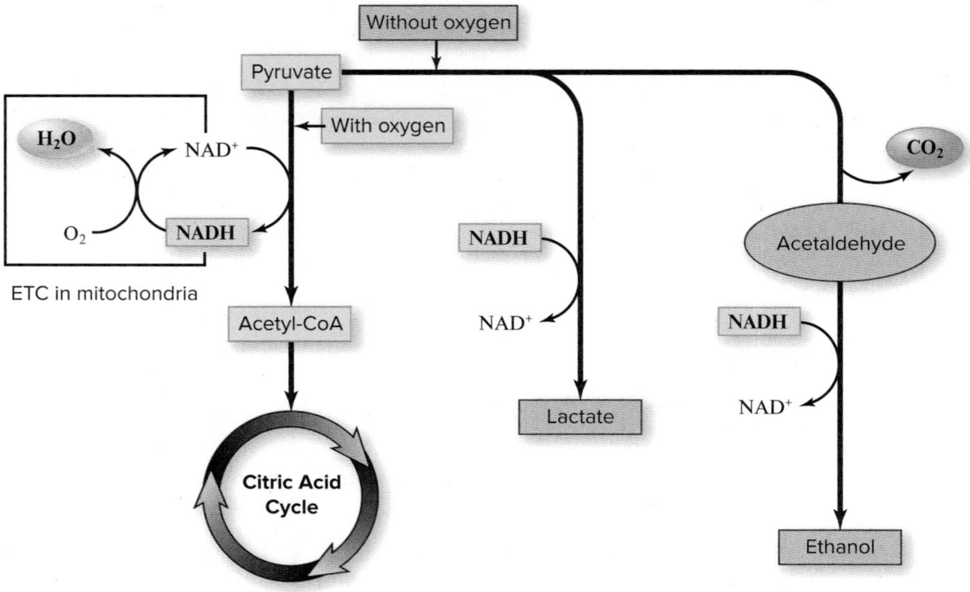

Figure 7.8 Meeting the need to regenerate NAD^+.

A brief history of glycolysis

Although far from ideal in terms of the amount of energy it releases, glycolysis does generate ATP. For more than a billion years during the anaerobic first stages of life on Earth, glycolysis was the primary way heterotrophic organisms generated ATP from organic molecules.

Why does glycolysis take place in modern organisms, given that its energy yield in the absence of oxygen is comparatively little? The answer is that evolution is an incremental process: change occurs by improving on past successes. In catabolic metabolism, glycolysis satisfied the one essential evolutionary criterion—it was an improvement. Cells that could not carry out glycolysis were at a competitive disadvantage, and only cells capable of glycolysis survived. Later improvements in catabolic metabolism built on this success. Metabolism evolved as one layer of reactions added to another. Nearly every present-day organism carries out glycolysis, as a metabolic memory of its evolutionary past.

Continuous Respiration Requires the Regeneration of NAD^+

LEARNING OBJECTIVE 7.2.2 Distinguish between aerobic respiration and fermentation.

Inspect for a moment the net reaction of the glycolytic sequence:

$$\text{glucose} + 2\ \text{ADP} + 2\ \text{P}_i + 2\ \text{NAD}^+ \longrightarrow 2\ \text{pyruvates} + 2\ \text{ATP} + 2\ \text{NADH} + 2\ \text{H}^+ + \text{H}_2\text{O}$$

You can see that three changes occur in glycolysis: (1) Glucose is converted into two molecules of pyruvate; (2) two molecules of ADP are converted into ATP via substrate-level phosphorylation; and (3) two molecules of NAD^+ are reduced to NADH. This leaves the cell with two problems: extracting the energy that remains in the two pyruvate molecules and regenerating NAD^+ to be able to continue glycolysis.

The need to regenerate NAD^+

As long as food molecules that can be converted into glucose are available, a cell can continually churn out ATP to drive its activities. In doing so, however, it accumulates NADH and depletes the pool of NAD^+ molecules. A cell does not contain a large amount of NAD^+, so for glycolysis to continue, NADH must be oxidized back to NAD^+—some molecule other than NAD^+ must ultimately accept the electrons taken from pyruvates and be reduced. Figure 7.8 shows how aerobic organisms regenerate NAD^+ by either aerobic respiration or fermentation. In section 7.7 we will consider anaerobic organisms.

Aerobic respiration

Oxygen has a high affinity for electrons, making it an excellent electron acceptor. Electrons are transferred through a series of membrane carriers, ultimately reducing oxygen and forming water. This process occurs in the mitochondria of eukaryotic cells in the presence of oxygen. Because air is rich in oxygen, this process is also referred to as *aerobic metabolism.* A significant amount of ATP is also produced.

Fermentation

When oxygen is unavailable, an organic molecule, such as acetaldehyde in wine fermentation, can accept electrons instead. In your muscles, pyruvate can accept electrons and be reduced to lactate under anaerobic conditions. This reaction plays an important role in the metabolism of most organisms, even those capable of aerobic respiration.

REVIEW OF CONCEPT 7.2

Glycolysis splits the 6-carbon sugar glucose into two 3-carbon molecules of pyruvate. This uses two ATP molecules and produces four molecules of ATP per glucose, for a net yield of two ATP. Oxidation reactions in glycolysis transfer electrons to NAD^+, producing two NADH. When oxygen is abundant, NADH is oxidized by aerobic respiration. When oxygen is absent, NAD^+ is regenerated by fermentation.

- *How many reactions in glycolysis result in ATP being produced?*

7.3 The Citric Acid Cycle Is the Oxidative Core of Cellular Respiration

The fate of the pyruvate produced by glycolysis depends on the presence or absence of O_2. In the presence of O_2, pyruvate can be further oxidized to produce acetyl coenzyme A, which is further oxidized in a series of reactions called the citric acid cycle. This is called aerobic respiration, as the electrons from these oxidations will eventually be transferred to O_2. In the absence of O_2, pyruvate, or a metabolite of pyruvate, is reduced and NADH is oxidized back to NAD^+.

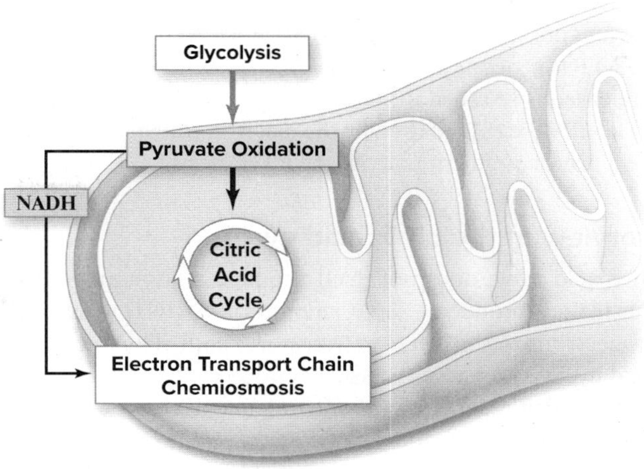

Pyruvate Is Converted to a 2-Carbon Acetyl Group for the Citric Acid Cycle

LEARNING OBJECTIVE 7.3.1 Explain how the oxidation of pyruvate links glycolysis with the rest of respiration.

In eukaryotic organisms, the extraction of additional energy from the pyruvate produced by glycolysis takes place exclusively inside mitochondria. Pyruvate is transported from the cytoplasm into the mitochondria by a carrier in the inner membrane. In prokaryotes, similar reactions take place in the cytoplasm and at the plasma membrane.

The cell harvests pyruvate's considerable energy in two steps. First, pyruvate is oxidized to produce a 2-carbon compound and CO_2, with the electrons transferred to NAD^+ to produce NADH. Next, the 2-carbon compound is oxidized to two molecules of CO_2 by the reactions of the citric acid cycle.

Pyruvate is oxidized in a "decarboxylation" reaction that cleaves off one of pyruvate's three carbons. This carbon departs as CO_2 (figure 7.9). The remaining 2-carbon compound, called an acetyl group, is then attached to coenzyme A; this entire molecule is called *acetyl-CoA.* In the reaction, a pair of electrons and one associated proton are transferred to the electron carrier NAD^+, reducing it to NADH, with a second proton donated to the solution.

This complex reaction involves three intermediate stages, catalyzed within mitochondria by a *multienzyme complex.* As chapter 6 noted, a multienzyme complex organizes a series of enzymatic steps so that the chemical intermediates do not diffuse away or undergo other reactions. Within the complex, component polypeptides pass the substrates from one enzyme to the next without releasing them. *Pyruvate dehydrogenase,* the complex of enzymes that removes CO_2 from pyruvate, is one of the largest enzymes known; it contains 60 subunits! The reaction can be summarized as

$$\text{pyruvate} + NAD^+ + \text{CoA} \rightarrow \text{acetyl-CoA} + NADH + CO_2 + H^+$$

The molecule of NADH produced is used later to produce ATP. The acetyl group is fed into the citric acid cycle, with the CoA being recycled for another oxidation of pyruvate. The citric acid cycle then completes the oxidation of the original carbons from glucose.

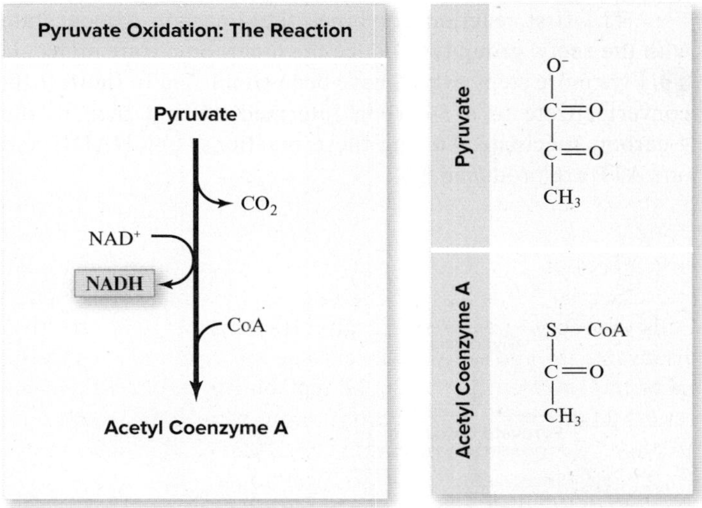

Figure 7.9 The oxidation of pyruvate. This complex reaction uses NAD^+ to accept electrons, reducing it to NADH. The product, acetyl coenzyme A (acetyl-CoA), feeds the acetyl unit into the citric acid cycle, and the CoA is recycled for another oxidation of pyruvate. NADH provides energetic electrons for the electron transport chain.

The Citric Acid Cycle Oxidizes 2-Carbon Acetyl Groups

LEARNING OBJECTIVE 7.3.2 Explain the fate of the electrons produced as products of the citric acid cycle.

Historically, the citric acid cycle has been known by multiple names: Krebs cycle, citric acid cycle, and TCA cycle. The first is for the codiscoverer of the sequence of reactions, and the next two for the key intermediate, citric acid, which is a tricarboxylic acid (TCA). Notice that under physiological conditions, citric acid will exist as its conjugate base citrate.

The reactions of the cycle oxidize 2-carbon units, which enter the cycle as acetyl-CoA. The acetyl-CoA comes from the oxidation of pyruvate, and the oxidation of fatty acids (refer to section 7.8). The acetyl group is added to a 4-carbon acid, oxaloacetate, to produce citrate. The reactions of the cycle convert citrate back to oxaloacetate, generating CO_2 and transferring electrons and protons to the electron carriers NADH and $FADH_2$. A single ATP is generated during the cycle as well, but most of the energy released is retained in the form of the electrons in NADH and $FADH_2$. These can be used by the electron transport chain to generate a proton gradient to drive ATP synthesis.

An overview of the citric acid cycle

The citric acid cycle does not itself produce a large amount of ATP, but it does produce a large amount of NADH, which can transfer electrons to the electron transport chain. The role of the citric acid cycle is to harvest these electrons by a series of oxidation reactions that regenerate the starting material.

The nine reactions of the citric acid cycle take in 2-carbon units in the form of acetyl-CoA (figure 7.10) and oxidize them, transferring electrons and protons to NADH and $FADH_2$.

The first reaction combines the 4-carbon oxaloacetate with the acetyl group to produce the 6-carbon citrate molecule ❶. Five more steps, which have been simplified in figure 7.10, convert citrate to a 5-carbon intermediate and then to the 4-carbon succinate. During these reactions, two NADH and one ATP are produced ❷.

Succinate undergoes three additional reactions, also simplified in the figure, to become oxaloacetate ❸. During these reactions, one more NADH is produced; in addition, a molecule of flavin adenine dinucleotide (FAD), another coenzyme, becomes reduced to $FADH_2$.

The Citric Acid Cycle Consists of Nine Reactions

LEARNING OBJECTIVE 7.3.3 Describe the nine reactions of the citric acid cycle, stating where in the cell they take place.

Figure 7.11 provides an overview of the nine citric acid cycle reactions: a 2-carbon group from acetyl-CoA enters the cycle at the beginning, and two CO_2 molecules, one ATP, and four pairs of electrons are produced.

Reaction 1: Condensation Citrate is formed by joining acetyl-CoA to oxaloacetate. This condensation reaction is irreversible, committing the 2-carbon acetyl group to the citric acid cycle. The reaction is inhibited when the cell's ATP concentration is high, and the reaction is stimulated when it is low. The result is that

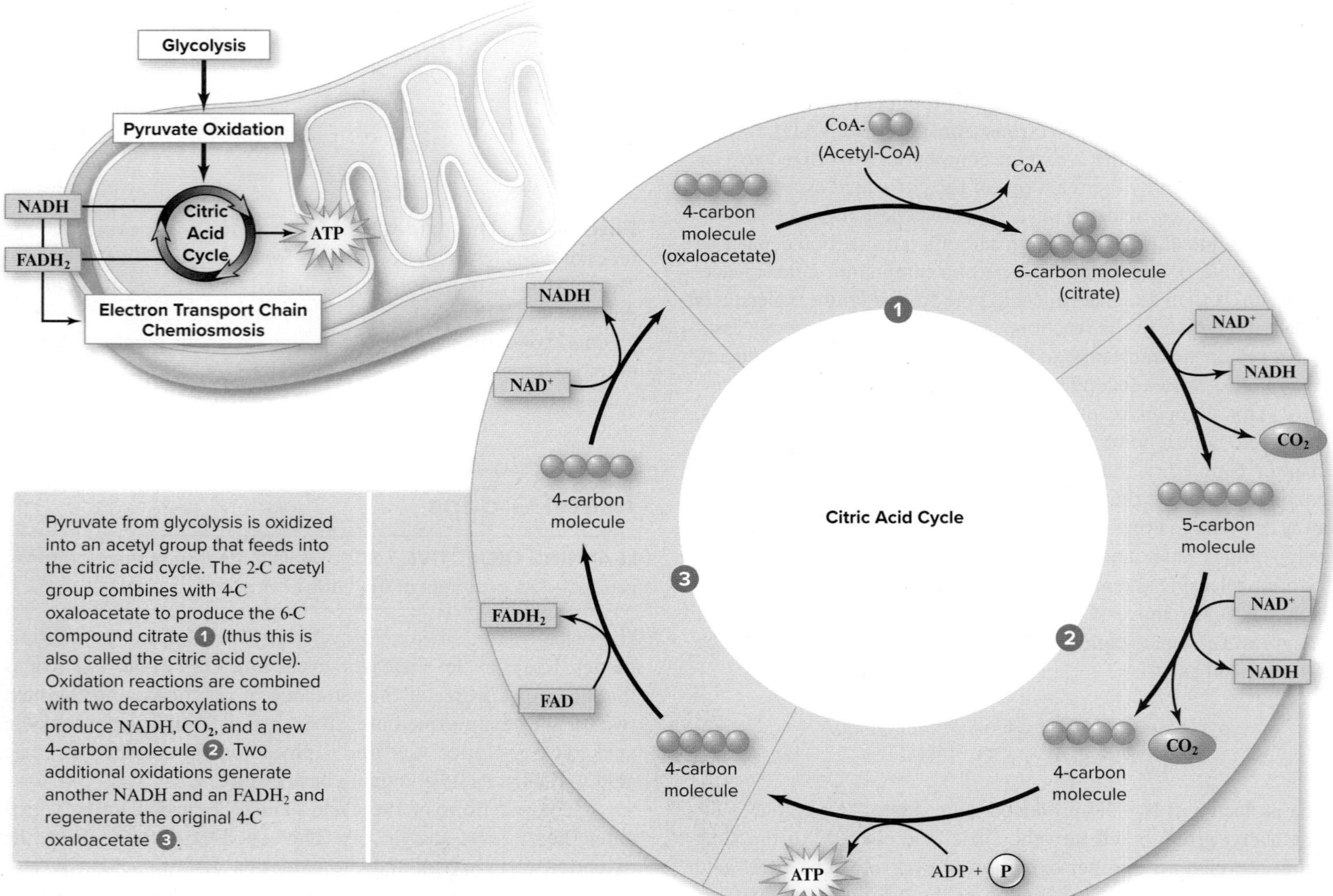

Pyruvate from glycolysis is oxidized into an acetyl group that feeds into the citric acid cycle. The 2-C acetyl group combines with 4-C oxaloacetate to produce the 6-C compound citrate ❶ (thus this is also called the citric acid cycle). Oxidation reactions are combined with two decarboxylations to produce NADH, CO_2, and a new 4-carbon molecule ❷. Two additional oxidations generate another NADH and an $FADH_2$ and regenerate the original 4-C oxaloacetate ❸.

Figure 7.10 How the citric acid cycle works.

Figure 7.11 The citric acid cycle. This series of reactions takes place within the matrix of the mitochondrion. For the complete breakdown of a molecule of glucose, the two molecules of acetyl-CoA produced by glycolysis and pyruvate oxidation each have to make a trip around the citric acid cycle. Follow the different carbons through the cycle, and notice the changes that occur in the carbon skeletons of the molecules and where oxidation reactions take place as they proceed through the cycle.

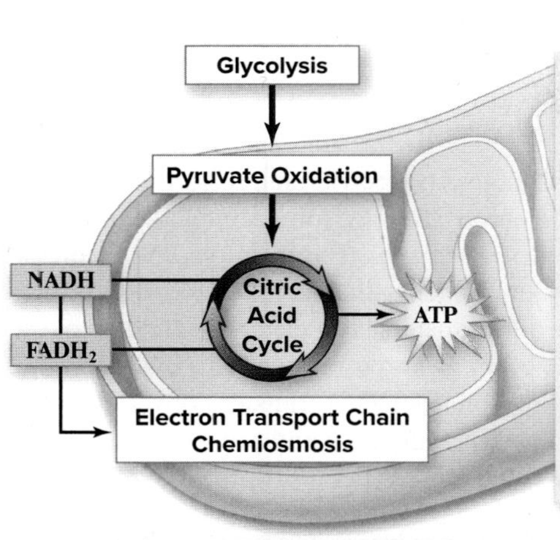

1. Reaction 1: Condensation
2–3. Reactions 2 and 3: Isomerization
4. Reaction 4: The first oxidation
5. Reaction 5: The second oxidation
6. Reaction 6: Substrate-level phosphorylation
7. Reaction 7: The third oxidation
8–9. Reactions 8 and 9: Regeneration of oxaloacetate and the fourth oxidation

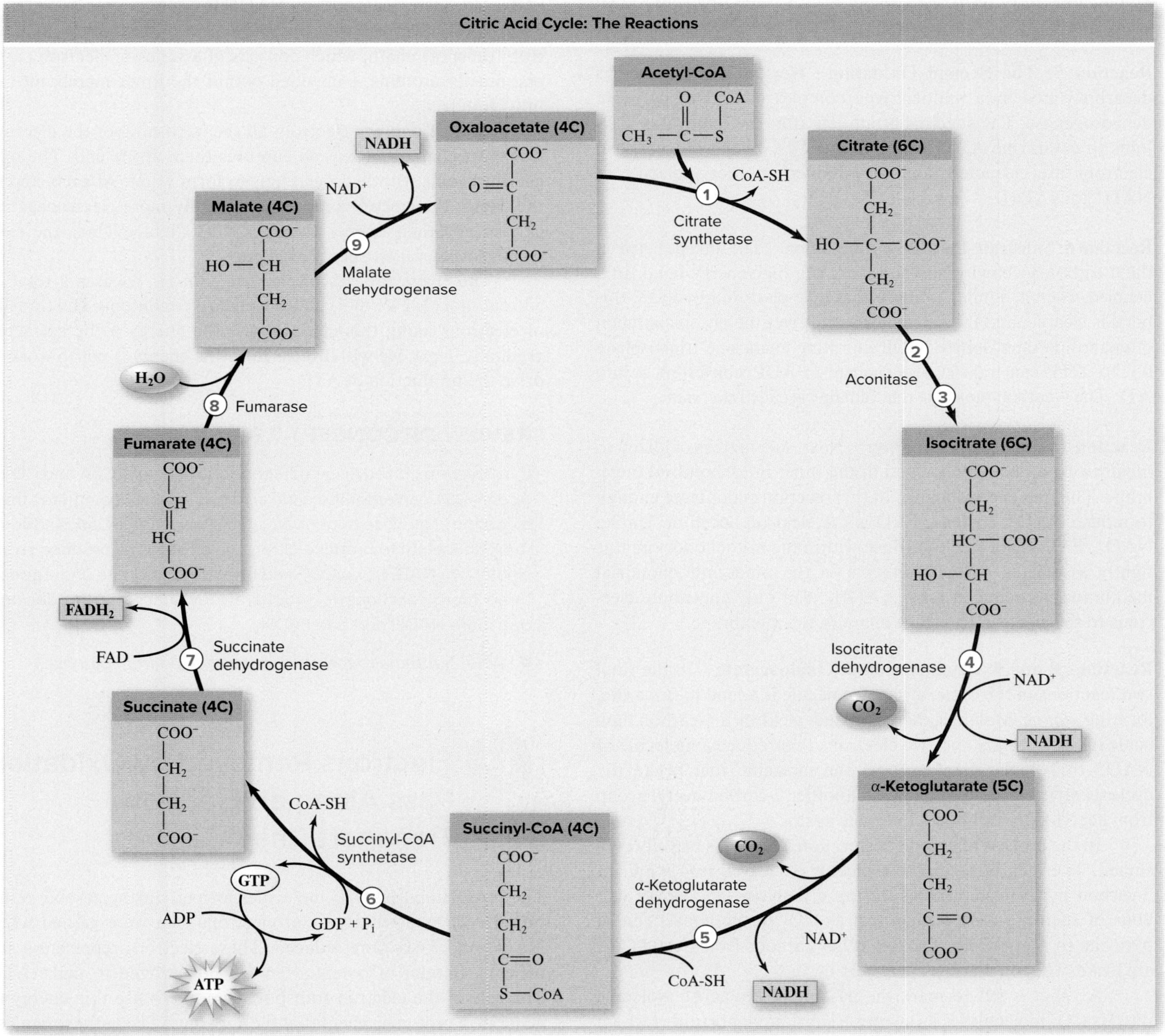

when the cell possesses ample amounts of ATP, the citric acid cycle shuts down and acetyl-CoA is channeled into fat synthesis.

Reactions 2 and 3: Isomerization Before the oxidation reactions can begin, the hydroxyl (—OH) group of citrate must be repositioned. This rearrangement is done in two steps: first, a water molecule is removed from one carbon; then water is added to a different carbon. As a result, an —H group and an —OH group change positions. The product is an isomer of citrate called *isocitrate.* This rearrangement facilitates the subsequent reactions.

Reaction 4: The First Oxidation In the first energy-yielding step of the cycle, isocitrate undergoes an oxidative decarboxylation reaction. First, isocitrate is oxidized, yielding a pair of electrons that reduce a molecule of NAD^+ to NADH. Then the oxidized intermediate is decarboxylated; the central carboxyl group splits off to form CO_2, yielding a 5-carbon molecule called *α-ketoglutarate.*

Reaction 5: The Second Oxidation Next, α-ketoglutarate is decarboxylated by a multienzyme complex similar to pyruvate dehydrogenase. The succinyl group left after the removal of CO_2 joins to coenzyme A, forming *succinyl-CoA.* In the process, two electrons are extracted, and they reduce another molecule of NAD^+ to NADH.

Reaction 6: Substrate-Level Phosphorylation The linkage between the 4-carbon succinyl group and CoA is a high-energy bond. In a coupled reaction similar to those that take place in glycolysis, this bond is cleaved, and the energy released drives the phosphorylation of guanosine diphosphate (GDP), forming guanosine triphosphate (GTP). GTP can transfer a phosphate to ADP, converting it into ATP. The 4-carbon molecule that remains is called *succinate.*

Reaction 7: The Third Oxidation Next, succinate is oxidized to *fumarate* by an enzyme located in the inner mitochondrial membrane. The free-energy change in this reaction is not large enough to reduce NAD^+. Instead, FAD is the electron acceptor. Unlike NAD^+, FAD is not free to diffuse within the mitochondrion; it is tightly associated with its enzyme in the inner mitochondrial membrane. Its reduced form, $FADH_2$, can only contribute electrons to the electron transport chain in the membrane.

Reactions 8 and 9: Regeneration of Oxaloacetate In the final two reactions of the cycle, a water molecule is added to fumarate, forming *malate.* Malate is then oxidized, yielding a 4-carbon molecule of *oxaloacetate* and two electrons that reduce a molecule of NAD^+ to NADH. Oxaloacetate, the molecule that began the cycle, is now free to combine with another 2-carbon acetyl group from acetyl-CoA and begin the cycle again.

In the process of aerobic respiration, glucose is entirely consumed. The 6-carbon glucose molecule is cleaved into a pair of 3-carbon pyruvate molecules during glycolysis. One of the carbons of each pyruvate is then lost as CO_2 in the conversion of pyruvate to acetyl-CoA. The two other carbons from acetyl-CoA are lost as CO_2 during the citric acid cycle.

All that is left to mark the passing of a glucose molecule into six CO_2 molecules is its energy, a substantial portion of which is preserved in four ATP molecules and in the reduced state of 12 electron carriers. Ten of these carriers are NADH molecules; the other two are $FADH_2$. The four ATP molecules are produced by substrate-level phosphorylation: two during glycolysis, and one for each turn of the citric acid cycle.

Following the electrons in the reactions

As you examine the changes in electric charge in the reactions that oxidize glucose, a good strategy for keeping the transfers clear is always to *follow the electrons.* For example, in glycolysis, an enzyme extracts two hydrogens—that is, two electrons and two protons—from glucose and transfers both electrons and one of the protons to NAD^+. The other proton is released as a hydrogen ion, H^+, into the surrounding solution. This transfer converts NAD^+ into NADH; that is, two electrons ($2e^-$) and one proton (H^+) are added to one positively charged NAD^+ to form NADH, which is electrically neutral.

The energy captured by NADH is not harvested all at once. The two electrons carried by NADH are passed along the electron transport chain, which consists of a series of electron carriers, mostly proteins, embedded within the inner membranes of mitochondria.

NADH delivers electrons to the beginning of the electron transport chain, and oxygen captures them at the end. The oxygen then joins with hydrogen ions to form water. At each step in the chain, the electrons move to a slightly more electronegative carrier, and their energy positions shift slightly. Thus, the electrons move *down* an energy gradient.

The entire process of electron transfer releases a total of 53 kcal/mol (222 kJ/mol) under standard conditions. The transfer of electrons along this chain allows the energy to be extracted gradually. Next, we will discuss how this energy is put to work to drive the production of ATP.

REVIEW OF CONCEPT 7.3

Pyruvate is oxidized in the mitochondria to acetyl-CoA and CO_2. Acetyl-CoA can enter the citric acid cycle, which completes the oxidation of glucose begun with glycolysis. Acetyl-CoA is added to oxaloacetate to produce citrate. Five reactions produce succinate, two NADH, two CO_2, and one ATP. Succinate undergoes three more reactions to regenerate oxaloacetate, producing one more NADH and one $FADH_2$.

- *What happens to the electrons removed from glucose?*

7.4 Electrons Removed by Oxidation Pass Along an Electron Transport Chain

The NADH and $FADH_2$ molecules formed during aerobic respiration each contain a pair of electrons that were gained when NAD^+ and FAD were reduced. These electrons represent a significant amount of potential energy not utilized to make ATP. The role of the electron transport chain is to use this energy to generate a proton gradient that the enzyme ATP synthase can use to drive the synthesis of ATP.

The Electron Transport Chain Produces a Proton Gradient

LEARNING OBJECTIVE 7.4.1 Describe the journey of an electron through the electron transport chain, identifying its final destination.

The NADH molecules carry their electrons to the inner mitochondrial membrane, where they transfer the electrons to a series of membrane-associated proteins collectively called the *electron transport chain.*

The first of the proteins to receive the electrons is a complex, membrane-embedded enzyme called **NADH dehydrogenase.** A carrier called *ubiquinone* then passes the electrons to a protein–cytochrome complex called the bc_1 complex. Each complex in the chain operates as a proton pump, driving a proton out across the membrane into the intermembrane space (figure 7.12).

The electrons are then carried by another carrier, *cytochrome c,* to the cytochrome oxidase complex. This complex uses four electrons to reduce a molecule of oxygen. Each oxygen atom then combines with two protons to form water:

$$O_2 + 4\ H^+ + 4\ e^- \rightarrow 2\ H_2O$$

In contrast to NADH, which contributes its electrons to NADH dehydrogenase, $FADH_2$, which is located in the inner mitochondrial membrane, feeds its electrons to ubiquinone, which is also in the membrane. Electrons from $FADH_2$ thus "skip" the first step in the electron transport chain.

The plentiful availability of a strong electron acceptor, oxygen, is what makes oxidative respiration possible. As you'll learn in chapter 8, the electron transport chain used in aerobic respiration is similar to, and may well have evolved from, the chain employed in photosynthesis.

Respiration takes place within the mitochondria present in nearly all eukaryotic cells. The internal compartment, or matrix, of a mitochondrion contains the enzymes that carry out the reactions of the citric acid cycle. As detailed in section 7.1, when electrons are transferred to NAD^+, protons (H^+) are produced. As the electrons harvested by oxidative respiration are passed along the electron transport chain, the energy they release transports protons out of the matrix and into the outer compartment called the intermembrane space.

Three transmembrane complexes of the electron transport chain in the inner mitochondrial membrane actually accomplish the proton transport (figure 7.12). The flow of highly energetic electrons induces a change in the shape of pump proteins, which causes them to transport protons across the membrane. The electrons contributed by NADH activate all three of these proton pumps, whereas those contributed by $FADH_2$ activate only two because of where they enter the chain. The overall effect is to increase the concentration of protons in the intermembrane space relative to the matrix.

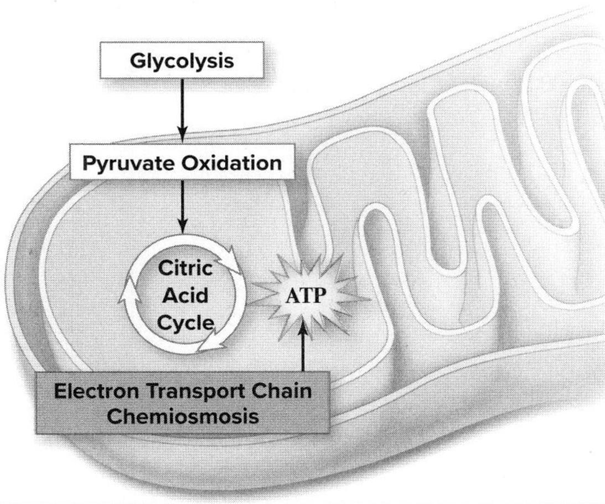

Figure 7.12 Electron transport and chemiosmosis. ***a.*** High-energy electrons removed from catabolized molecules are transported by the mobile electron carriers ubiquinone (Q) and cytochrome c (C) between three membrane complexes. Energy released by this process is used to pump protons out of the matrix and into the intermembrane space. The electrons are finally used to reduce oxygen, forming water. ***b.*** The electrochemical gradient of protons supplies energy for ATP synthesis by chemiosmosis.

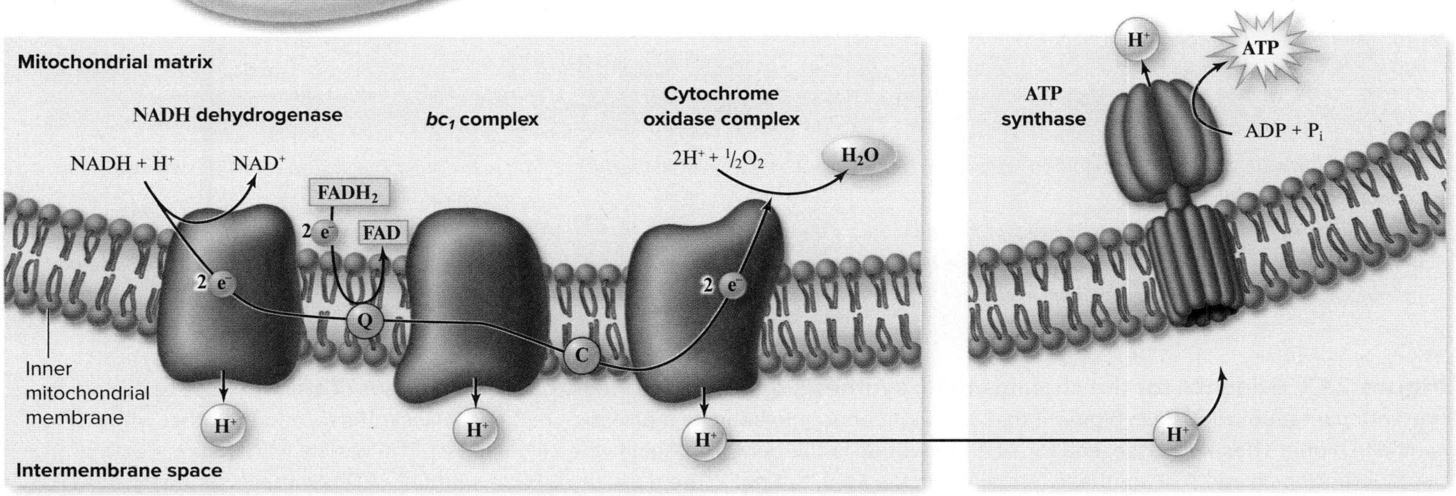

a. The electron transport chain

b. Chemiosmosis

Chemiosmosis Utilizes the Proton Gradient to Produce ATP

LEARNING OBJECTIVE 7.4.2 Describe the location, structure, and chemiosmotic function of ATP synthase.

Because the mitochondrial matrix is negative compared with the intermembrane space, positively charged protons are attracted to the matrix. The higher outer concentration of protons also tends to drive protons back in by diffusion, but because membranes are relatively impermeable to ions, this process occurs only very slowly. Most of the protons that reenter the matrix instead pass through ATP synthase, an enzyme channel that uses the energy of the gradient to catalyze the synthesis of ATP from ADP and P_i. Because the chemical formation of ATP is driven by a diffusion force similar to osmosis, this process is referred to as *chemiosmosis*. The newly formed ATP is transported by facilitated diffusion to the many places in the cell where enzymes require energy to drive endergonic reactions. This chemiosmotic mechanism for the coupling of electron transport and ATP synthesis was controversial when it was proposed. Over the years, experimental evidence accumulated to support this hypothesis (figure 7.13).

The energy released by the reactions of cellular respiration ultimately drives the proton pumps that produce the proton gradient. The proton gradient provides the energy required for the synthesis of ATP. Figure 7.14 summarizes the overall process.

ATP synthase is a molecular rotary motor

ATP synthase uses a fascinating molecular mechanism to perform ATP synthesis (figure 7.15). Structurally, the enzyme has a membrane-bound portion and a narrow stalk that connects the membrane portion to a knoblike catalytic portion.

SCIENTIFIC THINKING

Hypothesis: *ATP synthase enzyme uses a proton gradient to provide energy for phosphorylation reaction.*

Prediction: *The source of the proton gradient should not matter. A proton gradient formed by the light-driven pump bacteriorhodopsin should power phosphorylation in the light but not in the dark.*

Test: *Artificial vesicles are made with bacteriorhodopsin and ATP synthase, and ATP synthase alone. These are illuminated with light and assessed for ATP production.*

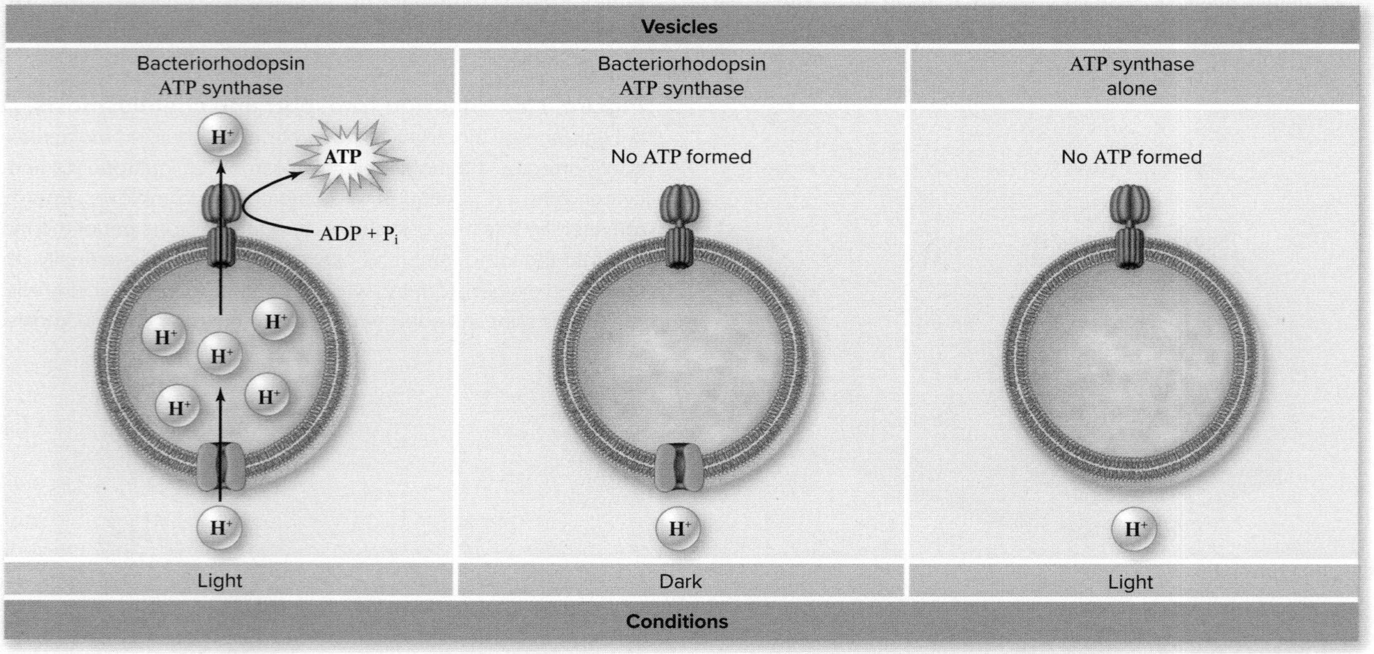

Result: *The vesicle with both bacteriorhodopsin and ATP synthase can form ATP in the light but not in the dark. The vesicle with ATP synthase alone cannot form ATP in the light.*

Conclusion: *ATP synthase is able to utilize a proton gradient for energy to form ATP.*

Further Experiments: *What other controls would be appropriate for this type of experiment? Why is this experiment a better test of the chemiosmotic hypothesis than the acid bath experiment in chapter 8 (refer to figure 8.16)?*

Figure 7.13 Evidence for the chemiosmotic synthesis of ATP by ATP synthase. Two kinds of artificial vesicles are made, one with bacteriorhodopsin (a pigment that uses light energy to activate a proton pump) and ATP synthase, and the other with ATP synthase alone. These are assessed for ATP production in the dark, and again when illuminated. The vesicle with both bacteriorhodopsin and ATP synthase forms ATP in the light, but not in the dark; the vesicle with only ATP synthase cannot form ATP even when illuminated. Conclusion: ATP synthase requires a proton gradient to form ATP.

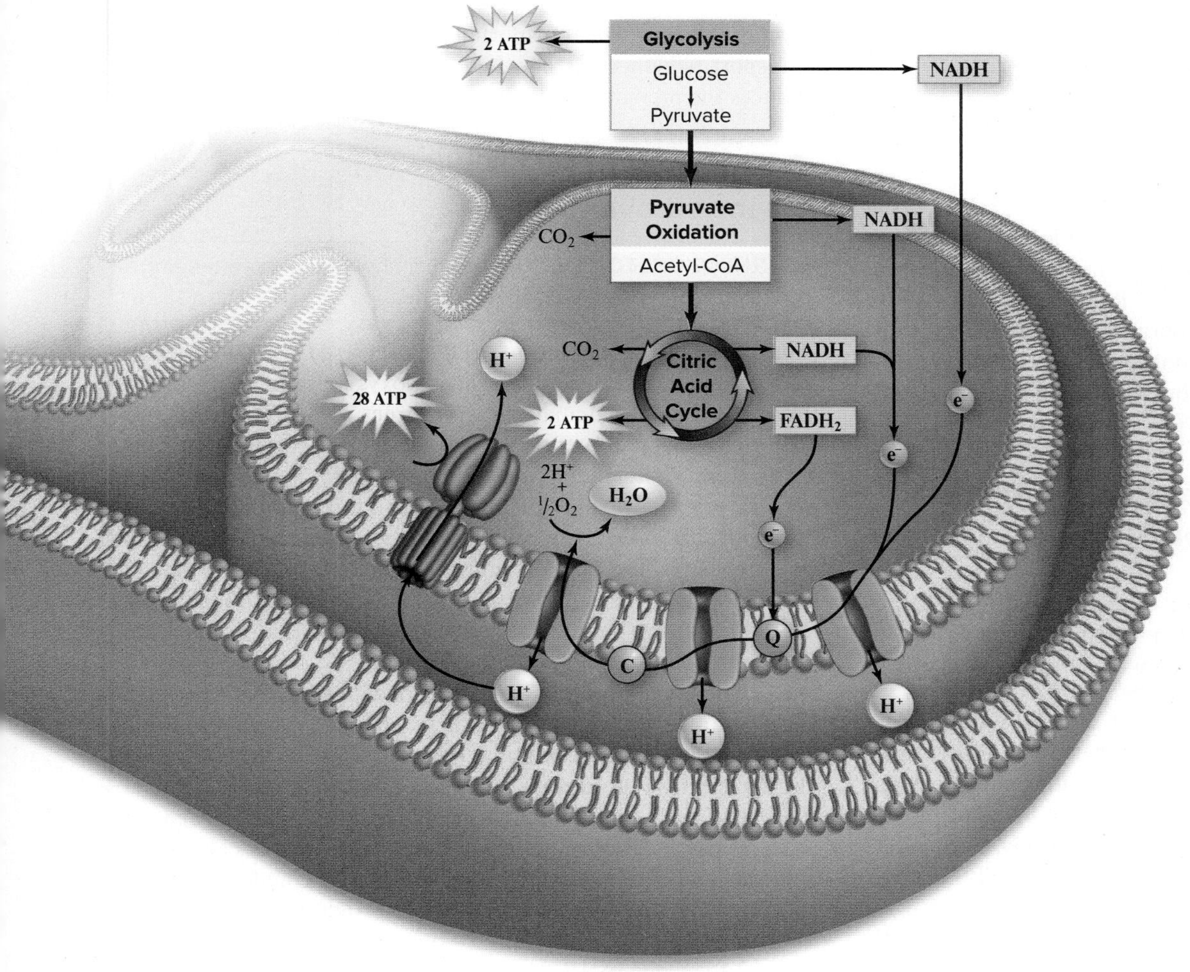

Figure 7.14 Overview of aerobic respiration in the mitochondria. The entire process of aerobic respiration is shown in cellular context. Glycolysis occurs in the cytoplasm with the pyruvate and NADH produced entering the mitochondria. Here, pyruvate is oxidized and fed into the citric acid cycle to complete the oxidation process. All the energetic electrons harvested by oxidations in the overall process are transferred by NADH and $FADH_2$ to the electron transport chain. The electron transport chain uses the energy released during electron transport to pump protons across the inner membrane. This creates an electrochemical gradient that contains potential energy. The enzyme ATP synthase uses this gradient to phosphorylate ADP to form ATP.

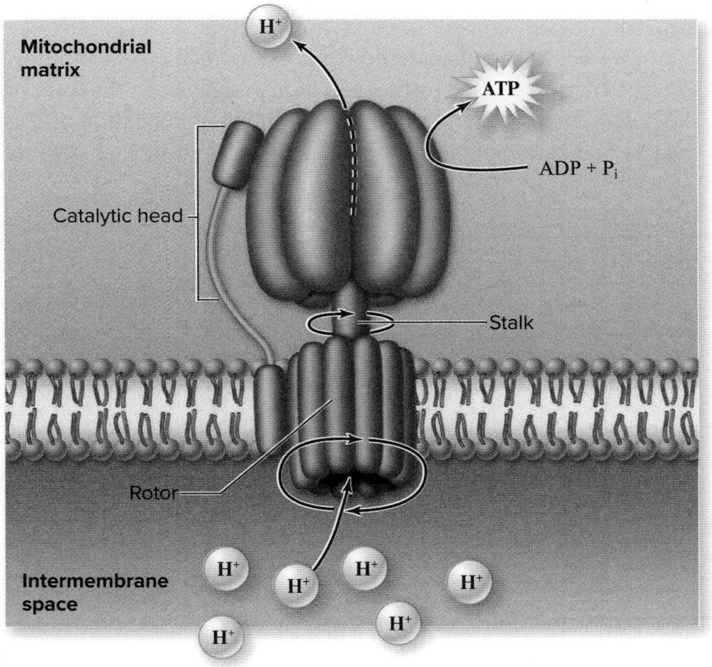

Figure 7.15 The ATP rotary engine. Protons move across the membrane down their concentration gradient. The energy released causes the rotor and stalk structures to rotate. This mechanical energy alters the conformation of the ATP synthase enzyme to catalyze the formation of ATP.

This complex can be dissociated into two subportions: the F_0 membrane-bound complex, and the F_1 complex composed of the stalk and a knob, or head domain.

The F_1 complex has enzymatic activity. The F_0 complex contains a channel through which protons move across the membrane down their concentration gradient. As they do so, their movement causes part of the F_0 complex and the stalk to rotate relative to the knob. The mechanical energy of this rotation is used to change the conformation of the catalytic domain in the F_1 complex.

Thus, the synthesis of ATP is achieved by a tiny rotary motor, the rotation of which is driven directly by a gradient of protons. The flow of protons is like that of water in a hydroelectric power plant. As the flow of water driven by gravity causes a turbine to rotate and generate electric current, the proton gradient produces the energy that drives the rotation of the ATP synthase generator.

REVIEW OF CONCEPT 7.4

The electron transport chain receives electrons from NADH and $FADH_2$ and passes them down the chain to oxygen, using the energy from electron transfer to pump protons across the membrane, creating an electrochemical gradient. The enzyme ATP synthase uses this gradient to drive the endergonic reaction of phosphorylating ADP to ATP.

- *How would poking a small hole in the outer membrane affect ATP synthesis?*

7.5 The Total Energy Yield of Aerobic Respiration Far Exceeds That of Glycolysis

Ever since the link was discovered between electron transport and the proton gradient used by ATP synthase, biochemists have attempted to determine the number of ATPs produced per NADH feeding electrons into electron transport. This number has proved to be surprisingly elusive. Early estimates were based on erroneous assumptions, but we now have both theoretical and calculated values that are in agreement.

The Theoretical Yield for Eukaryotes Is 30 Molecules of ATP per Glucose Molecule

LEARNING OBJECTIVE 7.5.1 Calculate the number of ATP produced by a cell via aerobic respiration.

The number of molecules of ATP produced by ATP synthase per molecule of glucose depends on the number of protons transported across the inner membrane and the number of protons needed per ATP synthesized. The number of protons transported per NADH and $FADH_2$ is 10 and 6 H^+, respectively. Each ATP synthesized requires 4 H^+, leading to $10/4 = 2.5$ ATP/NADH, and $6/4 = 1.5$ ATP/$FADH_2$.

Oxidizing glucose to pyruvate via glycolysis yields 2 ATP directly, and $2 \times 2.5 = 5$ ATP from NADH. The oxidation of pyruvate to acetyl-CoA yields another $2 \times 2.5 = 5$ ATP from NADH. Finally, the citric acid cycle produces 2 ATP directly, $6 \times 2.5 = 15$ ATP from NADH, and $2 \times 1.5 = 3$ ATP from $FADH_2$. Summing all of these leads to 32 ATP for respiration (figure 7.16).

This number is accurate for bacteria, but it does not hold for eukaryotes, because the NADH produced in the cytoplasm by glycolysis needs to be transported into the mitochondria by active transport, which costs one ATP per NADH transported. This reduces the predicted yield for eukaryotes to 30 ATP.

Calculation of P/O Ratios Has Changed over Time

LEARNING OBJECTIVE 7.5.2 Explain how our understanding of the P/O ratio has changed over time.

The value for the amount of ATP synthesized per O_2 molecule reduced is called the phosphate-to-oxygen ratio (P/O ratio). Both theoretical calculations and direct measurement of this value have been contentious issues. When theoretical calculations were first made, we lacked detailed knowledge of the respiratory chain and the mechanism for coupling electron transport to ATP synthesis. Because redox reactions occur at three sites for NADH and two sites for $FADH_2$, it was assumed that three molecules of ATP were produced per NADH and two per $FADH_2$. We now know that assumption was overly simplistic.

Understanding that a proton gradient is the link between electron transport and ATP synthesis changed the nature of the calculations. We need to know the number of protons pumped during electron transport: 10 H^+ per NADH, and 6 H^+ per $FADH_2$. Then we need to know the number of protons needed per ATP. Because ATP synthase is a rotary motor, this calculation depends on the number of binding sites for ATP and the number of protons required for rotation. We know that ATP synthase has three binding sites for ATP. If 12 protons are used per rotation, you get the value of 4 H^+ per ATP used in the previous calculation. Actual measurements of the P/O ratio have been problematic, but they now appear to be at most 2.5.

We can also calculate how efficiently respiration captures the free energy released by the oxidation of glucose in the form of ATP. The amount of free energy released by the oxidation of glucose is 686 kcal/mol, and the free energy stored in each ATP is

Figure 7.16 Theoretical ATP yield. The theoretical yield of ATP harvested from glucose by aerobic respiration totals 32 molecules. In eukaryotes this is reduced to 30 because it takes 1 ATP to transport each molecule of NADH that is generated by glycolysis in the cytoplasm into the mitochondria.

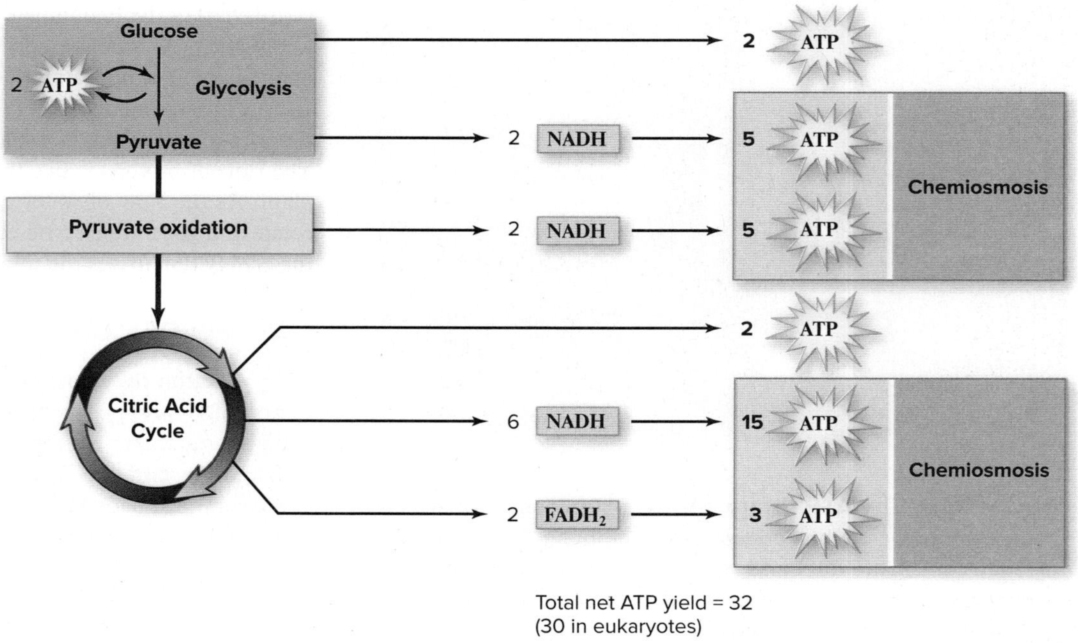

7.3 kcal/mol. Therefore, a eukaryotic cell harvests about (7.3 × 30)/686 = 32% of the energy available in glucose. (By comparison, a typical car converts only about 25% of the energy in gasoline into useful energy.)

The higher energy yield of aerobic respiration was one of the key factors that fostered the evolution of heterotrophs. As this mechanism for producing ATP evolved, nonphotosynthetic organisms became more effective at using respiration to extract energy from molecules derived from other organisms. As long as some organisms captured energy by photosynthesis, others could exist solely by feeding on them.

REVIEW OF CONCEPT 7.5

The number of ATP synthesized by chemiosmosis depends on the number of protons translocated by electron transport and the number used by ATP synthase. The maximum number of ATP per NADH is 2.5 (1.5 per $FADH_2$). This results in 32 ATP per glucose for prokaryotes and 30 for eukaryotes due to transport of cytoplasmic NADH into mitochondria. The amount of ATP per O_2 reduced is called the P/O ratio; this value has been contentious, but theoretical and measured values have converged.

- *What factors affect the yield of ATP by chemiosmosis?*

7.6 Aerobic Respiration Is Regulated by Feedback Inhibition

When cells possess plentiful amounts of ATP, the key reactions of glycolysis, the citric acid cycle, and fatty acid breakdown are inhibited, slowing ATP production. The regulation of these biochemical pathways by the level of ATP is an example of feedback inhibition. Conversely, when ATP levels in the cell are low, ADP levels are high, and ADP activates enzymes in the pathways of carbohydrate catabolism, to stimulate the production of more ATP.

There Are Two Key Points of Control

LEARNING OBJECTIVE 7.6.1 Identify the two key points at which cells can control the cellular respiration process.

Control of glucose catabolism occurs at two key points in the catabolic pathway, namely, at a point in glycolysis and at the beginning of the citric acid cycle (figure 7.17). The control point in glycolysis is the enzyme phosphofructokinase, which catalyzes the conversion of fructose phosphate to fructose bisphosphate. This is the first reaction of glycolysis that is not readily reversible, committing the substrate to the glycolytic sequence. ATP itself is an allosteric inhibitor (refer to chapter 6) of phosphofructokinase, as is the citric acid cycle intermediate citrate. High levels of both ATP and citrate inhibit phosphofructokinase. Thus, under conditions when ATP is in excess, or when the citric acid cycle is producing citrate faster than it is being consumed, glycolysis is slowed.

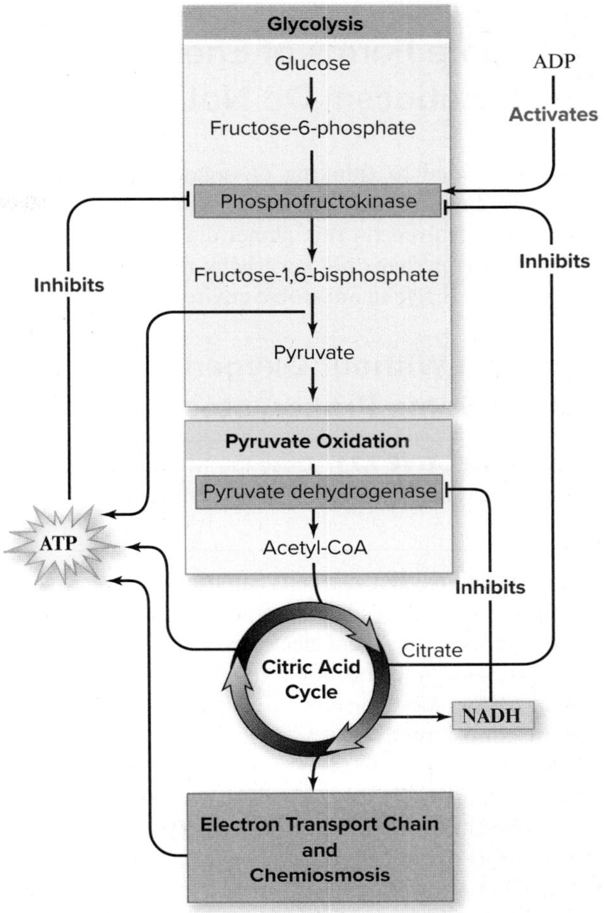

Figure 7.17 Control of glucose catabolism. The relative levels of ADP and ATP and key intermediates NADH and citrate control the catabolic pathway at two key points: the committing reactions of glycolysis and the citric acid cycle.

The main control point in the oxidation of pyruvate occurs at the committing step in the citric acid cycle with the enzyme pyruvate dehydrogenase, which converts pyruvate to acetyl-CoA. This enzyme is inhibited by high levels of NADH, a key product of the citric acid cycle.

Another control point in the citric acid cycle is the enzyme citrate synthetase, which catalyzes the first reaction, the conversion of oxaloacetate and acetyl-CoA into citrate. High levels of ATP inhibit citrate synthetase (as well as phosphofructokinase, pyruvate dehydrogenase, and two other citric acid cycle enzymes), slowing down the entire catabolic pathway.

REVIEW OF CONCEPT 7.6

Respiration is controlled by levels of ATP in the cell and levels of key intermediates in the process. The control point for glycolysis is the enzyme phosphofructokinase, which is inhibited by ATP or citrate (or both). The main control point in oxidation of pyruvate is the enzyme pyruvate dehydrogenase, inhibited by NADH.

- *How does feedback inhibition ensure economic production of ATP?*

7.7 Some Forms of Energy Metabolism Do Not Require O_2

We are accustomed to thinking of organisms as living in an oxygen-based environment, but many do not. In fact, the first organisms that evolved on this planet did not have any oxygen available. So it is perhaps not surprising that there are ways for living systems to thrive in anaerobic environments.

Respiration Without Oxygen Occurs in Some Prokaryotes

LEARNING OBJECTIVE 7.7.1 Describe two ways in which prokaryotes carry out respiration in the complete absence of oxygen.

In the presence of oxygen gas, cells can use the oxygen as an electron acceptor and produce large amounts of ATP. But even when no oxygen is present to accept electrons, some organisms can still respire *anaerobically,* using other inorganic molecules as final electron acceptors for an electron transport chain.

For example, many prokaryotes use sulfur, nitrate, carbon dioxide, or even inorganic metals as the final electron acceptor in place of oxygen (figure 7.18). The amount of free energy released using these other molecules as final electron acceptors is not as great as that released using oxygen, because these other molecules have a lower affinity for electrons. But even though less ATP is produced, the process is still respiration, not fermentation.

Methanogens use carbon dioxide

Among the heterotrophs that practice anaerobic respiration are Archaea, such as thermophiles and methanogens. **Methanogens** use carbon dioxide (CO_2) as the electron acceptor, reducing CO_2 to CH_4 (methane). The hydrogens are derived from organic molecules produced by other organisms. Methanogens are found in diverse environments, including soil and the digestive systems of ruminants such as cows.

Figure 7.18 Sulfur-respiring prokaryote. ***a.*** The micrograph shows the archaeal species *Thermoproteus tenax.* This organism can use elemental sulfur as a final electron acceptor for anaerobic respiration. ***b.*** *Thermoproteus* is often found in sulfur-containing hot springs such as the Norris Geyser Basin in Yellowstone National Park, shown here.

(a): Wolfgang Baumeister/Science Source; (b): National Park Service, photo by Jim Peaco

Sulfur bacteria use sulfate

Evidence of a second anaerobic respiratory process among primitive bacteria was found in a group of rocks about 2.7 billion years old, known as the Woman River iron formation. Organic material in these rocks is enriched for the light isotope of sulfur, ^{32}S, relative to the heavier isotope, ^{34}S. No known geochemical process produces such enrichment, but biological sulfur reduction does, in a process still carried out today by certain prokaryotes.

In this sulfate respiration, the prokaryotes derive energy from the reduction of inorganic sulfates (SO_4) to hydrogen sulfide (H_2S). The hydrogen atoms are obtained from organic molecules that have been produced by other organisms. These prokaryotes thus are similar to methanogens, but they use SO_4 as the oxidizing (electron-accepting) agent in place of CO_2.

The early sulfate reducers set the stage for the evolution of photosynthesis, creating an environment rich in H_2S. As discussed in chapter 8, the first form of photosynthesis harvested hydrogens from H_2S using the energy of sunlight.

Fermentations Donate Electrons Generated by Glycolysis to Organic Molecules Rather Than Oxygen

LEARNING OBJECTIVE 7.7.2 Define *fermentation,* and distinguish between ethanol and lactic acid fermentations.

In the absence of oxygen, cells that cannot utilize an alternative electron acceptor for respiration must rely exclusively on glycolysis to produce ATP. Under these conditions, the electrons generated by glycolysis are donated to organic molecules in a process called *fermentation.* This process recycles NAD^+, the electron acceptor that allows glycolysis to proceed.

Bacteria carry out more than a dozen kinds of fermentation reactions, often using pyruvate or a derivative of pyruvate to accept the electrons from NADH. Organic molecules other than pyruvate and its derivatives can be used as well; the important point is that the process regenerates NAD^+:

$$\text{organic molecule} + \text{NADH} \rightarrow \text{reduced organic molecule} + \text{NAD}^+$$

Often, the reduced organic compound is an organic acid—such as acetic acid, butyric acid, propionic acid, or lactic acid—or an alcohol.

Ethanol fermentation

Eukaryotic cells are capable of only a few types of fermentation. In one type, which occurs in yeast, the molecule that accepts electrons from NADH is derived from pyruvate, the end-product of glycolysis.

Yeast enzymes remove a terminal CO_2 group from pyruvate through decarboxylation, producing a 2-carbon molecule called acetaldehyde. The CO_2 released causes bread made with yeast to rise. The acetaldehyde accepts a pair of electrons from NADH, producing NAD^+ and ethanol (ethyl alcohol) (figure 7.19).

This type of fermentation is of great interest to humans, because it is the source of the ethanol in wine and beer. Ethanol is a by-product of fermentation that is actually toxic to yeast; as it approaches a concentration of about 12%, it begins to kill the yeast. That explains why naturally fermented wine contains only about 12% ethanol.

Lactic acid fermentation

Most animal cells regenerate NAD^+ in the absence of oxygen without decarboxylation. Muscle cells, for example, use the enzyme lactate dehydrogenase to transfer electrons from NADH back to the pyruvate that is produced by glycolysis.

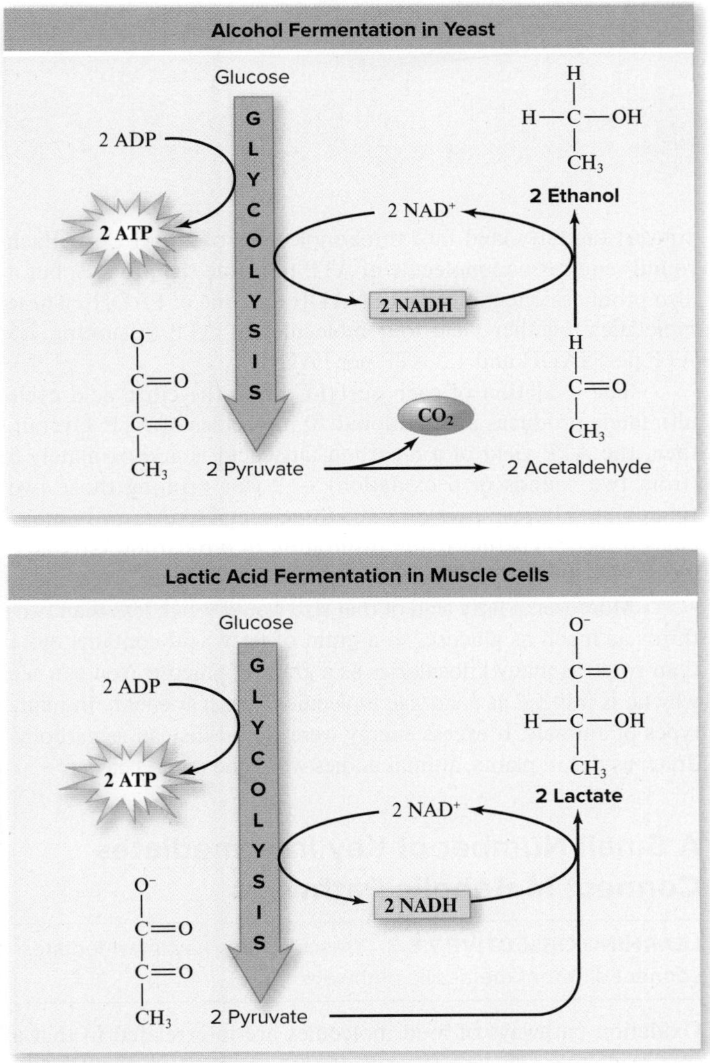

Figure 7.19 Fermentation. Yeasts carry out the conversion of pyruvate to ethanol. Muscle cells convert pyruvate into lactate, which is less toxic than ethanol. In each case, the reduction of a metabolite of glucose has oxidized NADH back to NAD^+ to allow glycolysis to continue under anaerobic conditions.

This reaction converts pyruvate into lactic acid and regenerates NAD^+ from NADH (figure 7.19). It therefore closes the metabolic circle, allowing glycolysis to continue as long as glucose is available.

Circulating blood removes excess lactate, the ionized form of lactic acid, from muscles, but when removal cannot keep pace with production, the accumulating lactic acid interferes with muscle function and contributes to muscle fatigue.

REVIEW OF CONCEPT 7.7

Nitrate, sulfur, and CO_2 are all used as terminal electron acceptors in anaerobic respiration of different organisms. Organic molecules can also accept electrons in fermentation reactions that regenerate NAD^+. Fermentation reactions produce a variety of compounds, including ethanol in yeast and lactic acid in humans.

- *In what kinds of ecosystems would you expect to find anaerobic respiration?*

7.8 Carbohydrates Are Not the Only Energy Source Used by Heterotrophs

Catabolism of Proteins Removes Amino Groups

LEARNING OBJECTIVE 7.8.1 Explain how cells extract energy from proteins.

Thus far, we have focused on the aerobic respiration of glucose, which organisms obtain from the digestion of carbohydrates or from photosynthesis. Organic molecules other than glucose, particularly proteins and fats, are also important sources of energy (figure 7.20).

Proteins are first broken down into their individual amino acids. The nitrogen-containing side group (the amino group) is then removed from each amino acid in a process called **deamination.** A series of reactions converts the carbon chain that remains into a molecule that enters glycolysis or the citric acid cycle. For example, alanine is converted into pyruvate, glutamate into α-ketoglutarate (figure 7.21), and aspartate into oxaloacetate. The reactions of cellular respiration then extract the high-energy electrons from these molecules and put them to work making ATP.

β Oxidation of Fatty Acids Produces Acetyl Groups

LEARNING OBJECTIVE 7.8.2 Calculate how many ATP can be produced by a fatty acid of a given length.

Fats are broken down into fatty acids plus glycerol. Long-chain fatty acids typically have an even number of carbons, and the many C–H bonds provide a rich harvest of energy. Fatty acids are

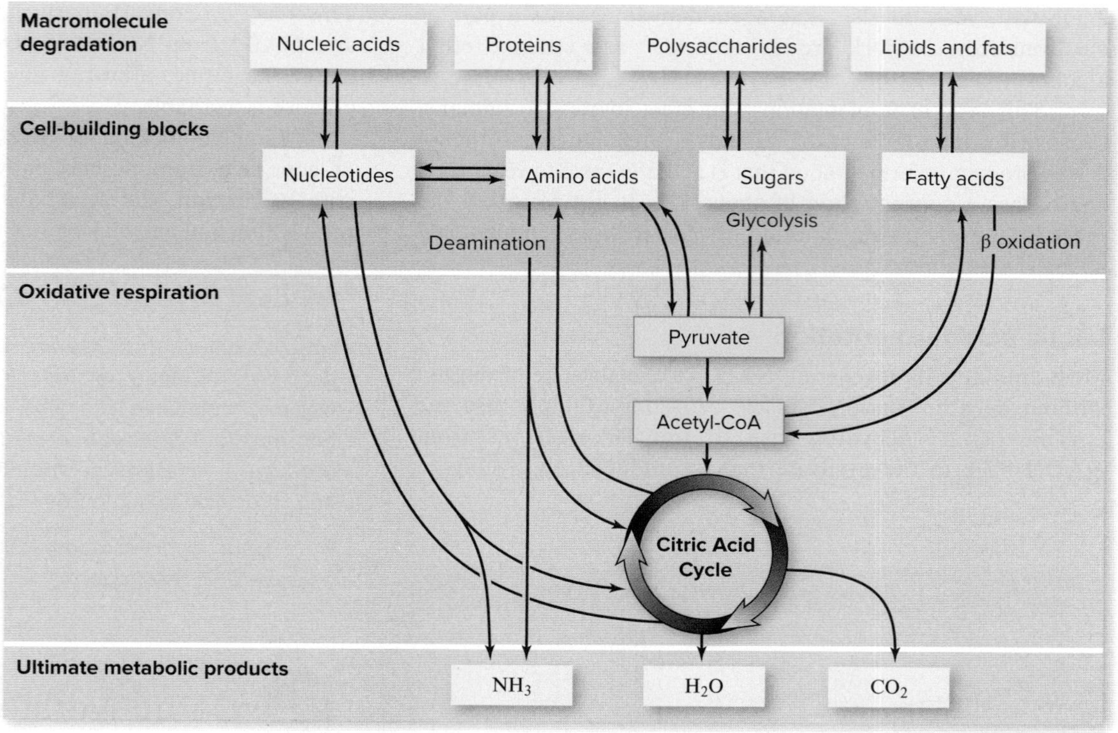

Figure 7.20 How cells extract chemical energy. All eukaryotes and many prokaryotes extract energy from organic molecules by oxidizing them. The first stage of this process, breaking down macromolecules into their constituent parts, yields little energy. The second stage, oxidative or aerobic respiration, extracts energy, primarily in the form of high-energy electrons, and produces water and carbon dioxide. Key intermediates in these energy pathways are also used for biosynthetic pathways, shown by reverse arrows.

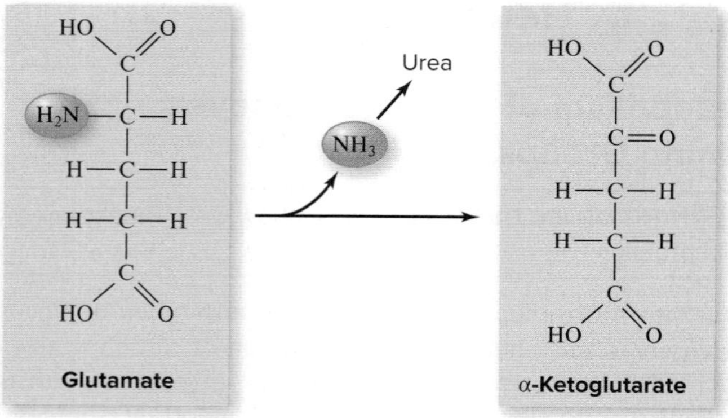

Figure 7.21 Deamination. After proteins are broken down into their amino acid constituents, the amino groups are removed from the amino acids to form molecules that participate in glycolysis and the citric acid cycle. For example, the amino acid glutamate becomes α-ketoglutarate, a citric acid cycle intermediate, when it loses its amino group.

oxidized in the matrix of the mitochondrion. Enzymes progressively remove 2-carbon acetyl groups from the terminus of each fatty acid, nibbling away at the end until the entire fatty acid is converted into acetyl groups (figure 7.22). Each acetyl group is combined with coenzyme A to form acetyl-CoA. This process is known as **β oxidation.** This process is oxygen-dependent, which explains why aerobic exercise burns fat but anaerobic exercise does not.

How much ATP does the catabolism of fatty acids produce? Let's compare a hypothetical 6-carbon fatty acid with the 6-carbon glucose molecule, which we've said yields about 30 molecules of ATP in a eukaryotic cell. Two rounds of β oxidation would convert the fatty acid into three molecules of acetyl-CoA. Each round requires one molecule of ATP to prime the process, but it also produces one molecule of NADH and one of $FADH_2$. These molecules together yield four molecules of ATP (assuming 2.5 ATP per NADH and 1.5 ATP per $FADH_2$).

The oxidation of each acetyl-CoA in the citric acid cycle ultimately produces an additional 10 molecules of ATP. Overall, then, the ATP yield of a 6-carbon fatty acid is approximately 8 (from two rounds of β oxidation) − 2 (for priming those two rounds) + 30 (from oxidizing the three acetyl-CoAs) = 36 molecules of ATP. Therefore, the respiration of a 6-carbon fatty acid yields 20% more ATP than the respiration of glucose.

Moreover, a fatty acid of that size would weigh less than two-thirds as much as glucose, so a gram of fatty acid contains more than twice as many kilocalories as a gram of glucose. You can see why fat is utilized as a storage molecule for excess energy in many types of animals. If excess energy were stored instead as carbohydrate, as it is in plants, animal bodies would be much heavier.

A Small Number of Key Intermediates Connect Metabolic Pathways

LEARNING OBJECTIVE 7.8.3 Describe how key intermediates connect different metabolic pathways.

Oxidation pathways of food molecules are interrelated in that a small number of key intermediates, such as pyruvate and acetyl-CoA, link the breakdown from different starting points. These key intermediates allow the interconversion of different types of molecules, such as sugars and amino acids (figure 7.20).

Cells can make glucose, amino acids, and fats, as well as get them from external sources. They use reactions similar to those that break down these substances. In many cases, the

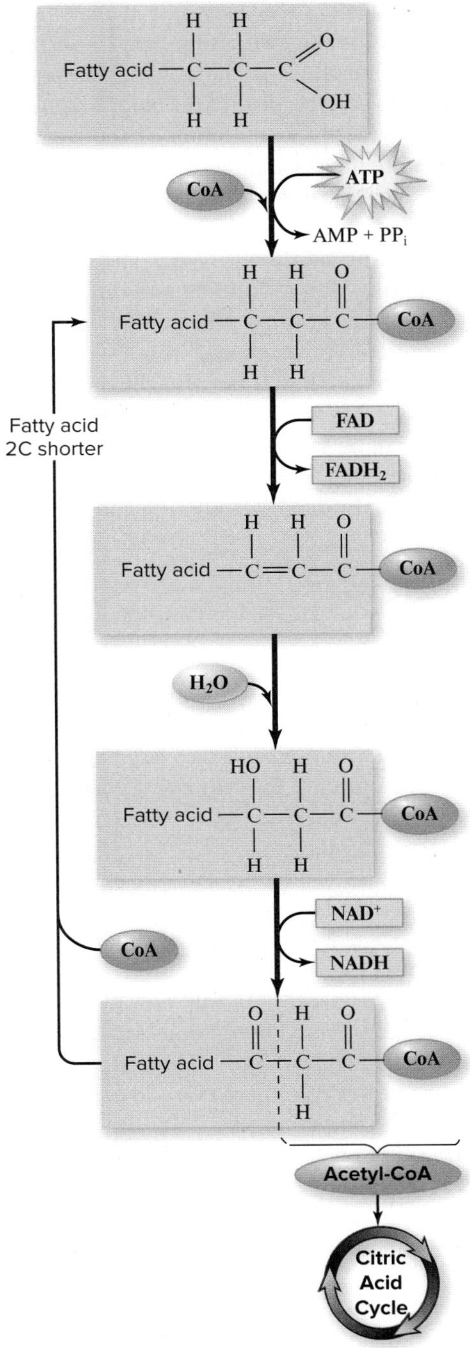

Figure 7.22 β oxidation. Through a series of reactions known as β oxidation, the last two carbons in a fatty acid combine with coenzyme A to form acetyl-CoA, which enters the citric acid cycle. The fatty acid, now two carbons shorter, enters the pathway again and keeps re-entering until all its carbons have been used to form acetyl-CoA molecules. Each round of β oxidation uses one molecule of ATP and generates one molecule each of $FADH_2$ and NADH.

reverse pathways even share enzymes if the free-energy changes are small. For example, gluconeogenesis, the process of making new glucose, uses all but three enzymes of the glycolytic pathway. Thus, much of glycolysis runs forward or backward, depending on the concentrations of the intermediates—with only three key steps having different enzymes for forward and reverse directions.

Acetyl-CoA has many roles

Many different metabolic processes generate acetyl-CoA. Not only does the oxidation of pyruvate produce it, but the metabolic breakdown of proteins, fats, and other lipids also generates acetyl-CoA. Indeed, almost all molecules catabolized for energy are converted into acetyl-CoA.

Acetyl-CoA has a role in anabolic metabolism as well. Units of two carbons derived from acetyl-CoA are used to build up the hydrocarbon chains in fatty acids. Acetyl-CoA produced from a variety of sources can therefore be channeled into fatty acid synthesis or into ATP production, depending on the organism's energy requirements. Which of these two options is taken depends on the level of ATP in the cell. When ATP levels are high, the oxidative pathway is inhibited, and acetyl-CoA is channeled into fatty acid synthesis. This explains why many animals (humans included) develop fat reserves when they consume more food than their activities require. Alternatively, when ATP levels are low, the oxidative pathway is stimulated, and acetyl-CoA flows into energy-producing oxidative metabolism.

REVIEW OF CONCEPT 7.8

Proteins can be broken into their constituent amino acids, which are deaminated and can enter metabolism at glycolysis or the citric acid cycle. Fats are broken into units of acetyl-CoA by β oxidation and then fed into the citric acid cycle. Many metabolic processes are used reversibly for anabolic and catabolic pathways. Key intermediates, such as pyruvate and acetyl-CoA, connect these processes.

- *Can fats be oxidized in the absence of O_2?*

How Do Swimming Fish Avoid Low Blood pH?

Animals that live in oxygen-poor environments, like worms living in the oxygen-free mud at the bottom of lakes, are not able to obtain energy from cellular respiration. Their cells lack the oxygen needed to act as a final electron acceptor for electron transport. Instead, these animals rely on glycolysis to obtain ATP and on regenerating NAD^+ by reducing pyruvate to lactic acid. Although the energy yield is much lower, this fermentative metabolism does not require oxygen. Even when oxygen is plentiful, the muscles of an active animal may use available oxygen, and so be forced to rely on glycolysis temporarily to generate the ATP for continued contraction.

This presents a particular problem for fish. Fish blood is much lower in carbon dioxide than yours is, and as a consequence, the amount of sodium bicarbonate acting as a buffer in fish blood is also quite low. Consider a trout, which needs to suddenly swim very fast to catch a mayfly for dinner. The vigorous swimming will cause its muscles to release large amounts of lactic acid into poorly buffered blood; this can severely disturb the blood's acid–base balance and thus impede the contraction of the trout's swimming muscles before it has captured its prey. One adaptation that helps fish solve this dilemma is that their muscles consist of two forms: red muscle that is primarily oxidative and white muscle that is primarily glycolytic. The red muscle can be activated during slow swimming and the white muscle during fast swimming to avoid predators or capture prey.

The graph presents the results of an experiment designed to explore how a trout solves this dilemma. In the experiment, a trout was made to swim vigorously for 15 minutes in a laboratory tank, then allowed a day's recovery. The lactic acid concentration in its blood was monitored periodically during swimming and recovery phases.

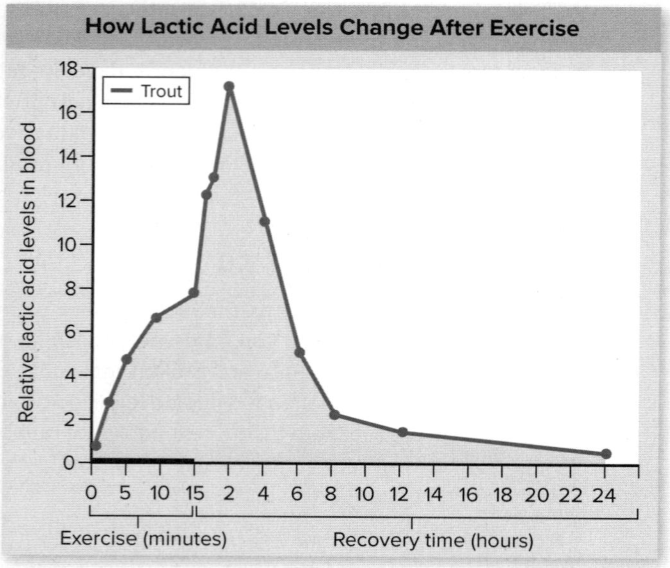
How Lactic Acid Levels Change After Exercise
Trout
Relative lactic acid levels in blood
0
2
4
6
8
10
12
14
16
18
0 5 10 15 2 4 6 8 10 12 14 16 18 20 22 24
Exercise (minutes)
Recovery time (hours)

Analysis

1. **Applying Concepts**
   a. ***Variable.*** What is the dependent variable?
   b. ***Recording data.*** Lactic acid levels are presented for both swimming and recovery periods. In what time units are the swimming data presented? The recovery data?
2. **Interpreting Data**
   a. What is the effect of exercise on the level of lactic acid in the trout's blood?
   b. Does the level of lactic acid change after exercise stops? How?
3. **Making Inferences** About how much of the total lactic acid created by vigorous swimming is released after this exercise stops? (*HINT:* Replot all points to the same scale on the *x*-axis and compare the areas under the curve.)
4. **Drawing Conclusions** Is this result consistent with the hypothesis that fish maintain blood pH levels by delaying the release of lactic acid from muscles? Why might this be beneficial to the fish?

Duane Raver, U.S. Fish and Wildlife Service

Retracing the Learning Path

CONCEPT 7.1 Cells Harvest Energy from Organic Compounds by Oxidation

7.1.1 The Potential Energy Stored in Organic Molecules Resides Largely in C—H Bonds Aerobic respiration uses oxygen as the final electron acceptor for redox reactions. Anaerobic respiration utilizes other inorganic molecules as electron acceptors, and fermentation uses organic molecules.

7.1.2 Electron Carriers Transfer Electrons from One Molecule to Another Electron carriers can be reversibly oxidized and reduced. For example, NAD^+ is reduced to NADH by acquiring two electrons; NADH is oxidized when it transfers these electrons to other molecules.

7.1.3 The Central Role of Respiration Is to Produce ATP The ultimate goal of cellular respiration is synthesis of ATP, which is used to power most of the cell's activities. Cells make ATP by two fundamentally different mechanisms: substrate-level phosphorylation and oxidative phosphorylation.

CONCEPT 7.2 Glycolysis Splits Glucose and Yields a Small Amount of ATP

7.2.1 Glycolysis Converts Glucose into Two Pyruvates and Yields Two ATP and Two NADH Glycolysis begins with priming reactions that add two phosphates to glucose; this is cleaved into two 3-carbon molecules of glyceraldehyde-3-phosphate (G3P). ATP is synthesized by substrate-level phosphorylation. Oxidation of G3P transfers electrons to NAD^+, yielding NADH. After four more reactions, the final product is two molecules of pyruvate. Glycolysis produces two net ATP, and two NADH.

7.2.2 Continuous Respiration Requires the Regeneration of NAD^+ In the presence of oxygen, NADH passes electrons to the electron transport chain. In the absence of oxygen, NADH passes the electrons to an organic molecule (fermentation).

CONCEPT 7.3 The Citric Acid Cycle Is the Oxidative Core of Cellular Respiration

7.3.1 Pyruvate Is Converted to a 2-Carbon Acetyl Group for the Citric Acid Cycle Pyruvate is oxidized to acetyl-CoA, also producing NADH and CO_2.

7.3.2 The Citric Acid Cycle Oxidizes 2-Carbon Acetyl Groups Each turn of the citric acid cycle produces 1 ATP, 3 NADH and 1 $FADH_2$. Acetyl-CoA is oxidized to $2CO_2$.

7.3.3 The Citric Acid Cycle Consists of Nine Reactions The first reaction is an irreversible condensation that produces citrate; it is inhibited by high levels of ATP. Two reactions rearrange citrate to isocitrate. This is followed by two oxidation reactions that reduce NAD^+ to NADH. A substrate-level phosphorylation produces GTP, and from that ATP. Another oxidation reduces FAD to $FADH_2$. The last two reactions regenerate oxaloacetate with a final oxidation that reduces NAD^+ to NADH.

CONCEPT 7.4 Electrons Removed by Oxidation Pass Along an Electron Transport Chain

7.4.1 The Electron Transport Chain Produces a Proton Gradient NADH is oxidized to NAD^+. A proton gradient forms as electrons move through electron carriers, ending with cytochrome oxidase where electrons, H^+, and O_2 form H_2O. Three protons are pumped into the intermembrane space.

7.4.2 Chemiosmosis Utilizes the Proton Gradient to Produce ATP Protons diffuse back into the mitochondrial matrix via the ATP synthase channel, which uses the energy of the proton gradient to synthesize ATP.

CONCEPT 7.5 The Total Energy Yield of Aerobic Respiration Far Exceeds That of Glycolysis

7.5.1 The Theoretical Yield for Eukaryotes Is 30 Molecules of ATP per Glucose Molecule As a glucose molecule is broken down to CO_2, some of its energy is preserved in 4 ATP, 10 NADH, and 2 $FADH_2$. Each NADH produces 2.5 ATP, and each $FADH_2$ produces 1.5 ATP. 2 ATP are used to transport NADH from glycolysis into the mitochondria.

7.5.2 Calculation of P/O Ratios Has Changed over Time The value for ATP synthesized per O_2 reduced has changed over time. It is now thought to be 2.5.

CONCEPT 7.6 Aerobic Respiration Is Regulated by Feedback Inhibition

7.6.1 There Are Two Key Points of Control Glucose catabolism is controlled by the concentration of ATP molecules and intermediates in the Krebs cycle.

CONCEPT 7.7 Some Forms of Energy Metabolism Do Not Require O_2

7.7.1 Respiration Without Oxygen Occurs in Some Prokaryotes Methanogens use carbon dioxide. Sulfur bacteria use sulfate.

7.7.2 Fermentations Donate Electrons Generated by Glycolysis to Organic Molecules Rather Than Oxygen Fermentation is the oxidation of NADH and reduction of an organic molecule. In yeast, pyruvate is decarboxylated, then reduced to ethanol. In animals, pyruvate is reduced directly to lactate.

CONCEPT 7.8 Carbohydrates Are Not the Only Energy Source Used by Heterotrophs

7.8.1 Catabolism of Proteins Removes Amino Groups

7.8.2 β Oxidation of Fatty Acids Produces Acetyl Groups Fatty acids are converted to acetyl groups by β oxidation. These acetyl groups feed into the citric acid cycle to be oxidized and generate NADH. If ATP is high, acetyl-CoA is converted into fatty acids.

7.8.3 A Small Number of Key Intermediates Connect Metabolic Pathways

Concept Overview

This Concept Overview diagrams the key concepts that were discussed in this chapter.

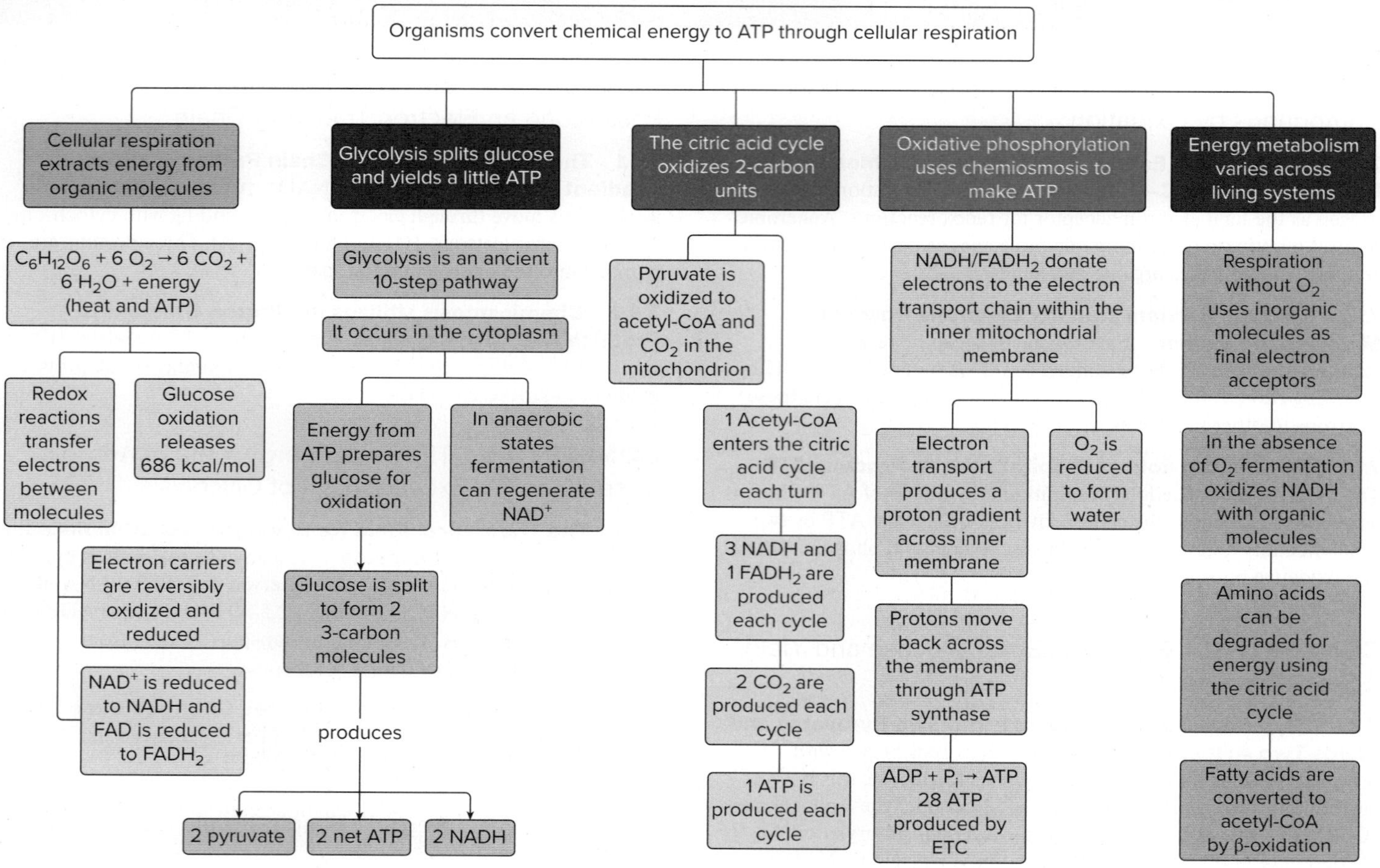

Assessing the Learning Path

Understand

1. What is the purpose of cellular respiration?
   a. To get O_2 into our bodies and to get CO_2 out of our bodies
   b. To use the energy of sunlight to synthesize sugars
   c. To make ATP to power cellular activities
   d. To hydrolyze ATP
2. In redox reactions,
   a. protons are transferred from one molecule to another.
   b. one substance loses electrons and the other gains electrons.
   c. a substance that loses electrons is reduced, while a substance that gains electrons is oxidized.
   d. Both b and c are correct.
3. ATP production during glycolysis
   a. requires an enzyme.
   b. occurs in the cytoplasm.
   c. occurs via substrate-level phosphorylation.
   d. All of the above
4. The reactions of the Krebs cycle occur in the
   a. inner membrane of the mitochondria.
   b. intermembrane space of the mitochondria.
   c. cytoplasm.
   d. matrix of the mitochondria.
5. In eukaryotic cells, the molecules that form the electron transport chain and ATP synthase are found in
   a. the matrix of the mitochondria.
   b. the inner mitochondrial membrane.
   c. the outer mitochondrial membrane.
   d. the intermembrane space.
6. The movement of electrons down the electron transport chain
   a. is used to make acetyl-CoA.
   b. generates CO_2.
   c. produces an electron gradient.
   d. pumps protons into the intermembrane space.

7. The P/O ratio
   a. is the amount of ATP synthesized by substrate-level phosphorylation.
   b. has changed over time due to our increased understanding of how ATP synthase works.
   c. is higher than originally thought because of our better understanding of chemiosmosis.
   d. All of the above
8. The production of the vast majority of the ATP molecules within the cells of your body is powered by electrons removed
   a. during the oxidation of pyruvate.
   b. during glycolysis.
   c. during the citric acid cycle.
   d. during passage through the electron transport chain.
9. When ATP is in excess within the cell, which of the following happens?
   a. Phosphofructokinase is inhibited.
   b. Citrate synthetase is inhibited.
   c. Pyruvate dehydrogenase is inhibited.
   d. All of the above
10. Which pathway is present in both fermentation and aerobic respiration?
   a. Citric acid cycle
   b. Electron transport
   c. Glycolysis
   d. Pyruvate oxidation
11. Which of the following is NOT true of β oxidation of fatty acids?
   a. The process occurs in the matrix of the mitochondrion.
   b. Acetyl groups are removed from fatty acid chain ends.
   c. Respiration of a 6-carbon fatty acid molecule yields fewer ATP than respiration of a 6-carbon glucose molecule.
   d. All of the above
12. Proteins
   a. do not contain enough potential energy to play a role in ATP production.
   b. are broken down and the products enter glycolysis or the citric acid cycle.
   c. are broken down into amino acids, which can then drive ATP synthase.
   d. yield more ATP (per mole) than fats.

Apply

1. When a yeast cell lacks oxygen (select all that apply),
   a. no ATP will be made.
   b. all ATP will be made during glycolysis.
   c. all ATP will be made by substrate-level phosphorylation.
   d. NADH is not produced.
2. If you started with 10 molecules of glucose, at the end of glycolysis you would have a net gain of ____ molecules of ATP and ____ molecules of NADH.
   a. 4; 2
   b. 2; 2
   c. 20; 20
   d. 40; unable to determine
3. Glycolysis yields two molecules of NADH. Where do the electrons carried by NADH come from?
   a. The enzymes that catalyze the reactions
   b. The covalent bonds of glucose
   c. The ATP that is used to "prime" glycolysis
   d. NAD^+
4. Starting with glycolysis and ending with complete oxidation of the molecule, which of the following are produced?
   a. 2 ATP, 4 CO_2, 2 $FADH_2$, 6 NADH
   b. 1 ATP, 2 CO_2, 1 $FADH_2$, 3 NADH
   c. 1 ATP, 6 CO_2, 1 FAD, 3 NAD^+
   d. 4 ATP, 6 CO_2, 2 $FADH_2$, 10 NADH
5. Dicyclohexylcarbodiimide (DCCD) is a chemical that prevents the flow of protons through ATP synthase. A cell treated with DCCD will die because
   a. glucose won't be converted to pyruvate.
   b. no ATP will be made by oxidative phosphorylation.
   c. electrons won't be passed down the electron transport chain and oxygen won't be reduced.
   d. a proton gradient won't be formed.
6. What would happen to the P/O ratio if 8 protons were needed for each rotation of ATP synthase instead of 12?
   a. The P/O ratio would increase.
   b. The P/O ratio would decrease.
   c. The P/O ratio would remain the same.
   d. It is not possible to tell, as protons have nothing to do with the P/O ratio.
7. Yeast cells that have mutations in genes that encode enzymes in glycolysis can still grow on glycerol. They are able to utilize glycerol because it
   a. enters glycolysis after the step affected by the mutation.
   b. can feed into the citric acid cycle and generate ATP via electron transport and chemiosmosis.
   c. can be utilized by fermentation.
   d. can donate electrons directly to the electron transport chain.
8. You discover a single-celled organism that lives in the rumen of sheep. The rumen is a part of the ruminant digestive system and is an O_2-free environment. Your new organism
   a. can produce ATP only via substrate-level phosphorylation.
   b. uses CO_2 as a final electron acceptor.
   c. produces ethanol as a by-product of glycolysis.
   d. will not be able to survive.
9. A 14-carbon fatty acid is catabolized by the process of β oxidation. The respiration of this compound yields how many ATP?
   a. 3
   b. 14
   c. 36
   d. 88

Synthesize

1. If the same amount of energy is released whether glucose is catabolized or burned, why doesn't glucose spontaneously burn?
2. Compare and contrast NAD^+ and FAD.
3. The electron carrier cytochrome *c* is one of many different cytochromes, but unlike the others, the sequence of the cytochrome *c* gene is nearly identical in all species. Indeed, among humans, no genetic disorder affecting cytochrome *c* has ever been reported. Why do you think this is so?
4. Estimates of the P/O ratio have been difficult to calculate. In eukaryotes, ADP must be transported into the mitochondria from the cytosol. How might this affect the P/O ratio in a eukaryotic cell versus a prokaryotic cell?
5. Soft drinks are artificially carbonated by the injection of CO_2 gas under pressure. Beer, on the other hand, is naturally carbonated. How does this natural carbonation occur?
6. Which of the following food molecules would generate the most ATP molecules, assuming that glycolysis and oxidative metabolism were both functioning optimally and that foods were consumed in equal amounts (e.g., the same number of moles): carbohydrates, proteins, or fats? Explain.

Sabrina Bekeschus Monteiro/EyeEm/Getty Images

8 Photosynthesis

Learning Path

- **8.1** Photosynthesis Uses Sunlight to Power the Synthesis of Organic Molecules
- **8.2** Experiments Revealed That Photosynthesis Is a Chemical Process
- **8.3** Pigments Capture Energy from Sunlight
- **8.4** Photosynthetic Pigments Are Organized into Photosystems
- **8.5** Energy from Sunlight Is Used to Produce a Proton Gradient
- **8.6** Carbon Fixation Incorporates CO_2 into Organic Molecules
- **8.7** Photorespiration Short-Circuits Photosynthesis

Concept Overview

This Concept Overview provides a diagram of the overarching concepts that are covered in this chapter. These concepts will be expanded upon in the Concept Overview at the end of the chapter.

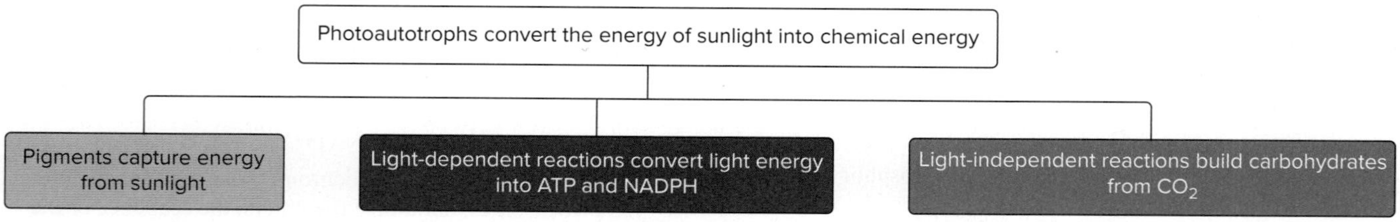

Photoautotrophs convert the energy of sunlight into chemical energy
Pigments capture energy from sunlight
Light-dependent reactions convert light energy into ATP and NADPH
Light-independent reactions build carbohydrates from CO_2

Introduction

The rich diversity of life that covers our Earth would be impossible without photosynthesis. Almost every oxygen atom in the air we breathe was once part of a water molecule, liberated by photosynthesis. All the energy released by the burning of coal, firewood, gasoline, and natural gas and by our bodies' burning of all the food we eat—all of this energy was first captured from sunlight by photosynthesis. It is vital, then, that we understand photosynthesis. Research may enable us to improve crop yields and land use, important goals in an increasingly crowded world. In chapter 7, we described how cells extract chemical energy from food molecules and use that energy to power their activities. In this chapter, we examine photosynthesis, the process by which organisms such as the aptly named sunflowers in the picture on the previous page capture energy from sunlight and use it to build food molecules rich in chemical energy.

8.1 Photosynthesis Uses Sunlight to Power the Synthesis of Organic Molecules

Life is powered by sunshine. The energy used by most living cells comes ultimately from the Sun and is captured by plants, algae, and bacteria through the process of photosynthesis.

Photosynthesis Has Light-Dependent and Light-Independent Reactions

LEARNING OBJECTIVE 8.1.1 Write the balanced equation for photosynthesis.

The diversity of life is possible only because our planet is awash in energy streaming Earthward from the Sun. In an hour on a sunny day, the earth receives radiant energy from the Sun equivalent to that needed to power 2,500 trillion 100-watt light bulbs! Global photosynthesis captures only about 1% of this huge supply of energy and this still provides the energy for most living systems.

Forms of photosynthesis

Photosynthesis occurs in a wide variety of organisms, and it comes in different forms. These include a form of photosynthesis that does not produce oxygen (anoxygenic) and a form that does (oxygenic). Anoxygenic photosynthesis is found in four different bacterial groups: purple bacteria, green sulfur bacteria, green nonsulfur bacteria, and heliobacteria. Oxygenic photosynthesis is found in cyanobacteria, seven groups of algae, and essentially all land plants. These two types of photosynthesis share similarities in the types of pigment molecules they use to trap light energy, but they differ in the arrangement and action of these pigment molecules.

Oxygenic photosynthesis, the photosynthesis of plants and algae, combines CO_2 and H_2O, producing glucose and O_2. It takes place primarily in the leaves and sometimes in the stems of land plants. Figure 8.1 illustrates how leaves are organized. As you learned in chapter 4, the cells of plant leaves contain organelles called chloroplasts, which are required for photosynthesis. No other structure in a plant cell is able to carry out photosynthesis.

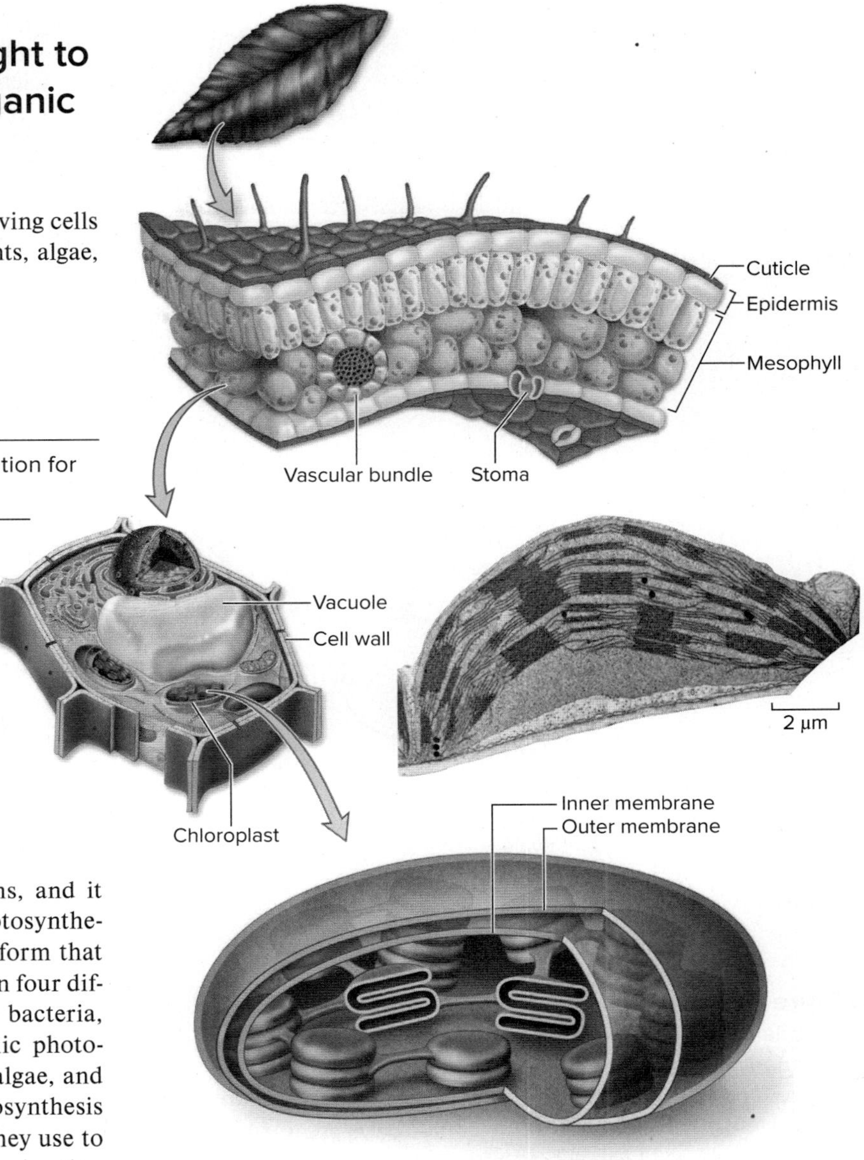
Cuticle
Epidermis
Mesophyll
Vascular bundle
Stoma
Vacuole
Cell wall
2 μm
Chloroplast
Inner membrane
Outer membrane

Figure 8.1 Journey into a leaf. A plant leaf possesses a thick layer of cells (the mesophyll) rich in chloroplasts. The inner membrane of the chloroplast is organized into flattened structures called thylakoid disks, which are stacked into columns called grana. The rest of the interior is filled with a semifluid substance called stroma.

(middle right): Courtesy Dr. Kenneth Miller, Brown University

The two stages of photosynthesis

Photosynthesis involves two sets of interrelated chemical reactions, illustrated in figure 8.2:

1. Reactions that use the energy in sunlight to make ATP, and reduce the electron carrier $NADP^+$ to NADPH. This also oxidizes water to provide electrons and produces O_2.
2. Reactions that use the ATP and NADPH to synthesize organic compounds using CO_2 from the air.

The first stage requires light and so these reactions are called the **light-dependent reactions.**

The second stage, the formation of organic molecules from CO_2, is called **carbon fixation.** This process takes place via a cyclic series of reactions called the Calvin cycle. As long as ATP and NADPH are available, the carbon fixation reactions can occur in either the presence or the absence of light, so these reactions are called the **light-independent reactions.** These reactions are also sometimes called the dark reactions, but this is misleading as they can take place in the presence of light.

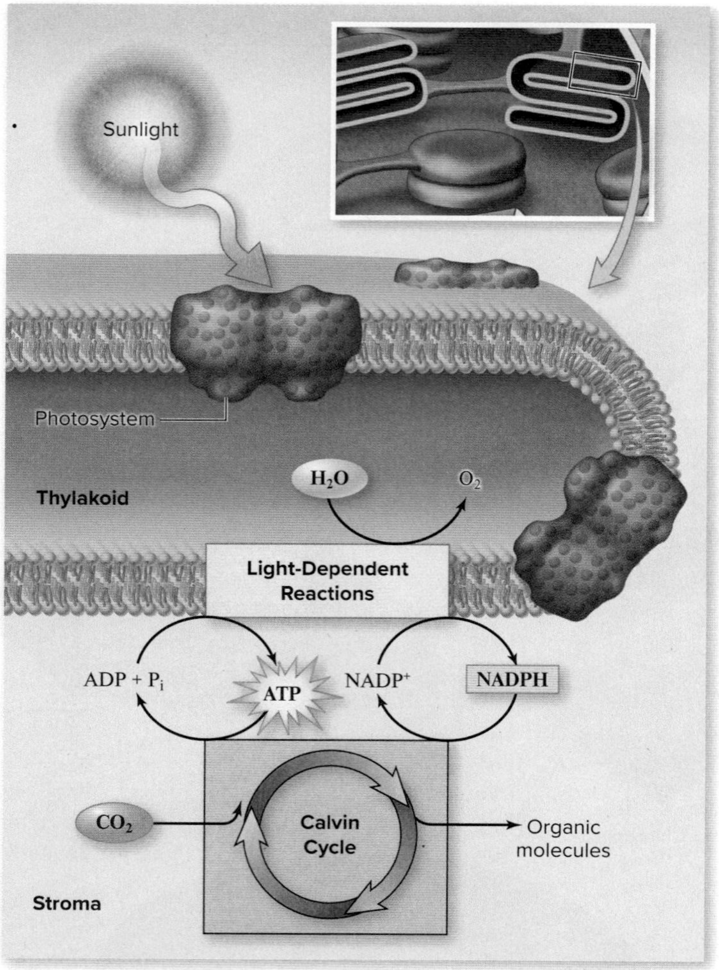

Figure 8.2 Overview of photosynthesis. Chlorophyll molecules are organized into photosystems. The light-dependent reactions begin when a chlorophyll molecule absorbs a photon of light. This light energy is used to generate ATP and NADPH. Electrons lost from chlorophyll are replaced by oxidizing water, producing O_2 as a by-product. The ATP and NADPH are used to reduce CO_2 using the Calvin cycle reactions in the stroma, producing organic molecules.

The following simple equation summarizes the overall process of photosynthesis:

$$\underset{\text{carbon dioxide}}{6\ CO_2} + \underset{\text{water}}{12\ H_2O} + \text{sunlight} \rightarrow \underset{\text{glucose}}{C_6H_{12}O_6} + \underset{\text{water}}{6\ H_2O} + \underset{\text{oxygen}}{6\ O_2}$$

You may notice that this equation is the reverse of the reaction for cellular respiration. In respiration, glucose is oxidized to CO_2 using O_2 as an electron acceptor. In photosynthesis, CO_2 is reduced to glucose using electrons gained from the oxidation of water. The oxidation of H_2O and the reduction of CO_2 require energy that is provided by light.

Remember that in chapter 6 we learned that all organisms constantly require energy to power endergonic reactions. The most common biochemical pathway to obtain this energy is cellular respiration. But the organic compounds used as a source of energy were originally synthesized by photosynthesis from CO_2 in the air, making us all dependent on photosynthesis as the ultimate source of energy. All of the O_2 used as a final electron acceptor in aerobic respiration also was produced by photosynthesis.

The Chloroplast Is a Photosynthetic Machine

LEARNING OBJECTIVE 8.1.2 Compare the structure of a chloroplast with the structure of a mitochondrion.

In chapter 7, you saw that a mitochondrion's complex structure of internal and external membranes contributes to its function. The same is true for the structure of the chloroplast.

The internal membrane of chloroplasts, called the thylakoid membrane, is a continuous phospholipid bilayer organized into flattened sacs, which are stacked on one another in columns called grana (singular, *granum*). The thylakoid membrane contains **chlorophyll** and other photosynthetic pigments for capturing light energy. The thylakoid membrane also contains the machinery to synthesize ATP by chemiosmosis.

Surrounding the thylakoid membrane system is a semiliquid substance called **stroma.** The stroma houses the enzymes needed to assemble organic molecules from CO_2 using energy from ATP coupled with reduction via NADPH. In the thylakoid membrane, photosynthetic pigments are clustered together to form **photosystems,** which act as large antennas, gathering the light energy harvested by many individual pigment molecules.

In sections 8.3 to 8.5, we will examine, in more detail, pigments and how they are organized into photosystems that function to produce ATP and NADPH using energy obtained from sunlight. We will then discuss how the ATP and NADPH can be used to power the incorporation of CO_2 into organic molecules (section 8.6). As background and context for this current understanding of photosynthesis, we first examine the experimental history of our basic understanding of the process.

REVIEW OF CONCEPT 8.1

Photosynthesis consists of light-dependent reactions that require sunlight, and others that convert CO_2 into organic molecules. The overall reaction is essentially the reverse of respiration and produces O_2 as a by-product. The thylakoid membrane is the site where photosynthetic pigments are clustered, allowing the passage of energy from one molecule to the next.

- *How is the structure of the chloroplast similar to that of the mitochondrion?*

8.2 Experiments Revealed That Photosynthesis Is a Chemical Process

Before considering the detailed chemistry and physics of photosynthesis, it is worth considering the long history of research on the process. This begins over 300 years ago, with curiosity about how plants manage to grow, increasing their organic mass. From this humble beginning we have built up a detailed picture of this amazing process that is critical for all life.

Soil Does Not Add Mass to Growing Plants

LEARNING OBJECTIVE 8.2.1 Demonstrate that plant mass is derived primarily from the air and not the soil.

From the time of the Greeks, plants were thought to obtain their food from the soil, literally sucking it up with their roots. The physician Jan Baptista van Helmont (1580–1644) thought of a simple way to test this idea.

He planted a small willow tree in a pot of soil, after first weighing the tree and the soil. The tree grew in the pot for several years, during which time van Helmont added only water. At the end of 5 years, the tree was much larger, its weight having increased by 74.4 kg. However, the soil in the pot weighed only 57 g less than it had 5 years earlier. With this experiment, van Helmont demonstrated that the substance of the plant was not produced only from the soil. He incorrectly concluded, however, that the water he had been adding mainly accounted for the plant's increased biomass.

A hundred years passed before the story became clearer. The key clue was provided by the scientist Joseph Priestly (1733–1804). On August 17, 1771, Priestly put a living sprig of mint into air in which a wax candle had burnt out. On August 27, he found that another candle could be burned in the same air. Somehow the vegetation seemed to have restored the air. Priestly found that although a mouse could not breathe candle-exhausted air, air "restored" by vegetation was not "at all inconvenient to a mouse." The key clue was that living vegetation adds something to the air.

How does vegetation "restore" air? Twenty-five years later, the physician Jan Ingenhousz (1730–1799) solved the puzzle. Working over several years, he reproduced and extended Priestly's results, demonstrating that air was restored only in the presence of sunlight and only by a plant's green leaves, not by its roots. He proposed that the green parts of the plant carry out a process (which we now call photosynthesis) that uses sunlight to split carbon dioxide into carbon and oxygen. He suggested that the oxygen was released as O_2 gas into the air, while the carbon atom combined with water to form carbohydrates. Other research refined his conclusions, and by the end of the 19th century the overall reaction for photosynthesis could be written as

$$CO_2 + H_2O + \text{light energy} \longrightarrow (CH_2O) + O_2$$

It turns out, however, that there's more to it than that. When researchers began to examine the process in more detail in the 20th century, the role of light proved to be unexpectedly complex.

Photosynthesis Is a Multistage Process in Which Some Steps Do Not Require Light

LEARNING OBJECTIVE 8.2.2 Demonstrate that a key portion of photosynthesis does not use light.

Ingenhousz's early equation for photosynthesis includes one factor we have not discussed: light energy. What role does light play in photosynthesis? At the beginning of the 20th century, the plant physiologist F. F. Blackman (1866–1947) began to address this question. In 1905, he came to the startling conclusion that photosynthesis is, in fact, a multistage process, only one portion of which uses light directly.

Blackman measured the effects of different light intensities, CO_2 concentrations, and temperatures on photosynthesis. His results are illustrated in figure 8.3. He found that as long as light intensity was relatively low, photosynthesis could be accelerated by increasing the amount of light, but not by increasing the temperature or CO_2 concentration (figure 8.3). At high light intensities, however, an increase in temperature or CO_2 concentration greatly accelerated photosynthesis.

Blackman concluded that photosynthesis consists of an initial set of what he called "light" reactions, which are largely independent of temperature but depend on light, and a second set of "dark" reactions (more properly called light-independent reactions), which seemed to be independent of light but limited by CO_2.

Do not be confused by Blackman's labels—the so-called dark reactions occur in the light (in fact, they require the products of the light-dependent reactions); his use of the word *dark* simply indicates that light is not *directly* involved in those reactions.

Blackman found that increased temperature increased the rate of the light-independent reactions, but only up to about 35°C. Higher temperatures caused the rate to decrease rapidly. Because many plant enzymes begin to be denatured at 35°C, Blackman concluded that enzymes must carry out the light-independent reactions.

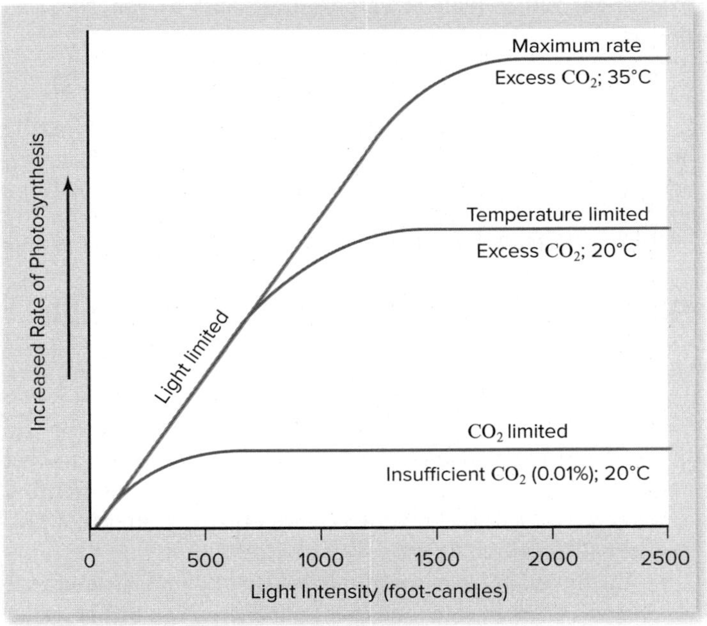

Figure 8.3 Discovery of the light-independent reactions. Blackman measured photosynthesis rates under differing light intensities, CO_2 concentrations, and temperatures. As this graph shows, light is the limiting factor at low light intensities, but temperature and CO_2 concentration are the limiting factors at higher light intensities. This implies the existence of reactions using CO_2 that involve enzymes.

Water Acts as an Electron Donor

LEARNING OBJECTIVE 8.2.3 Explain how photosynthesis generates O_2.

In the 1930s, C. B. van Niel (1897–1985), working at the Hopkins Marine Station at Stanford, discovered that purple sulfur bacteria do not release oxygen during photosynthesis; instead, they convert hydrogen sulfide (H_2S) into globules of pure elemental sulfur that accumulate inside them. The process van Niel observed was

$$CO_2 + 2\ H_2S + \text{light energy} \rightarrow (CH_2O) + H_2O + 2\ S$$

The striking parallel between this equation and Ingenhousz's equation led van Niel to propose that the generalized process of photosynthesis can be shown as

$$CO_2 + 2\ H_2A + \text{light energy} \rightarrow (CH_2O) + H_2O + 2\ A$$

In this equation, the substance H_2A serves as an electron donor. In photosynthesis performed by green plants, H_2A is water, whereas in purple sulfur bacteria, H_2A is hydrogen sulfide. The product, A, comes from the splitting of H_2A. Therefore, the O_2 produced during green plant photosynthesis results from splitting water, not carbon dioxide.

O_2 comes from water, not from CO_2

When isotopes came into common use in the early 1950s, van Niel's revolutionary proposal was tested. Investigators examined photosynthesis in green plants supplied with water containing heavy oxygen (^{18}O); they found that the ^{18}O label ended up in oxygen gas rather than in carbohydrate, just as van Niel had predicted:

$$CO_2 + 2\ H_2{}^{18}O + \text{light energy} \rightarrow (CH_2O) + H_2O + {}^{18}O_2$$

In algae and green plants, the carbohydrate typically produced by photosynthesis is glucose. The complete balanced equation for photosynthesis in these organisms thus becomes

$$6\ CO_2 + 12\ H_2O + \text{light energy} \rightarrow C_6H_{12}O_6 + 6\ H_2O + 6\ O_2$$

NADPH Is Used to Reduce Carbon Dioxide

LEARNING OBJECTIVE 8.2.4 Demonstrate that electrons from water are used to reduce $NADP^+$.

In his pioneering work on the light-dependent reactions, van Niel proposed that the H^+ ions and electrons generated by the splitting of water were used to convert CO_2 into organic matter in a process he called carbon fixation. Was he right?

In the 1950s, Robin Hill (1899–1991) demonstrated that van Niel was indeed right and that light energy can be harvested and used to generate reducing power. In his experiments, chloroplasts isolated from leaf cells were able to reduce a dye and release oxygen in response to light. Later experiments showed that the electrons released from water were transferred to $NADP^+$ and that illuminated chloroplasts deprived of CO_2 accumulate ATP. If CO_2 is introduced, neither ATP nor NADPH accumulates, and the CO_2 is assimilated into organic molecules.

These experiments are important for three reasons: first, they firmly demonstrate that photosynthesis in plants occurs within chloroplasts. Second, they show that the light-dependent reactions use light energy to reduce $NADP^+$ and to manufacture ATP. Third, they confirm that the ATP and NADPH from this early stage of photosynthesis are then used in the subsequent reactions to reduce carbon dioxide, forming simple sugars.

REVIEW OF CONCEPT 8.2

Early experiments indicated that plants produce oxygen in the presence of sunlight. Further experiments showed that there are both light-dependent and light-independent reactions. The light-dependent reactions produce O_2 from H_2O and generate ATP and NADPH. The light-independent reactions synthesize organic compounds from CO_2.

- *Where does the carbon in your body come from?*

8.3 Pigments Capture Energy from Sunlight

For plants to make use of the energy of sunlight, some biochemical structure must be present in chloroplasts and the thylakoids that can absorb this energy. Molecules that absorb light energy in the visible range are termed **pigments.** We are most familiar with them as dyes that impart a certain color to clothing or other materials. The color that we see is the color that is not absorbed—that is, it is reflected. To understand how plants use pigments to capture light energy, we must first review current knowledge about the nature of light.

Light Is a Form of Energy

LEARNING OBJECTIVE 8.3.1 Relate a photon's energy to its wavelength.

Light exhibits properties of both wave and particle. The wave nature of light produces an electromagnetic spectrum that differentiates light based on its wavelength. We are most familiar with the visible range of this spectrum because we can actually see it, but visible light is only a small part of the entire spectrum. Visible light can be divided into its separate colors by the use of a prism, which separates light based on wavelength.

A particle of light, termed a **photon,** acts as a discrete bundle of energy. We use the wave concept of light to understand different colors of light and the particle nature of light to understand the energy transfers that occur during photosynthesis. Thus, we will refer both to wavelengths of light and to photons of light throughout the chapter.

The energy in photons

The energy content of a photon is inversely proportional to the wavelength of the light: short-wavelength light contains photons

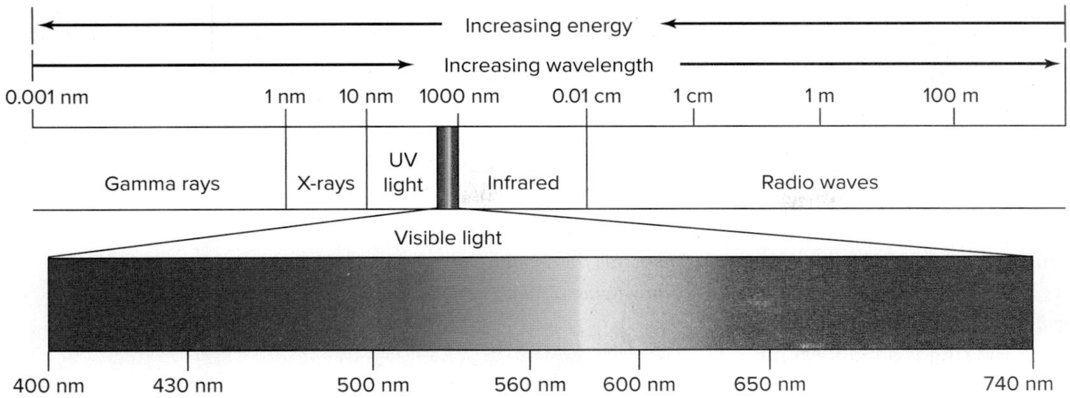

Figure 8.4 The electromagnetic spectrum. Light is a form of electromagnetic energy, conveniently thought of as a wave. The shorter the wavelength of light, the greater its energy. Visible light represents only a small part of the electromagnetic spectrum, between 400 and 740 nm.

of higher energy than long-wavelength light (figure 8.4). X-rays, which contain a great deal of energy, have very short wavelengths—much shorter than those of visible light.

A beam of light is able to remove electrons from certain materials, creating an electric current. This phenomenon is called the **photoelectric effect,** and it occurs when photons transfer energy to electrons. The strength of the photoelectric effect depends on the wavelength of light; that is, short wavelengths are generally much more effective than long ones in producing the photoelectric effect because they have more energy.

In photosynthesis, chloroplasts are acting as photoelectric devices: they absorb sunlight and transfer the excited electrons to a carrier. As we unravel the details of this process, it will become clear how this process traps energy and uses it to synthesize organic compounds.

Chlorophyll Is the Principal Photosynthetic Pigment

LEARNING OBJECTIVE 8.3.2 Relate the chlorophyll absorption spectra to the photosynthetic action spectrum.

When a photon strikes a molecule with the precise amount of energy needed to excite one of its electrons, the molecule will absorb the photon, raising that electron to a higher energy level. Whether the photon's energy is absorbed thus depends both on how much energy it carries (defined by its wavelength) and on the chemical nature of the molecule it hits. To boost an electron into a different energy level requires just the right amount of energy, just as reaching the next rung on a ladder requires you to raise your foot just the right distance. A specific atom, therefore, can absorb only certain photons of light—namely, those that correspond to the atom's available energy levels. As a result, each molecule has a characteristic **absorption spectrum,** the range and efficiency of photons it is capable of absorbing.

As mentioned earlier in this section, pigments absorb light in the visible range. Organisms have evolved a variety of pigments, but only two general types are used in green plant photosynthesis: chlorophylls and carotenoids. In some organisms, other molecules also absorb light energy.

Chlorophyll absorption spectra

Chlorophylls absorb photons within narrow energy ranges. Two kinds of chlorophyll in plants, chlorophyll *a* and chlorophyll *b,* preferentially absorb violet-blue and red light (figure 8.5). Neither of these pigments absorbs photons with wavelengths between about 500 and 600 nm; light of these wavelengths is reflected. When these reflected photons are subsequently absorbed by the retinal pigment in our eyes, we perceive them as green.

Chlorophyll *a* is the main photosynthetic pigment in plants and cyanobacteria and the only pigment that can act directly to convert light energy to chemical energy. **Chlorophyll *b,*** acting as an **accessory pigment,** or secondary light-absorbing pigment, complements the light absorption of chlorophyll *a*.

Chlorophyll *b* has an absorption spectrum shifted toward the green wavelengths. Because chlorophyll *b* can absorb

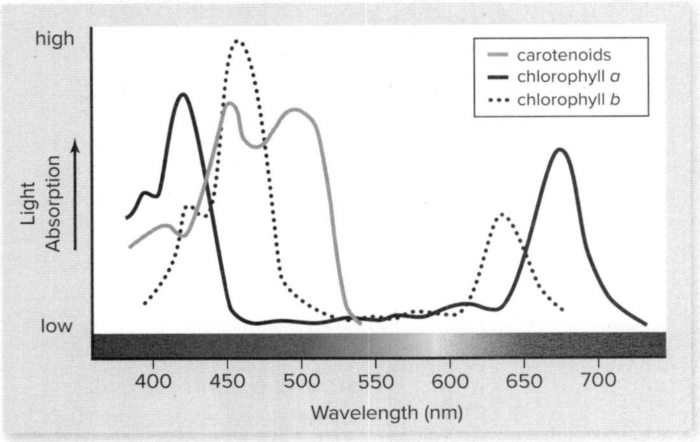

Figure 8.5 Absorption spectra for chlorophylls and carotenoids. The peaks represent wavelengths of light absorbed by the two common forms of photosynthetic pigment, chlorophylls *a* and *b,* and the carotenoids. Chlorophylls absorb predominantly violet-blue and red light in two narrow bands of the spectrum and reflect green light in the middle of the spectrum. Carotenoids absorb mostly blue and green light and reflect orange and yellow light.

green-wavelength photons that chlorophyll *a* cannot, channeling this energy to chlorophyll *a* greatly increases the proportion of the photons in sunlight that plants can harvest. A variety of other accessory pigments are found in plants, bacteria, and algae.

Structure of chlorophylls

Chlorophylls absorb photons by means of an excitation process analogous to the photoelectric effect. These pigments contain a complex ring structure, called a *porphyrin ring,* with alternating single and double bonds. At the center of the ring is a magnesium atom (figure 8.6).

Photons excite electrons in the porphyrin ring, which are then channeled away through the alternating carbon single- and double-bond system. Different small side groups attached to the outside of the ring alter the absorption properties of the molecule in the different kinds of chlorophyll (figure 8.6). The precise absorption spectrum is also influenced by the local microenvironment created by the association of chlorophyll with different proteins.

The **action spectrum** of photosynthesis—that is, the relative effectiveness of different wavelengths of light in promoting photosynthesis—corresponds to the absorption spectrum for chlorophylls. This was famously demonstrated in the classic experiment illustrated in figure 8.7. All plants, algae, and cyanobacteria use chlorophyll *a* as their primary pigments. This action spectrum can be broadened with the help of accessory pigments described later in this section.

SCIENTIFIC THINKING

Hypothesis: *All wavelengths of light are equally effective in promoting photosynthesis.*

Prediction: *Illuminating plant cells with light broken into different wavelengths by a prism will produce the same amount of O_2 for all wavelengths.*

Test: *A filament of algae immobilized on a slide is illuminated by light that has passed through a prism. Motile bacteria that require O_2 for growth are added to the slide.*

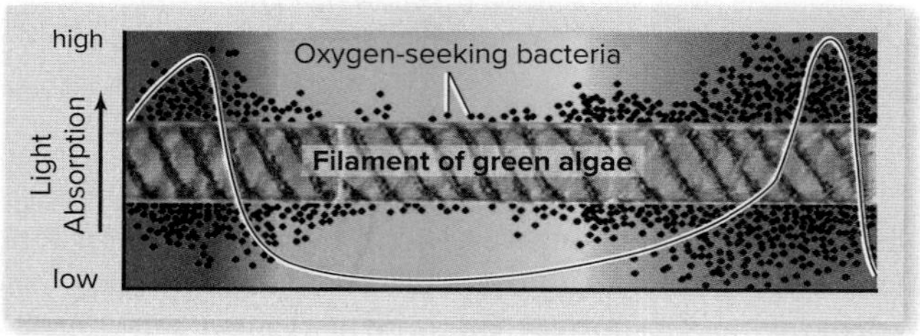

Result: *The bacteria move to regions of high O_2, or regions of most active photosynthesis. These are in the purple/blue and red regions of the spectrum.*

Conclusion: *All wavelengths are not equally effective at promoting photosynthesis. The most effective constitute the action spectrum for photosynthesis.*

Further Experiments: *How does the action spectrum relate to the various absorption spectra in figure 8.5?*

Figure 8.7 Determination of an action spectrum for photosynthesis.

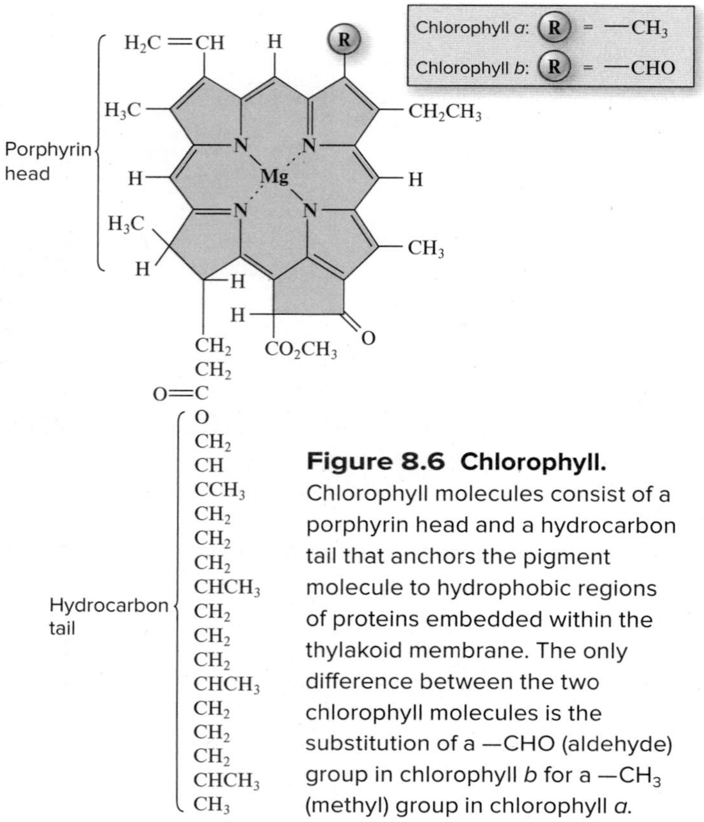

Figure 8.6 Chlorophyll. Chlorophyll molecules consist of a porphyrin head and a hydrocarbon tail that anchors the pigment molecule to hydrophobic regions of proteins embedded within the thylakoid membrane. The only difference between the two chlorophyll molecules is the substitution of a —CHO (aldehyde) group in chlorophyll *b* for a $—CH_3$ (methyl) group in chlorophyll *a*.

It is reasonable to ask why these photosynthetic organisms do not use a pigment like retinal (the pigment in our eyes), which has a broad absorption spectrum that covers the range of 500 to 600 nm. The most likely hypothesis involves *photoefficiency.* Although retinal absorbs a broad range of wavelengths, it does so with relatively low efficiency. Chlorophyll, in contrast, absorbs in only two narrow bands but does so with high efficiency. For this reason, plants and other photosynthetic organisms achieve far higher overall energy capture rates with chlorophyll than they would with other pigments.

Carotenoids Are Accessory Pigments

LEARNING OBJECTIVE 8.3.3 Explain the role of accessory pigments.

Carotenoids are pigment molecules that, like chlorophyll, consist of carbon rings linked to hydrocarbon chains, but in this case chains with alternating single and double bonds. Carotenoids can absorb photons with a wide range of energies, although they are not always highly efficient in transferring this energy. Carotenoids assist in photosynthesis by capturing energy from light composed of wavelengths that are not efficiently absorbed by chlorophylls (figure 8.8).

Carotenoids also perform a valuable role in scavenging free radicals. The oxidation–reduction reactions that occur in the

Maple leaf
in summer
Maple leaf
in autumn

Figure 8.8 Fall colors are produced by carotenoids and other accessory pigments. During the spring and summer, chlorophyll in leaves masks the presence of carotenoids and other accessory pigments. When cool fall temperatures cause leaves to cease manufacturing chlorophyll, the chlorophyll is no longer present to reflect green light, and the leaves reflect the orange and yellow light that carotenoids and other pigments do not absorb.

Rich Iwasaki/Stone/Getty Images

chloroplast can generate destructive free radicals. Carotenoids can act as general-purpose antioxidants to lessen damage. Thus, carotenoids have a protective role in addition to their role as light-absorbing molecules. This protective role is not limited to plant cells. Unlike the chlorophylls, carotenoids are found acting as antioxidants in many different kinds of organisms, including members of all three domains of life.

You may have heard that eating carrots, which are rich in β-carotene, can enhance vision. This would only be the case if you were deficient in vitamin A, which is synthesized from β-carotene. Vitamin A, in turn, is used to produce the pigment retinal, which is important for vertebrate vision (refer to chapter 33).

Phycobiliproteins are accessory pigments found in cyanobacteria and some algae. These pigments contain a system of alternating double bonds similar to those found in other pigments and molecules that transfer electrons. Phycobiliproteins can be organized to form another light-harvesting complex that can absorb green light, which is reflected by chlorophyll. These complexes are probably ecologically important to cyanobacteria, helping them to exist in low-light situations in oceans. In this habitat, green light remains; red and blue light have been absorbed by green algae closer to the surface.

REVIEW OF CONCEPT 8.3

Pigment molecules absorb light energy. An absorption spectrum describes the wavelengths of light absorbed most efficiently by a pigment. The color of a pigment results from the wavelengths it does not absorb, which we see. The main photosynthetic pigment is chlorophyll, which exists in two forms with slightly different absorption spectra. Accessory pigments have absorption spectra different from that of chlorophyll.

- *What is the difference between an action spectrum and an absorption spectrum?*

8.4 Photosynthetic Pigments Are Organized into Photosystems

Knowing that chlorophyll is the primary photosynthetic pigment leads to an important question: how are these pigments organized in chloroplasts? The answer to this question provided a surprise for plant biochemists studying photosynthesis.

The Rate of Photosynthesis Saturates at Low Light Intensity

LEARNING OBJECTIVE 8.4.1 Demonstrate the existence of photosystems in plant leaves.

One way to study the role that pigments play in photosynthesis is to measure the correlation between the output of photosynthesis and the intensity of illumination—that is, how much photosynthesis is produced by how much light. The results of this kind of experiment show a linear increase in photosynthetic output at low light intensities, but the rate of increase lessens at higher intensities, then plateaus at very high light intensity. This saturation occurs because all of the light-absorbing capacity of the plant is in use.

It is tempting to think that at saturation, all of a plant's pigment molecules are in use. In 1932, plant physiologists Robert Emerson and William Arnold set out to examine this question: at the saturation observed under high light intensity, have all chlorophyll molecules absorbed photons?

Finding an answer required being able to measure both photosynthetic output and the number of chlorophyll molecules present. The investigators could obtain both of these values using the unicellular alga *Chlorella*. In their experiments, Emerson and Arnold measured the oxygen yield of photosynthesis,

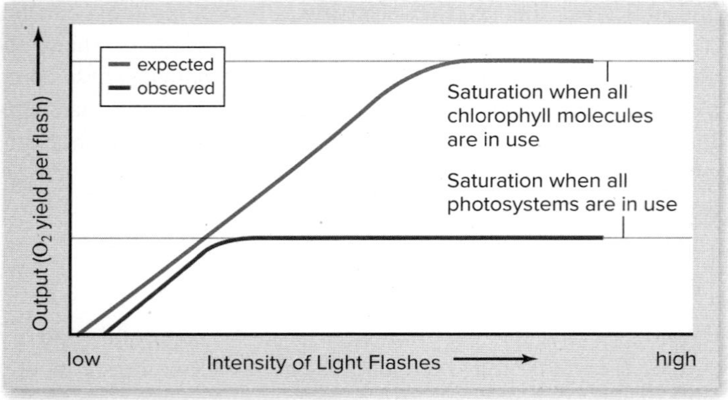

Figure 8.9 Saturation of photosynthesis. When photosynthetic saturation is achieved, further increases in intensity cause no increase in output. This saturation occurs far below the level expected for the number of individual chlorophyll molecules present. This led to the idea of organized photosystems, each containing many chlorophyll molecules. These photosystems saturate at a lower O_2 yield than that expected for the number of individual chlorophyll molecules.

illuminating a *Chlorella* culture with very brief pulses of light lasting only a few microseconds. Assuming the hypothesis of pigment saturation to be correct, they expected to find that as they increased the intensity of the light pulses, the yield per pulse would increase, until the system became saturated when every chlorophyll molecule had absorbed a photon. Then O_2 production could be compared directly with the number of chlorophyll molecules present in the culture.

That is not what happened. Instead, saturation was achieved much earlier, with only one molecule of O_2 per 2500 chlorophyll molecules (figure 8.9). This result, so very different from what was expected, led Emerson and Arnold to conclude that light is absorbed not by independent pigment molecules but by clusters of chlorophyll and accessory pigment molecules, which have come to be called *photosystems*. Light is absorbed by any one of hundreds of pigment molecules in a photosystem, and each pigment molecule transfers its excitation energy to a single molecule with a lower energy level than the others, called a *reaction center*. It was the saturation of these reaction centers, not the saturation of individual pigment molecules, that was observed by Emerson and Arnold.

Chlorophyll Pigments Are Organized to Efficiently Capture Light

LEARNING OBJECTIVE 8.4.2 Explain how the antenna complex and the reaction center function together in a photosystem.

In chloroplasts and all but one class of photosynthetic prokaryotes, light is captured by photosystems. Each photosystem contains a network of chlorophyll *a* molecules, and accessory pigments including chlorophyll *b* and carotenoids. These pigments are organized within a matrix of proteins embedded in the thylakoid membrane. The photosystem functions to channel absorbed light energy to a specific chlorophyll *a* molecule in the reaction center that uses this energy to transfer excited electrons out of the photosystem where they are used to drive the synthesis of ATP and organic molecules.

A photosystem thus consists of two closely linked components: (1) an antenna complex of hundreds of pigment molecules that gather photons and feed the captured light energy to the reaction center, and (2) a reaction center, consisting of one or more chlorophyll *a* molecules in a matrix of protein, that passes excited electrons out of the photosystem.

The antenna complex

The **antenna complex** is also called a light-harvesting complex, which accurately describes its role. This light-harvesting complex captures photons from sunlight and channels them to the reaction center chlorophylls.

In chloroplasts, light-harvesting complexes consist of a web of chlorophyll molecules linked together and held tightly in the thylakoid membrane by a matrix of proteins. Varying amounts of carotenoid accessory pigments may also be present. The protein matrix holds individual pigment molecules in orientations that are optimal for energy transfer.

The excitation energy resulting from the absorption of a photon passes from one pigment molecule to an adjacent molecule on its way to the reaction center (figure 8.10). After the transfer, the excited electron in each molecule returns to the low-energy level it had before the photon was absorbed. Consequently, it is energy, not the excited electrons themselves, that passes from one pigment molecule to the next. The antenna complex funnels the energy of many excited electrons to the reaction center.

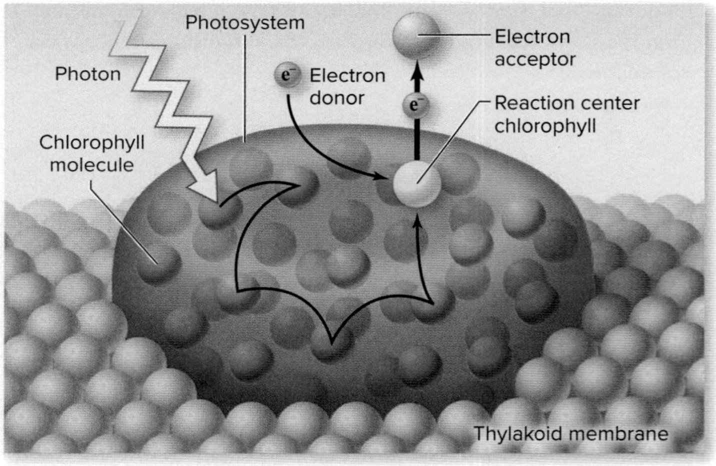

Figure 8.10 How the antenna complex works. Chlorophyll molecules of the antenna complex (dark green) absorb light energy, and transfer this energy to other chlorophyll molecules until the excitation energy reaches the reaction center. The reaction center chlorophyll (light green) uses the excitation energy to transfer an excited electron to an electron acceptor.

The reaction center

The **reaction center** is a transmembrane protein-pigment complex. The reaction center of purple photosynthetic bacteria is simpler than the one in chloroplasts but better understood. A pair of bacteriochlorophyll *a* molecules act as a trap for photon energy, passing an excited electron to an acceptor precisely positioned as its neighbor. Note that here in the reaction center, what is transferred is the excited electron itself, not just the energy, as was the case in the pigment–pigment transfers of the antenna complex. This difference allows the energy absorbed from photons to move away from the chlorophylls and is the key conversion of light into chemical energy.

Figure 8.11 shows the oxidation and reduction reactions that take place at the reaction center. The absorption of light by a reaction center chlorophyll excites an electron to a higher energy level, making this chlorophyll a strong electron donor. The energized electron is transferred to the primary acceptor, quinone, reducing it and making it a strong electron donor. A nearby weak electron donor passes a low-energy electron to the chlorophyll, reducing it. The quinone transfers its electrons to another acceptor, and the process is repeated.

In plant chloroplasts, water is used as the electron donor to chlorophyll. The oxidation of water produces molecular oxygen and two protons (H^+). Water is not readily oxidized, which indicates the very high oxidation potential produced by chlorophyll.

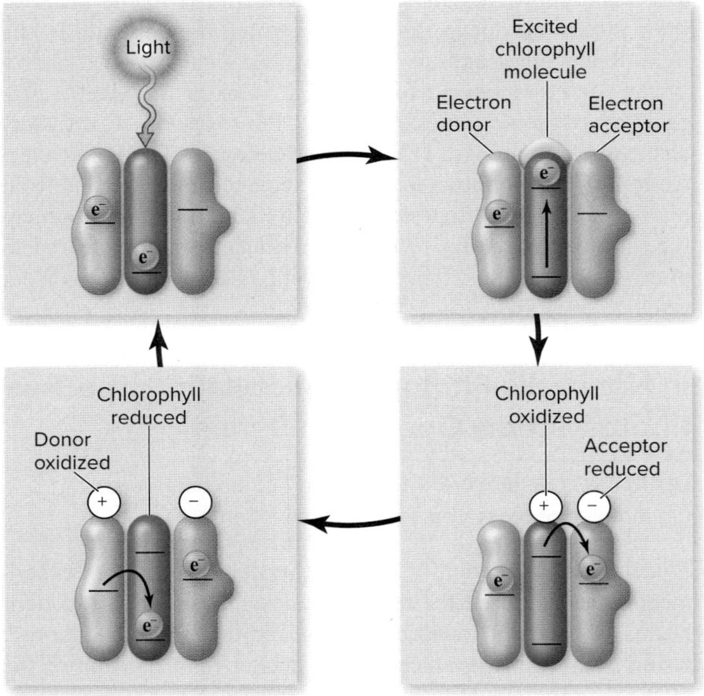

Figure 8.11 Converting light to chemical energy. A chlorophyll in the reaction center absorbs a photon, exciting an electron to a higher energy level. This electron is transferred to a primary electron acceptor, reducing it. The oxidized chlorophyll then replaces the lost electron by oxidizing another molecule. The source of the electron donor varies with the photosystem. The figure shows only the redox reactions and omits the thylakoid membrane where this would take place.

REVIEW OF CONCEPT 8.4

Chlorophylls and accessory pigments are organized into photosystems in the thylakoid membrane. A photosystem consists of an antenna complex for light harvesting and a reaction center where photochemical reactions occur. An excited electron is passed to an acceptor, transferring energy away from the chlorophylls. This is key to converting light into chemical energy.

- *Why were photosystems an unexpected finding?*

8.5 Energy from Sunlight Is Used to Produce a Proton Gradient

As you have learned, the light-dependent reactions of photosynthesis occur in membranes. In photosynthetic bacteria, the plasma membrane itself is the photosynthetic membrane. In many bacteria, the plasma membrane folds in on itself repeatedly to produce an increased surface area. In plants and algae, photosynthesis is carried out by chloroplasts, which are thought to be the evolutionary descendants of photosynthetic bacteria.

Light-Dependent Reactions Take Place in the Thylakoid Membrane

LEARNING OBJECTIVE 8.5.1 Describe the four stages of the light-dependent reactions.

The internal thylakoid membrane is highly organized and contains the structures involved in the light-dependent reactions. For this reason, the reactions are also referred to as the thylakoid reactions. The thylakoid reactions can be broken into four steps:

1. **Primary photoevent.** A photon of light is captured by a pigment. This primary photoevent excites an electron within the pigment.
2. **Charge separation.** This excitation energy is transferred to the reaction center, which transfers an energetic electron to an acceptor molecule, initiating electron transport.
3. **Electron transport.** The excited electrons are shuttled along a series of electron carrier molecules embedded within the photosynthetic membrane. This process is exergonic, with some of the energy released used to transport protons across the membrane, forming a proton gradient. The electrons are used to reduce the final acceptor, $NADP^+$, to NADPH.
4. **Chemiosmosis.** The proton gradient is a form of potential energy that can be used by an ATP synthase enzyme that catalyzes the reaction of ADP and P_i to form ATP. This enzyme is evolutionarily related to the mitochondrial enzyme with the same function.

These four steps make up the two stages of the light-dependent reactions described in section 8.1. It is important to remember that no matter how we divide these processes into parts to understand them, they are dynamic and continuous within a cell. We will consider both the evolution and the details of how this process occurs in the rest of the chapter.

Bacteria have a single photosystem

Photosynthetic pigment arrays are thought to have evolved more than 2 billion years ago (BYA) in bacteria similar to the purple and green bacteria alive today. In these bacteria, a single photosystem is used that generates ATP via electron transport. This process then returns the electrons to the reaction center. For this reason, it is called cyclic photophosphorylation. This process does not generate oxygen and is thus referred to as anoxygenic photosynthesis.

In the purple nonsulfur bacteria, peak absorption occurs at a wavelength of 870 nm (near-infrared, not visible to the human eye), and thus the reaction center pigment is called P_{870}. Absorption of a photon by chlorophyll P_{870} does not raise an electron to a high enough level to be passed to $NADP^+$; thus, these bacteria must generate reducing power in a different way.

When the P_{870} reaction center absorbs a photon, the excited electron is passed to an electron transport chain, which passes the electrons back to the reaction center, generating a proton gradient for ATP synthesis (figure 8.12). The proteins in the purple bacterial photosystem appear to be homologous to the proteins in the modern photosystem II.

In the green sulfur bacteria, peak absorption occurs at a wavelength of 840 nm (near-infrared, not visible to the human eye), and thus the reaction center pigment is called P_{840}. Excited electrons from this photosystem have enough energy to be passed to NADPH, or they can return to the chlorophyll by an electron transport chain similar to that in the purple bacteria. To replace electrons passed to NADPH, hydrogen sulfide is used as an electron donor. The proteins in the green sulfur bacterial photosystem appear to be homologous to the proteins in the modern photosystem I.

Because neither of these systems generates sufficient oxidizing power to oxidize H_2O, they are anoxygenic. This is in contrast to the linked photosystems of cyanobacteria and plant chloroplasts, which do generate the oxidizing power necessary to oxidize H_2O, allowing H_2O to serve as a source of both electrons and protons.

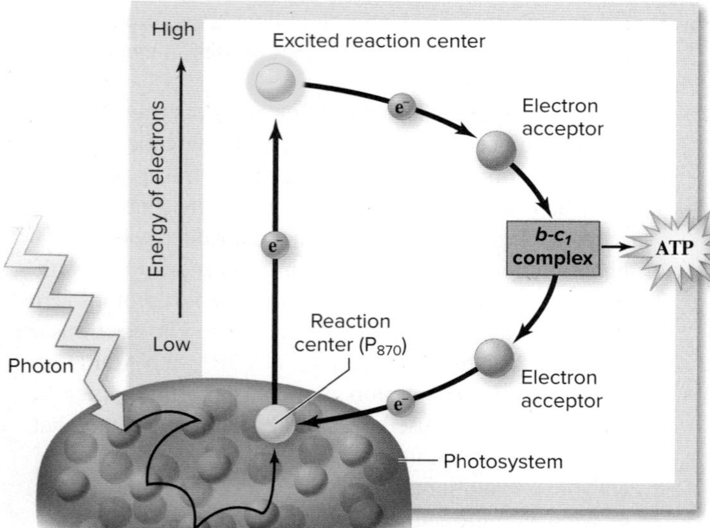

Figure 8.12 The path of an electron in purple nonsulfur bacteria. When a light-energized electron is ejected from the photosystem reaction center (P_{870}), it returns to the photosystem via a cyclic path that produces ATP but not NADPH.

Chloroplasts Utilize Two Connected Photosystems

LEARNING OBJECTIVE 8.5.2 Compare chloroplast photosystems with bacterial photosystems.

In contrast to the sulfur bacteria, plants have two linked photosystems. This overcomes the limitations of cyclic photophosphorylation by providing an alternative source of electrons from the oxidation of water. The oxidation of water also generates O_2—that is, it is oxygenic photosynthesis. Importantly, this noncyclic transfer of electrons also produces NADPH, which plays a key role in the biosynthesis of carbohydrates.

One photosystem, called **photosystem I,** has an absorption peak of 700 nm, so its reaction center pigment is called P_{700}. This photosystem functions in a way analogous to the photosystem found in the sulfur bacteria discussed earlier in this section. The other photosystem, called **photosystem II,** has an absorption peak of 680 nm, so its reaction center pigment is called P_{680}. This photosystem can generate an oxidation potential high enough to oxidize water. Working together, the two photosystems carry out a noncyclic transfer of electrons that is used to generate both ATP and NADPH (figure 8.13).

The photosystems were named I and II in the order of their discovery, not in the order in which they operate in the light-dependent reactions. In plants and algae, the two photosystems are specialized for different roles in the overall process of oxygenic photosynthesis. Photosystem I transfers electrons ultimately to $NADP^+$, producing NADPH. The electrons lost from photosystem I are replaced by electrons from photosystem II. Photosystem II, with its high oxidation potential, can oxidize water to replace the electrons transferred to photosystem I. Thus, there is an overall flow of electrons from water to NADPH.

In Noncyclic Photophosphorylation, the Two Photosystems Operate in Series

LEARNING OBJECTIVE 8.5.3 Differentiate between the functions of photosystem I and photosystem II.

Evidence for the action of two photosystems came from experiments that measured the rate of photosynthesis using two light beams of different wavelengths: one red and the other far-red. Using both beams produced a rate greater than the sum of the rates using individual beams of these wavelengths (figure 8.14). This surprising result, called the *enhancement effect,* can be explained by a mechanism involving two photosystems acting in series (that is, one after the other), one photosystem absorbing preferentially in the red, the other in the far-red.

Plants use photosystems II and I in series, first one and then the other, to produce both ATP and NADPH. This two-stage process is called **noncyclic photophosphorylation,** because the path of the electrons is not a circle—the electrons ejected from

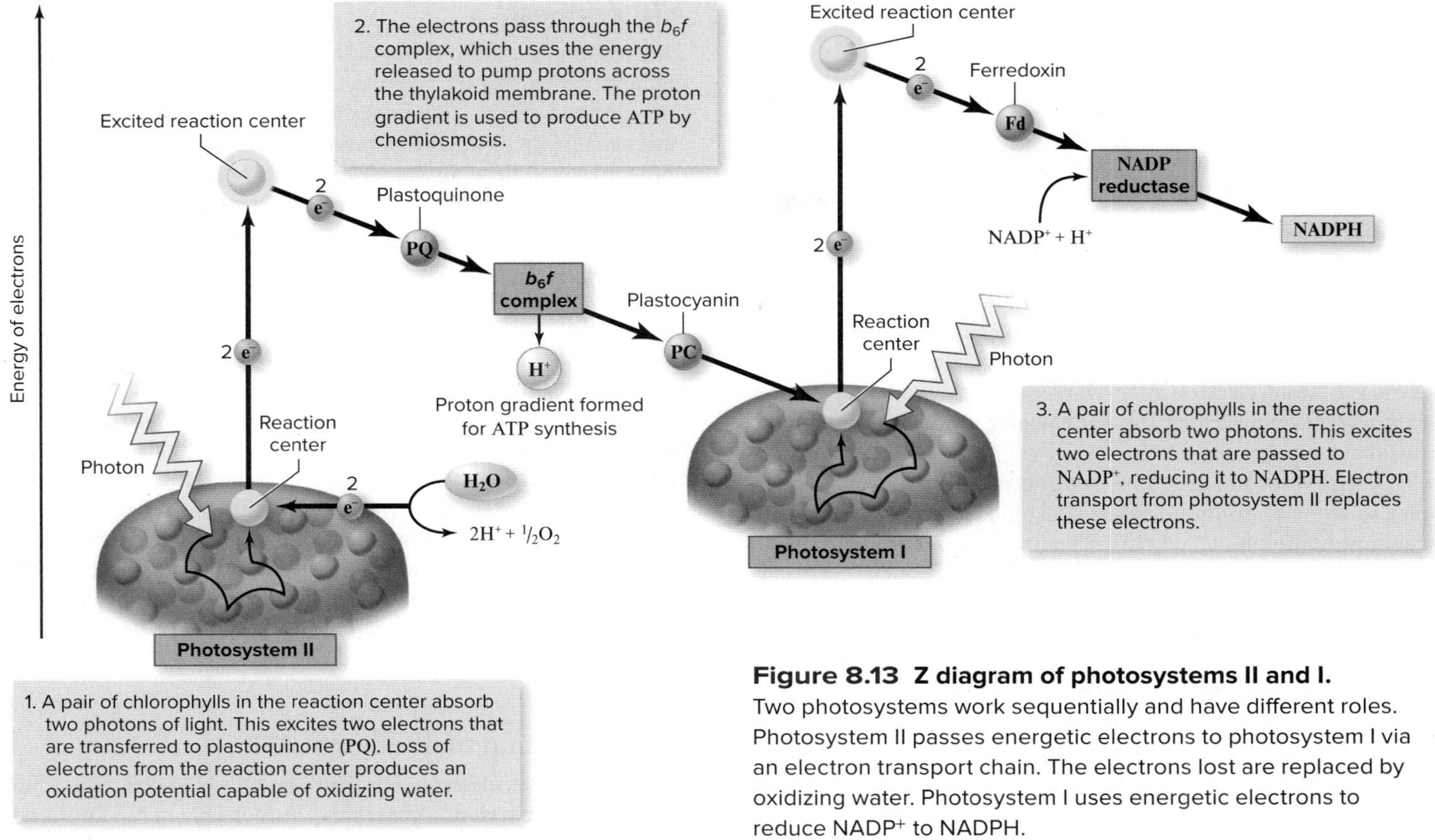

Energy of electrons
Excited reaction center
2. The electrons pass through the b_6f complex, which uses the energy released to pump protons across the thylakoid membrane. The proton gradient is used to produce ATP by chemiosmosis.
2 e⁻
Plastoquinone
PQ
b_6f complex
Plastocyanin
PC
H⁺
Proton gradient formed for ATP synthesis
2 e⁻
Reaction center
Photon
2 e⁻
H₂O
2H⁺ + ½O₂
Photosystem II
1. A pair of chlorophylls in the reaction center absorb two photons of light. This excites two electrons that are transferred to plastoquinone (PQ). Loss of electrons from the reaction center produces an oxidation potential capable of oxidizing water.
Excited reaction center
2 e⁻
Ferredoxin
Fd
NADP reductase
NADP⁺ + H⁺
NADPH
2 e⁻
Reaction center
Photon
Photosystem I
3. A pair of chlorophylls in the reaction center absorb two photons. This excites two electrons that are passed to NADP⁺, reducing it to NADPH. Electron transport from photosystem II replaces these electrons.

Figure 8.13 Z diagram of photosystems II and I. Two photosystems work sequentially and have different roles. Photosystem II passes energetic electrons to photosystem I via an electron transport chain. The electrons lost are replaced by oxidizing water. Photosystem I uses energetic electrons to reduce $NADP^+$ to NADPH.

the photosystems do not return to them but, rather, end up in NADPH. The photosystems are replenished with electrons obtained by splitting water.

The scheme shown in figure 8.13, called a *Z diagram,* illustrates the two electron-energizing steps, one catalyzed by each photosystem. The horizontal axis shows the progress of the light reactions and the relative positions of the complexes, and the vertical axis shows relative energy levels of electrons. The electrons originate from water, which holds on to its electrons very tightly (redox potential = +820 mV), and end up in NADPH, which holds its electrons much more loosely (redox potential = −320 mV).

Photosystem II acts first. High-energy electrons generated by photosystem II are used to synthesize ATP and are then passed to photosystem I to drive the production of NADPH. For every pair of electrons obtained from a molecule of water, one molecule of NADPH and slightly more than one molecule of ATP, on average, are produced.

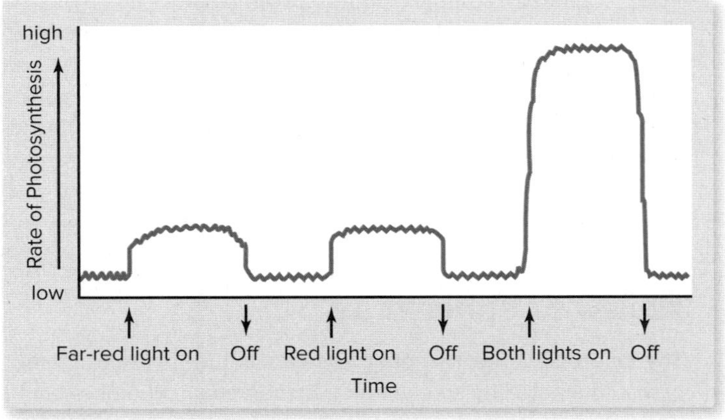

high
Rate of Photosynthesis
low
Far-red light on
Off
Red light on
Off
Both lights on
Off
Time

Figure 8.14 The enhancement effect. The rate of photosynthesis when red and far-red light are provided together is greater than the sum of the rates when each wavelength is provided individually. Photosynthesis is being carried out by two photochemical systems that act in series. One absorbs maximally in the far-red, the other in the red portion of the spectrum.

Photosystem II

The reaction center of photosystem II closely resembles the reaction center of purple bacteria. It consists of a core of 10 transmembrane protein subunits with electron transfer components and two P_{680} chlorophyll molecules arranged around this core. The light-harvesting antenna complex consists of molecules of chlorophyll *a* and accessory pigments bound to several protein chains. The reaction center of photosystem II differs from the reaction center of the purple bacteria in that it also contains four manganese atoms. These manganese atoms are essential for the oxidation of water.

Although the chemical details of the oxidation of water are not entirely clear, the outline is emerging. Four manganese atoms are bound in a cluster to reaction center proteins. Two water molecules are also bound to this cluster of manganese atoms. When the reaction center of photosystem II absorbs a photon, an electron in a P_{680} chlorophyll molecule is excited, and the P_{680} molecule transfers this electron to an acceptor. The oxidized P_{680} then removes an electron from a manganese atom. The oxidized manganese atoms, with the aid of reaction center proteins, remove electrons from oxygen atoms in the two water molecules. This process requires the reaction center to absorb four photons to complete the oxidation of two water molecules, producing one O_2 in the process.

The role of the b_6f complex

The primary electron acceptor for the light-energized electrons leaving photosystem II is a quinone molecule. The reduced quinone that results from accepting a pair of electrons (*plastoquinone*) is a strong electron donor; it passes the excited electron pair to a proton pump called the **b_6f complex** embedded within the thylakoid membrane.

Arrival of the energetic electron pair causes the b_6f complex to pump a proton into the thylakoid space. A small, copper-containing protein called plastocyanin then carries the electron pair to photosystem I.

Photosystem I

The reaction center of photosystem I consists of a core transmembrane complex consisting of 12 to 14 protein subunits with two bound P_{700} chlorophyll molecules. Energy is fed to it by an antenna complex consisting of chlorophyll *a* and accessory pigment molecules.

Photosystem I accepts an electron from plastocyanin into the "hole" created by the exit of a light-energized electron. The absorption of a photon by photosystem I boosts the electron leaving the reaction center to a very high energy level. The electrons are passed to an iron–sulfur protein called *ferredoxin*. Unlike photosystem II and the bacterial photosystem, the plant photosystem I does not rely on quinones as electron acceptors.

Photosystem I passes electrons to ferredoxin on the stromal side of the membrane (outside the thylakoid). The reduced ferredoxin carries an electron with very high potential. Two of them, from two molecules of reduced ferredoxin, are then donated to a molecule of $NADP^+$ to form NADPH. The reaction is catalyzed by the membrane-bound enzyme NADP *reductase.*

Because the reaction occurs on the stromal side of the membrane and involves the uptake of a proton in forming NADPH, it contributes further to the proton gradient established during photosynthetic electron transport. The function of the two photosystems is summarized in figure 8.15.

Photosynthetic Chemiosmosis Uses a Proton Gradient to Make ATP

LEARNING OBJECTIVE 8.5.4 Describe how a proton gradient acts as the energy source for ATP synthesis.

The b_6f complex of photosystem II pumps protons from the stroma into the thylakoid compartment. The splitting of water also adds protons to the thylakoid compartment. Because the thylakoid membrane is impermeable to protons, this creates an electrochemical gradient that can be used to synthesize ATP.

ATP synthase

The chloroplast has ATP synthase enzymes in the thylakoid membrane that form a channel, allowing protons to cross back out into the stroma. These channels protrude like knobs on the external surface of the thylakoid membrane. As protons pass out of the thylakoid through the ATP synthase channel, ADP is phosphorylated to ATP and released into the stroma (figure 8.15). The stroma contains the enzymes that catalyze the reactions of carbon fixation—the Calvin cycle reactions.

This mechanism is the same as that seen in the mitochondrial ATP synthase, and, in fact, the two enzymes are evolutionarily related. This similarity in generating a proton gradient by electron transport and using this gradient in generating ATP by chemiosmosis illustrates the similarities in structure and function between mitochondria and chloroplasts. Evidence for this chemiosmotic mechanism for photophosphorylation was actually discovered earlier (figure 8.16) and formed the background for the later experiments that revealed the existence of mitochondrial ATP synthase. The chemiosmotic mechanism was controversial when it was first proposed, and it stimulated much exploration into the mechanism of ATP synthesis in both mitochondria and chloroplasts.

The production of additional ATP

The passage of an electron pair from water to NADPH in noncyclic photophosphorylation generates one molecule of NADPH and slightly more than one molecule of ATP. But as you will learn in section 8.6, building organic molecules takes more energy than that—it takes 1.5 ATP molecules per NADPH molecule to fix carbon.

To produce the extra ATP, many plant species are capable of short-circuiting photosystem I, switching photosynthesis into a cyclic *photophosphorylation* mode, so that the light-excited electron leaving photosystem I is used to make ATP instead of NADPH. The energetic electrons are simply passed back to the b_6f complex, rather than passing on to $NADP^+$. The b_6f complex pumps protons into the thylakoid space, adding to the proton gradient that drives the chemiosmotic synthesis of ATP. The relative proportions of cyclic and noncyclic photophosphorylation in these plants determine the relative amounts of ATP and NADPH available for building organic molecules.

REVIEW OF CONCEPT 8.5

The chloroplast has two photosystems in the thylakoid membrane connected by an electron transport chain. Photosystem I passes an electron to NADPH. This electron is replaced by one from photosystem II. Photosystem II can oxidize water to replace the electron it has lost. A proton gradient is built up in the thylakoid space, then used to generate ATP by the ATP synthase enzyme.

- *If the thylakoid membrane were leaky to protons, would ATP still be produced? Would NADPH?*

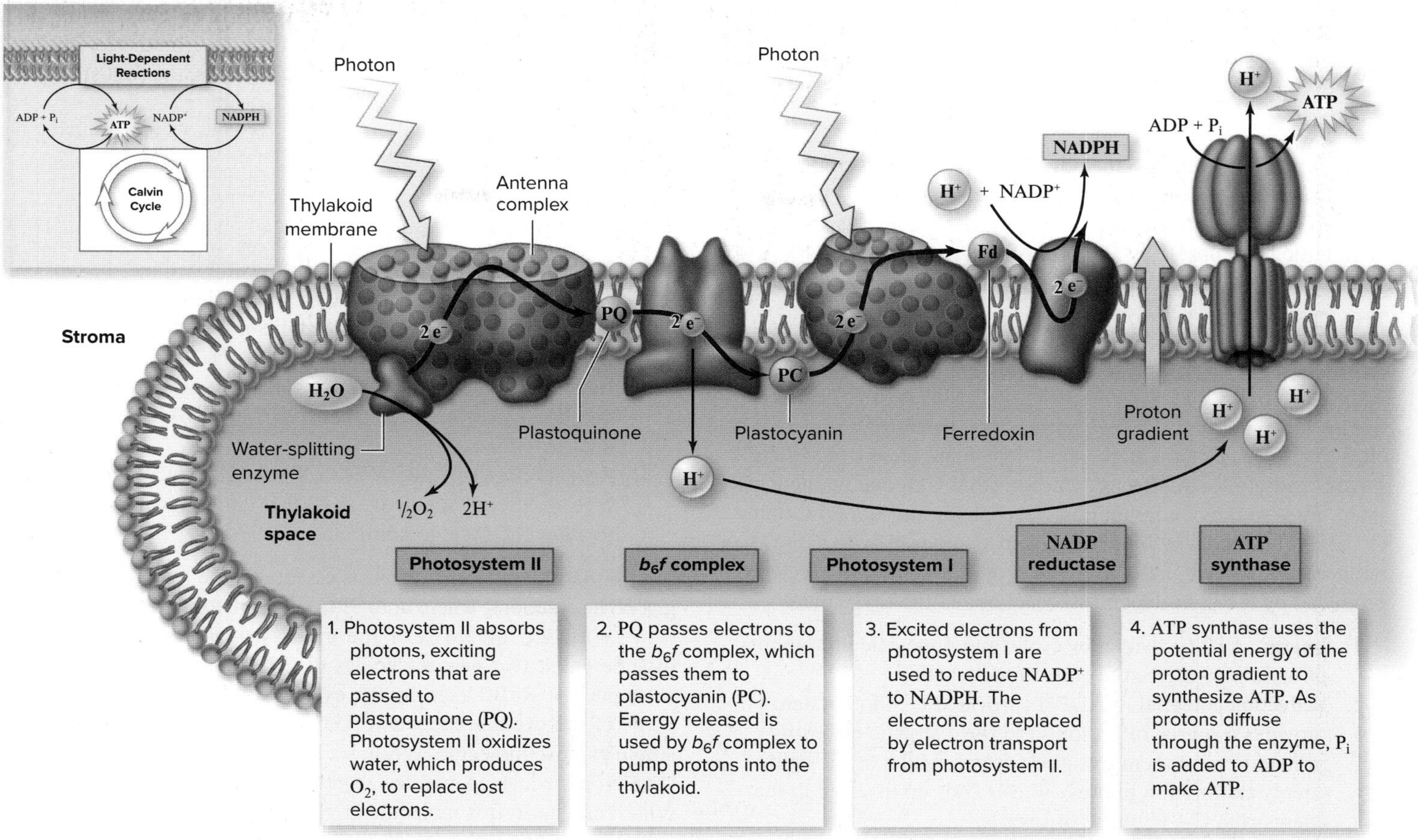
Light-Dependent Reactions
ADP + P_i
ATP
NADP+
NADPH
Calvin Cycle
Photon
Photon
Thylakoid membrane
Antenna complex
Stroma
H_2O
2 e-
PQ
2 e-
PC
2 e-
Fd
2 e-
H+ + NADP+
NADPH
ADP + P_i
H+
ATP
Water-splitting enzyme
Plastoquinone
Plastocyanin
Ferredoxin
Proton gradient
H+
H+
H+
H+
Thylakoid space
½O_2 2H+
Photosystem II
b_6f complex
Photosystem I
NADP reductase
ATP synthase
1. Photosystem II absorbs photons, exciting electrons that are passed to plastoquinone (PQ). Photosystem II oxidizes water, which produces O_2, to replace lost electrons.
2. PQ passes electrons to the b_6f complex, which passes them to plastocyanin (PC). Energy released is used by b_6f complex to pump protons into the thylakoid.
3. Excited electrons from photosystem I are used to reduce NADP+ to NADPH. The electrons are replaced by electron transport from photosystem II.
4. ATP synthase uses the potential energy of the proton gradient to synthesize ATP. As protons diffuse through the enzyme, P_i is added to ADP to make ATP.

Figure 8.15 The photosynthetic electron transport system and ATP synthase. The two photosystems are arranged in the thylakoid membrane joined by an electron transport system that includes the b_6f complex. These function together to create a proton gradient that is used by ATP synthase to synthesize ATP. Photosystem II can oxidize water to O_2, and photosystem I reduces $NADP^+$ to NADPH.

SCIENTIFIC THINKING

Hypothesis: *Photophosphorylation is coupled to electron transport by a proton gradient.*

Prediction: *If a proton gradient can be formed artificially, then isolated chloroplasts will phosphorylate ADP in the dark.*

Test: *Isolated chloroplasts are incubated in acid medium, then transferred in the dark to a basic medium to create an artificial proton gradient.*

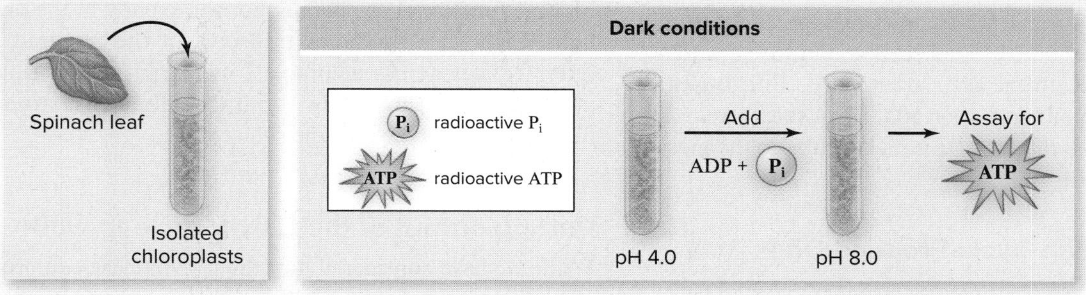
Spinach leaf
Isolated chloroplasts
Dark conditions
P_i radioactive P_i
ATP radioactive ATP
pH 4.0
Add
ADP + P_i
pH 8.0
Assay for
ATP

Result: *Isolated chloroplasts can phosphorylate ADP in the dark as assayed by the incorporation of radioactive PO_4 into ATP.*

Conclusion: *The energy from electron transport in the chloroplast is coupled to the phosphorylation of ADP by a proton gradient.*

Further Experiments: *How would the use of agents that inhibit electron transport affect this outcome? How could such agents be used as a further test of the hypothesis?*

Figure 8.16 The Jagendorf acid bath experiment.

8.6 Carbon Fixation Incorporates CO_2 into Organic Molecules

Consider all of the carbon in your body. Where did it come from? From your diet, but ultimately all of the carbon in your body was once CO_2 in the atmosphere. Photosynthesis converts CO_2 into an organic form usable by all organisms.

The Calvin Cycle Builds Organic Molecules

LEARNING OBJECTIVE 8.6.1 Diagram the action of rubisco in the Calvin cycle.

Now that we have explored the light-dependent reactions in detail, we are ready to learn how new carbon from CO_2 can be incorporated into organic compounds. Carbohydrates contain many C—H bonds and are highly reduced compared with CO_2. To build carbohydrates, cells use energy and a source of electrons produced by the light-dependent reactions of the thylakoids:

1. **Energy.** ATP (provided by cyclic and noncyclic photophosphorylation) drives the endergonic reactions.
2. **Reduction potential.** NADPH (provided by photosystem I) provides a source of protons and the energetic electrons needed to bind them to carbon atoms. Much of the light energy captured in photosynthesis ends up invested in the energy-rich C—H bonds of sugars.

We started this chapter noting that the overall reaction for photosynthesis was essentially the reverse of the overall reaction for cellular respiration. Like respiration, which includes the citric acid cycle, carbon fixation also involves a cycle of reactions.

This cycle was first described by Melvin Calvin and his colleagues in the 1950s, and thus the cycle is named the **Calvin cycle.** Since the first intermediate in the cycle is a 3-carbon acid, the process is also called **C_3 photosynthesis.** Notice that unlike the citric acid cycle, which degraded molecules, this cycle is building molecules.

The most important step in this process is the reaction of CO_2 with the organic molecule **ribulose 1,5-bisphosphate (RuBP)** to produce 3-phosphoglycerate (3PG). This reaction is catalyzed by the enzyme **ribulose bisphosphate carboxylase/oxygenase** (usually abbreviated **rubisco**). This is a large, 16-subunit enzyme found in the chloroplast stroma, which works slowly, processing only about three molecules of RuBP per second. Because it works so slowly, many molecules of rubisco are needed. In a typical plant leaf in a temperate climate, over 50% of all the protein is rubisco. Given the biomass of plants, this makes rubisco the most abundant protein on Earth.

Six Turns of the Calvin Cycle Produce One Molecule of Glucose

LEARNING OBJECTIVE 8.6.2 Describe how the Calvin cycle can produce a molecule of glucose.

Since we started our analysis of cellular respiration with glucose, we will analyze how the Calvin cycle can be used to make a molecule of glucose. Given the relationship to respiration, it should not be surprising that some of the intermediates in the Calvin Cycle are also intermediates in glycolysis, the first stage in the oxidation of glucose.

Phases of the cycle

The Calvin cycle can be divided into three phases (figure 8.17):

1. Carbon fixation: this is the reaction catalyzed by rubisco where CO_2 reacts with RuBP to form a transient 6-carbon intermediate that immediately splits into two molecules of the 3-carbon acid *3-phosphoglycerate* (*3PG*).
2. Reduction: The 3-phosphoglycerate is reduced to *glyceraldehyde-3-phosphate* (*G3P*) using ATP and NADPH from the light reactions.
3. Regeneration of RuBP. The G3P produced by reduction, with additional energy from ATP, is converted into RuBP to allow the cycle to continue.

Notice that the reduction phase is essentially the reverse of several glycolytic reactions, where both 3PG and G3P are intermediates.

With six full turns of the cycle, six molecules of carbon dioxide enter, two new molecules of G3P are produced, and six molecules of RuBP used by the cycle are regenerated. Thus, six turns of the cycle produce two G3P that can be used to make a single glucose molecule. While six turns of the cycle incorporate six CO_2 molecules to synthesize glucose, the six carbon atoms do not all end up in this molecule of glucose.

The net equation of the Calvin cycle is:

$$6\ CO_2 + 18\ ATP + 12\ NADPH + \text{water} \rightarrow$$
$$2\ \text{glyceraldehyde-3-phosphate} + 16\ P_i + 18\ ADP + 12\ NADP^+$$

To account for the carbon atoms necessary to make a molecule of glucose, 6 turns of the cycle incorporate 6 CO_2, and use 6 RuBP to make 12 G3P. 10 of the G3P are needed to make 6 RuBP to complete the cycle, leaving 2 G3P. These 2 G3P both represent the "new" carbon from carbon fixation, and can be used to make 1 molecule of the 6-carbon glucose.

While we call these reactions the light-independent reactions, we now know that light is required indirectly for different segments of the CO_2 reduction reactions. Five of the Calvin cycle enzymes–including rubisco–are light-activated; that is, they become functional or operate more efficiently in the presence of light. Light also promotes transport of the required 3-carbon intermediates across chloroplast membranes. And finally, light promotes the influx of Mg^{2+} into the chloroplast stroma, which further activates the enzyme rubisco.

Putting output of the Calvin cycle to work

While we have considered how the Calvin cycle can produce glucose, the primary output is G3P. Much of the G3P produced by the Calvin cycle is transported out of the chloroplast to the cytoplasm of the cell, where it is converted to glucose-1-phosphate and fructose-6-phosphate. These, in turn, are used to form sucrose, the major transport sugar in plants. Sucrose, commonly known as table sugar, is a disaccharide made of fructose and glucose.

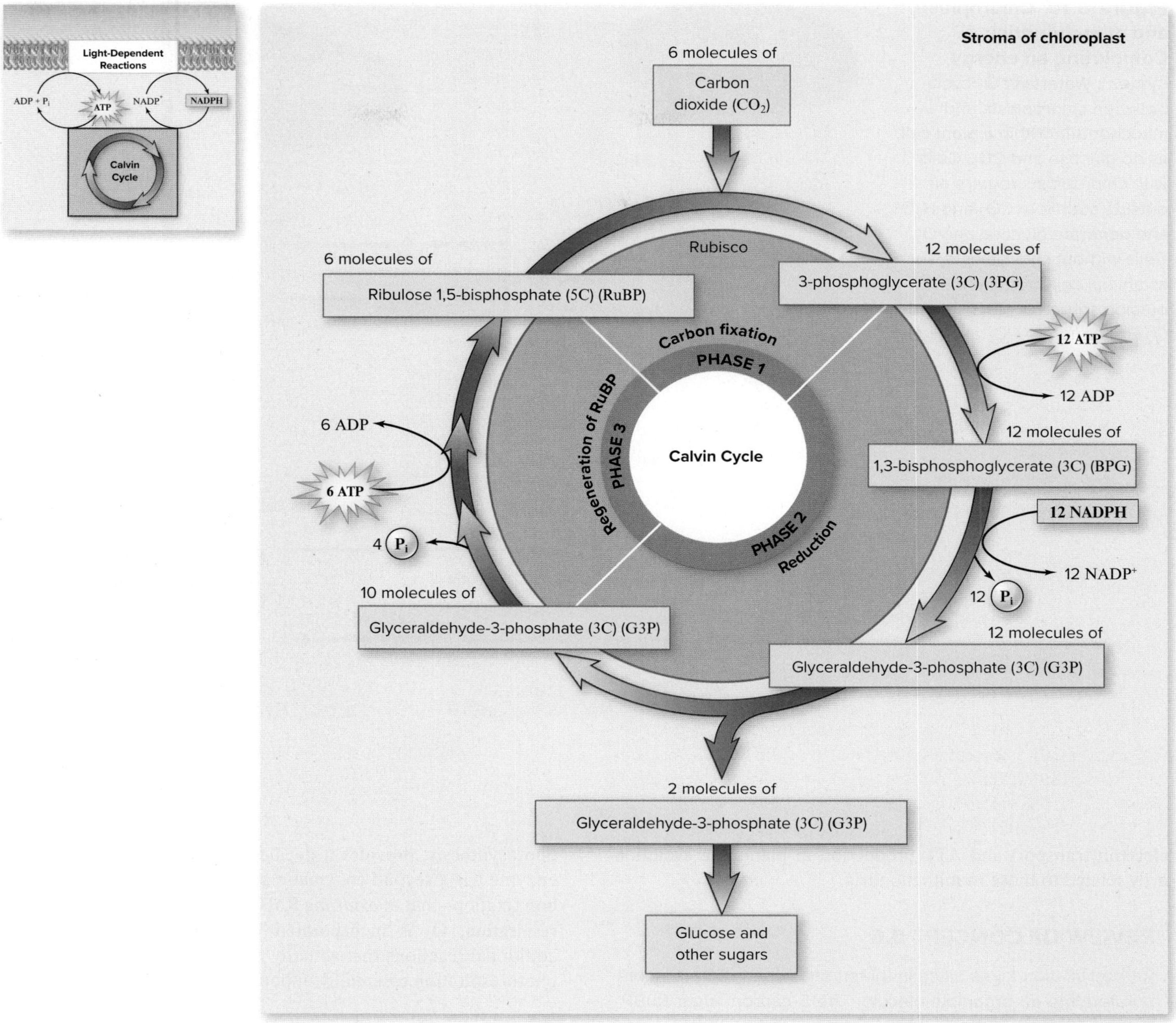

Figure 8.17 The Calvin cycle. The Calvin cycle accomplishes carbon fixation: converting inorganic carbon in the form of CO_2 into organic carbon in the form of carbohydrates. The cycle can be broken down into three phases: (1) carbon fixation, (2) reduction, and (3) regeneration of RuBP. For every six CO_2 molecules fixed by the cycle, a molecule of glucose can be synthesized from the products of the reduction reactions, G3P. The cycle uses the ATP and NADPH produced by the light reactions.

In times of intensive photosynthesis, G3P levels rise in the stroma of the chloroplast. As a consequence, some G3P in the chloroplast is converted to glucose-1-phosphate. This takes place in a set of reactions analogous to those occurring in the cytoplasm, by reversing several reactions similar to those of glycolysis. The glucose-1-phosphate is then combined into an insoluble polymer, forming long chains of starch stored as bulky starch grains in the cytoplasm. These starch grains represent stored glucose for later use.

The energy cycle

The energy-capturing metabolisms of the chloroplasts studied in this chapter and the mitochondria studied in chapter 7 are intimately related (figure 8.18). Photosynthesis uses the products of respiration as starting substrates, and respiration uses the products of photosynthesis as starting substrates. The production of glucose from G3P even uses part of the ancient glycolytic pathway, run in reverse. Also, the principal proteins involved in

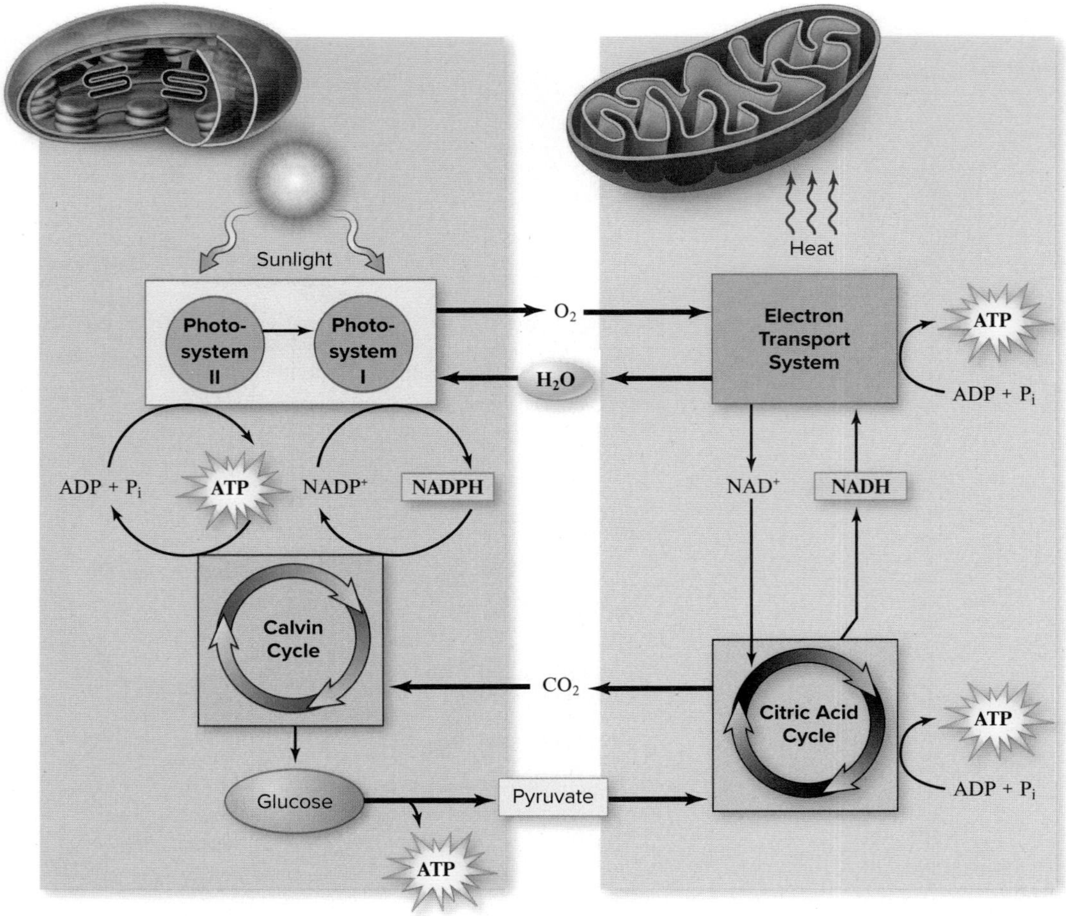

Figure 8.18 Chloroplasts and mitochondria: Completing an energy cycle. Water and O_2 cycle between chloroplasts and mitochondria within a plant cell, as do glucose and CO_2. Cells with chloroplasts require an outside source of CO_2 and H_2O and generate glucose and O_2. Cells without chloroplasts, such as animal cells, require an outside source of glucose and O_2 and generate CO_2 and H_2O.

electron transport and ATP production in plants are evolutionarily related to those in mitochondria.

REVIEW OF CONCEPT 8.6

Carbon fixation takes place in the stroma, where CO_2 is incorporated into an organic molecule. The 5-carbon sugar RuBP is combined with CO_2 in a reaction catalyzed by the enzyme rubisco. This produces 2 molecules of 3PG, which are then reduced to G3P. The Calvin cycle is divided into 3 phases: carbon fixation, reduction, and regeneration of RuBP. ATP and NADPH from the light reactions provide energy and electrons for the reduction reactions.

- *How does the Calvin cycle compare with glycolysis?*

8.7 Photorespiration Short-Circuits Photosynthesis

Evolution does not necessarily result in optimum solutions. Rather, it favors workable solutions that can be derived from features that already exist. Photosynthesis is no exception. Rubisco, the enzyme that catalyzes the key carbon-fixing reaction of photosynthesis, provides a decidedly suboptimal solution. This enzyme has a second enzymatic activity that interferes with carbon fixation—that of *oxidizing* RuBP. In this process, called **photorespiration,** O_2 is incorporated into RuBP, which undergoes additional reactions that actually release CO_2. By releasing CO_2, photorespiration essentially undoes carbon fixation.

C_4 Plants Incorporate CO_2 into a 4-Carbon Molecule

LEARNING OBJECTIVE 8.7.1 Compare and contrast C_3 plants with C_4 and CAM plants.

The carboxylation and oxidation of RuBP are catalyzed at the same active site on rubisco, and CO_2 and O_2 compete with each other at this site. Under normal conditions at 25°C, the rate of the carboxylation reaction is four times that of the oxidation reaction, meaning that under normal growing conditions, fully 20% of photosynthetically fixed carbon is lost to photorespiration.

This loss rises substantially as temperature increases, because under hot, arid conditions, specialized openings in the leaf called **stomata** (singular, *stoma;* figure 8.19) close to conserve water. This closing also cuts off the supply of CO_2 entering the leaf and does not allow O_2 to exit (figure 8.20). As a result, the low-CO_2 and high-O_2 conditions within the leaf favor photorespiration.

Plants that fix carbon using only C_3 photosynthesis (the Calvin cycle) are called **C_3 plants** (figure 8.21*a*). Other plants add CO_2 to phosphoenolpyruvate (PEP) to form a 4-carbon molecule. This reaction is catalyzed by the enzyme PEP *carboxylase.* This enzyme has two advantages over rubisco: it has a much greater affinity for CO_2 than rubisco, and it does not have oxidase activity.

The 4-carbon compound produced by PEP carboxylase undergoes further modification, only to be eventually decarboxylated. The CO_2 released by this decarboxylation is then used by rubisco in the Calvin cycle. This allows CO_2 to be pumped directly to the site of rubisco, which increases the local concentration of CO_2 relative to O_2, minimizing photorespiration. The 4-carbon compound produced by PEP carboxylase allows

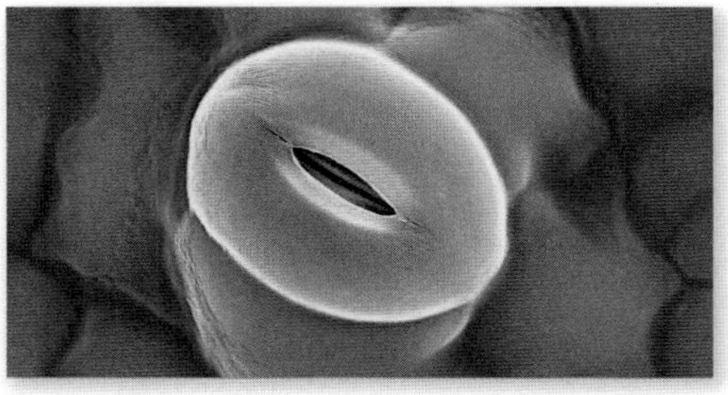

Figure 8.19 Stoma. A closed stoma in the leaf of a tobacco plant. Each stoma is formed from two guard cells whose shape changes with turgor pressure to open and close. Under dry conditions, plants close their stomata to conserve water.

Dr. Jeremy Burgess/Science Source

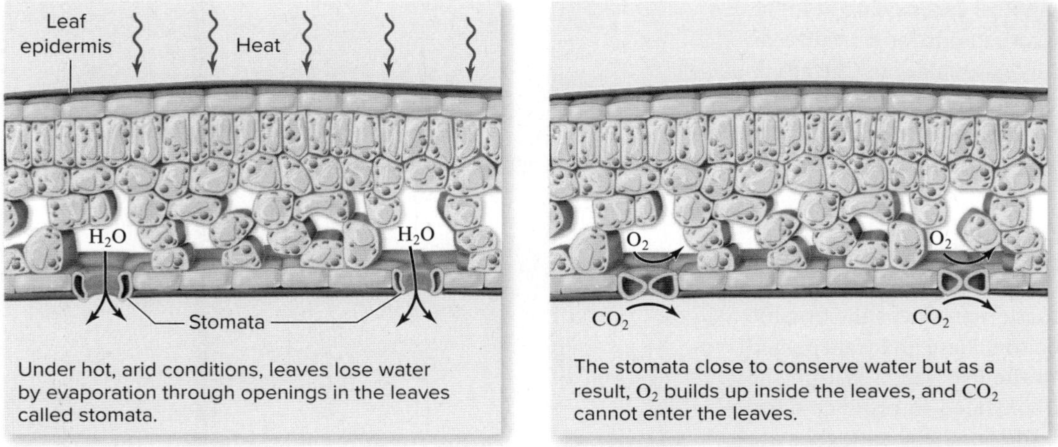

Figure 8.20 Conditions favoring photorespiration. In hot, arid environments, stomata close to conserve water, which also prevents CO_2 from entering and O_2 from exiting the leaf. The high-O_2/low-CO_2 conditions favor photorespiration.

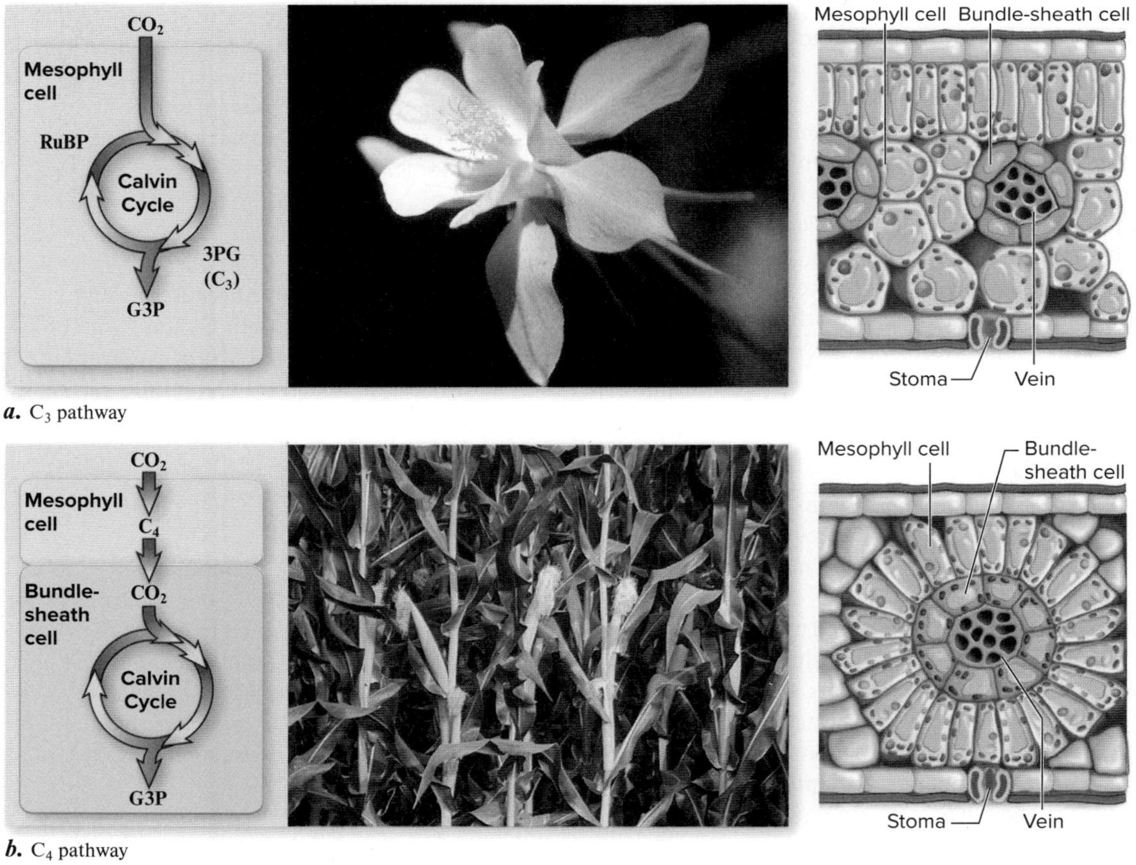

Figure 8.21 Comparison of C_3 and C_4 pathways of carbon fixation. ***a.*** The C_3 pathway uses the Calvin cycle to fix carbon. All reactions occur in mesophyll cells using CO_2 that diffuses in through stomata. ***b.*** The C_4 pathway incorporates CO_2 into a 4-carbon molecule of malate in mesophyll cells. This is transported to the bundle-sheath cells, where it is converted back into CO_2 and pyruvate, creating a high level of CO_2. This allows efficient carbon fixation by the Calvin cycle.

(a): Steven P. Lynch (b): Joseph Nettis/National Audubon Society Collection/Science Source

CO_2 to be stored in an organic form, to then be released in a different cell or at a different time, to keep the level of CO_2 high relative to O_2.

The reduction in the yield of carbohydrate as a result of photorespiration is not trivial. C_3 plants lose 25 to 50% of their photosynthetically fixed carbon in this way. The rate depends largely on temperature. In tropical climates, especially those in which the temperature is often above 28°C, the problem is severe, and it has a major effect on tropical agriculture.

The two main groups of plants that initially capture CO_2 using PEP carboxylase differ in how they maintain high levels of CO_2 relative to O_2. In **C_4 plants** (figure 8.21*b*), the capture of CO_2 occurs in one cell and the decarboxylation occurs in an adjacent cell. This represents a spatial solution to the problem of photorespiration. The second group, **CAM plants,** perform both reactions in the same cell but capture CO_2 using PEP carboxylase at night, then decarboxylate during the day. **CAM** stands for **crassulacean acid metabolism,** after the plant family Crassulaceae (the stonecrops, or hens-and-chicks), in which it was first discovered. This mechanism represents a temporal solution to the photorespiration problem.

C_4 plants

The C_4 plants include corn, sugarcane, sorghum, and a number of other grasses. These plants initially fix carbon using PEP carboxylase in mesophyll cells. This reaction produces the organic acid oxaloacetate, which is converted to malate and transported to bundle-sheath cells that surround the leaf veins. Within the bundle-sheath cells, malate is decarboxylated to produce pyruvate and CO_2 (figure 8.22). Because the bundle-sheath cells are impermeable to CO_2, the local level of CO_2 is high and carbon fixation by rubisco and the Calvin cycle is efficient. The pyruvate produced by decarboxylation is transported back to the mesophyll cells, where it is converted back to PEP, thereby completing the cycle.

The C_4 pathway, although it overcomes the problems of photorespiration, does have a cost. The conversion of pyruvate back to PEP requires breaking two high-energy bonds in ATP. Thus, each CO_2 transported into the bundle-sheath cells costs the equivalent of two ATP. To produce a single glucose, this requires 12 additional ATP compared with the Calvin cycle alone. Despite this additional cost, C_4 photosynthesis is advantageous in hot, dry climates where photorespiration would remove more than half of the carbon fixed by the usual C_3 pathway alone.

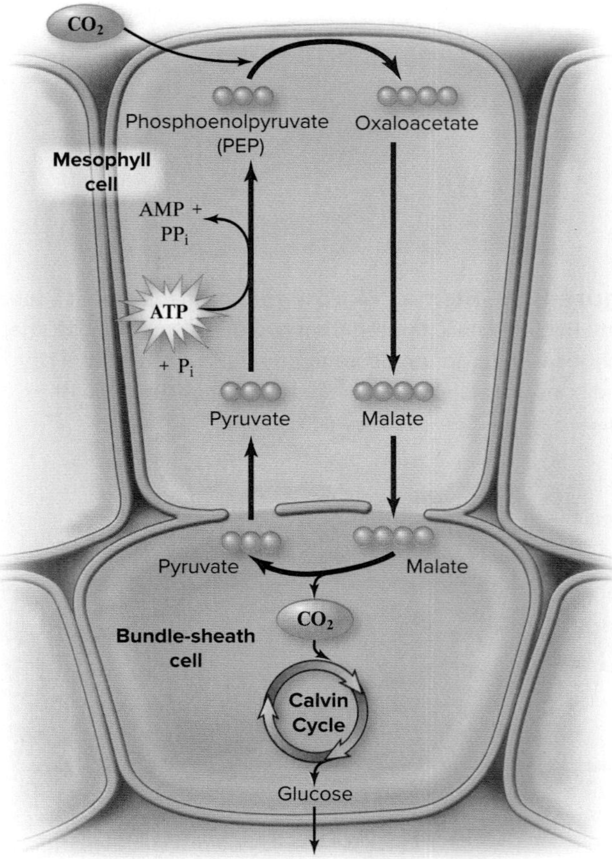

Figure 8.22 Carbon fixation in C_4 plants. The first molecule formed, oxaloacetate, is converted to malate, which moves into bundle-sheath cells, where it is decarboxylated back to CO_2 and pyruvate. This produces a high level of CO_2 in the bundle-sheath cells that can be fixed by the usual C_3 Calvin cycle with little photorespiration. The pyruvate diffuses back into the mesophyll cells, where it is converted back to PEP to be used in another C_4 fixation reaction.

The Crassulacean acid pathway

A second strategy to decrease photorespiration in hot regions has been adopted by the CAM plants. These include many succulent (water-storing) plants, such as cacti, pineapples, and some members of about two dozen other plant groups.

In these plants, the stomata open during the night and close during the day. This pattern of stomatal opening and closing is the reverse of that in most plants. CAM plants initially fix CO_2 using PEP carboxylase to produce oxaloacetate. The oxaloacetate is often converted into other organic acids, depending on the particular CAM plant. These organic compounds accumulate during the night and are stored in the vacuole. Then during the day, when the stomata are closed, the organic acids are decarboxylated to yield high levels of CO_2. These high levels of CO_2 drive the Calvin cycle and minimize photorespiration.

Like C_4 plants, CAM plants use both C_3 and C_4 pathways. They differ in that CAM plants use both of these pathways in the same cell: the C_4 pathway at night and the C_3 pathway during the day. In C_4 plants, the two pathways occur in different cells.

REVIEW OF CONCEPT 8.7

Rubisco can also oxidize RuBP under conditions of high O_2 and low CO_2. In plants that use only C_3 metabolism, up to 20% of fixed carbon is lost. Plants adapted to hot, dry environments are capable of storing CO_2 as a 4-carbon molecule and avoiding some of this loss; they are called C_4 plants. In CAM plants, CO_2 is fixed at night into a C_4 organic compound; in the daytime, this compound is used as a source of CO_2 for C_3 metabolism.

- *How do C_4 plants and CAM plants differ?*

Testing the Great Iron Dump Hypothesis

Greenhouse-induced global warming is caused by increasing levels of carbon dioxide, and other greenhouse gases, released into Earth's atmosphere by human activity. There are two ways to reduce these effects: reduce CO_2 emissions or remove existing CO_2 from the atmosphere. Reducing emissions appears to be the course preferred by many, but the approach to remove the excess CO_2 from the atmosphere can actually involve biology. Ocean ecologist John Martin once famously said, "Give me half a tanker of iron and I'll give you the next ice age." He was pointing out that Earth's oceans are rich in marine algae, their growth limited primarily by lack of iron (Fe); adding iron to the upper ocean could trigger an algal "bloom," its intense photosynthesis sequestering massive amounts of CO_2 in organic matter that would then fall to the ocean's bottom, where it could no longer react with Earth's atmosphere. Although there are many difficulties with this kind of scenario, it is sufficiently serious to merit experimental testing.

Martin's iron dump hypothesis is not an easy idea to test, as ocean currents quickly affect our ability to assess any results of Fe fertilization. At least 12 ocean iron dump (OID) experiments have been performed, but the results are equivocal. Recently, researchers have taken a novel approach. In the Southern Ocean near Antarctica, currents swirl, forming stable eddies within which the necessary measurements could be made.

Martin's iron dump hypothesis really has two parts: first, that an algal bloom occurs and, second, that the CO_2 sequestered into organic material would then fall downward into the deep ocean when the bloom died off. In this test, the researchers measured the amounts of organic carbon and chlorophyll in a 100-meter column of water, looking to see how they changed in the patch after Fe fertilization compared with those of surrounding waters. Some of the results obtained by the research team are presented in the two graphs. Red dots are from within the Fe-fertilized patch, and blue dots are from outside the fertilized patch.

The data shown plot particulate organic carbon (POC) versus time, and chlorophyll versus time. The measurements are averaged for a 100-meter column of water, which is the amount "mixed" with the surface water. The POC values are in mmol/m^3, and the chlorophyll values are in mg/m^3. Additional data, not shown, indicated that the changes in carbon and chlorophyll could be accounted for by changes in populations of diatom species.

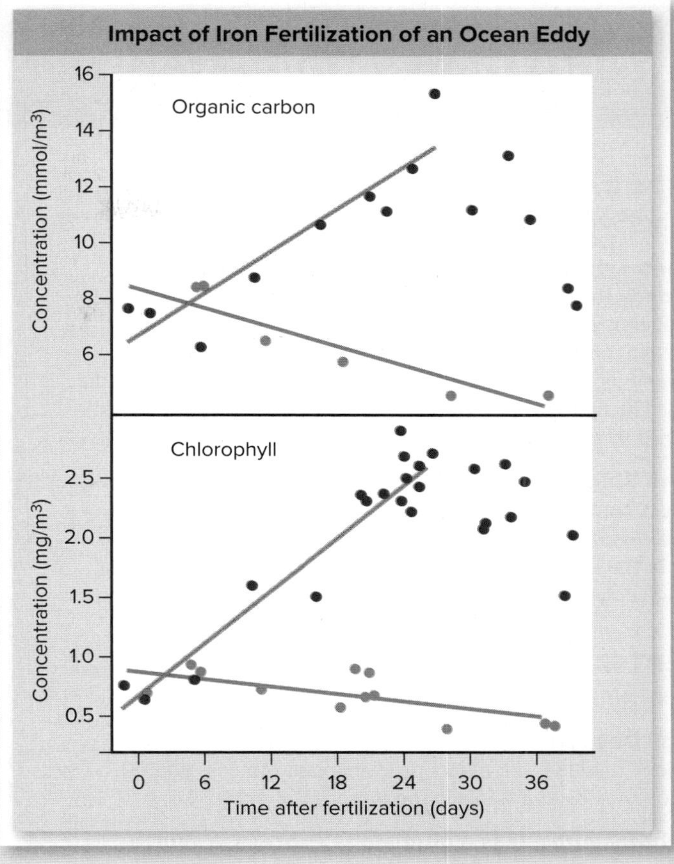

Analysis

1. **Applying Concepts**
   Variable. In the graph, what is the dependent variable?
2. **Interpreting Data**
   a. Does the concentration of organic carbon in seawater change in the weeks after Fe fertilization? Of chlorophyll?
   b. Over the course of the first 4 weeks after fertilization, is the change in organic carbon consistent? Of chlorophyll?
   c. Over the subsequent 2 weeks, is the change different? How so?
3. **Making Inferences** From these results, would it be reasonable to infer that the Fe fertilization resulted in an algal bloom at 100 meters ocean depth? Does this effect persist at this depth after 4 weeks?
4. **Drawing Conclusions**
   a. Do these results support the claim that lack of iron is limiting the growth of phytoplankton, and thus of photosynthesis, in certain areas of the oceans?
   b. Is Martin's ocean dump hypothesis verified?
5. **Further Analysis** How could you test the idea that any algal bloom was followed by a die-off and sinking of dead organisms? Does the fact that the major species involved in the bloom were diatoms have any bearing on the sinking hypothesis?

Retracing the Learning Path

CONCEPT 8.1 Photosynthesis Uses Sunlight to Power the Synthesis of Organic Molecules

8.1.1 Photosynthesis Has Light-Dependent and Light-Independent Reactions Photosynthesis combines CO_2 and H_2O, producing glucose and O_2. Photosynthesis has three stages: absorbing light energy, using this energy to synthesize ATP and NADPH, and using the ATP and NADPH to convert CO_2 to organic molecules. The first two stages consist of light-dependent reactions and the third stage of light-independent reactions.

8.1.2 The Chloroplast Is a Photosynthetic Machine Chloroplasts contain internal thylakoid membranes and a fluid matrix called stroma. The photosystems involved in energy capture are found in the thylakoid membranes, and enzymes for assembling organic molecules are in the stroma.

CONCEPT 8.2 Experiments Revealed That Photosynthesis Is a Chemical Process

8.2.1 Soil Does Not Add Mass to Growing Plants Early investigations revealed that plants produce plant biomass and O_2 from carbon dioxide and water in the presence of light.

8.2.2 Photosynthesis Is a Multistage Process in Which Some Steps Do Not Require Light The light-dependent reactions require light; the light-independent reactions occur in both daylight and darkness.

8.2.3 Water Acts as an Electron Donor The use of isotopes revealed the origins and fates of different molecules in photosynthetic reactions.

8.2.4 NADPH Is Used to Reduce Carbon Dioxide Carbon fixation requires ATP and NADPH, which are products of the light-dependent reactions.

CONCEPT 8.3 Pigments Capture Energy from Sunlight

8.3.1 Light Is a Form of Energy Light can remove electrons from some metals by the photoelectric effect, and in photosynthesis, chloroplasts act as photoelectric devices.

8.3.2 Chlorophyll Is the Principal Photosynthetic Pigment Chlorophyll *a* is the only pigment that can convert light energy into chemical energy. Chlorophyll *b* is an accessory pigment.

8.3.3 Carotenoids Are Accessory Pigments Accessory pigments absorb other wavelengths of light.

CONCEPT 8.4 Photosynthetic Pigments Are Organized into Photosystems

8.4.1 The Rate of Photosynthesis Saturates at Low Light Intensity Measurement of O_2 output led to the idea of photosystems—clusters of pigment molecules that channel energy to a reaction center.

8.4.2 Chlorophyll Pigments Are Organized to Efficiently Capture Light A photosystem is a network of chlorophyll *a,* accessory pigments, and proteins embedded in the thylakoid membrane. Pigment molecules of the antenna complex harvest photons and feed light energy to the reaction center. The reaction center is composed of two chlorophyll *a* molecules that pass an excited electron to an electron acceptor.

CONCEPT 8.5 Energy from Sunlight Is Used to Produce a Proton Gradient

8.5.1 Light-Dependent Reactions Take Place in the Thylakoid Membrane The light reactions can be broken down into four processes: primary photoevent, charge separation, electron transport, and chemiosmosis. An excited electron moves along a transport chain, producing a proton gradient, and eventually returns to the photosystem. In some bacteria, this can also produce NADPH.

8.5.2 Chloroplasts Utilize Two Connected Photosystems Photosystem I transfers electrons to $NADP^+$, reducing it to NADPH. Photosystem II replaces electrons lost by photosystem I. Electrons lost from photosystem II are replaced by electrons from oxidation of water, which also produces O_2.

8.5.3 In Noncyclic Photophosphorylation, the Two Photosystems Operate in Series Photosystem II and photosystem I are linked by an electron transport chain that pumps protons into the thylakoid space.

8.5.4 Photosynthetic Chemiosmosis Uses a Proton Gradient to Make ATP ATP synthase is a channel enzyme. As protons flow through the channel down their gradient, ADP is phosphorylated, producing ATP. Plants can make additional ATP by cyclic photophosphorylation.

CONCEPT 8.6 Carbon Fixation Incorporates, CO_2 into Organic Molecules

8.6.1 The Calvin Cycle Builds Organic Molecules The Calvin cycle, also known as C_3 photosynthesis, uses CO_2, ATP, and NADPH to build simple sugars.

8.6.2 Six Turns of the Calvin Cycle Produce One Molecule of Glucose The Calvin cycle occurs in three stages: carbon fixation via the enzyme rubisco's action on RuBP and CO_2; reduction of the resulting 3PG to G3P, using ATP and NADPH; and regeneration of RuBP. Six turns of the cycle fix enough carbon to produce two G3Ps, used to make one molecule of glucose.

CONCEPT 8.7 Photorespiration Short-Circuits Photosynthesis

Rubisco can catalyze the oxidation of RuBP, reversing carbon fixation. Dry, hot conditions tend to increase this reaction.

8.7.1 C_4 Plants Incorporate CO_2 into a 4-Carbon Molecule C_4 plants fix carbon by adding CO_2 to phosphoenolpyruvate to form oxaloacetate, which is converted to malate. Malate can be decarboxylated to pyruvate to increase the local concentration of CO_2 and minimize photorespiration. Carbon is fixed in one cell by the C_4 pathway; then CO_2 is released in another cell for the Calvin cycle (figure 8.22). CAM plants use the C_4 pathway during the day when stomata are closed, and the Calvin cycle at night in the same cell.

Concept Overview

This Concept Overview diagrams the key concepts that were discussed in this chapter.

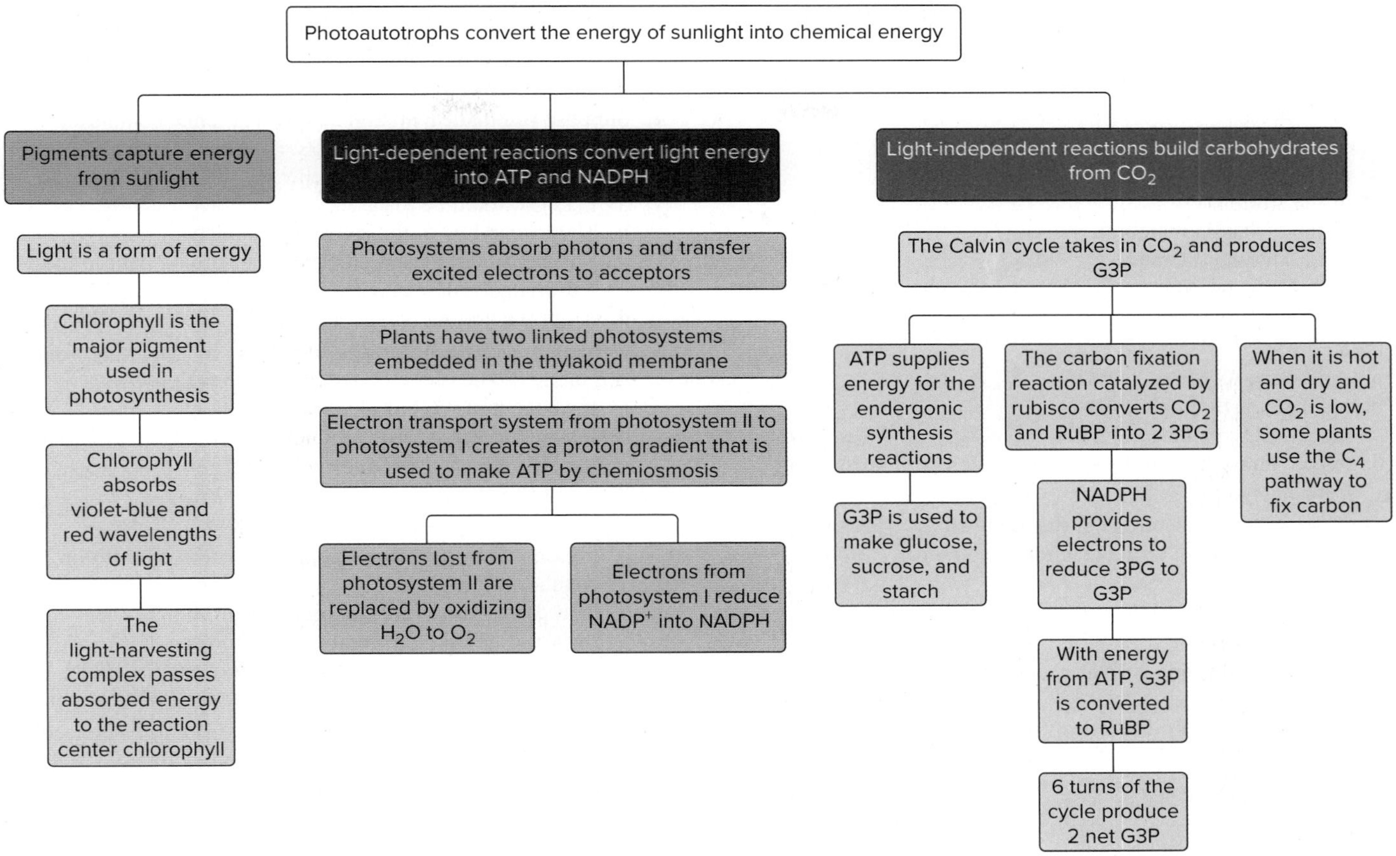

Assessing the Learning Path

Understand

1. Which of the following summarizes the overall process of photosynthesis?
   a. $6\ O_2 + C_6H_{12}O_6 + \text{light} \rightarrow 6\ CO_2 + 6\ H_2O$
   b. $6\ O_2 + C_6H_{12}O_6 \rightarrow 6\ CO_2 + 6\ H_2O + \text{ATP}$
   c. $6\ H_2O + 6\ CO_2 + \text{light} \rightarrow 6\ O_2 + C_6H_{12}O_6$
   d. $6\ CO_2 + 6\ H_2O + \text{light} \rightarrow 6\ O_2 + C_6H_{12}O_6 + \text{ATP} + \text{NADPH}$
2. The light-dependent reactions of photosynthesis
   a. produce ATP and NADPH and occur in the thylakoid membranes.
   b. produce glucose and occur in the stroma.
   c. produce ATP, reduce $NADP^+$, and occur in the stroma.
   d. utilize CO_2 and occur in the thylakoid membranes.
3. When a plant grows, its increase in biomass comes from
   a. the nutrients found in the soil.
   b. the water taken up by the roots.
   c. the carbon from CO_2 in the air.
   d. the photons in the sunlight.
4. The gas released as a product of photosynthesis comes from
   a. the breakdown of $C_6H_{12}O_6$.
   b. H_2O.
   c. CO_2.
   d. NADPH.
5. Which of the following statements about photons is NOT accurate?
   a. Photons are discrete bundles of light energy.
   b. Photons are found at all wavelengths of light.
   c. Energy from photons can be transferred to pigment molecules.
   d. All photons have the same amount of energy.
6. Chlorophyll *a*
   a. absorbs photons with wavelengths of 500–600 nm.
   b. reflects green light.
   c. has an absorption spectrum shifted toward green wavelengths.
   d. is considered an important accessory pigment.

7. How is a reaction center pigment in a photosystem different from a pigment in the antenna complex?
   a. The reaction center pigment is a chlorophyll molecule.
   b. The antenna complex pigment can only reflect light.
   c. The reaction center pigment loses an electron when it absorbs light energy.
   d. The antenna complex pigments are not attached to proteins.
8. In eukaryotes, photosystems are found
   a. in the cytosol.
   b. in the stroma.
   c. in the thylakoid membrane.
   d. in the thylakoid lumen.
9. In the light-dependent reactions of photosynthesis, photosystem I functions to ______, and photosystem II functions to ______.
   a. synthesize ATP; produce O_2
   b. reduce $NADP^+$; oxidize H_2O
   c. reduce CO_2; oxidize NADPH
   d. restore an electron to its reaction center; gain an electron from water
10. The overall flow of electrons in the light reactions is from
    a. antenna pigments to the reaction center.
    b. H_2O to CO_2.
    c. photosystem I to photosystem II.
    d. H_2O to NADPH.
11. Which of the following is NOT accurate about carbon fixation?
    a. It results in the production of CO_2.
    b. It requires NADPH to provide electrons and protons.
    c. It requires energy provided by the light-dependent reactions.
    d. It occurs in the stroma.
12. The overall process of photosynthesis
    a. results in the reduction of CO_2 and the oxidation of H_2O.
    b. results in the reduction of H_2O and the oxidation of CO_2.
    c. consumes O_2 and produces CO_2.
    d. produces O_2 from CO_2.
13. Which of the following is a similarity between C_4 and CAM plants?
    a. Both use PEP carboxylase to initially fix carbon.
    b. Both carry out carbon fixation and the Calvin cycle in separate cell types.
    c. Both carry out the Calvin cycle at night.
    d. All of the above

Apply

1. In theory, a well-watered plant in a normal atmosphere kept in total darkness could still manufacture glucose, if it were supplied with which molecules?
   a. CO_2 and H_2O
   b. O_2 and H_2O
   c. ATP and NADPH
   d. $FADH_2$, NADH, and CO_2
2. Illuminated chloroplasts deprived of CO_2
   a. cannot reduce $NADP^+$.
   b. accumulate NADPH and ATP.
   c. will not produce O_2.
   d. become ATP-depleted.
3. A plant that has normal chlorophyll but makes defective carotenoid pigments
   a. will not be able to absorb violet-blue and red light.
   b. will not produce free radicals.
   c. will not be able to carry out photosynthesis as efficiently as a plant with normal carotenoid.
   d. All of the above
4. Suppose you made a mutant cell with antenna complexes composed entirely of carotenoids. How would this cell compare with a normal plant cell?
   a. Less O_2 would be released.
   b. There would be more free-radical damage.
   c. The cell would be less efficient at absorbing light in the red/orange/yellow range.
   d. All of the above
5. If the thylakoid membrane became leaky to ions, what would you predict to be the result on the light reactions?
   a. It would stop ATP production.
   b. It would stop NADPH production.
   c. It would stop the oxidation of H_2O.
   d. All of the above
6. In chloroplast thylakoids,
   a. protons return passively to the stroma through proton channels.
   b. light causes the hydrogen ion concentration in the thylakoid to fall below that in the stroma.
   c. light leads to the flow of protons out of the thylakoid.
   d. ATP is formed when protons flow into the thylakoid lumen.
7. Six molecules of CO_2 enter the Calvin cycle and 12 molecules of G3P are made. Two of the G3P molecules are used to make sugar. What happens to the rest of the G3P?
   a. It is converted back to CO_2 and released from the plant.
   b. It is used to regenerate RuBP.
   c. It is converted to energy-storage molecules—either oil or starch, depending on the plant.
   d. It is used to make more ATP.
8. Why is PEP carboxylase more efficient than rubisco in hot, dry conditions?
   a. It works during the day, whereas rubisco can only work in the dark.
   b. It does not react with oxygen.
   c. It increases rates of photorespiration, thus leading to more sugar production.
   d. All of the above

Synthesize

1. Diagram the relationship between the reactants and the products of photosynthesis and cellular respiration, and identify in which organelles and organisms these reactions occur.
2. A key theme of biology is that form fits function. Explain how the structure of chlorophyll *a* makes it an excellent energy-harvesting molecule.
3. What is the advantage of having many pigment molecules in each photocenter for every reaction center chlorophyll? Why not couple every pigment molecule directly to an electron acceptor?
4. Compare and contrast the production of ATP during the light-dependent reactions and during cellular respiration.
5. Why are plants that consume 30 ATP molecules to produce 1 molecule of glucose (rather than the usual 18 molecules of ATP per glucose molecule) favored in hot climates but not in cold climates? What role does temperature play?

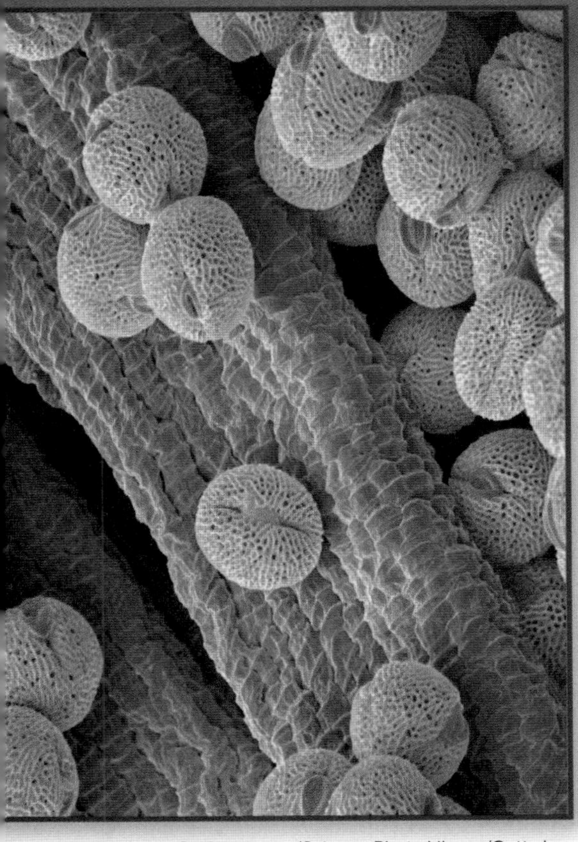

Steve Gschmeissner/Science Photo Library/Getty Images

9 Cell Communication

Learning Path

9.1 The Cells of Multicellular Organisms Communicate

9.2 Signal Transduction Begins with Cellular Receptors

9.3 Intracellular Receptors Respond to Signals by Regulating Gene Expression

9.4 Protein Kinase Receptors Respond to Signals by Phosphorylating Proteins

9.5 G Protein–Coupled Receptors Respond to Signals Through Effector Proteins

Concept Overview

This Concept Overview provides a diagram of the overarching concepts that are covered in this chapter. These concepts will be expanded upon in the Concept Overview at the end of the chapter.

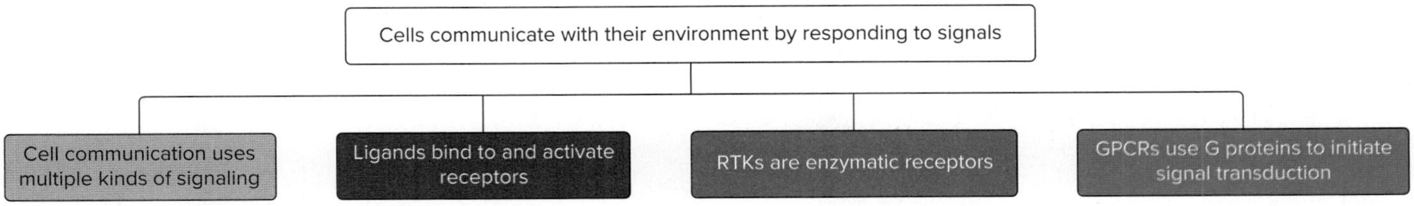
Cells communicate with their environment by responding to signals
Cell communication uses multiple kinds of signaling
Ligands bind to and activate receptors
RTKs are enzymatic receptors
GPCRs use G proteins to initiate signal transduction

Introduction

Spring is a time of rebirth and renewal. Trees that have appeared dead produce new leaves and buds, and flowers sprout from the ground. For people with seasonal allergies, this is not such a pleasant time. The pollen in the micrograph on the previous page and other allergens stimulate the immune system to produce histamine and other molecules that form cellular signals. These signals lead to the runny nose, itching, watery eyes, and other symptoms of the allergic reaction. We treat allergies using antihistamines, which act by blocking the receptor for the histamine signal. In this chapter, we will explore such signaling pathways in detail.

9.1 The Cells of Multicellular Organisms Communicate

Communication between cells is common in nature. Cell signaling occurs in all multicellular organisms, providing an indispensable mechanism for cells to influence one another. Effective signaling requires a signaling molecule, called a **ligand,** and a cellular protein that binds this ligand, called a **receptor protein.** The interaction of these two components initiates the process of *signal transduction,* which converts the information in the external signal into a cellular response (figure 9.1).

The cells of multicellular organisms use a variety of molecules as signals, including but not limited to peptides, large proteins, individual amino acids, nucleotides, and steroids and other lipids. Even dissolved gases such as nitric oxide (NO) are used as signals.

Signaling Mechanisms Vary with Distance from Source

LEARNING OBJECTIVE 9.1.1 Discriminate between methods of signaling based on distance from source to reception.

Any cell of a multicellular organism is exposed to a constant stream of signals. At any time, hundreds of different chemical signals may be present in the environment surrounding the cell. Each cell responds to only certain signals, however, and ignores the rest, like a person following the conversation of one or two individuals in a noisy, crowded room.

How does a cell "choose" which signals to respond to? This depends on the number and kind of receptor molecules found in the cell. A ligand will bind to a receptor that has a complementary shape, forming a receptor–ligand complex, much like an enzyme-substrate complex. Ligand binding induces a change in the receptor protein's structure, which initiates a signal transduction pathway to ultimately produce some kind of cellular response. A given cell responds only to those signaling molecules that can bind to the specific set of receptor proteins it possesses, and "ignores" any ligands that lack receptors.

Cells can communicate through any of four basic mechanisms, depending primarily on the distance between the signaling and responding cells (figure 9.2). These mechanisms are (1) direct contact, (2) paracrine signaling, (3) endocrine signaling, and (4) synaptic signaling.

In addition to using these four basic mechanisms, some cells actually send signals to themselves, secreting signals that bind to specific receptors on their own plasma membranes. This process, called *autocrine signaling,* is thought to play an important role in reinforcing developmental changes, and it is an important component of signaling in the immune system (refer to chapter 35).

Direct contact

As you learned in chapter 5, the surface of a eukaryotic cell is richly populated with proteins, carbohydrates, and lipids attached to and extending outward from the plasma membrane. When cells are very close to one another, some of the molecules on the plasma membrane of one cell can be recognized by receptors on the plasma membrane of an adjacent cell. Many of the important interactions between cells in early development occur by means of direct contact between cell surfaces. Cells also signal through gap junctions (figure 9.2*a*).

Paracrine signaling

Signal molecules released by cells can diffuse through the extracellular fluid to other cells. If those molecules are taken up by

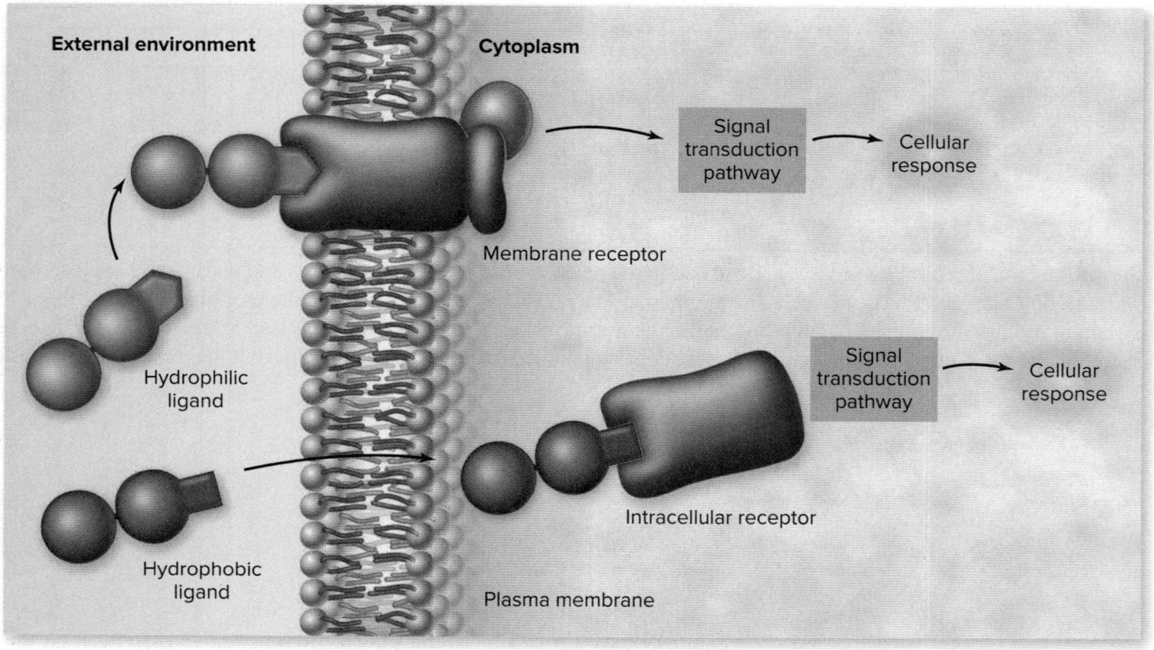
External environment
Cytoplasm
Signal transduction pathway
Cellular response
Membrane receptor
Hydrophilic ligand
Signal transduction pathway
Cellular response
Intracellular receptor
Hydrophobic ligand
Plasma membrane

Figure 9.1 Overview of cell signaling. Cell signaling involves a signal molecule called a ligand, a receptor, and a signal transduction pathway that produces a cellular response. The location of the receptor can be either intracellular, for hydrophobic ligands that can cross the membrane, or in the plasma membrane, for hydrophilic ligands that cannot cross the membrane.

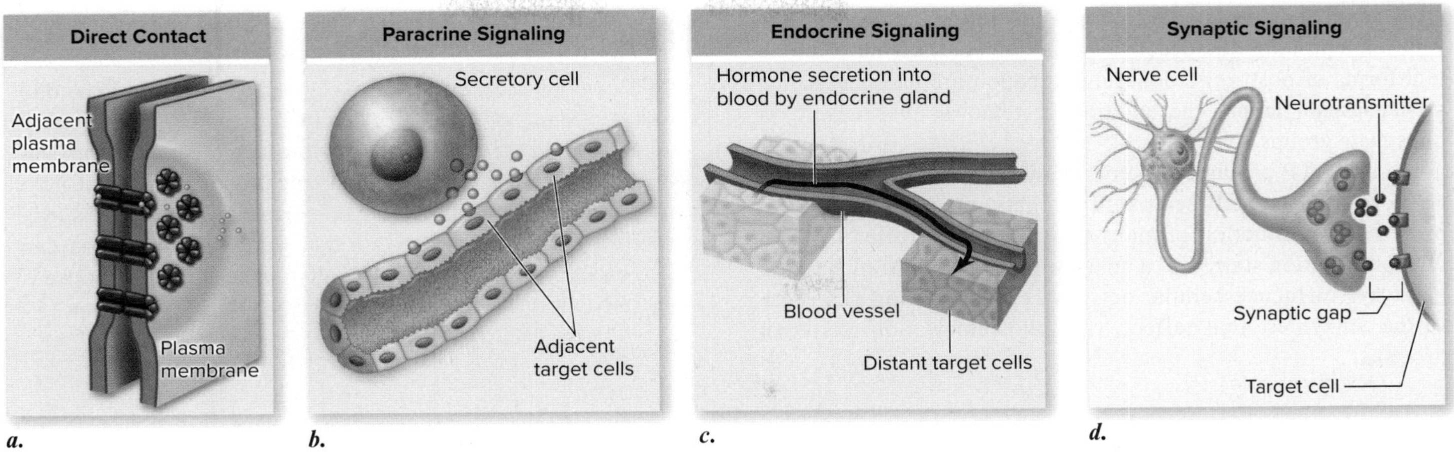

Figure 9.2 Four kinds of cell signaling. Cells communicate in four ways. ***a.*** Two cells in direct contact with each other may send signals across gap junctions. ***b.*** In paracrine signaling, secretions from one cell have an effect only on cells in the immediate area. ***c.*** In endocrine signaling, hormones are released into the organism's circulatory system, which carries them to the target cells. ***d.*** Chemical synapse signaling involves transmission of signal molecules, called neurotransmitters, from a neuron over a small synaptic gap to the target cell.

neighboring cells, destroyed by extracellular enzymes, or quickly removed from the extracellular fluid in some other way, their influence is restricted to cells in the immediate vicinity of the releasing cell. Signals with such short-lived, local effects are called **paracrine** signals (figure 9.2*b*). The immune response in vertebrates also involves paracrine signaling between immune cells (refer to chapter 35).

Endocrine signaling

A released signal molecule that remains in the extracellular fluid may enter the organism's circulatory system and travel widely throughout the body. These longer-lived signal molecules, which may affect cells very distant from the releasing cell, are called **hormones,** and this type of intercellular communication is known as **endocrine** signaling (figure 9.2*c*). Chapter 35 discusses endocrine signaling in detail.

Synaptic signaling

In animals, the cells of the nervous system provide rapid communication with distant cells. Their signal molecules, **neurotransmitters,** do not travel to the distant cells through the circulatory system as hormones do. Rather, the long, fiber-like extensions of nerve cells release neurotransmitters from their tips very close to the target cells (figure 9.2*d*). The association of a neuron and its target cell is called a **chemical synapse,** and this type of intercellular communication is called **synaptic signaling.** We will examine synaptic signaling more fully in chapter 33.

Signal Transduction Pathways Lead to Cellular Responses

LEARNING OBJECTIVE 9.1.2 Define signal transduction pathway.

These descriptive categories of signaling are useful, but tell us nothing about how cells respond to the signals. How cells respond to external signals is the field we call **signal transduction.** This process involves evolutionarily conserved signal transduction pathways that produce a particular cellular responses to a signal. These pathways can be activated by different signals and/or different receptors. This means that the same signal can produce a different response in different cells, and different signals can lead to the same response. This provides great flexibility in how cells in a multicellular organism respond to signals.

For example, multiple cell types respond to the hormone glucagon by mobilizing glucose. This response is part of how your body controls blood glucose (refer to chapter 35). In contrast, the hormone epinephrine has diverse effects on different cell types. We have all been startled or frightened by a sudden event. Your heart beats faster, you feel more alert, and you can even feel the hairs on your skin stand up. All of this is due in part to your body releasing the hormone epinephrine (also called adrenaline) into the bloodstream.

These distinct effects of epinephrine on different cell types depend on each cell having the same receptor for the hormone, but responding with different signal transduction pathways. In the liver, cells are stimulated to mobilize glucose; in heart muscle, cells contract more forcefully to increase blood flow; blood vessels respond by expanding in some areas and contracting in others to redirect blood flow to the liver, heart, and skeletal muscles. These different reactions depend on the fact that each cell type has an epinephrine receptor but different sets of response proteins.

Phosphorylation Is Key in Control of Protein Function

LEARNING OBJECTIVE 9.1.3 Differentiate between the activity of kinases and phosphatases.

Signal transduction pathways act to change the behavior or nature of a cell. This action may involve changing the composition of proteins in a cell, or altering the activity of cellular proteins. The activity of many proteins is affected by posttranslational modification, usually by the enzymatic addition of small molecules such as acetyl, methyl, or phosphate groups. These modifications will activate some proteins and deactivate others. Two of the most important forms of posttranslational modification are **phosphorylation** or **dephosphorylation,** the addition or removal, respectively, of phosphate groups.

The ATP produced by energy metabolism (refer to chapters 7 and 8) has another role: to donate phosphate groups to proteins. The function of many proteins is controlled by their phosphorylation state, and a major way that signal transduction pathways influence cellular activity is by activating or deactivating enzymes that affect the phosphorylation state of proteins.

Protein kinases

The class of enzyme that adds phosphate groups from ATP to proteins is called a *protein kinase.* These phosphate groups can be added to the three amino acids that have an OH as part of their R group, namely, serine, threonine, and tyrosine. We categorize protein kinases as either serine–threonine or tyrosine kinases based on the amino acids they modify (figure 9.3). Most cytoplasmic protein kinases fall into the serine–threonine kinase class.

Phosphatases

Part of the reason for the versatility of phosphorylation as a form of protein modification is that it is reversible. Another class of enzymes called **phosphatases** removes phosphate groups, reversing the action of kinases (figure 9.3). Thus, a protein activated by a kinase will be deactivated by a phosphatase, and a protein deactivated by a kinase will be activated by a phosphatase.

REVIEW OF CONCEPT 9.1

Cells communicate by chemical signals, or ligands, that bind to cellular receptors. Binding of ligand to receptor initiates signal transduction pathways that lead to a cellular response. Signaling varies with distance from source to reception of the signal, from direct contact to distant endocrine signals. Different cells may have the same response to one signal, and the same signal can also elicit different responses in different cells. One response is the phosphorylation of proteins.

- *How are receptor ligand interactions similar to enzyme substrate interactions?*

9.2 Signal Transduction Begins with Cellular Receptors

Understanding cell signaling requires first understanding the receptors themselves. For a cell to be able to respond to a specific signaling molecule, it must have a receptor that can bind to the signaling molecule. This specific binding interaction of receptor and ligand is an example of molecular recognition, where one molecule interacts with another based on their complementary shapes. This interaction causes subtle changes in the structure of the receptor, thereby activating it. This is the beginning of any signal transduction pathway.

Receptors can be differentiated based on their location and the chemical nature of their ligands. Hydrophobic ligands easily cross the membrane and thus bind to intracellular receptors. Hydrophilic ligands cannot easily cross the membrane and thus

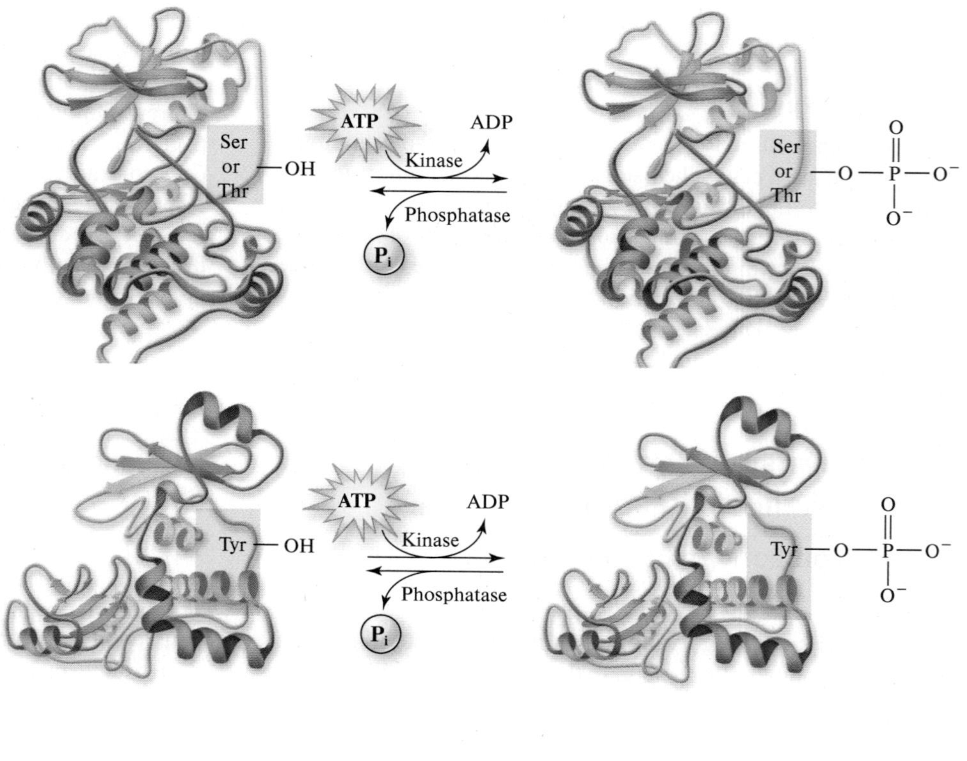

Figure 9.3 Phosphorylation of proteins. Many proteins are controlled by their phosphorylation state—that is, they are activated by phosphorylation and deactivated by dephosphorylation, or the reverse. The enzymes that add phosphate groups are called kinases. These form two classes, either serine–threonine kinases or tyrosine kinases, depending on the amino acid the phosphate is added to. The action of kinases is reversed by protein phosphatase enzymes.

bind to membrane receptors outside the cell (figure 9.1). Membrane receptors consist of transmembrane proteins that are in contact with both the cytoplasm and the extracellular environment.

Receptor Proteins Are Defined by Location and Mode of Response

LEARNING OBJECTIVE 9.2.1 Differentiate between different receptor types based on their cellular location.

When a receptor is a transmembrane protein, the ligand binds to the receptor outside of the cell and never actually crosses the plasma membrane. In this case, the receptor itself, not the signaling molecule, is responsible for information crossing the membrane. Membrane receptors can be categorized based on their structure and function.

Channel-linked receptors

Chemically gated ion channels are receptor proteins that allow the passage of ions (figure 9.4*a*). The receptors for many neurotransmitters are chemically gated ion channels. The channel is said to be chemically gated because it opens only when a chemical (the neurotransmitter) binds to it. The type of ion that flows across the membrane when a chemically gated ion channel opens depends on the shape and charge structure of the channel. Sodium, potassium, calcium, and chloride ions all have specific ion channels. The acetylcholine receptor found in muscle cell membranes functions as a Na^+ channel. When the receptor binds to acetylcholine, the channel opens, allowing Na^+ to flow into the muscle cell. This is a critical step linking the signal from a motor neuron to muscle cell contraction (refer to chapter 33).

Enzymatic receptors

Many cell-surface receptors either act as enzymes or are directly linked to enzymes (figure 9.4*b*). When a signal molecule binds to the receptor, it activates the enzyme. In almost all cases, these enzymes are **protein kinases,** enzymes that add phosphate groups to proteins. We discuss these receptors in detail in section 9.4.

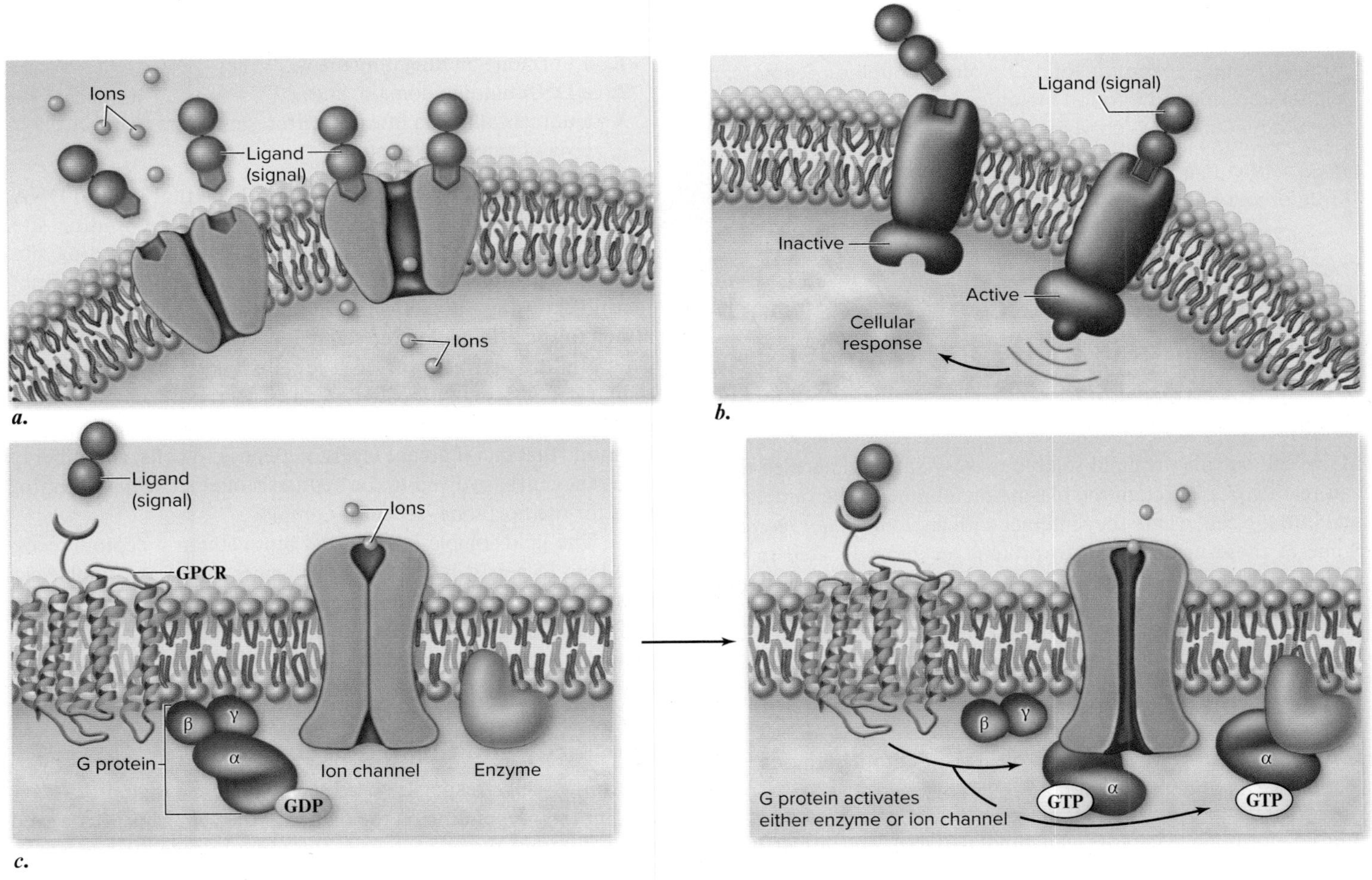

Figure 9.4 Cell-surface receptors. ***a.*** Chemically gated ion channels form a pore in the plasma membrane that can be opened or closed by chemical signals. They are usually selective, allowing the passage of only one type of ion. ***b.*** Enzymatic receptors bind to ligands on the extracellular surface. A catalytic region on their cytoplasmic portion transmits the signal across the membrane by acting as an enzyme in the cytoplasm. ***c.*** G protein–coupled receptors (GPCRs) bind to ligands outside the cell and to G proteins inside the cell. The G protein then activates an enzyme or ion channel, transmitting signals from the cell's surface to its interior.

G Protein–Coupled receptors

A third class of cell-surface receptors acts indirectly on enzymes or ion channels in the plasma membrane with the aid of an assisting protein, called a **G protein.** The G protein, which is so named because it binds the nucleotide *guanosine triphosphate (GTP),* can be thought of as being inserted between the receptors and the enzyme (effector). That is, the ligand binds to the receptor, activating it, which activates the G protein, which in turn activates the effector protein (figure 9.4*c*). These receptors are also discussed in detail in section 9.5.

Second messengers

Some enzymatic receptors and most G protein–coupled receptors utilize other substances to relay the message within the cytoplasm. These other substances, small molecules or ions called **second messengers,** alter the behavior of cellular proteins by binding to them and changing their shape. (The original external signal molecule is considered the "first messenger.") Two common second messengers are **cyclic adenosine monophosphate (cyclic AMP,** or **cAMP)** and calcium ions. The role of these second messengers will be explored in more detail in section 9.5.

REVIEW OF CONCEPT 9.2

Receptors are either intracellular receptors or membrane receptors. Membrane receptors include channel-linked receptors, enzymatic receptors, and G protein–coupled receptors. Signal transduction through membrane receptors can lead to the production of a second messenger inside the cell.

- *Would a hydrophobic molecule be expected to have an internal or a membrane receptor?*

9.3 Intracellular Receptors Respond to Signals by Regulating Gene Expression

Many cell signals are lipid-soluble or very small molecules that can readily pass through the plasma membrane of the target cell and into the cell, where they interact with an *intracellular receptor.* Some of these ligands bind to protein receptors located in the cytoplasm; others pass across the nuclear membrane as well and bind to receptors within the nucleus.

Steroid Hormone Receptors Affect Gene Expression

LEARNING OBJECTIVE 9.3.1 Explain how steroid hormone receptors can affect transcription.

Steroid hormone receptors can bind to DNA and affect gene expression. This kind of direct biochemical action does not involve a multistep pathway such as we will see later for membrane receptors. Turning on gene expression can lead to diverse and far-ranging effects, which are generally slower to manifest, but longer acting, than those of membrane receptors, which tend to have rapid effects that are short-lived.

Steroid hormones are a large class of compounds with a common nonpolar structure, which includes cortisol, estrogen, progesterone, and testosterone. Estrogen, progesterone, and testosterone are involved in sexual development and behavior (refer to chapter 36). Cortisol has varied effects depending on the target tissue, including the inhibition of white blood cells to control inflammation. This anti-inflammatory action is the basis of its major use in medicine.

Their nonpolar structure allows steroids to easily cross the membrane where they bind to a receptor in the cytoplasm. Formation of a hormone–receptor complex leads to a shift from the cytoplasm to the nucleus (figure 9.5). As the hormone–receptor complex functions in the nucleus, these receptors are often called *nuclear receptors.*

Steroid receptor action

The primary function of steroid hormone receptors, as well as receptors for a number of other small, lipid-soluble signal molecules such as vitamin D and thyroid hormone, is to act as regulators of gene expression (refer to chapter 16).

All of these receptors have similar structures; the genes that code for them appear to be the evolutionary descendants of a single ancestral gene. Because of their structural similarities, they are all part of the *nuclear receptor superfamily.*

Each of these receptors has three functional domains:

1. a hormone-binding domain,
2. a DNA-binding domain, and
3. a domain that can interact with coactivators to affect the level of gene transcription.

In its inactive state, the receptor typically cannot bind to DNA, because an inhibitor protein occupies the DNA-binding site. When the signal molecule binds to the hormone-binding site, the conformation of the receptor changes, releasing the inhibitor and exposing the DNA-binding site, allowing the receptor to attach to specific nucleotide sequences on the DNA (figure 9.5). This binding activates (or, in a few instances, suppresses) particular genes, usually located adjacent to the hormone-binding sequences. In the case of cortisol, which is a glucocorticoid hormone that can increase levels of glucose in cells, a number of different genes involved in the synthesis of glucose have binding sites for the hormone–receptor complex.

The lipid-soluble ligands that intracellular receptors recognize tend to persist in the blood far longer than water-soluble signals. Most water-soluble hormones break down within minutes, and neurotransmitters break down within seconds or even milliseconds. In contrast, a steroid hormone such as cortisol or estrogen persists for hours.

Specificity and the role of coactivators

The target cell's response to a lipid-soluble cell signal can vary enormously, depending on the nature of the cell. This is true even when different target cells have the same intracellular receptor. Given that the receptor proteins bind to specific DNA sequences, which are the same in all cells, this may seem puzzling. It is explained in part by the fact that the receptors act in concert with **coactivators,** and the number and nature of these molecules can differ from cell to cell. Thus, a cell's response depends not only on the receptors but also on the coactivators present.

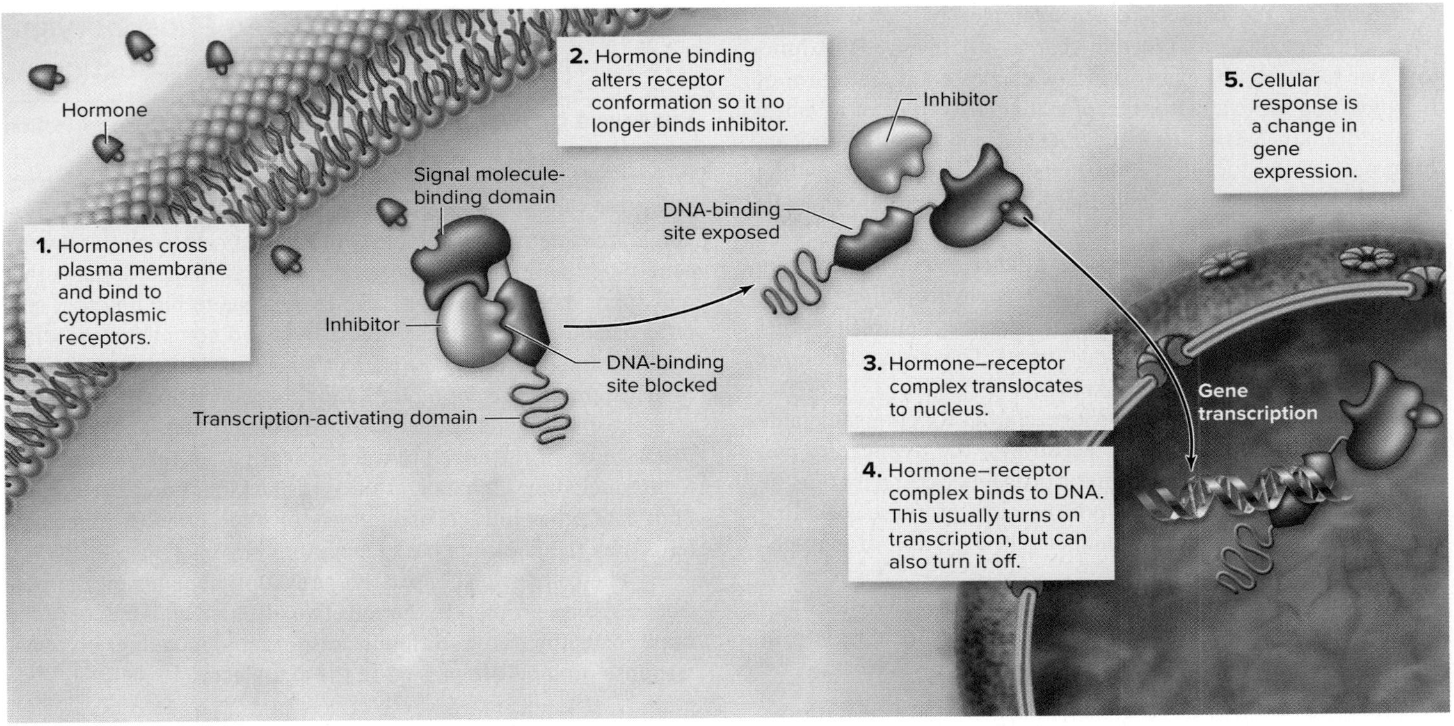

Figure 9.5 Intracellular receptors regulate gene transcription. Hydrophobic signaling molecules can cross the plasma membrane and bind to intracellular receptors. This starts a signal transduction pathway that produces changes in gene expression.

Other Intracellular Receptors Act as Enzymes

LEARNING OBJECTIVE 9.3.2 Differentiate between the action of NO and steroid hormones.

An example of a receptor acting as an enzyme is the nitric oxide (NO) receptor. This small gas molecule diffuses readily out of the cells where it is produced and into neighboring cells, where it binds to and activates guanylyl cyclase. This leads to the production of cyclic guanosine monophosphate (cGMP), an important signaling molecule with multiple effects. Among these effects is the relaxation of smooth muscle cells.

Signals from a nerve cell to epithelial cells in blood vessels lead to production of NO, which diffuses into adjacent smooth muscle. In the muscle cell, NO causes production of cGMP, leading to relaxation, expanding the vessel and increasing blood flow. This explains the use of nitroglycerin to treat the pain of angina caused by constricted blood vessels to the heart. Nitroglycerin is converted by cells to NO, which then acts to relax the blood vessels.

The drug sildenafil (better known as Viagra) also functions via this signal transduction pathway by binding to and inhibiting the enzyme cGMP phosphodiesterase, which breaks down cGMP. This keeps NO levels high by inhibiting degradation. The reason for Viagra's selective effect is that it binds to a form of cGMP phosphodiesterase found in cells in the penis. This allows relaxation of smooth muscle in erectile tissue, thereby increasing blood flow.

REVIEW OF CONCEPT 9.3

Hydrophobic signaling molecules can cross the membrane and bind to intracellular receptors. Steroid hormone receptors act by directly influencing gene expression. The receptor binds to the hormone, and the hormone–receptor complex moves into the nucleus, where it can affect gene expression. This may also involve a coactivator, so a cell's response to a hormone depends on the presence of both receptors and coactivators.

- *Would these types of intracellular receptors be fast-acting, or would they have effects of longer duration?*

9.4 Protein Kinase Receptors Respond to Signals by Phosphorylating Proteins

In section 9.1, you learned that protein kinases phosphorylate proteins to alter protein function and that the most common kinases act on the amino acids serine, threonine, and tyrosine. The **receptor tyrosine kinases (RTKs)** influence the cell cycle, cell migration, cell metabolism, and cell proliferation—virtually all aspects of the cell are affected by signaling through these receptors. Alterations to the function of these receptors and their signaling pathways can lead to cancers in humans and other animals.

Some of the earliest-known examples of cancer-causing genes, or oncogenes (discussed in chapter 10), involve RTK function. The avian erythroblastosis virus carries an altered form of the epidermal growth factor receptor that lacks most of its extracellular domain. When this virus infects a cell, the altered receptors produced are stuck in the "on" state. The continuous signaling from this receptor leads to cells that have lost the usual controls over growth.

Receptor tyrosine kinases are a large class of membrane receptors in animal cells that recognize hydrophilic ligands. Plants possess receptors with a similar overall structure and function, but they are serine–threonine kinases. These plant receptors have been named **plant receptor kinases.**

Because these receptors perform similar functions in plants and animals, but differ in their substrates, the duplication and divergence of both types of receptor probably occurred after the plant–animal divergence. The proliferation of these signaling molecules coincides with the independent evolution of multicellularity in each group.

In this section, we will concentrate on the RTK family of receptors, which has been extensively studied in a variety of animal cells.

Ligand Binding Causes RTKs to Phosphorylate Themselves, Activating Signal Transduction

LEARNING OBJECTIVE 9.4.1 Explain how autophosphorylation transmits a signal across the membrane.

Receptor tyrosine kinases are anchored in the membrane by a single transmembrane domain. They have an extracellular ligand-binding domain, as well as an intracellular kinase domain that adds phosphate groups to tyrosines. On ligand binding to a specific receptor, two of these receptor–ligand complexes associate together and phosphorylate each other, a process called autophosphorylation (figure 9.6). The autophosphorylation event transmits the extracellular signal across the membrane. The next step, propagation of the signal in the cytoplasm, can take a variety of forms, including activation of the tyrosine kinase domain to phosphorylate other intracellular targets or interaction of other proteins with the phosphorylated receptor.

The cellular response after activation depends on the possible response proteins in the cell. Two different cells can have the same receptor yet a different response, depending on what response proteins are present in the cytoplasm.

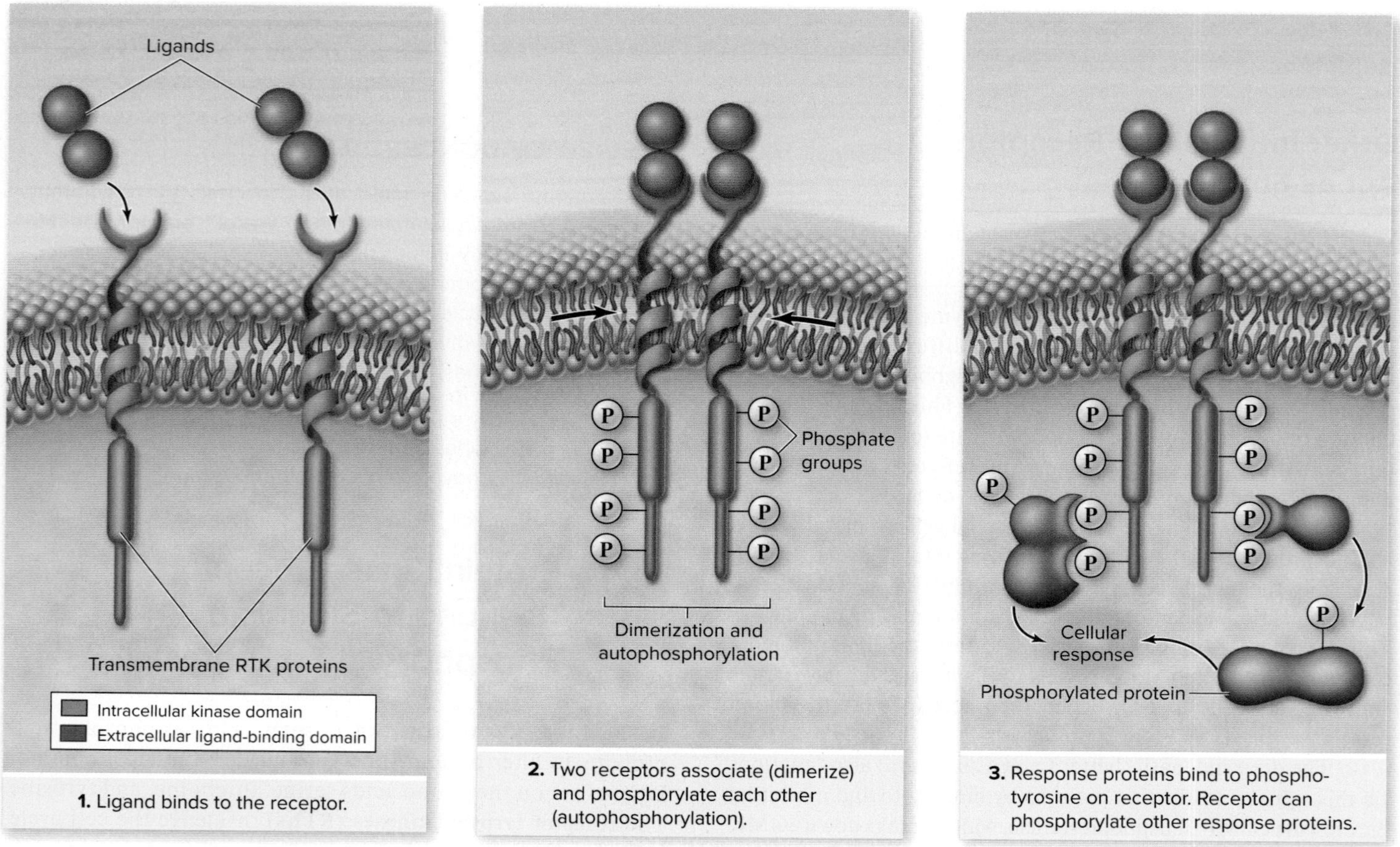

Figure 9.6 Activation of a receptor tyrosine kinase (RTK). These membrane receptors bind hormones or growth factors that are hydrophilic and cannot cross the membrane. The receptor is a transmembrane protein with an extracellular ligand-binding domain and an intracellular kinase domain. Signal transduction pathways begin with response proteins binding to phosphotyrosine on receptors and by receptor phosphorylation of response proteins.

Cellular Proteins Bind to Phosphotyrosine Domains to Transmit the Signal

LEARNING OBJECTIVE 9.4.2 Describe how protein–protein interactions transmit signals.

One way that the signal from the receptor can be propagated in the cytoplasm is via proteins that bind specifically to phosphorylated tyrosines in the receptor. When the receptor is activated, regions of the protein outside of the catalytic site are phosphorylated. This creates "docking" sites for proteins that bind specifically to phosphotyrosine.

The insulin receptor

The insulin receptor illustrates the role of docking proteins. Insulin is important in maintaining a constant level of blood glucose. Insulin acts to lower blood glucose by signaling through an RTK (figure 9.7). The insulin response protein binds to the phosphorylated receptor and passes the signal on to another response protein that activates glycogen synthase, which converts glucose to glycogen. Other response proteins inhibit the production of enzymes involved in making glucose and increase the number of glucose transporter proteins in the plasma membrane. Together these responses act to lower blood glucose.

Adapter proteins

Another class of proteins, **adapter proteins,** can also bind to phosphotyrosines. These proteins do not themselves participate in signal transduction but act as a link between the receptor and proteins that initiate downstream signaling events. For example, the Ras protein, discussed later in this section, is activated by adapter proteins binding to a receptor.

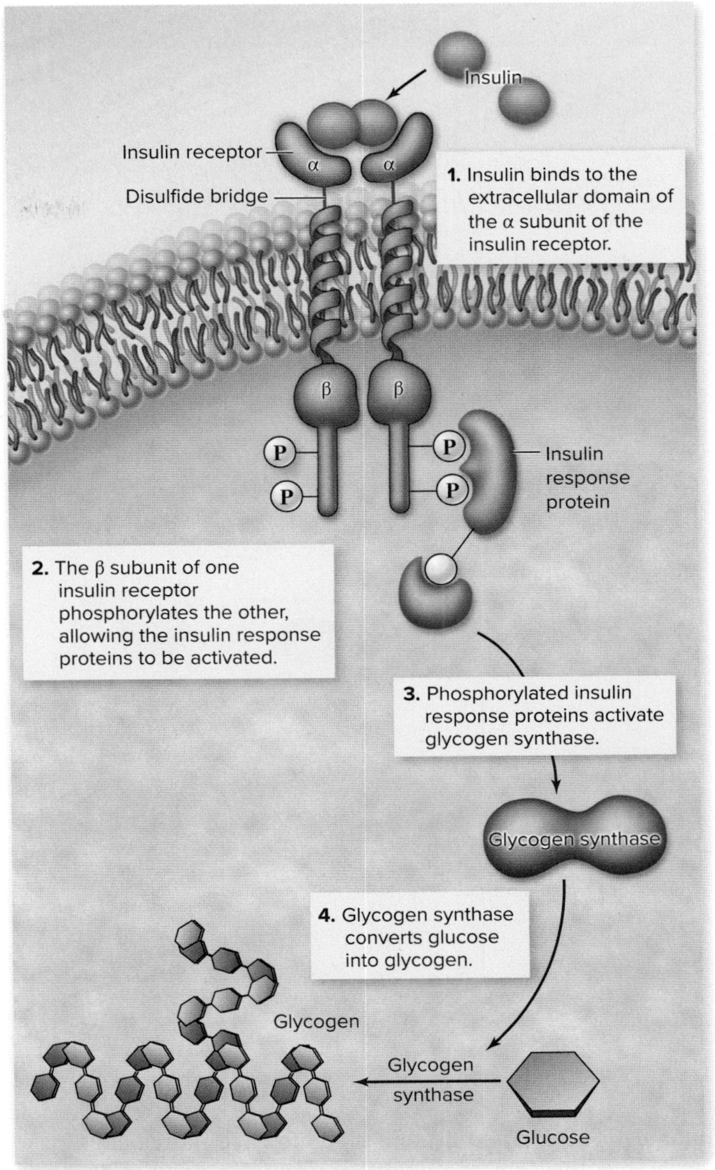

Figure 9.7 The insulin receptor. The insulin receptor is a receptor tyrosine kinase that initiates a variety of cellular responses related to glucose metabolism. One signal transduction pathway mediated by this receptor leads to the activation of the enzyme glycogen synthase. This enzyme converts glucose to glycogen.

Protein Kinase Cascades Can Amplify a Signal

LEARNING OBJECTIVE 9.4.3 Explain the role of protein kinase cascades in signal transduction.

One important class of cytoplasmic kinases are **mitogen-activated protein (MAP) kinases.** A *mitogen* is a chemical that stimulates cell division by activating the normal pathways that control division. The MAP kinases are activated by a signaling module called a *phosphorylation cascade* or a **kinase cascade.** This module is a series of protein kinases that phosphorylate each other in succession. The final step in the cascade is the activation by phosphorylation of MAP kinase itself (figure 9.8).

One function of a kinase cascade is to amplify the original signal. Because each step in the cascade is an enzyme, it can act on a number of substrate molecules. With each enzyme in the cascade acting on many substrates, this produces a large amount of the final product (figure 9.8). This allows a small number of initial signaling molecules to produce a large response.

The ultimate response from the MAP kinase cascade is usually to phosphorylate transcription factors that then activate gene expression (refer to chapter 16). In chapter 10, we will see how growth factors signal through a kinase cascade to control the process of cell division.

Scaffold proteins organize kinase cascades

The proteins in a kinase cascade need to act sequentially to be effective. One way the efficiency of this process can be increased is to organize them in the cytoplasm. Proteins called *scaffold proteins* can organize the components of a kinase cascade into a single protein complex, the ultimate in a signaling

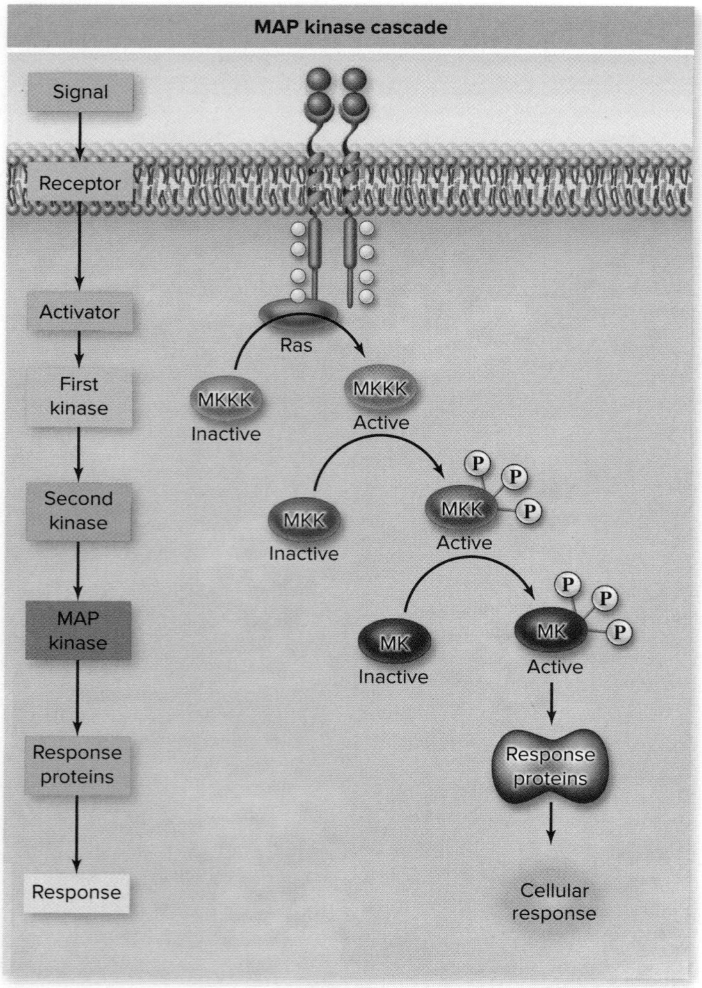
MAP kinase cascade
Signal
Receptor
Activator
First kinase
Second kinase
MAP kinase
Response proteins
Response
Ras
MKKK
Inactive
MKKK
Active
MKK
Inactive
MKK
Active
MK
Inactive
MK
Active
Response proteins
Cellular response

a.

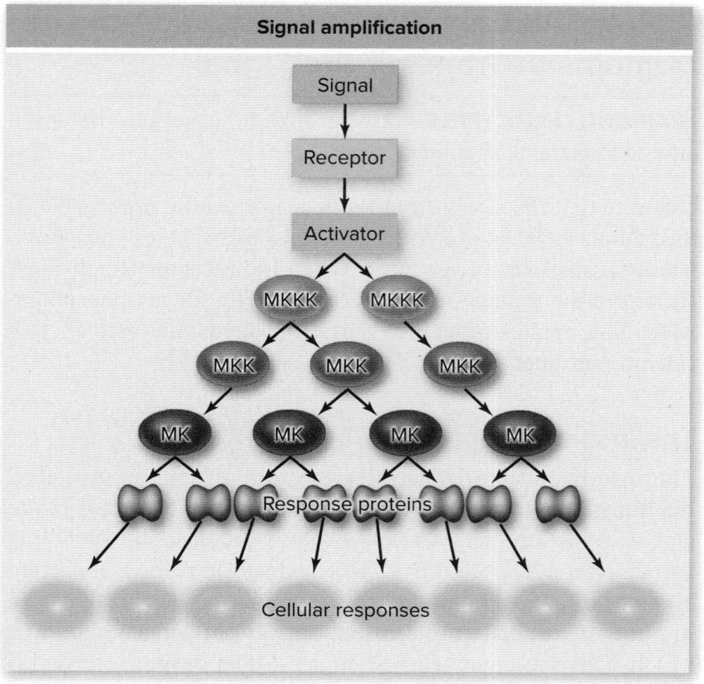
Signal amplification
Signal
Receptor
Activator
MKKK
MKKK
MKK
MKK
MKK
MK
MK
MK
MK
Response proteins
Cellular responses

b.

Figure 9.8 MAP kinase cascade leads to signal amplification. ***a.*** Phosphorylation cascade is shown as a flowchart on the left. The corresponding cellular events are shown on the right, beginning with the receptor in the plasma membrane. Each kinase is named starting with the last, the MAP kinase (MK), which is phosphorylated by a MAP kinase kinase (MKK), which is in turn phosphorylated by a MAP kinase kinase kinase (MKKK). The cascade is linked to the receptor protein by an activator protein. ***b.*** At each step, the enzymatic action of the kinase on multiple substrates leads to amplification of the signal.

module. The scaffold protein binds to each individual kinase such that they are spatially organized for optimal function (figure 9.9).

This kind of organization has advantages and disadvantages. Using a scaffold is more efficient than depending on diffusion to produce the appropriate order of events. This organization also allows the segregation of signaling modules in different cytoplasmic locations. The disadvantage of this kind of organization is that it reduces the amplification effect of the kinase cascade.

An example comes from mating behavior in budding yeast. Yeast cells respond to mating pheromones with changes in cell morphology and gene expression, mediated by a protein kinase cascade. The Ste5 protein is required for mating behavior, but it has no enzymatic activity. Instead, this protein interacts with all of the members of the kinase cascade and acts as a scaffold protein, organizing the cascade.

Figure 9.9 Kinase cascade can be organized by scaffold proteins. The scaffold protein binds to each kinase in the cascade, organizing them so that each substrate is next to its enzyme. This organization also sequesters the kinases from other signaling pathways in the cytoplasm.

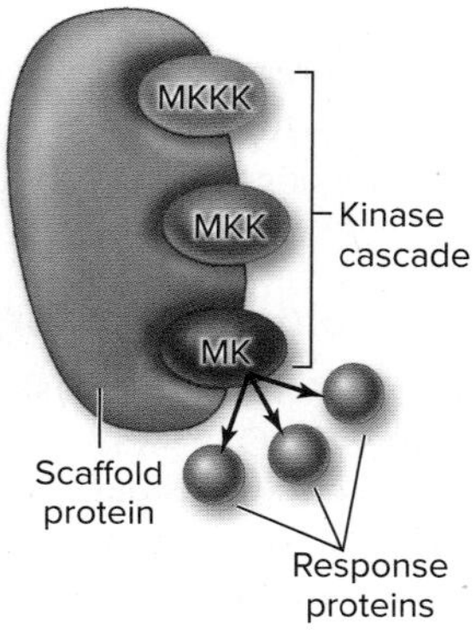
MKKK
MKK
MK
Kinase cascade
Scaffold protein
Response proteins

RTKs are inactivated by internalization

It is important that signaling pathways are not continuously active. Prolonged activation could render a cell unable to respond to other signals or cause it to respond inappropriately to a signal no longer relevant. RTKs are inactivated by two basic mechanisms–dephosphorylation and internalization. Internalization is by endocytosis with the receptor taken up in a vesicle where it can be degraded or recycled.

The enzymes in the kinase cascade are all controlled by dephosphorylation by phosphatase enzymes. This leads to termination of the response at the level of both the receptor and the response proteins.

Ras Is a Small G Protein That Acts as a Molecular Switch

LEARNING OBJECTIVE 9.4.4 Describe the role of Ras protein in signal transduction.

The link between RTKs and the MAP kinase cascade is a small GTP-binding protein (G protein) called Ras. Like all G proteins, Ras is active bound to GTP and inactive bound to GDP. The Ras protein is mutated in many human tumors, indicative of its central role in linking growth factor receptors to their cellular response. Ras was the first protein identified in a large superfamily of small G proteins with over 150 members in the human genome.

The roles of these small G proteins vary, affecting cell proliferation, the cytoskeleton, membrane transport, and nuclear transport. They are an example of how gene duplication and diversification allow evolution to create modular units with diverse functions. The common feature of all members of the family is to act as a molecular switch linking external signals to signal transduction pathways (figure 9.10).

The Ras switch is flipped by exchanging GDP for GTP, and by Ras hydrolyzing GTP to GDP. The switch is sensitive to signals that regulate the activation and deactivation of Ras. Guanine nucleotide exchange factors (GEFs) activate Ras by stimulating the exchange of GDP for GTP. Activated growth factor receptors bind to an adapter protein that acts as a GEF. The activated Ras protein then activates the first kinase in the MAP kinase cascade (refer to figure 9.8 and chapter 10).

The action of Ras is terminated by its intrinsic GTPase activity. This can be stimulated by a GTPase-activating protein (GAP), which provides the opportunity to fine-tune signaling based on the duration of Ras activity. The importance of these proteins is shown by mutations in GAPs that can lead to a predisposition for specific cancers, such as neurofibromatosis.

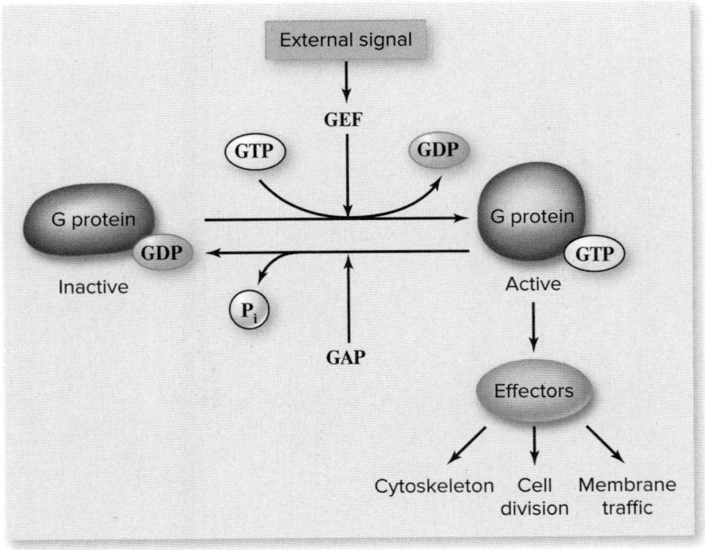

Figure 9.10 Small G proteins act as molecular switches. Small G proteins, such as Ras, link external signals to internal signal transduction pathways. External signals activate guanine nucleotide exchange (GEF) proteins, which activate the G protein. The G protein can be inactivated by its weak intrinsic GTPase activity, which can be stimulated by activating proteins (GAPs).

REVIEW OF CONCEPT 9.4

Receptor tyrosine kinases (RTKs) are membrane receptors that can phosphorylate tyrosine. When activated, they autophosphorylate, creating binding domains for other proteins that transmit the signal inside the cell. One signaling pathway involves the MAP kinase cascade, a series of kinases that each activate the next in the series. This ends by activating transcription factors to alter gene expression. Ras protein is a molecular switch that acts between the receptor and the kinase cascade.

- *Ras protein is mutated in many human cancers. How might these mutations affect Ras function?*

9.5 G Protein–Coupled Receptors Respond to Signals Through Effector Proteins

The single largest category of receptor type in animal cells is **G protein–coupled receptors (GPCRs),** so named because the receptors act by coupling with a G protein. These receptors bind diverse ligands, including histamine (mentioned in the introduction), ions, organic odorants, peptides, proteins, and lipids. Light-sensing receptors are also part of this family, so we could even count photons as "ligands."

This superfamily of proteins also has a characteristic structure, with seven transmembrane domains that anchor the receptors in the membrane (refer to chapter 5). The analysis of many animal genomes indicates that GPCRs are the largest gene family in most animals. They have an ancient origin, with duplication and divergence leading to a wide array of signaling pathways.

There are more than 850 genes encoding GPCRs in the human genome, with about half of these encoding olfactory receptors involved in the sense of taste and smell. The mouse genome contains more than 1000 different olfactory receptors. The GPCR gene family is subdivided into five subfamilies based on structure and function. In this section, we will concentrate on the basic mechanism of activation and some of the possible signal transduction pathways used by GPCRs.

G Proteins Link Receptors with Effector Proteins to Transmit the Signal

LEARNING OBJECTIVE 9.5.1 Describe how heterotrimeric G proteins are activated and inactivated.

The function of the G protein in signaling by GPCRs is to provide a link between a receptor that receives signals and effector proteins that produce cellular responses. The G protein functions as a switch that is turned on by the receptor. In its "on" state, the G protein activates effector proteins to cause a cellular response.

The main difference between the G proteins in GPCRs and the small G proteins described in section 9.4 is that these G proteins are composed of three subunits, called α, β, and γ. As a result, they are often called *heterotrimeric G proteins.* When a ligand binds to a GPCR and activates its associated G protein, the G protein exchanges GDP for GTP and dissociates into two parts, consisting of the G_α subunit bound to GTP and the G_β and G_γ subunits together ($G_{\beta\gamma}$). The signal can then be propagated by either the G_α or the $G_{\beta\gamma}$ components. The hydrolysis of bound GTP to GDP by G_α causes reassociation of the heterotrimer and restores the "off" state of the system (figure 9.11).

The effector proteins are usually enzymes. An effector protein might be a protein kinase that phosphorylates proteins to directly propagate the signal, or it may produce a second messenger to initiate a signal transduction pathway.

Effector Proteins Produce Multiple Second Messengers to Broadcast the Signal

LEARNING OBJECTIVE 9.5.2 Compare and contrast the action of different second messengers.

Often, the effector proteins activated by G proteins produce a second messenger. Two of the most common effectors are *adenylyl cyclase* and *phospholipase C,* which produce cAMP and IP_3 plus DAG, respectively (figure 9.12).

Cyclic AMP

All animal cells studied thus far use cAMP as a second messenger. When a signaling molecule binds to a GPCR that uses the enzyme **adenylyl cyclase** as an effector, a large amount of cAMP is produced within the cell (figure 9.12*a*). The cAMP then binds to and activates the enzyme protein kinase A (PKA), which adds phosphates to specific proteins in the cell (figure 9.13).

The effect of this phosphorylation on cell function depends on the identity of the cell and the proteins that are phosphorylated. In muscle cells, for example, PKA activates an enzyme necessary to break down glycogen and inhibits another enzyme necessary to synthesize glycogen. This leads to an increase in glucose available to the muscle. By contrast, in the kidney the action of PKA leads to the production of water channels that can increase the permeability of tubule cells to water.

Disruption of cAMP signaling can have a variety of effects. The symptoms of the disease cholera are due to altered cAMP levels in cells in the gut. The bacterium *Vibrio cholerae* produces a toxin that binds to a GPCR in the epithelium of the gut, causing it to be locked into an "on" state. This causes a large increase in intracellular cAMP that, in these cells, causes Cl^- ions to be transported out of the cell. Water follows the Cl^-, leading to the diarrhea and dehydration characteristic of the disease.

The molecule cAMP is also an extracellular signal. In the slime mold *Dictyostelium discoideum,* secreted cAMP acts as a signal for aggregation under conditions of starvation. Experiments have shown that the receptor for this signal is also a GPCR.

Inositol phosphates

A common second messenger is produced from the molecules called inositol phospholipids. These are inserted into the plasma membrane by their lipid ends and have the *inositol phosphate*

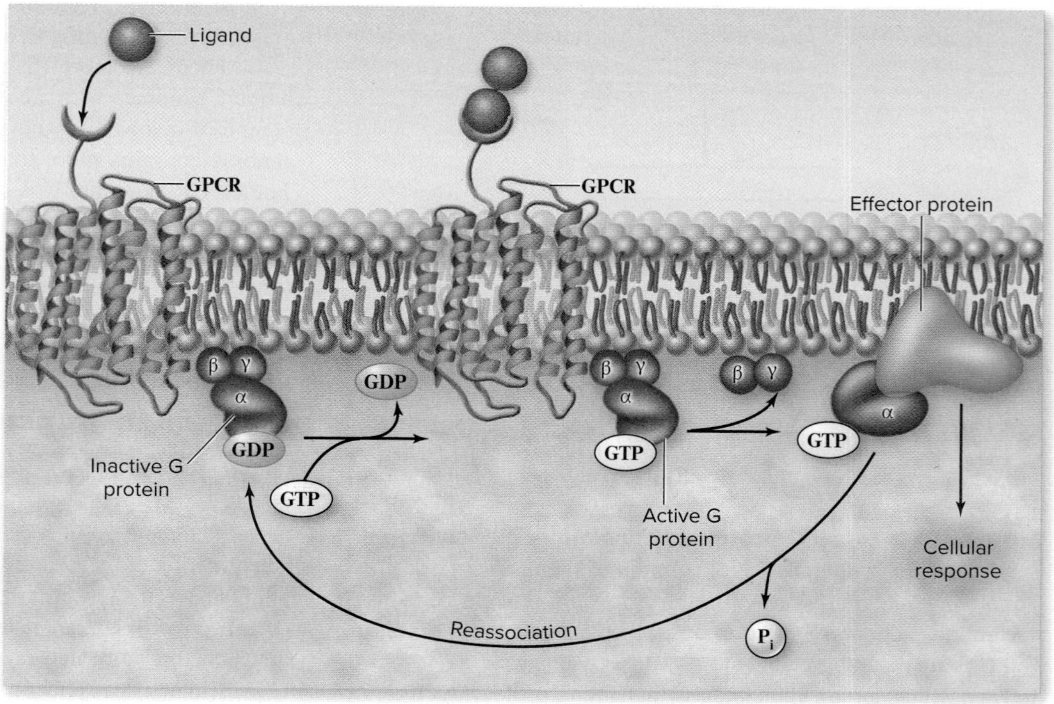

Figure 9.11 The action of G protein–coupled receptors. G protein–coupled receptors act through a heterotrimeric G protein that links the receptor to an effector protein. When ligand binds to the receptor, it activates an associated G protein, exchanging GDP for GTP. The active G protein complex dissociates into G_α and $G_{\beta\gamma}$. The G_α subunit (bound to GTP) is shown activating an effector protein. The effector protein may act directly on cellular proteins or produce a second messenger to cause a cellular response. G_α can hydrolyze GTP, inactivating the system, then reassociate with $G_{\beta\gamma}$.

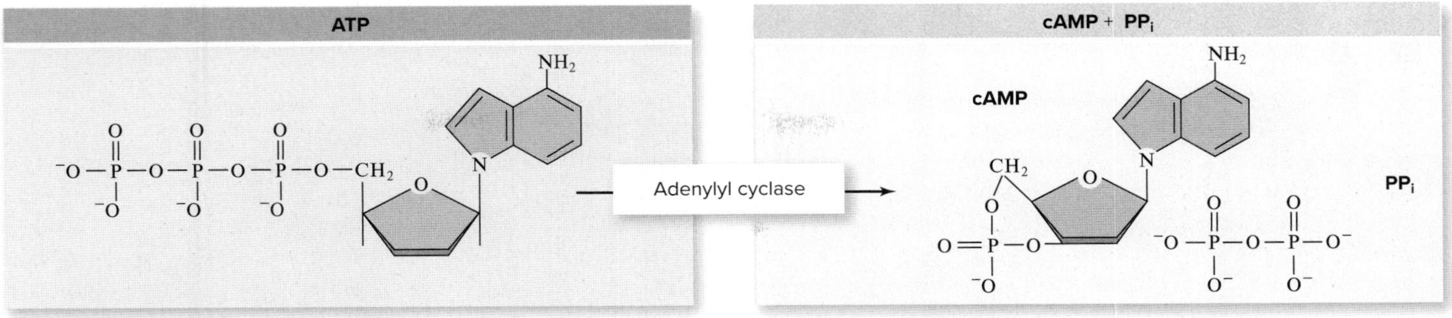

a.

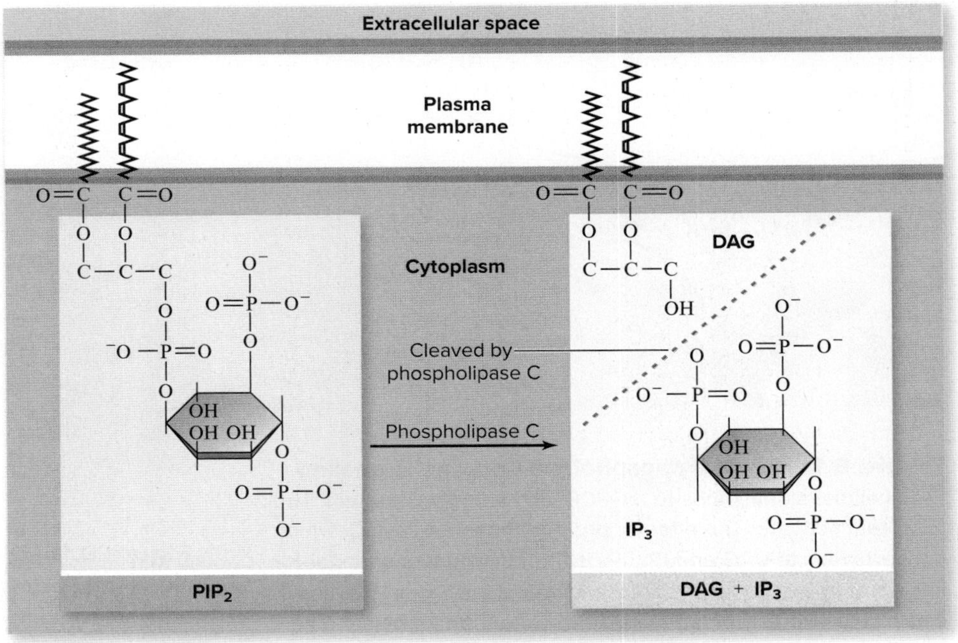

b.

Figure 9.12 Production of second messengers. Second messengers are signaling molecules produced within the cell. ***a.*** The nucleotide ATP is converted by the enzyme adenylyl cyclase into cyclic AMP, or cAMP, and pyrophosphate (PP_i). ***b.*** The inositol phospholipid PIP_2 is composed of two lipids and a phosphate attached to glycerol. The phosphate is also attached to the sugar inositol. This molecule can be cleaved by the enzyme phospholipase C to produce two different second messengers: DAG, made up of the glycerol with the two lipids, and IP_3, inositol trisphosphate.

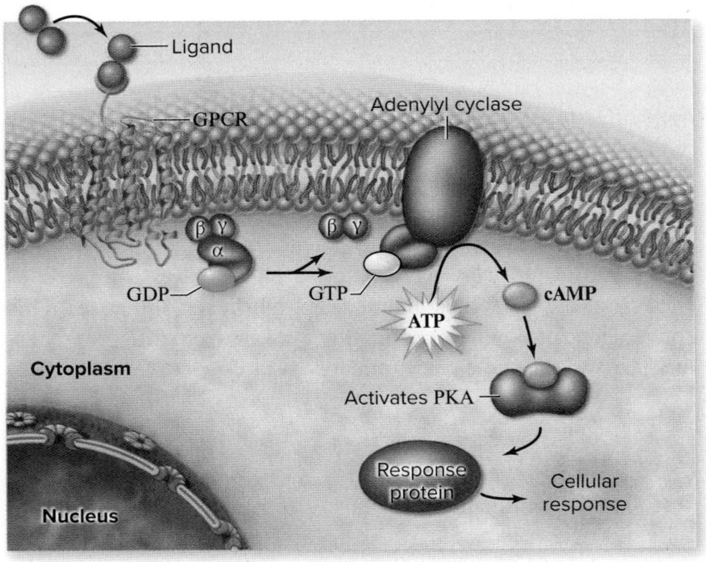

Figure 9.13 cAMP signaling pathway. Extracellular signal binds to a GPCR, activating a G protein. The G protein then activates the effector protein adenylyl cyclase, which catalyzes the conversion of ATP to cAMP. The cAMP then activates protein kinase A (PKA), which phosphorylates target proteins to cause a cellular response.

portion protruding into the cytoplasm. The most common inositol phospholipid is phosphatidylinositol-4,5-bisphosphate (PIP_2). This molecule is a substrate of the effector protein phospholipase C, which cleaves PIP_2 to yield **diacylglycerol (DAG)** and **inositol-1,4,5-trisphosphate (IP_3)** (figure 9.12*b*).

Both of these compounds then act as second messengers with a variety of cellular effects. DAG, like cAMP, can activate a protein kinase, in this case protein kinase C (PKC).

Calcium

Calcium ions (Ca^{2+}) serve widely as second messengers. Ca^{2+} levels inside the cytoplasm are normally very low (less than 10^{-7} M), whereas outside the cell and in the endoplasmic reticulum, Ca^{2+} levels are quite high (about 10^{-3} M). The endoplasmic reticulum has receptor proteins that act as ion channels to release Ca^{2+}. One of the most common of these receptors can bind the second messenger IP_3 to release Ca^{2+}, linking signaling through inositol phosphates with signaling by Ca^{2+} (figure 9.14).

The result of the outflow of Ca^{2+} from the endoplasmic reticulum depends on the cell type. For example, in skeletal muscle cells, Ca^{2+} stimulates muscle contraction, but in endocrine cells, it stimulates the secretion of hormones.

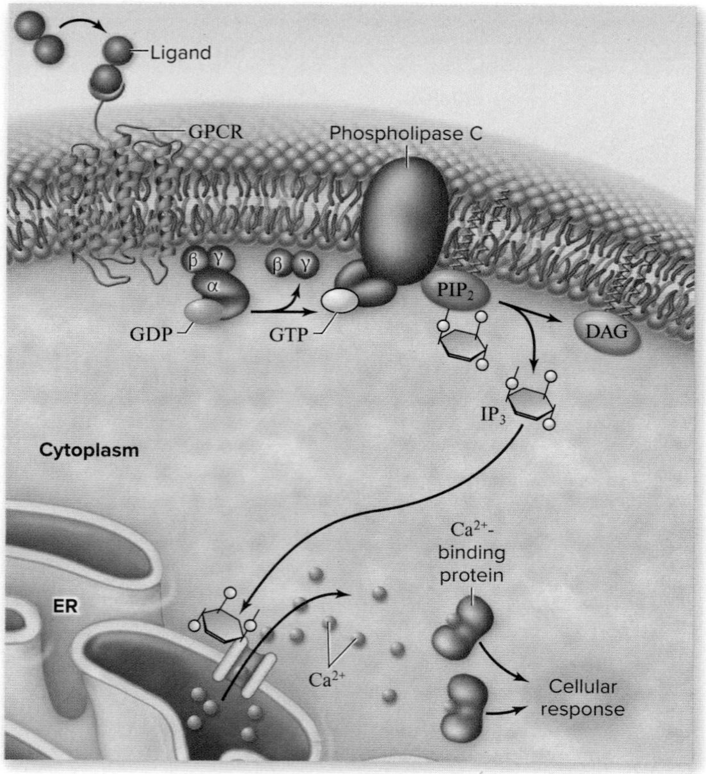

Figure 9.14 Inositol phospholipid and Ca^{2+} signaling. Extracellular signal binds to a GPCR, activating a G protein. The G protein activates the effector protein phospholipase C, which converts PIP_2 to DAG and IP_3. IP_3 is then bound to a channel-linked receptor on the endoplasmic reticulum (ER) membrane, causing the ER to release stored Ca^{2+} into the cytoplasm. The Ca^{2+} then binds to Ca^{2+}-binding proteins such as calmodulin and PKC to cause a cellular response.

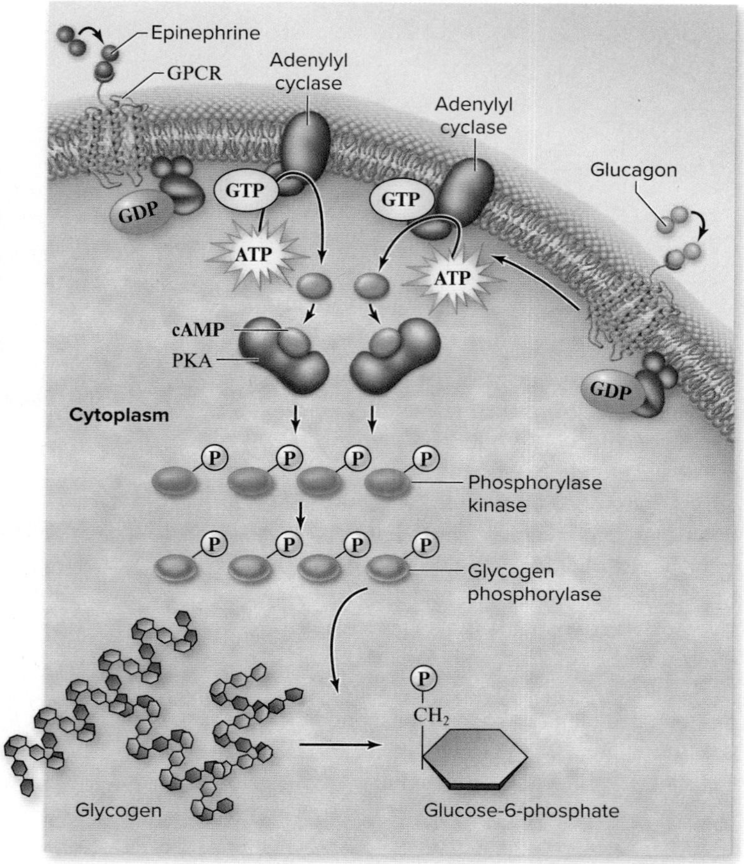

Figure 9.15 Different receptors can activate the same signaling pathway. The hormones glucagon and epinephrine both act through GPCRs. Each of these receptors acts via a G protein that activates adenylyl cyclase, producing cAMP. The activation of PKA begins a kinase cascade that leads to the breakdown of glycogen.

Different Signals Can Have the Same Effect, and the Same Signal Can Have Different Effects

LEARNING OBJECTIVE 9.5.3 Explain how signals can converge and diverge in cells.

The hormones glucagon and epinephrine can both stimulate liver cells to mobilize glucose. The reason these different signals have the same effect is that they both act by the same signal transduction pathway to stimulate the breakdown and inhibit the synthesis of glycogen.

The binding of either hormone to its receptor activates a G protein that stimulates adenylyl cyclase. The production of cAMP leads to the activation of PKA, which in turn activates another protein kinase called phosphorylase kinase. Activated phosphorylase kinase then activates glycogen phosphorylase, which cleaves off units of glucose-6-phosphate from the glycogen polymer (figure 9.15). The action of multiple kinases again leads to amplification, such that a few signaling molecules result in a large number of glucose molecules being released.

At the same time, PKA also phosphorylates the enzyme glycogen synthase, but in this case it inhibits the enzyme, thus preventing the synthesis of glycogen. In addition, PKA phosphorylates other proteins that activate the expression of genes encoding the enzymes needed to synthesize glucose. This convergence of signal transduction pathways from different receptors leads to the same result—glucose is mobilized.

Receptor subtypes

One signaling molecule can have different effects in different cells. This comes about if the receptors have multiple forms that produce different cellular responses. For example, the epinephrine receptor has nine different subtypes, or isoforms, each encoded by a different gene. They differ mainly in their cytoplasmic domains, which interact with G proteins. These different isoforms activate different G proteins, which activate different signal transduction pathways.

In heart muscle cells, one isoform of the epinephrine receptor activates adenylyl cyclase, leading to increased cAMP, which increases the rate and force of contraction. Smooth muscle cells in the intestine have a different isoform of the epinephrine receptor that inhibits adenylyl cyclase, which decreases cAMP. This produces the opposite result by relaxing the muscle.

GPCRs and RTKs can activate the same pathway

Different receptor types can affect the same signaling module. For example, RTKs were shown to activate the MAP kinase cascade, but GPCRs can also activate the same cascade. Similarly, the enzyme phospholipase C, which produces the second messengers DAG and IP_3, is activated by GPCR signaling, but it can also be activated by RTKs.

This cross-reactivity may appear to introduce complications into cell function, but, in fact, it provides the cell with an incredible amount of flexibility. Cells have a large, but limited, number of intracellular signaling modules, which can be turned on and off by different kinds of membrane receptors. This leads to signaling networks that interconnect possible cellular effectors with multiple incoming signals.

REVIEW OF CONCEPT 9.5

Signaling through GPCRs involves a receptor, a G protein, and an effector protein. G proteins are active bound to GTP and inactive bound to GDP. Ligand binding to the receptor activates the G protein, which then activates an effector protein. Adenylyl cyclase is an effector protein that produces the second messenger cAMP. The effector phospholipase C cleaves the inositol phosphates and results in the release of Ca^{2+} from the ER.

- *There are far more GPCRs than any other receptor type. What is a possible explanation for this?*

Why Measles Is a Respiratory Disease

Measles is a serious disease, infecting 10 million children each year, 120,000 of whom die of the infection. Measles is caused by the RNA paramyxovirus shown in the micrograph below. Because the virus is passed from one individual to another by sneezing, measles was long thought to be a respiratory disease, the virus replicating in the cells lining the respiratory tract before spreading through the air. Surprisingly, recent work has shown that the measles virus initially infects mobile immune system cells called macrophages in the airways. The infected macrophages cross the respiratory epithelium, transporting the infection to the lymph nodes, where the virus grows vigorously.

How does the legion of new virus particles produced in the lymph nodes of an infected person cross back into that person's airways, to disseminate out and infect other people when the infected person sneezes? New research has identified the highway used by the measles virus across the epithelial barrier.

Researchers added a green-fluorescent protein to a measles virus, then looked to see which cells of the respiratory tract fluoresced when exposed to the virus. The only cells affected were those with a particular cell-surface receptor called nectin-4. This adherens junction protein binds tightly to the measles virus attachment protein.

To test the hypothesis that nectin-4 is the cell-surface receptor used by the measles virus to enter the respiratory tract, the investigating team used small interfering RNA (siRNA) to shut off the *nectin-4* gene (siRNA is discussed in chapter 16).

First, they needed to demonstrate that their siRNA probe was specifically targeted to the *nectin-4* gene. In the graph, the amount of nectin-4 mRNA produced by cells treated with nectin-4 siRNA is compared with the amount produced by cells treated with another non-nectin-4 siRNA.

Then they needed to compare the number of viruses transmitted across the respiratory epithelium by epithelial cells treated with nectin-4 siRNA to the number transmitted by control cells treated with non-nectin-4 siRNA. Their results are seen in the graph.

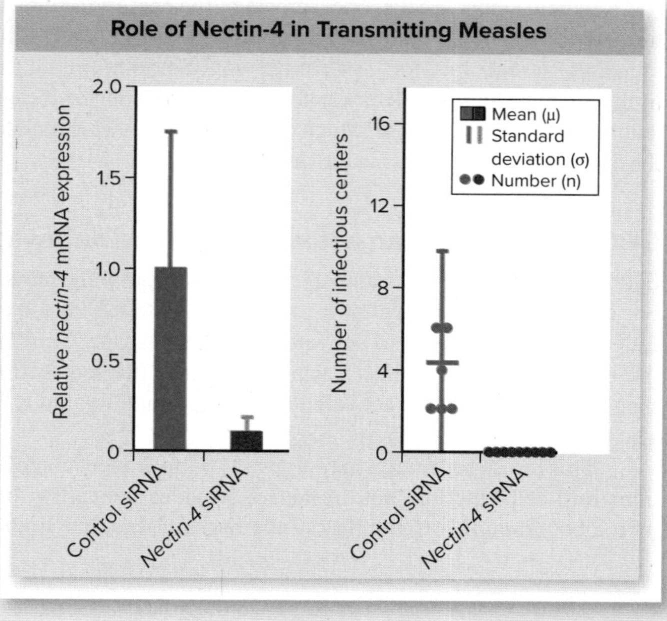

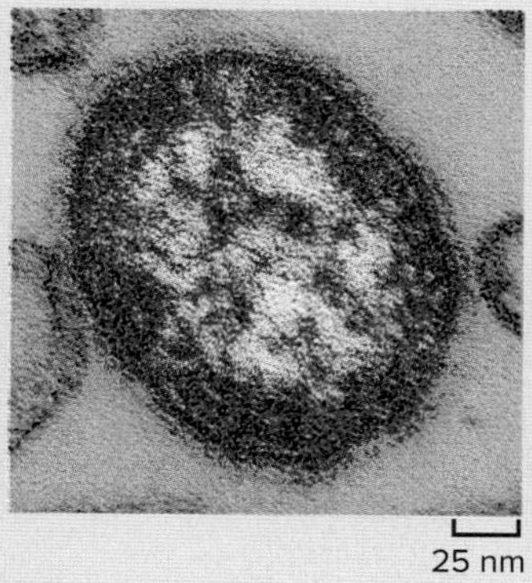

Cynthia Goldsmith/CDC

Analysis

1. **Applying Concepts**
   a. ***Dependent variable.*** In the graphs, what is the dependent variable?
   b. ***Relative expression.*** The *y*-axis variable in the left graph is measured relative to what? (*Hint*: The relative expression of the control is exactly 1.)
2. **Interpreting Data**
   a. Is the relative amount of mRNA transcribed from the *nectin-4* gene affected by nectin-4-specific siRNA? What percentage decrease is observed?
   b. What was the average number of infectious centers observed when control siRNA-treated cells were infected in 10 trials with measles virus? When nectin-4 siRNA-treated cells were infected in 10 trials?
3. **Making Inferences**
   a. ***Left graph.*** Would you infer from these results that treatment of respiratory epithelium cells with nectin-4 siRNA shuts off the *nectin-4* gene?
   b. ***Right graph.*** Would you infer from these results that nectin-4 siRNA inhibits measles virus infection of respiratory cells growing in culture?
4. **Drawing Conclusions**
   a. Does the silencing of the *nectin-4* gene block infection of respiratory epithelium cells by the measles virus?
   b. Is it reasonable to conclude that the nectin-4 cell-surface receptor provided a route for the measles virus across the respiratory epithelium?
5. **Further Analysis**
   The research team carried out other studies to verify that the measles virus physically crosses the epithelial barrier after binding to nectin-4 cell-surface receptors. What sort of experiments might you suggest they do to better establish this point?

Retracing the Learning Path

CONCEPT 9.1 The Cells of Multicellular Organisms Communicate

Cell communication requires signal molecules, called ligands, binding to specific receptor proteins, producing a cellular response.

9.1.1 Signaling Mechanisms Vary with Distance from Source Direct contact: Molecules on the plasma membrane of one cell contact the receptor molecules on an adjacent cell. Paracrine signaling: Short-lived signal molecules are released into the extracellular fluid and influence neighboring cells. Endocrine signaling: Long-lived hormones enter the circulatory system and are carried to target cells some distance away. Synaptic signaling: Short-lived neurotransmitters are released by neurons into the gap, called a synapse, between nerves and target cells.

9.1.2 Signal Transduction Pathways Lead to Cellular Responses Intracellular events initiated by a signaling event are called signal transduction. This converts the signal into a change in the cell.

9.1.3 Phosphorylation Is Key in Control of Protein Function Proteins can be controlled by phosphate added by kinase and removed by phosphatase enzymes. This control can be either positive or negative.

CONCEPT 9.2 Signal Transduction Begins with Cellular Receptors

9.2.1 Receptor Proteins Are Defined by Location and Mode of Response Receptors are broadly defined as intracellular or cell-surface receptors (membrane receptors). Membrane receptors are transmembrane proteins that transfer information across the membrane, but not the signal molecule. Channel-linked receptors are chemically gated ion channels that allow specific ions to pass through a central pore. Enzymatic receptors are enzymes activated by binding a ligand; these enzymes are usually protein kinases. G protein–coupled receptors interact with G proteins that control the function of effector proteins: enzymes or ion channels. Some enzymatic and most G protein–coupled receptors produce second messengers, to relay messages in the cytoplasm.

CONCEPT 9.3 Intracellular Receptors Respond to Signals by Regulating Gene Expression

Many cell signals are lipid-soluble and readily pass through the plasma membrane and bind to receptors in the cytoplasm or nucleus.

9.3.1 Steroid Hormone Receptors Affect Gene Expression Steroid hormones bind cytoplasmic receptors, then are transported to the nucleus. Thus, they are called nuclear receptors. Nuclear receptors have a hormone-binding, a DNA-binding, and a transcription-activating domain. This allows them to directly affect gene expression. A cell's response to a lipid-soluble signal depends on the hormone-receptor complex and other protein coactivators present.

9.3.2 Other Intracellular Receptors Act as Enzymes The receptor for nitric oxide activates the enzyme guanylyl cyclase, which produces cGMP. One action of this system is to relax smooth muscle cells.

CONCEPT 9.4 Protein Kinase Receptors Respond to Signals by Phosphorylating Proteins

Receptor kinases in plants and animals recognize hydrophilic ligands and influence the cell cycle, cell migration, cell metabolism, and cell proliferation. Because they are involved in growth control, alterations of receptor kinases and their signaling pathways can lead to cancer.

9.4.1 Ligand Binding Causes RTKs to Phosphorylate Themselves, Activating Signal Transduction The activated receptor can also phosphorylate other intracellular proteins.

9.4.2 Cellular Proteins Bind to Phosphotyrosine Domains to Transmit the Signal Adapter proteins can bind to phosphotyrosine and act as links between the receptors and downstream signaling events.

9.4.3 Protein Kinase Cascades Can Amplify a Signal Protein kinase enzymes can initiate a series of phosphorylation events that end in a cellular response. Each enzyme in the cascade can phosphorylate multiple copies of the next enzyme in the cascade, amplifying the signal as it progresses. Scaffold proteins and protein kinases form a single complex in which the enzymes act sequentially and are optimally functional. Receptors are inactivated by dephosphorylation or internalization. Enzymes in the cascade are inactivated by dephosphorylation.

9.4.4 Ras Is a Small G Protein That Acts as a Molecular Switch Small G proteins act as molecular switches linking external signals to signal transduction pathways. They are stimulated by guanine nucleotide exchange factors (GEFs) and their inactivation can be enhanced by factors that increase their GTPase activity (GAPs). These provide links to signal transduction pathways.

CONCEPT 9.5 G Protein–Coupled Receptors Respond to Signals Through Effector Proteins

G protein–coupled receptors function through activation of G proteins.

9.5.1 G Proteins Link Receptors with Effector Proteins to Transmit the Signal G proteins are active bound to GTP and inactive bound to GDP. Receptors promote exchange of GDP for GTP. The activated G protein dissociates into two parts, G_α and $G_{\beta\gamma}$, each of which can act on effector proteins. G_α also hydrolyzes GTP to GDP to inactivate the G protein.

9.5.2 Effector Proteins Produce Multiple Second Messengers to Broadcast the Signal Two common effector proteins are adenylyl cyclase and phospholipase C, which produce second messengers known as cAMP, and DAG and IP_3, respectively. Ca^{2+} is also a second messenger. Ca^{2+} release is triggered by IP_3 binding to channel-linked receptors in the ER. Ca^{2+} can bind to a cytoplasmic protein calmodulin, which in turn activates other proteins, producing a variety of responses.

9.5.3 Different Signals Can Have the Same Effect, and the Same Signal Can Have Different Effects Different GPCR receptors can converge to activate the same effector enzyme and thus produce the same second messenger. Receptor subtypes can lead to different effects in different cells. G protein–coupled receptors and receptor tyrosine kinases can activate the same signal transduction pathway. Both RTKs and GPCRs can activate MAP kinase cascades.

Concept Overview

This Concept Overview diagrams the key concepts that were discussed in this chapter.

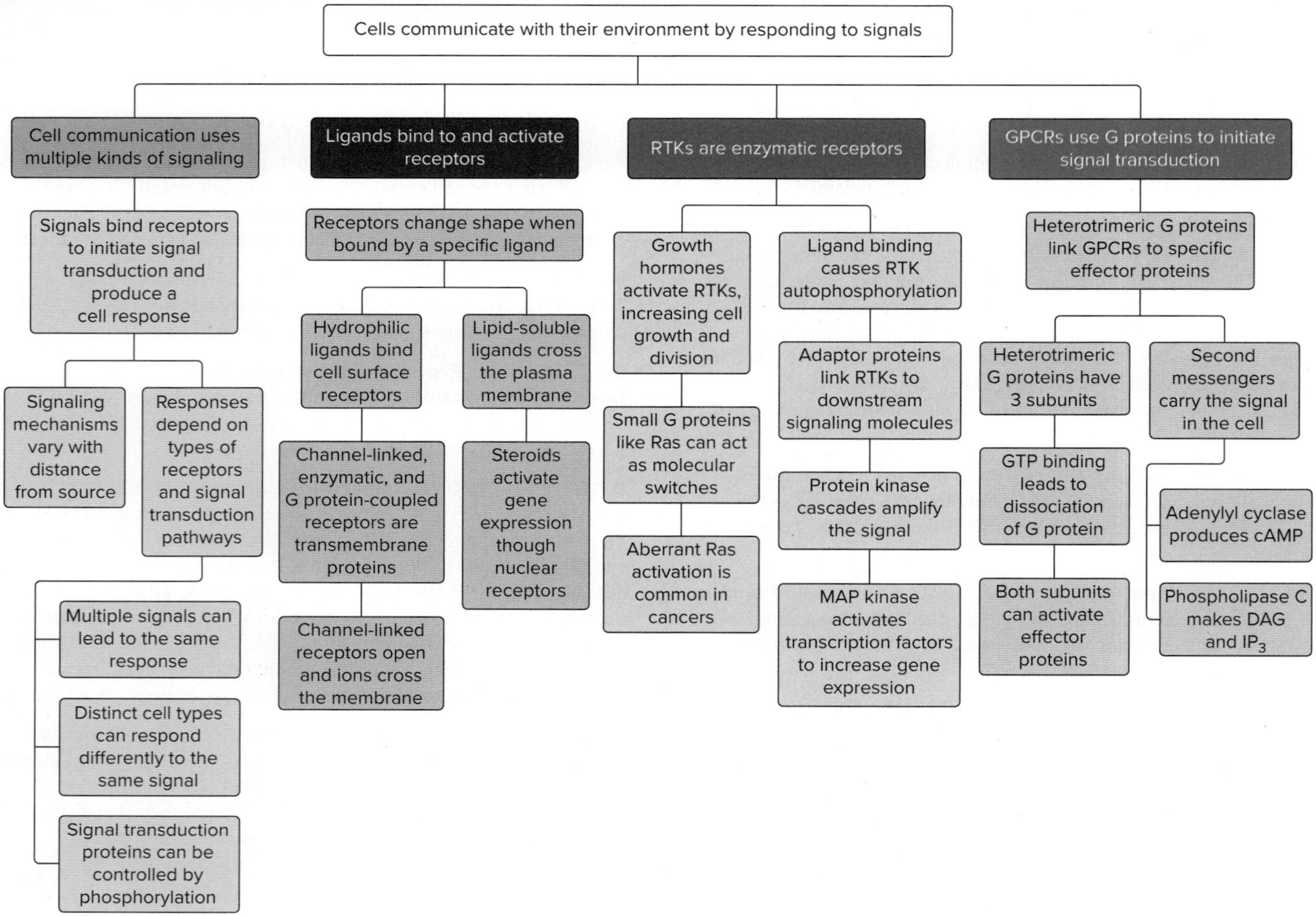

Cells communicate with their environment by responding to signals
Cell communication uses multiple kinds of signaling
Ligands bind to and activate receptors
RTKs are enzymatic receptors
GPCRs use G proteins to initiate signal transduction
Signals bind receptors to initiate signal transduction and produce a cell response
Signaling mechanisms vary with distance from source
Responses depend on types of receptors and signal transduction pathways
Multiple signals can lead to the same response
Distinct cell types can respond differently to the same signal
Signal transduction proteins can be controlled by phosphorylation
Receptors change shape when bound by a specific ligand
Hydrophilic ligands bind cell surface receptors
Lipid-soluble ligands cross the plasma membrane
Channel-linked, enzymatic, and G protein-coupled receptors are transmembrane proteins
Steroids activate gene expression though nuclear receptors
Channel-linked receptors open and ions cross the membrane
Growth hormones activate RTKs, increasing cell growth and division
Small G proteins like Ras can act as molecular switches
Aberrant Ras activation is common in cancers
Ligand binding causes RTK autophosphorylation
Adaptor proteins link RTKs to downstream signaling molecules
Protein kinase cascades amplify the signal
MAP kinase activates transcription factors to increase gene expression
Heterotrimeric G proteins link GPCRs to specific effector proteins
Heterotrimeric G proteins have 3 subunits
GTP binding leads to dissociation of G protein
Both subunits can activate effector proteins
Second messengers carry the signal in the cell
Adenylyl cyclase produces cAMP
Phospholipase C makes DAG and IP_3

Assessing the Learning Path

Understand

1. Signal transduction pathways
   a. are necessary for signals to cross the membrane.
   b. include the intracellular events stimulated by an extracellular signal.
   c. include the extracellular events stimulated by an intracellular signal.
   d. are found only in cases where the signal can cross the membrane.
2. The function of a __________ is to add phosphates to proteins, whereas a __________ functions to remove the phosphates.
   a. tyrosine; serine
   b. protein phosphatase; protein dephosphatase
   c. protein kinase; protein phosphatase
   d. receptor; ligand
3. Which of the following receptor types is NOT a membrane receptor?
   a. Channel-linked receptor
   b. Enzymatic receptor
   c. G protein–coupled receptor
   d. Steroid hormone receptor
4. A membrane receptor
   a. binds a ligand once it enters the cell.
   b. transmits a signal from the outside of the cell to the interior of the cell.
   c. is usually a peripheral membrane protein.
   d. generally binds hydrophobic ligands.
5. The action of steroid hormones is often longer-lived than that of peptide hormones. This is because they
   a. enter the cell and act as enzymes for a longer period of time.
   b. turn on gene expression to produce proteins that persist in the cell.

c. result in the production of second messengers that act directly on cellular processes.
d. stimulate G proteins that act directly on cellular processes.

6. Which of the following is NOT found in a protein that is a member of the nuclear receptor subfamily?
a. A DNA-binding domain
b. A coactivator-binding site
c. A hydrophobic tail that anchors it to the plasma membrane
d. A hormone-binding domain

7. What is the function of Ras during tyrosine kinase cell signaling?
a. It activates the opening of channel-linked receptors.
b. It is an enzyme that synthesizes second messengers.
c. It links the receptor protein to the MAP kinase pathway.
d. It phosphorylates other enzymes as part of a pathway.

8. Kinase cascades
a. can amplify the original signal presented to a cell.
b. are often organized by scaffold proteins.
c. can lead to several different cell responses from the same signal.
d. All of the above

9. Which of the following best describes the immediate effect of ligand binding to a G protein–coupled receptor?
a. The G protein trimer releases a GDP and binds a GTP.
b. The G protein trimer dissociates from the receptor.
c. The G protein trimer interacts with an effector protein.
d. The α subunit of the G protein becomes phosphorylated.

10. Which of the following is NOT a second messenger?
a. Adenylyl cyclase
b. Calcium
c. Phosphatidylinositol-4,5-bisphosphate
d. Cyclic AMP

11. The ion Ca^{2+} can act as a second messenger because it is
a. produced by the enzyme calcium synthase.
b. normally at a high level in the cytoplasm.
c. normally at a low level in the cytoplasm.
d. stored in the cytoplasm.

Apply

1. In what way(s) is a receptor protein similar to an enzyme?
a. They both speed up the rate of chemical reactions.
b. They both have the same primary structure.
c. They both have specific conformations that enable them to bind to specific molecules.
d. All of the above

2. When γ-aminobutyric acid binds to the membrane-bound GABA receptor, a pore opens, allowing Cl^- into the cell. The GABA receptor is
a. an enzymatic receptor.
b. a G protein–coupled receptor.
c. a channel-linked receptor.
d. a second messenger.

3. Thyroid hormone receptors regulate many processes in humans, including metabolism and heart rate. If a thyroid hormone receptor has a defective DNA-binding domain, what would be the effect of this defect?
a. Thyroid hormone would be unable to enter the cell.
b. Thyroid hormone would not be able to bind to the receptor.
c. Genes that regulate metabolism and heart rate would not be expressed correctly.
d. There would be no effect, as thyroid hormone would be able to bind directly to the DNA.

4. Which of the following defects in the PDGF signaling pathway could result in cancer?
a. A defect in which the PDGF receptor did not internalize
b. A defect in which the PDGF receptor was missing its internal domain
c. A defect in which a kinase in the pathway was unable to be activated
d. All of the above

5. In comparing small G proteins such as Ras and GPCR proteins, we can say that
a. both proteins have intrinsic GTPase activity that stops signaling.
b. both proteins are active bound to GTP.
c. Ras is active bound to GDP, and GPCRs are active bound to GTP.
d. Both a and b

6. You treat a cell with a phospholipase C inhibitor. Which of the following will occur?
a. Calcium will not be released from the endoplasmic reticulum.
b. Cyclic AMP will not be made.
c. G proteins will not be inactivated.
d. GTP will not be hydrolyzed.

Synthesize

1. Describe the common features found in all examples of cellular signaling discussed in this chapter. Provide examples to illustrate your answer.

2. Cortisol is a lipid-soluble hormone. Because it can move through cell membranes, does it elicit a response in all cells of the body? Explain.

3. Retinoic acid (RA, structure shown below) binds to the retinoic acid receptor (RAR). The expression of many *Hox* genes, which control early embryonic development, is affected by RAR. Based on the information given, what can you conclude about the structure of the RAR? If you created an antibody that mimics RA, what would happen to cell signaling and *Hox* gene expression?

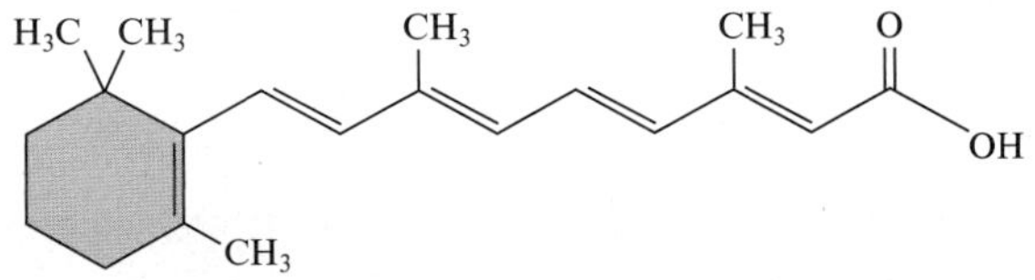

4. The fibroblast growth factor receptor 3 (FGFR3) is a receptor tyrosine kinase. FGFR3 binds to growth factors and regulates bone growth, development, and maintenance. Achondroplasia is a form of dwarfism in humans due to a defect in the gene for FGFR3. Individuals heterozygous for the *FGFR3* mutation (they produce half normal and half mutant FGFR3) have achondroplasia. Why would heterozygous individuals have achondroplasia?

5. Epinephrine (adrenaline) binds to G protein–coupled receptors and stimulates a signal transduction pathway that includes cyclic AMP. Our bodies' fight-or-flight response occurs when we are exposed to acute physical or mental stress and leads to an increase in heart rate, glucose release from glycogen stores, and awareness. The fight-or-flight response is mediated in part by epinephrine. Caffeine is a phosphodiesterase inhibitor; phosphodiesterases degrade cAMP. Explain how the ingestion of caffeine can cause our bodies to act as if they are in the fight-or-flight mode.

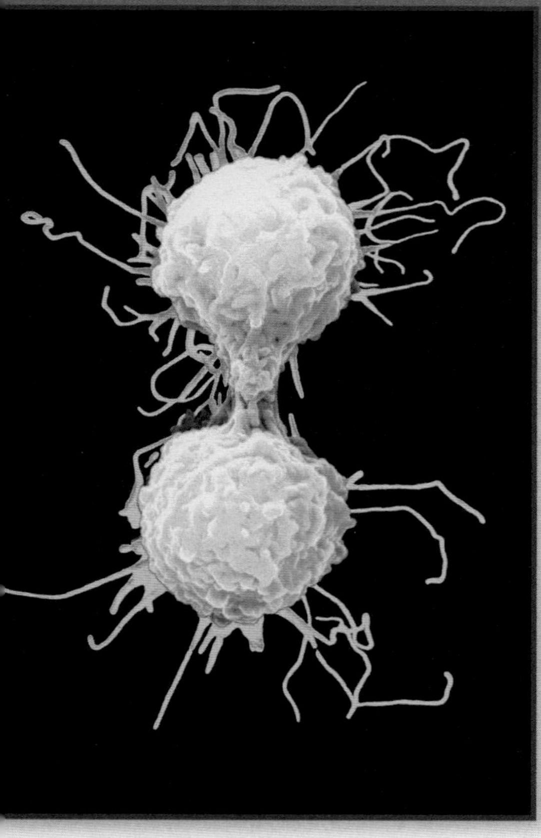

Stem Jems/Science Source

10 How Cells Divide

Learning Path

10.1 Bacterial Cell Division Is Clonal

10.2 Eukaryotes Have Large, Linear Chromosomes

10.3 The Eukaryotic Cell Cycle Is Complex and Highly Organized

10.4 During Interphase, Cells Grow and Prepare for Mitosis

10.5 In Mitosis, Chromosomes Segregate

10.6 Events of the Cell Cycle Are Carefully Regulated

10.7 Cancer Is a Failure of Cell-Cycle Control

Concept Overview

This Concept Overview provides a diagram of the overarching concepts that are covered in this chapter. These concepts will be expanded upon in the Concept Overview at the end of the chapter.

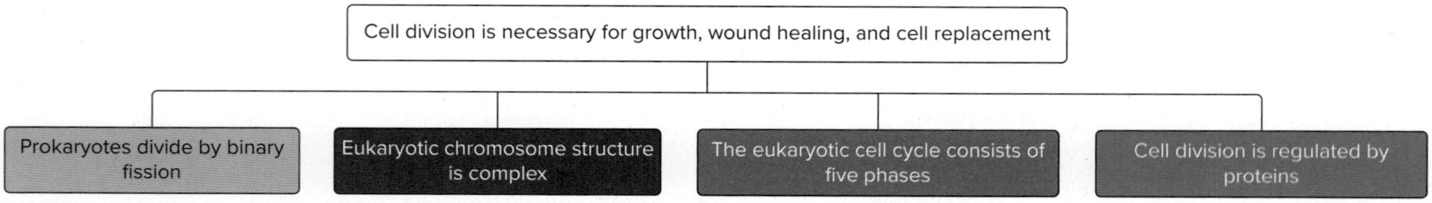
Cell division is necessary for growth, wound healing, and cell replacement
Prokaryotes divide by binary fission
Eukaryotic chromosome structure is complex
The eukaryotic cell cycle consists of five phases
Cell division is regulated by proteins

Introduction

From the smallest creature to the largest, all species produce offspring like themselves and pass on the hereditary information that makes them what they are. In this chapter, we examine how cells like the white blood cell shown in the micrograph on the previous page divide and reproduce. Cell division is necessary for the growth of organisms, for wound healing, and to replace cells that are lost regularly, such as those in your skin and in the lining of your gut. The process, which is direct in prokaryotes, is far more complex in eukaryotes, involving both the replication of chromosomes and their separation into daughter cells. Much of what we are learning about the causes of cancer relates to how cells control this process, a mechanism that in broad outline is the same in all eukaryotes.

10.1 Bacterial Cell Division Is Clonal

Bacteria divide as a way of reproducing themselves. Although bacteria exchange DNA, they do not have a sexual cycle like eukaryotes. Thus, all growth in a bacterial population is due to division to produce new cells. The reproduction of bacteria is clonal—that is, each cell produced by cell division is an identical copy of the original cell.

Binary Fission Divides Bacterial Cells in Two

LEARNING OBJECTIVE 10.1.1 Diagram the bacterial cell cycle.

In all organisms, cell division produces two new cells, each with the same genetic information as the original. Despite the host of differences between prokaryotes and eukaryotes, the essentials of the cell division process are the same: duplication and segregation of genetic information into daughter cells, and division of cellular contents. We begin by looking at the prokaryotic process, **binary fission,** as it occurs in bacteria.

Most bacteria have a genome made up of a single, circular DNA molecule. In spite of its apparent simplicity, the DNA molecule of the bacterium *Escherichia coli* is actually on the order of 500 times longer than the cell itself! Packaged very tightly to fit into the cell, the DNA is located in a region called the *nucleoid.* Although distinct from the cytoplasm around it, the nucleoid is not surrounded by a membrane.

The compaction and organization of the nucleoid are achieved by a class of proteins called structural maintenance of chromosome (SMC) proteins. These are ancient proteins that have diversified over evolutionary time to fulfill a variety of roles related to DNA organization in different lineages. The eukaryotic cohesin and condensin proteins discussed in section 10.2 are also SMC proteins.

During binary fission, the bacterial chromosome is replicated and the two products partitioned to each end of the cell, prior to the physical division of the cell. Binary fission begins with the replication of the bacterial DNA at a specific site—the origin of replication—and proceeds in both directions around the circular DNA to a specific site of termination (figure 10.1). As this occurs, the cell body proceeds to divide in two ("binary fission"), division occurring roughly at midcell.

Bacterial Cells Control Chromosome Partitioning and Septum Formation

LEARNING OBJECTIVE 10.1.2 Describe the events of DNA partitioning and cell fission.

One key feature of bacterial cell division is that replication and partitioning of the chromosome occur as a concerted process. In contrast, in eukaryotic cells DNA replication occurs early in division, and chromosome separation occurs much later.

For many years it was thought that newly replicated *E. coli* DNA molecules were passively segregated by attachment to and growth of the membrane as the cell elongated. To test this hypothesis, experiments followed the movement of the origin of

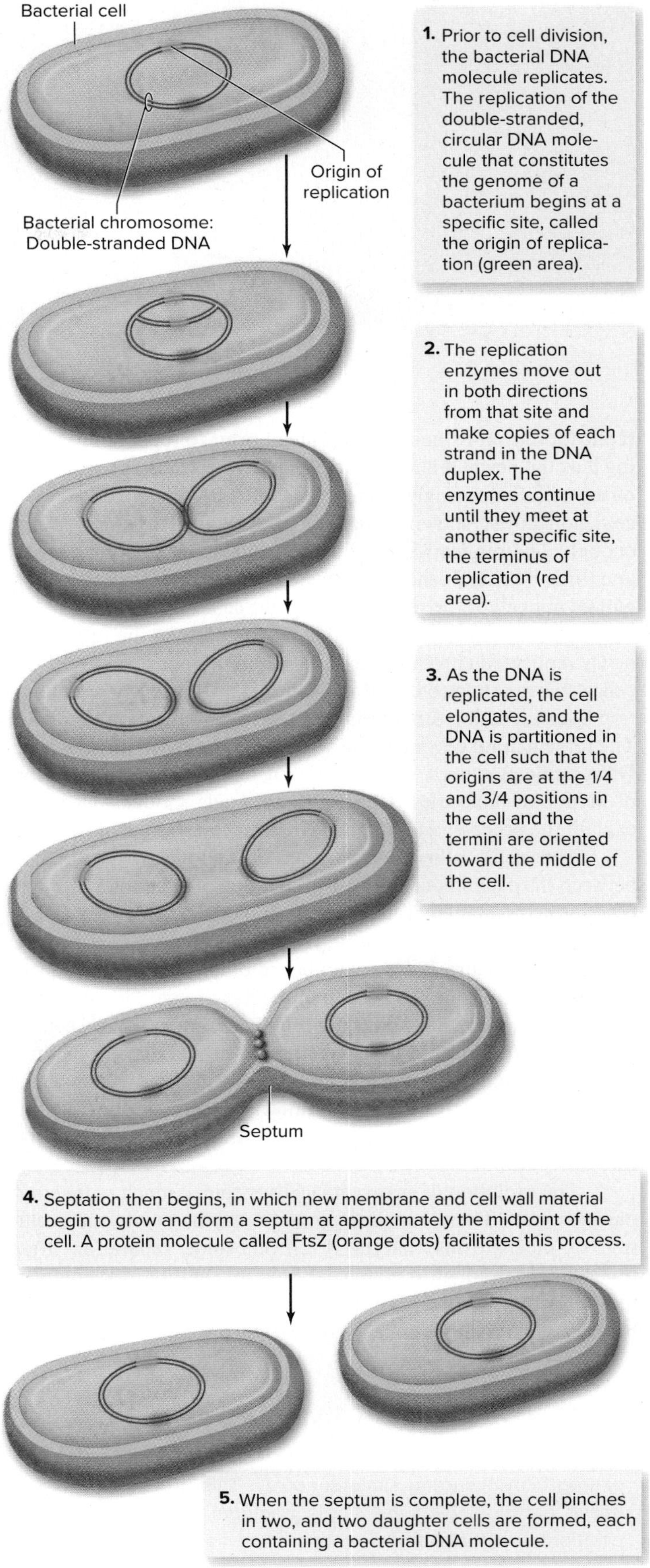

Bacterial cell
Origin of replication
Bacterial chromosome: Double-stranded DNA
1. Prior to cell division, the bacterial DNA molecule replicates. The replication of the double-stranded, circular DNA molecule that constitutes the genome of a bacterium begins at a specific site, called the origin of replication (green area).
2. The replication enzymes move out in both directions from that site and make copies of each strand in the DNA duplex. The enzymes continue until they meet at another specific site, the terminus of replication (red area).
3. As the DNA is replicated, the cell elongates, and the DNA is partitioned in the cell such that the origins are at the 1/4 and 3/4 positions in the cell and the termini are oriented toward the middle of the cell.
Septum
4. Septation then begins, in which new membrane and cell wall material begin to grow and form a septum at approximately the midpoint of the cell. A protein molecule called FtsZ (orange dots) facilitates this process.
5. When the septum is complete, the cell pinches in two, and two daughter cells are formed, each containing a bacterial DNA molecule.

Figure 10.1 Binary fission.

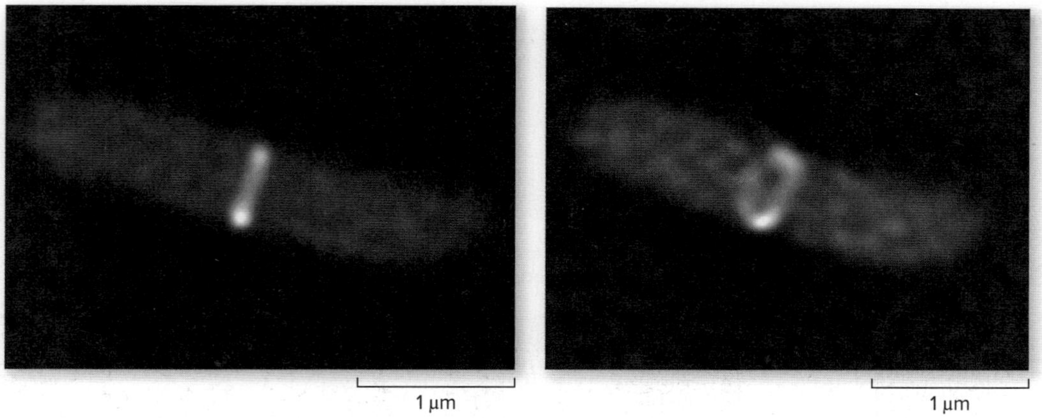

Figure 10.2 The FtsZ protein. In these dividing *E. coli* bacteria, the FtsZ protein is labeled with fluorescent dye to show its location during binary fission. The protein assembles into a ring at approximately the midpoint of the cell, where it facilitates septation and cell division. Bacteria carrying mutations in the *FtsZ* gene are unable to divide.
Courtesy of William Margolin

replication from its position at midcell prior to replication, tracking it as the newly replicated origins moved toward opposite ends of the cell. Surprisingly, this movement is faster than the rate of cell elongation, showing that membrane growth alone is not enough. The replication origins are captured at the one-quarter and three-quarter positions relative to the length of the cell, which will be midcell of the resulting daughter cells.

During replication, first the origin and then the rest of the newly replicated chromosomes are moved to opposite ends of the cell as two new nucleoids are assembled. The final event of replication is decatenation (untangling) of the final replication products. The force behind chromosome segregation has been attributed to DNA replication itself, to transcription, and to the polymerization of actin-like molecules. Currently, no single model appears to explain the process, and it may involve more than one of these.

The cell's other components are physically partitioned between the daughter cells by the production of a **septum.** This process, termed **septation,** usually occurs at the midpoint of the cell. It begins with the formation of a ring composed of many copies of the protein FtsZ (figure 10.2). A number of other proteins then accumulate, including ones embedded in the membrane. This structure contracts inward radially until the cell divides into two new cells.

The FtsZ protein is found in most prokaryotes, including archaea. It can form filaments and rings, and three-dimensional models show a high degree of similarity to eukaryotic tubulin. However, its role in bacterial division is quite different from the role of tubulin in mitosis.

The evolution of eukaryotic cells included much more complex genomes. These complex genomes may be due to the evolution of mechanisms that delay chromosome separation after replication. Although it is unclear how this ability to keep chromosomes together evolved, it does seem more closely related to binary fission than we once thought (figure 10.3).

REVIEW OF CONCEPT 10.1

Most bacteria divide by binary fission, with DNA replication and segregation occurring simultaneously. This process involves active partitioning of the single bacterial chromosome and positioning of the site of septation.

- *Would binary fission work as well if bacteria had many chromosomes?*

10.2 Eukaryotes Have Large, Linear Chromosomes

Chromosomes were first observed by the German embryologist Walther Flemming (1843–1905) in 1879, while he was examining the rapidly dividing cells of salamander larvae. Using a primitive light microscope, Flemming saw minute threads within their nuclei that appeared to be dividing lengthwise. Flemming called their division **mitosis**, based on the Greek word *mitos*, meaning "thread."

Chromosome Number Varies Among Species

LEARNING OBJECTIVE 10.2.1 Differentiate between haploid and diploid numbers of chromosomes in a species.

Chromosomes are found in the cells of all eukaryotes, although their numbers vary enormously between species. Organisms have been found with only a single pair of chromosomes, and others have more than 500 pairs, but most eukaryotes have between 10 and 50 chromosomes in their body cells.

When defining the number of different chromosomes in a species, we count the **haploid** (n) number of chromosomes, which is one complete set of chromosomes. For humans and many other species, the total number of chromosomes in a cell is called the **diploid** ($2n$) number, which is twice the haploid number. For humans, the haploid number is 23 and the diploid number is 46. Diploid chromosomes reflect the equal genetic contribution each parent makes to offspring. We refer to the maternal and paternal chromosomes as being **homologous**, and each one of the pair is termed a **homolog**.

Chromosomes vary in size, staining properties, the location of a visible constricted region called the centromere, and the relative length of the two arms on either side of the centromere (figure 10.4). The particular array of chromosomes an individual organism possesses is called its **karyotype**.

Chromatin Is Composed of DNA and Protein

LEARNING OBJECTIVE 10.2.2 Describe the structure of a nucleosome.

It is amazing that after more than 100 years and a variety of increasingly sophisticated methods of analysis, we still do not

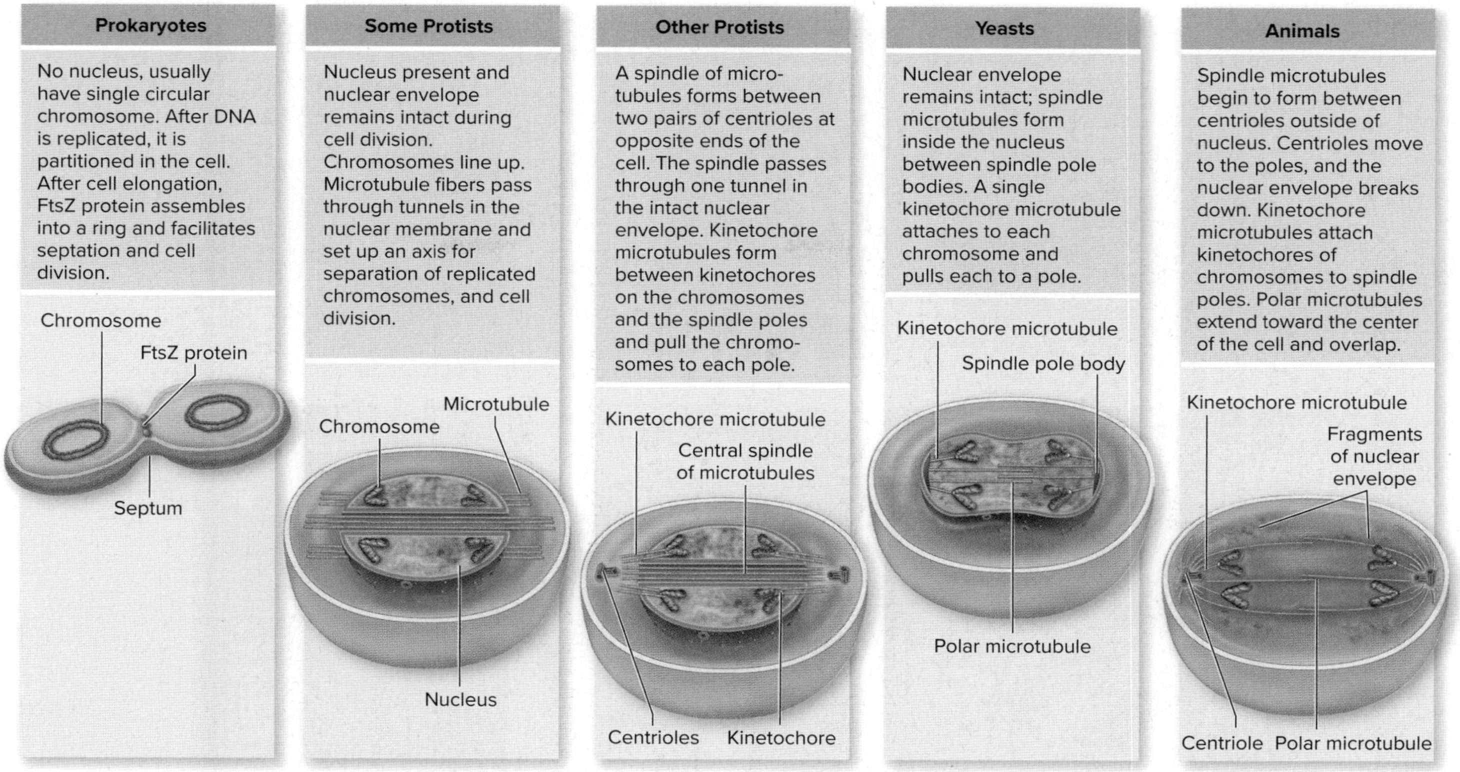
Prokaryotes
No nucleus, usually have single circular chromosome. After DNA is replicated, it is partitioned in the cell. After cell elongation, FtsZ protein assembles into a ring and facilitates septation and cell division.
Chromosome
FtsZ protein
Septum
Some Protists
Nucleus present and nuclear envelope remains intact during cell division. Chromosomes line up. Microtubule fibers pass through tunnels in the nuclear membrane and set up an axis for separation of replicated chromosomes, and cell division.
Microtubule
Chromosome
Nucleus
Other Protists
A spindle of microtubules forms between two pairs of centrioles at opposite ends of the cell. The spindle passes through one tunnel in the intact nuclear envelope. Kinetochore microtubules form between kinetochores on the chromosomes and the spindle poles and pull the chromosomes to each pole.
Kinetochore microtubule
Central spindle of microtubules
Centrioles
Kinetochore
Yeasts
Nuclear envelope remains intact; spindle microtubules form inside the nucleus between spindle pole bodies. A single kinetochore microtubule attaches to each chromosome and pulls each to a pole.
Kinetochore microtubule
Spindle pole body
Polar microtubule
Animals
Spindle microtubules begin to form between centrioles outside of nucleus. Centrioles move to the poles, and the nuclear envelope breaks down. Kinetochore microtubules attach kinetochores of chromosomes to spindle poles. Polar microtubules extend toward the center of the cell and overlap.
Kinetochore microtubule
Fragments of nuclear envelope
Centriole
Polar microtubule

Figure 10.3 A comparison of protein assemblies during cell division among different organisms. The prokaryotic protein FtsZ has a structure that is similar to that of the eukaryotic protein tubulin, the protein component of microtubules, which are the fibers eukaryotic cells use to construct the spindle apparatus used to separate chromosomes.

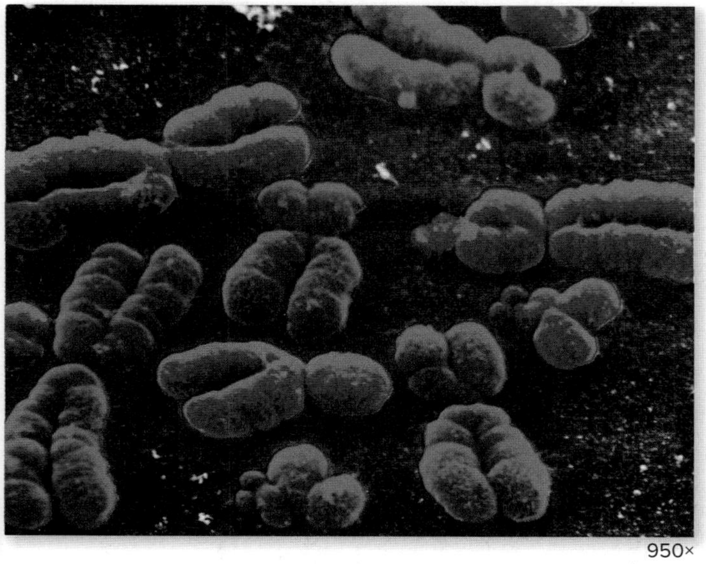
950×

Figure 10.4 Human chromosomes. This scanning electron micrograph shows human chromosomes as they appear immediately before nuclear division. Each DNA molecule has already replicated, forming identical copies held together at a visible constriction called the centromere. False color has been added to the chromosomes.

Biophoto Associates/Science Source

know the precise structure of chromosomes, and the details of how this structure changes over the course of the cell cycle. What has become clear is that the nucleus is much more organized than originally conceived.

Chromosomes are composed of **chromatin**, a complex of DNA and protein, usually about 40% DNA and 60% protein. A significant amount of RNA is also associated with chromosomes, because chromosomes are the sites of RNA synthesis. A representative human chromosome contains about 140 million (1.4×10^8) nucleotides in a single DNA molecule. If you were to stretch out this DNA strand, it would be about 4.3 cm (1.7 in.). Fitting this length of DNA into a cell nucleus is like cramming a string the length of a football field into a baseball—and that's only one chromosome! In the cell, however, the DNA is compacted, allowing it to fit into a much smaller space, and the level of compaction changes with the cell cycle as needed to move chromosomes about the cell.

Nucleosomes organize DNA

If we gently disrupt a eukaryotic nucleus and examine the DNA with an electron microscope, we find that it resembles a string of beads (figure 10.5). The beads are a structure we call the **nucleosome**, which is composed of DNA and **histone proteins.** Histones have a positive charge because of an abundance of the basic amino acids arginine and lysine. Thus they are strongly attracted

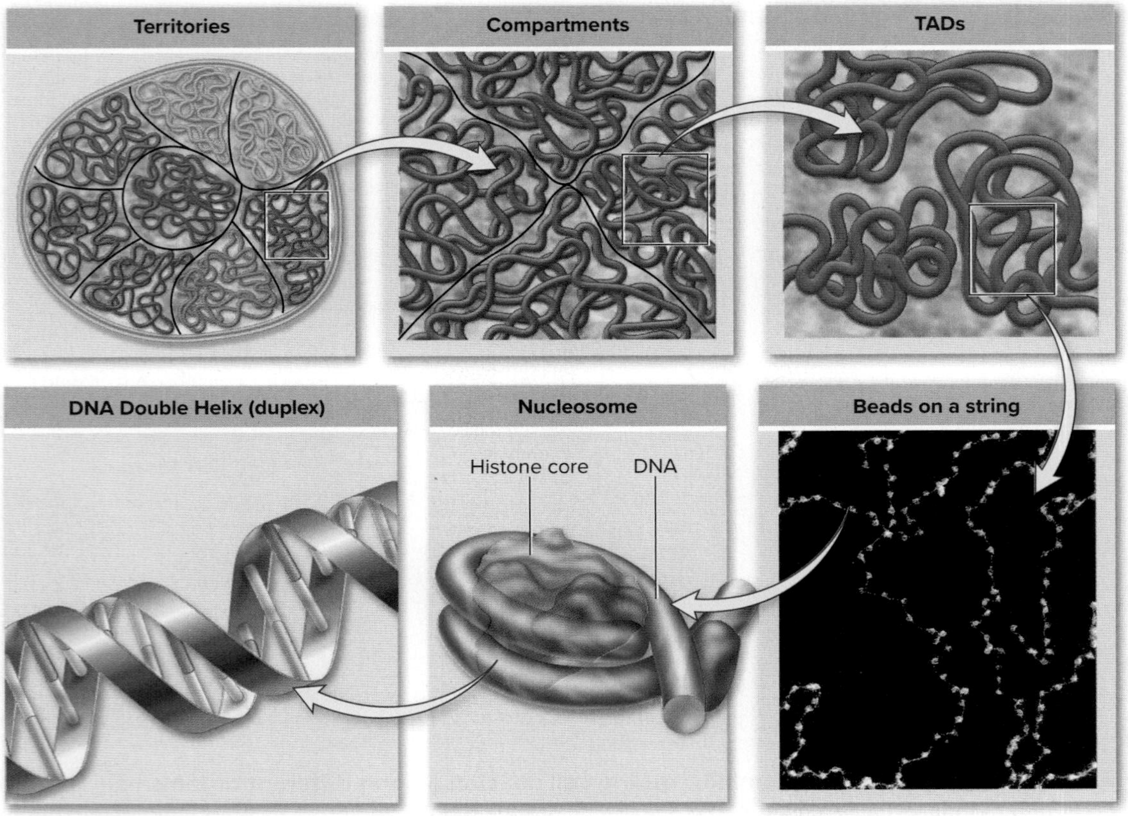

Figure 10.5 Organization of chromatin in the nucleus. Starting at the top left, the figure moves in a clockwise direction. The top row shows the organization of chromatin in the interphase nucleus. Individual chromosomes tend to be organized into territories that exclude other chromosomes. Within a chromosome, chromatin is organized into compartments that consist of either active or inactive regions. These in turn are formed by looped structures called TADs. The base of a loop is anchored by cohesin and CTCF. The DNA in TADs is organized into nucleosomes, which consist of DNA wrapped around a core particle of 8 histone proteins.
Don W. Fawcett/Science Source

to negatively charged DNA. Histone proteins are some of the most highly evolutionarily conserved proteins known, indicating both their age and their importance.

Nucleosomes contain four primary histones: H2A, H2B, H3, and H4. These associate together to form a large complex containing two of each core histone to form an octamer, which acts as a spool that DNA is wound around. Each histone octamer has 147 bp of DNA wrapped in 1.7 left-handed turns to form the nucleosome core particle, which has a diameter of about 10 nm. Each of these is separated by a variable length of DNA of 20–80 bp called linker DNA—thus the "beads on a string" appearance (figure 10.5). An additional histone, H1, can interact with the linker DNA, and the core particle histones also have protruding "tails" that are subject to posttraslational modifications, which facilitate interactions between nucleosomes.

If isolated DNA wrapped in nucleosomes is incubated under low-salt conditions with histone H1, it will form an even more compact structure with a diameter of 30 nm. This so-called "30 nm fiber" is formed either by coiling into a spiral called a solenoid, or by forming a zig-zag structure with the same diameter. The 30 nm fiber was long considered the usual state of chromatin in nondividing cells, but there is little actual evidence that this fiber exists in cells.

Chromatin Is Organized in the Nucleus

LEARNING OBJECTIVE 10.2.3 Explain how chromatin is organized in the nucleus.

New imaging techniques combined with high-throughput methods to assess interactions between genomic regions have revealed the organization of chromosomes in the eukaryotic nucleus. Chromatin is organized into territories, which contain individual chromosomes, and these are further organized into compartments within each chromosome. At the highest level of resolution, genomic regions interact to form **topologically associated domains (TADs)** (figure 10.5).

These TADs involve loops of DNA in nucleosomes that are anchored at their base by two proteins: a DNA-binding protein called **CTCF** (for CCCTC-binding factor) and the cohesin complex of proteins (discussed later). The loops may also involve interactions between nucleosomes. This organization may also affect the control of gene expression by bringing relatively distant regions close together, and keeping other regions compacted (refer to chapter 16).

Geneticists have recognized for years that some domains of chromatin, called **heterochromatin,** are not active, and other domains of chromatin, called **euchromatin,** are active. These names originally were based on staining patterns seen in the light microscope, but it now appears that this may extend to the kinds of organization found in compartments and TADs, with some being more condensed and inactive, and others more open and active.

Chromosome condensation changes during cell division

The state of chromosome condensation changes through the cell cycle. Between divisions, chromatin is relatively diffuse, and then is increasingly condensed to allow the separation of chromosomes during cell division, then decondensed again after division. If you have a mental image of a chromosome, it is likely the highly condensed "X-shaped" metaphase chromosome. This structure has been shown to be formed by organizing DNA on a protein scaffold. This scaffold was first visualized by removing histones from isolated metaphase chromosomes, which caused the DNA to form a massive tangle of spaghetti-like fibers of DNA surrounding an X-shaped scaffold of proteins.

For many years, the dominant model of chromosome condensation was of successive coiling to form the 30 nm fiber, which then was further compacted by radial looping from this scaffold of proteins. While models of this type have not been completely rejected, the only part with strong experimental support is the existence of a protein scaffold to organize chromatin.

One new model envisions condensin proteins organized along the scaffold, using energy from ATP hydrolysis to form regular loops of DNA in nucleosomes. This is then twisted over the course of the cell cycle to further compact the loops, with the number of loops also increasing with progress through prophase. In this model, DNA is wrapped into nucleosomes, and then interactions of linker histones, and the nucleosomes themselves can change the density of the "liquid DNA." While this may sound bizarre, it fits much of the new data on chromosome conformation.

Chromosome replication

When DNA is replicated, the single molecule in each chromosome becomes two identical DNA molecules held together at their centromere by **cohesins**. As each chromosome becomes more condensed, it becomes visible as the X-shaped structure in a karyotype. At this point, we still call this one chromosome, but it is composed of two **sister chromatids** (figure 10.6).

The fact that the products of replication are held together is critical to the division process. At the end of the division process, each cell must have a complete set of chromosomes. To insure this, the cell connects the replication products together at the centromere. The process of mitosis then separates all of these copies at the same time, ensuring that each daughter cell gets one copy of each chromosome.

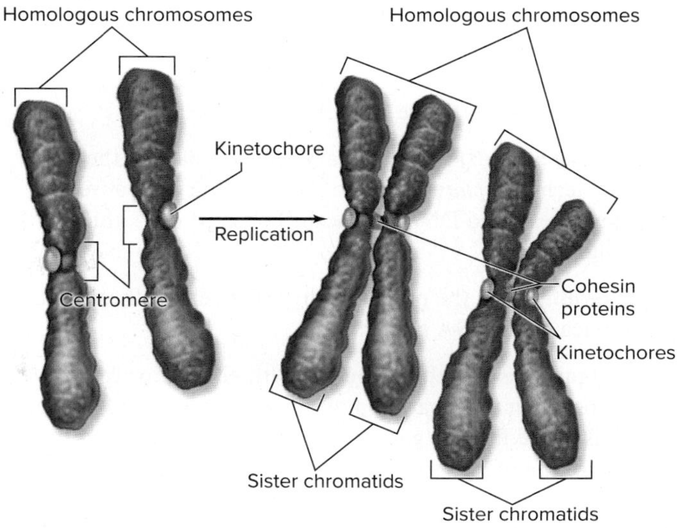

Figure 10.6 The difference between homologous chromosomes and sister chromatids. Homologous chromosomes are the maternal and paternal copies of the same chromosome. Sister chromatids are the two replicas of a single chromosome held together at their centromeres by cohesin proteins after DNA replication. The kinetochore (described in section 10.4) is composed of proteins found at the centromere that attach to microtubules during mitosis.

REVIEW OF CONCEPT 10.2

Chromosome number is highly variable between species. Diploid cells contain a maternal and a paternal copy, or homolog, for each chromosome. Chromatin is organized into nucleosomes consisting of DNA wrapped about histone proteins. Chromatin in the interphase nucleus is organized into compartments with individual chromosomes. These are organized into territories, which are further organized into topological domains (TADs). Chromosome condensation changes with the cell cycle. After chromosome replication, each homolog consists of two sister chromatids held together by proteins called cohesins at the centromere.

- *Is chromosome number related to organismal complexity?*

10.3 The Eukaryotic Cell Cycle Is Complex and Highly Organized

Eukaryotic genomes are larger and have a more complex organization than prokaryotes; this required radical changes in how replicated genomes are partitioned into daughter cells. The **cell cycle** requires the duplication of the genome, its accurate segregation, and the division of cellular contents.

The Cell Cycle Is Divided into Five Phases

LEARNING OBJECTIVE 10.3.1 Describe the events of the five stages of the eukaryotic cell cycle.

The cell cycle is divided into phases based on the key events of genome duplication, chromosome segregation, and distribution of cytoplasm. The cell cycle is usually diagrammed using the metaphor of a clock face (figure 10.7).

- **G_1 (gap phase 1)** is the primary growth phase of the cell. The term *gap phase* refers to its filling the gap between cytokinesis and DNA synthesis. For most cells, this is the longest phase.
- **S (synthesis)** is the phase in which the cell synthesizes a replica of its DNA genome.
- **G_2 (gap phase 2),** the second growth phase, involves preparation for separation of the newly replicated genome. This phase fills the gap between DNA synthesis and the beginning of mitosis. During this phase, microtubules begin to reorganize to form a spindle.

  G_1, S, and G_2 together constitute **interphase,** the portion of the cell cycle between cell divisions.
- **Mitosis** is the phase of the cell cycle in which the spindle apparatus assembles, binds to the chromosomes, and moves the sister chromatids apart. Mitosis is the essential step in the separation of the two daughter genomes. It is traditionally subdivided into five stages: prophase, prometaphase, metaphase, anaphase, and telophase.
- **Cytokinesis** is the phase of the cell cycle when the cytoplasm divides, creating two daughter cells. In animal cells, the microtubule spindle helps position a contracting ring of actin that constricts like a drawstring to pinch the cell in two. In cells with a cell wall, such as plant and fungal cells, a plate forms between the dividing cells.

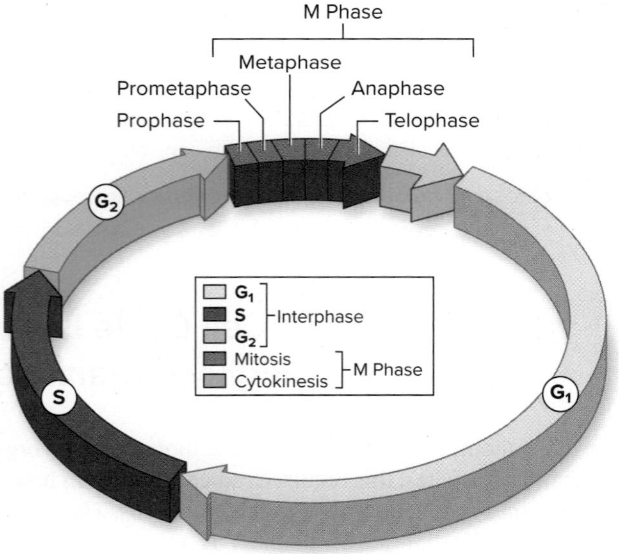

Figure 10.7 The cell cycle. The cell cycle is depicted as a circle. The first gap phase, G_1, involves growth and preparation for DNA synthesis. During S phase, a copy of the genome is synthesized. The second gap phase, G_2, prepares the cell for mitosis. During mitosis, replicated chromosomes are partitioned. Cytokinesis divides the cell into two cells with identical genomes.

Mitosis and cytokinesis together are usually referred to collectively as the M phase, to distinguish the dividing phase from interphase.

The duration of the cell cycle varies depending on cell type

The time it takes to complete a cell cycle varies greatly. Cells in animal embryos can complete their cell cycle in under 20 minutes. These cells simply divide their nuclei as quickly as they can replicate their DNA, without cell growth. Half of their cycle is taken up by S, half by M, and essentially none by G_1 or G_2.

Because mature cells require time to grow, their cycles are much longer than those of embryonic tissue. Typically, a dividing mammalian cell completes its cell cycle in about 24 hours, but some cells, such as certain cells in the human liver, have cell cycles lasting more than a year. During the cycle, growth occurs throughout the G_1 and G_2 phases, as well as during the S phase. The M phase takes only about an hour, a small fraction of the entire cycle.

Most of the variation in the length of the cell cycle between organisms or cell types occurs in the G_1 phase. Cells often pause in G_1 before DNA replication and enter a resting state called the **G_0 phase**; cells may remain in this phase for days to years before resuming cell division.

REVIEW OF CONCEPT 10.3

Cell division in eukaryotes is a complex process that involves five phases: a first gap phase (G_1); a DNA synthesis phase (S); a second gap phase (G_2); mitosis (M), during which chromatids are separated; and cytokinesis, in which a cell becomes two separate cells.

- *When during the cycle is a cell irreversibly committed to dividing?*

10.4 During Interphase, Cells Grow and Prepare for Mitosis

Interphase Includes the Synthesis and Gap Phases of the Cell Cycle

LEARNING OBJECTIVE 10.4.1 Describe the events that take place during interphase and how they affect the structure of the centromere after S phase.

Interphase is that portion of the cell cycle between cell divisions. The events that occur during interphase—the G_1, S, and G_2 phases—involve very important preparations for the successful completion of mitosis. During G_1, cells undergo the major portion of their growth. During the S phase, each chromosome replicates to produce two sister chromatids, which remain attached to each other at the centromere. In the G_2 phase, the chromosomes coil even more tightly.

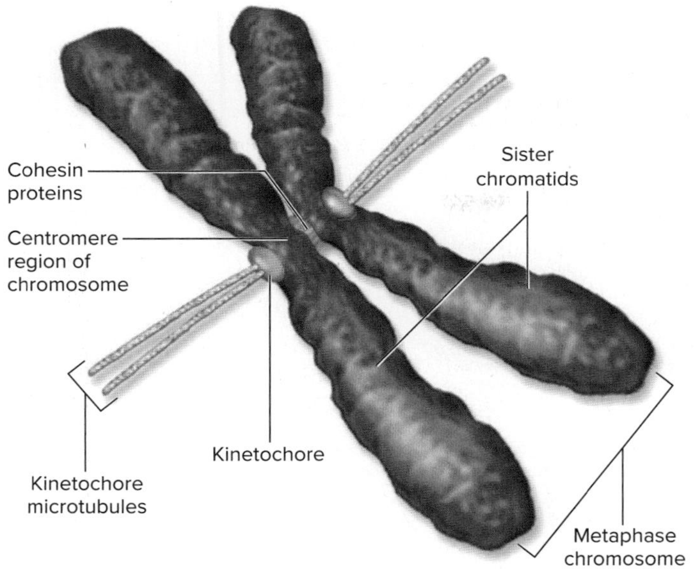

Figure 10.8 Kinetochores. Separation of sister chromatids during mitosis depends on microtubules attaching to proteins found in the kinetochore. These kinetochore proteins are assembled on the centromere of chromosomes. The centromeres of the two sister chromatids are held together by cohesin proteins.

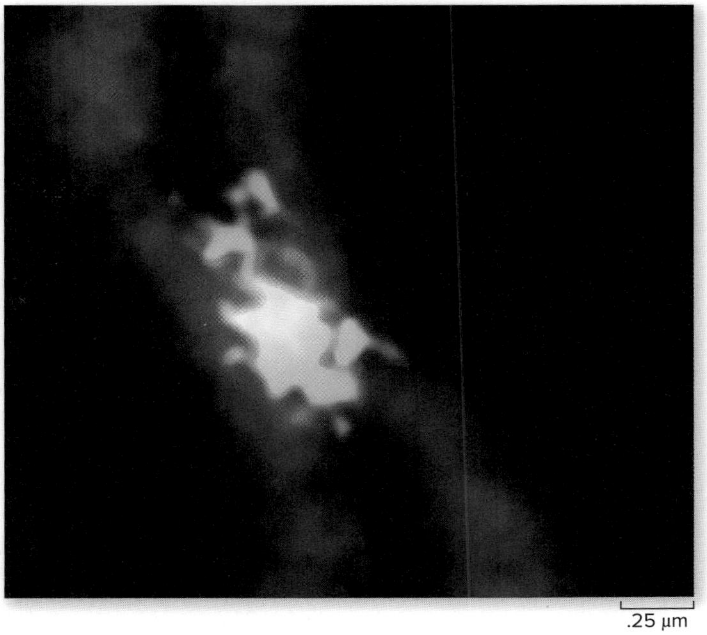

Figure 10.9 Proteins found at the centromere. In this image, chromatin, a cohesin protein, and a kinetochore protein have each been labeled with a different-colored fluorescent dye. Cohesin (green), which holds centromeres together, lies between the sister chromatids (blue). Each sister chromatid has its own kinetochore (red).

Courtesy of Pétér Lénárt and Jan-Michael Peters, IMP

The **centromere** is a point of constriction on the chromosome containing repeated DNA sequences that bind specific proteins. These proteins make up a disklike structure called the **kinetochore.** This disk functions as an attachment site for microtubules necessary to separate the chromosomes during cell division (figure 10.8). Each chromosome's centromere is located at a characteristic site along the length of the chromosome.

After the S phase, the newly synthesized sister chromatids appear to share a common centromere, but at the molecular level the DNA of the centromere has already replicated, so there are two complete DNA molecules. This means that you have two chromatids held together by cohesin proteins at the centromere, and each chromatid has its own set of kinetochore proteins (figure 10.9). In multicellular animals, most of the cohesin that holds sister chromatids together after replication is replaced by condensin as the chromosomes are condensed. This leaves the chromosomes still attached tightly at the centromere, but only loosely attached elsewhere.

A eukaryotic cell typically grows throughout interphase. The G_1 and G_2 segments of interphase are periods of active growth, during which proteins are synthesized and cell organelles are produced. However, the cell's DNA replicates only during the S phase of the cell cycle.

After the chromosomes have replicated in S phase, they remain dispersed, although cohesin proteins hold sister chromatids together. In G_2 phase, they begin the process of condensation, coiling ever more tightly. Also during G_2 phase, the cells begin to assemble the machinery they will later use to move the chromosomes to opposite poles of the cell. In animal cells, a pair of barrel-shaped organelles called **centrioles** replicate, producing one for each pole. These act as microtubule-organizing centers: surrounding each centriole is *pericentriolar material,* ring-shaped structures composed of tubulin that can nucleate the assembly of microtubules. Plants and fungi lack centrioles but still contain microtubule-organizing centers. All eukaryotic cells undertake an extensive synthesis of **tubulin,** the protein that forms microtubules.

REVIEW OF CONCEPT 10.4

During interphase, the cell grows; replicates chromosomes, organelles, and centrioles; and synthesizes components needed for mitosis, including tubulin. Cohesin proteins hold chromatids together at the centromere.

- *How would a mutation that deleted cohesin proteins affect cell division?*

10.5 In Mitosis, Chromosomes Segregate

Mitosis is one of the most dramatic and beautiful biological processes that we can easily observe. To better illustrate what happens, biologists have traditionally divided mitosis into five arbitrary phases. However, the actual process is dynamic and continuous, and not broken into discrete steps. This process is shown both schematically and in micrographs in figure 10.10.

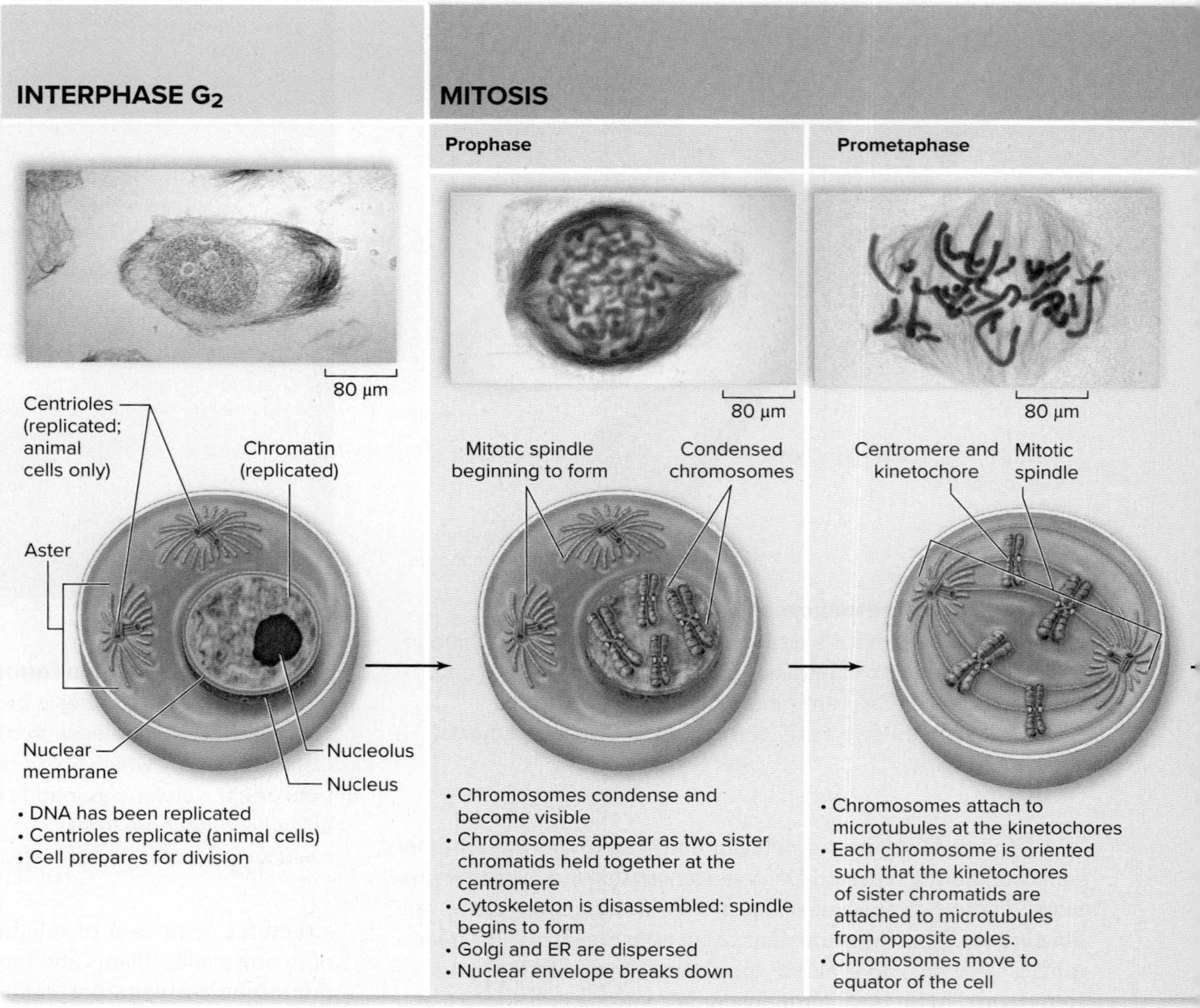

INTERPHASE G_2
MITOSIS
Prophase
Prometaphase
80 µm
Centrioles (replicated; animal cells only)
Chromatin (replicated)
Aster
Nuclear membrane
Nucleolus
Nucleus
• DNA has been replicated
• Centrioles replicate (animal cells)
• Cell prepares for division
80 µm
Mitotic spindle beginning to form
Condensed chromosomes
• Chromosomes condense and become visible
• Chromosomes appear as two sister chromatids held together at the centromere
• Cytoskeleton is disassembled: spindle begins to form
• Golgi and ER are dispersed
• Nuclear envelope breaks down
80 µm
Centromere and kinetochore
Mitotic spindle
• Chromosomes attach to microtubules at the kinetochores
• Each chromosome is oriented such that the kinetochores of sister chromatids are attached to microtubules from opposite poles.
• Chromosomes move to equator of the cell

Figure 10.10 Mitosis and cytokinesis. Mitosis is conventionally divided into five stages—prophase, prometaphase, metaphase, anaphase, and telophase—which together act to separate duplicated chromosomes. Mitosis is followed by cytokinesis, which divides the cell body into two separate cells. Photos depict mitosis and cytokinesis in a plant, the African blood lily (*Haemanthus katharinae*), with chromosomes stained blue and microtubules stained red. Drawings depict mitosis and cytokinesis in animal cells.

Waheeb Heneen, the Swedish University of Agricultural Sciences

During Prophase, the Mitotic Apparatus Forms

LEARNING OBJECTIVE 10.5.1 Describe how the mitotic apparatus forms during prophase.

The first stage of mitosis, **prophase,** is said to begin when the chromosome condensation initiated in G_2 phase reaches the point at which individual condensed chromosomes first become visible with the light microscope. Because the condensation process begun in G_2 continues throughout prophase, chromosomes that start prophase as minute threads appear quite bulky before its conclusion.

The spindle and centrioles

The assembly of the **spindle** apparatus that will later separate the sister chromatids occurs during prophase, replacing the normal microtubule structure of the cell that was disassembled in the G_2 phase. In animal cells, the two centriole pairs formed during G_2 phase begin to move apart early in prophase, forming between them an axis of microtubules referred to as spindle fibers. By the time the centrioles reach the opposite poles of the cell, they have established a bridge of microtubules, called the spindle apparatus, between them. In plant cells, a similar bridge of microtubular fibers forms between opposite poles of the cell, although centrioles are absent in plant cells.

In animal cell mitosis, the centrioles extend a radial array of microtubules toward the nearby plasma membrane when they reach the poles of the cell. This arrangement of microtubules is called an **aster.** Although the aster's function is not fully understood, it probably braces the centrioles against the membrane and stiffens the point of microtubular attachment during the retraction of the spindle. Plant cells, which have rigid cell walls, do not form asters.

Breakdown of the nuclear envelope

During the formation of the spindle apparatus, the nuclear envelope breaks down, and the endoplasmic reticulum reabsorbs its components. At this point, the microtubular spindle fibers extend completely across the cell, from one pole to the other. Their orientation determines the plane in which the cell will subsequently divide, through the center of the cell at right angles to the spindle apparatus.

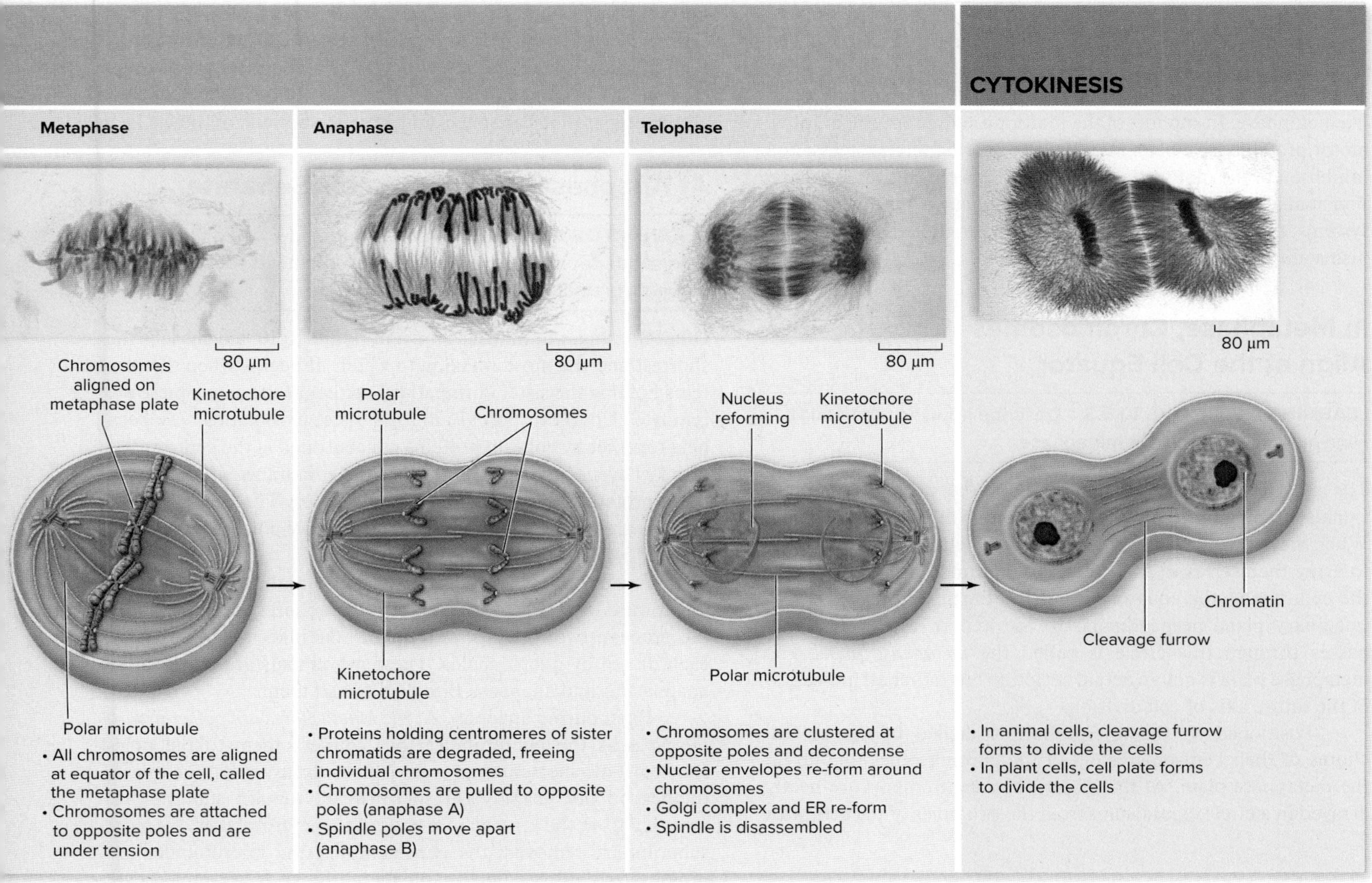
CYTOKINESIS
Metaphase
Anaphase
Telophase
80 µm
80 µm
80 µm
80 µm
Chromosomes aligned on metaphase plate
Kinetochore microtubule
Polar microtubule
• All chromosomes are aligned at equator of the cell, called the metaphase plate
• Chromosomes are attached to opposite poles and are under tension
Polar microtubule
Chromosomes
Kinetochore microtubule
• Proteins holding centromeres of sister chromatids are degraded, freeing individual chromosomes
• Chromosomes are pulled to opposite poles (anaphase A)
• Spindle poles move apart (anaphase B)
Nucleus reforming
Kinetochore microtubule
Polar microtubule
• Chromosomes are clustered at opposite poles and decondense
• Nuclear envelopes re-form around chromosomes
• Golgi complex and ER re-form
• Spindle is disassembled
Chromatin
Cleavage furrow
• In animal cells, cleavage furrow forms to divide the cells
• In plant cells, cell plate forms to divide the cells

In Prometaphase, Chromosomes Attach to the Spindle

LEARNING OBJECTIVE 10.5.2 Describe how chromosomes attach to the spindle during prometaphase.

The transition from prophase to **prometaphase** occurs following the disassembly of the nuclear envelope. During prometaphase the condensed chromosomes become attached to the spindle by their kinetochores. Each chromosome possesses two kinetochores, one attached to the centromere region of each sister chromatid (figure 10.8).

Microtubule attachment

As prometaphase continues, a second group of microtubules grows from the poles of the cell toward the centromeres. These microtubules are captured by the kinetochores on each pair of sister chromatids. This results in the kinetochores of each sister chromatid being connected to opposite poles of the spindle. This allows the proper separation, or **disjunction,** of the sister chromatids.

This bipolar attachment is critical to the process of mitosis; any mistakes in microtubule positioning can be disastrous. For example, the attachment of the kinetochores of both sister chromatids to the same pole leads to *nondisjunction* (a failure of the sister chromatids to separate): the two sisters will be pulled to the same pole and end up in the same daughter cell, with the other daughter cell missing that chromosome.

Movement of chromosomes to the cell center

Each chromosome is attached to the spindle by microtubules running from opposite poles to the kinetochores of sister chromatids. The chromosomes are being pulled simultaneously toward each pole, leading to a jerky motion that eventually pulls all of the chromosomes to the equator of the cell. At this point, the chromosomes are arranged at the equator with sister chromatids under tension and oriented toward opposite poles by their kinetochore microtubules.

The force that moves chromosomes has been of great interest since the process of mitosis was first observed. Two basic mechanisms have been proposed to explain this: (1) Assembly and disassembly of microtubules provides the force to move chromosomes, and (2) motor proteins located at the kinetochore and poles of the cell pull on microtubules to provide force. Data have been obtained that support both mechanisms.

In support of the microtubule-shortening proposal, isolated chromosomes can be pulled by microtubule disassembly. The spindle is a very dynamic structure, with microtubules being added to at the kinetochore and shortened at the poles, even during metaphase. In support of the motor protein proposal, multiple motor proteins have been identified as kinetochore proteins, and inhibition of the motor protein dynein slows chromosome separation at anaphase. Like many phenomena that we analyze in living systems, the answer is not a simple either/or choice. Both mechanisms are probably at work.

In Metaphase, Chromosomes Align at the Cell Equator

LEARNING OBJECTIVE 10.5.3 Describe how the chromatids align at the cell equator in metaphase.

The alignment of the chromosomes in the center of the cell signals the beginning of the third stage of mitosis, **metaphase.** When viewed with a light microscope, the chromosomes appear to array themselves in a circle along the inner circumference of the cell, just as the equator girdles the Earth (figure 10.11). An imaginary plane perpendicular to the axis of the spindle that passes through this circle is called the *metaphase plate.* The metaphase plate is not an actual structure but rather an indication of the future axis of cell division.

Positioned by the microtubules attached to the kinetochores of their centromeres, all of the chromosomes line up on the metaphase plate. At this point, their centromeres are neatly arrayed in a circle, equidistant from the two poles of the cell, with

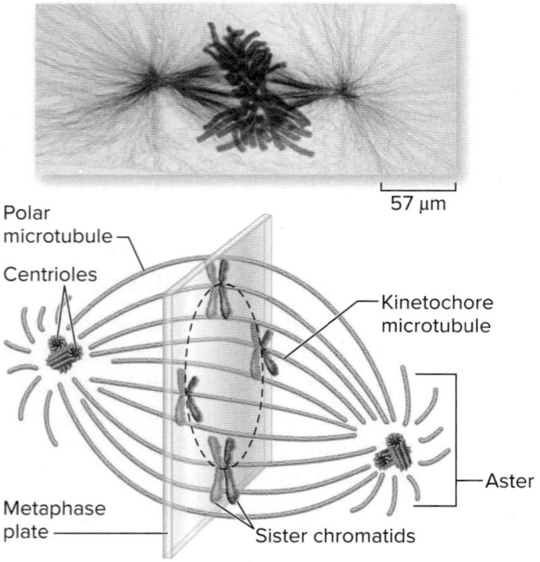

57 μm
Polar microtubule
Centrioles
Kinetochore microtubule
Aster
Metaphase plate
Sister chromatids

Figure 10.11 Metaphase. In metaphase, the chromosomes are arrayed at the midpoint of the cell. The imaginary plane through the equator of the cell is called the metaphase plate. As the spindle itself is a three-dimensional structure, the chromosomes are arrayed in a rough circle on the metaphase plate.

Waheeb Heneen, the Swedish University of Agricultural Sciences

microtubules extending back toward the opposite poles of the cell. The cell is prepared to properly separate sister chromatids, such that each daughter cell will receive a complete set of chromosomes. Thus, metaphase is really a transitional phase in which all the preparations are checked before the action continues.

At Anaphase the Chromatids Separate

LEARNING OBJECTIVE 10.5.4 Describe how, when the chromatids separate during anaphase, chromatid cohesion prevents premature separation.

Of all the stages of mitosis shown in figure 10.10, **anaphase** is the shortest and the most amazing to watch. It begins when the proteins holding the sister chromatids together at the centromere are removed. Up to this point in mitosis, sister chromatids have been held together by cohesin proteins concentrated at the centromere. The key event in anaphase, then, is the simultaneous removal of these proteins from all of the chromosomes. The details of this process are discussed in section 10.6 when we consider control of the cell cycle.

Freed from each other, the sister chromatids are pulled rapidly toward the poles to which their kinetochores are attached. In the process, two forms of movement take place simultaneously, each driven by microtubules. These movements are often called anaphase A and anaphase B to distinguish them.

First, during anaphase A, the *kinetochores are pulled toward the poles* as the microtubules that connect them to the poles shorten. This shortening process is not a contraction; the microtubules do not get any thicker. Instead, tubulin subunits are removed from the kinetochore ends of the microtubules. As more subunits are removed, the chromatid-bearing microtubules are progressively disassembled, and the chromatids are pulled ever closer to the poles of the cell.

Second, during anaphase B, the *poles move apart* as microtubular spindle fibers physically anchored to opposite poles slide past each other, away from the center of the cell. Because another group of microtubules attaches the chromosomes to the poles, the chromosomes move apart, too. If a flexible membrane surrounds the cell, it becomes visibly elongated.

When the sister chromatids separate in anaphase, the accurate partitioning of the replicated genome—the essential element of mitosis—is complete.

During Telophase the Nucleus Re-forms

LEARNING OBJECTIVE 10.5.5 Describe how the nucleus re-forms during telophase.

In **telophase,** the final phase of mitosis, the spindle apparatus disassembles as the microtubules are broken down into tubulin monomers that can be used to construct the cytoskeletons of the daughter cells. A nuclear envelope forms around each set of sister chromatids, which can now be called chromosomes because they are no longer attached at the centromere. The chromosomes soon begin to uncoil into the more extended form that permits gene expression. One of the early groups of genes expressed after the completion of mitosis is the rRNA genes, resulting in the reappearance of the nucleolus.

Telophase can be viewed as a reversal of the processes of prophase, bringing the cell back to the state of interphase. Mitosis is complete at the end of telophase. The eukaryotic cell has partitioned its replicated genome into two new nuclei positioned at opposite ends of the cell. Other cytoplasmic organelles, including mitochondria and chloroplasts (if present), were reassorted to areas that will later separate and become the cytoplasm of the daughter cells.

During Cytokinesis the Cytoplasm Is Divided

LEARNING OBJECTIVE 10.5.6 Compare cytokinesis in plant and animal cells.

Cell division is still not complete at the end of mitosis, because the division of the cell body proper has not yet begun. The final phase of the cell cycle, in which the cell actually divides, is called **cytokinesis.** It generally involves the cleavage of the cell body and cytoplasm into roughly equal halves.

In animal cells, a belt of actin pinches off the daughter cells

In animal cells and the cells of all other eukaryotes that lack cell walls, cytokinesis is achieved by means of a constricting belt of actin filaments. As these filaments slide past one another, the diameter of the belt decreases, pinching the cell and creating a **cleavage furrow** around the cell's circumference (figure 10.12*a*).

As constriction proceeds, the furrow deepens until it eventually slices all the way into the center of the cell. At this point, the cell is divided in two (figure 10.12*b*).

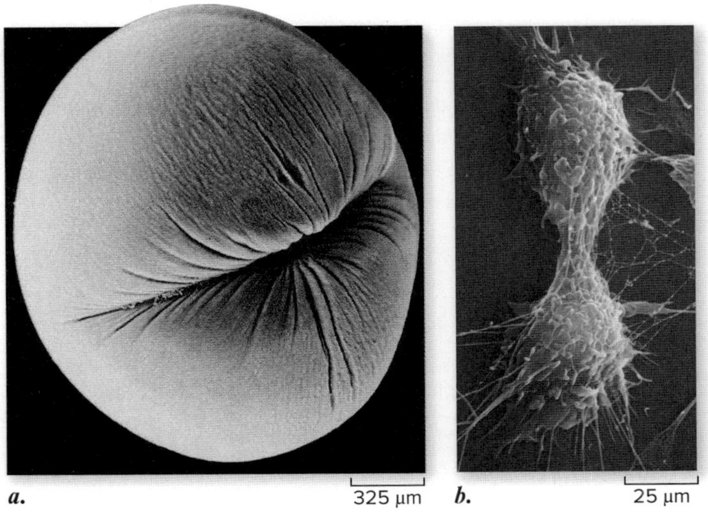

Figure 10.12 Cytokinesis in animal cells. ***a.*** A cleavage furrow forms around a dividing frog egg. ***b.*** The completion of cytokinesis in an animal cell. The two daughter cells are still joined by a thin band of cytoplasm occupied largely by microtubules.

(a): Don W. Fawcett/Science Source (b): Guenter Albrecht-Buehler, Northwestern University, Chicago

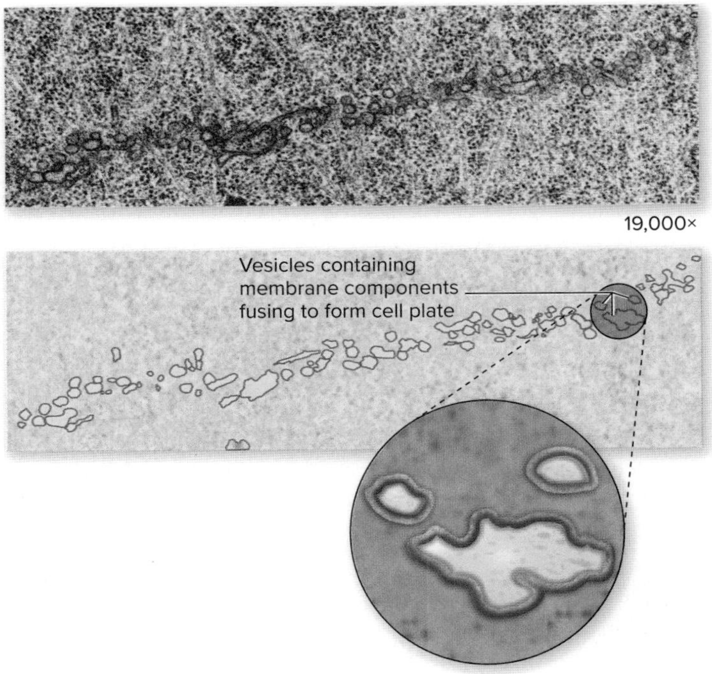

Figure 10.13 Cytokinesis in plant cells. In this photomicrograph and companion drawing, a cell plate is forming between daughter nuclei. The cell plate forms from the fusion of Golgi-derived vesicles. Once the plate is complete, there will be two cells.

Biophoto Associates/Science Source

In plant cells, a cell plate divides the daughter cells

Plant cell walls are far too rigid to be squeezed in two by actin filaments. Instead, these cells assemble membrane components in their interior, at right angles to the spindle apparatus. This expanding membrane partition, called a **cell plate,** continues to grow outward until it reaches the interior surface of the plasma membrane and fuses with it, effectively dividing the cell in two (figure 10.13). Cellulose is then laid down on the new membranes, creating two new cell walls. The space between the daughter cells becomes impregnated with pectins and is called a *middle lamella.*

In fungi and some protists, daughter nuclei are separated during cytokinesis

In most fungi and some groups of protists, the nuclear membrane does not dissolve, and as a result, all the events of mitosis occur entirely *within* the nucleus. Only after mitosis is complete in these organisms does the nucleus divide into two daughter nuclei; then, during cytokinesis, one nucleus goes to each daughter cell. This separate nuclear division phase of the cell cycle does not occur in plants, animals, or most protists.

After cytokinesis in any eukaryotic cell, the two daughter cells contain all the components of a complete cell. Whereas mitosis ensures that both daughter cells contain a full complement of chromosomes, no similar mechanism ensures that organelles such as mitochondria and chloroplasts are distributed equally between the daughter cells. But as long as at least one of each organelle is present in each cell, the organelles can later replicate to reach the number appropriate for that cell.

REVIEW OF CONCEPT 10.5

Mitosis is divided into phases: prophase, prometaphase, metaphase, anaphase, and telophase. The early phases involve restructuring the cell to create the microtubule spindle that pulls chromosomes to the equator of the cell in metaphase. Chromatids for each chromosome remain attached at the centromere by cohesin proteins. Chromatids are then pulled to opposite poles during anaphase when cohesin proteins are destroyed. The nucleus is re-formed in telophase.

Mitosis is followed by cytokinesis, which divides cell cytoplasm and organelles. In animal cells, actin pinches the cell in two; in plant cells, a cell plate forms in the middle of the dividing cell.

- *What would happen to a chromosome that lost cohesin protein between sister chromatids before metaphase?*

10.6 Events of the Cell Cycle Are Carefully Regulated

Controlling the process of cell division is critical both for individual cells and for multicellular organisms. In humans, loss of cell cycle control can lead to the disease cancer (refer to section 10.7). Biologists have used a variety of approaches to analyze this problem, uncovering factors that drive the cell cycle forward, and other factors that tend to rein in the division process. If we think about the process of cell division, there are two irreversible steps: DNA replication and the separation of the sister chromatids. Once DNA has been replicated, it cannot be unreplicated, and separated sister chromatids cannot be reattached. These two control points, and the decision to divide at all, are called checkpoints. At a checkpoint, the process can be assessed for accuracy and halted if there are errors, leading to a very high fidelity for the process. This also allows the cell cycle to respond to both internal and external signals, which are integrated at major checkpoints.

Cyclins and Cyclin-Dependent Kinases Control Key Steps of the Cell Cycle

LEARNING OBJECTIVE 10.6.1 Contrast the effects of cyclins and cyclin-dependent kinases on cell division.

Research uncovered cell-cycle control factors

The history of investigation into control of the cell cycle is instructive in two ways. First, it allows us to place modern observations into context; second, we can understand how biologists studying very different organisms often end up at the same place. The following brief history introduces three observations and then shows how they can be integrated into a single mechanism.

Observation 1. Discovery of MPF

Research on the activation of frog oocytes led to the discovery of a substance that was first called *maturation-promoting factor (MPF)*. Frog oocytes, which go on to become egg cells, become arrested near the end of their development at the G_2 stage before meiosis begins. They remain in this arrested state awaiting hormonal signaling to complete cell division.

Investigators found that cytoplasm taken from a variety of actively dividing cells could prematurely induce cell division when injected into oocytes (figure 10.14). These experiments indicated the presence of the positive regulator of cell-cycle progression in the cytoplasm of dividing cells, which the researchers called MPF. Cell fusion experiments done with mitotic and interphase cells also indicated the presence of a cytoplasmic positive regulator that could induce mitosis.

Further studies highlighted two key aspects of MPF. First, MPF activity varied during the cell cycle: low in early G_2, rising throughout this phase, and then peaking in mitosis. Second, the enzymatic activity of MPF involved the phosphorylation of

SCIENTIFIC THINKING

Hypothesis: *There are positive regulators of cell division.*

Prediction: *Frog oocytes are arrested in G_2 of meiosis I. They can be induced to mature (undergo meiosis) by progesterone treatment. If maturing oocytes contain a positive regulator of cell division, injection of cytoplasm should induce an immature oocyte to undergo meiosis.*

Test: *Oocytes are induced with progesterone, then cytoplasm from these maturing cells is injected into immature oocytes.*

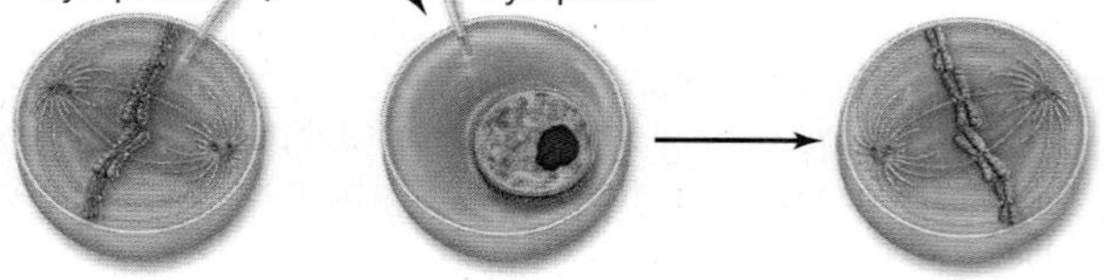
Remove cytoplasm
Inject cytoplasm
Progesterone-treated oocyte
Arrested oocyte
Oocyte in meiosis I

Result: *Injected oocytes progress from G_2 into meiosis I.*

Conclusion: *The progesterone treatment causes production of a positive regulator of maturation: Maturation Promoting Factor (MPF).*

Prediction: *If mitosis is driven by positive regulators, then cytoplasm from a mitotic cell should cause a G_1 cell to enter mitosis.*

Test: *M phase cells are fused with G_1 phase cells, then the nucleus from the G_1 phase cell is monitored microscopically.*

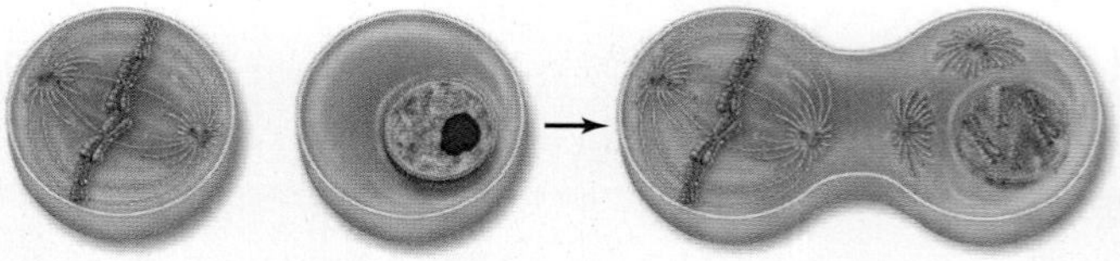
M phase cell
G1 phase cell
Fused cells

Conclusion: *Cytoplasm from M phase cells contains a positive regulator that causes a cell to enter mitosis.*

Further Experiments: *How can both of these experiments be rationalized? What would be the next step in characterizing these factors?*

Figure 10.14 Discovery of positive regulator of cell division.

proteins. This second point is not surprising, given the importance of phosphorylation as a reversible switch on the activity of proteins (refer to chapter 9). The first observations indicated that MPF itself was not always active but rather was being regulated with the cell cycle, while the second observation indicated the possible enzymatic mechanism of MPF activity.

Observation 2. Discovery of cyclins

Other researchers examined proteins produced during early cell divisions in sea urchin embryos and surf clams. They identified proteins that were produced in synchrony with the cell cycle and named them **cyclins.** Two forms of cyclin were found that cycled at slightly different times, reaching peaks in prophase of mitosis at the G_1/S and G_2/M boundaries. Despite much effort, no identified enzymatic activity was associated with these proteins. Their hallmark was the timing of their production rather than any intrinsic enzymatic activity.

Observation 3. Genetic control of the cell cycle

Geneticists using yeasts as model systems set out to determine the genes governing control of the cell cycle. By isolating mutants that were halted during division, they identified genes that were necessary for cell-cycle progression. These studies indicated that in yeast there were two critical control points: the commitment to DNA synthesis (called *START*) and the commitment to mitosis. One particular gene, named *cdc2,* was shown to be critical for passing both of these boundaries.

MPF is cyclin plus *cdc2*

All of these findings came together in an elegant fashion with the following three observations. First, the protein encoded by the *cdc2* gene was shown to be a protein kinase. Second, the purification and identification of MPF showed that it was composed of both a cyclin component and a kinase component (figure 10.15). Third, kinase was shown to be the Cdc2 protein!

The Cdc2 protein was the first identified **cyclin-dependent kinase (Cdk),** that is, a protein kinase enzyme that is active only when complexed with cyclin. This finding led to the renaming of MPF as *mitosis*-promoting factor, as its role was clearly more general than simply promoting the maturation of frog oocytes.

These Cdk enzymes are the key positive regulators of the cell cycle, often called the engines that drive cell division. The control of the cell cycle in higher eukaryotes is much more complex than the simple single-cycle engine of yeast, but the yeast model remains a useful framework for understanding more complex regulation.

The Cell Cycle Can Be Halted at Three Checkpoints

LEARNING OBJECTIVE 10.6.2 Distinguish the roles of the three key checkpoints in the eukaryotic cell cycle.

Although for clarity we have divided the eukaryotic cell cycle into arbitrary phases, the cell itself recognizes three phases, each marked by a **checkpoint** at which the cycle can be delayed or halted. The cell uses these three checkpoints both to assess its internal state and to integrate external signals (figure 10.16). The checkpoints are located at the G_1/S and G_2/M boundaries, as well as late in metaphase (the spindle checkpoint). Passage through these three checkpoints is controlled by the Cdk enzymes, as follows.

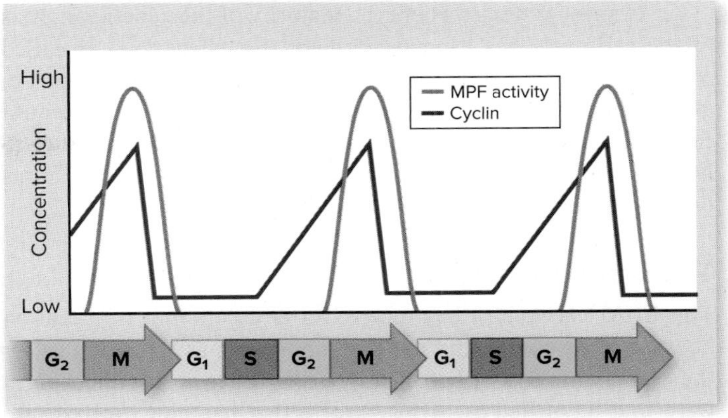

Figure 10.15 Correlation of MPF activity, amount of cyclin protein, and stages of the cell cycle. Cyclin concentration and MPF activity are plotted vs. stage of the cell cycle. Both change in a similar repeating pattern through the cell cycle. The reason for this correlation is that cyclin is actually one component of MPF, the other being a cyclin-dependent kinase (Cdk). Together, these act as a positive regulator of cell division.

The G_1/S checkpoint

The **G_1/S checkpoint** is the primary point at which the cell "decides" whether or not to divide. This checkpoint is therefore the primary point at which external signals can influence events of the cycle. It is the phase during which growth factors affect the cycle, as well as the phase that links cell division to cell growth and nutrition.

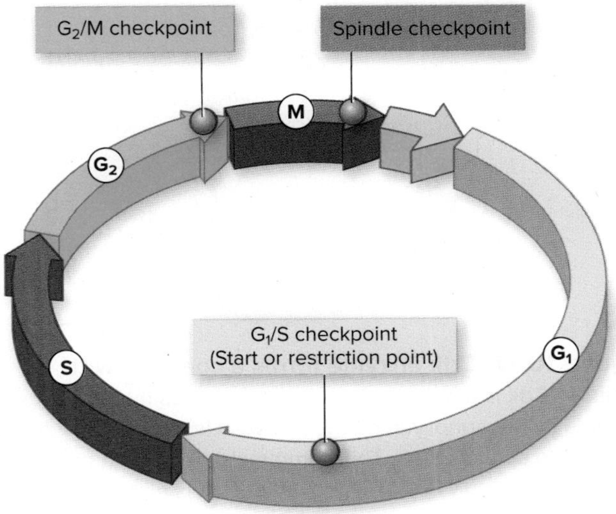

Figure 10.16 Control of the cell cycle. Cells use three key checkpoints to ensure that proper conditions have been achieved before initiating cell division.

In yeast systems, where the majority of the genetic analysis of the cell cycle has been performed, this checkpoint is called *START.* In animals, it is called the restriction point (*R* point). In all systems, once a cell has made this irreversible commitment to replicate its genome, it has committed itself to divide. Damage to DNA can halt the cycle at this point, as can starvation conditions or lack of growth factors.

The G_2/M checkpoint

The **G_2/M checkpoint** has received a large amount of attention because of its complexity and its importance as the stimulus for the events of mitosis. Historically, Cdks active at this checkpoint were first identified as MPFs, a term that has now evolved into **M-phase-promoting factor (MPF).**

Passage through this checkpoint commits the cell to mitosis and cytokinesis. This checkpoint assesses the success of DNA replication, and it can stall the cycle if DNA has not been accurately replicated. DNA-damaging agents result in arrest at this checkpoint as well as earlier, at the G_1/S checkpoint.

The spindle checkpoint

The **spindle checkpoint** ensures that all of the chromosomes are attached to the spindle in preparation for anaphase. The second irreversible step in the cell cycle is the separation of chromosomes during anaphase, and therefore it is critical that they are properly arrayed at the metaphase plate.

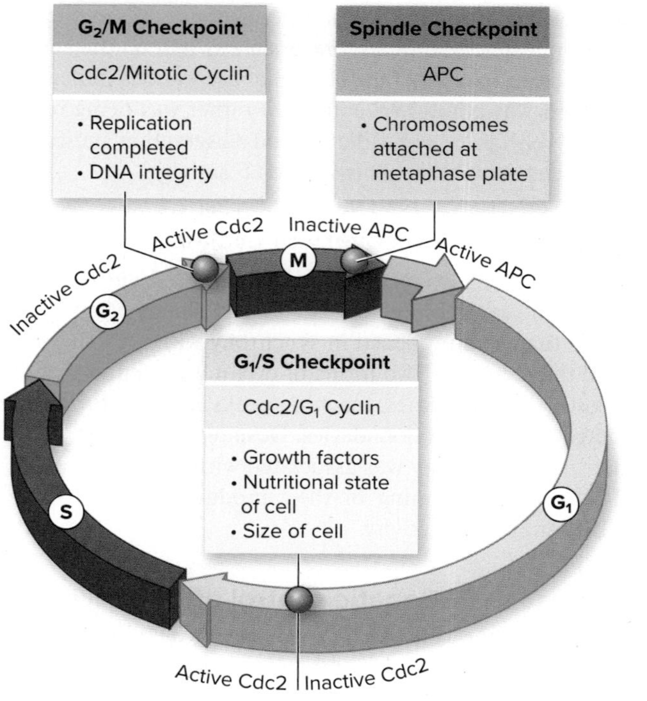

Figure 10.18 Checkpoints of the yeast cell cycle. The yeast cell cycle is the simplest that has been studied in detail. It is controlled at three main checkpoints and employs a single Cdk enzyme, called Cdc2. The Cdc2 enzyme partners with different cyclins to control the G_1/S and G_2/M checkpoints. The spindle checkpoint is controlled by the anaphase-promoting complex (APC).

Cyclin-Dependent Kinases Drive the Cell Cycle

LEARNING OBJECTIVE 10.6.3 Explain the role of cyclin-dependent kinases in mitosis.

The primary molecular mechanism of cell-cycle control is phosphorylation—the addition of a phosphate group to the amino acids serine, threonine, and tyrosine in proteins (refer to chapter 9). The enzymes that accomplish this phosphorylation are the Cdks (figure 10.17).

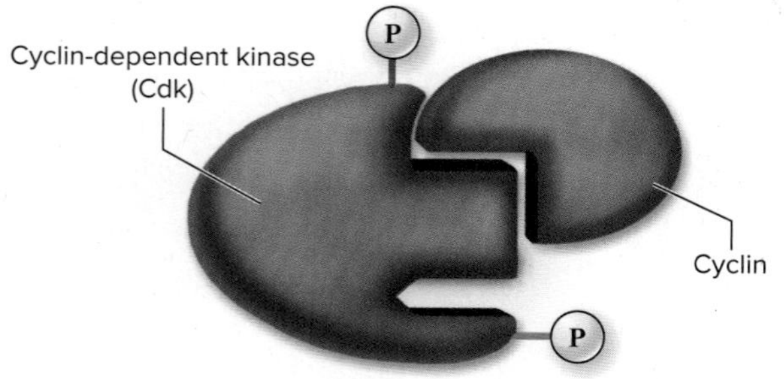

Figure 10.17 Cdk enzyme forms a complex with cyclin. Cdk is a protein kinase that activates numerous cell proteins by phosphorylating them. Cyclin is a regulatory protein required to activate Cdk. This complex is also called mitosis-promoting factor (MPF). The activity of Cdk is also controlled by the pattern of phosphorylation: phosphorylation at one site (represented by the red site) inactivates the Cdk, and phosphorylation at another site (represented by the green site) activates the Cdk.

Action of Cdks at the G_1/S checkpoint

The first important cell-cycle kinase was identified in fission yeast and named Cdc2 (now also called Cdk1). In yeast, this Cdk can partner with different cyclins at different points in the cell cycle (figure 10.18).

Even in the simplified cycle of the yeasts, we are left with the important question of what controls the activity of the Cdks during the cycle. For many years a common view was that cyclins drove the cell cycle—that is, that the periodic synthesis and destruction of cyclins acted as a clock. More recently, it has become clear that the Cdc2 kinase is also itself controlled by phosphorylation: phosphorylation at one site activates Cdc2, and phosphorylation at another site inactivates it (figure 10.17). Full activation of the Cdc2 kinase requires complexing with a cyclin and the appropriate pattern of phosphorylation.

As the G_1/S checkpoint is approached, the triggering signal in yeast appears to be the accumulation of G_1 cyclins. These form a complex with Cdc2 to create the active G_1/S Cdk, which phosphorylates a number of targets that bring about the increased enzyme activity for DNA replication.

Action of MPF at the G_2/M checkpoint

MPF and its role at the G_2/M checkpoint have been extensively analyzed in a number of different experimental systems. The control of MPF is sensitive to agents that disrupt or delay replication and to agents that damage DNA. It was once thought that MPF was controlled solely by the level of the M-phase-specific cyclins, but it has now become clear that this is not the case.

Although M-phase cyclin is necessary for MPF function, activity is controlled by inhibitory phosphorylation of the kinase component, Cdc2. The critical signal in this process is the removal of the inhibitory phosphates by a protein, phosphatase. This action forms a molecular switch based on positive feedback, because the active MPF further activates its own activating phosphatase.

The checkpoint assesses the balance of the kinase that adds inhibitory phosphates with the phosphatase that removes them. Damage to DNA acts through a complex pathway that includes damage sensing and a response to tip the balance toward the inhibitory phosphorylation of MPF. In section 10.7, we describe how some cancers overcome this inhibition.

Action of the anaphase-promoting complex at the spindle checkpoint

The molecular details of the sensing system at the spindle checkpoint are not clear. The presence of all chromosomes at the metaphase plate and the tension on the microtubules between opposite poles are both important. The signal is transmitted through the **anaphase-promoting complex,** also called the *cyclosome (APC/C).*

The function of the APC/C is to trigger anaphase itself. As described in section 10.5, the sister chromatids at metaphase are still held together by the protein complex cohesin. The APC does not act directly on cohesin but instead acts by marking for destruction a protein called *securin.* The securin protein acts as an inhibitor of another protease called *separase* that is specific for one component of the cohesin complex. Once inhibition is lifted, separase destroys cohesin.

This process has been analyzed in detail in budding yeast, where it has been shown that the separase enzyme specifically degrades a component of cohesin called Scc1. This leads to the release of the sister chromatids and results in their sudden movement toward opposite poles during anaphase.

In vertebrates, most cohesin is removed from the sister chromatids during chromosome condensation. At metaphase, the majority of the cohesin that remains on vertebrate chromatids is concentrated at the centromere (figure 10.9). The destruction of this cohesin explains the anaphase movement of chromosomes and the apparent "division" of the centromeres.

The APC/C has two main roles in mitosis: it activates the protease that removes the cohesins holding sister chromatids together, and it is necessary for the destruction of mitotic cyclins to drive the cell out of mitosis. The APC/C complex marks proteins for destruction by the proteasome, the organelle responsible for the controlled degradation of proteins (refer to chapter 4). The signal to degrade a protein is the addition of a molecule called *ubiquitin,* and the APC/C acts as a ubiquitin ligase. As we learn more about the APC/C and its functions, it is clear that the control of its activity is a key regulator of the cell cycle.

Control in multicellular eukaryotes

The major difference between more complex animals and single-celled eukaryotes such as fungi and protists is twofold: first, in animal cells, multiple Cdks control the cycle, as opposed to the single Cdk in yeasts; and second, animal cells respond to a greater variety of external signals than do yeasts, which respond primarily to signals necessary for mating.

In higher eukaryotes, there are more Cdk enzymes and more cyclins that can partner with these multiple Cdks, but their basic role is the same as in the yeast cycle. These more complex controls allow the integration of more input into control of the cycle. With the evolution of more complex forms of organization (tissues, organs, and organ systems), more complex forms of cell-cycle control evolved as well.

Multicellular organisms control cell proliferation so that only certain cells divide, and only at appropriate times. This ability to inhibit growth is seen in mammalian cells in tissue culture: cells will expand over a culture plate until they come in contact with other cells; then they stop dividing. If a sector of cells is cleared away, it will be refilled, but then cells stop dividing again.

How do cells sense the density of the cell culture around them? When cells come in contact, receptor proteins in the plasma membrane activate a signal transduction pathway that inhibits Cdk action. This prevents entry into the cell cycle.

Growth factors and the cell cycle

Growth factors act by triggering intracellular signaling systems. Fibroblasts, for example, possess numerous receptors on their plasma membranes for one of the first growth factors to be identified, **platelet-derived growth factor (PDGF).** The PDGF receptor is a receptor tyrosine kinase (RTK) that initiates a MAP kinase cascade to stimulate cell division (discussed in chapter 9).

PDGF was discovered when investigators found that fibroblasts would grow and divide in tissue culture only if the growth medium contained blood serum. Serum is the liquid that remains in blood after clotting; blood plasma, the liquid from which cells have been removed without clotting, would not work. The researchers hypothesized that platelets in the blood clots were releasing into the serum one or more factors required for fibroblast growth. Eventually, they isolated such a factor and named it PDGF.

Growth factors such as PDGF can override cellular controls that otherwise inhibit cell division. When a tissue is injured, a blood clot forms, and the release of PDGF triggers neighboring cells to divide, helping to heal the wound.

Over 300 different peptides that appear to have growth factor activity have been identified. These can be grouped into more than 30 families of related peptides. A specific cell-surface receptor recognizes each growth factor, its binding site fitting that growth factor precisely. These growth factor receptors often initiate MAP kinase cascades in which the final kinase enters the nucleus and activates transcription factors by phosphorylation (see figure 10.20). These transcription factors stimulate the production of G_1 cyclins and the proteins that are necessary for cell-cycle progression. Most animal cells need a combination of several different growth factors to overcome the various controls that inhibit cell division.

The G_0 phase

If cells are deprived of appropriate growth factors, they stop at the G_1 checkpoint of the cell cycle. With their growth and division arrested, they remain in this dormant G_0 phase.

The ability to enter G_0 accounts for the incredible diversity in cell cycle length in different tissues. Epithelial cells lining the human gut divide more than twice a day. By contrast, liver cells divide only once every year or two, spending most of their time in the G_0 phase. Mature neurons and muscle cells usually never leave G_0.

REVIEW OF CONCEPT 10.6

Cyclin proteins are produced in synchrony with the cell cycle. These proteins are complexed with cyclin-dependent kinases to drive the cell cycle. Three checkpoints exist in the cell cycle: the G_1/S checkpoint, the G_2/M checkpoint, and the spindle checkpoint. The cell cycle can be halted at these checkpoints if the process is not accurate. The anaphase-promoting complex/cyclosome (APC/C) triggers anaphase by lifting inhibition on a protease that removes cohesin holding chromatids together.

- *Would mutations in a Cdk be more likely to lead to no division or to uncontrolled division?*

10.7 Cancer Is a Failure of Cell-Cycle Control

The unrestrained, uncontrolled growth of cells in animals leads to the disease called **cancer.** Cancer is, in essence, a growth disorder of cells. It starts when an apparently normal cell begins to grow in an uncontrolled way. The result is a cluster of cells, called a **tumor,** that constantly expands in size. The clustered pink lung cells in the photo in figure 10.19 have begun to form a tumor. Malignant tumors are invasive, their cells able to break away from the tumor and spread to other areas of the body. Cells that leave a tumor and spread throughout the body, forming new tumors at distant sites, are called **metastases.**

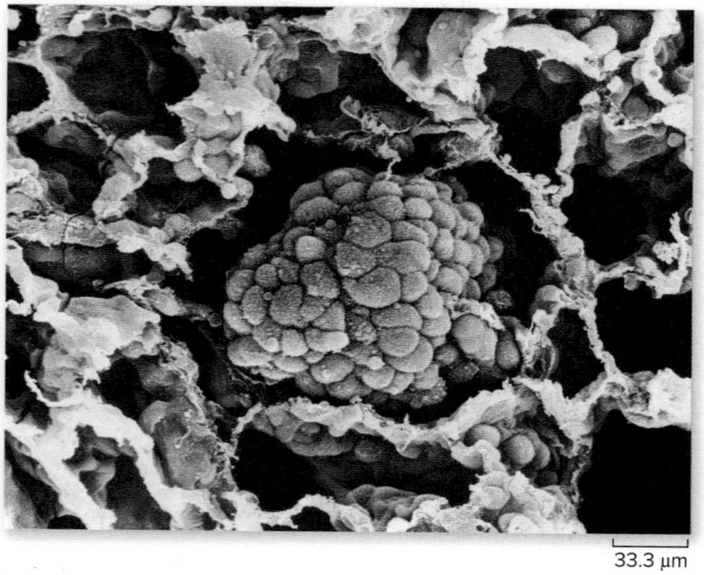

Figure 10.19 Lung cancer cells (300x). These cells are from a tumor located in an alveolus (air sac) of a human lung.
Moredun Scientific/Science Source

Mutations in Two Types of Genes Can Lead to Cancer

LEARNING OBJECTIVE 10.7.1 Distinguish between tumor-suppressor genes and oncogenes.

Put simply, cancer results from mutational damage to the genes encoding the proteins that regulate the cell cycle. This can be caused by anything that damages DNA: by chemicals like those in cigarette smoke, by environmental factors such as UV rays, or by random mutations that occur in somatic cells.

The search for cancer-causing genes produced two distinct classes of genes: **oncogenes** and **tumor-suppressor genes.** Oncogenes were discovered as genes carried by viruses that could cause cancer, the first being the *Src* gene, which is carried by the Rous sarcoma virus. When oncogenes are introduced into cells, they cause malignant transformation. The first tumor-suppressor gene was discovered as the cause of the familial cancer retinoblastoma. We will consider the similarities and differences between these two types of cancer genes.

Proto-oncogenes

The disease we call cancer is actually many different diseases, depending on the tissue affected. The common theme in all cases is the loss of control over the cell cycle. The discovery of oncogenes provided a simple genetic basis for this loss of control. The surprise came when it was discovered that most oncogenes have a cellular counterpart, so-called proto-oncogenes. The difference is that the alleles most people carry do not cause cancer, but mutation in these genes can lead to cancer. As only one allele needs to be mutated, these genes act in a dominant fashion genetically.

Signaling by growth factor receptors is a good example of how mutation in proto-oncogenes can lead to cancer. Some proto-oncogenes encode receptors for growth factors, and others encode proteins involved in signal transduction that act after the receptors. If a receptor for a growth factor becomes mutated such that it is permanently "on," the cell is no longer dependent on the presence of the growth factor for cell division. This is analogous to a light switch that is stuck on: the light will always be on. PDGF and EGF receptors both fall into the category of proto-oncogenes. Mutations in multiple points along these signaling pathways can lead to loss of control over the cell cycle. One simplistic way to think about this is an analogy to an automobile: oncogenes function as the accelerator, to push the cycle forward.

Tumor-suppressor genes

The second class of gene whose mutation can lead to cancer is the tumor-suppressor gene. In contrast to a proto-oncogene, both copies of a tumor-suppressor gene must lose function for the cancerous phenotype to develop; thus, these genes act in a recessive fashion genetically. In the automobile analogy, this makes tumor-suppressor genes the brakes.

The first tumor-suppressor gene identified was the **retinoblastoma susceptibility gene (*Rb*)** (figure 10.20), which predisposes individuals to a rare form of cancer that affects the retina

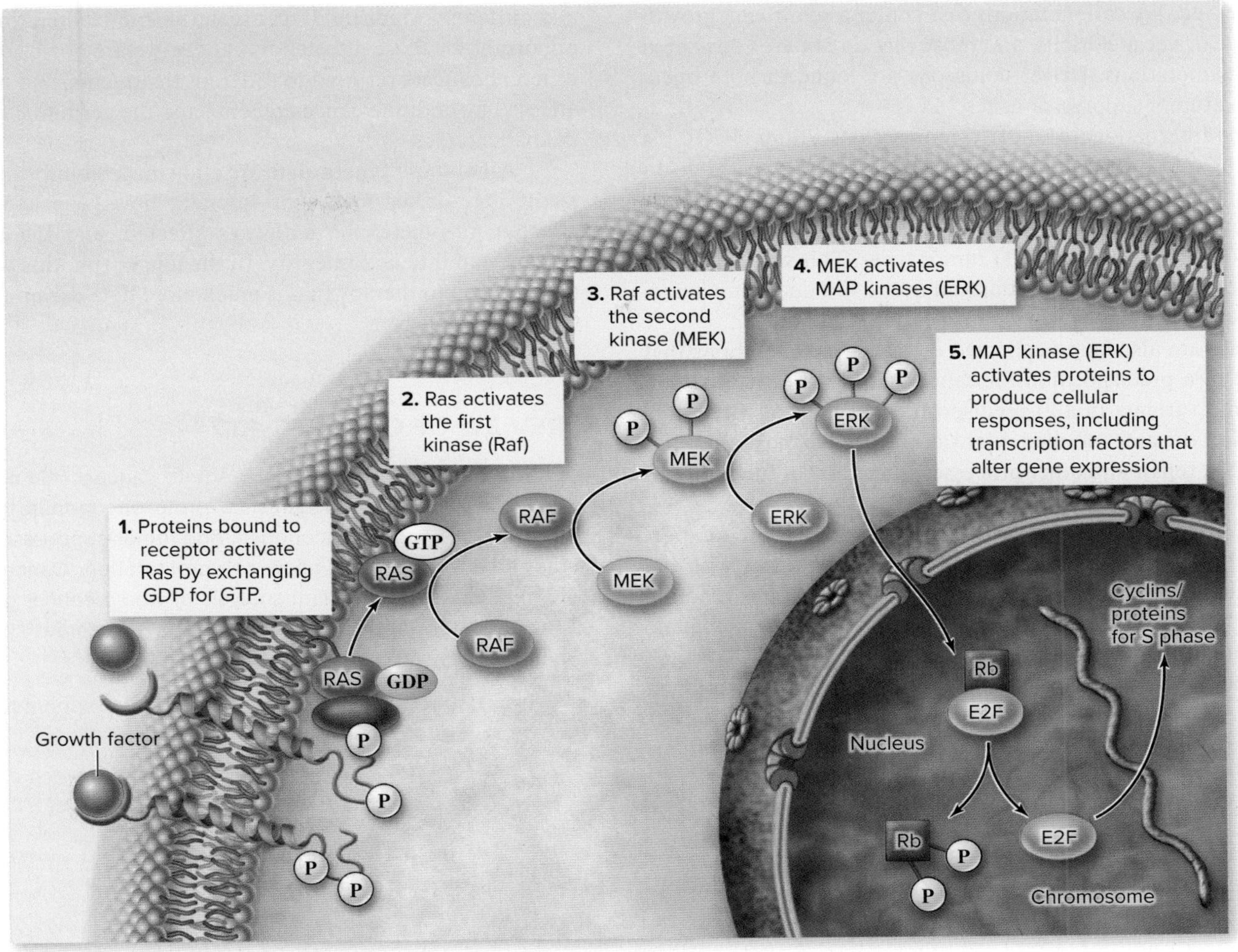

Figure 10.20 The cell-proliferation-signaling pathway. This pathway illustrates the roles of both proto-oncogenes and tumor-suppressor genes. Genes encoding proteins from the growth factor receptor to the MAP kinase cascade (described in chapter 9) can be mutated to cause inappropriate signaling (proto-oncogenes). The Rb protein binds to a transcription factor (E2F), preventing its signaling. When Rb is phosphorylated, E2F is released and growth-promoting proteins are synthesized. If Rb is inactivated, then these proteins can be produced in the absence of signals (tumor-suppressor gene).

of the eye. The role of the Rb protein in the cell cycle is to integrate signals from growth factors. The Rb protein is called a "pocket protein," because it has binding pockets for other proteins. Its role is therefore to bind important regulatory proteins and to prevent them from stimulating the production of cell-cycle-releasing proteins such as cyclins or Cdks.

The p53 gene

When we analyze mutations in human tumors, the gene with the largest number of mutations is *TP53*. This gene encodes the protein p53, which has been called the "guardian of the genome." The p53 protein is involved in the G_1/S checkpoint, where damaged DNA either causes p53 to halt the cell cycle, or if the damage is severe, to initiate the cell death program called apoptosis. This gives cells time to repair DNA, or if the damage is too great, removes them from the body.

Due to its obvious importance, *TP53* has been the subject of intensive research. This has revealed that p53 lies at the center of many intersecting signaling pathways and is important to integrate many aspects of physiology. The p53 protein responds to a variety of stressors, such as DNA damage, and affects a variety of transcriptional regulators. The action of p53 can affect the cell cycle and apoptosis, but also inflammation and even metabolism.

Whole-Genome Sequencing Is Shedding Light on Cancer Genetics

LEARNING OBJECTIVE 10.7.2 Explain how our view of cancer genetics has changed.

The classic model of cancer development is of a series of mutations that move a cell in a linear progression toward malignant transformation. This simplistic model is being refined by new data. Comparing DNA sequencing results from tumor cells and normal cells has allowed an unprecedented view of the genetic changes in cancer cells. This work has found an average of 30 to 60 genes mutated in common solid tumors, such as those from the colon, breast, brain, and pancreas. The majority of mutations are single-base substitutions, although more complex changes are also observed. We call mutations that confer a significant growth advantage driver mutations, and those that do not are considered passenger mutations. Driver mutations are found in both oncogenes and tumor-suppressor genes.

The International Cancer Genome Consortium (ICGC) is organizing and providing a data repository for this research. In the U.S. a project called The Cancer Genome Atlas (TCGA) is collecting sequence data from cancer cells derived from more than 10,000 cases, including 33 tumor types. This work has identified some 300 driver mutations and over 3400 oncogenic mutations, including many newly identified driver mutations.

Data are also being collected on germ-line alterations that can lead to a predisposition for cancer. This has identified 800 pathogenic variants in 100 predisposition genes, but these only affect 8% of the cases studied. Many of these predisposition genes affect DNA repair and other processes where loss of function can increase the frequency of mutations.

In addition to base substitutions, there is a broad spectrum of larger-scale genomic alterations, which varies with the type of tumor. It is not clear to what extent these larger genomic changes are causal, or whether they are the result of oncogenic transformation.

These data allow clinicians to classify tumors based on genetic lesions as well as tissue of origin and histology. It is clear that different cancers have different spectra of driver mutations and different kinds of larger-scale genomic alterations. More important for those affected by cancer, we are now learning how different cancers respond to different treatments, and how this is affected by both the cancer genome and the germ-line genome of the patient.

All of these genetic data are allowing a bioinformatic analysis of how driver mutations interact, how gene expression is affected, the metabolic pathways affected, and the effects on genome stability and integrity. In the longer run, this will enable an approach to therapy that is much more data driven than traditional approaches.

REVIEW OF CONCEPT 10.7

The loss of cell-cycle control leads to cancer. This occurs by two basic genetic mechanisms: proto-oncogenes that gain function to become oncogenes, and tumor-suppressor genes that lose function and allow cell proliferation. Cancer is due to multiple mutations, and analysis of whole cancer genomes has indicated that different cancers have different spectra of mutated genes.

- *How can you distinguish between a tumor-suppressor gene and a proto-oncogene?*

Do We Possess an Enzyme That Facilitates Cancerous Growth?

The DNA of all eukaryotic chromosomes is organized into chromatin by first wrapping DNA around a core of histone proteins to produce nucleosomes, then further coiling to produce the state of chromatin in interphase cells. Modifications to both DNA and histone proteins affect the expression of genes within a genomic region. One modification to histones that correlates with expression is acetylation at lysine residues. The converse of this is that loss of this acetylation is associated with gene silencing. Enzymes that can remove acetyl groups are called histone deacetylases. One class of histone deacetylases is the sirtuins, named for the prototype in yeast, *SIR2*, which is involved in silencing gene expression.

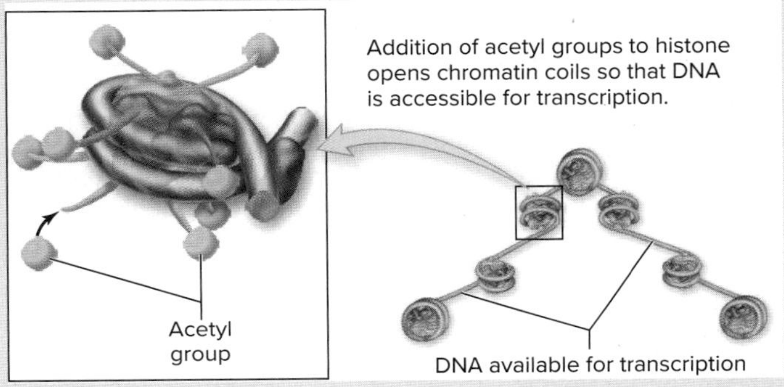
Addition of acetyl groups to histone opens chromatin coils so that DNA is accessible for transcription.
Acetyl group
DNA available for transcription

In mammals, some dozen sirtuins are known, and all are well characterized except one, SIRT-7. Focusing on this mysterious genomic expression regulator, researchers have begun to uncover its role in our lives.

Initial studies confirmed that SIRT-7 is found in cells bound to chromatin but revealed the interesting fact that SIRT-7 is highly selective, only removing an acetyl group from lysine 18 of histone 3, while having no activity at other histone acetylation sites.

What sorts of genes are covered by chromatin H3 units composed of acetylated lysine 18 histone? The researchers were able to detect 276 such SIRT-7 binding sites on human chromosomes. Many involved RNA processing, DNA metabolism, and transcription factors linked to tumor suppression, as well as genes found repressed in aggressive cancers. This result strongly suggests that SIRT-7 acts to release and maintain features essential for cancerous growth—that SIRT-7 is, in effect, a cancer-promoting gene.

To assess this possibility, the researchers monitored the growth of human cancer cells xenografted onto mice, comparing normal tumors (control) to knockdown tumors in which SIRT-7 activity had been depleted (S7KD). In each instance, cancerous growth was monitored over a period of 35 days, recording the volume of the tumors visible on the backs of the mice. The results obtained are seen in the graph.

Effect of SIRT-7 Protein on Tumor Growth

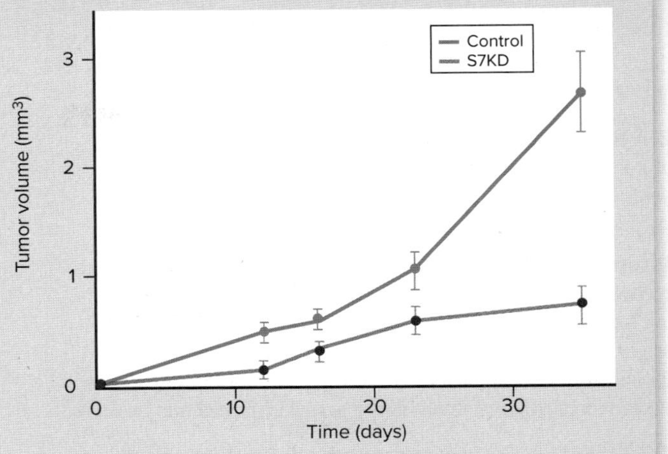
Control
S7KD
Tumor volume (mm³)
0
1
2
3
0
10
20
30
Time (days)

Analysis

1. **Applying Concepts**
   a. ***Variable.*** In the graph, what is the dependent variable?
   b. ***Error bars.*** What is the significance of the vertical lines passing through each point of the graph? What does it mean that the vertical line of a blue point does not overlap that of a red point measured at the same time interval? What would it mean if they did overlap? Explain.
2. **Interpreting Data**
   a. Does the control human tumor increase in size over the course of 35 days? Is the change continuous?
   b. Does the S7KD human tumor increase in size over the course of 35 days? Is the change continuous?
3. **Making Inferences** Comparing the growth of the S7KD tumors with the control tumors, would you say that they grew at the same rate over the 35 days of the experiment? If not, which tumor grew more rapidly? Is the rate of cancerous growth speeding up or slowing for either of the tumor types?
4. **Drawing Conclusions**
   a. Does depletion of SIRT-7 activity in the S7KD tumors decrease the rate of tumor growth?
   b. Is it reasonable to conclude from this result that SIRT-7 acts to unleash cancerous growth?
5. **Further Analysis** Given these data, if you were to look at a variety of tumors, what kinds of genetic changes in SIRT-7 activity would you expect to find? Is SIRT-7 likely to be up-regulated or down-regulated, and would you expect to find this in more or less aggressively growing tumors?

Retracing the Learning Path

CONCEPT 10.1 Bacterial Cell Division Is Clonal

10.1.1 Binary Fission Divides Bacterial Cells in Two Prokaryotic cell division is clonal, resulting in two identical cells. This involves duplication and segregation of the genome, and division of cellular contents.

10.1.2 Bacterial Cells Control Chromosome Partitioning and Septum Formation Bacterial DNA replication and chromosome partitioning is a concerted process. DNA replication is bidirectional with specific origin and termination sites. Chromosomes are segregated to opposite poles as they are replicated. A ring of FtsZ and other proteins accumulate at midcell, then contract inward to divide the cell into two cells.

CONCEPT 10.2 Eukaryotes Have Large, Linear Chromosomes

10.2.1 Chromosome Number Varies Among Species A complete genome consists of the haploid number of chromosomes. Many species are diploid with two copies, called homologs, of each chromosome.

10.2.2 Chromatin Is Composed of DNA and Protein Chromosomes are composed of chromatin, a complex of DNA and protein. The DNA of a single chromosome is a very long, double-stranded fiber, which must be compacted to fit into cells. DNA is wrapped around a core of eight histone proteins to form a nucleosome, which resemble beads on a string. This may be further coiled into a 30-nm fiber.

10.2.3 Chromatin Is Organized in the Nucleus Chromatin is organized in the nucleus into chromosome territories, which can be divided into compartments. Chromatin in compartments is organized into loops called topologically associated domains (TADs). These also correspond to expressed (euchromatin) and silent (heterochromatin) regions. Chromosome condensation changes over the cell cycle. Replicated chromosomes are attached at the centromere by cohesin proteins. After replication, a chromosome consists of two sister chromatids held together at the centromere.

CONCEPT 10.3 The Eukaryotic Cell Cycle Is Complex and Highly Organized

10.3.1 The Cell Cycle Is Divided into Five Phases The phases of the cell cycle are gap 1 (G_1), synthesis (S), gap 2 (G_2), mitosis, and cytokinesis. G_1, S, and G_2 are collectively called interphase, and mitosis and cytokinesis together are called M phase. The length of a cell cycle varies with age, cell type, and species. Cells can exit G_1 and enter a nondividing phase called G_0.

CONCEPT 10.4 During Interphase, Cells Grow and Prepare for Mitosis

10.4.1 Interphase Includes the Synthesis and Gap Phases of the Cell Cycle G_1, S, and G_2 are the three subphases of interphase. G_1 is the primary growth phase; during S phase, DNA synthesis occurs. G_2 phase occurs after S phase and before mitosis. The centromere binds proteins assembled into the kinetochore, where microtubules attach during mitosis.

CONCEPT 10.5 In Mitosis, Chromosomes Segregate

10.5.1 During Prophase, the Mitotic Apparatus Forms In prophase, chromosomes condense, the spindle is formed, and the nuclear envelope disintegrates. In animal cells, centriole pairs separate and migrate to opposite ends of the cell, establishing the axis of nuclear division.

10.5.2 In Prometaphase, Chromosomes Attach to the Spindle

10.5.3 In Metaphase, Chromosomes Align at the Cell Equator Chromatids of each chromosome are connected to opposite poles by kinetochore microtubules. They are held at the equator of the cell by the tension of being pulled toward opposite poles.

10.5.4 At Anaphase the Chromatids Separate At this point, cohesin proteins holding sister chromatids together at the centromeres are destroyed, and the chromatids are pulled to opposite poles. This movement is called anaphase A, and the movement of poles farther apart is called anaphase B.

10.5.5 During Telophase the Nucleus Re-forms Telophase reverses the events of prophase and prepares the cell for cytokinesis.

10.5.6 During Cytokinesis the Cytoplasm Is Divided In animal cells, a contractile belt of actin under the membrane pinches off the daughter cells. In plant cells, fusion of vesicles produces a new membrane in the middle of the cell to form the cell plate.

CONCEPT 10.6 Events of the Cell Cycle Are Carefully Regulated

10.6.1 Cyclins and Cyclin-Dependent Kinases Control Key Steps of the Cell Cycle Experiments showed that there are positive regulators of mitosis, and that there are proteins produced in synchrony with the cell cycle (cyclins). The positive regulators are cyclin-dependent kinases (Cdks). Cdks are complexes of a kinase and a regulatory molecule called cyclin. They phosphorylate proteins to drive the cell cycle.

10.6.2 The Cell Cycle Can Be Halted at Three Checkpoints Checkpoints are points at which the cell can assess the accuracy of the process and stop if need be. The G_1/S checkpoint is a commitment to divide; the G_2/M checkpoint ensures DNA integrity; and the spindle checkpoint ensures that all chromosomes are attached to spindle fibers, with bipolar orientation.

10.6.3 Cyclin-Dependent Kinases Drive the Cell Cycle Yeasts have one Cdk enzyme while vertebrates have many. The combination of G1 cyclin and Cdc2 kinase is the Cdk that triggers entry into S phase. The anaphase-promoting complex/cyclosome (APC/C) activates a protease that removes the choosiness holding together the centromeres of sister chromatids, allowing them to be separated. APC/C also degrades mitotic cyclins to exit mitosis. Growth factors stimulate cell division through a MAP kinase cascade, which results in the production of cyclins and activation of Cdks.

CONCEPT 10.7 Cancer Is a Failure of Cell-Cycle Control

10.7.1 Mutations in Two Types of Genes Can Lead to Cancer Mutations in proto-oncogenes have dominant, gain-of-function effects leading to cancer. Mutations in tumor-suppressor genes are recessive; loss of function of both copies leads to cancer.

10.7.2 Whole-Genome Sequencing Is Shedding Light on Cancer Genetics Driver mutations confer a growth advantage to cells. Over 300 driver mutations have been identified. Tumors can now be classified genetically, and different cancers respond to different treatments.

Concept Overview

This Concept Overview diagrams the key concepts that were discussed in this chapter.

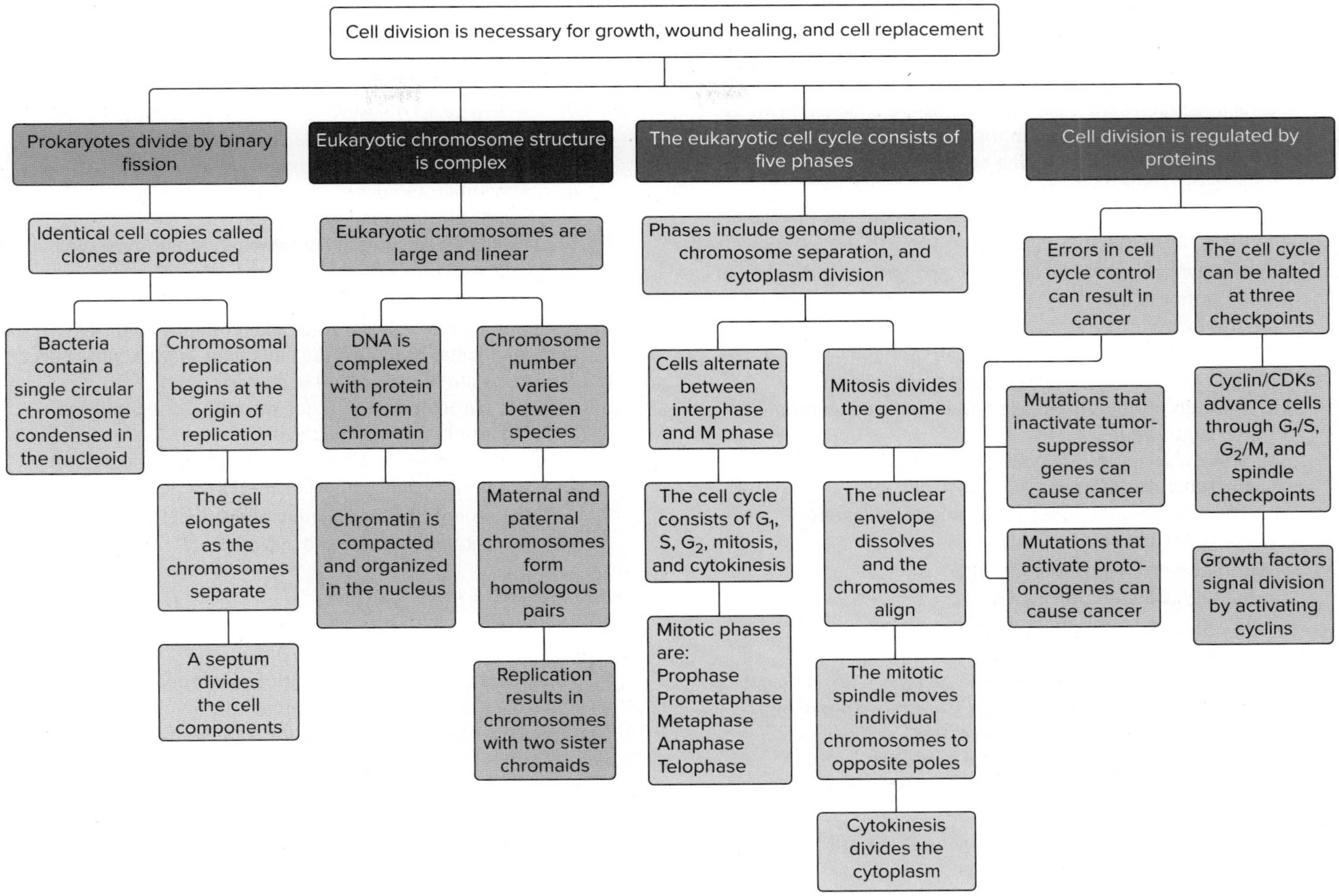

Assessing the Learning Path

Understand

1. Binary fission in prokaryotes does not require the
   a. replication of DNA.
   b. elongation of the cell.
   c. separation of daughter cells by septum formation.
   d. assembly of the nuclear envelope.
2. Chromatin is composed of
   a. RNA and protein.
   b. DNA and protein.
   c. sister chromatids.
   d. chromosomes.
3. What is a nucleosome?
   a. A region in the cell's nucleus that contains euchromatin
   b. A region of DNA wound around histone proteins
   c. A region of a chromosome made up of multiple loops of chromatin
   d. A 30-nm fiber found in chromatin
4. Just prior to cell division, homologous chromosomes
   a. are replicated.
   b. consist of two sister chromatids.
   c. become connected to each other.
   d. Both a and b
5. A cell receives a signal to divide. Which of the following is the correct order of steps it takes through the cell cycle?
   a. G_1, G_2, S, M
   b. G_1, S, G_2, interphase, M
   c. M, S, G_1, G_2
   d. G_1, S, G_2, M
6. How is the cell cycle (mitosis) similar to binary fission?
   a. Both are needed to produce genetically variable offspring.
   b. Both occur only in eukaryotes.
   c. Both involve DNA wrapped around histones.
   d. Both involve replication of DNA and separation of DNA equally into two daughter cells.

7. What is the role of cohesin proteins in cell division?
   a. They organize the DNA of the chromosomes into highly condensed structures.
   b. They hold the DNA of the sister chromatids together.
   c. They help the cell divide into two daughter cells.
   d. They connect microtubules and chromosomes.
8. A centromere
   a. is a region where sister chromatids are attached prior to their separation during mitosis.
   b. is a microtubule-organizing center.
   c. is the site of tubulin synthesis.
   d. is required for division in animal cells but not in cells of other eukaryotes.
9. In which stage of mitosis do duplicated chromosomes line up in the center of the cell?
   a. Prophase
   b. Telophase
   c. Anaphase
   d. Metaphase
10. Which of the following occurs during anaphase?
    a. Cohesin proteins are made and become concentrated at the centromere.
    b. Sister chromatids are pulled apart when kinetochore microtubules shorten.
    c. The cell elongates when polar microtubules shorten.
    d. All of the above
11. Maturation-promoting factor (MPF)
    a. activity is highest during G_1.
    b. is composed of cyclin and protein kinase.
    c. arrests division of frog oocytes.
    d. regulates the G_1/S checkpoint.
12. The anaphase-promoting complex triggers anaphase by
    a. causing polar microtubules to extend.
    b. signaling for the destruction of cohesion.
    c. increasing the concentration of mitotic cyclins in the cell.
    d. Both b and c
13. The normal function of the *TP53* gene in the cell is
    a. to act as a tumor-suppressor gene.
    b. to monitor the DNA for potential damage.
    c. to trigger the destruction of cells not capable of DNA repair.
    d. All of the above

Apply

1. A bacterial cell has a defect in the gene that encodes FtsZ protein. If the cell undergoes binary fission, what will be the result?
   a. Two daughter cells that are clones of each other
   b. One large cell with two chromosomes
   c. Two daughter cells, one that has two chromosomes and one that has no chromosome
   d. One large cell that looks genetically like the parent cell
2. The spotted skunk is $2n = 64$. Based on this information, you know that
   a. its haploid number is 64.
   b. each of its somatic cells has 64 chromosomes.
   c. it has 64 homologous chromosomes.
   d. its DNA has just been replicated.
3. Which phase of mitosis would be affected most by a mutation that inactivated condensin protein?
   a. Prophase
   b. Metaphase
   c. Anaphase
   d. Telophase
4. In some fungi, mitosis occurs without cytokinesis. What is the result?
   a. Two small cells with half the amount of DNA as the parent cell
   b. One large cell with the same amount of DNA as the parent cell
   c. One large cell with twice the DNA as the parent cell
   d. Two cells with the same amount of DNA as the parent cell
5. Drugs that affect microtubules have been used to create artificial polyploids in the laboratory. This is because disrupting microtubules
   a. prevents chromosome replication.
   b. allows DNA replication but prevents chromosome separation.
   c. would stop the cell cycle in G_1.
   d. would stop the cell cycle in G_2.
6. A mutant animal cell is formed that contains abnormal microfilaments. Assuming the cell could survive, what effect would this mutation have on cell division?
   a. Sister chromatids would not separate during mitosis.
   b. DNA would not be replicated.
   c. There would be no effect, as a cell plate could still form.
   d. Cytokinesis would not occur properly.
7. If DNA has not been completely or accurately replicated, at which checkpoint will the cell cycle arrest?
   a. G_1/S
   b. G_2/M
   c. Spindle
   d. Any of the above
8. You are trying to develop an anticancer drug. Which of the following might be a useful approach?
   a. A drug that shuts down tumor suppressors
   b. A drug that converts proto-oncogenes to oncogenes
   c. A drug that mimics a growth factor and binds to a mutant receptor
   d. A drug that stimulates metastasis

Synthesize

1. When bacterial cells divide, what mechanism ensures that each daughter cell receives one of the replicated genomes?
2. The plant *Haplopappus gracilis* has only two chromosomes, whereas the adder's tongue fern (*Ophioglossum* spp.) has 1262. There is much less variation in chromosome number among mammals. Can you suggest a reason for the wide variation in chromosome number among plants but not mammals?
3. If you could construct an artificial chromosome, what elements would you introduce into it, at a minimum, so that it could function normally in mitosis?
4. After cell division is complete, each daughter cell must have all of the organelles found in the parent cell. How are mitochondria and chloroplasts and the components of the endomembrane system divided up during cytokinesis?
5. Regulation of the cell cycle is very complex and involves multiple proteins. In yeast, the activity of the cyclin-dependent kinase Cdc2 is inhibited when it is phosphorylated by the Wee-1 kinase. What do you predict would be the phenotype of a *Wee-1* mutant yeast? What other genes could be altered in a Wee-1-deficient mutant strain that would make the cells act normally?
6. Despite all that we know about cancer today, some types of cancers are still increasing in frequency. Lung cancer among nonsmoking women is one of these. What reason(s) might there be for this increasing problem? Can you suggest a solution?

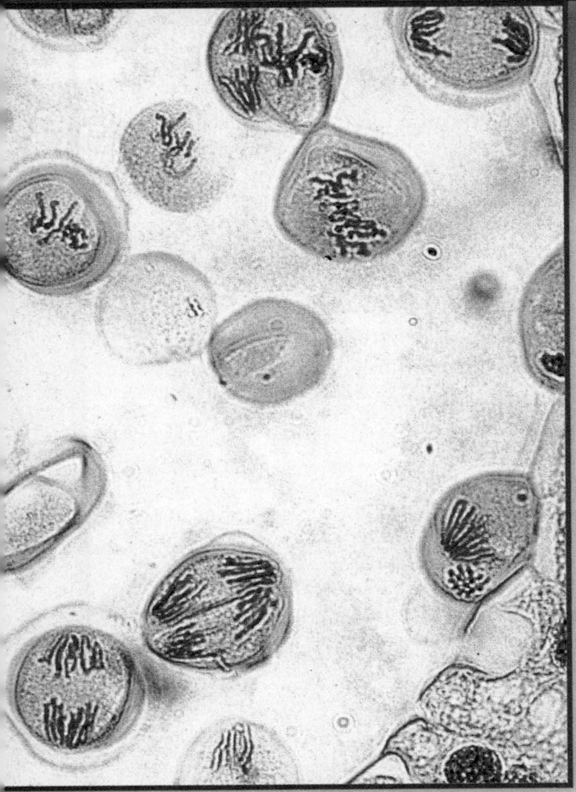

Steven P. Lynch

11 Sexual Reproduction and Meiosis

Learning Path

11.1 Sexual Reproduction Requires Meiosis

11.2 Meiosis Consists of Two Divisions with One Round of DNA Replication

11.3 The Process of Meiosis Involves Intimate Interactions Between Homologs

11.4 Meiosis Has Four Distinct Features

11.5 Genetic Variation Is the Evolutionary Consequence of Sex

Concept Overview

This Concept Overview provides a diagram of the overarching concepts that are covered in this chapter. These concepts will be expanded upon in the Concept Overview at the end of the chapter.

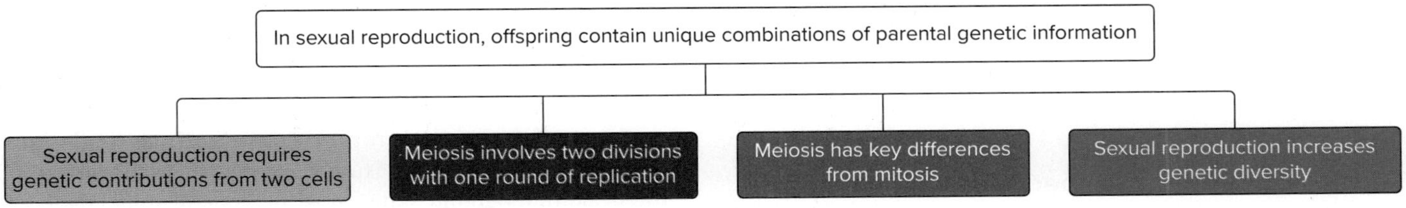
In sexual reproduction, offspring contain unique combinations of parental genetic information
Sexual reproduction requires genetic contributions from two cells
Meiosis involves two divisions with one round of replication
Meiosis has key differences from mitosis
Sexual reproduction increases genetic diversity

Introduction

Most animals and plants reproduce sexually. In humans, gametes of opposite sex unite to form a cell that, dividing repeatedly by mitosis, eventually gives rise to an adult body with some 100 trillion cells. The gametes that form the initial cell are the products of a special form of cell division called meiosis, visible in the photo on the previous page and the subject of this chapter. Meiosis is far more intricate than mitosis, and the details about it are not as well understood. The basic process, however, is clear. Also clear are the profound consequences of sexual reproduction: It plays a key role in generating the tremendous genetic diversity that is the raw material of evolution.

11.1 Sexual Reproduction Requires Meiosis

The essence of sexual reproduction is the merging of the genetic contribution of two cells from different individuals. This mode of reproduction results in evolutionary advantages that biologists have long recognized. However, we are only recently making progress on understanding the underlying mechanism that produces the elaborate behavior of chromosomes that occurs during meiosis, the process that underlies sexual reproduction. To begin, we briefly consider the history of meiosis and its relationship to sexual reproduction.

Gametes Contain Only One Copy of Each Chromosome

LEARNING OBJECTIVE 11.1.1 Compare the number of chromosomes in gametes and zygotes.

Fertilization joins two gametes

Only a few years after Walther Flemming's discovery of chromosomes in 1879, cytologist Edouard van Beneden was surprised to find different numbers of chromosomes in different types of cells in the roundworm *Ascaris.* Specifically, he observed that the **gametes** (eggs and sperm) each contained two chromosomes, but all of the nonreproductive cells, or **somatic cells,** of embryos and mature individuals each contained four.

From his observations, van Beneden proposed in 1883 that an egg and a sperm, each containing half the complement of chromosomes found in other cells, fuse to produce a single cell called a **zygote.** The zygote, like all of the cells ultimately derived from it, contains two copies of each chromosome. The fusion of gametes to form a new cell is called **fertilization,** or **syngamy.**

Meiosis reduces the number of chromosomes

It was clear even to early investigators that gamete formation must involve some mechanism that reduces the number of chromosomes to half the number found in other cells. If it did not, the chromosome number would double with each fertilization, and after only a few generations the number of chromosomes in each cell would become impossibly large. For example, in just 10 generations, the 46 chromosomes present in human cells would increase to over 47,000 (46×2^{10}).

The number of chromosomes does not explode in this way because of a special reduction division, **meiosis.** Meiosis occurs during gamete formation, producing **haploid** cells—cells with half the normal number of chromosomes (figure 11.1). The subsequent fusion of two of these cells to form a **diploid** cell—a cell with twice as many chromosomes as haploid cells—ensures a consistent chromosome number from one generation to the next. This reduction division process, the subject of this chapter, lies at the heart of sexual reproduction.

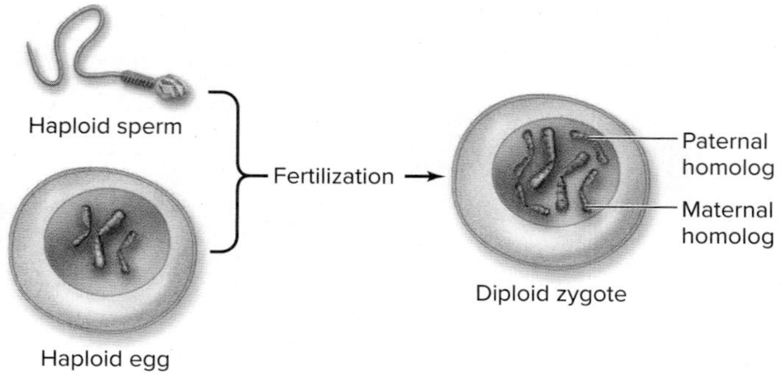

Figure 11.1 Diploid cells carry chromosomes from two parents. A diploid cell contains two versions of each chromosome, a maternal homolog contributed by the haploid egg of the mother, and a paternal homolog contributed by the haploid sperm of the father.

Sexual Life Cycles Alternate Between Haploid and Diploid

LEARNING OBJECTIVE 11.1.2 Differentiate between life cycles based on timing of meiosis and fertilization.

Meiosis and fertilization together constitute a cycle of reproduction. Figure 11.1 illustrates how two *haploid* cells, a sperm cell containing three chromosomes contributed by the father and an egg cell containing three chromosomes contributed by the mother, fuse to form a *diploid* zygote with six chromosomes.

Reproduction that involves this alternation of meiosis and fertilization is called **sexual reproduction.** Its outstanding characteristic is that offspring inherit chromosomes from *two* parents, as you saw in figure 11.1. You, for example, inherited 23 chromosomes from your mother (maternal homolog) and 23 from your father (paternal homolog).

The life cycles of all sexually reproducing organisms follow a pattern of alternation between diploid and haploid chromosome numbers, but there is variation in the pattern's timing. Many types of algae, for example, spend the majority of their life cycle in a haploid state. Most plants and some algae alternate between a multicellular haploid phase and a multicellular diploid phase (specific examples can be found in chapter 24). In most animals, by contrast, the diploid state dominates. The zygote first undergoes mitosis to produce diploid cells. Then, later in the life cycle, some of these diploid cells undergo meiosis to produce haploid gametes (figure 11.2).

Germ-line cells are set aside early in animal development

In animals, the single diploid zygote undergoes mitosis to give rise to somatic cells that form all of the cells in the adult body. The cells that will eventually undergo meiosis to produce gametes are set aside from somatic cells early in the course of development. These cells are referred to as **germ-line cells.** Both

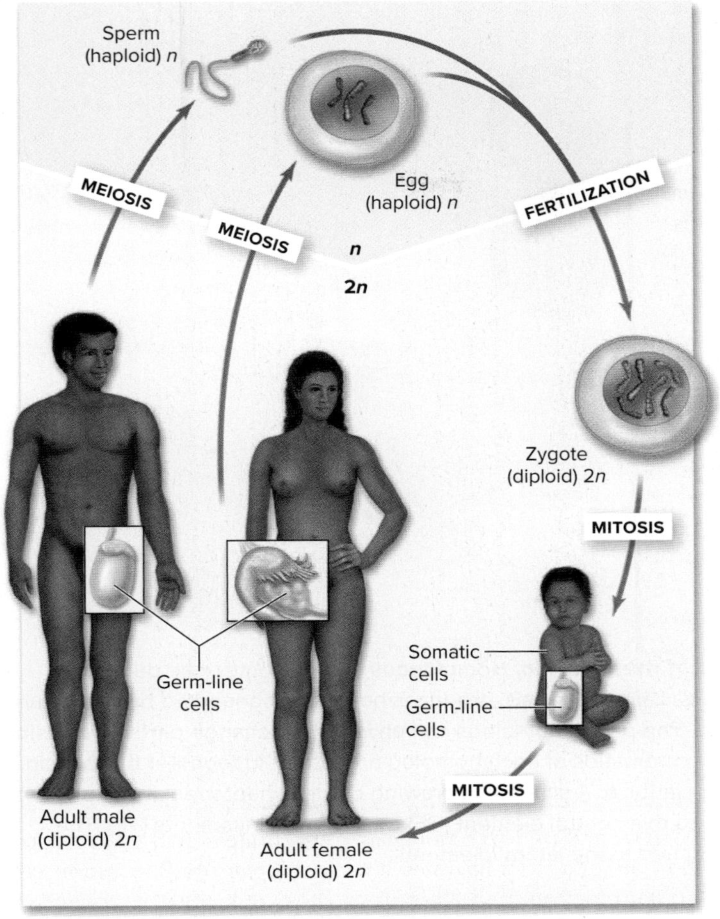

Figure 11.2 The sexual life cycle in animals. In animals, the zygote undergoes mitotic divisions and gives rise to all the cells of the adult body. Germ-line cells are set aside early in development and undergo meiosis to form the haploid gametes (eggs or sperm). The rest of the body cells are called somatic cells.

somatic cells and germ-line cells are diploid, but somatic cells undergo mitosis to form genetically identical, diploid daughter cells, and germ-line cells undergo meiosis to produce haploid gametes (figure 11.2).

Some organisms do not reproduce sexually and never produce gametes. Reproduction in these organisms is referred to as asexual reproduction. The cell division of yeasts we encountered in chapter 10 is an example of asexual reproduction, and some plants can reproduce asexually (refer to chapter 30).

REVIEW OF CONCEPT 11.1

Sexual reproduction involves the genetic contribution of two cells, each from a different individual. Meiosis produces haploid cells with half the number of chromosomes. Fertilization unites two haploid cells to restore the diploid state of the next generation. Only germ-line cells are capable of meiosis. All other cells in the body undergo only mitotic division.

- *Germ-line cells undergo meiosis, but how can the body maintain a constant supply of these cells?*

11.2 Meiosis Consists of Two Divisions with One Round of DNA Replication

Although the evolution of meiosis is not understood in detail, it clearly uses the same machinery as mitosis. In analyzing the process of meiosis, we will always look to comparisons with mitosis.

Homologous Pairing in Meiosis Leads to a Reductive Division

LEARNING OBJECTIVE 11.2.1 Describe the process of homologous pairing.

Although details vary among organisms, meiosis in all eukaryotes differs from mitosis in two principal respects.

1. Pairing of homologous chromosomes

Meiosis in a diploid organism consists of two rounds of division, called **meiosis I** and **meiosis II,** with each round consisting of prophase, metaphase, anaphase, and telophase stages. During early prophase I of meiosis, homologous chromosomes find each other and become closely associated, a process called pairing, or **synapsis** (figure 11.3*a*).

The synaptonemal complex

Homologous chromosomes become intimately associated during prophase I in an elaborate structure called the **synaptonemal complex,** joining the paired homologs. The synaptonemal complex consists of a central element connected by filaments to two sets of lateral elements that interact with the homologs. The structure of the synaptonemal complex appears similar in all systems that have been examined, although its exact function is unclear. A representation of the current model for this structure is shown in figure 11.3*b*. The result is that all four chromatids of the two homologs are closely associated during this phase of meiosis. This structure is also sometimes called a *tetrad* or *bivalent.*

Crossing over

Although homologs are paired in the synaptonemal complex, another process unique to meiosis occurs: **crossing over.** This process literally allows the homologs to exchange chromosomal arms, producing genetic recombination—alleles that were formerly on separate homologs are now found on the same homolog. The sites of crossing over are called **chiasmata** (singular, *chiasma*), and these sites of contact are maintained until anaphase I. The physical connection of homologs due to crossing over, as well as the continued joining of the sister chromatids by cohesin proteins, locks homologs together.

2. Reduction division

The most obvious distinction between meiosis and mitosis is that meiosis involves two successive divisions with no replication of

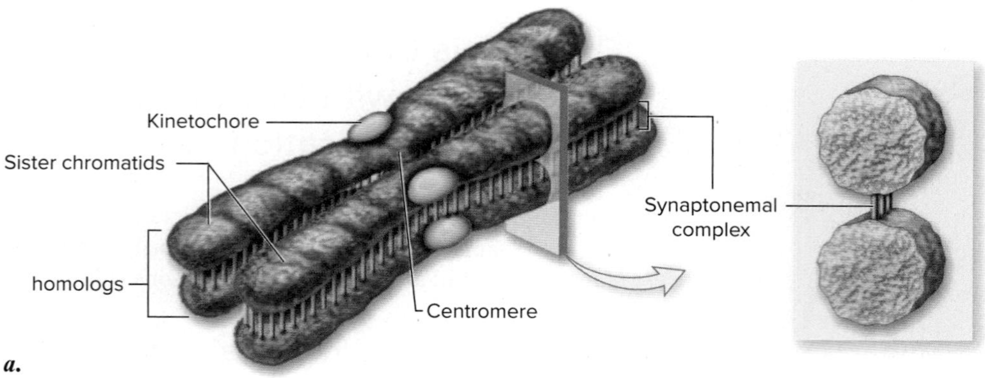

a.

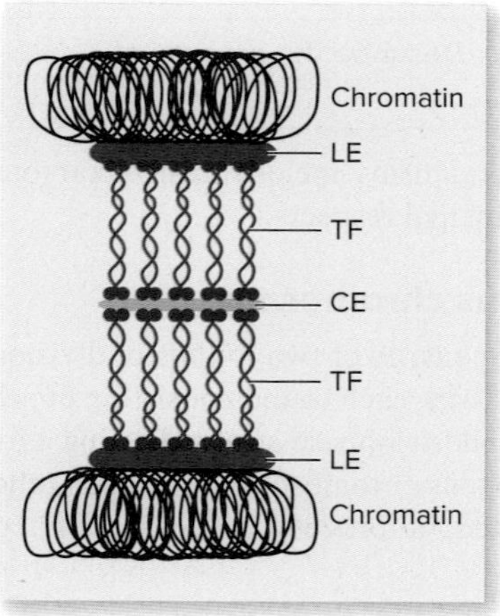

b.

Figure 11.3 Unique features of meiosis. ***a.*** Homologous chromosomes pair during prophase I of meiosis. This process, called synapsis, produces homologs connected by a structure called the synaptonemal complex. The paired homologs can physically exchange parts by crossing over, or recombination. The sister chromatids of each homolog are also held together by cohesin proteins, which are not shown for clarity. ***b.*** A schematic drawing of the synaptonemal complex. Two lateral elements (LE) are joined to a central element (CE) by transverse filaments (TF). The chromatin of each homolog is attached to the lateral elements.

genetic material between the two: DNA replication is suppressed between the two meiotic divisions. Homologous pairing leads to the first division being a reductive division: At the end of meiosis I, the number of chromosomes has been halved. The subsequent division without replication converts these cells into four cells, each with a complete haploid genome.

REVIEW OF CONCEPT 11.2

During prophase I, homologous chromosomes pair along their entire length. homologs may exchange chromosomal material at chiasmata. In anaphase I, homologs separate, reducing the number of chromosomes to the haploid number. This is followed without replication by a second division, in which sister chromatids separate to yield four haploid cells.

- *If sister chromatids separated at the first division, would meiosis still work?*

11.3 The Process of Meiosis Involves Intimate Interactions Between Homologs

The process of meiosis is complex, but we can understand it by following the behavior of chromosomes. The pairing of homologs that occurs during prophase I sets the stage for key later events.

Prophase I Sets the Stage for the Reductive Division

LEARNING OBJECTIVE 11.3.1 Describe the consequences of how homologous chromosomes pair in prophase I.

To understand the complex events of meiosis, we will track the behavior of chromosomes during each of its two divisions. The first meiotic division depends on each homologous pair behaving as a unit and not as individual chromosomes as they do in mitosis. This is accomplished by a complex set of processes that together join homologs until anaphase I, when they are separated.

During prophase I, homologous chromosomes pair

Meiotic cells have an interphase period that is similar to mitosis, with G_1, S, and G_2 phases. After interphase, germ-line cells enter meiosis I (figure 11.4). In prophase I, the DNA coils tighter, and individual chromosomes first become visible under the light microscope as a matrix of fine threads. Because DNA has already replicated before the onset of meiosis, each of these threads actually consists of two sister chromatids joined at their centromeres. In prophase I, homologous chromosomes become closely associated, exchange segments by crossing over, and later separate.

Synapsis

During interphase in germ-line cells, the ends of the chromatids appear to become attached to the nuclear envelope at specific sites. The sites to which homologs attach are adjacent, so that during prophase I the members of each homologous pair of chromosomes are positioned closely together. Homologous pairs then align closely side by side in the process called synapsis.

Synapsed homologs are joined along their entire length. The sister chromatids of each homolog are also joined together by cohesin proteins. This association of sister chromatids is called *sister chromatid cohesion.* Sister chromatid cohesion also occurs during mitosis, but in meiosis the cohesin complex contains a meiosis-specific cohesin that is important for later events.

The mechanism of crossing over

Crossing over involves a series of events in which DNA segments are broken and then rejoined at the same site, exchanging material between the homologs. The molecular events that break and rejoin homologs are controlled such that each chromosome arm usually has one or a few crossovers per meiosis. Human chromosomes typically have two or three.

After crossing over occurs, the synaptonemal complex breaks down. Homologous chromosomes are less tightly associated, but remain attached by chiasmata. At this point, each chromosome is organized into a bivalent with four chromatids: two homologs each with two sister chromatids.

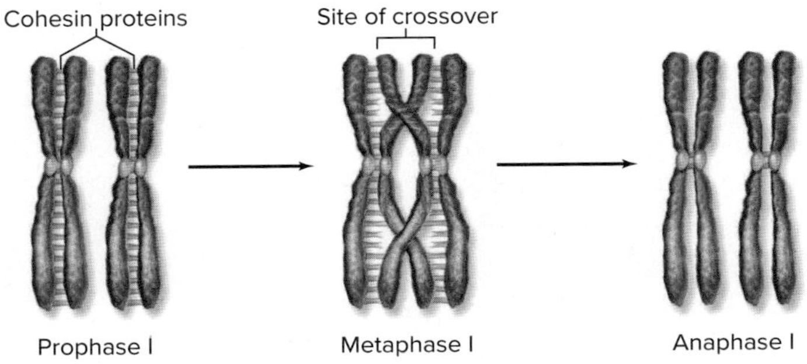

The four chromatids are held together in two ways: (1) The two sister chromatids of each homolog the products of DNA replication, are held together by cohesin proteins (sister chromatid cohesion); and (2) exchange of material by crossing over between homolog locks all four chromatids together.

This elaborate behavior of chromosome pairing is taking place during prophase I, but other key events also occur. The nuclear envelope is dispersed, along with the interphase structure of microtubules. These microtubules then re-form into a spindle, just as in mitosis.

During Metaphase I, Paired Homologs Align

LEARNING OBJECTIVE 11.3.2 Explain the importance of monopolar attachment of homolog pairs at metaphase I.

By the end of prophase I, the paired homologs are locked together as a bivalent. These bivalents capture spindle fibers and move to the center of the cell and align on the metaphase plate as paired homolog.

This alignment of homologs occurs because the kinetochores of sister chromatids for each homolog act as a unit to capture polar microtubules. Therefore microtubules from opposite poles are attached to homologs and not to sister chromatids, as occurs in mitosis. We call this monopolar attachment, as each homolog is attached to one pole instead of having its sister chromatids attached to opposite poles in the bipolar fashion of mitosis.

The molecular basis of monopolar attachment is not understood. One model with some support is that the centromere-kinetochore complex of sister chromatids is compacted during meiosis I, allowing them to function as a single unit.

This monopolar attachment of kinetochores of sister chromatids would be disastrous in mitosis, but it is critical to meiosis I. It produces tension on the paired homologs pulling them to the equator of the cell. In this way, each joined pair of homologs lines up on the metaphase plate (figure 11.4). The orientation of each pair on the spindle axis is random; either the maternal or the paternal homolog may be oriented toward a given pole.

During Anaphase I, Homologs Are Pulled to Opposite Poles

LEARNING OBJECTIVE 11.3.3 Compare the loss of cohesion between sister chromatids at the centromere and on the arms at anaphase I.

In anaphase I, the microtubules of the spindle fibers begin to shorten. As they shorten, they break the chiasmata and pull the centromeres toward the poles, dragging the chromosomes along with them.

Anaphase I comes about by the release of sister chromatid cohesion along the chromosome arms, but not at the centromeres. This release is thought to be the result of the destruction of meiosis-specific cohesin in a process analogous to anaphase in mitosis. The difference is that the destruction is inhibited at the centromeres by a mechanism that we have only recently begun to understand.

As a result of this release, the homologs are pulled apart, but the sister chromatids are not. Each homolog moves to one pole, taking both sister chromatids with it. When the spindle fibers have fully contracted, each pole has a complete haploid set of chromosomes consisting of one member of each homologous pair.

Because of the random orientation of homologous chromosomes on the metaphase plate, a pole may receive either the maternal or the paternal homolog from any chromosome pair. As a result, meiosis I results in the **independent assortment** of maternal and paternal chromosomes into the gametes.

Telophase I Completes Meiosis I

LEARNING OBJECTIVE 11.3.4 Identify the key event that occurs during telophase I.

By the beginning of telophase I, the chromosomes have segregated into two clusters, one at each pole of the cell. Now the nuclear membrane re-forms around each daughter nucleus.

Because each chromosome within a daughter nucleus had replicated before meiosis I began, each now contains two sister chromatids attached by a common centromere. Note that *the sister chromatids are no longer identical* because of the crossing over that occurred in prophase I; as you will see, this change has important implications for genetic variability.

As telophase I progresses, the cell is also divided into two cells by cytokinesis. This is similar to mitosis with animal cells using a cleavage furrow, and plant cells using a cell plate.

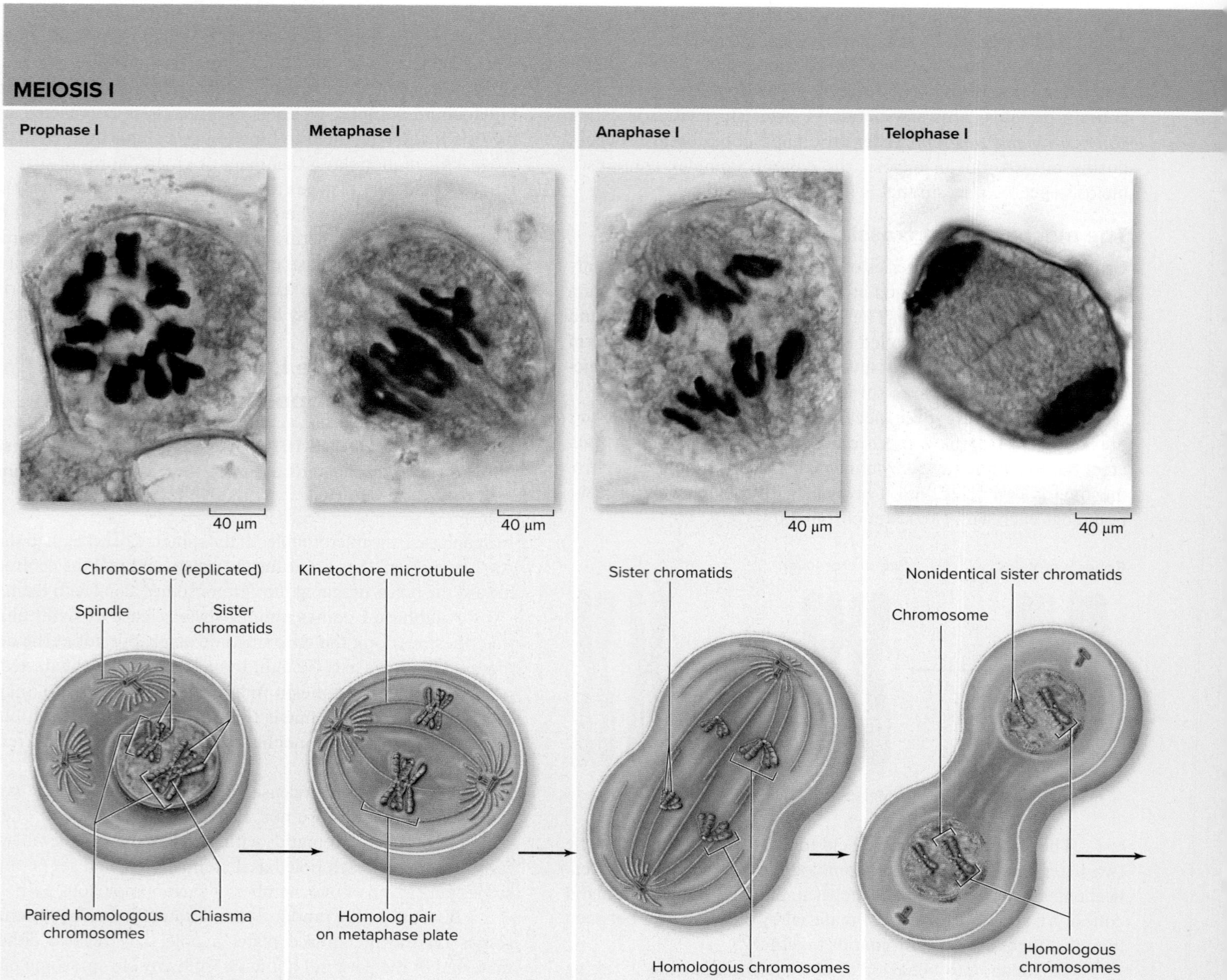
MEIOSIS I
Prophase I
Metaphase I
Anaphase I
Telophase I
40 μm
40 μm
40 μm
40 μm
Chromosome (replicated)
Spindle
Sister chromatids
Paired homologous chromosomes
Chiasma
Kinetochore microtubule
Homolog pair on metaphase plate
Sister chromatids
Homologous chromosomes
Nonidentical sister chromatids
Chromosome
Homologous chromosomes

In prophase I, the chromosomes begin to condense, and the spindle of microtubules begins to form. Each chromosome consists of two sister chromatids attached at the centromere. The cell illustrated here has four chromosomes, or two pairs of homologs. Homologous chromosomes pair along their entire length during synapsis. Crossing over occurs, forming chiasmata, which hold homologous chromosomes together.

In metaphase I, the pairs of homologous chromosomes align at the equator of the cell. Chiasmata keep homologous pairs together and microtubules from opposite poles attach to sister kinetochores of each homolog, producing tension. A kinetochore microtubule from one pole of the cell attaches to one homolog of a chromosome, while a kinetochore microtubule from the other cell pole attaches to the other homolog of a pair.

In anaphase I, kinetochore microtubules shorten, and homologous pairs are pulled apart. One duplicated homolog goes to one pole of the cell, and the other duplicated homolog goes to the other pole. Sister chromatids do not separate. This is in contrast to mitosis, in which duplicated homologs line up individually on the metaphase plate, and sister chromatids are pulled apart in anaphase.

In telophase I, the separated homologs form a cluster at each pole of the cell, and the nuclear envelope re-forms around each daughter cell nucleus. Cytokinesis may occur. The resulting two cells have half the number of chromosomes of the original cell: In this example, each nucleus contains two chromosomes (versus four in the original cell). Each chromosome consists of two sister chromatids, but sister chromatids are not identical because crossing over has occurred.

Figure 11.4 The stages of meiosis. Meiosis in plant cells (photos) and animal cells (drawings) is shown.

Ed Reschke/Photolibrary/Getty Images

MEIOSIS II

Prophase II | **Metaphase II** | **Anaphase II** | **Telophase II**

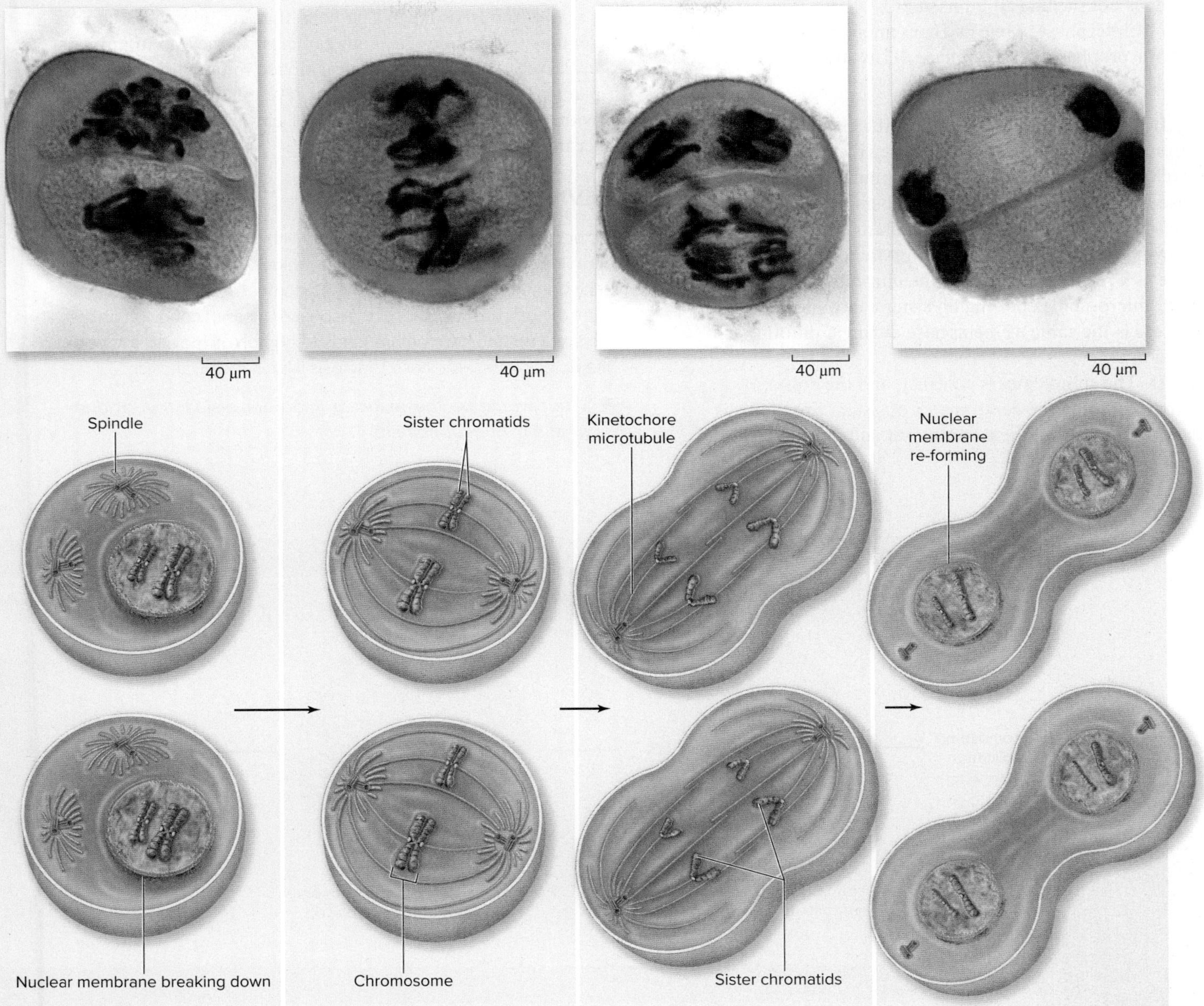
40 μm
40 μm
40 μm
40 μm
Spindle
Sister chromatids
Kinetochore microtubule
Nuclear membrane re-forming
Nuclear membrane breaking down
Chromosome
Sister chromatids

Prophase II

Following a brief interphase, with no S phase, meiosis II begins. During prophase II, a new spindle apparatus forms in each cell, and the nuclear envelope breaks down. In some species the nuclear envelope does not re-form in telophase I, obviating the need for nuclear envelope breakdown.

Metaphase II

In metaphase II, chromosomes consisting of sister chromatids joined at the centromere align along the metaphase plate in each cell. Now, kinetochore microtubules from opposite poles attach to kinetochores of sister chromatids, as in mitosis.

Anaphase II

When microtubules shorten in anaphase II, sister chromatids are pulled to opposite poles of the cells, as in mitosis.

Telophase II

In telophase II, the nuclear membranes re-form around four different clusters of chromosomes. After cytokinesis, four haploid cells result. No two cells are alike due to the random alignment of homologous pairs at metaphase I and crossing over during prophase I.

Meiosis II Is Like a Mitotic Division Without DNA Replication

LEARNING OBJECTIVE 11.3.5 Describe the events of meiosis II.

Typically, the period between meiosis I and meiosis II is brief and, critically, does not include an S phase. It is often called *interkinesis* instead of interphase. Meiosis II resembles a normal mitotic division with prophase II, metaphase II, anaphase II, and telophase II (figure 11.4).

Prophase II. At the two poles of the cell, the clusters of chromosomes enter a brief prophase II, each nuclear envelope breaking down as a new spindle forms.

Metaphase II. In metaphase II, spindle fibers from opposite poles bind to kinetochores of each sister chromatid, allowing each chromosome to migrate to the metaphase plate as a result of tension on the chromosomes from polar microtubules pulling on sister centromeres. This process is the same as metaphase during a mitotic division.

Anaphase II. The spindle fibers contract, and the cohesin complex joining the centromeres of sister chromatids is destroyed, splitting the centromeres and pulling the sister chromatids to opposite poles. This process is also the same as anaphase during a mitotic division.

Telophase II. Finally, the nuclear envelope re-forms around the four sets of daughter chromosomes. Cytokinesis then follows.

The final result of this division is four cells containing haploid sets of chromosomes. The cells that contain these haploid nuclei may develop directly into gametes, as they do in animals. Alternatively, they may themselves divide mitotically, as they do in plants, fungi, and many protists, eventually producing adult individuals with a haploid number of chromosomes, or a much larger number of haploid gametes.

REVIEW OF CONCEPT 11.3

Sister chromatid cohesion, combined with crossing over, connects homologous chromosomes during meiosis I. The centromere of each homolog shows monopolar attachment, leading to the alignment of homologous pairs at metaphase I. Loss of cohesion on the arms but not the centromere leads to homologs moving to opposite poles during anaphase I. Meiosis II is similar to mitosis.

- *What would be the result of improper disjunction at anaphase I? At anaphase II?*

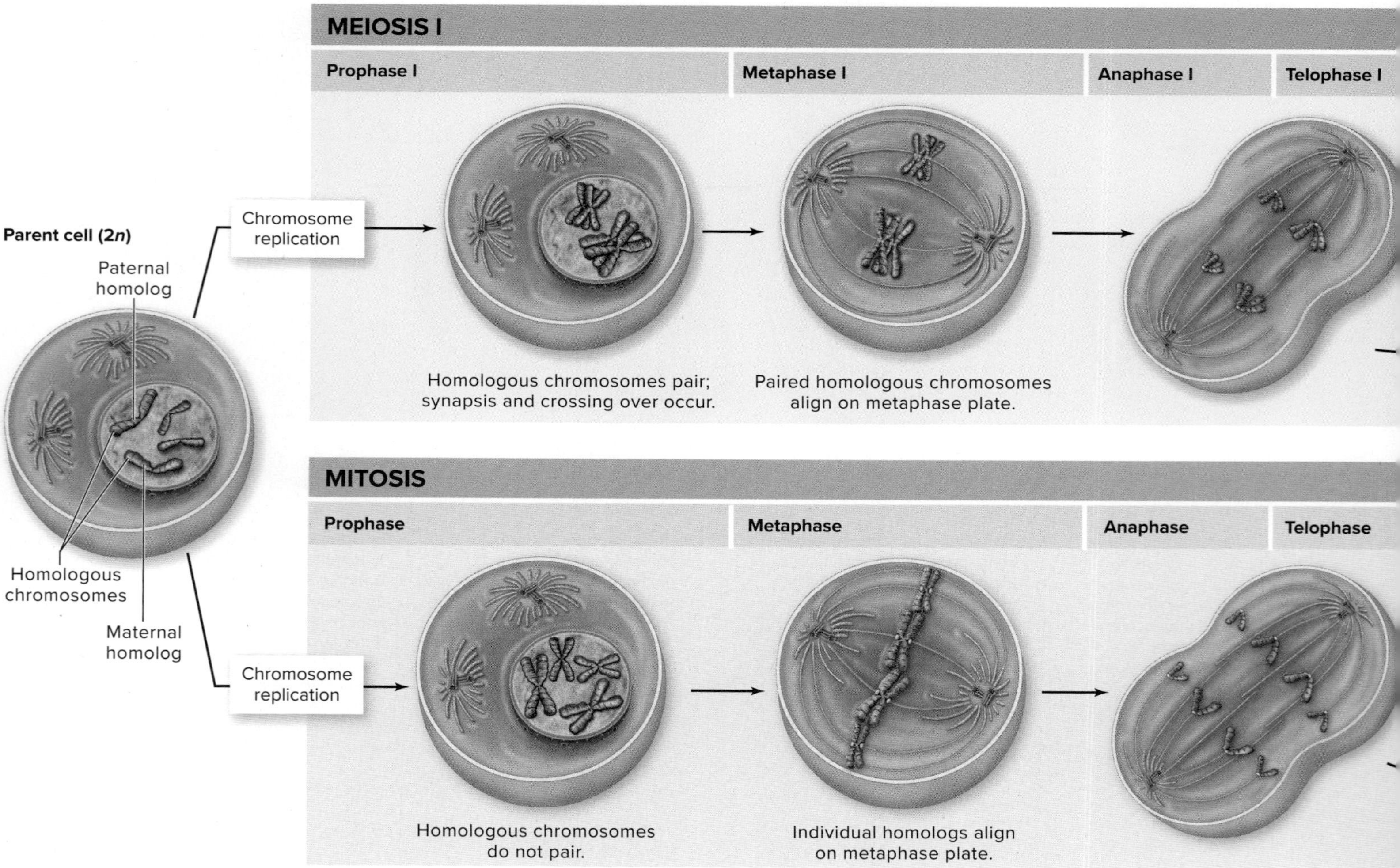
MEIOSIS I
Prophase I
Metaphase I
Anaphase I
Telophase I
Parent cell (2*n*)
Paternal homolog
Chromosome replication
Homologous chromosomes pair; synapsis and crossing over occur.
Paired homologous chromosomes align on metaphase plate.
MITOSIS
Prophase
Metaphase
Anaphase
Telophase
Homologous chromosomes
Maternal homolog
Chromosome replication
Homologous chromosomes do not pair.
Individual homologs align on metaphase plate.

11.4 Meiosis Has Four Distinct Features

The Behavior of Chromosomes Is Distinctly Different in Meiosis I

LEARNING OBJECTIVE 11.4.1 Discuss the molecular mechanisms responsible for the four distinct features of meiosis.

The key to understanding meiosis is understanding the differences between meiosis and mitosis. The basic machinery is the same in both processes, but the behavior of chromosomes is distinctly different during the first meiotic division (figure 11.5). Meiosis is characterized by four distinct features:

1. Homologous pairing and crossing over join maternal and paternal homologs during meiosis I.
2. Sister chromatids remain connected at the centromere and segregate together during anaphase of meiosis I.
3. Kinetochores of sister chromatids are attached to the same pole in meiosis I and to opposite poles in mitosis.
4. DNA replication is suppressed between the two meiotic divisions.

1. Homologous pairing is specific to meiosis

The pairing of homologs during prophase I of meiosis is the first deviation from mitosis, and this sets the stage for all of the subsequent differences (figure 11.6). How homologs find each other and become aligned is one of the great mysteries of meiosis. Some light has been shed on the mechanisms with the discovery of meiosis-specific cohesin proteins. In yeast, the protein Rec8 replaces the mitotic Scc1 protein as part of the cohesin complex. This component is the target of the separase enzyme (refer to chapter 10), and its replacement with a meiosis-specific version seems to be a common feature in systems analyzed to date.

The synaptonemal complex consists of seven proteins in mammals. The lateral elements consist of the proteins SYCP2 and SYCP3. The transverse filaments contain only the coiled-coil protein SYCP1, and the central element consists of four small proteins: SYCE1, SYCE2, SYCE3, and Tex12. Recent analysis indicates that, once thought not to be evolutionarily conserved, related proteins exist in the cnidarian *Hydra*, implying an ancient origin of these proteins. These proteins seem to have been replaced or to have diverged in flies and worms.

The recombination process requires homologous pairing, and many of the proteins involved are known. The recombination machinery is related to the machinery necessary for the repair of double-strand breaks in DNA. Recombination probably first evolved

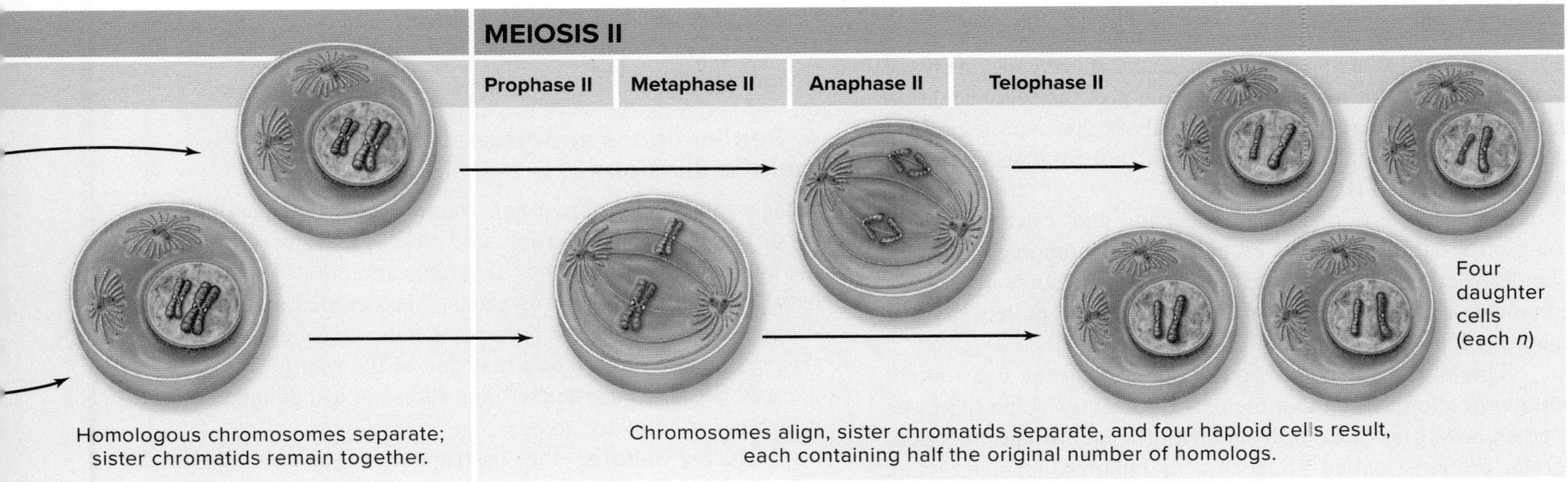

MEIOSIS II
Prophase II
Metaphase II
Anaphase II
Telophase II
Four daughter cells (each *n*)
Homologous chromosomes separate; sister chromatids remain together.
Chromosomes align, sister chromatids separate, and four haploid cells result, each containing half the original number of homologs.

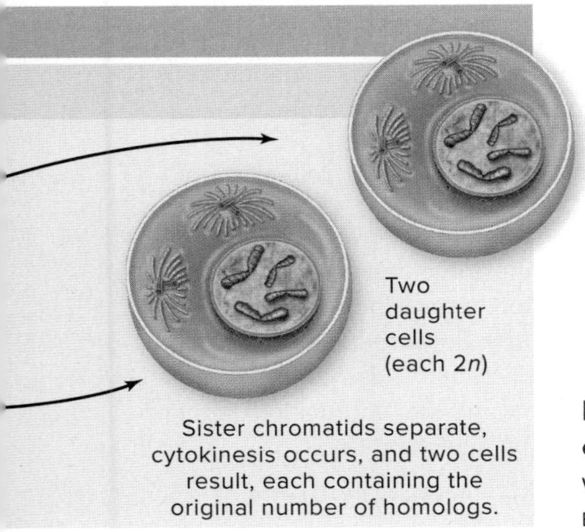

Two daughter cells (each 2*n*)
Sister chromatids separate, cytokinesis occurs, and two cells result, each containing the original number of homologs.

Figure 11.5 A comparison of meiosis and mitosis. Mitosis produces two daughter cells, each with the original diploid number of chromosomes. Meiosis produces four daughter cells, each with a complete haploid set of chromosomes, which is half the original diploid number.

Ed Reschke/Photolibrary/Getty Images

Figure 11.6 Alignment of chromosomes differs between meiosis I and mitosis. In metaphase I of meiosis I, the chiasmata and connections between sister chromatids hold homologous chromosomes together. The paired kinetochores of sister chromatids for each homolog are attached to microtubules from one pole. At anaphase I the connections between sister chromatids are broken, except for those at the centromere. This leads to homologs being pulled apart with sister chromatids remaining connected at the centromere. In contrast, in mitosis the centromeres for sister chromatids are connected to opposite poles at metaphase and sister chromatids are pulled apart.

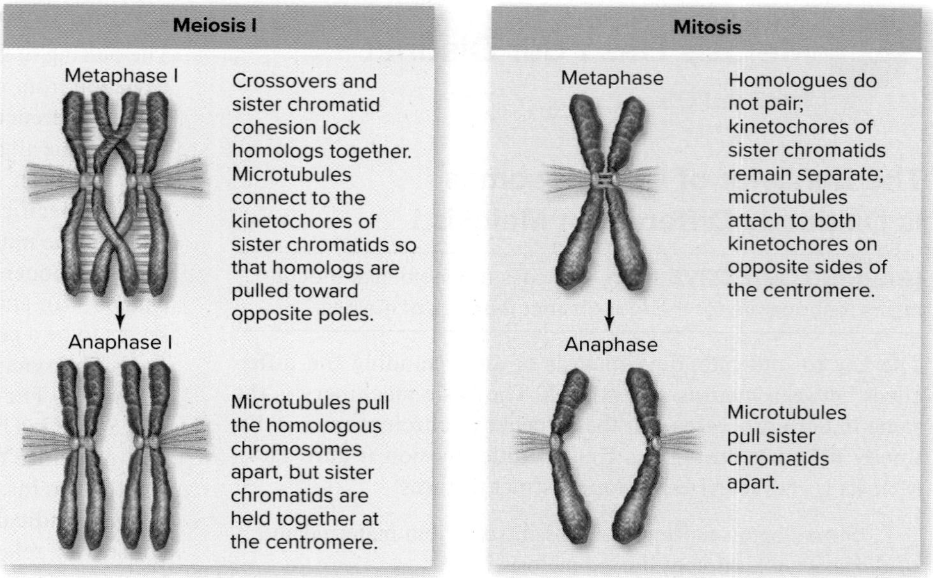

Meiosis I
Metaphase I
Crossovers and sister chromatid cohesion lock homologs together. Microtubules connect to the kinetochores of sister chromatids so that homologs are pulled toward opposite poles.
Anaphase I
Microtubules pull the homologous chromosomes apart, but sister chromatids are held together at the centromere.
Mitosis
Metaphase
Homologues do not pair; kinetochores of sister chromatids remain separate; microtubules attach to both kinetochores on opposite sides of the centromere.
Anaphase
Microtubules pull sister chromatids apart.

as a repair mechanism and was later co-opted for use in disjoining chromosomes. The importance of recombination for proper disjunction is clear from the observation that loss of function for recombination proteins results in higher levels of nondisjunction.

2. Centromeres of sister chromatids remain connected through meiosis I

Meiosis I is characterized by the segregation of homologs not sister chromatids, during anaphase. For this to occur, the centromeres of sister chromatids must move to the same pole (co-segregate) during anaphase I. This requires the removal of cohesin proteins from chromosome arms during meiosis I, but not from sister centromeres until meiosis II (figure 11.6).

Homologs are joined by chiasmata, and sister chromatid cohesion around the site of exchange holds homologs together. The destruction of Rec8 protein on the chromosome arms frees homologs from each other so that they can be pulled apart at anaphase I.

This process depends on the cohesins at the centromere being protected throughout meiosis I. Recently, some light was shed on how this is accomplished with the identification of conserved proteins, called Shugoshin (a Japanese term meaning "guardian spirit"), required for cohesin protection during meiosis I. Mice have two Shugoshins: Sgo-1 and Sgo-2. Depletion of Sgo-2 results in early sister chromatid separation. This leaves the problem of why Sgo-2 acts only at anaphase I and not at anaphase II. One hypothesis is that tension produced by anaphase II causes Sgo-2 to migrate from the centromere to the kinetochore.

3. Sister kinetochores are attached to the same pole during meiosis I

The co-segregation of sister centromeres during meiosis I requires that the kinetochores of sister chromatids be attached to the same pole. This attachment is in contrast to both mitosis (figure 11.6) and meiosis II, where sister kinetochores become attached to opposite poles.

The underlying basis of this monopolar attachment of sister kinetochores is unclear, but it seems to be based on structural differences between centromere–kinetochore complexes in meiosis I and in mitosis. Mitotic kinetochores seen in the electron microscope appear more recessed, making bipolar attachment more likely. Meiosis I kinetochores protrude more, making monopolar attachment easier. It is clear that both the maintenance of sister chromatid cohesion at the centromere and monopolar attachment are required for the segregation of homologs that distinguishes meiosis I from mitosis.

4. Replication is suppressed between meiotic divisions

After a mitotic division, a new round of DNA replication must occur before the next division. For meiosis to succeed in halving the number of chromosomes, this replication must be suppressed between the two meiotic divisions. The detailed mechanism of suppression of replication between meiotic divisions is unknown. One clue is the observation that one of the cyclins, cyclin B, is reduced between meiotic divisions but is not lost completely, as it is between mitotic divisions.

During mitosis, the destruction of mitotic cyclin is necessary for a cell to enter another division cycle. The result of persistence of cyclin B between meiotic divisions in germ-line cells is a failure to form the initiation complexes necessary for DNA replication to begin. This failure to form initiation complexes appears to be critical to suppressing DNA replication.

REVIEW OF CONCEPT 11.4

Meiosis is characterized by homolog pairing and crossing over; loss of sister chromatid cohesion in the arms but not at the centromere at the first division; sister kinetochores being attached to the same pole; and the suppression of DNA replication between the two meiotic divisions. These lead to homologs separating during meiosis I, then sister chromatids during meiosis II. The result is four unique haploid cells.

- *What features of meiosis are tied to the cell cycle?*

11.5 Genetic Variation Is the Evolutionary Consequence of Sex

Meiosis is a lot more complicated than mitosis. Why has evolution gone to so much trouble? Although our knowledge of how meiosis and sex evolved is sketchy, it is abundantly clear that meiosis and sexual reproduction have an enormous influence on how species continue to evolve today because of their ability to rapidly generate new genetic combinations. Three mechanisms make key contributions: independent assortment, crossing over, and random fertilization.

Three Mechanisms Increase Genetic Variability

LEARNING OBJECTIVE 11.5.1 Explain the ways in which meiosis increases genetic variability and why this is important.

Independent assortment

The reassortment of genetic material that takes place during meiosis helps to produce enormous genetic variation, and this variation has made possible the evolution of the incredible diversity of eukaryotic organisms over the past 1.5 billion years. To understand this, recall that most organisms have more than one pair of chromosomes, and that during metaphase I paired homologs align independently. This means that the number of possible alignments is equal to 2^n, where n is the haploid number. Consider an organism with a haploid number of 3: There are $2^3 = 8$ possible alignments of paired homologs (figure 11.7).

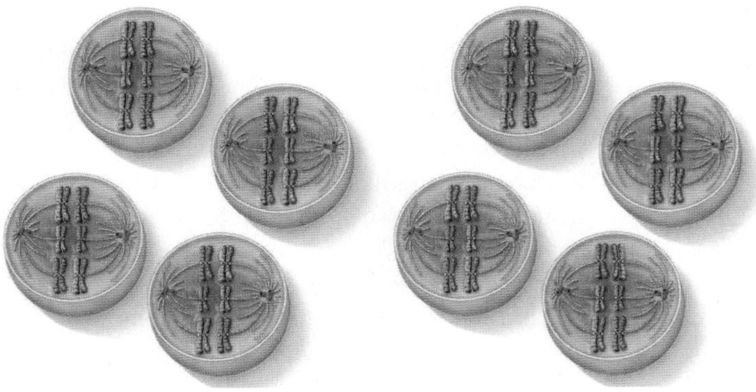

Figure 11.7 Random orientation of chromosomes on the metaphase plate. The number of possible chromosome orientations equals 2 raised to the power of the number of chromosome pairs. In this hypothetical cell with three chromosome pairs, eight (2^3) possible orientations exist. Each orientation produces gametes with different combinations of parental chromosomes.

Completion of meiosis will produce eight different types of gametes from this organism. Note that these eight alignments are actually four sets of complementary ones that will produce identical gametes. For instance, all blue homologues oriented to the "left" and red homologues oriented to the "right" produce the same gametes as the converse alignment of all blue "right" and all red "left." If we apply this logic to the human genome, with $n = 23$, we find that 2^{23} (more than 8 million) possible gametes can be produced.

Crossing over

The DNA exchange that occurs when the arms of nonsister chromatids cross over adds even more recombination to the independent assortment of chromosomes that occurs later in meiosis. Thus, the number of possible genetic combinations that can occur among gametes is virtually unlimited.

Random fertilization

Also, the zygote that forms a new individual is created by the fusion of two gametes, each produced independently, so fertilization squares the number of possible outcomes ($2^{23} \times 2^{23} =$ 70 trillion).

Importance of generating diversity

The pace of evolutionary change is quickened by genetic recombination, much of which results from sexual reproduction. Change is not always favored by selection, of course. Selection tends to preserve existing combinations of genes in asexually reproducing plants that live in especially demanding habitats. In vertebrates, by contrast, the evolutionary premium appears to have been on versatility, and sexual reproduction is the predominant mode of reproduction.

Whatever the forces that led to sexual reproduction, its evolutionary consequences have been profound. No genetic process generates diversity more quickly; as you will see in chapter 19, genetic diversity is the raw material of evolution, the fuel that drives it and determines its potential directions.

REVIEW OF CONCEPT 11.5

Meiosis generates genetic variability by independent assortment, crossing over, and random fertilization. Their ability to generate novel genetic forms has been of great evolutionary advantage to sexually reproducing organisms.

- *What features of meiosis lead to genetic variation?*

How Meiotic Recombination Matches Homolog Sequences in Crossing Over

A central event in meiosis is the recombination between homologous chromosomes that takes place during synapsis. Early in prophase, cohesin proteins form a lattice, called the synaptonemal complex, that brings homologous chromosomes into close contact. A nick is made in one of the two chromosomes by an enzyme called RecB, allowing that DNA duplex to become partially unwound. One of the short single strands of DNA that results then somehow finds its matching complementary sequence on the nearby homologous DNA duplex, forming a local brief triple-strand association. An exchange between strands at that point can then produce a genetic crossover, a physical exchange of segments between the homologous chromosomes.

The "somehow" is a matter of great interest to geneticists. How does the single-stranded DNA (ssDNA) produced by the initial nick locate its complementary sequence on the double-stranded DNA (dsDNA) of the homolog?

This feat is achieved by an unusual enzyme called RecA, shown as a ribbon diagram in the picture. This enzyme has binding sites for both ssDNA and dsDNA. Within the synaptonemal complex, a long string of RecA proteins binds to a newly formed ssDNA after the initial nick, wrapping around it tightly (six monomers of RecA per revolution around the DNA duplex) to form a long nucleoprotein filament. It is this RecA–ssDNA filament that searches for sequence similarity along the dsDNA of the nearby homologous chromosome. Only a double-stranded DNA sequence complementary to the ssDNA of the filament is able to bind the dsDNA binding sites of the filament's RecA proteins.

After decades of research, the mechanism of the ssDNA search process remains unknown. Recently, researchers have used single-molecule fluorescence microscopy to directly visualize what is happening. In one key experiment, they examined the influence of the length of the ssDNA within the filament on successful searching for complementary sequences in the homologous dsDNA. They measured the speed with which homologous pairing was achieved by ssDNA strands of three different lengths: Lengths of 162 nucleotides (too short to form loops) were compared with lengths of 430 (long enough to form a loop) and 1762 (long enough to form several loops). The results they obtained are presented in the graph.

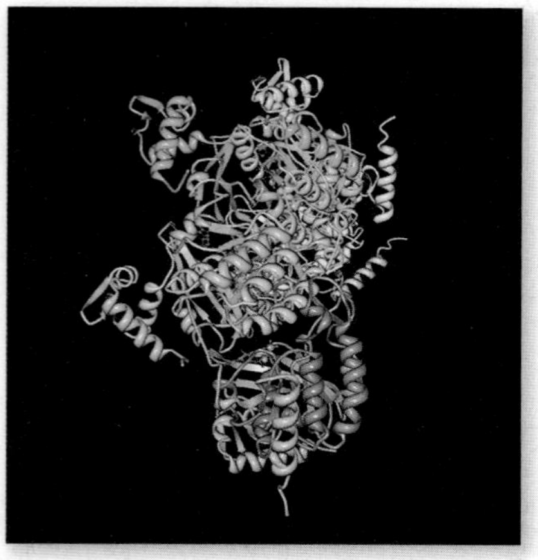

Image from the RCSB PDB (www.rcsb.org; H.M. Berman, et al. (2000), "The Protein Data Bank," *Nucleic Acids Research,* 28: 235–242).

PDB ID 3CMU: Chen, Z., Yang, H., Pavletich, N.P (2008) *Nature,* 453: 489–494

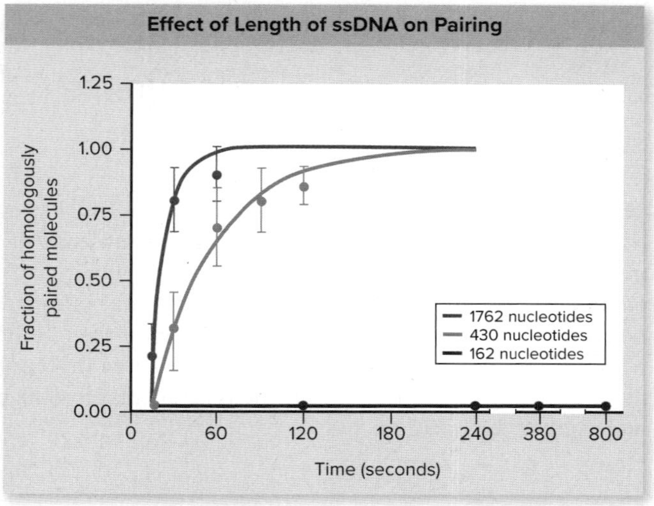

Analysis

1. **Applying Concepts**
   a. ***Variable.*** In the graph, what is the dependent variable?
   b. ***Fraction of homologously paired molecules.*** What is being paired with what?
2. **Interpreting Data**
   a. In these ssDNA–dsDNA binding experiments, does the fraction of ssDNA bound to dsDNA increase with time for 162 nucleotide ssDNA molecules? For 430 nucleotide ssDNA molecules? For 1762 nucleotide ssDNA molecules?
3. **Making Inferences**
   a. Which of the three ssDNA strand lengths achieves homologous pairing more rapidly? (*Hint:* How long does it take each of the three ssDNA strand lengths to achieve 50% homolog binding?)
   b. Is there any length of time in which 162 nucleotide strands of ssDNA can be expected to achieve 50% homolog binding?
4. **Drawing Conclusions**
   a. Is it fair to conclude that successful homolog binding requires an ssDNA strand longer than 162 nucleotides?
   b. Is it fair to conclude that the ability of the ssDNA to form loops is critical to its successful homologue binding?
5. **Further Analysis** If you wished to establish that the >430-nucleotide ssDNA length requirement indeed reflects the need to be able to form ssDNA loops, how could you go about attempting to test this hypothesis? Can you think of a way to investigate the possibility that successful homologue binding depends on the length of the dsDNA sequence for which the RecA–ssDNA filament is searching?

Retracing the Learning Path

CONCEPT 11.1 Sexual Reproduction Requires Meiosis

11.1.1 Gametes Contain Only One Copy of Each Chromosome Meiosis produces haploid ($1n$) eggs and sperm, each with half the normal number of chromosomes.

11.1.2 Sexual Life Cycles Alternate Between Haploid and Diploid During fertilization, or syngamy, fusion of two haploid gametes results in a diploid ($2n$) zygote. Meiosis and fertilization constitute a reproductive cycle in sexual organisms that alternates between diploid and haploid chromosome numbers. Somatic cells divide by mitosis and form the body of an organism. Cells that form haploid gametes by meiosis are called germ-line cells.

CONCEPT 11.2 Meiosis Consists of Two Divisions with One Round of DNA Replication

11.2.1 Homologous Pairing in Meiosis Leads to a Reductive Division Meiosis differs from mitosis in two important ways: homologous chromosomes pair during early prophase I, and meiosis involves two divisions with no DNA replication between them. Paired homologs are joined by the synaptonemal complex, and may exchange DNA by crossing over. The suppression of DNA replication between the two divisions results in a halving of the chromosome number.

CONCEPT 11.3 The Process of Meiosis Involves Intimate Interactions Between Homologs

11.3.1 Prophase I Sets the Stage for the Reductive Division In prophase I, homologous chromosomes align along their entire length. The sister chromatids are held together by cohesin proteins. Homologs exchange chromosomal material by crossing over, which assists in holding the homologues together during meiosis I. The nuclear envelope disperses and the spindle apparatus forms.

11.3.2 During Metaphase I, Paired Homologues Align Spindle fibers attach to the kinetochores of the homologues; the kinetochores of sister chromatids behave as a single unit. Homologues of each pair become attached by kinetochore microtubules to opposite poles, and homologous pairs move to the metaphase plate as a unit. The orientation of each homologous pair on the equator is random.

11.3.3 During Anaphase I, Homologues Are Pulled to Opposite Poles During anaphase I, the homologues of each pair are pulled to opposite poles as kinetochore microtubules shorten. Loss of sister chromatid cohesion on the arms but not at the centromeres allows homologues to separate. This is due to the loss of cohesin proteins on the arms but not at the centromere. At the end of anaphase I, each pole has a complete set of haploid chromosomes. Random orientation of homologous pairs at metaphase I results in the independent assortment of homologues.

11.3.4 Telophase I Completes Meiosis I During telophase I, the nuclear envelope re-forms around each daughter nucleus. This phase does not occur in all species. Cytokinesis may or may not occur after telophase I. Homologues are usually held together by chiasmata, but some systems segregate chromosomes without this.

11.3.5 Meiosis II Is Like a Mitotic Division Without DNA Replication A brief period with no DNA replication occurs after meiosis I. During meiosis II, cohesin proteins at the centromeres that hold sister chromatids together are destroyed, allowing each to migrate to opposite poles of the cell. The result of meiosis I and II is four cells, each containing haploid sets of chromosomes that are not identical.

CONCEPT 11.4 Meiosis Has Four Distinct Features

11.4.1 The Behavior of Chromosomes Is Distinctly Different in Meiosis I Homologous pairing is specific to meiosis. The proteins of the synaptonemal complex are conserved in structure but not sequence. There are meiosis-specific cohesin proteins involved in the differential loss on arms versus centromeres during meiosis. Sister chromatid cohesion is maintained through meiosis I but released in meiosis II. Shugoshin protein protects centromeric cohesin in anaphase I. Sister kinetochores are attached to the same pole during meiosis I. Kinetochores of sister chromatids must be attached to the same spindle fibers to segregate together. Replication is suppressed between meiotic divisions. This may be due to maintenance of some cyclin proteins that are degraded at the end of mitosis.

CONCEPT 11.5 Genetic Variation Is the Evolutionary Consequence of Sex

11.5.1 Three Mechanisms Increase Genetic Variability Sexual reproduction increases genetic variability through independent assortment in metaphase I, through crossing over in prophase I, and through random fertilization.

Concept Overview

This Concept Overview diagrams the key concepts that were discussed in this chapter.

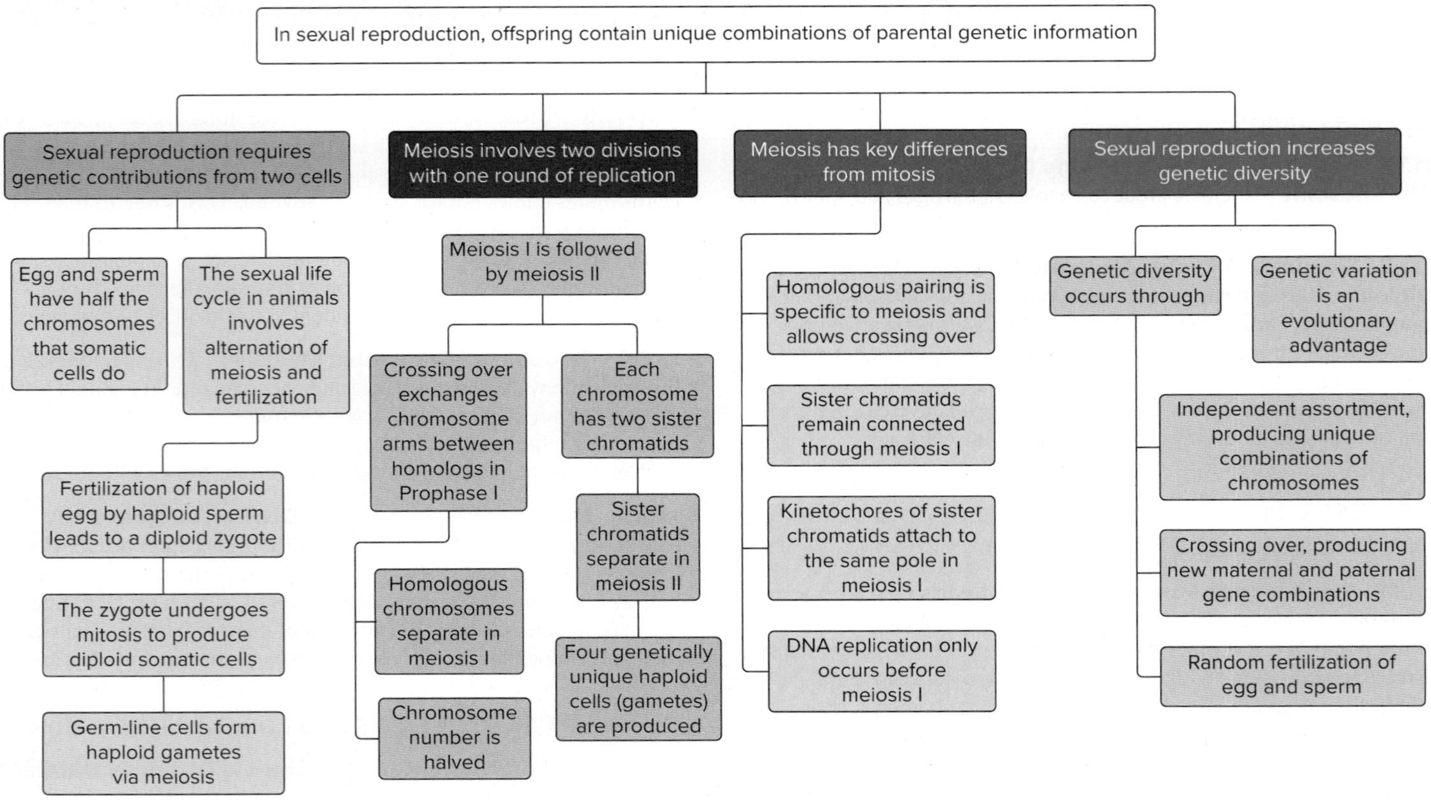

Assessing the Learning Path

Understand

1. Comparing somatic cells and gametes, somatic cells are
   a. diploid with half the number of chromosomes.
   b. haploid with half the number of chromosomes.
   c. diploid with twice the number of chromosomes.
   d. haploid with twice the number of chromosomes.
2. In the life cycle of most animals,
   a. there is an alternation between diploid and haploid chromosome numbers.
   b. gametes are produced by meiosis.
   c. the haploid stage is unicellular.
   d. All of the above
3. Synapsis occurs during
   a. mitosis.
   b. meiosis I.
   c. meiosis II.
   d. Both b and c
4. Which of the following is NOT true of crossing over?
   a. It occurs during prophase I.
   b. It occurs during prophase II.
   c. It occurs between homologs.
   d. Exchange of DNA occurs at the points of crossing over.
5. At metaphase I, the kinetochores of sister chromatids are
   a. attached to microtubules from the same pole.
   b. attached to microtubules from opposite poles.
   c. held together with cohesin proteins.
   d. not attached to any microtubules.
6. During anaphase I,
   a. sister chromatids separate and move to the poles.
   b. homologous chromosomes move to opposite poles.
   c. homologous chromosomes align at the middle of the cell.
   d. all chromosomes align independently at the cell's middle.
7. Anaphase I comes about because of
   a. release of sister chromatid cohesion along the chromosome arms.
   b. attachment of centromeres to microtubules originating from opposite poles.
   c. destruction of cohesin at the centromeres.
   d. release of sister chromatid cohesion at the centromere.
8. How are meiosis II and mitosis similar?
   a. DNA replicates before nuclear division occurs.
   b. Sister chromatids separate.
   c. Diploid daughter cells are produced.
   d. The chromosome number is halved.

9. Which of the following is NOT a distinct feature of meiosis?
   a. Pairing and exchange of genetic material between homologous chromosomes
   b. Attachment of sister kinetochores to spindle microtubules
   c. Movement of sister chromatids to the same pole
   d. Suppression of DNA replication
10. Crossing over
   a. seems to be important for correct disjunction.
   b. occurs while homologous chromosomes are paired.
   c. likely first evolved as a mechanism to repair double-stranded DNA breaks.
   d. All of the above
11. Which of the following give rise to genetic variation in sexually reproducing organisms?
   a. Crossing over, mutation, and fertilization
   b. Fertilization and crossing over
   c. Independent assortment, crossing over, mutation, and fertilization
   d. Crossing over and independent assortment

Apply

1. In human females, both meiotic divisions occur by a cytokinesis that is unequal, producing a large cell and a small polar body that is reabsorbed. In comparing male and female meiosis then, we can say that each meiosis produces
   a. 4 egg cells and 4 sperm cells.
   b. 1 egg cell and 4 sperm cells.
   c. 2 egg cells and 4 sperm cells.
   d. 2 egg cells and 2 sperm cells.
2. In at least some organisms, crossing over, or genetic exchange, can occur in the absence of the synaptonemal complex (SC). This implies that the SC may
   a. act to hold the homologs in place and not contain any enzymes involved.
   b. contain the enzymes necessary to break and rejoin chromosomes.
   c. contain enzymes, but they only act on sister chromatids not homologues.
   d. act to hold sister chromatids in place and not homologues.
3. Mutations that affect DNA repair often also affect the accuracy of meiosis. This is because
   a. the proteins involved in the repair of double-strand breaks are also involved in crossing over.
   b. the proteins involved in DNA repair are also involved in sister chromatid cohesion.
   c. DNA repair only occurs on condensed chromosomes such as those found in meiosis.
   d. cohesin proteins are also necessary for DNA repair.
4. A diploid cell with 6 chromosomes is undergoing meiosis. In meiosis I, a kinetochore microtubule does not attach to the kinetochore of one chromosome. At the end of meiosis, what will be the result of this error?
   a. Half haploid cells and half diploid cells
   b. Two cells with 4 chromosomes and two cells with 2 chromosomes
   c. One cell with 4 chromosomes, one cell with 2 chromosomes, and two cells with 3 chromosomes
   d. All diploid gametes
5. Structurally, meiotic cohesins have different components than mitotic cohesins. This leads to which functional difference?
   a. During metaphase I, the sister kinetochores become attached to the same pole.
   b. Centromeres remain attached during anaphase I of meiosis.
   c. Centromeres remain attached through both divisions.
   d. Centromeric cohesins are destroyed at anaphase I, and cohesins along the arms are destroyed at anaphase II.
6. You measure the amount of DNA in a diploid cell in the G_1 phase of the cell cycle. The cell completes meiosis I, and you measure the amount of DNA in one of the daughter cells. The amount of DNA in the daughter cell
   a. is double that of the G_1 cell.
   b. equals that of the G_1 cell.
   c. is one-half that of the G_1 cell.
   d. is one-quarter that of the G_1 cell.
7. Not considering crossing over, a $2n = 8$ organism could produce how many different types of gametes with respect to maternal and paternal homologs?
   a. 2
   b. 4
   c. 8
   d. 16

Synthesize

1. In many organisms, the haploid stage of the life cycle is dominant, with adult haploid individuals and only a brief diploid stage. No one would argue that the haploid individuals of these organisms are not alive. How, then, would you support or contest a statement that haploid human sperm or egg cells are not live individuals?
2. Diagram the process of meiosis for an imaginary cell with six chromosomes in a diploid cell.
   a. How many homologous pairs does the cell have? Draw this cell and label each homolog in a homologous pair as either maternal (M) or paternal (P).
   b. Draw this cell showing metaphase of meiosis I. Do all homologs from one parent line up on the same side of the cell?
   c. How would this picture differ if you were diagramming anaphase of meiosis II?
3. Individuals with Down syndrome have three copies of chromosome 21 in their somatic cells. Using what you know about meiosis and the sexual life cycle of animals, explain how this chromosomal anomaly can arise.
4. Mules are the offspring of the mating of a horse and a donkey. Mules are unable to reproduce. A horse has a total of 64 chromosomes, whereas donkeys have 62 chromosomes. Use your knowledge of meiosis to predict the diploid chromosome number of a mule. Propose a possible explanation for the inability of mules to reproduce.
5. Compare the processes of independent assortment and crossing over. Which process has the greatest influence on genetic diversity?

Peter Fakler/Alamy Stock Photo

12 Patterns of Inheritance

Learning Path

12.1 Experiments Carried Out by Mendel Explain Inheritance

12.2 Mendel's Principle of Segregation Accounts for 3:1 Phenotypic Ratios

12.3 Mendel's Principle of Independent Assortment Asserts That Genes Segregate Independently

12.4 Probability Allows Us to Predict the Results of Crosses

12.5 Extending Mendel's Model Provides a Clearer View of Genetics in Action

12.6 Genotype Dictates Phenotype by Specifying Protein Sequences

Concept Overview

This Concept Overview provides a diagram of the overarching concepts that are covered in this chapter. These concepts will be expanded upon in the Concept Overview at the end of the chapter.

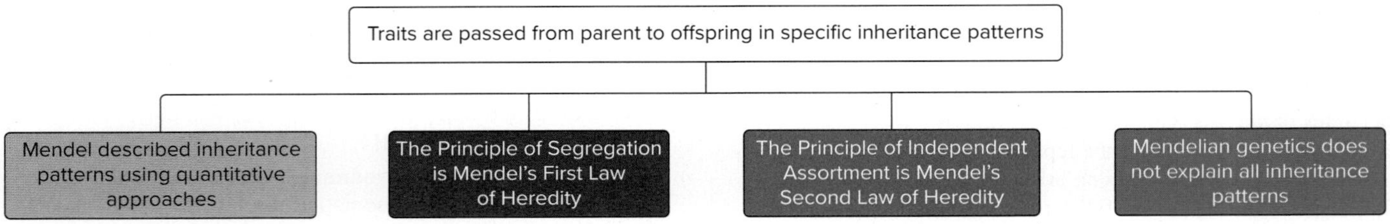

Traits are passed from parent to offspring in specific inheritance patterns
Mendel described inheritance patterns using quantitative approaches
The Principle of Segregation is Mendel's First Law of Heredity
The Principle of Independent Assortment is Mendel's Second Law of Heredity
Mendelian genetics does not explain all inheritance patterns

Introduction

In the pea pod on the previous page, you can see the shadowy outlines of seeds that will form part of the next generation of this pea plant. The seeds appear similar to one another, but the plants they produce may differ in significant ways. This is because the gametes that produced the seeds contribute chromosomes from both parents, in effect "shuffling the deck of cards" so that a progeny plant will have some characteristics from one parent and some from the other. About 150 years ago, Gregor Mendel first described this process, before anyone knew what genes or chromosomes were. We now understand the process of heredity in considerable detail. In this chapter, you will follow along as Mendel experiments with pea plants like the one pictured. Unlike researchers before him, Mendel carefully counted the number of each kind of pea plant his experiments produced, and in his results he found simple patterns. The theory he proposed to explain it has become one of the key principles of biology.

12.1 Experiments Carried Out by Mendel Explain Inheritance

It has long been recognized that we resemble members of our immediate family more than unrelated individuals, but there was no coherent explanation for this (figure 12.1). Prior to the 20th century, two concepts were the basis for most thinking about heredity. The first was that heredity occurs within species, and the second was that traits are transmitted directly from parents to offspring. Taken together, these ideas led to a view of inheritance as resulting from a blending of traits within fixed, unchanging species.

The Mystery of Heredity Was Solved in Stages

LEARNING OBJECTIVE 12.1.1 Describe the early experiments in hybridization.

Inheritance itself was viewed as traits being borne through fluid, usually identified as blood, that led to their blending in offspring. This older idea persists today in the use of the term "bloodlines" when referring to the breeding of domestic animals such as horses. However, this model leads to a paradox. If no variation enters a species from outside, and if the variation within each species blends in every generation, then all members of a species should soon have the same appearance. This is clearly not the case; individuals in a species differ from one another in traits that appear to be transmitted between generations.

Early hybridization experiments

There was a long history in 19th-century botany of producing hybrids from different strains within a species. The first such experiments were done by Josef Kölreuter in 1760 when he cross-fertilized (or crossed, for short) different strains of tobacco. When individuals from the hybrid generation were crossed, he observed great variation in the offspring, even seeing some resemblance to the original strains as well as to the hybrids, which contradicts the idea of direct transmission.

Throughout the 19th century, a number of other investigators followed up on Kölreuter's work. Among these was T. A. Knight, who crossed varieties of the garden pea, *Pisum sativum*. He crossed a variety with green seeds to another variety with yellow seeds to produce hybrids that all had yellow seeds. The original green and yellow varieties were both **true-breeding,** meaning that self-fertilization produces only one type. When he crossed the hybrids, he observed both yellow and green seeds. This result should have ended the concept of blending inheritance, but it did not.

Other investigators made similar observations, that alternate forms of a trait were being distributed among offspring. As we shall see, a modern version of this statement would be that the alternate forms of the trait were **segregating** among the progeny of a cross. This segregation of alternative forms of a trait provided the clue that led Gregor Mendel to his understanding of the nature of heredity. Within these deceptively simple results were the makings of a scientific revolution, although the process of segregation was not fully appreciated until the dawn of the 20th century.

Mendel's Experimental Design Was Quantitative, a Radical Change

LEARNING OBJECTIVE 12.1.2 Contrast Mendel's experimental design with those of earlier studies.

Born in 1822 to peasant parents in Austria, Gregor Mendel (figure 12.2) was educated in a monastery and went on to study science and mathematics at the University of Vienna, where he failed his examinations for a teaching certificate. He returned to the monastery and spent the rest of his life there, eventually

Figure 12.1 Heredity and family resemblance. Family resemblances are often strong—a visual manifestation of the mechanism of heredity.

Rob Marmion/Shutterstock

Figure 12.2 Gregor Mendel. The key to understanding the puzzle of heredity was solved by the monk Gregor Mendel, who cultivated pea plants in a garden alongside his monastery.

Leslie Holzer/Science Source

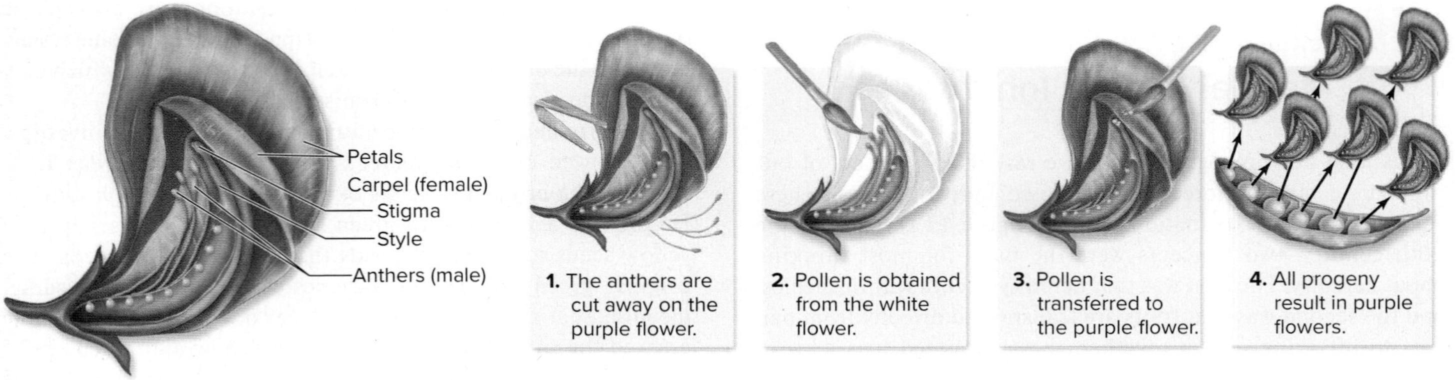
Petals
Carpel (female)
Stigma
Style
Anthers (male)
1. The anthers are cut away on the purple flower.
2. Pollen is obtained from the white flower.
3. Pollen is transferred to the purple flower.
4. All progeny result in purple flowers.

Figure 12.3 How Mendel conducted his experiments. In a pea plant flower, petals enclose both the male anther (containing pollen grains, which give rise to haploid sperm) and the female carpel (containing ovules, which give rise to haploid eggs). This ensures that self-fertilization will take place unless the flower is disturbed. Mendel collected pollen from the anthers of a white flower, then placed that pollen onto the stigma of a purple flower with anthers removed. This cross-fertilization yields all hybrid seeds that give rise to purple flowers. Using pollen from a purple flower to fertilize a white flower gives the same result.

becoming abbot. In the garden of the monastery, Mendel initiated his own series of experiments on plant hybridization. The results of these experiments would ultimately change our views of heredity irrevocably.

Mendel chose to use garden peas, as much work had already been done using this species. It was a good choice for numerous reasons, including that peas are small and easy to grow, and there were many varieties available. More important, it is also easy to do controlled crosses to produce hybrids, but because both male and female sexual organs are encased within a flower, they will self-fertilize if left alone (figure 12.3). Mendel initially examined 34 varieties. Then, for further study he selected lines that differed with respect to seven easily distinguishable traits, with clearly different alternate forms, such as flower color, with purple and white flowers.

There are a number of important differences between Mendel and his predecessors: he concentrated only on the traits of interest; he performed his crosses in a consistent and systematic manner; and, most important, he counted all of the offspring produced in his controlled crosses. Mendel used a simple format for all of his experiments, conducting his studies in three stages:

1. Plants of a given variety were allowed to self-cross for multiple generations to assure that the traits being studied were indeed true-breeding—that is, transmitted unchanged from generation to generation.
2. True-breeding varieties exhibiting alternative forms of traits were crossed. This also included reciprocal crosses: using pollen from a white-flowered plant to fertilize a purple-flowered plant, then using pollen from a purple-flowered plant to fertilize a white-flowered plant.
3. Finally, the hybrid offspring produced by these crosses were allowed to self-fertilize for several generations. This allowed Mendel to follow the inheritance of alternative forms of a trait for multiple generations. Most important, Mendel counted the numbers of offspring exhibiting each trait in each succeeding generation.

This quantification of results is what distinguished Mendel's research from that of earlier investigators, who only noted differences in a qualitative way. Mendel's mathematical analysis of experimental results led to a model of inheritance that is still valid today.

REVIEW OF CONCEPT 12.1

Prior to Mendel, concepts of inheritance did not form a consistent model. The dominant view was of blending inheritance, in which traits of parents were carried by fluid and "blended" in offspring. Plant hybridizers such as T. A. Knight had already observed characteristics in hybrids that seemed to change in second-generation offspring. Mendel's experiments were the first to quantify offspring using mathematical analysis.

- *Segregation of traits was observed before Mendel. What about his approach allowed him to formulate a complete model?*

12.2 Mendel's Principle of Segregation Accounts for 3:1 Phenotypic Ratios

Mendel studied seven traits in his experiments, each possessing alternative variants that differed from one another in ways that were easy to recognize and score. The traits he selected for his experiments involved the color or shape of the plant's flowers, seeds, pods, or adult form. A monohybrid cross is a cross that follows a single trait with two variations, such as white- and purple-colored flowers. We will first examine in detail Mendel's monohybrid crosses with the flower color trait.

Mendel Observed That Alternative Forms of a Trait Segregate in Crosses

LEARNING OBJECTIVE 12.2.1 Describe the outcome of Mendel's monohybrid crosses.

The F_1 generation exhibits only one of two traits

When Mendel crossed white-flowered peas to purple-flowered peas, the hybrid offspring all had purple flowers, not the lighter purple flowers that would be predicted by the blending inheritance hypothesis. These hybrid offspring are referred to as the **first filial generation,** or F_1. In a cross of white-flowered and purple-flowered plants, the F_1 offspring all had purple flowers, as had been reported by others before Mendel.

Mendel referred to the form of a trait expressed in the F_1 plants as **dominant,** and to the alternative form that was not expressed in the F_1 plants as **recessive.** For each of the seven pairs of contrasting traits that Mendel examined, one of the pair proved to be dominant and the other recessive.

The F_2 generation exhibits both traits in a 3:1 ratio

Mendel allowed the F_1 plants to self-fertilize to produce a second filial generation, or F_2. He observed both white- and purple-flowered plants—that is, the recessive trait that had disappeared in the F_1 reappeared in the F_2. When he actually counted the number of each type, he observed that among 929 F_2 individuals, there were 705 purple-flowered plants, and 224 white-flowered plants. If you express this as a percentage, there were 75.9% purple and 24.1% white. This is equivalent to a ratio of dominant:recessive very close to 3 dominant:1 recessive. This result was reproducible, and the outcome for reciprocal crosses was the same.

This striking regularity that Mendel observed was also independent of the trait he examined. The same numerical result was obtained with the other six characters he examined: of the F_2 individuals, ¾ exhibited the dominant form, and ¼ displayed the recessive form for each trait (figure 12.4). In other words, the dominant-to-recessive ratio among the F_2 plants was always close to 3:1.

There are two types of dominant plants in the F_2

Mendel went on to examine inheritance in his F_2 plants. He found that plants exhibiting the recessive form were always true-breeding. For example, self-fertilized white-flowered F_2 plants always produced white-flowered offspring. By contrast, ⅓ of the dominant, purple-flowered F_2 individuals (¼ of all F_2 offspring) were true-breeding, but ⅔ (½ of all F_2 offspring) were not. This latter class of purple-flowered F_2 produced a 3:1 ratio of dominant and recessive individuals in the third filial generation (F_3).

This result suggested that, for the entire sample, the 3:1 ratio that Mendel observed in the F_2 generation was really a disguised 1:2:1 ratio: ¼ true-breeding dominant individuals, ½ not-true-breeding dominant individuals, and ¼ true-breeding recessive individuals (figure 12.5).

Mendel's Principle of Segregation Explains Monohybrid Observations

LEARNING OBJECTIVE 12.2.2 Explain the principle of segregation.

From the simple framework of a monohybrid cross, a number of conclusions can be drawn. First, his initial hybrids did not have an intermediate appearance, as the hypothesis of blending

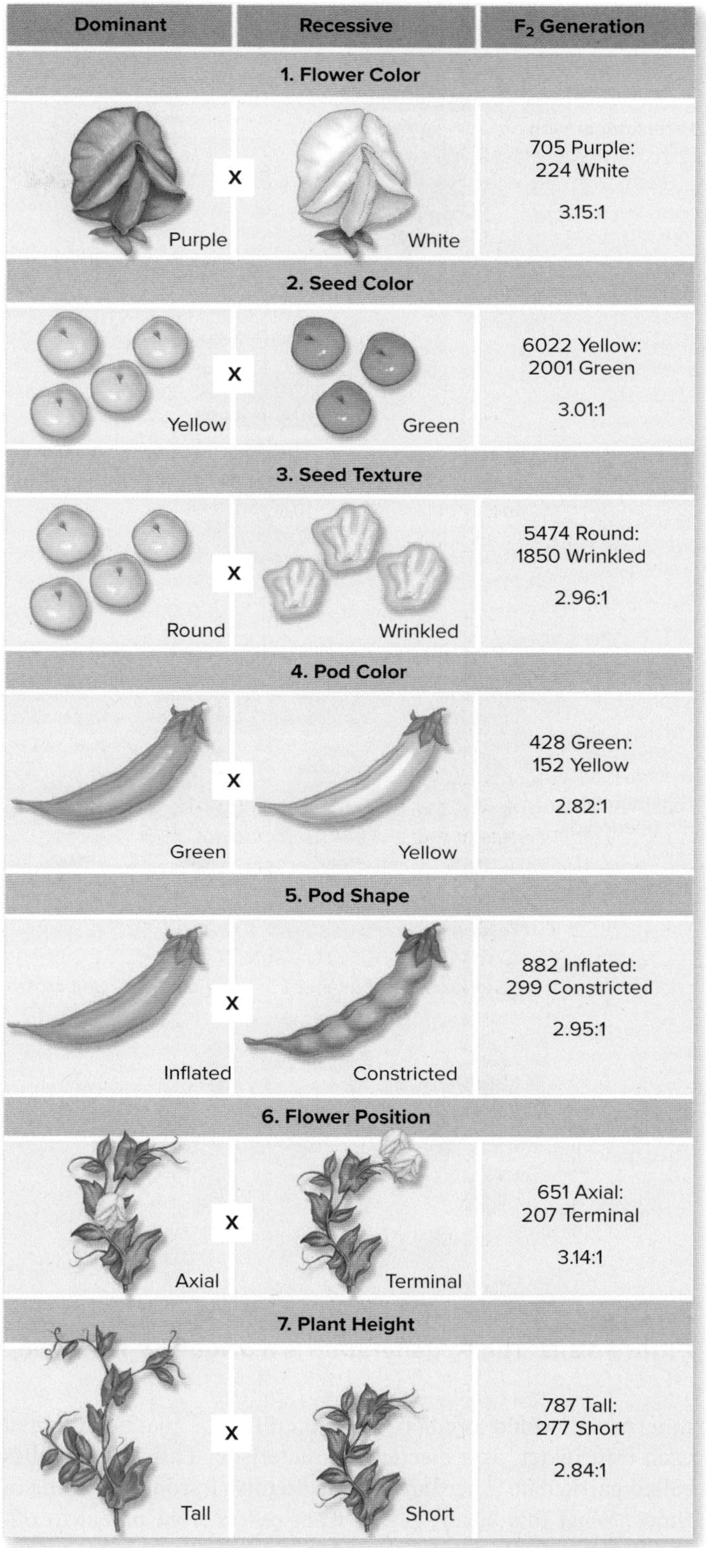

Figure 12.4 Mendel's seven traits. Mendel studied how differences among varieties of peas were inherited when the varieties were crossed. Similar experiments had been done before, but Mendel was the first to quantify the results and appreciate their significance. Results are shown for seven different monohybrid crosses. The F_1 generation is not shown in the figure.

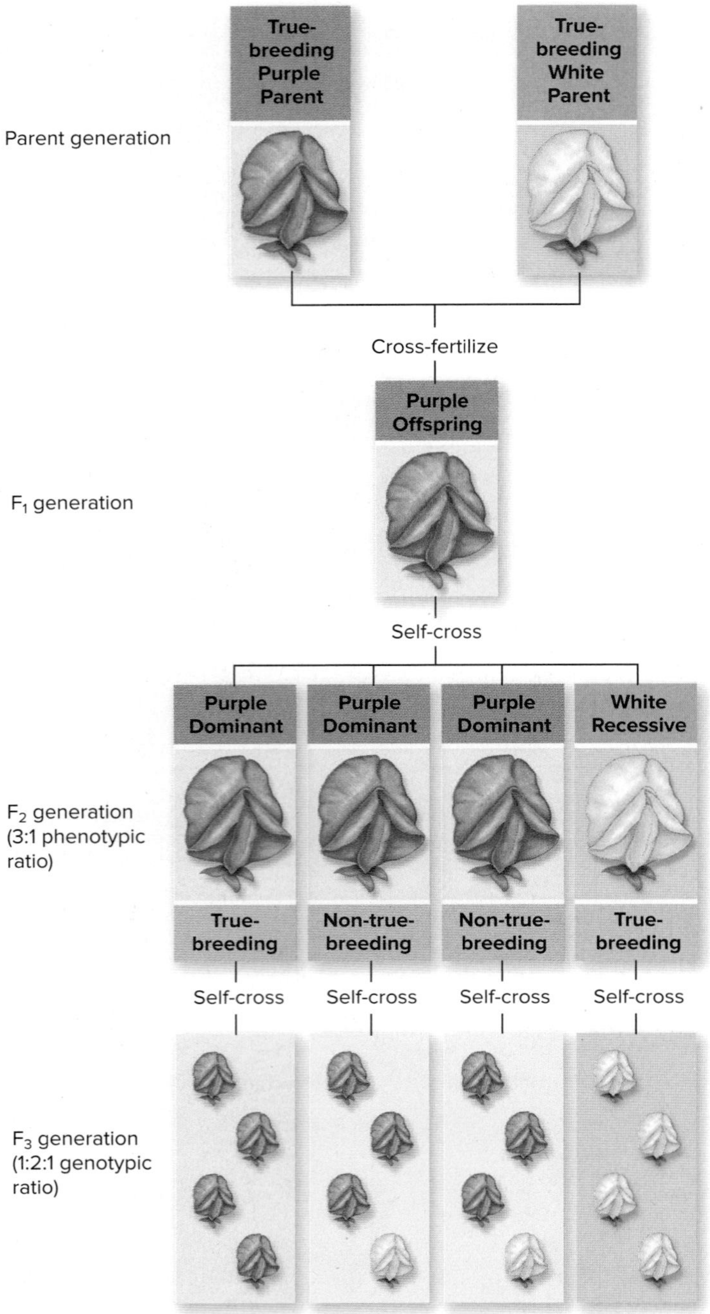

Parent generation
True-breeding Purple Parent
True-breeding White Parent
Cross-fertilize
Purple Offspring
F_1 generation
Self-cross
Purple Dominant
Purple Dominant
Purple Dominant
White Recessive
F_2 generation (3:1 phenotypic ratio)
True-breeding
Non-true-breeding
Non-true-breeding
True-breeding
Self-cross
Self-cross
Self-cross
Self-cross
F_3 generation (1:2:1 genotypic ratio)

Figure 12.5 The F_2 generation is a disguised 1:2:1 ratio.

inheritance would predict. Instead, different plants inherited each trait intact, as a discrete characteristic. This is sometimes called particulate inheritance, to distinguish it from a blending of fluids. What this means is that what passes from parent to offspring is a discrete entity, which Mendel called a "factor" but we now call a gene. We also now know that genes represent the information necessary to produce a particular trait.

Mendel also recognized that his factors, our genes, exist in different forms that produce different versions of the trait of interest, and for any trait, each parent provides one copy of the gene of interest. We now know that genes are on chromosomes, and that each individual is diploid, with one set of chromosomes from each parent.

He introduced the idea of dominance: that in a hybrid, only one form of a gene may be apparent. But, most important, the other form is not lost, but will appear in a predictable ratio in the next generation. We now call these different forms of a gene **alleles,** and they are segregated in the process of producing the next generation. To put this in more modern terms, during the process of gamete formation, alleles of a gene are segregated into individual gametes. The random combination of gametes from each parent then produces the observed ratios.

Genotype and phenotype

Geneticists now refer to the total set of alleles that an individual contains as the individual's **genotype.** The physical appearance or other observable characteristics of that individual, which result from an allele's expression, are together termed the individual's **phenotype.** In other words, the genotype is the blueprint, and the phenotype is the visible outcome in an individual.

This also allows us to present Mendel's ratios in more modern terms. The 3:1 ratio of dominant to recessive is the monohybrid phenotypic ratio. The 1:2:1 ratio of homozygous dominant to heterozygous to homozygous recessive is the monohybrid genotypic ratio. The genotypic ratio "collapses" into the phenotypic ratio because heterozygous and homozygous dominant individuals appear the same.

The Principle of Segregation

Mendel's model accounts for the ratios he observed in a neat and satisfying way. His main conclusion–that alternative alleles for a character segregate from each other during gamete formation and remain distinct–has since been verified in many other organisms. It is commonly referred to as the **Principle of Segregation.** It can be simply stated as: *The two alleles for a gene segregate during gamete formation and are rejoined at random, one from each parent, during fertilization.*

Although Mendel did not know this, we now know that the physical basis for allele segregation is the behavior of chromosomes during meiosis. As you saw in chapter 11, homologs for each chromosome disjoin during anaphase I of meiosis. The second meiotic division then produces gametes that contain only one homolog for each chromosome.

It is a tribute to Mendel that his analysis arrived at the correct scheme, even though he had no knowledge of the cellular mechanisms of inheritance; neither chromosomes nor meiosis had yet been described.

The Punnet Square Allows Symbolic Analysis

LEARNING OBJECTIVE 12.2.3 Explain the basis of the 3:1 Mendelian ratio using a Punnett square.

Mendel used symbols to represent traits, and then used this system to analyze his crosses. We still use this kind of analysis, although with slightly different symbols. Consider Mendel's cross of purple-flowered and white-flowered plants. We will assign the symbol P (uppercase) to the dominant allele, which produces purple flowers, and the symbol p (lowercase) to the

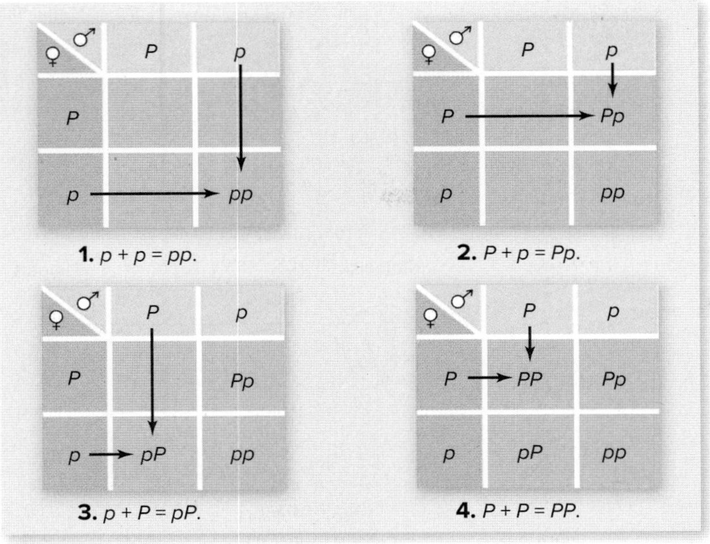

a.

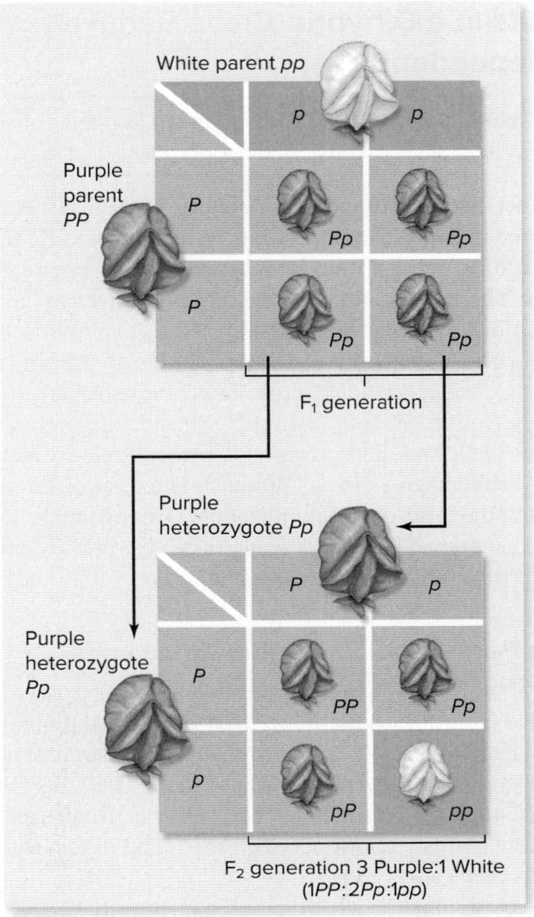

b.

Figure 12.6 Using a Punnett square to analyze Mendel's cross. ***a.*** To make a Punnett square, indicate the different female gametes from the cross along the side of a square and the different male gametes from the cross along the top. Each potential zygote is represented as the intersection of a vertical line and a horizontal line. ***b.*** In Mendel's cross of purple by white flowers, the F_1 are all purple, *Pp,* heterozygotes. These F_1 offspring make two types of gametes that can be combined to produce three kinds of F_2 offspring: *PP* homozygous dominant (purple); *Pp* heterozygous (also purple); and *pp* homozygous recessive (white). The phenotypic ratio is 3 purple:1 white. The genotypic ratio is 1 *PP*:2 *Pp*:1 *pp*.

recessive allele, which produces white flowers. We will also use the convention, throughout this book, that genotypes are shown as italic.

We can then symbolize the genotype of a homozygous recessive, white-flowered plant as *pp*. Similarly, the genotype of a homozygous dominant, purple-flowered plant would be *PP*. Finally, the genotype of a heterozygous, also purple-flowered plant, would be *Pp* (dominant allele first). Using these conventions and denoting a cross between two strains with X, we can symbolize Mendel's original purple X white cross a s *PP* X *pp*.

Because the white-flowered parent (*pp*) can produce only *p* gametes, and the true-breeding purple-flowered parent (*PP*) can produce only *P* gametes, only heterozygous *Pp* offspring will be produced in the F_1 generation. Because the *P* allele is dominant, all of these F_1 individuals will have purple flowers.

When F_1 individuals are allowed to self-fertilize, the *P* and *p* alleles segregate during gamete formation to produce both *P* gametes and *p* gametes, respectively. Gametes will randomly combine during fertilization to form F_2 individuals. The F_2 possibilities may be visualized in a simple diagram called a **Punnett square,** named after its originator, the geneticist R. C. Punnett (figure 12.6*a*). Analyzing Mendel's monohybrid cross with a Punnett square predicts an F_2 generation with ¾ purple-flowered plants and ¼ white-flowered plants, a phenotypic ratio of 3 purple:1 white (figure 12.6*b*).

REVIEW OF CONCEPT 12.2

Mendel's monohybrid crosses refute the idea of blending. One trait disappears in the F_1, then reappears in a predictable ratio in the F_2. The trait visible in the F_1 is called dominant, and the other trait is recessive. In the F_2, the ratio of dominant offspring to recessive is 3:1, and this represents a ratio of 1 homozygous dominant to 2 heterozygous to 1 homozygous recessive. The Principle of Segregation states that alleles segregate into different gametes. The physical basis for this is the separation of homologs during anaphase I of meiosis.

- *What fraction of tall F_2 plants are true-breeding?*

12.3 Mendel's Principle of Independent Assortment Asserts That Genes Segregate Independently

The Principle of Segregation explains the behavior of alternative forms of a single trait in a monohybrid cross. Mendel went on to ask if the inheritance of one trait, such as seed color, influences the inheritance of other traits, such as seed shape or flower color.

Traits in a Dihybrid Cross Behave Independently

LEARNING OBJECTIVE 12.3.1 Using a Punnett square, explain the genetic basis of a 9:3:3:1 dihybrid ratio.

Mendel first established a series of true-breeding lines of peas that differed in two of the seven traits he had studied. He then crossed contrasting pairs of the true-breeding lines to create F_1 heterozygotes. These heterozygotes are now doubly heterozygous, or dihybrid. Finally, he allowed the dihybrid F_1 plants to self-cross and produce an F_2 generation, and he counted all progeny types.

Consider Mendel's cross involving alternative seed shape alleles (round, *R*, and wrinkled, *r*) and alternative seed color alleles (yellow, *Y*, and green, *y*). Crossing round yellow (*RR YY*) with wrinkled green (*rr yy*) produces heterozygous F_1 individuals, all with the same dominant phenotype (round and yellow) and the same genotype (*Rr Yy*). Allowing these dihybrid F_1 individuals to self-fertilize produces an F_2 generation.

The F_2 generation exhibits four types of progeny in a 9:3:3:1 ratio

In analyzing these results, we first consider the number of possible phenotypes. We expect to see the two parental phenotypes: round yellow and wrinkled green. If the traits behave independently, then we can also expect one trait from each parent to produce plants with round green seeds and others with wrinkled yellow seeds.

Next consider what types of gametes the F_1 individuals can produce. Again, we expect the two types of gametes found in the parents: *RY* and *ry*. If the traits behave independently, then we can also expect the gametes *Ry* and *rY*. Using modern language, two genes each with two alleles can be combined four ways to produce these gametes: *RY*, *ry*, *Ry*, and *rY*.

We can then construct a Punnett square with these gametes. This is a 4 × 4 square with 16 possible outcomes. Filling in the Punnett square produces all possible offspring (figure 12.7). What are their phenotypes? You can see from the square that there are 9 round yellow, 3 wrinkled yellow, 3 round green, and 1 wrinkled green. This predicts a phenotypic ratio of 9:3:3:1.

Mendel's Principle of Independent Assortment explains dihybrid results

What did Mendel actually observe? From a total of 556 seeds from self-fertilized dihybrid plants, he observed the following results:

- 315 round yellow (signified *R__ Y__*, where the underscore indicates the presence of either allele),
- 108 round green (*R__ yy*),
- 101 wrinkled yellow (*rr Y__*), and
- 32 wrinkled green (*rr yy*).

These results are very close to a 9:3:3:1 ratio. (The expected 9:3:3:1 ratio from Mendel's 556 offspring would be 313:104:104:35.)

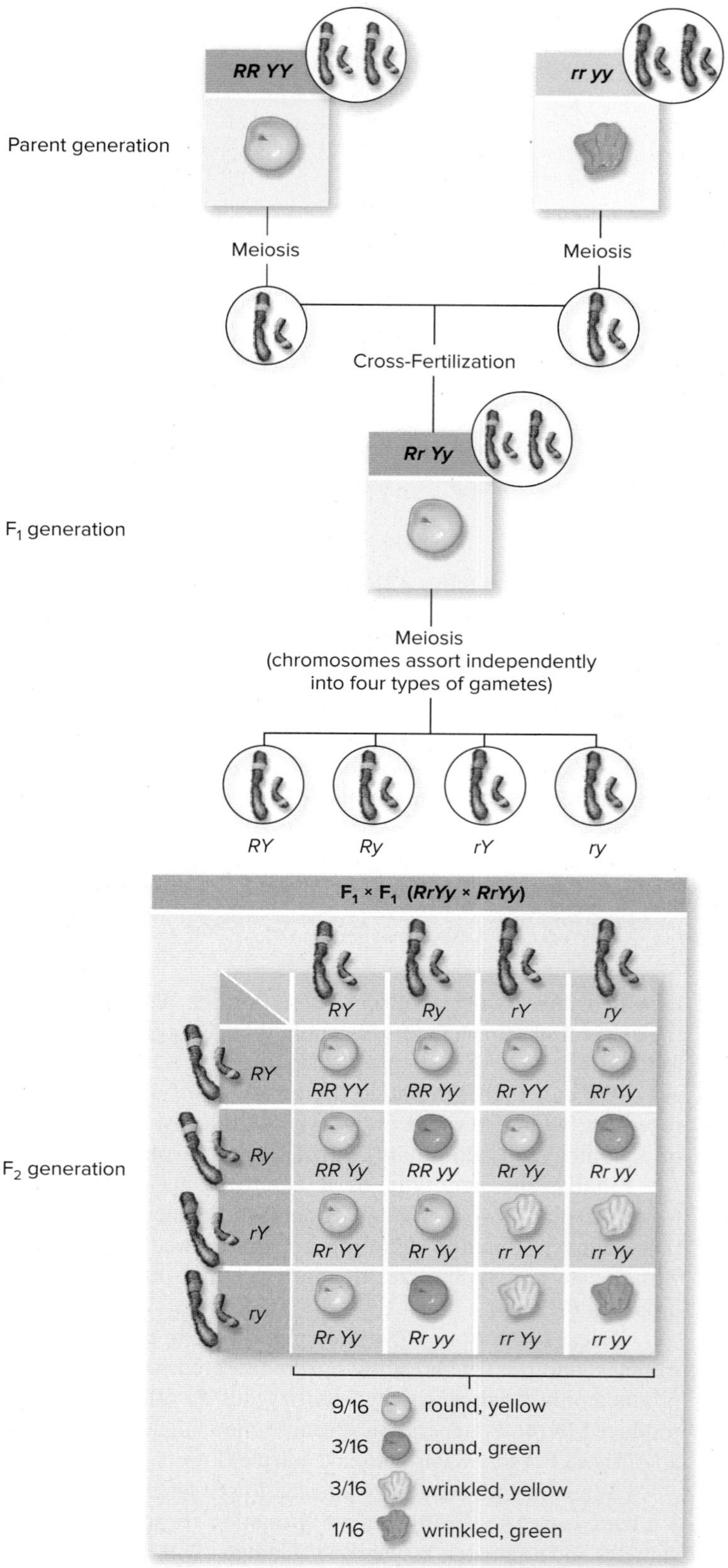

Figure 12.7 Analyzing a dihybrid cross. This Punnett square shows the results of Mendel's dihybrid cross between plants with round yellow seeds and plants with wrinkled green seeds. The ratio of the four possible combinations of phenotypes is predicted to be 9:3:3:1, the ratio that Mendel found.

Thus, Mendel concluded that the two traits were assorting independently of each other. Note that this *independent assortment* of different genes in no way alters the segregation of individual pairs of alleles for each gene. Round versus wrinkled seeds occur in a ratio of approximately 3:1 (423:133); so do yellow versus green seeds (416:140). Mendel obtained similar results for the other pairs of traits he tested.

We call this Mendel's Second Law of Heredity, or the **Principle of Independent Assortment.** This can also be stated simply: *In a cross, the alleles of a gene segregate independently of other genes.* In chapter 11 we saw that the different homologous chromosome pairs align independently during metaphase I of meiosis (refer to figure 11.7). This leads to the independent segregation of different genes.

REVIEW OF CONCEPT 12.3

Mendel's analysis of dihybrid crosses showed that the segregation of alleles for different traits is independent; this is known as Mendel's Principle of Independent Assortment. When individuals that differ in two traits are crossed and their progeny intercrossed, the result is four different types in a ratio of 9:3:3:1.

- *Which is more important in terms of explaining Mendel's laws, meiosis I or meiosis II?*

12.4 Probability Allows Us to Predict the Results of Crosses

Probability allows us to predict the likelihood of the outcome of random events. Because the behavior of different chromosomes during meiosis is independent, we can use probability to predict the outcome of crosses. This requires the use of only two simple rules.

Two Probability Rules Predict Cross Results

LEARNING OBJECTIVE 12.4.1 Apply the rule of addition and the rule of multiplication to genetic crosses.

Before we describe these rules and their uses, we need a definition. We say that two events are *mutually exclusive* if both cannot happen at the same time. The heads and tails of a coin flip are examples of mutually exclusive events. Notice that this is different from two consecutive coin flips where you can get two heads or two tails. In this case, each coin flip represents an *independent event.* The distinction between independent events and mutually exclusive events is the basis for our two rules.

The rule of addition

Consider a six-sided die instead of a coin: for any roll of the die, only one outcome is possible, and the possible outcomes are mutually exclusive. The probability of any particular number coming up is $\frac{1}{6}$. The probability of either of two different numbers is the sum of the individual probabilities, or restated as the **rule of addition:**

For two mutually exclusive events, the probability of either event occurring is the sum of the individual probabilities.

$$\text{Probability of rolling either a 2 or a 6 is } \tfrac{1}{6} + \tfrac{1}{6} = \tfrac{2}{6} = \tfrac{1}{3}$$

To apply this to our cross of heterozygous purple F_1 plants, four mutually exclusive outcomes are possible: *PP, Pp, pP,* and *pp.*

The probability of being heterozygous is the same as the probability of being either *Pp* or *pP*, or $\frac{1}{4}$ plus $\frac{1}{4}$, or $\frac{1}{2}$.

$$\text{Probability of } F_2 \text{ heterozygote} = \tfrac{1}{4}Pp + \tfrac{1}{4}pP = \tfrac{1}{2}$$

In the previous example, of 379 total offspring, we would expect $\frac{1}{2}$ of them, about 190, to be heterozygotes.

The rule of multiplication

The second rule, and by far the most useful for genetics, deals with the outcome of independent events. This is called the **product rule,** or **rule of multiplication,** and it states that the probability of two independent events both occurring is the *product* of their individual probabilities.

We can apply this to a monohybrid cross in which offspring are formed by gametes from each of two parents. Any particular outcome is the result of two independent events: the formation of two different gametes. Consider the purple F_1 parents from earlier. They are all *Pp* (heterozygotes), so the probability that a particular F_2 individual will be *pp* (homozygous recessive) is the probability of receiving a *p* gamete from the male ($\frac{1}{2}$) times the probability of receiving a *p* gamete from the female ($\frac{1}{2}$), or $\frac{1}{4}$:

$$\text{Probability of } pp \text{ homozygote} = \tfrac{1}{2}p \text{ (male parent)} \times \tfrac{1}{2}p \text{ (female parent)} = \tfrac{1}{4}pp$$

This is the basis for the Punnett square that we used before. Each cell in the square was the product of the probabilities of the gametes that contribute to the cell. We then use the addition rule to sum the probabilities of the mutually exclusive events that make up each cell.

We can use the result of a probability calculation to predict the number of homozygous recessive offspring in a cross between heterozygotes. For example, out of 379 total offspring, we would expect $\frac{1}{4}$ of them, about 95, to exhibit the homozygous recessive phenotype.

Analyzing a dihybrid cross

Probability analysis can be extended to the dihybrid case. We will use our example of seed shape and color from earlier. If the alleles affecting seed shape and seed color segregate independently, then the probability that a particular pair of alleles for seed shape would occur together with a particular pair of alleles for seed color is the product of the individual probabilities for each pair.

For example, the probability that an individual with wrinkled green seeds (*rr yy*) would appear in the F_2 generation would be equal to the probability of obtaining wrinkled seeds ($\frac{1}{4}$) times the probability of obtaining green seeds ($\frac{1}{4}$), or $\frac{1}{16}$.

$$\text{Probability of } rr\,yy = \tfrac{1}{4}\,rr \times \tfrac{1}{4}\,yy = \tfrac{1}{16}\,rr\,yy$$

Because of independent assortment, we can think of the dihybrid cross as consisting of two independent monohybrid crosses;

because these are independent events, the product rule applies. So, we can calculate the probabilities for each dihybrid phenotype:

Probability of round yellow ($R\_\_\ Y\_\_$) =
$\frac{3}{4}\ R\_\_ \times \frac{3}{4}\ Y\_\_ = \frac{9}{16}$

Probability of round green ($R\_\_\ yy$) =
$\frac{3}{4}\ R\_\_ \times \frac{1}{4}\ yy = \frac{3}{16}$

Probability of wrinkled yellow ($rr\ Y\_\_$) =
$\frac{1}{4}\ rr \times \frac{3}{4}\ Y\_\_ = \frac{3}{16}$

Probability of wrinkled green ($rr\ yy$) =
$\frac{1}{4}\ rr \times \frac{1}{4}\ yy = \frac{1}{16}$

The hypothesis that color and shape genes are independently sorted thus predicts that the F_2 generation will display a 9:3:3:1 phenotypic ratio—just as Mendel observed. In any dihybrid cross, these ratios can be applied to an observed total of offspring to predict the expected number in each phenotypic group. The underlying logic and the results are the same as those obtained using the Punnett square.

The Testcross Reveals Unknown Genotypes

LEARNING OBJECTIVE 12.4.2 Explain the outcome of a monohybrid testcross.

To test his model further, Mendel devised a simple and powerful procedure called the **testcross** (figure 12.8). In a testcross, an individual with unknown genotype is crossed with the homozygous recessive genotype—that is, the recessive parental variety. The contribution of the homozygous recessive parent can be ignored, because this parent can contribute only recessive alleles. Consider a purple-flowered pea plant. It is impossible to tell whether such a plant is homozygous or heterozygous simply by looking at it. To learn its genotype, you can perform a testcross to a white-flowered plant. In this cross, the two possible test plant genotypes will give different results (figure 12.8):

Alternative 1: Unknown individual is homozygous dominant (*PP*)
PP × *pp:* all offspring have purple flowers (*Pp*).
Alternative 2: Unknown individual is heterozygous (*Pp*) *Pp* × *pp:*
½ of offspring have white flowers (*pp*), and ½ have purple flowers (*Pp*).

Put simply, the appearance of the recessive phenotype in the offspring of a testcross indicates that the test individual is heterozygous for the gene in question.

TABLE 12.1 Dihybrid Testcross

Actual Genotype	Results of Testcross	
	Trait A	**Trait B**
AABB	Trait A breeds true.	Trait B breeds true.
AaBB	____	Trait B breeds true.
AABb	Trait A breeds true.	____
AaBb	____	____

For each pair of alleles Mendel investigated, he observed phenotypic F_2 ratios of 3:1 (figure 12.4) and testcross ratios of 1:1, just as his model had predicted. Testcrosses can also be used to determine the genotype of an individual when two genes are involved. Mendel often performed testcrosses to verify the genotypes of dominant-appearing F_2 individuals.

An F_2 individual exhibiting both dominant traits ($A\_\_\ B\_\_$) might have any of the following genotypes: *AABB, AaBB, AABb,* or *AaBb.* By crossing dominant-appearing F_2 individuals with homozygous recessive individuals (that is, $A\_\_\ B\_\_ \times aabb$), Mendel was able to determine whether either or both of the traits bred true among the progeny and so to determine the genotype of the F_2 parent (table 12.1).

REVIEW OF CONCEPT 12.4

The rule of addition states that the probability of either of two events occurring is the sum of their individual probabilities. The product rule states that the probability of two independent events both occurring is the product of their individual probabilities. These rules can be applied to predict the outcomes of genetic crosses. Crossing an individual with the dominant phenotype with a homozygous recessive individual, a testcross, will reveal its genotype.

- *In a testcross of Aa Bb CC by aa bb cc, what would be the expected phenotypic ratio?*

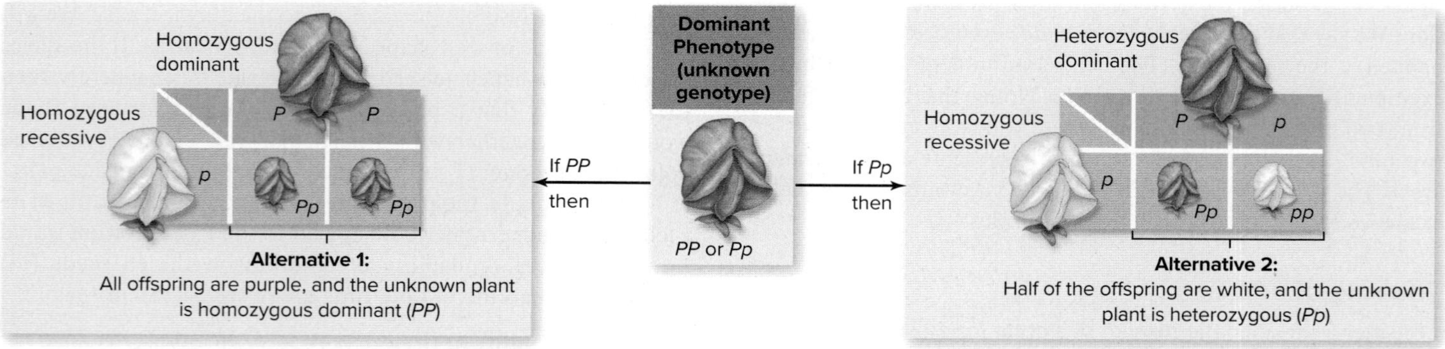

Figure 12.8 A testcross. To determine whether an individual exhibiting a dominant phenotype is homozygous or heterozygous for the dominant allele, Mendel crossed the individual with a plant that he knew to be homozygous recessive.

12.5 Extending Mendel's Model Provides a Clearer View of Genetics in Action

Although Mendel's results did not receive much notice during his lifetime, three other investigators independently rediscovered his pioneering paper in 1900, 16 years after his death. They came across it while searching the literature in preparation for publishing their own findings, which closely resembled those Mendel had presented more than 30 years earlier.

In the decades following the rediscovery of Mendel's ideas, many investigators set out to test them. However, scientists attempting to confirm Mendel's theory often had trouble obtaining the same simple ratios he had reported.

The reason Mendel's simple ratios were not obtained had to do with the traits that others examined. A number of assumptions are built into Mendel's model that are oversimplifications. These assumptions include that each trait is specified by a single gene with two alternative alleles; that there are no environmental effects; and that gene products act independently. The idea of dominance also hides a wealth of biochemical complexity. In this section, you'll see how Mendel's simple ideas can be extended to provide a more complete view of genetics.

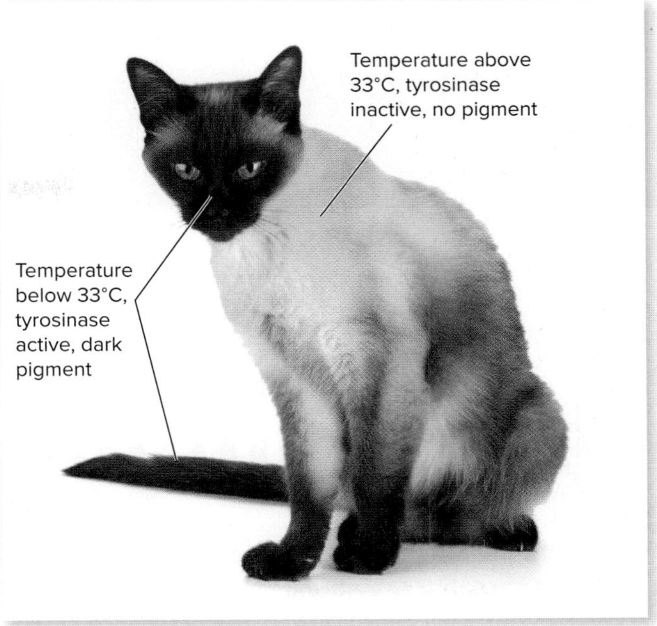
Temperature above 33°C, tyrosinase inactive, no pigment
Temperature below 33°C, tyrosinase active, dark pigment

Figure 12.9 Siamese cat. The pattern of coat color is due to an allele that encodes a temperature-sensitive form of the enzyme tyrosinase.
Evgeny Karandaev/Shutterstock

Phenotypes May Be Affected by the Environment

LEARNING OBJECTIVE 12.5.1 Explain how the environment might act to alter observed Mendelian ratios.

Perhaps the most obvious oversimplification is that the environment does not affect phenotype. Mendel ignored the environment, not unreasonably, as he performed his crosses under reasonably controlled conditions. Despite our attempts to control environmental conditions, it is clear that while genotype definitely produces phenotype, the environment lies in between. The production of different phenotypes for the same genotype due to environmental conditions is called **phenotypic plasticity.**

A particularly striking example involves sex determination in vertebrates. In birds and mammals, sex is determined by chromosomal constitution (refer to chapter 13), but many reptiles, and some fish, have so-called environmental sex determination. The incubation temperature for eggs determines the eventual sex of the offspring. What is most interesting about this observation is that the genes involved in the development of both male and female gonads are essentially the same in all vertebrates. This implies that the only differences are in the initiation of these developmental pathways. The striking pigment pattern seen in Siamese cats is due to an environmental effect on the enzyme tyrosinase, which catalyzes the first step in melanin synthesis. The cS allele in cats encodes a heat-sensitive variant of tyrosinase that is inactive in warmer central regions of the body, but active in colder peripheral regions (figure 12.9).

More Than One Gene Can Affect a Trait

LEARNING OBJECTIVE 12.5.2 Provide a genetic explanation of continuous variation.

Often, the relationship between genotype and phenotype is more complicated than a single allele producing a single trait. Most phenotypes do not reflect simple two-state alternatives like purple or white flowers.

Consider Mendel's crosses between tall and short pea plants. In reality, the "tall" plants have normal height, and the "short" plants are dwarfed by an allele at a single gene. But in most species of plants and animals, including humans, individual heights vary over a continuous range, rather than having discrete alternative values. This continuous distribution of a phenotype has a simple genetic explanation: more than one gene is at work. The mode of inheritance operating in this case is often called **polygenic inheritance.**

In reality, few phenotypes result from the action of only one gene. Instead, most characters reflect multiple additive contributions to the phenotype by several genes. When multiple genes act jointly to influence a character, such as height or weight, the character often shows a range of small differences. When these genes segregate independently, a gradation in the degree of difference can be observed when a group consisting of many individuals is examined (figure 12.10). We call this gradation **continuous variation,** and we call such traits **quantitative traits.** The greater the number of genes influencing a character, the more continuous the expected distribution of phenotypes.

This continuous variation in traits is similar to blending different colors of paint: combining one part red with seven parts

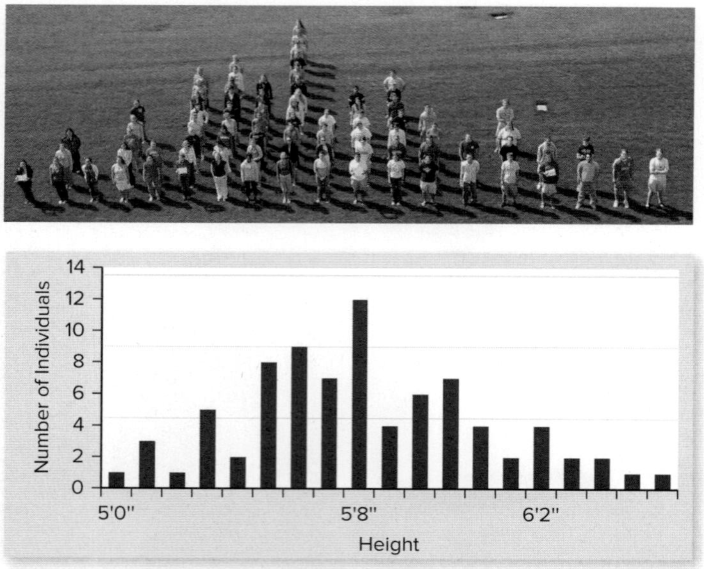

Figure 12.10 Height is a continuously varying trait. The photo and accompanying graph show variation in height among 83 students in Dr. Hyde's genetics classes from 2005 to 2007 at the University of Notre Dame.

David Hyde/Wayne Falda/McGraw Hill

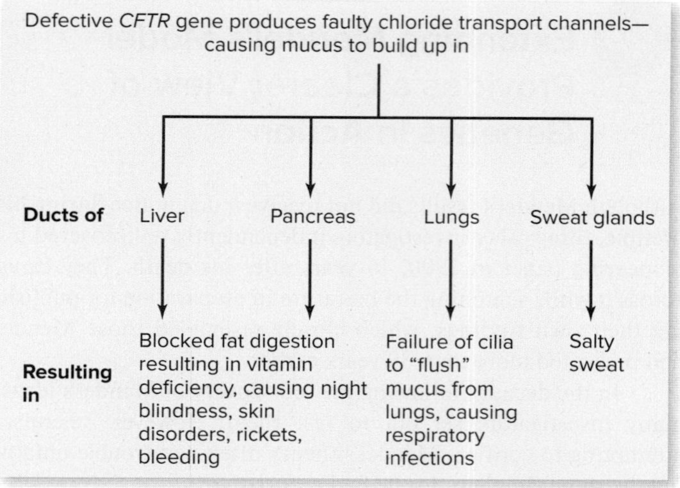

Figure 12.11 Pleiotropic effects of the cystic fibrosis gene. Cystic fibrosis (CF) is due to a single gene defect that prevents proper folding of a chloride ion channel in the plasma membrane. The mucus of CF patients loses water osmotically to cells as a result, becoming progressively thicker in many organs, leading to the listed symptoms.

white, for example, produces a much lighter shade of pink than does combining five parts red with three parts white. Different total amounts of red pigment in a quart of paint result in a continuum of shades, ranging from pure red to pure white.

Often, variations can be grouped into categories, such as different height ranges. Plotting the numbers in each height category produces a curve called a *histogram,* such as that shown in figure 12.10. The bell-shaped histogram approximates an idealized *normal distribution,* in which the central tendency is characterized by the mean and the spread of the curve indicates the amount of variation.

It is important to note that this kind of continuous variation can also be the result of environmental effects. Geneticists measure the variance observed in a phenotype, then use statistical techniques to partition this variance into environmental and genetic components. The variation in a phenotype that is due to genetic factors is called **heritability.** This can be measured and represents the amount of genetic variation that exists for a given trait in the population being analyzed.

A Single Gene Can Affect More Than One Trait

LEARNING OBJECTIVE 12.5.3 Explain the genetic basis of pleiotropic influences on inheritance.

Not only can more than one gene affect a single trait, but a single gene can affect more than one trait. Considering the complexity of biochemical pathways and the interdependent nature of organ systems in multicellular organisms, this should be no surprise.

An allele that has more than one effect on the phenotype is said to be **pleiotropic.** The pioneering geneticist Lucien Cuenot studied yellow fur in mice, a dominant trait, and found he was unable to obtain a pure-breeding yellow strain by crossing individual yellow mice with each other. Individuals homozygous for the yellow allele died, because the yellow allele was pleiotropic: one effect was yellow coat color, but another was a recessive lethal developmental defect.

A pleiotropic allele may be dominant with respect to one phenotypic consequence (yellow fur) and recessive with respect to another (lethal developmental defect). Pleiotropic effects are difficult to predict, because a gene that affects one trait often performs other, unknown functions.

Pleiotropic effects are characteristic of many inherited disorders in humans, including cystic fibrosis (figure 12.11) and sickle-cell anemia. In these disorders, multiple symptoms can be traced back to a single gene defect. Cystic fibrosis patients exhibit clogged blood vessels, overly sticky mucus, salty sweat, liver and pancreas failure, and several other symptoms (phenotypes). It is often difficult to deduce the nature of the primary defect from the range of a gene's pleiotropic effects. As it turns out, all these symptoms of cystic fibrosis are pleiotropic effects of a single mutation in a gene that encodes a chloride ion transmembrane channel. Ion channels such as these were discussed in detail in chapter 9.

Genes May Have More Than Two Alleles

LEARNING OBJECTIVE 12.5.4 Estimate the maximum number of alleles a gene may possess, and explain your estimate.

Mendel always looked at genes with two alternative alleles. Although any diploid individual can carry only two alleles for a gene, there may be more than two versions of a gene in a population. The example of ABO blood types in humans, described later in this section, involves an allelic series with three alleles.

If you think of a gene as a sequence of nucleotides in a DNA molecule, then the number of possible alleles is huge, because even a single change in any of the nucleotides could produce a new allele. In reality, more than two alleles usually exist for any gene in an outbreeding population, although the number of alleles observed for any gene is rarely more than a few dozen. The dominance relationships of these alleles are determined by observing the phenotypes for the various heterozygous combinations.

Dominance Is Not Always Complete

LEARNING OBJECTIVE 12.5.5 Explain how to distinguish between lack of dominance and incomplete dominance.

Mendel's observation of dominant and recessive traits can seem hard to explain in terms of modern biochemistry. For example, if a recessive trait is caused by the loss of function of an enzyme encoded by the recessive allele, then why should a heterozygote, with only half the activity of this enzyme, have the same appearance as a homozygous dominant individual? One answer is that if one functional allele provides the necessary level of enzyme activity for the relevant chemical reaction, dominance will be complete. If one functional allele does not provide sufficient enzyme activity, then you may see a different phenotype in the heterozygote.

Incomplete dominance

In **incomplete dominance,** the phenotype of the heterozygote is intermediate between those of the two homozygotes. For example, in a cross between red- and white-flowering Japanese four o'clock plants, all the F_1 offspring have pink flowers—indicating that neither red nor white flower color is dominant. Looking only at the F_1, we might conclude that this is a case of blending inheritance. But when two of the F_1 pink flowers are crossed, they produce red-, pink-, and white-flowered plants in a 1:2:1 ratio. In this case, the phenotypic ratio is the same as the genotypic ratio, because all three genotypes can be distinguished (figure 12.12).

We will consider human genetics in detail in chapter 13, but we will consider some examples involving dominance here. There is a gene that affects curly versus. straight hair, which displays incomplete dominance. A homozygous dominant individual has curly hair, a homozygous recessive individual has straight hair, and a heterozygous individual has wavy hair.

Codominance

Most genes in a population possess several different alleles, and often no single allele is dominant; instead, each allele has its own effect on the phenotype, and the heterozygote shows some aspect of the phenotype of both homozygotes. The alleles are said to be **codominant.**

Codominance can be distinguished from incomplete dominance by the appearance of the heterozygote. In incomplete dominance, the heterozygote is intermediate between the two homozygotes, whereas in codominance, some aspect of both alleles is seen in the heterozygote. One of the clearest human examples is found in the human blood groups.

The different phenotypes of human blood groups are based on the response of the immune system to proteins on the surface of red blood cells. In homozygotes, a single type of protein is found on the surface of cells, and in heterozygotes, two kinds of protein are found, leading to codominance.

SCIENTIFIC THINKING

Hypothesis: *The pink F_1 observed in a cross of red and white Japanese four o'clock flowers is due to failure of dominance and is not an example of blending inheritance.*
Prediction: *If pink F_1 are self-crossed, they will yield progeny the same as the Mendelian monohybrid genotypic ratio. This would be 1 red: 2 pink: 1 white.*
Test: *Perform the cross and count progeny.*

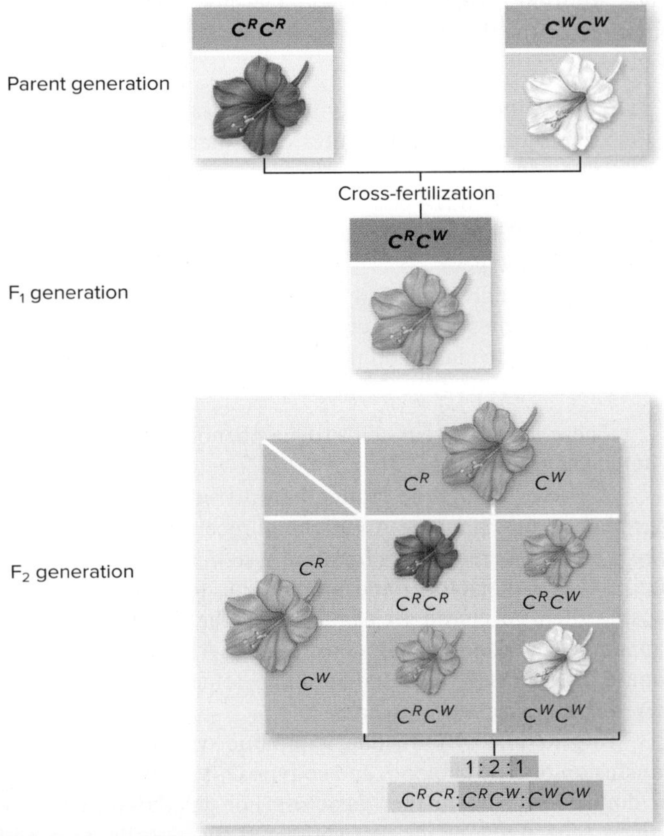

Result: *When this cross is performed, the expected outcome is observed.*
Conclusion: *Flower color in Japanese four o'clock plants exhibits incomplete dominance.*
Further Experiments: *How many offspring would you need to count to be confident in the observed ratio?*

Figure 12.12 Incomplete dominance. In a cross between a red-flowered (genotype C^RC^R) Japanese four o'clock and a white-flowered one (C^WC^W), neither allele is dominant. The heterozygous progeny have pink flowers and the genotype C^RC^W. If two of these heterozygotes are crossed, the phenotypes of their progeny occur in a ratio of 1:2:1 (red:pink:white).

The human ABO blood group system

The gene that determines ABO blood types encodes an enzyme that adds sugar molecules to proteins on the surface of red blood cells. These sugars act as recognition markers for the immune system (refer to chapter 35). The gene that encodes the enzyme, designated *I,* has three common alleles: I^A, whose product adds galactosamine; I^B, whose product adds galactose; and *i,* which codes for a protein that does not add sugar.

The three alleles of the *I* gene can be combined to produce six different genotypes. An individual heterozygous for the I^A and I^B alleles produces both forms of the enzyme and exhibits both galactose and galactosamine on red blood cells. Because both alleles are expressed simultaneously in heterozygotes, the I^A and I^B alleles are codominant. Both I^A and I^B are dominant over the *i* allele, because both I^A and I^B alleles lead to sugar addition, whereas the *i* allele does not. The different combinations of the three alleles produce four different phenotypes (figure 12.13):

1. Type A individuals add only galactosamine. They are either I^AI^A homozygotes or I^Ai heterozygotes (two genotypes).
2. Type B individuals add only galactose. They are either I^BI^B homozygotes or I^Bi heterozygotes (two genotypes).
3. Type AB individuals add both sugars and are I^AI^B heterozygotes (one genotype).
4. Type O individuals add neither sugar and are *ii* homozygotes (one genotype).

These four different cell-surface phenotypes are called the **ABO blood groups.**

A person's immune system can distinguish among these four phenotypes. If a type A individual receives a transfusion of type B blood, the recipient's immune system recognizes the "foreign" antigen (galactose) and attacks the donated blood cells, causing them to clump, or agglutinate. The same thing would happen if the donated blood were type AB. However, if the donated blood is type O, no immune attack occurs, because there are no galactose antigens.

In general, any individual's immune system can tolerate a transfusion of type O blood, and so type O is termed the "universal donor." Because neither galactose nor galactosamine is foreign to type AB individuals (whose red blood cells have both sugars), those individuals may receive any type of blood, and type AB is termed the "universal recipient." Nevertheless, matching blood is preferable for any transfusion.

Alleles	Blood Type	Sugars Exhibited	Donates and Receives
I^AI^A, I^Ai (I^A dominant to *i*)	A	Galactosamine	Receives A and O Donates to A and AB
I^BI^B, I^Bi (I^B dominant to *i*)	B	Galactose	Receives B and O Donates to B and AB
I^AI^B (codominant)	AB	Both galactose and galactosamine	Universal recipient Donates to AB
ii (*i* is recessive)	O	None	Receives O Universal donor

Figure 12.13 ABO blood groups illustrate both codominance and multiple alleles. There are three alleles of the *I* gene: I^A, I^B, and *i*. I^A and I^B are both dominant to *i* (see types A and B) but are codominant to each other (see type AB). The genotypes that give rise to each blood type are shown with the associated phenotypes in terms of sugars added to surface proteins and the body's reaction after a blood transfusion.

Gene Interactions May Alter Observed Genetic Ratios

LEARNING OBJECTIVE 12.5.6 Explain the genetic basis of a dihybrid phenotypic ratio of 9:7.

The last overly simple assumption in Mendel's model is that the products of genes do not interact. In fact, the product of one gene may influence the products of other genes. If gene products do not act independently, this may alter expected ratios in dihybrid crosses, even when alleles of the genes themselves are segregating independently. Given the interconnected nature of metabolism, it should not come as a surprise that many gene products are influenced by other gene products. For example, in a metabolic pathway, each gene is dependent on the action of genes that act earlier in the pathway. This can lead to different results in crosses than we have seen so far.

You have most likely encountered a common example already without knowing it: the different coat colors in Labrador retrievers (Labs). There are three coat colors in Labs: black Labs, brown (or chocolate) Labs, and yellow Labs. These 3 coat colors are primarily controlled by two genes: the *Brown* locus, with *B* or *b* alleles, and the *Extension* locus, with *E* or *e* alleles. Let's consider a cross between two individuals, both phenotypically black, and both heterozygous for Brown and Extension (*Bb Ee* × *Bb Ee*). In our dihybrid framework, we expect four phenotypic classes in the progeny of this cross. But we have only three kinds of Lab, so even before we see the results, we know they will deviate from expectation. The observed phenotypic ratio for this cross is 9 black:3 brown:4 yellow (figure 12.14).

Some background on mammalian pigmentation will shed light on this cross. Mammals make only one kind of pigment, but in two different forms. The pigment is melanin, and eumelanin is dark and appears black or brown, while pheomelanin appears yellow or red. The Brown gene affects the amount of eumelanin in hairs and skin such that the dominant B allele results in a black coat color, and the recessive b allele causes a brown coat color. This gene shows simple Mendelian inheritance: a monohybrid cross of *Bb* × *Bb* yields 3 black:1 brown.

The *Extension* gene encodes a receptor that controls the distribution of pigment in different tissues. The wild-type E allele results in uniform deposition of eumelanin in hair and skin, and leads to a black or brown coat depending on the genotype at the brown gene. The e allele is a recessive mutation that inactivates the receptor, and causes eumelanin to be deposited only in skin, and not in hairs. This results in a yellow coat color because of the presence of pheomelanin (controlled by another gene). The pheomelanin is not seen in black or brown Labs because it is obscured by the darker eumelanin. In the absence of eumelanin, the yellow to red pheomelanin produces the coat color. There are four yellow Labs in our cross because the genotype *BBee* and the two *Bbee* genotypes are all yellow, as well as the bbee genotype. Note that there is actually a difference among these yellow Labs as regards the color of their muzzle, which can be either black (*B_*) or brown (*bb*), so we are only considering the coat color in this analysis.

The action of one gene obscuring the effects of another gene is called **epistasis,** and in this case, the observed 9:3:4 ratio is really a 9:3:(3+1) ratio. In the case of Labs, because the distribution of pigment in the skin is different from that in the hairs of the coat,

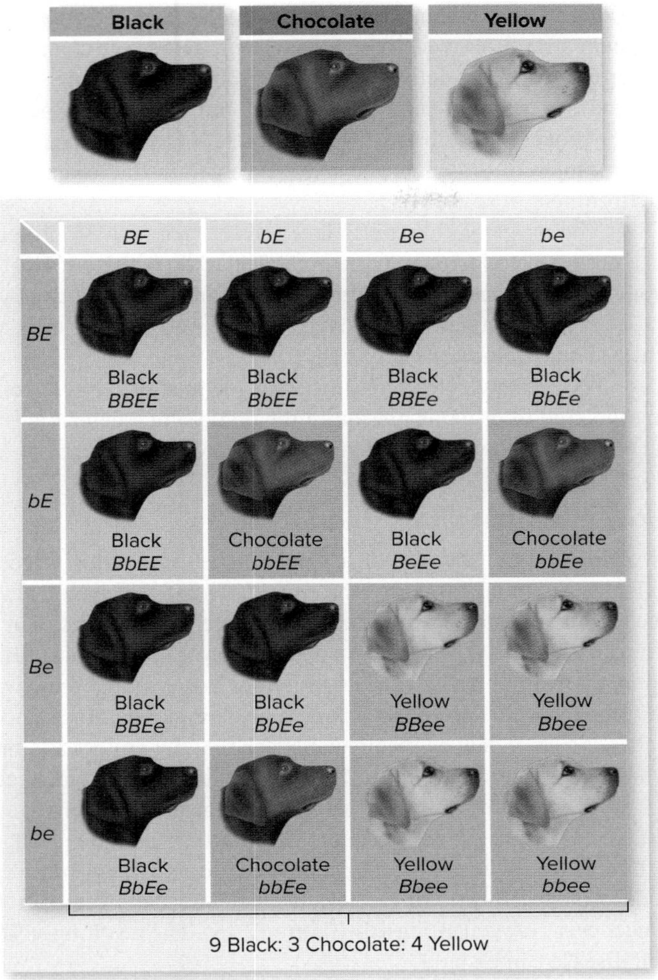

Figure 12.14 Labrador retriever coat color is due to epistasis. A cross of two black Labs produced the data shown in this Punnett square. The parent animals were each heterozygous for both the *Brown* locus and the *Extension* locus. The first affects the amount of pigment, and the second the distribution of pigment.

we actually can distinguish a difference in the color of the muzzle and footpads among the yellow animals: only the double recessive individual (*bbee*) would have a brown muzzle and foot pads.

Another example of epistasis comes from the analysis of kernel color in corn. Some varieties of corn have purple kernels due to the pigment anthocyanin, and others lack this pigment and are colorless. Two true-breeding colorless varieties are crossed, and this produces all colored offspring. This would be puzzling if only one gene were involved, but in this case, there are two genes that each encode a different enzyme needed to make anthocyanin. The two colorless varieties were each homozygous for a different gene needed to make anthocyanin. The colored offspring in turn were all heterozygous for both genes. If these colored offspring are self-crossed, they produce a ratio of 9 purple:7 colorless. This may sound strange, but it is another alteration of the usual Mendelian dihybrid ratio: 9:(3 + 3 + 1).

The examples in this section do not invalidate Mendel's model but, rather, extend it to provide a more complete version of inheritance. Table 12.2 summarizes a variety of genetic phenomena that can alter classical Mendelian expectations.

REVIEW OF CONCEPT 12.5

Mendel's model assumes that each trait is specified by one gene with only two alleles, that no environmental effects alter a trait, and that gene products act independently. All of these are oversimplifications. Traits produced by the action of multiple genes (polygenic inheritance) have continuous variation. One gene can affect more than one trait (pleiotropy). Genes may have more than two alleles, and these may not show simple dominance. The action of genes is not always independent, which can result in modified dihybrid ratios.

- *In mice, the dominant yellow allele of the agouti gene causes yellow fur, and is lethal when homozygous. What would you expect from a cross of yellow by yellow?*

TABLE 12.2 When Mendel's Laws/Results Might Not Be Observed

Genetic Occurrence	Definition	Examples
Polygenic inheritance	More than one gene can affect a single trait.	• Four genes are involved in determining eye color. • Human height
Pleiotropy	A single gene can affect more than one trait.	• A pleiotropic allele dominant for yellow fur in mice is recessive for a lethal developmental defect. • Cystic fibrosis • Sickle-cell anemia
Multiple alleles for one gene	Genes may have more than two alleles.	ABO blood types in humans
Dominance is not always complete.	• In incomplete dominance the heterozygote is intermediate. • In codominance no single allele is dominant, and the heterozygote shows some aspect of both homozygotes.	• Japanese four o'clocks • Human blood groups
Environmental factors	Genes may be affected by the environment.	Siamese cats
Gene interaction	Products of genes can interact to alter genetic ratios.	• The production of a purple pigment in corn • Coat color in mammals

12.6 Genotype Dictates Phenotype by Specifying Protein Sequences

Before we move on from Mendelian genetics, it is worth considering the actual nature of Mendel's factors, or genes. We will explore this in detail in later chapters, but for now we can sketch in broad strokes how a Mendelian trait is influenced by a particular gene, how a gene can be altered by mutation, and the potential long-term evolutionary consequences of such alterations to genes. We will use the protein hemoglobin as our example. Hemoglobin is a complex protein that in adults consists of four polypeptide chains: two α- and two β-chains.

How Genes Influence Traits

LEARNING OBJECTIVE 12.6.1 Explain how genotype determines phenotype.

From DNA to protein

Each body cell of an individual contains the same set of DNA molecules, called the genome of that individual. As you learned in chapter 3, DNA molecules are long chains of nucleotide subunits. There are four kinds of nucleotides (A, T, C, and G), and like an alphabet with four letters, the order of nucleotides determines the message encoded in the DNA of a gene.

The human genome contains 20,000 to 25,000 genes. The DNA of the human genome is subdivided into 23 pairs of chromosomes, each chromosome containing from 1000 to 2000 different genes. The genes for α- and β-globin are found in two clusters on chromosomes 16 and 11, respectively. These clusters contain different versions of the genes that are expressed at different developmental times.

The information in DNA is "read" by enzymes that create an RNA strand of the same sequence (except U is substituted for T). The RNA transcripts for α- and β-hemoglobin are exported to the cytoplasm, where they act as instructions for protein production by the ribosome. But in eukaryotic cells, the RNA transcript has more information than is needed, so it is first "edited" to remove unnecessary bits before it leaves the nucleus. For example, the initial RNA gene transcript encoding the β-subunit of the protein hemoglobin is 1660 nucleotides long; after "editing," the resulting "messenger" RNA is 1000 nucleotides long.

After an RNA transcript is edited, it leaves the nucleus as messenger RNA (mRNA) and is delivered to ribosomes in the cytoplasm. Each ribosome is a tiny protein-assembly plant, and it uses the sequence of the messenger RNA to determine the amino acid sequence of a particular polypeptide. In the case of β-hemoglobin, the messenger RNA encodes a polypeptide strand of 146 amino acids.

How proteins determine the phenotype

As we saw in chapter 3, polypeptide chains of amino acids spontaneously fold in water into complex three-dimensional shapes. Two α- and two β-hemoglobin polypeptides fold into compact, globular structures, which then associate together with four heme groups to form an active hemoglobin protein molecule that is present in red blood cells. The hemoglobin molecules bind oxygen (a process described fully in chapter 34) in the oxygen-rich environment of the lungs and release oxygen in the oxygen-poor environment of active tissues.

The oxygen-binding efficiency of the hemoglobin proteins in a person's bloodstream has a great deal to do with how well the body functions, particularly under conditions of strenuous physical activity, when delivery of oxygen to the body's muscles is the chief factor limiting the activity.

As a general rule, genes influence the phenotype by specifying the kind of proteins present in the body, which determines in large measure how that body functions.

How mutation alters phenotype

A change in the identity of a single nucleotide within a gene, called a mutation, can have a profound effect if the change alters the identity of the amino acid encoded there. When a mutation of this sort occurs, the new version of the protein may fold differently, altering or destroying its function. For example, how well the hemoglobin protein performs its oxygen-binding duties depends a great deal on the precise shape that the protein assumes when it folds. A change in the identity of a single amino acid can have a drastic effect on that final shape. In particular, a change in the sixth amino acid of β-hemoglobin from glutamic acid to valine causes the hemoglobin molecules to aggregate into stiff rods, which deform blood cells into a sickle shape that can no longer carry oxygen efficiently (figure 12.15). The resulting sickle-cell disease can be fatal.

Natural selection for alternative phenotypes leads to evolution

Because random mutations occur in all genes occasionally, populations usually contain several versions of a gene, usually all but one of them rare. Sometimes the environment changes in such a way that one of the rare versions functions better under the new conditions. When that happens, natural selection will favor the rare allele, which will then become more common. The sickle-cell version of the β-hemoglobin gene, rare throughout most of the world, is common in Central Africa, because heterozygous individuals obtain enough functional hemoglobin from their one normal allele to get along but are resistant to malaria, a deadly disease common there, due to their other sickle-cell allele.

REVIEW OF CONCEPT 12.6

Genes determine phenotypes by specifying the amino acid sequences, and thus the functional shapes, of the proteins that carry out cell activities. Mutations, by altering protein sequence, can change a protein's function and thus alter the phenotype in evolutionarily significant ways.

- *How does a single nucleotide change in the gene for β-hemoglobin lead to organ damage and death?*

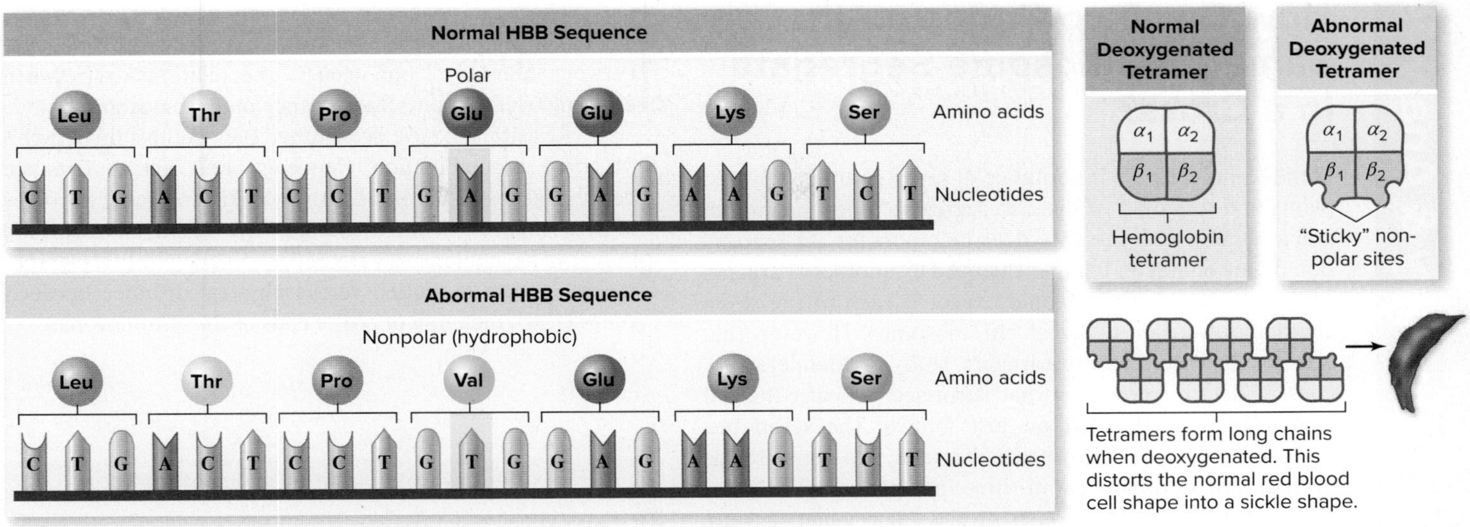

Figure 12.15 Sickle-cell anemia is caused by an altered protein. Hemoglobin is composed of a tetramer of two α-globin and two β-globin chains. The sickle-cell allele of the β-globin gene contains a single base change, resulting in the substitution of Val for Glu. This creates a hydrophobic region on the surface of the protein that is "sticky," leading to the association of tetramers into long chains that distort the shape of the red blood cells.

How Can Two Genes on the Same Chromosome Segregate in a Cross?

Mendel analyzed the inheritance of seven characters in pea plants and found all of them to segregate independently of one another. In no instance does he report the 3:1 segregation of one character being influenced by another character. This result is what you would expect if each of the seven characters were on different chromosomes. However, that was not the case. Later researchers, studying Mendel's varieties in detail, were somewhat surprised to learn that all seven genes are located on just four of the seven pea chromosomes—as shown in the diagram, two genes reside on the same chromosome in three instances. Pod shape is shown twice, on chromosomes #1 and #5, as geneticists are not sure which of these genes Mendel studied.

How, then, did Mendel observe all seven characters segregating independently? Well, for one thing, he didn't try all possible pairs. For example, he never tested pod shape versus plant height, the two genes located close to each other on chromosome #5. If he had, he would have found what other researchers have since discovered—that these two characters do not segregate independently in a dihybrid cross.

In another instance, however, Mendel did carry out dihybrid crosses between two characters located on the same chromosome. The flower color and seed color genes Mendel studied in garden peas are both located on chromosome #2—and yet Mendel observed them to segregate independently—that is, he observed 50% recombination among alleles, the same as random chance would predict.

These two genes are located very far apart, at opposite ends of chromosome #2. We know that crossing over occurs between homologs, but what is the relationship between crossing over and physical distance on chromosomes?

Researchers have determined the relationship between observed recombination frequency and actual distance between genes on chromosomes by analyzing genetic crosses of *Drosophila,* corn, and the fungus *Neurospora,* all organisms that have been intensively studied by geneticists. In the graph, recombination is plotted versus physical distance between genes (100% meaning opposite ends of the chromosome).

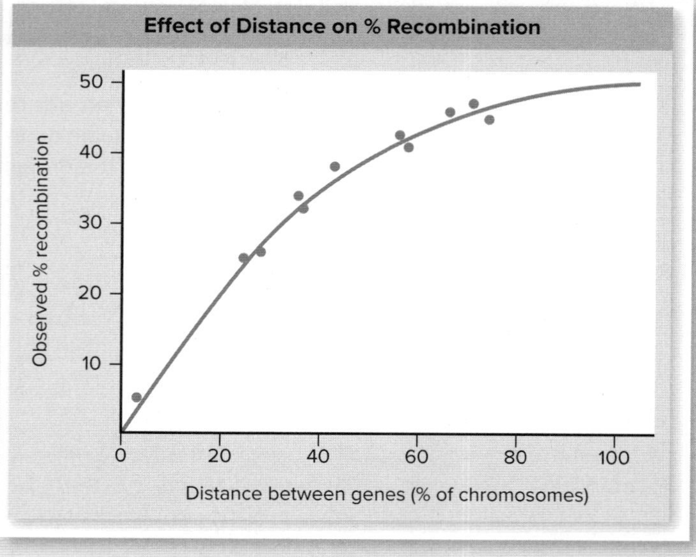
Effect of Distance on % Recombination
Observed % recombination
10
20
30
40
50
0
20
40
60
80
100
Distance between genes (% of chromosomes)

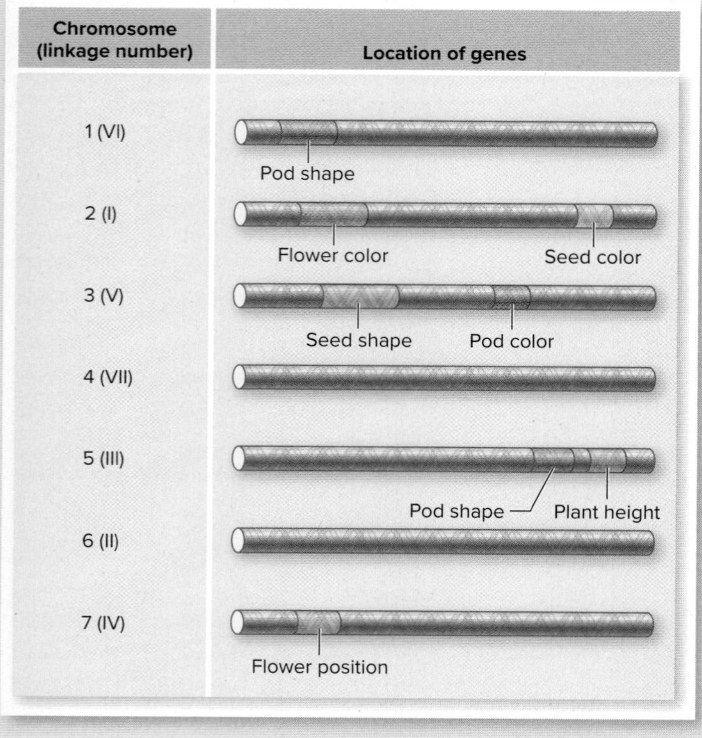
Chromosome (linkage number)
Location of genes
1 (VI)
Pod shape
2 (I)
Flower color
Seed color
3 (V)
Seed shape
Pod color
4 (VII)
5 (III)
Pod shape
Plant height
6 (II)
7 (IV)
Flower position

Analysis

1. **Applying Concepts**
   a. ***Variable.*** What is the dependent variable?
   b. ***Frequency.*** What is the highest frequency of recombination observed?
2. **Interpreting Data**
   a. Does the frequency of recombination increase as the distance between genes increases?
   b. Is the relationship linear?
3. **Making Inferences** Do genes that are located farther apart on a chromosome recombine more often? To what maximum value? Explain why recombination values greater than 50% are not observed. Can you imagine any circumstances when such values might be observed?
4. **Drawing Conclusions** Using the relationship established in the experimental curve, estimate how close together Mendel's flower and seed color genes are on chromosome #2, given that Mendel observed independent segregation (that is, 50% recombination) between them.

Retracing the Learning Path

CONCEPT 12.1 Experiments Carried Out by Mendel Explain Inheritance

12.1.1 The Mystery of Heredity Was Solved in Stages Plant breeders noticed that some forms of a trait disappear in one generation only to reappear later; they segregate rather than blend. Mendel studied heredity by crossing true-breeding garden peas that differed in easily scored alternative traits.

12.1.2 Mendel's Experimental Design Was Quantitative, a Radical Change Mendel's experiments involved reciprocal crosses between true-breeding pea varieties followed by one or more generations of self-fertilization. Mendel carefully counted the numbers of each type of progeny.

CONCEPT 12.2 Mendel's Principle of Segregation Accounts for 3:1 Phenotypic Ratios

12.2.1 Mendel Observed That Alternative Forms of a Trait Segregate in Crosses Mendel crossed two contrasting traits and counted offspring. All of the offspring exhibited one (dominant) trait, and none exhibited the other (recessive) trait. In the next generation, 25% were true-breeding dominant, 50% were not true-breeding dominant, and 25% were true-breeding recessive. This gives a 3 dominant:1 recessive phenotypic ratio.

12.2.2 Mendel's Principle of Segregation Explains Monohybrid Observations Traits are determined by discrete factors we now call genes, which exist in alternative forms we call alleles. Individuals carrying two identical alleles are homozygous, and individuals carrying different alleles are heterozygous. The genotype is the entire set of alleles of all genes possessed by an individual. The phenotype is an individual's appearance due to these alleles. The Principle of Segregation states that during gamete formation, alleles of a gene are segregated into different gametes. The physical basis of segregation is the separation of homologs during anaphase I of meiosis.

12.2.3 The Punnett Square Allows Symbolic Analysis Punnett squares are formed by placing the gametes from one parent along the top of the square, and the gametes from the other parent along the left side. The genotypes of zygotes are displayed as the blocks of the square.

CONCEPT 12.3 Mendel's Principle of Independent Assortment Asserts That Genes Segregate Independently

12.3.1 Traits in a Dihybrid Cross Behave Independently If parents differing in two traits are crossed, the F_1 will be all dominant. Each F_1 parent can produce four different gametes that can be combined to produce 16 possible outcomes in the F_2. This yields a phenotypic ratio of 9:3:3:1 of the four possible phenotypes. The Principle of Independent Assortment states that different traits segregate independently of one another. The physical basis of this is the independent behavior of different pairs of homologous chromosomes during meiosis I.

CONCEPT 12.4 Probability Allows Us to Predict the Results of Crosses

12.4.1 Two Probability Rules Predict Cross Results The rule of addition states that the probability of two independent events occurring is the sum of their individual probabilities. The product rule states that the probability of two independent events *both* occurring is the product of their individual probabilities. The product rule applies to dihybrid crosses and can be used to predict their outcomes.

12.4.2 The Testcross Reveals Unknown Genotypes In a testcross, an unknown genotype is crossed with a homozygous recessive genotype. The F_1 offspring will all be the same if the unknown genotype is homozygous dominant. The F_1 offspring will exhibit a 1:1 dominant:recessive ratio if the unknown genotype is heterozygous.

CONCEPT 12.5 Extending Mendel's Model Provides a Clearer View of Genetics in Action

12.5.1 Phenotypes May Be Affected by the Environment Genotype determines phenotype, but the environment, both internal and external, also influences the realized phenotype. For example, in Siamese cats a temperature-sensitive enzyme produces more pigment in the colder peripheral areas of the body.

12.5.2 More Than One Gene Can Affect a Trait Many traits, such as human height, reflect the multiple additive contributions by many genes, resulting in continuous variation.

12.5.3 A Single Gene Can Affect More Than One Trait A pleiotropic effect occurs when an allele affects more than one trait. These effects are difficult to predict.

12.5.4 Genes May Have More Than Two Alleles There may be more than two alleles of a gene in a population. Given the possible number of DNA sequences, this is not surprising.

12.5.5 Dominance Is Not Always Complete In incomplete dominance, heterozygotes have an intermediate phenotype; monohybrid genotypic and phenotypic ratios are the same. Codominant alleles each contribute to the phenotype of a heterozygote.

12.5.6 Gene Interactions May Alter Observed Genetic Ratios When one gene modifies the phenotypic expression of another gene, the interaction is an example of epistasis. Epistasis can modify Mendelian ratios in dihybrid crosses. In Labrador retrievers, coat color is due to interacting genes that affect the type and distribution of pigment. Crosses of doubly heterozygous individuals yield a 9 black:3 chocolate:4 yellow ratio.

CONCEPT 12.6 Genotype Dictates Phenotype by Specifying Protein Sequences

12.6.1 How Genes Influence Traits Genes determine phenotypes by specifying amino acid sequences, and thus the functional shapes, of proteins that carry out cell activities. Mutations alter protein sequences and change protein function and thus alter phenotype in an evolutionarily significant way.

Concept Overview

This Concept Overview diagrams the key concepts that were discussed in this chapter.

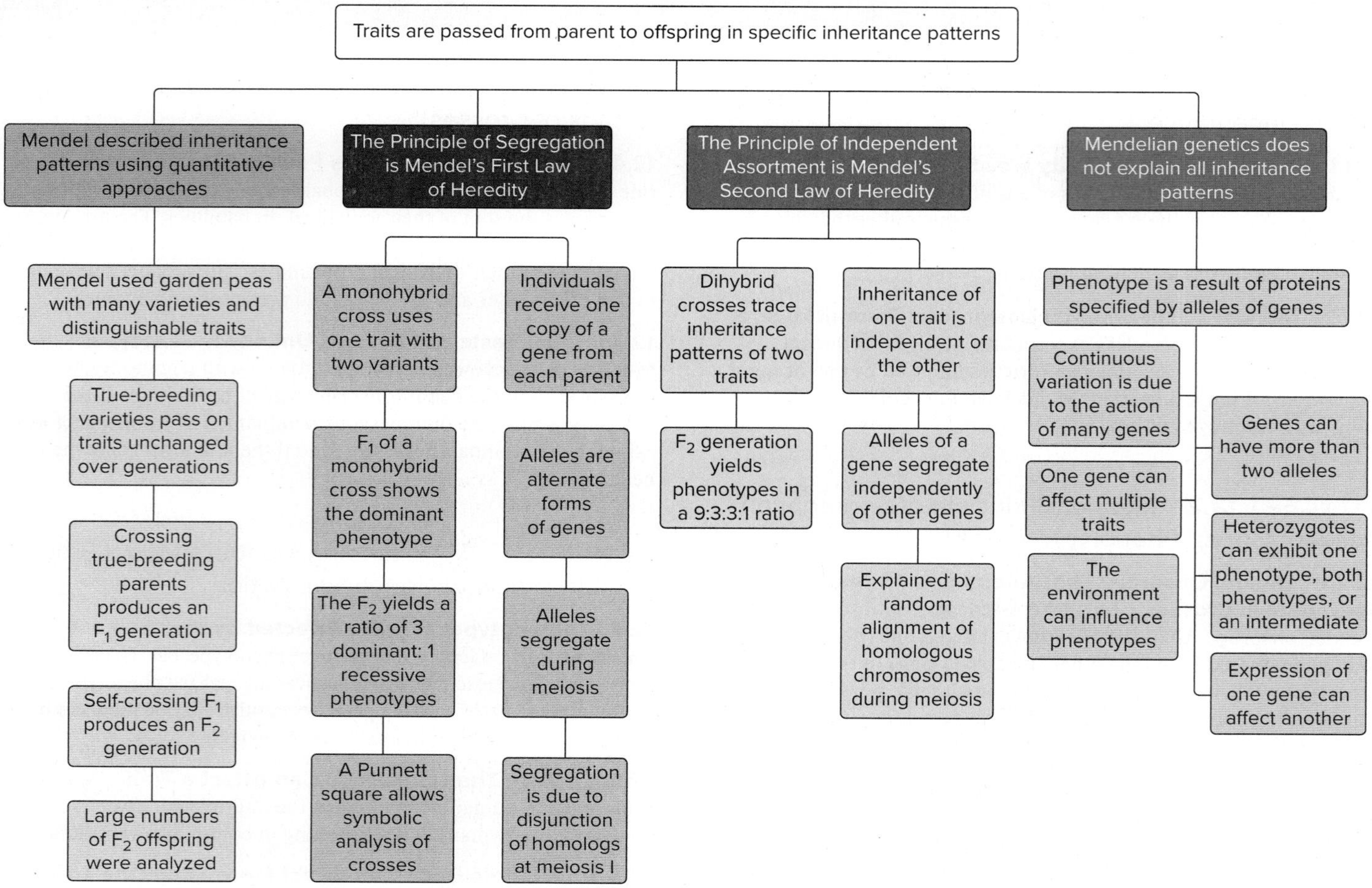

Assessing the Learning Path

Understand

1. What property distinguished Mendel's investigation from previous studies?
   a. Mendel used true-breeding pea plants.
   b. Mendel quantified his results.
   c. Mendel examined many different traits.
   d. Mendel examined the segregation of traits.
2. The F_1 generation of the monohybrid cross purple (*PP*) × white (*pp*) flower pea plants should
   a. all have white flowers.
   b. all have a light purple or blended appearance.
   c. all have purple flowers.
   d. have ¾ purple flowers and ¼ white flowers.
3. An organism's _____ is/are determined by its ________.
   a. genotype; phenotype
   b. phenotype; genotype
   c. alleles; phenotype
   d. genes; alleles
4. In all of Mendel's monohybrid crosses, the F_2 plants displayed a 3:1 ratio of dominant to recessive traits. Of those showing the dominant trait, what proportion were true-breeding?
   a. ¼
   b. ⅓
   c. ⅔
   d. None
5. A dihybrid cross between a plant with long, smooth leaves and a plant with short, hairy leaves produces a long, smooth F_1. If this F_1 is allowed to self-cross to produce an F_2, what would you predict for the ratio of F_2 phenotypes?
   a. 9 long, smooth:3 long, hairy:3 short, hairy:1 short, smooth
   b. 9 long, smooth:3 long, hairy:3 short, smooth:1 short, hairy
   c. 9 short, hairy:3 long, hairy:3 short, smooth:1 long, smooth
   d. 1 long, smooth:1 long, hairy:1 short, smooth:1 short, hairy
6. Of the following crosses, which is a testcross?
   a. *WW* × *WW*
   b. *WW* × *Ww*
   c. *Ww* × *ww*
   d. *Ww* × *W*

7. Consider a long, smooth F_2 from question 5. This plant's genotype
   a. must be homozygous for both leaf texture and length alleles.
   b. must be heterozygous for both leaf texture and length alleles.
   c. must be homozygous for one allele and heterozygous for the other allele.
   d. could be homozygous or heterozygous for both leaf texture and leaf length alleles.
8. Which is most important in explaining the idea behind Mendel's principle of independent assortment?
   a. Meiosis I
   b. Meiosis II
   c. Mitosis
   d. Both a and b are equally important.
9. To predict the outcome of independent events, it is useful to
   a. use the product rule.
   b. use the rule of addition.
   c. use the law of segregation.
   d. use a testcross.
10. Changing one nucleotide in a gene _____ changes the amino acid sequence of the protein and _____ changes the function of the protein.
    a. always; always
    b. always; sometimes
    c. sometimes; always
    d. sometimes; sometimes
11. Phenotypes such as height in humans, which show a continuous distribution, are usually the result of
    a. an alteration of dominance for multiple alleles of a single gene.
    b. the presence of multiple alleles for a single gene.
    c. the action of one gene on multiple phenotypes.
    d. the action of multiple genes on a single phenotype.
12. In epistasis,
    a. a portion of a chromosome is deleted.
    b. one gene masks the effect of another.
    c. only recessive traits are expressed.
    d. the behavior of linked genes is independent.

Apply

1. What is the probability of obtaining an individual with the genotype *bb* from a cross between two individuals with the genotype *Bb*?
   a. ½
   b. ¼
   c. ⅛
   d. 0
2. In a hypothetical animal, brown eyes (*B*) are dominant to pink eyes (*b*), and a solid coat (*S*) is dominant to a spotted coat (*s*). A true-breeding animal with brown eyes and a solid coat will produce which of the following gametes?
   a. *B, b, S, s*
   b. *Bb, BB, SS, Ss*
   c. *BS, bs*
   d. *BS, Bs, bS, bs*
3. A dihybrid cross between true-breeding animals with brown fur and two toes and others with white fur and three toes is allowed to proceed to the F_2 generation, where you see 9 brown two toes:3 brown three toes:3 white two toes:1 white three toes. Given this information, which of the following is true?
   a. Brown fur is dominant to white fur; two toes is dominant to three toes.
   b. White fur is dominant to brown fur; three toes is dominant to two toes.
   c. Brown fur is dominant to two toes; white fur is dominant to three toes.
   d. You can't determine which allele is dominant or recessive based on these data.
4. What would the phenotypic ratio be for a testcross of a doubly heterozygous brown eye, solid coat animal from question 2?
   a. 1 brown, solid:1 pink, spotted.
   b. 9 brown, solid:3 brown, spotted:3 pink, solid:1 pink, spotted.
   c. 1 brown, solid:1 brown, spotted:1 pink, solid:1 pink, spotted.
   d. 3 brown, solid:1 pink, spotted.
5. In beagles, glaucoma is inherited as an autosomal recessive trait. You breed two beagles that are heterozygous for glaucoma. What is the probability that both of their first two pups will have glaucoma?
   a. ¼
   b. ¾
   c. 1⁄16
   d. ½
6. Albinism is an autosomal recessive condition caused by a defect in the enzyme tyrosinase. A heat-sensitive tyrosinase is found in Siamese cats. Predict the outcome of crossing a Siamese cat to an animal homozygous for wild-type tyrosinase.
   a. All of the offspring would have a Siamese pigment pattern.
   b. All of the offspring would have a normal pigment pattern.
   c. Half of the offspring would be normal and half would be Siamese.
   d. There would be a ratio of 3 normal:1 Siamese.
7. The ABO blood group in humans is determined by multiple alleles. I^A and I^B are codominant and are both dominant over I^O. Consider a case in which a mother is type O and her newborn infant is type A. The possible phenotypes of the father are
   a. A, B, or AB.
   b. A or AB.
   c. O only.
   d. A or O.
8. You discover a new variety of plant with color varieties of purple and white. When you intercross these, the F_1 is a lighter purple. You consider that this may be an example of blending and self-cross the F_1. If Mendel is correct, what would you predict for the F_2?
   a. 1 purple:2 white:1 light purple
   b. 1 white:2 purple:1 light purple
   c. 1 purple:2 light purple:1 white
   d. 1 light purple:2 purple:1 white

Synthesize

1. Cuenot performed crosses with yellow mice and observed unusual results. When he crossed yellow to wild-type (agouti) mice, he observed one yellow to one agouti. When he crossed two yellow mice, he observed two yellow to one agouti. How can you explain these results? Is yellow dominant to agouti? What happened to one of the genotypes in the yellow × yellow cross?
2. In mammals, a variety of genes affect coat color. One of these is a gene with mutant alleles that results in the complete loss of pigment, or albinism. Another controls the type of dark pigment with alleles that lead to black or brown colors. The albinistic trait is recessive, and black is dominant to brown. Two black mice are crossed and yield 9 black:4 albino:3 brown. How would you explain these results?
3. You are studying a plant in which height appears to be a quantitative trait. To investigate this, you select the tallest and shortest plants, and cross tall to tall, and short to short for several generations. You then cross the tallest to the shortest and get an F_1 that is intermediate in height. When you intercross this F_1, you observe an F_2 with a distribution of heights that forms a bell curve. You find the frequency of the shortest plants is 1⁄64. If this is indeed a quantitative trait, how many genes are involved? What if the frequency of short plants was 1⁄256?

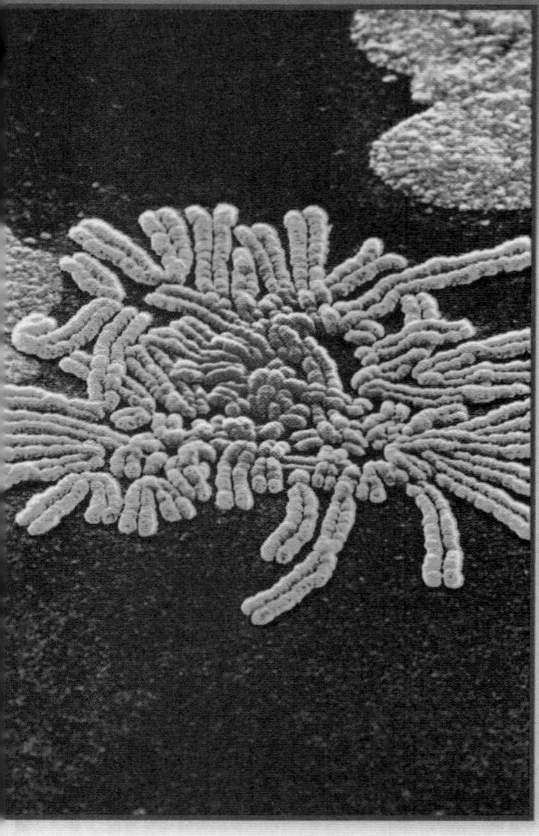

Adrian T Sumner/Science Source

13 The Chromosomal Basis of Inheritance

Learning Path

13.1 Sex Linkage and the Chromosomal Theory of Inheritance

13.2 There Are Two Major Exceptions to Chromosomal Inheritance

13.3 Some Genes Do Not Assort Independently: Linkage

13.4 Genetic Crosses Provide Data for Genetic Maps

13.5 Changes in Chromosome Number Can Have Drastic Effects

13.6 Inheritance in Humans Can Be Studied by Analyzing Pedigrees

Concept Overview

This Concept Overview provides a diagram of the overarching concepts that are covered in this chapter. These concepts will be expanded upon in the Concept Overview at the end of the chapter.

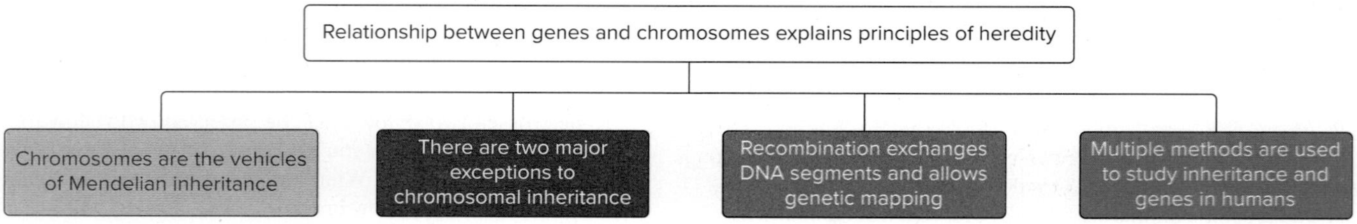
Relationship between genes and chromosomes explains principles of heredity
Chromosomes are the vehicles of Mendelian inheritance
There are two major exceptions to chromosomal inheritance
Recombination exchanges DNA segments and allows genetic mapping
Multiple methods are used to study inheritance and genes in humans

Introduction

Mendel's experiments provide a framework to understand heredity, but his work was largely ignored for 30 years. Then Mendel's work was simultaneously rediscovered at the turn of the 20th century by three different researchers who had come to similar conclusions. At the same time, detailed observations on the behavior of chromosomes in meiosis, seen in the picture on the previous page, were being made. The behavior of chromosomes during meiosis can explain Mendel's principles, and also leads to new and different approaches to the study of heredity. The ability to construct genetic maps of chromosomes is one of the most powerful tools of classical genetic analysis. In the 21st century, these tools have been refined and combined with genome-sequencing technologies to provide a view of the human genome that would have been considered science fiction as recently as the middle of the 20th century.

13.1 Sex Linkage and the Chromosomal Theory of Inheritance

A central role for chromosomes in heredity was first suggested in 1900 by Carl Correns, in one of the papers announcing the rediscovery of Mendel's work. Soon afterwards, observations that similar chromosomes paired with one another during meiosis led directly to the chromosomal theory of inheritance, first formulated by Walter Sutton in 1902. But while these observations can explain the patterns seen in Mendel's work, this does not prove that Mendel's factors (genes) are actually on chromosomes. This crucial connection came from early work on one of the most heavily studied genetic model systems known.

Morgan Correlated the Inheritance of a Trait with Sex Chromosomes

LEARNING OBJECTIVE 13.1.1 Demonstrate how white eye color in flies segregates with the X chromosome.

In 1910, Thomas Hunt Morgan, studying the fruit fly *Drosophila melanogaster,* discovered a mutant male fly with white eyes instead of red (figure 13.1). Morgan immediately set out to determine whether this new trait would show Mendelian inheritance. He crossed the mutant male to a normal red-eyed female to produce an F_1 with all red eyes, leading Morgan to conclude that red eye color was dominant over white.

Morgan then crossed the red-eyed flies from the F_1 generation with each other. Of the 4252 F_2 progeny Morgan examined, 782 (18%) had white eyes. Although the ratio of red eyes to white eyes in the F_2 progeny was greater than 3:1, the results nevertheless provided clear evidence that eye color segregates. However, something about the outcome was strange and totally unpredicted by Mendel's theory: all of the white-eyed F_2 flies were males (figure 13.2)!

One possible explanation was simply that for some unknown reason, white-eyed female flies are not viable. To test this idea, Morgan testcrossed the female F_1 progeny with the original white-eyed male. He observed white-eyed and red-eyed flies of both sexes in the expected 1:1:1:1 ratio (figure 13.3), which shows that white-eyed female flies are viable. This led Morgan in a more

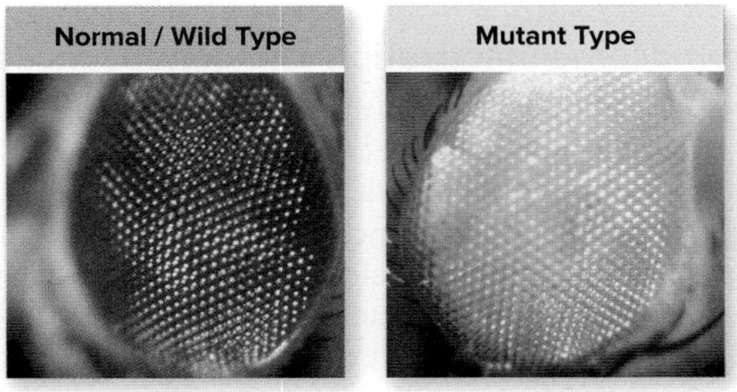
Normal / Wild Type
Mutant Type

Figure 13.1 Red-eyed (wild type) and white-eyed (mutant) *Drosophila*.

Photo Researchers/Science Source

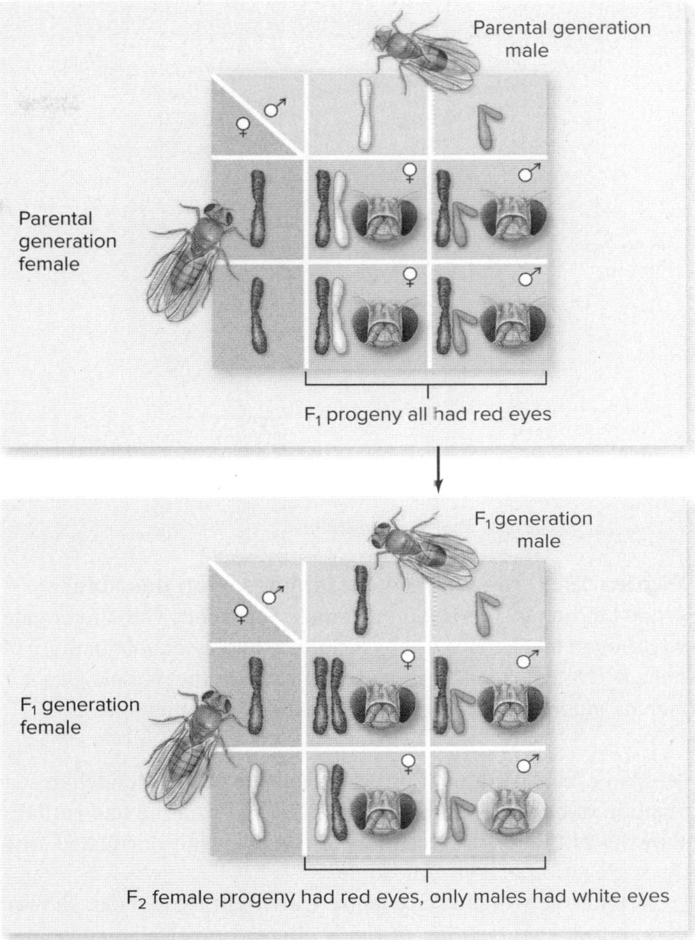
Parental generation male
Parental generation female
♀
♂
F_1 progeny all had red eyes
F_1 generation male
F_1 generation female
F_2 female progeny had red eyes, only males had white eyes

Figure 13.2 The chromosomal basis of sex linkage. Morgan crossed his white-eyed male fly to a red-eyed female. The F_1 flies all had red eyes, as expected for a recessive white-eye allele. In the F_2 generation Morgan observed both red and white-eyed flies—but unexpectedly, all of the white-eyed flies were male!

From Brian P. Chadwick and Huntington F. Willard, ""Multiple spatially distinct types of facultative heterochromatin on the human inactive X chromosome,"" PNAS, 101(50): 17450-17455, Fig. 3 ©2004 National Academy of Sciences, U.S.A.

interesting direction as he looked for an explanation in the actual chromosomes in males and females.

The gene for eye color is on the X chromosome

In *Drosophila,* the sex of an individual is determined by the number of copies it has of a particular chromosome, the **X chromosome.** Observations of *Drosophila* chromosomes revealed that female flies have two X chromosomes, but male flies have only one X chromosome. In males, the single X chromosome pairs in meiosis with a dissimilar partner called the **Y chromosome.** These two chromosomes are termed **sex chromosomes** because of their association with sex.

The solution to Morgan's puzzle is that the gene causing the white-eye trait in *Drosophila* resides only on the X chromosome—it is absent from the Y chromosome. (We now know that the Y chromosome in fruit flies carries few functional genes.) A trait determined by a gene on the X chromosome is said to be **sex-linked,** or **X-linked,** because it will segregate with the sex chromosomes. Because white is recessive, males carrying an X chromosome with this allele will always appear white-eyed.

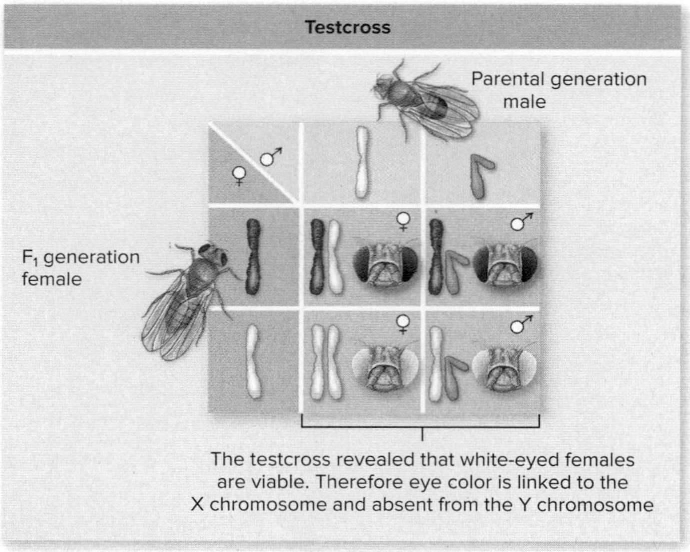

Figure 13.3 Are white-eyed females even possible? When Morgan test-crossed the female F_1 progeny with the original white-eyed male, the progeny included roughly equal numbers of male and female white-eyed flies. This shows that female flies are viable, and that eye color is linked to the X chromosome.

Morgan's result was a natural consequence of the Mendelian segregation of chromosomes (figure 13.2), and what he had actually shown was that the segregation of a visible trait correlated with the segregation of a chromosome.

One of Morgan's students, Calvin Bridges, later showed that aberrant segregation of the X chromosomes also correlated with aberrant segregation of the white gene. This is considered the final proof of genes being on chromosomes, as both normal and abnormal segregation of a chromosome pair could be perfectly correlated with normal and abnormal segregation of a trait.

Adult Body Cells of Mammals Have Only One Active X Chromosome

LEARNING OBJECTIVE 13.1.2 Explain the relationship between sex determination in mammals and the occurrence of dosage compensation.

The structure and number of sex chromosomes vary in different species. In the fruit fly, *Drosophila,* we have just learned that females are XX and males XY. This is also the case for mammals, including humans. However, in most birds, the male has two Z chromosomes, and the female has a Z and a W chromosome. We call the sex that is homozygous for a sex chromosome the homogametic sex, and the sex that carries different sex chromosomes the heterogametic sex. This has genetic consequences, as the unique sex chromosome in most systems has few active genes and does not carry the same genes as the shared sex chromosome. Thus recessive alleles will be expressed in the heterogametic sex, as we have just seen for white eyes in flies. We will examine the consequences for human genetics later in this chapter.

The "default" setting in mammalian embryonic development leads to female development. Some of the active genes on the Y chromosome are responsible for initiating male sexual development, which results in the masculinization of genitalia and secondary sex organs. Consequently, embryos with at least one Y chromosome will develop as male. In vertebrates, this form of sex determination is seen in mammals and birds and is called chromosomal sex determination. This is in contrast to some reptile species, where sex determination is controlled by environmental factors, which is called environmental sex determination.

Dosage compensation prevents doubling of sex-linked gene products

Although mammalian males have only one X chromosome and females have two, female cells do not produce twice the amount of the proteins encoded by genes on the X chromosome. Instead, one of the X chromosomes in females is inactivated early in embryonic development, shortly after the embryo's sex is determined. This inactivation is an example of **dosage compensation,** which ensures an equal level of expression from the sex chromosomes despite a differing number of sex chromosomes in males and females. In *Drosophila,* by contrast, dosage compensation is achieved by increasing the level of expression on the male X chromosome.

Which X chromosome is inactivated in mammalian females varies randomly from cell to cell. If a female is heterozygous for a sex-linked trait, some of the cells will express one allele and some the other. The inactivated X chromosome is highly condensed, making it visible as an intensely staining **Barr body,** visible in the photo, attached to the nuclear membrane.

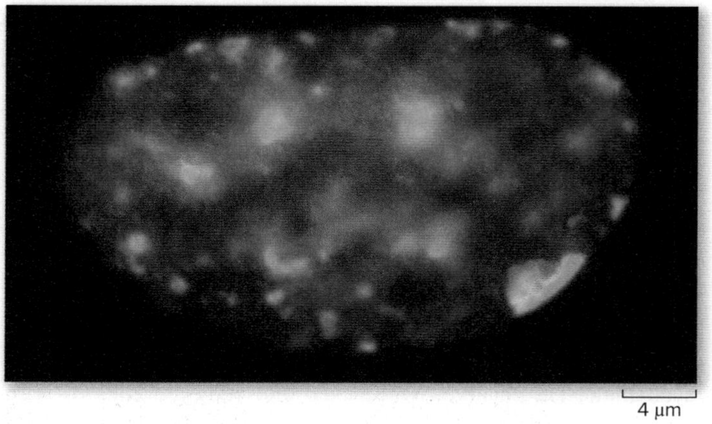

From Brian P. Chadwick and Huntington F. Willard, "Multiple spatially distinct types of facultative heterochromatin on the human inactive X chromosome," *PNAS*, 101(50): 17450–17455, Fig. 3

©2004 National Academy of Sciences, U.S.A.

X-chromosome inactivation and genetic mosaics

The X-chromosome inactivation mechanism of dosage compensation is found in all mammals. Females that are heterozygous for X-chromosome alleles are **genetic mosaics:** their individual cells may express different alleles, depending on which chromosome is inactivated. One example is the calico cat, a female that has a patchy distribution of dark fur, orange fur, and white fur (figure 13.4).

The dark fur and orange fur are due to heterozygosity for a gene on the X chromosome that determines pigment color. One allele results in dark fur, and another allele results in orange fur. Which of these colors is observed in any particular patch is due to inactivation of one X chromosome: if the chromosome

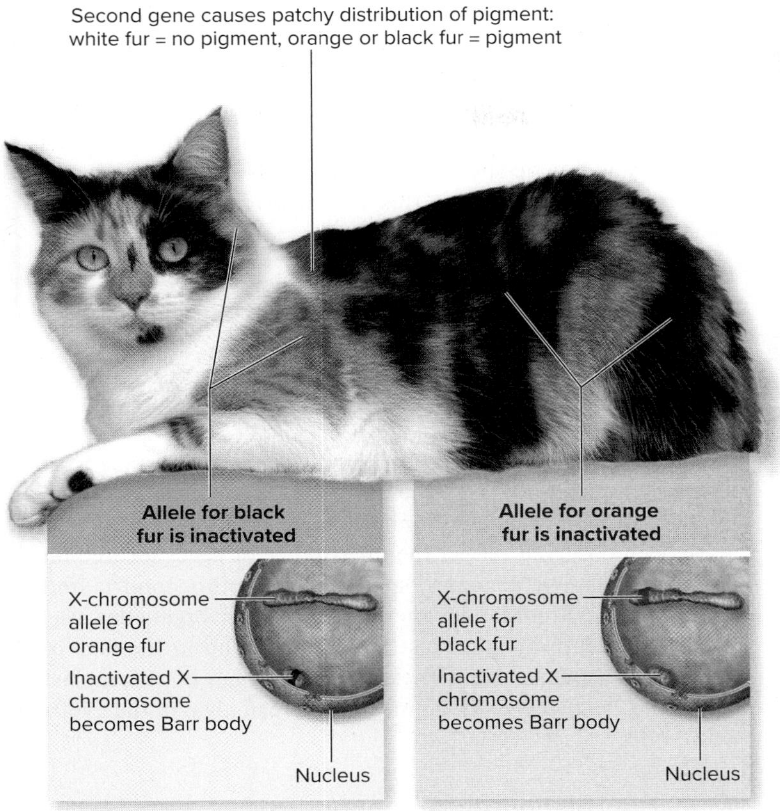
Second gene causes patchy distribution of pigment: white fur = no pigment, orange or black fur = pigment
Allele for black fur is inactivated
Allele for orange fur is inactivated
X-chromosome allele for orange fur
Inactivated X chromosome becomes Barr body
Nucleus
X-chromosome allele for black fur
Inactivated X chromosome becomes Barr body
Nucleus

Figure 13.4 A calico cat. The cat is heterozygous for alleles of a coat color gene that produce either black fur or orange fur. This gene is on the X chromosome, so the different-colored fur is due to inactivation of one X chromosome. The patchy distribution and white color are due to a second gene that is epistatic to the coat color gene and thus masks its effects.
cgbaldauf/Getty Images

containing the orange allele is inactivated, then the fur will be dark, and vice versa.

The patchy distribution of color, and the presence of white fur, is due to a second gene that is epistatic to the fur color gene (refer to chapter 12). That is, the presence of this second gene produces a patchy distribution of pigment, with some areas totally lacking pigment. In the areas that lack pigment, the effect of either fur color allele is masked. Thus, in this one animal we can see an excellent example of both epistasis and X inactivation.

REVIEW OF CONCEPT 13.1

Morgan showed that the trait for white eyes in *Drosophila* segregates with the sex of offspring. X and Y chromosomes also segregate with sex, which correlates the behavior of a trait with the behavior of chromosomes and supports idea that traits are carried on chromosomes. In mammals, males are XY, and therefore they exhibit recessive traits for alleles on the X chromosome. In mammalian females, one X chromosome in each cell becomes inactivated to balance the levels of gene expression. This random inactivation can lead to genetic mosaics.

- *What are the expectations for a cross in Drosophila of white-eyed females to red-eyed males?*

13.2 There Are Two Major Exceptions to Chromosomal Inheritance

By the late 20th century, geneticists were confident that they understood the basic mechanisms governing inheritance. It came as quite a surprise when mouse geneticists found an important exception to classical Mendelian genetics that appears to be unique to mammals.

Genomic Imprinting Depends on the Parental Origins of Alleles

LEARNING OBJECTIVE 13.2.1 Explain how genomic imprinting leads to non-Mendelian inheritance.

In **genomic imprinting,** the phenotype of a specific allele is expressed when the allele comes from one parent but not from the other. Genomic imprinting occurs during gamete formation, silencing a particular allele of a gene but not other alleles. The expression of the gene varies, depending on whether it passes through maternal or paternal germ lines. Some genes are inactivated in the paternal germ line and therefore are not expressed in the zygote. Other genes are inactivated in the maternal germ line, with the same result. This condition makes the zygote effectively haploid for an imprinted gene. The expression of variant alleles of imprinted genes depends on the parent of origin. A zygote expresses only one allele of an imprinted gene, that inherited from either the female or the male parent. The imprint is then transmitted to all body cells during development.

In each generation, the imprints received from parents are "erased" in that generation's gamete-producing cells. In this way, the choice of parents redefines the outcome in each generation, the gametes of each parent newly imprinted according to the sex of that parent. For each mammalian species that has been studied, the imprinted genes are always imprinted the same way: a gene imprinted to be expressed in female gametes is always imprinted this way, one generation to the next.

The mouse *igf 2* gene

One of the first imprinted genes to be identified was the gene for *insulin-like growth factor 2 (igf 2)* in the mouse. This gene encodes a growth factor that plays a critical role in prenatal development and growth. Healthy growth is impossible without it—but despite this, only the paternal allele is expressed. The discovery of the genomic imprinting of *igf 2* is described in figure 13.5, in which normal mice are crossed with dwarf mice homozygous for a recessive allele of the *igf 2* gene: the phenotypes of the heterozygous offspring (carrying one normal allele and one dwarf allele) are different, depending on which parent the mutant allele came from!

Prader–Willi and Angelman syndromes

An example of genomic imprinting in humans involves the two diseases Prader–Willi syndrome (PWS) and Angelman syndrome

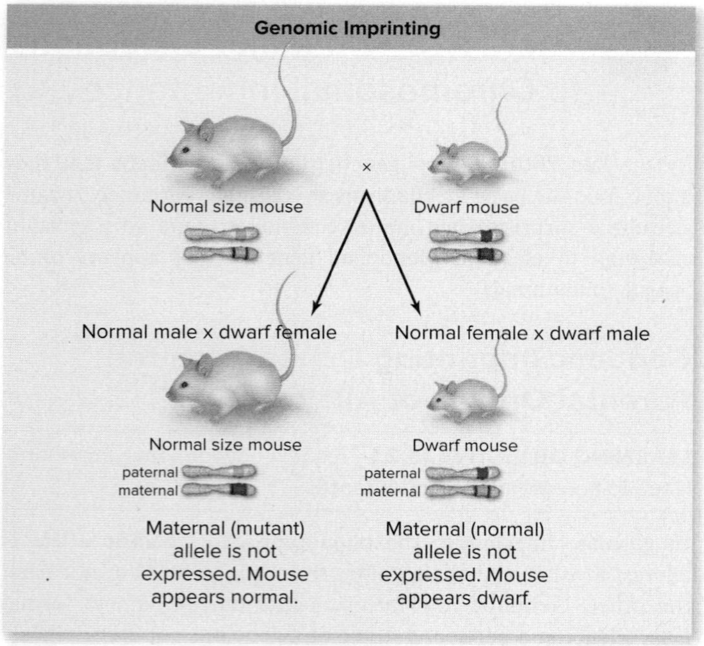

Figure 13.5 Genomic imprinting of the mouse *igf 2* gene.

(AS). The effects of PWS include respiratory distress, obesity, short stature, mild intellectual disability, and obsessive-compulsive behavior. The effects of AS include developmental delay, severe intellectual disability, hyperactivity, aggressive behavior, and inappropriate laughter.

Genetic studies have implicated a deletion of material on chromosome 15 for both disorders, and indeed the same deletion can cause either syndrome, depending on the parental origin of the deleted chromosome: if the chromosome with the deletion is paternally inherited, it causes PWS; if the chromosome with the deletion is maternally inherited, it causes AS.

The region of chromosome 15 that is lost is subject to imprinting, inactivating some genes. In PWS, genes are inactivated in the maternal germ line, such that deletion or other functional loss of paternally derived alleles produces the syndrome. The opposite is true for AS syndrome: genes are inactivated in the paternal germ line, such that loss of maternally derived alleles leads to the syndrome.

Genomic imprinting is an example of epigenetics

Genomic imprinting is actually an example of a more general phenomenon: **epigenetic inheritance.** An epigenetic change is defined as a mitotically and/or meiotically stable change in gene function that does not involve a change in DNA sequence. An example from section 13.1 is X-chromosome inactivation where an entire chromosome is silenced, and the effect is inherited through many mitotic divisions. In some mouse and rat models, this even includes the effects of maternal diet on F_2 animals.

Epigenetic mechanisms include changes in DNA methylation and histone modifications. Noncoding RNAs and nuclear organization have also been implicated. Alterations in chromatin structure and the accessibility of DNA seem to be a point of convergence for multiple epigenetic mechanisms (refer to chapter 16). In some well-studied cases, the pattern of imprinting that occurs in the male and female germ line is due to male- and female-specific patterns of DNA methylation and alterations to the proteins that are involved in chromosome structure.

Organellar Inheritance Involves Non-Nuclear DNA

LEARNING OBJECTIVE 13.2.2 Explain how mitochondrial and chloroplast DNA lead to non-Mendelian inheritance.

Genomic imprinting is not the only, or even the most common, pattern of non-Mendelian inheritance. Most instances that have been observed reflect the inheritance of genes located on DNA in organelle genomes, specifically in mitochondria and chloroplasts. Non-Mendelian inheritance via organelles was studied in depth by geneticist Ruth Sager, who in the face of universal skepticism constructed the first map of chloroplast genes in *Chlamydomonas,* a unicellular green alga, in the 1960s and 1970s.

Mitochondrial genes are inherited from the female parent

Mitochondria are usually inherited from only one parent, generally the mother. When a zygote is formed, it receives an equal contribution of the nuclear genome from each parent, but it gets all of its mitochondria from the egg cell, which contains a great deal more cytoplasm (and thus organelles). As the zygote goes on to divide by mitosis, these cytoplasmic mitochondria are partitioned randomly during cytokinesis into the two daughter cell cytoplasms. Because of this, the mitochondria in every cell of an adult organism can be traced back to the original maternal mitochondria present in the egg. This mode of uniparental (one-parent) inheritance from the mother is called **maternal inheritance.**

What genes in the mitochondrial genome can influence the eukaryotic cell? Not surprisingly, most mitochondrial genes encode subunits of the ATP synthetase enzyme and the protein complexes of the electron transport chain. Mutations in these genes that reduce the activity of these proteins can reduce the amount of ATP a cell produces, with severe metabolic consequences. Because the highest consumption of ATP occurs in the nervous system and muscle tissue, most inherited mitochondrial disorders affect these systems.

In humans, the disease Leber hereditary optic neuropathy (LHON) shows maternal inheritance. The genetic basis of this disease is a mutant allele for a subunit of NADH dehydrogenase. The mutant allele reduces the efficiency of electron flow in the electron transport chain in mitochondria (refer to chapter 7), in turn reducing ATP production. Some nerve cells in the optic system are particularly sensitive to reduced levels of ATP, causing degeneration of the optic nerve.

A mother with this disease will pass it on to all of her progeny, whereas a father with the disease will not pass it on to any of his progeny. Note that this condition differs from sex-linked inheritance because males and females are equally affected.

Chloroplast genes may also be passed on uniparentally

Like mitochondria, chloroplasts have their own DNA genomes that are inherited independently of meiosis, partitioned randomly by cytokinesis during mitosis. Thus, like mitochondria, the inheritance pattern of chloroplasts is also usually maternal (although paternal and biparental inheritance have been observed in some species).

Botanist Carl Correns first hypothesized in 1909 that chloroplasts were responsible for inheritance of variegation (mixed yellow or white patches on otherwise green leaves) in the plant commonly known as the four o'clock *(Mirabilis jalapa).* Variegation is quite common among plant species (figure 13.6). In all cases that have been examined, the offspring exhibit the variegation phenotype of the female parent, regardless of the male's phenotype.

The variegation is the result of mutations to genes in the chloroplast genome that control the production of chlorophyll or other pigments. Because the fertilized zygote may receive different alleles from its two parents, the subsequent mitoses will distribute the alleles randomly as development proceeds.

REVIEW OF CONCEPT 13.2

Genomic imprinting refers to inactivation of alleles depending on parental origin of alleles. This leads to differences in crosses depending on which parent is mutant, as seen in dwarf mice. The genomes of mitochondria and chloroplasts are inherited maternally.

- *How can you explain the lack of* mt^- *chloroplast DNA in* Chlamydomonas *zygotes from* mt^- *by* mt^+ *crosses?*

Figure 13.6 Variegated leaves in the ground elder *(Aegopodium podagraria).*

Asperra Images/Alamy Stock Photo

13.3 Some Genes Do Not Assort Independently: Linkage

Although Morgan's experiments with the white-eyed fly established that Mendel's genes reside on chromosomes, they leave an important question still unanswered: How can there be more independently segregating genes than chromosomes? Genes located on the same chromosome should segregate together. Yet Mendel's seven traits, located on only five of the pea plant chromosomes, assorted independently in Mendel's crosses. In fact, organisms generally have many more genes that assort independently than the number of their chromosomes. This means that independent assortment cannot be due only to the random alignment of chromosomes during meiosis.

Genetic Recombination Occurs Less Often Between Nearby Genes

LEARNING OBJECTIVE 13.3.1 Explain why recombination frequency is related to genetic distance.

The solution to this problem is found in an observation you first encountered in chapter 11: recombination, or crossing over, between homologs during meiosis I. In prophase I of meiosis, homologs appear to physically exchange material by crossing over (figure 13.7). As you learned in chapter 11, this is part of the mechanism that allows homologs, and not sister chromatids, to disjoin at anaphase I.

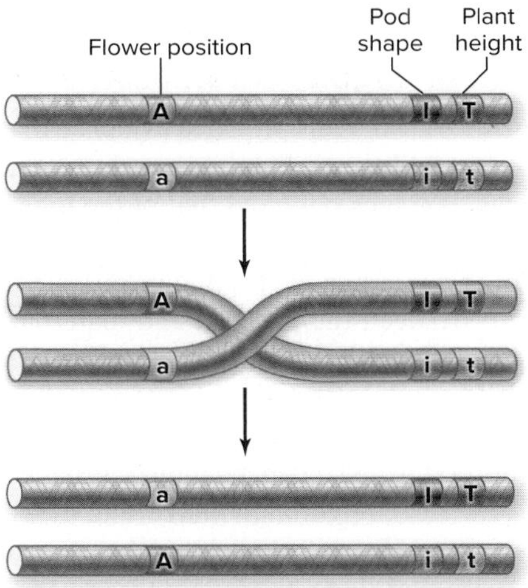

Figure 13.7 Linkage. Genes that are located farther apart on a chromosome, like the genes for flower position (*A*) and pod shape (*I*) in Mendel's peas, will assort independently because crossing over results in recombination of these alleles. Pod shape (*I*) and plant height (*T*), however, are positioned very near each other, such that crossing over usually would not occur. These genes are said to be linked and do not undergo independent assortment.

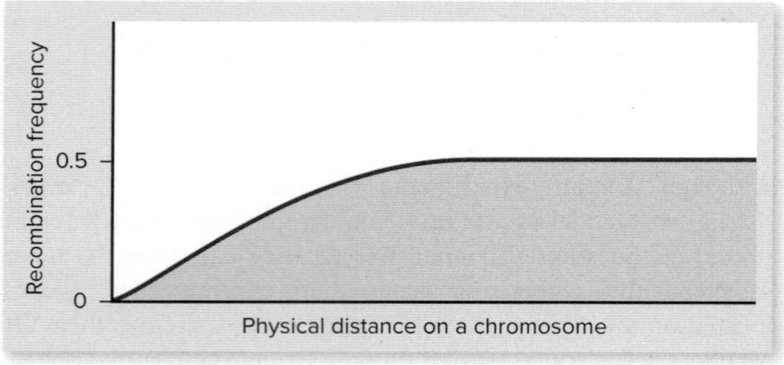

Figure 13.8 Relationship between true distance and recombination frequency. As distance between two genes on a chromosome increases, the probability of recombination between them also increases, to a maximum value of 0.5, with either combination equally likely. (This relationship was explored in the *Inquiry & Analysis* feature of chapter 12.)

Consider, first, two genes that are close together, like *I* and *T* in figure 13.7. Crossing over between these genes will be rare, and the *I* and *T*, and *i* and *t*, alleles (as shown) will usually segregate together. We call this **linkage,** and it will alter the expected outcomes of a cross as these two genes are not behaving independently. We will consider this in more detail in section 13.4. Now consider genes *A* and *I* in figure 13.5; they are much farther apart, so crossing over is much more likely between them. If a single crossover occurs, the alleles will be recombined, but if two crossovers occur, we are back to a parental configuration. If the genes are far enough apart, the number of odd and even crossovers will be about the same, and the genes will assort independently. This leads to a relationship between actual distance and the frequency of recombination, shown in figure 13.8.

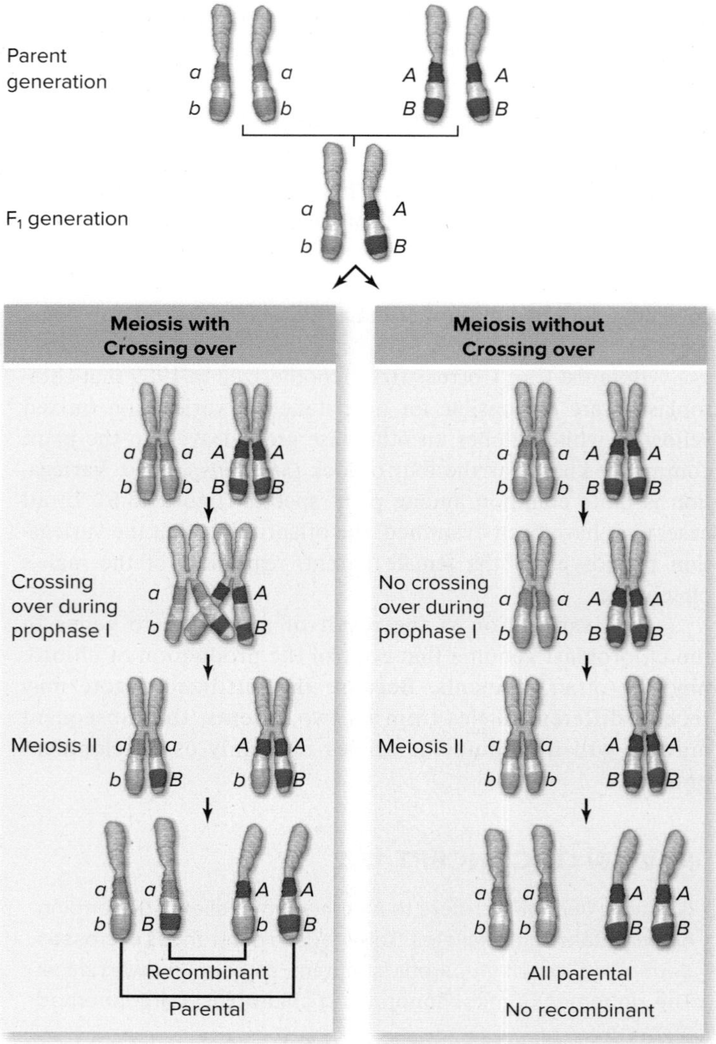

Figure 13.9 Crossing over exchanges alleles on homologs. When a crossover occurs between two loci, it leads to the production of recombinant chromosomes. When no crossover occurs, then the chromosomes will carry the parental combination of alleles.

Genetic recombination involves a physical exchange of homologous chromosome arms

In 1909, cytologist F. A. Janssens provided evidence that homologs interact during meiosis. Investigating chiasmata produced during amphibian meiosis, Janssens observed that of the four chromatids involved in each chiasma, two crossed each other and two did not. He suggested that this crossing of chromatids reflected a switch in chromosome arms between the paternal and maternal homologs, involving one chromatid in each homolog. It followed directly that crossing over would result in the recombination of genetic alleles (figure 13.9).

Experiments performed independently by geneticists Barbara McClintock and Harriet Creighton in maize, and by Curt Stern in *Drosophila,* provided evidence for this physical exchange of genetic material. The experiment done by Creighton and McClintock is detailed in figure 13.10. In this experiment, they used a chromosome with two alterations visible under a microscope: a knob on one end of the chromosome and an extension of the other end, making it longer. In addition to these visible markers, this chromosome also carried a gene that determines kernel color (colored or colorless) and a gene that determines kernel texture (waxy or starchy).

The long chromosome, which also had the knob, carried the dominant-colored allele for kernel color (*C*) and the recessive waxy allele for kernel texture (*wx*). Heterozygotes were constructed with this chromosome paired with a visibly normal chromosome carrying the recessive colorless allele for kernel color (*c*) and the dominant starchy allele for kernel texture (*Wx*) (figure 13.10). These plants appeared colored and starchy because they were heterozygous for both loci, and they were heterozygous for the two visibly distinct chromosomes.

These plants, heterozygous for both chromosomal and genetic markers, were test-crossed to colorless waxy plants with normal-appearing chromosomes. The progeny were analyzed for both physical recombination (using a microscope to observe chromosome appearance) and genetic recombination (by examining the phenotype of progeny). The results were striking: all of the progeny that were genetically recombinant (appear colored starchy or colorless waxy) also now had only one of the chromosomal markers. The researchers concluded that genetic recombination was accompanied by the physical exchange of chromosome arms.

SCIENTIFIC THINKING

Hypothesis: *Crossing over, or recombination, involves a physical exchange of genetic material.*

Prediction: *Recombination of visible differences in a chromosome should correlate with genetic recombination of alleles.*

Test: *In the cross shown, two visible chromosome markers (yellow extension marker, and green knob marker) have been combined with two genetic markers (kernel color and texture).*

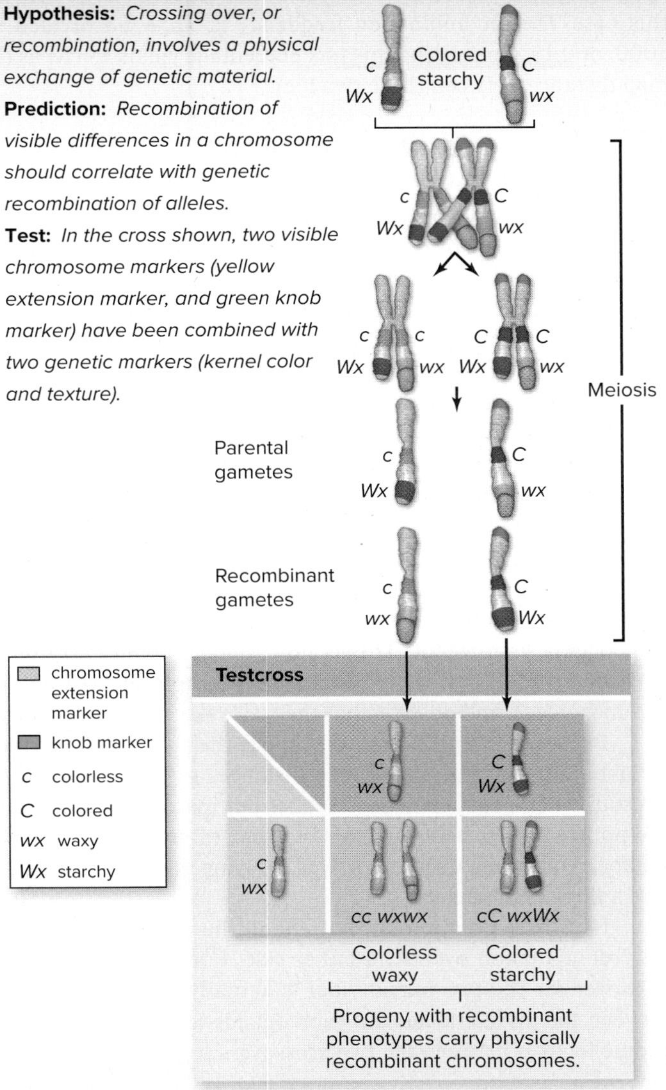

Colored
starchy
c
Wx
C
wx
Meiosis
Parental gametes
Recombinant gametes
Testcross
cc wxwx
cC wxWx
Colorless waxy
Colored starchy
Progeny with recombinant phenotypes carry physically recombinant chromosomes.
chromosome extension marker
knob marker
c colorless
C colored
wx waxy
Wx starchy

Result: *Genetically recombinant progeny also have physically recombinant chromosomes.*

Conclusion: *A physical exchange of genetic material accompanied genetic recombination.*

Further Experiments: *This experiment was performed using maize. What other genetic model system would you use to test this?*

Figure 13.10 The Creighton and McClintock experiment.

REVIEW OF CONCEPT 13.3

Crossing over during meiosis exchanges alleles on homologs and involves a physical exchange of chromosome arms. Genes that are close together tend not to experience crossing over as often as genes farther apart, and are said to be "linked."

- *How can you tell if two genes that assort independently are located on two different chromosomes or are located far apart on a single chromosome?*

13.4 Genetic Crosses Provide Data for Genetic Maps

The ability to map the location of genes on chromosomes using data from genetic crosses is one of the most powerful tools of genetics. The insight that allowed this technique, like many great insights, is so simple as to seem obvious in retrospect.

The Frequency of Recombination Allows Mapping of the Relative Positions of Genes on Chromosomes

LEARNING OBJECTIVE 13.4.1 Construct a genetic map using data from a testcross with linked genes.

Morgan had already suggested that the frequency of double-crossover recombinant progeny appeared to be greater for genes that were farther apart on the chromosome. An undergraduate in Morgan's laboratory, Alfred Sturtevant, extended this observation to its logical conclusion. Sturtevant first assumed that genes are arrayed in a linear order on chromosomes. If this is the case, he reasoned that as physical distance on a chromosome increases, so should the probability of recombination (a crossover event) occurring between the gene loci. Given this, the frequency of recombination observed in crosses between two genes could be used as a measure of the physical distance between them on a chromosome. This led Sturtevant to construct the first genetic map for a portion of the X chromosome in *Drosophila*.

Constructing maps from two-point crosses

The logic of genetic mapping is simple: the farther apart two genes are, the more frequently recombination should occur between them. In practice, this means measuring the **recombination frequency,** or the frequency of recombinant progeny in a testcross. The use of a testcross instead of Mendel's crosses simplifies analysis. In a testcross, the phenotypes of the progeny reflect the gametes produced by the doubly heterozygous F_1 individual. In the case of recombination, progeny that appear parental have not had a crossover, and progeny that appear recombinant have experienced a crossover between the two loci in question (figure 13.9).

Genes that are close together on a chromosome are said to be linked. We define linkage genetically as an excess of parental genotypes over recombinant genotypes. The frequency of recombination is defined as the number of recombinant progeny divided by the total number of progeny. This value is converted to a percentage, and each 1% of recombination represents one **map unit.** This unit has been named the centimorgan (cM) for T. H. Morgan, although it is also called simply a map unit (m.u.) as well.

Constructing genetic maps then becomes a process of performing testcrosses with doubly heterozygous individuals and counting progeny to determine percent recombination. This is illustrated with an example using a two-point cross. *Drosophila* homozygous for two mutations, vestigial wings (*vg*) and black body (*b*), are crossed to flies homozygous for the wild type, or normal alleles, of these genes (vg^+ b^+). The doubly heterozygous

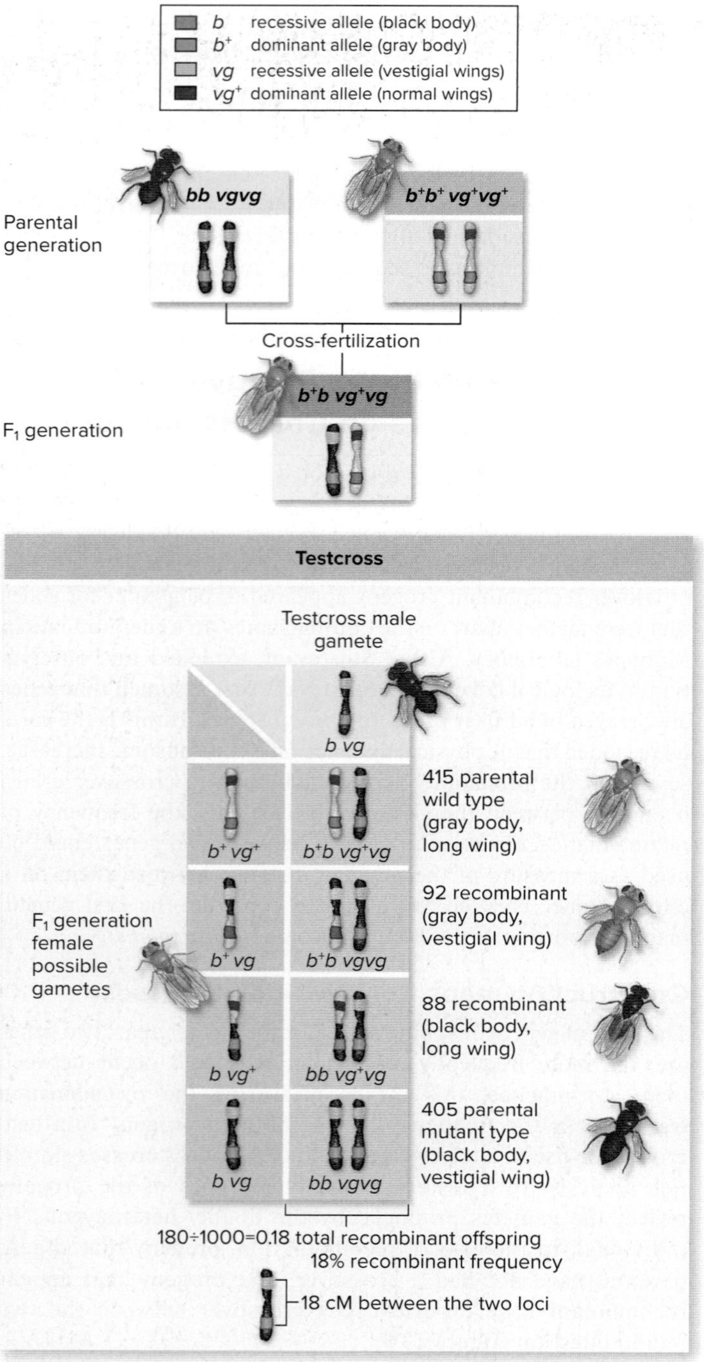

Figure 13.11 Two-point cross to map genes.

F_1 progeny are then test-crossed to homozygous recessive individuals (*vg b*/*vg b*), and progeny are counted (figure 13.11). Four different combinations of the two traits are possible, two parental and two recombinant. The numbers of each of the four types observed among the progeny are

vestigial wings, black body (*vg b*)	405 (parental)
long wings, gray body ($vg^+\ b^+$)	415 (parental)
vestigial wings, gray body ($vg\ b^+$)	92 (recombinant)
long wings, black body ($vg^+\ b$)	88 (recombinant)
Total progeny	1000

The sum of the recombinant progeny is divided by the total number of progeny to produce the recombination frequency. In this case, the recombination frequency is 92 + 88 divided by 1000, or 0.18. Converting this to a percentage yields 18 cM as the map distance between these two loci.

Three-point crosses can be used to put genes in order

There is a problem with constructing genetic maps by analyzing two-point crosses: as the distance separating loci increases, the probability of recombination occurring between them during meiosis also increases. But if homologs are far enough apart to undergo *two* crossovers between loci, the parental combination is restored! This leads to an underestimate of the true genetic distance, because not all the recombination events between the two distant loci are counted.

At long distances, multiple crossover events between loci become frequent. In this case, odd numbers of crossovers (1, 3, 5) produce recombinant gametes, whereas no crossover or an even number of crossovers (0, 2, 4) produces parental gametes. As you saw in figure 13.8, at large enough distances these frequencies are about equal, leading to the number of recombinant gametes being equal to the number of parental gametes, and the loci exhibit independent assortment! This is how Mendel could carry out a dihybrid cross of pod color and seed color, both located on pea chromosome 2, and observe that the traits assort independently.

Because multiple crossovers reduce the number of observed recombinant progeny, longer map distances are not accurate. As a result, when geneticists try to construct maps from a series of two-point crosses, determining the order of genes can be problematic. This problem can be solved by using three loci instead of two, a **three-point cross.**

In a three-point cross, the gene in the middle allows us to see recombination events on either side. For example, a double crossover for the two outside loci is actually a single crossover between the middle locus and each outside locus, so long as the middle locus is between the two crossovers (figure 13.12).

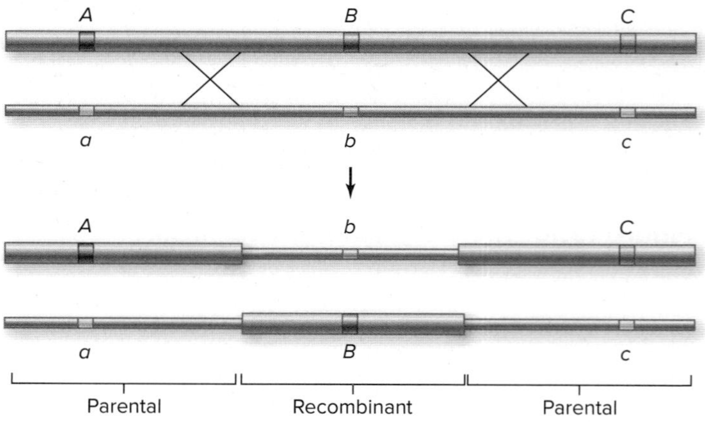

Figure 13.12 Use of a three-point cross to order genes. In a two-point cross, the outside loci appear parental for double crossovers. With the addition of a third locus, the two crossovers can still be detected, because the middle locus will be recombinant. This double crossover class should be the least frequent, so whatever locus has recombinant alleles in this class must be in the middle.

The probability of two crossovers is equal to the product of the probability of each individual crossover, each of which is relatively low. Therefore, in a three-point cross, the least frequent class contains offspring with two crossovers. Analyzing these individuals to determine which locus is recombinant will identify the locus that is in the middle of the three loci in the cross, as shown in figure 13.12.

In practice, geneticists use three-point crosses to determine the order of genes, then use data from the closest two-point crosses to determine distances. Longer distances are generated by simple addition of shorter distances. This avoids using inaccurate measures from two-point crosses between distant loci.

Genes Can Be Mapped in Humans

LEARNING OUTCOME 13.4.2 Explain the importance of molecular markers.

When genetic maps were first constructed in experimental organisms, the majority of genetic markers caused a visible phenotype, such as changes in eye color in flies. In humans, the equivalent genes cause disease states. These do not make good genetic markers both because of their low frequency in the population, and because few were known in the early days of human genetics. In the early 1980s, the number of markers for the human genome numbered in the hundreds.

Molecular markers

This situation changed with the identification of DNA differences in individuals that do not alter phenotype. One widely used marker is the **short tandem repeat (STR),** which is a short repeat of 2–4 bases that can differ in repeat number between individuals. These have been used in genetic mapping and also are the basis for some forensic DNA databases (refer to section 17.3). Molecular markers have evolved with technology and allowed the construction of a human genetic map that would have been unthinkable 30 years ago.

Data from sequencing human genomes allows an even finer level of analysis: identifying and mapping single differences between individuals. Any difference between individuals in populations is termed a polymorphism, so differences affecting a single base of a gene locus are called **single-nucleotide polymorphisms (SNPs).** There are now 600 million SNPs identified in the human genome. The vast majority of these are at a very low frequency in populations, with "only" about 10 million at a frequency of >0.5%. These have been identified and mapped by an international consortium, with the data residing in a public database.

The human genetic map

With these common molecular markers in hand, a dense genetic map was constructed as part of the Human Genome Project (refer to chapter 18). This genetic map provides a framework to locate genes of interest, that is, those that affect interesting phenotypes. Human genes can be mapped, similar to genetic mapping in model organisms, but using data derived from families instead of controlled crosses. The principle is the same—genetic distance is still proportional to recombination frequency—but the analysis requires the use of complex statistics and summing data from many families.

The power of genetic mapping is that it allows geneticists to find the genes responsible for specific phenotypes. Even before the Human Genome Project was completed, geneticists were finding genes that could be mutated to cause particular genetic diseases, including cystic fibrosis (recessive allele) and Huntington disease (dominant allele). Linkage analysis can identify rare causal variants in a population that have a large effect on phenotype. This limits gene discovery using linkage analysis to genetic disorders caused by a single gene, so-called monogenic disorders.

Genome-wide association studies

Although there are estimated to be some 7000 inherited diseases caused by rare variants of single genes, most human phenotypes, including many common diseases, are affected by many genes. We saw in the last chapter that complex traits are usually so-called quantitative traits that are due to the action of many genes. To address these traits, a different methodology was needed, and so the genome-wide association study, or GWAS, was born in 2005.

In these studies, it is possible to start with a complex disease phenotype, for example, coronary artery disease, then look for SNPs associated with this phenotype in a population. As the location of each SNP is known, you can then look for candidate genes in that region of the genome. Coronary artery disease is an interesting example, as a few variants were identified with relatively large effects. One variant found in both U.S. and European populations is the most common genetic risk factor and accounts for more than 10% of disease risk. But this and other variants with large effects do not account for a majority of the heritability, which is a measure of phenotypic variation due to genetic variation. There are literally thousands of variants with small effects that contribute to coronary artery disease, and account for the majority of the heritability. This pattern has held true for most complex diseases analyzed by GWAS.

REVIEW OF CONCEPT 13.4

The recombination of alleles on homologs can be used to construct genetic maps. Genes close together exhibit an excess of parental versus recombinant types in a testcross. The frequency of recombination types is used as a measure of genetic distance. Widely separated loci will have multiple crossovers between them, which can lead to independent assortment. Human genetic mapping uses statistical analysis of data from pedigrees. This is aided by molecular markers that do not cause a phenotype. Single nucleotide polymorphisms (SNP) can be used in association studies to map genes that affect complex traits.

- *If two genes are 10 cM apart, what fraction of gametes from a heterozygous individual will be parental?*

13.5 Changes in Chromosome Number Can Have Drastic Effects

The failure of homologs or sister chromatids to separate properly during meiosis is called **nondisjunction.** This failure leads to the gain or loss of a chromosome, a condition called **aneuploidy.** The frequency of aneuploidy in humans has been estimated at 7 to 10% of clinically recognized conceptions.

Nondisjunction Is a Failure of Meiotic Separation

LEARNING OBJECTIVE 13.5.1 Describe nondisjunction and its consequences in humans.

Nondisjunction of autosomes

The loss of a single chromosome is called **monosomy,** and the gain of a single chromosome is called **trisomy.** In humans, most monosomic zygotes do not survive embryonic development. In all but a few cases, trisomy in humans also leads to embryonic lethality. Aneuploidy is found in about only 0.3 of newborns, but it appears to be present in approximately 7–10% of clinically recognized pregnancies. However, data from spontaneous abortions imply aneuploidy levels as high as 35%.

Five of the smallest human autosomes—those numbered 13, 15, 18, 21, and 22—can be present as three copies and still allow the individual to survive, at least for a time. The presence of an extra chromosome 13, 15, or 18 causes severe developmental defects, and infants with such a genetic makeup usually die within a few months. In contrast, individuals who have an extra copy of chromosome 21, or, more rarely, chromosome 22, usually survive to adulthood. In these individuals, the maturation of the skeletal system is delayed, so they are often short with poor muscle tone. Their mental development is also affected.

The developmental defect produced by trisomy 21 (figure 13.13) was first described in 1866 by physician J. Langdon Down; for this reason, it is called **Down syndrome.** About 1 in every 750 children exhibits Down syndrome, and the frequency is comparable in all racial groups. Similar conditions also occur in chimpanzees and other related primates.

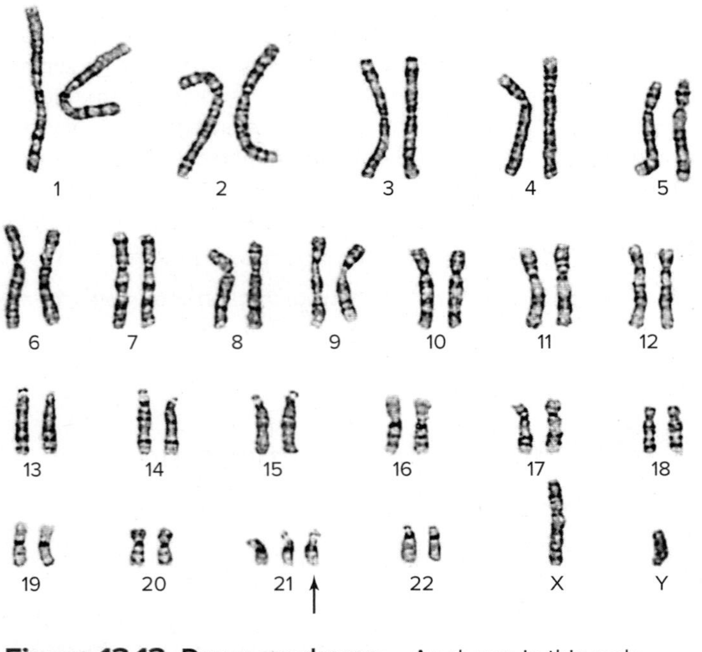

Figure 13.13 Down syndrome. As shown in this male karyotype, Down syndrome is associated with trisomy of chromosome 21 (arrow shows third copy of chromosome 21).

Karen Swisshelm/Colorado Genetics Laboratory

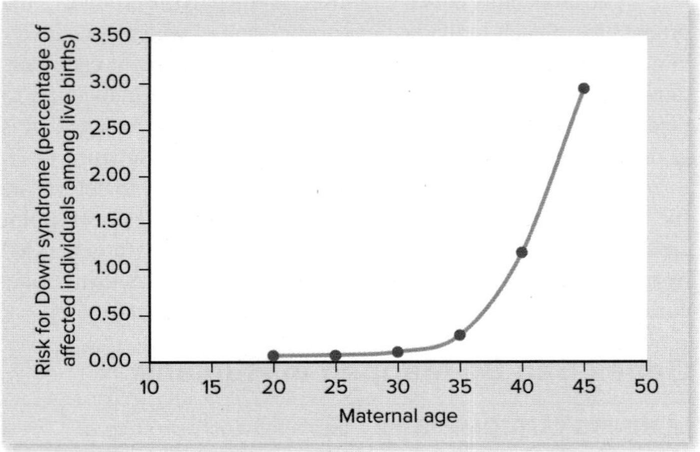

Figure 13.14 Correlation between maternal age and the incidence of Down syndrome. As women age, the chances they will bear a child with Down syndrome increase. After age 35, the frequency of Down syndrome rises rapidly.

Only a portion of chromosome 21 needs to be present in 3 copies to produce the clinical syndrome. In 3% of clinical cases, a portion of chromosome 21 containing the critical segment has been translocated to another chromosome (refer to chapter 15). This, along with the normal two copies of chromosome 21, produces the condition known as *translocation Down syndrome.*

There is an association between maternal age and trisomy. The risk for trisomy in clinically recognized pregnancies rises from 2–3% for females in their early 20s to 30–35% for females in their 40s. Figure 13.14 shows the relationship between maternal age and the risk of Down syndrome. Chromosomal abnormalities arise more frequently during female gametogenesis where gametes are set aside in early development, and arrested in meiosis I with homologs paired. While this is thought to be part of the mechanism for increased maternal aneuploidy, it is probably not the entire story. As we will learn in chapter 15, while aneuploidy is greater on the maternal side, most point mutations appear to arise in the paternal germ line.

Although maternal meiotic errors underlie the majority of cases of trisomy 21, about 10% appear to be due to paternal nondisjunction. There is also a small effect of increasing paternal age on the incidence of Down syndrome, especially when the mother is over 35. The underlying biology of this effect is unknown.

Nondisjunction of sex chromosomes

The gain or loss of a sex chromosome does not generally result in the severe developmental issues caused by autosomal aneuploidy.

X chromosome nondisjunction When X chromosomes fail to separate during meiosis, some of the gametes produced possess both X chromosomes, and so are XX gametes; the other gametes have no sex chromosome and are designated O.

If an XX gamete combines with an X gamete, the resulting XXX zygote develops as a female with one functional X chromosome and two Barr bodies. These individuals may be taller in stature, but otherwise show no visible phenotype from X chromosome trisomy.

If an XX gamete instead combines with a Y gamete, the effects are more serious. The resulting XXY zygote develops as a male but will often be infertile and may have weaker muscles and enlarged breasts. This condition, called *Klinefelter syndrome,* occurs in about 1 out of every 500 male births.

If an O gamete fuses with a Y gamete, the resulting OY zygote is nonviable and fails to develop further; humans cannot survive when they lack any X chromosome. But if an O gamete fuses with an X gamete, the XO zygote develops into a sterile female of short stature, with a webbed neck and sex organs that never fully mature during puberty. The mental abilities of an XO individual are often in the low-normal range. This condition, called *Turner syndrome,* occurs roughly once in every 5000 female births.

Y chromosome nondisjunction The Y chromosome can also fail to separate in meiosis, leading to the formation of YY gametes. When these gametes combine with X gametes, the XYY zygotes develop into fertile males with no visible phenotype from the extra Y chromosome. The frequency of the XYY genotype *(Jacob syndrome)* is about 1 per 1000 newborn males.

REVIEW OF CONCEPT 13.5

Nondisjunction during meiosis can result in aneuploid gametes with too few or too many chromosomes. While the condition is usually lethal, some aneuploid zygotes go on to develop, but often produce clinical syndromes.

- *During spermatogenesis, is there any difference in outcome between first- and second-division nondisjunction?*

13.6 Inheritance in Humans Can Be Studied by Analyzing Pedigrees

To study human heredity, scientists look at the results of crosses that have already been made. They study family trees, or **pedigrees,** to identify which relatives exhibit a trait. A pedigree is a consistent graphical representation of matings and offspring over multiple generations for a particular trait. In a pedigree, males are shown as squares, and females as circles. Those showing a trait, usually referred to as "affected" in the case of disease states, are shown with shaded squares or circles, and those unaffected are shown as unshaded. If an individual is known, or can be inferred, to be heterozygous, that individual is shown as half shaded.

To Analyze Human Pedigrees, Geneticists Ask Three Questions

LEARNING OBJECTIVE 13.6.1 Demonstrate how modes of inheritance can be analyzed using pedigrees.

We will use as an example the human trait of albinism. Individuals with albinism lack most or all of the pigment eumelanin. As a consequence, their hair and skin are usually completely white. In the United States, the frequency of albinism in the general population is about 1 in 20,000 individuals. There are multiple forms of albinism, with several genes that can be causal. There are also isolated populations that show elevated levels of albinism. On the Hopi reservation in Arizona, the incidence of one particular form of albinism is thought to be as high as 1 in 200. The pedigree in figure 13.15 is for a Hopi family with this form of albinism segregating.

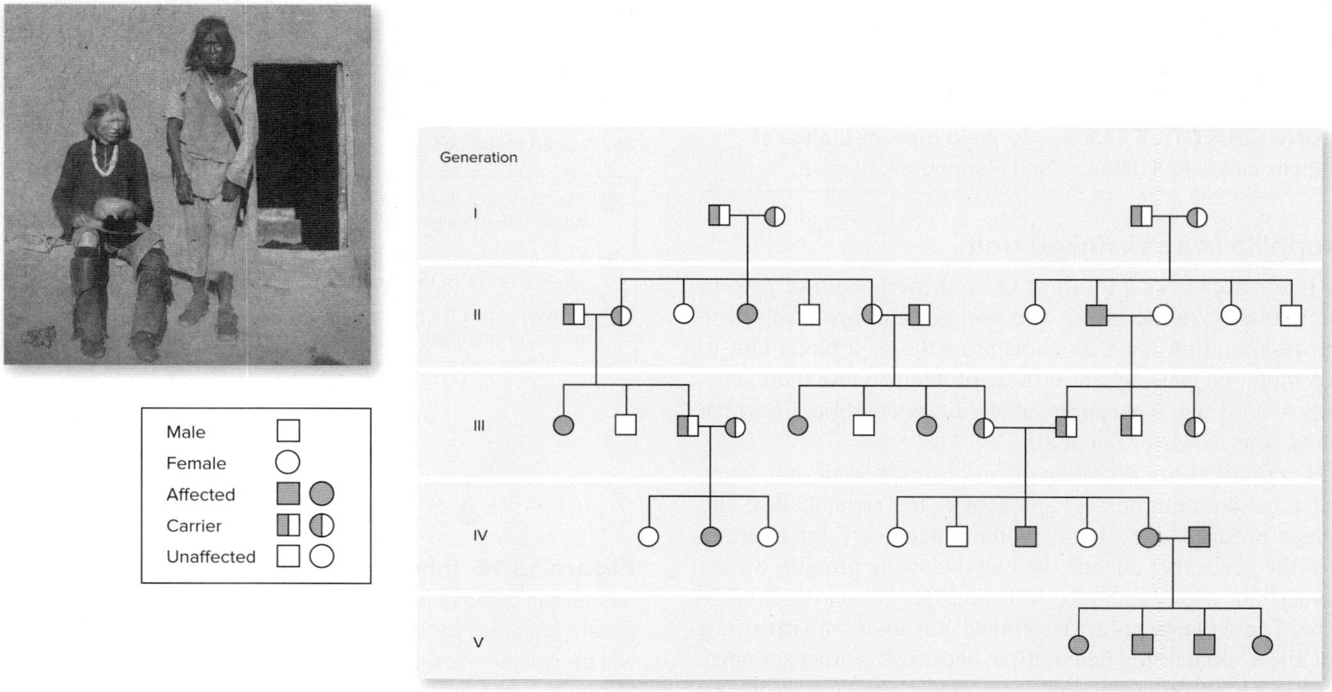
Male
Female
Affected
Carrier
Unaffected
Generation
I
II
III
IV
V

Figure 13.15 A pedigree of albinism. The pedigree shows the inheritance over five generations of a gene with a mutant allele that causes albinism, in a family of Hopi Indians. The solid green symbols indicate persons who are albino. In the photo, the individual on the left has albinism caused by the same allele.

Library of Congress Prints & Photographs Division [LC-DIG-stereo-1s00333]

To analyze a pedigree, try to fit a particular model of inheritance to the pedigree. If the pedigree does not allow you to rule out a model, it must be considered possible. In analyzing pedigrees, it is important to remember that disease-causing alleles are usually rare in the general population. If it is necessary to assume too many unrelated affected individuals, your model is probably wrong. The analysis can begin with simple questions:

1. Is the trait dominant or recessive? If the trait is dominant, every affected individual will have an affected parent, while recessive traits can arise in offspring from unaffected parents. Dominant traits also tend to appear in every generation.
2. Is the trait sex-linked or autosomal? If the trait is sex-linked, it is seen much more frequently in males; if it is autosomal, it appears in both sexes fairly equally. In sex-linked recessive traits, females can only be affected if their father is also affected, and males will tend to have affected grandfathers.
3. Is the trait determined by a single gene or by several? If the trait is determined by a single gene, then a ratio of 3:1 affected to unaffected offspring should be born to heterozygous parents, or about 25% affected overall. If two genes are involved, the trait would appear in far fewer individuals. While this is a general guide, the number are often too small for solid conclusions.

In the case of albinism shown in figure 13.15, the trait is clearly recessive, as all affected individuals, except those in generation V, have unaffected parents. It also appears to be autosomal, as there are roughly equal numbers of affected males and females. In addition, there are multiple affected females with unaffected fathers. Lastly, it appears to be single gene, as 8 of 28 offspring from heterozygous parents are affected..

Inherited Human Disorders Often Have Distinctive Pedigrees

LEARNING OBJECTIVE 13.6.2 Contrast the inheritance of hemophilia, sickle-cell disease, and Huntington disease.

Hemophilia is a sex-linked trait

Blood in a cut clots as a result of the polymerization of protein fibers circulating in the blood. A dozen proteins are involved in this process, and all must function properly for a blood clot to form. A mutation causing any of these proteins to lose their activity leads to a form of **hemophilia,** a hereditary condition in which the blood clots slowly or not at all.

Hemophilias are recessive disorders, expressed only when an individual does not possess any copy of the normal allele and so cannot produce one of the proteins necessary for clotting. Most of the genes that encode the blood-clotting proteins are on autosomes, but two (designated VIII and IX) are on the X chromosome. These two genes are sex-linked: any male who inherits a mutant allele will develop hemophilia, because his other sex chromosome is a Y chromosome that lacks any alleles of those genes.

The most famous instance of hemophilia, often called Royal hemophilia, is a sex-linked form that arose in the royal family of England. This hemophilia was caused by a mutation in gene IX that occurred in one of the parents of Queen Victoria of England (1819–1901). In the six generations since Queen Victoria, 10 of her male descendants have had hemophilia. The present British royal family has escaped the disorder because Queen Victoria's son King Edward VII did not inherit the defective allele, and all the subsequent rulers of England are his descendants. Three of Victoria's nine children did receive the defective allele, however, and they carried it by marriage into many of the other royal families of Europe.

Sickle-cell disease is a recessive trait

Sickle-cell disease is inherited as an autosomal recessive trait. This is shown in the pedigree in figure 13.16. Affected individuals have a defect in hemoglobin, the protein in red blood cells that carries oxygen. This defect impairs the ability of red blood cells to properly transport oxygen to tissues. The molecular nature of this defect is that the hemoglobin molecules tend to stick to one another, forming stiff, rodlike structures that deform the shape of the red blood cells. It is this characteristic "sickle" shape of red blood cells that gave the disease its name. As a result of their irregular shape, these cells have difficulty moving through the smallest blood vessels, where they tend to

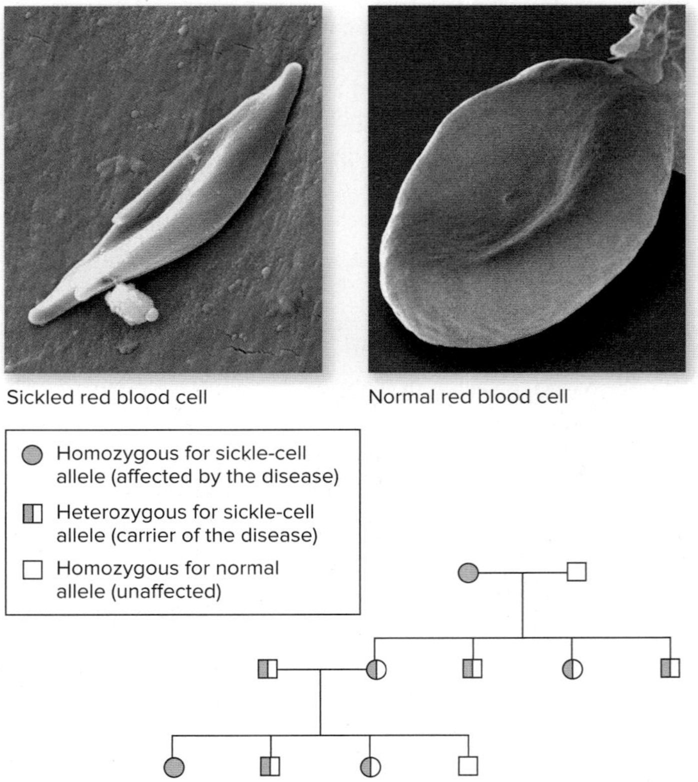

Figure 13.16 Inheritance of sickle-cell disease. Sickle-cell disease is a recessive autosomal disorder. If one parent is homozygous for the recessive trait, all of the offspring will be carriers (heterozygotes), like the F_1 generation of Mendel's testcross. A normal red blood cell is shaped like a flattened sphere. In individuals homozygous for the sickle-cell trait, many of the red blood cells have sickle shapes.

(a): Janice Haney Carr/CDC; (b): Steve Gschmeissner/Science Photo Library/Alamy Stock Photo accumulate and form clots. The resulting anemia produces a variety of symptoms.

The defective hemoglobin is due to a single amino acid change in ß-globin. The sickle-cell allele causes a substitution of valine for glutamic acid (shown in detail in figure 12.15). This amino acid substitution does not affect oxygen binding, but it does affect the surface of the protein. This simple change replaces a charged amino acid with a nonpolar one. Nonpolar amino acids tend to associate in a polar environment, and this aggregation leads to sickled red blood cells.

Individuals heterozygous for the sickle-cell allele are generally indistinguishable from individuals homozygous for the normal allele. However, some of their red blood cells show the sickling characteristic when they are exposed to low levels of oxygen. The allele responsible for sickle-cell disease is particularly common among people of African descent, because the sickle-cell allele is more common in Africa. Heterozygosity for the sickle-cell allele increases resistance to malaria, a common and serious disease in Central Africa. The interactions of sickle-cell disease and malaria are discussed further in chapter 19.

Huntington disease: A dominant trait

Not all hereditary disorders are recessive. **Huntington disease** is a hereditary condition caused by a dominant allele that causes the progressive deterioration of brain cells. Perhaps 1 in 24,000 individuals develops the disorder. Because the allele is dominant, every individual who carries the allele expresses the disorder. Nevertheless, the disorder persists in human populations because its symptoms usually do not develop until the affected individuals are more than 30 years old, and by that time most of those individuals have already had children. Consequently, as illustrated by the pedigree in figure 13.17, the allele is often transmitted before the lethal condition develops.

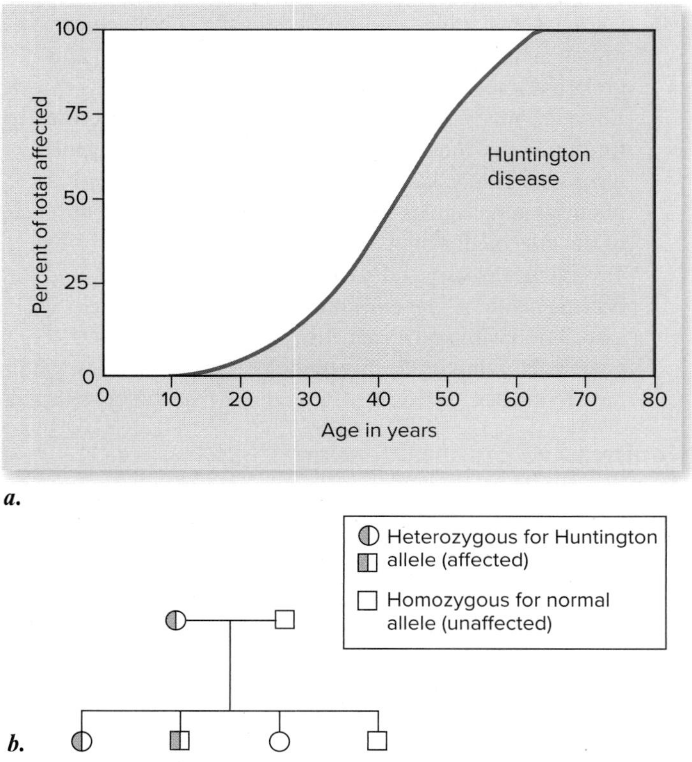

Figure 13.17 Huntington disease is a dominant genetic disorder. ***a.*** Because of the late age of onset of Huntington disease, the allele causing it persists despite being both dominant and fatal. ***b.*** Simple pedigree for the inheritance of the dominant Huntington allele.

Mutant alleles of many genes can produce human diseases

Table 13.1 provides additional examples of selected genetic diseases. This is obviously a very short list and raises the question of how many Mendelian phenotypes exist in humans. One way to answer this is to use the mouse, a fellow mammal, as a model system. Loss-of-function mutations were constructed in 900 mouse genes and analyzed for a phenotype. At least 30% of the mutants exhibited embryonic lethality, and the majority had at least one phenotype. This implies that the majority of human genes could potentially produce a phenotype when mutated.

With the proliferation of data on human genetic variation, there has also been a proliferation of databases to store this information. The oldest is Online Mendelian Inheritance in Man (OMIM). This effort was begun in 1966 with the publication of *Mendelian Inheritance in Man* by medical geneticist Victor McKusick. After multiple print editions, it was moved online and made publicly available in 1987. Summary statistics from 2020 indicate that there are 6771 phenotypes with a known molecular basis. From the perspective of genes, there are 4355 genes with mutations that cause phenotypes. Among these genes, many produce multiple phenotypes: 794 with 2 phenotypes, 293 with 3, and 238 with 4 or more phenotypes.

Genetic Counseling and Therapy

LEARNING OBJECTIVE 13.6.3 Describe three things geneticists examine in cells obtained by amniocentesis.

Although most genetic disorders cannot yet be cured, we are learning a great deal about them, and progress toward successful therapy is being made in many cases. However, in the absence of a cure, some parents may feel their only recourse is to try to avoid producing children with these conditions. The process of identifying parents at risk of producing children with genetic defects and of assessing the genetic state of early embryos is called **genetic counseling.** Genetic counseling can help prospective parents determine their risk of having a child with a genetic disorder and advise them on medical treatments or options if a genetic disorder is determined to exist in an unborn child.

High-risk pregnancies

If a genetic defect is caused by a recessive allele, how can potential parents determine the likelihood that they carry the allele? One way is through pedigree analysis, often employed as an aid in genetic counseling. By analyzing a person's pedigree, it is sometimes possible to estimate the likelihood that the person is a carrier for certain disorders. For example, if one of your relatives has been afflicted with a recessive genetic disorder such as cystic fibrosis, it is possible that you are a heterozygous carrier of the recessive allele for that disorder. When a pedigree analysis

TABLE 13.1 Some Important Genetic Disorders

Disorder	Symptom	Defect	Gene/Mode of Inheritance	Frequency Among Human Births
Cystic fibrosis	Mucus clogs lungs, liver, and pancreas.	Failure of chloride ion transport mechanism	Cystic fibrosis transmembrane conductance regulator (*CFTR*) Recessive	1/2,500 (Caucasians)
Sickle-cell anemia	Blood circulation is poor.	Abnormal hemoglobin molecules	β-globin (*HBB*) Recessive	1/600 (African Americans)
Tay–Sachs disease	Central nervous system deteriorates in infancy.	Defective enzyme (hexosaminidase A)	Hexosaminidase A (*HEXA*) Recessive	1/3,500 (Ashkenazi Jews)
Phenylketonuria	Brain fails to develop in infancy; it is treatable with dietary restriction.	Defective enzyme (phenylalanine hydroxylase)	Phenylalanine hydroxylase (*PAH*) Recessive	1/12,000
Hemophilia	Blood fails to clot.	Defective blood-clotting factor VIII	Coagulation factor VIII (*F8*) X-linked recessive	1/10,000 (Caucasian males)
Huntington disease	Brain tissue gradually deteriorates in middle age.	Production of an inhibitor of brain cell metabolism	Huntingtin (*HTT*) Dominant	1/24,000
Muscular dystrophy (Duchenne)	Muscles waste away.	Degradation of myelin coating of nerves, stimulating muscles	Dystrophin (*DMD*) X-linked recessive	1/3,700 (males)
Hypercholesterolemia	Excessive cholesterol levels in blood lead to heart disease.	Nonfunctional form of cholesterol cell-surface receptor	LDL receptor (*LDLR*) Dominant	1/500

indicates that both parents of an expected child have a significant probability of being heterozygous carriers of a recessive allele responsible for a serious genetic disorder, the pregnancy is said to be a high-risk pregnancy. In such cases, there is a significant probability that the child will exhibit the clinical disorder.

Genetic screening

When a pregnancy is determined to be high risk, many women elect to undergo **amniocentesis,** a procedure that permits the prenatal diagnosis of many genetic disorders. Figure 13.18*a* shows how an amniocentesis is performed. In the fourth month of pregnancy, a sterile hypodermic needle is inserted into the expanded uterus of the mother, and a small sample of the amniotic fluid bathing the fetus is removed. Within the fluid are free-floating cells derived from the fetus; once removed, these cells can be grown in cultures in the laboratory. During amniocentesis, the position of the needle and that of the fetus are usually observed by means of ultrasound.

Physicians have increasingly turned to another invasive procedure for genetic screening, called **chorionic villus sampling (CVS).** In this procedure, the physician removes cells from the chorion, a membranous part of the placenta that nourishes the fetus (figure 13.18*b*). This procedure can be used earlier in pregnancy (by the eighth week) and yields results much more rapidly than does amniocentesis, but it can increase the risk of miscarriage.

Genetic counselors look at three things in the cultures of cells obtained from amniocentesis or chorionic villus sampling:

1. **Chromosomal karyotype.** Analysis of the karyotype can reveal aneuploidy (extra or missing chromosomes) and gross chromosomal alterations.
2. **Enzyme activity.** In many cases it is possible to test directly for the proper functioning of enzymes involved in genetic disorders. The lack of the enzyme for breaking down phenylalanine signals PKU (phenylketonuria), the absence of the enzyme for breaking down gangliosides indicates Tay–Sachs disease, and so forth.
3. **Genetic markers.** Genetic counselors can look for an association with known genetic markers. For sickle-cell anemia, Huntington disease, and other diseases, investigators have found associated DNA alterations that can be detected.

Noninvasive prenatal testing

Amniocentesis and CVS are invasive procedures. The discovery that biomarkers found in maternal blood plasma could be correlated with Down syndrome now allows noninvasive screening for this trisomy. More recently, the discovery of fetal DNA in maternal blood plasma (cell-free DNA, or cfDNA) has led to new methods of noninvasive prenatal screening. At present these tests can be used to screen for trisomy of chromosomes 13, 18, and 21, as well as allow sex determination. Considerable effort is being put into expanding the scope of this type of screening. It should be

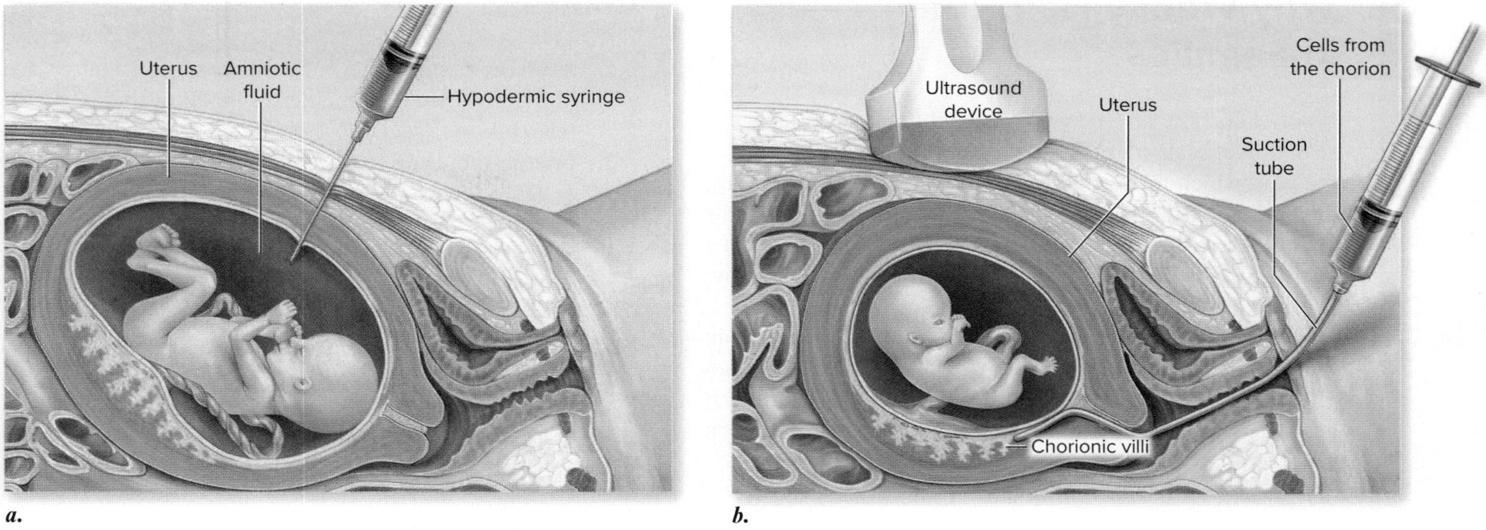

Figure 13.18 Two ways to obtain fetal cells. ***a.*** In amniocentesis, a needle is inserted into the amniotic cavity, and a sample of amniotic fluid, containing some free cells derived from the fetus, is withdrawn into a syringe. ***b.*** In chorionic villus sampling, cells are removed by suction with a tube inserted through the cervix. In each case, the cells can be grown in culture, then examined for karyotypes, and used in biochemical and genetic tests.

emphasized that this is a form of screening and is not at present diagnostic. A positive result indicates the need for more invasive techniques for a definitive diagnosis.

Parents using in vitro fertilization have access to a screening procedure known as **preimplantation genetic screening.** As the egg is fertilized outside the mother it can be allowed to divide until it contains eight cells, one of which is removed and tested for known genetic defects (figure 13.19). The remaining seven-cell embryo can develop normally, giving the parents the choice of identifying and implanting an embryo that is disease-free.

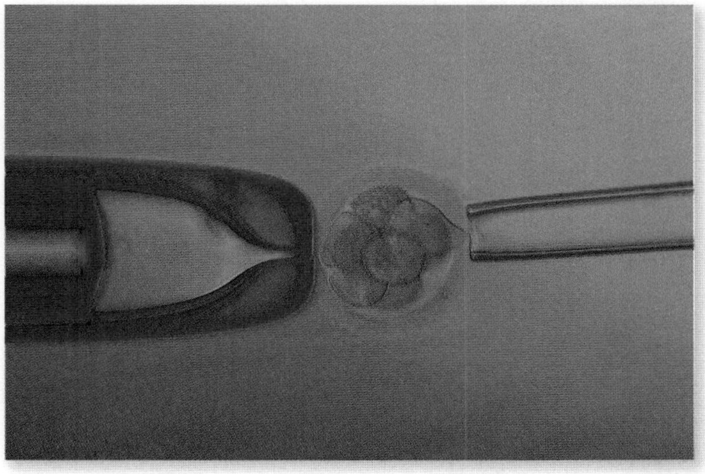

Figure 13.19 Preimplantation genetic screening. The photograph shows a human embryo at the eight-cell stage, just before one of the eight cells is to be extracted for genetic testing by researchers.

Pascal Goetgheluck/Science Source

REVIEW OF CONCEPT 13.6

Mutations in DNA that result in altered proteins can cause hereditary diseases. Pedigree studies and genetic testing may clarify the risk of disease.

- *Might mutations that do not alter proteins still cause hereditary disorders?*

Why Woolly Hair Runs in Families

Wooly hair is a rare congenital condition that affects the structure of scalp hair. The name refers to the appearance of the hair, which forms tight coils. The coils have an average diameter of 0.5 cm, and lie close together. This usually affects the entire scalp, but it can also be localized, with a patchy distribution. This syndrome also results in hair that is brittle and subject to breakage.

The syndrome can be apparent at birth, or appear in the first months of life. The growth rate of the hair is not affected, but it does not grow to be very long. It was first described in a European family in 1907, but has been observed in populations of European or Asian descent.

The woolly hair trait is rare, but it does appear to run in families. The extensive pedigree below (drawn curved so as to fit in the large families produced by the second and subsequent generations) records the incidence of woolly hair in five generations of a Norwegian family. Generations are indicated by the Roman numerals on the left, and as is the convention, affected individuals are indicated by solid symbols, with circles indicating females and squares indicating males. This pedigree has the information necessary for you to analyze potential modes of inheritance of this trait in this family.

Analysis

1. **Applying Concepts** In the diagram, how many individuals are documented? Are all of them related?
2. **Interpreting Data**
   a. Does the woolly hair trait appear in both sexes equally?
   b. Does every woolly-haired child have a woolly-haired parent?
   c. What percentage of the offspring born to a woolly-haired parent are also woolly haired?
3. **Making Inferences**
   a. Is woolly hair sex-linked or autosomal?
   b. Is woolly hair dominant or recessive?
   c. Is the woolly-hair trait determined by a single gene or by several?
4. **Drawing Conclusions**
   a. How many copies of the woolly-hair allele are necessary to produce a detectable change in a person's hair?
   b. Are there any woolly-hair homozygous individuals in the pedigree? Explain.

Pedigree of Woolly Hair Among a Norwegian Family

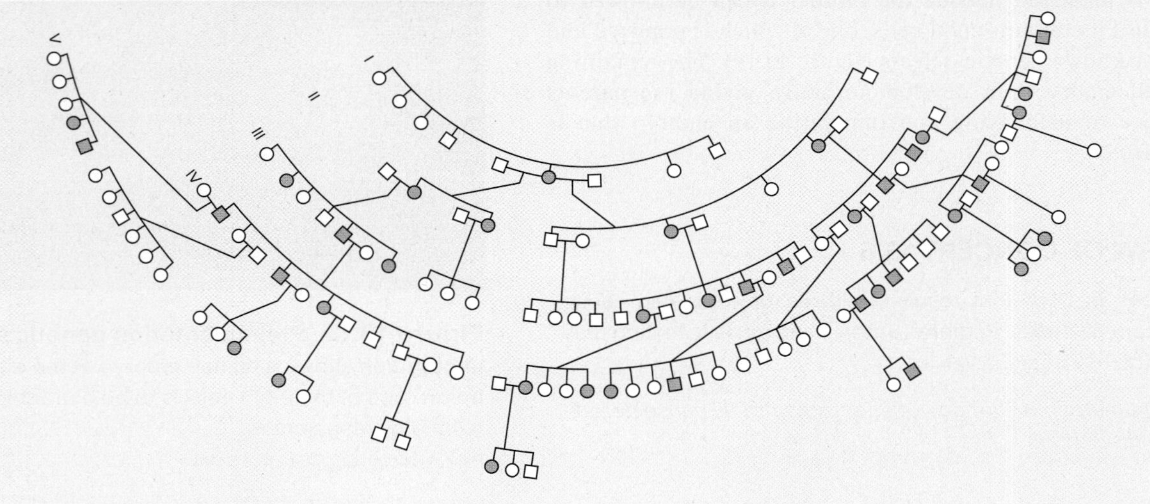

Retracing the Learning Path

CONCEPT 13.1 Sex Linkage and the Chromosomal Theory of Inheritance

13.1.1 Morgan Correlated the Inheritance of a Trait with Sex Chromosomes Morgan showed that the inheritance of eye color in *Drosophila* segregates with the X chromosome, a phenomenon termed sex-linked inheritance. This supports the idea that genes are on chromosomes. In flies and humans, males will show recessive traits on the X chromosome.

13.1.2 Adult Body Cells of Mammals Have Only One Active X Chromosome The homogametic sex has two similar chromosomes and the heterogametic sex carries a unique sex chromosome. In humans, the Y chromosome has very few genes. The *SRY* gene on the Y chromosome is responsible for male sexual development. An XY individual with a mutation in *SRY*, or someone who fails to respond to androgens, will develop as female. In fruit flies, males increase gene expression from their X chromosome, whereas in mammals, females randomly inactivate one X chromosome.

CONCEPT 13.2 There Are Two Major Exceptions to Chromosomal Inheritance

13.2.1 Genomic Imprinting Depends on the Parental Origin of Alleles In genomic imprinting, the expression of a gene depends on whether it passes through the maternal or paternal germ line. Imprinted genes appear to be inactivated by methylation. Imprinting is an example of epigenetics. Epigenetic changes are heritable through cell generations, but they do not involve a change in the DNA sequence.

13.2.2 Organellar Inheritance Involves Non-Nuclear DNA Mitochondria have their own genomes and are passed to offspring in the cytoplasm of the egg. This leads to maternal inheritance. Chloroplasts also have their own genomes. They are usually inherited maternally.

CONCEPT 13.3 Some Genes Do Not Assort Independently: Linkage

13.3.1 Genetic Recombination Occurs Less Often Between Nearby Genes Homologous chromosomes may exchange alleles by crossing over. This occurs by breakage and rejoining of chromosomes, shown in crosses of chromosomes carrying both visible and genetic markers.

CONCEPT 13.4 Genetic Crosses Provide Data for Genetic Maps

13.4.1 The Frequency of Recombination Allows Mapping of the Relative Positions of Genes on Chromosomes Genes close together on a single chromosome are said to be linked. The farther apart two linked genes are, the greater the frequency of recombination. This allows genetic maps to be constructed based on recombination frequency. Map units are expressed as the percentage of recombinant progeny. Multiple crossovers increase with longer distances and lead to a maximum recombination frequency of 50%, the same as independent assortment. Three-point crosses are used to order genes, as well as allow more accurate maps.

13.4.2 Genes Can Be Mapped in Humans Human linkage mapping is difficult because it requires multiple alleles segregating in a family. The process has been made easier by the use of identifiable molecular markers that do not cause a phenotype. Single-nucleotide polymorphisms (SNPs) can be used to detect differences between individuals for identification.

CONCEPT 13.5 Changes in Chromosome Number Can Have Drastic Effects

13.5.1 Nondisjunction Is a Failure of Meiotic Separation Nondisjunction is the failure of homologs or sister chromatids to separate during meiosis. The result is aneuploidy: monosomy or trisomy of a chromosome in the zygote. Most aneuploidies are lethal, but some, such as trisomy 21 in humans, result in viable offspring. Sex chromosome nondisjunction produces XX, YY, and O gametes. Fertilization yields viable XO, XXY, or XXX zygotes.

CONCEPT 13.6 Inheritance in Humans Can Be Studied by Analyzing Pedigrees

13.6.1 To Analyze Human Pedigrees, Geneticists Ask Three Questions The study of family trees can often reveal if an inherited trait is caused by a single gene, if that gene is located on the X chromosome, and if its mutant alleles are recessive.

13.6.2 Inherited Human Disorders Often Have Distinctive Pedigrees Mutations in blood-clotting factor IX, on the X chromosome, cause hemophilia. Inheritance of this is demonstrated by the European royal families. Sickle-cell disease is inherited as an autosomal recessive trait. Over 700 variants of hemoglobin structure have been characterized. Huntington disease is inherited as an autosomal dominant trait with late onset.

13.6.3 Genetic Counseling and Therapy Genetic defects in humans can be determined by pedigree analysis, amniocentesis, or chorionic villus sampling.

Concept Overview

This Concept Overview diagrams the key concepts that were discussed in this chapter.

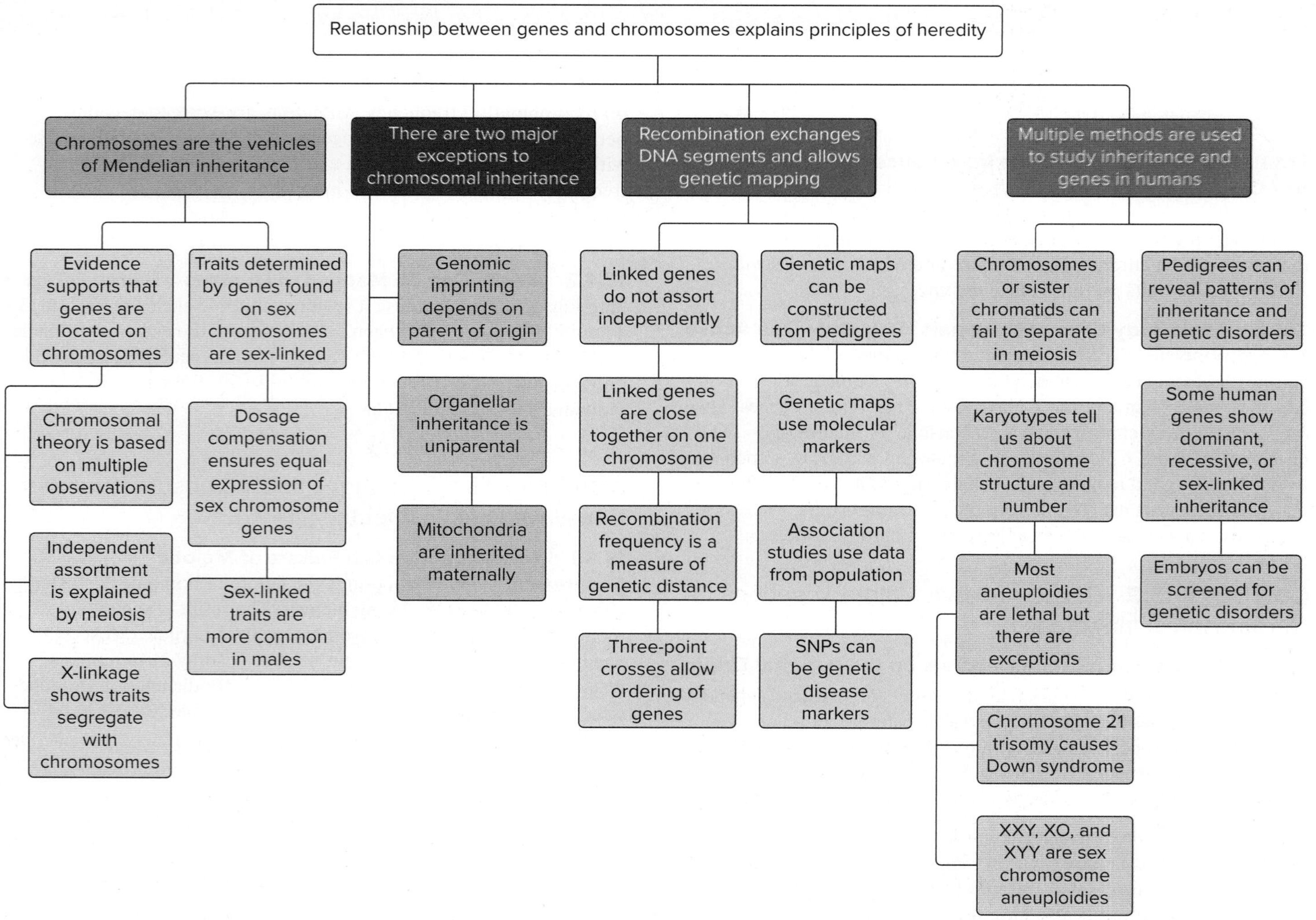

Assessing the Learning Path

Understand

1. Why is the white eye phenotype always observed in males carrying the white eye allele?
   a. Because the trait is dominant
   b. Because the trait is recessive
   c. Because the allele is located on the X chromosome and males have only one X chromosome
   d. Because the allele is located on the Y chromosome and only males have Y chromosomes
2. Dosage compensation is needed to
   a. balance expression from autosomes relative to sex chromosomes.
   b. balance expression from two autosomes in a diploid cell.
   c. balance expression of sex chromosomes in both sexes.
   d. inactivate female-specific autosomal chromosomes.
3. Genes that lie very close to each other on a chromosome
   a. segregate together.
   b. cross over frequently.
   c. assort independently.
   d. form chiasmata.
4. The map distance between two genes is determined by the
   a. recombination frequency.
   b. frequency of parental types.
   c. ratio of genes to length of a chromosome.
   d. ratio of parental to recombinant progeny.
5. As real genetic distance increases, the distance calculated by recombination frequency becomes an
   a. overestimate due to multiple crossovers that cannot be scored.
   b. underestimate due to multiple crossovers that cannot be scored.
   c. underestimate due to multiple crossovers adding to recombination frequency.
   d. overestimate due to multiple crossovers adding to recombination frequency.
6. Nondisjunction
   a. can occur in meiosis I or meiosis II.
   b. can be detected in a karyotype.
   c. can occur in sex chromosomes and autosomes.
   d. All of the above

7. Why is nondisjunction more common in oogenesis than in spermatogenesis?
   a. Eggs are formed daily, and this fast rate of production means more chances for problems.
   b. Egg development starts in the fetus, with oocytes arrested in meiosis I, which leads to meiotic errors later.
   c. Eggs are diploid and sperm are haploid.
   d. Oogenesis and spermatogenesis have equivalent rates of nondisjunction.
8. Which of the following CANNOT be determined from pedigree analysis?
   a. Whether a trait is dominant or recessive
   b. Whether a trait is sex-linked or autosomal
   c. The chance of future generations inheriting a trait
   d. The type of mutation leading to a disease state
9. The sickle-cell trait is recessive. This means
   a. an individual who is heterozygous makes all normal hemoglobin protein.
   b. an individual must have two abnormal copies of the hemoglobin gene to exhibit the trait.
   c. the condition is not very common in the human population.
   d. anyone affected by the disease must have a parent who is also affected.
10. How does maternal inheritance of mitochondrial genes differ from sex linkage?
    a. Mitochondrial genes do not contribute to the phenotype of an individual.
    b. Because mitochondria are inherited from the mother, only females are affected.
    c. Since mitochondria are inherited from the mother, females and males are equally affected.
    d. Mitochondrial genes must be dominant. Sex-linked traits are typically recessive.
11. An organism has an imprinted allele. Which of the following statements is NOT accurate about the imprinted allele?
    a. The imprint occurred during gamete formation in one of the organism's parents.
    b. The imprint will be passed to the organism's offspring.
    c. It is silenced.
    d. The imprint will be found in all body cells of the organism.

Apply

1. Color blindness is caused by a sex-linked recessive gene. If a woman whose father was color blind marries a man with normal color vision, what percentage of their children will be expected to be color blind?
   a. 100%
   b. 50%
   c. 25%
   d. 0%
2. What percentage of the sons of the couple described in the previous question will be color blind?
   a. 100%
   b. 50%
   c. 25%
   d. 0%
3. Mendel did not examine plant height and pod shape in his dihybrid crosses. The genes for these traits are very close together on the same chromosome. What would Mendel have found if he had studied these two traits in a dihybrid cross?
   a. The ratio in the F_2 generation would have been 9:3:3:1.
   b. The phenotypic ratio in the F_2 generation would have been 3:1, but the genotypic ratio would have been 1:2:1.
   c. The ratio would have been skewed from the expected value because of linkage.
   d. He would not have been able to set up a cross for these two traits, because they are found on the same chromosome.
4. In a plant, the genes for seed color and seed shape are located on the same chromosome. Yellow seeds are dominant to green seeds, and round seeds are dominant to wrinkled seeds. A plant heterozygous for both genes is test-crossed to a homozygous recessive plant, and the following data are obtained:

green, wrinkled	455
green, round	28
yellow, wrinkled	22
yellow, round	495

   How far apart are the two genes?
   a. 2.5 map units
   b. 5 map units
   c. 50 map units
   d. 95 map units
5. During the process of spermatogenesis, a nondisjunction event that occurs during the second division would be
   a. worse than the first division because all four meiotic products would be aneuploid.
   b. better than the first division because only two of the four meiotic products would be aneuploid.
   c. the same outcome as the first division, with all four products aneuploid.
   d. the same outcome as the first division, as only two products would be aneuploid.
6. Hypercholesterolemia is inherited in an autosomal dominant manner. Joe has hypercholesterolemia; his wife does not. They have a daughter with normal cholesterol levels. What is the chance that their next child will have the same phenotype as the daughter?
   a. 100%
   b. 50%
   c. No chance
   d. It is not possible to determine with the information provided.
7. Three of the traits that Mendel studied—round/wrinkled, tall/short, and purple/white—all segregate independently. True-breeding round tall purple is crossed to true-breeding wrinkled short white to produce an F_1 that is all round tall purple. If this F_1 is intercrossed, the probability of being round tall purple is
   a. $3/4$
   b. $9/16$
   c. $27/64$
   d. $81/256$
   e. $9/64$

Synthesize

1. In Morgan's testcross in figure 13.3, he obtained 129 red-eyed females, 132 red-eyed males, 88 white-eyed females, and 86 white-eyed males. This is a poor fit to the expected 1:1 ratio of red to white eyes. How might you account for this observation?
2. Is it possible to have a calico cat that is male? Why or why not?
3. Nondisjunction can also occur during mitosis. A human zygote is formed with two normal gametes (each $n = 23$). At the four-cell stage of embryonic development, nondisjunction occurs in a cell, resulting in some daughter cells with an additional copy of chromosome 21. What will be the phenotype of the individual who is born?
4. Female mice homozygous for the normal allele of *igf 2* are crossed to males heterozygous for normal and dwarf alleles. What proportion of offspring would you predict will be dwarf? If the genotypes of the parents were reversed, what would be the outcome? Explain your answers.
5. Most gene defects in humans that exhibit maternal inheritance seem to involve mental disorders or muscle function. Propose an explanation for this observation.

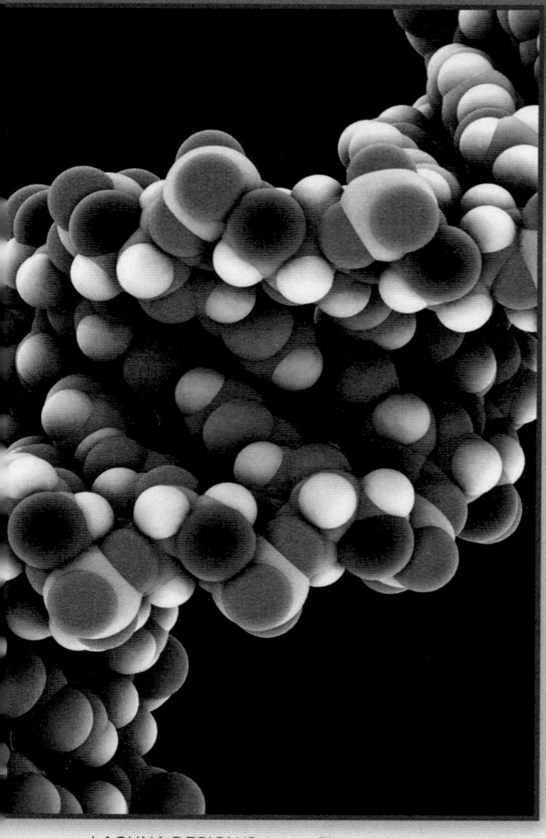

LAGUNA DESIGN/Science Photo Library/Alamy Stock Photo

14 DNA: The Genetic Material

Learning Path

14.1 DNA Is the Genetic Material

14.2 The DNA Molecule Is a Double Helix

14.3 Both Strands Are Copied During DNA Replication

14.4 Prokaryotes Organize the Enzymes Used to Duplicate DNA

14.5 Eukaryotic Chromosomes Are Large and Linear

14.6 Cells Repair Damaged DNA

Concept Overview

This Concept Overview provides a diagram of the overarching concepts that are covered in this chapter. These concepts will be expanded upon in the Concept Overview at the end of the chapter.

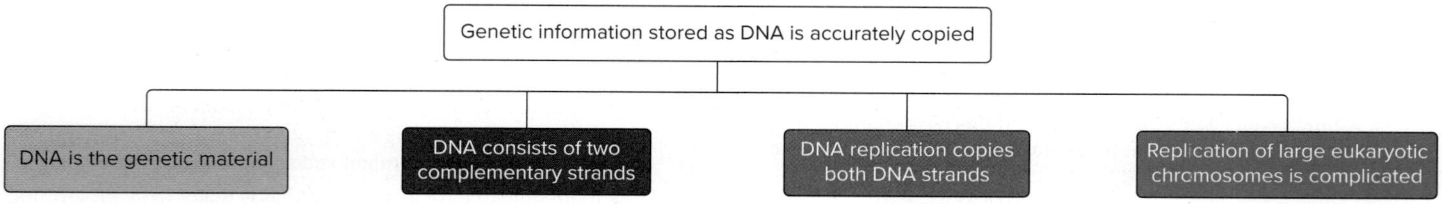
Genetic information stored as DNA is accurately copied
DNA is the genetic material
DNA consists of two complementary strands
DNA replication copies both DNA strands
Replication of large eukaryotic chromosomes is complicated

Introduction

The rediscovery of Mendel's work at the turn of the 20th century led to a period of rapid discovery of the genetic mechanisms detailed in the previous two chapters. One question not directly answered for more than 50 years was perhaps the simplest: What are genes actually made of? Genes were known to be on chromosomes, but they are complex structures composed of DNA, RNA, and protein. This chapter describes the chain of experiments that led to our current understanding of DNA, modeled in the picture on the previous page, and of the molecular mechanisms of heredity. These experiments are among the most elegant in science. The elucidation of the structure of DNA was the beginning of a molecular era, whose pace is only accelerating today.

14.1 DNA Is the Genetic Material

Mendel and the researchers who came after him unraveled the essential mystery of heredity: Hereditary traits are controlled by genes on chromosomes inherited from our parents. However, Mendel's work left a key question unanswered: What *is* a gene? Geneticists knew that chromosomes are composed primarily of both protein and DNA. It was possible to imagine that either of the two is the stuff that genes are made of—information might be stored in a sequence of different amino acids, or in a sequence of different nucleotides. But which?

DNA Transfer Produces Hereditary Transformation in Bacteria

LEARNING OBJECTIVE 14.1.1 Explain Griffith's transformation experiment.

The first clue came in 1928 with the work of microbiologist Frederick Griffith. Griffith was trying to make a vaccine that would protect against pneumonia, which was thought at the time to be caused by the bacterium *Streptococcus pneumoniae.* This bacterium has two forms: the normal, virulent form that causes pneumonia and a mutant, nonvirulent form that does not. The normal, virulent form of this bacterium is referred to as the S form, because its cells are encased in a polysaccharide capsule and so form smooth colonies on a culture dish. The mutant, nonvirulent form, which lacks an enzyme needed to manufacture the polysaccharide coat, is called the R form, because it forms rough colonies.

Griffith performed a series of simple experiments in which he infected mice with these bacteria, then monitored them for disease symptoms. Figure 14.1 details his experiments. When Griffith infected mice with the virulent S form of the bacterium, they died from pneumonia ①. However, when he infected similar mice with the nonvirulent R form, the mice showed no ill effects ②. If the virulent S form was heat-killed before injection, it did not cause pneumonia, showing that the coat itself is not sufficient to cause disease ③.

Finally ④, Griffith injected mice with a mixture containing dead S bacteria and live capsuleless R bacteria, each of which by itself did not harm the mice. Unexpectedly, many of them died! Furthermore, high levels of live virulent S bacteria were found in the lungs of the dead mice.

Somehow the information specifying the polysaccharide coat had passed from the dead virulent S bacteria to the live coatless R bacteria in the mixture, permanently transforming the coatless R bacteria into the virulent S variety. Griffith called this transfer of virulence from one cell to another **transformation.** It seemed clear from these results that genetic material was transferred between the cells—but the results gave no hint of how, or of what the material might be.

The Transforming Principle Is DNA

LEARNING OBJECTIVE 14.1.2 Describe how Avery's work demonstrated that DNA is the transforming principle.

The agent responsible for transforming *Streptococcus* went undiscovered until 1944, when in a classic series of experiments geneticists Oswald Avery and his coworkers Colin MacLeod and

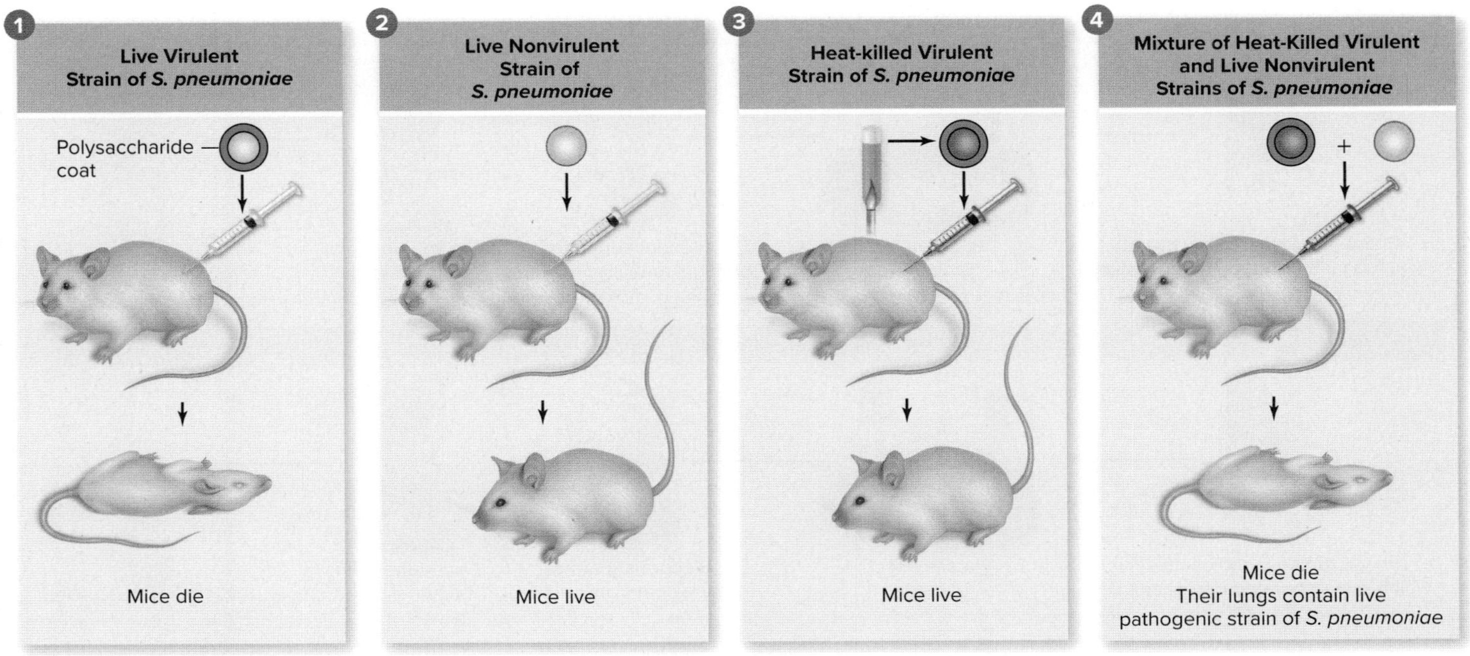

1
Live Virulent Strain of S. pneumoniae
Polysaccharide coat
Mice die
2
Live Nonvirulent Strain of S. pneumoniae
Mice live
3
Heat-killed Virulent Strain of S. pneumoniae
Mice live
4
Mixture of Heat-Killed Virulent and Live Nonvirulent Strains of S. pneumoniae
+
Mice die
Their lungs contain live pathogenic strain of S. pneumoniae

Figure 14.1 Griffith's experiment. Griffith was trying to make a vaccine against pneumonia and instead discovered transformation. ① Injecting live virulent bacteria into mice produced pneumonia. Injection of nonvirulent bacteria ② or heat-killed virulent bacteria ③ had no effect. ④ However, a mixture of heat-killed virulent and live nonvirulent bacteria produced pneumonia in the mice. This indicated that the genetic information for virulence was transferred from dead virulent cells to live nonvirulent cells, transforming them from nonvirulent to virulent.

Maclyn McCarty identified the substance responsible for transformation in Griffith's experiment.

They first prepared the mixture of dead S *Streptococcus* and live R *Streptococcus* that Griffith had used. Then they removed as much of the protein as they could from their preparation, eventually achieving 99.98% purity. They found that despite the removal of nearly all protein, the transforming activity was not reduced.

Moreover, the properties of this substance resembled those of DNA in several ways:

1. **Same chemistry as DNA.** When the purified principle was analyzed chemically, the elemental composition agreed closely with that of DNA.
2. **Same physical and chemical behavior as DNA.** When spun at high speeds in an ultracentrifuge, the transforming principle migrated to the same level (density) as DNA. In electrophoresis and other chemical and physical procedures, it also acted like DNA.
3. **Not affected by protein and lipid extraction.** Extracting proteins and lipids from the transforming principle did not reduce transforming activity.
4. **Not destroyed by protein- or RNA-digesting enzymes.** Protein-digesting enzymes did not affect the principle's transforming activity, nor did RNA-digesting enzymes.
5. **Destroyed by DNA-digesting enzymes.** Treatment of transforming principle with DNA-digesting enzymes destroyed all transforming activity.

The evidence of these experiments was overwhelming. Avery and coworkers concluded that "a nucleic acid of the deoxyribose type is the fundamental unit of the transforming principle"—in essence, that DNA is the hereditary material.

Virus Genes Are Made of DNA, Not Protein

LEARNING OBJECTIVE 14.1.3 Compare the findings of the Hershey–Chase experiment and the Avery experiment.

Avery's results were not widely accepted at first, because many biologists continued to believe that proteins were the genetic material. But in 1952 a simple experiment carried out by Alfred Hershey and Martha Chase was impossible to ignore (figure 14.2).

SCIENTIFIC THINKING

Hypothesis: *DNA is the genetic material in bacteriophage.*

Prediction: *The phage life cycle requires reprogramming the cell to make phage proteins. The information for this must be introduced into the cell during infection.*

Test: *Because DNA, but not protein, contains phosphate, and protein, but not DNA, contains sulfur, we can label each specifically with either radioactive phosphate (^{32}P for DNA) or radioactive sulfur (^{35}S for protein). This allows us to follow each molecule separately. Phage are grown on either ^{35}S or ^{32}P, then used to infect cells in two experiments. The phage heads are removed by brief agitation in a blender, and centrifuged to separate cells (pellet) from phage heads (supernatant), and each can be assayed for the radioactive labels.*

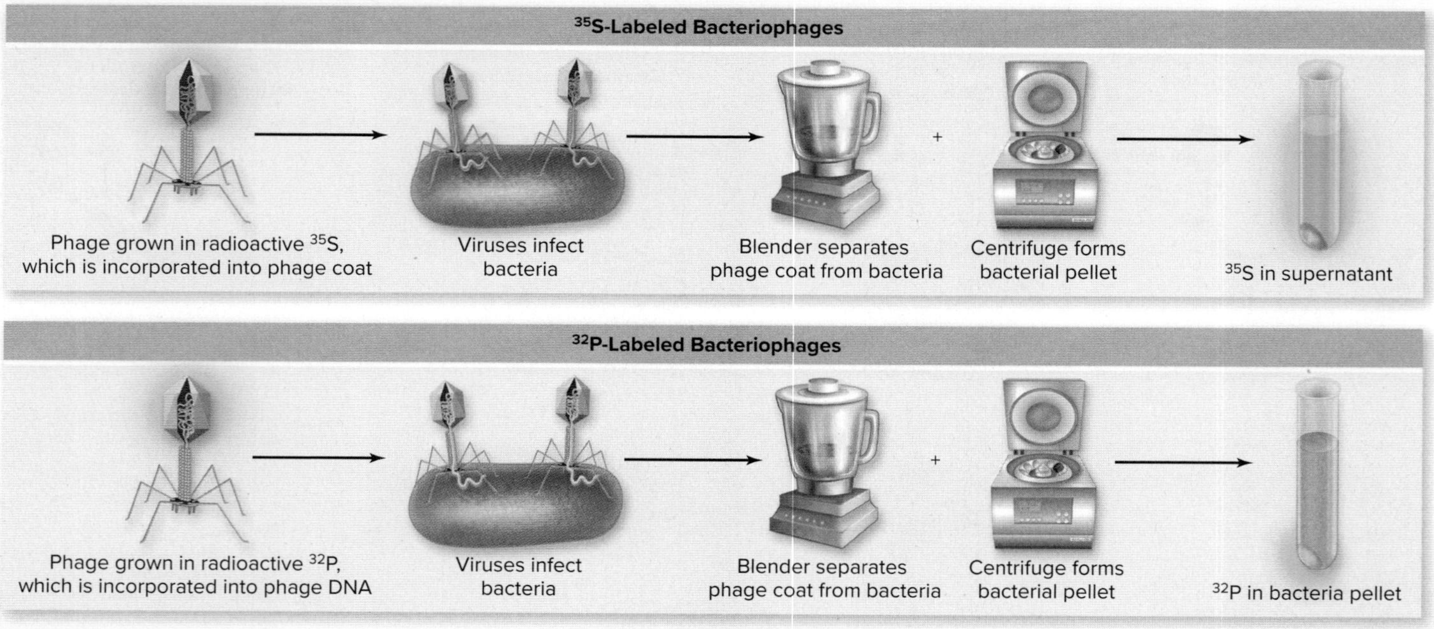

^{35}S-Labeled Bacteriophages
Phage grown in radioactive ^{35}S, which is incorporated into phage coat
Viruses infect bacteria
Blender separates phage coat from bacteria
+
Centrifuge forms bacterial pellet
^{35}S in supernatant
^{32}P-Labeled Bacteriophages
Phage grown in radioactive ^{32}P, which is incorporated into phage DNA
Viruses infect bacteria
Blender separates phage coat from bacteria
+
Centrifuge forms bacterial pellet
^{32}P in bacteria pellet

Result: *When the experiment is done, only ^{32}P makes it into the cell in any significant quantity.*

Conclusion: *Thus, DNA must be the molecule that is used to reprogram the cell.*

Further Experiments: *How does this experiment complement or extend the work of Avery on the identity of the transforming principle?*

Figure 14.2 Hershey–Chase experiment showed that DNA is genetic material for phage.

They studied the genes of viruses called **bacteriophages** ("bacteria eaters" in Latin) that infect bacteria. These viruses have a very simple structure: a core of DNA surrounded by a coat of protein. When these viruses infect a bacterial cell, they first bind to the cell's surface, then inject their genetic information into the cell. There it is expressed by the bacterial cell's gene expression machinery, leading to production of thousands of new viruses. The buildup of viruses eventually causes the cell to lyse, releasing progeny phage.

Because the T2 bacteriophage used by Hershey and Chase contained only DNA and protein, it provided the simplest possible system to differentiate the roles of DNA and protein. Hershey and Chase set out to identify the molecule that the phage injects into the bacterial cells. To do this, they needed a method to label both DNA and protein so that they could be distinguished from each other. Nucleotides contain phosphorus, but proteins do not, and some proteins contain sulfur, but DNA does not. Thus, the radioactive ^{32}P isotope can be used to label DNA specifically, and the radioactive isotope ^{35}S can be used to label proteins specifically. The two isotopes are easily distinguished based on the energy of the particles they emit when they decay.

Two experiments were performed (see figure 14.2). In one, viruses were grown on a medium containing ^{32}P, which was incorporated into DNA; in the other, viruses were grown on a medium containing ^{35}S, which was incorporated into T2 coat proteins. Each group of labeled viruses was then allowed to infect separate bacterial cultures.

After infection, the bacterial cell suspension was violently agitated in a blender to forcefully remove the infecting viral particles from the surfaces of the bacteria. This step ensured that only the part of the virus that had been injected into the bacterial cells—that is, the genetic material—would be detected when the cells were harvested.

Each bacterial suspension was then centrifuged to produce a pellet of cells for analysis. In the ^{32}P experiment, a large amount of radioactive phosphorus was found in the cell pellet, but in the ^{35}S experiment, very little radioactive sulfur was found in the pellet. Hershey and Chase deduced that DNA, not protein, constitutes the genetic information that viruses inject into bacteria.

REVIEW OF CONCEPT 14.1

Experiments with *Streptococcus* showed that information could be transferred from dead cells to live cells, transforming them. Purification of the transforming factor showed it was DNA. Experiments with bacteriophages using labeled DNA and protein confirmed this.

- *Why was protein an attractive candidate for the genetic material?*

14.2 The DNA Molecule Is a Double Helix

Chemist Friedrich Miescher discovered DNA in 1869, only four years after Mendel's work was published—although it is unlikely that Miescher knew of Mendel's experiments.

DNA Is a Polymer of Nucleotides

LEARNING OBJECTIVE 14.2.1 Identify the four DNA nucleotides.

Miescher extracted a white substance from the nuclei of human cells and fish sperm. The proportion of nitrogen and phosphorus in the substance was different from that found in any other known constituent of cells, which convinced Miescher that he had discovered a new biological substance. He called this substance "nuclein," because it seemed to be specifically associated with the nucleus. Miescher's nuclein was slightly acidic and came to be called **nucleic acid.**

DNA's components were known, but its three-dimensional structure was not

Although the three-dimensional structure of the DNA molecule was a mystery, the components of nucleic acids were known (figure 14.3):

1. A 5-carbon sugar
2. A phosphate (PO_4) group
3. A nitrogen-containing (nitrogenous) base. The base may be a **purine** (adenine, A, or guanine, G), a two-ringed structure; or a **pyrimidine** (thymine, T, or cytosine, C), a single-ringed structure. RNA contains the pyrimidine uracil (U) in place of thymine.

We need a way to be able to unambiguously refer to any carbon atom in a nucleotide. The convention used is that the numbers referring to carbons in the sugar include a prime symbol (′) to distinguish them from the numbers that refer to atoms in the bases. The ribose sugars found in nucleic acids consist of a five-membered ring with four carbon atoms and an oxygen atom. As illustrated in figure 14.3, the carbon atoms are numbered 1′ to 5′, proceeding clockwise from the oxygen atom.

The phosphate group is attached to the 5′ carbon atom of the sugar, and bases are attached to the 1′ carbon. In addition, a free hydroxyl (—OH) group is attached to the 3′ carbon atom. This means each nucleotide has a 5′ phosphate end, and a 3′ hydroxyl end.

Nucleotide monomers are joined together by a dehydration reaction involving the 5′ phosphate of one nucleotide with the 3′ hydroxyl of another nucleotide. This linkage is called a **phosphodiester bond** because the phosphate group is now linked to the two sugars by means of a pair of ester bonds (figure 14.4). Literally hundreds of millions of nucleotides are joined together via these linkages to form the nucleic acid polymers in your DNA.

Polynucleotide strands of DNA or RNA can be thought of as starting with a 5′ phosphate group and ending with a 3′ hydroxyl group. Therefore, every DNA and RNA molecule has an intrinsic polarity, and we can refer unambiguously to each end of the molecule. By convention, the sequence of bases is usually written in the 5′-to-3′ direction.

DNA Is Not a Simple, Repeating Polymer

LEARNING OBJECTIVE 14.2.2 State Chargaff's findings on relative abundances of the four bases.

The early studies of DNA structure, carried out before highly sensitive chemical analysis was possible, suggested that all four types

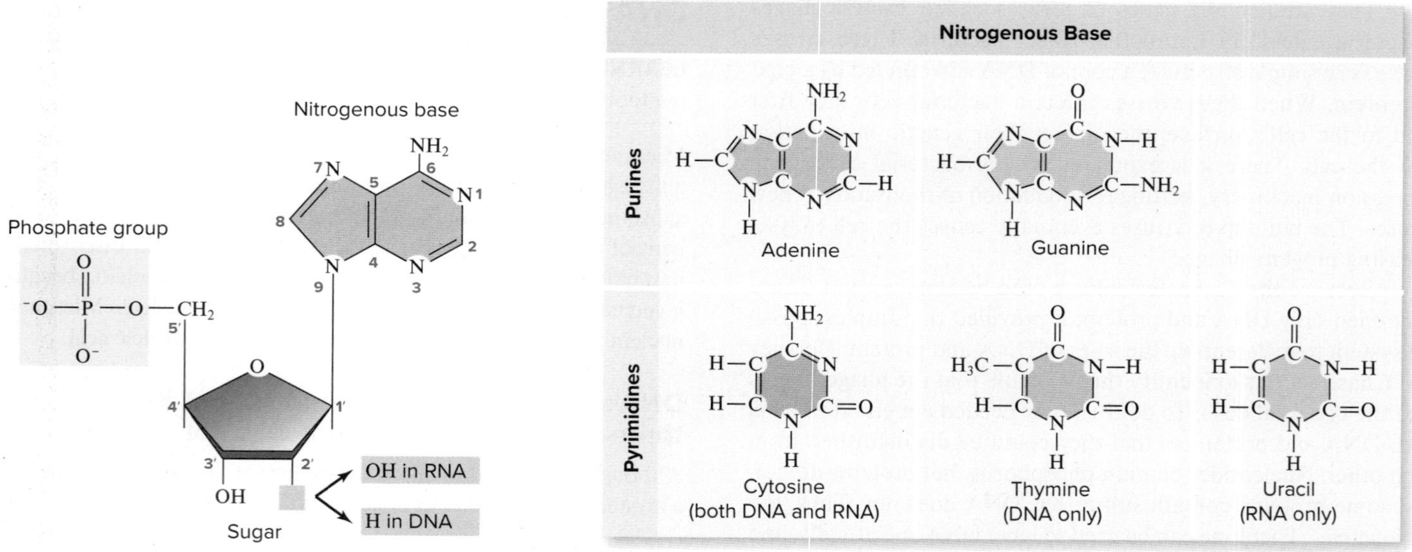
Nitrogenous base
Phosphate group
Sugar
OH in RNA
H in DNA
Nitrogenous Base
Purines
Adenine
Guanine
Pyrimidines
Cytosine
(both DNA and RNA)
Thymine
(DNA only)
Uracil
(RNA only)

Figure 14.3 Nucleotide subunits of DNA and RNA. The nucleotide subunits of DNA and RNA have three components: a 5-carbon sugar (deoxyribose in DNA and ribose in RNA); a phosphate group; and a nitrogenous base (either a purine or a pyrimidine).

of nucleotides were present in roughly equal amounts. This result, which was erroneous, led to the widely accepted but mistaken "tetranucleotide hypothesis" that DNA was a simple structural polymer with a four-base sequence that never varied, like CGATCGATCGAT.

A careful study carried out by Erwin Chargaff showed that the nucleotide composition of DNA molecules varied in complex ways, depending on the source of the DNA. This strongly suggested that DNA was not a simple repeating polymer and that it might have the information-encoding properties genetic material requires. Despite DNA's complexity, however, Chargaff observed an important underlying regularity in the ratios of the bases found in native DNA: *the amount of adenine present in DNA always equals the amount of thymine, and the amount of guanine always equals the amount of cytosine.* Two important findings from this work are often called **Chargaff's rules:**

1. The proportion of A always equals that of T, and the proportion of G always equals that of C (A = T and G = C).
2. The relative proportions of A/T and G/C vary widely among species.

As mounting evidence indicated that DNA molecules store the hereditary information, investigators began to puzzle over how such a seemingly simple molecule could carry out such a complex coding function.

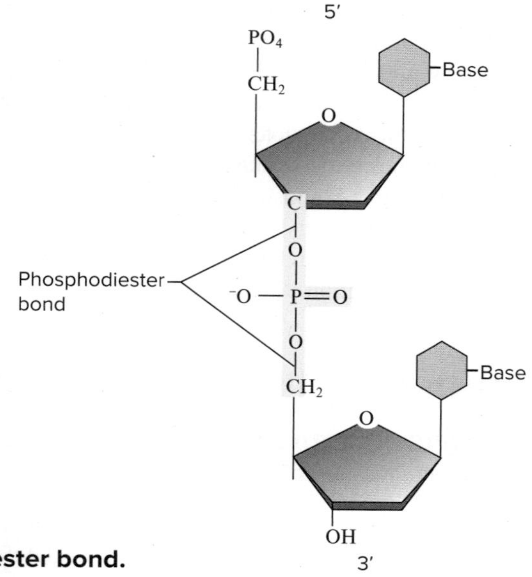
5′
PO_4
CH_2
Base
Phosphodiester bond
Base
3′

Figure 14.4
A phosphodiester bond.

X-Ray Diffraction Patterns Suggest DNA Has a Helical Shape

LEARNING OBJECTIVE 14.2.3 Explain the importance of Franklin's X-ray diffraction picture.

The techniques of modern physics soon provided more direct information about the possible structure of DNA. Chemists Maurice Wilkins and Rosalind Franklin (figure 14.5*a*) used the technique of X-ray diffraction to analyze DNA. In X-ray diffraction, a molecule is bombarded with a beam of X-rays. The rays are bent, or diffracted, by the molecules they encounter, and the diffraction pattern is recorded on photographic film. The patterns resemble the ripples created by tossing a rock into a smooth lake. When analyzed mathematically, the pattern can yield information about the three-dimensional structure of a molecule.

X-ray diffraction works best on substances that can be prepared as perfectly regular crystalline arrays, but in the 1950s it was impossible to obtain true crystals of natural DNA. However, Maurice Wilkins learned how to prepare uniformly oriented DNA fibers, and with graduate student Ray Gosling he succeeded in obtaining the first crude diffraction information

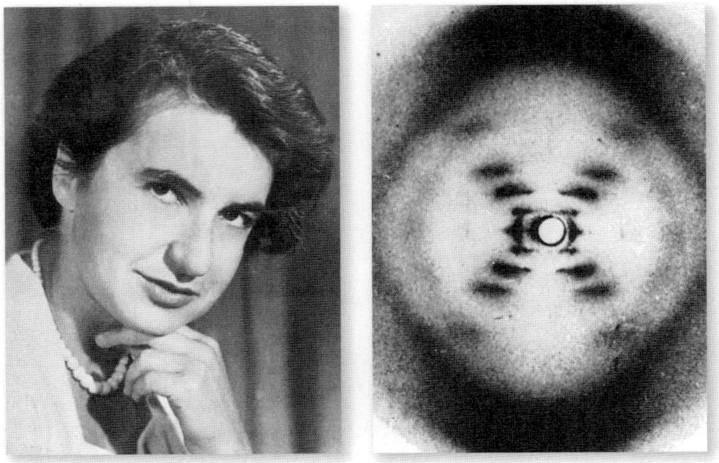

a. *b.*

Figure 14.5 Rosalind Franklin's X-ray diffraction patterns. ***a.*** Rosalind Franklin. ***b.*** This famous X-ray diffraction photograph of DNA fibers, made in March 1952 by Rosalind Franklin, was shown to Watson and Crick by Wilkins in 1953.
(a) National Library of Medicine/Science Source; (b) Omikron/Science Source

Figure 14.6 The DNA double helix. James Watson (left) and Francis Crick (right) deduced the structure of DNA in early 1953 from Chargaff's rules, knowing the proper tautomeric forms of the bases and using Franklin's diffraction studies.
Barrington Brown/Science Source

on natural DNA in 1950. Their early X-ray photos suggested that the DNA molecule has the shape of a helix, or corkscrew. Rosalind Franklin perfected the Wilkins approach over the next two years, her DNA diffraction patterns taking ever more clearly the form of a cross (figure 14.5*b*). This was a key result, as the clarity of her photographs both confirmed that DNA was a helix, and allowed calculation of the dimensions of the molecule, indicating a diameter of about 2 nm and a complete helical turn every 3.4 nm.

Tautomeric forms of bases

One piece of chemical evidence not revealed by the X-ray diffraction patterns was the form of the bases themselves. Because of the alternating double and single bonds in nitrogenous bases, when in solution they actually exist in equilibrium between two different structural forms. Such alternative structural forms are called *tautomers.* The predominant form of the bases when in solution contains keto (C═O) and amino groups ($—NH_2$), but the prominent biochemistry texts of the time actually illustrated the alternative, and incorrect, enol (C—OH) and imino (═NH) forms. The difference is critical, as the two alternative forms have very different hydrogen-bonding possibilities. Legend has it that Watson learned the correct form while having lunch with a biochemist friend.

DNA Is a Double Helix

LEARNING OBJECTIVE 14.2.4 Illustrate Watson and Crick's proposed structure for the DNA molecule.

Learning informally of Franklin's results before they were published in 1953, James Watson and Francis Crick, two young investigators at Cambridge University, quickly worked out a likely structure for the DNA molecule (figure 14.6), which we now know was substantially correct. Watson and Crick did not perform a single experiment themselves related to DNA structure; rather, they built detailed molecular models based on the data discussed earlier in this section.

The key to the model was their understanding that each DNA molecule is actually made up of *two* chains of nucleotides that wrap around one another—a double helix.

The phosphodiester backbone

The two strands of the double helix are made up of long polymers of nucleotides, and as described earlier in this section, each strand is made up of repeating sugar and phosphate units joined by phosphodiester bonds (figure 14.7). We call this the *phosphodiester backbone* of the molecule. The two strands of the backbone are then wrapped about a common axis, forming a double helix (figure 14.8). The helix is often compared to a spiral staircase, in which the two strands of the double helix are the handrails on the staircase.

Complementarity of bases

Watson and Crick proposed that the two strands were held together by hydrogen bonds between bases on opposite strands. The nature of these hydrogen bonds results in specific base-pairs: adenine (A) can form two hydrogen bonds with

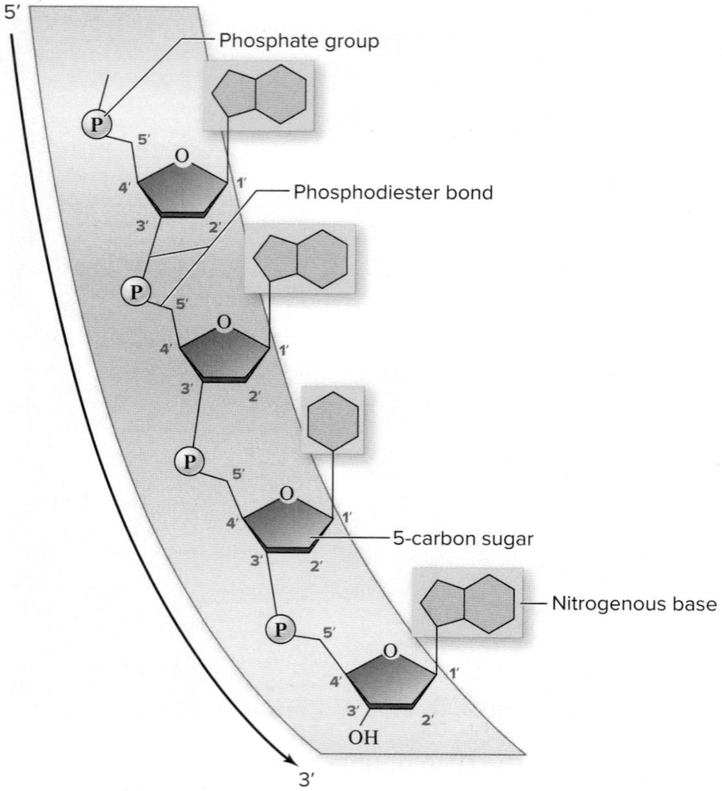

Figure 14.7 Structure of a single strand of DNA. The phosphodiester backbone is composed of alternating sugar and phosphate groups. The bases are attached to each sugar.

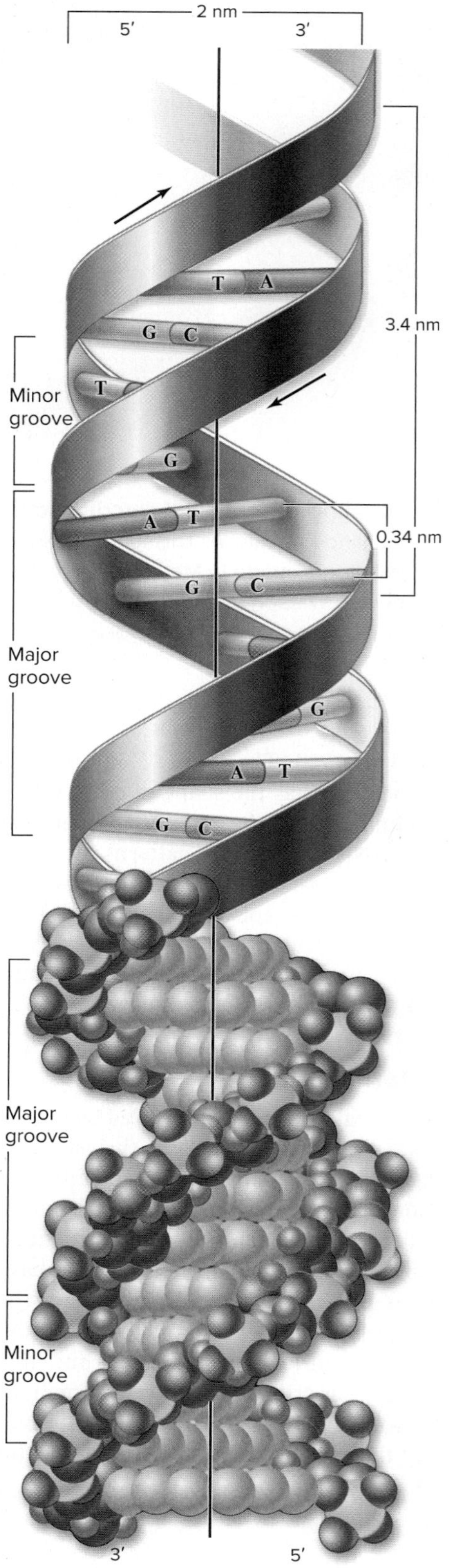

Figure 14.8 The double helix. Shown with the phosphodiester backbone as a ribbon on top and a space-filling model on the bottom. The bases protrude into the interior of the helix, where they hold it together by base-pairing. The backbone forms two grooves, the larger major groove and the smaller minor groove.

thymine (T) in an A–T base-pair, and guanine (G) can form three hydrogen bonds with cytosine (C) in a G–C base-pair (figure 14.9).

Note that each base-pair combines a two-ringed purine with a single-ringed pyrimidine, which means that the diameter of each base-pair is the same. A purine–purine base-pair would be wider, and a pyrimidine–pyrimidine base-pair would be narrower. In either case, this would distort the double helix formed by base-pairs.

We call this pattern of base-pairing *complementary,* and this is a critical feature of all aspects of DNA function. The two strands of the DNA molecule are not identical, but each strand specifies the other because of complementarity of base-pairs. This means that if you know one strand of a DNA double helix, you know the other strand, even though they are not the same.

For instance, the DNA strand:

5′ AGGCTATGCTAAGCTCCTTAA 3′

will have the complementary sequence:

3′ TCCGATACGATTCGAGGAATT 5′

We will see this principle of complementarity in action as we learn about DNA replication and expression in section 14.3.

The Watson–Crick model also explains Chargaff's results: the reason that the amount of A equals T, and the amount of G equals C, is that A base-pairs with T, and G base-pairs with C. The relative amounts of A/T and G/C can vary, but the amount of A must be equal to T, and the amount of G must be equal to C because of complementary base-pairing.

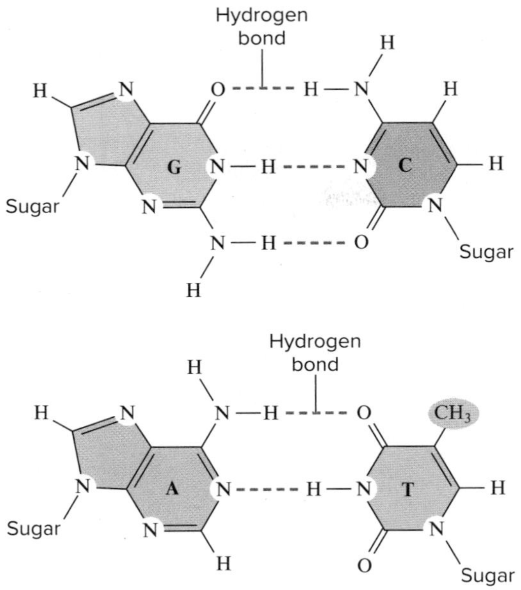

Figure 14.9 Base-pairing holds strands together. The hydrogen bonds that form between A and T and between G and C are shown with dashed lines. These produce AT and GC base-pairs that hold the two strands together. This always pairs a purine with a pyrimidine, keeping the diameter of the double helix constant.

Antiparallel configuration

As we learned earlier in this section, a single phosphodiester strand has an inherent polarity, meaning that one end terminates in a 3′ OH and the other end terminates in a 5′ PO_4. Strands are thus referred to as having either a 5′-to-3′ or a 3′-to-5′ polarity. Two strands could be put together in two ways: with the polarity the same in each (parallel) or with the polarity opposite (antiparallel). Native double-stranded DNA always has the antiparallel configuration, with one strand running 5′ to 3′ and the other running 3′ to 5′, as shown in figure 14.8.

The Watson–Crick DNA molecule

In the Watson and Crick model, a DNA molecule consists of two phosphodiester strands wrapped about the same helical axis, with the bases extending to the interior of the helix. The sequences of bases on the two strands are complementary, so they form base-pairs that hold the strands together. Each strand is polar, with a 5′ and a 3′ end, and these are oriented in the opposite (antiparallel) direction (figure 14.9).

Although each individual hydrogen bond is low-energy, the sum of thousands, or millions, of such bonds makes a very stable molecule. This also means that a DNA molecule can have regions where the helix can be "opened up" without affecting the stability of the entire molecule, which is critical for its function.

Although the Watson–Crick model provided a rational structure for DNA, researchers had to answer further questions about how DNA could be replicated, a crucial step in cell division, and about how cells could repair damaged or otherwise altered DNA. We explore these questions in the rest of this chapter. (In chapter 15, we will continue with the genetic code and the connection between the code and protein synthesis.)

REVIEW OF CONCEPT 14.2

Chargaff showed that the amount of adenine is equal to that of thymine, and the amount of guanine is equal to that of cytosine, in DNA. Structural studies by Franklin and Wilkins indicated that DNA forms a helix. Watson and Crick's model consists of two antiparallel strands wrapped about a common helical axis. The strands are held together by hydrogen bonds between the bases: adenine pairs with thymine and guanine pairs with cytosine. This makes the strands complementary to each other.

- *Why was information about the proper tautomeric form of the bases critical?*

14.3 Both Strands Are Copied During DNA Replication

The accurate replication of DNA prior to cell division is a crucial function of the cell cycle. Research has revealed that this process requires the participation of a large number of cellular proteins. Before geneticists could begin to sort out these details, however, they first needed to gain a clearer idea of the general mechanism.

DNA Replication Is Semiconservative

LEARNING OBJECTIVE 14.3.1 Relate the results of the Meselson–Stahl experiment to possible modes of DNA replication.

The Watson–Crick model of DNA immediately suggested that the basis for copying the genetic information is complementarity: one chain of the DNA molecule may have any conceivable base sequence, but this sequence completely determines the sequence of its partner in the duplex.

In order to accurately replicate a DNA molecule, the sequence of parental strands must be accurately duplicated in the daughter strands. That is, one parental helix with two strands must yield two daughter helices, each with two strands that complement each other—four strands in all.

Three models of DNA replication are possible (figure 14.10):

1. In a *conservative model,* both strands of the parental duplex would remain intact (conserved), and both strands of the new DNA duplex would contain all-new nucleotides.
2. In a *semiconservative model,* one strand of the parental duplex remains intact in daughter strands (semiconserved), with a new complementary strand built for each parental strand consisting of new nucleotides. Daughter strands would consist of one parental strand and one newly synthesized strand.

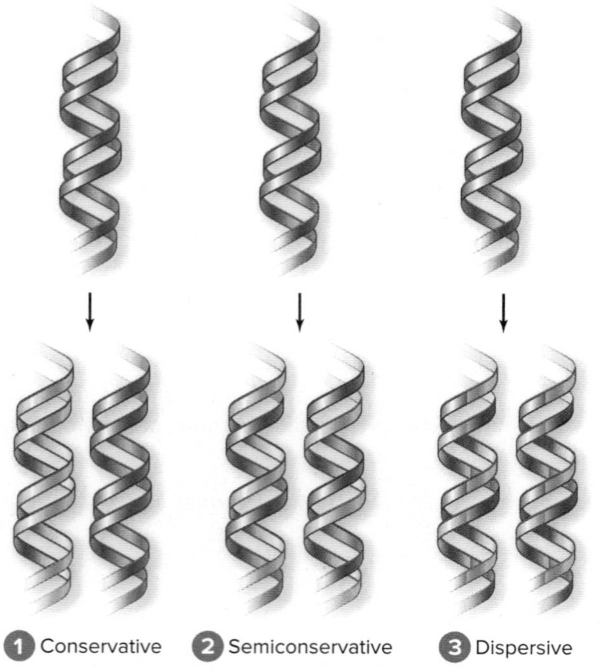

Figure 14.10 Three possible models for DNA replication. 1 The conservative model produces one entirely new molecule and conserves the old. 2 The semiconservative model produces two hybrid molecules of old and new strands. 3 The dispersive model produces hybrid molecules with each strand a mixture of old and new.

3 In a *dispersive model,* copies of DNA would consist of mixtures of parental and newly synthesized strands; that is, the new DNA would be dispersed throughout each strand of both daughter molecules after replication.

Notice that these three models suggest different mechanisms of replication, without specifying molecular details of the process.

The Meselson–Stahl experiment

The three models for DNA replication were evaluated in 1958 by geneticist Matthew Meselson and molecular biologist Franklin Stahl. To distinguish between these models, they labeled DNA and then followed the labeled DNA through two rounds of replication (figure 14.11).

The label Meselson and Stahl used was a heavy isotope of nitrogen (^{15}N), not a radioactive label. DNA molecules containing ^{15}N have a greater density than those containing the common ^{14}N isotope. The tool they used was the ultracentrifuge, which spins so fast it can be used to separate DNA molecules that have different densities.

Bacteria were grown in a medium containing ^{15}N, which became incorporated into the bases of the bacterial DNA. After several generations, the DNA of these bacteria was denser than that of bacteria grown in a medium containing the normally available ^{14}N. They then transferred the bacteria from the ^{15}N medium to ^{14}N medium and collected the DNA at various time intervals.

The DNA for each interval was dissolved in a solution containing a heavy salt, cesium chloride. This solution was spun at very high speeds in an ultracentrifuge. The enormous centrifugal forces caused cesium ions to migrate toward the bottom of the

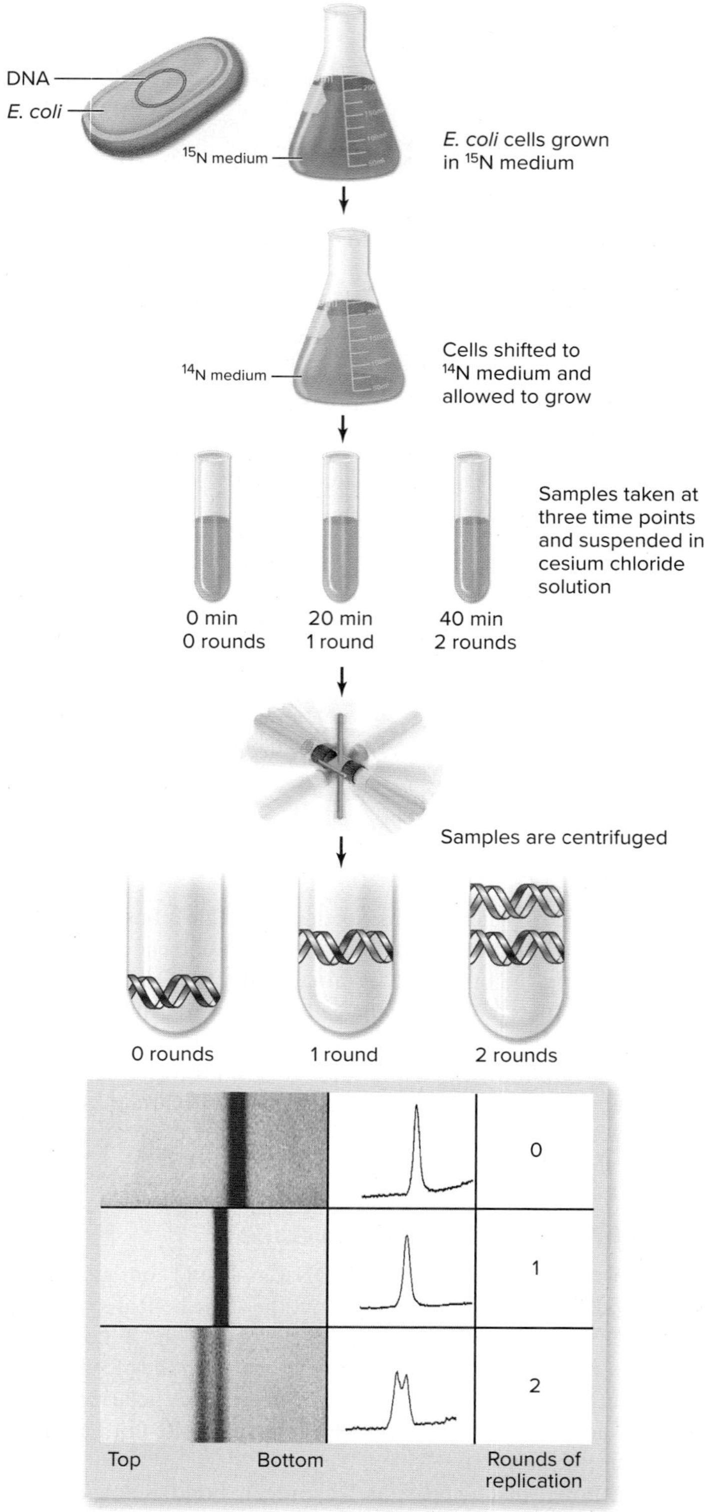

Figure 14.11 The Meselson–Stahl experiment. Bacteria grown in heavy ^{15}N medium are shifted to light ^{14}N medium and grown for two rounds of replication. Samples are taken at time points corresponding to zero, one, and two rounds of replication and centrifuged in cesium chloride to form a gradient. The actual data are shown at the bottom with the interpretation of semiconservative replication shown schematically.

Photo: Meselson, M., Stahl, F., (1958) "The replication of DNA in Escherichia coli," PNAS, 44(7):671-682, Fig. 4a; Source: M. Meselson and F.W. Stahl, PNAS, 44(1958): 671.

centrifuge tube, creating a gradient of cesium concentration, and thus of density. Each DNA strand floated or sank in the gradient until it reached the point at which its density exactly matched the density of the cesium at that location. Because ^{15}N strands are denser than ^{14}N strands, they migrated farther down the tube.

The DNA collected immediately after the transfer of bacteria to new ^{14}N medium was all of one density, equal to that of ^{15}N DNA alone. However, after the bacteria completed a first round of DNA replication, the density of their DNA had decreased to a value intermediate between ^{14}N DNA alone and ^{15}N DNA. After the second round of replication, two density bands of DNA were observed: one intermediate and one equal to that of ^{14}N DNA (figure 14.11).

Interpretation of the Meselson–Stahl findings

Meselson and Stahl compared their experimental data with the results that would be predicted on the basis of the three models:

1. The conservative model was not consistent with the data, because after one round of replication, two densities should have been observed: DNA strands would either be all-heavy (parental) or all-light (daughter). This model is rejected.
2. The semiconservative model was consistent with all observations: After one round of replication, a single density would be predicted, because all DNA molecules would have a light strand and a heavy strand. After two rounds of replication, half of the molecules would have two light strands, and half would have a light strand and a heavy strand—and so two densities would be observed. Therefore, the results support the semiconservative model.
3. The dispersive model was consistent with the data from the first round of replication, because in this model, every DNA helix would consist of strands that are mixtures of ½ light (new) and ½ heavy (old) molecules. But after two rounds of replication, the dispersive model would still yield only a single density; DNA strands would be composed of ¾ light and ¼ heavy molecules. Instead, two densities were observed. Therefore, this model is also rejected.

The basic mechanism of DNA replication is semiconservative. At the simplest level, then, DNA is replicated by opening up a DNA helix and making copies of both strands to produce two daughter helices, each consisting of one old strand and one new strand.

Semiconservative DNA replication requires a template, nucleotides, and enzymes

DNA replication requires three things: something to copy (the parental DNA molecules serve as a template), something to do the copying (enzymes copy the template), and building blocks to assemble into the copy (nucleoside triphosphates).

A number of enzymes work together to accomplish the task of assembling a new strand, but the enzyme that actually matches the existing DNA bases with complementary nucleotides and then links the nucleotides together to make the new strand is **DNA polymerase** (figure 14.12). As we will discuss in section 14.4, all

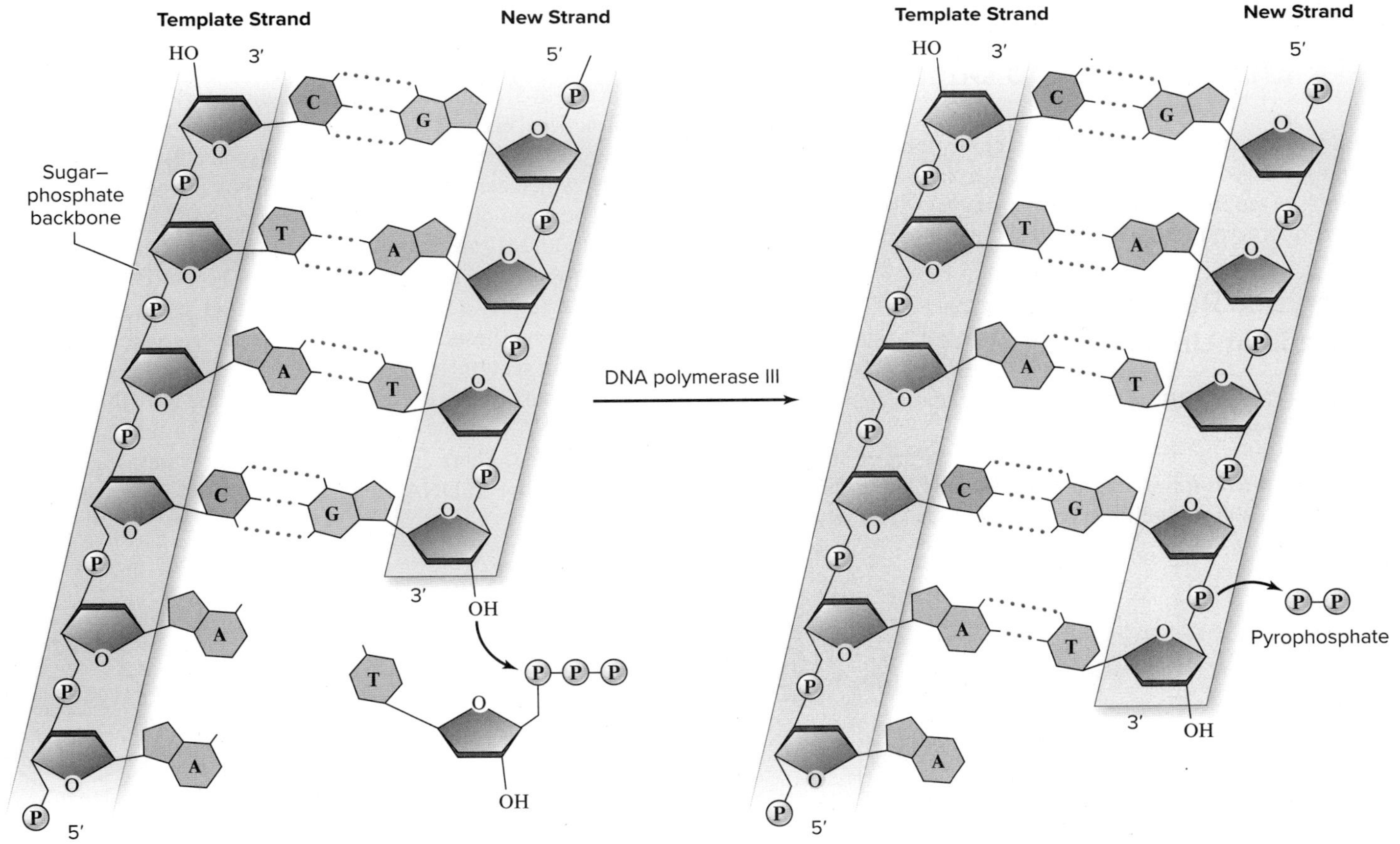

Figure 14.12 Action of DNA polymerase. DNA polymerases add nucleotides to the 3′ end of a growing chain. The nucleotide added depends on the base that is in the template strand. Each new base must be complementary to the base in the template strand. With the addition of each new nucleoside triphosphate, two of its phosphates are cleaved off as pyrophosphate.

DNA polymerases that have been examined have several common features. They all add new bases to the 3′ end of existing strands. That is, they synthesize in a 5′-to-3′ direction by extending a strand base-paired to the template. All DNA polymerases also require a *primer* to begin synthesis; they cannot begin without a strand of RNA or DNA base-paired to the template. RNA polymerases do not have this requirement, so they usually synthesize the primers.

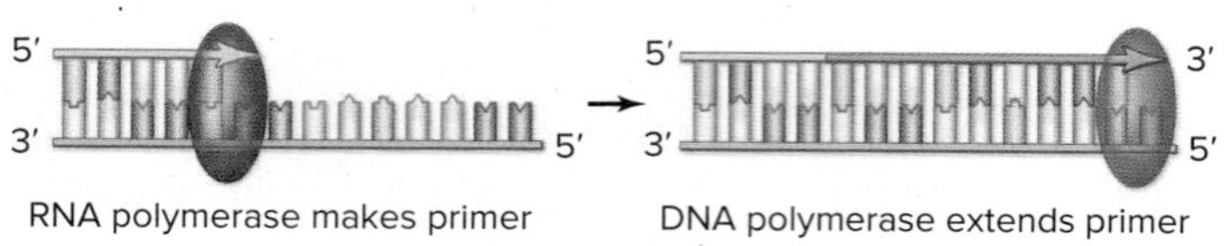

REVIEW OF CONCEPT 14.3

Meselson and Stahl demonstrated that the basic mechanism of DNA replication is semiconservative: Each new DNA helix is composed of one old strand and one new strand. The process of replication requires a template, nucleoside triphosphates, and the enzyme DNA polymerase, which synthesize DNA in a 5′-to-3′ direction from a primer, usually RNA.

- *What would the results be if the DNA were denatured prior to separation by ultracentrifugation?*

14.4 Prokaryotes Organize the Enzymes Used to Duplicate DNA

To build up a more detailed picture of replication, we first concentrate on prokaryotic replication using *E. coli* as a model. We can then look at eukaryotic replication primarily in how it differs from the prokaryotic system.

Prokaryotic Replication Starts and Ends at Unique Sites

LEARNING OBJECTIVE 14.4.1 Describe the enzymes used to synthesize DNA.

Replication in *E. coli* initiates at a specific site, the origin (called *oriC*), and ends at a specific site, the terminus. The sequence of *oriC* consists of repeated nucleotides that bind an initiator protein, and an AT-rich sequence that can be opened easily during initiation of replication. (A–T base-pairs have only two hydrogen bonds, compared with the three hydrogen bonds in G–C base-pairs.)

After initiation, replication proceeds bidirectionally from this unique origin to the unique terminus (figure 14.13). We call the DNA controlled by an origin a **replicon.** In this case, the chromosome plus the origin forms a single replicon.

E. coli has at least three different DNA polymerases

As mentioned in section 14.3, *DNA polymerase* refers to a class of enzymes that use a DNA template to assemble a new complementary strand. The first DNA polymerase isolated in *E. coli* was **DNA polymerase I (Pol I).** At first, investigators assumed this polymerase was all that was required for DNA replication. Later, a mutant was isolated with no Pol I activity but that could still replicate its chromosome. Two additional polymerases were isolated from this strain of *E. coli,* **DNA polymerase II (Pol II)** and **DNA polymerase III (Pol III).** As with all known DNA polymerases, all three of these enzymes synthesize polynucleotide strands only in the 5′-to-3′ direction and require a primer.

In addition to adding nucleotides to a growing DNA strand, some polymerases can also remove nucleotides, or act as a nuclease. Enzymes that act as nucleases are classified as either **endonucleases** (cut DNA internally) or **exonucleases** (remove nucleotides from the end of DNA). DNA Pol I, Pol II, and Pol III all have 3′-to-5′ exonuclease activity, which serves as a proofreading function, because it allows the enzyme to "back up" and remove a mispaired base. DNA Pol I also has 5′-to-3′ exonuclease activity, which can be used to remove RNA primers.

The three different polymerases have different roles in the cell. Pol I and III are both involved in DNA replication, with Pol III being the main replication enzyme. The Pol II enzyme does not appear to play a role in DNA replication, but it is involved in new DNA synthesis needed for some DNA repair processes, described in section 14.6.

For many years, these three polymerases were thought to be the only DNA polymerases in *E. coli,* but several new ones have been identified. There are now five known polymerases, although not all are active in DNA replication.

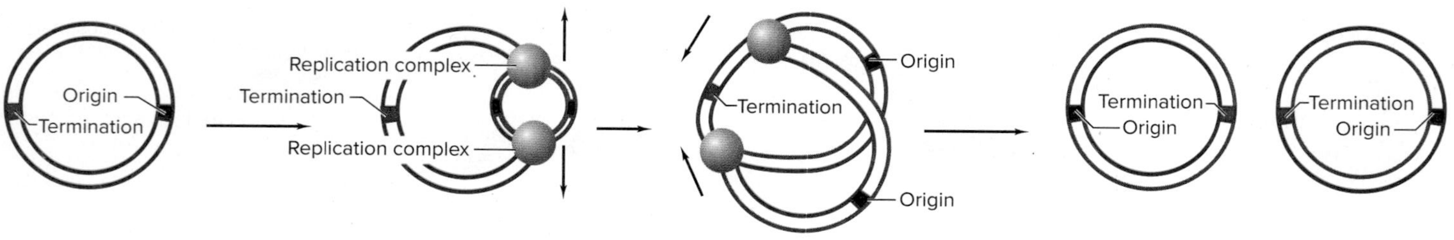

Figure 14.13 Replication is bidirectional from a unique origin. Replication initiates from a unique origin. Two separate replication complexes containing all of the necessary enzymes are loaded onto the origin and initiate synthesis in the opposite directions on the chromosome. These two complexes continue in opposite directions until they come to a unique termination site.

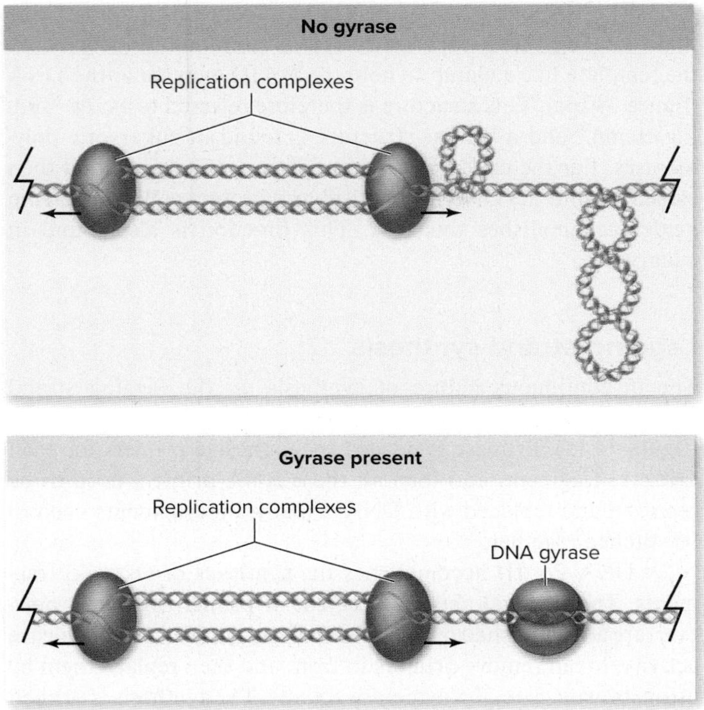

Figure 14.14 Unwinding the helix causes torsional strain. Unwinding the helix produces torsional strain in front of the replication fork. This causes the double helix to coil in space (supercoiling) as shown in the top panel. DNA gyrase, acting in front of the replication fork, removes supercoils (bottom panel).

Unwinding DNA requires energy and causes torsional strain

Although most DNA polymerases can unwind DNA during synthesis, replication is more efficient if the helix is unwound ahead of the polymerase. An enzyme with DNA-unwinding activity is called a **helicase.** This process uses energy from ATP hydrolysis and can progressively unwind DNA, forming single strands. These single strands are not stable, because the hydrophobic bases are exposed to water. Cells solve this problem with another protein, single-strand-binding protein (SSB), that will coat exposed single strands.

The unwinding of the two strands introduces torsional strain in the DNA molecule. Imagine two rubber bands twisted together. If you now unwind the rubber bands, what happens? The rubber bands, already twisted about each other, will further coil in space. When this happens with a DNA molecule, it is called **supercoiling** (figure 14.14). The branch of mathematics that studies how forms twist and coil in space is called *topology,* and therefore we describe this coiling of the double helix as the *topological state* of DNA. This state describes how the double helix itself coils in space. You have already encountered an example of this coiling with DNA wrapped about histone proteins in the nucleosomes of eukaryotic chromosomes (refer to chapter 10).

Enzymes that can alter the topological state of DNA are called **topoisomerases.** Topoisomerase enzymes act to relieve the torsional strain caused by unwinding and to prevent this supercoiling from happening. **DNA gyrase** is the topoisomerase involved in DNA replication.

Replication Is Semidiscontinuous

LEARNING OBJECTIVE 14.4.2 Explain why DNA synthesis is not continuous on both strands.

In section 14.2, DNA was described as being antiparallel—meaning that one strand runs in the 3′-to-5′ direction, and its complementary strand runs in the 5′-to-3′ direction. The antiparallel nature of DNA combined with the nature of the polymerase enzymes puts constraints on the replication process. Because polymerases can synthesize DNA in only one direction, and the two DNA strands run in opposite directions, polymerases on the two strands must be synthesizing DNA in opposite directions (figure 14.15).

The requirement of DNA polymerases for a primer means that on one strand primers need to be added as the helix is opened up, as shown in figure 14.15. This means that one strand can be synthesized in a continuous fashion from an initial primer, but the other strand must be synthesized in a discontinuous fashion with multiple priming events and short sections of DNA being assembled. The strand that is continuous is called the **leading strand,** and the strand that is discontinuous is the **lagging strand.** DNA fragments synthesized on the lagging strand are named **Okazaki fragments** in honor of the molecular biologists Reiji and Tsuneko Okazaki, who first experimentally demonstrated discontinuous synthesis. This implies a need for even more enzymatic activity on the lagging strand, as is described next.

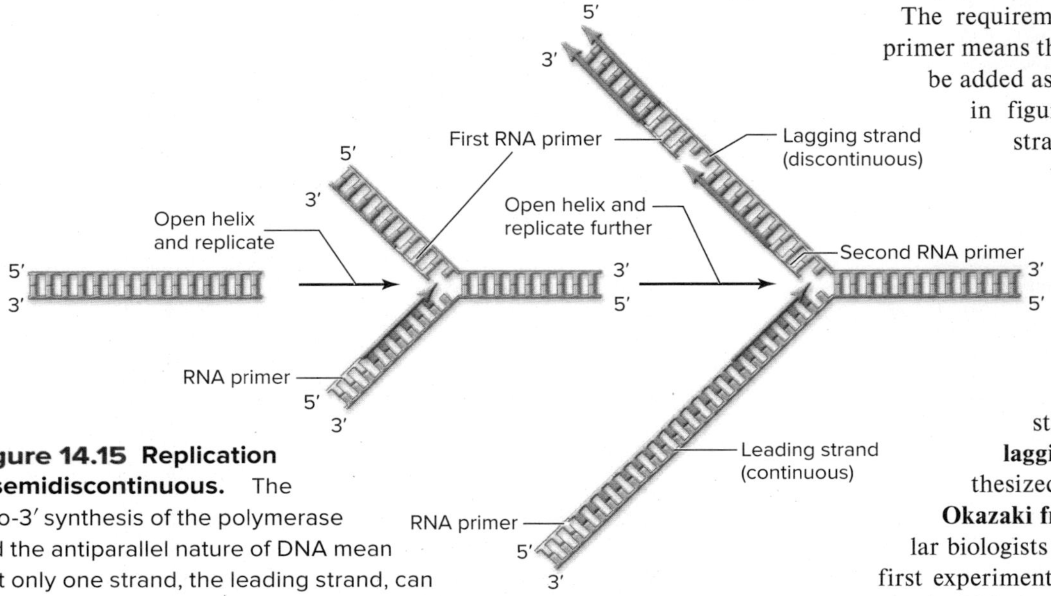

Figure 14.15 Replication is semidiscontinuous. The 5′-to-3′ synthesis of the polymerase and the antiparallel nature of DNA mean that only one strand, the leading strand, can be synthesized continuously. The other strand, the lagging strand, must be made in pieces, each with its own primer.

Synthesis occurs at the replication fork

The partial opening of a DNA helix to form two single strands produces a forked structure, which is called the **replication fork.** All of the enzymatic activities that we have discussed, plus a few more, are found at the replication fork. Synthesis on the leading strand and synthesis on the lagging strand proceed in different ways, however.

Priming

The primers required by DNA polymerases during replication are synthesized by the enzyme *DNA primase.* This enzyme is an RNA polymerase that synthesizes short stretches of RNA 10 to 20 base-pairs (bp) long that function as primers for DNA polymerase. Later on, the RNA primer is removed and replaced with DNA.

Leading-strand synthesis

Synthesis on the leading strand is relatively simple. A single priming event is required, and then the strand can be extended indefinitely by the action of DNA Pol III. If the enzyme remains attached to the template, it can synthesize around the entire circular *E. coli* chromosome.

The ability of a polymerase to remain attached to the template is called *processivity.* The Pol III enzyme is a large, multisubunit enzyme that has high processivity due to the action of one subunit of the enzyme, called the ß *subunit* (figure 14.16*a*).

The ß subunit is made up of two identical protein chains that come together to form a circle. This circle can be loaded onto the template like a clamp to hold the Pol III enzyme to the DNA (figure 14.16*b*). This structure is therefore referred to as the "sliding clamp," and a similar structure is found in eukaryotic polymerases. For the clamp to function, it must be opened and then closed around the DNA. A multisubunit protein called the clamp loader accomplishes this task. This function is also found in eukaryotes.

Lagging-strand synthesis

The discontinuous nature of synthesis on the lagging strand requires the cell to do much more work than on the leading strand (figure 14.15). Primase is needed to synthesize primers for each Okazaki fragment, and then all these RNA primers need to be removed and replaced with DNA. Finally, the fragments need to be stitched together.

DNA Pol III accomplishes the synthesis of Okazaki fragments. The removal and replacement of primer segments, however, are accomplished by DNA Pol I. Using its 5′-to-3′ exonuclease activity, it can remove primers in front and then replace them by using its usual 5′-to-3′ polymerase activity. The synthesis is primed by the previous Okazaki fragment, which is composed of DNA and has a free 3′ OH that can be extended.

This leaves only the last phosphodiester bond to be formed where synthesis by Pol I ends. This is done by **DNA ligase,** which seals this "nick," eventually joining the Okazaki fragments into complete strands. All of this activity on the lagging strand is summarized in figure 14.17.

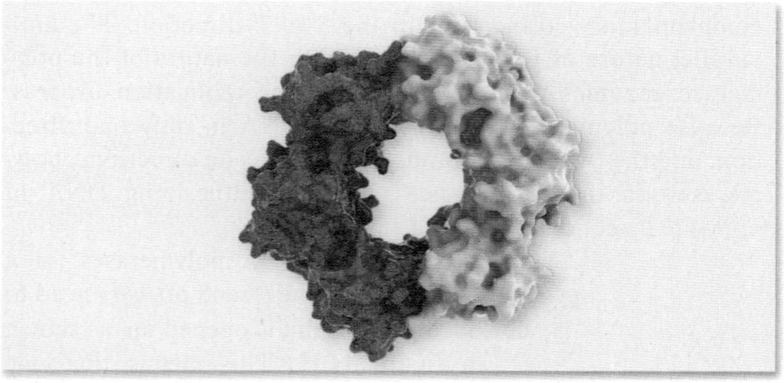

a.

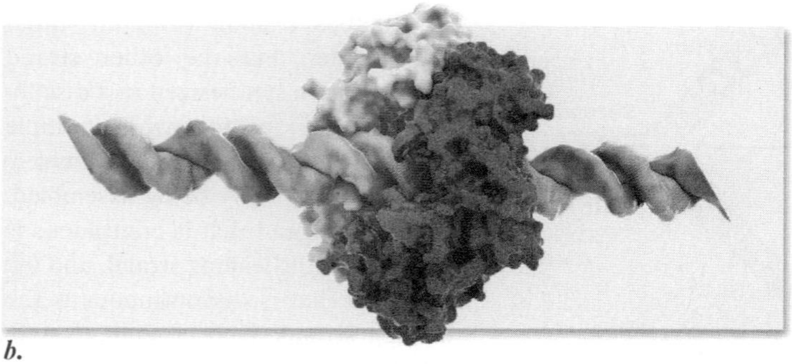

b.

Figure 14.16 The DNA polymerase sliding clamp. ***a.*** The β subunit forms a ring that can encircle DNA. ***b.*** The β subunit is shown attached to the DNA. This forms the "sliding clamp" that keeps the polymerase attached to the template.

(a, b): Ramon Andrade 3Dciencia/SPL/Science Source

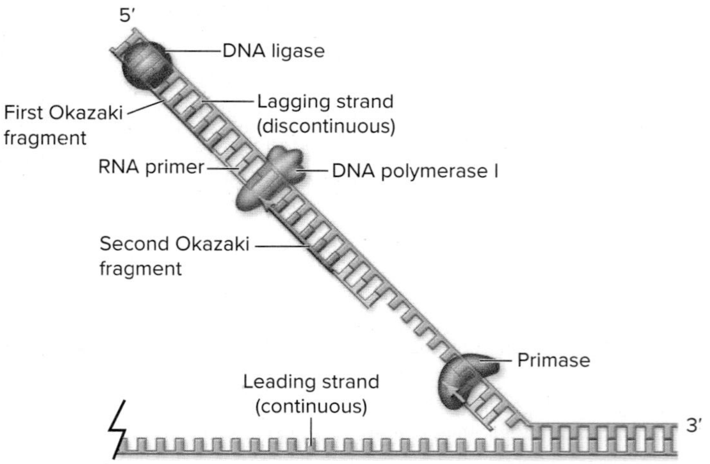

Figure 14.17 Lagging-strand synthesis. The action of primase synthesizes the primers needed by DNA polymerase III (not shown). These primers are removed by DNA polymerase I using its 5′-to-3′ exonuclease activity, then extending the adjacent Okazaki fragment to replace the RNA. The nick between Okazaki fragments after primer removal is sealed by DNA ligase.

Termination

Termination occurs at a specific site roughly opposite *oriC* on the circular chromosome. The last stages of replication produce two daughter molecules that are intertwined like two rings in a chain. These intertwined molecules are unlinked by the same enzyme that relieves torsional strain at the replication fork–DNA gyrase.

Bacteria Have a DNA Replication Organelle

LEARNING OBJECTIVE 14.4.3 Diagram the functioning of the bacterial DNA replication organelle.

The enzymes involved in bacterial DNA replication (table 14.1) form a macromolecular assembly called the **replisome.** This assembly can be thought of as the "replication organelle," just as the ribosome is the organelle that synthesizes protein. The replisome has two main subcomponents: the *primosome* and a complex of two DNA Pol III enzymes, one for each strand. The primosome is composed of primase and helicase, along with a number of accessory proteins. The need for constant priming on the lagging strand explains the need for the primosome complex as part of the replisome.

The two Pol III complexes include two synthetic core subunits, each with its own ß subunit. The entire replisome complex is held together by a number of proteins that include the clamp loader. The clamp loader is required to periodically load a ß subunit on the lagging strand and to transfer the Pol III to this new ß subunit (figure 14.18).

TABLE 14.1 DNA Replication Enzymes of *E. coli*

Protein	Role	Size (kDa)	Molecules per Cell
Helicase	Unwinds the double helix	300	20
Primase	Synthesizes RNA primers	60	50
Single-strand-binding protein	Stabilizes single-stranded regions	74	300
DNA gyrase	Relieves torque	400	250
DNA polymerase III	Synthesizes DNA	≈900	20
DNA polymerase I	Erases primer and fills gaps	103	300
DNA ligase	Joins the ends of DNA segments; DNA repair	74	300

Even given the difficulties with lagging-strand synthesis, the two Pol III enzymes in the replisome are active on both leading and lagging strands simultaneously. How can the two strands be synthesized in the same direction when the strands are antiparallel? The model first proposed, still with us in some form,

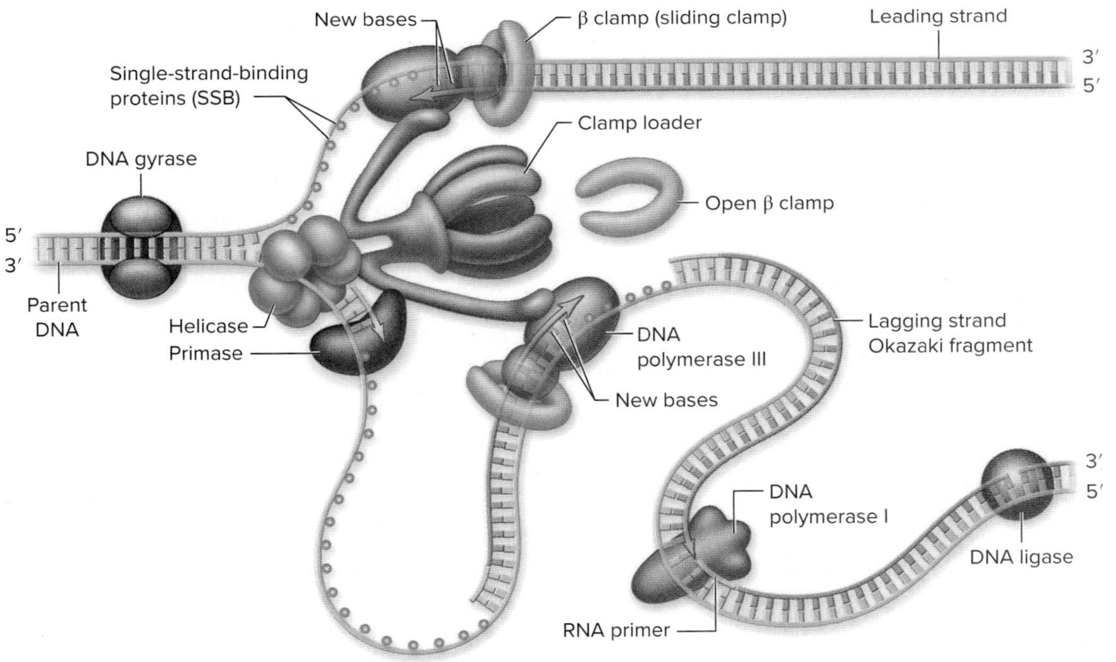

Figure 14.18 The replication fork. A model for the structure of the replication fork with two polymerase III enzymes held together by a large complex of accessory proteins. These include the "clamp loader," which loads the β subunit sliding clamp periodically on the lagging strand. The polymerase III on the lagging strand periodically releases its template and reassociates along with the β clamp. The loop in the lagging-strand template allows both polymerases to move in the same direction despite DNA being antiparallel. Primase, which makes primers for the lagging-strand fragments, and helicase are also associated with the central complex. Polymerase I removes primers, and ligase joins the fragments together.

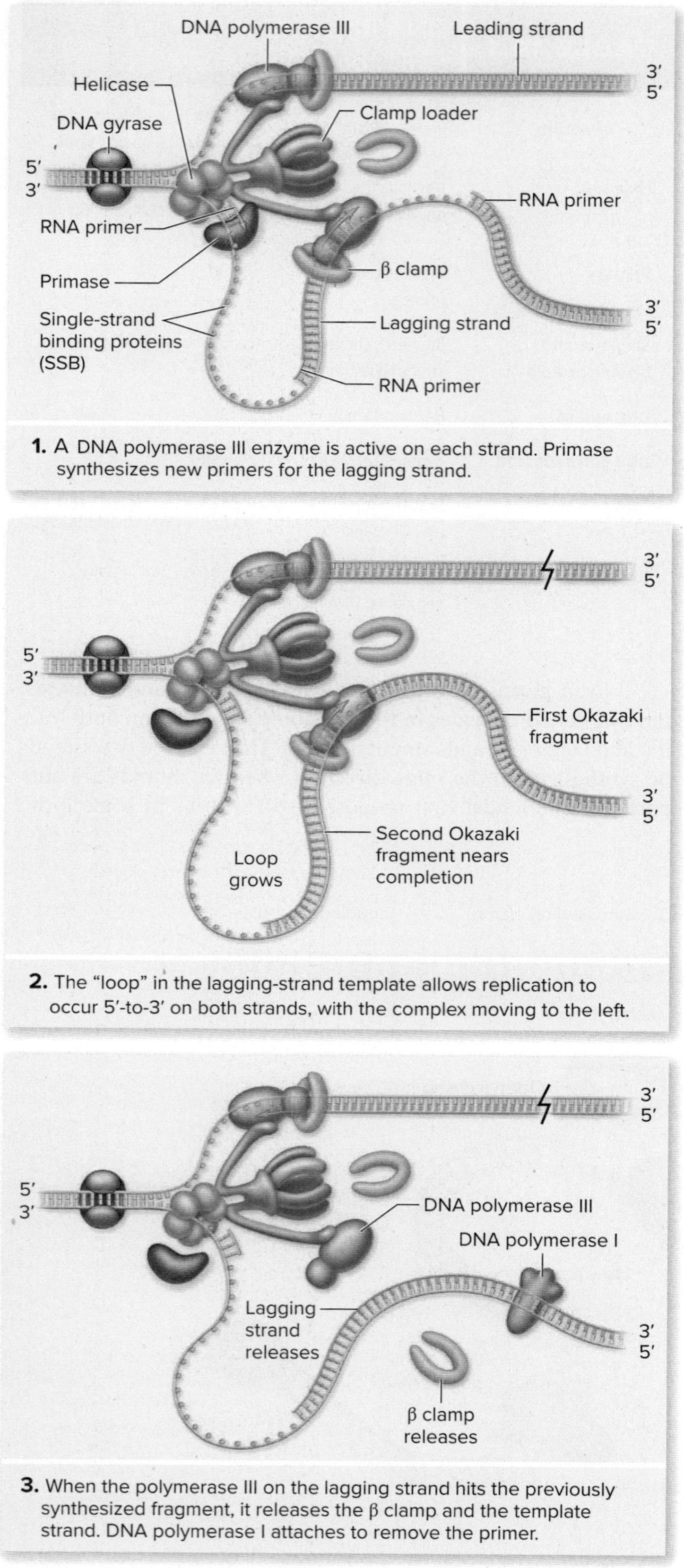

1. A DNA polymerase III enzyme is active on each strand. Primase synthesizes new primers for the lagging strand.

2. The "loop" in the lagging-strand template allows replication to occur 5′-to-3′ on both strands, with the complex moving to the left.

3. When the polymerase III on the lagging strand hits the previously synthesized fragment, it releases the β clamp and the template strand. DNA polymerase I attaches to remove the primer.

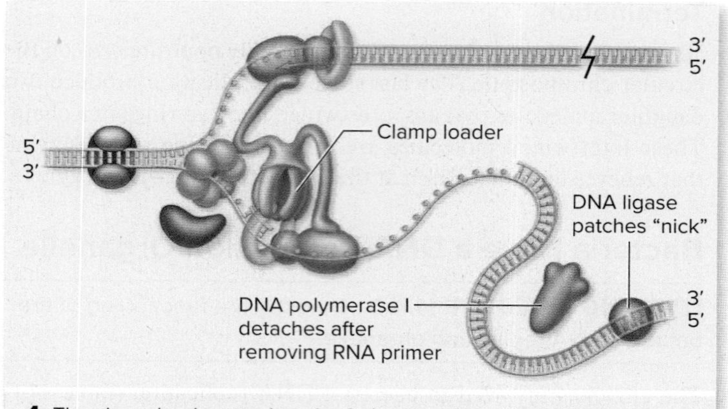

4. The clamp loader attaches the β clamp and transfers this to polymerase III, creating a new loop in the lagging-strand template. DNA ligase joins the fragments after DNA polymerase I removes the primers.

5. After the β clamp is loaded, the DNA polymerase III on the lagging strand adds bases to the next Okazaki fragment.

Figure 14.19 DNA synthesis by the replisome. The semidiscontinuous synthesis of DNA is illustrated in stages using the model from figure 14.18.

involves a loop formed in the lagging strand, so that the polymerases can move in the same direction. Current evidence also indicates that this replication complex is probably stationary, with the DNA strand moving through it like thread in a sewing machine, rather than the complex moving along the DNA strands. This stationary complex also pushes the newly synthesized DNA outward, which may aid in chromosome segregation. This process is summarized in figure 14.19.

REVIEW OF CONCEPT 14.4

E. coli has three DNA polymerases, two of which are used during replication. DNA strands are separated at the replication fork, where a massive complex, the replisome, is assembled. This contains DNA polymerase III, primase, helicase, and other proteins. The antiparallel structure of DNA and 5′-to-3′ synthesis of polymerases lead to discontinuous replication on one strand. DNA polymerase I removes primers and replaces them with DNA on this strand; DNA ligase joins Okazaki fragments.

- *How do the functions of the two polymerases differ during replication?*

14.5 Eukaryotic Chromosomes Are Large and Linear

Eukaryotic replication is complicated by two main factors: the larger amount of DNA organized into multiple chromosomes, and the linear structure of the chromosomes. This process requires new enzymatic activities only for dealing with the ends of chromosomes; otherwise, the basic enzymology is the same.

Replicating Very Large Chromosomes Creates New Problems

LEARNING OBJECTIVE 14.5.1 Compare eukaryotic DNA replication with prokaryotic DNA replication.

Eukaryotic replication uses multiple origins

The sheer amount of DNA and how it is packaged constitute a problem for eukaryotes (figure 14.20). Most eukaryotes have multiple chromosomes, each of which is larger than the *E. coli* chromosome. If only a single unique origin existed for each chromosome, the length of time necessary for replication would be prohibitive. This problem is solved by the use of multiple origins of replication for each chromosome, resulting in multiple *replicons.*

The origins are not as sequence-specific as *oriC,* and their recognition seems to depend on chromatin structure as well as on sequence. The number of origins used can also be adjusted during the course of development, so that early on, when cell divisions need to be rapid, more origins are activated. Each origin must be used only once per cell cycle.

Figure 14.20 DNA of a single human chromosome. This chromosome has been relieved of most of its packaging proteins, leaving the DNA in its native form. The residual protein scaffolding appears as the dark material in the lower part of the micrograph.
Don W. Fawcett/Science Source

The eukaryotic replication fork is more complex

Prior to S phase, helicase complexes are loaded onto possible replication origins, but not activated. Then, during S phase a subset of these are activated, and the rest of the replisome assembled. The action of the replisome differs from that in *E. coli* in a few ways we will highlight.

Priming is accomplished by a complex of *DNA polymerase α* and primase. This complex synthesizes primers consisting of RNA and a short stretch of DNA. There are also different polymerases for the leading and lagging strands. *DNA polymerase epsilon* (*Pol ε*) is responsible for leading-strand synthesis and *DNA polymerase delta* (*Pol δ*) synthesizes the lagging strand. The sliding clamp subunit that allows the enzyme complex to stay attached to the template is called PCNA (for proliferating cell nuclear antigen). The name comes from its identification as an antibody-inducing protein in proliferating (dividing) cells. The PCNA sliding clamp forms a trimer, but has similar structure and function to the β subunit sliding clamp.

Synthesis on the lagging strand differs most from related events in prokaryotic DNA replication. The Okazaki fragments are significantly shorter: 200 bp instead of 1000 bp. As each fragment reaches the previous fragment, synthesis continues to displace the strand containing the previous primer. The displaced primer is removed by a nuclease, and the remaining fragments are joined by ligase.

Archaeal and eukaryotic replication proteins are evolutionarily related

The replisome, which can replicate both strands of DNA simultaneously, requires a variety of functions, including DNA unwinding (helicase), synthesis (primase and polymerase), and processivity factors (sliding clamps and their loaders). There is clear functional conservation of these activities across all domains of life, but major components appear to have evolved independently in bacteria and in archaea/eukarya.

Members of the DNA polymerase family of enzymes have similar structure and homologous synthetic domains. They are differentiated based on nonhomologous regions into seven subfamilies. The subfamily used by eukaryotes and archaea is the same, but different from the one used by bacteria. Replicative

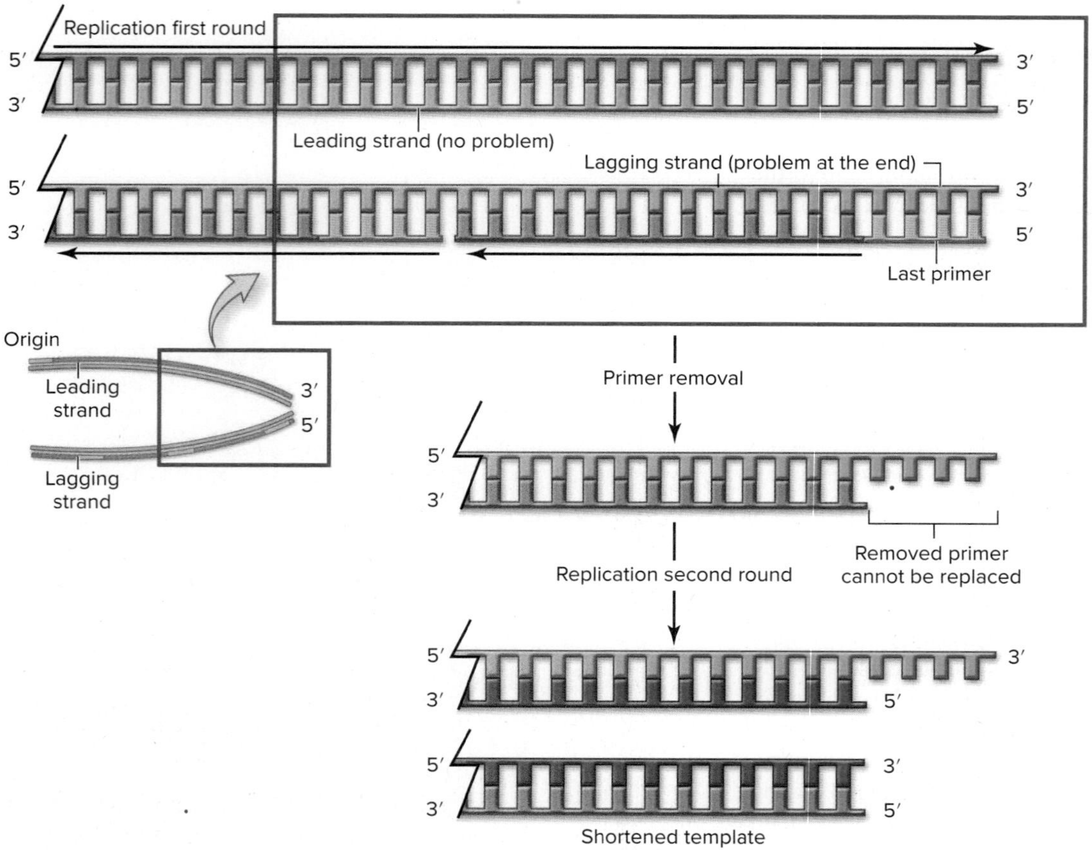

Figure 14.21 Replication of the end of linear DNA. Only one end is shown, for simplicity; the problem exists at both ends. The leading strand can be completely replicated, but the lagging strand cannot be finished. When the last primer is removed, it cannot be replaced. During the next round of replication, when this shortened template is replicated, it will produce a shorter chromosome.

helicases all have a similar structure, forming a ring around a single strand of DNA, but the bacterial and archaeal/eukaryal enzymes are not homologous and even move in the opposite direction along DNA. Lastly, the primase component of the replisome is also not homologous between bacteria and archaea/eukarya. The one function that is clearly homologous in all domains of life is the processivity factors: sliding clamps and clamp loaders.

The ends of eukaryotic chromosomes have specialized structures called telomeres

Specialized structures called **telomeres** are found on the ends of all eukaryotic chromosomes. These structures protect the ends of chromosomes from nucleases and are necessary to maintain the integrity of linear chromosomes. Although these telomeres are composed of specific DNA sequences, they are not made by the replication complex.

A problem with how replication ends

The very structure of a linear chromosome creates a problem in replicating the ends. The directionality of polymerases, combined with their requirement for a primer, creates this problem.

Consider a simple linear molecule like the one in figure 14.21. Replicating right to the 5′ end of the leading-strand template poses no problem: When the polymerase reaches this end, synthesizing in the 5′-to-3′ direction, it eventually runs out of template and is finished.

But on the other strand's end, the 3′ end of the lagging strand, removal of the last primer on this end leaves a gap. This gap cannot be primed, meaning that the polymerase complex cannot finish this end properly. The result would be a gradual shortening of chromosomes with each round of cell division (figure 14.21).

The structure of telomeres provided the answer to this problem. Molecular biologist Elizabeth Blackburn found that telomeric regions are composed of several thousand repeats of the sequence TTAGGG. She also found that the telomeric region of chromosomes is substantially shorter in somatic cells than in germ-line cells. She speculated that this is due to a loss of DNA in telomeres during DNA replication.

How do germ-line cells, dividing continuously for decades, avoid this? Blackburn and collaborator Jack Szostak proposed that cells possess a telomere-lengthening enzyme, and in 1984 Blackburn's graduate student Carol Greider isolated the enzyme called telomerase. This enzyme has an internal RNA it uses as a template instead of DNA (figure 14.22). The enzyme synthesizes short DNA sequences complementary to the internal RNA, producing the repeated sequences observed in telomeres. The other strand of these repeats is synthesized by the replication machinery using the telomerase-generated sequence as a template. Lastly, telomerase is active in germ cells, but not in somatic cells.

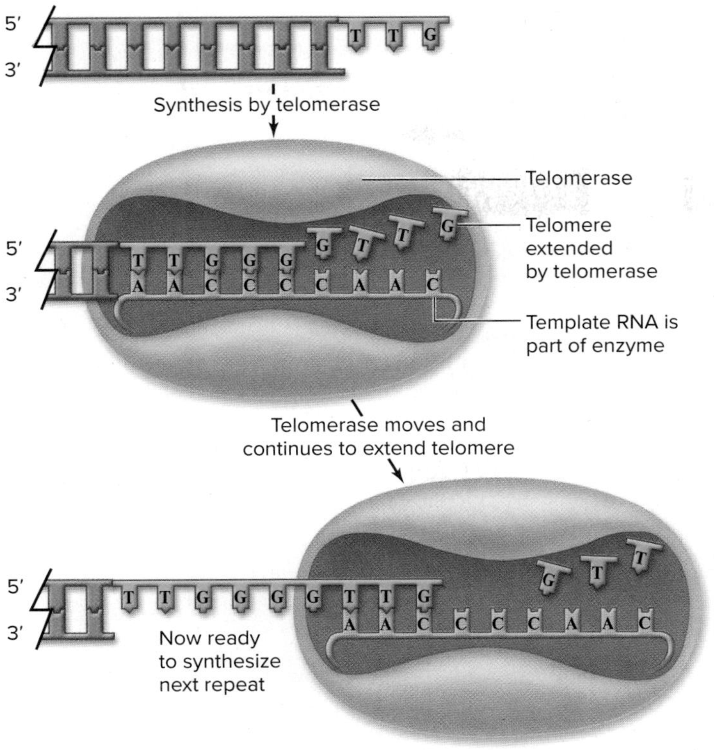

Figure 14.22 Action of telomerase. Telomerase contains an internal RNA, which the enzyme uses as a template to extend the DNA of the chromosome end. Multiple rounds of synthesis by telomerase produce repeated sequences. This single strand is completed by normal synthesis, using it as a template (not shown).

Telomerase, aging, and cancer

In the absence of telomerase activity, there is a gradual shortening of chromosomes. In humans, telomerase activity is high during embryonic and childhood development, but it is low in adult somatic tissue. The exception to this are cells such as lymphocytes that must divide in adults. This inhibition of telomerase in adults results from a lack of expression of the telomerase gene.

Experiments with mice lacking telomerase activity provided direct evidence for chromosome shortening. These mice appear to be normal for up to six generations, but they show steadily decreasing telomere length, which eventually leads to nonviable offspring. This finding indicates a relationship between cell senescence (aging) and telomere length.

Normal cells undergo only a specified number of divisions when grown in culture. This limit is at least partially based on telomere length. Experiments in which telomerase is introduced into fibroblasts in culture support this relationship. The life span of these cells is increased relative to controls with no added telomerase. Interestingly, these cells do not show the hallmarks of malignant cells, indicating that activation of telomerase alone does not make cells malignant.

Telomerase does play a role in cancer, however. Cancer cells could not continue to divide indefinitely if their chromosomes were being continually shortened. The telomerase gene is activated in cancer cells, allowing them to maintain chromosome ends. But this is only one aspect of conditions that allow them to escape normal growth controls.

REVIEW OF CONCEPT 14.5

Eukaryotic replication is complicated by a large amount of DNA organized into chromosomes, and by the linear nature of chromosomes. Eukaryotes speed replication by using multiple origins of replication. The ends of linear chromosomes cannot be completely replicated. Specialized ends, the telomeres, are synthesized by the enzyme telomerase using an internal RNA template.

- *What might be the result of abnormal shortening of telomeres or a lack of telomerase activity?*

14.6 Cells Repair Damaged DNA

As you learned in section 14.4, many DNA polymerases have 3′-to-5′ exonuclease activity that allows "proofreading" of added bases. This action increases the accuracy of replication, but errors still occur. Without additional error correction mechanisms, cells would accumulate errors at an unacceptable rate, leading to high levels of deleterious or lethal mutations.

Cells Are Constantly Exposed to DNA-Damaging Agents

LEARNING OBJECTIVE 14.6.1 Compare and contrast the different forms of DNA repair.

In addition to errors in DNA replication, cells are constantly exposed to agents that can damage DNA. Agents that damage DNA can lead to mutations, and any agent that increases the number of mutations above background levels is called a **mutagen.** The number of potentially mutagenic agents that organisms encounter is huge. Sunlight itself includes radiation in the UV range that is mutagenic. Ozone normally screens out much of the harmful UV radiation in sunlight, but some remains. The relationship between sunlight and mutations is shown clearly by the increase in skin cancer in regions of the southern hemisphere that are underneath a seasonal "ozone hole."

Organisms also may encounter chemical mutagens in their diet or surroundings. When a simple test was designed to detect mutagens, screening of possible sources indicated an amazing diversity of mutagens in our environment and food sources. As a result, consumer products are now screened to reduce the load of mutagens we are exposed to.

DNA repair restores damaged DNA

Cells cannot escape exposure to mutagens, but systems have evolved that enable cells to repair some of the DNA damage they create. These DNA repair systems are vital to cells' continued existence, whether a cell is a free-living, single-celled organism or part of a complex, multicellular organism.

The importance of DNA repair is indicated by the multiplicity of repair systems that have been discovered and characterized. All cells that have been examined show multiple pathways for repairing damaged DNA and for reversing errors that occur during replication. These systems are not perfect, but they do reduce the mutational load on organisms to an acceptable level. In the rest of this section, we illustrate the action of DNA repair by concentrating on two examples drawn from these multiple repair pathways.

Mismatch repair reverses replication errors

Mutations due to DNA replication errors occur even less often than predicted for a proofreading DNA polymerase. This is due to the action of mismatch repair (MMR), which removes incorrect bases that have been incorporated, replacing them with the correct base by copying the template strand. For this system to work, it must be able to distinguish between the template strand and the newly synthesized strand. In *E. coli,* this involves methylation of the base adenine in the sequence 5′ GATC 3′. This sequence is a palindrome, so the complement (3′ CTAG 5′) is the same 5′ to 3′, and the A in each strand is methylated. The methylase responsible does not act immediately after replication, leaving a short window for the MMR system to be able to identify the newly synthesized strand.

Photorepair: A specific repair mechanism

Photorepair is specific for one particular form of damage caused by UV light, namely the *thymine dimer* (figure 14.23). Thymine dimers are formed from adjacent thymine bases in DNA ❶ by a photochemical reaction to UV light that causes the thymines to react, covalently linking them together, creating a thymine dimer ❷.

Repair of these thymine dimers can be accomplished by multiple pathways, including photorepair. In photorepair, an enzyme called a *photolyase* absorbs light in the visible range and uses this energy to cleave the thymine dimer ❸. This action restores the two thymines to their original state ❹. Photorepair does not occur in cells deprived of visible light.

The photolyase enzyme is an ancient repair system found in both prokaryotes and eukaryotes. For as long as cells have existed on Earth, they have been exposed to UV light and its potential to damage DNA.

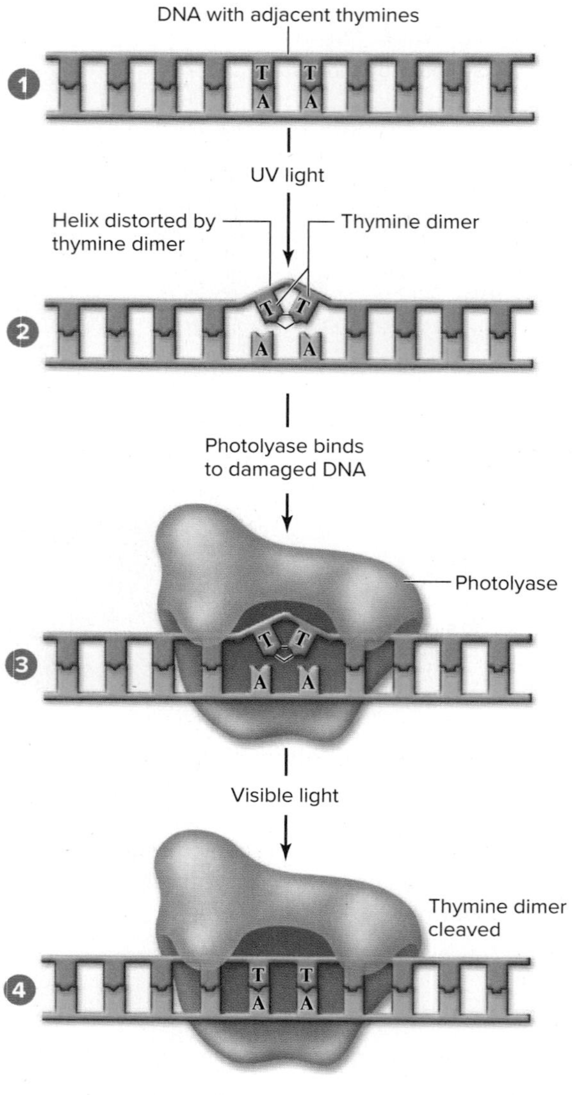

Figure 14.23 Repair of thymine dimer by photorepair. UV light can catalyze a photochemical reaction to form a covalent bond between two adjacent thymines, thereby creating a thymine dimer. A photolyase enzyme recognizes the damage and binds to the thymine dimer. The enzyme absorbs visible light and uses the energy to cleave the thymine dimer.

Excision repair: A nonspecific repair mechanism

A common form of nonspecific repair is **excision repair.** In this pathway, a damaged region is removed, or excised, and is then replaced by DNA synthesis (figure 14.24). In *E. coli,* this action is accomplished by proteins encoded by the *uvr A, B,* and *C* genes.

Excision repair follows three steps: ❶ recognition of damage, ❷ removal of the damaged region, and ❸ resynthesis using the information on the undamaged strand as a template. Recognition and excision are accomplished by the UvrABC complex. The UvrABC complex binds to damaged DNA and then cleaves a single strand on either side of the damage, removing it. In the synthesis stage, DNA Pol I or Pol II replaces the damaged DNA. This restores the original information in the damaged strand by using the information in the complementary strand.

DNA repair can be error-free or error-prone

None of the forms of DNA repair described introduce errors into DNA, but there is an error-prone repair pathway. It may seem

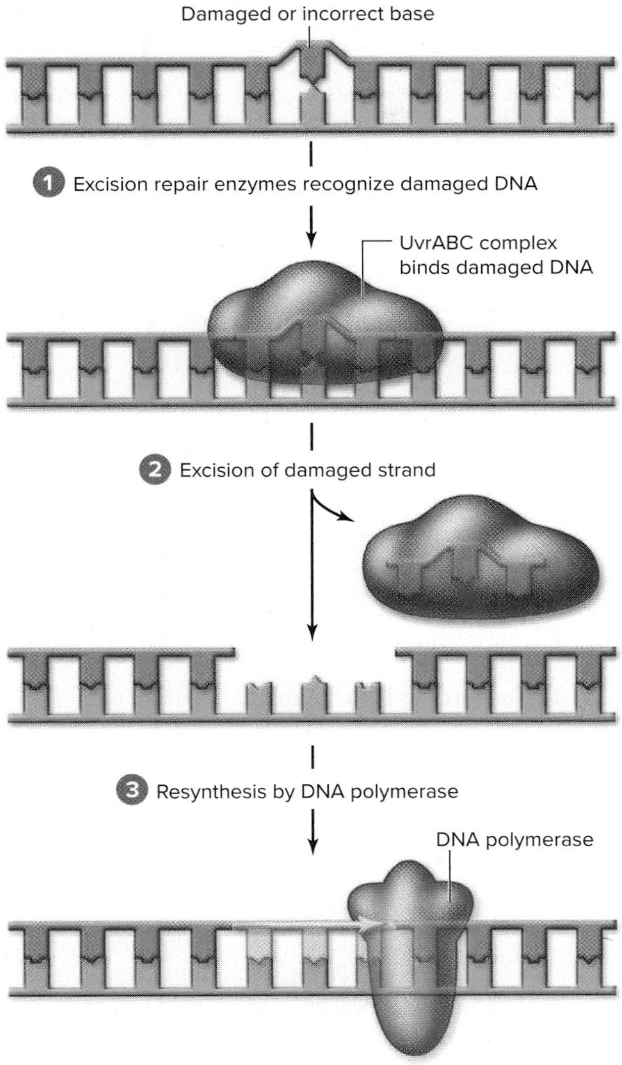
Damaged or incorrect base
1 Excision repair enzymes recognize damaged DNA
UvrABC complex
binds damaged DNA
2 Excision of damaged strand
3 Resynthesis by DNA polymerase
DNA polymerase

Figure 14.24 Repair of damaged DNA by excision repair. Damaged DNA is recognized by the Uvr complex, which binds to the damaged region and removes it. Synthesis by DNA polymerase replaces the damaged region. DNA ligase finishes the process (not shown).

strange to have an error-prone pathway, but it can be thought of as a last-ditch effort to save a cell that has been exposed to such massive damage that it has overwhelmed the error-free systems. In fact, this system in *E. coli* is part of what is called the "SOS response."

Cells can also repair damage that produces breaks in DNA. These systems use enzymes related to those that are involved in recombination during meiosis (refer to chapter 11). This system requires a second copy of a chromosome, usually a homologue, for repair. This does not introduce errors, but can change information in a chromosome as homologues are not identical. The enzymes involved in this process are thought to have been co-opted by evolution for recombination during meiosis.

REVIEW OF CONCEPT 14.6

DNA repair is critical to remove replication errors and to reverse the effects of DNA-damaging agents. Mismatch repair occurs immediately after replication to remove errors. Repair systems can be either specific or nonspecific. Photorepair is specific, removing thymine dimers caused by UV light. Excision repair is nonspecific, removing and replacing different kinds of damage.

- *Could a cell survive with no form of DNA repair?*

Are Mutations Random or Directed by the Environment?

Once biologists appreciated that Mendelian traits were, in fact, alternative versions of DNA sequences, which resulted from mutations, a very important question arose and needed to be answered: Are mutations random events that might happen anywhere on the DNA in a chromosome, or are they directed to some degree by the environment? For example, do the mutagens in cigarettes damage DNA at random locations, or do they preferentially seek out and alter specific sites such as those regulating the cell cycle?

This key question was addressed and answered in an elegant, deceptively simple experiment carried out in 1943 by two of the pioneers of molecular genetics, Salvadore Luria and Max Delbrück. They chose to examine a particular mutation that occurs in laboratory strains of the bacterium *E. coli*. These bacterial cells are susceptible to T1 viruses, tiny chemical parasites that infect, multiply within, and kill the bacteria. If 10^5 bacterial cells are exposed to 10^{10} T1 viruses, and the mixture is spread on a culture dish, not one cell grows—every single *E. coli* cell is infected and killed. However, if you repeat the experiment using 10^9 bacterial cells, lots of cells survive! When tested, these surviving cells prove to be mutants, resistant to T1 infection. The question is, did the T1 virus cause the mutations, or were they present all along, too rare to be present in a sample of only 10^5 cells but common enough to be present in 10^9 cells?

To answer this question, Luria and Delbrück devised a simple experiment they called a "fluctuation test," illustrated here. Five cell generations are shown for each of four independent bacterial cultures, all tested for resistance in the fifth generation. If the T1 virus causes the mutations (top row), then each culture will have more or less the same number of resistant cells, with only a little fluctuation (that is, variation among the four). If, on the other hand, mutations are spontaneous and therefore equally likely to occur in any generation, then bacterial cultures in which the T1-resistance mutation occurs in earlier generations will possess far more resistant cells by the fifth generation than cultures in which the mutation occurs in later generations, resulting in wide fluctuation among the four cultures. The table presents the data obtained for 20 individual cultures.

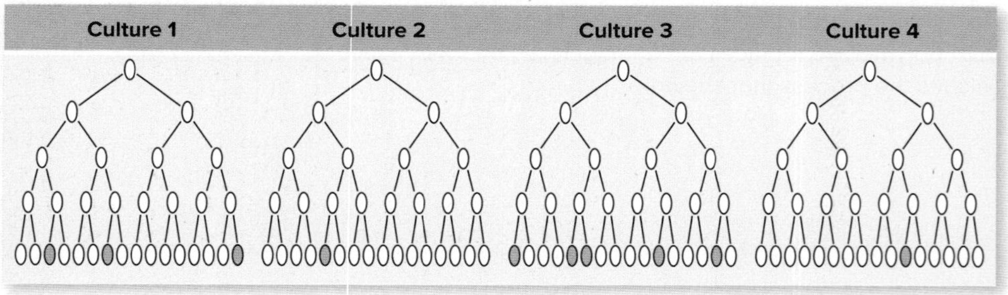

a.

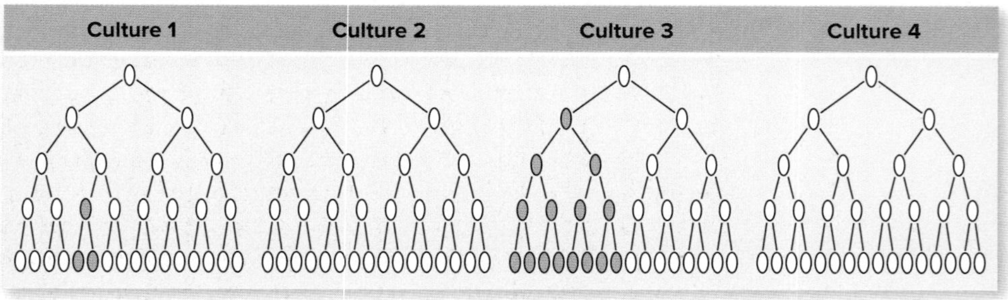

b.

Number of Bacteria Resistant to T1 Virus

Culture number	Resistant colonies found	Culture number	Resistant colonies found
1	1	11	107
2	0	12	0
3	3	13	0
4	0	14	0
5	0	15	1
6	5	16	0
7	0	17	0
8	5	18	64
9	0	19	0
10	6	20	35

Analysis

1. **Applying Concepts** Is there a dependent variable in this experiment? Explain.
2. **Interpreting Data** What is the mean number of T1-resistant colonies found in the 20 individual cultures?
3. **Making Inferences**
   a. Comparing the 20 individual cultures, do the cultures exhibit similar numbers of T1-resistant bacterial cells?
   b. Which of the two alternative outcomes illustrated, (a) or (b), is more similar to the outcome obtained by Luria and Delbrück in this experiment?
4. **Drawing Conclusions** Are these data consistent with the hypothesis that the mutation for T1 resistance among *E. coli* bacteria is caused by exposure to T1 virus? Explain.

Retracing the Learning Path

CONCEPT 14.1 DNA Is the Genetic Material

14.1.1 DNA Transfer Produces Hereditary Transformation in Bacteria Nonvirulent *S. pneumoniae* could take up an unknown substance from a virulent strain and become virulent.

14.1.2 The Transforming Principle Is DNA The transforming substance could be inactivated by DNA-digesting enzymes, but not by protein-digesting enzymes.

14.1.3 Virus Genes Are Made of DNA, Not Protein Radioactive labeling showed that the infectious agent of phage is its DNA, and not its protein.

CONCEPT 14.2 The DNA Molecule Is a Double Helix

14.2.1 DNA Is a Polymer of Nucleotides The nucleotide building blocks for DNA contain deoxyribose and the bases adenine (A), guanine (G), cytosine (C), and thymine (T). Phosphodiester bonds are formed between the 5′ phosphate of one nucleotide and the 3′ hydroxyl of another nucleotide.

14.2.2 DNA Is Not a Simple, Repeating Polymer Chargaff found equal amounts of adenine and thymine, and of cytosine and guanine, in DNA.

14.2.3 X-Ray Diffraction Patterns Suggest DNA Has a Helical Shape X-ray diffraction studies by Franklin and Wilkins indicated that DNA has a helical structure.

14.2.4 DNA Is a Double Helix DNA consists of two antiparallel polynucleotide strands wrapped about a common helical axis. These are held together by hydrogen bonds forming specific base-pairs (AT and GC). The two strands are complementary.

CONCEPT 14.3 Both Strands Are Copied During DNA Replication

14.3.1 DNA Replication Is Semiconservative Semiconservative replication uses each strand of a DNA molecule to specify the synthesis of a new strand. Meselson and Stahl used density-labeled DNA to show the products are composed of one new and one old strand. DNA replication requires a template, nucleotides, and a polymerase enzyme. All new DNA molecules are produced by DNA polymerase copying a template. Polymerases all synthesize new DNA in the 5′-to-3′ direction and require a primer.

CONCEPT 14.4 Prokaryotes Organize the Enzymes Used to Duplicate DNA

14.4.1 Prokaryotic Replication Starts and Ends at Unique Sites The *E. coli* origin has AT-rich sequences that are easily opened. The chromosome and its origin form a replicon. *E. coli* has at least three different DNA polymerases. Some DNA polymerases have exonuclease activity. Pol I, II, and III all have 3′-to-5′ exonuclease activity that can remove mispaired bases. Pol I can remove bases in the 5′-to-3′ direction, important to removing RNA primers. Unwinding DNA requires energy and causes torsional strain. DNA helicase uses energy from ATP to unwind DNA. The torsional strain introduced is removed by the enzyme DNA gyrase.

14.4.2 Replication Is Semidiscontinuous Replication is discontinuous on one strand. The continuous strand is called the leading strand, and the discontinuous strand is called the lagging strand. Synthesis occurs at the replication fork. At the fork, synthesis on the leading strand requires a single primer, and the polymerase is clamped to the template by the β subunit. On the lagging strand, DNA primase adds primers periodically, and DNA Pol III synthesizes the Okazaki fragments. DNA Pol I removes primer segments, and DNA ligase joins the fragments.

14.4.3 Bacteria Have a DNA Replication Organelle The replisome consists of two copies of Pol III, DNA primase, DNA helicase, and a number of accessory proteins. It moves in one direction by creating a loop in the lagging strand, allowing the antiparallel template strands to be copied in the same direction.

CONCEPT 14.5 Eukaryotic Chromosomes Are Large and Linear

14.5.1 Replicating Very Large Chromosomes Creates New Problems Multiple origins of replication are needed for DNA to be able to replicate in the time available in S phase. The eukaryotic primase synthesizes a short stretch of RNA and then switches to making DNA. This primer is extended by the main replication polymerase. The sliding clamp subunit is a protein called PCNA. The replication proteins of archaea, including the sliding clamp, clamp loader, and DNA polymerases, are more similar to those of eukaryotes than those of prokaryotes. Linear chromosomes have specialized ends called telomeres. They are made by telomerase using an internal RNA as a template. Adult cells lack telomerase activity.

CONCEPT 14.6 Cells Repair Damaged DNA

14.6.1 Cells Are Constantly Exposed to DNA-Damaging Agents Errors from replication and damage by agents such as UV light and chemical mutagens can lead to mutations. Without repair mechanisms, cells would accumulate mutations. Repair can either be specific, like photorepair, or nonspecific, using a single mechanism to repair many types of damage. Excision repair is nonspecific, removing and replacing a damaged region of DNA.

Concept Overview

This Concept Overview diagrams the key concepts that were discussed in this chapter.

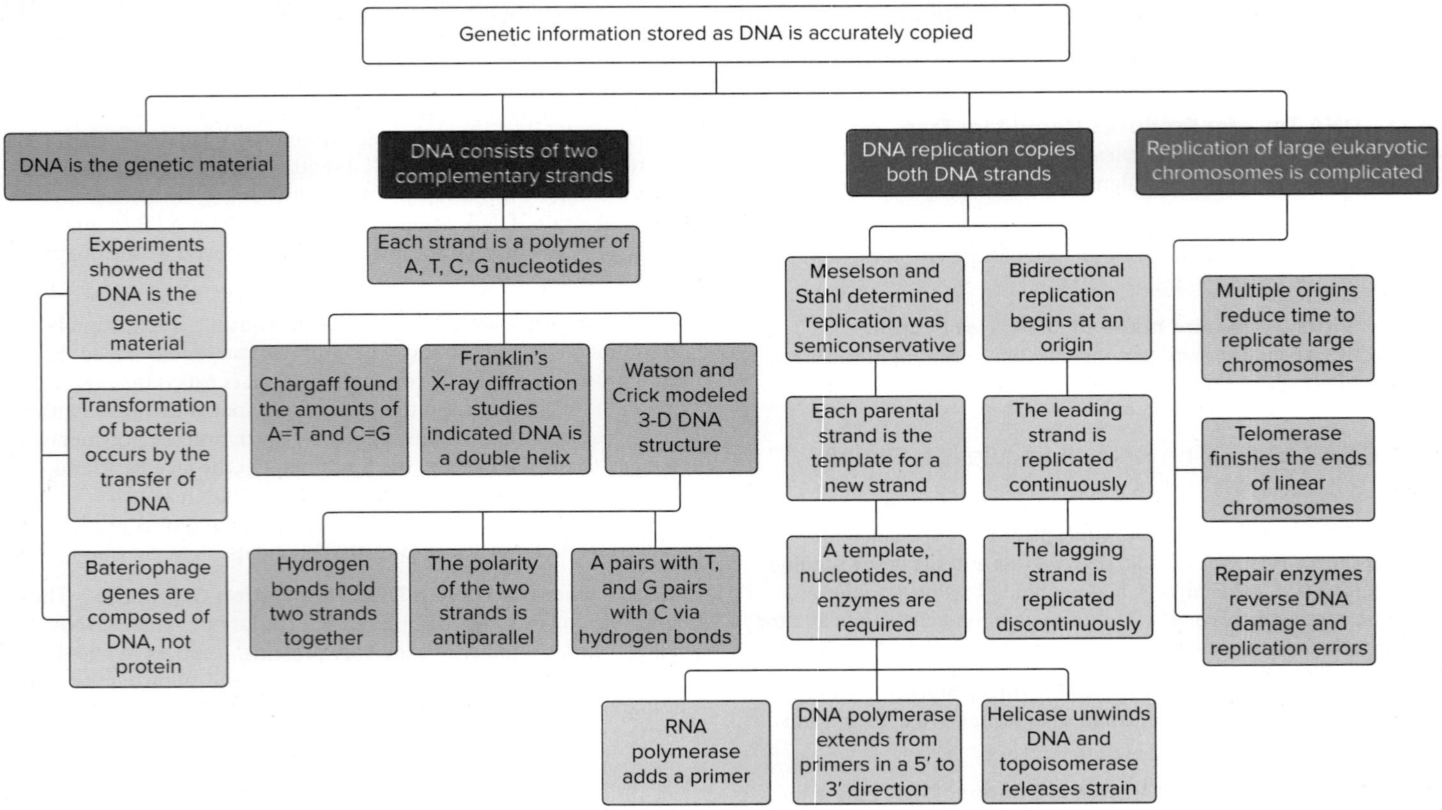

Assessing the Learning Path

Understand

1. What was the key observation made by Griffith in his experiments using live and heat-killed pathogenic bacteria?
   a. Bacteria with a smooth coat can kill mice.
   b. Bacteria with a rough coat are not lethal.
   c. DNA is the genetic material.
   d. Genetic material can be transferred from dead to live bacteria.
2. Hershey and Chase used radioactive phosphorus and sulfur to
   a. label DNA and protein with the same molecule.
   b. differentially label DNA and protein.
   c. identify the transforming principle.
   d. Both b and c
3. The bonds that hold two complementary strands of DNA together are
   a. hydrogen bonds.
   b. peptide bonds.
   c. ionic bonds.
   d. phosphodiester bonds.
4. Analysis of Rosalind Franklin's X-ray diffraction data showed
   a. which bases are complementary to each other.
   b. that DNA is a helix with a 2-nm diameter.
   c. which tautomeric forms of bases are found in DNA.
   d. Both a and b
5. Which of the following is NOT part of the Watson–Crick model of the structure of DNA?
   a. DNA is composed of two strands.
   b. The two DNA strands are oriented in parallel (5′-to-3′).
   c. Purines bind to pyrimidines.
   d. DNA forms a double helix.
6. The basic mechanism of DNA replication is semiconservative with two new molecules,
   a. each with new strands.
   b. one with all new strands and one with all old strands.
   c. each with one new and one old strand.
   d. each with a mixture of old and new strands.
7. The Meselson and Stahl experiment used a density label to be able to
   a. determine the directionality of DNA replication.
   b. differentially label DNA and protein.
   c. distinguish between newly replicated and old strands.
   d. distinguish between replicated DNA and RNA primers.
8. Which of the following statements about DNA polymerases is NOT accurate?
   a. They are enzymes.
   b. They make DNA polymers by adding nucleotides to the 3′ end of a growing chain.

c. They require ATP for addition of each nucleotide to a growing strand.
d. They require a template.

9. Which of the following does NOT occur during DNA replication?
   a. Complementary base-pairing between old and new strands
   b. Production of short segments joined by a ligase enzyme
   c. Polymerization in the 3′-to-5′ direction
   d. Use of an RNA primer
10. How is eukaryotic DNA replication different from bacterial DNA replication?
   a. They use entirely different sets of enzymes to copy their DNA.
   b. Eukaryotes lack lagging strands, because their DNA polymerases can synthesize DNA in either direction.
   c. Bacteria have circular chromosomes with one origin of replication, whereas eukaryotes have linear chromosomes with multiple origins of replication.
   d. Topoisomerase is needed for bacterial DNA replication, but not for eukaryotic DNA replication.
11. Telomeres
   a. are found in prokaryotic and eukaryotic cells.
   b. shorten with each round of DNA replication in most somatic cells.
   c. contain essential genes.
   d. are A–T-rich regions where DNA replication starts in eukaryotes.
12. Which of the following is an enzyme involved in excision repair?
   a. Photolyase
   b. DNA polymerase III
   c. Endonuclease
   d. Telomerase
13. Thymine dimers
   a. generally result from mistakes in DNA replication.
   b. occur when two thymines are mistakenly base-paired with each other.
   c. can be repaired by photolyase.
   d. Both a and b

Apply

1. To isolate the factor that allowed R cells to become S cells, Avery and coworkers fractionated dead S cells and then treated solutions of the fractionate as follows:

   Solution A—treated with chemicals that remove carbohydrate capsule

   Solution B—treated with enzyme that degrades proteins

   Solution C—treated with enzyme that degrades RNA

   Solution D—treated with enzyme that degrades DNA

   Each solution was then added separately to R cells. Which solution(s) would be unable to transform R cells to S cells?
   a. Solutions A, B, and C
   b. Solutions B and C
   c. Solution D
   d. All solutions—live S cells are needed for this experiment
2. Suppose that, in analyzing DNA from your own cells, you are able to determine that 15% of the nucleotide bases it contains are thymine. What percentage of the bases are cytosine?
   a. 15%
   b. 35%
   c. 60%
   d. 85%
3. From a hospital patient afflicted with a mysterious illness, you isolate and culture cells and then purify DNA from the culture. You find that the DNA sample obtained from the culture contains two quite different kinds of DNA: One is double-stranded human DNA and the other is single-stranded virus DNA. You analyze the base composition of the two purified DNA preparations, with the following results:

   Tube #1 22.1% A : 27.9% C : 29.7% G : 22.1% T
   Tube #2 31.3% A : 31.3% C : 18.7% G : 18.7% T

   Which of the two tubes contains single-stranded virus DNA?
4. If replication were conservative instead of semiconservative, what pattern of banding would Meselson and Stahl have seen after two rounds of DNA replication?
   a. Only a thick, heavy band
   b. A heavy band, a light band, and an intermediate band
   c. A heavy band and a thick, light band
   d. Heavy and light bands of equal intensity
5. If the two strands of DNA were arranged in a parallel and not antiparallel fashion, and if nothing else was different, how would this affect DNA replication?
   a. Replication on both strands would be like the lagging strand.
   b. Replication would be the same.
   c. Replication on both strands would be like the leading strand, but replication could not be bidirectional.
   d. Replication on both strands would be like the leading strand, and replication would be bidirectional.
6. If the activity of DNA ligase were removed from replication, this would have a greater effect on
   a. synthesis on the lagging strand versus the leading strand.
   b. synthesis on the leading strand versus the lagging strand.
   c. priming of DNA synthesis versus actual DNA synthesis.
   d. photorepair of DNA versus DNA replication.
7. How does telomerase solve the end-replication problem of eukaryotes?
   a. It attaches to DNA polymerase, giving it the ability to make DNA without a primer.
   b. It can build a DNA strand in the 3′-to-5′ direction.
   c. It circularizes the DNA so that there are no ends.
   d. It uses an intrinsic RNA template to build DNA.
8. In excision repair, DNA polymerase can make DNA without the need for primase. How?
   a. The DNA polymerase used in excision repair has its own primase activity.
   b. The 3′-OH is provided by existing DNA adjacent to the DNA that was removed.
   c. The DNA polymerase used in excision repair can build DNA without an existing 3′-OH.
   d. None of the above

Synthesize

1. The discovery that DNA is the genetic material was an experimental journey rather than a flash of insight. Highlighting individual experiments, use this journey to defend the statement attributed to Sir Isaac Newton in 1676 that scientists build new ideas by "standing on the shoulders of giants."
2. In the Meselson–Stahl experiment, a control experiment was done to show that the hybrid bands after one round of replication were, in fact, two complete strands, one heavy and one light. Using the same experimental setup as detailed in the text, how can this be addressed?
3. Why do you think it is important that the sugar–phosphate backbone of DNA is held together by covalent bonds, but the cross-bridges between the two strands are held together by hydrogen bonds?
4. Enzyme function is critically important for the proper replication of DNA. Predict the consequence of a loss of function for each of the following enzymes.
   a. DNA gyrase
   b. DNA polymerase III
   c. DNA ligase
   d. DNA polymerase I

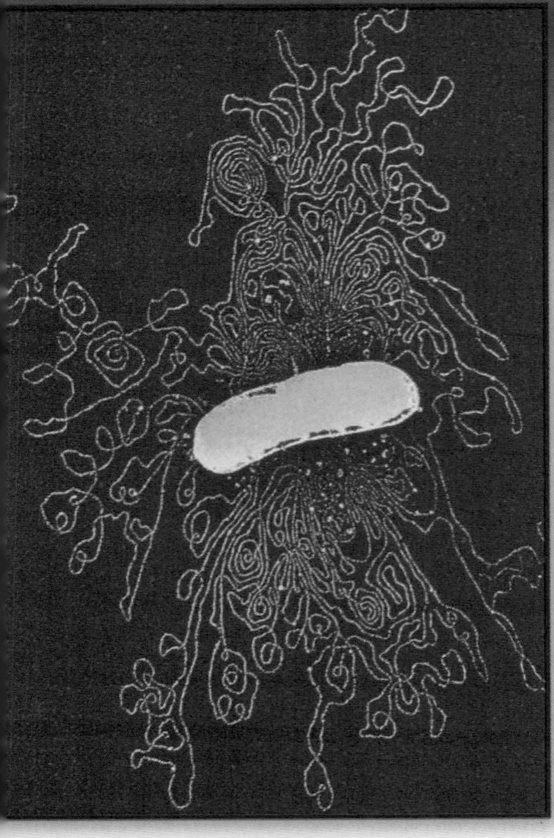

Dr. Gopal Murti/Science Source

15 Genes and How They Work

Learning Path

15.1 Experiments Have Revealed the Nature of Genes

15.2 The Genetic Code Relates Information in DNA and Protein

15.3 Prokaryotes Exhibit All the Basic Features of Transcription

15.4 Eukaryotes Use Three Polymerases and Extensively Modify Transcripts

15.5 Eukaryotic Genes May Contain Noncoding Sequences

15.6 The Ribosome Is the Machine of Protein Synthesis

15.7 The Process of Translation Is Complex and Energy-Expensive

15.8 Mutations Are Heritable Changes in Genetic Material

Concept Overview

This Concept Overview provides a diagram of the overarching concepts that are covered in this chapter. These concepts will be expanded upon in the Concept Overview at the end of the chapter.

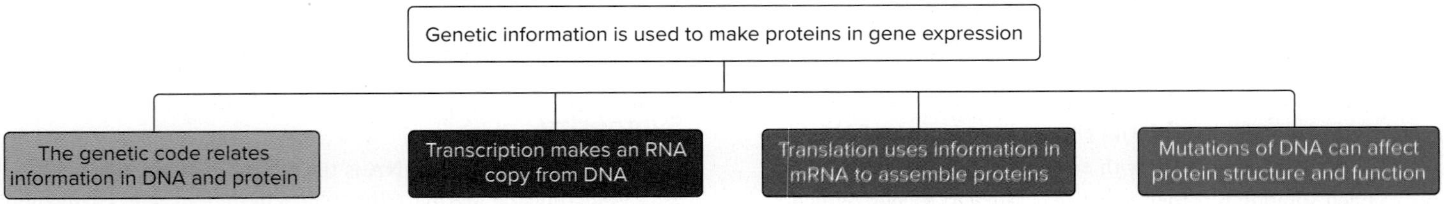
Genetic information is used to make proteins in gene expression
The genetic code relates information in DNA and protein
Transcription makes an RNA copy from DNA
Translation uses information in mRNA to assemble proteins
Mutations of DNA can affect protein structure and function

Introduction

The experimental journey that led to the discovery of DNA as the genetic material opened the door to the new field of molecular genetics—the study of genes and how they work. A burst of experiments soon revealed how molecules of DNA, like the one you see streaming out of the bacterial cell in the micrograph on the previous page, use the information in their nucleotide sequence to produce particular proteins. Later experiments have focused on how the DNA sequence controls which proteins are produced, and when. In this chapter, we will examine how the proteins of prokaryotes and eukaryotes are synthesized from the information in DNA. In subsequent chapters, we will explore how this process is regulated, and how that regulation has evolved among multicellular organisms.

15.1 Experiments Have Revealed the Nature of Genes

For many years after Mendel, it was not clear how genes influence an organism's phenotype. A clue came in 1902, when physician Archibald Garrod noted that certain diseases seemed to be more prevalent in particular families. He examined several generations of these families and found that some diseases behaved as though they were due to simple recessive alleles. Garrod concluded that these disorders were Mendelian traits, resulting from changes in the hereditary information in an ancestor of the affected families.

Garrod investigated four diseases, including alkaptonuria, in detail. Alkaptonuria patients produce urine that containes homogentisic acid (alkapton). This substance oxidizes rapidly when exposed to air, turning the urine black. Normally, homogentisic acid is metabolized and does not accumulate. Garrod concluded that alkaptonuric patients lack an enzyme necessary for this metabolism. He speculated that other inherited diseases might also reflect enzyme deficiencies.

Beadle and Tatum Showed That Genes Specify Enzymes

LEARNING OBJECTIVE 15.1.1 Describe the evidence supporting the "one-gene/one-polypeptide" hypothesis.

Garrod was ahead of his time, as a direct connection between Mendelian alleles and enzymes was not established for almost 40 years. In 1941 the link between traits followed in crosses and proteins was demonstrated by a series of experiments by geneticists George Beadle and Edward Tatum. They used the then-new technology of inducing mutations by X-rays to create a series of mutants that behaved in a Mendelian fashion for detailed study. Their work is summarized in figure 15.1.

Neurospora crassa, the bread mold

Beadle and Tatum chose the bread mold *Neurospora crassa* for their experiments. This fungus can be grown readily in the laboratory on a defined medium consisting of only a carbon source (glucose), a vitamin (biotin), plus inorganic salts. This is called minimal medium, because it represents the minimal requirements to support growth. Any cells that can grow on minimal medium are able to synthesize all necessary biological molecules.

When Beadle and Tatum exposed spores to X-rays, they expected to find some mutants that would be unable to grow on minimal medium. These so-called **nutritional mutants** result from damage to genes encoding functions necessary to make important biological molecules.

Nutritional mutants

To identify nutritional mutants, Beadle and Tatum first grew cultures on rich medium, then placed subcultures of individual fungal cells onto minimal medium. This identified any cells that had

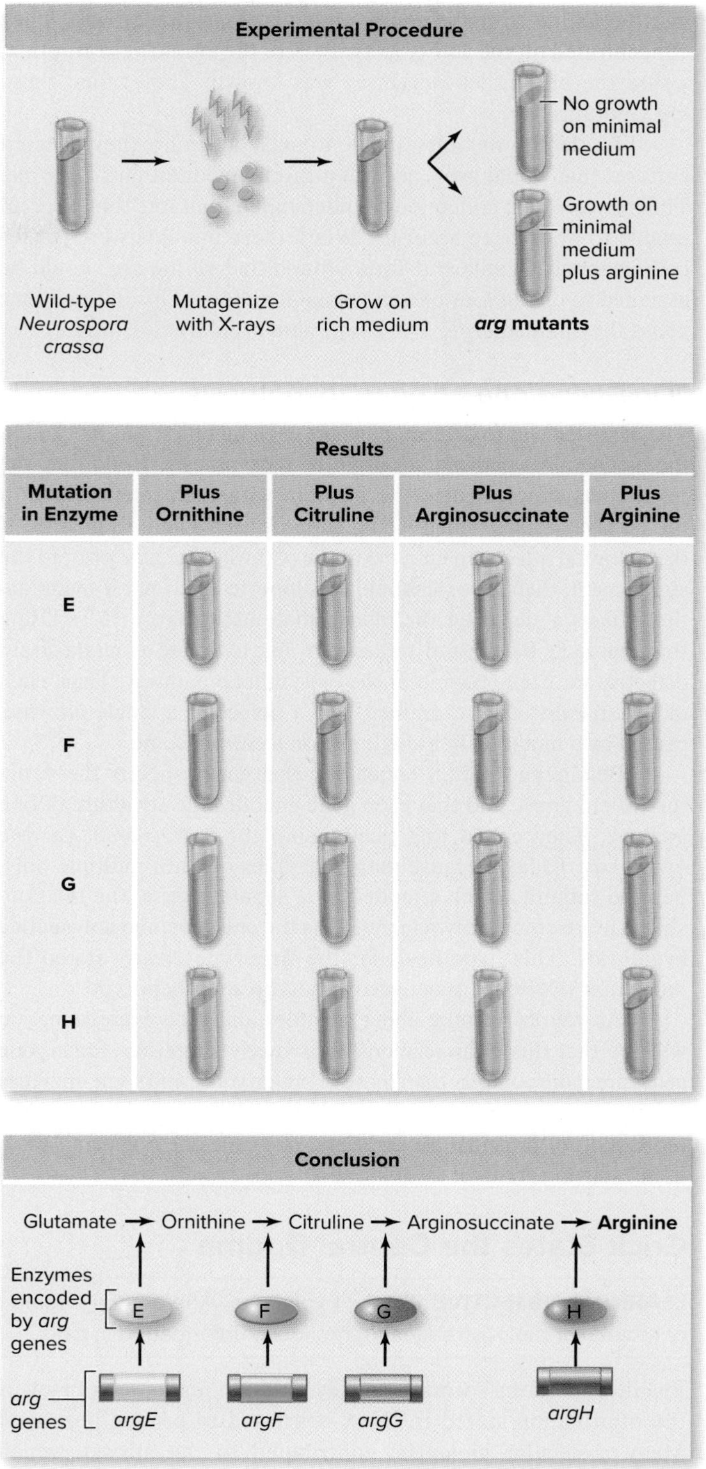

Experimental Procedure
No growth on minimal medium
Growth on minimal medium plus arginine
Wild-type Neurospora crassa
Mutagenize with X-rays
Grow on rich medium
arg mutants
Results
Mutation in Enzyme
Plus Ornithine
Plus Citruline
Plus Arginosuccinate
Plus Arginine
E
F
G
H
Conclusion
Glutamate → Ornithine → Citruline → Arginosuccinate → Arginine
Enzymes encoded by arg genes
E
F
G
H
arg genes
argE
argF
argG
argH

Figure 15.1 The Beadle and Tatum experiment. Wild-type *Neurospora* were mutagenized with X-rays to produce mutants deficient in the synthesis of arginine (top panel). The specific defect in each mutant was identified by growing on medium supplemented with intermediates in the biosynthetic pathway for arginine (middle panel). A mutant will grow only on media supplemented with an intermediate produced after the defective enzyme in the pathway for each mutant. The enzymes in the pathway can then be correlated with genes on chromosomes (bottom panel).

lost the ability to make compounds necessary for growth. They concentrated on the ability to synthesize the amino acid arginine, where the biosynthetic pathway was known. They called these *arg*⁻ mutants.

To identify mutants unable to make arginine, they selected cultures that would grow only on minimal medium plus arginine. This resulted in a collection of independent mutants that were all unable to synthesize arginine. When these mutants were genetically mapped, Beadle and Tatum found that all the *arg*⁻ could be assigned to one of four chromosomal locations. This defined four genes they named *argE, argF, argG,* and *argH* (figure 15.1).

"One-gene/one-enzyme" hypothesis

The next step was to determine where each mutation was blocked in the biochemical pathway for arginine biosynthesis. To do this, the researchers supplemented the medium with each intermediate in the pathway to see which would support each mutant's growth. If the mutation affects an enzyme in the pathway that acts prior to the supplement, then growth should be supported—but not if the mutation affects a step after the intermediate used (figure 15.1). Using this approach, Beadle and Tatum were able to isolate a mutant strain defective for each enzyme in the biosynthetic pathway. Thus, each of the mutants they examined had a defect in a single enzyme, caused by a mutation at a single site on a chromosome.

Beadle and Tatum concluded that genes specify the structure of enzymes, and that each gene encodes the structure of one enzyme. They called this relationship the *one-gene/one-enzyme hypothesis.* Today, because many enzymes contain multiple polypeptide subunits, each encoded by a separate gene, the relationship is more commonly referred to as the **one-gene/one-polypeptide hypothesis.** This hypothesis for the first time clearly stated the molecular relationship between genotype and phenotype.

As you learn more about genomes and gene expression, you will see that this clear relationship is overly simplistic. Eukaryotic genes are more complex than those of prokaryotes, and some enzymes are composed, at least in part, of RNA, itself an intermediate in the production of proteins. Nevertheless, one gene/one polypeptide is a useful starting point for thinking about gene expression.

Crick States the Central Dogma

LEARNING OBJECTIVE 15.1.2 Explain how the central dogma of molecular biology relates to the flow of information in cells.

Beadle and Tatum's work left a key question unanswered: How is the information stored in DNA converted to protein enzymes? Many molecular biologists contributed to the answer, which involves the other nucleic acid, RNA. Information passes in one direction from the gene (DNA) to an RNA copy of the gene, which directs the sequential assembly of a chain of amino acids into a protein (figure 15.2). Stated briefly,

$$\text{DNA} \rightarrow \text{RNA} \rightarrow \text{protein}$$

This **central dogma,** first articulated by one of the discoverers of mRNA, Francis Crick, provides an intellectual framework that describes information flow in biological systems. We call the DNA-to-RNA step **transcription,** because it produces an exact copy of the DNA, much as a legal transcription contains the exact words of a court proceeding. The RNA-to-protein step is termed

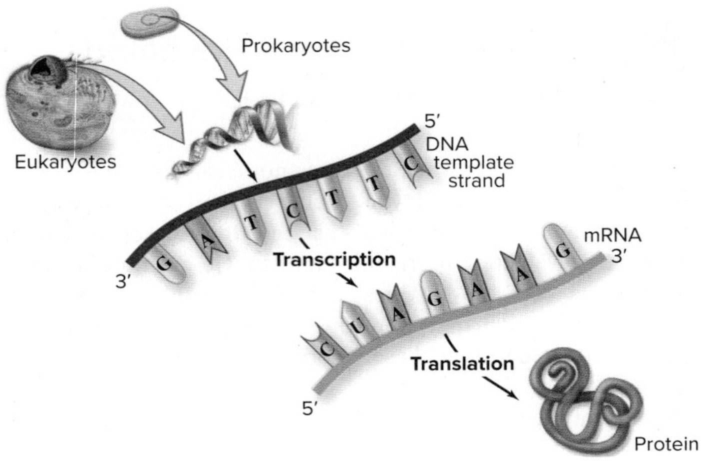

Figure 15.2 The central dogma of molecular biology. DNA is transcribed to make mRNA, which is translated to make a protein.

translation, because it requires translating from the nucleic acid to the protein "languages."

Since the original formulation of the central dogma, a class of viruses called **retroviruses** was discovered that can convert their RNA genome into a DNA copy, using the viral enzyme **reverse transcriptase.** This conversion violates the direction of information flow of the central dogma, and the discovery has forced an updating of the possible flow of information to allow for this sort of "reverse" flow from RNA to DNA.

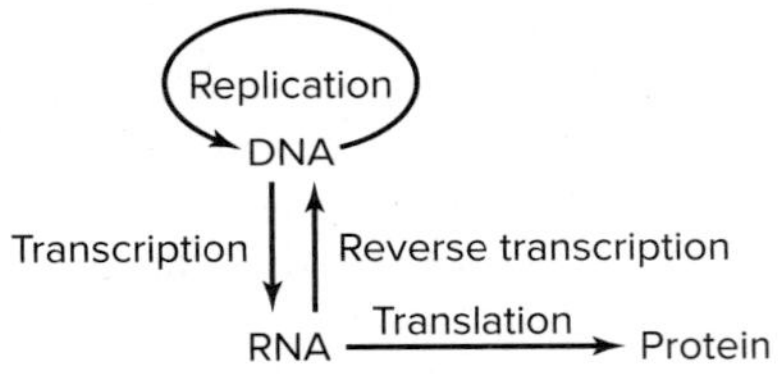

Transcription makes an RNA copy of DNA

The process of transcription produces an RNA copy of the information in DNA. That is, transcription is the DNA-directed synthesis of RNA by the enzyme RNA polymerase (figure 15.3). This process uses the principle of complementarity, described in chapter 14, to use DNA as a template to make RNA.

Because DNA is double-stranded and RNA is single-stranded, only one of the two DNA strands needs to be copied. We call the strand that is copied the **template strand.** The RNA transcript's sequence is complementary to the template strand. The strand of DNA not used as a template is called the **coding strand.** It has the same sequence as the RNA transcript, except that U (uracil) in the RNA is T (thymine) in the DNA-coding strand. Another naming convention for the two strands of the DNA is to call the coding strand the sense strand, as it has the same "sense" as the RNA. The template strand would then be the antisense strand.

Coding (sense) 5′–TCAGCCGTCAGCT–3′ ⎤ DNA
Template (antisense) 3′–AGTCGGCAGTCGA–5′ ⎦
Transcription ↓
Coding 5′–UCAGCCGUCAGCU–3′ mRNA

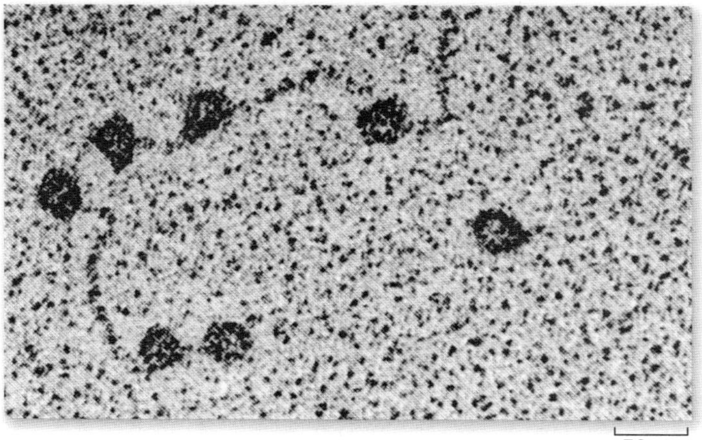

Figure 15.3 RNA polymerase. In this electron micrograph, the dark circles are RNA polymerase molecules synthesizing RNA from a DNA template.

From R.C. Williams, PNAS, 74(1977): 2313

The RNA transcript used to direct the synthesis of polypeptides is termed **messenger RNA (mRNA).** Its name reflects its use by the cell to carry the DNA message to the ribosome for processing.

Translation uses information in RNA to synthesize proteins

The process of translation is, by necessity, much more complex than transcription. In this case, RNA cannot be used as a direct template for a protein because there is no complementarity—that is, a sequence of amino acids cannot be aligned to an RNA template based on any kind of "chemical fit." Molecular geneticists suggested that some kind of adapter molecule must exist that can interact with both RNA and amino acids, and *transfer RNA (tRNA)* was found to fill this role. This need for an intermediary adds a level of complexity to the process that is not present in either DNA replication or transcription of RNA.

RNA has multiple roles in gene expression

All RNAs are synthesized from a DNA template by transcription. Gene expression involves multiple kinds of RNA, each with different roles in the overall process. Here is a brief summary of these roles, which are described in detail throughout the chapter.

Messenger RNA. Even before the details of gene expression were unraveled, geneticists recognized that there must be an intermediate form of the information in DNA that can be transported out of the eukaryotic nucleus to the cytoplasm, where proteins are made on ribosomes. This hypothesis was called the "messenger hypothesis" and the RNA molecules called **messenger RNA (mRNA).**

Ribosomal RNA. The class of RNA found in ribosomes is called **ribosomal RNA (rRNA).** There are multiple forms of rRNA, each critical to the function of the ribosome; rRNA is found in both ribosomal subunits.

Transfer RNA. The intermediary adapter molecule between mRNA and amino acids is **transfer RNA (tRNA).** Transfer RNA molecules have amino acids covalently attached to one end and an anticodon that can base-pair with an mRNA codon at the other. The tRNAs act to interpret information in mRNA and to help position the amino acids on the ribosome.

Small nuclear RNA. Small nuclear RNAs (snRNAs) are part of the machinery that is involved in nuclear processing of eukaryotic "pre-mRNA." We discuss this splicing reaction later, in section 15.5.

SRP RNA. In eukaryotes, in which some proteins are synthesized by ribosomes on the rough endoplasmic reticulum (RER), this process is mediated by the **signal recognition particle,** or **SRP,** described in section 15.7. The SRP contains both RNA and proteins.

Small RNAs. This class of RNA includes both micro-RNA (miRNA) and small interfering RNA (siRNA). These are involved in the control of gene expression, discussed in chapter 16.

REVIEW OF CONCEPT 15.1

Garrod showed that altered enzymes can cause metabolic disorders. Beadle and Tatum demonstrated that each gene encodes a unique enzyme. Genetic information flows from DNA (genes) to protein (enzymes) using messenger RNA as an intermediate. Transcription converts information in DNA into an RNA transcript, and translation converts this information into protein. There are multiple forms of RNA with different functions; these include mRNA, tRNA (adapter), and rRNA (in ribosomes), as well as snRNA, SRP RNA, and miRNA.

- *Why do cells need an adapter molecule, such as tRNA, between RNA and protein?*

15.2 The Genetic Code Relates Information in DNA and Protein

How does a sequence of nucleotides in a DNA molecule specify the sequence of amino acids in a polypeptide? The answer to this essential question came in 1961, through an experiment led by Francis Crick and biologist Sydney Brenner. That experiment was so elegant and the result so critical to understanding the genetic code that we describe it here in detail.

Crick and Brenner Learn How the Code Is Read

LEARNING OBJECTIVE 15.2.1 Predict the results of deleting or adding one, two, or three DNA bases.

Crick and Brenner reasoned that the genetic code consisted of a series of nucleotides, grouped into blocks of information called **codons,** each codon corresponding to an amino acid in the encoded protein. How many nucleotides per codon? They reasoned that the most likely number was three, for the simple reason that using only two nucleotides in each codon (with four DNA nucleotides G, C, T, and A) can produce only 4^2, or 16, different codons—not enough to code for 20 amino acids. However, using three nucleotides results in 4^3, or 64, different combinations of three, more than enough.

Do genetic messages include spaces?

Crick and Brenner recognized that the sequence of codons in a gene could be arranged in two quite different ways. If the information in the genetic message were separated by spaces, then altering any single word would not affect the entire sentence. In contrast, if all of the words were run together but read in groups of three, then any alteration not in groups of three would alter the entire sentence. These two ways of using information in DNA imply different methods of translating the information into protein.

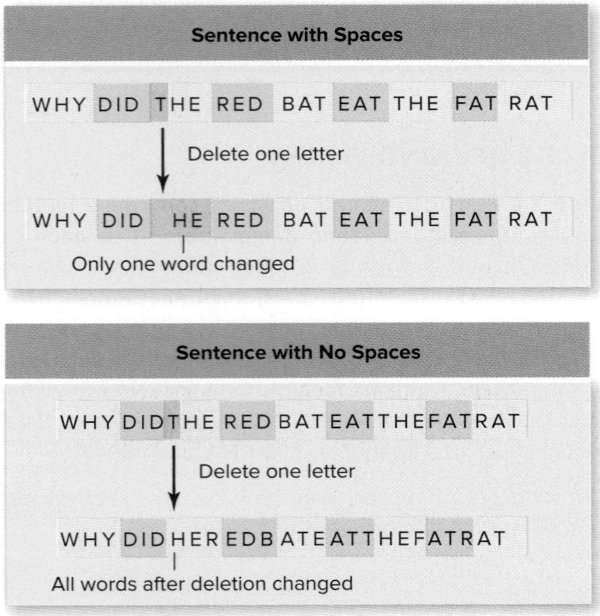

The Crick–Brenner experiment

To choose between these alternative mechanisms, Crick and Brenner used a chemical to create mutations with single-base insertions or deletions from a viral DNA molecule. When they made a single addition or deletion, the **reading frame** of the genetic message shifted, and the downstream gene was transcribed as nonsense. They then showed that combining an insertion with a deletion restored function, even though either one individually displayed loss of function. In this case, only the region between the insertion or deletion was altered. By choosing a region of the gene that encoded a part of the protein not critical to function, this small change did not cause a change in phenotype.

When they combined two deletions near each other, the genetic message frame-shifted, altering all of the amino acids after the deletion. However, when they made three deletions, the protein after the deletions was normal. They obtained the same results with additions of one, two, and three nucleotides.

Crick and Brenner concluded that the genetic code is read in groups of three nucleotides (a triplet code) and is read continuously without punctuation between the three-nucleotide units (figure 15.4).

These experiments indicate the importance of the reading frame for the genetic message. Because there is no punctuation, the reading frame established by the first codon in the sequence determines how all subsequent codons are read. We now call the kinds of mutations that Crick and Brenner used **frameshift mutations,** because they alter the reading frame of the genetic message.

SCIENTIFIC THINKING

Hypothesis: *The genetic code is read in groups of three bases.*

Prediction: *If the genetic code is read in groups of three, then deletion of one or two bases would shift the reading frame after the deletion. Deletion of three bases, however, would produce a protein with a single amino acid deleted but no change downstream.*

Test: *Single-base deletion mutants are collected, each of which exhibits a mutant phenotype. Three of these deletions in a single region are combined to assess the effect of deletion of three bases.*

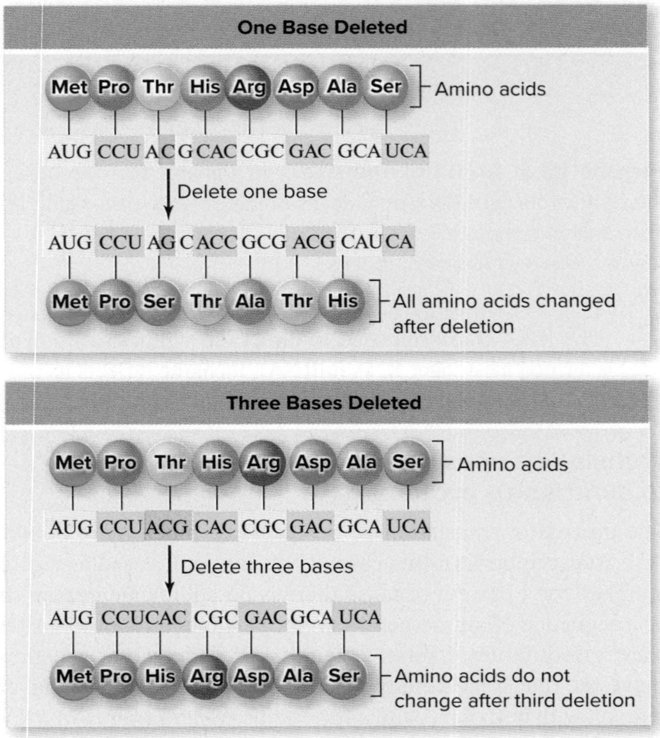

Result: *The combination of three deletions does not have the same drastic effect as the loss of one or two bases.*

Conclusion: *The genetic code is read in groups of three.*

Further Experiments: *If you also had mutants with one single-base addition, what would be the effect of combining a deletion and an addition?*

Figure 15.4 The genetic code is a continuous triplet code.

Nirenberg and Khorana Decipher the Code

LEARNING OBJECTIVE 15.2.2 Describe the features of the genetic code.

The assignment of each of the 64 possible codons to specific amino acids was a highlight of 20th-century biochemistry. This decryption of the genetic code depended on two related technologies: (1) cell-free biochemical systems that would support protein synthesis from defined RNAs, and (2) the ability to produce synthetic, defined RNAs for use in cell-free systems.

From 1961 to 1966, work performed primarily in biochemist Marshall Nirenberg's laboratory opened the door to elucidating the genetic code. Nirenberg's group discovered that adding a synthetic RNA with only uracil nucleotides (polyU) to their cell-free systems produced only polyphenylalanine. They concluded that the codon UUU must encode the amino acid phenylalanine. The researchers then used enzymes to synthesize defined three-base sequences that were tested for their ability to bind defined tRNA/ribosome complexes. This so-called triplet-binding assay allowed the tentative identification of 54 of the 64 possible triplets.

During this same time period, the organic chemist H. Gobind Khorana (figure 15.5) pioneered the organic synthesis of RNA molecules of defined sequence. These synthetic RNAs could then be used in cell-free systems. This allowed the conclusive determination of all 64 possible three-nucleotide sequences—the full genetic code (table 15.1). In addition to his work on synthetic RNA, Khorana actually performed experiments that anticipated the technique known as the polymerase chain reaction. Khorana and Nirenberg shared the Nobel prize in physiology and medicine with Robert Holley, who determined the first tRNA structure, in 1968.

Figure 15.5 H. Gobind Khorana. Best known for contributions to deciphering the genetic code, Khorana was a pioneer in the chemical synthesis of nucleic acids.

Science History Images/Alamy Stock Photo

TABLE 15.1 The Genetic Code

First Letter	SECOND LETTER: U		C		A		G		Third Letter
U	UUU	Phe Phenylalanine	UCU	Ser Serine	UAU	Tyr Tyrosine	UGU	Cys Cysteine	U
	UUC		UCC		UAC		UGC		C
	UUA	Leu Leucine	UCA		**UAA**	"Stop"	**UGA**	"Stop"	A
	UUG		UCG		**UAG**	"Stop"	UGG	Trp Tryptophan	G
C	CUU	Leu Leucine	CCU	Pro Proline	CAU	His Histidine	CGU	Arg Arginine	U
	CUC		CCC		CAC		CGC		C
	CUA		CCA		CAA	Gln Glutamine	CGA		A
	CUG		CCG		CAG		CGG		G
A	AUU	Ile Isoleucine	ACU	Thr Threonine	AAU	Asn Asparagine	AGU	Ser Serine	U
	AUC		ACC		AAC		AGC		C
	AUA		ACA		AAA	Lys Lysine	AGA	Arg Arginine	A
	AUG	Met Methionine; "Start"	ACG		AAG		AGG		G
G	GUU	Val Valine	GCU	Ala Alanine	GAU	Asp Aspartate	GGU	Gly Glycine	U
	GUC		GCC		GAC		GGC		C
	GUA		GCA		GAA	Glu Glutamate	GGA		A
	GUG		GCG		GAG		GGG		G

A codon consists of three nucleotides read in the sequence shown. For example, ACU codes for threonine. The first letter, A, is in the First Letter column; the second letter, C, is in the Second Letter column; and the third letter, U, is in the Third letter column. Many amino acids are specified by more than one codon. For example, threonine is specified by four codons, which differ only in the third nucleotide (ACU, ACC, ACA, and ACG).

The code is degenerate but specific

Some obvious features of the code jump out of table 15.1. First, 61 of the 64 possible codons are used to specify amino acids. Three codons, UAA, UGA, and UAG, are reserved for another function: they signal "stop" and are known as **stop codons.** The only other form of "punctuation" in the code is that AUG not only encodes the amino acid methionine (Met) but also is used to signal "start" and is therefore the **start codon.**

It is clear that 61 codons are more than enough to encode 20 amino acids. That leaves lots of extra codons. One way to deal with this abundance would be to use only 20 of the 61 codons, but that is not what cells do. In reality, all 61 codons are used, making the code **degenerate,** which means that some amino acids are specified by more than one codon. The reverse, however, in which a single codon would specify more than one amino acid, is never found.

The code is practically universal, but not quite

The genetic code is the same in almost all organisms. The universality of the genetic code is among the strongest evidence that all living things share a common evolutionary heritage. Because the code is universal, genes can be transferred from one organism to another and can be successfully expressed in their new host (figure 15.6). This universality of gene expression is central to many of the advances of genetic engineering discussed in chapter 17.

In 1979 investigators began to determine the complete nucleotide sequences of the mitochondrial genomes in humans, cattle, and mice. It came as something of a shock when these investigators learned that the genetic code used by these mammalian mitochondria was not quite the same as the "universal code" that had become so familiar to biologists.

In the mitochondrial genomes, what should have been a stop codon, UGA, was instead read as the amino acid tryptophan; AUA was read as methionine rather than as isoleucine; and AGA and AGG were read as stop codons rather than as arginine.

Figure 15.6 Transgenic pig. The piglet on the right is a conventional piglet. The piglet on the left was engineered to express a gene from jellyfish that encodes green fluorescent protein. The color of this piglet's nose and hooves is due to expression of this introduced gene. Such transgenic animals indicate the universal nature of the genetic code.
Steve Morse/University of Missouri Extension

Furthermore, minor differences from the universal code have also been found in the genomes of chloroplasts and in ciliates (certain types of protists).

Thus, it appears that the genetic code is not quite universal. Some time ago, presumably after they began their endosymbiotic existence, mitochondria and chloroplasts began to read the code differently, particularly the portion associated with "stop" signals.

REVIEW OF CONCEPT 15.2

The genetic code is a triplet code with no spaces. Sixty-one codons specify amino acids, 1 of which also codes for "start," and 3 codons encode "stop," for 64 total. Because some amino acids have more than one codon, the code is degenerate. Each codon encodes only one amino acid.

- *What would be the outcome if a codon specified more than one amino acid?*

15.3 Prokaryotes Exhibit All the Basic Features of Transcription

We begin an examination of gene expression by describing the process of transcription in prokaryotes. The later description of eukaryotic transcription will concentrate on their differences from prokaryotes.

Stages of Prokaryotic Transcription

LEARNING OBJECTIVE 15.3.1 Describe the transcription process in bacteria, identifying its unique features.

The single **RNA polymerase** of prokaryotes exists in two forms: the *core polymerase* and the *holoenzyme.* The core polymerase can synthesize RNA using a DNA template, but it cannot initiate synthesis accurately. The holoenzyme can accurately initiate synthesis.

The core polymerase is composed of four subunits: two identical α subunits, a β subunit, and a β' subunit (figure 15.7*a*). The two α subunits help to hold the complex together and can bind to regulatory molecules. The active site of the enzyme is formed by the β and β' subunits, which bind to the DNA template and the ribonucleotide triphosphate precursors.

The *holoenzyme* is formed by the addition of a σ (sigma) subunit to the core polymerase. Sigma's ability to recognize specific signal sequences in DNA allows the holoenzyme to locate the beginning of genes, critical to its function. Note that initiation of mRNA synthesis does not require a primer, in contrast to DNA replication.

Initiation

Accurate initiation of transcription requires two sites on DNA: one called a **promoter,** which forms a recognition and binding site for the RNA polymerase, and the actual **start site.** The polymerase also needs a signal to end transcription, which we call a **terminator.** We refer to the region from promoter to terminator as a **transcription unit.**

The action of the polymerase moving along the DNA can be thought of as analogous to water flowing in a stream. We can speak of sites on the DNA as being "upstream" or "downstream" of the start site. We can also use this comparison to form a simple system for numbering bases in DNA to refer to positions in the transcription unit. The first base transcribed is called **+1,** and this numbering continues downstream until the last base is transcribed. Any bases upstream of the start site receive negative numbers, starting at **−1.**

The promoter is a short sequence found upstream of the start site and is therefore not transcribed by the polymerase. Two six-base sequences are common to bacterial promoters: one is located 35 nucleotides upstream of the start site (−35), and the other is located 10 nucleotides upstream of the start site (−10; figure 15.7*b*). These two sites provide the promoter with asymmetry; they indicate not only the site of initiation but also the direction of transcription.

The binding of RNA polymerase to the promoter is the first step in transcription. Promoter binding is controlled by the σ subunit of the RNA polymerase holoenzyme, which recognizes the −35 sequence in the promoter and positions the RNA polymerase at the correct start site, oriented to transcribe in the correct direction.

Once bound to the promoter, the RNA polymerase begins to unwind the DNA helix at the −10 site (figure 15.7*b*). The polymerase covers a region of about 75 bp but unwinds only about 12 to 14 bp.

Elongation

In prokaryotes, the transcription of the RNA chain usually starts with ATP or GTP. One of these forms the 5′ end of the chain, which grows 5′-to-3′ as ribonucleotides are added to the 3′ end. As the RNA polymerase molecule leaves the promoter region, the σ factor is no longer required, although it may remain in association with the enzyme.

This process of leaving the promoter, called *clearance,* or *escape,* involves more than just synthesizing the first few nucleotides of the transcript and moving on. The enzyme made strong contacts with the DNA of the promoter region during initiation, and these contacts must be broken to enable the enzyme's progressive movement down the template. The enzyme goes through conformational changes during this clearance stage, and it subsequently contacts less of the DNA than it does during the initial promoter binding.

The region containing the RNA polymerase, the DNA template, and the growing RNA transcript is called the **transcription bubble** because it contains a locally unwound "bubble" of DNA (figure 15.8). The polymerase makes a variety of contacts with DNA, including about 18 bp downstream of the active site. The bubble itself is 12–14 nucleotides (nt), including an 8–10 bp RNA–DNA hybrid. Finally about 5 nt of the RNA contact the enzyme in an exit channel.

The transcription bubble created by RNA polymerase moves down the bacterial DNA at a constant rate of about 50 nt/sec, with the growing RNA moving out the exit channel. After the

Figure 15.7 Bacterial RNA polymerase and transcription initiation. ***a.*** RNA polymerase has two forms: core polymerase and holoenzyme. ***b.*** The σ subunit of the holoenzyme recognizes promoter elements at −35 and −10 and binds to the DNA. The helix is opened at the −10 region, and transcription begins at the start site at +1.

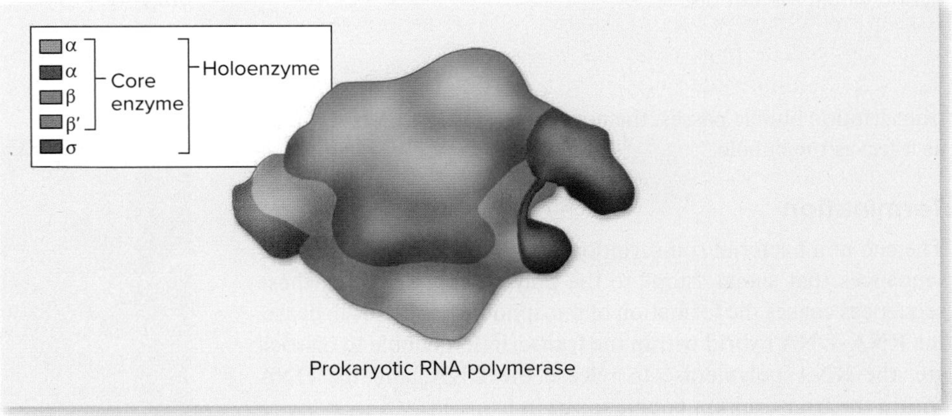

a.

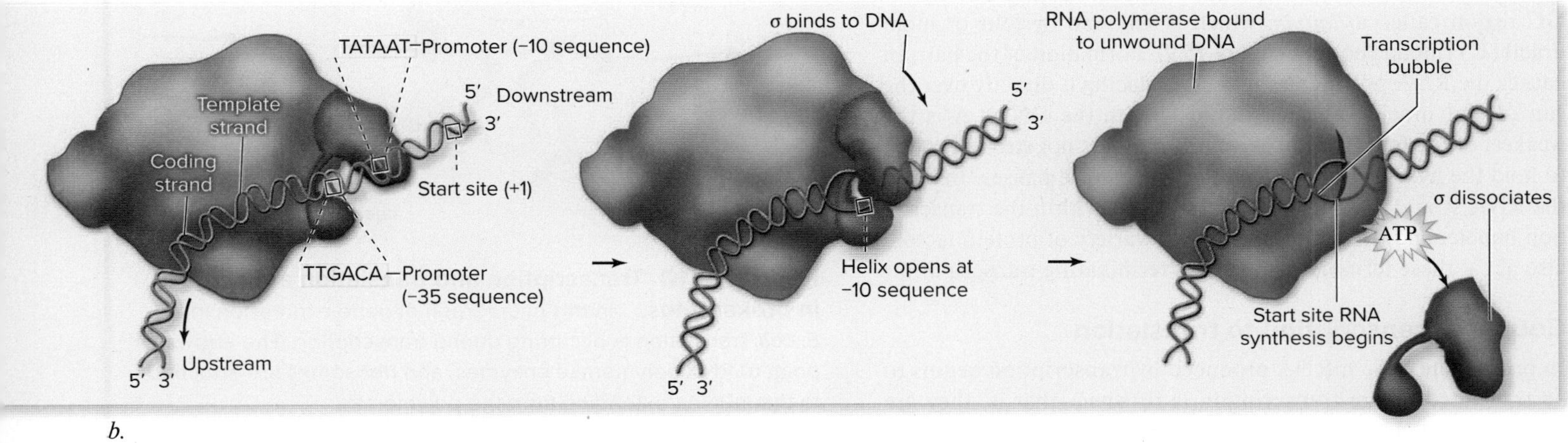

b.

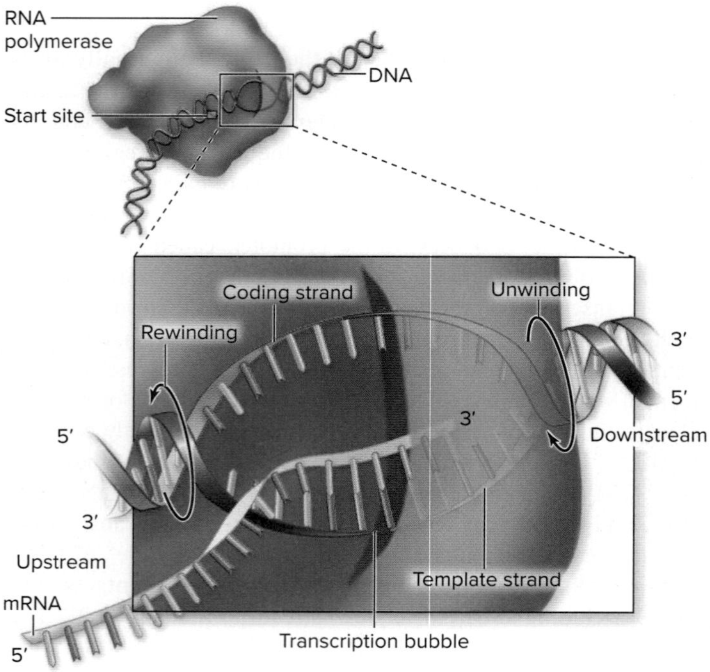

Figure 15.8 Model of a transcription bubble. The DNA duplex is unwound by the RNA polymerase complex, rewinding at the end of the bubble. One of the strands of DNA functions as a template, and nucleotide building blocks are added to the 3′ end of the growing RNA. There is a short region of RNA–DNA hybrid within the bubble.

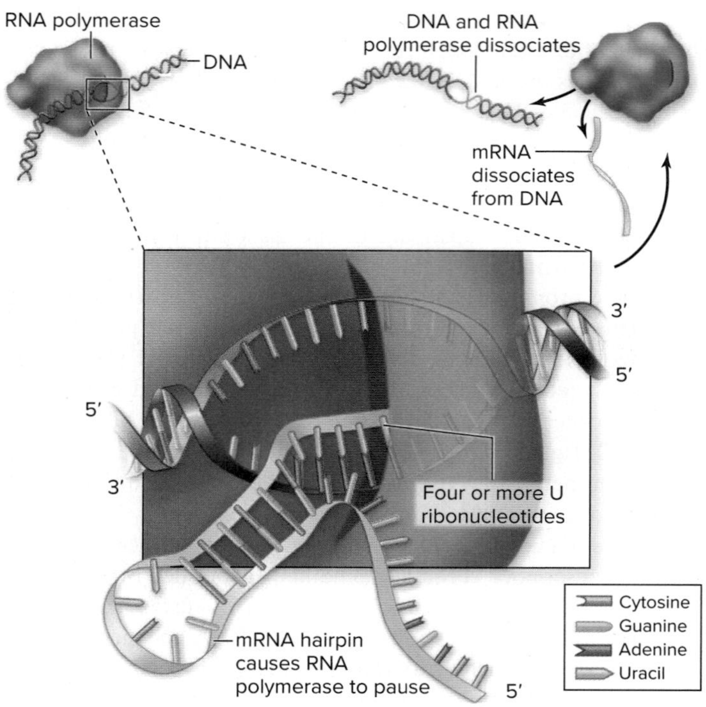

Figure 15.9 Bacterial transcription terminator. The self-complementary G–C region forms a double-stranded stem with a single-stranded loop called a hairpin. The stretch of U's forms a less stable RNA–DNA hybrid that falls off the enzyme.

transcription bubble passes, the now-transcribed DNA is rewound as it leaves the bubble.

Termination

The end of a bacterial transcription unit is marked by terminator sequences that signal "stop" to the polymerase. Reaching these sequences causes the formation of phosphodiester bonds to cease, the RNA-DNA hybrid within the transcription bubble to dissociate, the RNA polymerase to release the DNA, and the DNA within the transcription bubble to rewind.

The simplest terminators consist of a series of G-C base-pairs followed by a series of A-T base-pairs. The RNA transcript of this stop region can form a double-stranded structure in the GC region called a *hairpin,* which is followed by four or more uracil (U) ribonucleotides (figure 15.9). Formation of the hairpin causes the RNA polymerase to pause, placing it directly over the run of four uracils. The pairing of U with the DNA's A is the weakest of the four hybrid base-pairs, and it is not strong enough to hold the hybrid strands when the polymerase pauses. Instead, the RNA strand dissociates from the DNA within the transcription bubble, and transcription stops. A variety of protein factors also act at these terminators to aid in terminating transcription.

Coupling transcription to translation

In prokaryotes, the mRNA produced by transcription begins to be translated before transcription is finished—that is, they are *coupled* (figure 15.10). As soon as a 5′ end of the mRNA becomes

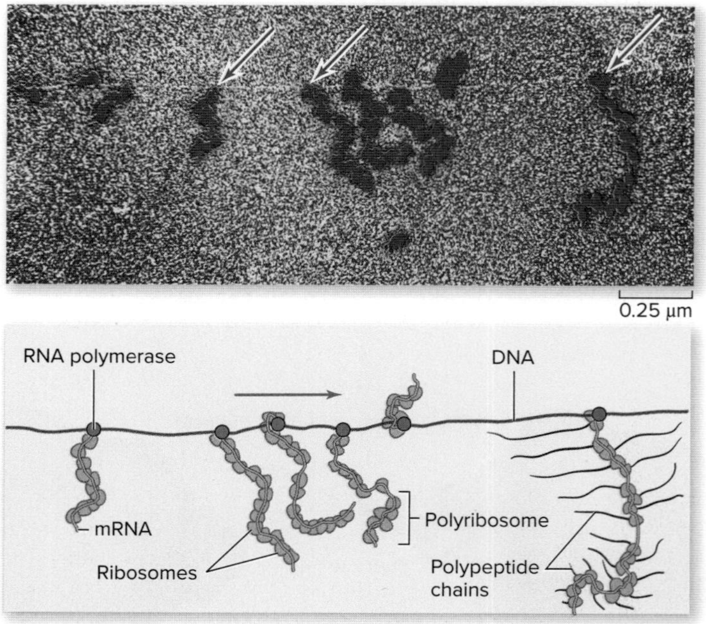

Figure 15.10 Transcription and translation are coupled in prokaryotes. In this micrograph of gene expression in *E. coli,* translation is occurring during transcription. The arrows point to RNA polymerase enzymes, and ribosomes are attached to the mRNAs extending from the polymerase.

Professor Oscar Miller/Science Source

available, ribosomes are loaded onto this to begin translation. This coupling cannot occur in eukaryotes, because transcription occurs in the nucleus and translation occurs in the cytoplasm.

Another difference between prokaryotic and eukaryotic gene expression is that the mRNA produced in prokaryotes may contain multiple genes. Prokaryotic genes are often organized such that genes encoding related functions are clustered together and transcribed onto the same mRNA molecule. This grouping of functionally related genes is referred to as an **operon.** An operon is a single transcription unit that encodes multiple enzymes. When genes are clustered in this fashion, functionally related genes can be regulated together by controlling transcription, a topic that we return to in chapter 16.

REVIEW OF CONCEPT 15.3

RNA polymerase in bacteria has two forms: a core polymerase and a holoenzyme. Holoenzyme can recognize promoter sequences and initiate transcription. During elongation the core enzyme adds RNA nucleotides based on the template. The enzyme stops at terminator sites and the transcript is released. In prokaryotes, translation begins before transcription is finished, so the processes are coupled.

- *Yeast are unicellular organisms like bacteria; would you expect them to have the same transcription/translation coupling?*

15.4 Eukaryotes Use Three Polymerases and Extensively Modify Transcripts

The basic mechanism of transcription by RNA polymerase is similar in all systems. In fact, the RNA polymerases of bacteria, archaea, and eukaryotes appear to be descended from the same ancestral enzyme. However, the details of RNA production in eukaryotes differ enough to merit consideration.

Eukaryotes Have Three RNA Polymerases

LEARNING OBJECTIVE 15.4.1 Explain how the three eukaryotic RNA polymerases differ in their functions.

An obvious difference between bacteria and eukaryotes is that bacteria have a single RNA polymerase, but eukaryotes have three distinct RNA polymerases, with different cellular roles. The enzyme **RNA polymerase I** transcribes rRNA, **RNA polymerase II** transcribes mRNA and some small nuclear RNAs, and **RNA polymerase III** transcribes tRNA and some other small RNAs. Together, these three enzymes accomplish all transcription in eukaryotic cells.

The three eukaryotic RNA polymerases require different control elements in the DNA to allow each polymerase to recognize where to initiate transcription. Each polymerase recognizes a different promoter structure.

RNA polymerase I promoters

RNA polymerase I promoters at first puzzled biologists, because comparisons of rRNA genes between species showed no similarities outside the coding region. The current view is that these promoters are also specific for each species, which explains the lack of conserved sequences in cross-species comparisons.

RNA polymerase II promoters

The RNA polymerase II promoters are the most complex of the three types, probably a reflection of the huge diversity of genes that are transcribed by this polymerase. When the first eukaryotic genes were isolated, many had a sequence called the **TATA box** upstream of the start site. This sequence was similar to the prokaryotic −10 sequence, and it was assumed that the TATA box was the primary promoter element. With the sequencing of entire genomes, many more genes have been analyzed, and this assumption has proved too simple. It has been replaced by the idea of a "core promoter" that can be composed of a number of different elements, including the TATA box. Additional control elements allow for tissue-specific and developmental time-specific expression, as discussed in chapter 16.

RNA polymerase III promoters

Promoters for RNA polymerase III also were a source of surprise for biologists examining the control of eukaryotic gene expression. A common technique for analyzing regulatory regions is to make successive deletions from the 5′ end of genes until enough is deleted to abolish specific transcription. This works well with prokaryotes, in which the regulatory regions are always found at the 5′ end of genes. But in the case of eukaryotic tRNA genes, the 5′ deletions have no effect on expression! The promoters were found to actually be internal to the gene itself. This has not proved to be the case for all polymerase III genes, but it appears to be for most.

Polymerase II Initiation and Termination Requires Other Factors

LEARNING OBJECTIVE 15.4.2 Contrast initiation of transcription in eukaryotes and prokaryotes.

The initiation at RNA polymerase II promoters is analogous to prokaryotic initiation, but eukaryotes use a host of general **transcription factors** instead of a single factor for promoter recognition. The transcription factors interact with RNA polymerase II to form an initiation complex at the promoter (figure 15.11) that is necessary for transcription to occur. Other specific transcription factors control the level of transcription in different tissues and at different developmental times, as we will explore in chapter 16.

During early elongation, the polymerase may pause within 50 bp of initiation. Global screens of promoter occupancy indicate that a majority of genes in multicellular organisms may harbor a paused polymerase. It is not yet clear how much of this is due to abortive short transcripts and how much to paused polymerases that can restart. Release from pausing involves a complex interplay of promoter elements, elongation factors, and nucleosome positioning.

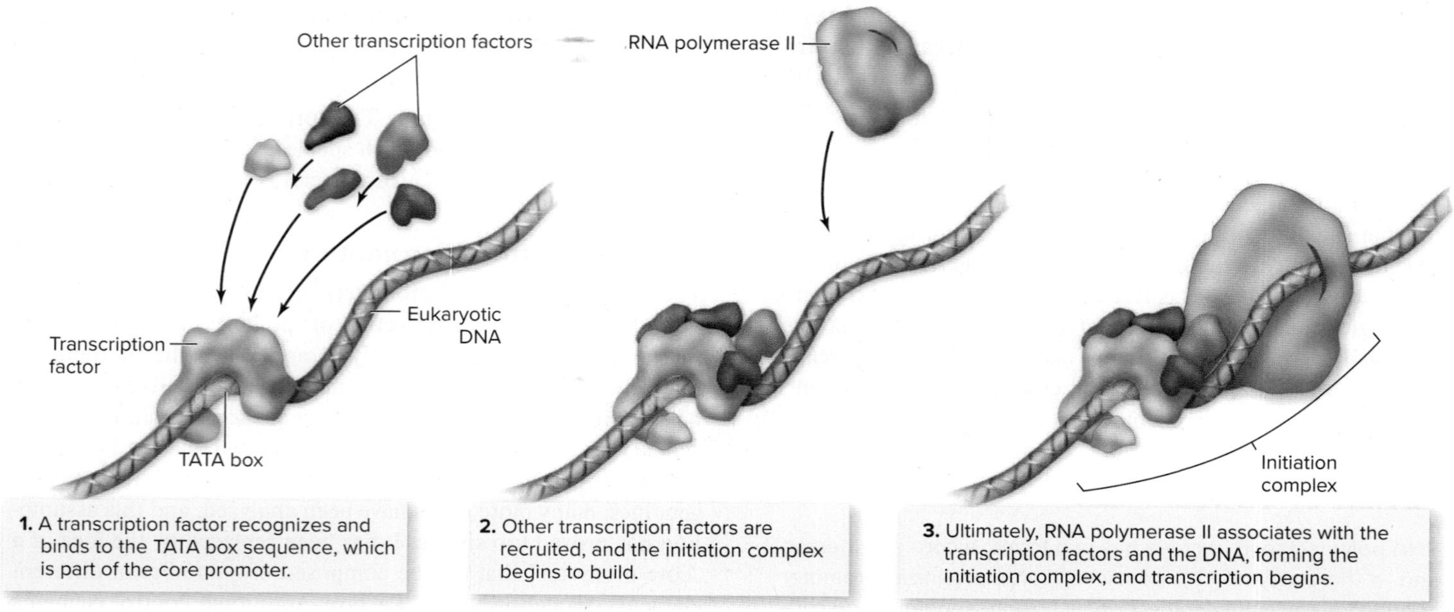

Figure 15.11 Eukaryotic initiation complex. Unlike transcription in prokaryotic cells, in which the RNA polymerase recognizes and binds to the promoter, eukaryotic transcription requires the binding of transcription factors to the promoter before RNA polymerase II binds to the DNA. The association of transcription factors and RNA polymerase II at the promoter is called the initiation complex.

The RNA polymerase recruits elongation factors through its *carboxyl terminal domain* (*CTD*). This domain contains repeated units of amino acids that can be phosphorylated. The phosphorylated CTD binds to elongation factors to form the transcription elongation complex. The phosphorylated CTD also helps recruit the modifying enzymes discussed in the next section.

The termination of transcription for RNA polymerase II also differs from prokaryotic termination. Although Pol II termination sites exist, they are not well defined, and the end of the mRNA is also not even formed by RNA polymerase II because of modifications to the primary transcript described in section 5.5.

Eukaryotic Transcripts Are Modified

LEARNING OBJECTIVE 15.4.3 Describe how eukaryotic RNA transcripts are modified.

A primary difference between prokaryotes and eukaryotes is the fate of the transcript itself. Prokaryotes translate the mRNA during transcription, but eukaryotes extensively modify the transcript in the nucleus before its translation in the cytoplasm. We call the RNA synthesized by RNA polymerase II the **primary transcript,** which is processed to produce the **mature mRNA.**

The 5′ cap

When the transcript reaches about 20 nt, it is modified by the addition of GTP to the 5′ PO_4^- group, forming what is known as the **5′ cap** (figure 15.12). This cap is joined to the transcript by its 5′ end and is the only such 5′-to-5′ bond found in nucleic acids. The G in the GTP is also modified by the addition of a methyl group, so it is often called a *methyl-G cap.* The cap is added while transcription is still in progress. This cap protects the 5′ end of the mRNA from degradation and participates in translation initiation.

The 3′ poly-A tail

Unlike in prokaryotes, in eukaryotes the end of the transcript is not the end of the mRNA. The eukaryotic transcript is cleaved downstream of a specific site prior to the termination site called the polyadenylation signal sequence (AAUAAA). A series of adenine (A) residues, called the **3′poly-A tail,** is added after this cleavage by the enzyme poly-A polymerase. Thus, the end of the mRNA is not created by RNA polymerase II (see figure 15.12).

The enzyme poly-A polymerase is part of a complex that recognizes the polyadenylation signal sequence, cleaves the transcript, then adds 100–200 A's to the end. The poly-A tail appears to play a role in the stability of mRNAs by protecting them from degradation (refer to chapter 16).

Splicing of primary transcripts

Eukaryotic genes may contain noncoding sequences that have to be removed to produce the final mRNA. This process, called pre-mRNA splicing, is accomplished by a macromolecular machine called the *spliceosome.* This complex topic is discussed in section 15.5.

REVIEW OF CONCEPT 15.4

Eukaryotes have three RNA polymerases: RNA Pol I, II, and III. Each synthesizes a different RNA and recognizes its own promoter. The RNA Pol I promoter is species-specific. The Pol II promoter is complex but often includes a TATA box. The Pol III promoter is internal to the gene, not at the 5′ end. Pol II is responsible for mRNA synthesis. The primary transcript is modified with a 5′ cap and a 3′ poly-A tail consisting of 100–200 adenines. Noncoding regions are removed by splicing.

- *Does the complexity of the eukaryotic genome require three polymerases?*

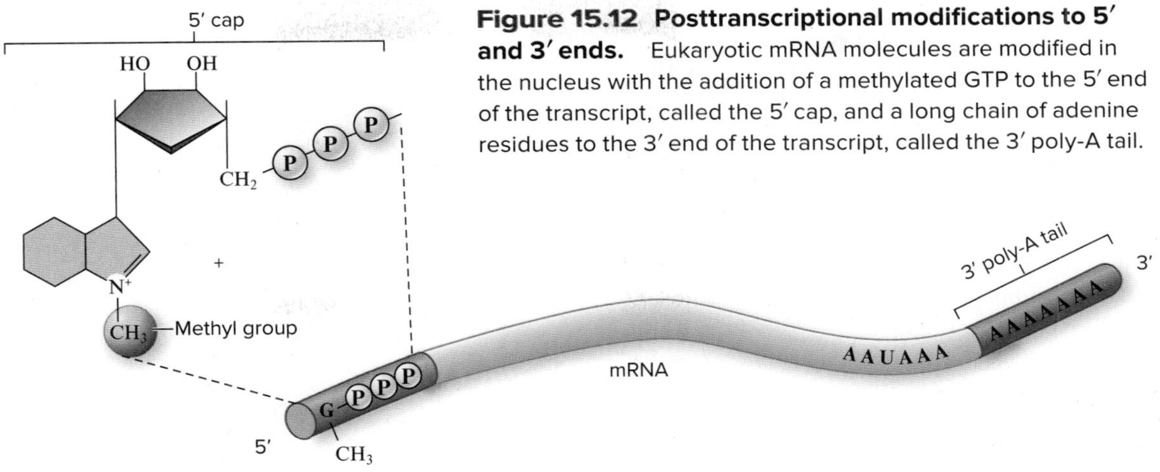

Figure 15.12 Posttranscriptional modifications to 5′ and 3′ ends. Eukaryotic mRNA molecules are modified in the nucleus with the addition of a methylated GTP to the 5′ end of the transcript, called the 5′ cap, and a long chain of adenine residues to the 3′ end of the transcript, called the 3′ poly-A tail.

15.5 Eukaryotic Genes May Contain Noncoding Sequences

The first genes successfully isolated were prokaryotic genes found in *E. coli* and its viruses. A clear picture of the nature and some of the control of gene expression emerged from these systems before any eukaryotic genes were isolated. It was assumed that although details would differ, the outline of gene expression in eukaryotes would be similar.

The world of biology was in for a shock with the isolation of the first genes from eukaryotic organisms: they appeared to contain within them sequences that were not represented in the mRNA! It is hard to exaggerate how unexpected this finding was. A basic tenet of molecular biology based on *E. coli* was that a gene is *colinear* with its protein product, that is, the sequence of bases in the gene corresponds to the sequence of bases in the mRNA, which in turn corresponds to the sequence of amino acids in the protein.

In the case of eukaryotes, it turns out that many genes are interrupted by sequences not represented in the mRNA and the protein. We call the noncoding DNA that interrupts the sequence of the gene "intervening sequences," or **introns,** and we call the coding sequences **exons,** because they are expressed (figure 15.13).

RNA Splicing Is Carried Out by the Spliceosome

LEARNING OBJECTIVE 15.5.1 Explain how the spliceosome processes a primary transcript.

It is still true that the mature eukaryotic mRNA is colinear with its protein product, but a eukaryotic gene that contains introns is not. Imagine looking at an interstate highway from a satellite. Scattered randomly along the thread of concrete would be cars, some moving in clusters, others individually; most of the road would be bare. That is what a eukaryotic gene is like—scattered exons embedded within much longer sequences of introns.

In humans, only 1 to 1.5% of the genome is devoted to the exons that encode proteins; 24% is devoted to the noncoding introns within which these exons are embedded.

The splicing reaction

The obvious question is, How do eukaryotic cells deal with the noncoding introns? The answer is that the primary transcript

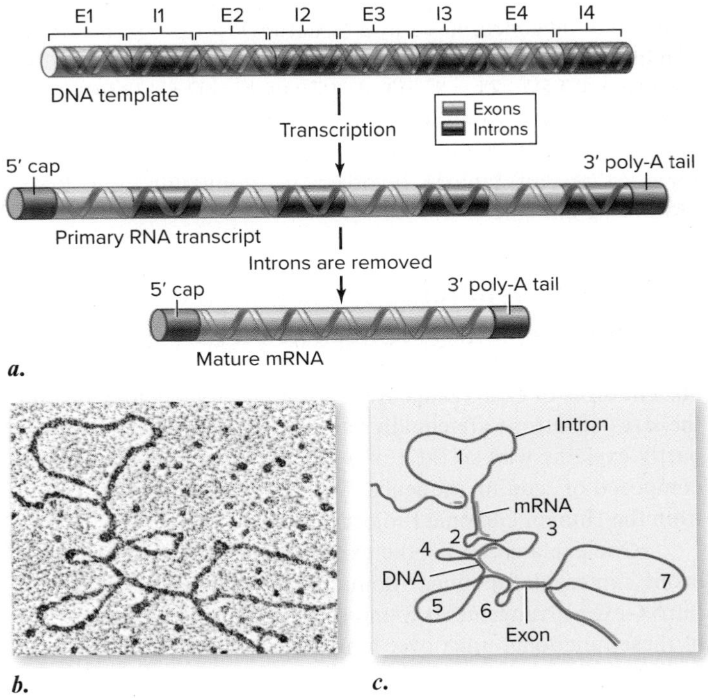

Figure 15.13 Eukaryotic genes contain introns and exons. ***a.*** Eukaryotic genes contain sequences that form the coding sequence called exons and intervening sequences called introns. ***b.*** An electron micrograph showing hybrids formed with the mRNA and the DNA of the ovalbumin gene, which has seven introns. Introns within the DNA sequence have no corresponding sequence in the mRNA and thus appear as seven loops. ***c.*** A schematic drawing of the micrograph.

(b) Dr. Bert O'Malley

is cut and put back together to produce the mature mRNA. The latter process is referred to as **pre-mRNA splicing,** and it occurs in the nucleus prior to the export of the mRNA to the cytoplasm.

The intron–exon junctions are recognized by **small nuclear ribonucleoprotein particles,** called **snRNPs** (pronounced "snurps"). The snRNPs are complexes composed of snRNA and protein. These snRNPs then cluster together with other associated proteins to form a larger complex called the **spliceosome,** which is responsible for the splicing, or removal, of the introns.

For splicing to occur accurately, the spliceosome must be able to recognize intron–exon junctions. Introns all begin with the same two-base sequence and end with another two-base sequence that tags them for removal. In addition, within the intron is a conserved A nucleotide, called the *branch point,* which is important for the splicing reaction (figure 15.14).

The splicing process begins with cleavage of the 5′ end of the intron. This 5′ end becomes attached to the 2′ OH of the branch point A, forming a branched structure called a *lariat* due to its resemblance to the loop of rope in a cowboy's lariat, as illustrated in figure 15.14. The 3′ end of the first exon is then used to displace the 3′ end of the intron, joining the two exons together and releasing the intron as a lariat.

The processes of transcription and RNA processing do not occur in a linear sequence but, rather, are all part of a concerted process that produces the mature mRNA. This appears to be coordinated by a structural feature of the largest subunit of RNA polymerase II, called the carboxyl terminal domain (CTD). The CTD consists of seven amino acids repeated many times (52 repeats in humans). The CTD contains serines that can be phosphorylated, and it acts as a platform to recruit factors involved in elongation and RNA-modifying enzymes.

Distribution of introns

No rules govern the number of introns per gene or the sizes of introns and exons. Some genes have no introns; others may have 50. The sizes of exons range from a few nucleotides to 7500, and the sizes of introns are equally variable. The presence of introns partly explains why so little of a eukaryotic genome is actually composed of "coding sequences" (refer to chapter 18 for results from the Human Genome Project).

One explanation for the existence of introns suggests that exons represent functional domains of proteins, and that the intron–exon arrangements found in genes represent the shuffling of these functional units over long periods of evolutionary time. This hypothesis, called *exon shuffling,* was proposed soon after the discovery of introns and has been the subject of much debate over the years.

The recent flood of genomic data has shed light on this issue by allowing statistical analysis of the placement of introns and of intron–exon structure. This analysis has provided support for the exon shuffling hypothesis for many genes; however, it is also clearly not universal, because not all proteins show this kind of pattern. It is possible that introns do not have a single origin, and therefore cannot be explained by a single hypothesis.

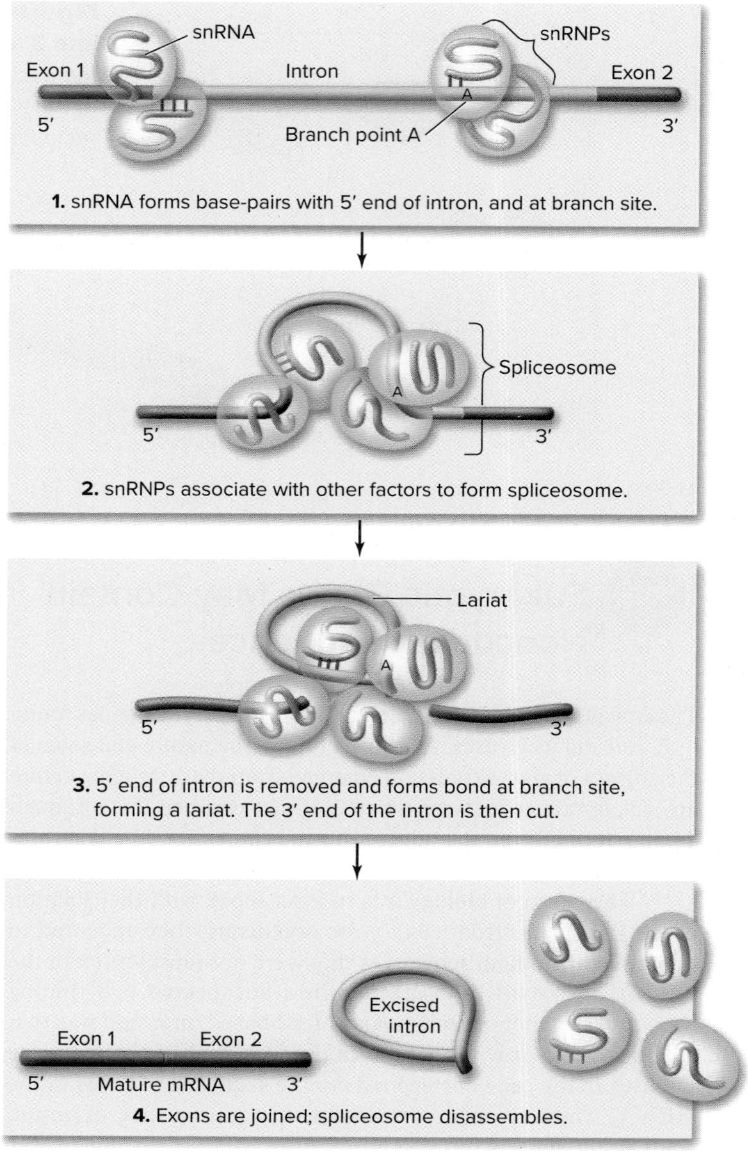

snRNA
snRNPs
Exon 1
Intron
Exon 2
5′
Branch point A
3′
1. snRNA forms base-pairs with 5′ end of intron, and at branch site.
Spliceosome
5′
3′
2. snRNPs associate with other factors to form spliceosome.
Lariat
5′
3′
3. 5′ end of intron is removed and forms bond at branch site, forming a lariat. The 3′ end of the intron is then cut.
Excised intron
Exon 1
Exon 2
5′
Mature mRNA
3′
4. Exons are joined; spliceosome disassembles.

Figure 15.14 Pre-mRNA splicing by the spliceosome. Particles called snRNPs contain snRNA that interacts with the 5′ end of an intron and with a branch site internal to the intron. Several snRNPs come together with other proteins to form the spliceosome. As the intron forms a loop, the 5′ end is cut and linked to a site near the 3′ end of the intron. The intron forms a lariat, which is excised, and the exons are spliced together. The spliceosome then disassembles and releases the spliced mRNA.

Alternative Splicing Can Produce Multiple Transcripts from the Same Gene

LEARNING OBJECTIVE 15.5.2 Explain how eukaryotes can produce many more proteins than they have genes.

We call all of the RNAs produced from a genome the **transcriptome,** and all the proteins produced from a genome the **proteome.** If every gene were spliced to include all exons, then the number of genes, transcripts, and proteins would all be the same. Unfortunately, it is not so simple: a single primary transcript can be spliced into different mRNAs by using different sets of exons, a process called **alternative splicing.**

Before the human genome project (refer to chapter 18), it was estimated that humans had 100,000 genes. The current number is down to around 20,000 genes that encode proteins, but more than 80,000 protein-encoding transcripts have been identified, which implies a high level of alternative splicing.

Using high-throughput systems to sequence cellular RNAs, more than 200,000 transcripts have been detected, leading to estimates of up to 95% of human genes producing multiple splice products. Estimates of the average number of transcripts per gene vary from as many as 10 to as few as 1.5 per gene. A conservative estimate is 85,000 possible protein-encoding transcripts, but only a minority of these are experimentally verified.

Other observations indicate that 80% of protein-coding genes appear to have a single predominant transcript, although this predominant transcript and minor transcripts can switch with different cell and tissue types. So, while there are well-characterized examples of alternative splicing producing different functional proteins, the overall biological significance remains unclear.

It is clear that the transcriptome is much more diverse than one gene equals one transcript, but it is not clear to what extent alternative splicing contributes to proteome diversity. The actual size of the proteome is also not yet clear, although progress is being made using mass spectrometry (refer to chapter 18).

REVIEW OF CONCEPT 15.5

Unlike prokaryotes, genes in eukaryotes are interrupted. They consist of exons, or expressed regions, and introns, or intervening sequences. The introns are removed by the spliceosome in a process that joins two exons. Alternative splicing can generate different mRNAs, and thus different proteins, from the same gene.

- *What advantages would alternative splicing confer on an organism?*

15.6 The Ribosome Is the Machine of Protein Synthesis

RNA is the key actor in a cell's translation of its genetic message. The translation of the nucleotide sequence of DNA into the amino acid sequence of a protein requires the participation of mRNA, rRNA, tRNA, and a host of other factors. Even the formation of the peptide bond in the protein assembly process turns out to be an RNA-catalyzed process. The amino acid assembly process had been traditionally assumed to be catalyzed by proteins within the ribosome, with the rRNA acting as a scaffold to properly position the proteins. Powerful X-ray diffraction studies have recently revealed the reverse to be true: it is the ribosome's RNA, not its proteins, that catalyzes the joining together of amino acids.

Critical to the process of translation is the interaction of the ribosomes with tRNA. To understand this, we first examine the structure of tRNA, the tRNA adapter molecule, and the ribosome itself.

tRNA Is a Bifunctional Molecule

LEARNING OBJECTIVE 15.6.1 Describe how the two ends of a tRNA differ functionally.

tRNA

Transfer RNA is a bifunctional molecule that must be able to interact with both mRNA and amino acids. The structure of tRNAs is highly conserved in all living systems. The tRNA molecule can be folded into a two-dimensional cloverleaf type of structure, as a result of intramolecular base-pairing that produces double-stranded regions. This basic structure is then folded in space to form an L-shaped three-dimensional molecule that has two functional ends: the **acceptor stem** and the **anticodon loop** (figure 15.15).

The acceptor stem is the 3′ end of the molecule, which always ends in 5′ CCA 3′. The amino acid is attached to this end of the molecule. The anticodon loop is the bottom loop of the cloverleaf, and it can base-pair with codons in mRNA.

Activating Enzymes Attach Amino Acids to tRNA

LEARNING OBJECTIVE 15.6.2 Explain why activating enzymes are said to be the cell's translators of the genetic code.

For protein synthesis to proceed, each amino acid must be attached to a tRNA with the correct anticodon. This covalent attachment is accomplished by the action of **activating enzymes,** more formally called **aminoacyl-tRNA synthetases.** One of these activating enzymes is present for each of the 20 common amino acids.

The charging reaction

The aminoacyl-tRNA synthetases must be able to recognize specific tRNA molecules as well as their corresponding amino acids. Although 61 codons code for amino acids, there are actually not 61 tRNAs in cells, although the number varies from species to species. Therefore, some aminoacyl-tRNA synthetases must be able to recognize more than one tRNA—but each recognizes only a single amino acid.

The reaction catalyzed by the activating enzymes is called the tRNA **charging reaction,** and the product is an amino acid joined to a tRNA, now called a *charged tRNA.* An ATP molecule provides energy for this endergonic reaction. The charged tRNA produced by the reaction is an activated intermediate that can undergo the peptide-bond-forming reaction without an additional input of energy.

The charging reaction joins the acceptor stem of tRNA to the carboxyl terminus of an amino acid (figure 15.16). Keeping this directionality in mind is critical to understanding the function of the ribosome, because each peptide bond will be formed between the amino group of one amino acid and the carboxyl group of another amino acid.

The correct attachment of amino acids to tRNAs is important, because the ribosome does not verify this attachment. Ribosomes can only ensure that the codon–anticodon pairing is correct. In an elegant experiment, cysteine was converted chemically to alanine after the charging reaction, when the amino acid

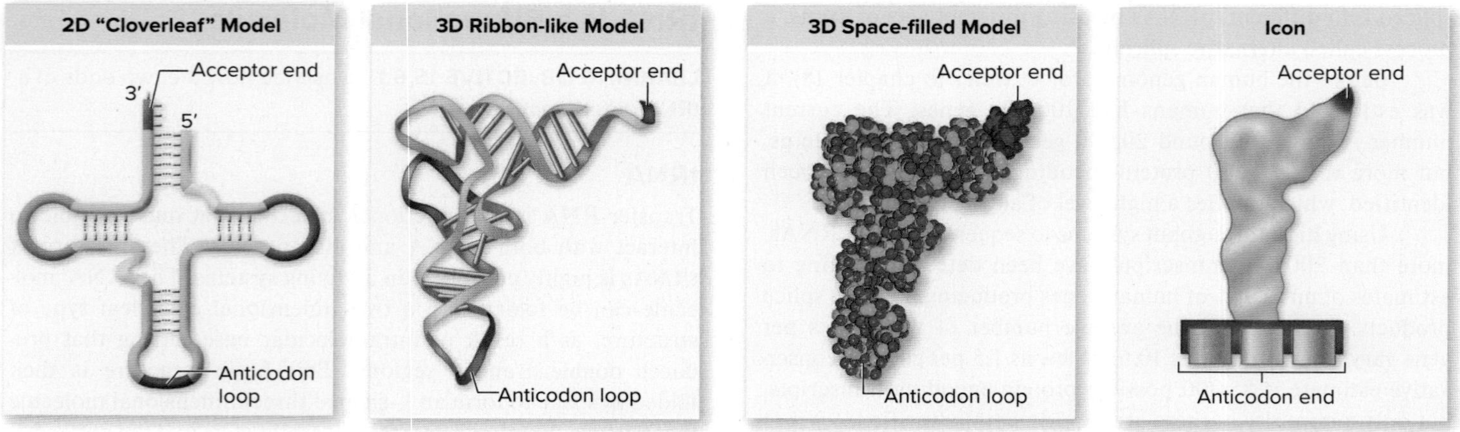

Figure 15.15 The structure of tRNA. Base-pairing within the molecule creates three stem-and-loop structures in a characteristic cloverleaf shape. The loop at the bottom of the cloverleaf contains the anticodon sequence, which can base-pair with codons in the mRNA. Amino acids are attached to the free, single-stranded –OH end of the acceptor stem. In its final three-dimensional structure, the loops of tRNA are folded into the final L-shaped structure.

(3D Space-filled model): Image of PDB ID 2TRA (Westhof, E., Dumas, P., Moras, D. (1988) Restrained refinement of two crystalline forms of yeast aspartic acid and phenylalanine transfer RNA crystals. *Acta Crystallogr.*, Sect.A 44: 112-123) created by John Beaver using ProteinWorkshop, a product of the RCSB PDB, and built using the Molecular Biology Toolkit developed by John Moreland and Apostol Gramada (mbt.sdsc.edu). The MBT is financed by grant GM63208

was already attached to tRNA. When this charged tRNA was used in an in vitro protein synthesis system, alanine was incorporated in the place of cysteine, showing that the ribosome cannot "proofread" the amino acids attached to tRNA.

In a very real sense, therefore, it is the activating enzymes that translate the genetic code, their charging reactions matching tRNA anticodon to amino acid. Afterward, amino acids are incorporated into a peptide based solely on the tRNA anticodon's complementary interaction with the mRNA.

The Ribosome Has Multiple tRNA-Binding Sites

LEARNING OBJECTIVE 15.6.3 Differentiate between the functions of different tRNA-binding sites on the ribosome.

The synthesis of any biopolymer can be broken down into initiation, elongation, and termination. In the case of translation, or protein synthesis, all three of these steps take place on the

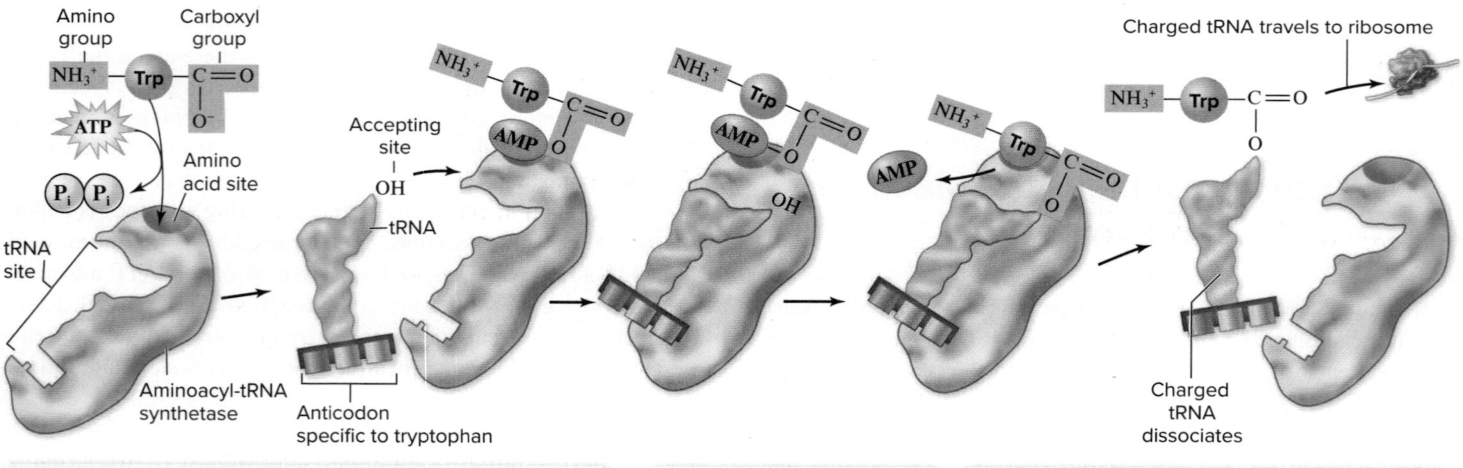

1. In the first step of the reaction, the amino acid is activated. The amino acid reacts with ATP to produce an intermediate with the carboxyl end of the amino acid attached to AMP. The two terminal phosphates (pyrophosphates) are cleaved from ATP in this reaction.
2. The amino acid-AMP complex remains bound to the enzyme. The tRNA next binds to the enzyme.
3. The second step of the reaction transfers the amino acid from AMP to the tRNA, producing a charged tRNA and AMP. The charged tRNA consists of a specific amino acid attached to the 3′ acceptor stem of its RNA.

Figure 15.16 tRNA charging reaction. There are 20 different aminoacyl-tRNA synthetase enzymes, each specific for one amino acid, such as tryptophan (Trp). The enzyme must also recognize and bind to the tRNA molecules with anticodons specifying that amino acid, ACC for tryptophan. The reaction uses ATP and produces an activated intermediate that will not require further energy for peptide bond formation.

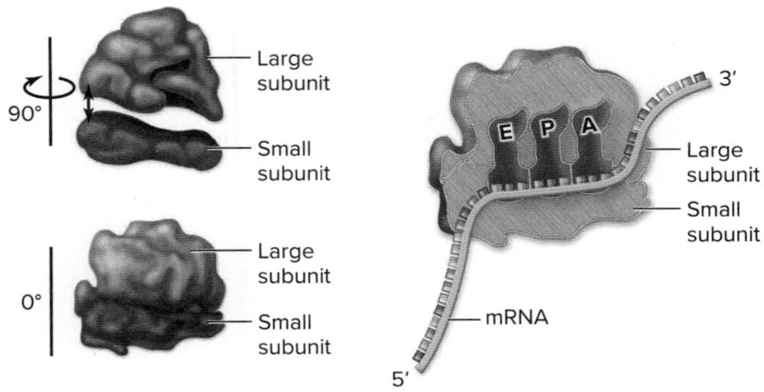

Figure 15.17 Ribosomes have two subunits. Ribosome subunits come together and apart as part of a ribosome cycle. The smaller subunit fits into a depression on the surface of the larger one. Ribosomes have three tRNA-binding sites: aminoacyl site (A), peptidyl site (P), and exit site (E).

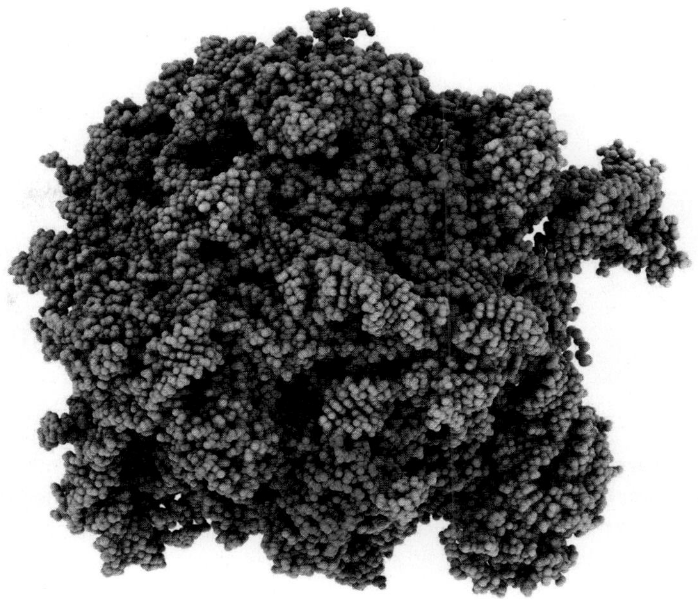

Figure 15.18 3-D structure of eukaryotic ribosome. The complete atomic structure of the yeast large ribosomal subunit is shown. The ribosomal RNA is beige, blue, and pale green; all other colors are different ribosomal proteins. The faces of each ribosomal subunit are lined with rRNA such that their interaction with tRNAs, amino acids, and mRNA all involve rRNA. Proteins are absent from the active site but abundant everywhere on the surface. The proteins stabilize the structure by interacting with adjacent RNA strands.

Science Photo Library/age fotostock

ribosome—a large, macromolecular assembly consisting of rRNA and proteins.

For the ribosome to function, it must be able to bind to at least two charged tRNAs at once so that a peptide bond can be formed between their amino acids, as described in the previous overview. The bacterial ribosome contains three binding sites, summarized in figure 15.17:

- The **A site** (aminoacyl) binds to the tRNA carrying the next amino acid to be added.
- The **P site** (peptidyl) binds to the tRNA attached to the growing peptide chain.
- The **E site** (exit) binds to the tRNA that previously carried the growing peptide chain.

Transfer RNAs move through these sites successively during the process of elongation. Relative to the mRNA, the sites are arranged 5′ to 3′ in the order E, P, and A. The incoming charged tRNAs enter the ribosome at the A site, transit through the P site, and then leave via the E site.

The ribosome has both decoding and enzymatic functions

The two functions of the ribosome involve (1) decoding the transcribed message and (2) forming peptide bonds. The decoding function resides primarily in the small subunit of the ribosome. The formation of peptide bonds requires the enzyme **peptidyl transferase,** which resides in the large subunit.

Our view of the ribosome has changed dramatically over time. Initially, molecular biologists assumed that the proteins in the ribosome carried out these decoding and catalytic functions. Now this view has been revised. The ribosome is seen instead as an assembly of rRNAs, held in place by proteins, that carry out the key chemical reactions. The faces of the two subunits that interact with each other are lined with rRNA, and the parts of both subunits that interact with mRNA, tRNA, and amino acids are also primarily rRNA (figure 15.18). It is now thought that the peptidyl transferase activity resides in an rRNA in the large subunit.

REVIEW OF CONCEPT 15.6

One end of tRNA can bond with amino acids, and one end can base-pair with mRNA. The tRNA charging reaction joins the carboxyl end of an amino acid to the 3′ acceptor stem of its tRNA. This reaction is catalyzed by 20 aminoacyl-tRNA synthetases, one for each amino acid. The ribosome has three binding sites for tRNA, one for the tRNA attached to the growing peptide (P site), one for the next charged tRNA (A site), and one for the previous tRNA (E site). The ribosome has both a decoding function and an enzymatic function. Function is now thought to reside primarily in the rRNA.

- *What would be the effect on translation of a mutant tRNA that had an anticodon complementary to a stop codon?*

15.7 The Process of Translation Is Complex and Energy-Expensive

The process of translation is one of the most complex and energy-expensive tasks that cells perform, although the basic process is simple: an mRNA molecule is threaded through a ribosome, where tRNAs carrying amino acids interact with the mRNA by base-pairing with the mRNA's codons. The ribosome positions

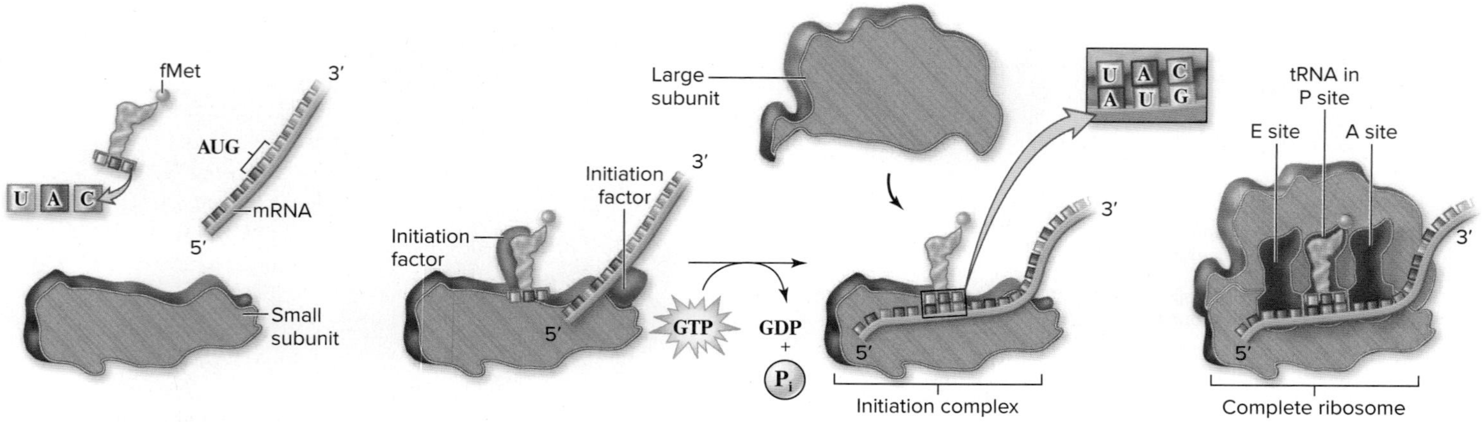

Figure 15.19 Initiation of translation. In prokaryotes, initiation factors play key roles in positioning the small ribosomal subunit, the initiator $tRNA^{fMet}$, and the mRNA. When the $tRNA^{fMet}$ is positioned over the first AUG codon of the mRNA, the large ribosomal subunit binds, forming the E, P, and A sites where successive tRNA molecules bind to the ribosomes, and polypeptide synthesis begins.

the amino acids such that peptide bonds can be formed between each new amino acid and the growing polypeptide.

Initiation Requires Accessory Factors

LEARNING OBJECTIVE 15.7.1 Contrast initiation in prokaryotes and eukaryotes.

Prokaryotic initiation

In prokaryotes, the **initiation complex** includes a special **initiator tRNA** molecule charged with a chemically modified methionine, N-*formylmethionine.* The initiator tRNA is shown as $tRNA^{fMet}$. The initiation complex also includes the small ribosomal subunit and the mRNA strand (figure 15.19). The small subunit is positioned correctly on the mRNA due to a conserved sequence in the 5′ end of the mRNA called the **ribosome-binding sequence (RBS)** that is complementary to the 3′ end of a small subunit rRNA. Also called the Shine–Dalgarno sequence for the discoverers, John Shine and Lynn Dalgarno, it is usually found about 8 bp upstream of the AUG in bacterial and archaeal mRNAs.

As mentioned when considering the genetic code, the start codon is AUG. The ribosome usually uses the first AUG it encounters in an mRNA strand to signal the start of translation. A number of initiation factors mediate this interaction of the ribosome, mRNA, and $tRNA^{fMet}$ to form the initiation complex. These factors are involved in initiation only and are not part of the ribosome.

Once the complex of mRNA, initiator tRNA, and small ribosomal subunit is formed, the large subunit is added, and translation can begin. With the formation of the complete ribosome, the initiator tRNA is bound to the P site with the A site empty.

Eukaryotic initiation

Initiation in eukaryotes is similar, although it differs in two important ways. First, in eukaryotes the initiating amino acid is methionine rather than *N*-formylmethionine. Second, the initiation complex is far more complicated than in prokaryotes, containing nine or more protein factors, many consisting of several subunits. Eukaryotic mRNAs also lack an RBS. The small subunit instead binds initially to the mRNA by binding to the 5′ cap at the end of the modified mRNA.

Elongation Adds Successive Amino Acids

LEARNING OBJECTIVE 15.7.2 Indicate in what order the A, E, and P sites of a ribosome are occupied by each tRNA.

When the entire ribosome is assembled around the initiator tRNA and mRNA, the second charged tRNA can be brought to the ribosome and bind to the empty A site. This requires an **elongation factor** called **EF-Tu,** which binds to the charged tRNA and to GTP.

A peptide bond can then form between the amino acid of the initiator tRNA and the newly arrived charged tRNA in the A site. The geometry of this bond relative to the two charged tRNAs is critical to understanding the process. Remember that an amino acid is attached to a tRNA by its carboxyl terminus. The peptide bond is formed between the amino end of the incoming amino acid (in the A site) and the carboxyl end of the growing chain (in the P site; see figure 15.21).

The addition of successive amino acids is a series of events that occur in a cyclic fashion. Figure 15.20 shows the details of the elongation cycle.

1. **Matching tRNA anticodon with mRNA codon.** Each new charged tRNA comes to the ribosome bound to EF-Tu and GTP. The charged tRNA binds to the A site if its anticodon is complementary to the mRNA codon in the A site.

   After binding, GTP is hydrolyzed, and EF-Tu-GDP dissociates from the ribosome, where it is recycled by another factor. This two-step binding and hydrolysis of GTP is thought to increase the accuracy of translation.

2. **Peptide bond formation.** Peptidyl transferase, located in the large subunit, catalyzes the formation of a peptide bond between the amino group of the amino acid in the A site and the carboxyl group of the growing chain (figure 15.21).

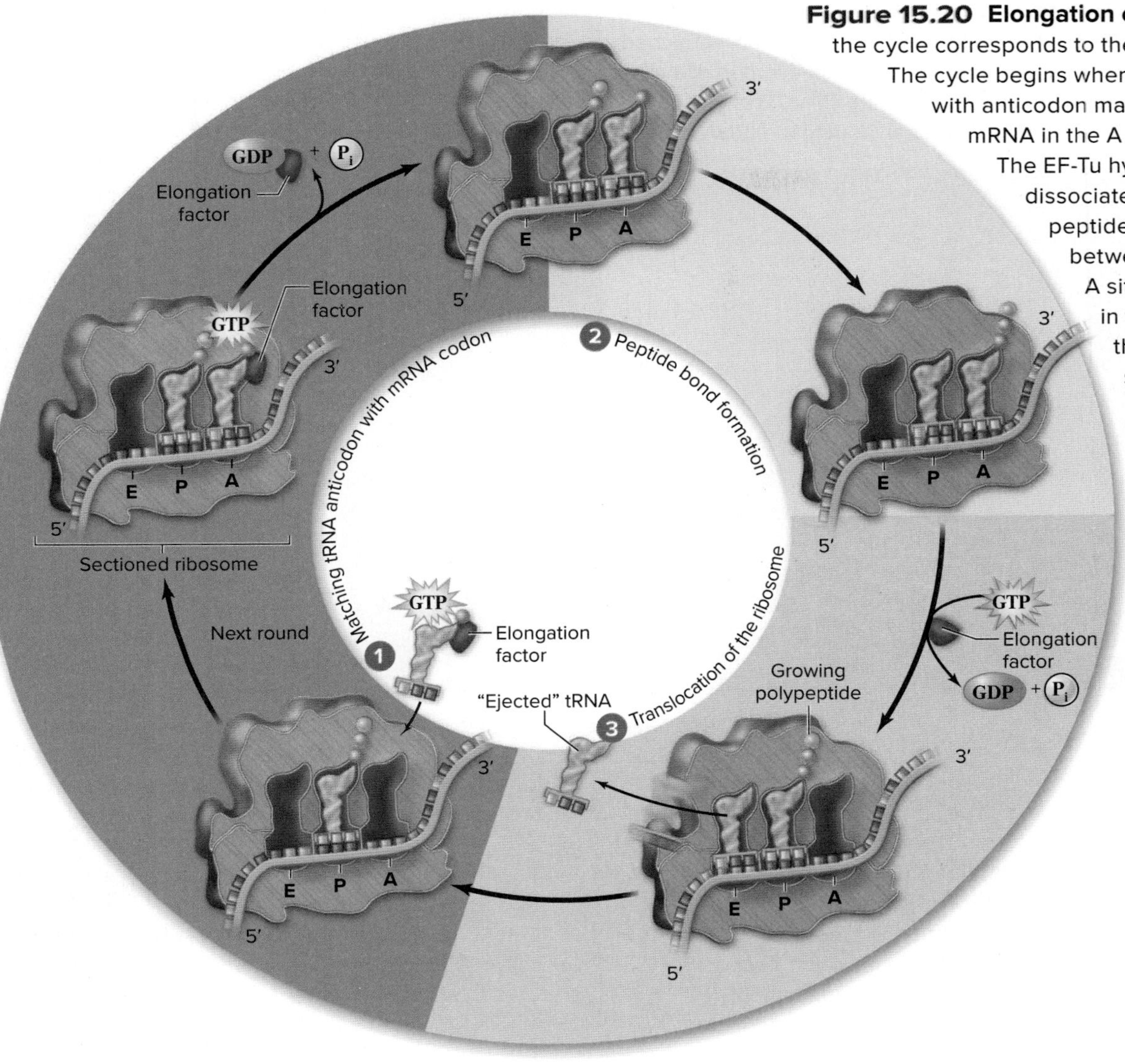

Figure 15.20 Elongation cycle. Numbering of the cycle corresponds to the numbering in the text. The cycle begins when a new charged tRNA with anticodon matching the codon of the mRNA in the A site arrives with EF-Tu. The EF-Tu hydrolyzes GTP and dissociates from the ribosome. A peptide bond is formed between the amino acid in the A site and the growing chain in the P site, transferring the growing chain to the A site and leaving the tRNA in the P site empty. Ribosome translocation requires another elongation factor and GTP hydrolysis. This moves the tRNA in the A site into the P site, the next codon in the mRNA into the A site, and the empty tRNA into the E site.

This also breaks the bond between the growing chain and the tRNA in the P site, leaving it empty (no longer charged). The overall result of this is to transfer the growing chain to the tRNA in the A site.

3. **Translocation of the ribosome.** After the peptide bond has been formed, the ribosome moves relative to the mRNA and the tRNAs. The next codon in the mRNA shifts into the A site, and the tRNA with the growing chain moves to the P site. The uncharged tRNA formerly in the P site is now in the E site, and it will be ejected in the next cycle. This translocation step requires the accessory factor EF-G and the hydrolysis of another GTP.

This elongation cycle continues with each new amino acid added. The ribosome moves down the mRNA in a 5′-to-3′ direction, reading successive codons. The tRNAs move through the ribosome in the opposite direction, from the A site to the P site and finally the E site, before they are ejected as empty tRNAs, which can be charged with another amino acid and then used again.

Wobble pairing

As mentioned, there are fewer tRNAs than codons. This situation is a result of the fact that the pairing between the 3′ base of the codon and the 5′ base of the anticodon is less stringent than normal. In some tRNAs, the presence of modified bases with less accurate pairing in the 5′ position of the anticodon enhances this flexibility. This effect is referred to as **wobble pairing,** because these tRNAs can "wobble" a bit on the mRNA, so that a single tRNA can recognize more than one base possibility in the third position of the mRNA codon.

Termination requires additional accessory factors

Elongation continues in this fashion until a chain-terminating stop codon is reached (for example, UAA in figure 15.22). These stop codons do not bind to tRNA; instead, they are recognized by release factors, proteins that release the newly made polypeptide from the ribosome.

Release of a polypeptide from the final tRNA and dissociation of the ribosome conclude the process of gene expression.

Figure 15.21 Peptide bond formation. Peptide bonds are formed between a "new" charged tRNA in the A site and the growing chain attached to the tRNA in the P site. The bond forms between the amino group of the new amino acid and the carboxyl group of the growing chain. This breaks the bond between the growing chain and its tRNA, transferring the tRNA to the A site as the new amino acid remains attached to its tRNA.

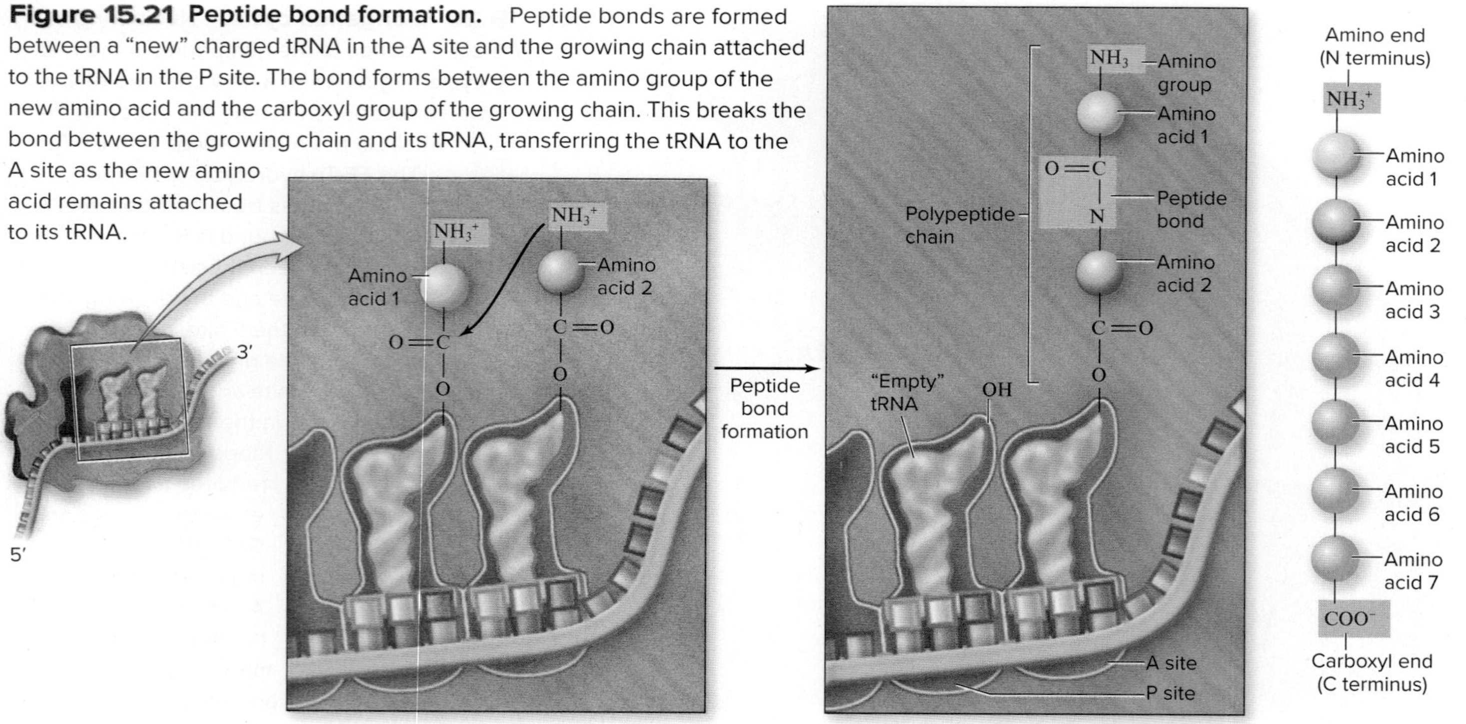

Information from a gene is now represented in a polypeptide. The information in DNA was converted to RNA, which was further processed, then finally translated on the ribosome to produce a polypeptide. This complex process can seem overwhelming when first encountered, so a graphical summary of eukaryotic gene expression is presented in figure 15.23. In addition, differences between prokaryotes and eukaryotes are highlighted in table 15.2.

In Eukaryotes, Proteins May Be Targeted to the ER

LEARNING OBJECTIVE 15.7.3 Compare translation on the RER to that in the cytoplasm.

In eukaryotes, translation can occur either in the cytoplasm or on the RER. Proteins that are translated on the RER are targeted there, based on their own initial amino acid sequence. The ribosomes found on the RER are actively translating and are not permanently bound to the ER.

A polypeptide that starts with a short series of amino acids called a **signal sequence** is specifically recognized and bound by a cytoplasmic complex of proteins called the *signal recognition particle (SRP).* The complex of signal sequence and SRP is in turn recognized by a receptor protein in the ER membrane. The binding of the ER receptor to the signal sequence/SRP complex holds the ribosome engaged in translation of the protein on the ER membrane, a process called *docking* (figure 15.24).

As the protein is assembled, it passes through a channel formed by the docking complex and into the interior ER

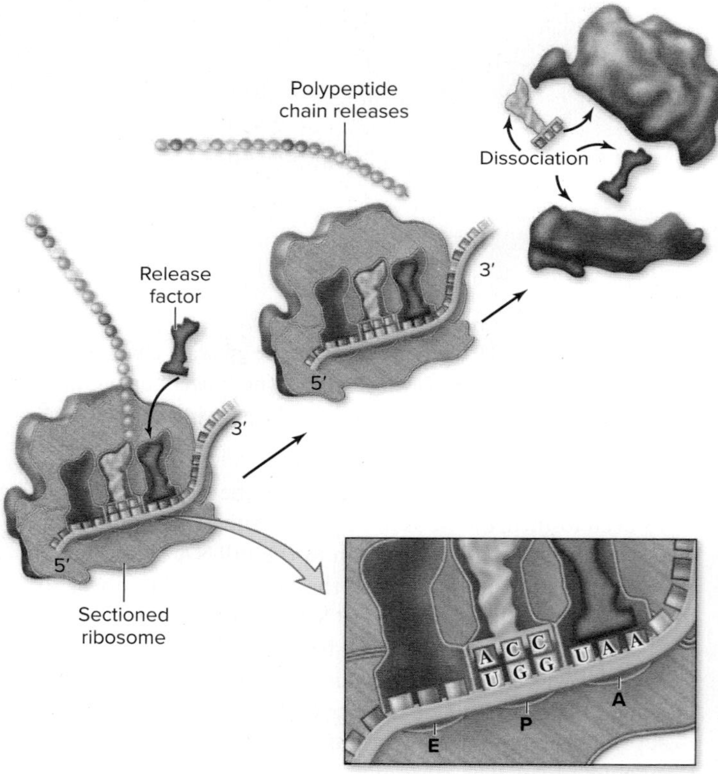

Figure 15.22 Termination of protein synthesis. There is no tRNA with an anticodon complementary to any of the three termination signal codons. When a ribosome encounters a termination codon, it stops translocating. A specific protein release factor facilitates the release of the polypeptide chain by breaking the covalent bond that links the polypeptide to the P site tRNA.

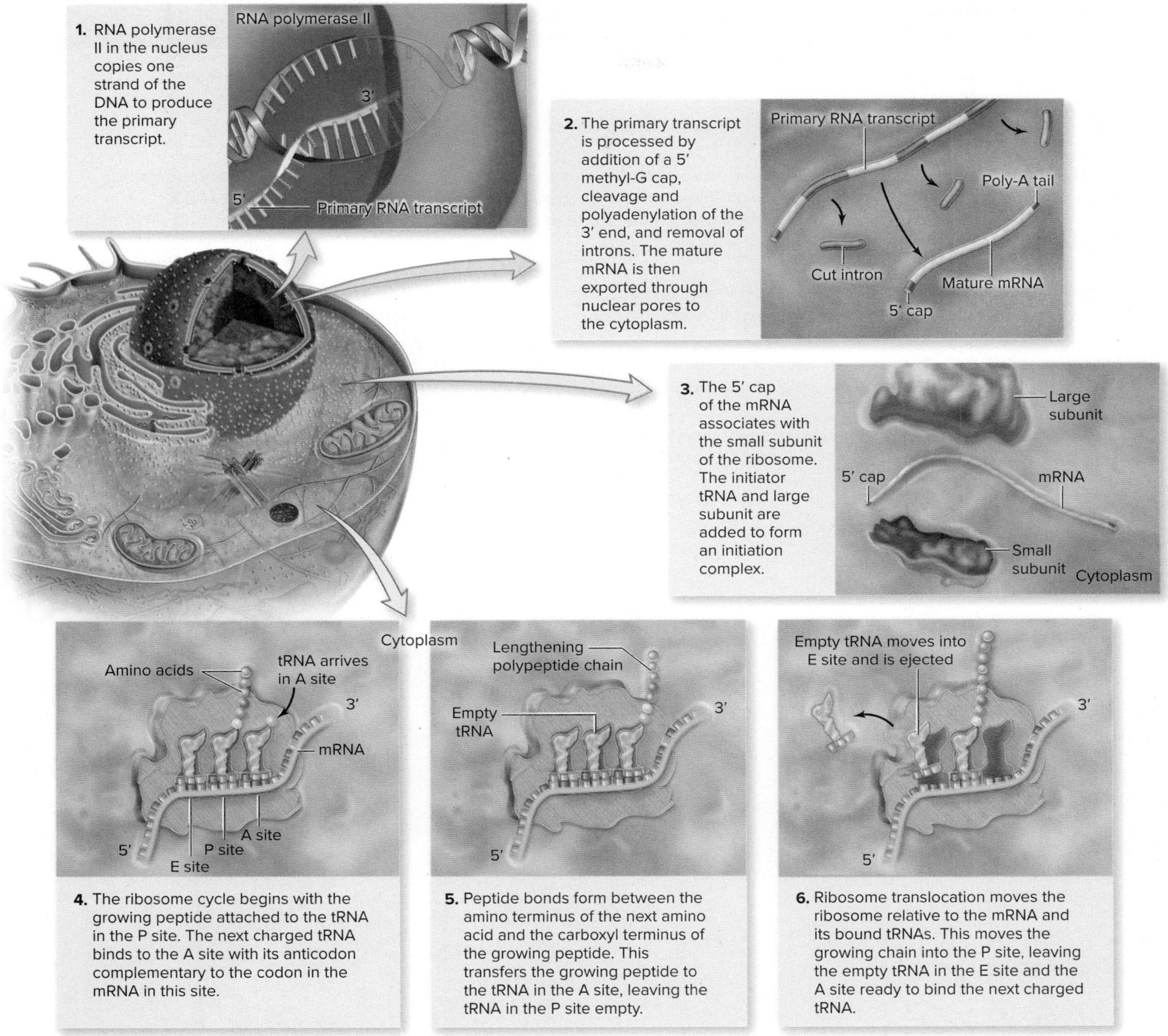

1. RNA polymerase II in the nucleus copies one strand of the DNA to produce the primary transcript.
RNA polymerase II
3′
5′
Primary RNA transcript
2. The primary transcript is processed by addition of a 5′ methyl-G cap, cleavage and polyadenylation of the 3′ end, and removal of introns. The mature mRNA is then exported through nuclear pores to the cytoplasm.
Primary RNA transcript
Poly-A tail
Cut intron
Mature mRNA
5′ cap
3. The 5′ cap of the mRNA associates with the small subunit of the ribosome. The initiator tRNA and large subunit are added to form an initiation complex.
Large subunit
5′ cap
mRNA
Small subunit
Cytoplasm
Cytoplasm
Amino acids
tRNA arrives in A site
3′
mRNA
A site
P site
E site
5′
4. The ribosome cycle begins with the growing peptide attached to the tRNA in the P site. The next charged tRNA binds to the A site with its anticodon complementary to the codon in the mRNA in this site.
Lengthening polypeptide chain
Empty tRNA
3′
5′
5. Peptide bonds form between the amino terminus of the next amino acid and the carboxyl terminus of the growing peptide. This transfers the growing peptide to the tRNA in the A site, leaving the tRNA in the P site empty.
Empty tRNA moves into E site and is ejected
3′
5′
6. Ribosome translocation moves the ribosome relative to the mRNA and its bound tRNAs. This moves the growing chain into the P site, leaving the empty tRNA in the E site and the A site ready to bind the next charged tRNA.

Figure 15.23 An overview of gene expression in eukaryotes.

compartment, the cisternal space. This is the basis for the docking metaphor—the ribosome is not actually bound to the ER itself, but with the newly synthesized protein entering the ER, the ribosome is like a boat tied to a dock with a rope.

Once within the ER cisternal space, or lumen, the newly synthesized protein can be modified by the addition of sugars (glycosylation) and transported by vesicles to the Golgi apparatus (refer to chapter 4). This is the beginning of the protein-trafficking pathway that can lead to other intracellular targets, to incorporation into the plasma membrane, or to release outside of the cell itself.

REVIEW OF CONCEPT 15.7

During initiation the small ribosomal subunit binds to mRNA and a charged initiator tRNA. The elongation cycle involves bringing in new charged tRNAs to the ribosome's A site, forming peptide bonds between amino acids, and translocating the ribosome along the mRNA chain. The tRNAs transit through the ribosome from A to P to E sites. In eukaryotes, a signal sequence in newly forming polypeptides targets them to the RER where they are translocated into the cisternal space during synthesis.

- *What stages of translation require energy?*

TABLE 15.2 Differences Between Prokaryotic and Eukaryotic Gene Expression

Characteristic	Prokaryotes	Eukaryotes
Introns	No introns, although some archaeal genes possess them.	Most genes contain introns.
Number of genes in mRNA	Several genes may be transcribed into a single mRNA molecule. Often these have related functions and form an operon, which helps coordinate regulation of biochemical pathways.	Only one gene per mRNA molecule; regulation of pathways accomplished in other ways.
Site of transcription and translation	No membrane-bounded nucleus; transcription and translation are coupled.	Transcription in nucleus; mRNA is transported to the cytoplasm for translation.
Initiation of translation	Begins at AUG codon preceded by special sequence that binds the ribosome.	Begins at AUG codon preceded by the 5′ cap (methylated GTP) that binds the ribosome.
Modification of mRNA after transcription	None; translation begins before transcription is completed. Transcription and translation are coupled.	A number of modifications while the mRNA is in the nucleus: introns are removed and exons are spliced together; a 5′ cap is added; a poly-A tail is added.

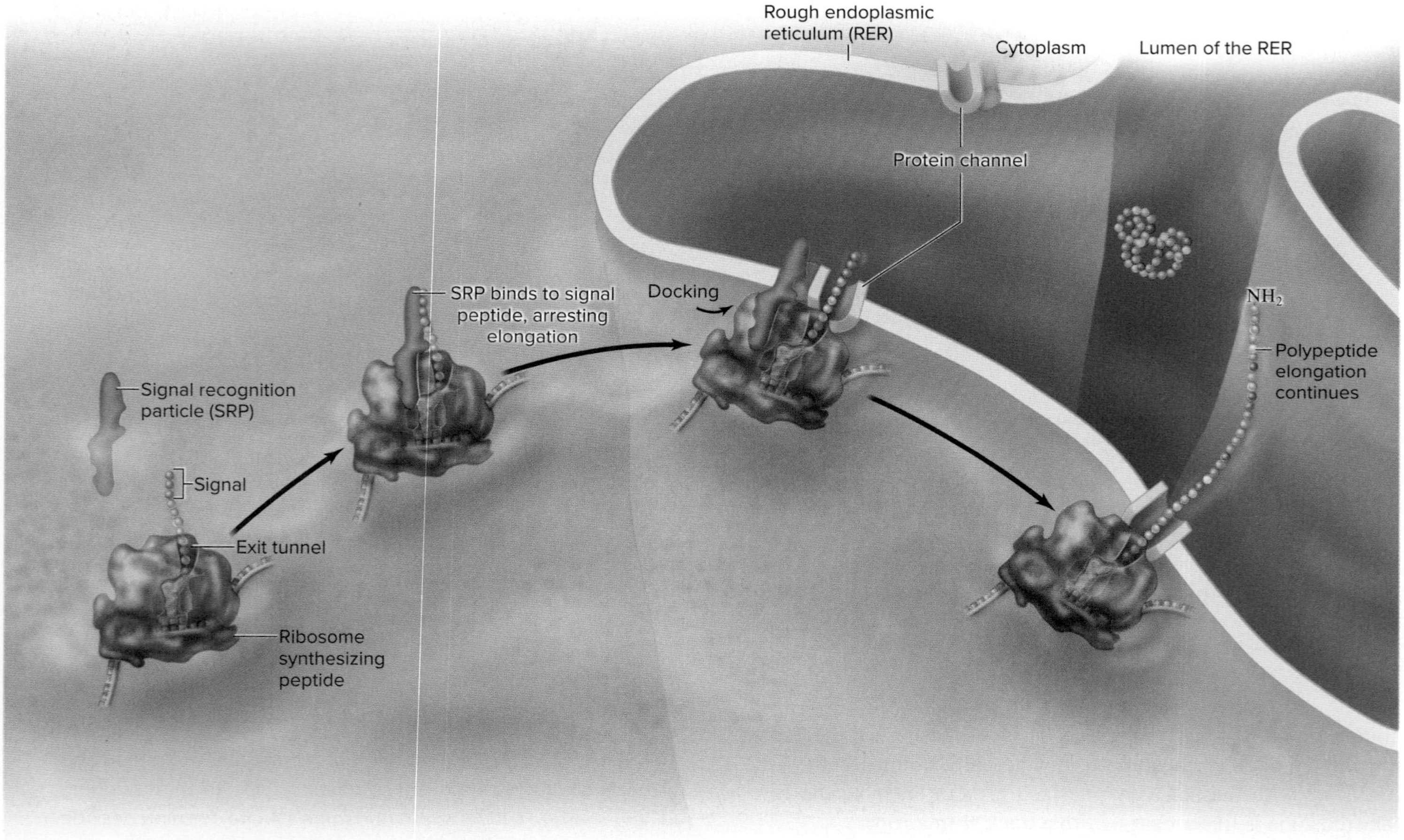

Figure 15.24 Synthesis of proteins on RER. Proteins that are synthesized on RER arrive at the ER because of sequences in the peptide itself. A signal sequence in the amino terminus of the polypeptide is recognized by the signal recognition particle (SRP). This complex docks with a receptor associated with a channel in the ER. The peptide passes through the channel into the lumen of the ER as it is synthesized.

15.8 Mutations Are Heritable Changes in Genetic Material

A mutation is a heritable change in the genetic material. A change as simple as altering a single base can result in an amino acid substitution that can lead to a debilitating clinical phenotype. This concept was introduced in chapter 12 using the example of sickle-cell anemia (refer to figure 12.15). In the sickle-cell allele, a single A is changed to a T, causing a glutamic acid to be replaced with a valine. The nonpolar valine causes the β-chains to aggregate into polymers, which consequently alters the shape of the blood cells, leading to anemia.

Genetic differences between individuals in a population are called polymorphisms, and they arise by mutation. In this section, we will consider the different kinds of changes that can occur in a genome.

Point Mutations Affect a Single Site in the DNA

LEARNING OBJECTIVE 15.8.1 Contrast the different kinds of point mutations.

A mutation that alters a single base is termed a **point mutation,** which leads to **single-nucleotide variation (SNV),** in populations; as these are polymorphisms in populations, they are also called **single-nucleotide polymorphisms (SNP).** The replacement of one base for another is called a base substitution mutation (figure 15.25).

Missense mutations

Because the genetic code is degenerate, a base substitution mutation may not change the encoded amino acid. If it does not, we say the mutation is *synonymous*, or silent (figure 15.25*b*). If it changes an amino acid, we say the mutation is *nonsynonymous.* We also refer to nonsynonymous changes as **missense mutations,** as the "sense" of the codon involved has been changed (figure 15.25*c*). We further classify nonsynonymous changes as conservative or nonconservative. Conservative changes replace an amino acid with a chemically similar one (nonpolar for nonpolar, for example), while nonconservative changes replace an amino acid with a chemically different one, as in our sickle-cell example.

Nonsense mutations

A mutation that changes a codon from one that encodes an amino acid to a stop codon is called a **nonsense mutation** because the codon no longer "makes sense" to the translation apparatus (figure 15.25*d*). The stop codon will cause translation to terminate prematurely, producing a truncated protein. The length of the resulting protein depends on the location of the new stop codon in the gene.

Small insertions or deletions (indels)

The gain or loss of 1–50 bp is called an *indel* (for "insertion/deletion"). Since codons consist of three bases, insertions or deletions that do not occur in multiples of three will shift the reading frame

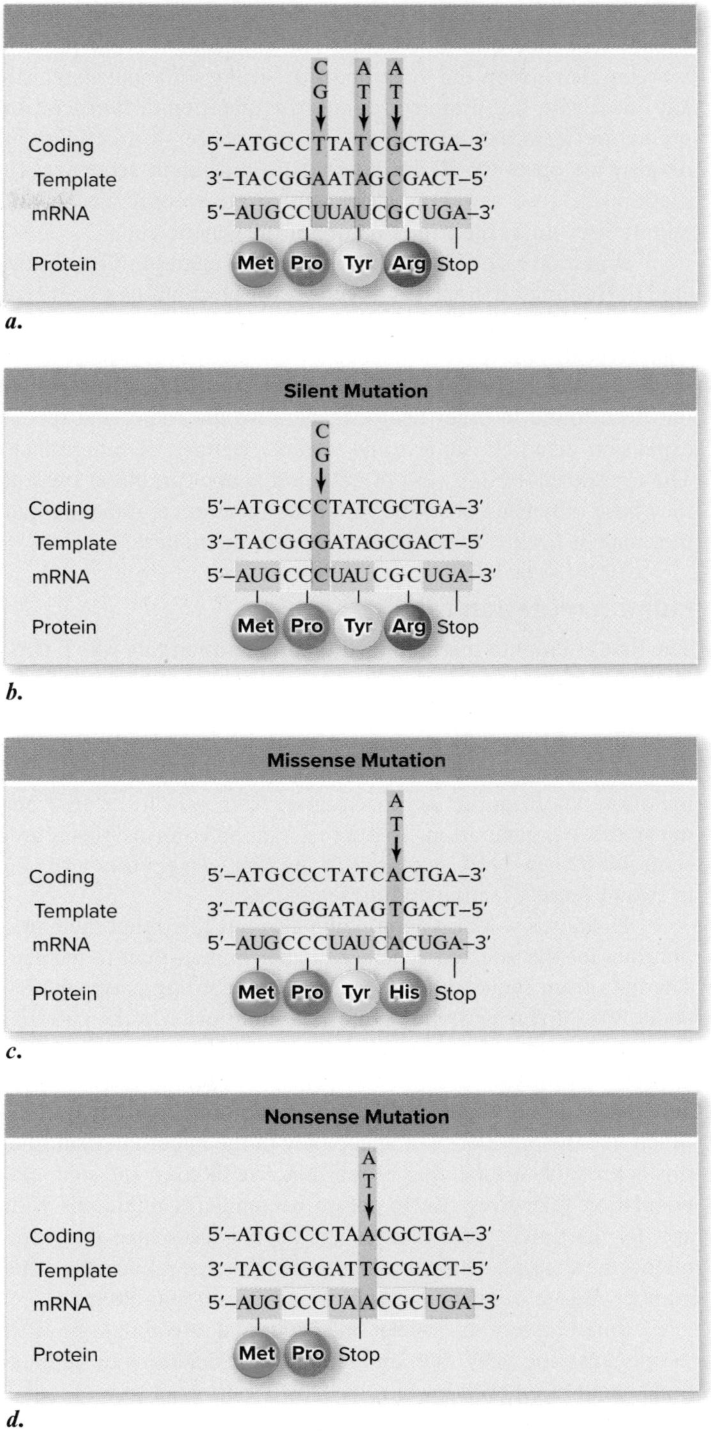

Figure 15.25 Types of mutations. ***a.*** A hypothetical gene is shown with encoded mRNA and protein. Arrows above the gene indicate sites of mutations described in the rest of the figure. ***b.*** Silent mutation. A change in the third position of a codon is often silent due to degeneracy in the genetic code. In this case, T/A to C/G mutation does not change the amino acid encoded (proline). ***c.*** Missense mutation. The G/C to A/T mutation changes the amino acid encoded from arginine to histidine. ***d.*** Nonsense mutation. The T/A to A/T mutation produces a UAA stop codon in the mRNA.

in the mRNA. Such a **frameshift mutation** randomizes the downstream sequence of amino acids. The consequences for protein function depend on the location of the frameshift, but they usually inactivate the protein. Frameshift mutations often lead to premature termination, as 3 in 64 codons are stop codons, or roughly a stop every 20 amino acids in a random sequence. In section 15.2, we saw how Crick and Brenner used frameshift mutations to infer the triplet nature of the genetic code.

A special case of insertions is a class of mutation first discovered in Huntington disease. The gene involved contains a sequence of 3 bases that are repeated, called a **trinucleotide repeat** (**TNR**) or **triplet repeat.** This repeat sequence is expanded in the disease allele relative to the normal allele. This kind of triplet repeat expansion has been found in other neurodegenerative disorders, and repeat expansion disorders share some unusual features of inheritance. The prevalence of this kind of mutation is unclear, but at present they have only been observed in humans and mice, implying that they may be limited to vertebrates, or even mammals.

Human mutation rate

The first attempt to measure the human mutation rate was J. B. S. Haldane's pioneering work using the prevalence of hemophilia in the population. We have now advanced to using DNA sequencing technology to directly measure a mutation rate by sequencing parent–offspring trios (trio = mom, dad, child). Estimates of mutation rate from this approach range from $1.0-1.8 \times 10^{-8}$ per nucleotide per generation. To put this in simpler terms, this represents 44–82 new SNV mutations in the average genome, with one to two of these affecting coding sequences.

Haldane's work on hemophilia also indicated that the mutation rate for the affected gene was higher in men than in women. The data from sequencing trios also show a strong paternal bias: about 80% of new germ-line point mutations arise in the paternal genome. This rate also increases with increasing paternal age. It is estimated from this work that for every one-year increase in the father's age at conception, there are 2 additional new mutations. Much like the prevalence of aneuploidy in the maternal germ line, this is probably a reflection of reproductive biology: the stem cell population that gives rise to sperm accumulates mutations with age. By the time a male reaches age 20, it is estimated that spermatogenic stem cells have undergone 160 genome replications; this number rises to 610 genome replications in a 40-year-old male.

Small insertions or deletions occur at a rate about one-fifth to one-tenth the SNV rate, and much larger deletions or duplications occur at an even lower rate. This is estimated to be about 1 large rearrangement per 42 births. Some of the mobile genetic elements that litter our genome (refer to chapter 18) can still move. The rate for mobile element insertion appears to be on the order of about 1 per 20 births.

Chromosomal Mutations Change the Number or Position of Large DNA Segments

LEARNING OBJECTIVE 15.8.2 Compare the different kinds of chromosomal mutations.

Point mutations affect a single site in a chromosome, but more extensive changes that affect more than 50 bp are called **structural variation.** Many human cancers are associated with chromosomal abnormalities, so these can be clinically relevant. We will consider the different kinds of chromosomal mutations and their possible effects.

Copy number variation (CNV)

When genomic rearrangements result in differences in the number of copies of a particular genomic region, it is called **copy number variation** (**CNV**). This can be the result of loss of DNA by **deletion,** or the gain of DNA by **duplication** (figure 15.26*a, b*) or **insertion.** A special case of possible copy number variation is the insertion of mobile genetic elements (refer to chapter 18), or mobile element insertion (MEI).

The amount of DNA that can be deleted without drastic consequences is dependent on the region of the genome, but large deletions are usually lethal. One human syndrome caused by a deletion is cri-du-chat, which is French for "cry of the cat" after the sound made by children with this syndrome. Cri-du-chat syndrome is caused by a large deletion from the short arm of chromosome 5. It usually results in early death, although many affected individuals do not have an abbreviated lifespan, but can have effects in multiple physiological systems.

The duplication of a chromosomal region may or may not lead to phenotypic consequences. Effects depend upon the location of the "breakpoints" where the duplication occurred. If the duplicated region does not lie within a gene, there may be no effect. If the duplication occurs next to the original region, it is termed a tandem duplication. Some gene families, such as the globin genes that encode hemoglobin, arose as the result of tandem duplication.

Balanced rearrangements

Material in chromosomes can become inverted or transferred between chromosomes, events called **inversions** and **reciprocal translocations,** respectively (figure 15.26 *c, d*). In each case, the total amount of genetic material remains the same, but it has been rearranged. More complex rearrangements can also occur, but they are even more rare.

An inversion results when a segment of a chromosome is broken in two places, reversed, and put back together. An inversion may not have an effect on phenotype if the sites where the inversion occurs do not break within a gene. In fact, although humans all have the "same" genome, the order of genes in all individuals in a population is not precisely the same due to inversions that occur in different lineages.

If a piece of one chromosome is broken off and joined to another chromosome, we call this a translocation. Translocations are complex because they cause problems during meiosis, when two different chromosomes try to pair with each other, which can lead to unbalanced translocation chromosomes. Translocations can also move genes such that their expression is affected. Two forms of leukemia have been shown to be associated with translocations that move oncogenes into regions of a chromosome where they are expressed inappropriately in blood cells.

Mutations and evolution

Without a continuous source of new variation, evolution would not occur. However, mutations are much more often detrimental

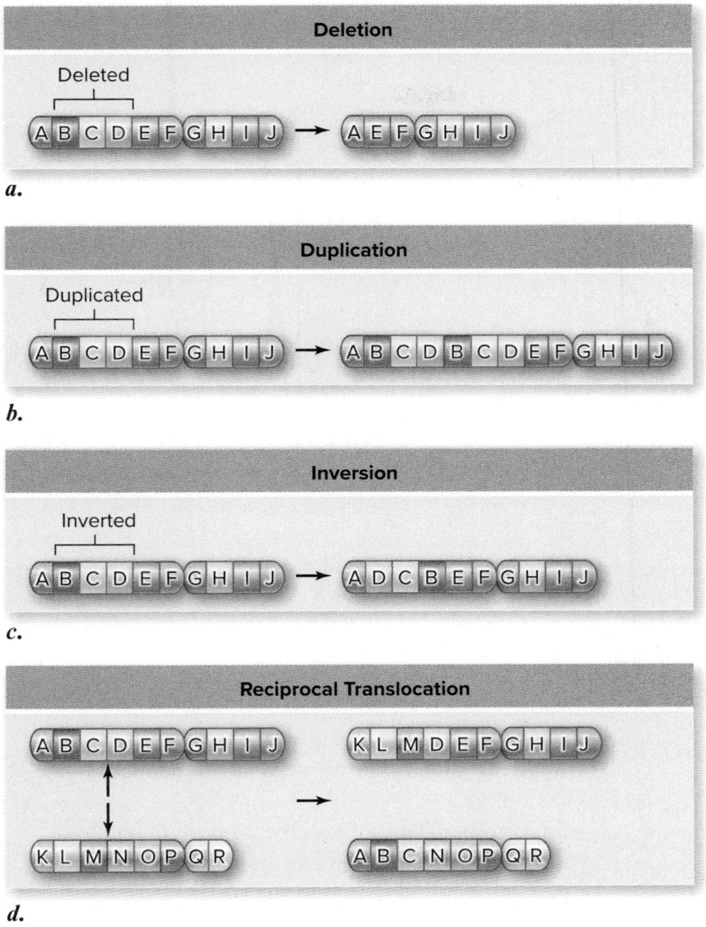

Figure 15.26 Chromosomal mutations. Larger-scale changes in chromosomes are also possible. Material can be deleted ***(a)***, duplicated ***(b)***, or inverted ***(c)***. Translocations occur when one chromosome is broken and becomes part of another chromosome. This often occurs where both chromosomes are broken and exchange material, an event called a reciprocal translocation ***(d)***.

than beneficial to an individual. A delicate balance must exist between the amount of new variation that arises in a species and the health of individuals in the species. The larger-scale alteration of chromosomes has also been important in evolution, although its role is poorly understood. It is clear that gene families arise by the duplication of an ancestral gene, followed by the functional divergence of the duplicated copies. It is also clear that even among closely related species, the number and arrangements of genes on chromosomes can differ. There have even been whole-genome duplications in lineages.

REVIEW OF CONCEPT 15.8

Point mutations (single-base changes, additions, or deletions) include missense mutations that substitute one amino acid for another, nonsense mutations that halt transcription, and frameshift mutations that alter the reading frame. Triplet repeat expansion is the abnormal duplication of a codon with each round of cell division. The average human genome has 44–82 new point mutations. Larger genetic changes occur at a much lower rate. Mutations affecting chromosomes include deletions, duplications, inversions, and translocations.

- *Is a silent mutation most likely in the first, second, or third base of a codon?*

Building Proteins in a Test Tube

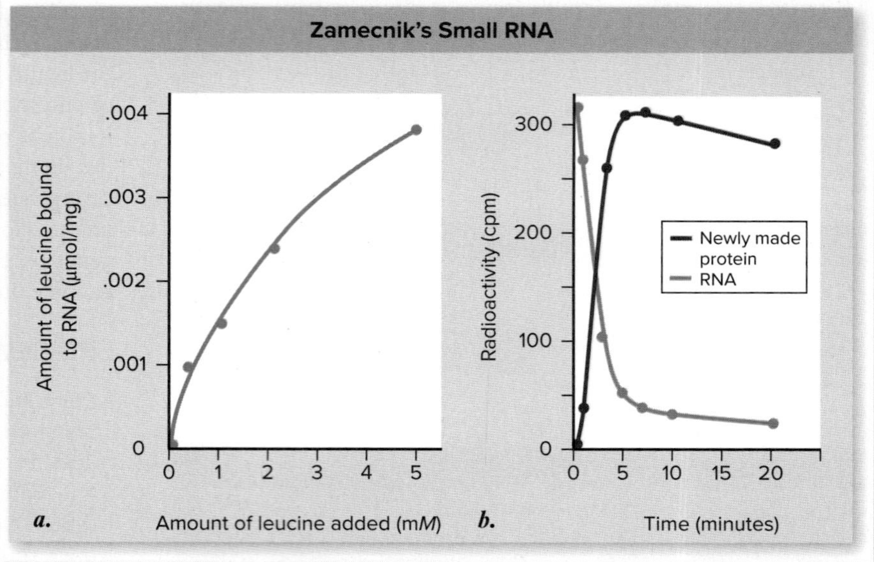

The complex mechanisms used by cells to build proteins were not discovered all at once. Our understanding came slowly, accumulating through a long series of experiments, each telling us a little bit more. To gain some sense of the incremental nature of this experimental journey, and to appreciate the excitement that each step gave, it is useful to step into the shoes of an investigator back when little was known and the way forward was not clear.

The shoes we will step into are those of Paul Zamecnik, an early pioneer in protein synthesis research. Working with colleagues at Massachusetts General Hospital in the early 1950s, Zamecnik first asked the most direct of questions: Where in the cell are proteins synthesized? To find out, the researchers injected radioactive amino acids into rats. After a few hours, the labeled amino acids could be found as part of newly made proteins in the livers of the rats. And in the livers that were removed and checked only minutes after injection, radioactive-labeled proteins were found associated only with small particles in the cytoplasm. Composed of protein and RNA, these particles, later named ribosomes, had been discovered years earlier by electron microscope studies of cell components. This experiment identified them as the sites of protein synthesis in the cell.

After several years of trial-and-error tinkering, Zamecnik and his colleagues had worked out a "cell-free" protein synthesis system that would lead to the synthesis of proteins in a test tube. It included ribosomes, mRNA, and ATP to provide energy. It also included a collection of required soluble "factors" isolated from homogenized rat cells that somehow worked with the ribosome to get the job done. When Zamecnik's team characterized these required factors, they found most of them to be proteins, as expected, but also unexpectedly present in the mix was a small RNA.

To determine the role of this small RNA, the team performed several experiments. In the first, they added various amounts of ^{14}C-leucine (the radioactively labeled amino acid leucine) to a cell-free system containing the soluble factors, ribosomes, and ATP. After incubating the mixture with ^{14}C-leucine, they isolated the small RNA from the mixture assayed it for radioactivity. The results are presented in graph a.

In a second experiment, they mixed the ^{14}C-leucine–small RNA complex generated by the first experiment with cell extracts capable of protein synthesis. They then looked to see where the radioactive label went. In this case, they analyzed both newly made protein, and the small RNA. The results are shown in graph b.

Analysis

EXPERIMENT A, shown in graph *a*

1. **Applying Concepts** What is the dependent variable?
2. **Interpreting Data** Does the amount of leucine added to the test tube have an effect on the amount of leucine found bound to the small RNA?
3. **Making Inferences** Is the amount of leucine bound to small RNA proportional to the amount of leucine added to the mixture?
4. **Drawing Conclusions** Can you reasonably conclude from this result that the amino acid leucine is binding to the small RNA?

EXPERIMENT B, shown in graph *b*

1. **Applying Concepts** What is the dependent variable?
2. **Interpreting Data**
   a. If radioactivity is monitored for 20 minutes after the addition of the radioactive leucine–small RNA complex to the cell extract, what happens to the level of radioactivity in the small RNA (blue)?
   b. Over the same period, what happens to the level of radioactivity in the newly made protein (red)?
3. **Making Inferences** Is the same amount of radioactivity being lost from the small RNA that is being gained by the newly made protein?
4. **Drawing Conclusions**
   a. Is it reasonable to conclude that the small RNA is donating its amino acid to the growing protein? (*Hint:* Consider the different kinds of RNA you learned about in this chapter.)
   b. If you were to isolate the protein made in this experiment after 20 minutes, which amino acids would be radioactively labeled? Explain.

Retracing the Learning Path

CONCEPT 15.1 Experiments Have Revealed the Nature of Genes

Garrod found that alkaptonuria is due to an altered enzyme.

15.1.1 Beadle and Tatum Showed That Genes Specify Enzymes *Neurospora* mutants unable to synthesize arginine were found to lack specific enzymes. Beadle and Tatum advanced the "one-gene/one-polypeptide" hypothesis.

15.1.2 Crick States the Central Dogma The central dogma of molecular biology describes the flow of information in cells from DNA through RNA to make protein. We call the DNA strand copied to mRNA the template (antisense) strand; the other, the coding (sense) strand. Transcription makes an RNA copy of DNA. Translation uses information in RNA to synthesize proteins. RNA has multiple roles in gene expression.

CONCEPT 15.2 The Genetic Code Relates Information in DNA and Protein

15.2.1 Crick and Brenner Learn How the Code Is Read Crick and Brenner showed that the code is triplet and nonoverlapping. This established the concept of reading frame.

15.2.2 Nirenberg and Khorana Decipher the Code A three-base codon specifies an amino acid. There are three "stop" codons and one "start" codon, which also encodes methionine; 61 codons encode the 20 amino acids. Some amino acids have more than one codon, but each codon specifies only one amino acid. The code is practically universal, but not quite. In some mitochondrial and protist genomes, a stop codon is read as an amino acid.

CONCEPT 15.3 Prokaryotes Exhibit All the Basic Features of Transcription

15.3.1 Stages of Prokaryotic Transcription The single prokaryotic RNA polymerase exists in two forms: core polymerase, which can synthesize mRNA; and holoenzyme, core plus σ factor, which can accurately initiate synthesis. Initiation requires a start site and a promoter. The promoter is upstream of the start site, and binding of RNA polymerase holoenzyme to its -35 region positions the polymerase properly. Transcription proceeds in the $5'$-to-$3'$ direction. The transcription bubble contains RNA polymerase, the locally unwound DNA template, and the growing mRNA transcript. Terminators consist of complementary sequences that form a double-stranded hairpin loop where the polymerase pauses. Prokaryotic transcription is coupled to translation. In prokaryotes, translation begins while mRNAs are still being transcribed.

CONCEPT 15.4 Eukaryotes Use Three Polymerases and Extensively Modify Transcripts

15.4.1 Eukaryotes Have Three RNA Polymerases RNA polymerase I transcribes rRNA; polymerase II transcribes mRNA and some snRNAs; polymerase III transcribes tRNA. Each polymerase has its own promoter.

15.4.2 Polymerase II Initiation and Termination Requires Other Factors RNA polymerase II promoters require a host of transcription factors. The end of the mRNA is modified after transcription.

15.4.3 Eukaryotic Transcripts Are Modified After transcription, a methyl-GTP cap is added to the $5'$ end of the transcript. A poly-A tail is added to the $3'$ end. Noncoding internal regions are also removed by splicing.

CONCEPT 15.5 Eukaryotic Genes May Contain Noncoding Sequences

Coding DNA (an exon) is interrupted by noncoding introns. These introns are removed by splicing.

15.5.1 RNA Splicing Is Carried Out by the Spliceosome snRNPs recognize intron–exon junctions and recruit spliceosomes. The spliceosome ultimately joins the $3'$ end of the first exon to the $5'$ end of the next exon.

15.5.2 Alternative Splicing Can Produce Multiple Transcripts from the Same Gene

CONCEPT 15.6 The Ribosome Is the Machine of Protein Synthesis

15.6.1 tRNA Is a Bifunctional Molecule

15.6.2 Activating Enzymes Attach Amino Acids to tRNA The tRNA charging reaction attaches the carboxyl terminus of an amino acid to the $3'$ end of the correct tRNA and requires one ATP.

15.6.3 The Ribosome Has Multiple tRNA-Binding Sites Ribosomes hold tRNAs and mRNA in position for a ribosomal enzyme to form peptide bonds. Charged tRNAs first bind to the A site, move to the P site bound to the peptide, then exit from the E site with no amino acid.

CONCEPT 15.7 The Process of Translation Is Complex and Energy-Expensive

15.7.1 Initiation Requires Accessory Factors In prokaryotes, initiation-complex formation is aided by the ribosome-binding sequence (RBS) of mRNA, complementary to a small subunit. Eukaryotes use the $5'$ cap for the same function.

15.7.2 Elongation Adds Successive Amino Acids As the ribosome moves $5'$ to $3'$ along the mRNA, new amino acids from charged tRNAs are added to the growing peptide. This process requires one ATP for each new charged tRNA, and another ATP to move the ribosome after each peptide bond. Stop codons are recognized by termination factors.

15.7.3 In Eukaryotes, Proteins May Be Targeted to the ER In eukaryotes, proteins with a signal sequence in their amino terminus bind to the SRP, and this complex docks on the ER.

CONCEPT 15.8 Mutations Are Heritable Changes in Genetic Material

15.8.1 Point Mutations Affect a Single Site in the DNA Base substitutions exchange one base for another, and frameshift mutations involve the addition or deletion of a base. Triplet repeat expansion mutations can cause genetic diseases.

15.8.2 Chromosomal Mutations Change the Number or Position of Large DNA Segments Chromosomal mutations include additions, deletions, inversions, and translocations.

Concept Overview

This Concept Overview diagrams the key concepts that were discussed in this chapter.

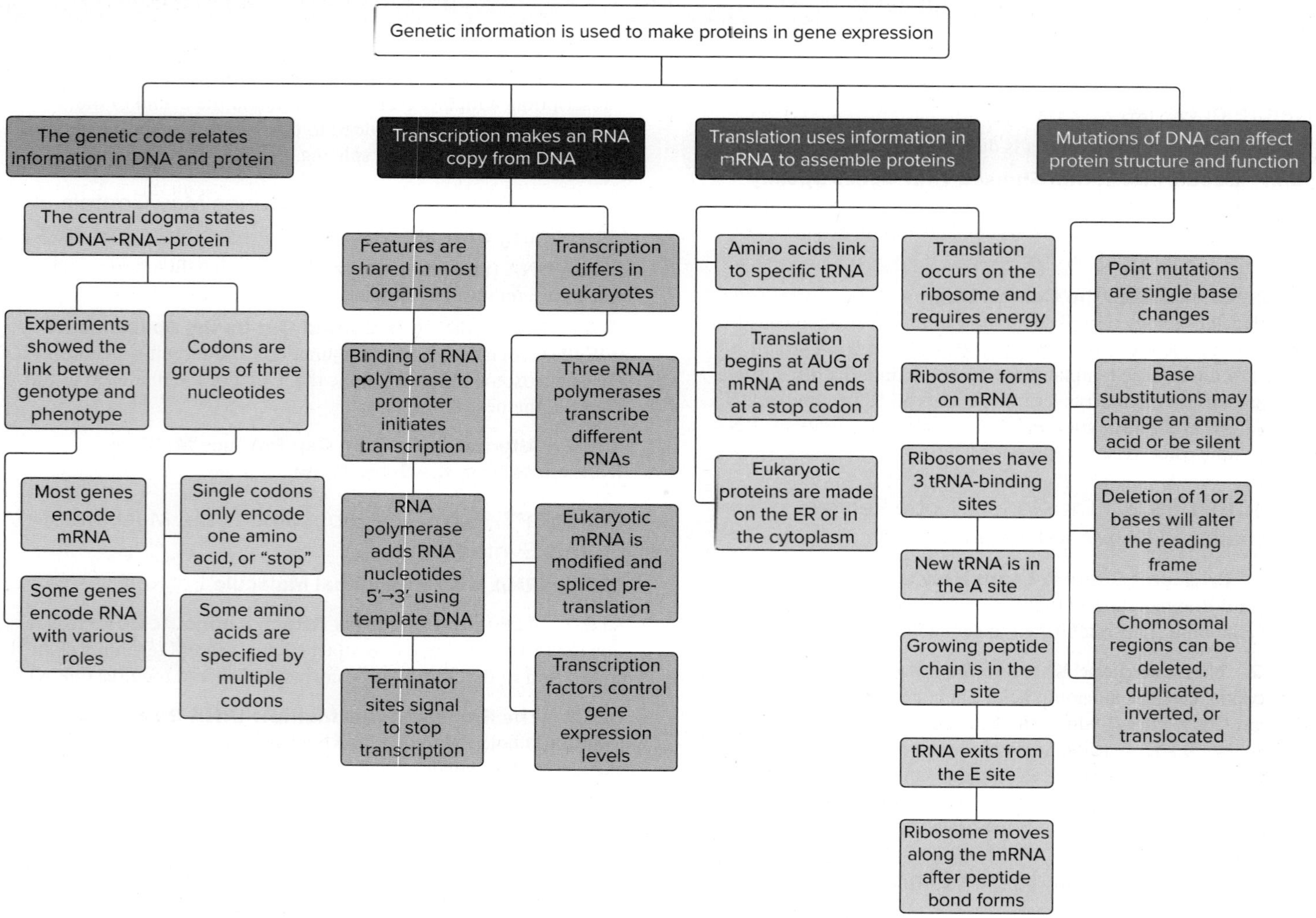

Assessing the Learning Path

Understand

1. Which of the following RNA molecules is NOT correctly matched with its function?
   a. mRNA—carries genetic message from DNA to protein-synthesizing machinery
   b. tRNA—connects genetic message to amino acids needed in protein being built
   c. snRNA—localizes ribosomes to the rough ER
   d. rRNA—structural component of ribosomes
2. In the genetic code, one codon
   a. consists of three bases.
   b. specifies a single amino acid.
   c. specifies more than one amino acid.
   d. Both a and b
3. Which amino acid is specified by the codon CUC? (Consult table 15.1.)
   a. Lysine
   b. Alanine
   c. Glutamate
   d. Leucine
4. RNA polymerase is similar to DNA polymerase in that
   a. both enzymes require a primer.
   b. both enzymes make a nucleic acid in the 5′-to-3′ direction.
   c. both enzymes require helicase to expose the template strand.
   d. All of the above
5. Which of the following functions as a "stop" signal for a prokaryotic RNA polymerase?
   a. A specific sequence of bases called a terminator
   b. The poly-A site
   c. Addition of a 5′ cap
   d. A region of the mRNA that can base-pair to form a hairpin

6. Eukaryotic transcription differs from prokaryotic in that
   a. eukaryotes have only one RNA polymerase.
   b. eukaryotes have three RNA polymerases.
   c. prokaryotes have three RNA polymerases.
   d. Both a and c
7. Which of the following is NOT true about poly-A tails?
   a. They are made by RNA polymerase II; each gene has a long stretch of T's that marks its end.
   b. They are added to the RNA transcript in the nucleus.
   c. They protect mRNA from degrading.
   d. They are found at the 3′ end of eukaryotic mRNA.
8. Introns
   a. are coded in the DNA.
   b. are added to the RNA after transcription is complete.
   c. are removed from RNA before it leaves the nucleus.
   d. Both a and c
9. How are tRNAs charged?
   a. By aminoacyl-tRNA synthetases
   b. By ribosomes
   c. By peptidyl transferase
   d. By randomly attaching to amino acids from a cytosolic pool
10. The ribosome
   a. is composed of two subunits.
   b. is a complex of RNA and protein.
   c. has tRNA- and mRNA-binding sites.
   d. All of the above
11. In the initiation stage of prokaryotic translation, where does $tRNA^{fMet}$ bind?
   a. At the start codon in the P site
   b. At the start codon in the A site
   c. At the start codon in the E, or entrance, site
   d. At the mRNA-binding site
12. Codons are found in which of the following types of RNA?
   a. snRNA
   b. mRNA
   c. tRNA
   d. rRNA
13. A nonsense mutation
   a. results in large-scale change to a chromosome.
   b. will lead to the premature termination of transcription.
   c. will lead to the premature termination of translation.
   d. is the same as a transversion.

Apply

1. Beadle and Tatum isolated mutants in the fungus *Neurospora* that were unable to synthesize arginine. One of these mutants would only grow on media supplemented with arginine and not on any of the intermediates in the pathway. This strain
   a. must have a mutation in the first enzyme in the biosynthetic pathway.
   b. must have a mutation in the last enzyme in the biosynthetic pathway.
   c. could have a mutation in any enzyme in the pathway.
   d. must have a nonsense mutation
2. Which would you predict to have the largest effect on the primary structure of a protein, adding one nucleotide near the beginning of an mRNA or deleting one nucleotide near the end of an mRNA?
   a. Deleting one nucleotide near the end of the mRNA.
   b. Additions and deletions should have the same effect.
   c. Neither would have an effect on primary structure, but both would affect tertiary structure.
   d. Adding one nucleotide near the beginning of the mRNA.
3. How would mutations in the −10 region of a promoter that changed A/T base-pairs to G/C base-pairs affect gene expression?
4. A gene has a mutation in which the AAUAAA site just upstream of the termination site is altered. Which of the following would be the most likely consequence of this mutation?
   a. The mRNA would not have a cap.
   b. The mRNA could not leave the nucleus.
   c. The mRNA would be short-lived.
   d. The mRNA would not position itself correctly on the ribosome.
5. A mutation that deleted only the 5′ end of an intron would likely result in
   a. no effect on the mRNA.
   b. a longer mRNA because the intron would not be removed.
   c. a shorter mRNA because more material would be removed during splicing.
   d. a frameshift in the mRNA where the end of the intron meets the next exon.
6. A tRNA has the anticodon 5′ UAA 3′. That tRNA will be charged with which amino acid?
   a. None, because that is a stop codon
   b. Ile (isoleucine)
   c. Asn (asparagine)
   d. Leu (leucine)
7. Insulin is a peptide hormone secreted from pancreatic cells that produce it. If pancreatic cells produced insulin with a defective signal sequence, the insulin
   a. would not bind to SRP and would not be directed to the ER.
   b. would be made on ribosomes that were permanently attached to the ER.
   c. would have no effect on the function of the protein.
   d. would be secreted from the cell.

Synthesize

1. It is widely accepted that RNA polymerase has no proofreading capacity. Would you expect high or low levels of error in transcription compared with DNA replication? Why do you think it is more important for DNA polymerase than for RNA polymerase to proofread?
2. Frameshift mutations often result in truncated proteins. Explain this observation based on the genetic code.
3. You are provided with a sample of aardvark DNA. As part of your investigation of this DNA, you transcribe mRNA from the DNA and purify it. You then separate the two strands of the DNA and analyze the base composition of each strand, and of the mRNA transcripts. You obtain the following results:

	A	G	C	T	U
DNA strand #1	19.1	26.0	31.0	23.9	0
DNA strand #2	24.2	30.8	25.7	19.3	0
mRNA	19.0	25.9	30.8	0	24.3

   Which strand of the DNA is the "sense" strand that serves as the template for mRNA synthesis?
4. Describe how each of the following mutations will affect the final protein product (protein begins with start codon). Name the type of mutation.
   Original template strand:
   a. 3′ - CGTTACCCGAGCCGTACGATTAGG - 5′
   b. 3′ - CGTTACCCGAGCCGTAACGATTAGG - 5′
   c. 3′ - CGTTACCCGATCCGTACGATTAGG - 5′
   d. 3′ - CGTTACCCGAGCCGTTCGATTAGG - 5′

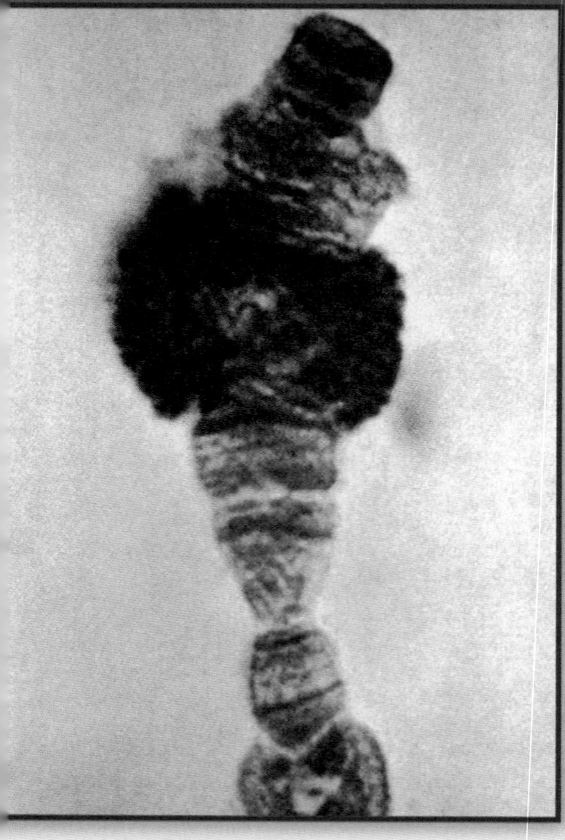

Science Source

16 Control of Gene Expression

Learning Path

- **16.1** All Organisms Control Expression of Their Genes
- **16.2** Regulatory Proteins Control Genes by Interacting with Specific DNA Nucleotide Sequences
- **16.3** Prokaryotes Regulate Their Genes in Clusters
- **16.4** Transcription Factors Control Gene Transcription in Eukaryotes
- **16.5** Chromatin Structure Affects Gene Expression
- **16.6** Eukaryotic Genes Are Also Regulated After Transcription
- **16.7** Gene Regulation Determines How Cells Will Develop

Concept Overview

This Concept Overview provides a diagram of the overarching concepts that are covered in this chapter. These concepts will be expanded upon in the Concept Overview at the end of the chapter.

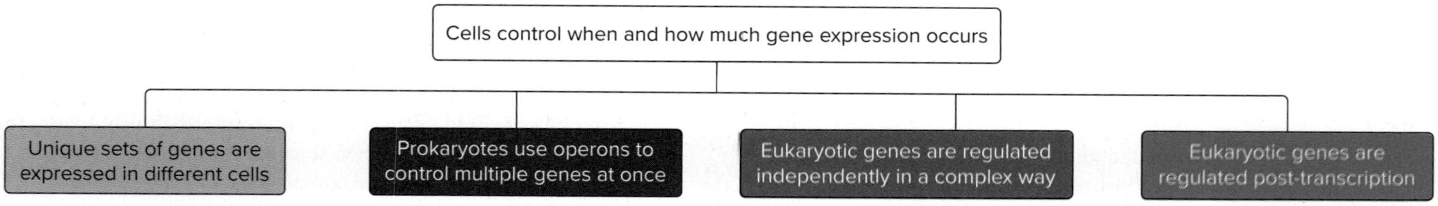

Cells control when and how much gene expression occurs
Unique sets of genes are expressed in different cells
Prokaryotes use operons to control multiple genes at once
Eukaryotic genes are regulated independently in a complex way
Eukaryotic genes are regulated post-transcription

Introduction

In a symphony, various instruments play their own parts at different times; the musical score determines which instruments play when. Similarly, in an organism, different genes are expressed at different times, with a "genetic score," written in regulatory regions of the DNA, determining which genes are active when. The picture on the previous page shows the expanded "puff" of a Drosophila chromosome, which represents genes that are being actively expressed. We have considered the machinery of gene expression; now we turn to how cells control this expression.

16.1 All Organisms Control Expression of Their Genes

Control of gene expression is essential to all organisms. In prokaryotes, it allows the cell to take advantage of changing environmental conditions. In multicellular eukaryotes, it is critical for directing development and maintaining homeostasis.

Unicellular eukaryotes also use different control mechanisms from those of prokaryotes. All eukaryotes have a membrane-bounded nucleus, use similar mechanisms to condense DNA into chromosomes, and have the same gene expression machinery, all of which differ from those of prokaryotes.

Control of Gene Expression Can Occur at Many Levels

LEARNING OBJECTIVE 16.1.1 Identify the point at which control of gene expression usually occurs.

You learned in chapter 15 that gene expression is the conversion of genotype to phenotype—the flow of information from DNA to produce functional proteins that control cellular activities. This traditional view of gene expression includes the idea that the control of the process occurs primarily at the initiation of transcription. Although this view remains valid, evidence is accumulating that both the extent of the genome transcribed and the control of this transcription in multicellular organisms are more complex than anticipated.

In this chapter, we will investigate the control of initiation of transcription in some detail because of its importance, and because it is still the best-studied mechanism for the control of gene expression. With this framework in place, we will also consider how chromatin structure affects gene expression and how control can be exerted posttranscriptionally as well. The latter topic will lead us into the exciting new world of regulatory RNA molecules.

RNA polymerase is key to transcription, and it must have access to the DNA helix and must be capable of binding to the gene's promoter for transcription to begin. **Regulatory proteins** act by modulating the ability of RNA polymerase to bind to the promoter. This process of controlling the access of RNA polymerase to a promoter is common to both prokaryotes and eukaryotes, but the details differ greatly, as you will see.

These regulatory proteins bind to specific nucleotide sequences on the DNA that are usually only 10–15 nt in length. (Even a large regulatory protein has a "footprint," or binding area, of only about 20 nt.) Hundreds of these regulatory sequences have been characterized, and each provides a binding site for a specific protein that is able to recognize the sequence. Binding of the protein either *blocks* transcription by getting in the way of RNA polymerase or *stimulates* transcription by facilitating the binding of RNA polymerase to the promoter.

Control Strategies Differ Between Prokaryotes and Eukaryotes

LEARNING OBJECTIVE 16.1.2 Compare strategies for control of gene expression in prokaryotes and eukaryotes.

Control of gene expression is accomplished very differently in prokaryotes than it is in eukaryotes. Prokaryotic cells have been shaped by evolution to grow and divide as rapidly as possible, enabling them to exploit transient resources. Proteins in prokaryotes turn over rapidly, allowing these organisms to respond quickly to changes in their external environment by changing patterns of gene expression.

In prokaryotes, the primary function of gene control is to adjust the cell's activities to its immediate environment. Changes in gene expression alter which enzymes are present in response to the quantity and type of available nutrients and the amount of oxygen. Almost all of these changes are fully reversible, allowing the cell to adjust its enzyme levels up or down in response to environmental changes.

The cells of multicellular organisms, in contrast, have been shaped by evolution to be protected from transient changes in their immediate environment. Most of them experience fairly constant conditions. Indeed, *homeostasis*—the maintenance of a constant internal environment—is considered by many to be the hallmark of multicellular organisms. Cells in such organisms respond to signals in their immediate environment (such as growth factors and hormones) by altering gene expression, and in doing so they participate in regulating the body as a whole.

Some of these changes in gene expression compensate for changes in the physiological condition of the body. Others mediate the decisions that produce the body, ensuring that the correct genes are expressed in the right cells at the right time during development. Later chapters deal with the details, but for now we can simplify by saying that the growth and development of multicellular organisms entail a long series of biochemical reactions, each catalyzed by a specific enzyme. Once a particular developmental change has occurred, these enzymes cease to be active, lest they disrupt the events that must follow.

To produce this sequence of enzymes, genes are transcribed in a carefully prescribed order, each for a specified period of time, following a fixed genetic program that may even lead to programmed cell death **(apoptosis).** The onetime expression of the genes that guide a developmental program is fundamentally different from the reversible metabolic adjustments prokaryotic cells make in response to the environment. In all multicellular organisms, changes in gene expression within particular cells serve the needs of the whole organism, rather than the survival of individual cells.

REVIEW OF CONCEPT 16.1

Gene expression is usually controlled at the level of transcription initiation. Regulatory proteins bind to specific DNA sequences and affect the binding of RNA polymerase to promoters. Prokaryotes regulate gene expression to adjust the cell's activities to the environment. Multicellular eukaryotes regulate gene expression to maintain internal homeostasis.

- *Would the control of gene expression in a unicellular eukaryote such as yeast be more like that of humans or like that of* E. coli*?*

16.2 Regulatory Proteins Control Genes by Interacting with Specific DNA Nucleotide Sequences

The ability of regulatory proteins to bind to *specific* DNA sequences provides the basic tool of gene regulation—the key ability that makes transcriptional control possible. To understand how cells control gene expression, it is first necessary to gain a clear picture of this molecular recognition process.

DNA-Binding Motifs Interact with Specific DNA Sequences Without Unwinding the Helix

LEARNING OBJECTIVE 16.2.1 Describe the common features of DNA-binding motifs.

Regulatory proteins bind to specific sequences of bases without actually unwinding the DNA helix. When a protein binds to the surface of DNA, the edges of the bases are exposed in two grooves in the molecule: a larger major groove and a smaller minor groove (refer to figure 14.8 for review). Within the deeper major groove, the nucleotides' hydrogen bond donors and acceptors are accessible. The pattern created by these chemical groups is unique for each of the four base-pair arrangements, allowing proteins nestled in the groove to read the sequence of bases (figure 16.1).

Protein–DNA recognition is critical to all DNA functions, including replication, gene expression, and the control of these activities. When the structure of a large number of DNA-binding proteins was determined, a small number of **DNA-binding motifs** were identified (motifs are described in chapter 3). These motifs arose relatively early in evolution and have been reused with alterations in a wide variety of proteins. These DNA-binding motifs share the property of interacting with specific sequences of bases, usually through the major groove of the DNA helix.

DNA-binding motifs are the key structure within the DNA-binding domain of these proteins. This domain is a functionally distinct part of the protein necessary to bind to DNA in a sequence-specific manner. Regulatory proteins must also be able to interact with the transcription apparatus, which is accomplished by a different regulatory domain.

Two proteins that share the same DNA-binding domain do not necessarily bind to the same DNA sequence. The similarities in the DNA-binding motifs appear in their three-dimensional structure, not in the specific contacts they make with DNA. We will consider four of the best-known motifs to provide a sense of how DNA-binding proteins interact with DNA.

The helix-turn-helix motif

The most common DNA-binding motif is the helix-turn-helix, constructed from two α-helical segments of the protein linked by a short, nonhelical segment, the "turn" (figure 16.2*a*). This was one of the first such motifs discovered, and is found in hundreds of DNA-binding proteins.

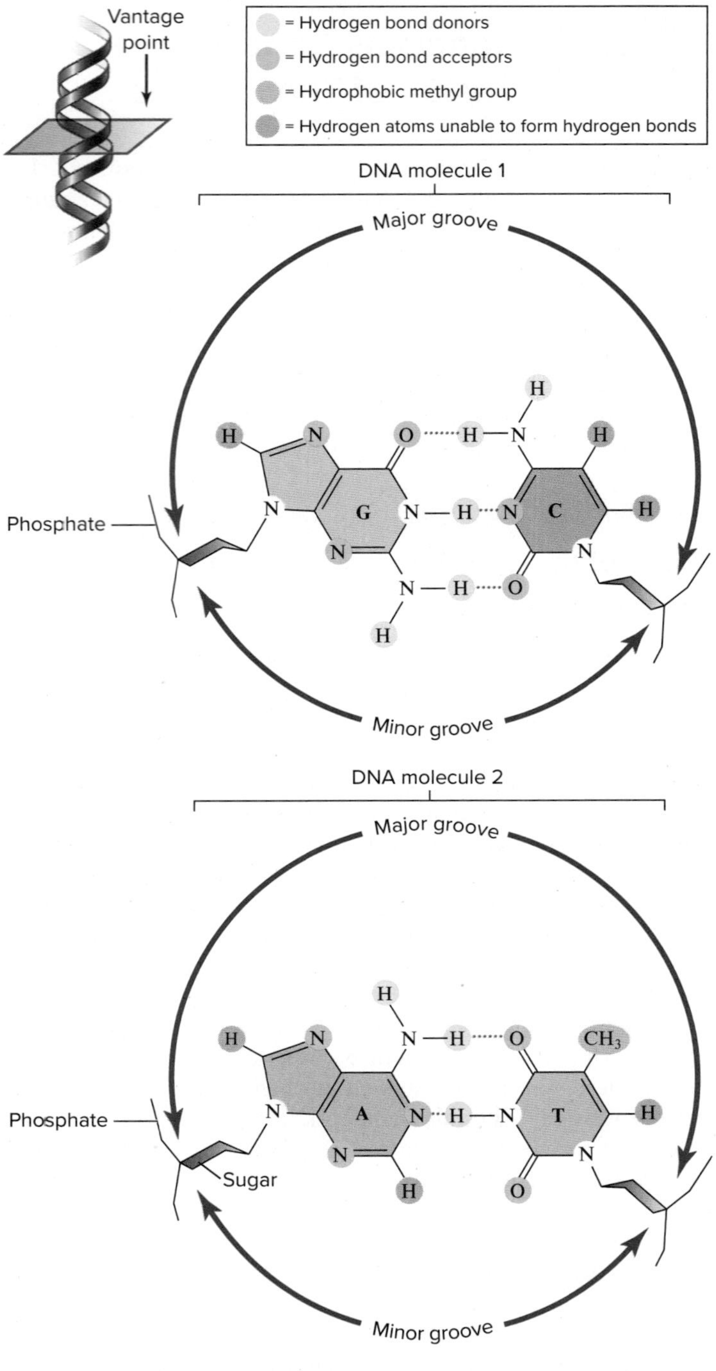

Figure 16.1 Reading the major groove of DNA. Looking down into the major groove of a DNA helix, we can see the edges of the bases protruding into the groove. Each of the four possible base-pair arrangements (two are shown here) extends a unique set of chemical groups into the groove, indicated in this diagram by differently colored circles. A regulatory protein can identify the base-pair arrangement by this characteristic signature.

The structure of a helix-turn-helix motif reveals how it interacts with the major groove of DNA. The helical segments of the motif interact with one another such that they are held at roughly right angles. One of the helices, the recognition helix, fits into the major groove, and the other interacts with the outside of the DNA molecule to help position the recognition helix.

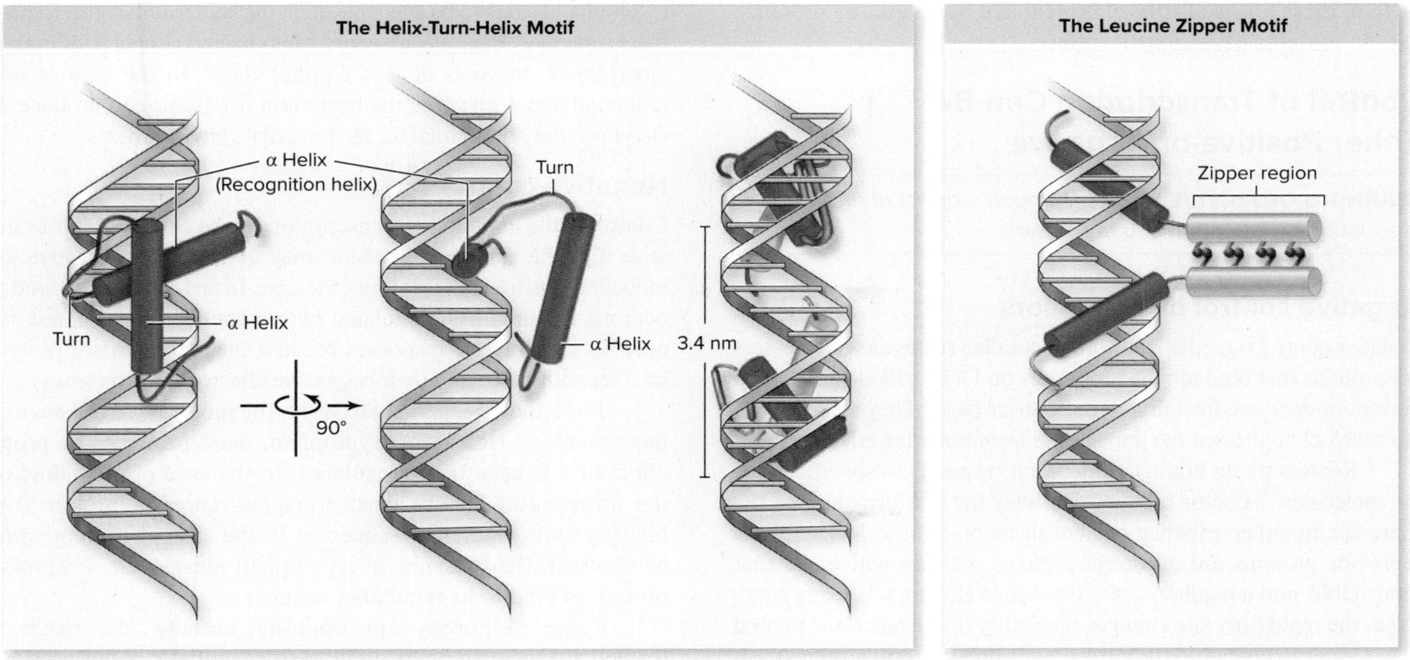

a. *b.*

Figure 16.2 Major DNA-binding motifs. Two different DNA-binding motifs are pictured interacting with DNA. ***a.*** The helix-turn-helix motif binds to DNA using one α helix, the recognition helix, to interact with the major groove. The other helix positions the recognition helix. Proteins with this motif are usually dimers, with two identical subunits, each containing the DNA-binding motif. The two copies of the motif (red) are separated by 3.4 nm, precisely the spacing of one turn of the DNA helix. ***b.*** The leucine zipper acts to hold two subunits in a multisubunit protein together, thereby allowing α-helical regions to interact with DNA.

The DNA-regulatory sequences recognized by helix-turn-helix proteins occur in symmetrical pairs. The proteins contain two helix-turn-helix motifs separated by 3.4 nm, which is the distance required for one turn of the DNA helix. Having two DNA-binding sites strengthens the interaction between protein and DNA.

The homeodomain motif

A class of helix-turn-helix motif, the **homeodomain,** plays a critical role in development in many eukaryotic organisms, including humans. Mutations in *Drosophila* called homeotic mutations cause one body part to be replaced by another. Analysis of homeotic genes led to the discovery of a 60-amino-acid domain called the homeodomain. The most conserved part of the homeodomain contains a recognition helix of a helix-turn-helix motif. This was the first indication that developmental mechanisms are ancient and shared across great phylogenetic distances. At the level of our genes, we are much closer to a fly than you might imagine.

The zinc finger motif

A different kind of DNA-binding motif uses one or more zinc atoms to coordinate its binding to DNA. This motif is called a zinc **finger motif** because a zinc atom links an α-helical segment to a β-sheet segment (refer to chapter 3) where the helical segment fits into the major groove of DNA. This type of motif often occurs in clusters, the β sheets spacing the helical segments so that each helix contacts the major groove. The effect is like a hand wrapped around the DNA with the fingers lying in the major groove.

The leucine zipper motif

Another DNA-binding motif is formed by the interaction of different protein subunits. This motif is created where a region on one subunit containing several hydrophobic amino acids (usually leucines) interacts with a similar region on the other subunit. This interaction holds the two subunits together at those regions, while the rest of the subunits remain separated. Called a **leucine zipper,** this structure has the shape of a Y, with the two arms of the Y being helical regions that fit into the major groove of DNA (figure 16.2*b*). Because the two subunits can contribute quite different helical regions to the motif, leucine zippers allow for great flexibility in controlling gene expression.

REVIEW OF CONCEPT 16.2

The DNA helix has a major groove and a minor groove that allow regulatory proteins to interact with bases without unwinding the DNA. These proteins have DNA-binding motifs, which often include α-helical segments. These motifs are active in DNA binding, whereas other domains interact with the transcription machinery.

- *What would be the effect of a mutation in a helix-turn-helix protein that altered the spacing of the two helices?*

16.3 Prokaryotes Regulate Their Genes in Clusters

Control at the level of transcription initiation can be either positive or negative. **Positive control** increases the frequency of initiation, and **negative control** decreases the frequency of initiation.

Each of these forms of control is mediated by regulatory proteins, but the proteins have opposite effects.

Control of Transcription Can Be Either Positive or Negative

LEARNING OBJECTIVE 16.3.1 Compare control of enzyme production by induction and repression.

Negative control by repressors

Negative control is mediated by proteins called **repressors.** Repressors are proteins that bind to regulatory sites on DNA called **operators** to prevent or decrease the initiation of transcription. They act as a kind of roadblock to prevent the polymerase from initiating effectively.

Repressors do not act alone; each responds to specific effector molecules. Effector binding can alter the conformation of the repressor to either enhance or abolish its binding to DNA. These repressor proteins are allosteric proteins with an active site that binds DNA and a regulatory site that binds effectors. Effector binding at the regulatory site changes the ability of the repressor to bind DNA (refer to chapter 6 for more details on allosteric proteins).

Positive control by activators

Positive control is mediated by another class of regulatory, allosteric proteins called *activators* that can bind to DNA and stimulate the initiation of transcription. These activators enhance the binding of RNA polymerase to the promoter to increase the frequency of transcription initiation. Activators are the logical and physical opposites of repressors. Effector molecules can either enhance or decrease activator binding.

Adjusting gene expression in response to environmental changes

Changes in the environments that bacteria and archaea encounter often result in changes in gene expression. In general, genes encoding proteins involved in catabolic pathways (breaking down molecules) are controlled in a manner opposite from genes encoding proteins involved in anabolic pathways (building up molecules). In the discussion that follows, we will use lactose utilization as a model for catabolic functions, and tryptophan biosynthesis as a model for anabolic functions.

As mentioned in chapter 15, prokaryotic genes are often organized into **operons,** multiple genes that are part of a single transcription unit having a single promoter. Genes that are involved in the same metabolic pathway are often organized in this fashion. The proteins necessary for the utilization of lactose are encoded by the ***lac* operon,** and the proteins necessary for the synthesis of tryptophan are encoded by the ***trp* operon.**

Induction and repression

If a bacterium encounters lactose, it begins to make the enzymes necessary to utilize lactose. When lactose is not present, however, there is no need to make these proteins. Thus, we say that the synthesis of the proteins is *induced* by the presence of lactose. **Induction** therefore occurs when enzymes for a certain pathway are produced in response to a substrate.

When tryptophan is available in the environment, a bacterium will not synthesize the enzymes necessary to make tryptophan. If tryptophan ceases to be available, then the bacterium begins to make these enzymes. **Repression** occurs when bacteria capable of making biosynthetic enzymes do not produce them. In the case of both induction and repression, the bacterium is adjusting to produce the enzymes that are optimal for its immediate environment.

Negative control

Control of the initiation of transcription can be either positive or negative. On the surface, repression may appear to be negative and induction positive, but this is not the case. In fact, both the *lac* and *trp* operons are negatively regulated by repressor proteins. These two operons have inverse responses because the two repressor proteins interact with DNA and their respective effectors in inverse ways.

For either mechanism to work, the molecule in the environment, such as lactose or tryptophan, must produce the proper effect on the gene being regulated. In the case of *lac* induction, the presence of lactose must *prevent* a repressor protein from binding to its regulatory sequence. In the case of *trp* repression, by contrast, the presence of tryptophan must *cause* a repressor protein to bind to its regulatory sequence.

These responses are opposite because the needs of the cell are opposite in anabolic versus catabolic pathways. Each pathway is examined in detail to show how protein–DNA interactions allow the cell to respond to environmental conditions.

Induction of the *lac* Operon Is Negatively Regulated by the *lac* Repressor

LEARNING OBJECTIVE 16.3.2 Explain how the *lac* operon is regulated based on the availability of lactose.

The *lac* operon consists of the genes that encode functions necessary to utilize lactose: β-galactosidase *(lacZ),* lactose permease *(lacY*), and lactose transacetylase *(lacA),* plus the regulatory regions necessary to control the expression of these genes (figure 16.3). In addition, the gene for the *lac* repressor *(lacI)* is linked to the rest of the *lac* operon and is thus considered part of the operon, although it has its own promoter. The arrangement of the control regions upstream of the coding region is typical of most prokaryotic operons, although the linked repressor is not.

Action of the repressor

Initiation of transcription of the *lac* operon is controlled by the *lac* repressor. The repressor binds to the operator, which is adjacent to the promoter (figure 16.4*a*). This binding prevents RNA polymerase from binding to the promoter. This DNA binding is sensitive to the presence of lactose: the repressor binds DNA in the absence of lactose, but not in the presence of lactose.

Interaction of repressor and inducer

In the absence of lactose, the *lac* repressor binds to the operator, and the operon is repressed (figure 16.4*a*). The effector that controls the DNA binding of the repressor is a metabolite of lactose, allolactose, which is produced when lactose is available. Allolactose binds to the repressor, altering its conformation so that it no longer can bind to the operator (figure 16.4*b*). The operon is now induced. Because allolactose allows induction of the operon, it is usually called the inducer.

As the level of lactose falls, allolactose concentrations decrease, making allolactose no longer available to bind to the

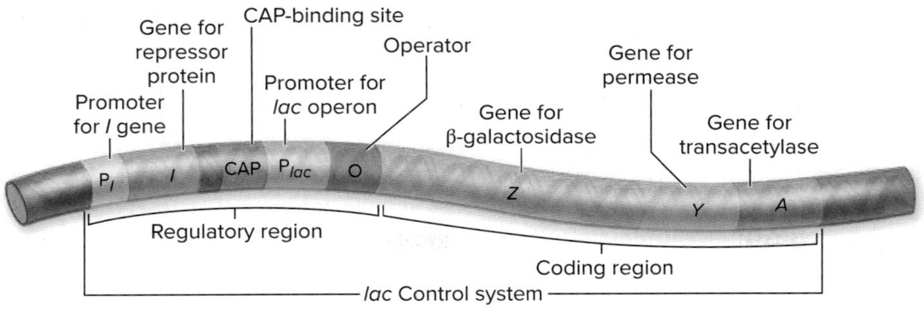

Figure 16.3 The *lac* region of the *Escherichia coli* chromosome. The *lac* operon consists of a promoter, an operator, and three genes (*lac Z, Y,* and *A*) that encode proteins required for the metabolism of lactose. In addition, there is a binding site for the catabolite activator protein (CAP), which affects RNA polymerase binding to the promoter. The *I* gene encodes the repressor protein, which can bind the operator and block transcription of the *lac* operon.

repressor, which in turn allows the repressor to bind to DNA again. Thus, this system of negative control by the *lac* repressor and its inducer, allolactose, allows the cell to respond to changing levels of lactose in the environment.

Even in the absence of lactose, the *lac* operon is expressed at a very low level. When lactose becomes available, it is transported into the cell, and enough allolactose is produced that induction of the operon can occur.

Expression of the *lac* Operon Is Also Affected by Glucose Levels

LEARNING OBJECTIVE 16.3.3 Explain how glucose affects the production of lactose-utilizing enzymes.

Glucose repression is a mechanism for the preferential use of glucose in the presence of other sugars such as lactose. If bacteria are

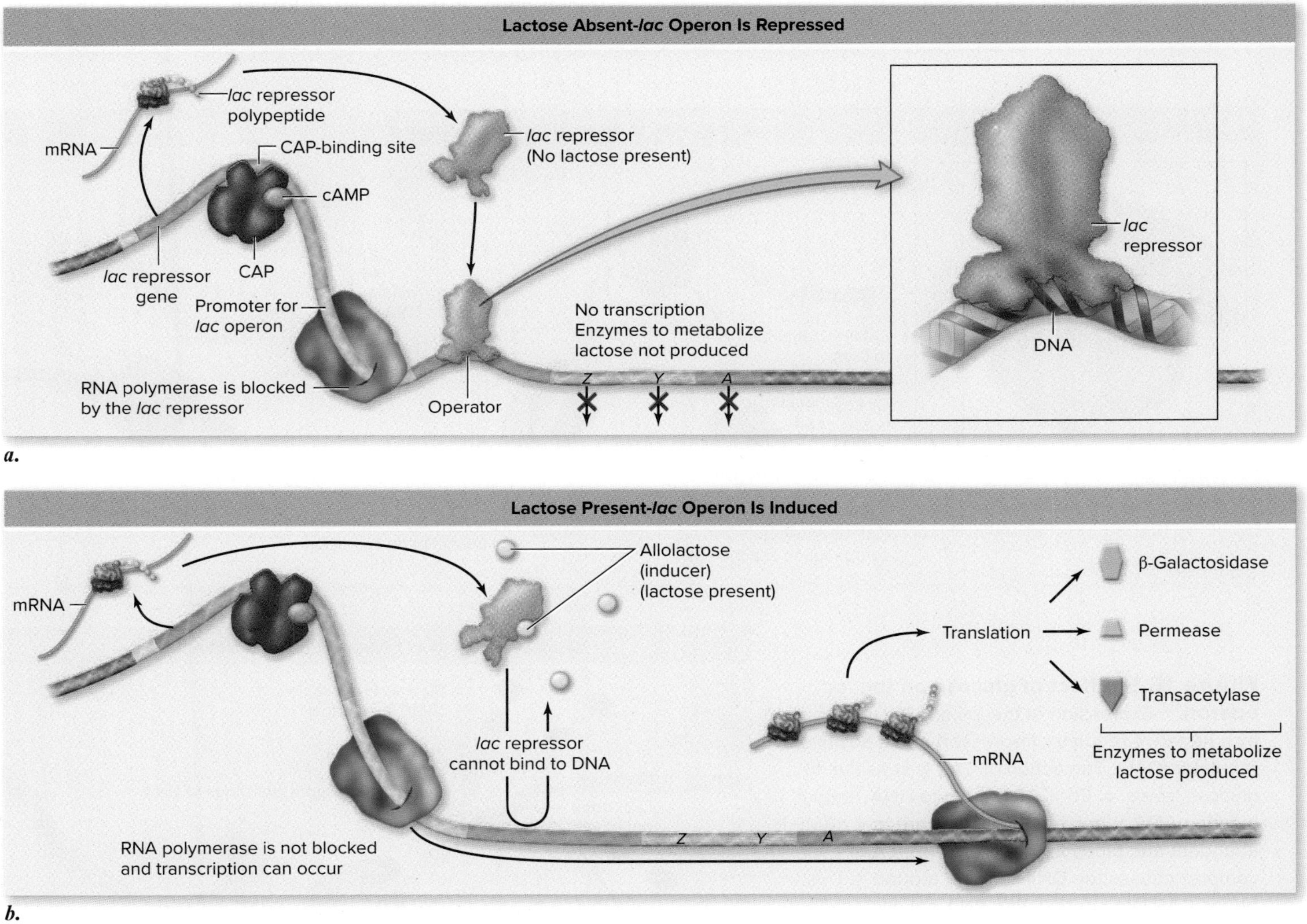

Figure 16.4 Induction of the *lac* operon. ***a.*** In the absence of lactose, the *lac* repressor binds to DNA at the operator site, thus preventing transcription of the operon. When the repressor protein is bound to the operator site, the *lac* operon is shut down (repressed). ***b.*** The *lac* operon is transcribed (induced) when CAP is bound and when the repressor is not bound. Allolactose binding to the repressor alters the repressor's shape so it cannot bind to the operator site and block RNA polymerase activity.

grown in the presence of both glucose and lactose, the *lac* operon is not induced. As glucose is used up, the *lac* operon is induced, allowing lactose to be used as an energy source.

Despite the name *glucose repression,* this mechanism involves an activator protein that can stimulate transcription from multiple catabolic operons, including the *lac* operon. This activator, **catabolite activator protein (CAP),** is an allosteric protein with cAMP as an effector. This protein is also called **cAMP response protein (CRP)** because it binds cAMP, but we will use the name CAP to emphasize its role as a positive regulator. CAP alone does not bind to DNA, but binding of the effector cAMP to CAP changes its conformation such that it can bind to DNA (figure 16.5). The level of cAMP in cells is reduced in the presence of glucose so that no stimulation of transcription from CAP-responsive operons takes place.

The CAP–cAMP system was long thought to be the sole mechanism of glucose repression. But more recent research has indicated that the presence of glucose inhibits the transport of lactose into the cell. This deprives the cell of the *lac* operon inducer, allolactose, allowing the repressor to bind to the operator. This mechanism, called **inducer exclusion,** is now thought to be the main form of glucose repression of the *lac* operon.

Given that inducer exclusion occurs, the role of CAP in the absence of glucose seems superfluous. But, in fact, the positive control of CAP–cAMP is necessary, because the promoter of the *lac* operon alone is not efficient in binding RNA polymerase. This inefficiency is overcome by the action of the positive control of the CAP–cAMP activator. Thus, the highest levels of expression occur in the absence of glucose and the presence of lactose. In this case, the presence of the activator and the absence of the repressor combine to produce the highest levels of expression (figure 16.5).

The *trp* Operon Is Negatively Regulated by the *trp* Repressor

LEARNING OBJECTIVE 16.3.4 Explain how the *trp* operon is regulated by levels of tryptophan.

Like the *lac* operon, the *trp* operon consists of a series of genes that encode enzymes involved in the same biochemical pathway. In the case of the *trp* operon, these enzymes are necessary for synthesizing tryptophan. The regulatory region that controls transcription of these genes is located upstream of the genes. The *trp* operon is controlled by a repressor encoded by a gene

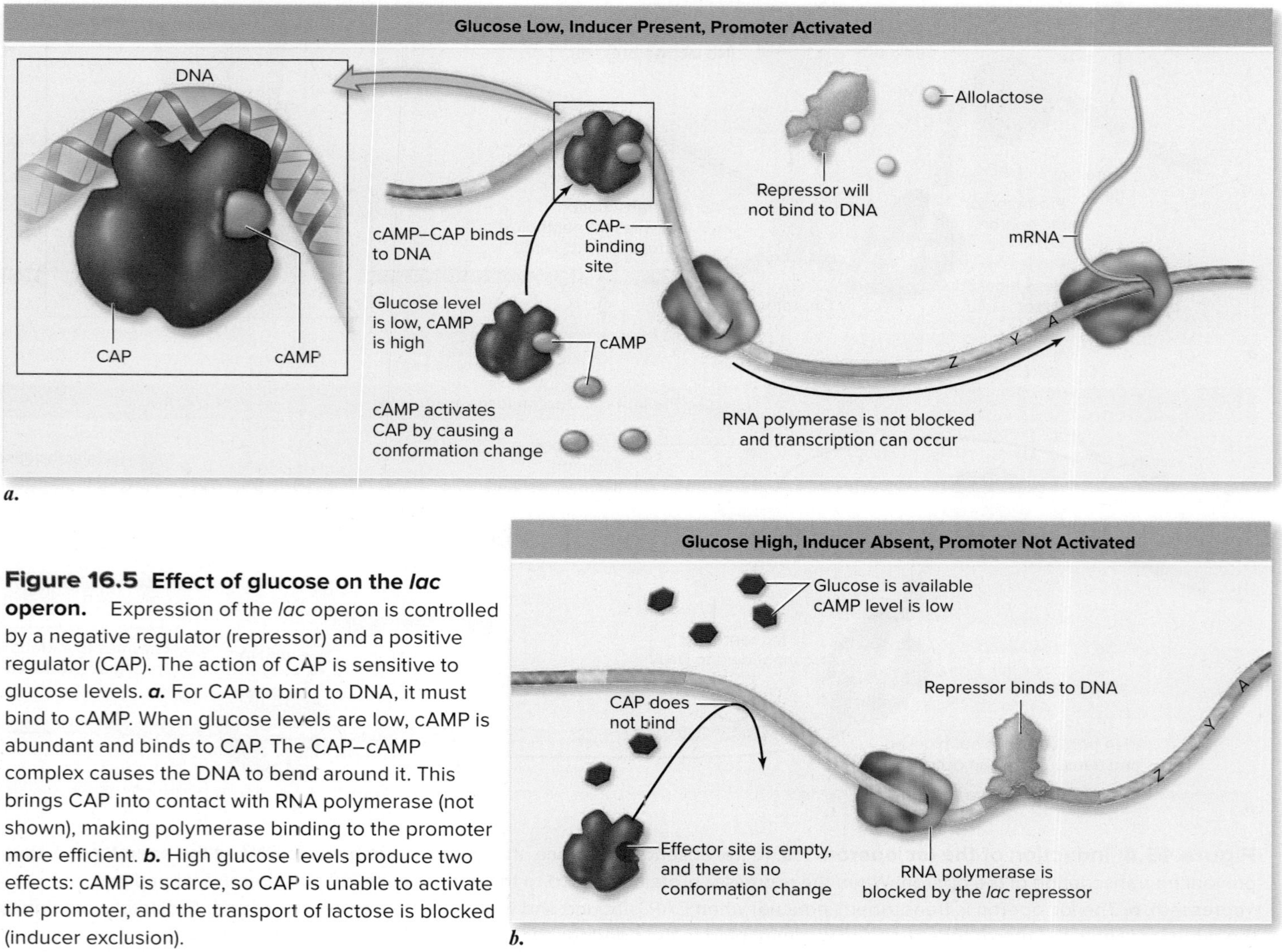

Figure 16.5 Effect of glucose on the *lac* operon. Expression of the *lac* operon is controlled by a negative regulator (repressor) and a positive regulator (CAP). The action of CAP is sensitive to glucose levels. ***a.*** For CAP to bind to DNA, it must bind to cAMP. When glucose levels are low, cAMP is abundant and binds to CAP. The CAP–cAMP complex causes the DNA to bend around it. This brings CAP into contact with RNA polymerase (not shown), making polymerase binding to the promoter more efficient. ***b.*** High glucose levels produce two effects: cAMP is scarce, so CAP is unable to activate the promoter, and the transport of lactose is blocked (inducer exclusion).

located outside the *trp* operon. The *trp* operon is continuously expressed in the absence of tryptophan and is not expressed in the presence of tryptophan.

The *trp* repressor is a helix-turn-helix protein that binds to the operator site located adjacent to the *trp* promoter (figure 16.6). In the absence of tryptophan, the *trp* repressor does not bind to its operator, allowing expression of the operon and production of the enzymes necessary to make tryptophan.

When levels of tryptophan rise, then tryptophan (the *corepressor*) binds to the repressor and alters its conformation, allowing it to bind to its operator. Binding of the repressor–corepressor complex to the operator prevents RNA polymerase from binding to the promoter. The actual change in repressor structure due to tryptophan binding is an alteration of the orientation of a pair of helix-turn-helix motifs that allows their recognition helices to fit into adjacent major grooves of the DNA.

When tryptophan is present and bound to the repressor and this complex is bound to the operator, the operon is said to be *repressed*. As tryptophan levels fall, the repressor alone cannot bind to the operator, allowing expression of the operon. In this state, the operon is said to be **derepressed,** distinguishing this state from induction (figure 16.6).

The key to understanding how both induction and repression can be due to negative regulation is knowledge of the behavior of repressor proteins and their effectors. In induction, the repressor alone can bind to DNA, and the inducer prevents DNA binding. In

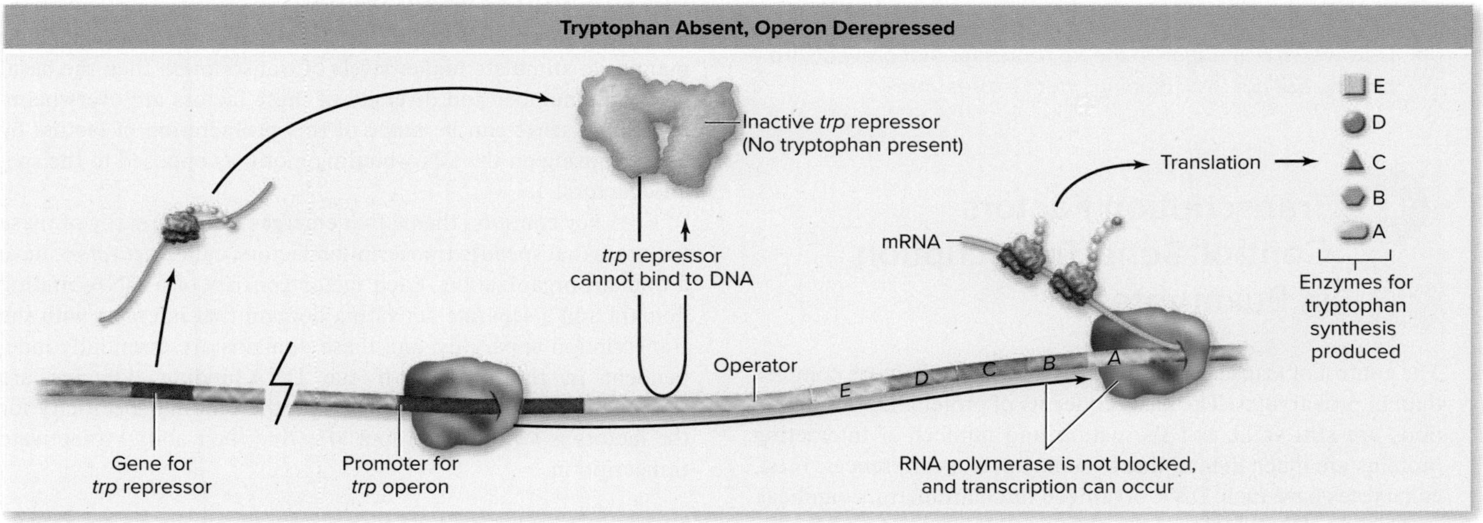

a.

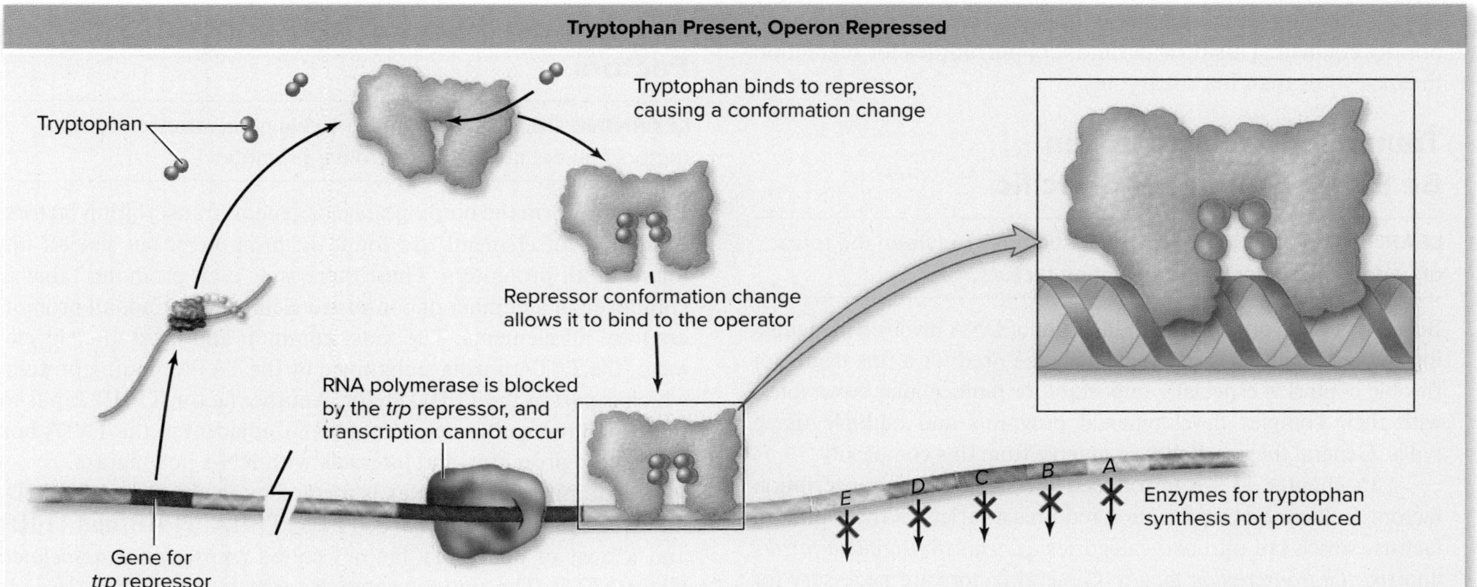

b.

Figure 16.6 How the *trp* operon is controlled. The tryptophan operon encodes the enzymes necessary to synthesize tryptophan. ***a.*** The tryptophan repressor alone cannot bind to DNA. The promoter is free to function, and RNA polymerase transcribes the operon. ***b.*** When tryptophan is present, it binds to the repressor, altering its conformation so it now binds DNA. The tryptophan–repressor complex binds tightly to the operator, preventing RNA polymerase from initiating transcription.

the case of repression, the repressor binds DNA only when bound to the corepressor. Induction and repression are excellent examples of how interactions of molecules can affect their structures and how molecular structure is critical to function.

REVIEW OF CONCEPT 16.3

Turning on gene expression in response to a substrate is called induction, and turning off gene expression in response to levels of a metabolite is called repression. The *lac* operon is induced in response to lactose. This involves negative control mediated by a repressor protein that binds DNA in the absence of lactose; repressor bound to the inducer allolactose can no longer bind DNA. The *trp* operon is also negatively regulated by a repressor. In this case, the repressor only binds DNA when also bound to tryptophan.

- *How would a mutation in the* trp *repressor that prevents* trp *binding but not DNA binding affect expression?*

16.4 Transcription Factors Control Gene Transcription in Eukaryotes

The control of transcription in eukaryotes is much more complex than in prokaryotes. The basic concepts of protein–DNA interactions are still valid, but the nature and number of interacting proteins are much greater due to some obvious differences. First, eukaryotes have their DNA organized into chromatin, complicating protein–DNA interactions considerably. Second, eukaryotic transcription occurs in the nucleus, and translation occurs in the cytoplasm; in prokaryotes, these processes are spatially and temporally coupled. This provides more opportunities for regulation in eukaryotes than in prokaryotes.

Transcription Factors Can Be Either General or Specific

LEARNING OBJECTIVE 16.4.1 Distinguish between the roles of general and specific transcription factors.

Because of these differences, the amount of DNA involved in regulating eukaryotic genes is much greater. The need for a fine degree of flexible control is especially important for multicellular eukaryotes, with their complex developmental programs and multiple tissue types. General themes, however, emerge from this complexity.

In chapter 15 we introduced the concept of transcription factors. Eukaryotic transcription requires a variety of these protein factors, which fall into two categories: *general transcription factors* and *specific transcription factors.* General factors are necessary for transcription to occur at all, and specific factors increase the level of transcription in certain cell types or in response to signals.

General transcription factors

Transcription of RNA polymerase II templates (the majority being genes that encode protein products) requires more than just RNA polymerase II to initiate transcription. A host of general transcription factors are also necessary to recruit RNA polymerase II to a promoter and assemble an initiation complex for productive initiation. These factors are required for transcription to occur, but they do not increase the rate above this basal rate.

General transcription factors are named with letter designations that follow the abbreviation TFII, for "transcription factor RNA polymerase II" (also true of polymerase I and III factors, except that these would be named TFI and TFIII, respectively). These general factors interact with other accessory factors and function to recognize a promoter. The RNA polymerase II enzyme interacts with the general factors to form an initiation complex that is competent to initiate transcription.

Specific transcription factors

Specific transcription factors act in a tissue- or time-dependent manner to stimulate higher levels of transcription than the basal level. The number and diversity of these factors are overwhelming. Some sense can be made of this proliferation of factors by concentrating on the DNA-binding motif, as opposed to the specific factors.

A key common theme that emerges from the study of these factors is that specific transcription factors, called *activators,* have a domain organization. Each factor consists of a DNA-binding domain and a separate activating domain that interacts with the transcription apparatus, and these domains are essentially independent in the protein. If the DNA-binding domains are "swapped" between different factors, the binding specificity for the factors is switched without affecting their ability to activate transcription.

Promoters and Enhancers Are Binding Sites for Transcription Factors

LEARNING OBJECTIVE 16.4.2 Explain how transcription factors can act at a distance from a promoter.

Promoters form the binding sites for general transcription factors. A number of elements are found in promoters, but not all are found in all promoters. Thus, there is a "core promoter" that is composed of a number of conserved elements, but not all promoters have all elements. The most common and most ancient element, the TATA box, is recognized by the TATA-binding protein, which is part of the TFIID factor. Another factor, TFIIB, binds to the TFIIB recognition element (BRE), adjacent to the TATA box in the core promoter, and interacts with RNA polymerase.

An initiation complex is formed by the binding of TFIID, followed by binding of TFIIE, TFIIF, TFIIA, TFIIB, and TFIIH and a host of accessory factors called *transcription-associated factors, TAFs.* The *initiation complex* that results (figure 16.7) is much more complex than the bacterial equivalent, yet even this complex is only capable of basal levels of transcription; high levels require the action of other specific factors.

In eukaryotes, promoters are the binding sites for general transcription factors, which then mediate RNA polymerase II binding to the promoter. In contrast, the holoenzyme portion of

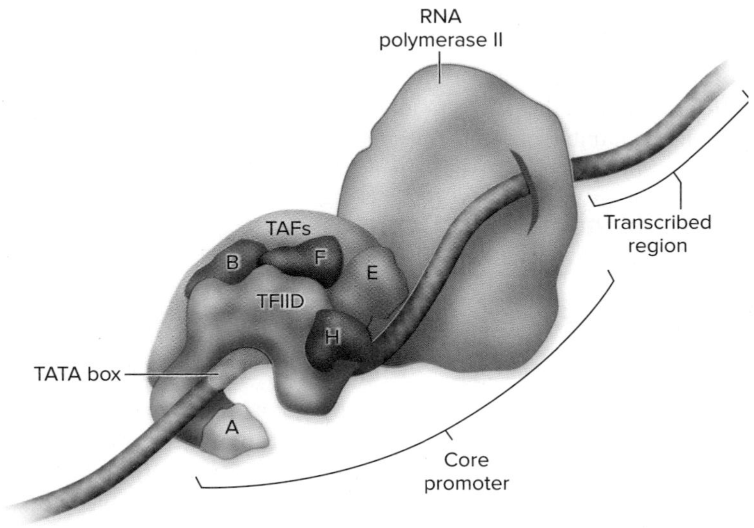

Figure 16.7 Formation of a eukaryotic initiation complex. The general transcription factor, TFIID, binds to the TATA box and is joined by the other general factors, TFIIE, TFIIF, TFIIA, TFIIB, and TFIIH. This complex is added to by a number of transcription-associated factors (TAFs) that together recruit the RNA Pol II molecule to the core promoter.

the RNA polymerase of prokaryotes can directly recognize a promoter and bind to it.

Enhancers were originally defined as DNA sequences necessary for high levels of transcription that can act independently of position or orientation. This was counterintuitive to molecular biologists conditioned by prokaryotic systems to expect control regions to be immediately upstream of the coding region. We now know that enhancers are the binding sites for specific transcription factors, and their ability to act over large distances is due to DNA looping to bring the enhancer close to the promoter.

Although more important in eukaryotic systems, this looping was first demonstrated using prokaryotic DNA-binding proteins (figure 16.8).

This fits well with the emerging view of the organization of chromatin in the nucleus (refer to chapter 10).

Topologically associated domains (*TADs*) are structures formed by chromosome loops that bring together physically distant regions on chromosomes. These TADs appear to correlate with regions that are being actively transcribed, and would allow an activator bound to an enhancer to be in close contact with transcription factors bound to a distant promoter (figure 16.9). This kind of chromatin organization is likely much more important than physical distance on a chromosome.

Coactivators and Mediators Link Transcription Factors to RNA Polymerase II

LEARNING OBJECTIVE 16.4.3 Contrast the roles of coactivators and transcription factors.

Other factors specifically mediate the action of transcription factors. These *coactivators* and *mediators* are also necessary for activation of transcription by the transcription factor. They act by

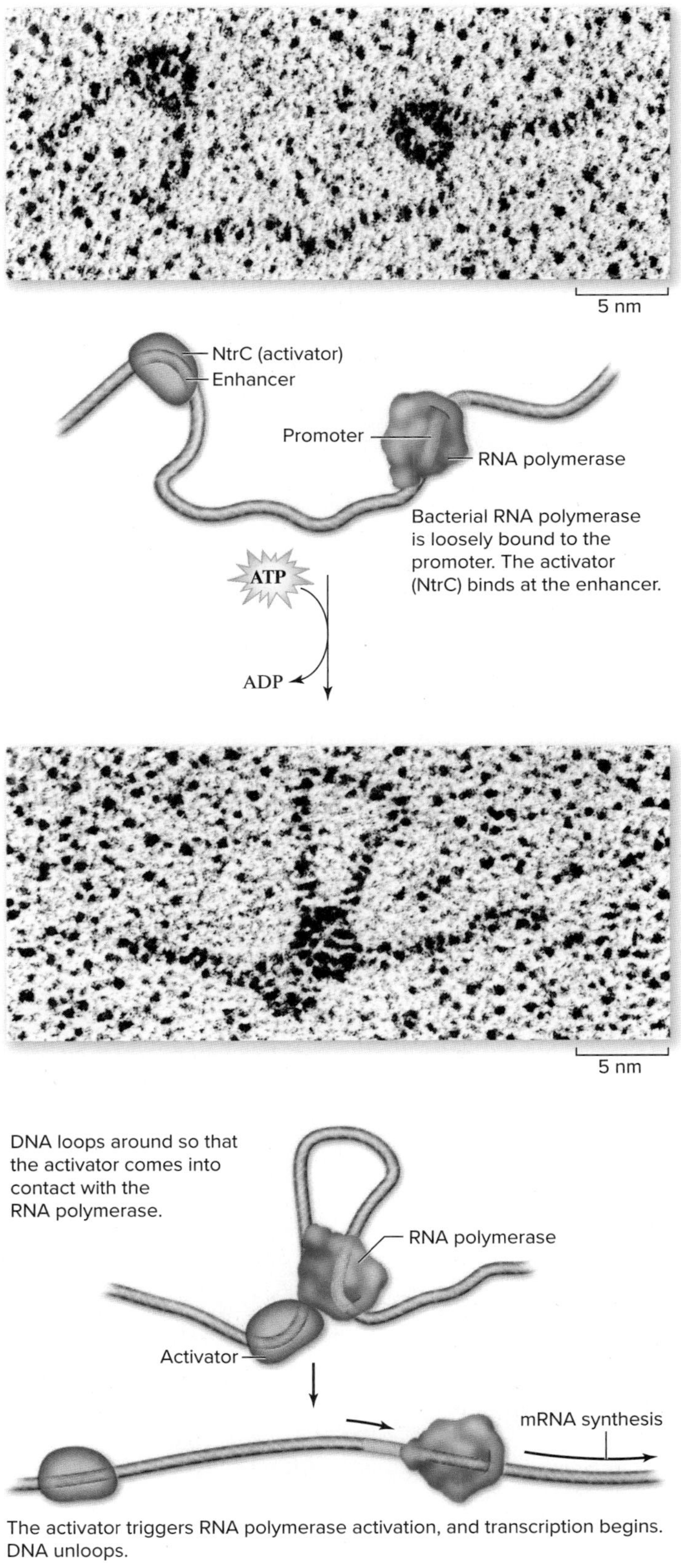

Figure 16.8 DNA looping caused by proteins. When the bacterial activator NtrC binds to an enhancer, it causes the DNA to loop over to a distant site where RNA polymerase is bound, thereby activating transcription. Although such enhancers are rare in prokaryotes, they are common in eukaryotes.

(micrographs): Courtesy of Dr. Harrison Echols and Dr. Sydney Kustu

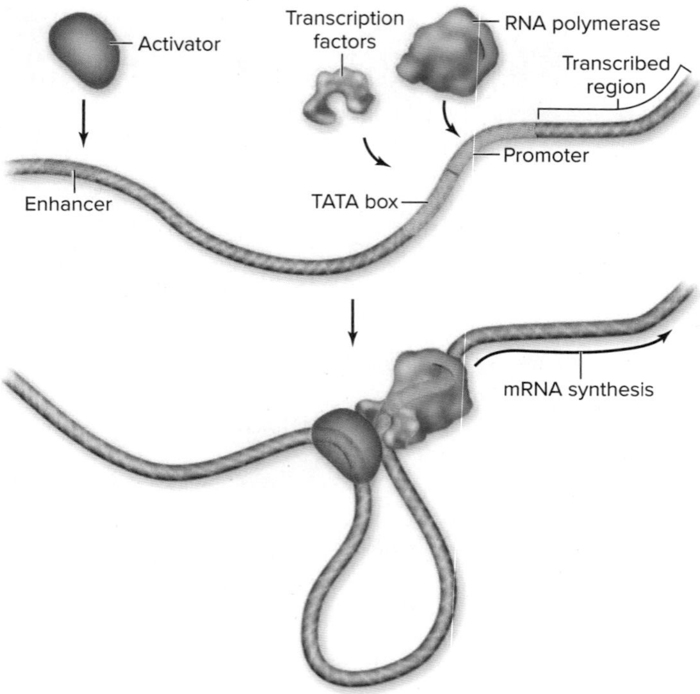

Activator
Transcription factors
RNA polymerase
Transcribed region
Promoter
Enhancer
TATA box
mRNA synthesis

Figure 16.9 How enhancers work. The enhancer site is located far away from the gene being regulated. Binding of an activator (gray) to the enhancer allows the activator to interact with the transcription factors (blue) associated with RNA polymerase, stimulating transcription.

binding the transcription factor and then binding to another part of the transcription apparatus. Mediators are essential to the function of some transcription factors, but not all transcription factors require them. The number of coactivators is much smaller than the number of transcription factors because the same coactivator can be used with multiple transcription factors.

The Transcription Complex Integrates Many Levels of Control

LEARNING OBJECTIVE 16.4.4 Describe the interactions of the components of the eukaryotic transcription complex.

Although a few general principles apply to a broad range of situations, nearly every eukaryotic gene—or group of genes with coordinated regulation—represents a unique case. Virtually all genes that are transcribed by RNA polymerase II need the same suite of general factors to assemble an initiation complex, but the assembly of this complex and its ultimate level of transcription depend on specific transcription factors that in combination make up the **transcription complex** (figure 16.10).

The assembly of this transcription complex is also influenced by chromatin organization, and regulators that affect chromatin structure. The formation of TADs can bring distant regions close

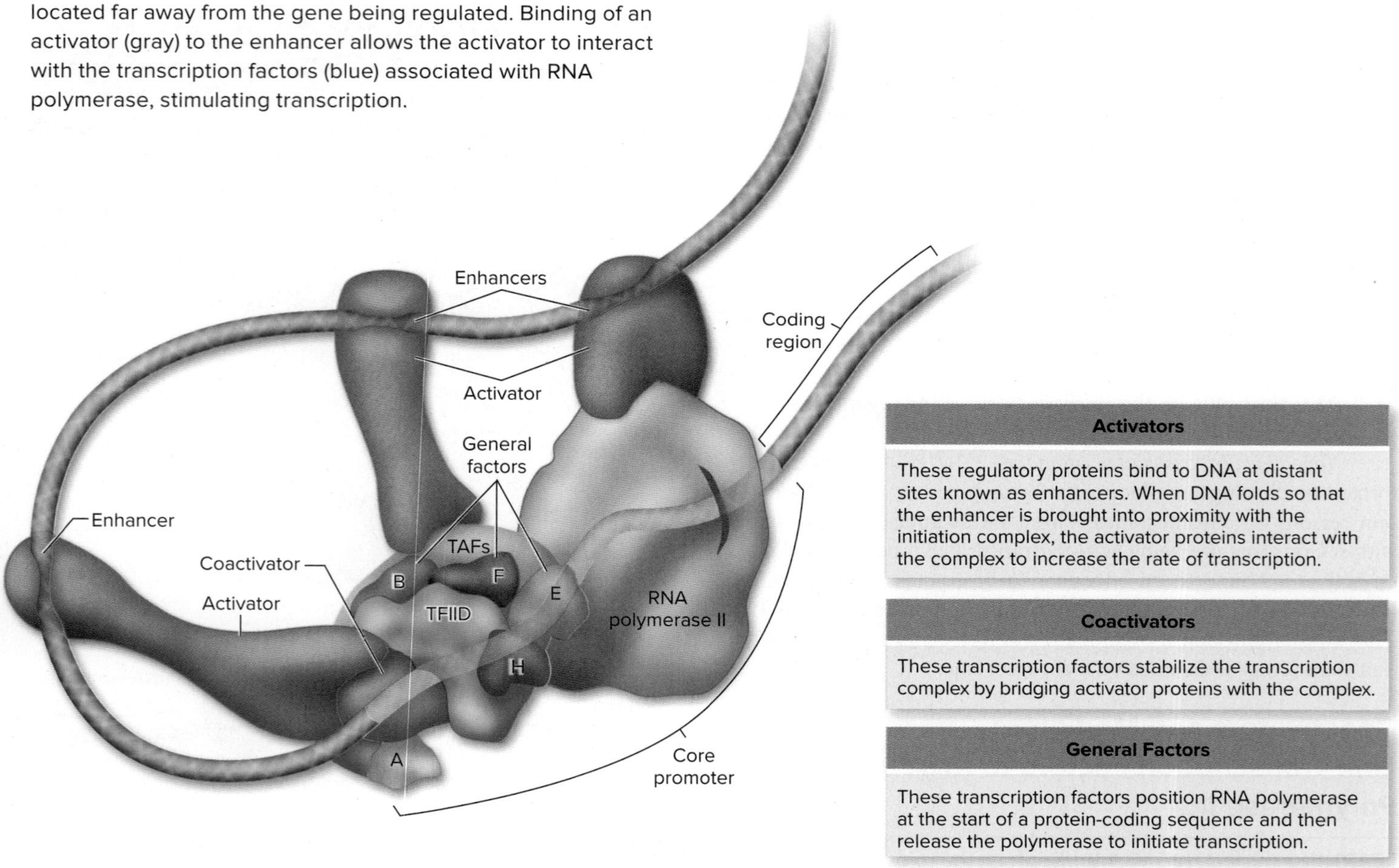

Enhancers
Coding region
Activator
General factors
Enhancer
TAFs
Coactivator
B
F
E
Activator
TFIID
RNA polymerase II
H
A
Core promoter
Activators
These regulatory proteins bind to DNA at distant sites known as enhancers. When DNA folds so that the enhancer is brought into proximity with the initiation complex, the activator proteins interact with the complex to increase the rate of transcription.
Coactivators
These transcription factors stabilize the transcription complex by bridging activator proteins with the complex.
General Factors
These transcription factors position RNA polymerase at the start of a protein-coding sequence and then release the polymerase to initiate transcription.

Figure 16.10 Interactions of various factors within the transcription complex. All specific transcription factors bind to enhancer sequences that may be distant from the promoter. These proteins can then interact with the initiation complex by DNA looping to bring the factors into proximity with the initiation complex. As detailed in the text, some transcription factors, called activators, can directly interact with the RNA polymerase II or the initiation complex, whereas others require additional coactivators.

together, and create regions of active chromatin. The complexity of this organization and the number of actors necessary provide great flexibility in how cells can respond to the many signals they receive that affect transcription, allowing integration of these signals.

REVIEW OF CONCEPT 16.4

In eukaryotes, initiation requires general transcription factors that bind to the promoter and recruit RNA polymerase II to form an initiation complex. General factors produce the basal level of transcription. Specific transcription factors, which bind to enhancer sequences, can increase the level of transcription. Enhancers can act at a distance, because DNA can loop, bringing an enhancer and a promoter closer together. Additional coactivators and mediators link certain specific transcription factors to RNA polymerase II.

- *What would be the effect of a mutation that resulted in the loss of a general transcription factor versus the loss of a specific factor?*

16.5 Chromatin Structure Affects Gene Expression

In a diploid human cell, about 6×10^9 bp of DNA is wrapped into 3×10^7 nucleosomes. Further compaction forms the chromatin that fits about 2 m of DNA into a 5- to 10-µm nucleus (refer to chapter 10). This obviously complicates the process of transcription. The way the cell deals with this conundrum is to selectively modulate the structure of chromatin by modifying DNA and histones, to allow the DNA transactions necessary for life. These alterations to chromatin are thought to be the basis for **epigenetics**—heritable changes in phenotype not due to changes in DNA sequence.

DNA Methylation, Histone Modification, and Noncoding RNA All Affect Chromatin Structure

LEARNING OBJECTIVE 16.5.1 Describe the role of methylation in gene regulation.

One definition of an epigenetic alteration is that it must persist in the absence of the initiating stimulus and must be inherited through cell division. Alterations to chromatin structure can satisfy this definition, although the mechanisms of mitotic inheritance are not yet clear. Chromatin structure is affected by a wide variety of modifications to histones as well as DNA methylation, and even noncoding RNAs.

DNA methylation

DNA methylation was the first modification of chromosome structure shown to act epigenetically. The addition of a methyl group to cytosine by a methylase enzyme creates 5-methylcytosine, but this change has no effect on its base-pairing with guanine (figure 16.11). High levels of DNA methylation correlate with inactive genes, and the allele-specific gene expression seen in genomic imprinting (refer to chapter 13) is mediated at least in part by DNA methylation.

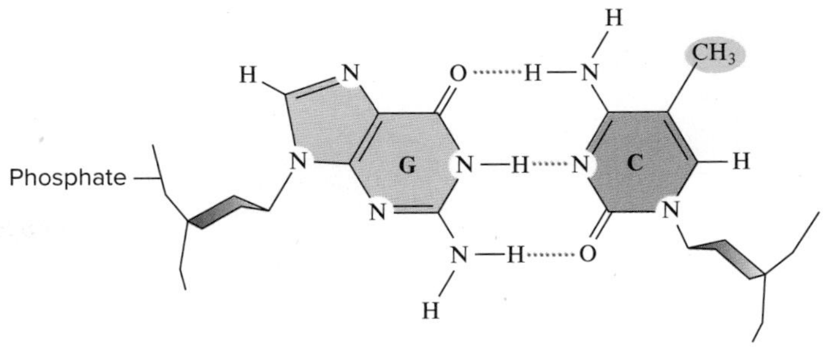

Figure 16.11 DNA methylation. Cytosine is methylated, creating 5-methylcytosine. Because the methyl group (green) is positioned to the side, it does not interfere with the hydrogen bonds of a G—C base-pair, but it can be recognized by proteins.

Methylation state can be maintained through cell division by well-known mechanisms. The semiconservative replication of DNA methylated on both strands produces hemimethylated DNA, which becomes fully methylated by the action of a maintenance methylase. In humans, DNMT1 is the maintenance methylase. Other methylases can initiate the process, but they do not need to remain active.

X-chromosome inactivation

Mammalian females inactivate one X chromosome to equalize X-chromosome expression in both sexes. First proposed by geneticist Mary Lyon in 1961, this process continues to be an area of active research. It also represents the first example of a long noncoding RNA acting to regulate gene expression. The region of the chromosome that initiates the inactivation process contains a gene called X-inactivation specific transcript *(Xist),* which does not encode a protein. The process of inactivation is quite complex, but a streamlined version is that *Xist* RNA "coats" the entire inactive X chromosome and acts to recruit the chromatin regulator Polycomb repressive complex (PRC2). This leads to the histone modifications associated with inactive chromatin.

Histone modification

Histone modification is complex because nucleosomes consist of four histones, each of which has multiple possible modifications. Modifications include acetylation and methylation of lysine as well as phosphorylation of serine, threonine, and tyrosine. In general, acetylation, especially of H3, is correlated with active sites of transcription, both in regulatory regions and in the transcribed region of the gene itself. But methylation of the same histone (H3) can have the opposite effect, depending on the lysine methylated.

Proteins Can Alter Chromatin Structure

LEARNING OBJECTIVE 16.5.2 Describe how alteration of chromatin structure can affect gene expression.

Some transcription activators alter chromatin structure

Some eukaryotic transcription activators seem to interact directly with the initiation complex or with coactivators that themselves interact with the initiation complex, as described in section 16.4.

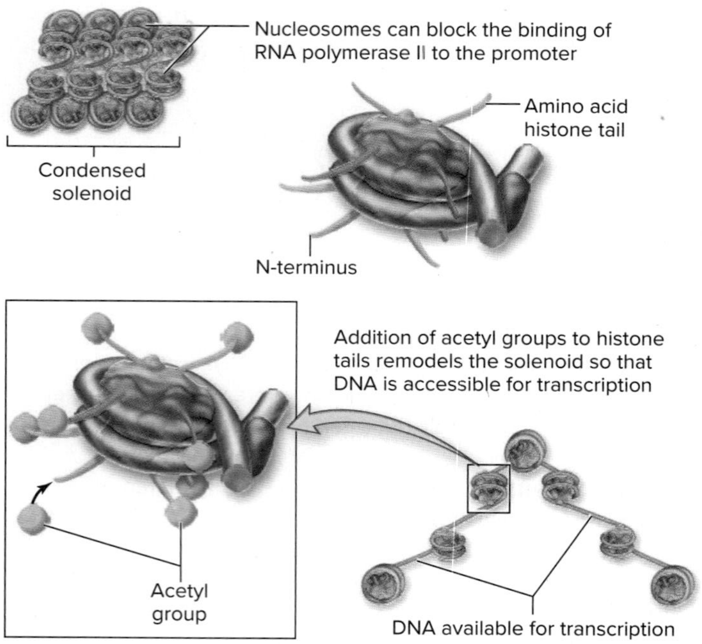
Nucleosomes can block the binding of RNA polymerase II to the promoter
Condensed solenoid
Amino acid histone tail
N-terminus
Addition of acetyl groups to histone tails remodels the solenoid so that DNA is accessible for transcription
Acetyl group
DNA available for transcription

Figure 16.12 Histone modification affects chromatin structure. DNA in eukaryotes is organized first into nucleosomes and then into higher-order chromatin structures. The histones that make up the nucleosome core have amino tails that protrude. These amino tails can be modified by the addition of acetyl groups. The acetylation alters the structure of chromatin, making it accessible to the transcription apparatus.

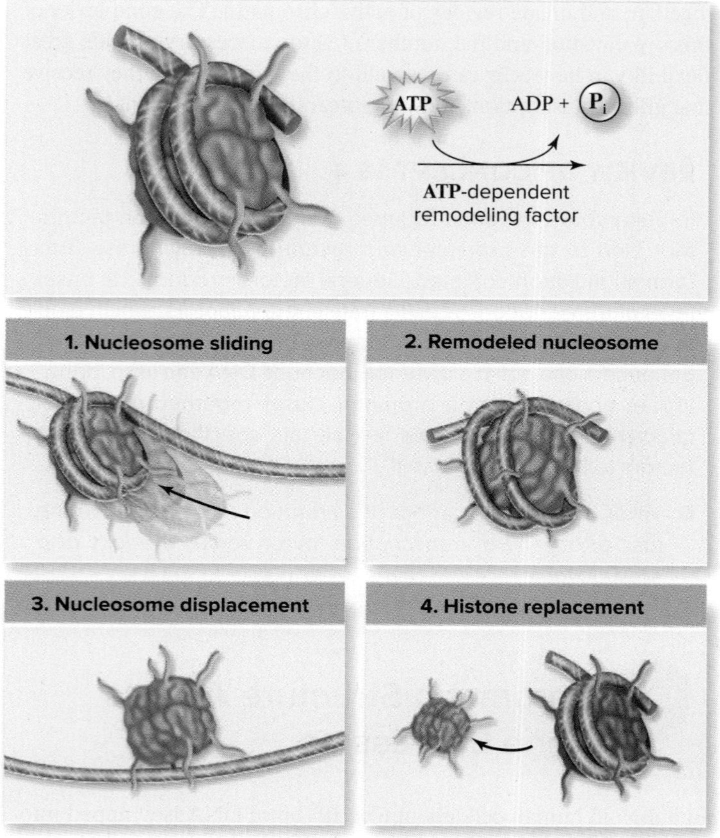
ATP
ADP + P_i
ATP-dependent remodeling factor
1. Nucleosome sliding
2. Remodeled nucleosome
3. Nucleosome displacement
4. Histone replacement

Figure 16.13 Function of ATP-dependent remodeling factors. ATP-dependent remodeling factors use the energy from ATP to alter chromatin structure. They can (**1**) slide nucleosomes along DNA to reveal binding sites for proteins; (**2**) create a remodeled state of chromatin where the DNA is more accessible; (**3**) completely remove nucleosomes from DNA; and (**4**) replace histones in nucleosomes with variant histones.

But in other cases, coactivators are enzymes called **histone acetylases (HATs).** This acetylation can increase transcription by removing higher-order chromatin structure that prevents transcription (figure 16.12). Some corepressors are also enzymes, but these are **histone deacetylases (HDACs),** which remove acetyl groups from histones, making chromatin less accessible.

These observations have led to the idea of a "histone code," analogous to the genetic code. This histone code is thought to underlie the control of chromatin structure and, thus, of access of the transcription machinery to DNA.

Chromatin-remodeling complexes also change chromatin structure

Although the details of higher-order chromatin structure are unclear, it is clear that this structure can affect gene expression. So-called **chromatin-remodeling complexes** contain enzymes that modify histones and DNA, and alter chromatin structure directly.

One class of remodeling factors are the ATP-dependent chromatin-remodeling factors. These function as molecular motors that use energy from ATP hydrolysis to alter the relationships between histones and DNA. They can catalyze four different changes in histone/DNA binding (figure 16.13): (1) nucleosome sliding along DNA, which changes the position of a nucleosome on the DNA; (2) creation of a remodeled state where DNA is more accessible; (3) removal of nucleosomes from DNA; and (4) replacement of histones with variant histones. These functions all make DNA more accessible to regulatory proteins that in turn affect gene expression.

The process of transcription actually requires altering chromatin so RNA polymerase can move through nucleosomes. The elongation complex contains a histone chaperone called FACT (facilitates chromatin transcription) that helps remove nucleosomes in front of the polymerase and return them after transcription.

REVIEW OF CONCEPT 16.5

Chromatin structure affects eukaryotic gene expression and is the basis for epigenetic effects. DNA methylation is part of the basis of genomic imprinting, and the *Xist* RNA mediates X-chromosome inactivation. Transcription activators can act as histone acetylases to convert inactive to active chromatin. Chromatin-remodeling complexes can change chromatin structure by affecting nucleosome type and positioning, to remodel chromatin to make it more accessible to regulatory proteins.

- *Genes that are turned on in all cells are called "housekeeping" genes. Explain the idea behind this name.*

16.6 Eukaryotic Genes Are Also Regulated After Transcription

The separation of transcription in the nucleus and translation in the cytoplasm in eukaryotes allows regulation not seen in prokaryotes. As prokaryotes were first studied in detail, these regulatory mechanisms were initially viewed as less important. It is now clear that posttranscriptional regulation is a central aspect in the control of eukaryotic gene expression. In this section, we will consider how gene expression is regulated after transcription, beginning with regulation by small RNAs.

Small RNAs Act After Transcription to Control Gene Expression

LEARNING OBJECTIVE 16.6.1 Describe the role of small RNAs in regulating gene expression.

The process of development is a series of coordinated events that require changes in gene expression. Thus the study of development often provides insight into the control of gene expression. The analysis of mutants that change developmental timing in the nematode *Caenorhabditis elegans* led to the discovery of small RNAs that affect gene expression. The *lin-14* mutant is a so-called heterochronic mutant, which changes developmental timing. This gene encodes a transcription factor that is expressed at a high level in L1 stage larvae; then expression is reduced in later stages. This is controlled by another gene, *lin-4*. The surprise from this work was that the *lin-4* gene encodes a small RNA of 61 nt and not a protein. This small RNA is further processed to form an only 22-nt RNA that is complementary to sequences in the 3′ untranslated region of the *lin-14* mRNA. This small RNA was shown to act as a translational repressor of the *lin-14* mRNA (figure 16.14). This kind of regulatory RNA is now called a **micro-RNA**, or **mi-RNA**.

Research begun in plants led to the discovery that double-stranded RNA can silence gene expression. This phenomenon is mediated by another class of small RNAs called **small interfering RNAs**, or **siRNAs**. These siRNAs can be introduced experimentally, be derived from invading viruses, or even be encoded in the genome. The analysis of siRNA and miRNA has revealed cellular machinery that uses small RNAs to control gene expression.

Since its discovery, gene silencing by small RNAs has been a source of great interest, both for its experimental uses and as an explanation for posttranslational control of gene expression. Additional kinds of small RNAs have been discovered, but we will confine ourselves to the two classes of miRNA and siRNA, as these are well established and illustrate the RNA-silencing machinery.

miRNA genes

The discovery of the role of miRNAs in gene expression initially appeared to be confined to nematodes, because the *lin-4* gene did not have any obvious homologs in other systems. Seven years later, a second gene, *let-7*, was discovered in the same pathway in *C. elegans*. The *let-7* gene also encoded a 22-nt RNA that could influence translation. In this case, homologs for *let-7* were immediately found in both *Drosophila* and humans.

SCIENTIFIC THINKING

Hypothesis: The reduced expression of *lin-14* in L2 is due to action of the *lin-4* miRNA. This requires the region of the *lin-14* mRNA to be complementary to the *lin-4* miRNA.

Prediction: If the *lin-4* complementary region of the *lin-14* gene is necessary for correct regulation, then a reporter gene containing this region should show high expression in L1 and low expression in L2.

Test: In transgenic worms (*Caenorhabditis elegans*), expression of the reporter gene, β-galactosidase, produces a blue color. Two versions of this reporter gene were constructed:

1. The β-galactosidase gene with the *lin-14* 3′ untranslated region complementary to *lin-4* (construction shown below)
2. The β-galactosidase gene with a control 3′ untranslated region not complementary to *lin-4*.

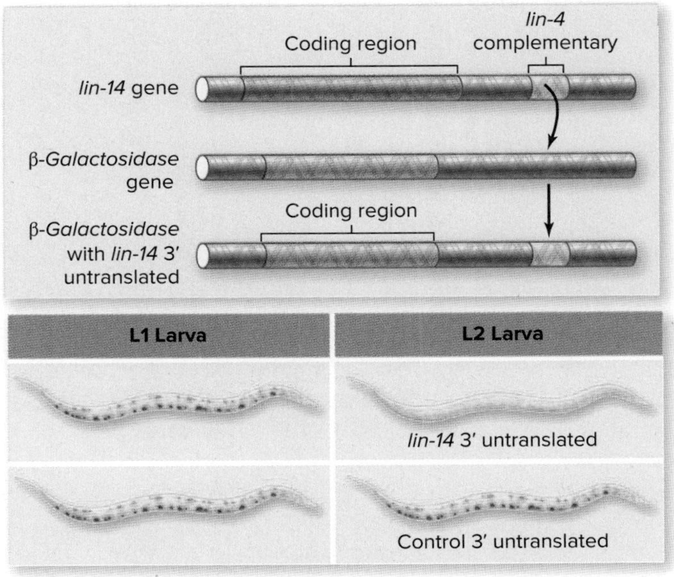

Result:

1. Transgenic worms with reporter gene plus *lin-14* 3′ untranslated region show expression in L1 but not L2 stage larvae. This is the pattern expected for the *lin-14* gene, which is controlled by *lin-4*.
2. Transgenic worms with reporter gene with control 3′ untranslated region do not show expression pattern expected for control by *lin-4*.

Conclusion: The 3′ untranslated region from *lin-14* is sufficient to turn off gene expression in L2 larvae.

Further Experiments: What expression pattern would you predict for these constructs in a mutant that lacks *lin-4* function?

Figure 16.14 Control of *lin-14* gene expression. The expression of the *lin-14* gene is high in L1 and then low in L2. This is controlled by the *lin-4* gene product, which is a small RNA.

As more miRNAs were discovered in different organisms, miRNA gene discovery turned to computer searching and high-throughput methods such as microarrays and new next-generation sequencing. A database devoted to miRNAs currently lists 1917 known human miRNA sequences. These may affect the regulation of up to one-third of human genes.

Genes for miRNA are found in a variety of locations, including the introns of expressed genes, and they are often clustered with multiple miRNAs in a single transcription unit. They are also found in regions of the genome that were previously considered transcriptionally silent. This finding is particularly exciting, because other work looking at transcription across animal genomes has found that much of what we thought was transcriptionally silent is actually not.

miRNA biogenesis and function

The production of a functional miRNA begins in the nucleus and ends in the cytoplasm with an approximately 22-nt RNA that functions to repress gene expression (figure 16.15). The initial transcript of an miRNA gene occurs by RNA polymerase II producing a transcript called the pri-miRNA. The region of this transcript containing the miRNA can fold back on itself and base-pair to form a stem-and-loop structure. This is cleaved in the nucleus by a nuclease called Drosha, which trims the miRNA to just the stem-and-loop structure, which is now called the pre-miRNA. This pre-miRNA is exported from the nucleus through a nuclear pore bound to the protein exportin 5. Once in the cytoplasm, the pre-miRNA is further cleaved by another nuclease called Dicer to produce a short, double-stranded RNA containing the miRNA. The miRNA is loaded into a complex of proteins called an RNA-induced silencing complex (RISC). The RISC includes the RNA-binding protein Argonaute (Ago), which interacts with the miRNA. The complementary strand is removed either by a nuclease or during the loading process.

At this point, the RISC is targeted to repress the expression of other genes based on sequence complementarity to the miRNA. The complementary region is usually in the 3′ untranslated region of genes, and the result can be cleavage of the mRNA or inhibition of translation. It appears that in animals the inhibition of translation is more common than the cleavage of the mRNA, although the precise mechanism of this inhibition is still unclear. In plants, the cleavage of the mRNA by the RISC is common and seems to be related to the more precise complementarity found between plant miRNAs and their targets than that found in animal systems.

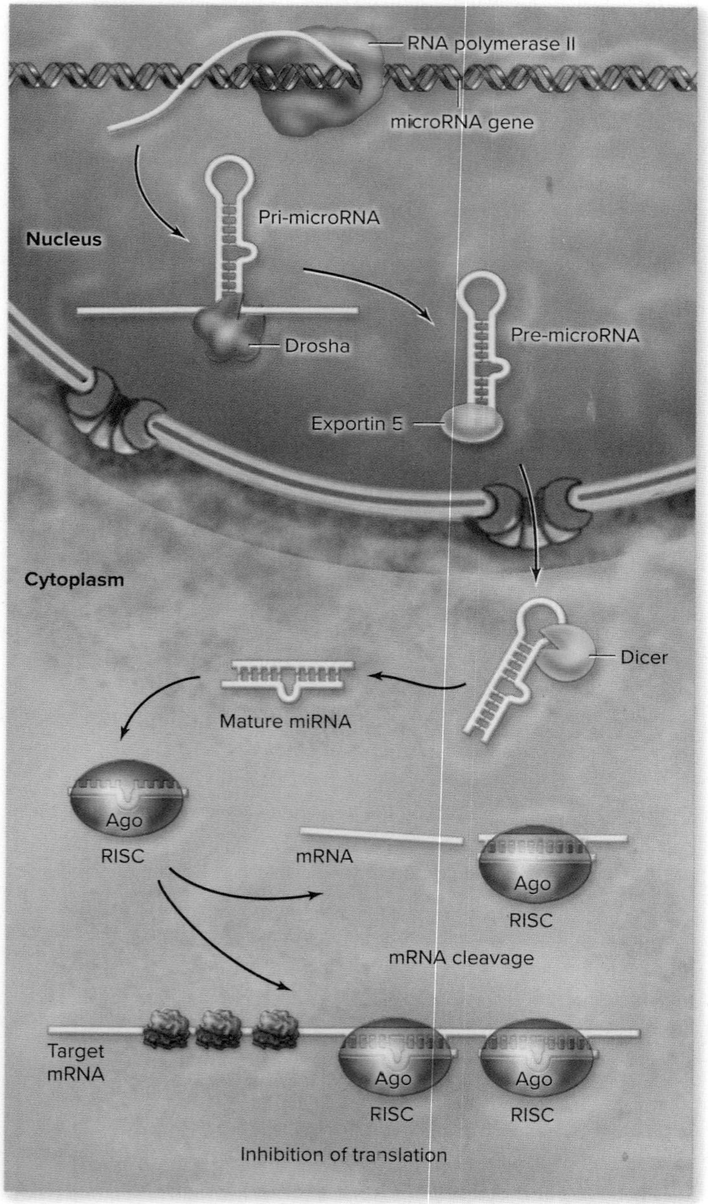

Figure 16.15 Biogenesis and function of miRNA. Genes for miRNAs are transcribed by RNA polymerase II to produce a pri-miRNA. This is processed by the Drosha nuclease to produce the pre-miRNA, which is exported from the nucleus bound to export factor exportin 5. Once in the cytoplasm, the pre-miRNA is processed by Dicer nuclease to produce the mature miRNA. The miRNA is loaded into a RISC, which can either cleave target mRNAs or inhibit translation of target mRNAs.

RNA interference

Small-RNA-mediated gene silencing has been known for a number of years. Some confusion arose in the nomenclature in this area, because work in different systems led to a profusion of names. However, RNA interference, cosuppression, and posttranscriptional gene silencing all act through similar biochemical mechanisms. The term *RNA interference (RNAi)* is currently the most commonly used and involves the production of siRNAs.

The production of siRNAs is similar to that of miRNAs, except that they arise from a long piece of double-stranded RNA (figure 16.16). This can be either a very long region of self-complementarity or a combination of two complementary RNAs. These long, double-stranded RNAs are processed by Dicer to yield multiple siRNAs that are loaded into an Ago containing RISC. The siRNAs usually have near-perfect complementarity to their target mRNAs, and the result is cleavage of the mRNA by the siRNA containing RISC.

The source of the double-stranded RNA to produce siRNAs can be either from the cell or from outside the cell. From the cell itself, genes can produce RNAs with long regions of self-complementarity that fold back to produce a substrate for Dicer in the cytoplasm. They can also arise from repeated regions of the genome that contain transposable elements. Exogenous double-stranded RNAs can be introduced experimentally or by infection with a virus.

Distinguishing miRNAs and siRNAs

The biogenesis of both miRNA and siRNA involves cleavage by Dicer and incorporation into a RISC complex. The main thing that distinguishes these small RNAs is their targets: miRNAs

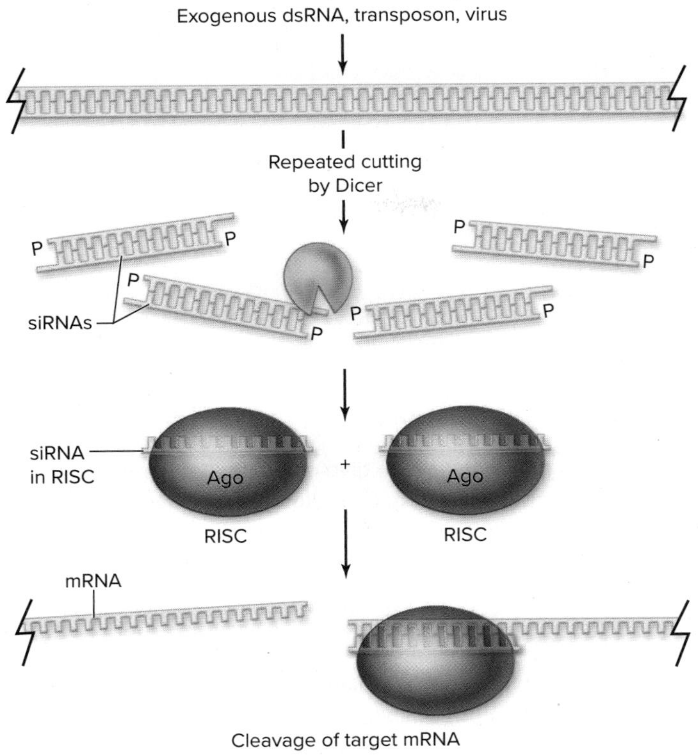

Exogenous dsRNA, transposon, virus
Repeated cutting by Dicer
P
siRNAs
siRNA in RISC
Ago
RISC
+
mRNA
Cleavage of target mRNA

Figure 16.16 Biogenesis and function of siRNA. siRNAs can arise from a variety of sources that all produce long, double-stranded regions of RNA. The double-stranded RNA is processed by Dicer nuclease to produce a number of siRNAs, each of which is loaded onto its own RISC. The RISC then cleaves target mRNA.

tend to repress genes different from their origin, whereas endogenous siRNAs tend to repress the genes they were derived from. Additionally, siRNAs are used experimentally to turn off the expression of genes. This takes advantage of the cell's RNA-silencing machinery to turn off genes complementary to an introduced double-stranded RNA.

The two classes of small RNA have other differences. When multiple species are examined, miRNAs tend to be evolutionarily conserved, but siRNAs do not. Although the biogenesis for both is similar in terms of the nucleases involved, the actual structures of the double-stranded RNAs are not the same. The transcripts of miRNA genes form stem-and-loop structures containing the miRNA, but the double-stranded RNAs generating siRNAs may be bimolecular, or very long stem-loops. These longer, double-stranded regions lead to multiple siRNAs, whereas only a single miRNA is generated from a pre-miRNA.

Small RNAs Can Mediate Heterochromatin Formation

LEARNING OBJECTIVE 16.6.2 Describe how RNA silencing may act to alter chromatin structure.

RNA-silencing pathways have also been implicated in the formation of heterochromatin in fission yeast, plants, and *Drosophila*. In fission yeast, centromeric heterochromatin formation is driven by siRNAs produced by the action of the Dicer nuclease. This heterochromatin formation also involves modification of histone proteins and thus connects RNA interference with chromatin-remodeling complexes in this system. It is not yet clear how widespread this phenomenon is.

Plants are interesting in this regard because they have a variety of small RNA species. The RNA interference pathway in plants is more complex than that in animals, with multiple forms of Dicer nuclease proteins and Argonaute RNA-binding proteins. One class of endogenous siRNA can lead to heterochromatin formation by DNA methylation and histone modification.

Small RNAs have a protective role

The observation that viral RNA can be degraded via the RNA-silencing pathway may point toward the evolutionary origins of small RNAs. A related observation is that RNA silencing can control the action of transposons as well. In both mice and fruit flies, genetic evidence supports the involvement of the RNA interference machinery in the germ line where a specific class of small RNA appears to be involved in silencing transposons during spermatogenesis and oogenesis. Thus, the origins of this mechanism may be an ancient pathway for protection of the genome from assault from both within and without. The conservation of key proteins suggests that the ancestor to all eukaryotes had some form of RNA-silencing pathway.

Alternative Splicing Can Produce Multiple Proteins from One Gene

LEARNING OBJECTIVE 16.6.3 Describe how alternative splicing can produce tissue-specific gene expression.

The high frequency of alternative splicing discussed in chapter 15 indicates nearly 100,000 transcripts produced from the human genome. Elucidating the functional significance of this large number of transcripts will require analysis of specific genes. We will briefly consider two examples.

The pattern of splicing can change during different stages of development or in different tissues. An example of developmental differences is found in *Drosophila*, in which sex determination is the result of a complex series of alternative splicing events that differ in males and females.

A human example of tissue-specific alternative splicing involves proteins produced by the thyroid gland and the hypothalamus. These two organs produce two distinct hormones: *calcitonin* and *CGRP* (*calcitonin-gene-related peptide*) from one gene. Calcitonin controls calcium uptake and the balance of calcium in tissues such as bones and teeth. CGRP is involved in a number of neural and endocrine functions. Despite their different physiological roles, they are produced from the same primary transcript (figure 16.17). Whether calcitonin or CGRP is produced depends on splicing factors that differ in the thyroid and the hypothalamus.

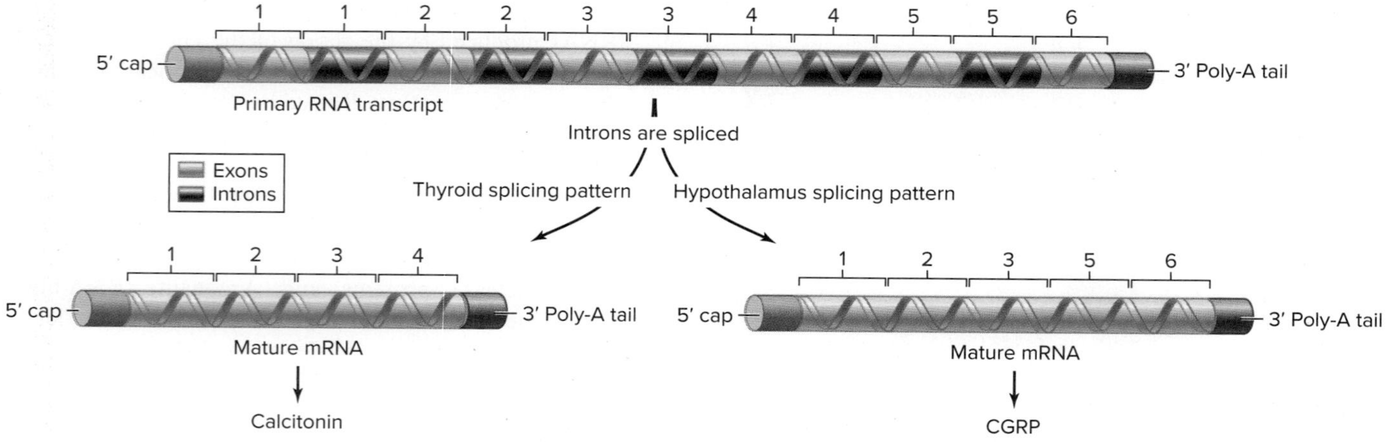

Figure 16.17 Alternative splicing. Many primary transcripts can be spliced in different ways to give rise to multiple mRNAs. In this example, in the thyroid the primary transcript is spliced to contain four exons encoding the protein calcitonin. In the hypothalamus the fourth exon, which contains the poly-A site used in the thyroid, is skipped and two additional exons are added to encode the protein calcitonin gene-related peptide (CGRP).

A functional analysis of every gene in the human genome is unrealistic at present. Another approach is to analyze many proteins at the same time using high-throughput mass spectrometry. This proteomics approach holds great promise, but at the moment there are conflicting results as to how many transcripts are translated.

RNA Editing Alters mRNA Base Sequences After Transcription

LEARNING OBJECTIVE 16.6.4 Explain how editing of RNA transcripts can affect gene expression.

In some cases, the editing of mature mRNA transcripts can produce an altered mRNA that is not truly encoded in the genome—an unexpected possibility. RNA editing was first discovered as the insertion of uracil residues into some RNA transcripts in protozoa, and it was thought to be an anomaly.

RNA editing of a different sort has since been found in mammalian species, including humans. In this case, the editing involves chemical modification of a base to change its base-pairing properties, usually by deamination. For example, both deamination of cytosine to uracil and deamination of adenine to inosine have been observed (inosine pairs as G would during translation).

Apolipoprotein B

The human protein apolipoprotein B is involved in the transport of cholesterol and triglycerides. The gene that encodes this protein, *apoB,* is large and complex, consisting of 29 exons scattered across almost 50 kilobases (kb) of DNA.

The protein exists in two isoforms: a full-length APOB100 form and a truncated APOB48 form. The truncated form is due to an alteration of the mRNA that changes a codon for glutamine into a stop codon. Furthermore, this editing occurs in a tissue-specific manner; the edited form appears only in the intestine, whereas the liver makes only the full-length form. The full-length APOB100 form is part of the low-density lipoprotein (LDL) particle that carries cholesterol. High levels of serum LDL are thought to be a major predictor of atherosclerosis in humans. It does not appear that editing has any effect on the levels of the intestine-specific transcript.

Initiation of Translation Can Also Be Controlled

LEARNING OBJECTIVE 16.6.5 Describe how gene expression can be regulated at the level of translation.

Processed mRNA transcripts exit the nucleus through the nuclear pores (described in chapter 4). The passage of a transcript across the nuclear membrane is an active process that requires the transcript to be recognized by receptors lining the interior of the pores. Specific portions of the transcript, such as the poly-A tail, appear to play a role in this recognition.

There is little hard evidence that gene expression is regulated at this point, although it could be. On average, about 10% of primary transcripts consist of exons that will make up mRNA sequences, but only about 5% of the total mRNA produced as primary transcript ever reaches the cytoplasm. This observation suggests that about half of the exons in primary transcripts never leave the nucleus, but it is unclear whether the disappearance of this mRNA is selective. The translation of a processed mRNA transcript by ribosomes in the cytoplasm involves a complex of proteins called *translation factors.* In at least some cases, gene expression is regulated by modification of one or more of these factors. In other instances, **translation**

repressor proteins shut down translation by binding to the beginning of the transcript so that it cannot attach to the ribosome.

In humans, the production of ferritin (an iron-storing protein) is normally shut off by a translation repressor protein called aconitase. Aconitase binds to a 30-nt sequence at the beginning of the ferritin mRNA, forming a stable loop to which ribosomes cannot bind. When iron enters the cell, the binding of iron to aconitase causes the aconitase to dissociate from the ferritin mRNA, freeing the mRNA to be translated and increasing ferritin production 100-fold.

Protein and mRNA Levels Are Controlled

LEARNING OBJECTIVE 16.6.6 Describe how eukaryotic cells control mRNA and protein degradation.

If all of the proteins produced by a cell during its lifetime remained in the cell, this would cause a variety of problems. Eukaryotic cells deal with this by controlling the lifetime of both mRNA and proteins.

The ubiquitin–proteasome pathway

Eukaryotic cells selectively mark proteins for destruction. This includes proteins that are misfolded, are nonfunctional, or are produced cyclically, as in cell-cycle proteins. The signal for degradation is the attachment of a ubiquitin molecule. Ubiquitin, so named because it is found in essentially all eukaryotic cells (that is, it is ubiquitous), is a 76-amino-acid protein that can exist as an isolated molecule or in longer chains that are attached to other proteins. The longer chains are added to proteins in a stepwise fashion by the enzyme ubiquitin ligase. These polyubiquitinated proteins are now marked for degradation by a macromolecular machine called the proteasome.

The process of ubiquitination followed by degradation by the proteasome is called the ubiquitin–proteasome pathway. It functions as a cycle in that the ubiquitin added to proteins is not itself destroyed in the proteasome. As the proteins are degraded, the ubiquitin chain itself is simply cleaved back into ubiquitin units that can then be reused (figure 16.18). The proteasome is also an example of compartmentalization on a very small scale, in that proteins marked for destruction are moved into a large complex, which isolates them from the rest of the cytoplasm.

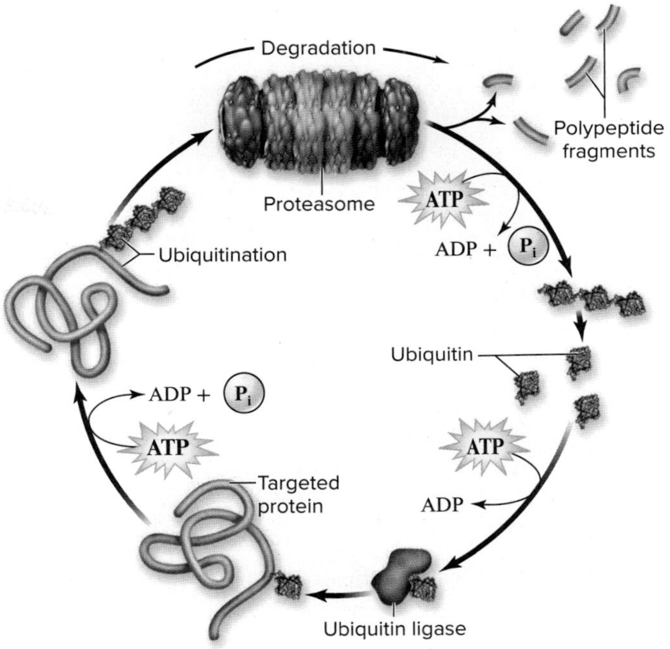

Figure 16.18 Degradation by the ubiquitin–proteasome pathway. Proteins are first ubiquitinated, then enter the proteasome, where they are degraded. In the proteasome, the polyubiquitin is removed and converted back to single ubiquitin molecules that can be reused.

mRNA stability

The stability of mRNA transcripts in the cell cytoplasm also affects protein turnover. Unlike prokaryotic mRNA transcripts, which typically have a half-life of about 3 min, eukaryotic mRNA transcripts can be very stable. For example, β-globin gene transcripts have a half-life of over 10 hr, an eternity in the fast-moving metabolic life of a cell.

The transcripts encoding regulatory proteins and growth factors, however, are usually much less stable, with half-lives of less than 1 hr. The stability of these particular transcripts is affected by specific sequences near their 3′ ends. A sequence of A and U nucleotides near the 3′ poly-A tail of a transcript promotes removal of the tail. Loss of the poly-A tail leads to rapid degradation by 3′-to-5′ RNA exonucleases. Loss of the poly-A tail also stimulates decapping enzymes that remove the 5′ cap, leading to degradation by 5′-to-3′ RNA exonucleases.

Other mRNA transcripts have sequences near their 3′ ends that are recognition sites for endonucleases, which cause these transcripts to be digested quickly. The short half-lives of the mRNA transcripts of many regulatory genes are critical to their function, allowing levels of these proteins to change rapidly.

A review of various methods of posttranscriptional control of gene expression is provided in figure 16.19.

REVIEW OF CONCEPT 16.6

Small RNAs control gene expression by either selective degradation of mRNA, inhibition of translation, or alteration of chromatin structure. Multiple mRNAs can be formed from a single gene via alternative splicing, which can be tissue- or developmentally specific. The sequence of an mRNA transcript can also be altered by RNA editing.

- *How could the phenomenon of RNA interference be used in drug design?*

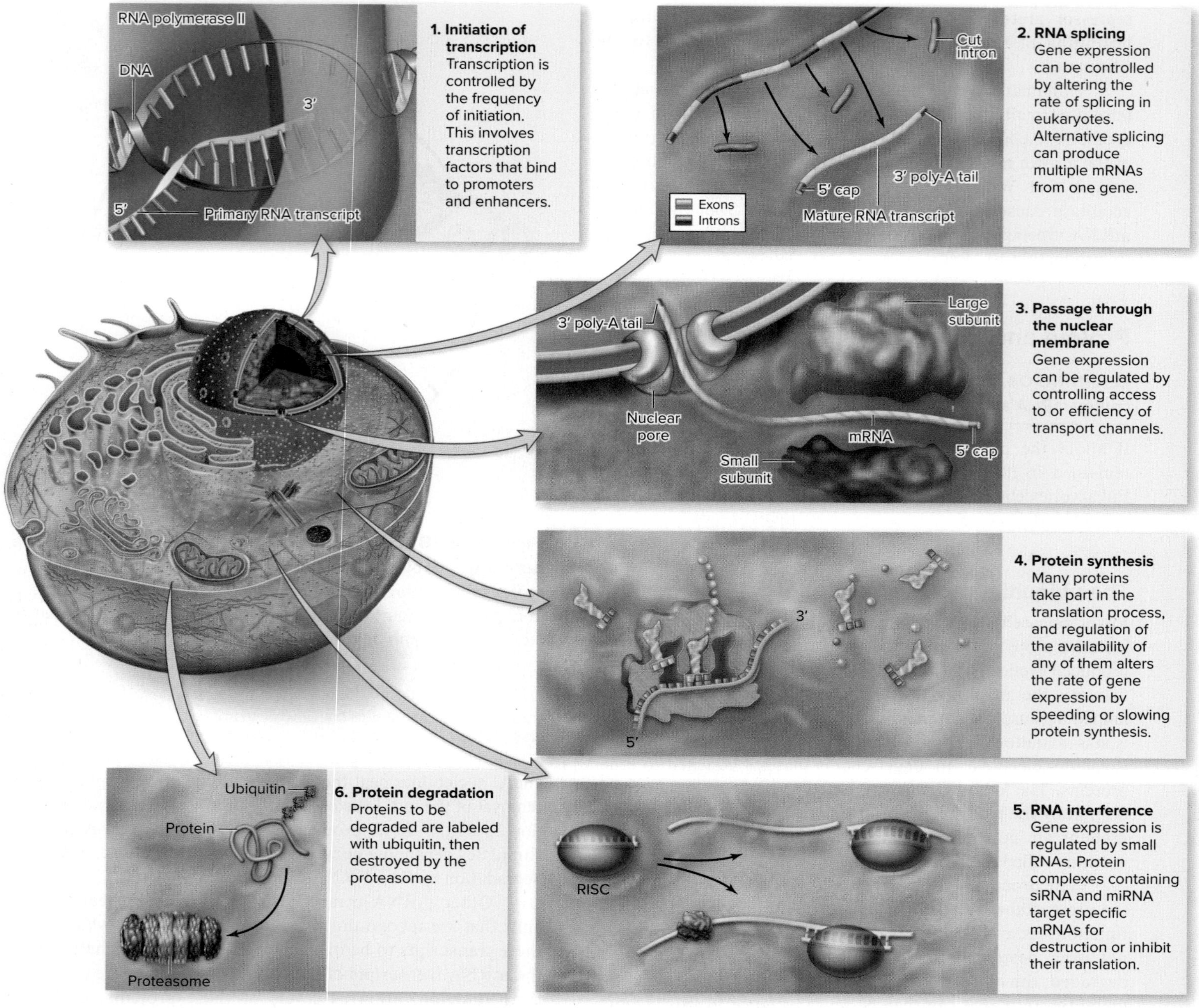
RNA polymerase II
DNA
3′
5′
Primary RNA transcript
1. Initiation of transcription
Transcription is controlled by the frequency of initiation. This involves transcription factors that bind to promoters and enhancers.
Cut intron
5′ cap
3′ poly-A tail
Exons
Introns
Mature RNA transcript
2. RNA splicing
Gene expression can be controlled by altering the rate of splicing in eukaryotes. Alternative splicing can produce multiple mRNAs from one gene.
3′ poly-A tail
Large subunit
Nuclear pore
mRNA
5′ cap
Small subunit
3. Passage through the nuclear membrane
Gene expression can be regulated by controlling access to or efficiency of transport channels.
3′
5′
4. Protein synthesis
Many proteins take part in the translation process, and regulation of the availability of any of them alters the rate of gene expression by speeding or slowing protein synthesis.
Ubiquitin
Protein
Proteasome
6. Protein degradation
Proteins to be degraded are labeled with ubiquitin, then destroyed by the proteasome.
RISC
5. RNA interference
Gene expression is regulated by small RNAs. Protein complexes containing siRNA and miRNA target specific mRNAs for destruction or inhibit their translation.

Figure 16.19 Mechanisms for control of gene expression in eukaryotes.

16.7 Gene Regulation Determines How Cells Will Develop

The mechanisms that control eukaryotic gene expression are critical for the development of multicellular organisms, with complex tissues and organs formed from different cell types. During development, cells become different from one another by expressing different subsets of genes at different times and in different locations in the growing embryo. We now explore some of the mechanisms that lead to differential gene expression during development.

Determination Commits a Cell to a Particular Developmental Pathway

LEARNING OBJECTIVE 16.7.1 Describe the progressive nature of cell determination.

The human body contains around 300 major types of differentiated cells. These can be distinguished by the proteins they

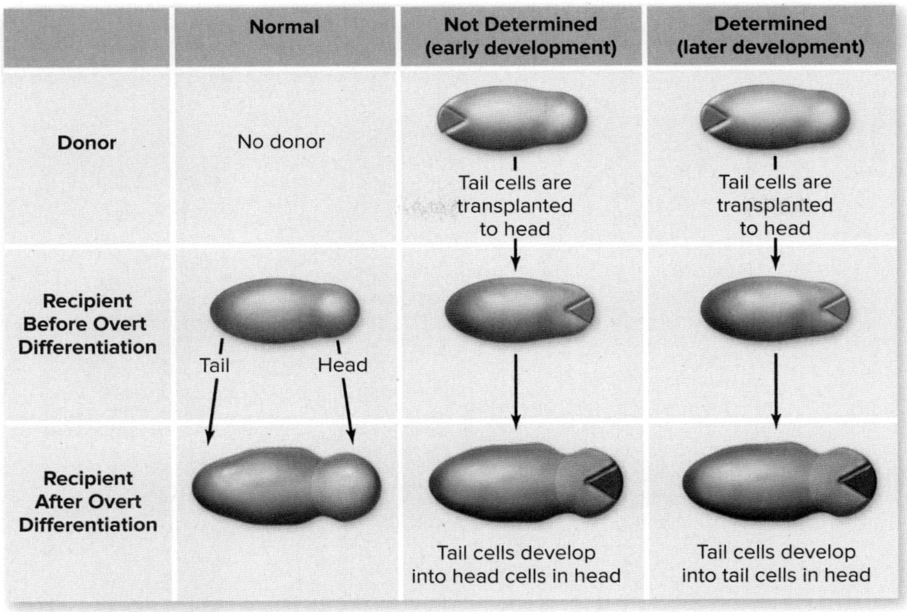

Normal
Not Determined (early development)
Determined (later development)
Donor
No donor
Tail cells are transplanted to head
Tail cells are transplanted to head
Recipient Before Overt Differentiation
Tail
Head
Recipient After Overt Differentiation
Tail cells develop into head cells in head
Tail cells develop into tail cells in head

Figure 16.20 The standard test for determination. The gray ovals represent embryos at early stages of development. The cells to the right normally develop into head structures, whereas the cells to the left usually form tail structures. If prospective tail cells from an early embryo are transplanted to the opposite end of a host embryo, they develop according to their new position into head structures. These cells are not determined. At later stages of development, the tail cells are determined, because they now develop into tail structures after transplantation into the opposite end of a host embryo.

synthesize, their morphologies, and their specific functions. The total number of different human cells is being catalogued by a project called the Human Cell Atlas. In this effort, cells are distinguished primarily by patterns of gene expression. This is revealing a large diversity of subtypes within known cell types.

Before overt differentiation takes place, cells make a molecular decision to become a particular cell type. This process is called cell **determination,** and it commits a cell to a particular developmental pathway.

Tracking determination

Determination is often not visible in the cell and can be detected only experimentally. The standard experiment to test whether a cell or group of cells is determined is to move the donor cell(s) to a different location in a host (recipient) embryo. If the cells of the transplant develop into the same type of cell as they would have if left undisturbed, then they are judged to be already determined (figure 16.20).

The time course of determination can be determined by transplantation experiments. For example, a cell in the prospective brain region of an amphibian embryo at the early gastrula stage (refer to chapter 36) has not yet been determined; if transplanted elsewhere in the embryo, it will develop according to the site of transplant. By the late gastrula stage, however, additional cell interactions have occurred, determination has taken place, and the cell will develop as neural tissue no matter where it is transplanted.

The molecular basis of determination

Cells initiate developmental changes by using transcription factors to change patterns of gene expression. One effect of activating these transcription factor genes is often to reinforce their own activation. This leads to a "permanent" change in gene expression. These developmental changes in gene expression also involve epigenetic changes that are inherited through cell divisions to maintain the expression pattern.

Cells in which a set of regulatory genes has been activated may not actually undergo differentiation until sometime later, when other factors interact with the regulatory protein and cause it to activate still other genes. Nevertheless, once the initial "switch" is thrown, the cell is fully committed to its future developmental path.

Cells become committed to a particular developmental fate by two basic mechanisms: cytoplasmic determinants, and cell-to-cell interactions. Cytoplasmic determinants are deposited on the maternal side during oogenesis, and cell-to-cell interactions can occur at any time after the zygote begins mitotic divisions..

Determination Can Be Triggered by Cytoplasmic Factors

LEARNING OBJECTIVE 16.7.2 Explain the role of cytoplasmic determinants in determination.

Many invertebrate embryos provide good visual examples of cell determination through the differential inheritance of **cytoplasmic determinants.** Tunicates are marine invertebrates, and most adults have simple, saclike bodies that are attached to the underlying substratum. Tunicates are placed in the phylum Chordata, however, due to the characteristics of their swimming, tadpole-like larval stage, which has a dorsal nerve cord and notochord (figure 16.21*a*). The muscles that move the tail develop on either side of the notochord.

In many tunicate species, colored pigment granules become asymmetrically localized in the egg following fertilization and subsequently segregate to the tail muscle cell progenitors during cleavage (figure 16.21*b*). When these pigment granules are shifted experimentally into other cells that normally do not develop into muscle, the fate of those cells is changed and they become muscle cells. Thus, the molecules that

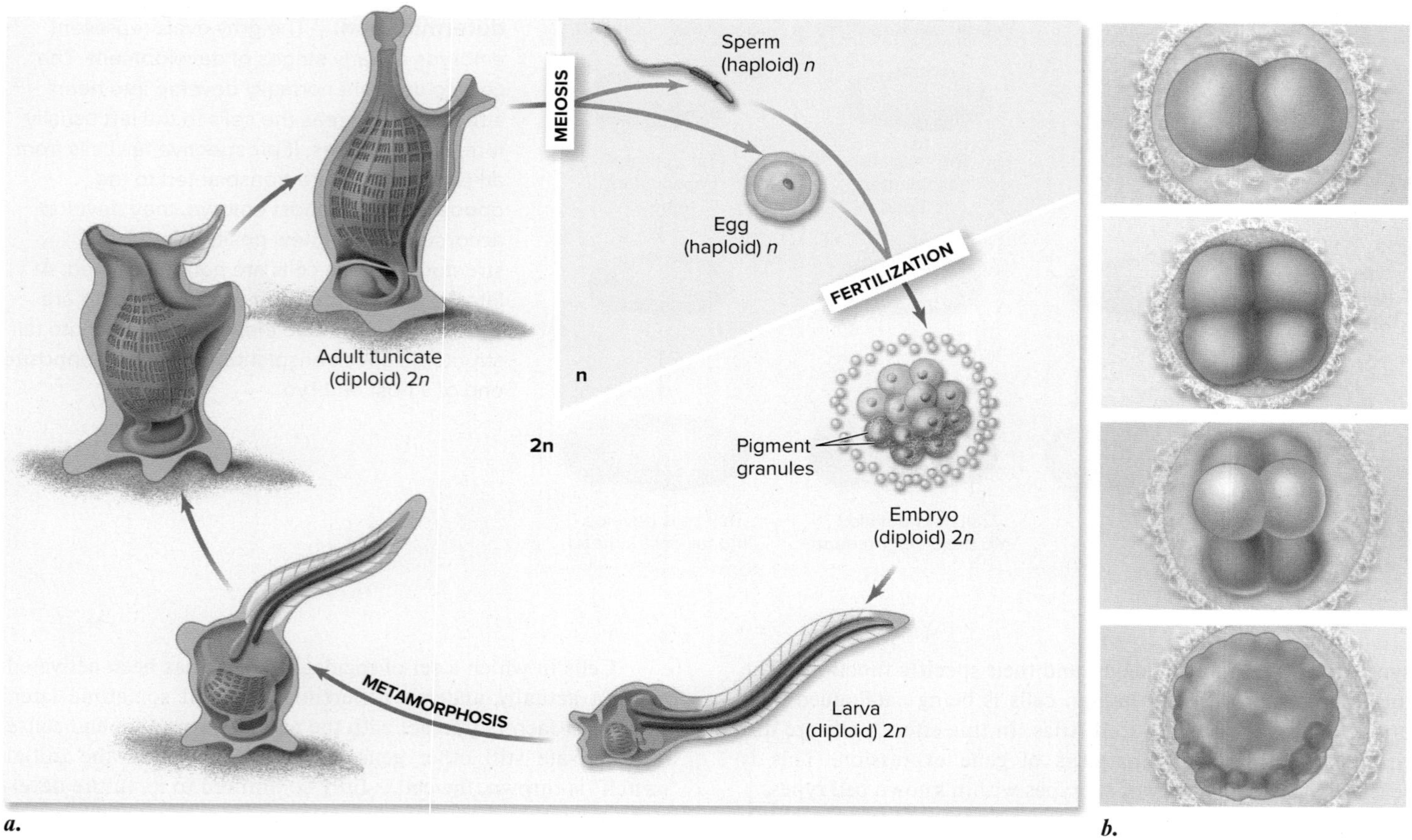

Figure 16.21 Muscle determinants in tunicates. ***a.*** The life cycle of a solitary tunicate. Muscle cells that move the tail of the swimming tadpole are arranged on either side of the notochord and nerve cord. The tail is lost during metamorphosis into the sedentary adult. ***b.*** The egg of the tunicate *Styela* contains bright yellow pigment granules. These become asymmetrically localized in the egg following fertilization, and cells that inherit the yellow granules during cleavage will become the larval muscle cells. Embryos at the 2-cell, 4-cell, 8-cell, and 64-cell stages are shown. The tadpole tail will grow out from the lower region of the embryo in the bottom panel.

flip the switch for muscle development appear to be associated with the pigment granules.

The next step is to determine the identity of the molecules involved. Experiments indicate that the female parent provides the egg with mRNA encoded by the *macho-1* gene. The elimination of *macho-1* function leads to a loss of tail muscle in the tadpole, and the misexpression of *macho-1* mRNA leads to the formation of additional (ectopic) muscle cells from nonmuscle lineage cells. The *macho-1* gene product has been shown to be a transcription factor that can activate the expression of several muscle-specific genes.

Induction Can Change the Fate of a Cell

LEARNING OBJECTIVE 16.7.3 Contrast the role of induction with the role of cytoplasmic determinants.

In chapter 9 we examined a variety of ways by which cells communicate with one another. We can demonstrate the importance of cell–cell interactions in development by separating the cells of an early frog embryo and allowing them to develop independently.

Isolated blastomeres from one pole of the embryo (the "animal pole") become ectoderm, and blastomeres from the opposite pole (the "vegetal pole") become endoderm. Neither of the two types of isolated cells will become mesoderm. However, if animal-pole and vegetal-pole cells are placed next to each other, some of the animal-pole cells develop as mesoderm. The interaction with endodermal cells changes the fate of ectodermal cells. A change in cell fate caused by interaction with an adjacent cell is called induction. Signaling molecules from the inducing cells cause changes in gene expression in the target cells—in this case, some of the animal-pole cells.

Another example of inductive cell interactions is the formation of the notochord and mesenchyme, a specific tissue, in tunicate embryos. Muscle, notochord, and mesenchyme all arise from mesodermal cells that form at the vegetal margin of the 32-cell-stage embryo. These prospective mesodermal cells receive signals from the underlying endodermal precursor cells that lead to the formation of notochord and mesenchyme (figure 16.22).

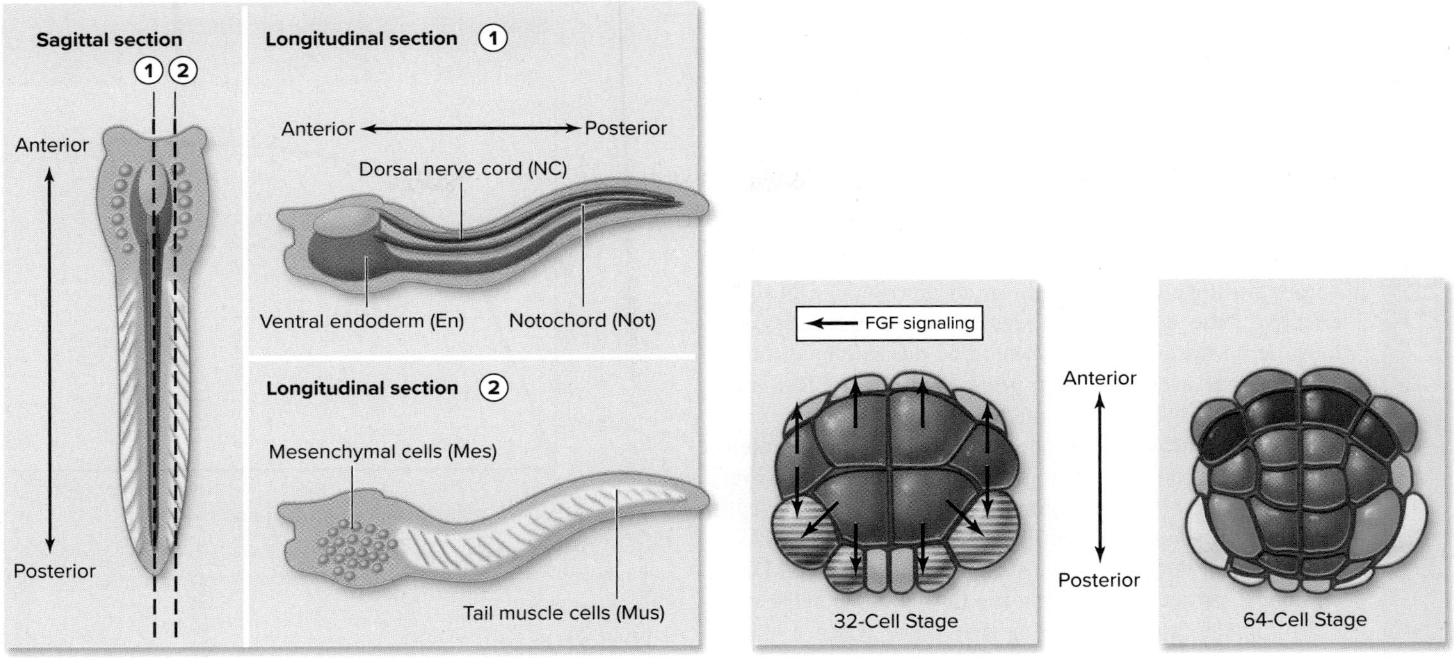

Sagittal section
1 2
Anterior
Posterior
Longitudinal section 1
Anterior
Posterior
Dorsal nerve cord (NC)
Ventral endoderm (En)
Notochord (Not)
Longitudinal section 2
Mesenchymal cells (Mes)
Tail muscle cells (Mus)
FGF signaling
32-Cell Stage
Anterior
Posterior
64-Cell Stage

a. *b.* *c.*

Figure 16.22 Inductive interactions contribute to cell fate specification in tunicate embryos. ***a.*** Internal structures of a tunicate larva. To the left is a sagittal section through the larva, with dotted lines indicating two longitudinal sections. Section 1, through the midline of a tadpole, shows the dorsal nerve cord (NC), the underlying notochord (Not), and the ventral endoderm cells (En). Section 2, a more lateral section, shows the mesenchymal cells (Mes) and the tail muscle cells (Mus). ***b.*** View of the 32-cell stage, looking up at the endoderm precursor cells. FGF secreted by these cells is indicated with light green arrows. Only the surfaces of the marginal cells that directly border the endoderm precursor cells bind FGF signal molecules. Note that the posterior vegetal blastomeres also contain the *macho-1* determinants (red and white stripes). ***c.*** Cell fates have been fixed by the 64-cell stage. Colors are as in ***a.*** Cells on the anterior margin of the endoderm precursor cells become notochord and nerve cord, respectively, whereas cells that border the posterior margin of the endoderm cells become mesenchyme and muscle cells, respectively.

This chemical signal is a member of the *fibroblast growth factor (FGF)* family. It induces the overlying marginal zone cells to differentiate into either notochord (anterior) or mesenchyme (posterior). The FGF receptor on the marginal zone cells is a receptor tyrosine kinase that signals through a MAP kinase cascade (refer to chapter 9). This results in gene expression leading to differentiation. Thus, cell fate depends on both FGF signaling and the Macho-1 muscle determinant discussed earlier. Cells with Macho-1 become muscle without FGF, and mesenchyme with FGF. Cells with no Macho-1 become nerve chord without FGF, and notochord with FGF.

REVIEW OF CONCEPT 16.7

Cell differentiation is preceded by determination, in which cells are committed to a developmental fate but not differentiated. Differential inheritance of cytoplasmic factors can cause determination, as can interactions between neighboring cells (induction). Inductive changes are mediated by signaling molecules.

- *How could you distinguish whether a cell becomes determined by induction or by cytoplasmic factors?*

How Do Transcription Factors Find Specific Regulatory Sites on DNA?

If you think about it, one of the greatest challenges in controlling gene expression is the daunting "needle-in-a-haystack" task facing repressor proteins and transcription factors, each of which must find its specific binding site among millions of other sites on chromosomal DNA. For effective gene control, the regulatory site discovery rate must be far faster than what would be possible by diffusion alone. What are regulatory proteins doing to find their DNA-binding sites quickly and accurately?

Researchers have proposed that regulatory proteins and transcription factors facilitate regulatory site discovery by combining three-dimensional diffusion through the cytoplasm with one-dimensional diffusion along DNA—in essence, after encountering a strand of chromosomal DNA, regulatory proteins slide along the DNA double helix, reading its nucleotide sequence as they pass. When a sliding regulatory protein passes over a DNA sequence that fits its particular shape, the protein stops sliding and binds to the site.

To determine if this sliding process actually occurs in living bacteria, a research team chose to examine the well-studied *lac* operon of *E. coli*, focusing on the binding of the *lac* repressor protein to the *lac* operator site.

The researchers used a yellow fluorescent protein-labeled version of the *lac* repressor protein to carry out single-molecule imaging in real time in living bacterial cells. Binding of this repressor protein to *lac* operator sites on the DNA could be detected as fixed spots against a background of freely diffusing molecules. The association rate of repressor to binding site was then calculated from the average number of spots per cell as a function of time, after removal of the inducer started the discovery process.

The key question the researchers sought to answer was whether the *lac* repressor actually slides along the DNA, and if so, how far. To find out, they made strains of *E. coli* with two *lac* operator sequences separated by different distances and measured how fast the *lac* repressor protein found either site. The results they obtained are presented in the graph.

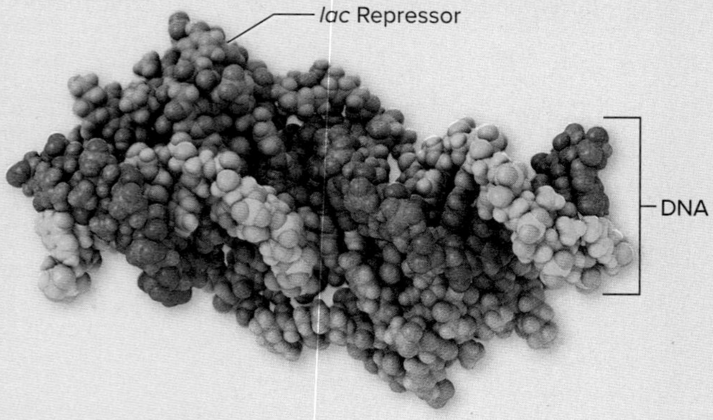

lac Repressor
DNA

Laguna Design/Science Source

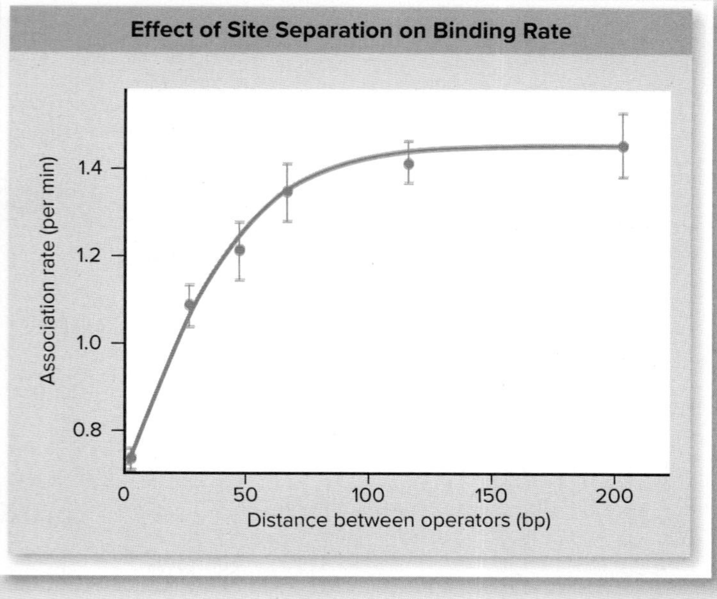

Effect of Site Separation on Binding Rate
Association rate (per min)
1.4
1.2
1.0
0.8
0
50
100
150
200
Distance between operators (bp)

Analysis

1. **Applying Concepts**
   a. ***Variable.*** In the graph, what is the dependent variable?
2. **Interpreting Data**
   a. Is the discovery time—the average time it takes a *lac* repressor protein to associate with a *lac* operator site—affected by the distance between operators?
   b. Is there a minimum separation between the two operator sites, below which binding by the repressor protein becomes less effective? How much less effective? (*Hint:* For each data point, estimate the ratio of its association rate to the maximal association rate.)
3. **Making Inferences**
   a. Does the observation that the distance separating two operators affects how quickly an operator site is bound by a repressor imply that the repressor is moving along the DNA? Explain.
   b. What would you infer from the fact that less effective operator site binding is seen when two operator sites are closer? (*Hint:* What would you expect to happen if two operator sites overlapped? Does the answer to question 2b shed any light on this question?)
4. **Drawing Conclusions** Is it reasonable to conclude from these results that the *lac* repressor protein is sliding along bacterial DNA?
5. **Further Analysis** If the *lac* repressor slides on DNA, then binding the transcription factor TetR next to the operator should block the repressor's sliding to the operator from that side (refer to illustration). When researchers placed TetR there to test this prediction, they found the rate of repressor binding reduced by a factor of 1.8. Is this result consistent with *lac* repressor sliding?

Retracing the Learning Path

CONCEPT 16.1 All Organisms Control Expression of Their Genes

16.1.1 Control of Gene Expression Can Occur at Many Levels Transcription is controlled by regulatory proteins that modulate the ability of RNA polymerase to bind to the promoter. These may either block transcription or stimulate it.

16.1.2 Control Strategies Differ Between Prokaryotes and Eukaryotes Prokaryotes change gene expression to adapt to different environmental conditions. Eukaryotes control gene expression to maintain homeostasis and during development.

CONCEPT 16.2 Regulatory Proteins Control Genes by Interacting with Specific DNA Nucleotide Sequences

16.2.1 DNA-Binding Motifs Interact with Specific DNA Sequences Without Unwinding the Helix A DNA double helix exhibits a major groove and a minor groove; bases in the major groove are accessible to regulatory proteins. A region of the regulatory protein that can bind to the DNA is termed a DNA-binding motif.

CONCEPT 16.3 Prokaryotes Regulate Their Genes in Clusters

16.3.1 Control of Transcription Can Be Either Positive or Negative Negative control is mediated by proteins called repressors that prevent transcription. Positive control is mediated by a class of proteins called activators that stimulate transcription.

16.3.2 Induction of the *lac* Operon Is Negatively Regulated by the *lac* Repressor The *lac* operon is induced when the effector (allolactose) binds to the repressor, altering its conformation such that it no longer binds DNA.

16.3.3 Expression of the *lac* Operon Is Also Affected by Glucose Levels Maximal expression of the *lac* operon requires positive control by catabolite activator protein (CAP) complexed with cAMP. When glucose is low, cAMP is high.

16.3.4 The *trp* Operon Is Negatively Regulated by the *trp* Repressor The *trp* operon is repressed when tryptophan, acting as a corepressor, binds to the repressor, and this complex binds DNA. This prevents expression in the presence of excess *trp.*

CONCEPT 16.4 Transcription Factors Control Gene Transcription in Eukaryotes

16.4.1 Transcription Factors Can Be Either General or Specific General transcription factors are needed to assemble the transcription apparatus and recruit RNA polymerase II at the promoter. Specific factors act in a tissue- or time-dependent manner to stimulate higher rates of transcription.

16.4.2 Promoters and Enhancers Are Binding Sites for Transcription Factors General factors bind to the promoter. Specific factors bind to enhancers, which may be distant from the promoter but can be brought closer by DNA looping.

16.4.3 Coactivators and Mediators Link Transcription Factors to RNA Polymerase II Some transcription factors interact with mediators or coactivators.

16.4.4 The Transcription Complex Integrates Many Levels of Control The transcription complex contains all factors necessary for transcription.

CONCEPT 16.5 Chromatin Structure Affects Gene Expression

16.5.1 DNA Methylation, Histone Modification, and Noncoding RNA All Affect Chromatin Structure Epigenetic effects are due to alterations of chromatin. DNA methylation is part of the basis of genomic imprinting, and *Xist* RNA mediates X-chromosome inactivation.

16.5.2 Proteins Can Alter Chromatin Structure Some activators are histone acetylase enzymes that convert inactive into active chromatin. Chromatin-remodeling complexes change chromatin structure with enzymes that move, reposition, and transfer nucleosomes.

CONCEPT 16.6 Eukaryotic Genes Are Also Regulated After Transcription

16.6.1 Small RNAs Act After Transcription to Control Gene Expression RNA interference is mediated by siRNAs formed by cleavage of double-stranded RNA by Dicer. An RNA-induced silencing complex (RISC) consists of siRNA bound to an Argonaute protein. The RISC can cleave mRNA or inhibit translation. Another regulatory small RNA, miRNA, is produced by the action of Drosha and Dicer. These also form a RISC.

16.6.2 Small RNAs Can Mediate Heterochromatin Formation In fission yeast, *Drosophila,* and plants, RNA interference pathways lead to the formation of heterochromatin.

16.6.3 Alternative Splicing Can Produce Multiple Proteins from One Gene Alternative splicing produces multiple proteins from one gene. This can be tissue-specific or occur during development.

16.6.4 RNA Editing Alters mRNA Base Sequences After Transcription

16.6.5 Initiation of Translation Can Also Be Controlled Translation factors may be modified to control initiation; translation repressor proteins can prevent binding to the ribosome.

16.6.6 Protein and mRNA Levels Are Controlled An mRNA transcript is relatively stable, but it may carry targets for enzymes that degrade it more quickly as needed by the cell.

CONCEPT 16.7 Gene Regulation Determines How Cells Will Develop

16.7.1 Determination Commits a Cell to a Particular Developmental Pathway Cells become determinate to a developmental pathway prior to differentiating. This is not visible but can be tracked experimentally. This arises by differential inheritance of cytoplasmic factors or cell–cell interactions.

16.7.2 Determination Can Be Triggered by Cytoplasmic Factors In tunicates, determination of tail muscle cells depends on the presence of the *macho-1* transcription factor, deposited in the egg cytoplasm during gamete formation.

16.7.3 Induction Can Change the Fate of a Cell Induction occurs when one cell type produces signal molecules that induce gene expression in neighboring target cells. In tunicates, signaling by the growth factor FGF induces mesoderm development.

Concept Overview

This Concept Overview diagrams the key concepts that were discussed in this chapter.

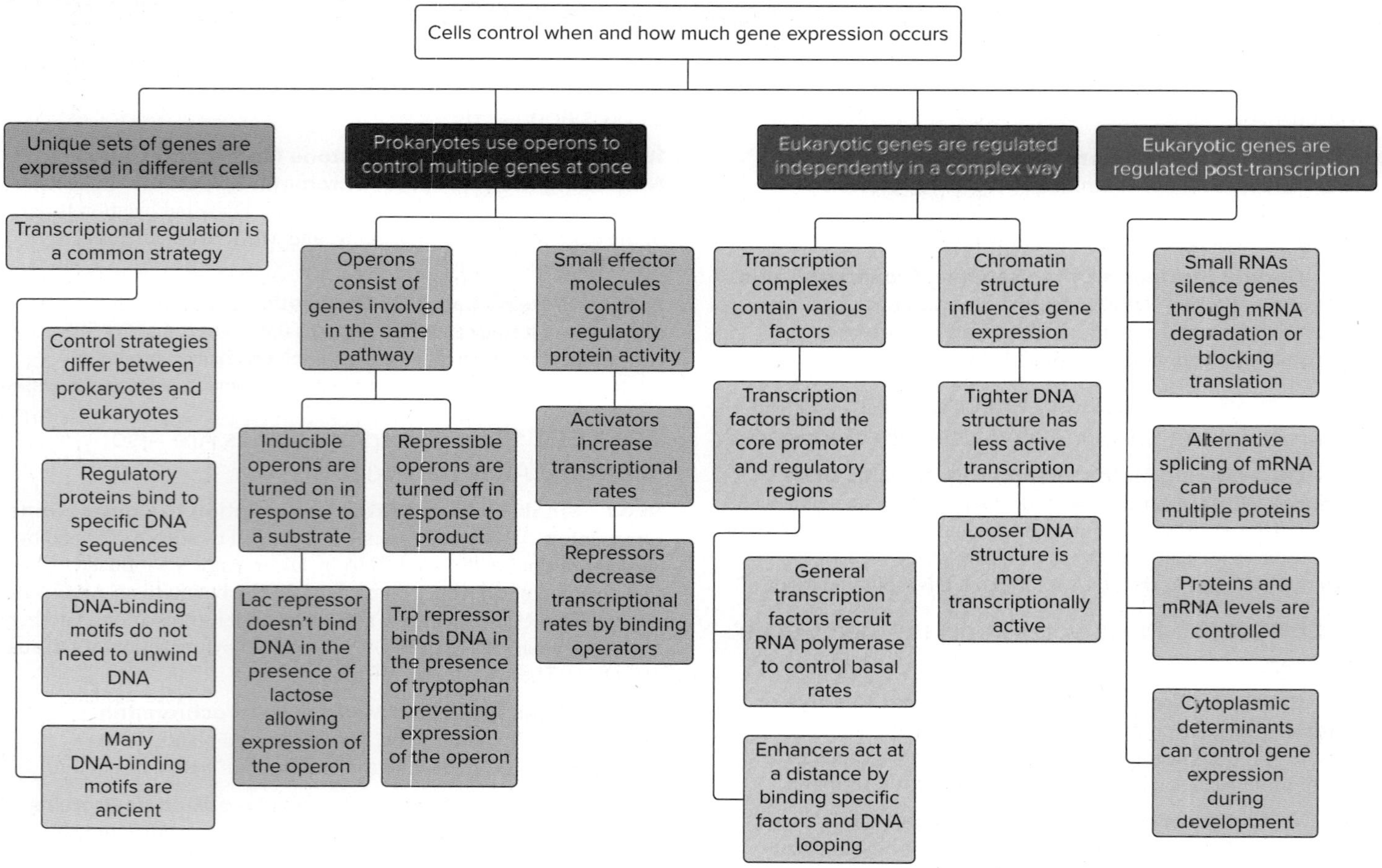

Assessing the Learning Path

Understand

1. In prokaryotes, control of gene expression usually occurs at the
   a. splicing of pre-mRNA into mature mRNA.
   b. initiation of translation.
   c. initiation of transcription.
   d. All of the above
2. In *E. coli*, induction in the *lac* operon and repression in the *trp* operon are both examples of
   a. negative control by a repressor.
   b. positive control by a repressor.
   c. negative control by an activator.
   d. positive control by an activator.
3. The *lac* repressor, the *trp* repressor, and CAP are all
   a. negative regulators of transcription.
   b. positive regulators of transcription.
   c. allosteric proteins that bind to DNA and an effector.
   d. proteins that can bind DNA or other proteins.
4. In the *trp* operon, the repressor binds to DNA
   a. in the absence of *trp*.
   b. in the presence of *trp*.
   c. in either the presence or the absence of *trp*.
   d. only when *trp* is needed in the cell.
5. In eukaryotes, binding of RNA polymerase to a promoter requires the action of
   a. specific transcription factors.
   b. general transcription factors.
   c. repressor proteins.
   d. inducer proteins.
6. Eukaryotic transcription differs from prokaryotic transcription in that it
   a. occurs in the cytosol.
   b. is initiated with the binding of a transcription factor to the TATA box.
   c. is dependent upon enhancers that lie between the promoter and the gene.
   d. All of the above
7. In eukaryotic DNA packaged into chromatin, 5-methylcytosine
   a. does not gain access to promoters.
   b. forms a base-pair with adenine.
   c. is not recognized by DNA polymerase.
   d. correlates with inactive genes.

8. ATP-dependent chromatin-remodeling complexes can do all of the following EXCEPT
   a. change the position of a nucleosome on DNA.
   b. render DNA less accessible.
   c. remove nucleosomes from DNA.
   d. replace some histones with variant forms of histone.
9. Regulation by small RNAs and alternative splicing are similar in that both
   a. act after transcription.
   b. act via RNA–protein complexes.
   c. regulate the transcription machinery.
   d. Both a and b
10. Control of gene expression in eukaryotes includes all of the following EXCEPT
   a. methylation of histone-packaged DNA.
   b. regulation by transcription factors.
   c. stabilization of mRNA transcripts by siRNA.
   d. alternative RNA splicing.
11. During development, cells become
   a. differentiated before they become determined.
   b. determined before they become differentiated.
   c. determined by the loss of genetic material.
   d. differentiated by the loss of genetic material.

Apply

1. If prospective tail cells from an early embryo are transplanted to the opposite end of a host embryo and still develop into a tail,
   a. the prospective tail cells were determined.
   b. the prospective tail cells were not determined.
   c. the prospective tail cells did not differentiate.
   d. the prospective tail cells responded to new positional signals.
2. A homeodomain protein found in human cells has a defect in its recognition helix. What will be the result of this defect?
   a. The protein will not bind correctly into the minor groove of the DNA molecule.
   b. Development will not proceed correctly.
   c. The zinc atom needed for maintaining the tertiary structure of the protein will not be able to complex with the protein.
   d. Both a and b
3. If there is a mutation in the operator sequence of the *lac* operon such that the repressor cannot bind, then
   a. the enzymes that degrade lactose will always be expressed.
   b. the enzymes that degrade lactose will never be expressed, even when lactose is present.
   c. the enzymes that degrade lactose will be expressed only when lactose is present.
   d. None of the above
4. What would be the result of a mutation in the *trp* repressor that altered its conformation such that it did not bind to *trp*?
   a. No tryptophan would be made.
   b. Tryptophan would be made all the time.
   c. There would be no effect on the regulation of the *trp* operon.
   d. CAP protein would not be made.
5. An animal has a gene for the protein PDQ. The TATA box in the PDQ is mutated so that TFIID cannot bind. What will be the effect of that mutation?
   a. PDQ will be made at a very fast rate.
   b. RNA polymerase II will not bind.
   c. PDQ will not be expressed.
   d. Both b and c
6. Tumor-suppressor proteins are essential for normal cell-cycle control. Which of the following would be a potential anticancer (antitumor) drug?
   a. A drug that removes acetyl groups from histones
   b. A drug that methylates tumor-suppressor genes
   c. A drug that removes methyl groups from tumor-suppressor genes
   d. Both a and b
7. Telomerase is an enzyme that replicates the ends of chromosomes and keeps the chromosomes from shortening with each cell division. If you used RNA interference to silence the telomerase gene in cancer cells,
   a. the telomerase gene would not be transcribed.
   b. the telomerase gene would not be translated.
   c. chromosomes would become shorter.
   d. Both b and c
8. What is the common theme in cell determination by induction or cytoplasmic determinants?
   a. The activation of transcription factors
   b. The activation of cell division
   c. A change in gene expression
   d. Both a and c
9. A tissue that makes an inducer is transplanted to another site in an embryo. The tissue adjacent to the inducer will
   a. develop into the usual structure
   b. develop into the induced tissue
   c. develop into a mixture of usual and induced tissues
   d. develop randomly
10. Which statement about induction is NOT true?
   a. A single cell cannot act as an inducer.
   b. Induction triggers distantly located target cells to develop in a certain way.
   c. A tissue that makes an inducer can also itself be induced.
   d. One group of cells can induce an adjacent group of cells.

Synthesize

1. Why would two proteins with the same DNA-binding domain not be able to bind to the same DNA sequence?
2. You have isolated a series of mutants that cause continuous expression of the Lac operon. You have the wild type alleles for these mutants on plasmids that can be introduced into cells to make them diploid for these genes. How can these be used to determine which mutants affect DNA-binding sites in the operon, and which affect proteins that bind to this region?
3. Within chromatin, strings of nucleosomes are coiled into a series of ever-more-compact, higher-level structures. What effect do you imagine this higher-level organization has on gene expression? Explain.
4. The eukaryotic gene encoding the protein catalyzing the breakdown of unobtanium has three introns. How many different proteins could be made by alternative splicing of the pre-mRNA from this gene?
5. Do you imagine humans might regulate embryonic development by possessing genes that influence where in an egg specific cytoplasmic determinants are located? Explain how such cytoplasmic control might work.

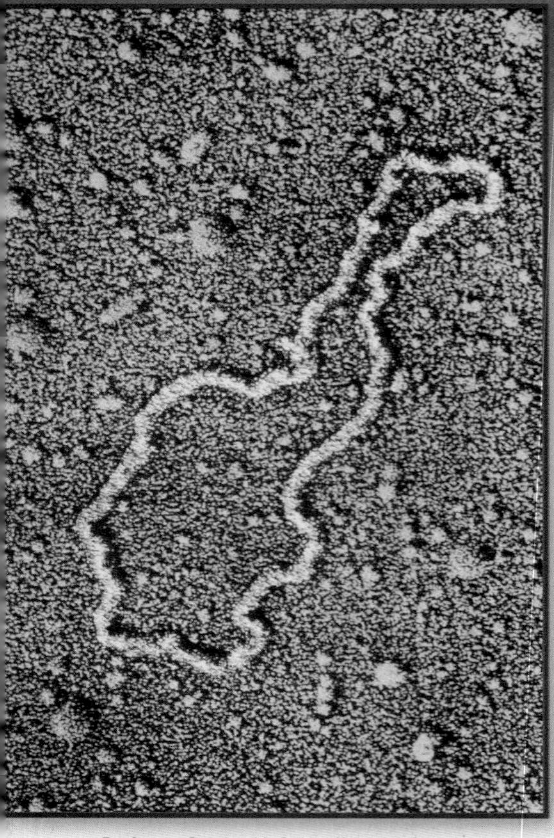

Professor Stanley N. Cohen/Science Source

17 Biotechnology

Learning Path

- **17.1** Enzymes Allow the Creation of Recombinant Molecules In Vitro
- **17.2** The Polymerase Chain Reaction Is Used to Amplify Specific DNA Sequences
- **17.3** Molecular Tools Allow Us to Analyze and Modify Genetic Variation
- **17.4** Transgenic Organisms Are Used to Analyze Gene Function
- **17.5** Genetic Tools Are Changing Modern Medicine
- **17.6** Genetic Engineering Is Used in Industry and Agriculture

Concept Overview

This Concept Overview provides a diagram of the overarching concepts that are covered in this chapter. These concepts will be expanded upon in the Concept Overview at the end of the chapter.

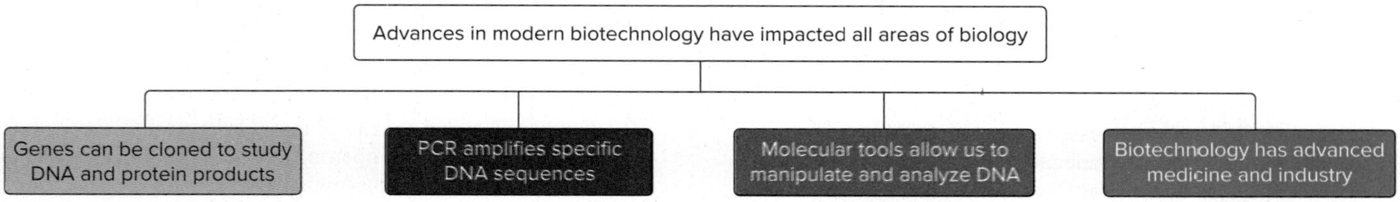

Advances in modern biotechnology have impacted all areas of biology
Genes can be cloned to study DNA and protein products
PCR amplifies specific DNA sequences
Molecular tools allow us to manipulate and analyze DNA
Biotechnology has advanced medicine and industry

Introduction

Biotechnology involves manipulating biological systems to generate useful products or to improve environmental, medical, or agricultural processes. Biotechnology is not a new human endeavor, however: the use of commercial yeast to produce leavened bread dates to ancient Egypt and fermentation to make wine probably predates this by several thousand years. Modern biotechnology combines recent discoveries in molecular biology and genetics with more traditional biotechnologies such as selective breeding and hybridization. Modern biotechnology would not be possible without advances made in other disciplines. Progress in computer science accelerated genomics, and discoveries in chemical engineering moved nanobiotechnology out of the realm of science fiction.

In this chapter, we explore biotechnology by first considering advances in molecular biology and genetics such as the use of plasmids and recombinant DNA techniques. Then, we consider how biotechnology can be used to solve problems in environmental science, medicine, and agriculture.

17.1 Enzymes Allow the Creation of Recombinant Molecules In Vitro

Biotechnology is not necessarily a discipline of the 21st century. For example, the initial domestication of dogs from wolves by our hunter-gatherer ancestors some 15,000 years ago, and their subsequent selective breeding, has produced the 180 existing dog breeds. Later, our agriculturalist ancestors began to domesticate crops, and the development of hybrid corn in the early 20th century substantially increased yield. Recent discoveries in genetics and molecular biology have ushered in a modern era of biotechnological innovation and discovery. The ability to isolate and manipulate DNA revolutionized biotechnology, accelerated rates of discovery, and made possible novel applications of biotechnology. The construction of **recombinant DNA,** a single DNA molecule made from different genetic sources, began in the mid-1970s.

In 1972, Paul Berg created the first recombinant DNA molecule when he inserted viral DNA into bacterial DNA. Building on Berg's work in 1973, Herbert Boyer and Stanley Cohen introduced genes from the toad *Xenopus laevis* into bacteria and showed that the genes could be passed from generation to generation, and that they were expressed in the bacteria. This conclusively demonstrated that the genes from slowly reproducing animals could be replicated and expressed in much more rapidly growing bacteria. In the following three decades, the discoveries of Berg, Boyer, and Cohen would become the basis of biotechnologies as advanced as whole-genome sequencing and the production of recombinant human insulin in bacteria.

Restriction Endonucleases Cleave DNA at Specific Sites

LEARNING OBJECTIVE 17.1.1 Describe how restriction endonucleases are used to make recombinant DNA.

Early cloning experiments were some of the first to utilize new tools called **restriction enzymes,** or **endonucleases.** Arising from basic research on how the host range of a bacterial virus can be restricted, these enzymes recognize specific sequences of DNA and act as a nuclease to cleave the DNA. Originally identified by geneticist Werner Arber studying the host restriction phenomenon, the first enzyme that cleaved DNA at its recognition site was isolated by microbiologist Hamilton O. Smith from the bacterium *Haemophilus influenza* and was thus named *Hin*dIII. Arber, Smith, and microbiologist Daniel Nathans shared the 1978 Nobel Prize in Physiology or Medicine for this work. The discovery of restriction endonucleases is important for two reasons: first, they can cut DNA into specific fragments, and second, this cutting ability can be applied to genome mapping (refer to chapter 18).

How restriction endonucleases work

There are three types of restriction enzymes, but only type II enzymes cleave at specific sequences. These enzymes recognize a specific DNA sequence, ranging from 4 bases to 12 bases, and cleave the DNA at a specific base within this sequence (figure 17.1). The recognition sites for most type II enzymes are linguistic palindromes: a word or phrase that reads the same forward and in reverse, such as the name Hannah. The palindromic DNA sequence reads the same from 5′ to 3′ on one strand as it does on the complementary strand.

Given this kind of sequence, cutting the DNA at the same base on either strand produces staggered cuts, which leave "sticky ends," or overhangs. These short, unpaired sequences are the same for any DNA that is cut by this enzyme. This allows DNA molecules from different sources to be joined together easily due to the short regions of complementarity in the overhangs (figure 17.1). Although less common, some type II restriction enzymes can cut both strands in the same position, producing blunt, not sticky, ends. Blunt-cut ends can be joined with other blunt-cut ends.

Gel Electrophoresis Separates DNA Fragments

LEARNING OBJECTIVE 17.1.2 Explain how DNA fragments can be separated with gel electrophoresis and why this is useful.

To create a piece of recombinant DNA, two or more fragments of DNA must be joined together. Before this can happen, the fragments of DNA must be purified, or isolated. A common technique to separate different fragments of cut DNA so that they can be isolated is gel electrophoresis. This technique takes advantage of the negative charge on DNA by using an electrical field to separate DNA molecules based on size.

The gel, which is made of either agarose or polyacrylamide and spread thinly on supporting material, provides a three-dimensional matrix that separates molecules based on size (figure 17.2). The gel is submerged in a buffer solution containing ions that can carry current and is subjected to an electrical field. The strong negative charges from the phosphate groups in the DNA backbone cause it to migrate toward the positive pole. The gel acts as a sieve to separate DNA molecules based on size: the larger the molecule, the more slowly it will move through the gel matrix. In the gel, smaller molecules migrate farther than larger ones. The DNA in gels can be visualized using a fluorescent dye that binds to DNA. Gels can be made that are used solely to analyze the sizes of DNA fragments, or so that individual fragments of DNA can be cut from the gel, purified, and prepared for further use.

DNA Fragments Are Combined to Make Recombinant Molecules

LEARNING OBJECTIVE 17.1.3 Describe the role of DNA ligase in recombinant DNA molecule construction.

The next step in constructing a recombinant DNA molecule is to combine pieces of DNA from different sources into a single molecule. The recombinant DNA can be stably replicated in a host organism such as the common gut bacterium *Escherichia coli.* The most common strategy to create a stable recombinant DNA molecule is to add a fragment of DNA of interest—for example, a gene

Figure 17.1 Many restriction endonucleases produce DNA fragments with sticky ends. The restriction endonuclease *Eco*RI always cleaves the sequence 5′–GAATTC–3′ between G and A. Because the same sequence occurs on both strands, both are cut. However, the two sequences run in opposite directions on the two strands. As a result, single-stranded tails called "sticky ends" are produced that are complementary to each other. These complementary ends can then be joined to a fragment from another DNA that is cut with the same enzyme. These two molecules can then be joined by DNA ligase to produce a recombinant molecule.

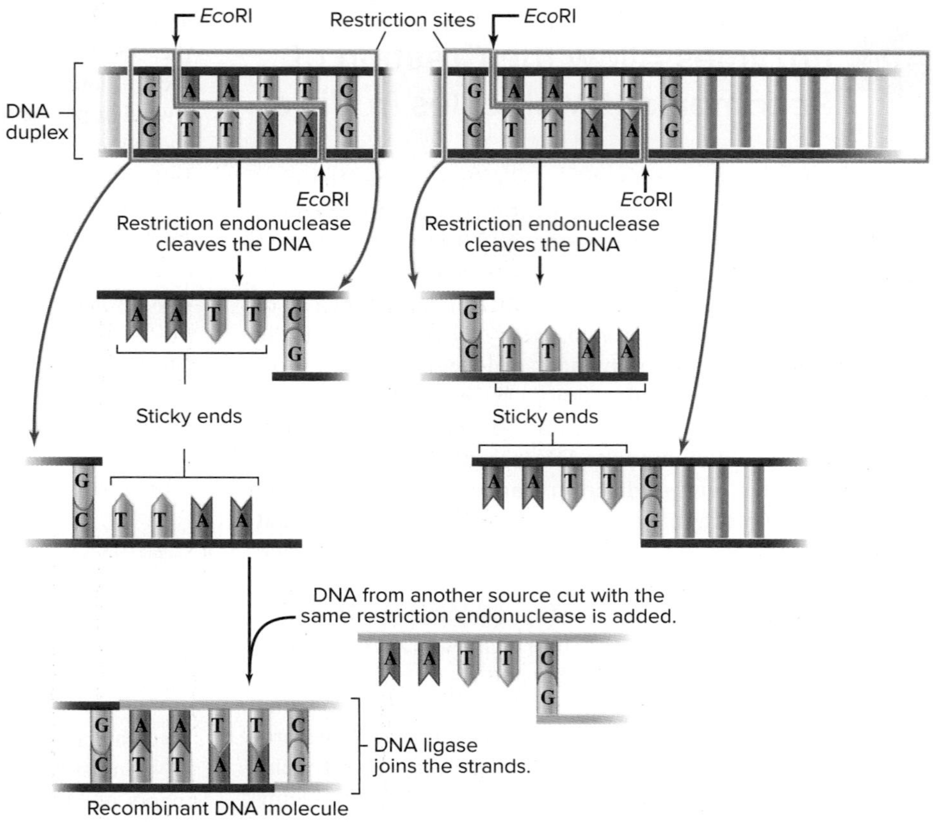

sequence to be studied—into a bacterial plasmid. The recombinant plasmid can then be introduced to *E. coli,* where it will be replicated and maintained.

Joining fragments of DNA using DNA ligase

DNA fragments that have been cut by restriction endonucleases and purified using an agarose gel can be joined together using DNA ligase. DNA ligase catalyzes the formation of a phosphodiester bond between the 5′ phosphate group of one strand of DNA and the 3′ hydroxyl group of another strand of DNA. This is the same reaction that is used by the DNA ligase that joins Okazaki fragments together during lagging-strand DNA synthesis in the cell (refer to chapter 14).

If two fragments of cut DNA have sticky ends, then the short complementary overhangs (as shown in figure 17.1) will hold the two strands of DNA together while DNA ligase forms the phosphodiester bond. Certain DNA ligases can also join DNA fragments with blunt ends.

In figure 17.3, a piece of foreign DNA is being introduced into a plasmid that has been cut by a specific restriction endonuclease. The piece of foreign DNA has also been cut by the same restriction endonuclease and has overhanging ends that are complementary to the cut ends in the plasmid. The sticky ends will hold the two molecules together and will allow ligase to form phosphodiester bonds to join them, creating a recombinant plasmid. The whole process of producing recombinant DNA molecules is often called *molecular cloning.*

Reverse Transcriptase Makes DNA from RNA

LEARNING OBJECTIVE 17.1.4 Describe how the different polymerase activities of reverse transcriptase are involved in the synthesis of cDNA from mRNA.

Although the "central dogma" of molecular biology indicates that the flow of information is from DNA to RNA to protein, there are several instances in biology when DNA can be made using the information in RNA. Reverse transcriptase, first identified in a class of viruses called retroviruses, can use RNA as a template to synthesize a DNA molecule (refer to chapter 23).

Retroviruses, like the human immunodeficiency virus (HIV), have RNA genomes but must convert the RNA to DNA when they infect cells. Reverse transcriptase is an RNA-dependent DNA polymerase; this means that it uses the information in RNA to make a complementary strand of DNA. Reverse transcriptase is also a DNA-dependent DNA polymerase, which means it can make new DNA using the information contained in a DNA strand. The DNA made from the information in mRNA is called **complementary DNA (cDNA).** The creation of cDNA from mRNAs of eukaryotic cells is a major use of reverse transcriptase (figure 17.4). When cDNA is made from mature eukaryotic mRNAs, the function of just the coding regions of a gene (the exons) can be studied.

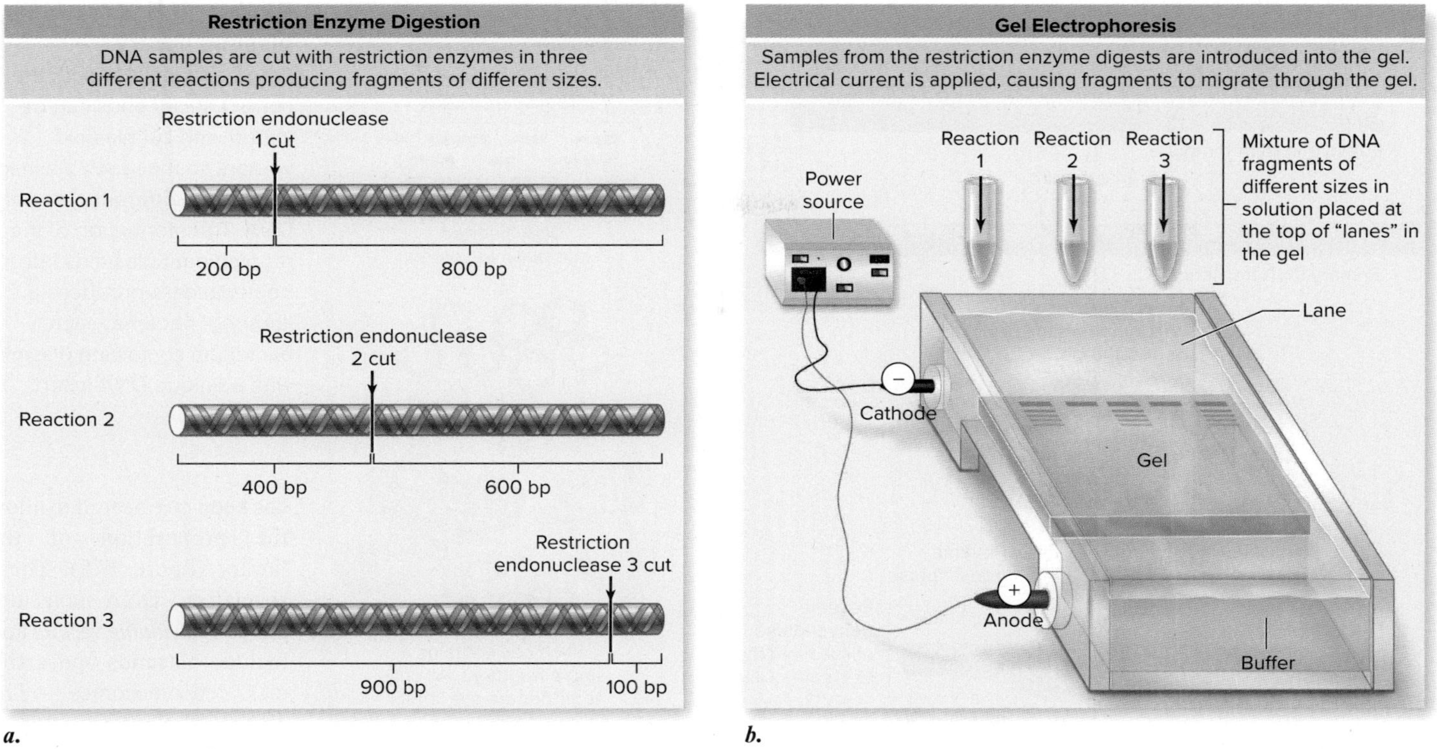
Restriction Enzyme Digestion
DNA samples are cut with restriction enzymes in three different reactions producing fragments of different sizes.
Restriction endonuclease 1 cut
Reaction 1
200 bp
800 bp
Restriction endonuclease 2 cut
Reaction 2
400 bp
600 bp
Restriction endonuclease 3 cut
Reaction 3
900 bp
100 bp
a.
Gel Electrophoresis
Samples from the restriction enzyme digests are introduced into the gel. Electrical current is applied, causing fragments to migrate through the gel.
Reaction 1
Reaction 2
Reaction 3
Mixture of DNA fragments of different sizes in solution placed at the top of "lanes" in the gel
Power source
Lane
Cathode
Gel
Anode
Buffer
b.

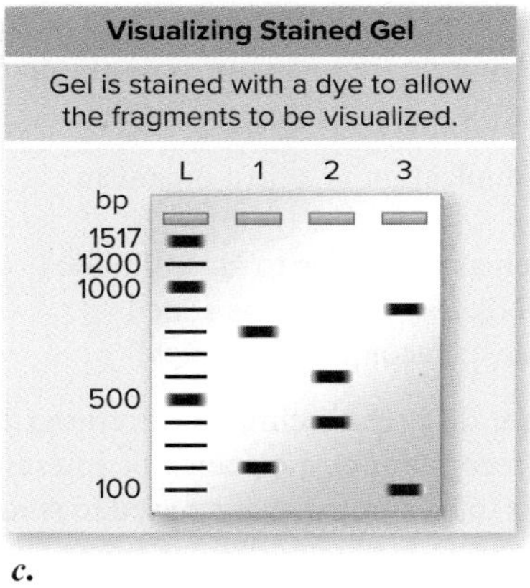
Visualizing Stained Gel
Gel is stained with a dye to allow the fragments to be visualized.
L
1
2
3
bp
1517
1200
1000
500
100
c.

Figure 17.2 Gel electrophoresis separates DNA fragments based on size. ***a.*** Three restriction enzymes are used to cut DNA into specific pieces, depending on each enzyme's recognition sequence. ***b.*** The fragments are loaded into a gel (agarose or polyacrylamide), and an electric current is applied. The DNA fragments migrate through the gel based on size, with larger fragments moving more slowly. ***c.*** This results in a pattern of fragments separated based on size, with the smaller fragments migrating farther than larger ones. A series of fragments of known sizes produces a "ladder" so that sizes of fragments of unknown size can be estimated (bp = base-pairs; L = ladder; 1, 2, 3 = fragments from piece of DNA cut with restriction endonucleases 1, 2, and 3, respectively).

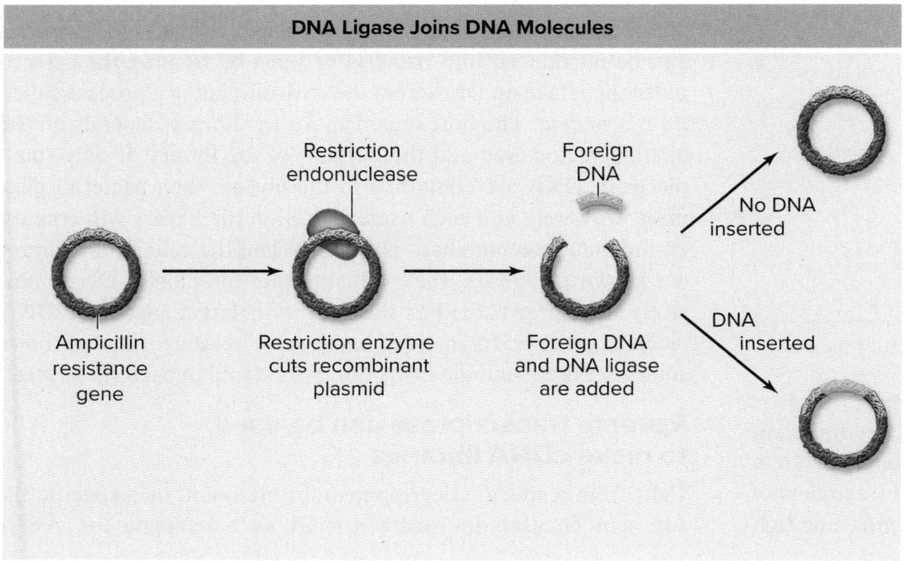
DNA Ligase Joins DNA Molecules
Restriction endonuclease
Foreign DNA
No DNA inserted
DNA inserted
Ampicillin resistance gene
Restriction enzyme cuts recombinant plasmid
Foreign DNA and DNA ligase are added

Figure 17.3 Molecular cloning with vectors.
Plasmids are cut using restriction endonucleases at a specific site, and foreign DNA and DNA ligase are added. If foreign DNA is successfully incorporated into the plasmid, a recombinant plasmid is created. In a process called transformation, the plasmids can be introduced into bacteria. Spreading transformed cells on a medium containing the antibiotic ampicillin selects for plasmid-containing cells.

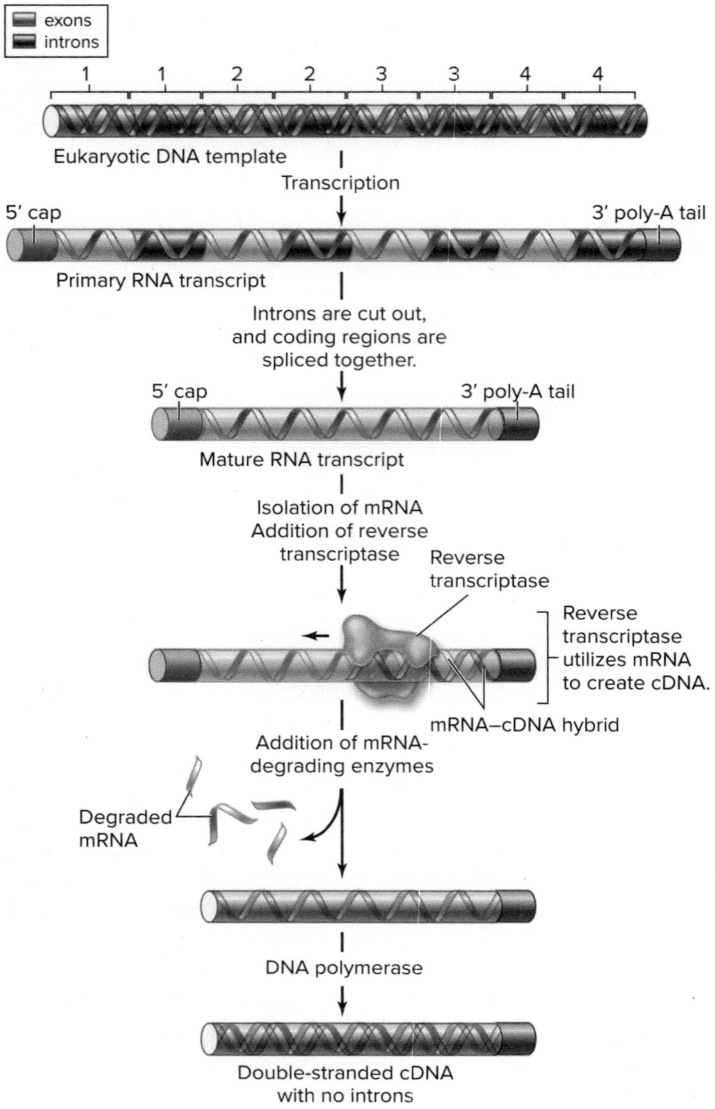

Figure 17.4 Making cDNA from mRNA using reverse transcription. A mature mRNA transcript is usually much shorter than the gene producing it, due to the removal of introns. mRNA is isolated from the cytoplasm of a cell, and used by the enzyme reverse transcriptase as a template to make a DNA strand complementary to the mRNA. The newly made strand of DNA is the template for the DNA polymerase activity of reverse transcriptase, producing cDNA—a double-stranded DNA version of the mature mRNA.

DNA Libraries Are Collections of Recombinant DNA Molecules

LEARNING OBJECTIVE 17.1.5 Describe the construction and uses of recombinant DNA libraries.

A DNA library is a collection of many different recombinant DNA molecules that can be stably maintained and replicated in a suitable host organism. To create a library, different fragments of DNA of interest are spliced together with a DNA molecule that

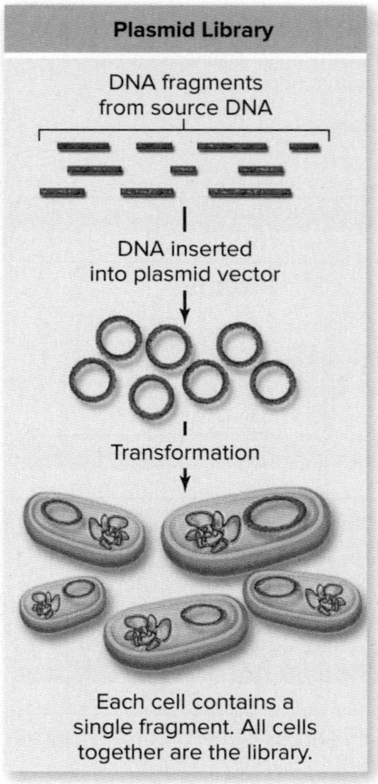

Figure 17.5 Creating a plasmid library. Fragments of DNA with the correct sticky ends can be joined with cut plasmid vectors so that each plasmid contains a different piece of DNA. Transformation of the recombinant plasmids into bacterial cells produces a library of bacteria, each bacterium containing plasmid with a unique DNA insert.

has been engineered to allow the propagation of the library (figure 17.5). These specialized DNA molecules are called *cloning vectors* and include plasmids and artificial chromosomes. The choice of cloning vector to use when making a library depends on what the library will be used for and the sizes of the pieces of DNA to be cloned. Cloning vectors have several important features:

1. A sequence that allows replication in a host organism (such as *E. coli* or yeast)
2. A selectable marker, such as resistance to an antibiotic
3. Sequences that allow DNA fragments to be added (restriction endonuclease or recombination sites)

Once a library has been made, different techniques are used to identify and isolate specific recombinant molecules of interest. Libraries can produce proteins to be studied, can be used to study genome structure, or can be used to isolate particular genes.

Constructed libraries can be stored by being introduced into an appropriate host organism (figure 17.5). Growing the host cells will then "amplify" the library and produce many copies of recombinant molecules that can be isolated or analyzed. Libraries can be introduced into bacteria or yeast by treating the cells to make them take up DNA from the environment in a process called *transformation*. The host organism for the library depends on the cloning vector used and the purpose of the library. If only small pieces of DNA are contained in the library, then bacterial plasmids work well, and each bacterial cell in the library will contain an individual recombinant plasmid. When the cells in the library are grown in culture, they will replicate the cloned DNAs passively during growth. For libraries with larger pieces of DNA (such as genomic fragments), bacterial or yeast artificial chromosomes are used, and the DNA library is stored in bacteria or yeast.

Reverse transcriptase can be used to make cDNA libraries

Cells from a specific developmental time point, or a specific tissue, can be used to isolate mRNA as a template for cDNA synthesis (figure 17.4). The resulting cDNAs can then be utilized to construct a library that represents only the genes expressed at a specific time point or in a specific tissue. Genomic libraries made from different cells or tissues of an organism all contain similar sequences, but this is not true for cDNA libraries. A brain cDNA library will contain many brain-specific sequences not represented in a fibroblast cDNA library, which in turn will contain fibroblast-specific sequences not represented in the brain cDNA library.

Because a collection of cDNAs made from a tissue reflects the mRNA present in that tissue, comparisons of the collections of cDNAs produced from different tissues can be informative. For example, comparisons made between cDNA libraries from different cell types might provide information about the relationships between the genes being expressed, the proteome, and cell structure and function.

REVIEW OF CONCEPT 17.1

Type II restriction endonucleases cleave DNA at specific sites. Gel electrophoresis uses an electrical field to separate DNA fragments according to size. DNA ligase will join DNA fragments from different sources to produce recombinant DNA. Reverse transcription converts mRNA into cDNA. DNA libraries are stable collections of recombinant DNA molecules used for many kinds of genetic analyses.

- *Why is it important to be able to convert mRNA into DNA?*

17.2 The Polymerase Chain Reaction Is Used to Amplify Specific DNA Sequences

The development of the **polymerase chain reaction (PCR)** in 1993 accelerated the construction of recombinant DNA molecules and ushered in a new era of genetic engineering. PCR mimics the processes of DNA replication to produce millions of copies of a DNA sequence (a process called amplification) without the need for molecular cloning. A major benefit of producing DNA fragments using PCR is that the exact sequence of DNA to be amplified can be chosen. When DNA molecules are cloned into vectors based on producing DNA fragments by restriction endonuclease digestion, the DNA fragments available are limited by the location of restriction endonuclease sites.

To amplify a DNA sequence using PCR, two 20- to 25-nt-long single-stranded DNA sequences called *primers* are chemically synthesized. One primer is complementary to one strand of the DNA at one end of the chosen sequence, and the other primer is complementary to the opposite strand at the other end of the chosen sequence. If the primers are allowed to bind to their complementary DNA sequences, DNA polymerase can extend the two primers so that two new strands of DNA are produced, in a process similar to DNA replication in cells. These two strands of DNA are complementary and contain the original primer binding sites. Repeating this process over and over results in an exponential increase in the number of DNA molecules. A DNA sequence up to about 10,000 bp long can be amplified this way.

PCR Mimics DNA Replication

LEARNING OBJECTIVE 17.2.1 Relate the process of DNA replication to PCR.

PCR mimics the processes of DNA replication, and an individual reaction is cycled through a series of steps in which each step is analogous to a step in DNA replication:

1. **Denaturation.** Heat is used to separate strands of double-stranded DNA.
2. **Annealing of primers.** Primers provide the 3′ OH required for elongation by DNA polymerase.
3. **Synthesis.** DNA polymerase makes new DNA.

New DNA molecules are synthesized in an exponential manner when these steps are cyclically repeated; a 25-cycle PCR produces 2^{25} molecules of DNA from a single target molecule! When PCR was initially developed, the steps were done manually using a mixture containing the DNA to be amplified (called the template DNA) and the primers specific for the template sequence. The mixture was heated to 95°C to separate the two strands of DNA, it was cooled to a temperature at which the primers could anneal to the template DNA, and then a polymerase was added to synthesize new DNA. Unfortunately, when the mixture was heated in the next cycle, the DNA polymerase was denatured and new polymerase had to be added. Because a typical PCR involves 25 to 30 cycles, this made PCR tedious and labor-intensive.

PCR was made practical by two innovations. First, a thermostable DNA polymerase called **Taq polymerase** was isolated from the thermophilic bacterium *Thermus aquaticus*. This overcame the need to add new DNA polymerase during every cycle of PCR because the Taq polymerase is stable at the denaturation temperature. The second innovation was the development of machines with heating blocks that can be accurately and rapidly cycled over large temperature ranges.

A typical PCR is now set up in a single tube containing the template DNA, the primers, nucleotides, and a thermostable DNA polymerase. Although Taq polymerase is still commonly used, other, more accurate thermostable polymerases are also used. The tube is placed into a PCR machine that can quickly cycle between the denaturation temperature, the primer annealing temperature, and the DNA synthesis temperature (figure 17.6). Twenty-five cycles of PCR can be performed in as little as an hour. These developments have made PCR one of the most common techniques for generating specific DNA molecules for a variety of uses.

Reverse Transcription PCR Makes Amplified DNA from mRNA

LEARNING OBJECTIVE 17.2.2 Describe the benefits of RT-PCR and how it can be used to quantify mRNA levels.

In addition to using a variety of types of DNA (plasmids, genomic DNA) as a template, PCR can also be performed on cDNA made from mRNA. Isolated mRNA is incubated with reverse transcriptase and the appropriate primer(s); the

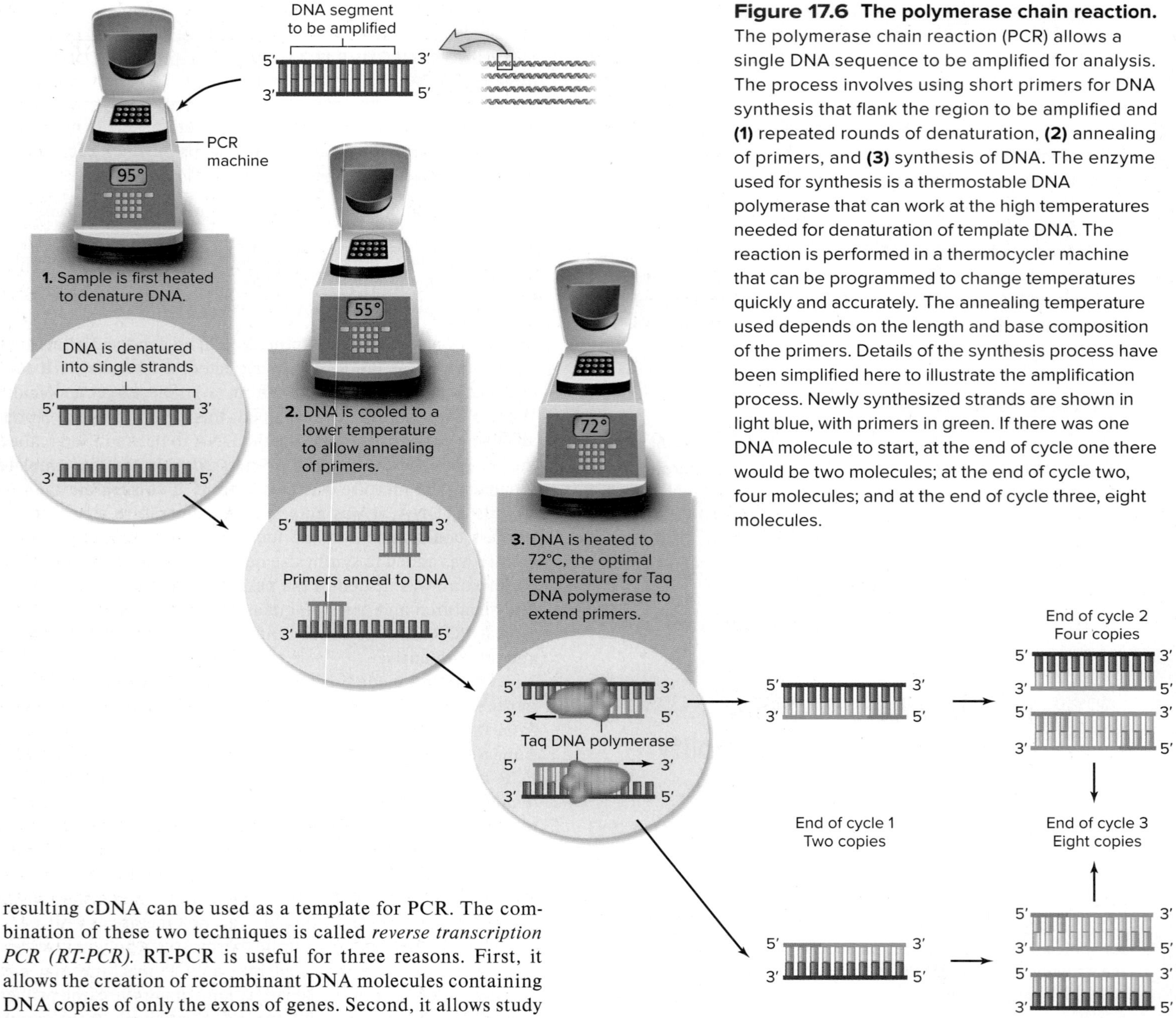

Figure 17.6 The polymerase chain reaction. The polymerase chain reaction (PCR) allows a single DNA sequence to be amplified for analysis. The process involves using short primers for DNA synthesis that flank the region to be amplified and **(1)** repeated rounds of denaturation, **(2)** annealing of primers, and **(3)** synthesis of DNA. The enzyme used for synthesis is a thermostable DNA polymerase that can work at the high temperatures needed for denaturation of template DNA. The reaction is performed in a thermocycler machine that can be programmed to change temperatures quickly and accurately. The annealing temperature used depends on the length and base composition of the primers. Details of the synthesis process have been simplified here to illustrate the amplification process. Newly synthesized strands are shown in light blue, with primers in green. If there was one DNA molecule to start, at the end of cycle one there would be two molecules; at the end of cycle two, four molecules; and at the end of cycle three, eight molecules.

resulting cDNA can be used as a template for PCR. The combination of these two techniques is called *reverse transcription PCR (RT-PCR).* RT-PCR is useful for three reasons. First, it allows the creation of recombinant DNA molecules containing DNA copies of only the exons of genes. Second, it allows study of the structure and function of gene products. Third, it can be used to determine relative levels of gene expression in cells and tissues.

RT-PCR can be quantified to determine mRNA abundance

Cells often respond to changes in environmental conditions by changing gene expression. Analysis of gene expression involves accurate quantification of cellular mRNA levels. One of the fastest and easiest ways to measure relative changes in gene expression is using reverse transcription quantitative PCR (RT-qPCR). This involves isolating mRNA, using reverse transcriptase to convert this to cDNA, then using PCR to amplify specific cDNAs using gene-specific primers. The amount of DNA produced can be measured in real time by the PCR machine. For this reason, quantitative PCR is also known as real-time PCR. Two common techniques are used, both of which depend on the use of fluorescent dyes.

The simplest way to quantify the amount of DNA made during the PCR part of RT-qPCR is to add a dye to the PCR reaction that binds nonspecifically to DNA. If DNA is present, the dye will bind to the DNA and fluoresce when illuminated by a laser. After each cycle of PCR, a laser in the PCR machine illuminates the reaction and detectors measure the amount of fluorescence. As the amount of DNA increases with each PCR cycle, more dye becomes bound to the DNA, and the fluorescence increases. The data can be output to a computer to plot a graph of fluorescence versus the number of PCR cycles.

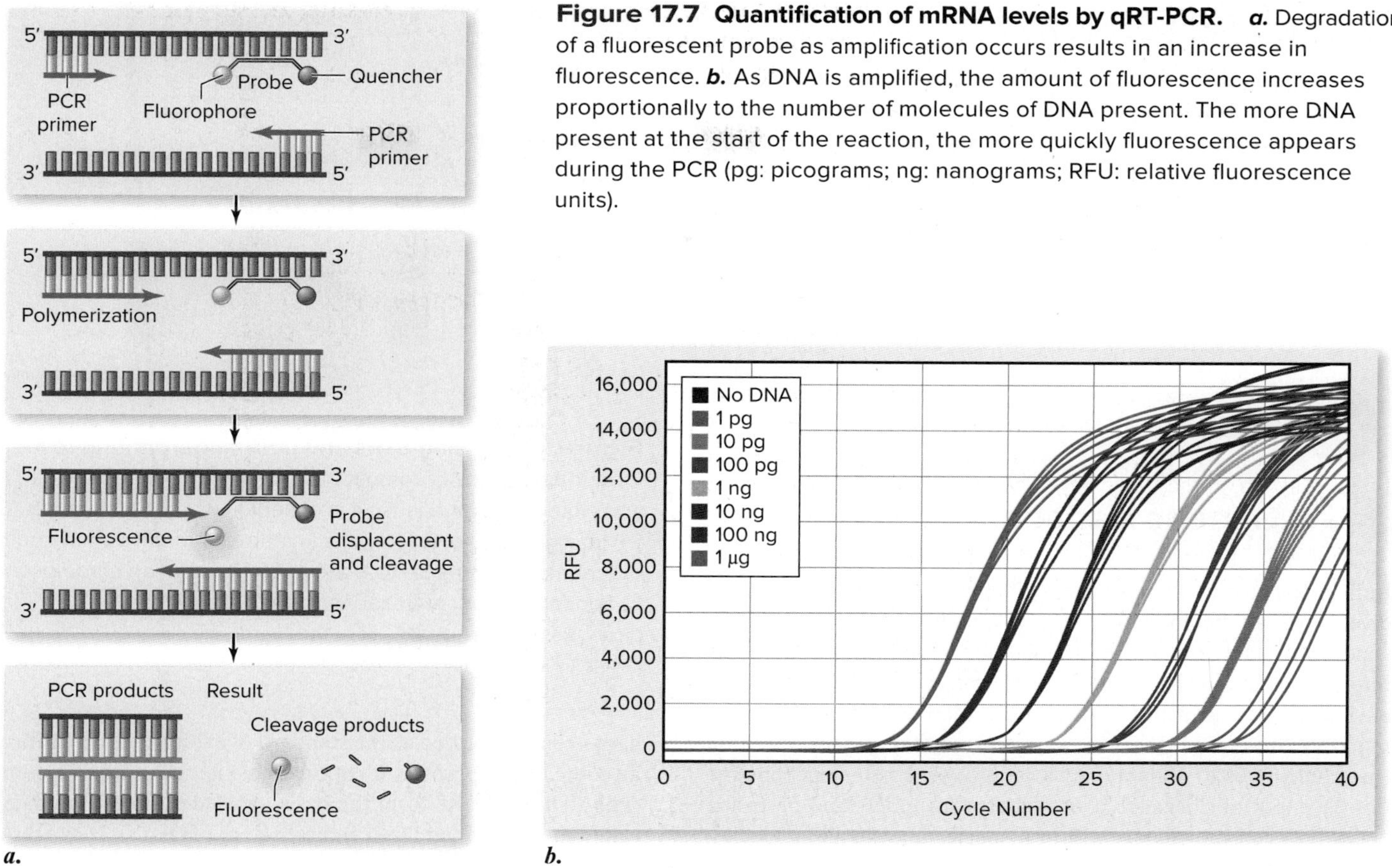

Figure 17.7 Quantification of mRNA levels by qRT-PCR. ***a.*** Degradation of a fluorescent probe as amplification occurs results in an increase in fluorescence. ***b.*** As DNA is amplified, the amount of fluorescence increases proportionally to the number of molecules of DNA present. The more DNA present at the start of the reaction, the more quickly fluorescence appears during the PCR (pg: picograms; ng: nanograms; RFU: relative fluorescence units).

A more accurate way to quantify the amount of PCR product is to introduce a fluorescently labeled probe into the qPCR reaction. This 20- to 30-nt-long probe is complementary to a short region of the template cDNA and has a fluorescent molecule (fluorophore) added to one end. When illuminated, the fluorophore emits fluorescence that can be measured by the PCR machine. A so-called quencher molecule added to the other end of the probe absorbs the light emitted by the fluorophore. As long as the fluorophore and quencher are attached to the same probe, they are close enough that fluorescence from the fluorophore is quenched.

If the probe is included in the PCR reaction, it can be used to detect the amount of DNA produced. After the denaturation step, when the reaction is cooled to the primer annealing temperature, the qPCR probe also anneals to one of the template strands between the two PCR primers. As the polymerase extends the PCR primers, it will run into the annealed probe on one strand. As the Taq polymerase has an exonuclease activity (refer to chapter 14), it can degrade the probe and complete copying the template. When the probe is degraded, the fluorophore is released from the quencher, and its fluorescence can be detected (figure 17.7*a*). The PCR cycles continue and with every round, there will be an increase in fluorescence that is proportional to the amount of DNA being made. The more template (cDNA made from mRNA) there is to begin with, the faster the fluorescent signal will increase (figure 17.7*b*).

PCR Is the Basis of Many Sequencing Technologies

LEARNING OBJECTIVE 17.2.3 Describe the benefits of using PCR-based methods for next-generation sequencing.

It took 13 years, from 1990 to 2003, to obtain a finished version of the sequence of the human genome at an estimated cost of $2.7 billion. Today, a human genome can be sequenced in as little as three days at a cost close to $1000. DNA sequencing requires that many copies of fragments of the genome be obtained. For the initial Human Genome Project, this involved cloning small fragments of genomic DNA into cloning vectors that were engineered to facilitate the manual sequencing of DNA (refer to chapter 18). Clones then had to be grown in bacteria, and DNA isolated for sequencing, all of which is both labor-intensive and time-consuming.

PCR overcame the need to use cloning vectors to obtain enough DNA to be sequenced. Many new sequencing technologies, called next-generation sequencing (NGS), use variations of PCR to produce the DNA that will be sequenced (discussed in chapter 18). Without PCR, sequencing genomes quickly and cheaply would not be possible. The value of being able to resequence genomes quickly and cheaply is that it allows comparison between genotypes and the identification of genetic variations that might be responsible for disease, for responses to certain drugs, or for general health.

REVIEW OF CONCEPT 17.2

The polymerase chain reaction (PCR) is used to rapidly amplify specific DNA sequences from small amounts of starting material. RT-PCR allows DNA copies of expressed mRNAs to be amplified, and RT-qPCR allows the relative quantification of specific mRNAs. The use of PCR in next-generation sequencing has contributed to a reduction in time and cost to sequence a genome.

- *How are PCR and DNA replication in cells similar? How do they differ?*

17.3 Molecular Tools Allow Us to Analyze and Modify Genetic Variation

Individuals in sexually reproducing populations are phenotypically diverse, yet all members of a species usually have the same set of genes. Most diversity within populations is produced by complex interactions among genetic variation, epigenetic differences, and environmental variability. Advances in manipulating and analyzing DNA have produced new ways to study the genetic variation that contributes to phenotypic variation. Analyzing genetic variation is important because it contributes to understanding differences between individuals within a species and between species. Recent advances in DNA technology also allow us to create or repair mutant alleles as a way to investigate gene function.

Forensics Uses DNA Fingerprinting to Identify Individuals

LEARNING OBJECTIVE 17.3.1 Describe how genetic information can be used to identify an unknown individual.

It is sometimes useful to be able to identify an individual based on a small amount of tissue or bodily fluids. This occurs during the investigation of crimes, as well as in the identification of victims of catastrophic events. Although more sensitive techniques have been developed (refer to chapter 18), forensics continues to use DNA fingerprinting as a primary technique. This is at least in part due to the large databases that have been assembled by law enforcement organizations.

DNA fingerprinting takes advantage of short, repeated sequences that vary among individuals. Short tandem repeats (STRs), typically 2 to 4 nt long, are not part of coding or regulatory regions of genes and mutate over generations so that the length of the repeats varies. Over time, populations of individuals become polymorphic for these molecular markers. A combination of these markers can be used as DNA "fingerprints" in criminal investigations and other identification applications (figure 17.8).

This is another example of the flexibility and utility of the PCR process. Primers are designed to flank a region known to contain an STR. These primers can then be used to amplify the STR-containing region from very small amounts of starting material. The analysis can also be done using the same technology used for automated sequencing (refer to chapter 18). Since 1997, 13 STRs have been established as the standard of evidence for identification in court, and approved identification kits are available commercially. The 13 STRs form the basis of a federal profiling database called the Combined DNA Index System (CODIS). New DNA fingerprints can be compared with those of known individuals in CODIS.

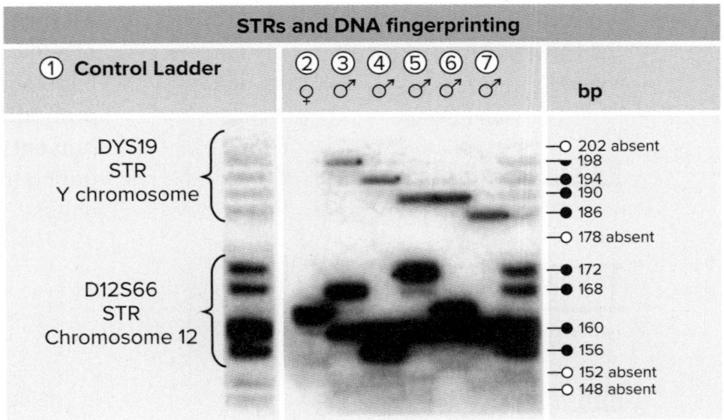

Figure 17.8 Using STRs and DNA fingerprinting to identify individuals. A Y-chromosome STR distinguishes between men and women *(top)*, because it is absent in women (column 2). The STR occurs in different lengths in men, depending on the number of repeats (columns 3–7). A second STR found on chromosome 12 appears in both men and women *(bottom)*.

Lutz Roewer, Department of Forensic Genetics, Institute of Legal Medicine and Forensic Sciences, Charité - Universitätsmedizin Berlin

DNA fingerprinting has also been used to exonerate wrongly convicted individuals who had been imprisoned for years before DNA analysis was widely available, as well as to identify individuals after catastrophes. After the September 11, 2001, attacks on the World Trade Centers in New York, DNA fingerprinting was the only means for identifying some of the victims. After a devastating earthquake in the Republic of Haiti in 2010, DNA fingerprinting was used to reunite children with their families.

PCR Can Be Used to Create Point Mutations in Specific DNA Sequences

LEARNING OBJECTIVE 17.3.2 Explain how we can construct mutations in genes in vitro.

Changes to one or a few nucleotides in a genome may have no effect on phenotype; the change(s) could easily be in noncoding regions such as introns or intergenic regions. Any change to the genome that results in a phenotypic change is recognized as a mutation. For years, geneticists created mutations by randomly using chemicals (mutagens), but now mutations at specific sequences in the genome can be created in vitro. Studying mutations allows us to better understand the normal function of the altered DNA. One important use of this approach is to replace a wild-type gene with a mutant copy to test the function of the mutated gene.

PCR-based, site-specific mutagenesis

Remember that the amplification of a region of DNA by PCR involves a template molecule, which is often a gene or cDNA copy of a gene, and a pair of primers used to initiate DNA replication of the template. If the sequence of one of the primers is made to be slightly different from the template sequence, then as DNA amplification occurs, most of the DNA molecules made in the PCR will end up with a specific mutation at a specific site. In this way, different mutations can be introduced into a gene or cDNA, and the effects of those mutations can be analyzed in a variety of ways, some of which we explore in this section.

PCR-based random mutagenesis

As we previously discussed, one of the polymerases used in PCR is called Taq polymerase. Unlike the polymerases used to replicate DNA in cells, Taq polymerase is not able to correct errors it makes during DNA synthesis. Taq polymerase will randomly introduce a mutation approximately every 45,000 nt synthesized. Changing the chemical conditions of the PCR can further increase the mutation rate so that by performing repeated rounds of PCR, randomly mutagenized cloned genes or cDNA can be obtained. The effects of these mutations on gene product function can then be analyzed in a variety of ways.

RNA Interference Can Reduce the Level of a Gene Product

LEARNING OBJECTIVE 17.3.3 Explain the mechanism of RNA interference.

Although the ability to make and analyze mutants in vitro is extremely powerful, in many experimental systems it is not possible to introduce these constructs back into a living cell to assess their effects on phenotype. The technique of RNA interference (RNAi) allows researchers to reduce the amount of a gene product in cells. This is the equivalent of a mutation that eliminates the function of a gene product. RNAi acts to degrade or block translation of a specific mRNA in cells, which in turn reduces the level of the encoded protein. This technique takes advantage of systems in cells that have evolved to use small RNAs to control the level of gene expression posttranscriptionally (refer to chapter 16 for details).

Using RNAi to reduce or eliminate the production of a specific protein requires the production of a short, double-stranded RNA complementary to the mRNA that encodes the protein. A variety of techniques exist to create and introduce double-stranded RNA into cells or organisms. RNAi then depends on the cellular mechanisms described in chapter 16. For any organism for which we have a complete genome sequence, a library of short, double-stranded RNA molecules can be created to target every gene in the genome. These libraries can be used to systematically reduce the expression of individual proteins or combinations of proteins so that their roles in cells can be investigated.

New Technologies Allow Direct Editing of Genomes

LEARNING OBJECTIVE 17.3.4 Compare and contrast different strategies for direct genome editing.

PCR-based mutagenesis allows the in vitro alteration of gene sequences and RNAi allows the reduction of gene products. These both provide indirect ways to investigate gene function in living cells. The next step is to directly edit genes in living cells. For research, this allows a direct assessment of how changing a protein affects its function, and for the clinic, it offers the possibility to alter genetic variants that cause genetic diseases. While this was once the province of science fiction, the **CRISPR (clustered regularly interspaced short palindromic repeats)/Cas9** system makes this not only possible, but relatively straightforward.

The CRISPR/Cas9 system

Much as the study of host restriction in bacteria led to the discovery of restriction endonucleases, the study of how some bacteria become "immune" to infection by a virus led to the discovery of the CRISPR/Cas9 system. Some bacteria appear to have a kind of adaptive immunity that involves a complex system that has evolved to incorporate part of the viral genome into the bacteria's genome. This viral DNA can then be used to target future infection by a virus with the same DNA sequence. A bacterium uses the viral DNA to transcribe an RNA that is used as a "guide" by a nuclease (Cas9) to target DNA with the same sequence.

Once inside a cell and associated with one another, a guide RNA and the Cas9 protein are able to seek out and associate with a gene that matches the guide RNA sequence (figure 17.9*a*). Cas9 is a nuclease and can cut the DNA with which it is associated. Inaccurate repair of the cut DNA can result in small deletions or insertions to the genomic sequence, which can result in loss of gene function. Alternatively, if linear, double-stranded DNA is available, homology-directed repair can accurately repair the cut. By providing linear, double-stranded DNA with a mutation, changes to the sequence of the gene can be made. Alternatively, if a defective allele of a gene is targeted, a mutation can be replaced with a normal gene sequence. This technique can inactivate the genome of the human immunodeficiency virus (HIV) that is integrated into cells grown in vitro and has the potential to treat certain genetic diseases.

Gene expression can be regulated using this system as well. Gene expression can be activated or repressed by constructing a fusion between Cas9 lacking nuclease activity and a transcriptional activator or repressor, respectively. It is also possible to create fusions between fluorescent proteins and nuclease-deficient Cas9 proteins (figure 17.9*b*). This could simplify the detection and analysis of oncogenes that are amplified in certain cancers.

REVIEW OF CONCEPT 17.3

Individuals can be identified through DNA fingerprinting. A specific set of STR loci makes up the CODIS database. PCR can be used to create site-specific or random mutations in DNA. RNA interference can be used to reduce levels of gene products. Genome-editing techniques can be used to remove gene function and to introduce new alleles in the chromosome.

- *What advantages do CRISPR/Cas9 systems offer over conventional molecular cloning?*

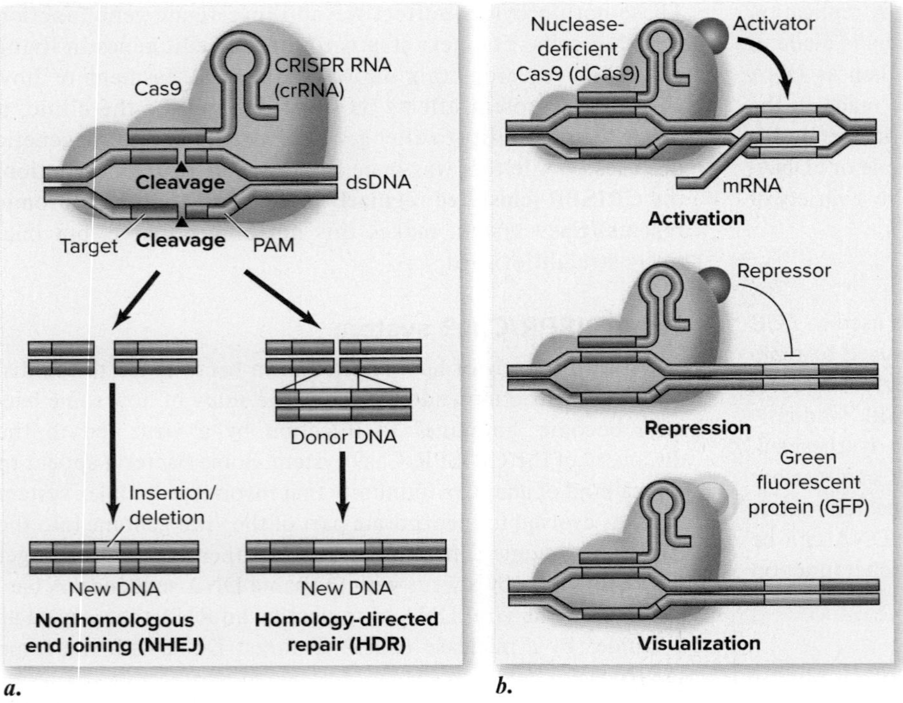

Figure 17.9 The CRISPR/Cas9 genome editing system. ***a.*** A sequence-specific guide RNA 9 (crRNA) targets Cas9 to a target sequence, just upstream from a sequence called a protospacer adjacent motif (PAM), where Cas9 cuts the target sequence. Repair can be inaccurate, resulting in gene inactivation, or it can be accurate in the presence of a donor DNA sequence that results in replacement of the DNA target sequence with the donor sequence. ***b.*** Recombinant DNA technology can be used to target Cas9 to a gene and turn it on (activation), turn it off (inactivation), or reveal its location in the genome.

Source: New England BioLabs Inc., www.neb.com/tools-and-resources/feature-articles/crispr-cas9-and-targeted-genome-editing-a-new-era-in-molecular-biology

17.4 Transgenic Organisms Are Used to Analyze Gene Function

Transgenic organisms contain a gene from a different species, called a *transgene*, which has been incorporated into the genome through genetic engineering. In addition to adding genes to an organism's genome, it is also possible to remove and change genes within the genome. Genetically modified organisms are organisms that have been genetically altered by techniques other than conventional breeding. The construction of genetically modified organisms has revealed the function of many genes and has applications in medicine, environmental biology, and agriculture.

The Universal Nature of the Genetic Code Allows Transgenesis

LEARNING OBJECTIVE 17.4.1 Explain how the universal nature of the genetic code allows transgenesis.

A human gene can be placed into a mouse genome, and the cells of the mouse will express the gene. Similarly, a human gene can be placed into the genome of an *E. coli* bacterium and it will make the encoded protein. This is because the cells of organisms of different species interpret genetic information identically. The genetic code used to decipher the information in a gene is the same code used in bacteria, archaea, and eukaryotes. For example, the amino acid phenylalanine will always be coded for by the codon UUU; this is true if we look at the genetic code in humans or in *E. coli*. There are minor variations to the genetic code, however. For example, the genetic code of mitochondria is slightly different from the genetic code used to interpret nuclear genes.

Removal or Addition of Genes Can Reveal Function

LEARNING OBJECTIVE 17.4.2 Compare the processes for making knockout and knockin mice and describe the purposes these mice serve.

The techniques for making recombinant DNA can be combined with other techniques to create organisms that have genes removed or genes added. In some cases, a gene can be removed and then replaced with a version that has been genetically altered. The study of these animals and plants allows scientists to better understand the relationships between genotype and phenotype. By introducing mutant alleles into organisms, scientists can study the effects of specific mutations in disease processes in the whole organism.

Knockout mice

A **knockout** animal is one that has had a gene inactivated so that the function of the gene is lost. It is possible to inactivate genes in a number of different experimental organisms. The most relevant to human genetics is the mouse, which has a rich genetic history. The effect of knocking out a gene can be assessed in the adult mouse—or if the gene is essential for survival, the developmental stage requiring the gene's function can be identified. A streamlined description of the steps in creating a knockout mouse are outlined as follows and illustrated in figure 17.10:

1. The cloned gene is disrupted by replacing part of it with a marker gene using recombinant DNA techniques. The marker gene codes for resistance to the antibiotic neomycin in bacteria, which allows mouse cells to survive when grown in a medium containing the related drug G418. The construction is done such that the marker gene is flanked by the DNA normally flanking the gene of interest in the chromosome.
2. The interrupted gene is introduced into **embryonic stem cells (ES cells).** These cells are derived from early embryos

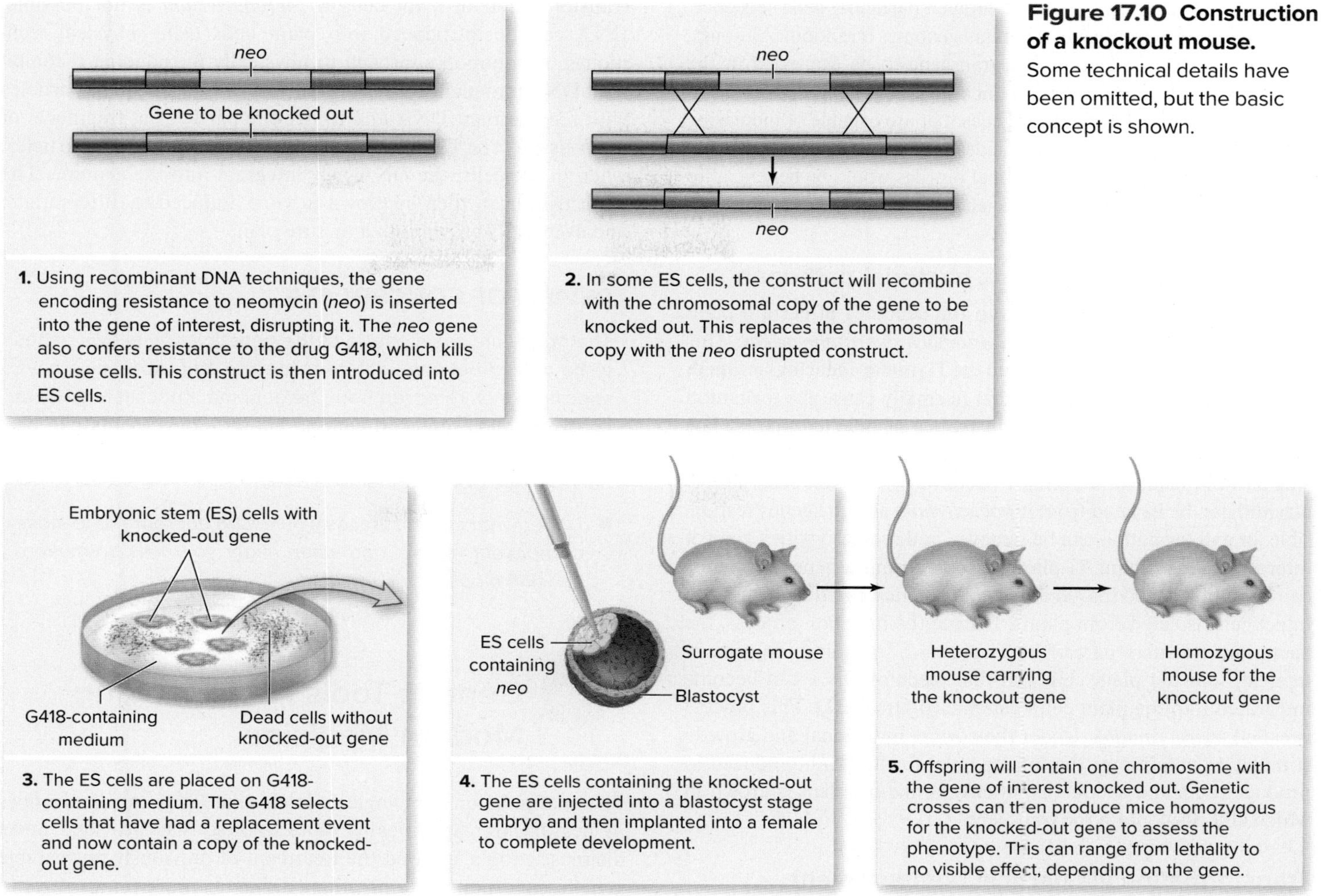
neo
Gene to be knocked out
1. Using recombinant DNA techniques, the gene encoding resistance to neomycin (neo) is inserted into the gene of interest, disrupting it. The neo gene also confers resistance to the drug G418, which kills mouse cells. This construct is then introduced into ES cells.
neo
neo
2. In some ES cells, the construct will recombine with the chromosomal copy of the gene to be knocked out. This replaces the chromosomal copy with the neo disrupted construct.
Embryonic stem (ES) cells with knocked-out gene
G418-containing medium
Dead cells without knocked-out gene
3. The ES cells are placed on G418-containing medium. The G418 selects cells that have had a replacement event and now contain a copy of the knocked-out gene.
ES cells containing neo
Surrogate mouse
Blastocyst
4. The ES cells containing the knocked-out gene are injected into a blastocyst stage embryo and then implanted into a female to complete development.
Heterozygous mouse carrying the knockout gene
Homozygous mouse for the knockout gene
5. Offspring will contain one chromosome with the gene of interest knocked out. Genetic crosses can then produce mice homozygous for the knocked-out gene to assess the phenotype. This can range from lethality to no visible effect, depending on the gene.

Figure 17.10 Construction of a knockout mouse. Some technical details have been omitted, but the basic concept is shown.

and can develop into different adult tissues. In these cells, the gene can recombine with the chromosomal copy of the gene based on the flanking DNA. This is the same kind of recombination used to map genes (refer to chapter 13). The knockout gene with the drug resistance gene does not have an origin of replication, and thus it will be lost if no recombination occurs. Cells are grown in medium containing G418 to select for recombination events. (Only those containing the marker gene can grow in the presence of G418.)

3. ES cells containing the knocked-out allele are injected into a mouse embryo early in its development, which is then implanted into a pseudopregnant female (a female that has been mated with a vasectomized male and as a result has a receptive uterus). Some of the cells in the pups born to this female have one of the two alleles for the gene of interest knocked out and, so, these animals are chimeras. The chimeric mice are mated with normal mice and some of the offspring will be completely heterozygous. In some cases, there may be a phenotype that can be detected and analyzed if heterozygosity affects any biological function. If the mutation is recessive and not lethal, heterozygotes can then be crossed to generate homozygous mice. The homozygous mice can be analyzed for phenotypes.

Knockin mice

Knockin mice have a normal allele replaced with an allele that has a specific genetic alteration. The ability to eliminate alleles provides insight into a complete loss of gene function, a so-called null allele, but does not allow the analysis of alleles that alter but do not completely remove function. Specific mutations can be constructed in vitro, then introduced into the mouse to assess the effects of the alterations. This allows an investigator to use naturally occurring alleles, targeted mutations based on other information, or any other desired alteration. Mutations can be introduced that result in complete loss of function, partial loss of function, or even gain of function of the gene product.

Plants Require Different Techniques to Be Genetically Modified

LEARNING OBJECTIVE 17.4.3 Describe how transgenic plants are created.

Genetically modified plants are created using techniques that are considerably different from those employed to create genetically modified animals. Transgenic plants can be created by introducing foreign DNA into plant cells by electroporation, physical bombardment, chemical treatment, and bacterial transfer. Whichever technique is used, it is not generally possible to target a particular gene sequence, and integration of transgenes into plant genomes is random. Although this allows the introduction of a transgene to be expressed in the plant, it is a problem for knocking out a specific gene. It is desirable to introduce into crop species transgenes that are capable of increasing resistance to disease, frost, herbicides, and drought; also, creating crop plants with modified nutritional profiles would be beneficial to populations where nutritional deficiency is a problem.

Transforming plants with a bacterial pathogen

Transgenes can be transferred into the genomes of certain plant species from the plant pathogen *Agrobacterium tumefaciens.* This bacterium harbors a plasmid called the **Ti (tumor-inducing) plasmid.** The Ti plasmid contains genes that normally cause the formation of a plant tumor—called a gall—in tissues of infected plants. The plasmid also contains genetic sequences responsible for transferring part of the plasmid from the bacterium into plant cells. The Ti plasmid can be isolated from *Agrobacterium,* and the genes responsible for gall formation can be removed and replaced with a gene of interest. Recombinant Ti plasmids containing a gene of interest can be reintroduced into *Agrobacterium,* which can then be used to infect cells isolated from plants. During the infection process, the bacterium transfers part of its Ti plasmid, including the gene of interest, into the plant cell, and that piece of DNA can become integrated into the plant cell's genome (figure 17.11). This process is called *transformation.* Under the correct nutritional and growth-stimulating conditions, the infected plant cells can be induced to produce roots and shoots. Eventually, a mature plant is grown in which all cells contain the transgene.

Transformation via particle bombardment

Agrobacterium cannot infect all plant species, so not all plant species can be transformed using the method just described. When transformation of plant cells by *Agrobacterium* is not possible, DNA can be introduced into plant cells using physical techniques. A common approach to physically introducing recombinant DNA into plant cells is to coat gold or tungsten nanoparticles with recombinant DNA and to fire the particles at fragments of plant tissue. The DNA is carried into cells on the tiny particles; when in the cells, the DNA may integrate into the genome. The plant tissue can then be grown in vitro, induced to differentiate, and eventually grown into a mature plant.

REVIEW OF CONCEPT 17.4

The nearly universal nature of the genetic code allows genes to be moved between different species. Knockout mice have specific alleles removed from the genome. Knockin mice have specific alleles replaced with an altered copy. Plants can be transformed with Ti plasmid DNA from *Agrobacterium* or by nanoparticle bombardment.

- *Under what circumstances would it be appropriate to make a knockout mouse, and when might you need to make a knockin mouse?*

17.5 Genetic Tools Are Changing Modern Medicine

The manipulation of biological systems has led to major advances in health care. Specifically, biotechnology has provided novel diagnostic tools, allowed the treatment of previously untreatable genetic conditions and infectious diseases, and given us ways to make novel medicines. The application of biotechnology to medicine is not a new endeavor; however, advances in genetics,

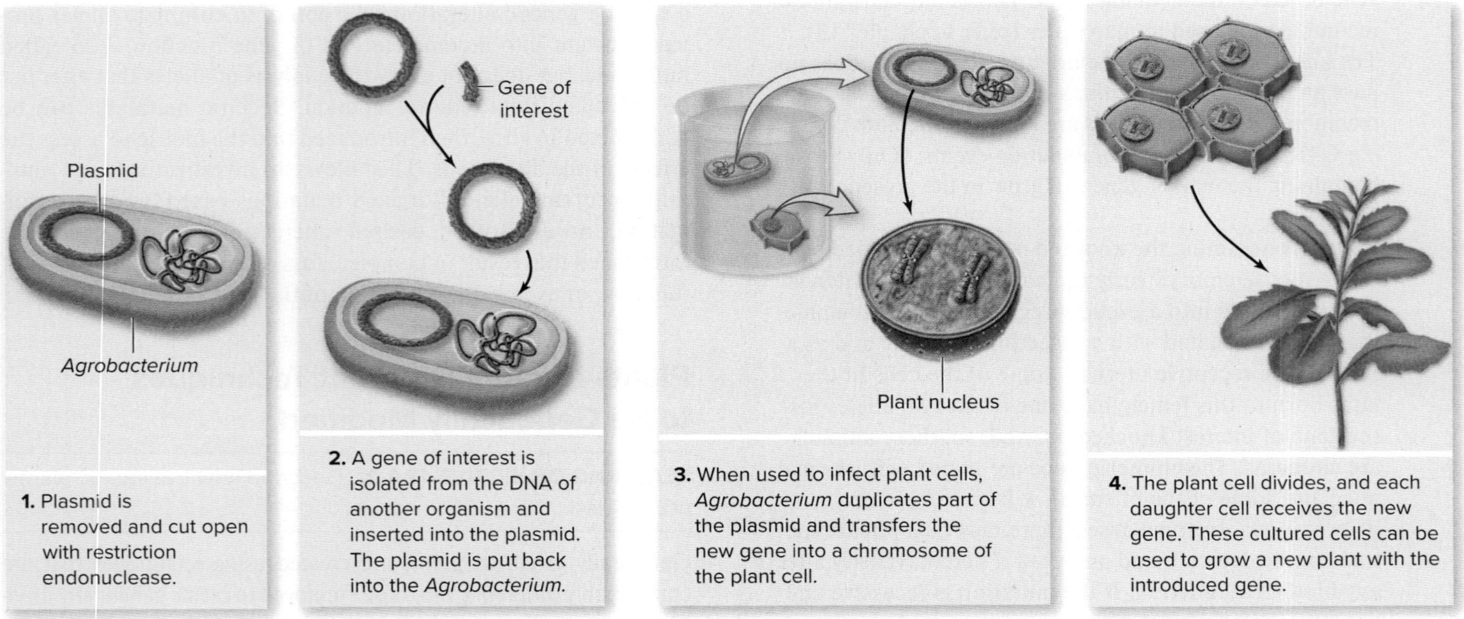

Plasmid
Agrobacterium
1. Plasmid is removed and cut open with restriction endonuclease.
Gene of interest
2. A gene of interest is isolated from the DNA of another organism and inserted into the plasmid. The plasmid is put back into the Agrobacterium.
Plant nucleus
3. When used to infect plant cells, Agrobacterium duplicates part of the plasmid and transfers the new gene into a chromosome of the plant cell.
4. The plant cell divides, and each daughter cell receives the new gene. These cultured cells can be used to grow a new plant with the introduced gene.

Figure 17.11 Creating transgenic plants using *Agrobacterium* transformation. Steps in the generation of transgenic plants using *Agrobacterium tumefaciens* transformation.

genomics, and recombinant DNA technologies have accelerated the rate of discovery in the last 20 years.

Important Proteins Can Be Produced Using Recombinant DNA

LEARNING OBJECTIVE 17.5.1 Describe how genetic engineering can be used to overcome the challenges of expressing human proteins in bacteria.

Diseases that are due to the body's inability to produce a specific protein can be treated by providing the missing protein. Traditionally, these are isolated from animal tissues, but recombinant DNA allows us to express proteins in bacteria and other systems. The first medical product to be produced via recombinant DNA was insulin, which is produced in the pancreas, then transported via the bloodstream. Insulin is one of the hormones involved in regulating glucose levels (refer to chapter 35), and low levels of insulin cause blood sugar to rise. High blood sugar is a symptom of type I diabetes, which can be managed with daily insulin injections. The production of insulin using recombinant DNA is instructive because it shows both the promise of this technology, and also the challenges.

In human cells, insulin is expressed as a single mRNA that is translated into a polypeptide that is cleaved into two shorter polypeptides that are then linked by disulfide bonds (figure 17.12*a*). Using recombinant DNA, a cDNA derived from the coding region of the insulin gene can be expressed in *E. coli*, and used to produce recombinant protein. However, *E. coli* will not process this protein as in human cells. To get around this, two cDNAs, which encode the two polypeptides that make up active insulin protein, are used instead of a single full-length cDNA. Each of these cDNAs is spliced into a separate plasmid vector that can be expressed in *E. coli*.

Proteins isolated from two strains of *E. coli* expressing the individual insulin polypeptides can then be purified and mixed together to make a functional insulin molecule (figure 17.12*b*). When purified of other contaminating bacterial proteins, the insulin reaches very high concentrations. Unfortunately, at these high concentrations, the insulin molecules stick together, impairing their function. To solve this last problem, specific codons in the insulin cDNA sequences were altered so that the resulting proteins don't stick together. This allows the production of high concentrations of highly effective insulin.

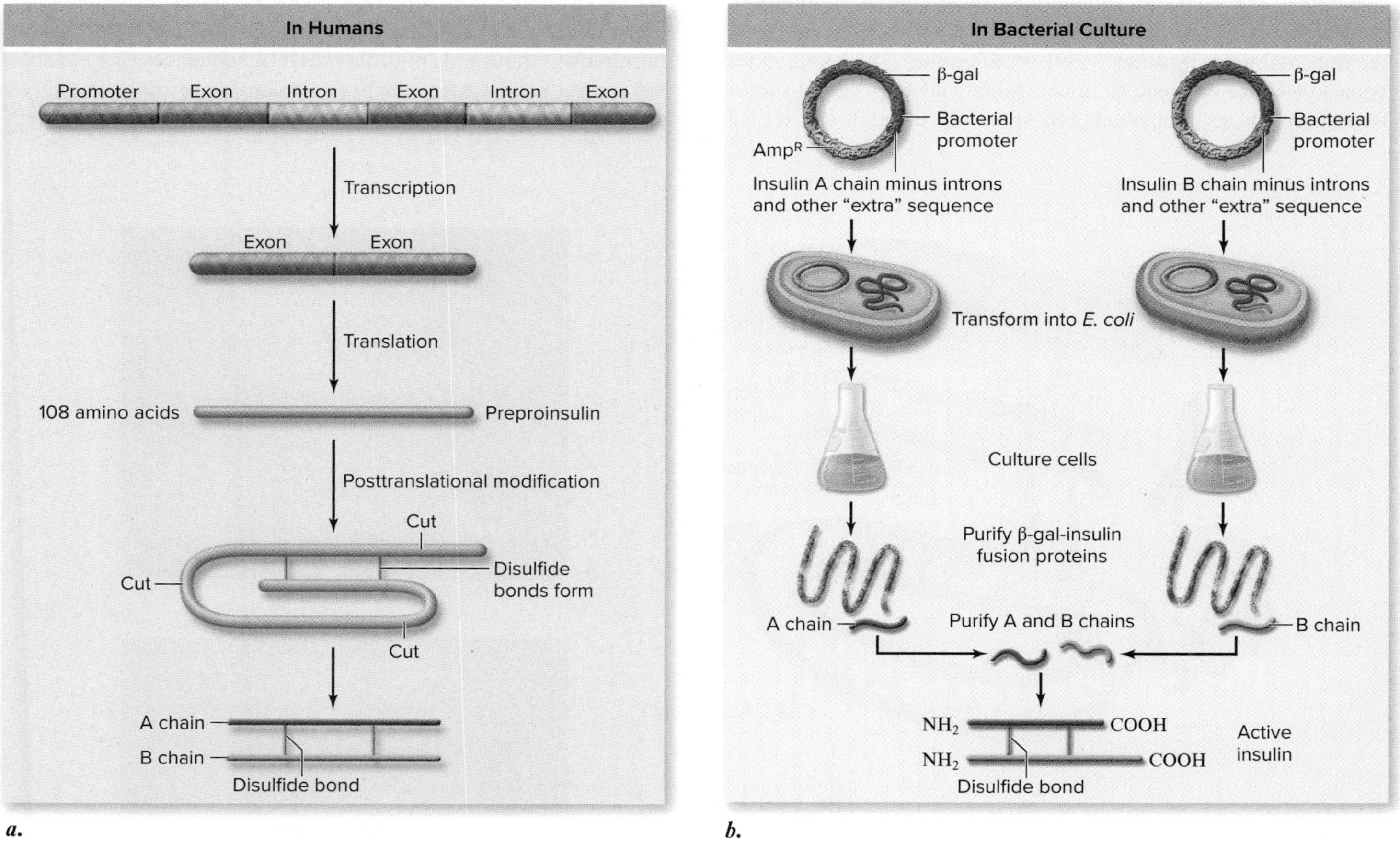

Figure 17.12 Making insulin in genetically engineered *E. coli*. ***a.*** In human cells, one preproinsulin polypeptide is processed posttranslationally into insulin chains A and B that associate via disulfide bonds to form mature insulin. ***b.*** Two cDNAs corresponding to the gene sequences for insulin chain A and chain B are cloned into plasmids and introduced into different *E. coli*. Cultures of the two types of *E. coli* express either chain A or chain B. Chains A and B are purified from the different *E. coli* and, when mixed, associate into active, mature insulin.

Fluorescent In Situ Hybridization Can Detect Gross Chromosomal Rearrangements

LEARNING OBJECTIVE 17.5.2 Explain how FISH can be used to detect gross chromosomal rearrangements.

Fluorescent in situ hybridization (FISH) takes advantage of the ability of DNA to be reversibly denatured and renatured. Many techniques in molecular biology use hybridization to detect a specific DNA in a complex mixture. Hybridization uses a probe, a relatively short piece of DNA identical to the sequence you wish to locate, that is labeled with either a fluorescent or a radioactive label. Both probe and target DNA are denatured and mixed together. When conditions are restored to allow renaturation, the probe will find any complementary sequences in the mixture and form a hybrid. In FISH, the target DNA is a metaphase spread of chromosomes, or cells with intact chromosomes. This allows the detection of gross chromosomal abnormalities, such as large deletions, inversions, duplications, and translocations (refer to chapter 15 for descriptions).

Aneuploidy and chromosomal abnormalities are associated with many forms of cancer. In about 25% of invasive breast cancers, a gene called *HER2* has been duplicated. Instead of the usual two copies expected in a diploid cell, there can be hundreds of copies. The HER2 protein is a receptor that activates signal-transduction pathways leading to cell proliferation. Therefore, extra copies of *HER2* can lead to uncontrolled cell growth and tumor formation. Drugs that block cell signaling through the HER2 protein can be used to treat this kind of tumor. However, these drugs are only effective in treating tumors caused by *HER2* duplications, making data on the copy number of *HER2* genes vital for clinicians. These data allow treatment to be personalized based on a patient's specific genetic profile.

FISH can be used to detect HER2 in cells taken from a tumor biopsy. Tumor cells are collected and treated to expose the chromosomes. A fluorescently labeled *HER2* probe is used for hybridization to identify *HER2* DNA in the chromosome (figure 17.13*a*). When cells are illuminated with UV light, which causes the probe to fluoresce, *HER2* sequences are visible.

In normal cells, or in tumor cells from breast cancers with only the diploid complement of *HER2* genes, there would be just two small dots of fluorescence seen under the microscope (figure 17.13*b*). If the breast cancer cells contain more than two copies of the gene, then there will be multiple, much larger dots of fluorescence (figure 17.13*c*). This is an example of personalized medicine, where drugs are administered only when necessary.

Gene Chips Can Identify Genetic Markers for Disease

LEARNING OBJECTIVE 17.5.3 Compare and contrast FISH and gene chip technologies for diagnosing disease.

Gene chips, also known as DNA microarrays, are collections of hundreds to thousands of different DNA sequences that are spotted onto a solid surface such as a microscope slide. The DNA sequences are usually short, chemically synthesized pieces of

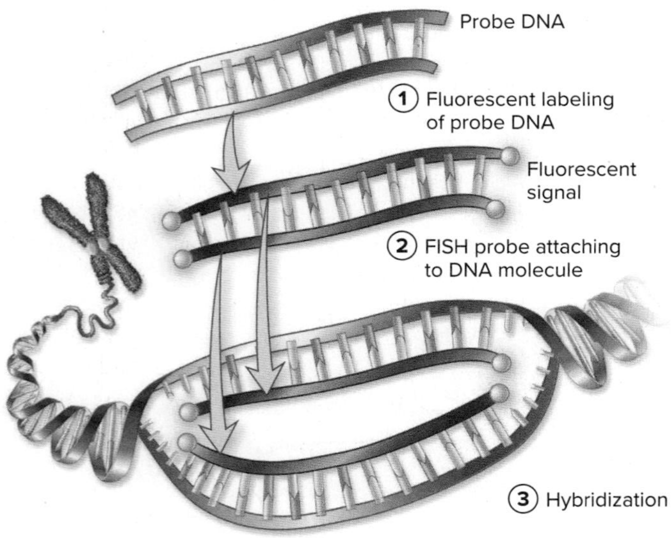

a.

b.

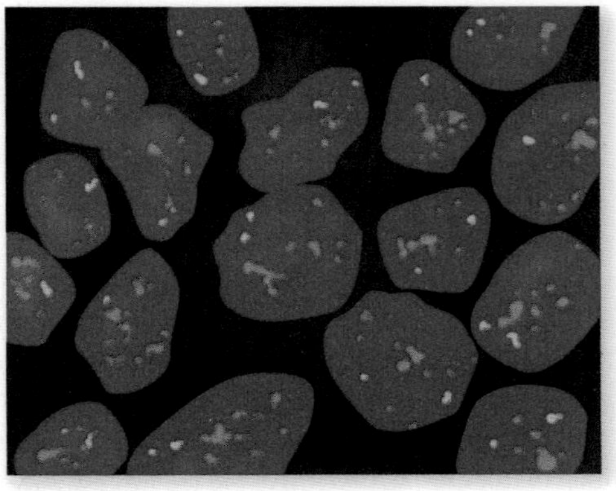

c.

Figure 17.13 Fluorescent in situ hybridization. ***a.*** Fluorescent probes bind to specific DNA sequences in chromosomes. ***b.*** Only two spots of fluorescence are visible in the nuclei of some cancerous cells, indicating that *HER2* is not amplified. ***c.*** Multiple large fluorescent spots in the tumor cells indicates amplification of *HER2*.

DNA or short sequences of DNA amplified by PCR or RT-PCR. By using different gene chips, and different experimental conditions, a variety of kinds of questions can be assessed.

A disease marker, or biomarker, is a biological molecule found in affected individuals, and not in unaffected individuals. Biomarkers can be diagnostic of a disease state, or the progress of a disease, and help monitor the response to therapy. If a specific gene is only expressed in affected individuals, then the mRNA encoded by that gene can be a biomarker. For example, individuals with some autoimmune diseases or chronic inflammatory conditions express different sets of genes in immune cells than healthy individuals. Gene chips can be used to monitor gene expression, including the relative levels of expression for different genes.

Gene chips are also used to genotype SNPs in an individual, with literally hundreds of thousands of SNPs on a single chip. This allows the GWAS studies discussed in chapter 13 to collect very large data sets quickly and cheaply. If you have your DNA analyzed by any of the companies that offer to provide you information on your ancestry, they are not sequencing your entire genome; they are using microarrays with thousands to millions of informative SNPs.

Taken together, technologies such as FISH and gene chips are changing the way that medicine is administered. The more that can be learned about the specific underlying causes of a disease on an individual basis, the more specific and personalized the health care provided. These technologies face some challenges, however. First, they require great skill to be successfully used; second, they are relatively expensive; and third, not all diseases are caused or influenced by factors specific to the individual.

Gene and Stem Cell Therapy and Gene Editing Are All Moving into the Clinic

LEARNING OBJECTIVE 17.5.4 Compare and contrast gene therapy, stem cell therapy, and gene editing.

From the earliest days of "genetic engineering," biologists envisioned using this technology to restore or replace defective genes in humans. The first clinical trial in the U.S. of gene therapy was in 1990 for a rare inherited form of severe immune deficiency. This trial used autologous (from the patient) T cells to which a normal copy of a defective gene had been added. The trial was a partial success, and since that first trial, more than 2200 clinical trials have occurred worldwide, with the majority (63%) in the U.S. Despite this, there have been only a few gene therapy products approved.

While gene therapy targets genetic diseases, stem cell therapy seeks to treat degenerative diseases, or tissue damage from injury, by replacing damaged tissue. The most common use of stem cells has been bone marrow transplantation, which is now used for 50,000 patients annually worldwide. Bone marrow contains stem cells that produce the different types of blood cells, and transplantation is used to treat a number of blood disorders, including various types of leukemia and anemia. This technique has also been successful in stopping progression of multiple sclerosis in some patients because it results in a complete reprogramming of the immune system.

The majority of approved stem cell therapies use **mesenchymal stem cells.** These stem cells can differentiate into cell types that form connective tissues such as bone and cartilage. These therapies can promote connective tissue formation and repair, and reduce inflammation. Stem cell therapies for the treatment of the tissue damage caused by heart attacks are also moving through clinical trials. Despite some successes, stem cell therapies face significant hurdles before they can realize the potential promised by researchers.

These two kinds of therapy can be extended by new genome editing technology (section 17.4). In the most far-reaching form of this therapy, induced pluripotent stem cells (refer to chapter 36) derived from a patient can have defective genes restored, before reintroduction back into the patient. This is already being considered for a number of genetic diseases that result in blindness. Genetic diseases that affect a specific tissue are all potential targets for this methodology. Work is ongoing for treatment of diabetes, and of blood disorders such as sickle-cell disease.

This kind of technology can also be used to treat infectious disease by targeting cellular receptors used by infectious agents to gain entry into cells. In a 2014 clinical trial, researchers disrupted the CCR5 gene in patient-derived $CD4^+$ T cells. Infusion of these cells showed some promise in lessening the effects of HIV infection (refer to chapter 23). There are now multiple clinical trials using CRISPR/Cas9 systems to edit the CCR5 gene in T cells.

Because of ethical issues, including the anticipated high cost of such therapy, the most important use of this technology may be for research. Cells taken from patients suffering from specific diseases can be used to create stem cells that represent a kind of "disease-in-a-dish." These cells can be used for research into disease mechanisms, and potential drug therapies.

New Vaccine Technologies Are Helping to Fight the COVID Pandemic

LEARNING OBJECTIVE 17.5.5 Differentiate between new vaccine technologies and traditional vaccines.

Traditional vaccines for viral pathogens have used two main technologies: inactivated virus and weakened (or attenuated) virus. The first uses the actual virus, but chemically treated such that it is not infectious. Weakened viruses are made by passaging a virus through cells, accumulating mutations, until the virus can still replicate, but does not cause disease.

New vaccine technologies fall into three categories: nucleic acid vaccines, viral vector vaccines, and subunit (recombinant protein) vaccines. The common thread in all of these is that they use a variety of technologies to produce a single protein, or even part of a protein, to be used as an antigen to stimulate an immune response (refer to chapter 35). This requires significant knowledge about the infectious agent, and the immune response to it. For example, with the COVID pandemic of 2021, while an enormous amount of research was devoted to the SARS-CoV-2 virus, this effort was also aided by work on the original SARS virus.

We will use SARS-CoV-2 to illustrate how these technologies work. It was known from work on SARS that the antigen that stimulated the strongest immune response was the spike (S) protein, which binds to the ACE2 receptor protein on cells. The DNA and mRNA vaccines both encode this S protein and are intended to enter cells and cause the production of S protein, which will then stimulate the immune system. Subunit vaccines skip this step and consist of just the S protein, or the S protein attached to the surface of a nanoparticle. Lastly, the viral vector vaccines consist of other innocuous human viruses that contain the gene for the S protein. These are usually the same virus vectors that are used in gene therapy experiments.

By the end of 2020, there were almost 200 candidate vaccines based on these technologies being developed for SARS-CoV-2, and two mRNA vaccines had received approval for emergency use. By the end of 2021, the two mRNA vaccines were approved by the FDA, and another vaccine that uses an adenovirus vector to produce the spike protein was also approved for emergency use. Worldwide, a variety of vaccines that use all of the approaches discussed, including the traditional methods, are being used.

Infectious Disease Agents Are Detected by Diagnostic Testing

LEARNING OBJECTIVE 17.5.6 Describe the different kinds of diagnostic tests used for infectious agents.

Traditionally diseases were diagnosed based on analysis of symptoms, including relevant biomarkers. This is increasingly being replaced by direct testing for the actual agent responsible. There are three kinds of tests now in common use: nucleic acid amplification tests, antigen tests, and antibody tests. Nucleic acid amplification tests (NAATs) detect the genome of the infectious agent using PCR, either conventional PCR for DNA genomes, or RT-PCR for RNA genomes (section 17.2). Antigen and antibody tests both use antibody technology (refer to chapter 35), although they address different questions.

It is possible to make antibodies against a range of viruses and bacteria, which can be used in immunoassays to detect pathogens in patient samples. An antibody against a specific virus can be coated into a plastic well on a microtiter plate (figure 17.14). When an unknown sample is added to the well, if the virus is present, it will be bound by the antibody. Bound virus can be detected using a second antibody against the virus that has been linked to a fluorescent molecule, or an enzyme that produces a colored product from a suitable substrate. The amount of fluorescence or colored product made by the enzyme is proportional to the amount of virus present. This same technology is used to detect antibodies against an infectious agent, which gives us knowledge of who has been infected rather than finding current infections.

During the COVID pandemic, NAATs have been the primary diagnostic test because of their sensitivity, even detecting viral RNA in asymptomatic individuals. As SARS-CoV-2 is an RNA virus, these use RT-PCR, which requires more handling, and skilled technicians and can take time to get results. This has led to the development of rapid antigen tests, which are less sensitive, but very rapid and simple to perform. The difference in sensitivity depends on the comparisons, but rapid antigen tests are at least 100 times less sensitive. However, they appear to be able to detect virus in individuals who are infectious, and as they can provide an answer in 10–15 minutes as opposed to hours to days for NAAT, they are a valuable tool.

REVIEW OF CONCEPT 17.5

Recombinant DNA technology allows the mass production of human gene products. FISH is used to detect chromosomal rearrangements, including those involved in cancers. This diagnostic information allows personalization of treatment. Gene chips are used to determine patterns of gene expression for many genes, or detect many SNPs, at a time. Gene therapy, stem cell therapy, and gene editing are being used for therapy. New vaccine technologies are changing how we prevent infectious diseases. Diagnostic tests using molecular tools are now common.

- *Describe how genetic engineering techniques would be used to produce a specific protein product.*

17.6 Genetic Engineering Is Used in Industry and Agriculture

Possibly the greatest impact of biotechnology has been in agriculture and other industries. Many applications of biotechnology in industrial and environmental sciences focus on recycling waste

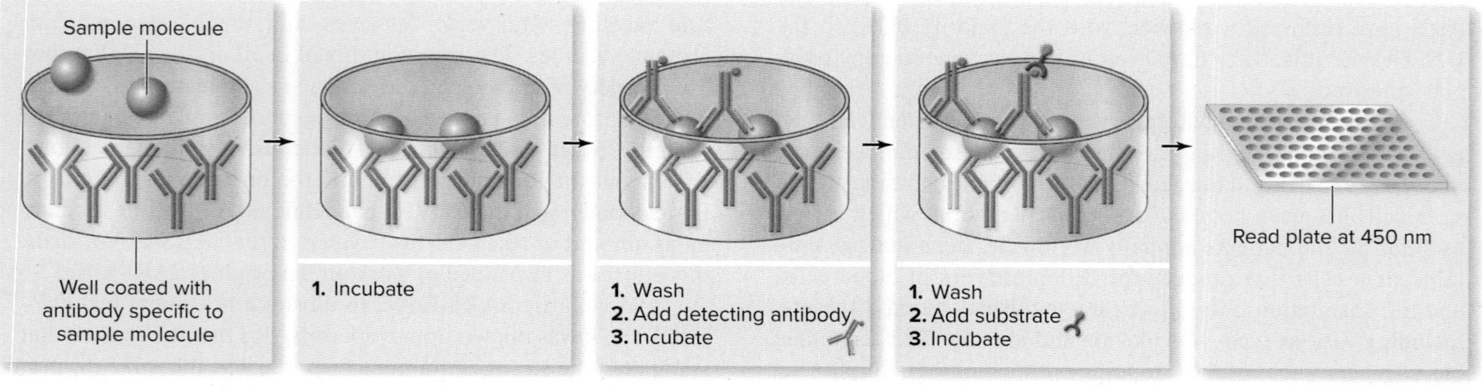

Figure 17.14 Detecting virus using an immunoassay. Steps in an immunoassay to detect viral antigens in a sample. Sample wells are coated with antibodies specific to proteins found in a particular pathogen. If a sample contains the virus, it is bound by the antibodies. The virus can be detected by another fluorescently tagged antibody.

materials, or using waste from one process as input to another process. For example, there is currently much interest in using biological processes to generate renewable fuel sources.

In agriculture, transgenic crops are being created that resist disease, are tolerant of herbicides and drought, and have improved nutritional quality. Plants are also being used to produce pharmaceuticals, and domesticated animals are being genetically modified to produce biologically active compounds.

Algae Can Be Used to Manufacture Biofuels

LEARNING OBJECTIVE 17.6.1 Describe the benefits of biofuel production from algae.

Biofuels are derived from biomass made from recently fixed carbon sources, and not "ancient" biomass such as fossil fuels. Common sources of biomass for biofuel generation are crop plants and algae. There are two common forms of liquid biofuel: ethanol from the fermentation of plant carbohydrates and biodiesel produced from the chemical transformation of plant or algal lipids. A number of species of algae can be used to produce biodiesel; however, all are microalgae with cell diameters in the range of 3 to 30 μm.

Biofuels produced from microalgae have several advantages over petroleum-based fossil fuels. They are renewable as long as carbon dioxide can be fixed into biomass by photosynthesis. They are environmentally friendly, as the carbon in the fuel is derived from atmospheric carbon dioxide, a greenhouse gas that contributes to climate change. The production of biodiesel from microalgae can also be combined with wastewater treatment or carbon dioxide capture systems of fossil-fuel-burning industrial plants.

The simplest system to culture microalgae is the use of large, open ponds. Although relatively cheap to construct and maintain, these are subject to contamination by environmental microorganisms or algal predators, and they require large areas of land. More expensive, but more easily controlled, are self-contained culture systems that circulate algal cultures through water with dissolved nutrients, and a light harvesting system (figure 17.15). These photobioreactors can even be fed with wastewater, or use carbon dioxide from industrial processes, for carbon for fixation by photosynthesis. Photosynthesis produces reduced carbon that can be metabolically converted to acetyl-coA, the building block of fatty acids.

Some species of microalgae are heterotrophic and use sugars from the culture medium to form biomass for growth or the production of lipids. Other species simultaneously use both heterotrophic and photoautotrophic nutritional strategies, producing very high concentrations of lipid in their cells. Under conditions of stress, the algal cells can be composed of 40 to 50% triacylglycerides.

Harvesting the lipids from the algae is usually the most expensive part of biodiesel production. Once the lipids have been isolated from the algae, chemical treatments separate the glycerol from the fatty acid chains in the triglyceride. Further refinement yields a biodiesel product.

Microorganisms Can Degrade Hydrocarbons in the Environment

LEARNING OBJECTIVE 17.6.2 Describe the factors that limit microbial degradation of hydrocarbons in natural environments.

Our dependence on hydrocarbon-derived fuels—whether petroleum- or biofuel-based—has an inevitable and serious consequence: those fuels can enter ecosystems and cause environmental damage. Certain hydrocarbons are persistent in terrestrial, aquatic, and marine ecosystems for decades and can accumulate on and in organisms, leading to death.

Hydrocarbons that persist in the environment can dissipate slowly by the evaporation of volatile compounds, by dissolution and dispersion in watery environments, and by oxidation in the presence of light. Importantly, certain microorganisms are capable of metabolizing hydrocarbon pollutants. The degradation or metabolism of hydrocarbon pollutants by microorganisms is called *bioremediation*. The rate at which these microorganisms metabolize hydrocarbons is affected by a number of variables, including temperature, the physical and chemical properties of the pollutant, oxygen availability, pH, salinity, other microorganisms present, and, perhaps most critically, nutrient availability.

Lab studies have shown that the addition of certain nutrients and other chemical additives to a polluted environment can increase the rate of bioremediation of hydrocarbon pollutants. Unfortunately, the increases in decontamination observed in the environment are not as great as those in the lab. An interesting possibility to increase the rate of decontamination by

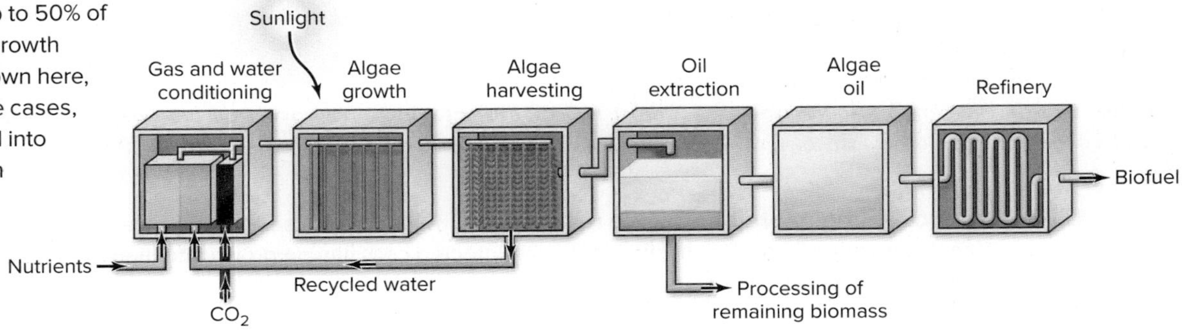

Figure 17.15 Using microalgae to make biofuel. Microalgae grown under certain conditions accumulate up to 50% of their biomass as lipids. Growth can be in ponds or, as shown here, photobioreactors. In some cases, carbon dioxide can be fed into the photobioreactors from industrial plants. Lipids harvested from the algae are processed and refined into biofuel.

microorganisms is genetic engineering to optimize the pathways involved in hydrocarbon uptake and catabolism.

Bioremediation of hydrocarbons is an important part of a hydrocarbon contamination cleanup plan. When metabolized by microorganisms, hydrocarbons end up being metabolized into biomass or catabolized into carbon dioxide, so no additional toxic compounds are released into the environment.

Herbicide-Resistant Crops Allow for No-Till Planting

LEARNING OBJECTIVE 17.6.3 Describe the benefits of creating transgenic crops.

Tilling is the process in which soils are turned over to prepare them for planting. This process removes weeds, which compete with crop plants for nutrients, and improves irrigation. There is an economic cost to tilling, however, because it requires expensive equipment and takes time to perform. Weeds can also be removed from crop fields by the applications of herbicides, but most common herbicides are not particularly discriminating, killing a variety of plant species. Soils that are not tilled have greater levels of organic and inorganic nutrients that promote growth, often retain more water, and are less prone to soil erosion. If crops are planted without the need for tilling, farms can be more efficient—which can translate into economic gains.

Agriculturally important broadleaf plants such as soybeans have been genetically engineered to be resistant to **glyphosate,** a powerful, biodegradable herbicide that kills most actively growing plants (figure 17.16). Glyphosate works by inhibiting an enzyme called 5-enolpyruvylshikimate-3-phosphate (EPSP) synthase, which plants require to make aromatic amino acids. Animals obtain aromatic amino acids from their diet, so inhibiting this pathway does not affect animals living in and around crops, or in waterways receiving crop runoff. To make glyphosate-resistant plants, scientists used a Ti plasmid to insert extra copies of the EPSP synthase gene into plants. These engineered plants produce 20 times the normal level of EPSP synthase, enabling them to synthesize aromatic amino acids and grow despite glyphosate's inhibition of the enzyme (figure 17.16).

These advances are of great interest to farmers, because a crop resistant to glyphosate does not have to be weeded through tilling. Because glyphosate is a broad-spectrum herbicide, farmers don't need to use a variety of herbicides, most of which kill only a few kinds of weeds. Furthermore, unlike many other herbicides, glyphosate breaks down readily in the environment.

Six important crop plants have been modified to be glyphosate-resistant: maize (corn), cotton, soybeans, canola, sugarbeets, and alfalfa. The use of glyphosate-resistant soy has been especially popular, accounting for 60% of the global area of transgenic crops worldwide. In the United States, 90% of the soy currently grown is transgenic. Other commercially important crops,

SCIENTIFIC THINKING

Hypothesis: *Petunias can be made tolerant to the herbicide glyphosate when their cells overexpress EPSP synthase.*

Prediction: *Transgenic petunia plants with a chimeric EPSP synthase gene with strong promoter will be glyphosate tolerant.*

Test:

1. *Use restriction enzymes and ligase to "paste" the cauliflower mosaic virus promoter (35S) to the EPSP synthase gene and insert the construct in Ti plasmids.*
2. *Transform* Agrobacterium *with the recombinant plasmid.*
3. *Infect petunia cells and regenerate plants. Regenerate uninfected plants as controls.*
4. *Challenge plants with glyphosate.*

Result: *Glyphosate kills control plants, but not transgenic plants.*

Conclusion: *Additional EPSP synthase provides glyphosate tolerance.*

Further Experiments: *The transgenic plants are tolerant, but not resistant (note bleaching at shoot tip). How could you determine if additional copies of the gene would increase tolerance? Can you think of any downsides to expressing too much EPSP synthase in petunia?*

Figure 17.16 Genetically engineered herbicide resistance. Ti plasmid is used to introduce the EPSP synthase gene into *Agrobacterium,* which subsequently infects cultured plant cells. Infected cells can be grown into mature plants. Tolerant plants are resistant to the effects of glyphosate.

Rob Horsch/Monsanto Company

such as barley, rice, and wheat, have been made resistant to a range of other herbicides.

Bt Crops Are Resistant to Some Insect Pests

LEARNING OBJECTIVE 17.6.4 Explain how Bt crops are resistant to certain pests.

Many commercially important plants are sensitive to insect pests, and the usual defense against this has been insecticides. Over 40% of the chemical insecticides used today are targeted against boll weevils, bollworms, and other insects that eat cotton plants. Scientists have produced plants that are resistant to insect pests, removing the need to use many externally applied insecticides.

The approach is to insert into crop plants genes encoding proteins that are harmful to the pests but harmless to other organisms. The most commonly used protein is a toxin produced by the soil bacterium *Bacillus thuringiensis* (Bt toxin). When insects ingest Bt toxin, endogenous enzymes convert it into an insect-specific toxin, causing paralysis and death. Because these enzymes are not found in other animals, the protein is harmless to them.

Many of the same crops that have been modified for herbicide resistance have also been modified for insect resistance using the Bt toxin. Bt maize is the second most common genetically modified (GM) crop, representing 14% of global area of GM crops in nine countries. The global distribution of these crops is similar to that of the herbicide-resistant relatives.

Given the popularity of both of these types of crop modifications, it is not surprising that they have also been combined, to produce so-called 'stacked' GM crops in species such as maize and cotton. Stacked crops now represent 9% of the global area of GM crops.

Golden Rice Shows the Potential of Transgenic Crops

LEARNING OBJECTIVE 17.6.5 Describe the social and economic benefits of transgenic crops such as Golden Rice.

Golden Rice is a transgenic cultivar of rice, genetically modified to express β-carotene, the precursor of vitamin A. Ingested β-carotene can be converted to vitamin A in the body to alleviate the symptoms of vitamin A deficiency. The World Health Organization (WHO) estimates that vitamin A deficiency affects between 140 and 250 million preschool children worldwide, 250,000 to 500,000 of whom become blind. The deficiency is especially severe in developing countries where the major food is rice.

Golden Rice is named for its distinctive color, imparted by the presence of β-carotene in the endosperm (the outer layer of rice that has been milled). Rice does not normally make β-carotene in endosperm tissue, but it does produce a precursor, geranylgeranyl diphosphate (GGPP), which can be converted by three enzymes—phytoene synthase, phytoene desaturase, and lycopene β-cyclase—to β-carotene. Genes for these three enzymes were altered using recombinant DNA technologies to be expressed in endosperm and introduced into rice to complete the biosynthetic pathway producing β-carotene in endosperm (figure 17.17).

This case of genetic engineering is interesting for two reasons. First, it introduces a new biochemical pathway into tissue of the transgenic plants. Second, expression of β-carotene in the endosperm could not have been achieved by conventional breeding, as no known rice cultivar produces these enzymes in endosperm. A second-generation rice cultivar that makes much higher levels of β-carotene has also been produced by using the gene for phytoene synthase from maize in place of the original daffodil gene.

Golden Rice was originally constructed in a public facility in Switzerland and made available for free with no commercial incentives. Since its creation, Golden Rice has been improved by public groups and industry scientists; these improved versions are also being made available without commercial strings attached.

Transgenic Crops Raise a Number of Social Issues

LEARNING OBJECTIVE 17.6.6 Evaluate issues on each side of the transgenic crop debate.

The adoption of transgenic crops has been resisted in some places for a variety of reasons. Some people have wondered about the safety of these crops for human consumption, the likelihood of

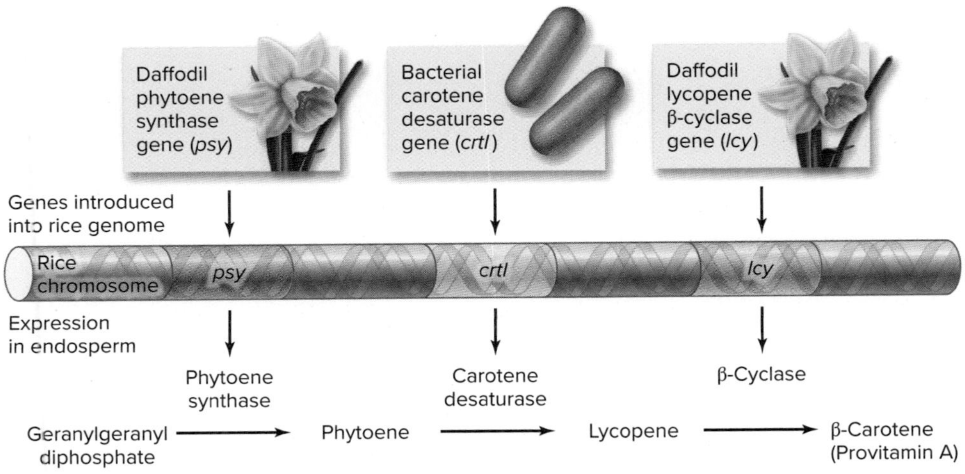

Figure 17.17 Construction of Golden Rice. Three genes were added to the rice genome to allow the synthesis of β-carotene in endosperm. The sources of the genes and the pathway for synthesis of β-carotene are shown. The result is Golden Rice, which contains enriched levels of β-carotene in endosperm.

introduced genes moving into wild relatives, and the possible loss of biodiversity associated with these crops.

On the side promoting the use of transgenic crops are the multinational companies using these technologies to produce seeds for the crops. Also on this side are groups noting the benefits of transgenic crops like Golden Rice in feeding the developing world and preventing nutritional deficiencies. On the opposing side are a variety of political and social organizations that raise concerns for the ecological and health implications of genetically modified foods. Scientists can be found on both sides of the debate.

The controversy originally centered on the safety of consuming foods containing introduced genes. In the United States, this issue has been "settled" for the crops already mentioned, and a large amount of transgenic soy and maize is consumed in this country. There remains a perceived risk of allergic reactions and longer-term effects, although there is no documentation of adverse effects on human health.

Also of concern is the potential for the spread of the transgenes from a transgenic crop to a nontransgenic crop. The amount of cross-pollination between plants of the same species varies. Soybeans tend to self-pollinate with limited outcrossing. Corn, however, freely outcrosses, and so genes from transgenic plants move more frequently to nontransgenic corn plants. This is a concern for organic farmers with fields close to transgenic corn fields, because current regulations for organic certification exclude transgenic crops. The amount of outcrossing depends on the proximity of crops, wind speed and direction, and temperature and humidity. For plants that depend on pollinators for reproduction, the range of species the pollinator visits and the distance it travels are also factors.

Gene flow can also occur between related species through hybridization. In the United States, at least 15 weedy species have hybridized with crop plants. For two of the major U.S. crops, corn and soybean, no closely related weedy species reproduce with them, so this is an unlikely mechanism for acquiring herbicide resistance. Sunflowers, however, were first domesticated in the United States, and the risk of a transgene moving into a weedy population is a concern. One study revealed that 10 to 33% of wild sunflowers had hybridized with domesticated varieties.

In the case of herbicide-resistant transgenic crop plants, the repeated application of a single herbicide over a number of years creates selective pressure on weedy species. Over time, weeds with resistance to the herbicide will increase in frequency, and the herbicide becomes ineffective. Almost 200 weedy species worldwide have acquired resistance to one or more herbicides. Management practices that reduce herbicide use and vary the herbicides used can slow the appearance of herbicide resistance.

REVIEW OF CONCEPT 17.6

Industrial biotechnology uses advances in genomics, microbiology, genetics, and computing to solve problems of nonrenewable resource use. Microalgae can be used to produce biofuel. Nutrient supplementation of contaminated environments can stimulate hydrocarbon breakdown. Herbicide resistance, pathogen protection, nutritional enhancement, and drug production have been targets of agricultural genetic engineering. Controversy regarding the use of transgenic plants has centered on the potential for unforeseen effects on human health and on the environment.

- *Grasses resistant to glyphosate have been constructed but never released as a product. What might keep a grass seed company from releasing such a grass seed?*

Does Bt Cotton Promote Biological Forms of Pest Control?

Until recently, more pesticide was applied worldwide to cotton crops than to all other crops combined. Most of these chemicals were targeted at the cotton bollworm—a hardy pest that, uncontrolled, devastates cotton fields. Over the past 16 years, however, this picture has begun to change, with widespread planting of genetically engineered cotton that expresses the insect-specific endotoxin Bt. As shown in the photo, Bt cotton (on the right) is not attacked by the cotton bollworm, but nonengineered cotton (on the left) is.

Because the bollworm does not attack Bt cotton, there is no need to apply pesticide. Since 1996, there have been vast plantings of Bt cotton. In 2011, more than 6.6×10^7 hectares were planted worldwide, with a drastic decrease in insecticide use. This decrease is expected to increase the population numbers of other insects, which in the past have been killed by these chemicals.

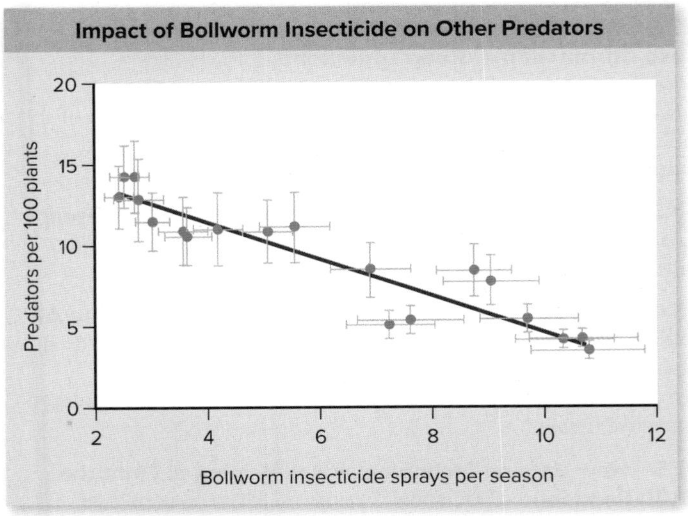

Rob Horsch/Monsanto Corporation

Generalist insect and arachnid predators, escaping cotton-targeted pesticides, should become available to help control insect pests such as aphids in common neighboring crops, including corn, peanuts, and soybeans. This sort of damping down of pest levels by other insects is termed biocontrol services.

But does this really happen? Does a decrease in cotton bollworm insecticide spraying actually lead to an increase in the number of generalist insect predators? To see, investigators in China sampled predators from 2001 to 2011 in Bt and non-Bt cotton fields at 36 sites in six provinces of northern China. When insecticides were not sprayed on the fields, there was no difference in numbers of predator insects between Bt and non-Bt cotton. So what happens when plants are sprayed with cotton bollworm insecticide? The results are presented in the graph. Each blue dot represents the average result over 10 years for one field, with error bars indicated. Bt cotton fields were typically sprayed two to three times a year; non-Bt cotton fields, as often as 11 times a year.

Analysis

1. **Applying Concepts**
   a. ***Variable.*** In the graph, what is the dependent variable?
   b. ***Error bars.*** When the vertical or horizontal error bars of two points overlap, what does this indicate? When they do not overlap? Explain.
2. **Interpreting Data**
   a. What is the relationship between predators per 100 plants and frequency of sprays per season?
   b. When the error bars overlap there can be no statistically significant difference between the mean values (the blue dots). Scientists set statistical significance to values of, for example, $p < 0.05$. Why is this approach problematic when interpreting data?
3. **Making Inferences** The red line is a best fit line through the data points. Approximately how big a difference on predators per 100 plants does a threefold increase in spraying frequency have (for example, compare three sprayings with nine sprayings.) How confident are you in this difference in spraying effectiveness?
4. **Drawing Conclusions** Is it reasonable to conclude from these results that a decrease in spraying of cotton bollworm insecticide leads to an increase in population numbers of generalized predator insects?
5. **Further Analysis** Of course, the next step was for the researchers to see if the increase in numbers of predator insects was indeed promoting biocontrol services by reducing aphids and other insect pests on neighboring noncotton crops. How would you have recommended that they go about doing this?

Retracing the Learning Path

CONCEPT 17.1 Enzymes Allow the Creation of Recombinant Molecules in Vitro

17.1.1 Restriction Endonucleases Cleave DNA at Specific Sites Different DNA fragments produced by the same type II enzyme can be joined.

17.1.2 Gel Electrophoresis Separates DNA Fragments DNA fragments can be separated based on size through a gel matrix driven by an electrical field.

17.1.3 DNA Fragments Are Combined to Make Recombinant Molecules Recombinant DNA is formed by joining DNA from different sources using DNA ligase.

17.1.4 Reverse Transcriptase Makes DNA from RNA Reverse transcriptase uses an mRNA template to make a DNA copy we call cDNA.

17.1.5 DNA Libraries Are Collections of Recombinant DNA Molecules DNA libraries hold pieces of DNA, which can be stored and replicated in a host organism.

CONCEPT 17.2 The Polymerase Chain Reaction Is Used to Amplify Specific DNA Sequences

17.2.1 PCR Mimics DNA Replication The polymerase chain reaction (PCR) uses cycles of heating, cooling, and DNA synthesis to amplify short regions of DNA between two primers.

17.2.2 Reverse Transcription PCR Makes Amplified DNA from mRNA PCR can be used on cDNA to amplify only the coding sequence of a gene. This can be used to quantify the amount of specific mRNAs in a sample.

17.2.3 PCR Is the Basis of Many Sequencing Technologies The cost of sequencing genomes has dropped with new DNA sequencing technologies, many relying on PCR.

CONCEPT 17.3 Molecular Tools Allow Us to Analyze and Modify Genetic Variation

17.3.1 Forensics Uses DNA Fingerprinting to Identify Individuals DNA fingerprinting uses short tandem repeats (STRs) that vary in number between individuals in a population. The CODIS database stores data from 13 different STR loci.

17.3.2 PCR Can Be Used to Create Point Mutations in Specific DNA Sequences PCR can be used to create site-specific mutations by using mismatched primers. Error-prone DNA polymerases can create random mutations.

17.3.3 RNA Interference Can Reduce the Level of a Gene Product Short sequences of double-stranded RNA can be used to reduce expression of specific genes.

17.3.4 New Technologies Allow Direct Editing of Genomes Genes can be directly edited using the CRISPR/Cas9 system. This allows the editing of specific genes in a living cell.

CONCEPT 17.4 Transgenic Organisms Are Used to Analyze Gene Function

17.4.1 The Universal Nature of the Genetic Code Allows Transgenesis The universal nature of the genetic code allows genes from one organism to be expressed in other organisms.

17.4.2 Removal or Addition of Genes Can Reveal Function Function can be assessed by removing genes in knockout mice, or replacing them with a variant in knockin mice.

17.4.3 Plants Require Different Techniques to Be Genetically Modified Transgenic plants can be created by transformation with *Agrobacterium* and a recombinant Ti plasmid or by particle bombardment. Targeted gene knockouts are not currently possible.

CONCEPT 17.5 Genetic Tools Are Changing Modern Medicine

17.5.1 Important Proteins Can Be Produced Using Recombinant DNA Bacteria and yeast that produce human insulin have been created using recombinant DNA. This technology has also increased the potency of this insulin.

17.5.2 Fluorescent In Situ Hybridization Can Detect Gross Chromosomal Rearrangements FISH uses fluorescent DNA probes to identify genes in chromosomes, allowing detection of chromosomal alterations. Identification of amplifications of the *HER2* gene can help design personalized treatments.

17.5.3 Gene Chips Can Identify Genetic Markers for Disease Gene chips allow the analysis of gene expression in diseased and healthy states. These data may aid in clinical interventions.

17.5.4 Gene and Stem Cell Therapy and Gene Editing Are All Moving into the Clinic Bone marrow transplants have been used to treat some blood cancers, anemias, and some cases of multiple sclerosis. Mesenchymal stem cells can be used to repair some damaged connective tissues.

17.5.5 New Vaccine Technologies Are Helping to Fight the COVID Pandemic Traditional vaccines use killed or attenuated viruses. New technologies use DNA, mRNA, viral vectors, or protein fragments to stimulate the immune system.

17.5.6 Infectious Disease Agents Are Detected by Diagnostic Testing PCR is used to detect DNA or RNA from pathogens. Antigen tests detect pathogen proteins using antibodies, and antibody testing can identify previously infected individuals.

CONCEPT 17.6 Genetic Engineering Is Used in Industry and Agriculture

17.6.1 Algae Can Be Used to Manufacture Biofuels Lipids harvested from microalgae can be converted into biodiesel. Some microalgae can produce up to 50% of their biomass in lipids.

17.6.2 Microorganisms Can Degrade Hydrocarbons in the Environment Bacterial species can be used to degrade toxic hydrocarbons in aquatic, terrestrial, and marine environments.

17.6.3 Herbicide-Resistant Crops Allow for No-Till Planting Herbicide-resistant crops are widely used in the United States, reducing both the need for tilling and fossil fuel consumption.

17.6.4 Bt Crops Are Resistant to Some Insect Pests The development of insect-resistant Bt plants reduces the use of insecticides.

17.6.5 Golden Rice Shows the Potential of Transgenic Crops Golden Rice produces a precursor for vitamin A. Dietary use of Golden Rice can reduce blindness in countries where rice is a staple crop.

17.6.6 Transgenic Crops Raise a Number of Social Issues Transgenic plants raise concerns about allergic reactions, but none have been observed. The spread of transgenes into noncultivated plants is being followed closely. Up to 200 weedy species worldwide have acquired resistance to herbicides.

Concept Overview

This Concept Overview diagrams the key concepts that were discussed in this chapter.

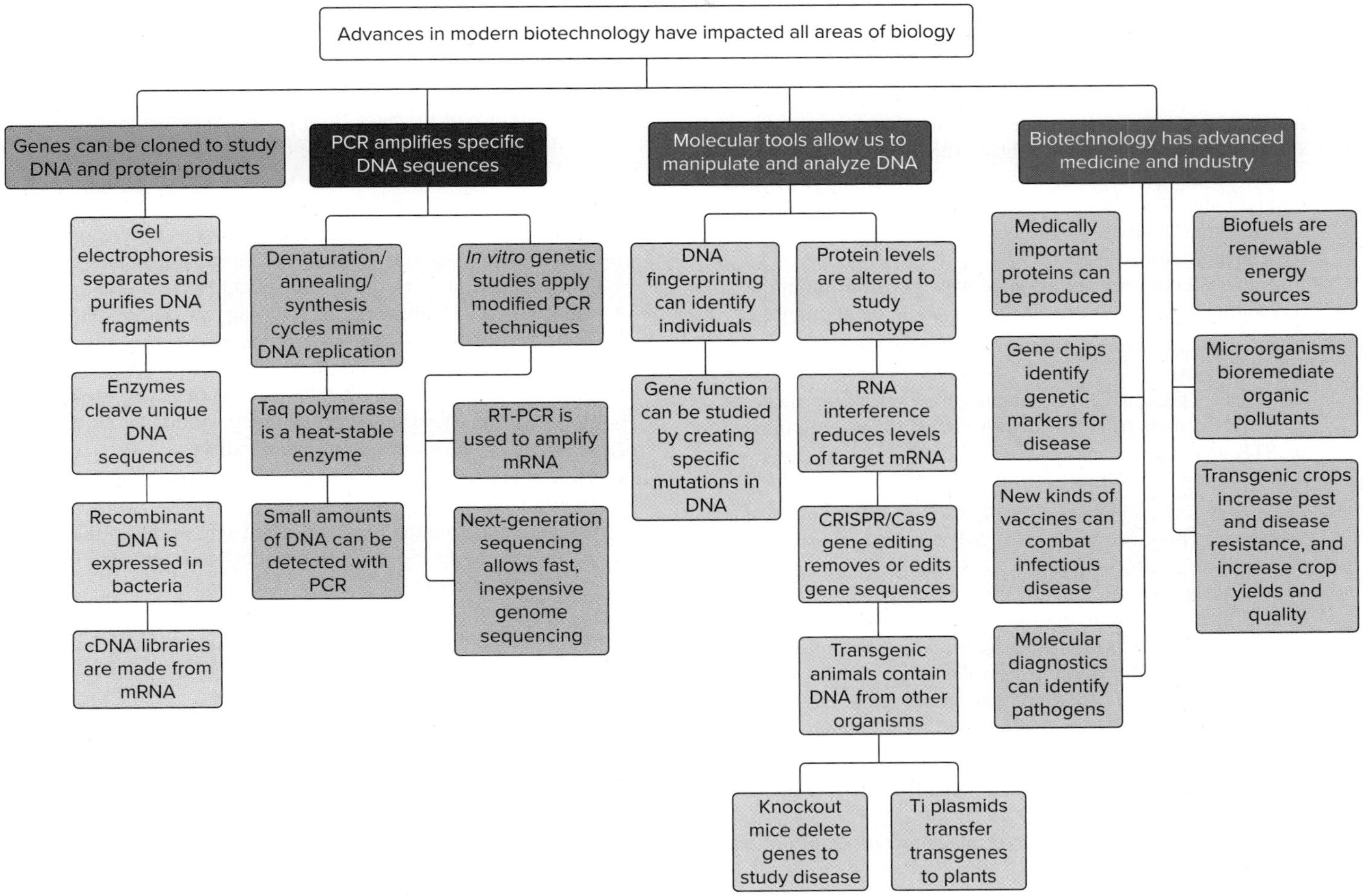

Assessing the Learning Path

Understand

1. The palindromic recognition sequence for the restriction enzyme *Bam*H1 is 5′ GGATCC 3′. Remember, the other strand of DNA will have the complementary set of bases and will read the same as this sequence, but just "backwards." The enzyme cuts between the first and second Gs in the sequence. What is the "sticky-end" sequence when the enzyme cuts double-stranded DNA?
   a. 5′ GATCC 3′
   b. 5′ CCATG 3′
   c. 5′ GATC 3′
   d. 5′ CATG 3′
2. What is the basis of separation of different DNA fragments by gel electrophoresis?
   a. The positive charge on DNA
   b. The size of the DNA fragments
   c. The sequence of the fragments
   d. The presence of a dye in the gel
3. DNA replication and PCR share some similarities. What function is common to the primer that initiates DNA replication in cells and to the DNA primers used in PCR?
   a. Both are DNA.
   b. Both are RNA.
   c. Both provide a 5′ phosphate so that DNA synthesis can be started.
   d. Both provide a 3′ hydroxyl group so that DNA synthesis can be started.

4. Which of the following statements is accurate for PCR, but not for DNA replication in your cells?
   a. Primers are required.
   b. DNA polymerase is stable at high temperatures.
   c. Ligase is essential.
   d. Nucleotides are necessary.
5. The universal nature of the genetic code means that
   a. all organisms have only one codon to specify each amino acid.
   b. all organisms use the same codons to specify the same amino acids.
   c. all codons specify the same amino acid.
   d. there is no redundancy in the genetic code.
6. You study a gene known to be important in DNA repair. You have several variations of a gene in which different nucleotide substitutions produce proteins that seem to have subtle differences in the DNA bases they repair. What kind of transgenic mouse would allow you to study the effects of these different alleles in vivo?
   a. Knockout mouse
   b. Knockin mouse
   c. Either a knockout or a knockin would be appropriate.
   d. Neither a knockout nor a knockin would be appropriate.
7. If you wanted to express a human gene in bacteria, you would most likely use
   a. a piece of genomic DNA containing the human gene of interest.
   b. a cDNA made from the mRNA coding the gene of interest.
   c. STR analysis.
   d. CRISPR/Cas9 to transform the bacteria.
8. FISH analysis of a breast tumor biopsy for *HER2* gene reveals two spots of fluorescence in cells. What conclusion do these data support about the tumor?
   a. It is likely to contain high levels of the HER2 mRNA.
   b. It is likely to contain high levels of the HER2 protein.
   c. The *HER2* gene is amplified, and treatments targeting the HER2 protein may not be suitable.
   d. The *HER2* gene is not amplified, so particular treatments targeting the HER2 protein may not be suitable.
9. What is a Ti plasmid?
   a. A vector that can transfer recombinant genes into plant genomes
   b. A vector that can be used to produce recombinant proteins in yeast
   c. A vector that is specific to cereal plants like rice and corn
   d. A vector that is specific to embryonic stem cells

Apply

1. If a PCR is started using 10 pieces of template DNA, how many pieces of DNA would there be after 10 cycles?
   a. About 100
   b. About 1000
   c. About 10,000
   d. About 10^{10}
2. You set up four qRT-PCR reactions (A to D), all containing the same amount of total RNA. When the reactions are complete, you see that you have 10,000 units of fluorescence after 20 cycles for sample A, after 25 cycles for sample B, after 18 cycles for sample C, and after 22 cycles for sample D. Which of the samples had the most copies of target mRNA at the start of the qRT-PCR?
   a. Sample A
   b. Sample B
   c. Sample C
   d. Sample D
3. A piece of DNA is digested with the restriction enzyme *Eco*RI. A control shows that the enzyme is active and works as expected. The piece of DNA cut is known to have three *Eco*RI cut sites in it. When the digest is analyzed by gel electrophoresis, there are three clearly visible pieces of DNA on the gel. Which of the following best explains the data?
   a. The DNA is a plasmid.
   b. The DNA is a linear piece of DNA.
   c. The enzyme cuts at additional sites besides its known cut site.
   d. The DNA was not isolated from the same organism that the enzyme was obtained from.
4. If you created a line of knockout mice in which the mice were heterozygous for a knocked-out allele, how could you get a mouse that is homozygous for the knocked-out allele?
   a. Make a new knockout line and cross your first line with the new line.
   b. Cross your line with the mouse line you started with to make your knockout line.
   c. Cross two of the mice from the line together and determine if any of the offspring are homozygous.
   d. Use RNAi to eliminate RNA produced from the remaining intact allele in the heterozygote.
5. Assuming you crossed two mice heterozygous with respect to a knocked-out allele (for example, both mice had an intact allele and a removed allele), under what circumstances would you not expect to get 25% of the offspring having two knocked-out alleles?
   a. If the gene is essential for early development
   b. If the gene is on chromosome 4
   c. If all the offspring from the heterozygous cross are female
   d. All of the above would probably not result in 25% of the offspring having two knocked-out alleles.
6. How might genetic engineering be used to improve the yields of biofuel from microalgae?
   a. By stimulating the production of triglycerides in the algae
   b. By increasing the efficiency of rubisco in the algae
   c. By reducing the surface area of the algae
   d. Both a and b
7. A researcher wants to purify a recombinant protein that is known to be glycosylated only in the Golgi apparatus. What would be a major problem of trying to make such a protein in bacteria such as *E. coli?*
   a. The protein will be too large for bacterial expression.
   b. The protein will be recognized as foreign by the bacteria and destroyed.
   c. Bacteria lack Golgi so the protein will not be glycosylated and may be inactive.
   d. Bacteria cannot phosphorylate proteins so the expressed protein would be inactive.

Synthesize

1. The N-*myc* gene is sometimes amplified in a type of brain cancer called neuroblastoma. How might you use PCR and/or RT-qPCR to determine if this gene is amplified in a particular brain cancer sample and if there is also an increase in mRNA levels due to any amplification?
2. If you needed to reduce the expression of a gene in cultured animal cells using RNA interference or the CRISPR/Cas9 system, which one would you select to use, and why?
3. You make a particular knockout mouse, but find that you cannot obtain a mouse homozygous for the deleted gene. What might you conclude and why? How can you analyze this gene further?
4. Many human proteins, such as hemoglobin, are only functional as an assembly of multiple subunits, which occurs within the endoplasmic reticulum and Golgi apparatus of a eukaryotic cell. Discuss the limitations, if any, of the large-scale production of genetically engineered hemoglobin.

18 Genomics

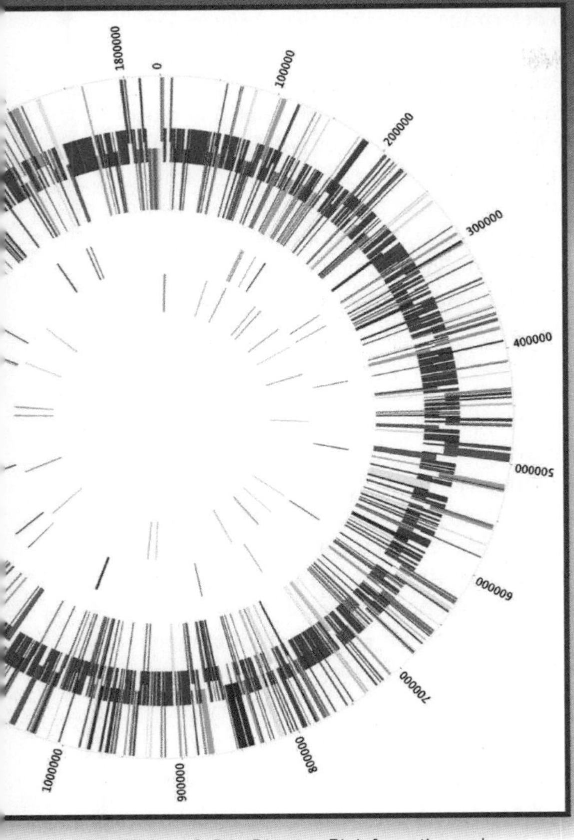

1800000
0
100000
200000
300000
400000
500000
600000
700000
800000
900000
1000000

William C. Ray, Director, Bioinformatics and Computational Biology Division, Biophysics Program, The Ohio State University

Learning Path

18.1 Mapping Identifies and Locates Functional Elements in Genomes

18.2 The Modernization of DNA Sequencing Has Accelerated Discovery

18.3 Genome Projects Reveal Insights into Medicine and Agriculture

18.4 Genome Annotation Assigns Functional Information to Genomes

18.5 Genome Comparisons Provide Information About Genomic Structure and Function

18.6 Comparative Genomics Informs Evolutionary Biology

Concept Overview

This Concept Overview provides a diagram of the overarching concepts that are covered in this chapter. These concepts will be expanded upon in the Concept Overview at the end of the chapter.

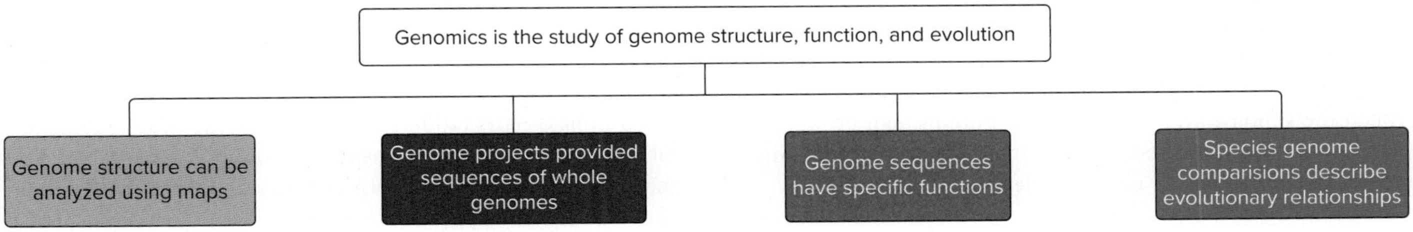

Genomics is the study of genome structure, function, and evolution
Genome structure can be analyzed using maps
Genome projects provided sequences of whole genomes
Genome sequences have specific functions
Species genome comparisions describe evolutionary relationships

Introduction

The pace of discovery in biology in the last 30 years has been dramatic. Starting with the isolation of the first genes in the mid-1970s, researchers obtained the first complete genome sequence, that of the bacterium *Haemophilus influenza* (shown in the diagram on the previous page, with genes of similar function similarly colored), by the mid-1990s. A complete draft sequence of the human genome had been completed by the turn of the 21st century. Put another way, science moved from cloning a single gene, to determining the sequence of a million base-pairs in 20 years, to determining the sequence of 3 billion base-pairs in five more years, to now being able to sequence 20 billion base-pairs in a few hours for as little as $1000. In chapter 17 you learned about biotechnology, and in this chapter you will learn how some of those technologies have been applied to the analysis of whole genomes. Genomics integrates classical and molecular genetics with biotechnology to study the structure, function, and relatedness of genomes.

18.1 Mapping Identifies and Locates Functional Elements in Genomes

We use maps to find locations, and depending on the nature of the location, we may use different types of maps. In genomics, a gene can be located to a chromosome, to a subregion of a chromosome, and finally to a precise location in the sequence of chromosomal DNA. A map produced at the DNA sequence level requires knowing the entire sequence of the genome, something that was once technologically impossible. Knowing the entire sequence of a genome is useless, however, without other information, such as what parts of the genome affect particular phenotypes. This sort of information resides in genetic maps. Genetic maps are analogous to geographical maps: finding a single gene within the sequence of the human genome is like trying to find a house on a map of the world.

Genomics Uses Many Approaches to Analyze Entire Genomes

LEARNING OBJECTIVE 18.1.1 Discriminate between genetic maps and physical maps.

We construct different kinds of maps of genomes from different kinds of data in order to perform different kinds of analyses. Fundamentally, there are two kinds of maps: genetic maps and physical maps. Both of these types of maps use **genetic markers,** which can be any detectable differences between individuals. **Genetic maps** are abstract maps that place the relative location of genetic markers on chromosomes based on recombination frequency between markers (refer to chapter 13). **Physical maps** precisely position genetic markers in the genome, with the ultimate physical map being the complete DNA sequence of a genome. Ultimately, genomics unifies many different types of maps, or views, of the genome to facilitate large-scale analysis.

Although not technically the study of the genome, the relationship between the genome and the expressed RNAs, and between the genome and the expressed proteins, can be studied. Genomic studies have shed light on evolutionary relationships, genetic susceptibility to disease, and development.

Genetic Maps Provide Relative Distances Between Genetic Markers

LEARNING OBJECTIVE 18.1.2 Describe the importance and limitations of genetic maps.

Genetic maps, also called linkage maps, use the process of recombination during meiosis, which alters the linkage relationships of alleles on chromosomes (refer to chapter 13). These genetic markers can be genes, as detected by phenotypic differences, or differences in DNA sequences that can be detected by PCR or restriction endonuclease digestion, as described in chapter 17. Classical genetic mapping uses controlled crosses to determine the number of recombinant progeny for specific pairs of markers. Obviously, this is not practical in humans, so we use preexisting data, in the form of family relationships and pedigrees, and we use statistical analysis to determine recombination frequencies.

Distances on a genetic map are measured in centimorgans (cM) where 1 cM corresponds to 1% recombination frequency between two loci. Two markers 1 cM apart are relatively close together, as there is only a 1% chance that they will be separated during meiosis. Early human genetic maps consisted of linked genes for disease susceptibility that had been mapped to specific chromosomes, but no overall map. This did not change until the advent of molecular markers that did not cause detectable phenotypes (refer to chapter 13). The human genetic map is now quite dense, with a marker roughly every 1 cM.

Genetic maps are very important, because they locate genetic traits on a chromosome, but they also have limitations. First, the distribution of crossover, or recombination, events is not random. In fact, extensive data now indicate the presence of hot spots for recombination in humans, mice, and most other organisms examined. Additionally, recombination frequencies can vary between individuals and are affected by chromatin structure. All of this means that the relationship between genetic distance (in centimorgans) and actual physical distance (in base-pairs [bp]) varies across the genome. But with a complete genome sequence and a dense genetic map, they can still be superimposed to provide complementary views of the genome.

Physical Maps Provide Absolute Distances Between Genetic Markers

LEARNING OBJECTIVE 18.1.3 Compare and contrast the different types of physical maps.

Unlike a genetic map, which provides the relative position of a marker in the genome, a physical map provides the absolute location of a marker. The first physical maps used the early tools of molecular biology: restriction enzymes. As technologies became more sophisticated, so did the nature of physical maps.

The ultimate form of a physical map is the placement of many genetic markers on the complete DNA sequence of a genome. Distances between markers on a physical map are measured in base-pairs (1000 bp equal 1 kilobase [kb]). Segments of DNA can be physically mapped without knowing the DNA sequence or whether the DNA encodes any genes. In fact, many genome-sequencing projects begin by constructing physical maps. There are three general types of physical maps: (1) restriction maps, constructed using restriction endonucleases; (2) chromosome maps (created using techniques discussed in chapter 17); and (3) sequence-tagged site (STS) maps.

Restriction maps

Restriction maps are generally not suitable for DNA molecules greater than about 50 kb, which means that they are only useful for some organelles and some viral genomes. The first physical maps were created by cutting DNA with single restriction enzymes and with combinations of restriction enzymes (figure 18.1). The analysis of the patterns of fragments generated is used to generate a map.

Restriction mapping can be applied to larger genomes if the DNA is first fragmented into smaller pieces. In terms of larger pieces of DNA, this process is repeated and then used to put the

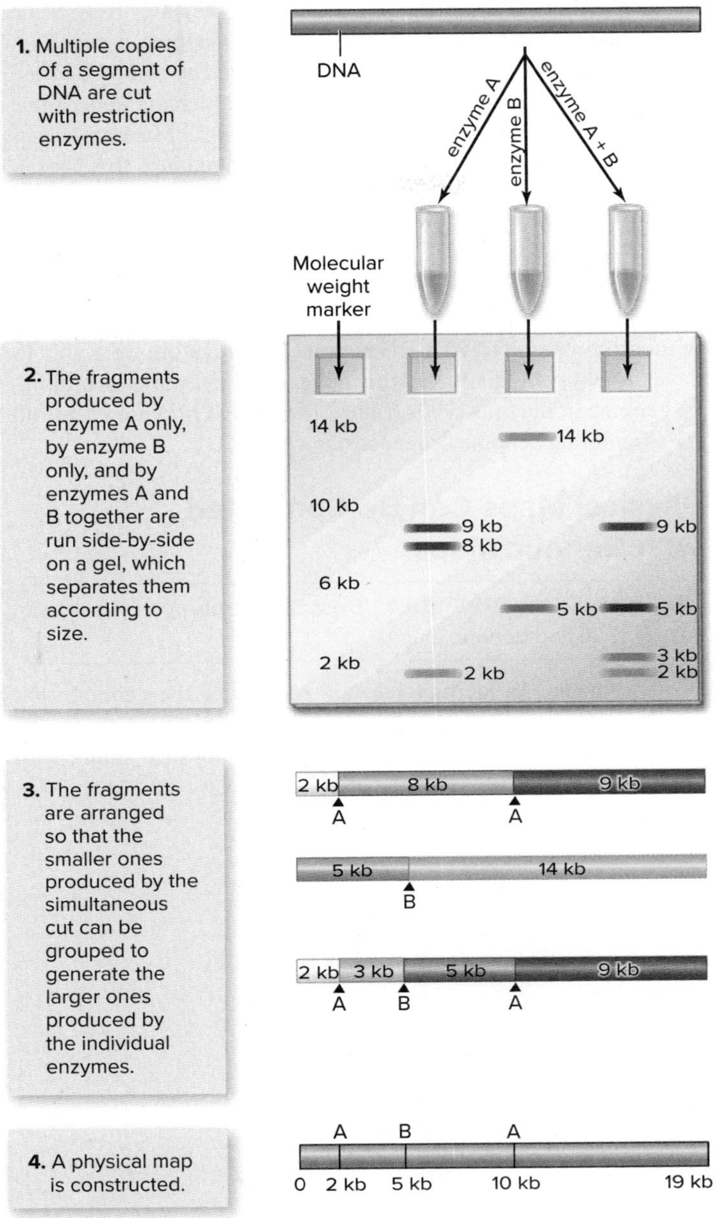
1. Multiple copies of a segment of DNA are cut with restriction enzymes.
DNA
enzyme A
enzyme B
enzyme A + B
Molecular weight marker
2. The fragments produced by enzyme A only, by enzyme B only, and by enzymes A and B together are run side-by-side on a gel, which separates them according to size.
14 kb
10 kb
6 kb
2 kb
9 kb
8 kb
2 kb
14 kb
5 kb
9 kb
5 kb
3 kb
2 kb
3. The fragments are arranged so that the smaller ones produced by the simultaneous cut can be grouped to generate the larger ones produced by the individual enzymes.
2 kb
8 kb
9 kb
A
A
5 kb
14 kb
B
2 kb
3 kb
5 kb
9 kb
A
B
A
4. A physical map is constructed.
A
B
A
0
2 kb
5 kb
10 kb
19 kb

Figure 18.1 Restriction enzymes can be used to create a physical map. DNA is digested with two different restriction enzymes singly and in combination. The DNA fragments are separated based on size using gel electrophoresis. The location of sites can be deduced by comparing the sizes of fragments from the individual reactions with the combined reaction.

pieces back together, based on size and overlap, into a contiguous segment of the genome, called a **contig.**

Chromosome maps

Cytologists studying chromosomes with light microscopes found that by using different stains, they could produce reproducible patterns of bands on the chromosomes. In this way, they could identify all of the chromosomes and divide them into subregions based on banding pattern. The use of different stains allows for the construction of a cytological map of the entire genome. These maps break each chromosome into a p arm and a q arm and then break these into regions and subregions based on banding patterns. All later maps have been correlated with this low-resolution map. These large-scale physical maps are like a map of an entire country in that they encompass the whole genome but do so at low resolution.

Cytological maps are used to characterize chromosomal abnormalities associated with human diseases, such as chronic myelogenous leukemia. In this disease, a reciprocal translocation occurs between chromosome 9 and chromosome 22 (figure 18.2*a*), resulting in an altered form of tyrosine kinase that is always turned on, causing white blood cell proliferation and, consequently, leukemia.

Fluorescent in situ hybridization, which we discussed in chapter 17, can also be used to generate chromosomal maps (figure 18.2*b*). Initially, the level of resolution of FISH was approximately 1 megabase (Mb); this means that two points on a chromosome closer than 1 Mb could not be distinguished as

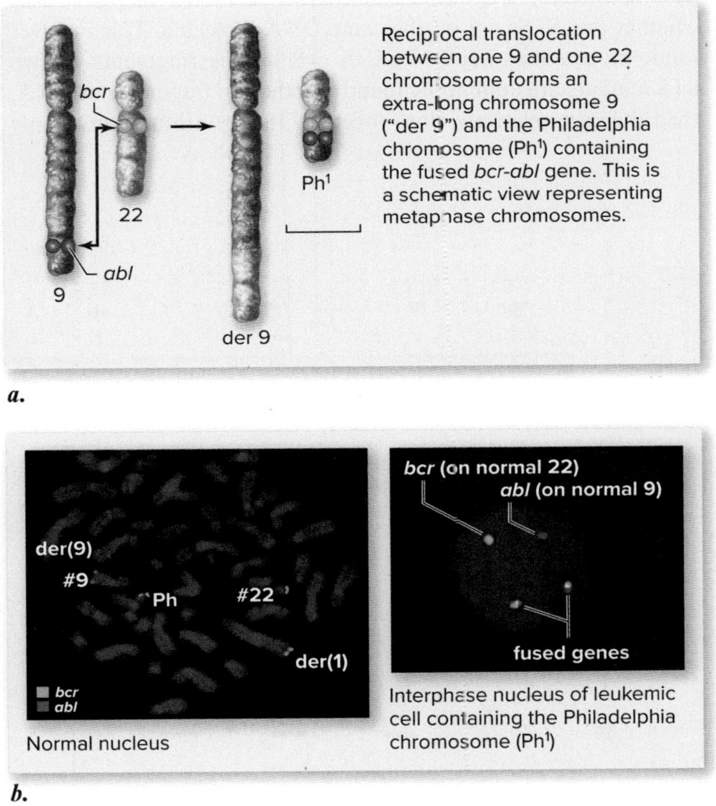
bcr
abl
9
22
der 9
Ph1
Reciprocal translocation between one 9 and one 22 chromosome forms an extra-long chromosome 9 ("der 9") and the Philadelphia chromosome (Ph1) containing the fused bcr-abl gene. This is a schematic view representing metaphase chromosomes.
a.
der(9)
#9
Ph
#22
der(1)
bcr
abl
Normal nucleus
bcr (on normal 22)
abl (on normal 9)
fused genes
Interphase nucleus of leukemic cell containing the Philadelphia chromosome (Ph1)
b.

Figure 18.2 Use of fluorescence in situ hybridization (FISH) to correlate cloned DNA with cytological maps. ***a.*** Karyotype of human chromosomes showing the translocation between chromosomes 9 and 22. ***b.*** FISH (refer to chapter 17) using a *bcr* (green) and *abl* (red) probe. The yellow color indicates the fused genes (red plus green fluorescence combined). The *abl* gene and the fused *bcr-abl* gene both encode a tyrosine kinase, but the fused gene is always expressed.

©W. Achkar, A. Wafa, H. Mkrtchyan, F. Moassass and T. Liehr, "Novel complex translocation involving 5 different chromosomes in a chronic myeloid leukemia with Philadelphia chromosome: a case report," Molecular Cytogenetics, 2: 21 (2009). doi:10.1186/1755-8166-2-21 separate points. This limitation made FISH only really useful for determining whether a particular DNA sequence was found on a particular chromosome. Technical advances in the 1990s allowed the use of interphase chromosomes as well as metaphase chromosomes, which improved the resolution of FISH to 25 kb. This advance meant the technique could be used for finer mapping.

Sequence-tagged site maps

Restriction mapping allows relatively high-throughput, technically easy, high-resolution mapping of relatively small DNA molecules. FISH, on the other hand, provides low-resolution mapping of large DNA molecules and is technically challenging. Sequence-tagged site (STS) mapping provides a powerful alternative that combines the best of restriction mapping with the best of mapping via FISH. STS mapping allows the rapid construction of high-resolution physical maps of large DNA molecules with few technical challenges.

An STS is a short, unique stretch of genomic DNA that can be amplified by PCR (refer to chapter 17). STS mapping asks whether two STSs are on the same DNA molecule. This involves randomly fragmenting DNA into overlapping fragments. If two STS markers are commonly found together on fragments of DNA, then they are relatively close together. If they are not commonly found together on fragments of DNA, then they are likely to be farther apart (figure 18.3). Much as genetic mapping is based on the frequency with which markers are separated by recombination, STS mapping is based on the frequency with which markers can be separated by breaking DNA into fragments. Markers closer together will be separated less frequently than markers that are farther apart. The farther apart two markers are, the more likely it is that a break can occur between them.

Fragments of DNA can be pieced together using the STSs by identifying overlapping regions in fragments. Because of the high density of STSs in the human genome and the relative ease of identifying an STS in a DNA clone, investigators were able to develop physical maps on the huge scale of the 3.2 billion-bp genome in the mid-1990s (figure 18.3). STSs provide a scaffold for assembling genome sequences.

Physical Maps Can Be Correlated with Genetic Maps

LEARNING OBJECTIVE 18.1.4 Describe the purpose of correlating physical and genetic maps.

Genetic maps can identify relevant regions of the genome that affect specific phenotypes, but they tell us nothing about the nature of these genes. The correlation of physical maps with

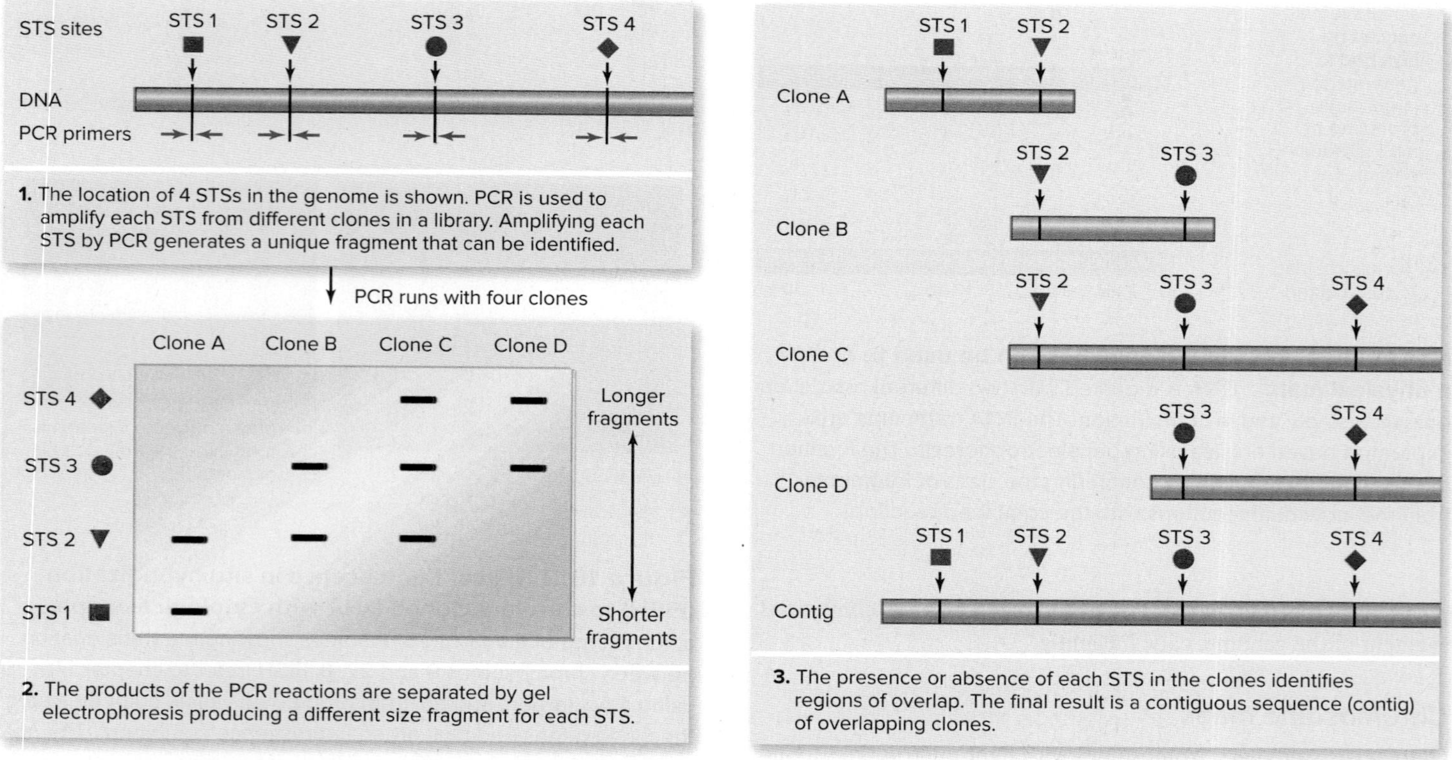

Figure 18.3 Creating a physical map with sequence-tagged sites. The presence of landmarks called sequence-tagged sites, or STSs, in the human genome made it possible to begin creating a physical map large enough in scale to provide a foundation for sequencing the entire genome. **(1)** Primers (green arrows) that recognize unique STSs are added to cloned DNA, followed by DNA amplification via PCR. **(2)** PCR products are separated based on size on a DNA gel, and the STSs contained in each clone are identified. **(3)** Cloned DNA segments are aligned based on STSs to create a contig.

genetic maps allows us to find the sequence of genes that have been mapped genetically.

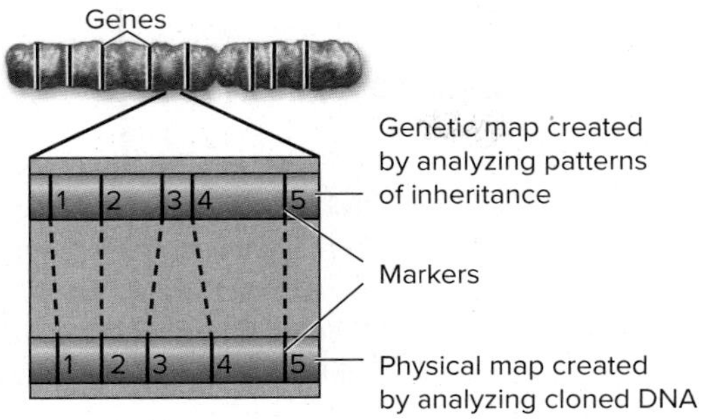
Genes
1
2
3
4
5
1
2
3
4
5
Genetic map created by analyzing patterns of inheritance
Markers
Physical map created by analyzing cloned DNA

The problem in finding genes is that the resolution of genetic maps is at present not nearly as good as the resolution of the genome sequence. In the human genome, markers that are 1 cM apart may be as much as a million base-pairs apart.

Because the markers used to construct genetic maps are now primarily molecular markers, they can be easily located within a genome sequence. Similarly, any gene that has been cloned can be placed within the genome sequence and can be mapped genetically. This provides an automatic correlation between the two maps. Genes that have been mapped genetically can, in theory, be readily located within the DNA sequence of a genome. In practice, this can be challenging, as the relationship between genetic distance and physical distance varies across the genome. So 1 cM of genetic distance can actually be composed of different numbers of base-pairs in different regions of a genome.

REVIEW OF CONCEPT 18.1

Maps of genomes can be either physical maps or genetic maps. Physical maps include cytological maps of chromosome banding or restriction maps. Genetic maps are correlated with physical maps by using DNA markers such as sequence-tagged sites (STSs) unique to each genome. Although genetic maps position markers, such as genes, relative to one another, physical maps position them at a specific genetic location on a sequence of DNA. Genetic and physical maps are used to aid in the sequencing phase of genome projects.

- *How can it be that two markers on a genetic map are 10 cM apart, and two markers in a different part of the same genetic map are 12 cM apart, but both sets of markers are separated by the same amount of DNA?*

18.2 The Modernization of DNA Sequencing Has Accelerated Discovery

The ultimate physical map is the base-pair sequence of an entire genome. Large-scale, high-throughput DNA sequencing has made whole-genome sequencing realistic for smaller labs and biotechnology companies. Conceptually, all approaches to sequencing DNA adapt DNA replication to an in vitro environment. Older approaches use modified nucleotides to stop or pause replication and limit the replication to just one strand of the DNA. The rapid advances in genomics have been enabled by newer and much faster automated sequencing technologies.

Dideoxy Terminator Sequencing Remains Important in Genome Sequencing

LEARNING OBJECTIVE 18.2.1 Describe the process of dideoxy terminator sequencing.

In the last 10 years, huge improvements have been made to the technologies used to sequence DNA. In some situations, however, the original DNA sequencing methods are useful. One original method for sequencing DNA developed by Fred Sanger in the 1970s mimics DNA replication. This technique depends on the use of **dideoxynucleotides** in DNA sequencing reactions. The dideoxynucleotides act as chain terminators and halt replication when they are incorporated. All DNA nucleotides lack a hydroxyl group at the 2′ carbon of the sugar, but dideoxynucleotides also have no 3′ OH group. Recall from chapter 14 that each additional base is added to the growing 3′ end of a strand of DNA by a reaction with the OH group of the previous nucleotide to the triphosphate of the next nucleotide (refer to figure 14.12). A dideoxynucleotide can be incorporated, but then cannot be added to, as it lacks the 3′ OH, and thus terminates the growing chain.

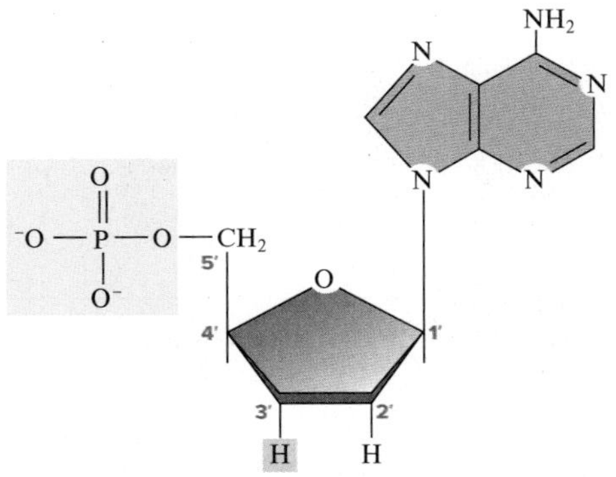
NH2
N
N
N
N
O
-O
P
O
CH2
O-
5'
O
4'
1'
3'
2'
H
H

Dideoxy terminator sequencing was originally performed manually; the experimenter would perform four separate reactions, each with a single dideoxynucleotide and the four deoxynucleotides. When a dideoxynucleotide is incorporated, it will terminate synthesis at a specific base. Thus, each reaction produces a set of nested fragments, which each end in a specific base, and the four reactions together represent fragments that terminate at each possible base. The fragments are separated by high-resolution gel electrophoresis, which can separate fragments differing in length by a single base. This produces a ladder of fragments that allows the sequence to be read from bottom (smallest fragment) to top (longest fragment).

To automate this, each dideoxynucleotide is labeled with a fluorescent dye, and all four are added to a single DNA synthesis reaction. As before, the random incorporation of dideoxynucleotides produces a set of fragments that end in specific bases, but in this case, each base is represented by a color. The dye-labeled nested fragments are separated in a capillary tube by electrophoresis and as each molecule passes a laser, it fluoresces. The nucleotide

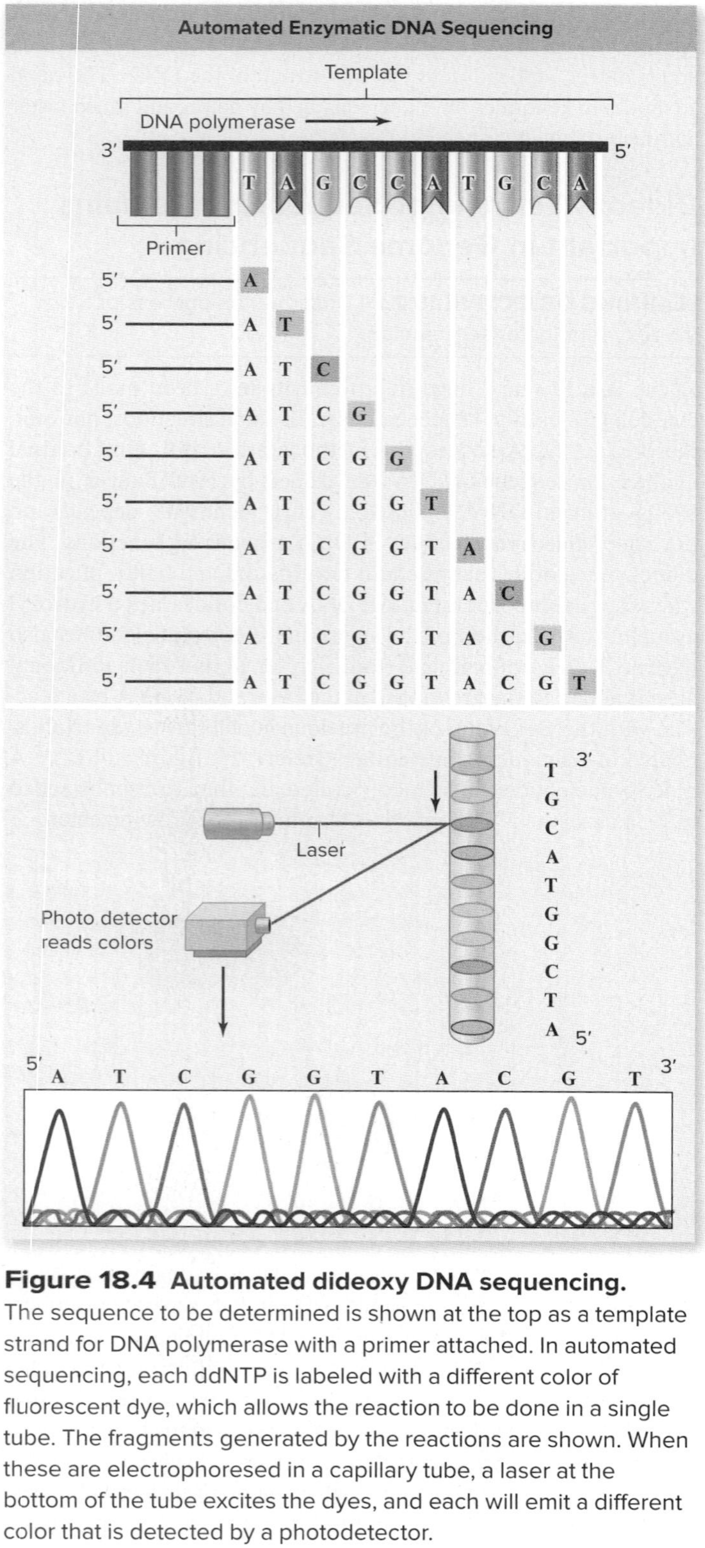

Figure 18.4 Automated dideoxy DNA sequencing. The sequence to be determined is shown at the top as a template strand for DNA polymerase with a primer attached. In automated sequencing, each ddNTP is labeled with a different color of fluorescent dye, which allows the reaction to be done in a single tube. The fragments generated by the reactions are shown. When these are electrophoresed in a capillary tube, a laser at the bottom of the tube excites the dyes, and each will emit a different color that is detected by a photodetector.

terminating the DNA strand is interpreted based on the color of the fluorescence and automatically reported to a computer (figure 18.4). Although this approach is still labor-intensive, automation allows thousands of sequencing reactions to be performed in parallel. This means it is possible to sequence nearly 900 kb of DNA in a single day. In some ways, this was the first step in creating the massively parallel sequencing platforms that characterized the next generation of sequencing technologies.

Next-Generation Sequencing Uses Massively Parallel Technologies to Increase Throughput

LEARNING OBJECTIVE 18.2.2 Compare and contrast dideoxy terminator sequencing with Illumina next-generation sequencing.

Next-generation sequencing (NGS) technologies first appeared 12 or so years ago. Since then, there have been huge improvements in the speed of sequencing, in the cost per sequenced base, and in the length of sequence read per sequencing reaction. The result is a staggering increase in the number of genomes sequenced and resequenced in the past 10 years. In the future, this abundance of data could be translated into discoveries that reveal novel evolutionary relationships, improve the personalization of health care, and improve diagnostic medicine.

There are numerous NGS technologies in the market, but all have some common features. First, unlike dideoxy sequencing, DNA can be sequenced without a need to first construct a genomic library by conventional cloning (refer to chapter 17). Second, instead of thousands of reactions being prepared and analyzed simultaneously, up to millions of sequencing reactions can be performed simultaneously; it is this feature of the technology that has led to the term *massively parallel.* Third, the time-consuming electrophoresis previously required is not needed. Instead, sequencing reactions occur simultaneously in solution and are read directly by the sequencing equipment (figure 18.5). Because of the increase in sequencing speed and the reduction in cost, some bacterial genomes can be sequenced in as little as a few hours!

Although sequencing times and cost have dropped dramatically, a drawback of some of the technologies is the amount of DNA sequence—called *read length*—produced per sequencing reaction. Read length varies significantly between the different technologies and ranges from 35 bases up to a rare example of 20,000 bases. Given the variability in the read length, different technologies are suitable for different applications; some will be good for genome sequencing, whereas others will be more appropriate for diagnostic medicine. Most of the common technologies produce read lengths in the range of 50 to 1000 bp. Shorter read lengths greatly complicate the assembly of the individual pieces of sequence into a finished genome. Because of this problem, novel algorithms for genome assembly have been created.

In 2014, it was reported that a human genome had been sequenced for $1000. This represents a cost reduction of nearly 10,000-fold relative to the initial cost of sequencing a human-sized genome approximately 10 years before that.

Sequenced Genome Fragments Are Assembled into Complete Sequences

LEARNING OBJECTIVE 18.2.3 Compare and contrast shotgun and clone-contig methods of genome sequencing and assembly.

The most complex task in a large-scale sequencing project is the assembly of individual pieces of sequence information into a complete genome. A genome is first fragmented into overlapping pieces, which are sequenced but then must be reassembled by matching short overlapping regions. The shorter the sequenced

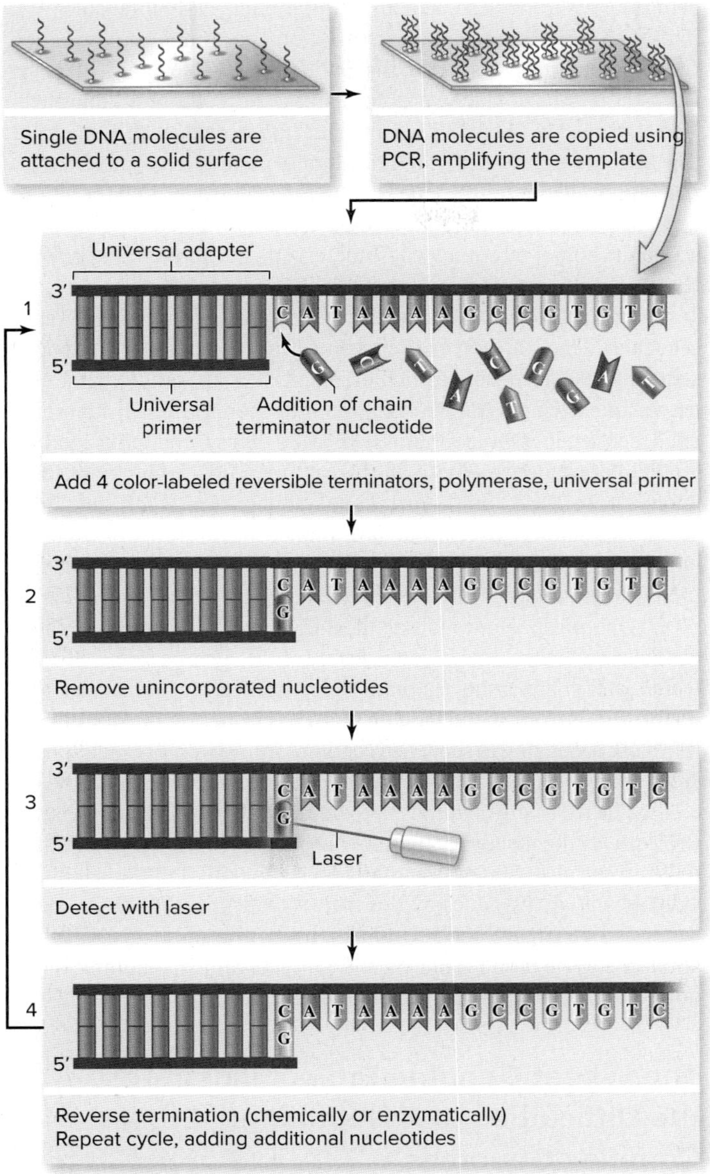

Figure 18.5 Illumina next-generation DNA sequencing. In Illumina next-generation sequencing, the template DNA molecules are attached to a solid surface and amplified by PCR to create islands of DNA. Four fluorescently labeled terminator nucleotides are then added, and after one is incorporated into new DNA, synthesis stops. The color of the incorporated nucleotide is read, and the nucleotide is chemically altered so that another nucleotide can be added. This process of single-nucleotide addition is repeated to read the sequence of the DNA in each island.

fragments, and the more repetitive DNA present in the genome, the harder it is to piece together the puzzle. Two strategies can be employed: (1) Assemble portions of a chromosome first, and then figure out how the bigger pieces fit together **(clone-contig assembly)** or (2) try to assemble all the pieces at once, instead of in a stepwise fashion **(shotgun assembly).**

Shotgun assembly

A benefit of shotgun assembly is that it does not rely on any genetic or physical maps. This means that no prior information about the sequenced genome is required. The first genome ever sequenced, the genome of the bacterium *Haemophilus influenzae,* was sequenced and then assembled using the shotgun approach. The genome of the bacterium was broken into fragments approximately 1.5 to 2 kb in length, which were cloned into suitable vectors (refer to chapter 17). The ends of the fragments were sequenced and then, based on overlaps in the sequence, the pieces of DNA sequence could be stitched back together to form the whole 1.8-Mb genome (figure 18.6*a*). This strategy rarely gives a completely assembled genome sequence. A variety of other approaches are used to close the gaps that become obvious only when the sequences have been assembled.

Clone–contig assembly

Clone-contig assembly methods (figure 18.6*b*) are used to assemble the pieces of sequenced DNA from more complex genomes such as those of eukaryotes. Unlike shotgun assembly, clone-contig assembly requires a physical map to build the completely

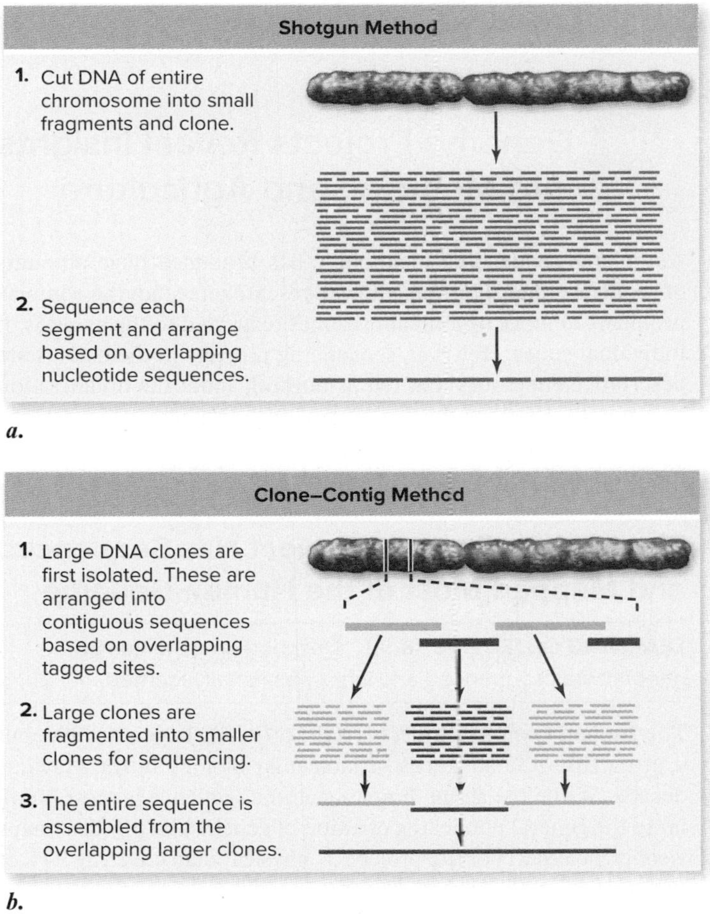

Figure 18.6 Comparison of sequencing and genome assembly methods. ***a.*** In the shotgun method, the entire genome is fragmented into small clones and sequenced. Computer algorithms assemble the final DNA sequence based on overlapping nucleotide sequences. ***b.*** The clone–contig method uses large clones assembled into overlapping regions by STSs. Once assembled, these can be fragmented into smaller clones for sequencing.

reassembled genomic sequence. A genome is fragmented into larger fragments—approximately 1 to 1.5 Mb. Then, those fragments are mapped into a larger contiguous segment of DNA, called a **contig,** using genetic markers such as STSs or restriction endonuclease sites. The fragments of DNA 1 to 1.5 Mb in size can be sequenced and assembled using the shotgun approach (figure 18.6*a*).

REVIEW OF CONCEPT 18.2

Because of the enormous size of some genomes, sequencing requires the use of automated sequencers running many samples in parallel. Two approaches have been developed for whole-genome sequencing and assembly: one that uses clones already aligned by physical mapping (clone–contig sequencing and assembly) and one that involves sequencing random clones and using a computer to assemble the final sequence (shotgun sequencing and assembly). In either case, significant computing power is necessary to assemble a final sequence.

- *What are the pros and cons of the different approaches to whole-genome sequencing and assembly?*

18.3 Genome Projects Reveal Insights into Medicine and Agriculture

Automated-sequencing technology has produced huge amounts of sequence data. This has allowed researchers studying complex problems to move beyond approaches restricted to the analysis of individual genes. However, sequencing projects in themselves are descriptive endeavors that tell us nothing about the organization of genomes, let alone the function of gene products and how they may be interrelated. Research using data from genome projects has produced both insight and perplexity.

The Human Genome Project Has Sequenced and Mapped Most of the Human Genome

LEARNING OBJECTIVE 18.3.1 Describe how the human genome was sequenced and what was initially learned.

The Human Genome Project (HGP) officially began in 1990, but it arose from the advances in molecular biology in the previous decade. While the actual human genome sequence is most familiar to the general public, the creation of genetic and physical maps were critical parts of the project. A physical map covering 94% of the genome, with markers about every 200 kb, was completed by 1995. A genetic map with markers about every 1.6 cM was completed in 1996. Since these early maps, the number of markers and resolution has increased significantly. The project drove advances in automation that increased the speed and reduced the cost of sequencing. Ultimately, this formed the basis for the clone-contig and shotgun approaches used by the HGP.

In 1998 a competitor from the private sector emerged when Craig Venter, who founded The Institute for Genomic Research, claimed that they could sequence the genome faster and cheaper using just shotgun sequencing and better assembly software. Although there is some truth to this, Venter's company, now called Celera Genomics, also had the publicly available physical maps from the HGP, so they were not using a pure shotgun approach. The approach worked for the bacterium *Haemophilus influenzae* but this genome is orders of magnitude smaller, and less complex, than the human genome. In June 2000, a *draft sequence* of the entire genome was jointly announced by the HGP and Celera. This draft sequence contained many gaps and errors, but represented a functional, albeit incomplete, version of the human genome.

In 2003, the draft sequence was replaced by the first **reference sequence**. The difference between these is that the reference sequence has fewer sequencing errors, fewer gaps, and greater coverage, and attempts to take variation into account. The initial reference sequence still contained 150,000 gaps, and consisted of 2.69 Gb of sequence. These data are curated by the Genome Reference Consortium, an international consortium, with reference sequences named GRCh plus a number. We are up to GRCh38, which consists of 3.1 Gb of sequence and only 350 gaps, and also includes 360 regions with alternative loci.

One obvious question about the reference genome is, Whose genome is it? In fact, the original source of DNA came from 20 volunteers. Later analysis showed that 70% of the sequence is from a single donor, 23% is from 10 individuals, and 7% is from over 50 DNA libraries. Work is ongoing to sequence genomes from diverse populations to create a "pan genome" that better reflects actual human diversity.

The first significant result to come out of the Human Genome Project was the discovery that the number of genes, long estimated to be around 100,000, was actually only about 20,000. This is about 1.5 times as many genes as the fruit fly, and nearly half as many genes as rice (figure 18.7). Humans, mice, and puffer fish all have about the same number of genes. Clearly, organismal complexity is not a simple function of either genome size or gene number.

The Wheat Genome Project Illustrates the Difficulties of Assembling Complex Genomes

LEARNING OBJECTIVE 18.3.2 Explain why sequencing the human genome was a relatively simple endeavor compared with sequencing the wheat genome.

Wheat, *Triticum aestivum*, is used to feed about 30% of the world's population. Better understanding of wheat genomics can inform breeding practices to increase yield and improve drought and disease resistance. It can also provide insight into the evolution of wheat species. Given expanding populations, dwindling water resources, and global climate change, applying modern genetics to the improvement of crop plants will be critical to maintain secure global food supplies.

Genetically, wheat is an allohexaploid (refer to chapter 21) produced by two hybridization events. This means the wheat genome contains the contributions of three diploid ancestors designated A, B, and D, each containing about 5 billion bp of DNA. The modern wheat genome has 21 chromosomes, 7 from each ancestral genome, and an estimated size of 16 Gb. There are two main challenges to sequencing and assembling the wheat genome: (1) its large size, and (2) a large amount of repetitive DNA,

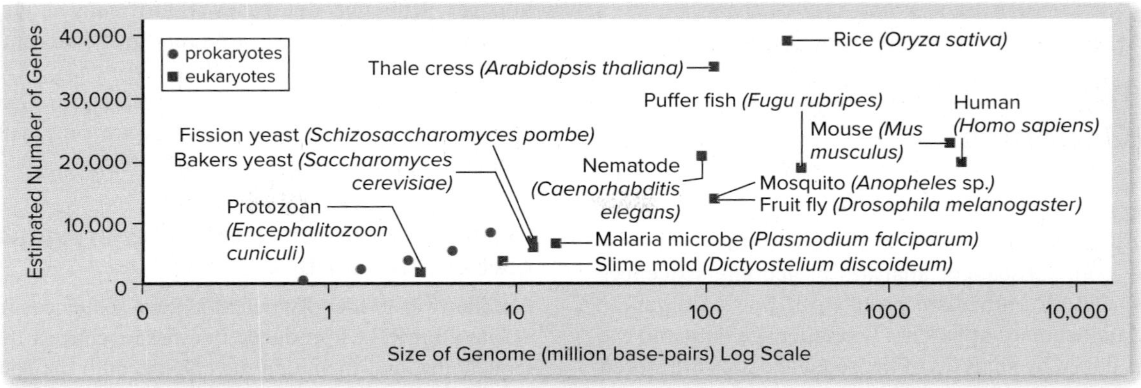

Figure 18.7 Size and complexity of genomes. In general, eukaryotic genomes are larger and have more genes than prokaryotic genomes, although the size of the organism is not the determining factor. The mouse genome is larger than the human genome, and the rice genome is larger than both the human and the mouse genomes.

including nearly identical sequences scattered throughout due to its polyploid nature.

A first attempt to sequence the wheat genome in 2012 assembled only 5.4 Gbp, or about a third of the genome. A second attempt published 2 years later attempted to sequence a chromosome at a time, yielding a genome assembly of 10.2 Gbp, or about two-thirds of the genome. In early 2017, a third attempt was produced that covered 78% of the genome. An additional problem with all 3 of these assemblies was that the length of contigs (refer to section 18.2) was relatively short. Finally, later in 2017, the most complete sequence yet was published that covers 15.3 Gbp, with contigs more than 10 times longer than the previous best effort. This was accomplished by combining newer sequencing technology that produces very long reads, but with a high frequency of errors, with multiple runs with more accurate short reads.

A similar approach was used to produce the first sequence of *Aegilops tauschii*, the species that contributed the "D genome" in the last hybridization event leading to modern wheat. The *Ae. tauschii* genome has a high degree of dispersed repeated sequences, including transposable elements that make up 84% of the genome. The *Ae. tauschii* genome appears to be accumulating large-scale rearrangements faster than other cereal genomes, probably by recombination between these repetitive elements.

Cancer Genome Projects Seek the Genetic Basis of Cancer

LEARNING OBJECTIVE 18.3.3 Describe the potential benefits of the cancer genome project.

Cancer kills approximately 1500 people a day in the United States. As average life span increases, this rate will likely continue to increase. Many years of study have revealed the outline of the genetic basis of cancer: more than one somatic mutation is necessary to convert a normal cell to a cancer cell—it is a progressive series of mutations. We have also identified two categories of genes involved: oncogenes that can cause cancer with gain-of-function mutations, or inappropriate activation, and tumor-suppressor genes that lead to cancer with loss-of-function mutations. These can also be thought of as positive, "accelerator"-type mutations and negative, "brake"-type mutations (refer to chapter 10).

Genomics has helped to fill out this simple framework by comparing tumor genomes with the genomes of matched normal tissues. At this point, more than 5000 tumors and matched normal-tissue samples, representing more than 20 tumor types, have been sequenced. This has led to the identification of genes found associated with one or more tumor types. Mutations that are found in a tumor genome are divided into "driver" mutations, which affect progress toward cancer, and "passenger" mutations, which accumulate but do not lead to cancer. More than 200 different genes have been identified as potential drivers, including both oncogenes and tumor-suppressor genes, and this is not yet a complete catalog. These genes affect signal-transduction pathways, DNA repair pathways, the control of gene expression, chromatin structure, and even metabolism. Some genes are mutated in multiple tumor types, and different tumors may have from only a few to hundreds of mutations.

This avalanche of information has both clarified our understanding and provided new problems for analysis. There is no simple model in which mutations in a set of genes always lead to a specific type of cancer, but patterns do begin to emerge. It is also clear that even specific types of tumors have heterogeneity. In the long run, characterizing each type of tumor will help in both diagnosis and the design of treatments based on the genotype of the patient and the mutations in each patient's tumor.

REVIEW OF CONCEPT 18.3

The Human Genome Project cost $2.7 billion and took 13 years to complete. By comparison, the much larger and more complex wheat genome could be sequenced and assembled at a fraction of that cost and in half as much time. The sequencing and assembly of large, complex, highly repetitive genomes are technologically challenging. Cancer genomics hopes to identify sets of mutations that could indicate predisposition to cancer, predict disease progression, or anticipate responsiveness to particular treatments.

- *Explain why complex genomes, such as the wheat genome, are technically harder to sequence than relatively simple genomes.*

18.4 Genome Annotation Assigns Functional Information to Genomes

The genome sequencing and assembly stages of a genome project produce vast amounts of data. Add to these data the additional data generated from subsequent analyses of genome sequences and structures, and an immediate problem of how to organize and store the data becomes apparent. The sequence data and the information produced by analyses of the sequence data are stored in a variety of searchable, often publicly available, databases.

Although laborious to produce, the genome sequence itself is of little intrinsic interest. More important are the number and kind of sequences present in the genome and how these sequences contribute to phenotype. The assignment of this kind of information to a genome sequence is called **genome annotation.**

Genome Annotation Assigns Function to DNA Sequences in Genomes

LEARNING OBJECTIVE 18.4.1 Explain how genomes are annotated and the purpose of annotation.

An annotation is a record attached to a DNA sequence held in a database that identifies the DNA sequence as a gene. The record can contain information about the gene, its structure and function, and the product it ultimately produces (RNA or protein). The annotation may also include evidence about how the assignment of gene function was made. The overarching goal is to attempt to add biologically relevant information to the DNA sequences in the genome.

Genome annotation is largely an automated process that is checked as needed by experts. The annotation process often begins by using computer algorithms to look for regions of DNA that are likely to contain a gene. The algorithms may look for sequences that contain regulatory regions such as promoters, or they may search for a start codon followed by codons that encode amino acids, and eventually a stop codon. Sequences such as this are called **open reading frames (ORFs).** It is also possible to search for splicing signals to predict intron–exon structure and consequently to predict the sequence and structure of an RNA that might be produced, and thus the amino acid sequence coded for by the putative gene.

Inferring function across species: The BLAST algorithm

Once a potential gene or gene product has been identified, it is possible to infer function by finding a similar sequence in other species. The tool that makes this possible is a search algorithm called the Basic Local Alignment Search Tool (BLAST). BLAST searching can also be automated to speed up the annotation process. Using a networked computer, potential gene or gene product sequences are submitted to the BLAST server and a search is conducted for similar sequences already deposited into the sequence databases. Significant sequence similarity with other genes or gene products in a different species could suggest conservation of function.

Using computer programs to search for genes, to compare genomes, and to assemble genomes are only a few of the new genomics approaches called bioinformatics.

Expressed sequence tags identify genes that are transcribed

Another way to identify potential gene sequences in genomes is to isolate all mRNAs produced in a tissue, convert this to cDNA for sequencing, and then map the cDNAs onto the genomic sequence (figure 18.8). This can be simplified by sequencing just one or both ends of as many cDNAs as possible. These short sections of cDNA have been named **expressed sequence tags (ESTs).** These ESTs are another form of STS to include in physical maps, and also can be used in BLAST searches for similar sequences in other species.

ESTs have been used to identify 87,000 cDNAs in different human tissues. About 80% of these cDNAs were previously unknown. The estimated 20,000 genes of the human genome can result in these 87,000 different cDNAs because of *alternative splicing* (refer to chapter 15).

Genomes Contain Both Coding and Noncoding Sequences

LEARNING OBJECTIVE 18.4.2 Compare and contrast the types of DNA found in genomes.

DNA sequences that are used to produce a protein are called coding sequences (CDSs). This leaves all the sequences that do not result in the synthesis of polypeptides as noncoding sequences. Genome annotation identifies the distribution of coding sequences in the genome, and subsequent experimentation can reveal the function of those genes and their products. In some cases, the function of noncoding DNA is known; for

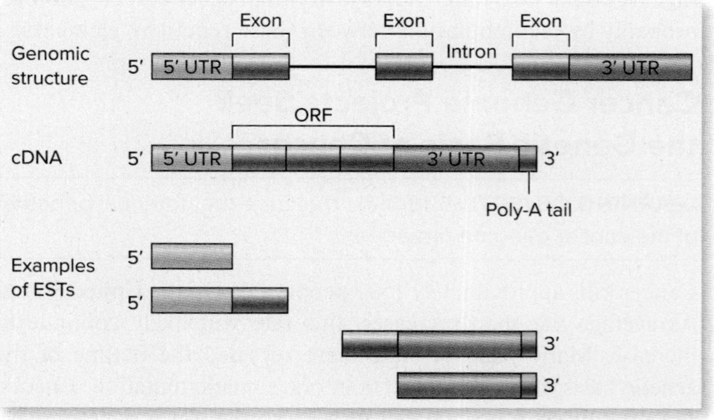

Figure 18.8 ESTs can be used to map expressed genes onto the genome. Sequencing the ends of ESTs produces sequence data that can be aligned with predicted cDNA structure and mapped onto corresponding genomic structures. By sequencing many ESTs, complete genomic structures can be determined.

example, genes that produce functional RNA molecules such as tRNAs, rRNAs, and miRNAs are functionally well characterized. On the other hand, there is much noncoding DNA that is less well characterized.

Noncoding DNA in eukaryotes

A notable characteristic of eukaryotic genomes is the amount of noncoding DNA they possess. The Human Genome Project revealed a particularly startling picture: each of your cells contains nearly 2 m of DNA, but only about 2.5 cm of that DNA is coding sequence! Nearly 99% of the DNA in your cells is noncoding DNA.

Genes coding for proteins are scattered in clumps among the noncoding DNA in the human genome. Seven major types of noncoding DNA have been described in the human genome (table 18.1 shows the composition of the human genome, including noncoding DNA):

Noncoding DNA within genes. As discussed in chapter 15, a human gene is made up of numerous pieces of coding DNA (exons) interspersed with lengths of noncoding DNA (introns). Introns constitute about 24% of the human genome, whereas the exons constitute less than 1.5%.

Structural DNA. Some regions of the chromosomes remain highly condensed, tightly coiled, and untranscribed. These regions, called *constitutive heterochromatin,* tend to be localized around the centromere or near the ends of chromosomes, at telomeres.

Simple sequence repeats. There are two forms of tandem repeat variation that differ in size and are scattered within the genome. **Short tandem repeats (STR)** consist of repeat units of 2-6 bp. **Variable number tandem repeats (VNTR)** contain repeat units of 7-49 bp. Together they make up about 3% of the genome.

Segmental duplications. These are blocks of genomic sequences of 10,000 to 300,000 bp that have duplicated and moved either within a chromosome or to a nonhomologous chromosome.

Pseudogenes. These inactive genes may have lost function because of mutation. There are almost 15,000 characterized pseudogenes in the human genome.

Transposable elements. Forty-five percent of the human genome consists of DNA sequences that can move from one place in the genome to another. Some of these sequences code for proteins that allow their movement, but many do not. Because of the significance of these elements, they are described in this section.

microRNA genes. Hidden within the noncoding DNA is a mechanism for controlling gene expression. Specifically, DNA that was once considered "junk" has been shown to encode microRNAs (miRNAs), which are processed after transcription to lengths of 21 to 23 nt but are never translated. About 10,000 unique miRNAs have been identified that are complementary to one or more mature mRNAs. These miRNAs regulate some of the complex developmental processes in eukaryotes by down-regulating translation.

Long, noncoding RNA. In addition to the many small RNAs such as microRNAs that are not translated into protein but serve a regulatory role, tens of thousands of longer, noncoding RNAs likely regulate gene expression. This recently discovered, hidden world of regulatory networks reveals a new level of complexity in the precise control of gene expression. Long, noncoding RNAs have important roles in physiology and development. Only approximately 200 long, noncoding RNAs have been functionally characterized, but the biological functions, if any, of the others remain unclear.

TABLE 18.1 Classes of DNA Sequences Found in the Human Genome

Class	Description
Protein-encoding genes	Translated portions of the approximately 20,000 genes scattered about the chromosomes
Introns	Noncoding DNA that constitutes the great majority of each human gene
Segmental duplications	Regions of the genome that have been duplicated
Pseudogenes (inactive genes)	Sequences that have characteristics of a gene but are not functional genes
Structural DNA	Constitutive heterochromatin, localized near centromeres and telomeres
Simple sequence repeats	Stuttering repeats of a few nucleotides, such as CGG, repeated thousands of times
Transposable elements	21%: Long interspersed elements (LINEs), which are active transposons 13%: Short interspersed elements (SINEs), which are active transposons 8%: Retrotransposons, which contain long terminal repeats (LTRs) at each end 3%: DNA transposon fossils
Noncoding RNA	RNA that does not encode a protein but has regulatory functions, many of which are currently unknown

Transposable elements: Mobile DNA

Transposable elements, also called *transposons* and *mobile genetic elements,* are sequences of DNA able to move from one location in the genome to another. Transposable elements move around in different ways. In some cases, the transposon is duplicated, and the duplicated DNA moves to a new place in the genome. When this happens, the number of copies of the transposon increases. Other types of transposons are excised without duplication and insert themselves elsewhere in the genome.

Human chromosomes contain four types of transposable elements. About 21% of the genome consists of **long interspersed elements (LINEs).** These ancient and very successful elements are about 6000 bp long, and they encode the enzymes needed for transposition. LINEs encode a reverse transcriptase enzyme that can make a cDNA copy of the transcribed LINE RNA. The result is a double-stranded DNA fragment that can reinsert into the genome rather than undergo translation into a protein. Since these elements use an RNA intermediate, they are called retrotransposons.

Short interspersed elements (SINEs) are similar to LINEs, but they do not encode the transposition enzymes and so cannot move without using the transposition machinery of LINEs. Nested within the genome's LINEs are over half a million copies of a SINE element called Alu (named for a restriction enzyme that cuts within the sequence). The Alu SINE is 300 bp and represents 10% of the human genome. An Alu SINE can use the enzymes of the LINE of which it is a part to jump to a new chromosome location. Alu sequences can also jump into coding DNA sequences, causing mutations.

Two other sorts of transposable elements are also found in the human genome. First, about 8% of the human genome is made from a type of retrotransposon called a **long terminal repeat (LTR).** Although the transposition mechanism is a bit different from that of LINEs, LTRs also use reverse transcriptase to create double-stranded copies of themselves that can integrate into the cell's genome. Second, dead transposons occupy approximately 3% of the genome; these are transposons inactivated by mutation that can no longer move.

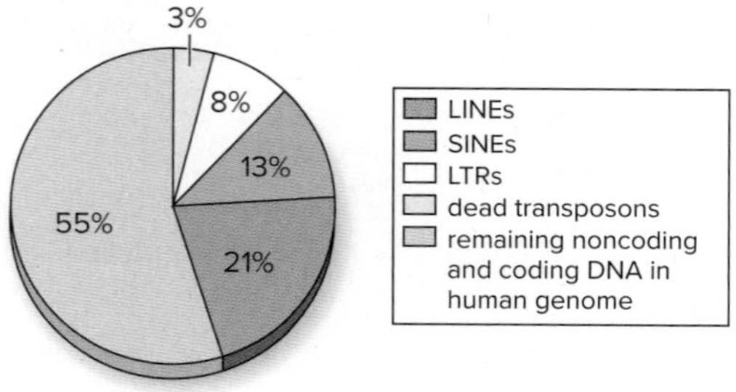

The ENCODE Project Seeks to Identify All Functional Elements in the Human Genome

LEARNING OBJECTIVE 18.4.3 Evaluate the conclusions of the ENCODE project.

The presence of noncoding DNA in genomes has been known for many years, but the biological importance of this DNA remained unclear. Since no function was assigned, this noncoding DNA was often referred to as "junk DNA." Extensive analyses of the noncoding DNA have led some to suggest that the term "junk DNA" should be reevaluated. It may be best to reframe the DNA of no known function as "nonfunctional DNA."

The Encyclopedia of DNA Elements (ENCODE) project is a collaborative effort to identify all functional elements in the human genome. The primary conclusion of its work was that 80% of the DNA sequences in the human genome are functional. Of that 80%, approximately 62% are defined as transcriptionally active, although the majority of the transcribed, or potentially transcribed, RNAs have no known function. There has been much debate in the scientific community regarding what these numbers actually mean in terms of biological function, and the ENCODE consortium has revised its estimate of functionality to somewhere between 20 and 80%.

ENCODE's definition of function

ENCODE's definition of a functional element includes any DNA sequence that results in protein production, that is transcribed, or that has a distinct and reproducible biochemical signature (figure 18.9).

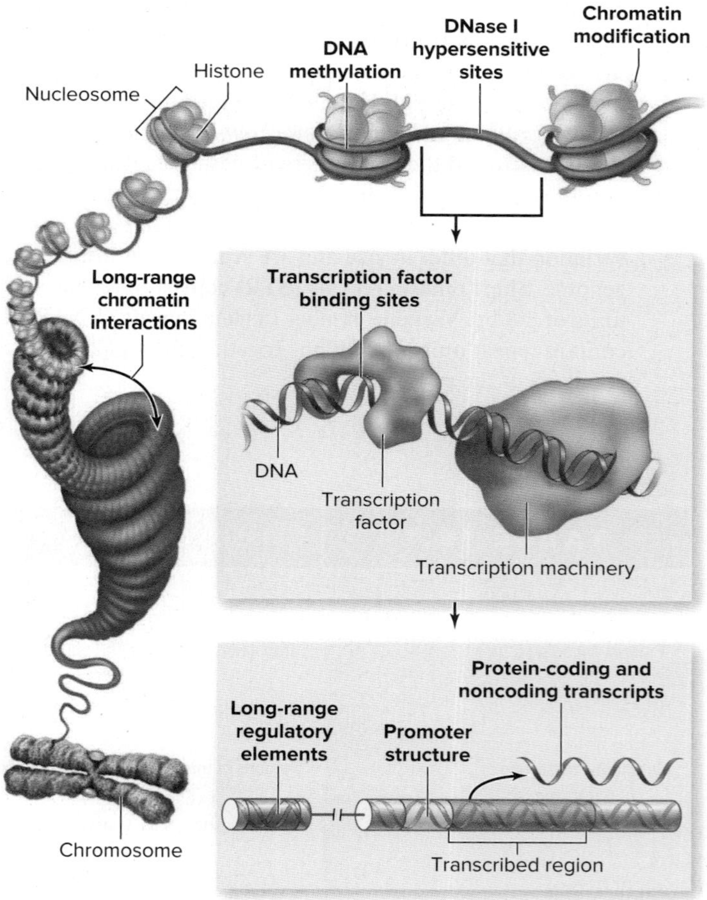

Figure 18.9 Functional DNA elements defined by ENCODE. ENCODE uses a biochemical signature definition of function. Biochemical signatures include methylation patterns of DNA sequences; modifications of histones in chromatin; sensitivity to DNase I, which suggests regions of transcriptional activity; sites of transcription factor binding; sequences identified as promoters; production of coding or noncoding transcripts; long-range regulatory elements such as enhancers; and long-range chromatin interactions.

A biochemical signature appears when a protein binds to a DNA sequence, when chromatin bends to form long-range interactions, or when a DNA sequence exhibits a particular pattern of histone modification. Given that there is likely significant difference between these functionalities in different cell types, ENCODE derived its 80% value by testing the genomes of 144 different cultured human cell lines for functional elements.

Some criticism has been directed at ENCODE for its use of cultured cell lines and stem cell lines. These cell lines may not be reflective of the situation in the genome of cells of an organism and may be more transcriptionally active than cells in tissues. Critics also claim that there are likely to be significant differences between functional elements in different developmental stages of an organism and under different environmental conditions.

Biological function

The functional definition used by ENCODE is not necessarily synonymous with biological function. A sequence of DNA that is transcribed—and therefore biochemically functional, according to ENCODE—may not produce an RNA with a function necessary for the organism. Also, just because one or several transcribed RNAs have a demonstrated biological effect, this does not mean that all transcribed RNAs have biological effects. A natural extension of the definition of functional element could include any DNA sequence that is replicated; given such a biochemical definition, 100% of the genome would be functional, as 100% of the genome is replicated during cell division.

Selected-effect function

ENCODE's conclusions have also drawn fire from some evolutionary biologists. Most evolutionary biologists, and many biologists in general, subscribe to a selected-effect definition of function. This definition of function requires that the function of a characteristic be the one selected for by purifying selection. For example, the selected function of the mammalian lung is gas exchange, and selective pressures have shaped that function. The rising and falling of the chest during breathing is caused by the inflation and deflation of the lungs, but that rising and falling action has not been selected as the lung's function.

If a genetic sequence has a particular function, this will eventually be destroyed by the accumulation of random mutations over time, unless this is counteracted by natural selection. The only way that a functional sequence is protected from destruction is if purifying selection acts to resist the accumulation of mutations that degrade the function. Many biologists would argue that only DNA sequences under purifying selection should be considered functional. Such sequences are more likely to be conserved in a specific lineage and are more likely to be conserved in related species. Based on these criteria, comparative genomics studies suggest that as little as 5 to 15% of the genome is functional.

Clearly, there are conflicting perspectives on the topic of the functionality of DNA sequences in the human genome, and ENCODE's calculation of 80% functionality remains debatable. Nevertheless, the technologies and means of data collection advanced by the ENCODE effort significantly advance our potential for understanding functions of DNA sequences in genomes. Similarly, ENCODE's data regarding the biochemical functions of noncoding DNA provide an impressive map that can be used by others to explore the biological relevance of the identified functional elements in the human genome.

REVIEW OF CONCEPT 18.4

Once a genome has been sequenced, assembled, and deposited into an appropriate database, functional elements of the genome are annotated with useful information. Functional elements can be identified in a variety of ways, including by inference using the BLAST program. Only a fraction of the DNA in a genome contains information for making proteins; the remaining noncoding DNA may have function. Because of ambiguity in the definition of function, conclusions of what percentage of a genome is composed of functional DNA require critical evaluation.

- *What explanation could you suggest, based on principles of natural selection, for the many repeated transposable elements in the human genome?*

18.5 Genome Comparisons Provide Information About Genomic Structure and Function

Comparative genomics uses information from one genome to learn about a second genome. For example, the known function of a gene in one organism's genome can be used to hypothesize about the function of a similar gene in a related organism's genome. It turns out that 60% of the genes involved in triggering human cancer are conserved in the fruit fly genome. The functions of those genes can be analyzed in the fruit fly, which is significantly easier than analyzing them in humans. Comparative genomics also reveals information about the relatedness of organisms, about how different organisms perform similar biological functions, about what makes different species different, and even about the minimum number of genes it would take to build a functional cell—an engineering feat of interest to synthetic biologists.

Functional genomics is an extension of genomics that investigates the relationship between genotype and phenotype. The phenotype of an organism is determined by the pattern of expression of genes and by the interaction of the organism with the environment. For example, by comparing the gene expression pattern in a healthy cell with the gene expression pattern in a diseased cell, it is possible to learn which genes might be involved in the disease process.

Comparative Genomics Reveals Conserved Regions in Genomes

LEARNING OBJECTIVE 18.5.1 Describe the value of using comparative genomics to learn about the properties of genomes.

One of the striking lessons learned from the sequence of the human genome is how similar humans are to other organisms. More than half of the genes of *Drosophila* have human counterparts. Among mammals, the similarities are even greater. Humans have only 300 genes that have no counterpart in the mouse genome. The use of comparative genomics to ask evolutionary questions is also a field of great promise. The comparison of the many prokaryotic genomes already sequenced indicates a greater degree of lateral gene transfer than was previously suspected.

Comparative genomics based on synteny

Comparative genomics allows for a large-scale approach to comparing genomes by taking advantage of synteny. *Synteny* refers to the conserved arrangements of segments of DNA in related genomes. Physical mapping techniques can be used to look for synteny in genomes that have not been sequenced. Comparisons with the sequenced, syntenous segment in another species can provide information about the unsequenced genome.

To illustrate this, consider rice and its grain relatives, corn, sugarcane, and wheat. Even though these plants diverged more than 50 MYA, the chromosomes of rice, corn, wheat, and other grass crops show extensive synteny (figure 18.10). In a genomic sense, "rice is wheat," and the information gleaned from the physical and genetic maps of corn and rice genomes undoubtedly aided the sequencing of the wheat genome.

By understanding the rice and corn genomes at the level of their DNA sequences, the identification and isolation of genes from grains with larger genomes should be much easier. DNA sequence analysis of cereal grains could be valuable for identifying genes associated with disease resistance, crop yield, nutritional quality, and growth capacity.

Functional Genomics Reveals Gene Function at the Genome Level

LEARNING OBJECTIVE 18.5.2 Explain how functional genomics is used to learn about the functions of genomes.

To fully understand how a genome contributes to the structure and function of an organism, we need to characterize the products of the genome—the RNA molecules and the proteins. This information is essential to understanding cell biology, physiology, development, and evolution. Functional genomics allows connections to be drawn between the genotype and the phenotype of the organism.

Functional genomics uses a range of high-throughput techniques to learn about the products of a genome, when they are produced, and how they change during development or in response to the environment. Functional genomics can be broken into three separate, but related, categories: (1) the study of all the RNA molecules produced by a genome (the transcriptome); (2) the study of the proteins produced from the genome (the proteome); and (3) the

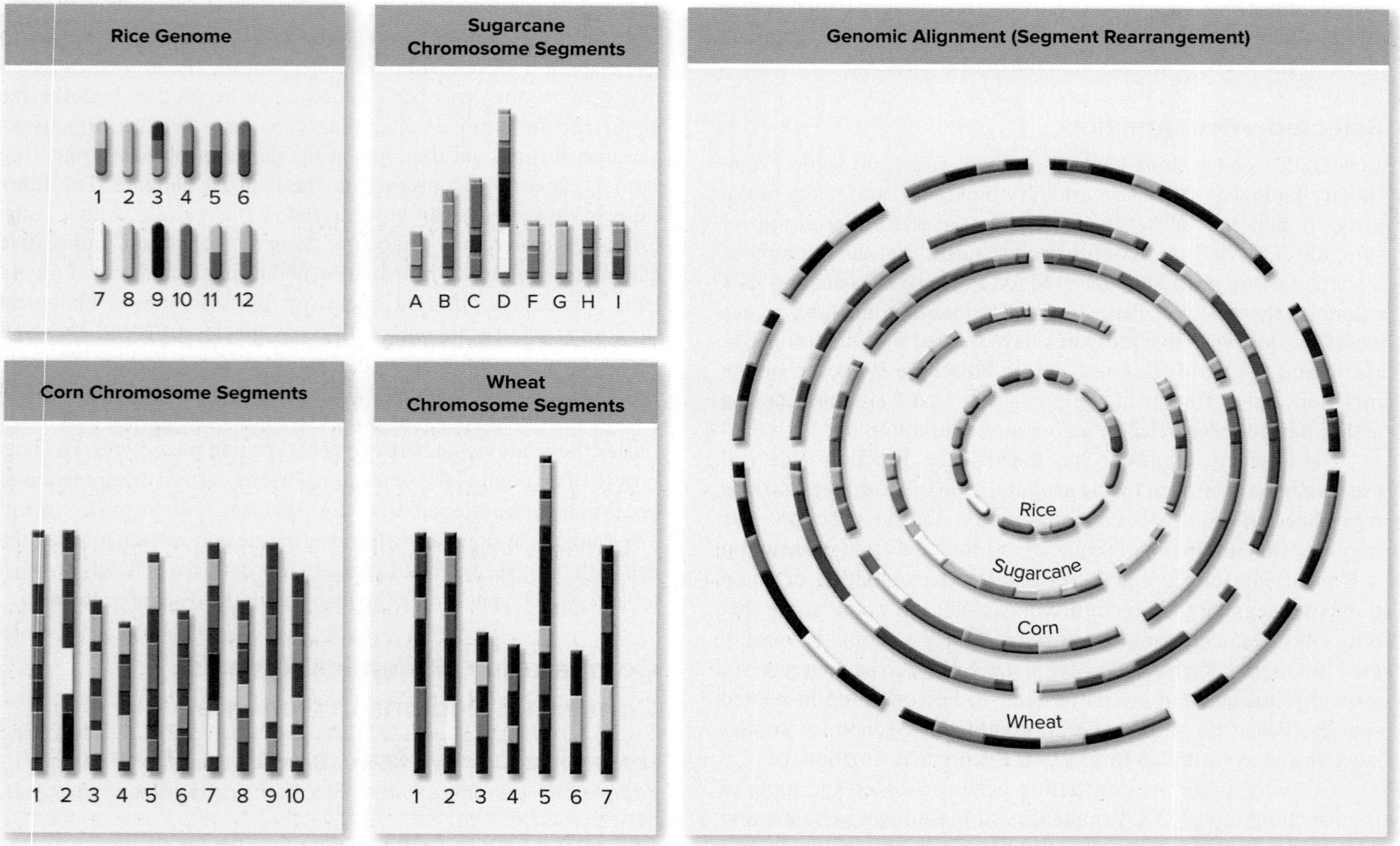

Figure 18.10 Grain genomes are rearrangements of similar chromosome segments. Shades of the same color represent pieces of DNA that are conserved among the different species but have been rearranged. By splitting the individual chromosomes of major grass species into segments and rearranging the segments, researchers have found that the genome components of rice, sugarcane, corn, and wheat are highly conserved. This implies that the order of the segments in the ancestral grass genome has been rearranged by recombination as the grasses have evolved.

SCIENTIFIC THINKING

Hypothesis: *Flowers and leaves will express some of the same genes.*

Prediction: *When mRNAs isolated from* Arabidopsis *flowers and leaves are used as probes on an* Arabidopsis *genome microarray, the two different probe sets will hybridize to both common and unique sequences.*

Test:

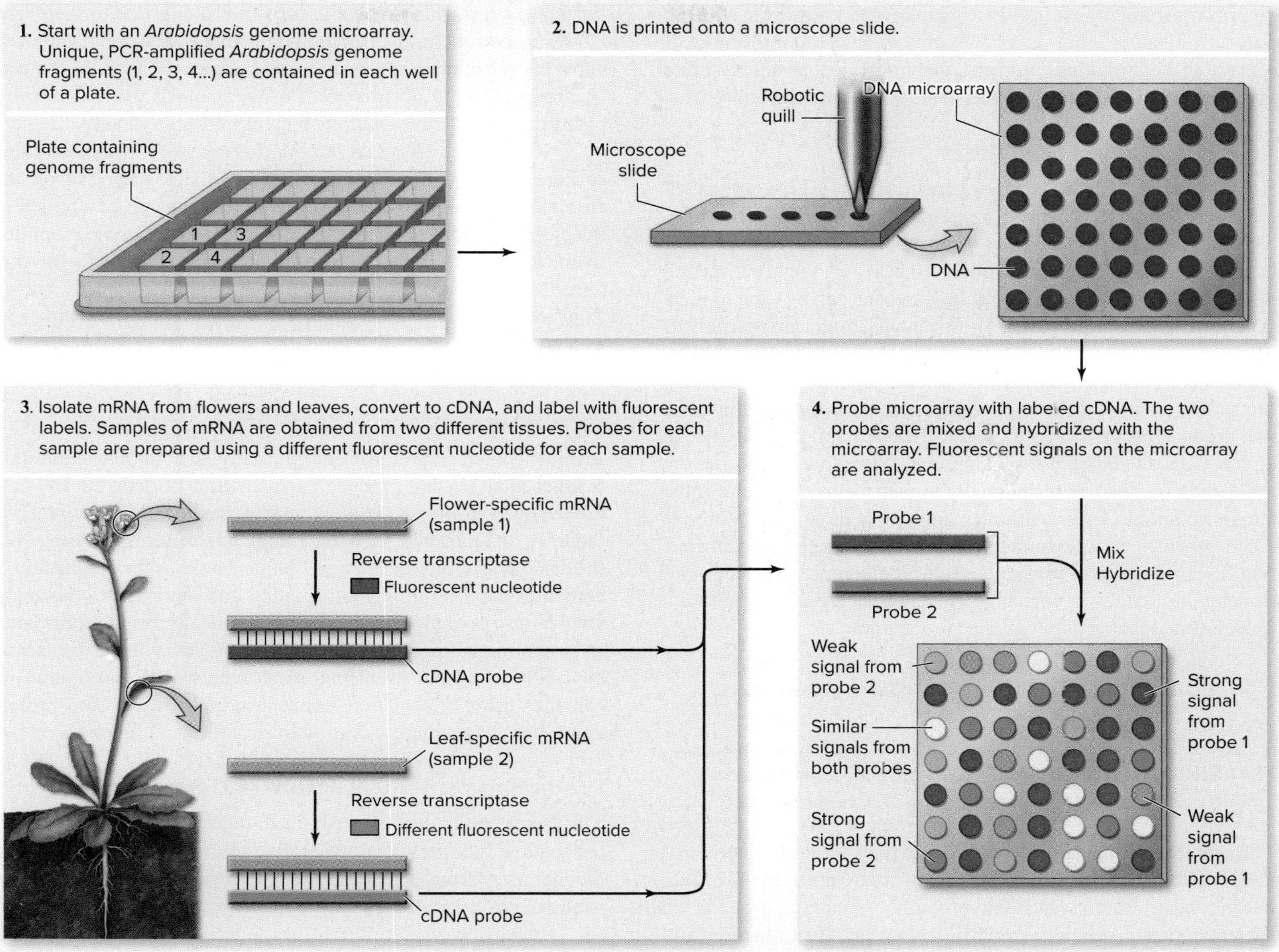

1. Start with an *Arabidopsis* genome microarray. Unique, PCR-amplified *Arabidopsis* genome fragments (1, 2, 3, 4...) are contained in each well of a plate.
Plate containing genome fragments
1
3
2
4
2. DNA is printed onto a microscope slide.
Robotic quill
DNA microarray
Microscope slide
DNA
3. Isolate mRNA from flowers and leaves, convert to cDNA, and label with fluorescent labels. Samples of mRNA are obtained from two different tissues. Probes for each sample are prepared using a different fluorescent nucleotide for each sample.
Flower-specific mRNA (sample 1)
Reverse transcriptase
Fluorescent nucleotide
cDNA probe
Leaf-specific mRNA (sample 2)
Reverse transcriptase
Different fluorescent nucleotide
cDNA probe
4. Probe microarray with labeled cDNA. The two probes are mixed and hybridized with the microarray. Fluorescent signals on the microarray are analyzed.
Probe 1
Probe 2
Mix
Hybridize
Weak signal from probe 2
Similar signals from both probes
Strong signal from probe 2
Strong signal from probe 1
Weak signal from probe 1

Result: *Yellow spots represent sequences that hybridized to cDNA from both flowers and leaves. Red spots represent genes expressed only in flowers. Green spots represent genes expressed only in leaves.*

Conclusion: *Some* Arabidopsis *genes are expressed in both flowers and leaves, but there are genes expressed in flowers but not leaves and leaves but not flowers.*

Further Experiments: *How could you use microarrays to determine whether the genes expressed in both flowers and leaves are housekeeping genes or are unique to flowers and leaves?*

Figure 18.11 Microarrays.

study of the interactions and products of interactions between proteins. We consider proteomics later in this section, but first we summarize several of the technologies used to study the transcriptome.

DNA microarrays

DNA microarrays were created for the analysis of gene expression at the whole-genome level. The arrays allow questions to be asked about how gene expression patterns change during development, in response to environmental stimuli, and even during the development of disease. This kind of analysis requires accurate annotation of a genome and the use of this knowledge to construct an array containing all of the coding sequences (figure 18.11).

Microarrays called gene chips were discussed in chapter 17 as a tool in medical diagnostics. Chips can be used in hybridization experiments with labeled mRNA from different sources. This gives a high-level view of genes that are active and inactive in a variety of conditions or states. Arrays can also be prepared so that the DNA dot contains variations of the DNA sequence so that only transcripts from genes with single-nucleotide polymorphisms are detected.

Microarrays have some limitations, however. The sequence of a genome must be known to prepare a microarray with appropriate sequences. Also, if a microarray is being prepared to look for rare polymorphic transcripts in a sample, then knowledge of those polymorphisms is required for the array to be created. Because of some of these limitations, and given technological advances, other approaches have become popular in studying the transcriptome.

RNA sequencing (RNA-seq)

High-throughput methods to sequence cellular RNAs are called RNA-seq. RNA-seq provides a snapshot of all the mRNA in a sample but does so by directly sequencing cDNA using NGS technologies. RNA-seq can be modified based on the experimental questions being asked. For example, RNA-seq can be used to identify intron–exon boundaries, map single-nucleotide polymorphisms to unsequenced genomes, and determine the contribution of different alleles to phenotype when alleles of a gene are present. RNA-seq suffers from the problem that in many cases it is not the RNA that produces a phenotype—rather, it is the protein. The presence of an RNA detected by RNA-seq does not take into account that the RNA may be posttranscriptionally regulated in a way that interferes with protein levels. Although analysis of the transcriptome of a cell provides a large amount of potentially important information, it can be critical to also consider the complete collection of proteins in a cell at a particular time. These studies constitute a field of functional genomics called **proteomics.**

Proteomics Catalogs Proteins Encoded by the Genome

LEARNING OBJECTIVE 18.5.3 Describe the relationships among the genome, the transcriptome, and the proteome.

The proteome is much harder to analyze than either the genome or the transcriptome, because proteins are harder to analyze than nucleic acids. Proteins must fold properly to function; proteins are subject to posttranslational modifications that affect function; and a single gene can produce multiple proteins by alternative splicing (figure 18.12). All of this makes identifying the relationships between genomes, transcriptomes, and proteomes a difficult task.

Many high-throughput techniques have been developed to study the proteome. The technique used depends on the specific research question being asked. One technique becoming more common for analyzing the proteome is *mass spectroscopy (mass spec).* Mass spec determines the charge-to-mass ratio of a molecule, which with small molecules allows for unambiguous identification. For many years, this was not used with proteins due to their size. However, new techniques combine purification and ionization to allow the use of mass spec to analyze complex mixtures of proteins (figure 18.13). One advantage of this technique is that it can reveal posttranslational modifications such as phosphorylation.

Proteomics also uses the microarray technology described earlier in this section. Rather than DNA being spotted on the chip, antibodies recognizing specific proteins can be applied to the chip. If a mixture of proteins isolated from a sample is applied to an antibody chip, then the antibodies will bind to specific proteins in the sample. Using this approach, specific proteins, known as *biomarkers,* can rapidly be identified in a complex sample. As well as being useful in a research setting, this technique has potential for disease diagnosis and screening.

Bioinformatics and proteomics

Just like genome and transcriptome analysis, high-throughput proteomics analyses can produce huge amounts of data, but the data are largely meaningless without approaches to study and interpret them. At the interface between computer science and biology lies the discipline of bioinformatics. **Bioinformatics** is the application of computer programming, mathematics, and modeling to the analysis of large sets of biological data. The application of bioinformatics to proteomics allows the rapid identification of proteins discovered by high-throughput profiling experiments, the annotation of genomes using information from mass spectroscopy and protein chips, and the prediction of protein structure and function.

Predicting protein structure and function

The use of new computer methods to quickly identify and characterize large numbers of proteins distinguishes traditional protein biochemistry from proteomics. As with genomics, the challenge is one of scale.

Ideally, we would like to use a gene sequence to infer the structure and function of the encoded protein. At present, we can do this only by comparing a new protein sequence with the sequences of already identified proteins. Inferences about

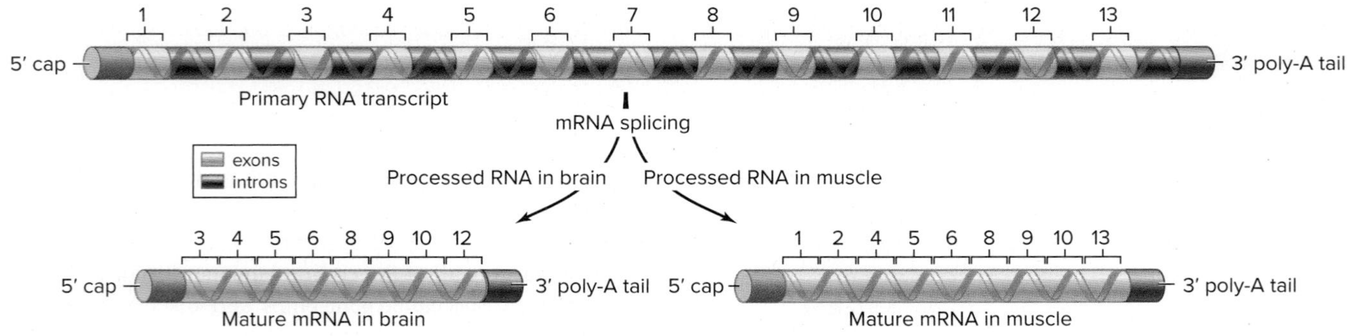

Figure 18.12 Alternative splicing can result in the production of different mRNAs from the same coding sequence. In some cells, exons can be excised along with neighboring introns, resulting in different proteins. Alternative splicing explains why approximately 20,000 human genes can code for three to four times as many proteins.

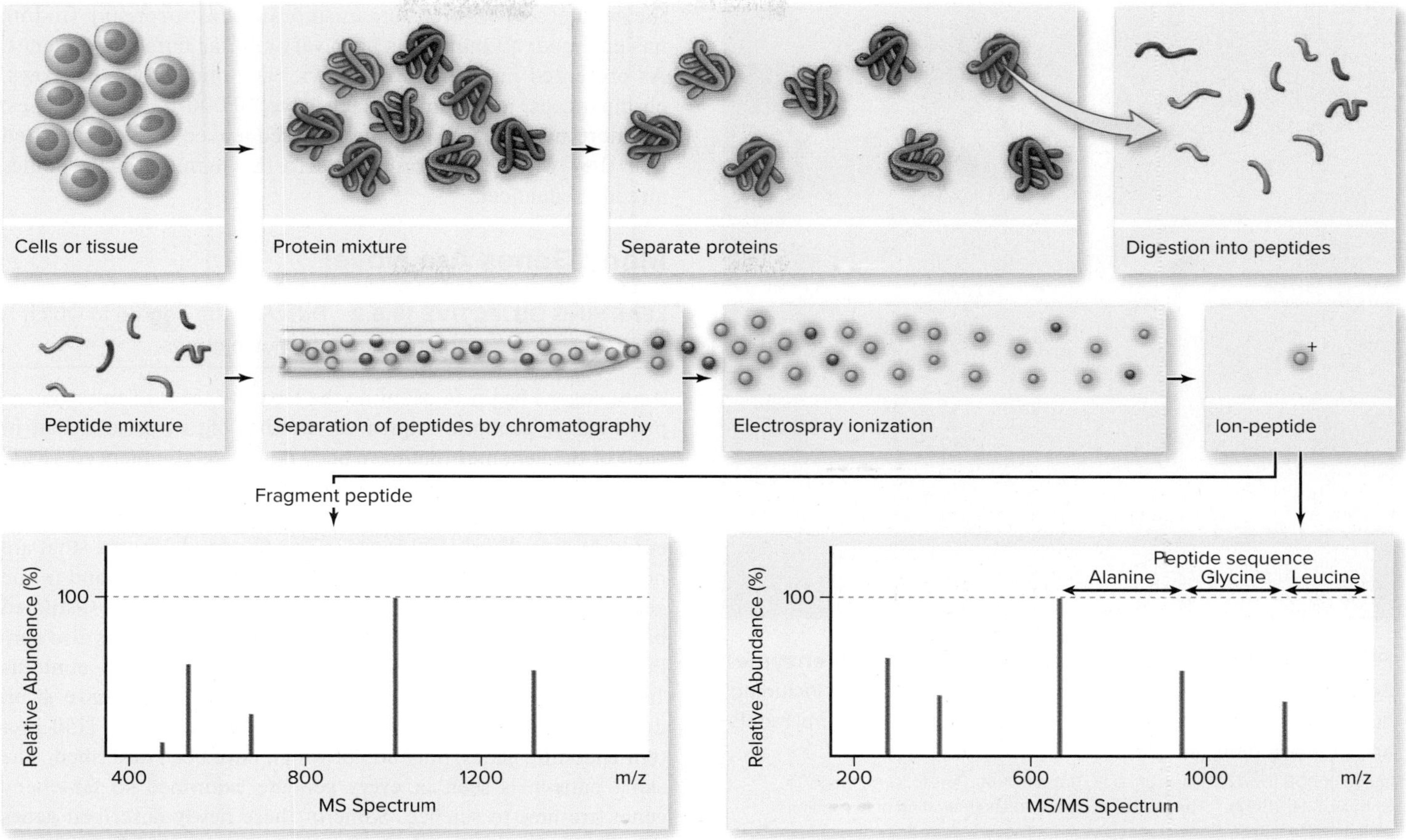

Figure 18.13 Mass spectroscopy can be used to analyze proteins. Proteins can be isolated from cells or tissues and partially separated using gels or biochemical techniques. Partially purified or individual proteins are digested into small peptides by proteases and ionized. The charged peptides can be separated by mass in a spectrometer. Computer algorithms can match the peptide fragments with predicted digestion patterns to tentatively identify proteins. Alternatively, individual peptides can be further fragmented and amino acid composition can be determined by mass spectroscopy. (MS spectrum: abundance of an ion as a function of mass-to-charge ratio of the ion; MS/MS spectrum: abundance of an ion as a function of mass-to-charge ratio of the ion when the ions are the product of tandem ionization events; m/z: the mass of an ion divided by the charge of an ion, commonly referred to as the mass-to-charge ratio)

structure and function are sometimes possible due to the fact that similar sequences of amino acids will produce a similar structure. However, it is also true that similar structures can be produced from very different sequences of amino acids. In these cases, similar functions can be inferred only if structural information is available.

Inference of function from structure is an even more challenging task. Proteins with similar sequences and structures often have similar functions, but there are many examples where this is not the case. Proteins with significantly different structures perform many similar functions. It is helpful if we have information about the evolutionary history of a protein. We can be more confident in our inferences about shared structures and functions if we are comparing proteins derived from a common ancestor.

Until recently, it has been very difficult to predict the variety of ways in which chains of amino acids could fold into secondary and tertiary structures (an example of predicted protein structure is shown in figure 18.14). One type of computer algorithm attempts to predict protein structure by first looking for stretches of amino acids that have characteristics of a protein domain. Then, the algorithm takes short fragments from the protein domain and compares them to known folds in proteins whose structures are known. A computer program then uses these data to build a model of how all the fragments might be arranged to produce specific structures. When the same structure appears independently from different simulations, there is a higher probability that the structure is correct.

REVIEW OF CONCEPT 18.5

Comparisons of different genomes allow geneticists to infer structural, functional, and evolutionary relationships between genes and proteins, as well as relationships between species. Functional genomics provides tools to begin to understand the functions of the genes in a genome. DNA microarrays and RNA-seq can be used to learn about the transcriptome, but there are pros and cons to the two approaches. Proteomics involves analyses of the proteins produced by a genome under specific conditions. Although it is possible to make some predictions about protein structure and function, the analyses are complex and require powerful computers.

- *Why is establishment of a species' transcriptome an important step in studying its proteome?*

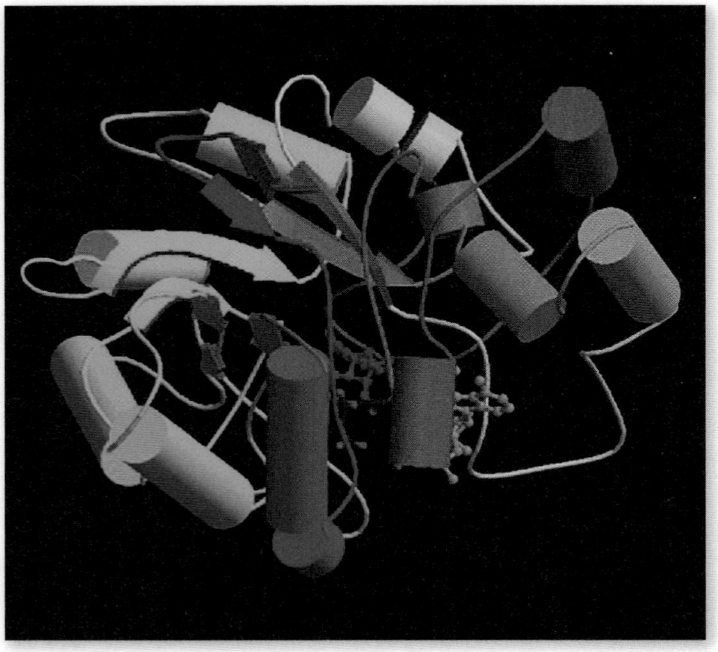

Figure 18.14 Computer-generated model of an enzyme. Searchable databases contain known protein structures, including human aldose reductase, shown here. Secondary structural motifs are shown in different colors.

Image of PDB ID 1AZ2 (Harrison, D.H., Bohren, K.M., Petsko, G.A., Ringe, D., Gabbay, K.H., (1997), ""The alrestatin double-decker: binding of two inhibitor molecules to human aldose reductase reveals a new specificity determinant,"" Biochemistry, 36 (51)

18.6 Comparative Genomics Informs Evolutionary Biology

Comparing genomes (entire DNA sequences) of different species provides a powerful new tool for biologists, as well as insights into phylogenetic relationships in the tree of life. The genomic sequences that are already completed give exciting clues of what is to come.

All Mammals Have Comparably Sized Genomes

LEARNING OBJECTIVE 18.6.1 Compare the number of genes in the different mammalian genomes.

Comparing the number of genes in the genomes of organisms suggests that, very roughly speaking, more-complex organisms have more genes. Figure 18.15 shows that insects (*Drosophila* and mosquito) have twice as many genes as single-celled organisms like bacteria and yeasts, and about half as many as mammals.

Perhaps the most surprising finding of the HGP, however, was the small number of genes required to encode a human being—only about 20,000 to 25,000 genes. In comparisons of the genomes of different organisms, another surprising finding has emerged: all mammals have very similar numbers of genes. As presented in figure 18.15, cows, rats, mice, dogs, monkeys, chimpanzees, and humans each have 20,000 to 25,000 genes. As more mammalian genomes are sequenced, it is anticipated that they will be of similar size to the mammalian genomes already sequenced.

Many Genes Are Novel

LEARNING OBJECTIVE 18.6.2 Describe the degree to which genomics is revealing previously unknown genes.

As biologists begin to examine the treasure trove of information provided by genomic sequences, another big surprise is that in each of the genomes that have been completely sequenced so far, there are large numbers of unfamiliar protein-encoding genes. The genome of the prokaryote *Aeropyrum pernix,* for example, contains more than 1500 genes—57% of its total genome—that are not found in any other organism. Some 4000 genes found in the genome of *Mycobacterium tuberculosis,* one of the best-studied bacteria, fall into the same category. Eukaryote genomes also contain many unexpected genes. Human chromosome 6 contains 1557 protein-encoding genes, only 772 of which have been described. Similarly, human chromosome 7 contains 1150 protein-encoding genes, only 605 of which have been described. The same pattern is seen in every genome examined so far—many genes are new to science. Some of these newly described genes resemble other genes whose functions are known, but it is not known what these particular proteins are doing in the organism. Despite centuries of examination by biologists, it seems that there remains much we don't yet understand.

Large Differences in Genome Sizes Can Arise Through Genome Duplication

LEARNING OBJECTIVE 18.6.3 Explain why some organisms have far more genes than their complexity would seem to predict.

Even a casual look at figure 18.15 reveals three glaring exceptions to the general finding that more-complex organisms have more genes—cottonwood trees, rice, and pufferfish all have far more genes than their organismal complexity would suggest. A pufferfish has 40% more genes than a human! Is the pufferfish more complex than a human? Given these organisms' relative complexities, it seems that other factors must be at play to explain the difference in the sizes of their genomes. When the genomes of cottonwood trees, rice, and pufferfish are examined, the reason for their high gene number is found to be the result of whole-genome duplication, rather than the addition of new genes.

We can look at the pufferfish more closely to better understand how this happens. The last common ancestor of pufferfish and humans was a primitive bony fish that lived some 230 MYA. The descendants of this long-extinct fish evolved into two distinct lineages, the ray-finned fishes (including the pufferfish) and the lobe-finned fishes (including the ancestors of humans). Sometime early in the ray-finned fish lineage, the entire genome duplicated.

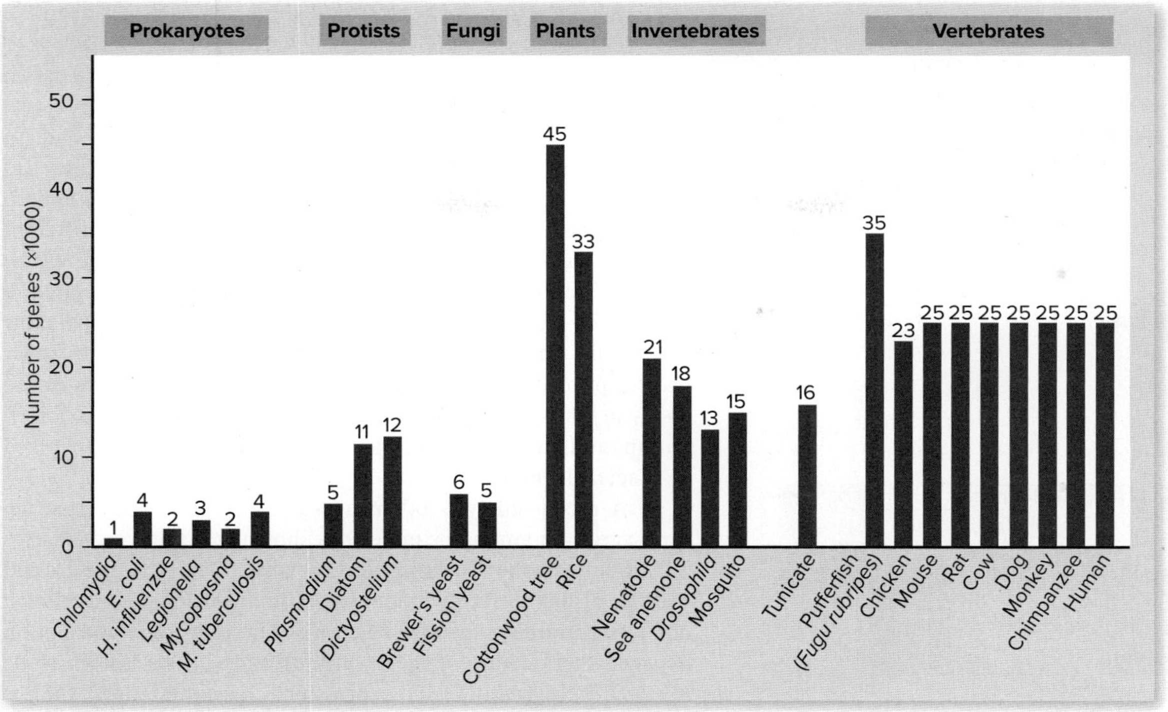

Figure 18.15 Comparing genome size. All mammals have the same-size genome, 20,000 to 25,000 protein-encoding nuclear genes. The unexpectedly larger sizes of the plant and pufferfish genomes are thought to reflect whole-genome duplications rather than increased complexity.

When the positions of more than 6000 pufferfish genes are compared with the positions of corresponding genes in the human genome, one chromosomal region in humans matches *two* in pufferfish, across the entire genome. Illustrated in figure 18.16 (with duplicated genes highlighted in color), this pattern is a clear reflection of the whole-genome duplication among ray-finned fishes. Many similar duplications arose during the evolution of the plants.

Key Genes Tend to Be Conserved

LEARNING OBJECTIVE 18.6.4 Explain why sequences, like the *HOX* genes, are highly conserved within genomes.

Once the discovery of DNA sequencing made it possible to compare DNA sequences from different organisms, it became clear that certain sequences are widespread among the kingdoms of life. Multiple copies of a gene called the *HOX* gene occur in the genomes of all animals. *HOX* genes play a key role in guiding development of the body plan, determining the number and orientation of body segments. Changes in four of these *HOX* genes have been shown to account in large part for the major differences in the bodies of crustaceans and insects. Related *HOX*-like gene sequences have been found in plants and yeasts, and even in prokaryotes. Apparently this sequence evolved very early in the history of life and was of such importance that it has been conserved virtually unchanged in animals, fungi, and plants for hundreds of millions of years.

How common are such highly conserved genes? As more entire genomes become sequenced, it will become possible to address this question. Already it is clear that the number will be large. Only about one-third of the genes in cottonwood trees and rice appear to be in some sense "plant" genes, not found in any animal or fungal genome sequenced so far. These include the many thousands of genes involved in photosynthesis. Two-thirds of the plant genomes are devoted to genes similar to those found in animal and fungi genomes, particularly genes involved in basic intermediary metabolism and in genome replication and repair.

Rates of Evolution Vary Greatly

LEARNING OBJECTIVE 18.6.5 Compare the rates of genomic change in major groups of organisms.

Comparison of rodent (mouse and rat) and primate (human and chimpanzee) genomes reveals that since the time of the last common ancestor in the mouse and human lineages, about 75 MYA, rodent DNA has mutated about twice as fast as primate DNA. This is a fascinating observation in search of an explanation. The difference in generation time between mice and humans (the average time between two successive generations, or the time needed for offspring to become parents) could account for some of this difference, as mice have much shorter generation times and would have had more opportunities to mix and match genomic components during meiosis.

The insects are the most species-rich and morphologically diverse animal group on Earth. Two important insect genomes that have been sequenced are the fruit fly *Drosophila* and the mosquito *Anopheles.* These organisms are separated by approximately 250 million years of evolution, and they appear to have evolved more rapidly over that interval than have vertebrates. The extent of the similarity between these two insects is equivalent to that between humans and pufferfish, which diverged 450 MYA.

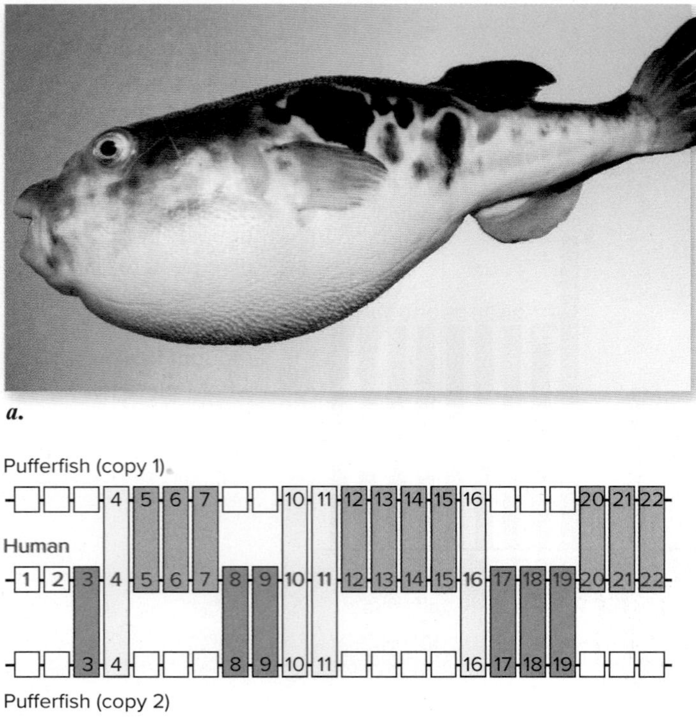

Figure 18.16 Detecting whole-genome duplications. ***a.*** The genome of the pufferfish has, in its ancestry, undergone a whole-genome duplication. ***b.*** Human and pufferfish genes are shown. The middle section shows human genes, numbered 1 through 22, that are also found in the same position in pufferfish. Some of them, in yellow, have related genes found in two different chromosomal locations in pufferfish (labeled as copy 1 and copy 2). Others have relatives only on copy 1 (red) or copy 2 (blue), the other copy having been lost or diversified.

(a) B. Venkatesh

What Makes Us Human?

LEARNING OBJECTIVE 18.6.6 Compare human and other primate genomes.

Genome projects have been completed or are under way for the majority of extant primates. Among these, the chimpanzee, *Pan troglodytes,* and the bonobo, *Pan paniscus,* share the most recent common ancestor with humans. Comparisons among these genomes reveal 1.4% divergence in single-site substitutions between humans and chimps, and 1.5% between humans and bonobos. There is another 1.5% of the genome that differs in small insertions and deletions (indels) between humans and chimps, and a similar percentage between humans and bonobos.

Comparing all sequenced primates indicates that the prevalence of transposable elements observed in humans is also true for other primates. About 50% of all primate genomes consist of various transposons. Species-specific insertions of these elements vary considerably in different lineages, and different elements have expanded in some lineages more than others. In humans, Alu insertions have expanded almost twice as fast as in the chimp and bonobo genomes. As these elements can also facilitate deletions or duplications, they can have effects on genome evolution outside of insertion events.

The majority of protein-coding genes found in humans have homologs among both great apes and Old World apes. Despite this, some gene families have experienced lineage-specific expansion and contraction in different lineages. Genes in the major histocompatibility complex (MHC) are expanded in macaques relative to humans, perhaps related to differential exposure to pathogens. Humans and chimps show interesting lineage-specific gains and losses of the zinc finger transcription factor family of genes. The effect of these differences in transcription factors on patterns of gene expression in chimps and humans is not clear, but might potentially be involved in species differences.

A recent analysis of lineage-specific differences in structural variation among primates has shown that there are close to 18,000 human-specific structural variants. These include around 12,000 different fixed human-specific insertions, and almost 6000 fixed human-specific deletions. There is some correlation between the location of these variants and regions known to be involved in regulating gene expression. One tantalizing example is the deletion of a region thought to regulate two cell-cycle genes active in cortical neurons. As one human-specific morphological difference is an enlarged brain, this kind of alteration could lead to more cell division during brain development.

We can also examine differences in gene expression in humans and other primates. RNA sequencing (RNA-Seq; refer to chapter 17) has been used to measure mRNA from single cells from cerebral organoids. These organoids are cultured from pluripotent stem cells that are induced to differentiate into organized neural tissue, including radial glial cells and excitatory neurons. Comparing chimp and human cerebral organoids, investigators found 383 genes up-regulated in radial glial cells and 220 genes up-regulated in excitatory neurons. Conversely, 285 genes in radial glial cells, and 165 genes in excitatory neurons, were down-regulated. A subset of these genes proved to be associated with human-specific structural variations.

These expression studies are complemented by the analysis of a gene thought to be involved in speech. The *FOXP2* gene encodes a transcription factor expressed in the brain. Individuals heterozygous for a mutant allele of *FOXP2* show difficulties in learning and sequencing the facial movements necessary to produce speech, but not impaired language comprehension. The function of *FOXP2* is necessary for the development of brain circuits in the neocortex, basal ganglia, and cerebellum, which are involved in language and fine motor control.

Analysis of the evolution of this gene has only added to interest in its role. In comparisons between humans and mice, *FOXP2* is among the most conserved 5% of genes. Despite this conservation, the rate of amino acid substitution in FOXP2 protein in humans is increased 50-fold over the average rate in the genome. It is also found in a region of the genome that has little archaic DNA. There are three amino acid substitutions in the human protein compared to the protein in mice. Of these, two occurred after humans diverged from chimps and bonobos. The FOXP2 protein is

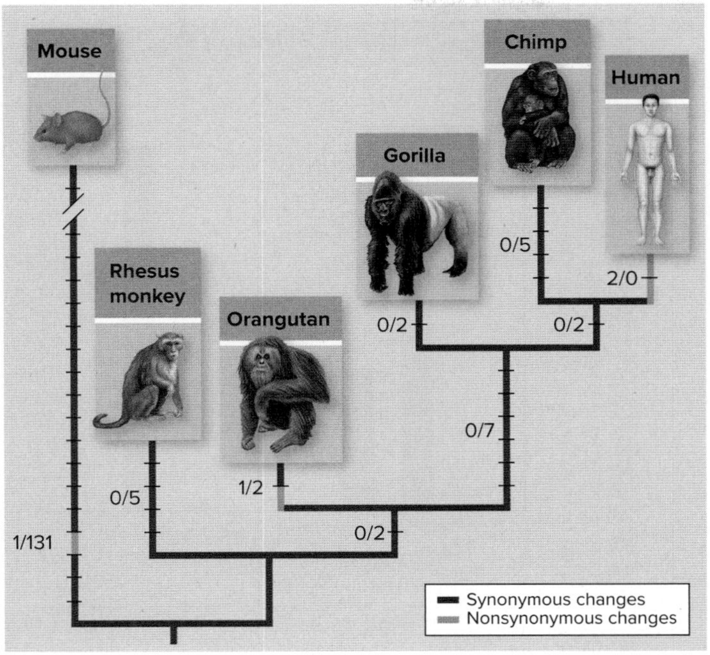

Figure 18.17 Evolution of *FOXP2*. Beige bars represent synonymous nucleotide substitutions, and brown bars represent nonsynonymous changes. Numbers indicate nonsynonymous over synonymous nucleotide substitutions unique to each lineage. Humans have two amino acid substitutions not found in other primates.

identical in all nonhuman primates examined, with the exception of a single amino acid substitution in the orangutan.

The downstream targets of the FOXP2 transcription factor have not been characterized, but there are interesting hints that its role in communication may not be limited to humans. Song birds with the level of FOXP2 protein reduced by RNA interference show disturbed development of song variability, especially learning by imitation. Mice communicate via squeaks, with lost young mice emitting high-pitched squeaks. *FOXP2* mutations leave mice squeakless, and transgenic baby mice carrying the human *FOXP2* squeak differently. Both downstream targets of FOXP2 protein and its role in other species are areas of active research that should shed further light on the role of FOXP2 in human evolution.

REVIEW OF CONCEPT 18.6

Comparison of different genomes allows us to infer structural, functional, and evolutionary relationships between genes and proteins and the relationships between species. Increased complexity and chromosomal–genome duplication lead to larger genomes. Key genes tend to be conserved, whereas novel genes and noncoding DNA are added. Genomes evolve at different rates, but closer relatives have more similar genomes.

- *What functional groups of genes would you expect to share with a plant?*

Recombination Is Directed Away from the Functional Mouse Genome

Meiotic recombination is the hallmark of eukaryotic genetics, responsible for much of the diversity found in the animal and plant kingdoms. Recombination can be a disruptive process, with function sometimes being lost if recombination occurs within a gene. Mice and other mammals appear to have evolved a particularly clever way to avoid such problems, however. Meiotic crossing over begins with the introduction of a double-stranded break in DNA, and in mice like the laboratory mouse shown below, the location of this break depends on the activity of a meiosis-specific histone, methyltransferase enzyme, called PRDM9.

PRDM9 creates preferred recombination sites by binding to specific DNA sequences that are present at many places on the mouse genome, acting as a flag to attract the enzyme that creates the double-stranded breaks. This labeling by PRDM9 is not required for recombination. Knockout mice with a deletion of the *PRDM9* gene have the same level of meiotic recombination as wild-type mice with a functional *PRDM9* gene. Rather, it seems that PRDM9 is directing the positioning of the double-stranded breaks to specific sites on the mouse genome.

This raises the interesting possibility that mice may be using PRDM9 to direct double-stranded breaks, and thus meiotic recombination, away from key areas of the genome that might be disrupted by the process.

To test this hypothesis, researchers posed a simple question: Is recombination directed away from the functional genome in wild-type mice possessing active PRDM9? They defined the functional genome in the simplest way, as that portion of the genome that is transcribed, and scored double-stranded recombinational breaks as affecting the functional genome if the double-stranded breaks are associated with transcription start sites (TSSs). The results of their investigation are presented in the histograms above. Red bars represent recombination affecting transcription start sites and, so, the functional genome. Green bars represent recombination not affecting the transcribed genome.

David Hosking/Science Source

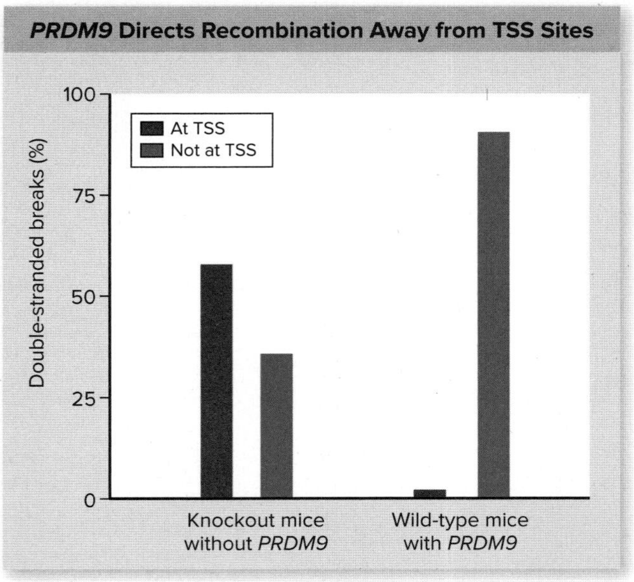

PRDM9 Directs Recombination Away from TSS Sites
Double-stranded breaks (%)
100
75
50
25
0
At TSS
Not at TSS
Knockout mice without *PRDM9*
Wild-type mice with *PRDM9*

Analysis

1. **Applying Concepts**
   a. ***Variable.*** In the graph, what is the dependent variable?
   b. ***Percentage.*** A value of 75 represents what proportion of the total number of double-stranded breaks analyzed in the study?
2. **Interpreting Data**
   a. Among knockout mice lacking any PRDM9 activity, is there any difference in the relative proportion of recombination occurring at TSSs versus that occurring away from TSSs?
   b. Among wild-type mice with PRDM9 activity, is there any difference in the relative proportion of recombination occurring at TSSs versus that occurring away from TSSs?
3. **Making Inferences**
   a. Among the knockout mice lacking PRDM9 activity, what proportion of recombination affects the transcribed (functional) genome?
   b. Among wild-type mice with PRDM9 activity, what proportion of recombination affects the transcribed (functional) genome?
4. **Drawing Conclusions** Is it reasonable to conclude from these results that in mice the gene *PRDM9* acts to direct the enzyme initiating meiotic recombination (by making a double-stranded break in DNA) away from transcription start sites, and thus away from the functional genome?
5. **Further Analysis** These results seem to suggest that in wild-type mice many non-TSS DNA sites are subject to recombination that would otherwise not be subject to recombination in the absence of the *PRDM9* gene. How might the researchers go about learning more about the nature of these sites? Do you imagine they might be evolutionarily important? Explain.

Retracing the Learning Path

CONCEPT 18.1 Mapping Identifies and Locates Functional Elements in Genomes

18.1.1 Genomics Uses Many Approaches to Analyze Entire Genomes Genomics uses genetic and physical maps to assign genetic landmarks, called markers, to positions in the genome.

18.1.2 Genetic Maps Provide Relative Distances Between Genetic Markers Linkage mapping uses recombination frequency to determine the relative position of genetic markers in a genome.

18.1.3 Physical Maps Provide Absolute Distances Between Genetic Markers Physical maps include restriction, cytological, and STS maps. These all identify the actual location of markers in the sequence of a genome.

18.1.4 Physical Maps Can Be Correlated with Genetic Maps Any gene that can be cloned can be placed within the genome sequence, but genetic distances will vary across the physical map.

CONCEPT 18.2 The Modernization of DNA Sequencing Has Accelerated Discovery

18.2.1 Dideoxy Terminator Sequencing Remains Important in Genome Sequencing Despite advances in sequencing technologies, dideoxy terminator sequencing, which uses altered nucleotides to terminate DNA synthesis, is still important.

18.2.2 Next-Generation Sequencing Uses Massively Parallel Technologies to Increase Throughput Next-gen sequencers eliminate the need to clone DNA for sequencing. Isolated DNA is sequenced in many parallel reactions, producing massive amounts of data. This has reduced the time and cost of genome sequencing.

18.2.3 Sequenced Genome Fragments Are Assembled into Complete Sequences Shotgun sequencing and assembly produces a genome sequence from short pieces of overlapping DNA. The clone–contig method uses physical maps and larger DNA fragments to assemble a genome sequence.

CONCEPT 18.3 Genome Projects Reveal Insights into Medicine and Agriculture

18.3.1 The Human Genome Project Has Sequenced and Mapped Most of the Human Genome The race to sequence the human genome led to rapid advances in technology and early completion of the project. The human genome contains significantly fewer genes than predicted.

18.3.2 The Wheat Genome Project Illustrates the Difficulties of Assembling Complex Genomes Large, highly repetitive genomes are difficult to sequence and assemble. Complex genome structures require the use of the clone–contig approach.

18.3.3 Cancer Genome Projects Seek the Genetic Basis of Cancer Identifying similarities and differences in cancer genomes and epigenomes should improve diagnostics and allow improved therapy.

CONCEPT 18.4 Genome Annotation Assigns Functional Information to Genomes

18.4.1 Genome Annotation Assigns Function to DNA Sequences in Genomes Genes are identified as open reading frames in DNA sequences. Search tools can compare these to existing genes to help infer function. Identification of ESTs supports the annotation process.

18.4.2 Genomes Contain Both Coding and Noncoding Sequences Noncoding DNA in eukaryotes makes up about 99% of DNA. Almost half of the human genome consists of transposable elements. Noncoding DNA includes micro-RNAs, long noncoding RNAs, and pseudogenes.

18.4.3 The ENCODE Project Seeks to Identify All Functional Elements in the Human Genome The ENCODE consortium predicts that 20 to 80% of the human genome is functional. Critics argue that this view of function is too broad and ignores evolutionary constraints on genome function.

CONCEPT 18.5 Genome Comparisons Provide Information About Genomic Structure and Function

18.5.1 Comparative Genomics Reveals Conserved Regions in Genomes More than half of the genes of *Drosophila* have human counterparts. Comparative genomics can be exploited to more rapidly sequence novel genomes and often relies on synteny between related genomes. *Synteny* is the conserved arrangements of segments of DNA in related genomes.

18.5.2 Functional Genomics Reveals Gene Function at the Genome Level Functional genomics uses high-throughput approaches and bioinformatics to analyze gene function and gene products. DNA microarrays and RNA-seq provide ways to study the transcriptome.

18.5.3 Proteomics Catalogs Proteins Encoded by the Genome Proteomics characterizes all of the proteins produced by a cell. This is complicated by posttranslational modifications and variant proteins produced by the alternative splicing of genes.

CONCEPT 18.6 Comparative Genomics Informs Evolutionary Biology

18.6.1 All Mammals Have Comparably Sized Genomes In general, genome size increases with organismal complexity. Mammalian genomes appear to have roughly the same number of genes.

18.6.2 Many Genes Are Novel In every genome sequenced, there are a large number of previously unknown protein-encoding genes.

18.6.3 Large Differences in Genome Sizes Can Arise Through Genome Duplication Both whole chromosomes and entire genomes have been duplicated. The range of genome sizes varies much more than the number of genes.

18.6.4 Key Genes Tend to Be Conserved Plants, animals, and fungi have approximately 70% of their genes in common.

18.6.5 Rates of Evolution Vary Greatly Rodents have evolved twice as fast as humans, and insects more rapidly than either. Short generation time may partly explain this.

18.6.6 What Makes Us Human? Most of the difference may be in the timing and tissue location of gene expression. Small evolutionary changes in the FOXP2 protein and its expression may have led to human speech.

Concept Overview

This Concept Overview diagrams the key concepts that were discussed in this chapter

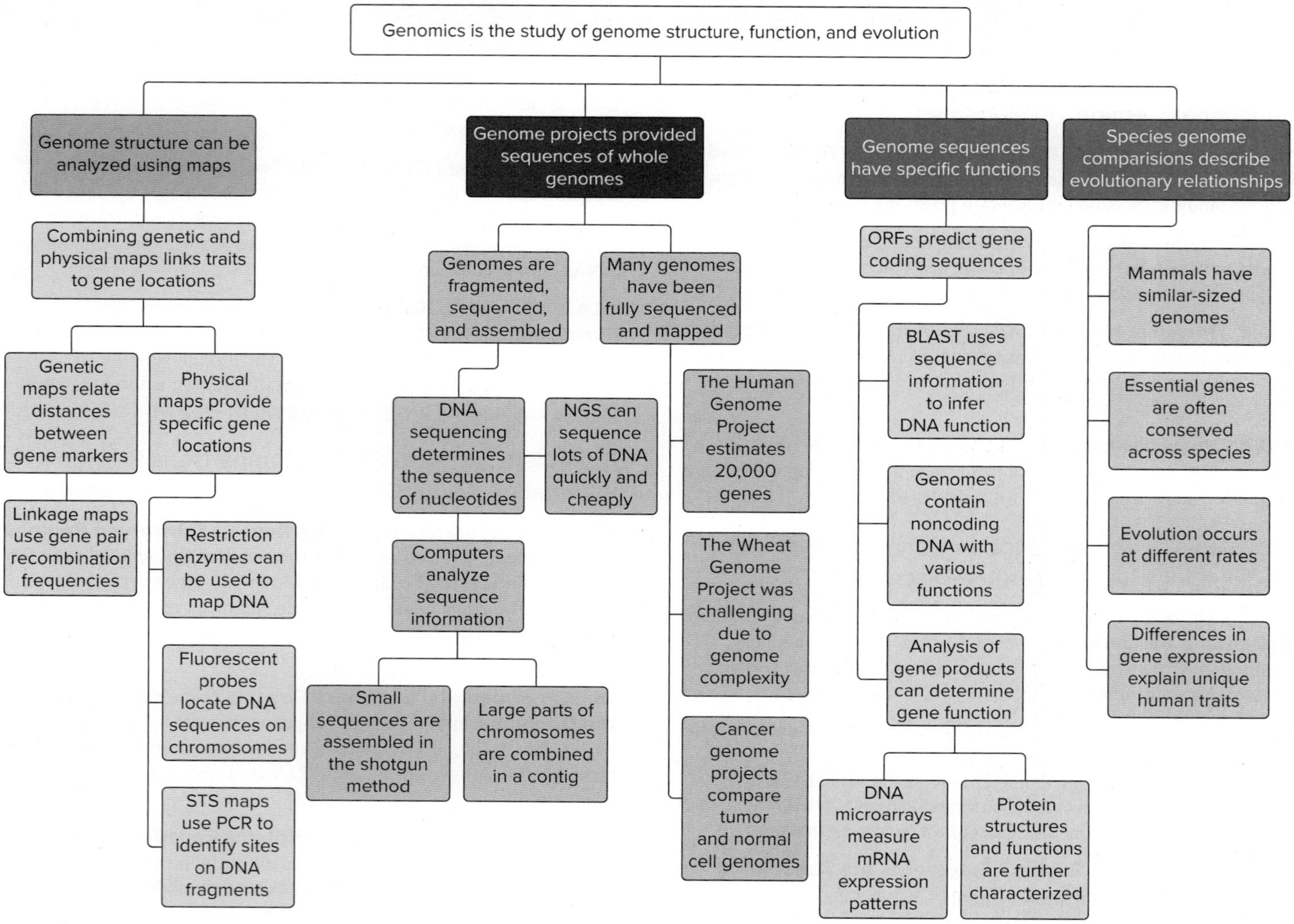
Genomics is the study of genome structure, function, and evolution
Genome structure can be analyzed using maps
Combining genetic and physical maps links traits to gene locations
Genetic maps relate distances between gene markers
Physical maps provide specific gene locations
Linkage maps use gene pair recombination frequencies
Restriction enzymes can be used to map DNA
Fluorescent probes locate DNA sequences on chromosomes
STS maps use PCR to identify sites on DNA fragments
Genome projects provided sequences of whole genomes
Genomes are fragmented, sequenced, and assembled
Many genomes have been fully sequenced and mapped
DNA sequencing determines the sequence of nucleotides
NGS can sequence lots of DNA quickly and cheaply
Computers analyze sequence information
Small sequences are assembled in the shotgun method
Large parts of chromosomes are combined in a contig
The Human Genome Project estimates 20,000 genes
The Wheat Genome Project was challenging due to genome complexity
Cancer genome projects compare tumor and normal cell genomes
Genome sequences have specific functions
ORFs predict gene coding sequences
BLAST uses sequence information to infer DNA function
Genomes contain noncoding DNA with various functions
Analysis of gene products can determine gene function
DNA microarrays measure mRNA expression patterns
Protein structures and functions are further characterized
Species genome comparisions describe evolutionary relationships
Mammals have similar-sized genomes
Essential genes are often conserved across species
Evolution occurs at different rates
Differences in gene expression explain unique human traits

Assessing the Learning Path

Understand

1. Two genetic markers are separated by 2.4 cM on a genetic map. What does this mean?
   a. If the chromosome is stretched out, the two markers are physically 2.4 cm apart along the chromosome.
   b. There is a 2.4% chance they will be separated by recombination during meiosis.
   c. They will be expressed together in a cell 2.4% of the time due to their proximity.
   d. None of the above
2. What is the main difference between dideoxy terminator sequencing and next-generation sequencing (NGS)?
   a. NGS cannot sequence GC-rich genomes, but dideoxy terminator sequencing can.
   b. Dideoxy terminator sequencing is faster and cheaper than NGS.
   c. In NGS but not dideoxy terminator sequencing, nucleotides are fluorescently labeled.
   d. In NGS, many sequencing reactions are carried out simultaneously.
3. Which kind of genome assembly approach would be best for a complex, highly repetitive genome?
   a. Shotgun assembly, because it uses a genetic map to align sequences
   b. Clone–contig assembly, because a physical map can be used to locate markers in the DNA
   c. Shotgun assembly, because it is cheaper for complex genomes
   d. It wouldn't really matter because complex, repetitive genomes cannot be sequenced.

4. Why is genome assembly a challenge in large, polyploid genomes where there has been extensive genomic duplication?
   a. The repetitive and duplicated nature of the DNA makes it hard to know to which chromosome a given sequence of DNA belongs.
   b. Repetitive DNA cannot be sequenced.
   c. It is hard to isolate the DNA from the cells of these kinds of species.
   d. In polyploid species, the chromosomes are often tangled and hard to separate for sequencing.
5. What is a BLAST search?
   a. A mechanism for aligning consensus regions during whole-genome sequencing
   b. A search for similar gene sequences from other species
   c. A method of screening a DNA library
   d. A method of identifying ORFs
6. What is a proteome?
   a. The collection of all genes encoding proteins
   b. The collection of all proteins encoded by the genome
   c. The collection of all proteins present in a cell
   d. The amino acid sequence of a protein
7. A transcriptome is
   a. all of the RNA present in a cell or tissue.
   b. all of the nontranscribed sequences in the genome.
   c. less complex and smaller than the genome.
   d. all of the protein-coding genes present in a cell or tissue.

Apply

1. You are in the early stages of a genome-sequencing project and have mapped sequenced fragments of DNA using STSs. Use the following STSs to make a contig.
   Clone 1: –M–E–N–L–   Clone 3: –Z–L–N–
   Clone 2: –R–Q–M–   Clone 4: –Q–R–P
   a. M–E–N–L–R–Q–P–Z
   b. P–R–Q–M–E–N–L–Z
   c. E–L–M–N–P–Q–R–Z
   d. Z–E–L–M–R–Q–P–N
2. If it was possible to sequence the genome of cells as they progressed from being normal to being completely tumorigenic, which kind of genetic changes do you think would appear first?
   a. Passenger mutations
   b. Driver mutations
   c. Mutations in noncoding regions
   d. Recessive mutations in only one allele of a gene
3. ENCODE defines function in a biochemical way. Which of the following genomics approaches could increase the accuracy of its estimate of genome functionality in humans?
   a. Comparing sequences with other organisms to see if they are functionally conserved
   b. Performing RNA-seq to determine if genes are transcribed or not
   c. Performing shotgun assembly versus clone–contig assembly
   d. Analyzing differences in gene expression under different conditions using a DNA microarray
4. You are trying to map a set of six markers associated with a particular disease. You have determined recombination frequencies for four of the markers, but despite your efforts you cannot establish a recombination frequency for two of them. Which of the following could explain your difficulty?
   a. The two markers are so close together in genetic space that they cannot be separated by recombination.
   b. The two markers are on the same chromosome.
   c. The two markers are so far apart in genetic space that they cannot be separated by recombination.
   d. Both a and b
5. You cut a linear piece of DNA with the restriction enzyme *Eco*RI. You obtain 3 fragments sized at 5 kb, 6 kb, and 7 kb. You cut the same piece of DNA with the restriction enzyme *Hin*dIII and obtain two fragments of 2 kb and 16 kb. You cut the same piece of DNA with both enzymes and obtain fragments of 2 kb, 3 kb, 6 kb, and 7 kb. Which of the *Eco*RI fragments was cut with *Hin*dIII?
   a. The 5-kb fragment
   b. The 6-kb fragment
   c. The 7-kb fragment
   d. It is impossible to determine.
6. You take another piece of linear DNA and cut it with the restriction enzyme *Eco*RV. You get fragments sized at 5 kb, 6 kb, and 7 kb. You cut the same piece of DNA with the restriction enzyme *Pvu*II and obtain two fragments of 2 kb and 16 kb. You cut the same piece of DNA with both enzymes and get fragments sized 2 kb, 5 kb, and 6 kb. How big is the piece of DNA you digested?
   a. 15 kb
   b. 12 kb
   c. 18 kb
   d. Cannot be determined from the data
7. Consider again the data described in question 6. In which *Eco*RV fragment does *Pvu*II cut?
   a. The 5-kb fragment
   b. The 6-kb fragment
   c. The 7-kb fragment
   d. It is impossible to determine.
8. Refresh your memory about FISH shown in figure 18.2 (also refer to chapter 17). Imagine that a piece of chromosomal DNA had been duplicated rather than translocated from one chromosome to another. That cell divides with the duplication and enters the next cell cycle. How many dots would a FISH analysis of cells show with a single duplication event if the cell was in G2? Assume the cell is diploid.
   a. One dot   c. Three dots
   b. Two dots   d. Four dots

Synthesize

1. What problems do you think exist for the assembly of genomes that contain a lot of repetitive sequences and gene duplications?
2. The Human Genome Project discovered many new genes. Mutations in these genes have not been observed. Does this mean their functions are not important for survival? What factors might prevent such mutations being observed in human populations?
3. The *FOXP2* gene is associated with speech in humans. It is also found in chimpanzees, gorillas, orangutans, Rhesus macaques, and even mice, yet none of these mammals speak. Develop a hypothesis that explains why *FOXP2* supports speech in humans but not in other mammals.

Tetra Images/Getty Images

19 Genes Within Populations

Learning Path

19.1 Natural Populations Exhibit Genetic Variation

19.2 Frequencies of Alleles Can Change

19.3 Five Agents Are Responsible for Evolutionary Change

19.4 Selection Can Act on Traits Affected by Many Genes

19.5 Natural Selection Can Be Studied Experimentally

19.6 Fitness Is a Measure of Evolutionary Success

19.7 Evolutionary Processes Sometimes Maintain Variation

19.8 Sexual Selection Determines Reproductive Success

Concept Overview

This Concept Overview provides a diagram of the overarching concepts that are covered in this chapter. These concepts will be expanded upon in the Concept Overview at the end of the chapter.

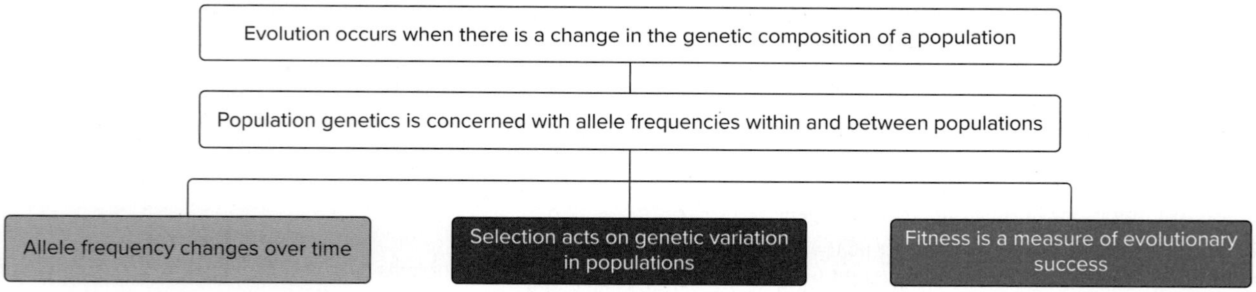

Evolution occurs when there is a change in the genetic composition of a population
Population genetics is concerned with allele frequencies within and between populations
Allele frequency changes over time
Selection acts on genetic variation in populations
Fitness is a measure of evolutionary success

Introduction

Individuals in most populations vary greatly in phenotype, as the wild horses shown above illustrate; these differences are often the result of genetic differences among individuals. Often, the particular characteristics of an individual have an important bearing on its survival, on its chances to reproduce, and on the success of its offspring. Evolution is driven by such factors, as different alleles rise and fall in populations. These deceptively simple matters lie at the core of evolutionary biology, which is the topic of this chapter and chapters 20 and 21.

19.1 Natural Populations Exhibit Genetic Variation

With this chapter we begin our treatment of evolution. The word *evolution* is widely used in the natural and social sciences. It refers to how an entity—be it a social system, a gas, or a planet—changes through time. Although development of the modern concept of evolution in biology can be traced to Darwin's landmark work *On the Origin of Species,* the word *evolution* never appears in the first five editions of his book. Rather, Darwin used the phrase "descent with modification." Darwin's words capture well the essence of biological evolution: through time, species accumulate differences; as a result, descendants differ from their ancestors. In this way, new species arise from existing ones.

Genetic Variation Is the Raw Material of Evolution

LEARNING OBJECTIVE 19.1.1 Differentiate between evolution by natural selection and the inheritance of acquired characteristics.

Darwin was not the first to propose a theory of evolution. He followed a long line of earlier naturalists who deduced that the many kinds of organisms around us must have been produced by a process of evolution. Unlike his predecessors, however, Darwin proposed natural selection as the mechanism of evolution. Natural selection produces evolutionary change when some individuals in a population possess certain inherited characteristics and then produce more surviving offspring than individuals lacking these characteristics. As a result, the population gradually comes to include more and more individuals with the advantageous characteristics. In this way, the population evolves and becomes better adapted to its local circumstances.

A rival theory, championed by the biologist Jean-Baptiste Lamarck, has not stood the test of time. Lamarck proposed that evolution occurred by the **inheritance of acquired characteristics.** That is, changes acquired during an individual's life are passed on to its offspring. For example, he proposed that ancestral giraffes tended to stretch their necks to feed on tree leaves, and this extension of the neck was passed on to subsequent generations, leading to the long-necked giraffe (figure 19.1*a*). In Darwin's theory, by contrast, variation is not created by experience but reflects preexisting genetic differences that affect neck length (figure 19.1*b*). Giraffes with long necks can exploit resources not available to those with shorter necks, so they will leave more offspring, and thus make a greater genetic contribution to the population.

Darwin understood that a key assumption of his theory of evolution by natural selection was that a population contains genetic alternatives, so that selection could favor one over another. Darwin himself considered this requirement a key internal contradiction of his theory, as the theory seemed, on the one hand, to predict that natural selection would quickly act to cleanse a population of all but the most favorable variant alleles—and yet, on the other hand, the same theory requires that such variants persist in a population to provide the raw material for future natural selection. This highlights the importance of variation but leaves open the question of how much variation exists in natural populations.

Figure 19.1 Two ideas of how giraffes might have evolved long necks.

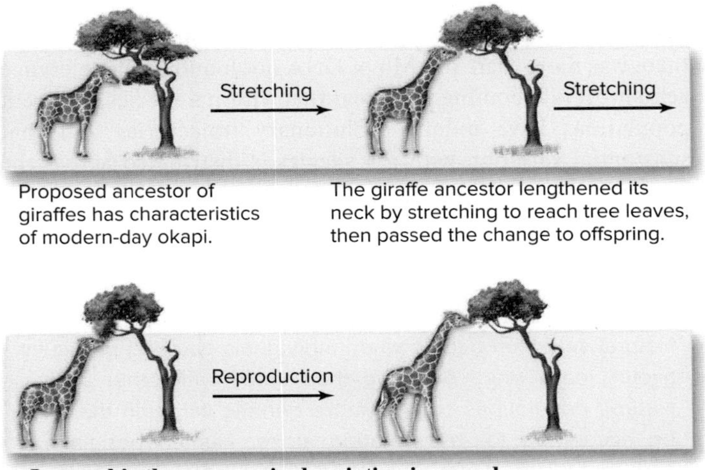
Stretching
Stretching
Proposed ancestor of giraffes has characteristics of modern-day okapi.
The giraffe ancestor lengthened its neck by stretching to reach tree leaves, then passed the change to offspring.
Reproduction

***a.* Lamarck's theory: acquired variation is passed on to descendants.**

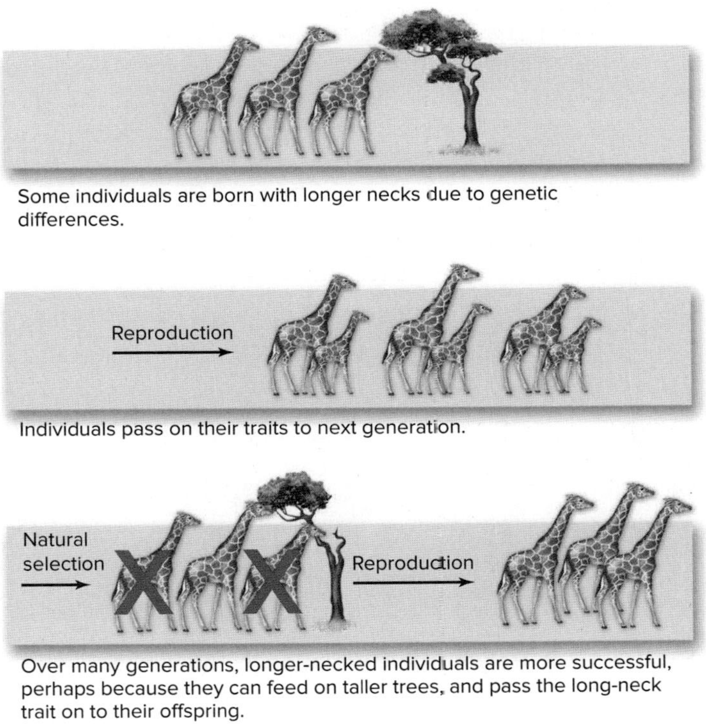
Some individuals are born with longer necks due to genetic differences.
Reproduction
Individuals pass on their traits to next generation.
Natural selection
Reproduction
Over many generations, longer-necked individuals are more successful, perhaps because they can feed on taller trees, and pass the long-neck trait on to their offspring.

***b.* Darwin's theory: natural selection, or genetically based variation, leads to evolutionary change.**

Natural Populations Contain Ample Genetic Variation

LEARNING OBJECTIVE 19.1.2 Describe the extent of genetic variation in populations.

Evolution can result from any process that causes a change in the genetic composition of a population. We cannot talk about evolution, therefore, without also considering **population genetics,** the study of the properties of genes in populations. It is best to start by looking at the genetic variation present among individuals within a species. This is the raw material available for evolutionary change.

Many loci in a particular population have more than one allele at frequencies significantly greater than would occur due to mutation alone. The question has always been how we can analyze and measure this variation. The simplest case is morphological variation, which is visible (figure 19.2). Classically, we define any allele that exists at a frequency of 1% or higher in a population as a **polymorphism.**

Enzyme polymorphism

Initial attempts to measure the amount of genetic variation concentrated on loci that encode enzymes, which could be separated electrophoretically, then visualized by staining for enzyme activity.

Figure 19.2 Polymorphic variation. This natural population of lupines *(Lupinus perennis)* exhibits considerable variation in flower color. Individual differences are inherited and passed on to offspring.

Rosita So Image/Getty Images

We now know that most populations of insects and plants are polymorphic at more than half of their enzyme-encoding loci. Vertebrates are somewhat less polymorphic. Heterozygosity, the probability that a randomly selected gene will be heterozygous in a randomly selected individual, is about 15% in *Drosophila* and other invertebrates, between 5 and 8% in vertebrates, and around 8% in outcrossing plants. Values of heterozygosity tend to be lower than the frequency of polymorphic loci because many common alleles in a population are homozygous.

Variation in humans

The population that has been analyzed most extensively for genetic variation is, in fact, our own. The availability of a high-quality reference genome sequence for comparison, and the interest in medically relevant genetic variation, has fueled large-scale studies of human genetic variation. This has been accelerated by the use of next-generation sequencing technologies that reveal genetic variation on a scale that is truly staggering.

By the time you read this, more than 100,000 human genomes will have been partially or entirely sequenced (refer to chapter 18). This has revealed an enormous amount of genetic variation, where any specific individual will have around 3% of their genome varying from the reference genome.

As well as extensive variation within human populations, these studies are also documenting substantial variation between populations. This kind of work is in progress looking at genomes from specifically Asian and specifically African populations, and in one case a Dutch population. The early results from these forays into human genetic variation are significant and surprising. Not only do these populations have additional genetic variation not previously seen, but they have entire genomic regions that are not represented in the reference genome, which comes from individuals in North America. We have long known that African populations have greater genetic diversity than European and Asian populations. This is a result of human history: our species originated in Africa, and the populations there represent the "oldest" human genomes. A recent study that involved sequencing over 900 African genomes, from different regions of Africa, showed that taking all of these genomes into account, there was more than 300 Mb of DNA not found in the reference genome. It is becoming very clear that within a species, different populations have unique evolutionary trajectories and that substantial variation within a species is distributed across the species' range.

REVIEW OF CONCEPT 19.1

Natural selection occurs when individuals carrying particular alleles leave more offspring than those with other alleles. Natural populations contain more genetic variation than can be accounted for by mutation alone. Human populations have extensive genetic variation with a pattern that fits an expanding population.

- *Why is genetic variation in a population necessary for evolution to occur?*

19.2 Frequencies of Alleles Can Change

Genetic variation within natural populations was a puzzle to Darwin and his contemporaries in the mid-1800s. The idea that genetic segregation during meiosis could produce variation in progeny was not understood. Although Mendel's work was published during this period, his results were ignored until their rediscovery at the turn of the 20th century. Darwin reasoned that selection should always favor an optimal form and thus reduce variation. Moreover, the theory of blending inheritance was widely accepted. Blending inheritance predicts that new genetic variation will be diluted in subsequent generations.

The Hardy–Weinberg Principle Characterizes Populations at Equilibrium

LEARNING OBJECTIVE 19.2.1 Describe the characteristics of a population in Hardy–Weinberg equilibrium.

Following the rediscovery of Mendel's research, biologists studying heredity were initially confused about why, after many generations, a population didn't come to be composed solely of individuals with the dominant phenotype. In 1908, two people solved the puzzle of why genetic variation persists: mathematician Godfrey H. Hardy and physician Wilhelm Weinberg. The conclusion they came to independently was that the original proportions of the genotypes in a population will remain constant from generation to generation, as long as the following assumptions are met:

1. No mutation takes place.
2. No genes are transferred to or from other sources (no immigration or emigration takes place).
3. Random mating is occurring.
4. The population size is very large.
5. No selection occurs.

If these assumptions are met, the genotypes' proportions will not change and the population is said to be in **Hardy–Weinberg equilibrium.**

The Hardy–Weinberg equation with two alleles

In algebraic terms, the Hardy-Weinberg principle is written as an equation. Consider a population of 100 cats in which 84 are black and 16 are white (figure 19.3). The frequencies of the two phenotypes would be 0.84 (or 84%) black and 0.16 (or 16%) white. Based on these phenotypic frequencies, can we deduce the underlying frequency of genotypes?

If we assume that the white cats are homozygous recessive for an allele we designate as *b*, and the black cats are either homozygous dominant *BB* or heterozygous *Bb*, we can calculate the **allele frequencies** of the two alleles in the population from the proportion of black and white individuals, assuming that the population is in Hardy-Weinberg equilibrium.

Let the letter p designate the frequency of the *B* allele and the letter q the frequency of the alternative allele. Because there are only two alleles, p plus q must always equal 1 (that is, the total population). In addition, we know that the sum of the three genotype frequencies must also equal 1. If the frequency of the *B* allele is p, then the probability that an individual will have two *B* alleles is simply the probability that each of its alleles is a *B*. The probability of two events happening independently is the product of the probability of each event; in this case, the probability that the individual received a *B* allele from its father is p, and the probability the individual received a *B* allele from its mother is also p, so the probability that both happened is $p \times p = p^2$ (figure 19.3). By the same reasoning, the probability that an individual will have two *b* alleles is q^2.

What about the probability that an individual will be a heterozygote? There are two ways this could happen: the individual could receive a *B* from its father and a *b* from its mother, or vice versa. The probability of the first case is $p \times q$ and the probability of the second case is $q \times p$. Because the result in either case is that the individual is a heterozygote, the probability of that outcome is the sum of the two probabilities, or $2pq$.

Thus, if a population is in Hardy-Weinberg equilibrium with allele frequencies of p and q, then the probability that an individual will have each of the three possible genotypes is $p^2 + 2pq + q^2$. You may recognize this as the *binomial expansion:*

$$(p + q)^2 = p^2 + 2pq + q^2$$

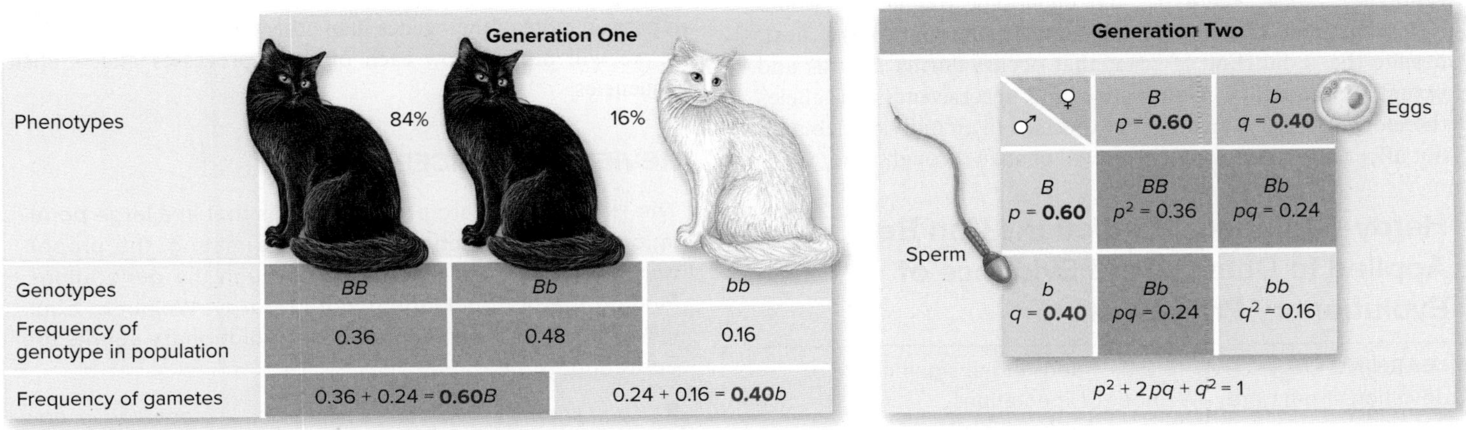

Figure 19.3 The Hardy–Weinberg equilibrium. In the absence of factors that alter them, the frequencies of gametes, genotypes, and phenotypes remain constant generation after generation.

Finally, we may use these probabilities to predict the distribution of genotypes in the population, again assuming that the population is in Hardy–Weinberg equilibrium. If the probability that any individual is a heterozygote is $2pq$, then we would expect the proportion of heterozygous individuals in the population to be $2pq$; similarly, the frequency of *BB* and *bb* homozygotes would be expected to be p^2 and q^2.

Let us return to our example. Remember that 16% of the cats are white. If white is a recessive trait, then this means that such individuals must have the genotype *bb*. If the frequency of this genotype is $q^2 = 0.16$ (the frequency of white cats), then q (the frequency of the *b* allele) $= 0.4$. Because $p + q = 1$, therefore, p, the frequency of allele *B*, would be $1.0 - 0.4 = 0.6$ (remember, the frequencies must add up to 1). We can now easily calculate the expected **genotype frequencies:** Homozygous dominant *BB* cats would make up the p^2 group, and the value of $p^2 = (0.6)^2 = 0.36$, or 36 homozygous dominant *BB* individuals in a population of 100 cats. The heterozygous cats have the *Bb* genotype and would have the frequency corresponding to $2pq$, or $(2 \times 0.6 \times 0.4) = 0.48$, or 48 heterozygous *Bb* individuals.

Using the Hardy–Weinberg equation to predict frequencies in subsequent generations

Examine figure 19.3 again. We will trace genetic reassortment during sexual reproduction and uncover how it affects the frequencies of the *B* and *b* alleles during the next generation.

In constructing this Punnett-square-like diagram, we have assumed that the union of sperm and egg in these cats is random, so that all combinations of *b* and *B* alleles occur. The alleles are therefore mixed randomly and are represented in the next generation in proportion to their original occurrence. Each individual egg or sperm in each generation has a 0.6 chance of receiving a *B* allele ($p = 0.6$) and a 0.4 chance of receiving a *b* allele ($q = 0.4$).

In the next generation, therefore, the chance of combining two *B* alleles is p^2, or 0.36 (that is, 0.6×0.6), and approximately 36% of the individuals in the population will continue to have the *BB* genotype. The frequency of *bb* individuals is q^2 (0.4×0.4) and so will continue to be about 16%, and the frequency of *Bb* individuals will be $2pq$ $(2 \times 0.6 \times 0.4)$, or on average, 48%.

Phenotypically, if the population size remained at 100 cats, we would still have approximately 84 black individuals (with either *BB* or *Bb* genotypes) and 16 white individuals (with the *bb* genotype). Allele, genotype, and phenotype frequencies would have remained unchanged from one generation to the next, despite the reshuffling of genes that occurs during meiosis and sexual reproduction. Dominance and recessiveness of alleles therefore affect how an allele is expressed in an individual, but do not affect how allele frequencies will change through time.

Hardy–Weinberg Predictions Can Be Applied to Data to Find Evidence of Evolutionary Processes

LEARNING OBJECTIVE 19.2.2 Interpret the significance of deviations from Hardy–Weinberg expectations.

The lesson to take home from the example of black and white cats in figure 19.3 is that if all five of the assumptions listed earlier hold true, the allele and genotype frequencies will not change from one generation to the next. But in reality, most populations in nature will not fit all five assumptions. The primary utility of this method is to determine whether some evolutionary process or processes are operating in a population and, if so, to suggest hypotheses about what they may be.

Suppose, for example, that the observed frequencies of the *BB, bb,* and *Bb* genotypes in a different population of cats are 0.6, 0.2, and 0.2, respectively. We can calculate the allele frequencies for *B* as follows: 60% (0.6) of the cats have two *B* alleles, 20% have one, and 20% have none. This means that the average number of *B* alleles per cat is 1.4 $[(0.6 \times 2) + (0.2 \times 1) + (0.2 \times 0) = 1.4]$. Because each cat has two alleles for this gene, the frequency is $1.4/2.0 = 0.7$. Similarly, you should be able to calculate that the frequency of the *b* allele $= 0.3$.

If the population were in Hardy–Weinberg equilibrium, then, according to the equation earlier in this section, the frequency of the *BB* genotype would be $0.7^2 = 0.49$, lower than it really is. Similarly, you can calculate that there are fewer heterozygotes and more *bb* homozygotes than expected; clearly, the population is not in Hardy–Weinberg equilibrium.

What could cause such an excess of homozygotes and deficit of heterozygotes? A number of possibilities exist, including (1) natural selection favoring homozygotes over heterozygotes, (2) individuals choosing to mate with genetically similar individuals (because *BB* × *BB* and *bb* × *bb* matings always produce homozygous offspring, but only half of *Bb* × *Bb* produce heterozygous offspring, such mating patterns would lead to an excess of homozygotes), or (3) an influx of homozygous individuals from outside populations (or conversely, emigration of heterozygotes to other populations). By detecting a lack of Hardy–Weinberg equilibrium, we can generate potential hypotheses, which we can then investigate directly.

Changes in allele frequencies between generations also indicate that one of the assumptions is not met. Suppose, for example, that the frequency of *b* is 0.53 in one generation and 0.61 in the next. Again, there are a number of possible explanations: for example, (1) selection favoring individuals with *b* over those with *B*, (2) immigration of *b* into the population or emigration of *B* out of the population, or (3) high rates of mutation that more commonly occur from *B* to *b* than vice versa. Another possibility is that the population is small, and that the change represents the random fluctuations that result because, simply by chance, some individuals pass on more of their genes than others. In the rest of this chapter, we will discuss how each of these processes affects allele frequencies.

REVIEW OF CONCEPT 19.2

The Hardy–Weinberg principle states that in a large population with no selection and random mating, the proportion of alleles does not change through the generations. Finding that a population is not in Hardy–Weinberg equilibrium indicates that one or more evolutionary agents are operating.

- *If you know the genotype frequencies in a population, how can you determine whether the population is in Hardy–Weinberg equilibrium?*

19.3 Five Agents Are Responsible for Evolutionary Change

The five assumptions of the Hardy-Weinberg principle identify the five agents that can lead to evolutionary change in populations. They are mutation, gene flow, nonrandom mating, genetic drift in small populations, and natural selection. Any one of these five agents may bring about changes in allele or genotype proportions, causing the population to deviate from Hardy-Weinberg equilibrium.

Mutation Can Change Allele Frequencies

LEARNING OBJECTIVE 19.3.1 Describe how mutation can cause a population to deviate from Hardy–Weinberg equilibrium.

Mutation from one allele to another can obviously change the proportions of particular alleles in a population. Mutation rates are generally so low that they have little effect on the Hardy-Weinberg proportions of common alleles. A typical gene mutates about once per 100,000 cell divisions. Because this rate is so low, other evolutionary processes are usually more important in determining how allele frequencies change.

Nonetheless, mutation is the ultimate source of genetic variation and thus makes evolution possible (figure 19.4*a*). It is important to remember, however, that the likelihood of a particular mutation occurring is not affected by natural selection; that is, mutations do not occur more frequently in situations in which they would be favored by natural selection (a concept explored in the Inquiry & Analysis feature of chapter 14).

Gene Flow Occurs When Alleles Move Between Populations

LEARNING OBJECTIVE 19.3.2 Illustrate how migration can cause deviations from Hardy–Weinberg equilibrium.

Gene flow is the movement of alleles from one population to another. It can be a powerful agent of change. Sometimes gene flow is obvious, as when an animal physically migrates from one place to another. If the characteristics of the newly arrived individual differ from those of the animals already there, and if the newcomer is adapted well enough to the new area to survive and mate successfully, the genetic composition of the receiving population may be altered.

Other important kinds of gene flow are not as obvious. These subtler movements include the drifting of gametes or the immature stages of plants or marine animals from one place to another (figure 19.4*b*). Pollen, the male gamete of flowering plants, is often carried great distances by insects and other animals that visit flowers. Seeds may also blow in the wind or be carried by animals to new populations far from their place of origin.

Consider two populations initially different in allele frequencies: in population 1, $p = 0.2$ and $q = 0.8$; in population 2, $p = 0.8$ and $q = 0.2$. Gene flow will tend to bring the rarer allele into each population. Thus, allele frequencies will change from generation to generation, and the populations will not be in Hardy-Weinberg equilibrium. Only when allele frequencies reach 0.5 for both alleles in both populations will equilibrium be attained (assuming that rates of gene flow are the same in both directions). This example illustrates the important concept that gene flow tends to homogenize allele frequencies among populations.

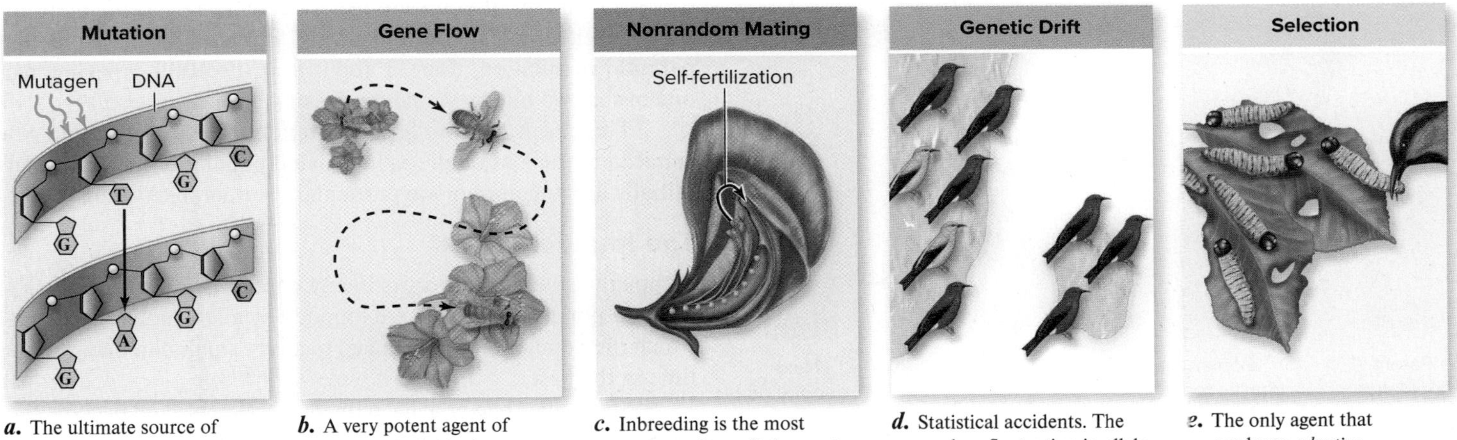

a. The ultimate source of variation. Individual mutations occur so rarely that mutation alone usually does not change allele frequency much.

b. A very potent agent of change. Individuals or gametes move from one population to another.

c. Inbreeding is the most common form. It does not alter allele frequency but reduces the proportion of heterozygotes.

d. Statistical accidents. The random fluctuation in allele frequencies increases as population size decreases.

e. The only agent that produces *adaptive* evolutionary changes.

Figure 19.4 Five agents of evolutionary change. ***a.*** Mutation. ***b.*** Gene flow. ***c.*** Nonrandom mating. ***d.*** Genetic drift. ***e.*** Selection.

Nonrandom Mating Shifts Genotype Frequencies

LEARNING OBJECTIVE 19.3.3 Discuss how nonrandom mating can lead to deviations from Hardy–Weinberg equilibrium.

Individuals with certain genotypes sometimes mate with one another more commonly than would be expected on a random basis, a phenomenon known as *nonrandom mating* (figure 19.4*c*). **Assortative mating,** in which phenotypically similar individuals mate, is a type of nonrandom mating that causes the frequencies of particular genotypes to differ greatly from those predicted by the Hardy–Weinberg principle.

Assortative mating does not change the frequency of the individual alleles but, rather, increases the proportion of homozygous individuals. For example, populations of self-fertilizing plants consist primarily of homozygous individuals. Inbreeding (mating between relatives) is the most common form of assortative mating in animals, although any tendency for phenotypically similar animals to breed more often with each other will increase the proportion of homozygotes, because phenotypically similar individuals are likely to be genetically similar. By contrast, **disassortative mating,** in which phenotypically different individuals mate more often, produces an excess of heterozygotes.

Genetic Drift May Alter Allele Frequencies in Small Populations

LEARNING OBJECTIVE 19.3.4 Demonstrate how genetic drift can have a larger effect on small populations.

In small populations, frequencies of particular alleles may change drastically by chance alone. Such changes in allele frequencies occur randomly, as if the frequencies were drifting from their values. These changes are thus known as **genetic drift** (figure 19.4*d*). For this reason, a population must be large to be in Hardy–Weinberg equilibrium.

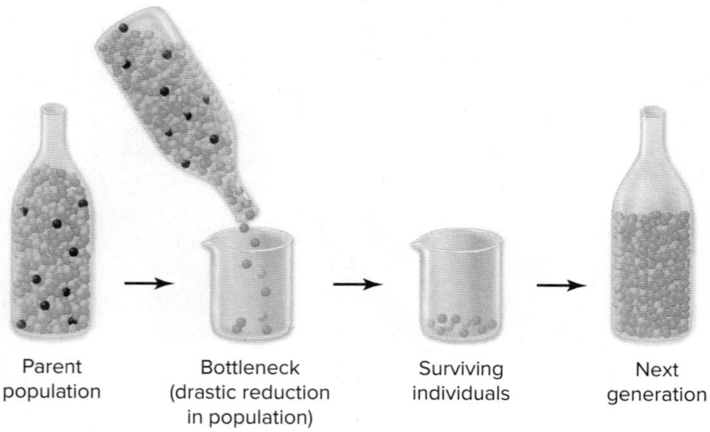

Parent population
Bottleneck (drastic reduction in population)
Surviving individuals
Next generation

Figure 19.5 Genetic drift: A bottleneck effect. The parent population contains roughly equal numbers of green and yellow individuals and a small number of red individuals. By chance, the few remaining individuals that contribute to the next generation are mostly green, and the red allele has disappeared entirely. The bottleneck occurs because so few individuals form the next generation, as might happen after an epidemic or a catastrophic storm.

If the gametes of only a few individuals form the next generation, the alleles they carry may by chance not be representative of the parent population from which they were drawn, as illustrated in figure 19.5. In this example, a small number of individuals are removed from a bottle containing many. By chance, most of the individuals removed are green, so the new population has a much higher population of green individuals than the parent generation had.

A pair of small populations that are isolated from each other may come to differ strongly as a result of genetic drift, even if the forces of natural selection are the same for both. Because of genetic drift, sometimes harmful alleles may increase in frequency in small populations, despite selective disadvantage, and favorable alleles may be lost even though they are selectively advantageous. It is interesting to realize that humans have lived in small groups for much of the course of their evolution; consequently, genetic drift may have been a particularly important factor in the evolution of our species.

Larger populations also experience the effect of genetic drift, but to a lesser extent than smaller populations—the magnitude of genetic drift is inversely related to population size. However, large populations may have been much smaller in the past, and genetic drift may have greatly altered allele frequencies at that time. Imagine a population containing only two alleles of a gene, *B* and *b*, in equal frequency (that is, $p = q = 0.50$). In a large Hardy-Weinberg population, the genotype frequencies are expected to be 0.25 *BB*, 0.50 *Bb*, and 0.25 *bb*. If only a small sample of individuals produces the next generation, large deviations in these genotype frequencies can occur simply by chance.

Suppose, for example, that from a population in which *B* and *b* are equally represented (that is, $p = q = 0.5$), four individuals formed the next generation, and that by chance they were two *Bb* heterozygotes and two *BB* homozygotes—that is, the allele frequencies in the next generation would be $p = 0.75$ and $q = 0.25$. In fact, if you were to replicate this experiment 1000 times, each time randomly drawing four individuals from the parental population, then in about 8 of the 1000 experiments, one of the two alleles would be missing entirely.

This result leads to an important conclusion: genetic drift can lead to the loss of alleles in isolated populations. Alleles that initially are uncommon are particularly vulnerable (figure 19.5).

The founder effect

Although genetic drift occurs in any population, it is particularly likely in populations that were founded by a few individuals or in which the population was reduced to a very small number at some time in the past.

Sometimes one or a few individuals disperse and become the founders of a new, isolated population at some distance from their place of origin. These pioneers are not likely to carry all the alleles present in the source population. Thus, some alleles may be lost from the new population, and others may change drastically in frequency. In some cases, previously rare alleles in the source population may be a significant fraction of the new population's genetic endowment. This phenomenon, in which the allele frequencies of the founding individuals is different from the allele frequencies in the population from which they came, is called the **founder effect.**

Founder effects are not rare in nature. Many self-pollinating plants start new populations from a single seed. Founder effects have been particularly important in the evolution of organisms on distant oceanic islands, such as the Hawaiian and Galápagos Islands. Most of the organisms in such areas probably derive from one or a few initial founders.

In a similar way, isolated human populations begun by relatively few individuals are often dominated by genetic features characteristic of their founders. Amish populations in the United States, for example, were established from relatively few individuals and have unusually high frequencies of a number of conditions, such as polydactylism (the presence of a sixth finger).

The bottleneck effect

Even if organisms do not move from place to place, occasionally their populations may be drastically reduced in size. This may result from flooding, drought, epidemic disease, and other natural forces or from changes in the environment. Just as we have discussed with founder effects, the few surviving individuals may by chance have allele frequencies different from the entire population, and some alleles may be lost entirely (natural selection may also be responsible for change in allele frequencies, but here we are referring to random changes resulting from small population size). The resulting alterations and loss of genetic variability have been termed the **bottleneck effect.**

The genetic variation of some living species appears to be severely depleted, probably as the result of a bottleneck effect in the past. For example, the northern elephant seal, which breeds on the western coast of North America and nearby islands, was nearly hunted to extinction in the 19th century and was reduced to a single population containing perhaps no more than 20 individuals on the island of Guadalupe off the coast of Baja, California. As a result of this bottleneck, the species has lost almost all of its genetic variation, even though the seal populations have rebounded and now number in the tens of thousands and breed in locations as far north as Vancouver Island.

Selection Favors Some Phenotypes over Others

LEARNING OBJECTIVE 19.3.5 Explain what conditions are required for natural selection to occur.

As Darwin pointed out, some individuals leave behind more progeny than others, and the rate at which they do so is affected by their phenotype. We describe the results of this process as **selection** when individuals with one phenotype leave, on average, more surviving offspring in the next generation than individuals with an alternative phenotype (figure 19.4*e*). In *artificial selection,* a breeder selects for the desired characteristics. In *natural selection,* environmental conditions determine which individuals in a population produce the most offspring.

For natural selection to occur and to result in evolutionary change, three conditions must be met:

1. **Variation must exist among individuals in a population.** Natural selection works by favoring individuals with some traits over individuals with alternative traits. If no variation exists, natural selection cannot operate.
2. **Variation among individuals must result in differences in the number of offspring surviving in the next generation.** This is the essence of natural selection. Because of their phenotype or behavior, some individuals are more successful than others in producing offspring.
3. **Variation must be genetically inherited.** For natural selection to result in evolutionary change, the selected differences must have a genetic basis. Not all variation has a genetic basis—often, genetically identical individuals may be phenotypically quite distinctive if they grow up in different environments. When phenotypically different individuals do not differ genetically, differences in the number of their offspring will not alter the genetic composition of the population in the next generation, and thus no evolutionary change will have occurred.

It is important to remember that natural selection and evolution are not the same—the two concepts often are incorrectly equated. Natural selection is a process, whereas evolution is the historical record, or outcome, of change through time. Natural selection (the process) can lead to evolution (the outcome), but natural selection is only one of several processes that can do so. Moreover, natural selection can occur without producing evolutionary change; only if variation is genetically based will natural selection lead to evolution.

Selection to avoid predators

A common result of evolution driven by natural selection is that populations become better adapted to their environment. Many of the most dramatic documented instances of adaptation involve genetic changes that decrease the probability of capture by a predator. The caterpillar larvae of the common sulphur butterfly, *Colias eurytheme,* usually exhibit a pale green color, providing excellent camouflage against the alfalfa plants on which they feed. An alternative bright yellow color morph is kept at very low frequency, because this color renders the larvae highly visible on the food plant, making it easier for bird predators to see them (figure 19.4*e*).

One of the most dramatic examples of background matching involves ancient lava flows in the deserts of the American Southwest. In these areas, the black rock formations produced when the lava cooled contrast starkly with the surrounding bright glare of the desert sand. Populations of many species of animals occurring on these rocks—including lizards, rodents, and a variety of insects—are dark in color, whereas sand-dwelling populations in surrounding areas are much lighter (figure 19.6). Predation is the likely cause for these differences in color. Laboratory studies have confirmed that predatory birds such as owls are adept at picking out individuals on backgrounds to which they are not adapted.

A similar phenomenon occurs in Finland, where the tawny owl (*Strix aluco*) is either brown or gray. In the past, gray owls far outnumbered brown ones. Researchers studied a population for 30 years, individually marking each bird to determine how long it lived. In years when there was more snow, brown owls died at much higher rates, probably because they were more visible against the snow to their predators. But the climate is changing, and Finland is getting much less snow than it used to. The researchers discovered that the less snow there is in a year, the smaller the advantage to the gray owls. In recent years, survival of the two types has been equal, and the population is now evenly split between brown and gray

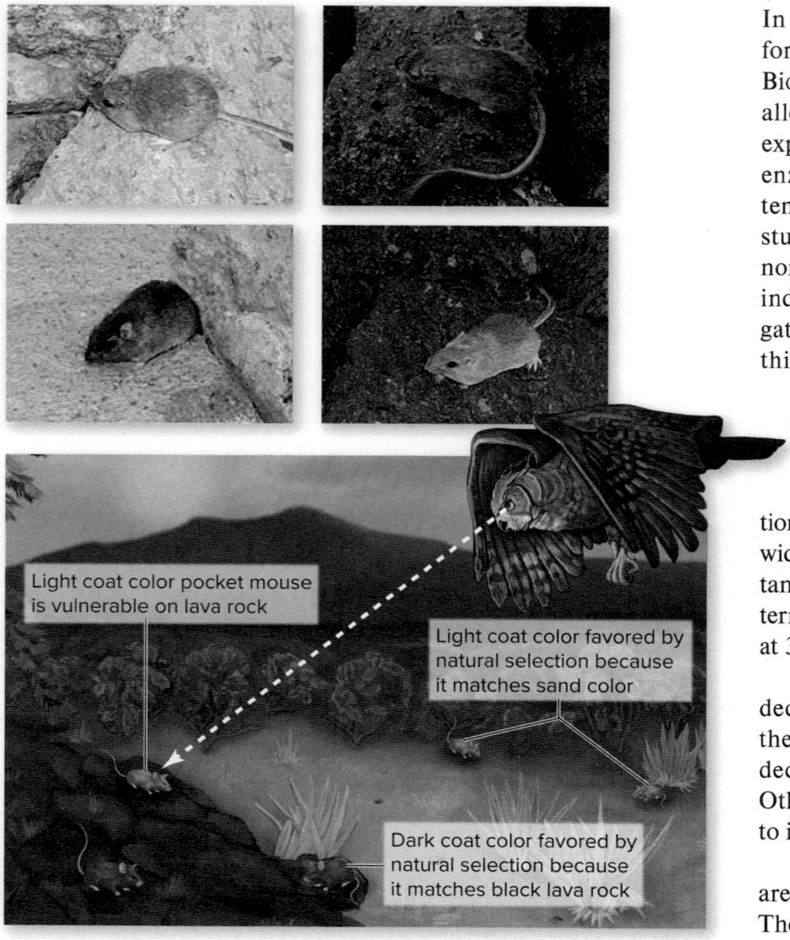
Light coat color pocket mouse is vulnerable on lava rock
Light coat color favored by natural selection because it matches sand color
Dark coat color favored by natural selection because it matches black lava rock

Figure 19.6 Pocket mice from the Tularosa Basin of New Mexico whose color matches their background. Black lava formations are surrounded by desert, and selection favors coat color in pocket mice that matches their surroundings. Genetic studies indicate that the differences in coat color are the result of small differences in the DNA of alleles of a single gene.

(photos): Michael W. Nachman, Hopi E. Hoekstra, and Susan L. D'Agostino, "The genetic basis of adaptive melanism in pocket mice," PNAS, Vol. 100, No. 9, April 29, 2003, pp. 5268-5273. ©National Academy of Sciences, U.S.A. Used with permission.

owls. Humans are changing the world in many ways. Evolutionary biologists are being kept busy studying how populations will respond to changing natural selection pressures.

Selection to match climatic conditions

Climate also affects populations directly: individuals need to be able to survive in the temperature and moisture conditions where they live. Many studies of climate-based selection have focused on genes encoding enzymes, because enzyme function varies with temperature, allowing investigators to study whether different enzyme alleles are most fit in different environments.

Often, investigators find that enzyme allele frequencies vary with latitude so that, for instance, one allele might be more common in northern populations but progressively less common at more southern locations. A superb example comes from studies of the mummichog (*Fundulus heteroclitus*), a minnowlike fish that ranges along the eastern coast of North America. In this fish, geographical variation occurs in allele frequencies for the gene that produces the enzyme lactate dehydrogenase. Biochemical studies show that the enzymes formed by these alleles function differently at different temperatures, thus explaining their geographical distributions. The form of the enzyme more frequent in the north is a better catalyst at low temperatures than is the enzyme from the south. Moreover, studies indicate that at low temperatures, individuals with the northern allele swim faster, and presumably survive better, than individuals with the alternative allele. This classic field investigation is explored in the Inquiry & Analysis feature at the end of this chapter.

Selection for pesticide and microbial resistance

A particularly clear example of selection in natural populations is provided by studies of pesticide resistance in insects. The widespread use of insecticides has led to the rapid evolution of resistance in more than 500 pest species. The cost of this evolution, in terms of crop losses and increased pesticide use, has been estimated at 3 to 8 billion dollars per year.

In the housefly, the resistance allele of the *pen* gene decreases the uptake of insecticide, whereas resistance alleles of the *kdr* and *dld-r* genes decrease the number of target sites, thus decreasing the binding ability of the insecticide (figure 19.7). Other resistance alleles enhance the ability of the insects' enzymes to identify and detoxify insecticide molecules.

Another important example is antibiotic resistance. Bacteria are currently winning their evolutionary struggle with humans. There are now pathogenic bacterial strains with resistance to multiple antibiotics. The obvious lesson is that any agent that targets a cellular protein will lead to the evolution of resistance.

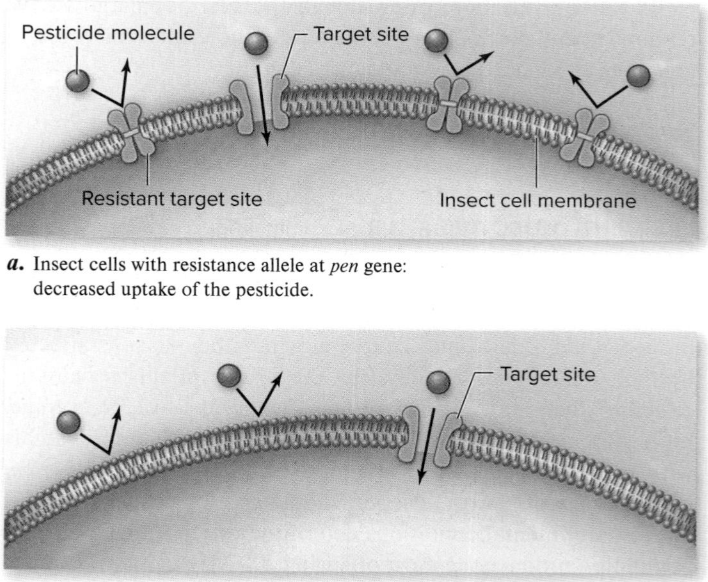
Pesticide molecule
Target site
Resistant target site
Insect cell membrane
Target site

a. Insect cells with resistance allele at *pen* gene: decreased uptake of the pesticide.

b. Insect cells with resistance allele at *kdr* gene: decreased number of target sites for the pesticide.

Figure 19.7 Selection for pesticide resistance. Resistance alleles at genes such as *pen* and *kdr* allow insects to be more resistant to pesticides. Insects that possess these resistance alleles have become more common through selection.

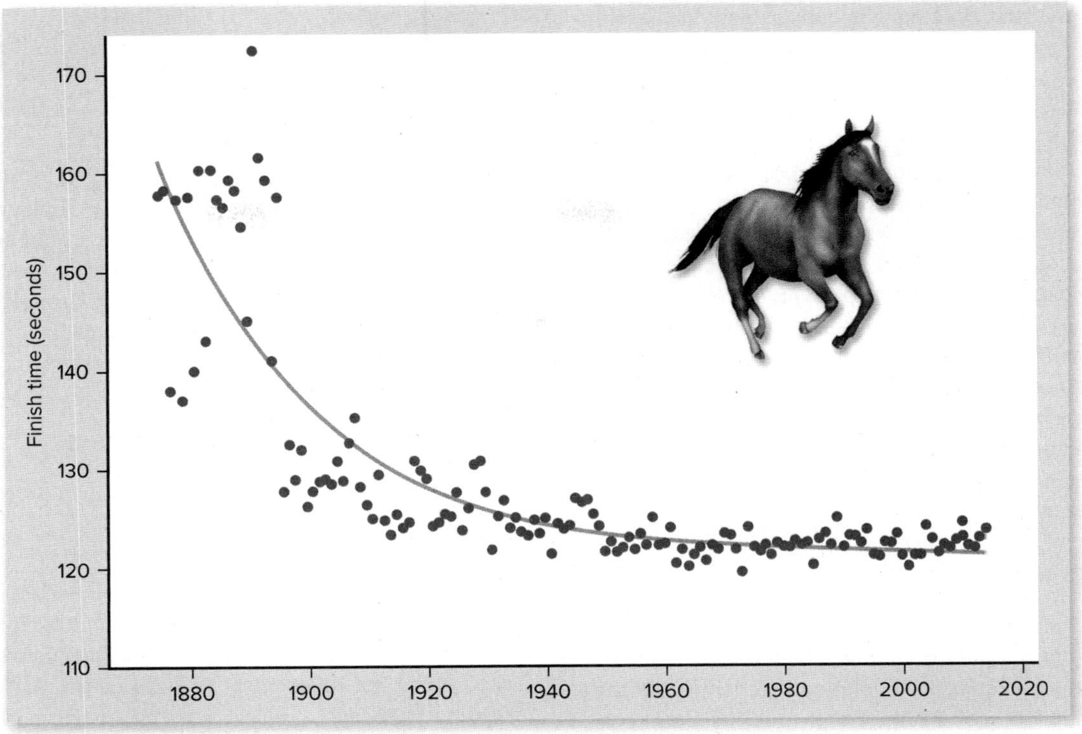

Figure 19.8 Selection for increased speed in racehorses is no longer effective. Kentucky Derby winning speeds have not improved significantly since 1950.

There Are Limits to What Selection Can Accomplish

LEARNING OBJECTIVE 19.3.6 Explain how selection is limited by genetics.

Although selection is the most powerful of the principal agents of genetic change, there are limits to what it can accomplish. These limits result from (1) multiple phenotypic effects of alleles, (2) lack of genetic variation upon which selection can act, and (3) interactions between genes.

Multiple phenotypic effects

Alleles often affect multiple aspects of a phenotype, a phenomenon called *pleiotropy*. These multiple effects tend to set limits on how much a phenotype can be altered. In chickens, the same gene affects the size of a hen's comb and the rate at which she lays eggs—alleles that produce large combs also lead to faster egg production. As a result of this linkage, selection for hens that lay many eggs but have a small comb would be very difficult.

Lack of genetic variation

Over 80% of the gene pool of the thoroughbred horses racing today goes back to 31 ancestors from the late 18th century. Despite intense directional selection on thoroughbreds, their performance times have not improved for more than 60 years (figure 19.8). Decades of intense selection presumably have removed variation from the population and now little genetic variation remains, with the result that further evolutionary change is not possible.

In some cases, phenotypic variation for a trait may never have had a genetic basis. The compound eyes of insects are made up of hundreds of visual units, termed ommatidia (figure 19.9). In some individuals, the left eye contains more ommatidia than the right. In other individuals, the right eye contains more than the left. However, despite intense selection experiments in the laboratory, scientists have never been able to produce a line of fruit flies that consistently has more ommatidia in the left eye than in the right. The reason is that separate genes do not exist for the left and right eyes. Rather, the same genes affect both eyes, and differences in the number of ommatidia result from differences that occur as the eyes are formed in the development process. Thus, despite the existence of phenotypic variation, no underlying genetic variation is available for selection to favor.

Interactions between genes

Often, the effect of selection on an allele depends on the genotype of other genes. If a population is polymorphic for a second gene,

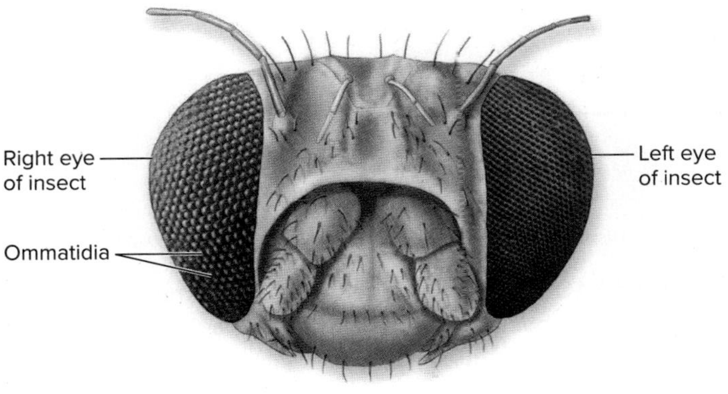

Figure 19.9 Phenotypic variation in insect ommatidia. In some individuals, the number of ommatidia in the left eye is greater than the number in the right.

then selection on the first gene may be constrained, because different alleles are favored in different individuals of the same population.

Studies on bacteria illustrate how selection on alleles for one gene can depend on which alleles are present at other genes. In *E. coli,* two biochemical processes exist to break down gluconate, using enzymes produced by different genes. One process uses the enzyme 6-PGD, whose gene possesses several alleles. When the common allele for the gene that codes for the other gluconate breakdown enzyme is present, selection does not favor one allele over another at the *6-PGD* gene. In some *E. coli,* however, an alternative allele at the second gene occurs, encoding an enzyme that is not functional. The bacteria with this alternative allele are forced to rely on only the 6-PGD pathway, and in this case selection favors one *6-PGD* allele over another. Thus, the outcome of natural selection on the *6-PGD* gene depends on which allele is present at the second gene locus.

REVIEW OF CONCEPT 19.3

Five factors can bring about deviation from the predicted Hardy–Weinberg genotype frequencies. Of these, only selection regularly produces adaptive evolutionary change, but the genetic constitution of populations, and thus the course of evolution, can also be affected by mutation, gene flow, nonrandom mating, and genetic drift.

- *How does each of these processes cause populations to vary from Hardy–Weinberg equilibrium?*

19.4 Selection Can Act on Traits Affected by Many Genes

In nature, many—perhaps most—phenotypic traits are affected by more than one gene. Such traits usually display a continuous pattern of variation among individuals in a population, rather than occurring as discrete alternatives. In such cases, selection operates on all the genes, influencing most strongly those that make the greatest contribution to the phenotype. How selection changes the population depends on which genotypes are favored.

Disruptive Selection Removes Intermediate Phenotypes

LEARNING OBJECTIVE 19.4.1 Describe the evolutionary outcome of disruptive selection.

In some situations, selection acts to eliminate intermediate types, a phenomenon called **disruptive selection** (figure 19.10*a*). A clear example is the different beak sizes of the African black-bellied seedcracker finch *Pyrenestes ostrinus* (figure 19.11). Populations of these birds contain individuals with large and small beaks, but very few individuals with intermediate-size beaks.

As their name implies, these birds feed on seeds, and the available seeds fall into two size categories: large and small. Only large-beaked birds can open the tough shells of large seeds,

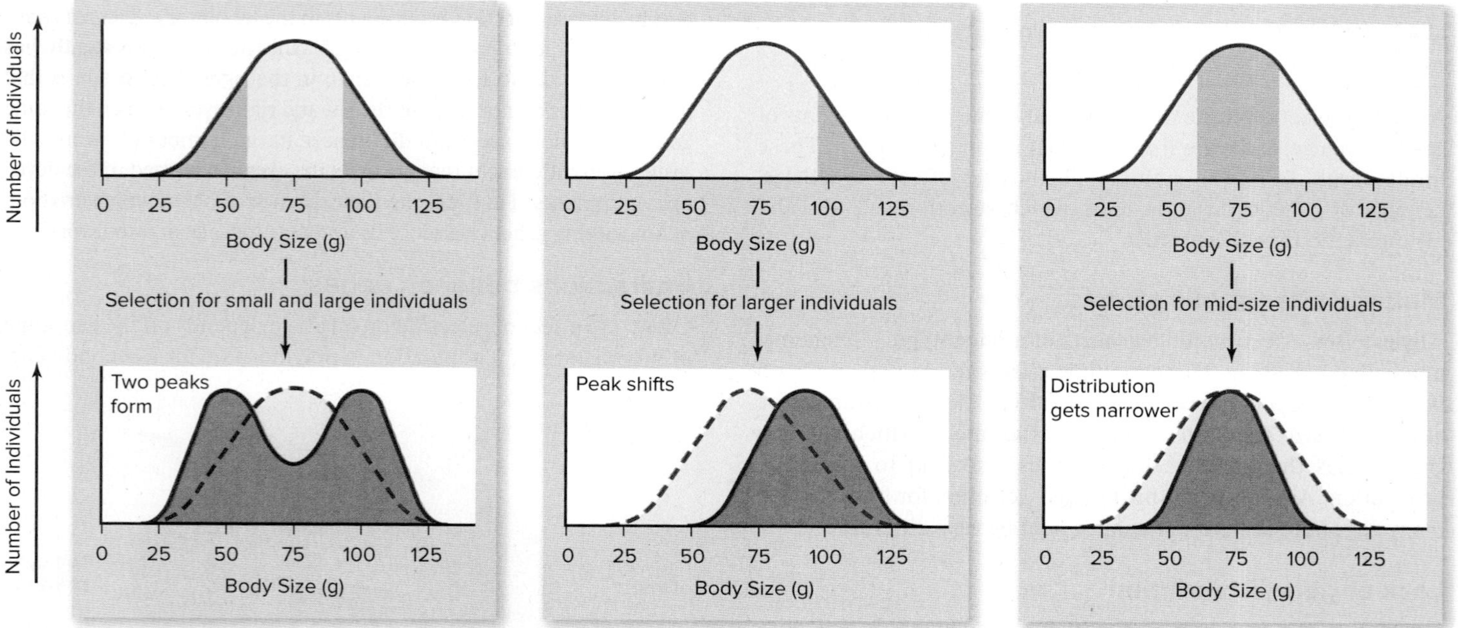

a. Disruptive selection *b.* Directional selection *c.* Stabilizing selection

Figure 19.10 Three kinds of selection. The top panels show the populations before selection has occurred (under the solid red line). Within the population, those favored by selection are shown in light brown. The bottom panels indicate what the populations would look like in the next generation. The dashed red lines are the distribution of the original population and the solid dark brown lines are the true distribution of the population in the next generation. ***a.*** In disruptive selection, individuals in the middle of the range of phenotypes of a certain trait are selected against, and the extreme forms of the trait are favored. ***b.*** In directional selection, individuals concentrated toward one extreme of the array of phenotypes are favored. ***c.*** In stabilizing selection, individuals with midrange phenotypes are favored, with selection acting against both ends of the range of phenotypes.

SCIENTIFIC THINKING

Question: *Does disruptive selection promote differences in beak size in the African black-bellied seedcracker finches (*Pyrenestes ostrinus*)?*

Field Study: *Capture, measure, and release birds in a population. Follow the birds through time to determine how long each lives.*

Result: *Large- and small-beaked birds have higher survival rates than birds with intermediate-sized beaks.*

Prediction: *What would happen if the distribution of seed size and hardness in the environment changed?*

Figure 19.11 Disruptive selection for large and small beaks. Differences in beak size in the black-bellied seedcracker finch of west Africa are the result of disruptive selection.

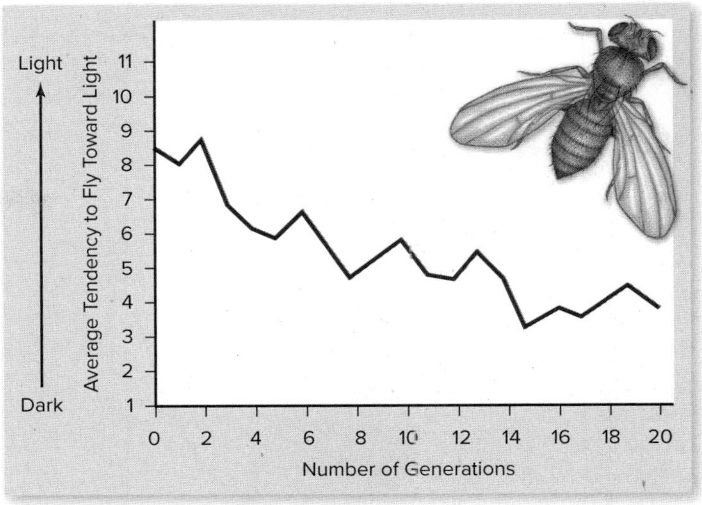

Figure 19.12 Directional selection for negative phototropism in *Drosophila*. Flies that moved toward light were discarded, and only flies that moved away from light were used as parents for the next generation. This procedure was repeated for 20 generations, producing substantial evolutionary change.

whereas birds with the smaller beaks are more adept at handling small seeds. Birds with intermediate-size beaks are at a disadvantage with both seed types—they are unable to open large seeds and too clumsy to efficiently process small seeds. Consequently, selection acts to eliminate the intermediate phenotypes, in effect partitioning (or "disrupting") the population into two phenotypically distinct groups.

Directional Selection Eliminates Phenotypes at One End of a Range

LEARNING OBJECTIVE 19.4.2 Describe the evolutionary outcome of directional selection.

When selection acts to eliminate one extreme from an array of phenotypes, the genes promoting this extreme become less frequent in the population and may eventually disappear. This form of selection is called **directional selection** (figure 19.10*b*). Thus, in the *Drosophila* population illustrated in figure 19.12, eliminating flies that move toward light causes the population over time to contain fewer individuals with alleles promoting such behavior. If you were to pick an individual at random from a later generation of flies, there is a smaller chance that the fly would spontaneously move toward light than if you had selected a fly from the original population. Artificial selection has changed the population in the direction of being less attracted to light. Directional selection often occurs in nature when the environment changes; one example is the widespread evolution of pesticide resistance discussed in section 19.3.

Stabilizing Selection Favors Individuals with Intermediate Phenotypes

LEARNING OBJECTIVE 19.4.3 Contrast the effects of directional, disruptive, and stabilizing selection.

When selection acts to eliminate both extremes from an array of phenotypes, the result is to increase the frequency of the already common intermediate type. This form of selection is called **stabilizing selection** (figure 19.10*c*). In effect, selection is operating to prevent change away from this middle range of values. Selection does not change the most common phenotype of the population but, rather, makes it even more common by eliminating extremes. Many examples are known. In humans, infants with intermediate weight at birth have the highest survival rate (figure 19.13). In ducks and chickens, eggs of intermediate weight have the highest hatching success.

REVIEW OF CONCEPT 19.4

In disruptive selection, extreme forms increase; in stabilizing selection, intermediates increase, whereas in disruptive selection, they decrease. Directional selection shifts frequencies toward one end and may eventually eliminate alleles entirely.

- *If a population of squirrels evolved to have larger jaws that allowed them to eat larger seeds, what kind of selection could be acting?*

19.5 Natural Selection Can Be Studied Experimentally

Evolutionary biology is not entirely an observational science. Darwin was right about many things, but one area in which he was mistaken concerns the pace at which evolution occurs. Darwin

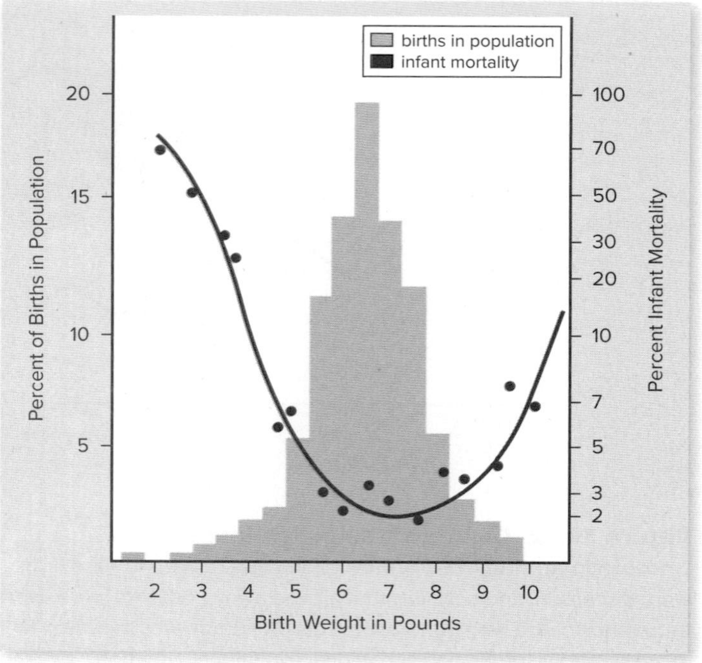

Figure 19.13 Stabilizing selection for birth weight in humans. The death rate among babies (red curve, right *y*-axis) is lowest at an intermediate birth weight; both smaller and larger babies have a greater tendency to die than those around the most frequent weight (tan area, left *y*-axis) of between 7 and 8 pounds. Recent medical advances have reduced mortality rates for small and large babies.

thought that evolution occurred at a very slow, almost imperceptible pace. But in recent years, many case studies have demonstrated that in some circumstances evolutionary change can occur rapidly, and in these instances experimental studies can be devised to test evolutionary hypotheses.

Guppy Color Variation in Different Environments Illustrates Natural Selection

LEARNING OBJECTIVE 19.5.1 Recount how laboratory and field experiments with guppies demonstrated the ongoing action of natural selection.

Although laboratory studies on fruit flies and other organisms have been common for more than 50 years, scientists have only recently started conducting experimental studies of evolution in nature. One excellent example of how evolutionary biologists today are combining detailed investigations in the lab with rigorous experiments in the field concerns research on the guppy, *Poecilia reticulata.*

The guppy is a popular aquarium fish because of its bright coloration and prolific reproduction. In nature these guppies are found in small streams in northeastern South America and in many mountain streams on the nearby island of Trinidad. One interesting feature of several of the streams is that they have waterfalls. Amazingly, guppies and some other fish are capable of colonizing portions of the stream above the waterfall.

The killifish is a particularly good colonizer; apparently on rainy nights it will wriggle out of the stream and move through the damp leaf litter. Guppies are not so proficient at moving through damp leaves but are good at swimming upstream. During flood seasons, rivers can overflow their banks, creating secondary channels that move through the forest. On these occasions, guppies may be able to swim upstream in the secondary channels and invade the pools above waterfalls.

By contrast, some other species of fishes are not capable of either kind of dispersal and thus are only found in streams below the first waterfall. One species whose distribution is restricted by waterfalls is the pike cichlid, a voracious predator that feeds on other fish, including guppies.

Because of these barriers to dispersal, guppies can be found in two very different environments. In pools just below the waterfalls, predation by the pike cichlid is a substantial risk and rates of survival are relatively low. But in similar pools just above the waterfall, the only predator present is the killifish, which rarely preys on guppies.

Guppy populations above and below waterfalls exhibit many differences. In the high-predation pools, guppies exhibit drab coloration. Moreover, they tend to reproduce at a younger age and attain relatively smaller adult sizes. Male fish above the waterfall, in contrast, are colorful (figure 19.14), mature later, and grow to larger sizes.

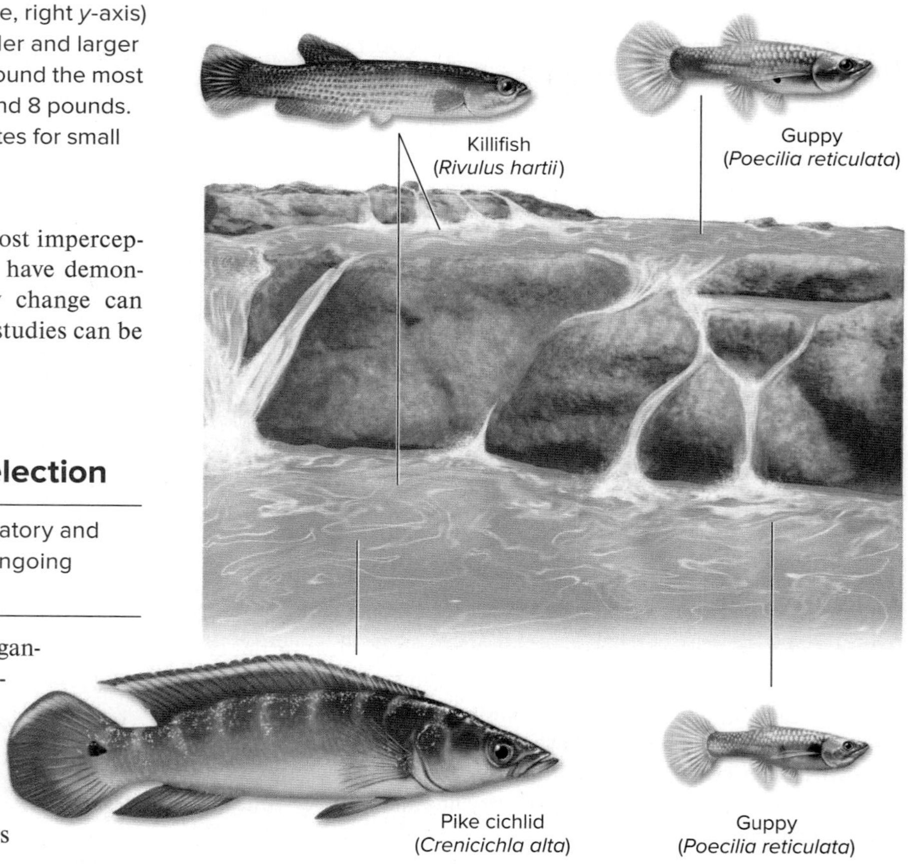

Figure 19.14 The evolution of protective coloration in guppies. In pools below waterfalls where predation is high, male guppies are drab in color. In the absence of the highly predatory pike cichlid (*Crenicichla alta*) in pools above waterfalls, male guppies are much more colorful and attractive to females. The killifish is also a predator, but it only rarely eats guppies. The evolution of these differences in guppies can be experimentally tested.

These differences suggest the operation of natural selection. In the low-predation environment, males display gaudy colors and spots that they use to attract females. Moreover, larger males are most successful at holding territories and mating with females, and larger females lay more eggs. Thus, in the absence of predators, larger and more colorful fish may have produced more offspring, leading to the evolution of those traits.

In pools below the waterfall, natural selection favors different traits. Colorful males are likely to attract the attention of the pike cichlid, and high predation rates mean that most fish are short-lived. Individuals that are more drab and shunt energy into early reproduction, rather than growth to a larger size, are therefore likely to be favored by natural selection.

Experimentation Reveals the Agent of Selection

LEARNING OBJECTIVE 19.5.2 Contrast laboratory and field experiments on selection on guppies.

Although the differences between guppies living above and below the waterfalls suggest evolutionary responses to differences in the strength of predation, alternative explanations are possible. Perhaps, for example, only very large fish are capable of crawling past the waterfall to colonize pools. If this were the case, then a founder effect would occur in which the new population was established solely by individuals with genes for large size. The only way to rule out such alternative possibilities is to conduct a controlled experiment.

The laboratory experiment

The first experiments were conducted in large pools in laboratory greenhouses. At the start of an experiment, a group of 2000 guppies was divided equally among 10 large pools. Six months later, pike cichlids were added to four of the pools and killifish to another four, with the remaining two pools left to serve as "no-predation" controls.

Fourteen months later (which corresponds to 10 guppy generations), the scientists compared the populations. The guppies in the killifish and control pools were indistinguishable—brightly colored and large. In contrast, the guppies in the pike cichlid pools were smaller and drab in coloration (figure 19.15).

These results established that predation can lead to rapid evolutionary change, but do these laboratory experiments reflect what occurs in nature?

The field experiment

To find out whether the laboratory results were an accurate reflection of natural processes, the scientists located two streams that had guppies in pools below a waterfall, but not above it. As in other Trinidadian streams, the pike cichlid was present in the lower pools, but only the killifish was found above the waterfalls.

The scientists then transplanted guppies to the upper pools and returned at several-year intervals to monitor the populations. Despite originating from populations in which predation levels were high, the transplanted populations rapidly evolved the traits characteristic of low-predation guppies: they matured late, attained greater size, and had brighter colors. The control populations in the lower pools, by contrast, continued to be drab and to mature early and at a smaller size. Laboratory analysis confirmed that the variations between the populations were the result of genetic differences.

These results demonstrate that substantial evolutionary change can occur in just a few years. The results give strong support to the theory of evolution by natural selection.

SCIENTIFIC THINKING

Question: *Does the presence of predators affect the evolution of guppy color?*

Hypothesis: *Predation on the most colorful individuals will cause a population to become increasingly dull through time. Conversely, in populations with few or no predators, increased color will evolve.*

Experiment: *Establish laboratory populations of guppies in large pools with or without predators.*

Result: *The populations with predators evolved to have fewer spots, while the populations in pools without predators evolved more spots.*

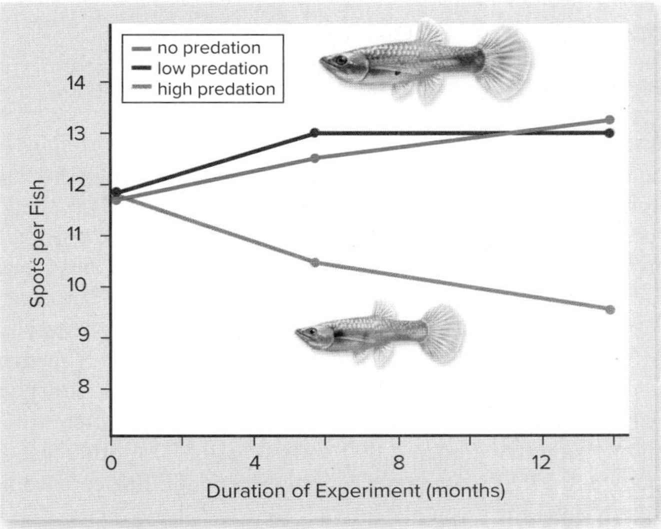

Interpretation: *Why does color increase in the absence of predators? How would you test your hypothesis?*

Figure 19.15 Evolutionary change in spot number. Guppy populations raised for 10 generations in low-predation or no-predation environments in laboratory greenhouses evolved a greater number of spots, whereas selection in more dangerous environments, such as the pools with the highly predatory pike cichlid, led to less-conspicuous fish. The same results are found in field experiments conducted in pools above and below waterfalls.

REVIEW OF CONCEPT 19.5

Much of evolutionary theory is derived from observation, but experiments are possible in natural settings. Studies have revealed that traits can shift in populations in a relatively short time. Experiments on coloration in guppies in the laboratory and in natural settings show genetic changes under selective pressure.

- *What type of selection is observed in the guppy predation experiments?*

19.6 Fitness Is a Measure of Evolutionary Success

Selection occurs when individuals with one phenotype leave more surviving offspring in the next generation than individuals with an alternative phenotype. Evolutionary biologists quantify reproductive success as **fitness,** the number of surviving offspring left in the next generation.

A Phenotype with Greater Fitness Usually Increases in Frequency

LEARNING OBJECTIVE 19.6.1 Define evolutionary fitness.

Fitness is a relative concept; the most fit phenotype is simply the one that produces, on average, the greatest number of offspring. Suppose, for example, that in a population of toads, two phenotypes exist: green and brown. Suppose further that green toads leave, on average, 4.0 offspring in the next generation, but brown toads leave only 2.5. By custom, the most fit phenotype is assigned a fitness value of 1.0, and the fitness values of other phenotypes are expressed in relative proportion to 1. In this case, the fitness of the green phenotype would be $4.0/4.0 = 1.000$, and the fitness of the brown phenotype would be $2.5/4.0 = 0.625$. The difference in fitness would therefore be $1.000 - 0.625 = 0.375$. A difference in fitness of 0.375 is quite large; natural selection in this case would strongly favor the green phenotype.

If differences in color have a genetic basis, then we would expect evolutionary change to occur; the frequency of green toads should be substantially greater in the next generation.

Fitness May Consist of Many Components

LEARNING OBJECTIVE 19.6.2 List the three principal components of fitness.

Although selection is often characterized as "survival of the fittest," differences in survival are only one component of fitness.

Even if no differences in survival occur, selection may operate if some individuals are more successful than others in attracting mates. In many territorial animal species, for example, large males mate with many females, and small males rarely get to mate. Selection with respect to mating success is termed *sexual selection*.

In addition, the number of offspring produced per reproductive event is also important. Large female frogs and fish lay more eggs than do smaller females, and thus they may leave more offspring in the next generation.

Fitness is therefore a combination of survival, mating success, and number of offspring per mating. Selection favors phenotypes with the greatest fitness, but predicting fitness from a single component can be tricky, because traits favored for one component of fitness may be at a disadvantage for others. As an example, in water striders, larger females lay more eggs per day. Thus, natural selection at this stage favors large size. However, larger females also die at a younger age and thus have fewer opportunities to reproduce than smaller females. Overall, the two opposing directions of selection cancel each other out, and the intermediate-size females leave the most offspring in the next generation (figure 19.16).

REVIEW OF CONCEPT 19.6

Fitness is defined by an organism's reproductive success relative to other members of its population. This success is determined by how long it survives, how often it mates, and how many offspring it produces per mating. Relative fitness assigns numerical values to different phenotypes relative to the most fit phenotype.

- *Is one of these factors always the most important in determining reproductive success? Explain.*

19.7 Evolutionary Processes Sometimes Maintain Variation

The amount of genetic variation in a population is determined by the relative strength of different evolutionary forces acting on it. Sometimes these forces act together to change allele or genotype frequencies; in other cases they work in opposition. In addition, natural selection sometimes acts to maintain variation by favoring alleles when they are rare or by favoring heterozygotes.

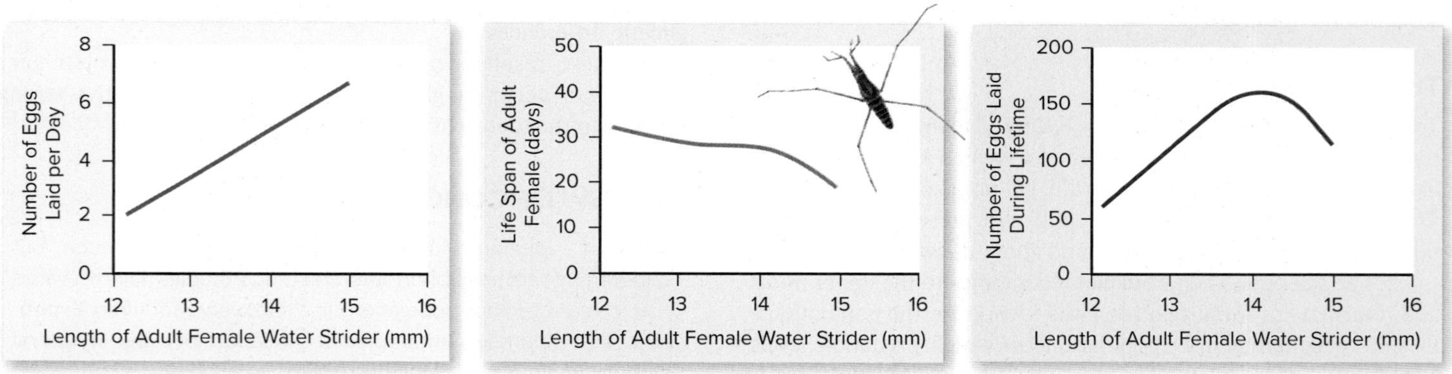

Figure 19.16 Body size and egg-laying in water striders. Larger female water striders lay more eggs per day (left panel), but they also survive for a shorter period of time (center panel). As a result, intermediate-size females produce the most offspring over the course of their entire lives.

Mutation and Genetic Drift May Counter the Influence of Natural Selection

LEARNING OBJECTIVE 19.7.1 Demonstrate how mutation and genetic drift may act to counter natural selection.

In theory, if allele *B* mutates to allele *b* at a high enough rate, allele *b* can be maintained in the population, even if natural selection strongly favors allele *B*. In nature, however, mutation rates are rarely high enough to counter the effects of natural selection.

The effect of natural selection also may be countered by genetic drift. Both of these processes often act to remove variation from a population. But selection is a nonrandom process that operates to increase the representation of alleles that enhance survival and reproductive success, whereas genetic drift is a random process in which any allele may increase. Thus, in some cases drift may lead to a decrease in the frequency of an allele that is favored by selection. In some extreme cases, drift may even lead to the loss of a favored allele from a population.

Remember, however, that the magnitude of drift is inversely related to population size; consequently, natural selection is expected to overwhelm drift except when populations are very small.

Gene Flow May Promote or Constrain Evolutionary Change

LEARNING OBJECTIVE 19.7.2 Examine how gene flow can interact with natural selection to affect a population.

Gene flow can be either a constructive or a constraining force. On the one hand, gene flow can spread a beneficial mutation that arises in one population to other populations. On the other hand, gene flow can impede adaptation within a population by the continual inflow of inferior alleles from other populations.

Consider two populations of a species that live in different environments. In this situation, natural selection might favor different alleles—*B* and *b*—in the two populations. In the absence of other evolutionary processes such as gene flow, the frequency of *B* would be expected to reach 100% in one population and 0% in the other. However, if gene flow occurred between the two populations, then the less favored allele would continually be reintroduced into each population. As a result, the frequency of the alleles in the populations would reflect a balance between the rate at which gene flow brings the inferior allele into a population and the rate at which natural selection removes it.

A classic example of gene flow opposing natural selection occurs on abandoned mine sites in Great Britain. Although mining activities ceased hundreds of years ago, the concentration of heavy-metal ions in the soil of mine tailings is still much greater than in surrounding soils. Large concentrations of heavy metals are generally toxic to plants, but alleles at certain genes confer the ability to grow on soils high in heavy metals. The ability to tolerate heavy metals comes at a price, however; individuals with the resistance allele exhibit lower growth rates on nonpolluted soil. Consequently, we would expect the resistance allele to occur with a frequency of 100% on mine sites and 0% elsewhere.

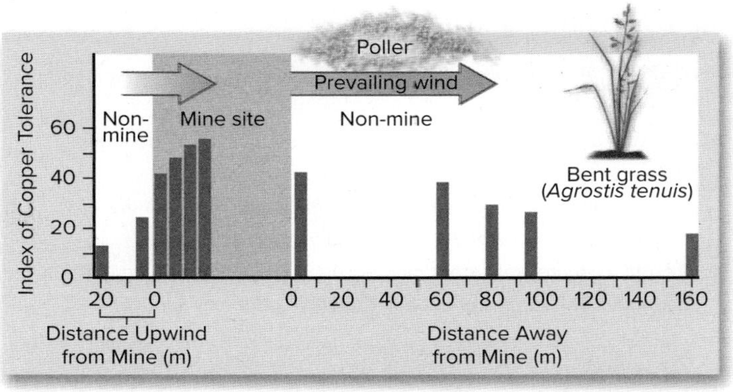

Figure 19.17 Degree of copper tolerance in grass plants on and near ancient mine sites. Individuals with tolerant alleles have decreased growth rates on unpolluted soil. Thus, we would expect copper tolerance to be 100% on mine sites and 0% on non-mine sites. However, prevailing winds blow pollen containing nontolerant alleles onto the mine site and tolerant alleles beyond the site's borders. The amount of pollen received decreases with distance, which explains the changes in levels of tolerance. The index of copper tolerance is calculated as the growth rate of a plant on soil with high concentrations of copper relative to growth rate on soils with low levels of copper; the higher the index, the more tolerant the plant is of heavy-metal pollution.

Heavy-metal tolerance in plants growing on mine tailings has been studied intensively in the slender bent grass *Agrostis tenuis,* in which the resistance allele occurs at intermediate levels on tailings, rather than at 100% (figure 19.17). Why not 100%? The explanation relates to the reproductive system of this grass, in which pollen, the floral equivalent of sperm, is dispersed by the wind. As a result, pollen grains—and the alleles they carry—can move great distances, leading to levels of gene flow between mine sites and unpolluted areas high enough to counteract the effects of natural selection.

Frequency-Dependent Selection May Favor Either Rare or Common Phenotypes

LEARNING OBJECTIVE 19.7.3 Describe the effect of frequency-dependent selection.

So far, we have discussed natural selection as a process that removes variation from a population by favoring one allele over others at a gene locus. However, in some circumstances selection can do exactly the opposite, actually maintain population variation. In these instances, the amount that selection favors a phenotype depends on how commonly or uncommonly it occurs within the population, a phenomenon termed **frequency-dependent selection.**

Negative frequency-dependent selection

In negative frequency-dependent selection, rare phenotypes are favored by selection. Assuming a genetic basis for phenotypic variation, such selection will have the effect of making rare alleles more common, thus maintaining variation.

Negative frequency-dependent selection often occurs when animals or people searching for something form a "search image."

SCIENTIFIC THINKING

Question: *Does negative frequency-dependent selection maintain variation in a population?*

Hypothesis: *Fish may disproportionately capture water boatmen (a type of aquatic insect) with the most common color.*

Experiment: *Place predatory fish in different aquaria with the different frequencies of the color types in each aquarium.*

Result: *Fish prey disproportionately on the common color in each aquarium. The rare color in each aquarium generally survives best.*

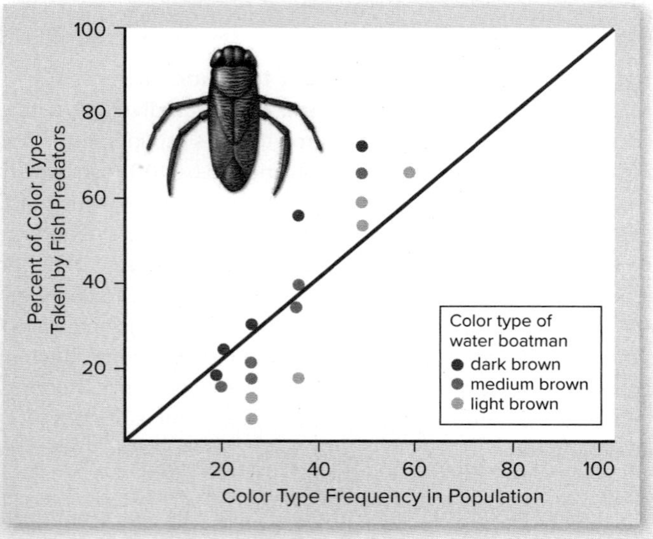

Figure 19.18 Frequency-dependent selection.

That is, they become particularly adept at picking out certain objects. In just this way, predators may form a search image for common prey phenotypes, with the result that rare forms are preyed upon less frequently.

An example is fish predation on the water boatman, an insect that occurs in three different colors. Experiments indicate that each of the color types is preyed upon disproportionately when it is the most common one (figure 19.18).

Another cause of negative frequency dependence is resource competition. If genotypes differ in their resource requirements, as occurs in many plants, then the rarer genotype will have fewer competitors, and when resources are equally abundant, the rarer genotype will be at an advantage relative to the more common genotype.

Positive frequency-dependent selection

Positive frequency-dependent selection has the opposite effect. By favoring common forms, it tends to eliminate variation from a population. For example, predators don't always select common individuals. In some cases, "oddballs" stand out from the rest and attract attention.

The strength of selection should change through time as a result of frequency-dependent selection. In negative frequency-dependent selection, rare genotypes should become increasingly common, and their selective advantage will decrease correspondingly. Conversely, in positive frequency dependence, the rarer a genotype becomes, the greater the chance it will be selected against.

In Oscillating Selection, the Favored Phenotype Changes as the Environment Changes

LEARNING OBJECTIVE 19.7.4 Define oscillating selection, and explain how it influences the amount of genetic variation in a population.

In some cases, selection favors one phenotype at one time and another phenotype at another time, a phenomenon called **oscillating selection.** If selection repeatedly oscillates in this fashion, the effect is to maintain genetic variation in the population.

One example, discussed in chapter 21, concerns the medium ground finch of the Galápagos Islands. In times of drought, the supply of small, soft seeds is depleted, but there are still enough large seeds around. Consequently, birds with big bills are favored. However, when wet conditions return, the ensuing abundance of small seeds favors birds with smaller bills.

Oscillating selection and frequency-dependent selection are similar, because in both cases the strength of selection changes through time. But it is important to recognize that they are not the same: in oscillating selection, the fitness of a phenotype does not depend on its frequency; rather, environmental changes lead to the oscillation in selection. In contrast, in frequency-dependent selection, it is the change in frequencies themselves that leads to the changes in fitness of the different phenotypes.

Heterozygotes May Exhibit Greater Fitness Than Homozygotes

LEARNING OBJECTIVE 19.7.5 Explain how heterozygous advantage can affect allele frequencies in a population.

If heterozygotes are favored over homozygotes **(heterozygous advantage),** then natural selection favors individuals with copies of both alleles and thus works to maintain both alleles in the population. Some evolutionary biologists think that heterozygous advantage is pervasive and can explain the high levels of polymorphism observed in natural populations. Others, however, think that it is relatively rare.

The best-documented example of heterozygous advantage is sickle-cell anemia, a hereditary disease affecting hemoglobin in humans. Individuals with sickle-cell anemia exhibit symptoms of severe anemia with abnormal red blood cells that are irregular in shape. Red blood cells containing the abnormal hemoglobin have a long sickle shape (figure 19.19).

The average incidence of the *S* allele in central African populations is about 0.12, far higher than that found among Americans of African descent. From the Hardy–Weinberg principle, you can calculate that 1 in 5 central African individuals are heterozygous at the *S* allele, and 1 in 100 are homozygous and develop the fatal form of the disorder. People who are homozygous for the sickle-cell allele almost never reproduce, because they usually die before they reach reproductive age.

Why, then, is the *S* allele not eliminated from the central African population by selection rather than being maintained at such high levels? As it turns out, one of the leading causes of illness and death in central Africa, especially among young children, is malaria. People who are heterozygous for the sickle-cell allele (and thus do not suffer from sickle-cell anemia) are much less susceptible to malaria. The reason is that when the parasite that causes malaria, *Plasmodium falciparum,* enters a red blood cell, it causes extremely low oxygen tension in the cell, which leads to sickling in cells of individuals either homozygous or heterozygous for the sickle-cell allele (but not in individuals that do not have the sickle-cell allele). Such cells are quickly filtered out of the bloodstream by the spleen, thus eliminating the parasite. (The spleen's filtering effect is what leads to anemia in persons homozygous for the sickle-cell allele, because large numbers of red blood cells become sickle-shaped and are removed; in the case of heterozygotes, only those cells containing the *Plasmodium* parasite sickle, whereas the remaining cells are not affected, and thus anemia does not occur.)

Consequently, even though most homozygous recessive individuals die at a young age, the sickle-cell allele is maintained at high levels in these populations because it is associated with resistance to malaria in heterozygotes and, for reasons not yet fully understood, with increased fertility in female heterozygotes. Figure 19.19 shows the overlap between regions where sickle-cell anemia is found and those where malaria is prevalent.

For people living in areas where malaria is common, having the sickle-cell allele in the heterozygous condition has adaptive value. However, in the United States, the environment does not place a premium on resistance to malaria because the disease is now essentially absent in North America. Consequently, no adaptive value counterbalances the ill effects of the disease; in this nonmalarial environment, selection is acting to eliminate the *S* allele. As a result, 1 in 375 Americans of African descent, many of whose ancestors lived in Central Africa, develops sickle- cell anemia, far fewer than in central Africa.

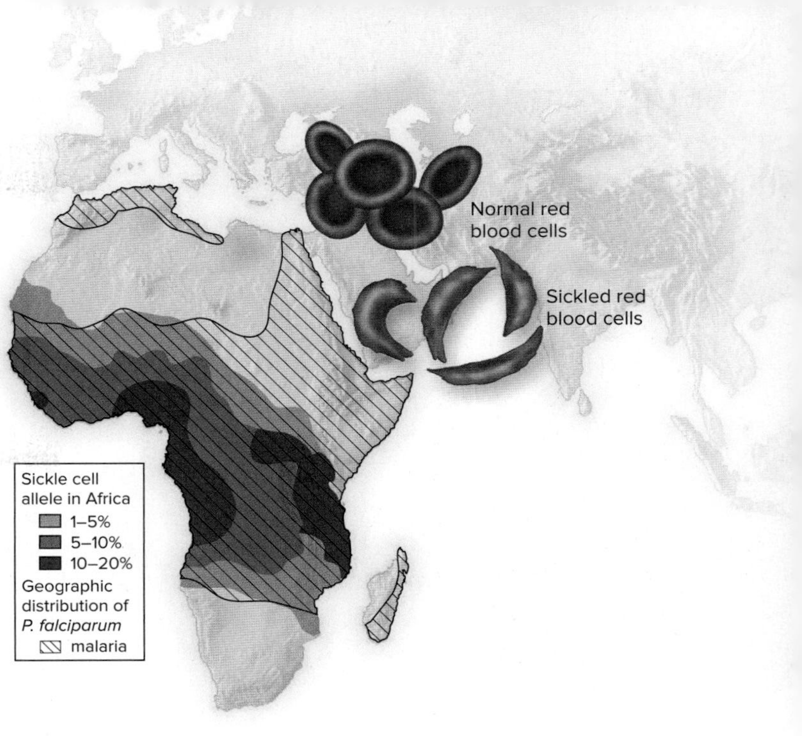

Figure 19.19 Frequency of sickle-cell allele and distribution of *Plasmodium falciparum* malaria. The red blood cells of people homozygous for the sickle-cell allele collapse into sickled shapes when the oxygen level in the blood is low. The distribution of the sickle-cell allele in Africa coincides closely with that of *P. falciparum* malaria.

REVIEW OF CONCEPT 19.7

Allele frequencies sometimes reflect a balance between opposing processes. Gene flow, for example, may increase some alleles, whereas natural selection decreases them. When several processes are involved, observed frequencies depend on the relative strength of the processes. Selection can maintain variation in a number of ways. Negative frequency-dependent selection favors rare phenotypes. Oscillating selection favors different phenotypes at different times. In some cases, heterozygotes have a selective advantage that may act to retain deleterious alleles.

- *How would genetic variation in a population change if heterozygotes had the lowest fitness? Under what circumstances might evolutionary processes operate in the same direction, and what would be the outcome?*

19.8 Sexual Selection Determines Reproductive Success

Energetic costs of reproduction appear to have been critically important to behavioral differences between females and males. Ecological factors such as the way food resources, nest sites, and members of the opposite sex are spatially distributed in the environment, as well as disease, are important in the evolution of reproductive decisions.

The Sexes Often Have Different Reproductive Strategies

LEARNING OBJECTIVE 19.8.1 Explain parental investment and the prediction it makes about mate choice.

Males and females have the common goal of improving the quantity and quality of offspring they produce, but usually they differ in the way they attempt to maximize fitness. Such a difference in reproductive behavior is clearly revealed in mate choice. Darwin was the first to observe that females often do not mate with the first male they encounter but, instead, seem to evaluate a male's quality and then decide whether to mate. Peahens prefer to mate with peacocks that have more eyespots on their elaborate tail feathers (figure 19.20*b, c*). Similarly, female frogs prefer to mate with males having more acoustically complex, and thus attractive, calls. This behavior, called mate choice, is well known in many invertebrate and vertebrate species.

Males are much less selective in choosing a mate than females are. Why should this be? Many of the differences in reproductive strategies between the sexes can be understood by comparing the parental investment made by males and females. **Parental investment** refers to the energy and time each sex spends ("invests") in producing and rearing offspring; it is, in effect, an estimate of the energy expended by males and females in each reproductive event.

a. *b.*

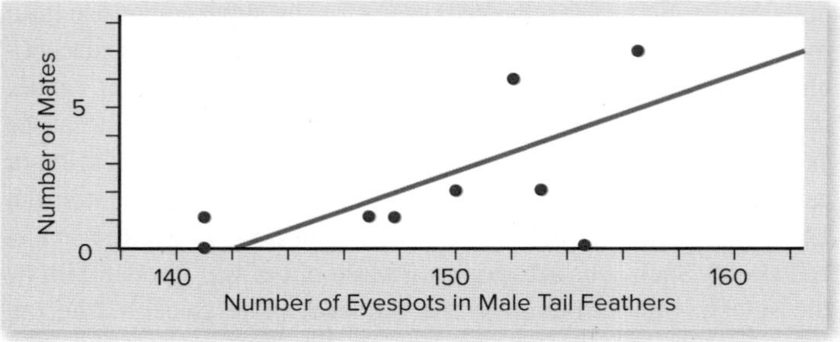

c.

Figure 19.20 Products of sexual selection. Attracting mates with long feathers is common in bird species such as (***a***) the raggiana bird of paradise (*Paradisaea raggiana*) and (***b***) the peacock (*Pavo cristatus*), which show pronounced sexual dimorphism. ***c.*** Female peahens prefer to mate with males having greater numbers of eyespots in their tail feathers.

(a): Bruce Beehler/Science Source; (b): Carole_R/Flickr Flash/Getty Images

Numerous studies have shown that females generally have a higher parental investment. One reason is that eggs are much larger than sperm—195,000 times larger, in humans! Eggs contain proteins and lipids in the yolk and other nutrients for the developing embryo, but sperm are little more than mobile DNA packages. In some groups of animals, females are responsible for gestation and lactation, costly reproductive functions only they can carry out.

The consequence of such inequalities in reproductive investment is that the sexes face very different selective pressures. Because any single reproductive event is relatively inexpensive for males, they can best increase their fitness by mating with as many females as possible. By contrast, each reproductive event for females is much more costly, and the number of eggs that can be produced often limits reproductive success. For this reason, a female should be choosy, trying to pick the male that can provide the greatest benefit to her offspring and thus improve her fitness.

These conclusions hold only when female reproductive investment is much greater than that of males. In species with biparental care, males may contribute equally to the cost of raising young; in this case, the degree of mate choice should be more equal between the sexes.

In some cases, male investment exceeds that of females. For example, male Mormon crickets transfer a protein-containing packet (a spermatophore) to females during mating. Almost 30% of a male's body weight is made up by the spermatophore, which provides nutrition for the female and helps her develop her eggs. As we might expect from our model of mate choice, in this case it is the females that compete with one another for access to males, which are the choosy sex. Indeed, males are quite selective, favoring heavier females. Heavier females have more eggs; thus, males that choose larger females leave more offspring (figure 19.21).

In many species, including seahorses and a number of birds and insects, males care for eggs and the developing young. As with Mormon crickets, these males are often choosy, and females compete for mates.

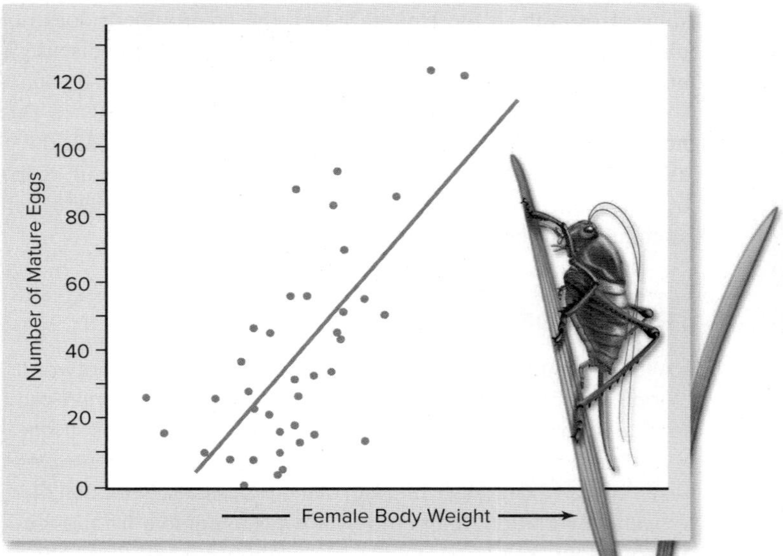

Figure 19.21 Male mate choice can increase fitness. Male Mormon crickets (*Anabrus simplex*) choose heavier females as mates, and larger females have more eggs. Thus, male mate selection increases fitness.

Sexual Selection Occurs Through Mate Competition and Mate Choice

LEARNING OBJECTIVE 19.8.2 Describe how sexual selection affects the evolution of secondary sex characteristics.

As we've discussed, the reproductive success of an individual is determined by how long the individual lives, how frequently it mates, and how many offspring it produces per mating. The second of these factors, competition for mates, is termed **sexual selection.** Some people consider sexual selection to be distinguishable from natural selection, but others understand it as a subset of natural selection, just one of the many factors affecting an organism's fitness.

Sexual selection involves both **intrasexual selection,** or competitive interactions between members of one sex ("the power to conquer other males in battle," as Darwin put it), and **intersexual selection,** which is another name for mate choice ("the power to charm"). Sexual selection leads to the evolution of structures used in combat with other males, such as a deer's antlers and a ram's horns, as well as ornaments used to "persuade" members of the opposite sex to mate, such as long tail feathers and bright plumage. These traits are called **secondary sex characteristics.**

Selection strongly favors any trait that confers greater ability in mate competition. Larger body size is a great advantage if dominance is important, as it is in territorial species. Males may thus be considerably larger than females. Such differences between the sexes are referred to as **sexual dimorphism.** In other species, structures used for fighting, such as horns, antlers, and large canine teeth, have evolved to be larger in males because of the advantage they give in intrasexual competition.

Sometimes **sperm competition** occurs between the sperm of different males if females mate with multiple males. This type of competition, which occurs after mating, has selected for sperm-transfer organs designed to remove the sperm of a prior mating, large testes to produce more sperm per mating, and sperm that hook themselves together to swim more rapidly. These traits enhance the likelihood of fertilizing an egg.

Intrasexual selection

In many species, individuals of one sex—usually males—compete with one another for the opportunity to mate. Competition can occur for a territory in which females feed or bear young. Males may also directly compete for the females themselves. A few successful males may engage in an inordinate number of matings, but most males do not mate at all. For example, elephant seal males control territories on breeding beaches, and a few dominant males do most of the breeding (figure 19.22). On one beach, for example, 8 males impregnated 348 females, but the remaining males mated rarely, if at all.

Intersexual selection

Intersexual selection concerns the active choice of a mate. Mate choice has both direct and indirect benefits.

Direct Benefits of Mate Choice In some cases, the benefits of mate choice are obvious. If males help raise offspring, females benefit by choosing the male that can provide the best care—the better the parent, the more offspring she is likely to rear. In other species, males provide no care but maintain territories that provide food, nesting sites, and predator refuges. In red deer, males that hold territories with the highest-quality grasses mate with the most females. In this case, there is a direct benefit for a female mating with such a territory owner: she feeds with little disturbance on quality food.

Indirect Benefits of Mate Choice In many species, however, males provide no direct benefits of any kind to females. In such cases, it is not intuitively obvious what females have to gain by being "choosy." Moreover, what could be the benefit of choosing a male with an extremely long tail or a complex song?

A number of theories have been proposed to explain the evolution of such preferences. One idea is that females choose the male that is the healthiest or oldest. Large males, for example, have probably been successful at living long, acquiring a lot of food, and resisting parasites and disease. In other species, features other than size may indicate a male's condition. In guppies and some birds, the brightness of a male's color reflects the quality of his diet and overall health. Females may gain two benefits from mating with the healthiest males. First, healthy males are less likely to be carrying diseases, which might be transmitted to the female during mating. Second, to the extent that the male's success in living long and prospering is the result of his genetic makeup, the female will be ensuring that her offspring receive good genes from their father.

Several experimental studies in fish and moths have examined whether female mate choice leads to greater reproductive success. In these experiments, females in one group were allowed to choose males, whereas males were randomly mated to a different group of females. Offspring of females that chose their mates were more vigorous and survived better than offspring from females given no choice, which suggests that females preferred males with a better genetic makeup.

A variant of this theory goes one step further. In some cases, females prefer mates with traits that appear to be detrimental to survival (figure 19.20*c*). The long tail of the peacock is a hindrance in flying and makes males more vulnerable to predators. Why should females prefer males with such traits? The **handicap hypothesis** states that only genetically superior mates can survive with such a handicap. By choosing a male with the largest handicap, the female is ensuring that her offspring will

Figure 19.22 Male northern elephant seals (*Mirounga angustirostris*) compete for mates. Male elephant seals fight with one another for possession of territories. Only the largest males can hold territories, which contain many females.
Cathy & Gordon ILLG

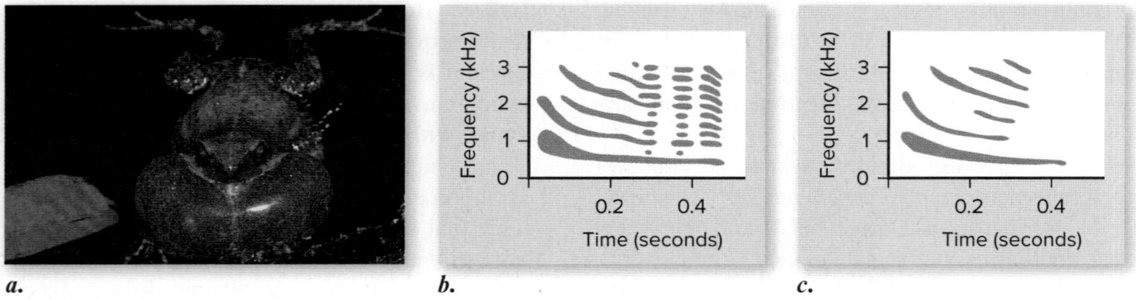

Figure 19.23 Male Túngara frog (*Physalaemus pustulosus*) calling. Female frogs of several species in the genus *Physalaemus* prefer males that include a "chuck" in their call. However, only males of the Túngaru frog (***a***) produce such calls (***b***); males of other species do not (***c***).

W. Perry Conway/Getty Images

receive these quality genes. Of course, the male offspring will also inherit the genes for the handicap. For this reason, evolutionary biologists are still debating the merit of this hypothesis.

Alternative Theories About the Evolution of Mate Choice Some courtship displays appear to have evolved from a predisposition in the female's sensory system to respond to certain stimuli. For example, females may be better able to detect particular colors or sounds at a certain frequency, and thus be attracted to such signals. **Sensory exploitation** involves the evolution in males of a signal that "exploits" these preexisting biases. For example, if females are particularly adept at detecting red objects, then red coloration may evolve in males as part of a courtship display.

To understand the evolution of courtship calls, consider the vocalizations of the Túngara frog (figure 19.23). Unlike related species, males include a short burst of sound, termed a "chuck," at the end of their calls. Recent research suggests that not only are females of this species particularly attracted to calls of this sort but so are females of related species, even though males of these species do not produce "chucks."

A great variety of other hypotheses have been proposed to explain the evolution of mating preferences. Many of these hypotheses may be correct in some circumstances, but none seems capable of explaining all of the variation in mating behavior in the animal world. This is an area of vibrant research, with new discoveries appearing regularly.

REVIEW OF CONCEPT 19.8

The sex with greater parental investment tends to exhibit mate choice. Females or males can be selective, depending on the energy and time they devote to parental care. Mate competition and mate choice can lead to sexual selection. Both direct and indirect benefits may explain why mate choice increases fitness; sometimes mate choice evolves because of predispositions in the sensory system.

- *Pipefish males incubate young in a brood pouch. Which sex would you expect to show mate choice? Why?*

Does Natural Selection Act to Maintain Enzyme Polymorphism?

The essence of Darwin's theory of evolution by natural selection is that, in nature, selection favors some gene alternatives over others. Many studies of natural selection have focused on genes encoding enzymes, because populations in nature tend to possess many alternative alleles of their enzymes (a phenomenon called *enzyme polymorphism*). Often, investigators have investigated how weather influences which alleles are more common in natural populations. A particularly nice example of such a study was carried out on a fish, the mummichog *(Fundulus heteroclitus)*, a kind of minnow that lives in coastal waters along the east coast of North America. Researchers in the laboratory of Richard Koehn at the State University of New York at Stony Brook studied allele frequencies of the gene encoding the enzyme lactate dehydrogenase, which catalyzes the conversion of pyruvate to lactate. As you learned in chapter 7, this reaction is a key step in energy metabolism, particularly when oxygen is in short supply. There are two common alleles of lactate dehydrogenase in these fish populations, with allele *a* being a better catalyst at lower temperatures than allele *b*.

In an experiment, investigators sampled the frequency of allele *a* in 41 fish populations located over 14 degrees of latitude, from Jacksonville, Florida (31° N), to Bar Harbor, Maine (44° N). Annual mean water temperatures change 1°C per degree change in latitude. The survey was designed to test a prediction of the hypothesis that natural selection acts on this enzyme polymorphism. If it does, then you would expect that allele *a*, producing a better "low-temperature" enzyme, would be more common in the colder waters of the more northern latitudes. The graph presents the results of this survey. The points on the graph are derived from the pie chart data such as that shown for 20 populations in the map. The blue line on the graph is the line that best fits the data.

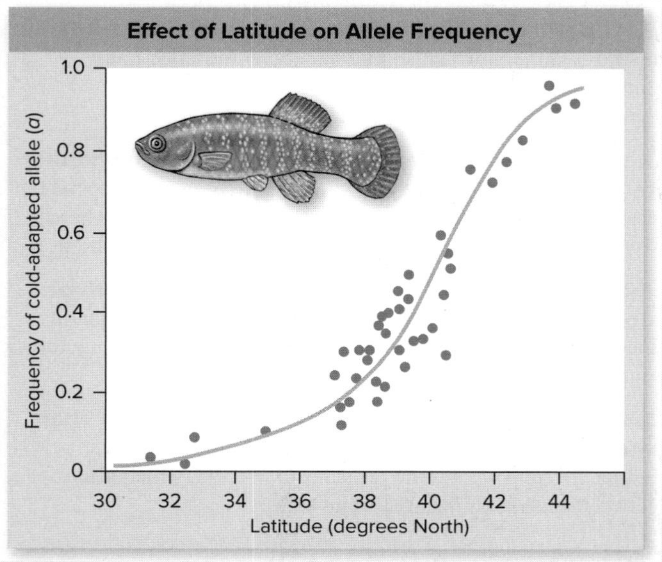
Effect of Latitude on Allele Frequency
Frequency of cold-adapted allele (a)
1.0
0.8
0.6
0.4
0.2
0
30
32
34
36
38
40
42
44
Latitude (degrees North)

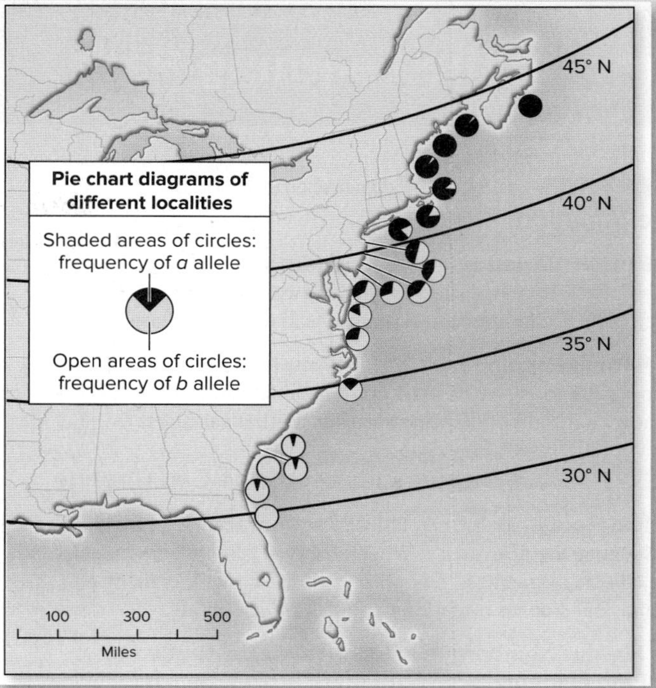
45° N
40° N
35° N
30° N
Pie chart diagrams of different localities
Shaded areas of circles: frequency of a allele
Open areas of circles: frequency of b allele
100
300
500
Miles

Analysis

1. **Applying Concepts**
   a. ***Variable.*** In the graph, what is the dependent variable?
   b. ***Reading pie charts.*** In the fish population located at 35° N latitude, what is the frequency of the *a* allele? Locate this point on the graph.
   c. ***Analyzing a continuous variable.*** Compare the frequency of allele *a* among fish captured in waters at 44° N latitude with the frequency among fish captured at 31° N latitude. Is there a pattern? Describe it.
2. **Interpreting Data** At what latitude do fish populations exhibit the greatest variability in allele *a* frequency?
3. **Making Inferences**
   a. Are fish populations in cold waters at 44° N latitude more or less likely to contain heterozygous individuals than fish populations in warm waters at 31° N latitude? What causes this difference, or lack of it?
   b. Where along this latitudinal gradient in the frequency of allele *a* would you expect to find the highest frequency of heterozygous individuals? Why?
4. **Drawing Conclusions** Are the differences in population frequencies of allele *a* consistent with the hypothesis that natural selection is acting on the alleles encoding this enzyme? Explain.
5. **Further Analysis** If you were to release fish captured at 32° N into populations located at 44° N, so that the local population now had equal frequencies of the two alleles, what would you expect to happen in future generations? How might you test this prediction?

Retracing the Learning Path

CONCEPT 19.1 Natural Populations Exhibit Genetic Variation

19.1.1 Genetic Variation Is the Raw Material of Evolution Darwin proposed that evolution of species occurs by the process of natural selection. Other processes can also lead to evolutionary change.

19.1.2 Natural Populations Contain Ample Genetic Variation For a population to evolve, it must contain genetic variation. DNA testing shows natural populations have substantial variation.

CONCEPT 19.2 Frequencies of Alleles Can Change

19.2.1 The Hardy–Weinberg Principle Characterizes Populations at Equilibrium When observed genotype frequencies match calculated frequencies, a population is in Hardy–Weinberg equilibrium. This implies evolutionary forces are not acting.

19.2.2 Hardy–Weinberg Predictions Can Be Applied to Data to Find Evidence of Evolutionary Processes If genotype frequencies are not in Hardy–Weinberg equilibrium, then evolutionary processes must be at work.

CONCEPT 19.3 Five Agents Are Responsible for Evolutionary Change

19.3.1 Mutation Can Change Allele Frequencies Mutations are the ultimate source of genetic variation. Mutation usually is not responsible for deviations from Hardy–Weinberg equilibrium.

19.3.2 Gene Flow Occurs When Alleles Move Between Populations Gene flow can introduce genetic variation and homogenize allele frequencies in populations.

19.3.3 Nonrandom Mating Shifts Genotype Frequencies Assortative mating, when similar individuals tend to mate, increases homozygosity; disassortative mating increases heterozygosity.

19.3.4 Genetic Drift May Alter Allele Frequencies in Small Populations Genetic drift refers to random shifts in allele frequency. Its effects may be severe in small populations.

19.3.5 Selection Favors Some Phenotypes over Others For natural selection to occur, genetic variation must exist, it must result in differential reproductive success, and it must be inheritable.

19.3.6 There Are Limits to What Selection Can Accomplish Pleiotropic genes set limits on how a phenotype can be altered. Intense selection pressure may remove the genetic variation required for selection. The fitness of an allele of one gene may vary depending on the genotype at other genes.

CONCEPT 19.4 Selection Can Act on Traits Affected by Many Genes

19.4.1 Disruptive Selection Removes Intermediate Phenotypes When intermediate phenotypes are at a disadvantage, a population may exhibit a bimodal trait distribution.

19.4.2 Directional Selection Eliminates Phenotypes at One End of a Range Directional selection tends to shift the mean value of the population toward the favored end of the distribution.

19.4.3 Stabilizing Selection Favors Individuals with Intermediate Phenotypes Stabilizing selection eliminates both extremes, increasing the frequency of intermediate types.

CONCEPT 19.5 Natural Selection Can Be Studied Experimentally

19.5.1 Guppy Color Variation in Different Environments Illustrates Natural Selection Guppies subject to predation are less colorful and smaller.

19.5.2 Experimentation Reveals the Agent of Selection Guppies in natural populations subject to different predators were shown to undergo color change over generations.

CONCEPT 19.6 Fitness Is a Measure of Evolutionary Success

19.6.1 A Phenotype with Greater Fitness Usually Increases in Frequency Fitness is defined as the reproductive success of an individual. Usually, the phenotype with highest relative fitness increases in frequency in the next generation, assuming that phenotypic differences are the result of genetic differences.

19.6.2 Fitness May Consist of Many Components Reproductive success is determined by how long an individual survives, how often it mates, and surviving offspring per reproductive event.

CONCEPT 19.7 Evolutionary Processes Sometimes Maintain Variation

19.7.1 Mutation and Genetic Drift May Counter the Influence of Natural Selection In theory, high mutation rates can oppose natural selection, but this rarely happens. Genetic drift also can work counter to natural selection.

19.7.2 Gene Flow May Promote or Constrain Evolutionary Change Gene flow can spread beneficial mutations, but an influx of low fitness alleles can impede adaptation to an environment.

19.7.3 Frequency-Dependent Selection May Favor Either Rare or Common Phenotypes Negative frequency-dependent selection favors rare phenotypes and maintains variation, whereas positive selection for the common phenotype reduces variation.

19.7.4 In Oscillating Selection, the Favored Phenotype Changes as the Environment Changes If environmental change is cyclical, selection favors first one phenotype, then another, maintaining variation.

19.7.5 Heterozygotes May Exhibit Greater Fitness Than Homozygotes Sickle-cell anemia exhibits heterozygote advantage. Heterozygotes are resistant to malaria, which is a selective advantage in malaria-prone areas.

CONCEPT 19.8 Sexual Selection Determines Reproductive Success

19.8.1 The Sexes Often Have Different Reproductive Strategies One sex may be choosier than the other, often depending on the degree of parental investment.

19.8.2 Sexual Selection Occurs Through Mate Competition and Mate Choice Intrasexual selection acts by competition within a sex, usually involving males. Intersexual selection is usually mate choice by females and can have direct or indirect benefits.

Concept Overview

This Concept Overview diagrams the key concepts that were discussed in this chapter

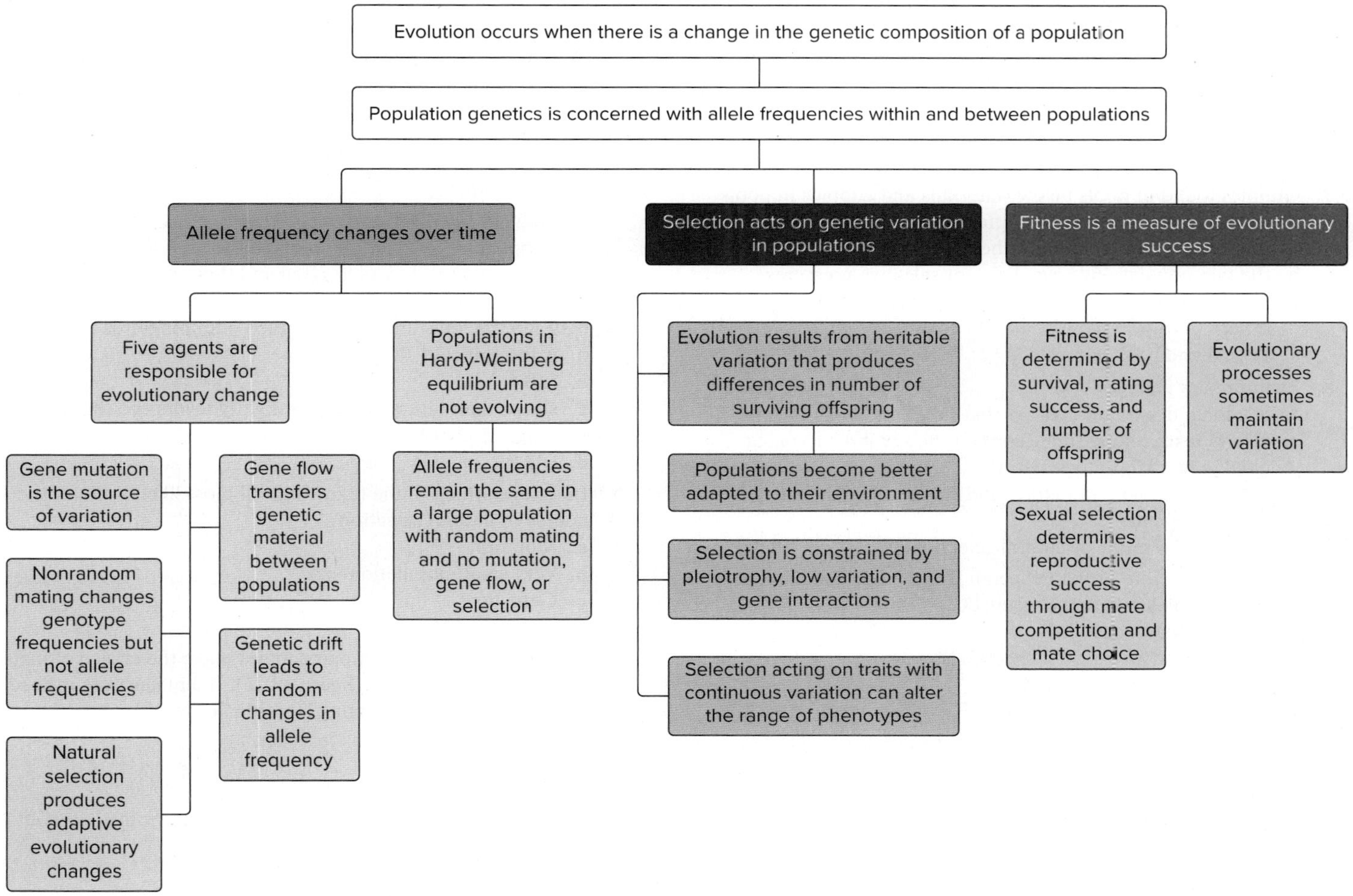

Assessing the Learning Path

Understand

1. Genetic variation
   a. is created by natural selection.
   b. must be present for natural selection to act.
   c. is lower in sexually reproducing organisms than in asexually reproducing organisms.
   d. is lower in organisms that have many heterozygous loci than in organisms with few.

2. In the equation $p^2 + 2pq + q^2$, the heterozygote is represented by
   a. p^2.
   b. $2pq$.
   c. q^2.
   d. $p^2 + q^2$.

3. Which of the following agents are NOT responsible for evolutionary change?
   a. Mutation
   b. Migration
   c. Assortative mating
   d. Random mating

4. Gene flow
   a. occurs when a population is in Hardy–Weinberg equilibrium.
   b. reduces genetic variation between populations.
   c. often leads to speciation.
   d. leads to increased rates of mutation.
5. In disruptive selection,
   a. intermediate forms of a trait increase in frequency.
   b. intermediate forms of a trait decrease in frequency.
   c. one extreme form of a trait is favored.
   d. allele frequencies remain unchanged.
6. Which type of natural selection is most likely to lead to speciation (generation of new species)?
   a. Directional selection
   b. Stabilizing selection
   c. Disruptive selection
   d. Any type of selection that reduces genetic variation in a population
7. Guppies found in pools below waterfalls and guppies in pools above waterfalls have evolved differently because
   a. there is abundant gene flow between the populations.
   b. the two populations have different selective pressures.
   c. natural selection has acted in one population but not the other.
   d. Both a and b
8. Evolutionary change
   a. is a long, drawn-out process that takes millions of years.
   b. always takes a long time (eons) in nature but can occur quickly in a laboratory setting.
   c. can occur rapidly in nature, and this can be confirmed by lab experiments.
   d. is solely the result of natural selection.
9. Cheetah A is more fit than cheetah B. Based on this statement, you know that cheetah A
   a. is bigger than cheetah B.
   b. has more offspring than cheetah B.
   c. is stronger than cheetah B.
   d. has evolved more than cheetah B.
10. The elaborate tail feathers of a male peacock evolved because they
    a. improve the reproductive success of males and females.
    b. improve male survival.
    c. reduce survival.
    d. None of the above

Apply

1. In a population of red (dominant allele) or white flowers in Hardy–Weinberg equilibrium, the frequency of red flowers is 91%. What is the frequency of the red allele?
   a. 9%
   b. 30%
   c. 91%
   d. 70%
2. In the population described in question 1, which proportion of the population is heterozygous for the red allele?
   a. 9%
   b. 21%
   c. 42%
   d. 58%
3. If you came across a population of plants and discovered a surprisingly high level of homozygosity, what would you predict about their mating system?
   a. Their pollen is dispersed by wind.
   b. They probably reproduce asexually.
   c. They are predominantly self-fertilizing.
   d. They are predominantly outcrossing.
4. Genetic drift and natural selection can both lead to rapid rates of evolution. However,
   a. genetic drift works fastest in large populations.
   b. only genetic drift leads to adaptation.
   c. natural selection requires genetic drift to produce new variation in populations.
   d. both processes of evolution can be slowed by gene flow.
5. Farmers have bred hogs to be leaner (have less fat in their meat) over time. This is an example of
   a. directional selection.
   b. disruptive selection.
   c. artificial selection.
   d. Both a and c
6. What would happen to average birth weight if over the next several years advances in medical technology reduced infant mortality rates of large babies to equal those of intermediate-size babies (figure 19.13)? Assume that differences in birth weight have a genetic basis.
   a. Over time, average birth weight would only increase.
   b. Over time, average birth weight would only decrease.
   c. Both a and b
   d. None of the above
7. In a population of hummingbirds, those with long tongues leave an average of two offspring and those with short tongues leave an average of two offspring. The difference in fitness between the long- and short-tongued hummingbirds
   a. is insignificant.
   b. is 0.5.
   c. shows that natural selection slightly favors the short-tongued phenotype.
   d. is 2.0.
8. In nature, which of the following will most likely counter the course of natural selection?
   a. Mutation
   b. Very small population size
   c. Gene flow
   d. All of the above
9. Among Mormon cricket pairs, males make the choice of mate, selecting larger females (figure 19.21). If females were instead to do the choosing, selecting larger males would
   a. increase fitness.
   b. decrease fitness.
   c. have no effect on fitness.
   d. affect fitness in a way that cannot be determined without more information.

Synthesize

1. Can Darwinian evolution by natural selection occur among genetically identical clones? Explain your reasoning.
2. If the frequency of individuals homozygous for a recessive allele were reduced from 1 in 1000 to 1 in 10,000, what would be the expected change in the proportion of individuals heterozygous for the allele?
3. If you found that a series of small islands were each occupied by a genetically distinct population of land snails, what kind of evidence would you gather to attempt to determine whether the differences had resulted from natural selection or from genetic drift?
4. It is often said that "fitness is relative." What does this mean?
5. Can you suggest an evolutionary reason that many vertebrate reproductive groups are composed of one male and numerous females, rather than the reverse?

Aneese/iStock/Getty Images

20 The Evidence for Evolution

Learning Path

- **20.1** The Beaks of Darwin's Finches Provide Evidence of Natural Selection
- **20.2** Peppered Moths and Industrial Melanism Illustrate Natural Selection in Action
- **20.3** Human-Initiated Artificial Selection Is Also a Powerful Agent of Change
- **20.4** Fossils Provide Direct Evidence of Evolution
- **20.5** Anatomical Evidence for Evolution Is Extensive and Persuasive
- **20.6** Genes Carry a Molecular Record of the Evolutionary Past
- **20.7** Natural Selection Favors Convergent Evolution in Similar Environments
- **20.8** Addressing Common Criticisms of Evolutionary Theory

Concept Overview

This Concept Overview provides a diagram of the overarching concepts that are covered in this chapter. These concepts will be expanded upon in the Concept Overview at the end of the chapter.

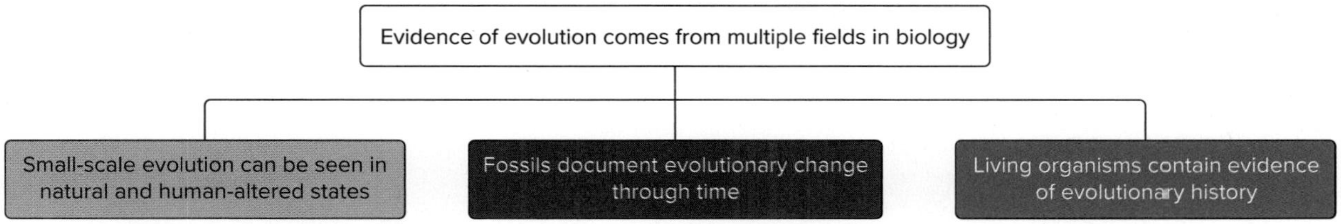

Introduction

As we discussed in chapter 1, when Darwin proposed his revolutionary theory of evolution by natural selection, little actual evidence existed to bolster his case. Instead, Darwin relied on observations of the natural world, logic, and results obtained by breeders working with domestic animals. Since his day, however, the evidence for Darwin's theory has become overwhelming.

The case is built upon two pillars: first, evidence that natural selection can produce evolutionary change, and second, evidence from the fossil and molecular records that evolution has occurred. In addition, information from many different areas of biology—fields as different as anatomy, molecular biology, and biogeography—is interpretable scientifically only as being the outcome of evolution.

20.1 The Beaks of Darwin's Finches Provide Evidence of Natural Selection

As you learned in chapter 19, a variety of processes can produce evolutionary change. Most evolutionary biologists, however, agree with Darwin's thinking that natural selection is the primary process responsible for evolution. Although we cannot travel back through time, modern-day evidence allows us to test hypotheses about how evolution proceeds, and it confirms the power of natural selection as an agent of evolutionary change. This evidence comes from both the field and the laboratory and from both natural and human-altered situations.

Galápagos Finches Exhibit Variation Related to Food Gathering

LEARNING OBJECTIVE 20.1.1 Describe the different feeding adaptations in Darwin's finches.

Darwin's finches are a classic example of evolution by natural selection. When he visited the Galápagos Islands off the coast of Ecuador in 1835, Darwin collected 31 specimens of finches from three islands. Not being an expert on birds, Darwin, had trouble identifying the specimens and believed from an examination of their bills that his collection contained wrens, "gross-beaks," and blackbirds.

Upon Darwin's return to England, ornithologist John Gould informed Darwin that the species he had collected were not closely related to different English birds, but rather were a closely related group of distinct species, all similar to one another except for their bills. In all, 14 species are now recognized.

The diversity of Darwin's finches is illustrated in figure 20.1. Ground finches feed on seeds that they crush in their powerful beaks, whereas species with smaller and narrower bills, such as the warbler finch, eat insects. Some species eat fruit and buds, and others feed on cactus fruits and the insects they attract. Some populations of the sharp-beaked ground finch even include "vampires" that can creep up on seabirds and use their sharp beaks to pierce the seabirds' skin and drink their blood. Perhaps most remarkable are the tool users: woodpecker finches that will pick up a twig, cactus spine, or leaf stalk, trim it with their bills, and then poke it into dead branches to pry out grubs.

The correspondence between the beaks of the finch species and their food sources suggested to Darwin that natural selection had shaped them. In *The Voyage of the Beagle,* published several

Large ground finch (*Geospiza magnirostris*)
Warbler finch (*Certhidea olivacea*)
Vegetarian tree finch (*Platyspiza crassirostris*)
Cactus finch (*Geospiza scandens*)
Woodpecker finch (*Cactospiza pallida*)

Figure 20.1 Darwin's finches. These species show differences in bills and feeding habits among Darwin's finches. This diversity arose when an ancestral species of finch colonized the islands and diversified into habitats lacking other types of small birds. The bills of several species resemble those of different families of birds on the mainland. For example, the warbler finch has a beak very similar to that of warblers, to which it is not closely related.

years after his return to England, Darwin wrote, "Seeing this gradation and diversity of structure in one small, intimately related group of birds, one might really fancy that from an original paucity of birds in this archipelago, one species has been taken and modified for different ends."

Modern Research Has Verified Darwin's Selection Hypothesis

LEARNING OBJECTIVE 20.1.2 Explain how climatic variation drives evolutionary change in the medium ground finch.

Darwin's observations suggest that differences among species in beak size and shape have evolved as the species adapted to use different food resources, but can this hypothesis be tested? In chapter 19, you learned that evolution by natural selection requires genetic variation in a population, and that this variation must lead to differences in lifetime reproductive success.

The key to successfully testing Darwin's proposal proved to be patience. For more than 40 years, starting in 1973, Peter and Rosemary Grant of Princeton University and their students have studied the medium ground finch, *Geospiza fortis,* on a tiny island in the center of the Galápagos called Daphne Major. These finches feed preferentially on small, tender seeds produced in abundance by plants in wet years. The birds resort to larger, drier seeds, which are harder to crush, only when small seeds become depleted during long periods of dry weather, when plants produce few seeds.

The Grants quantified beak shape among the medium ground finches of Daphne Major by carefully measuring beak depth (height of beak, from top to bottom, at its base) on individual birds. Measuring many birds every year, they were able to assemble for the first time a detailed portrait of evolution in action. The Grants found that not only did a great deal of variation in beak depth exist among members of the population, but the average beak depth changed from one year to the next in a predictable fashion.

During droughts, plants produced few seeds, and all available small seeds were quickly eaten, leaving large seeds as the major remaining source of food. As a result, birds with deeper, more powerful beaks survived better because they were better able to break open these large seeds (figure 20.2*a*). Conversely, in particularly wet years, plants flourished, producing an abundance of small seeds; as a result, small-beaked birds were favored, and beak depth decreased greatly.

Could these changes in beak dimension reflect the action of natural selection? An alternative possibility might be that the changes in beak depth do not reflect changes in gene frequencies but, rather, are simply a response to diet—for example, perhaps crushing large seeds causes a growing bird to develop a larger beak.

To rule out this possibility, the Grants measured the relation of parent beak size to offspring beak size, examining many broods over several years. The depth of the beak was very similar between parents and offspring regardless of environmental conditions (figure 20.2*b*), suggesting that the differences among individuals in beak size reflect genetic differences, and therefore that the year-to-year changes in average beak depth represent evolutionary change resulting from natural selection.

Natural Selection Can Produce Adaptive Radiations in New Environments

LEARNING OBJECTIVE 20.1.3 Describe how the diversity of Darwin's finches arose in the Galápagos Islands.

Darwin believed that each Galápagos finch species had adapted to the particular foods and other conditions on the

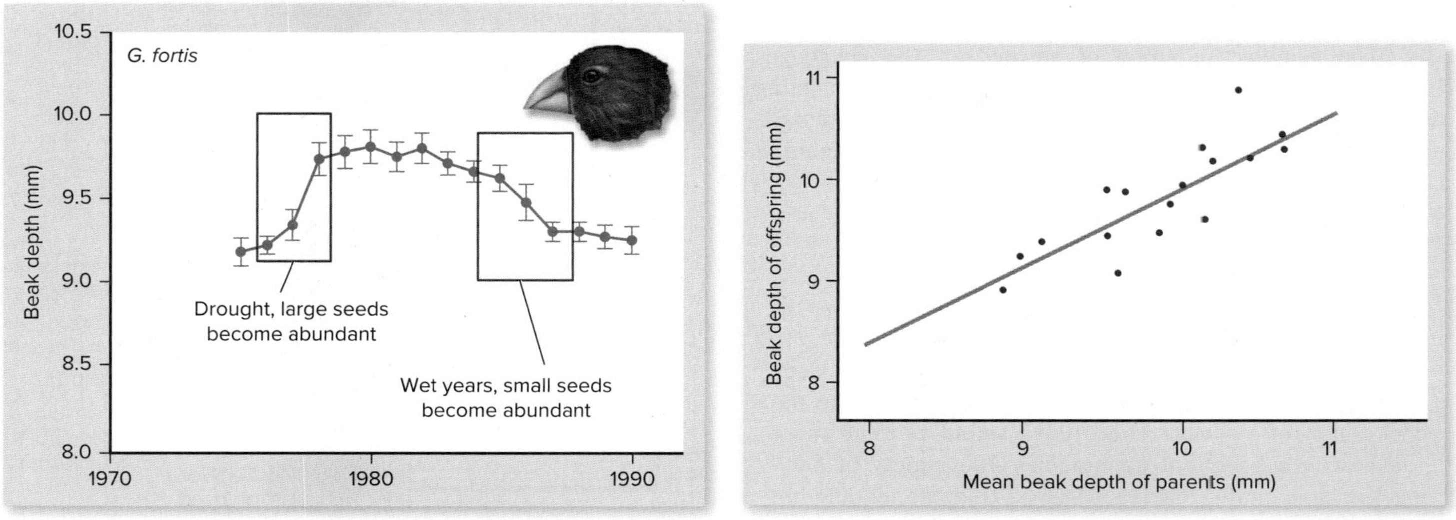

Figure 20.2 Evidence that natural selection alters beak shape in the medium ground finch, *Geospiza fortis.* ***a.*** In dry years, when only large, tough seeds are available, the mean beak depth increases. In wet years, when many small seeds are available, mean beak depth decreases. ***b.*** Beak depth is inherited by offspring from their parents. Like many quantitative traits (refer to chapter 12, for a discussion of quantitative traits), beak depth is probably determined by many genes and, on average, offspring tend to have a beak depth equal to the mean of their parents' beak depth.

island it inhabited. Because the islands presented different opportunities, a cluster of species resulted. Presumably, the ancestor of Darwin's finches reached these newly formed islands before other land birds, so that when it arrived, all of the niches occupied by birds on the mainland were unoccupied. A **niche** is what a biologist calls the way a species makes a living—the biological conditions (other organisms) and physical conditions (climate, food, shelter, etc.) organisms interact with as they attempt to survive and reproduce. As the new arrivals to the Galápagos moved into vacant niches and adopted new lifestyles, they were subjected to diverse sets of selective pressures. Under these circumstances, the ancestral finches rapidly split into a series of populations, some of which evolved into separate species.

A group of species evolving to occupy a variety of different habitats within a region is called an **adaptive radiation.** Figure 20.3 shows how the 14 species of Darwin's finches on the Galápagos Islands and Cocos Island are thought to have evolved. The ancestral population, indicated by the base of the brackets, migrated to the islands about 2 MYA and underwent adaptive radiation, giving rise to the 14 different species.

The descendants of the original finches that reached the Galápagos Islands now occupy many different kinds of habitats on the islands. The 14 species that inhabit the Galápagos Islands and Cocos Island occupy four types of niches, which can be specified by bird type:

1. **Ground finches.** Most of the ground finches have stout beaks suitable for feeding on seeds. As you have discovered, the size of their beaks is related to the size of the seeds they eat.
2. **Tree finches.** Tree finches have narrower beaks that are suitable for feeding on insects.
3. **Vegetarian finch.** The very heavy bill of this species is used to wrench buds from branches.
4. **Warbler finches.** Warbler finches have small, slender beaks used to capture elusive insects on leaves.

The adaptive radiation of finches on the Galápagos Islands has been the subject of extensive research, including detailed comparisons of species at the DNA level. This remarkable example of diversity creation is examined in more detail in chapter 21.

REVIEW OF CONCEPT 20.1

Among Darwin's finches, natural selection has been responsible for changes in the shape of the beak corresponding to characteristics of the available food supply. Because beak morphology is a heritable trait, a beak better suited to the distribution of available seed types would become more common in subsequent generations. The variety of habitats available in the Galápagos Islands selected for different types of finches, eventually producing the 14 species that Darwin encountered.

- *Suppose that the act of eating hard seeds caused birds to develop bigger beaks. Would this lead to an evolutionary increase in beak size after a drought?*

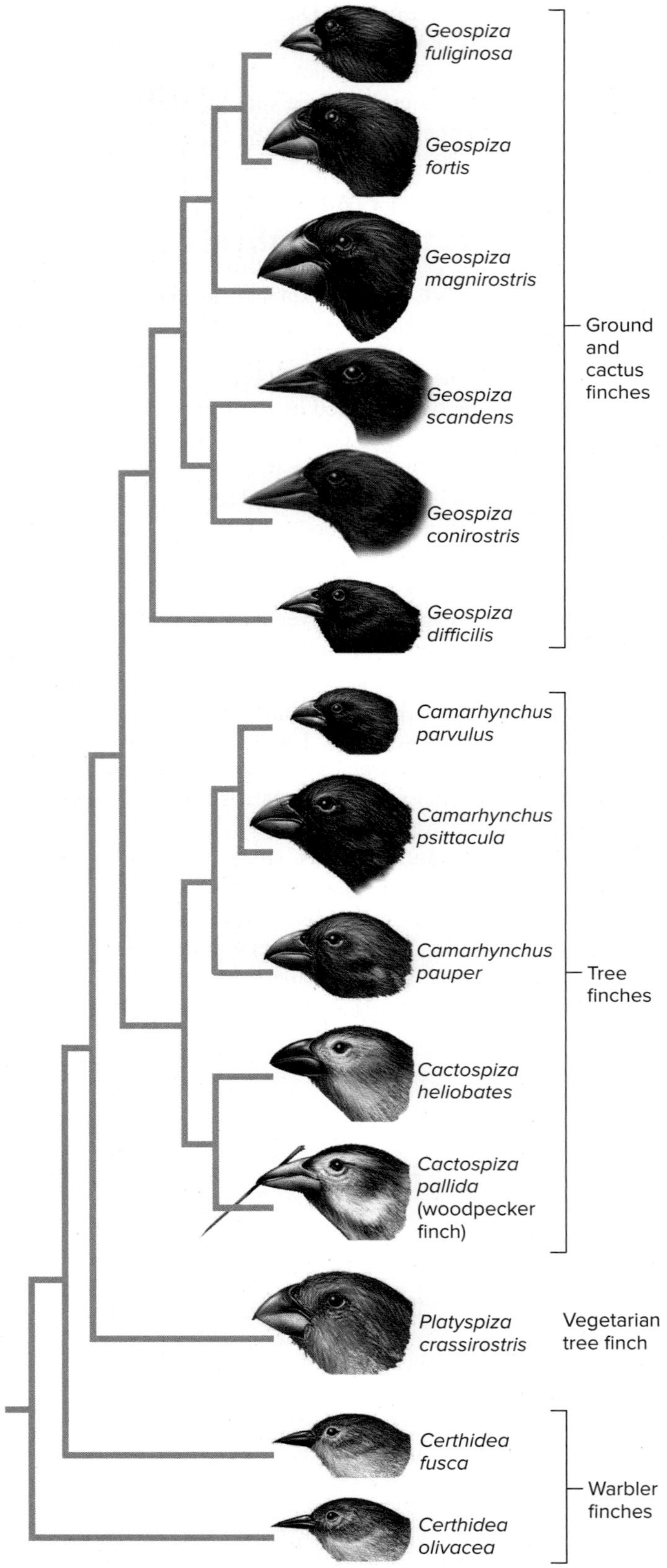

Figure 20.3 An evolutionary tree of Darwin's finches. This family tree was constructed by comparing DNA of the 14 species. Their position at the base of the finch tree suggests that warbler finches were among the first adaptive types to evolve in the Galápagos.

20.2 Peppered Moths and Industrial Melanism Illustrate Natural Selection in Action

When the environment changes, natural selection often may favor different traits in a species. One classic example concerns the peppered moth, *Biston betularia.* Adults come in a range of shades, from light gray with black speckling (hence the name "peppered" moth) to jet black (melanic).

Light-Colored Moths Are Less Frequent in Polluted Areas

LEARNING OBJECTIVE 20.2.1 Explain the relationship between altered environment and evolution in peppered moths.

Recent molecular genetics studies have demonstrated that all-black moths are the descendants of a single mutation. This dominant allele was present but very rare in populations before 1850. From that time on, dark individuals increased in frequency in moth populations near industrialized centers until they made up almost 100% of these populations. In industrialized regions where the dark moths were common, the tree trunks were darkened almost black by the soot of pollution, which also killed many of the light-colored lichens on tree trunks.

Why did dark moths gain a survival advantage around 1850? In 1896 an amateur moth collector named J. W. Tutt proposed what became the most commonly accepted hypothesis explaining the decline of the light-colored moths. He suggested that peppered forms were more visible to predators on sooty trees that have lost their lichens. Consequently, birds ate the peppered moths resting on the trunks of trees during the day. The black forms, in contrast, had an advantage because they were camouflaged (figure 20.4).

Although Tutt initially had no evidence, ecologist Bernard Kettlewell tested the hypothesis in the 1950s by releasing equal numbers of dark and light individuals into two sets of woods in England: one near heavily polluted Birmingham and the other in unpolluted Dorset. Kettlewell then set up lights in the woods to attract moths to traps to find out how many of both kinds of moths survived. To evaluate his results, he had marked the released moths with a dot of paint on the underside of their wings, where birds could not see it.

In the polluted area near Birmingham, Kettlewell recaptured only 19% of the light moths but 40% of the dark ones. This indicated that dark moths survived better in these polluted woods, where tree trunks were dark. In the relatively unpolluted Dorset woods, Kettlewell recovered 12.5% of the light moths but only 6% of the dark ones. This result indicated that where the tree trunks were still light-colored, light moth survival was twice as great as that of dark moths.

Kettlewell later solidified his argument by placing moths on trees and filming birds looking for food. Sometimes the birds actually passed right over a moth that was the same color as its background.

Kettlewell's finding that birds more frequently detect moths whose color does not match their background has subsequently been confirmed in many field studies. Recently, for example, an enormous six-year study by researcher Michael Majerus involving the release of 4864 moths solidly confirmed Kettlewell's findings. Conducted in an unpolluted forest, the study found that dark-colored moths disappeared at a rate 10% higher than the rate for light-colored moths. Majerus's data provide clear evidence implicating camouflage and bird predation in the rise and fall of melanism in moths.

When environmental conditions reverse, so does selection pressure

In industrialized areas throughout Eurasia and North America, dozens of other species of moths have evolved in the same way as the peppered moth. The term **industrial melanism** refers to the phenomenon in which darker individuals come to predominate over lighter ones. In the second half of the 20th century, with the widespread implementation of pollution controls, the trend toward melanism began reversing for many species of moths throughout the northern continents.

In England, the air pollution that promoted industrial melanism began to reverse following enactment of the Clean Air Act

Figure 20.4 Tutt's hypothesis explaining industrial melanism. These photographs show preserved specimens of the peppered moth (*Biston betularia*) placed on trees. Tutt proposed that the dark melanic variant of the moth is more visible to predators on unpolluted trees (left), whereas the light "peppered" moth is more visible to predators on bark blackened by industrial pollution (right).

(peppered moth): IanRedding/Shutterstock; (blackened bark): The Natural History Museum/Alamy Stock Photo in 1956. Beginning in 1959, the *Biston* population at Caldy Common outside Liverpool has been sampled each year. The frequency of the melanic (dark) form dropped from a high of 93% in 1959 to a low of 15% in 1995 (figure 20.5).

The drop correlates well with a significant drop in air pollution, particularly with a lowering of the levels of sulfur dioxide and suspended particulates, both of which act to darken trees. The drop is consistent with a 15% selective disadvantage acting against moths with the dominant melanic allele, very much in line with Majerus's more recent result.

Interestingly, the same reversal of melanism occurred in the United States. Of 576 peppered moths collected at a field station near Detroit from 1959 to 1961, 515 were melanic, a frequency of 89%. The American Clean Air Act, passed in 1963, led to significant reductions in air pollution. Resampled in 1994, the Detroit field station peppered moth population had only 15% melanic moths (figure 20.5). The moth populations in Liverpool and Detroit, both part of the same natural experiment, exhibit strong evidence for natural selection.

The agent of selection may be difficult to pin down

Although the evidence for natural selection in the case of the peppered moth is strong, Tutt's hypothesis about the agent of selection is currently being refined. Researchers have noted that the recent selection against melanism does not appear to correlate with changes in tree lichens.

At Caldy Common, the light form of the peppered moth began to increase in frequency long before lichens began to reappear on the trees. At the Detroit field station, the lichens never changed significantly as the dark moths first became dominant and then declined over a 30-year period. In fact, investigators have not been able to find peppered moths on Detroit trees at all, whether covered with lichens or not. Some evidence suggests the moths rest on leaves in the treetops during the day, but no one is sure. Could poisoning by pollution rather than predation by birds be the agent of natural selection on the moths? Perhaps—but to date, only predation by birds is backed by experimental evidence.

Researchers supporting the bird predation hypothesis point out that a bird's ability to detect moths may depend less on the presence or absence of lichens, and more on other ways in which the environment is darkened by industrial pollution. Pollution tends to cover all objects in the environment with a fine layer of particulate dust, which tends to decrease how much light is reflected by surfaces. In addition, pollution has a particularly severe effect on birch trees, which are light in color. Both effects would tend to make the environment darker, and thus would favor darker moths by protecting them from predation by birds.

Despite this lingering uncertainty over how the agent of selection is acting, the overall pattern is clear. Kettlewell's and Majerus's experiments establish indisputably that selection favors dark moths in heavily polluted habitats and light moths in areas with cleaner air. The increase and subsequent decrease in the frequency of melanic moths, correlated with levels of pollution independently on two continents, demonstrates clearly that this selection drives evolutionary change.

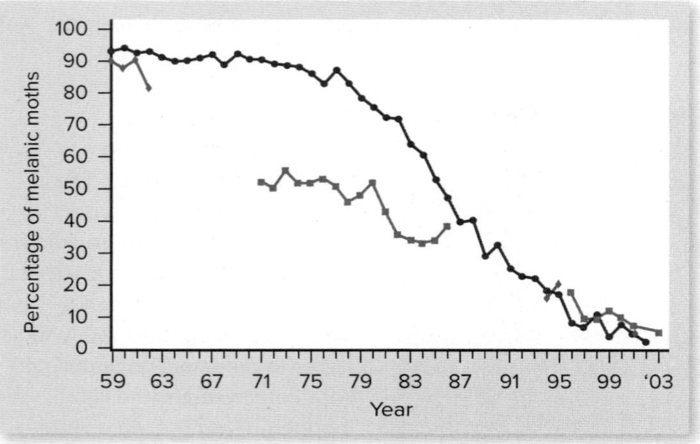

Figure 20.5 Selection against melanism. The red circles indicate the frequency of melanic *Biston betularia* moths at Caldy Common in England. Green diamonds indicate frequencies of melanic *B. betularia* in Michigan, and the blue squares indicate corresponding frequencies in Pennsylvania.

REVIEW OF CONCEPT 20.2

Natural selection has favored the dark form of the peppered moth in areas subject to severe air pollution, perhaps because on darkened trees they are less easily seen by moth-eating birds. As pollution has abated, selection has in turn shifted to favor the light form. Although selection is clearly occurring, further research is required to understand whether predation by birds is the agent of selection.

- *How would you test the idea that predation by birds is the agent of selection on moth coloration?*

20.3 Human-Initiated Artificial Selection Is Also a Powerful Agent of Change

Humans have imposed selection upon plants and animals since the dawn of civilization. Just as in natural selection, artificial selection operates by favoring individuals with certain phenotypic traits, allowing them to reproduce and pass their genes on to the next generation. Assuming that phenotypic differences are genetically determined, this directional selection should lead to evolutionary change, and indeed it has.

Experimental Selection Can Produce Changes in Populations

LEARNING OBJECTIVE 20.3.1 Contrast the processes of artificial and natural selection.

Artificial selection, imposed in laboratory experiments, agriculture, and the domestication process, has produced substantial change in almost every case in which it has been applied. This success is strong proof that selection is an effective evolutionary process.

With the rise of genetics as a field of science in the 1920s and 1930s, researchers began conducting experiments to test the hypothesis that selection can produce evolutionary change. A favorite subject was the laboratory fruit fly, *Drosophila melanogaster.* Geneticists have imposed selection on just about every conceivable aspect of the fruit fly—including body size, eye color, growth rate, life span, and exploratory behavior—with a consistent result: selection for a trait leads to strong and predictable evolutionary response.

In one classic experiment, scientists selected for fruit flies with many bristles (stiff, hairlike structures) on their abdomens. At the start of the experiment, the average number of bristles was 9.5. Each generation, scientists picked out the 20% of the population with the greatest number of bristles and allowed them to reproduce, thus establishing the next generation. After 86 generations of this directional selection, the average number of bristles had quadrupled, to nearly 40! In another experiment, fruit flies in one population were selected for high numbers of bristles, while fruit flies in the other cage were selected for low numbers of bristles. Within 35 generations, the populations did not overlap at all in range of variation (figure 20.6).

SCIENTIFIC THINKING

Question: *Can artificial selection lead to substantial evolutionary change?*

Hypothesis: *Strong directional selection will quickly lead to a large shift in the mean value of the population.*

Experiment: *In one population, every generation select the 20% of the population with the most bristles and allow them to reproduce to form the next generation. In the other population, do the same with the 20% with the smallest number of bristles.*

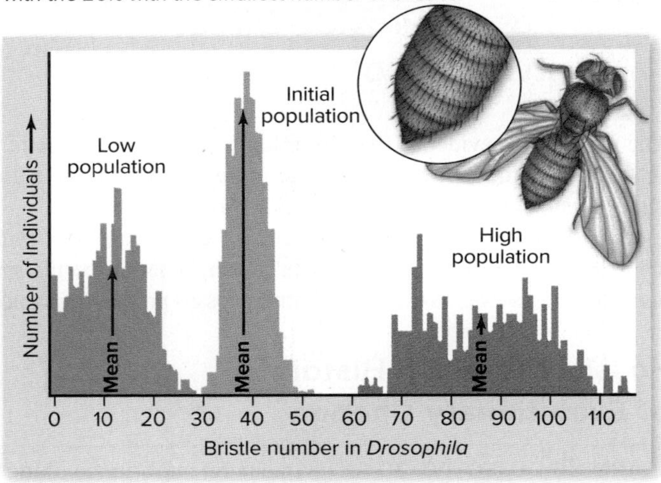

Result: *After 35 generations, mean number of bristles has changed substantially in both populations.*

Interpretation: *Note that at the end of the experiment, the range of variation lies outside the range discovered in the initial population. Selection can move a population beyond its original range because mutation and recombination continuously introduce new variation into populations.*

Figure 20.6 Artificial selection can lead to rapid and substantial evolutionary change.

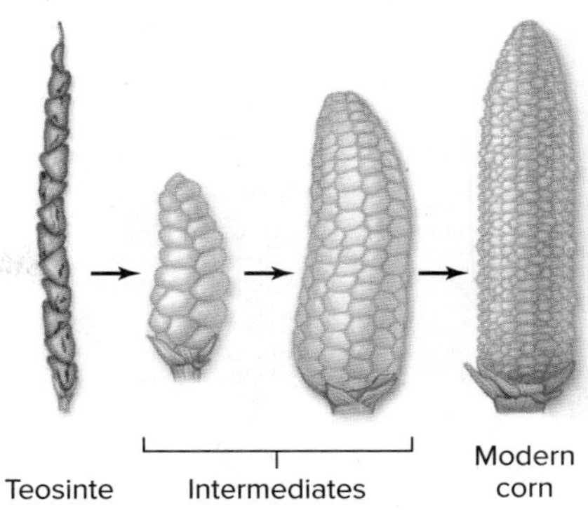

Figure 20.7 Corn looks very different from its ancestor. Teosinte, which can be found today in a remote part of Mexico, is very similar to the ancestor of modern corn. Artificial selection has transformed it into the form we know today.

Similar experiments have been conducted on a wide variety of other laboratory organisms. For example, by selecting for rats that were resistant to tooth decay, in less than 20 generations scientists were able to increase the average time for onset of decay from barely over 100 days to more than 500 days.

Agricultural selection has led to extensive modification of crops and livestock

Familiar livestock, such as cattle and pigs, and crops, such as corn and strawberries, are greatly different from their wild ancestors (figure 20.7). These differences have resulted from generations of human selection for such desirable traits as greater milk production and larger corn ear size.

An experiment with corn demonstrates the ability of artificial selection to rapidly produce major change in crop plants. In 1896 agricultural scientists began selecting for the oil content of corn kernels, which initially was 4.5%. Just as in the fruit fly experiments, the top 20% of all individuals were allowed to reproduce. By 1986, at which time 90 generations had passed, average oil content of the corn kernels had increased approximately 450%.

Domesticated breeds have arisen from artificial selection

Human-imposed selection has produced a great variety of breeds of cats, dogs (figure 20.8), pigeons, and other domestic animals. Some breeds have been developed for particular purposes. Greyhound dogs, for example, resulted from selection for maximal running ability, resulting in an animal with long legs, a long tail for balance, an arched back to increase stride length, and great muscle mass. By contrast, the odd proportions of the ungainly dachshund resulted from selection for dogs that could enter narrow holes in pursuit of badgers. Other breeds have been selected primarily for their appearance, such as the many colorful breeds of pigeons and cats.

Domestication also has led to unintentional selection for some traits. In recent years, as part of an attempt to domesticate the silver fox, Russian scientists chose the most docile animals in each generation and allowed them to reproduce. Within 40 years, most foxes were exceptionally tame, not only allowing themselves to be petted but also whimpering to get attention and sniffing and licking their caretakers (figure 20.9). In many respects, they had become no different from domestic dogs.

It was not only their behavior that changed, however. These foxes also began to exhibit other traits found in some dog breeds,

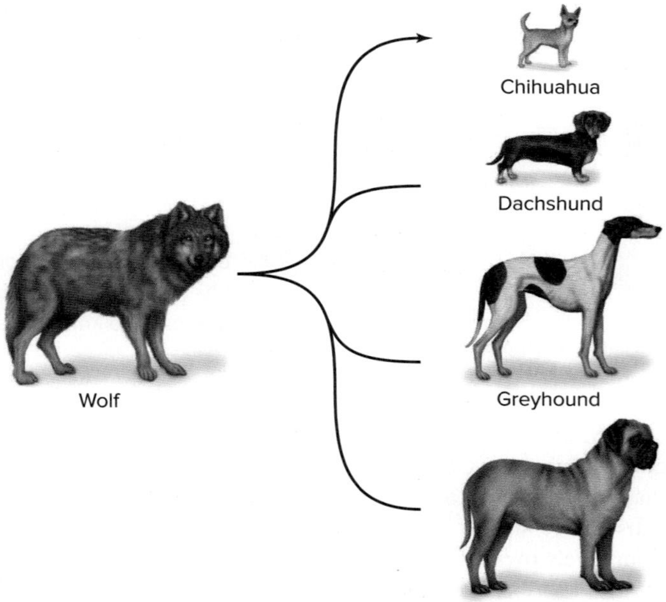

Figure 20.8 Breeds of dogs. The differences among dog breeds are greater than the differences displayed among wild species of canids.

Figure 20.9 Domesticated foxes. After 40 years of selectively breeding the tamest individuals, artificial selection has produced silver foxes that not only are as friendly as domestic dogs but also exhibit many physical traits in dog breeds.

Artyom Geodakyan/ITAR-TASS News Agency/Alamy Stock Photo

such as different color patterns, floppy ears, curled tails, and shorter legs and tails. Presumably, the genes responsible for docile behavior either affect these traits as well or are closely linked to the genes for these other traits (the phenomena of pleiotropy and linkage, which are discussed in chapters 12 and 13).

Can selection produce major evolutionary changes?

Given that we can observe the results of selection operating over a relatively short time, most scientists think that natural selection is the process responsible for the evolutionary changes documented in the fossil record. Some critics of evolution accept that selection can lead to changes within a species, but they contend that such changes are relatively minor in scope and not equivalent to the substantial changes documented in the fossil record. In other words, it is one thing to change the number of bristles on a fruit fly or the size of an ear of corn, but quite another to produce an entirely new species.

This argument does not fully appreciate the extent of change produced by artificial selection. Consider, for example, the existing breeds of dogs, all of which have been produced in the last few thousand years after wolves were first domesticated, perhaps 15,000 years ago. If the various dog breeds did not exist and a paleontologist found fossils of animals similar to dachshunds, greyhounds, mastiffs, and chihuahuas, there is no question that they would be considered different species. Indeed, the differences in size and shape exhibited by these breeds are greater than those between members of different genera in the family Canidae—such as coyotes, jackals, foxes, and wolves—which have been evolving separately for 5 to 10 million years. Consequently, the claim that artificial selection produces only minor changes is clearly incorrect. If selection operating over a period of only 15,000 years can produce such substantial differences, it should be powerful enough, over the course of many millions of years, to produce the diversity of life around us today.

REVIEW OF CONCEPT 20.3

In artificial selection, humans choose which plants or animals to mate in an attempt to affect specific traits. Rapid and substantial results can be obtained over a very short time, often in a few generations. From this, we can understand that natural selection is capable of producing major evolutionary change.

- *In what circumstances might artificial selection fail to produce a desired change?*

20.4 Fossils Provide Direct Evidence of Evolution

The most direct evidence that evolution has occurred is found in the fossil record. Today we have a far more complete understanding of this record than was available in Darwin's time.

Fossils are the preserved remains of once-living organisms. They include specimens preserved in amber, Siberian permafrost, and dry caves, as well as the more common fossils preserved as rocks.

Fossils Present a History of Evolutionary Change

LEARNING OBJECTIVE 20.4.1 Explain the importance of the discovery of transitional fossils.

Rock fossils are created when three events occur. First, the organism must become buried in sediment; then, the calcium in bone or other hard tissue must mineralize; and finally, the surrounding sediment must eventually harden to form rock.

The process of fossilization occurs only rarely. Usually, animal or plant remains decay or are scavenged before the process can begin. In addition, many fossils occur in rocks that are inaccessible to scientists. When they do become available, they are often destroyed by erosion and other natural processes before they can be collected. As a result, only a very small fraction of the species that have ever existed (estimated by some to be as many as 500 million) are known from fossils. Nonetheless, the fossils that have been discovered are sufficient to provide detailed information on the course of evolution through time.

The age of fossils can be estimated by rates of radioactive decay

By dating the rocks in which fossils occur, we can get an accurate idea of how old the fossils are. In Darwin's day, rocks were dated by their position with respect to one another (*relative dating*); rocks in deeper strata are generally older. Knowing the relative positions of sedimentary rocks and the rates of erosion of different kinds of sedimentary rocks in different environments, geologists of the 19th century derived a fairly accurate idea of the relative ages of rocks.

Today geologists take advantage of radioactive decay to establish the age of rocks (*absolute dating*). Many types of rock, such as the igneous rocks formed when lava cools, contain radioactive elements such as uranium-238. These isotopes transform at a precisely known rate into nonradioactive forms. For example, for U^{238} the *half-life* (that is, the amount of time needed for one-half of the original amount to be transformed) is 4.5 billion years. Once a rock is formed, no additional radioactive isotopes are added. Therefore, by measuring the ratio of the radioactive isotope to its derivative "daughter" isotope (figure 20.10), geologists can determine the age of the rock. If a fossil is found between two layers of rock, each of which can be dated, then the age at which the fossil formed can be determined.

When fossils are arrayed according to their age, from oldest to youngest, they often provide evidence of successive evolutionary change. At the largest scale, the fossil record documents the course of life through time, from the origin of first prokaryotic and then eukaryotic organisms, through the evolution of fishes, the rise of land-dwelling organisms, the reign of the dinosaurs, and on to the origin of humans. In addition, the fossil record shows the waxing and waning of biological diversity through time, such as the periodic mass extinctions that have reduced the number of living species.

Fossils document evolutionary transitions

Given the low likelihood of fossil preservation and recovery, it is not surprising that there are gaps in the fossil record. Nonetheless, intermediate forms are often available to illustrate how the major transitions in life occurred.

Undoubtedly the most famous of these is the oldest known bird, *Archaeopteryx* (meaning "ancient feather"), which lived around 165 MYA (figure 20.11). This specimen is clearly intermediate between birds and dinosaurs. Its feathers, similar in many respects to those of birds today, clearly reveal that it is a bird. Nonetheless, in many other respects—for example, possession of teeth, a bony tail, and other anatomical characteristics—it is indistinguishable from some carnivorous dinosaurs. Indeed, it is so similar to these dinosaurs that several specimens lacking preserved feathers were misidentified as dinosaurs and lay in the wrong natural history museum cabinet for several decades before the mistake was discovered!

Archaeopteryx reveals a pattern commonly found in intermediate fossils—rather than being intermediate in every trait, such fossils usually exhibit some traits like those of their ancestors and others like those of their descendants. In other words, traits evolve at different rates and different times; expecting an intermediate form to be intermediate in every trait would not be correct.

The first *Archaeopteryx* fossil was discovered in 1859, the year Darwin published *On the Origin of Species.* Since then, paleontologists have continued to fill in the gaps in the fossil record. Today the fossil record is far more complete, particularly among the vertebrates; fossils have been found linking all the major groups.

In recent years, spectacular discoveries have closed some of the major remaining gaps in our understanding of vertebrate

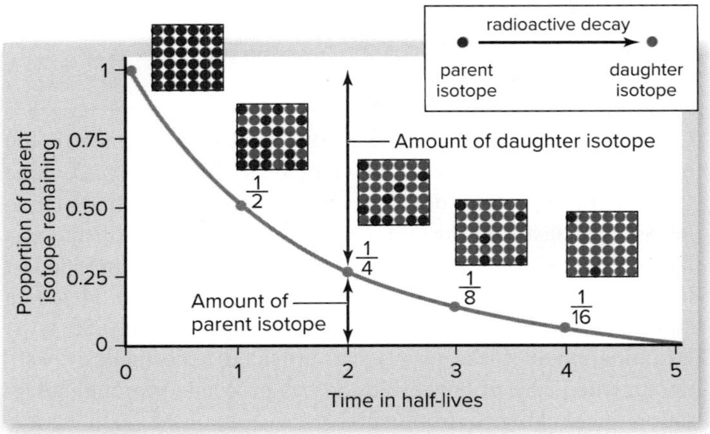

Figure 20.10 Radioactive decay. Radioactive elements decay at a known rate, called their half-life. After one half-life, one-half of the original amount of parent isotope has transformed into a nonradioactive daughter isotope. After each successive half-life, one-half of the remaining amount of parent isotope is transformed.

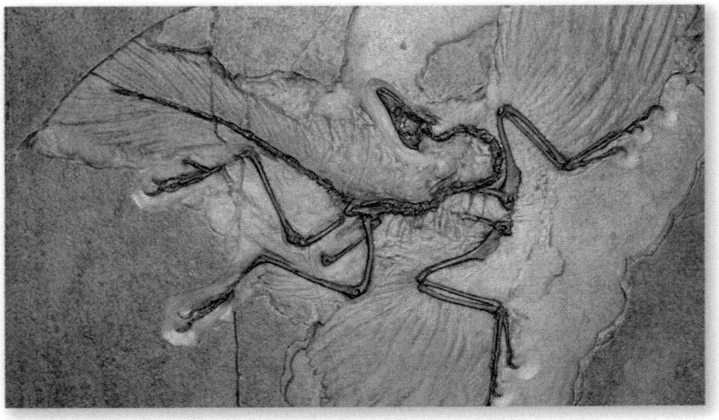

Figure 20.11 Fossil of *Archaeopteryx*, the first bird. The remarkable preservation of this specimen reveals soft parts usually not preserved in fossils; the presence of feathers makes it clear that *Archaeopteryx* was a bird, despite the presence of many dinosaurian traits.

Jason Edwards/National Geographic/Getty Images

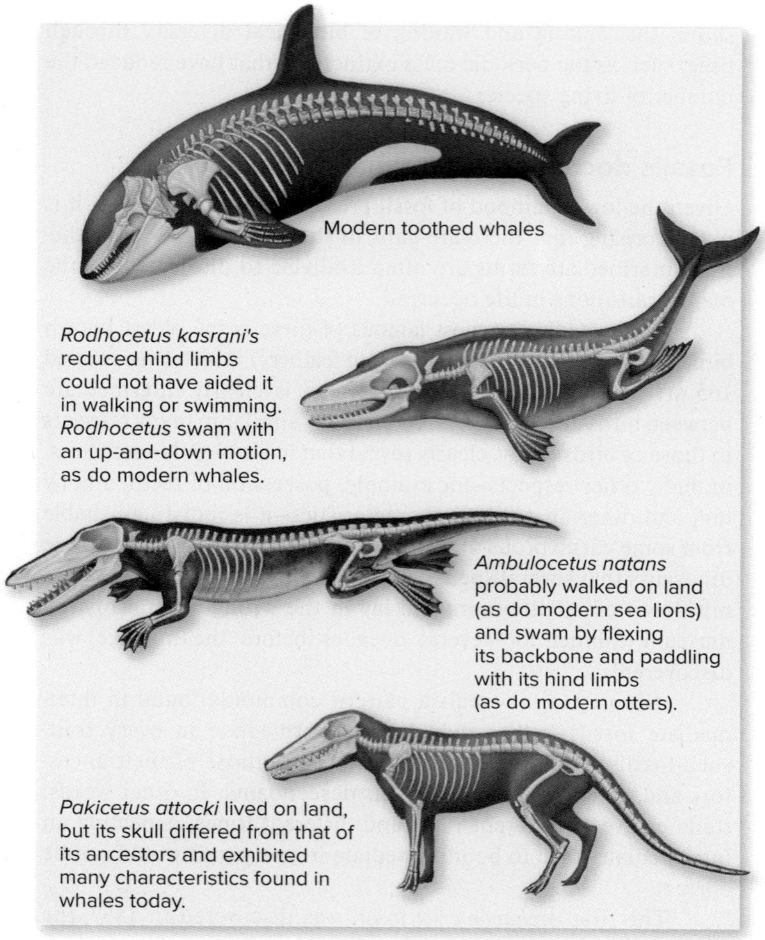

Figure 20.12 Whale "missing links." The recent discoveries of *Ambulocetus, Rodhocetus,* and *Pakicetus* have filled in the gaps between whales and their hoofed mammal ancestors. The features of *Pakicetus* illustrate that intermediate forms are not intermediate in all characteristics; rather, some traits evolve before others. In the case of the evolution of whales, changes occurred in the skull prior to evolutionary modification of the limbs. All three fossil forms occurred in the Eocene period, 45–55 MYA.

evolution. For example, four-legged aquatic mammals have been discovered that provide important insights into the evolution of whales and dolphins from land-dwelling, hoofed ancestors (figure 20.12). Similarly, fossil snakes with legs have shed light on the evolution of snakes, which are descended from lizards that gradually became more and more elongated, with the simultaneous reduction and eventual disappearance of the limbs. Another recent discovery is *Tiktaalik*, a species that bridged the gap between fish and the first land-living vertebrates.

On a finer scale, evolutionary change within some types of animals is known in exceptional detail. For example, about 200 MYA, oyster shells evolved from having small, curved shells to having larger, flatter ones, with progressively flatter fossils observed in the fossil record over a period of 12 million years. A host of other examples illustrate similar records of successive change. The demonstration of this successive change is one of the strongest lines of evidence that evolution has occurred.

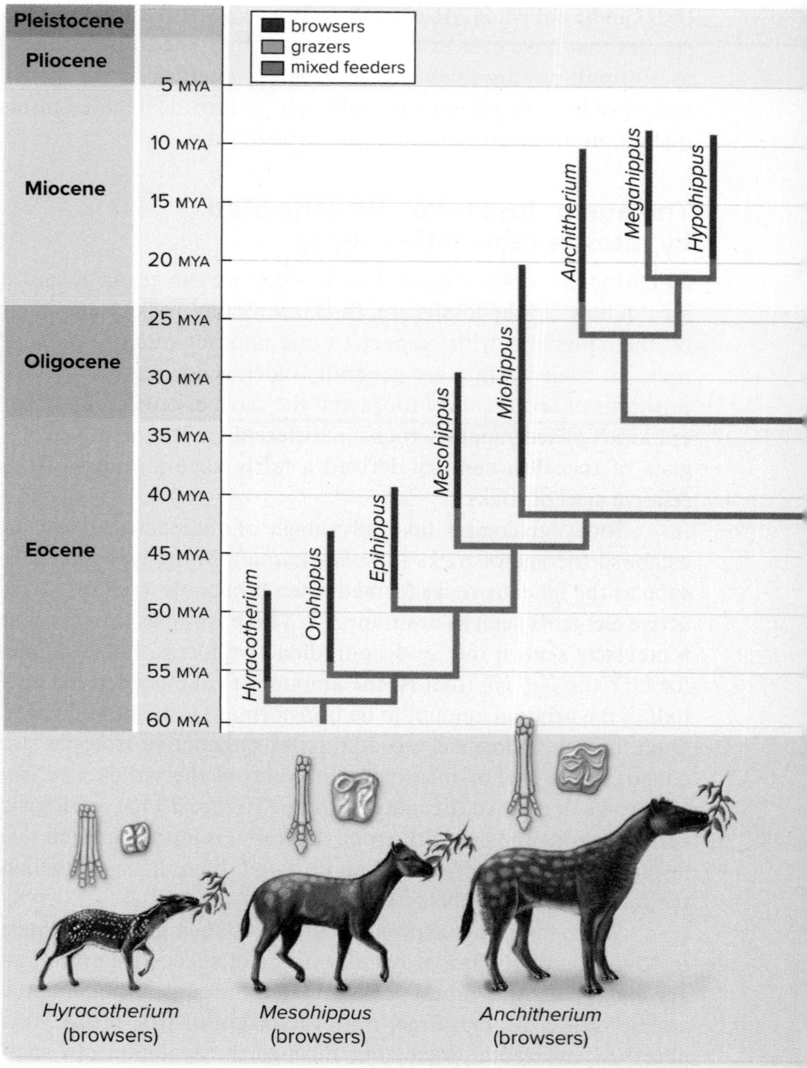

The Fossil Record Provides Clear Evidence for the Evolution of Horses

LEARNING OBJECTIVE 20.4.2 Describe the patterns of evolution in the horse.

One of the most studied cases in the fossil record concerns the evolution of horses. Modern-day members of the family Equidae include horses, zebras, donkeys, and asses, all of which are large, long-legged, fast-running animals adapted to living on open grasslands. These species, all classified in the genus *Equus,* are the last living descendants of a long lineage that has produced 34 genera since its origin in the Eocene period, approximately 55 MYA. Examination of these fossils has provided a particularly well-documented case of how evolution has proceeded through adaptation to changing environments.

The first horse

The earliest known members of the horse family, species in the genus *Hyracotherium,* didn't look much like modern-day horses at all. Small, with short legs and broad feet, these species occurred

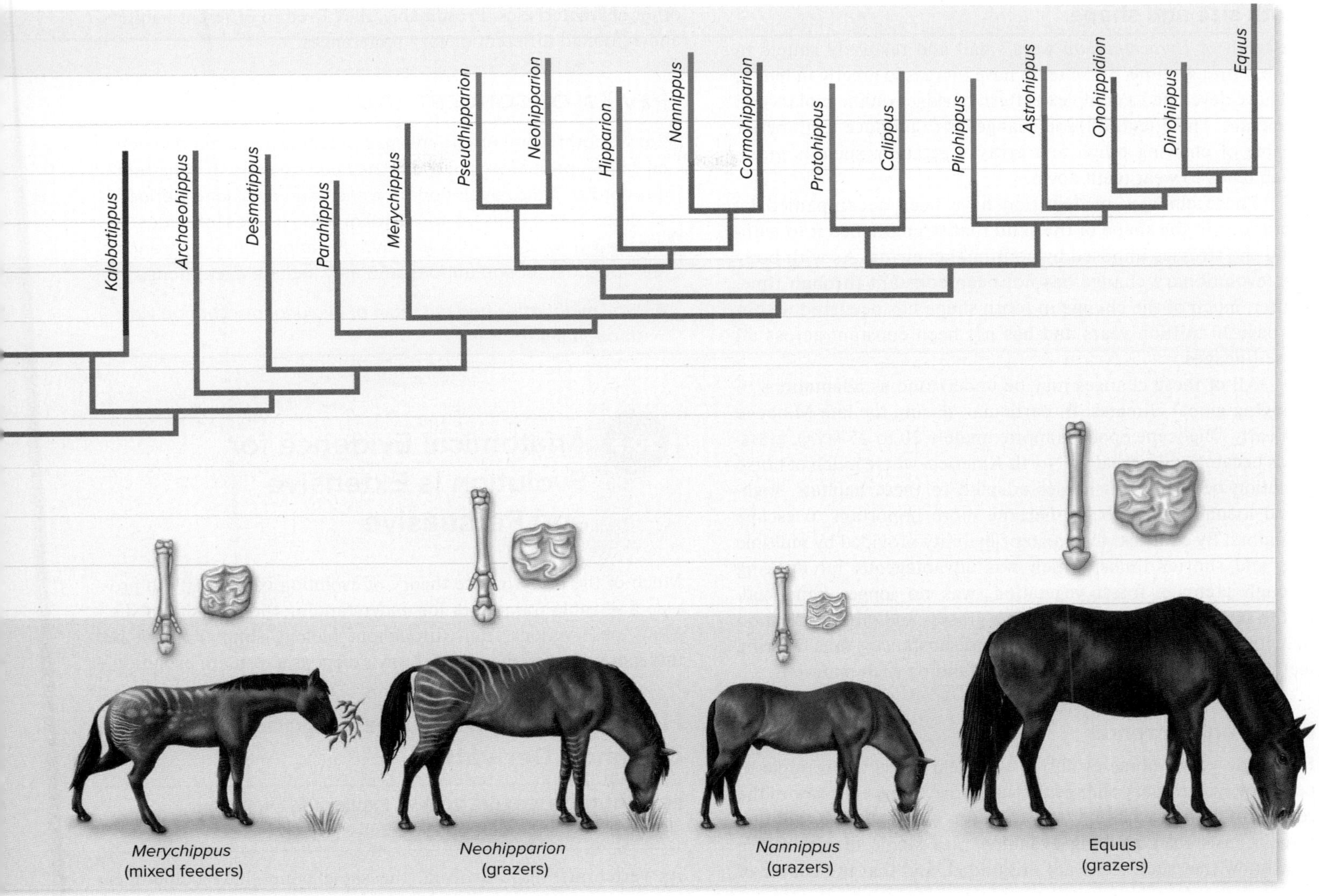

Figure 20.13 Evolutionary change in body size of horses. Lines indicate evolutionary relationships of the horse family. Horse evolution is more like a bush than a single-trunk tree; diversity was much greater in the past than it is today. In general, there has been a trend toward larger size, more-complex molar teeth, and fewer toes, but this trend has exceptions. For example, a relatively recent form, *Nannippus,* evolved in the opposite direction, toward decreased size.

in wooded habitats, where they probably browsed on leaves and herbs and escaped predators by dodging through openings in the forest vegetation. The evolutionary path from these diminutive creatures to the workhorses of today has involved changes in a variety of traits, including size, toe reduction, and tooth size and shape (figure 20.13).

Changes in size

The first species of horses were as big as a large house cat. By contrast, modern equids can weigh more than 500 kg. Examination of the fossil record reveals that horses changed little in size for their first 30 million years, but since then, a number of different lineages have exhibited rapid and substantial increases. However, trends toward decreased size were also found in some branches of the equid evolutionary tree, as revealed, for example, by *Nannippus.*

Toe reduction

The feet of modern horses have a single toe enclosed in a tough, bony hoof. By contrast, *Hyracotherium* had four toes on its front feet and three on its hind feet. Rather than hooves, these toes were encased in fleshy pads like those of dogs and cats.

Examination of fossils clearly shows the transition through time: a general increase in length of the central toe, development of the bony hoof, and reduction and loss of the other toes (figure 20.13). As with body size, these trends occurred concurrently on several different branches of the horse evolutionary tree and were not exhibited by all lineages.

At the same time that toe reduction was occurring, these horse lineages were evolving changes in the length and skeletal structure of their limbs, leading to animals capable of running long distances at high speeds.

Tooth size and shape

The teeth of *Hyracotherium* were small and relatively simple in shape. Through time, horse teeth have increased greatly in length and have developed a complex pattern of ridges on the molars and premolars. The effect of these changes is to produce teeth better capable of chewing tough and gritty vegetation, such as grass, which tends to wear teeth down.

These changes in dentition have been accompanied by alterations in the shape of the skull that strengthened it to withstand the stresses imposed by continual chewing. As with body size, evolutionary change has not been constant through time. Rather, much of the change in tooth shape has occurred within the past 20 million years and has not been constant across all horse lineages.

All of these changes may be understood as adaptations to changing global climates. In particular, during the late Miocene and early Oligocene epochs (approximately 20 to 25 MYA), grasslands became widespread in North America, where much of horse evolution occurred. As horses adapted to these habitats, high-speed locomotion probably became more important to escape predators. By contrast, the greater flexibility provided by multiple toes and shorter limbs, which was advantageous for ducking through complex forest vegetation, was no longer beneficial. At the same time, horses were eating grasses and other vegetation that contained more grit and other hard substances, thus favoring teeth and skulls better suited for withstanding such materials.

Evolutionary trends

For many years, horse evolution was held up as an example of constant evolutionary change through time. Some even saw in the record of horse evolution evidence for a progressive, guiding force, consistently pushing evolution in a single direction. We now know that such views are misguided, and that the course of evolutionary change over millions of years is rarely so simple.

Rather, the fossils demonstrate that even though overall trends have been evident in a variety of characteristics, evolutionary change has been far from constant and uniform through time. Instead, rates of evolution have varied widely, with long periods of little observable change and some periods of great change. Moreover, when changes happen, they often occur simultaneously in multiple lineages of the horse evolutionary tree.

Finally, even when a trend exists, exceptions, such as the evolutionary decrease in body size exhibited by some lineages, are not uncommon. These patterns are usually discovered for any group of plants and animals for which we have an extensive fossil record; human evolution is one such example, which we discuss in chapter 28.

Horse diversity

One reason that horse evolution was originally conceived of as linear through time may be that modern horse diversity is relatively limited. For this reason, it is easy to mentally picture a straight line from *Hyracotherium* to modern-day *Equus.* But today's limited horse diversity—only one surviving genus—is unusual. In fact, at the peak of horse diversity in the Miocene epoch, 13 genera of horses could be found in North America alone. These species differed in body size and in a wide variety of other characteristics. Presumably, they lived in different habitats and exhibited different dietary preferences.

REVIEW OF CONCEPT 20.4

Fossils form when an organism is preserved in a matrix such as amber, permafrost, or rock. They can be used to construct a record of major evolutionary transitions over long periods of time. The extensive fossil record for horses provides a detailed view of evolutionary diversification, although trends are not constant and uniform and may include exceptions.

- *Why might rates and direction of evolutionary change vary through time?*

20.5 Anatomical Evidence for Evolution Is Extensive and Persuasive

Much of the power of the theory of evolution is its ability to provide a sensible framework for understanding the diversity of life. Many observations from throughout biology simply cannot be understood in any meaningful way except as a result of evolution.

Homologous Structures Suggest Common Derivation

LEARNING OBJECTIVE 20.5.1 Explain the evolutionary significance of homologous structures.

As vertebrates have evolved, the same bones have sometimes been put to different uses. Yet the bones are still recognizable, their presence betraying their evolutionary past. For example, the forelimbs of vertebrates are all **homologous structures**—structures with different appearances and functions that all derived from the same body part in a common ancestor.

Figure 20.14 shows how the bones of the forelimb have been modified in different ways for different mammals. Why should these very different structures be composed of the same bones—a single upper forearm bone, a pair of lower forearm bones, several small carpals, and one or more digits? If evolution had not occurred, this would indeed be a riddle. But when we consider that all of these animals are descended from a common ancestor, it is easy to understand that natural selection has modified the same initial starting blocks to serve very different purposes.

Early Embryonic Development Shows Similarities in Some Groups

LEARNING OBJECTIVE 20.5.2 Describe how patterns of early development provide evidence of evolution.

Some of the strongest anatomical evidence supporting evolution comes from comparisons of how organisms develop. Embryos of

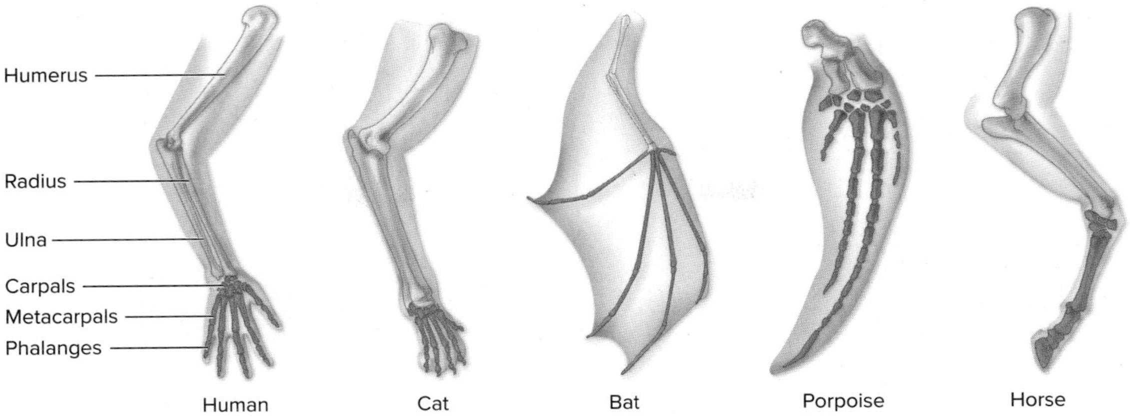
Humerus
Radius
Ulna
Carpals
Metacarpals
Phalanges
Human
Cat
Bat
Porpoise
Horse

Figure 20.14 Homology of the bones of the forelimbs of mammals. Although these structures show considerable differences in form and function, the same basic bones are present in the forelimbs of humans, cats, bats, porpoises, and horses.

different types of vertebrates, for example, often are similar early on but become more different as they develop. Early in their development, vertebrate embryos possess pharyngeal pouches, which develop into different structures. In humans, for example, they become various glands and ducts; in fish, they turn into gill slits. At a later stage, every human embryo has a long tail, the vestige of which we carry to adulthood as the coccyx at the end of our spine. Human fetuses even possess a fine fur (called *lanugo*) during the fifth month of development.

Similarly, although most frogs go through a tadpole stage, some species develop directly and hatch out as little, fully formed frogs. However, the embryos of these species still exhibit tadpole features, such as the presence of a tail, which disappear before the froglet hatches (figure 20.15).

These relict developmental forms suggest strongly that our development has evolved, with new instructions modifying ancestral developmental patterns.

Some Structures Are Imperfectly Suited to Their Use

LEARNING OBJECTIVE 20.5.3 Illustrate how imperfect design is evidence for natural selection.

There is a popular myth that organisms are perfectly adapted to their environments. In fact, most organisms are not perfectly adapted to their environments, because they are the result of natural selection and not purposeful design. Natural selection is limited to the variation present in a population. For example, most animals with long necks have many neck vertebrae for enhanced flexibility: geese have up to 25, and plesiosaurs, the long-necked marine reptiles from the Age of Dinosaurs, had as many as 76. By contrast, almost all mammals have only 7 neck vertebrae, even the giraffe. In the absence of variation in vertebra number, selection led to an evolutionary increase in vertebra size to produce the long neck of the giraffe.

An excellent example of an imperfect design is the eye of vertebrate animals, in which the photoreceptors face backward, toward the wall of the eye (figure 20.16*a*). As a result, the nerve fibers extend not backward, toward the brain, but forward into the eye chamber, where they slightly obstruct light. Moreover, these fibers bundle together to form the optic nerve, which exits through a hole at the back of the eye, creating a blind spot.

By contrast, the eyes of mollusks—such as squid and octopuses—are better arranged: the photoreceptors face forward, and the nerve fibers exit at the back, neither obstructing light nor creating a blind spot (figure 20.16*b*).

Such examples illustrate that natural selection is like a tinkerer, working with whatever material is available to craft a workable solution, rather than like an engineer, who can design and build the best possible structure for a given task. Workable, but imperfect, structures such as the vertebrate eye are an expected outcome of evolution by natural selection.

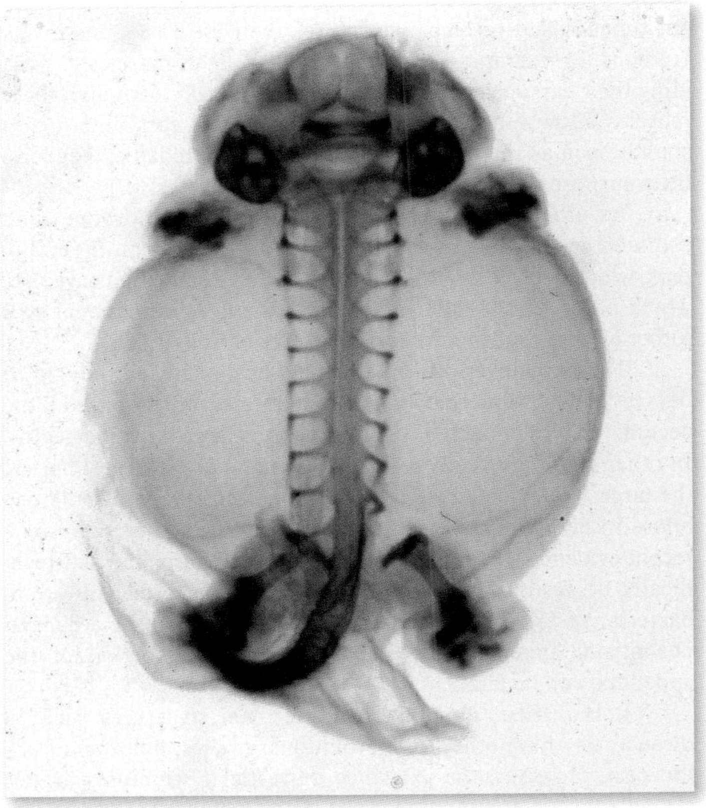

Figure 20.15 Developmental features reflect evolutionary ancestry. Some species of frogs have lost the tadpole stage. Nonetheless, tadpole features first appear and then disappear during development in the egg.

James Hanken, Museum of Comparative Zoology, Harvard University, Cambridge

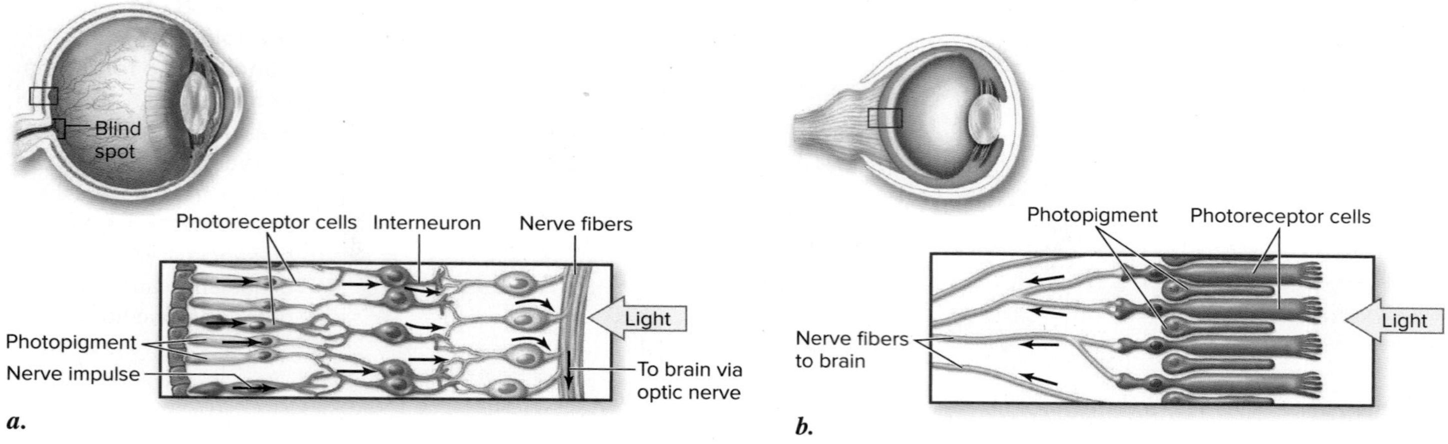

Figure 20.16 The eyes of vertebrates and mollusks. ***a.*** Photoreceptors of vertebrates point backward, whereas (***b***) those of mollusks face forward. As a result, vertebrate nerve fibers pass in front of the photoreceptor; where they bundle together and exit the eye, a blind spot is created. Mollusks' eyes have neither of these problems.

Vestigial Structures Can Be Explained as Holdovers from the Past

LEARNING OBJECTIVE 20.5.4 Explain the evolutionary significance of vestigial structures.

Many organisms possess **vestigial structures** that have no apparent function but resemble structures their ancestors possessed. Humans, for example, possess a complete set of muscles for wiggling their ears, just as many other mammals do. Although these muscles allow other mammals to move their ears to pinpoint sounds such as the movements or growl of a predator, they have little purpose in humans other than amusement.

As other examples, boa constrictors have hip bones and rudimentary hind legs. Manatees have fingernails on their flippers, which evolved from legs (figure 20.17). Some blind cave fish, which never see the light of day, have small, nonfunctional eyes (other blind cave fish have lost their eyes entirely).

The human vermiform appendix has long been thought to be vestigial; it represents the degenerate terminal part of the cecum, the blind pouch or sac in which the large intestine begins. In other mammals, such as mice, the cecum is the largest part of the large intestine and functions in storage—usually of bulk cellulose in herbivores. Although greatly diminished in size, some recent evidence suggests that the human appendix may not technically be vestigial because it may harbor beneficial intestinal bacteria. Vestigial or not, the human appendix can be a dangerous organ: appendicitis, which results from infection of the appendix, can be fatal.

It is difficult to understand vestigial structures such as these as anything other than evolutionary relics, holdovers from the past. However, the existence of vestigial structures argues strongly for the common ancestry of the members of the groups that share them, regardless of how different those groups have subsequently become.

All of these anatomical lines of evidence—homology, development, and imperfect and vestigial structures—are readily understandable as a result of descent with modification—that is, evolution.

REVIEW OF CONCEPT 20.5

Comparisons of the anatomy of different living animals often reveal evidence of shared ancestry. In cases of homology, the same organ has evolved to carry out different functions. In other cases, an organ is still present, usually in diminished form, even though it has lost its function altogether; such an organ or structure is termed vestigial.

- *How might homologous and vestigial structures be explained other than as a result of evolutionary descent with modification?*

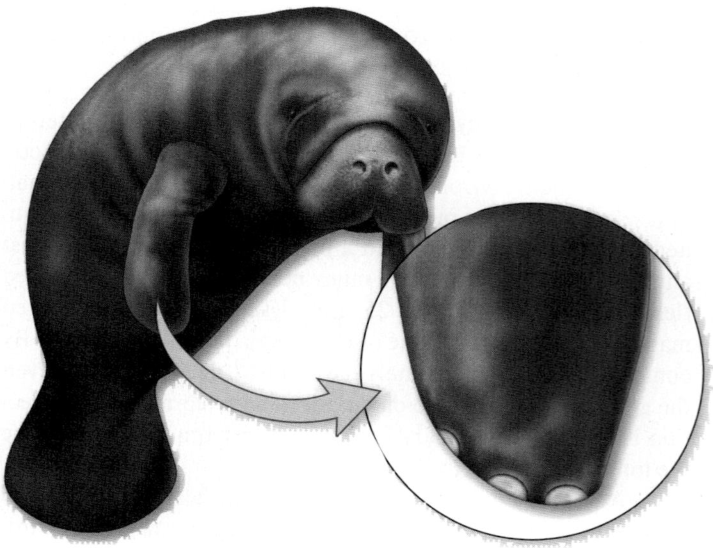

Figure 20.17 Vestigial structures. The flippers of the West Indian manatee (*Trichechus manatus*), a relative of the elephant and descended from a terrestrial mammal, have retained nails, even though the manatee never leaves the water.

20.6 Genes Carry a Molecular Record of the Evolutionary Past

Traces of our evolutionary past are also evident at the molecular level. We possess the same set of color vision genes as those of our ancestors, only more complex, and we use the pattern formation genes during early development that all animals share. Indeed, the fact that organisms have evolved from a series of simpler ancestors implies that a record of evolutionary change is present in the cells of each of us, in our DNA.

Darwin's Theory Predicts the Continual Accumulation of Gene Differences

LEARNING OBJECTIVE 20.6.1 Describe the molecular evidence that evolution has occurred.

According to evolutionary theory, new alleles arise from older ones by mutation and come to predominance through favorable selection. A series of evolutionary changes thus implies a continual accumulation of genetic changes in the DNA. From this you can observe that evolutionary theory makes a clear prediction: organisms that are more distantly related should have accumulated a greater number of evolutionary differences than two species that are more closely related.

This prediction is now subject to direct test. The sequencing of complete genomes of a wide variety of species allows us to directly compare organisms at the DNA level. The result is clear: for a broad array of vertebrates, the more distantly related two organisms are, the greater their genomic difference. This research was described in chapter 1 (refer to figure 1.14).

This pattern of divergence is clear found at the protein level. Comparison of the hemoglobin amino acid sequence of different species with the human sequence in figure 20.18 reveals that species more closely related to humans have fewer differences in the amino acid structure of their hemoglobin. Macaques, primates closely related to humans, have fewer differences from humans (only 8 different amino acids) than do more distantly related mammals such as dogs (which have 32 different amino acids). Nonmammalian terrestrial vertebrates differ even more, and marine vertebrates are the most different of all. Again, the prediction of evolutionary theory is strongly confirmed.

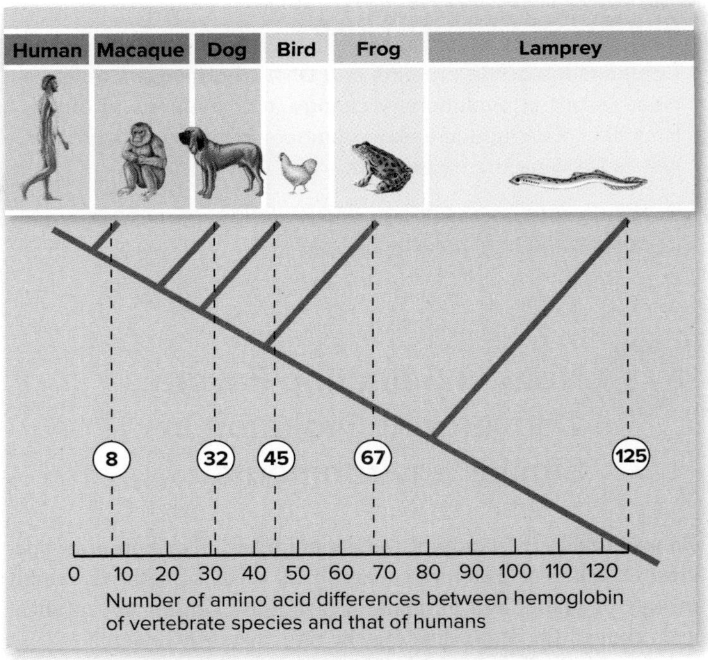

Figure 20.18 Molecules reflect evolutionary divergence. The greater the evolutionary distance from humans (evolutionary tree is based on the fossil record), the greater the number of amino acid differences in the vertebrate hemoglobin polypeptide.

Molecular clocks

This same pattern is found when the DNA sequence of an individual gene is compared over a much broader array of organisms. One well-studied case is the mammalian *cytochrome c* gene (cytochrome c is a protein that plays a key role in oxidative metabolism). Figure 20.19 compares the time when two species diverged (x-axis) to the number of differences in their *cytochrome c* gene (y-axis). The relationship between nucleotide substitutions and time is a linear one, implying that the rate of evolution is constant. This led to the idea of a molecular clock, which ticks away, marking evolutionary time. It turns out that not all genes show a constant rate of evolution, nor is the rate the same for all genes. Nonetheless, molecular clocks can be useful when we can combine molecular data with the fossil record.

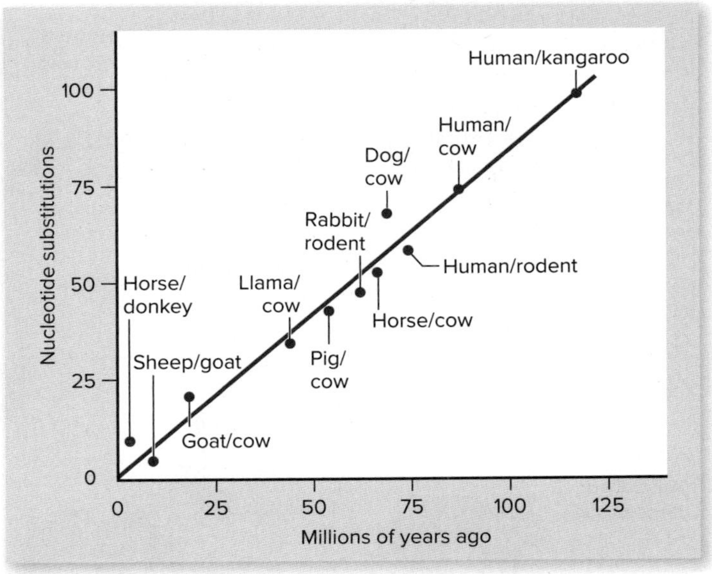

Figure 20.19 The molecular clock of *cytochrome c*. When the time since each pair of organisms presumably diverged is plotted against the number of nucleotide differences in *cytochrome c*, the result is a straight line, suggesting that the *cytochrome c* gene is evolving at a constant rate.

REVIEW OF CONCEPT 20.6

Comparisons of the proteins and DNA of organisms reveal a clear record of evolutionary change, the genomes of organisms accumulating increasing numbers of changes over time, just as Darwin's theory predicts.

- *Why might evolutionary accumulation of changes not correspond to a molecular clock?*

20.7 Natural Selection Favors Convergent Evolution in Similar Environments

Biogeography, the study of the geographical distribution of species, reveals that different geographical areas sometimes exhibit groups of plants and animals of strikingly similar appearance, even though the organisms may be only distantly related.

Convergent Evolution Is a Widespread Phenomenon

LEARNING OBJECTIVE 20.7.1 Explain the principle of convergent evolution.

It is difficult to explain so many similarities as being the result of coincidence. Instead, natural selection appears to have favored parallel evolutionary adaptations in similar environments. Because selection in these instances has tended to favor changes that made the two groups more alike, their phenotypes have converged. This form of evolutionary change is referred to as **convergent evolution.**

Marsupials and placentals demonstrate convergence

In the best-known case of convergent evolution, two major groups of mammals—marsupials and placentals—have evolved in very similar ways in different parts of the world. Marsupials are a group in which the young are born in a very immature condition and held in a pouch until they are ready to emerge into the outside world. In placentals, by contrast, offspring are not born until they can safely survive in the external environment (with varying degrees of parental care).

Australia separated from the other continents more than 70 MYA; at that time, both marsupials and placental mammals had evolved, but in different places. In particular, only marsupials occurred in Australia. As a result of this continental separation, besides humans and species recently brought there by humans, the only placental mammals in Australia today are bats and a few colonizing rodents (which arrived relatively recently), and Australia is dominated by marsupials.

What are the Australian marsupials like? To an astonishing degree, they resemble the placental mammals living today on the other continents (figure 20.20). The similarity between some individual members of these two sets of mammals argues strongly that they are the result of convergent evolution, similar forms having evolved in different, isolated areas because of similar selective pressures in similar environments.

When species interact with the environment in similar ways, they often are exposed to similar selective pressures, and they therefore frequently develop the same evolutionary adaptations. Consider, for example, fast-moving marine predators (figure 20.21). The hydrodynamics of moving through water require a

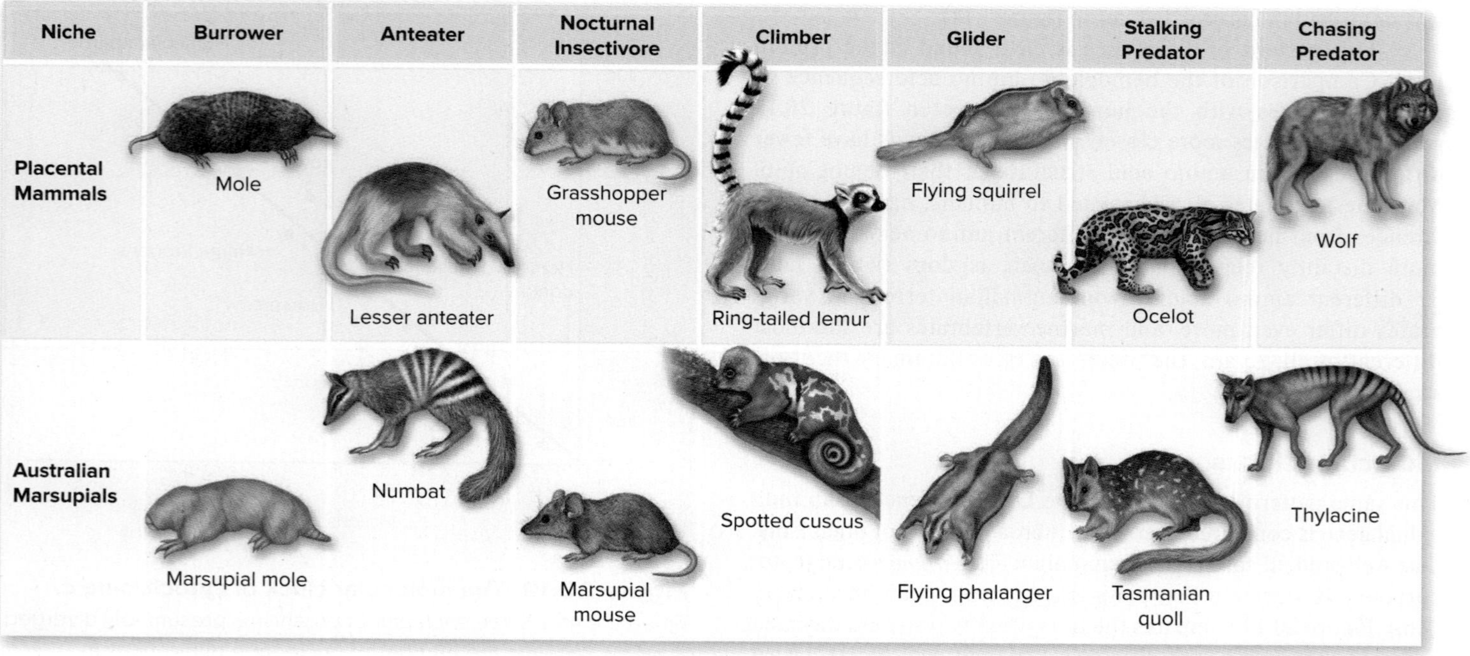

Figure 20.20 Convergent evolution. Many marsupial species in Australia resemble placental mammals occupying similar ecological niches elsewhere in the rest of the world. Marsupials evolved in isolation after Australia separated from other continents.

Figure 20.21 Convergence among fast-swimming predators. Fast movement through water requires a streamlined body form, which has evolved numerous times.

streamlined body shape to minimize friction. It is no coincidence that dolphins, sharks, and tuna—among the fastest of marine species—all evolved to have the same basic shape. We can infer as well that ichthyosaurs—marine reptiles that lived during the Age of Dinosaurs—exhibited a similar lifestyle.

Island trees exhibit a similar phenomenon. Most islands are covered by trees (or were until the arrival of humans). Careful inspection of these trees, however, reveals that they are not closely related to the trees with which we are familiar. Although they have all the characteristics of trees, such as being tall and having a tough outer covering, many island trees are members of plant families that exist elsewhere only as flowers, shrubs, or other small bushes. For example, on many islands the native trees are members of the sunflower family.

Why do these plants evolve into trees on islands? Probably because seeds from trees rarely make it to isolated islands. As a result, those species that do manage to colonize distant islands face an empty ecological landscape upon arrival. In the absence of other treelike plants, natural selection often favors individual plants that can capture the most sunlight for photosynthesis, and the result is the evolution of similar treelike forms on islands throughout the world.

Convergent evolution even occurs in humans. People in most populations stop producing lactase, the enzyme that digests milk, at some point in childhood. However, individuals in African and European populations that raise cattle produce lactase throughout their lives. DNA analysis indicates that this has been accomplished by the incorporation of different mutations in Africa and Europe, which indicates that the populations have independently (convergently) acquired this adaptation.

Biogeographical Studies Document Evolutionary Divergence

LEARNING OBJECTIVE 20.7.2 Demonstrate how the biogeographical distribution of plant and animal species on islands provides evidence of evolutionary diversification.

Darwin made several important observations during his voyage around the world. He noted that many islands are missing plants and animals, such as frogs and land mammals, that are common on continents. Accidental human introductions have proved that these species can survive on these islands, so lack of suitable habitat is not the cause. In addition, the species present on islands often have diverged from their continental relatives and sometimes—as with Darwin's finches and the island trees just discussed—occupy ecological niches used by other species on continents. Finally, island species usually are closely related to species on nearby continents, even though the continental environment may not be similar to that on the island.

The explanation for this phenomenon proves to be powerful evidence for natural selection. Darwin's observations concerned islands of a type called oceanic islands: those that have never been connected to the mainland. So-called continental islands, which were once connected to a continent, do not show these unusual patterns of flora and fauna. Consider the British Isles, or Japan, where the distribution of species resembles that in Europe or Asia, respectively.

The species that occur on oceanic islands arrived by dispersing across the water; dispersal from nearby areas is more likely, though long-distance colonization does occur occasionally. Species that can fly, float, or swim are the most likely colonizers. Some, like frogs, are particularly vulnerable to dehydration in salt water, making them unlikely colonizers. The absence of many types of plants and animals provides opportunity for those that make it; as a result, colonizers often evolve into many species exhibiting great ecological and morphological diversity.

Geographical proximity is not always a good predictor of evolutionary relationships, however. Earth's continents are constantly moving because of the process known as continental drift. Although this drift is slow, on the order of several centimeters per year, the configuration of the continents has changed considerably over geologic time. As a result, closely related species that at one time occurred near each other may now be separated by thousands of miles. Many examples occur on the southern continents, which were last united as the supercontinent Gondwana more than 100 MYA. One such example is the southern beech tree, which is found in Chile, as well as Australia and New Zealand. In cases such as this, Earth's geological and evolutionary history must be considered jointly to make sense of geographical distributions.

REVIEW OF CONCEPT 20.7

Convergence is the evolution of similar forms in different lineages when exposed to similar selective pressures. The biogeographical distribution of species often reflects the outcome of evolutionary diversification with closely related species in nearby areas.

- *Why does convergent evolution occur, and why might species occupying similar environments in different localities sometimes not exhibit it?*

20.8 Addressing Common Criticisms of Evolutionary Theory

In the century and a half since he proposed it, Darwin's theory of evolution by natural selection has become nearly universally accepted by biologists, but it has been a source of controversy among some members of the general public. Often, this controversy arises from a lack of accurate information about the nature of evolutionary theory. It can also be based on literal interpretations of religious texts that do not represent scientific thinking.

Seven Objections Have Been Raised to Darwin's Theory

LEARNING OBJECTIVE 20.8.1 Characterize the criticisms of evolutionary theory and list counterarguments that can be made.

Here we discuss seven principal objections that critics raise to the teaching of evolution as biological fact, along with some answers that scientists present in response.

1. **Evolution is not solidly demonstrated.** "Evolution is just a theory," Darwin's critics point out, as though *theory* means a lack of knowledge, or some kind of guess.

   Scientists, however, use the word *theory* in a very different sense than the general public does. Theories are the solid ground of science—that about which we are most certain. Few of us doubt the theory of gravity because it is "just a theory."
2. **There are no fossil intermediates.** "No one ever saw a fin on the way to becoming a leg," critics claim, pointing to the many gaps in the fossil record in Darwin's day.

   Since that time, however, many fossil intermediates in vertebrate evolution have indeed been found. A clear line of fossils now traces the transition between hoofed mammals and whales, between reptiles and mammals, between dinosaurs and birds, and between apes and humans. The fossil evidence of evolution between major forms is compelling.
3. **The intelligent design argument.** "The organs of living creatures are too complex for a random process to have produced—the existence of a clock is evidence of the existence of a clockmaker."

   Evolution by natural selection is not a random process—quite the contrary. By favoring those variations that lead to the highest reproductive fitness, natural selection is a nonrandom process that can construct highly complex organs by incrementally improving them from one generation to the next.

   For example, the intermediates in the evolution of the mammalian ear have been found in fossils, and many intermediate "eyes" are known in various invertebrates. These intermediate forms arose because they have value—being able to detect light slightly is better than not being able to detect it at all. Complex structures such as eyes evolved as a progression of slight improvements. Moreover, inefficiencies of certain designs, such as the vertebrate eye and the existence of vestigial structures, do not support the idea of an intelligent designer.
4. **Evolution violates the Second Law of Thermodynamics.** "A jumble of soda cans doesn't by itself jump neatly into a stack—things become more disorganized due to random events, not more organized."

   Biologists point out that this argument ignores what the second law really says: disorder increases in a closed system, which the Earth most certainly is not. Energy continuously enters the biosphere from the Sun, fueling life and all the processes that organize it.
5. **Proteins are too improbable.** "Hemoglobin has 141 amino acids. The probability that the first one would be leucine is 1/20, and that all 141 would be the ones they are by chance is $(1/20)^{141}$, an impossibly rare event."

   This argument illustrates a lack of understanding of probability and statistics—probability cannot be used to argue backward. The probability that a student in a classroom has a particular birthdate is 1/365; arguing this way, the probability that everyone in a class of 50 would have the birthdates that they do is $(1/365)^{50}$, and yet there the class sits, all with their actual birthdates.
6. **Natural selection does not imply evolution.** "No scientist has come up with an experiment in which fish evolve into frogs and leap away from predators."

   Can we extrapolate from our understanding that natural selection produces relatively small changes that are observable in populations *within* species to explain the major differences observed *between* species? Most biologists who have studied the problem think so. The differences between breeds produced by artificial selection—such as chihuahuas, mastiffs, and greyhounds—are more distinctive than the differences between some wild species, and laboratory selection experiments sometimes create forms that cannot interbreed and thus would in nature be considered different species. Thus, production of radically different forms has indeed been observed, repeatedly. To object that evolution still does not explain major differences, such as those between fish and amphibians, simply takes us back to point number 2. These changes take millions of years, and they are clearly evident in the fossil record.
7. **The irreducible complexity argument.** Because each part of a complex cellular mechanism, such as blood clotting, is essential to the overall process, the intricate machinery of the cell cannot be explained by evolution from simpler stages.

   The error in this argument is that each part of a complex molecular machine evolves as part of the whole system. Natural selection can act on a complex system because, at every stage of its evolution, the system functions. Parts that improve function are added. Subsequently, other parts may be modified or even lost, so that parts that were not essential when they first evolved become essential. In this way, an "irreducible, complex" structure can evolve by natural selection. The same process works at the molecular level.

   For example, snake venom initially evolved as enzymes to increase the ability of snakes to digest large prey items, which were captured by biting the prey and then constricting them with coils. Subsequently, the digestive enzymes evolved to become increasingly lethal. Rattlesnakes kill large prey by injecting them with venom, letting them go, and then tracking them down and eating them after they die. To do so, they have evolved extremely toxic venom; highly modified, syringe-like front teeth; and many other characteristics. Take away the fangs or the venom and the rattlesnakes can't feed—what initially evolved as nonessential parts are now indispensable; irreducible complexity has evolved by natural selection.

The mammalian blood-clotting system similarly has evolved from much simpler systems. The core clotting system evolved at the dawn of the vertebrates more than 500 MYA, and it is found today in primitive fishes such as lampreys. One hundred million years later, as vertebrates continued to evolve, proteins were added to the clotting system, making it sensitive to substances released from damaged tissues. Fifty million years later, a third component was added, triggering clotting by contact with the jagged surfaces produced by injury. At each stage, as the clotting system evolved to become more complex, its overall performance came to depend on the added elements. Thus, blood clotting has become "irreducibly complex" as a result of Darwinian evolution.

Statements that various structures could not have been built by natural selection have repeatedly been made over the past 150 years. In many cases, after detailed scientific study, the likely path by which such structures have evolved has been discovered.

REVIEW OF CONCEPT 20.8

Darwin's theory of evolution is controversial to some in the general public. Objections are often based on a misunderstanding of the theory. In scientific usage, a hypothesis is a preliminary idea, whereas a theory is an explanation that fits available evidence and has withstood rigorous testing.

- *Why is evolution a theory and not a hypothesis?*

Experiments on Corn Demonstrate That Selection Can Cause Evolutionary Change

Darwin developed his ideas about evolution through natural selection from his intimate knowledge of the practices of agriculturists and animal breeders. Darwin was particularly familiar with the world of pigeon breeders, who had developed a great variety of distinctive breeds, all descended from the humble rock dove.

Paul D Stewart/Science Source

We now know this practice as artificial selection, a method still used today in agriculture and pet breeding (new breeds of dogs and cats are regularly developed), as well as in many laboratory experiments. A typical study takes, say, 100 individuals and selects the top few percent (let's say 20%) with the highest value of the trait being selected for, such as wing size or plant height. Those individuals are then allowed to mate with each other, and the whole process is performed again on their offspring in the next generation. Such selection, implemented generation after generation, can rapidly lead to substantial change, producing phenotypes vastly different from those present the initial population.

One long-running example is the Illinois Long-Term Selection Experiment on Corn, begun in 1896 and continuing to this day. Every year, University of Illinois experimenters measured the oil content in corn plants and chose seeds from 20% of those plants with the highest content of oil in their kernels to plant for the next generation. Very quickly, oil content increased and; it now has quadrupled from its original 5% level. Conversely, a parallel selection experiment chose the individuals with the lowest oil content; that line, too, evolved quickly, so much so that the experiment had to be discontinued because the oil content level became so low that it could not be accurately measured. In 1947, a new line was created from the High Oil Selection experiment, selecting for low oil content. In this line, the trend toward higher oil content immediately reversed course. When this line was then selected for higher oil content, it reversed back, yet again favoring high oil content.

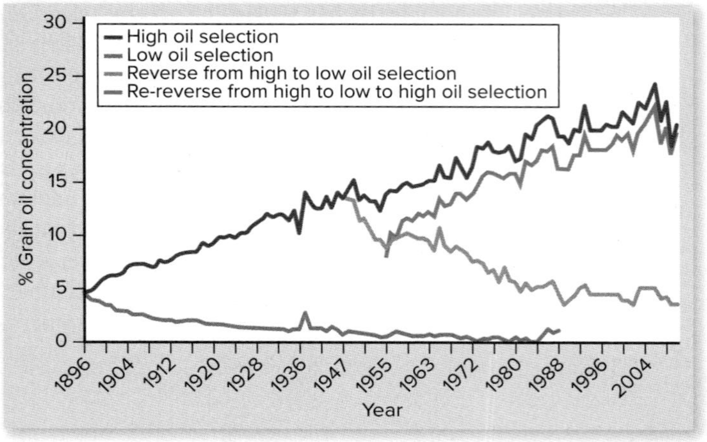

Analysis

1. **Applying Concepts**
   a. ***Variable.*** In the graph, what is the dependent variable?
   b. ***Reading a Line.*** Does oil content always change in the same direction from one year to the next, either within or between selection treatments?
   c. ***Comparing Lines.*** Are there consistent trends through time in each treatment?
2. **Interpreting Data**
   a. Which evolves faster, corn selected for greater oil content or that selected for lesser content?
   b. Since 1954, there have been two experimental lines in which high oil content is being selected. How similar are the evolutionary responses in these lines?
3. **Making Inferences**
   a. How many years of data do you think are necessary to figure out how the populations are responding to a selection treatment?
   b. How long do you think the reversed high-to-low oil selection treatment will take before the oil content is so low that it can't be measured?
4. **Drawing Conclusions** At the start of the experiment, the average oil content was 5% and the range was about 4 to 6%. What does the degree of change in this experiment tell us about the power of selection?

Retracing the Learning Path

CONCEPT 20.1 The Beaks of Darwin's Finches Provide Evidence of Natural Selection

20.1.1 Galápagos Finches Exhibit Variation Related to Food Gathering The correspondence between beak shape and a beak's use in obtaining food suggested to Darwin that finch species had diversified and adapted to eating different foods.

20.1.2 Modern Research Has Verified Darwin's Selection Hypothesis Natural selection acts on variation in beak morphology, favoring larger-beaked birds during extended droughts and smaller-beaked birds during long periods of heavy rains.

20.1.3 Natural Selection Can Produce Adaptive Radiations in New Environments The Galápagos finches adapted to eating a variety of foods.

CONCEPT 20.2 Peppered Moths and Industrial Melanism Illustrate Natural Selection in Action

20.2.1 Light-Colored Moths Are Less Frequent in Polluted Areas In polluted areas where soot built up on tree trunks, the darker form of the peppered moth became more common. In unpolluted areas, lighter forms remained predominant. Experiments suggested that predation by birds was the cause. In the last 40 years, pollution has decreased in many areas and the frequency of lighter moths has rebounded. Whether bird predation is the agent of selection has been questioned. Regardless, changes in frequency of the two morphs over time indicate that natural selection has acted on moth coloration.

CONCEPT 20.3 Human-Initiated Artificial Selection Is Also a Powerful Agent of Change

20.3.1 Experimental Selection Can Produce Changes in Populations Laboratory experiments in directional selection have shown substantial evolutionary change in controlled populations. Agricultural selection has led to extensive modification of crops and livestock. Many crops and animal breeds are substantially different from their wild ancestors. These experiments support the idea that natural selection could have created the Earth's diversity of life over millions of years.

CONCEPT 20.4 Fossils Provide Direct Evidence of Evolution

20.4.1 Fossils Present a History of Evolutionary Change The history of life on Earth can be traced through the fossil record. In recent years, new fossil discoveries have provided a more detailed understanding of major evolutionary transitions.

20.4.2 The Fossil Record Provides Clear Evidence for the Evolution of Horses The fossil record indicates that horses have evolved from small, forest-dwelling animals to the large and fast plains-dwelling species alive today. Over the course of 50 million years, evolution has not been constant and uniform. Change has been rapid at some times, slow at others.

CONCEPT 20.5 Anatomical Evidence for Evolution Is Extensive and Persuasive

20.5.1 Homologous Structures Suggest Common Derivation Homologous structures may have different appearances and functions even though derived from the same common ancestral body part.

20.5.2 Early Embryonic Development Shows Similarities in Some Groups Embryonic development shows similarity among species whose adult phenotypes are very different. Species that have lost features present in ancestral forms often develop and then lose those features during development.

20.5.3 Some Structures Are Imperfectly Suited to Their Use Natural selection can influence only the variation present in a population; evolution often results in workable but imperfect structures, such as the vertebrate eye.

20.5.4 Vestigial Structures Can Be Explained as Holdovers from the Past The existence of vestigial structures supports the concept of common ancestry among organisms that share them.

CONCEPT 20.6 Genes Carry a Molecular Record of the Evolutionary Past

20.6.1 Darwin's Theory Predicts the Continual Accumulation of Gene Differences More closely related species have more similar DNA and proteins.

CONCEPT 20.7 Natural Selection Favors Convergent Evolution in Similar Environments

20.7.1 Convergent Evolution Is a Widespread Phenomenon Convergent evolution may occur in species or populations exposed to similar selective pressures. Marsupial mammals in Australia have converged upon features of their placental counterparts elsewhere. Other examples include hydrodynamic streamlining in marine species.

20.7.2 Biogeographical Studies Document Evolutionary Divergence Island species usually are closely related to species on nearby continents, even if the environments are different. Early island colonizers often evolve into diverse species because competing species are scarce.

CONCEPT 20.8 Addressing Common Criticisms of Evolutionary Theory

20.8.1 Seven Objections Have Been Raised to Darwin's Theory Darwin's theory of evolution by natural selection is almost universally accepted by biologists. Many criticisms have been made both historically and recently, but most stem from a lack of understanding of scientific principles, the theory's actual content, or the time spans involved in evolution.

Concept Overview

This Concept Overview diagrams the key concepts that were discussed in this chapter.

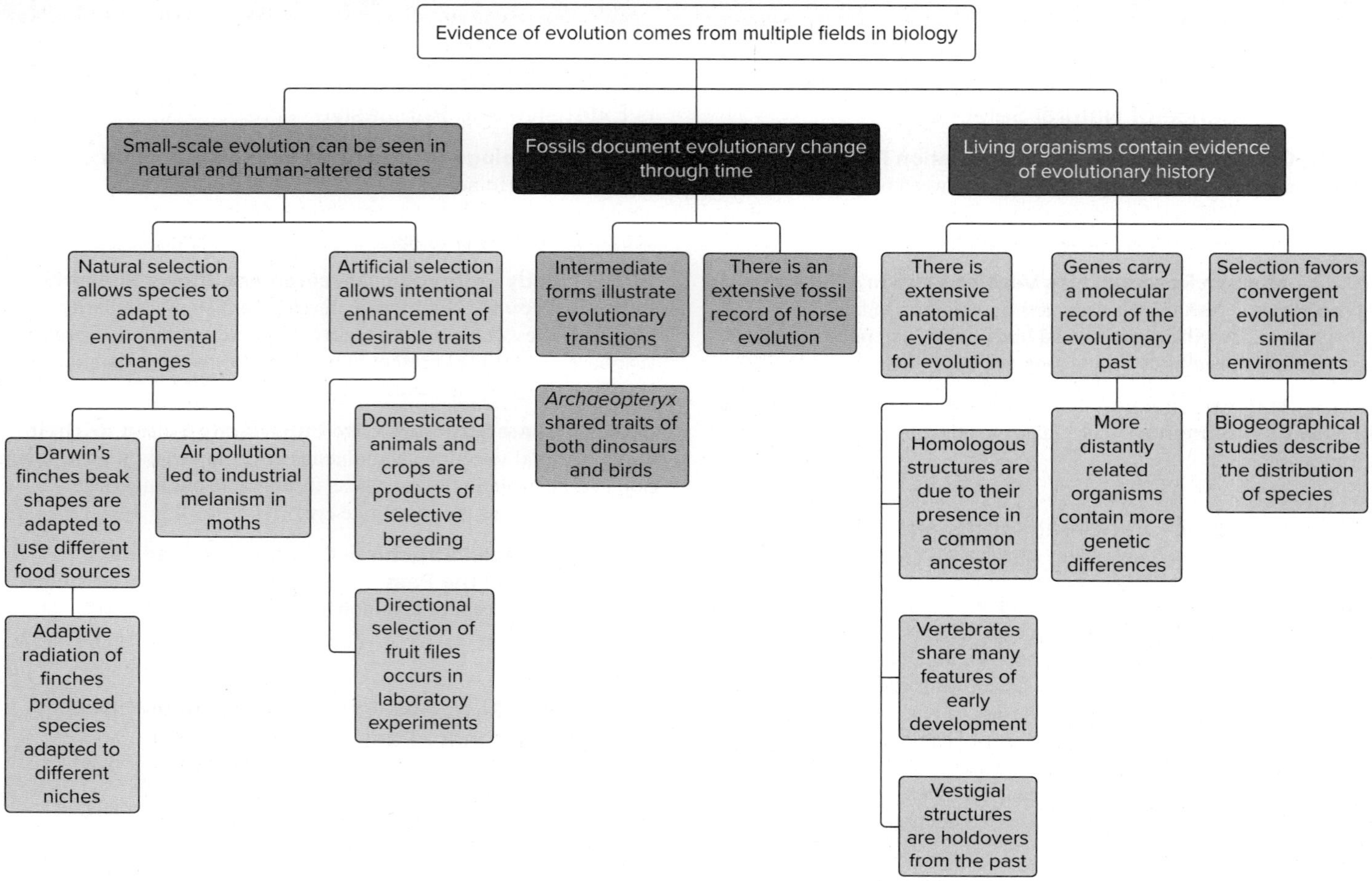

Assessing the Learning Path

Understand

1. Darwin's finches are a noteworthy case study of evolution by natural selection because evidence suggests
   a. they are descendants of many different species that colonized the Galápagos.
   b. they radiated from a single species that colonized the Galápagos.
   c. they are more closely related to mainland species than to one another.
   d. None of the above

2. When Kettlewell released equal numbers of light-colored and dark melanic moths in the industrial areas near Birmingham, and later recaptured moths there,
   a. he captured only melanic moths.
   b. he recaptured a greater percentage of dark moths than light moths.
   c. he recaptured a greater percentage of light moths than dark moths.
   d. he recaptured equal proportions of dark and light moths.

3. Starting with a wild mustard species, humans have developed cauliflower, broccoli, kale, and cabbage. Which of the following statements is supported by this fact?
   a. Natural selection never worked on the mustard plant.
   b. There is no natural variation in the wild mustard plant.
   c. There was enough natural variation in the mustard plant that humans were able to exaggerate certain features by artificial selection.
   d. Both a and b

4. Gaps in the fossil record
   a. demonstrate our inability to date geologic sediments.
   b. are expected, since the probability that any organism will fossilize is extremely low.
   c. have not been filled in as new fossils have been discovered.
   d. weaken the theory of evolution.

5. The evolution of modern horses (*Equus*) is best described as
   a. the constant change and replacement of one species by another over time.
   b. a complex history of lineages that changed over time, with many going extinct.

c. a simple history of lineages that have always resembled extant horses.
d. None of the above

6. Homologous structures
a. are structures in two or more species that originate as the same structure in a common ancestor.
b. are structures that look the same in different species.
c. cannot serve different functions in different species.
d. must serve different functions in different species.

7. Possession of fine fur in 5-month-old human embryos indicates
a. that the womb is cold at that point in pregnancy.
b. that humans evolved from a hairy ancestor.
c. that hair is a defining feature of mammals.
d. that some parts of the embryo grow faster than others.

8. The *cytochrome c* molecular clock
a. measures the rate of evolution of oxidative metabolism.
b. measures the rate of evolutionary change of *cytochrome c*.
c. runs at a constant rate.
d. All of the above

9. The amino acid sequences of two species are compared, and few differences are found. This suggests that these species
a. are hybridizing and will become one species.
b. have identical genomes.
c. have similar evolutionary histories.
d. have both acquired the same mutations.

10. Convergent evolution
a. is an example of stabilizing selection.
b. depends on natural selection to independently produce similar phenotypic responses in different species or populations.
c. occurs only on islands.
d. is expected when different lineages are exposed to vastly different selective environments.

11. The irreducible complexity argument states that
a. cells are impossibly complex unless invented by an intelligent agent.
b. cells are too complex to ever be understood.
c. natural selection cannot operate on a complex system.
d. None of the above

Apply

1. Which of the following conditions need NOT be met for evolution by natural selection to occur in a population?
a. Variation must be genetically transmissible to the next generation.
b. Variants in the population must have a differential effect on lifetime reproductive success.
c. Variation must be detectable by the opposite sex.
d. Variation must exist in the population.

2. If Kettlewell monitored a mildly polluted region where tree trunks were a gray color, what would be a reasonable hypothesis with regard to the moth population?
a. There would be few light- and dark-colored moths, and many moths of intermediate color.
b. There would be no dark- or light-colored moths.
c. There would be mostly light-colored moths.
d. There would be equal numbers of dark-, light-, and intermediate-colored moths.

3. Artificial selection experiments in the laboratory, such as the one in figure 20.6, are an example of
a. stabilizing selection.
b. negative frequency-dependent selection.
c. directional selection.
d. disruptive selection.

4. A change in toe number in the evolution of horses was advantageous because
a. it allowed them to stand for long times, chewing grass.
b. it gave animals improved speed to escape predators.
c. it provided great flexibility to navigate through forest shrubbery.
d. it made them taller so that they could eat from higher trees.

5. Many beetle species have nonfunctional wings hidden under a hard shell. Which of the following is ACCURATE?
a. The ancestors of these beetles could probably fly.
b. The hidden wings are vestigial structures.
c. They are the result of convergent evolution.
d. Both a and b

6. You are examining the DNA sequence of a particular gene in Darwin's finches. In which of the following pairs would you expect to find the most differences in sequence? Refer to figure 20.3.
a. *Geospiza fuliginosa* and *Geospiza fortis*
b. *Cactospiza pallida* and *Cactospiza heliobates*
c. *Cactospiza heliobates* and *Camarhynchus parvulus*
d. *Cactospiza pallida* and *Platyspiza crassirostris*

7. Cacti and euphorbs are both succulent desert plants. Cacti, found in the western hemisphere, have spines that are modified leaves; euphorbs, found in the eastern hemisphere, have thorns that are modified branches. Cacti spines and euphorb thorns are
a. homologous.
b. analogous.
c. the result of convergent evolution.
d. Both b and c

Synthesize

1. On figure 20.2*b*, draw the relationship between offspring beak depth and parent beak depth, assuming that there is no genetic basis to beak depth in the medium ground finch.
2. What can you conclude from the fact that the frequency of melanic moths decreased to the same degree in Caldy Common in England as in Michigan?
3. Refer to figure 20.6. In this experiment on artificial selection, one population of *Drosophila* was selected for low numbers of bristles and the other for high numbers. Note that not only did the means of the populations change greatly in 35 generations, but also all individuals in both experimental populations lay outside the range of the initial population. What would the result of this experiment have been if only flies with high numbers of bristles had been allowed to breed?
4. The ancestor of horses was a small, many-toed animal that lived in forests, whereas today's wild horses are large animals with a single hoof that live on open plains. A series of intermediate fossils illustrate how this transition has occurred, and for this reason, many old treatments of horse evolution portrayed it as a steady increase through time in body size accompanied by a steady decrease in toe number. Why is this interpretation incorrect?
5. Other than the eye, can you suggest a human structure only imperfectly suited to its use?
6. Can you suggest a reasonable scientific explanation of the pattern of genomic variation in figure 20.8 other than evolution over time?
7. The thylacine, also called the Tasmanian wolf, became extinct only recently (the last individual died in a zoo in 1936). What would you accept as evidence that the marsupial thylacine is an example of evolutionary convergence with mammalian wolves?
8. In a courtroom in 2005, biologist Ken Miller criticized the claims of intelligent design. After noting that 99.9% of the organisms that have ever lived on Earth are now extinct, he said that "an intelligent designer who designed things, 99.9% of which didn't last, certainly wouldn't be very intelligent." Evaluate Miller's criticism.

JohnMernick/iStock/Getty Images

21 The Origin of Species

Learning Path

21.1 The Biological Species Concept Highlights Reproductive Isolation

21.2 Natural Selection May Reinforce Reproductive Isolation

21.3 Natural Selection and Genetic Drift Play Key Roles in Speciation

21.4 Speciation Is Influenced by Geography

21.5 Adaptive Radiation Requires Both Speciation and Habitat Diversity

21.6 The Pace of Evolution Varies

21.7 Speciation and Extinction Have Molded Biodiversity Through Time

Concept Overview

This Concept Overview provides a diagram of the overarching concepts that are covered in this chapter. These concepts will be expanded upon in the Concept Overview at the end of the chapter.

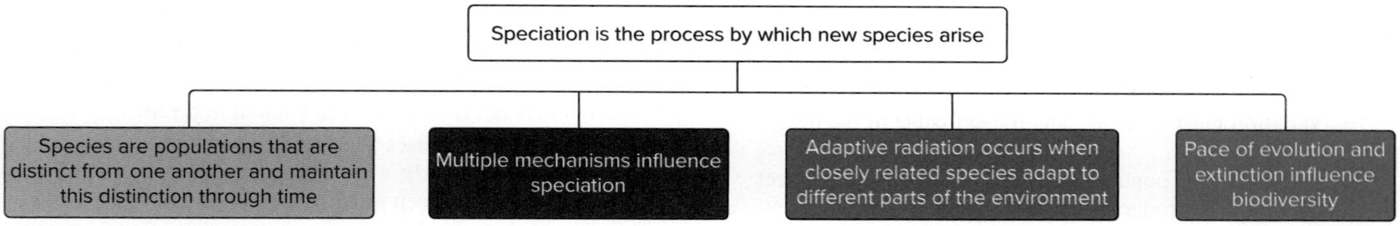

Speciation is the process by which new species arise
Species are populations that are distinct from one another and maintain this distinction through time
Multiple mechanisms influence speciation
Adaptive radiation occurs when closely related species adapt to different parts of the environment
Pace of evolution and extinction influence biodiversity

Introduction

Although Darwin titled his book *On the Origin of Species*, he never actually discussed what he referred to as that "mystery of mysteries"—how one species gives rise to another. Rather, his argument concerned evolution by natural selection—that is, how one species evolves through time to adapt to its changing environment. Although an important mechanism of evolutionary change, the process of adaptation does not explain how one species becomes another, a process we call *speciation*. As we will discuss, adaptation may be involved in the speciation process, but it does not have to be.

Before we can discuss how one species gives rise to another, we need to understand exactly what a species is. Even though the definition of a species is of fundamental importance to evolutionary biology, this issue has still not been completely settled and is the subject of considerable research and debate.

21.1 The Biological Species Concept Highlights Reproductive Isolation

Any concept of a species must account for two phenomena: the distinctiveness of species that occur together at a single locality and the connection that exists among different populations belonging to the same species.

The Biological Species Concept Focuses on the Ability to Exchange Genes

LEARNING OBJECTIVE 21.1.1 Explain the basis for the biological species concept.

Sympatric species inhabit the same locale but remain distinct

Put out a birdfeeder on your balcony or in your backyard and you will attract a wide variety of birds (especially if you include different kinds of foods). In the midwestern United States, for example, you might routinely find cardinals, blue jays, downy woodpeckers, house finches—even hummingbirds in the summer.

Although it might take a few days of careful observation, you would soon be able to readily distinguish among the many different species. The reason is that species that occur together (termed **sympatric**) are distinctive entities that are phenotypically different, utilize different parts of the habitat, and behave differently. This observation is generally true not only for birds but also for most other types of organisms.

Occasionally, two species occur together that appear to be nearly identical. In such cases, we need to go beyond visual similarities. When other aspects of the phenotype are examined, such as the mating calls or the chemicals exuded by each species, they usually reveal great differences. In other words, even though we might have trouble distinguishing them, the organisms themselves have no such difficulties.

Populations of a species exhibit geographical variation

Within a single species, individuals in populations that occur in different areas may be distinct from one another. Such groups of distinctive individuals may be classified as **subspecies** (the vague term *race* has a similar connotation but is no longer commonly used). In areas where these populations occur close to one another, individuals often exhibit combinations of features characteristic of both populations (figure 21.1). In other words, even though geographically distant populations may appear distinct, they are usually connected by intervening populations that are intermediate in their characteristics.

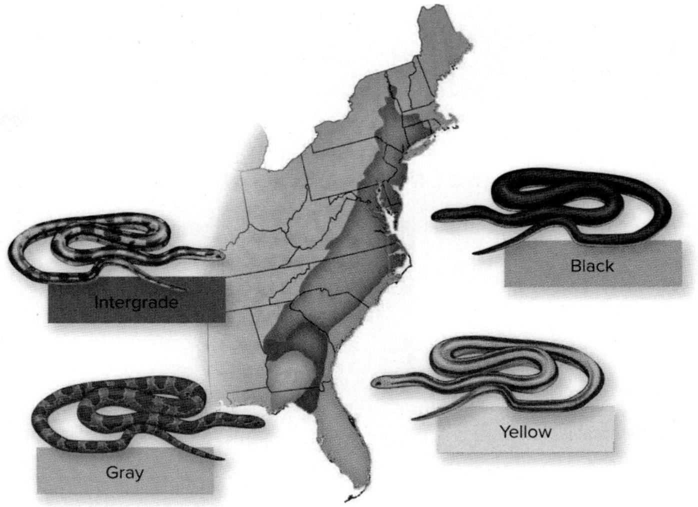

Figure 21.1 Geographic variation in the eastern rat snake, *Pantherophis alleghaniensis*. Although populations at the eastern, western, and northern ends of the species' range are phenotypically quite distinctive from one another, they are connected by populations that are phenotypically intermediate.

What can account for both the distinctiveness of sympatric species and the connectedness of geographically separate populations of the same species? One obvious possibility is that species exchange genetic material only with other members of the same species. If sympatric species commonly exchanged genes, which they generally do not, they would rapidly lose any distinctions. This would occur because the gene pools (all of the alleles present in a species) of the different species would become homogenized by intermixing. Conversely, the ability of members of geographically distinct populations to interbreed and share genes may keep these populations integrated as a single species.

Based on these ideas, in 1942 the evolutionary biologist Ernst Mayr set forth the **biological species concept,** which defines *species* as "groups of actually or potentially interbreeding natural populations which are reproductively isolated from other such groups."

In other words, the biological species concept says that a species is composed of populations whose members mate with each other and produce fertile offspring—or would do so if they came into contact. Conversely, populations whose members do not mate with each other or who cannot produce fertile offspring are said to be *reproductively isolated* and, therefore, not members of the same species.

What causes reproductive isolation? If organisms cannot interbreed or cannot produce fertile offspring, they clearly belong to different species. However, some populations that are considered separate species can interbreed and produce fertile offspring, but they ordinarily do not do so under natural conditions. They are still considered reproductively isolated, in that genes from one species generally will not enter the gene pool of the other.

Table 21.1 summarizes the steps at which barriers to successful reproduction may occur. Such barriers are termed **reproductive isolating mechanisms** because they prevent genetic exchange between species. We will discuss examples of these next, beginning with those that prevent the formation of zygotes, which are called **prezygotic isolating mechanisms.** Mechanisms that prevent the proper functioning of zygotes after they form are called **postzygotic isolating mechanisms.**

TABLE 21.1 Reproductive Isolating Mechanisms

Mechanism		Description
PREZYGOTIC ISOLATING MECHANISMS		
Ecological isolation		Species occur in the same area, but they occupy different habitats and rarely encounter each other.
Behavioral isolation		Species differ in their mating rituals.
Temporal isolation		Species reproduce in different seasons or at different times of the day.
Mechanical isolation		Anatomical differences between species prevent mating.
Prevention of gamete fusion	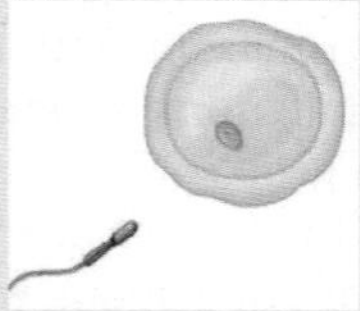	Gametes of one species function poorly with the gametes of another species or within the reproductive tract of another species.
POSTZYGOTIC ISOLATING MECHANISMS		
Hybrid inviability or infertility		Hybrid embryos do not develop properly, hybrid adults do not survive in nature, or hybrid adults are sterile or have reduced fertility.

Prezygotic Isolating Mechanisms Prevent the Formation of a Zygote

LEARNING OBJECTIVE 21.1.2 Distinguish among the various forms of prezygotic isolating mechanisms.

Mechanisms that prevent formation of a zygote include ecological or environmental isolation, behavioral isolation, temporal isolation, mechanical isolation, and prevention of gamete fusion.

Ecological isolation

Even if two species occur in the same area, they may utilize different portions of the environment and thus not hybridize because they do not encounter each other. For example, lions and tigers can produce hybrid offspring in zoos, but even though the ranges of the two species overlapped in India until about 150 years, no natural hybrids were ever discovered. Because of their ecological and behavioral differences, lions and tigers rarely came into direct contact with each other, even though their ranges overlapped over thousands of square kilometers.

In another example, the ranges of two toads, *Bufo woodhousei* and *B. americanus,* overlap in some areas. Although these two species can produce viable hybrids, they usually do not interbreed because they utilize different portions of the habitat for breeding. *B. woodhousei* prefers to breed in streams, and *B. americanus* breeds in rainwater puddles.

Similar situations occur among plants. Two species of oaks occur widely in California: the valley oak, *Quercus lobata,* and the scrub oak, *Q. dumosa.* The valley oak, a graceful deciduous tree that can be as tall as 35 m, occurs in the fertile soils of open grassland on gentle slopes and valley floors. In contrast, the scrub oak is an evergreen shrub, usually only 1 to 3 m tall, which often forms the kind of dense scrub known as chaparral. The scrub oak is found on steep slopes in less fertile soils. Hybrids between these different oaks do occur and are fully fertile, but they are rare. The sharply distinct habitats of their parents limit their occurrence together, and there is little intermediate habitat where the hybrids might flourish.

Behavioral isolation

Many species have elaborate courtship and mating rituals (discussed more in chapter 37), which may affect sexual selection in these groups (refer to chapter 19). Many related species of organisms, such as birds, differ in their courtship rituals, which tends to keep these species distinct in nature even if they inhabit the same places (figure 21.2). For example, mallard and pintail ducks are perhaps the two most common freshwater ducks in North America. In captivity they produce completely fertile hybrid offspring, but in nature they nest side by side and only rarely hybridize.

Sympatric species mate preferentially with members of their own species using a variety of forms of communication. Every mode of communication imaginable may be used by some species. Differences in visual signals are common, but many animals rely more on other sensory cues such as sound or scent for communication. Frogs, birds, and a variety of insects use sound to attract mates. Predictably, sympatric species of these animals produce different calls. Similarly, lacewings produce "songs" when they vibrate their abdomens against the surface on which they are sitting, and sympatric species produce different vibration patterns (figure 21.3).

Other species rely on the detection of chemical signals, called **pheromones.** The use of pheromones in moths has been particularly well studied. When female moths are ready to mate, they emit a pheromone that males can detect at great distances. Sympatric species differ in the pheromone they produce: either they use different chemical compounds or, if

Figure 21.2 Differences in courtship rituals can isolate related bird species. These Galápagos blue-footed boobies select their mates only after an elaborate courtship display. This male is lifting his feet in a ritualized high-step that shows off his bright blue feet. The display behavior of the two other species of boobies that occur in the Galápagos is very different, as is the color of their feet.

Rene Baars/Shutterstock

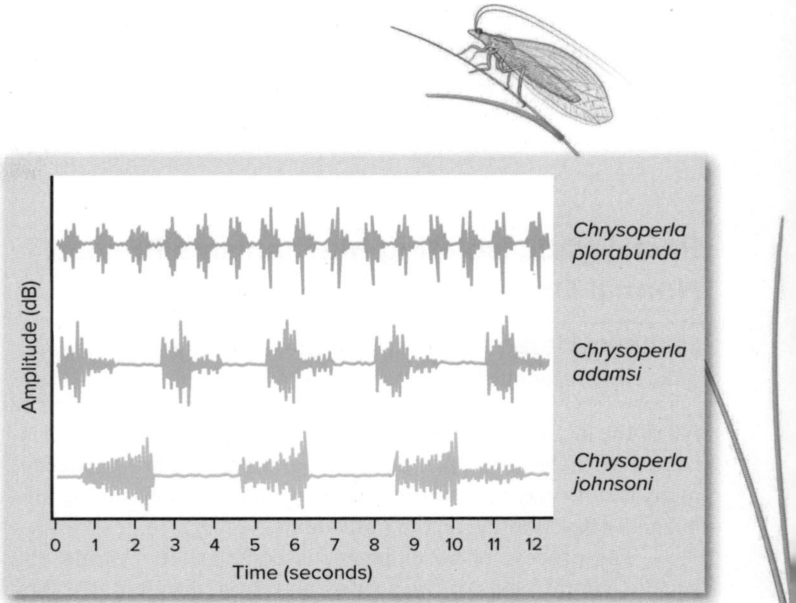

Figure 21.3 Differences in courtship song of sympatric species of lacewings. Lacewings are small insects that rely on auditory signals produced by moving their abdomens to vibrate the surface on which they are sitting to attract mates. As these recordings indicate, the vibration patterns produced by sympatric species differ greatly. Females, which detect the calls as they are transmitted through solid surfaces such as branches, are able to distinguish calls of different species and respond only to individuals producing their own species' call.

using the same compounds, the proportions used are different. Laboratory studies indicate that males are remarkably adept at distinguishing the pheromones of their own species from those of other species or even from synthetic compounds that are similar, but not identical, to that of their own species.

Some species even use electroreception. African and South Asian electric fish independently have evolved specialized organs in their tails that produce electrical discharges and electroreceptors on their skins to detect them. These discharges are used to communicate in social interactions; field experiments indicate that males can distinguish between signals produced by their own and other species, probably on the basis of differences in the timing of the electrical pulses.

Temporal isolation

Lactuca graminifolia and *L. canadensis,* two species of wild lettuce, grow together along roadsides throughout the southeastern United States. Hybrids between these two species are easily made experimentally and are completely fertile. But these hybrids are rare in nature because *L. graminifolia* flowers in early spring and *L. canadensis* flowers in summer. When their blooming periods overlap, as happens occasionally, the two species do form hybrids, which may become locally abundant.

Many species of closely related amphibians have different breeding seasons that prevent hybridization. For example, four species of frogs of the genus *Rana* occur together in most of the eastern United States, but hybrids are rare, because the peak breeding time is different for each of them.

Mechanical isolation

Structural differences prevent mating between some related species of animals. Aside from such obvious features as size, the structure of the male and female copulatory organs may be incompatible. In many insect and other arthropod groups, the sexual organs, particularly those of the male, are so diverse that they are used as a primary basis for distinguishing species.

Similarly, flowers of related species of plants often differ significantly in their proportions and structures. Some of these differences limit the transfer of pollen from one plant species to another. For example, bees may carry the pollen of one species on a certain place on their bodies; if this area does not come into contact with the receptive structures of the flowers of another plant species, the pollen is not transferred.

Prevention of gamete fusion

In animals that shed gametes directly into water, the eggs and sperm derived from different species may not attract or fuse with one another. Many land animals may not hybridize successfully because the sperm of one species functions so poorly within the reproductive tract of another that fertilization never takes place. In plants, the growth of pollen tubes may be impeded in hybrids between different species. In both plants and animals, isolating mechanisms such as these prevent the union of gametes, even following successful mating.

Postzygotic Isolating Mechanisms Prevent Normal Development into Fertile Adults

LEARNING OBJECTIVE 21.1.3 Differentiate between postzygotic and prezygotic isolating mechanisms.

All of the factors we have discussed so far tend to prevent hybridization. If hybrid matings do occur and zygotes are produced, many factors may still prevent those zygotes from developing into normally functioning, fertile individuals.

The process of development is complex. In hybrids, the genetic complements of two species may be so different that they cannot function together normally in embryonic development. For example, hybridization between sheep and goats usually produces embryos that die in the earliest developmental stages.

The leopard frogs (*Rana pipiens* complex) of the eastern United States are a group of similar species, assumed for a long time to constitute a single species (figure 21.4). Careful examination, however, revealed that although the frogs appear similar, successful mating between them is rare because of problems that occur as the fertilized eggs develop. Many of the hybrid combinations cannot be produced even in the laboratory.

Examples of this kind, in which similar species have been recognized only as a result of hybridization experiments, are common in plants. Sometimes the hybrid plant embryos can be removed at an early stage and grown in an artificial medium. When these hybrids are supplied with extra nutrients or other supplements that compensate for their weakness or inviability, they may complete their development normally.

Even when hybrids survive the embryo stage, they still may not develop normally. If the hybrids are less physically fit than their parents, they will almost certainly be eliminated in nature. Even if a hybrid is vigorous and strong, as in the case of the mule, which is a hybrid between a female horse and a male donkey, it may still be sterile and thus incapable of contributing to succeeding generations.

Hybrids may be sterile because the development of sex organs is abnormal, because the chromosomes derived from the respective parents cannot pair properly during meiosis, or due to a variety of other causes.

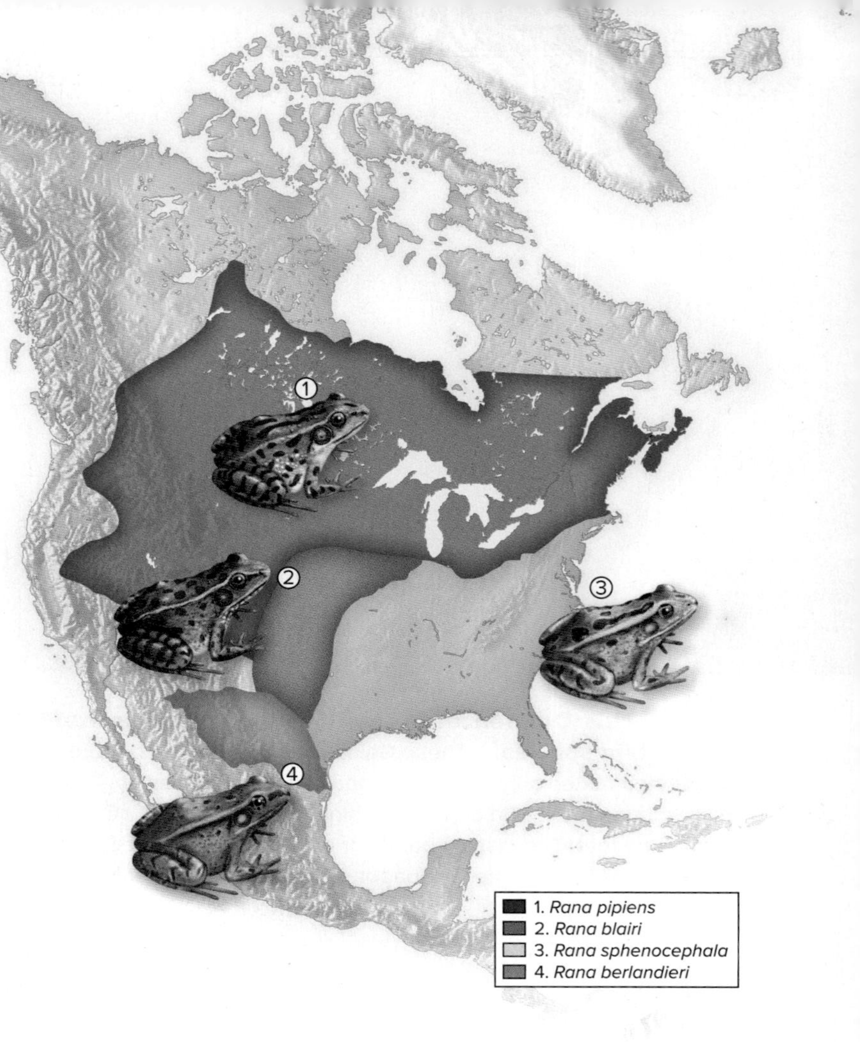

1
2
3
4
1. Rana pipiens
2. Rana blairi
3. Rana sphenocephala
4. Rana berlandieri

Figure 21.4 Postzygotic isolation in leopard frogs. These four species resemble one another closely in their external features. Their status as separate species first was suspected when hybrids between some pairs of these species were found to produce defective embryos in the laboratory. Subsequent research revealed that the mating calls of the four species differ substantially, indicating that the species have both pre- and postzygotic isolating mechanisms.

The Biological Species Concept Does Not Explain All Observations

LEARNING OBJECTIVE 21.1.4 Explain the weaknesses of the biological species concept.

The biological species concept is an effective way to explain the existence of species in nature. Nonetheless, it does have some weaknesses and does not take into account all observations. One criticism of the biological species concept concerns the extent to which all species truly are reproductively isolated. By definition, under the biological species concept, different species should not interbreed and produce fertile offspring. In recent years, biologists have observed more hybridization than expected in populations that seem to coexist as distinct biological entities.

Botanists have always been aware that plant species often undergo substantial amounts of hybridization. More than 50% of California plant species included in one study, for example, were not well defined by genetic isolation. This coexistence without genetic isolation can be long-lasting: fossil data show that balsam poplars and cottonwoods have been phenotypically distinct for 12 million years, but they also have routinely produced hybrids throughout this time. Consequently, many botanists have long felt that the biological species concept applies only to animals.

New evidence, however, increasingly indicates that hybridization is not all that uncommon in animals, either. In recent years, many cases of substantial hybridization between animal species have been documented. One recent survey indicated that almost 10% of the world's 9500 bird species are known to have hybridized in nature.

The Galápagos finches provide a particularly well-studied example. Three species on the island of Daphne Major—the medium ground finch, the cactus finch, and the small ground finch—are clearly distinct morphologically, and they occupy different ecological niches. Studies over the past 20 years by Peter and Rosemary Grant found that, on average, 2% of the medium ground finches and 1% of the cactus ground finches mated with other species every year. Furthermore, hybrid offspring appeared to be at no disadvantage in terms of survival or subsequent reproduction. This is not a trivial amount of genetic exchange, and one might expect to find that the species coalesce into one genetically variable population—but the species are maintaining their distinctiveness.

These observations do not mean that hybridization is rampant throughout the animal world. Most bird species do not hybridize, and probably even fewer experience significant amounts of hybridization. Still, hybridization is common enough to cast doubt on whether reproductive isolation is the only force maintaining the integrity of species.

Natural selection and the ecological species concept

The ecological species concept proposes that the distinctions among species are maintained by natural selection. The idea is that each species has adapted to its own specific part of the environment. Stabilizing selection, described in chapter 19, then maintains the species' adaptations. Hybridization has little effect, because alleles introduced into one species' gene pool from other species are quickly eliminated by natural selection.

The problem with this view is that the interaction between gene flow and natural selection can have many outcomes, as discussed in chapter 19. Strong selection can overwhelm any effects of gene flow, but gene flow can also prevent less successful alleles from being eliminated from a population. So, as a general explanation, an ecological species concept may not be more useful than the biological species concept, although it may be a more successful description for some organisms or habitats.

Other weaknesses of the biological species concept

The biological species concept has been criticized for other reasons as well. One problem is that reproductive interactions cannot be studied in fossils, making the concept inapplicable to extinct species.

Another issue is that it can be difficult to apply the concept to populations that are geographically separated in nature. Because individuals of these populations do not encounter each other, it is not possible to observe whether they would interbreed naturally.

Although experiments can determine whether fertile hybrids can be produced, this information is not enough. Many species that coexist without interbreeding in nature will readily hybridize in the artificial settings of the laboratory or zoo. Consequently, evaluating whether such populations constitute different species is ultimately a judgment call. In addition, the concept is more limited than its name would imply. Many organisms are asexual and reproduce without mating. Reproductive isolation therefore has no meaning for such organisms.

For these reasons, a variety of other ideas have been put forward to establish criteria for defining species. Many of these are specific to a particular type of organism, and none has universal applicability. In reality, there may be no single explanation for what maintains the identity of species. Given the incredible variation evident in plants, animals, and microorganisms in all aspects of their biology, it would not be surprising to find that different processes are operating in different organisms.

In addition, some scientists have turned from emphasizing the processes that maintain species distinctions to examining the evolutionary history of populations. These phylogenetic species concepts are currently a topic of great debate and are discussed further in chapter 22.

REVIEW OF CONCEPT 21.1

Species are populations of organisms distinct from others. The biological species concept defines species based on their ability to interbreed. Reproductive isolating mechanisms prevent successful interbreeding between species. The ecological species concept relies on adaptation and natural selection as forces for maintaining separation of species.

- *How does the ability to exchange genes explain why sympatric species remain distinct but geographical populations of one species remain connected?*

21.2 Natural Selection May Reinforce Reproductive Isolation

One of the oldest questions in the field of evolution is, How does one ancestral species become divided into two descendant species (a process termed cladogenesis)? If species are defined by the existence of reproductive isolation, then the process of speciation is identical to the evolution of reproductive isolating mechanisms.

Selection May Act to Strengthen Isolating Mechanisms

LEARNING OBJECTIVE 21.2.1 Explain how natural selection can reinforce reproductive isolation.

The formation of species is a continuous process, and as a result, two populations may be only partially reproductively isolated. For example, because of behavioral or ecological differences, individuals of two populations may be more likely to mate with members of their own population, and yet between-population matings may still occur. If mating occurs and fertilization produces a zygote, postzygotic barriers may also be incomplete: developmental problems may result in lower embryo survival or reduced fertility, but some individuals may survive and reproduce.

Figure 21.5 Reinforcement in European flycatchers. The pied flycatcher (*Ficedula hypoleuca*) and the collared flycatcher (*F. albicollis*) appear very similar when they occur alone. However, in places where the two species occur sympatrically (indicated by the yellow color on the map), they have evolved differences in color and pattern, which allow individuals to choose mates from their own species and thus avoid hybridizing.

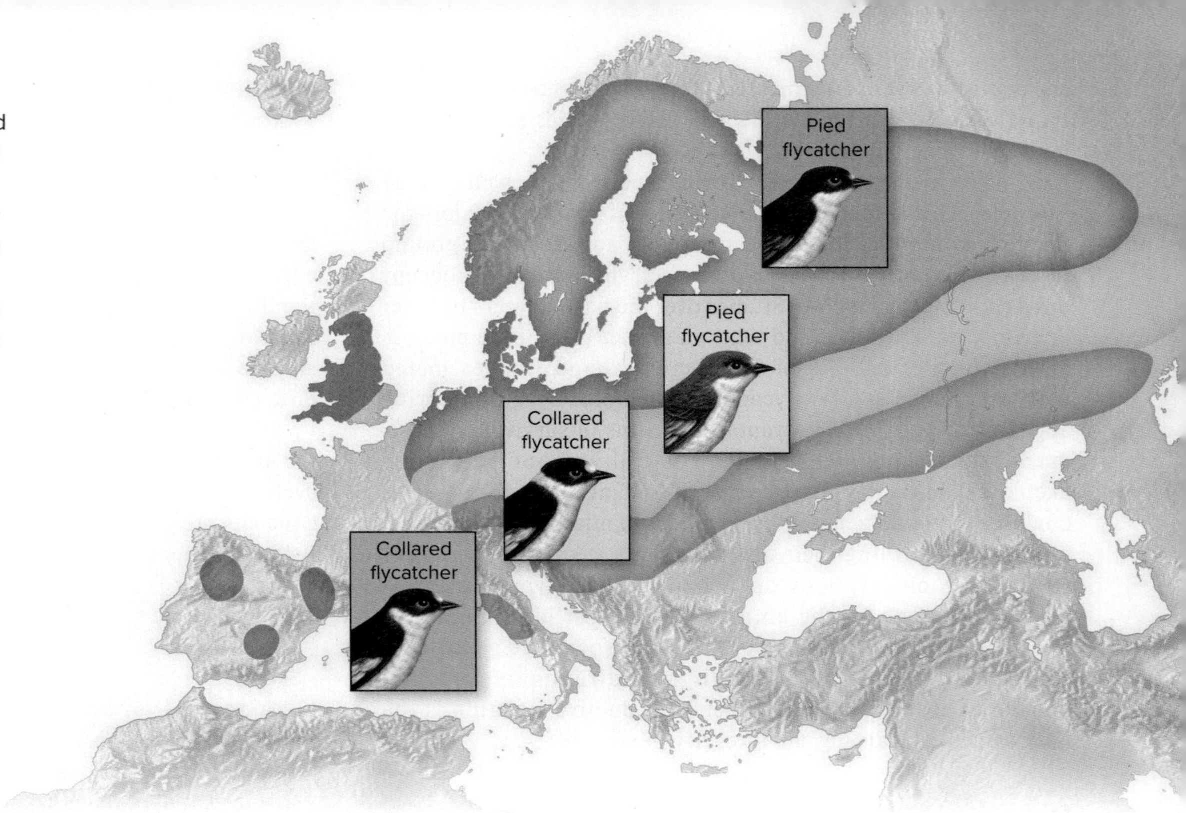

What happens when two populations come into contact thus depends on the extent to which isolating mechanisms have already evolved. If isolating mechanisms have not evolved at all, then the two populations will interbreed freely, and whatever other differences have evolved between them should disappear over the course of time, as genetic exchange homogenizes the populations. Conversely, if the populations are completely reproductively isolated, then no genetic exchange will occur, and the two populations will remain different species.

How reinforcement can complete the speciation process

The intermediate state, in which reproductive isolation has partially evolved but is not complete, is perhaps the most interesting situation. If the hybrids are partly sterile, or not as well adapted to the existing habitats as their parents, they will be at a disadvantage. Selection would favor any alleles in the parental populations that prevented hybridization, because matings within species would produce more successful offspring.

The result would be the continual improvement of prezygotic isolating mechanisms until the two populations were completely reproductively isolated. This process is termed **reinforcement,** because initially incomplete isolating mechanisms are reinforced by natural selection until they are completely effective.

An example of reinforcement is provided by pied and collared flycatchers. Throughout much of eastern and central Europe, these two bird species are geographically separated **(allopatric)** but are very similar in color (figure 21.5). However, in the Czech Republic and Slovakia, the two species occur together and occasionally hybridize, producing offspring that usually have very low fertility. At those sites, the species have evolved to look very different from each other, and birds prefer to mate with individuals with their own species' coloration. In contrast, birds from the allopatric populations prefer the allopatric color pattern. As a consequence of the color differences, where the species are sympatric the rate of hybridization is extremely low. These results indicate that when populations of the two species came into contact, natural selection led to the evolution of differences in color patterns, resulting in the evolution of behavioral, prezygotic isolation.

How gene flow may counter speciation

Reinforcement is not inevitable, however. When incompletely isolated populations come together, gene flow immediately begins to occur between them. Although hybrids may be inferior, they are not completely inviable or infertile—if they were, the species would already be completely reproductively isolated. When these surviving hybrids reproduce with members of either population, they serve as a conduit of genetic exchange from one population to the other, and the two populations tend to lose their genetic distinctiveness. Thus, a race ensues: can complete reproductive isolation evolve before gene flow erases the differences between the populations? Experts disagree on the likely outcome, but many consider reinforcement to be the much less common outcome.

REVIEW OF CONCEPT 21.2

Natural selection may favor the evolution of increased prezygotic reproductive isolation between sympatric populations. This phenomenon is termed reinforcement, and it may lead to populations becoming completely reproductively isolated. In contrast, however, genetic exchange between populations may decrease genetic differences among populations, thus preventing speciation from occurring.

- *How might the initial degree of reproductive isolation affect the probability that reinforcement will occur when two populations come into sympatry?*

21.3 Natural Selection and Genetic Drift Play Key Roles in Speciation

What role does natural selection play in the speciation process? Certainly, the process of reinforcement is driven by natural selection, favoring the evolution of complete reproductive isolation. But reinforcement may not be common. In situations where reinforcement is not occurring, does natural selection play a role in the evolution of reproductive isolating mechanisms?

Both Random Changes and Adaptation Can Lead to Isolation and Speciation

LEARNING OBJECTIVE 21.3.1 Compare how natural selection and genetic drift affect speciation.

Random changes may cause reproductive isolation

Populations may diverge due to random causes. Genetic drift, founder effects, and population bottlenecks all may lead to changes in traits that cause reproductive isolation.

For example, in the Hawaiian Islands, closely related species of *Drosophila* often differ greatly in their courtship behavior. Colonization of new islands by these fruit flies probably involved a founder effect, in which one or a few flies—perhaps only a single pregnant female—were blown by strong winds to the new island. Changes in courtship behavior between ancestor and descendant populations may be the result of such founder events.

Given enough time, any two isolated populations will diverge because of genetic drift (remember that even large populations experience drift, but at a lower rate than small populations). In some cases, this random divergence may affect traits responsible for reproductive isolation, and speciation may occur.

Adaptation can lead to speciation

Although random processes are acting, in many cases natural selection probably plays a role in the speciation process. As populations of a species adapt to different circumstances, they likely accumulate many differences that lead to reproductive isolation. It is important to realize that in cases like this, contrary to what occurs during reinforcement, natural selection isn't acting to favor those individuals that avoid hybridizing with the other species. Rather, increased reproductive isolation evolves as an incidental consequence of the evolution of distinctive adaptations in the different populations. For example, if one population of flies adapts to wet conditions and another to dry ones, then natural selection will favor a variety of corresponding differences in physiological and sensory traits. These differences may promote ecological and behavioral isolation that could cause hybrids between the two populations to be poorly adapted to either habitat.

Selection on mating behavior might also incidentally lead to the evolution of reproductive isolation. For example, male *Anolis* lizards court females by extending a colorful flap of skin, called a *dewlap,* located under their throats (figure 21.6). The ability of one lizard to see the dewlap of another lizard depends on both the color of the dewlap and the environment of the lizards. A light-colored dewlap is most effective in reflecting light in a dim forest, whereas dark colors are more apparent in the bright glare of open habitats. As a result, when these lizards occupy new habitats, natural selection favors evolutionary change in dewlap color because males whose dewlaps cannot be seen will not attract many mates. But the lizards also distinguish members of their own species from other species by the color of the dewlap. Adaptive change in mating signals in new environments could therefore have the incidental consequence of producing reproductive isolation from populations in the ancestral environment.

Laboratory experiments on fruit flies and other fast-reproducing organisms, using isolated populations, allow measurements of the factors involved in reproductive isolation. These experiments indicate that genetic drift by itself can lead to some degree of reproductive isolation, but in general, reproductive

Figure 21.6 Dewlaps of different species of Caribbean *Anolis* lizards. Males use their dewlaps in both territorial and courtship displays. Coexisting species almost always differ in their dewlaps, which are used in species recognition. Darker-colored dewlaps, such as those of the two species on the left, are easier to see in open habitats, whereas lighter-colored dewlaps, like those of the two species on the right, are more visible in shaded environments.

(all): Jonathan Losos

isolation evolves more rapidly when the populations are forced to adapt to an altered laboratory environment, such as temperature or food type changes. Although natural selection in the experiment does not directly favor traits because they lead to reproductive isolation, the incidental effect of adaptive divergence is that populations in different environments become reproductively isolated. For this reason, some biologists believe that the term *isolating mechanisms* is misguided, because it implies that the traits evolved specifically for the purpose of genetically isolating a species, which in most cases—except reinforcement—is probably incorrect.

REVIEW OF CONCEPT 21.3

Genetic drift refers to randomly generated changes in a population's genetic makeup. Isolated populations will eventually diverge because of genetic drift. Adaptation to different environments may also lead to populations becoming reproductively isolated from each other.

- *How can complete reproductive isolation evolve without reinforcement occurring?*

21.4 Speciation Is Influenced by Geography

Thus far, we have considered reproductive isolation from a purely biological perspective. This is actually too simplistic, as geography can affect the process. Populations that are not in the same place are by definition isolated.

Allopatric Speciation Takes Place When Populations Are Geographically Isolated

LEARNING OBJECTIVE 21.4.1 Compare allopatric species with sympatric species.

Speciation is a two-part process. Initially identical populations must diverge, and reproductive isolation must evolve to maintain these differences. The difficulty with this process, as discussed previously, is that the homogenizing effect of gene flow between populations is constantly acting to erase any differences that may arise, by either genetic drift or natural selection. Gene flow only occurs between populations that are in contact, however, and populations can become geographically isolated for a variety of reasons (figure 21.7). Consequently, evolutionary biologists have long recognized that speciation is much more likely in geographically isolated populations.

Ernst Mayr was the first biologist to demonstrate that geographically separated, or *allopatric,* populations appear much more likely to have evolved substantial differences leading to speciation. Marshaling data from a wide variety of organisms and localities, Mayr made a strong case for allopatric speciation as the primary means of speciation.

For example, the little paradise kingfisher varies little throughout its wide range in New Guinea, despite the great variation in the island's topography and climate. By contrast, isolated populations on nearby islands are strikingly different from one another and from the mainland population (figure 21.8). Thus, geographical isolation seems to have been an important prerequisite for the evolution of differences between populations.

Many other examples indicate that speciation can occur under allopatric conditions. Because we would expect isolated populations to diverge over time by either drift or selection,

a. *b.* *c.*

Figure 21.7 Populations can become geographically isolated for a variety of reasons. ***a.*** Colonization of remote areas by one or a few individuals can establish populations in a distant place. ***b.*** Barriers to movement can split an ancestral population into two isolated populations. ***c.*** Extinction of intermediate populations can leave the remaining populations isolated from one another.

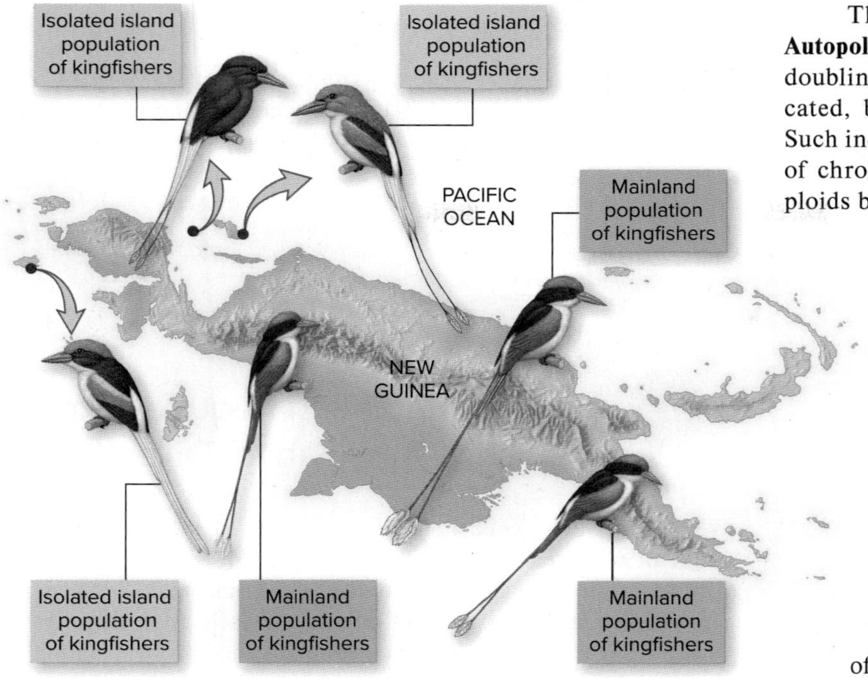

Figure 21.8 Phenotypic differentiation in the little paradise kingfisher, *Tanysiptera hydrocharis*, in New Guinea. Isolated island populations (left) are quite distinctive, showing variation in tail feather structure and length, plumage coloration, and bill size, whereas kingfishers on the mainland (right) show little variation.

this result is not surprising. Rather, the more intriguing question becomes, Is geographical isolation *required* for speciation to occur?

Sympatric Speciation Occurs without Geographical Separation

LEARNING OBJECTIVE 21.4.2 Explain the conditions required for sympatric speciation to occur.

For decades, biologists have debated whether one species can split into two at a single locality, without the two new species ever having been geographically separated. Investigators have suggested that this sympatric speciation could occur either instantaneously or over the course of multiple generations. Although most of the hypotheses suggested so far are highly controversial, one mechanism of immediate sympatric speciation is common in plants: polyploidy.

Instantaneous speciation through polyploidy

Instantaneous sympatric speciation occurs when an individual arises that is reproductively isolated from all other members of its species. In most cases, a mutation that would cause an individual to differ greatly from others of its species would have many adverse pleiotropic side effects, and the individual would not survive. One exception, often found in plants, is the process of **polyploidy,** which produces individuals with more than two sets of chromosomes.

There are two ways that polyploid individuals can arise. **Autopolyploidy** occurs within a single species, usually by a doubling of chromosomes. This can occur when DNA is replicated, but cytokinesis does not occur during cell division. Such individuals, called *tetraploids* because they have four sets of chromosomes, can self-fertilize or mate with other tetraploids but cannot mate and produce fertile offspring with normal diploids. This is because the tetraploid species produce diploid gametes that when combined with normal haploid gametes produce triploid offspring (having three sets of chromosomes). Triploids are sterile, because the odd number of chromosomes prevents proper pairing during meiosis.

A more common type of polyploid speciation is **allopolyploidy,** which happens when two species hybridize prior to a chromosome-doubling event (figure 21.9). The resulting offspring, having one copy of the chromosomes of each species, is usually infertile because the chromosomes do not pair correctly in meiosis. However, such individuals are often otherwise healthy. Sometimes, the chromosomes of such an individual spontaneously double, as just described for autopolyploidy. Consequently, the resulting tetraploid has two copies of each set of chromosomes and pairing during meiosis is no longer a problem. As a result, such tetraploids are able to interbreed with other similar tetraploids, and a new species has been created.

It is estimated that about half of the approximately 260,000 species of plants have a polyploid episode in their history, including many of great commercial importance, such as bread wheat, cotton, tobacco, sugarcane, bananas, and potatoes. Speciation by polyploidy is also known to occur in a variety of animals, including insects, fish, and salamanders, although much more rarely than in plants.

Sympatric speciation by disruptive selection

Some investigators believe that sympatric speciation can occur over the course of multiple generations through the process of disruptive selection. As noted in chapter 19, disruptive selection can cause a population to contain individuals exhibiting two different phenotypes.

One might think that if selection were strong enough, these two phenotypes would evolve over a number of generations into different species. But before the two phenotypes could become different species, they would have to evolve reproductive isolating mechanisms. Initially, the two phenotypes would not be reproductively isolated at all, and genetic exchange between individuals of the two phenotypes would tend to prevent genetic divergence in mating preferences or other isolating mechanisms. As a result, the two phenotypes would be retained as polymorphisms within a single population. For this reason, most biologists consider sympatric speciation of this type to be a rare event.

In recent years, however, a number of cases have appeared that are difficult to interpret in any way other than as sympatric speciation. An example occurs on Lord Howe Island, a small (16 km^2), remote island in the Pacific Ocean 600 km

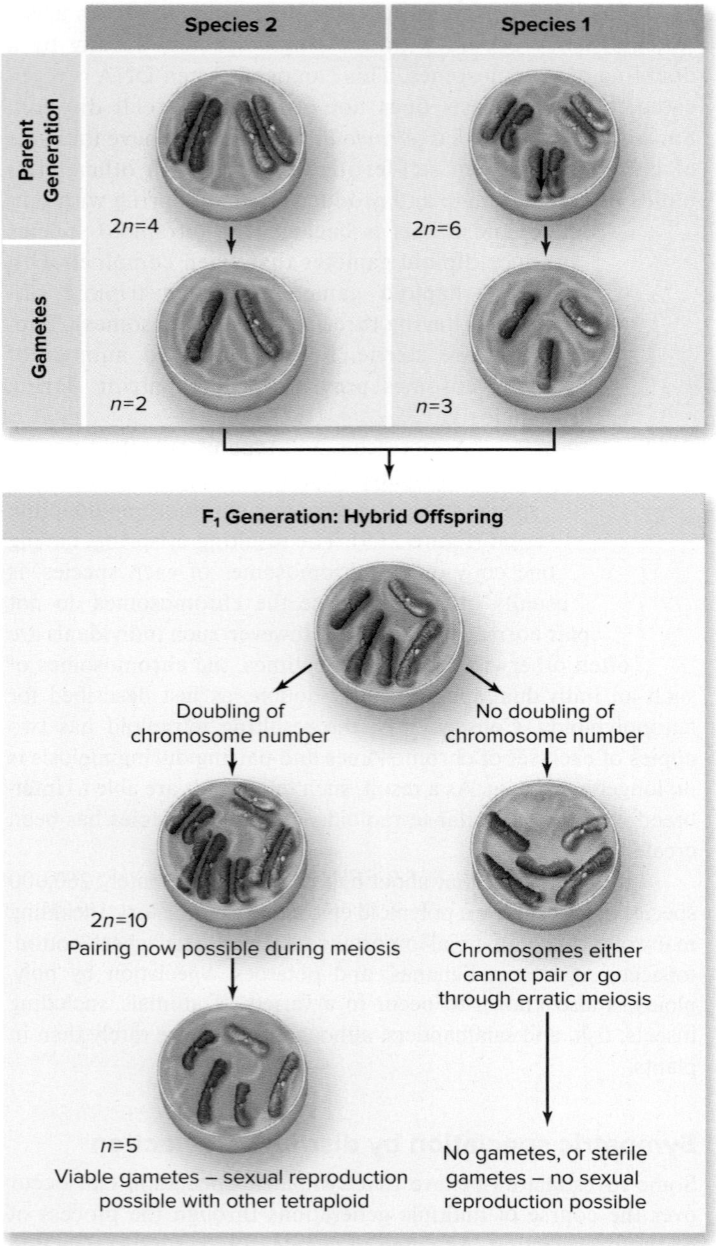

Figure 21.9 Allopolyploid speciation. Hybrid offspring from parents with different numbers of chromosomes often cannot reproduce sexually. Sometimes the number of chromosomes in such hybrids doubles to produce a tetraploid individual, which can undergo meiosis and reproduce with similar tetraploid individuals.

from the nearest large landmass—Australia. The palm genus *Howea* contains two species, both found on Lord Howe Island and adapted to living on different types of soil. Given the small size of the island and that the pollen of these trees is dispersed by the wind, little opportunity would have existed for divergence of two populations in isolation from each other. The most reasonable explanation is that the ancestral *Howea* species colonized the island and subsequently underwent sympatric speciation.

REVIEW OF CONCEPT 21.4

Sympatric speciation occurs without geographical separation, whereas allopatric speciation occurs in geographically isolated populations. Polyploidy and disruptive selection are two ways by which a single species may undergo sympatric speciation.

- *How do polyploidy and disruptive selection differ as ways in which sympatric speciation can occur?*

21.5 Adaptive Radiation Requires Both Speciation and Habitat Diversity

We say that these collections of closely related species resulted from an **adaptive radiation.** Darwin himself observed this in the finches of the Galápagos. One of the most visible manifestations of evolution is the existence of groups of closely related species that have recently evolved from a common ancestor by adapting to different parts of the environment.

Key Innovations Can Lead to Adaptive Radiations

LEARNING OBJECTIVE 21.5.1 Describe how a key innovation can lead to adaptive radiation.

These adaptive radiations are particularly common in situations where a species occurs in an environment with few other species and many available resources. One example is the creation of new islands through volcanic activity, such as the Hawaiian and Galápagos Islands. Another example is a catastrophic event leading to the extinction of most other species.

Adaptive radiation can also result when a new trait, called a **key innovation,** evolves within a species, allowing it to use resources or other aspects of the environment that were previously inaccessible to it. Classic examples of key innovation leading to adaptive radiation are the evolution of lungs in fish and of wings in birds and insects, both of which allowed descendant species to diversify and adapt to many newly available parts of the environment.

Adaptive radiation requires both speciation and adaptation to different habitats. A classic model postulates that a species colonizes multiple islands in an archipelago. Speciation subsequently occurs allopatrically, and then the newly arisen species colonize other islands, producing multiple species per island (figure 21.10).

Character Displacement May Drive Sympatric Species Divergence

LEARNING OBJECTIVE 21.5.2 Explain how character displacement may promote sympatric speciation.

Adaptation to new habitats can occur either during the allopatric phase, as the species respond to different environments on the

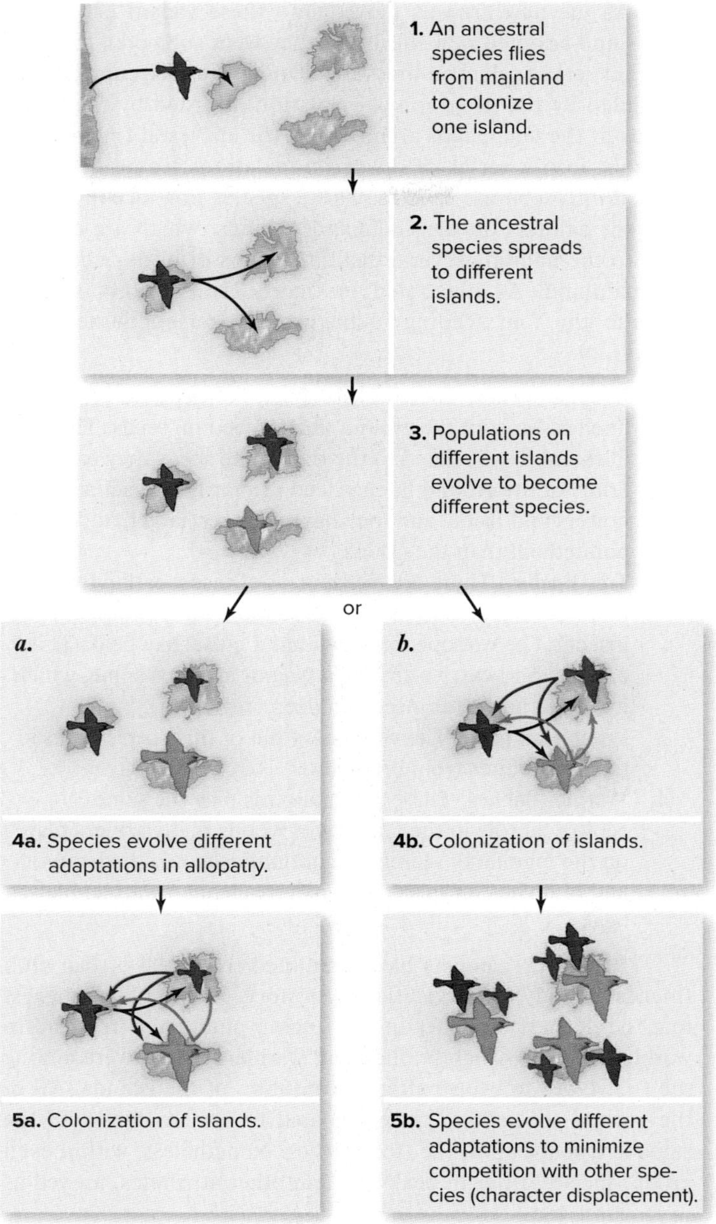
1. An ancestral species flies from mainland to colonize one island.
2. The ancestral species spreads to different islands.
3. Populations on different islands evolve to become different species.
or
a.
b.
4a. Species evolve different adaptations in allopatry.
4b. Colonization of islands.
5a. Colonization of islands.
5b. Species evolve different adaptations to minimize competition with other species (character displacement).

Figure 21.10 Classic model of adaptive radiation on island archipelagoes. ***(1)*** An ancestral species colonizes an island in an archipelago. Subsequently, the population colonizes other islands ***(2),*** after which the populations on the different islands speciate in allopatry ***(3).*** Then some of these new species colonize other islands, leading to local communities of two or more species. Adaptive differences can evolve either when species are in allopatry in response to different environmental conditions ***(4a,5a)*** or as a result of ecological interactions between species ***(4b,5b)*** by the process of character displacement.

different islands, or after two species become sympatric. In the latter case, this adaptation may be driven by selective pressures to minimize competition for available resources with other species. This process is termed **character displacement.** In character displacement, natural selection in each species favors those individuals that use resources not used by the other species. Because those individuals will have greater fitness, whatever traits cause the differences in resource use will increase in frequency (assuming that a genetic basis exists for these differences), and over time the species will diverge (figure 21.11).

An alternative possibility is that adaptive radiation occurs through repeated instances of sympatric speciation, producing a suite of species adapted to different habitats. These scenarios continue to be hotly debated.

SCIENTIFIC THINKING

Question: *Does competition for resources cause character displacement?*

Hypothesis: *Competition with similar species will cause natural selection to promote evolutionary divergence.*

Experiment: *Place a species of fish in a pond with another, similar fish species and measure the form of selection. As a control, place a population of the first species in a pond without the second species. Note that the size of food that these fish eat is related to the size of the fish.*

Result: *In the pond with two species, directional selection favors those individuals that have phenotypes most dissimilar from the other species, and thus are most different in resource use. Directional selection does not occur in the control population.*

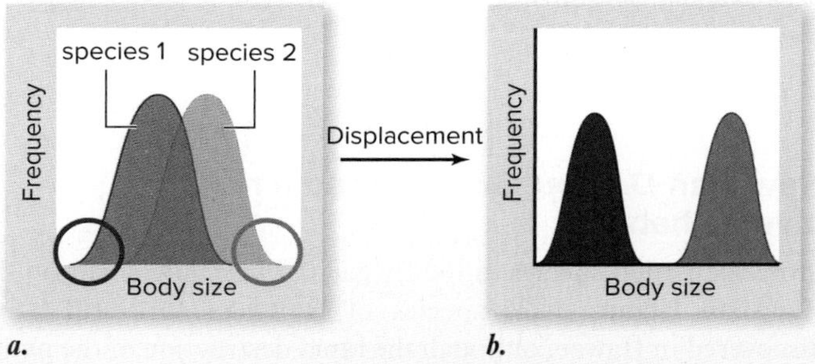
species 1
species 2
Frequency
Body size
Displacement
Frequency
Body size
a.
b.

Interpretation: *Would you expect character displacement to occur if resources were unlimited?*

Figure 21.11 Character displacement. ***a.*** Two species are initially similar and thus overlap greatly in resource use, as might happen if the two species are similar in size (in many species, body size and food size are closely related). Individuals in each species that are most different from the other species (circled) will be favored by natural selection, because they will not have to compete with the other species. For example, the smallest individuals of one species and the largest of the other would not compete with the other species for food and thus would be favored. ***b.*** As a result, the species will diverge in resource use and minimize competition between the species.

Adaptive Radiation Is Very Well Documented

LEARNING OBJECTIVE 21.5.3 Describe examples of adaptive radiation.

In the following paragraphs, we discuss four exemplary cases of adaptive radiation.

a. b.

Figure 21.12 Hawaiian *Drosophila*. The hundreds of species that have evolved on the Hawaiian Islands are extremely variable in appearance; note in particular the differences in head shape and body color between these two species. ***a.*** *Drosophila heteroneura.* ***b.*** *Drosophila silvestris.*

(a and b) Kevin T. Kaneshiro

Hawaiian *Drosophila* exploited a rich, diverse habitat

More than 1000 species in the fly genus *Drosophila* occur on the Hawaiian Islands. New species of *Drosophila* are still being discovered in Hawaii, although the rapid destruction of the native vegetation is making the search more difficult.

Aside from their sheer number, Hawaiian *Drosophila* species are unusual because of their incredible diversity of morphological and behavioral traits (figure 21.12). When their ancestors first reached these islands, they encountered many "empty" habitats that other kinds of insects and other animals occupied elsewhere. As a result, the species have adapted to all manners of fruit fly life and include predators, parasites, and herbivores, as well as species specialized for eating the detritus in leaf litter and the nectar of flowers. The larvae of various species live in rotting stems, fruits, bark, leaves, or roots or feed on sap. No comparable diversity of *Drosophila* species is found anywhere else in the world.

The great diversity of Hawaiian species is a result of the geologic history of these islands. New islands have continually arisen from the sea in this region. It appears that as they have done so, they have been invaded successively by the various *Drosophila* groups present on the older islands. New species thus have evolved as new islands have been colonized.

In addition, the Hawaiian Islands are among the most volcanically active islands in the world. Periodic lava flows have created many patches of habitat within an island surrounded by a "sea" of barren rock. These land islands are termed *kipukas. Drosophila* populations isolated in these kipukas often undergo speciation. In these ways, rampant speciation combined with ecological opportunity has led to an unparalleled diversity of insect life.

Darwin's finch species adapted to use different food types

The diversity of finches on the Galápagos Islands was noted by Darwin, as mentioned in chapter 20. He surmised that the ancestor of these finches reached the newly formed Galápagos Islands from South America before other land birds did. Many of the types of habitats that small birds use on the mainland had not yet been occupied on the new islands.

As the new arrivals moved into these vacant ecological niches and adopted new lifestyles, they were subjected to many different sets of selective pressures. Under these circumstances, and aided by the geographical isolation afforded by the many islands of the Galápagos archipelago, the ancestral finches rapidly split into a series of diverse populations, some of which evolved into separate species. These species now occupy many different habitats on the Galápagos Islands, which are comparable to the habitats several distinct groups of birds occupy on the mainland. As illustrated in figure 21.13, the 14 species fall into the four feeding niche groups you encountered in chapter 20:

1. **Ground finches.** There are six species of *Geospiza* ground finches. Most of the ground finches feed on seeds. The size of their bills is related to the size of the seeds they eat. Some of the ground finches feed primarily on cactus flowers and fruits, and they have a longer, larger, and more pointed bill than the others.
2. **Tree finches.** There are five species of insect-eating tree finches. Four species have bills suitable for feeding on insects. The woodpecker finch has a chisel-like beak. This unusual bird carries around a twig or a cactus spine, which it uses to probe for insects in deep crevices.
3. **Vegetarian finch.** The very heavy bill of this species is used to wrench buds from branches.
4. **Warbler finches.** These unusual birds play the same ecological role in the Galápagos woods that warblers play on the mainland, searching continuously over the leaves and branches for insects. They have slender, warbler-like beaks.

Recently, scientists have examined the DNA of Darwin's finches to study their evolutionary history. These studies suggest that the deepest branches in the finch evolutionary tree lead to warbler finches, which implies that warbler finches were among the first types to evolve after colonization of the islands. All of the ground species are closely related to one another, and the same is true for all of the tree finches. Nonetheless, within each group, species differ in beak size and other attributes, as well as in resource use.

Field studies, conducted in conjunction with those discussed in chapter 20, demonstrate that ground species compete for resources. The differences between species likely resulted from character displacement as initially similar species diverged to minimize competitive pressures.

Lake Victoria cichlid fishes diversified very rapidly

Lake Victoria is an immense, shallow, freshwater sea about the size of Switzerland in the heart of equatorial East Africa. Until recently, the lake was home to an incredibly diverse collection of over 450 species of cichlid fishes.

Geologically recent radiation. The cluster of cichlid species appears to have evolved recently and quite rapidly. By sequencing the *cytochrome b* gene in many of the lake's fish, scientists have been able to estimate that the first cichlids entered Lake Victoria only 200,000 years ago.

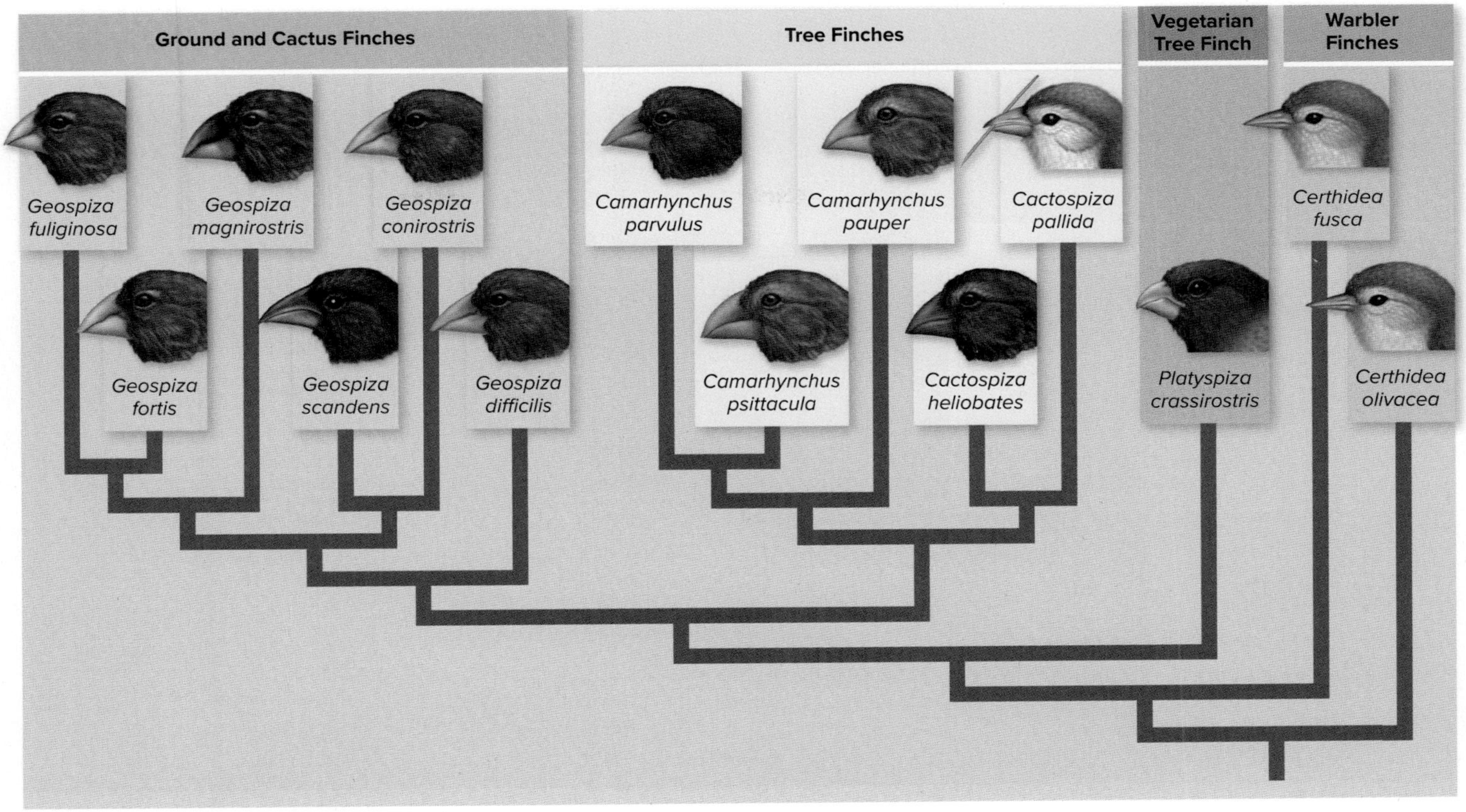

Figure 21.13 An evolutionary tree of Darwin's finches. This evolutionary tree, derived from examination of DNA sequences, suggests that warbler finches are an early offshoot. Ground and tree finches subsequently diverged, and then species within each group specialized to use different resources. Recent studies have shown, surprisingly, that the two warbler finches are not each other's closest relatives. Rather, *Certhidea fusca* is more closely related to the remaining Darwin's finches than it is to *C. olivacea.*

Dramatic changes in water level encouraged species formation. As the lake rose, it flooded new areas and opened up new habitats. Many of the species may have originated after the lake dried down 14,000 years ago, isolating local populations in small lakes until the water level rose again.

Cichlid diversity. Cichlids are small, perchlike fishes ranging from 5 to 25 cm in length, and the males come in many varieties of colors. The ecological and morphological diversity of these fish is remarkable, particularly given the short span of time over which they have evolved.

We can gain some sense of the vast range of types by looking at how different species eat. There are mud biters, algae scrapers, leaf chewers, snail crushers, zooplankton eaters, insect eaters, prawn eaters, and fish eaters. Snail shellers pounce on slow-crawling snails and spear their soft parts with long, curved teeth before the snail can retreat into its shell. Scale scrapers rasp slices of scales off other fish. There are even cichlid species that are "pedophages," eating the young of other cichlids.

Cichlid fish have a remarkable key innovation that may have been instrumental in their evolutionary radiation: they carry a second set of functioning jaws (see inset in figure 21.14*a*). This trait occurs in many other fish, but in cichlids it is greatly enlarged. The ability of these second jaws to manipulate and process food has freed the oral jaws to evolve for other purposes, and the result has been the incredible diversity of ecological roles filled by these fish (figure 21.14).

Abrupt extinction in the last several decades. Recently, much of the cichlid diversity has disappeared. In the 1950s, the Nile perch, a large commercial fish with a voracious appetite, was introduced into Lake Victoria. Since then, it has spread through the lake, eating its way through the cichlids.

By 1990 many of the open-water cichlid species, as well as others living in rocky shallow regions, had become extinct. Over 70% of all the named Lake Victoria cichlid species had disappeared, as had untold numbers of species that had yet to be described.

New Zealand alpine buttercups underwent speciation in glacial habitats

Adaptive radiations such as those we have described in Hawaiian *Drosophila,* Galápagos finches, and cichlid fishes seem to have been favored by periodic isolation. A clear example of the role periodic isolation plays in species formation is evident in the alpine buttercups that grow among the glaciers of New Zealand (figure 21.15).

More species of alpine buttercups grow on the two main islands of New Zealand than in all of North and South America combined. The evolutionary mechanism responsible for this

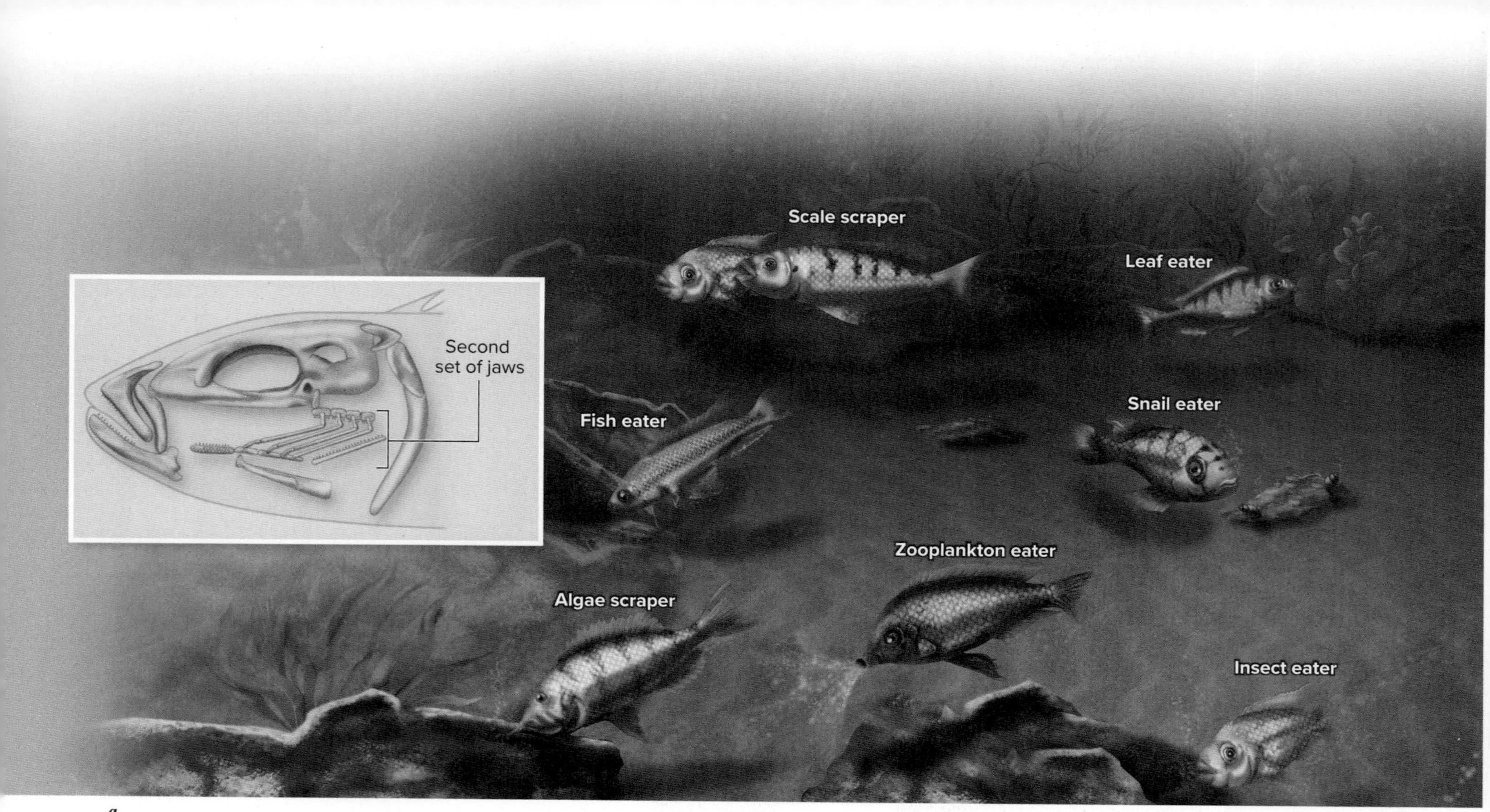

a.

b.

Figure 21.14 Cichlid fishes of Lake Victoria. ***a.*** These fishes have evolved adaptations to use a variety of different habitats. The enlarged second set of jaws located in the throat of these fish has provided evolutionary flexibility, allowing oral jaws to be modified in many ways. ***b.*** A difference in two genes is responsible for a short snout in *Labeotropheus fuelleborni* and a long snout in *Maylandia zebra*.
(b, left) Roberto Nistri/Alamy Stock Photo; (b, right) Frank Hecker/Alamy Stock Photo

diversity is recurrent isolation associated with the recession of glaciers.

The 14 species of alpine buttercups occupy five distinctive habitats within glacial areas:

- *snowfields*—rocky crevices among outcrops in permanent snowfields at 2130- to 2740-m elevation,
- *snowline fringe*—rocks at lower margins of snowfields between 1220 and 2130 m,
- *stony debris*—slopes of exposed loose rocks at 610 to 1830 m,
- *sheltered situations*—shaded by rock or shrubs at 305 to 1830 m, and
- *boggy habitats*—sheltered slopes and hollows, poorly drained tussocks at elevations between 760 and 1525 m.

Buttercup speciation and diversification have been promoted by repeated cycles of glacial advance and retreat. As the glaciers retreat up the mountains, populations become isolated on mountain peaks, permitting speciation (figure 21.15). In the next glacial advances, these new species can expand throughout the mountain range, coming into contact with their close relatives. In this way, one initial species could give rise to many descendants. Moreover, on isolated mountaintops during glacial retreats, species have convergently evolved to occupy similar habitats; these distantly related but ecologically similar species

snowfield
snowline fringe
stony debris
sheltered
boggy
a.
Glaciers recede
Glaciation
Glaciers link alpine zones into one continuous range.
Mountain populations become isolated, permitting divergence and speciation.
Alpine zones are reconnected. Separately evolved species come back into contact.
b.

Figure 21.15 New Zealand alpine buttercups (genus *Ranunculus*). Periodic glaciation encouraged species formation among alpine buttercups in New Zealand. ***a.*** Fourteen species of alpine *Ranunculus* grow among the glaciers and mountains of New Zealand. ***b.*** The formation of extensive glaciers during the Pleistocene epoch linked the alpine zones (white) of many mountains together. When the glaciers receded, these alpine zones were isolated from one another, only to become reconnected with the advent of the next glacial period. During periods of isolation, populations of alpine buttercups diverged in the isolated habitats.

(*Ranunculus glacialis*): Bob Gibbons/Alamy Stock Photo; (*Ranunculus acris*): Jim Harding/agefotostock; (*Ranunculus buchananii*): Colin Harris/Light Touch Images/Alamy Stock Photo; (*Ranunculus lyallii*): Cephas Picture Library/Alamy Stock Photo; (*Ranunculus insignis*): Brian Enting/Science Source

have then been brought back into contact in subsequent glacial advances.

REVIEW OF CONCEPT 21.5

Adaptive radiation occurs when a species diversifies, producing descendant species that are adapted to use many different parts of the environment. Adaptive radiation may occur under conditions of recurrent isolation, which increases the rate at which speciation occurs, and by occupation of areas with few competitors and many types of available resources, such as on volcanic islands. The evolution of a key innovation may also allow adaptation to parts of the environment that previously couldn't be utilized.

- *In contrast to the archipelago model, how might an adaptive radiation proceed in a case of sympatric speciation by disruptive selection?*

21.6 The Pace of Evolution Varies

We have discussed the manner in which speciation may occur, but we haven't yet considered the relationship between speciation and the evolutionary change that occurs within a species. Two hypotheses, *gradualism* and *punctuated equilibrium,* have been advanced to explain the relationship.

There Are Two Distinct Modes of Evolutionary Change

LEARNING OBJECTIVE 21.6.1 Compare stasis, gradual evolutionary change, and punctuated equilibrium.

Gradualism is the accumulation of small changes

For more than a century after the publication of *On the Origin of Species,* the standard view was that evolution occurred very slowly. Such change would be nearly imperceptible from generation to generation but would accumulate such that, over the course of thousands and millions of years, major changes could occur. This view is termed **gradualism** (figure 21.16*a*).

Punctuated equilibrium is long periods of stasis followed by relatively rapid change

Gradualism was challenged in 1972 by paleontologists Niles Eldredge of the American Museum of Natural History in New York and Stephen Jay Gould of Harvard University, who argued that species experience long periods of little or no evolutionary change (termed **stasis**), punctuated by bursts of evolutionary change occurring over

Figure 21.16 Two views of the pace of macroevolution. ***a.*** Gradualism suggests that evolutionary change occurs slowly through time and is not linked to speciation, whereas ***b.*** punctuated equilibrium surmises that phenotypic change occurs in bursts associated with speciation, separated by long periods of little or no change.

geologically short time intervals. They called this phenomenon **punctuated equilibrium** (figure 21.16*b*) and argued that these periods of rapid change occur only during the speciation process.

Initial criticism of the punctuated equilibrium hypothesis focused on whether rapid change could occur over short periods of time. As we have explored in chapters 19 and 20, however, when natural selection is strong, rapid and substantial evolutionary change can occur. A more difficult question involves the long periods of stasis: Why would species exist for thousands, or even millions, of years without changing?

Although a number of possible reasons have been suggested, most researchers now believe that a combination of stabilizing and oscillating selection is responsible for stasis. If the environment does not change over long periods of time, or if environmental changes oscillate back and forth, then stasis may occur for long periods. One factor that may enhance this stasis is the ability of species to shift their ranges; for example, during the ice ages, when the global climate cooled, the geographical ranges of many species shifted southward, so that the species continued to experience similar environmental conditions.

Evolution may include both types of change

Eldredge and Gould's proposal prompted a great deal of research. The fossil record shows that some well-documented groups, such as African mammals, clearly have evolved gradually, not in spurts. Other groups, such as marine bryozoa, seem to show the irregular pattern of evolutionary change predicted by the punctuated equilibrium model. It appears, in fact, that gradualism and punctuated equilibrium are two ends of a continuum. Although some groups appear to have evolved solely in a gradual manner and others only in a punctuated mode, many other groups show evidence of both gradual and punctuated episodes at different times in their evolutionary history.

The idea that speciation is necessarily linked to phenotypic change has not been supported, however. On the one hand, it is now clear that speciation can occur without substantial phenotypic change. For example, many closely related salamander species are nearly indistinguishable. On the other hand, it is also clear that phenotypic change can occur within species in the absence of speciation.

REVIEW OF CONCEPT 21.6

Gradualism is the accumulation of almost imperceptible changes that eventually result in major differences. Punctuated equilibrium proposes that long periods of stasis are interrupted (punctuated) by periods of rapid change. Evidence for both gradualism and punctuated equilibrium has been found in different groups. *Stasis* refers to a period in which little or no evolutionary change occurs. Stasis may result from stabilizing or oscillating selection.

- *Could evolutionary change be punctuated in time (that is, rapid and episodic) but not linked to speciation?*

21.7 Speciation and Extinction Have Molded Biodiversity Through Time

Biological diversity has increased vastly since the Cambrian period, but the trend has been far from consistent. After a rapid rise, diversity reached a plateau for about 200 million years, but since then it has risen steadily. Because changes in the number of species reflect the rate of origin of new species relative to the rate at which existing species disappear, this long-term trend reveals that speciation has, in general, surpassed extinction.

Five Mass Extinctions Occurred in the Distant Past

LEARNING OBJECTIVE 21.7.1 Define mass extinction, and identify when major mass extinctions occurred.

Nonetheless, speciation has not always outpaced extinction. In particular, interspersed in the long-term increase in species diversity have been a number of sharp declines, termed **mass extinctions.** Five major mass extinctions have been identified, the most severe one occurring at the end of the Permian period, approximately 250 MYA (figure 21.17). At that time, more than half of all plant and animal families and as much as 96% of all species may have perished.

Although not as drastic as the Permian extinction, the most famous and well-studied mass extinction occurred at the end of

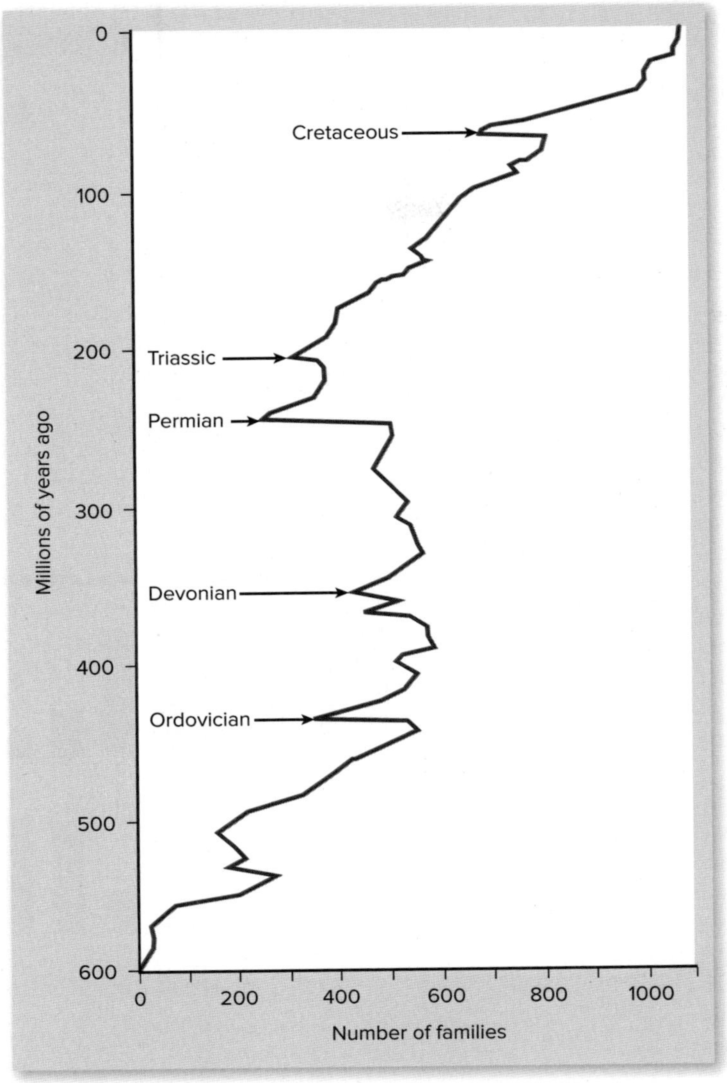

Figure 21.17 Biodiversity through time. The taxonomic diversity of families of marine animals has increased since the Cambrian period, although occasional dips have occurred. The fossil record is most complete for marine organisms, because they are more readily fossilized than terrestrial species. Families are shown, rather than species, because many species are known from only one specimen, thus introducing error into estimates of the time of extinction. Arrows indicate the five major mass extinction events.

the Cretaceous period (66 MYA), when the dinosaurs (except for birds, which are one type of dinosaur—refer to chapter 28) and a variety of other organisms became extinct. There is a great deal of support for the hypothesis that this event was triggered by a large asteroid impact, perhaps causing global forest fires and obscuring the Sun for months by throwing particles into the air. The cause of other mass extinction events is less certain. Some scientists suggest that asteroids may have played a role in at least some of the other mass extinction events; other hypotheses implicate global climate change, massive volcanic eruptions, and other causes.

One important result of mass extinctions is that not all groups of organisms are affected equally. For example, in the extinction at the end of the Cretaceous, not only most dinosaurs but also marine and flying reptiles, as well as ammonites (a type of mollusk), went extinct. Marsupials, flowering plants, birds, and some forms of plankton were greatly reduced in diversity. In contrast, turtles, crocodilians, and amphibians seemed to have been unscathed. Why some groups were harder hit than others is not clear, but one hypothesis suggests that survivors were those animals that could shelter underground or in water, and that either could scavenge or required little food in the cool temperatures that resulted from the blockage of sunlight.

A consequence of mass extinctions is that previously dominant groups may perish, thus changing the course of evolution. This is certainly true of the Cretaceous extinction. During the Cretaceous period, placental mammals were a minor group composed of species that were mostly no larger than a house cat. When the dinosaurs, which had dominated the world for more than 100 million years, disappeared at the end of this period, the placental mammals underwent a significant adaptive radiation. It is humbling to think that humans might never have arisen, had that asteroid not struck Earth 66 MYA.

As the world around us illustrates today, species diversity does rebound after mass extinctions, but this recovery is not rapid. Examination of the fossil record indicates that rates of speciation do not immediately increase after an extinction pulse but, rather, take about 10 million years to reach their maximum. The cause of this delay is not clear, but it may be a result of the time required for ecosystems to recover, and for the processes of speciation and adaptive diversification to begin. Consequently, species diversity may require 10 million years, or even much longer, to attain its previous level.

A Sixth Mass Extinction Is Under Way

LEARNING OBJECTIVE 21.7.2 Evaluate the contention that we are in the midst of a mass extinction today.

The number of species in the world in recent times is greater than it has ever been. Unfortunately, that number is decreasing at an alarming rate due to human activities.

Some estimate that as much as one-fourth of all species will become extinct in the near future, a rate of extinction not found on Earth since the Cretaceous mass extinction. Moreover, the rebound in species diversity may be even slower than those that followed previous mass extinction events. Instead of the ecologically impoverished but energy-rich environment that existed after previous mass extinction events, a large proportion of the world's resources will be taken up already by human activities, leaving few resources available for adaptive radiation.

REVIEW OF CONCEPT 21.7

The number of species has increased through time, although not at a constant rate. Five major extinction events have substantially, though briefly, reduced the number of species. Diversity rebounds, but the recovery is not rapid, and the groups making up that diversity are not the same as those that existed before the extinction event. Unfortunately, humans are currently causing a sixth mass extinction event.

- *In what ways is the current mass extinction event different from those that occurred in the past?*

Are "Ecotypes" Incipient Species?

One of the most important cohesive forces that hold a species together is the interbreeding that occurs among its members. However, populations in nature are often widely separated, with little chance for local populations to interbreed and share genes with other populations. Such species, when distributed over a broad area, are often broken by their environments into geographical subunits, natural selection acting on different subunits in different ways.

In classic studies done a half-century ago, ecologists Jens Clausen, David Keck, and William Hiesey of Stanford University's Carnegie Institute examined local populations of the yarrow plant along an east–west transect (a transect is a straight line on a map) across California from the high Sierras to the Pacific. They found that local populations of these grassy plants differed in many heritable traits such as height and growing season, and that these variations were not simply the consequence of growing under differing environmental conditions: when grown together in a common garden, plants continued to exhibit the differences for generations. Natural selection had created genetically distinct subpopulations of the yarrow species in different portions of the transect, subpopulations that the researchers called "ecotypes."

Are ecotypes incipient species in the process of diverging? This question was addressed by famous evolutionary biologist Theodosius Dobzhansky in a study of natural populations of the wild fruit fly *Drosophila pseudoobscura*. A wild cousin of the famous laboratory fruit fly *D. melanogaster*, this fly is widely distributed in the western United States. The third chromosome of *D. pseudoobscura* exhibits a variety of rearrangements such as inversion or translocations (refer to chapter 15). Dobzhansky surveyed the frequencies of several of these third-chromosome rearrangements among *D. pseudoobscura* populations collected from 12 localities on an east–west transect running along the United States–Mexico border. Results for the three most frequent rearrangements are presented.

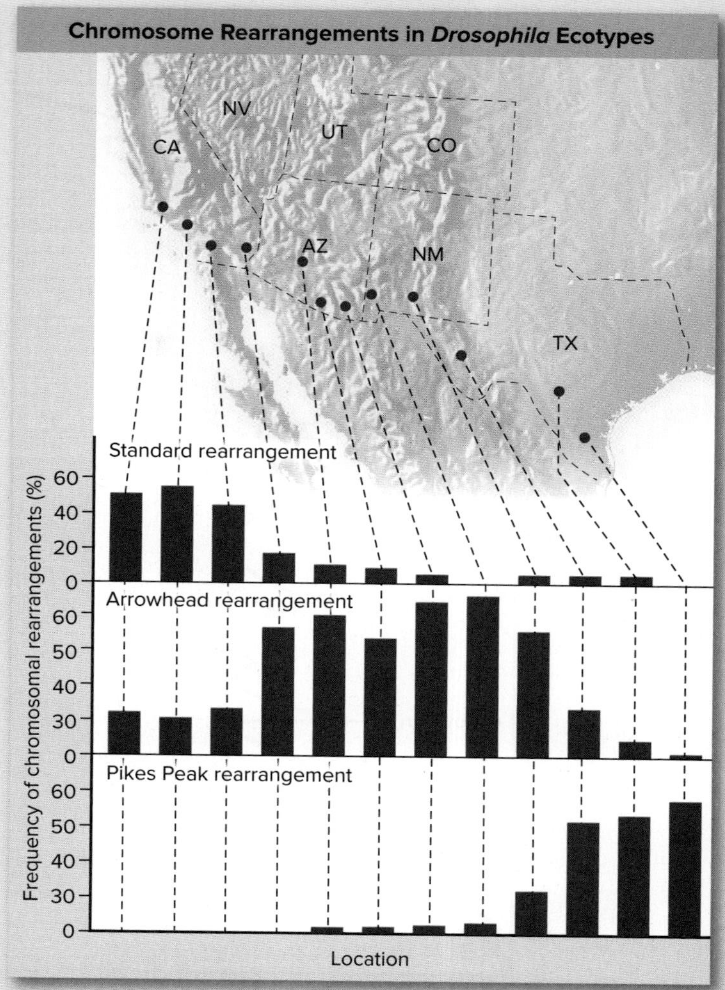

Analysis

1. **Applying Concepts**
   a. ***Variable.*** In the histograms, what is (are) the dependent variable(s)?
   b. ***Comparing categories.*** Comparing the three histograms, how many populations possess all three third-chromosome rearrangements? In how many populations is one of the three rearrangements present in more than half of the population's individuals?
2. **Interpreting Data**
   a. In Texas, what is the most common third-chromosome rearrangement? In California? In between the two?
   b. Along an east–west transect, is there any significant pattern to the frequency of the Standard rearrangement? The Arrowhead rearrangement? The Pikes Peak rearrangement?
3. **Making Inferences** Are the patterns in third-chromosome rearrangement frequencies the same for the three rearrangements? Locate the population with the highest frequency of each third-chromosome rearrangement. Place them on the east–west transect. Do the three rearrangements group into different subpopulations of the species? If so, do the three subpopulations overlap along this transect?
4. **Drawing Conclusions** Are the three subpopulations of *Drosophila pseudoobscura* from Texas, Arizona/New Mexico, and California genetically distinct? How do they differ?
5. **Further Analysis** How would you measure the degree of gene exchange among the three subpopulations?

Retracing the Learning Path

CONCEPT 21.1 The Biological Species Concept Highlights Reproductive Isolation

21.1.1 The Biological Species Concept Focuses on the Ability to Exchange Genes Biological species are defined as populations that interbreed, or can, to produce fertile offspring. Reproductive isolating mechanisms prevent genetic exchange between species. Sympatric species inhabit the same locale but remain distinct. Populations that appear distinct are called subspecies. When these populations occur close to each other, there are usually individuals with features of both populations.

21.1.2 Prezygotic Isolating Mechanisms Prevent the Formation of a Zygote Prezygotic isolating mechanisms prevent a viable zygote from being created. These include ecological, behavioral, temporal, and mechanical isolation and prevention of gamete fusion.

21.1.3 Postzygotic Isolating Mechanisms Prevent Normal Development into Fertile Adults Postzygotic isolating mechanisms prevent a zygote from developing into a viable and fertile individual.

21.1.4 The Biological Species Concept Does Not Explain All Observations The ecological species concept focuses on the role of natural selection and differences among species in their ecological requirements. No one species concept can explain all the diversity of life.

CONCEPT 21.2 Natural Selection May Reinforce Reproductive Isolation

21.2.1 Selection May Act to Strengthen Isolating Mechanisms If populations that have evolved only partial reproductive isolation come into contact, natural selection can increase isolation, a process termed reinforcement. Gene flow between populations can also homogenize them, preventing speciation.

CONCEPT 21.3 Natural Selection and Genetic Drift Play Key Roles in Speciation

21.3.1 Both Random Changes and Adaptation Can Lead to Isolation and Speciation In small populations, genetic drift may cause populations to diverge. This may lead to the population becoming reproductively isolated. Adaptation to different situations or environments may incidentally lead to reproductive isolation. Natural selection can also directly select for traits that increase reproductive isolation.

CONCEPT 21.4 Speciation Is Influenced by Geography

21.4.1 Allopatric Speciation Takes Place When Populations Are Geographically Isolated Geographically isolated populations are much more likely to evolve into separate species, because no gene flow occurs. Most speciation probably occurs in allopatry.

21.4.2 Sympatric Speciation Occurs Without Geographical Separation Sympatric speciation can occur in two ways. One is polyploidy, which instantly creates a new species. Disruptive selection also may cause one species to divide into two.

CONCEPT 21.5 Adaptive Radiation Requires Both Speciation and Habitat Diversity

21.5.1 Key Innovations Can Lead to Adaptive Radiations The evolution of a new trait that allows individuals to use previously inaccessible parts of the environment may also trigger an adaptive radiation.

21.5.2 Character Displacement May Drive Sympatric Species Divergence In character displacement, selection favors individuals who use resources not used by other species.

21.5.3 Adaptive Radiation Is Very Well Documented Darwin's finch species adapted to use different food types. Fourteen species in four genera have evolved to exploit four different habitats based on type of food. Cichlids in Lake Victoria underwent rapid radiation to form 450 species, although 70% are now extinct. Periodic isolation by glaciers has led to 14 species of alpine buttercups in distinct habitats.

CONCEPT 21.6 The Pace of Evolution Varies

21.6.1 There Are Two Distinct Modes of Evolutionary Change Scientists generally agree that evolutionary change occurs on a continuum, with gradualism and punctuated change being the extremes. Historically, scientists took the view that speciation occurred gradually through very small, cumulative changes. The punctuated equilibrium hypothesis contends that not only is change rapid and episodic, it is associated only with the speciation process.

CONCEPT 21.7 Speciation and Extinction Have Molded Biodiversity Through Time

21.7.1 Five Mass Extinctions Occurred in the Distant Past Mass extinctions led to dramatic decreases in species diversity. Five such mass extinctions occurred in the distant past due to asteroids hitting the Earth and global climate change, among other events.

21.7.2 A Sixth Mass Extinction Is Under Way Humans are currently causing a sixth mass extinction.

Concept Overview

This Concept Overview diagrams the key concepts that were discussed in this chapter.

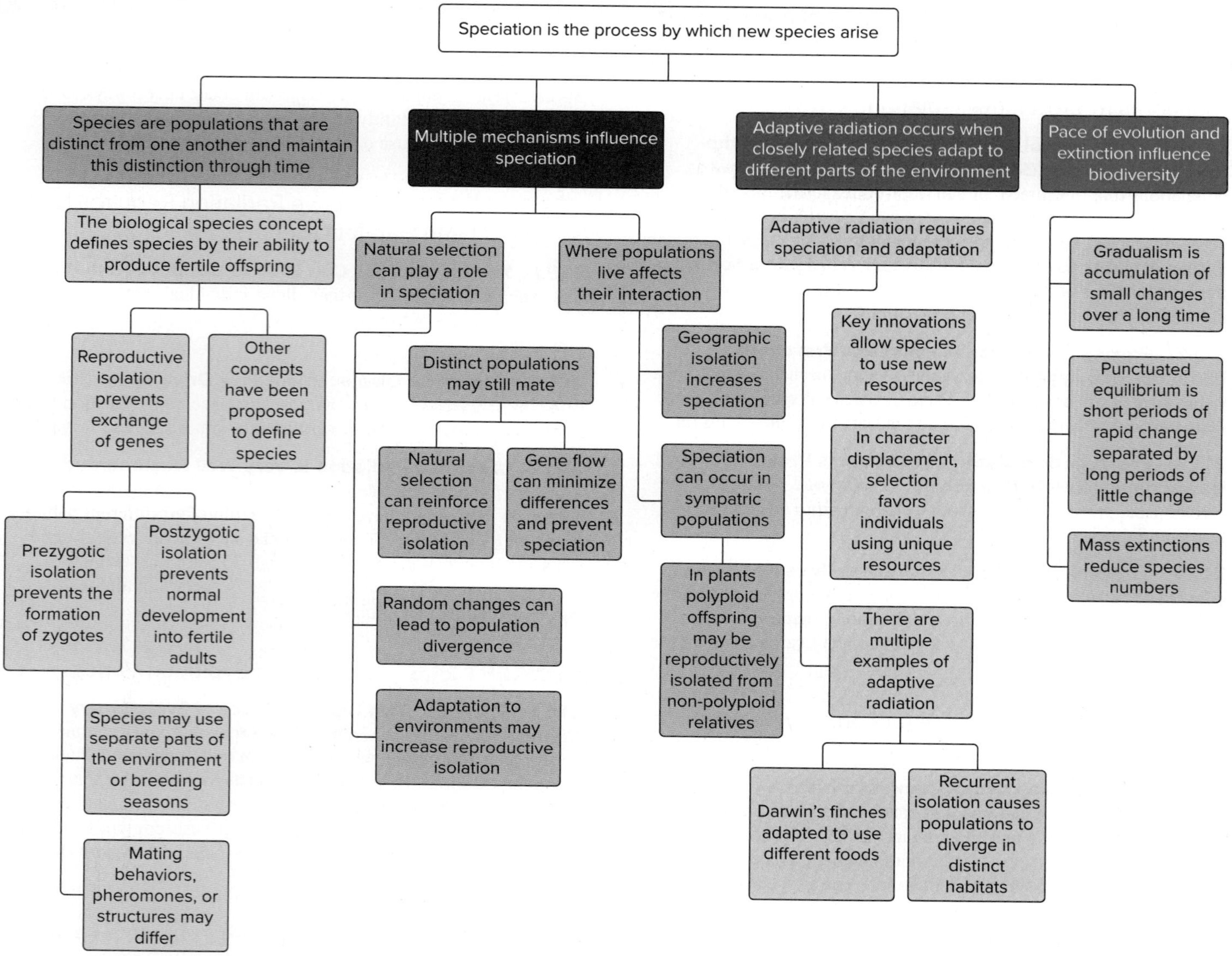

Assessing the Learning Path

Understand

1. Prezygotic isolating mechanisms include all of the following EXCEPT
   a. hybrid sterility.
   b. courtship rituals.
   c. habitat separation.
   d. seasonal reproduction.
2. A key element of the biological species concept is
   a. homologous isolation.
   b. divergent isolation.
   c. convergent isolation.
   d. reproductive isolation.
3. Natural selection can
   a. enhance the probability of speciation.
   b. enhance reproductive isolation.
   c. act against hybrid survival and reproduction.
   d. All of the above
4. The process of ___________ continually improves the prezygotic isolating mechanisms until complete reproductive isolation is achieved.
   a. hybridization
   b. reinforcement
   c. convergence
   d. allopatry
5. Which of the following does NOT lead to a hereditary change in one or more traits that may lead to reproductive isolation?
   a. Genetic drift in small populations
   b. Founder effects
   c. Geographical isolation
   d. Bottlenecks in population size

6. The immediate genetic divergence of a small population that colonized a geographically isolated habitat would be due to
   a. natural selection.
   b. a founder effect.
   c. postzygotic isolation.
   d. adaptation.
7. Allopatric speciation
   a. is less common than sympatric speciation.
   b. involves geographical isolation of some kind.
   c. is the only kind of speciation that occurs in plants.
   d. requires polyploidy.
8. Speciation by allopolyploidy
   a. takes a long time.
   b. is common in birds.
   c. leads to reduced numbers of chromosomes.
   d. occurs after hybridization between two species.
9. What drove the adaptive radiation that produced the diversity of finch species on the Galápagos Islands?
   a. Competition for scarce food resources
   b. The need for nesting space
   c. Predation from iguanas
   d. Genetic drift
10. Gradualism and punctuated equilibrium are
   a. two ends of the continuum of the rate of evolutionary change over time.
   b. mutually exclusive views about how all evolutionary change takes place.
   c. mechanisms of reproductive isolation.
   d. None of the above
11. Following a mass extinction, a rebound in the number of species occurs due to
   a. low levels of speciation.
   b. adaptive radiations.
   c. high rates of gene flow.
   d. mutations induced by the mass extinction event.

Apply

1. Scrub oak and valley oak, both found in California, do not form many hybrids in nature because
   a. they do not occur together.
   b. hybrids are not as well suited to the habitats where the two oak species occur together.
   c. they are pollinated by different insects.
   d. they produce flowers in different seasons.
2. Natural selection can lead to speciation
   a. by causing small populations to diverge more than large populations.
   b. because the evolutionary changes that two populations acquire while adapting to different habitats may have the effect of making them reproductively isolated.
   c. by favoring the same evolutionary change in multiple populations.
   d. by favoring intermediate phenotypes.
3. In laboratory experiments, reproductive isolation evolves more rapidly when
   a. populations experience uniform environments.
   b. populations are forced to adapt to different environments.
   c. populations are kept so small that genetic drift occurs.
   d. populations experience bottlenecks in population size.
4. If reinforcement is weak and hybrids are not completely infertile,
   a. genetic divergence between populations may be overcome by gene flow.
   b. speciation will occur 100% of the time.
   c. gene flow between populations will be impossible.
   d. the speciation will be more likely than if hybrids were completely infertile.
5. Character displacement
   a. arises through competition and natural selection, favoring divergence in resource use.
   b. arises through competition and natural selection, favoring convergence in resource use.
   c. does not promote speciation.
   d. reduced speciation rates in Galápagos finches.
6. The tremendous diversity of Lake Victoria would not have been possible without
   a. the large range in body size of the ancestral fish population.
   b. the development of lungs in cichlids.
   c. the mating rituals of cichlids.
   d. the secondary set of jaws present in cichlids.
7. Which statement about speciation is NOT true?
   a. Reproductive isolation may develop slowly.
   b. Among plants, it is often the result of polyploidy.
   c. Among animals, it usually requires a physical barrier.
   d. Formation of a new species always takes thousands of years.
8. What type of fossil evidence would support evolution via punctuated equilibrium?
   a. A series of fossils with slight variations between them
   b. A fossil record showing a near-constant state of change
   c. A long period of stability in the fossil record followed by the rapid appearance of new forms
   d. Fossils cannot be used to demonstrate evolutionary changes.
9. Compare biodiversity now to that in the time of the dinosaurs.
10. How does the modern-day mass extinction differ from previous ones?
   a. The number of species is not decreasing.
   b. The dominant life form is not going extinct.
   c. The extinct forms may reappear.

Synthesize

1. Natural selection can lead to the evolution of prezygotic isolating mechanisms, but not postzygotic isolating mechanisms. Explain.
2. Refer to figure 21.5. In Europe, pied and collared flycatchers are dissimilar in sympatry but very similar in allopatry, consistent with character divergence in coloration. In this case, there is no competition for ecological resources as in other cases of character divergence discussed. How would you explain these observations?
3. How would you experimentally distinguish between isolation mechanisms that are the incidental consequence of adaptive divergence and those selected for in order to generate a new species?
4. When two partially differentiated populations of a single species come into contact with each other after a period of isolation, their differences may increase or decrease. What conditions favor each outcome, and what are the consequences in each case?
5. Sulphur butterfly species of the genus *Colias* are frequently divided into local populations between which there is little gene exchange. These local populations typically look alike, exhibiting little morphological variation from one population to the next. Design field and lab studies that would determine what evolutionary forces are acting to maintain this similarity.

Part V The Diversity of Life

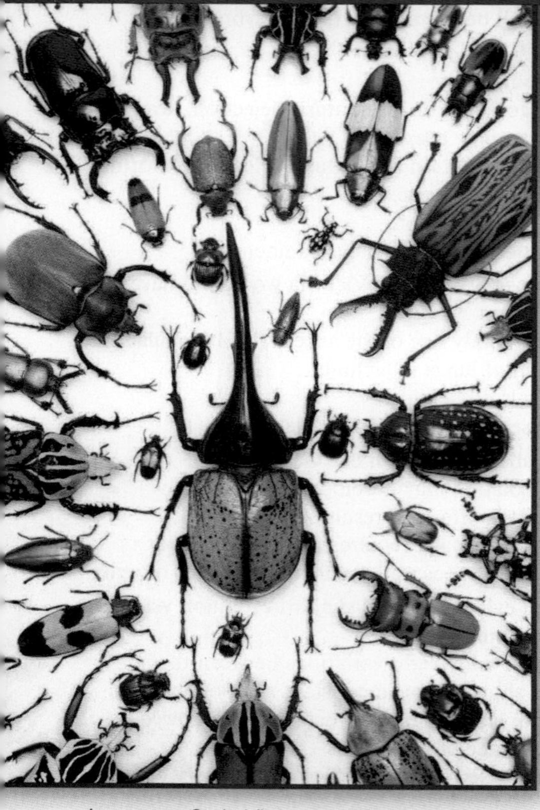

Imagemore Co, Ltd./Imagemore/Getty Images

22 Systematics and Phylogeny

Learning Path

22.1 Systematics Reconstructs Evolutionary Relationships

22.2 Cladistics Focuses on Traits Derived from a Common Ancestor

22.3 Classification Is a Labeling Process, Not an Evolutionary Reconstruction

22.4 Taxonomy Attempts to Classify Organisms in an Evolutionary Context

22.5 The Largest Taxa Are Domains

Concept Overview

This Concept Overview provides a diagram of the overarching concepts that are covered in this chapter. These concepts will be expanded upon in the Concept Overview at the end of the chapter.

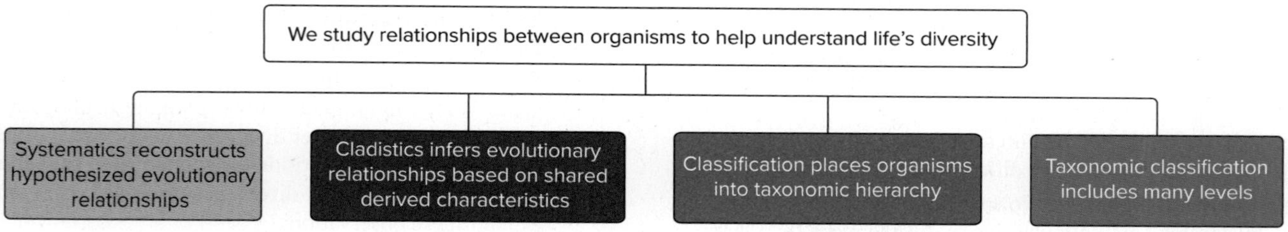

We study relationships between organisms to help understand life's diversity
Systematics reconstructs hypothesized evolutionary relationships
Cladistics infers evolutionary relationships based on shared derived characteristics
Classification places organisms into taxonomic hierarchy
Taxonomic classification includes many levels

Introduction

All organisms share many biological characteristics. They are composed of one or more cells, carry out metabolism and transfer energy with ATP, and encode hereditary information in DNA. Yet there is also a tremendous diversity of life, ranging from bacteria and amoebas to blue whales and sequoia trees. For generations, biologists have tried to group organisms based on shared characteristics. The most meaningful groupings are based on the study of evolutionary relationships among organisms. New methods for constructing evolutionary trees and a sea of molecular sequence data are leading to improved evolutionary hypotheses to explain life's diversification.

22.1 Systematics Reconstructs Evolutionary Relationships

One of the great challenges of modern science is to understand the history of ancestor–descendant relationships that unites all forms of life on Earth, from the earliest single-celled organisms to the complex organisms we find around us today. If the fossil record were perfect, we could trace the evolutionary history of species and examine how each arose and proliferated; however, as discussed in chapter 20, the fossil record is far from complete. Although it answers many questions about life's diversification, it leaves many others unsettled.

Branching Diagrams Depict Evolutionary Relationships

LEARNING OBJECTIVE 22.1.1 Recognize what a phylogeny represents.

Given the imperfections of the fossil record, scientists must utilize additional types of evidence to establish the best hypothesis of evolutionary relationships. Bear in mind that the outcomes of such studies *are* hypotheses, and as such, they require further testing. All hypotheses may be disproved by new data, leading to the formation of better, more accurate scientific ideas.

The reconstruction and study of evolutionary relationships is called **systematics.** By looking at the similarities and differences between species, systematists can construct an evolutionary tree, or **phylogeny,** which represents a hypothesis about patterns of relationship among species.

Darwin envisioned that all species were descended from a single common ancestor, and that the history of life could be depicted as a branching tree (figure 22.1). In Darwin's view, the twigs of the tree represent existing species. As one works down the tree, the joining of twigs and branches reflects the pattern of common ancestry back in time to the single common ancestor of all life. The process of descent with modification from common ancestry results in all species being related in this branching, hierarchical fashion, and their evolutionary history can be depicted using branching diagrams, or phylogenetic trees. Figure 22.1*b* shows how evolutionary relationships are depicted with a branching diagram. Humans and chimpanzees are descended from a common ancestor and are each other's closest living relative (the position of this common ancestor is indicated by the node labeled 1). Humans, chimps, and gorillas share an older common ancestor (node 2), and all great apes share a more distant common ancestor (node 3).

One key to interpreting a phylogeny is to look at how recently species share a common ancestor, rather than looking at the

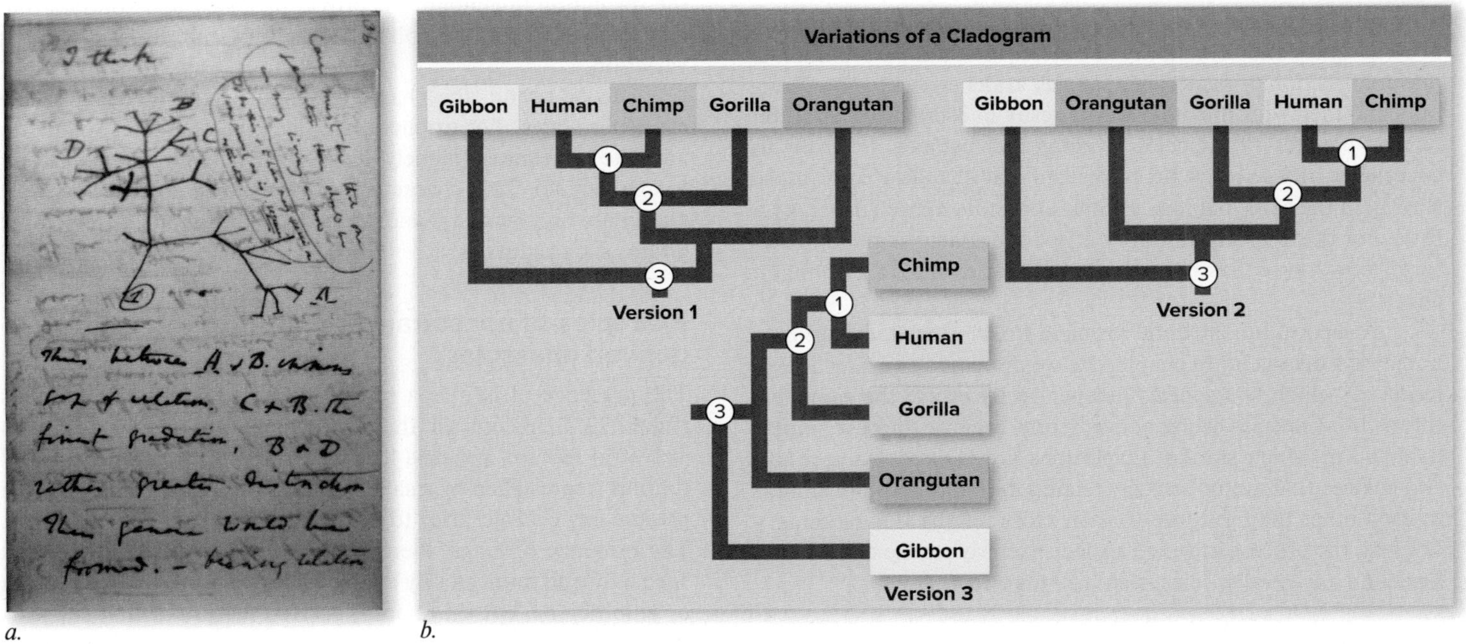

a. *b.*

Figure 22.1 Phylogenies depict evolutionary relationships. ***a.*** A drawing from one of Darwin's notebooks, written in 1837 as he developed his ideas that led to *On the Origin of Species.* Darwin viewed life as a branching process akin to a tree, with species on the twigs, and evolutionary change represented by the branching pattern displayed by a tree as it grows. ***b.*** An example of a phylogeny. Humans and chimpanzees are more closely related to each other than they are to any other living species. This is apparent because they share a common ancestor (the node labeled 1) that was not an ancestor of other species. Similarly, humans, chimpanzees, and gorillas are more closely related to one another than any of them is to orangutans, because they share a common ancestor (node 2) that was not ancestral to orangutans. Node 3 represents the common ancestor of all apes. Note that these three versions convey the same information despite the differences in arrangement of species and orientation.

(a): Letz/SIPA/Newscom

arrangement of species across the top of the tree. If you compare the three versions of the phylogeny of figure 22.1*b,* you find that the relationships are the same: regardless of where they are positioned, chimpanzees and humans are still more closely related to each other than to any other species.

Moreover, even though humans are placed next to gibbons in version 1 of figure 22.1*b,* the pattern of relationships still indicates that humans are more closely related (that is, share a more recent common ancestor) to gorillas and orangutans than to gibbons. Phylogenies are also sometimes displayed on their side, rather than upright (figure 22.1*b,* version 3), but this arrangement also does not affect its interpretation.

Similarity May Not Accurately Predict Evolutionary Relationships

LEARNING OBJECTIVE 22.1.2 Explain the relationship between phenotypic similarity and evolutionary history.

We might expect that the more time that has passed since two species diverged from a common ancestor, the more different they would be. Early systematists relied on this reasoning and constructed phylogenies based on overall similarity. If, in fact, species evolved at a constant rate, then the amount of divergence between two species would be a function of how long they had been diverging, and thus phylogenies based on degree of similarity would be accurate. As a result, we might think that chimps and gorillas are more closely related to each other than either is to humans.

But as chapter 21 revealed, evolution can occur very rapidly at some times and very slowly at others. In addition, evolution is not unidirectional—sometimes species' traits evolve in one direction, and then back the other way (a result of oscillating selection; refer to chapter 19). Species invading new habitats are likely to experience new selective pressures and may change greatly; those staying in the same habitats as their ancestors may change only a little. For this reason, similarity is not necessarily a good predictor of how long it has been since two species shared a common ancestor.

A second fundamental problem exists as well: evolution is not always divergent. In chapter 20, we discussed convergent evolution, in which two species independently evolve the same features. Often, species evolve convergently, because they use similar habitats, in which similar adaptations are favored. As a result, two species that are not closely related may end up more similar to each other than they are to their close relatives. Evolutionary reversal, the process in which a species re-evolves the characteristics of an ancestral species, also has this effect.

REVIEW OF CONCEPT 22.1

Systematics is the study of evolutionary relationships. Phylogenies, or phylogenetic trees, are graphic representations of relationships among species. Similarity of organisms alone does not necessarily correlate with their relatedness, because evolutionary change is not constant in rate and direction.

- *Why might a species be most phenotypically similar to a species that is not its closest evolutionary relative?*

22.2 Cladistics Focuses on Traits Derived from a Common Ancestor

Because phenotypic similarity may be misleading, most systematists no longer construct their phylogenetic hypotheses solely on this basis. Rather, they distinguish similarity among species that is inherited from the most recent common ancestor of an entire group, which is called **derived,** from similarity that arose prior to the common ancestor of the group, which is termed *ancestral.* In this approach, termed **cladistics,** only **shared derived characters** are considered informative in determining evolutionary relationships.

Cladistics Requires That Character States Be Identified as Ancestral or Derived

LEARNING OBJECTIVE 22.2.1 Differentiate between ancestral and derived characters.

To employ the method of cladistics, systematists first gather data on a number of characters for all the species in the analysis. Characters can be any aspect of the phenotype, including morphology, physiology, behavior, and DNA. As chapter 18 shows, the revolution in genomics is providing a vast body of data that may revolutionize our ability to identify and study character variation.

To be useful, the characters should exist in recognizable **character states.** For example, consider the character "teeth" in amniote vertebrates (namely, birds, reptiles, and mammals; refer to chapter 28). This character has two states: presence in most mammals and reptiles, and absence in birds and a few other groups such as turtles.

Examples of ancestral versus derived characters

The presence of hair is a shared derived feature of mammals (figure 22.2); in contrast, the presence of lungs in mammals is an ancestral feature, because it is also present in amphibians and reptiles (represented by a salamander and a lizard) and therefore presumably evolved prior to the common ancestor of mammals. The presence of lungs, therefore, does not tell us that mammal species are all more closely related to one another than to reptiles or amphibians, but the shared, derived feature of hair suggests that all mammal species share a common ancestor that existed more recently than the common ancestor of mammals, amphibians, and reptiles.

To return to the question concerning the relationships of humans, chimps, and gorillas, a number of morphological and DNA characters exist that are derived and shared by chimps and humans, but not by gorillas or other great apes. These characters suggest that chimps and humans diverged from a common ancestor (figure 22.1*b,* node 1) that existed more recently than the common ancestor of gorillas, chimps, and humans (node 2).

Determination of ancestral versus derived

Once the data are assembled, the first step in a manual cladistic analysis is to **polarize** the characters—that is, to determine whether particular character states are ancestral or derived. To polarize the character "teeth," for example, systematists must determine which state—presence or absence—was exhibited by the most recent common ancestor of this group.

Usually the fossils available do not represent the most recent common ancestor—or we cannot be confident that they do. As a result, the method of *outgroup comparison* is used to assign character polarity. To use this method, a species or group of species that is closely related to, but not a member of, the group under study is designated as the **outgroup.**

When the group under study exhibits multiple character states, and one of those states is exhibited by the outgroup, then that state is considered to be ancestral and other states are considered to be derived. However, outgroup species also evolve from their ancestors, so the outgroup species will not always exhibit the ancestral condition. For this reason, polarity assignments are most reliable when the same character state is exhibited by several different outgroups. In the preceding example, teeth are generally present in the nearest outgroups of amniotes—amphibians and fish—as well as in many species of amniotes themselves. Consequently, the presence of teeth in mammals and reptiles is considered ancestral, and their absence in birds and turtles is considered derived.

Construction of a cladogram

Once all characters have been polarized, systematists use this information to construct a **cladogram,** which depicts a hypothesis of evolutionary relationships. Species that share a common ancestor, as indicated by the possession of shared derived characters, are said to belong to a **clade.** Clades are thus evolutionary units and refer to a common ancestor and all of its descendants. A derived character shared by clade members is called a **synapomorphy** of that clade. Figure 22.2*b* illustrates that a simple cladogram is a nested set of clades, each characterized by its own synapomorphies. For example, amniotes are a clade for which the evolution of an amniotic membrane is a synapomorphy. Within that clade, mammals are a clade, with hair as a synapomorphy, and so on.

Ancestral states are also called **plesiomorphies,** and shared ancestral states are called **symplesiomorphies.** In contrast to synapomorphies, symplesiomorphies are not informative about phylogenetic relationships.

Consider, for example, the character state "presence of a tail," which is exhibited by lampreys, sharks, salamanders, lizards, and tigers. Does this mean that tigers are more closely related to—and shared a more recent common ancestor with—lizards and sharks than to apes and humans, their fellow mammals? The answer, of course, is no: because symplesiomorphies reflect character states inherited from a distant ancestor, they do not imply that species exhibiting that state are closely related.

Homoplasy Complicates Cladistic Analysis

LEARNING OBJECTIVE 22.2.2 Contrast informative shared derived characters from noninformative ones.

In real-world cases, phylogenetic studies are rarely as simple as the examples we have shown so far. The reason is that in some cases, the same character has evolved independently in several

Traits: Organism	Jaws	Lungs	Amniotic Membrane	Hair	No Tail	Bipedal
Lamprey	0	0	0	0	0	0
Shark	1	0	0	0	0	0
Salamander	1	1	0	0	0	0
Lizard	1	1	1	0	0	0
Tiger	1	1	1	1	0	0
Gorilla	1	1	1	1	1	0
Human	1	1	1	1	1	1

a.

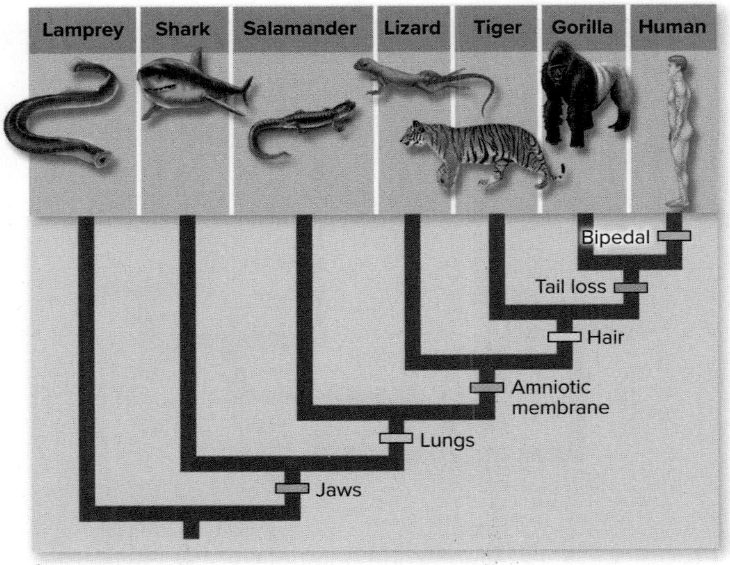

b.

Figure 22.2 A cladogram. ***a.*** Morphological data for a group of seven vertebrates are tabulated. A "1" indicates possession of the derived character state, and a "0" indicates possession of the ancestral character state (note that the derived state for character "no tail" is the absence of a tail; for all other traits, absence of the trait is the ancestral character state). ***b.*** A tree, or cladogram, diagrams the relationships among the organisms based on the presence of derived characters. The derived characters between the cladogram branch points are shared by all organisms above the branch points and are not present in any below them. The outgroup (in this case, the lamprey) does not possess any of the derived characters.

species. These characters would be categorized as shared derived characters, but they would be false signals of a close evolutionary relationship. In addition, derived characters may sometimes be lost as species within a clade re-evolve to the ancestral state.

Homoplasy refers to a shared character state that has not been inherited from a common ancestor exhibiting that character state. Homoplasy can result from convergent evolution or from evolutionary reversal. For example, adult frogs do not have a tail. Thus, absence of a tail is a synapomorphy that unites not only gorillas and humans but also frogs. However, frogs have neither an amniotic membrane nor hair, both of which are synapomorphies for clades that contain gorillas and humans.

In cases when there are conflicts among the characters, systematists rely on the **principle of parsimony,** which favors the hypothesis that requires the fewest assumptions. As a result, the phylogeny that requires the fewest evolutionary events is considered the best hypothesis of phylogenetic relationships (figure 22.3). In the example just stated, therefore, grouping frogs with salamanders is favored because it requires only one instance of homoplasy (the multiple origins of taillessness), whereas a phylogeny in which frogs were most closely related to humans and gorillas would require two homoplastic evolutionary events (the loss of both amniotic membranes and hair in frogs).

The examples presented so far have all involved morphological characters, but systematists increasingly use DNA sequence data to construct phylogenies because of the large number of characters that can be obtained through sequencing. Cladistics analyzes sequence data in the same manner as any other type of data: character states are polarized by reference to the sequence of an outgroup, and a cladogram is constructed that minimizes the amount of character evolution required (figure 22.4).

Other Phylogenetic Methods Work Better Than Cladistics in Some Situations

LEARNING OBJECTIVE 22.2.3 Discuss the drawbacks of molecular clocks for timing evolutionary events.

If characters evolve from one state to another at a slow rate compared with the frequency of speciation events, then the principle of parsimony works well in reconstructing evolutionary relationships. In this situation, the principle's underlying assumption—that shared derived similarity is indicative of recent common ancestry—is usually correct. In recent years, however, systematists have realized that some characters evolve so rapidly that the principle of parsimony may be misleading.

Rapid rates of evolutionary change and homoplasy

Of particular interest is the rate at which some parts of the genome evolve. Although the fraction of our genome that is "junk DNA" is controversial, much of our DNA does not appear to be functionally constrained. The rate of evolution of new character states can be quite high in these regions, due to genetic drift, and the altered states not being removed by natural selection.

Moreover, because only four character states are possible for any nucleotide base (A, C, G, or T), there is a high probability that two species will independently evolve the same derived

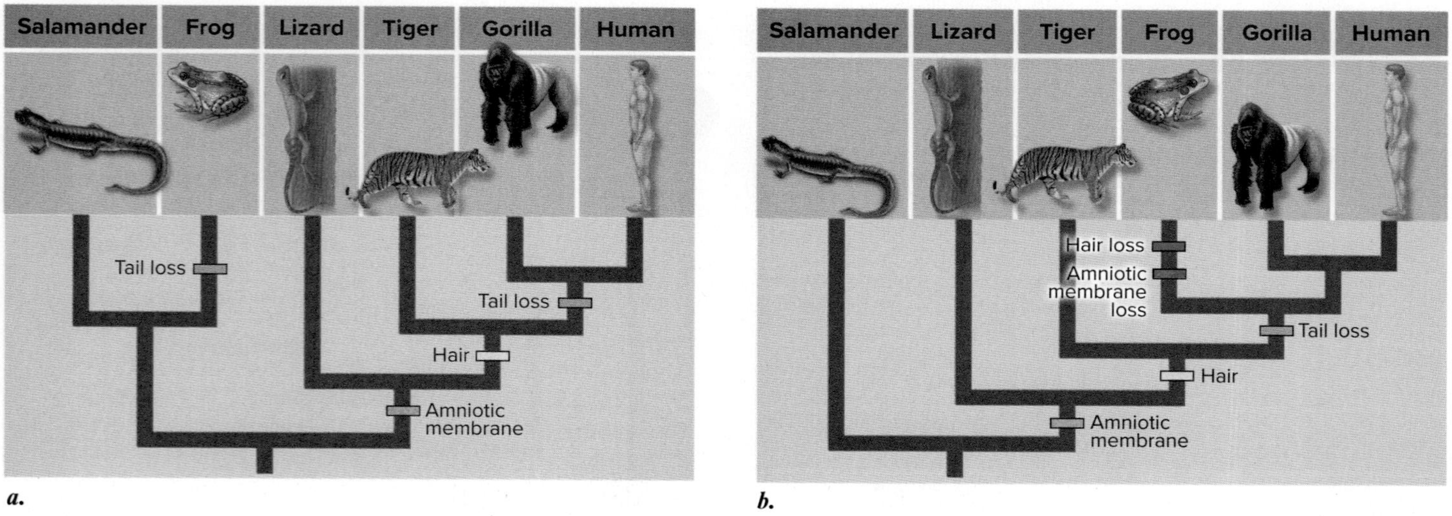

Figure 22.3 Parsimony and homoplasy. ***a.*** The placement of frogs as closely related to salamanders requires that tail loss evolved twice, an example of homoplasy. ***b.*** If frogs are closely related to gorillas and humans, then tail loss only had to evolve once. However, this arrangement would require two additional evolutionary changes: frogs would have had to have lost the amniotic membrane and hair (alternatively, hair could have evolved independently in tigers and the clade of humans and gorillas; this interpretation would require two changes, hair evolving twice, the same number of changes shown in the figure, in which hair evolved only once but then was lost in frogs). Based on the principle of parsimony, the cladogram that requires the fewest number of evolutionary changes is favored; in this case, the cladogram in (***a***) requires four changes, whereas that in (***b***) requires five, so (***a***) is considered the preferred hypothesis of evolutionary relationships.

	DNA Sequence									
Site	1	2	3	4	5	6	7	8	9	10
Species A	G	C	A	T	A	G	G	C	G	T
Species B	A	C	A	G	C	C	G	C	A	T
Species C	G	C	A	T	A	G	G	T	G	T
Species D	A	C	A	T	C	G	G	T	G	G
Outgroup	A	T	A	T	C	C	G	T	A	T

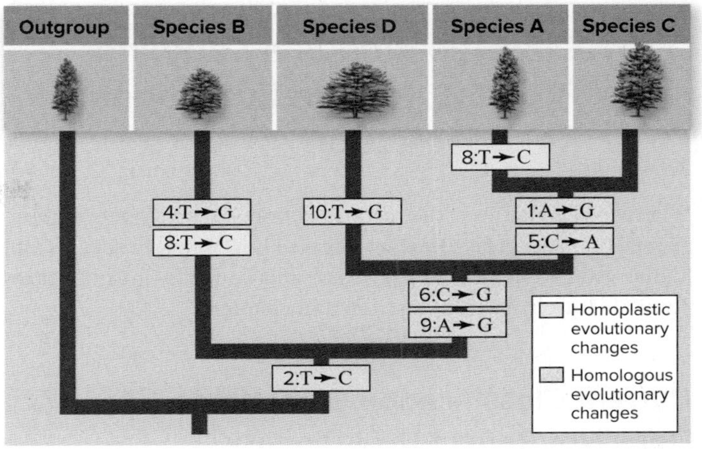

Figure 22.4 Cladistic analysis of DNA sequence data. Sequence data are analyzed just like any other data. The most parsimonious interpretation of the DNA sequence data requires nine evolutionary changes. Each of these changes is indicated on the phylogeny. Change in site 8 is homoplastic: species A and B independently evolved from thymine to cytosine at that site.

character state at any particular base position. If such homoplasy dominates the character data set, then the assumptions of the principle of parsimony are violated, and as a result, phylogenies inferred using this method are likely to be inaccurate.

Statistical approaches

Because evolution can sometimes proceed rapidly, systematists in recent years have been exploring other methods based on statistical approaches, such as maximum likelihood, to infer phylogenies. These methods start with an assumption about the rate at which characters evolve and then fit the data to these models to derive the phylogeny that best accords (that is, is "maximally likely") with these assumptions.

One advantage of these methods is that different assumptions of rate of evolution can be used for different characters. If some DNA characters evolve more slowly than others—for example, because they are constrained by natural selection—then the methods can employ different models of evolution for the different characters. This approach is more effective than parsimony in dealing with homoplasy when rates of evolutionary changes are high.

The molecular clock

In general, cladograms such as the one in figure 22.2 only indicate the order of evolutionary branching events; they do not contain information about the timing of these events. In some cases, however, branching events can be timed, either by reference to fossils or by making assumptions about the rate at which characters change. One widely used but controversial method is the **molecular clock,** which states that the rate of evolution of a molecule is constant through time. In this model, divergence in DNA can be used to calculate the times at which branching events have occurred. To make such estimates, the timing of one or more divergence events must be confidently estimated. For example, the fossil record may indicate that two clades diverged from a common ancestor at a particular time. Alternatively, the timing of separation of two clades may be estimated from geologic events that likely led to their divergence, such as the rise of a mountain that now separates the two clades. With this information, the amount of DNA divergence separating two clades can be divided by the length of time separating the two clades, which produces an estimate of the rate of DNA divergence per unit of time (usually, per million years). Assuming a molecular clock, this rate can then be used to date other divergence events in a cladogram.

Although the molecular clock appears to hold true for a variety of individual genes (see figure 20.19), in most instances the data indicate that rates of evolution have not been the same through time across all branches in an evolutionary tree. Thus, although *cytochrome c* may have evolved at a particular constant rate, other genes in the same organisms may have evolved at quite different rates, or not in a clocklike manner. For this reason, though molecular clocks provide convincing evidence of persistent evolutionary influence on particular genes, dates of evolutionary divergence derived from such data must be treated cautiously. Recently, methods have been developed to date evolutionary events without assuming that molecular evolution has been clocklike. These methods hold great promise for providing more reliable estimates of evolutionary timing.

REVIEW OF CONCEPT 22.2

Cladistics uses shared derived character states to produce groups. Derived characters are distinguished from ancestral ones using comparison to a closely related outgroup. A clade contains all descendants of a common ancestor. Cladograms are hypothetical representations of evolutionary relationships based on derived character states. Homoplasies may give a false picture of these relationships.

- *Why is cladistics more successful at inferring phylogenetic relationships in some cases than in others?*

22.3 Classification Is a Labeling Process, Not an Evolutionary Reconstruction

Whereas systematics is the reconstruction and study of evolutionary relationships, **classification** refers to how we place species and higher groups—genus, family, class, and so forth—into the taxonomic hierarchy we first discussed in chapter 1.

Current Classification Sometimes Does Not Reflect Evolutionary Relationships

LEARNING OBJECTIVE 22.3.1 Differentiate among monophyletic, paraphyletic, and polyphyletic groups.

Systematics and traditional classification are not always congruent; to understand why, we need to consider how species may be grouped based on their phylogenetic relationships. A **monophyletic** group includes the most recent common ancestor of the group and all of its descendants. By definition, a clade is a monophyletic group. A **paraphyletic** group includes the most recent common ancestor of the group, but not all its descendants, and a **polyphyletic** group does not include the most recent common ancestor of all members of the group (figure 22.5).

Taxonomic hierarchies are based on shared traits, and ideally they should reflect evolutionary relationships. Traditional taxonomic groups, however, do not always fit well with new understanding of phylogenetic relationships. For example, birds have historically been placed in the class Aves, and dinosaurs have been considered part of the class Reptilia. But recent phylogenetic advances make clear that birds evolved from dinosaurs. The last common ancestor of all birds and a dinosaur was a meat-eating dinosaur, as presented in figure 22.5.

Therefore, having two separate monophyletic groups, one for birds and one for reptiles (including dinosaurs and crocodiles, as well as lizards, snakes, and turtles), is not possible based on phylogeny. And yet the terms *Aves* and *Reptilia* are so familiar and well established that suddenly referring to birds as a type of dinosaur, and thus a type of reptile, is difficult for some people. Nonetheless, biologists increasingly refer to birds as a type of dinosaur and hence a type of reptile.

Situations like this are not uncommon. Another example concerns the classification of plants. Traditionally, three major groups were recognized: green algae, bryophytes, and vascular plants (figure 22.6). However, recent research reveals that neither the green algae nor the bryophytes constitute monophyletic groups. Rather, some bryophyte groups are more closely related to vascular plants than they are to other bryophytes, and some green algae are more closely related to bryophytes and vascular plants than they are to other green algae. As a result, systematists no longer recognize green algae or bryophytes as evolutionary groups, and the classification system has been changed to reflect evolutionary relationships.

The Phylogenetic Species Concept Focuses on Shared Derived Characters

LEARNING OBJECTIVE 22.3.2 Discuss the phylogenetic species concept and its drawbacks.

In chapter 21, you learned about a number of different ideas concerning what determines whether two populations belong to the same species. The biological species concept (BSC) defines species as groups of interbreeding populations that are reproductively isolated from other groups. In recent years, a new phylogenetic perspective has been applied to the question of species concepts. Advocates of the **phylogenetic species concept (PSC)** propose that the term *species* should be applied to groups of populations that have been evolving independently of other groups of populations. Moreover, they suggest that phylogenetic analysis is the way to identify such species. In this view, a species is a population or set of populations characterized by one or more shared derived characters.

This approach solves two of the problems with the BSC that were discussed in chapter 21. First, the BSC cannot be applied to allopatric populations, because scientists cannot determine whether individuals of the populations would interbreed and produce fertile offspring if they ever came together. The PSC solves this problem: instead of trying to predict what will happen in the future if allopatric populations ever come into contact, the PSC looks to the past to determine whether a population (or groups of populations) has evolved independently for a long enough time to develop its own derived characters.

Second, the PSC can be applied equally well to both sexual and asexual species, in contrast to the BSC, which deals only with sexual forms.

The PSC also has drawbacks

The PSC is controversial, however, for several reasons. First, some critics contend that it will lead to the recognition of every slightly differentiated population as a distinct species. In Missouri, for example, open, desert-like habitat patches called glades are distributed throughout much of the state. These glades contain a variety of warmth-loving species of plants and animals that do not occur in the forests that separate the glades. Glades have been isolated from one another for a few thousand years, allowing enough time for populations on each glade to evolve differences in some rapidly evolving parts of the genome. Does that mean that each of the hundreds, if not thousands, of Missouri glades contains its own species of lizards, grasshoppers, and scorpions? Some scientists argue that if one took the PSC to its logical extreme, that is exactly what would result.

A second problem is that species may not always be monophyletic, contrary to the definition of some versions of the phylogenetic species concept. Consider, for example, a species composed of five populations, with evolutionary relationships like those indicated in figure 22.7. Suppose that population C becomes isolated and evolves differences that make it qualify as a species by any concept (for example, reproductively isolated, ecologically differentiated). But this distinction would mean that the remaining populations, which might still be perfectly capable of exchanging genes, would be paraphyletic,

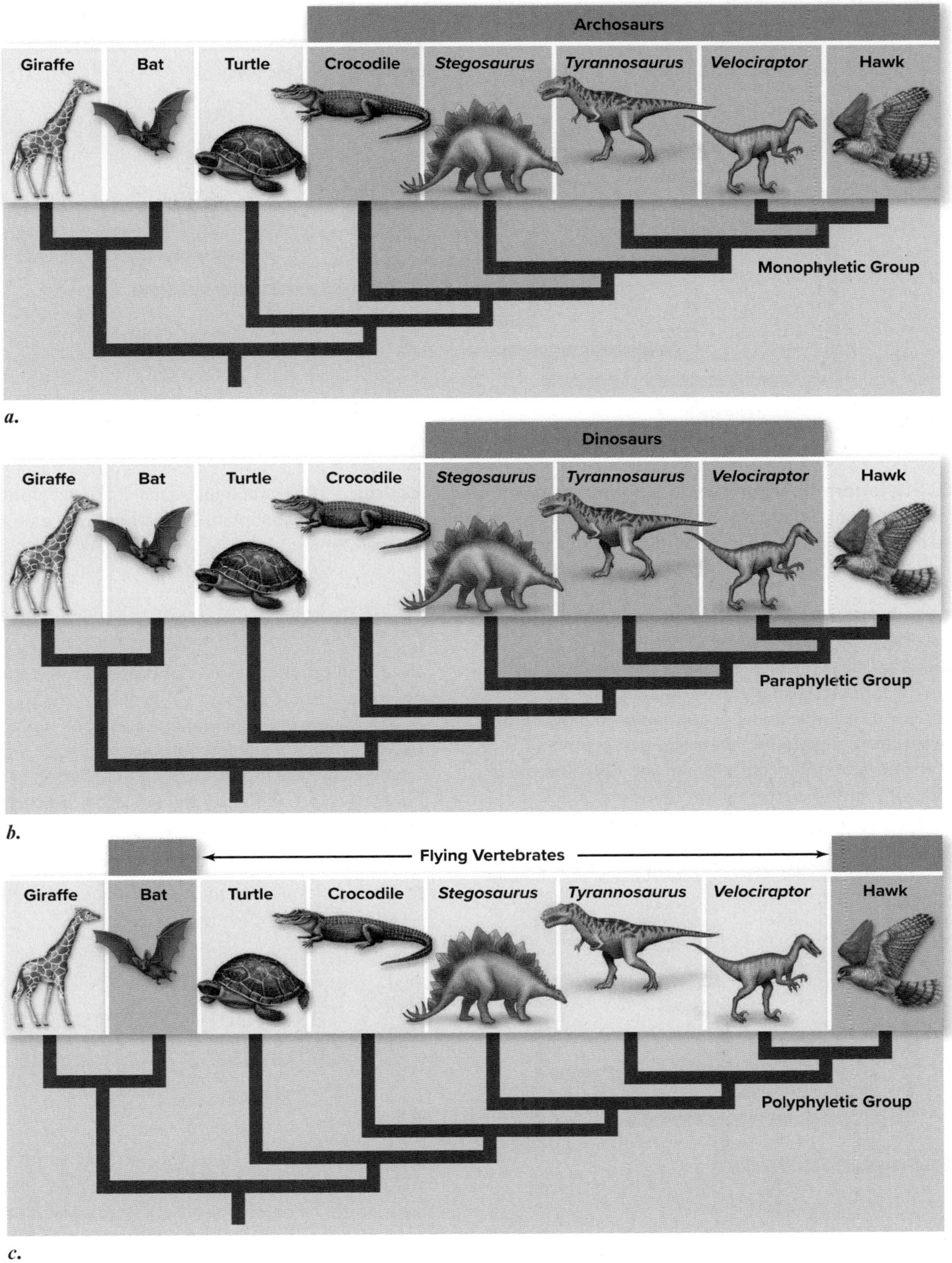

Figure 22.5 Monophyletic, paraphyletic, and polyphyletic groups. ***a.*** A monophyletic group consists of the most recent common ancestor and all of its descendants. For example, the name "Archosaurs" is given to the monophyletic group that includes crocodiles, *Stegosaurus, Tyrannosaurus, Velociraptor,* and hawks. ***b.*** A paraphyletic group consists of the most recent common ancestor and some of its descendants. For example, some, but not all, taxonomists traditionally give the name "dinosaurs" to the paraphyletic group that includes *Stegosaurus, Tyrannosaurus,* and *Velociraptor.* This group is paraphyletic because one descendant of the most recent ancestor of these species, the bird, is not included in the group. Other taxonomists include birds within the Dinosauria because *Tyrannosaurus* and *Velociraptor* are more closely related to birds than to other dinosaurs. ***c.*** A polyphyletic group does not contain the most recent common ancestor of the group. For example, bats and birds could be classified in the same group, which we might call "flying vertebrates," because they have similar shapes, anatomical features, and habitats. However, their similarities reflect convergent evolution, not common ancestry.

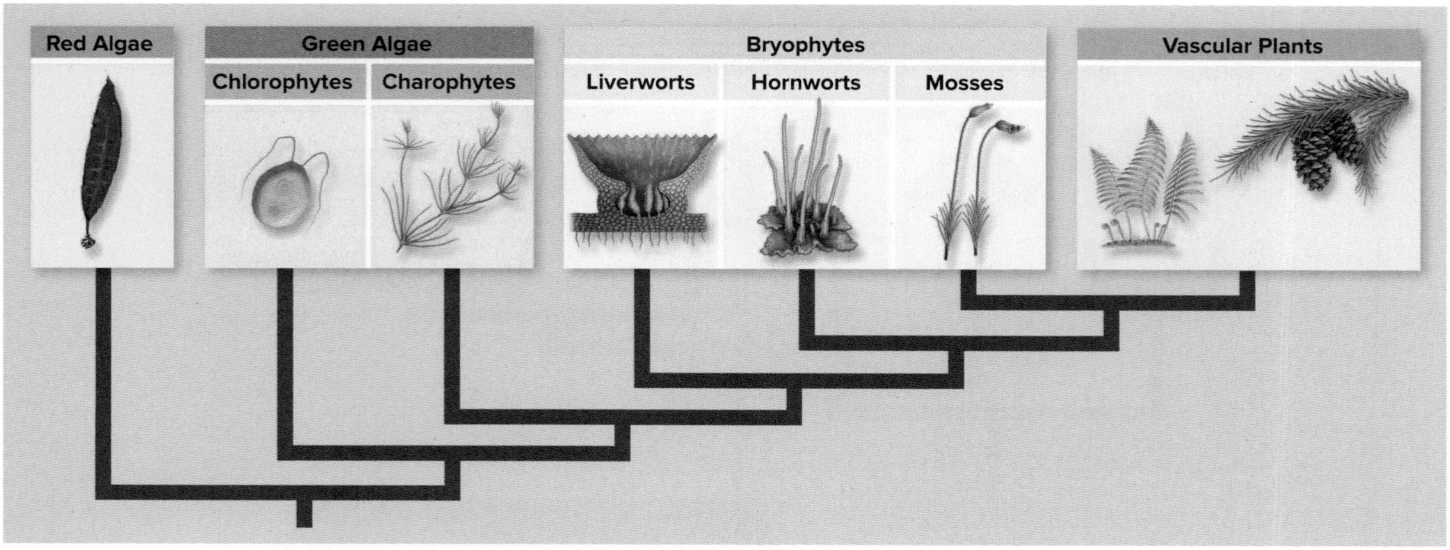

Figure 22.6 Phylogenetic information transforms plant classification. The traditional classification included two groups that we now realize are not monophyletic: the green algae and bryophytes. For this reason, plant systematists have developed a new classification of plants that does not include these groups (discussed in chapter 26).

rather than monophyletic. Such situations probably occur often in the natural world.

Phylogenetic species concepts, of which there are many different permutations, are increasingly used, but they are also contentious for the reasons just discussed. Evolutionary biologists are trying to find ways to reconcile the historical perspective of the PSC with the process-oriented perspective of the BSC and other species concepts.

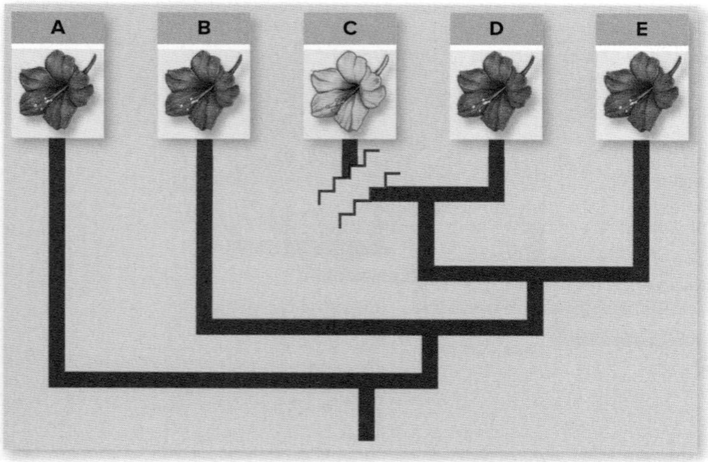

Figure 22.7 Paraphyly and the phylogenetic species concept. The five populations initially were all members of the same species, with their historical relationships indicated by the cladogram. Then, population C evolved in some ways to become greatly differentiated ecologically and reproductively from the other populations. By all species concepts, this population would qualify as a different species. However, the remaining four species do not form a clade; they are paraphyletic because population C has been removed and placed in a different species. This scenario may occur commonly in nature, but most versions of the phylogenetic species concept do not recognize paraphyletic species.

REVIEW OF CONCEPT 22.3

By definition, a clade is monophyletic. A paraphyletic group contains the most recent common ancestor, but not all its descendants; a polyphyletic group does not contain the most recent common ancestor of all members. The phylogenetic species concept focuses on the possession of shared derived characters, in contrast to the biological species concept, which emphasizes reproductive isolation. The PSC solves some problems of the BSC but has difficulties of its own.

- *Under the biological species concept, is it possible for a species to be polyphyletic?*

22.4 Taxonomy Attempts to Classify Organisms in an Evolutionary Context

People have known from the earliest times that differences exist between organisms. Early humans learned that some plants could be eaten but others were poisonous. Some animals could be hunted or domesticated; others were dangerous hunters themselves. In this section, we review formal scientific classification.

Taxonomy Is a Quest for Identity and Relationships

LEARNING OBJECTIVE 22.4.1 Explain how taxonomists name and group organisms.

More than 2000 years ago, the Greek philosopher Aristotle formally categorized living things as either plants or animals. The

Greeks and Romans expanded this simple system and grouped animals and plants into basic units such as cats, horses, and oaks. Eventually, these units began to be called **genera** (singular, *genus*), the Latin word for "groups." Starting in the Middle Ages, these names began to be systematically written down in Latin, the language used by scholars at that time. Thus, cats were assigned to the genus *Felis,* horses to *Equus,* and oaks to *Quercus.*

Linnaeus instituted the use of binomial names

Until the mid-1700s, whenever biologists wanted to refer to a particular kind of organism, which they called a species, they added a series of descriptive terms to the name of the genus; this was a polynomial, or "many names," system.

A much simpler system of naming organisms stemmed from the work of the Swedish biologist Carolus Linnaeus (1707–1778). In the 1750s, Linnaeus used the polynomial names *Apis pubescens, thorace subgriseo, abdomine fusco, pedibus posticis glabris utrinque margine ciliates* to denote the European honeybee. But as a kind of shorthand, he also included a two-part name for the honeybee; he designated it *Apis mellifera.* These two-part names, or **binomials,** have become our standard way of designating species. You have already encountered many binomial names in earlier chapters.

Taxonomy is the science of classifying living things. A group of organisms at a particular level in a classification system is called a *taxon* (plural, *taxa*). By agreement among taxonomists throughout the world, no two organisms can have the same scientific name. The scientific name of an organism is the same anywhere in the world; this avoids the confusion caused by common names (figure 22.8).

Also by agreement, the first word of the binomial name is the genus to which the organism belongs. This word is always capitalized. The second word is called the specific epithet and is not capitalized. The genus and specific epithet together constitute the species name (or scientific name) and are written in italics—for example, *Homo sapiens.* Once a genus has been used in the body of a text, it is often abbreviated in later uses. For example, the dinosaur *Tyrannosaurus rex* becomes *T. rex.*

Figure 22.8 Common names make poor labels. In North America, the common name "bear" brings a clear image to mind, but the image is very different for someone in Australia.
(left): Moodboard/Image Source; (right): John White Photos/Moment/Flickr/Getty Images

Taxonomic hierarchies have limitations

Named species are organized into larger groups based on shared characteristics. As discussed in section 22.2, sound evolutionary hypotheses can be constructed when organisms are grouped based on derived characters, not ancestral characters. Early taxonomists were not aware that the distinction between derived and ancestral characters could make a difference; as a result, many hierarchies are now being re-examined. As the phylogenetic and systematic revolution continues, other limitations of the original levels of taxonomic organization, called the *Linnaean taxonomy,* are being revealed.

The Linnaean hierarchy

In the decades following Linnaeus, taxonomists began to group organisms into larger, more inclusive categories. Genera with similar characters were grouped into a cluster called a **family,** and similar families were placed into the same **order** (figure 22.9). Orders with common properties were placed into the same **class,** and classes with similar characteristics into the same **phylum** (plural, *phyla*). Finally, the phyla were assigned to one of several great groups, the **kingdoms.** These kingdoms include two kinds of prokaryotes (Archaea and Bacteria), a largely unicellular group of eukaryotes (Protista), and three multicellular groups (Fungi, Plantae, and Animalia). As you will learn in chapter 24, the protists are not a monophyletic group, and the term *kingdom* is still used in classification systems, but other groupings are more appropriate for showing evolutionary relationships.

In addition, an eighth level of classification, called a **domain,** is frequently used. Biologists recognize three domains, which will be discussed in section 22.5. The names of the taxonomic units from the genus level and higher are capitalized.

The categories at the different levels may include many, a few, or only one taxon. For example, there is only one living genus in the family Ornithorhynchidae (*Ornithorhynchus*, containing the species *O. anatinus*, the duck-billed platypus), but there are several living genera of Fagaceae (the birch family). To someone familiar with classification or having access to the appropriate reference books, each taxon implies both a set of characteristics and a group of organisms belonging to the taxon.

To return to the example of the European honeybee, we can analyze the bee's taxonomic classification as follows:

1. **Species level:** *Apis mellifera,* meaning honey-bearing bee
2. **Genus level:** *Apis,* a genus of bees
3. **Family level:** Apidae, a bee family. All members of this family are bees—some solitary, some living in colonies as *A. mellifera* does.
4. **Order level:** Hymenoptera, a grouping that includes bees, wasps, ants, and sawflies—all of which have wings with membranes
5. **Class level:** Insecta, a very large class that comprises animals with three major body segments, three pairs of legs attached to the middle segment, and wings

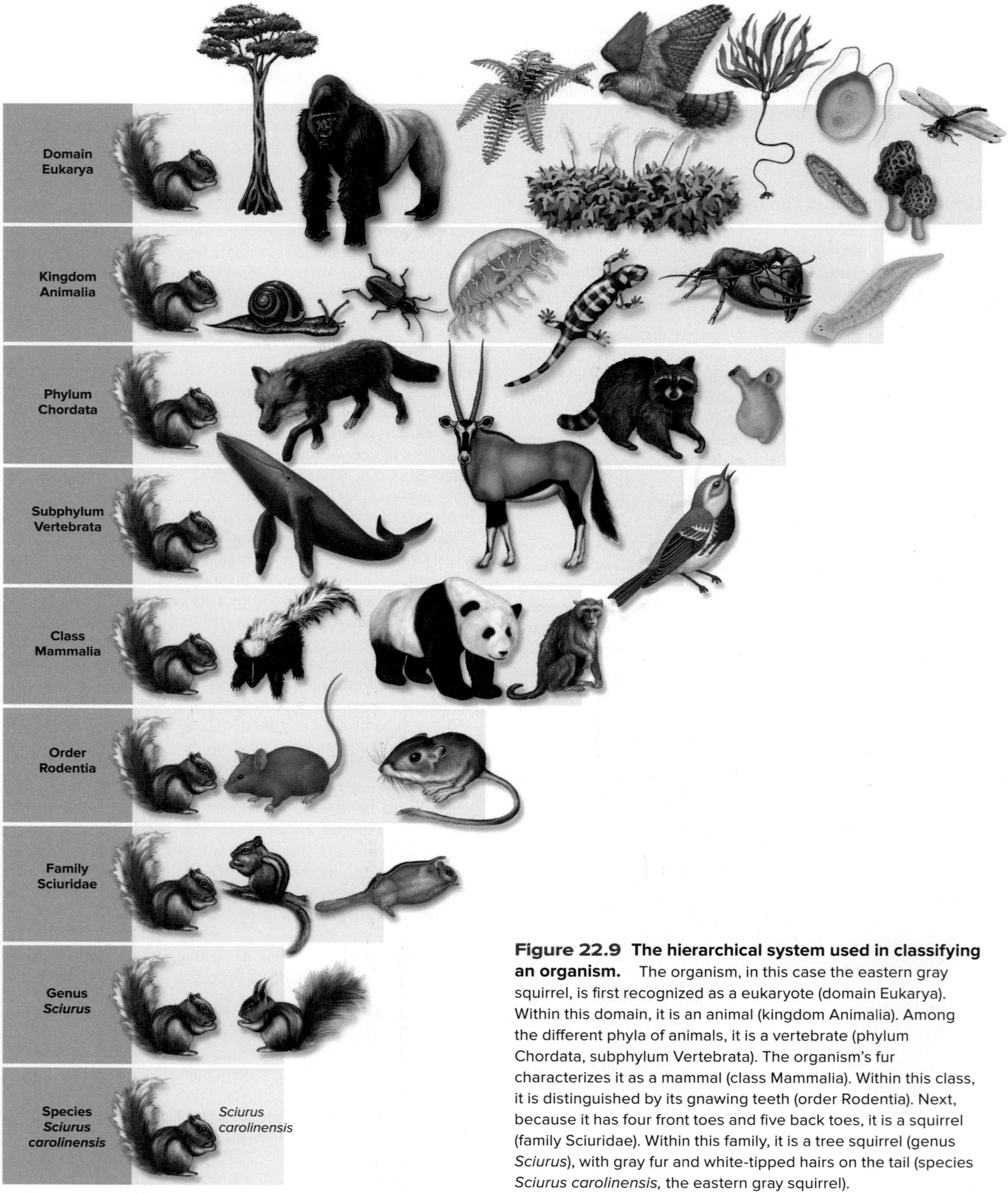

Figure 22.9 The hierarchical system used in classifying an organism. The organism, in this case the eastern gray squirrel, is first recognized as a eukaryote (domain Eukarya). Within this domain, it is an animal (kingdom Animalia). Among the different phyla of animals, it is a vertebrate (phylum Chordata, subphylum Vertebrata). The organism's fur characterizes it as a mammal (class Mammalia). Within this class, it is distinguished by its gnawing teeth (order Rodentia). Next, because it has four front toes and five back toes, it is a squirrel (family Sciuridae). Within this family, it is a tree squirrel (genus *Sciurus*), with gray fur and white-tipped hairs on the tail (species *Sciurus carolinensis,* the eastern gray squirrel).

6. **Phylum level:** Arthropoda. Animals in this phylum have a hard exoskeleton made of chitin and jointed appendages.
7. **Kingdom level:** Animalia. The animals are multicellular heterotrophs with cells that lack cell walls.

Limitations of the hierarchy

In sections 22.1 and 22.2, we discussed the modern phylogenetic approach, which distinguishes relationships between different species based on evolutionary history. New phylogenies, based on molecular data, reveal that the Linnaean hierarchy does not always reflect evolutionary history. It was never intended to, so this should not be surprising. This discovery is leading to new evolutionary hypotheses that can be tested against rapidly accumulating molecular data.

One problem with the Linnaean system is that many higher taxonomic ranks (for example, Reptilia) are not monophyletic and therefore do not represent natural groups. A common ancestor and all of its descendants form a natural group that results from descent from that common ancestor, but any other type of group (paraphyletic or polyphyletic) is an artificial group created by taxonomists.

In addition, Linnaean ranks, as currently recognized, are not equivalent in any meaningful way. For example, two families may not represent clades that originated at the same time. One family may have diverged 70 million years before another family, and therefore these families have had vastly different amounts of time to diverge and develop evolutionary adaptations. Two groups that diverged from a common ancestor at the same time may be given different ranks. Thus, comparisons using Linnaean categories may be misleading. It is much better to use hypotheses of phylogenetic relationships in such instances.

One result of all these differences is that families demonstrate different degrees of biological diversity. Here's one example. It is difficult to say that the legume family, with 16,000 species, represents the same level of taxonomic organization as the cat family, with only 36 species. The differences across a single rank—whether class, order, or family—limit the usefulness of taxonomic hierarchies in making evolutionary predictions.

REVIEW OF CONCEPT 22.4

By convention, a species is given a binomial name. The first part of the name identifies the genus, and the second part the individual species. The Linnaean taxonomic hierarchy groups species into genera, then families, orders, classes, phyla, and kingdoms. Traditional classification systems are based on similar traits, but because they include a mix of derived and ancestral traits, they do not necessarily take into account evolutionary relationships.

- *What can you infer about evolutionary relationships by comparing a taxonomic hierarchy for a squirrel and a fox? What questions remain unanswered?*

22.5 The Largest Taxa Are Domains

In this section, we examine the largest groupings of organisms: kingdoms and domains. The earliest classification systems recognized only two kingdoms of living things: animals and plants. But as biologists discovered microorganisms and learned more about other multicellular organisms, they added kingdoms in recognition of certain fundamental differences. The six-kingdom system was first proposed by Carl Woese of the University of Illinois (figure 22.10*b*).

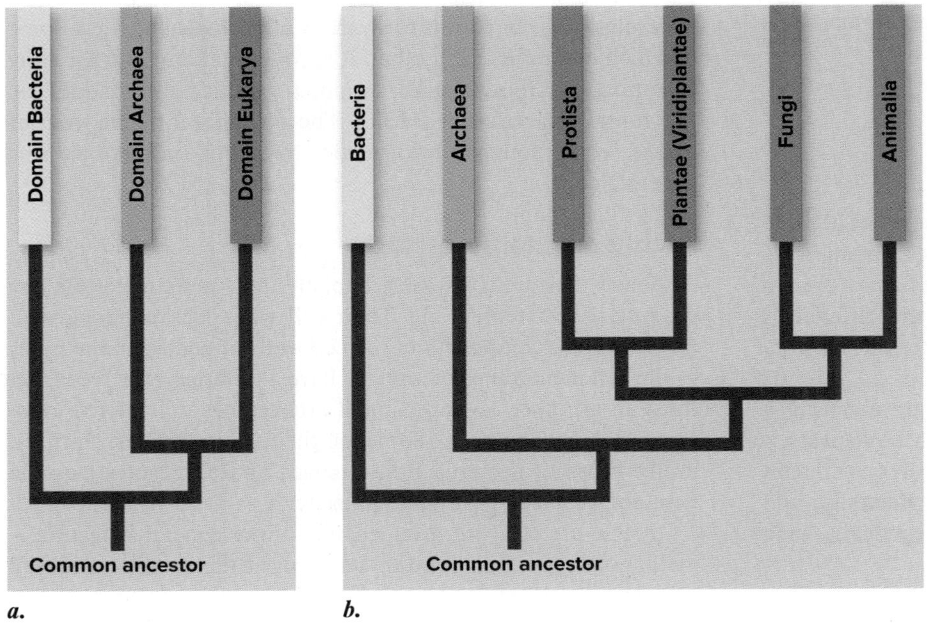

Figure 22.10 Different approaches to classifying living organisms. ***a.*** Bacteria and Archaea are so distinct that they have been assigned to separate domains distinct from the Eukarya. Members of the domain Bacteria are thought to have diverged early from the evolutionary line that gave rise to the archaea and eukaryotes. ***b.*** Eukarya are grouped into four kingdoms, but these, especially the protists, are not necessarily monophyletic groups.

Molecular Data Support the Existence of Three Domains

LEARNING OBJECTIVE 22.5.1 List examples showing that the three domains of life are monophyletic, but the five kingdoms are not.

We currently recognize five kingdoms

The six-kingdom system included Protista as a Kingdom, which many biologists no longer find tenable. That leaves five kingdoms that are recognized by the majority of biologists. Three of the kingdoms consist of eukaryotic organisms. The two most familiar are Animalia and Plantae, which only contain organisms that are multicellular during most of their life cycle. The kingdom Fungi contains multicellular forms and single-celled yeasts.

Fundamental differences divide these three kingdoms. Plants are mainly stationary, but some have motile sperm; most fungi lack motile cells; animals are mainly motile or mobile. Animals ingest their food, plants manufacture it, and fungi digest and absorb it by means of secreted extracellular enzymes.

The large number of eukaryotes that do not fit any of the three eukaryotic kingdoms are arbitrarily grouped together as the *protists* (refer to chapter 24). Most protists are unicellular or, in the case of some algae, have a unicellular phase in their life cycle. This group reflects the current controversy between taxonomic and phylogenetic approaches. The protists are a paraphyletic group, containing several nonmonophyletic adaptive lineages with distinct evolutionary origins.

The remaining two kingdoms, **Archaea** and **Bacteria,** consist of prokaryotic organisms, which are vastly different from all other living things (refer to chapter 23). Archaea are a diverse group that includes the methanogens, a variety of extremophiles, and that differs greatly from Bacteria, the other prokaryotic domain.

The three domains probably are monophyletic

As biologists have learned more about the Archaea, it has become increasingly clear that this group is very different from all other organisms. When the full genomic DNA sequences of an archaean and a bacterium were first compared in 1996, the differences proved striking. Archaea are as different from bacteria as bacteria are from eukaryotes.

Recognizing this, biologists are increasingly adopting a classification of living organisms that recognizes three **domains,** a taxonomic level higher than kingdom. Archaea are in one domain **(domain Archaea),** bacteria in a second **(domain Bacteria),** and eukaryotes in the third **(domain Eukarya).** Phylogenetically, each of these domains forms a clade.

In the remainder of this section, we preview the major characteristics of the three domains.

Bacteria Are More Numerous Than Any Other Organism

LEARNING OBJECTIVE 22.5.2 List the distinctive characteristics of bacteria.

Bacteria are the most abundant organisms on Earth. There are more living bacteria in your mouth than there are mammals living on Earth. Although too tiny to see with the unaided eye, bacteria play critical roles throughout the biosphere. Some extract from the air all the nitrogen used by organisms, and they play key roles in cycling carbon and sulfur. Much of the world's photosynthesis is carried out by bacteria. In contrast, certain bacteria are also responsible for many forms of disease. Understanding bacterial metabolism and genetics is a critical part of modern medicine.

Bacteria are highly diverse, and the evolutionary links among species are not well understood. Although taxonomists disagree about the details of bacterial classification, most recognize 12 to 15 major groups of bacteria. Comparisons of the nucleotide sequences of ribosomal RNA (rRNA) molecules are beginning to reveal how these groups are related to one another and to the other two domains.

Archaea May Live in Extreme Environments

LEARNING OBJECTIVE 22.5.3 Distinguish between bacteria and archaea.

The archaea seem to have diverged very early from the bacteria and are more closely related to eukaryotes than to bacteria (figure 22.10). This conclusion comes largely from comparisons of genes that encode ribosomal RNAs.

Horizontal gene transfer in microorganisms

Comparing whole-genome sequences from microorganisms has led evolutionary biologists to a variety of phylogenetic trees, some of which contradict each other. It appears that during their early evolution, microorganisms swapped genetic information via horizontal gene transfer (HGT). The potential for gene transfer makes constructing phylogenetic trees for microorganisms very difficult.

Archaean characteristics

Although they are a diverse group, all archaea share certain key characteristics (table 22.1). Their cell walls lack peptidoglycan (an important component of the cell walls of bacteria); the lipids in the cell membranes of archaea have a different structure from those in all other organisms; and archaea have distinctive ribosomal RNA sequences. Some of their genes possess introns, unlike those of bacteria. Both archaea and eukaryotes lack the peptidoglycan cell wall found in bacteria.

The archaea are grouped into three general categories—methanogens, extremophiles, and nonextreme archaea—based primarily on the environments in which they live or on their specialized metabolic pathways. The word *extreme* refers to our

TABLE 22.1 Features of the Three Domains of Life

Feature	Archaea	Bacteria	Eukarya
Amino acid that initiates protein synthesis	Methionine	Formylmethionine	Methionine
Introns	Present in some genes	Absent	Present
Membrane-bounded organelles	Few	Few	Present
Membrane lipid structure	Branched	Unbranched	Unbranched
Nuclear envelope	Absent	Absent	Present
Number of different RNA polymerases	Several	One	Several
Peptidoglycan in cell wall	Absent	Present	Absent
Response to the antibiotics streptomycin and chloramphenicol	Growth not inhibited	Growth inhibited	Growth not inhibited

current environment. When archaea first appeared on the scene, their now extreme habitats may have been typical.

Methanogens obtain their energy by using hydrogen gas (H_2) to reduce carbon dioxide (CO_2) to methane gas (CH_4). They are strict anaerobes, poisoned by even traces of oxygen. They live in swamps, marshes, and the intestines of mammals. Methanogens release about 2 billion tons of methane gas into the atmosphere each year.

Extremophiles are able to grow under conditions that seem extreme to us. These conditions include temperature, salt, pH, and pressure extremes. There are both cold-adapted archaea in glacier and alpine lakes and thermophiles living in temperatures ranging from 60° to 80°C in hot springs. Halophiles live in salty environments such as the Dead Sea and Great Salt Lake, actually requiring salinity of 15 to 20%. Archaea tolerant to pH also range from acidic (pH = 0.7) to basic (pH = 11) extremes. Pressure-tolerant archaea are found at great depths in the ocean, where they require pressures of more than 300 atmospheres (atm) to survive and can tolerate up to 800 atm.

Nonextreme archaea are found in "normal" terrestrial, marine, and aquatic environments. With the advent of large-scale environmental DNA sequencing, this group has proved to be much larger than expected. Entire new phyla of archaea have been identified, although not cultured, based on the analysis of DNA sequence data. This includes a large group with the closest affinity to eukaryotes based on genome analysis.

Eukaryotes Exhibit Extensive Compartmentalization

LEARNING OBJECTIVE 22.5.4 Distinguish between prokaryotes and eukaryotes.

The progenitor for all life now on earth arose in the Archean eon. We hypothesize the existence of a last universal common ancestor, or LUCA. The descendants of LUCA include the three domains Eubacteria, Archaea, and Eukarya. The relationships between these three domains are still contentious, although the most common model has Eukarya branching from within Archaea. A group of deep-sea archaea called Asgard appear to be the closest prokaryotic relatives to modern eukaryotes.

Roots of the eukaryotic tree

Despite a large amount of work by many investigators, the roots of the eukaryotic tree remain problematic. There are issues with the traditional kingdoms, especially the Protista, but can we use molecular genetic data to discover evolutionarily significant groups in the eukaryotic domain? Using this approach, it does appear that the deepest branches form five so-called supergroups: **Excavata, SAR** (a monophyletic clade consisting of Stramenopila, Alveolata, and Rhizaria), **Archaeplastida, Amoebozoa,** and **Opisthokonta.** These groups are based on both morphological similarities and molecular analysis. They appear to form monophyletic clades, although this has not been definitively shown for all groups. The SAR group combines the older supergroup called Chromalveolata (Stramenopiles and Alveolates) with Rhizaria. These groups are shown in figure 22.11 and described in more detail in chapter 24. Here we will consider the key evolutionary events supporting this incredible diversity of life evolving in response to an ever-changing environment.

Endosymbiosis and the origin of eukaryotes

The hallmark of eukaryotes is complex cellular organization, highlighted by an extensive endomembrane system that subdivides the eukaryotic cell into functional compartments (refer to chapter 4). Not all cellular compartments, however, are derived from the endomembrane system.

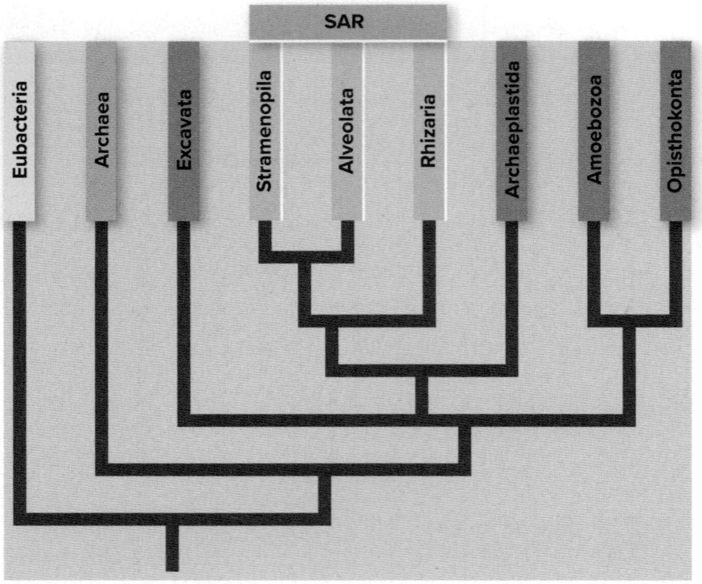

Figure 22.11 Five supergroups have been identified within the domain Eukarya, one of three domains of life on Earth. The SAR group is a monophyletic clade that joins the older supergroup Chromalveolata with Rhizaria.

With few exceptions, modern eukaryotic cells possess the energy-producing organelles termed *mitochondria,* and photosynthetic eukaryotic cells possess *chloroplasts,* the energy-harvesting organelles. Mitochondria and chloroplasts are both believed to have entered early eukaryotic cells by a process called **endosymbiosis,** which is discussed in more detail in chapter 24.

Mitochondria are the descendants of relatives of purple sulfur bacteria and the parasitic *Rickettsia* that were incorporated into eukaryotic cells early in the history of the group. Chloroplasts are derived from cyanobacteria (figure 22.12). The red and green algae acquired their chloroplasts by directly engulfing a cyanobacterium. The brown algae most likely engulfed red algae to obtain chloroplasts.

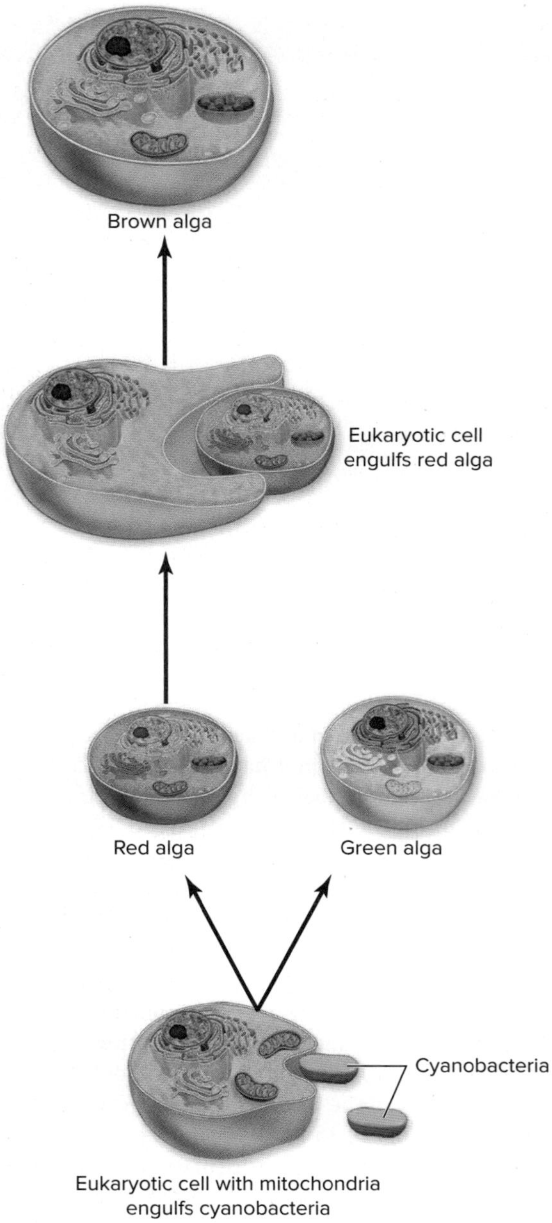

Figure 22.12 All chloroplasts are monophyletic. The same cyanobacteria were engulfed by multiple hosts that were ancestral to the red and green algae. Brown algae share the same ancestral chloroplast DNA but most likely gained it by engulfing red algae.

Key characteristics of the eukaryotes

Although eukaryotic organisms are extraordinarily diverse, they share characteristics that distinguish them from prokaryotes. Eukaryotic cells show extensive compartmentalization, primarily through an internal membrane system, and membrane-bounded organelles. Multicellularity evolved within eukaryotes, although there are unicellular groups, and sexual reproduction also evolved within eukaryotes.

Compartmentalization Although prokaryotes do have some specialized compartments, eukaryotes are characterized by the extensive use of internal membranes to divide the cell into different functional compartments. This provides evolutionary opportunities for increased specialization within the cell and the evolution of different specialized cell types. The nuclear membrane, found only in eukaryotes, also allows for increased complexity. In eukaryotes, RNA transcripts from nuclear DNA are processed and transported across the nuclear membrane into the cytosol, where translation occurs. The physical separation of transcription and translation in eukaryotes adds additional levels of control to the process of gene expression.

Multicellularity The unicellular body plan has been tremendously successful: unicellular prokaryotes and eukaryotes make up about half of the biomass on Earth. But a single cell has limits. The evolution of multicellularity allowed organisms to deal with their environments in novel ways through differentiation of cell types into tissues and organs.

True multicellularity, where the activities of individual cells are coordinated and the cells themselves are in contact, occurs only in eukaryotes and is one of their major characteristics. Bacteria and some protists form colonial aggregates of many cells, but the cells in the aggregates have little differentiation or integration of function.

Multicellularity appears to have evolved multiple times. The red, brown, and green algae have each independently attained multicellularity. One lineage of multicellular green algae was the ancestor of the plants (refer to chapters 24 and 26), and most taxonomists now place its members in the green plant kingdom, the Viridiplantae.

Fungi and animals also illustrate the multiple origins of multicellularity. Although fungi and animals are more closely related than either is to plants, they arose from unicellular protist ancestors with different characteristics. As you will discover in chapters 25 and 27, the groups that seem to have given rise to each of these kingdoms are still in existence.

Sexual Reproduction Another major characteristic of eukaryotic species as a group is sexual reproduction. Although some interchange of genetic material occurs in bacteria, it is certainly not a regular, predictable mechanism in the same sense that sex is in eukaryotes. Sexual reproduction allows greater genetic diversity through the processes of meiosis and crossing over, as you learned in chapter 11.

In many of the unicellular phyla of protists, sexual reproduction occurs only occasionally. The first eukaryotes were probably haploid; diploids seem to have arisen on a number of separate occasions by the fusion of haploid cells, which then eventually divided by mitosis.

REVIEW OF CONCEPT 22.5

The six kingdoms are not necessarily based on common lineage; kingdom Protista, for example, is not a monophyletic group. The three domains, however, do appear to be monophyletic. Bacteria and archaea are tiny but numerous unicellular organisms that lack internal compartmentalization. Eukaryotic cells are highly compartmentalized, and they have acquired mitochondria and chloroplasts by endosymbiosis.

- *What is the relationship between the six supergroups and the three domains?*

What Causes New Forms to Arise?

Biologists once presumed that new forms—genera, families, and orders—arose most often during times of massive geologic disturbance, stimulated by the resulting environmental changes. But does such a relationship exist? An alternative hypothesis was proposed by evolutionist George Simpson in 1953. He proposed that diversification followed new evolutionary innovations, "inventions" that permitted an organism to occupy a new "adaptive zone." After a burst of new orders that define the major groups, subsequent specialization would lead to new genera.

The early bony fishes, typified by the sturgeon (bottom fish image), had feeble jaws and long, sharklike tails. They dominated the Devonian (the Age of Fishes), to be succeeded in the Triassic (the period when dinosaurs appeared) by fishes like the gar pike, with a shorter, more powerful jaw that improved feeding and a shortened, more maneuverable tail that improved locomotion. They were in turn succeeded by teleost fishes like the perch, with an even better tail for fast, maneuverable swimming and a complex mouth with a mobile upper jaw that slides forward as the mouth opens.

This history allows a clear test of Simpson's hypothesis. Was the appearance of these three orders followed by a burst of evolution as Simpson predicts, the new innovations in feeding and locomotion opening wide the door of opportunity? If so, many new genera should occur in the fossil record soon after the appearance of each new order. If not, the pattern of when new genera appear should not track the appearance of new orders.

The graph shows the evolutionary history of the class Osteichthyes, the bony fishes, since they first appeared in the Silurian some 420 MYA.

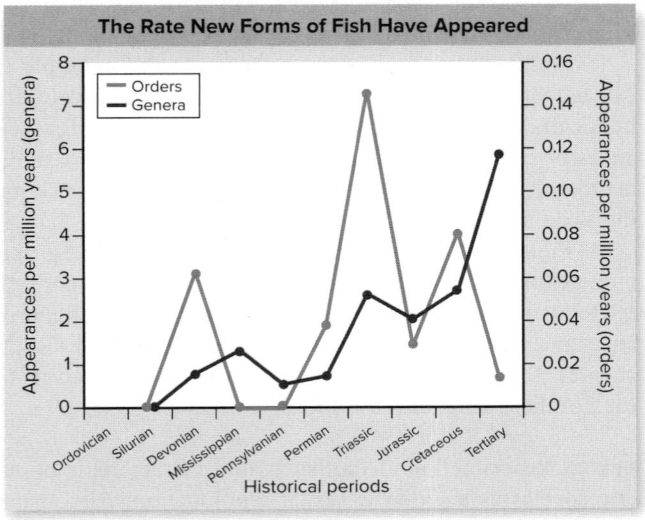

Perch

Gar pike

Sturgeon

Analysis

1. **Applying Concepts** In the graph, what is the dependent variable?
2. **Interpreting Data** Three great innovations in jaw and tail occur during the history of the bony fishes, producing the superorders represented by sturgeons, then gars, and then teleost fishes. In what period did each innovation occur?
3. **Making Inferences** Do bursts of new genera appear at these same three times, or later?
4. **Drawing Conclusions** Do the data presented in the graph support Simpson's hypothesis? Explain.
5. **Further Analysis** If you were to plot on the graph the rate at which new families of fishes appeared, what general pattern would you expect to find, relative to new orders, if Simpson is right? Explain.

Retracing the Learning Path

CONCEPT 22.1 Systematics Reconstructs Evolutionary Relationships

22.1.1 Branching Diagrams Depict Evolutionary Relationships Systematics is the study of evolutionary relationships, which are depicted on branching evolutionary trees, called phylogenies.

22.1.2 Similarity May Not Accurately Predict Evolutionary Relationships The rate of evolution can vary among species and can even reverse direction. Closely related species can therefore be dissimilar in phenotypic characteristics. Conversely, convergent evolution results in distantly related species being phenotypically similar.

CONCEPT 22.2 Cladistics Focuses on Traits Derived from a Common Ancestor

22.2.1 Cladistics Requires That Character States Be Identified as Ancestral or Derived Derived character states are those that differ from the ancestral condition. Character polarity is established using an outgroup comparison in which the outgroup consists of closely related species or a group of species, relative to the group under study. Character states exhibited by the outgroup are assumed to be ancestral, and other character states are considered derived. A cladogram is a graphically represented hypothesis of evolutionary relationships.

22.2.2 Homoplasy Complicates Cladistic Analysis *Homoplasy* refers to a shared character state, such as wings of birds and wings of insects, that has not been inherited from a common ancestor. Cladograms are constructed based on the principle of parsimony, which indicates that the phylogeny requiring the fewest evolutionary changes is accepted as the best working hypothesis.

22.2.3 Other Phylogenetic Methods Work Better Than Cladistics in Some Situations When evolutionary change is rapid, other methods, such as statistical approaches and the use of the molecular clock, are sometimes more useful.

CONCEPT 22.3 Classification Is a Labeling Process, Not an Evolutionary Reconstruction

22.3.1 Current Classification Sometimes Does Not Reflect Evolutionary Relationships A monophyletic group consists of the most recent common ancestor and all of its descendants. A paraphyletic group consists of the most recent common ancestor and some of its descendants. A polyphyletic group does not contain the most recent ancestor of the group. Some currently recognized taxa are not monophyletic, such as reptiles, which are paraphyletic with respect to birds.

22.3.2 The Phylogenetic Species Concept Focuses on Shared Derived Characters The phylogenetic species concept emphasizes the possession of shared derived characters, whereas the biological species concept focuses on reproductive isolation. Many versions of this concept recognize only species that are monophyletic. The phylogenetic species concept (PSC) also has drawbacks. Two main criticisms of the PSC are that it can create too many species via impractical distinctions and that the PSC definition of a group may not always apply as selection proceeds.

CONCEPT 22.4 Taxonomy Attempts to Classify Organisms in an Evolutionary Context

22.4.1 Taxonomy Is a Quest for Identity and Relationships Taxonomy is the science of assigning organisms to a particular level of classification called a taxon. Taxonomic hierarchies are organized by domain, kingdom, phylum, class, order, family, genus, and species. Linnaeus instituted the use of binomial names. He devised a system of giving individual species unique names, beginning with the capitalized genus, followed by a species name. These names are italicized. Taxonomic hierarchies have limitations. Traditional classifications are limited because they are based on similar traits and do not take into account evolutionary relationships.

CONCEPT 22.5 The Largest Taxa Are Domains

22.5.1 Molecular Data Support the Existence of Three Domains Domain Eukarya contains the kingdoms Protista, Plantae, Fungi, and Animalia; the other two domains, Bacteria and Archaea, each contain only prokaryotes.

22.5.2 Bacteria Are More Numerous Than Any Other Organism Bacteria are the most abundant and diverse organisms on Earth; they consist of 12 to 15 major groups, but their evolutionary links are unclear.

22.5.3 Archaea May Live in Extreme Environments Archaea are prokaryotes that are more closely related to eukaryotes than to bacteria. They are grouped into methanogens, extremophiles, and nonextreme archaea.

22.5.4 Eukaryotes Exhibit Extensive Compartmentalization Eukaryote cells have compartmentalized organelles and other structures. Eukaryotes acquired mitochondria and chloroplasts by endosymbiosis. Many eukaryotes are multicellular, and most undergo sexual reproduction.

Concept Overview

This Concept Overview diagrams the key concepts that were discussed in this chapter.

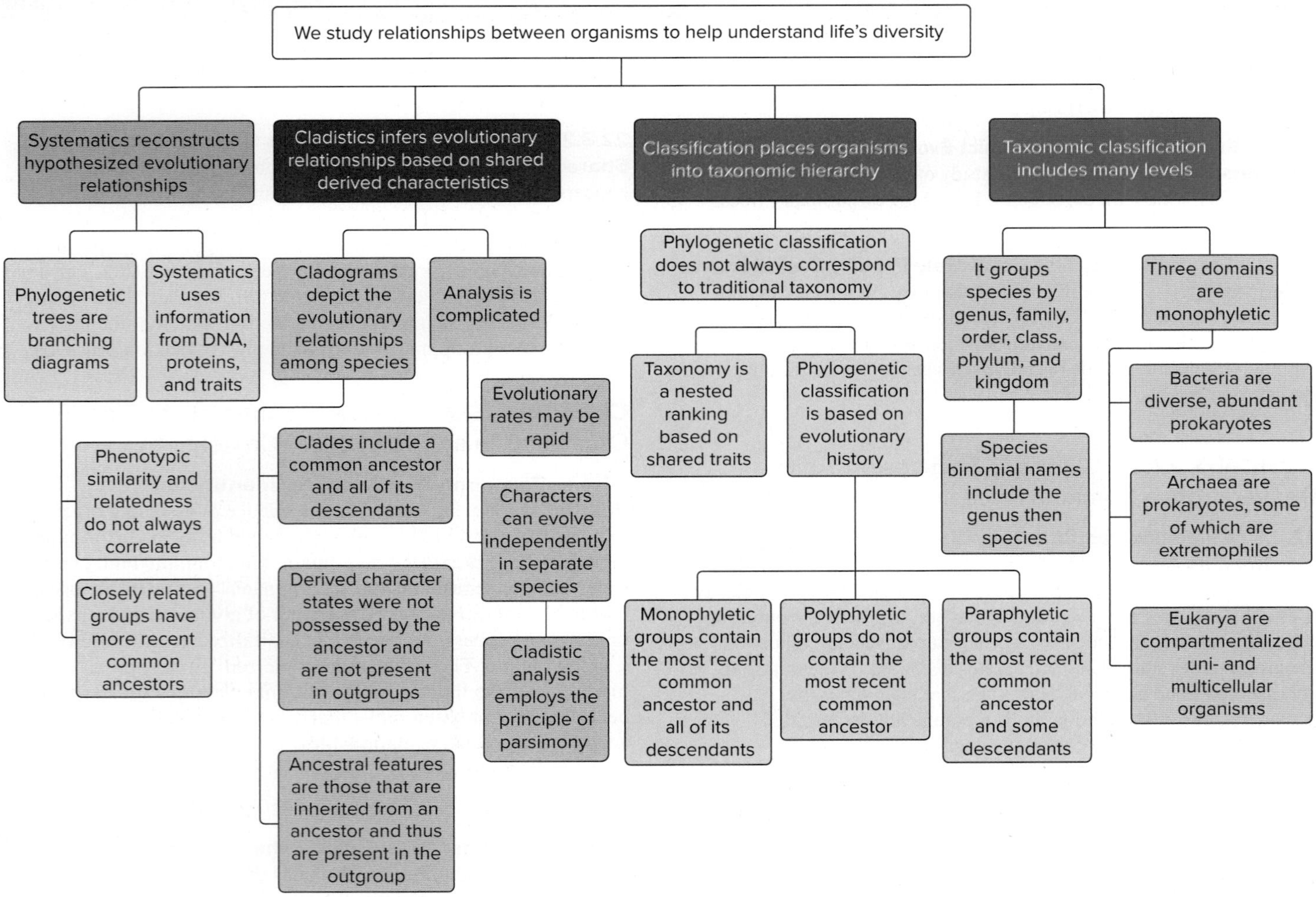

Assessing the Learning Path

Understand

1. The evolutionary relationships between organisms, and their relationships to other species, are its
   a. taxonomy.
   b. phylogeny.
   c. ontogeny.
   d. systematics.
2. Overall similarity of phenotypes may not always reflect evolutionary relationships
   a. due to convergent evolution.
   b. because of variation in rates of evolutionary change of different kinds of characters.
   c. due to homoplasy.
   d. due to all of the above.
3. Which statement accurately describes the relationship between humans and chimpanzees?
   a. Humans are descendants of chimpanzees.
   b. Chimpanzees are descendants of humans.
   c. Humans and chimpanzees share a common ancestor.
   d. Humans and chimpanzees have no relationship.
4. Cladistics
   a. is based on overall similarity of phenotypes.
   b. requires distinguishing similarity due to inheritance from a common ancestor from other reasons for similarity.
   c. is not affected by homoplasy.
   d. None of the above
5. The principle of parsimony
   a. helps evolutionary biologists distinguish among competing phylogenetic hypotheses.
   b. does not require that the polarity of traits be determined.
   c. is a way to avoid having to use outgroups in a phylogenetic analysis.
   d. cannot be applied to molecular traits.

6. Rapid rates of character change relative to the rate of speciation pose a problem for cladistics because
   a. the frequency with which distantly related species evolve the same derived character state may be high.
   b. evolutionary reversals may occur frequently.
   c. homoplasy will be common.
   d. All of the above
7. In a paraphyletic group,
   a. all species are more closely related to each other than they are to a species outside the group.
   b. evolutionary reversal is common.
   c. polyphyly also usually occurs.
   d. some species are more closely related to species outside the group than they are to some species within the group.
8. Species recognized by the phylogenetic species concept
   a. sometimes also would be recognized as species by the biological species concept.
   b. are sometimes paraphyletic.
   c. are characterized by symplesiomorphies.
   d. are more frequent in plants than in animals.
9. Humans, chimpanzees, and gorillas constitute a __________ group.
   a. monophyletic
   b. paraphyletic
10. The wolf, domestic dog, and red fox are all in the same family, Canidae. The scientific name for the wolf is *Canis lupis*, the domestic dog is *Canis familiaris*, and the red fox is *Vulpes vulpes*. This means that
   a. the red fox is in the same family as but a different genus from dogs and wolves.
   b. the dog is in the same family as but a different genus from red foxes and wolves.
   c. the wolf is in the same family as but a different genus from dogs and red foxes.
   d. all three organisms are in different genera.
11. Which of the following events occurred first in eukaryotic evolution?
   a. Endosymbiosis and mitochondria evolution
   b. Endosymbiosis and chloroplast evolution
   c. Compartmentalization and formation of the nucleus
   d. Formation of multicellular organisms
12. Given your understanding of phylogenetics, where would you place viruses in the tree of life?
   a. Archaea
   b. Fungi
   c. Bacteria
   d. None of the above

Apply

1. Organisms that are closer together on a cladogram
   a. are in the same family.
   b. constitute an outgroup.
   c. share a more recent common ancestor than those organisms that are farther apart.
   d. share fewer derived characters than organisms that are farther apart.
2. In order to determine polarity for different states of a character,
   a. there must be a fossil record of the groups in question.
   b. genetic sequence data must be available.
   c. an appropriate name for the taxonomic group must be selected.
   d. an outgroup must be identified.
3. Parsimony suggests that parental care in birds, crocodiles, and some dinosaurs
   a. evolved independently multiple times by convergent evolution.
   b. evolved once in an ancestor common to all three groups.
   c. is a homoplastic trait.
   d. is not a homologous trait.
4. Dating divergences with molecular clock data must be done cautiously, because
   a. the same change may have evolved independently multiple times by convergent evolution.
   b. rates of evolution may not have been constant across all branches of an evolutionary tree.
   c. molecular clocks run at different rates for different genes.
   d. the molecular clock is not a homologous trait.
5. A taxonomic group that contains species that have similar phenotypes due to convergent evolution is
   a. paraphyletic.
   b. monophyletic.
   c. polyphyletic.
   d. a good cladistic group.
6. Phylogenetic classification emphasizes key traits. Such traits often allow us to distinguish species immediately. Most of us are very familiar with dogs and cats, which are common household pets. What key traits can you think of that would always distinguish a dog from a cat?
7. The forelimb of a bird and the forelimb of a rhinoceros
   a. are homologous and symplesiomorphic.
   b. are not homologous but are symplesiomorphic.
   c. are homologous and synapomorphic.
   d. are not homologous but are synapomorphic.
8. If two organisms are in the same class, then they should
   a. belong to the same genus.
   b. both be members of the same order.
   c. belong to the same family.
   d. both be in the same phylum.
9. The three domains probably are
   a. paraphyletic.
   b. monophyletic.
   c. polyphyletic.
   d. homeoplastic.

Synthesize

1. Your friend wants to know what the big deal is—everyone knows that a rose is a rose, so why bother with the fancy Latin name *Rosa odorata*? What do you tell him?
2. Why do high rates of evolutionary change and a limited number of character states cause problems for parsimony analyses?
3. Birds and mammals have four-chambered hearts, whereas most living reptiles have three-chambered hearts. Such a fundamental difference suggests birds and mammals should be placed in the same clade, yet biologists now usually lump birds in with the reptile clade. Evaluate their decision to do so.
4. As noted in this chapter, cladistics is a widely utilized method of systematics, and our classification system (taxonomy) is increasingly becoming reflective of our knowledge of evolutionary relationships. Using birds as an example, discuss the advantages and disadvantages of recognizing them as reptiles versus as a group separate from and equal to reptiles.
5. Can you think of any alternatives to convergence to explain the presence of wings in birds and bats? What types of data might be used to test these hypotheses?

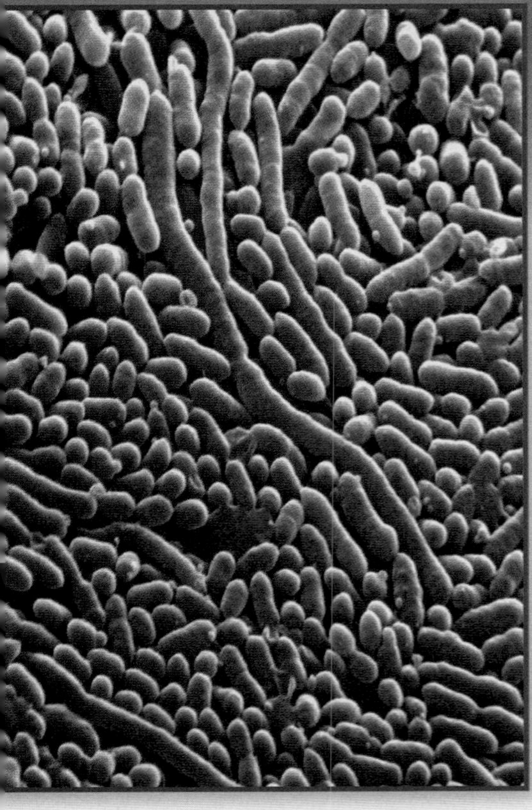

David M. Phillips/Science Source

23 Prokaryotes and Viruses

Learning Path

- **23.1** Prokaryotes Are the Most Ancient Organisms
- **23.2** Prokaryotes Have an Organized but Simple Structure
- **23.3** The Genetics of Prokaryotes Focuses on DNA Transfer
- **23.4** Prokaryotic Metabolism Is Diverse
- **23.5** Bacteria Cause Important Human Diseases
- **23.6** Viruses Are Not Organisms
- **23.7** Bacterial Viruses Infect by DNA Injection
- **23.8** Animal Viruses Infect by Endocytosis

Concept Overview

This Concept Overview provides a diagram of the overarching concepts that are covered in this chapter. These concepts will be expanded upon in the Concept Overview at the end of the chapter.

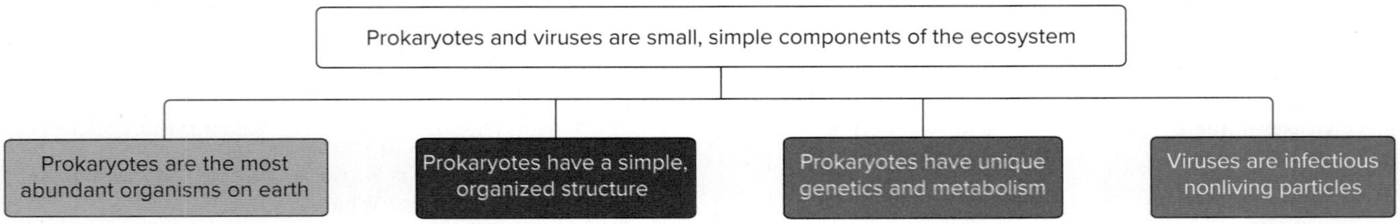

Introduction

Prokaryotes are the smallest and simplest of all organisms. Each is composed of a single cell that is too small to be seen by the naked eye and lacks internal compartments. Viewed with a microscope, they appear as seen in the micrograph on the previous page, with few notable external features, and typically in large numbers. There are actually 10 times as many prokaryotic cells as human cells in your body. Prokaryotes are thought to be the most ancient of living organisms, eukaryotes having arisen from them only after billions of years of life on Earth. Prokaryotic photosynthesis, for example, is thought to have been the source for the oxygen in the ancient Earth's atmosphere, and it still contributes significantly to oxygen production today. Indeed, the diverse eukaryotic organisms that currently live on Earth depend for their existence on prokaryotes, which make possible many of the essential functions of Earth's ecosystems. Thus, we start our study of biological diversity with an overview of prokaryotes—organisms essential to understanding all life on Earth, past and present.

We conclude the chapter with a brief look at viruses. Viruses are not organisms, but strands of nucleic acid encased in a protein coat that are not capable of independent life. All organisms appear to be infected by viruses, which can reproduce only within the cells of organisms. Viruses can have a major effect on the health of the organisms they infect, and they are responsible for many important diseases in plants and animals.

23.1 Prokaryotes Are the Most Ancient Organisms

Life originated on Earth over 3.5 BYA. What little we know of early life has come from tiny fossil microbes called **microfossils** collected from very old rock. Microfossils are fossilized forms of microscopic life. Many microfossils are small (1–2 μm in diameter) and appear to be single-celled, lack external appendages, and have little evidence of internal structure (figure 23.1). Thus, microfossils seem to resemble present-day prokaryotes.

Microfossils Indicate That the First Cells Were Probably Prokaryotic

LEARNING OBJECTIVE 23.1.1 Describe the basic features of archaea and bacteria.

The oldest known microfossils are 3.5 billion years old. The claim that these microfossils are the remains of living organisms is supported by isotopic data and by spectroscopic analysis that indicates they do contain complex carbon molecules. Whether these microscopic structures are true fossil cells is still controversial, and the identity of the prokaryotic groups represented by the various microfossils is still unclear. Arguments have been made for various bacteria, including cyanobacteria (described in section 23.2), being the microfossils in question, but definitive interpretation is difficult.

Today prokaryotes are the most abundant forms of life on Earth. Although thousands of different kinds of prokaryotes are recognized, many thousands more await proper identification. New molecular techniques have allowed scientists to identify and study microorganisms without culturing them. As a result, microbiologists have discovered thousands of new species that were never discovered or characterized because they could not be maintained in culture.

It is estimated that only 1 to 10% of all prokaryotic species are known and characterized, leaving 90 to 99% unknown and undescribed. Every place microbiologists look, new species are being discovered, often altering the way we think about prokaryotes. In the 1970s and 1980s, a new type of prokaryote was identified and analyzed that eventually led to the division of prokaryotes into two groups: the *Archaea* (formerly called Archaebacteria) (figure 23.2*a*) and the *Bacteria* (sometimes also called Eubacteria; figure 23.2*b*).

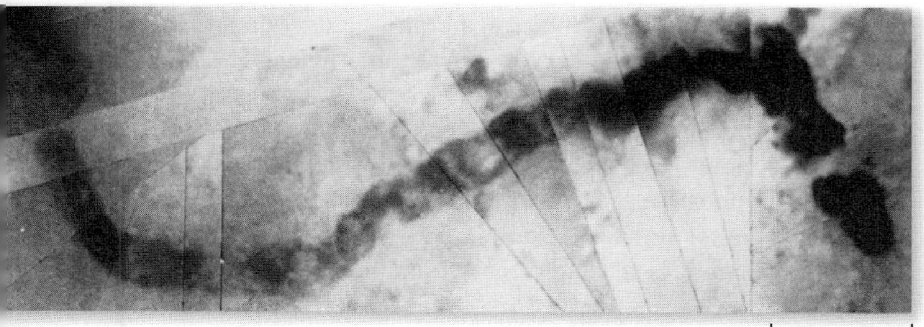

Figure 23.1 Evidence of bacterial fossils. Rocks between approximately 1 billion and 3.5 billion years old have tiny fossils resembling bacterial cells embedded within them.
J. William Schopf, UCLA

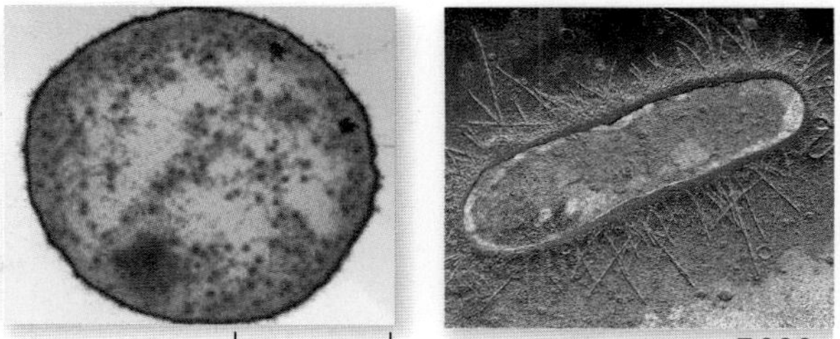

Figure 23.2 The two kinds of prokaryotes. Two of the three domains of life are prokaryotes. ***a.*** Archaea often live in extreme environments (*Methanococcoides burtonii*). ***b.*** Bacteria are as different from archaea as they are from eukaryotes (*Escherichia coli*).
(a): SPL/Science Source (b): BSIP SA/Alamy Stock Photo

Archaea and bacteria are the oldest, structurally simplest, and most abundant forms of life. They share a prokaryotic cellular organization with no membrane-bounded nucleus or endomembrane system. Prokaryotes were abundant for over a billion years before the first eukaryotes evolved. Early photosynthetic bacteria (cyanobacteria) altered the Earth's atmosphere by producing oxygen, which stimulated extreme bacterial and eukaryotic diversity. Prokaryotes are ubiquitous and live everywhere eukaryotes do, but they also thrive in places where no eukaryote could live.

Bacteria and archaea have been found in deep-sea caves, in volcanic rims, and deep within glaciers. They have even been recovered living beneath 435 m of ice in Antarctica! The first archaea characterized were extremophiles, living in hot springs that would cook other organisms, in hypersaline environments that would dehydrate others' cells, and in atmospheres rich in otherwise toxic gases such as methane or hydrogen sulfide. More recent surveys of microbes using modern sequencing technology have greatly expanded both the range and diversity of archaean life.

These harsh environments may be similar to the conditions present on the early Earth when life first began. It is likely that prokaryotes evolved to dwell in these harsh conditions early on and have retained the ability to exploit these areas as the rest of the atmosphere has changed.

Prokaryotes Are Fundamentally Different from Eukaryotes

LEARNING OBJECTIVE 23.1.2 Differentiate between prokaryotes and eukaryotes.

Two of the three domains of life are prokaryotes (figure 23.3). Prokaryotes differ from eukaryotes in numerous important features. These differences represent some of the most fundamental distinctions that separate any groups of organisms.

Unicellularity. With a few exceptions, prokaryotes are single-celled. In some types, individual cells adhere to one another within a matrix and form filaments; however, the

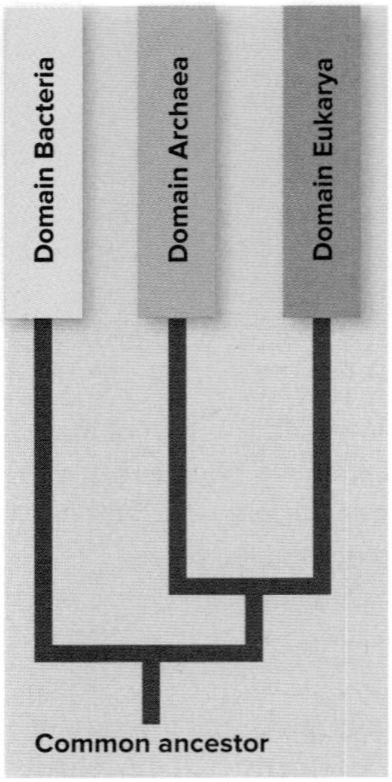

Figure 23.3 The three domains of life. The two prokaryotic domains, Archaea and Bacteria, are not closely related, though both are prokaryotes. In many ways, archaea more closely resemble eukaryotes than they resemble bacteria. This tree is based on rRNA sequences.

cells retain their individuality. Cyanobacteria, in particular, form such associations, but their cytoplasm is not interconnected. These filaments do have a common cell wall, however, making it difficult to isolate single cells.

In their natural environments, most bacteria appear to be capable of forming a complex community of different species called a **biofilm.** Although not a multicellular organism, a biofilm is more resistant to environmental stressors than is a simple colony of a single type of microbe.

Cell size. As new species of prokaryotes are discovered, investigators are finding that prokaryotic cells vary greatly in size, by as much as five orders of magnitude. The largest bacterial cells currently characterized are from *Thiomargarita namibia.* A single cell from this species is up to 750 μm across and is visible to the naked eye. Most prokaryotic cells are only 1 μm or less in diameter, and most eukaryotic cells are well over 10 times bigger. However, this generality is misleading, because there are very small eukaryotes and very large prokaryotes.

Chromosomes. Eukaryotic cells have a membrane-bounded nucleus containing linear chromosomes made up of both nucleic acids and histone proteins. Prokaryotes do not have membrane-bounded nuclei; instead, most have a single circular chromosome consisting of DNA and histone-like proteins in a *nucleoid* region of the cell. Exceptions include *Vibrio cholerae,* which has two circular chromosomes. Many prokaryotic cells also have accessory DNA molecules called plasmids.

Cell division and genetic recombination. Eukaryotic cells divide by mitosis, which involves spindles composed of microtubules. Prokaryotic cells divide mainly by binary fission (refer to chapter 10), which is also a form of asexual reproduction. True sexual reproduction occurs only in eukaryotes and requires the production of haploid gametes that fuse to form a diploid zygote (refer to chapter 11). Prokaryotes do have mechanisms that can transfer genetic material between cells. These mechanisms are collectively called *horizontal gene transfer.*

Internal compartmentalization. Prokaryotes do not have extensive membrane-bound organelles like eukaryotes; however, some species have structures that can isolate specific metabolic pathways, or provide storage. These structures may be bounded by a lipid bilayer or monolayer, or by a protein shell. The plasma membrane can also be extensively invaginated to isolate processes like cellular respiration or photosynthesis.

Flagella. Prokaryotic flagella are composed of a single protein fiber and not the 9 + 2 arrangement of microtubules found in eukaryotic flagella. Bacterial flagella have a twist, are rigid, and rotate to drive cell movement; eukaryotic flagella, on the other hand, have a whiplike motion. The flagella of bacteria and archaea have similar function but actually have independent evolutionary origins.

Metabolic diversity. Only one kind of photosynthesis occurs in eukaryotes, and it involves the release of oxygen. Photosynthetic bacteria have two basic patterns of photosynthesis: *oxygenic,* producing oxygen, and *anoxygenic,* non-oxygen-producing. Anoxygenic photosynthesis involves the formation of products such as sulfur and sulfate instead of oxygen.

Prokaryotic cells can also be *chemolithotrophic,* meaning that they use the energy stored in chemical bonds of inorganic molecules to synthesize carbohydrates; eukaryotes are not capable of this metabolic process.

Archaea Have Distinct Cell Architecture and Metabolism

LEARNING OBJECTIVE 23.1.3 Explain how archaea differ from bacteria and eukaryotes.

Archaea and bacteria both lack a membrane-bounded nucleus and the endomembrane system found in eukaryotes, but they are distinctly different cell types. The original argument for three domains of life was based on the analysis of rRNA sequences, but archaea have proven to have unique structural and metabolic features. Once thought to be found only in extreme environments, archaea have now been found essentially everywhere.

Archaeal cell structure. The most obvious structural difference in archaea concerns their membrane lipids, which differ from both bacterial and eukaryotic membrane lipids. The phospholipid structure described in chapter 5 holds for eukaryotes and bacteria but not for archaea. Archaeal membrane lipids are composed of nonpolar hydrocarbons called isoprenoids, not fatty acid chains, and these are joined to glycerol phosphate by ether linkages, not ester linkages (figure 23.4*a*). The stereoisomer of glycerol phosphate is also different from that in bacteria and eukaryotes, which use glycerol-3-phosphate, whereas archaea use glycerol-1-phosphate. Archaeal membrane lipids can be branched, can include cyclic compounds, and may even be organized into tetraethers that can be used to form a monolayer membrane (figure 23.4*b*).

The unique lipid composition may aid membrane stability in some extremophile archaea. The differences in membrane

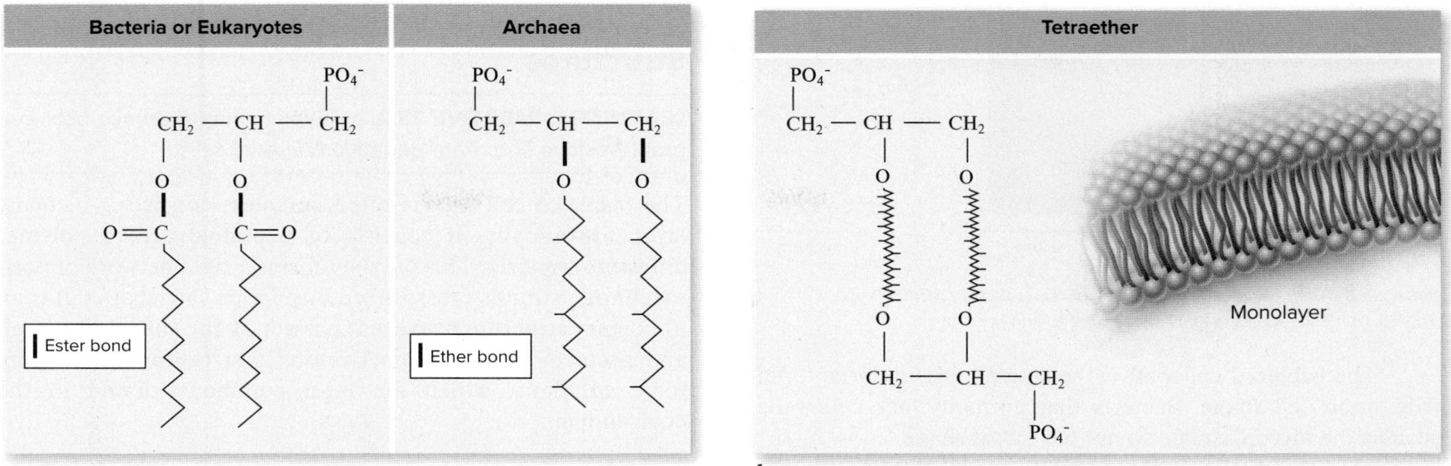

a. *b.*

Figure 23.4 Archaea membrane lipids. ***a.*** Archaeal membrane phospholipids differ from bacterial and eukaryotic membrane lipids in three ways: different stereoisomers of glycerol phosphate, isoprenoid hydrocarbons in archaea and fatty acids in bacteria and eukaryotes, and ether bonds in archaea instead of ester bonds. ***b.*** Archaeal membrane phospholipids can also be tetraethers instead of diethers. These may also include ringed structures (not shown) and be branched.

lipids between archaea and eukaryotes present an evolutionary conundrum in that these groups are considered to be more closely related than either is to bacteria. The basis for this so-called "lipid divide" is an active area of research for those interested in the origins of modern cells.

The other distinct structural feature of archaea is the lack of peptidoglycan in their cell walls. Archaeal cells have cell walls with diverse structure, including pseudomurein, which is similar to peptidoglycan. Most archaea also have a layer outside the cell wall, or even replacing it, called the **S-layer.** Although some bacteria have an analogous structure, it appears to be dispensable. The archaeal S-layer is composed of glycoproteins that form a rigid paracrystalline surface surrounding the cell, which can function to reinforce the plasma membrane and provide the cell structure.

Lastly, the archaeal flagella have proved to represent a third kind of flagella. Despite structural and functional similarities to bacterial flagella, they are not homologous.

Archaeal genetics and metabolism. Most archaea have a single origin of replication, like bacteria, but the proteins involved in initiation are more similar to eukaryotes. The same is true for gene expression: archaea show transcription–translation coupling, but the enzymatic machinery is more like that in eukaryotes.

Metabolically, methanogenesis seems to be limited to archaeal species, meaning all organically produced methane comes from archaea. There are also related species that can anaerobically oxidize methane. Ammonia oxidation was once thought to be only bacterial, but ammonia-oxidizing archaeal (AOA) species now appear to be major contributors to the global nitrogen cycle. The aerobic oxidation of ammonia, which produces N_2, appears to be another process limited to archaea. The advent of metagenomic studies of uncultured microbes should provide new insights into metabolic features among archaea.

REVIEW OF CONCEPT 23.1

The earliest microfossils are controversial, but there is evidence for life at least 3.5 BYA. Prokaryotes lack both a membrane-bounded nucleus and diverse organelles, and reproduce by binary fission. Archaea have unique membrane lipids with different nonpolar hydrocarbons, and ether linkages joined to a different stereoisomer of glycerol phosphate. The archaeal cell wall is also distinct, lacking peptidoglycan, and surrounded by an S-layer.

- *What features distinguish archaea from both bacteria and eukaryotes?*

23.2 Prokaryotes Have an Organized but Simple Structure

Prokaryotic cells are relatively simple, but they can be categorized based on cell shape. They also have some variations in structure that give them different staining properties for certain dyes. Other features are found in some types of cells but not in others.

Prokaryotes Have Three Basic Shapes: Bacilli, Cocci, and Spirilla

LEARNING OBJECTIVE 23.2.1 Describe the three basic shapes of prokaryotes.

Although it is an oversimplification, it is useful to divide bacteria based on easily definable morphologies. Most prokaryotes exhibit one of three basic shapes: rod-shaped, often called a *bacillus* (plural, *bacilli*); *coccus* (plural, *cocci*), spherical- or ovoid-shaped; and *spirillum* (plural, *spirilla*), long and helical-shaped, also called *spirochetes.*

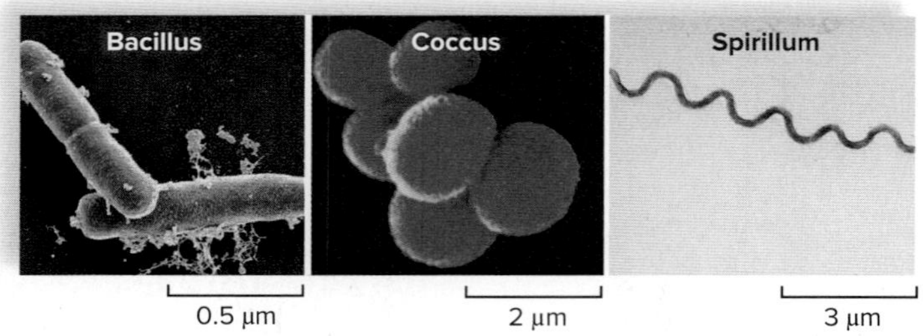

(bacillus): Dr. Gary Gaugler/SPL/Science Source; (coccus): CNRI/Science Source; (spirillum): Centers for Disease Control and Prevention

The bacterial cell wall is the single most important contributor to cell shape. Bacteria that normally lack cell walls, such as the mycoplasmas, do not have a set shape.

As diverse as their shapes may be, prokaryotic cells also have many different methods to allow movement through their environment. A *flagellum* or several flagella may be found on the outer surface of many prokaryotic cells. These structures are used to propel the organisms in a fluid environment. Some rod-shaped and spherical bacteria form colonies, adhering end-to-end after they have divided, forming chains. Some bacterial cells change into stalked structures or grow long, branched filaments. Some filamentous bacteria are capable of a gliding motion on solid surfaces, often combined with rotation around a longitudinal axis.

Bacteria Have a Variety of External Structures

LEARNING OBJECTIVE 23.2.2 Explain the difference between gram-positive and gram-negative bacterial cells.

The bacterial cell wall is often complex, consisting of many layers. Minimally, it consists of peptidoglycan, a polymer unique to bacteria. This polymer forms a rigid network of polysaccharide strands cross-linked by peptide side chains. It is an important structure because it maintains the shape of the cell and protects the cell from swelling and rupturing in hypotonic solutions, which are most commonly found in the environment.

Gram-positive and gram-negative bacteria

Two types of bacteria can be identified using a staining process called the **Gram stain,** hence their names. **Gram-positive** bacteria have a thicker peptidoglycan wall and stain a purple color, whereas the more common **gram-negative** bacteria contain less peptidoglycan and do not retain the purple-colored dye. These gram-negative bacteria can be stained with a red counterstain and then appear dark pink (figure 23.5).

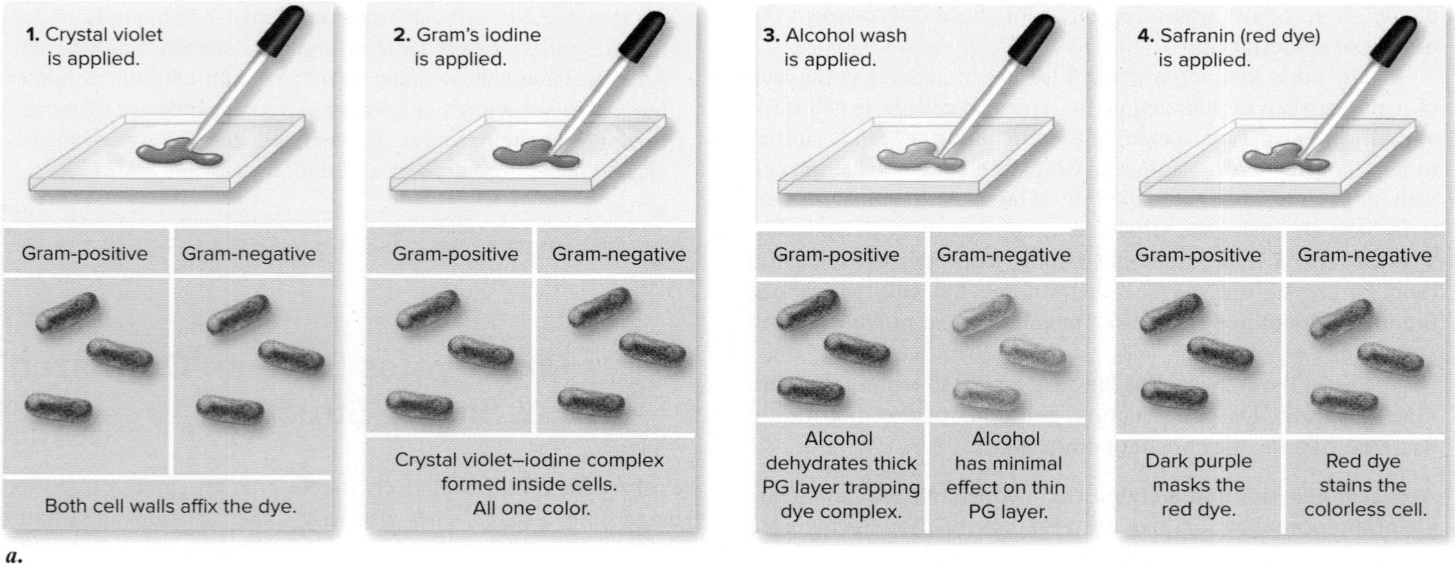

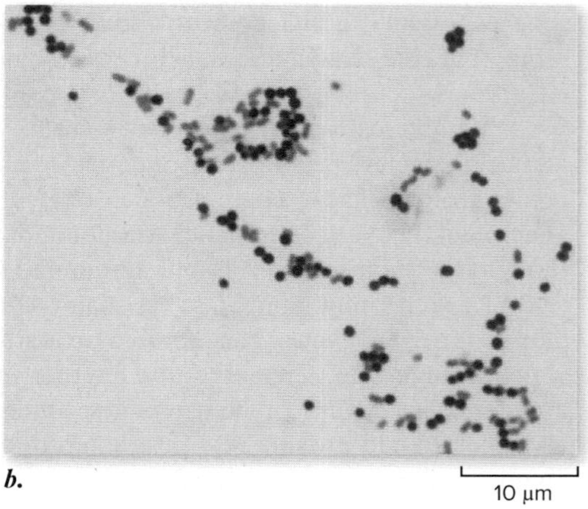

Figure 23.5 The Gram stain. ***a.*** The thick peptidoglycan (PG) layer encasing gram-positive bacteria traps crystal violet dye, so the bacteria appear purple in a Gram-stained smear, named after the bacteriologist Hans Christian Gram (1853–1938), who developed the technique. Because gram-negative bacteria have much less peptidoglycan (located between the plasma membrane and an outer membrane), they do not retain the crystal violet dye and so exhibit the red counterstain (usually a safranin dye). ***b.*** A micrograph showing the results of a Gram stain with both gram-positive and gram-negative cells.

(b): Lisa Burgess/McGraw Hill

In the gram-positive bacteria, the peptidoglycan forms a thick, complex network around the outer surface of the cell. This network also contains lipoteichoic and teichoic acid, which protrudes from the cell wall. In the gram-negative bacteria, a thin layer of peptidoglycan is sandwiched between the plasma membranes and a second outer membrane (figure 23.6). The outer membrane contains large molecules of **lipopolysaccharide,** lipids with polysaccharide chains attached. The outer membrane layer makes gram-negative bacteria resistant to many antibiotics that interfere with cell-wall synthesis in gram-positive bacteria. For example, penicillin acts to inhibit the cross-linking of peptidoglycan in a gram-positive cell wall, killing growing bacterial populations.

The capsule

In some bacteria, an additional gelatinous layer, the **capsule,** surrounds the cell wall. A capsule enables a prokaryotic cell to adhere to surfaces and to other cells and, most important, to evade an immune response by interfering with recognition by phagocytic cells. Therefore, a capsule often contributes to the ability of bacteria to cause disease.

Bacterial flagella and pili

Many kinds of prokaryotes have slender, rigid, helical flagella composed of the protein **flagellin** (figure 23.7). These flagella range from 3 to 12 μm in length and are very thin—only 10 to 20 nm thick. They are anchored in the cell wall and spin like a propeller, moving the cell through a liquid environment. Bacterial cells that have lost the genes for flagellin are not able to swim.

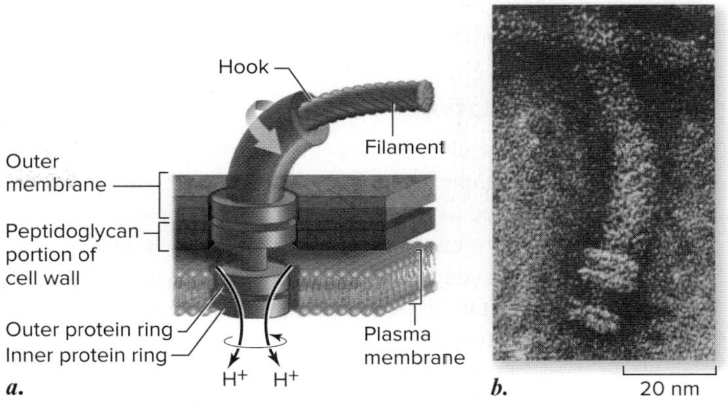

Figure 23.7 The flagellar motor of a gram-negative bacterium. ***a.*** A protein filament, composed of the protein flagellin, is attached to a protein rod that passes through a sleeve in the outer membrane and through a hole in the peptidoglycan layer to rings of protein anchored in the cell wall and plasma membrane, like rings of ball bearings. The rod rotates when the inner protein ring attached to the rod turns with respect to the outer ring fixed to the cell wall. The inner ring is an H^+ ion channel, a proton pump that uses the flow of protons into the cell to power the movement of the inner ring past the outer one. The membrane wall anchor of the flagellum is called the basal body. ***b.*** Electron micrograph of bacterial flagellum.

(b): Julius Adler

Pili (singular, *pilus*) are other hairlike structures that occur on the cells of some gram-negative prokaryotes. They are shorter than prokaryotic flagella and about 7.5 to 10 nm thick. Pili are more

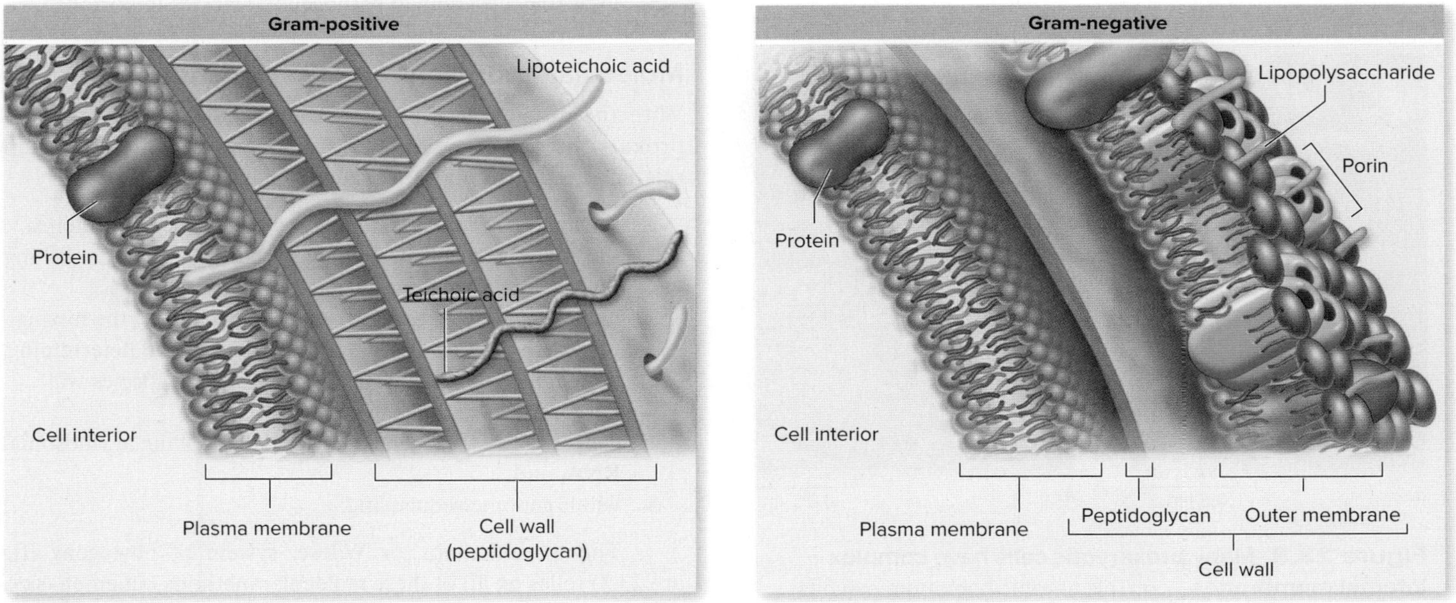

Figure 23.6 The structure of gram-positive and gram-negative cell walls. The gram-positive cell wall is much simpler, composed of a thick layer of cross-linked peptidoglycan chains. Molecules of lipoteichoic acid and teichoic acid are also embedded in the wall and exposed on the surface of the cell. The gram-negative cell wall is composed of multiple layers. The peptidoglycan layer is thinner than in gram-positive bacteria and is surrounded by an additional membrane composed of lipopolysaccharide. Porin proteins form aqueous pores in the outer membrane. The space between the outer membrane and peptidoglycan is called the periplasmic space.

important in adhesion than in movement, and they have a role in the exchange of genetic information (discussed in section 23.3).

Endospore formation

Some prokaryotes are able to form **endospores,** developing a thick wall around their genome and a small portion of the cytoplasm when they are exposed to environmental stress. These endospores are highly resistant to environmental stress, especially heat, and when environmental conditions improve, they can germinate and return to normal cell division to form new individuals after decades or even centuries.

The Bacterial Cell Interior Is Organized

LEARNING OBJECTIVE 23.2.3 Characterize the internal structure of prokaryotic cells.

While less complex than eukaryotes, prokaryotes do have organized substructure. There are no organelles common to all prokaryotes, but particular species do have specific organelles that provide specialized functions. For example, the *magnetosome,* found in bacteria that can move along a magnetic field, consist of spherical membranes containing iron oxide crystals. Both bacterial and archaeal species may also have internal membrane-bound storage structures, and some bacteria also contain cellular compartments bounded by a semipermeable protein shell. These **bacterial microcompartments (BMCs)** range in size from 40 to 400nm, and act to isolate a specific metabolic process, or to store a particular substance. While the structure is quite different, these compartments are functional analogs of eukaryotic organelles. Prokaryotes can also have infoldings of the plasma membrane that serve to segregate metabolic reactions (figure 23.8).

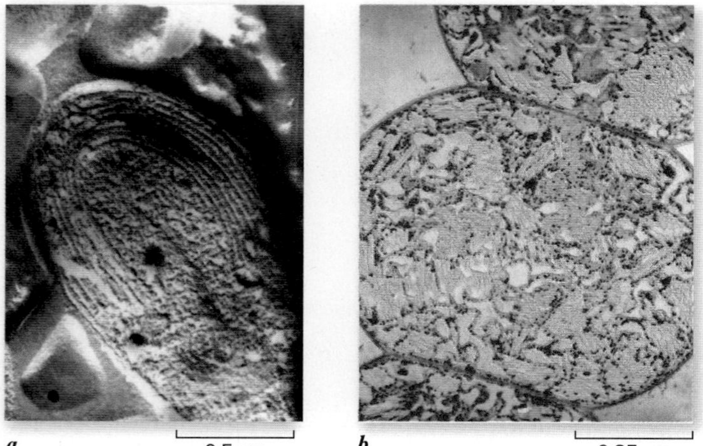

Figure 23.8 Many prokaryotic cells have complex internal membranes. ***a.*** This aerobic bacterium exhibits extensive respiratory membranes (long, dark curves that hug the cell wall) within its cytoplasm, not unlike those seen in mitochondria. ***b.*** This cyanobacterium has thylakoid-like membranes (ripple-like shapes along the edges and in the center) that provide a site for photosynthesis.

(a): Dr. W.J. Ingledew/Science Source; (b): Biophoto Associates/Science Source

While lacking a membrane-bounded nucleus, prokaryotic genomes are organized into a compact structure called a nucleoid. Prokaryotes also may have extrachromosmal DNAs that provide nonessential functions that may provide a selective advantage, such as antibiotic resistance.

Prokaryotic ribosomes differ in protein and RNA content and are smaller than those of eukaryotes. Antibiotics such as tetracycline and chloramphenicol that can bind to prokaryotic ribosomes and block protein synthesis do not bind to eukaryotic ribosomes.

Most Prokaryotes Have Not Been Characterized

LEARNING OBJECTIVE 23.2.4 Explain the methods used to classify prokaryotes.

Traditional methods of classification based on morphology are not easily applied to prokaryotes. Instead, biochemical and metabolic characteristics are used for overall classification schemes comparable to those used for other organisms.

Early approaches to classification

Early systems for classifying prokaryotes relied on differential stains such as the Gram stain and differences in the observable phenotype of the organism. Key characteristics once used in classifying prokaryotes were

1. photosynthetic or nonphotosynthetic,
2. motile or nonmotile,
3. unicellular or colony-forming or filamentous,
4. formation of spores or division by transverse binary fission, and
5. importance as human pathogens or lack of importance.

Molecular approaches to classification

With the development of genetic and molecular approaches, prokaryotic classifications may help reflect true evolutionary relatedness. Molecular approaches include

1. the analysis of the amino acid sequences of key proteins,
2. the analysis of nucleic acid–base sequences by establishing the percent of guanine (G) and cytosine (C),
3. nucleic acid hybridization, which is essentially the mixing of single-stranded DNA from two species and determining the amount of base-pairing (closely related species will have more bases pairing),
4. gene and RNA sequencing, especially looking at ribosomal RNA, and
5. whole-genome sequencing.

The three-domain, or Woese, system of phylogeny (figure 23.3) relies on all of these molecular methods but emphasizes the comparison of rRNA sequences to establish the evolutionary relatedness of all organisms. The rRNA sequences were chosen for their high degree of evolutionary conservation to address questions about these most ancient splits in the tree of life.

Based on these sorts of molecular data, several groupings of prokaryotes have been proposed. The most widely accepted is that presented in *Bergey's Manual of Systematic Bacteriology,* second edition, which is published in five volumes (figure 23.9). At the same time, large-scale sequencing of randomly sampled collections of bacteria shows an incredible amount of diversity. Although it has always been challenging to assign bacteria to species, these new data indicate that the vast majority of bacteria have never been cultured and studied in any detail. The field is in a state of flux as attempts are being made to define the nature of bacterial species.

REVIEW OF CONCEPT 23.2

The three basic shapes of prokaryotes are rod-shaped, spherical, and spiral-shaped. Bacteria have a peptidoglycan cell wall, which is the basis for the Gram stain. Gram-positive bacteria have a thick cell wall relative to gram-negative species. Many also have an external capsule, and some have flagella and pili. Some species can form heat-resistant endospores. Prokaryotic cells do not have membrane-bounded organelles but may have an organized interior, including infolding of the plasma membrane. Prokaryotic DNA is localized in a nucleoid region. Classification is aided by DNA analysis, but most prokaryotes remain unidentified.

- *What would be the simplest method to determine whether two bacteria belong to the same species?*

23.3 The Genetics of Prokaryotes Focuses on DNA Transfer

In sexually reproducing populations, traits are transferred from parent to child in what is called vertical transmission. Prokaryotes do not reproduce sexually, but they can exchange DNA between different cells, in horizontal transmission. This horizontal gene transfer can be by **conjugation,** mediated by plasmids; by **transduction,** mediated by viruses; and by **transformation,** the direct uptake of DNA.We will concentrate on bacterial systems, primarily *E. coli,* but these processes appear to be acting across all prokaryotes.

Conjugation Depends on the Presence of a Conjugative Plasmid

LEARNING OBJECTIVE 23.3.1 Describe how conjugation may be used to map the genes of bacteria.

Plasmids often encode functions that confer an advantage to the cell, but are not essential functions. Some plasmids can be transferred from one cell to another via conjugation. The best-known transmissible plasmid is called the **F plasmid** (the "F" stands for "fertility factor"). Cells containing F plasmids are termed F^+ cells, and cells that lack the F plasmid are F^- cells. The F plasmid encodes proteins necessary for transfer, but uses cellular machinery for replication.

F plasmid transfer

The F plasmid contains a DNA replication origin and encodes protein subunits that assemble on the surface of the bacterial cell to form a hollow pilus that is necessary for plasmid transfer (figure 23.10*a*). Transfer begins when the F plasmid binds to a site on the interior of the F^+ cell near the pilus, now called a *conjugation bridge.* Then, by a process called *rolling-circle replication,* the F plasmid begins to copy its DNA at the binding point. As it is replicated, the displaced single strand of the plasmid passes into the other cell. There, a complementary strand is added, creating a new, stable F plasmid (figure 23.10*b*).

DNA transfer by Hfr cells

The F plasmid can also integrate into the host chromosome by recombining with it. This process is similar to the recombination (crossing over) between homologs that occurs during meiosis in eukaryotes (refer to chapter 13). Recombination requires regions of homology, which in this case are regions called insertion sequences (IS) that exist in both the F plasmid and the *E. coli* chromosome. Recombination between the F plasmid and the chromosome integrates the plasmid into the chromosome (figure 23.11). This process uses host proteins and depends on the presence of IS elements.

A cell with an integrated F plasmid is called an Hfr cell (for "high frequency of recombination"), because F plasmid can now transfer chromosomal DNA. Because the origin of transfer is in the middle of the integrated plasmid, usually only a portion of the chromosome and not all of the F plasmid is transferred. It takes around 100 minutes to transfer the entire chromosome, and the conjugation bridge usually does not survive that long. This leads to transfer of portions of donor chromosome that can replace regions of the recipient chromosome with two recombination events (figure 23.12).

The original *E. coli* genetic map was constructed by taking advantage of the incomplete nature of Hfr gene transfer. As the transfer is progressive, the farther from the origin of transfer, the later the transfer. This allowed mapping based on the time of entry of different genes (figure 23.13). Genes are followed by using a donor with wild-type alleles that replace mutant alleles in the recipient, with mating interrupted at time points. These experiments have shown that the genetic map of E. coli is circular, with minutes as the units of the map, and the entire chromosome is 100 minutes long.

The F plasmid can also excise itself by reversing the integration process (see figure 23.11). If excision is inaccurate, the F plasmid can pick up some chromosomal DNA in the process to create an F′ plasmid. Normal transfer of this F′ will produce a cell that is diploid for the material carried by the plasmid. Such **partial diploids** were used in classical bacterial genetics for a variety of purposes.

Transfer of antibiotic resistance

One of the reasons that antibiotic resistance is such a huge problem is that many conjugative plasmids that can transfer DNA have picked up antibiotic resistance genes. These so-called resistance plasmids, or **R plasmids,** can spread antibiotic resistance rapidly, sometimes even between different species. R plasmids often acquire antibiotic resistance genes through transposable elements (described in chapter 18). As these elements move back and forth between a chromosome and plasmids, they can transfer antibiotic resistance genes. If a conjugative plasmid picks up these genes, the bacterium carrying it has a selective advantage in the presence of antibiotics. This kind of horizontal gene transfer can

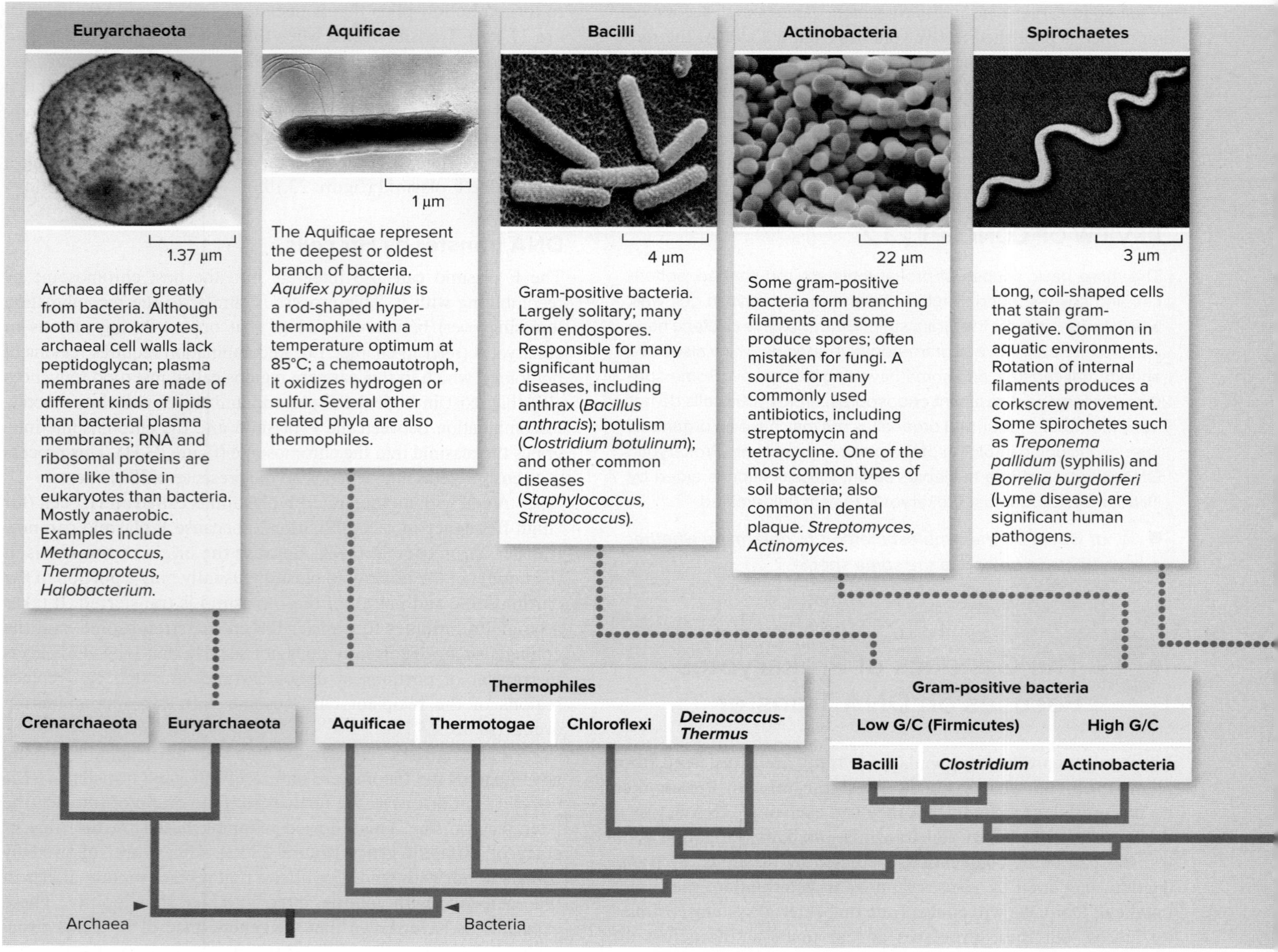
Euryarchaeota
1.37 μm
Archaea differ greatly from bacteria. Although both are prokaryotes, archaeal cell walls lack peptidoglycan; plasma membranes are made of different kinds of lipids than bacterial plasma membranes; RNA and ribosomal proteins are more like those in eukaryotes than bacteria. Mostly anaerobic. Examples include *Methanococcus, Thermoproteus, Halobacterium.*
Aquificae
1 μm
The Aquificae represent the deepest or oldest branch of bacteria. *Aquifex pyrophilus* is a rod-shaped hyper-thermophile with a temperature optimum at 85°C; a chemoautotroph, it oxidizes hydrogen or sulfur. Several other related phyla are also thermophiles.
Bacilli
4 μm
Gram-positive bacteria. Largely solitary; many form endospores. Responsible for many significant human diseases, including anthrax (*Bacillus anthracis*); botulism (*Clostridium botulinum*); and other common diseases (*Staphylococcus, Streptococcus*).
Actinobacteria
22 μm
Some gram-positive bacteria form branching filaments and some produce spores; often mistaken for fungi. A source for many commonly used antibiotics, including streptomycin and tetracycline. One of the most common types of soil bacteria; also common in dental plaque. *Streptomyces, Actinomyces.*
Spirochaetes
3 μm
Long, coil-shaped cells that stain gram-negative. Common in aquatic environments. Rotation of internal filaments produces a corkscrew movement. Some spirochetes such as *Treponema pallidum* (syphilis) and *Borrelia burgdorferi* (Lyme disease) are significant human pathogens.
Thermophiles
Crenarchaeota
Euryarchaeota
Aquificae
Thermotogae
Chloroflexi
Deinococcus-Thermus
Gram-positive bacteria
Low G/C (Firmicutes)
High G/C
Bacilli
Clostridium
Actinobacteria
Archaea
Bacteria

also transfer genes involved in virulence. The pathogenic *E. coli* O157:H7 (responsible for hemorrhagic gastroenteritis) arose by transfer of virulence genes in this way.

Viruses Transfer DNA by Transduction

LEARNING OBJECTIVE 23.3.2 Explain how general transducing phage arise.

Horizontal transfer of DNA can also be mediated by bacteriophage. In **generalized transduction,** virtually any gene can be transferred between cells; in **specialized transduction,** only a few genes are transferred.

Generalized transduction

Generalized transduction is really an accident of the biology of some types of lytic phage (see section 23.7). In these viruses, the viral genome is packaged into new phage heads by a mechanism called "headful packaging." DNA in long repeating units of viral genome is stuffed into a phage head until it is full; then it is cleaved and another head "stuffed." If the packaging machinery starts with a piece of chromosomal DNA, then it will fill viral heads with this DNA. When these viral particles go on to infect another cell, they will transfer this chromosomal DNA, and it can be incorporated by homologous recombination, as we already seen (figure 23.12).

Generalized transduction has also been used for mapping purposes in *E. coli,* although the logic is different from that in conjugation. In transduction, the closer together two genes are, the more likely it is that they will be transferred in a single transduction event. Prior to the advent of recombinant DNA technology, transduction was the best method to construct specialized strains of *E. coli.*

Specialized transduction

Specialized transduction is limited to lysogenic phage (see section 23.7). When phage lambda (λ) infects an *E. coli* cell and becomes a

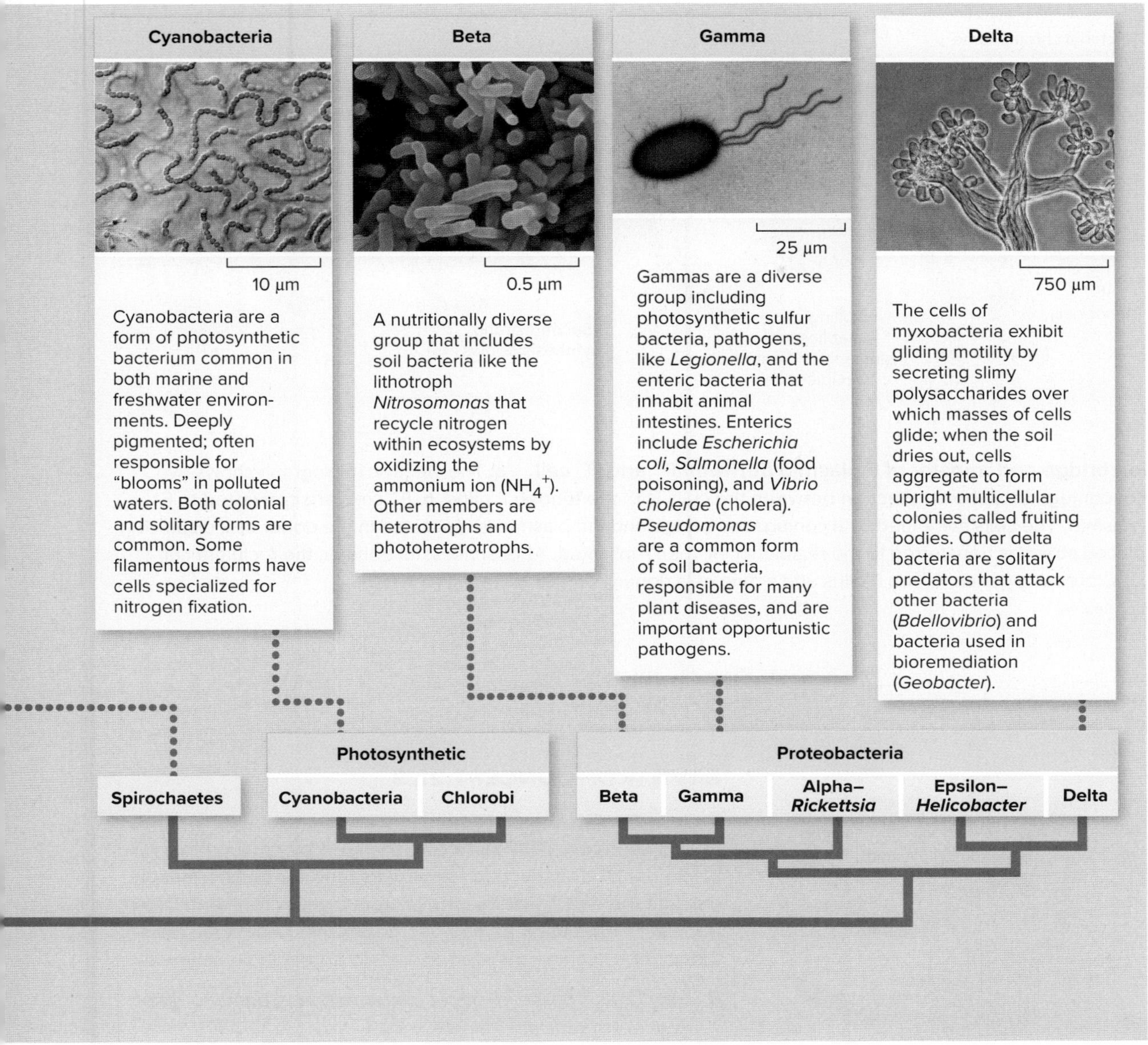
Cyanobacteria
10 µm
Cyanobacteria are a form of photosynthetic bacterium common in both marine and freshwater environments. Deeply pigmented; often responsible for "blooms" in polluted waters. Both colonial and solitary forms are common. Some filamentous forms have cells specialized for nitrogen fixation.
Beta
0.5 µm
A nutritionally diverse group that includes soil bacteria like the lithotroph Nitrosomonas that recycle nitrogen within ecosystems by oxidizing the ammonium ion (NH_4^+). Other members are heterotrophs and photoheterotrophs.
Gamma
25 µm
Gammas are a diverse group including photosynthetic sulfur bacteria, pathogens, like Legionella, and the enteric bacteria that inhabit animal intestines. Enterics include Escherichia coli, Salmonella (food poisoning), and Vibrio cholerae (cholera). Pseudomonas are a common form of soil bacteria, responsible for many plant diseases, and are important opportunistic pathogens.
Delta
750 µm
The cells of myxobacteria exhibit gliding motility by secreting slimy polysaccharides over which masses of cells glide; when the soil dries out, cells aggregate to form upright multicellular colonies called fruiting bodies. Other delta bacteria are solitary predators that attack other bacteria (Bdellovibrio) and bacteria used in bioremediation (Geobacter).
Spirochaetes
Photosynthetic
Cyanobacteria
Chlorobi
Proteobacteria
Beta
Gamma
Alpha–Rickettsia
Epsilon–Helicobacter
Delta

Figure 23.9 Some major clades of prokaryotes. The classification adopted here was that used in *Bergey's Manual of Systematic Bacteriology,* second edition, 2001. G/C refers to %G/C in genome.

(euryarchaeota): SPL/Science Source; (aquificae): Prof. Dr. R. Rachel and Prof. Dr. K. O. Stetter, University of Regensburg, Regensburg, Germany; (bacilli): Andrew Syred/Science Source; (actinobacteria): Microfield Scientific Ltd/Science Source; (spirochaetes): Alfred Pasieka/Science Source; (cyanobacteria): Don Rubbelke/McGraw Hill; (beta, gamma): Dennis Kunkel Microscopy/Science Source; (delta): Prof. Dr. Hans Reichenbach, Helmholtz Centre for Infection Research, Braunschweig

lysogen, its genome is integrated into the chromosome as a prophage. This integration event is similar to the integration of the F plasmid, except that λ integrates into a specific site using phage proteins. The prophage encodes the functions necessary to excise itself and undergo lytic growth. If this excision event is imprecise, it will take chromosomal DNA with it, producing a specialized transducing phage. These specialized transducing phages carry only genes adjacent to the integration site, unlike generalized transducing phage that can carry any part of the bacterial chromosome.

Transformation Is the Uptake of DNA Directly from the Environment

LEARNING OBJECTIVE 23.3.3 Distinguish between natural and artificial transformation.

Transformation occurs when one bacterial cell is damaged and releases fragmented DNA into the surrounding environment. This DNA can be taken up by another cell and incorporated into its genome by homologous recombination, as in conjugation and transduction (see figure 23.12). When the uptake occurs under natural conditions, it is termed natural transformation. Some species of both gram-positive and gram-negative bacteria exhibit natural transformation, although by different mechanisms.

The proteins involved in the process of natural transformation are all encoded by the bacterial chromosome. This implies that natural transformation is a mechanism of DNA exchange that evolved as part of normal cellular machinery. The transfer of chromosomal DNA by either conjugation or transduction are more accidents of plasmid or phage biology, respectively.

When transformation is accomplished in the laboratory, it is called artificial transformation. Artificial transformation is critical for gene cloning and DNA manipulation (refer to chapter 17).

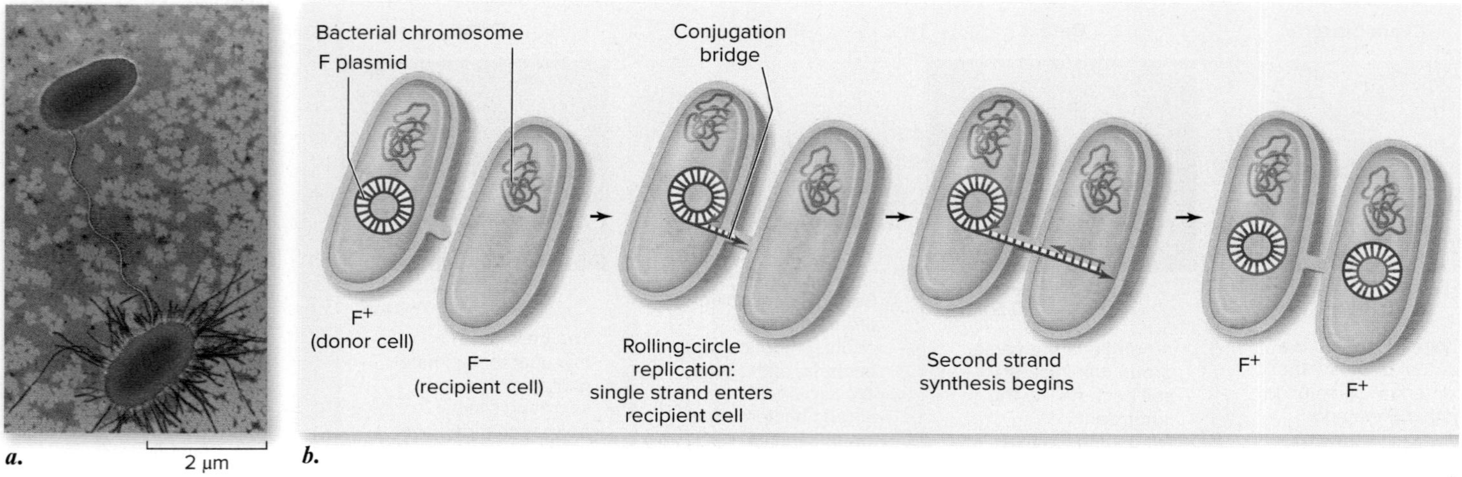

Figure 23.10 Conjugation bridge and transfer of F plasmid between F$^+$ and F$^-$ cell. ***a.*** The electron micrograph shows two *E. coli* cells caught in the act of conjugation. The connection between the cells is the extended F pilus. ***b.*** F$^-$ cells are converted to F$^+$ cells by the transfer of the F plasmid. The cells are joined by a conjugation bridge and the plasmid is replicated in the donor cell, displacing one parental strand. The displaced strand is transferred to the recipient cell, then replicated. After successful transfer, the recipient cell becomes an F$^+$ cell capable of expressing genes for the F pilus and acting as a donor.

(a): Dennis Kunkel Microscopy/Science Source

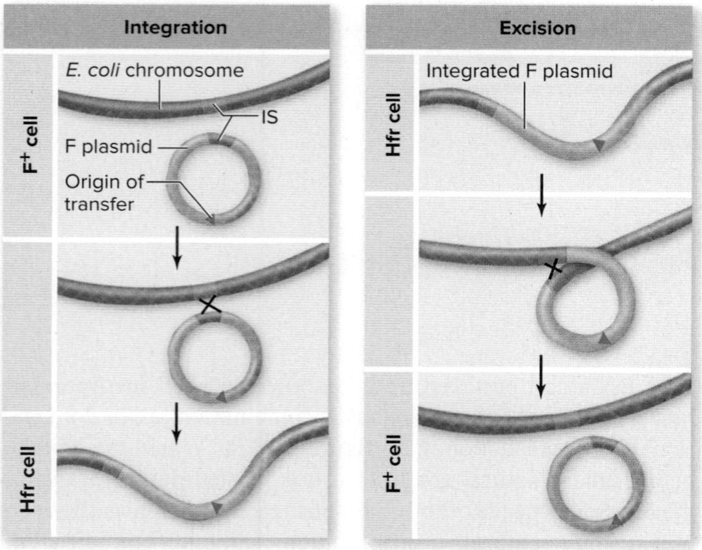

Figure 23.11 Integration and excision of F plasmid. The F plasmid contains short insertion sequences (IS) that also exist in the chromosome. This allows the plasmid to pair with the chromosome, and a single recombination event between two circles leads to a larger circle. This integrates the plasmid into the chromosome, creating an Hfr cell, as shown on the left. The process is reversible, because the IS sequences in the integrated plasmid can pair, and now a recombination event will return the two circles and convert the Hfr back to an F$^+$ cell, as shown on the right.

REVIEW OF CONCEPT 23.3

Prokaryotic DNA exchange is horizontal, from donor cell to recipient cell. DNA can be exchanged by conjugation via plasmids, by transduction via viruses, and by transformation through the direct uptake of DNA from the environment. DNA transfer can be used to map bacterial genes. Antibiotic resistance genes can be transferred, rapidly spreading resistance.

- *How does transfer of genetic information in bacteria differ from eukaryotic sex?*

23.4 Prokaryotic Metabolism Is Diverse

The variation seen in prokaryotes manifests itself most noticeably in biochemical rather than morphological diversity. Wide variation has been found in the types of metabolism prokaryotes exhibit, especially in the means by which they acquire energy and carbon.

Prokaryotes Include Both Autotrophs and Heterotrophs

LEARNING OBJECTIVE 23.4.1 Compare the different ways that prokaryotes acquire carbon and energy.

Like all living organisms, prokaryotes' two most basic needs are for energy and carbon. We categorize organisms into autotrophs

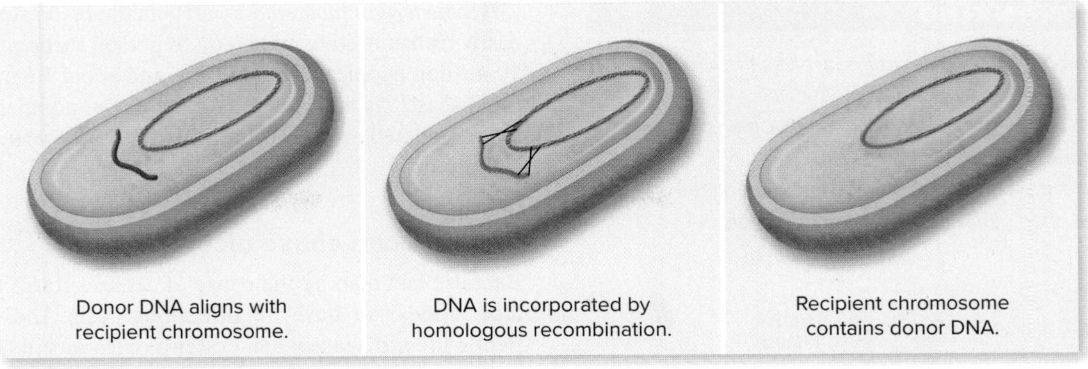
Donor DNA aligns with recipient chromosome.
DNA is incorporated by homologous recombination.
Recipient chromosome contains donor DNA.

Figure 23.12 Integration of transferred DNA. Fragments of donor DNA can be transferred by conjugation, transduction, or transformation. This donor DNA can replace the corresponding region in the recipient chromosome by two homologous recombination events.

and heterotrophs, based on how they acquire carbon and energy. These can then be further broken down into four metabolic lifestyles.

Photoautotrophs can acquire energy from the Sun, and convert carbon dioxide to organic form (fix carbon). Prokaryotes exhibit a variety of forms of photosynthesis ranging from cyanobacteria, which function similarly to green plants, and produce oxygen (refer to chapter 8), to sulfur bacteria that use light energy and H_2S as an electron donor, producing elemental sulfur. These so-called purple and green sulfur bacteria have a simpler, single photosystem form of photosynthesis that does not produce oxygen (is anoxygenic).

Some archaeal species exhibit an even simpler form of photosynthesis. This involves a single protein, bacteriorhodopsin, that uses energy from light to translocate protons across a membrane. This then provides a proton motive force for ATP synthesis. Recent surveys of microbial diversity in marine ecosystems have found a new relative of the rhodopsin family called proteorhodopsin. This raises the possibility that photosynthesis in marine systems may be more widespread and complex than previously thought.

Unlike the purple and green sulfur bacteria, the purple and green nonsulfur bacteria are **photoheterotrophs.** These species use light energy but obtain carbon from organic molecules, such as carbohydrates or alcohols that have been produced by other organisms.

Chemolithoautotrophs obtain energy by oxidizing inorganic substances. Nitrifiers oxidize ammonia or nitrite to obtain energy, producing the nitrate that is taken up by plants. This process of nitrification is essential for both marine and terrestrial ecosystems. There are both bacterial and archaeal species that can oxidize ammonia.

Other chemolithoautotrophs oxidize sulfur, hydrogen gas, and other inorganic molecules. On the dark ocean floor at depths of 2500 m, entire ecosystems subsist on prokaryotes that oxidize hydrogen sulfide as it escapes from thermal vents.

Most prokaryotes are actually much like you and are classified as **chemoheterotrophs.** That is, they obtain both carbon and energy from organic molecules. This group includes decomposers and most pathogens.

REVIEW OF CONCEPT 23.4

Prokaryotes have both autotrophic and heterotrophic species. Photoautotrophs use light as an energy source; chemolithoautotrophs oxidize inorganic compounds. Photoheterotrophs use light as an energy source and organic compounds as carbon sources. Chemoheterotrophs use organic compounds for both energy and carbon.

- *Why is metabolism a better way than morphology to characterize prokaryotes?*

23.5 Bacteria Cause Important Human Diseases

In the early 20th century, before the discovery and widespread use of antibiotics, infectious diseases killed nearly 20% of all U.S. children before they reached the age of five. Sanitation and antibiotics considerably improved the situation (table 23.1).

A Wide Variety of Diseases Are Caused by Bacteria

LEARNING OBJECTIVE 23.5.1 Describe several important human diseases caused by bacteria.

Bacteria have many different methods to spread through a susceptible population. Tuberculosis and many other bacterial diseases of the respiratory tract are mostly spread through the air in droplets of mucus or saliva. Diseases such as typhoid fever, cholera, and dysentery are spread by fecal contamination of food or water. Lyme disease and Rocky Mountain spotted fever are spread to humans by tick vectors. Many important sexually transmitted diseases are bacterial.

SCIENTIFIC THINKING

Hypothesis: *Conjugation using Hfr strains involves the linear transfer of information from donor to recipient cell.*

Prediction: *If there is a linear transfer of information, then different markers should appear in a time sequence.*

Test: *Mating strains are agitated at time points to break the conjugation bridge, then plated to determine genotype.*

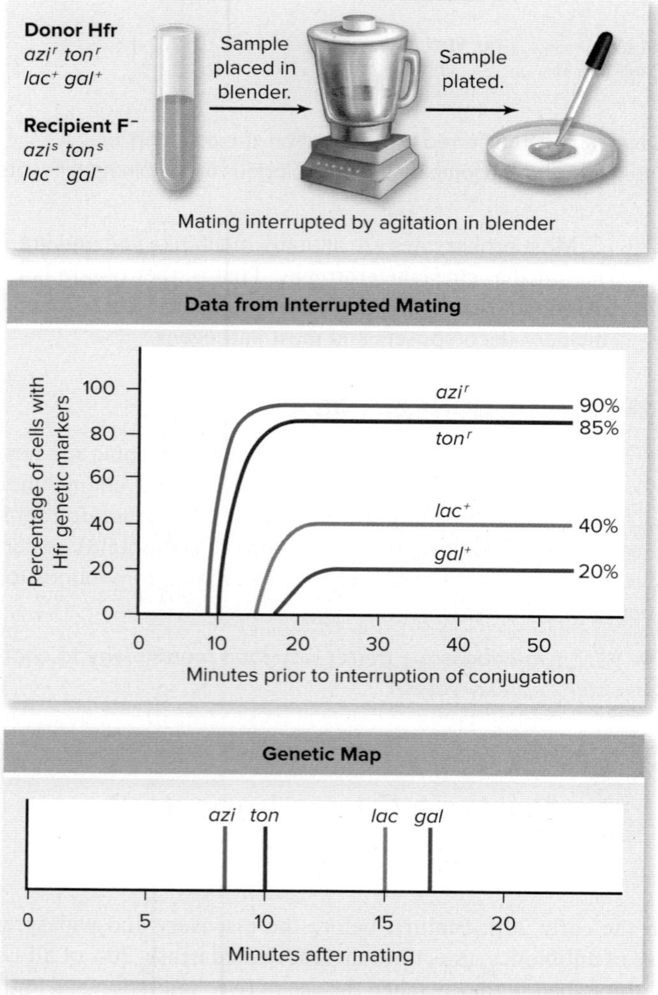

Donor Hfr
azi^r ton^r
lac^+ gal^+
Recipient F^−
azi^s ton^s
lac^− gal^−
Sample placed in blender.
Sample plated.
Mating interrupted by agitation in blender
Data from Interrupted Mating
Percentage of cells with Hfr genetic markers
100
80
60
40
20
0
azi^r
ton^r
lac^+
gal^+
90%
85%
40%
20%
0
10
20
30
40
50
Minutes prior to interruption of conjugation
Genetic Map
azi
ton
lac
gal
0
5
10
15
20
Minutes after mating

Result: *The different genes from the donor strain appear in a linear time sequence.*

Conclusion: *The transfer of genetic information is linear. This sequence can be used to construct a genetic map ordering the genes on the chromosomes.*

Further Experiments: *Can other methods of DNA exchange also be used for genetic mapping?*

Figure 23.13 Interrupted mating experiment allows construction of genetic map.

Tuberculosis has infected humans for all of recorded history

Tuberculosis (TB) is second only to HIV/AIDS as the cause of death worldwide from an infectious agent. The TB bacillus (*Mycobacterium tuberculosis*) afflicts the respiratory system and is easily transmitted from person to person through the air. In 2017, 10 million people contracted TB, and about 1.6 million died from the disease. As much as 25% of the global population is estimated to be infected with TB, but only 5 to 15% will develop disease symptoms.

Bacteria can cause ulcers

Bacteria can also be the cause of disease states that, on the surface, appear to have no infectious basis. Individuals develop peptic ulcer disease when crater-like lesions in the gastrointestinal tract are exposed to peptic acid. Ulcers can be caused by drugs, as well as by some tumors of the pancreas that cause an oversecretion of peptic acid. In 1982, a bacterium named *Campylobacter pylori* (now named *Helicobacter pylori*) was isolated from gastric juices. Over the years, evidence has accumulated that this bacterium is actually the causative agent in the majority of cases of peptic ulcer disease. Antibiotic therapy can now eliminate *H. pylori,* treating the cause of the disease, and not just the symptoms.

Bacterial biofilms are involved in tooth decay

Bacteria and other organisms form mixed cultures on surfaces that are extremely difficult to treat. On teeth, this biofilm, or plaque, consists largely of bacterial cells in a polysaccharide matrix. Plaque is a complex environment consisting of many different species, most of which are not necessarily bad. Tooth decay, or dental caries, is caused by certain bacteria, primarily *Streptococcus sobrinus* and *S. mutans,* fermenting simple sugars to lactic acid. The acid enhances breakdown of the hydroxyapatite that makes tooth enamel hard. As the enamel degenerates, the remaining soft matrix of the tooth is vulnerable to infection.

REVIEW OF CONCEPT 23.5

Many human diseases, including tuberculosis, streptococcal and staphylococcal infection, and sexually transmitted diseases, are due to bacteria. The causative agent of most peptic ulcers is *Helicobacter pylori,* an inhabitant of the digestive tract. Bacteria are responsible for tooth decay as well as many STDs, including gonorrhea, syphilis, and chlamydia.

- *Why is infection by most pathogens not fatal?*

23.6 Viruses Are Not Organisms

Viruses are not considered organisms, because they lack many of the features associated with life, including cellular structure and independent metabolism or replication. Viral particles are called virions, and they are generally described as active or inactive. Because of their disease-producing potential, however, viruses are important biological entities. Viruses cause such diseases as AIDS, severe acute respiratory syndrome (SARS), and hemorrhagic fever, and some cause certain forms of cancer.

TABLE 23.1 Important Human Bacterial Diseases

Disease	Pathogen	Vector/Reservoir	Epidemiology
Anthrax	*Bacillus anthracis*	Animals, including processed hides	Bacterial infection that can be transmitted through contact or ingestion; rare except in sporadic outbreaks; may be fatal
Botulism	*Clostridium botulinum*	Improperly prepared food	Contracted through ingestion or contact with wound; produces acute toxic poison; can be fatal
Chlamydia	*Chlamydia trachomatis*	Humans, sexually transmitted disease (STD)	Urogenital infections with possible spread to eyes and respiratory tract; increasingly common over the past 20 years
Cholera	*Vibrio cholerae*	Human feces, plankton	Causes severe diarrhea that can lead to death by dehydration; a major killer in times of crowding and poor sanitation
Dental caries	*Streptococcus mutans, Streptococcus sabrinus*	Humans	A dense collection of these bacteria on the surface of teeth leads to secretion of acids that destroy minerals in tooth enamel
Diphtheria	*Corynebacterium diphtheriae*	Humans	Acute inflammation and lesions of respiratory membranes; spread through respiratory droplets; vaccine available
Gonorrhea	*Neisseria gonorrhoeae*	Humans only	STD, on the increase worldwide, but left untreated it can be fatal
Hansen disease (leprosy)	*Mycobacterium leprae*	Humans, feral armadillos	Chronic infection of the skin; worldwide about 10–12 million infected; spread through contact with infected individuals
Lyme disease	*Borrelia burgdorferi*	Ticks, deer, small rodents	Spread through bite of infected tick; lesion followed by malaise, fever, fatigue, pain, stiff neck, and headache
Peptic ulcers	*Helicobacter pylori*	Humans	Originally thought to be caused by stress or diet, most peptic ulcers now appear to be caused by this bacterium
Plague	*Yersinia pestis*	Fleas of wild rodents: rats and squirrels	Killed 25% of European population in the 14th century; endemic in wild rodent populations of western United States today
Pneumonia	*Streptococcus, Mycoplasma, Chlamydia, Haemophilus*	Humans	Acute infection of the lungs; often fatal without treatment; vaccine for streptococcal pneumonia available
Tuberculosis	*Mycobacterium tuberculosis*	Humans, badgers	Acute bacterial infection of the lungs, lymph, and meninges; incidence is on the rise, including antibiotic-resistant strains
Typhoid fever	*Salmonella typhi*	Humans	Systemic disease of worldwide incidence; fewer than 500 cases a year in the U.S.; spread through contaminated water or foods; vaccines available
Typhus	*Rickettsia typhi*	Lice, rat fleas, humans	Historically, a major killer in times of crowding and poor sanitation; transmitted through the bite of infected lice and fleas

Viruses Are Strands of Nucleic Acids Encased in a Protein Coat

LEARNING OBJECTIVE 23.6.1 Describe the different structural forms of viruses.

All viruses have the same basic structure—a core of nucleic acid surrounded by protein. This structure lacks cytoplasm, and it is not a cell. Individual viruses contain only a single type of nucleic acid, either DNA or RNA. The DNA or RNA genome may be linear or circular, and single-stranded or double-stranded. Nearly all viruses form a protein sheath, or **capsid,** around their nucleic acid core (figure 23.14). The capsid is composed of one to a few different protein molecules repeated many times. The repeating units are called capsomeres.

Most viruses come in one of two simple shapes, helical or icosahedral. Most animal viruses are icosahedral, the design of a geodesic dome that maximizes internal capacity. Some viruses, like the T-even bacteriophage, are complex, with an icosahedral head and helical axis. Large animal viruses, like the flu virus, may have a regular but polymorphic shape.

In several viruses, specialized enzymes are stored with the nucleic acid, inside the capsid. One example is reverse transcriptase, which is required for retroviruses to complete their cycle and is not found in the host. This enzyme is needed early in the infection process and is carried within each virion.

Many animal viruses have an *envelope* around the capsid that is rich in proteins, lipids, and glycoprotein molecules. The lipids found in the envelope are derived from the host cell; however, the proteins in a viral envelope are generally virally encoded.

Viruses replicate by taking over host machinery

Viruses can reproduce only when they enter cells. A virus is simply a set of instructions, the viral genome, that can trick the cell's replication and metabolic enzymes into making copies of the virus. When they are outside a cell, viral particles are called *virions* and are metabolically inert. Viruses lack the machinery necessary for protein synthesis as well as most, if not all, of the enzymes for nucleic acid replication. In addition, viruses lack the amino acid and nucleotide monomers necessary to synthesize proteins and nucleic acids. Inside cells, the

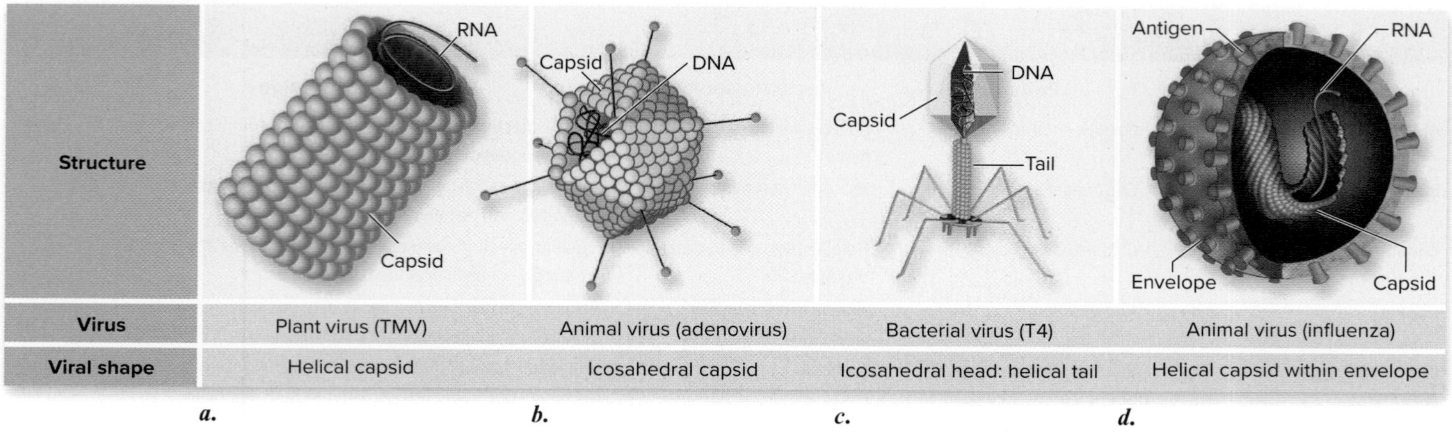

Figure 23.14 Structure of virions. Viruses are characterized as helical, icosahedral, binal, or polymorphic, depending on their symmetry. ***a.*** The capsid may have helical symmetry as in the tobacco mosaic virus (TMV). TMV infects plants and consists of 2130 identical protein molecules (green) that form a cylindrical coat around the single strand of RNA (red). ***b.*** The capsid of icosahedral viruses has 20 facets made of equilateral triangles. ***c.*** Bacteriophage come in a variety of shapes, but binal symmetry is exclusively seen in phage such as the T4 phage of *E. coli*. ***d.*** Viruses such as the Influenza virus can also have an envelope surrounding the capsid. This gives the virus a polymorphic shape.

virus hijacks the transcription and translation systems to produce viral proteins.

Viruses have been found to infect every kind of organism studied by biologists. They infect fungal cells and protists, as well as prokaryotes, animals, and plants. However, each type of virus can replicate in only a very limited number of cell types. This host specificity depends on the presence of receptors on the host cell surface. Bacterial viruses do not infect human cells because they lack the appropriate receptors.

Viral genomes exhibit great variation

Viruses vary greatly in size (figure 23.15), type of nucleic acid, and number of genome strands (table 23.2). Some viruses, including those that cause flu, measles, and AIDS, possess RNA

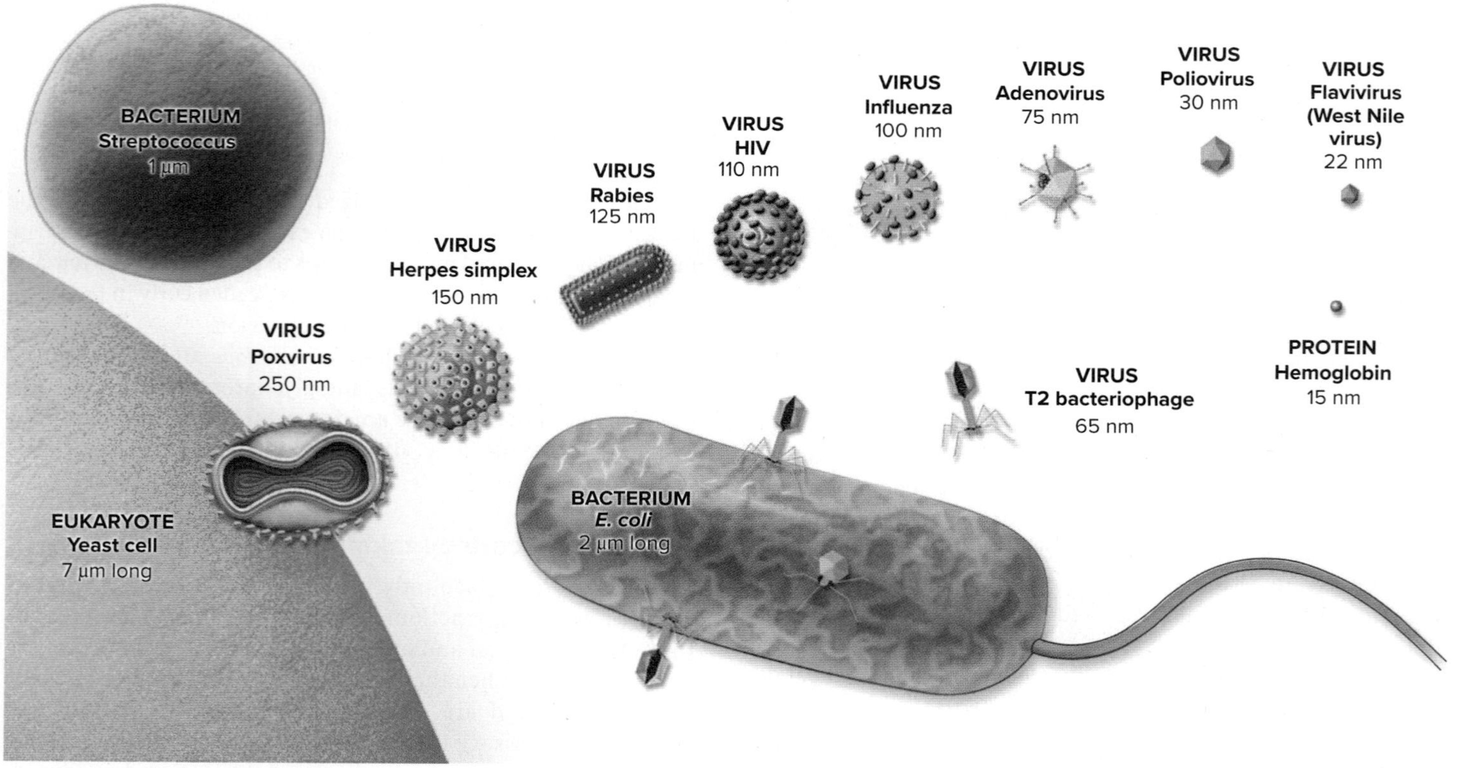

Figure 23.15 Viruses vary in size and shape. Note the dramatic differences in the size of a eukaryotic yeast cell, prokaryotic bacterial cells, and the many different viruses.

TABLE 23.2 Important Human Viral Diseases

Disease	Pathogen	Genome	Vector/Epidemiology
Chicken pox	Varicella-zoster virus	Double-stranded DNA	Spread through contact with infected individuals; rarely fatal; vaccine approved in U.S. in early 1995; exhibits latency, leading to shingles, which usually occurs in older people.
Hepatitis B (viral)	Hepadnavirus	Double-stranded DNA	Highly infectious through contact with infected body fluids; vaccine available; no cure; can be fatal
Herpes	Herpes simplex virus	Double-stranded DNA	HSV-1 spread primarily by skin-to-skin contact with cold sores. HSV-2 is genital form also spread by skin-to-skin contact. Very prevalent worldwide; no cure; both forms exhibit latency.
Mononucleosis	Epstein–Barr virus	Double-stranded DNA	Can cause extreme fatigue and flulike symptoms that can persist. Spread via bodily fluids, especially saliva. Infection may increase likelihood of some rare cancers.
Cervical and penile cancer	Human papillomaviruses 16/18	Double-stranded DNA	HPV is the most common sexually transmitted infection in the United States. HPV types 16 and 18 can cause cervica and penile cancers; other subtypes can cause genital warts. A safe and effective vaccine is available for HPV 16/18.
AIDS	HIV	(+) Single-stranded RNA (two copies)	Destroys immune defenses, resulting in death by opportunistic infection or cancer; for the year 2017, WHO estimated 1.8 million new infections and 940,000 deaths
Polio	Enterovirus	(+) Single-stranded RNA	Acute viral infection of the CNS that can lead to paralysis and is often fatal; close to being eliminated worldwide
Yellow fever	Flavivirus	(+) Single-stranded RNA	Spread from individual to individual by mosqu to bites; if untreated, has a peak mortality rate of 60%
Ebola	Filoviruses	(–) Single-stranded RNA	Acute hemorrhagic fever; virus attacks connective tissue, leading to massive hemorrhaging and death; peak mortality is 50–90%
Influenza	Influenza viruses	(–) Single-stranded RNA (eight segments)	Periodic pandemics (50 million died in 1918 pandemic) due to antigen reassortment in avian species, pigs, and humans
Measles	Paramyxoviruses	(–) Single-stranded RNA	Extremely contagious through contact with infected individuals; vaccine available; childhood disease; more dangerous to adults and infants younger than 12 months old
SARS	SARS-CoV	(+) Single-stranded RNA	Identified in early 2003; spread via aerosolic droplets of saliva. Causes severe acute respiratory syndrome with roughly 3% mortality.
Rabies	Rhabdovirus	(–) Single-stranded RNA	Acute viral encephalomyelitis transmitted by the bite of an infected animal; fatal if untreated
Zika virus disease	Zika virus	(+) single-stranded RNA	Vectored by *Aedes* mosquitoes. Can be transmitted from pregnant mother to fetus. Can cause fever, rash, headache. Complications in pregnancy can cause microcephaly. No vaccine exists.
COVID-19	SARS-CoV-2	(+) single-stranded RNA	First reported in late 2019 from a cluster of cases of Wuhan province, China. Causes fever, dry cough, fatigue, and often loss of taste or smell. Caused a global pandemic starting in 2020.

genomes. Most RNA viruses are single-stranded and are replicated and assembled in the cytosol of infected eukaryotic cells. RNA virus replication is error-prone, leading to high rates of mutation. This makes them difficult targets for the host immune system, vaccines, and antiviral drugs.

In single-stranded RNA viruses, if the genome has the same base sequence as the mRNA used to produce viral proteins, then the genomic RNA can serve as the mRNA. Such viruses are called *positive-strand viruses* (*+ssRNA viruses*). In contrast, if the genome is complementary to the viral mRNA, then the virus is called a *negative-strand virus (–ssRNA viruses).*

A special class of RNA viruses, called *retroviruses,* have an RNA genome that is reverse-transcribed into DNA by the enzyme *reverse transcriptase.* The DNA fragments produced by reverse transcription are often integrated into a host's chromosomal DNA. *Human immunodeficiency virus (HIV),* the agent that causes *acquired immune deficiency syndrome (AIDS),* is a retrovirus. (We describe HIV in detail in section 23.8.)

Other viruses, such as the viruses causing smallpox and herpes, have DNA genomes. Most DNA viruses are double-stranded, and their DNA is replicated in the nucleus of eukaryotic host cells.

Giant viruses

One common practical definition of viruses was that they were "filterable agents," infectious agents that could pass through a 500-nm filter. The first virus to fail this centuries-old test was Mimivirus, originally misidentified as a bacterium that infected the unicellular protist *Acanthamoeba*. Mimivirus has a virion 750 nm in diameter and a genome of more than 1 megabases (Mb). A number of other so-called giant viruses were isolated, with the current largest being Pithovirus, which was revived from 30,000-year-old Siberian ice in 2014. With virions 1.5 μm long, Pithovirus is larger than some bacteria, and even than some eukaryotic cells.

Giant viruses challenge assumptions about the distinctions between viruses and cells in other ways. The Mimivirus genome contains genes central to the translation process, normally not found in a virus. The presence of part of the translation machinery implies they could once have been autonomous. These viruses also encode virtually all of the DNA replication machinery, and they replicate in the cytoplasm, further blurring the line between a virus and an intracellular parasitic bacterium.

REVIEW OF CONCEPT 23.6

Viruses have a very simple structure that includes a nucleic acid genome encased in a protein coat. Viruses replicate by taking over a host's cell systems and are thus obligate intracellular parasites. Viruses show diverse genomes that are composed of DNA or RNA, which may be single- or double-stranded; most DNA viruses are double-stranded.

- *Why can't viruses replicate outside of a cell?*

23.7 Bacterial Viruses Infect by DNA Injection

Bacterial Viruses Exhibit Two Reproductive Cycles

LEARNING OBJECTIVE 23.7.1 Distinguish between the lytic and lysogenic cycles in bacteriophage.

The usual result of viral infection is production and release of new virus particles, usually killing the cell. This release of viruses then allows infection of new cells, or horizontal transmission of the virus. This kind of infection cycle is called a lytic cycle, because the virus usually causes the cells to rupture, or lyse. Some bacterial viruses, however, can also enter a latent phase, called the lysogenic cycle, after the initial infection. These latent viruses are then transmitted vertically–through cell division.

The lytic cycle

The lytic cycle of virus reproduction is illustrated in figure 23.16. The basic steps of a lytic bacteriophage cycle are similar to those of a nonenveloped animal virus. We will use the lytic cycle as an example of a viral replication cycle. Although the basic outline of this infective cycle is common to most viruses, the details vary, as do the viruses themselves.

The first step is called **attachment** (or adsorption), in which the virus contacts the cell and becomes specifically bound to the cell. This step limits the host range of the virus, because it binds to specific proteins on the surface of the cell. Different phage may target different parts of the outer surface of a bacterial cell. The next step is called **penetration** and results in the release of the viral genome into the host. This has been studied in detail in the binal phage, such as T4. Once contact is established, the tail contracts, and the tail tube passes through an opening that appears in the base plate, piercing the bacterial cell wall. The viral genome is literally injected into the host cytoplasm.

Once inside the bacterial cell, in the **synthesis** phase, the virus takes over the cell's replication and protein synthesis machinery in order to synthesize viral components. During the **assembly** phase, viral components are put together to produce mature virus particles. In the **release** phase, mature virus particles are released, either through the action of enzymes that lyse the host cell or by budding through the host cell wall. The time between attachment and the formation of new viral particles is called an eclipse period, because if a cell is lysed at this point, few, if any, active virions can be released.

The lysogenic cycle

Bacteriophage that are capable of latent infection do this by integrating their nucleic acid into the genome of the infected host cell. This integration allows a virus to be replicated along with the host cell's DNA as the host divides. These viruses are called **temperate,** or **lysogenic,** phage. The DNA segment that is integrated into a host cell's genome is called a **prophage,** and the resulting cell is called a **lysogen** (figure 23.17, right).

In a lysogenic phage, the expression of its genome is repressed (refer to chapter 16) by a viral regulatory protein. However, in times of cell stress the prophage can be derepressed and the viral genome excised. The viral genome then is in the same state as early infection, and the lytic cycle leads to formation of viral particles and lysis of the cell.

REVIEW OF CONCEPT 23.7

Bacteriophage are viruses that infect bacteria. They have two major types of infection cycle: the lytic cycle, which results in immediate death of the host, and the lysogenic cycle, in which the virus becomes part of the host genome. This viral genome is then transmitted vertically by cell division. Under certain conditions, the lysogenic phage can switch to the lytic cycle.

- *What kinds of mutations would make a host cell immune to infection by a particular phage?*

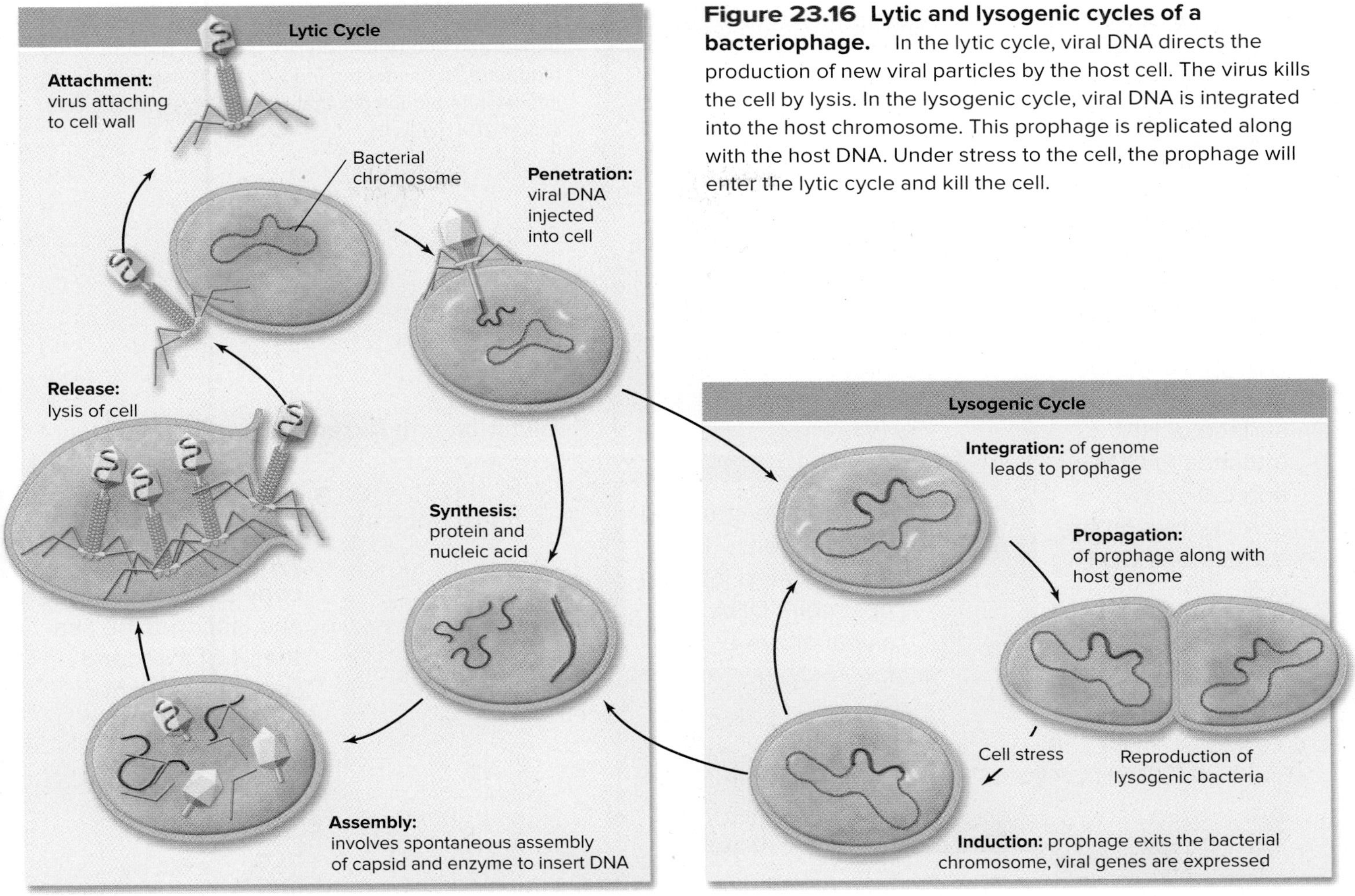
Lytic Cycle
Attachment:
virus attaching
to cell wall
Bacterial
chromosome
Penetration:
viral DNA
injected
into cell
Release:
lysis of cell
Synthesis:
protein and
nucleic acid
Assembly:
involves spontaneous assembly
of capsid and enzyme to insert DNA
Lysogenic Cycle
Integration: of genome
leads to prophage
Propagation:
of prophage along with
host genome
Cell stress
Reproduction of
lysogenic bacteria
Induction: prophage exits the bacterial
chromosome, viral genes are expressed

Figure 23.16 Lytic and lysogenic cycles of a bacteriophage. In the lytic cycle, viral DNA directs the production of new viral particles by the host cell. The virus kills the cell by lysis. In the lysogenic cycle, viral DNA is integrated into the host chromosome. This prophage is replicated along with the host DNA. Under stress to the cell, the prophage will enter the lytic cycle and kill the cell.

23.8 Animal Viruses Infect by Endocytosis

A diverse array of viruses occurs among animals. A good way to gain a general idea of the characteristics of these viruses is to look at one animal virus in detail. Here we examine the virus responsible for the fatal viral disease, *acquired immune deficiency syndrome (AIDS).*

AIDS Is Caused by the Animal Virus HIV

LEARNING OBJECTIVE 23.8.1 Describe how the HIV virus infects human cells.

The disease now known as AIDS was first reported in the United States in 1981, although a few dozen people in the United States had likely died of AIDS prior to that time and had not been diagnosed. Frozen plasma samples and estimates based on evolutionary speed and current diversity of HIV strains trace the origins of HIV in the human population to Africa in the 1950s. It was not long before the infectious agent, a retrovirus, was identified by laboratories in France. Study of HIV revealed it to be closely related to a chimpanzee virus (simian immunodeficiency virus, SIV), suggesting a recent host expansion to humans from chimpanzees in central Africa.

HIV infection compromises the host's immune system

The HIV virus targets cells that are critical to the human immune response. Immune cells are characterized based on the proteins they display on their surface. HIV targets cells that express the antigen CD4, thus $CD4^+$ cells. The specific cell type infected by HIV is the helper T cell, which is critical to regulating the human immune response. The action of the T helper cell is described in chapter 35.

HIV infects and kills the $CD4^+$ cells until very few are left. Without these crucial immune system cells, the body cannot mount a defense against invading bacteria or viruses. AIDS patients die of infections that a person with a healthy immune system could fight off. These infections, called *opportunistic infections,* usually do not cause disease and are part of the progression from HIV infection to AIDS.

Clinical symptoms typically do not begin to develop until after a long latency period, generally 8 to 10 years after the initial infection with HIV. Some individuals, however, may develop symptoms in as few as two years. During latency, HIV particles are not in circulation, but the virus can be found integrated within the genome of macrophages and $CD4^+$ T cells as a provirus (equivalent to a prophage in bacteria).

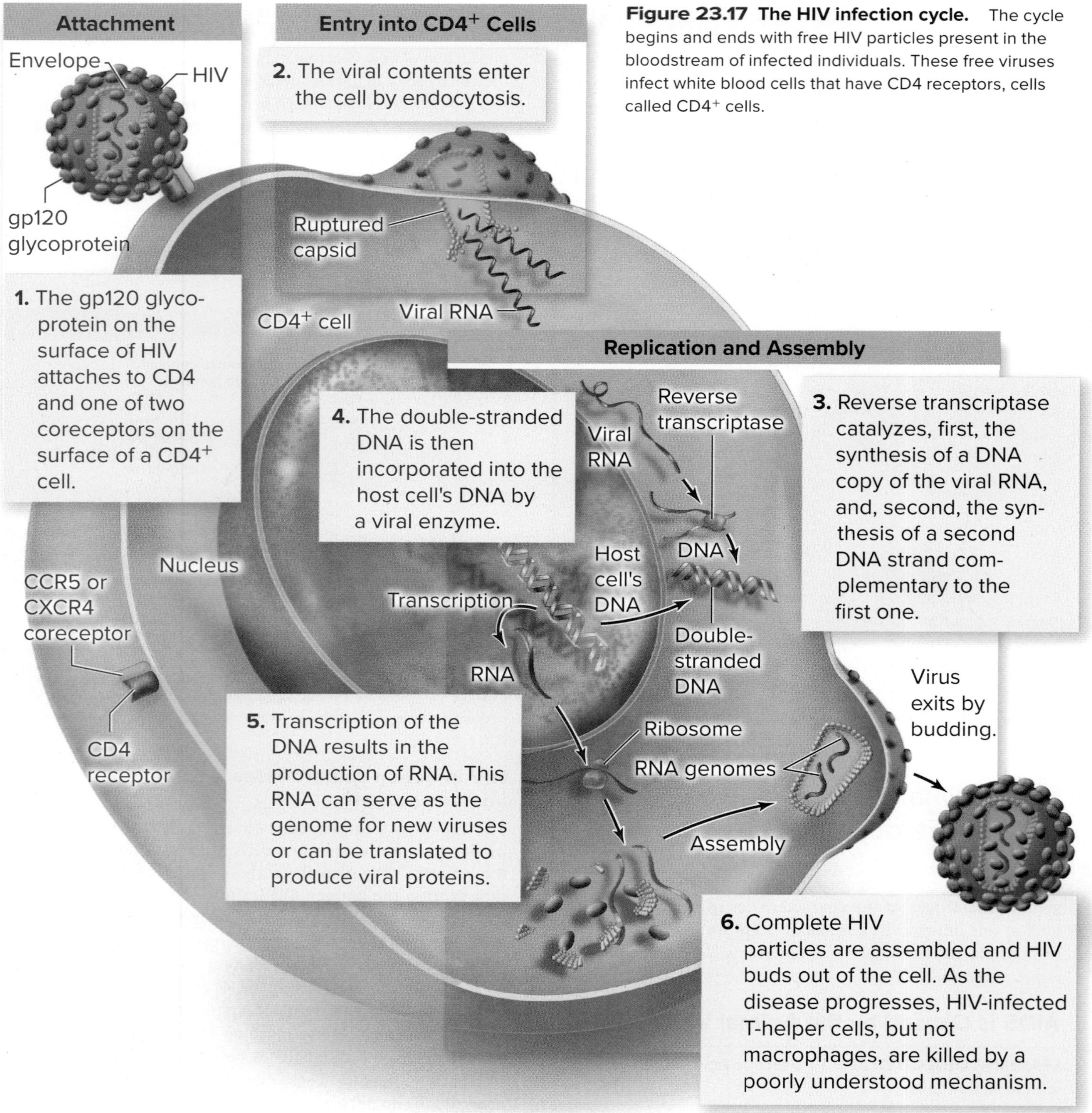
Attachment
Envelope
HIV
gp120 glycoprotein
1. The gp120 glycoprotein on the surface of HIV attaches to CD4 and one of two coreceptors on the surface of a CD4+ cell.
CCR5 or CXCR4 coreceptor
CD4 receptor
Nucleus
Entry into CD4+ Cells
2. The viral contents enter the cell by endocytosis.
Ruptured capsid
Viral RNA
CD4+ cell
Replication and Assembly
Reverse transcriptase
Viral RNA
3. Reverse transcriptase catalyzes, first, the synthesis of a DNA copy of the viral RNA, and, second, the synthesis of a second DNA strand complementary to the first one.
4. The double-stranded DNA is then incorporated into the host cell's DNA by a viral enzyme.
Host cell's DNA
DNA
Double-stranded DNA
Transcription
RNA
5. Transcription of the DNA results in the production of RNA. This RNA can serve as the genome for new viruses or can be translated to produce viral proteins.
Ribosome
RNA genomes
Assembly
Virus exits by budding.
6. Complete HIV particles are assembled and HIV buds out of the cell. As the disease progresses, HIV-infected T-helper cells, but not macrophages, are killed by a poorly understood mechanism.

Figure 23.17 The HIV infection cycle. The cycle begins and ends with free HIV particles present in the bloodstream of infected individuals. These free viruses infect white blood cells that have CD4 receptors, cells called $CD4^+$ cells.

HIV infects key immune system cells

The way in which HIV infects humans provides a good example of how animal viruses replicate (figure 23.17). Most other viral infections follow a similar course, although the details of entry and replication differ in individual cases.

Attachment

When HIV is introduced into the human bloodstream, the virus particles circulate throughout the body but infect only $CD4^+$ cells. How does a virus such as HIV recognize a target cell? Recall from chapter 4 that every kind of cell in the human body has a specific array of cell-surface glycoprotein markers that identify them to other, similar cells. Invading viruses take advantage of this to bind to specific cell types. Each HIV particle possesses a glycoprotein called gp120 on its surface that precisely fits the cell-surface marker protein CD4 on the surfaces of the immune system macrophages and T cells. Macrophages, another type of white blood cell, are infected first. Because macrophages commonly interact with $CD4^+$ T cells, this may be one way that the T cells are infected. Several coreceptors, including the CCR5 receptor, which is mutated in HIV-immune individuals, also significantly affect the likelihood of viral entry into cells.

Entry of virus

After docking onto the CD4 receptor of a cell, HIV requires a coreceptor such as CCR5 to pull itself across the cell membrane. After gp120 binds to CD4, it goes through a conformational change that allows it to then bind the coreceptor. Receptor binding is thought to ultimately result in fusion of the viral and target cell membranes and entry of the virus through a fusion pore. The coreceptor, CCR5, is hypothesized to have been used by the smallpox virus.

Replication

Once inside the host cell, the HIV particle sheds its protective coat. This leaves viral RNA floating in the cytoplasm, along with the reverse transcriptase enzyme that was also within the virion. Reverse transcriptase synthesizes a double strand of DNA complementary to the virus RNA, often making mistakes and introducing new mutations. This double-stranded DNA then enters the nucleus along with a viral enzyme that incorporates the viral DNA into the host cell's DNA. After a variable period of dormancy, the HIV provirus directs the host cell's machinery to produce many copies of the virus.

As is the case with most enveloped viruses, HIV does not directly rupture and kill the cells it infects. Instead, the new viruses are released from the cell by *budding,* a process much like exocytosis. HIV synthesizes large numbers of viruses in this way, challenging the immune system over a period of years. In contrast, naked viruses, those lacking an envelope, generally lyse the host cell in order to exit. Some enveloped viruses may produce enzymes that damage the host cell enough to kill it or may produce lytic enzymes as well.

Evolution of HIV during infection

During an infection, HIV is constantly replicating and mutating. The reverse transcriptase enzyme is less accurate than DNA polymerases, leading to a high mutation rate. Eventually, by chance, variants in the gene for gp120 arise that cause the gp120 protein to alter its second-receptor partner. This new form of gp120 protein will bind to a different second receptor—for example, CXCR4—instead of CCR5. During the early phase of an infection, HIV primarily targets immune cells with the CCR5 receptor. Eventually, the virus mutates to infect a broader range of cells. Ultimately, infection results in the destruction and loss of critical helper T cells.

This destruction of T cells blocks the body's immune response and leads directly to the onset of AIDS. Most deaths due to AIDS are not a direct result of HIV, but are from other diseases that usually do not harm an individual with a fully functional immune system.

Zoonotic Events Can Cause Human Pandemics

LEARNING OBJECTIVE 23.8.2 Explain how zoonotic events can give rise to pandemics.

When a pathogen moves from a wild animal to humans it is called zoonotic spillover. This kind of movement from vertebrate animals to humans is thought to be the origin of up to 75% of human pathogens. It is certainly the most common source of new infectious diseases, and the origin of pandemics both ancient and modern.

Influenza

Influenza virus has been one of the most lethal pathogens in human history. Flu viruses are segmented –ssRNA viruses with 11 genes contained on 8 genomic segments. An individual flu virus resembles a sphere studded with spikes composed of two kinds of protein: hemagglutinin (H) and neuraminidase (N) (figure 23.18). Hemagglutinin is involved in the virus binding to the surface of respiratory cells, and neuraminidase is involved in the release of viral particles. Different strains of flu virus, called subtypes, are differentiated by their H–N composition. Flu viruses are currently classified into 13 H subtypes and 9 N subtypes and are given names such as H1N1. Each subtype requires a different vaccine to protect against infection.

The viral genome is replicated by an RNA-dependent RNA polymerase, which is error prone. This leads to a high mutation rate, and mutations in either the H or N proteins can produce variants that make a vaccine ineffective. This so-called **antigenic drift** is the reason we need yearly vaccination. As antigenic drift

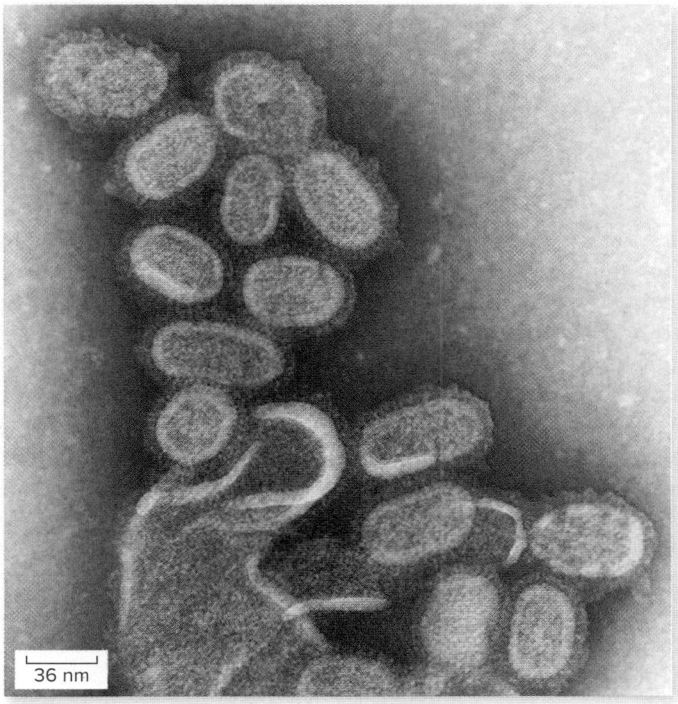

Figure 23.18 H3N2 influenza virus. Thousands of Americans die of this virus every year.

Cynthia Goldsmith/Centers for Disease Control and Prevention is a random process, it is impossible to predict which viral strains will emerge each year. Scientists at the Centers for Disease Control and Prevention (CDC) use information on circulating strains to decide which strains to use in next year's flu vaccines.

The segmented nature of the influenza genome causes a larger problem: **antigenic shift.** This arises when the virus packages its genome into new virions. If a cell is infected by multiple flu viruses, the different genomic segments can be recombined to produce novel combinations that can become pandemic viruses. This can even involve viruses from different species, primarily birds, pigs, and humans. Antigenic shifts in H–N combinations were responsible for the three major flu pandemics in the 20th century. The "Spanish flu" of 1918, A(H1N1), killed 50 million people worldwide. The Asian flu of 1957, A(H2N2), killed over 100,000 Americans. The Hong Kong flu of 1968, A(H3N2), infected 50 million people in the United States alone, of whom 70,000 died.

Three conditions are necessary for a pandemic: (1) The new strain must have a novel combination of H and N proteins, so human population will have no significant immunity; (2) the new strain must be able to replicate in human cells and cause death; and (3) the new strain must be efficiently transmitted between humans. The new strain need not be deadly to every infected person in order to produce a pandemic—the H1N1 flu of 1918 had an overall mortality rate estimated to be 2.5% yet killed 50 million people. Why did so many die? Because so much of the world's population was infected.

The first pandemic of the 21st century started in February of 2009, and by mid-May 11,000 people in 41 countries had been affected. Modern genomics provide us a unique view into how this virus arose. In the 1990s, three flu viruses infected pigs: a swine flu virus, an avian flu virus, and a human H3N2 virus. Recombination produced a new virus that circulated in swine herds, but could not infect humans. Pigs harboring the new virus were infected by an avian-like influenza virus, and this reassortment created a new virus called H1N1/09, which could now infect humans.

Coronaviruses

Coronaviruses are enveloped, +ssRNA viruses. A wide variety of coronaviruses appear to circulate in bats, and from these have jumped to other vertebrates, and to humans. Coronaviruses can infect a variety of vertebrate animals and occasionally they spill over from their nonhuman hosts into humans, where they can spread rapidly to cause epidemics or pandemics. There are seven coronaviruses that infect humans, four of which have long been circulating in human populations, and cause roughly 40% of what we call the common cold. The other three coronaviruses, SARS-CoV, MERS-CoV, and SARS-CoV-2, cause more serious respiratory disease and are all pathogens that have recently moved into humans. Coronaviruses are spread by close contact via respiratory droplets expelled during coughing or sneezing, but may also be spread by contaminated surfaces.

SARS-CoV first appeared in southern China in late 2002 and spread to create an epidemic in about 24 countries before being contained in 2003. No cases of SARS have been reported since 2004. MERS-CoV first appeared in the middle east in 2012, with additional outbreaks in South Korea in 2015 and Saudi Arabia in 2018. Both SARS-CoV and MERS-CoV cause severe respiratory disease often leading to pneumonia. Fortunately, despite being new pathogens, neither of these diseases became pandemic.

A third new coronavirus, SARS-CoV-2, is the causative agent of coronavirus disease 2019 (**COVID-19**), first identified in the Wuhan province of China in January 2020. This virus does not cause as severe a respiratory disease as either SARS or MERS, but spreads through human populations much more readily. The virus spread globally, causing about 1 million infections in the three months from January to March 2020. By June 2020, the number of infections globally reached 10 million, causing about 500,000 deaths. By January 2022, the World Health Organization reported over 300 million infections and 5.5 million deaths globally. Toward the end of 2020, vaccines began to be available, including two mRNA vaccines (refer to chapter 17) that provide approximately 90% protection against infection.

SARS-CoV-2 infects respiratory epithelia by binding, via its surface spike proteins, to the ACE2 protein found on the surface of cells in the respiratory tract (figure 23.19). Once inside the cell, the viral genome is replicated and expressed, and new virions are assembled. New, infectious virions are released from the cell by exocytosis and can infect nearby cells, spreading the infection.

Individuals infected with SARS-CoV-2 exhibit a range of morbidities from asymptomatic to severe respiratory distress. Some individuals develop neurological symptoms such as dizziness and headache, and loss of taste and smell. In individuals who clear the infection and recover in 2–3 weeks, there can be long-term complications from infection: heart palpitations, carditis, impaired lung function, hair loss, and skin rashes. This spectrum of longer-term symptoms is now referred to as *long COVID.*

The infection-to-fatality ratio (IFR), which expresses the percentage of infected individuals that die from the disease, is highly variable for COVID-19. The most obvious factor is the age of the infected individual. IFR ranges from 0.002% at age 10 to 15% at age 85, with a sharp increase in deaths starting at about age 65 (1.4%). This is also complicated by other risk factors from preexisting conditions, and it appears that males suffer more severe disease and death than females.

While SARS-CoV-2 virus replication is not as error-prone as either Influenza A or HIV due to a proofreading function, it is less accurate than DNA replication. This means that the mutation rate is still high enough to continually produce new viral variants. This has complicated how we deal with the ongoing pandemic, as the protection from vaccines is based on the original strain, and has not been as effective against all new variants. The World Health Organization (WHO) tracks and names new variants using Greek letters, and determines which are considered "variants of concern."

REVIEW OF CONCEPT 23.8

The HIV retrovirus enters cells via membrane fusion. The virus primarily infects human $CD4^+$ T cells, ultimately resulting in massive cell death, compromising the immune system of infected individuals. Zoonosis is the movement of a pathogen from another vertebrate species to humans. Influenza infects a variety of vertebrates, and has caused periodic pandemics in humans. Coronaviruses have circulated in humans for years, but the new SARS-CoV-2 virus caused the worst pandemic of the 21st century.

- *How does the biology of the HIV virus complicate vaccine development?*

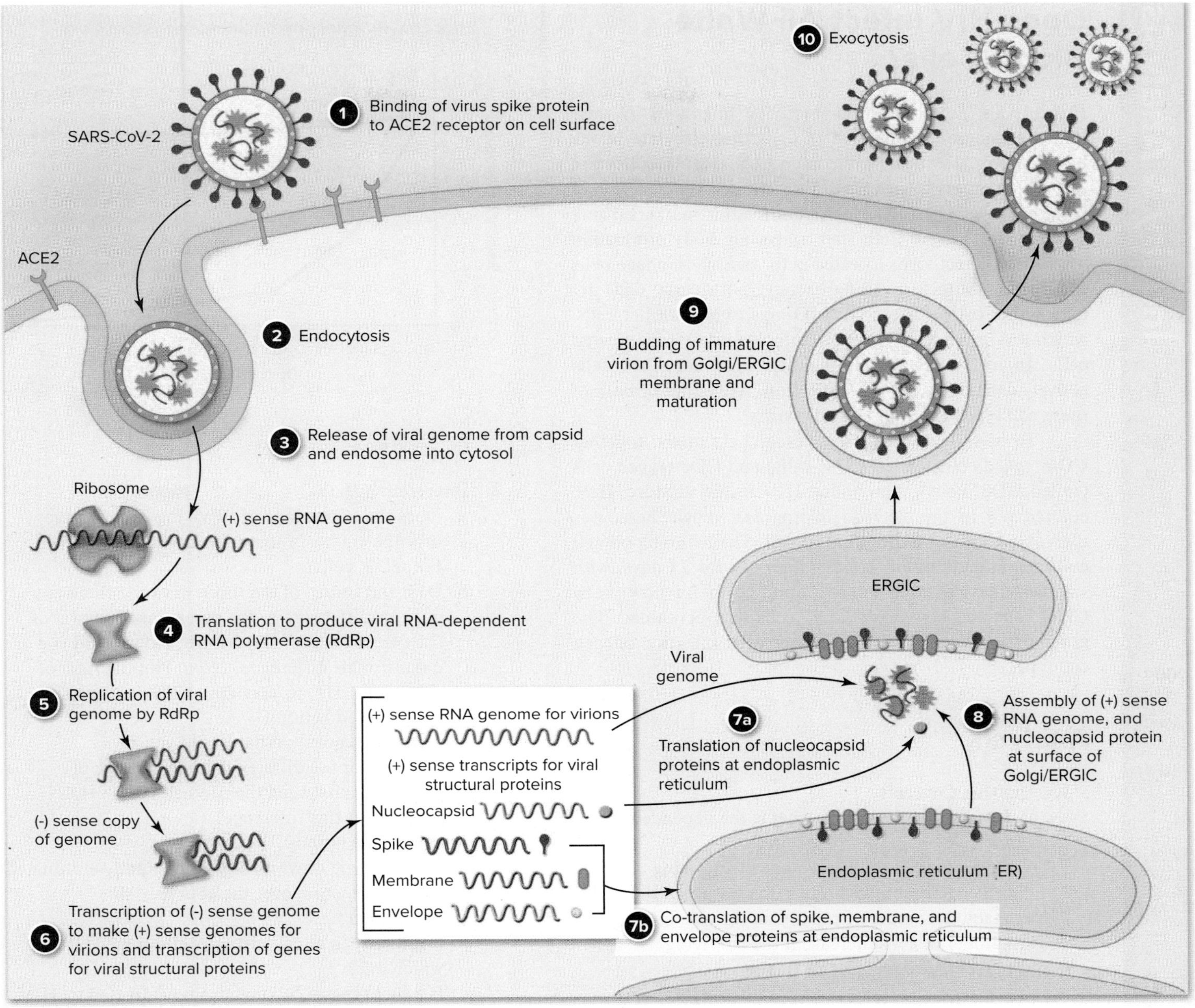

Figure 23.19 SARS-CoV-2 infection and replication mechanism. (1) Virions attach to ACE2 receptors on the host cell surface; (2) virions enter the cell by receptor-mediated endocytosis; (3) virions are released from endosomes and the viral genome is released into the cytoplasm; (4) translation by host ribosomes produces RNA-dependent RNA polymerase (RdRp); (5) RdRp makes (−) sense copies of the viral genome; (6) (−) sense RNAs can then be copied into thousands of (+) sense RNA genomes; (7) (+) sense RNAs are used to make viral proteins in the cytoplasm and on the ER; (8) viral genomes are assembled with the N protein to form capsids at the membrane of the endoplasmic reticulum–Golgi intermediate compartment (ERGIC), which contains envelope proteins; (9) immature virions bud from the surface of the ERGIC; and (10) virions mature in vesicles and are exocytosed.

Does HIV Infect All White Blood Cells?

Humans are protected from microbial infections by their immune system, a collection of cells that circulate in the blood. Loosely called "white blood cells," this collection actually contains a variety of different cell types. Some of them possess CD4 cell-surface identification markers (think of them as ID tags). Cells that trigger antibody production when they detect virus-infected cells and macrophage cells that initially attack invading bacteria both carry CD4 ID tags. Other cells possess CD8 ID tags, such as killer cells, which are immune cells that bore holes into virus-infected cells. In an AIDS patient, neither CD4 nor CD8 cells actively defend against HIV infection. Are either or both of these cell types killed by the HIV virus?

To investigate this issue, researchers mixed together CD4-tagged cells (called $CD4^+$ cells) and CD8-tagged cells (called $CD8^+$ cells), then added HIV to the mixture. HIV, colored red in the electron micrograph shown here, was then able to infect either kind of cell. The white blood cell culture was monitored at 5-day intervals for 25 days, with samples taken at each interval and scored for how many $CD4^+$ cells and how many $CD8^+$ cells they contained. The graph presents the survival over time after infection of each cell type.

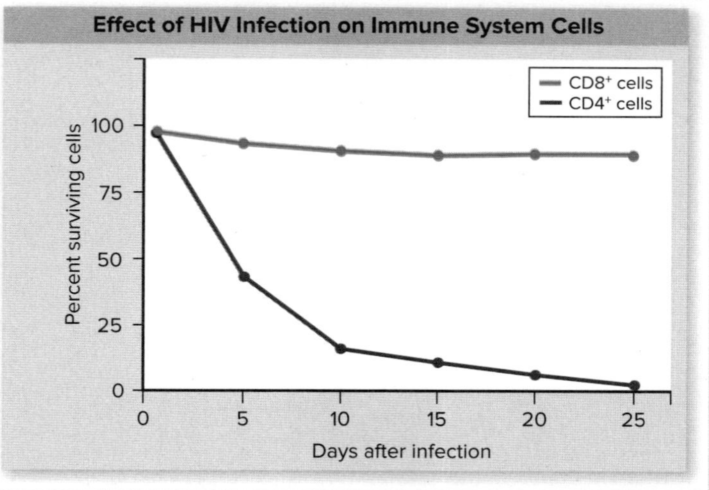

Analysis

1. **Applying Concepts**
   a. ***Variable.*** In the graph, what is the dependent variable?
   b. ***Percentage.*** If the percentage of surviving cells decreases, what does this say about the absolute number of cells? Can the absolute number of cells increase if the surviving percentage decreases? Explain.

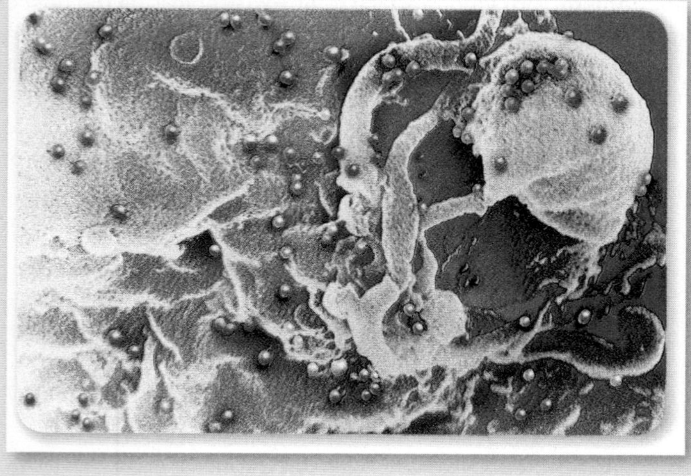

CDC/Science Source

2. **Interpreting Data**
   a. Does the percentage of surviving cells change over the course of three weeks for $CD4^+$ cells? For $CD8^+$ cells?
   b. Over the course of the three weeks, is there any obvious difference in the percentage survival of the two cell types? Describe it. How would you quantify this difference? (*Hint:* Plot the *ratio* of surviving $CD4^+$ to surviving $CD8^+$ cells versus days after infection.)
3. **Making Inferences** What would you say is responsible for the difference in percentage of surviving cells between the two cell types? How might you test this inference?
4. **Drawing Conclusions**
   a. Is either type of white blood cell totally eliminated by HIV infection over the course of this experiment?
   b. Is either type of cell virtually eliminated? If so, which one?
   c. Is either type of cell not strongly affected by HIV infection? If so, which one? Can you think of a reason the percentage of surviving cells of this cell type changes at all? How might you test this hypothesis?
5. **Further Analysis** Neither $CD4^+$ nor $CD8^+$ cells actively defend AIDS patients. If one of them is not eliminated by HIV, why do you suppose it ceases to defend HIV-infected AIDS patients? Can you think of a way to investigate this possibility?

Retracing the Learning Path

CONCEPT 23.1 Prokaryotes Are the Most Ancient Organisms

23.1.1 Microfossils Indicate That the First Cells Were Probably Prokaryotic The oldest microfossils are 3.5 billion years old. Prokaryotes, bacteria and archaea, are the most abundant life forms, and are found in all environments.

23.1.2 Prokaryotes Are Fundamentally Different from Eukaryotes Prokaryotic features include unicellularity; small, circular DNA; division by binary fission; lack of internal compartmentalization; a singular flagellum; and metabolic diversity.

23.1.3 Archaea Have Distinct Cell Architecture and Metabolism Archaea have unique membrane lipids with isoprenoids, and not fatty acids, joined by ether linkages to a different isomer of glycerol phosphate. Their cell wall lacks peptidoglycan, and is surrounded by an S-layer. Archaeal DNA replication and gene expression are more like eukaryotes than bacteria. Some species can oxidize ammonia, and only archaea are capable of methanogenesis.

CONCEPT 23.2 Prokaryotes Have an Organized but Simple Structure

23.2.1 Prokaryotes Have Three Basic Shapes: Bacilli, Cocci, and Spirilla

23.2.2 Bacteria Have a Variety of External Structures Bacteria are either gram-positive or gram-negative based on cell wall structure. Gram-positive cell walls have a thick layer of peptidoglycan while gram negative have thin layer, and an outer membrane with lipopolysaccharide. Some bacteria have a gelatinous layer, the capsule. Many bacteria have a slender, rigid, helical flagellum composed of flagellin, which rotates to drive movement.

23.2.3 The Bacterial Cell Interior Is Organized In prokaryotes, invaginated regions of the plasma membrane function in respiration and photosynthesis. The nucleoid region contains a compacted, circular DNA with no bounding membrane.

23.2.4 Most Prokaryotes Have Not Been Characterized Nine clades of prokaryotes have been identified, but many bacteria have not been studied.

CONCEPT 23.3 The Genetics of Prokaryotes Focuses on DNA Transfer

23.3.1 Conjugation Depends on the Presence of a Conjugative Plasmid DNA can tranferred by conjugative plasmids, like the F plasmid in *E. coli.* The plasmid can be transferred, or if it is integrated into the chromosome, host DNA can be transferred. Transferred donor DNA is integrated into recipient genome by recombination.

23.3.2 Viruses Transfer DNA by Transduction Generalized transduction occurs when viruses package host DNA and transfer it on subsequent infection. Specialized transduction is limited to lysogenic phage.

23.3.3 Transformation Is the Uptake of DNA Directly from the Environment Transformation occurs when cells take up DNA from the surrounding medium.

CONCEPT 23.4 Prokaryotic Metabolism Is Diverse

23.4.1 Prokaryotes Include Both Autotrophs and Heterotrophs Photoautotrophs carry out photosynthesis and obtain carbon from carbon dioxide. Chemolithoautotrophs obtain energy by oxidizing inorganic substances. Photoheterotrophs use light for energy but obtain carbon from organic molecules. Chemoheterotrophs, the largest group, obtain carbon and energy from organic molecules.

CONCEPT 23.5 Bacteria Cause Important Human Diseases

23.5.1 A Wide Variety of Diseases Are Caused by Bacteria Tuberculosis has infected humans for all of recorded history. The potentially dangerous sexually transmitted diseases gonorrhea, syphilis, and chlamydia are caused by bacteria. Most stomach ulcers are caused by infection with *Helicobacter pylori.*

CONCEPT 23.6 Viruses Are Not Organisms

23.6.1 Viruses Are Strands of Nucleic Acids Encased in a Protein Coat Viral genomes can be DNA or RNA and are classified as DNA viruses, RNA viruses, or retroviruses. Most viruses have a protein capsid around their nucleic acid core. Many animal viruses have an outer envelope composed of virally encoded proteins and host cell lipids. Viruses vary in size and come in two simple shapes: helical (rodlike) or icosahedral (spherical). The DNA or RNA viral genome may be linear or circular, single- or double-stranded. RNA viruses may have multiple RNA molecules (segmented) or only one (nonsegmented). Retroviruses contain RNA that is converted to DNA.

CONCEPT 23.7 Bacterial Viruses Infect by DNA Injection

23.7.1 Bacterial Viruses Exhibit Two Reproductive Cycles The lytic cycle kills the host cell, whereas the lysogenic cycle incorporates the virus into the host genome as a prophage. Steps in infection include attachment, injection of DNA, macromolecular synthesis, assembly of new phage, and release of progeny phage.

CONCEPT 23.8 Animal Viruses Infect by Endocytosis

23.8.1 AIDS Is Caused by the Animal Virus HIV HIV targets macrophages and $CD4^+$ lymphocytes. The loss of these cell types compromises the ability to fight opportunistic infections. The viral protein gp120 binds to the cell-surface protein $CD4^+$ triggering receptor-mediated endocytosis. Replicated viruses bud off infected cells by exocytosis.

23.8.2 Zoonotic Events Can Cause Human Pandemics Influenza type A has caused periodic pandemics due to antigenic shift, where genome segments from different viruses are mixed. Coronaviruses cause respiratory illness, and a new coronavirus caused the COVID-19 pandemic

Concept Overview

This Concept Overview diagrams the key concepts that were discussed in this chapter.

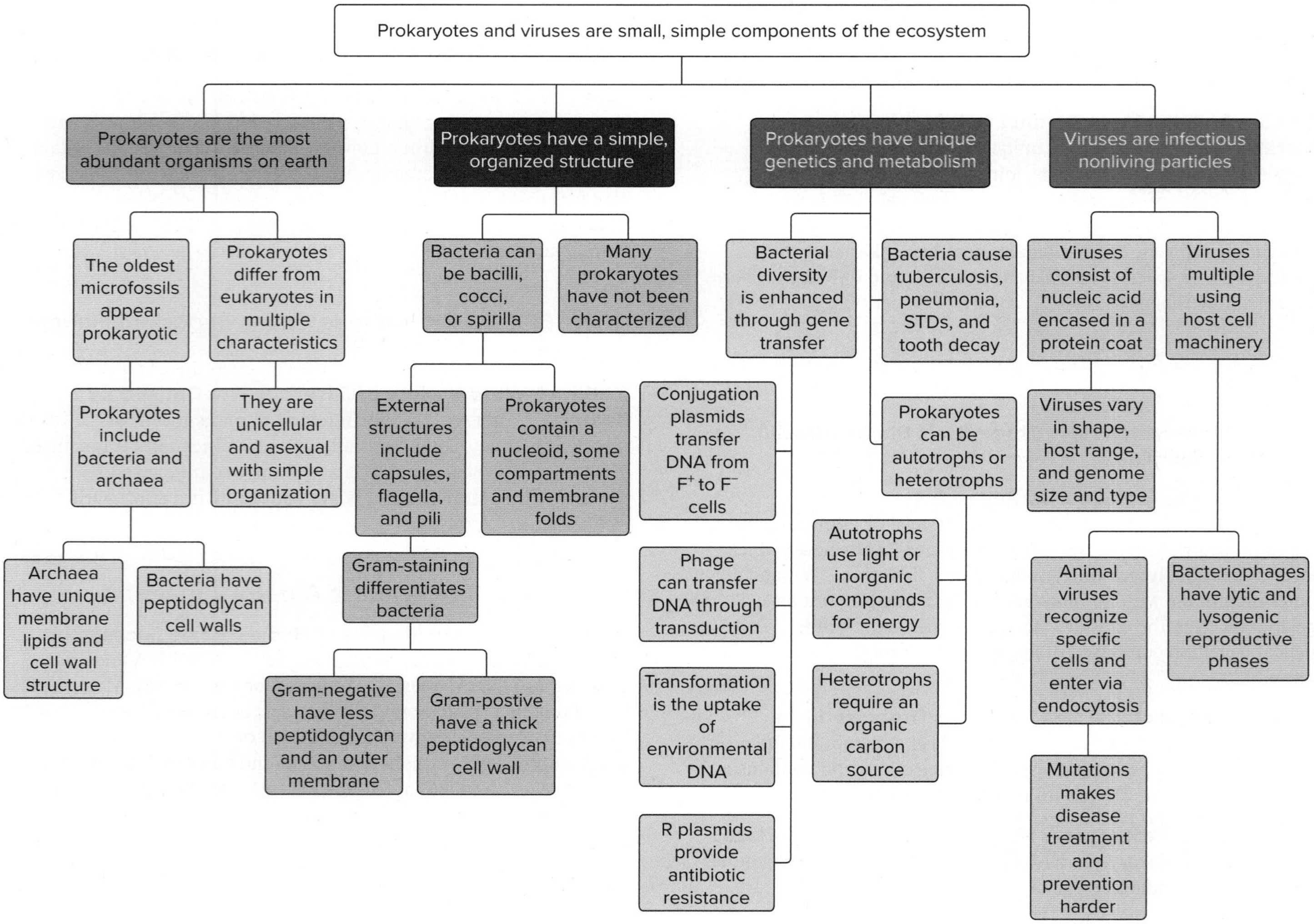

Assessing the Learning Path

Understand

1. Prokaryotic cellular organization is characterized by all of the following EXCEPT
   a. lipid bilayer plasma membranes.
   b. cell walls.
   c. a nucleus.
   d. ribosomes.

2. Which of the following characteristics is unique to the archaea?
   a. A fluid mosaic model of plasma membrane structure
   b. The use of an RNA polymerase during gene expression
   c. Ether-linked phospholipids
   d. A single origin of DNA replication

3. Which of the following is NOT a body form commonly found among the bacteria?
   a. Rod-shaped
   b. Spherical
   c. Plate-shaped
   d. Helical
4. The mechanisms of DNA exchange in prokaryotes share the feature of
   a. vertical transmission of information.
   b. horizontal transfer of information.
   c. requiring cell contact.
   d. the presence of a plasmid in one cell.
5. A cell that can use energy from the sun and CO_2 as a carbon source is a
   a. photoautotroph.
   b. chemolithoautotroph.
   c. photoheterotroph.
   d. chemoheterotroph.
6. Which of the following is common in animal viruses but not in bacteriophage?
   a. DNA
   b. Capsid
   c. Envelope
   d. Icosahedral shape
7. A membrane filter has pores in it that can be used to sort microbes by size. Which pore size would be most effective for removing bacteria but not viruses?
   a. 1 μm
   b. 300 nm
   c. 15 nm
   d. 1 nm
8. Which of the following is NOT part of the infection cycle of a lytic virus?
   a. Macromolecular synthesis
   b. Attachment to the host cell
   c. Assembly of progeny virus
   d. Integration into the host genome
9. Prior to entry, the __________ glycoprotein of the HIV virus recognizes the ______ receptor on the surface of the macrophage.
   a. CCR5; gp120
   b. CXCR4; CCR5
   c. CD4; CCR5
   d. gp120; CD4
10. A pandemic disease
    a. spreads rapidly over a large portion of the world.
    b. has a very high mortality rate.
    c. can be caused only by a virus.
    d. cannot be treated.

Apply

1. Bacterial and archaeal flagella, which both differ from eukaryotic flagella, do not appear to be homologous despite similar structure and function. From this we can infer
   a. that bacteria and archaea are more closely related than either is to eukaryotes.
   b. that bacteria and eukaryotes are more closely related than either is to archaea.
   c. that archaea and eukaryotes are more closely related than either is to bacteria
   d. nothing about the evolutionary relationships of these 3 groups
2. There have been no archaeal pathogens identified to date. If we were to find an archaeal pathogen, antibiotics like penicillin would
   a. not be effective.
   b. be effective.
   c. be effective if the archaea did not have an S-layer.
   d. be effective if the archaea did have an S-layer.
3. What do prokaryotic genetics tell us about how prokaryotes evolve?
   a. It implies that prokaryotic evolution occurs by vertical gene transfer.
   b. The transfer of DNA from cell to cell leads to horizontal gene transfer.
   c. Because they do not have sex, evolution does not act in prokaryotes.
   d. Evolution occurs by horizontal gene transfer, but very slowly.
4. The cell wall in both gram-positive and gram-negative cells is
   a. composed of phospholipids.
   b. a target for antibiotics that affect peptidoglycan synthesis.
   c. composed of peptidoglycan.
   d. surrounded by a membrane.
5. Compare and contrast photosynthesis among the archaea.
6. Ulcers and tooth decay do not appear related but, in fact, both
   a. are due to eating particular kinds of foods.
   b. are caused by viral infection.
   c. are caused by environmental factors.
   d. can be due to bacterial infection.
7. The reverse transcriptase enzyme is found in the virion of retroviruses. How is this related to the retroviral infection cycle?
   a. A viral RNA genome needs to be converted to DNA for gene expression.
   b. The viral genome is a single-stranded DNA that needs to be made double-stranded.
   c. The genome cannot be translated, so the enzyme is necessary to make a DNA copy.
   d. The enzyme uses the RNA genome to make an mRNA for translation.
8. Bacterial and animal viruses are similar in that they both
   a. have only DNA as genetic material.
   b. have only RNA as genetic material.
   c. require host functions for some aspect of their life cycle.
   d. do not require any host proteins.
9. Poliovirus attacks nerve cells, hepatitis attacks the liver, and AIDS attacks white blood cells. How does each virus know which cell to attack?
   a. They enter all cells but only reproduce in certain cells.
   b. They key in on certain cell-surface molecules characteristic of each cell type.
   c. The virus in each case is derived from that kind of cell.
   d. The cell types practice virus-specific phagocytosis.

Synthesize

1. Why do you think a biofilm is more resistant to antibiotics than is a laboratory culture?
2. Capsules enable prokaryote cells to adhere to surfaces and other cells. Can you suggest a reason capsules, often found in bacteria, are not common among archaea?
3. In what ways does conjugation create bacterial cells that are at least partially diploid? Is this a stable condition?
4. Explain how chemolithoautotrophs can exist in deep-sea vents where there is no light or free oxygen.
5. Most biologists believe that viruses evolved following the origin of the first cells. Defend or critique this concept.
6. Despite significant effort, we have been unable to produce a vaccine for HIV. While we have a vaccine for influenza, yearly vaccination is required. How does the biology of these two viruses affect vaccine design?

Stephen Durr

24 Protists

Learning Path

- **24.1** Protists, the First Eukaryotes, Arose by Endosymbiosis
- **24.2** Protists Are a Very Diverse Group
- **24.3** The Rough Outlines of Protist Phylogeny Are Becoming Clearer
- **24.4** Excavata Are Flagellated Protists Lacking Mitochondria
- **24.5** SAR: Stramenopiles and Alveolates Exhibit Secondary Endosymbiosis
- **24.6** SAR: Rhizaria Have Silicon Exoskeletons or Limestone Shells
- **24.7** Archaeplastida Are Descended from a Single Endosymbiosis Event
- **24.8** Amoebozoa and Opisthokonta Are Closely Related

Concept Overview

This Concept Overview provides a diagram of the overarching concepts that are covered in this chapter. These concepts will be expanded upon in the Concept Overview at the end of the chapter.

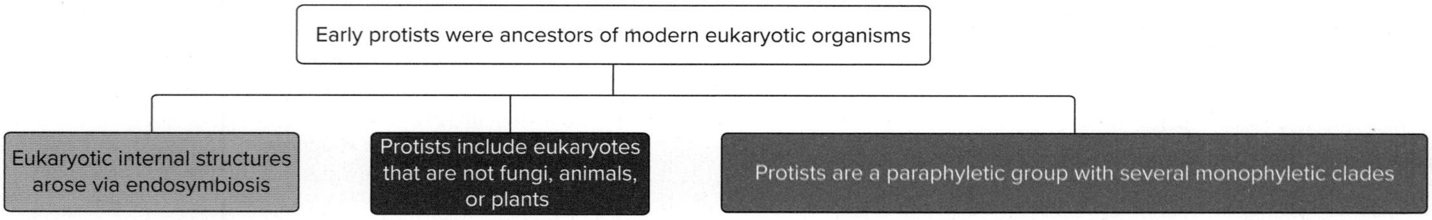
Early protists were ancestors of modern eukaryotic organisms
Eukaryotic internal structures arose via endosymbiosis
Protists include eukaryotes that are not fungi, animals, or plants
Protists are a paraphyletic group with several monophyletic clades

Introduction

For more than half of the history of life on Earth, all life was microscopic. For more than 2 billion years, the largest organisms in existence were unicellular prokaryotes. These prokaryotes lacked internal membranes, except for invaginations of surface membranes in photosynthetic bacteria.

The first evidence of a different kind of organism is found in tiny fossils in rock 1.5 billion years old. These fossil cells are much larger than bacteria (up to 10 times larger) and contain internal membranes and what appear to be small, membrane-bounded structures. The complexity and diversity of form among these single cells are astonishing. The step from relatively simple to quite complex cells marks one of the most important events in the evolution of life, the appearance of the eukaryote. Protists are a diverse group of eukaryotes that are not classified as animals, plants, or fungi. Although the group may seem a little strange at first glance, its members play important roles in human health, in ecology, and in nutrient cycling in the environment.

24.1 Protists, the First Eukaryotes, Arose by Endosymbiosis

Eukaryotic cells are distinguished from prokaryotes by the presence of a cytoskeleton and compartmentalization that includes a nuclear envelope and organelles. Exactly how large, complex eukaryotic cells arose will likely never be known. Scientists agree, however, that several key events happened. The loss of a rigid cell wall allowed membranes to fold inward, increasing surface area, and membrane flexibility made it possible for one cell to engulf another.

Fossil Evidence Dates the Origins of Eukaryotes

LEARNING OBJECTIVE 24.1.1 Describe the earliest evidence of eukaryotes.

Indirect chemical evidence suggests that eukaryotes may have arisen as long ago as 2.7 billion years, but no fossils as yet support such an early appearance. In rocks about 1.5 billion years old, we begin to see the first microfossils that are noticeably different in appearance from the earlier, simpler forms, none of which were more than 6 μm in diameter (figure 24.1). These cells are much larger than those of prokaryotes and have internal membranes and thicker walls.

These microfossils mark a major event in the evolution of life: a new kind of organism had appeared. These new cells are called eukaryotes. Cells must be either prokaryotic or eukaryotic.

In this chapter, the origins of eukaryotic cell internal structure are considered. Keep in mind that horizontal gene transfer is thought to have occurred frequently while eukaryotic cells were evolving. Eukaryotic cells evolved not only through horizontal gene transfer but also through infolding membranes and engulfing other cells. Today's eukaryotic cell is the result of cutting and pasting of DNA and exchange of organelles between species.

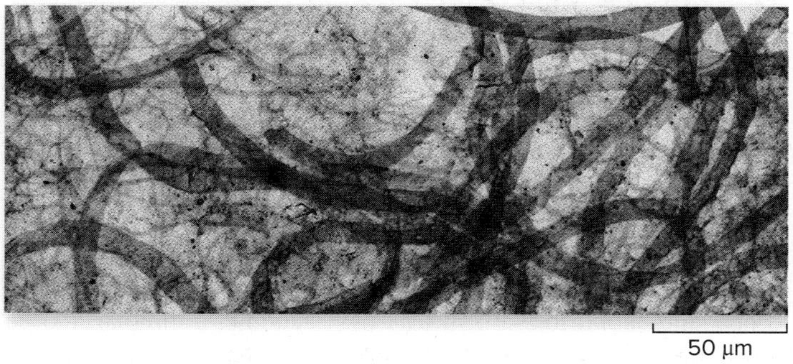

Figure 24.1 Early eukaryotic fossil. Fossil algae that lived in Siberia 1 billion years ago (BYA).

Andrew H. Knoll/Harvard University

The nucleus and ER arose from membrane infoldings

Many prokaryotes have infoldings of their outer membranes, which extend into the cytoplasm and serve as passageways to the surface. The network of internal membranes in eukaryotes is called the endoplasmic reticulum (ER). The nuclear envelope, an extension of the ER network that isolates and protects the nucleus, is thought to have evolved from these membrane infoldings (figure 24.2).

Mitochondria Evolved from Engulfed Aerobic Bacteria

LEARNING OBJECTIVE 24.1.2 Illustrate how endosymbiosis relates to the evolution of mitochondria.

Bacteria that live within other cells and perform specific functions for their host cells are called *endosymbiotic bacteria*. Their widespread presence in nature led biologist Lynn Margulis in the early 1970s to champion the theory of endosymbiosis, which was first proposed by Konstantin Mereschkowsky in 1905. *Endosymbiosis* means "living together in close association."

Endosymbiosis, a concept that is now widely accepted, suggests that a critical stage in the evolution of eukaryotic cells involved endosymbiotic relationships with prokaryotic organisms. According to this theory, energy-producing bacteria may have come to reside within larger bacteria, eventually evolving into what we now know as mitochondria (figure 24.3). Possibly the original host cell was anaerobic with hydrogen-dependent metabolic pathways. The symbiont had a form of respiration that produced H_2.

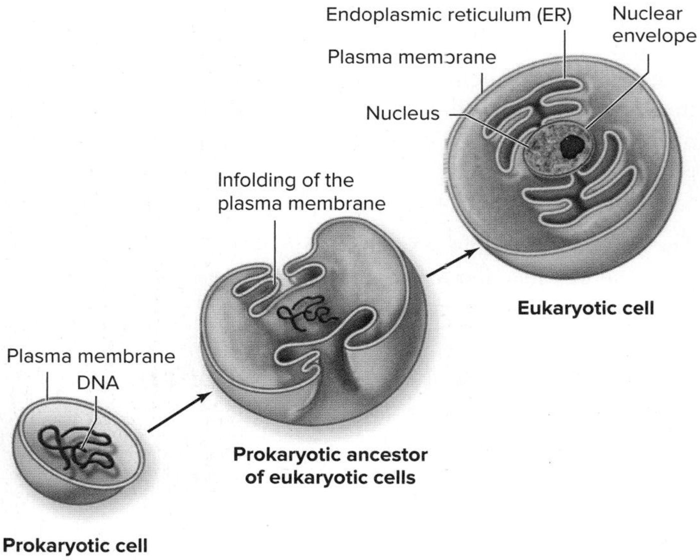

Figure 24.2 Origin of the nucleus and endoplasmic reticulum. Many prokaryotes today have infoldings of the plasma membrane. The eukaryotic internal membrane system, called the endoplasmic reticulum (ER), and the nuclear envelope may have evolved from such infoldings, encasing the DNA of prokaryotic cells that gave rise to eukaryotic cells.

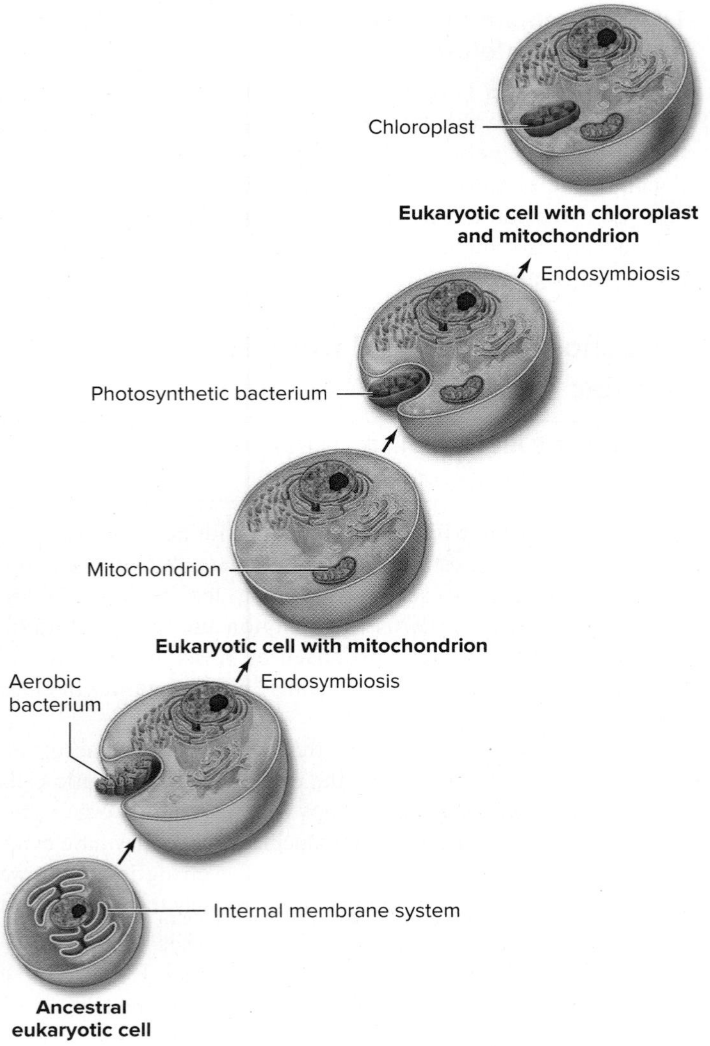
Chloroplast
Eukaryotic cell with chloroplast and mitochondrion
Endosymbiosis
Photosynthetic bacterium
Mitochondrion
Eukaryotic cell with mitochondrion
Aerobic bacterium
Endosymbiosis
Internal membrane system
Ancestral eukaryotic cell

Figure 24.3 The theory of endosymbiosis. Scientists propose that ancestral eukaryotic cells, which already had an internal system of membranes, engulfed aerobic bacteria, which then became mitochondria in the eukaryotic cell. Chloroplasts also originated this way, with eukaryotic cells engulfing photosynthetic bacteria.

The host depended on the symbiont for H_2 under anaerobic conditions and was able later to adapt to an O_2-rich atmosphere using the symbiont's respiratory pathways.

Chloroplasts Evolved from Engulfed Photosynthetic Bacteria

LEARNING OBJECTIVE 24.1.3 Explain the origin of chloroplasts.

Chloroplasts—the photosynthetic organelles of plants and algae—probably evolved when photosynthetic bacteria were engulfed by other, larger bacteria (figure 24.3). The history of chloroplast evolution is an example of the care that must be taken in phylogenetic studies. All chloroplasts are likely derived from a single line of cyanobacteria, but the organisms that host these chloroplasts are not monophyletic. This apparent paradox is resolved by

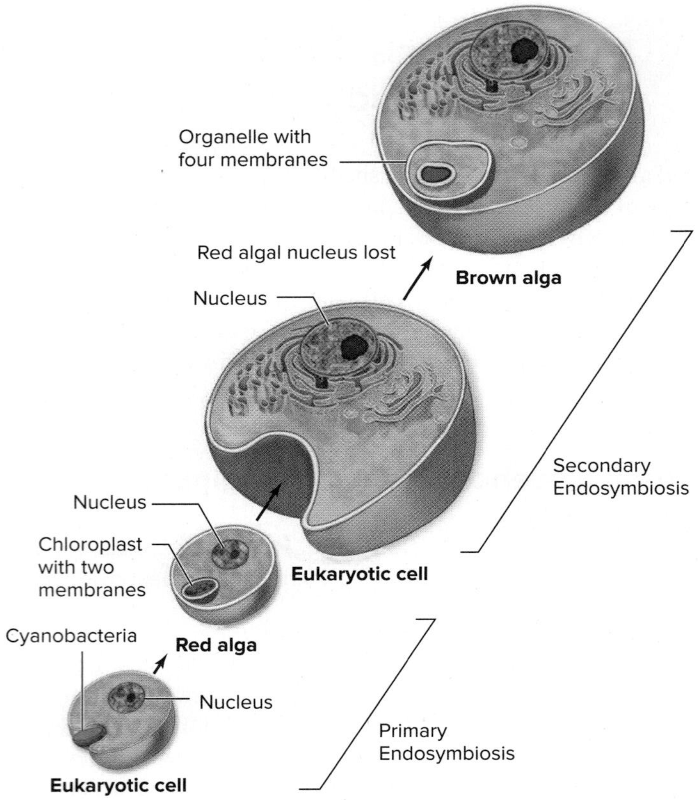
Organelle with four membranes
Red algal nucleus lost
Brown alga
Nucleus
Secondary Endosymbiosis
Nucleus
Chloroplast with two membranes
Eukaryotic cell
Cyanobacteria
Red alga
Nucleus
Primary Endosymbiosis
Eukaryotic cell

Figure 24.4 Endosymbiotic origins of chloroplasts in red and brown algae.

considering the possibility of secondary, and even tertiary, endosymbiosis. Red and green algae both obtained their chloroplasts by engulfing photosynthetic cyanobacteria. The brown algae most likely obtained their chloroplasts by engulfing one or more red algae, a process called **secondary endosymbiosis** (figure 24.4).

A phylogenetic tree based only on chloroplast gene sequences from red and green algae reveals an incredibly close evolutionary relationship. This tree is misleading, however, because it is not possible to tell just from these data how much the two algal lines had diverged at the time they engulfed the same line of cyanobacteria. Morphological and chemical traits are more helpful than chloroplast gene sequences in sorting out the evolutionary relationships between red and green algae.

Endosymbiosis is supported by a range of evidence

The fact that we now observe so many symbiotic relationships lends general support to the endosymbiotic theory. Even stronger support comes from the observation that present-day organelles such as mitochondria and chloroplasts contain their own DNA, which is remarkably similar in size and character to the DNA of bacteria. During the billion and a half years in which mitochondria have existed as endosymbionts inside eukaryotic cells, most of their genes have been transferred to the chromosomes of the host cells. Each mitochondrion still has its own genome, a circular, closed molecule of DNA similar to that found in bacteria. The chromosome has genes that encode the proteins essential for oxidative metabolism. These genes are transcribed within the mitochondrion, using mitochondrial ribosomes that are smaller than those of eukaryotic cells, very much like bacterial ribosomes in size and structure. Many antibiotics that inhibit protein synthesis in bacteria also inhibit protein synthesis in mitochondria and chloroplasts, but not ribosomes in the cytoplasm. The bacterial origin of chloroplasts and mitochondria is further supported by the observation that they, like bacteria, replicate by binary fission, not by mitosis.

Mitosis Evolved in Eukaryotes

LEARNING OBJECTIVE 24.1.4 Explain why mitosis is not believed to have evolved all at once.

The mechanisms of mitosis and cytokinesis, now so common among eukaryotes, did not evolve all at once. Traces of very different, and possibly intermediate, mechanisms survive today in some of the eukaryotes. In fungi and in some groups of protists, for example, the nuclear membrane does not dissolve, as it does in plants, animals, and most other protists, in which mitosis is confined to the nucleus. When mitosis is complete in these organisms, the nucleus divides into two daughter nuclei, and only then does the rest of the cell divide. We do not know whether mitosis without nuclear membrane dissolution represents an intermediate step on the evolutionary journey or simply a different way of solving the same problem.

REVIEW OF CONCEPT 24.1

Eukaryotes are organisms that contain a nucleus and other membrane-bounded organelles. Endoplasmic reticulum and the nuclear membrane are believed to have evolved from infoldings of the outer membranes. According to the endosymbiont theory, mitochondria and chloroplasts evolved from engulfed bacteria that remained intact. Mitochondria and chloroplasts have their own DNA, which is similar to that of prokaryotes. Mitosis did not evolve all at once; different mechanisms persist in different organisms.

- *What evidence supports the endosymbiont theory?*

24.2 Protists Are a Very Diverse Group

Protists are the most ancient eukaryotes and are classified on the basis of what they are not: they are not fungi, plants, or animals. In all other respects, they are highly variable, with no unifying features. Many are unicellular, like the *Vorticella* you see in figure 24.5, but there are numerous colonial and multicellular groups. Most are microscopic, but some are as large as trees.

Protists Are Eukaryotes That Are Not Fungi, Animals, or Plants

LEARNING OBJECTIVE 24.2.1 Describe the features that distinguish protists from other eukaryotes.

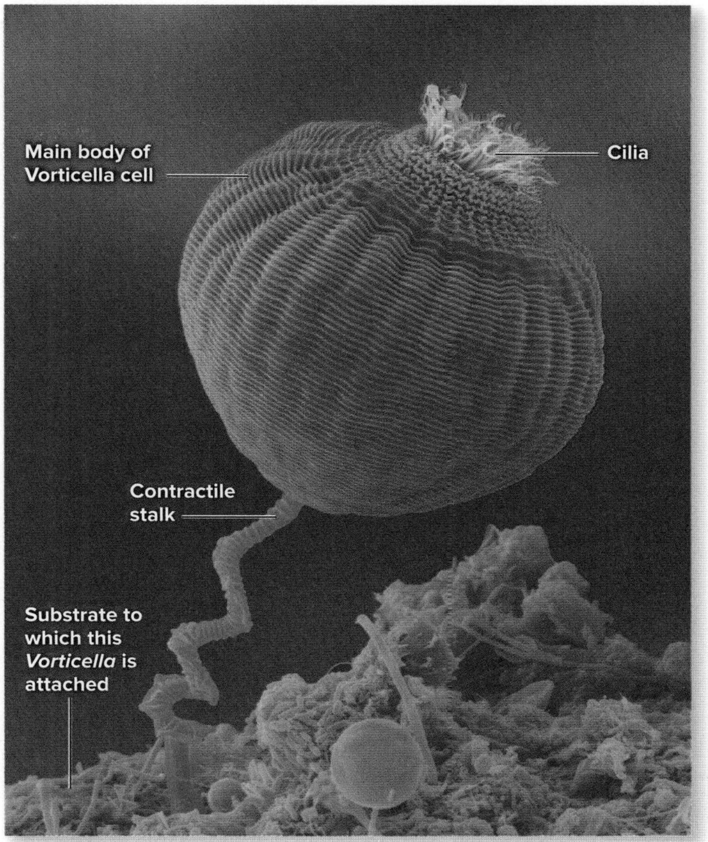

Figure 24.5 A unicellular protist. The protists are a catch all classification for many different groups of unicellular organisms, such as this *Vorticella* (a ciliate of the phylum Alveolata), which is heterotrophic, feeding on bacteria, and has a contractible stalk.
Dennis Kunkel Microscopy/Science Source

Protists possess a varied array of cell surfaces. Some protists, such as amoebas, are surrounded only by their plasma membrane. All other protists have a plasma membrane with an extracellular matrix (ECM) deposited on the outside of the membrane. Some ECMs form strong cell walls; for instance, diatoms and foraminifera secrete shells of silica.

Locomotor organelles

Protists move using diverse mechanisms. Many protists wave one or more flagella to move themselves through the water, whereas others use banks of short, flagella-like structures called cilia to create water currents for their feeding or propulsion. Among amoebas, large, blunt extensions of the cell body called pseudopodia (Greek, meaning "false feet") are the chief means of locomotion. Other protists extend thin, branching protrusions, sometimes supported by axial rods of microtubules. These so-called axopodia can be extended or retracted. Because the tips can adhere to adjacent surfaces, the cell can move by a rolling motion, shortening the axopodia in front and extending those in the rear.

Cyst formation

Some protists, even those with delicate cell surfaces, can survive and reproduce asexually in harsh conditions by forming cysts.

A cyst is a dormant form of a cell with a resistant outer covering in which cell metabolism is virtually shut down. Not all cysts are comparably resistant to hostile environments, however. Vertebrate parasitic amoebas, for example, form cysts that are quite resistant to gastric acid but will not tolerate desiccation or high temperature.

Nutrition

Protists can be heterotrophic or autotrophic. The heterotrophs obtain energy from organic molecules synthesized by other organisms. Some autotrophic protists are photosynthetic, whereas others are chemoautotrophic.

Some heterotrophic protists are *phagotrophs*; these are organisms that ingest particles of food into vesicles called food vacuoles or phagosomes. Lysosomes fuse with the food vacuoles, introducing enzymes that digest the food particles within. Digested molecules are absorbed across the vacuolar membrane.

Another example of the protists' tremendous nutritional flexibility is seen in *mixotrophs*, protists that are both phototrophic and heterotrophic.

Multicellularity

A single cell has its limits. As a cell becomes larger, its volume increases more rapidly than does its surface area. The cell surface cannot perform the functions that a high-volume cell needs at a certain cell size. The evolution of multicellular individuals solved this problem. Multicellularity is a condition in which an organism is composed of many cells, permanently associated with one another, that integrate their activities. Besides solving the surface-area-to-volume problem, a key advantage of multicellularity is that it allows specialization—distinct types of cells, tissues, and organs, each with different functions, can be differentiated within an individual's body. With such functional "division of labor," a multicellular organism can possess cells devoted specifically to protecting the body, others to moving it about, still others to seeking mates and prey, and yet others to carrying out a host of other activities.

Colonies. A colonial organism is a collection of cells that are permanently associated but with little or no integration of cell activities. Many protists form colonial assemblies, consisting of many cells with little differentiation or integration. In some protists, the distinction between colonial and multicellular is blurred. For example, in the green algae *Volvox,* shown in figure 24.6, individual motile cells aggregate into a hollow ball of cells that moves by the coordinated beating of the flagella of the individual cells. Although a few cells near the rear of the moving colony are reproductive cells; most are relatively undifferentiated.

Multicellular Individuals. True multicellularity, in which the activities of the individual cells are coordinated and the cells themselves are in contact, occurs only in eukaryotes and is one of their defining characteristics. Three groups of protists have independently attained true but simple multicellularity—the brown algae (phylum Phaeophyta), green algae (phylum Chlorophyta), and red algae (phylum Rhodophyta). In multicellular organisms, individuals are composed of many specialized cells that interact with one another and coordinate their activities.

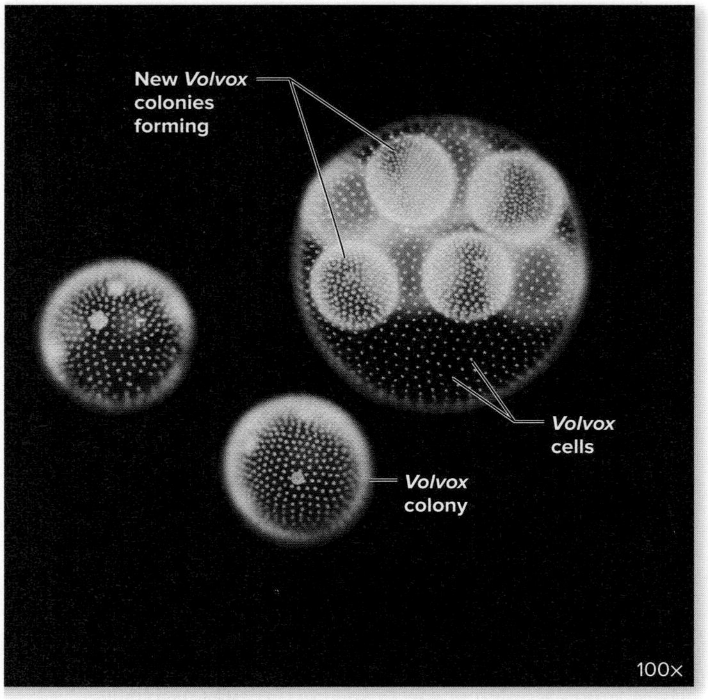

Figure 24.6 A colonial protist. Individual, motile, unicellular green algae are united in the protist *Volvox* (phylum Chlorophyta) as a hollow colony of cells that moves by the beating of the flagella of its individual cells. Some species of *Volvox* have cytoplasmic connections between the cells that help coordinate colony activities. The complex *Volvox* colony has many of the properties of multicellular life.
Stephen Durr

Reproduction

Protists typically reproduce asexually, although some have an obligate sexual reproductive phase and others undergo sexual reproduction at times of stress, including during food shortages.

Asexual Reproduction. **Asexual reproduction** involves mitosis, but the process often differs from the mitosis in multicellular animals. For example, the nuclear membrane often persists throughout mitosis, with the mitotic spindle forming within the nucleus.

In some species, a cell simply splits into nearly equal halves after mitosis; in some species, particularly ciliate species, this process is nearly always called binary fission. In some cases, however, division is not equal and the daughter cell is considerably smaller than its parent and then grows to adult size—a type of cell division called **budding.** In *schizogony,* common among some protists, cell division is preceded by several nuclear divisions. This allows cytokinesis to produce several individuals almost simultaneously.

Sexual Reproduction. Most eukaryotic cells also possess the ability to reproduce sexually, something prokaryotes cannot do. Meiosis is a major evolutionary innovation that arose in ancestral protists and allows for the production of haploid cells from diploid cells. **Sexual reproduction** is the process of producing offspring by fertilization, the union of two haploid

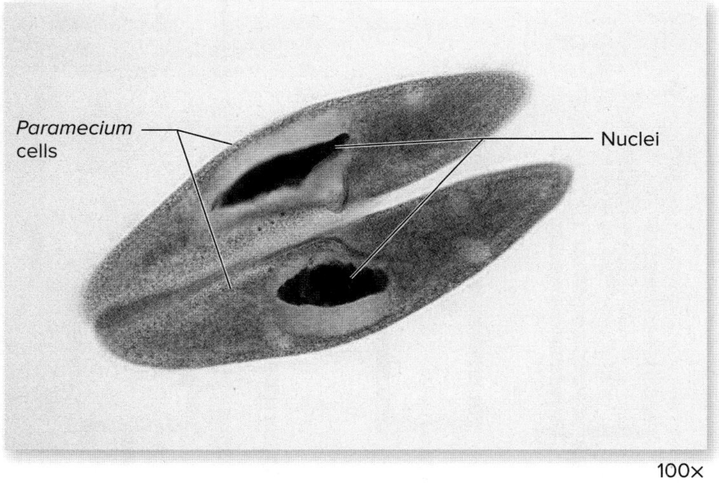

Figure 24.7 Sexual reproduction among paramecia. In sexual reproduction, two mature cells fuse in a process called conjugation and exchange haploid nuclei.
Ed Reschke/Stone/Getty Images

cells. The great advantage of sexual reproduction is that it allows for frequent genetic recombination, which contributes to the variation that is the starting point of evolution. Most eukaryotes reproduce sexually only in times of stress, like the paramecia undergoing conjugation in figure 24.7. The evolution of meiosis and sexual reproduction contributed significantly to the tremendous explosion of diversity among the eukaryotes.

REVIEW OF CONCEPT 24.2

All protists have plasma membranes, but other cell-surface components, such as deposited extracellular matrix (ECM), are highly variable. Protists mainly use flagella or pseudopodial movement to propel themselves. Phototrophic protists carry out photosynthesis and phagotrophs ingest food particles. Sexual reproduction is common under conditions of stress, but asexual reproduction is the rule in most groups. Multicellular organisms likely arose from colonial protists.

- *What are the advantages of movement by pseudopodia?*

24.3 The Rough Outlines of Protist Phylogeny Are Becoming Clearer

The origin of eukaryotes, which began with ancestral protists, is one of the most significant events in the evolution of life. It has led to an explosion of diversity, which continues to this day. As a matter of convenience, taxonomists placed all eukaryotes, except plants, fungi, and animals, into one group, the Protists. Some 200,000 or so distantly related organisms were joined into one paraphyletic group comprising 15 major phyla. We will keep this grouping for convenience in discussing the protists, but also consider other organizational schema.

Monophyletic Clades Have Been Identified Among the Protists

LEARNING OBJECTIVE 24.3.1 Describe how the 15 major phyla of protists are related.

Eukaryotes diverged rapidly in a world that was shifting from anaerobic to aerobic conditions. We may never be able to completely sort out the relationships among different lineages during this major evolutionary transition. Molecular methods provide insight into the evolutionary relationships among protists. Molecular systematics now allows us to sort out the roots of the entire eukaryotic tree. One hypothesis for this complete eukaryotic tree is presented in figure 24.8 with all known eukaryotes grouped into five supergroups:

Excavata. Named after a groove on one side of the cell body in some forms, this supergroup contains three major monophyletic clades. Two (Diplomonads and Parabasalids) have modified mitochondria; the third (Euglenozoa) has structurally unique flagella.

SAR. Stramenopiles and alveolates were formerly grouped as Chromalveolata. They are now grouped with Rhizaria to form the supergroup SAR, which is currently thought to be monophyletic. Stramenopiles and alveolates are largely photosynthetic. They include diatoms, dinoflagellates, and ciliates and appear to have arisen from a secondary symbiosis event. Rhizaria includes forams and radiolarians. Despite their many differences they are grouped together.

Archaeplastida. The red and green algae of this group contain related photosynthetic plastids. Plants arose from a green alga.

Amoebozoa. This group includes free-living amoebas and social amoebas (slime molds).

Opisthokonta. The fungi and animals of this group have been shown to be closely related. This group also includes the ancestors of animals (choanoflagellates).

Massive amounts of sequence data are being generated from diverse protist lineages with the goal of untangling the early branching of the eukaryotic tree. This area of research is fast moving and includes significant areas of controversy. Although not all protist lineages can be placed on this phylogenetic tree with full confidence, the arrangement presented in figure 24.8 presents a simplified version of current views. This exemplifies the challenges and excitement of the changes currently sweeping through taxonomy and phylogeny, which were explored briefly in chapter 22.

REVIEW OF CONCEPT 24.3

The paraphyletic group called protists contains the ancestors of plants, fungi, and animals. Current polyphyletic molecular studies suggest that protists fall into 15 monophyletic clades, clustered in five supergroups.

- *Why are the protists considered to be a paraphyletic group?*

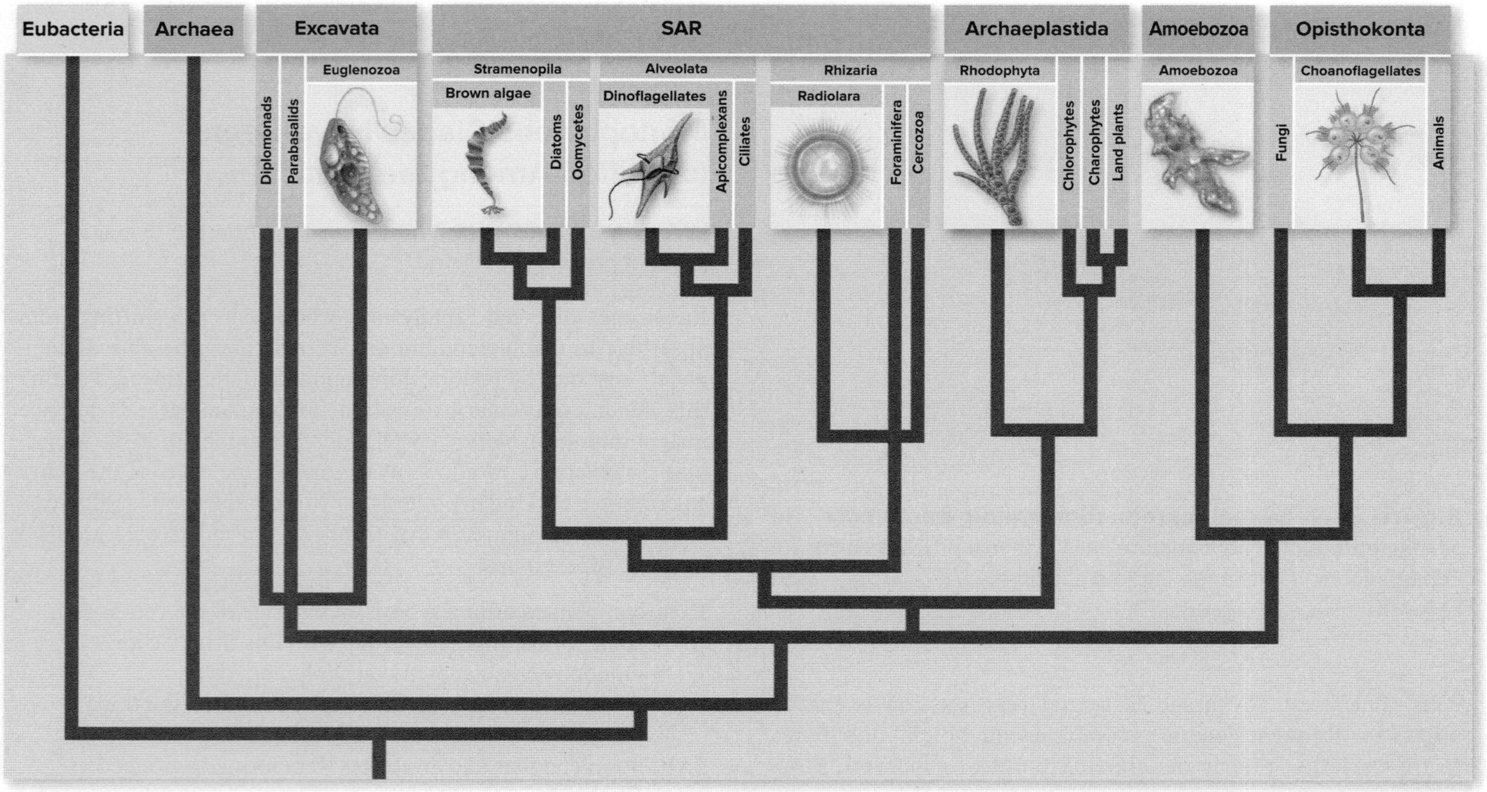

Figure 24.8 Eukaryotic evolutionary relationships. Analyses of molecular data group the eukaryotes into the five supergroups shown on the top level of the figure. The phylogeny has been simplified to reflect the lack of consensus over deeper branching relationships. Within the eukaryotes, plants, fungi, and animals are monophyletic clades, but protists are paraphyletic. Protist lineages are shaded in gray.

24.4 Excavata Are Flagellated Protists Lacking Mitochondria

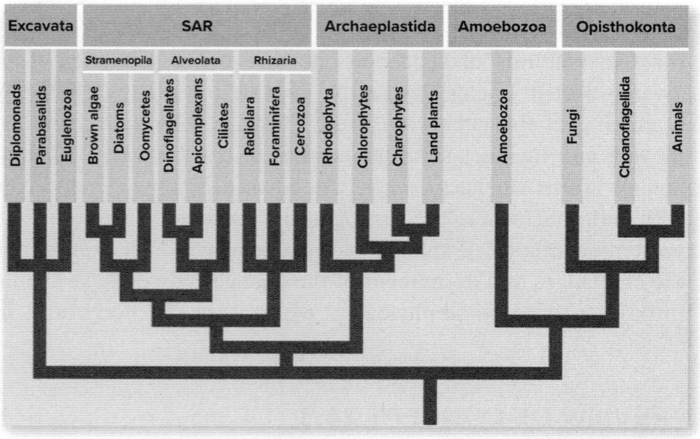

Diplomonads Have Two Nuclei

LEARNING OBJECTIVE 24.4.1 List the main features of diplomonads.

The supergroup Excavata is composed of three monophyletic clades, the *diplomonads,* the *parabasalids,* and the Euglenozoa. The name *Excavata* refers to a groove down one side of the cell body in some groups.

Diplomonads are unicellular and move with flagella. Members of this group have two nuclei. The human parasite *Giardia intestinalis* is a diplomonad (figure 24.9). *Giardia* can pass from human to human via contaminated water and cause diarrhea. Their nuclei contain mitochondrial genes, implying that *Giardia* evolved from aerobes. Electron micrographs of *Giardia* cells stained with mitochondrial-specific antibodies reveal **mitochondrion-related organelles (MROs)** called mitosomes. Thus, *Giardia* is unlikely to represent an early protist.

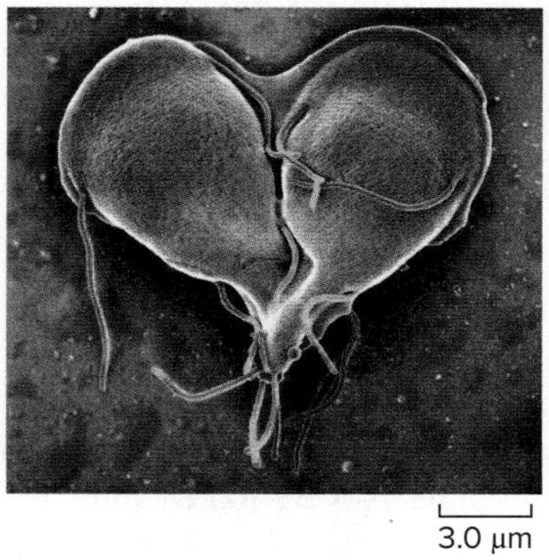

Figure 24.9 ***Giardia intestinalis.*** This parasitic diplomonad lacks a mitochondrion.

Dr. Stan Erlandsen/Centers for Disease Control and Prevention

Parabasalids Have Undulating Membranes

LEARNING OBJECTIVE 24.4.2 List the main features of parabasalids and give an example.

Parabasalids contain an intriguing array of species. Some live in the gut of termites and digest cellulose, the main component of the termite's wood-based diet. The symbiotic relationship includes another level, because these parabasalids also have a symbiotic relationship with bacteria that aid in the digestion of cellulose. The persistent activity of these three symbiotic organisms from three different kingdoms can lead to the collapse of a home built of wood or recycle tons of fallen trees in a forest. Another parabasalid, *Trichomonas vaginalis,* causes a sexually transmitted disease in humans.

Parabasalids have undulating membranes that assist in locomotion (figure 24.10). Like diplomonads, parabasalids also use flagella to move and lack classic mitochondria but possess MROs.

Euglenozoa Are Free-Living Eukaryotes with Anterior Flagella and Often Chloroplasts

LEARNING OBJECTIVE 24.4.3 Describe the distinguishing feature of euglenoids and kinetoplastids.

Euglenozoa diverged early and were among the earliest free-living eukaryotes to possess mitochondria. A number of the Euglenozoa have acquired chloroplasts via endosymbiosis, and this is one of their distinguishing features. None of the algae are closely related to Euglenozoa, a reminder that endosymbiosis is widespread. About one-third of the approximately 40 genera of Euglenozoa have chloroplasts and are fully autotrophic; the others are heterotrophic and ingest their food.

Euglena: The best-known euglenoid

The best-known phylum of Euglenozoa are the euglenoids. Individual euglenoids range from 10 to 500 μm long and vary greatly in form. Interlocking proteinaceous strips arranged in a helical pattern form a flexible structure called a *pellicle,* which lies within the plasma membrane of the euglenoids. Because its pellicle is flexible, a euglenoid is able to change its shape.

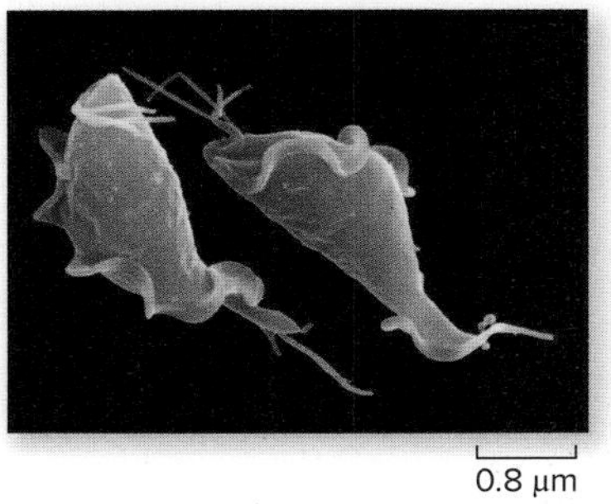
0.8 μm

Figure 24.10 Undulating membrane characteristic of parabasalids. Vaginitis can be caused by this parasite species, *Trichomonas vaginalis.*
David M. Phillips/Science Source

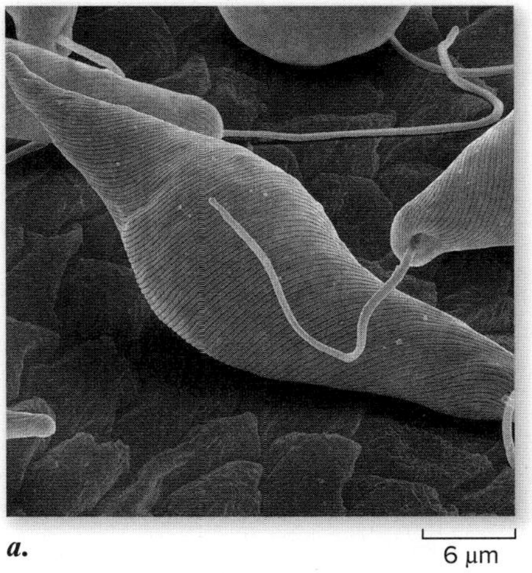
a.
6 μm

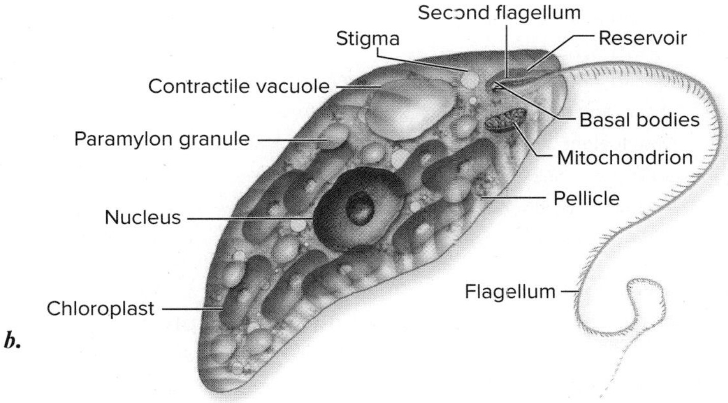
b.
Second flagellum
Stigma
Reservoir
Contractile vacuole
Basal bodies
Paramylon granule
Mitochondrion
Pellicle
Nucleus
Flagellum
Chloroplast

Figure 24.11 Euglenoids. ***a.*** Micrograph of *Euglena gracilis.* ***b.*** Diagram of *Euglena.* Paramylon granules are areas where food reserves are stored.
(a): Andrew Syred/Science Source

Reproduction in this phylum occurs by mitotic cell division. The nuclear envelope remains intact throughout the process of mitosis. No sexual reproduction is known to occur in this group.

In *Euglena* (figure 24.11), two flagella are attached at the base of a flask-shaped opening called the *reservoir,* which is located at the anterior end of the cell. One of the flagella is long and has a row of very fine, short, hairlike projections along one side. A second, shorter flagellum is located within the reservoir but does not emerge from it. Contractile vacuoles collect excess water from all parts of the organism and empty it into the reservoir; this apparently helps regulate the osmotic pressure within the organism. The stigma, which is also found in the green algae (phylum Chlorophyta), helps these photosynthetic organisms move toward light.

Cells of *Euglena* contain numerous small chloroplasts. These chloroplasts, like those of the green algae and plants, contain chlorophylls *a* and *b,* together with carotenoids. Although the chloroplasts of euglenoids differ somewhat in structure from those of green algae, they probably had a common origin. *Euglena*'s photosynthetic pigments are light-sensitive (figure 24.12). It seems likely that euglenoid chloroplasts ultimately evolved from a symbiotic relationship through ingestion of green algae.

Trypanosomes: Disease-causing kinetoplastids

Another major group within the *Euglenozoa* is the *kinetoplastids.* The name *kinetoplastid* refers to a unique, single mitochondrion in each cell. The mitochondria contain an unusual DNA

SCIENTIFIC THINKING

Hypothesis: Euglena *cells do not retain photosynthetic pigments in a dark environment.*

Prediction: *Photosynthetic pigments will be degraded when light-grown* Euglena *cells are transferred to the dark and new pigment will not be produced.*

Test: *Grow* Euglena *under normal light conditions. Transfer the culture to two flasks. Take a sample from each flask and measure the amount of photosynthetic pigments in each. Maintain one flask in the light and transfer the other to the dark. After several days, extract the photosynthetic pigments from each flask, and compare amounts with each other and with initial levels.*

Grow culture of *Euglena* under light.
Partition into two flasks.
Take a sample from each flask and quantify photosynthetic pigment.
Put one flask in dark, and expose the other to light.
Allow growth for several days.
Quantify photosynthetic pigment in each flask.

Result: *Photosynthetic pigment levels are lower in the flask grown in the dark than in the flask grown in light. Pigment levels in the dark-grown flask are lower than at the beginning of the experiment. Pigment levels in the light-grown flask are unchanged.*

Conclusion: *The hypothesis is supported. Keeping* Euglena *in the dark resulted in a loss of photosynthetic pigment. Pigments were degraded in the dark-grown flask.*

Further Experiments: *Transfer dark-grown flasks back to the light and measure changes in pigment levels over time. Are original pigment levels restored after growth in light?*

Figure 24.12 Effect of light on *Euglena* photosynthetic pigments.

structure called kDNA composed of interlocked minicircles and maxicircles. The maxicircles encode proteins, and the minicircles encode guide RNAs involved in editing maxicircle-encoded RNAs. This group also has modified peroxisomes called glycosomes that are unique in containing metabolic enzymes, including some glycolytic enzymes.

Parasitism has evolved multiple times within the kinetoplastids. This includes trypanosomes, which cause many serious human diseases, the most familiar being trypanosomiasis, also known as African sleeping sickness, which causes extreme lethargy and fatigue (figure 24.13).

Leishmaniasis, which is transmitted by sand flies, is a trypanosomic disease that causes skin sores and in some cases can affect internal organs, leading to death. In 2020 the WHO reported about 220,000 new cases of leishmaniasis globally. Estimates based on under-reporting suggest, however, that the global burden is closer to 1.3 million new cases each year. Deaths each year are on the order of 20,000 to 30,000. The rise in leishmaniasis in South America correlates with the movement of infected individuals from rural to urban environments, where there is a greater chance of spreading the parasite.

Chagas disease is caused by *Trypanosoma cruzi*. About 6 to 7 million people are infected worldwide, mostly in Latin America. The disease is spread mainly by contact with the feces of blood-sucking triatomine bugs, which are passive vectors. The bugs live in the cracks of poorly constructed housing, making this a problem in poorer areas. Blood transfusions have also increased the spread of the infection. Chagas disease can lead to severe cardiac and digestive problems in humans and domestic animals.

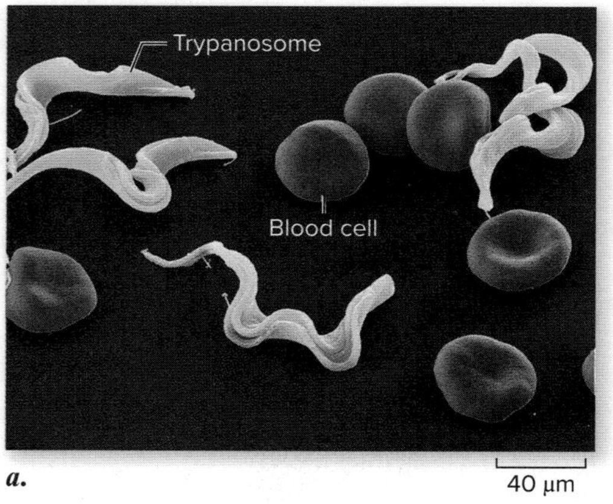

Trypanosome
Blood cell
40 μm

a.

b.

Figure 24.13 A kinetoplastid. ***a.*** *Trypanosoma* among red blood cells. The anterior flagella and undulating, changeable shape of the trypanosomes are visible in this photomicrograph. ***b.*** A tsetse fly, shown here sucking blood from a human arm, can carry trypanosomes.

(a): Eye of Science/Science Source
(b): Image Quest Marine/Alamy Stock Photo

Control of trypanosomes is especially difficult because of some interesting genetic features of these organisms. For example, tsetse-fly-transmitted trypanosomes have evolved an elaborate mechanism for repeatedly shifting the antigens in their protective glycoprotein coat, thus evading the antibodies produced by their hosts (refer to chapter 35).

In the guts of the flies that spread them, trypanosomes are noninfective. When they are ready to transfer to the skin or bloodstream of their host, trypanosomes migrate to the salivary glands and acquire the thick coat of glycoprotein antigens that protect them from the host's antibodies. Later, when they are taken up by a tsetse fly, the trypanosomes again shed their coats.

The production of vaccines against such a system is complex, but work is ongoing. Releasing sterilized flies to impede the reproduction of populations is also being tried to control the fly population. Traps made of dark cloth and scented like cows, but poisoned with insecticides, have likewise proved effective.

The recent sequencing of the genomes of the three kinetoplastids revealed a core of common genes in all three. The devastating toll of all three on human health and life could be alleviated by the development of a single drug targeted at one or more of the core proteins shared by the three parasites.

REVIEW OF CONCEPT 24.4

Diplomonads lack classic mitochondria but have mitochondrial-related organelles (MROs). They are unicellular, have two nuclei, and move with flagella. Parabasalids also have MROs and use flagella and undulating membranes for locomotion. The Euglenozoa, among the earliest protists to contain mitochondria, include both phototrophs and heterotrophs. The kinetoplastids contain a single mitochondrion with two types of DNA and the ability to edit RNA with RNA guides. Trypanosomes are disease-causing kinetoplastids.

- *How does a contractile vacuole regulate osmotic pressure in a* Euglena *cell?*

24.5 SAR: Stramenopiles and Alveolates Exhibit Secondary Endosymbiosis

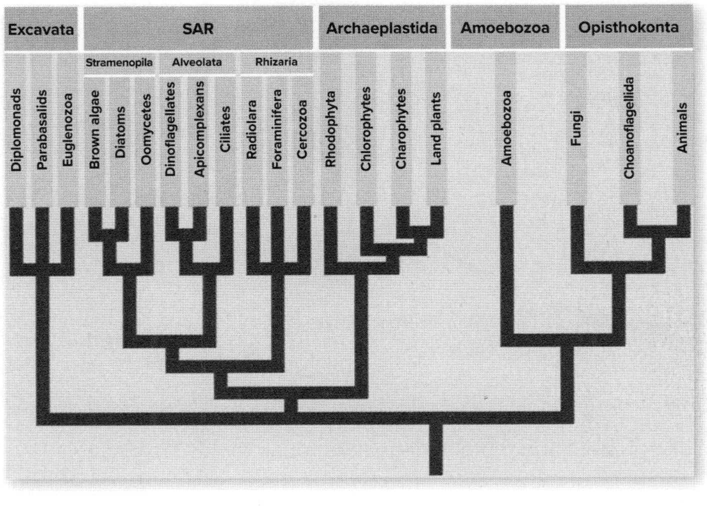

The **Stramenopila** and **Alveolata** are each thought to form monophyletic clades within the SAR supergroup. The DNA evidence no longer supports a monophyletic clade uniting these two groups, but the SAR supergroup joining them with Rhizaria does appear, at present, to be monophyletic. Given the similarities between Stramenopila and Alveolata, we will consider them together. Both groups are largely photosynthetic and are thought to have originated over a billion years ago when an ancestor of both groups engulfed a single-celled photosynthetic red alga. Because red algae originated by primary endosymbiosis, this event is referred to as secondary endosymbiosis; it yields a chloroplast with four membranes instead of two.

Stramenopila Have Fine Hairs

LEARNING OBJECTIVE 24.5.1 Describe the distinguishing feature of brown algae, diatoms, and water molds.

Stramenopiles include *brown algae, diatoms,* and the *oomycetes* (water molds). The name *stramenopila* refers to unique, fine hairs (figure 24.14) found on the flagella of members of this group, although a few species have lost the hairs during evolution.

Brown algae include large seaweeds

Brown algae are the most conspicuous seaweeds in many northern regions (figure 24.15). The life cycle of the brown algae is marked by an alternation of generations between a multicellular sporophyte (diploid) and a multicellular gametophyte (haploid) (figure 24.16). Some sporophyte cells go through meiosis and produce spores. These spores germinate and undergo mitosis to produce the large individuals we recognize, such as the kelps. The gametophytes are often much smaller, filamentous individuals, being perhaps only a few centimeters wide.

Even in an aquatic environment, transport of nutrients throughout an organism can be a challenge for the very large brown algal species. Distinctive transport cells that stack one

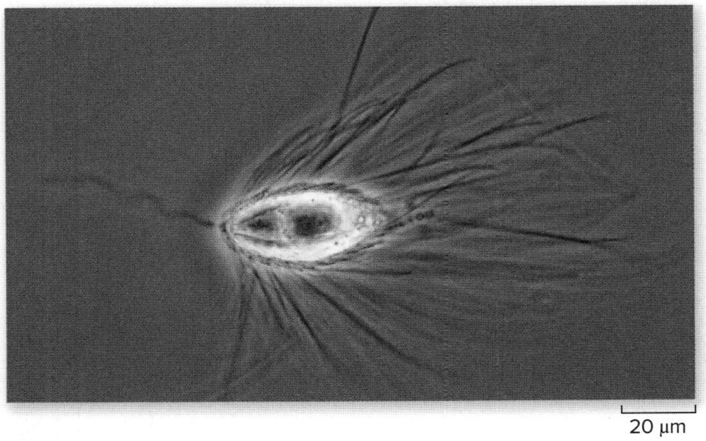

Figure 24.14 Stramenopiles have very fine hairs on their flagella.

Michele Bahr and D.J. Patterson, used under license

Figure 24.15 Brown alga. The giant kelp, *Macrocystis pyrifera,* grows in relatively shallow water along the coasts throughout the world and provides food and shelter for many different kinds of organisms.
Ethan Daniels/Shutterstock

upon the other enhance transport within some species. However, even though the large kelp look like plants, it is important to realize that they do not contain complex tissues, such as xylem, that are found in plants.

Diatoms are unicellular organisms with double shells

Diatoms, members of the phylum Chrysophyta, are photosynthetic, unicellular organisms with unique double shells made of opaline silica, which are often strikingly marked (figure 24.17). The shells of diatoms are like small boxes with lids, one half of the shell fitting inside the other. Their chloroplasts, containing chlorophylls *a* and *c,* as well as carotenoids, resemble those of the brown algae and dinoflagellates. Diatoms produce a unique carbohydrate called chrysolaminarin.

Some diatoms move by using two long grooves, called *raphes,* which are lined with vibrating fibrils (figure 24.18). The exact way in which these organisms move remains unclear but may involve the ejection of mucopolysaccharide streams from the raphe that propel the diatom. Pencil-shaped diatoms can slide back and forth over each other, creating an ever-changing shape.

Oomycetes, the "water molds," have some pathogenic members

All oomycetes are either parasites or saprotrophs (organisms that live by feeding on dead organic matter). At one time, these

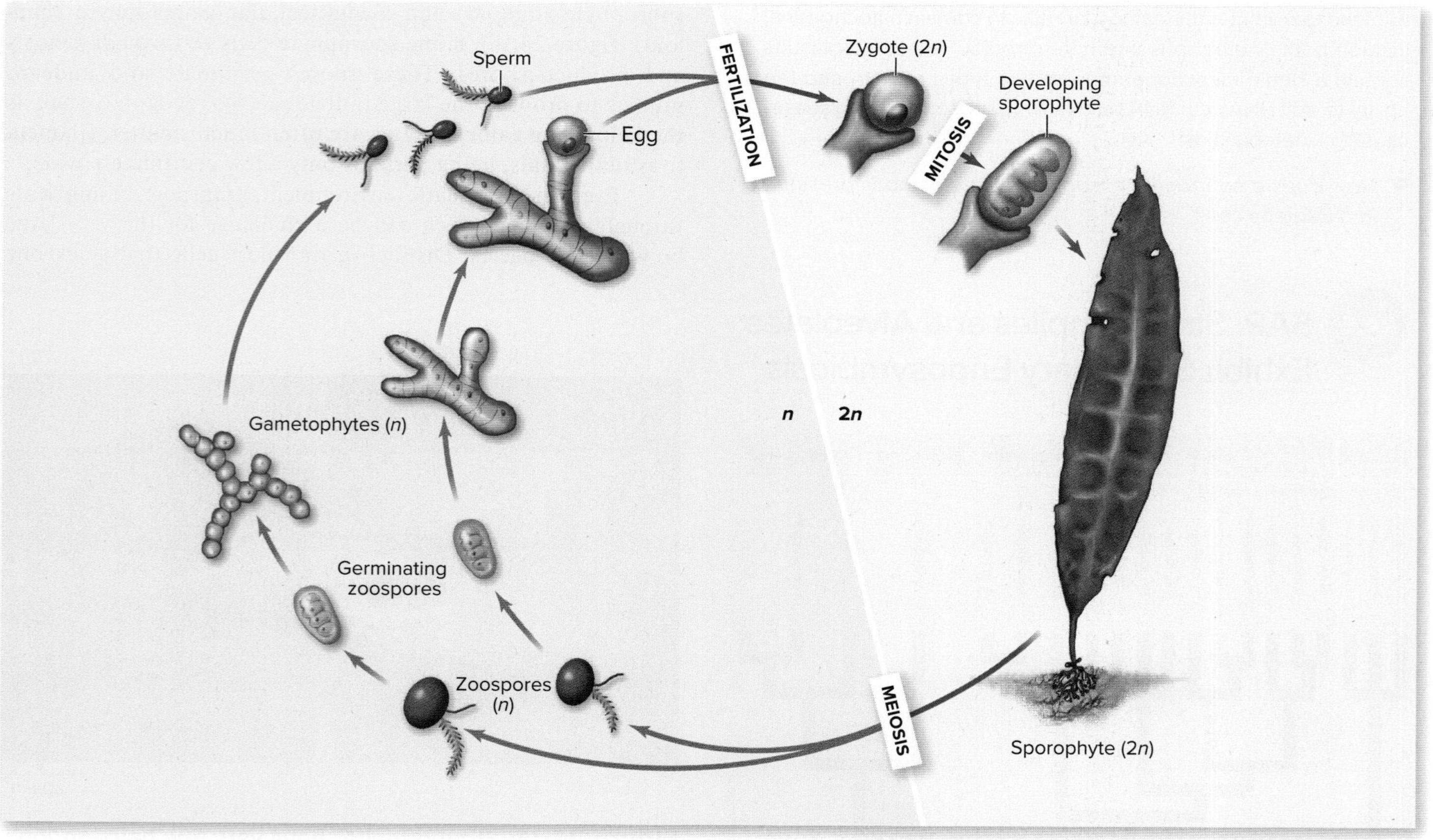

Figure 24.16 Life cycle of *Laminaria,* a brown alga. Multicellular haploid and diploid stages are found in this life cycle, although the male and female gametophytes are relatively small.

Figure 24.17 Diatoms. These different radially symmetrical diatoms have unique two-part silica shells.
Dennis Kunkel Microscopy/Science Source

organisms were considered fungi, which is the origin of the term *water mold* and why their name contains *-mycetes.*

They are distinguished from other protists by the structure of their motile spores, or zoospores, which bear two unequal flagella, one pointed forward and the other backward. Zoospores are produced asexually in a sporangium. Sexual reproduction involves the formation of male and female reproductive organs that produce gametes. Most oomycetes are found in water, but their terrestrial relatives are plant pathogens.

Phytophthora infestans caused the late blight of potatoes in Ireland between 1845 and 1849. The blight, along with the politics of the time, resulted in the Great Famine. During the famine, about 400,000 people starved to death or died of diseases complicated by starvation, and about 2 million Irish immigrated to the United States and elsewhere.

Another oomycete, *Saprolegnia,* is a fish pathogen that can cause serious losses in fish hatcheries. When these fish are released into lakes, the pathogen can infect amphibians and kill millions of amphibian eggs at a time at certain locations. This pathogen is thought to contribute to the phenomenon of amphibian decline.

Alveolata Have Submembrane Vesicles

LEARNING OBJECTIVE 24.5.2 Describe the distinguishing feature of dinoflagellates, apicomplexans, and ciliates.

The group Alveolata contains three major subgroups: dinoflagellates, apicomplexans, and ciliates, all of which have a common lineage but diverse modes of locomotion. One trait common to all three subgroups is the presence of flattened vesicles called alveoli (hence the name *Alveolata*) stacked in a continuous layer below their plasma membranes (figure 24.19). The precise function of the alveoli is not clear. They may function in membrane transport, similar to Golgi bodies, or perhaps to regulate the cell's ion concentration.

Dinoflagellates are photosynthesizers with distinctive features

Most dinoflagellates are photosynthetic unicells with two flagella. Dinoflagellates live in both marine and freshwater environments. Some dinoflagellates are luminous and contribute to the twinkling or flashing effects seen in the sea at night, especially in the tropics.

The flagella, protective coats, and biochemistry of dinoflagellates are distinctive, and the dinoflagellates do not appear to be directly related to any other phylum. Plates made of a cellulose-like material, often encrusted with silica, encase the dinoflagellate cells (figure 24.20). Grooves at the junctures of these plates usually house the flagella, one encircling the cell like a belt, and the other perpendicular to it. By beating in their grooves, these flagella cause the dinoflagellate to spin as it moves.

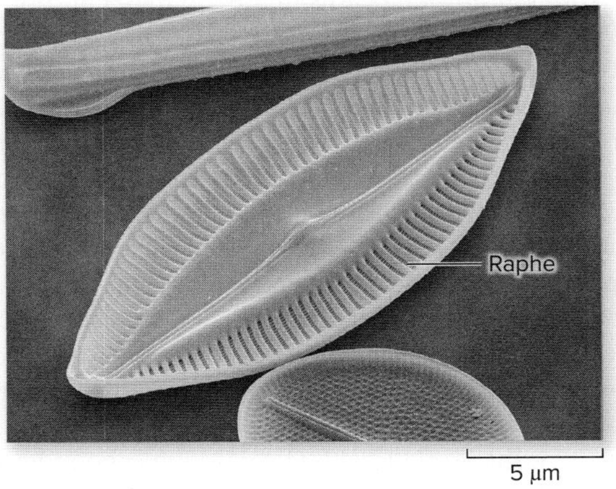

Figure 24.18 Diatom raphes are lined with fibrils that aid in locomotion.
Andrew Syred/Science Source

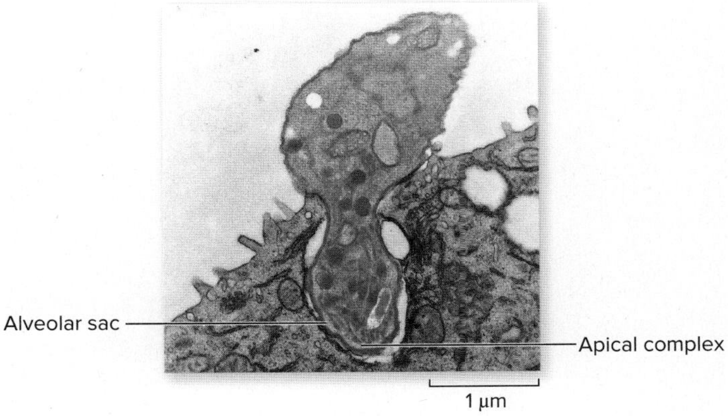

Figure 24.19 Alveoli are a continuum of vesicles just below the plasma membrane of dinoflagellates, apicomplexans, and ciliates. The apical complex of apicomplexans forces the parasite into host cells.
Vern Carruthers, David Elliott

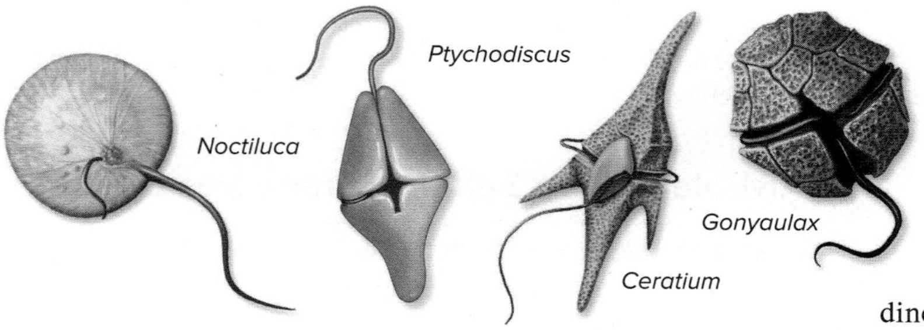

Figure 24.20 Some dinoflagellates. *Noctiluca,* which lacks the heavy cellulose armor characteristic of most dinoflagellates, is one of the bioluminescent organisms that cause the waves to sparkle in warm seas. In the other three genera, the shorter, encircling flagellum is seen in its groove, with the longer one projecting away from the body of the dinoflagellate (not drawn to scale).

Most dinoflagellates have chlorophylls *a* and *c,* in addition to carotenoids, so that the biochemistry of their chloroplasts resembles that of the diatoms and the brown algae. Possibly this lineage acquired such chloroplasts by forming endosymbiotic relationships with members of those groups.

The poisonous and destructive "red tides" that occur frequently in coastal areas are often associated with great population explosions, or "blooms," of dinoflagellates, whose pigments color the water (figure 24.21). Red tides have a profound, detrimental effect on the fishing industry worldwide. Some 20 species of dinoflagellates produce powerful toxins that inhibit the diaphragm and cause respiratory failure in many vertebrates. When the toxic dinoflagellates are abundant, many fishes, birds, and marine mammals may die.

Although sexual reproduction does occur under starvation conditions, dinoflagellates reproduce primarily by asexual cell division. Asexual cell division relies on a unique form of mitosis in which the permanently condensed chromosomes divide within a permanent nuclear envelope. After the numerous chromosomes duplicate, the nucleus divides into two daughter nuclei.

Figure 24.21 Red tide. Although small in size, huge populations of dinoflagellates can color the sea and release toxins into the water. In this case, the dinoflagellates are green.
Visual&Written SL/Alamy Stock Photo

Also, the dinoflagellate chromosome is unique among eukaryotes in that the DNA is not generally complexed with histone proteins. In all other eukaryotes, the chromosomal DNA is complexed with histones to form nucleosomes, structures that represent the first order of DNA packaging in the nucleus (refer to chapter 10). How dinoflagellates maintain distinct chromosomes with a small amount of histones remains a mystery.

Apicomplexans include the malaria parasite

Apicomplexans are spore-forming parasites of animals. They are called apicomplexans because of a unique arrangement of fibrils, microtubules, vacuoles, and other cell organelles at one end of the cell, termed an *apical complex* (figure 24.19). The apical complex is a cytoskeletal and secretory complex that enables the apicomplexan to invade its host. The best-known apicomplexan is the malarial parasite *Plasmodium.*

***Plasmodium* and Malaria.** *Plasmodium* glides inside the red blood cells of its host with amoeboid-like contractility. Like other apicomplexans, *Plasmodium* has a complex life cycle involving sexual and asexual phases and alternation between different hosts—in this case, mosquitoes *(Anopheles gambiae)* and humans (figure 24.22). Even though *Plasmodium* has mitochondria, it grows best in a low-O_2, high-CO_2 environment.

In the period from 2000 to 2015, new cases of malaria fell by 37% globally and by 42% in Africa, the region with the highest burden of malaria cases. Malaria control involves primarily vector control, and progress in this area has focused on the use of insecticide-treated mosquito nets (ITNs). In sub-Saharan Africa, the percentage of the population that had access to ITNs had increased to 55% compared with only 2% in 2000. The use of ITNs combined effectively with indoor residual spraying (IRS) helps control vectors worldwide. Unfortunately, from 2018–2019 the number of new malaria cases held steady at 228 and 229 million cases, respectively. Similarly, the number of malaria deaths was relatively unchanged in the same period, with 411 million deaths in 2018 and 409 million deaths in 2019.

Treatment of malaria sufferers has also improved with artemisinin-based combination therapies, in which the relatively new drug artemisinin is combined with traditional antimalarial drugs. As the mode of action of artemisinin is quite different from that of the conventional drugs, the combination therapy has been used to combat drug resistance. As with all such efforts to control insect pests and to use drugs to control infections in humans, the biggest problem is the appearance of parasite resistance. There have been reports of artemisinin resistance in Asia, and resistance to the insecticides used on nets and indoor spraying is also possible.

In 2021, the World Health Organization endorsed a vaccine targeting the malarial parasite *Plasmodium falciparum*. The vaccine, called Mosquirix, is the first vaccine targeting a parasite. Although only moderately effective, the vaccine is expected to save the lives of tens of thousands of children each year, most in sub-Saharan Africa.

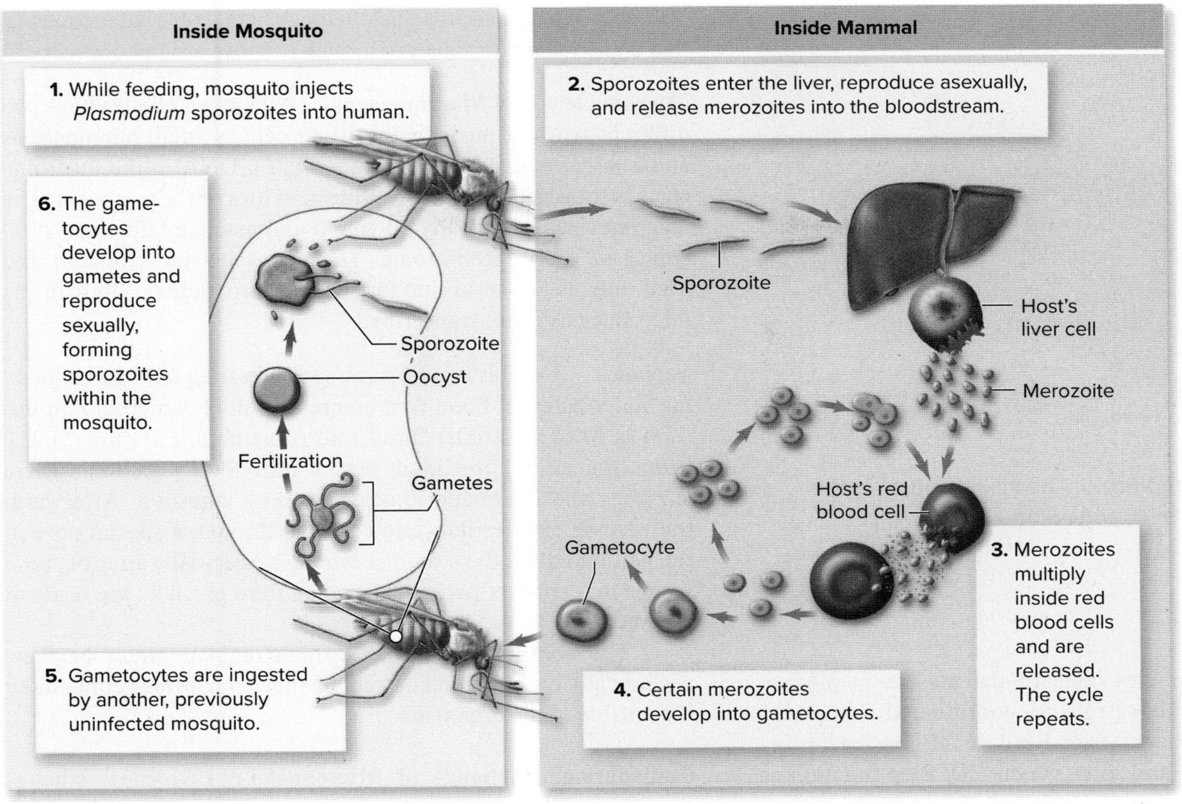

Inside Mosquito
Inside Mammal
1. While feeding, mosquito injects Plasmodium sporozoites into human.
2. Sporozoites enter the liver, reproduce asexually, and release merozoites into the bloodstream.
Sporozoite
Host's liver cell
Merozoite
6. The gametocytes develop into gametes and reproduce sexually, forming sporozoites within the mosquito.
Sporozoite
Oocyst
Fertilization
Gametes
Host's red blood cell
Gametocyte
3. Merozoites multiply inside red blood cells and are released. The cycle repeats.
5. Gametocytes are ingested by another, previously uninfected mosquito.
4. Certain merozoites develop into gametocytes.

Figure 24.22 The life cycle of *Plasmodium*. *Plasmodium,* the apicomplexan that causes malaria, has a complex life cycle that alternates between mosquitoes and mammals.

Gregarines. Gregarines are another group of apicomplexans that use their distinctive apical complex to attach themselves in the intestinal epithelium of arthropods, annelids, and mollusks. Most of the gregarine body, aside from the apical complex, is in the intestinal cavity, and nutrients appear to be obtained through the apicomplex attachment to the cell (figure 24.23).

Toxoplasma. Using its apical complex, *Toxoplasma gondii* invades the epithelial cells of the human gut. Most individuals infected with the parasite mount an immune response, preventing any permanent damage. In the absence of a fully functional immune system, however, *Toxoplasma* can damage the brain (figure 24.24), heart, and skeletal tissues, in addition to gut and lymph tissue, during extended infections. Individuals with AIDS are particularly susceptible to *Toxoplasma* infection. If a pregnant woman touches a cat litter box, *Toxoplasma* parasites from the cat can, if ingested, cross the placental barrier and harm the developing fetus with an immature immune system.

Ciliates are characterized by their mode of locomotion

As the name indicates, most **ciliates** feature large numbers of cilia (tiny, beating hairs). These heterotrophic, unicellular protists are 10 to 3000 μm long. Their cilia are usually arranged either in longitudinal rows or in spirals around the cell. Cilia

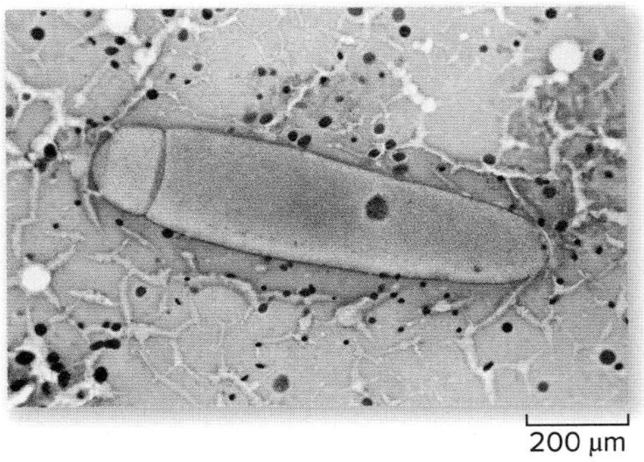

200 μm

Figure 24.23 Gregarine entering a cell.
De Agostini Picture Library/Getty Images

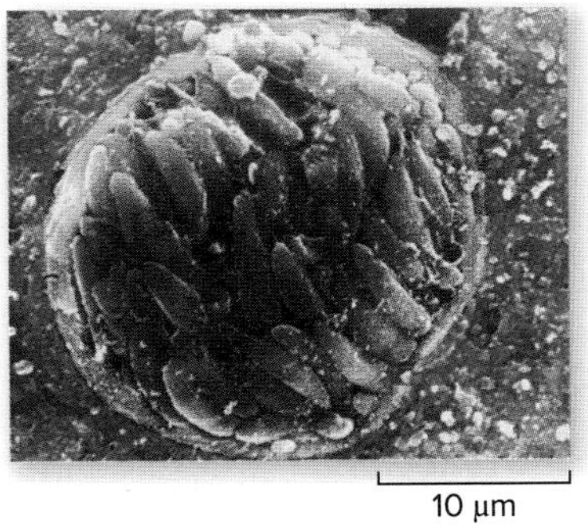

10 μm

Figure 24.24 Micrograph of a cyst filled with *Toxoplasma*. *Toxoplasma* (red in the micrograph) can enter the brain and form cysts filled with slowly replicating parasites.
Prof. David J.P. Ferguson, Oxford University

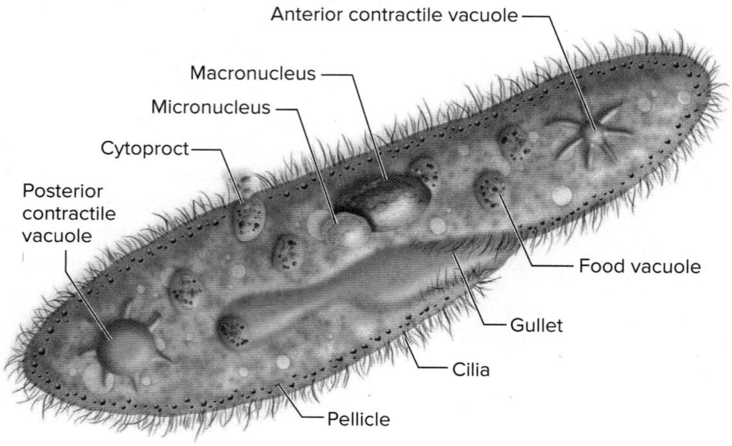
Anterior contractile vacuole
Macronucleus
Micronucleus
Cytoproct
Posterior contractile vacuole
Food vacuole
Gullet
Cilia
Pellicle

Figure 24.25 ***Paramecium.*** The main features of this ciliate include cilia, two nuclei, and numerous specialized organelles.

are anchored to microtubules beneath the plasma membrane (refer to chapter 4), and they beat in a coordinated fashion. In some groups, the cilia have specialized functions, becoming fused into sheets, spikes, and rods, which may then function as mouths, paddles, teeth, or feet.

The ciliates have a pellicle, a tough but flexible outer covering, that enables them to squeeze through or move around obstacles.

Micronucleus and Macronucleus. All known ciliates have two different types of nuclei within their cells: a small **micronucleus** and a larger **macronucleus** (figure 24.25). DNA in the macronucleus is transcribed for the routine activities of the organism. The macronucleus is typically polyploid and may contain up to 1000 copies of each chromosome. The micronucleus is diploid and used only as the germ line for sexual reproduction. DNA in the micronucleus is not transcribed.

Vacuoles. Ciliates form vacuoles for ingesting food and regulating water balance. Food first enters the gullet, which in *Paramecium* is lined with cilia fused into a membrane (figure 24.25). From the gullet, the food passes into food vacuoles, where enzymes and hydrochloric acid aid in its digestion. Afterward, the vacuole empties its waste contents through a special pore in the pellicle called the *cytoproct,* which is essentially an exocytotic vesicle that appears periodically when solid particles are ready to be expelled.

The contractile vacuoles, which regulate water balance, periodically expand and contract as they empty their contents to the outside of the organism.

Conjugation: Exchange of Micronuclei. Like most ciliates, *Paramecium* undergoes **conjugation** (figure 24.26), a sexual process

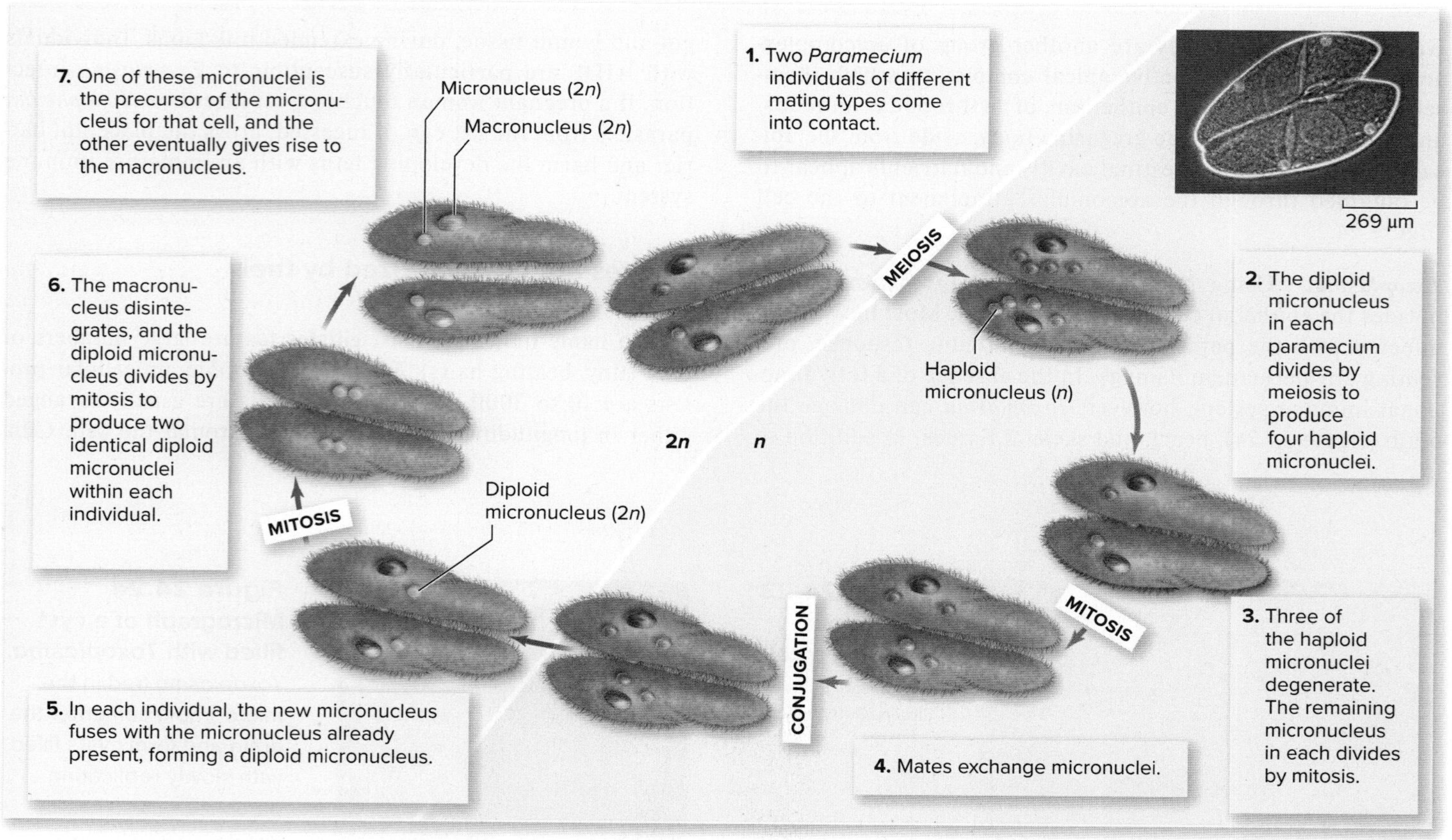
1. Two *Paramecium* individuals of different mating types come into contact.
269 μm
MEIOSIS
Haploid micronucleus (*n*)
2. The diploid micronucleus in each paramecium divides by meiosis to produce four haploid micronuclei.
MITOSIS
3. Three of the haploid micronuclei degenerate. The remaining micronucleus in each divides by mitosis.
CONJUGATION
4. Mates exchange micronuclei.
Diploid micronucleus (2*n*)
5. In each individual, the new micronucleus fuses with the micronucleus already present, forming a diploid micronucleus.
MITOSIS
6. The macronucleus disintegrates, and the diploid micronucleus divides by mitosis to produce two identical diploid micronuclei within each individual.
2*n*
n
Micronucleus (2*n*)
Macronucleus (2*n*)
7. One of these micronuclei is the precursor of the micronucleus for that cell, and the other eventually gives rise to the macronucleus.

Figure 24.26 Life cycle of *Paramecium.* In sexual reproduction, two mature cells fuse in a process called conjugation.

(top right photo): M. I. Walker/Science Source

in which two individual cells remain attached to each other for up to several hours.

Paramecia have multiple mating types. Only cells of two different genetically determined mating types can conjugate. Meiosis in the micronuclei produces several haploid micronuclei, and the two partners exchange a pair of their micronuclei through a cytoplasmic bridge between them.

In each conjugating individual, the new micronucleus fuses with one of the micronuclei already present in that individual, resulting in the production of a new diploid micronucleus. After conjugation, the macronucleus in each cell disintegrates, and the new diploid micronucleus undergoes mitosis, thus giving rise to two new identical diploid micronuclei in each individual.

One of these micronuclei becomes the precursor of the future micronuclei of that cell, while the other micronucleus undergoes multiple rounds of DNA replication, becoming the new macronucleus. This complete segregation of the genetic material is unique to the ciliates and makes them ideal organisms for the study of certain aspects of genetics.

REVIEW OF CONCEPT 24.5

Most members of the Stramenopila have fine hairs on their flagella. Brown algae are large seaweeds that undergo an alternation of generations. Diatoms are unicellular with silica in their cell walls, which forms a shell with two halves. Oomycetes are unique in the production of zoospores that bear two unequal flagella. The Alveolata have a flattened stacked vesicle layer below the plasma membrane. Dinoflagellates have paired flagella and swim with a spinning motion. Apicomplexans are spore-forming parasites. Ciliates are unicellular, heterotrophic protists that use cilia for feeding and locomotion.

- *How could you distinguish between the sporophyte and the gametophyte of a brown alga?*

24.6 SAR: Rhizaria Have Silicon Exoskeletons or Limestone Shells

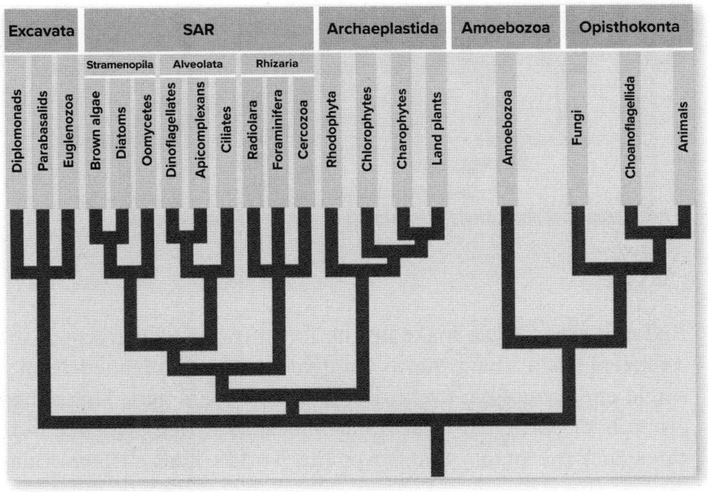

The Rhizaria has been combined with the Stramenopila and the Alveolata to form the supergroup SAR, which is currently thought to be monophyletic. The Rhizaria include two monophyletic groups—radiolarians and forams—and a third has been proposed, the cercozoans. These three groups are morphologically quite different from one another, and until recently they had not been grouped together or linked to a particular branch of the protist phylogenic tree. As with much of our rapidly changing picture of protist phylogeny, this grouping will undoubtedly be refined as new information comes to light.

Actinopoda Have Silicon Internal Skeletons

LEARNING OBJECTIVE 24.6.1 Describe the pseudopodia of radiolarians.

Many Rhizaria have amorphous shapes, with protruding pseudopods that are constantly changing form. Roughly characterized as amoeboid, these same pseudopod-shaped bodies are also found in other protist supergroups. One group of Rhizaria, however, has more distinct structures. Members of the phylum Actinopoda, often called **radiolarians,** secrete glassy exoskeletons made of silica. These skeletons give the unicellular organisms a distinct shape, exhibiting either bilateral or radial symmetry. The shells of different species form many elaborate and beautiful shapes, with pseudopods extruding outward along spiky projections of the skeleton (figure 24.27). Microtubules support these cytoplasmic projections.

Foraminifera Fossils Created Huge Limestone Deposits

LEARNING OBJECTIVE 24.6.2 Describe the exterior features of Foraminifera, and give an example.

Members of the phylum Foraminifera are heterotrophic marine protists. They range in diameter from about 20 µm to several centimeters. They resemble tiny snails and can form 3-m-deep layers in marine sediments. Characteristic of the group are pore-studded shells (called *tests*) composed of organic materials usually reinforced with grains of calcium carbonate, sand, or even plates from shells of echinoderms or spicules (minute needles of calcium carbonate) from sponge skeletons.

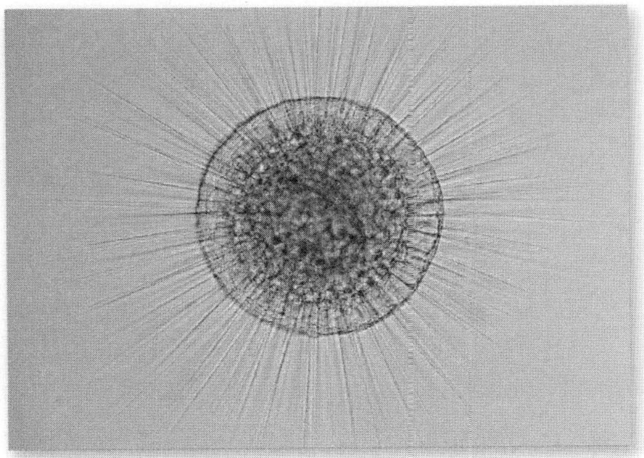

Figure 24.27 ***Actinosphaerium*** **with needle-like pseudopods.**
Wim Van Egmond/SPL/Science Source

Depending on the building materials they use, Foraminifera may have shells of very different appearance. Some of them are brilliantly colored red, salmon, or yellow-brown.

Most Foraminifera live in sand or are attached to other organisms, but two families consist of free-floating planktonic organisms. Their tests may be single-chambered but are more often multichambered, and some have a spiral shape resembling that of a tiny snail. Thin cytoplasmic projections called *podia* emerge through openings in the tests (figure 24.28). Podia are used for swimming, gathering materials for the tests, and feeding. Foraminifera consume a wide variety of small organisms.

The life cycles of Foraminifera are extremely complex, involving alternation between haploid and diploid generations. Foraminifera have contributed massive accumulations of their tests to the fossil record for more than 200 million years. Because of the excellent preservation of their tests and the striking differences among them, forams are very important as geologic markers. The pattern of occurrence of different forams is often used as a guide in searching for oil-bearing strata. Many limestones all over the world, including the famous White Cliffs of Dover in southern England, are rich in forams (figure 24.29).

Cercozoans Feed in Many Ways

LEARNING OBJECTIVE 24.6.3 Explain how you would identify a cercozoan.

Identified by metagenomic analyses (refer to chapter 18), cercozoans are a large group of amoeboid and flagellated protists that feed in an unusually wide variety of ways. Some are predatory heterotrophs that feed on bacteria, fungi, and other protists, while others are photosynthetic autotrophs. Still others can both ingest bacteria and also perform photosynthesis.

Figure 24.28 A representative of the Foraminifera. Podia, thin cytoplasmic projections, extend through pores in the calcareous test, or shell, of this living foram.

O. Roger Anderson, Columbia University, Lamont-Doherty Earth Observatory

Figure 24.29 White Cliffs of Dover. The limestone that forms these cliffs is composed almost entirely of fossil shells of protists, including Foraminifera.

Markus Keller/imageBROKER/age fotostock

REVIEW OF CONCEPT 24.6

The Rhizaria contain two disparate groups, the radiolarians with glassy exoskeletons and the forams with rocky shells. A third group has been proposed based on molecular similarities.

- *If forams are encased in limestone shells, how do they feed?*

24.7 Archaeplastida Are Descended from a Single Endosymbiosis Event

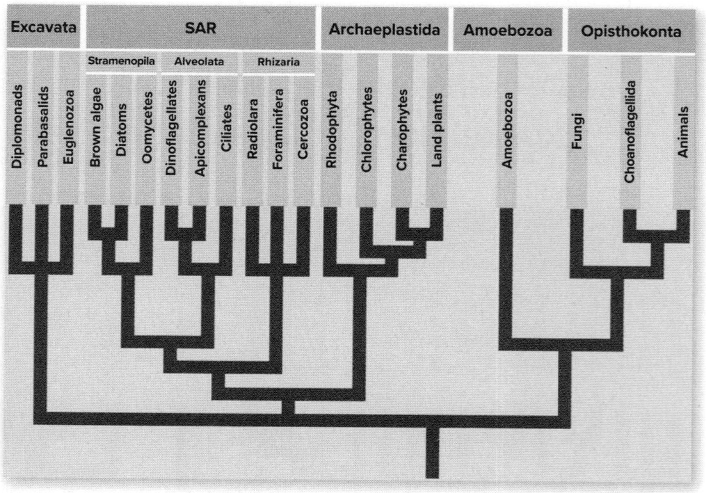

Red and green algae make up the fourth protist supergroup, the Archaeplastida; the group is thought to have descended from a single endosymbiotic event over a billion years ago. This supergroup is of particular importance, because strong evidence indicates that the plants that cover the world's land surface today evolved from the green algae.

Rhodophyta Are Photosynthetic, Multicellular Marine Algae

LEARNING OBJECTIVE 24.7.1 List the major characteristics of red algae.

There are some 6000 described species of **Rhodophyta,** the red algae, living in the world's oceans. Although the origin of Rhodophyta has been a source of controversy, genomic comparisons indicate very early eukaryotic origins and a common ancestry with green algae. Molecular comparisons of the chloroplasts within red and green algae also support a single endosymbiotic origin for both.

The red color of Rhodophyta (*rhodos* is Greek for "red") is due to the presence of an accessory photosynthetic pigment called phycoerythrin, which masks the green of chlorophyll. Phycoerythrin, along with the accessory pigments phycocyanin and allophycocyanin, are arranged within structures called *phycobilisomes.* These allow the algae to absorb blue and green light, which penetrate relatively deeply into ocean waters and allow red algae to live at great depths.

The red algae range in size from microscopic, single-celled organisms to multicellular, "seaweed"-like *Schizymenia borealis* with blades as long as 2 m (figure 24.30). Most are multicellular, and they are the most common algae in tropical coastal waters. They have many commercial uses. Sushi rolls are wrapped in nori, a multicellular red alga of the genus *Porphyra.* Red algal polysaccharides are also used commercially to thicken ice cream and cosmetics.

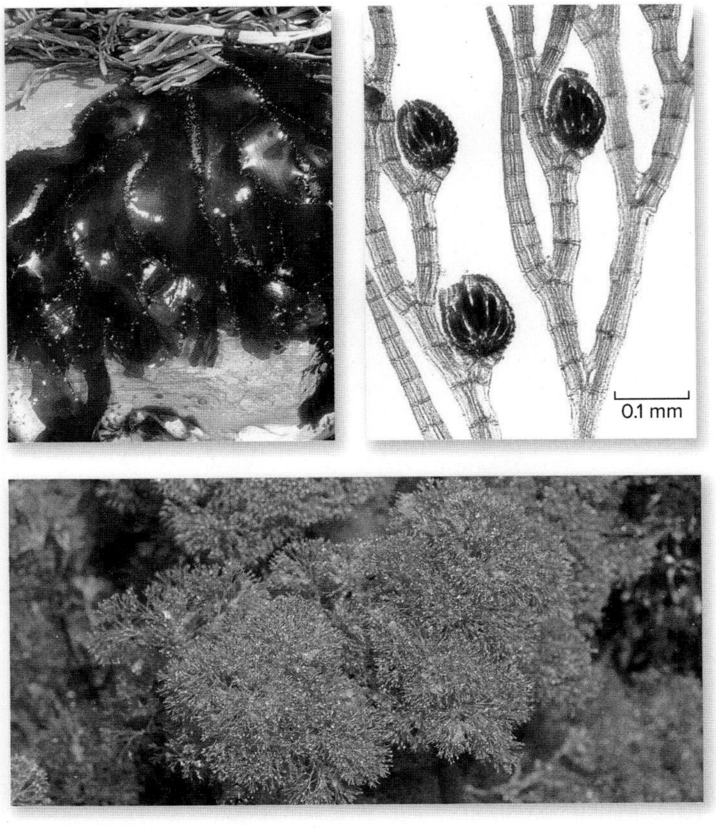

Figure 24.30 Red algae come in many forms and sizes. Three examples of red algae, illustrating diversity of form.
(top photos): Steven P. Lynch; (bottom): Premaphotos/Alamy Stock Photo

Chlorophyta Are Unusually Diverse Green Algae

LEARNING OBJECTIVE 24.7.2 List the major characteristics of green algae.

Green algae have two distinct lineages: the **chlorophytes,** discussed here, and another lineage, the *streptophytes,* which contains the Charophytes that gave rise to the land plants (refer to figure 26.1). The chlorophytes are of special interest here because of their unusual diversity and lines of specialization. The chlorophytes have an extensive fossil record dating back 900 million years, and they closely resemble land plants, especially in their chloroplasts, which are biochemically quite similar to those of plants, containing chlorophylls *a* and *b* as well as an array of carotenoids.

Many chlorophytes are unicellular

Early green algae probably resembled *Chlamydomonas reinhardtii,* diverging from land plants over 1 BYA (figure 24.31). Individuals are microscopic (usually less than 25 μm long), green, and rounded, and they have two flagella at the anterior end. They are soil dwellers that move rapidly in water by beating their flagella in opposite directions. Most individuals of *Chlamydomonas* are haploid. *Chlamydomonas* reproduces asexually as well as sexually, but because it is always unicellular, the life cycle is not haplodiplontic (figure 24.31).

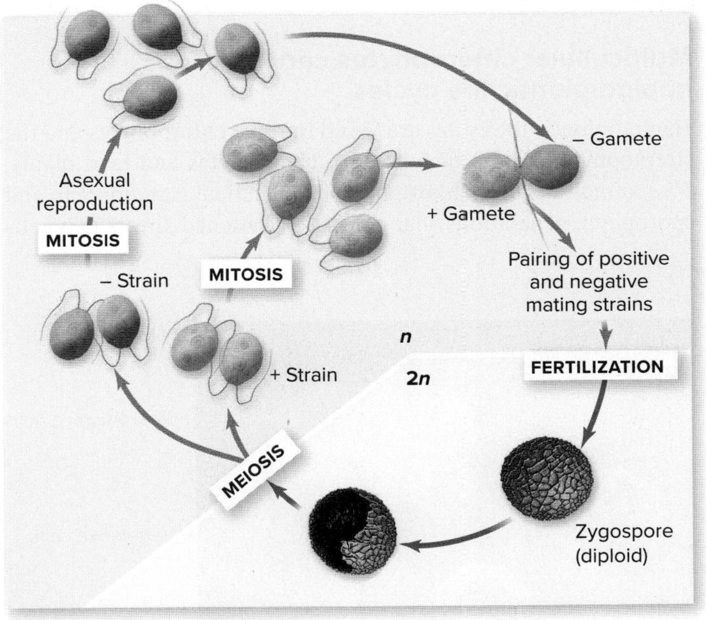

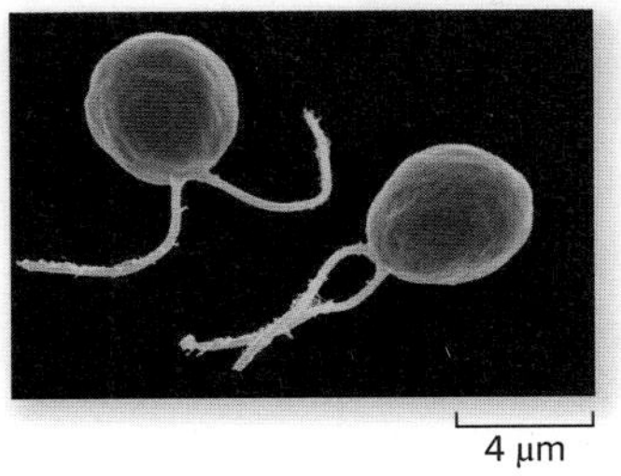

Figure 24.31 ***Chlamydomonas*** **life cycle.** This single-celled chlorophyte has both asexual and sexual reproduction. Unlike in plants, gamete fusion is not followed by mitosis.
Aaron J. Bell/Science Source

Several lines of evolutionary specialization have been derived from organisms such as *Chlamydomonas,* including the evolution of nonmotile, unicellular green algae. *Chlamydomonas* is capable of retracting its flagella and settling down as an immobile unicellular organism if the pond in which it lives dries out. Some common algae found in soil and bark, such as *Chlorella,* are essentially like *Chlamydomonas* in this trait, but they do not have the ability to form flagella.

Genome-sequencing projects are providing new insights into the evolution of plants. The 6968 protein families predicted from the *Chlamydomonas* genome were compared to proteins predicted by a red algal genome and two plant (streptophyte) genomes (moss and *Arabidopsis*). Of these proteins, only 172 are unique to plants. Analysis of these conserved proteins among the many branches of the plant phylogenetic tree will provide insights into plant evolution.

Colonial chlorophytes have some cell specialization

Multicellularity arose many times in the eukaryotes. Colonial chlorophytes provide examples of cellular specialization, an aspect of multicellularity. A line of specialization from cells like those of *Chlamydomonas* concerns the formation of motile, colonial organisms. In these genera of green algae, the *Chlamydomonas*-like cells retain some of their individuality.

The most elaborate of these organisms is *Volvox* (figure 24.32), a hollow sphere made up of a single layer of 500 to 60,000 individual cells, each cell having two flagella. Only a small number of the cells are reproductive. Some reproductive cells may divide asexually, bulge inward, and give rise to new colonies that initially remain within the parent colony.

Multicellular chlorophytes can have haplodiplontic life cycles

Haplodiplontic life cycles are found in some chlorophytes and the streptophytes, which include both charophytes and land plants. *Ulva,* a multicellular chlorophyte, has identical gametophyte and sporophyte generations that consist of flattened sheets two cells thick (figure 24.33). Unlike the charophytes, none of the ancestral chlorophytes gave rise to land plants.

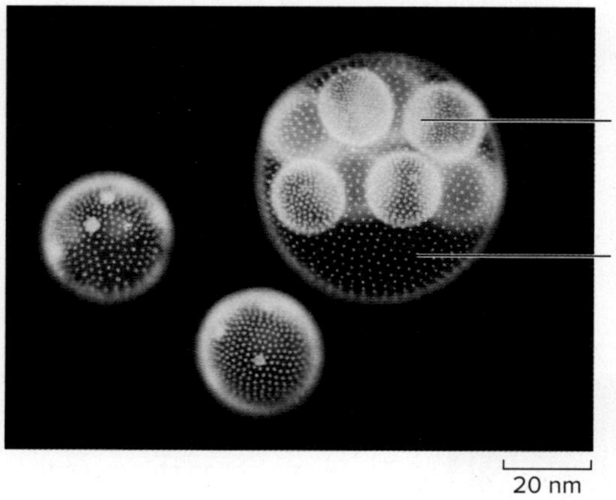

Figure 24.32 ***Volvox.*** This chlorophyte forms a colony where some cells specialize for reproduction.
Stephen Durr

Charophytes Are the Closest Relatives to Plants

LEARNING OBJECTIVE 24.7.3 Explain why charophytes are considered the closest relatives of plants.

Charophytes, a clade of streptophytes, are also green algae, and they are distinguished from chlorophytes by their close phylogenetic relationship to the plants. Charophytes have haplontic life cycles, indicating that the evolution of a diplontic embryo and haplodiplontic life cycle occurred after the move onto land.

Identifying which of the charophyte clades is sister (most closely related) to the land plants puzzled biologists for a long

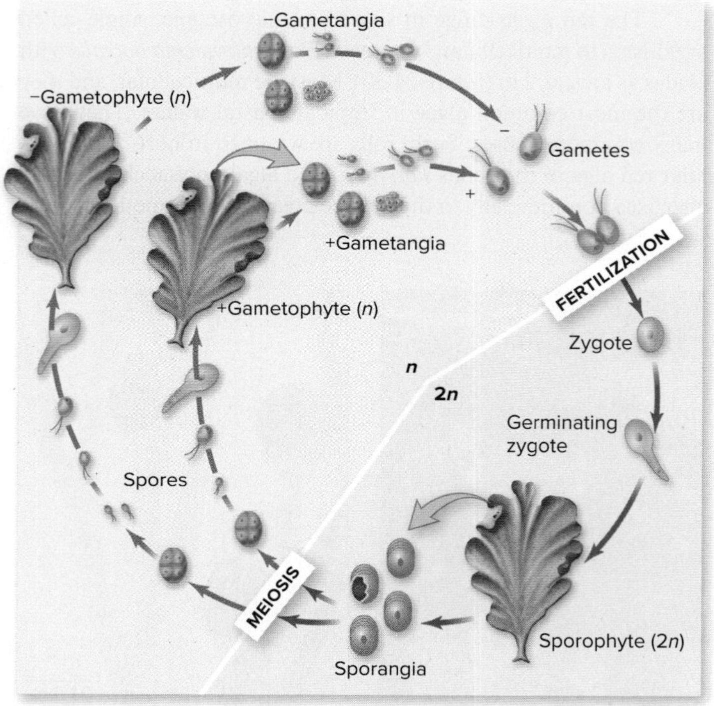

Figure 24.33 Life cycle of *Ulva.* This chlorophyte alga has a haplodiplontic life cycle. The gametophyte and sporophyte are multicellular and identical in appearance.
Dr. Diane S. Littler

time. The charophyte algae fossil record is scarce. Currently, the molecular evidence from rRNA and DNA sequences favors the charophytes as the green algal clade within the streptophytes that gave rise to plants.

The two candidate Charophyta clades have been the Charales, with about 300 species, and the Coleochaetales, with about 30 species (figure 24.34). Both lineages are primarily freshwater algae, but the Charales are huge, relative to the microscopic Coleochaetales. Both clades have similarities to land plants. *Coleochaete* and its relatives have cytoplasmic linkages between cells called *plasmodesmata,* which are found in land plants. The species *Chara* in the Charales undergoes mitosis and cytokinesis like land plant cells. Sexual reproduction in both relies on a large, nonmotile egg and flagellated sperm. These gametes are more similar to those of land plants than to those of many charophyte relatives. Both charophyte clades form green mats around the edges of freshwater ponds and marshes. One species must have successfully inched its way onto land through adaptations to drying.

REVIEW OF CONCEPT 24.7

Red algae vary greatly in size and produce accessory pigments that give them a red color. They lack centrioles and flagella, and they typically reproduce using an alternation of generations. The chlorophytes have chloroplasts very similar to those of plants. Specializations in this group include the evolution of nonmotile, unicellular species that can tolerate drying and of colonial organisms that exhibit a degree of cell specialization. Charophytes are the group of green algae that are thought to be sisters of today's plants, based on a wide variety of morphological and molecular evidence.

- *What major barrier must be overcome for sexual reproduction of green algae to be possible on land?*

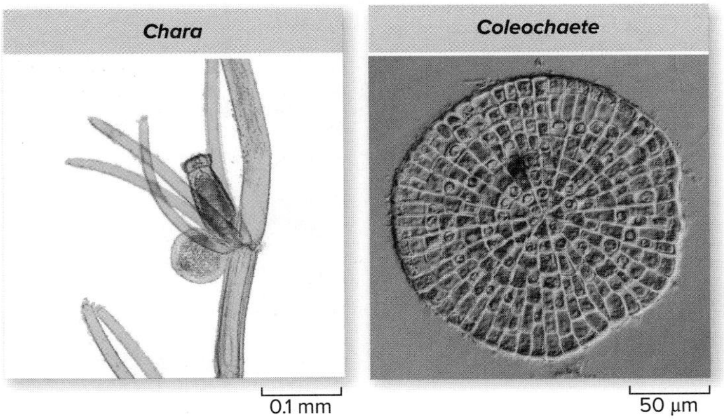

Figure 24.34 ***Chara,*** **a member of the Charales, and** ***Coleochaete,*** **a member of the Coleochaetales, represent the two clades most closely related to land plants.**

(chara): Steven P. Lynch; (coleochaete): Lee W. Wilcox

24.8 Amoebozoa and Opisthokonta Are Closely Related

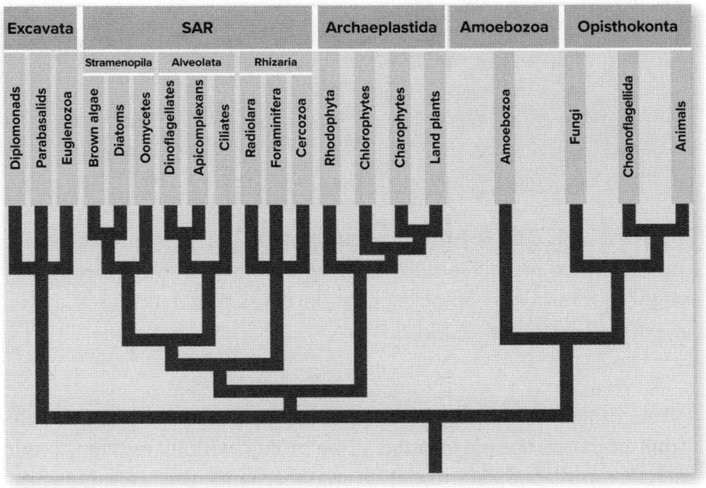

Amoebas move from place to place by means of their pseudopods, which are projections of cytoplasm that extend and pull the amoeba forward or engulf food particles. An amoeba puts a pseudopod forward and then flows cytoplasm into it (figure 24.35). Microfilaments of actin and myosin similar to those found in muscles are associated with these movements. The pseudopods can form at any point on the cell body, so the amoeba can move in any direction. The amoebas in the Amoebozoa supergroup are most closely related to Opisthokonta, which include the close relatives of fungi and animals. Members of these two supergroups have a single flagellum, in contrast with two or more in the other supergroups. Amoebozoa and Opisthokonta are collectively referred to as Unikonts, with *uni* referring to the single flagellum; however, they are distinct supergroups.

Amoebozoa Are the Plasmodial and Cellular Slime Molds

LEARNING OBJECTIVE 24.8.1 Distinguish between cellular and plasmodial slime molds.

The Amoebozoa, or slime molds, are but one of several groups of protists that assume an amoeboid form. Like water molds, the

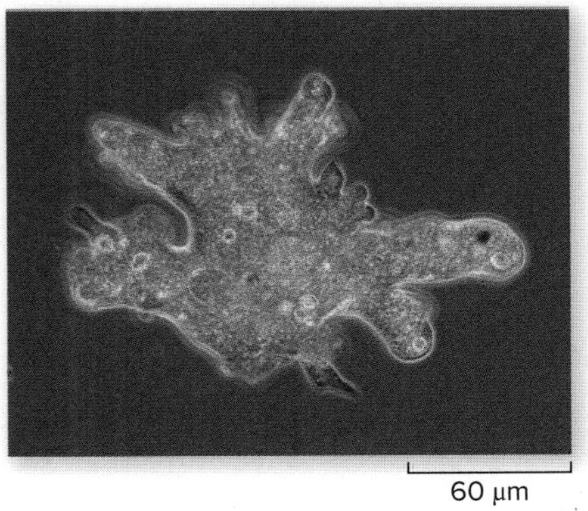

Figure 24.35 ***Amoeba proteus.*** The projections are pseudopods; an amoeba moves by flowing cytoplasm into them.

Melba Photo Agency/Alamy Stock Photo

Figure 24.36 A plasmodial protist. This multinucleate pretzel slime mold, *Hemitrichia serpula,* moves about in search of the bacteria and other organic particles that it ingests.
Eye of Science/Science Source

Figure 24.37 Sporangia of a plasmodial slime mold. These *Arcyria* sporangia are found in the phylum Myxomycota.
Eye of Science/Science Source

slime molds were once considered fungi. There are two quite distinct lineages: the plasmodial slime molds, which are huge, single-celled, multinucleate, oozing masses, and the cellular slime molds, in which single cells combine into masses and differentiate, creating an early model of multicellularity.

Plasmodial slime molds

Plasmodial slime molds stream along as a **plasmodium,** a non-walled, multinucleate mass of cytoplasm that resembles a moving mass of slime (figure 24.36). This form is called the *feeding phase,* and the plasmodia may be orange, yellow, or another color.

Plasmodia show a back-and-forth streaming of cytoplasm that can easily be seen using a light microscope. They are able to pass through the mesh in cloth or simply flow around or over obstacles. As they move, they engulf and digest bacteria, yeasts, and other small particles of organic matter.

A multinucleated *Plasmodium* cell undergoes mitosis synchronously, with the nuclear envelope breaking down, but only at late anaphase or telophase. Centrioles are absent.

When either food or moisture is in short supply, the plasmodium migrates relatively rapidly to a new area. Here it stops moving and either forms a mass in which spores differentiate or divides into a large number of small mounds, each of which produces a single, mature **sporangium,** the structure in which spores are produced. These sporangia are often beautiful and extremely complex in form (figure 24.37). The spores are highly resistant to unfavorable environmental influences and may remain viable for years if kept dry.

Cellular slime molds

The cellular slime molds have become an important group for the study of cell differentiation because of their relatively simple developmental systems (figure 24.38). The individual organisms behave

Figure 24.38 Development in *Dictyostelium discoideum*, a cellular slime mold. Dictyostelium discoideum spores (top) germinate into amoebas. Scarcity of food triggers the amoebas to aggregate, eventually forming a mound which differentiates into a crawling phototactic "slug" (bottom). Culmination is a process of differentiation that results in the production of a stalked structure supporting a spore-producing fruiting body (left).

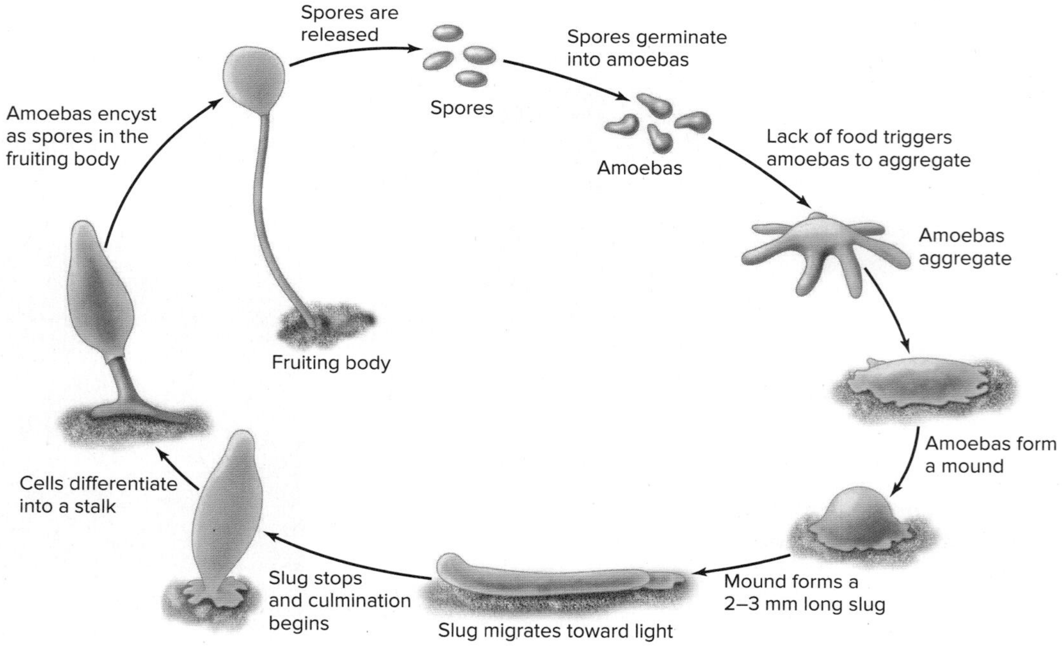
Spores are released
Spores
Spores germinate into amoebas
Amoebas
Lack of food triggers amoebas to aggregate
Amoebas aggregate
Amoebas form a mound
Mound forms a 2–3 mm long slug
Slug migrates toward light
Slug stops and culmination begins
Cells differentiate into a stalk
Fruiting body
Amoebas encyst as spores in the fruiting body

as separate amoebas, moving through the soil and ingesting bacteria. When food becomes scarce, the individuals aggregate to form a moving "slug." Cyclic adenosine monophosphate (cAMP) is sent out in pulses by some of the cells, and other cells move in the direction of the cAMP to form the slug. In the cellular slime mold *Dictyostelium discoideum,* this slug goes through morphogenesis to make stalk and spore cells. The spores then go on to form a new amoeba if they land in a moist habitat.

Choanoflagellates Are Likely Animal Ancestors

LEARNING OBJECTIVE 24.8.2 Describe the evolutionary significance of the choanoflagellates.

At structural and molecular levels, the Choanoflagellates are the protists most similar to sponges, the most ancient of animals. Choanoflagellates have a single emergent flagellum surrounded by a funnel-shaped, contractile collar composed of closely placed filaments, a structure that is exactly matched in the sponges, which are animals. These protists feed on bacteria strained out of the water by their collar. Colonial forms resemble freshwater sponges (figure 24.39).

The close relationship of choanoflagellates to animals is further demonstrated by the strong homology between a cell-surface protein (a tyrosine kinase receptor; refer to chapter 9) found in choanoflagellates and one found in sponges.

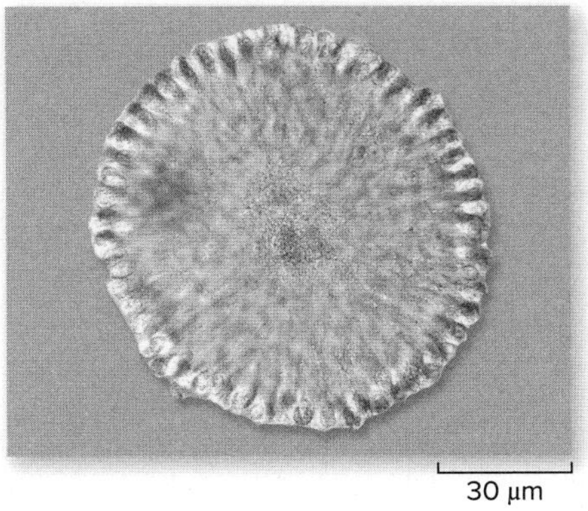

Figure 24.39 Colonial choanoflagellates resemble their close animal relatives, the sponges.
David Patterson for Maple Ferryman Pty Ltd.

REVIEW OF CONCEPT 24.8

Like other amoebas, slime molds move with the aid of pseudopods. Plasmodial slime molds consist of a single large, multinucleate cell, while cellular slime molds are multicellular. Choanoflagellates are the closest relatives of animals. Colonial forms are similar to freshwater sponges, and both organisms have a homologous cell-surface receptor.

- *What other types of studies might connect choanoflagellates with sponges?*

Defining a Treatment Window for Malaria

Although malaria kills more people each year than any other infectious disease, the combination of mosquito control and effective treatment has virtually eliminated this disease from the United States. In 1941, more than 4000 Americans died of malaria; in the year 2010, by contrast, only 5 people died of malaria contracted in the United States.

The key to controlling malaria has come from understanding its life cycle. The first critical advance came in 1897 in a remote field hospital in Secunderabad, India, when physician Ronald Ross observed that hospital patients who did not have malaria were more likely to develop the disease in the open wards (those without screens or netting) than in wards with closed windows or screens. Observing closely, he saw that patients in the open wards were being bitten by mosquitoes of the genus *Anopheles*. Dissecting mosquitoes that had bitten malaria patients, he found the *Plasmodium* parasite. Newly hatched mosquitoes that had not yet fed, when allowed to feed on malaria-free blood, did not acquire the parasite. Ross reached the conclusion that mosquitoes were spreading the disease from one person to another, passing along the parasite while feeding. In every country where it has been possible to eliminate the *Anopheles* mosquitoes, the incidence of malaria has plummeted.

The second critical advance came with the development of drugs to treat malaria victims. The British had discovered in India in the mid-1800s that a bitter substance called quinine taken from the bark of cinchona trees was useful in suppressing attacks of malaria. The boys in the photograph are being treated with an intravenous solution of quinine. Quinine also reduces the malarial fever during attacks, but it does not cure the disease.

Today physicians instead use the synthetic drugs chloroquine and primaquine, which are much more effective than quinine, with fewer side effects. Unlike quinine, these two drugs can cure patients completely, because they attack and destroy one of the phases of the *Plasmodium* life cycle, the merozoites released into the bloodstream several days after infection—but only if the drugs are administered soon enough after the bite that starts the infection.

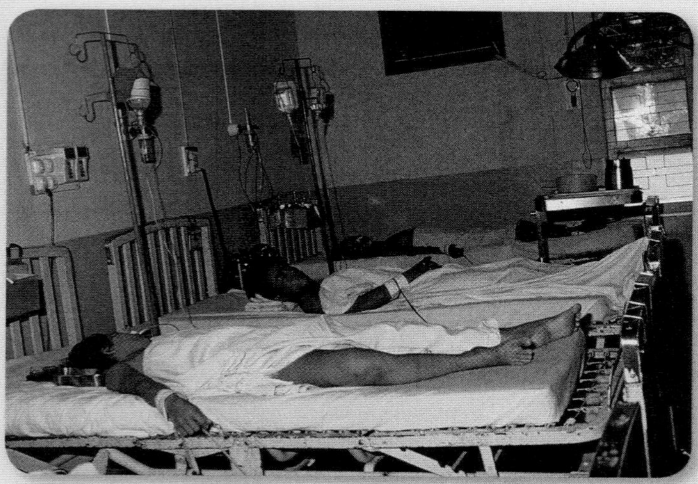

Sue Ford/Science Source

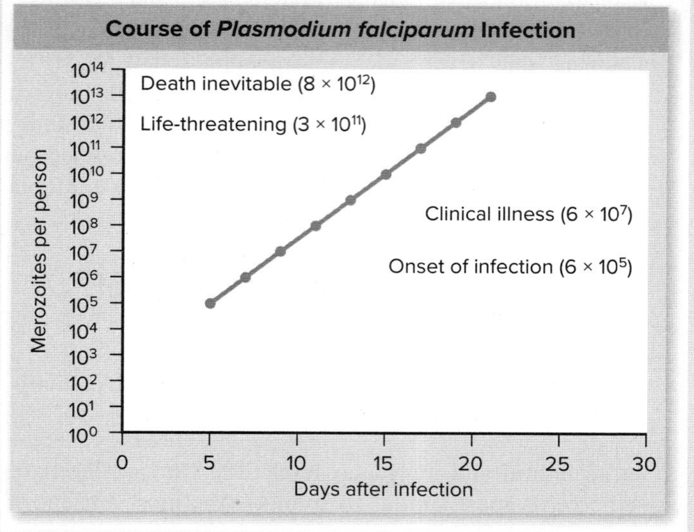
Course of *Plasmodium falciparum* Infection
Death inevitable (8×10^{12})
Life-threatening (3×10^{11})
Clinical illness (6×10^{7})
Onset of infection (6×10^{5})
Merozoites per person
10^{0} 10^{1} 10^{2} 10^{3} 10^{4} 10^{5} 10^{6} 10^{7} 10^{8} 10^{9} 10^{10} 10^{11} 10^{12} 10^{13} 10^{14}
0 5 10 15 20 25 30
Days after infection

To determine the time frame for successful treatment, doctors have carefully studied the time course of a malarial infection. The graph presents what they have found. Numbers of merozoites are presented on the *y*-axis on a log scale—each step reflects a 10-fold increase in numbers. The infection becomes life-threatening if 1% of red blood cells become infected, and death is almost inevitable if 20% of red blood cells are infected.

Analysis

1. **Applying Concepts** In the graph, what is the dependent variable? What is the independent variable?
2. **Making Inferences**
   a. How long after infection (that is, initial infection by sporozoites) is it before the liver releases merozoites into the bloodstream? Before the disease becomes life-threatening? Before death is inevitable?
   b. How long does it take merozoites to multiply 10-fold?
   c. Between days 5 and 20, how many orders of magnitude change in the number of merozoites per person are there?
3. **Drawing Conclusions** After the first appearance of clinical illness symptoms, for how many days can the disease be treated before it becomes life-threatening? Before treatment has little or no chance of saving the patient's life?

Retracing the Learning Path

CONCEPT 24.1 Protists, the First Eukaryotes, Arose by Endosymbiosis

24.1.1 Fossil Evidence Dates the Origins of Eukaryotes Although eukaryotes may have arisen earlier, the fossil evidence of their appearance dates back to 1.5 BYA.

24.1.2 Mitochondria Evolved from Engulfed Aerobic Bacteria The theory of endosymbiosis states that ancestral eukaryotic cells engulfed aerobic bacteria, which became mitochondria.

24.1.3 Chloroplasts Evolved from Engulfed Photosynthetic Bacteria Chloroplasts are believed to have arisen when ancestral eukaryotic cells engulfed photosynthetic bacteria.

24.1.4 Mitosis Evolved in Eukaryotes Mechanisms of mitosis vary among organisms, suggesting that the process did not evolve all at once.

CONCEPT 24.2 Protists Are a Very Diverse Group

24.2.1 Protists Are Eukaryotes That Are Not Fungi, Animals, or Plants Protists mainly use flagella or pseudopods for locomotion, although many other means of propulsion are found. Nutritional strategies include phototrophy, heterotrophy, and mixotrophy—the use of both. Protists can reproduce asexually by mitosis, budding, or schizogony. They may also carry out sexual reproduction.

CONCEPT 24.3 The Rough Outlines of Protist Phylogeny Are Becoming Clearer

24.3.1 Monophyletic Clades Have Been Identified Among the Protists Molecular comparisons have begun to clarify protist classification. This research is also shedding light on the roots of the eukaryotic tree. All eukaryotes fall into one of five supergroups. Plants are thought to have evolved from one of them, fungi and animals from another.

CONCEPT 24.4 Excavata Are Flagellated Protists Lacking Mitochondria

24.4.1 Diplomonads Have Two Nuclei Diplomonads are unicellular, move with flagella, and have two nuclei.

24.4.2 Parabasalids Have Undulating Membranes Parabasalids use flagella and undulating membranes for locomotion.

24.4.3 Euglenozoa Are Free-Living Eukaryotes with Anterior Flagella and Often Chloroplasts Euglenoids can produce chloroplasts to carry out photosynthesis in the light. They contain a pellicle and move via anterior flagella.

CONCEPT 24.5 SAR: Stramenopiles and Alveolates Exhibit Secondary Endosymbiosis

24.5.1 Stramenopila Have Fine Hairs Brown algae typically are large seaweeds that undergo an alternation of generations, producing gametophyte and sporophyte stages. Diatoms have silica in their cell walls. Each diatom produces two overlapping, glassy shells that fit together. Oomycetes, the water molds, are parasitic and are unique in the production of asexual spores (zoospores) that bear two unequal flagella.

24.5.2 Alveolata Have Submembrane Vesicles Dinoflagellates have pairs of flagella arranged so that they swim with a spinning motion. Blooms of dinoflagellates can cause toxic red tides. Apicomplexans are spore-forming animal parasites. They have a unique arrangement of organelles at one end of the cell, called the apical complex, which is used to invade the host. Ciliates are unicellular, heterotrophic protists that use numerous cilia for feeding and propulsion. Each cell has a macronucleus and a micronucleus.

CONCEPT 24.6 SAR: Rhizaria Have Silicon Exoskeletons or Limestone Shells

24.6.1 Actinopoda Have Silicon Internal Skeletons

24.6.2 Foraminifera Fossils Created Huge Limestone Deposits The Foraminifera are heterotrophic marine protists with pore-studded shells primarily formed by deposit of calcium carbonate.

24.6.3 Cercozoans Feed in Many Ways

CONCEPT 24.7 Archaeplastida Are Descended from a Single Endosymbiosis Event

24.7.1 Rhodophyta Are Photosynthetic, Multicellular Marine Algae Red algae produce accessory pigments that may give them a red color. These algae range in size from microscopic single-celled organisms to large seaweeds.

24.7.2 Chlorophyta Are Unusually Diverse Green Algae Unicellular chlorophytes include *Chlamydomonas*, which has two flagella, and *Chlorella*, which has no flagella and reproduces asexually. *Volvox* is a colonial green alga with some specialized cells.

24.7.3 Charophytes Are the Closest Relatives to Plants Both candidate Streptophyta clades—Charales and Coleochaetales—exhibit plasmodesmata. They also undergo mitosis and cytokinesis like terrestrial plants.

CONCEPT 24.8 Amoebozoa and Opisthokonta Are Closely Related

24.8.1 Amoebozoa Are the Plasmodial and Cellular Slime Molds All slime molds can aggregate to form a moving "slug" that produces spores.

24.8.2 Choanoflagellates Are Likely Animal Ancestors Colonial choanoflagellates are structurally very similar to freshwater sponges, and molecular similarities have also been found.

Concept Overview

This Concept Overview diagrams the key concepts that were discussed in this chapter.

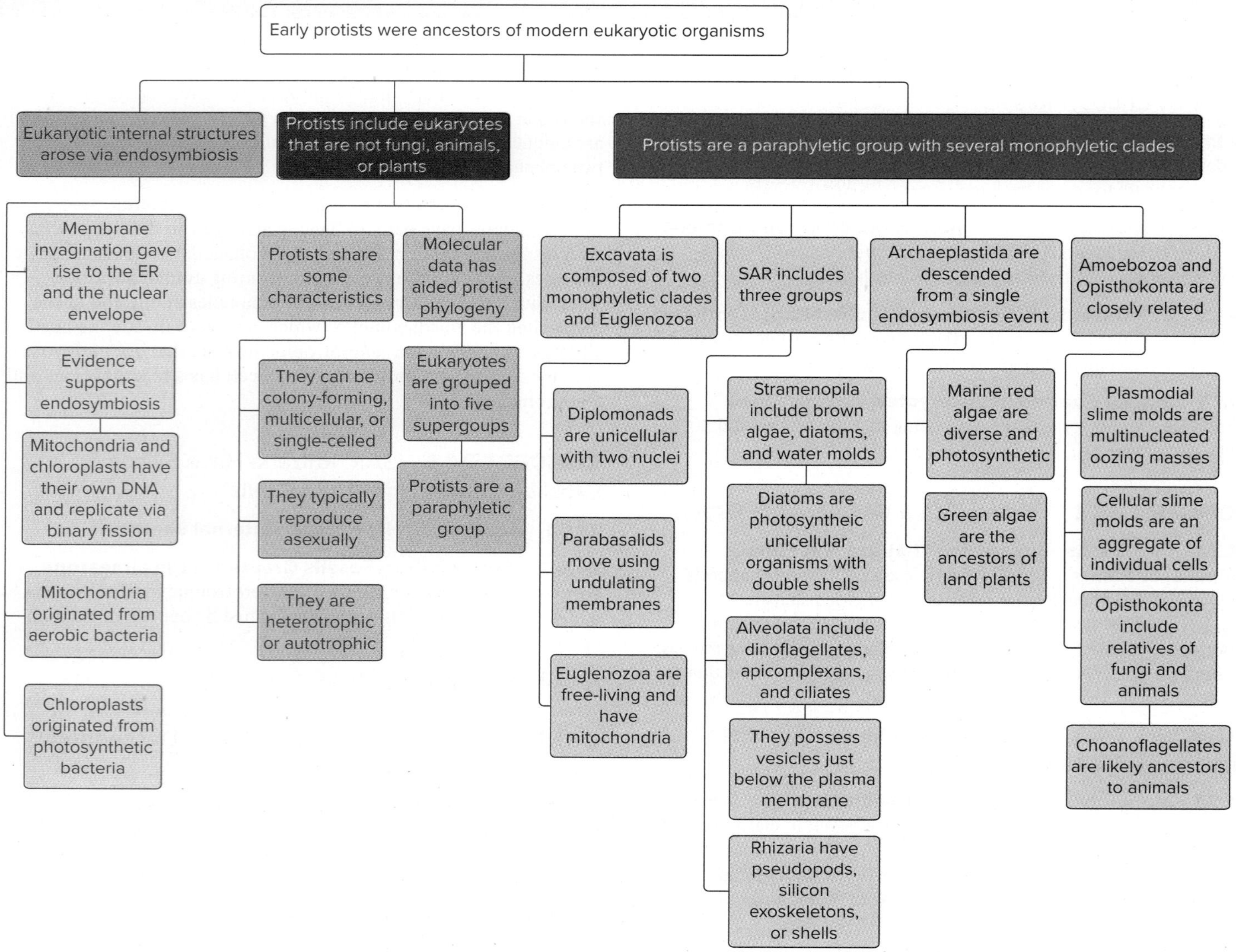

Assessing the Learning Path

Understand

1. Fossil evidence of eukaryotes dates back to
   a. 2.5 BYA.
   b. 1.5 BYA.
   c. 2.5 MYA.
   d. 1.5 MYA.
2. One piece of evidence supporting the endosymbiotic theory for the origin of eukaryotic cells is that
   a. eukaryotic cells have internal membranes.
   b. mitochondria and chloroplasts have their own DNA.
   c. the nuclear membrane resembles the organelle membranes.
   d. mitochondria have silica in their cell walls.
3. Which of the following is NOT a protist mode of locomotion?
   a. Waving cilia
   b. Extending pseudopodia
   c. Shortening axopodia
   d. Gliding on extruded slime
4. All protists possess
   a. cell walls.
   b. functional chloroplasts.
   c. multicellular organization.
   d. nuclei.

5. Among protists, a mixotroph is both
   a. phototrophic and heterotrophic.
   b. unicellular and colonial.
   c. mobile and sessile.
   d. asexual and sexual.
6. Protists do not include
   a. algae.
   b. ciliates.
   c. apicomplexans.
   d. fungi.
7. The protists that share the most recent common ancestor to plants are found in the
   a. Rhizaria.
   b. Archaeplastida.
   c. Amoebozoa.
   d. Chromalveolata.
8. Which of these protist supergroups has given rise to two multicellular kingdoms?
   a. SAR
   b. Amoebozoa
   c. Opisthokonta
   d. Archaeoplastida
9. Both diplomonads and parabasalids
   a. contain chloroplasts.
   b. have multinucleate cells.
   c. lack mitochondria.
   d. have silica in their cell walls.
10. Stramenopila are
   a. tiny flagella.
   b. large cilia.
   c. small hairs on flagella.
   d. pairs of large flagella.
11. Kelp, which sometimes forms large, underwater forests, is actually composed of multicellular protists called
   a. chlorophyta.
   b. red algae.
   c. brown algae.
   d. dinoflagellates.
12. The Actinopoda do NOT
   a. have hard calcium carbonate shells.
   b. have amoeboid shapes in many groups.
   c. have glassy silicon exoskeletons.
   d. have spiky, needle-like pseudopods.
13. Which protists are used as geologic markers and are used to search for petroleum?
   a. Radiolarians
   b. Forams
   c. Diatoms
   d. Myxomycotes
14. Amoebas, forams, and radiolarians move using their
   a. cytoplasm.
   b. flagella.
   c. cilia.
   d. setae.
15. *Chlamydomonas* is unlike plants in that in *Chlamydomonas*
   a. gamete fusion is not followed by mitosis.
   b. only chlorophyll *a* is found.
   c. phycoerythrin masks the green of chlorophyll.
   d. both haploid and diploid phases of the life cycle occur.
16. The sporophyte form of the protist *Ulva*
   a. produces gametes.
   b. contains haploid sperm.
   c. produces diploid spores.
   d. is the product of fertilization.
17. Which of the following is most likely the ancestor of animals?
   a. Trypanosomes
   b. Diplomonads
   c. Ciliates
   d. Choanoflagellates

Apply

1. Analyze the following statements and choose the one that most accurately supports the endosymbiotic theory.
   a. Mitochondria rely on mitosis for replication.
   b. Chloroplasts contain DNA but translation does not occur in chloroplasts.
   c. Vacuoles have double membranes.
   d. Antibiotics that inhibit protein synthesis in bacteria can have the same effect on mitochondria.
2. *Giardia* lack mitochondria, implying that
   a. they do not carry out oxidative respiration.
   b. their cell nuclei do not contain mitochondrial genes.
   c. they are a primitive form of protist.
   d. they are not a primitive form of protist.
3. If a cell contains a pellicle, it
   a. can change shape readily.
   b. is shaped like a sphere.
   c. is shaped like a torpedo.
   d. must have a contractile vacuole.
4. Protists that form aggregates, have cellulose cell walls, and are heterotrophic are probably
   a. ciliates.
   b. choanoflagellates.
   c. radiolarians.
   d. slime molds.
5. Examine the life cycle of cellular slime molds. Which feature affords the greatest advantage for surviving food shortages?
   a. Cellular slime molds produce spores when starved.
   b. Cellular slime molds are saprobes.
   c. A diet of bacteria ensures there will never be a shortage of food.
   d. Cellular slime molds use cAMP to guide each other to food sources.
6. When the protein-encoding genes of *Chlamydomonas* are compared with red algae and plant genomes,
   a. most of the *Chlamydomonas* genes are unique.
   b. most of the red algae genes are unique.
   c. most of the plant genes are unique.
   d. most of the genes are common to all three groups.
7. Based on what you've learned about the protists in this chapter, what seems to be the major barrier to classifying protists?
   a. A lack of a fossil record
   b. An absence of diversity
   c. Many shared traits
   d. Lack of molecular data upon which to base phylogenetic analyses
8. You believe you have identified a new protist species. It has a pair of nuclei, expresses a gene responsible for producing a protein used to make flagella, and has several genes involved in aerobic metabolism. In an attempt to classify the potentially new organism, you start by eliminating groups of protists. Which group would you eliminate from the list below?
   a. Diplomonads
   b. Parabasilids
   c. Apicomplexans
   d. All of the above would best be eliminated.
9. Which of the following might be expected as a symptom in a malaria patient based on the life cycle of the parasite?
   a. Increased sugar in the urine
   b. An increase in skin infections
   c. Irregular heartbeat
   d. Anemia
10. A 25-year-old pregnant student has been admitted to the emergency department with acute respiratory failure. The student's medical notes show that she recently returned from a trip to the northeast coast of the United States on a visit to an elderly relative who has 22 cats and owns a fishery. The relative's partner recently returned from a trip to Latin America and is suffering from Chagas disease. What protist is most likely responsible for the student's condition?
   a. *Trypanosoma cruzi*
   b. *Toxoplasma*
   c. *Plasmodium falciparum*
   d. A dinoflagellate

Synthesize

1. In plants, the mitochondrial, chloroplast, and nuclear genomes all contain ribosomal RNA genes. Would you expect the ribosomal RNA genes of a plant's mitochondria to be more like those of its chloroplasts or more closely related to its nuclear ribosomal RNA genes? Explain.
2. Would you classify *Volvox* (see figure 24.6) as multicellular? What distinguished the multicellularity of brown, red, and green algae from the multicellular organization of a *Volvox* colony?
3. Modern taxonomy relies heavily on phylogenetic data. In the past, taxonomists often used a morphological species concept, in which species were defined based on similarities in growth form. Give an example to show how a morphological species concept would group a set of protists differently than a phylogenetic species concept would.
4. The flagellated protist *Giardia* does not have true mitochondria but does have tiny organelles called mitosomes. What features would you look for to test the hypothesis that mitosomes are derived from mitochondria?
5. Three methods have been used to try to eradicate malaria. One is to eliminate the mosquito vectors of the parasite, the second is to kill the parasites after they enter the human body, and the third is to develop a vaccine against the parasite, allowing the human immune system to provide protection from the disease. Which do you suppose is the most promising in the long run? Why?
6. Radiolarians have exterior glassy shells of silica, whereas forams have exterior stony shells made of calcium carbonate (the stuff of limestone). Why do you imagine these protists utilize this hard-exterior strategy?
7. If plants were derived from green algae, why don't taxonomists classify green algae as plants?
8. List the instances in which multicellularity has arisen among the protists. Can you suggest a common theme among these instances favoring this evolutionary development?

David Clapp/Getty Images

25 Fungi

Learning Path

25.1 Fungi Have Unique Reproductive and Nutritional Strategies

25.2 Fungi Have an Enormous Ecological Impact

25.3 Fungi Are Important Plant and Animal Pathogens

25.4 Fungi Are Taxonomically Diverse

25.5 Microsporidia Are Unicellular Parasites

25.6 Chytridiomycota and Relatives: Fungi with Zoospores

25.7 Zygomycota Produce Zygotes

25.8 Glomeromycota Are Asexual Plant Symbionts

25.9 Basidiomycota Are the Mushroom Fungi

25.10 Ascomycota Are the Most Diverse Phylum of Fungi

Concept Overview

This Concept Overview provides a diagram of the overarching concepts that are covered in this chapter. These concepts will be expanded upon in the Concept Overview at the end of the chapter.

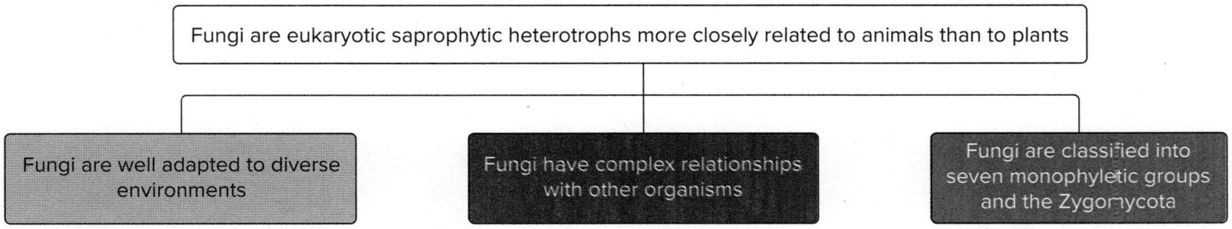

Fungi are eukaryotic saprophytic heterotrophs more closely related to animals than to plants
Fungi are well adapted to diverse environments
Fungi have complex relationships with other organisms
Fungi are classified into seven monophyletic groups and the Zygomycota

Introduction

Fungi are often believed to be more closely related to plants than they are to animals. In fact, fungi share a more recent common ancestor with animals and are therefore more closely related to them. Their nutritional strategy of absorbing externally digested nutrients that are rapidly distributed throughout the organism allows for rapid growth and adaptation to a wide range of ecological niches. Fungi are found everywhere on Earth—from the tropics to the tundra and in both terrestrial and aquatic environments—and have a profound influence on ecology and human health. Fungi made it possible for plants to colonize land by associating with rootless stems of plants to aid in the uptake of nutrients and water. Fungi's continued relationship with plants places them at the center of the carbon cycle, where they act as catalysts of cellulose and lignin degradation, rapidly mobilizing plant-fixed carbon into forms useful to other organisms.

25.1 Fungi Have Unique Reproductive and Nutritional Strategies

We usually become aware of fungi when we see them growing in our lawns, on trees, or on bread left out on a counter too long. Mycologists estimate, however, that there may be as many as 1.5 million species of fungi. Most of those fungi are multicellular; however, one group—the yeasts—are unicellular.

Because some of the more obvious fungal structures, like the mushroom pictured on the previous page, grow from the ground up, some people assume fungi to be more related to plants than to animals. In fact, fungi share several characteristics with animals, despite being very different from animals. Fungi have evolved strategies that allow them to exist in environments as diverse as oceans and deserts and from the Arctic to the equator. Fungi use a variety of life strategies to succeed in diverse ecological niches: some are mutualistic, some are parasitic, and some are free living.

The Body of a Fungus Is a Mass of Connected Hyphae

LEARNING OBJECTIVE 25.1.1 Compare fungal tissues with animal and plant tissues.

Fungi grow in two basic ways: as unicellular forms called yeasts and as multicellular, threadlike forms called mycelia. Mycelia are aggregations of slender, filamentous structures called **hyphae.** Hyphae consist of single cells joined end-to-end, which are collectively surrounded by a common cell wall. Individual cells can be separated from one another by a perforated wall called a septum, which allows the flow of cytoplasm, making them effectively one large cell (figure 25.1).

The streaming of cytoplasm from cell to cell in a hypha is an adaptation that allows fungi to grow rapidly when nutrients and moisture are abundant. The molecules needed to build cellular components can be quickly moved to the growing tips of the hyphae, allowing very rapid growth rates; you may have seen mushrooms appear overnight in a lawn after rain. Some types of fungi have hyphae that lack septa entirely and are one giant cell containing multiple nuclei; these structures are called coenocytic hyphae.

The mycelium

A mass of connected hyphae is called a **mycelium** (plural, *mycelia*). A mycelium is the main body of multicellular fungi that grows into, on, or through soil, wood, or tissues of host organisms (figure 25.2). Mycelia can be huge, with hyphae being many meters long. Some mycelial mats can spread across acres of ground. Because of the threadlike nature of hyphae, a mycelium has a high surface-area-to-volume ratio. This means that nutrient absorption by hyphae is very efficient, which also contributes to the high growth rates of fungi under ideal conditions.

When growth conditions are not optimal, reproductive structures can form from dense masses of hyphae. The spores produced by these structures are resistant to adverse conditions, such as desiccation, and germinate when conditions become suitable. These reproductive structures are most obvious in the forms of mushrooms, toadstools, and puffballs and have been used historically to classify the fungi into groups.

Cell walls contain chitin

Unlike the cell walls of plants and many protists, fungi have cell walls composed primarily of chitin. Chitin is very similar to cellulose, except it is made from a form of glucose called *N*-acetylglucosamine. Chitin is also found in the exoskeletons of mollusks, insects, and crustaceans (refer to chapter 27). This shared feature is part of the evidence that fungi are more closely related to animals than they are to plants. Additional evidence comes from the observation that both fungi and animals store excess glucose in the form of glycogen; plants, on the other hand, store excess glucose as starch.

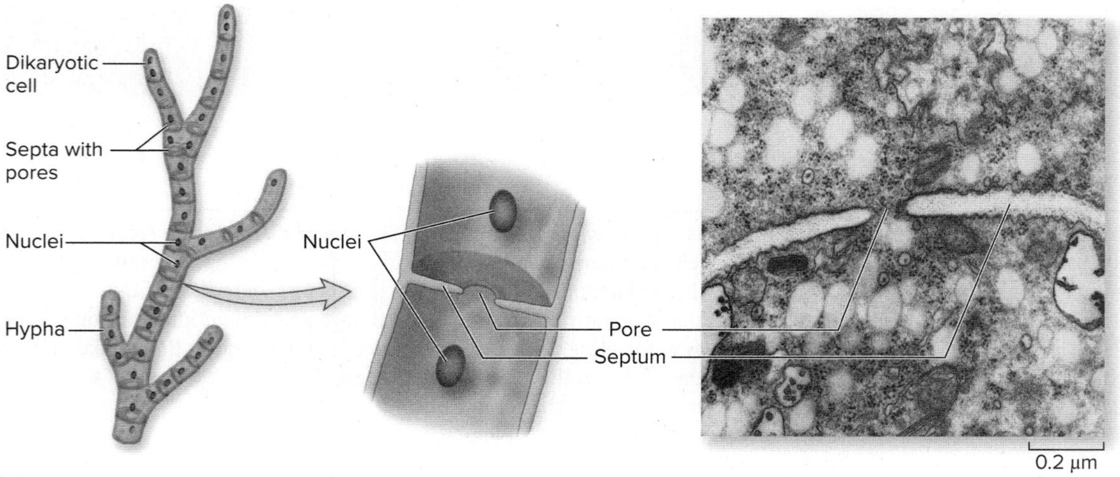

Figure 25.1 Hyphal structure showing a septum between cells. This transmission electron micrograph of a section through a hypha of an ascomycete shows a pore through which the cytoplasm streams.

Biophoto Associates/Science Source

Hyphae
10 μm
Mycelium

Figure 25.2 Fungal mycelium. This mycelium, composed of hyphae, is growing through leaves on a forest floor in Maryland. The dense mat of hyphae has a massive surface area relative to its volume and is adapted to absorb nutrients.

(inset): Nano Creative/Science Source; (right): Darlyne A. Murawski/Getty Images

Some cells (or hyphae) contain multiple nuclei

Fungi are different from most animals and plants in that each cell (or hypha) can house one, two, or more nuclei. A hypha with only one nucleus is **monokaryotic;** a cell with two nuclei is **dikaryotic.** In a dikaryotic cell, the two haploid nuclei exist independently. Dikaryotic hyphae have some of the genetic properties of diploids, because both genomes may be transcribed.

Sometimes many nuclei intermingle in the common cytoplasm of a fungal mycelium, which can lack distinct cells. If a dikaryotic or multinucleate hypha has nuclei that are derived from two genetically distinct individuals, the hypha is called **heterokaryotic.** Hyphae whose nuclei are genetically similar to one another are called **homokaryotic.**

Reproduction in Fungi Reflects Their Unusual Body Organization

LEARNING OBJECTIVE 25.1.2 Compare cell division in fungi with cell division in higher eukaryotes.

Fungal reproduction is influenced by the way fungal structures are organized in several key respects.

Cell division does not always follow mitosis

Most higher eukaryotic cell cycles are characterized by the replication and segregation of the genetic material followed by cytokinesis. This makes the cell the unit of division. In many filamentous fungi, however, the nucleus is the unit of division, and cytokinesis does not occur. In many fungal mitoses, the nuclear envelope does not break down as it does in higher eukaryotes, and the mitotic spindle forms within the nucleus. Except in chytrids, mitotic spindles are formed using spindle pole bodies and centrioles are absent. In some fungi, septa may form similarly to how new walls are formed when plant cells divide. Alternatively, the formation of septa may be suppressed, leading to multinucleate cells.

Fungi can reproduce both sexually and asexually

Many fungi are capable of reproducing using both sexual and asexual mechanisms. When a fungus reproduces sexually, two haploid hyphae of compatible mating types may come together and fuse.

The Dikaryon Stage. In animals, plants, and some fungi, the fusion of two haploid cells during reproduction immediately results in a diploid cell ($2n$). But in other fungi—namely, basidiomycetes and ascomycetes—an intervening dikaryotic stage ($1n + 1n$) occurs before the parental nuclei fuse and form a diploid nucleus. In ascomycetes, this **dikaryon** stage is brief, occurring in only a few cells of the sexual reproductive structure. In basidiomycetes, however, it can last for most of the life of the fungus, including both the mycelial and sexual spore-producing structures.

Reproductive structures

Some fungi produce specialized reproductive structures that make spores. For example, the mushrooms and puffballs that can be seen above ground are spore-forming structures. These structures are made from dense collections of hyphae that arise from an underground mycelium. Unlike in a mycelium, however, the hyphae forming the reproductive structures have cells separated by intact septa, and there is no cytoplasmic streaming. Fungi form four general types of reproductive structure: motile zoospores, zygosporangia, basidia, and asci.

Spores

Spores are the most common means of reproduction among fungi and are resistant to drying out, heating, and freezing. They may form as a result of either asexual or sexual processes, and they are often dispersed by the wind, insects, or other small animals. When spores land in a suitable place, they germinate, giving rise to a hypha that can grow into a new fungal mycelium. A few fungal phyla retain ancestral flagella and have motile zoospores. These organisms tend to live in wet or moist environments.

Because spores are very small, between 2 and 75 μm in diameter (figure 25.3), they can remain suspended in the air for a long time. Many of the fungi that cause diseases in plants and animals are spread rapidly by such means. Several species of fungus—most notably the molds—produce spores that can trigger allergic reactions. Although some of these molds grow outside, some are often found indoors in damp environments such as bathrooms and kitchens.

Fungi Are Adapted for Efficient Nutrient Absorption

LEARNING OBJECTIVE 25.1.3 Relate the structure of fungi to their nutritional strategy.

Fungi, like animals, are heterotrophs. This means that they obtain carbon and energy from the breakdown and oxidation of organic, reduced compounds. However, unlike animals, which use internal digestion, fungi secrete digestive enzymes and then absorb the digested organic molecules. In fungi that act as decomposers, the material to be degraded is commonly dead plants and animals; however, parasitic fungi will invade the living cells of their hosts and absorb nutrients directly.

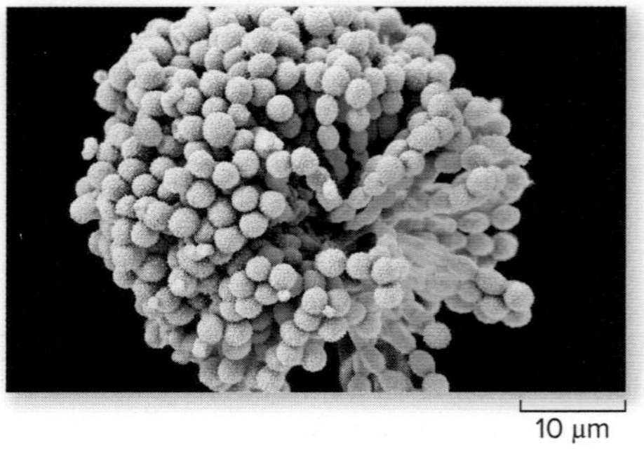
10 µm

Figure 25.3 Fungal spores. Scanning electron micrograph of fungal spores from Aspergillus.
Eye of Science/Science Source

The structure of fungi reflects their nutritional strategy. Unicellular fungi (yeasts) have the greatest surface-area-to-volume ratio of any fungus, which means they can absorb nutrients very efficiently and grow rapidly. The extensive threadlike networks of hyphae in mycelia also provide an enormous surface area for enzyme secretion and nutrient digestion. The rapid growth seen in many fungi is in part due to the efficient breakdown and absorption of nutrients.

Many fungi are able to break down the cellulose in wood, cleaving the bonds between glucose subunits and then absorbing the glucose molecules as nutrients. Most fungi also digest lignin, an insoluble organic compound that strengthens plant cell walls and is found in wood and bark. The specialized metabolic pathways of fungi allow them to obtain nutrients from woody plants and from an extraordinary range of organic compounds, including tiny roundworms called nematodes (figure 25.4*a*). In the environment, the majority of dead plants and plant-based materials are decomposed by fungi. Without this activity, ecosystems would become overwhelmed with dead plant material.

The mycelium of the edible oyster mushroom *Pleurotus ostreatus* (figure 25.4*b*) secretes a substance that paralyzes nematodes that feed on the fungus. When the worms become sluggish and inactive, the fungal hyphae envelop and penetrate their bodies. The fungus then secretes digestive juices and absorbs the nematode's nutritious contents.

Pleurotus usually grows within living trees or on old stumps, obtaining the bulk of its glucose through the enzymatic digestion of cellulose and lignin from plant cell walls. The nematodes it consumes apparently serve mainly as a source of nitrogen—a nutrient often in short supply in biological systems. Other fungi are even more active predators than *Pleurotus,* snaring, trapping, or firing projectiles into nematodes, rotifers, and other small animals on which they prey.

Because of their ability to break down almost any carbon-containing compound, fungi have potential uses in bioremediation (refer to chapter 17). For example, they have the potential to clean up contaminated forest litter, soil, or water.

REVIEW OF CONCEPT 25.1

A fungus consists of a mass of hyphae termed a mycelium. Hyphal cells have walls containing the polysaccharide chitin and store glucose in glycogen. Cell division in fungi divides the nucleus but not always the hypha itself. Sexual reproduction may occur when hyphae of two different mating types fuse. Dikaryon stages of the life cycle can occur when haploid nuclei do not fuse but instead share cytoplasm. Spores are produced sexually or asexually and are spread by wind or animals. Fungi secrete digestive enzymes externally and then absorb the products of the digestion. Secretion and absorption are maximized by high surface-area-to-volume ratios. Fungi can break down many different kinds of organic compound.

- *What differentiates fungi from animals?*

25.2 Fungi Have an Enormous Ecological Impact

Fungi, along with bacteria, are the principal decomposers in the biosphere. They break down organic materials and return the substances locked in those molecules to circulation in the ecosystem. Fungi can break down cellulose and lignin, an insoluble organic compound that is a major constituent of wood. By breaking down such substances, fungi release carbon, nitrogen, and phosphorus from the bodies of living or dead organisms and make them available to other organisms.

Fungi Are Involved in a Range of Symbioses

LEARNING OBJECTIVE 25.2.1 Distinguish between different kinds of symbiotic relationships involving fungi.

In addition to their role as decomposers, fungi have formed relationships, called symbioses, with a variety of organisms. In some cases, the relationship is essential for survival of the fungus and it is an

Fungal loop
Fungus
Nematode
a.
400 µm
b.

Figure 25.4 Carnivorous fungi. ***a.*** Fungus obtaining nutrients from a nematode. ***b.*** The oyster mushroom *Pleurotus ostreatus* not only decomposes wood but also immobilizes nematodes, which the fungus uses as a source of nitrogen.
(a): Carolina Biological/Medical Images/DIOMEDIA
(b): L. West/Science Source

obligate symbiosis. In other cases, the fungus can survive without its symbiotic partner and the relationship is a **facultative symbiosis.** Both kinds of symbiosis can be found in a group of closely related fungi. Three kinds of symbiotic relationship can be identified:

- **Pathogens** and **parasites** gain resources from their host and have a negative effect on the host that can lead to death. Pathogens always cause disease, whereas parasites, except in extreme cases, usually do not.
- **Commensal** symbiosis benefits one partner but does not harm the other.
- **Mutualistic** symbiosis benefits both partners.

Endophytes live inside plants and may protect plants from parasites

Some species of fungi live inside plants, obtaining nutrients from them; in some cases, they produce compounds that protect the plant against herbivory or that increase the plant's nutrient absorption. The relationship between these endophytic fungi and a plant can be parasitic, commensal, or mutualistic. In some cases, the nature of the symbiotic relationship can change, depending on interactions with other organisms or in response to the environment.

One way to assess whether an endophyte is enhancing the health of its host plant is to grow plots of plants with and without an endophyte. An experiment with perennial ryegrass, *Lolium perenne,* demonstrated that it is more resistant to aphid feeding when an endophytic fungus, *Neotyphodium,* is present (figure 25.5).

Lichens Are an Example of Symbiosis Between Different Kingdoms

LEARNING OBJECTIVE 25.2.2 Describe the ecological roles played by lichens.

Lichens (figure 25.6) are symbiotic associations between a fungus and a photosynthetic partner. Although many lichens are excellent examples of mutualism, some fungi are parasitic on their photosynthetic host.

Composition of a lichen

Ascomycetes are the fungal partners in all but about 20 of the approximately 15,000 species of lichens that we think exist. Most of the visible body of a lichen consists of its fungus, but between the filaments of that fungus are cyanobacteria, green algae, or sometimes both.

Specialized fungal hyphae penetrate or envelop the photosynthetic cell walls within the partner and transfer nutrients directly to that partner. Although fungi penetrate the cell wall, they do not penetrate the plasma membrane. Biochemical signals sent out by the fungus apparently direct its cyanobacterial or green algal component to produce metabolic substances that it does not produce when growing independently of the fungus.

The fungi in lichens are unable to grow normally without their photosynthetic partners, and the fungi protect their partners from strong light and desiccation. When fungal components of lichens are isolated from their photosynthetic partner, they survive, but grow very slowly.

SCIENTIFIC THINKING

Hypothesis: *Endophytic fungi can protect their host from herbivory.*

Prediction: *There will be fewer aphids* (Rhopalosiphum padi, *an herbivore) on perennial ryegrass* (Lolium perenne) *infected with endophytic fungi than on uninfected ryegrass.*

Test: *Place five adult aphids on each pot of 2-week-old grass plants with and without endophytic fungi. Place pots in perforated bags and grow for 36 days. Count the number of aphids in each pot.*

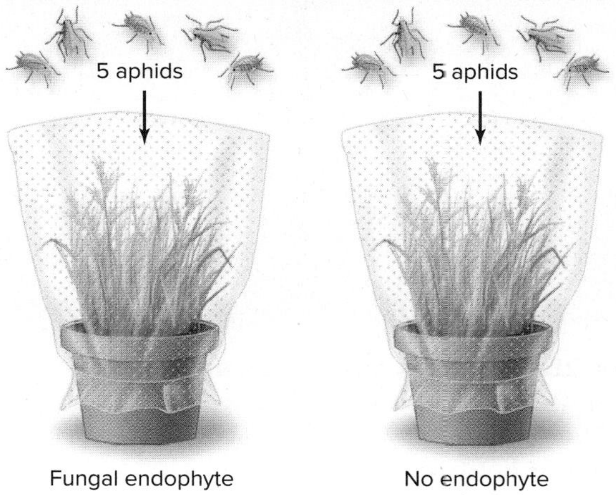

Result: *Significantly more aphids were found on the uninfected grass plants.*

Conclusion: *Endophytic fungi protect host plants from herbivory.*

Further Experiments: *How do you think the fungi protect the plants from herbivory? If they secrete chemical toxins, could you use this basic experimental design to test specific fungal compounds?*

Figure 25.5 Effect of the fungal endophyte *Neotyphodium* on the aphid population living on perennial ryegrass (*Lolium perenne*).

(left): Nigel Cattlin/Alamy Stock Photo; (right): B. Borrell Casals/Getty Images

Figure 25.6 Lichens growing on a rock.
Perry Mastrovito/Corbis/Getty Images

Ecology of lichens

The durable construction of the fungus combined with the photosynthetic properties of its partner allow lichens to grow anywhere it is relatively dry with a stable substrate for growth. Lichens are resilient to drying and can survive extended periods without water, remaining capable of growth if provided with moisture. In harsh, exposed areas, lichens are often the first colonists, breaking down rock to produce new soil and setting the stage for colonization by other organisms.

Because many lichens absorb aerosolic compounds directly across their surfaces, they often accumulate environmental pollutants to high levels. Those pollutants can affect growth and, so, some lichens are good indicators of air quality. Lichens are generally absent in and around cities because of automobile and industrial pollution, but some are adapted to these conditions. As pollution decreases, lichen populations tend to increase.

Mycorrhizae Are Fungi Associated with Roots of Plants

LEARNING OBJECTIVE 25.2.3 Compare and contrast arbuscular mycorrhizae and ectomycorrhizae.

About 90% of all plant families have species with roots that form obligate symbiotic relationships with fungi called **mycorrhizae.** These symbioses are so pervasive in plants that these fungi probably amount to about 15% of the total weight of the world's plant roots. The fungi in mycorrhizal associations act as extensions of the plant root system. The fungal hyphae dramatically increase the surface area of the plant root that is in contact with the soil, resulting in much more efficient absorption of nutrients such as nitrogen, phosphorus, and trace elements. The plant, on the other hand, supplies sugars to the fungus, so the symbiosis is mutualistic.

There are two types of mycorrhizae: arbuscular mycorrhizae and ectomycorrhizae (figure 25.7):

- **Arbuscular mycorrhizae** have hyphae that penetrate the outer cells of the plant root. They grow through the plant cell walls and make contact with the root cell's plasma membrane via highly branched structures called arbuscules.

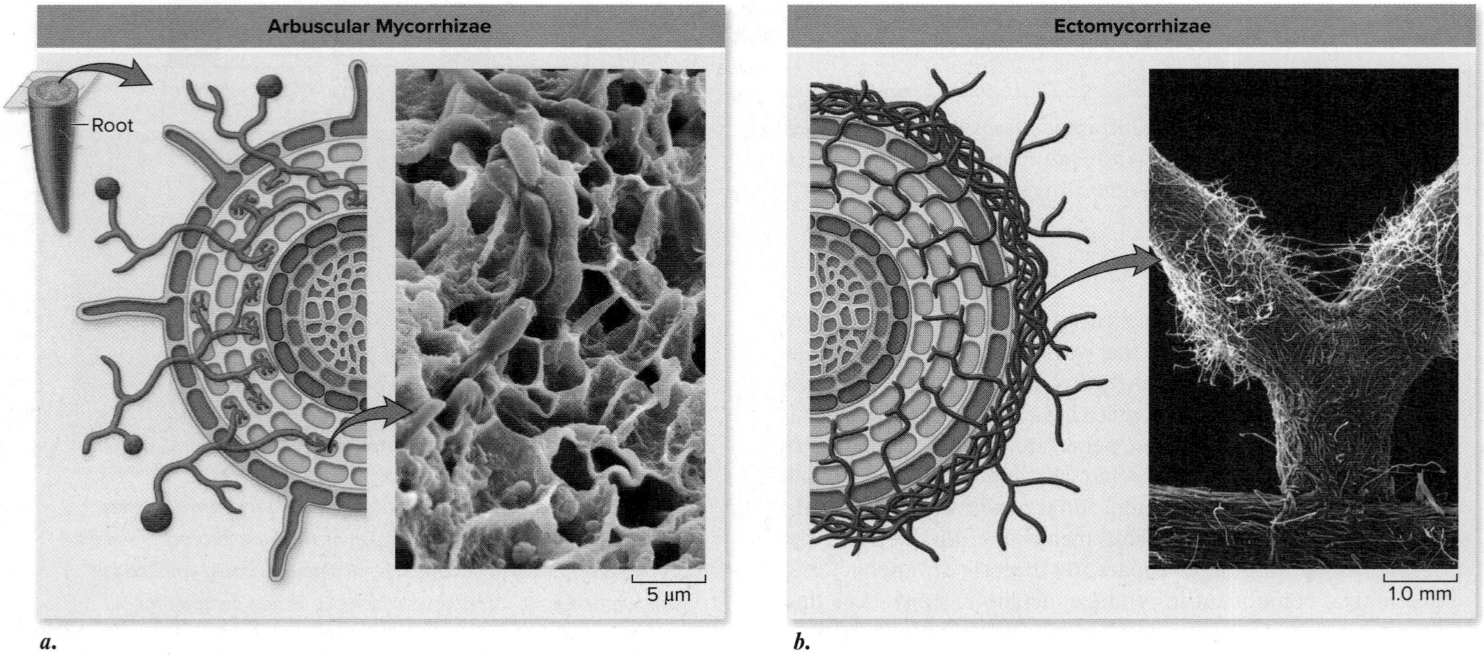

Figure 25.7 Arbuscular mycorrhizae and ectomycorrhizae. ***a.*** In arbuscular mycorrhizae, fungal hyphae penetrate the root cell wall of plants, making close contact with the cell's plasma membrane but not penetrating it. ***b.*** Ectomycorrhizae, shown here on the roots of a *Eucalyptus* tree, grow around and extend between the root cells but never penetrate the cell wall.
(a): Eye of Science/Science Source; (b): H.B. Massicotte, R.L. Peterson and L.H. Melville

- **Ectomycorrhizae** have hyphae that grow around and surround, but do not penetrate, the cell walls of the root cells.

In both cases, the mycelium extends far out into the soil so that the fungus is interacting more closely with the soil via its thread-like hyphae than can root hairs. A single root may associate with many fungal species, creating a complex set of symbiotic relationships between different species and different kingdoms.

Arbuscular mycorrhizae

Arbuscular mycorrhizae are far more common than ectomycorrhizae and are associated with about 80% of all plant species (figure 25.7*a*). The fungal partner in arbuscular mycorrhizae is always a glomeromycete. The glomeromycetes are associated with more than 200,000 species of plants; however, because none produce an aboveground reproductive structure like a mushroom, it is difficult to determine the exact number of species.

The earliest fossil plants often show arbuscular mycorrhizal roots, and this kind of association was likely the first type to arise. Such associations may have played an important role in allowing plants to grow in the infertile soils present when plants began to colonize land. Plants that form mycorrhizal associations are particularly successful in infertile soils, and it seems reasonable that mycorrhizal associations helped the earliest plants succeed on such soils. The closest living relatives of early vascular plants surviving today continue to depend strongly on mycorrhizae.

Ectomycorrhizae

Ectomycorrhizae (figure 25.7*b*) are associated with fewer plant species than are arbuscular mycorrhizae. Most ectomycorrhizal hosts are forest trees, such as pines, oaks, and birches. Although fewer plant species are involved in ectomycorrhizal symbioses, they are of agricultural interest due to some partner plants being important timber species. The majority of fungal species in ectomycorrhizae are basidiomycetes, the fruiting bodies (mushrooms), that can sometimes appear close to host trees.

Most ectomycorrhizal fungi are not restricted to a single species of plant, and most ectomycorrhizal plants form associations with many ectomycorrhizal fungi. Different combinations of fungus have different effects on the physiological characteristics of the plant and its ability to survive under different environmental conditions. Regardless of the species of fungus involved, their primary function is to extract nitrogen and other nutrients from soil and to pass them to the plant host.

Some Fungi Form Mutual Symbioses with Animals

LEARNING OBJECTIVE 25.2.4 Explain why the relationship between leaf-cutter ants and certain fungi is an obligate symbiosis.

Numerous fungus–animal symbioses have been identified. For example, ruminant animals contain certain fungi in their gut that can release nutrients by degrading plant materials with high cellulose and lignin content. In return, the fungus gets to live in the nutrient-rich rumen of the host animal.

A notable fungus–animal symbiosis occurs between leaf-cutter ants and some species of basidiomycete. These ants, members of the phylogenetic tribe Attini, have an obligate symbiosis with specific fungi, which they have domesticated and farm in an underground garden. The ants provide leaves to the fungi as a nutrient source and protect the fungus from pathogens and other predators (figure 25.8). The fungi are the ants' food source.

Figure 25.8 Ant–fungal symbiosis. Ants delivering leaf fragments to their fungal garden. The fungi degrade the plant material and use the nutrients for growth. The ants then eat the fungus.
Scott Camazine/Science Source

Depending on the species of ant, the ant nest can be as small as a golf ball or as large as 50 cm in diameter and many feet deep. Some nests are inhabited by millions of leaf-cutter ants that maintain fungal gardens. These social insects have a caste system, and different ants have specific roles. Traveling on trails as long as 200 m, leaf-cutter ants search for foliage for their fungi. A colony of ants can defoliate an entire tree in a day.

REVIEW OF CONCEPT 25.2

Fungi are the primary decomposers in ecosystems. Several types of symbiotic relationships between fungi and plants have evolved. Endophytes live inside plant tissues and may offer protection from parasites and herbivores. Lichens are a complex symbiosis between fungi and cyanobacteria or green algae. Mycorrhizal associations between fungi and plant roots are mutually beneficial. Fungi have also coevolved with animals in mutualistic relationships.

- *Why might it be argued that leaf-cutter ants are parasitic on the fungus they farm?*

25.3 Fungi Are Important Plant and Animal Pathogens

Fungi can destroy crops and create significant health problems for humans and other animals. A major problem in treatment and prevention is that fungi are eukaryotes, as are plants and animals. Understanding how fungi are different from plants and animals may lead to better treatments for fungal infections.

Fungal Infestation Can Harm Plants and Those Who Eat Them

LEARNING OBJECTIVE 25.3.1 Explain why treating fungal infections in animals is particularly difficult.

Fungi cause many diseases in plants and are responsible for billions of dollars in agricultural losses every year. In addition to causing plant disease, fungi can also spoil food products that have been harvested. Moreover, some fungi secrete toxic substances into the foods they are infecting.

Aflatoxins are produced by *Aspergillus* species that grow on corn, peanuts, and cotton seed (figure 25.9*a*). Aflatoxins can cause severe kidney and nervous system damage in animals, including humans. Aflatoxin B1 is one of the most carcinogenic compounds known, being nearly 100 times more carcinogenic than some of the compounds in cigarette smoke.

In contrast, corn smut is a maize fungal disease that is harmful to the plant but not to animals that consume it (figure 25.9*b*). Corn smut is caused by the basidiomycete *Ustilago maydis* and is edible.

Fungal infections are difficult to treat in humans and other animals

Human and animal diseases can also be caused by fungus. Some common diseases, such as ringworm (which is not a worm but a fungus), athlete's foot, nail fungus, and oral and vaginal candidiasis can be treated with topical antimycotics and in some cases with oral medication. However, these infections can be difficult to treat because of the close phylogenetic relationship between fungi and animals. When a pathogen is closely related to its host, it can be difficult to find drugs that target the biology of the pathogen but which do not interfere with the biology of the host. When two groups of organisms are well separated phylogenetically, as are bacteria and animals, it is easier to find drugs that discriminate between the biology of the two groups.

An example of a parasitic fungal–animal symbiosis is **chytridiomycosis,** first identified in 1998 as an emerging infectious disease of amphibians. Amphibian populations have been declining worldwide for over three decades. The decline correlates with the presence of the chytrid *Batrachochytrium dendrobatidis* in the skin of the infected amphibian (figure 25.10). Sick and dead frogs were more likely than healthy frogs to have flask-like structures encased in their skin, which proved to be associated with chytrid spore production.

It is possible that the pathogenicity of *B. dendrobatidis* is related to changes in symbiotic relationships between the amphibian partner and a variety of bacteria that live on the amphibian's skin. Some studies have shown that the bacteria produce antifungal compounds that combat the growth of the fungus. Changes to the normal microbiota on the skin may lead to uncontrollable fungal growth. Other studies have linked the increase in chytridiomycosis to increased pesticide use, which suppresses amphibian immunity to infection.

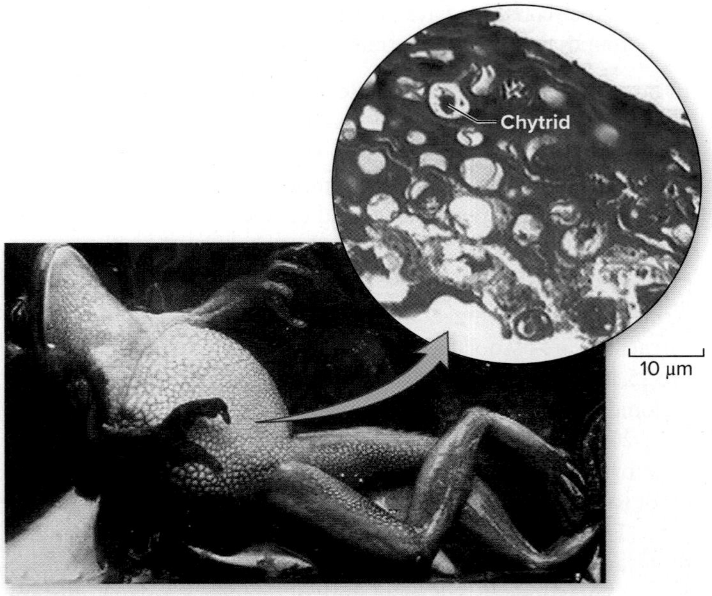

Figure 25.10 Frog killed by chytridiomycosis. Lesions formed by the chytrid can be seen on the abdomen of this frog.
(left): School of Biological Sciences, University of Canterbury, New Zealand; (inset): Dr. Peter Daszak

Figure 25.9 Maize (corn) fungal infections. ***a.*** *Ustilago maydis* infections of maize are a delicacy in Mexican cuisine. ***b.*** A photomicrograph of *Aspergillus flavus* conidia. *Aspergillus flavus* infects maize and can produce aflatoxins that are harmful to animals.
(a): Scott Camazine/Alamy Stock Photo; (b): BSIP SA/Alamy Stock Photo

REVIEW OF CONCEPT 25.3

Fungal infections can cause disease or death in both plants and animals, either by direct infection or by secretion of toxins. Treatment of fungal disease and parasitism in animals is made difficult by the close relationship between fungi and animals; drugs that are damaging to the fungus may also affect the host.

- *What is likely to be the most common mechanism for the spread of fungal disease?*

25.4 Fungi Are Taxonomically Diverse

Fossils and molecular data indicate that animals and fungi last shared a common ancestor close to 670 MYA, probably a *nucleariid* protist, based on DNA analysis of multiple genes. The oldest fungal fossils resemble extant members of the genus *Glomus* that arose within the Glomeromycota.

Fungal Taxonomy Is Undergoing Rapid Change

LEARNING OBJECTIVE 25.4.1 List the major phyla of fungi.

A complete and accurate phylogeny for the fungi remains to be determined. Based primarily on characteristics of meiosis, four fungal phyla were identified: Chytridiomycota ("chytrids"), Zygomycota ("zygomycetes"), Ascomycota ("ascomycetes"), and Basidiomycota ("basidiomycetes"). Subsequent molecular analyses showed that chytrids and zygomycetes are not monophyletic. This led to the "chytrids" being subdivided into three phyla: Chytridiomycota, Blastocladiomycota, and Neocallimastigomycota. The relationships within the zygomycetes have yet to be resolved. In 2007, mycologists agreed on seven monophyletic phyla: Microsporidia, Blastocladiomycota, Neocallimastigomycota, Chytridiomycota, Glomeromycota, Basidiomycota, and Ascomycota (figure 25.11 and table 25.1). The Microsporidia are sister to all other fungi, but there is disagreement as to whether they are true fungi.

REVIEW OF CONCEPT 25.4

Fungi are descended from nucleariid protists. Molecular analysis and features of meiosis divide the fungi into seven monophyletic phyla, plus the Zygomycota.

- *In respect to phylogenetics, what differentiates the Zygomycota from other fungal phyla?*

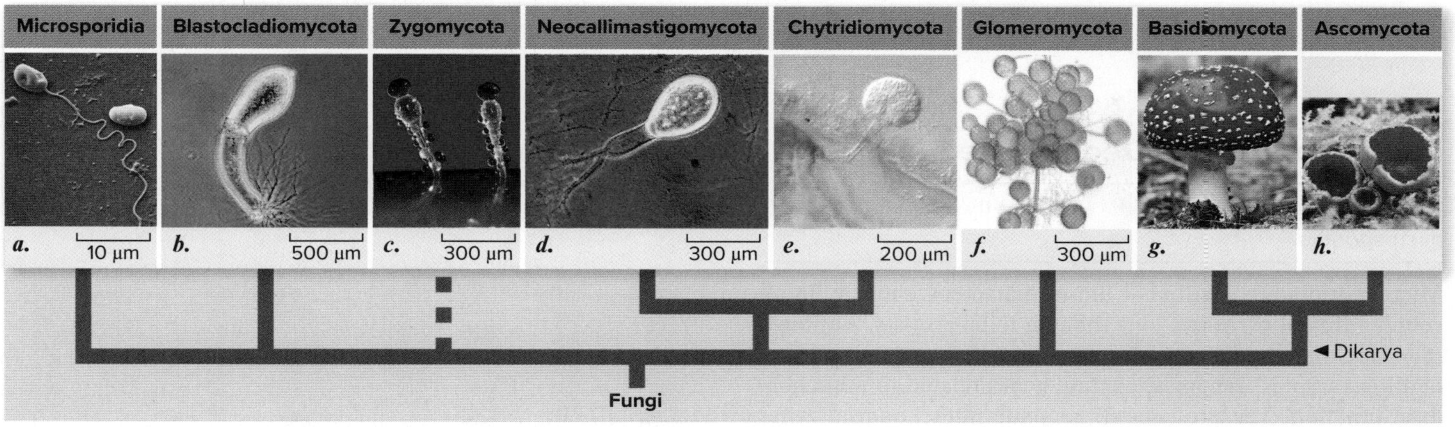

Figure 25.11 The major phyla of fungi. All phyla, except Zygomycota, are monophyletic. ***a.*** Microsporidia, including *Encephalitozoon cuniculis,* are animal parasites. ***b.*** *Allomyces arbuscula,* a water mold, is a blastocladiomycete. ***c.*** *Pilobolus,* a zygomycete, grows on animal dung and on culture medium. ***d.*** Neocallimastigomycota, including *Piromyces communis,* decompose cellulose in the rumens of herbivores. ***e.*** Some chytrids, including members of the genus *Rhizophydium,* parasitize green algae. ***f.*** Spores of *Glomus intraradices,* a glomeromycete associated with roots. ***g.*** *Amanita muscaria,* the fly agaric, is a toxic basidiomycete. ***h.*** The cup fungus *Cookeina tricholoma* is an ascomycete from the rainforest of Costa Rica.

(a): Dr. Ronny Larsson; (b): Don Barr/Mycological Society of America; (c): Carolina Biological/Medical Images/DIOMEDIA; (d) Contributed by Don Barr, Mycological Society of America; (e): Dr. Yuuji Tsukii; (f): Yolande Dalpé, Agriculture and Agri-Food Canada; (g): Ondrej83/Shutterstock; (h): Mantonature/Getty Images

TABLE 25.1 FUNGI

Group	Typical Examples	Key Characteristics	Approximate Number of Living Species
Chytridiomycota	*Allomyces*	Aquatic, flagellated fungi that produce haploid gametes in sexual reproduction or diploid zoospores in asexual reproduction	1,000
Zygomycota	*Rhizopus, Pilobolus*	Multinucleate hyphae lack septa, except for reproductive structures; fusion of hyphae leads directly to formation of a zygote in zygosporangium, in which meiosis occurs just before it germinates; asexual reproduction is most common	1,050
Glomeromycota	*Glomus*	Form arbuscular mycorrhizae; multinucleate hyphae lack septa; reproduce asexually	150
Ascomycota	Truffles, morels	In sexual reproduction, ascospores are formed inside a sac called an ascus; asexual reproduction is also common	45,000
Basidiomycota	Mushrooms, toadstools, rusts	In sexual reproduction, basidiospores are borne on club-shaped structures called basidia; asexual reproduction occurs occasionally	22,000

25.5 Microsporidia Are Unicellular Parasites

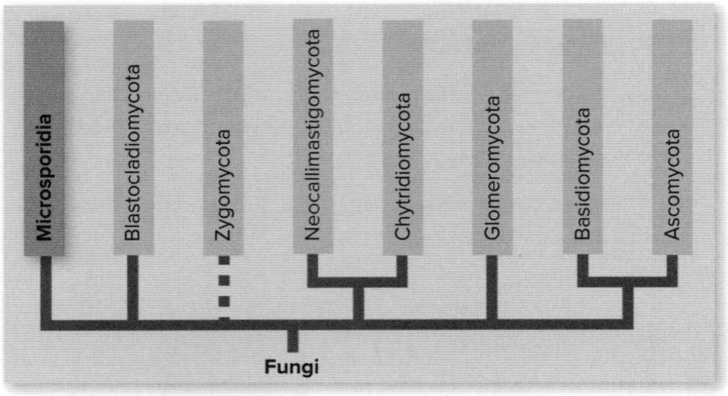

Microsporidia Are Degenerate Fungi

LEARNING OBJECTIVE 25.5.1 Describe the characteristics of microsporidia.

Microsporidia are small (1–4 μm in diameter), obligate, intracellular animal parasites, long thought to be protists. Although they lack mitochondria, their genomes, which are the smallest of all eukaryotes' genomes, contain genes related to mitochondrial function. The presence of genomic mitochondrial genes led to the hypothesis that ancestors of the microsporidia had mitochondria. Consistent with this is the observation that greatly reduced, mitochondrion-derived organelles can be found in microsporidia. These observations, combined with phylogenies derived from analyses of new sequence data, resulted in the microsporidia being moved from the protists to the fungi.

Whereas microsporidia parasitize many groups of animals, including mammals, crustaceans, fish, and birds, most parasitic species cause disease in insects. The infection with microsporidia of mosquitoes bearing *Plasmodium* spp. has even been proposed to control the spread of malaria. Infection usually occurs via the gastrointestinal (GI) tract. Ingested spores absorb water, swell, and then rupture in the GI tract. Rupture of the spores leads to the violent extrusion of a tightly coiled structure called the polar tube (figure 25.12). The polar tube penetrates an epithelial cell, and the infectious contents of the spore are passed into the host cell. In humans, microsporidia rarely cause disease; however, in immunocompromised individuals such as people with AIDS infection can lead to severe disease.

REVIEW OF CONCEPT 25.5

Microsporidia lack mitochondria, but they possess genes related to mitochondrial function and contain a mitochondrion-related organelle. As obligate parasites, microsporidia can cause diseases in animals, including humans.

- *How would you distinguish a microsporidian from the parasitic protistan* Plasmodium?

25.6 Chytridiomycota and Relatives: Fungi with Zoospores

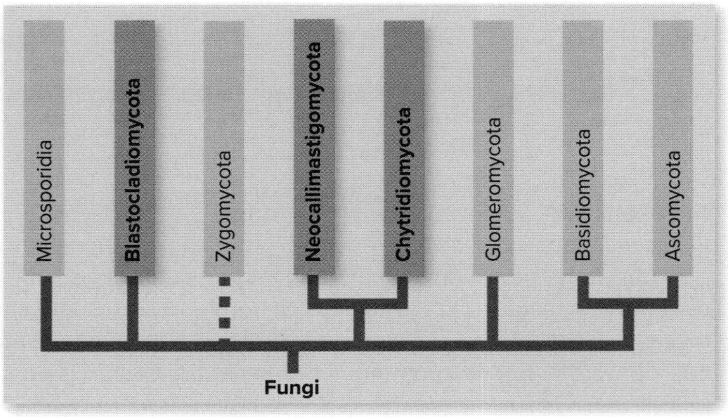

Until recently, members of the Chytridiomycota, the Blastocladiomycota, and the Neocallimastigomycota were grouped together in a single phylum—the Chytridiomycota—due to their common characteristic of flagellated zoospores. Flagella have been lost in

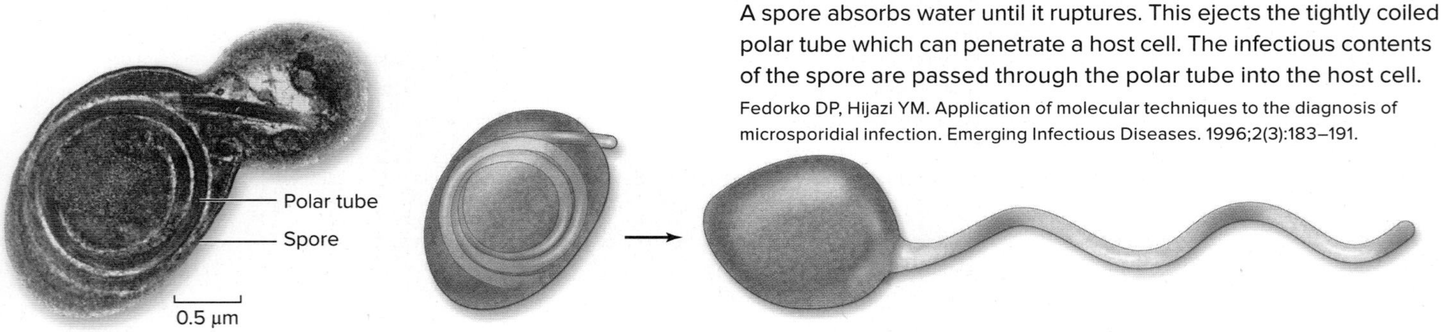

Figure 25.12 Polar tube of a microsporidian infects cells. A spore absorbs water until it ruptures. This ejects the tightly coiled polar tube which can penetrate a host cell. The infectious contents of the spore are passed through the polar tube into the host cell.

Fedorko DP, Hijazi YM. Application of molecular techniques to the diagnosis of microsporidial infection. Emerging Infectious Diseases. 1996;2(3):183–191.

all other fungal lineages except for the microsporidia. Recent molecular analyses have separated the single paraphyletic Chytridiomycota into three monophyletic phyla.

Chytrids Are Closely Related to Ancestral Fungi

LEARNING OBJECTIVE 25.6.1 Describe the characteristics of chytrids.

The close relationship of Chytridiomycota with basal lineages of the fungi combined with their aquatic ecology provide insight into the origins of aquatic fungi. The production of motile zoospores is a distinguishing character of these fungi. The word *chytrid* has its origin in the Greek word *chytridion*, meaning "little pot," referring to the structure that releases the flagellated zoospores (figure 25.13).

Chytrids include *Batrachochytrium dendrobatidis*, which has been implicated in the die-off of amphibians (refer to section 25.3). Other chytrids have been identified as plant pathogens.

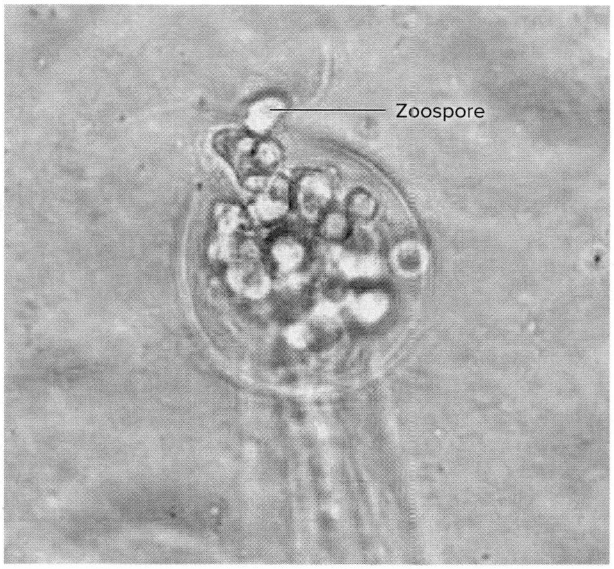

Figure 25.13 Zoospore release. The potlike structure (*chytridion* in Greek) containing the zoospores gives chytrids their name.

Contributed by Daniel Wubah, Mycological Society of America

Blastocladiomycetes Are the Only Fungi to Exhibit Alternation of Generations

LEARNING OBJECTIVE 25.6.2 Describe the defining characteristics of blastocladiomycetes.

The Blastocladiomycota are found in both aquatic and terrestrial environments. The fungi in the single class and single order of this phyla are parasites of plants, invertebrates, algae, and in one case, other fungi. Possibly the most distinguishing feature of blastocladiomycetes is their alternation of multicellular haploid and diploid life cycles.

Because of the ease with which some members of the blastocladiomycetes can be grown in culture, they are sometimes used in teaching laboratories to demonstrate the alternation of generations that is seen in some plant reproductive cycles. As in these cycles, the alternation of generations in members of the blastocladiomycota alternates a diploid ($2n$), spore-producing sporophyte with a haploid, gamete-producing gametophyte. As in plants, both generations are multicellular, with the sporophyte producing spores by meiosis and the gametophyte producing gametes by mitosis (figure 25.14).

Neocallimastigomycetes Anaerobically Digest Cellulose in Ruminant Herbivores

LEARNING OBJECTIVE 25.6.3 *Describe the defining characteristics of Neocallimastigomycetes.*

Within the rumens of mammalian herbivores, **neocallimastigomycetes** enzymatically digest the cellulose and lignin of the plant biomass in the animals' grassy diet. Sheep, cows, kangaroos, and elephants all depend on these fungi to obtain sufficient nutrients and calories. The cellulose-degrading neocallimastigomycete population of ruminants is critical when these animals are fed diets that are otherwise very difficult to digest. These fungi lack true mitochondria and instead use hydrogenosomes, which are thought to be derived from mitochondria, to produce ATP.

Organisms in genus *Neocallimastix* can survive on cellulose alone. Genes encoding digestive enzymes such as cellulase made their way into the *Neocallimastix* genomes via horizontal gene transfer from bacteria. Due to the anaerobic growth of these fungi and their abilities to degrade cellulose at relatively high temperatures, they have potential uses in biotechnology for biofuel production.

REVIEW OF CONCEPT 25.6

Three closely related phyla, Chytridiomycota, Blastocladiomycota, and Neocallimastigomycota, have flagellated zoospores. *Chytrid* refers to the potlike shape of the structure releasing the zoospores, and one species of chytrid is responsible for declining amphibian populations. Blastocladiomycetes are the only fungi to be characterized by a reproductive strategy involving the alternation of generations. Neocallimastigomycetes acquired cellulases from bacteria. They aid ruminant animals in digesting cellulose from plants and have potential for biofuel production.

- *What three features differentiate blastocladiomycetes from microsporidians?*

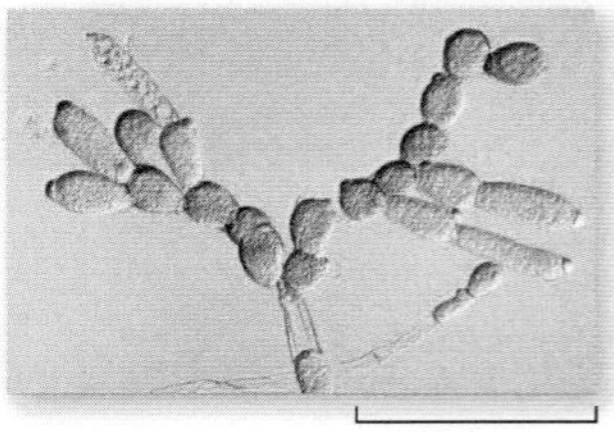

a. 100 µm

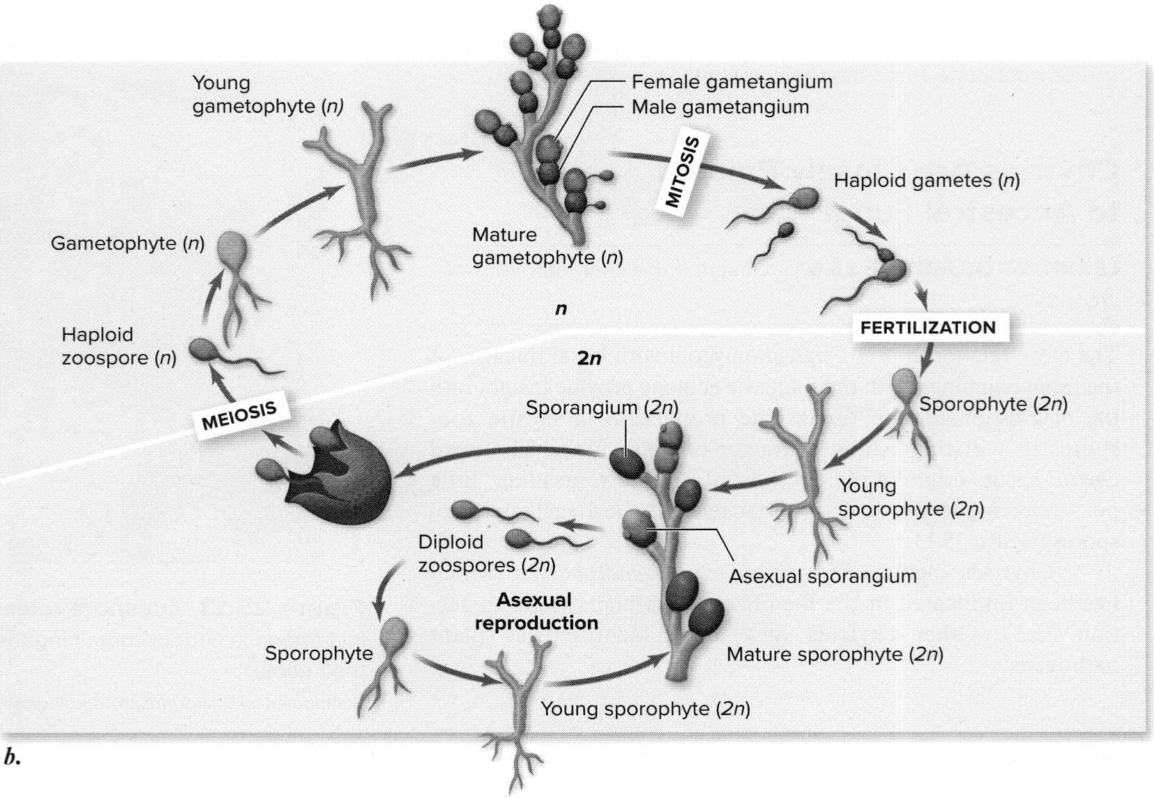

b.

Figure 25.14 ***Allomyces,* a blastocladiomycete that grows in the soil.** ***a.*** The spherical sporangia can produce either diploid zoospores via mitosis or haploid zoospores via meiosis. ***b.*** Life cycle of an *Allomyces* species, which has both haploid and diploid multicellular stages (alternation of generations).

(a): Carolina Biological/Medical Images/DIOMEDIA

25.7 Zygomycota Produce Zygotes

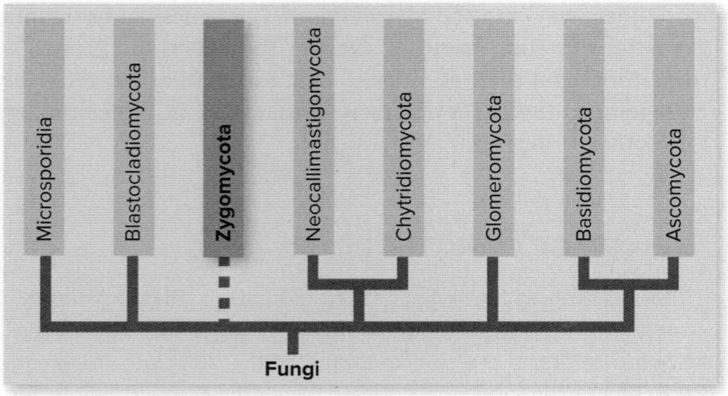

In Sexual Reproduction, Zygotes Form Inside a Zygosporangium

LEARNING OBJECTIVE 25.7.1 Describe the defining characteristics of zygomycetes.

Zygomycetes (phylum Zygomycota) include only about 1050 named species, but they are incredibly diverse and are some of the fastest-growing fungi. Among them are some of the more common bread molds (figure 25.15), as well as a variety of species found on decaying organic material such as strawberries and other fruits. Some members of this group can produce and modify steroids and, so, have use in biotechnology. Others have found uses in the production of foods such as tempeh, where they are used to partially degrade protein in soybeans to increase their digestibility. A few members are pathogens of plants and animals, including immunocompromised humans.

Zygomycetes have predominantly coenocytic hyphae, and they form diploid zygote nuclei during sexual reproduction using structures called gametangia and zygosporangia. They are also notable because their cell walls contain a modified form of chitin called chitosan.

When conditions are unfavorable, **chemotropic** hyphae of opposite mating types attract one another, and sexual reproduction can occur. Attraction between hyphae of different mating types leads to gametangia formation. Gametangia, containing multiple haploid nuclei, fuse and are isolated from the rest of the hyphae by complete septa.

Inside the fused gametangia, haploid nuclei fuse in a process called karyogamy. The fused gametangia containing diploid nuclei develop into a structure called a **zygosporangium** (figure 25.15*b*). Inside the zygosporangium, zygospores develop, producing thick coats that help the spores survive unfavorable growth conditions.

Zygosporangia are sensitive to physical damage and easily burst to release many haploid spores. Upon germination of released spores, haploid hyphae grow into a new mycelium.

a.

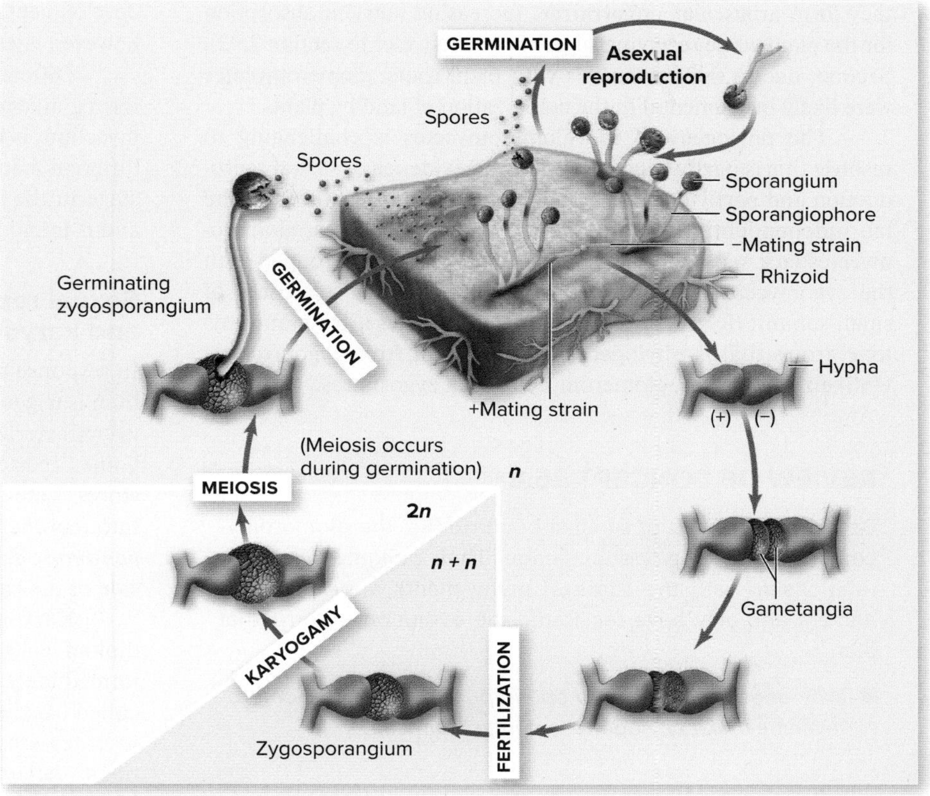

b.

Figure 25.15 ***Rhizopus*****, a zygomycete that grows on simple sugars.** This fungus is often found on moist bread or fruit. ***a.*** The dark, spherical, spore-producing sporangia are on hyphae about 1 cm tall. The rootlike hyphae (rhizoids) anchor the sporangia. ***b.*** Life cycle of *Rhizopus*. The Zygomycota group is named for the zygosporangia characteristic of *Rhizopus*. The (+) and (–) denote mating types.

(a): Carolina Biological/Medical Images/DIOMEDIA

Asexual Reproduction Is More Common

LEARNING OBJECTIVE 25.7.2 Describe the structure and functioning of sporangiophores.

Asexual reproduction occurs much more frequently than sexual reproduction in the zygomycetes. During asexual reproduction, hyphae produce clumps of erect stalks, called **sporangiophores.** The tips of the sporangiophores form sporangia, which are separated by septa. Thin-walled haploid spores are produced within the sporangia. These spores are shed above the food substrate, in a position where they may be picked up by the wind and dispersed to a new food source. Under favorable conditions, haploid hyphae grow from germinating spores. If conditions become unfavorable, then sexual reproduction can occur.

REVIEW OF CONCEPT 25.7

Zygomycetes are named for the production of diploid zygote nuclei via karyogamy. Zygomycete hyphae are coenocytic, forming septa only where gametangia or sporangia are produced. Sexual reproduction produces zygosporangia, which contain zygospores that are able to withstand harsh conditions. Asexual reproduction makes haploid spores in aerial structures called sporangia and is more common than sexual reproduction.

- *Under what conditions would you expect a zygomycete to produce zygospores rather than haploid spores?*

25.8 Glomeromycota Are Asexual Plant Symbionts

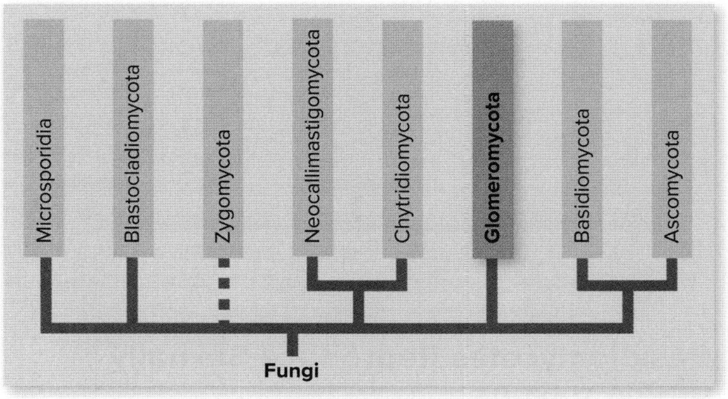

Glomeromycetes Facilitated the Invasion of Land by Plants

LEARNING OBJECTIVE 25.8.1 *Describe the ecological importance of the glomeromycetes.*

Despite being a small group of fungi with only 150 or so known species, the glomeromycetes are important for two related reasons. First, they are obligate symbionts with plant roots, where they form **arbuscular mycorrhizae,** increasing nutrient absorption for the plant while receiving sugars in return (refer to section 25.2). Second, due to their symbiosis with plant roots, glomeromycetes were likely instrumental in the colonization of land by plants.

The phylogeny of the glomeromycetes is challenging to resolve; this is partly because there is no evidence of sexual reproduction and partly because these fungi cannot be cultured in the lab independently of plant roots. Like zygomycetes, glomeromycetes lack septae in their hyphae and were once grouped with the zygomycetes. However, comparisons of DNA sequences of small-subunit rRNAs reveal that glomeromycetes are a monophyletic group that is phylogenetically distinct from zygomycetes. Unlike zygomycetes, glomeromycetes lack zygospores.

REVIEW OF CONCEPT 25.8

Based on analyses of small-subunit rRNAs, the glomeromycetes are a monophyletic lineage. Their obligate symbiotic relationship with the roots of many plants appears to be ancient and may have facilitated the evolution of terrestrial plants.

- *Why does the inability to obtain pure cultures of glomeromycetes make phylogenetic analysis difficult?*

25.9 Basidiomycota Are the Mushroom Fungi

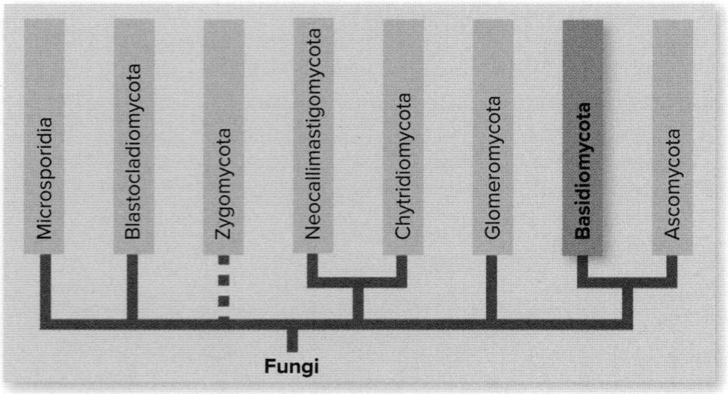

Basidiomycetes Reproduce Sexually Within Basidia

LEARNING OBJECTIVE 25.9.1 Distinguish primary from secondary mycelia in a basidiomycete.

The **basidiomycetes** (phylum Basidiomycota) are a large group of fungi with over 30,000 species, and include the common edible forms of portobello and white button mushrooms. Basidiomycetes include not only the mushrooms, toadstools, and puffballs, but also many important plant pathogens (figure 25.16*a*).

The majority of the life cycle is spent as monokaryotic hyphae growing as a primary mycelium on or through a substrate. Early in development, the hyphae of the primary mycelium are coenocytic; however, later in development, septa form to separate the nuclei.

Different mating types of monokaryotic hyphae may fuse to form a dikaryotic, or secondary, mycelium (figure 25.16*b*). Such a mycelium is heterokaryotic, with two nuclei representing the two different mating types between each pair of septa. This dikaryotic stage in the life cycle is a common characteristic of many fungi and is found in both the ascomycetes and the basidiomycetes.

Sexual reproduction involves basidia and karyogamy

In response to particular environmental cues, a secondary mycelium is triggered to enter the sexual reproductive cycle. This cycle is characterized by the formation of dense, aboveground fruiting bodies (basidiocarps) easily recognized as mushrooms or toadstools. Late in fruiting body development, some hyphae develop into reproductive, club-shaped structures called **basidia** (singular, *basidium*). Basidia are usually found on gills that line the underside of the basidiocarp.

Karyogamy occurs within basidia, giving rise to the only diploid cells of the life cycle (figure 25.16*b*). Meiosis occurs immediately after karyogamy to produce four haploid spores called basidiospores. In most members of this phylum, the basidiospores are borne at the end of the basidia on slender projections called sterigmata. Released basidiospores are easily distributed by wind and, under suitable conditions, germinate to produce monokaryotic hyphae that form a primary mycelium.

REVIEW OF CONCEPT 25.9

The primary mycelium consists of monokaryotic hyphae resulting from spore germination. Fusion of hyphae leads to the formation of a dikaryotic mycelium. Fruiting bodies form under certain conditions, and karyogamy in basidia leads to the formation of haploid basidiospores. Basidiospore germination produces hyphae, forming a monokaryotic primary mycelium.

- *What distinguishes a dikaryotic cell from a diploid cell?*

25.10 Ascomycota Are the Most Diverse Phylum of Fungi

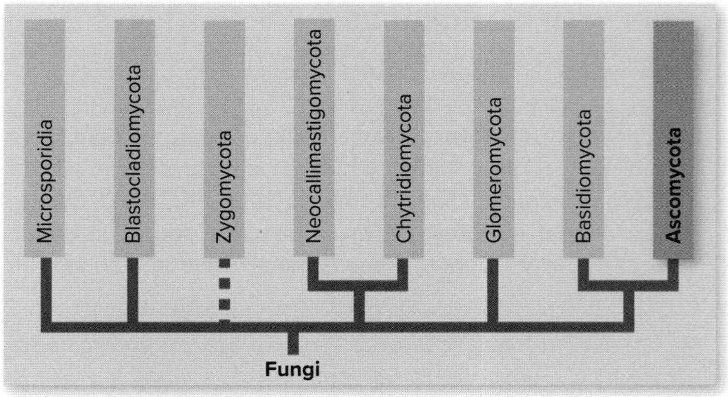

a.

Figure 25.16 Basidiomycetes.
a. The death cap mushroom, *Amanita phalloides*. When eaten, these mushrooms are usually fatal. ***b.*** Life cycle of a basidiomycete. The basidia are the reproductive structures and in this example are found on gills on the underside of the basidiocarp.
(a): Alexandra Lowry/Science Source

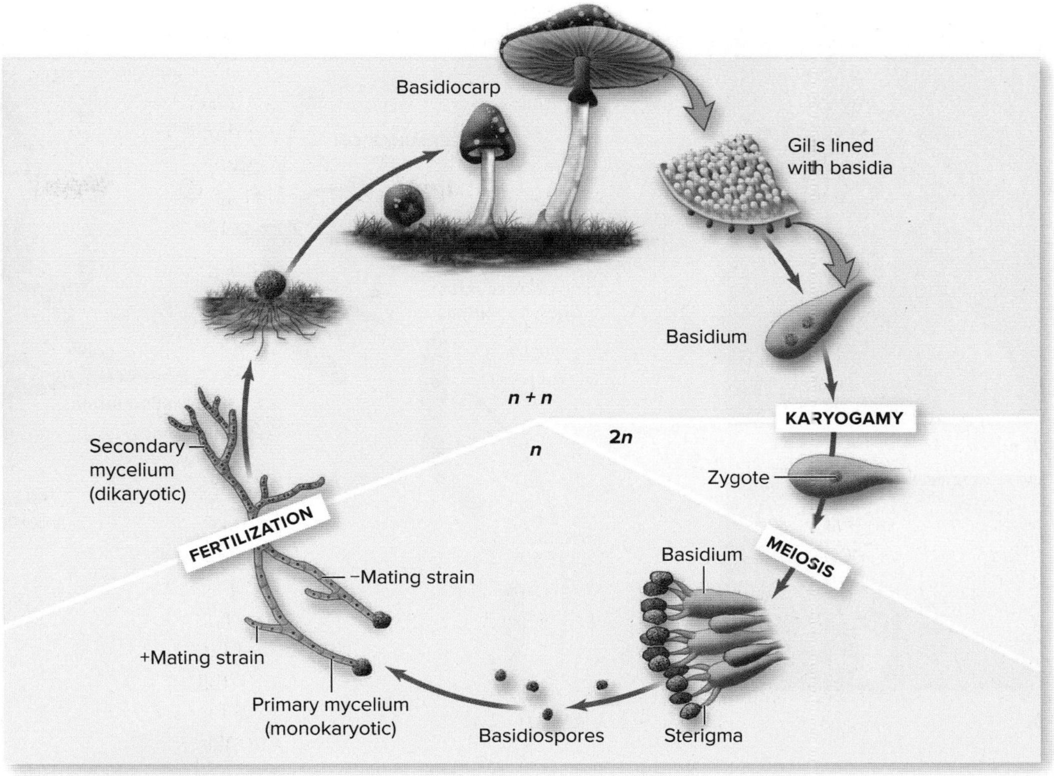

b.

The phylum **Ascomycota** contains about 75% of the known fungi and is incredibly diverse. Among the ascomycetes are such familiar and economically important fungi as bread yeasts, common molds, morels (figure 25.17*a*), cup fungi (figure 25.17*b*), and truffles. Also included in this phylum are many plant pathogens, including those causing chestnut blight and Dutch elm disease. Maybe the most famous ascomycete, *Penicillium chrysogenum*, produces the antibiotic penicillin.

Numerous semisynthetic and fully synthetic forms have been derived from the original penicillin, penicillin G. The antibiotic works by interfering with cell-wall synthesis in gram-positive bacteria. Despite serious antibiotic resistance in some pathogens, the penicillin family of antibiotics remains important in the treatment of a wide range of infectious diseases. A close relative of the penicillins, the cephalosporins, are produced by the mold *Cephalosporium* and also work by interfering with cell-wall synthesis but, importantly, work against certain gram-negative species as well as gram-positive species.

Asexual Reproduction Occurs Within Conidiophores

LEARNING OBJECTIVE 25.10.1 Distinguish between conidia and basidiospores.

Asexual reproduction is common in the ascomycetes and results in the production of **conidia** (singular, *conidium*). These are asexual spores cut off by septa at the ends of modified hyphae called conidiophores. Conidia are easily dispersed by airflow, water, or animals and allow for the rapid colonization of a new food source. The hyphae produced from the germination of conidia are divided by septa, but the septa are perforated, and cytoplasm flows along the length of each hypha.

Sexual Reproduction Occurs Within an Ascus

LEARNING OBJECTIVE 25.10.2 Describe the process of sexual reproduction in an ascomycete.

Sexual reproduction is initiated when two compatible hyphae make contact. One hypha produces an antheridium, and the other produces an ascogonium; their fusion results in the mixing of the cytoplasm from each hypha. Hyphae containing pairs of nuclei are then produced, which can form a fruiting body, called an ascocarp, with the primary mycelium (figure 25.17*c*).

Certain hyphae in the ascocarp develop into structures called asci (singular, *ascus*), after which these fungi are named. Karyogamy results in the production of a diploid nucleus at the tip of forming asci. In mature asci, the diploid nucleus undergoes meiosis to produce four haploid cells, each of which then undergoes a single mitosis to produce eight haploid ascopores.

In many ascomycetes, the ascus becomes turgid at maturity, eventually bursting. Under suitable conditions, ascospores can germinate to produce haploid hyphae that can grow vegetatively into mycelia, producing conidia to reproduce asexually, or antheridia and ascogonia to reproduce sexually.

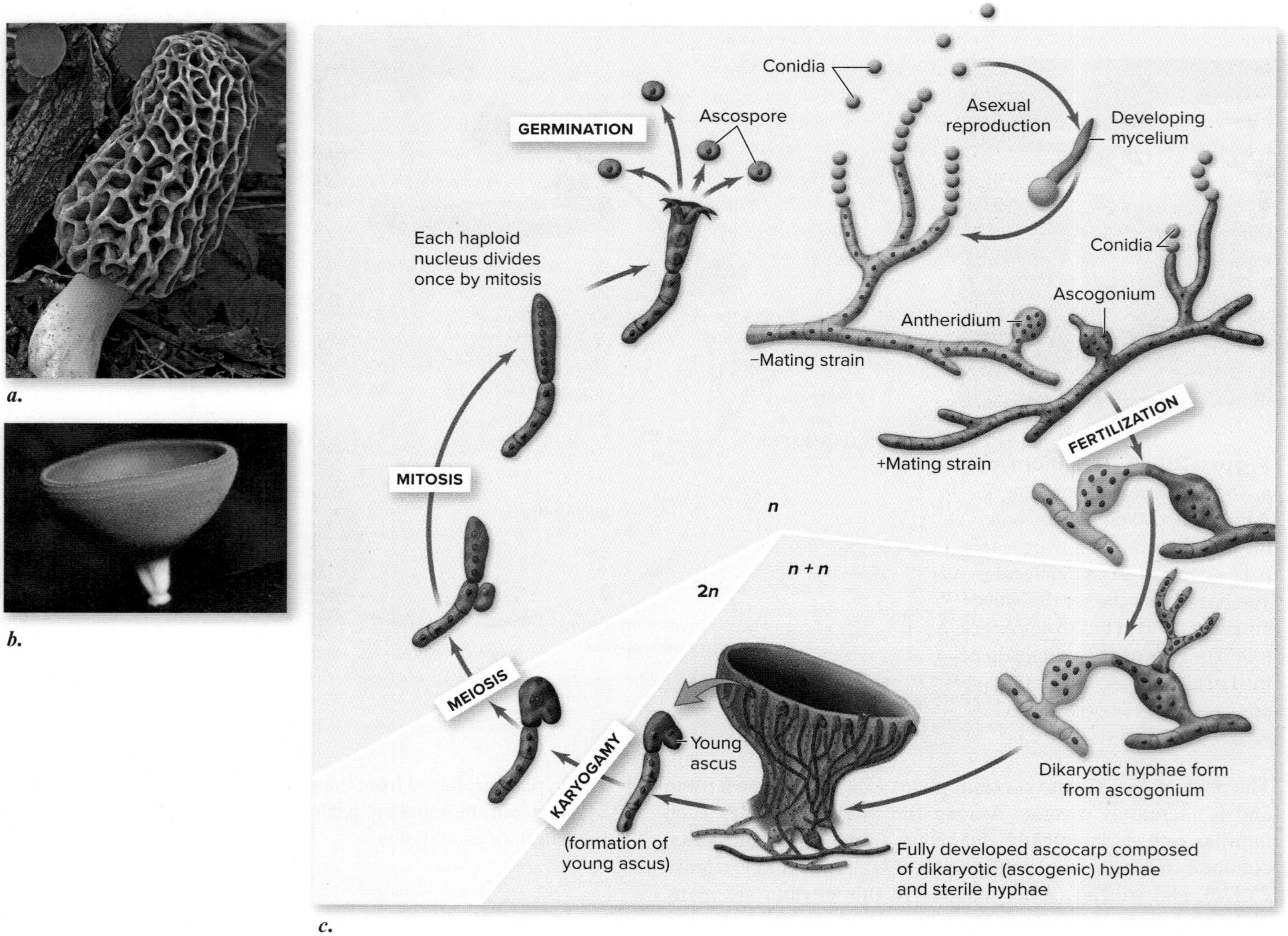

Figure 25.17 Ascomycetes. ***a.*** The morel, *Morchella esculenta,* is an edible ascomycete that appears in early spring. ***b.*** A cup fungus. ***c.*** Life cycle of an ascomycete. Haploid ascospores form within the ascus.

(a): Kenneth M. Highfill/Science Source; (b): Ed Reschke/Getty Images

Some Ascomycetes Are Unicellular Yeasts

LEARNING OBJECTIVE 25.10.3 Describe the commercial importance of unicellular ascomycetes.

Most yeasts are ascomycetes with a single-celled lifestyle. Yeast reproduction is usually asexual and takes place by cell fission or budding, when a smaller cell forms from a larger one (figure 25.18). Sometimes two yeast cells of opposite mating type fuse to form a dikaryon. This cell may then function as an ascus, with karyogamy followed immediately by meiosis. The resulting ascospores germinate into new asexually reproducing, haploid yeast cells.

The ability of yeasts to ferment carbohydrates, breaking down glucose to produce ethanol and carbon dioxide, is exploited in the production of bread, beer, and wine. Many different strains of yeast have been domesticated and selected for these processes, using the sugars in rice, barley, wheat, and corn. Wild yeasts—those that occur naturally in the areas where wine is made—were important in wine making historically, but domesticated cultured yeasts are normally used now.

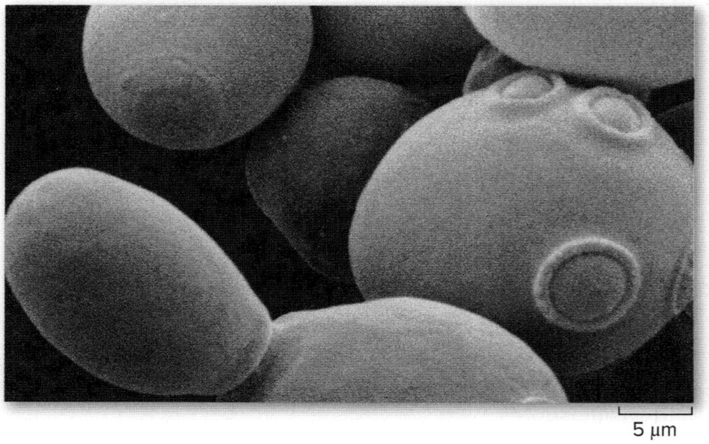

Figure 25.18 Budding in *Saccharomyces.* As shown in this scanning electron micrograph, the cells tend to hang together in chains, a feature that calls to mind the derivation of single-celled yeasts from multicellular ancestors.

David Scharf/Science Source

The ability to ferment simple sugars, found in abundance in fruits as the fall arrives, likely evolved in response to competition for those sugars from other microorganisms. Organisms only able to use oxygen as a terminal electron acceptor would be at a competitive disadvantage over organisms able to break down abundant environmental sugars in the absence of oxygen using fermentation. Yeasts' ability to metabolize sugars in the absence of oxygen would give them a growth advantage over bacteria that would otherwise compete for the resource. Also, the alcohol produced by the yeasts is likely to inhibit bacterial growth, providing another selective advantage.

The most important yeast in baking, brewing, and winemaking is *Saccharomyces cerevisiae.* This yeast has been used by humans throughout recorded history. Yeast is also employed as a nutritional supplement, because it contains high levels of B vitamins and because about 50% of yeast is protein.

REVIEW OF CONCEPT 25.10

Ascomycetes undergo karyogamy within a characteristic saclike structure, the ascus. Meiotic division of a diploid nucleus produces ascospores in the ascus. Yeasts within this group generally reproduce asexually by budding but can also form ascospores in an ascus-like structure. Ascomycetes include both beneficial forms used as foods and in the production of foods, and harmful forms responsible for diseases and spoilage.

- *Coccidioidomycosis is caused by inhaling spores; it often occurs in farmworkers in the southwestern United States. What would help prevent this disease?*

Inquiry & Analysis

Are Chytrids Killing the Frogs?

As you learned in section 25.3, chytrid fungi are thought to play a major role in several amphibian extinctions. Awareness of the possible role of chytrids began in Queensland (the northeastern portion of Australia) in 1993, when a mass die-off of frogs was reported. Many different kinds of frogs were affected, and entire populations were wiped out. In the rainforests of northern Queensland, populations of the sharp-nosed torrent frog *(Taudactylus acutirostris)* were found to be so seriously affected as to be in danger of extinction. Captive colonies were set up at James Cook University and at the Melbourne and Taronga zoos in an attempt to preserve the species. Unfortunately, these attempts failed. Every frog in the colonies died.

Using scanning electron microscopy, researchers examined the skin of sick frogs and saw what is shown in the electron micrograph below. The normally smooth surface of the frog's skin was covered in protruding spherical structures.

The protrusions were zoosporangia: asexual reproductive structures of a chytrid fungus. One is shown in the close-up in the photos. Each zoosporangium is roughly spherical, with one or more small, projecting tubes. Millions of tiny zoospores develop in each zoosporangium. When the plug blocking the tip of a tube disappears, the spores are discharged onto the surface of adjacent skin cells, or into the water, where their flagella allow them to swim until they encounter another host. When one of the zoospores contacts the skin of another frog, it attaches and forms a new zoosporangium in the subsurface layer of the skin, renewing the infection cycle.

The chytrids turned out to be *Batrachochytrium dendrobatidis*. Chytrids are typically found in water and soil; this was the first reported case of them infecting a vertebrate host.

The initial scanning electron micrograph results were compelling evidence that chytrids had caused the mass die-off of frogs in Queensland. To provide more direct evidence, experiments were carried out in which the ability of the chytrid fungus to kill frogs was directly assessed.

In one typical experiment, some frogs were exposed to chytrids and others were not. After three weeks, all the frogs were examined for shed skin, a clinical sign of chytrid infection and disease. The results are shown in the two pie charts.

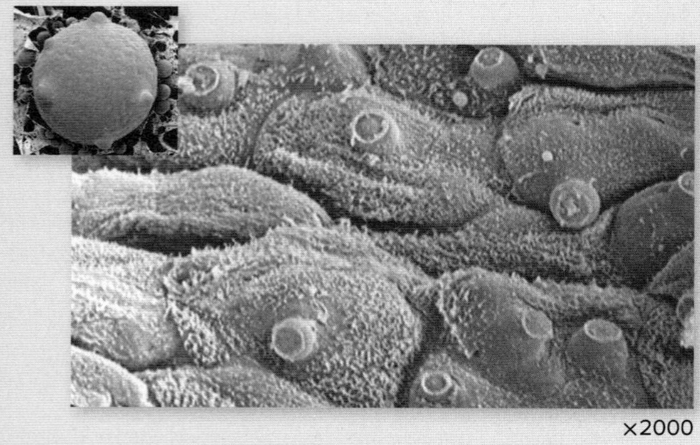
×2000

Lee Berger, Alex D. Hyatt, Rick Speare, Joyce E. Longcore, "Life cycle stages of the amphibian chytrid *Batrachochytrium dendrobatidis*," *Diseases of Aquatic Organisms*, 68: 51–63, 2005.

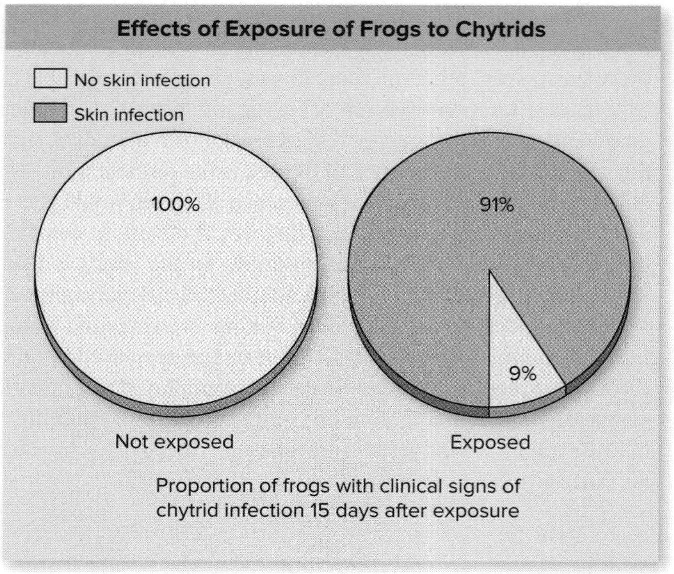
Effects of Exposure of Frogs to Chytrids
No skin infection
Skin infection
100%
91%
9%
Not exposed
Exposed
Proportion of frogs with clinical signs of chytrid infection 15 days after exposure

Analysis

1. **Applying Concepts** In this study, what are the dependent and independent variables?
2. **Interpreting Data** What is the incidence of skin infection in frogs not exposed versus in those exposed to the chytrid?
3. **Making Inferences** Is there any association between exposure to *B. dendrobatidis* and development of skin infection?
4. **Drawing Conclusions** What is the impact of exposure to chytrids upon the likelihood of infection?
5. **Further Analysis**
   a. Many kinds of frogs and salamanders are dying all over the world. Does this experiment suggest a way to determine the general susceptibility of amphibians to chytrid infection?
   b. A few frog die-offs have occurred in the past, but none have been nearly this serious. Do you think *B. dendrobatidis* is a new species, or could changing environmental conditions be related to the epidemic? Discuss.

Retracing the Learning Path

CONCEPT 25.1 Fungi Have Unique Reproductive and Nutritional Strategies

25.1.1 The Body of a Fungus Is a Mass of Connected Hyphae Fungi have hyphal cells with walls containing chitin. A mass of connected hyphae is called a mycelium. Hyphae can be continuous and multinucleate (coenocytic) or divided into long chains of cells separated by cross-walls called septa. A hypha with one nucleus is monokaryotic; a hypha with two nuclei is dikaryotic.

25.1.2 Reproduction in Fungi Reflects Their Unusual Body Organization The nucleus, not the cell, is the relevant unit of reproduction. The mitotic spindle forms inside the nuclear envelope, which does not break down and re-form. Fungi can reproduce sexually by the fusion of hyphae from two compatible mating types. Spores can form by either asexual or sexual reproduction.

25.1.3 Fungi Are Adapted for Efficient Nutrient Absorption Fungi obtain their nutrients by secreting enzymes for external digestion and then absorbing the products. Fungal structures for the absorption of nutrients have high surface-area-to-volume ratios.

CONCEPT 25.2 Fungi Have an Enormous Ecological Impact

25.2.1 Fungi Are Involved in a Range of Symbioses Fungi can be pathogenic or parasitic, commensal or mutualistic.

25.2.2 Lichens Are an Example of Symbiosis Between Different Kingdoms A lichen is composed of a fungus, usually an ascomycete, along with cyanobacteria, green algae, or both.

25.2.3 Mycorrhizae Are Fungi Associated with Roots of Plants Arbuscular mycorrhizae are common and involve glomeromycetes; ectomycorrhizae are primarily found in forest trees and involve basidiomycetes and a few ascomycetes.

25.2.4 Some Fungi Form Mutual Symbioses with Animals Some ants grow "farms" of fungi by providing plant material. The ants consume the fungi as a source of nutrients.

CONCEPT 25.3 Fungi Are Important Plant and Animal Pathogens

25.3.1 Fungal Infestation Can Harm Plants and Those Who Eat Them Infestation of agriculturally valuable crops causes billions of dollars of damage annually. Fungi that infect plants can produce chemicals that make food unpalatable, carcinogenic, or poisonous.

CONCEPT 25.4 Fungi Are Taxonomically Diverse

25.4.1 Fungal Taxonomy Is Undergoing Rapid Change Fungi form seven monophyletic phyla, and much remains to be resolved about the correct classification of some fungi.

CONCEPT 25.5 Microsporidia Are Unicellular Parasites

25.5.1 Microsporidia Are Degenerate Fungi Microsporidia are small, obligate, cellular parasites lacking mitochondria. Some species can cause potentially severe disease in humans.

CONCEPT 25.6 Chytridiomycota and Relatives: Fungi with Zoospores

25.6.1 Chytrids Are Closely Related to Ancestral Fungi The ecology and evolutionary history of the Chytridiomycota place them close to the common ancestor of fungi. Chytrids have been implicated in the die-off of amphibians.

25.6.2 Blastocladiomycetes Are the Only Fungi to Exhibit Alternation of Generations Parasitizing plants, animals, and other fungi, blastocladiomycetes alternate multicellular haploid and diploid structures in their life cycle.

25.6.3 Neocallimastigomycetes Anaerobically Digest Cellulose in Ruminant Herbivores The ability of the neocallimastigomycetes to degrade plant material makes them attractive to some biotechnology companies.

CONCEPT 25.7 Zygomycota Produce Zygotes

25.7.1 In Sexual Reproduction, Zygotes Form Inside a Zygosporangium Zygomycetes all produce a diploid zygote. In sexual reproduction, fusion of the haploid nuclei of gametangia produces diploid zygote nuclei, which undergo meiosis to form zygospores.

25.7.2 Asexual Reproduction Is More Common Sporangia produce haploid spores that are airborne; bread mold is a common example of a zygomycete.

CONCEPT 25.8 Glomeromycota Are Asexual Plant Symbionts

25.8.1 Glomeromycetes Facilitated the Invasion of Land by Plants Glomeromycete hyphae form intracellular associations with plant roots and are called arbuscular mycorrhizae.

CONCEPT 25.9 Basidiomycota Are the Mushroom Fungi

25.9.1 Basidiomycetes Reproduce Sexually Within Basidia The primary mycelium is monokaryotic, but different mating types may fuse to form the heterokaryotic secondary mycelium. The basidiocarp is the visible reproductive structure. Karyogamy occurs within the basidia of the basidiocarp, giving rise to a diploid cell. Meiosis then results in four haploid basidiospores that germinate to produce a primary mycelium.

CONCEPT 25.10 Ascomycota Are the Most Diverse Phylum of Fungi

25.10.1 Asexual Reproduction Occurs Within Conidiophores Asexual reproduction is very common and occurs by means of conidia formed at the end of modified hyphae called conidiophores.

25.10.2 Sexual Reproduction Occurs Within an Ascus Karyogamy occurs only in the ascus and results in a diploid nucleus. Meiosis and mitosis produce eight haploid nuclei in walled ascospores.

25.10.3 Some Ascomycetes Are Unicellular Yeasts Unicellular ascomycetes, yeasts, are exploited in the commercial production of breads, beers, and wines.

Concept Overview

This Concept Overview diagrams the key concepts that were discussed in this chapter.

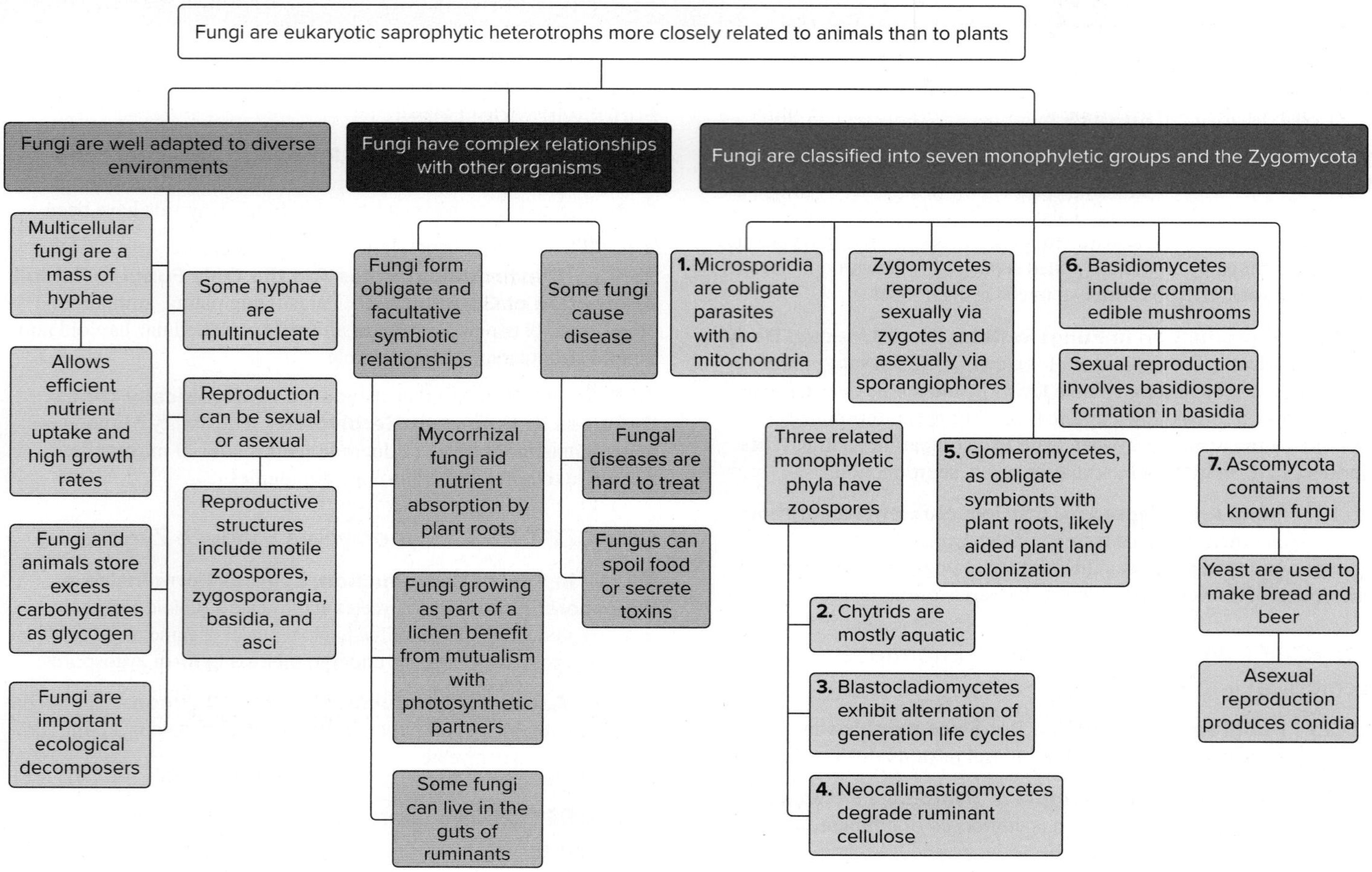

Assessing the Learning Path

Understand

1. Mycelia composed of hyphae support the high rates of growth seen in many fungi because they
   a. have cell walls made of chitin, which supports nutrient uptake.
   b. have very thin cell walls, which allow for efficient nutrient absorption.
   c. have a high surface-area-to-volume ratio to increase rates of nutrient absorption.
   d. are more easily hidden in soils and other substrates and are thus poorly predated.
2. What kind of relationship exists between a fungus and a green alga, if the fungus cannot grow without the alga, takes nutrients from the alga, but returns nothing?
   a. Obligate parasite
   b. Obligate mutualist
   c. Facultative pathogen
   d. Facultative mutualist
3. How could a normal chytrid population on the skin of a frog suddenly become pathogenic?
   a. By a second fungus also infecting the frog, and the two fungi working together to produce toxins that each alone cannot make
   b. By changes in the chytrid that make it grow faster and more aggressively on the skin
   c. By changing the pH of the water in which the frog lives
   d. By interfering with the normal bacterial skin population
4. Neocallimastigomycetes can digest the cellulose that is found in plant cell walls and live inside the digestive tract of many herbivores. Of what kind of symbiosis is this an example?
   a. Parasitism
   b. Mutualism
   c. Commensalism
5. Zygospores have thick coats to
   a. protect them from adverse conditions such as drying out.
   b. help them attach to the bodies of insects and other animals that distribute them.
   c. act as a source of nutrition while they germinate into hyphae.
   d. reduce predation by small invertebrates such as nematodes.
6. Glomeromycota are obligate plant symbionts, meaning that they
   a. harm the plant.
   b. grow best when associated with a plant.
   c. cannot live without a plant symbiont.
   d. are parasites.

7. How could you differentiate between a primary and a secondary mycelium?
   a. By the number of nuclei per cell in the mature hyphae
   b. By the diameter of the hyphae in the mycelium
   c. By whether or not the hyphae contain mitochondria
   d. By the presence of asci in the secondary but not the primary mycelium
8. Mitosis in multicellular fungi differs from that seen in other multicellular organisms in that
   a. the spindle apparatus is formed outside the nucleus.
   b. there is no DNA replication between mitotic divisions.
   c. the nuclear envelope does not break down.
   d. centrioles regulate the formation of microtubules.

Apply

1. The cell walls of fungi contain chitin, which is a form of cellulose used in making a sugar with nitrogen in it. If you isolated chitin from cell walls of fungi and analyzed the types of atoms present, which elements would you find?
   a. Nitrogen, carbon, hydrogen, and oxygen
   b. Nitrogen, sulfur, phosphorus, and oxygen
   c. Carbon, hydrogen, and oxygen
   d. Sulfur, oxygen, phosphorus, and selenium
2. An examination of a fungal structure shows that it is a dense collection of septate hyphae with spores present. This is most likely to be a
   a. nutrient-absorbing structure.
   b. reproductive structure.
   c. structure involved in the formation of gametes.
   d. structure used to parasitize cells of insects.
3. Why would a microsporidian with a mutation that prevented its polar tube from extruding not be able to establish an infection?
   a. The cell wall would break and the parasitic cell would die.
   b. The spore could not swell and rupture.
   c. The parasite could not attach to cells.
   d. Spore contents could not enter cells.
4. In the figure 25.5, what is the dependent variable?
   a. Presence or absence of endophytes
   b. Length of time plants are exposed to aphids
   c. Number of aphids
   d. Species of plant being used
5. Why would an antifungal treatment that targeted translation not be a good idea for use in animals when some antibiotics used on bacterial infections of animals interfere with translation?
   a. Fungi are phylogenetically closer to animals than bacteria are to animals, so the drug would not be selectively toxic due to the similarities in ribosomes and translation.
   b. The drug would be unlikely to be absorbed by the fungus, so could never get into the cytoplasm where translation occurs; this is not a problem in bacteria.
   c. Stopping translation in fungi would not affect fungi as it would affect bacteria, because fungi grow much more slowly than do bacteria.
   d. The animal cells would degrade the antifungal compound, rendering it useless.
6. Why might it be difficult to accurately classify all fungi into correct phylogenetic groups?
   a. They have unusual growth characteristics, which makes analysis difficult.
   b. If they can't be grown in the lab, then it is hard to obtain DNA to sequence specific genes.
   c. They have such varied reproductive strategies that it is hard to find patterns of similarity.
   d. They are small and hard to visualize microscopically.
7. If you were to compare three masses of hyphae of unknown origin, which characteristics would be useful in determining that one was an ascomycete?
   a. Presence of coenocytic hyphae
   b. Presence of gametangia
   c. Chitinous cell walls
   d. a and b
   e. b and c
8. Which two structures are functionally analogous with respect to nuclear fusion when comparing the Glomeromycota and the Basidiomycota?
   a. Asci and basidia
   b. Gametangia and basidia
   c. Gametangia and gills
   d. Sporangiophore and basidiocarp
9. In which part of a yeast's reproductive cycle would it be easiest to analyze the effects of mutations?
   a. The haploid stage
   b. The diploid stage
   c. Either stage
10. Cephalosporin antibiotics target cell wall synthesis in certain gram-negative and gram-positive species of bacteria. Why might the antibiotic not prevent growth in fungi that make it?
   a. Fungi use a form of gene editing to make the drug ineffective.
   b. Fungi lack the genes that are targeted by the drug.
   c. Fungi have very different molecules in their cell walls.
   d. Fungi break the drug down as they make it so they are not harmed by it.
11. Which of the following fungi or fungal structures would you expect to grow fastest?
   a. A yeast with a diameter of 10 μm and a single diploid genome of 13 Mbp in ideal growth conditions
   b. A cylindrical hypha 400 μm long and 2 μm wide with a single diploid genome of 13 Mbp in ideal growth conditions
   c. A yeast with a diameter of 10 μm and a single haploid genome of 6.5 Mbp in ideal growth conditions
   d. A conidiospore of 3 μm diameter that landed on a dry surface
12. Which of the following kinds of bacteria would be best adapted to compete with a sugar-fermenting yeast growing on an apple that has fallen from a tree in late summer in Dubuque, Iowa?
   a. One that can effectively grow anaerobically using fructose and starch as electron donors
   b. One that can metabolize ethanol
   c. One that can produce spores
   d. a and b
   e. b and c

Synthesize

1. Your friend has athlete's foot and is unhappy with how long it is taking to treat. Your friend asks you why antibiotics such as penicillin (derived from a fungus) are so much more effective in treating bacterial infections of humans than fungicides are in treating human fungal infections. What do you tell your friend?
2. What might be some problems with using microsporidia as a biological means to control malaria?
3. If you were able to find a way to culture glomeromycotes independently of plant roots, why would this be of interest to an agricultural biotechnology company?

Geofile/Doug Sherman

26 Plants

Learning Path

26.1 Land Plants Are Multicellular Autotrophs Adapted to Terrestrial Life

26.2 Bryophytes Have a Dominant Gametophyte Generation

26.3 Seedless Vascular Plants Evolved Roots, Stems, and Leaves

26.4 Lycophytes Have a Dominant Sporophyte Generation

26.5 Pterophytes Are Ferns and Their Relatives

26.6 Seed Plants Were a Key Step in Plant Evolution

26.7 Gymnosperms Are Plants with “Naked Seeds”

26.8 Angiosperms Are Flowering Plants

Concept Overview

This Concept Overview provides a diagram of the overarching concepts that are covered in this chapter. These concepts will be expanded upon in the Concept Overview at the end of the chapter.

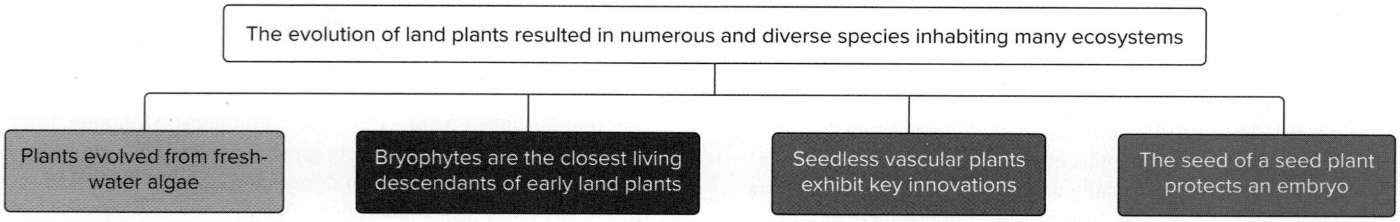

The evolution of land plants resulted in numerous and diverse species inhabiting many ecosystems
Plants evolved from fresh-water algae
Bryophytes are the closest living descendants of early land plants
Seedless vascular plants exhibit key innovations
The seed of a seed plant protects an embryo

Introduction

The colonization of land by plants fundamentally altered the history of life on Earth. A terrestrial environment offers abundant CO_2 and solar radiation for photosynthesis. But for at least 500 million years, the lack of water and more intense ultraviolet radiation on land confined the ancestors of land plants to aquatic environments. Evolutionary innovations for reproduction, structural support, and prevention of water loss were key in the adaptation of plants to terrestrial life. Numerous evolutionary solutions to the challenges of life on land have resulted in over 300,000 species of plants organized into 10 monophyletic taxa. Found in virtually every terrestrial habitat, from forests to alpine tundra and from agricultural fields to deserts, plants affect almost every aspect of our lives. They are used to provide pharmaceuticals, food, fuels, building materials, and clothing. This chapter explores the evolutionary history and life strategies of land plants from the earliest land colonizers, the liverworts, to the most recently diverged group, the flowering plants.

26.1 Land Plants Are Multicellular Autotrophs Adapted to Terrestrial Life

As we learned in chapter 24, the phylogenetic relationships among protists have been revised. We now know that green algae and the plants that dominate the land today shared a common protist ancestor about 1 BYA.

Plants Evolved from Freshwater Algae

LEARNING OBJECTIVE 26.1.1 Describe the relationship between green algae and plants.

A single species of freshwater algae gave rise to the entire terrestrial plant lineage. Given the innovations required to adapt to life on land, it is not surprising that the transition from water occurred only once. The identity of the ancestral alga remains unclear; however, close relatives classified as charophytes exist in freshwater lakes today.

The green algae split into two major groups: the chlorophytes, which never made it to land, and the charophytes, which are sister to all land plants. Collectively, the charophytes and the land plants are called the streptophytes (figure 26.1). The transition from water to land, and diversification into different terrestrial niches, required the following innovations:

- Ways to limit water loss and to control gas exchange
- Ways to limit damage to DNA by ultraviolet light
- Ways to transport water through the body of the plant
- Tissues to resist the effects of wind and gravity
- Ways to protect and distribute reproductive structures
- Life cycles that promote genetic diversity

Plants Have Adapted to Terrestrial Life

LEARNING OBJECTIVE 26.1.2 Explain how plants adapted to the challenges of life on land.

Living in water has several advantages over living on land:

1. Water is abundantly available and easily absorbed.
2. Water filters much harmful UV light.
3. Water provides support.
4. Reproductive cells are easily distributed when released into water.

On land, water is limited and easily lost by evaporation. As an adaptation to living on land, most plants are protected from dessication by a **cuticle.** The cuticle is made from waxy, water-impermeable substances that reduce water loss due to evaporation. However, the relatively impermeable cuticle limits the

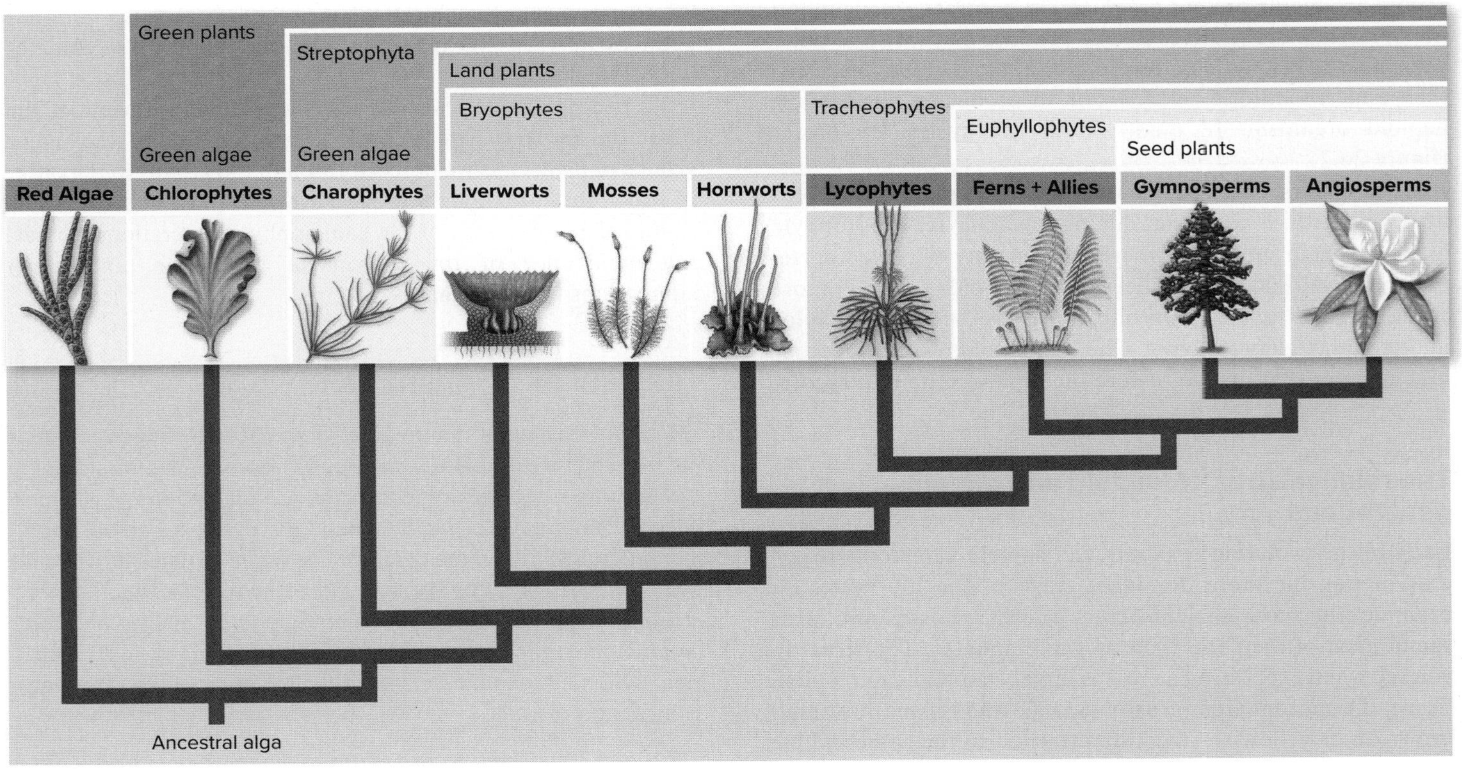

Figure 26.1 **Plant phylogeny.**

exchange of gases needed for photosynthesis and cellular respiration. The evolution of **stomata** (singular, *stoma*), tiny, mouth-shaped openings on the plant surface, allows gas exchange to occur. Stomata can be closed to limit water loss under certain conditions (refer to chapter 31).

Due to their nutritional strategy, plants have an absolute requirement for light. Sunlight contains wavelengths of light used by plants for photosynthesis in addition to wavelengths of light (UV) that are damaging to DNA. In water, much of the harmful UV is filtered out, and damage to DNA is significantly reduced. On land, however, plants must have other ways to protect DNA from the effects of UV light. One innovation is the production of pigments in land plants that absorb harmful UV light, protecting the genome from damage.

All land plants have both haploid and diploid generations. In a haploid genome, any deleterious alleles will manifest in the phenotype or be lethal to the plant. There has been an evolutionary shift toward a dominant diploid generation, which allows greater genetic variability to persist in plants.

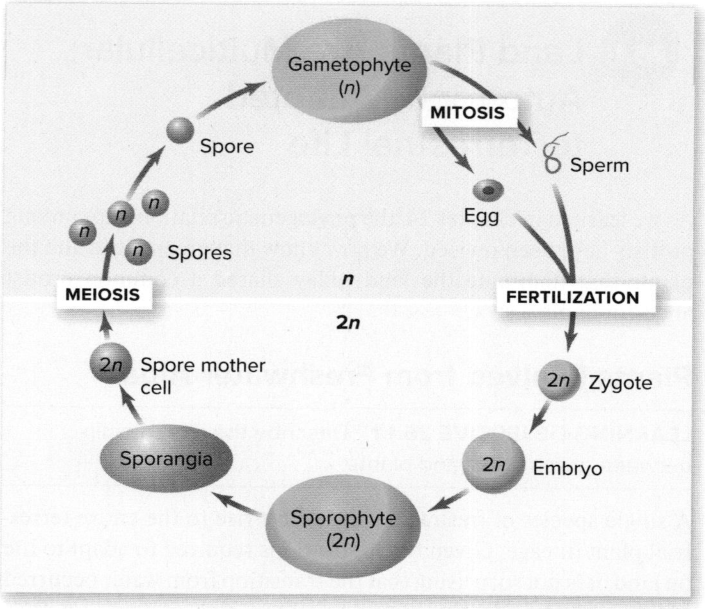

Figure 26.2 A generalized multicellular plant life cycle. Note that both haploid and diploid individuals can be multicellular. Also, spores are produced by meiosis, whereas gametes are produced by mitosis.

Plants Exhibit Alternation of Generations

LEARNING OBJECTIVE 26.1.3 Distinguish between a sporophyte and a gametophyte.

Many organisms, including humans, have a diplontic life cycle: meiotic cell division produces haploid gametes, which then fuse at fertilization to produce a diploid zygote. The zygote undergoes mitotic cell division to produce a multicellular, zygote-producing diploid organism. In contrast, in a haplontic life cycle, haploid gametes fuse to produce a zygote that undergoes meiotic cell division to produce haploid cells. These cells then undergo mitotic cell division to produce a multicellular haploid organism. In plants, both life cycles are present (figure 26.2).

In plants, fertilization produces a diploid zygote that develops into a spore-producing **sporophyte.** Sporophytes make spores in sporangia where **spore mother cells (sporocytes)** undergo meiosis to produce four haploid spores. Spores germinate and undergo mitotic cell division to form haploid multicellular **gametophytes.** The gametophytes produce haploid gametes that fuse to restore the diploid state in a zygote.

The relative sizes of haploid and diploid generations vary

Early in the evolution of plants, the haploid gametophyte generation was the more dominant stage of the life cycle. For example, the more ancient plants, such as mosses and ferns, spend most of their life cycle as gametophytes. The structure you see when examining a moss is largely haploid gametophyte tissue; the sporophyte is greatly reduced and is usually a small, brownish or yellowish structure attached to the gametophyte. The sporophyte is usually nutritionally dependent on the gametophyte and cannot exist independently.

In a member of a more recently evolved plant phylum, such as a conifer or a flowering plant like a daffodil, the most visible stage in the life cycle is the sporophyte. Regardless of which generation is most prominent and long-lived, gametophyte size is limited in all plants. Different plants produce gametes in ways that maximize the likelihood these gametes will meet.

In this chapter, we will use the general overview of the plant life cycle to consider the characteristics of the major groups of plants. As we proceed through the groups in the order in which they are thought to have evolved, you will see a reduction in the dominance of the gametophyte generation, increasing dominance of the sporophyte generation, and many adaptations to life on land.

REVIEW OF CONCEPT 26.1

A single freshwater green alga successfully invaded land. Its descendants, the plants, developed reproductive strategies, vasculature, stomata, and cuticles as adaptations to life on land. Plants have a haplodiplontic life cycle: a haploid form alternates with a diploid form in a single organism. Diploid sporophytes produce haploid spores by meiosis. Each spore can develop into a haploid gametophyte by mitosis and the gametophyte form produces haploid gametes, again by mitosis. When the gametes fuse, the diploid sporophyte is formed once more.

- *What distinguishes gamete formation in plants from gamete formation in humans?*

26.2 Bryophytes Have a Dominant Gametophyte Generation

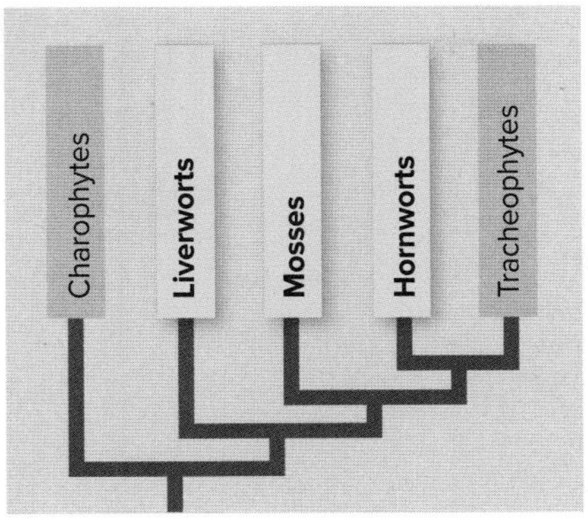

Bryophytes are the closest living descendants of the first land plants. Three monophyletic groups make up the bryophytes: the liverworts, the mosses, and the hornworts. Due to their lack of water-conducting tracheids, they are considered to be nonvascular plants.

Bryophytes Exhibit Many Adaptations for Living on Land

LEARNING OBJECTIVE 26.2.1 Describe adaptations of bryophytes for living on land.

The 24,700 or so bryophyte species have simple body plans but are well adapted to a range of terrestrial environments, including even deserts and arctic regions. Despite several similarities with their more modern relatives, such as the gymnosperms and angiosperms, the bryophytes have several distinguishing characteristics.

Most bryophytes are short; few exceed 7 cm in height. Height is limited because of a lack of supporting vascular tissue and roots. Although bryophytes have primitive water-conducting cells, they are not reinforced with lignin and do not provide structural support as tracheids do in the vascular plants. Unlike all other plants, bryophytes lack roots. Instead, they have primitive anchoring systems called rhizoids that can absorb water.

The gametophytes of bryophytes are photosynthetic and the dominant generation in the life cycle. The greatly reduced sporophytes are attached to the gametophytes, and they depend in varying degrees on the gametophytes for nutrition. This is in contrast to modern plants, like angiosperms, in which the sporophyte is the dominant generation in the life cycle.

Like ferns and some other vascular plants, bryophytes require water (such as rainwater) to reproduce sexually, a characteristic consistent with their aquatic origins. It is not surprising, therefore, that bryophytes are abundant in moist places, both in the tropics and in temperate regions.

Liverworts Are an Ancient Phylum

LEARNING OBJECTIVE 26.2.2 Describe the roles of gametophyte and sporophyte in liverwort reproduction.

The Old English word *wyrt* means "plant" or "herb." The most familiar **liverworts** (phylum Hepaticophyta) have flattened gametophytes with lobes resembling those of the liver—hence the name "liverwort" (figure 26.3). They constitute about 20% of the species of the phylum; the other 80% are leafy and superficially resemble mosses. The gametophytes lie flat and close to the ground, instead of growing erect, and are anchored using single-celled rhizoids that grow from the lower epidermis.

Haploid spores germinate to produce male or female gametophytes called thalli (singular, *thallus*). In some liverworts, gametes are produced in aerial, umbrella-like structures. Water splashing onto male gametangia can transfer sperm to female gametangia, where fertilization occurs. The resulting diploid embryo develops into a tiny sporophyte, dependent on the gametophyte for nutrition. In the sporophyte, meiosis produces spores, which are released and distributed by wind or water. Under favorable conditions, spores germinate into new thalli. Some liverworts can also reproduce asexually by releasing fragments of tissue from structures called gemmae, found on the surface of the thallus.

Mosses and Hornworts Have Distinct Evolutionary Innovations

LEARNING OBJECTIVE 26.2.3 Explain the relationship between moss gametophytes and sporophytes.

Unlike other bryophytes, the gametophytes of mosses typically consist of small, leaflike structures arranged spirally or alternately around a stemlike axis (figure 26.4); the axis is anchored to its substrate by means of rhizoids. Each rhizoid consists of several cells that absorb water, but not nearly the volume of water that is absorbed by a vascular plant root.

Figure 26.3 A common liverwort, *Marchantia* (phylum Hepaticophyta). The microscopic sporophytes are formed by fertilization within the tissues of the umbrella-shaped structures that arise from the surface of the flat, green, creeping gametophyte.
Steven P. Lynch

Figure 26.4 A hair-cup moss, *Polytrichum* (phylum Bryophyta). The leaflike structures belong to the gametophyte. The brownish stalks with a podlike sporangium at each tip are sporophytes.

Richard H. Gross/McGraw Hill

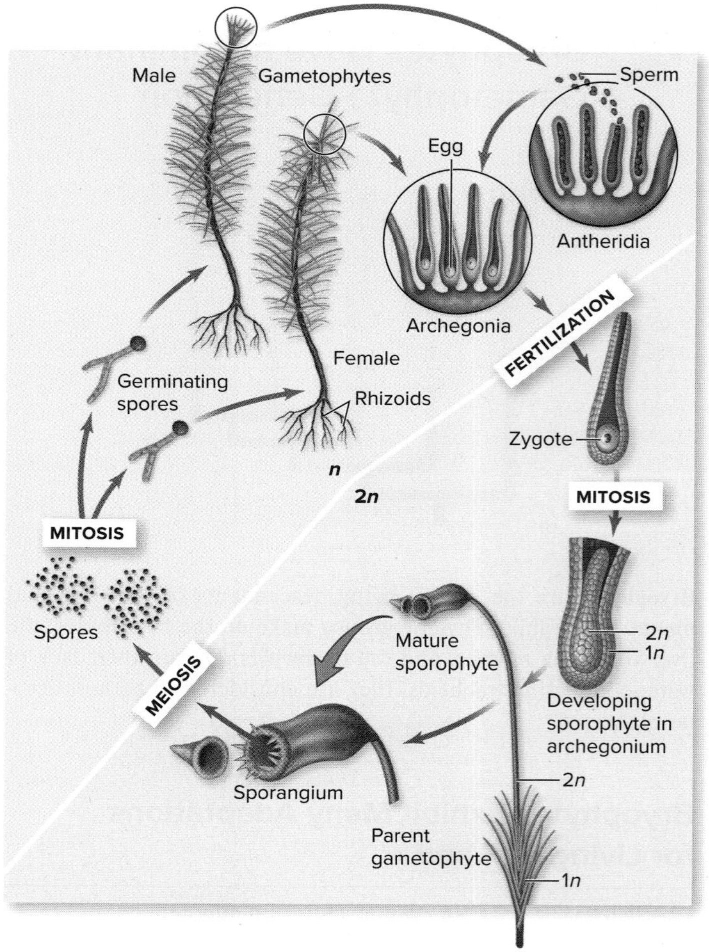

Figure 26.5 Life cycle of a typical moss. The majority of the life cycle of a moss is in the haploid state. The leafy gametophyte is photosynthetic, but the smaller sporophyte is not and is nutritionally dependent on the gametophyte.

Leaflike structures in mosses have little in common with leaves of vascular plants, except for the superficial appearance of the green, flattened blade and slightly thickened midrib that runs lengthwise down the middle. The leaflike structures are only one cell layer thick (except at the midrib), lacking vasculature and stomata, and all the cells are haploid.

Water may rise up a strand of specialized cells in the center of a moss gametophyte axis; however, most moves upward by capillary action along the surface of the plant. Water is absorbed directly across the surface of the plant.

Moss reproduction

Multicellular gametangia are formed at the tips of the leafy gametophytes (figure 26.5). Female gametangia **(archegonia)** may develop either on the same gametophyte as the male gametangia **(antheridia)** or on separate plants. A single egg is produced in the swollen lower part of an archegonium, whereas numerous sperm are produced in an antheridium.

When sperm are released from an antheridium, they swim with the aid of flagella through a film of dew or rainwater to the archegonia. One haploid sperm unites with a haploid egg, forming a diploid zygote. The zygote divides by mitotic cell division and develops into the sporophyte, a slender, basal stalk with a swollen capsule, the sporangium, at its tip. As the sporophyte develops, its base is embedded in gametophyte tissue, its nutritional source.

The sporangium is often cylindrical or club-shaped. Spore mother cells within the sporangium undergo meiosis, each producing four haploid spores. In many mosses at maturity, the top of the sporangium pops off, and the spores are released. A spore that lands in a suitable damp location may germinate and grow into a threadlike structure, which branches to form rhizoids and "buds" that grow upright. Each bud develops into a new gametophyte plant consisting of a leafy axis.

Moss distribution

In the Arctic and the Antarctic, mosses are the most abundant plants. The greatest diversity of moss species, however, is found in the tropics. Many mosses are able to withstand prolonged periods of drought, although mosses are not usually found in deserts.

Most mosses are highly sensitive to air pollution and are rarely found in abundance in or near cities or other areas with high levels of air pollution. Some mosses, such as the peat mosses *(Sphagnum),* can absorb up to 25 times their weight in water and are valuable commercially as a soil conditioner or as a fuel when dry.

The moss genome

Mosses can survive dessication—an adaptive trait in the early colonization of land (figure 26.6). Desiccation tolerance and

SCIENTIFIC THINKING

Hypothesis: *Desiccation tolerance genes in moss and flowering plants first appeared in a common ancestor.*

Prediction: *The late embryogenesis abundant* (LEA) *protein gene, a desiccation tolerance gene from flowering plants, will be expressed in moss plants when they experience severe water loss.*

Test: *Isolate RNA from moss plants that have not been water stressed (control), have been dehydrated to 84% water loss, and have been dehydrated to 95% water loss. Load a gel with equal amounts of RNA from each treatment. Probe the gel with a cDNA sequence for the* LEA *gene that is labeled.*

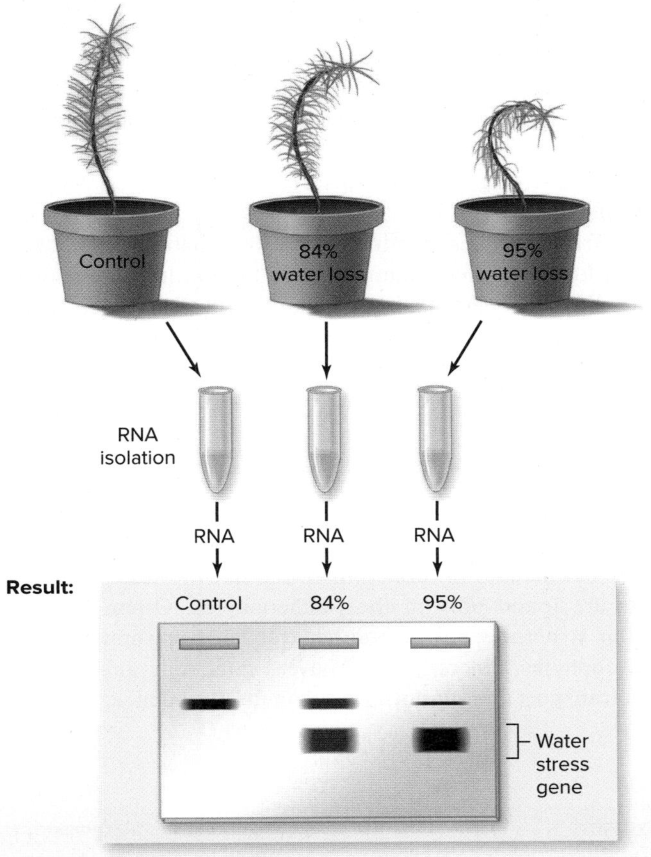

Conclusion: *Moss and flowering plants share a gene that is expressed under water stress conditions.*

Further Experiments: *Are other stress genes shared by bryophytes and flowering plants? Repeat the experiment with other stress-induced genes.*

Figure 26.6 Moss and flowering plants share desiccation tolerance genes.

phylogenetic position led researchers to sequence the genome of the moss *Physcomitrella patens.* Although the moss genome is a single genome bracketed by *Chlamydomonas* and the flowering plant *Arabidopsis,* many evolutionary hints are hidden within it. Evidence indicates the loss of genes associated with an aquatic life, including those for flagellar arms, from the genomes of the flowering plants. Genes associated with tolerance of terrestrial stresses, including temperature and water availability, are absent in *Chlamydomonas* and present in moss. The genome data add rich sets of traits to be used in phylogenetic analyses.

Figure 26.7 Hornworts (phylum Anthocerotophyta). Hornwort sporophytes are indicated in this photo. Unlike the sporophytes of other bryophytes, most hornwort sporophytes are photosynthetic.

Lee W. Wilcox

Hornworts developed stomata

The origin of **hornworts** (phylum Anthocerotophyta) is a puzzle. They are most likely among the earliest land plants, yet the earliest hornwort fossil spores date from the Cretaceous period (65 to 145 MYA), when angiosperms were emerging.

The small hornwort sporophytes resemble tiny green broom handles or horns, rising from filmy gametophytes usually less than 2 cm in diameter (figure 26.7). The sporophyte base is embedded in gametophyte tissue, from which it derives some of its nutrition. However, the sporophyte has stomata to regulate gas exchange, is photosynthetic, and provides much of the energy needed for growth and reproduction. Hornwort cells usually have a single large chloroplast.

REVIEW OF CONCEPT 26.2

The bryophytes exhibit adaptations to terrestrial life. All bryophytes exhibit alternation of generations and have dominant gametophyte generations. Moss adaptations include rhizoids to anchor the moss body and to absorb water, as well as water-conducting tissues. Hornworts developed stomata that can open and close to regulate gas exchange. With the exception of the hornworts, the sporophyte is completely dependent on the gametophyte for nutrition.

- *What might account for the abundance of mosses in the Arctic and Antarctic?*

26.3 Seedless Vascular Plants Evolved Roots, Stems, and Leaves

Several key innovations were important in the success of the first tracheophytes: the ability to produce more spores than their ancestors, a dominant sporophyte generation, the ability to better regulate gas exchange, and the appearance of vascular tissue throughout the body of the plant. As leaves became flatter and bigger, and tracheophytes could grow taller, photosynthetic potential increased significantly.

Vasculature Distributes Nutrients and Provides Support

LEARNING OBJECTIVE 26.3.1 Distinguish between xylem and phloem.

After the first plants with vasculature (networks of fluid conducting vessels or structures in the body of a plant) appeared approximately 420 MYA, they spread rapidly with little competition from existing plants. These primitive vascular plants lacked true roots or leaves and consisted of little more than a branching stemlike structure. Because sporangia could be produced at the tips of each branch, however, these plants were able to produce many more spores than the bryophytes already present. As other vascular plants evolved, more complex arrangements of sporangia followed.

In addition to being able to produce more spores than their competitors, vascular plants were successful colonizers of the land due to their efficient water- and food-conducting vascular tissues. These tissues consist of strands of specialized cylindrical or elongated cells that form a network throughout a plant, extending from near the tips of the roots, through the stems, and into true leaves, defined by the presence of vascular tissue in the blade. One type of vascular tissue, **xylem,** conducts water and dissolved minerals upward from the roots; another type of tissue, **phloem,** conducts sucrose and hormones throughout the plant. Vascular tissue enabled the tracheophytes to grow taller and larger. Vasculature develops in the sporophyte, but (with a few exceptions) not in the gametophyte. Vascular tissue was a key evolutionary advance (figure 26.8). A cuticle and stomata are also characteristic of vascular plants.

The Three Clades of Vascular Plants Include Seven Extant Phyla

LEARNING OBJECTIVE 26.3.2 Explain the evolutionary significance of roots, leaves, and stems.

Three clades of vascular plants, the **tracheophytes,** exist today: (1) lycophytes (club mosses), (2) pterophytes (ferns and their relatives), and (3) seed plants. Advances in molecular systematics have changed how we view the evolutionary history of vascular plants. Whisk ferns and horsetails were long believed to be distinct phyla that were transitional between bryophytes and vascular plants. Phylogenetic evidence now shows they are the closest living relatives to ferns, and they are grouped as pterophytes.

Tracheophytes dominate terrestrial habitats everywhere, except for the highest mountains and the tundra. The haplodiplontic life cycle persists, but the gametophyte has been reduced in size relative to the sporophyte during the evolution of tracheophytes. A similar reduction in multicellular gametangia has also occurred.

Stems evolved before roots

Fossils of early vascular plants reveal stems, but no roots or leaves. The earliest vascular plants had transport cells in their stems, but the lack of roots limited the size of these plants. True roots are found only in the tracheophytes. Other, somewhat similar structures enhance either transport or support in nontracheophytes, but only roots have a dual function—providing both transport and support. Based on fossil evidence, lycophytes

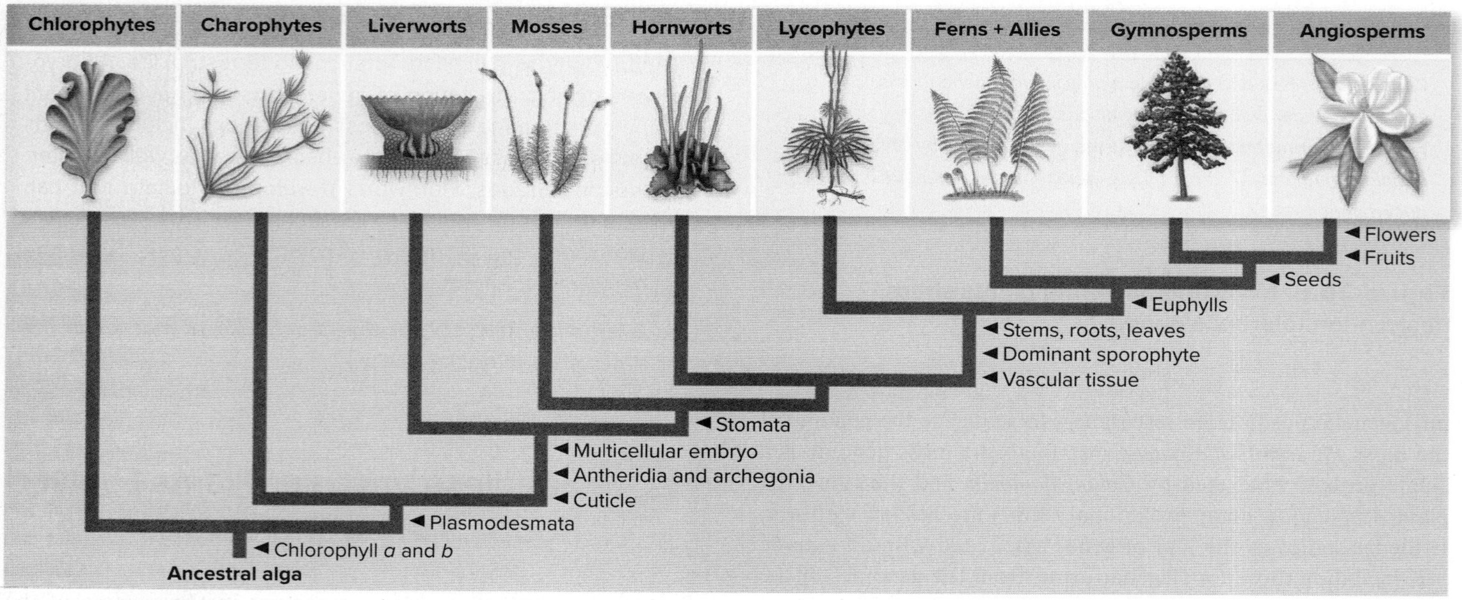

Figure 26.8 Major plant innovations.

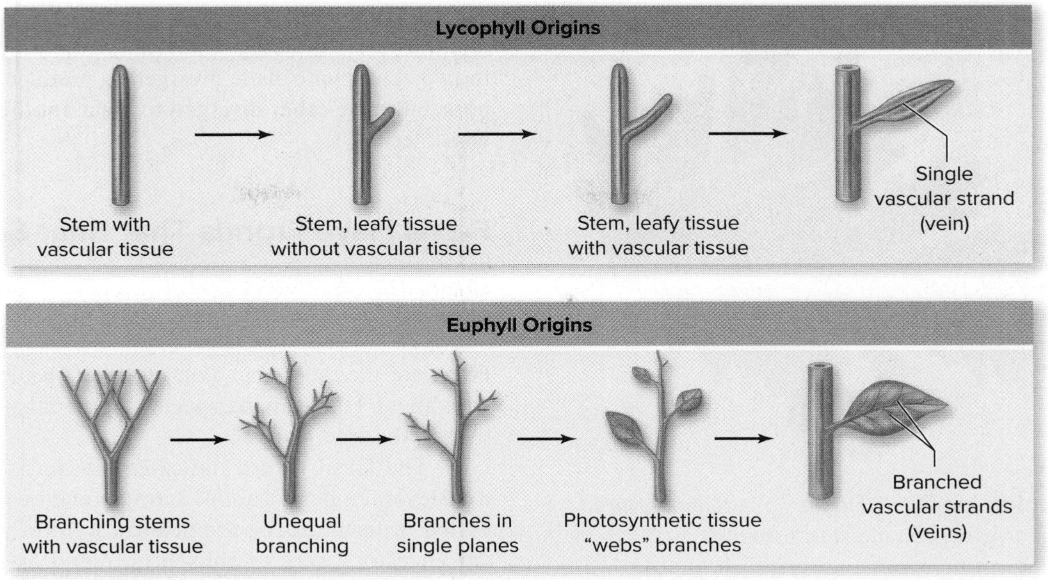

Figure 26.9 Evolution of leaves.

diverged from other tracheophytes before roots appeared. It appears that roots evolved at least twice.

Leaves evolved more than once

Leaves increased the surface area of the sporophyte, which enhanced their photosynthetic capacity. Lycophytes have single vascular strands supporting relatively small leaves called lycophylls. True leaves, called euphylls, are found only in ferns and seed plants, which have distinct origins from lycophylls (figure 26.9). Lycophylls may have resulted from vascular tissue penetrating small, leafy protuberances on stems. Euphylls most likely arose from branching stems that became webbed with leaf tissue.

Although beneficial for photosynthesis, larger leaves increase leaf temperature, which can be lethal. Stomatal openings in the leaf enhance the evaporation of water out of the leaf, thereby cooling it. The density of stomata on leaf surfaces correlates with CO_2 concentration, as the stomatal openings are essential for gas exchange. As the atmospheric CO_2 levels dropped, plants could not obtain sufficient CO_2 for photosynthesis. In the low-CO_2 atmosphere, natural selection favored plants with higher stomatal densities. Higher stomatal densities favored larger leaves with a photosynthetic advantage that did not overheat.

REVIEW OF CONCEPT 26.3

Most tracheophytes have well-developed vascular tissues, including tracheids, that enable efficient delivery of water and nutrients throughout the plant. The evolution of vasculature allowed plants to become larger and to exploit environments where water is not abundant. Tracheophytes also exhibit specialized roots, stems, leaves, cuticles, and stomata.

- *Why would vascular tissue be prevalent in the sporophyte, but not the gametophyte generation?*

26.4 Lycophytes Have a Dominant Sporophyte Generation

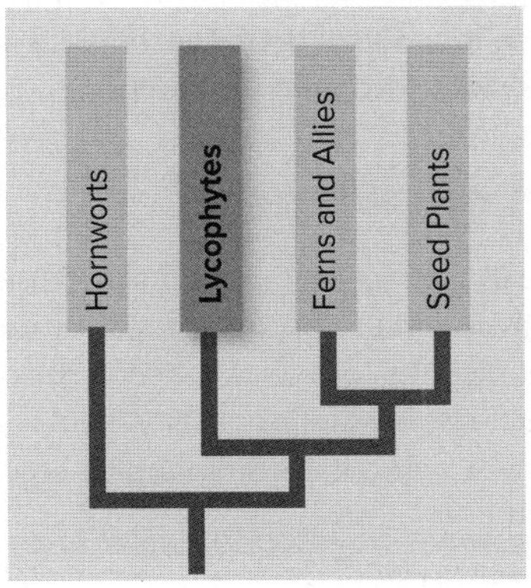

The earliest vascular plants lacked seeds. Members of four phyla of living vascular plants also lack seeds, as do at least three other phyla known only from fossils. Although not covered in this section, five extant phyla of tracheophytes produce seeds; we focus on those plants in later sections. As we explore the adaptations of the vascular plants, we focus on both reproductive strategies and the advantages of increasingly complex transport systems.

Club Mosses Are the Only Extant Lycophytes

LEARNING OBJECTIVE 26.4.1 Explain the features that differentiate lycophytes from bryophytes.

The lycophytes (club mosses) are relic species of an ancient past when vascular plants first evolved (figure 26.10). They are the sister group to all vascular plants. Several genera of club mosses, some of them large trees, became extinct about 270 MYA, and today all lycophytes are herblike. Lycophytes have worldwide distribution, being most abundant in the tropics and moist temperate regions.

Figure 26.10 A club moss. *Selaginella moellendorffii*'s sporophyte generation grows on moist forest floors.

Dr. Jody Banks, Purdue University

Members of the 12 to 13 genera and about 1150 living species of club mosses superficially resemble true mosses, but once their internal vascular structure and reproductive processes became known, it was clear that they are unrelated to mosses. The diploid sporophyte is the dominant life stage; sporophytes have leafy stems that are seldom more than 30 cm long. It is believed that genetic diversity can be increased when the diploid life stage is more long-lived than the haploid life stage due to masking of deleterious alleles in diploids.

REVIEW OF CONCEPT 26.4

Lycophytes are basal to all other vascular plants. Although they superficially resemble bryophytes, they contain tracheid-based vascular tissues but lack vascularized leaves. Their reproductive cycle is like that of other vascular plants, with the diploid sporophyte being the dominant life stage.

- *What events might have contributed to the extinction of large club mosses 270 MYA?*

26.5 Pterophytes Are Ferns and Their Relatives

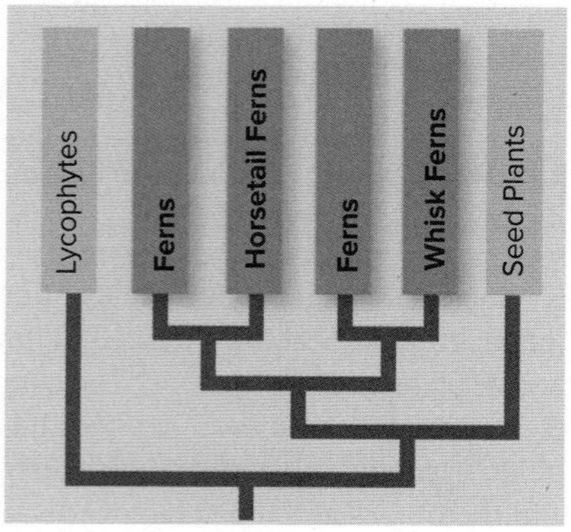

The phylogenetic relationships among ferns and their near relations are still being resolved. A common ancestor gave rise to two clades: one clade diverged to produce a line of ferns and horsetails; the other diverged to yield another line of ferns and whisk ferns.

Ferns Have Fronds That Bear Sori

LEARNING OBJECTIVE 26.5.1 Describe fern morphology and reproduction.

Ferns are the most abundant group of seedless vascular plants, with about 11,500 living species. Ferns may be the closest relatives to the seed plants.

The fossil record indicates that ferns originated during the Devonian period about 350 MYA and became abundant and varied in form during the next 50 million years. Their apparent ancestors were established on land as much as 375 MYA. Rainforests and swamps of lycopsid and fern trees growing in the eastern United States and Europe over 300 MYA formed the coal currently being mined. Today ferns flourish in a wide range of habitats throughout the world; however, about 75% of the species occur in the tropics.

The conspicuous sporophytes may be less than a centimeter in diameter (as in small aquatic ferns such as *Azolla*) or more than 24 m tall, with leaves up to 5 m or longer in the tree ferns (figure 26.11). The sporophytes and the much smaller gametophytes, which rarely reach 6 mm in diameter, are both photosynthetic.

The fern life cycle (figure 26.12) differs from that of a moss primarily in the much greater development, independence, and dominance of the fern's sporophyte. The fern sporophyte is structurally more complex than the moss sporophyte, having vascular tissue and well-differentiated roots, stems, and leaves. The gametophyte, however, lacks the vascular tissue found in the sporophyte.

Figure 26.11 A tree fern (phylum Pterophyta) in the forests of Malaysia. The ferns are by far the largest group of seedless vascular plants.

NHPA/Photoshot/Avalon

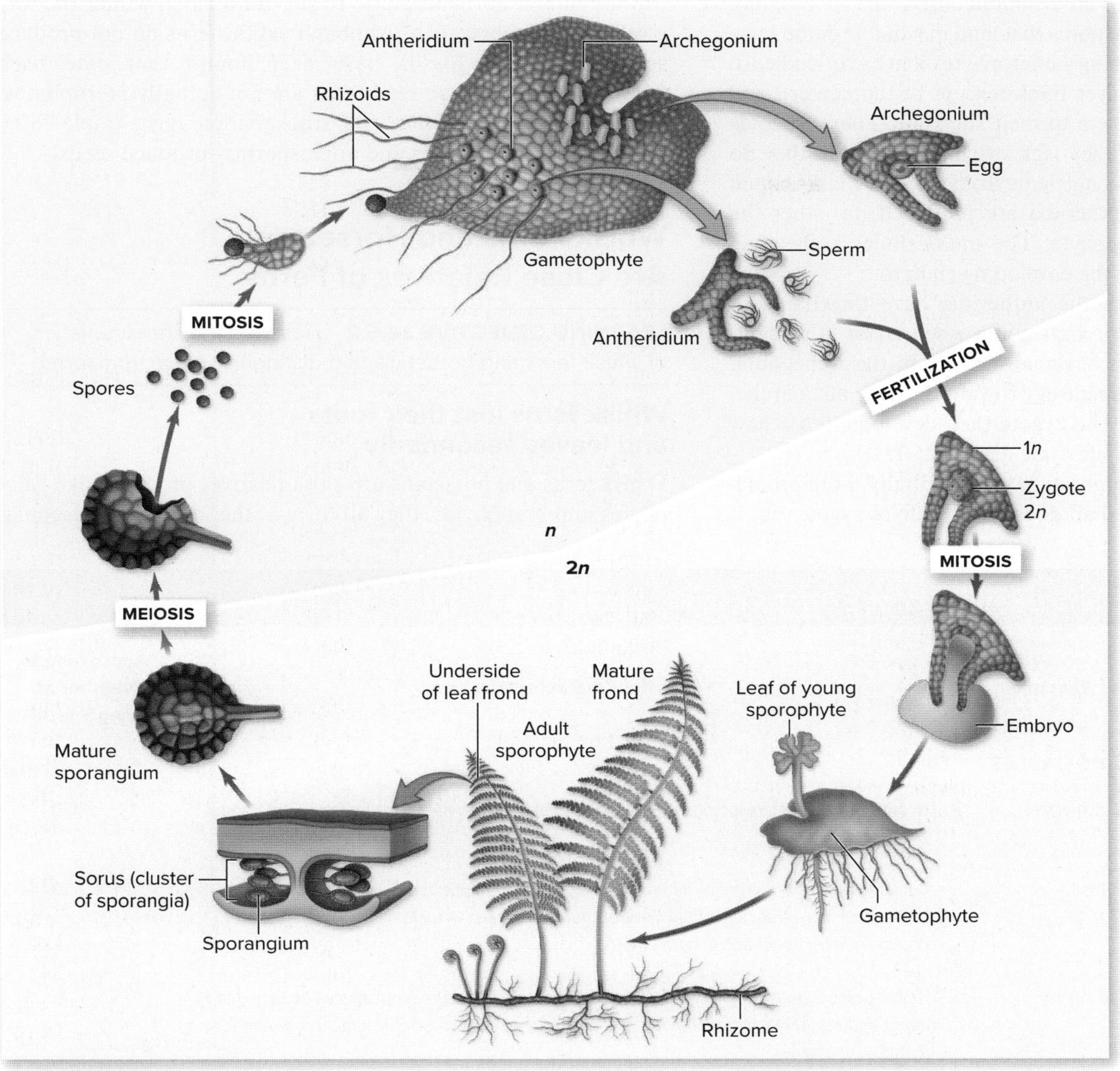

Figure 26.12 Life cycle of a typical fern. Both the gametophyte and the sporophyte are photosynthetic and can live independently. Water is necessary for fertilization. Sperm are released on the underside of the gametophyte and swim in moist soil to neighboring gametophytes. Spores are dispersed by wind.

Fern morphology

Fern sporophytes have rhizomes. Leaves, referred to as *fronds,* usually develop at the tip of the rhizome as tightly rolled-up coils ("fiddleheads") that unroll and expand (figure 26.13).

Many fronds are highly dissected and feathery, making the ferns that produce them prized as ornamental garden plants. Some ferns, such as *Marsilea,* have fronds that resemble a four-leaf clover, but *Marsilea* fronds still begin as coiled fiddleheads. Other ferns produce a mixture of photosynthetic fronds and nonphotosynthetic reproductive fronds that tend to be brownish in color.

Fern reproduction

Ferns produce distinctive sporangia, usually in clusters called **sori** (singular, *sorus*), typically on the underside of the fronds. Sori are often protected during their development by a transparent, umbrella-like covering. Diploid spore mother cells in each sporangium produce haploid spores by meiosis.

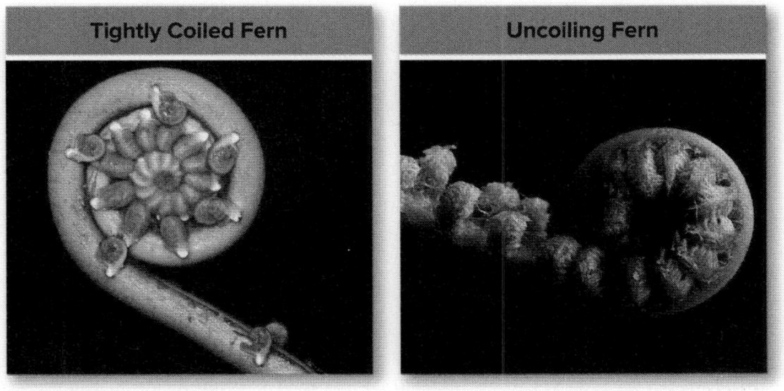

Figure 26.13 Fern "fiddlehead." Fronds of ferns develop in a coil and slowly unfold, including the tree fern fronds in these photos. Fiddleheads of some species of fern are considered a delicacy in several cuisines, but other species contain secondary compounds linked to stomach cancer.

(coiled fern): Zoonar GmbH/Alamy Stock Photo; (uncoiling fern): Ed Reschke

At maturity, the spores are catapulted from the sporangium by a snapping action. Those spores that land in suitable damp locations may germinate, producing gametophytes that are often heart-shaped, are only one cell layer thick (except in the center), and have rhizoids that anchor them to their substrate. These rhizoids are not true roots, because they lack vascular tissue, but they do aid in transporting water and nutrients from the soil. Flask-shaped archegonia and globular antheridia are produced on either the same or a different gametophyte. The multicellular archegonia provide some protection for the developing embryo.

The sperm formed in the antheridia have flagella, with which they swim toward the archegonia when water is present, often in response to a chemical signal secreted by the archegonia. One sperm unites with the single egg toward the base of an archegonium, forming a zygote. The zygote then develops into a new sporophyte, completing the life cycle (figure 26.12).

The developing fern embryo has substantially more protection from the environment than does a charophyte zygote, but it cannot enter a dormant phase to survive a harsh winter the way a seed plant embryo can. Although extant ferns do not produce seeds, seed fern fossils have been found that date back 365 million years. The seed ferns are not actually pterophytes, but gymnosperms. Of the seven tracheophyte phyla (table 26.1), only two—gymnosperms and angiosperms—produce seeds.

Whisk Ferns and Horsetails Are Close Relatives of Ferns

LEARNING OBJECTIVE 26.5.2 Describe the characteristics of whisk ferns and horsetails that distinguish them from ferns.

Whisk ferns lost their roots and leaves secondarily

Whisk ferns and horsetails are close relatives of ferns. Like lycophytes and bryophytes, they all form antheridia and archegonia.

TABLE 26.1 The Seven Phyla of Extant Vascular Plants

Phylum	Examples	Key Characteristics	Approximate Number of Living Species
SEEDLESS VASCULAR PLANTS			
Lycophyta	Club mosses	Homosporous or heterosporous; sperm motile; external water necessary for fertilization; about 12–13 genera	1,150
Pterophyta	Ferns	Primarily homosporous (a few heterosporous); sperm motile; external water necessary for fertilization; leaves uncoil as they mature; sporophytes and virtually all gametophytes are photosynthetic; about 365 genera	11,500
	Horsetails	Homosporous; sperm motile; external water necessary for fertilization; stems ribbed, jointed, either photosynthetic or nonphotosynthetic; leaves scalelike in whorls, nonphotosynthetic at maturity; one genus	15
	Whisk ferns	Homosporous; sperm motile; external water necessary for fertilization; no differentiation between root and shoot; no leaves; one of two genera has scalelike extensions, the other, leaflike appendages	6
SEED PLANTS			
Coniferophyta	Conifers (pines, spruces, firs, redwood, and others)	Heterosporous seed plants; sperm not motile and conducted to the egg by a pollen tube; leaves mostly needle-like or scalelike; tree shrubs; about 50 genera; many produce seeds in cones	601
Cycadophyta	Cycads	Heterosporous; sperm flagellated and motile but confined within a pollen tube that grows to the vicinity of the egg; palmlike plants with pinnate leaves; secondary growth slow compared with that of the conifers; 10 genera; seeds in cones	206
Gnetophyta	Gnetophytes	Heterosporous; sperm not motile and conducted to egg by a pollen tube; the only gymnosperms with vessels; trees, shrubs, vines; three very diverse genera *(Ephedra, Gnetura, Wehvitschia)*	65
Ginkgophyta	*Ginkgo*	Heterosporous; sperm flagellated and motile but conducted to the vicinity of the egg by a pollen tube; deciduous tree with fan-shaped leaves that have evenly forking veins; seeds resemble a small plum with fleshy, foul-smelling outer covering; one genus	1
Anthophyta	Flowering plants (angiosperms)	Heterosporous; sperm not motile and conducted to egg by a pollen tube; seeds enclosed within a fruit; leaves greatly varied in size and form; herbs, vines, shrubs, trees; about 14,000 genera	250,000

Water is required for the process of fertilization, during which the flagellated sperm swim to and unite with the eggs. In contrast, most seed plants have nonflagellated sperm.

In whisk ferns, which are found in the tropics and subtropics, the sporophytic generation consists merely of evenly forking green stems anchored to a substrate with a rhizome but lacking true roots and leaves (figure 26.14). Currently, systematists believe that whisk ferns lost leaves and roots when they diverged from others in the fern lineage.

The gametophytes of whisk ferns are essentially colorless and are less than 2 mm in diameter, but they can be up to 18 mm long. They form symbiotic associations with fungi, which provide nutrients. Some develop elements of vascular tissue and have the distinction of being the only gametophytes known to do so.

Horsetails have jointed stems with brushlike leaves

The 15 living species of horsetails are all homosporous. They constitute a single genus, *Equisetum.* Fossil forms of *Equisetum* extend back 300 million years to an era when some of their relatives were treelike. Today, they are widely scattered around the world, mostly in damp places. Some that grow among the coastal redwoods of California may reach a height of 3 m, but most are less than a meter tall (figure 26.15).

Horsetail sporophytes consist of ribbed, jointed, photosynthetic stems that arise from branching underground rhizomes with roots at their nodes. A whorl of nonphotosynthetic, scalelike leaves emerges at each node. The hollow stems have silica deposits in the epidermal cells of the ribs, and the interior parts of the stems have two sets of vertical, tubular canals. The larger outer canals, which alternate with the ribs, contain air, while the smaller inner canals opposite the ribs contain water.

Figure 26.15 A horsetail, *Equisetum telmateia.* This species forms two kinds of erect stems; one is green and photosynthetic, and the other, which terminates in a spore-producing "cone," is mostly light brown.

Stephen P. Parker/Science Source

Figure 26.14 A whisk fern. Whisk ferns have no roots or leaves. The green, photosynthetic stems have yellow sporangia attached.

Michael Szonyi/imageBROKER/Alamy Stock Photo

REVIEW OF CONCEPT 26.5

Ferns and their relatives have a large and conspicuous sporophyte with vascular tissue. Many have well-differentiated roots, stems, and leaves (fronds). The gametophyte generation is small and lacks vascular tissue.

- *What would be the advantage of silica deposits in stems, as are found in horsetails?*

26.6 Seed Plants Were a Key Step in Plant Evolution

The history of the land plants is filled with evolutionary innovations allowing the ancestors of aquatic algae to colonize harsh and varied terrestrial terrains. Early innovations made survival on land possible. Later innovations drove a radiation of plant life that continues to change the landscape and the atmosphere, and that supports diverse animal life.

The Seed Protects the Embryo

LEARNING OBJECTIVE 26.6.1 List the evolutionary advantages of seeds.

Seed-producing plants have come to dominate the terrestrial landscape over the last several hundred million years. Much of the remarkable success of seed plants, both gymnosperms and angiosperms, can be attributed to the evolution of the seed, an innovation that protects and provides food for delicate embryos. Seeds allow embryos to remain dormant, germinating after a harsh winter or an extremely dry season has passed. Fruits, a later innovation, provided extra protection and enhanced the dispersal of embryos across a broader landscape.

Ancestral seed plants first appeared about 305 to 465 MYA and preceded the appearance of gymnosperms and angiosperms. Seed plants appear to have evolved from spore-bearing plants known as **progymnosperms.** Progymnosperms shared several features with modern gymnosperms, including secondary vascular tissues (which allow for an increase in girth later in development). Some progymnosperms had leaves. Their reproduction was very simple, and it is unclear which particular group of progymnosperms gave rise to seed plants.

From an evolutionary and ecological perspective, the seed represents an important advance. The embryo is protected by an extra layer or two of sporophyte tissue called the **integument,** creating the **ovule** (figure 26.16). Within the ovule, the megasporangium divides by meiosis to produce a haploid megaspore. The megaspore produces the egg that combines with the sperm, resulting in the zygote. Seeds also contain a food supply for the developing embryo.

During development, the integuments harden to produce the seed coat. In addition to protecting the embryo from drought, the seed coat allows the seed to be easily dispersed. Perhaps even more significantly, the presence of seeds introduces into the life cycle a dormant phase, which allows the embryo to survive until environmental conditions are favorable for further growth.

A pollen grain is the male gametophyte

Seed plants produce two kinds of gametophytes—male and female—each of which consists of just a few cells. Pollen grains, multicellular male gametophytes, are conveyed to the egg in the female gametophyte by wind or by a pollinator. In some seed plants, the sperm moves toward the egg through a growing **pollen tube.** This eliminates the need for external water through which sperm swim. In contrast to the seedless plants, the whole male gametophyte, rather than just the sperm, moves to the female gametophyte.

A female gametophyte forms within the protection of the integuments, collectively forming the ovule. In angiosperms, the ovules are completely enclosed within additional diploid sporophyte tissue. The ovule and the surrounding protective tissue are called the ovary. The ovary develops into the fruit.

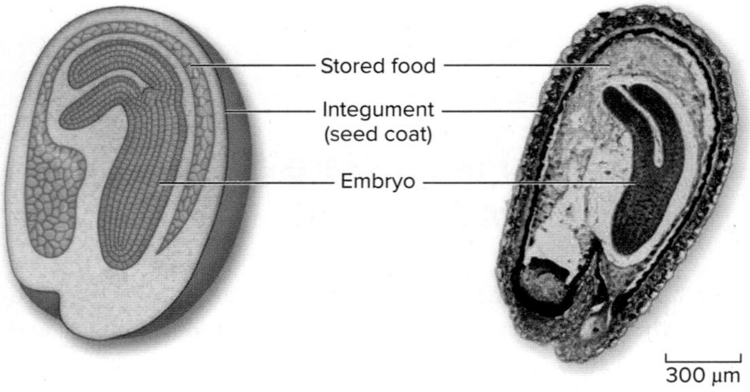

Stored food
Integument
(seed coat)
Embryo
300 μm

Figure 26.16 Cross section of an ovule.
Biology Media/Science Source

REVIEW OF CONCEPT 26.6

A common ancestor that had seeds gave rise to the gymnosperms and the angiosperms. Seeds protect the embryo, aid in dispersal, and can allow for an extended pause in the life cycle. Seed plants produce multicellular male and female gametophytes; the male gametophyte is a pollen grain, which is carried to the female gametophyte by wind or other means. The sperm is within the pollen grain.

- *Why is water not essential for fertilization in seed plants?*

26.7 Gymnosperms Are Plants with "Naked Seeds"

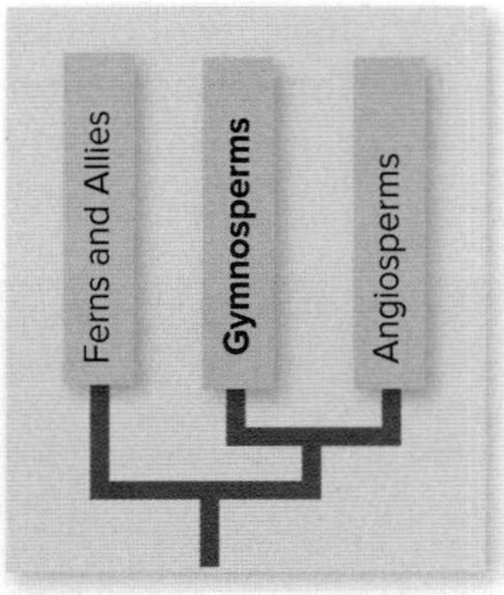

Ferns and Allies
Gymnosperms
Angiosperms

There are four groups of living **gymnosperms:** coniferophytes, cycadophytes, gnetophytes, and ginkgophytes. They all lack the flowers and fruits characteristic of angiosperms. In all of them, the ovule, which becomes a seed, rests exposed on a scale (a modified shoot or leaf) and is not completely enclosed by sporophyte tissues at the time of pollination. The name *gymnosperm* literally means "naked seed." Although the ovules are naked at the time of pollination, at maturity the seeds of gymnosperms are sometimes enclosed by sporophyte tissues.

Details of reproduction vary somewhat in gymnosperms, and their forms vary greatly. For example, cycads and *Ginkgo* have motile sperm, whereas conifers and gnetophytes have sperm with no flagella. All sperm are, however, carried within a pollen tube. The female cones range from tiny, woody structures weighing less than 25 g and having a diameter of a few millimeters to massive structures produced in some cycads, weighing more than 45 kg and growing to lengths of more than a meter.

Conifers Are the Largest Gymnosperm Phylum

LEARNING OBJECTIVE 26.7.1 Explain why conifer reproduction favors forest formation.

The most familiar gymnosperms are **conifers** (phylum Coniferophyta), which include pines (figure 26.17), spruces, firs, cedars, hemlocks, yews, larches, and cypresses. The coastal redwood *(Sequoia sempervirens),* a conifer native to northwestern California and southwestern Oregon, is the tallest living vascular plant; it can attain a height of nearly 100 m (300 ft). Another conifer, the bristlecone pine (*Pinus longaeva)* of the eastern White Mountains of California, is the oldest reported living tree; one specimen is reported as being about 5065 years old.

Conifers are found in the colder temperate and sometimes drier regions of the world. Various species are sources of timber, paper, resin, taxol (used to treat some kinds of cancer), and other economically important products.

Pines exemplify the conifer genus

More than 100 species of pines exist today, all native to the northern hemisphere, although the range of one species does extend a little south of the equator. Pines and spruces, which belong to the same family, are members of the vast coniferous forests that extend from just below the arctic tundra to the temperate deciduous forests and prairies to their south. During the past century, pines have been extensively planted in the southern hemisphere.

Pine morphology

Pines have tough, needle-like leaves produced mostly in clusters of two to five. Among the conifers, only pines have clustered leaves. The leaves have two adaptations for reducing water loss: a thick cuticle and recessed stomata. This strategy is important, because many of the trees grow in areas where the topsoil is frozen for part of the year, relative humidity is low, and winds are often high, all of which make obtaining and retaining water difficult.

Figure 26.17 Conifers. Longleaf pines, *Pinus palustris,* in Florida are representative of the Coniferophyta, the largest phylum of gymnosperms.
Scott Nodine/Alamy Stock Photo

The leaves and other parts of the sporophyte have canals into which surrounding cells secrete resin. The resin deters insect and fungal attacks. The resin of certain pines is harvested commercially for its volatile liquid portion, called turpentine, and for the solid rosin, which is used on the bows of stringed instruments such as violins. The wood of pines lacks some of the more rigid cell types found in other trees, and it is considered a "soft" rather than a "hard" wood. The thick bark of pines is an adaptation for surviving fires and subzero temperatures. Some cones actually depend on fire to open them, releasing seeds to reforest burned areas.

Reproductive structures

All seed plants produce two types of spores that give rise to two types of gametophytes (figure 26.18). The male gametophytes (pollen grains) of pines develop from microspores, which are produced in male cones that develop in clusters of 30 to 70, typically at the tips of the lower branches; there may be hundreds of such clusters on any single tree.

The male pine cones generally are 1 to 4 cm long and consist of small, papery scales arranged in a spiral or in whorls. A pair of microsporangia form as sacs within each scale. Numerous microspore mother cells in the microsporangia undergo meiosis, each becoming four microspores. The microspores develop into four-celled pollen grains with a pair of air sacs that give them added buoyancy when released into the air. A single cluster of male pine cones may produce more than a million pollen grains.

Female pine cones typically are produced on the upper branches of the same tree that produces male cones. Female cones are larger than male cones, and their scales become woody.

Two ovules develop toward the base of each scale. Each ovule contains a megasporangium called the **nucellus.** The nucellus itself is completely surrounded by a thick layer of cells called the integument, which has a small opening (the **micropyle**) toward one end. One of the layers of the integument later becomes the seed coat. A single megaspore mother cell within each megasporangium undergoes meiosis, becoming a row of four megaspores. Three of the megaspores break down, but the remaining one slowly develops into a female gametophyte. The female gametophyte at maturity may consist of thousands of cells, with two to six archegonia formed at the micropylar end. Each archegonium contains an egg so large it can be seen without a microscope.

Fertilization and seed formation

Female cones usually take two or more seasons to mature. At first, they may be reddish or purplish in color, but they soon turn green, and during the first spring the scales spread apart. While the scales are open, pollen grains carried by the wind drift down between them, some catching in sticky fluid oozing out of the

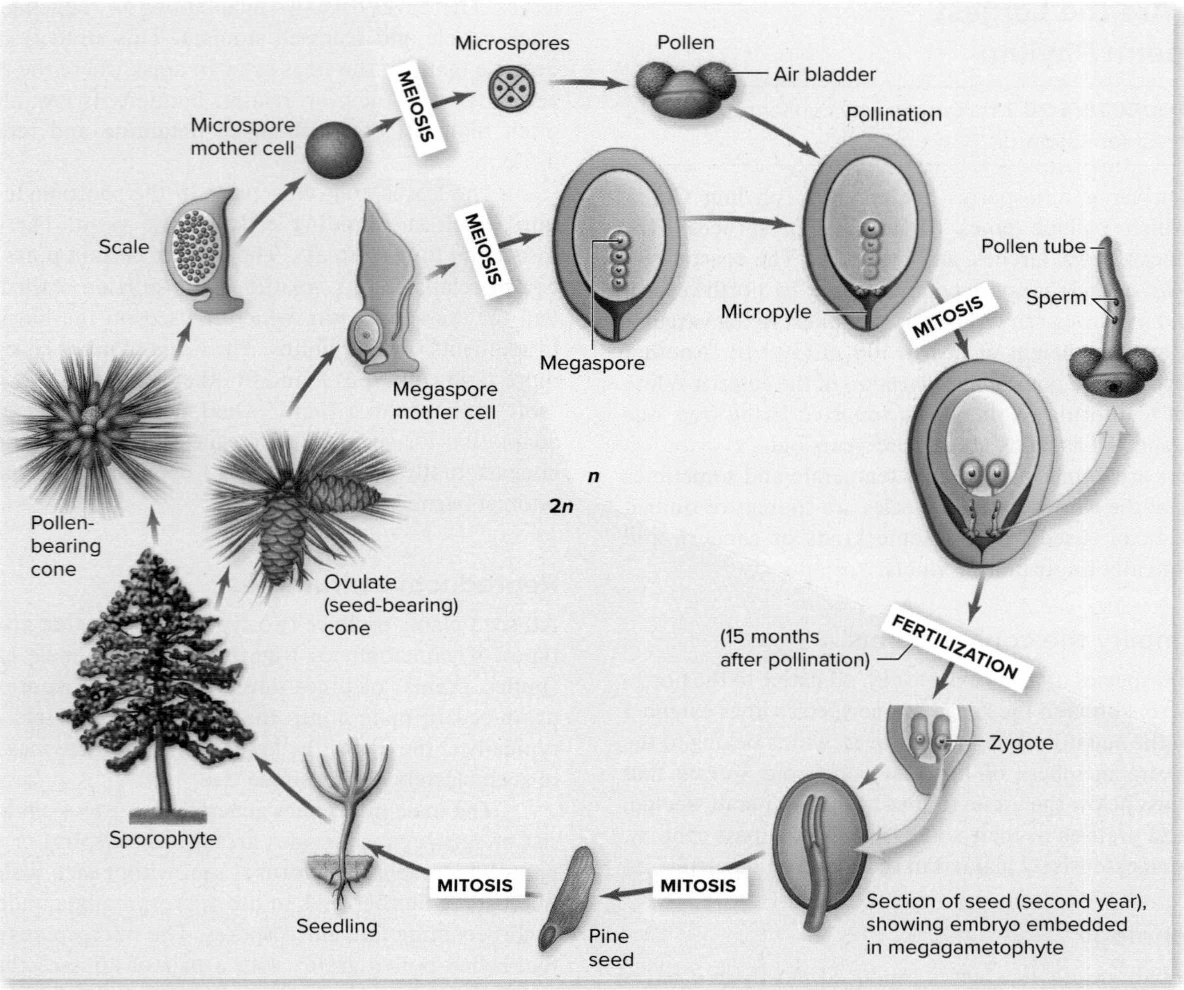

Figure 26.18 Life cycle of a typical pine. The male and female gametophytes are dramatically reduced in size in seed plants. Wind generally disperses the male gametophyte (pollen), which produces sperm. Pollen tube growth delivers the sperm to the egg on the female cone. Additional protection for the embryo is provided by the integument, which develops into the seed coat.

micropyle. The pollen grains within the sticky fluid are slowly drawn down through the micropyle to the top of the nucellus, and the scales close shortly thereafter.

The archegonia and the rest of the female gametophyte are not mature until about a year later. While the female gametophyte is developing, a pollen tube emerges from a pollen grain at the bottom of the micropyle and slowly digests its way through the nucellus to the archegonia. During growth of the pollen tube, one of the pollen grain's four cells, the generative cell, divides by mitosis, with one of the resulting two cells dividing once more. These last two cells function as sperm. The germinated pollen grain with its two sperm is the mature male gametophyte, a very limited haploid phase compared with fern gametophytes.

About 15 months after pollination, the pollen tube reaches an archegonium and discharges its contents into it. One sperm unites with the egg, forming a zygote. The other sperm and cells of the pollen grain degenerate. The zygote develops into an embryo within the seed. After seed dispersal and germination, the young sporophyte of the next generation develops into a tree.

Cycads, Gnetophytes, and Ginkgophytes Are Three Minor Phyla of Gymnosperms

LEARNING OBJECTIVE 26.7.2 Describe the three nonconifer gymnosperms.

Cycads resemble palms but are not flowering plants

Cycads (phylum Cycadophyta) are slow-growing gymnosperms of tropical and subtropical regions. The sporophytes of most of the 206 known species resemble palm trees (figure 26.19*a*) with trunks that can attain heights of 15 m or more. Unlike palm trees, which are flowering plants, cycads produce cones and have a life cycle similar to that of pines.

The female cones, which develop upright among the leaf bases, are huge in some species and can weigh up to 45 kg. The sperm of cycads, although formed within a pollen tube, are released within the ovule to swim to an archegonium. These sperm are the largest sperm cells among all living organisms.

Figure 26.19 Three phyla of gymnosperms. ***a.*** A cycad, *Cycas circinalis.* ***b.*** *Welwitschia mirabilis* represents one of the three genera of gnetophytes. ***c.*** Maidenhair tree, *Ginkgo biloba,* the only living representative of the phylum Ginkgophyta.

(*Cycas circinalis*): Luca Invernizzi Tetto/age fotostock; (*Welwitschia mirabilis*): Juan Carlos Muñoz/age fotostock; (*Ginkgo biloba*): Nancy Nehring/iStockphoto/Getty Images

Several species of cycads are facing extinction in the wild and soon may exist only in botanical gardens.

Gnetophytes have xylem vessels

There are three genera and about 65 living species of gnetophytes (phylum Gnetophyta). They are the only gymnosperms with vessels in their xylem. **Vessels** are a particularly efficient conducting cell type that is a common feature in flowering plants.

The members of the three genera differ greatly from one another in form. One of the most bizarre of all plants is *Welwitschia* (figure 26.19*b*), which occurs in the Namib and Mossamedes deserts of southwestern Africa. The stem is shaped like a large, shallow cup that tapers into a taproot below the surface. It has two strap-shaped, leathery leaves that grow continuously from their base, splitting as they flap in the wind. The reproductive structures of *Welwitschia* are conelike, appear toward the bases of the leaves around the rims of the stems, and are produced on separate male and female plants.

More than half of the gnetophyte species are in the genus *Ephedra,* which is common in arid regions of the western United States and Mexico. Species are found on every continent except Australia. The plants are shrubby, with stems that superficially resemble those of horsetails, being jointed and having tiny, scale-like leaves at each node. Male and female reproductive structures may be produced on the same or different plants.

The drug ephedrine, widely used in the treatment of respiratory problems, was in the past extracted from Chinese species of *Ephedra,* but it has now been largely replaced with synthetic preparations (pseudoephedrine). Because ephedrine found in herbal remedies for weight loss was linked to strokes and heart attacks, it was withdrawn from the market in April 2004. Sales restrictions were placed on pseudoephedrine-containing products in 2006 because they can be used to manufacture the illegal drug methamphetamine.

The best-known species of the third genus, *Gnetum,* is a tropical tree, but most species are vinelike. All species have broad leaves similar to those of angiosperms. One *Gnetum* species is cultivated in Java for its tender shoots, which are cooked as a vegetable.

Ginkgo biloba is the only extant ginkgophyte

The fossil record indicates that members of the ginkgophytes (phylum Ginkgophyta) were once widely distributed, particularly in the northern hemisphere; today only one living species, *Ginkgo biloba,* remains (figure 26.19*c*). This tree, which sheds its leaves in the fall, was first encountered by Europeans in cultivation in Japan and China; it apparently no longer exists in the wild.

Like the sperm of cycads, those of *Ginkgo* have flagella (figure 26.20). The ginkgo is **dioecious**—that is, the male and female reproductive structures are produced on separate trees. The fleshy outer coverings of the seeds of female ginkgo plants exude the foul smell of rancid butter, caused by the presence of butyric and isobutyric acids. As a result, male plants vegetatively propagated from shoots are preferred for cultivation in the West. Because of its beauty and resistance to air pollution, *Ginkgo* is commonly planted along city streets.

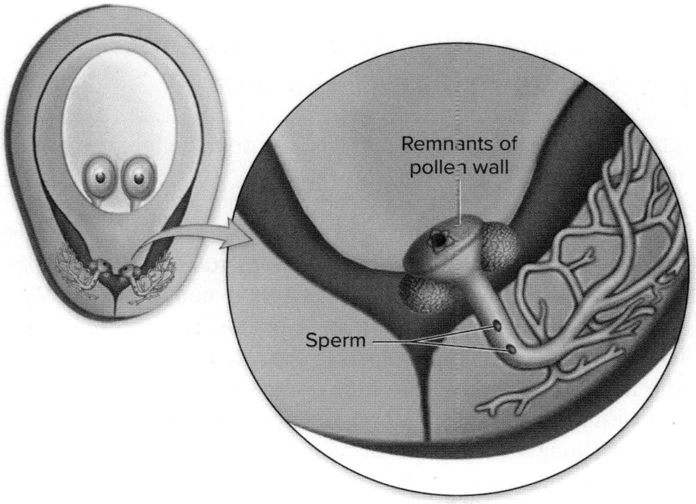

Figure 26.20 ***Ginkgo*** **pollen tube growth.** The *Ginkgo* pollen tube grows intercellularly in the ovule tissue, forming a highly branched structure. The enlarged basal end contains the two flagellated sperm; when it ruptures, the two sperm swim to the eggs within the female gametophyte.

REVIEW OF CONCEPT 26.7

Gymnosperms are mostly cone-bearing seed plants. In gymnosperms, the ovules are not completely enclosed by sporophyte tissue at pollination, and thus these plants have "naked seeds." The four groups of gymnosperms are conifers, cycads, gnetophytes, and ginkgophytes.

- *What adaptation do conifers exhibit to capture wind-borne pollen?*

26.8 Angiosperms Are Flowering Plants

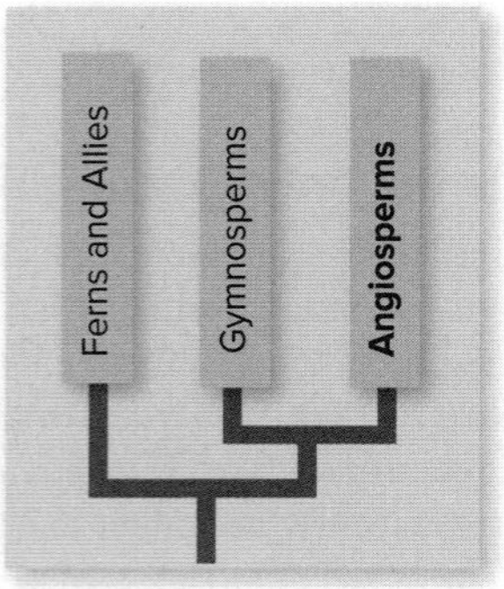

The 250,000 known species of flowering plants are called **angiosperms,** because their ovules, unlike those of gymnosperms, are enclosed within diploid tissues at the time of pollination. The *carpel,* a modified leaf that encapsulates seeds, develops into the fruit, a unique angiosperm feature. Although some gymnosperms, including the yew (*Taxus* spp.), have fleshlike tissue around their seeds, it is of a different origin and not a true fruit.

Flowers and Other Adaptations Drove the Radiation of Angiosperms

LEARNING OBJECTIVE 26.8.1 List the defining features of the angiosperms.

Recent fossil evidence, along with molecular sequence data, suggest that basal angiosperms first appeared sometime around the Upper Jurassic to Lower Cretaceous period, about 208–145 MYA. After their initial appearance, they diversified and spread rapidly. The speed with which new species appeared perplexed Darwin, who found the rapid nature of the group's evolution in contradiction with his belief that "nature does not make a leap."

Although gymnosperms and angiosperms share many features, the angiosperms have distinctive characteristics (table 26.2). Most notable is the presence of the flower organ in the angiosperms. Flowers are species-specific and attract animals such as insects, birds, and bats to the plant.

In the course of harvesting nutritious nectar, many animals collect pollen and transfer it to the reproductive structures of other flowers of the same species. Many animals have co-evolved with the angiosperms they feed upon. The presence of flowers results in a more efficient transfer of pollen, which in turn increased the diversity of the angiosperms.

TABLE 26.2 Dlstinctive Angiosperm Characteristics

Characteristic	Significance
Flowers	Improved efficiency of pollination; variety of breeding systems promotes genetic diversity
Reduced gameotophytes	Faster seed production; reduced resource allocation to gametophyte generation
Carpels (and fruit)	Protection of the ovule; dispersal
Endosperm	Provides nutrition to developing embryo
Vasculature	Vessel elements allow more efficient water transport than do the tracheids of gymnosperms

Flowers House the Gametophyte Generation

LEARNING OBJECTIVE 26.8.2 Describe the structure of an angiosperm flower.

Flowers are considered to be modified stems bearing modified leaves. Regardless of their size and shape, they all share certain features (figure 26.21*a*). Each flower originates as a **primordium** that develops into a bud at the end of a stalk called a **pedicel.** The pedicel expands slightly at the tip to form the **receptacle,** to which the remaining flower parts are attached.

Flower morphology

The other flower parts typically are attached in circles called **whorls.** The outermost whorl is composed of **sepals.** Most flowers have three to five sepals, which are green and somewhat leaflike. The next whorl consists of **petals** that are often colored, attracting pollinators such as insects, birds, and some small mammals. The petals, which also commonly number three to five, may be separate, fused together, or missing altogether in wind-pollinated flowers.

The third whorl consists of **stamens** and is collectively called the **androecium.** This whorl is where the male gametophytes, pollen, are produced. Each stamen consists of a pollen-bearing **anther** and a stalk called a **filament,** which may be missing in some flowers.

At the center of the flower is the fourth whorl, called the **gynoecium,** where the small female gametophytes are housed; the gynoecium consists of one or more **carpels.** The first carpel is believed to have been formed from a leaflike structure with ovules along its margins (figure 26.22). Primitive flowers can have several to many separate carpels, but in most flowers, two to several carpels are fused together. Such fusion is present in an orange sliced in half; each segment represents one carpel.

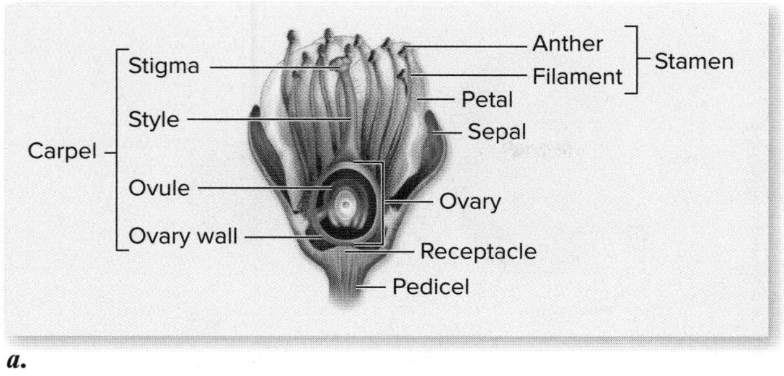

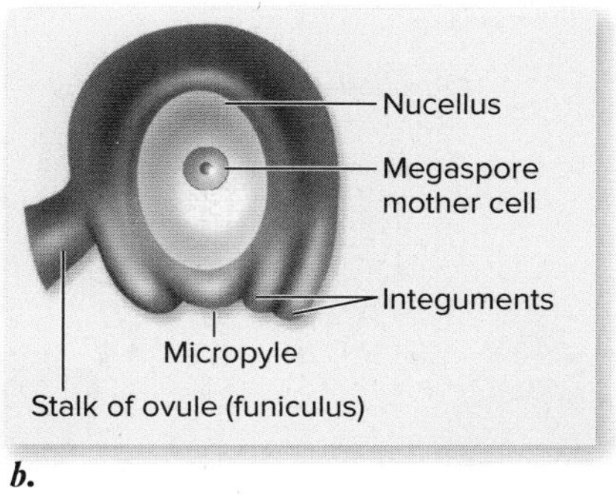

Figure 26.21 Diagram of an angiosperm flower. ***a.*** The main structures of the flower are labeled. ***b.*** Details of an ovule. The ovary as it matures will become a fruit; as the ovule's outer layers (integuments) mature, they will become a seed coat.

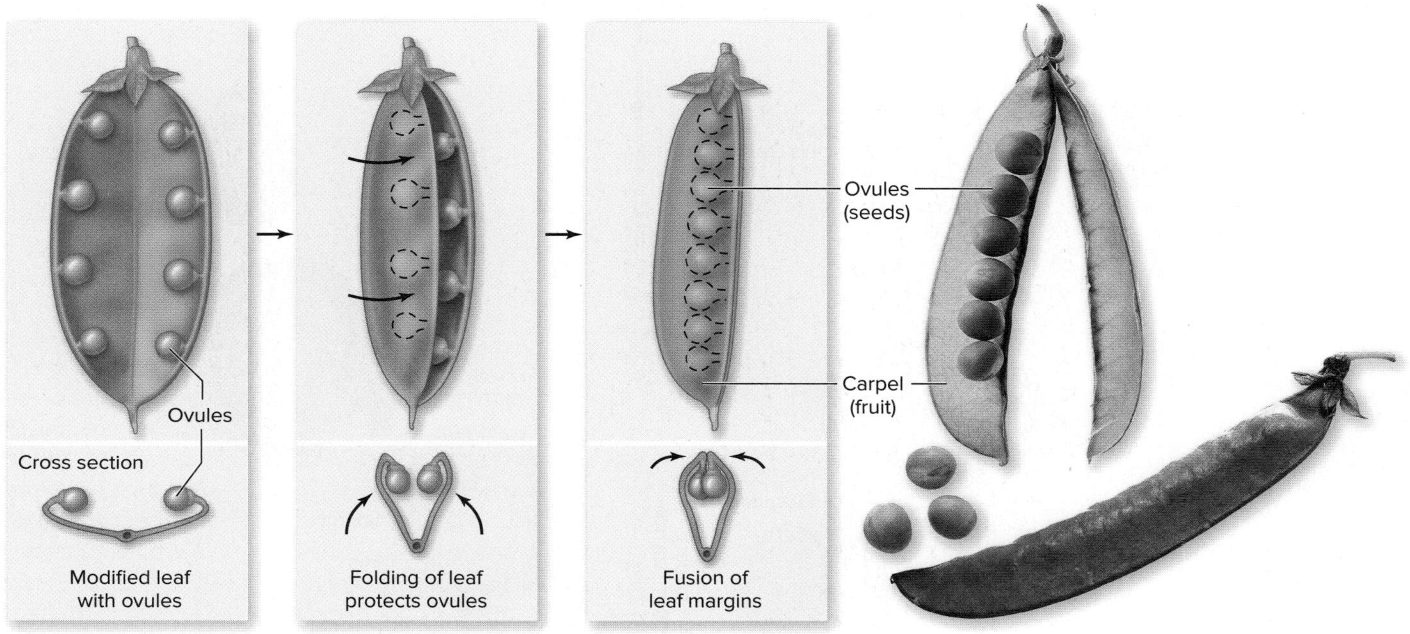

Figure 26.22 Evolution of fruit. Unlike the ovule of gymnosperms, the ovule of angiosperms is surrounded by diploid tissue derived from leaves. This tissue is called the carpel and develops into the fruit. The leaf origins of carpels are still visible in the pea pod.

Goodshoot/Alamy Stock Photo

Structure of the carpel

A carpel has three major regions (figure 26.21*a*). The **ovary** is the swollen base, which contains from one to hundreds of ovules; the ovary later develops into a **fruit.** The tip of the carpel is called a **stigma.** Most stigmas are sticky or feathery, causing pollen grains that land on them to adhere. Typically, a neck or stalk called a **style** connects the stigma and the ovary; in some flowers, the style is very short or even missing.

Many flowers have nectar-secreting glands called *nectaries,* often located toward the base of the ovary. Nectar is a fluid containing sugars, amino acids, and other molecules that attracts insects, birds, and other animals to flowers.

Most species use flowers to attract pollinators and reproduce

Eudicots (about 175,000 species) include the great majority of familiar angiosperms—almost all kinds of trees and shrubs, snapdragons, mints, peas, sunflowers, and other plants. Monocots (about 65,000 species) include the lilies, grasses, cattails, palms, agaves, yuccas, orchids, and irises and share a common ancestor with the eudicots. Some of the monocots, including maize, rely on wind rather than pollinators to reproduce.

The Angiosperm Life Cycle Includes Double Fertilization

LEARNING OBJECTIVE 26.8.3 Describe double fertilization and its outcome.

During development of a flower bud, a single megaspore mother cell in the ovule undergoes meiosis, producing four megaspores (figure 26.23). In most flowering plants, three of the megaspores soon disappear; the nucleus of the remaining megaspore divides mitotically, and the cell slowly expands until it becomes many times its original size.

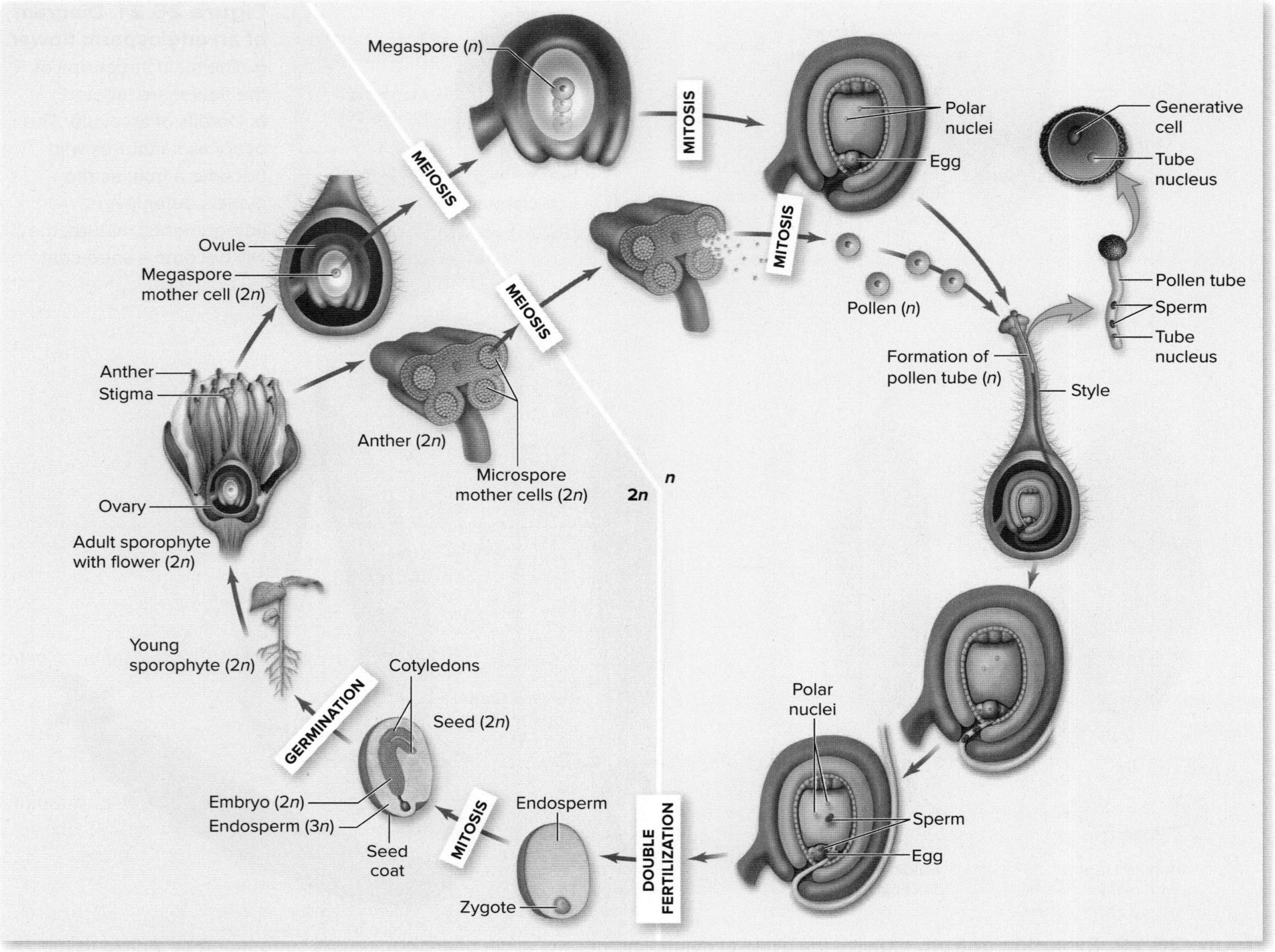

Figure 26.23 Life cycle of a typical angiosperm. As in pines, external water is no longer required for fertilization. In most species of angiosperms, animals carry pollen to the carpel. The outer wall of the carpel forms the fruit, which often entices animals to disperse the seed.

The female gametophyte

While the expansion of the megaspore is occurring, each of the daughter nuclei divides twice, resulting in eight haploid nuclei arranged in two groups of four. At the same time, two layers of the ovule, the integuments, differentiate and become the *seed coat* of a seed. The integuments, as they develop, form the micropyle, a small gap or pore at one end (figure 26.21*b*).

One nucleus from each group of four migrates toward the center, where they function as polar nuclei. Polar nuclei may fuse together, forming a single diploid nucleus, or they may form a single cell with two haploid nuclei. Cell walls also form around the remaining nuclei. In the group closest to the micropyle, one cell functions as the egg; the other two nuclei are called synergids. At the other end, the three cells are now called antipodals; they have no apparent function and eventually break down and disappear.

The large sac with eight nuclei in seven cells is called an **embryo sac;** it constitutes the female gametophyte. Although it is completely dependent on the sporophyte for nutrition, it is a multicellular haploid individual.

Pollen production

While the female gametophyte is developing, a similar but less complex process takes place in the anthers (figure 26.23). Most anthers have patches of tissue (usually four) that eventually become chambers lined with nutritive cells. The tissue in each patch is composed of many diploid microspore mother cells that undergo meiosis more or less simultaneously, each producing four microspores.

The four microspores at first remain together as a quartet, or tetrad, and the nucleus of each microspore divides once; in most species, the microspores of each quartet then separate. At the same time, a two-layered wall develops around each microspore. As the microspore-containing anther continues to mature, the wall between adjacent pairs of chambers breaks down, leaving two larger sacs. At this point, the binucleate microspores have become pollen grains.

The outer pollen grain wall layer often becomes beautifully sculptured, and it contains chemicals that may react with others in a stigma to signal whether development of the male gametophyte should proceed to completion. The pollen grain has areas called *apertures,* through which a pollen tube may later emerge.

Pollination and the male gametophyte

Pollination is simply the transfer of pollen from its source (an anther) to a receptive area (the stigma of a flowering plant). Most pollination takes place between flowers of different plants and is brought about by insects, wind, water, gravity, bats, and other animals. In as many as one-quarter of all angiosperms, however, a pollen grain may be deposited directly on the stigma of its own flower, and self-pollination occurs. Pollination may or may not be followed by *fertilization,* depending on the genetic compatibility of the pollen grain and the flower on whose stigma it has landed.

If the stigma is receptive, the pollen grain's dense cytoplasm absorbs substances from the stigma and bulges through an aperture. The bulge develops into a pollen tube that responds to chemical and mechanical stimuli that guide it to the embryo sac. It follows a diffusion gradient of the chemicals and grows down through the style and into the micropyle. The pollen tube usually takes several hours to two days to reach the micropyle, but the journey may take up to a year. Pollen tube growth is more rapid in angiosperms than in gymnosperms.

One of the pollen grain's two cells, the *generative cell,* lags behind. Its nucleus divides in the pollen grain or in the pollen tube, producing two sperm cells. Unlike sperm in mosses, ferns, and some gymnosperms, the sperm of flowering plants have no flagella. At this point, the pollen grain with its tube and sperm has become a mature male gametophyte.

Double fertilization and seed production

As the pollen tube enters the embryo sac, it destroys a synergid in the process and then discharges its contents. Both sperm are functional, and an event called **double fertilization** follows. One sperm unites with the egg and forms a zygote, which develops into an embryo sporophyte plant. The other sperm and the two polar nuclei unite, forming a triploid primary endosperm nucleus.

The primary endosperm nucleus begins dividing rapidly and repeatedly, becoming triploid endosperm tissue that may soon consist of thousands of cells. Endosperm tissue can become an extensive part of the seed in grasses such as corn, and it provides nutrients for the embryo in most flowering plants.

Until recently, the nutritional, triploid endosperm was believed to be the ancestral state in angiosperms. A recent analysis of extant, basal angiosperms revealed that diploid endosperms were also common. The female gametophyte in these species has four, not eight, nuclei. At the moment, it is unclear whether diploid or triploid endosperms are the more primitive.

Germination and growth of the sporophyte

As mentioned in section 26.6, a seed may remain dormant for many years, depending on the species. When environmental conditions become favorable, the seed undergoes germination, and the young sporophyte plant emerges. Again depending on the species, the sporophyte may grow and develop for many years before becoming capable of reproduction, or it may quickly grow and produce flowers in a single growing season.

Chapter 30 presents a more detailed description of plant reproduction.

REVIEW OF CONCEPT 26.8

Angiosperms are characterized by ovules that at pollination are enclosed within an ovary at the base of a carpel, a structure unique to the phylum; a fruit develops from the ovary. Evolutionary innovations of angiosperms include flowers to attract pollinators, fruits to protect embryos and aid in their dispersal, and double fertilization, which provides endosperm to help nourish the embryo.

- *What advantage does an angiosperm gain by producing a fruit eaten by animals?*

How Does Arrowgrass Tolerate Salt?

Plants grow almost everywhere on Earth, thriving in many places where exposure, drought, and other severe conditions challenge their survival. In deserts, a common stress is the presence of high levels of salt in the soils. Soil salinity is also a problem for millions of acres of abandoned farmland, because the accumulation of salt from irrigation water restricts growth. Why does excess salt in the soil present a problem for a plant? For one thing, the high levels of sodium ions taken up by the roots are toxic. For another, a plant's roots cannot obtain water when growing in salty soil. Osmosis (the movement of water molecules to areas of higher solute concentrations; refer to section 31.1) causes water to be drawn out of the roots by the soil's high levels of salt. And yet, some plants do grow in these soils. How do they manage?

To investigate this, researchers have studied seaside arrowgrass *(Triglochin maritima),* the plant shown in the photo below. Arrowgrass plants are able to grow in very salty seashore soils, where few other plants survive. How do they do it? Researchers found that their roots do not take up salt and, so, do not accumulate toxic levels of salt.

However, the arrowgrass still has to prevent its root cells from losing water to the surrounding salty soil. How do the roots achieve osmotic balance? In an attempt to find out, researchers grew arrowgrass plants in nonsalty soil for two weeks, then transferred them to one of several soils that differed in salt level. After 10 days, shoots were harvested and analyzed for amino acids, because accumulating amino acids could be one way that the cells increase intracellular solute concentrations to maintain osmotic balance. Results are presented in the graph.

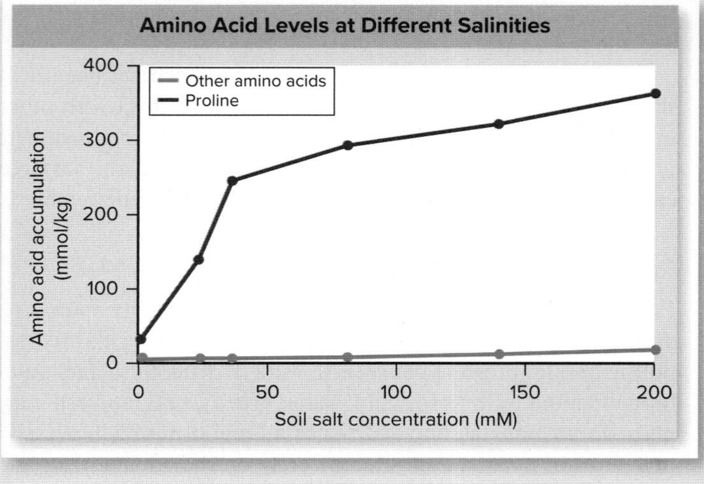

Analysis

1. **Applying Concepts**
   a. What do the abbreviations "mM" and "mmol/kg" mean?
   b. What concentration would you get if you dissolved 4 mmol of solute in 100 mL of solvent?
2. **Interpreting Data**
   a. In beach soils with a salt concentration of 35 mM, roughly how much proline has accumulated in the roots after 10 days? How much of other amino acids?
   b. Describe the relationship between soil salt concentration and accumulation of proline in the plants at 10 days.
3. **Making Inferences**
   a. In general, what is the effect of soil salt concentration on an arrowgrass plant's accumulation of the amino acid proline? Of other amino acids?
   b. Is the effect of salt on proline accumulation the same at lower salt levels (below 50 mM) as at higher salt levels (above 50 mM)?
4. **Drawing Conclusions** Are these results consistent with the hypothesis that arrowgrass accumulates proline to achieve osmotic balance with salty soils?
5. **Further Analysis** What do you think might account for the different rates of proline accumulation in low-salt and high-salt soils? Can you think of a way to test this hypothesis?

blickwinkel/Alamy Stock Photo

Retracing the Learning Path

CONCEPT 26.1 Land Plants Are Multicellular Autotrophs Adapted to Terrestrial Life

26.1.1 Plants Evolved from Freshwater Algae All plants arose from a single freshwater green algal species. The charophytes are the sister clade of plants.

26.1.2 Plants Have Adapted to Terrestrial Life Land plants have multicellular haploid and diploid phases. A waxy cuticle, stomata, and ways to survive harmful UV light enhance survival.

26.1.3 Plants Exhibit Alternation of Generations Plants have a haplodiplontic life cycle with multicellular diploid sporophytes and multicellular haploid gametophytes.

CONCEPT 26.2 Bryophytes Have a Dominant Gametophyte Generation

26.2.1 Bryophytes Exhibit Many Adaptations for Living on Land Bryophytes consist of three distinct clades: liverworts, mosses, and hornworts. Bryophytes do not have true roots or tracheids, but they do have conducting cells to move water and nutrients.

26.2.2 Liverworts Are an Ancient Phylum The gametophyte of some liverworts is flattened and has lobes that resemble those of the liver. They produce upright structures that contain the gametangia.

26.2.3 Mosses and Hornworts Have Distinct Evolutionary Innovations Mosses exhibit alternation of generations and have widespread distribution.

CONCEPT 26.3 Seedless Vascular Plants Evolved Roots, Stems, and Leaves

26.3.1 Vasculature Distributes Nutrients and Provides Support The evolution of tracheids in the sporophytes allowed more efficient vascular systems to develop.

26.3.2 The Three Clades of Vascular Plants Include Seven Extant Phyla The vascular plants found today exist in three clades: lycophytes, pterophytes, and seed plants.

CONCEPT 26.4 Lycophytes Have a Dominant Sporophyte Generation

26.4.1 Club Mosses Are the Only Extant Lycophytes Lycophyte ancestors were the earliest vascular plants and were among the first plants to have a dominant sporophyte generation.

CONCEPT 26.5 Pterophytes Are Ferns and Their Relatives

26.5.1 Ferns Have Fronds That Bear Sori The pterophytes require water for fertilization and are seedless. The leaves of ferns, called fronds, develop as tightly rolled coils that unwind to expand. Sporangia called sori develop on the underside of the fronds. The gametophyte can often live independently.

26.5.2 Whisk Ferns and Horsetails Are Close Relatives of Ferns Scalelike leaves of horsetail sporophytes emerge in a whorl. The stems have silica deposits in epidermal cells of their ribs.

CONCEPT 26.6 Seed Plants Were a Key Step in Plant Evolution

26.6.1 The Seed Protects the Embryo Seeds are resistant structures that protect the embryo from desiccation. The gametophytes of seed plants consist of only a few cells. Pollen grains are male gametophytes containing sperm cells. The female gametophyte develops within an ovule that forms the seed.

CONCEPT 26.7 Gymnosperms Are Plants with "Naked Seeds"

26.7.1 Conifers Are the Largest Gymnosperm Phylum Gymnosperms have ovules that are not completely enclosed at the time of pollination. They produce male and female cones. In male cones, microspore mother cells in the microsporangia give rise to microspores that then develop into four-celled pollen grains, the male gametophytes. In female cones, a megasporangium produces a megaspore mother cell that becomes four megaspores, one of which develops into a female gametophyte. Upon pollination, a pollen tube emerges from the pollen grain and grows through the nucellus. Two sperm cells migrate through the tube, and one unites with the egg.

26.7.2 Cycads, Gnetophytes, and Ginkgophytes Are Three Minor Phyla of Gymnosperms Cycads resemble palms but are not flowering plants. Gnetophytes have xylem vessels. Only one species of the ginkgophytes remains.

CONCEPT 26.8 Angiosperms Are Flowering Plants

26.8.1 Flowers and Other Adaptations Drove the Radiation of Angiosperms Angiosperms are distinct because they have flowers, their ovules are enclosed within diploid tissue at the time of fertilization, and they form fruits. These features make them the dominant plant life on Earth.

26.8.2 Flowers House the Gametophyte Generation Flowers are modified stems that bear modified leaves. Flower parts are organized into four whorls: sepals, petals, androecium, and gynoecium. Most species use flowers to attract pollinators. Nectar and scent attract animal pollinators, which carry pollen from one flower to another; some angiosperms are wind-pollinated.

26.8.3 The Angiosperm Life Cycle Includes Double Fertilization After landing on a receptive stigma, a pollen grain develops a pollen tube that grows toward the embryo sac. Two sperm pass through this tube. One fuses with the egg to form a zygote, and the other fuses with polar bodies to form a triploid endosperm that develops to nourish the embryo.

Concept Overview

This Concept Overview diagrams the key concepts that were discussed in this chapter.

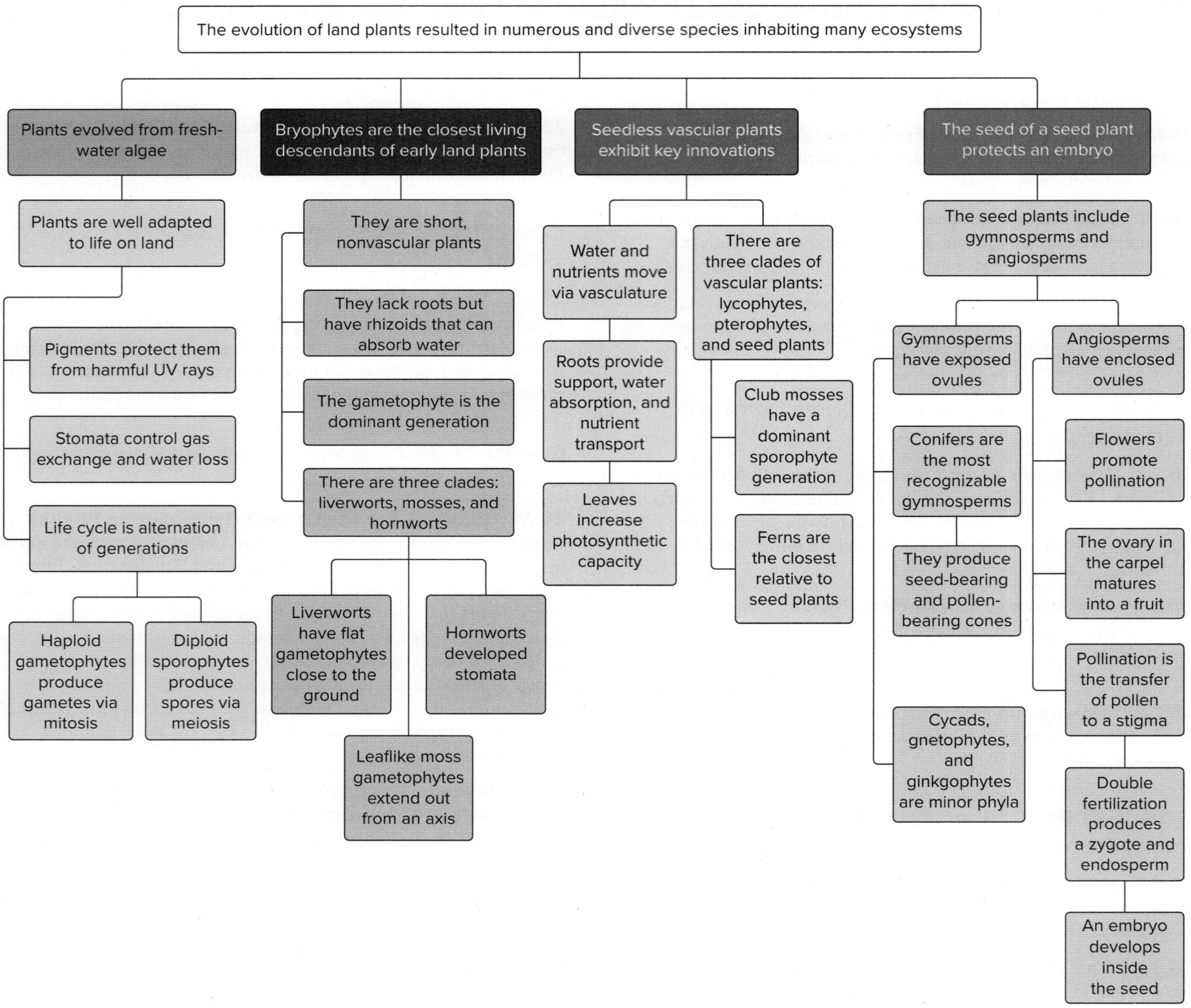

The evolution of land plants resulted in numerous and diverse species inhabiting many ecosystems
Plants evolved from fresh-water algae
Bryophytes are the closest living descendants of early land plants
Seedless vascular plants exhibit key innovations
The seed of a seed plant protects an embryo
Plants are well adapted to life on land
Pigments protect them from harmful UV rays
Stomata control gas exchange and water loss
Life cycle is alternation of generations
Haploid gametophytes produce gametes via mitosis
Diploid sporophytes produce spores via meiosis
They are short, nonvascular plants
They lack roots but have rhizoids that can absorb water
The gametophyte is the dominant generation
There are three clades: liverworts, mosses, and hornworts
Liverworts have flat gametophytes close to the ground
Hornworts developed stomata
Leaflike moss gametophytes extend out from an axis
Water and nutrients move via vasculature
Roots provide support, water absorption, and nutrient transport
Leaves increase photosynthetic capacity
There are three clades of vascular plants: lycophytes, pterophytes, and seed plants
Club mosses have a dominant sporophyte generation
Ferns are the closest relative to seed plants
The seed plants include gymnosperms and angiosperms
Gymnosperms have exposed ovules
Conifers are the most recognizable gymnosperms
They produce seed-bearing and pollen-bearing cones
Cycads, gnetophytes, and ginkgophytes are minor phyla
Angiosperms have enclosed ovules
Flowers promote pollination
The ovary in the carpel matures into a fruit
Pollination is the transfer of pollen to a stigma
Double fertilization produces a zygote and endosperm
An embryo develops inside the seed

Assessing the Learning Path

Understand

1. Why does living on land require adaptations to protect against UV light?
   a. Because of the increase in heat associated with higher levels of UV light
   b. Because UV light damages DNA and is mutagenic
   c. Because UV light promotes evaporation of water
   d. Both a and b
   e. Both b and c

2. What does it mean that plants alternate generations during their reproductive cycles?
   a. When a mature diploid plant dies, its offspring will be haploid, and this pattern repeats generation after generation.
   b. All plants have a unicellular generation followed by a multicellular generation.
   c. A haploid multicellular gametophyte produces gametes that unite to form a spore, producing diploid

multicellular sporophytes, and these generations alternate.
d. Sperm and egg are produced by a sporophyte; they unite to make a gametophyte that gives rise again to a sporophyte.

3. In what way(s) is a rhizoid like a root?
a. It provides structural support for vertical growth.
b. It anchors the root to a substrate.
c. It absorbs water.
d. a and b
e. b and c

4. A major innovation of land plants is embryo protection. How is a moss embryo protected from desiccation?
a. By the seed
b. By the antheridium
c. By the archegonium
d. By the lycophyll

5. Which of the following plant structures is NOT matched to its correct function?
a. Stomata–allow gas transfer
b. Tracheids–allow the movement of water and minerals
c. Cuticle–prevents excessive gas loss
d. All of the above are matched correctly.

6. Which feature of vascular plants was the first to evolve?
a. Vascular tissue
b. Seeds
c. Flowers
d. Stems, roots, and leaves

7. Why aren't club mosses classified as bryophytes?
a. They have tracheids.
b. They produce seeds.
c. They have flowers.
d. They aren't photosynthetic.

8. One characteristic that separates ferns from higher, more complex vascular plants is that ferns do not have
a. alternation of generations in their life cycle.
b. seeds.
c. a cuticle to prevent desiccation.
d. a vascular system.

9. Which of the following adaptations allow plants to pause their life cycle until environmental conditions are optimal?
a. Stomata
b. Phloem and xylem
c. Seeds
d. Flowers

10. In seeds, the endosperm helps with
a. fertilization.
b. nourishment.
c. protection.
d. dispersal.

11. In a pine tree, the microspores and megaspores are produced by the process of
a. fertilization.
b. mitosis.
c. meiosis.
d. fusion.

12. Which phylum of gymnosperms has the least number of species?
a. Coniferophyta
b. Gnetophyta
c. Cycadophyta
d. Ginkgophyta

13. What separates the angiosperms from other seed plants?
a. A vascular system
b. Wind dispersal of pollen
c. Ovules not completely covered by the sporophyte
d. Fruits and flowers

14. In double fertilization, one sperm produces a diploid __________, and the other produces a triploid __________.
a. zygote; primary endosperm
b. primary endosperm; microspore
c. antipodal; zygote
d. polar nucleus; zygote

15. A plant has a dominant gametophyte generation, rhizoids, and stomata. Which of these features would best identify the plant as a hornwort?
a. The presence of stomata
b. The dominant gametophyte generation
c. The presence of rhizoids
d. All of these are unique to hornworts.

Apply

1. Which is the correct comparison of what happens to a spore mother cell as it gives rise to a spore and what happens to a spore as it gives rise to a gametophyte?
a. The spore mother cell and the spore both go through meiosis.
b. The spore mother cell and the spore both go through mitosis.
c. The spore mother cell goes through mitosis, and the spore goes through meiosis.
d. The spore mother cell goes through meiosis, and the spore goes through mitosis.

2. Mosses do not reach a large size, because
a. they lack chlorophyll.
b. they do not have specialized vascular tissue to transport water very high.
c. moss photosynthesis does not take place at a very fast rate.
d. alternation of generations does not allow the plant to grow very tall before reproduction.

3. How can a plant without roots obtain sufficient nutrients from the soil?
a. It cannot; all land plants have roots.
b. Mycorrhizal fungi associate with the plant and assist with the transfer of nutrients.
c. Charophytes associate with the plant and assist with the transfer of nutrients.
d. It relies on its xylem in the absence of a root.

4. What is the best reason most extant lycophytes live in moist areas such as forest floors?
a. They absorb water across their surface and thus need to be in close contact with water.
b. They lack a cuticle and so are susceptible to dessication.
c. They have motile sperm that need to swim through water for fertilization to occur.
d. They need an abundant water supply, because they are poor photosynthesizers.

5. An example of a seedless vascular plant is
a. a hornwort.
b. a club moss.
c. a pine tree.
d. a horsetail.

6. Which of the following is NOT associated with a male portion of a plant?
a. Megaspore
b. Antheridium
c. Pollen grain
d. Microspore

7. Which of the following gymnosperms possess a form of vascular tissue that is similar to that found in the angiosperms?
a. Cycads
b. Gnetophytes
c. Ginkgophytes
d. Conifers

8. Which innovation likely contributed to the tremendous success of angiosperms?
a. Homospory
b. Fruits that attract animal dispersers
c. Cones that protect the seed
d. Dominant gametophyte generation

9. Following fertilization, the primary endosperm nucleus is ______; this is because a ______ sperm unites with ______ haploid polar nucleus/nuclei.
   a. diploid; haploid; two
   b. triploid; diploid; one
   c. triploid; haploid; two
   d. triploid; diploid; two

Synthesize

1. Compare and contrast the adaptations of plants and fungi to a terrestrial existence. Which group of organisms has been more successful and why?
2. In all plants, from bryophytes to angiosperms, the zygote is nourished with substances provided by the maternal gametophyte as it develops into the multicellular embryo and eventually the mature sporophyte. Would it then be appropriate to term all plants "embryophytes"?
3. In the late Carboniferous period (the "Age of Coal"), much of North America was lowland covered by shallow seas or swamps. Tall, slender lycophyte trees dominated the environment, growing to heights of 35 m. Their compressed remains are what we know today as coal. Experimentally, how might you confirm that coal is largely composed of lycophytes?
4. How can a dominant diploid sporophyte generation be considered an adaptive advantage when being compared with a dominant haploid gametophyte generation?
5. Why do ferns require free water to complete their life cycle, when angiosperms do not? What are the reasons for the difference?
6. Why is the ability to form seeds always associated with a separation of the gametophyte generation into two distinct kinds, megagametophytes and microgametophytes? How has this relationship changed during the course of evolution?
7. In New Zealand, large gymnosperms are far more common than large angiosperms. Under what conditions might gymnosperms have an evolutionary advantage over angiosperms?
8. The relationship between flowering plants and pollinators is often used as an example of coevolution. Many flowering plant species have flower structures that are adaptive to a single species of pollinator. What are the pros and cons of using such a specialized relationship?

Simon Murrell/Cultura/Getty Images

27 Animal Diversity

Learning Path

27.1 The Diversity of Animal Body Plans Arose by a Series of Evolutionary Innovations

27.2 Molecular Data Are Clarifying the Animal Phylogenetic Tree

27.3 True Tissue Evolved in Simple Animals

27.4 Flatworms and Rotifers Are Very Simple Bilaterians

27.5 Mollusks and Annelids Are the Largest Groups of Lophotrochozoans

27.6 Lophophorates Are Very Simple Marine Organisms

27.7 Nematodes and Arthropods Are Both Large Groups of Ecdysozoans

27.8 Deuterostomes Are Composed of Echinoderms and Chordates

Concept Overview

This Concept Overview provides a diagram of the overarching concepts that are covered in this chapter. These concepts will be expanded upon in the Concept Overview at the end of the chapter.

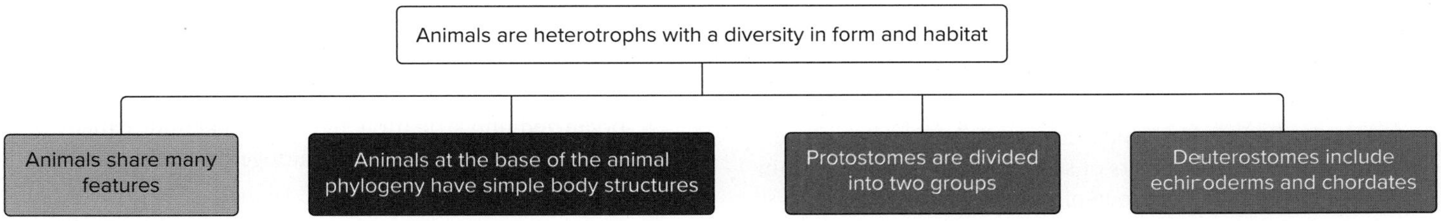

Animals are heterotrophs with a diversity in form and habitat
Animals share many features
Animals at the base of the animal phylogeny have simple body structures
Protostomes are divided into two groups
Deuterostomes include echinoderms and chordates

Introduction

We now explore the great diversity of modern animals, the result of a long evolutionary history. Found in almost every habitat, they bewilder us with their diversity in form, habitat, behavior, and lifestyle. About a million and a half species have been described. Our exploration of the great diversity of animals starts with the morphologically simplest members—sponges, jellyfish, and some of the worms. The major organization of the animal body—the basic body plan from which all the rest of the animals evolved—first evolved in these animals. In this chapter, we will explore the diversity of animals lacking a backbone (invertebrates), then cover the vertebrates in chapter 28. By far the most diverse group of animals is the arthropods. Two-thirds of all named animal species are arthropods, 80% of them insects, like the locust on the preceding page. There are seven times as many species of beetles as there are vertebrates!

27.1 The Diversity of Animal Body Plans Arose by a Series of Evolutionary Innovations

In this section, we will introduce the diversity of animals by considering some common features, then the evolutionary innovations that led to the diversity of animal body plans. This will provide a framework for understanding modern animal phylogeny and forms the organization for our tour of invertebrate diversity.

Animals are so diverse that few criteria fit them all. But one does apply: all animals are eaters, or consumers, and not producers like plants. Other criteria, such as that animals are mobile (can move about), have exceptions. Taken together, the universal characteristics and other features of major importance exhibited by most species are convincing evidence that animals are monophyletic. This conclusion has been confirmed when DNA data has been used to build phylogenies of all living organisms.

Animals Share Many Features

LEARNING OBJECTIVE 27.1.1 Describe some common features of animals.

Any list of shared features is necessarily incomplete, but in general the following are the common characteristics of animals.

- **Heterotrophy.** All animals are heterotrophs—that is, they obtain energy and organic molecules by ingesting other organisms. Unlike autotrophic plants and algae, animals cannot construct organic molecules from CO_2. Some animals (herbivores) consume autotrophs; other animals (carnivores) consume heterotrophs; some animals (omnivores) consume both autotrophs and heterotrophs; and still others (detritivores) consume decomposing organisms.
- **Multicellularity.** All animals are multicellular; many have complex bodies. Unicellular heterotrophic organisms are now considered members of several different clades within the large and diverse group of protists, discussed in chapter 24.
- **No cell walls.** Animal cells differ from those of other multicellular organisms: they lack rigid cell walls and are usually quite flexible. The many cells of animal bodies are held together by extracellular frames of structural proteins such as collagen. Other proteins form unique intercellular junctions between cells.
- **Active movement.** Although single-celled organisms are able to travel from place to place, most animals move in more complex ways. This is due to the evolution of nerve and muscle tissues. A form of movement unique to animals is flying, as seen in the butterfly (phylum Arthropoda) shown in figure 27.1*a*. Although some animals cannot move from place to place (they are sessile) or do so rarely or slowly (they are sedentary), they have muscles or muscle fibers that allow parts of their bodies to move. Sponges, however, have little capacity for movement.
- **Diversity of form.** Animals vary greatly in form, ranging in size from organisms too small to see to enormous whales and giant squids. Almost all animals lack a backbone—they are called invertebrates, like the millipede (phylum Arthropoda) in figure 27.1*b*. Of the 1.5 million known living animal species, fewer than 60,000 have a backbone—they are referred to as vertebrates.
- **Diversity of habitat.** Animals are grouped into 35 to 40 phyla, most with members that occur only in the sea. Members of fewer phyla occur in fresh water, and members of still fewer occur on land. Three phyla that are successful in the marine environment—Arthropoda, Nematoda, and Chordata—also dominate animal life on land.
- **Sexual reproduction.** Most animals reproduce sexually. Animal eggs, which are not mobile, are much larger than the small, usually flagellated sperm. In animals, cells formed in meiosis function as gametes. These haploid cells fuse directly with each other to form the zygote. Consequently, there is no counterpart among animals to the alternation of generations characteristic of plants.
- **Embryonic development.** An animal zygote first undergoes a series of mitotic divisions, called cleavage, and as in the dividing frog's egg in figure 27.1*c*, cleavage produces a ball of cells, the blastula. Embryos of most kinds of animals develop into a larva, which looks unlike the adult of the species, lives in a different habitat, and eats different sorts of food. A larva undergoes metamorphosis, a radical reorganization, to transform into the adult body form.
- **Tissues.** The cells of all animals, except sponges, are organized into structural and functional units called tissues—collections of cells that together are specialized to perform specific tasks.

The Evolution of Tissue and Symmetrical Bodies Was a Critical Early Innovation

LEARNING OBJECTIVE 27.1.2 Compare and contrast radial and bilateral symmetry.

A typical sponge is asymmetrical, growing as an irregular mass, with no tissue. Virtually all other animals have tissue, as well as a definite shape and symmetry that can be defined along an imaginary axis drawn through the animal's body. The two main types of symmetry are radial and bilateral (note that many animals are not perfectly symmetrical, but close enough that they are considered symmetrical).

Figure 27.1 Some characteristics shared by animals. ***a.*** Many, but not all, animals are able to move from place to place. Other organisms are also able to move, but flying is unique to animals. ***b.*** Animals exhibit a wide array of different sizes and forms, but the vast majority of animals are invertebrates, lacking a backbone, such as this millipede. ***c.*** Animals undergo a process of development beginning with a series of cell divisions called cleavage, producing a multicelled structure called a blastula.

(a): Lisa Thornberg/Getty Images; (b): Edward S. Ross; (c): Carolina Biological/Medical Images/DIOMEDIA

Evolution of tissue

The zygote (a fertilized egg) has the capability of giving rise to all the kinds of cells in an animal's body. We say that it is totipotent. During embryonic development, cells become specialized to carry out particular functions. In all animals except sponges, the process is irreversible: the descendants of a differentiated cell remain differentiated.

A sponge cell that has specialized to serve one function (such as lining the cavity where feeding occurs) can lose the special attributes that serve that function and change to serve another function (such as being a gamete). Thus, a sponge cell can dedifferentiate and redifferentiate. Cells of all other animals are organized into tissues, each of which is characterized by cells of particular morphology and capability. But their competence to dedifferentiate prevents sponge cells from forming clearly defined tissues (and therefore organs, which are composed of tissues). The evolution of specialized tissues was a key innovation—a trait that fosters evolutionary diversification.

Radial symmetry

Radial symmetry is exhibited by members of the phylum Cnidaria (jellyfish, sea anemones, and corals; the *C* of *Cnidaria* is silent). Its parts are arranged in such a way that any longitudinal plane passing through the central axis divides the organism into halves that are approximately mirror images. A pie, for example, is radially symmetrical, and so is a sea anemone (figure 27.2*a*).

Bilateral symmetry

The bodies of most animals other than sponges and cnidarians exhibit **bilateral symmetry,** in which the body has right and left halves that are mirror images of each other. The clade containing animals with this body plan is termed the Bilateria. The sagittal plane defines these halves (figure 27.2*b*). In addition to the left and right halves, a bilaterally symmetrical body has dorsal and ventral portions, which are divided by the frontal plane, and anterior (front) and posterior (rear) ends, which are divided by the transverse plane.

Sometimes, it is not easy to decide whether an organism should be considered bilaterally or radially symmetrical. For example, in echinoderms (sea stars and their relatives), adults have five axes of symmetry (pentaradial symmetry), but the larvae are bilaterally symmetrical. In this case, examination of animal phylogeny indicates that the pentaradial symmetry of the adult is an evolutionarily derived condition from a bilaterally symmetrical ancestral condition.

Bilateral symmetry constitutes a major evolutionary advance in the animal body plan. Bilaterally symmetrical animals have the ability to move through the environment in a consistent direction (typically, anterior first)—a feat that is difficult for radially symmetrical animals. That capacity for directional movement is associated with the grouping of nerve cells into a brain, with sensory structures at the anterior end of the body. This concentration of nervous tissue at the anterior end, which appears to have occurred early in evolution, is called **cephalization.** Much of the layout of the nervous system in bilaterally symmetrical

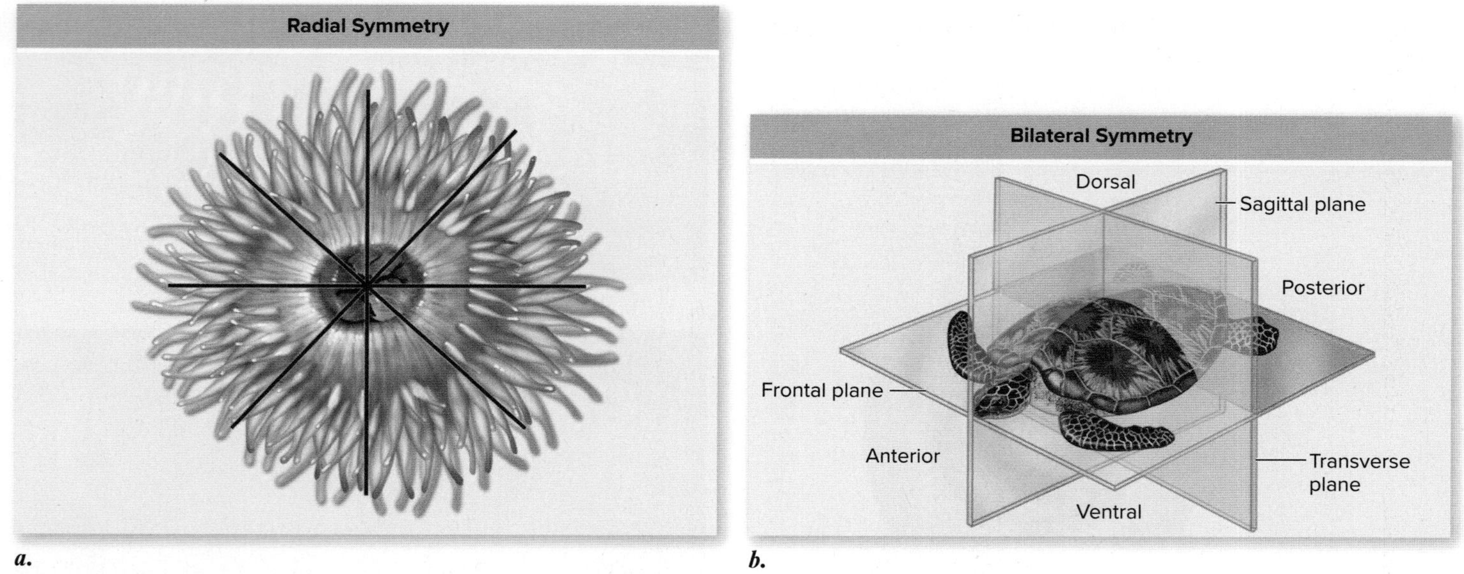

Figure 27.2 A comparison of radial and bilateral symmetry. ***a.*** Radially symmetrical animals, such as this sea anemone (phylum Cnidaria), can be bisected into equal halves by any longitudinal plane that passes through the central axis. ***b.*** Bilaterally symmetrical animals, such as this turtle (phylum Chordata), can be bisected into equal halves only in one plane (the sagittal plane).

animals is centered on one or more major longitudinal nerve cords that transmit information from the anterior sense organs and brain to the rest of the body.

A Body Cavity Made the Development of Advanced Organ Systems Possible

LEARNING OBJECTIVE 27.1.3 Describe the types of body cavities found in animals.

In the process of embryonic development, the cells of most types of animals organize into three layers (called germ layers): an outer ectoderm, an inner endoderm, and an intermediate mesoderm. Animals with three embryonic cell layers are said to be triploblastic. During maturation from the embryo, certain organs and organ systems develop from each germ layer. The ectoderm gives rise to the outer covering of the body and the nervous system; the endoderm gives rise to the digestive system, including the intestine; and the skeleton and muscles develop from the mesoderm (refer to chapter 36). Cnidarians are diploblastic, with only two layers—the endoderm and the ectoderm—and they lack organs. Sponges lack germ layers altogether; they, of course, have no tissues or organs. All bilaterians are triploblastic.

Body cavities

A key innovation in the body plan of some bilaterians was a body cavity isolated from the exterior of the animal. This is different from the digestive cavity, which is open to the exterior. A fluid-filled body cavity accommodated the evolution of complex organ systems by providing support, by aiding the efficient distribution of materials, and by fostering complex developmental interactions. In most animals, the fluid is liquid, but in vertebrates it is gas—a human's body cavity filling with liquid is a life-threatening condition. A few types of bilaterians (the acoelomates) have no body cavity. In these animals, the space between tissues that develop from the mesoderm and those that develop from the endoderm is filled with cells and connective tissue (figure 27.3).

Body cavities appear to have evolved multiple times in the Bilateria. In some animals, a body cavity (the **pseudocoelom**) develops embryologically between the mesoderm and endoderm and thus occurs in the adult between tissues derived from the mesoderm and those derived from endoderm. Animals with this type of body cavity are termed pseudocoelomates. Although the word *pseudocoelom* means "false coelom," this is a true body space and characterizes many groups of animals. A **coelom** is a cavity that develops entirely within the mesoderm (figure 27.3). The coelom is surrounded by a layer of epithelial cells derived from the mesoderm and termed the peritoneum.

Animals can have open or closed circulatory systems

In many small animals, nutrients and oxygen are distributed and wastes are removed by fluid in the body cavity. Most larger animals, in contrast, have a **circulatory system,** a network of vessels that carry fluids to and from the parts of the body distant from the sites of digestion (gut) and gas exchange (gills or lungs). The circulating fluid carries nutrients and oxygen to the tissues and removes wastes, including carbon dioxide, by diffusion between the circulatory fluid and the other cells of the body.

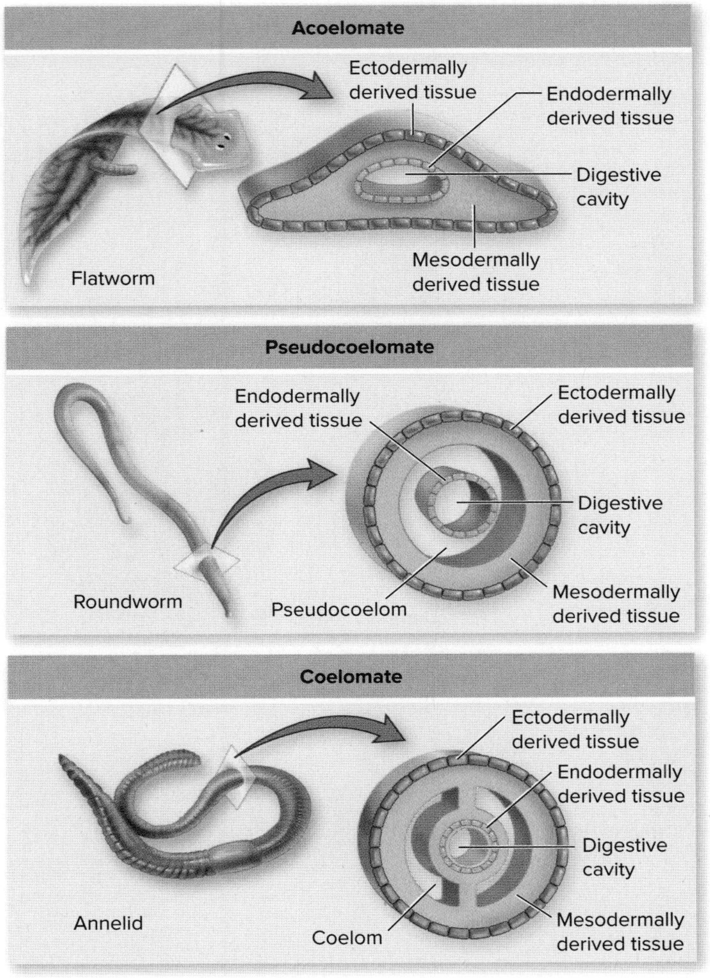

Acoelomate
Ectodermally derived tissue
Endodermally derived tissue
Digestive cavity
Mesodermally derived tissue
Flatworm
Pseudocoelomate
Endodermally derived tissue
Ectodermally derived tissue
Digestive cavity
Roundworm
Pseudocoelom
Mesodermally derived tissue
Coelomate
Ectodermally derived tissue
Endodermally derived tissue
Digestive cavity
Annelid
Coelom
Mesodermally derived tissue

Figure 27.3 Three body plans for bilaterally symmetrical animals. Acoelomates, such as flatworms, have no body cavity between the digestive tract (derived from the endoderm) and the musculature layer (derived from the mesoderm). Pseudocoelomates have a body cavity, the pseudocoelom, between tissues derived from the endoderm and those derived from the mesoderm. Coelomates have a body cavity, the coelom, that develops entirely within tissues derived from the mesoderm and, so, is lined on both sides by tissue derived from the mesoderm.

In an **open circulatory system,** the blood passes from vessels into sinuses, mixes with body fluid that bathes the cells of tissues, then reenters vessels in another location. In a **closed circulatory system,** the blood is entirely confined to blood vessels, so it is physically separated from other body fluids. Blood moves through a closed circulatory system faster and more efficiently than it does through an open system; open systems are typical of animals that are relatively inactive and, so, do not have a high demand for oxygen. In small animals, blood can be pushed through a closed circulatory system by the animal's movements. In larger animals, the body musculature does not provide enough force, so the blood must be propelled by contraction of one or more hearts, which are specialized muscular parts of the blood vessels.

Early Developmental Differences Divide Bilaterians into Protostomes and Deuterostomes

LEARNING OBJECTIVE 27.1.4 Compare and contrast protostome and deuterostome development.

The process of embryonic development in animals is discussed fully in chapter 36. Briefly, development of a bilaterally symmetrical animal begins with mitotic cell divisions (cleavages) of the egg, which lead to the formation of a hollow ball of cells, which subsequently indents to form a two-layered ball. The internal space that is created through such indentation is the **archenteron** (literally, the "primitive gut"); it communicates with the outside by a **blastopore.**

Protostomes and deuterostomes

In a **protostome,** the mouth of the adult animal develops from the blastopore or from an opening near the blastopore (*protostome* means "first mouth"–the first opening becomes the mouth). Protostomes include most bilaterians, including flatworms, nematodes, mollusks, annelids, and arthropods. In some protostomes, both mouth and anus form from the embryonic blastopore; in other protostomes, the anus forms later in another region of the embryo. Two outwardly dissimilar groups–the echinoderms and the chordates–together with a few other small phyla, constitute the **deuterostomes,** in which the mouth of the adult animal does not develop from the blastopore. The deuterostome blastopore gives rise to the organism's anus, and the mouth develops from a second pore that arises later in development (*deuterostome* means "second mouth").

Cleavage patterns

The cleavage pattern relative to the embryo's polar axis determines how the resulting cells lie with respect to one another. Two patterns of cleavage characterize major animal clades. In one clade of protostomes, the Lophotrochozoa, each new cell cleaves off at an angle oblique to the polar axis. As a result, a new cell nestles into the space between the older ones in a closely packed array. This pattern is called **spiral cleavage,** because a line drawn through a sequence of dividing cells spirals outward from the polar axis (figure 27.4, *top*). Spiral cleavage is characteristic of annelids, mollusks, nemerteans, and related phyla.

By contrast, in all deuterostomes and, convergently, in a few protostomes such as brachiopods, the cells divide parallel to and at right angles to the polar axis. As a result, the pairs of cells from each division are positioned directly above and below one another, a process that gives rise to a loosely packed ball. This pattern is called **radial cleavage,** because a line drawn through a sequence of dividing cells describes a radius outward from the polar axis (figure 27.4, *bottom*).

Determinate versus indeterminate development

Many protostomes exhibit **determinate development,** in which the type of tissue each embryonic cell will form in the adult is determined early, in many lineages even before cleavage begins, when

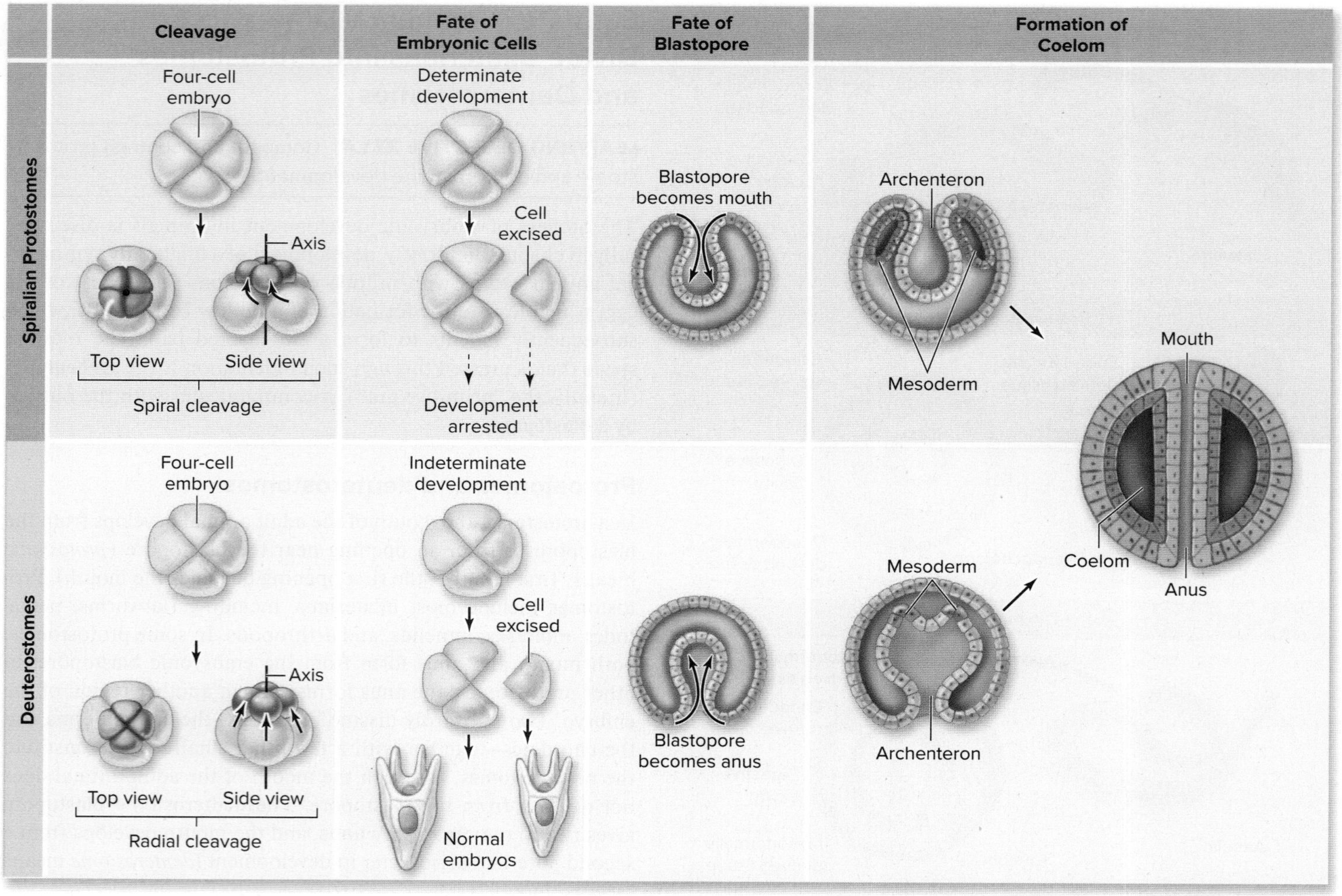

Figure 27.4 Embryonic development in protostomes and deuterostomes. In lophotrochozoans, embryonic cells cleave in a spiral pattern and exhibit determinate development; the blastopore becomes the animal's mouth, and the coelom originates from a split among endodermal cells. In deuterostomes, embryonic cells cleave radially and exhibit indeterminate development; the blastopore becomes the animal's anus, and the coelom originates from an invagination of the archenteron.

the molecules that act as developmental signals are localized in different regions of the egg. Consequently, the cell divisions that occur after fertilization segregate molecular signals into different daughter cells, specifying the fate of even the very earliest embryonic cells. Each embryonic cell is destined to occur only in particular parts of the adult body, so if the cells are separated, development cannot proceed.

Deuterostomes, conversely, display **indeterminate development.** The first few cell divisions of the zygote produce identical daughter cells. If the cells are separated, any one can develop into a complete organism, because the molecules that signal the embryonic cells to develop differently are not segregated into different cells until later in the embryo's development. (This is how identical twins are formed.) Thus, each cell remains totipotent, and its fate is not determined for several cleavages.

Formation of the coelom

The coelom arises within the mesoderm. In protostomes, cells simply move apart from one another to create an expanding coelomic cavity within the mass of mesodermal cells. In deuterostomes, groups of cells pouch off the end of the archenteron (literally, the "primitive gut"), the hollow in the center of the developing embryo that is lined with endoderm.

The consistency of deuterostome development and its distinctiveness from that of the protostomes suggest that it evolved once, in the ancestor of the deuterostome phyla. The mode of development in protostomes is more diverse, but because of the distinctiveness of spiral development, scientists infer that it also evolved once, in the common ancestor to all lophotrochozoan phyla. As we will see, our changing understanding of animal phylogeny supports these conclusions.

Segmentation Has Evolved Multiple Times

LEARNING OBJECTIVE 27.1.5 Explain how segmentation is an important innovation.

Segmented animals consist of a series of linearly arrayed compartments that typically look alike (see figures 27.16, 27.20, and 27.28),

TABLE 27.1 Animal Phyla with the Most Species

Phylum	Typical Examples	Key Characteristics	Approximate Number of Named Species
Arthropoda (arthropods)	Insects, crabs, spiders, scorpions, centipedes	Chitinous exoskeleton covers segmented, coelomate body. With paired, jointed appendages; many types of insects have wings. They occupy marine, terrestrial, and freshwater habitats.	1,000,000
Mollusca (mollusks)	Snails, clams, octopuses, slugs	Coelomate body of many mollusks is covered by one or more shells secreted by a part of the body termed the mantle. Many kinds possess a unique rasping tongue, a radula. Members occupy marine, terrestrial, and freshwater habitats.	110,000
Chordata (chordates)	Mammals, fish, reptiles, amphibians	Each coelomate individual possesses a notochord, a dorsal nerve cord, pharyngeal slits, and a postanal tail at some stage. In vertebrates, the notochord is replaced by the spinal column. Members occupy marine, terrestrial, and freshwater habitats.	56,000
Platyhelminthes (flatworms)	Planarians, tapeworms, flukes	These are unsegmented, acoelomate, bilaterally symmetrical worms. Digestive cavity has only one opening; tapeworms lack a gut. Many species are parasites of medical and veterinary importance. Members occupy marine, terrestrial, and freshwater habitats.	20,000
Nematoda (roundworms)	*Ascaris*, pinworms, hookworms	Pseudocoelomate, unsegmented, bilaterally symmetrical worms; tubular digestive tract has mouth and anus. Members occupy marine, terrestrial, and freshwater habitats; some are important parasites of plants and animals, including humans.	25,000 (though some think the number may be much greater)
Annelida (segmented worms)	Earthworms, tube worms, leeches	Segmented, bilaterally symmetrical, coelomate worms with a complete digestive tract; most have bristles (chaetae) on each segment. They occupy marine, terrestrial, and freshwater habitats.	22,000
Cnidaria (cnidarians)	Jellyfish, corals, sea anemones	Radially symmetrical, acoelomate body has tissues but no organs. Mouth opens into a simple digestive sac and is surrounded by tentacles armed with stinging capsules (nematocysts). The very few nonmarine species live in fresh water.	10,000
Echinodermata (echinoderms)	Sea stars, sea urchins, sand dollars	Adult body is pentaradial (fivefold) in symmetry. Water–vascular system is a coelomic space; endoskeleton is of calcium carbonate plates. Many can regenerate lost body parts. They are exclusively marine.	7,000
Porifera (sponges)	Sponges	Asymmetrical bodies make defining “an individual” difficult. Body lacks tissues or organs. Channels open to the outside and internal cavities are lined with food-filtering, flagellated cells (choanocytes). Most species are marine.	7,000
Bryozoa (moss animals)	Sea mats, sea moss	The only exclusively colonial phylum; each colony comprises small, coelomate individuals (zooids) connected by an exoskeleton. A ring of ciliated tentacles (lophophore) surrounds the mouth of each zooid.	4,500
SOME IMPORTANT ANIMAL PHYLA WITH FEWER SPECIES			
Rotifera (wheel animals)	Rotifers	Small pseudocoelomates with a complete digestive tract, including a set of complex jaws. Cilia at the anterior end beat so they resemble a revolving wheel.	2,000
Nemertea (ribbon worms)	*Lineus*	Protostome worms notable for their fragility. Long, extensible proboscis occupies a coelomic space; most marine, but some live in fresh water, and a few are terrestrial.	900
Tardigrada (water bears)	*Hypsibius*	Microscopic protostomes with five body segments and four pairs of clawed legs; an individual lives a week or less but can enter a state of suspended animation for many decades. Occupy marine, freshwater, and terrestrial habitats.	800
Brachiopoda (lamp shells)	*Lingula*	Protostomous animals encased in two shells; a ring of ciliated tentacles (lophophore) surrounds the mouth.	300
Onychophora (velvet worms)	*Peripatus*	Segmented protostomous worms resemble tardigrades, with a chitinous soft exoskeleton and unsegmented appendages.	110
Ctenophora (sea walnuts)	Comb jellies, sea walnuts	Gelatinous, almost transparent, often bioluminescent marine animals; they have eight bands of cilia. They are the largest animals that use cilia for locomotion and have a complete digestive tract with anal pore.	100
Chaetognatha (arrow worms)	*Sagitta*	Small, bilaterally symmetrical, transparent marine worms with a fin along each side, powerful bristly jaws, and lateral nerve cords; it is uncertain if they are coelomates.	100

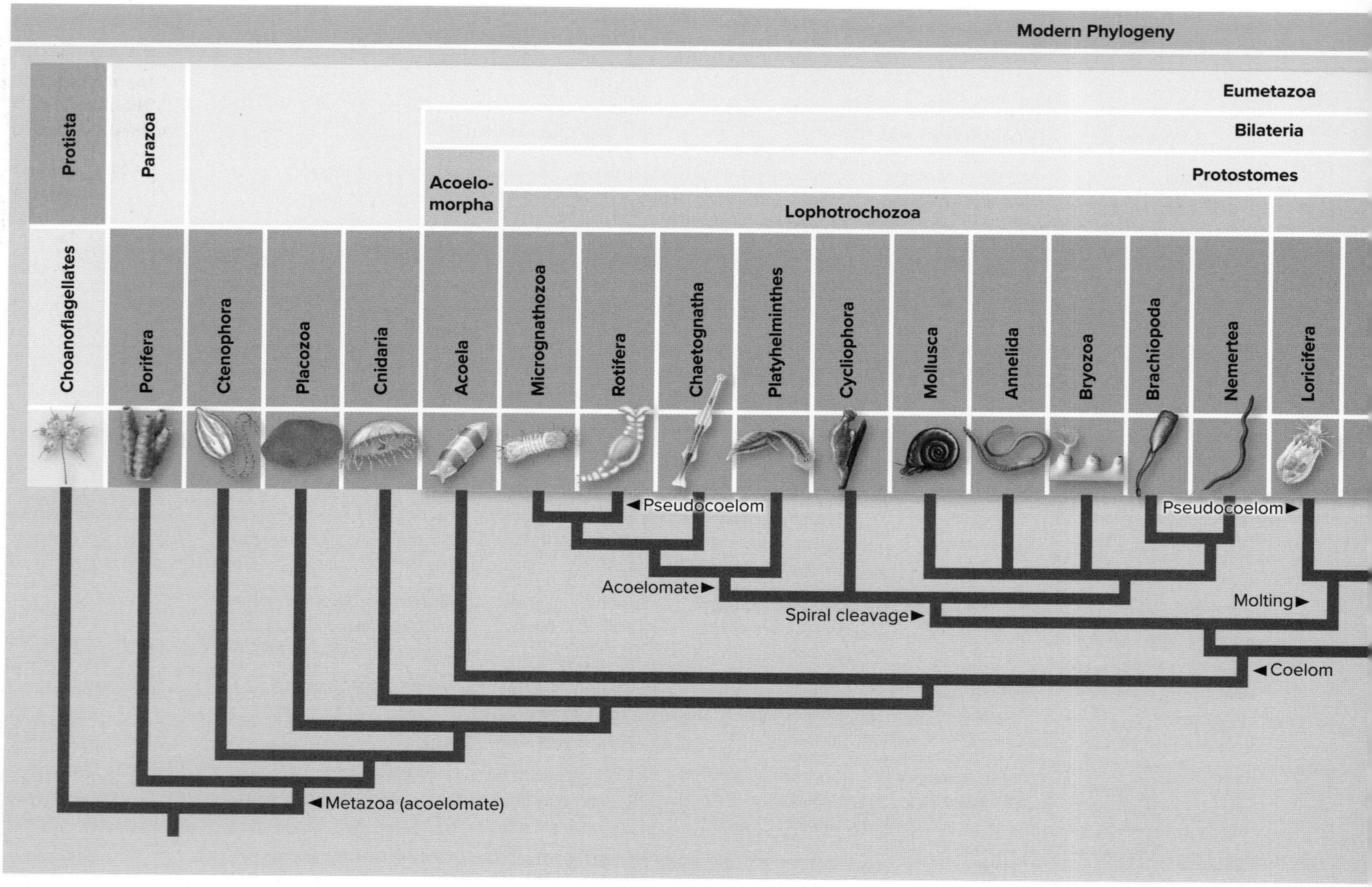

Figure 27.5 Proposed revision of the animal tree of life. A phylogeny of many of the 35–40 phyla reflects a consensus based on interpretation of anatomical and developmental data as well as results derived from molecular phylogenetic studies. The position of ctenophores is currently being debated, with some studies suggesting that it is the sister taxon to other animals, rather than Porifera occupying that position.

at least early in development, but that may have specialized functions. During early development, segments first are obvious in the mesoderm but later are reflected in the ectoderm and endoderm. Two advantages result from early embryonic segmentation:

1. In highly segmental animals, such as earthworms (phylum Annelida), each segment may develop a more or less complete set of adult organ systems. Because these are redundant systems, damage to any one segment need not be fatal.
2. Locomotion is more efficient when individual segments can move semi-independently. Because partitions isolate the segments, each can contract or expand autonomously. Therefore, a long body can move in ways that are often quite complex.

Segmentation underlies the organization of body plans of the most morphologically complex animals. In some adult arthropods, the segments are fused, but segmentation is usually apparent in embryological development. In vertebrates, the backbone and muscle blocks are segmented, although segmentation is often disguised in the adult form.

Previously, zoologists considered that true segmentation was found only in annelids, arthropods, and chordates, but segmentation is now recognized to be more widespread. Animals such as onychophorans (velvet worms), tardigrades (water bears), and kinorhynchs (mud dragons) are also segmented.

REVIEW OF CONCEPT 27.1

Animals are distinguished on the basis of symmetry, tissues, type of body cavity, sequence of embryonic development, and segmentation. Bilateral animals have bodies with a left and a right side, which are mirror images. A body cavity forms in most bilaterians. A pseudocoelom develops between the mesoderm and endoderm; a coelom develops entirely within mesoderm. Protostomes develop the mouth prior to the anus; deuterostomes develop the mouth after the anus has formed. Segmentation allows redundant systems and efficient locomotion.

- *How is cephalization related to body symmetry?*

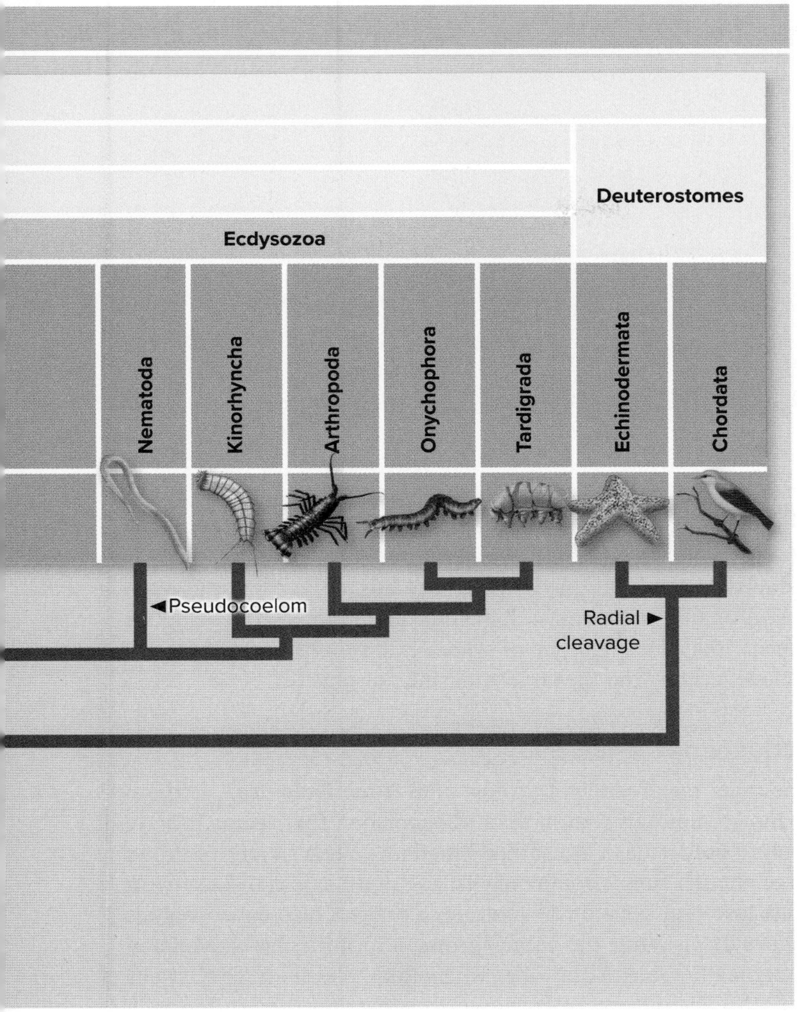
Deuterostomes
Ecdysozoa
Nematoda
Kinorhyncha
Arthropoda
Onychophora
Tardigrada
Echinodermata
Chordata
◄Pseudocoelom
Radial ►
cleavage

27.2 Molecular Data Are Clarifying the Animal Phylogenetic Tree

Multicellular animals, or metazoans, are traditionally divided into 35 to 40 phyla. There is little disagreement among biologists about the placement of most animals in phyla, although zoologists disagree on the status of some, particularly those with few members. The diversity of animals is obvious in table 27.1, which describes the key characteristics of 17 of the phyla.

Molecular Data Are Revising Animal Phylogeny

LEARNING OBJECTIVE 27.2.1 Explain how our understanding of animal phylogeny has changed.

Traditionally the phylogeny of animals has been inferred using features of anatomy and aspects of embryological development, from which a broad consensus emerged over the last century concerning the main branches of the animal tree of life. In the past 20 years, gene sequence data have accumulated at an accelerating pace for all animal groups, leading to some rethinking of classification schemes. Phylogenies developed from different molecules sometimes suggest quite different evolutionary relationships among the same groups of animals, but combining data from multiple genes and even entire genomes has resolved the relationships of most phyla.

Although they differ from one another in some respects, phylogenies incorporating molecular data or based entirely on them share some deep structure with the traditional animal tree of life. Figure 27.5 is a summary of animal phylogeny developed from morphological, molecular, life-history, and other types of relevant data. Some parts of this phylogeny are not firmly established, and new studies are constantly appearing, often with somewhat different conclusions. It is an exciting time to be a systematist, but shifts in understanding of relationships among groups of animals can be frustrating to some! Like any scientific idea, a phylogeny is a hypothesis, open to being revised in light of additional data.

One consistent result is that Porifera (sponges) constitutes a monophyletic group. Some systematists had considered sponges to comprise two (or three) groups that are not particularly closely related, but molecular data support what had been the majority view, that phylum Porifera is monophyletic. And, as mentioned in section 27.1, all animals are found to be monophyletic.

Among remaining animals, termed Eumetazoa, molecular data are in accord with the traditional view that cnidarians (hydras, sea jellies, and corals) branch off the tree before the origin of animals with bilateral symmetry, the Bilateria.

Our understanding of the phylogeny of the deuterostome branch of Bilateria has not changed much, but our understanding of the phylogeny of protostomes has been drastically altered by molecular data. Molecular analysis has made clear that annelids and arthropods, which had been considered closely related because both exhibit segmentation, are, in fact, not; instead, they belong to separate clades. Nematodes and arthropods, which previously were not thought to be closely related, but which share the feature of molting, are now thought to be part of the same clade. Molecular sequence data can help test our ideas of which morphological features reveal evolutionary relationships best; in this case, molecular data allowed us to see that, contrary to our hypothesis, segmentation seems to have evolved convergently, but molting did not.

Molecular data are changing and clarifying our view of animal phylogeny. It is clear that animals are monophyletic, and within animals, sponges are also monophyletic—relationships that were uncertain. Although these data will not resolve all phylogenetic issues, they have changed our view of relationships in protostomes.

Traditional morphology-based phylogeny focused on body cavities

This new understanding of animal phylogeny has led to a rethinking of how body cavities have evolved. Zoologists previously inferred that the first animals lacked a body cavity (were acoelomate), that some of their descendants evolved a pseudocoelom, and that some pseudocoelomate descendants evolved the coelom. This view was so widespread that classification systems were based on the state of the coelom. However, as you learned in chapter 21, evolution rarely occurs in such a linear and directional way. Our understanding of animal phylogeny now makes it clear that the state of the internal cavity has evolved many more times than previously realized, and consequently, that it is not a reliable character to infer phylogenetic relationships.

In particular, a coelom appears to have evolved just once, in the ancestor of the clade comprising protostomes and deuterostomes. Subsequently, one clade within the Lophotrochozoa reverted secondarily to the acoelomate condition, and the pseudocoelomate condition arose several times from coelomate ancestors within the protostome clade. As a result, all deuterostomes have coeloms, but the condition in protostomes is mixed. In addition, although the acoelomate condition is the ancestral state for animals, some animals have lost the body space, becoming acoelomate secondarily.

Protostomes Are Divided into Lophotrochozoans and Ecdysozoans

LEARNING OBJECTIVE 27.2.2 Contrast lophotrochozoans and ecdysozoans.

Two major clades of protostomes are recognized as having evolved independently since ancient times: the **lophotrochozoans,** named for their larval and feeding structures, and the **ecdysozoans,** named for their trait of periodic molting (or ecdysis) that accompanies growth of the organism.

Lophotrochozoa

Lophotrochozoan animals grow by gradual addition of mass to the body. Most live in water and propel themselves through it using cilia or contractions of the body musculature. One prominent group is the flatworms (phylum **Platyhelminthes**), animals with a simple body, no circulatory or respiratory system, but a complicated reproductive system. This group includes marine and freshwater planarians as well as the parasitic flukes and tapeworms.

Many lophotrochozoan groups have a type of free-living larva known as a trochophore, and some have a feeding structure termed a lophophore, a horseshoe-shaped crown of ciliated tentacles around the mouth, used in filter feeding (figure 27.6). The phyla characterized by a lophophore are Bryozoa and Brachiopoda. Lophophorate animals are sessile (anchored in place).

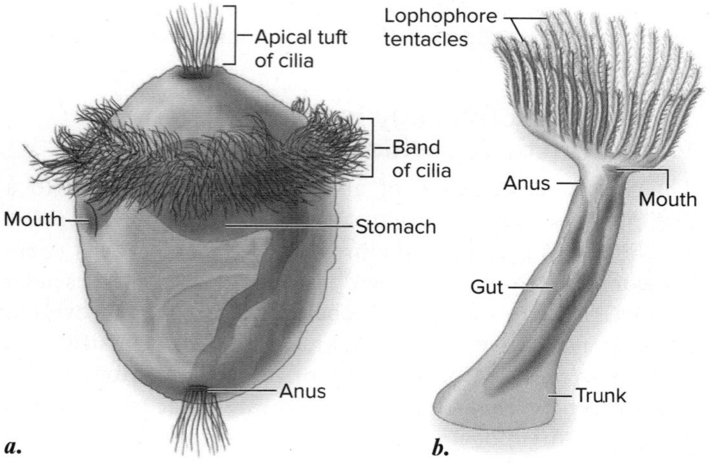

Figure 27.6 Trochophore (*a*) and lophophore (*b*).

Figure 27.7 A damselfly, a type of insect, molting.
TommyIX/Getty Images

Among the lophotrochozoans with a trochophore are phyla Mollusca and Annelida. Mollusks are unsegmented, with a reduced coelom. This phylum includes animals as diverse as octopuses, snails, and clams. Annelids are segmented coelomate worms, the most familiar of which is the earthworm, but also include leeches and the largely marine polychaetes.

Ecdysozoa

The other major clade of protostomes is the **Ecdysozoa.** Ecdysozoans are animals that molt, a phenomenon that seems to have evolved only once in the animal kingdom. When an animal grows large enough that it completely fills its hard external skeleton, it must lose that skeleton by molting, a process also called **ecdysis** (figure 27.7). While the animal grows, it forms a new exoskeleton underneath the existing one. In molting, the body first swells until the existing exoskeleton cracks open and is shed. After molting that skeleton, the animal inflates the soft, new one, expanding it using body fluids (and, in many insects and spiders, air as well). When the new one hardens, it is larger than the molted one was and has room for growth. Thus, rather than being continuous, as in other animals, the growth of ecdysozoans occurs in steps.

Two phyla not previously thought to be related, nematodes and arthropods, are both now assigned to the Ecdysozoa. Arthropoda contains the largest number of described species of any phylum. Both of these phyla contain one of the model organisms used in laboratory studies that have informed much of our current understanding of genetics and development: the fruit fly *Drosophila melanogaster* and the roundworm *Caenorhabditis elegans.*

REVIEW OF CONCEPT 27.2

Scientists have traditionally defined phyla based on tissues, symmetry, presence or absence of a coelom, and development stages. Molecular data have led to a reassessment of protostomes, contrasting lophotrochozoan organisms, whose body size simply increases, with ecdysozoans, which must molt in order to grow larger.

- *Why are phylogenies constructed with phenotypic characteristics like presence or absence of the coelom sometimes mistaken?*

27.3 True Tissue Evolved in Simple Animals

Systematists traditionally divided the kingdom Animalia (also termed Metazoa) into two main branches. Parazoa ("near animals") comprises animals that, for the most part, lack definite symmetry and do not possess tissues. These are the sponges, phylum Porifera. Because they are so different in so many ways from other animals, some scientists inferred that sponges were not closely related to other animals, which would mean that what we consider animals had two separate origins. Eumetazoa ("true animals") are animals that have a definite shape and symmetry. All have tissues, and most have organs and organ systems. Now most systematists agree that Parazoa and Eumetazoa are descended from a common ancestor whose closest living relative is the choanoflagellates. This implies that animal life had a single origin. In the past, phylogenies constructed with molecular data placed Parazoa at the base of the animal tree of life, but some recent studies indicate that ctenophores occupy that position.

Porifera Lack Symmetry and Specialized Tissue

LEARNING OBJECTIVE 27.3.1 Explain the function of choanocytes.

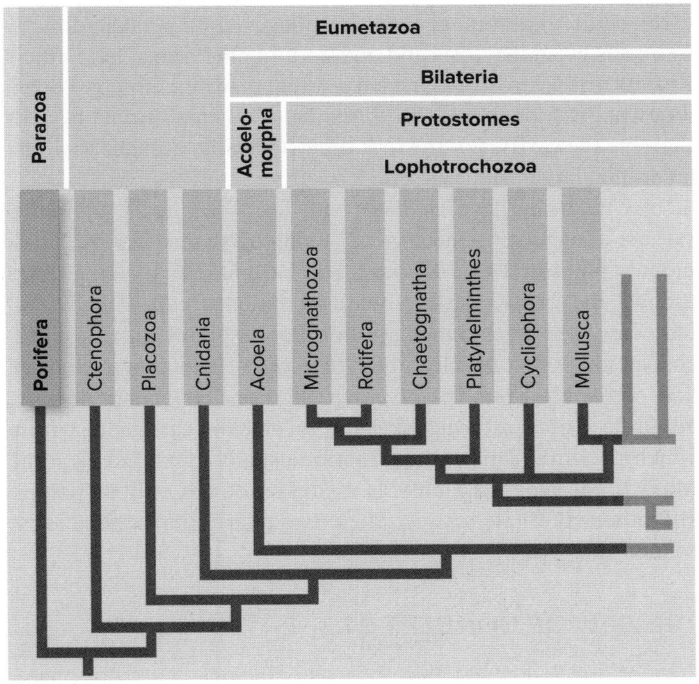

Sponges, members of the phylum **Porifera,** are the simplest animals. Most sponges are asymmetrical, and although some of their cells are highly specialized, they are not organized into tissues. However, sponge cells do possess a key property of animal cells: cell recognition. For example, when a sponge is passed through a fine silk mesh, individual cells separate and then reaggregate on the other side to re-form the sponge. Clumps of cells disassociated from a sponge can give rise to entirely new sponges. Molecular analysis has also shown that some of the molecules involved in cell adhesion exist in the sponge genome, although they do not form the specialized junctions seen in true tissue.

Nearly 7000 species of sponges live in the sea, and perhaps 150 species live in fresh water. Marine sponges occur at all depths and may be among the most abundant animals in the deepest part of the oceans. Although some sponges are small (no more than a few millimeters across), some may reach 2 m or more in diameter.

As is true of many marine invertebrate animals, larval sponges are free-swimming. After a sponge larva attaches to an appropriate surface, it metamorphoses into an adult and remains attached to that surface for the rest of its life. Thus, adult sponges are sessile; that is, they are anchored, immobile, on rocks or other submerged objects (figure 27.8*a*).

The cells that compose a sponge include a layer of choanocytes, a layer of epithelial cells, and amoeboid cells in the protein-rich matrix called **mesohyl** between the two layers. The body of the sponge is perforated by tiny holes. The name of the phylum, Porifera, refers to this system of pores. Unique flagellated cells called **choanocytes,** or collar cells, line the body cavity of the sponge (figure 27.8*b*). The beating of the flagella of the many choanocytes draws water in through the pores. Why all this moving of water? The sponge is a "filter feeder." The beating of each choanocyte's flagellum draws water through its collar, made of small, hairlike projections resembling a picket fence. Any food particles in the water, such as protists and tiny animals, are trapped in the fence. The choanocytes of sponges very closely resemble protist choanoflagellates (refer to chapter 24).

Whether a sponge should be considered colonial is an illustration of the limitations of human language. A colony of invertebrate animals, such as coral, is generally defined as a group of individuals that are physically connected (and may be physiologically connected as well), all having been produced by asexual reproduction (such as budding or dividing) from a single progenitor that arose by sexual reproduction. Nearly all sponges grow by multiplying the number of flagellated chambers connected to a single osculum, but whether these units can be considered "individuals" is debatable.

Cnidarians Possess Both Symmetry and True Tissue

LEARNING OBJECTIVE 27.3.2 Describe the features of cnidarians.

All animals other than sponges have both symmetry and tissues. Two taxa exhibit radial symmetry in which their bodies are organized around a central axis, like the petals of a daisy. Radial symmetry offers advantages to these animals, as their bodies—attached to a surface or free-floating—don't pass through the environment but, rather, interact with it on all sides. These two phyla are **Cnidaria** (pronounced ni-DAH-ree-ah) and **Ctenophora** (pronounced ten-NO-fo-rah). Cnidaria includes jellyfish, hydra (figure 27.9), corals, and sea anemones. Ctenophora is a minor phylum of comb jellies; although they resemble Cnidarians with their gelatinous, medusa-like form, they are

a.

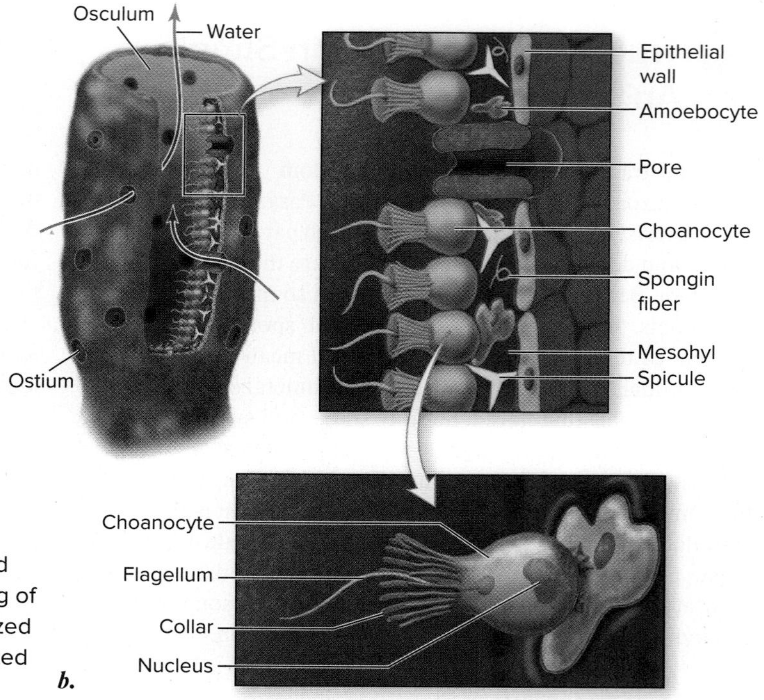

b.

Figure 27.8 Phylum Porifera: Sponges. ***a.*** *Aplysina longissima.* This beautiful bright orange and purple, elongated sponge is found on deep coral reefs. ***b.*** Diagrammatic drawing of the simplest type of sponge. The sponge body has no organized tissues, and most are not symmetrical. They do have specialized cells, illustrated here.

(a): Andrew J. Martinez/Science Source

structurally more complex. The bodies of all other animals with tissues are marked by a fundamental bilateral symmetry.

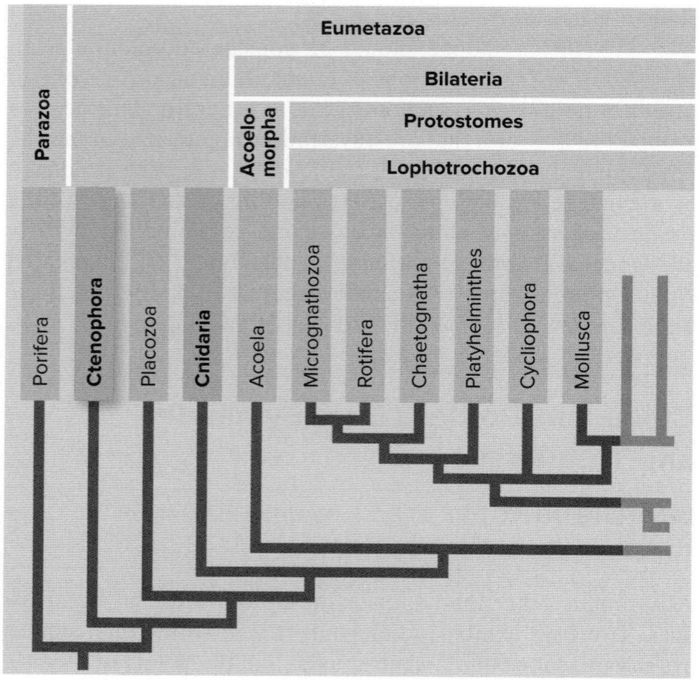

Cnidarians (phylum Cnidaria) are carnivores that capture prey such as fishes and shellfish with tentacles that ring their mouths. These tentacles, and sometimes the body surface, bear unique stinging cells called **nematocytes.** Within each nematocyte is a small but powerful harpoon called a **nematocyst,** which cnidarians use to spear their prey and then draw back the harpooned prey. The nematocyte builds up a very high internal osmotic pressure and uses it to push the nematocyst outward so explosively that the barb can penetrate the hard shell of a crab.

A major evolutionary innovation that arose in these phyla is extracellular digestion of food. In sponges, food is taken directly into cells by endocytosis and digested. In cnidarians, food enters a cavity, the gastrovascular cavity, where digestive enzymes break it down; the products of this digestion are then absorbed by cells that line the cavity. For the first time, it became possible to digest an animal larger than oneself.

Cnidarians have two basic body forms (figure 27.10). Medusae are free-floating, gelatinous, umbrella-shaped forms. Their mouths point downward, with a ring of tentacles hanging down around the edges. Medusae are commonly called "jellyfish" because of their gelatinous interior. Polyps are cylindrical, pipe-shaped forms that usually attach to a rock. *Hydra,* sea anemones, and corals are examples of polyps. For shelter and protection, corals deposit an external "skeleton" of calcium carbonate, within which they live. This is the structure usually identified as coral. Many cnidarians exist only as medusae, others only as polyps; still others alternate between these two phases during the course of their life cycles.

REVIEW OF CONCEPT 27.3

Sponges, phylum Porifera, possess multicellularity but have neither tissue-level development nor body symmetry. Sponge choanocyte cells have flagella that beat to circulate water through the sponge body. Members of phylum Cnidaria have both symmetry and tissues. They are carnivores, and their nematocysts are harpoons used in defense and prey capture.

- *What features of a sponge make it seem to be a colony, and what features make it seem to be a single organism?*

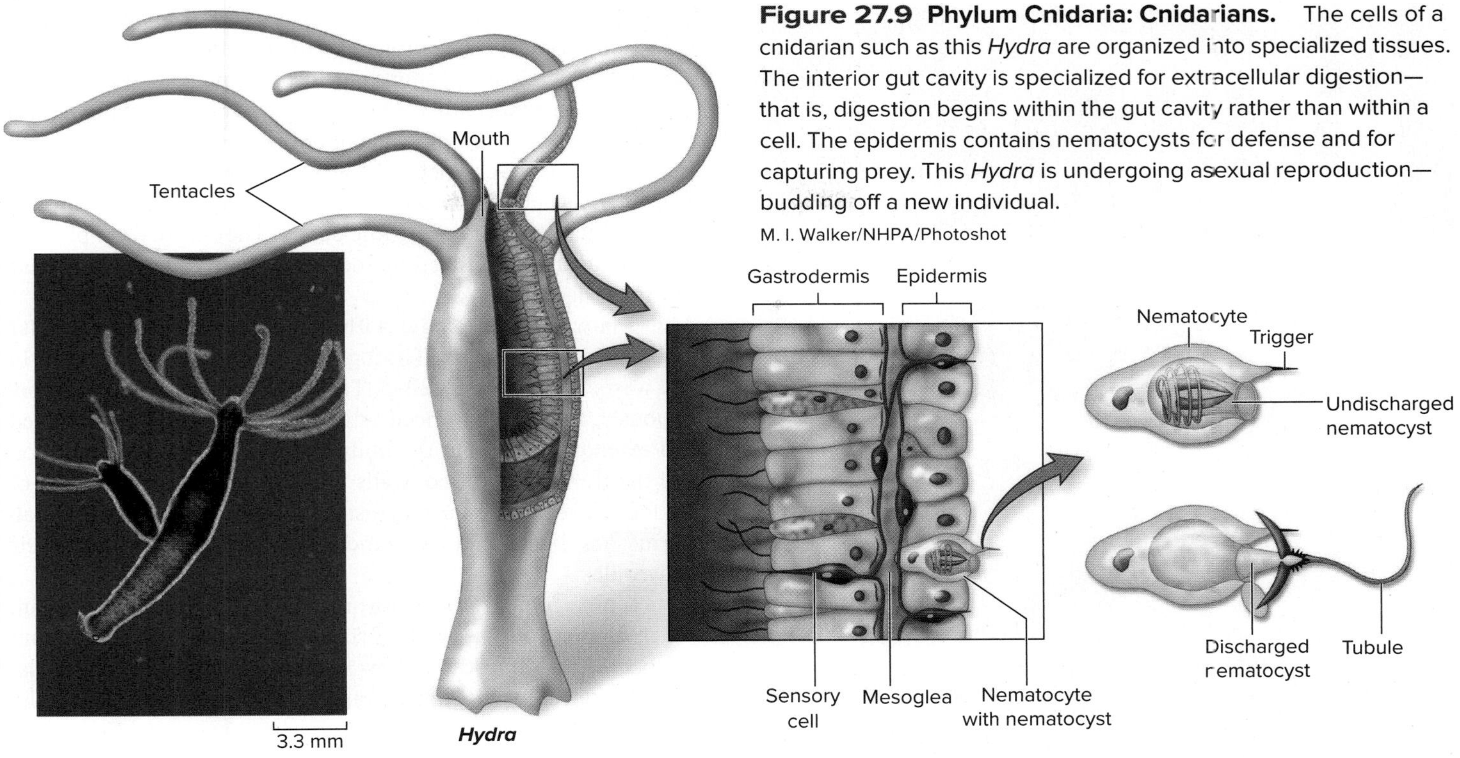

Figure 27.9 Phylum Cnidaria: Cnidarians. The cells of a cnidarian such as this *Hydra* are organized into specialized tissues. The interior gut cavity is specialized for extracellular digestion—that is, digestion begins within the gut cavity rather than within a cell. The epidermis contains nematocysts for defense and for capturing prey. This *Hydra* is undergoing asexual reproduction—budding off a new individual.

M. I. Walker/NHPA/Photoshot

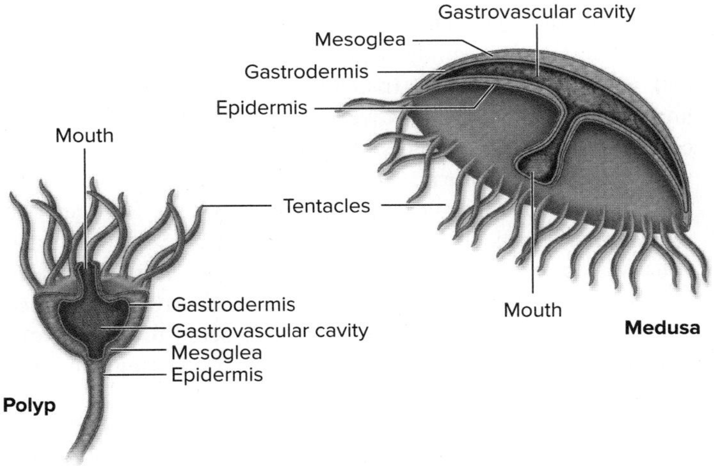

Figure 27.10 Two body forms of cnidarians: The polyp and the medusa.

27.4 Flatworms and Rotifers Are Very Simple Bilaterians

Bilateria is characterized by a key transition in the animal body plan—bilateral symmetry—which allowed animals to achieve high levels of internal body specialization, such as the concentration of sensory structures at the anterior. Bilateria is divided into two clades. One clade comprises the protostomes and deuterostomes, discussed in this chapter and in chapter 28; the other clade is the acoel flatworms, which are not discussed in detail.

Flatworms Are Bilaterally Symmetrical and Have Lost a True Coelom

LEARNING OBJECTIVE 27.4.1 Describe the features of platyhelminthes.

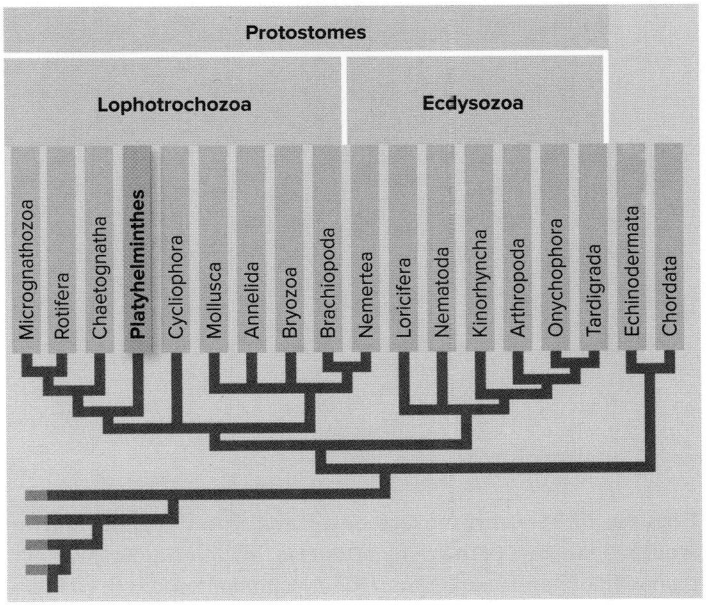

The phylum **Platyhelminthes** (pronounced plat-ee-hel-MIN-theeze) consists of some 20,000 species. These ciliated, soft-bodied animals are flattened dorsoventrally, the anatomy that gives them the name flatworms. Flatworm bodies are solid, aside from an incomplete digestive cavity (figure 27.11). Although among the

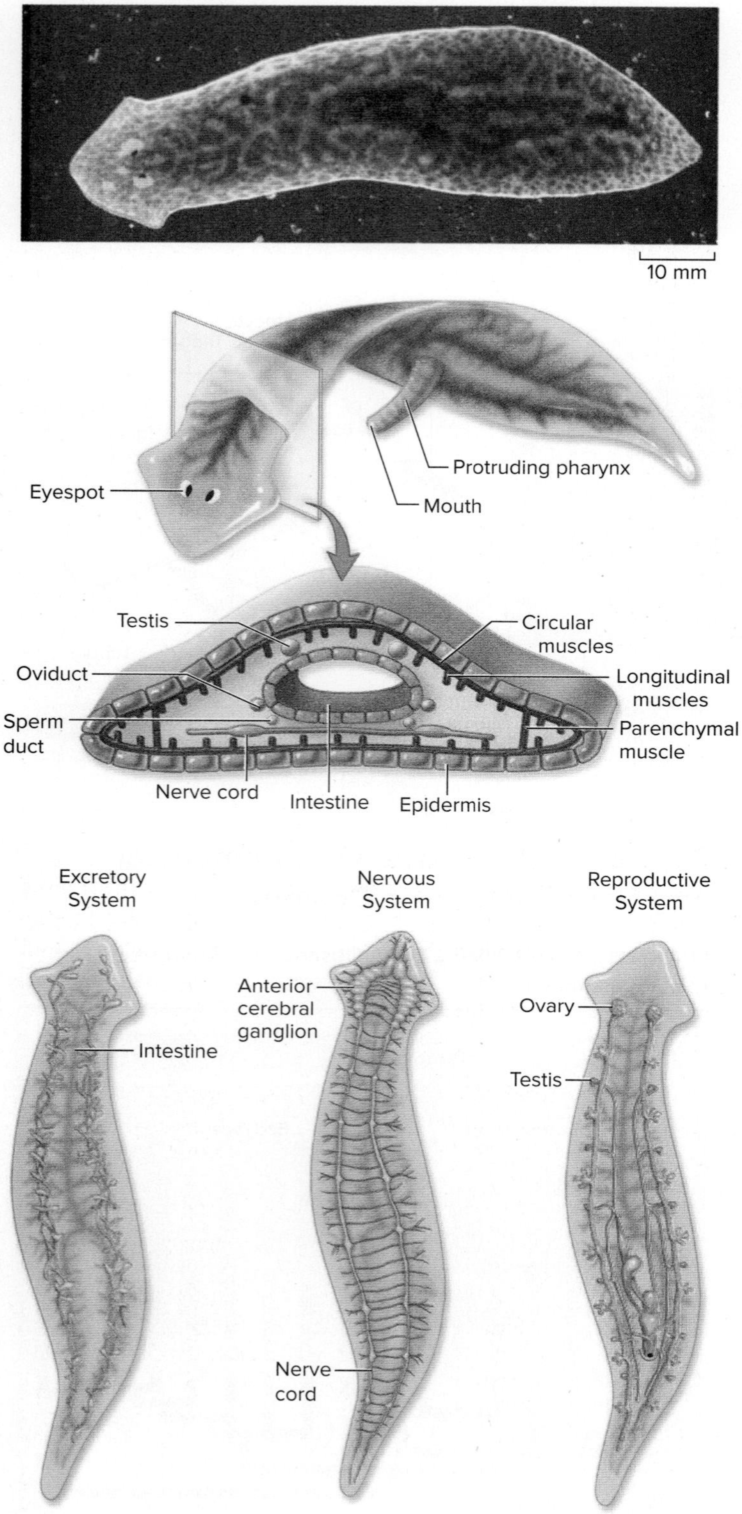
10 mm
Eyespot
Protruding pharynx
Mouth
Testis
Circular muscles
Oviduct
Longitudinal muscles
Sperm duct
Parenchymal muscle
Nerve cord
Intestine
Epidermis
Excretory System
Nervous System
Reproductive System
Anterior cerebral ganglion
Intestine
Ovary
Testis
Nerve cord

Figure 27.11 Architecture of a flatworm. A photo and an idealized diagram of the genus *Dugesia,* the familiar freshwater planarian of ponds and rivers. Upper schematic shows a whole animal and a transverse section through the anterior part of the body. Schematics below show the digestive, central nervous, and reproductive systems.

(a): M. I. Walker/NHPA/Photoshot

morphologically simplest of bilaterally symmetrical animals, they have some complex structures, like their reproductive apparatus, and they have the most complex life cycles among animals.

Most species of flatworms are parasitic, occurring within the bodies of many other kinds of animals. Other flatworms are free-living in a wide variety of marine and freshwater habitats, as well as moist places on land. Free-living flatworms eat various small animals and bits of organic debris. They move from place to place by means of ciliated epithelial cells concentrated on their ventral surfaces.

Flatworms have a gut with only one opening, a muscular tube called the pharynx. Through this opening, food is taken in and waste material is expelled. Thus, flatworms cannot feed continuously, as more advanced animals can. The gut is branched and extends throughout the body, functioning in both digestion and the transport of food. Cells that line the gut engulf most of the food particles by phagocytosis and digest them. Parasitic flatworms lack digestive systems and absorb their food through their body walls.

Unlike cnidarians, flatworms have an excretory system, which consists of a network of fine tubules that runs throughout the body. Cilia line the hollow centers of bulblike **flame cells,** which are located on the side branches of the tubules. Cilia in the flame cells move water and excretory substances into the tubules and then out through exit pores located between the epidermal cells. Flame cells were so named because of the flickering movements of the tuft of cilia within them. They play a major role in regulating the body's water balance.

Like sponges, cnidarians, and ctenophorans, flatworms lack a circulatory system. Instead, flatworms have thin bodies and highly branched digestive cavities, so that all flatworm cells are within diffusion distance of oxygen and food. The nervous system of flatworms is very simple: a longitudinal nerve cord that acts as a simple central nervous system. Between two longitudinal cords are cross-connections, so the flatworm nervous system resembles a ladder. Free-living flatworms also have eyespots on their heads. These are inverted, pigmented cups containing light-sensitive cells connected to the nervous system. The eyespots enable the worms to distinguish light from dark.

The reproductive systems of flatworms are complex. Most flatworms are hermaphroditic, with each individual containing both male and female sexual structures. In some parasitic flatworms, there is a complex succession of distinct larval forms. Some genera of flatworms are also capable of asexual regeneration; when a single individual is divided into two or more parts, each part can regenerate an entirely new flatworm.

Rotifers Are Tiny and Lack a True Coelom

LEARNING OBJECTIVE 27.4.2 Describe the features of rotifers.

Rotifers (phylum **Rotifera**) are bilaterally symmetrical, unsegmented pseudocoelomates (figure 27.12). Several features suggest that their ancestors may have resembled flatworms.

At 50 to 500 μm long, rotifers are smaller than some ciliate protists. But they have complex bodies with three cell layers, highly developed internal organs, and a complete gut. An extensive pseudocoelom acts as a hydrostatic skeleton; the cytoskeleton

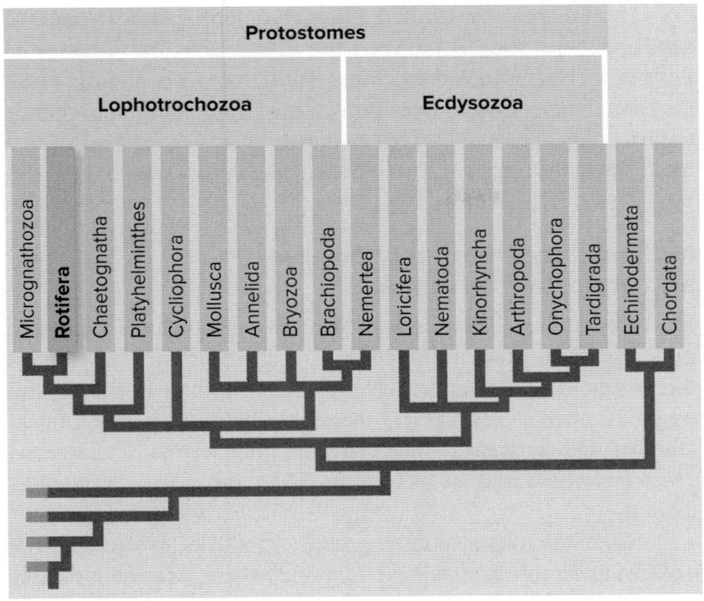

provides rigidity. A rotifer has a rigid external covering, but its body can lengthen and shorten greatly, because the posterior part is tapered, so it can fold up like a telescope. Many have adhesive toes, used for clinging to vegetation and other such objects.

Diversity and distribution

About 2,000 species are known, most of which occur in fresh water; a few rotifers live in soil, the capillary water in cushions of mosses, and the ocean. The life span of a rotifer is typically no longer than one or two weeks, but some species can survive in a desiccated, inactive state on the leaves of plants; when rain falls, the rotifers become active and feed in the film of water that temporarily covers the leaf.

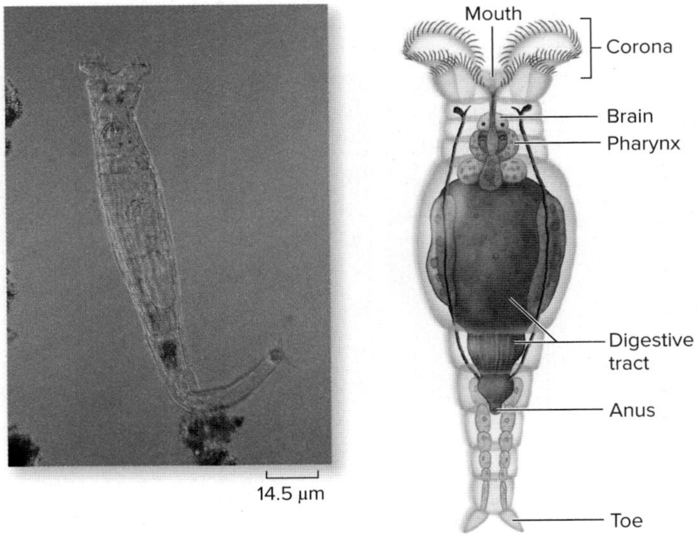

Figure 27.12 Phylum Rotifera. Microscopic in size (***a***), rotifers are smaller than some ciliate protists yet have complex internal organs (***b***).

(a): MELBA PHOTO AGENCY/Alamy Stock Photo

Food gathering

The corona, a conspicuous ring of cilia at the anterior end (figure 27.12), is the source of the common name "wheel animals" for rotifers, because the beating cilia make it appear that a wheel is rotating around the head of the animal. The corona is used for locomotion, but its cilia also sweep food into the rotifer's mouth. Once food is swallowed, it is crushed with a complex jaw in the pharynx.

REVIEW OF CONCEPT 27.4

Flatworms are compact, bilaterally symmetrical animals that lack a body cavity. Most have a blind digestive cavity with only one opening; they also have an excretory system consisting of flame cells within tubules that run throughout the body. Many are free-living, but some cause human diseases. Rotifers are extremely small but with highly complex body structure; the name "wheel animals" comes from the apparent motion of their beating cilia.

- *Does the anatomy of a tapeworm relate to its way of life?*

27.5 Mollusks and Annelids Are the Largest Groups of Lophotrochozoans

One of the realignments of protostome phylogeny was the finding that annelids, segmented worms, are more closely related to mollusks than to arthropods. Segmentation was once thought to join arthropods and annelids, but it now appears to have arisen independently in the two lineages.

Mollusks Are a Large and Diverse Group

LEARNING OBJECTIVE 27.5.1 Differentiate between different types of mollusks.

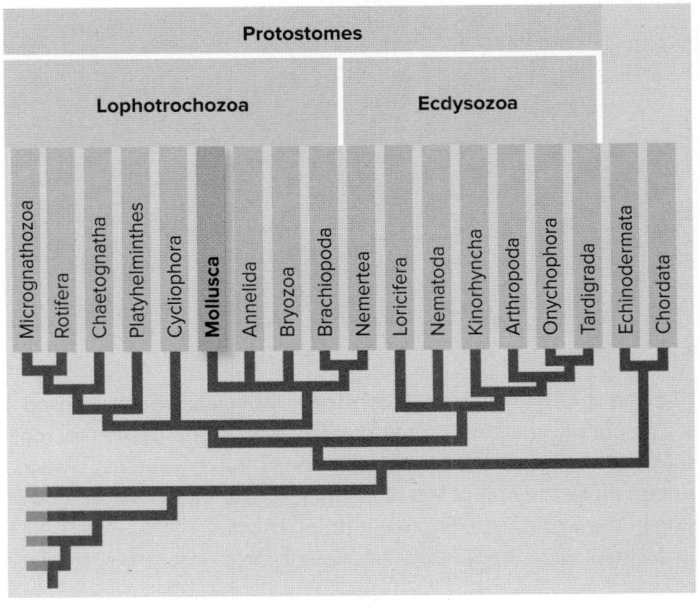

The mollusks (phylum **Mollusca**) are the second largest animal phylum, with over 110,000 species (figure 27.13). Mollusks, although mostly marine, are found nearly everywhere.

The mollusk body is composed of three distinct parts: a head-foot, a central section called the visceral mass that contains the body's organs, and a **mantle** (figure 27.14). The foot of a mollusk is muscular and may be adapted for locomotion, attachment, food capture, or various combinations of these functions. The mantle is a heavy fold of tissue wrapped around the visceral mass like a cape, with the respiratory organs (gills or lungs) positioned on its inner surface, like the lining of a coat.

There are seven or eight recognized classes of mollusks, the three most common of which are the gastropods (snails and slugs), bivalves (clams, oysters, scallops, and mussels), and cephalopods (octopuses, squids, and nautiluses; figure 27.13). All terrestrial mollusks are gastropods, in which the mantle secretes a hard shell. Bivalves secrete a two-part shell with a hinge, and they filter-feed by drawing water into their shell. Cephalopods have a large brain and a modified mantle cavity that creates a jet propulsion system.

Mollusks were among the first animals to develop an efficient excretory system. Tubular structures called **nephridia** (a type of kidney) gather wastes from the coelom to discharge into the mantle cavity. Mollusks also have a circulatory system, a network of vessels that carries fluids, oxygen, and food molecules to all parts of the body. The circulating fluid is usually pushed through the circulatory system by contraction of one or more muscular hearts.

One of the characteristic features of gastropods and cephalopods is the **radula,** a rasping, tonguelike organ. With rows of pointed, backward-curving teeth, the radula is used by some snails to scrape algae off rocks. The small holes seen in oyster shells are produced by gastropods that have bored holes to kill the oyster and extract its body.

Pearls are formed when a foreign object, such as a grain of sand, becomes lodged between the mantle and the inner shell layer of a bivalve, including clams and oysters. The mantle coats the foreign object with layer upon layer of shell material to reduce the irritation. The shell serves primarily as protection, with some mollusks withdrawing into their shells when threatened.

Mollusks are extremely diverse—and important to humans

Mollusks range in size from almost microscopic to huge. Although most measure a few millimeters to centimeters in their largest dimension, the giant squid may grow to more than 15 m long and weigh as much as 250 kg. It is therefore one of the heaviest invertebrates. Other large mollusks are the giant clams of the genus *Tridacna,* which may be as long as 1.5 m and weigh as much as 270 kg.

Like all major animal groups, mollusks evolved in the oceans, and most groups have remained there. Marine mollusks are widespread and many are abundant. Snails and slugs have invaded freshwater and terrestrial habitats, and freshwater mussels live in lakes and streams (the flat foot of a snail or slug allows it to crawl, but the foot of clams, mussels, and other bivalved mollusks is adapted to digging, so they cannot move about on land). Some places where terrestrial mollusks live, such as crevices of desert rocks, may appear dry, but if mollusks live there, the habitat has at least a temporary supply of water.

Mollusks—including oysters, clams, scallops, mussels, octopuses, and squids—are an important source of food for humans. They are also economically significant in other ways. For example, the material called mother-of-pearl (nacre), which is used for jewelry and other decorative objects, and formerly for buttons, comes from mollusk shells, most notably that of the abalone. Mollusks can also be pests. Bivalves called shipworms burrow through wood exposed to the sea, damaging boats, docks,

a. *b.* *c.* *d.*

Figure 27.13 Mollusk diversity. Mollusks exhibit a broad range of variation. ***a.*** The flame scallop, *Lima scabra,* is a filter feeder. ***b.*** The common octopus, *Octopus vulgaris,* is one of the few mollusks dangerous to humans. Strikingly beautiful, it is equipped with a sharp beak that can deliver a poisonous bite! ***c.*** Nautiluses, such as this chambered nautilus, *Nautilus pompilius,* have been around since before the age of the dinosaurs. ***d.*** The banana slug, *Ariolimax columbianus,* native to the Pacific Northwest, is the second largest slug in the world, attaining a length of 25 cm.

(a): Comstock Images/Getty Images; (b): Juniors Bildarchiv GmbH/Alamy Stock Photo; (c): Douglas Faulkner/Science Source; (d): Mike Anich/age fotostock/SuperStock

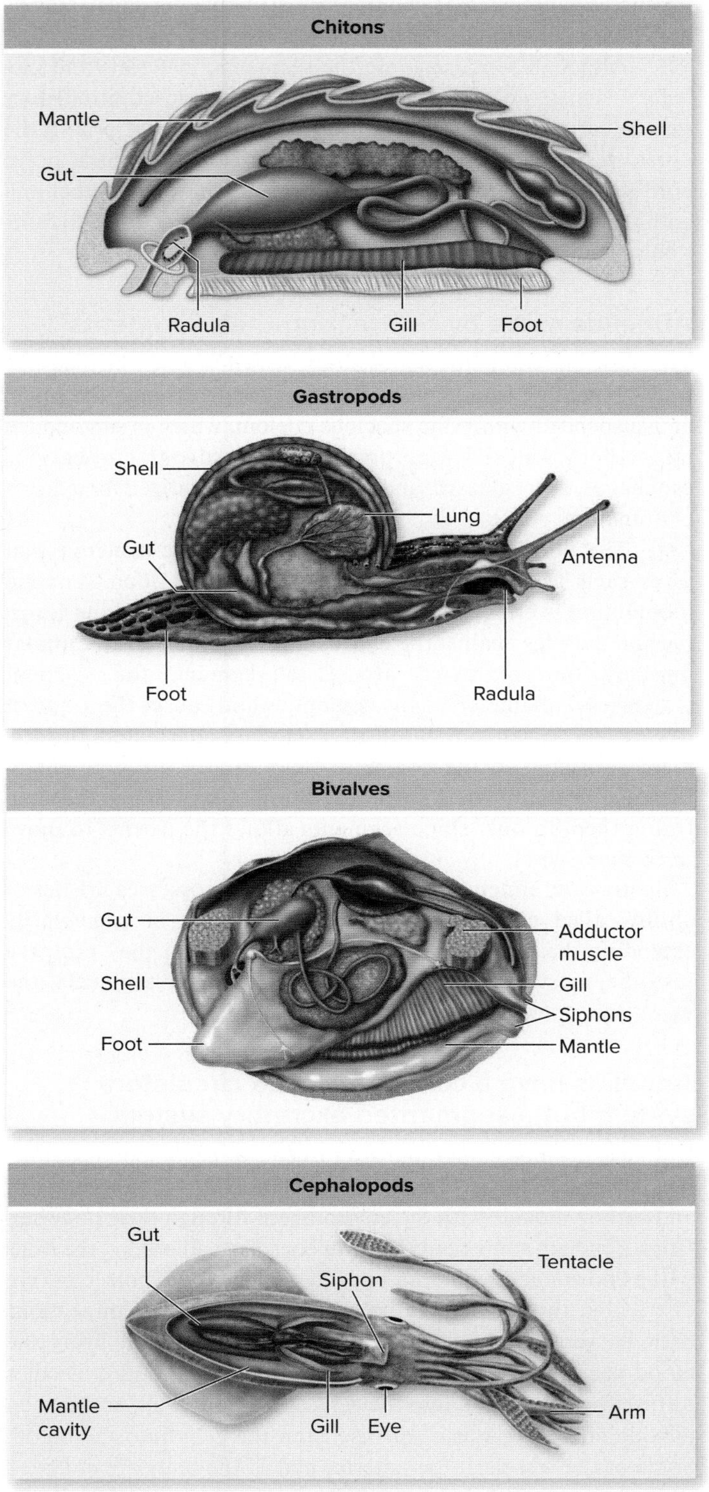
Chitons
Mantle
Gut
Shell
Radula
Gill
Foot
Gastropods
Shell
Lung
Gut
Antenna
Foot
Radula
Bivalves
Gut
Adductor muscle
Shell
Gill
Siphons
Foot
Mantle
Cephalopods
Gut
Tentacle
Siphon
Mantle cavity
Gill
Eye
Arm

Figure 27.14 Body plans of some mollusks.

and pilings. The zebra mussel *(Dreissena polymorpha)* has recently invaded many North American freshwater ecosystems. Many slugs and snails damage flowers, vegetable gardens, and crops. Other mollusks serve as hosts to the larval stages of many serious parasites.

The Annelid Body Is Composed of Ringlike Segments

LEARNING OBJECTIVE 27.5.2 Explain how circular and longitudinal muscles facilitate moving a segmented body.

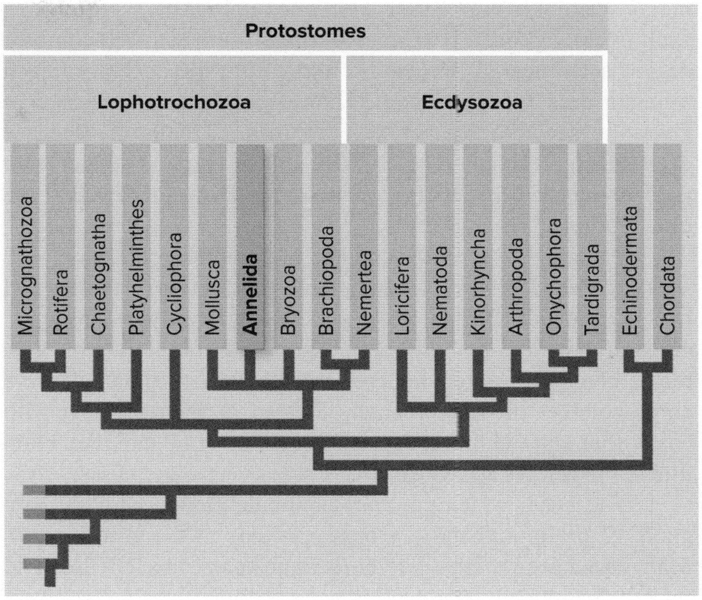
Protostomes
Lophotrochozoa
Ecdysozoa
Micrognathozoa
Rotifera
Chaetognatha
Platyhelminthes
Cycliophora
Mollusca
Annelida
Bryozoa
Brachiopoda
Nemertea
Loricifera
Nematoda
Kinorhyncha
Arthropoda
Onychophora
Tardigrada
Echinodermata
Chordata

Segmentation, the building of a body from a series of repeated units, has evolved multiple times. Worms of the phylum **Annelida** (figure 27.15) are segmented, with a body composed of repeating units resembling a stack of coins. One advantage of a segmented body is that the development and function of individual segments or groups of segments can differ. For example, some segments may be specialized for reproduction, whereas others are adapted for locomotion or excretion.

The majority of annelids are marine, but they can also be found in freshwater, and in soil. They range in length from

Figure 27.15 A polychaete annelid. *Nereis virens* is a wide-ranging, predatory, marine polychaete worm equipped with feathery parapodia for movement and respiration, as well as jaws for hunting. You may have purchased *Nereis* as fishing bait!
Derrick Alderman/Alamy Stock Photo

0.5 mm to the 3 m long Australian earthworm. The basic body plan of an annelid is a tube within a tube: the digestive tract is suspended within the coelom, which is itself a tube running from mouth to anus (figure 27.16). The body segments of an annelid are visible as a series of ringlike structures running the length of the body, looking like a stack of doughnuts. The segments are divided internally from one another by partitions. In each of the cylindrical segments, the excretory and locomotor organs are repeated. The body fluid within the coelom of each segment creates a hydrostatic (liquid-supported) skeleton that gives the segment rigidity, like an inflated balloon. Because each segment is separate, each is able to expand or contract independently. This lets the worm's body move in ways that are quite complex.

The anterior (front) segments of annelids contain the sensory organs. Elaborate eyes with lenses and retinas have evolved in some annelids. One anterior segment contains a well-developed cerebral ganglion, or brain. The digestive tract, circulatory system, and nervous system are connected between segments. Nerve cords connect the nerve centers located in each segment with each other and the brain (figure 27.16). The brain can then coordinate the worm's activities.

The head, which contains a well-developed cerebral ganglion, or brain, and sensory organs occurs at the anterior end. Many species have eyes, which in some species have lenses and retinas. Technically, the head is not a segment, nor is the posterior end of the worm, the pygidium. In embryonic development, the head and tail form first, and then segments form between them; if a worm is cut in pieces, generally only those parts containing either head or tail can regenerate the missing parts and the middle bits just die.

Internally, the segments are divided from one another by partitions called septa, just as bulkheads separate the compartments of a submarine. Each segment has a pair of excretory organs, a ganglion, and locomotory structure; in most marine annelids, each also has a set of reproductive organs.

Although septa separate the segments, materials and biological signals do pass between segments. A closed circulatory system carries blood the length of the animal, anteriorly in the dorsal vessel and posteriorly in the ventral one. A ventral nerve cord connects the ganglia in each segment with one another and with the brain. These neural connections allow the worm to function as a unified and coordinated organism.

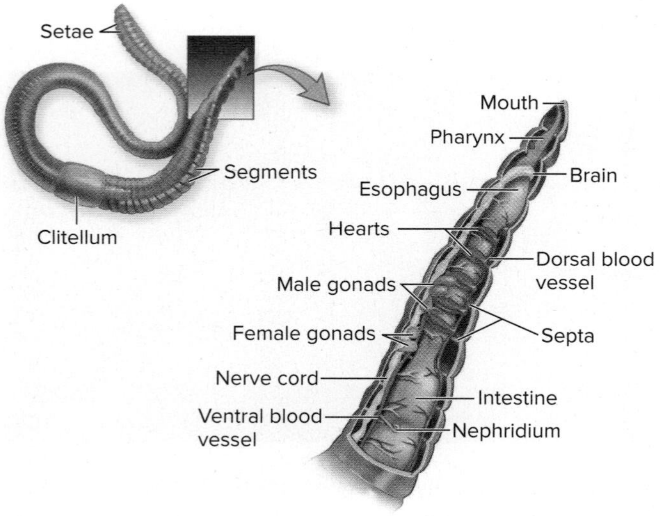

Setae
Segments
Clitellum
Mouth
Pharynx
Brain
Esophagus
Hearts
Dorsal blood vessel
Male gonads
Female gonads
Septa
Nerve cord
Intestine
Ventral blood vessel
Nephridium

Figure 27.16 Diagram of the body plan of an annelid. The earthworm body plan is based on repeated body segments. Segments are separated internally from each other by septa.

Annelids move by contracting their segments

The basic annelid body plan is a tube within a tube, the digestive tract—extending from mouth to anus—passing through the septa, and suspended within the spacious coelom, which is surrounded by the body wall. Each portion of the digestive tract—pharynx, esophagus, crop, gizzard, and intestine—is specialized for a different function.

The coelomic fluid creates a hydrostatic skeleton that gives each segment rigidity, like an inflated balloon. Annelid locomotion is effected by contraction of the circular and longitudinal muscles against the hydrostatic skeleton. When circular muscles are contracted around a segment, the segment decreases in diameter, so the coelomic fluid causes the segment to elongate. When longitudinal muscles are contracted, the segment shortens, so the coelomic fluid causes the segment to increase in diameter. Alternating these contractions and confining them to only some segments allows the worms to move in complex ways.

In most annelid groups, each segment possesses bristles of chitin called **chaetae** (or setae—singular, *seta* or *chaeta*). By extending the chaetae in some segments so that they protrude into the substrate and retracting them in other segments, the worm can extend its body, but not slip.

Annelids have a common closed circulatory system but a segmented excretory system

Unlike arthropods and mollusks (except for cephalopods), annelids have a closed circulatory system. Annelids exchange oxygen and carbon dioxide with the environment through their body surfaces, although some nonterrestrial ones have gills along the sides of the body or at the anterior end. Gases (and food molecules) are distributed throughout the body in blood vessels. Connections between ventral and dorsal vessels in each segment bring the blood near enough to each cell that oxygen and food molecules diffuse from the blood into the cells of the body wall, and carbon dioxide and other wastes diffuse from the cells into the blood. Some of the vessels at the anterior end of the body are enlarged and heavily muscular, serving as hearts that pump the blood.

The excretory system of annelids consists of ciliated, funnel-shaped nephridia like those of mollusks. Each segment has a pair of nephridia that collect wastes and transport them out of the body by way of excretory tubes. Some polychaetes have protonephridia like the flame cells of planarians.

Segmentation underlies the body organization of all complex coelomate animals, not only annelids but also arthropods (crustaceans, spiders, and insects) and chordates (lancelets, tunicates, and vertebrates, like you).

REVIEW OF CONCEPT 27.5

Mollusks and annelids are the largest groups of lophotrochozoans. There are three main categories of mollusks: gastropods, bivalves, and cephalopods. Although varied in appearance, these all have the same basic body design. Annelids are segmented worms. Each segment has its own excretory and locomotor elements; circular and longitudinal muscles in segments cause the body to extend and contract, respectively.

- *Why might a radula be unnecessary in a bivalve?*

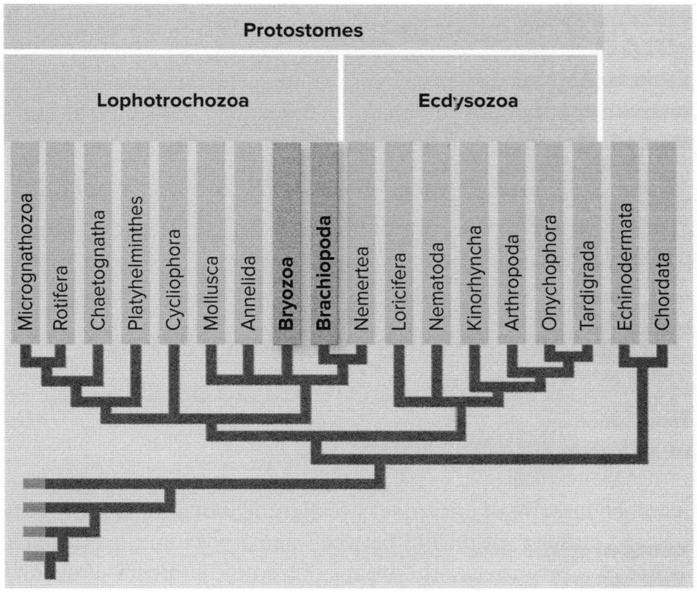

27.6 Lophophorates Are Very Simple Marine Organisms

Two phyla of mostly marine animals—**Bryozoa** and **Brachiopoda**—are characterized by a lophophore, a circular or U-shaped ridge around the mouth bearing one or two rows of ciliated tentacles into which the coelom extends. The lophophore functions as a surface for gas exchange, and the cilia of the lophophore guide the organic detritus and plankton on which the animal feeds to the mouth. Because of the lophophore, bryozoans and brachiopods have been considered related to one another, but recent evidence indicates that brachiopods and nemerteans are sister taxa, and whether Bryozoa is the sister taxon of this clade is uncertain. As a result, lophophores may have evolved convergently in brachiopods and bryozoans.

Bryozoans Are the Only Exclusively Colonial Animals

LEARNING OBJECTIVE 27.6.1 Describe how bryozoans obtain food.

Bryozoans are small—usually less than 0.5 mm long—and live in colonies that look like patches of moss on the surfaces of submerged objects (figure 27.17). Their common name, "moss-animals," is a direct translation of the Latin word *bryozoa*. The digestive system is U-shaped, with the anus opening near the mouth, as in many sessile animals.

The 4500 species of bryozoans include both marine and freshwater forms. Each individual bryozoan—a zooid—secretes a tiny, chitinous chamber called a zoecium (plural, *zoecia*), which is attached to rocks or other substrates such as the leaves of marine plants and algae. Calcium carbonate is deposited in the wall of a zoecium in many marine bryozoans, and in early geologic times bryozoans formed reefs just as corals do today. A zooid can divide or bud to asexually create another zooid beside the existing one, so one wall of the new zooid's zoecium is shared with that of the existing one; this expanding group of zoecia constitutes a colony. Individuals in the colony communicate chemically through pores between the zoecia. Not all zoecia of a colony are identical; some are specialized for functions such as feeding, reproduction, or defense.

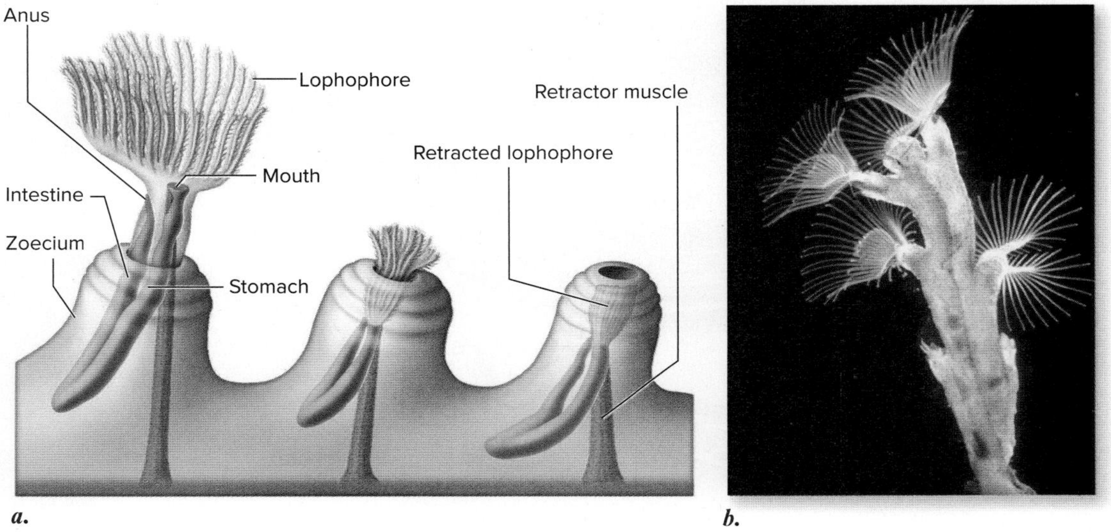

Figure 27.17 Bryozoans (phylum Bryozoa). ***a.*** This drawing depicts a small portion of a colony of the freshwater bryozoan genus *Plumatella,* which grows on rocks. The individual at the left has a fully extended lophophore. The tiny individuals disappear into the zoecium when disturbed. ***b.*** *Plumatella repens,* another freshwater bryozoan.

Hecker/Sauer/age fotostock

REVIEW OF CONCEPT 27.6

Brachiopods and bryozoans are primarily marine animals. Both have a lophophore for feeding. Bryozoans are colonial and have both marine and freshwater forms.

- *Bryozoans live attached to a surface. Why do we consider them animals and not plants?*

27.7 Nematodes and Arthropods Are Both Large Groups of Ecdysozoans

Another surprise of the modern realignment of animal phylogeny was the closer relationship between nematodes and arthropods. These two very successful groups, based on numbers of species and individuals, are both now recognized as clades within the ecdysozoans (molting animals).

Nematodes Consist of Many Different Kinds of Roundworm

LEARNING OBJECTIVE 27.7.1 Describe the characteristic features of nematodes.

Nematodes (phylum Nematoda) are bilaterally symmetrical, unsegmented worms (figure 27.18). They are covered by a flexible, thick cuticle, which they shed as they grow by molting. Their muscles constitute a layer beneath the epidermis and extend along

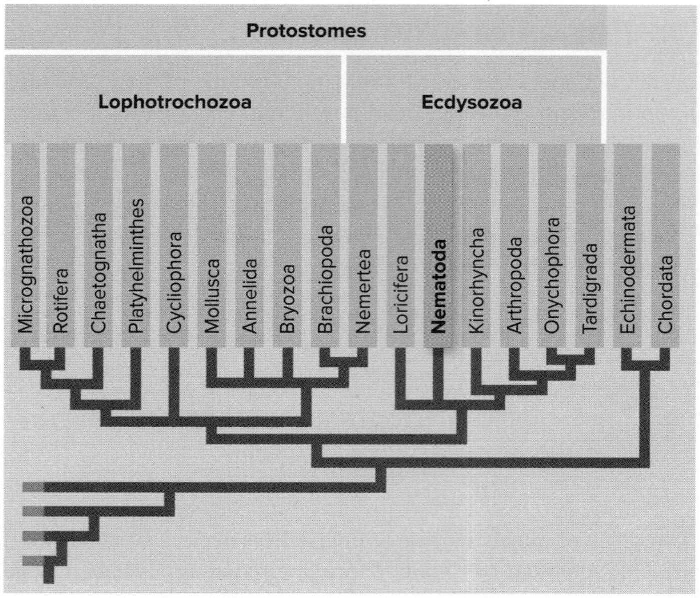

the length of the worm, rather than encircling its body. These longitudinal muscles attach to the outer layer of the body and pull against the cuticle and the pseudocoelom, which forms a type of fluid skeleton. When nematodes move, their bodies whip about from side to side.

Lacking specialized respiratory organs, nematodes exchange oxygen and carbon dioxide through their cuticles. Nematodes possess a well-developed digestive system and feed on a diversity of food sources. Near the mouth, at the anterior end, are hairlike sensory structures. The mouth may be equipped with piercing organs called stylets. Food passes into the mouth as a result of the sucking action produced by the rhythmic contraction

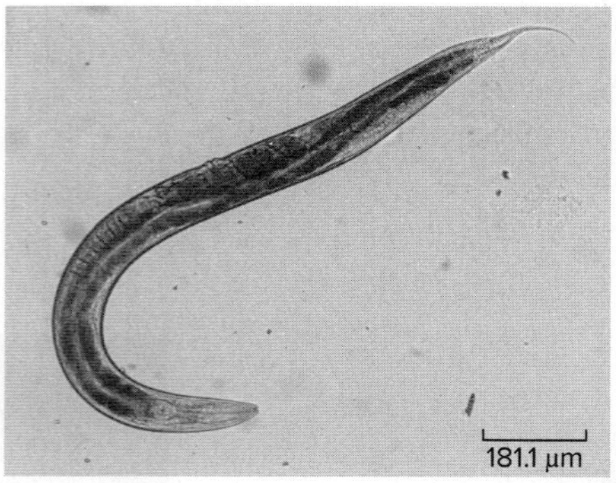

Figure 27.18 Phylum Nematoda: Roundworms. Roundworms such as this male nematode possess a pseudocoelum between the gut and the body wall. It allows nutrients to circulate throughout the body and prevents organs from being deformed by muscle movements.

lostkabab/Shutterstock

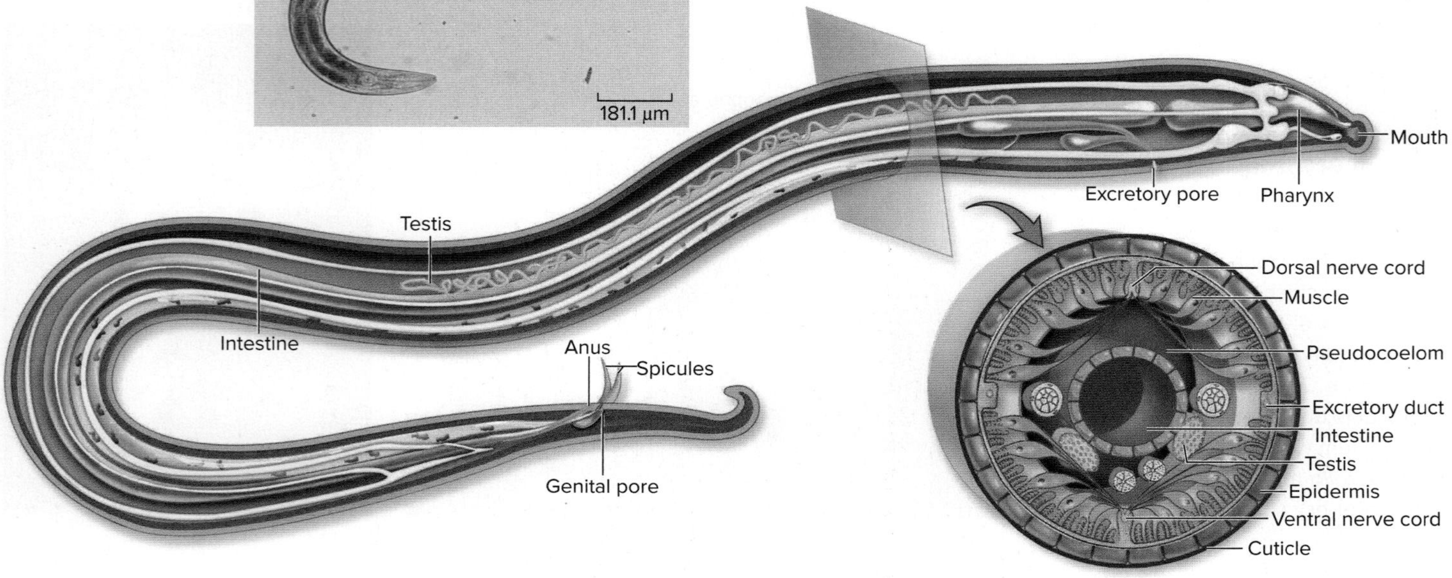

of a muscular pharynx and continues through the intestine (figure 27.18). Some of the water with which the food has been mixed is reabsorbed near the end of the digestive tract, and material that has not been digested is eliminated through the anus.

Nematodes completely lack flagella or cilia, even on sperm cells. Reproduction in nematodes is sexual, with sexes usually separate. Their development is simple, and the adults consist of very few cells. For this reason, nematodes have become extremely important subjects for genetic and developmental studies. The 1-mm-long *Caenorhabditis elegans* matures in only three days, its body is transparent, and it has only 959 cells. It was the first animal whose complete developmental cellular anatomy was known, and is the first animal whose genome (97 million DNA bases encoding over 21,000 different genes) has been fully sequenced.

Nematode-caused human diseases

About 50 species of nematodes, including several that are rather common in the United States, regularly parasitize human beings. Hookworms, most of the genus *Necator,* can be common in southern states. By sucking blood through a person's intestinal wall, they can produce anemia.

The most serious and common nematode-caused disease in temperate regions is trichinosis. Worms of the genus *Trichinella* live in the small intestine of some mammals, especially pigs and bears. Fertilized female nematodes burrow through the intestinal wall and release as many as 1500 live young. These are transported through the lymphatic system to muscles throughout the body, where they mature to form highly resistant cysts. Eating undercooked pork or bear carrying these cysts transmits the worm to humans. Fatal infections are rare. Trichinosis used to be common in the United States, but now, only 11 cases are reported, on average, every year.

It is estimated that pinworms, *Enterobius vermicularis,* infect about 30% of children and 16% of adults in the United States. Adult pinworms live in the human rectum, where they usually cause nothing more serious than itching of the anus; large numbers, however, can lead to prolapse of the rectum. The worms can easily be killed with medication.

Some nematode-caused diseases are extremely serious in the tropics. Filariasis is caused by several species of nematodes that infect at least 250 million people worldwide. Filarial worms of some species live in the circulatory system. The larval filarial worms are transmitted by an intermediate host, typically a bloodsucking insect such as a mosquito.

Arthropods Are the Most Successful of All Animals

LEARNING OBJECTIVE 27.7.2 List the four classes of insects.

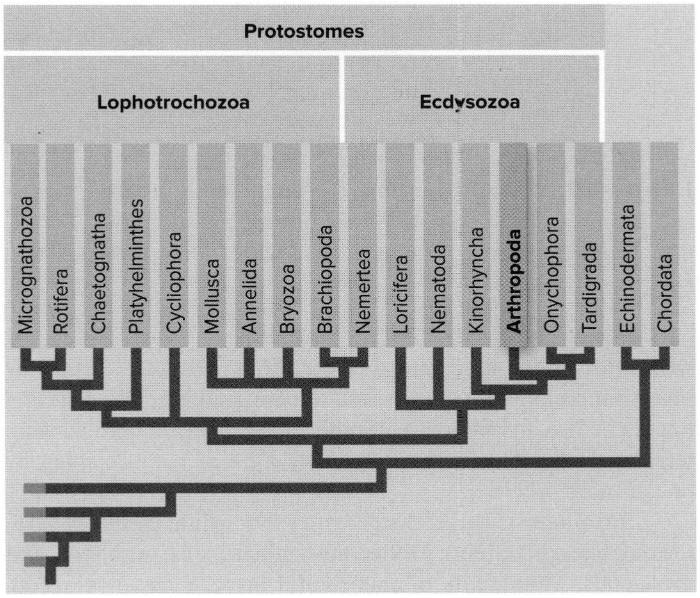

Arthropods are by far the most successful of all animals (table 27.2). Well over 1 million species—about two-thirds of all the named species on Earth—are members of the phylum Arthropoda (figure 27.19). About 200 million individual insects are alive at any time for each human! Insects and other arthropods occur, usually in great numbers, in every habitat on the planet, but there are few marine insects. Most members of the phylum are small, generally a few millimeters in length, but adults range in size from about 80 μm long (some parasitic mites) to 3 m across (Japanese spider crabs).

Arthropods are of enormous economic importance, affecting all aspects of human life. They pollinate crops and are valuable as food for humans and other animals but they also compete

TABLE 27.2 Major Groups of the Phylum Arthropoda

Class	Characteristics	Members
Chelicerata	Anterior appendages (chelicera) specialized as pincers or fangs.	Spiders, mites, ticks, scorpions, daddy long-legs, horseshoe crabs
Crustacea	Mouthparts are mandibles; appendages are biramous ("two-branched"); the head has two pairs of antennae.	Lobsters, crabs, shrimps, isopods, barnacles
Hexapoda	Mouthparts are mandibles; the body consists of three regions: a head with one pair of antennae, a thorax, and an abdomen; appendages are uniramous.	Insects (beetles, bees, flies, grasshoppers, butterflies, termites), springtails
Myriapoda	Mouthparts are mandibles; the body consists of a head with one pair of antennae, and numerous segments, each bearing paired uniramous appendages.	Centipedes, millipedes

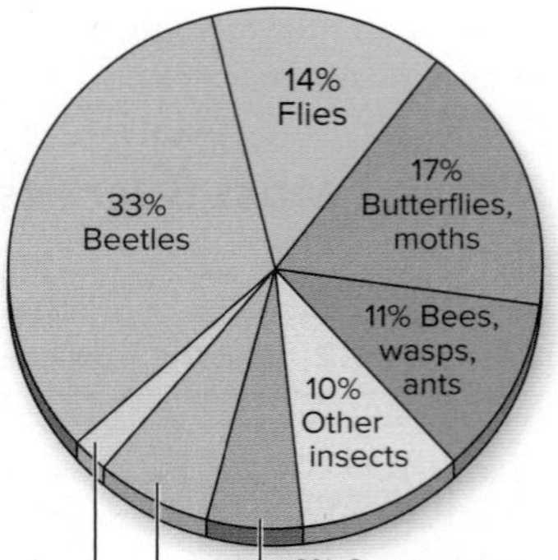

Figure 27.19 Arthropods are a successful group. About two-thirds of all named species are arthropods. About 80% of all arthropods are insects, and about half of the named species of insects are beetles.

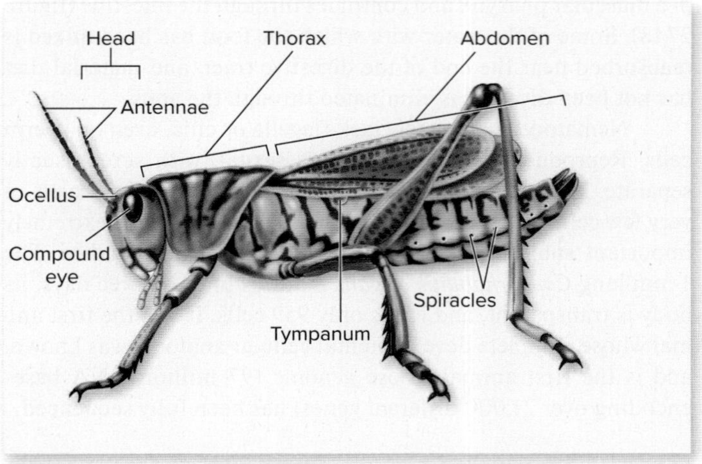

a.

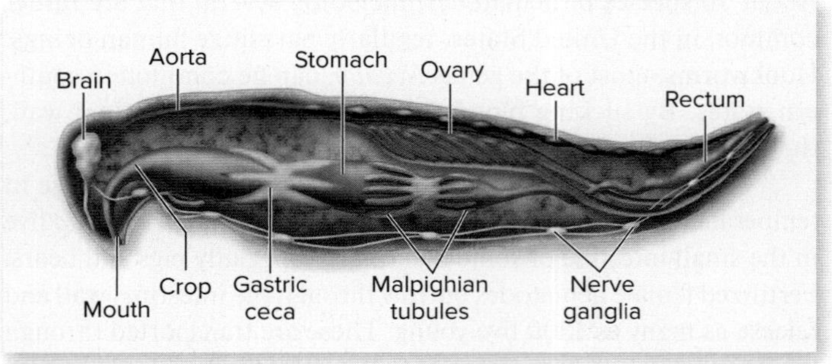

b.

Figure 27.20 A grasshopper (order Orthoptera). This grasshopper illustrates the major structural features of the insects, the arthropod group with the greatest number of species. ***a.*** External anatomy. ***b.*** Internal anatomy.

with humans for food and damage crops. Diseases spread by insects and ticks strike every kind of plant and animal, including human beings. Insects are by far the most important herbivores in terrestrial ecosystems: virtually every kind of plant is eaten by one or more species.

Taxonomists currently recognize four extant classes (a fifth, the trilobites, is extinct): chelicerates, crustaceans, hexapods, and myriapods. Mouthparts of chelicerates are chelicerae (pincers), whereas those of the other three classes are **mandibles** (biting jaws). Mandibles are inferred to have arisen (probably from a pair of limbs) in the common ancestor of crustaceans, hexapods, and myriapods, which means that these groups are more closely related to one another than any of them is to chelicerates.

The Arthropod Body Exhibits Three Key Features

LEARNING OBJECTIVE 27.7.3 Explain the advantages and disadvantages of an exoskeleton.

Arthropods exhibit a segmented body with a rigid exoskeleton and jointed appendages. They have an open circulatory system with a longitudinal muscular heart near the dorsal surface (figure 27.20). The nervous system consists of a segmented ganglion along the animal's ventral surface with fused dorsal ganglia forming a brain. The segmented nature of the system allows ganglia in each segment to control much of the animal's activity.

The advantages of segmentation were discussed in section 27.1. A hard exoskeleton confers protection against predators but also restricts motion. Joints in the appendages maintain protection while providing some flexibility. With this system, arthropods have developed many efficient modes of locomotion, both in the oceans and on land.

Segmentation

In members of some classes of arthropods, many body segments look alike. In others, the segments are specialized into functional groups, or **tagmata** (singular, *tagma*), such as the head, thorax, and abdomen of an insect (figure 27.20). The fusion of segments, known as tagmatization, is of central importance in the evolution of arthropods. Typically, the segments can be distinguished during larval development, but fusion in development obliterates them. All arthropods have a distinct head; in many crustaceans and chelicerates, head and thorax fuse to form the cephalothorax, or prosoma.

An exoskeleton

The tough external skeleton, or exoskeleton, is made of chitin and protein. In most animals, the skeleton provides a surface for muscle attachment, support for the body, and protection against physical forces. The arthropod exoskeleton protects against water loss, which was a powerful advantage in insects colonizing land. Chitin is chemically similar to cellulose, the dominant structural component of plants, and shares with it properties of toughness and flexibility.

An exoskeleton has inherent limitations, however. As arthropods increase in size, their exoskeletons must get disproportionately thick to bear the pull of the muscles. If beetles were as large as eagles, or crabs the size of cows, the exoskeleton would be so thick that the animal would be unable to move its great weight. Most terrestrial arthropods weigh only a few grams, but aquatic ones can be heavier, because water, being denser than air, provides more support. Another limitation is that an exoskeleton requires arthropods to periodically undergo ecdysis. The anterior and posterior regions of the digestive tract as well as the compound eyes are covered with cuticle and therefore are also shed at ecdysis. The animal is especially vulnerable during molting while the exoskeleton is soft.

Jointed appendages

The name *arthropod* means "jointed feet"; all arthropods have jointed appendages. Appendages may be modified into antennae, mouthparts of various kinds, or legs.

One advantage of jointed appendages is that they can be extended and retracted by bending. Imagine how difficult life would be if your arms and legs could not bend. In addition, joints serve as a fulcrum, or stable point, making leverage possible. A small muscle force on a lever can produce a large movement, just as extending your lower arm takes advantage of the fulcrum of the elbow. A small contraction distance in your muscles moves your hand through a large arc.

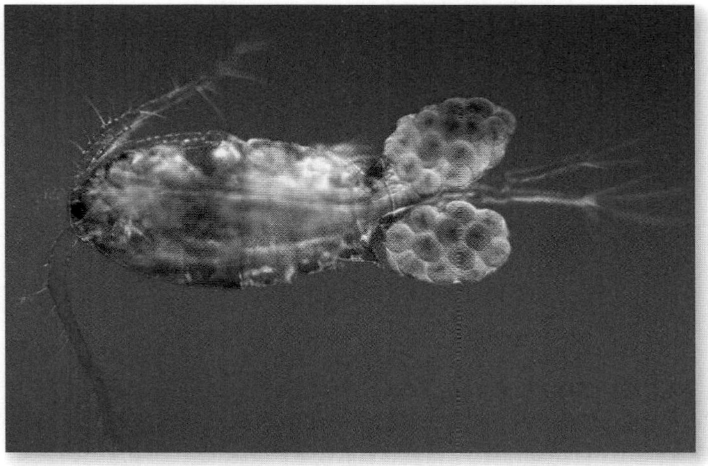

Figure 27.22 Freshwater crustacean. A copepod with attached eggs. The order Copepoda is an important component of plankton. Most are a few millimeters long.
Roland Birke/Getty Images

There Are Four Classes of Arthropods

LEARNING OBJECTIVE 27.7.4 Compare and contrast the four classes of arthropods.

Class Chelicerata: Spiders, mites, and ticks

In the class Chelicerata, with some 70,000 species, the most anterior appendages, called **chelicerae** (singular, *chelicera*), may function as fangs or pincers. The body of a chelicerate is divided into two tagmata: the anterior prosoma, which bears all the appendages, and the posterior opisthosoma, which contains the reproductive organs. Chelicerates include familiar, largely terrestrial arthropods, such as 35,000 named species of spiders (figure 27.21) and 35,000 species of ticks and mites, as well as scorpions and daddy long-legs. Although most live on land, 4000 known species of mite and 1 species of spider live in freshwater habitats, and a few mites live in the sea. Exclusively marine groups of chelicerates are horseshoe crabs and sea spiders. In addition to a pair of chelicerae, a chelicerate has four pairs of walking legs. Most chelicerates are carnivorous, but mites are largely herbivorous. Aside from the daddy long-legs, which can ingest small particles, most cannot consume solid food. They subsist on liquids, including solid food, which they liquefy by injecting with digestive enzymes and then suck up with the muscular pharynx.

Class Crustacea: Crabs, shrimps, and lobsters

The crustaceans (class Crustacea) comprise some 35,000 species of largely marine organisms, such as crabs, shrimps, lobsters, and barnacles. However, some groups, such as crayfish, occur in fresh water, and some crabs and copepods (figure 27.22) are among the most abundant multicellular organisms on Earth. Only a small number are terrestrial, including pillbugs and some sand fleas. Some crustaceans (such as lobsters and crayfish) are valued as food for humans; planktonic crustaceans (such as krill) are the primary food of baleen whales and many smaller marine animals.

A typical crustacean has three tagmata; the anteriormost two—the cephalon and thorax—may fuse to form the cephalothorax (figure 27.23). Most crustaceans have two pairs of antennae, three pairs of appendages for chewing and manipulating food, and various pairs of legs. Crustacean appendages, with the possible exception of the first pair of antennae, are biramous ("two-branched"). Crustaceans differ from hexapods, but resemble myriopods, in

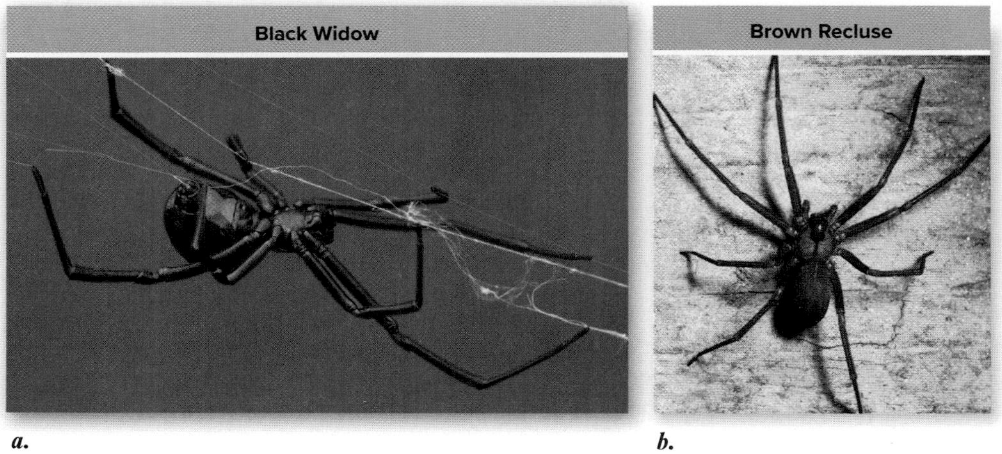

Figure 27.21 Two common venomous spiders. ***a.*** The southern black widow, *Latrodectus mactans*. ***b.*** The brown recluse, *Loxosceles reclusa*. Both species are common throughout temperate and subtropical North America.
(a): Mark Kostich/E+/Getty Images
(b): S. Camazine/K. Visscher/Science Source

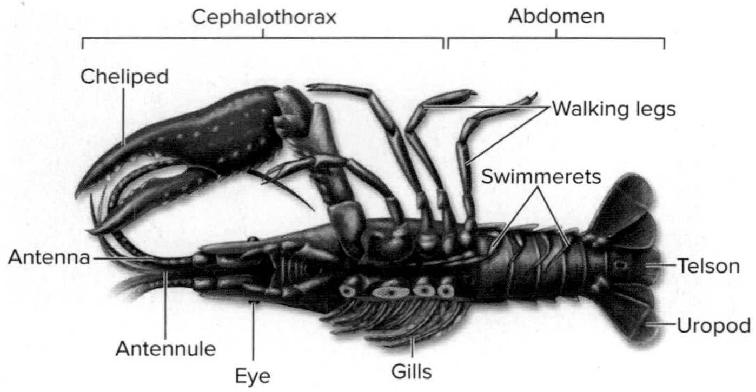

Figure 27.23 Decapod crustacean. Ventral view of a lobster, *Homarus americanus,* with some of its principal features labeled.

having appendages on their abdomen as well as their thorax. They are the only arthropods with two pairs of antennae.

Large crustaceans have feathery gills for respiration near the bases of their legs (figure 27.23). Oxygen extracted from the gills is distributed through the circulatory system. In smaller crustaceans, gas exchange takes place directly through the thinner areas of the cuticle or the entire body.

The **nauplius** larva is characteristic of Crustacea, providing evidence that all members of this diverse group descended from a common ancestor that had a nauplius in its life cycle. The sessile barnacles, with their shell-like exoskeleton, had been thought to be related to mollusks until they were discovered to have a nauplius larva.

Class Hexapoda: Insects

The insects, members of class Hexapoda, have six legs, as their name implies—and wings. Insects are by far the largest group of animals on Earth, in terms of number of species and number of individuals. Insects live in every habitat on land and in fresh water, but very few have invaded the sea. More than half of all named animal species are insects, and the actual proportion may be higher, because millions of forms await discovery, classification, and naming.

Approximately 90,000 described species of insects occur in the United States and Canada alone; the actual number probably approaches 125,000. Many suburban gardens may have 1500 or more species. A typical single hectare of lowland tropical forest is estimated to be inhabited by as many as 41,000 species of insects! Approximately a billion billion (10^{18}) individual insects are alive at any one time. A glimpse into the enormous diversity of insects is presented in figure 27.24.

During the course of their development, many insects undergo **metamorphosis.** For those such as grasshoppers, in which immature individuals are quite similar to adults, a series of molts results in an individual gradually getting bigger and more developed; this is termed simple metamorphosis. Those such as moths and butterflies have a wormlike larval stage, a resting stage called a pupa or chrysalis, during which metamorphosis occurs, and then a final molt into the adult form or imago; this is termed complete metamorphosis.

Class Myriapoda: Centipedes and millipedes

The body of a centipede (subclass Chilopoda) and millipede (subclass Diplopoda) consists of a head region posterior to which are numerous, more or less similar segments. Nearly all segments of a centipede have one pair of appendages, and nearly all segments of a millipede have two pairs of appendages. Each segment of a millipede is a simple tagma derived evolutionarily from two ancestral segments, which explains why millipedes have twice as many legs per segment as centipedes. Although the name *centipede* implies an animal with 100 legs and the name *millipede* one with 1000, adult centipedes usually have far fewer than 100 legs (most have 15, 21, or 23 pairs), and most adult millipedes have 100 or fewer.

Centipedes, with some 3000 species known, are carnivorous, feeding mainly on insects. The appendages of the first trunk segment are modified into a pair of venom-delivering fangs. The venom may be toxic to humans, and although extremely painful, centipede bites are never fatal. In contrast, most millipedes are herbivores, feeding mainly on decaying vegetation such as leaf litter and rotting logs. Many millipedes can roll their bodies into a flat coil or sphere to defend themselves. More than 12,000 species of millipedes have been named.

Figure 27.24 Insect diversity. Although there are around 30 insect orders, most of the described species are in only four. Examples of three of these are shown. ***a.*** Luna moth, *Actias luna* ***b.*** Robber fly, *Efferia sp.* ***c.*** European rhinoceros beetle, *Oryctes nasicornis*. Not pictured is the order Hymenoptera, which includes social insects such as bees and wasps.

(a): Cleveland P. Hickman; (b): nofilm2011/Shutterstock; (c): Horia Bogdan/Shutterstock

REVIEW OF CONCEPT 27.7

Nematodes and arthropods are the two largest groups of ecdysozoans. Nematodes are unsegmented worms that have longitudinal, not circular, muscle. Some nematodes are parasites of animals and plants. Arthropods are segmented animals with exoskeletons and jointed appendages. The four living classes are Chelicerata, Crustacea, Hexapoda (insects), and Myriopoda. These are distinguished by differences in mouthparts, segmental fusions, and numbers of appendages.

- *What would explain why the largest arthropods are found in marine environments?*

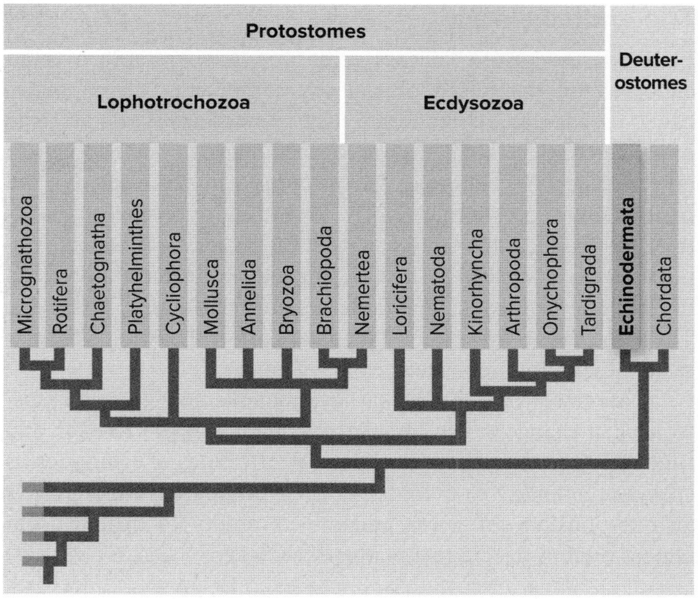

27.8 Deuterostomes Are Composed of Echinoderms and Chordates

Deuterostomes consist of fewer phyla and species than protostomes and are more uniform in many ways, despite great differences in appearance. Echinoderms such as sea stars, and chordates such as humans, share a mode of development that is evidence of their evolution from a common ancestor and that separates them clearly from protostomes.

Echinoderms Are Ancient and Unmistakable

LEARNING OBJECTIVE 27.8.1 Describe the basic body plan for echinoderms.

Members of the exclusively marine phylum Echinodermata are characterized by deuterostome development and an endoskeleton. The endoskeleton is composed of hard, calcium carbonate plates that lie just beneath the delicate skin. The term *echinoderm,* which means "spiny skin," refers to the spines or bumps that occur on these plates in many species. Echinoderms include sea stars, sea cucumbers, sea urchins and sand dollars, sea lilies and feather stars, and brittle stars (figure 27.25). Although an excellent fossil record extends back into the Cambrian period, the origin of echinoderms remains unclear. They are thought to have evolved from bilaterally symmetrical ancestors, because echinoderm larvae are bilaterally symmetrical.

In many echinoderms, the oral surface faces the substratum, although in sea cucumbers the animal's axis is horizontal, so the animal crawls oral surface foremost, and in crinoids (sea lilies and feather stars) the oral surface is located opposite the substrate.

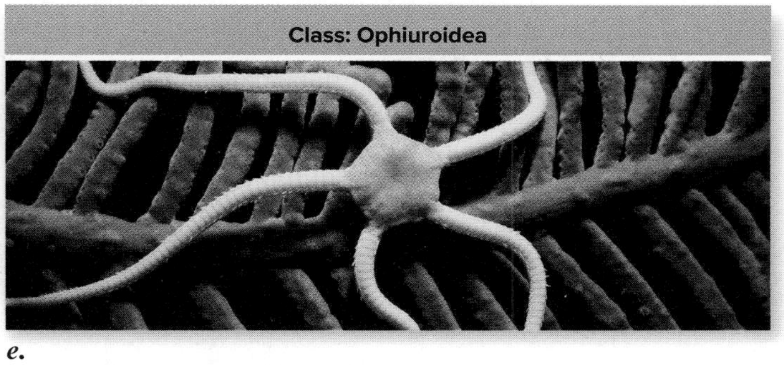

Figure 27.25 Diversity in echinoderms. ***a.*** Sea star, *Pisaster ochraceus.* ***b.*** Warty sea cucumber, *Parastichopus parvimensis,* Philippines. ***c.*** Sea urchin of the genus *Heterocentrotus,* Papahanaumokuakea Marine National Monument, Hawaii. ***d.*** Orange feather star, *Cenolia trichoptera,* from Indonesia. ***e.*** Gaudy brittle star, *Ophioderma ensiferum,* Grand Turk Island, Caribbean Sea.

(a): NOAA National Estuarine Research Reserve Collection; (b): Randy Morse/GoldenStateImages.com; (c): NOAA's Sanctuaries Collection; (d): Source: NOAA Okeanos Explorer Program, INDEX-SATAL 2010; (e): Jeff Rotman/Science Source

The endoskeleton

Echinoderms have a delicate epidermis that stretches over an endoskeleton composed of calcium carbonate (calcite) plates called **ossicles.** In echinoderms such as asteroids (sea stars), the individual skeletal elements are loosely joined to one another. In others, especially echinoids (sea urchins and sand dollars), the ossicles abut one another tightly, forming a rigid shell (called a test). In sea cucumbers, by contrast, the ossicles are widely scattered, so the body wall is flexible. The ossicles in certain portions of the body of some echinoderms are perforated by pores. Through these pores extend tube feet, part of the water-vascular system that is a unique feature of this phylum.

Members of this phylum have mutable collagenous tissue, which can change in texture from tough and rubbery to weak and fluid. This amazing tissue accounts for attributes of echinoderms such as the ability to autotomize (cast off) parts. This tissue is also responsible for a sea cucumber's ability to change from almost rigid to flaccid in a matter of seconds.

The water–vascular system

The water-vascular system is radially organized. From the ring canal, which encircles the animal's esophagus, a radial canal extends into each branch of the body (figure 27.26). Water enters the water-vascular system through a sievelike plate, the **madreporite,** which in most echinoderms is on the animal's surface, and flows to the ring canal through a stone canal, so named because it is reinforced by calcium carbonate. Each radial canal, in turn, extends through short side branches into the hollow tube feet (figure 27.26*b*). In some echinoderms, each tube foot has a sucker at its end; in others, suckers are absent. At the base of each tube foot in most types of echinoderms is a muscular sac, the **ampulla.** When the ampulla contracts, the fluid, prevented from entering the radial canal by a one-way valve, is forced into the tube foot, thus extending it. Contraction of longitudinal muscles on one side of the tube foot wall causes the tube foot to bend; relaxation of the muscles in the ampulla and contraction of all the longitudinal muscles in the tube foot force the fluid back into the ampulla.

In asteroids and echinoids, concerted action of a very large number of small, individually weak tube feet causes the animal to move across the seafloor. The tube feet around the mouth of a holothurian are used in feeding. In crinoids, tube feet that arise from the branches of the arms, which extend from the margins of an upward-directed cup, are used in capturing food from the surrounding water. Ophiuroid tube feet are pointed and specialized for feeding.

Gas exchange in echinoderms is through the body surface and tube feet. In addition, a holothurian has paired respiratory trees, which branch off the hindgut. Water is drawn into them and exits from them through the anus. In an asteroid, one of the coelomic spaces branches into protrusions from the epidermis called papulae through which gas exchange occurs.

All Vertebrates Are Chordates

LEARNING OBJECTIVE 27.8.2 Describe the defining features of chordates.

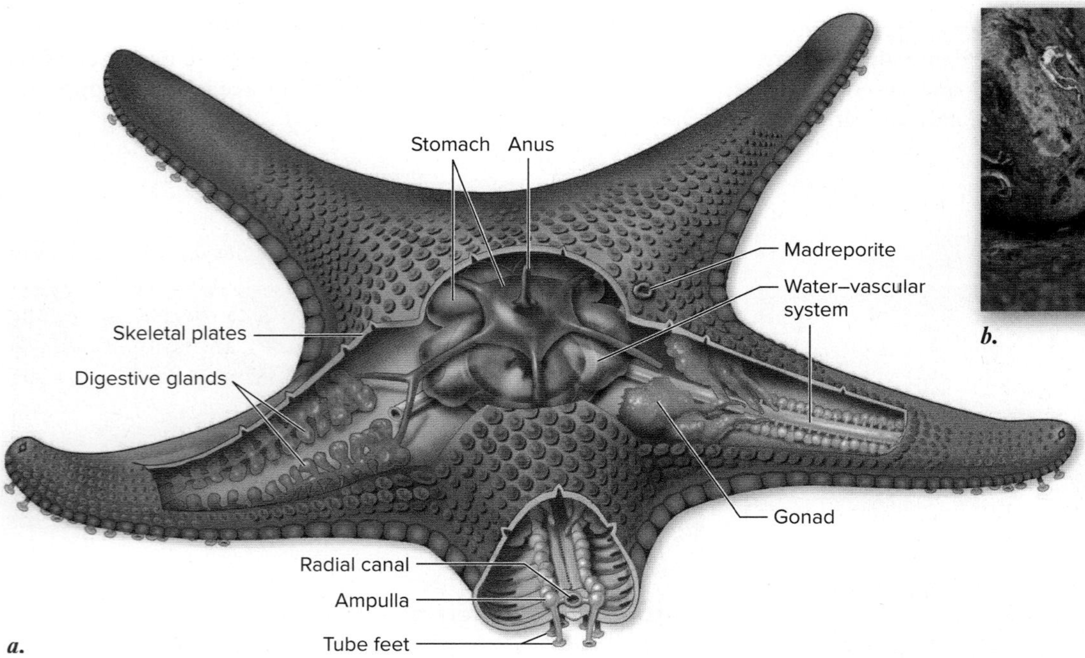

a.

b.

Figure 27.26 Phylum Echinodermata. ***a.*** Echinoderms, such as sea stars (class Asteroidea), are coelomates with a deuterostome pattern of development and an endoskeleton made of calcium carbonate plates. The water–vascular system of an echinoderm is shown in detail. Radial canals transport liquid to the tube feet. As the ampulla in each tube foot contracts, the tube foot extends and can attach to the substrate. When the muscles in the tube feet contract, the tube foot bends, pulling the animal forward. ***b.*** Extended nonsuckered tube feet of the sea star *Luidia magnifica.*

(b): Dan Bagur/Shutterstock

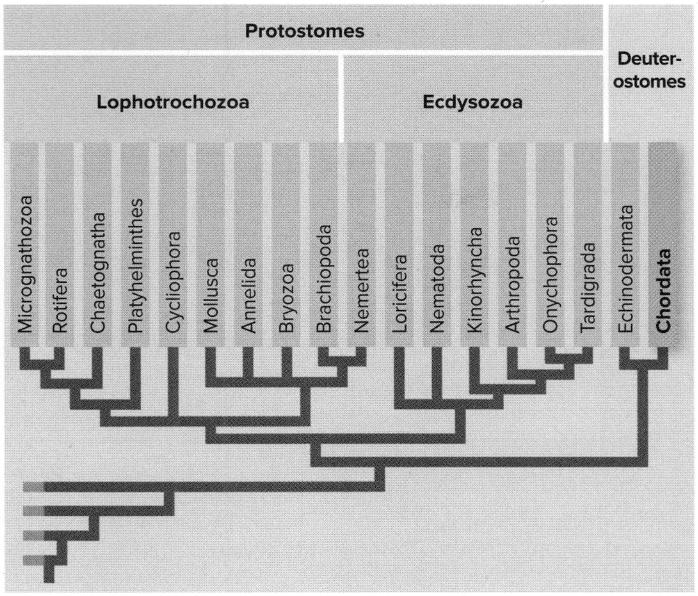

Chordates (phylum Chordata) are the only other major phylum of deuterostomes. There are some 56,000 species of chordates, a phylum that includes all vertebrates: fishes, amphibians, reptiles, birds, and mammals.

Four features characterize the chordates and have played an important role in the evolution of the phylum (figure 27.27):

1. A single, hollow nerve cord runs just beneath the dorsal surface of the animal. In vertebrates, the dorsal nerve cord differentiates into the brain and spinal cord.
2. A flexible rod, the notochord, forms on the dorsal side of the archenteron in the early embryo and is present at some developmental stage in all chordates. The notochord is located just below the nerve cord. The notochord may persist in some chordates; in others it is replaced during embryonic development by the vertebral column that forms around the nerve cord.
3. Pharyngeal slits connect the pharynx, a muscular tube that links the mouth cavity and esophagus with the external environment. In terrestrial vertebrates, the slits do not connect to the outside and are better termed pouches. Pharyngeal pouches are present in the embryos of all vertebrates. They become slits, open to the outside in animals with gills. The presence of these structures in all vertebrate embryos is evidence of their aquatic ancestry.
4. Chordates have a postanal tail that extends beyond the anus, at least during their embryonic development. Nearly all other animals have a terminal anus.

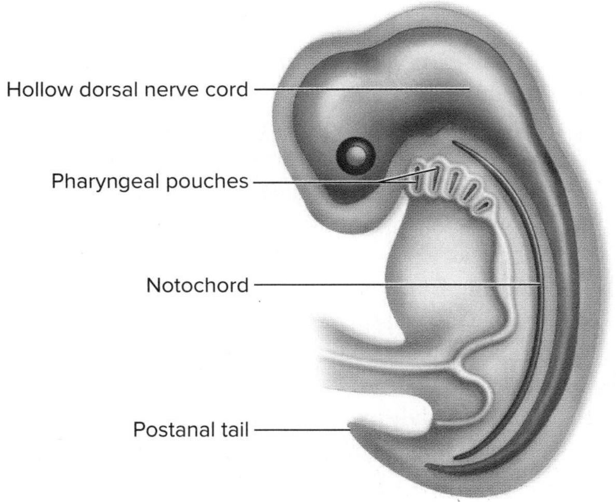

Figure 27.27 The four principal features of the chordates, as shown in a generalized embryo.

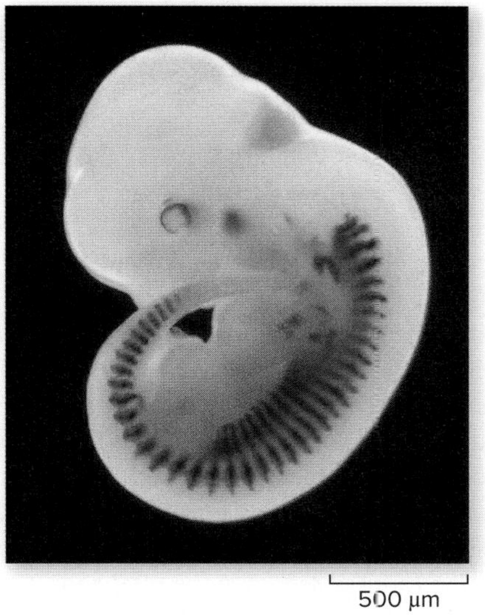

Figure 27.28 A mouse embryo. At 11.5 days of development, the mesoderm is already divided into segments called somites (stained dark in this photo), reflecting the fundamentally segmented nature of all chordates.

Eric N. Olson, Ph.D./The University of Texas MD Anderson Cancer Center

All chordates have all four of these characteristics at some time in their lives. For example, humans as embryos have pharyngeal pouches, a dorsal nerve cord, a postanal tail, and a notochord. As adults, the nerve cord remains, and the notochord is replaced by the vertebral column. All but one pair of pharyngeal pouches are lost; this remaining pair forms the Eustachian tubes that connect the throat to the middle ear. The postanal tail regresses, forming the tailbone (coccyx). Chordate muscles are arranged in segmented blocks that affect the basic organization of the chordate body and can often be clearly seen in embryos of this phylum (figure 27.28). Most chordates have an internal skeleton against which the muscles work. We will consider the chordates, and especially the vertebrates, in detail in chapter 28.

REVIEW OF CONCEPT 27.8

Echinoderms are marine deuterostomes with endoskeletons. They are characterized by pentaradial symmetry in the adult. The water–vascular system and tube feet that act as suction cups aid in movement and feeding. Chordates are deuterostomes characterized by a hollow dorsal nerve cord, a notochord, pharyngeal pouches, and a postanal tail at some point in their development. All vertebrates are chordates.

- *What distinguishes a chordate from an echinoderm?*

Punctuated Equilibrium: Evaluating a Case History

Biologists have long argued over the rate at which evolution occurs. Some organisms appear to have evolved gradually (gradualism), whereas in others evolution seems to have occurred in spurts (punctuated equilibrium). There is evidence of both patterns in the fossil record. Perhaps the most famous claim of punctuated equilibrium has been made by researchers studying the fossil record of marine bryozoans. Bryozoans are microscopic aquatic animals that form branching colonies. You encountered them in section 27.6 as lophophorates. The fossil record is particularly well documented for Caribbean bryozoan species of the genus *Metrarabdotos,* whose fossil record extends back more than 15 million years without interruption (a fossil record is the total collection of fossils of that particular kind of organism known to science). The sketches below of four different *Metrarabdotos* species indicate the diversity of form within this genus.

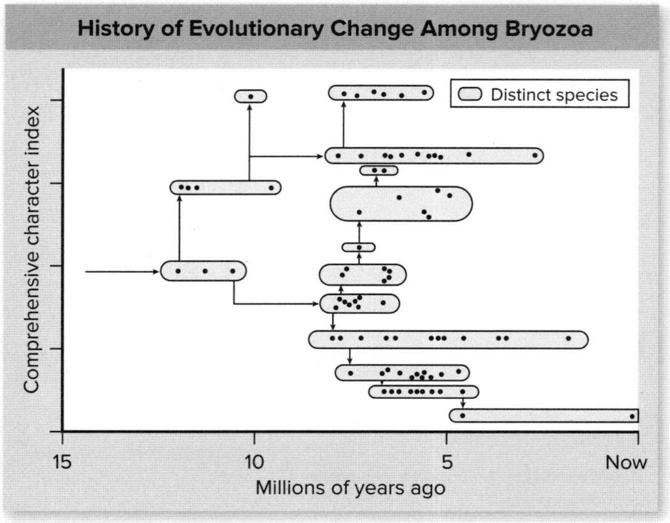

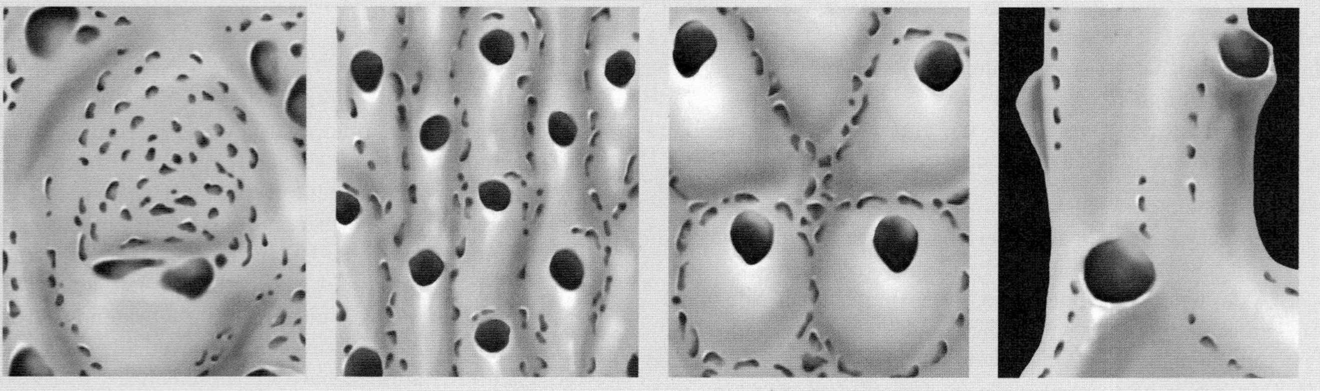

The graph displays an analysis of the *Metrarabdotos* fossil record. This uses a technique called a character index, which is based on assigning numbers to specific morphological traits. Different characteristics are measured and assigned quantitative values, and the character index is determined by adding together the individual character values that apply to the specimen. In this case, researchers formulated a comprehensive character index based on a broad array of bryozoan traits. The closer the character indices are for two specimens, the more closely related they are. Each fossil was measured for all of the traits, and the index number for each fossil was plotted on the graph as a black dot. Each cluster of dots within an oval represents a distinct species.

Analysis

1. **Applying Concepts**
   a. ***Variable.*** In the diagram, is there a dependent variable? If so, what is it?
   b. ***Analyzing diagrams.*** How many different species are included in the study illustrated by this diagram? How many of these are extinct?
2. **Interpreting Data**
   a. For each species, estimate how long that species survives in the fossil record. For simplicity, a species found only once should be assigned a duration of 1 million years. What is the average evolutionary duration of a *Metrarabdotos* species?
   b. Create a histogram of your species-duration estimates (place the duration times on the *x*-axis and the number of species on the *y*-axis). What general statement can be made regarding the distribution of *Metrarabdotos* species durations?
3. **Making Inferences**
   a. How many of the species exhibit variation in the comprehensive character index?
   b. How does the magnitude of this variation within species compare with the variation seen between species?
4. **Drawing Conclusions** Does major evolutionary change, as measured by significant changes in this comprehensive character index, occur gradually or in occasional bursts?
5. **Further Analysis** Plot the number of *Metrarabdotos* species versus date (millions of years ago), in increments of 1 million years. Characterize the result. What do you suppose is responsible for this? How would you go about assessing this possibility?

Retracing the Learning Path

CONCEPT 27.1 The Diversity of Animal Body Plans Arose by a Series of Evolutionary Innovations

27.1.1 Animals Share Many Features These include multicellularity, heterotrophy, diversity of form and habitat, embryonic development, and movement.

27.1.2 The Evolution of Tissue and Symmetrical Bodies Was a Critical Early Innovation Complex organisms have organs made of tissues composed of cells with characteristic form and function. Sponges lack tissue and symmetry, but other animals are radially or bilaterally symmetrical. Animals more complex than cnidarians have bilateral symmetry.

27.1.3 A Body Cavity Made the Development of Advanced Organ Systems Possible Most bilaterian animals possess a body cavity other than the gut. A coelom is a cavity that lies within tissues derived from mesoderm. A pseudocoelom lies between tissues derived from mesoderm and the gut. The coelom evolved only once. Larger animals have a circulatory system to transport nutrients and gases. This system can be open or closed.

27.1.4 Early Developmental Differences Divide Bilaterians into Protostomes and Deuterostomes In protostomes, the mouth develops from or near the blastopore. Protostomes have determinate development, and many have spiral cleavage. In deuterostomes, the anus develops from the blastopore. Deuterostomes have indeterminate development and radial cleavage.

27.1.5 Segmentation Has Evolved Multiple Times Segmentation, which evolved multiple times, allows for efficient and flexible movement. Segmentation underlies the organization of the body plan of morphologically complex animals.

CONCEPT 27.2 Molecular Data Are Clarifying the Animal Phylogenetic Tree

27.2.1 Molecular Data Are Revising Animal Phylogeny Biologists agree on the placement of most animals into 35 to 40 phyla, but relationships between these phyla have been debated. Using molecular data in phylogenetic analyses modified the traditional view. The protostome branch of the tree has been altered more dramatically than the deuterostome branch.

27.2.2 Protostomes Are Divided into Lophotrochozoans and Ecdysozoans Lophotrochozans grow by gradual addition of mass and undergo spiral cleavage. Ecdysozoans grow by molting an external skeleton.

CONCEPT 27.3 True Tissue Evolved in Simple Animals

27.3.1 Porifera Lack Symmetry and Specialized Tissue The sponges, phylum Porifera, have a loose body organization. Sponges lack tissues and organs and a definite symmetry, but they do have a complex multicellularity.

27.3.2 Cnidarians Possess Both Symmetry and True Tissue Cnidarians are carnivorous and radially symmetrical with distinct tissue but no organs. They have two body forms: a sessile polyp and a free-floating medusa. Cnidarians have extracellular digestion and stinging capsules called nematocysts.

CONCEPT 27.4 Flatworms and Rotifers Are Very Simple Bilaterians

27.4.1 Flatworms Are Bilaterally Symmetrical and Have Lost a True Coelom Free-living flatworms, phylum Platyhelminthes, move by muscles and ciliated epithelial cells. They also exhibit a head and an incomplete gut. Flatworms have an excretory system containing a network of tubules with flame cells.

27.4.2 Rotifers Are Tiny and Lack a True Coelom Rotiferans are tiny lophotrochozoans with complex form. They are either free-swimming or sessile.

CONCEPT 27.5 Mollusks and Annelids Are the Largest Groups of Lophotrochozoans

27.5.1 Mollusks Are a Large and Diverse Group Mollusks are bilaterally symmetrical at some point in their lives. They use a muscular foot for locomotion, attachment, and food capture. All mollusks except bivalves have a radula, a rasplike structure used in feeding.

27.5.2 The Annelid Body Is Composed of Ringlike Segments The annelid body is segmented with duplicate organs and a closed circulatory system. Segments are separated by septa and are connected by a ventral nerve cord with an anterior brain region.

CONCEPT 27.6 Lophophorates Are Very Simple Marine Organisms

27.6.1 Bryozoans Are the Only Exclusively Colonial Animals Each individual zooid produces a chitinous chamber called a zoecium that attaches to substrates and other colony members.

CONCEPT 27.7 Nematodes and Arthropods Are Both Large Groups of Ecdysozoans

27.7.1 Nematodes Consist of Many Different Kinds of Roundworm Nematodes are ecdysozoans that reproduce sexually and exhibit sexual dimorphism. Nematodes exchange gases through their cuticle and have a well-developed digestive system.

27.7.2 Arthropods Are the Most Successful of All Animals Well over 1 million species of arthropods have been named, making up two-thirds of all named species.

27.7.3 The Arthropod Body Exhibits Three Key Features Arthropods are segmented with an exoskeleton. The exoskeleton is molted during ecdysis, allowing the arthropod to grow. Jointed appendages may be modified into mouth parts, antennae, or legs.

27.7.4 There Are Four Classes of Arthropods Classes Chelicerata, Crustacea, Hexapoda, and Myriapoda differ based on the structure of mouth parts, segments, and appendages. The hexapods are the largest and most diverse group and include insects. Crustacea includes crabs, shrimp, and lobster.

CONCEPT 27.8 Deuterostomes Are Composed of Echinoderms and Chordates

27.8.1 Echinoderms Are Ancient and Unmistakable Echinoderms are deuterostomes with pentaradial symmetry, an endoskeleton covered by an epidermis, and a water-vascular system.

27.8.2 All Vertebrates Are Chordates Chordates possess a dorsal nerve chord, a notochord, pharyngeal pouches, and a postanal tail. Chordate mesoderm is organized into segmented blocks during development.

Concept Overview

This Concept Overview diagrams the key concepts that were discussed in this chapter.

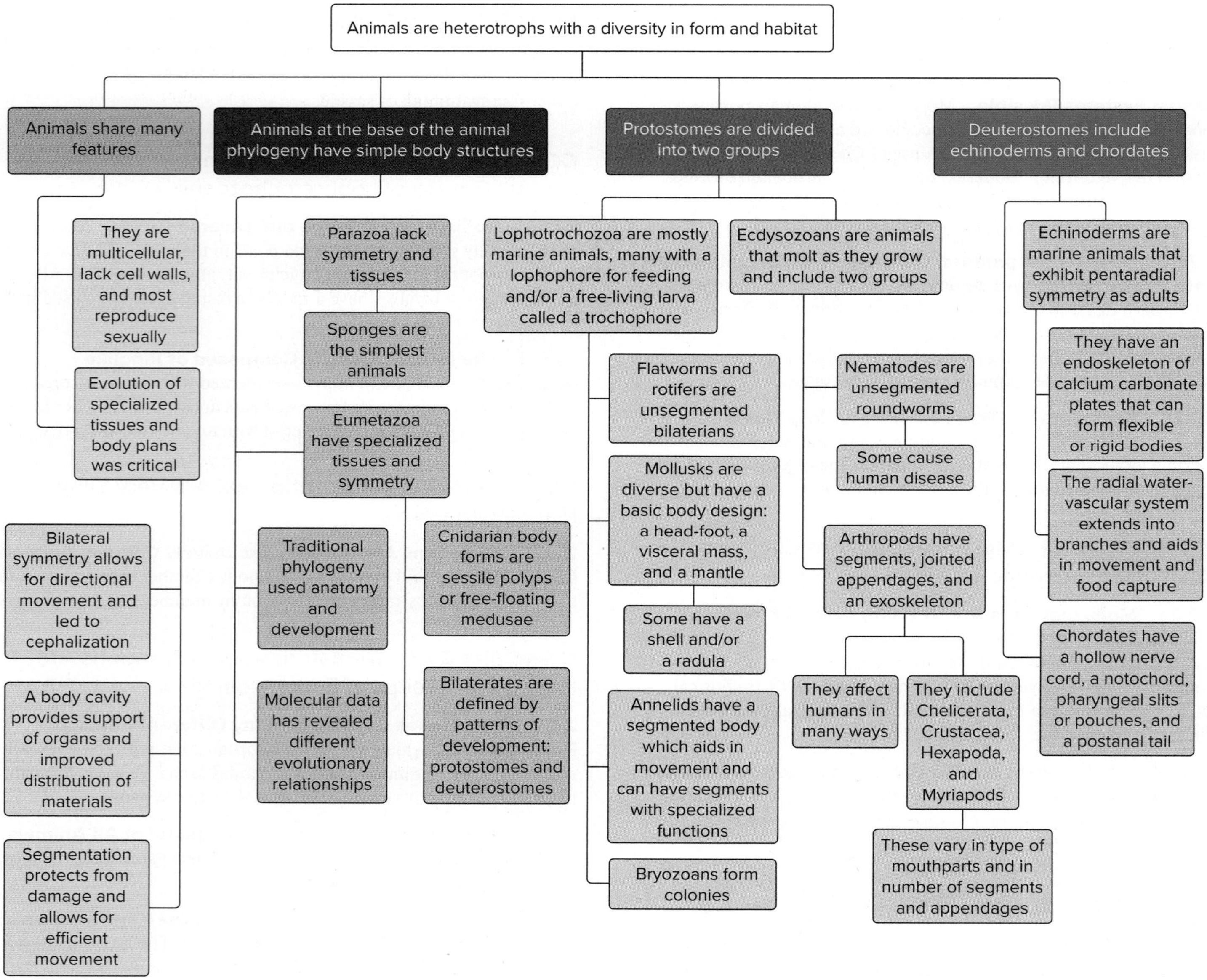

Assessing the Learning Path

Understand

1. Animals are unique in the fact that they possess ________ for movement and ________ for conducting signals between cells.
   a. brains; muscles
   b. muscle tissue; nervous tissue
   c. limbs; spinal cords
   d. flagella; nerves
2. Animal cell walls are composed of
   a. chitin.
   b. cellulose.
   c. peptidoglycans.
   d. silicon dioxide.
   e. None of the above; animal cells lack cell walls.
3. Body cavities differ from digestive cavities in that
   a. they are not connected to the outside.
   b. they are always surrounded by ectoderm.
   c. they are always filled with fluid.
   d. they are only found in asymmetrical organisms.
4. With regard to classification of the animals, the study of which of the following is changing our understanding of the organization of the kingdom?
   a. Molecular systematics
   b. Origin of tissues
   c. Patterns of segmentation
   d. Evolution of morphological characteristics
5. Molting is a key feature of
   a. lophotrochozoans.
   b. ecdysozoans.
   c. mollusks.
   d. platyhelminths.
6. Sponges exhibit which of the following characteristics?
   a. They have true tissue.
   b. They are bilaterally symmetrical.
   c. They are radially symmetrical.
   d. They are asymmetrical.
7. Cnidarians possess
   a. bilateral symmetry.
   b. extracellular digestion.
   c. internal skeletons.
   d. a complex nervous system.
8. Rotifers are named for the wheel of cilia at the top of their bodies. What is its function?
   a. Digestion
   b. Locomotion
   c. Excretion
   d. Respiration
9. The ________ of a mollusk is an efficient excretory structure.
   a. nephridium
   b. radula
   c. ctenidium
   d. veliger
10. The only exclusively colonial animals are
    a. bryozoans.
    b. brachiopods.
    c. annelids.
    d. ctenophores.
11. In terms of numbers of species, the most successful phylum on the planet is the
    a. Mollusca.
    b. Arthropoda.
    c. Echinodermata.
    d. Annelida.
12. Which of the following classes of arthropod possesses chelicerae?
    a. Chilopoda
    b. Crustacea
    c. Hexapoda
    d. Chelicerata
13. Based on embryonic development, which of the following phyla is the closest to the chordates?
    a. Annelida
    b. Arthropoda
    c. Echinodermata
    d. Mollusca
14. Which of the following statements regarding all species of chordates is false?
    a. Chordates are deuterostomes.
    b. A notochord is present in the embryo.
    c. The notochord is surrounded by bone or cartilage.
    d. All possess a postanal tail during embryonic development.

Apply

1. All animals have which of the following characteristics?
   a. Body symmetry
   b. Tissues
   c. Multicellularity
   d. Body cavity
2. In modern phylogenetic analysis of the animals, the protostomes are divided into two major groups based on what characteristic?
   a. Their symmetry
   b. Having a head
   c. Their ability to molt
   d. The presence or absence of vertebrae
3. Choanocytes of sponges bear a striking resemblance to the ________, members of the Unikonta group of protists.
   a. nuclearia
   b. stramenophytes
   c. choanoflagellates
   d. charophytes
4. Which of the following characteristics is NOT seen in the phylum Platyhelminthes?
   a. Cephalization
   b. Segmentation
   c. A body cavity
   d. Bilateral symmetry
5. To which of the following groups would a species that does not molt, possesses a coelom, and has a trochophore larva belong?
   a. Arthropods
   b. Nematodes
   c. Mollusks
   d. Echinoderms
6. Serial segmentation is a key characteristic of which of the following phyla?
   a. Mollusca
   b. Brachiopoda
   c. Bryozoa
   d. Annelida
7. A paleontologist discovers a thick layer of fossilized zoecia on rocks. What does this say about the rocks when the animals were alive?
   a. The rocks must have been under water.
   b. The rocks were exposed to lava flows.
   c. The rocks were part of a mountain that eroded.

8. Which of the following characteristics is NOT found in the arthropods?
   a. Jointed appendages
   b. Segmentation
   c. Closed circulatory system
   d. Segmented ganglia
9. Features that unite chordates and echinoderms as deuterostomes include
   a. bilaterally symmetrical adult forms.
   b. radially symmetrical adult forms.
   c. a distinct form of development.
   d. a dorsal nerve cord.
10. Which of the following structures is NOT a component of the water-vascular system of an echinoderm?
   a. Ossicles
   b. Ampullae
   c. Radial canals
   d. Madreporites

Synthesize

1. Animals first evolved in the sea. List the major groups that have successfully invaded land. What do they have in common that the noninvading groups lack?
2. In the new phylogeny, arthropods and nematodes are both Ecdysozoa. Does this mean the coelom evolved more than once?
3. Compare sponges with cnidarians. How do food particles and waste products enter and leave the two kinds of organisms? How might the similarity be explained?
4. Does the lack of a digestive system in tapeworms indicate that they are a primitive, ancestral form of Platyhelminthes? Explain your answer.
5. Compare terrestrial mollusks to terrestrial annelids. Why do you think terrestrial mollusks and annelids are less diverse than terrestrial arthropods?
6. Discuss the similarities and differences between bryozoans and brachiopods.
7. In the rainforest, you discover a new species that is terrestrial, has determinate development, molts during its lifetime, and possesses jointed appendages. To which phylum of animals should it be assigned?
8. Do you think it is their internal skeletons that allow vertebrates to be so much larger than other animals? Explain.

Ingram Publishing/SuperStock

28 Vertebrates

Learning Path

28.1 Nonvertebrate Chordates Do Not Form Bone

28.2 Almost All Chordates Are Vertebrates

28.3 Fishes Are the Earliest and Most Diverse Vertebrates

28.4 Amphibians Are Moist-Skinned Descendants of the Early Tetrapods

28.5 Reptiles Are Fully Adapted to Terrestrial Living

28.6 Birds Are Essentially Flying Reptiles

28.7 Mammals Are the Least Diverse of Vertebrates

28.8 Primates Include Lemurs, Monkeys, Apes, and Humans

Concept Overview

This Concept Overview provides a diagram of the overarching concepts that are covered in this chapter. These concepts will be expanded upon in the Concept Overview at the end of the chapter.

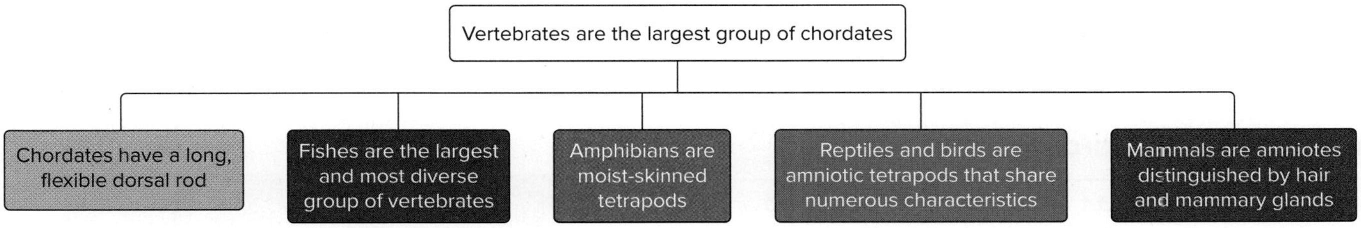
Vertebrates are the largest group of chordates
Chordates have a long, flexible dorsal rod
Fishes are the largest and most diverse group of vertebrates
Amphibians are moist-skinned tetrapods
Reptiles and birds are amniotic tetrapods that share numerous characteristics
Mammals are amniotes distinguished by hair and mammary glands

Introduction

Members of the phylum Chordata exhibit great changes in the endoskeleton from that seen in echinoderms. As discussed in chapter 27, the endoskeleton of echinoderms is functionally similar to the exoskeleton of arthropods—a hard shell with muscles attached to its inner surface. Chordates employ a very different kind of endoskeleton, one that is truly internal. Members of the phylum Chordata are characterized by a flexible rod that develops along the back of the embryo. Muscles attached to this internal rod allowed early chordates to swing their bodies from side to side, swimming through the water. This key evolutionary advance, attaching muscles to an internal element, started chordates along an evolutionary path that led to the vertebrates—and, for the first time, to truly large animals.

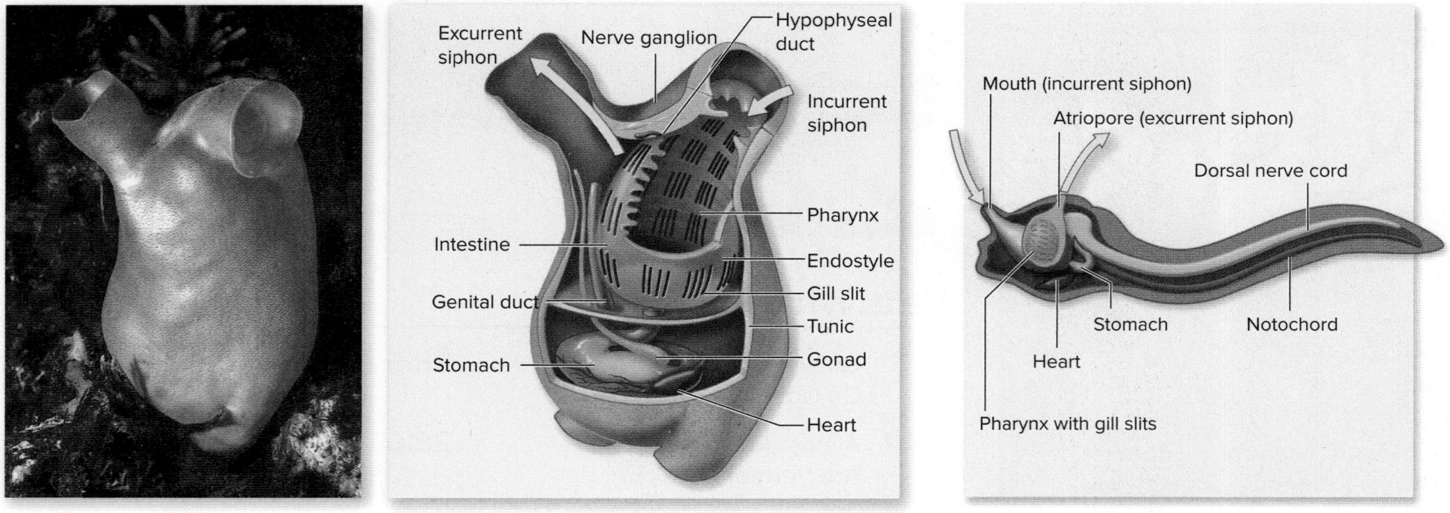

Figure 28.1 Tunicates (phylum Chordata, subphylum Urochordata). ***a.*** The sea peach, *Halocynthia aurantium,* like other tunicates, does not move as an adult but, rather, is firmly attached to the seafloor. ***b.*** Diagram of the structure of an adult tunicate. ***c.*** Diagram of the structure of a larval tunicate, showing the characteristic tadpole-like form. Larval tunicates resemble the postulated common ancestor of the chordates.

(a): Rick Harbo

28.1 Nonvertebrate Chordates Do Not Form Bone

Chordates (phylum **Chordata**) are one of two major phyla of deuterostomes. Echinoderms, members of the other major phylum, are so different they were covered in detail in chapter 27. Phylum Chordata can be divided into three subphyla. Two of these, Urochordata and Cephalochordata, are nonvertebrate; the third subphylum is Vertebrata, including fishes, amphibians, reptiles, birds, and mammals. The nonvertebrate chordates do not form vertebrae or other bones, and in the case of the urochordates, their adult form is much different from what we expect of a chordate.

Tunicates Have Chordate Larval Forms

LEARNING OBJECTIVE 28.1.1 Describe the nonvertebrate chordates.

The tunicates (subphylum Urochordata) are a group of about 1250 species of marine animals. Most of them are immobile as adults, with only the larvae having a notochord and nerve cord. The tadpole-like larvae of tunicates plainly exhibit all of the basic characteristics of chordates and mark the tunicates as having the most primitive combination of features found in any chordate (figure 28.1*c*). The larvae remain free-swimming for only a few days before settling to the bottom and attaching themselves to a suitable substrate by means of a sucker. Their adult form is greatly different from what we expect chordates to look like. As adults, they exhibit neither a major body cavity nor visible signs of segmentation (figure 28.1*a, b*). Most species occur in shallow waters, but some are found at great depths.

Lancelets Have Chordate Adult Forms

LEARNING OBJECTIVE 28.1.2 Distinguish between lancelets and tunicates.

Lancelets (subphylum Cephalochordata) resemble a small, two-edged surgical knife. These scaleless chordates, a few centimeters long, occur widely in shallow water throughout the oceans of the world. There are about 23 species of this subphylum. In lancelets, the notochord runs the entire length of the body and persists throughout the animal's life.

Lancelets spend most of their time partly buried in sandy or muddy substrates, with only their anterior ends protruding (figure 28.2). Lancelets filter-feed on microscopic plankton, creating a current by beating cilia that line the pharynx.

Figure 28.2 Lancelets. Two lancelets, *Branchiostoma lanceolatum,* partly buried in shell gravel, with their anterior ends protruding. The muscle segments are clearly visible.

Heather Angel/Natural Visions

REVIEW OF CONCEPT 28.1

Nonvertebrate chordates have notochords but no vertebrae or bones. Tunicates have chordate larval forms but drastically different adult forms. Lancelets do not change body form as adults.

- *How do lancelets and tunicates differ from vertebrates?*

28.2 Almost All Chordates Are Vertebrates

The largest clade within the chordates by far is the vertebrates. There are more species of vertebrates than any phylum other than arthropods and mollusks.

Vertebrates Have Vertebrae, a Distinct Head, and Other Features

LEARNING OBJECTIVE 28.2.1 Distinguish vertebrates from other chordates.

Vertebrates (subphylum Vertebrata) are chordates with a spinal column. The name *vertebrate* comes from the individual bony or cartilaginous segments, called vertebrae, that make up the spine. Vertebrates differ from the tunicates and lancelets in several important respects (figure 28.3):

Vertebral column. In all vertebrates except the earliest diverging fishes, the notochord is replaced during embryonic development by a vertebral column. The column is a series of bony or cartilaginous vertebrae that enclose and protect the dorsal nerve cord like a sleeve.

Head. Vertebrates have a distinct and well-differentiated head with three pairs of well-developed sensory organs; the brain is encased within a protective box—the skull, or cranium, made of bone or cartilage.

In addition, vertebrates differ from other chordates in other important respects (figure 28.4):

Neural crest. A unique group of embryonic cells called the neural crest contributes to the development of many vertebrate structures. These cells develop on the crest of the neural tube as it forms by invagination and pinching together of the neural plate (chapter 36 gives a detailed account).

Internal organs. Internal organs characteristic of vertebrates include a liver, kidneys, and endocrine glands. The ductless endocrine glands secrete hormones that help regulate many of the body's functions. All vertebrates have a heart and a closed circulatory system. Vertebrate circulatory and excretory functions differ markedly from those of other animals.

Endoskeleton. The endoskeleton of most vertebrates is made of cartilage or bone. Cartilage and bone are specialized tissues containing fibers of the protein collagen compacted together (chapter 32 includes a more detailed account). The advantage of bone over chitin for a skeleton is that bone is a dynamic, living tissue that is strong without being brittle. The vertebrate endoskeleton makes possible the great size and extraordinary powers of movement that characterize this group.

REVIEW OF CONCEPT 28.2

Vertebrates are characterized by a vertebral column and a distinct head. Other distinguishing features are the development of a neural crest, a closed circulatory system, specialized organs, and a bony or cartilaginous endoskeleton that has the strength to support larger body size and powerful movements.

- *In what ways would an exoskeleton limit the size of an organism?*

28.3 Fishes Are the Earliest and Most Diverse Vertebrates

Over half of all vertebrates are fishes. The most diverse vertebrate group, fishes provided the evolutionary base for invasion of land. In many ways, early tetrapods can be viewed as transitional—"fish out of water."

Fishes Exhibit Five Key Characteristics

LEARNING OBJECTIVE 28.3.1 Describe the evolutionary innovations of fishes.

From whale sharks 18 m long to tiny gobies no larger than your fingernail, fishes vary considerably in size, shape, color, and appearance (figure 28.5). Some live in freezing arctic seas; others are found in warm, freshwater lakes; and still others spend a lot of time entirely out of water. However varied, all fishes have important characteristics in common:

1. **Vertebral column.** Fish have an internal skeleton, with a bony or cartilaginous spine surrounding the dorsal nerve cord, and a bony or cartilaginous skull encasing the brain. Exceptions are the jawless hagfish and lampreys. In hagfish, a cartilaginous skull is present, but vertebrae only occur in the larvae (and only in some species) and then disappear; the notochord persists and provides support. In lampreys, a cartilaginous skeleton and notochord are present, but rudimentary cartilaginous vertebrae also surround the notochord in places.
2. **Jaws and paired appendages.** Fishes other than lampreys and hagfish have jaws and paired appendages, features that are also seen in tetrapods. Jaws allowed these fish to capture larger and more active prey. Most fishes have two pairs of fins: a pair of pectoral fins at the shoulder and a pair of pelvic fins at the hip. In the lobe-finned fish, these pairs of fins became jointed.
3. **Internal gills.** Fishes are water-dwelling creatures and must extract oxygen dissolved in the water around them. They do this by directing a flow of water through their

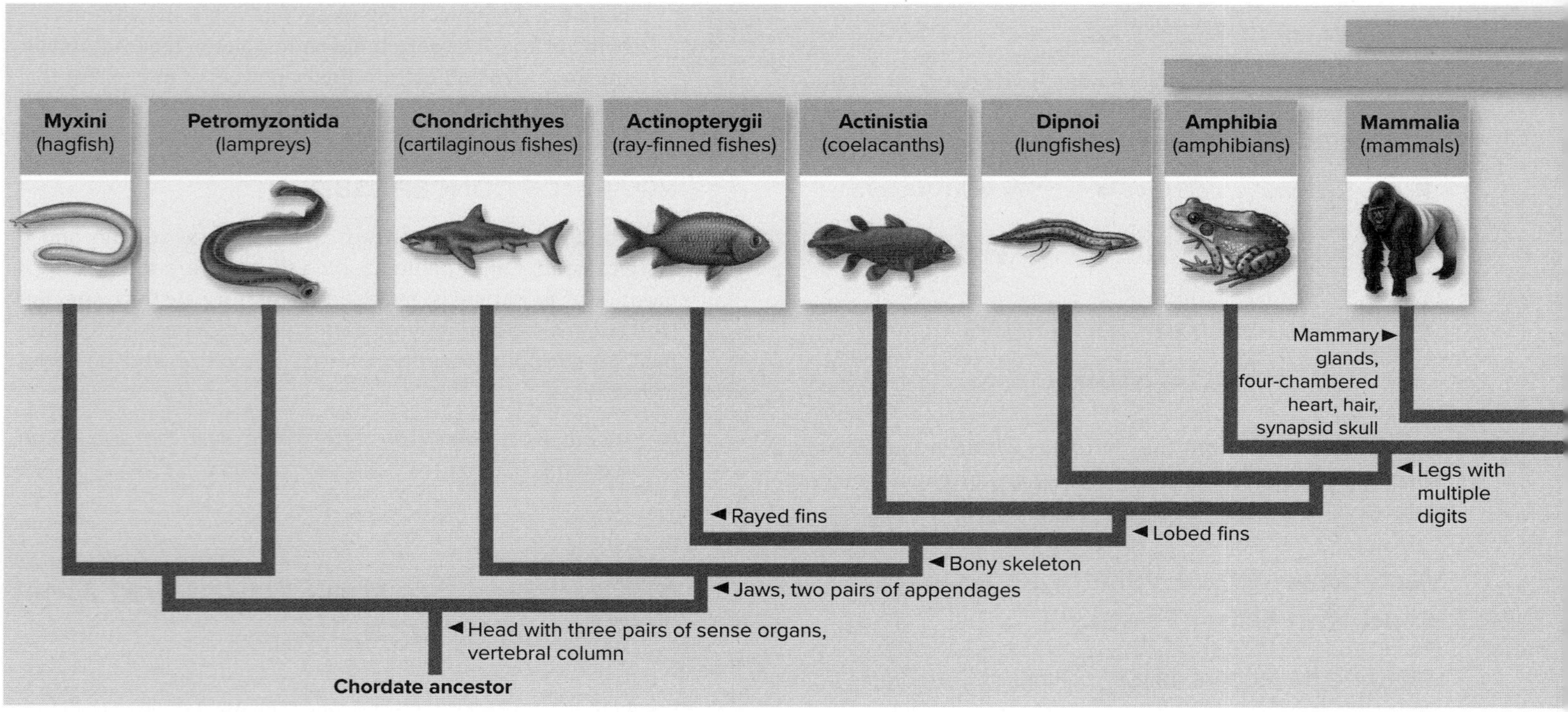

Figure 28.3 Phylogeny of the living vertebrates. Some of the key characteristics that evolved among the vertebrate groups are shown in this phylogeny.

mouths and across their gills (refer to chapter 34). The gills are composed of fine filaments of tissue that are rich in blood vessels.

4. **Single-loop blood circulation.** Blood is pumped from the heart to the gills. From the gills, the oxygenated blood passes to the rest of the body, then returns to the heart. The heart is a muscular tube-pump made of two chambers that contract in sequence.
5. **Nutritional deficiencies.** Fishes are unable to synthesize the aromatic amino acids (phenylalanine, tryptophan, and tyrosine; refer to chapter 3), so they must consume them in their foods. This inability has been inherited by all of their vertebrate descendants.

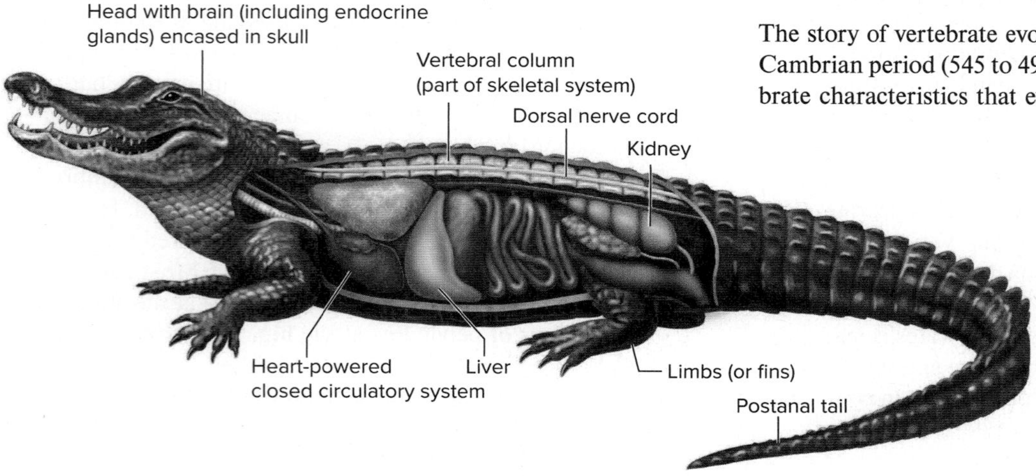

Figure 28.4 Major characteristics of vertebrates. Adult vertebrates are characterized by an internal skeleton of cartilage or bone, including a vertebral column and a skull. Several other internal and external features are characteristic of vertebrates.

The First Fishes Lacked Jaws

LEARNING OBJECTIVE 28.3.2 Describe the evolution of jaws in early fishes.

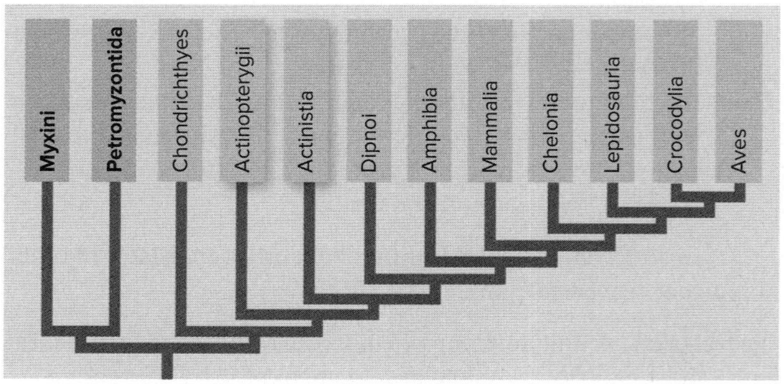

The story of vertebrate evolution started in the ancient seas of the Cambrian period (545 to 490 MYA). Figure 28.3 shows the key vertebrate characteristics that evolved subsequently. Wriggling through

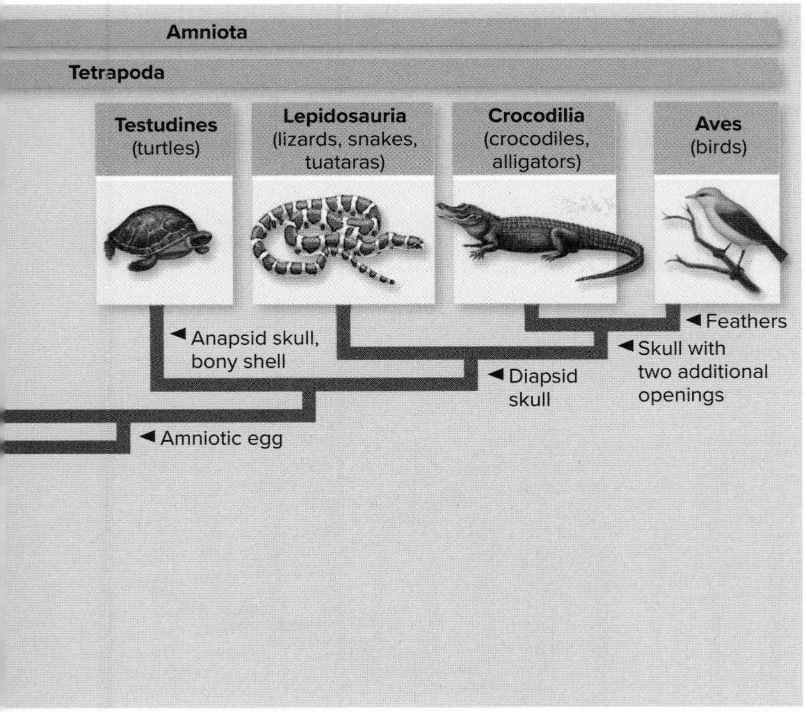

Amniota
Tetrapoda
Testudines
(turtles)
Lepidosauria
(lizards, snakes,
tuataras)
Crocodilia
(crocodiles,
alligators)
Aves
(birds)
Feathers
Anapsid skull,
bony shell
Skull with
two additional
openings
Diapsid
skull
Amniotic egg

the water, jawless and toothless, the first fishes sucked up small food particles from the ocean floor like miniature vacuum cleaners. Most were less than a foot long, respired with gills, and had no paired fins; they did have a head and a tail to push them through the water.

For 50 million years, during the Ordovician period (490 to 438 MYA), these simple fishes were the only vertebrates. By the end of this period, fish had developed fins to help them swim and

Figure 28.5 Fish. Fish are the most diverse vertebrates and include more species than all other kinds of vertebrates combined. Top, ribbon eel, *Rhinomuraena quaesita;* bottom left, leafy sea-dragon, *Phycodurus eques;* bottom right, yellowfin tuna, *Thunnus albacares.*

(a): Andre Seale/age fotostock/SuperStock; (b): Ian D M Robertson/Shutterstock; (c): Shane Gross/Shutterstock

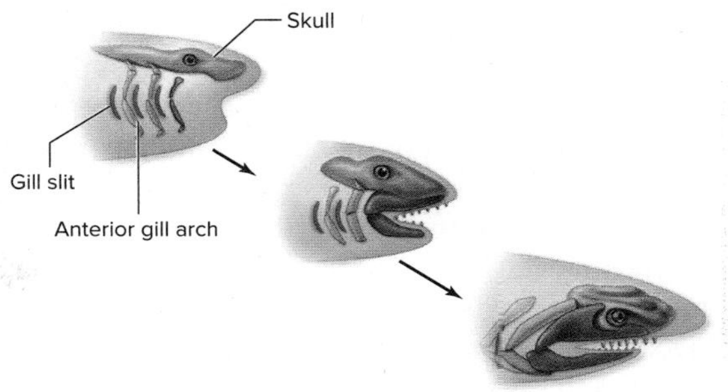

Skull
Gill slit
Anterior gill arch

Figure 28.6 Evolution of the jaw. Jaws evolved from the anterior gill arches of ancient jawless fishes.

massive shields of bone for protection. Jawed fishes first appeared during the Silurian period (438 to 408 MYA), and along with them came a new mode of feeding.

The first fishes did not have jaws; instead, they had only a mouth at the front end of the body that could be opened to take in food. Two groups survive today as hagfish (class Myxini) and lampreys (class Petromyzontida).

Another group were the *ostracoderms* (meaning "shell-skinned"). Only their head-shields were made of bone; their elaborate internal skeletons were constructed of cartilage.

Evolution of the jaw

A fundamentally important evolutionary advance that occurred in the late Silurian period was the development of jaws. Jaws evolved from the most anterior of a series of arch-supports made of cartilage, which reinforced the tissue between gill slits to hold the slits open (figure 28.6). Armored fishes called placoderms and spiny fishes called acanthodians both had jaws.

This transformation was not as radical as it might at first appear. Each gill arch was formed by a series of several cartilages (which later evolved to become bones) arranged somewhat in the shape of a V turned on its side, with the point directed outward. Imagine the fusion of the front pair of arches at top and bottom, with hinges at the points, and you have the primitive vertebrate jaw. The top half of the jaw is not attached to the skull directly except at the rear. Teeth developed on the jaws from modified scales on the skin that lined the mouth (figure 28.6).

Sharks, with Cartilaginous Skeletons, Became Top Predators

LEARNING OBJECTIVE 28.3.3 Explain how a lateral line system works.

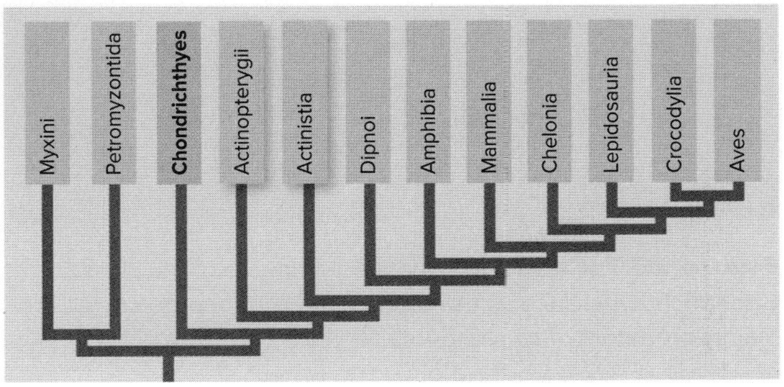

Myxini
Petromyzontida
Chondrichthyes
Actinopterygii
Actinistia
Dipnoi
Amphibia
Mammalia
Chelonia
Lepidosauria
Crocodylia
Aves

At the end of the Devonian period, these early armored fishes disappeared, replaced by sharks and bony fishes in the early Devonian, 400 MYA. In these fishes, the jaw was improved even further, with the upper part of the first gill arch behind the jaws being transformed into a supporting strut or prop, joining the rear of the lower jaw to the rear of the skull. This allowed the mouth to open much wider than was previously possible.

During the Carboniferous period (360 to 280 MYA), sharks became the dominant predators in the sea. Sharks, as well as skates and rays (all in the class Chondrichthyes), have a skeleton made of cartilage, like early fishes, but it is "calcified," strengthened by granules of calcium carbonate deposited in the outer layers of cartilage. The result is a very light and strong skeleton. Streamlined, with paired fins and a light, flexible skeleton, sharks are superior swimmers (figure 28.7). Their pectoral fins are particularly large, jutting out stiffly, like airplane wings—and that is how they function, adding lift to compensate for the downward thrust of the tail fin. Some early sharks reached enormous size.

The evolution of teeth

Sharks were among the first vertebrates to develop teeth. These teeth evolved from rough scales on the skin and are not set into the jaw as human teeth are; rather, they sit atop it. They are not firmly anchored and are easily lost. In a shark's mouth, the teeth are arrayed in up to 20 rows; the teeth in front do the biting and cutting, and behind them other teeth grow and wait their turn. When a tooth breaks or is worn down, a replacement from the next row moves forward. A single shark may eventually use more than 20,000 teeth in its lifetime.

A shark's skin is covered with tiny, toothlike scales, giving it a rough, "sandpaper" texture. Like the teeth, these scales are constantly replaced throughout the shark's life.

The lateral line system

Sharks, as well as bony fishes, possess a fully developed **lateral line system** consisting of a series of sensory organs that project into a canal beneath the surface of the skin. The canal runs the length of the fish's body and is open to the exterior through a series of sunken pits. Movement of water past the fish forces water through the canal. The pits are oriented so that some are stimulated no matter what direction the water moves. Details of the lateral line system's function are described in chapter 33.

Figure 28.7 Chondrichthyes. Members of the class Chondrichthyes, such as this tiger shark, *Galeocerdo cuvier,* are mainly predators or scavengers.

Alastair Pollock Photography/Flickr Open/Getty Images

Bony Fishes Dominate Today's Seas

LEARNING OBJECTIVE 28.3.4 Explain the importance of a swim bladder.

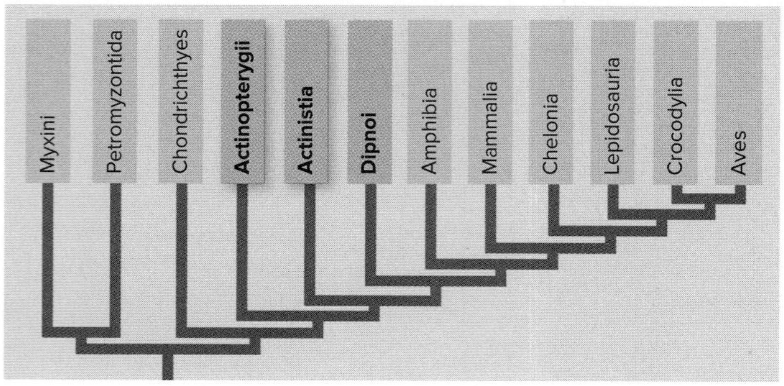

Bony fishes evolved at the same time as sharks, some 400 MYA, but took quite a different evolutionary road. Instead of gaining speed through lightness, as sharks did, bony fishes adopted a heavy internal skeleton made completely of bone.

Bone is very strong, providing a base against which very strong muscles can pull. Not only is the internal skeleton ossified, but so is the outer covering of plates and scales. Most bony fishes have highly mobile fins, very thin scales, and completely symmetrical tails (which keep the fish on a straight course as it swims through the water). Bony fishes are the most species-rich group of fishes, indeed of all vertebrates. There are several dozen orders containing more than 30,000 living species.

The remarkable success of the bony fishes has resulted from a series of significant adaptations that have enabled them to dominate life in the water. These include the swim bladder and the gill cover (figure 28.8).

Swim bladder

Although bones are heavier than cartilaginous skeletons, most bony fishes are still buoyant because they possess a **swim bladder,** a gas-filled sac that allows them to regulate their buoyant density and so remain suspended at any depth in the water effortlessly. Sharks, by contrast, must move through the water or sink, because lacking a swim bladder, their bodies are denser than water.

In most of today's bony fishes, the swim bladder is an independent organ that is filled and drained of gases, mostly nitrogen and oxygen, internally. How do bony fishes manage this remarkable trick? It turns out that the gases are harvested from the blood by a unique gland that discharges the gases into the bladder when more buoyancy is required. To reduce

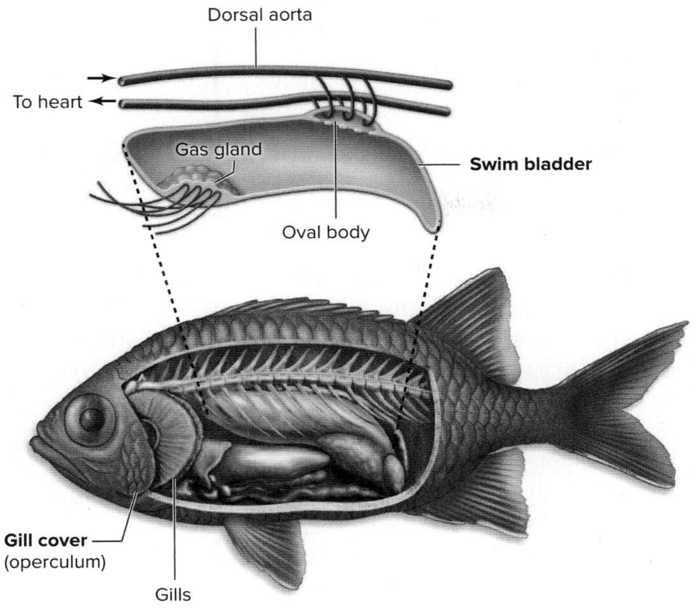

Figure 28.8 Diagram of a swim bladder. The bony fishes use this structure, which evolved as a dorsal outpocketing of the pharynx, to control their buoyancy in water.

buoyancy, gas is reabsorbed into the bloodstream through a structure called the oval body. A variety of physiological factors control the exchange of gases between the bloodstream and the swim bladder.

Gill cover

Most bony fishes have a hard plate called an **operculum,** which covers the gills on each side of the head. Flexing the operculum permits bony fishes to pump water over their gills. The gills are suspended in the pharyngeal slits that form a passageway between the pharynx and the outside of the fish's body. When the operculum is closed, it seals off the exit.

When the mouth is open, closing the operculum increases the volume of the mouth cavity, so that water is drawn into the mouth. When the mouth is closed, opening the operculum decreases the volume of the mouth cavity, forcing water past the gills to the outside. Using this very efficient bellows, bony fishes can pass water over the gills while remaining stationary in the water. That is what a goldfish is doing when it seems to be gulping in a fish tank.

The evolutionary path to land ran through the lobe-finned fishes

Three major groups of bony fish are the ray-finned fishes (class Actinopterygii), the lungfish (class Dipnoi), and the lobe-finned fishes (class Actinistia). Lobe-finned fishes evolved 390 MYA, shortly after the first bony fishes appeared. Only eight species survive today, two species of coelacanth and six species of lungfish. Although rare today, lobe-finned fishes played an important part in the evolutionary story of vertebrates. Tetrapods almost certainly evolved from the lobe-finned fishes (figure 28.9).

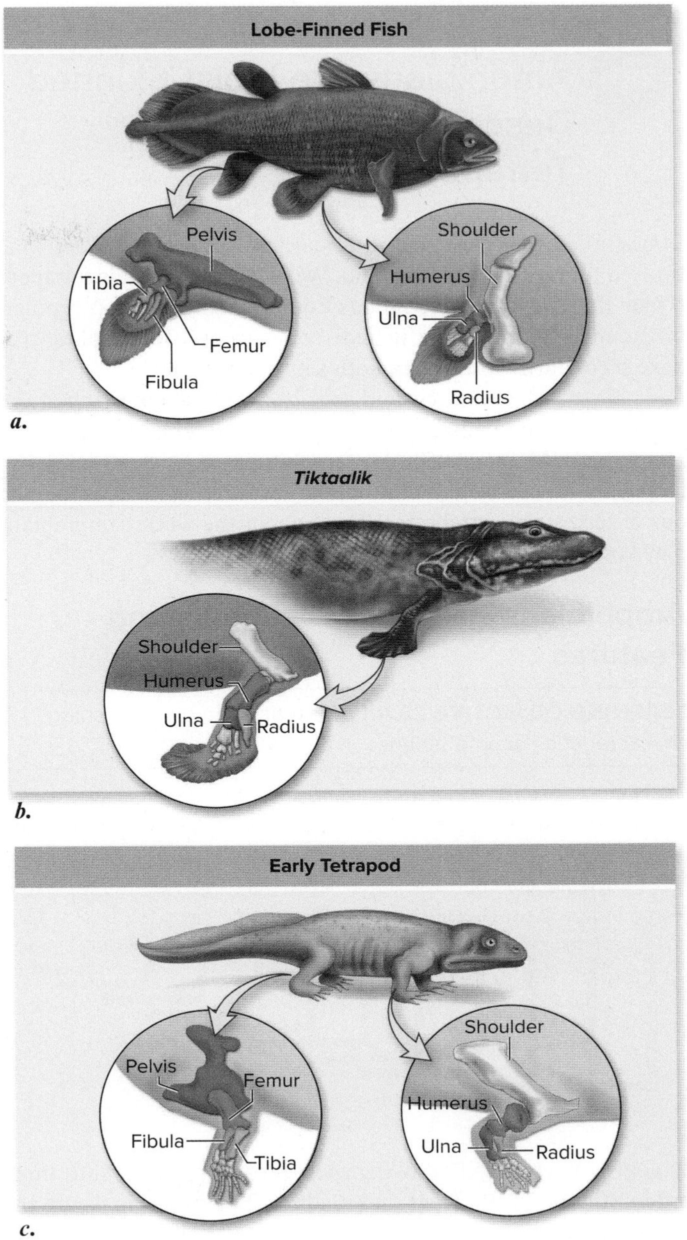

Figure 28.9 Comparison between limbs of a lobe-finned fish, *Tiktaalik,* and an early amphibian. ***a.*** A lobe-finned fish. Some of these animals could probably move on land. ***b.*** *Tiktaalik.* The shoulder and limb bones are like those of an amphibian, but the fins are like those of a lobe-finned fish. ***c.*** An early tetrapod. The legs of such an animal could function better for movement on land.

REVIEW OF CONCEPT 28.3

Fishes are generally characterized by the possession of a vertebral column, jaws, paired appendages, a lateral line system, internal gills, and single-loop circulation. Cartilaginous fishes have lightweight skeletons and were among the first vertebrates to develop teeth. The very successful bony fishes have unique characteristics such as swim bladders and gill covers, as well as ossified skeletons. One type of bony fish, the lobe-finned fish, gave rise to the ancestors of amphibians.

- *Are fish a monophyletic group?*

28.4 Amphibians Are Moist-Skinned Descendants of the Early Tetrapods

Lobe-finned fish gave rise to a species that left the water and evolved four legs to walk on land. We call these animals **tetrapods** ("four limbs"). Living tetrapods comprise amphibians, reptiles, birds, and mammals and include some groups, like snakes, that have secondarily lost one or both sets of limbs.

Frogs, salamanders, and caecilians, the damp-skinned vertebrates, are one clade of Tetrapoda, the amphibians (class Amphibia). Most present-day amphibians are small and live largely unnoticed by humans, but they are among the most numerous of terrestrial vertebrates. Throughout the world, amphibians play key roles in terrestrial food chains.

Amphibians Have Five Distinguishing Features

LEARNING OBJECTIVE 28.4.1 Describe the distinguishing characteristics of amphibians.

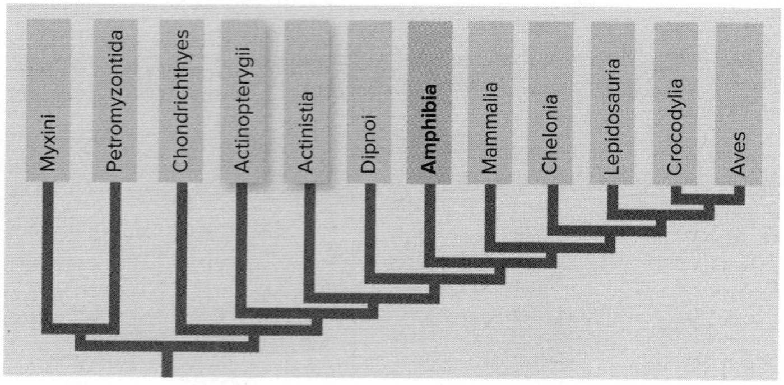

Biologists have classified living species of amphibians into three orders (figure 28.10): frogs and toads in 22 families make up the order Anura ("without a tail"); salamanders and newts in 9 families make up the order Caudata ("visible tail"); and 6 families of wormlike, nearly blind organisms called caecilians that live in the tropics make up the order Apoda ("without legs"). These amphibians have several key characteristics in common:

1. **Legs.** Frogs and most salamanders have four legs and can move about on land quite well. Legs were one of the key adaptations to life on land. Caecilians lost their legs during the course of adapting to a burrowing existence.
2. **Lungs.** Most amphibians possess a pair of lungs, although the internal surfaces have much less surface area than do reptilian or mammalian lungs. Amphibians breathe by lowering the floor of the mouth to suck in air, then raising it back to force the air down into the lungs.
3. **Cutaneous respiration.** Frogs, salamanders, and caecilians all supplement the use of lungs by respiring through their skin, which is kept moist and provides an extensive surface area.
4. **Pulmonary veins.** After blood is pumped through the lungs, two large veins called pulmonary veins return the aerated blood to the heart for repumping. In this way, aerated blood is pumped to the tissues at a much higher pressure.
5. **Partially divided heart.** A dividing wall helps prevent aerated blood from the lungs from mixing with nonaerated blood being returned to the heart from the rest of the body. The blood circulation is thus divided into two separate paths: pulmonary and systemic. The separation is imperfect, however, because no dividing wall exists in one chamber of the heart, the ventricle (refer to chapter 34).

Several other specialized characteristics are shared by all present-day amphibians. In all three orders, there is a zone of weakness between the base and the crown of the teeth. They also have a peculiar type of sensory rod cell in the retina of the eye called a "green rod," the function of which is unknown.

Diversification of the tetrapods

By moving onto land, tetrapods were able to utilize many resources and access many habitats. Tetrapods first became common during the Carboniferous period (360 to 280 MYA).

a. *b.* *c.*

Figure 28.10 Class Amphibia. ***a.*** A red-eyed tree frog, *Agalychnis callidryas* (order Anura). ***b.*** An adult tiger salamander, *Ambystoma tigrinum* (order Caudata). ***c.*** A caecilian, *Caecilia tentaculata* (order Apoda).

(a): Irina Kozorog/Shutterstock; (b): Suzanne L. Collins & Joseph T. Collins/Science Source; (c): Jany Sauvanet/Science Source

Fourteen families are known from the early Carboniferous, nearly all of them aquatic or semiaquatic, such as *Ichthyostega*. By the late Carboniferous, much of North America was covered by low-lying tropical swamplands, and a great variety of tetrapods thrived in this wet terrestrial environment.

In the early Permian period that followed (280 to 248 MYA), a remarkable change occurred among tetrapods—they began to leave the marshes for dry uplands. Many of these terrestrial tetrapods had bony plates and armor covering their bodies and grew to be very large, some as big as a pony. Both their large size and the complete covering of their bodies indicate that these tetrapods did not use the skin respiratory system of present-day amphibians but instead had an impermeable, leathery skin to prevent water loss. Consequently, they must have relied entirely on their lungs for respiration. By the mid-Permian period, most tetrapods were terrestrial or semiterrestrial.

Amniotes, one of the clades within the Tetrapoda, evolved in the Carboniferous. By the end of the Permian period, one amniote group, therapsids, had become common and ousted nonamniote tetrapods from terrestrial habitats.

Modern Amphibians Belong to Three Groups

LEARNING OBJECTIVE 28.4.2 Contrast the three major orders of living amphibians.

One clade of aquatic nonamniote tetrapods survived the rise of the therapsids and the subsequent extinction that occurred at the end of the Age of the Dinosaurs. During the Tertiary period (66 to 2 MYA), these moist-skinned amphibians accomplished a highly successful invasion of wet habitats all over the world, and today there are over 7000 species of amphibians in the orders Anura (frogs), Caudata (salamanders), and Apoda (caecilians).

Order Anura: Frogs and toads

Frogs and toads, amphibians without tails, live in a variety of environments, from deserts and mountains to ponds and puddles (figure 28.10*a*). Frogs have smooth, moist skin; a broad body; and long hind legs that make them excellent jumpers. Most frogs live in or near water and go through an aquatic tadpole stage before metamorphosing into frogs; however, some tropical species that don't live near water bypass this stage and hatch out as little froglets. Unlike frogs, toads have a dry, bumpy skin and short legs, and they are well adapted to dry environments.

Most frogs and toads return to water to reproduce, laying their eggs directly in water. Their eggs lack watertight external membranes and would dry out quickly on land. Eggs are fertilized externally and hatch into swimming larval forms called tadpoles. Tadpoles live in the water, where they generally feed on algae. After considerable growth, the tadpole undergoes metamorphosis into an adult frog.

Order Caudata: Salamanders

Salamanders have elongated bodies, long tails, and smooth, moist skin (figure 28.10*b*). They typically range in length from a few inches to a foot, although giant Asiatic salamanders of the genus *Andrias* are as much as 1.5 m long and weigh up to 33 kg. Most salamanders live in moist places, such as under stones or logs, or among the leaves of tropical plants. Some salamanders live entirely in water.

Like anurans, many salamanders go through a larval stage before metamorphosing into adults. However, unlike anurans, in which the tadpole is strikingly different from the adult frog, larval salamanders are quite similar to adults, although most live in water and have external gills and gill slits, which disappear at metamorphosis.

Order Apoda: Caecilians

Caecilians, members of the order Apoda, are a highly specialized group of tropical burrowing amphibians (figure 28.10*c*). These legless, wormlike creatures average about 30 cm long but can be up to 1.3 m long. They have very small eyes and many are blind. They resemble worms but have jaws with teeth. They eat worms and other soil invertebrates. Fertilization is internal.

REVIEW OF CONCEPT 28.4

Amphibians—including frogs and toads, salamanders, and caecilians—are generally characterized by legs, lungs, cutaneous respiration, and a more complex and divided circulatory system. All of these features developed as adaptations to life on land. Most species rely on a water habitat for reproduction. Although some early forms reached the size of a pony, modern amphibians are generally quite small.

- *What challenges did tetrapods overcome to make the transition to living on land?*

28.5 Reptiles Are Fully Adapted to Terrestrial Living

If we think of amphibians as a first draft of a manuscript about survival on land, then amniotes—the clade that comprises reptiles, mammals, and birds—are the finished book. For each of the five key challenges of living on land, amniotes (named for the amniotic egg they all possess) improved on the innovations of amphibians. The arrangement of legs evolved to support the body's weight more effectively, allowing amniote bodies to be bigger and to run. Lungs and heart became more efficient. The skin was covered with dry plates or scales—followed later by the evolution of hair in mammals and feathers in birds—to minimize water loss, and watertight coverings evolved for eggs.

Amniotes Exhibit Three Key Characteristics

LEARNING OBJECTIVE 28.5.1 Describe the significance of the evolution of the amniotic egg.

All living **amniotes** share certain fundamental characteristics, features they retain from the time when they replaced nonamniote tetrapods as dominant terrestrial vertebrates. Among the most important are the following:

1. **Amniotic eggs.** Amphibians' eggs must be laid in water or a moist setting to avoid drying out. Ancestral amniotes laid watertight eggs that contained a food source (the yolk) and a series of four membranes: the yolk sac, the amnion, the allantois, and the chorion (figure 28.11). Each membrane plays a role in making the egg an independent life-support system. All modern reptiles, as well as birds and mammals, show exactly this same pattern of membranes within the egg. This is true regardless of whether the egg is covered with a calcium shell and laid, as seen in all birds, many reptiles, and a few mammals, or the egg is retained within the body and the young are born alive, as in some reptiles and almost all mammals.

   The outermost membrane of the egg is the **chorion,** which lies just beneath the porous shell. It allows the exchange of respiratory gases but retains water. The **amnion** encases the developing embryo within a fluid-filled cavity. The **yolk sac** provides food from the yolk for the embryo via blood vessels connecting to the embryo's gut. The **allantois** surrounds a cavity into which waste products from the embryo are excreted.
2. **Dry skin.** Most surviving amphibians have moist skin and must remain in moist places to avoid drying out. Reptiles have dry, watertight skin. A layer of scales covers their bodies, preventing water loss. These scales develop as surface cells fill with keratin, the same protein that forms claws, hair, and bird feathers.
3. **Thoracic breathing.** Amphibians breathe by squeezing their throat to pump air into their lungs; this limits their breathing capacity to the volume of their mouths. Reptiles developed pulmonary breathing, expanding and contracting the rib cage to suck air into the lungs and then force it out. The capacity of this system is limited only by the volume of the lungs.

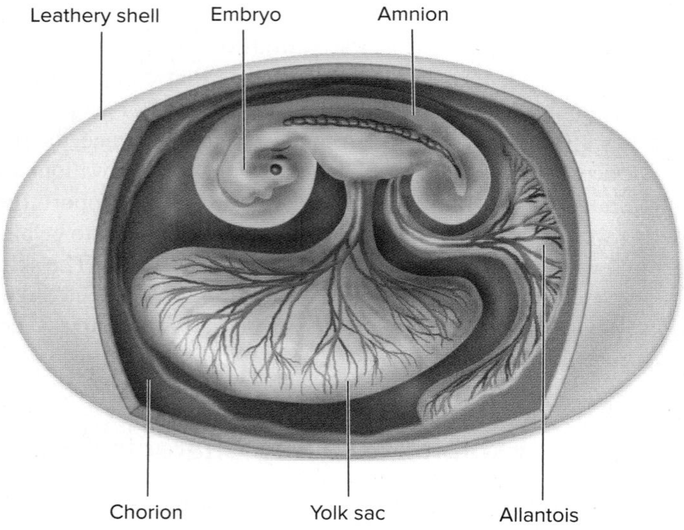

Leathery shell
Embryo
Amnion
Chorion
Yolk sac
Allantois

Figure 28.11 The watertight egg. The amniotic egg is perhaps the most important feature that allows reptiles to live in a wide variety of terrestrial habitats.

The Amniota Contains the Synapsids and Diapsids

LEARNING OBJECTIVE 28.5.2 Distinguish between synapsids and diapsids.

The amniote clade contains two major groups. One, the Synapsida, gave rise to mammals, whereas birds and reptiles belong to the other, the Diapsida.

Synapsids

An important feature of amniote classification is the presence and number of openings behind the eyes (figure 28.12). Amniotes' jaw muscles were anchored to these holes, which allowed them to bite more powerfully. The first group to rise to dominance were the **synapsids,** whose skulls had a single temporal hole behind the opening for each eye.

Pelycosaurs, an important group of early synapsids, were dominant for 50 million years and made up 70% of all land vertebrates; some species weighed as much as 200 kg. With long, sharp, "steak-knife" teeth, these pelycosaurs were the first land vertebrates to kill prey their own size.

About 250 MYA, pelycosaurs were replaced by another type of synapsid, the therapsids. Some evidence indicates that they may have been endotherms and perhaps even possessed hair. This would have permitted therapsids to be far more active than other vertebrates of that time, when winters were cold and long.

For 20 million years, therapsids were the dominant land vertebrate, until they were largely replaced 230 MYA by another group of amniotes, the diapsids, the group that includes

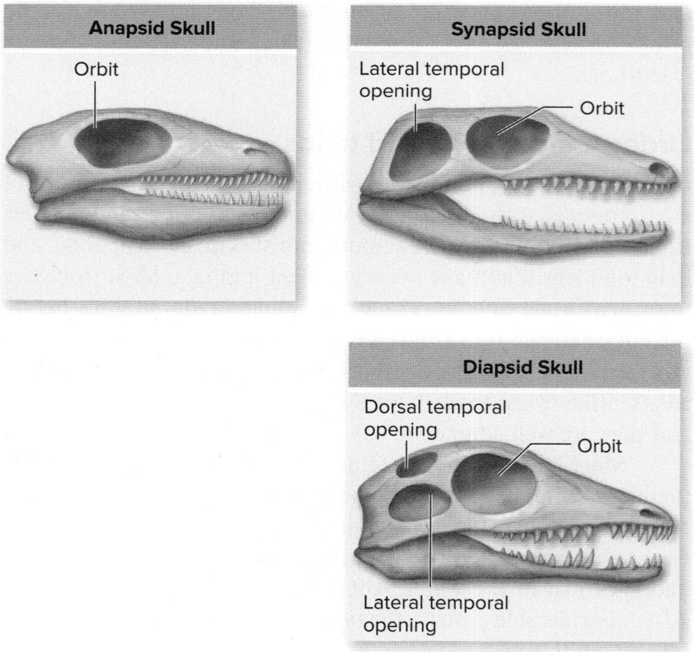

Anapsid Skull
Orbit
Synapsid Skull
Lateral temporal opening
Orbit
Diapsid Skull
Dorsal temporal opening
Orbit
Lateral temporal opening

Figure 28.12 Skulls of amniote groups. Amniote groups are distinguished by the number of holes on the side of the skull behind the eye orbit: 0 (anapsids), 1 (synapsids), or 2 (diapsids). Turtles are the only living anapsids, although there are several extinct groups.

reptiles and birds. Most therapsids became extinct 170 MYA, but one group survived and has living descendants today—the mammals.

Archosaurs

Diapsids have skulls with two holes on each side of the head. A variety of diapsids occurred in the Triassic period (248 to 213 MYA), but one group, the archosaurs, were of particular evolutionary significance, because they gave rise to crocodiles, pterosaurs, dinosaurs, and birds.

Among the early archosaurs were the largest animals the world had seen and the first land vertebrates to be bipedal—to stand and walk on two feet. By the end of the Triassic period, however, one archosaur group had risen to prominence: the dinosaurs.

Dinosaurs evolved about 220 MYA. Unlike previous bipedal diapsids, their legs were positioned directly underneath their bodies (figure 28.13). This design placed the weight of the body directly over the legs, which allowed dinosaurs to run with great speed and agility. Subsequently, a number of types of dinosaur evolved enormous size and reverted to a four-legged posture to support their massive weight. Other types—the theropods—became the most fearsome predators the Earth had ever seen, and one theropod line evolved to become birds. Dinosaurs went on to become the most successful of all land vertebrates, dominating for more than 150 million years. All dinosaurs, except their bird descendants, became extinct rather abruptly 66 MYA, apparently as a result of an asteroid's impact.

Important characteristics of modern reptiles

As you might imagine from the structure of the amniotic egg, reptiles and other amniotes do not practice external fertilization, as most amphibians do. Sperm would be unable to penetrate the membrane barriers protecting the egg. Instead, the male places sperm inside the female, where sperm fertilizes the egg before the protective membranes are formed. This is called internal fertilization.

Figure 28.13 Mounted skeleton of *Afrovenator*. This bipedal carnivore was about 30 feet long and lived in Africa about 130 MYA.

Didier Dutheil/Getty Images

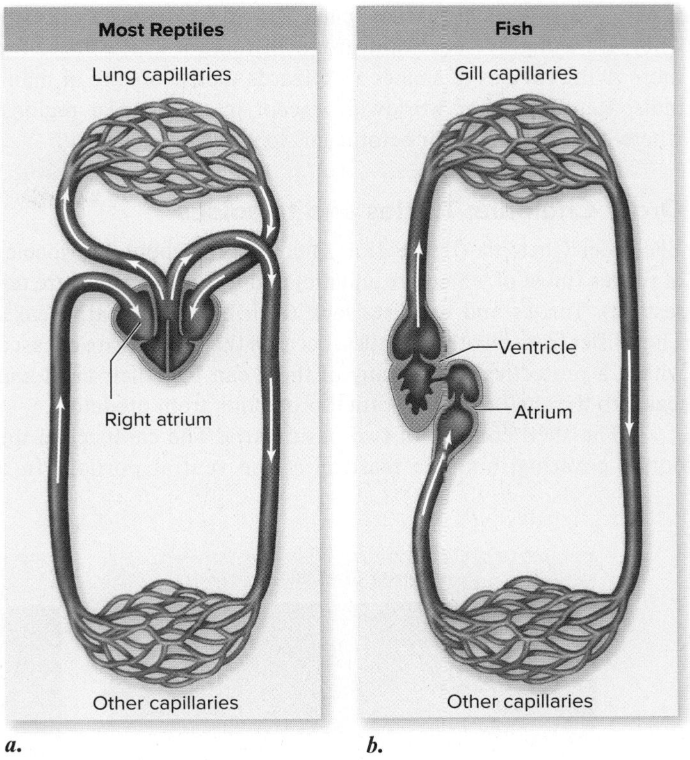

Figure 28.14 A comparison of reptile and fish circulation. ***a.*** In most reptiles, oxygenated blood (red) is repumped after leaving the lungs, and circulation to the rest of the body remains vigorous. ***b.*** The blood in fishes flows from the gills directly to the rest of the body, resulting in slower circulation.

The circulatory system of reptiles is an improvement over that of fish and amphibians, providing oxygen to the body more efficiently (figure 28.14 and refer to chapter 34). The improvement is achieved by extending the septum within the heart from the atrium partway across the ventricle. This septum creates a partial wall that tends to lessen the mixing of oxygen-poor blood with oxygen-rich blood within the ventricle. In crocodiles, the septum completely divides the ventricle, creating a four-chambered heart, just as it does in birds and mammals (and probably did in later dinosaurs).

Modern Reptiles Belong to Four Orders

LEARNING OBJECTIVE 28.5.3 Describe the characteristics of the major groups of living reptiles.

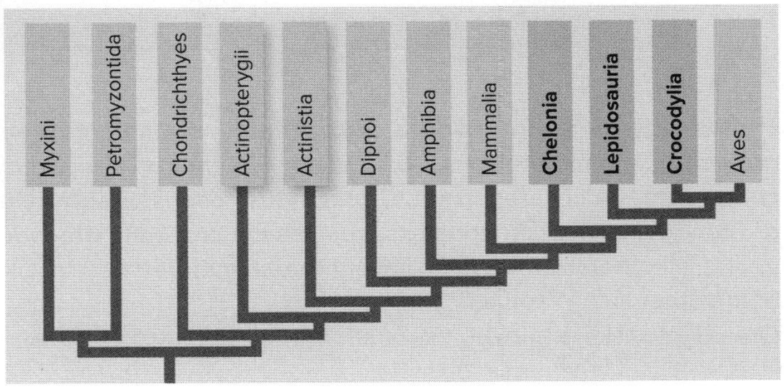

Over 7000 species of reptiles (class Reptilia) now live on Earth. They are a highly successful group in today's world; there are more living species of snakes and lizards than there are of mammals. Reptiles occur worldwide except in the coldest regions, where it is impossible for ectotherms to survive.

Order Chelonia: Turtles and tortoises

The order Chelonia (figure 28.15*a*) consists of about 250 species of turtles (most of which are aquatic) and tortoises (which are terrestrial). Turtles and tortoises lack teeth but have sharp beaks. They differ from all other reptiles, because their bodies are encased within a protective shell. Many of them can pull their head and legs into the shell as well, for total protection from predators.

The shell consists of two basic parts. The carapace is the dorsal covering, and the plastron is the ventral portion. In a fundamental commitment to this shell architecture, the vertebrae and ribs of most turtle and tortoise species are fused to the inside of the carapace. All of the support for muscle attachment comes from the shell.

Although marine turtles spend their lives at sea, they must return to land to lay their eggs. Many species migrate long distances to do this. Atlantic green turtles *(Chelonia mydas)* migrate from their feeding grounds off the coast of Brazil to Ascension Island in the middle of the South Atlantic—a distance of more than 2000 km—to lay their eggs on the same beaches where they themselves hatched.

Order Rhynchocephalia: Tuataras

Today the order Rhynchocephalia contains only one species of tuatara, a large, lizard-like animal about half a meter long

Figure 28.15 Living orders of reptiles. ***a.*** Chelonia. The Indian tent turtle, *Pangshura tentoria*. The domed shell provides protection against predators. ***b.*** Tuatara, *Sphenodon punctatus*. The sole living members of the ancient group Rhynchocephalia. ***c.*** Squamata. A collared lizard, *Crotaphytus collaris*, is shown left, and a smooth green snake, *Liochlorophis vernalis*, on the right. ***d.*** Crocodylia. Most crocodilians, such as the crocodile *Crocodylus acutus*, resemble birds and mammals in having four-chambered hearts; all other living reptiles have three-chambered hearts. Like birds, crocodiles are more closely related to dinosaurs than to any of the other living reptiles.

(a): sabirmallick/iStockphoto/Getty Images; (b, c left): Jonathan Losos; (c right): Rod Planck; (d): Judd Patterson/National Park Service (figure 28.15*b*). New Zealand is the only place in the world where this endangered species is found. The limited diversity of modern rhynchocephalians belies their rich evolutionary past: in the Triassic period, rhynchocephalians experienced a great adaptive radiation, producing many species that differed greatly in size and habitat.

An unusual feature of the tuatara is the inconspicuous "third eye" on the top of its head, called a parietal eye. Concealed under a thin layer of scales, the eye has a lens and a retina and is connected by nerves to the brain. Rhynchocephalians are the closest relatives of snakes and lizards, with whom they form the group Lepidosauria.

Order Squamata: Lizards and snakes

The order Squamata (figure 28.15*c*) includes 3800 species of lizards and about 3000 species of snakes. A distinguishing characteristic of this order is the presence of paired copulatory organs in the male. In addition, changes to the morphology of the head and jaws allow greater strength and mobility. Most lizards and snakes are carnivores, preying on insects and small animals, and these improvements in jaw design have made a major contribution to their evolutionary success.

Snakes, which evolved from a lizard ancestor, are characterized by the lack of limbs, movable eyelids, and external ears, as well as a great number of vertebrae (sometimes more than 300). Limblessness has actually evolved more than a dozen times in lizards; snakes are simply the most extreme case of this evolutionary trend.

Common lizards include iguanas, chameleons, geckos, and anoles. Most are small, measuring less than a foot in length. The largest lizards belong to the monitor family. The largest of all monitor lizards is the Komodo dragon of Indonesia, which reaches 3 m in length and can weigh more than 100 kg. Snakes also vary in length from only a few inches to more than 10 m.

Many lizards and snakes rely on agility and speed to catch prey and elude predators. Only two species of lizard are venomous, the Gila monster of the southwestern United States and the beaded lizard of western Mexico. Similarly, most species of snakes are nonvenomous. Of the 13 families of snakes, only 4 contain venomous species: the elapids (cobras, kraits, and coral snakes); the sea snakes; the vipers (adders, bushmasters, rattlesnakes, water moccasins, and copperheads); and some colubrids (African boomslang and twig snake).

Order Crocodylia: Crocodiles and alligators

The order Crocodylia is composed of 28 species of large, primarily aquatic reptiles (figure 28.15*d*). In addition to crocodiles and alligators, the order includes the less familiar caimans and gavials. Although all crocodilians are similar in appearance today, much greater diversity existed in the past, including species that were entirely terrestrial and others that achieved a total length in excess of 50 feet.

Crocodiles are largely nocturnal animals that live in or near water in tropical or subtropical regions of Africa, Asia, and the Americas. The American crocodile *(Crocodylus acutus)* is found in southern Florida, in Cuba, and throughout tropical Central America. Nile crocodiles *(Crocodylus niloticus)* and estuarine crocodiles *(Crocodylus porosus)* can grow to enormous size and are responsible for many human fatalities each year.

There are only two species of alligators: one living in the southern United States *(Alligator mississippiensis)* and the other a rare endangered species living in China *(Alligator sinensis)*. Caimans, which resemble alligators, are native to Central America. Gharials, or gavials, are a group of fish-eating crocodilians with long, slender snouts that live only in India and Burma.

All crocodilians are carnivores. They generally hunt by stealth, waiting in ambush for prey and then attacking ferociously. Their bodies are well adapted for this form of hunting, with eyes on top of their heads and their nostrils on top of their snouts, so that they can see and breathe while lying quietly submerged in water. They have enormous mouths, studded with sharp teeth, and very strong necks. A valve in the back of the mouth prevents water from entering the air passage when a crocodilian feeds under water.

In many ways, crocodiles resemble birds far more than they do other living reptiles. For example, crocodiles build nests and care for their young (traits they share with birds and at least some dinosaurs), and they have a four-chambered heart, as birds do. Most biologists agree that birds are, in fact, the direct descendants of dinosaurs, which means that crocodiles and birds are more closely related to each other than either is to lizards and snakes.

REVIEW OF CONCEPT 28.5

Reptiles have a hard, scaly skin that minimizes water loss, thoracic breathing, and an enclosed, amniotic egg that does not need to be laid in water. Synapsids had a single hole in the skull behind the eyes and included ancestors of mammals; diapsids had two holes and are ancestors of modern reptiles and birds. The four living orders of reptiles include the turtles and tortoises, tuataras, lizards and snakes, and crocodiles.

- *How do reptile eggs differ from those of amphibians?*

28.6 Birds Are Essentially Flying Reptiles

Today, birds (class **Aves**) are the most diverse of all terrestrial vertebrates, with 28 orders containing a total of 166 families and about 8600 species. The success of birds lies in the development of a structure unique in the animal world: the feather. Developed from reptilian scales, feathers are the ideal adaptation for flight, serving as lightweight airfoils that are easily replaced if damaged (unlike the vulnerable skin wings of bats and the extinct pterosaurs).

Key Characteristics of Birds Are Feathers and a Lightweight Skeleton

LEARNING OBJECTIVE 28.6.1 Describe the key characteristics of birds.

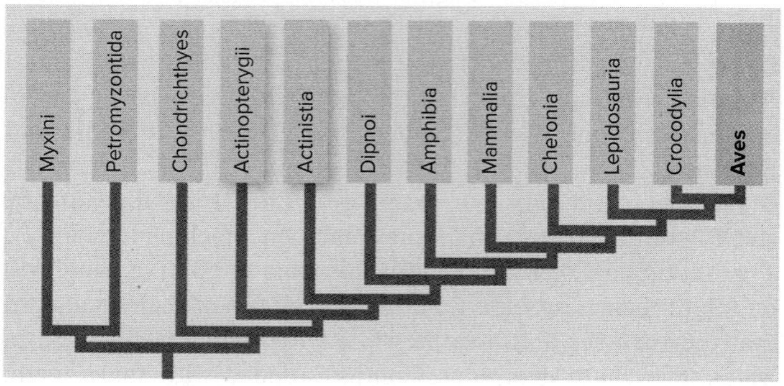

Modern birds lack teeth and have only vestigial tails, but they retain many reptilian characteristics. For instance, birds lay amniotic eggs. Also, reptilian scales are present on the feet and lower legs of birds. Two primary characteristics distinguish birds from living reptiles:

1. **Feathers.** Feathers are modified reptilian scales made of keratin, just like hair and scales. Feathers serve two functions: providing lift for flight and conserving heat. The structure of feathers combines maximum flexibility and strength with minimum weight (figure 28.16).

   Feathers develop from tiny pits in the skin called follicles. In a typical flight feather, a shaft emerges from the follicle, and pairs of vanes develop from its opposite sides. At maturity, each vane has many branches called barbs. The barbs, in turn, have many projections called barbules, which are equipped with microscopic hooks. These hooks link the barbs to one another, giving the feather a continuous surface and a sturdy but flexible shape.

   Like scales, feathers can be replaced. Among living animals, feathers are unique to birds. Recent fossil finds suggest that some dinosaurs may have had feathers.

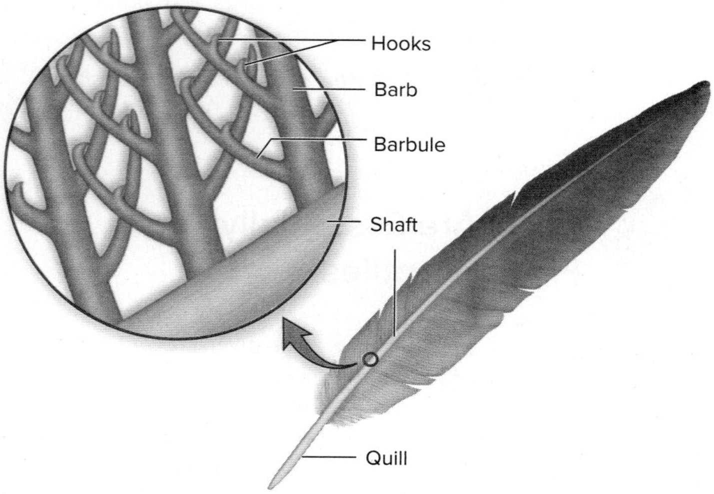

Figure 28.16 A feather. The enlargement shows how the secondary branches and barbs of the vanes are linked together by microscopic barbules.

2. **Flight skeleton.** The bones of birds are thin and hollow. Many of the bones are fused, making the bird skeleton more rigid than a reptilian skeleton. The fused sections of backbone and of the shoulder and hip girdles form a sturdy frame that anchors muscles during flight. The power for active flight comes from large breast muscles, which can make up 30% of a bird's total body weight. These muscles stretch down from the wing and attach to the breastbone, which is greatly enlarged and bears a prominent keel for muscle attachment. Breast muscles also attach to the fused collarbones that form the so-called wishbone. No other living vertebrates have a fused collarbone or a keeled breastbone.

Birds Arose About 150 MYA

LEARNING OBJECTIVE 28.6.2 Explain why some consider birds to be one type of reptile.

A 150-million-year-old fossil of the first known bird, *Archaeopteryx* (figure 28.17), was found in 1862 in a limestone quarry in Germany, the impression of its feathers stamped clearly into the rocks. The skeleton of *Archaeopteryx* shares many features with those of small theropod dinosaurs. About the size of a crow, *Archeopteryx* had a skull with teeth. Very few of its bones were fused to one another, and its bones are thought to have been solid, not hollow like a bird's. Also, it had a long, reptilian tail and no enlarged breastbone such as modern birds use to anchor flight muscles. Finally, the skeletal structure of the forelimbs was nearly identical to that of theropods.

Because of its many dinosaur features, several *Archaeopteryx* fossils were originally classified as *Compsognathus,* a small theropod dinosaur of similar size—until feathers were discovered on the fossils. What makes *Archaeopteryx* distinctly avian is the presence of feathers on its wings and tail.

The remarkable similarity of *Archaeopteryx* to *Compsognathus* has led almost all paleontologists to conclude that *Archaeopteryx* is the direct descendant of dinosaurs—indeed, that today's birds are "feathered dinosaurs." Recent discoveries of fossils of feathered dinosaurs in China lend strong support to this inference. The dinosaur *Caudipteryx,* for example, is clearly intermediate between *Archaeopteryx* and dinosaurs, having large feathers on its tail and arms but also many features of dinosaurs like *Velociraptor* (figure 28.17). Because the arms of *Caudipteryx* were too short to use as wings, feathers probably didn't evolve for flight but instead served as insulation, much as fur does for mammals.

Flight is an ability certain kinds of dinosaurs achieved as they evolved longer arms. We call these dinosaurs birds. Despite their close affinity to dinosaurs, birds exhibit three evolutionary novelties: feathers, hollow bones, and physiological mechanisms, such as superefficient lungs, that permit sustained, powered flight.

By the early Cretaceous period, only a few million years after *Archaeopteryx* lived, a diverse array of birds had evolved, with many of the features of modern birds. Fossils in Mongolia, Spain, and China discovered within the last few years reveal a diverse collection of toothed birds with the hollow bones and breastbones necessary for sustained flight (figure 28.18). Other fossils reveal highly specialized, flightless diving birds. These diverse birds shared the skies with pterosaurs for 70 million years until the flying reptiles went extinct at the end of the Cretaceous.

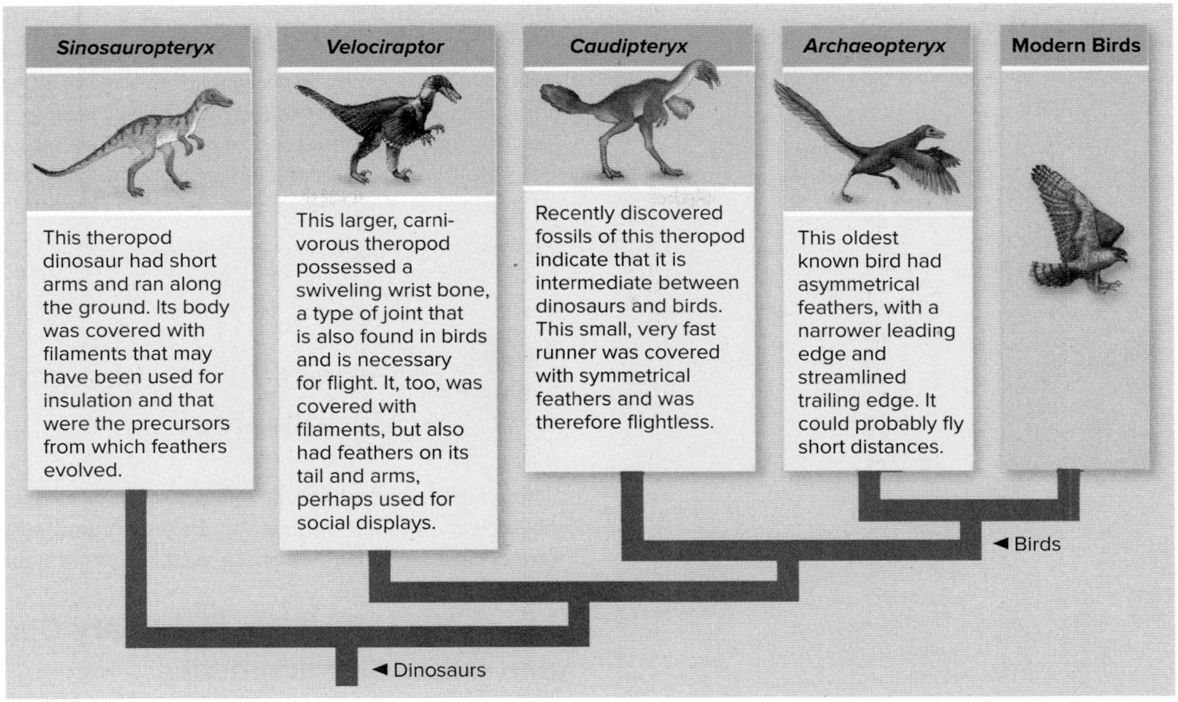
Sinosauropteryx
Velociraptor
Caudipteryx
Archaeopteryx
Modern Birds
This theropod dinosaur had short arms and ran along the ground. Its body was covered with filaments that may have been used for insulation and that were the precursors from which feathers evolved.
This larger, carnivorous theropod possessed a swiveling wrist bone, a type of joint that is also found in birds and is necessary for flight. It, too, was covered with filaments, but also had feathers on its tail and arms, perhaps used for social displays.
Recently discovered fossils of this theropod indicate that it is intermediate between dinosaurs and birds. This small, very fast runner was covered with symmetrical feathers and was therefore flightless.
This oldest known bird had asymmetrical feathers, with a narrower leading edge and streamlined trailing edge. It could probably fly short distances.
Birds
Dinosaurs

Figure 28.17 The evolutionary path to the birds. Almost all paleontologists now accept the theory that birds are the direct descendants of theropod dinosaurs.

Because the impression of feathers is rarely fossilized and modern birds have hollow, delicate bones, the fossil record of birds is incomplete. Relationships among the 166 families of modern birds are mostly inferred from studies of anatomy and degree of DNA similarity.

Modern Birds Are Diverse but Share Several Characteristics

LEARNING OBJECTIVE 28.6.3 Explain the adaptations birds have to cope with the energetic demands of flight.

The most ancient living birds appear to be the flightless birds, such as the ostrich. Ducks, geese, and other waterfowl evolved next, in the early Cretaceous, followed by a diverse group of woodpeckers, parrots, swifts, and owls. The largest of the bird orders, Passeriformes, evolved in the mid-Cretaceous and comprises 60% of species alive today. Overall, there are 28 orders of birds, the largest consisting of over 5000 species (figure 28.19).

You can tell a great deal about the habits and food of a bird by examining its beak and feet. For instance, carnivorous birds such as owls have curved talons for seizing prey and sharp beaks for tearing apart their meal. The beaks of ducks are flat for shoveling through mud, and the beaks of finches are short, thick seed-crushers.

Many adaptations enabled birds to cope with the heavy energy demands of flight, including respiratory and circulatory adaptations and endothermy.

Efficient respiration

Flight muscles consume an enormous amount of oxygen during active flight. The reptilian lung has a limited internal surface area, not nearly enough to absorb all the oxygen needed. Mammalian lungs have a greater surface area, but bird lungs satisfy this challenge with a radical redesign.

When a bird inhales, the air goes past the lungs to a series of air sacs located near and within the hollow bones of the back; from there, the air travels to the lungs and then to a set of anterior air sacs before being exhaled. Because air passes all the way through the lungs in a single direction, gas exchange is highly efficient. Respiration in birds is described in more detail in chapter 34.

Efficient circulation

The revved-up metabolism needed to power active flight also requires very efficient blood circulation, so that the oxygen captured by the lungs can be delivered to the flight muscles quickly. In the heart of most living reptiles, oxygen-rich blood coming from the lungs mixes with oxygen-poor blood returning from the body, because the wall dividing the ventricle into two chambers is not complete. In birds, the wall dividing the ventricle is complete, and the two blood circulations do not mix—thus, flight muscles receive fully oxygenated blood.

In comparison with reptiles and most other vertebrates, birds have a rapid heartbeat. A hummingbird's heart beats about 600 times a minute, and an active chickadee's heart beats 1000 times a minute. In contrast, the heart of the large, flightless ostrich averages 70 beats per minute.

Endothermy

Birds, like mammals, are endothermic (an example of convergent evolution). Many paleontologists believe that the dinosaurs from which birds evolved were endothermic as well. Birds maintain body temperatures significantly higher than those of most mammals, ranging from 40° to 42°C (human body temperature is 37°C). Feathers provide excellent insulation, helping to conserve body heat.

Figure 28.18 A fossil bird from the early Cretaceous. *Confuciusornis* had long tail feathers. Some fossil specimens of this species lack the long tail feathers, suggesting that this trait was present in only one sex, as in some modern birds.
Layne Kennedy/Getty Images

The high temperatures maintained by endothermy permit metabolism in the bird's flight muscles to proceed at a rapid pace, to provide the ATP necessary to drive rapid muscle contraction.

REVIEW OF CONCEPT 28.6

Birds have the greatest diversity of species of all terrestrial vertebrates. *Archaeopteryx,* the oldest fossil bird, exhibited many traits shared with theropod dinosaurs. Key features of birds are feathers and a lightweight, hollow skeleton; additional features include auxiliary air sacs and a four-chambered heart.

- *What traits do birds share with reptiles?*

28.7 Mammals Are the Least Diverse of Vertebrates

There are about 5400 living species of mammals (class Mammalia), fewer than the number of fishes, amphibians, reptiles, or birds. Most large, land-dwelling vertebrates are mammals.

Mammals Have Hair, Mammary Glands, and Other Characteristics

LEARNING OBJECTIVE 28.7.1 Describe the characteristics of mammals.

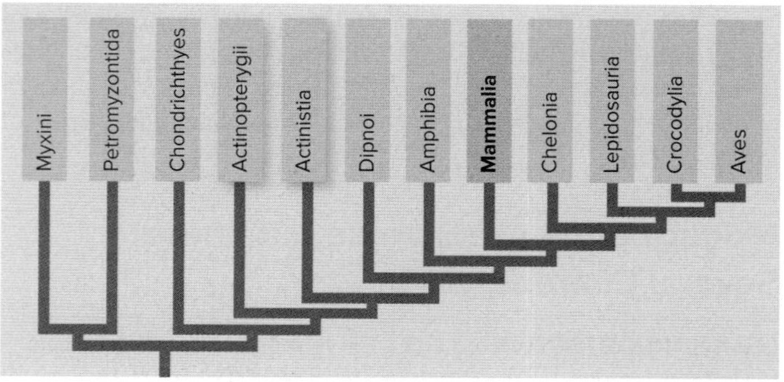

a. *b.* *c.* *d.*

Figure 28.19 Diversity of Passeriformes, the largest order of birds. ***a.*** Variable sunbird, *Cinnyris venustus.* ***b.*** Indigo bunting, *Passerina cyanea.* ***c.*** Blue Jay, *Cyanocitta cristata.* ***d.*** Blue-winged Pitta, *Pitta moluccensis.*
(a): Aubrey Stoll/Getty Images; (b): Dave Menke/USFWS; (c): Frank Miles/USFWS; (d): Myron Tay/Getty Images

When we look out over an African plain, we see the big mammals—the lions, zebras, gazelles, and antelopes. But the typical mammal is not that large. Of the 5400 species of mammals, 4000 are rodents, bats, shrews, or moles.

Mammals are distinguished from all other classes of vertebrates by two fundamental characteristics—hair and mammary glands—and are marked by two other notable features, endothermy and a placenta:

1. **Hair.** All mammals have hair. Even apparently hairless whales and dolphins grow sensitive bristles on their snouts. An individual mammalian hair is a long filament that extends like a stiff thread from a bulblike foundation beneath the skin, which is known as a hair follicle. The filament is composed mainly of dead cells filled with the fibrous protein keratin. The evolution of fur enabled mammals to invade the colder climates that ectothermic reptiles do not inhabit. Mammals typically maintain body temperatures higher than their surroundings, and the dense undercoat of many mammals reduces loss of body heat.

   Another function of hair is camouflage. The coloration and pattern of a mammal's coat usually match its background. Hairs also function as sensory structures. The whiskers of cats and dogs are stiff hairs that are very sensitive to touch. Finally, hair can serve as a defensive weapon. Porcupines and hedgehogs protect themselves with long, sharp, stiff hairs called quills.
2. **Mammary glands.** Female mammals possess mammary glands that can secrete milk. Newborn mammals, born without teeth, suckle this milk as their primary food. Even baby whales are nursed by their mother's milk. Milk is a very high-calorie food (human milk has 750 kcal per liter), important because of the high energy needs of a rapidly growing newborn. About 50% of the energy in the milk comes from fat.
3. **Endothermy.** Endothermy is a crucial adaptation that has allowed mammals to be active at any time of the day or night and to colonize severe environments, from deserts to ice fields. Also, more efficient blood circulation provided by the four-chambered heart and more efficient respiration provided by the *diaphragm* (refer to chapter 34) make higher metabolic rates possible.
4. **Placenta.** In most mammal species, females carry their developing young internally in a uterus, nourishing them through the placenta, and give birth to live young. The placenta is a specialized organ that brings the bloodstream of the fetus into close contact with the bloodstream of the mother (figure 28.20). Food, water, and oxygen can pass across from mother to child, and wastes can pass over to the mother's blood to be eliminated.

In addition to these main characteristics, the mammalian lineage gave rise to several other adaptations in certain groups. These include specialized teeth, the ability of grazing animals to digest plants, hooves and horns made of keratin, and adaptations for flight in bats.

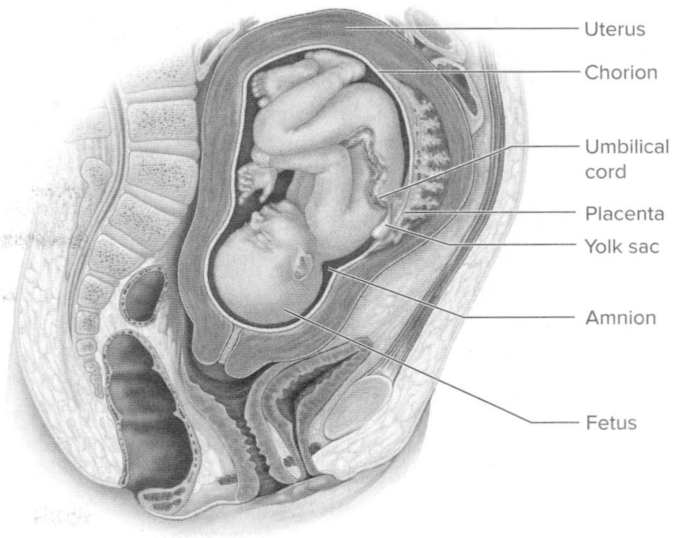
Uterus
Chorion
Umbilical cord
Placenta
Yolk sac
Amnion
Fetus

Figure 28.20 The placenta. The placenta is characteristic of the largest group of mammals, the placental mammals. It evolved from membranes in the amniotic egg. The umbilical cord evolved from the allantois. The chorion, or outermost part of the amniotic egg, forms most of the placenta itself. The placenta serves as the provisional lungs, intestine, and kidneys of the fetus, without ever mixing maternal and fetal blood.

Specialized teeth

Mammals have different types of teeth that are highly specialized to match particular eating habits. It is usually possible to determine a mammal's diet simply by examining its teeth. A dog's long canine teeth, for example, are well suited for biting and holding prey, and some of its premolar and molar teeth are triangular and sharp for ripping off chunks of flesh.

In contrast, large herbivores such as deer lack canine teeth; instead, a deer clips off mouthfuls of plants with flat, chisel-like incisors on its lower jaw. The deer's molars are large and covered with ridges to effectively grind and break up tough plant tissues.

Digestion of plants

Most mammals are herbivores, eating mostly (or only) plants. Cellulose forms the bulk of a plant's body and is a major source of food for mammalian herbivores. Mammals lack the enzymes, however, to degrade cellulose. Herbivorous mammals rely on a mutualistic partnership with bacteria in their digestive tracts that have the necessary cellulose-splitting enzymes.

Mammals such as cows, buffalo, antelope, goats, deer, and giraffes have huge, four-chambered fermentation vats derived from the esophagus and stomach. The first chamber is the largest and holds a dense population of cellulose-digesting bacteria. Chewed plant material passes into this chamber, where the bacteria attack the cellulose. The material is then digested further in the other three chambers.

Rodents, horses, rabbits, and elephants, by contrast, have relatively small stomachs; they digest plant material in their large intestine, as a termite does. The bacteria that actually carry out cellulose digestion live in a pouch, called the **cecum,** that branches from the end of the small intestine.

Development of hooves and horns

Keratin, the protein of hair, is also the structural building material in claws, fingernails, and hooves. Hooves are specialized keratin pads on the toes of horses, cows, sheep, and antelope. The pads are hard and horny, protecting the toe and cushioning it from impact. The horns of cattle, sheep, and antelope are composed of a core of bone surrounded by a sheath of keratin. The bony core is attached to the skull, and the horn is not shed.

Deer antlers are made not of keratin but of bone. Male deer grow and shed a set of antlers each year. While growing during the summer, antlers are covered by a thin layer of skin known as velvet.

Flying mammals: Bats

Bats are the only mammals capable of powered flight (figure 28.21). Like the wings of birds and pterosaurs, bat wings are modified forelimbs. The bat wing is a leathery membrane of skin and muscle stretched over the bones of four fingers. The edges of the membrane attach to the side of the body and to the hind leg. When resting, most bats prefer to hang upside down by the claws on their toes.

After rodents, bats are the second largest order of mammals. They have been a particularly successful group, because many species have been able to utilize a food resource that most birds do not use: night-flying insects.

How do bats navigate in the dark? Late in the 18th century, the biologist Lazzaro Spallanzani showed that a blinded bat could fly without crashing into things and still capture insects. Clearly, bats were using a sense other than vision to navigate in the dark. When Spallanzani plugged the ears of a bat, it was unable to navigate and collided with objects. Spallanzani concluded that bats "hear" their way through the night world.

Figure 28.21 Gambian epauletted fruit bat, *Epomophorus gambianus*. Bats are the only mammal capable of true flight.

Ivkuzmin/iStock/360/Getty Images

Modern Mammals Are Placed into Three Groups

LEARNING OBJECTIVE 28.7.2 Compare the three groups of living mammals.

Mammals have been around since the time of the dinosaurs, about 220 MYA. At the end of the Cretaceous period, 65 MYA, the dinosaurs and numerous other land and marine animals became extinct, but mammals survived, perhaps because they lived in burrows underground and were able to survive by scavenging and eating seeds. In the Tertiary period (lasting from 66 MYA to 2 MYA), mammals rapidly diversified, taking over many of the ecological niches once dominated by dinosaurs.

Mammals reached their maximum diversity late in the Tertiary period, about 15 MYA. At that time, tropical conditions existed over much of the world. During the last 15 million years, world climates have changed, and the area covered by tropical habitats has decreased, causing a decline in the total number of mammalian species.

For 155 million years, while the dinosaurs flourished, mammals were a minor group of small insectivores and herbivores. The earliest evolving mammals were members of the subclass **Prototheria.** Most prototherians were small and resembled modern shrews. All prototherians laid eggs, as did their synapsid ancestors. The only prototherians surviving today are the monotremes.

The other major mammalian group is the subclass **Theria.** Therians are viviparous (that is, their young are born alive). The two living therian groups are marsupials, or pouched mammals (including kangaroos, opossums, and koalas), and the placental mammals (dogs, cats, humans, horses, and most other mammals).

Monotremes: Egg-laying mammals

The duck-billed platypus *(Ornithorhynchus anatinus)* and four extant species of echidna are the only living **monotremes** (figure 28.22*a*). Among living mammals, only monotremes lay shelled eggs. The structure of their shoulder and pelvis is more similar to that of the early reptiles than to that of any other living mammal. Also like reptiles, monotremes have a cloaca, a single opening through which feces, urine, and reproductive products leave the body.

Despite the retention of some reptilian features, monotremes have the diagnostic mammalian characters: a single bone on each side of the lower jaw, fur, and mammary glands. Young monotremes drink their mother's milk after they hatch from eggs. Females lack well-developed nipples; instead, the milk oozes onto the mother's fur, and the babies lap it off with their tongues.

The platypus, found only in Australia, lives much of its life in the water and is a good swimmer. It uses its bill much as a duck does, rooting in the mud for worms and other soft-bodied animals. However, unlike ducks, the platypus has electroreceptors in its bill that can detect the electrical discharges produced by muscle contractions in its prey, helping it to locate its next meal. Echidnas, found in Australia and New Guinea, also have electroreceptors on their beaks. In addition, they have very strong, sharp claws, which they use for burrowing and digging. The echidna probes with its snout for insects, especially ants and termites.

Monotremes
a.
Marsupials
b.
Placental Mammals
c.

Figure 28.22 Today's mammals. ***a.*** Monotremes: the short-nosed echidna, *Tachyglossus aculeatus* (left), and the duck-billed platypus, *Ornithorhynchus anatinus* (right). ***b.*** Marsupials: the red kangaroo, *Macropus rufus* (left), and the opossum, *Didelphis virginiana* (right). ***c.*** Placental mammals: the lion, *Panthera leo* (left), and the bottle-nosed dolphin, *Tursiops truncatus* (right).

(a left): Sir Francis Canker Photography/Getty Images; (a right): Dave Watts/Alamy Stock Photo; (b left): Craig Dingle/E+/Getty Images; (b right): W. Perry Conway/Ramble/Getty Images; (c left): Jack Weinberg/Image Source/Getty Images; (c right): Corbis/Alamy Stock Photo

Marsupials: Pouched mammals

The major difference between **marsupials** (figure 28.22*b*) and other mammals is their pattern of embryonic development. In marsupials, a fertilized egg is surrounded by chorion and amniotic membranes, but no shell forms around the egg as it does in monotremes. During most of its early development, the marsupial embryo is nourished by an abundant yolk within the egg. Shortly before birth, a short-lived placenta forms from the chorion membrane. Soon after, sometimes within eight days of fertilization, the embryonic marsupial is born. It emerges tiny and hairless, and it crawls into the marsupial pouch, where it latches onto a mammary-gland nipple and continues its development.

Marsupials evolved shortly before placental mammals, about 125 MYA. Today most species of marsupials live in Australia and South America, areas that have undergone long periods of geographical isolation. Marsupials in Australia and New Guinea have diversified to fill ecological positions occupied by placental mammals elsewhere in the world. The placental mammals in Australia and New Guinea today arrived relatively recently and include some introduced by humans. The only marsupial found in North America is the Virginia opossum *(Didelphis virginiana),* which migrated north from Central America within the last 3 million years.

Placental mammals

A placenta that nourishes the embryo throughout its development forms in the uterus of placental mammals (figure 28.22*c*). Most species of mammals living today, including humans, are in this group. Of the 19 orders of living mammals, 17 are placental mammals (although some scientists recognize 4 or more orders of marsupials, rather than 1). They range in size from 1.5-g pygmy shrews to 100,000-kg whales.

Early in the course of embryonic development, the placenta forms. Both fetal and maternal blood vessels are abundant in the placenta, and substances can be exchanged efficiently between the bloodstreams of mother and offspring (figure 28.20). The fetal placenta is formed from the membranes of the chorion and allantois. In placental mammals, unlike in marsupials, the young undergo a considerable period of development before they are born.

REVIEW OF CONCEPT 28.7

Mammals are the only animals with hair and mammary glands. Other mammalian specializations include endothermy, the placenta, a tooth design suited to diet, and specialized sensory systems. Today three subgroups of mammals are recognized: monotremes, which lay eggs; marsupials, which feed embryonic young in a marsupial pouch; and placental mammals, in which the placenta nourishes the embryo throughout its development.

- *What features found in both mammals and birds are examples of convergent evolution?*

28.8 Primates Include Lemurs, Monkeys, Apes, and Humans

Primates are the mammalian group that includes our own species. Primates evolved two distinct features that allowed them to succeed as arboreal (tree-dwelling) insectivores:

1. **Grasping fingers and toes.** Unlike the clawed feet of tree shrews and squirrels, primates have grasping hands and

feet that enable them to grip limbs, hang from branches, seize food, and, in some primates, use tools. The first digit in many primates, the thumb, is opposable, and at least some of the digits have nails.

2. **Binocular vision.** Unlike the eyes of shrews and squirrels, which sit on either side of the head, the eyes of primates are shifted forward to the front of the face. This produces overlapping binocular vision, which lets the brain judge distance precisely—important to an animal moving through the trees and trying to grab or pick up food items.

The Anthropoid Lineage Led to the Earliest Humans

Figure 28.23
A tarsier. This tarsier, *Tarsius,* native to tropical Asia, shows the characteristic features of primates: grasping fingers and toes and binocular vision.
Magdalena Biskup Travel Photography/Getty Images

LEARNING OBJECTIVE
28.8.1 Distinguish among the major groups of primates.

Other mammals have binocular vision—for example, carnivorous predators—but only primates have both binocular vision and grasping hands, making them particularly well adapted to their arboreal environment.

Tarsiers, lemurs, and lorises used to be considered **prosimians** ("before monkeys"), but it is now realized that this group is paraphyletic, with tarsiers being more closely related to monkeys and apes than they are to lemurs and lorises (figure 28.23). Tarsiers, lorises, and some lemurs are small and nocturnal, but other lemur species are larger and diurnal. Lemurs occur only on the island of Madagascar, where they experienced an adaptive radiation in the absence of competition from monkeys, which do not occur there.

Anthropoids

Anthropoids include monkeys, apes, and humans. Anthropoids are almost all diurnal—that is, active during the day—feeding mainly on fruits and leaves. Natural selection favored many changes in eye design, including color vision, that were adaptations to daytime foraging. An expanded brain governs the improved senses, with the braincase forming a larger portion of the head.

Anthropoids, like the diurnal lemurs, live in groups with complex social interactions. They tend to care for their young for prolonged periods, allowing for a long childhood of learning and brain development.

About 30 MYA some anthropoids migrated to South America. Their descendants, known as the New World monkeys (figure 28.24*a*), are easy to identify: all are arboreal; they have flat, spreading noses; and many of them grasp objects with long, prehensile tails.

Anthropoids that remained in Africa gave rise to two lineages: the Old World monkeys (figure 28.24*b*) and the clade containing apes and humans (figure 28.24*c*). Old World monkeys include ground-dwelling as well as arboreal species. None of them have prehensile tails, their nostrils are close together, their noses point downward, and some have toughened pads of skin on their rumps for prolonged sitting.

Hominids

The **hominids** include the apes and humans and their direct ancestors. The living apes consist of the gibbon (genus *Hylobates*), orangutan *(Pongo),* gorilla *(Gorilla),* and chimpanzee *(Pan).* Apes have larger brains than monkeys, and they lack tails. With the exception of the gibbon, which is small, all living apes are larger than any monkey. Apes exhibit the most adaptable behavior of any mammal except human beings. Once widespread in Africa and Asia, apes are rare today, living in relatively small areas. No apes ever occurred in North or South America.

Studies of ape DNA have explained a great deal about how the living apes evolved. The Asian apes evolved first. The line of apes leading to gibbons diverged from other apes

a. *b.* *c.*

Figure 28.24 Anthropoids. ***a.*** New World monkey: the squirrel monkey, *Saimiri oerstedii.* ***b.*** Old World monkey: the mandrill, *Mandrillus sphinx.* ***c.*** Hominids: human, *Homo sapiens* (left), and gorilla, *Gorilla gorilla* (right).
(a): Tier und Naturfotografie/J & C Sohns/age fotostock; (b): blickwinkel/Alamy Stock Photo; (c left): VisualCommunications/E+/Getty Images; (c right): Erni/Shutterstock

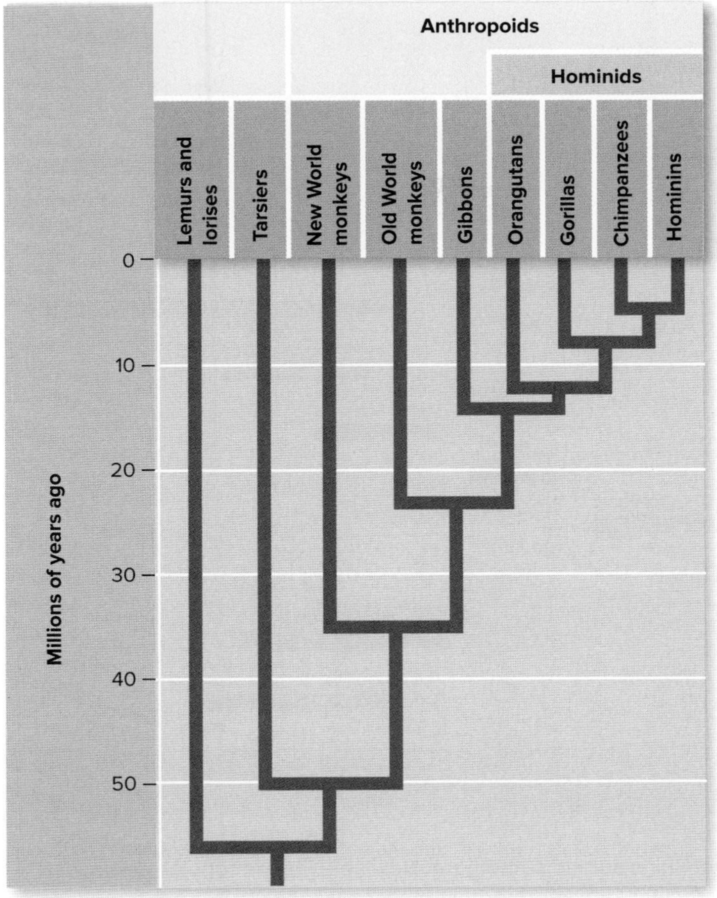

Figure 28.25 A primate evolutionary tree. Lemurs, lorises, and tarsiers used to be considered a group, the prosimians, which we now realize to be paraphyletic. Similarly, apes constitute a paraphyletic group because some apes are more closely related to nonape species (hominins) than they are to other apes.

about 15 MYA, whereas orangutans split off about 10 MYA (figure 28.25).

The African apes evolved more recently. These apes are the closest living relatives to humans. Based on genetic differences, scientists estimate that gorillas diverged from the line leading to chimpanzees and humans some 8 MYA. The taxonomic group "apes" is a paraphyletic group; some apes are more closely related to humans than they are to other apes. For this reason, taxonomists have placed humans, orangs, gorillas, and chimps (that is, all apes except gibbons) in the Hominidae. We refer to the members of this group as hominids.

Hominins

Soon after the gorilla lineage diverged, the ancestor of humans split off from the chimpanzee line to begin the evolutionary journey leading to humans; we refer to all species more closely related to humans than they are to chimps as hominins. Because this split was so recent, few genetic differences between humans and chimpanzees have had time to evolve. For example, a human hemoglobin molecule differs from its chimpanzee counterpart in only a single amino acid. In general, humans and chimpanzees exhibit a level of genetic similarity normally found between closely related species of the same genus! In fact, when geneticists sequenced the genome of the chimpanzee, they found that it was 99% identical to the human genome.

Comparing hominids

The common ancestor of hominids is thought to have been an arboreal climber. Much of the subsequent evolution of the hominids reflected different approaches to locomotion. Hominins became bipedal, walking upright; in contrast, the other hominids evolved knuckle-walking, supporting their weight on the dorsal sides of their fingers (monkeys, by contrast, walk using the palms of their hands).

Humans depart from other hominids in several areas of anatomy related to bipedal locomotion. Because humans walk on two legs, our vertebral column is more curved than an ape's, and the human spinal cord exits from the bottom, rather than the back, of the skull. The human pelvis has become broader and more bowl-shaped, with the bones curving forward to center the weight of the body over the legs. The hip, knee, and foot have all changed proportions.

Being bipedal, humans carry much of the body's weight on the lower limbs, which constitute 32 to 38% of the body's weight and are longer than the upper limbs; human upper limbs do not bear the body's weight and make up only 7 to 9% of human body weight. African apes walk on all fours, with both the upper and the lower limbs bearing the body's weight; in gorillas, the longer upper limbs account for 14 to 16% of body weight, the somewhat shorter lower limbs for about 18%.

Australopithecines Were Early Hominins

LEARNING OBJECTIVE 28.8.2 Describe the role of bipedalism in the evolution of early hominins.

Five to 10 MYA, the world's climate began to get cooler, and the great forests of Africa were largely replaced with savannas and open woodland. In response to these changes, a new kind of hominid was evolving, one that was bipedal. These new hominids are classified as hominins—that is, of the human line.

The major groups of hominins include three to seven species of the genus *Homo* (scientists disagree on the exact number); seven species of the older, smaller-brained genus *Australopithecus;* and several even older lineages (figure 28.26). In every case where the fossils allow a determination to be made, the hominins are bipedal—the hallmark of hominin evolution.

In recent years, anthropologists have found a remarkable series of early hominin fossils extending as far back as 6 to 7 million years. Often displaying a mixture of primitive and modern traits, these fossils have thrown the study of early hominins into turmoil. Although the inclusion of these fossils among the hominins seems warranted, only a few specimens have been discovered, and they do not provide enough information to determine with certainty their relationships to australopithecines and

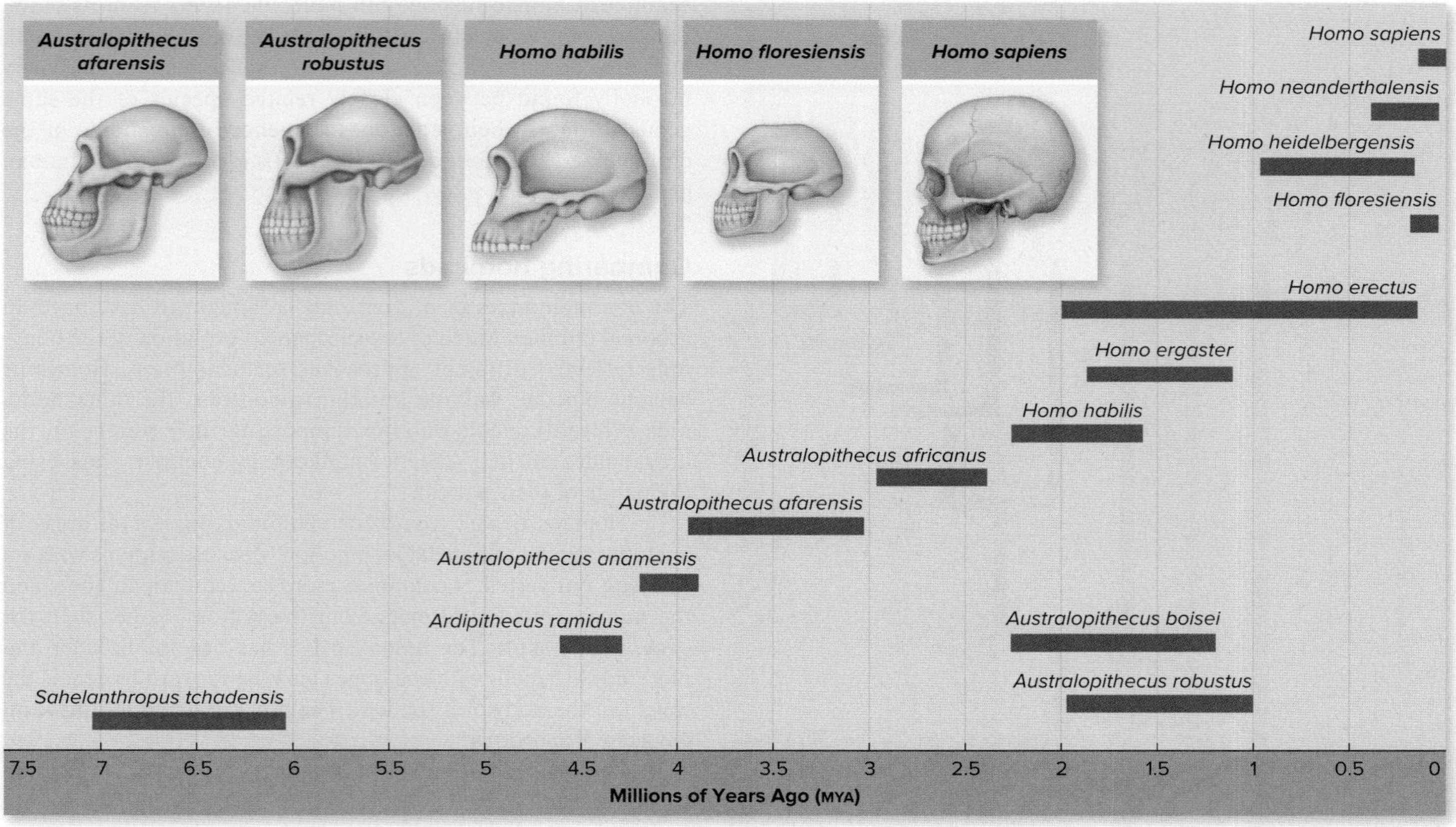

Figure 28.26 Hominin fossil history. Most, but not all, fossil hominins are indicated here. New species are regularly being discovered, though great debate sometimes exists about whether a particular specimen represents a new species.

humans. The search for additional early hominin fossils continues.

Early australopithecines

Our knowledge of australopithecines is based on hundreds of fossils, all found in South and East Africa (except for one specimen from Chad in West Africa). Australopithecines may have lived over a much broader area of Africa, but rocks of the proper age that might contain fossils are not exposed elsewhere. The evolution of hominins seems to have begun with an initial radiation of numerous species. The seven species identified so far provide ample evidence that australopithecines were a diverse group.

These early hominins weighed about 18 kg and were about 1 m tall. Their dentition was distinctly hominin, but their brains were no larger than those of apes, generally 500 cubic centimeters (cm^3) or less. *Homo* brains, by comparison, are usually larger than 600 cm^3; modern *H. sapiens* brains average 1350 cm^3.

The structure of australopithecine fossils clearly indicates that they walked upright. Evidence of bipedalism includes a set of some 69 hominin footprints found at Laetoli, East Africa. Two individuals, one larger than the other, walked upright side-by-side for 27 m, their footprints preserved in a layer of 3.7-million-year-old volcanic ash. Importantly, the big toe is not splayed out to the side as in a monkey or ape, indicating that these footprints were clearly made by hominins.

Bipedalism

The evolution of bipedalism marks the beginning of hominins. Bipedalism seems to have evolved as australopithecines left dense forests for grasslands and open woodland.

Whether larger brains or bipedalism evolved first was a matter of debate for some time. One school of thought proposed that hominin brains enlarged first, and then hominins became bipedal. Another school of thought saw bipedalism as a precursor to larger brains, arguing that bipedalism freed the forelimbs to manufacture and use tools, leading to the evolution of bigger brains. Recently, fossils unearthed in Africa have settled the debate. These fossils demonstrate that bipedalism extended back 4 million years; the knee joint, pelvis, and leg bones all exhibit the hallmarks of an upright stance. Substantial brain expansion, on the other hand, did not appear until roughly 2 MYA. In hominin evolution, upright walking clearly preceded large brains.

The reason bipedalism evolved in hominins remains a matter of controversy. Ideas include that walking upright is faster and uses less energy; that an upright posture permits hominins to pick fruit from trees and see over tall grass; that being upright reduces the body surface exposed to the Sun's rays; and that bipedalism frees the forelimbs of males to carry food back to females, encouraging pair-bonding. All of these suggestions have their proponents, and none is universally accepted, leaving the origin of bipedalism unknown.

The Genus *Homo* Arose Roughly 2 MYA

LEARNING OBJECTIVE 28.8.3 Contrast australopithicenes and the genus *Homo*.

The first humans (genus *Homo*) evolved from australopithecine ancestors about 2 MYA. The exact ancestor has not been clearly identified but is commonly thought to be *Australopithecus afarensis.* Only within the last 30 years have a significant number of fossils of early *Homo* been uncovered. With intensive field exploration, and new fossil finds, our picture of the base of the human evolutionary tree is being clarified. The following historical account will undoubtedly be supplanted by future discoveries, but it provides a good example of science at work.

The first human: *Homo habilis*

In the early 1960s, stone tools were found scattered among hominin bones close to the site where *A. boisei* had been unearthed. Although the fossils were badly crushed, painstaking reconstruction of the many pieces suggested a skull with a brain volume of about 680 cm^3, larger than the australopithecine range of 400 to 550 cm^3. Because of its association with tools, this early human was called *Homo habilis,* meaning "handy man." Partial skeletons discovered in 1986 indicate that *H. habilis* was small in stature, with arms longer than its legs and a skeleton much like that of *Australopithecus.* Because of its general similarity to australopithecines, many researchers at first questioned whether this fossil was human.

Out of Africa: *Homo erectus*

Our picture of what early *Homo* was like lacks detail, because it is based on only a few specimens. We have much more information about the species that replaced it, *Homo erectus* and *H. ergaster,* which are sometimes considered a single species.

These species were a lot larger than *Homo habilis*—about 1.5 m tall. They had a large brain, about 1000 cm^3, and walked erect. Their skull had prominent brow ridges and, like modern humans, a rounded jaw. Most interesting of all, the shape of the skull interior suggests that *H. erectus* was able to talk.

Far more successful than *H. habilis, H. erectus* quickly became widespread and abundant in Africa, Asia, and Europe. A social species, *H. erectus* lived in tribes of 20 to 50 people, often dwelling in caves. They successfully hunted large animals, butchered them using flint and bone tools, and cooked them over fires—a site in China contains the remains of horses, bears, elephants, and rhinoceroses.

Homo erectus survived for nearly two million years, longer than any other species of human. These very adaptable humans disappeared in Africa only about 500,000 years ago as modern humans were emerging. Interestingly, they survived even longer in Asia, until 100,000 years ago.

New additions to the human family: *Homo floresiensis* and *Homo naledi*

The world was stunned in 2004 with the announcement of the discovery of fossils of a new human species from the tiny Indonesian island of Flores (figure 28.27). *Homo floresiensis* was notable for its diminutive stature; standing only a meter tall, and with a brain size of just 380 cm^3, the species was quickly nicknamed "the Hobbit" after the diminutive characters in J. R. R. Tolkien's books. Just as surprising was the age of the fossils, the youngest of which was originally thought to be only 15,000 years old. More recent research, however, has pushed the age of *H. floresiensis* back to 60,000 to 100,000 years ago.

Despite its recency, a number of skeletal features suggest to most scientists that *H. floresiensis* is not most closely related to *H. sapiens*. Rather, the prevailing view is that *H. floresiensis* was most closely related to *H. erectus*, or even to the long-extinct *H. habilis*.

Why *H. floresiensis* evolved to such small size is unknown, although a number of experts have pointed to the phenomenon of "island dwarfism," in which mammal species evolve to be much smaller on islands. Indeed, *H. floresiensis* coexisted with and preyed on a miniature species of elephant that also lived on Flores, but which also has gone extinct. These findings have rekindled interest in explaining why island dwarfism occurs.

Another recently described species is *Homo naledi,* found in a cave in South Africa. This new species possesses a surprising mix of features, some typical of *Homo*, such as the shape of the feet and teeth; yet the curved fingers and shape of the legs suggest that it was a species that climbed a lot, and its brain was quite small, more akin to australopithecines than to *Homo.* In other words, it doesn't seem to fit the story of a steady progression from

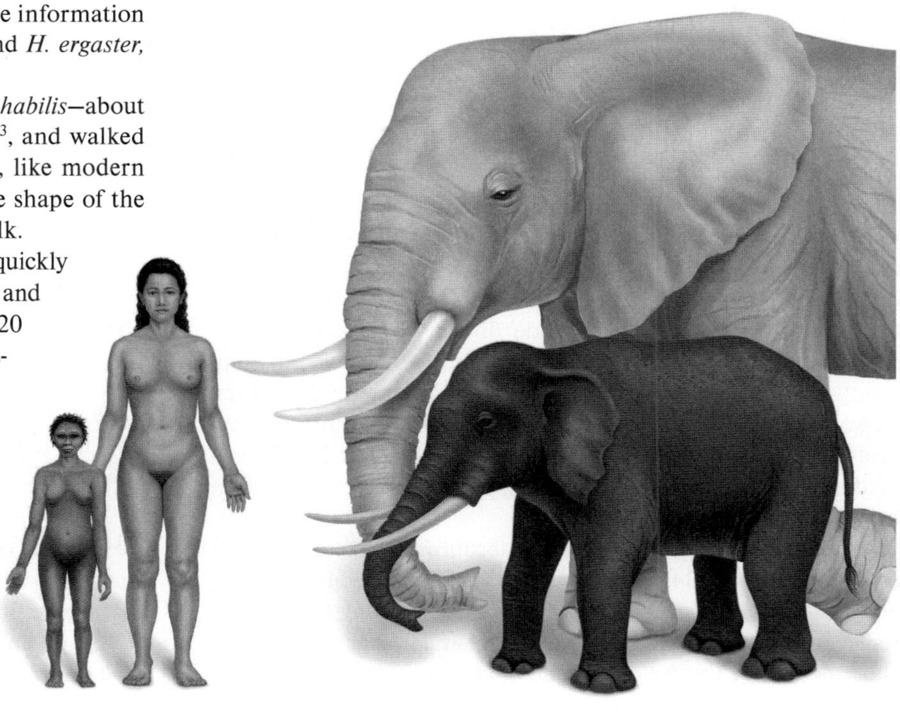

Figure 28.27 ***Homo floresiensis.*** This diminutive species (compare the modern human female on the right with the female *H. floresiensis* on the left) occurred on the small island of Flores in what is now Indonesia. *H. floresiensis* preyed upon a dwarf species of elephant, *Stegodon sondaari,* which also occurred on Flores (compare with the larger African elephant, *Loxodonta african,* in gray).

Australopithecus to *H. erectus* to modern humans. In some regards this is not surprising. We know that even when evolutionary trends occur, as with general increase in horse body and tooth size (refer to figure 20.13), exceptions usually exist as well: evolution tends to be more like a bush than a straight-line linear progression.

A second wrinkle to the *H. naledi* puzzle emerged in 2017 when further research revealed that the fossils are only about 300,000 years old. This means that *H. naledi* may have coexisted with the modern humans. We can only speculate about how the two species may have interacted and whether our ancestors may have played a role in *H. naledi's* disappearance.

Humans

The evolutionary journey entered its final phase when humans first appeared in Africa about 700,000 years ago. Investigators who focus on human diversity denote three species: *Homo heidelbergensis, H. neanderthalensis,* and *H. sapiens.* Other investigators lump the three species into one, *H. sapiens* ("wise man").

The oldest human, *Homo heidelbergensis,* is known from a 700,000-year-old fossil from Ethiopia. Although it coexisted with *H. erectus* in Africa, *H. heidelbergensis* has more advanced anatomical features, including a bony keel running along the midline of the skull, a thick ridge over the eye sockets, and a large brain. Also, its forehead and nasal bones are very much like those of *H. sapiens.*

As *H. erectus* was becoming rarer, about 130,000 years ago, a new species of human arrived in Europe from Africa. *Homo neanderthalensis* likely branched off the ancestral line leading to modern humans more than 500,000 years ago. Compared with modern humans, Neanderthals were stocky and powerfully built. Their skulls were massive, with protruding faces; heavy, bony ridges over the brows; and larger braincases.

Neanderthals and early modern humans

The Neanderthals (classified by many as a separate species, *Homo neanderthalensis*) were named after the Neander Valley of Germany, where their fossils were first discovered in 1856. Rare at first outside of Africa, they became progressively more abundant in Europe and Asia; by 70,000 years ago, they had become common. The Neanderthals made diverse tools and lived in huts or caves. They took care of their injured and sick and commonly buried their dead, often placing food, weapons, and even flowers with the bodies. Such attention to the dead strongly suggests that they believed in a life after death. This is the first evidence of the symbolic thinking characteristic of modern humans (figure 28.28).

Fossils of *H. neanderthalensis* abruptly disappear from the fossil record about 34,000 years ago and are replaced by fossils of early modern *H. sapiens.* We can only speculate why this sudden replacement occurred, but it was complete all over Europe in a short period. A variety of evidence indicates that early modern humans came from Africa—fossils of essentially modern aspect but as much as 100,000 years old have been found there. Early modern humans seem to have replaced the Neanderthals completely in the Middle East by 40,000 years ago, and then spread across Europe, coexisting with the Neanderthals for several thousand years.

Figure 28.28 Early modern human art. Rhinoceroses are among the animals depicted in this reproduction of the remarkable cave painting found in 1995 near Vallon-Pont d'Arc, France.
Delmarty/Andia/Alamy Stock Photo

DNA analysis reveals Neanderthals to be quite distinct from early modern humans. Interestingly, modern human populations from Europe, Asia, and the Americas, but not Africa, carry on average 2.5% Neanderthal DNA. This indicates that humans and Neanderthals interbred, after humans migrated out of Africa. Although an individual has only 2.5% of Neanderthal DNA in his or her genome, if we accumulate these sequences over hundreds of human genomes, Neanderthal DNA representing up to 1 Gb of sequence can be identified in our genomes. Collectively, we may carry 20 to 35% of the Neanderthal genome. Neanderthal DNA content varies across the genome, being least common on the X chromosome. This has been interpreted to represent negative selection, and possibly reduced male fertility.

DNA analysis also reveals the existence of a previously unknown human population, the Denisovans. A finger bone found in a cave in Siberia yielded ancient DNA that was not modern human but also not Neanderthal. The entire genome has been sequenced from this bone, and although people in Europe, Asia, Africa, and the Americas lack Denisovan DNA, up to 4 to 6% of Denisovan DNA is found in some people of the South Pacific (Melanesians).

This analysis of ancient DNA supports the outline of the "out of Africa" theory of human origins, but it also indicates that there was considerable mixing of humans with other populations after the initial migration. Further genome sequencing of ancient DNA will continue to provide a window on human evolution.

Our own species: *Homo sapiens*

Homo sapiens is the only surviving species of the genus *Homo,* and indeed the only surviving hominin. Some of the best fossils of *H. sapiens* are 20 well-preserved skeletons with skulls found in a cave near Nazareth in Israel. Modern dating techniques estimate these humans to be between 90,000 and 100,000 years old. The skulls are modern in appearance and size with high, short braincases; vertical foreheads with only slight brow ridges; and a cranial capacity of roughly 1550 cm^3. Our evolution has been marked by a progressive increase in brain size, distinguishing us from other animals in several ways. First, humans are able to make and

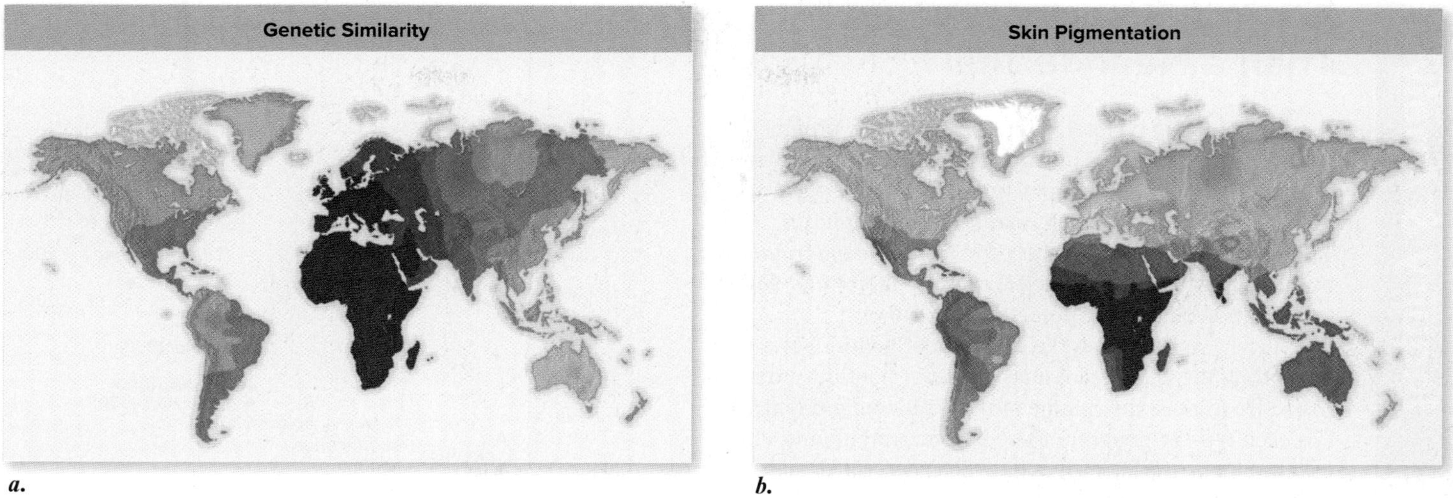

Figure 28.29 Patterns of genetic variation in human populations differ from patterns of skin color variation. ***a.*** Genetic variation among *Homo sapiens.* The more similar areas are in color, the more similar they are genetically based on many enzyme and blood group genetic loci. ***b.*** Similarity among *Homo sapiens* based on skin color. The color of an area represents the skin pigmentation of the people native to that region.

use tools more effectively than any other animal. Second, although not the only animal capable of conceptual thought, humans have refined and extended this ability until it has become the hallmark of our species. Finally, we use symbolic language and can, with words, shape concepts out of experience and transmit that accumulated experience from one generation to another.

Humans have undergone what no other animal ever has: extensive cultural evolution. Through culture, we have found ways to change and mold our environment, rather than changing evolutionarily in response to the environment's demands.

Human races

Human beings, like all other species, have differentiated in their characteristics as they have spread throughout the world, an example of the phenomenon of geographic variation discussed in chapter 21. Local populations in one area often appear significantly different from those that live elsewhere. For example, many northern Europeans have fair hair, light-colored skin, and blue eyes, whereas many Africans have black hair, dark skin, and brown eyes. These traits may play a role in adapting the particular populations to their environments. Blood groups may be associated with immunity to diseases more common in certain geographical areas, and dark skin shields the body from the damaging effects of ultraviolet radiation, which is much stronger in the tropics than in temperate regions.

Despite these geographic differences, there is no question that all humans belong to a single species, *H. sapiens*. The number of groups into which the human species might logically be divided has long been a point of contention. Historically, anthropologists have divided people into as many as 30 "races," or as few as 3: Caucasoid, Negroid, and Oriental. Native Americans, Bushmen, and Aborigines are examples of particularly distinctive populations that are sometimes regarded as distinct groups.

The problem with classifying people or other organisms in this fashion is that the characteristics used to define the races are usually not well correlated with one another, so the determination of race is arbitrary. Humans are visually oriented; consequently, we have relied on visual cues—primarily skin color—to define races. However, when other types of characteristics, such as blood groups, are examined, patterns of variation correspond very poorly with visually determined racial classes. Indeed, if we sort the human species into subunits based on overall genetic similarity, the groupings are very different from those based on visual features (figure 28.29).

Those characteristics that are differentiated among populations, such as skin color, represent classic examples of the antagonism between gene flow and natural selection. When selection is strong enough, as it is for dark coloration in tropical regions, populations can differentiate even in the presence of gene flow. However, gene flow will still ensure that populations are relatively homogeneous for genetic variation at other loci.

Relatively little of the variation in the human species represents differences between the described races. Indeed, one study calculated that only 8% of all genetic variation among humans could be accounted for as differences that exist among racial groups. Racial categories do a very poor job in describing the vast majority of human genetic variation. For this reason, most biologists reject human racial classifications as reflecting patterns of biological differentiation in the human species.

REVIEW OF CONCEPT 28.8

Primates include lemurs, monkeys, apes, and humans. Primates have grasping fingers and toes and binocular vision. The evolution of bipedal locomotion in hominins led to modification of the spine, pelvis, and limbs. Several species of *Homo* evolved in Africa, and some migrated from there to Europe and Asia. Analysis of ancient DNA indicates humans interbred with Neanderthals and Denisovans. Conventional human races are not supported by analysis of variation.

- *Which of these groups is monophyletic: lemurs, monkeys, apes, hominins?*

Are You Smarter Than a Neanderthal?

As noted in this chapter, brain size became progressively larger as hominins evolved. Interestingly, Neanderthal fossils (*left* in photo below) typically have larger brains than fossils of modern humans (*right* in photo), about 1650 cubic centimeters (cm^3) for *Homo neanderthalensis* versus about 1500 cm^3 for *H. sapiens*. Does this suggest that Neanderthals were smarter than us?

The graph explores the evolution of hominin brain size by plotting the age of each major type of hominin versus its brain size (that is, the volume of the skull cranium's interior). For each type of hominin, there is some variation in cranial volume among the fossils that have been described, and a typical value is presented (the number in parentheses by each point). The value for *H. neanderthalensis,* for example, is plotted as a typical 1650 cm^3, even though a skull found in the Amud Cave of Israel is 90 cm^3 larger. Some paleontologists consider *H. ergaster* to be a variant of *H. erectus,* and *H. heidelbergensis* and *H. neanderthalensis* to be variants of *H. sapiens.* However, for the sake of this analysis, the "splitters" view is presented. Although the question remains controversial, many anthropologists now feel that *H. neanderthalensis* and *H. sapiens* are separate species, both descended from *H. heidelbergensis* (however it is named).

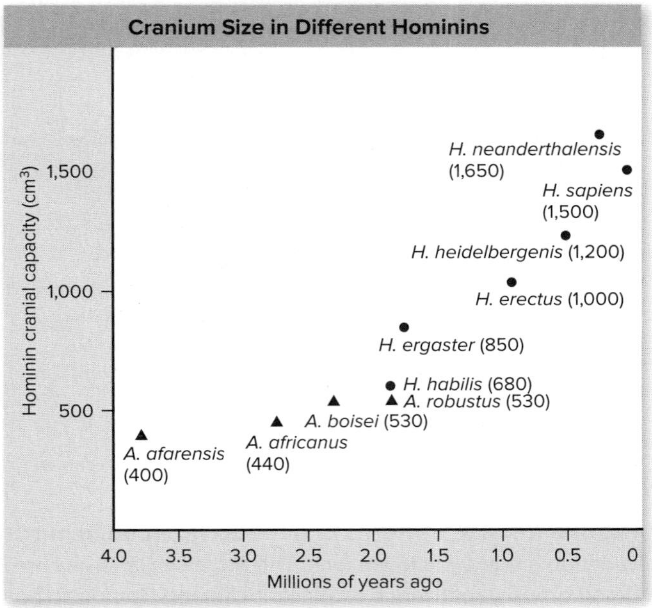
Cranium Size in Different Hominins
Hominin cranial capacity (cm³)
1,500
1,000
500
H. neanderthalensis (1,650)
H. sapiens (1,500)
H. heidelbergenis (1,200)
H. erectus (1,000)
H. ergaster (850)
H. habilis (680)
A. robustus (530)
A. boisei (530)
A. africanus (440)
A. afarensis (400)
4.0
3.5
3.0
2.5
2.0
1.5
1.0
0.5
0
Millions of years ago

Analysis

1. **Applying Concepts** In the graph, what is the dependent variable?
2. **Interpreting Data**
   a. Which human species of *Homo* has the biggest brain? The smallest?
   b. Which australopithecine has the biggest brain? The smallest?
   c. Does any australopithecine have a brain as large as a human's?
3. **Making Inferences**
   a. Over 2 million years, does the brain size of australopithecines change? If so, how much? What percent increase is this?
   b. Over 2 million years, does the brain size of humans change? If so, how much? What percent increase is this?
4. **Drawing Conclusions**
   a. Does brain size appear to have evolved faster in the genus *Homo* than in the genus *Australopithecus*? If so, how much faster?
   b. Given the clear and undisputed larger brain size of Neanderthals, and the conclusion you drew in question 4a, are you able to further conclude that Neanderthals were smarter than today's humans?
5. **Further Analysis** What key unverified assumption does the conclusion in question 4b depend upon? If you do not accept this further conclusion, why, then, do you think brain size has evolved as rapidly as it has in the genus *Homo?*

John Reader/Science Source

Retracing the Learning Path

CONCEPT 28.1 Nonvertebrate Chordates Do Not Form Bone

28.1.1 Tunicates Have Chordate Larval Forms Tunicates have a swimming larval form exhibiting all the features of a chordate, but their adult form is sessile and baglike.

28.1.2 Lancelets Have Chordate Adult Forms Lancelets have chordate features throughout life, but as adults they lack bones and have no distinct head.

CONCEPT 28.2 Almost All Chordates Are Vertebrates

28.2.1 Vertebrates Have Vertebrae, a Distinct Head, and Other Features In vertebrates, a vertebral column encloses and protects the dorsal nerve cord. The distinct, well-differentiated head carries sensory organs. Vertebrates have specialized organs and a bony or cartilaginous endoskeleton.

CONCEPT 28.3 Fishes Are the Earliest and Most Diverse Vertebrates

28.3.1 Fishes Exhibit Five Key Characteristics Modern fish have a vertebral column of bone or cartilage, jaws, paired appendages, internal gills, and a closed circulatory system.

28.3.2 The First Fishes Lacked Jaws Hagfish and lampreys are the only surviving fish with a mouth but no jaws.

28.3.3 Sharks, with Cartilaginous Skeletons, Became Top Predators Sharks, rays, and skates are cartilaginous fishes. Sharks are streamlined for fast swimming. They evolved teeth that enabled them to grab, kill, and devour prey. The lateral line system of sharks and bony fishes detects changes in pressure waves.

28.3.4 Bony Fishes Dominate Today's Seas Bony fishes belong to one of three groups (1) the ray-finned fishes (Actinopterygii), (2) lungfish (Dipnoi), or (3) the lobe-finned fishes (Actinistia). Ray-finned fishes have fins stiffened with bony parallel rays. The lobe-finned fishes have muscular lobes with bones connected by joints. The fossil *Tiktaalik* represents an intermediate between these fishes and tetrapods.

CONCEPT 28.4 Amphibians Are Moist-Skinned Descendants of the Early Tetrapods

28.4.1 Amphibians Have Five Distinguishing Features Amphibian adaptations include legs, lungs, cutaneous respiration, pulmonary veins, and a partially divided heart.

28.4.2 Modern Amphibians Belong to Three Groups The Anura (frogs and toads) lack tails as adults; many have a larval tadpole stage. The Caudata (salamanders) have tails as adults and larvae similar to the adult form. The Apoda (caecilians) are legless.

CONCEPT 28.5 Reptiles Are Fully Adapted to Terrestrial Living

28.5.1 Amniotes Exhibit Three Key Characteristics Amniotes possess a watertight amniotic egg; dry, watertight skin; and thoracic breathing. Modern reptiles and other amniotes practice internal fertilization and are ectothermic.

28.5.2 The Amniota Contains the Synapsids and Diapsids Synapsids gave rise to the therapsids that became the mammalian line. Diapsids gave rise to modern reptiles and the birds.

28.5.3 Modern Reptiles Belong to Four Orders The four orders of reptiles are Chelonia (turtles and tortoises), Rhynchocephalia (tuataras), Squamata (lizards and snakes), and Crocodylia (crocodiles and alligators).

CONCEPT 28.6 Birds Are Essentially Flying Reptiles

28.6.1 Key Characteristics of Birds Are Feathers and a Lightweight Skeleton The feather is a modified reptilian scale. Feathers provide lift in gliding or flight and conserve heat. The lightweight skeleton of birds is an adaptation to flight.

28.6.2 Birds Arose About 150 MYA Birds evolved from theropod dinosaurs. Feathers probably first arose to provide insulation, only later being modified for flight.

28.6.3 Modern Birds Are Diverse but Share Several Characteristics In addition to the key characteristics, birds have efficient respiration and circulation and are endothermic.

CONCEPT 28.7 Mammals Are the Least Diverse of Vertebrates

28.7.1 Mammals Have Hair, Mammary Glands, and Other Characteristics Mammals are distinguished by hair and by mammary glands, which provide milk to feed the young. Mammals are also endothermic. Mammals evolved from therapsids (synapsids) about 220 MYA and reached maximum diversity about 15 MYA.

28.7.2 Modern Mammals Are Placed into Three Groups The monotremes lay shelled eggs. In marsupials, an embryo completes development in a pouch. Placental mammals produce a placenta in the uterus to nourish the embryo.

CONCEPT 28.8 Primates Include Lemurs, Monkeys, Apes, and Humans

28.8.1 The Anthropoid Lineage Led to the Earliest Humans Primates share two innovations: grasping fingers and toes, and binocular vision. Anthropoids include monkeys, apes and humans. The Hominidae includes humans, chimps, gorillas, and orangutans. All species more closely related to humans than to chimps are called hominins.

28.8.2 Australopithecines Were Early Hominins The distinguishing characteristics of hominids are upright posture and bipedal locomotion.

28.8.3 The Genus *Homo* Arose Roughly 2 MYA Common features of early *Homo* species include a larger body and brain size. *Homo sapiens* is the only extant species. Humans exhibit conceptual thought, tool use, and symbolic language.

Concept Overview

This Concept Overview diagrams the key concepts that were discussed in this chapter.

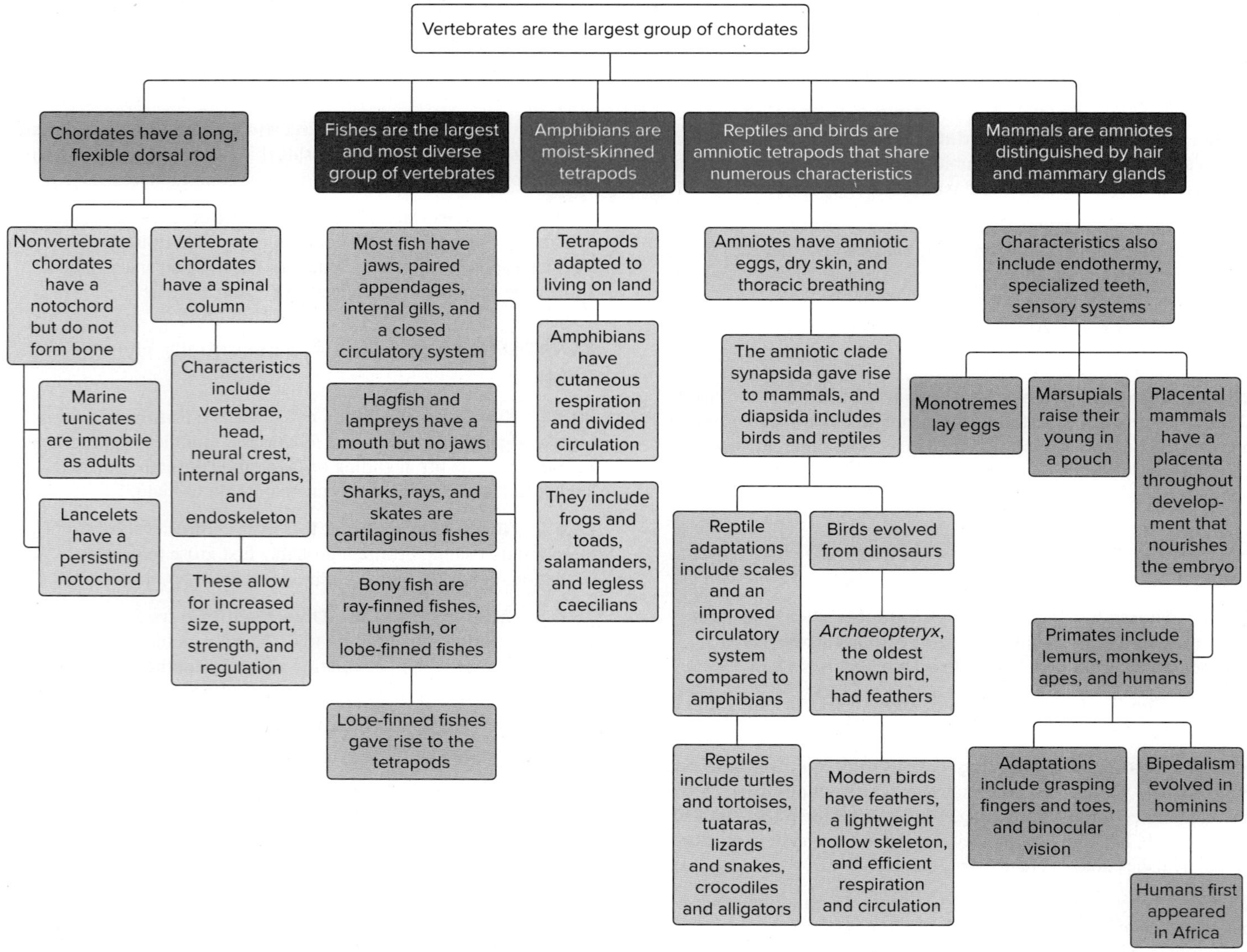

Assessing the Learning Path

Understand

1. Which of the following statements regarding all species of chordates is false?
   a. Chordates are deuterostomes.
   b. A notochord is present in the embryo.
   c. The notochord is surrounded by bone or cartilage.
   d. All possess a postanal tail during embryonic development.
2. A cephalochordate lacks
   a. segmentation.
   b. a dorsal nerve cord.
   c. a bony structure to protect the nerve cord.
   d. cartilage.
3. Vertebrates differ from other chordates in all of the following respects EXCEPT for the presence of
   a. a neural crest.
   b. a head.
   c. an adult notochord.
   d. an endoskeleton.
4. All vertebrates
   a. are capable of maintaining an internal temperature.
   b. possess waterproof, keratinized skin.
   c. have a completely closed circulatory system.
   d. develop jaws with teeth as they mature.
5. All fish species, living and extinct, share all of the following characteristics EXCEPT
   a. gills.
   b. jaws.

c. an internal skeleton with dorsal nerve cord.
d. a single-loop circulatory system.

6. The first fishes lacked jaws. Jaws evolved from
a. ear bones.
b. gill arches.
c. modified skin scales.
d. small plates of bone.

7. Sharks were among the first vertebrates to evolve all of the following EXCEPT
a. a calcified endoskeleton.
b. teeth.
c. the lateral line system.
d. a swim bladder.

8. Why was the evolution of the pulmonary veins important for tetrapods?
a. To move oxygen to and from the lungs
b. To increase the metabolic rate
c. For increased blood circulation to the brain
d. None of the above

9. In the Paleozoic era, the first vertebrate animal group to live successfully on land was the
a. tetrapods.
b. dinosaurs.
c. fishes.
d. therapsids.

10. Adaptations in amniotes did NOT include
a. an amniotic egg.
b. a layer of scales on the skin.
c. middle ear bones.
d. modifications to the respiratory system.

11. Waste products are stored in the
a. amnion.
b. chorion.
c. yolk sac.
d. allantois.

12. Which of the following evolutionary adaptations allows the birds to become efficient at flying?
a. Structure of the feather
b. High metabolic temperatures
c. Increased respiratory efficiency
d. All of the above

13. All of the following contributed to the birds' ability to cope with the heavy energy demands of flight EXCEPT
a. efficient respiration.
b. endothermy.
c. efficient circulation.
d. efficient digestion.

14. A characteristic unique to most species of mammals and no other vertebrate is
a. an amniotic egg.
b. endothermy.
c. middle ear bones.
d. hair.

15. All of the following are therians EXCEPT
a. kangaroos.
b. opossums.
c. echidnas.
d. humans.

16. Anthropoids are primates, which include all of the following EXCEPT
a. monkeys.
b. apes.
c. lemurs.
d. humans.

17. Hominids include all of the following EXCEPT
a. mandrills.
b. gorillas.
c. humans.
d. chimpanzees.

Apply

1. A key distinction between tunicates and lancets is
a. the presence of a neural crest.
b. filter feeding.
c. an adult notochord.
d. a cilia-lined pharynx.

2. During embryonic development, neural crest cells
a. form the spinal cord.
b. migrate to various locations in the embryo.
c. form the adult notochord.
d. form the embryonic brain.

3. Chondrichthyes (sharks) and Osteichthyes (bony fishes) have evolved anatomical solutions to increase swimming speed and maneuverability. Which modification is not found in Osteichthyes?
a. A lateral line system
b. Buoyancy control through swim bladders
c. A light internal skeleton made of cartilage
d. An operculum

4. Which of the following is the closest relative of lungfish?
a. Hagfish
b. Sharks
c. Ray-finned fish
d. Mammals

5. In order for tetrapods to be successful on land, they had to develop which of the following?
a. A more efficient swim bladder
b. Cutaneous respiration and lungs
c. Water-tight skin
d. Shelled eggs

6. How are a crocodile and a hawk similar?
a. Both are synapsids.
b. Both are homoeothermic.
c. Both have four-chambered hearts.
d. Both have unkeratinized skin.

7. Both birds and mammals share the physiological characteristic of endothermy. How do these animals maintain a high body temperature?
a. They live in warm environments.
b. They have high metabolic rates.
c. They run or fly, which produces heat.
d. They eat a lot.

8. The reason that both birds and crocodilians build nests might be that they
a. are both warm-blooded.
b. both eat fish.
c. both inherited the trait from a common ancestor.
d. both lay eggs.

9. Mammals began to diversify, with large forms evolving,
a. after the Cretaceous extinction.
b. during the Jurassic.
c. at the same time that large-bodied dinosaurs evolved.
d. after the great Permian extinction.

10. The fact that monotremes lay eggs
a. indicates that they are more closely related to some reptiles than they are to some mammals.
b. is a plesiomorphic trait.
c. demonstrates that the amniotic egg evolved multiple times.
d. is a result of ectothermy.

11. Although many mammals have binocular vision, the anatomical adaptation that sets primates apart from these other mammals is
a. a prehensile tail.
b. opposable digits on hands.
c. a large brain.
d. hair-covered skin.

Synthesize

1. Of the animal phyla, which were the only three to successfully populate terrestrial habitats in large numbers of species and individuals? What about these three favored living on land?
2. Do all vertebrates have a bony backbone? Explain.
3. Some fishes spend much of their time out of water. How do you think they are able to manage this?
4. The fossils of skates are first seen around 248 MYA, some 200 million years after sharks first appear in the fossil record. What evidence would you accept that skates are, in fact, close relatives of sharks?
5. Tragically, despite their great evolutionary success, amphibians appear to be undergoing a precipitous decline in recent years. What features might make amphibians particularly vulnerable to human modification of their environment?
6. Some people state that the dinosaurs have not "gone extinct" and are with us today. What evidence can be used to support this statement?
7. Among terrestrial vertebrates, flight has evolved three times. Among what class of terrestrial vertebrates has flight never evolved? Can you think of a reason why not?
8. Shrews eat insects, kangaroos eat plants or fungi, and lions are meat eaters. What differences would you expect to find in the teeth of these three kinds of mammals?
9. Exactly how would you distinguish between a cat and a dog (be specific)?
10. Once common in Australia, a huge, flightless bird called *Genyornis newtoni* is now extinct. Isotopic dating of its fossilized egg shells indicates that no *Genyornis* eggs are younger than 50,000 years. Humans colonized Australia 50,000 years ago. What types of evidence would be needed to support the hypothesis that humans hunted this flightless bird to extinction?

Susan Singer

29 Plant Form

Learning Path

29.1 Meristems Articulate the Body Plan

29.2 Plants Contain Three Main Tissues

29.3 Roots Have Four Growth Zones

29.4 Stems Provide Support for Aboveground Organs

29.5 Leaves Are a Plant's Photosynthetic Organs

Concept Overview

This Concept Overview provides a diagram of the overarching concepts that are covered in this chapter. These concepts will be expanded upon in the Concept Overview at the end of the chapter.

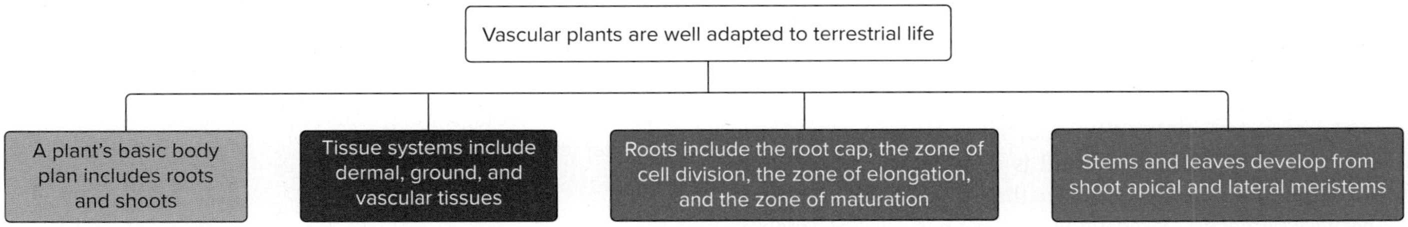

Vascular plants are well adapted to terrestrial life
A plant's basic body plan includes roots and shoots
Tissue systems include dermal, ground, and vascular tissues
Roots include the root cap, the zone of cell division, the zone of elongation, and the zone of maturation
Stems and leaves develop from shoot apical and lateral meristems

Introduction

Although the similarities between a cactus, an orchid, and a hardwood tree might not be obvious at first sight, these and most other plants share a similar basic structure. This unity is reflected in how plants are constructed, and in how they grow, develop, manufacture, and transport their food. This chapter addresses the question of how a vascular plant is "built." We will focus on the cells, tissues, and organs that compose the mature plant body. The basic body plan of a plant—above-ground shoots and below-ground roots—is established during embryogenesis. In this chapter we explore the carefully regulated developmental processes that lead from embryogenesis to the adult plant form.

29.1 Meristems Articulate the Body Plan

Among modern vascular plants, the presence of organs such as large, broad leaves, flowers, stems, and roots reflects adaptation to the demands of a terrestrial existence. For example, obtaining water is a major challenge for organisms living on land, and roots are adapted for water absorption from the soil.

Although the development of organs may be precisely controlled in modern plants, many aspects of leaf, stem, and root development are flexible and dynamic. Plants exhibit **indeterminate growth**; they add structures to their bodies throughout their lives in response to environmental change. This chapter emphasizes the unifying aspects of plant form, using the flowering plants as a model.

Vascular Plants Have Roots and Shoots

LEARNING OBJECTIVE 29.1.1 Distinguish between the functions of roots and shoots.

Fundamentally, a vascular plant is built from roots that anchor the plant in the soil, and a stem (called the shoot), from which leaves grow (figure 29.1). Roots and shoots grow at their tips, which are called apices (singular, **apex**).

The **root system** anchors the plant and penetrates the soil, from which it absorbs water and ions crucial for the plant's nutrition. Root systems are often extensive, and growing roots can exert great force, moving soil as they grow. Roots evolved later than the shoot system as an adaptation to living on land.

The **shoot system** consists of a stem and its leaves. Stems serve as a scaffold for positioning the leaves, the principal sites of photosynthesis. The arrangement, size, and other features of the leaves are critically important in a plant's photosynthetic efficiency. Flowers, other reproductive organs, and ultimately fruits and seeds are also formed on the shoot.

The repeating unit of the vegetative shoot consists of the internode, node, leaf, and axillary bud, but not reproductive structures (figure 29.1). An axillary bud is a lateral shoot apex that allows the plant to branch or replace the main shoot if it is eaten or otherwise damaged. A vegetative axillary bud has the capacity to re-create the development of the primary shoot. When the plant has shifted to the reproductive phase of development, axillary buds can produce flowers or floral shoots.

Roots and Shoots Are Composed of Tissue Systems

LEARNING OBJECTIVE 29.1.2 Describe the three types of tissue in a vascular plant.

Roots, shoots, and leaves all contain three basic types of tissue: dermal, ground, and vascular tissue. Because each of these tissues extends through the root and shoot systems, they are called **tissue systems.**

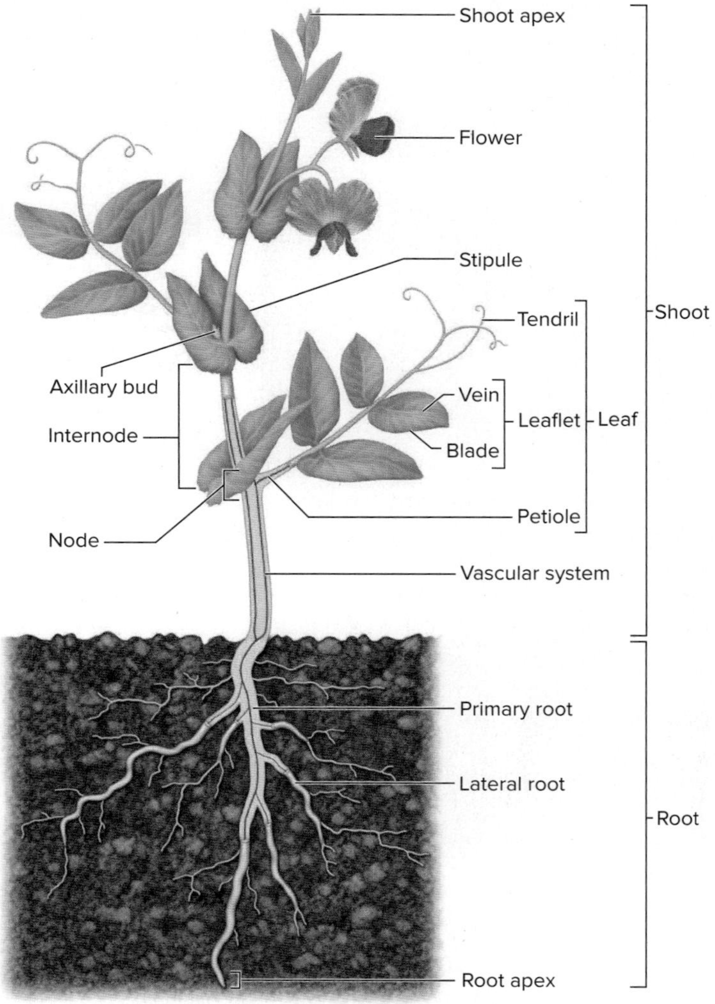

Figure 29.1 Diagram of a plant body. Branching root and shoot systems create the plant's architecture. Each root and shoot has an apex that extends growth. Leaves are initiated at the nodes of the shoot, which also contain axillary buds that can remain dormant, grow to form lateral branches, or make flowers. A leaf can be a simple blade or consist of multiple parts, as shown here. Roots, shoots, and leaves are all connected with vascular (conducting) tissue.

Plant cell types, found in different tissue systems, can be distinguished by the size of their vacuoles, by whether they are living or not at maturity, and by the thickness of secretions found in their cellulose cell walls, a distinguishing feature of plant cells (refer to chapter 4 to review cell structure). Some cells have only a primary cell wall of cellulose, synthesized by enzymes near the cell membrane. Microtubules align within the cell and determine the orientation of the cellulose fibers (figure 29.2*a*). Cells that support the plant body have more heavily reinforced cell walls with multiple layers of cellulose. Cellulose layers are laid down at angles to adjacent layers, like plywood, which enhances the strength of the cell wall (figure 29.2*b*).

Plant cells contribute to three tissue systems: **dermal tissue, ground tissue,** and **vascular tissue** (table 29.1). These tissues and their functions are described in detail in section 29.2.

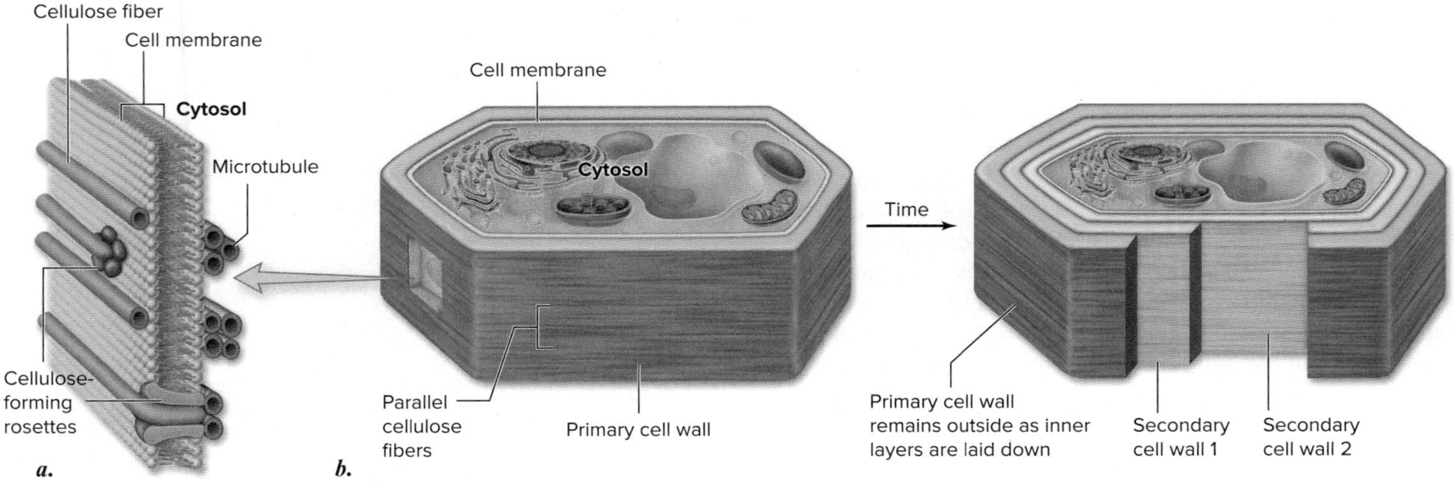

Figure 29.2 Synthesis of a plant cell wall. ***a.*** Cellulose is a glucose polymer that is produced at the cellulose-forming rosettes in the cell membrane to form the cell wall. Cellulose fibers are laid down parallel to microtubules inside the cell membrane. Additional substances that strengthen and waterproof the cell wall are added to the cell wall in some cell types. ***b.*** Some cells extrude additional layers of cellulose, increasing the mechanical strength of the wall. Because new cellulose is produced at the cell membrane, the oldest layers of cellulose are on the outside of the cell wall. All cells have a primary cell wall. Additional layers of cellulose and lignin contribute to the secondary cell wall.

TABLE 29.1 The Three Tissue Systems of Plants

Tissue System	Function
Dermal	Composed primarily of the epidermis, a protective outer covering of the plant
Ground	Several cell types that function in storage, support, photosynthesis, and secretion
Vascular	Cell types for conducting fluids and dissolved substances through the plant body

Meristems Elaborate the Body Plan Throughout the Plant's Life

LEARNING OBJECTIVE 29.1.3 Describe the types and functions of meristems.

When a seed germinates, only a tiny portion of the adult plant exists. Although embryo cells can undergo division and differentiation to form many cell types, the fate of most adult cells is more restricted. Further development of the plant body depends on the activities of meristems, specialized cells found in shoot and root apices, as well as other parts of the plant.

Overview of meristems

Meristems are clusters of small cells with dense cytoplasm and large nuclei that act as stem cells do in animals. That is, one cell divides to give rise to two cells, of which one remains a meristem cell, while the other undergoes differentiation and contributes to the plant body (figure 29.3). In this way, the population of meristem cells is continually renewed. Genetic evidence suggests that animal stem cells and plant meristem cells may also share some common pathways of gene expression. Extension of both root and shoot takes place as a result of repeated cell divisions and subsequent elongation of the cells produced by the **apical meristems.** In some vascular plants, including shrubs and most trees, **lateral meristems** produce an increase in root and shoot diameter.

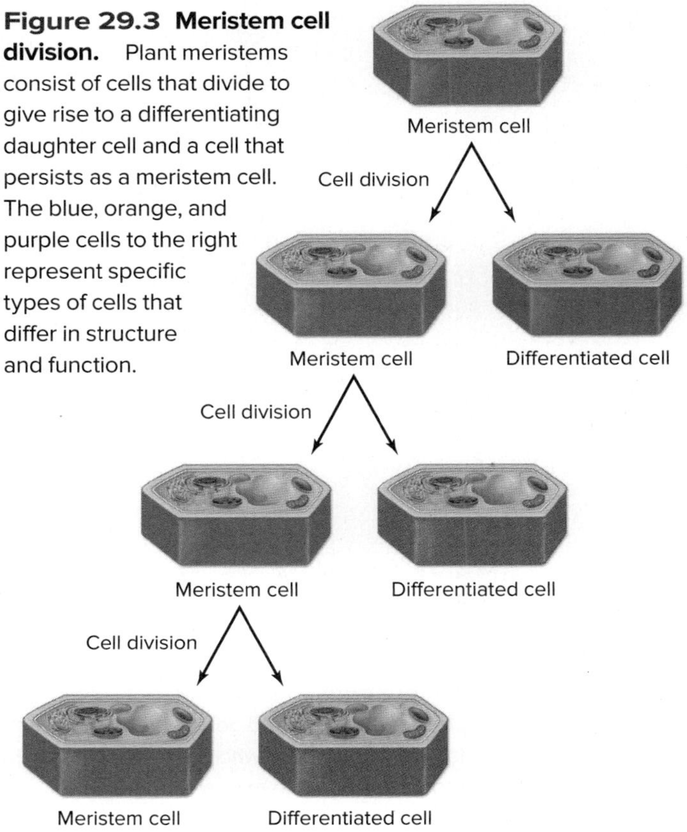

Figure 29.3 Meristem cell division. Plant meristems consist of cells that divide to give rise to a differentiating daughter cell and a cell that persists as a meristem cell. The blue, orange, and purple cells to the right represent specific types of cells that differ in structure and function.

Apical meristems

Apical meristems are located at the tips of stems and roots (figure 29.4). During periods of growth, the cells of apical meristems divide and continually add more cells at the tips. Tissues derived from apical meristems are called **primary tissues,** and the extension of the root and stem forms what is known as the **primary plant body.** The primary plant body comprises the young, soft shoots and roots of a tree or shrub, or the entire plant body in some plants.

Both root and shoot apical meristems are composed of delicate cells that need protection (figure 29.4). The root apical meristem is protected by the root cap, which is described in section 29.3. Root cap cells are produced by the root meristem and are sloughed off and replaced as the root pushes through the soil.

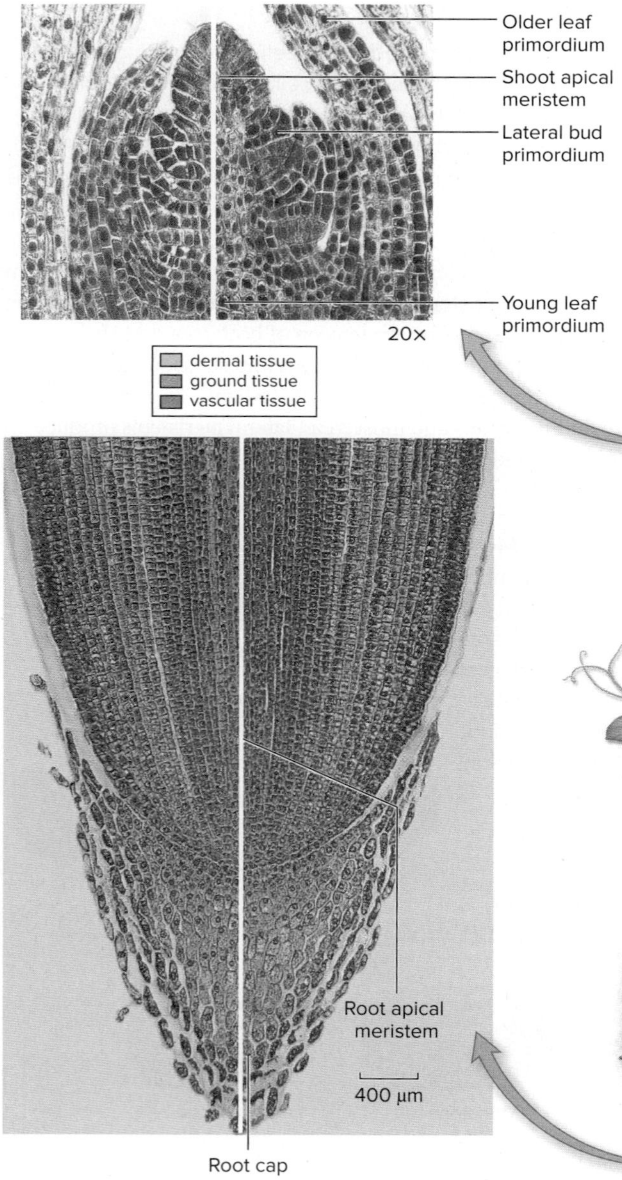

Older leaf primordium
Shoot apical meristem
Lateral bud primordium
Young leaf primordium
20×
dermal tissue
ground tissue
vascular tissue
Root apical meristem
400 μm
Root cap

Figure 29.4 Apical meristems. Shoot and root apical meristems extend the plant body above and below the ground. Leaf primordia protect the fragile shoot meristem, whereas the root meristem produces a protective root cap in addition to new root tissue.

(shoot): Steven P. Lynch; (root): Garry DeLong/Getty Images

TABLE 29.2	Types and Functions of Primary Meristems
Type of Meristem	**Function**
Protoderm	Produces the epidermis, a protective outer covering of the plant
Procambium	Produces primary vasculature for water and mineral transport (primary xylem) and organic nutrient transport (primary phloem)
Ground meristem	Produces ground tissue such as that used in support (e.g., sclerenchyma)
Intercalary meristem (only some plant groups)	Produces tissues for rapid internode growth or for tissue replacement in response to herbivory

In contrast, leaf primordia protect the growing shoot apical meristem, which is particularly susceptible to desiccation because of its exposure to air and sun.

The apical meristem gives rise to the three tissue systems by first initiating primary meristems: **protoderm, procambium,** and **ground meristem.** In some plants, such as corn, intercalary meristems are also present (table 29.2).

Lateral meristems

Many herbaceous plants—plants with fleshy, not woody, stems—exhibit only primary growth, but others also exhibit **secondary growth,** which may result in a substantial increase of diameter. Secondary growth is accomplished by the lateral meristems—peripheral cylinders of meristematic tissue within the stems and roots that increase the girth (diameter) of gymnosperms and most angiosperms. Lateral meristems form from ground tissue that is derived from apical meristems (figure 29.5). Secondary growth does not occur in monocots.

Although secondary growth increases girth in many nonwoody plants, its effects are most dramatic in woody plants, which have two lateral meristems. Within the bark of a woody stem is the **cork cambium**—a lateral meristem that contributes to the outer bark of the tree. Just beneath the bark is the **vascular cambium**—a lateral meristem that produces secondary vascular tissue. The vascular cambium forms between the xylem and phloem in vascular bundles, adding secondary vascular tissue to both of its sides.

Secondary xylem is the main component of wood. Secondary phloem is very close to the outer surface of a woody stem. Removing the bark of a tree damages the phloem and may eventually kill the tree. Tissues formed from lateral meristems, which comprise most of the trunk, branches, and older roots of trees and shrubs, are known as **secondary tissues** and are collectively called the **secondary plant body.**

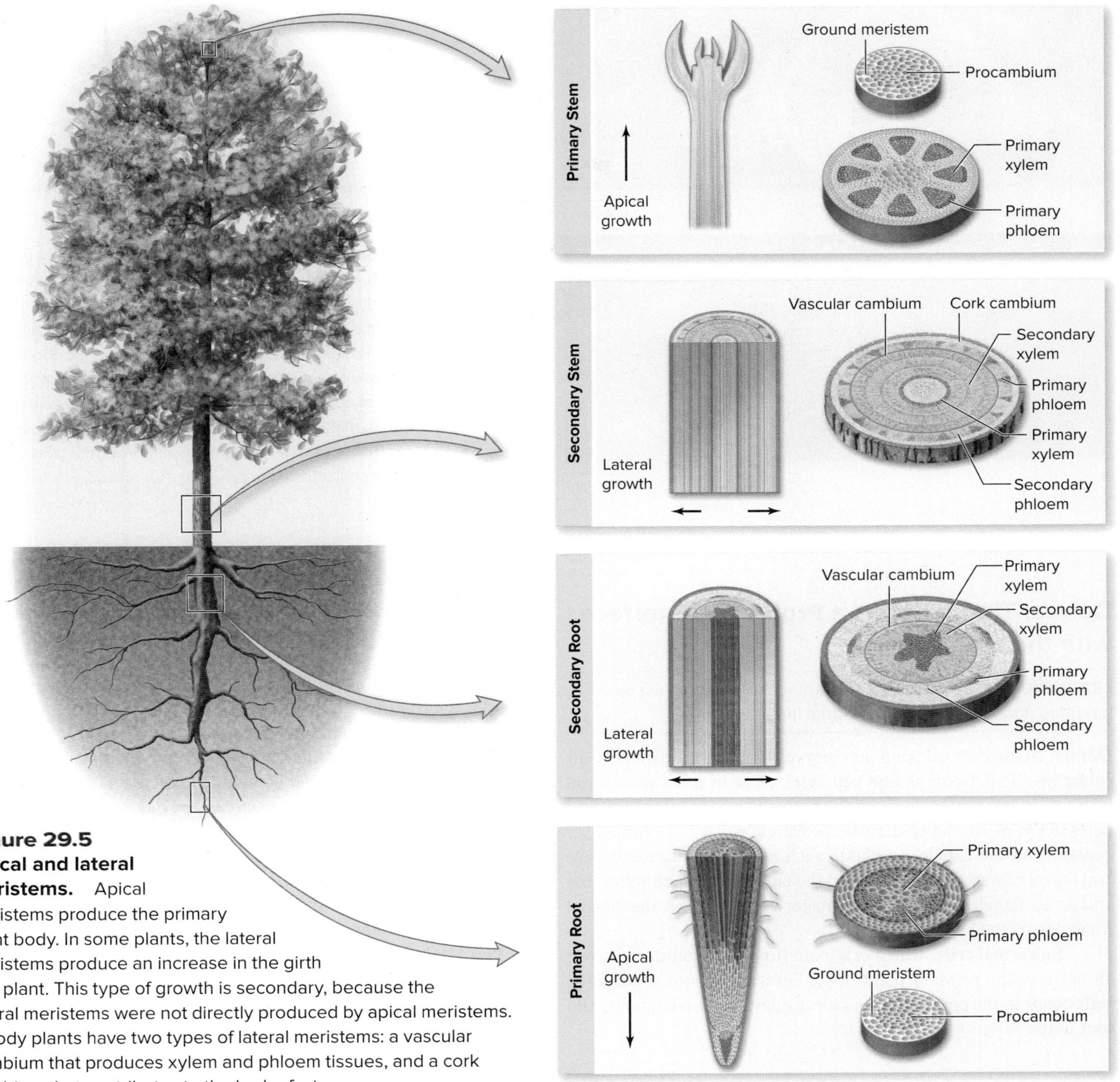

Figure 29.5
Apical and lateral meristems. Apical meristems produce the primary plant body. In some plants, the lateral meristems produce an increase in the girth of a plant. This type of growth is secondary, because the lateral meristems were not directly produced by apical meristems. Woody plants have two types of lateral meristems: a vascular cambium that produces xylem and phloem tissues, and a cork cambium that contributes to the bark of a tree.

REVIEW OF CONCEPT 29.1

The root system anchors plants and absorbs water and nutrients. The shoot system, consisting of stems, leaves, and flowers, carries out photosynthesis and sexual reproduction. The three general types of tissue are dermal, ground, and vascular tissue. Primary growth is produced by apical meristems at the tips of roots and shoots; secondary growth is produced by lateral meristems and increases girth.

- *Why are both primary and secondary growth necessary in a woody plant?*

29.2 Plants Contain Three Main Tissues

Three main categories of tissue can be identified in the vascular plant body. These are (1) dermal tissue on external surfaces, which serves a protective function; (2) ground tissue, which forms several different internal tissue types and can participate in photosynthesis, serve a storage function, or provide structural support; and (3) vascular tissue, which conducts water and nutrients.

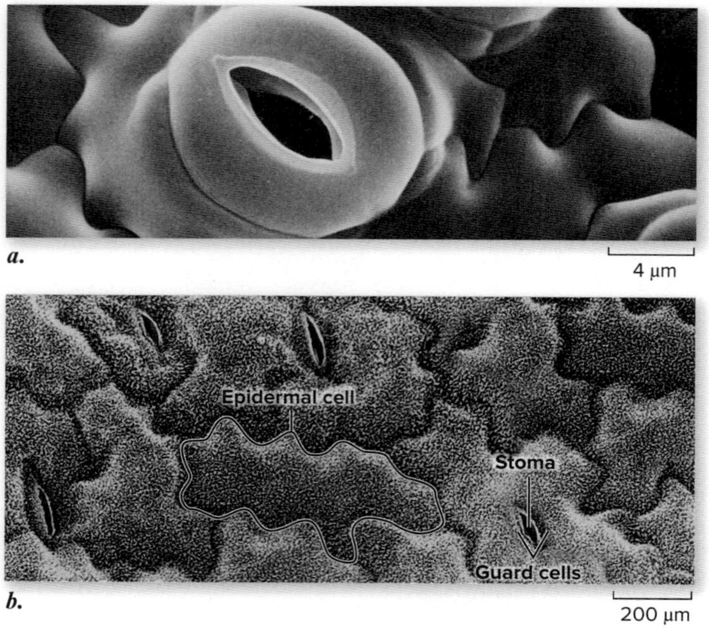
a.
4 µm
Epidermal cell
Stoma
Guard cells
b.
200 µm

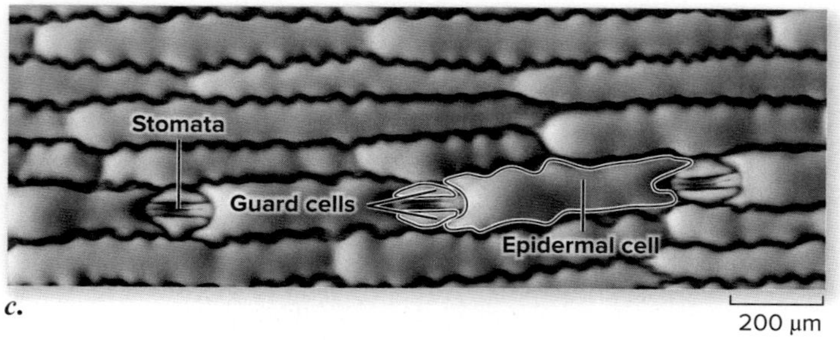
Stomata
Guard cells
Epidermal cell
c.
200 µm

Figure 29.6 Stomata. ***a.*** A stoma is the space between two guard cells, which regulate the size of the opening. Stomata are evenly distributed within the epidermis of monocots and eudicots, but the patterning is quite different. ***b.*** A pea (eudicot) leaf with a random arrangement of stomata. ***c.*** A maize (corn, a monocot) leaf with stomata evenly spaced in rows. These photomicrographs also show the variety of cell shapes in plants. Some plant cells are boxlike, as seen in maize **(*c*)**, and others are irregularly shaped, as seen in the jigsaw-puzzle shapes of the pea epidermal cells **(*b*)**.
(a, b): Dr. Jeremy Burgess/SPL/Science Source; (c): Patricia Goggin

Dermal Tissue Forms a Protective Interface with the Environment

LEARNING OBJECTIVE 29.2.1 Explain how dermal tissue provides adaptations for a terrestrial lifestyle.

Dermal tissue derived from an embryo or apical meristem forms **epidermis.** This tissue is one cell layer thick in most plants and forms the outer protective covering of the plant. In young, exposed parts of the plant, the epidermis is covered with a fatty **cutin** layer constituting the **cuticle.** In plants such as desert succulents, several layers of wax may be added to the cuticle to reduce water loss and protect against ultraviolet damage. In some cases, the dermal tissue forms the bark of trees.

Epidermal cells, which originate from the protoderm, cover all parts of the primary plant body. Several types of specialized cells occur in the epidermis, including guard cells, trichomes, and root hairs.

Guard cells

Guard cells are paired, sausage-shaped cells flanking a **stoma** (plural, *stomata*), an epidermal pore found on the surfaces of leaves and sometimes on other parts of the plant (figure 29.6*a*). Guard cells, unlike other epidermal cells, contain chloroplasts.

The movement of oxygen and carbon dioxide, as well as the diffusion of water vapor, occurs almost exclusively through stomata. Stomatal density ranges from 500 to 100,000 stomata per square centimeter of leaf surface, depending on the plant species. In many plants, stomata are more numerous on the lower side of the leaf than on the upper—a factor that helps minimize water loss. Some plants have stomata only on the lower epidermis, and a few, such as water lilies, have them only on the upper epidermis to maximize gas exchange.

An asymmetrical division of an epidermal cell produces a guard cell and a subsidiary cell that aids in the opening and closing of the stoma. The patterning of these asymmetrical divisions that results in stomatal distribution has intrigued developmental biologists (figure 29.6*b, c*).

Research on mutants that are unable to correctly position stomata is providing information on the timing of stomatal initiation and the kind of intercellular communication that triggers guard cell formation. For example, the *too many mouths (tmm)* mutation that occurs in *Arabidopsis* disrupts the normal pattern of cell division that spatially separates stomata (figure 29.7). Investigations of this and other stomatal patterning genes have revealed a coordinated network of cell–cell communication (refer to chapter 9) that informs cells of their position relative to other cells and determines cell fate. The *TMM* gene encodes a membrane-bound receptor that is part of a signaling pathway controlling asymmetrical cell division.

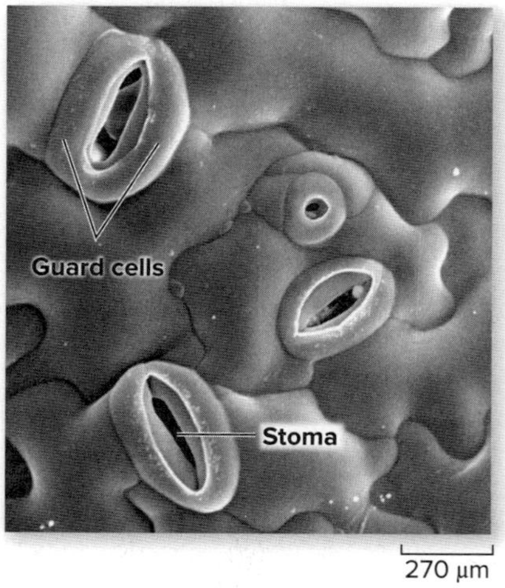
Guard cells
Stoma
270 µm

Figure 29.7 The *too many mouths* stomatal mutant. This *Arabidopsis* mutant plant lacks an essential signal for spacing stomata. Usually, a differentiating guard cell pair inhibits differentiation of a nearby cell into a guard cell; however, here two stomata are positioned too close together (middle right pair).
Jessica Lucas & Fred Sack

Trichomes

Trichomes are single-celled or multicellular hairlike outgrowths of the epidermis that occur frequently on stems, leaves, and reproductive organs (figure 29.8). Trichomes keep leaf surfaces cool and reduce evaporation by covering stomatal openings. They also protect leaves from high light intensities and ultraviolet radiation and can buffer against temperature fluctuations. In some species, they have been shown to sometimes regulate plant growth and shape by sticking flower plants together to help reproduction. If a leaf appears "fuzzy" or "woolly," like the leaves of tomato plants, it is probably covered with trichomes that can be seen clearly with a microscope under low magnification. Trichomes can vary greatly in form; some consist of a single cell, whereas others are multicellular. Some are glandular, often secreting sticky or toxic substances to deter herbivory.

Genes that regulate trichome development have been identified, including *GLABRA2*. When trichome-initiating proteins reach threshold levels relative to trichome-inhibiting proteins, *GLABRA2* expression causes an epidermal cell to become a trichome. Signals from this trichome cell prevent neighboring cells from expressing trichome-promoting genes.

Root hairs

Root hairs, which are tubular extensions of individual epidermal cells, occur in a zone just behind the tips of young, growing roots (figure 29.9). Because a root hair is simply an extension of an epidermal cell and not a separate cell, no wall isolates the hair from the rest of the cell. Root hairs keep the root in close contact with the surrounding soil particles and greatly increase the root's surface area, which maximizes water and nutrient absorption.

GLABRA2 is also involved in root hair development. However, unlike in leaf epidermal cells in which *GLABRA2* triggers the development of a trichome, in roots *GLABRA2* is responsible for the development of non-root-hair cells. In root epidermal cells where *GLABRA2* is not expressed, root hairs form. As a root grows, the extent of the root hair zone remains roughly constant as root hairs at the older end slough off while new ones are produced at the apex. Most of the absorption of water and minerals occurs through root hairs, especially in herbaceous plants.

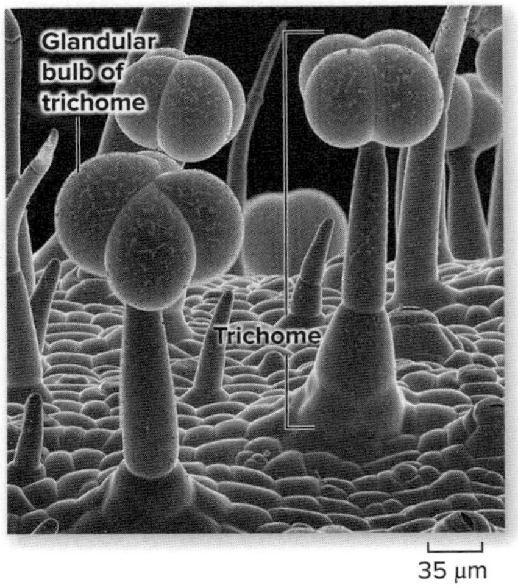

Figure 29.8 Trichomes. The trichomes with tan, bulbous tips on this tomato plant are glandular trichomes. These trichomes secrete substances that can literally glue insects to the trichome.
Andrew Syred/Science Source

Figure 29.9 Root hairs. Root hair cells are a type of epidermal cell; they increase the surface area of the root to enhance water and mineral uptake.
Nigel Cattlin/Alamy Stock Photo

Ground Tissue Cells Perform Many Functions, Including Storage, Photosynthesis, and Support

LEARNING OBJECTIVE 29.2.2 Compare and contrast the different kinds of ground tissue.

Ground tissue consists primarily of thin-walled parenchyma cells that function in storage, photosynthesis, and secretion. Other ground tissue, composed of collenchyma cells and sclerenchyma cells, provides support and protection.

Parenchyma

Parenchyma cells are the most common type of plant cell. They have large vacuoles and thin walls, and they are initially (but briefly) more or less spherical. These cells, which have living protoplasts, push up against each other shortly after they are produced, however, and assume other shapes, often ending up with 11 to 17 sides.

Parenchyma cells may live for many years. They function in the storage of food and water, photosynthesis, and secretion. They are the most abundant cells of primary tissues and may also occur, to a much lesser extent, in secondary tissues (figure 29.10*a*). Most parenchyma cells have only primary walls, which are walls laid down while the cells are still

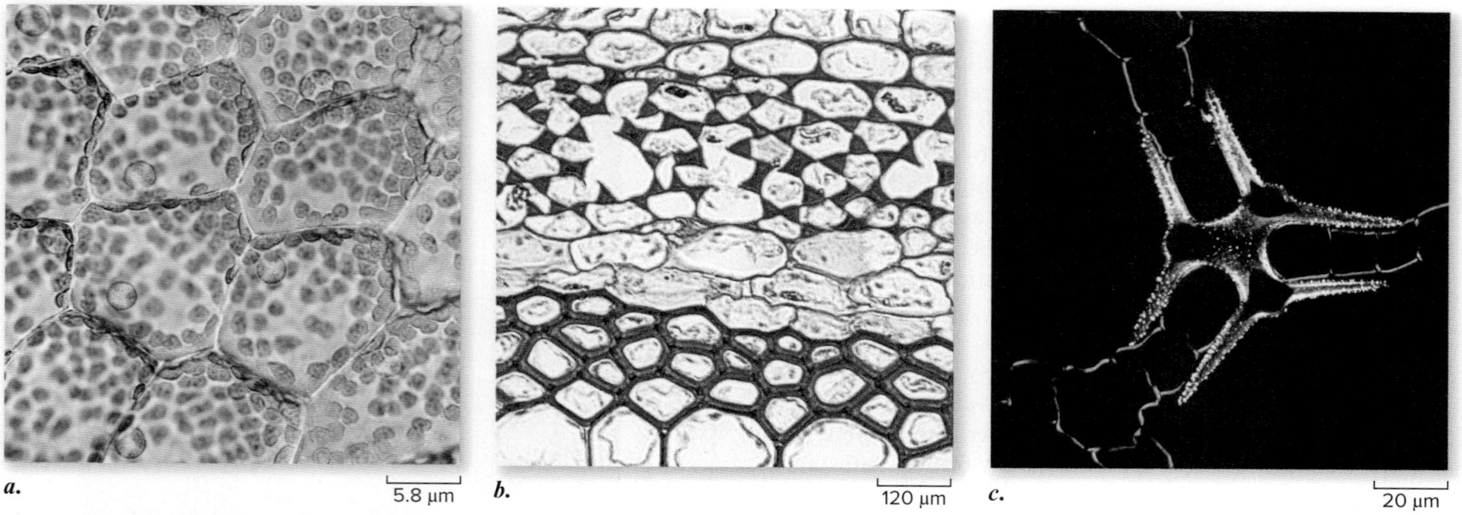
a.
5.8 µm
b.
120 µm
c.
20 µm

Figure 29.10 The three types of ground tissue. ***a.*** Parenchyma cells. Only primary cell walls are seen in this cross section of parenchyma cells from grass. ***b.*** Collenchyma cells. Thickened side walls are seen in this cross section of collenchyma cells from the stem of a cucumber (*Cucumis sativus*). In other kinds of collenchyma cells, the thickened areas may occur at the corners of the cells or in other kinds of strips. ***c.*** Sclereids. Clusters of sclereids ("stone cells"), stained red in this preparation. The surrounding thin-walled cells, stained green, are parenchyma. Sclereids are one type of sclerenchyma tissue, which also contains fibers.

(a, c): Lee W. Wilcox; (b): Steven P. Lynch

maturing. Parenchyma are less specialized than other plant cells, although many variations occur with special functions, such as nectar and resin secretion or storage of latex, proteins, and metabolic wastes.

Parenchyma cells have functional nuclei and are capable of dividing, and they usually remain alive after they mature; in some plants (for example, cacti), they may live to be over 100 years old. The majority of cells in fruits such as apples are parenchyma. Some parenchyma contain chloroplasts, especially in leaves and in the outer parts of herbaceous stems. Such photosynthetic parenchyma tissue is called chlorenchyma.

Collenchyma

If you have ever had celery "strings" trapped in your teeth, you are familiar with tough, flexible **collenchyma cells.** Like parenchyma cells, collenchyma cells have living protoplasts and may live for many years. These cells, which are usually a little longer than wide, have walls that vary in thickness (figure 29.10*b*).

Flexible collenchyma cells provide support for plant organs, allowing them to bend without breaking. They often form strands or continuous cylinders beneath the epidermis of stems or leaf petioles (stalks) and along the veins in leaves. Strands of collenchyma provide much of the support for stems in the primary plant body.

Sclerenchyma

Sclerenchyma cells have tough, thick walls. Unlike collenchyma and parenchyma, they usually lack living protoplasts at maturity. Their secondary cell walls are often impregnated with **lignin,** a highly branched polymer that makes cell walls more rigid. Lignin is an important component in wood. Cell walls containing lignin are said to be lignified. Lignin is common in the walls of plant cells that have a structural or mechanical function, such as those found in wood.

Sclerenchyma is present in two general types: fibers and sclereids. Fibers are long, slender cells that are usually grouped together in strands. Linen, for example, is woven from strands of sclerenchyma fibers that occur in the phloem of flax (*Linum* spp.) plants. Sclereids vary in shape but are often branched. They may occur singly or in groups; they are not elongated but may have many different forms, including that of a star. The gritty texture of a pear is caused by groups of sclereids throughout the soft flesh of the fruit (figure 29.10*c*). Sclereids are also found in hard seed coats. Both of these tough, thick-walled cell types strengthen the tissues in which they are found.

Vascular Tissues Conduct Water and Nutrients Throughout the Plant

LEARNING OBJECTIVE 29.2.3 Distinguish between xylem and phloem.

As mentioned earlier, vascular tissue includes two types of conducting tissues: (1) xylem, which conducts water and dissolved minerals and (2) phloem, which conducts a solution of carbohydrates. The phloem also transports hormones, amino acids, and other substances necessary for plant growth. Xylem and phloem differ in structure as well as in function.

Xylem

Xylem, the principal water-conducting tissue of plants, usually contains a combination of vessels and tracheids. Vessels are continuous tubes formed from dead, hollow, cylindrical cells

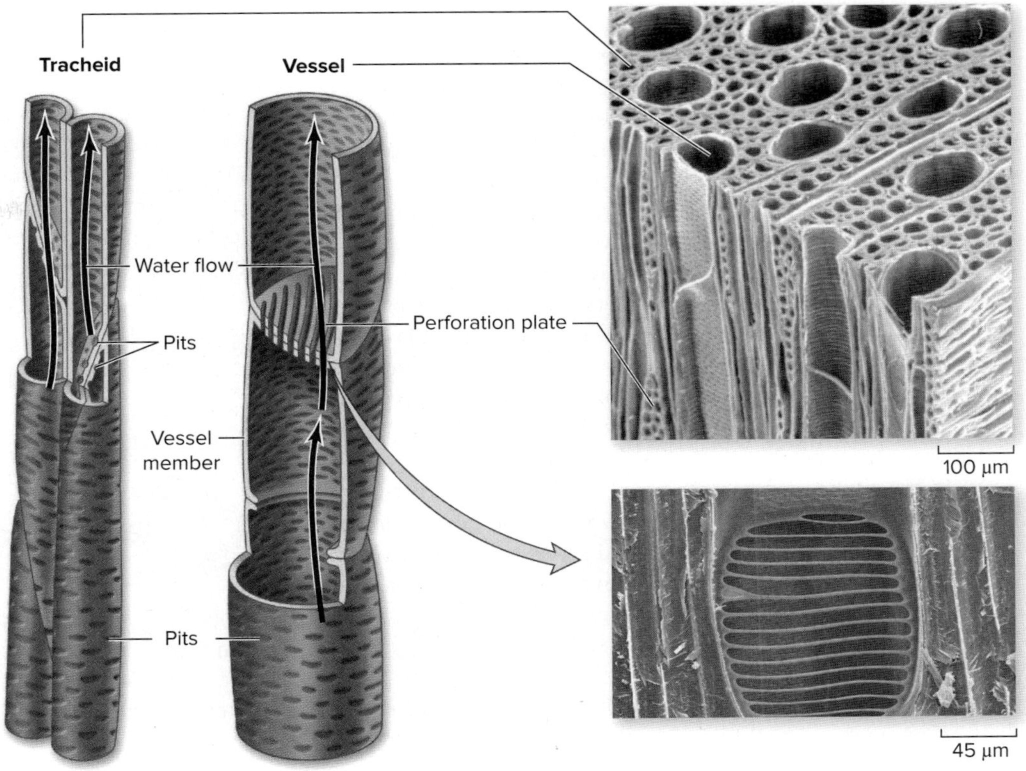
Tracheid
Vessel
Water flow
Pits
Vessel member
Pits
Perforation plate
100 μm
45 μm

Figure 29.11 Comparison between tracheids and vessel members. In tracheids, the water passes from cell to cell by means of pits. In vessel members, water moves by way of perforation plates (as seen in the lower photomicrograph in this figure). In gymnosperm wood, tracheids both conduct water and provide support; in most kinds of angiosperms, vessels are present in addition to tracheids. These two types of cells conduct water, and fibers provide additional support. The wood of red maple, *Acer rubrum,* contains both tracheids and vessels, as seen in the top electron micrograph in this figure.

(top): NC Brown Center for Ultrastructure Studies, SUNY, College of Environmental Science and Forestry, Syracuse, NY; (bottom): USDA Forest Service, Forest Products Laboratory, Madison, WI

arranged end-to-end. Tracheids are dead cells that taper at the ends and overlap one another (figure 29.11). Primary xylem is derived from the procambium produced by the apical meristem. Secondary xylem is formed by the vascular cambium, a lateral meristem. Wood consists of accumulated secondary xylem.

Vessels evolved independently multiple times; in some plants (for example, the conifers), tracheids are the only water-conducting cells present. Water passes in an unbroken stream through the xylem from the roots up through the shoot and into the leaves. When the water reaches the leaves, much of it diffuses in the form of water vapor into the intercellular spaces and out of the leaves into the surrounding air, mainly through the stomata. This diffusion of water vapor from a plant is known as **transpiration** (refer to chapter 31). In addition to conducting water, dissolved minerals, and inorganic ions such as nitrates and phosphates throughout the plant, xylem provides support for the plant body.

Vessel members tend to be shorter and wider than tracheids. When viewed with a microscope, they resemble soda cans with both ends removed. Both vessel members and tracheids have thick, lignified secondary walls and no living protoplasts at maturity. Lignin is produced by the cell and secreted to strengthen the cellulose cell walls before the protoplast dies, leaving only the cell wall.

Tracheids contain pits, which are small areas, round to elliptical in shape, where no secondary wall material has been deposited. The pits of adjacent cells occur opposite one another; the continuous stream of water flows through these pits from tracheid to tracheid. In contrast, vessel members, which are joined end to end, may be almost completely open or have bars or strips of wall material across the open ends (figure 29.11). Vessels appear to conduct water more efficiently than do the overlapping strands of tracheids.

In addition to conducting cells, xylem typically includes fibers and parenchyma cells. It is likely that some types of fibers have evolved from tracheids, becoming specialized for strengthening rather than conducting fluids. The parenchyma cells, which are usually produced in horizontal rows called rays by special ray initials (meristematic cells) of the vascular cambium, function in lateral conduction and food storage.

In cross sections of woody stems and roots, the rays can be seen radiating out from the center of the xylem like the spokes of a wheel. Fibers are abundant in some kinds of wood, such as oak (*Quercus* spp.), and the wood is correspondingly dense and heavy. The arrangements of these and other kinds of cells in the xylem make it possible to identify most plant genera and many species from their wood alone.

Over 2000 years ago, paper as we recognize it today was made in China by mashing herbaceous plants in water and separating out a thin layer of phloem fibers on a screen. Not until the third century of the common era did the secret of making paper make its way out of China. Today the ever-growing demand for paper is met by extracting xylem fibers from relatively soft woods such as those from spruce. The lignin-rich cell walls yield brown paper, which is often bleached. In addition, tissues from many other plants, including kenaf and hemp, have been developed as sources of paper. U.S. paper currency is 75% cotton and 25% flax.

Phloem

Phloem, which is located toward the outer part of roots and stems, is the principal food-conducting tissue in vascular plants. If a plant is girdled—by removing a substantial strip of bark down to the vascular cambium around the entire stem's circumference—the plant eventually dies from starvation of the roots.

Food conduction in phloem is carried out through two kinds of elongated cells: sieve cells and sieve-tube members. Gymnosperms, ferns, and horsetails have only sieve cells; most angiosperms have sieve-tube members. Both types of cells have clusters of pores known as sieve areas because the cell walls resemble sieves. Sieve areas are more abundant on the overlapping ends of the cells and connect the protoplasts of adjoining sieve cells and sieve-tube members. Both of these types of cells are living, but most sieve cells and all sieve-tube members lack a nucleus at maturity.

In sieve-tube members, some sieve areas have larger pores and are called sieve plates (figure 29.12). Sieve-tube members occur end to end, forming longitudinal series called sieve tubes. Sieve cells are less specialized than sieve-tube members, and the pores in all of their sieve areas are roughly of the same diameter. Sieve-tube members are more specialized and presumably more efficient than sieve cells.

Each sieve-tube member is associated with an adjacent, specialized parenchyma cell known as a companion cell. Companion cells apparently carry out some of the metabolic functions needed to maintain the associated sieve-tube member. In angiosperms, a common initial cell divides asymmetrically to produce a sieve-tube member cell and its companion cell. Companion cells have all the components of normal parenchyma cells, including nuclei, and numerous plasmodesmata (cytoplasmic connections between adjacent cells) connect their cytoplasm with that of the associated sieve-tube members.

Sieve cells in nonflowering plants have albuminous cells that function as companion cells. Unlike a companion cell, an albuminous cell is not necessarily derived from the same mother cell as its associated sieve cell. Fibers and parenchyma cells are often abundant in phloem.

REVIEW OF CONCEPT 29.2

Dermal tissue protects a plant from its environment and produces specialized cells such as guard cells, trichomes, and root hairs. Ground tissue serves several functions, including storage (parenchyma cells), photosynthesis (specialized parenchyma called chlorenchyma), and structural support (collenchyma and sclerenchyma). Vascular tissue carries water through the xylem (primarily vessels) and nutrients through the phloem (primarily sieve-tube members).

- *Contrast the structure and function of mature vessels and sieve-tube members.*

29.3 Roots Have Four Growth Zones

Roots have a simpler pattern of organization and development than stems, so we will consider them first. Keep in mind, however, that roots evolved after shoots and are a major innovation for terrestrial living.

Roots Are Adapted for Growing Underground and Absorbing Water and Solutes

LEARNING OBJECTIVE 29.3.1 Describe the four regions of a typical root.

Four regions are commonly recognized in developing roots: the root cap, the zone of cell division, the zone of elongation, and the zone of maturation (figure 29.13). The boundaries between the last three regions are not clearly defined in the plant root.

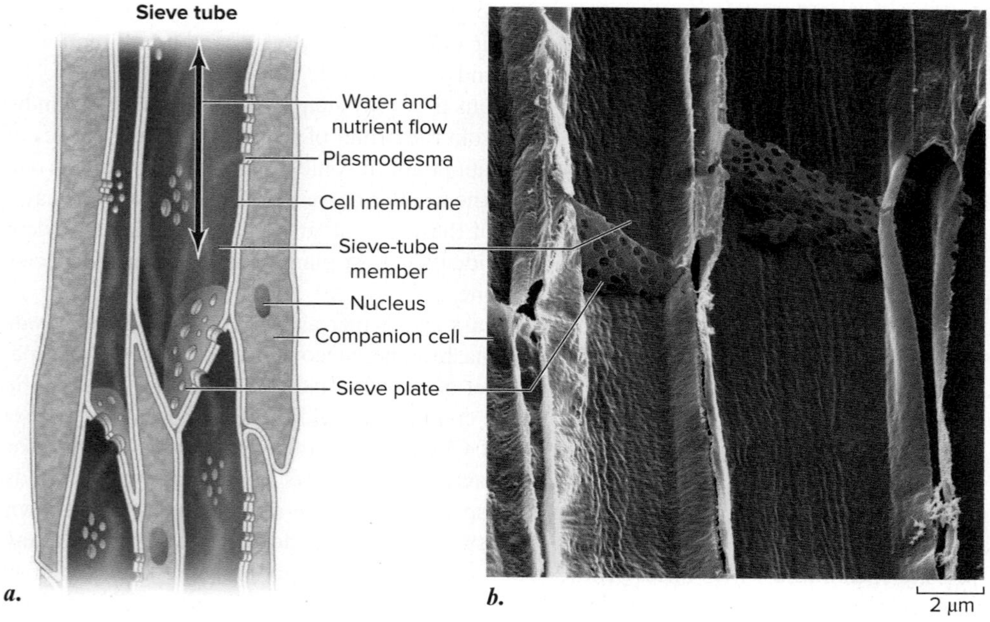
Sieve tube
Water and nutrient flow
Plasmodesma
Cell membrane
Sieve-tube member
Nucleus
Companion cell
Sieve plate
a.
b.
2 μm

Figure 29.12 A sieve-tube member. ***a.*** Sieve-tube member cells are stacked, with sieve plates forming the connection. The narrow cell with the nucleus at the right of the sieve-tube member is a companion cell. This cell nourishes the sieve-tube members, which have plasma membranes but no nuclei. ***b.*** Looking up into sieve plates in maize phloem reveals the perforations through which sucrose and hormones move.

(b): Kage Mikrofotografie/Medical Images/DIOMEDIA

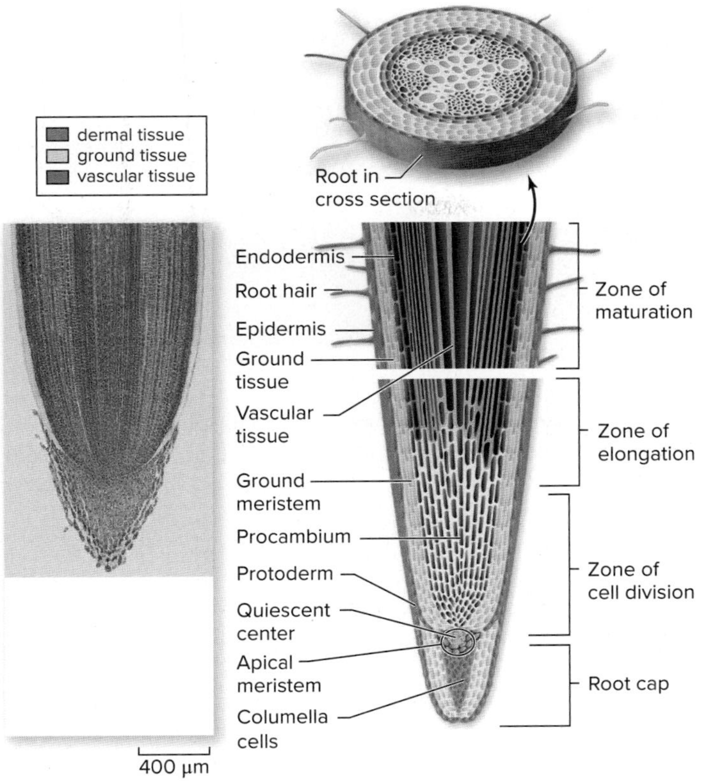
dermal tissue
ground tissue
vascular tissue
Root in cross section
Endodermis
Root hair
Epidermis
Ground tissue
Vascular tissue
Ground meristem
Procambium
Protoderm
Quiescent center
Apical meristem
Columella cells
Zone of maturation
Zone of elongation
Zone of cell division
Root cap
400 μm

Figure 29.13 Root structure. A root tip in corn, *Zea mays.*
Garry DeLong/Getty Images

When cells of the root apical meristem divide, daughter cells that end up on the tip end of the root become root cap cells. Daughter cells that end up on the nontip end pass through the three other zones before they finish differentiating.

The root cap

The **root cap** has no structural equivalent in shoots. Functionally, however, leaf primordia protect the shoot apical meristem just as the root cap protects the root apical meristem. The root apical meristem is composed of two types of cells: the inner columella cells, and the outer, lateral root cap cells, which are continuously replenished by the root apical meristem. In some plants with larger roots, the root cap is quite obvious. Its main function is to protect the delicate tissues behind it as growth pushes the root through abrasive soil particles.

Cells in the outer root cap secrete a slimy substance produced by Golgi bodies. The root cap cells, which have an average life of less than a week, are constantly being replaced from the inside, forming a mucilaginous lubricant that eases the root through the soil. The slimy mass also provides a medium for the growth of beneficial nitrogen-fixing bacteria in the roots of plants such as legumes. A new root cap is produced when an existing one is artificially or accidentally removed from a root.

The root cap also functions in the perception of gravity, or geotropism. The columella cells are highly specialized, with the endoplasmic reticulum in the periphery and the nucleus located at either the middle or the top of the cell. Columella cells lack large vacuoles but contain **amyloplasts** (plastids with starch grains) that collect on the sides of cells facing the pull of gravity. When a potted plant is placed on its side, the amyloplasts drift or tumble down to the side nearest the source of gravity, and the root bends in that direction.

Experiments where lasers are used to kill individual columella cells have shown that it takes only two columella cells to sense gravity. The precise mechanism of **gravitropism** is unknown, but some evidence indicates that calcium ions in the amyloplasts influence the distribution of the growth hormone auxin. Because bending has been seen in the absence of auxin, multiple signaling mechanisms probably influence **gravitropism**.

The zone of cell division

The apical meristem is located in the center of the root tip in the area protected by the root cap. Most of the activity in this **zone of cell division** takes place toward the edges of the meristem, where the cells divide every 12 to 36 hours, often coordinately, reaching a peak of division once or twice a day.

Most of the cells are essentially cuboidal, with small vacuoles and proportionately large, centrally located nuclei. These rapidly dividing cells are daughter cells of the apical meristem. A group of cells in the center of the root apical meristem, termed the quiescent center, divide only very infrequently. The presence of the quiescent center makes sense if you think about a solid ball expanding—the outer surface would have to increase far more rapidly than the very center.

The apical meristem daughter cells soon subdivide into the three primary tissues: protoderm, procambium, and ground meristem. Genes have been identified in the relatively simple root of *Arabidopsis* that regulate the patterning of these tissue systems. The patterning of these tissues begins in this zone, but the anatomical and morphological expression of this patterning is not fully revealed until the cells reach the zone of maturation.

For example, the *WEREWOLF (WER)* gene is required for the patterning of the two root epidermal cell types, those with and those without root hairs (figure 29.14). Plants with a mutation in *WER* have an excess of root hairs, because *WER* is needed to prevent root hair development in nonhair epidermal cells. Similarly, the *SCARECROW (SCR)* gene is necessary in ground cell differentiation (figure 29.15). A ground meristem cell undergoes an asymmetrical cell division that gives rise to two nested cylinders of cells from one if *SCR* is present. The outer cell layer becomes ground tissue and serves a storage function. The inner cell layer forms the endodermis, which regulates the intercellular flow of water and solutes into the vascular core of the root. The *scr* mutant, in contrast, forms a single layer of cells that have both endodermal and ground cell traits.

SCR illustrates the importance of the orientation of cell division. If a cell's relative position changes because of a mistake in cell division or the death of another cell, the cell develops according to its new position. The fate of most plant cells is determined by their position relative to other cells.

The zone of elongation

In the **zone of elongation,** roots lengthen because the cells produced by the primary meristems become several times longer

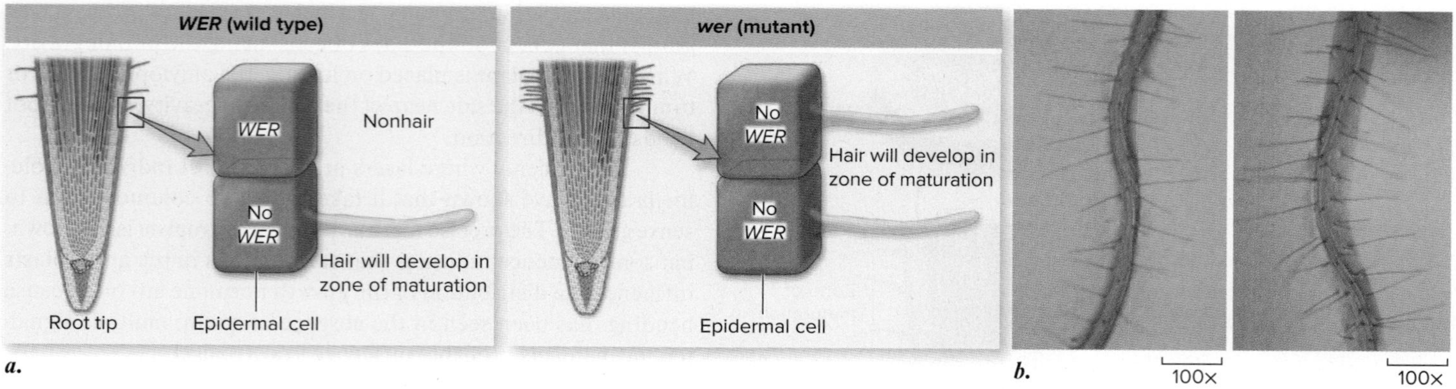

Figure 29.14 Tissue-specific gene expression. ***a.*** The *WEREWOLF* gene of *Arabidopsis* is expressed in some, but not all, epidermal cells and suppresses root hair development. The *wer* mutant is covered with root hairs. ***b.*** On the left is a root from a wild-type plant and on the right is a root from a mutant *wer* plant, containing more root hairs.

(b): ©John Schiefelbein

while their width increases only slightly. The small vacuoles merge and grow until they occupy 90% or more of the volume of each cell. It is metabolically efficient for the bulk of the volume of a cell to be filled with vacuole versus cytoplasm. Cytoplasm is metabolically "expensive" to make, but vacuoles are not. When growth is rapid, as in these cells, it makes sense to use the least expensive way to fill the space—the vacuole. No further increase in cell size occurs above the zone of elongation. The mature parts of the root, except for increasing in girth, remain stationary for the life of the plant.

The zone of maturation

The cells that have elongated in the zone of elongation become differentiated into specific cell types in the **zone of maturation** (figure 29.13). The cells of the root surface cylinder mature into *epidermal cells,* which have a very thin cuticle, and include both root hair and nonhair cells. Although the root hairs are not visible until this stage of development, their fate was established much earlier, as you saw with the expression patterns of *WER* (figure 29.14).

Root hairs can provide over 37,000 cm^2 of surface area for a root. This large increase in surface area greatly increases the absorptive capacity of the root. Symbiotic bacteria that fix atmospheric nitrogen into a form usable by legumes enter the plant via root hairs and "instruct" the plant to create a nitrogen-fixing nodule around them (refer to chapter 31).

Parenchyma cells are produced by the ground meristem immediately to the interior of the epidermis. This tissue, called the **cortex,** may be many cell layers wide and stores food. As just described, the inner boundary of the cortex differentiates into a single-layered cylinder of **endodermis,** after an asymmetrical cell division regulated by *SCR* (figure 29.16). Endodermal primary walls are impregnated with suberin, a fatty substance that is impervious to water. The suberin is produced in bands, called **Casparian strips,** that surround each adjacent endodermal cell wall

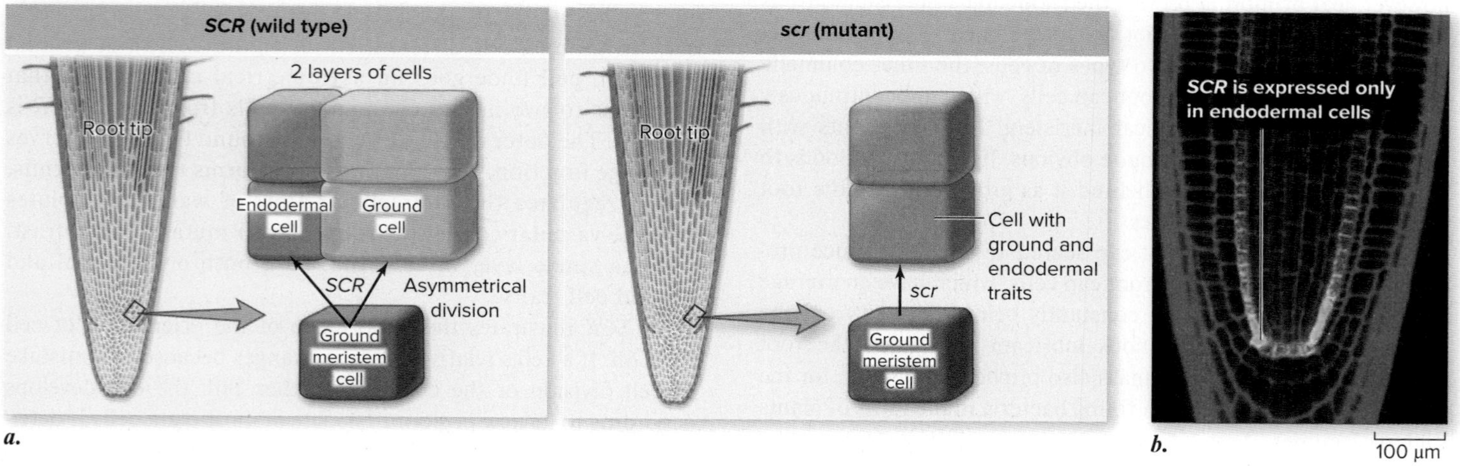

Figure 29.15 *SCARECROW* regulates asymmetrical cell division. ***a.*** *SCR* is needed for an asymmetrical cell division leading to the differentiation of daughter cells into endodermal and ground cells. ***b.*** The *SCR* promoter was attached to a gene coding for a green fluorescent protein to find out exactly where in the wild-type root *SCR* is expressed. *SCR* is expressed only in the endodermal cells, not in the ground cells.

Jee Jung and Philip Benfey

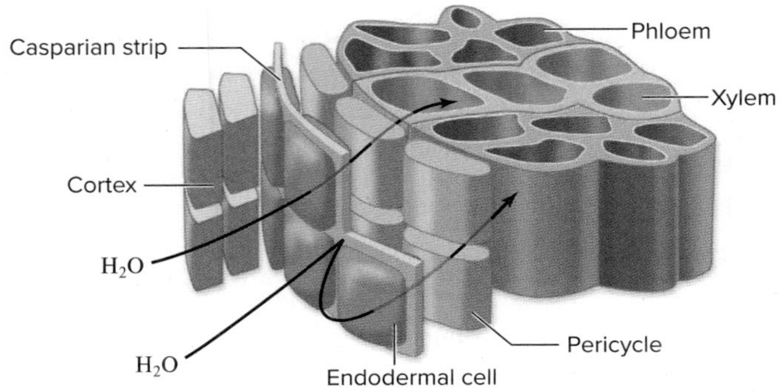
Casparian strip
Phloem
Xylem
Cortex
H_2O
H_2O
Pericycle
Endodermal cell

Figure 29.16 Cross sections of the zone of maturation of roots. Both monocot and eudicot roots have a Casparian strip, as seen in the cross section of greenbriar (*Smilax*), a monocot, and buttercup (*Ranunculus*), a eudicot. The Casparian strip is a water-proofing band that forces water and minerals to pass through the plasma membranes rather than through the spaces in the cell walls.

(top left): Carolina Biological/Medical Images/DIOMEDIA; (top right, bottom right): George S. Ellmore; (bottom left): Lee W. Wilcox

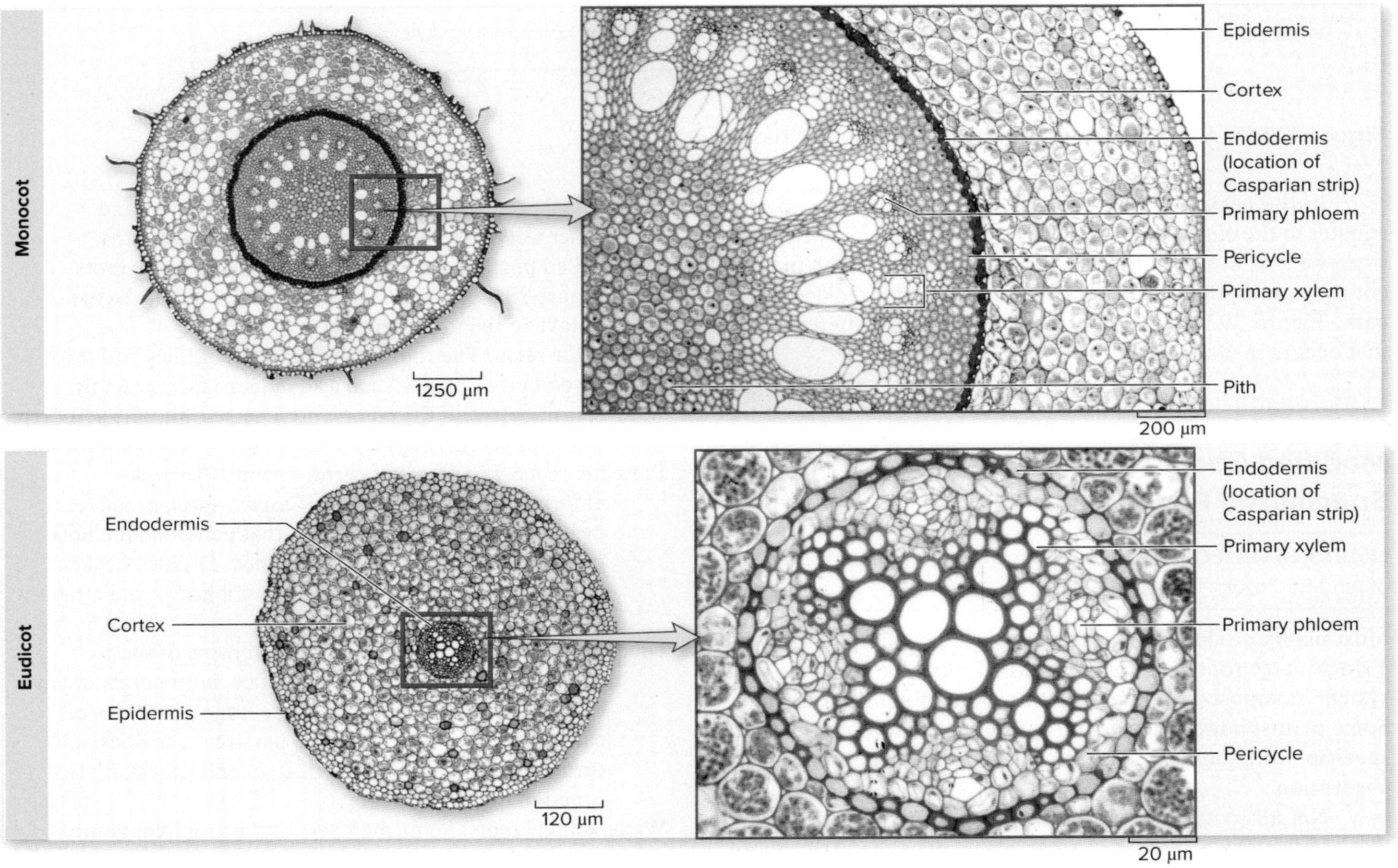
Monocot
1250 μm
Epidermis
Cortex
Endodermis (location of Casparian strip)
Primary phloem
Pericycle
Primary xylem
Pith
200 μm
Eudicot
Endodermis
Cortex
Epidermis
120 μm
Endodermis (location of Casparian strip)
Primary xylem
Primary phloem
Pericycle
20 μm

perpendicular to the root's surface. These strips block transport between cells. The two surfaces that are parallel to the root surface are the only way into the vascular tissue of the root, and the selectively permeable nature of the plasma membrane controls what passes through. Plants with an *scr* mutation lack this waterproof Casparian strip.

All the tissues interior to the endodermis are collectively referred to as the **stele.** Immediately adjacent and interior to the endodermis is a cylinder of parenchyma cells known as the **pericycle.** Pericycle cells divide, even after they mature, and can give rise to lateral (branch) roots or, in eudicots, to the two lateral meristems, the vascular cambium and the cork cambium.

The water-conducting cells of the primary xylem are differentiated as a solid core in the center of young eudicot roots. In a cross section of a eudicot root, the central core of primary xylem often is somewhat star-shaped, having from two to several radiating arms that point toward the pericycle (figure 29.16). In monocot (and a few eudicot) roots, the primary xylem is in discrete vascular bundles arranged in a ring, which surrounds parenchyma cells, called *pith,* at the very center of the root. Primary phloem, composed of cells involved in food conduction, is differentiated in discrete groups of cells adjacent to the xylem in both eudicot and monocot roots.

In eudicots and other plants with secondary growth, part of the pericycle and the parenchyma cells between the phloem patches and the xylem become the root vascular cambium, which starts producing secondary xylem to the inside and secondary phloem to the outside. Eventually, the secondary tissues acquire the form of concentric cylinders. The primary phloem, cortex, and epidermis become crushed and are sloughed off as more secondary tissues are added.

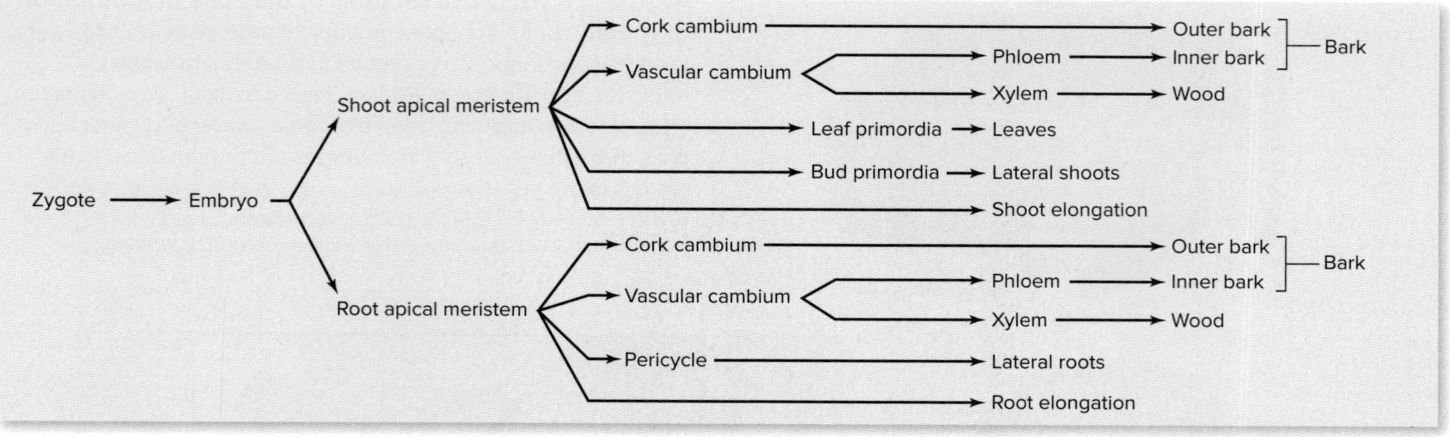

Figure 29.17 Stages in the differentiation of plant tissues.

In the pericycle of woody plants, the cork cambium contributes to the outer bark, which will be discussed in more detail when we look at stems. In the case of secondary growth in eudicot roots, everything outside the stele is lost and replaced with bark. Figure 29.17 summarizes the process of differentiation that occurs in plant tissue.

Modified Roots Accomplish Specialized Functions

LEARNING OBJECTIVE 29.3.2 Describe the functions of modified roots.

Most plants produce either a taproot system, characterized by a single large root with smaller branch roots, or a fibrous root system, composed of many smaller roots of similar diameter. Some plants, however, have intriguing root modifications with specific functions in addition to those of anchorage and absorption.

Not all roots are produced by preexisting roots. Any root that arises along a stem or in some place other than the root of the plant is called an **adventitious** root. For example, climbing plants such as ivy produce roots from their stems; these can anchor the stems to tree trunks or to a brick wall. Adventitious root formation in ivy depends on the developmental stage of the shoot. When the shoot enters the adult phase of development, it is no longer capable of initiating these roots. Following are the functions of modified roots.

Prop roots. Some monocots, such as corn, produce thick adventitious roots from the lower parts of the stem. These roots grow down to the ground and brace the plants against wind (figure 29.18*a*); corn plants lacking prop roots are easily blown over.

Aerial roots. Plants such as epiphytic orchids, which are attached to tree branches and grow unconnected to the ground, have roots that extend into the air (figure 29.18*b*). Some aerial roots have a thickened epidermis to reduce water loss. Aerial roots may also be photosynthetic.

Pneumatophores. Some plants that grow in swamps and other wet places may produce spongy outgrowths called pneumatophores from their underwater roots (figure 29.18*c*). The pneumatophores facilitate oxygen uptake in the roots beneath.

Contractile roots. The roots from the bulbs of lilies and from several other plants, such as dandelions, contract by spiraling to pull the plant a little deeper into the soil each year, until they reach an area of relatively stable temperature.

Parasitic roots. The stems of certain plants that lack chlorophyll, such as dodder (*Cuscuta* spp.), produce peglike roots, called haustoria, that penetrate the host plants around which they are twined. Dodder weakens plants and can spread disease when it grows and attaches to several plants.

Food-storage roots. The xylem of branch roots of sweet potatoes and similar plants produce, at intervals, many extra parenchyma cells that store large quantities of carbohydrates. Carrots, beets, parsnips, radishes, and turnips have combinations of stem and root that also function in food storage.

Water-storage roots. Some members of the pumpkin family (Cucurbitaceae), especially those that grow in arid regions, produce water-storage roots that can weigh over 50 kg (figure 29.18*d*).

Buttress roots. Species of fig and other tropical trees produce huge buttress roots toward the base of the trunk, which provide considerable stability (figure 29.18*e*).

REVIEW OF CONCEPT 29.3

The root cap protects the root apical meristem and helps sense gravity to give growth directionality. New cells formed in the zone of cell division grow in length in the zone of elongation. Cells differentiate in the zone of maturation, and root hairs appear there. Root hairs greatly increase the absorptive surface area of roots. Modified roots allow plants to carry out many additional functions, including bracing, aeration, and storage of nutrients and water.

- *Why do you suppose root hairs are not formed in the region of elongation?*

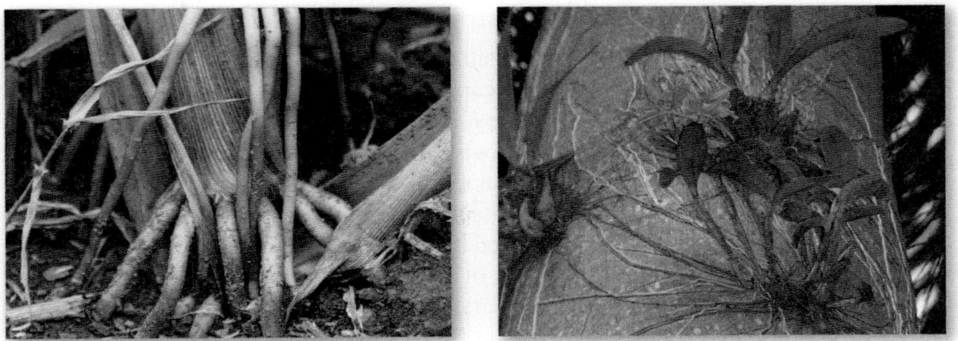

a. *b.*

Figure 29.18 Five types of modified roots. ***a.*** Maize (corn) prop roots originate from the stem and keep the plant upright. ***b.*** Epiphytic orchids attach to trees far above the tropical soil. Their roots are adapted to obtain water from the air rather than the soil. ***c.*** Pneumatophores (*foreground*) of the swam cypress (*Taxodium distichum*) are spongy outgrowths from the roots below. ***d.*** A water-storage root weighing over 25 kg (60 pounds). ***e.*** Buttress roots of a tropical fig tree.

(a): NokHoOkNoi/iStock/Getty Images; (b): F. Scholz/picture alliance/Arco Images/Newscom; (c): Mark Newman/FLPA/age fotostock; (d): Dennis Albert; (e): Patjo/Shutterstock

c. *d.* *e.*

29.4 Stems Provide Support for Aboveground Organs

The supporting structure of a vascular plant's shoot system is the mass of stems that extends from the root system below ground into the air, often reaching a great height. Rigid stems capable of rising upward against gravity are an ancient adaptation that allowed plants to move into terrestrial ecosystems.

Stems Carry Leaves and Flowers, and Support Their Weight

LEARNING OBJECTIVE 29.4.1 Compare the internal structure of eudicot and monocot stems.

Like roots, stems contain three types of plant tissue. Stems also undergo growth from cell division in apical and lateral meristems. The stem may be thought of as an axis from which other stems or organs grow. The shoot apical meristems are capable of producing these new stems and organs.

External stem structure

The shoot apical meristem initiates stem tissue and intermittently produces primordia that are capable of developing into leaves, other shoots, or even flowers (figure 29.19). Leaves may be arranged in a spiral around the stem, or they may be in pairs opposite or alternate to one another; they also may occur in whorls (circles) of three or more (figure 29.20). The spiral arrangement is the most common, with sequential leaves placed approximately 137.5° apart. This angle relates to the golden mean, a mathematical ratio found in nature. The angle of coiling in shells of some gastropods is the same. In plants, this pattern of leaf arrangement, called **phyllotaxy,** may optimize the exposure of leaves to the sun by ensuring that each leaf is shaded as little as possible by those above it.

The point of leaf attachment to the stem is called a **node** and the region of stem between two nodes is called an **internode.** Most leaves have a flattened blade, and some have a petiole (stalk). The angle between a leaf's petiole (or blade) and the stem is called an **axil.** An **axillary bud** is produced in each axil. This bud is a product of the primary shoot apical meristem, and it is itself a shoot apical meristem. Axillary buds frequently develop into branches with leaves, or they may form flowers.

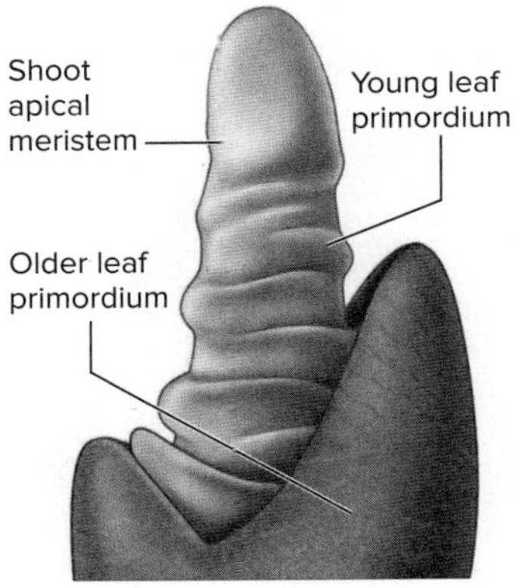
Shoot
apical
meristem
Young leaf
primordium
Older leaf
primordium

Figure 29.19 A shoot apex. Scanning electron micrograph of the apical meristem of wheat (*Triticum*).

Courtesy of J.H. Troughton, L.A. Donaldson and Callaghan Innovation, NZ

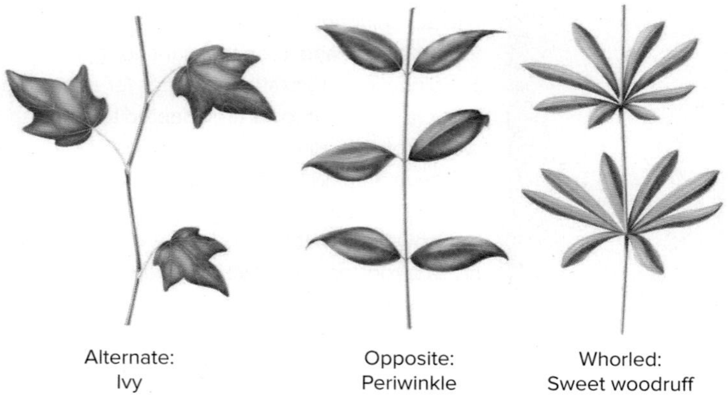
Alternate:
Ivy
Opposite:
Periwinkle
Whorled:
Sweet woodruff

Figure 29.20 Types of leaf arrangements. The three common types of leaf arrangements are alternate, opposite, and whorled.

Neither monocots nor herbaceous eudicot stems produce a cork cambium. The stems in these plants are usually green and photosynthetic, with at least the outer cells of the cortex containing chloroplasts. Herbaceous stems commonly have stomata and may have various types of trichomes (hairs).

Woody stems can persist over a number of years and develop distinctive markings in addition to the original organs that form (figure 29.21). Terminal buds usually extend the length of the shoot system during the growing season. Some buds, such as those of geraniums, are unprotected, but most buds of woody plants have protective winter bud scales that drop off, leaving tiny bud scale scars as the buds expand.

Some twigs have tiny scars of a different origin. A pair of butterfly-like appendages called stipules (part of the leaf) develops at the base of some leaves. The stipules can fall off and leave stipule scars. When the leaves of deciduous trees drop in the fall, they leave leaf scars with tiny bundle scars, marking where vascular connections existed. The shapes, sizes, and other features of leaf scars can be distinctive enough to identify deciduous plants in winter, when they lack leaves.

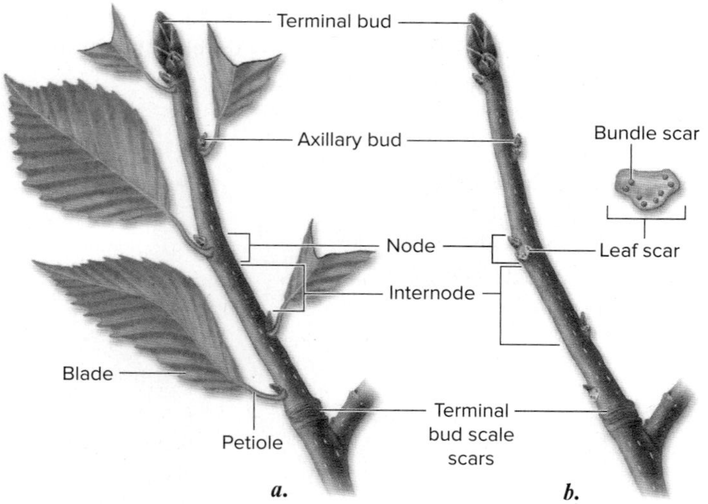
Terminal bud
Axillary bud
Bundle scar
Node
Leaf scar
Internode
Blade
Petiole
Terminal bud scale scars
a.
b.

Figure 29.21 A woody twig. ***a.*** In summer. ***b.*** In winter.

Internal stem structure

A major distinguishing feature between monocot and eudicot stems is the organization of the vascular tissue system (figure 29.22). Most monocot vascular bundles are scattered throughout the ground tissue system, whereas eudicot vascular tissue is arranged in a ring with internal ground tissue (pith) and external ground tissues (cortex). The arrangement of vascular tissue is directly related to the ability of the stem to undergo secondary growth. In eudicots, a vascular cambium may develop between the primary xylem and the primary phloem (figure 29.23). In many ways, this is a connect-the-dots game in which the vascular cambium connects the ring of primary vascular bundles. There is no logical way to connect primary monocot vascular tissue that would allow a uniform increase in girth. Monocots do not have secondary growth due to their lack of vascular cambium.

Rings in the stump of a tree reveal annual patterns of vascular cambium growth; cell size varies, depending on growth conditions (figure 29.24). Large cells form under favorable conditions such as abundant rainfalls. Rings of smaller cells mark the seasons where growth is limited. In woody eudicots and gymnosperms, a second cambium, the cork cambium, arises in the outer cortex (occasionally in the epidermis or phloem); it produces box-like cork cells to the outside and may produce parenchyma-like phelloderm cells to the inside.

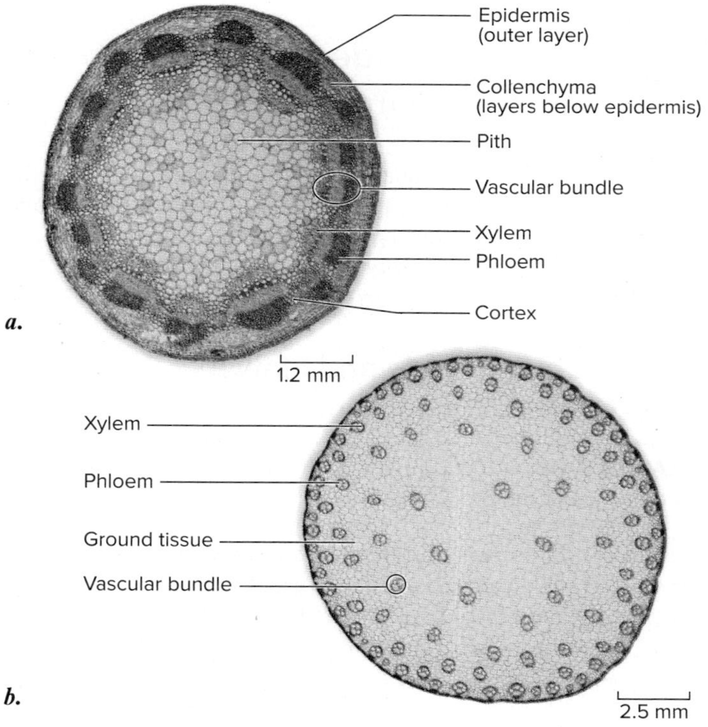
Epidermis
(outer layer)
Collenchyma
(layers below epidermis)
Pith
Vascular bundle
Xylem
Phloem
Cortex
a.
1.2 mm
Xylem
Phloem
Ground tissue
Vascular bundle
b.
2.5 mm

Figure 29.22 Stems. Transverse sections of a young stem in ***(a)*** a eudicot, the common sunflower (*Helianthus annus*), in which the vascular bundles are arranged around the outside of the stem; and ***(b)*** a monocot, corn (*Zea mays*), with characteristically scattered vascular bundles.

(both): Ed Reschke/Stone/Getty Images

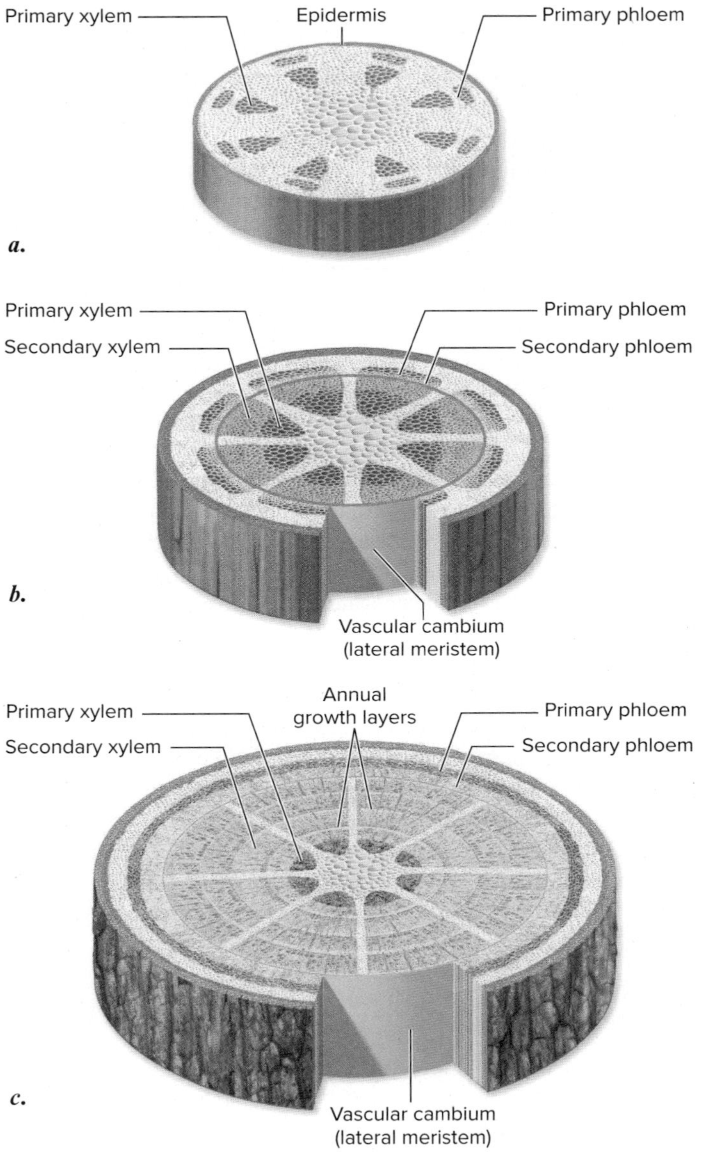
Primary xylem
Epidermis
Primary phloem
a.
Primary xylem
Secondary xylem
Primary phloem
Secondary phloem
b.
Vascular cambium (lateral meristem)
Annual growth layers
Primary xylem
Secondary xylem
Primary phloem
Secondary phloem
c.
Vascular cambium (lateral meristem)

Figure 29.23 Secondary growth. ***a.*** Before secondary growth begins in eudicot stems, primary tissues continue to elongate as the apical meristems produce primary growth. ***b.*** As secondary growth begins, the vascular cambium produces secondary tissues, and the stem's diameter increases. ***c.*** In this four-year-old stem, the secondary tissues continue to widen, and the trunk has become thick and woody. Note that the vascular cambium forms a cylinder that runs axially (up and down) in the roots and shoots that have this tissue.

The cork cambium, cork, and phelloderm are collectively referred to as the periderm (figure 29.25). Cork tissues, the cells of which become impregnated with water-repellent suberin shortly after they are formed and which then die, constitute the outer bark. The cork tissue cuts off water and food to the epidermis, which dies and sloughs off. In young stems, gas exchange between stem tissues and the air takes place through stomata, but as the cork cambium produces cork, it also produces patches of unsuberized cells beneath the stomata. These unsuberized cells, which permit gas exchange to continue, are called lenticels (figure 29.26).

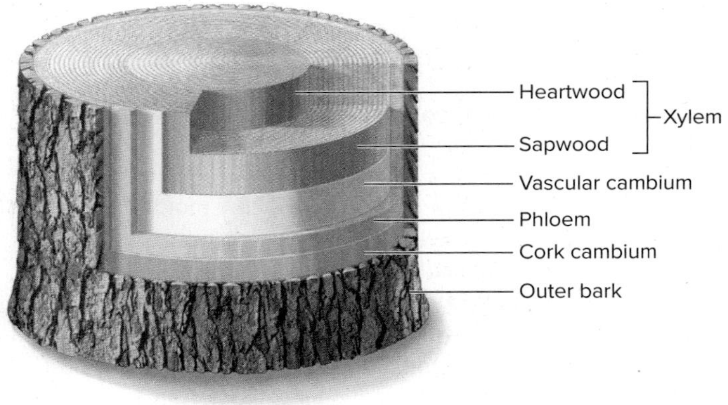
Heartwood
Sapwood
Xylem
Vascular cambium
Phloem
Cork cambium
Outer bark

Figure 29.24 Tree stump. The vascular cambium produces rings of xylem (sapwood and nonconducting heartwood) and phloem, and the cork cambium produces the cork.

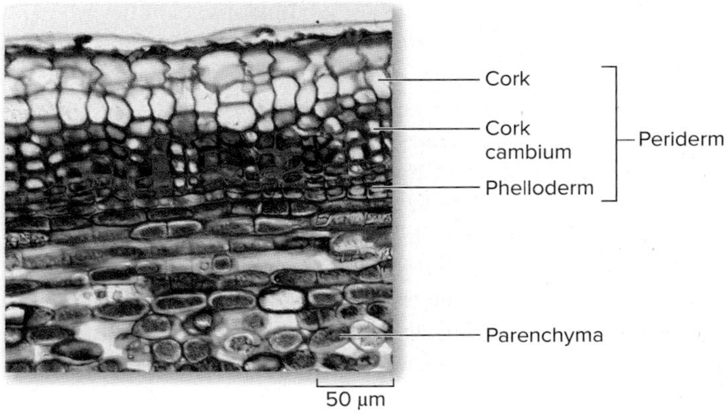
Cork
Cork cambium
Phelloderm
Periderm
Parenchyma
50 μm

Figure 29.25 Section of periderm. An early stage in the development of periderm in cottonwood, *Populus* sp.

Ed Reschke

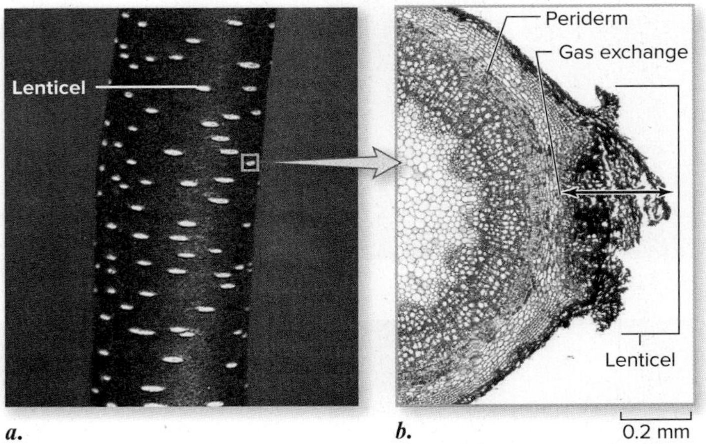
Lenticel
Periderm
Gas exchange
Lenticel
a.
b.
0.2 mm

Figure 29.26 Lenticels. ***a.*** Lenticels, the numerous small, pale, raised areas shown here on cherry tree bark (*Prunus cerasifera*), allow gas exchange between the external atmosphere and the living tissues immediately beneath the bark of woody plants. ***b.*** Transverse section through a lenticel in a stem of elderberry, *Sambucus canadensis*.

(a): Ed Reschke; (b): Steven P. Lynch

Modified Stems Carry Out Vegetative Propagation and Store Nutrients

LEARNING OBJECTIVE 29.4.2 List three functions of modified stems.

Although most stems grow upright, some have modifications that serve special purposes, including natural vegetative propagation. In fact, the widespread artificial vegetative propagation of plants, both commercial and private, frequently involves cutting modified stems into segments, which are then planted, producing new plants. As you become acquainted with the following modified stems, keep in mind that stems have leaves at nodes, with internodes between the nodes, and buds in the axils of the leaves, whereas roots have no leaves, nodes, or axillary buds.

Bulbs. Onions, lilies, and tulips have swollen underground stems that are really large buds with adventitious roots at the base (figure 29.27*a*). Most of a bulb consists of fleshy leaves attached to a small, knoblike stem. For most bulbs, next year's foliage comes from the tip of the shoot apex, which is protected by storage leaves.

Corms. Crocuses, gladioluses, and other popular garden plants produce corms that superficially resemble bulbs. Cutting a corm in half, however, reveals no fleshy leaves. Instead, almost all of a corm consists of stem, with a few papery, brown, nonfunctional leaves on the outside and adventitious roots below.

Rhizomes. Perennial grasses, ferns, bearded iris, and many other plants produce rhizomes, which typically are horizontal stems that grow underground, often close to the surface (figure 29.27*b*). Each node has an inconspicuous, scalelike leaf with an axillary bud; much larger photosynthetic leaves may be produced at the rhizome tip. Adventitious roots are produced throughout the length of the rhizome, mainly on the lower surface.

Runners and stolons. Strawberry plants produce horizontal stems with long internodes that usually grow along the surface of the ground. Several runners radiate out from a single plant (figure 29.27*c*). Some biologists reserve the term *stolon* for a stem with long internodes (but no roots) that grows underground, as seen in potato plants. A potato itself, however, is another type of modified stem—a tuber.

Tubers. In potato plants, carbohydrates may accumulate at the tips of rhizomes, which swell, becoming tubers; the rhizomes die after the tubers mature (figure 29.27*d*). The "eyes" of a potato are axillary buds formed in the axils of scalelike leaves. These leaves soon drop off; the tiny ridge adjacent to each "eye" of a mature potato is a leaf scar. Crop potatoes are propagated vegetatively from "seed potatoes." A tuber is cut up into pieces that contain at least one eye, and these pieces are planted.

Tendrils. Many climbing plants, such as grapes and English ivy, produce modified stems known as tendrils that twine around supports and aid in climbing (figure 29.27*e*). Some other tendrils, such as those of peas and pumpkins, are actually modified leaves or leaflets.

Cladophylls. Cacti and several other plants produce flattened, photosynthetic stems called cladophylls that resemble leaves (figure 29.27*f*). In cacti, the spines are modified leaves (refer to section 29.5).

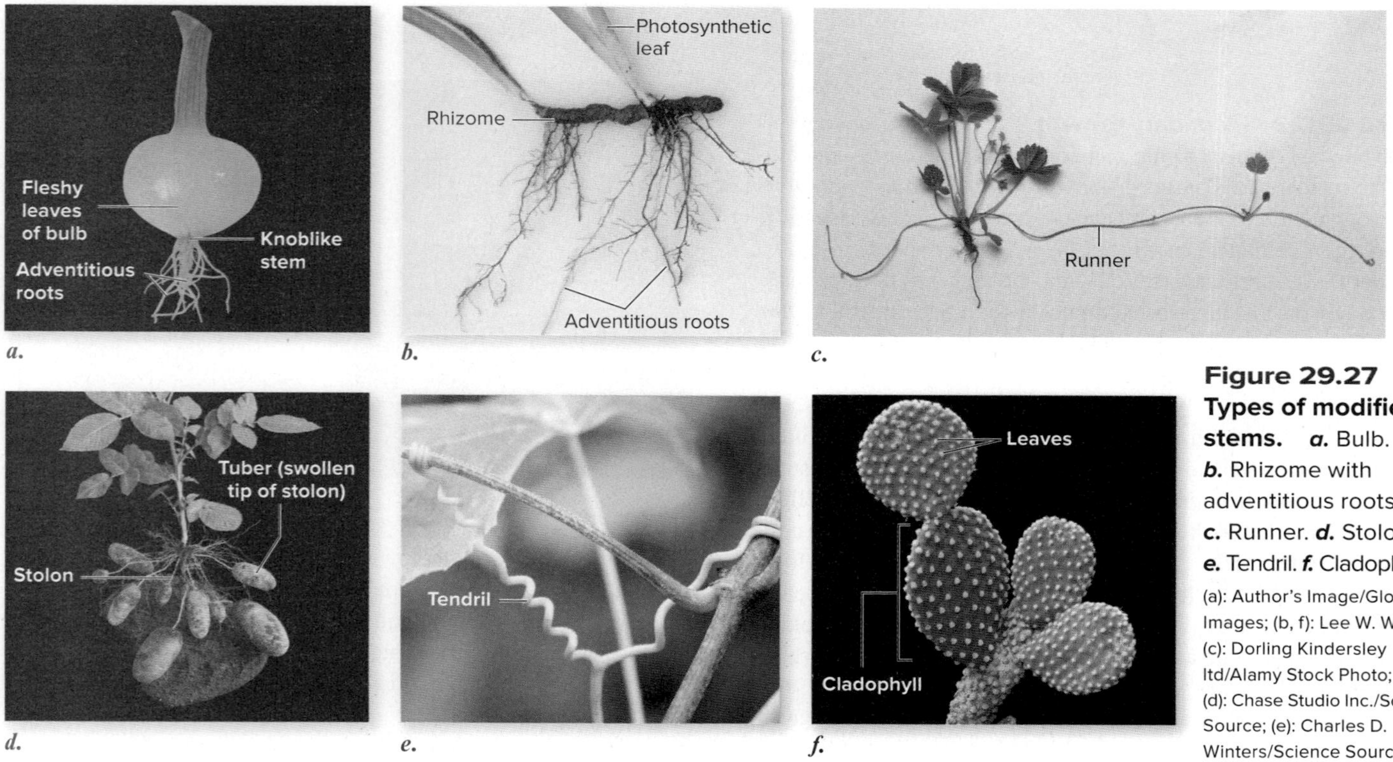
Fleshy leaves of bulb
Knoblike stem
Adventitious roots
a.
Photosynthetic leaf
Rhizome
Adventitious roots
b.
Runner
c.
Tuber (swollen tip of stolon)
Stolon
d.
Tendril
e.
Leaves
Cladophyll
f.

Figure 29.27 Types of modified stems. ***a.*** Bulb. ***b.*** Rhizome with adventitious roots. ***c.*** Runner. ***d.*** Stolon. ***e.*** Tendril. ***f.*** Cladophyll.

(a): Author's Image/Glow Images; (b, f): Lee W. Wilcox; (c): Dorling Kindersley ltd/Alamy Stock Photo; (d): Chase Studio Inc./Science Source; (e): Charles D. Winters/Science Source

REVIEW OF CONCEPT 29.4

Shoots grow from apical and lateral meristems. Axillary buds may develop into branches, flowers, or leaves. In monocots, vascular tissue is evenly spaced throughout the stem ground tissue; in eudicots, vascular tissue is arranged in a ring with inner and outer ground tissues. Some plants produce modified stems for support, vegetative reproduction, or nutrient storage.

- *Why don't stems produce the equivalent of root caps?*

29.5 Leaves Are a Plant's Photosynthetic Organs

Leaves are the principal sites of photosynthesis on land. They contribute to the atmospheric oxygen used by many forms of life, and they produce the carbohydrates consumed by heterotrophs. Leaves are determinate structures; their growth stops at maturity. The forms, sizes, internal structures, and arrangements of leaves on a stem differ greatly among plant species and reflect adaptations to different environments.

External Leaf Structure Reflects Vascular Morphology

LEARNING OBJECTIVE 29.5.1 Distinguish between a simple and a compound leaf.

Leaves are an extension of the shoot apical meristem and stem development. Experiments in which very young leaf primordia are isolated from fern and coleus plants have shown that when primordia first emerge, they are not committed to being leaves. If the primordia are young enough, they will form an entire shoot rather than a leaf. The positioning of leaf primordia and the initial cell divisions occur before those cells are committed to the leaf developmental pathway.

Leaves fall into two morphological groups, which may reflect differences in evolutionary origin. A **microphyll** is a leaf with one vein branching from the vascular cylinder of the stem and not extending the full length of the leaf. Microphylls are generally small and are commonly found in the Lycophyta. Most plants have leaves called **megaphylls,** which have several to many veins.

Most eudicot leaves have a flattened **blade** and a slender stalk, the **petiole.** The flattening of the leaf blade reflects a shift from radial symmetry to dorsal–ventral (top–bottom) symmetry. Leaf flattening increases the surface for capturing light and maximizes photosynthetic rate.

In addition, leaves may have a pair of **stipules,** which are outgrowths at the base of the petiole. The stipules, which may be leaflike or modified as spines (as in the black locust, *Robinia pseudo-acacia*) or glands (as in the purple-leaf plum tree *Prunus cerasifera*), vary considerably in size from the microscopic to almost half the size of the leaf blade. Grasses and other monocot leaves usually lack a petiole; these leaves tend to sheathe the stem toward the base.

a.

b.

Figure 29.28 Eudicot and monocot leaves. ***a.*** The leaves of eudicots, such as this African violet relative from Sri Lanka, have netted, or reticulate, veins. ***b.*** Those of monocots, such as this cabbage palmetto, have parallel veins. The eudicot leaf has been cleared with chemicals and stained with a red dye to make the veins show more clearly.
(a): Michal Durinik/Shutterstock; (b): Steven P. Lynch

Veins, the vascular bundles in leaves, consist of both xylem and phloem and are distributed throughout the leaf blades. The main veins are parallel in most monocot leaves; the veins of eudicots, on the other hand, often form an intricate network (figure 29.28).

Leaf blades come in a variety of forms, from oval to deeply lobed to having separate leaflets. In **simple leaves** (figure 29.29*a*), such as those of lilacs or birch trees, the blades are undivided, but simple leaves may have teeth, indentations, or lobes of various sizes, as in the leaves of maples and oaks.

In **compound leaves** (figure 29.29*b*), such as those of ashes, box elders, and walnuts, the blade is divided into *leaflets.* The relationship between the development of compound and simple leaves is an open question. Two explanations are being debated: (1) A

a. *b.*

Figure 29.29 Simple versus compound leaves. ***a.*** A simple leaf, its margin deeply lobed, from the oak tree (*Quercus robur*). ***b.*** A pinnately compound leaf, from a black walnut (*Juglans nigra*). A compound leaf is associated with a single lateral bud, located where the petiole is attached to the stem.
(a): Gusto/Science Source; (b): Dorling Kindersley ltd/Alamy Stock Photo

compound leaf is a highly lobed simple leaf; or (2) a compound leaf utilizes a shoot development program, and each leaflet was once a leaf.

Internal Leaf Structure Regulates Gas Exchange and Evaporation

LEARNING OBJECTIVE 29.5.2 Compare the mesophyll of a monocot leaf with that of a eudicot leaf.

The entire surface of a leaf is covered by a transparent epidermis, and most of these epidermal cells have no chloroplasts. As described in section 29.2, the epidermis has a waxy cuticle, and different types of glands and trichomes may be present. Also, the lower epidermis (and occasionally the upper epidermis) of most leaves contains numerous mouth-shaped stomata flanked by guard cells (figure 29.30).

The tissue between the upper and lower epidermis is called **mesophyll.** Mesophyll is interspersed with veins of various sizes.

Most eudicot leaves have two distinct types of mesophyll. Closest to the upper epidermis are one to several (usually two) rows of tightly packed, barrel-shaped to cylindrical chlorenchyma cells (parenchyma with chloroplasts), which constitute the palisade mesophyll (figure 29.31). Some plants, including species of *Eucalyptus,* have leaves that hang down, rather than extend horizontally. They have palisade mesophyll on both sides of the leaf. Such an arrangement presumably maximizes light capture.

Nearly all eudicot leaves have loosely arranged, spongy mesophyll cells between the palisade mesophyll and the lower epidermis, with many air spaces throughout the tissue. The interconnected intercellular spaces, along with the stomata, function in gas exchange and the passage of water vapor from the leaves.

The mesophyll of monocot leaves often is not differentiated into palisade and spongy layers, and there is often little distinction between the upper and lower epidermis. Instead, cells surrounding the vascular tissue are distinctive and are the site of carbon fixation. This anatomical difference often correlates with a modified photosynthetic pathway, C_4 photosynthesis, which maximizes the amount of CO_2 relative to O_2 to reduce energy loss through photorespiration (refer to chapter 8). Leaf anatomy reflects a careful balance between maximizing light capture, allowing efficient gas exchange, staying cool, and transporting photosynthetic products to the rest of the plant.

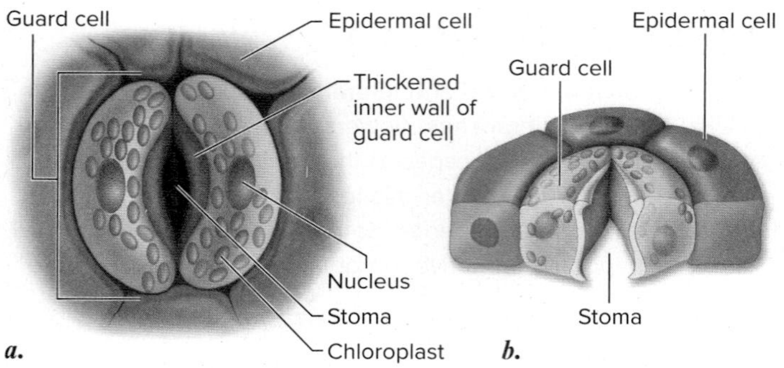

Figure 29.30 A stoma. ***a.*** Surface view. ***b.*** View in cross section.

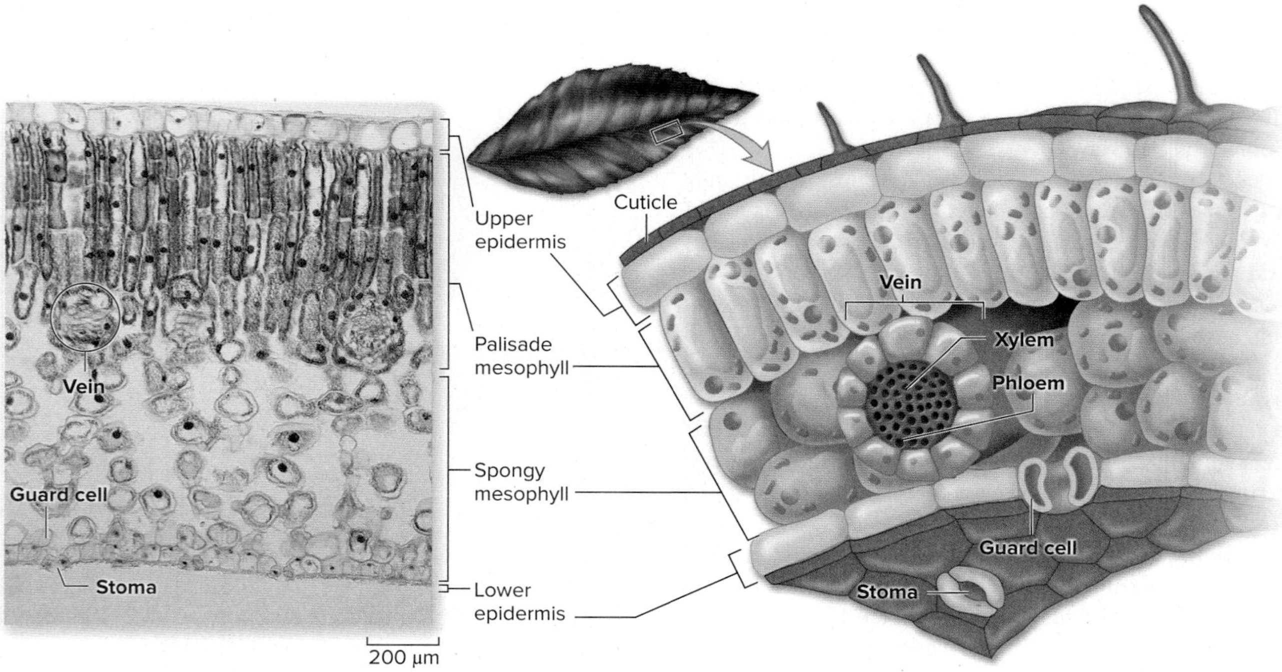

Figure 29.31 A leaf in cross section. Transection of a leaf showing the arrangement of palisade and spongy mesophyll, a vascular bundle or vein, and the epidermis with paired guard cells flanking the stoma.

Ed Reschke/Stone/Getty Images

Modified Leaves Are Highly Versatile Organs

LEARNING OBJECTIVE 29.5.3 Describe the functions of modified leaves.

As plants colonized a wide variety of environments, from deserts to lakes to tropical rainforests, plant organ modifications arose that would adapt the plants to their specific habitats. Leaves, in particular, have evolved some remarkable adaptations. A brief discussion of a few of these modifications follows:

Floral leaves (bracts). Poinsettias and dogwoods have relatively inconspicuous, small, greenish yellow flowers. However, both produce large, modified leaves called *bracts* (red, pink, or white in poinsettias and white or pink in dogwoods). These bracts surround the true flowers and perform the same function as showy petals. In other plants, bracts can be small and inconspicuous.

Spines. The leaves of many cacti and other plants are modified as *spines* (figure 29.27*f*). In cacti, having less leaf surface reduces water loss, and the sharp spines may deter predators. Spines should not be confused with *thorns,* which are modified stems, or the prickles on raspberries, which are simply outgrowths from the epidermis.

Reproductive leaves. Several plants, notably *Kalanchoë,* produce tiny but complete plantlets along their margins. Each plantlet, when separated from the leaf, is capable of growing independently into a full-sized plant. The "walking" fern *(Asplenium rhizophyllum)* produces new plantlets at the tips of its fronds.

Window leaves. Several genera of plants growing in arid regions produce cone-shaped leaves with transparent tips. The leaves often become mostly buried in sand blown by the wind, but the transparent tips, which have a thick epidermis and cuticle, admit light to the hollow interiors. This allows photosynthesis to take place beneath the surface of the ground.

Shade leaves. Leaves produced in the shade, where they receive little sunlight, tend to be larger in surface area but thinner and with less mesophyll than leaves on the same tree receiving more direct light. Environmental signals can have a major effect on development.

Insectivorous leaves. Almost 200 species of flowering plants are known to have leaves that trap insects; some plants digest the insects' soft parts. Plants with insectivorous leaves often grow in acidic swamps that are deficient in essential nutrients or that contain nutrients in forms not readily available to the plants. Their needs are met, however, by the supplementary absorption of nutrients from captured animals such as small insects.

REVIEW OF CONCEPT 29.5

Leaves come in a range of forms. A simple leaf is undivided, whereas a compound leaf has a number of separate leaflets. Monocots typically produce leaves with parallel veins, whereas those of eudicots are netted. Mesophyll cells carry out photosynthesis; in monocots, mesophyll is undifferentiated, whereas in eudicots it is divided into palisade and spongy mesophyll. Stomata facilitate the exchange of gases needed for photosynthesis. Leaves may be modified for reproduction, protection, water conservation, uptake of nutrients, and even the trapping of insects.

- *Why would a plant with vertically oriented leaves produce palisade, but not spongy, mesophyll cells?*

Is There a Unified Theory Relating Size to Metabolic Rate?

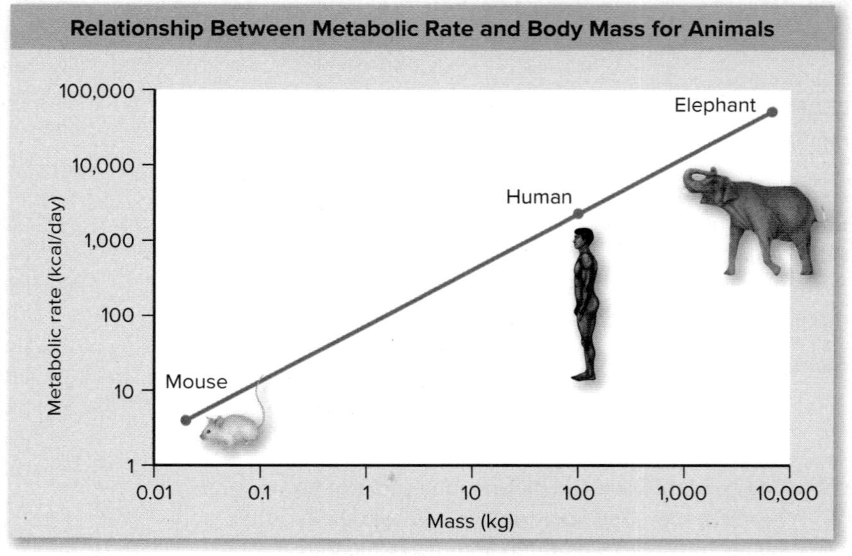

For many years, biologists have accepted a unified metabolic theory, called Kleiber's law, that relates the size of an organism to its metabolic rate. In 1932, the agricultural biologist Max Kleiber observed that metabolic rates of mammals and birds scale as the 3/4 power of body mass. This means that the logarithm of the basal metabolic rate of organisms—as different as mice, humans, and elephants—varies with body mass as a straight line with a slope of 0.75 (see the upper graph).

In 1960, Kleiber's 3/4 slope was shown to apply across six orders of magnitude in the body size of mammals, reptiles, fish, insects, and even unicellular organisms. This is an extraordinarily broad metabolic pattern, and it is thought to reflect surface-to-volume relationships.

But does Kleiber's law apply to plants? Until 2006, there were too few direct measures of whole-plant respiration to confirm or reject the 3/4 scaling prediction. In 2006, plant physiologist Peter Reich carried out direct measurements of whole-plant respiration for some 500 individual plants of 43 species. The plants selected for the study ranged across six orders of magnitude in their body mass, from one of the tallest living plants to plants smaller than a fingernail.

The results of Reich's study are presented in the lower graph. The metabolic rates of the plants are measured directly as whole-plant respiration rates (kcal/day). The green line represents the 3/4 scaling line predicted by Kleiber's law. The blue line shows the relationship for plants growing in nutrient-poor natural soils. The red line shows the relationship for the same array of plant species growing in nutrient-rich greenhouse soils.

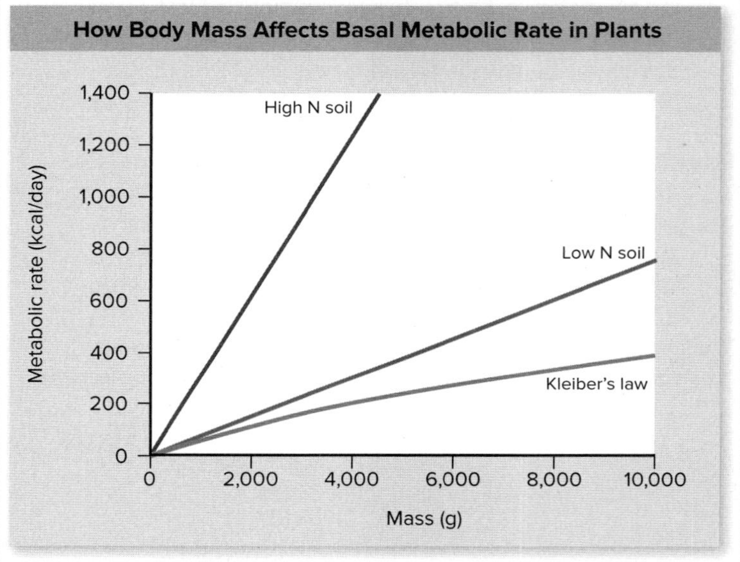

Analysis

1. **Applying Concepts** In the graphs, what is the dependent variable?
2. **Interpreting Data**
   a. Are Reich's respiration rate measurements (lower graph) plotted on a logarithmic or a linear scale?
   b. Plotted as they are, the 3/4 scaling relationship of Kleiber's law, which is linear on a logarithmic scale, shows a curve bending slightly to the right. Do the two plant lines also curve, or are they linear?
3. **Making Inferences**
   a. What is the slope of the plot for low-nitrogen plants? (*Hint:* If you double plant mass from 4000 g to 8000 g, how much does the respiration rate change?)
   b. What is the slope of the plot for high-nitrogen plants? (*Hint:* If you double plant mass from 2000 g to 4000 g, how much does the respiration rate change?)
   c. Do plants metabolize more rapidly in nitrogen-rich greenhouse soils or in nitrogen-poor natural soils, or is there no difference?
4. **Drawing Conclusions**
   a. Compare the slopes of the plots of plants growing in natural soils with the slopes of those growing in greenhouse soils. Does the rate of plant growth influence the scaling relationship?
   b. A straight line on Reich's plot would indicate a strictly proportional (linear) relationship between whole-plant respiration and body mass, but a curve such as that seen in the green line would indicate a nonlinear proportional relationship. Based on Reich's data as plotted here, do plants obey Kleiber's law?

Retracing the Learning Path

CONCEPT 29.1 Meristems Articulate the Body Plan

29.1.1 Vascular Plants Have Roots and Shoots The root system is primarily below ground; roots anchor the plant and absorb water and minerals. The shoot system is above ground and provides support for leaves and flowers.

29.1.2 Roots and Shoots Are Composed of Tissue Systems The three types of tissues are dermal tissue, ground tissue, and vascular tissue.

29.1.3 Meristems Elaborate the Body Plan Throughout the Plant's Life Apical meristems are located on the tips of stems and near the tips of roots. Lateral meristems are found in plants that exhibit secondary growth. They increase the diameter of a stem or root.

CONCEPT 29.2 Plants Contain Three Main Tissues

29.2.1 Dermal Tissue Forms a Protective Interface with the Environment Dermal tissue is primarily the epidermis, which is usually one cell thick and is covered with a fatty or waxy cuticle to reduce water loss. Guard cells in the epidermis control water loss through stomata. Root hairs are outgrowths of epidermal cells that increase the absorptive area of roots.

29.2.2 Ground Tissue Cells Perform Many Functions, Including Storage, Photosynthesis, and Support Ground tissue is mainly composed of parenchyma cells, which function in storage, photosynthesis, and secretion. Collenchyma cells provide flexible support, and sclerenchyma cells provide rigid support.

29.2.3 Vascular Tissues Conduct Water and Nutrients Throughout the Plant Xylem tissue conducts water through dead cells called tracheids and vessel elements. Phloem tissue conducts nutrients such as dissolved sucrose through living cells called sieve-tube members and sieve cells.

CONCEPT 29.3 Roots Have Four Growth Zones

29.3.1 Roots Are Adapted for Growing Underground and Absorbing Water and Solutes Developing roots have four regions: (1) the root cap, which protects the root; (2) the zone of cell division, which contains the apical meristem; (3) the zone of elongation, which extends the root through the soil; and (4) the zone of maturation, in which cells become differentiated.

29.3.2 Modified Roots Accomplish Specialized Functions Most plants produce either a taproot system containing a single large root with smaller branch roots, or a fibrous root system composed of many small roots. Adventitious roots may be modified for support, stability, acquisition of oxygen, storage of water and food, or parasitism of a host plant.

CONCEPT 29.4 Stems Provide Support for Aboveground Organs

29.4.1 Stems Carry Leaves and Flowers, and Support Their Weight Leaves are attached to stems at nodes. The axil is the area between the leaf and the stem, and an axillary bud develops in axils of eudicots. The vascular bundles in stems of monocots are randomly scattered; in eudicots the bundles are arranged in a ring. Vascular cambium develops between the inner xylem and the outer phloem, allowing secondary growth in dicots.

29.4.2 Modified Stems Carry Out Vegetative Propagation and Store Nutrients Bulbs, corms, rhizomes, runners and stolons, tubers, tendrils, and cladophylls are examples of modified stems. The tubers of potatoes are both a food source and a means of propagating new plants.

CONCEPT 29.5 Leaves Are a Plant's Photosynthetic Organs

29.5.1 External Leaf Structure Reflects Vascular Morphology Leaves are the principal sites of photosynthesis. The arrangement, form, size, and internal structure can vary greatly among environments. Vascular bundles are parallel in monocots but form a network in eudicots. The leaves of most eudicots have a flattened blade and a slender petiole; monocots usually do not have a petiole. Leaf blades may be simple or compound (divided into leaflets).

29.5.2 Internal Leaf Structure Regulates Gas Exchange and Evaporation The tissues of the leaf include the epidermis with guard cells, vascular tissue, and mesophyll in which photosynthesis takes place. The mesophyll in eudicot leaves has a horizontal orientation, partitioned into palisade cells near the upper surface and spongy cells near the lower surface. The mesophyll of monocot leaves is often not differentiated.

29.5.3 Modified Leaves Are Highly Versatile Organs Leaves vary greatly in form and are adapted to serve many different functions. Leaves may be modified for reproduction, protection, storage, mineral uptake, or even the trapping of insects in carnivorous plants.

Concept Overview

This Concept Overview diagrams the key concepts that were discussed in this chapter.

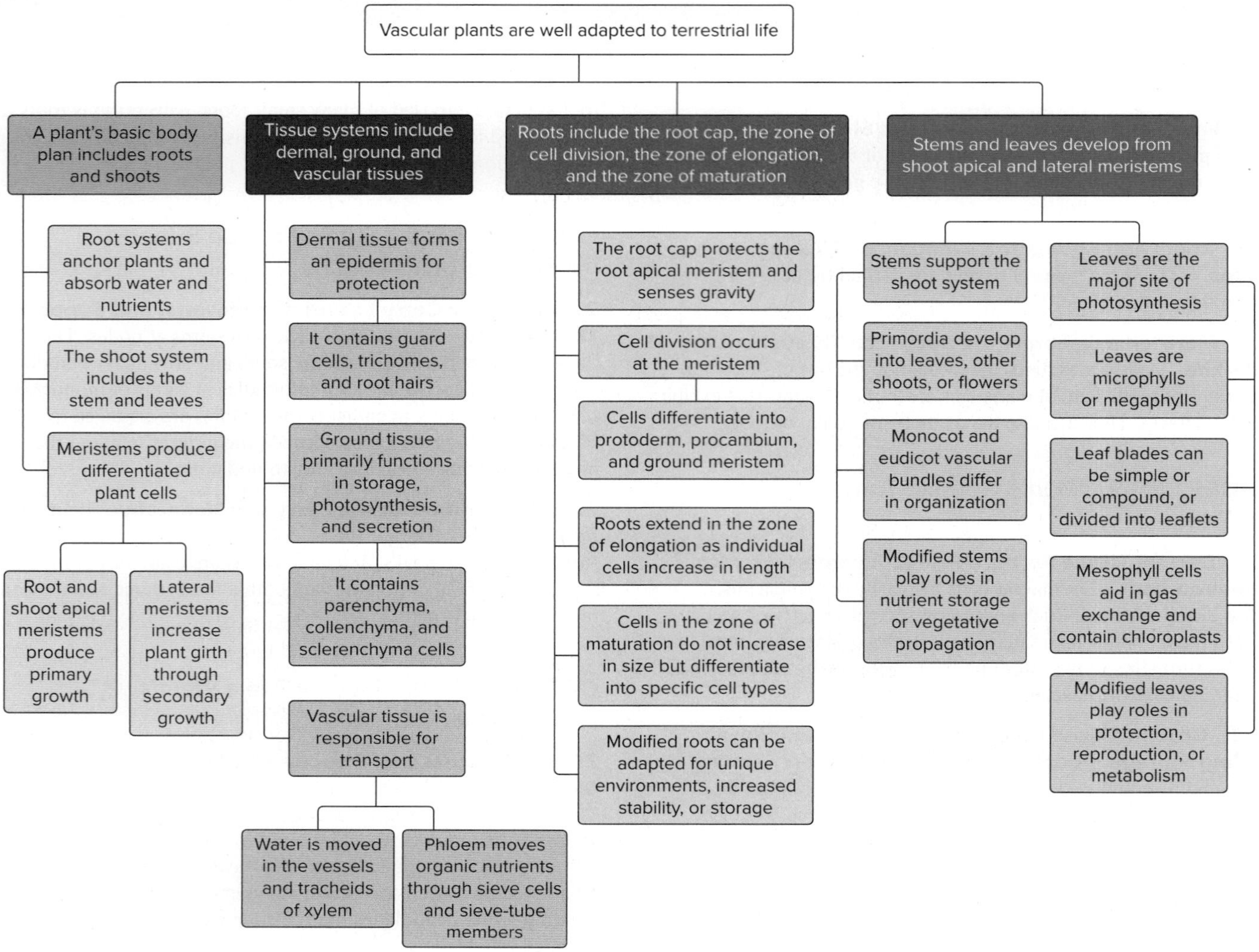

Assessing the Learning Path

Understand

1. A unique feature of plants is indeterminate growth. Indeterminate growth is possible because
   a. meristematic regions for primary growth occur throughout the entire plant body.
   b. all cell types in a plant often give rise to meristematic tissue.
   c. meristematic cells continually replace themselves.
   d. all cells in a plant continue to divide indefinitely.
2. One function of stomata is to
   a. allow carbon dioxide absorption.
   b. repel insects and other herbivores.
   c. support leaf tissue.
   d. allow water and dissolved minerals to be absorbed.
3. The food-conducting cells in an oak tree are called
   a. tracheids.
   b. vessels.
   c. companion cells.
   d. sieve-tube members.
4. Root hairs form in the zone(s) of
   a. cell division.
   b. elongation.
   c. maturation.
   d. cell division, elongation, and maturation.
5. The regulation of water flow laterally between the vascular tissue and cell layers in the outer portion of the root is performed by the
   a. periderm.
   b. endodermis.
   c. pericycle.
   d. xylem.
6. Which of the following is NOT a modified stem?
   a. A tuber
   b. A rhizome
   c. A stolon
   d. A bract
7. Palisade and spongy parenchyma are typically found in the mesophyll of
   a. monocots.
   b. eudicots.
   c. monocots and eudicots
   d. neither monocots nor eudicots.

8. A bract is
   a. an inconspicuous flower.
   b. a leaf.
   c. a spine.
   d. a petal.
9. Plant organs form by
   a. cell division in gamete tissue.
   b. cell division in meristematic tissue.
   c. cell migration into the appropriate position in the tissue.
   d. elimination of chromosomes in the precursor cells.
10. Roots differ from stems in that roots lack
   a. vessel elements.
   b. nodes.
   c. an epidermis.
   d. ground tissue.
11. Leaves are an extension of
   a. meristematic tissue.
   b. the axillary bud.
   c. the shoot apical meristem.
   d. the microphyll.
12. In vascular plant leaves, gases enter and leave the plant through pores called
   a. stomata.
   b. meristems.
   c. trichomes.
   d. lenticels.
13. You can determine the age of an oak tree by counting the annual rings of ________________ formed by the ______________.
   a. primary xylem; apical meristem
   b. secondary phloem; vascular cambium
   c. dermal tissue; cork cambium
   d. secondary xylem; vascular cambium

Apply

1. Trees close to water sources can be killed due to the feeding habits of beaver. The beaver may not fell the tree, but the tree may be killed when the beaver feeds on the bark and deeper plant tissues of the tree trunk. To which tissue is the beaver most likely doing damage that causes the tree to die?
   a. Vascular cambium
   b. Cork
   c. Cork cambium
   d. Primary phloem
   e. Bark
2. A substance toxic to a plant has entered the soil in which the plant is growing. Which of the following plant structures will afford the plant the most protection against the substance?
   a. The vascular cambium
   b. The secondary xylem
   c. The Casparian strip
   d. The pericycle
3. When a plant is grazed by an herbivore, like a giraffe grazes a tree, which structure is damaged that most impacts plant growth?
   a. Root apical meristems
   b. Shoot apical meristems
   c. Cork cambium
   d. Vascular cambium
4. What phenotype would you expect to see in a plant that had a mutation in the *GLABRA2* gene such that the gene was expressed at much higher levels than normal?
   a. If the plant was a tomato plant, the leaves would have fewer trichomes than would the leaf of a normal tomato plant.
   b. Roots would have more root hairs in the mutant than in a normal plant.
   c. If the plant produced tubers, tuber formation would be promoted at a greater frequency than it would in a normal plant.
   d. Internodes would be longer in the stem than they would in the normal plant.
5. Which of the following animal cells or tissues is *most* functionally analogous to collenchyma?
   a. Epithelial tissue
   b. Striated muscle tissue
   c. Cartilage
   d. Nerve tissue
6. Which of the following animal structures is *most* functionally analogous to phloem?
   a. Alveoli
   b. Capillaries
   c. Bones
   d. The small intestine
7. Lignin is *most* functionally analogous to which of the following proteins?
   a. A cell surface receptor on a lung epithelial cell which receives signals to promote cell division
   b. A protein that stimulates the formation of cilia on epithelial cells of the upper respiratory tract
   c. Tubulin, the protein which is used to build microtubules
   d. Collagen
8. The upper and lower surfaces of all of the leaves on a vascular plant have been painted with a thin layer of water-impermeable nail varnish. Which of the following effects on the plant might you anticipate?
   a. Water levels in the soil around the plant's roots would not change except for water added or water lost due to evaporation from the soil surface.
   b. Sugar transport through the plant would shift from the phloem to the xylem.
   c. Photosynthetic rate would increase to compensate.
   d. Growth at the apical meristems would halt immediately and growth at the lateral meristems would be stimulated.
9. If you attached the promoter for the *GLABRA2* gene to a fluorescent protein and expressed the gene in plants, where would you definitely expect to see fluorescence?
   a. Epidermal tissue
   b. Phloem
   c. Vascular cambium
   d. Cork cambium

Synthesize

1. A friend just returned from a family trip to northern Michigan, where he visited a sugar maple farm where maple syrup is harvested. On the maple trees, harvesters make just one relatively small cut all the way through the bark (or two cuts on larger trees) and hang a bucket beneath to catch the sap. Why, your friend asks, don't they just make a cut completely around the tree and collect much more sap, much faster?
2. If you were to relocate the pericycle of a plant root to the epidermal layer, how would it affect root growth?
3. If you were given an unfamiliar vegetable, how could you use an examination of its external features and a microscopic cross section to tell if it was a root or a stem?
4. Fifteen years ago, your parents hung a swing from the lower branch of a large tree growing in your yard. When you go and sit in it today, you realize it is exactly the same height off the ground as it was when you first sat in it 15 years ago. Why is the swing not any higher off the ground than it was 15 years ago?
5. Why do you think commercial potatoes are not grown from seeds?

Steven P. Lynch/McGraw Hill

30 Flowering Plant Reproduction

Learning Path

- **30.1** Reproduction Starts with Flowering
- **30.2** Flowers Attract Pollinators
- **30.3** Fertilization Leads to Embryogenesis
- **30.4** Seeds Protect Angiosperm Embryos
- **30.5** Fruits Promote Seed Dispersal
- **30.6** Germination Begins Seedling Growth
- **30.7** Plant Life Spans Vary Widely
- **30.8** Asexual Reproduction Is Common Among Flowering Plants

Concept Overview

This Concept Overview provides a diagram of the overarching concepts that are covered in this chapter. These concepts will be expanded upon in the Concept Overview at the end of the chapter.

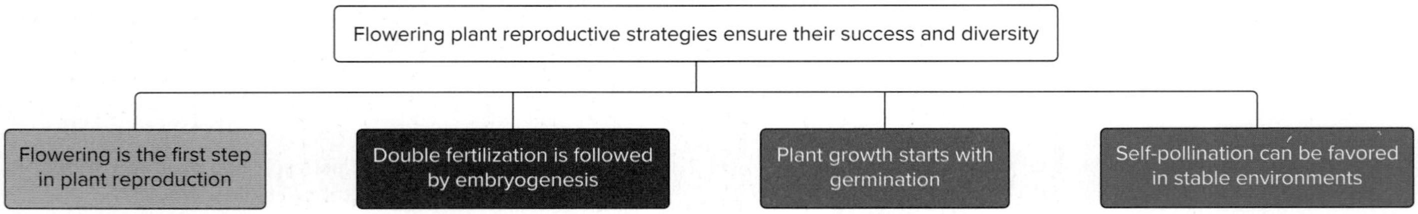

Flowering plant reproductive strategies ensure their success and diversity
Flowering is the first step in plant reproduction
Double fertilization is followed by embryogenesis
Plant growth starts with germination
Self-pollination can be favored in stable environments

Introduction

The remarkable evolutionary success of flowering plants can be linked to their reproductive strategies. In this chapter, we explore the reproductive strategies of the angiosperms and how their unique features—flowers and fruits—have contributed to their success. This is, in part, a story of coevolution between plants and animals that ensures greater genetic diversity by widely dispersing pollen and embryos in seeds. Once an egg is fertilized, the timing and directionality of each cell division is carefully orchestrated during plant development. This orchestration is critical because plant cells cannot migrate as do animal cells. Cell differentiation to produce specialized cell types is controlled by the position of one cell relative to another. The developing embryo is fragile, and many protective structures have evolved since plants first colonized land. Following the emergence of a seedling from the soil, plant organs develop throughout the plant's life, which can range from a single season to hundreds of years long.

30.1 Reproduction Starts with Flowering

In chapter 26, we discussed that angiosperms represent an evolutionary innovation with their production of flowers and fruits. In this section, we describe the additional changes that occur in a vegetative plant to produce the elaborate structures associated with flowering (figure 30.1).

Flowering Is Carefully Regulated

LEARNING OBJECTIVE 30.1.1 Describe how flowering is regulated.

Flowering and subsequent seed production are carefully regulated, because reproduction is a metabolically expensive process. Producing seeds under poor conditions can waste a lot of nutritional resources if seeds are not viable. Flowering usually occurs when light, water abundance, and temperature are optimal for the production of sugars needed for seed production. Flowering is usually preceded by a period of vegetative growth when leaves and the structures supporting leaves are produced and photosynthesis can occur optimally.

Carefully regulated processes determine when and where flowers will form, and many species flower synchronously as a way to increase reproductive success. Once plants are competent to reproduce, a combination of factors—including light, temperature, and both stimulatory and inhibitory internal signals—determines when a flower is produced. These signals turn on genes that specify formation of the floral organs—sepals, petals, stamens, and carpels. Once cells have instructions to become a specific floral organ, other developmental signals lead to the production of flower parts.

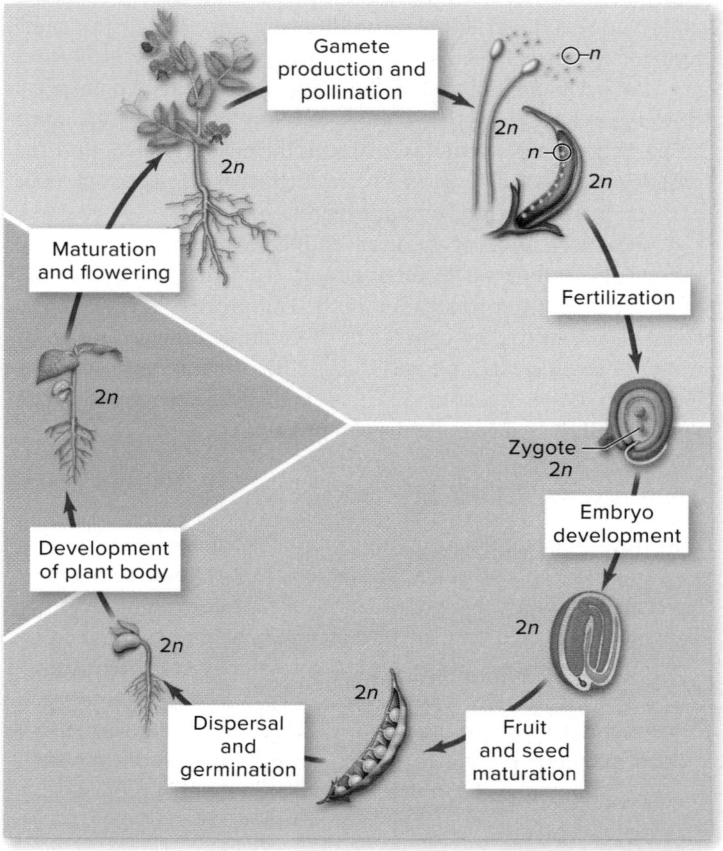

Figure 30.1 Life cycle of a flowering plant (*Angiosperm*).

The transition to flowering competence is called phase change

Phase change is the developmental change that occurs when a plant goes from being incapable to being capable of producing flowers. At germination, most plants cannot flower, even if all the necessary environmental cues are present. Internal developmental changes allow plants to become competent to respond to external or internal signals (or both) that trigger flowering. Even though a plant has reached the adult stage of development, it may still not produce reproductive structures. Other factors may be necessary to trigger flowering.

Four genetically controlled pathways regulate flowering: (1) the light-dependent pathway, (2) the temperature-dependent pathway, (3) the gibberellin-dependent pathway, and (4) the autonomous pathway. Plants can rely primarily on one pathway, but all four pathways can be present.

The environment can promote or repress flowering, but in some cases it can have little influence. For example, increasing light duration can be a signal that long summer days have arrived in a temperate climate and that conditions are favorable for reproduction. In other cases, plants depend on light to accumulate sufficient amounts of sucrose to fuel reproduction, but flower independently of day length.

Temperature can also be a signal for flowering. The dependence of a shoot on a period of chilling to trigger flowering is called **vernalization.** Vernalization assures that plants in temperate climates do not flower late in the summer or during the autumn when conditions are not ideal for seed production. After winter, seed production would be favored by the more ideal growth conditions of spring and summer. Assuming that regulation of reproduction first arose in more constant tropical environments, many of the day-length and temperature controls would have evolved as plants colonized more temperate climates.

A Complete Flower Has Four Whorls of Parts

LEARNING OBJECTIVE 30.1.2 Relate flower structure to flower function.

The flower not only houses the haploid generations that will produce gametes but also functions to increase the probability that fertilization occurs.

The diversity of angiosperms is partly due to the evolution of a great variety of floral phenotypes, which may enhance the effectiveness of pollination. Floral organs are thought to have evolved from leaves. In some early angiosperms, these organs maintain the spiral developmental pattern often found in leaves. The trend has been toward four distinct whorls of parts. A complete flower has four whorls (calyx, corolla, androecium, and gynoecium; figure 30.2). An incomplete flower lacks one or more of the whorls.

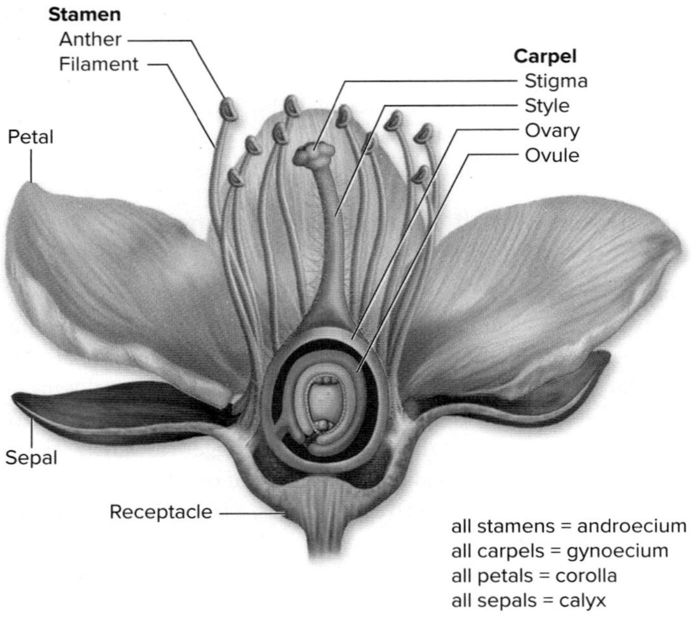
Stamen
Anther
Filament
Petal
Carpel
Stigma
Style
Ovary
Ovule
Sepal
Receptacle
all stamens = androecium
all carpels = gynoecium
all petals = corolla
all sepals = calyx

Figure 30.2 A complete angiosperm flower.

Flower morphology

In both complete and incomplete flowers, the **calyx** usually constitutes the outermost whorl; it consists of flattened appendages, called sepals, that protect the flower in the bud. The petals collectively make up the **corolla,** which may be fused. Many petals function to attract pollinators. Although these two outer whorls of floral organs are not involved directly in gamete production or fertilization, their structures enhance reproductive success.

Male structures. *Androecium* is a collective term for all the stamens (male structures) of a flower. Stamens are specialized structures that bear the angiosperm microsporangia. Similar structures bear the microsporangia in the pollen cones of gymnosperms. Most living angiosperms have stamens with filaments ("stalks") that are slender and often threadlike; four microsporangia are evident at the apex in a swollen portion, the **anther.**

Female structures. *Gynoecium* is a collective term for all the female parts of a flower. In most flowers, the gynoecium consists of a single carpel or two or more fused carpels. Single or fused carpels are often referred to as simple or compound pistils, respectively. Examples of flowers with a compound pistil include tomatoes and oranges. Buttercups and stonecups have flowers with several to many separate pistils, each formed from a single carpel.

The **ovary** is the pistil's swollen lower portion that protects ovules (which develop into seeds). The ovary narrows toward the top into a slender structure called a style. Atop the style is a pollen-receptive structure called the stigma. Sometimes the stigma is divided, with the number of stigma branches indicating how many carpels compose the particular pistil.

Carpels are essentially rolled floral leaves with ovules along the margins. It is possible that the first carpels were leaf blades that folded longitudinally; the leaf margins, which had hairs, did not actually fuse until the fruit developed, but the hairs interlocked and were receptive to pollen. Evidence indicates that during the course of evolution, the hairs became localized into a stigma, a style was formed, and the fusing of the carpel margins ultimately resulted in a pistil. In many modern flowering plants, the carpels have become highly modified and are not visually distinguishable from one another unless the pistil is cut open.

Gametes Are Produced in the Gametophytes of Flowers

LEARNING OBJECTIVE 30.1.3 Compare gamete production in animals with gamete production in flowering plants.

Reproductive success depends on uniting the gametes (egg and sperm) found in the embryo sacs and pollen grains of flowers (figure 30.3). As you learned in chapter 26, plant sexual life cycles are characterized by an *alternation of generations,* in which a diploid sporophyte generation gives rise to a haploid gametophyte generation. In angiosperms, the multicellular gametophyte generation is very small and is completely enclosed within the tissues of the parent sporophyte. The male gametophytes, or microgametophytes, are **pollen grains** (figure 30.4). The female gametophyte, or megagametophyte, is the **embryo sac.** Pollen grains and the embryo sac both are produced in separate, specialized structures of the angiosperm flower.

Like animals, angiosperms have separate structures for producing male and female gametes, but the reproductive organs of angiosperms differ from those of animals in two ways. First, both male and female structures usually occur together in the same individual flower. Second, angiosperm reproductive structures are not permanent parts of the adult individual. Angiosperm flowers and reproductive organs develop seasonally. In some cases, reproductive structures are produced only once, and the parent plant dies. The germ line in angiosperms is not set aside early on but forms quite late during phase change.

The formation of the male and female gametophytes, and the events leading up to fertilization, were discussed in detail in chapter 26 (refer to section 26.8). The process is reviewed in figure 30.3 within the context of the complete angiosperm life cycle. In section 30.2, we will discuss the various ways that pollen grains arrive at a stigma. This sets the stage to discuss fertilization and the subsequent development of the embryo.

REVIEW OF CONCEPT 30.1

In a flowering plant life cycle, fertilization produces an embryo in a seed. The embryo develops into a plant that flowers, to again produce gametes. Flowers consist of four concentric whorls: calyx, corolla, androecium, and gynoecium. Male gametes are produced in pollen grains, and female gametes are produced in egg sacs. The microspore mother cells in flowers undergo meiosis to produce microspores, which undergo mitosis to produce microgametophytes (pollen grains). Megaspore mother cells undergo a similar process to produce megaspores, which result in megagametophytes (embryo sacs).

- *What is the main evolutionary advantage of the flower?*

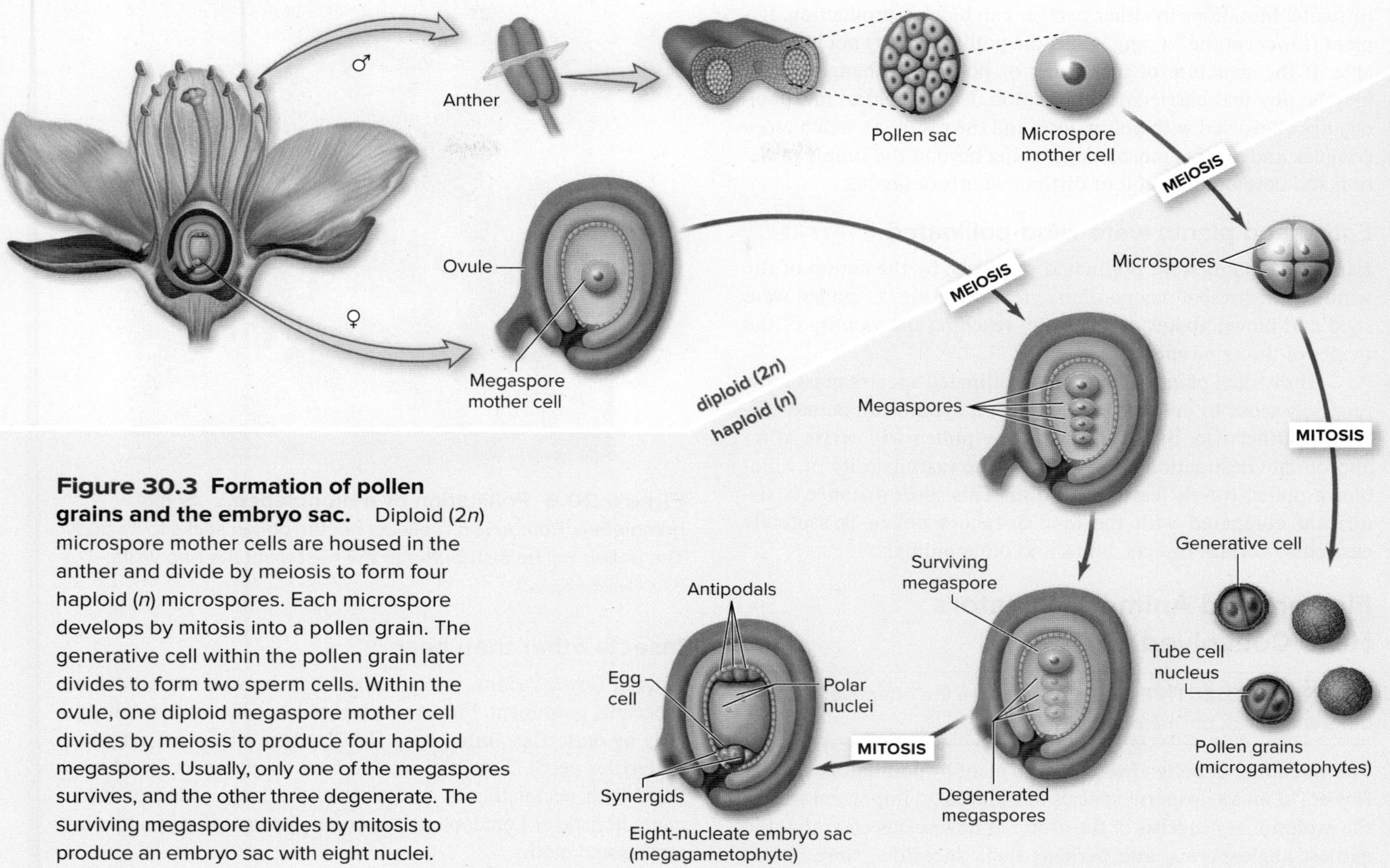

Figure 30.3 Formation of pollen grains and the embryo sac. Diploid (2*n*) microspore mother cells are housed in the anther and divide by meiosis to form four haploid (*n*) microspores. Each microspore develops by mitosis into a pollen grain. The generative cell within the pollen grain later divides to form two sperm cells. Within the ovule, one diploid megaspore mother cell divides by meiosis to produce four haploid megaspores. Usually, only one of the megaspores survives, and the other three degenerate. The surviving megaspore divides by mitosis to produce an embryo sac with eight nuclei.

Figure 30.4 Pollen grains. ***a.*** The Easter lily, *Lilium candidum*. ***b.*** A plant of the sunflower family, *Hyoseris longiloba*.

(both): ©L. DeVos/Free University of Brussels

30.2 Flowers Attract Pollinators

Pollination and fertilization are different events involving different processes. Pollination is the process by which pollen is placed on the stigma. Fertilization is the unification of haploid gametes to produce a diploid zygote. We begin this section by considering the events of pollination.

When pollen from a flower's anther pollinates the same flower's stigma, the process is called *self-pollination.* When pollen from the anther of one flower pollinates the stigma of a different flower, the process is termed *cross-pollination,* or *outcrossing.* Pollen may be carried to a flower by wind or by animals.

Pollen May Reach a Flower in Many Ways

LEARNING OBJECTIVE 30.2.1 Compare the various ways pollen may reach a flower.

Pollination in angiosperms does not involve direct contact between the pollen grain and the ovule. When pollen reaches the stigma, it germinates, and a pollen tube grows down through the stigma and style, carrying the sperm nuclei to the embryo sac. After double fertilization (discussed later in this section) takes place, development of the embryo and endosperm begins. The seed matures within the ripening fruit; eventually, the germination of the seed initiates another life cycle.

In many angiosperms, successful pollination depends on the regular attraction of **pollinators,** such as insects, birds, and other animals, which transfer pollen between plants of the same species. When animals disperse pollen, they perform the same function for flowering plants that they do for themselves when they actively search out mates.

The relationship between plant and pollinator can be quite intricate. Mutations in either partner can block reproduction. If a plant flowers at the "wrong" time, the pollinator may not be available. If the structure of the flower or pollinator changes, there may be physical barriers to pollination. Clearly, floral morphology has coevolved with pollinators, and the result is a much more complex and diverse morphology, going beyond the simple initiation and development of four distinct whorls of organs.

Early seed plants were wind-pollinated

Early seed plants were pollinated passively, by the action of the wind. As in present-day conifers, great quantities of pollen were shed and blown about, occasionally reaching the vicinity of the ovules of the same species.

Individual plants of any wind-pollinated species must grow relatively close to one another for such a system to operate efficiently. Otherwise, the chance that any pollen will arrive at an appropriate destination is very small. The vast majority of wind-blown pollen travels less than 100 m. This short distance is significant compared with the long distances pollen is routinely carried by certain insects, birds, and other animals.

Flowers and Animal Pollinators Have Coevolved

LEARNING OBJECTIVE 30.2.2 Compare the effectiveness of different animal pollinators.

The spreading of pollen from plant to plant by pollinators visiting flowers of an angiosperm species has played an important role in the evolutionary success of the group. It now seems clear that the earliest angiosperms, and perhaps their ancestors, were insect-pollinated; the coevolution of insects and plants has been important for both groups for more than 100 million years.

Bees

Among insect-pollinated angiosperms, the most numerous groups are those pollinated by bees (figure 30.5). Like most insects, bees initially locate sources of food by odor and then orient themselves on a flower or group of flowers by the flower's shape, color, and texture.

Some bees collect nectar, which is a source of food for adult bees and occasionally for larvae. Most of the approximately 20,000 species of bees visit flowers to obtain pollen, which is used to provide food to bee larvae while they complete their development.

Except for a few hundred species of social and semisocial bees and about 1000 species parasitic in nests of other bees, the great majority of bees—at least 18,000 species—are solitary. Solitary bees in temperate regions characteristically produce only a single generation a year. Many are active as adults for only a few weeks a year.

Many solitary bees use the flowers of a particular group of plants almost exclusively as sources of larval food. The highly constant relationships of such bees with specific flowers may lead to coevolutionary modifications in both the flowers and the bees. For example, the time of day flowers open may correlate with when bees appear; the mouthparts of the bees may become elongated in relation to tubular flowers; or the bees' pollen-collecting apparatuses may be adapted to the anthers of plants they visit. These relationships provide an efficient mechanism of pollination for the flowers, as well as a constant source of food for the bees.

Figure 30.5 Pollination by a bumblebee. As this bumblebee, *Bombus* sp., collects nectar, pollen sticks to its body. The pollen will be distributed to the next plant the bee visits.
IT Stock/age fotostock

Insects other than bees

Among flower-visiting insects other than bees, a few groups are especially prominent. Flowers such as phlox, which are visited regularly by butterflies, often have flat "landing platforms" on which butterflies perch. They also tend to have long, slender floral tubes filled with nectar that is accessible to the long, coiled proboscis characteristic of Lepidoptera, the order of insects that includes butterflies and moths.

Flowers such as jimsonweed *(Datura stramonium),* evening primrose *(Oenothera biennis),* and others visited regularly by moths are often white, yellow, or some other pale color; they also tend to be heavily scented, making the flowers easy to locate at night.

Birds

Several interesting groups of plants are regularly visited and pollinated by birds, especially the hummingbirds of North and South America (figure 30.6) and the sunbirds of Africa. Such plants must produce large amounts of nectar, because birds will not continue to visit flowers if they do not find enough food to maintain themselves. But flowers producing large amounts of nectar have no advantage in being visited by insects, because an insect could obtain its energy requirements from a single flower and would not cross-pollinate the flower. How are these different selective forces balanced in flowers that are "specialized" for hummingbirds and sunbirds?

The answer involves the evolution of flower color. Ultraviolet light is highly visible to insects. Carotenoids are yellow or orange pigments responsible for the colors of many flowers, including sunflowers and mustard. Carotenoids reflect both in the yellow range and in the ultraviolet range, the mixture resulting in a distinctive color called "bee's purple." Such yellow flowers may also be marked in distinctive ways normally invisible to us but highly visible to bees and other insects (figure 30.7).

In contrast, red does not stand out as a distinct color to most insects, but it is a very conspicuous color to birds. To most insects, the red upper leaves of poinsettias look just like the other

Figure 30.6 Hummingbirds and flowers. A Violet Sabrewing hummingbird (*Campylopterus hemileucurus*) extracts nectar from a Red Button Ginger flower (*Costus woodsonii*), in Costa Rica.

Adam Jones/Getty Images

Figure 30.7 How a bee sees a flower. ***a.*** The yellow flower of *Ludwigia peruviana* (Peruvian primrose) photographed in normal light and **(*b*)** with a filter that selectively transmits ultraviolet light. The outer sections of the petals reflect both yellow and ultraviolet, a mixture of colors called "bee's purple"; the inner portions of the petals reflect yellow only and therefore appear dark in the photograph that emphasizes ultraviolet reflection. To a bee, this flower appears as if it has a conspicuous central bull's-eye.

(both): Thomas Eisner, Cornell University

leaves of the plant. Consequently, even though the flowers produce abundant supplies of nectar and attract hummingbirds, insects tend to bypass them. Thus, the red color both signals to birds the presence of abundant nectar and makes that nectar as inconspicuous as possible to insects.

Other animal pollinators

Other animals, including bats and small rodents, may aid in pollination. The signals here are also species specific. As an example, the saguaro cactus (*Carnegiea gigantea*) of the Sonoran Desert is pollinated by bats that feed on nectar at night, as well as by birds and insects.

These animals may also assist in dispersing the seeds and fruits that result from pollination. Monkeys are attracted to orange and yellow, and thus can be effective in dispersing fruits of this color in their habitats.

Some Flowering Plants Continue to Use Wind Pollination

LEARNING OBJECTIVE 30.2.3 Compare animal and wind pollination.

A number of groups of angiosperms are wind-pollinated. Among these groups are oaks, birches, cottonwoods, grasses, sedges, and nettles. The flowers of these plants are small, greenish, and odorless; their corollas are reduced or absent. Such flowers often are grouped together in fairly large numbers and may hang down in tassels that wave about in the wind and shed pollen freely.

Many wind-pollinated plants have stamen- and carpel-containing flowers separated between individuals or physically separated on a single individual. Maize is a good example, with pollen-producing tassels at the top of the plant and axillary shoots with female flowers lower down. Separation of pollen-producing and ovule-bearing flowers is a strategy that greatly promotes outcrossing, because pollen from one flower must land on a different flower for fertilization to have any chance of occurring. Some wind-pollinated plants, especially trees and shrubs, flower in the spring, before the development of their leaves can interfere with the wind-borne pollen.

Self-Pollination Is Favored in Stable Environments

LEARNING OBJECTIVE 30.2.4 Explain why self-pollination may be favored.

Thus far, we have considered examples of pollination that tend to lead to outcrossing, which is as advantageous for plants as it is for eukaryotic organisms. Nevertheless, self-pollination also occurs among angiosperms, particularly in temperate regions. Most self-pollinating plants have small, relatively inconspicuous flowers that shed pollen directly onto the stigma, sometimes even before the bud opens.

You might reasonably ask why many self-pollinated plant species have survived if outcrossing is as important genetically for plants as it is for animals. Biologists propose two basic reasons for the frequent occurrence of self-pollinated angiosperms:

1. Self-pollination is advantageous under certain ecological circumstances, because self-pollinators do not need to be visited by animals to produce seed. As a result, self-pollinated plants expend less energy in producing pollinator attractants, such as expensive nectar, and can grow in areas where the kinds of insects or other animals that might visit them are absent or very scarce—as in the Arctic or at high elevations.
2. In genetic terms, self-pollination produces offspring that are more uniform than those that result from outcrossing.

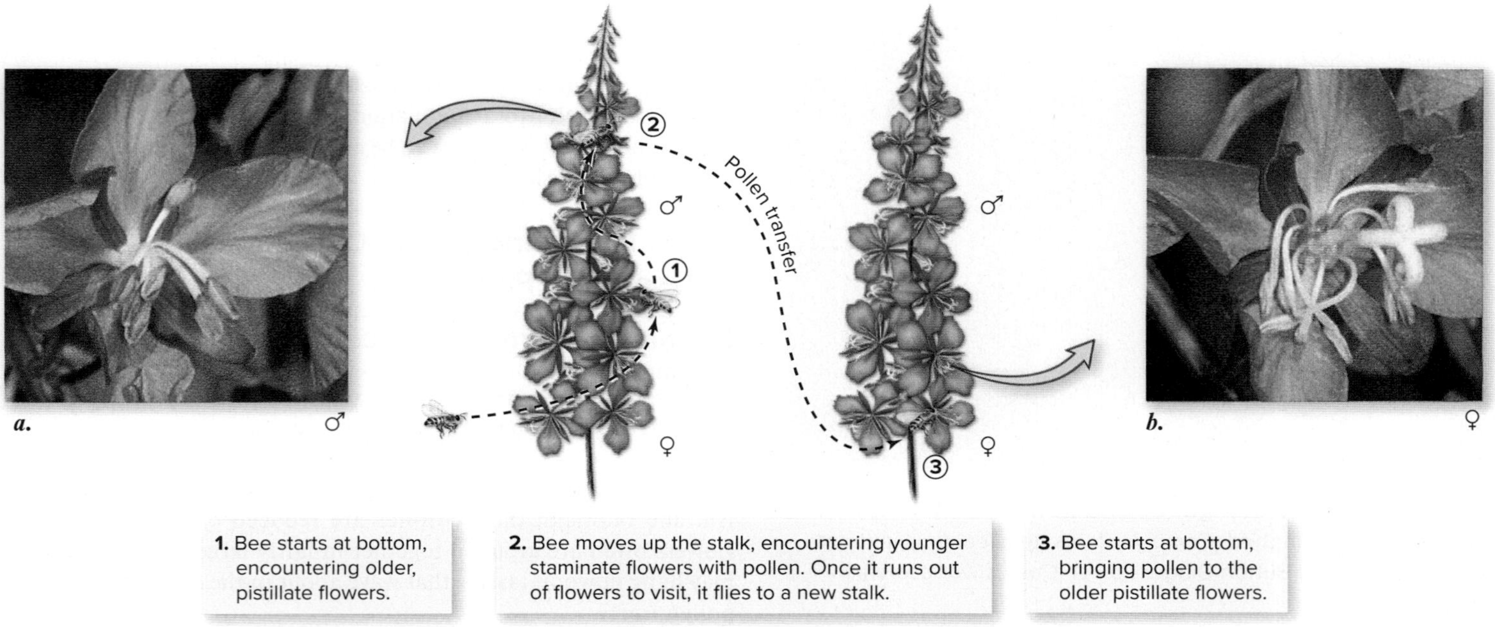
a.
♂
♂
♀
Pollen transfer
①
②
③
♂
♀
b.
♀
1. Bee starts at bottom, encountering older, pistillate flowers.
2. Bee moves up the stalk, encountering younger staminate flowers with pollen. Once it runs out of flowers to visit, it flies to a new stalk.
3. Bee starts at bottom, bringing pollen to the older pistillate flowers.

Figure 30.8 Dichogamy in the fireweed, *Chamerion angustifolium*. More than 200 years ago, fireweed was one of the first plant species to have its pollination described. First, the anthers shed pollen, and then the style elongates above the stamens while the four lobes of the stigma curl back and become receptive. Consequently, flowers are functionally staminate at first, becoming pistillate about two days later. The flowers open progressively up the stem, so that the lowest are visited first. Working up the stem, the bees encounter pollen-shedding, staminate-phase flowers, and become covered with pollen, which they then carry to the lower, functionally pistillate flowers of another plant. Shown here are flowers in **(*a*)** the staminate phase and **(*b*)** the pistillate phase.

(both): Steven P. Lynch/McGraw Hill

Because meiosis is involved, recombination still takes place and therefore the offspring will not be identical to the parent. However, such offspring may contain high proportions of individuals well adapted to particular habitats.

Self-pollination in normally outcrossing species tends to produce large numbers of ill-adapted individuals, because it brings together deleterious recessive alleles—but some of these combinations may be highly advantageous in particular habitats. In these habitats, it may be advantageous for the plant to continue self-pollinating indefinitely.

Several Evolutionary Strategies Promote Outcrossing

LEARNING OBJECTIVE 30.2.5 Describe three evolutionary strategies that promote outcrossing.

Outcrossing, as we have stressed, is critically important for the adaptation and evolution of all eukaryotic organisms, with a few exceptions. Many flowers contain both stamens and pistils (figure 30.8), increasing the likelihood of self-pollination. One general strategy to promote outcrossing, therefore, is to separate stamens and pistils. Another strategy involves self-incompatibility, which prevents self-fertilization.

Separation of male and female structures in space or in time

In a number of species—for example, willows and some mulberries—staminate and pistillate flowers may occur on separate plants. Such plants, which produce only ovules or only pollen, are called dioecious, meaning "two houses." These plants clearly cannot self-pollinate and must rely exclusively on outcrossing. In other kinds of plants, such as oaks, birches, corn (maize), and pumpkins, separate male and female flowers may both be produced on the same plant. Such plants are called **monoecious,** meaning "one house" (figure 30.9). In monoecious plants, the separation of pistillate and staminate flowers, which may mature at different times, greatly enhances the probability of outcrossing.

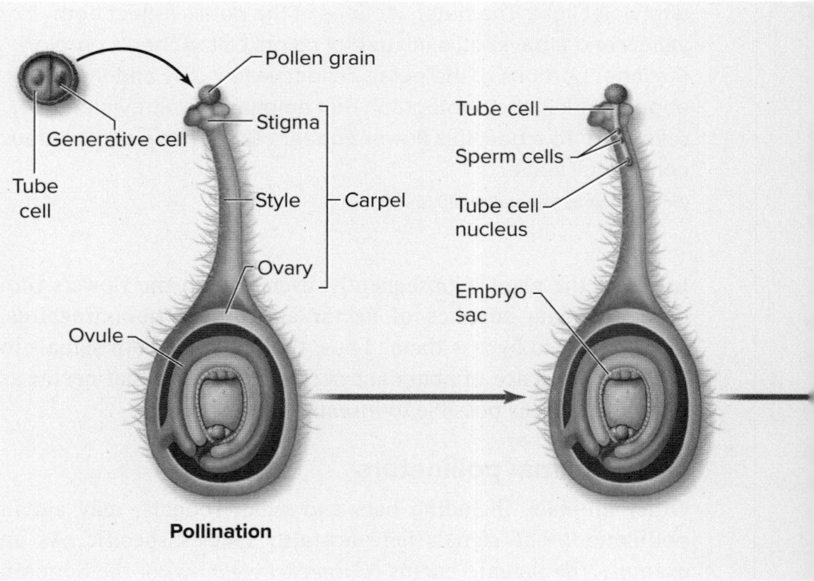
Pollen grain
Generative cell
Tube cell
Stigma
Style
Carpel
Ovary
Ovule
Tube cell
Sperm cells
Tube cell nucleus
Embryo sac
Pollination

Figure 30.9 Staminate and pistillate flowers of a birch, *Betula* sp. Birches are monoecious; their staminate flowers hang down in long, yellowish tassels, and their pistillate flowers mature above the tassels into clusters of small, brownish, conelike structures.
Guenter Fischer/Imagebroker/Alamy Stock Photo

Even if, as is usually the case, functional stamens and pistils are both present in each flower of a particular plant species, these organs may reach maturity at different times. Plants in which this occurs are called **dichogamous.** If the stamens mature first, shedding their pollen before the stigmas are receptive, the flower is effectively staminate at that time. Once the stamens have finished shedding pollen, the stigma or stigmas may become receptive, and the flower may become essentially pistillate (figure 30.8). This separation in time has the same effect as if individuals were dioecious; the outcrossing rate is significantly increased.

Self-incompatibility

Even when a flower's stamens and stigma mature at the same time, genetic self-incompatibility, which is widespread in flowering plants, increases outcrossing. Self-incompatibility results when the pollen and stigma recognize each other as being genetically related, and pollen tube growth is blocked.

Angiosperms Undergo Double Fertilization

LEARNING OBJECTIVE 30.2.6 Describe the process of double fertilization.

Fertilization in angiosperms is a complex process in which two sperm cells participate in a process called double fertilization. Double fertilization produces (1) a diploid zygote and (2) triploid endosperm tissue that nourishes the embryo.

Once a pollen grain has been spread by wind, by animals, or through self-pollination, it adheres to the sticky, sugary substance that covers the stigma and begins to grow a pollen tube that pierces the style (figure 30.10). The pollen tube, nourished by the sugary substance, grows until it reaches the ovule in the ovary. Meanwhile, the generative cell within the pollen grain tube cell divides to form two sperm cells.

Figure 30.10 The formation of the pollen tube and double fertilization. When pollen lands on the stigma of a flower, the pollen tube cell grows toward the embryo sac, forming a pollen tube. While the pollen tube is growing, the generative cell divides to form two sperm cells. When the pollen tube reaches the embryo sac, it enters one of the synergids and releases the sperm cells. In a process called double fertilization, one sperm cell nucleus fuses with the egg cell to form the diploid (2*n*) zygote, and the other sperm cell nucleus fuses with the two polar nuclei to form the triploid (3*n*) endosperm nucleus.

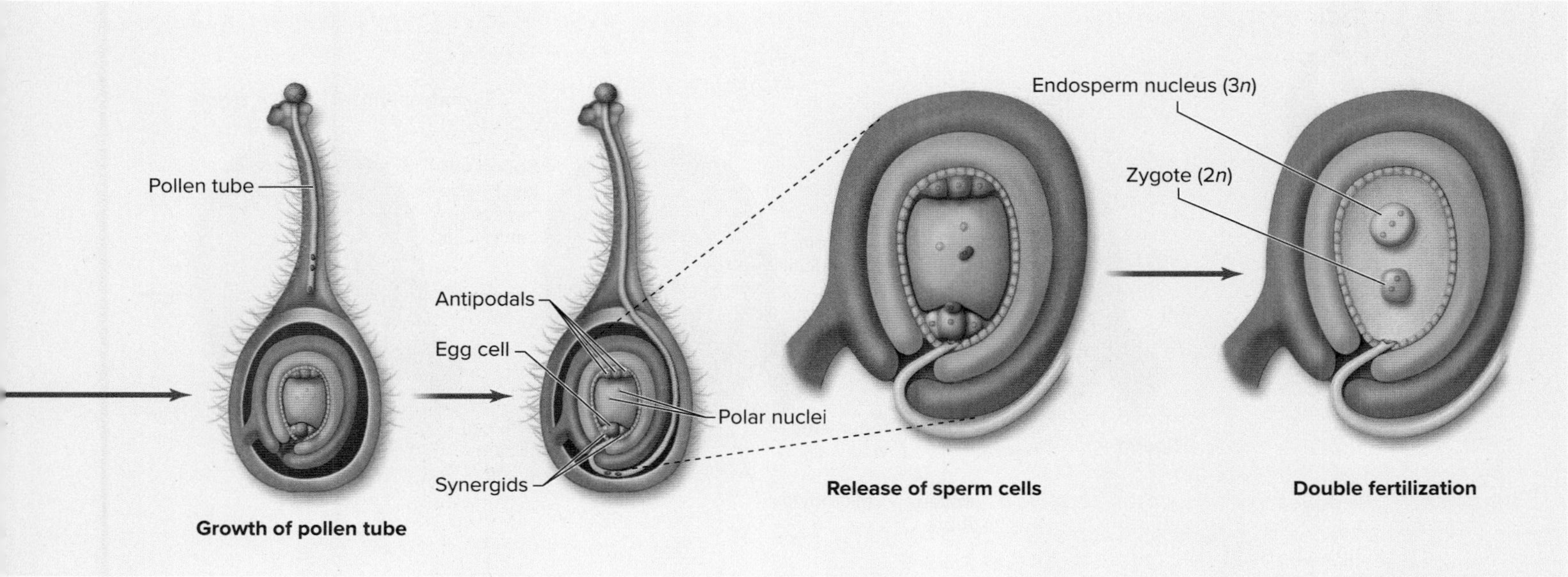

The pollen tube eventually reaches the embryo sac in the ovule. At the entry to the embryo sac, one of the nuclei flanking the egg cell degenerates, and the pollen tube enters that cell. The tip of the pollen tube bursts and releases the two sperm cells. One of the sperm cells fertilizes the egg cell, forming a zygote. The other sperm cell fuses with the two polar nuclei located at the center of the embryo sac, forming the triploid primary endosperm nucleus. The primary endosperm nucleus eventually develops into the endosperm. The endosperm is a food supply for the developing embryo and also provides nourishment during seed germination.

Once fertilization is complete, the embryo develops as its cells divide numerous times by mitosis. Meanwhile, protective tissues enclose the embryo, resulting in the formation of the seed. The seed, in turn, is enclosed in another structure, called the fruit. These typical angiosperm structures evolved in response to the need for seeds to be dispersed over long distances to ensure genetic variability.

REVIEW OF CONCEPT 30.2

Pollen may be carried by wind, insects, or birds or other animals. Self-pollination may be favored when pollinators are absent or when plants are adapted to a stable environment. Mechanisms to promote outcrossing include the production of separate male and female flowers, maturation of male flowers at a different time from that of female flowers, and genetically controlled self-incompatibility. Double fertilization produces a diploid embryo and triploid endosperm that provides nutrition.

- *Are all offspring of a self-pollinating plant identical?*

30.3 Fertilization Leads to Embryogenesis

Embryo development begins once the egg cell is fertilized. The growing pollen tube from a pollen grain enters the angiosperm embryo sac through one of the synergids, releasing two sperm cells (figure 30.11). One sperm cell fertilizes the central cell with

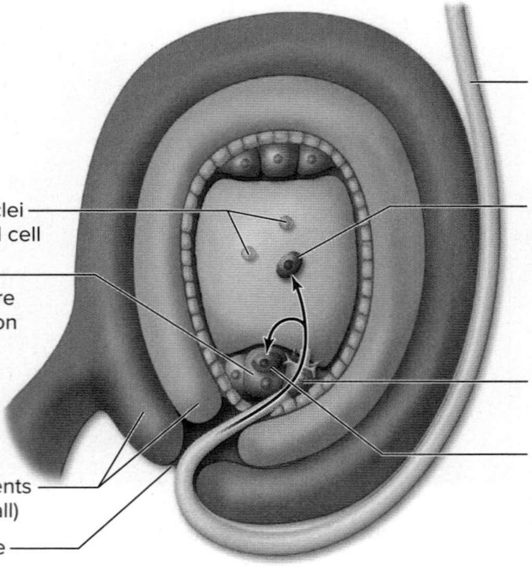

Figure 30.11 Fertilization triggers embryogenesis. The egg cell, within the embryo sac, is fertilized by one sperm cell released from the pollen tube. The second sperm cell fertilizes the central cell and initiates endosperm development. This diagram shows sperm just before fertilization.

Figure 30.12 Stages of development in a eudicot angiosperm embryo. The very first cell division is asymmetrical. Differentiation begins almost immediately after fertilization.

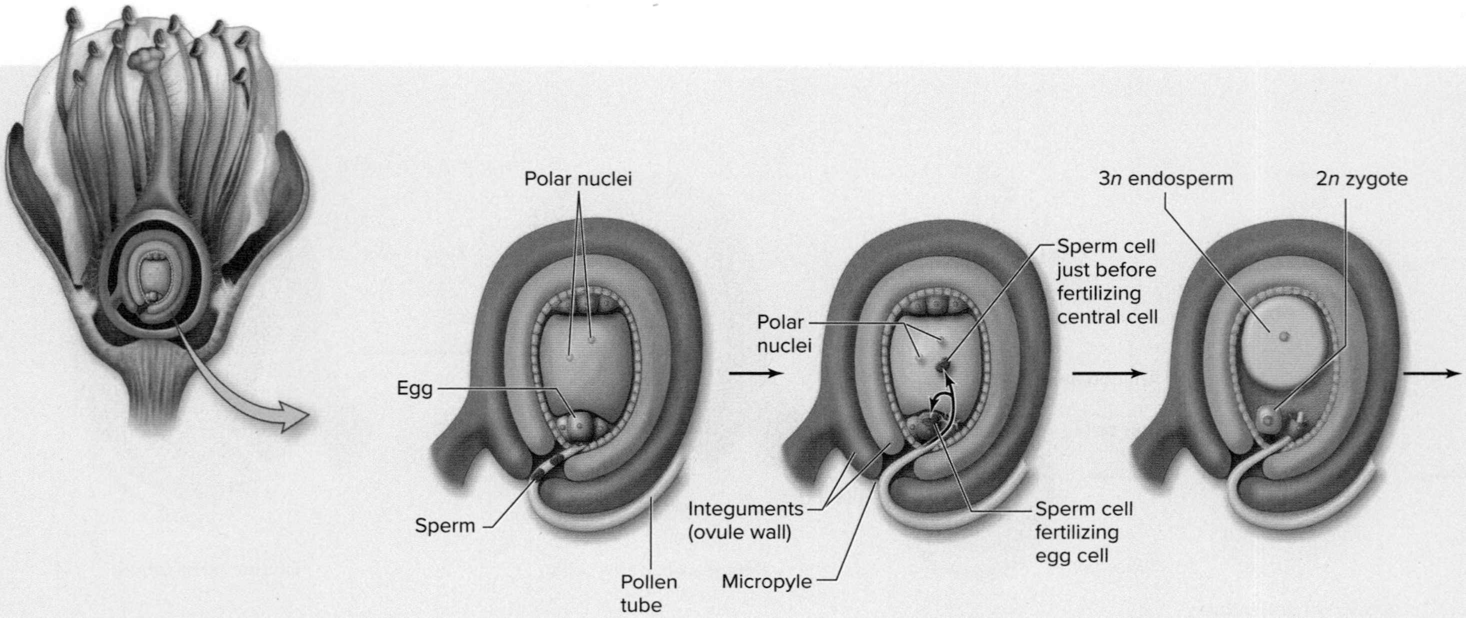

its polar nuclei, and the resulting cell division produces a nutrient source, the **endosperm,** for the embryo. The other sperm cell fertilizes the egg to produce a zygote, and cell division soon follows, creating the **embryo.**

A Single Cell Divides to Produce a Three-Dimensional Body

LEARNING OBJECTIVE 30.3.1 Discuss the role of the suspensor in embryo development.

The first division of the zygote (fertilized egg) in a flowering plant is asymmetrical and generates cells with two different fates (figure 30.12). One daughter cell is small, with dense cytoplasm. That cell, which is destined to become the embryo, begins to divide repeatedly in different planes, forming a ball of cells. The other, larger daughter cell divides repeatedly, forming an elongated structure called a **suspensor,** which links the embryo to the nutrient tissue of the seed. The suspensor also provides a route for nutrients to reach the developing embryo. The root–shoot axis also forms at this time; cells near the suspensor are destined to form a root, whereas those at the other end of the axis ultimately become a shoot.

A Simple Body Plan Emerges During Embryogenesis

LEARNING OBJECTIVE 30.3.2 Describe how three tissue systems arise in the embryo.

In plants, three-dimensional shape and form arise by regulating the amount and the pattern of cell division. A vertical axis (root–shoot axis) becomes established at a very early stage, as does a radial axis (inner–outer axis). Although the first cell division gives rise to a single row of cells, cells soon begin dividing in different directions, producing a three-dimensional, solid ball of cells. The root–shoot axis lengthens as cells divide. New cell walls form perpendicular to the root–shoot axis, stacking new cells along the root–shoot axis to lengthen the embryo in this direction.

Apical meristems, the actively dividing cell regions at the tips of roots and shoots, establish the root–shoot axis in the globular stage, from which the three basic tissue systems arise: *dermal, ground,* and *vascular* tissue (refer to chapter 29). These tissues are organized radially around the root–shoot axis. Both the shoot and the root meristems are apical meristems, but their formation is controlled independently by different sets of genes.

Formation of the three tissue systems

Three basic tissues, called *primary meristems,* differentiate while the plant embryo is still a ball of cells (called the globular stage). Unlike in animal development where cell migration is critical in embryogenesis, cells do not migrate during plant embryo development. The protoderm consists of the outermost cells in a plant embryo and will become *dermal tissue* (refer to chapter 29). These cells almost always divide with their cell plate perpendicular to the body surface, thus creating a single outer layer of cells. Dermal tissue protects the plant from desiccation. Stomata that open and close to facilitate gas exchange and minimize water loss are derived from dermal tissue.

A ground meristem gives rise to the bulk of the embryonic interior, consisting of *ground tissue* cells that eventually function in food and water storage. Finally, procambium at the core of the embryo will form the future vascular tissue, which is responsible for water and nutrient transport.

Morphogenesis

The globular stage gives rise to a heart-shaped embryo with two bulges in one group of angiosperms (the eudicots) and a ball with a bulge on a single side in another group (the monocots). These bulges

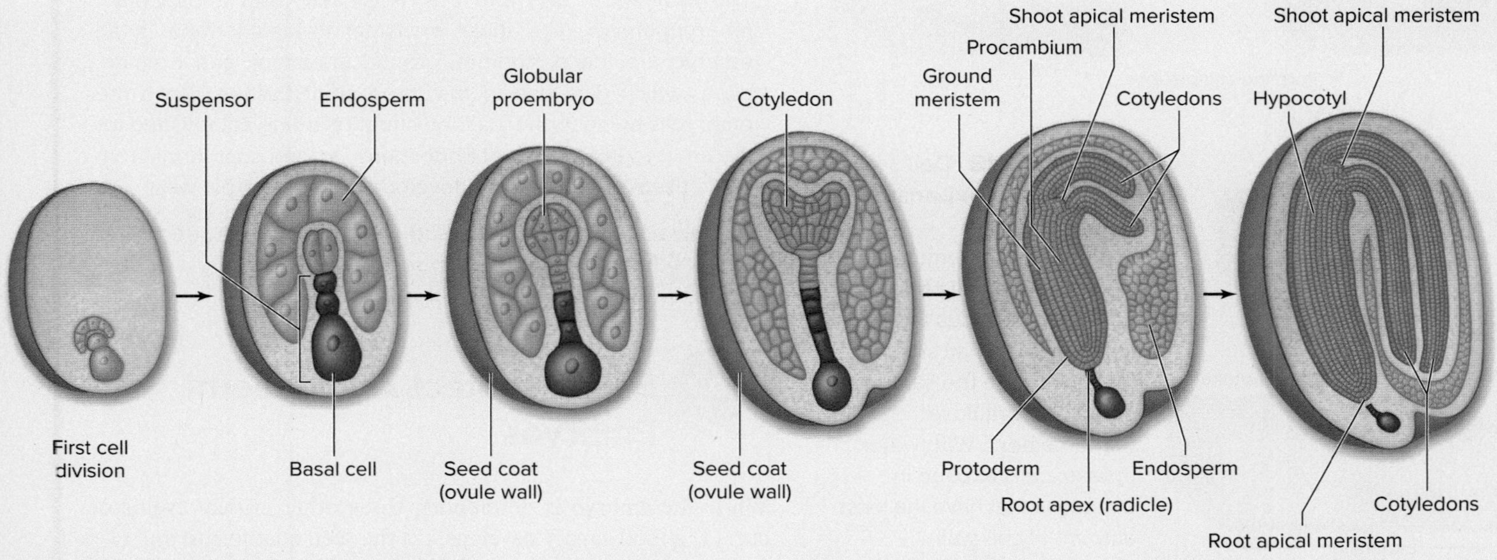

are **cotyledons** ("first leaves") and are produced by the embryonic cells, not by the shoot apical meristem that begins forming during the globular stage. This process, called morphogenesis (generation of form), results from changes in planes and rates of cell division.

Because plant cells cannot migrate, the form of a plant body is largely determined by the plane in which its cells divide. It is also controlled by changes in cell shape as cells expand by absorbing water after they form (figure 30.13). A structure called the **cell plate,** which corresponds to where chromosomes will line up during metaphase of mitosis, determines the direction of division. Microtubules and actin determine where the cell plate forms. Plant hormones and other factors influence the orientation of bundles of microtubules on the interior of the plasma membrane. These microtubules also guide cellulose deposition as the cell wall forms around the outside of a new cell, where four of the six sides are reinforced more heavily with cellulose; the cell tends to expand and grow in the direction of the two sides having less reinforcement (figure 30.13*b*).

Much is being learned about morphogenesis at the cellular level from mutants that are able to divide but cannot control their plane of cell division or the direction of cell expansion. The lack of root meristem development in *hobbit* mutants is just one such example. As the procambium begins differentiating in the root, a critical division parallel to the root's surface is regulated by the gene *WOODEN LEG* (*WOL;* figure 30.14). Without that division, the cylinder of cells that would form phloem is missing. Only xylem forms in the vascular tissue system, giving the root a "wooden leg."

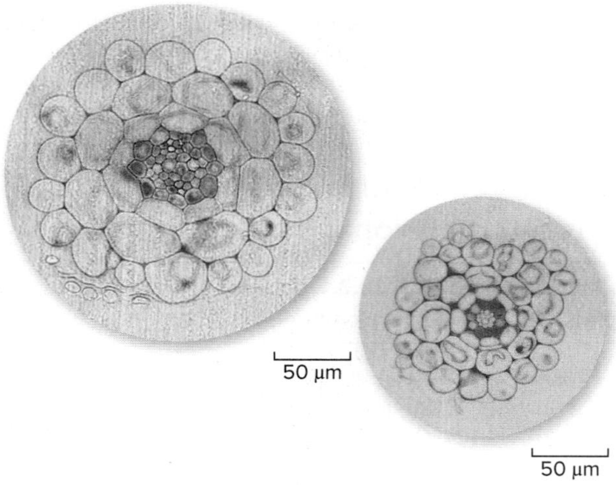

50 μm
50 μm

Figure 30.14 ***WOODEN LEG* is needed for phloem development.** The *wol* mutant (right) has less vascular tissue than wild-type *Arabidopsis* (left), and all of it is xylem.

(both): A. P. Mahonen

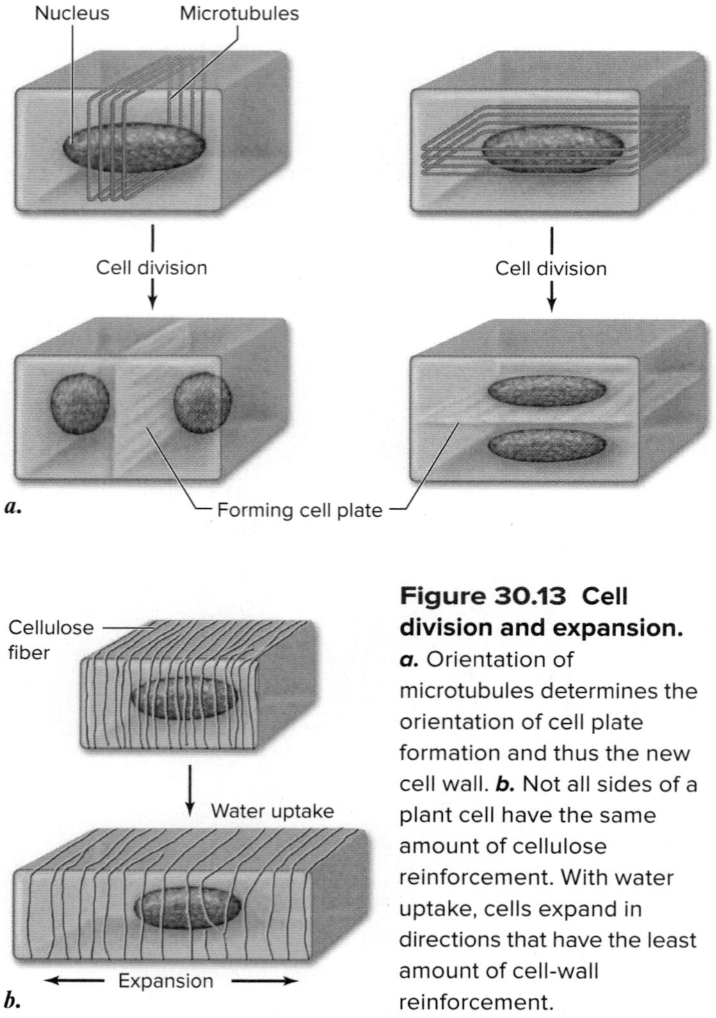

Nucleus
Microtubules
Cell division
Cell division
a.
Forming cell plate
Cellulose fiber
Water uptake
Expansion
b.

Figure 30.13 Cell division and expansion. ***a.*** Orientation of microtubules determines the orientation of cell plate formation and thus the new cell wall. ***b.*** Not all sides of a plant cell have the same amount of cellulose reinforcement. With water uptake, cells expand in directions that have the least amount of cell-wall reinforcement.

Early in embryonic development, most cells can give rise to a wide range of cell and organ types, including leaves. As development proceeds, the cells with multiple potentials are mainly restricted to the meristem regions. Many meristems have been established by the time embryogenesis ends and the seed becomes dormant. After germination, apical meristems continue adding cells to the growing root and shoot tips. Apical meristem cells of corn, for example, divide every 12 hours, producing half a million cells per day in an actively growing corn plant. Lateral meristems can cause an increase in the girth of some plants, whereas intercalary meristems in the stems of grasses allow for elongation.

REVIEW OF CONCEPT 30.3

The root–shoot axis and the radial axis form during plant embryogenesis. The three *meristem* tissues formed in an embryo are the protoderm, ground meristem, and procambium, which give rise to the three adult tissues. While the embryo is being formed, a food supply is being established for the embryo in the form of endosperm; a seed coat forms from ovule tissues; and the fruit develops from the carpel wall.

- *How does the nutritive tissue of a gymnosperm seed differ from that of an angiosperm seed?*

30.4 Seeds Protect Angiosperm Embryos

While the embryo is developing, three other critical events occur: (1) a food supply develops; (2) the seed coat forms; and (3) a fruit grows, eventually surrounding the seed. These events are important in the nourishment and protection of the embryo and in dispersal of the seed that forms.

Food Reserves Form During Embryogenesis

LEARNING OBJECTIVE 30.4.1 Describe how food reserves develop in the embryo.

In angiosperms, double fertilization produces endosperm for nutrition; in gymnosperms, the megagametophyte is the food source (refer to chapter 26). The seed coat is the result of the differentiation of ovule tissue (from the parental sporophyte) to form a hard, protective covering around the embryo. The seed then enters a dormant phase, signaling the end of embryogenesis. In angiosperms, the fruit develops from the carpel wall surrounding the ovule. Fruit development and seed germination are addressed in sections 30.5 and 30.6, respectively. In this section, we focus on nutritional reserves that form during embryogenesis.

Throughout embryogenesis, starch, lipids, and proteins are synthesized. The seed-storage proteins are so abundant that the genes coding for them were the first cloning targets for plant molecular biologists. Providing nutritional resources in a seed is part of the evolutionary trend toward enhancing embryo survival.

In angiosperms, the sporophyte transfers nutrients via the suspensor. (This is in contrast to gymnosperms, in which the suspensor serves only to push the embryo closer to the megagametophytic nutrient source.) This transfer of nutrients happens concurrently with the development of the endosperm, which is present only in angiosperms (although double fertilization has been observed in the gymnosperm *Ephedra*). Endosperm formation varies with species and may be extensive or minimal.

The form that endosperm takes also varies considerably: Endosperm in coconut includes the liquid "milk"; however, in corn the endosperm is solid. In popping corn, the endosperm expands with heat to form the white edible part of the popped corn. In peas and beans, the endosperm is used up during embryo development, and nutrients are stored in thick, fleshy cotyledons (figure 30.15).

Because the photosynthetic machinery is built in response to light, stored nutrients are critical to early embryonic growth. The germinating sporophyte will utilize cellular respiration to extract needed energy from nutrients stored in the seed until the sporophyte is capable of photosynthesis. A seed buried too deeply in the soil will use up all its reserves in cellular respiration before reaching the surface and sunlight.

The Seed Coat Protects the Embryo

LEARNING OBJECTIVE 30.4.2 Explain how seeds help to ensure the survival of a plant's offspring.

Early in the development of an angiosperm embryo, a profoundly important event occurs: the embryo stops developing. In many plants, development of the embryo is arrested soon after the meristems and cotyledons differentiate. Then, the integuments—the outer cell layers of the ovule—develop into a relatively impermeable **seed coat,** which encloses the seed with its dormant embryo and stored food (figure 30.16).

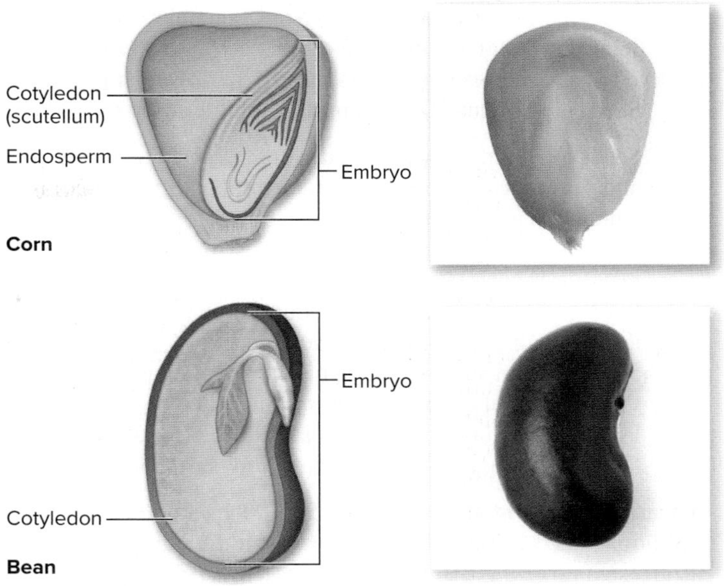

Figure 30.15 Endosperm in maize and the bean. The maize kernel has endosperm that is still present at maturity, but the endosperm in the bean has disappeared. The bean embryo's cotyledons take over food-storage functions.

(corn): Somchai Som/Shutterstock; (bean): Metta image/Alamy Stock Photo

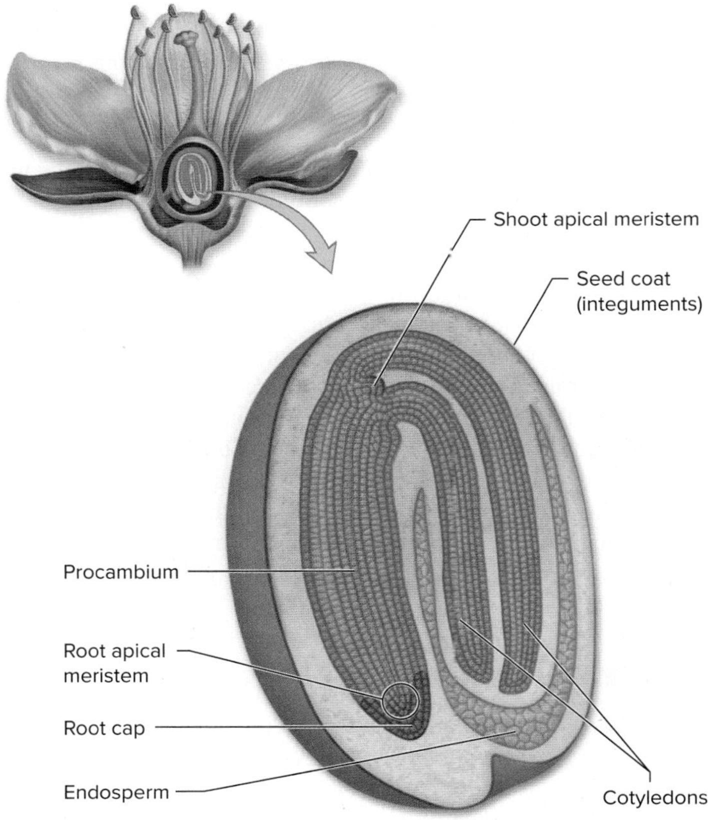

Figure 30.16 Seed development. The integuments of this mature angiosperm ovule are forming the seed coat. Note that the two cotyledons have grown into a bent shape to fill the tight confines of the seed. In some embryos, the shoot apical meristem will have already initiated a few leaf primordia as well.

The seed is a vehicle for dispersing the embryo to distant sites. Being encased in the protective layers of a seed allows a plant embryo to survive in environments that might kill a mature plant.

Seeds are an important adaptation in at least four ways:

1. Seeds maintain dormancy under unfavorable conditions and postpone development until better conditions arise. If conditions are marginal, a plant can "afford" to have some seeds germinate, because some of those that germinate may survive, while others remain dormant.
2. Seeds provide maximum protection to the young plant at its most vulnerable stage of development.
3. Seeds contain stored food that allows a young plant to grow and develop before photosynthetic activity begins.
4. Perhaps most important, seeds are adapted for dispersal, facilitating the migration of plants into new habitats.

A mature seed contains only about 5 to 20% water. Under these conditions, the seed and the young plant within it are very stable. Its arrested growth is primarily due to the progressive and severe desiccation of the embryo and the associated reduction in metabolic activity. Germination cannot take place until water and oxygen reach the embryo. Seeds of some plants have been known to remain viable for hundreds or, in rare instances, thousands of years.

a.

b.

Figure 30.17 Fire induces seed release in some pines. Fire can destroy adult jack pines but stimulate growth of the next generation. ***a.*** The cones of a jack pine are tightly sealed and cannot release the seeds protected by the scales. ***b.*** High temperatures lead to the release of the seeds.

(a): Ed Reschke; (b): NPS Photo by Don Despain

Specialized Seed Adaptations Improve the Odds That Offspring Will Survive

LEARNING OBJECTIVE 30.4.3 Discuss the role of environmental conditions in seed germination in some plants.

Specific adaptations often help ensure that seeds will germinate only under appropriate conditions. Some seeds lie within tough cones that do not open until they are exposed to the heat of a fire (figure 30.17). This strategy causes the seed to germinate in an open, fire-cleared habitat where nutrients are relatively abundant, having been released from plants burned in the fire.

Seeds of other plants germinate only when inhibitory chemicals leach from their seed coats, thus guaranteeing their germination when sufficient water is available. Still other seeds germinate only after they pass through the intestines of birds or mammals or are regurgitated by them, which both weakens the seed coats and ensures dispersal. Sometimes seeds of plants thought to be extinct in a particular area germinate under unique or improved environmental circumstances, and the plants may then reestablish themselves.

REVIEW OF CONCEPT 30.4

The seed coat originates from the integuments and encloses the embryo and stored nutrients. The four advantages conferred by seeds are dormancy, protection of the embryo, nourishment, and a method of dispersal. Fire, heavy rains, or passage through an animal's digestive tract may be required for germination in some species.

- *What type of seed dormancy would you expect to find in trees living in climates with cold winters?*

30.5 Fruits Promote Seed Dispersal

Survival of angiosperm embryos depends on fruit development as well as seed development. Fruits are most simply defined as mature ovaries (carpels). During seed formation, the flower ovary begins to develop into fruit (figure 30.18). In some cases, pollen landing on the stigma can initiate fruit development, but more frequently the coordination of fruit, seed coat, embryo, and endosperm development follows fertilization.

Fruits Are Adapted for Seed Dispersal

LEARNING OBJECTIVE 30.5.1 Identify the structures that develop into fruit.

It is possible for fruits to develop without seed development. Commercial bananas, for example, have aborted seed development but do produce mature, edible ovaries. Commercially grown bananas are propagated asexually, because no embryo develops.

Fruits form in many ways and exhibit a wide array of adaptations for dispersal. Three layers of ovary wall, also called the *pericarp,* can have distinct fates, which account for the diversity of fruit types, from fleshy to dry and hard. The differences among some of the fruit types are shown in figure 30.19.

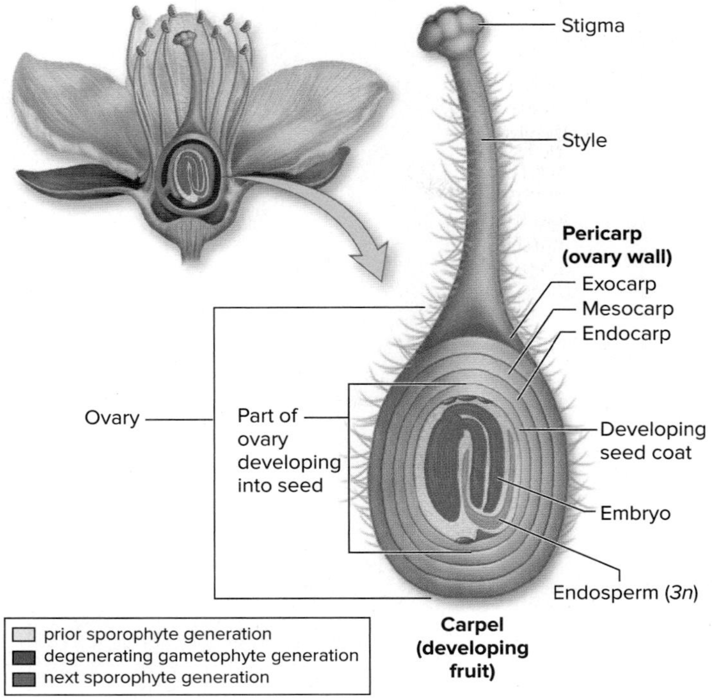

Stigma
Style
Pericarp (ovary wall)
Exocarp
Mesocarp
Endocarp
Ovary
Part of ovary developing into seed
Developing seed coat
Embryo
Endosperm (3n)
Carpel (developing fruit)
prior sporophyte generation
degenerating gametophyte generation
next sporophyte generation

Figure 30.18 Fruit development. The carpel (specifically, the ovary) wall is composed of three layers: the exocarp, mesocarp, and endocarp. One, some, or all of these layers develop to contribute to the recognized fruit in different species. The seed matures within this developing fruit.

Developmentally, fruits are fascinating organs that contain three genotypes. The fruit and seed coat are from the prior sporophyte generation. Remnants of the gametophyte generation that produced the egg are found in the developing seed, and the embryo represents the next sporophyte generation.

Fruits allow angiosperms to colonize large areas

Not only do fruits form in multiple ways, but they also exhibit a wide array of specialized dispersal methods. Fruits with fleshy coverings, often shiny black or bright blue or red, are normally dispersed by birds or other vertebrates (figure 30.20*a*). Like red flowers, red fruits signal an abundant food supply. By feeding on these fruits, birds and other animals may carry seeds from place to place and thus transfer plants from one suitable habitat to another. Such seeds require a hard seed coat to resist the stomach acids and digestive enzymes of the animals that eat and transport them.

Fruits with hooked spines, like those of burrs (figure 30.20*b*), are typical of several genera of plants that occur in the northern deciduous forests. Such fruits are often disseminated by mammals, including humans, when they hitch a ride on fur or clothing.

Other fruits, including those of maples, elms, and ashes, have wings that aid in their distribution by the wind. Orchids have minute, dustlike seeds, which are likewise blown away by the wind. The dandelion provides another familiar example of a fruit type that is wind-dispersed (figure 30.20*c*).

Coconuts and other plants that characteristically occur on or near beaches are regularly spread by floating in water (figure 30.20*d*). This sort of dispersal is especially important in the colonization of distant island groups, such as the Hawaiian Islands.

It has been calculated that the seeds of about 175 angiosperms, nearly one-third from North America, must have reached Hawaii to have evolved into the roughly 970 species found there today. Some of these seeds blew through the air, others were transported on the feathers or in the guts of birds, and still others floated across the Pacific.

REVIEW OF CONCEPT 30.5

As a seed develops, the pericarp layers of the ovary wall develop into the fruit. A berry has a fleshy pericarp; a legume has a dry pericarp that opens to release seeds; the outer layers of a drupe pericarp are fleshy; and a samara is a dry structure with a wing. Animals often distribute the seeds of fleshy fruits and fruits with spines or hooks. Wind disperses lightweight seeds and samara forms.

- *What features of fruits might encourage animals to eat them?*

30.6 Germination Begins Seedling Growth

In suitable conditions, the embryo emerges from its previously desiccated state, utilizes food reserves, and resumes growth. Although **germination** is a process characterized by several stages, it is often defined as the emergence of the **radicle** (first root) through the seed coat.

External Signals and Conditions Trigger Germination

LEARNING OBJECTIVE 30.6.1 Describe how seed germination occurs.

Germination begins when a seed absorbs water and its metabolism resumes. The amount of water a seed can absorb is phenomenal, and osmotic pressure creates a force strong enough to break the seed coat. At this point, it is important that oxygen be available to the developing embryo, because plants, like animals, require oxygen for cellular respiration. Few plants produce seeds that germinate successfully under water, although some, such as rice, have evolved a tolerance to anaerobic conditions.

Even though a dormant seed may have absorbed a full supply of water and may be respiring, expressing genes, and apparently carrying on normal metabolism, it may fail to germinate without an additional signal from the environment. This signal may be light of the correct wavelength and intensity, a series of cold days, or simply the passage of time at temperatures appropriate for germination. The seeds of many plants will not germinate unless they have been **stratified**—held for periods of time at low temperatures. This phenomenon prevents the seeds of plants that grow in seasonally cold areas from germinating until they have passed the winter, thus protecting their tender seedlings from harsh, cold conditions.

Legumes
Split along two carpel edges (sutures) with seeds attached to edges; peas, beans. Unlike fleshy fruits, the three tissue layers of the ovary do not thicken extensively. The entire pericarp is dry at maturity.
Pericarp
Stigma
Seed
Style
Samaras
Not split and with a wing formed from the outer tissues; maples, elms, ashes.
Pericarp
Seed
Multiple Fruits
Individual flowers form fruits around a single stem. The fruits fuse as seen with pineapple.
Main stem
Pericarp of individual flower
True Berries
The entire pericarp is fleshy, although there may be a thin skin. Berries have multiple seeds in either one or more ovaries. The tomato flower had four carpels that fused. Each carpel contains multiple ovules that develop into seeds.
Outer pericarp
Fused carpels
Seed
Drupes
Single seed enclosed in a hard pit; peaches, plums, cherries. Each layer of the pericarp has a different structure and function, with the endocarp forming the pit.
Pericarp
Exocarp (skin)
Mesocarp
Endocarp (pit)
Seed
Aggregate Fruits
Derived from many ovaries of a single flower; strawberries, blackberries. Unlike tomato, these ovaries are not fused and covered by a continuous pericarp.
Sepals of a single flower
Ovary
Seed

Figure 30.19 Examples of some kinds of fruits. Legumes and samaras are examples of dry fruits. Legumes open to release their seeds; samaras do not. Drupes and true berries are simple, fleshy fruits; they develop from a flower with a single pistil composed of one or more carpels. Aggregate and multiple fruits are compound, fleshy fruits; they develop from flowers with more than one pistil or from more than one flower.

(legumes): Goodshoot/Alamy Stock Photo; (samaras): Damann/Shutterstock; (multiple fruits): David A. Tietz/McGraw Hill; (true berries): D. Hurst/Alamy Stock Photo; (drupes): lynx/iconotec/Glowimages; (aggregate fruits): Alex Coan/Shutterstock

a.
b.
c.
d.

Figure 30.20 Seed dispersal strategies. ***a.*** The red berries of this honeysuckle, *Lonicera hispidula,* are highly attractive to birds. After eating the fruits, birds may carry the seeds they contain for great distances. ***b.*** The fruits of *Cenchrus incertus* have spines that adhere readily to any passing animal. ***c.*** False dandelion, *Pyrrhopappus carolinianus,* has "parachutes" that widely disperse the fruits in the wind. ***d.*** This fruit of the coconut palm, *Cocos nucifera,* is sprouting on a sandy beach. Coconuts have become established on other islands by drifting there on the waves.

(a): Rainyclub/Shutterstock; (b): Steven P. Lynch; (c): Brian Jackson/123RF; (d): John Kaprielian/Science Source

Germination can occur over a wide temperature range (5° to 30°C), although certain species may have relatively narrow optimum ranges. Some seeds will not germinate even under the best conditions. In some species, a significant fraction of a season's seeds remain dormant for an indeterminate length of time, providing a gene pool of great evolutionary significance to the future plant population. The presence of ungerminated seeds in the soil of an area is referred to as the **seed bank.**

Nutrient reserves sustain the growing seedling

Germination occurs when all internal and external requirements are met. Germination and early seedling growth require the utilization of metabolic reserves stored as starch in amyloplasts (colorless plastids) and protein bodies. Fats and oils, also stored in some kinds of seeds, can readily be digested during germination to produce glycerol and fatty acids, which yield energy through cellular respiration. Depending on the kind of plant, any of these reserves may be stored in the embryo or in the endosperm to be used as a food source during germination.

In the kernels of cereal grains, the single cotyledon is modified into a relatively massive structure called the **scutellum.** The abundant food stored in the scutellum is used up first during germination. Later, while the seedling is becoming established, the scutellum serves as a nutrient conduit from the endosperm to the rest of the embryo.

The utilization of stored starch by germinating plants is one of the best examples of how hormones modulate plant development (figure 30.21). The embryo produces gibberellic acid, a hormone, that signals the outer layer of the endosperm, called the **aleurone,** to produce α-amylase. This enzyme is responsible for breaking down the endosperm's starch, primarily amylose, into sugars that are passed by the scutellum to the embryo.

When the Seedling Becomes Oriented in the Environment, Photosynthesis Begins

LEARNING OBJECTIVE 30.6.2 Contrast the pattern of shoot emergence in the bean (dicot) with that in maize (monocot).

As the sporophyte pushes through the seed coat, it orients with the environment so that the root grows down and the shoot grows up. New growth comes from delicate meristems that are protected from environmental stresses. The shoot becomes photosynthetic, and the postembryonic phase of growth and development begins. Figure 30.22 shows the process of germination and subsequent development of the plant body in eudicots and monocots.

The emerging shoot and root tips are protected by additional tissue layers in the monocots—the *coleoptile* surrounding the shoot, and the *coleorhiza* surrounding the radicle. Other protective strategies include having a bent shoot emerge so tissues with more robust cell walls push through the soil.

The emergence of the embryonic root and shoot from the seed during germination varies widely from species to species. In most plants, the root emerges before the shoot appears and anchors the young seedling in the soil. In plants such as peas, the cotyledons may be held below ground; in other plants, such as beans, radishes, and onions, the cotyledons are held above ground. The cotyledons may

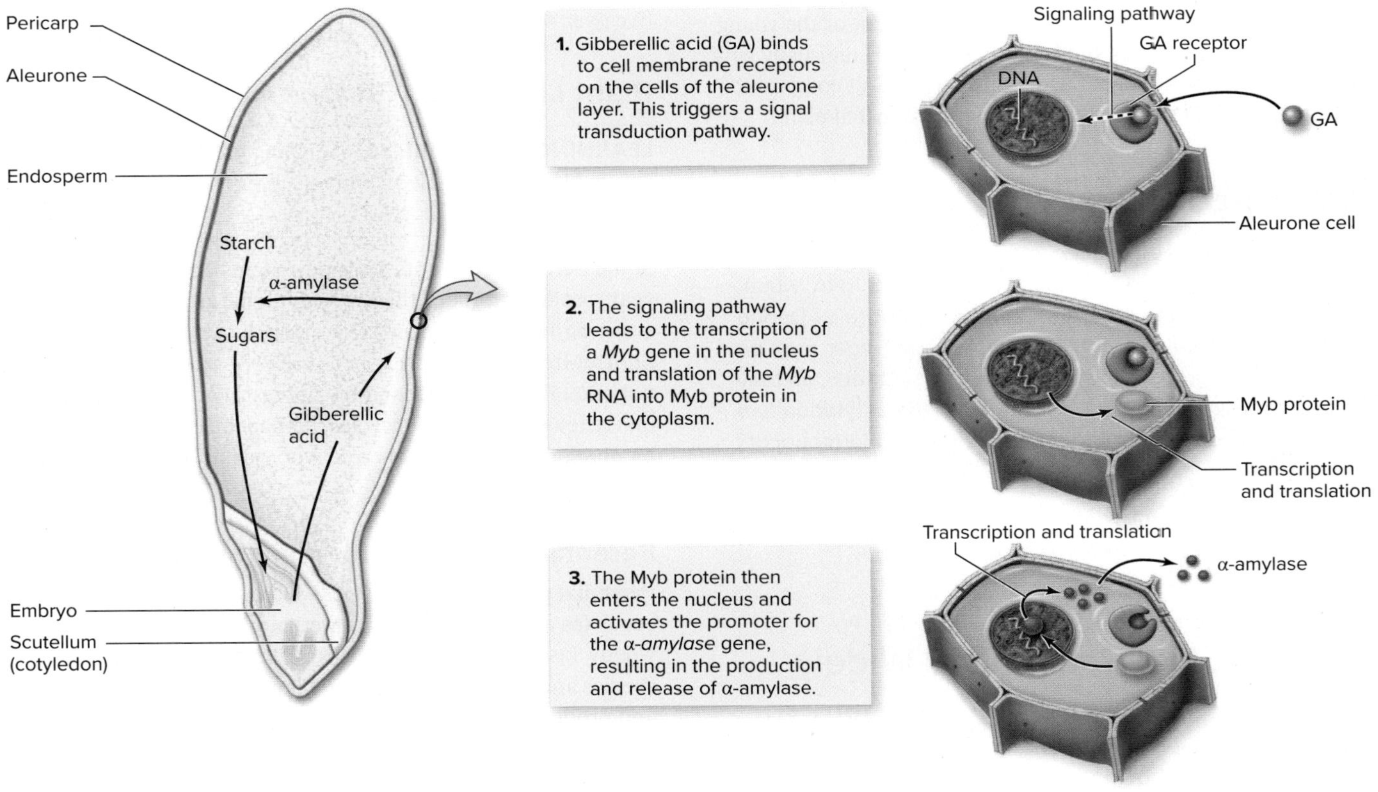

Figure 30.21 Hormonal regulation of seedling growth.

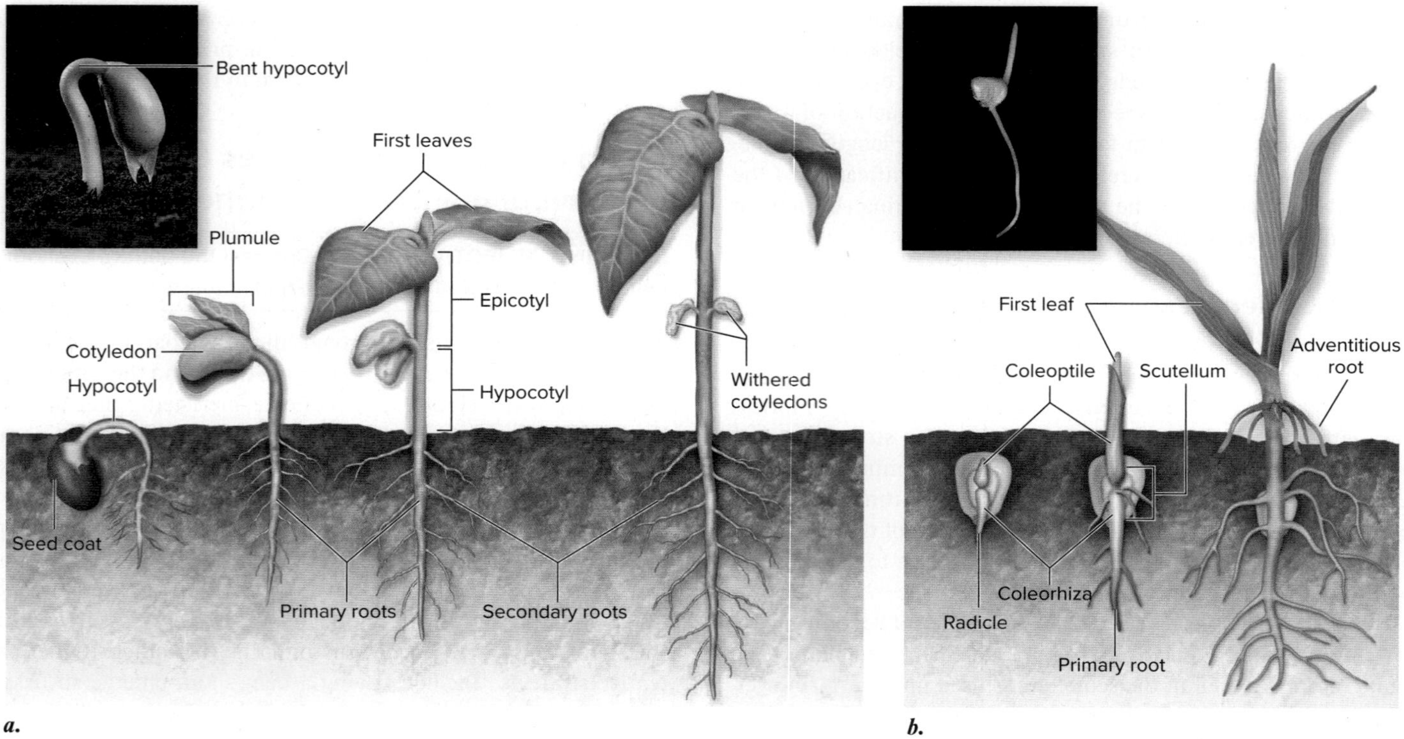

Figure 30.22 Germination. The stages shown are for **(*a*)** a eudicot, the common bean (*Phaseolus vulgaris*), and **(*b*)** a monocot, maize (*Zea mays*). Note that the bending of the hypocotyl (region below the cotyledons) protects the delicate bean shoot apex as it emerges through the soil. Maize radicles are protected by a layer of tissue called the coleorhiza, in addition to the root cap found in both the bean and maize. A sheath of cells called the coleoptile, rather than a hypocotyl tissue, protects the emerging maize shoot tip.
(a): Nigel Cattlin/Alamy Stock Photo; (b): Martin Shields/Alamy Stock Photo

become green and contribute to the nutrition of the seedling as it becomes established, or they may shrivel relatively quickly. The period from the germination of the seed to the establishment of the young plant is critical for the plant's survival; the seedling is unusually susceptible to disease and drought during this period. Soil composition and pH can also affect the survival of a newly germinated plant.

REVIEW OF CONCEPT 30.6

During germination, the seed and embryo take up water, increase respiration, and synthesize protein and RNA. Metabolic reserves in seeds include starch, fats, and oils. During seedling emergence, the cotyledons and seed coat may be pulled out of the ground and become photosynthetic, as they do in eudicots such as beans. Alternatively, the cotyledon and seed coat may remain in the ground, as they do in monocots such as maize.

- *What might be an advantage of retaining a seed in the ground during seedling emergence?*

30.7 Plant Life Spans Vary Widely

Once established, plants' life spans vary widely, depending on the species. Life span may or may not correlate with reproductive strategy. Woody plants, which have extensive secondary growth, nearly always live longer than herbaceous plants, which have limited or no secondary growth. Bristlecone pine, for example, can live over 4000 years.

Plant Life Spans Fit One of Three Patterns

LEARNING OBJECTIVE 30.7.1 Distinguish between herbaceous plants, that may be annuals, biennials, or perennials, and woody perennials.

Some herbaceous plants send new stems above the ground every year, producing them from woody underground structures. Others germinate and grow, flowering just once before they die. Shorter-lived plants rarely become very woody, because there is not enough time for secondary tissues to accumulate. Depending on the length of their life cycles, herbaceous plants may be annual, biennial, or perennial, whereas woody plants are generally perennial (figure 30.23).

Perennial plants live for many years

Perennial plants continue to grow year after year and may be herbaceous (as are many woodland, wetland, and prairie wildflowers) or woody (as are trees and shrubs). The majority of vascular plant species are perennials. Perennial plants in general are able to flower and produce seeds and fruit for numerous growing seasons.

Herbaceous perennials rarely experience any secondary growth in their stems; the stems die each year after a period of relatively rapid growth and food accumulation. Food is often

a.

b.

Figure 30.23 Annual and perennial plants. Plants live for very different lengths of time. ***a.*** Desert annuals complete their entire life span in a few weeks, flowering just once. ***b.*** Some trees, such as the giant redwood (*Sequoiadendron giganteum*), which occurs in scattered groves along the western slopes of the Sierra Nevada in California, live 2000 years or more.

(a): Anthony Arendt/Alamy Stock Photo; (b): SuperStock/Alamy Stock Photo

stored in the plants' roots or underground stems, which can become quite large in comparison with their less substantial aboveground parts.

Trees and shrubs are either *deciduous,* with all the leaves falling at one time of year and the plants remaining bare for a period, or *evergreen,* with some leaves dropping throughout the year and the plants never appearing completely bare. In northern temperate regions, conifers are the most familiar evergreens, but in tropical and subtropical regions, most angiosperms are evergreen, except where there is severe seasonal drought. In these areas, many angiosperms are deciduous, losing their leaves during the drought and thus conserving water.

Annual plants grow, reproduce, and die in a single year

Annual plants grow, flower, and form fruits and seeds within one growing season and die when the process is complete. Many crop plants, including corn, wheat, and soybeans, are annuals. Annuals generally grow rapidly under favorable conditions and in proportion to the availability of water or nutrients. The lateral meristems of some annuals, such as sunflowers or giant ragweed, do produce some secondary tissues for support, but most annuals are entirely herbaceous.

Annuals typically die after flowering once; the developing flowers or embryos use hormonal signaling to reallocate nutrients, so the parent plant literally starves to death. The process that leads to the death of a plant is called **senescence.**

Biennial plants follow a two-year life cycle

Biennial plants, which are much less common than annuals, have life cycles that take two years to complete. During the first year, biennials store the products of photosynthesis in underground storage organs. During the second year of growth, flowering stems are produced using energy stored in the underground parts of the plant. Certain crop plants, including carrots, cabbage, and beets, are biennials, but these plants generally are harvested for food during their first season, before they flower. They are grown for their leaves or roots, not for their fruits or seeds.

REVIEW OF CONCEPT 30.7

Woody perennials produce secondary growth, but herbaceous perennials typically do not. Perennial plants continue to grow year after year, whereas annual plants die after one growing season. During the first year of a biennial plant life cycle, food is produced and stored in underground storage organs. During the second year of growth, the stored energy is used to produce flowering stems.

- *What are the advantages and disadvantages of a biennial life cycle compared to an annual cycle?*

30.8 Asexual Reproduction Is Common Among Flowering Plants

Self-pollination reduces genetic variability, but asexual reproduction results in genetically identical individuals because only mitotic cell divisions occur. In the absence of meiosis, individuals that are highly adapted to a relatively unchanging environment persist for the same reasons that self-pollination is favored. If conditions change dramatically, there will be less variation in the population for natural selection to act on, and the species may be less likely to survive.

Asexual reproduction is also used in agriculture and horticulture to propagate a particularly desirable plant with traits that would be altered by sexual reproduction or even by self-pollination. Most roses and potatoes, for example, are vegetatively (asexually) propagated.

Apomixis Involves Development of Diploid Embryos

LEARNING OBJECTIVE 30.8.1 Define *apomixis.*

In certain plants, including some citruses, certain grasses (such as Kentucky bluegrass), and dandelions, the embryos in the seeds may be produced asexually from the parent plant. This kind of asexual reproduction is known as **apomixis.** Seeds produced in this way give rise to individuals that are genetically identical to their parents.

Although these plants reproduce by cloning diploid cells in the ovule, they also gain the advantage of seed dispersal, an adaptation usually associated with sexual reproduction. Asexual reproduction in plants is far more common in harsh or marginal environments, where there is little leeway for variation. For example, a greater proportion of asexual plants occur in the Arctic than in temperate regions.

In Vegetative Reproduction, New Plants Arise from Nonreproductive Tissues

LEARNING OBJECTIVE 30.8.2 List examples of plant parts involved in vegetative reproduction.

In a very common form of asexual reproduction called vegetative reproduction, new plant individuals are simply cloned from parts of adults (figure 30.24). The forms of vegetative reproduction in plants are many and varied.

Runners, or stolons. Some plants reproduce by means of *runners* (also called stolons)—long, slender stems that grow along the surface of the soil. In the cultivated strawberry, leaves, flowers, and roots are produced at every other node on the runner. Just beyond each second node, the tip of the runner turns up and becomes thickened. This produces first adventitious roots and then a new shoot.

Rhizomes. Underground horizontal stems, or *rhizomes,* are also important reproductive structures, particularly in grasses and sedges. Rhizomes invade areas near the parent plant, and each node can give rise to a new flowering shoot. The noxious character of many weeds results from this type of growth pattern, and many garden plants, such as irises, are propagated almost entirely from rhizomes. Corms and bulbs are vertical, underground stems. Tubers are also stems specialized for storage and reproduction. Tubers are the terminal storage portion of a rhizome.

Suckers. The roots of some plants—for example, cherry, apple, raspberry, and blackberry—produce *suckers,* or sprouts, which give rise to new plants. When the root of a dandelion is broken, as it may be if one attempts to pull it from the ground, each root fragment may give rise to a new plant.

Figure 30.24 Vegetative reproduction. Small plants arise from notches along the leaves of the houseplant *Kalanchoë daigremontiana.* The plantlets can fall off and grow into new plants, an unusual form of vegetative reproduction.
Jerome Wexler/Science Source

Adventitious plantlets. In a few plant species, even the leaves are reproductive. One example is the houseplant *Kalanchoë daigremontiana* (figure 30.24), familiar to many people as the "maternity plant." The common names of this plant are based on the fact that numerous plantlets arise from meristematic tissue located in notches along the leaves.

Plants Can Be Cloned from Isolated Cells in the Laboratory

LEARNING OBJECTIVE 30.8.3 Outline the steps involved in protoplast regeneration.

Whole plants can be cloned by regenerating plant cells or tissues on nutrient medium with growth hormones. This is another form of asexual reproduction. Cultured leaf, stem, and root tissues can undergo organogenesis in culture and form roots and shoots. In some cases, individual cells can also give rise to whole plants in culture.

Individual cells can be isolated from tissues with enzymes that break down cell walls, leaving behind the *protoplast,* a plant cell enclosed only by a plasma membrane. Plant cells have greater developmental plasticity than most vertebrate animal cells, and many, but not all, cell types in plants maintain the ability to generate organs or an entire organism in culture. Consider the limited number of adult stem cells in vertebrates and the challenges associated with cloning discussed in chapter 36.

When single plant cells are cultured, wall regeneration takes place. Cell division follows to form an undifferentiated mass of cells called a callus (figure 30.25). Once a callus is formed, whole plants can be produced in culture. Whole-plant development can go through an embryonic stage or can start with the formation of a shoot or root. Often, the plant cell being cultured is the product of a protoplast fusion event in which cells are combined to produce genetically modified plants.

Tissue culture has many agricultural and horticultural applications. Virus-free raspberries and sugarcane can be propagated by culturing meristems, which are generally free of viruses, even in an infected plant. As with other forms of asexual reproduction, genetically identical individuals can be propagated.

REVIEW OF CONCEPT 30.8

In apomixis, embryos are produced by mitosis rather than fertilization; in contrast, asexual vegetative reproduction occurs from vegetative plant parts. Examples include runners, stolons, rhizomes, suckers, and adventitious plant parts. In the laboratory, protoplasts are produced by isolating cells and removing the cell walls. Inducing mitosis results in a cluster of undifferentiated cells called a callus, which can then be stimulated to differentiate into a plant.

- *Under what conditions would vegetative reproduction benefit survival?*

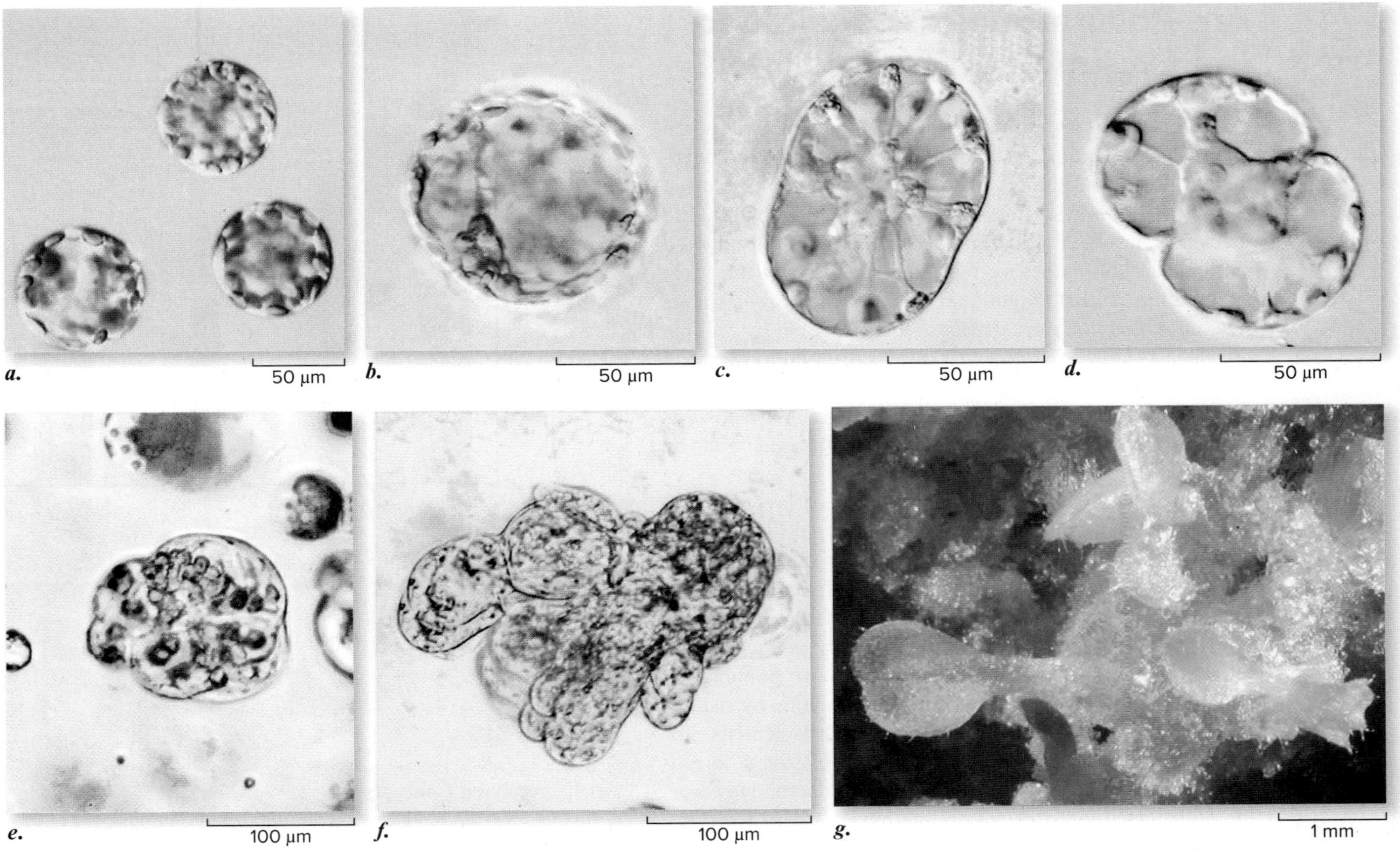

Figure 30.25 Protoplast regeneration. The relevant stages of regeneration of plantlets from protoplasts via cell colony and callus of tobacco. ***a.*** Freshly isolated leaf protoplasts of *Nicotiana tabacum* cv.petit havana. ***b.*** Cell wall formation (day 2). ***c.*** Preparing for first cell division (day 2 to 3). ***d.*** Completion of first cell division (day 2 to 3). ***e.*** Formation of microcolony (day 7). ***f.*** Colony of approx. 50 cells (day 9). ***g.*** Shoot regeneration (day 20). Rooting of regenerated shoots to give whole plants can be easily achieved.

(all): Hans-Ulrich Koop

Inquiry & Analysis

Are Pollinators Responsible for the Evolution of Flower Color?

Evolution results from many types of interactions between organisms, including predator–prey relationships, competition, and mate selection. An important type of coevolution among plants and animals involves flowering plants and their pollinators. Pollinators need flowers for food, and plants need pollinators for reproduction. It is logical, then, to hypothesize that evolutionary changes in flower shape, size, odor, and color are driven to a large extent by pollinators such as bees. We know that insects respond to variation in flower traits by visiting flowers with certain features, but few studies have been carried out to predict and then evaluate the response of plant populations to selection by pollinators.

In wild radish populations, honeybees preferentially visit yellow and white flowers, whereas syrphid flies prefer pink flowers. Rebecca Irwin and Sharon Strauss at the University of California, Davis, studied the response of wild radish flower color to selection by pollinators in natural populations. They compared the frequency of four flower colors (yellow, pink, white, and bronze) in two populations of wild radishes. Bees created the first population by visiting flowers based on their color preferences. The scientists produced the second population by hand-pollinating wild radish flowers with no discrimination by flower color. However, the pollen used for artificial pollinations contained varying proportions of each type of plant, based on the frequencies of each in the wild. In the graph shown here, you can see the distribution of flower colors in these two populations. The blue bars indicate the number of plants with yellow, white, pink, and bronze flowers in the population generated by the bee pollination. The red bars show the number of each type of flower in the population generated by researcher pollination, with no selection for flower color.

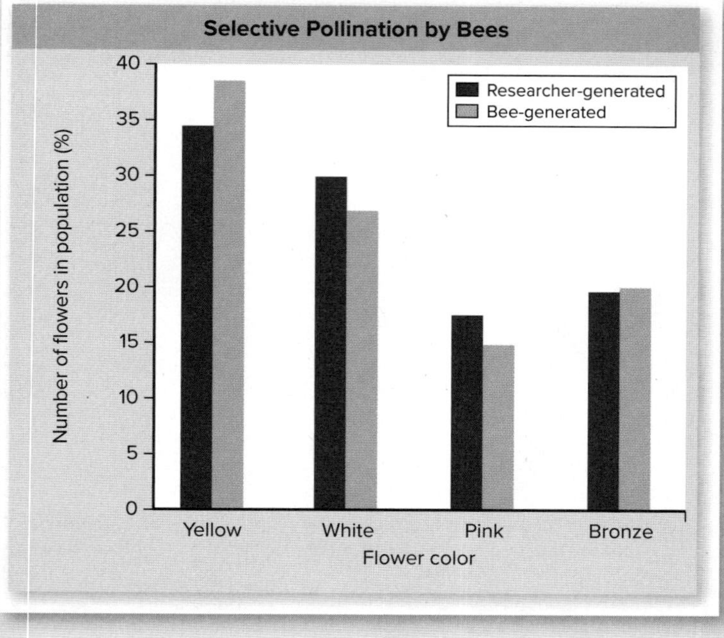

An uncommon visitor: A syrphid fly is visiting a yellow flower, usually preferred by bees.

David M. Phillips/Science Source

Analysis

1. **Applying Concepts**
   a. ***Variable.*** In the graph, what is the dependent variable?
   b. ***Reading a bar graph.*** Does this graph reflect data on flower colors from syrphid fly pollinations?
2. **Interpreting Data**
   a. Which flower color was most common?
   b. Which flower color(s) did the bees preferentially appear to visit?
3. **Making Inferences**
   a. Which type of insect do you suppose was most abundant in the region where this study was carried out, and why?
   b. Because there were pink flowers in these populations, can you say that syrphid flies *had* to be present in the area?
4. **Drawing Conclusions**
   a. Does it appear that insects are influencing the evolution of flower color in wild radish populations?
   b. Why did the researcher-pollinated population of flowers exhibit a pattern of flower colors similar to that of the bee-pollinated population? Were the researchers showing some experimental bias in their color selections?
5. **Further Analysis** Assume the wild radish population in this study is visited again in 10 years, and although a slight increase in yellow-flowered plants is observed, the proportion is not as high as would be predicted by this study. Provide some explanation for a slower-than-expected increase in the yellow-flowered plants.

Retracing the Learning Path

CONCEPT 30.1 Reproduction Starts with Flowering

30.1.1 Flowering Is Carefully Regulated Phase change, the shift from the inability of a plant to flower to flowering competence, is genetically regulated. In some cases, the environment (light and temperature) plays important roles in the regulation of flowering.

30.1.2 A Complete Flower Has Four Whorls of Parts The four floral organs are the calyx, corolla, androecium, and gynoecium. Incomplete flowers lack one or more whorls.

30.1.3 Gametes Are Produced in the Gametophytes of Flowers Meiosis in the anthers produces microspores, which will produce pollen grains, the male gametophytes. Meiosis in the ovules produces megaspores, which will produce embryo sacs, the female gametophytes.

CONCEPT 30.2 Flowers Attract Pollinators

30.2.1 Pollen May Reach a Flower in Many Ways Wind pollination is passive and does not carry pollen long distances.

30.2.2 Flowers and Animal Pollinators Have Coevolved Animal pollinators provide an efficient transfer of pollen over long distances. The flowers produce scents and visual cues for pollinators.

30.2.3 Some Flowering Plants Continue to Use Wind Pollination Wind-pollinated species have male and female flowers on separate individuals or separate parts of each individual.

30.2.4 Self-Pollination Is Favored in Stable Environments Plants adapted to a stable environment benefit from having uniform progeny that are likely to be more successful than those arising from cross-pollination.

30.2.5 Several Evolutionary Strategies Promote Outcrossing The two main strategies are separation of male and female structures in time and space and self-incompatibility.

30.2.6 Angiosperms Undergo Double Fertilization Double fertilization produces a diploid zygote and a triploid endosperm that provides nourishment to the zygote.

CONCEPT 30.3 Fertilization Leads to Embryogenesis

30.3.1 A Single Cell Divides to Produce a Three-Dimensional Body The first cell divisions of the zygote produce two cell populations: one will become the embryo and the other the suspensor, which will connect the embryo to the nutritive tissue of the seed. The root-shoot axis also forms early during development.

30.3.2 A Simple Body Plan Emerges During Embryogenesis Shoot and root apical meristems develop, and protoderm, ground meristem, and procambium differentiate; these will become the three types of tissue in an adult plant.

CONCEPT 30.4 Seeds Protect Angiosperm Embryos

30.4.1 Food Reserves Form During Embryogenesis While the embryo is being formed, a food supply is being established for the embryo. In angiosperms, this consists of the endosperm produced by double fertilization; in gymnosperms, the megagametophyte is the food source.

30.4.2 The Seed Coat Protects the Embryo Seeds help to ensure the survival of the next generation by maintaining dormancy during unfavorable conditions, protecting the embryo, providing food for the embryo, and providing a means for dispersal.

30.4.3 Specialized Seed Adaptations Improve the Odds That Offspring Will Survive Prior to germination, the seed must become permeable so that water and oxygen can reach the embryo.

CONCEPT 30.5 Fruits Promote Seed Dispersal

30.5.1 Fruits Are Adapted for Seed Dispersal A fruit is the mature ovary of an angiosperm. Plants produce many types of fruit, with the developmental fate of the pericarp determining the type of fruit formed. Fruits contain tissues from parent and offspring sporophyte and gametophyte. Dispersal mechanisms include ingestion and transport by animals and birds and dispersal by wind or water.

CONCEPT 30.6 Germination Begins Seedling Growth

30.6.1 External Signals and Conditions Trigger Germination A seed must absorb water to germinate. Abundant oxygen is necessary to support the high respiration rate of a germinating seed. Environmental signals are often needed for germination. Examples include light of a certain wavelength, an appropriate temperature, and stratification (a period of chilling).

30.6.2 When the Seedling Becomes Oriented in the Environment, Photosynthesis Begins In most plants, the root emerges before the shoot appears, anchoring the young seedling. A seedling enters the postembryonic phase of growth and development when the emerging shoot becomes photosynthetic.

CONCEPT 30.7 Plant Life Spans Vary Widely

30.7.1 Plant Life Spans Fit One of Three Patterns Perennial plants live for many years. Annual plants grow, reproduce, and die in a single year. Biennial plants follow a two-year life cycle: growth in year one, flowering and seed production in year two.

CONCEPT 30.8 Asexual Reproduction Is Common Among Flowering Plants

30.8.1 Apomixis Involves Development of Diploid Embryos Some plants can produce seeds in a process called apomixis. Seeds made in this way produce plants that are genetically identical to their parents.

30.8.2 In Vegetative Reproduction, New Plants Arise from Nonreproductive Tissues Vegetative parts such as runners, rhizomes, suckers, and adventitious plantlets may give rise to new individual clones.

30.8.3 Plants Can Be Cloned from Isolated Cells in the Laboratory Removing the cell wall creates a protoplast. This divides to produce a callus, which can differentiate into a complete plant.

Concept Overview

This Concept Overview diagrams the key concepts that were discussed in this chapter.

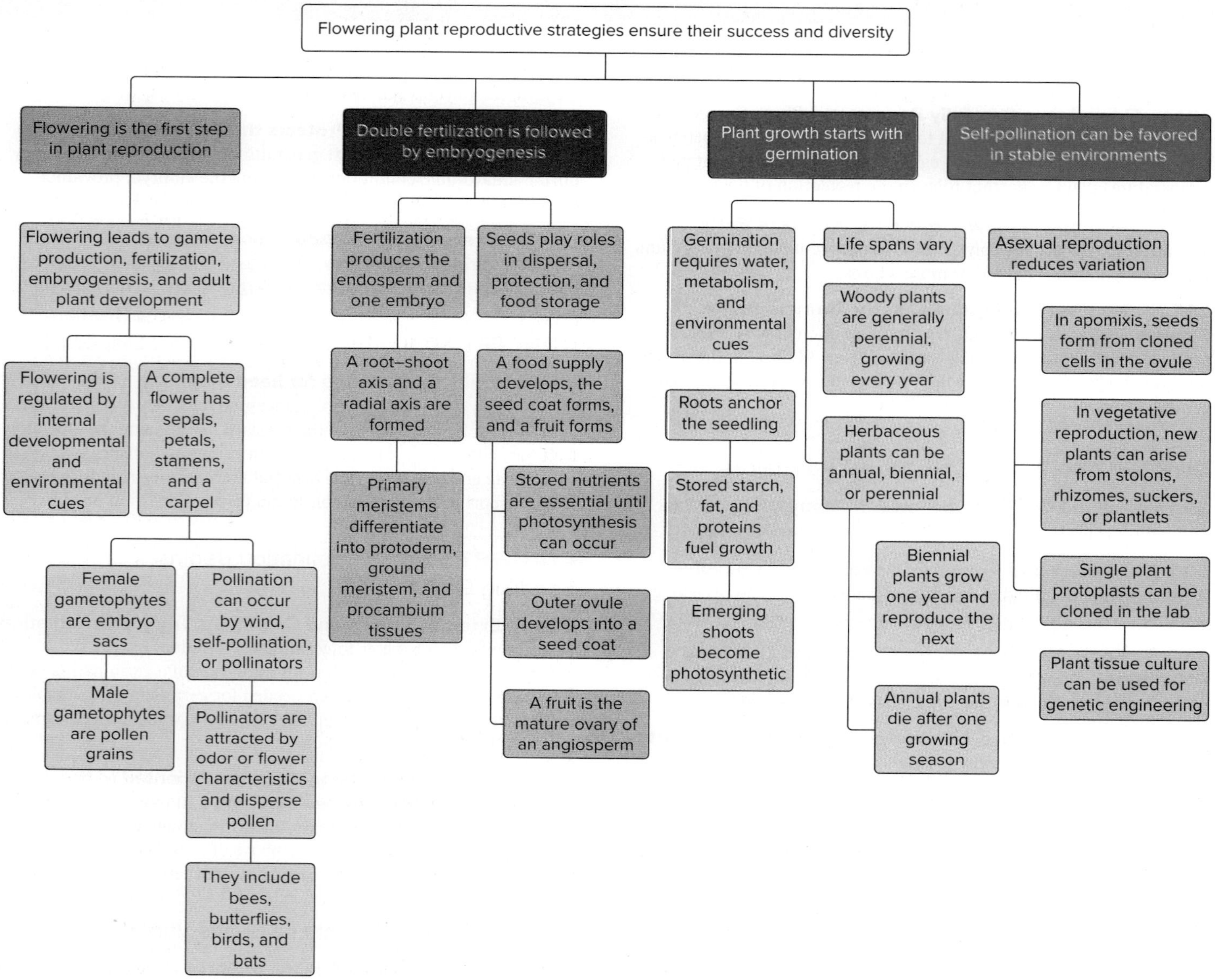

Assessing the Learning Path

Understand

1. Which of the following is NOT a component of a flower?
   a. Sepal
   b. Stamen
   c. Carpel
   d. Bract
2. What is the ploidy of pollen grains?
   a. Haploid, even though they are made by mitosis
   b. Diploid, because they are the products of mitosis
   c. Diploid, because they are the products of meiosis
   d. Haplodiploid, because of the alternation of generations seen in angiosperms
3. After the first mitotic division of the zygote, the larger of the two cells becomes the
   a. embryo.
   b. endosperm.
   c. suspensor.
   d. micropyle.
4. Endosperm is produced by the union of
   a. a central cell with a sperm cell.
   b. a sperm cell with a synergid cell.
   c. an egg cell with a sperm cell.
   d. a suspensor with an egg cell.

5. Why are seeds an important evolutionary improvement over spores?
   a. They contain little water.
   b. They can remain dormant until conditions are right for germination.
   c. Seeds such as beans, corn, and rice are important food resources.
   d. Seeds can be far larger than spores.
6. Fruits develop from which flower structure?
   a. Pericarp (ovary wall)
   b. Developing seed coat
   c. Mesocarp only
   d. Mature endosperm
7. The shoot tip of an emerging maize seedling is protected by the
   a. hypocotyl.
   b. epicotyl.
   c. coleoptile.
   d. plumule.
8. Perennial plants are
   a. always herbaceous.
   b. always woody.
   c. either herbaceous or woody.
   d. neither herbaceous nor woody.
9. Asexual reproduction is likely to be most common in which ecosystem?
   a. Tropical rainforest
   b. Temperate grassland
   c. Arctic tundra
   d. Deciduous forest
10. Underground horizontal stems are called
   a. runners.
   b. stolons.
   c. rhizomes.
   d. tubers.
11. When pollen from the anther of one flower pollinates the stigma of a different flower, the process is called
   a. stigmatization.
   b. outcrossing.
   c. self-pollination.
   d. wind pollination.
12. The integuments of an ovule will develop into the
   a. embryo.
   b. endosperm.
   c. fruit.
   d. seed coat.
13. Embryo growth in the seed is arrested because
   a. the seed coat prevents further tissue expansion.
   b. severe desiccation occurs.
   c. stored food is depleted.
   d. temperatures are too low.
14. During seed germination, which hormone produces the signal for the aleurone to begin starch breakdown?
   a. Abscisic acid
   b. Ethylene
   c. Gibberellic acid
   d. Auxin
15. The emerging shoot of a monocot is protected by a tissue layer not present in dicots called the
   a. coleorhiza.
   b. hypocotyl.
   c. coleoptile.
   d. root tip.
16. Senescence is
   a. plant death.
   b. reproductive growth.
   c. pollination.
   d. the accumulation of storage reserves.

Apply

1. How would a loss-of-function mutation in a gene required for mitosis during gametogenesis affect gamete production? If it helps, refer to figure 30.3 to work out the answer.
   a. Microspore mother cells and megaspores would not be produced.
   b. Microspores would not be produced.
   c. Megaspores would not be produced.
   d. Neither an embryo sac nor pollen grains would be produced.
2. How would a loss-of-function mutation in the *α-amylase* gene affect seed germination?
   a. The seed could not absorb water.
   b. The embryo would starve.
   c. The seed coat would not rupture.
   d. The seed would germinate prematurely.
3. A plant lacking the *WOODEN LEG* gene will likely
   a. be incapable of transporting water to its leaves.
   b. lack xylem and phloem.
   c. be incapable of transporting sugars made by photosynthesis.
   d. All of the above
4. At what point in development would you expect a mutation in a gene regulating the formation of the vertical plant axis (root–shoot axis) to have an effect?
   a. When the first mitotic division of the zygote occurs
   b. Around the time that the cotyledons begin to form
   c. After the first mitotic division of the zygote, but relatively early in embryogenesis
   d. When the radicle emerges from the seed
5. Which of the following structures is/are the gametophyte generation in angiosperms?
   a. Embryo sacs
   b. Pollen grains
   c. Megaspore
   d. a and b
   e. b and c
6. An exam question asks you to explain how seeds increase the reproductive success of angiosperms. You and your friend from class disagree, and he insists that seeds don't increase reproductive success because they get eaten by birds and other animals. What would be a good counterargument to his point?
   a. Seeds eaten by animals are protected from fire, because animals flee fires.
   b. Seeds absorb water in the GI tracts of animals that have eaten them, and that causes them to germinate when expelled from the animal.
   c. Animals carry the seeds in their GI tracts to new environments, where the plant may be successful and produce new offspring away from competition.
   d. Some animals can't eat all the seeds they collect, so they bury them. They bury them at just the right depth, and that promotes their germination.
7. A classmate is once again arguing with you about an exam question. Your classmate believes that fruits contain only one type of genotype, that of the next sporophyte generation in the form of the developing embryo. Why is she only partially correct?
   a. The fruit is from the previous sporophyte generation.
   b. Remnants of the embryo sac are from the gametophyte generation and found in the seed.
   c. The carpel is the male gametophyte generation.
   d. Both a and b
   e. Both b and c
8. In a highly stable environment where there is little variation in conditions, plants may use asexual reproduction instead of sexual reproduction. Why would this make sense in such an environment?
   a. Absolutely no genetic variation is required, because no adaptation is required.
   b. The large amounts of genetic variation produced by meiosis and sexual reproduction are not required when there is little variation to adapt to.
   c. It is less metabolically expensive to reproduce asexually in these environments.
   d. Asexual reproduction produces much more densely clustered populations, which favors success in this kind of environment.

9. Why is a recessive mutation that interferes with mitosis more likely to influence formation of the antipodal cells than megaspore formation?
   a. Embryo sacs, where you find antipodal cells, are diploid, and therefore a recessive mutation would be masked by a dominant allele.
   b. Antipodal cells are haploid, and a recessive mutation in the megaspore that gives rise to the embryo sac could not be masked by a dominant allele.
   c. Meiosis, not mitosis, produces the antipodal cells where mitosis and not meiosis produces the megaspore.
   d. Two of the above
10. In metazoans, mate choice is common. A peacock puts on an elaborate display to attract females, and females presumably pick and mate with the fittest males. In which of the following processes of plant reproduction could mate choice be said to occur by plants?
   a. The formation of flowers of different colors by different species of plants
   b. Embryo sac formation in the archegonium
   c. Self-incompatibility
   d. Seed germination

Synthesize

1. Why is vernalization an evolutionary advantage to a flowering plant in a temperate climate?
2. When two sperm are released from a pollen tube into the embryo sac, what do you think prevents both from fertilizing the egg?
3. Why do you suppose the cones of jack pines are tightly sealed, preventing seed release and therefore germination? How is the next jack pine generation achieved?
4. As you are eating an apple one day, you decide that you'd like to save the seeds and plant them. You do so, but they fail to germinate. Why might the seeds not have germinated? How could you improve your chance of germinating the seeds?
5. Some plants flower once and die, whereas others flower many times. What are the relative advantages of the two strategies?

Richard Rowan/Science Source

31 The Living Plant

Learning Path

31.1 Water Moves Through Plants Based on Potential Differences

31.2 Roots Absorb Minerals and Water

31.3 Xylem Transports Water from Root to Shoot

31.4 Transpiration Rate Reflects Environmental Conditions

31.5 Plants Are Adapted to Water Stress

31.6 Phloem Transports Organic Molecules

31.7 Plants Require a Variety of Nutrients

31.8 Plants Use Hormones to Regulate Growth

31.9 Plant Growth Is Responsive to Light

31.10 Plant Growth Is Sensitive to Gravity

Concept Overview

This Concept Overview provides a diagram of the overarching concepts that are covered in this chapter. These concepts will be expanded upon in the Concept Overview at the end of the chapter.

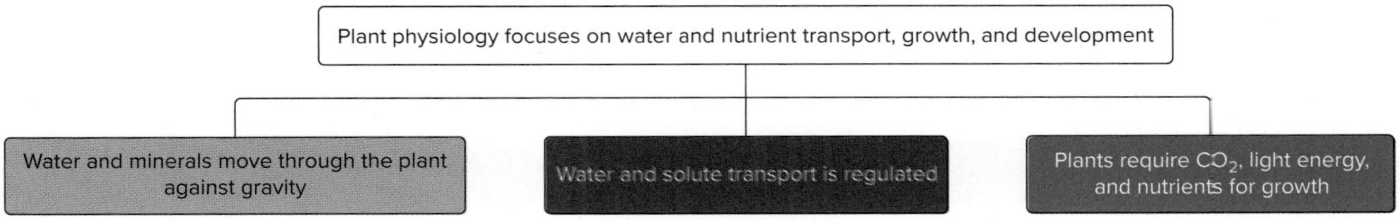
Plant physiology focuses on water and nutrient transport, growth, and development
Water and minerals move through the plant against gravity
Water and solute transport is regulated
Plants require CO_2, light energy, and nutrients for growth

Introduction

This chapter focuses on the daily life of a vascular plant—on those activities that allow it to grow and flourish. A plant faces three major challenges: maintaining water and nutrient balance, creating sufficient structural support for upright growth, and adapting growth to a changing environment. Vascular systems transport water, minerals, and organic molecules between roots and shoots, often over great distances—particularly in tall trees like the one pictured on the previous page. To remain healthy, a plant needs various inorganic nutrients, which it acquires from the soil. The lack of an important nutrient may slow the plant's growth, make the plant more susceptible to disease, or even kill it. Like all organisms, a plant senses and responds to its environment. The plant's survival and growth are critically influenced by abiotic factors it cannot avoid, including water, wind, gravity, and light. In this chapter, we explore how a plant senses these factors to elicit an appropriate physiological, growth, or developmental response. Hormones, keyed in many ways to the environment, play an important role in the internal signaling that results in a plant's responses to the environment.

31.1 Water Moves Through Plants Based on Potential Differences

Water is required for many physiological processes that occur in leaves. Leaves may be many meters above the roots where water and nutrients are extracted from the soil in which the plant grows. How does water move against gravity over great distances from root to shoot?

Water Potential Regulates Movement of Water Through the Plant Body

LEARNING OBJECTIVE 31.1.1 Use water potential to predict the movement of water.

In plants, water and any solutes it contains can move through porous cell walls, through plasmodesmata (connections between cells), across plasma membranes, and through the interconnected vasculature (figure 31.1*a*). Water first enters the roots and then moves to the xylem, the innermost vascular tissue of plants. Water moves against gravity through the xylem due to several factors, and much of that water exits through the stomata in the leaves. The products of photosynthesis can move both upward and downward in the plant body (figure 31.2).

Local changes result in long-distance movement of materials

Water molecules and dissolved minerals travel the greatest distances in the xylem. Once water enters the xylem in the roots of a redwood, for example, it can move upward as much as 100 m. Most of the force moving the water is "pulling" caused by transpiration—evaporation of water from inside of leaves via stomata. A pulling force is generated because water molecules stick to each other (cohesion) and to the walls of the tracheid or xylem vessel (adhesion) due to hydrogen bonding. The result is a stable column of water that can move water molecules great distances, including upwards against the force exerted on the water by gravity.

The movement of water at the cellular level also plays a significant role in bulk water transport in the plant, although over much shorter distances. Water can diffuse slowly across plasma membranes; however, charged ions and organic compounds depend on facilitated diffusion or active transport to enter or exit cells (refer to chapter 5). ATP-dependent proton pumps often fuel active transport by creating a proton gradient across a membrane. This gradient can be used in a variety of ways, including transporting sucrose, as shown in figure 31.1*b*. Unequal concentrations of solutes (such as ions and organic molecules) drive water movement across membranes via the process of osmosis. Using a quantitative approach to osmosis, it is possible to predict which way water will move across a selectively permeable membrane.

The forces that act on water to drive its movement within a plant are explained in terms of potentials. Potentials are a way of representing free energy (the potential to do work; refer to chapter 6). **Water potential** (Ψ_w, pronounced "psi-W") is used to predict the direction of water movement from one place to another, often across a selectively permeable cell membrane. Water will move from a cell or solution with higher water potential to a cell or solution with lower water potential. Water potential is measured in units of pressure called **megapascals (MPa)**. If you turn on your kitchen or bathroom faucet full blast, the water pressure should be between 0.2 and 0.3 MPa (30 and 45 psi).

Movement of water by osmosis

If a plant cell is placed into a **hypotonic** solution, water moves into the cell by the process of **osmosis** (refer to chapter 5). As water moves

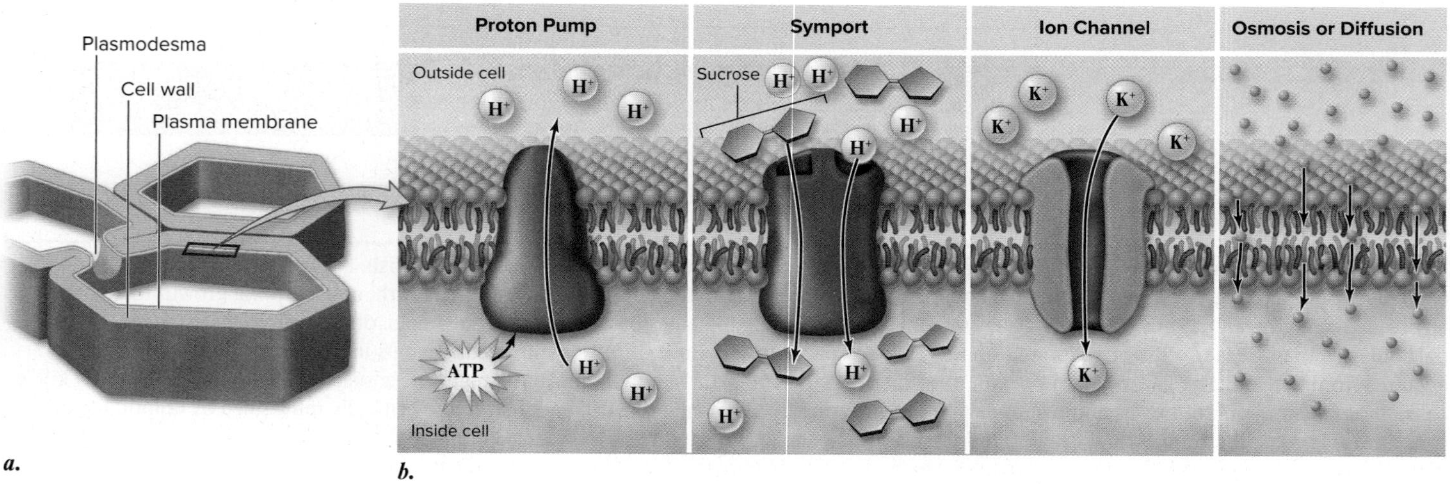

Figure 31.1 Transport between cells. ***a.*** Relationship between cell wall and membrane showing a plasmodesma through which solutes and water can diffuse to move from cell to cell. ***b.*** Solutes can also move between cells by active transport across membranes, by facilitated diffusion using transporters and channels, or by passive diffusion. Details of membrane transport are found in chapter 5.

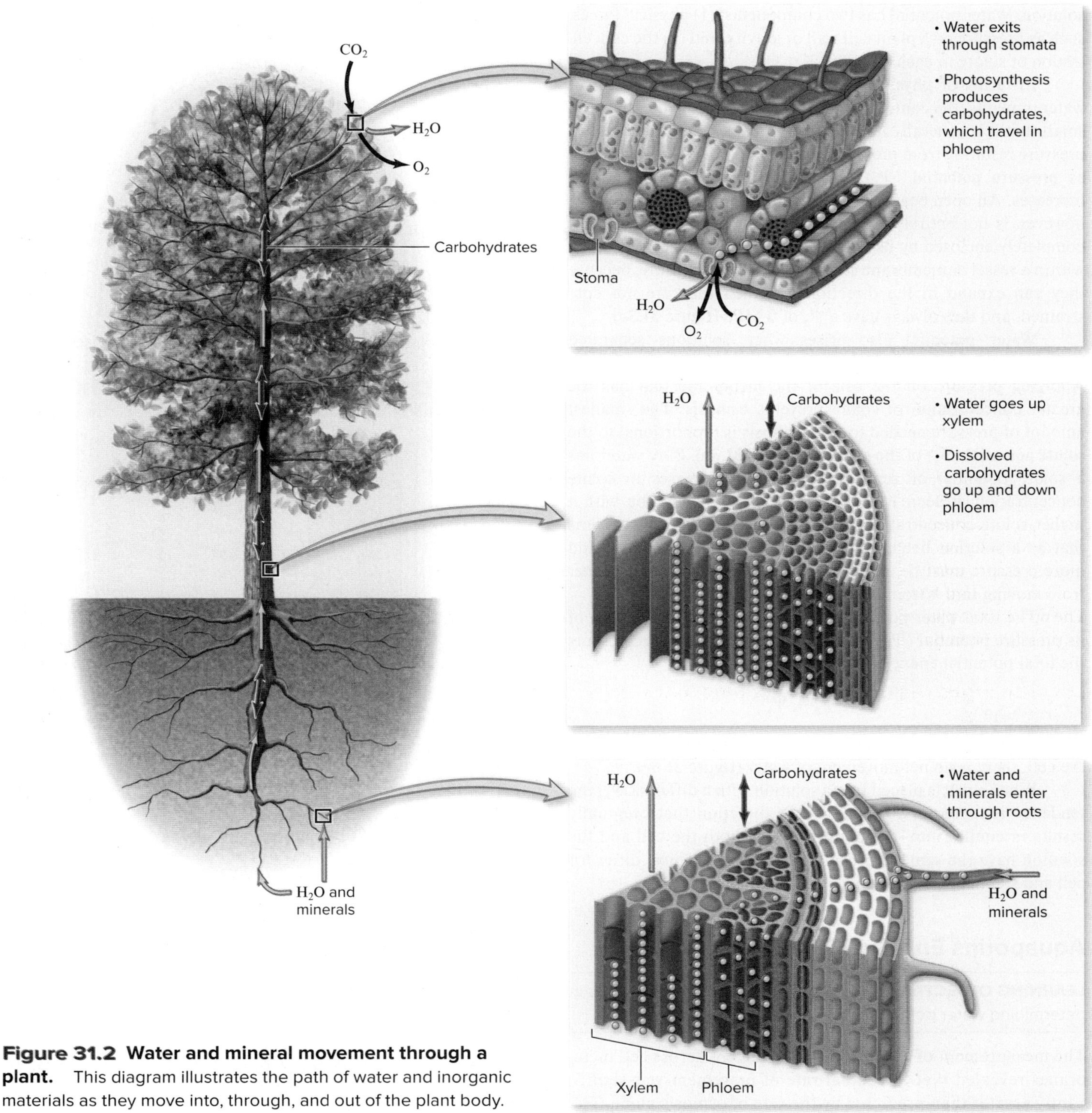

Figure 31.2 Water and mineral movement through a plant. This diagram illustrates the path of water and inorganic materials as they move into, through, and out of the plant body.

into the cell, it expands and its membrane presses against the cell wall, making it turgid because of the cell's increased turgor pressure. By contrast, if a cell is placed into a **hypertonic** solution, water leaves the cell and turgor pressure drops—the effects of this water movement can be seen when plants "wilt." The cell membrane pulls away from the cell wall as the volume of the cell shrinks. This process is called **plasmolysis,** and if the cell loses too much water, it will die.

Even a tiny change in cell volume causes a large change in turgor pressure.

Calculation of water potential

Changes in turgor pressure can be predicted more accurately by calculating the water potential of the cell and the surrounding solution. Water potential has two components: (1) physical forces, such as pressure on a plant cell wall or gravity, and (2) the concentration of solute in each solution.

In terms of physical forces that affect the movement of water in plants, the contribution of gravity to water potential is so small that it is generally not included in calculations. The turgor pressure resulting from pressure against the cell wall is referred to as **pressure potential** (Ψ_p). As turgor pressure increases, Ψ_p increases. An open beaker of water containing dissolved sucrose, however, is not entirely enclosed in the way that a plant cell is completely enclosed by its wall. Solutions that are not contained within a vessel or membrane cannot have turgor pressure, because they can expand in the direction in which they are not constrained, and they always have a Ψ_p of 0 MPa (figure 31.3*a*).

Water potential also arises when solutions separated by a membrane have different **tonicity** and osmosis occurs. Applying pressure on the side of the membrane that has the greater concentration of solute prevents osmosis. The smallest amount of pressure needed to stop osmosis is proportional to the **solute potential** (Ψ_s) of the solution (figure 31.3*b*). Pure water has a solute potential of zero. As a solution increases in solute concentration, it decreases in Ψ_s (< 0 MPa). A solution with a higher solute concentration has a more negative Ψ_s. This means that as a solution becomes increasingly hypertonic, more and more pressure must be applied to that solution to prevent water from moving into it from a hypotonic solution.

The total water potential (Ψ_w) of a plant cell is the sum of its pressure potential (Ψ_p) and solute potential (Ψ_s); it represents the total potential energy of the water in the cell:

$$\Psi_w = \Psi_p + \Psi_s$$

When the Ψ_w inside the cell equals that of a solution outside the cell, there is no net movement of water (figure 31.3*c*).

When a cell is placed into a solution with a different Ψ_w, the tendency is for water to move in the direction that eventually results in equilibrium—the point at which both the cell and the solution have the same Ψ_w. The Ψ_p and Ψ_s values may differ for cell and solution, but the sum ($= \Psi_w$) should be the same.

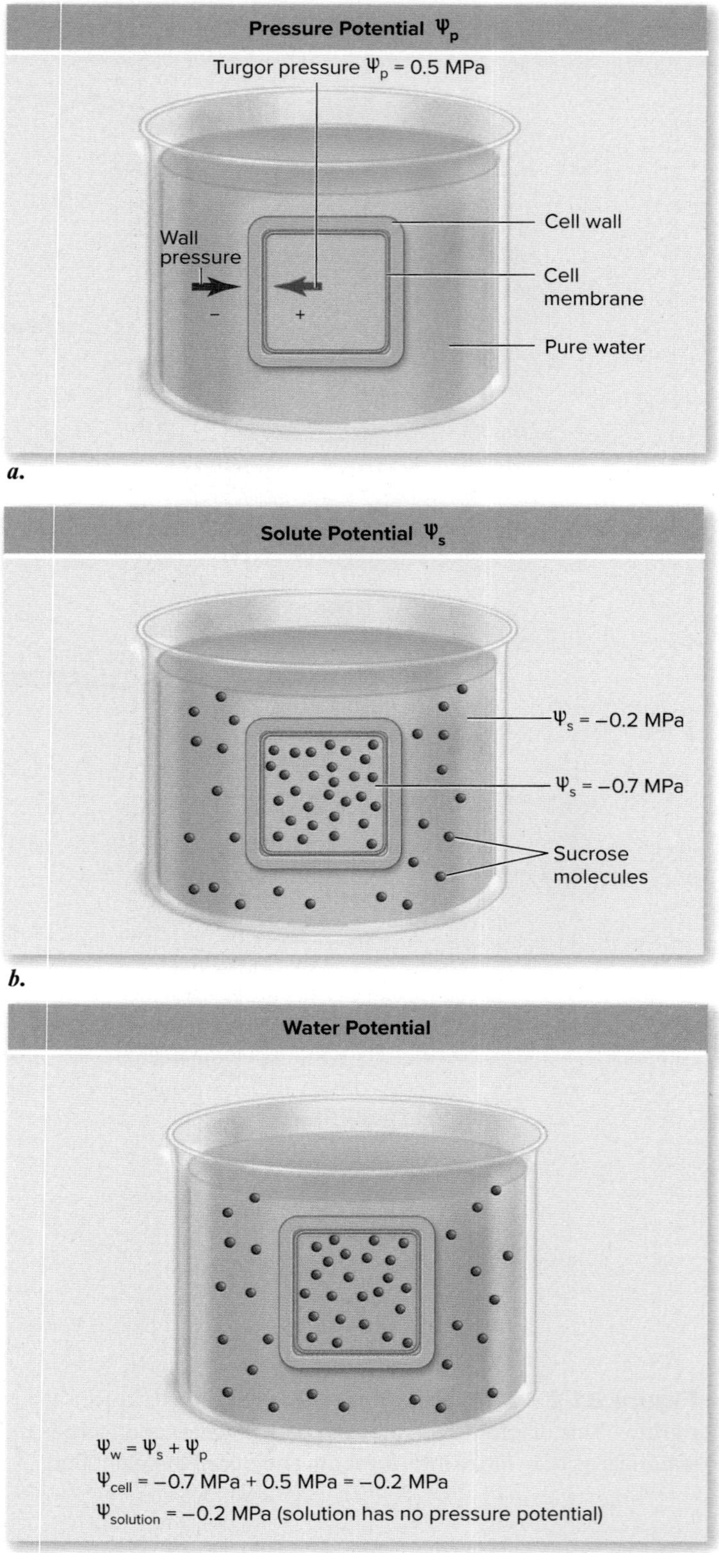

Figure 31.3 Determining water potential. ***a.*** Cell walls exert pressure in the opposite direction of cell turgor pressure. ***b.*** Using the given solute potentials, predict the direction of water movement based only on solute potential. ***c.*** Total water potential is the sum of Ψ_s and Ψ_p. Because the water potential inside the cell (0.2 MPa) equals that of the solution (0.2 MPa), there is no net movement of water. Note that the pressure potential of the solution surrounding the cell is 0 MPa because it is an open container.

Aquaporins Enhance Osmosis

LEARNING OBJECTIVE 31.1.2 Explain the role of aquaporins in determining water potential.

The measurement of rates of water movement across cell membranes revealed that the actual rate of movement was significantly greater than predicted by the rate of osmosis alone. The rate of water movement across membranes is enhanced by channels through which water can move called aquaporins (figure 31.4; refer to chapter 5). These transport channels occur in both plants and animals; in plants they exist in both vacuole and plasma membranes and allow for bulk flow of water across the membrane.

At least 30 different genes code for aquaporin-like proteins in *Arabidopsis.* Aquaporins speed up osmosis, but they do not change the direction of water movement. They are important in maintaining water balance within a cell and play a major role in moving water into the xylem.

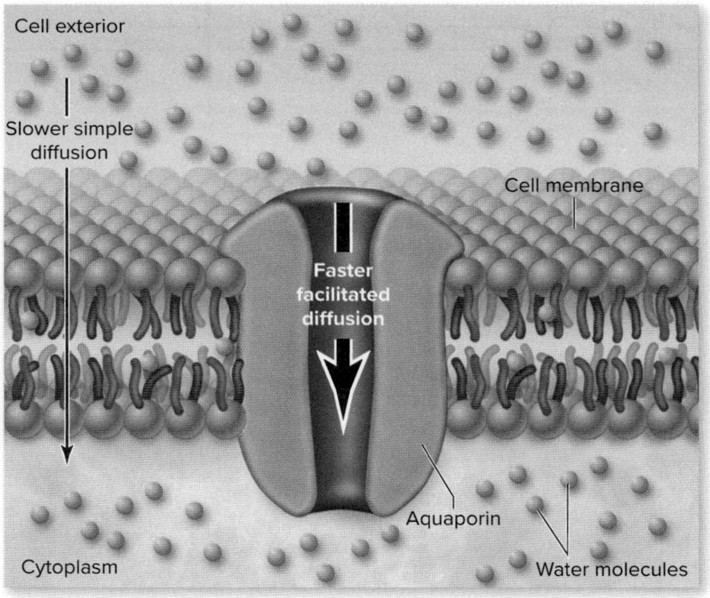

Figure 31.4 Aquaporins. Aquaporins are water-selective pores in the plasma membrane that increase the rate of osmosis by facilitating the diffusion of water across cellular membranes.

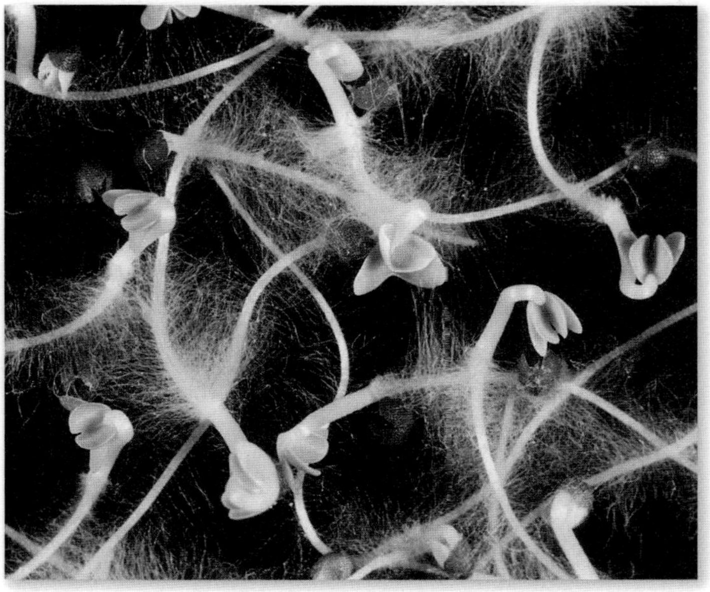

Figure 31.5 Water and minerals move into roots in regions rich with root hairs.
Dan Guravich/Science Source

Water potential and pressure gradients are foundational concepts for understanding local and long-distance transport in plants. We will explore this in detail throughout the chapter.

REVIEW OF CONCEPT 31.1

Transpiration is the evaporation of water from leaves via stomata, which creates a pulling force on water in xylem. Water potential is the sum of the pressure potential and the solute potential; water moves from an area of high water potential to an area of low water potential. Aquaporins increase the rate of osmosis by facilitating the diffusion of water across membranes.

- *Explain how physical pressure and solute concentration contribute to water potential.*

31.2 Roots Absorb Minerals and Water

Most water absorbed by a plant enters through the region of the root with root hairs (figure 31.5). Recall from chapter 29 that root hairs are extensions of root epidermal cells located just behind the tips of growing roots that increase the surface area for absorption of water and nutrients.

The surface area for the absorption of water and minerals is further increased in many species of plants due to interactions between roots and mycorrhizal fungi (refer to chapter 25). These fungi extend the absorptive net far beyond that of root hairs and are particularly helpful in the uptake of phosphorus in the soil.

There Are Three Transport Routes into Roots

LEARNING OBJECTIVE 31.2.1 Describe the routes water can take from soil to the plant vasculature.

Once absorbed through root hairs, water and minerals must move across cell layers until they reach the vascular tissues; water and dissolved ions then enter the xylem and move throughout the plant.

Water and minerals can follow three pathways to the vascular tissue of the root (figure 31.6). The **apoplast route** includes movement through the cell walls and the space between cells. Transport through the apoplast avoids membrane transport. In the **symplast route,** movement is through the continuum of cytoplasm between connected cells. Once molecules are inside a cell, they can move between cells through plasmodesmata without crossing a plasma membrane. The **transmembrane route** involves membrane transport between cells and across the membranes of vacuoles within cells. Of the three routes, the transmembrane route allows cells the ability to carefully control what solutes enter and exit the vasculature and, ultimately, be transported around the plant. These three routes are not mutually exclusive, and molecules can change pathways at any time, until reaching the endodermis of the root.

Transport Through the Endodermis Is Selective

LEARNING OBJECTIVE 31.2.2 Describe the function of Casparian strips.

Eventually, on their journey toward the root vasculature, molecules reach the endodermis. Any further passage through the

Figure 31.6 Transport routes between cells.

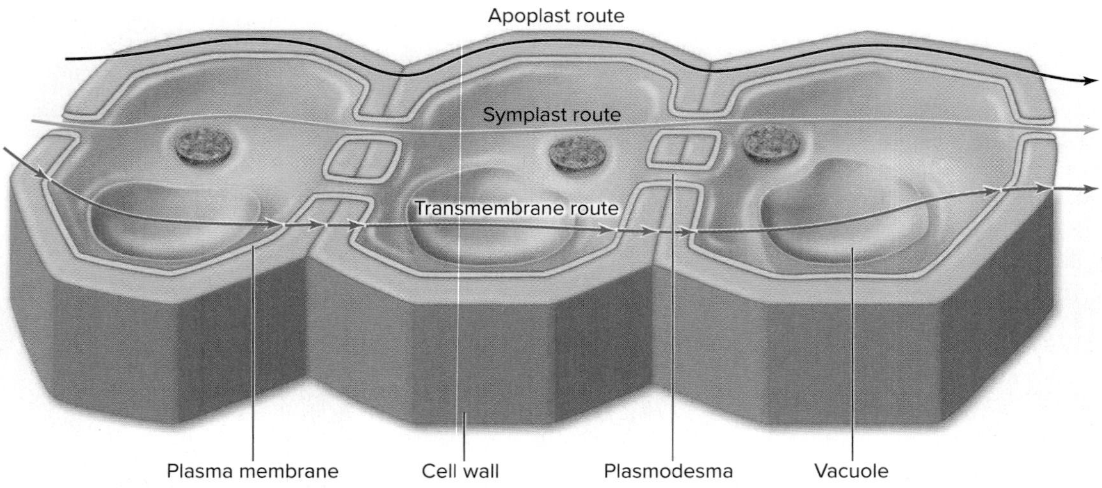

Apoplast route
Symplast route
Transmembrane route
Plasma membrane
Cell wall
Plasmodesma
Vacuole

cell walls is blocked by the Casparian strips. These are interconnected rings of waterproof **suberin** found in all walls of the ring of endodermal cells; they prevent water and dissolved nutrients from passing through the endodermis into the stele. Molecules must pass through the plasma membranes and protoplasts of the endodermal cells to reach the xylem (figure 31.7). The endodermis, with its unique structure, along with the cortex and epidermis, controls water and nutrient flow to the xylem. This helps to regulate water potential, limits the leakage of water out of the root, and restricts the movement of toxins into the

Figure 31.7 The pathways of mineral transport in roots. Minerals are absorbed at the surface of the root primarily by root hairs. In passing through the cortex, minerals must either follow the cell walls and the spaces between them or go directly through the plasma membranes and the protoplasts of the cells, passing from one cell to the next by way of the plasmodesmata. When they reach the endodermis, however, their further passage through the cell walls is blocked by the Casparian strips, and they must pass through the membrane and protoplast of an endodermal cell before they can reach the xylem.

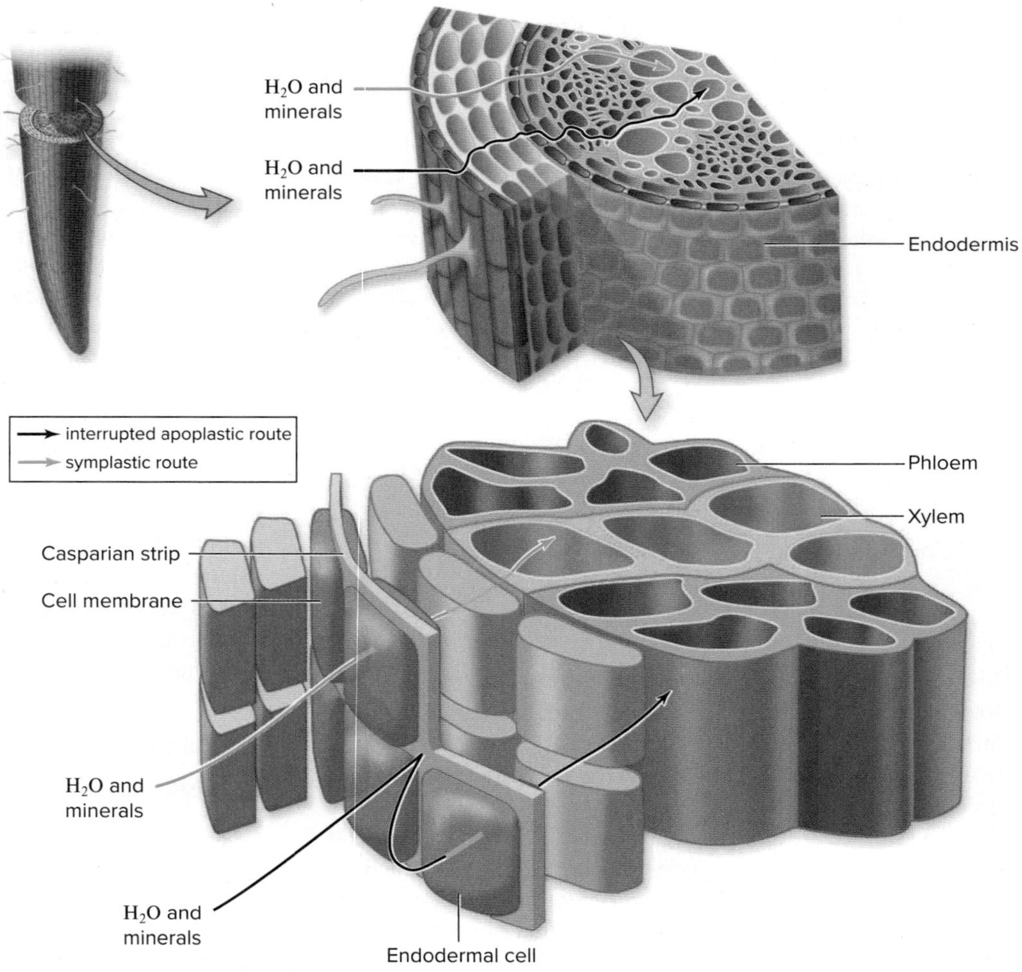

H_2O and minerals
H_2O and minerals
Endodermis
interrupted apoplastic route
symplastic route
Phloem
Xylem
Casparian strip
Cell membrane
H_2O and minerals
H_2O and minerals
Endodermal cell

vasculature where they would be transported to other parts of the plant.

Because the mineral ion concentration is usually much lower in soil water than it is inside the plant, primary (ATP-powered) active transport is required to accumulate ions inside the plant. The plasma membranes of endodermal cells contain a variety of protein transport channels, through which proton pumps transport specific ions against concentration gradients. Once inside the stele, the ions, which are plant nutrients, are transported via the xylem throughout the plant.

REVIEW OF CONCEPT 31.2

Water and minerals move into roots from the soil, particularly through root hairs. The three water transport routes are the apoplast route between cells, the symplast route through plasmodesmata, and the transmembrane route across cell membranes. Transport through the endodermis is selective due to the Casparian strip.

- *What properties of the cell membrane allow it to act as a selective barrier?*

31.3 Xylem Transports Water from Root to Shoot

A Water Potential Gradient from Roots to Shoots Enables Transport

LEARNING OBJECTIVE 31.3.1 Describe the conditions that produce root pressure.

The aqueous solution that passes through the membranes of endodermal cells enters the plant's vascular tissues and moves into the tracheids and vessel members of the xylem. As ions are transported into the root from the soil, the tonicity of root cells increases. The increased tonicity results in movement of water into the root cells and subsequently water potential and turgor pressure increase in those cells (figure 31.8).

Root pressure is present even when transpiration is low or not occurring

Root pressure is caused by the accumulation of ions in the roots at times when transpiration from the leaves is very low or absent. Because transpiration rates are low at night, root pressure often drops overnight. The accumulation of ions results in an increasingly high ion concentration within the cells, which in turn draws more water into the root by osmosis. Ion transport further decreases the Ψ_s of the roots. The result is movement of water into the plant and up the xylem columns despite the absence of transpiration.

Under certain circumstances, root pressure is so strong that water will ooze out of a cut plant stem for hours or even days. When root pressure is very high, it can force water up to the leaves, where it may be lost in a process known as **guttation.** Guttation cannot move water upward over great distances or at rapid speeds. The water moved to the leaves via guttation is not lost through the stomata but, instead, exits through special groups of cells located near the ends of small veins.

Transpiration drives xylem transport

Root pressure alone is insufficient to explain xylem transport. Transpiration provides the main force for moving water and dissolved solutes from roots to leaves.

Water moves from the soil into the plant only if water potential of the soil is greater than in the root. Too much fertilizer or drought conditions lower the Ψ_w of the soil and limit water flow into the plant. Water in a plant moves along a Ψ_w gradient from the soil (where the Ψ_w may be close to zero under wet conditions) to successively more negative water potentials in the roots, stems, leaves, and atmosphere (figure 31.8).

Evaporation of water in a leaf creates negative pressure or tension in the xylem, which literally pulls water up the stem from the roots. The strong pressure gradient between leaves and the atmosphere cannot be explained by evaporation alone. As water diffuses from the xylem of tiny, branching veins in a leaf, it forms a thin film along mesophyll cell walls. If the surface of the air-water interface is fairly smooth (flat), the water potential is higher than if the surface becomes rippled.

The driving force for transpiration is the humidity gradient from 100% relative humidity inside the leaf to much less than 100% relative humidity outside the leaf. Molecules diffusing from the xylem replace evaporating water molecules. As the rate of evaporation increases, diffusion cannot replace all the water molecules. The film is pulled back into the cell walls and becomes rippled rather than smooth. The change increases the pull on the column of water in the xylem, concurrently increasing the rate of transpiration.

Vessels and Tracheids Accommodate Bulk Flow

LEARNING OBJECTIVE 31.3.2 Explain the effect of cavitation on the flow of water in the xylem.

Water has an inherent **tensile strength** that arises from the cohesion of its molecules, which is a result of hydrogen bonds between water molecules (refer to chapter 2). These two factors are the basis of the cohesion–tension theory of the bulk flow of water in the xylem. The tensile strength of a column of water varies inversely with the diameter of the column; that is, the narrower the column, the greater the tensile strength of the column. Because plant tracheids and vessels are very narrow, the tensile strength of the water column can overcome the destabilizing effect of the pull of gravity. The water molecules also adhere to the sides of the tracheid or xylem vessels, further stabilizing the long column of water.

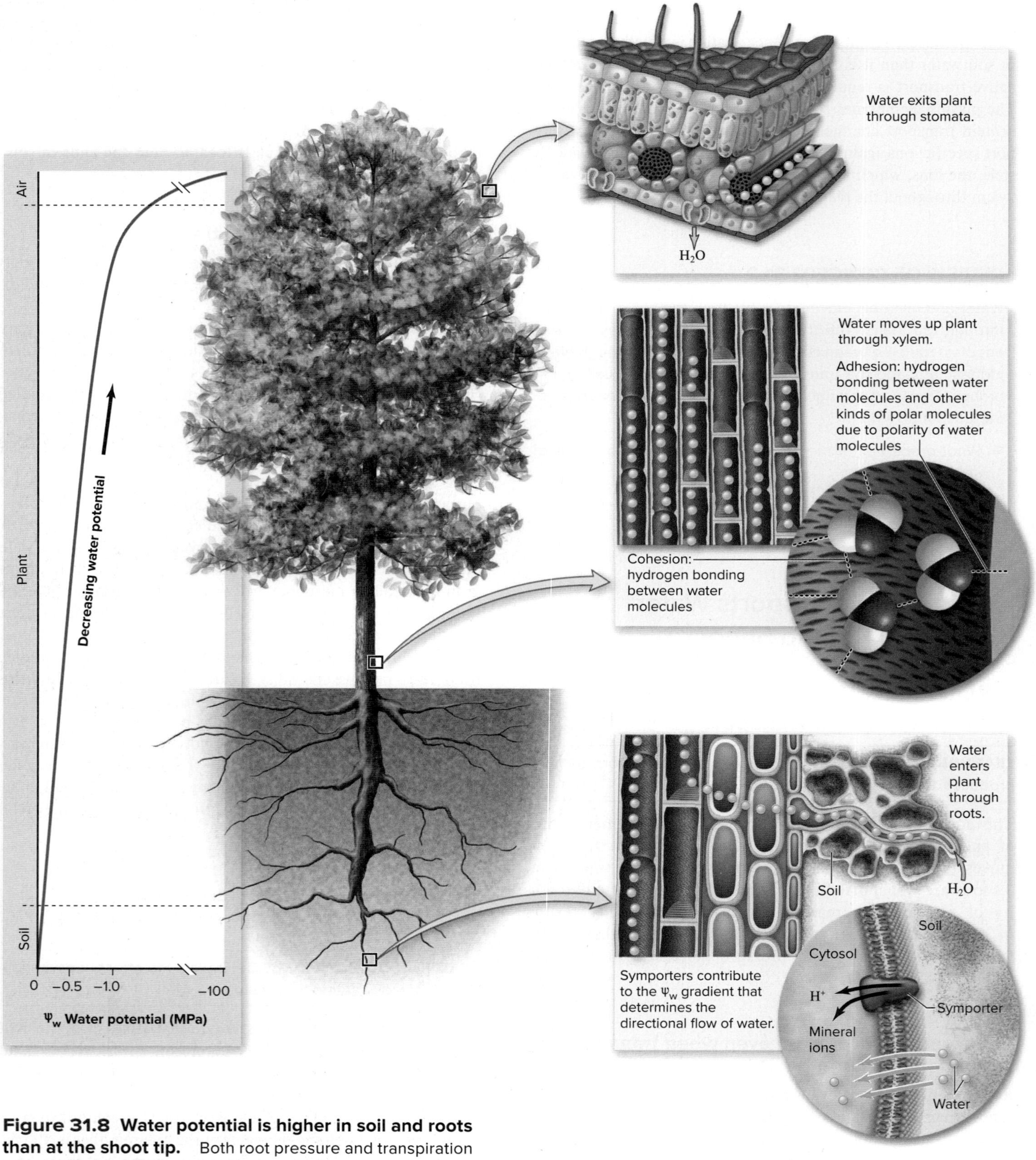

Air
Plant
Soil
Decreasing water potential
0
−0.5
−1.0
−100
Ψ_w Water potential (MPa)
Water exits plant through stomata.
H_2O
Water moves up plant through xylem.
Adhesion: hydrogen bonding between water molecules and other kinds of polar molecules due to polarity of water molecules
Cohesion: hydrogen bonding between water molecules
Water enters plant through roots.
Soil
H_2O
Soil
Cytosol
H^+
Symporter
Mineral ions
Water
Symporters contribute to the Ψ_w gradient that determines the directional flow of water.

Figure 31.8 Water potential is higher in soil and roots than at the shoot tip. Both root pressure and transpiration cause additional water to move upward in the xylem and to enter the plant through the roots. Water potential drops substantially in the leaves due to transpiration.

Given that a narrower column of water has greater tensile strength, it is intriguing that vessels, having diameters that are larger than those of tracheids, are found in so many plants. The difference in diameter has a larger effect on the mass of water in the column than on the tensile strength of the column. The volume of liquid moving in a column per second is proportional to r^4, where r is the radius of the column, at constant pressure. A 2-fold increase in radius would result in a 16-fold increase (2^4, or $2 \times 2 \times 2 \times 2$) in the volume of liquid moving through the column. Given equal cross-sectional areas of xylem, a plant with larger-diameter vessels can move more water up its stems than a plant with narrower tracheids.

The effect of cavitation

The tensile strength of a water column depends on its being unbroken. Air bubbles introduced into the column when a vessel is broken or cut would cause the continuity and the cohesion to fail. A gas-filled bubble can expand and block a tracheid or vessel, a process called **cavitation.** Cavitation stops water transport and can lead to dehydration and death of part or all of a plant (figure 31.9).

Anatomical adaptations can compensate for the problem of cavitation, including the presence of alternative pathways that can be used if one path is blocked. Individual tracheids and vessel members are connected to other tracheids or vessels by pits in their walls, and air bubbles are generally larger than these openings. In this way, bubbles cannot pass through the pits to further block transport. Freezing or deformation of cells can also cause small bubbles of air to form within xylem cells, especially with seasonal temperature changes. Cavitation is one reason older xylem often stops conducting water and is why it is a good idea to cut the bottom few centimeters of stems from fresh flowers before placing them in a vase. It is in this region that cavitation is most likely to occur.

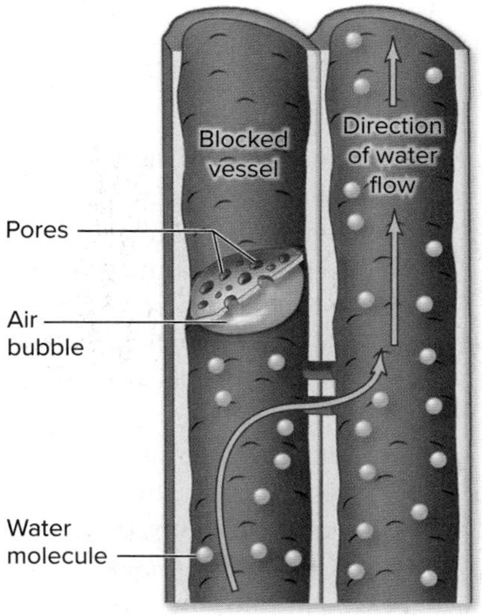

Figure 31.9 Cavitation. An air bubble can break the tensile strength of the water column. Bubbles are larger than pits and can block transport to the next tracheid or vessel. Water drains to surrounding tracheids or vessels.

Mineral transport

Tracheids and vessels are essential for the bulk transport of minerals. Ultimately, the minerals that are actively transported into the roots are relocated through the xylem to other metabolically active parts of the plant. Phosphorus, potassium, nitrogen, and sometimes iron may be abundant in the xylem during certain seasons. In many plants, this pattern of ionic concentration helps conserve these essential nutrients, which may move from mature deciduous parts such as leaves and twigs to areas of active growth, namely, meristem regions.

Keep in mind that minerals that are relocated via the xylem must move with the generally upward flow through the xylem. Not all minerals can reenter xylem once they leave. For example, calcium, an essential nutrient, cannot be transported elsewhere once it has been deposited in a particular part of the plant.

REVIEW OF CONCEPT 31.3

The general upward movement of water and minerals through a plant is due to the combined forces produced by root pressure and transpiration. Guttation occurs when root pressure is high but transpiration is low. Water's high tensile strength results from its cohesive and adhesive properties. This allows water to be pulled up the xylem by transpiration. Cavitation, which stops water movement, results from a bubble in the water transport system that breaks cohesion.

- *How might a plant overcome loss of water flow through a xylem vessel or tracheid due to cavitation?*

31.4 Transpiration Rate Reflects Environmental Conditions

More than 90% of the water taken in by the roots of a plant is ultimately lost to the atmosphere. Water moves from the tips of veins into mesophyll cells, and from the surface of these cells it evaporates into pockets of air in the leaf. As discussed in chapter 29, these intercellular spaces are in contact with the air outside the leaf via the stomata.

Stomata Open and Close to Balance H_2O and CO_2 Needs

LEARNING OBJECTIVE 31.4.1 Explain how guard cells regulate the opening of stomata.

Water is essential for plant metabolism but is continuously being lost to the atmosphere. At the same time, photosynthesis requires a supply of carbon dioxide from the atmosphere. Plants, therefore, face two somewhat conflicting requirements: the need to reduce water loss and the need to absorb carbon dioxide. The cuticle and stomata are structural adaptations that allow plants to overcome these environmental challenges.

The rate of transpiration depends on environmental conditions such as humidity, temperature, and wind speed. Transpiration from leaves decreases at night, when stomata are closed and the vapor pressure gradient between the leaf and the atmosphere is less. During the day, sunlight increases the temperature of the leaf, while transpiration cools the leaf through evaporative cooling.

On a short-term basis, closing the stomata can control water loss. This occurs in many plants when they are subjected to water stress. However, CO_2 cannot diffuse across the waxy cuticle into the leaf, so the stomata must be open at least part of the time so that CO_2 can enter. As CO_2 enters the intercellular spaces, it dissolves in water before entering the plant's cells, where it is used in photosynthesis. The gas dissolves mainly in water on the walls of the intercellular spaces below the stomata. The continuous stream of water that reaches the leaves from the roots keeps these walls moist.

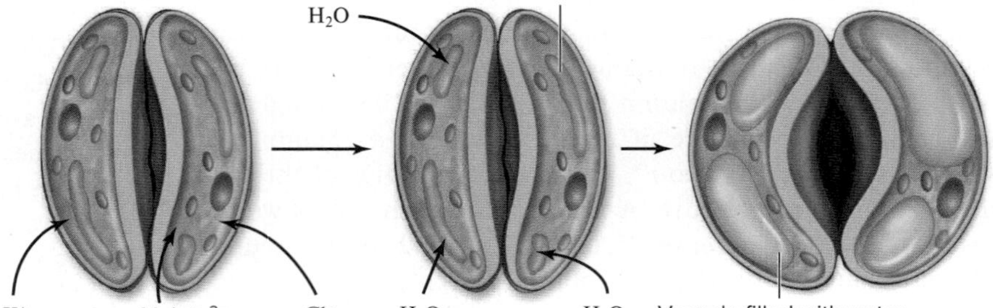

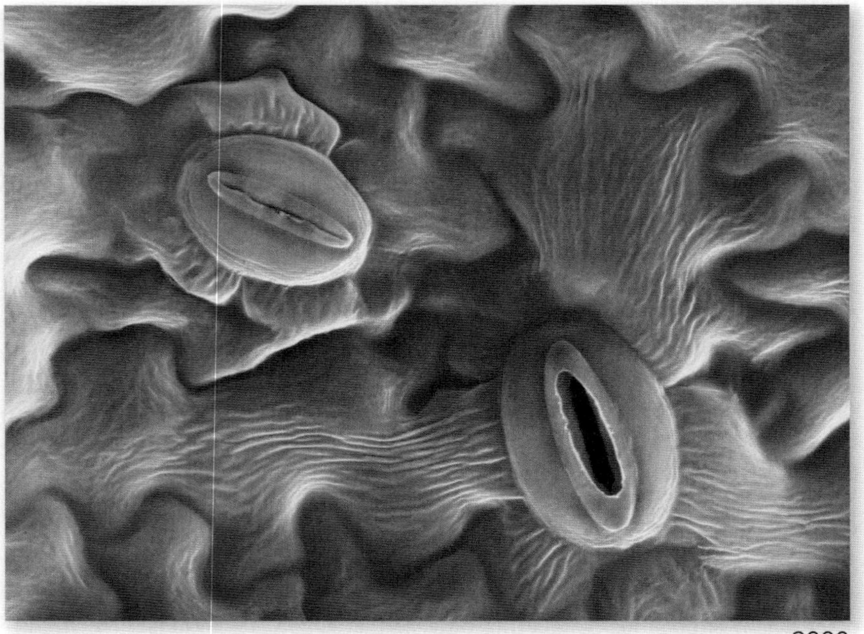

Figure 31.10 How a stoma opens. ***a.*** When H^+ ions are pumped from guard cells, K^+ and Cl^- ions move in, and the guard cell turgor pressure increases as water enters by osmosis. The increased turgor pressure causes the guard cells to bulge, with the thick walls on the inner side causing each guard cell to bow outward, thereby opening the stoma. ***b.*** Scanning electron micrograph of open and closed stomata.

(b): Power and Syred/Science Source

Turgor pressure in guard cells causes stomata to open and close

The two sausage-shaped guard cells on each side of a stoma stand out from other epidermal cells not only because of their shape but also because they are the only epidermal cells containing chloroplasts. Their distinctive wall construction, which is thicker on the inside and thinner elsewhere, results in them bending when they become turgid.

Turgor in guard cells results from the active uptake of potassium (K^+), chloride (Cl^-), and malate ions. As solute concentration increases, water potential decreases in the guard cells, and water enters via osmosis. As a result, these cells accumulate water and become turgid, opening the stomata (figure 31.10). Ions are moved into the guard cells by a light-activated ATP-driven proton pump.

The guard cells of many plant species regularly become turgid in the morning, when photosynthesis occurs, and lose turgor in the evening, regardless of the availability of water. During the course of a day, sucrose accumulates in the photosynthetic guard cells. The active pumping of sucrose out of guard cells in the evening may lead to loss of turgor and close the guard cell.

Environmental factors affect transpiration rates

Transpiration rates increase with temperature and wind speed, because water molecules evaporate more quickly. As humidity increases, the water potential difference between the leaf and the atmosphere decreases, but even at 95% relative humidity in the atmosphere, the vapor pressure gradient can sustain full transpiration. When a whole plant wilts because insufficient water is available, the guard cells may lose turgor, and as a result the stomata may close. Fluctuations in transpiration rate are tempered by opening or closing stomata.

Several pathways regulate stomatal opening and closing. **Abscisic acid (ABA),** a plant hormone discussed in section 31.8, plays an important role in allowing K^+ to pass rapidly out of guard cells, causing the stomata to close in response to drought. ABA binds to receptor sites in the plasma membranes of guard cells, triggering a signaling pathway that opens K^+, Cl^-, and malate ion channels. Turgor pressure decreases as water loss follows, and the guard cells close (figure 31.11).

CO_2 concentration, light, and temperature also affect stomatal opening. When CO_2 concentration is high, the guard cells

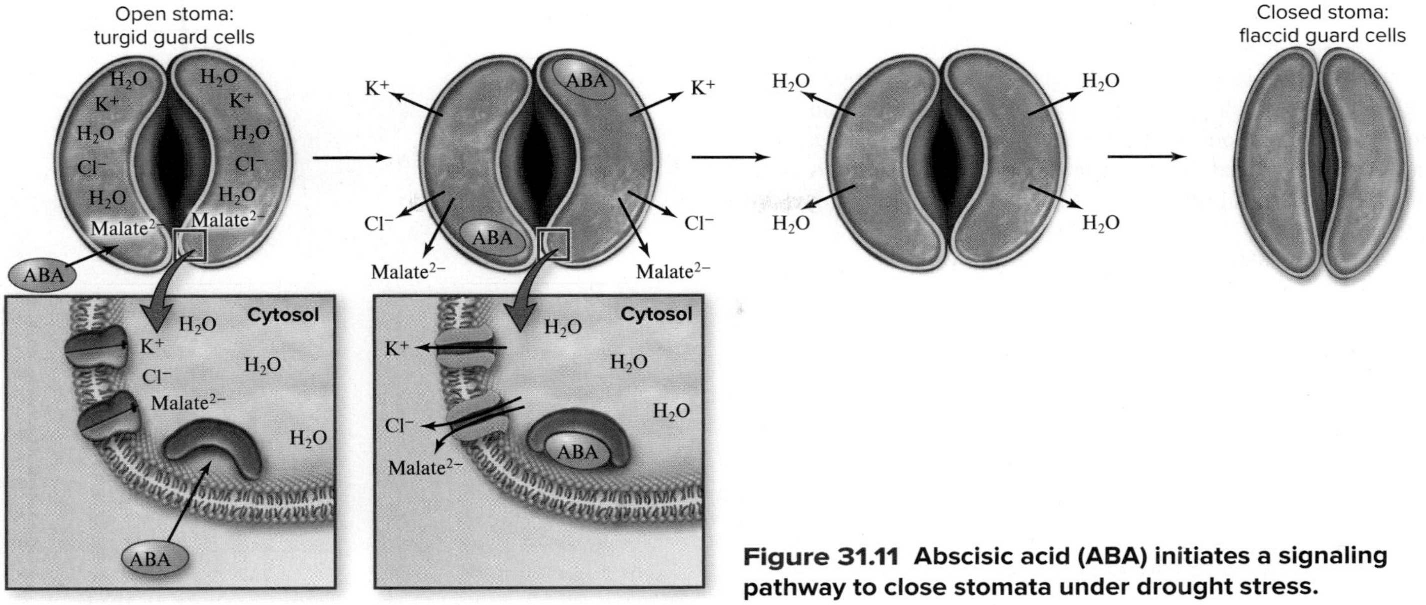

Figure 31.11 Abscisic acid (ABA) initiates a signaling pathway to close stomata under drought stress.

of many plant species are triggered to decrease the stomatal opening. Additional CO_2 is not needed at such times, and water is conserved when the guard cells are closed. As light intensity increases, leaf temperature rises. Opening the stomata allows the increased evaporation of water, which in turn cools the plant surface.

Stomata may also close when the temperature exceeds 30° to 34°C and water stress is likely to be high. To ensure sufficient gas exchange, the stomata open when it is dark and the temperature has dropped. Some plants are able to collect CO_2 at night in a modified form to be utilized in photosynthesis during daylight hours (refer to chapter 8).

REVIEW OF CONCEPT 31.4

Stomatal opening is controlled by guard cells, which change shape with changes in water content. Abscisic acid controls the movement of potassium ions, which affects osmosis. Stomata close when a plant is under water stress, but they open when carbon dioxide is needed and transpiration does not cause excess water loss.

- *Why is it critical that carbon dioxide dissolve in water upon entering plants?*

31.5 Plants Are Adapted to Water Stress

Because plants cannot simply move when water availability or salt concentrations change, adaptations to the environment allow plants to cope with diverse environmental conditions, including drought and high soil salinity.

Plant Adaptations to Drought Include Limiting Water Loss

LEARNING OBJECTIVE 31.5.1 Describe how plants are adapted to limit water loss.

Many mechanisms for controlling the rate of water loss have evolved in plants. Regulating the opening and closing of stomata provides an immediate response. Morphological adaptations provide longer-term solutions to drought periods. For example, some plants become dormant during dry times of the year; other plants limit transpiration by shedding their leaves. Such deciduous plants are common in areas that periodically experience dramatic changes in water availability. In a broad sense, annual plants conserve water when conditions are unfavorable simply by going into "dormancy" as seeds.

Thick, hard leaves, many of which have relatively few stomata—often with stomata only on the lower side of the leaf—lose water far more slowly than large, pliable leaves with abundant stomata. Leaves covered with masses of woolly-looking trichomes (hairs) reflect more sunlight and thereby reduce the heat load on the leaf and the demand for transpiration for evaporative cooling.

Plant Responses to Flooding Include Short- and Long-Term Adaptations

LEARNING OBJECTIVE 31.5.2 Describe how flooding affects plant growth.

Plants can also receive too much water: flooding rapidly depletes available oxygen in the soil and interferes with the transport of minerals and carbohydrates in the roots. Oxygen deprivation is among the most significant problems, because it leads to decreased cellular respiration. Standing water has much less oxygen than moving water, and standing-water flooding is generally more harmful to a plant. Flooding that occurs when a plant is dormant is much less harmful than flooding when it is growing actively.

Physical changes that occur in the roots as a result of oxygen deprivation may halt the flow of water through the plant. Paradoxically, even though the roots of a plant may be standing in water, its leaves may be drying out. Plants can respond to flooded conditions by forming larger lenticels (which facilitate gas exchange) and adventitious roots that reach above flood level for gas exchange.

Section of pneumatophore
Lenticel
O_2
O_2 transported to submerged portions of plants

Figure 31.12 How mangroves get oxygen to their submerged parts. The black mangrove (*Avicennia germinans*) grows in areas that are commonly flooded, and much of each plant is usually submerged. However, modified roots called pneumatophores supply the submerged portions of the plant with oxygen, because these roots emerge above the water and have large lenticels.

Mark Boulton/Science Source

Plant Adaptations to High Salt Concentration Include Elimination

LEARNING OBJECTIVE 31.5.3 Describe plant adaptations to a salty environment.

Plants such as mangroves that grow in areas normally flooded with salt water must not only provide a supply of oxygen to their submerged parts but also control their salt balance. The salt must be excluded, actively secreted, or diluted as it enters. The black mangrove (*Avicennia germinans*) has long, spongy, air-filled roots that emerge above the mud. These roots, called pneumatophores, have large lenticels on their above-water portions through which oxygen enters; the oxygen is then transported to the submerged roots (figure 31.12). In addition, the succulent leaves of some mangrove species contain large quantities of water, which dilute the salt that reaches them. Many plants that grow in such conditions also either secrete large quantities of salt or block salt uptake at the root level.

Some plants are **halophytes** ("salt lovers") and can tolerate soils with high salt concentrations. They do this by producing high concentrations of organic molecules within their roots to alter the water potential gradient between the soil and the root so that water enters the root.

REVIEW OF CONCEPT 31.5

Adaptations to drought include dormancy, leaf loss, and leaves that minimize water loss. When plants are exposed to flooding, oxygen deprivation leads to lower cellular respiration rates, impedance of mineral and carbohydrate transport, and changes in hormone levels. If a plant is exposed to a salty environment, it may exclude the salt from uptake, secrete it after it has been taken up, or dilute it.

- *Why are flooded plants in danger of oxygen deprivation when photosynthesis produces oxygen?*

31.6 Phloem Transports Organic Molecules

Because leaves are the main site of photosynthesis, plants need to be able to transport the products, sugars, from leaves to roots and shoots in the entire plant body. Unlike the unidirectional flow in the xylem, the phloem conducts sugar water throughout the plant body.

Leaves are the main sites of photosynthesis; however, plants need the products of photosynthesis at many other locations throughout the plant body. This means that, unlike water in xylem, which is always moved upward, sugars can move through phloem bidirectionally.

Organic Molecules Are Transported Up and Down the Shoot

LEARNING OBJECTIVE 31.6.1 Describe the process of translocation.

Most sugars made in leaves and other photosynthetic tissues travel through the phloem to the rest of the plant. This process, known as **translocation,** provides carbohydrates for the roots and other actively growing parts of the plant. Carbohydrates (like starch), which are concentrated in storage organs (like tubers), can be converted into soluble molecules like sucrose and transported through the phloem. In this section, we discuss how carbohydrates and other nutrient-rich fluids, collectively called **sap,** are moved through the plant body.

The movement of sugars and other substances can be followed in phloem using radioactive labels. Radioactive carbon from carbon dioxide ($^{14}CO_2$) can be incorporated into glucose by photosynthesis. Glucose molecules are used to make the disaccharide sucrose, which is transported in the phloem. Experiments that track the movement of the radioactive sucrose through a plant have shown that sucrose moves bidirectionally in the phloem.

Aphids, a group of insects that extract plant sap for food, have been valuable tools in understanding translocation. Aphids insert their stylets (piercing mouthparts) into phloem cells of leaves and stems to obtain the abundant sugars there. When a feeding aphid is removed by cutting its stylet, the liquid from the phloem continues to flow through the detached mouthpart and is thus available in pure form for analysis (figure 31.13). If the water is evaporated out of phloem to leave just the solid matter, the sugar is about 10 to 25% of the weight of the solids. Using aphids to obtain the critical samples and radioactive tracers to mark them, plant biologists have demonstrated that substances in phloem can move as fast as 50 to 100 cm/h.

Phloem also transports plant hormones. Changes in the environment can result in the rapid translocation of hormones in the plant. Messenger RNA can also move through the phloem, providing a mechanism for a kind of long-distance communication among cells. In addition, phloem carries other molecules, such as a variety of sugars, amino acids, organic acids, proteins, and ions.

Turgor Pressure Differences Drive Bulk Flow in the Phloem

LEARNING OBJECTIVE 31.6.2 Explain how sugars move through the plant body using the pressure–flow model of phloem transport.

The most widely accepted model explaining how sugars move through the phloem is called the **pressure–flow model.** Dissolved carbohydrates are loaded into the phloem at a ***source,*** and flow from the source to the ***sink,*** where they are unloaded. Carbohydrate sources include photosynthetic tissues, such as the mesophyll of leaves. Food-storage tissues, such as the cortex of roots, can be either sources or sinks. Sinks are any highly metabolically active tissues, such as those found at the growing tips of roots and stems, and in developing fruits. Also, because sources and sinks can change through time as needs change, the direction of phloem flow can change.

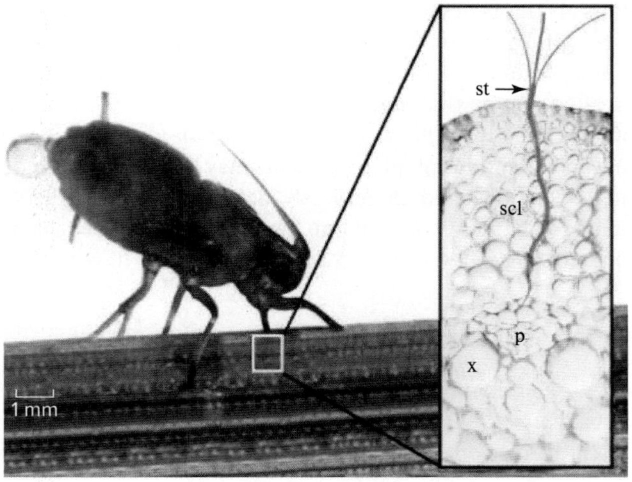

Figure 31.13 Feeding on phloem. Aphids feed on the food-rich contents of the phloem, which they extract through their piercing mouthparts, called stylets.

Reproduced with permission from Plants In Action, http://plantsinaction.science.uq.edu.au, published by the Australian Society of Plant Scientists

In a process known as **phloem loading,** sugars (mostly sucrose) enter the phloem in the smallest veins at the source. Some sucrose travels from mesophyll cells to the companion and sieve cells via the symplast. Much of the sucrose arrives at the sieve cell through apoplastic transport and is moved across the membrane using secondary transport, employing a sucrose/H^+ symporter (refer to chapter 5). Companion cells and parenchyma cells next to the sieve tubes provide the ATP energy to drive this transport. Unlike vessels and tracheids, sieve cells must be alive to participate in active transport.

Bulk flow occurs in the sieve tubes without additional energy requirements. Because of the difference between the water potential in the sieve tubes and that in the nearby xylem cells, water flows into the sieve tubes by osmosis. Turgor pressure in the sieve tubes thus increases, and this pressure drives the fluid throughout the plant's system of sieve tubes. At the sink, sucrose and hormones are actively removed from the sieve tubes, and water follows by osmosis. The turgor pressure at the sink drops, causing a mass flow from the stronger pressure at the source to the weaker pressure at the sink (figure 31.14). Most of the water at the sink then diffuses back into the xylem, where it may be either recirculated or lost through transpiration.

REVIEW OF CONCEPT 31.6

Translocation is the movement of dissolved material throughout a plant through the phloem. Sap in the phloem contains sucrose and other sugars, hormones, mRNA, amino acids, organic acids, proteins, and ions. According to the pressure–flow model, carbohydrates loaded into sieve tubes create a difference in water potential. As a result, water enters the tubes, creating pressure to move fluid through the phloem.

- *What is the key difference between the fluid in xylem and the fluid in phloem?*

31.7 Plants Require a Variety of Nutrients

The major source of plant nutrition is the fixation of atmospheric carbon dioxide (CO_2) into simple sugars using light energy. CO_2 enters through the stomata while oxygen (O_2), a waste product of photosynthesis, exits the plant through the stomata. Oxygen is used in cellular respiration to support growth and maintenance in the plant.

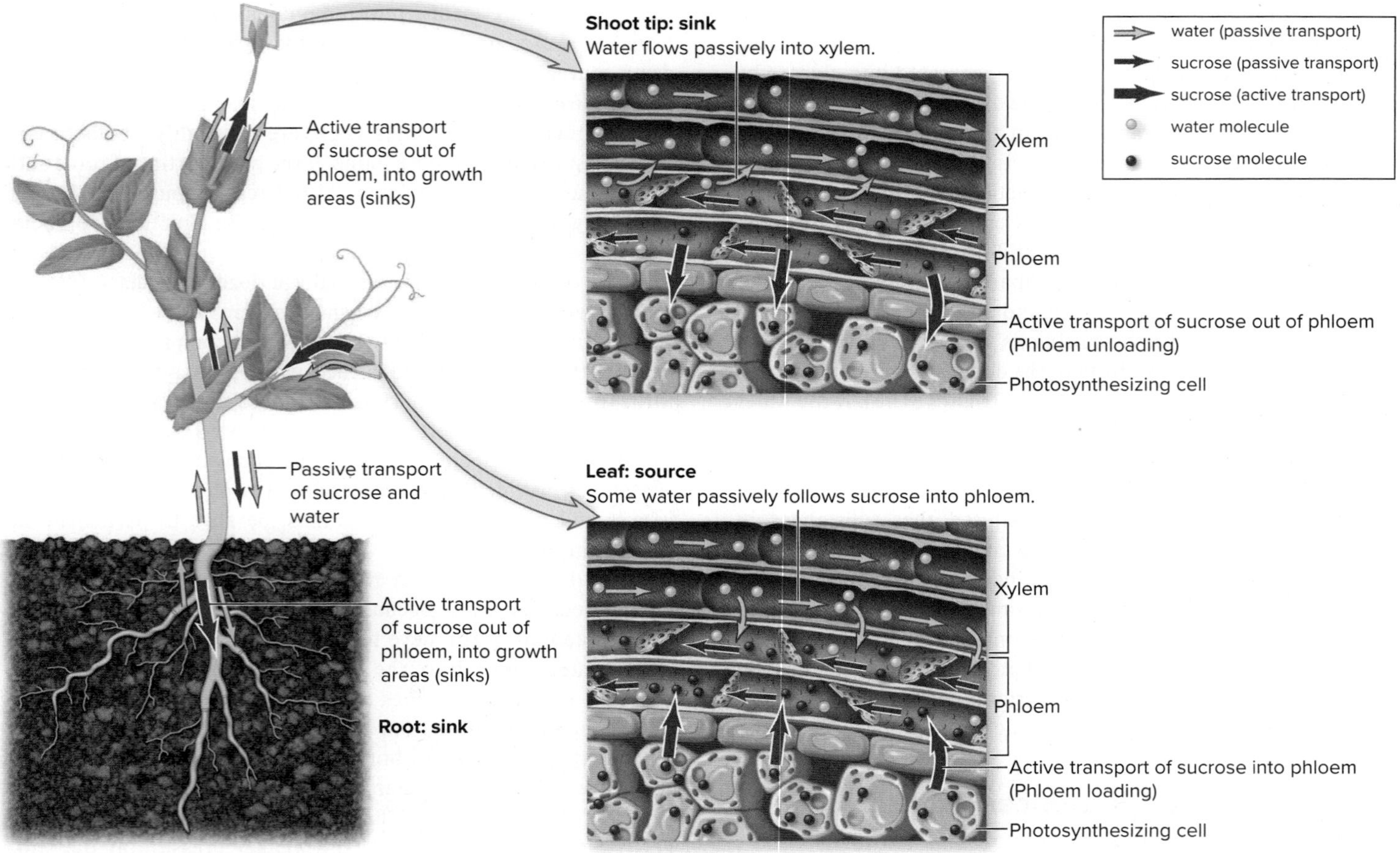
Shoot tip: sink
Water flows passively into xylem.
water (passive transport)
sucrose (passive transport)
sucrose (active transport)
water molecule
sucrose molecule
Active transport of sucrose out of phloem, into growth areas (sinks)
Xylem
Phloem
Active transport of sucrose out of phloem (Phloem unloading)
Photosynthesizing cell
Passive transport of sucrose and water
Leaf: source
Some water passively follows sucrose into phloem.
Active transport of sucrose out of phloem, into growth areas (sinks)
Root: sink
Xylem
Phloem
Active transport of sucrose into phloem (Phloem loading)
Photosynthesizing cell

Figure 31.14 Diagram of mass flow. In this diagram, red dots represent sucrose molecules and blue dots symbolize water molecules. After moving from the mesophyll cells of a leaf or another part of the plant into the conducting cells of the phloem, the sucrose molecules are transported to other parts of the plant by mass flow and are unloaded where they are required.

Plants Require Nine Macronutrients and Seven Micronutrients

LEARNING OBJECTIVE 31.7.1 Distinguish between macronutrients and micronutrients.

CO_2 and light energy are necessary, but not sufficient, to meet all the metabolic needs of a plant. Plants require a large number of inorganic nutrients as well. Some of these are **macronutrients,** which plants need in relatively large amounts, and others are **micronutrients,** required in trace amounts.

The nine macronutrients are carbon, oxygen, and hydrogen—the three elements found in all organic compounds—plus nitrogen (essential for amino acids), potassium, calcium, magnesium (the center of the chlorophyll molecule), phosphorus, and sulfur. Each of these nutrients approaches or, in the case of carbon, may greatly exceed 1% of the dry weight of a healthy plant.

The seven micronutrient elements—chlorine, iron, manganese, zinc, boron, copper, and molybdenum—constitute from less than one part per million to several hundred parts per million in most plants. A deficiency of any one can have severe effects on plant growth (figure 31.15).

Nutritional requirements are assessed by growing plants in hydroponic cultures in which the plant roots are suspended in aerated nutrient solutions. For the purposes of testing, the solutions contain all the necessary nutrients in the right proportions, but with certain known or candidate nutrients left out. The plants are then allowed to grow and are studied for altered growth patterns and leaf coloration that might indicate a need for the missing element (figure 31.16). This example will give an idea of how small the needed quantities of micronutrients may be: the standard dose of molybdenum added to seriously deficient soils in Australia amounts to about 34 g (about one handful) per hectare (a square 100 meters on a side—about 2.5 acres), once every 10 years!

REVIEW OF CONCEPT 31.7

Plants require nine macronutrients in relatively large amounts and seven micronutrients in trace amounts. Plants are grown in controlled hydroponic solutions to determine which nutrients are required for growth.

- *Why would a lack of magnesium in the soil limit food production? (Hint: look at a diagram of a cholorophyll molecule.)*

Figure 31.15 Mineral deficiencies in plants. ***a.*** Healthy tomato plant. ***b.*** Manganese-deficient leaves. ***c.*** Copper-deficient tomato leaf. ***d.*** Phosphorus deficiency in tomato.

(a): McPhoto/SHU/Insadco/age fotostock; (b): Nigel Cattlin/Alamy Stock Photo; (c): Dave Bevan/Garden World Images/age fotostock; (d): Allen Barker

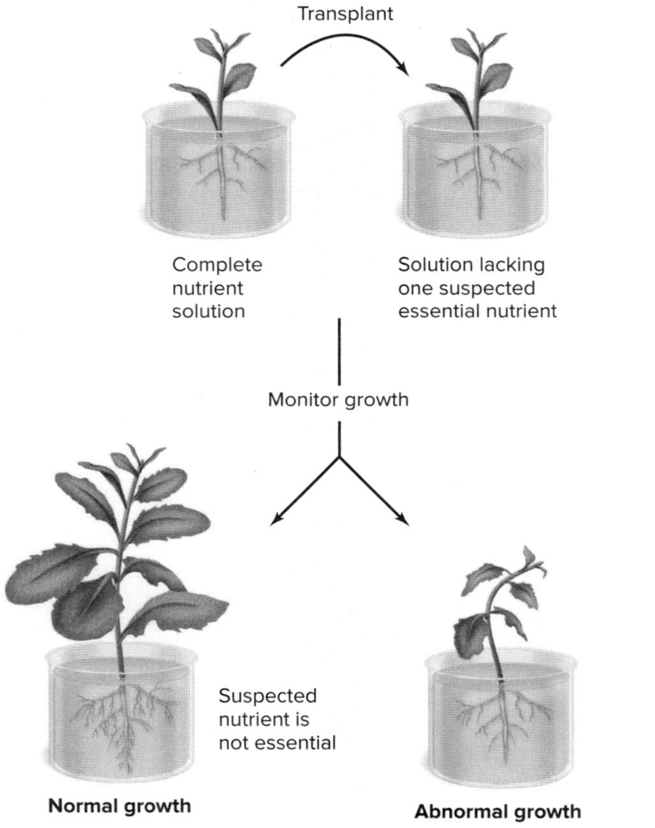

Figure 31.16 Identifying nutritional requirements of plants. A seedling is first grown in a complete nutrient solution. The seedling is then transplanted to a solution that lacks one nutrient thought to be essential. The growth of the seedling is studied for the presence of symptoms indicative of abnormal growth, such as discolored leaves or stunting. If the seedling's growth is normal, the nutrient that was left out may not be essential; if the seedling's growth is abnormal, the nutrient that is lacking is essential for growth.

31.8 Plants Use Hormones to Regulate Growth

Sensory responses, such as responses to light and gravity, that alter morphology rely on complex physiological networks. Many internal signaling pathways involve plant hormones, which are the focus of this section. Hormones are involved in responses to the environment, as well as in internal regulation of development.

The Hormones That Guide Plant Growth Are Responsive to the Environment

LEARNING OBJECTIVE 31.8.1 Describe the properties of plant hormones.

Hormones are information-containing chemical substances produced in small, often minute, quantities in one part of an organism, which are transported to another part where they bring about physiological or developmental responses. The response caused by a hormone is influenced by both the type of hormone and the tissue that receives the information in the hormone.

In animals, most hormones are produced at specific sites, most commonly in organs such as glands. In plants, hormone production is not restricted to a dedicated tissue or organ but, instead, occurs in tissues that also carry out more obvious functions. Seven major kinds of plant hormones have been identified: auxins, cytokinins, gibberellins, brassinosteroids, oligosaccharins, ethylene, and abscisic acid. Current research is focused on the biosynthesis of hormones and characterization of the hormone receptors involved in signal transduction pathways. Much of the molecular basis of hormone function remains poorly understood.

Because hormones are involved in so many aspects of plant function and development, we have integrated examples of hormone activity with specific aspects of plant biology throughout the text. In this section, our goal is to give a brief overview of these hormones.

Auxin Allows Elongation and Organizes the Body Plan

LEARNING OBJECTIVE 31.8.2 Describe how auxin controls plant growth.

More than a century ago, an organic substance known as auxin was the first plant hormone to be discovered. **Auxin** increases the plasticity of plant cell walls and is involved in elongation of stems. The discovery of auxin and its role in plant growth is an elegant example of thoughtful experimental design and is summarized here for that reason.

Discovery of auxin

Later in his life, Charles Darwin became increasingly devoted to the study of plants. In 1881, he and his son Francis reported their experiments on the responses of growing plants to light—the responses that came to be known as phototropisms (discussed in more detail in section 31.9).

The Darwins made a number of observations showing that plants can grow toward light and that the effect depends on the light hitting the tip of a seedling, but not the lower parts of the shoot. They hypothesized that when seedlings are illuminated from one side, they bend toward the light in response to an "influence" transmitted downward from its source at the tip of the shoot.

Some 30 years later, the plant physiologists Peter Boysen-Jensen and Arpad Paal independently showed that the substance causing the shoots to bend is a diffusible chemical. They found that if the tip of a germinating grass seedling was cut off and then replaced, with a small block of agar separating it from the rest of the seedling, the seedling still grew normally.

This use of agar, which can absorb diffusible substances, was taken a step further in 1926 by plant physiologist Frits Went. He cut off the tips of normally illuminated oat seedlings and set them on agar (figure 31.17). He similarly removed the tips of oat seedlings that had been grown in the dark. Went then placed agar from illuminated seedlings off-center on the tops of the decapitated dark-grown seedlings. Despite not having being exposed to the light themselves, the tipless seedlings bent away from the side with the agar block.

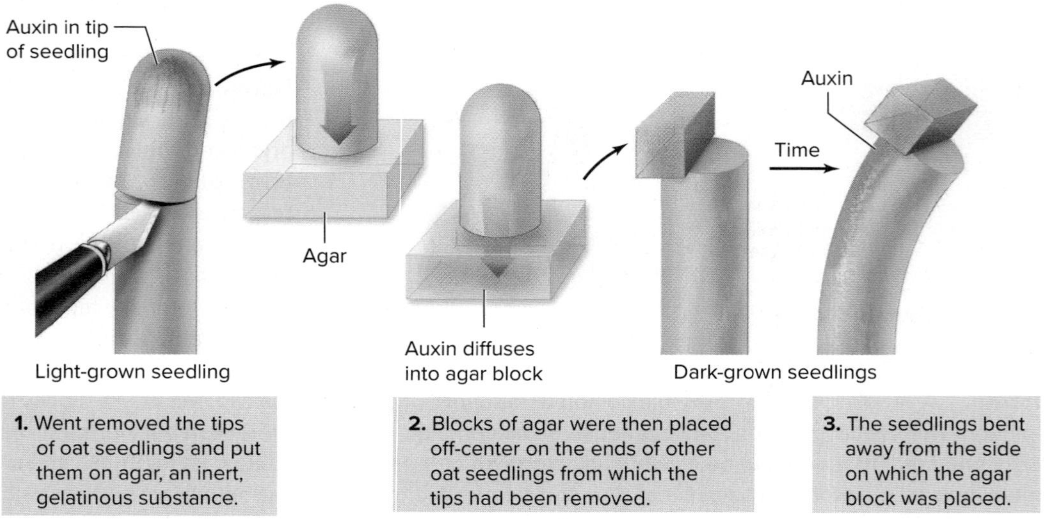

Auxin in tip of seedling
Agar
Auxin diffuses into agar block
Auxin
Time
Light-grown seedling
Dark-grown seedlings
1. Went removed the tips of oat seedlings and put them on agar, an inert, gelatinous substance.
2. Blocks of agar were then placed off-center on the ends of other oat seedlings from which the tips had been removed.
3. The seedlings bent away from the side on which the agar block was placed.

Figure 31.17 Frits Went's experiment. Went concluded that a substance he named auxin promotes the elongation of the cells and that it accumulates on the side of an oat seedling away from the light.

As a negative control, Went put blocks of pure agar on the decapitated stem tips and noted either no effect or a slight bending toward the side where the agar blocks were placed. Finally, Went cut sections out of the lower portions of the light-grown seedlings and placed these sections on the tips of decapitated, dark-grown oat seedlings and again observed no effect.

These experiments showed that a diffusible substance in the tips of light-grown oat seedlings can make seedlings bend. This bending is due to a signaling molecule causing cells on the side of the agar block to elongate more than on the other side of the seedling. Went named this substance auxin.

This research provides a basis for understanding the phototropic response. The oat seedlings bent toward the light because of differences in the auxin concentrations on the two sides of the shoot (figure 31.18). The side of the shoot that was in the shade had more auxin, which promoted cell elongation on this side, bending the plant toward the light.

The effects of auxin

Auxin acts to promote growth and elongation, and environmental signals can influence the distribution of auxin in the plant. How does the environment, specifically light, exert this influence? The light can affect the production, destruction, or distribution of auxin within the plant.

Plant biologist Winslow Briggs added a wrinkle to the agar block type experiments by placing a thin sheet of transparent mica vertically between the half of the shoot oriented toward the light and the half of the shoot oriented away from it. The mica blocks movement of auxin, and these experiments showed that the effect of light is to promote movement of auxin from the illuminated side of the tip to the shaded side. This experiment showed that it is the distribution of auxin within the seedling shoot that causes bending. When auxin is distributed preferentially to the shaded side of the shoot, cells on that side elongate to cause bending away from shade, toward the light.

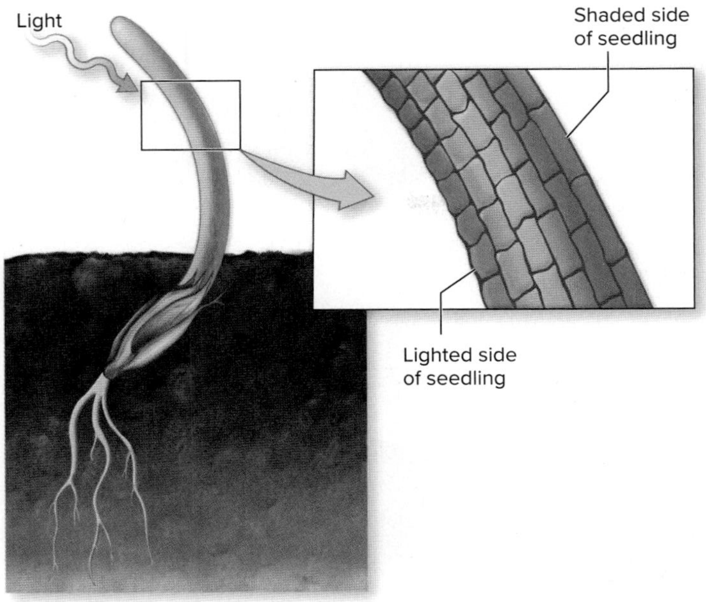

Figure 31.18 Auxin causes cells on the shaded side to elongate. Plant cells that are in the shade have more auxin and grow faster than cells on the illuminated side, causing the plant to bend toward light. Further experiments showed exactly why there is more auxin on the shaded side of a plant.

The effects of auxin are numerous and varied. Auxin promotes the activity of the vascular cambium and the vascular tissues. Also, auxin is present in pollen in large quantities and plays a key role in the development of fruits. Synthetic auxins are used commercially for the same purpose. Fruits will normally not develop if fertilization has not occurred and seeds are not present, but frequently they will develop if auxin is applied. Pollination may trigger auxin release in some species, leading to fruit development even before fertilization has taken place.

How auxin works

In spite of the long history of research, auxin's molecular basis of action has been difficult to work out. The chemical structure of the most common auxin, indoleacetic acid (IAA), resembles that of the amino acid tryptophan, from which it is probably synthesized by plants (figure 31.19).

An auxin-binding protein (ABP1) was identified two decades ago, but its role in auxin response is still unclear. Mutants that lack ABP1 do not survive embryogenesis, because cell elongation is inhibited and the basic body plan is not organized. But the ABP1 mutant cells divide, indicating that part of the auxin pathway functions.

More recently, two families of proteins that mediate rapid, auxin-induced changes in gene expression have been identified: the auxin response factors (ARFs) and the Aux/IAA proteins. The ARF proteins act as transcription factors to regulate transcription, and Aux/IAA proteins inhibit this action. However, neither of these is the actual auxin receptor, which was finally identified in 2005. This protein, called transport inhibitor response protein 1 (TIR1), binds to auxin directly. This finding led to a model that incorporates TIR1, ARFs, and Aux/IAA proteins. In this model, auxin binding to TIR1 leads to the degradation of Aux/IAA by interacting with a protein complex called SCF. The complex of auxin/TIR1/SCF stimulates the degradation of Aux/IAA by the ubiquitin pathway (refer to chapter 4). With Aux/IAA no longer inhibiting the ARFs, this allows auxin-induced gene expression. This model is detailed in figure 31.20.

Unlike with animal hormones, a specific signal is not sent to specific cells, eliciting a predictable response. Most likely, multiple auxin reception sites are present. Auxin is also unique among the plant hormones in that it is transported toward the base of the plant. Two families of genes have been identified in *Arabidopsis* that are involved in auxin transport. For example, one family of proteins (the PINs) are involved in the top-to-bottom transport of auxin, while two other proteins function in the root tip to regulate the growth response to gravity (described in section 31.10).

One of the direct effects of auxin is an increase in the plasticity of the plant cell wall, but this effect works only on young cell walls lacking extensive secondary cell-wall formation and may or may not involve rapid changes in gene expression. The acid growth hypothesis provides a model linking auxin to cell-wall expansion. According to this hypothesis, auxin causes responsive cells to actively transport hydrogen ions from the cytoplasm into the cell-wall space. This decreases

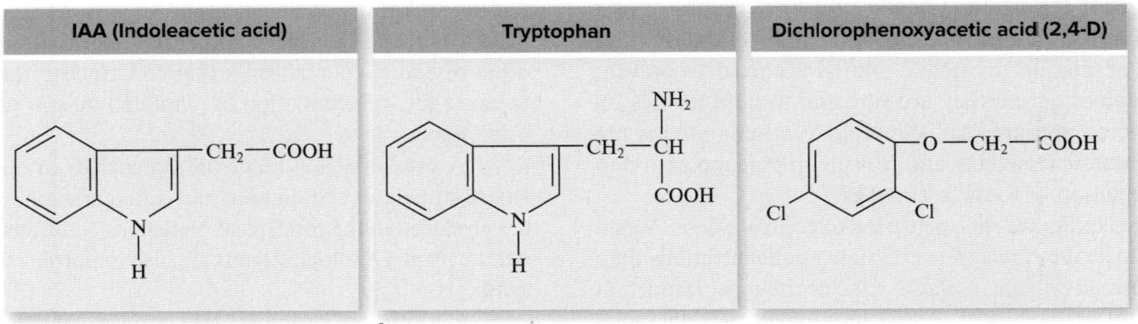

a. *b.* *c.*

Figure 31.19 Auxins. ***a.*** Indoleacetic acid (IAA), the principal naturally occurring auxin. ***b.*** Tryptophan, the amino acid from which plants probably synthesize IAA. ***c.*** Dichlorophenoxyacetic acid (2,4-D), a synthetic auxin, is a widely used herbicide.

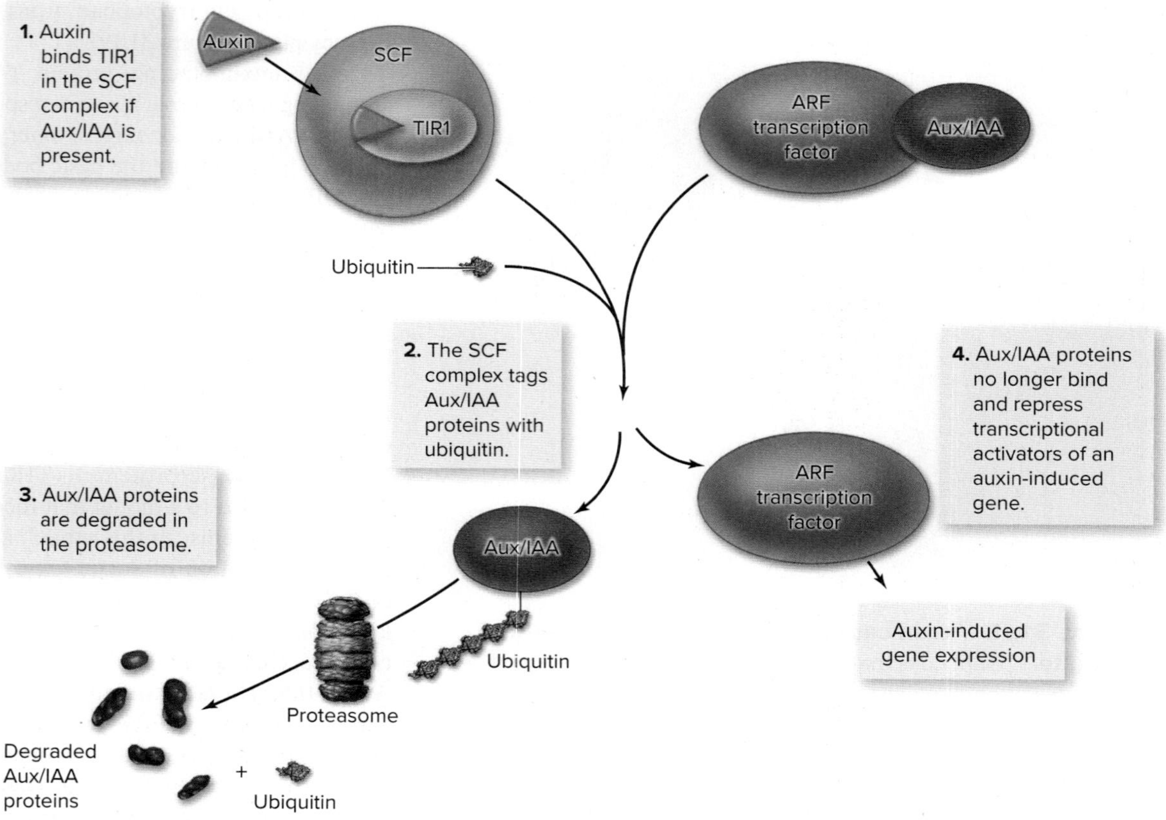

1. Auxin binds TIR1 in the SCF complex if Aux/IAA is present.
Auxin
SCF
TIR1
ARF transcription factor
Aux/IAA
Ubiquitin
2. The SCF complex tags Aux/IAA proteins with ubiquitin.
4. Aux/IAA proteins no longer bind and repress transcriptional activators of an auxin-induced gene.
3. Aux/IAA proteins are degraded in the proteasome.
ARF transcription factor
Aux/IAA
Auxin-induced gene expression
Ubiquitin
Proteasome
Degraded Aux/IAA proteins
+
Ubiquitin

Figure 31.20 Auxin regulation of gene expression. Auxin activates a ubiquitin pathway that releases ARF transcription factors from repression by Aux/IAA proteins. The result is auxin-induced gene expression.

the pH, which activates enzymes that can break the bonds between cell-wall fibers.

This hypothesis has been experimentally supported in several ways. Buffers that prevent cell-wall acidification block cell expansion. And other compounds that release hydrogen ions from the cell can also cause cell expansion. Finally, the movement of hydrogen ions has been observed in response to auxin treatment. The snapping of the Venus flytrap is postulated to involve an acid growth response that allows cells to expand in just 0.5 sec and close the trap.

Synthetic auxins

Synthetic auxins, such as naphthalene acetic acid (NAA) and indolebutyric acid (IBA), have many uses in agriculture and horticulture. One of their most important uses is based on their prevention of abscission. Synthetic auxins are used to prevent fruit drop in apples before they are ripe and to hold berries on holly that is being prepared for shipping. Synthetic auxins are also used to promote flowering and fruiting in pineapples and to induce the formation of roots in cuttings.

Synthetic auxins are routinely used to control weeds. When used as herbicides, they are applied in higher concentrations than IAA would normally occur in plants. One of the most important synthetic auxin herbicides is 2,4-dichlorophenoxyacetic acid, known as 2,4-D (figure 31.19*c*). It kills weeds in grass lawns by selectively eliminating broad-leaf dicots.

A related herbicide, 2,4,5-trichlorophenoxyacetic acid, known as 2,4,5-T, was widely used until its ban for most uses in the United States in 1979. The manufacture of 2,4,5-T results in contamination with minute amounts of dioxin, which causes liver and lung diseases, leukemia, miscarriages, and birth defects in laboratory animals.

Cytokinins Stimulate Cell Division and Differentiation

LEARNING OBJECTIVE 31.8.3 Describe how auxins and cytokinins influence plant form.

Cytokinins comprise another group of naturally occurring growth hormones in plants. Botanist Gottlieb Haberlandt showed that an unknown chemical found in various tissues of vascular plants would cause parenchyma cells to become meristematic. In other research, coconut milk that contained cytokinins was used to cause the differentiation of plant cell masses grown in culture into organs.

A **cytokinin** is a plant hormone that, in combination with auxin, stimulates cell division and differentiation. Most cytokinins are produced in the root apical meristems and transported throughout the plant. Developing fruits are also important sites of cytokinin synthesis.

Cytokinins promote the growth of lateral buds into branches. Conversely, cytokinins inhibit the formation of lateral roots, whereas auxins promote their formation. As a consequence of these relationships, the balance between cytokinins and auxin largely determines a plant's form.

Gibberellins Enhance Plant Growth and Nutrient Utilization

LEARNING OBJECTIVE 31.8.4 Describe how gibberellins affect plant growth.

Gibberellins are named after the fungus *Gibberella fujikuroi*. This fungus grows parasitically on rice plants and causes them to grow abnormally tall. The plant pathologist Eiichi Kurosawa investigated bakanae ("foolish seedling") disease in the 1920s. He grew *Gibberella* in culture and obtained a substance that, when applied to rice plants, produced bakanae (figure 31.21). This substance was isolated and its structural formula identified by Japanese chemists in 1939.

These molecules, initially nothing more than a curiosity, are in fact members of a class of more than 100 plant hormones. All are acidic and are usually abbreviated GA (for gibberellic acid), with a different subscript (GA_1, GA_2, and so forth) to distinguish each one.

Gibberellins, which are synthesized in the apical portions of stems and roots, have important effects on stem elongation. The elongation effect is enhanced if auxin is also present. Gibberellins also affect a number of other aspects of plant growth and development.

Ethylene Induces Fruit Ripening and Aids Plant Defenses

LEARNING OBJECTIVE 31.8.5 Describe how ethylene induces fruit ripening.

Long before its role as a plant hormone was appreciated, the gaseous hydrocarbon **ethylene** (C_2H_4) was known to defoliate plants when it leaked from gaslights in old-fashioned streetlamps. Ethylene is, however, a natural product of plant metabolism that, in minute amounts, interacts with other plant hormones to regulate plant growth and development.

Figure 31.21 Effects of gibberellins. One of two identical pea plants (right) was treated with gibberellic acid. 48 hours later a significant effect can be seen on the upward growth of the treated plant.
Omikron/Science Source

When auxin is transported downward from shoot apical meristem, it stimulates the production of ethylene in the tissues around the lateral buds and thus prevents their growth. Ethylene also suppresses stem and root elongation, probably in a similar way. An ethylene receptor has been identified and characterized, and it appears to have evolved early in the evolution of photosynthetic organisms, sharing features with environmental-sensing proteins identified in bacteria.

Ethylene plays a major role in fruit development. At first, auxin, which is produced in significant amounts in pollinated flowers and developing fruits, stimulates ethylene production; this in turn hastens fruit ripening. Complex carbohydrates are broken down into simple sugars, chlorophylls are broken down, cell walls become soft, and the volatile compounds associated with flavor and scent in ripe fruits are produced.

One of the first observations that led to the recognition of ethylene as a plant hormone was the premature ripening in bananas produced by gases coming from oranges. Such relationships have led to major commercial uses of ethylene. For example, tomatoes are often picked green and artificially ripened later by the application of ethylene. Ethylene is widely used to speed the ripening of lemons and oranges as well. Carbon dioxide has the opposite effect of arresting ripening; fruits are often shipped in an atmosphere high in carbon dioxide.

Abscisic Acid Suppresses Growth and Induces Dormancy

LEARNING OBJECTIVE 31.8.6 Describe the major roles of abscisic acid.

Abscisic acid (ABA) appears to be synthesized mainly in mature green leaves, fruits, and root caps. The hormone earned its name because applications of it appear to stimulate fruit dropping (abscission) in cotton, but there is little evidence that it plays an important role in this process. Ethylene is actually the hormone that promotes senescence and abscission.

ABA probably induces the formation of winter buds—dormant buds that remain through the winter (figure 31.22). The conversion of leaf primordia into bud scales follows. Like ethylene, ABA may also suppress the growth of dormant lateral buds. It appears that ABA, by suppressing growth and elongation of buds, can counteract some of the effects of gibberellins; it also promotes senescence by counteracting auxin.

ABA plays a role in seed dormancy and is antagonistic to gibberellins during germination. ABA levels rise during embryogenesis. As maize embryos develop in the kernels on the cob, ABA is necessary to induce dormancy and prevent premature germination, called vivipary. It is also important in controlling the opening and closing of stomata.

ABA is found in all plants and has apparently been functioning as a growth-regulating substance since early in the evolution of the plant kingdom. Relatively little is known about the exact nature of its physiological and biochemical effects, but

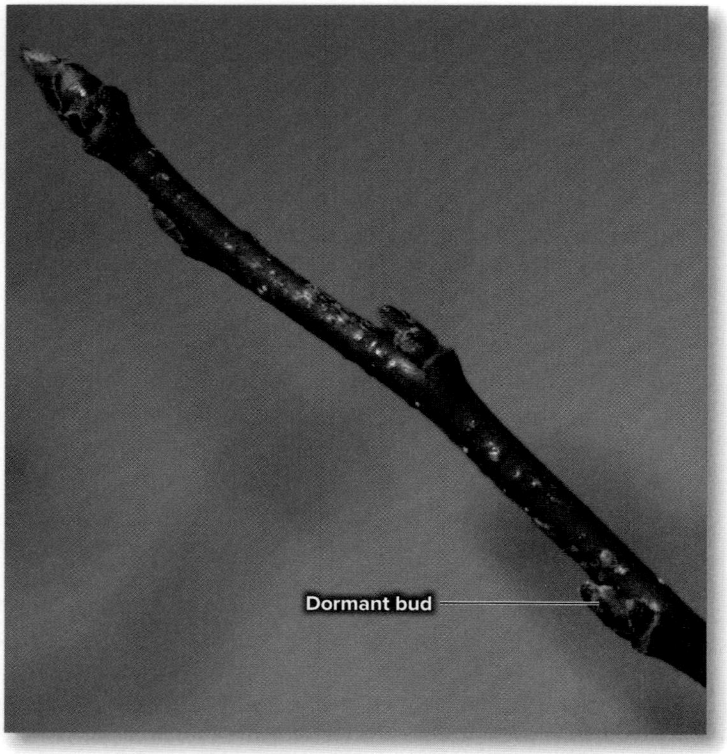

Figure 31.22 Effect of abscisic acid on dormancy. ABA plays a role in the formation of these winter buds of an American basswood. As long as ABA remains present, these buds will remain dormant for the winter, with bud scales—modified leaves—protecting the buds from desiccation.

Evelyn Jo Johnson/McGraw Hill

these effects are very rapid—often taking place within a minute or two—and, therefore, they must be at least partly independent of gene expression.

REVIEW OF CONCEPT 31.8

Hormones are chemicals produced in small quantities in one region of the plant and then transported to another region, where they cause a physiological or developmental response. Both auxins and cytokinins are produced in meristems and promote growth; however, auxins stimulate growth by cell elongation, whereas cytokinins stimulate cell division. In contrast, abscisic acid inhibits growth and promotes dormancy.

- *What methods could you use to test whether abscisic acid produced in root caps can affect bud growth in stems?*

31.9 Plant Growth Is Responsive to Light

In chapter 8, we covered the details of photosynthesis, the process by which plants convert light energy into chemical potential energy. We described pigments, molecules that are capable of absorbing light energy, and you learned that chlorophylls are the primary pigment molecules in photosynthesis. Plants also contain other pigments, and one of the functions of these other pigments is to detect light and to mediate plant responses to light.

Phytochrome Facilitates Expression of Light-Response Genes

LEARNING OBJECTIVE 31.9.1 Compare the pigments phytochrome and chlorophyll.

Several environmental factors, including light, can initiate seed germination, flowering, and other critical developmental events in the life of a plant. **Photomorphogenesis** is nondirectional, light-triggered development. It can result in complex changes in form, including flowering.

Unlike photomorphogenesis, **phototropisms** are directional growth responses to light. Both photomorphogenesis and phototropisms compensate for the plant's inability to move away from unfavorable environmental conditions.

Phytochrome is a pigment present in all groups of plants and in a few genera of green algae, but not in other protists, bacteria, or fungi. Phytochrome-based systems probably evolved among the green algae and were present in the common ancestor of the plants.

The phytochrome molecule exists in two interchangeable forms: the first form, $\mathbf{P_r}$, absorbs red light at a wavelength of 660 nm; the second, $\mathbf{P_{fr}}$, absorbs far-red light at 730 nm. Sunlight has more red than far-red light. P_r is biologically inactive; it is converted into P_{fr}, the active form, when red photons are present. P_{fr} is converted back into P_r when far-red photons are available. In other words, biological processes that are affected by phytochrome occur when P_{fr} is present. When most of the P_{fr} has been replaced by P_r, the P_{fr}-dependent processes will not occur (figure 31.23).

The amount of P_{fr} is also regulated by degradation. Ubiquitin is a protein that tags P_{fr} for transport to the **proteasome,** a multi-protein complex that degrades ubiquitin-tagged proteins. The proteasome has a channel in the center, and as proteins pass through, they are degraded into amino acids that can be recycled to build other proteins (refer to chapter 4). The process of tagging and recycling P_{fr} is carefully regulated to maintain the amounts of phytochrome that a cell needs.

The phytochrome protein consists of two parts: a smaller part that is sensitive to light, called the chromophore, and a larger portion, called the apoprotein. The apoprotein facilitates expression of light-response genes. Over 2500 genes, 10% of the *Arabidopsis* genome, are involved in biological responses that begin with a conformational change of phytochrome in response to red light. Phytochromes are involved in numerous signaling pathways that lead to gene expression.

Phytochrome is found in the cytoplasm; however, when P_r is converted to P_{fr}, it can move to the nucleus. Once in the nucleus, P_{fr} binds with other proteins to form a transcription complex, which can turn on the expression of light-regulated genes (figure 31.24).

Phytochrome also works through protein kinase–signaling pathways. When phytochrome is converted to the P_{fr} form, the protein kinase domain of the apoprotein may phosphorylate

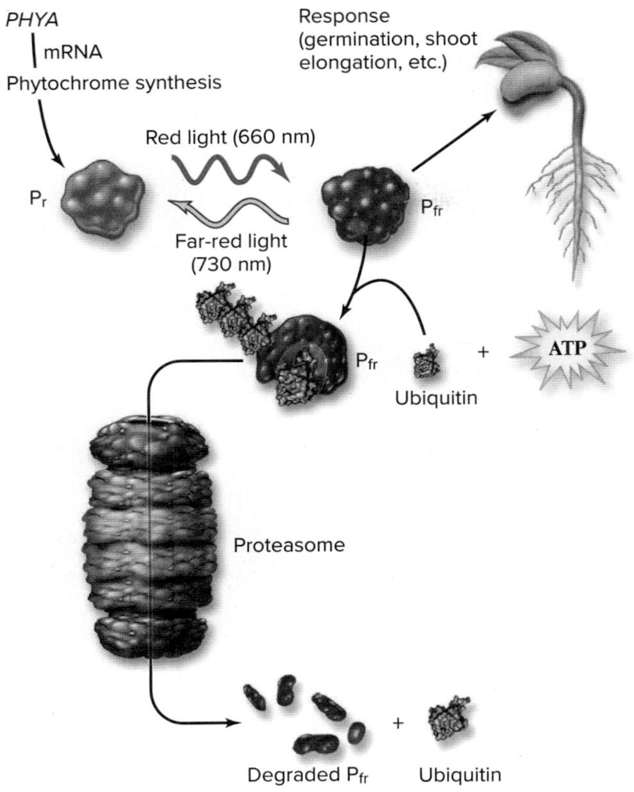

Figure 31.23 How phytochrome works. *PHYA* is one of the five *Arabidopsis* phytochrome genes. When exposed to red light, P_r changes to P_{fr}, the active form that elicits a response in plants. P_{fr} is converted to P_r when exposed to far-red light. The amount of P_{fr} is regulated by protein degradation. The protein ubiquitin tags P_{fr} for degradation in the proteasome.

serines in the amino terminus of the phytochrome itself (autophosphorylation), or it may phosphorylate the serine of another protein involved in light signaling. Phosphorylation initiates a signaling cascade that can activate transcription factors and lead to the transcription of light-regulated genes.

Although phytochrome is involved in multiple signaling pathways, it does not directly initiate the expression of that 10% of the *Arabidopsis* plant genome. Rather, phytochrome initiates expression of master regulatory genes that manage the complex interactions leading to photomorphogenesis and phototropisms.

Although we refer to phytochrome as a single molecule here, several different phytochromes have been identified that appear to have specific functions.

Many Growth Responses Are Linked to Phytochrome Action

LEARNING OBJECTIVE 31.9.2 Describe growth responses influenced by phytochromes.

Phytochrome is involved in a number of plant growth responses, including seed germination, shoot elongation, and detection of plant spacing.

Seed germination

Seed germination is inhibited by far-red light and stimulated by red light in many plants. Because chlorophyll absorbs red light strongly but does not absorb far-red light, light filtered through the green leaves of canopy trees above a seed contains a reduced

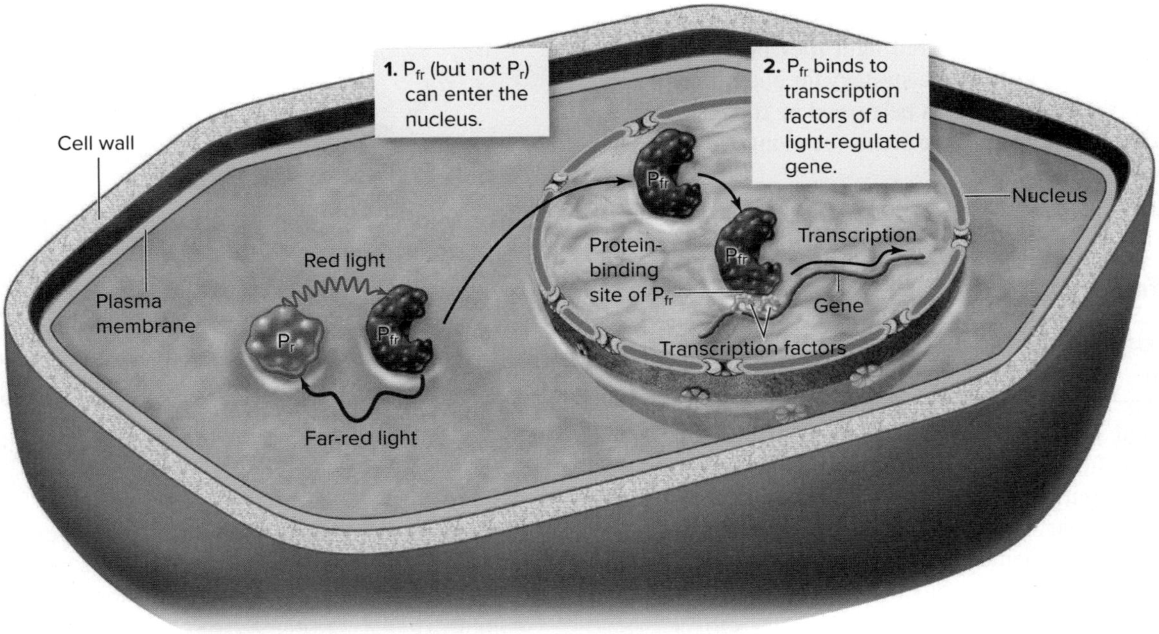

Figure 31.24 P_{fr} enters the nucleus and regulates gene expression.

amount of red light. The far-red light inhibits seed germination by converting P_{fr} into the biologically inactive P_r form.

Consequently, seeds on the ground under deciduous plants, which lose their leaves in winter, are more apt to germinate in the spring after the leaves have decomposed and the seeds are exposed to direct sunlight and a greater amount of red light. This adaptation greatly improves the chances that seedlings will become established before leaves on taller plants shade the seedlings and reduce sunlight available for photosynthesis.

Shoot elongation

Elongation of the shoot in an etiolated seedling (one that is pale and slender from having been kept in the dark) is caused by a lack of red light. The morphology of such plants becomes normal when they are exposed to red light, increasing the amount of P_{fr}.

Etiolation is an energy conservation strategy to help plants growing in the dark reach the light before they die. They don't become green until light becomes available, and they divert energy to internode elongation. This strategy is useful for seedlings when they have sprouted underground or under leaf cover.

The de-etiolated *(det2) Arabidopsis* mutant has a poor etiolation response; seedlings fail to elongate in the dark (figure 31.25). The *det2* mutants are defective in an enzyme necessary for biosynthesis of a brassinosteroid hormone, leading researchers to propose that brassinosteroid hormones play a role in plant responses to light via phytochrome.

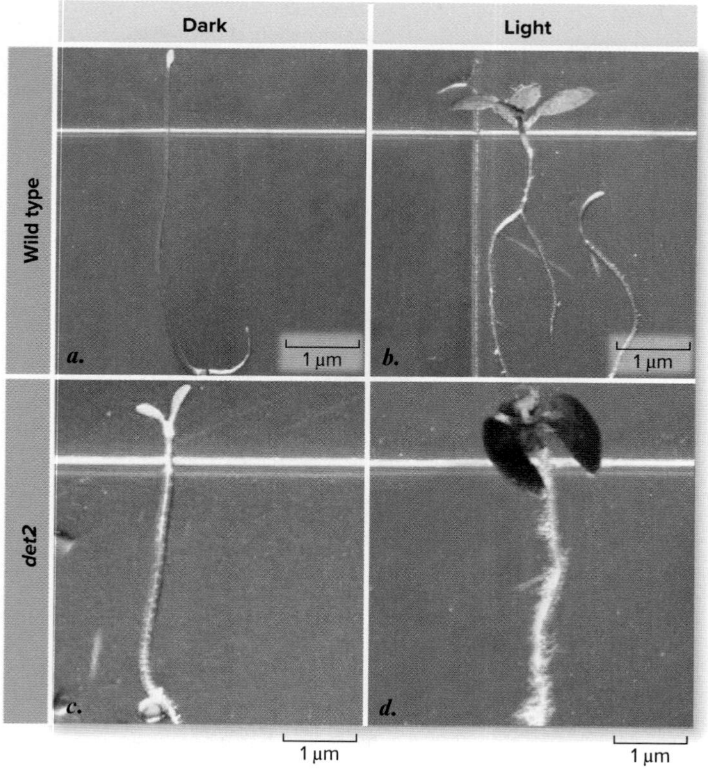

Figure 31.25 Etiolation is regulated by light and the *DET2* gene in *Arabidopsis*. *Arabidopsis* plants wild type for *DET2* grown in the (***a***) dark become etiolated, while those grown under (***b***) normal light do not etiolate and produce leaves. *Arabidopsis det2* mutants fail to etiolate when grown in the (***c***) dark and appear relatively normal under (***d***) conditions of normal light.

(a-d): Niko Geldner, UNIL

Detection of plant spacing

Red and far-red light also signal plant spacing. Again, leaf shading increases the amount of far-red light relative to red light. Plants somehow measure the amount of far-red light reflected to them from neighboring plants. The closer together plants are, the more far-red relative to red light they perceive and the more likely they are to grow tall, a strategy for outcompeting others for light. If their perception is perturbed by the placement of a light-blocking collar around the stem, the elongation response no longer occurs.

Light Affects Directional Growth

LEARNING OBJECTIVE 31.9.3 Describe the mechanism of phototropism.

The directional growth responses of a plant to light are called phototropisms (figure 31.26). Different plants' phototropisms contribute to the variety of plant forms seen within a single species. Tropisms are particularly intriguing because they challenge us to connect environmental signals with cellular perception of the signal, transduction into biochemical pathways, and ultimately an altered growth response.

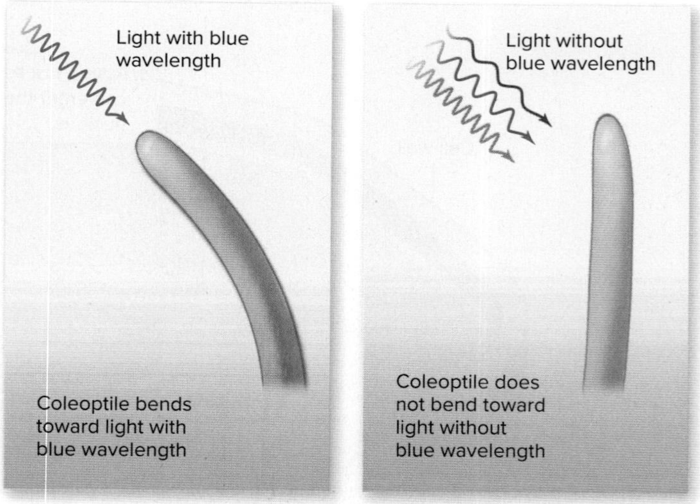

Figure 31.26 Phototropism. Oat coleoptiles growing toward light with blue wavelengths. Colors indicate the color of light shining on coleoptiles. Arrows indicate the direction of light.

Positive phototropism in shoots

Phototropic responses include the bending of growing stems and other plant parts toward sources of light with blue wavelengths (460-nm range; figure 31.26). In general, stems are positively phototropic, growing toward a light source, but most roots do not respond to light or, in exceptional cases, bend very slightly away from light.

The phototropic reactions of stems are clearly of adaptive value, giving plants greater exposure to available light. They are also important in determining the development of plant organs and, therefore, the appearance of the plant. Individual leaves may also display phototropic responses; the position of leaves is important to the photosynthetic efficiency of the plant.

Blue-light receptors

The recent identification of blue-light receptors in plants provided a connection between the light signal and the phototropic response. A blue-light receptor phototropin 1 (phot1) was identified through the characterization of a nonphototropic mutant. The phot1 protein is a plant receptor kinase (refer to chapter 9) that is activated by light. Currently, only the early steps in this signal transduction are understood.

Blue light is perceived by proteins called phototropins. In *Arabidopsis* there are two phototropins: phot1 and phot2. Both phototropins are membrane-bound serine/threonine protein kinase receptors (refer to chapter 9). They presumably cause coleoptile bending by phosphorylating proteins that ultimately result in a change in the distribution of the plant growth hormone auxin.

Circadian Clocks Are Independent of Light but Are Entrained by Light

LEARNING OBJECTIVE 31.9.4 Explain how light entrains circadian cycles of plant growth.

Circadian rhythms are intrinsically programmed physiological routines that run for about 24 hours. They are particularly common among eukaryotes and have been found in plants, animals, fungi, and some types of bacteria. Although they are an inherent type of cyclical process, they can be modified (entrained) or reset by external factors such as light. If you have ever experienced jet lag after a long flight across several time zones, your circadian rhythm has been disrupted and only "resets" after becoming entrained to the new time zone.

For example, plants kept in darkness will continue the circadian cycle, but the cycle's period may gradually move away from the actual day–night cycle, becoming desynchronized. In the natural environment, the cycle is entrained to a daily cycle through the action of phytochrome and blue-light photoreceptors. Circadian rhythms are an adaptive advantage to plants, because they allow coordination between certain physiological processes such as leaf movement and light intensity, which is dependent on time of day.

REVIEW OF CONCEPT 31.9

Plants grow and develop in response to environmental signals. Phytochrome, a red-light receptor, acts through pathways that control gene expression. Phytochrome influences seed germination, shoot elongation, and other growth. Phototropism is directional growth in response to light and is controlled by a blue-light receptor. Circadian rhythms are 24-hour cycles entrained to the day–night cycle that are adaptations to maximize growth.

- *Why would it be an advantage to have both phytochromes and chlorophylls as pigments?*

31.10 Plant Growth Is Sensitive to Gravity

When a potted plant is tipped over and left in place, the shoot bends and grows upward (figure 31.27). The same thing happens when a storm knocks plants over in a field. These are examples of **gravitropism,** the response of a plant to the gravitational field of the Earth. Because plants also grow in response to light, separating out phototropic effects is important in the study of gravitropism.

Plants Align with the Gravitational Field

LEARNING OBJECTIVE 31.10.1 Describe how plants may perceive and respond to gravity.

Gravitropic responses are present at germination, when the root grows down and the shoot grows up. Why does a shoot have a

Figure 31.27 Plant response to gravity. This plant was placed horizontally and allowed to grow for seven days. Note how the shoot is now growing upward: a negative gravitational response.

Ray F. Evert

negative gravitropic response (growth away from gravity), whereas a root has a positive one? Auxins play a primary role in gravitropic responses, but they may not be the only way gravitational information is sent through the plant.

Opportunities to do experiments in space have accelerated research in the area of gravitropism. Analysis of gravitropic mutants is also adding to our understanding of gravitropism. Scientists propose that four general steps lead to a gravitropic response:

1. Gravity is perceived by the cell.
2. A mechanical signal is transduced into a physiological signal in the cell that perceives gravity.
3. The physiological signal is transduced inside the cell and externally to other cells.
4. Differential cell elongation occurs, affecting cells in the "up" and "down" sides of the root or shoot.

The exact steps involved in how a gravitropic effect is elicited in plants are unclear. In shoots, gravity is sensed along the length of the stem in the endodermal cells that surround the vascular tissue (figure 31.28*a*), and signaling occurs toward the outer epidermal cells. In roots, the cap is the site of gravity perception, and a signal must trigger differential cell elongation and division in the elongation zone (figure 31.28*b*).

In both shoots and roots, amyloplasts, plastids that contain starch, sink toward the center of the gravitational field and thus may be involved in sensing gravity. Auxin evidently plays a role in transmitting a signal from the gravity-sensing cells that contain amyloplasts and the site where growth occurs. The link between amyloplasts and auxin is not fully understood.

Stems bend away from a center of gravity

Increased auxin concentration on the lower side in stems causes the cells in that area to grow more than the cells on the upper side. The result is a bending upward of the stem against the force of gravity—in other words, a negative gravitropic response. Such differences in hormone concentration have not been as well documented in roots. Nevertheless, the upper sides of roots oriented horizontally grow more rapidly than the lower sides, causing the root ultimately to grow downward; this phenomenon is known as a positive gravitropic response.

Two *Arabidopsis* mutants, *scarecrow (scr)* and *short root (shr),* were identified because they caused abnormal root phenotypes, but they also affect shoot gravitropism (figure 31.29). Both genes are needed for normal endodermal development. Without a fully functional endodermis, stems lack a normal gravitropic response. These endodermal cells carry amyloplasts in the stems, and in the mutants, stem endodermis fails to differentiate and produce gravity-sensing amyloplasts.

Roots bend toward a center of gravity

In roots, the gravity-sensing cells are located in the root cap, and the cells that actually undergo asymmetrical growth are in the distal elongation zone, which is closest to the root cap. How the information is transferred over this distance is an intriguing question. Auxin may be involved, but when auxin transport is suppressed, a gravitropic response still occurs in the distal elongation zone. Some type of electrical signaling involving membrane polarization has been hypothesized, and this idea was tested in space. So far, the exact mechanism remains unclear.

The growing number of auxin mutants affecting roots confirms that auxin has an essential role in root gravitropism, even if it may not be the long-distance signal between the root cap and the elongation zone. Mutations that affect both influx and efflux of auxin can eliminate the gravitropic response by altering the directional transport of this hormone.

Given normal gravitropic responses, it is a little bewildering that in tropical rainforests the roots of some plants grow up the stems of neighboring plants, instead of exhibiting the normal positive gravitropic responses typical of other roots. It appears that rainwater dissolves nutrients, both while passing

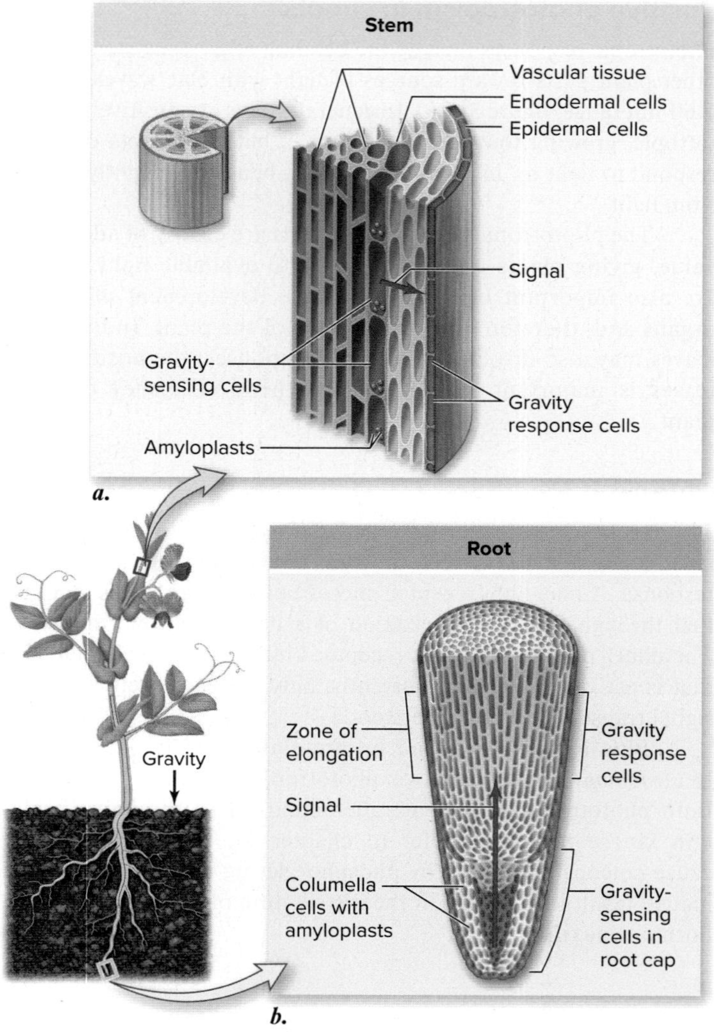

Figure 31.28 Sites of gravity sensing and response in roots and shoots.

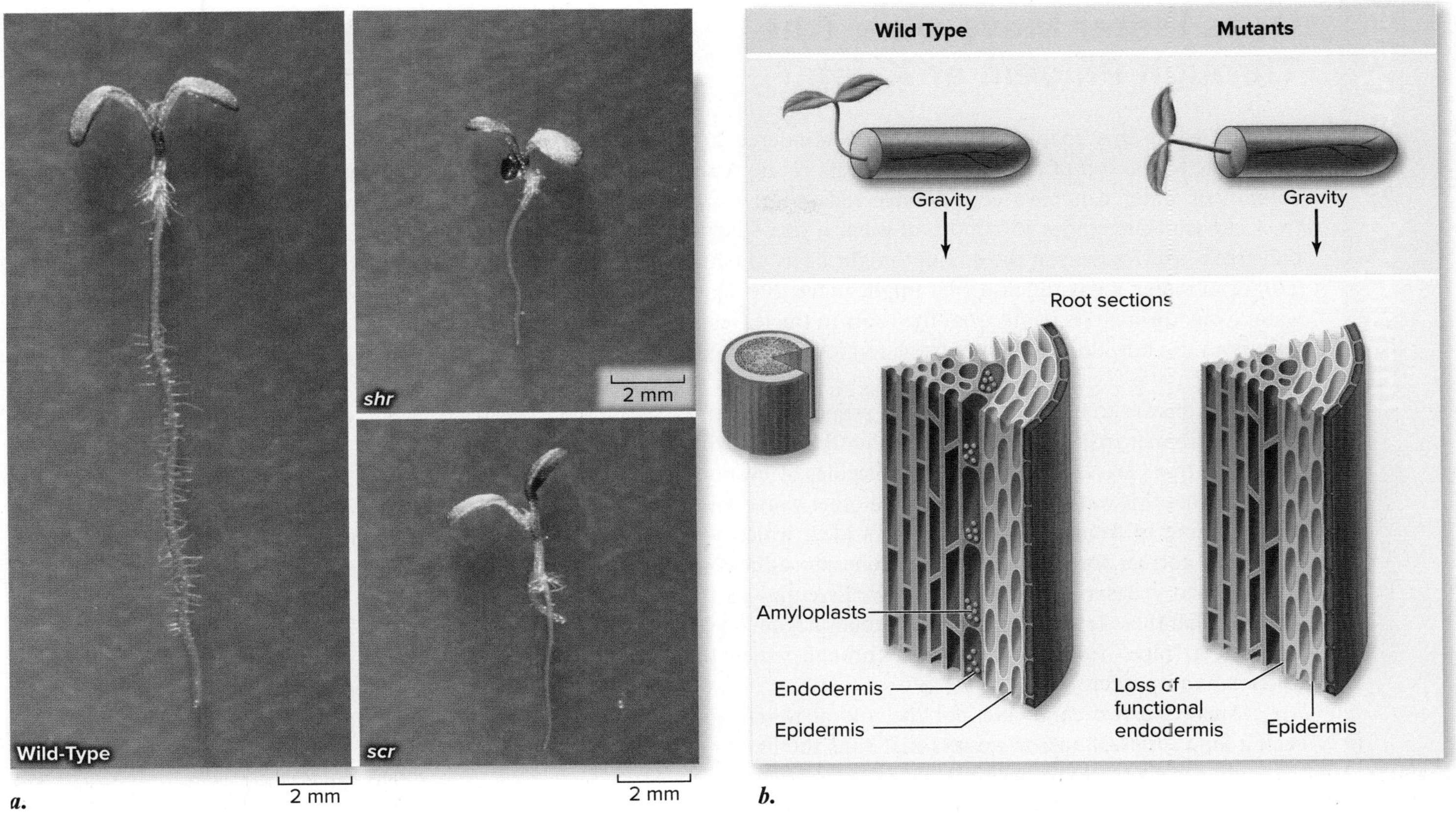

Figure 31.29 Amyloplasts in stem endoderm are needed for gravitropism. ***a.*** The *scr* and *shr* mutants of *Arabidopsis* have abnormal root development—they lack a fully differentiated endodermal layer. ***b.*** The endodermal defect extends into the stem, eliminating the positive gravitropic response of wild-type stems.

(Left, Top right, Bottom): ©Jee Jung and Philip Benfey

through the lush upper canopy of the forest and subsequently while trickling down the tree trunks. This water is a more reliable source of nutrients for the roots than the nutrient-poor rainforest soils in which the plants are anchored. Explaining this observation in terms of current hypotheses is a challenge. It has been proposed that roots are more sensitive to auxin than are shoots and that auxin may inhibit growth on the lower side of a root, resulting in a positive gravitropic response. Perhaps in these tropical plants, the sensitivity to auxin in roots is reduced.

REVIEW OF CONCEPT 31.10

Gravitropism is the response of a plant to gravity. In endodermis cells of shoots and root cap cells of roots, amyloplasts settle to the bottom, allowing the plant to sense the direction of gravitational pull. In response, cells on the lower side of stems and the upper side of roots grow faster than other cells, causing stems to grow upward and roots to grow downward.

- *What would happen to a plant growing under weightless conditions, such as in an orbiting spacecraft?*

Does Water Move Up a Tree Through Phloem or Xylem?

Before reading this chapter, you may have wondered how water gets to the top of a tree, high above its roots. A tall column of water can have considerable mass. Very large trees can transpire about 100,000 L of water a year—during that tree's growing season the amount might be as much as 270 kg of water a day. Such a tree might move 200 kg of water a day through its trunk from the roots to the leaves in its canopy. So how does a plant manage to move that weight of water?

The answer to this puzzle was first proposed by biologist Otto Renner in 1911. He suggested that dry air moving across the tree's leaves captured water molecules by evaporation, and that this water was replaced with other water molecules coming in from the roots. Renner's idea, which was essentially correct, forms the core of the cohesion–adhesion–tension theory described in this chapter. Essential to the theory is that there is an unbroken water column from leaves to roots, a "pipe" from top to bottom through which the water can move freely.

There are two candidates for the role of water pipe, each a long series of narrow vessels that runs the length of the stem of a tree. As you have learned, these two vessel systems are called xylem and phloem. In principle, either xylem or phloem could provide the plumbing through which water moves up a tree trunk or other stem. How did scientists work out which it is?

An elegant experiment demonstrates which of these vessel systems carries water up a tree stem. A section of a stem was placed in water containing the radioactive potassium isotope ^{42}K. A piece of wax paper was carefully inserted between the xylem and the phloem in a 23-cm section of the stem to prevent any lateral transport of water between xylem and phloem.

After enough time had elapsed to allow water movement up the stem, the 23-cm section of the stem was removed and cut into six segments, and the amounts of ^{42}K were measured both in the xylem and in the phloem of each segment, as well as in the stem immediately above and below the 23-cm section. The amount of radioactivity recorded provided a direct measure of the amount of water that had moved up from the bottom of the stem through either the xylem or the phloem.

The results are presented in the graph.

SuperStock/Alamy Stock Photo

Analysis

1. **Interpreting Data**
   a. In which segments of the stem is there more than 25 ppm ^{42}K in both xylem and phloem? In which parts of the stem is there less than 25 ppm ^{42}K in the phloem?
2. **Making Inferences**
   a. In the 23-cm section, is more ^{42}K found in xylem or in phloem? What might you conclude from this?
   b. Above and below the 23-cm section, is more ^{42}K found in xylem or in phloem? How would you account for this? (*Hint:* These sections did not contain the wax paper barrier that prevented lateral transport between xylem and phloem.)
   c. Within the 23-cm section, the phloem in segments 1 and 6 contains more ^{42}K than interior segments. What best accounts for this?
   d. Is it fair to infer that water could move through either xylem or phloem vessel systems?
3. **Drawing Conclusions** Does water move up a stem through phloem or through xylem? Explain.

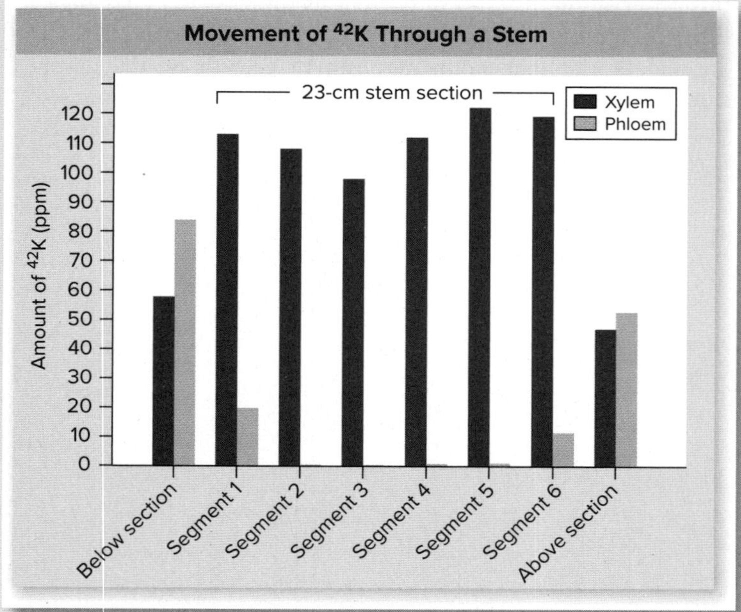

Retracing the Learning Path

CONCEPT 31.1 Water Moves Through Plants Based on Potential Differences

31.1.1 Water Potential Regulates Movement of Water Through the Plant Body The major force for water transport in a plant is the pulling of water by transpiration. Water moves from an area of high water potential to an area of low water potential.

31.1.2 Aquaporins Enhance Osmosis Aquaporins are water channels in plasma membranes that facilitate rapid diffusion of water across membranes.

CONCEPT 31.2 Roots Absorb Minerals and Water

31.2.1 There Are Three Transport Routes into Roots The three water transport routes are the apoplast route between cells, the symplast route through plasmodesmata, and the transmembrane route across cell membranes.

31.2.2 Transport Through the Endodermis Is Selective Casparian strips in the endoderm force water and nutrients to move across the cell membranes, allowing selective flow to the xylem.

CONCEPT 31.3 Xylem Transports Water from Root to Shoot

31.3.1 A Water Potential Gradient from Roots to Shoots Enables Transport Water moves into roots when soil water potential is greater than in roots. Evaporation from leaves creates negative water potential that pulls water upward through the xylem.

31.3.2 Vessels and Tracheids Accommodate Bulk Flow The volume of water that can be transported by a xylem vessel or tracheid is a function of its diameter. Cavitation occurs when a gas bubble forms in a water column and water movement stops.

CONCEPT 31.4 Transpiration Rate Reflects Environmental Conditions

31.4.1 Stomata Open and Close to Balance H_2O and CO_2 Needs More than 90% of the water absorbed by the roots is lost by evaporation through open stomata. Stomata close at high temperatures or when carbon dioxide concentrations increase.

CONCEPT 31.5 Plants Are Adapted to Water Stress

31.5.1 Plant Adaptations to Drought Include Limiting Water Loss Regulating stomatal opening is a short-term response; morphological adaptations are long-term responses.

31.5.2 Plant Responses to Flooding Include Short- and Long-Term Adaptations Prolonged exposure to standing water is more damaging than prolonged exposure to flowing water. Adaptive responses to flooding involve an increase in lenticel size and adventitious root formation.

31.5.3 Plant Adaptations to High Salt Concentration Include Elimination Pneumatophores allow plants to obtain oxygen for root tissues in flooded environments. Halophytic plants dilute, exclude, or secrete excessively salty fluids in which they grow.

CONCEPT 31.6 Phloem Transports Organic Molecules

31.6.1 Organic Molecules Are Transported Up and Down the Shoot Translocation is the bidirectional movement of carbohydrates through the phloem of a plant. Other molecules, such as hormones and mRNA molecules, can also move through the phloem.

31.6.2 Turgor Pressure Differences Drive Bulk Flow in the Phloem In a photosynthetic leaf, active transport of sugars into the phloem reduces water potential. As water moves into the phloem, pressure drives the contents to nonphotosynthetic tissue, where sugar is unloaded.

CONCEPT 31.7 Plants Require a Variety of Nutrients

31.7.1 Plants Require Nine Macronutrients and Seven Micronutrients Macronutrient elements required are C, O, H, N, K, Ca, Mg, P, and S. Micronutrient elements are Cl, Fe, Mn, Zn, B, Cu, and Mb. Hydroponics can be used to identify nutrients that limit plant growth.

CONCEPT 31.8 Plants Use Hormones to Regulate Growth

31.8.1 The Hormones That Guide Plant Growth Are Responsive to the Environment Hormones are produced in one part of a plant and transported to another part, where they bring about physiological or developmental responses.

31.8.2 Auxin Allows Elongation and Organizes the Body Plan Auxins are produced in apical meristems and affect gene expression. Auxins promote stem elongation and cell division, inhibit leaf abscission, and induce ethylene production.

31.8.3 Cytokinins Stimulate Cell Division and Differentiation Cytokinins are commonly produced in the root apical meristem and in developing fruits. In combination with auxin they stimulate cell division and differentiation. The balance between cytokinins and auxins regulates the form of a plant.

31.8.4 Gibberellins Enhance Plant Growth and Nutrient Utilization The giberellins are a large family of plant hormones synthesized in the apical parts of roots and stems; they play important roles in plant development, including stem elongation.

31.8.5 Ethylene Induces Fruit Ripening and Aids Plant Defenses Ethylene is a gas that controls leaf, flower, and fruit abscission, promotes fruit ripening, and suppresses stem and root elongation.

31.8.6 Abscisic Acid Suppresses Growth and Induces Dormancy Abscisic acid inhibits bud growth and the effects of other hormones, induces seed dormancy, and controls stomatal closure.

CONCEPT 31.9 Plant Growth Is Responsive to Light

31.9.1 Phytochrome Facilitates Expression of Light-Response Genes Phytochrome is activated by absorbing red light and inactivated by absorbing far-red light. The active form enters the nucleus to form a transcription complex, leading to expression of light-regulated genes.

31.9.2 Many Growth Responses Are Linked to Phytochrome Action Phytochrome is involved in seed germination, shoot elongation, and detection of plant spacing.

31.9.3 Light Affects Directional Growth Phototropisms are directional growth responses of stems toward blue light.

31.9.4 Circadian Clocks Are Independent of Light but Are Entrained by Light Circadian rhythms are daily cycles controlled by phytochrome and blue-light photoreceptors.

CONCEPT 31.10 Plant Growth Is Sensitive to Gravity

31.10.1 Plants Align with the Gravitational Field Cells in plants perceive gravity when amyloplasts are pulled downward, generating a physiological signal, causing cell elongation in other cells.

Concept Overview

This Concept Overview diagrams the key concepts that were discussed in this chapter.

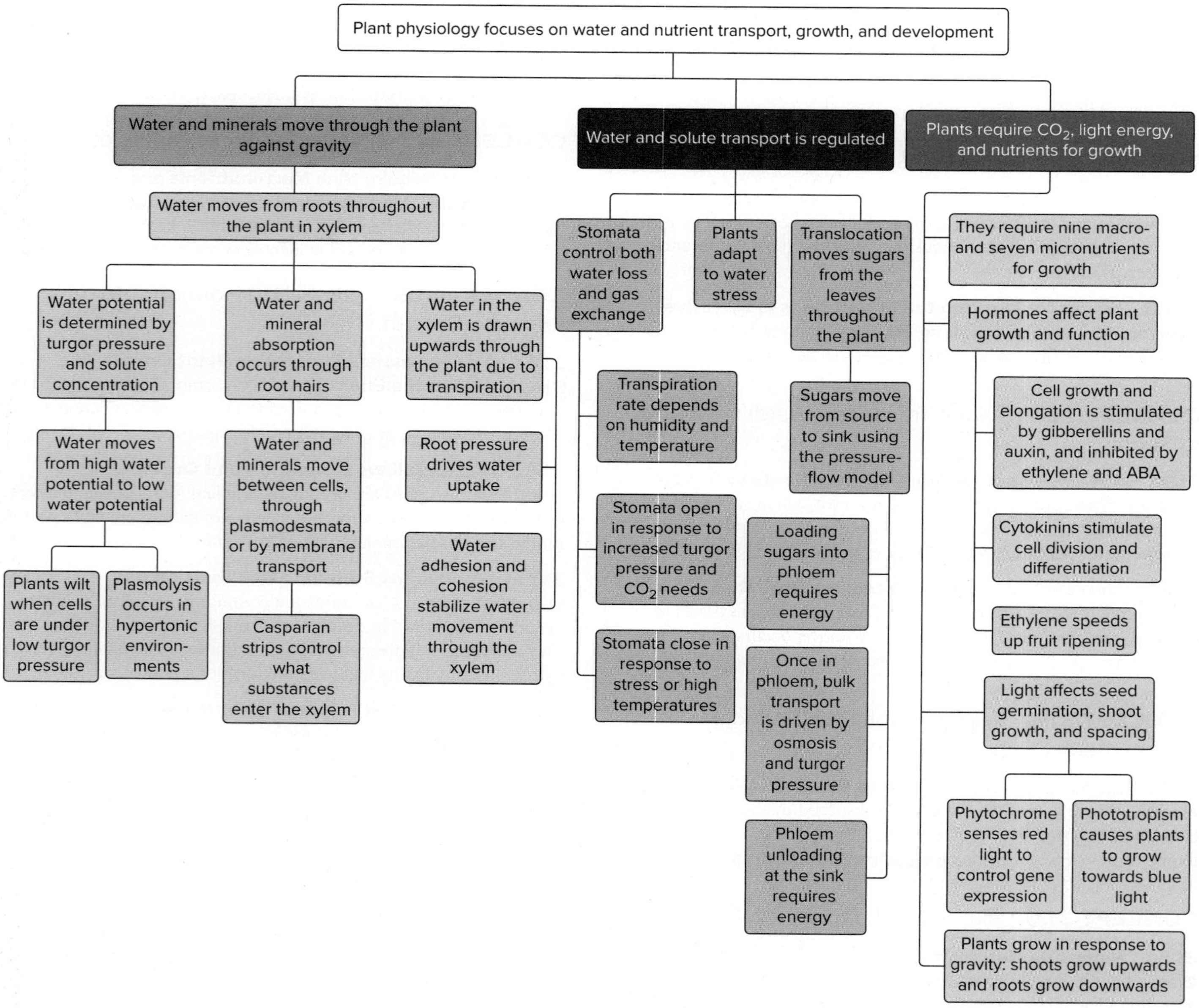

Assessing the Learning Path

Understand

1. The water potential of a plant cell is the
   a. sum of the membrane potential and gravity.
   b. difference between membrane potential and gravity.
   c. sum of the pressure potential and solute potential.
   d. difference between pressure potential and solute potential.
2. Water movement through cell walls is
   a. via the apoplast route.
   b. via the symplast route.
   c. via the transmembrane route.
   d. Both a and b
   e. Both b and c
3. How does the Casparian strip ensure that toxins taken up by root hairs do not enter the stele and get distributed throughout the plant via the vasculature?
   a. By forcing water to enter companion cells, which actively export toxins
   b. By forcing water to enter a cell wall that can filter toxins
   c. By forcing water to cross a selectively permeable cell membrane
   d. By stimulating the efflux of toxins from the xylem of the vasculature
4. If a mutation in an ion pump prevented solutes from being moved into cells of the root, how would such a mutation affect root pressure?
   a. Root pressure would increase.
   b. Root pressure would decrease.
   c. Nothing; root pressure is not related to the movement of ions into the cells of the roots.
5. Stomata open when guard cells
   a. take up potassium and then absorb water.
   b. lose potassium and then absorb water.
   c. take up potassium and then lose water.
   d. cotransport sugars and sodium, then lose water.
6. According to the pressure–flow theory, which of the following is the source for carbohydrates?
   a. Shoots
   b. Roots
   c. Leaves
   d. Stem
7. Auxin promotes plant growth toward a light source by
   a. increasing cell division rates on the shaded side of the stem.
   b. shortening the cells on the light-exposed side of the stem.
   c. causing cells on the shaded side of the stem to elongate.
   d. decreasing the rate of cell division on the light-exposed side of the stem.
8. Which of the following is stimulated by blue light?
   a. Seed germination
   b. Detection of plant spacing
   c. Phototropism
   d. Shoot elongation
9. Nondirectional light-triggered development is mediated by
   a. phytochrome.
   b. SCF.
   c. auxin.
   d. ubiquitin.
10. The tensile strength of a column of water
    a. varies directly with the diameter of the column.
    b. varies directly with the height of the column.
    c. varies inversely with the diameter of the column.
    d. varies inversely with the height of the column.
11. Which of the following is NOT an adaptation to a highly saline environment?
    a. Secretion of salts
    b. Lowering of root water potential
    c. Exclusion of salt
    d. Production of pneumatophores
12. Stems bend away from gravity because of
    a. increased auxin on the upper side.
    b. decreased auxin on the upper side.
    c. increased auxin on the lower side.
    d. decreased auxin on the lower side.

Apply

1. What will happen if a cell with a solute potential of −0.4 MPa and a pressure potential of 0.2 MPa is placed in a chamber filled with pure water that is pressurized to 0.5 MPa?
   a. Water will move out of the cell.
   b. Water will move into the cell.
   c. There will be no net movement of water into or out of the cell.
   d. The cell will be crushed.
   e. The cell will burst.
2. Which of the following would be a consequence of removing the Casparian strip?
   a. Water and mineral nutrients would not be able to reach the xylem.
   b. There would be less selectivity as to what passed into the xylem.
   c. Water and mineral nutrients would be lost from the xylem back into the soil.
   d. Water and mineral nutrients would no longer be able to pass through the cell walls of the endodermis.
3. If you could override the control mechanisms that open stomata and force them to remain closed, what would you expect to happen to the plant?
   a. Sugar synthesis would likely slow down.
   b. Water transport would likely slow down.
   c. Both a and b
   d. Neither a nor b
4. If you exposed seeds to a series of red-light versus far-red-light treatments, which of the following exposures would result in seed germination?
   a. Red; far-red
   b. Far-red; red
   c. Red; far-red; red; far-red; red; far-red; red; far-red
5. What would be the effect on stomata if the ABA receptor on guard cell stomata was mutated so that it could no longer bind ABA?
   a. Stomata would be stuck open due to efflux of K^+, Cl^-, and malate^{2-} into guard cells.
   b. Stomata would be stuck closed due to influx of water into guard cells.
   c. Stomata would be stuck open because binding of ABA promotes stomata closure.
   d. Stomata would be stuck closed because binding of ABA promotes water influx.

6. What do you predict would happen if you repeated the Frits Went experiment (figure 31.17) but also illuminated the side of the seedling on which the agar block is placed?
   a. Nothing; the source of the auxin is on the other side, and so the auxin cannot be redistributed due to the illumination.
   b. The seedling would bend away from the light more.
   c. The seedling would bend toward the light and straighten up, possibly even bending more toward the light than away.
7. What might happen if you repeated the Frits Went experiment using a mutant plant that had a mutation in the *Aux/IAA* gene, and that mutation prevented ubiquitin being added to the Aux/IAA protein by SCF/TIR1?
   a. You would get no bending of the coleoptile, assuming bending is dependent on gene expression in response to auxin.
   b. You would get the same kind of bending as seen in the original experiment, assuming bending is based on gene expression in response to auxin.
   c. You would get more extreme bending than seen in the original experiment, assuming bending is dependent on gene expression in response to auxin.
8. You have isolated a new gene controlling gravitropism in *Arabidopsis*. You have named the gene *tribblestones*. Which of the following processes could the gene be involved in?
   a. Auxin synthesis in the shoot apical meristem
   b. Starch synthesis
   c. The uptake of malate ions by guard cells
   d. Cytokinin hormone reception in the cuticle

Synthesize

1. If you feed a houseplant with fertilizer too often, it may look wilted despite the soil being wet. Explain what has happened in terms of water potential.
2. If a mutation increased the radius of a xylem vessel threefold, how would the movement of water through the plant be affected? Be specific and try to calculate the change in water movement based on the change in the diameter of the xylem.
3. Compare how halophytes and sharks meet the challenge of living in a hyperosmotic environment.
4. Aphids extract food from a plant's phloem with a piercing mouthpart called a stylet. The plant's transport tissues are well below the surface. How does a hungry aphid avoid piercing xylem?
5. Describe an experiment to determine the amount of boron needed for the normal growth of tomato seedlings.
6. You are given seed of a plant with a mutation in the protein kinase domain of phytochrome. Would you expect to see any red-light-mediated responses when you germinated the seed? Explain your answer.

Part VII Animal Form and Function

Stockbyte/Getty Images

32 The Animal Body and How It Moves

Learning Path

32.1 The Vertebrate Body Has a Hierarchical Organization

32.2 Epithelial Tissue Covers Body Surfaces

32.3 Nerve Tissue Conducts Signals Rapidly

32.4 Connective Tissue Supports the Body

32.5 Muscle Tissue Powers the Body's Movements

32.6 Skeletal Systems Anchor the Body's Muscles

32.7 Vertebrate Endoskeletons Are Made of Bone

32.8 Muscles Contract Because Their Myofilaments Slide

32.9 Animal Locomotion Takes Many Forms

Concept Overview

This Concept Overview provides a diagram of the overarching concepts that are covered in this chapter. These concepts will be expanded upon in the Concept Overview at the end of the chapter.

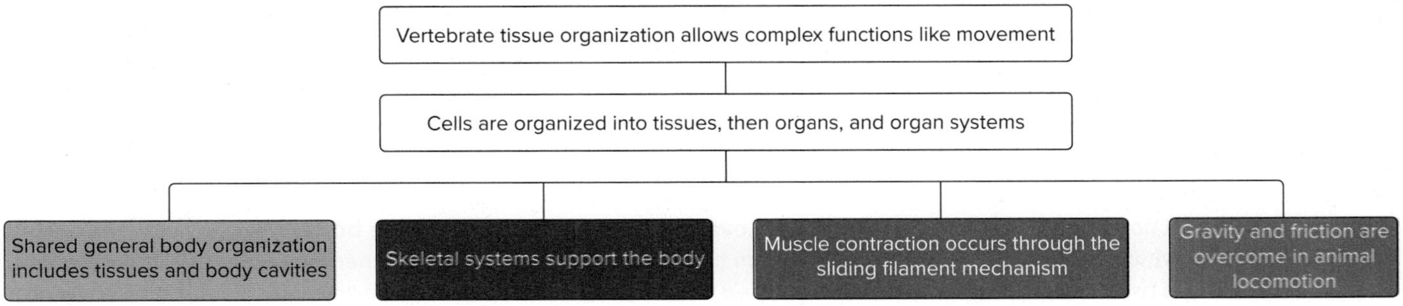
Vertebrate tissue organization allows complex functions like movement
Cells are organized into tissues, then organs, and organ systems
Shared general body organization includes tissues and body cavities
Skeletal systems support the body
Muscle contraction occurs through the sliding filament mechanism
Gravity and friction are overcome in animal locomotion

Introduction

The organization of cells into tissues was an important evolutionary innovation in multicellular animals. Tissues in turn are organized into the organs and organ systems vital to complex animals like yourself. When people think of animals, they may think of a cat or a dog, perhaps a fish in an aquarium, or even an owl like the one pictured on the previous page. Despite obvious differences, a dog, a fish, and a bird are all vertebrates with a similar basic body plan and similar tissues and organs. In this chapter we begin a detailed consideration of the biology of animals, concentrating on vertebrates, but including other examples for comparison. We begin with the different types of tissues and go on to examine how the vertebrate body can move. Complex movements of the body are a hallmark of vertebrates. This involves a semirigid skeletal system, joints that act as hinges, and a muscular system to pull on this skeleton. When the great horned owl in the picture takes off in flight, its wings exert force on the air, literally pushing against it. Similarly, when you run, your feet push against the ground, shoving you forward. The complex system of muscle and bone that allows these movements is a marvel of evolution.

32.1 The Vertebrate Body Has a Hierarchical Organization

The vertebrate body has four levels of organization: (1) cells, (2) tissues, (3) organs, and (4) organ systems (figure 32.1). All animals are composed of many different cell types. There is a more than 150-year history of "counting" cell types in humans, but we still cannot give an exact number. It is estimated that more than 300 cell types arise during development and exist in the adult body. An international effort is under way to construct a Human Cell Atlas based on molecular phenotypes of cells.

Vertebrate Bodies Are Formed from Four Tissues

LEARNING OBJECTIVE 32.1.1 List the kinds of tissues found within the vertebrate body.

Tissues are groups of cells of a single type and function

An important early developmental event produces the three fundamental embryonic tissue types: *endoderm, mesoderm,* and *ectoderm* (refer to chapter 36). These three **germ layers** will give rise to all adult tissue types. Table 32.1 provides a simplified list of the fate of these germ layers. A tissue in the adult is an organized group of cells with similar features and the same developmental origin.

In adult vertebrates, there are four kinds of *primary tissues:* (1) **epithelial,** (2) **nerve,** (3) **connective,** and (4) **muscle.** Each type is discussed in a separate section of this chapter.

Organs and organ systems provide specialized functions

Organs are body structures composed of several different types of tissues that form a structural and functional unit. One example is the heart, which contains cardiac muscle, connective tissue, and epithelial tissue. Nerve tissue connects the brain and spinal cord to the heart and helps regulate the heartbeat.

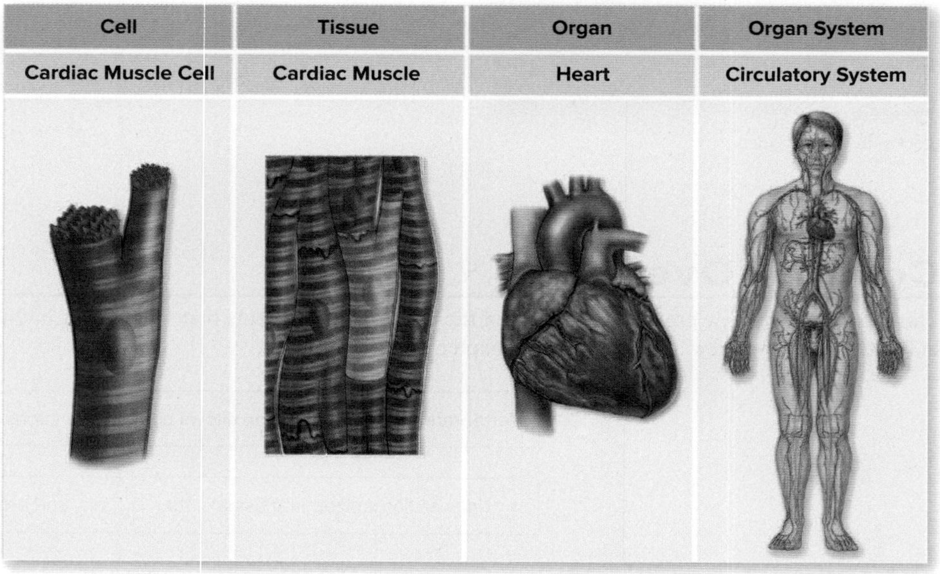

Figure 32.1 Levels of organization within the body. Similar cell types operate together and form tissues. Tissues functioning together form organs such as the heart, which is composed primarily of cardiac muscle with a lining of epithelial tissue. An organ system consists of several organs working together to carry out a function for the body. An example of an organ system is the circulatory system, which consists of the heart, blood vessels, and blood.

An **organ system** is a group of organs that cooperate to perform the major activities of the body. For example, the circulatory system is composed of the heart and blood vessels (refer to chapter 34). These organs cooperate in the transport of blood and help distribute substances about the body. The vertebrate body contains 11 principal organ systems.

The General Body Plan of Vertebrates Is a Tube Within a Tube

LEARNING OBJECTIVE 32.1.2 Describe how body cavities are organized in vertebrates.

The bodies of all vertebrates have the same general architecture. The body plan is essentially a tube suspended within a tube. The inner tube is the digestive tract, a long tube that travels from the mouth to the anus. An internal skeleton made of jointed bones or cartilage that grows as the body grows supports the outer tube, which forms the main vertebrate body. The outermost layer of the vertebrate body is the integument, or skin, and its many accessory organs and parts—hair, feathers, scales, and sweat glands.

TABLE 32.1	Developmental Fates of Vertebrate Primary Germ Layers
Ectoderm	Epidermis of skin, nervous system, sense organs
Mesoderm	Skeleton, muscles, blood vessels, heart, blood, gonads, kidneys, dermis of skin
Endoderm	Lining of digestive and respiratory tracts, liver, pancreas, thymus, thyroid

Vertebrates have both dorsal and ventral body cavities

Inside the main vertebrate body are two identifiable cavities. The *dorsal body cavity* forms within a bony skull and a column of bones, the vertebrae. The skull surrounds the brain, and within the stacked vertebrae is a channel that contains the spinal cord.

The *ventral body cavity* is much larger and extends anteriorly from the area bounded by the rib cage and vertebral column and posteriorly to the area contained within the ventral body muscles (the abdominals) and the pelvic girdle. In mammals, a sheet of muscle, the diaphragm, breaks the ventral body cavity anteriorly into the *thoracic cavity,* which contains the heart and lungs, and posteriorly into the *abdominopelvic cavity,* which contains many organs, including the stomach, intestines, liver, kidneys, reproductive organs, and urinary bladder (figure 32.2*a*).

Recall from the discussion of the animal body plan in chapter 27 that a coelom is a fluid-filled body cavity completely formed within the embryonic mesoderm layer of some animals (vertebrates included). The coelom is present in vertebrates, but compared with that in invertebrates it is constricted, folded, and subdivided. The mesodermal layer that lines the coelom extends from the body wall to envelop and suspend several organs within the ventral body cavity (figure 32.2*b*). In the abdominopelvic cavity, the coelomic space is the *peritoneal cavity.*

In the thoracic cavity, the heart and lungs occupy and greatly constrict the coelomic space. The thin space within mesodermal layers around the heart is the *pericardial cavity,* and the two thin spaces around the lungs are the *pleural cavities* (see figure 32.2*b*).

REVIEW OF CONCEPT 32.1

The body's cells are organized into tissues, which are organized into organs and organ systems. Tissue types include epithelial, connective, muscle, and nerve tissue. Mammals have a dorsal and a ventral cavity, which is divided by the diaphragm into thoracic and abdominopelvic cavities. The adult coelom subdivides into the peritoneal, pericardial, and pleural cavities.

- *Can an organ be made of more than one tissue?*

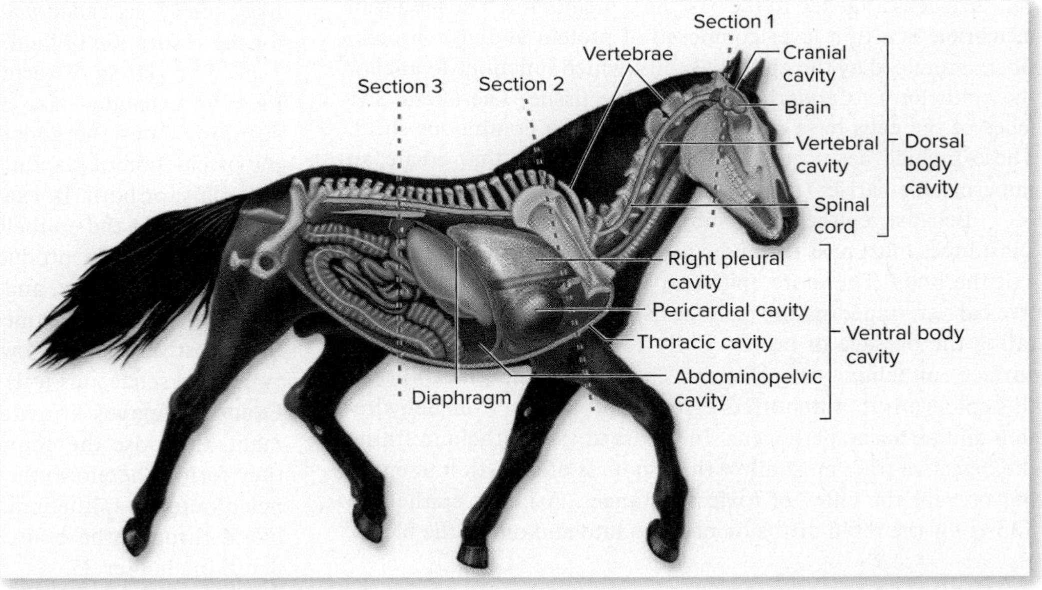

a.

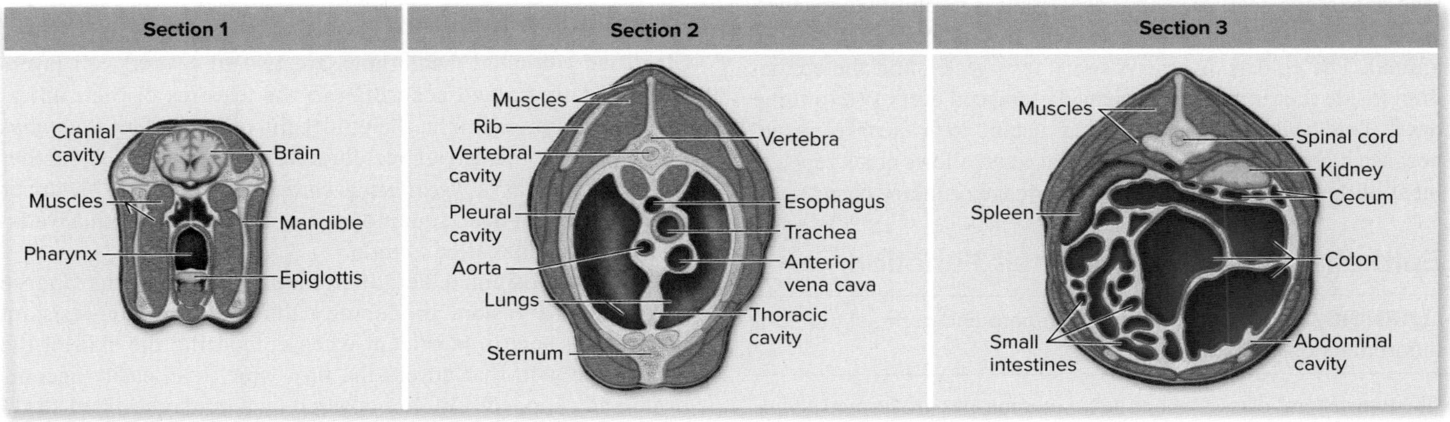

b.

Figure 32.2 Architecture of the vertebrate body. ***a.*** All vertebrates have dorsal and ventral body cavities. The dorsal cavity divides into the cranial (contains the brain) and vertebral (contains the spinal cord) cavities. In mammals, a muscular diaphragm divides the ventral cavity into the thoracic and abdominopelvic cavities. ***b.*** Cross sections through three body regions show the relationships among body cavities, major organs, and coeloms (pericardial, pleural, and peritoneal cavities).

32.2 Epithelial Tissue Covers Body Surfaces

An epithelial membrane, or **epithelium** (plural, *epithelia*), covers every surface of the vertebrate body. Epithelial tissues provide the linings and coverings of organs and body cavities, they constitute the outer surface of bodies (skin), and they are the primary tissues found in glands. Together with connective tissues, they form membranes. Epithelial tissues can arise from all three embryonic germ layers: endoderm, mesoderm, and ectoderm. For example, in vertebrates the epidermis of the skin arises from ectoderm, but the epithelial lining of blood vessels arises from mesoderm.

Epithelium Covers Body Surfaces

LEARNING OBJECTIVE 32.2.1 Describe the structure and function of an epithelium.

Epithelial tissue is organized into connected sheets of cells. The cells have an apical surface that is exposed either to the outside environment, such as in skin, or internally, to the lumen of tubes like blood vessels and the digestive tract. The opposite, or basal, cell surface is attached to the basement membrane. The **basement membrane** is a thin layer composed of protein and glycoprotein fibers produced by the epithelial cells, which functions to anchor the epithelium to underlying connective tissue. The lateral surfaces of the cells have connections to form a continuous sheet. These include desmosomes and tight junctions that create an impermeable barrier (refer to figure 4.26).

Because epithelial membranes cover all body surfaces, substances must pass through an epithelium in order to enter or exit the body. Therefore epithelial membranes provide a selective barrier, impeding the passage of some substances and facilitating the passage of others. Fully 15% of your body weight is surface epithelium, or epidermis. The relative impermeability of this epidermis to water offers essential protection from dehydration and airborne pathogens. In contrast, the epithelium lining the digestive tract must allow the products of digestion to enter, but prevent the entry of toxic substances. A lung's epithelium allows for the rapid diffusion of gases into and out of the blood.

Epithelial regeneration

One remarkable feature of epithelial tissue is its ability to regenerate tissue, constantly replacing its cells throughout the life of the animal. The epidermis renews every two weeks, and the epithelium inside the stomach is completely replaced every two to three days. This ability to regenerate is useful in a surface tissue, because it constantly renews the surface and allows quick replacement of the protective layer, should damage or injury occur.

Epithelial Types Reflect Their Function

LEARNING OBJECTIVE 32.2.2 Compare and contrast the different kinds of epithelia.

The two general classes of epithelial membranes are termed *simple* (single layer of cells) and *stratified* (multiple layers of cells). These classes are further subdivided into squamous, cuboidal, and columnar, based on the shape of the cells: *squamous cells* are flat, *cuboidal cells* are about as wide as they are tall, and *columnar cells* are taller than they are wide (table 32.2).

Simple epithelium

Simple epithelial membranes are one cell thick. A simple squamous epithelium is composed of epithelial cells that have a flattened shape when viewed in cross section. Examples of such membranes are those that line the lungs and blood capillaries, where the thin, delicate nature of these membranes permits the rapid movement of molecules (such as the diffusion of gases).

A *simple cuboidal epithelium* lines kidney tubules and several glands. In the case of glands, these cells are specialized for secretion.

A *simple columnar epithelium* lines the airways of the respiratory tract and the inside of most of the gastrointestinal tract, among other locations. Interspersed among the columnar epithelial cells of mucous membranes are numerous *goblet cells,* which are specialized to secrete mucus. The columnar epithelial cells of the respiratory airways contain cilia on their apical surface (the surface facing the lumen, or cavity), which move mucus and dust particles toward the throat. In the small intestine, the apical surface of the columnar epithelial cells forms fingerlike projections called *microvilli,* which increase the surface area for the absorption of food.

The glands of vertebrates form from invaginated epithelia. The expanded size of both cuboidal and columnar cells accommodates the added intracellular machinery needed for the production of glandular secretions, the active absorption of materials, or both. In **exocrine glands,** the connection between the gland and the epithelial membrane remains as a duct. The duct channels the product of the gland to the surface of the epithelial membrane, and thus to the external environment (or to an interior compartment that opens to the exterior, such as the digestive tract). A few examples of exocrine glands include sweat and sebaceous (oil) glands, as well as the salivary glands. **Endocrine glands** are called ductless glands. During development, they lose the connection to the epithelium from which they form. Therefore, the hormones they secrete are not channeled onto an epithelium, but enter capillaries and circulate in blood through the body. We will cover these glands in more detail in chapter 35.

Stratified epithelium

Stratified epithelial membranes are two to several cell layers thick and are named according to the features of their apical cell layers. For example, the epidermis is a *stratified squamous epithelium.* In terrestrial vertebrates, the epidermis is further characterized as a *keratinized epithelium,* because its upper layer consists of dead squamous cells and is filled with a water-resistant protein called *keratin.*

The deposition of keratin in the skin increases in response to repeated abrasion, producing calluses. The water-resistant property of keratin is evident when comparing the skin of the face to the skin that covers the lips, which can easily become dried and chapped. Lips are covered by a nonkeratinized stratified squamous epithelium.

TABLE 32.2 Epithelial Tissue

SIMPLE EPITHELIUM

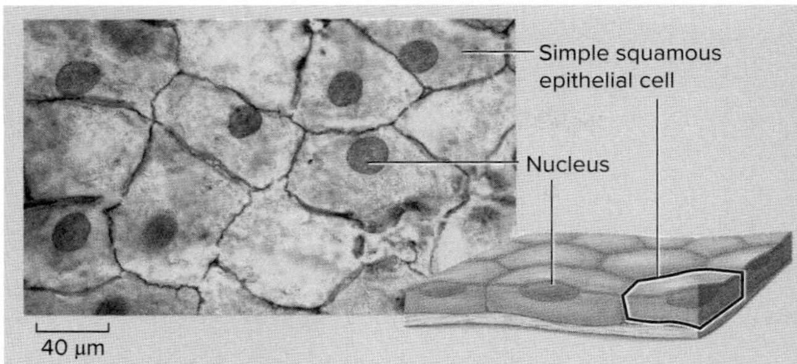
Simple squamous epithelial cell
Nucleus
40 µm

Ed Reschke

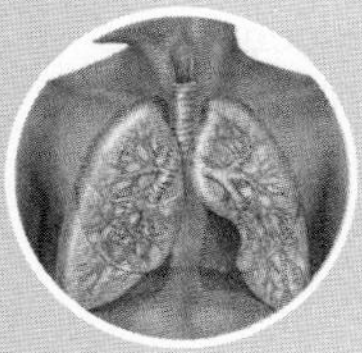

Squamous

Typical Location

Lining of lungs, capillary walls, and blood vessels

Function

Cells form thin layer across which diffusion can readily occur

Characteristic Cell Types

Epithelial cells

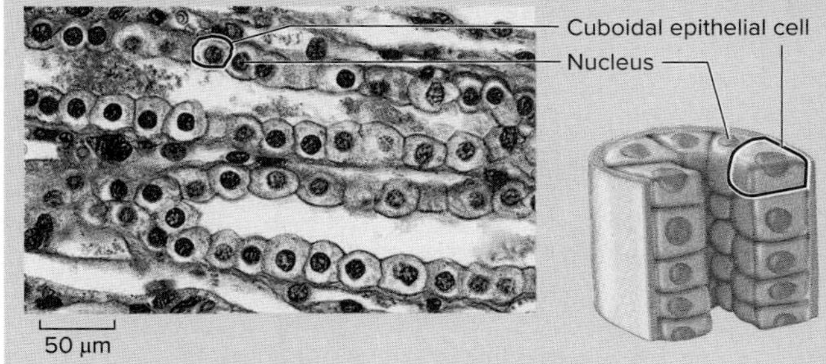
Cuboidal epithelial cell
Nucleus
50 µm

©Victor P. Eroschenko

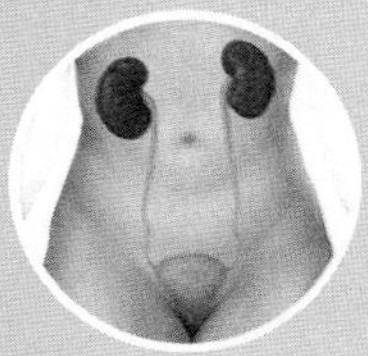

Cuboidal

Typical Location

Lining of some glands and kidney tubules; covering of ovaries

Function

Cells rich in specific transport channels; functions in secretion and absorption

Characteristic Cell Types

Gland cells

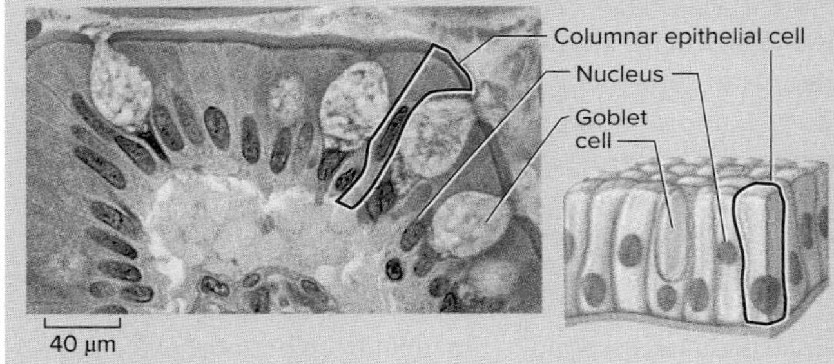
Columnar epithelial cell
Nucleus
Goblet cell
40 µm

Ed Reschke

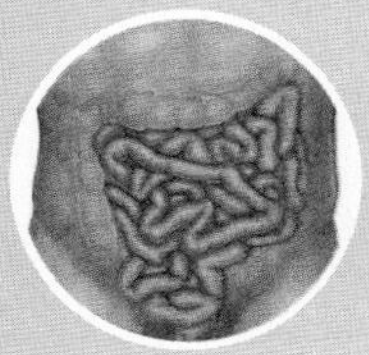

Columnar

Typical Location

Surface lining of stomach, intestines, and parts of respiratory tract

Function

Thicker cell layer; provides protection and functions in secretion and absorption

Characteristic Cell Types

Epithelial cells

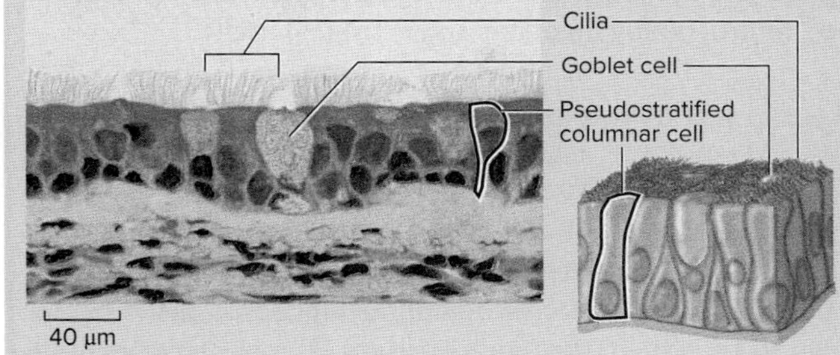
Cilia
Goblet cell
Pseudostratified columnar cell
40 µm

©Victor P. Eroschenko

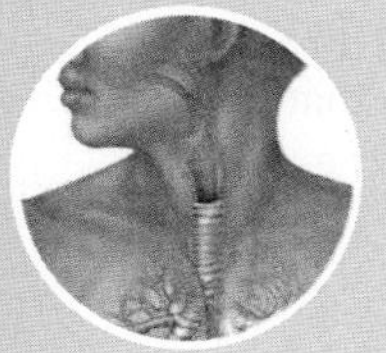

Pseudostratified Columnar

Typical Location

Lining of parts of the respiratory tract

Function

Secretes mucus; dense with cilia that aid in movement of mucus; provides protection

Characteristic Cell Types

Gland cells; ciliated epithelial cells

STRATIFIED EPITHELIUM

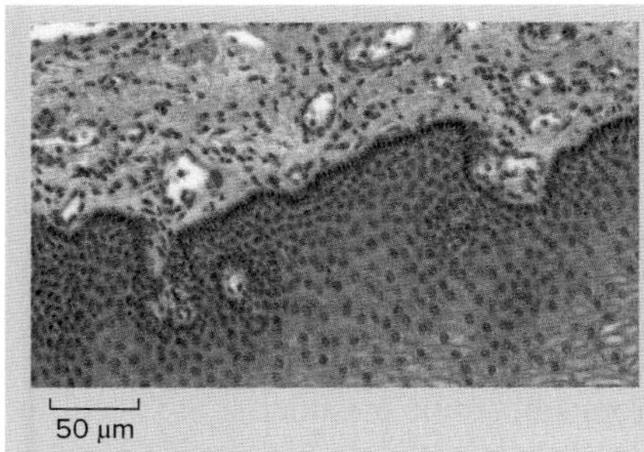
50 µm

Al Telser/McGraw Hill

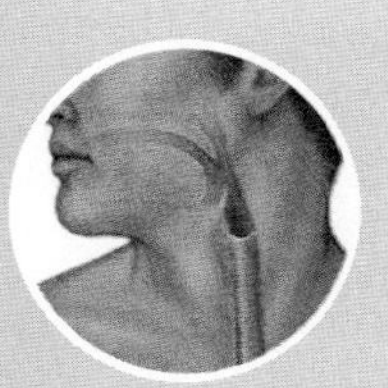

Squamous

Typical Location

Outer layer of skin; lining of mouth

Function

Tough layer of cells; provides protect on

Characteristic Cell Types

Epithelial cells

REVIEW OF CONCEPT 32.2

Epithelial tissues cover all body surfaces both inside and outside. Glands are formed from epithelial tissue. Epithelia form sheets of cells with a basal surface attached to connective tissue and an apical surface that is free. Epithelia may be specialized for protection or for transport and secretion. Simple epithelium has a single cell layer and is classified as squamous, cuboidal, columnar, or pseudostratified. A stratified epithelium contains two or more cell layers.

- *How does the epithelium in a gland function differently from that in the lining of the gut?*

32.3 Nerve Tissue Conducts Signals Rapidly

The second major class of vertebrate tissue is nerve tissue. Its cells include neurons and their supporting cells, called neuroglia. Neurons are specialized to produce and conduct electrochemical events, or impulses. We will examine the function of both neurons and the nervous system in detail in chapter 33.

Neurons Are the Primary Cells of the Nervous System

LEARNING OBJECTIVE 32.3.1 Describe the basic structure of neurons and their supporting cells.

Most **neurons** consist of three parts: a cell body, dendrites, and an axon (figure 32.3). The *cell body* of a neuron contains the nucleus. *Dendrites* are thin, highly branched extensions that receive incoming stimulation and conduct electrical impulses to the cell body. The *axon* is a single extension of cytoplasm that conducts impulses away from the cell body. Axons and dendrites can be quite long. For example, the cell bodies of neurons that control the muscles in your feet lie in the spinal cord, and their axons may extend over a meter to your feet.

Neuroglia provide support for neurons

Neuroglia do not conduct electrical impulses but, instead, support and insulate neurons and eliminate foreign materials in and around neurons. In many neurons, neuroglial cells associate with the axons and form an insulating covering, a *myelin sheath,* produced by successive wrapping of the membrane around the axon. Gaps in the myelin sheath, known as *nodes of Ranvier,* serve as sites for accelerating an impulse (refer to chapter 33).

Two divisions of the nervous system coordinate activities

The nervous system is divided into the **central nervous system (CNS),** which includes the brain and spinal cord, and the **peripheral nervous system (PNS),** which includes *nerves* and *ganglia.* Nerves consist of axons in the PNS that are bundled together in much the same way that wires are bundled together in a cable. Ganglia are collections of neuron cell bodies. The role of the CNS is to integrate, interpret, and respond to the input from both the external and internal environments. The PNS carries signals to the CNS from a wide variety of sensory neurons, and carries signals away from the CNS to control bodily functions and activate muscles.

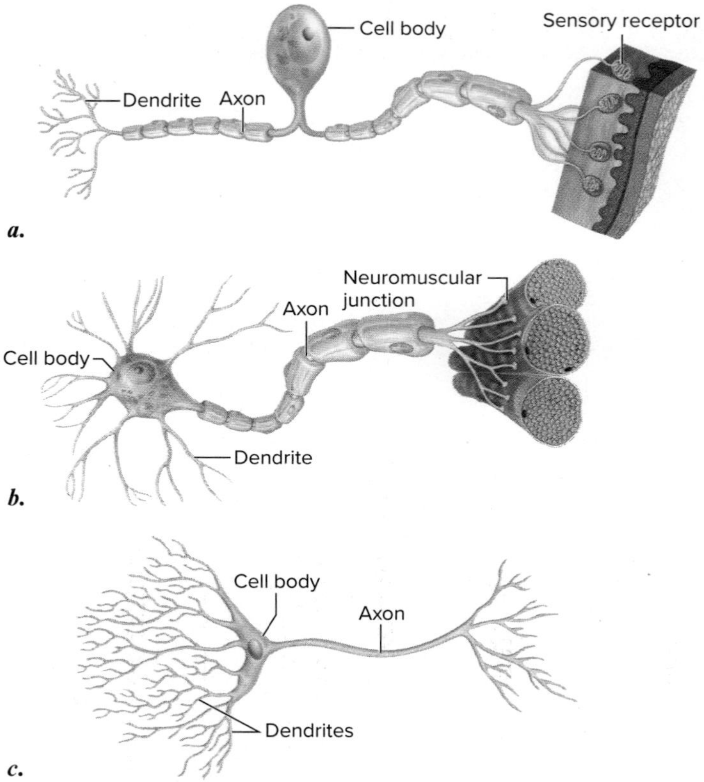

Figure 32.3 The three basic types of neurons. ***a.*** Sensory neurons receive information both internally and externally. ***b.*** Motor neurons stimulate muscles and glands. ***c.*** Interneurons integrate information by conducting signals between neurons.

REVIEW OF CONCEPT 32.3

Nerve tissue is composed of neurons and neuroglia. Neurons include a cell body with a nucleus, dendrites that receive incoming signals, and axons that conduct impulses away from the cell body. Neuroglia have support functions, and provide insulation to axons. The nervous system consists of central and peripheral components.

- *Surface-area-to-volume ratio limits cell size. How do neurons reach up to a meter in length in spite of this?*

32.4 Connective Tissue Supports the Body

Connective tissues derive from embryonic mesoderm and occur in many different forms. We divide these various forms into two major classes: *connective tissue proper,* which further divides into loose and dense connective tissues, and **special connective tissues,** which include cartilage, bone, and blood.

Connective Tissue Proper May Be Either Loose or Dense

LEARNING OBJECTIVE 32.4.1 Differentiate the structure and function of loose and dense connective tissue.

At first glance, it may seem odd that such diverse tissues are in the same category, yet all connective tissues share a common structural feature: they all have abundant extracellular material, because their cells are spaced widely apart. This extracellular material is called the **matrix** of the tissue. In bone, the matrix contains crystals that make the bones hard; in blood, the matrix is plasma, the fluid portion of the blood. The matrix itself consists of protein fibers and **ground substance,** the fluid material between cells and fibers containing a diverse array of proteins and polysaccharides.

During the development of both loose and dense connective tissues, cells called fibroblasts produce and secrete the extracellular matrix. Loose connective tissue contains other cells as well, including mast cells and macrophages—cells of the immune system.

Loose connective tissue

Loose connective tissue consists of cells scattered within a matrix that contains a large amount of ground substance. This gelatinous material is strengthened by a loose scattering of protein fibers such as collagen, which supports the tissue by forming a meshwork (figure 32.4); elastin, which makes the tissue elastic; and reticulin, which helps support the network of collagen. The flavored gelatin of certain desserts consists primarily of extracellular material extracted from the loose connective tissues of animals.

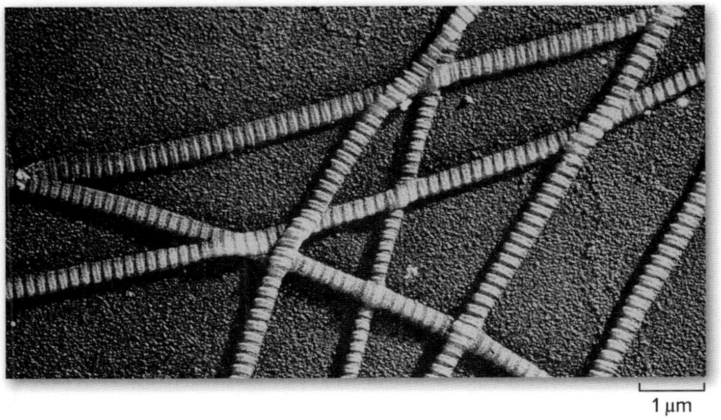

Figure 32.4 Collagen fibers. These fibers, shown under an electron microscope, are composed of many individual collagen strands and can be very strong under tension.
J. Gross/Biozentrum, University of Basel/Science Source

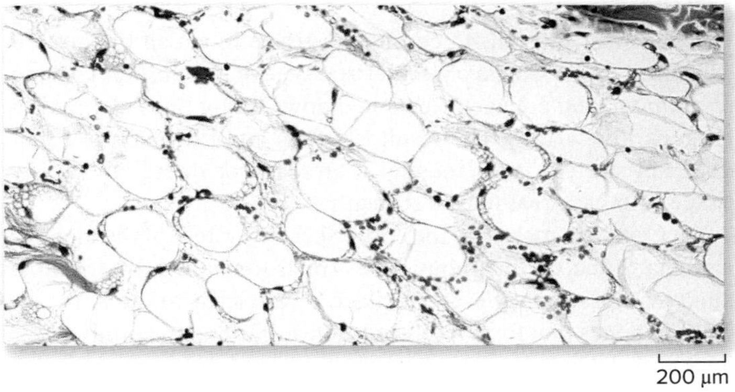

Figure 32.5 Adipose tissue. Fat is stored in globules of adipose tissue, a type of loose connective tissue. As a person gains or loses weight, the size of the fat globules swells or shrinks. Losing weight does not decrease a person's number of fat cells.
Biophoto Associates/Science Source

Adipose cells, more commonly termed fat cells, are important for nutrient storage, and they occur in loose connective tissue. In certain areas of the body, including under the skin, in bone marrow, and around the kidneys, these cells can develop in large groups, forming **adipose tissue** (figure 32.5).

Each adipose cell contains a droplet of triglycerides within a storage vesicle. When needed for energy, the adipose cell hydrolyzes its stored triglyceride and secretes fatty acids into the blood for oxidation by the cells of the muscles, liver, and other organs. The number of adipose cells in an adult is generally fixed, although there is some small turnover of these cells. When a person gains weight, the cells become larger, and when weight is lost, the cells shrink.

Dense connective tissue

Dense connective tissue, with less ground substance, contains tightly packed collagen fibers, making it stronger than loose connective tissue. It consists of two types: regular and irregular. The collagen fibers of *dense regular connective tissue* line up in parallel, like the strands of a rope. This is the structure of tendons, which bind muscle to bone, and ligaments, which bind bone to bone.

In contrast, the collagen fibers of *dense irregular connective tissue* have many different orientations. This type of connective tissue produces the tough coverings that package organs, such as the capsules of the kidneys and adrenal glands. It also covers muscle, nerves, and bones.

Special Connective Tissues Have Unique Characteristics

LEARNING OBJECTIVE 32.4.2 Describe cartilage, bone, and blood tissue.

The special connective tissues—cartilage, bone, and blood—each have unique cells and matrices that allow them to perform their specialized functions.

Cartilage

Cartilage is a specialized connective tissue in which the ground substance forms from a characteristic type of glycoprotein, called *chondroitin,* and collagen fibers laid down along lines of stress in long, parallel arrays. The result is a firm and flexible tissue that does not stretch, is far tougher than loose or dense connective tissue, and has great tensile strength.

Cartilage makes up the entire skeletal system of the modern agnathans and cartilaginous fishes (refer to chapter 28). In most adult vertebrates, however, cartilage is restricted to the joint surfaces of bones that form freely movable joints and to certain other locations. In humans, for example, the tip of the nose, the outer ear, the intervertebral disks of the backbone, the larynx, and a few other structures are composed of cartilage.

Chondrocytes, the cells of cartilage, live within spaces called **lacunae** within the cartilage ground substance. These cells remain alive even though there are no blood vessels within the cartilage matrix; they receive oxygen and nutrients by diffusion through the cartilage ground substance from surrounding blood vessels. This diffusion can occur only because the cartilage matrix is well hydrated and not calcified, as is bone.

Bone

Bone cells, or **osteocytes,** remain alive even though the extracellular matrix becomes hardened with crystals of calcium phosphate. Blood vessels travel through central canals into the bone, providing nutrients and removing wastes. Osteocytes extend cytoplasmic processes toward neighboring osteocytes through tiny canals, or *canaliculi.* Osteocytes communicate with the blood vessels in the central canal through this cytoplasmic network.

In the course of fetal development, the bones of vertebrate fins, arms, and legs, among other appendages, are first "modeled" in cartilage. The cartilage matrix then calcifies at particular locations, so that the chondrocytes are no longer able to obtain oxygen and nutrients by diffusion through the matrix. Living bone replaces the dying and degenerating cartilage. Bone is described in more detail in section 32.7, when we consider how bone and muscle function together to enable complex movement.

Blood

We classify *blood* as a connective tissue because it contains abundant extracellular material, the fluid plasma. The cells of blood are *erythrocytes,* or red blood cells, and *leukocytes,* or white blood cells. Blood also contains platelets, or *thrombocytes,* which are fragments of a type of bone marrow cell. We discuss blood more fully in chapter 34.

All connective tissues have similarities

Although the descriptions of the types of connective tissue suggest numerous different functions for these tissues, they have some similarities. As mentioned, connective tissues originate as embryonic mesoderm, and they all contain abundant extracellular material called matrix; however, the extracellular matrix material is different in different types of connective tissue. Embedded within the extracellular matrix of each tissue type are varieties of cells, each with specialized functions.

REVIEW OF CONCEPT 32.4

Connective tissues are characterized by extracellular materials forming a matrix between loosely organized cells. Connective tissue proper is either loose or dense. Special connective tissues have a unique extracellular matrix. Cartilage has a matrix of organic materials, bone has calcium crystals, and blood has a fluid called plasma.

- *Why is blood considered connective tissue?*

32.5 Muscle Tissue Powers the Body's Movements

Muscles are the motors of the vertebrate body. The characteristic that makes muscle cells unique is the relative abundance and organization of actin and myosin filaments within them. Although these filaments form a fine network in all eukaryotic cells, where they contribute to movement of materials within the cell, they are far more abundant and organized in muscle cells, which are specialized for contraction.

Vertebrates Possess Three Kinds of Muscle

LEARNING OBJECTIVE 32.5.1 Compare and contrast the three kinds of muscle and muscle cells.

Vertebrates possess three kinds of muscle: *smooth, skeletal,* and *cardiac* (table 32.3). Skeletal and cardiac muscles are also known as *striated muscles,* because their cells appear to have transverse stripes when viewed in longitudinal section under the microscope. The contraction of each skeletal muscle is under voluntary control, whereas the contraction of cardiac and smooth muscles is generally involuntary.

Smooth muscle is found in most organs

Smooth muscle was the earliest form of muscle to evolve, and it is found throughout most of the animal kingdom. In vertebrates, smooth muscle occurs in the organs of the internal environment, or *viscera,* and is also called *visceral muscle.* Smooth muscle tissue is arranged into sheets of long, spindle-shaped cells, each cell containing a single nucleus. In vertebrates, muscles of this type line the walls of many blood vessels. In other smooth muscle tissues, such as those in the wall of the digestive tract, the muscle cells themselves may spontaneously initiate electrical impulses, leading to a slow, steady contraction of the tissue.

Skeletal muscle moves the body

Skeletal muscles are usually attached to bones by tendons, so that their contraction causes the bones to move at their joints. A skeletal muscle is made up of numerous, very long muscle cells called **muscle fibers,** which have multiple nuclei. The fibers lie parallel

TABLE 32.3 Muscle Tissue

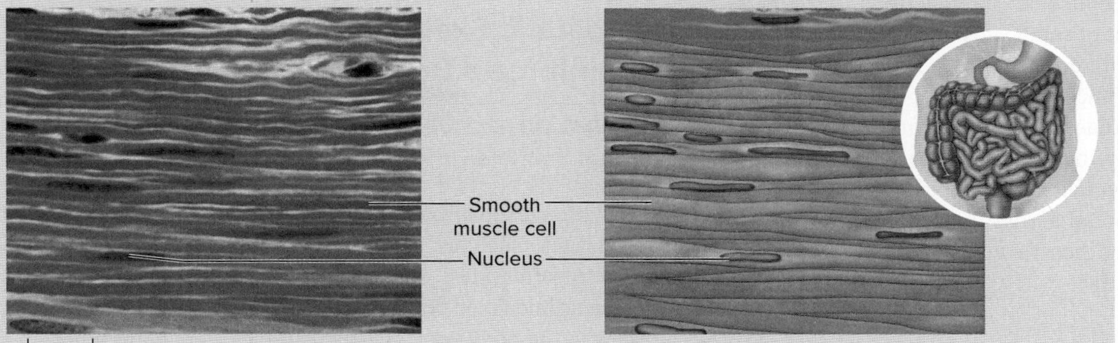
Smooth muscle cell
Nucleus
40 µm

Ed Reschke

Smooth Muscle

Typical Location

Walls of blood vessels; hollow organs such as the stomach and uterus; and intestines

Function

Powers rhythmic, involuntary contractions commanded by the central nervous system

Characteristic Cell Types

Smooth muscle cells

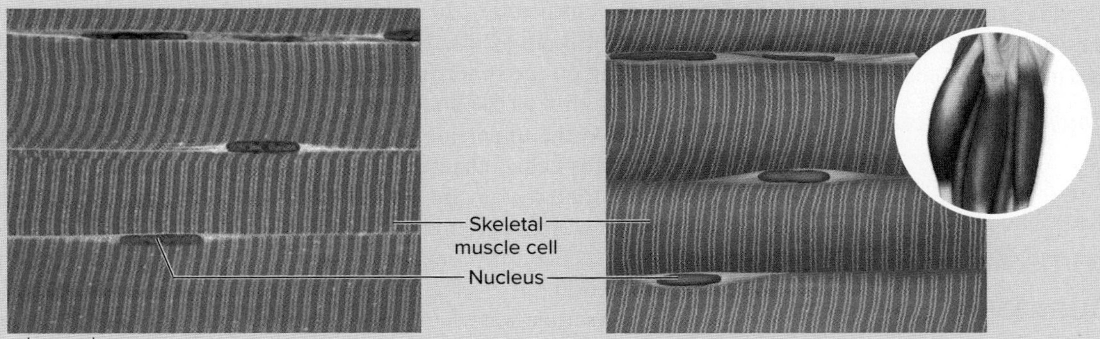
Skeletal muscle cell
Nucleus
100 µm

Ed Reschke

Skeletal Muscle

Typical Location

Voluntary muscles

Function

Powers walking, lifting, talking, and all other voluntary movement

Characteristic Cell Types

Skeletal muscle cells

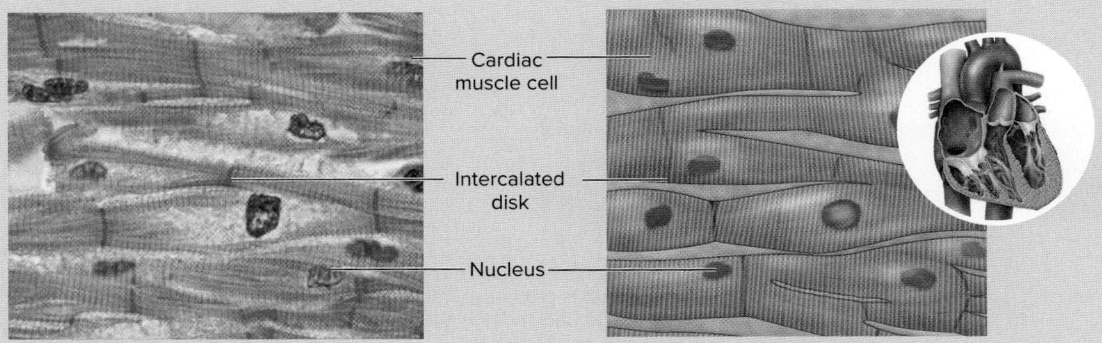
Cardiac muscle cell
Intercalated disk
Nucleus
40 µm

Ed Reschke

Cardiac Muscle

Typical Location

Walls of heart

Function

Highly interconnected cells; promotes rapid spread of signal initiating contraction

Characteristic Cell Types

Cardiac muscle cells

to each other within the muscle and are connected to the tendons on the ends of the muscle. Each skeletal muscle fiber is stimulated to contract by a motor neuron.

The nervous system controls the overall strength of a skeletal muscle contraction by controlling the number of motor neurons that fire and, therefore, the number of muscle fibers stimulated to contract. Each muscle fiber contracts by means of substructures called **myofibrils** containing highly ordered arrays of actin and myosin myofilaments. These filaments give the muscle fiber its striated appearance. We will examine the molecular details of contraction in section 32.8.

Skeletal muscle fibers are formed by the fusion of several cells, end to end. This embryological development explains why a mature muscle fiber contains many nuclei.

The heart is composed of cardiac muscle

The hearts of vertebrates are made up of striated muscle cells arranged very differently from the fibers of skeletal muscle. Instead of having very long, multinucleate cells running the length of the muscle, **cardiac muscle** consists of smaller, interconnected cells, each with a single nucleus. The interconnections between adjacent cells appear under the microscope as lines where gap junctions link adjacent cells. Gap junctions have openings that permit the movement of small substances and ions from one cell to another. These interconnections enable the cardiac muscle cells to form a single functioning unit.

Certain specialized cardiac muscle cells can generate electrical impulses spontaneously, but the nervous system usually regulates the rate of impulse activity. The impulses generated by the specialized cell groups spread across the gap junctions from cell to cell, synchronizing the heart's contraction. Chapter 34 describes this process more fully.

REVIEW OF CONCEPT 32.5

Muscles are the motors of the body; they are able to contract to change their length. Muscle tissue is of three types: smooth, skeletal, and cardiac. Smooth muscles provide a variety of visceral functions. Skeletal muscles enable the vertebrate body to move. Cardiac muscle forms a muscular pump, the heart.

- *Why is it important that cardiac muscle cells have gap junctions?*

32.6 Skeletal Systems Anchor the Body's Muscles

Muscles have to pull against something to produce the changes that cause movement. This necessary form of supporting structure is called a skeletal system. Zoologists recognize three types of skeletal systems in animals: hydrostatic skeletons, exoskeletons, and endoskeletons.

Hydrostatic Skeletons Use Water Pressure to Reinforce a Body Wall

LEARNING OBJECTIVE 32.6.1 Describe how animals with a hydrostatic skeleton move about.

Hydrostatic skeletons are found primarily in soft-bodied terrestrial invertebrates, such as earthworms and slugs, and soft-bodied aquatic invertebrates, such as jellyfish and squids. In these animals, a fluid-filled central cavity is encompassed by two sets of muscles in the body wall: circular muscles, which are repeated in segments and run the length of the body, and longitudinal muscles, which oppose the action of the circular muscles.

Muscles act on the fluid in the body's central space, which constitutes the hydrostatic skeleton. As locomotion begins (figure 32.6), the anterior circular muscles contract, pressing on the inner fluid and forcing the front of the body to become thin as the body wall in this region extends forward.

On the underside of a worm's body are short, bristle-like structures called chaetae. When circular muscles act, the chaetae of that region are pulled up close to the body and lose contact with the ground. Circular-muscle activity is passed backward, segment by segment, to create a backward wave of contraction.

As this wave continues, the anterior circular muscles now relax, and the longitudinal muscles take over, thickening the front end of the worm and allowing the chaetae to protrude and regain contact with the ground. The chaetae now prevent that body section from slipping backward. This locomotion

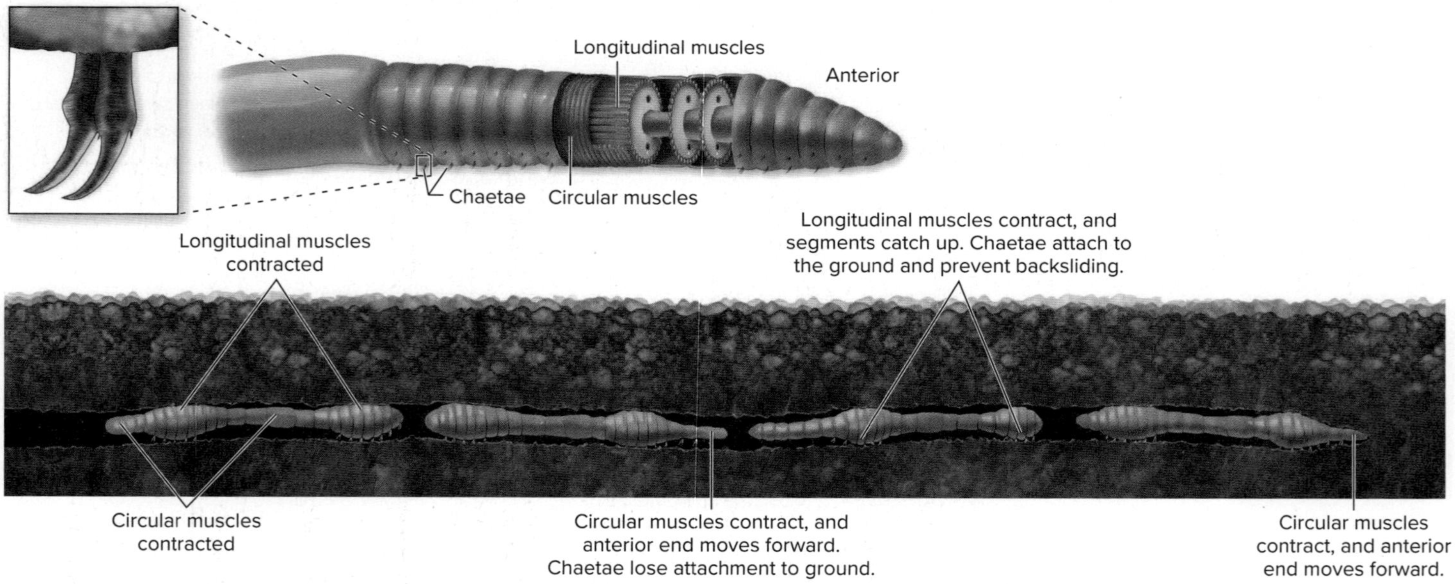

Figure 32.6 Locomotion in earthworms. The hydrostatic skeleton of the earthworm uses muscles to move fluid within the segmented body cavity, changing the shape of the animal. When circular muscles contract, the pressure in the fluid rises. At the same time, the longitudinal muscles relax, and the body becomes longer and thinner. When the longitudinal muscles contract and the circular muscles relax, the chaetae of the worm's lower surface extend to prevent backsliding. A wave of circular-muscle contractions followed by longitudinal-muscle contractions down the body produces forward movement.

process proceeds as waves of circular-muscle contraction are followed by waves of longitudinal-muscle effects.

Exoskeletons Consist of a Rigid Outer Covering

LEARNING OBJECTIVE 32.6.2 Discuss the limitations of exoskeletons.

An **exoskeleton** is a rigid, hard case that surrounds the body. Arthropods, such as crustaceans and insects, have exoskeletons made of the polysaccharide *chitin,* also found in the cell walls of fungi and some protists.

A chitinous exoskeleton resists bending and thus acts as the skeletal framework of the body; it also protects the internal organs and provides attachment sites for the muscles, which lie inside the exoskeletal casing (figure 32.7*a*). But in order to grow, the animal must periodically molt, shedding the exoskeleton. The animal is vulnerable to predation until the new (slightly larger) exoskeleton forms.

Exoskeletons have other limitations as well. The chitinous framework is not as strong as a bony, internal one. This fact by itself would set a limit for insect size, but there is a more important factor: insects breathe through openings in their body that lead into tiny tubes, and as insect size increases beyond a certain limit, the ratio between the inside surface area of the tubes and the volume of the body exceeds the capacity of this sort of respiratory system.

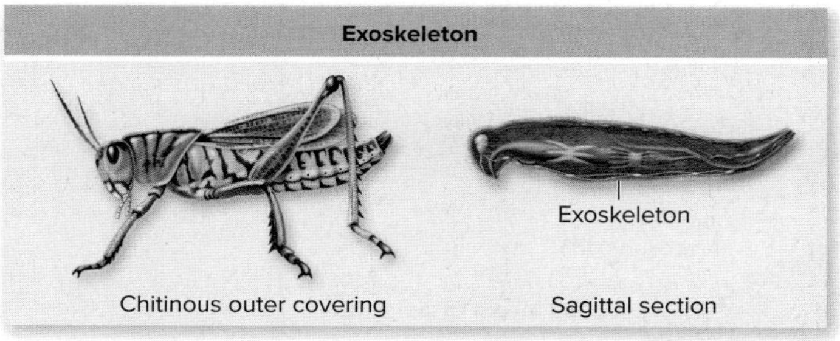

Exoskeleton
Exoskeleton
Chitinous outer covering
Sagittal section

a.

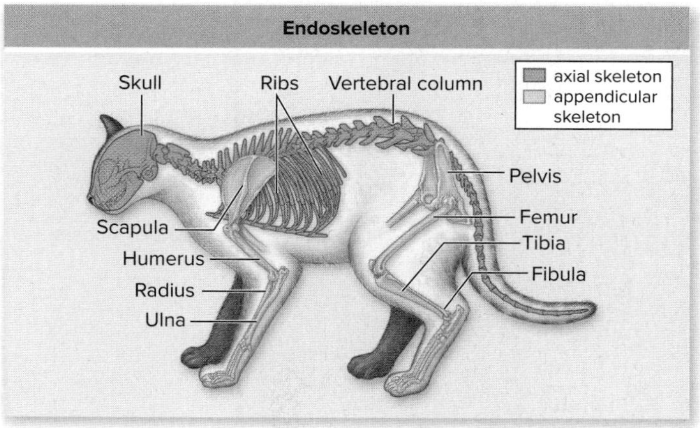

Endoskeleton
Skull
Ribs
Vertebral column
axial skeleton
appendicular skeleton
Pelvis
Femur
Tibia
Fibula
Scapula
Humerus
Radius
Ulna

b.

Figure 32.7 Exoskeleton and endoskeleton. ***a.*** The hard, tough outer covering of an arthropod, such as this grasshopper, is its exoskeleton and is composed of chitin. ***b.*** Vertebrates, such as this cat, have endoskeletons formed of bone and cartilage. Some of the major bony features are labeled.

Endoskeletons Are Composed of Hard Internal Structures

LEARNING OBJECTIVE 32.6.3 Compare endoskeletons to exoskeletons.

Endoskeletons, found in vertebrates and echinoderms, are rigid internal skeletons that form the body's framework and offer surfaces for muscle attachment. Echinoderms, such as sea urchins and sand dollars, have skeletons made of calcite, a crystalline form of calcium carbonate. This calcium compound is different from that in bone, which is based on calcium phosphate.

Vertebrate skeletal tissues

The vertebrate endoskeleton (figure 32.7*b*) includes fibrous dense connective tissue along with the more rigid special connective tissues, cartilage or bone. Cartilage is strong and slightly flexible, a characteristic important in such functions as padding the ends of bones where they come together in a joint. Although some large, active animals such as sharks have totally cartilaginous skeletons, bone is the main component in most vertebrate skeletons. Bone is much stronger than cartilage and much less flexible.

Unlike chitin, both cartilage and bone are living tissues. Bone, particularly, can have high metabolic activity, especially if bone cells are present throughout the matrix, a common condition. Bone, and to some extent cartilage, can change and remodel itself in response to injury or to physical stresses.

REVIEW OF CONCEPT 32.6

Movement with a hydrostatic skeleton uses muscle contraction to put pressure on body fluids. Invertebrate exoskeletons consist of hard chitin that is shed and renewed for growth. Endoskeletons are composed of fibrous dense connective tissue along with cartilage or mineralized bone.

- *What limitations does an exoskeleton impose on terrestrial invertebrates?*

32.7 Vertebrate Endoskeletons Are Made of Bone

Bone is a hard but resilient tissue that is unique to vertebrate animals. This connective tissue first appeared over 520 MYA and is now found in all vertebrates except cartilaginous fishes.

Bone Tissue Is Highly Structured

LEARNING OBJECTIVE 32.7.1 Compare the structure of different parts of a long bone.

Figure 32.8 shows a typical long bone, the humerus. It is covered by a dense fibrous connective tissue, the **periosteum**. At either end the bone has an **epiphysis** that is covered in articular (hyaline) cartilage. The shaft of the bone is called the **diaphysis**. The epiphyses and the diaphysis fuse at the **epiphyseal plates,** often called the growth plates. They are so named because this is the region that contains cartilage where new bone is deposited during growth. This allows for the bone to increase in length. In humans these cartilaginous plates transform into bone and are fused when growth ceases, usually between the ages of 18 and 20.

A longitudinal section reveals that there are two types of bone: the outer **compact** bone and the inner **spongy** (medullary) bone. Compact bone is densely packed with an extracellular matrix of calcium phosphate and collagen with no gaps. This provides strength and support in the skeleton. Spongy bone is arranged in branching plates called **trabeculae** that afford resistance to compressive forces while reducing weight. It has a porous, or "spongy," appearance. The medullary cavity houses bone marrow, which can be red because of red blood cells, or yellow because of fat cells. Red marrow is important for the production of blood cells, and is found in the epiphyses of long bones. Based upon the mode of locomotion and metabolic demands, the bones in all vertebrates vary in the proportion of compact bone, spongy bone, and marrow. For example, birds have thin layers of compact bone and little, if any, marrow. This evolutionary innovation helps reduce weight, making flight more efficient by making it less energetically demanding.

Bone was introduced in section 32.5 as a connective tissue. Recall that mature bone cells are called **osteocytes.** In compact bone, osteocytes are found in **lacunae** ("little lakes") embedded in the extracellular matrix and are arranged in concentric circles around a central canal containing blood vessels and nerves. The osteocytes exchange nutrients and waste through a network of tiny canals, **canaliculi.** The combination of osteocytes encircling a central canal is termed an osteon.

Osteons extend the entire length of the bone. Central canals are connected by perforating canals that run in a perpendicular fashion. In spongy bone, the osteocytes are not arranged in osteons, but they also employ canaliculi for communication and exchange.

In human long bones, bone marrow can be yellow, due to fat cells, or red due to red blood cells. The epiphysis of long bones contain red marrow and the marrow cavity contains yellow marrow.

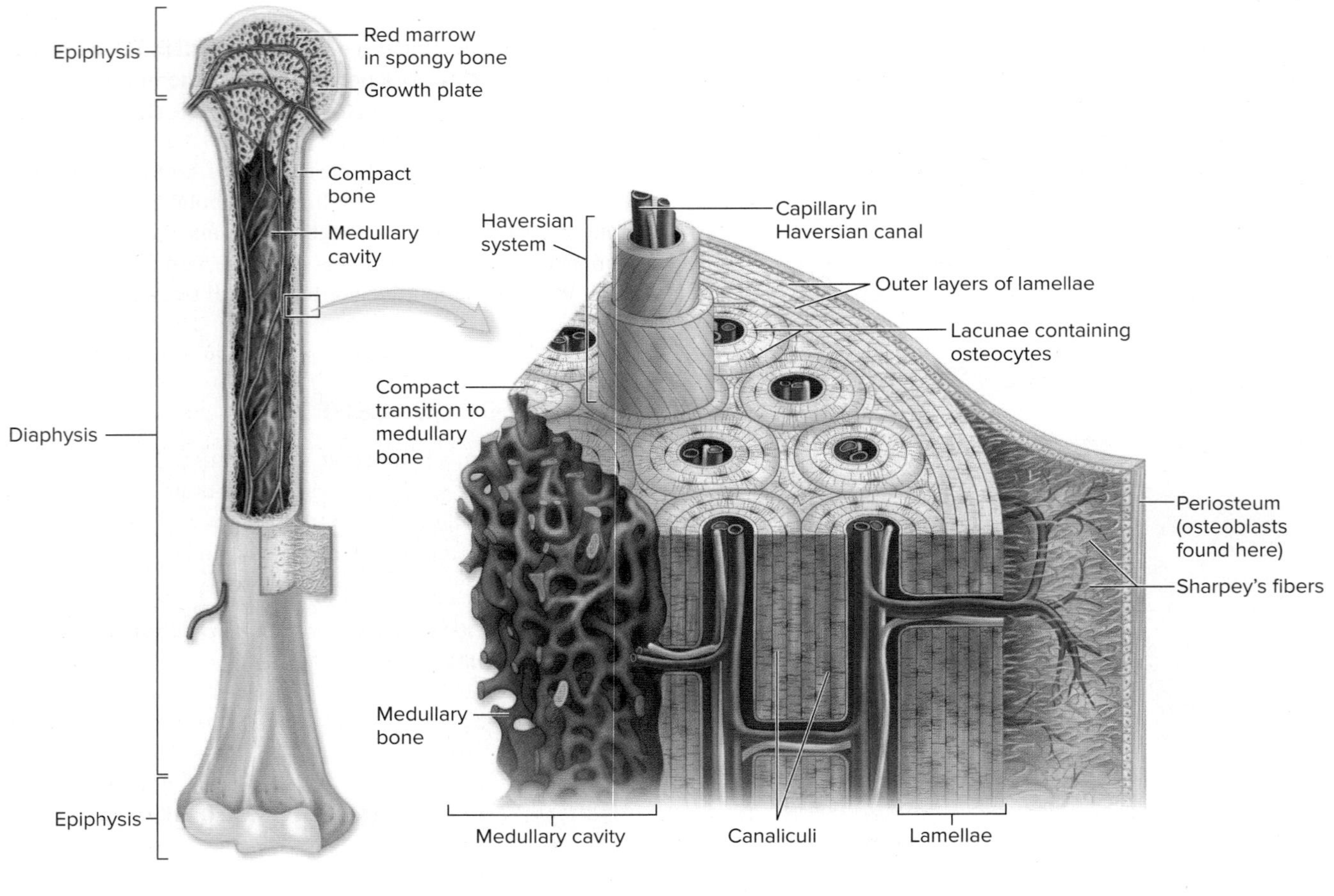

Figure 32.8 The structure of bone. A mammalian humerus is partly opened to show its interior on the left. A section has been removed and magnified on the right to show the difference in structure between the outer compact bone and the inner spongy bone in the medullary cavity. Details of basic layers, Haversian canals, and osteocytes in lacunae can be seen here.

Bones Can Be Classified by Two Modes of Development

LEARNING OBJECTIVE 32.7.2 Compare intramembranous and endochondral development.

Bone tissue itself can be of several types classified in a few different ways. The most common system is based on the way in which bone develops.

Intramembranous development

In intramembranous development, bones form within a layer of connective tissue. Many of the flat bones that make up the exterior of the skull and jaw are intramembranous.

Typically, the site of the intramembranous bone-to-be is in a designated region in the dermis of the skin. During embryonic development, the dermis is formed largely of **mesenchyme**—a loose tissue consisting of undifferentiated mesenchyme cells and other cells that have arisen from them—along with collagen fibers. Some of the undifferentiated mesenchyme cells differentiate to become specialized cells called **osteoblasts.** These osteoblasts arrange themselves along the collagenous fibers and begin to secrete the enzyme alkaline phosphatase, which causes calcium phosphate salts to form in a crystalline configuration called *hydroxyapatite.* The crystals merge along the fibers to encase them.

The crystals give the bone its hardness; however, without the resilience afforded by collagen's stretching ability, bone would be rigid but dangerously brittle. Typical bones have roughly equal volumes of collagen and hydroxyapatite, but hydroxyapatite contributes about 65% to the bone's weight.

As the osteoblasts continue to make bone crystals, some become trapped in the bone matrix and undergo dramatic changes in shape and function, becoming osteocytes (figure 32.8). They reside in the lacunae, where the canaliculi permit contact of the starburst-like extensions of each osteocyte with those of its neighbors. This allows many cells within bones to engage in intercellular communication.

As an intramembranous bone grows, it requires alterations of shape. Imagine that you were modeling with clay and wanted to make a tiny clay bowl larger. Simply adding clay to the outside would not work; you would need to remove clay from the inside as well. As bone grows, it must undergo a remodeling process, with matrix being added in some regions and removed in others. This is where osteoclasts come in. These cells form by the fusion of monocytes, a type of white blood cell, to form large, multinucleate cells. Their function is to break down the bone matrix.

Endochondral development

Bones that form through endochondral development are typically those that are deeper in the body and form its architectural framework. Examples include the vertebrae, the ribs, the bones of the shoulder and pelvis, the long bones of the limbs, and the most internal of the skull bones. Endochondral bones begin as tiny, cartilaginous models that have the rough shape of the bones that eventually will be formed. Bone development of this kind consists of adding bone to the outside of the cartilaginous model, while replacing the interior cartilage with bone.

Bone added to the outside of the model is produced in the fibrous sheath that envelopes the cartilage. This sheath is tough and made of collagen fibers, but it also contains undifferentiated mesenchyme cells. Osteoblasts arise and sort themselves out along the fibers in the deepest part of the sheath. Bone is then formed between the sheath and the cartilaginous matrix. This process is somewhat similar to what occurs in the dermis in the production of intramembranous bone.

As the outer bone is formed, the interior cartilage begins to calcify. The calcium source for this process seems to be the cartilage cells themselves. As calcification continues, the inner cartilaginous tissue breaks down into pieces of debris. Blood vessels from the sheath, now called the periosteum, force their way through the outer bony jacket, thus entering the interior of the cartilaginous model, and remove the debris. Again, trapped osteoblasts transform into osteocytes, and osteoclasts for bone remodeling arise from cell fusions in the same manner as occurs in intramembranous bone. Growth in bone thickness occurs by the adding of additional bone layers just beneath the periosteum.

Growth in length usually ceases in humans by late adolescence. Although growth of the bone length is curtailed at this time, growth in width is not. The diameter of the shaft can be enhanced by bone addition just beneath the periosteum throughout an individual's life.

Bone Remodeling Allows Bone to Respond to Use or Disuse

LEARNING OBJECTIVE 32.7.3 Explain how bone remodeling occurs.

It is easy to think of bones as being inert, especially because we rarely encounter them except as the skeletons of dead animals. But just as muscles, skin, and other body tissues may change depending on the stresses of the environment, bone also is a dynamic tissue that can change with demands made on it.

Mechanical stresses such as compression at joints, the forces of muscles on certain portions and features of a bone, and similar effects may all be remodeling factors that shape the bone not only during its embryonic development but after birth as well. Depending on the directions and magnitudes of forces impinging on it, a bone may thicken; the surface features to which muscles, tendons, or ligaments attach may change in size and shape; even the direction of the tiny, bony struts that make up spongy bone may be altered.

Exercise and frequent use of muscles for a particular task change more than just the muscles; blood vessels and fibrous connective tissue increase, and the skeletal frame becomes more robust through bone thickening and enhancement.

The phenomenon of remodeling is easiest to describe in a long bone. Small forces may not have much of an effect on the bone, but larger ones—if frequent enough—can initiate remodeling (figure 32.9). In the example shown, larger compressive forces may tend to bend a bone, even if the bend is imperceptible to the eye. This bending stress promotes bone formation that thickens the bone. As the bone becomes thicker, the amount of bending is reduced (figure 32.9*c*). Further bone addition will eventually prevent significant bending

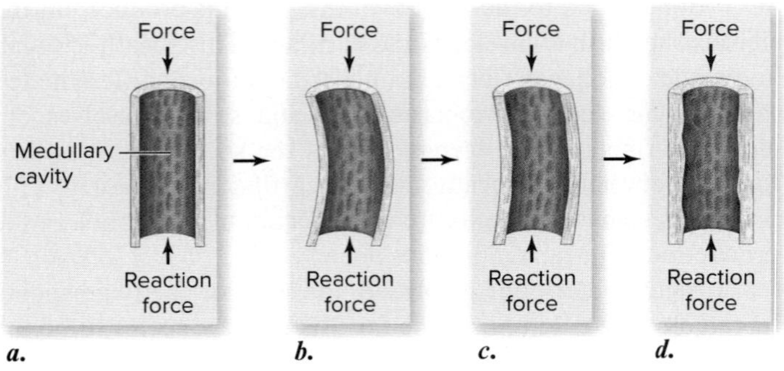

Figure 32.9 Model of stress and remodeling in a long bone. This figure shows a diagrammatic section of a long bone, such as a leg bone. The section is placed under a load or force, which causes a reaction force from the ground the leg is standing upon. ***a.*** Under a mild compressive load, the bone does not bend. ***b.*** If the load is large enough, and the bone is not sufficiently thick, the bone will bend (the bending shown is exaggerated for clarity). ***c.*** Osteoblasts are signaled by the stresses in the bending section to produce additional bone. As the bone becomes thicker, the degree of bending is reduced. ***d.*** When sufficient bone is added to prevent significant bending, the production of new osteoblasts stops and no more bone is added.

(figure 32.9*d*). At this point, bone addition stops, a form of negative feedback.

This phenomenon has important medical implications. Osteoporosis, which is characterized by a loss of bone mineral density, is a debilitating and potentially life-threatening ailment that afflicts more than 25 million people in the United States. Osteoporosis affects primarily postmenopausal women, but also those suffering from malnutrition and a number of diseases. One treatment is a regimen of weight-lifting to stimulate bone deposition.

Bones Move Relative to One Another at Joints

LEARNING OBJECTIVE 32.7.4 Describe the function of the different kinds of joints.

Movements of the endoskeleton are powered by the skeletal musculature. The skeletal movements that respond to muscle action occur at **joints,** or articulations, where one bone meets another. Each movable joint within the skeleton has a characteristic range of motion. Four basic joint movement patterns can be distinguished: *ball-and-socket, hinge, gliding,* and *combination.*

Ball-and-socket joints are like those of the hip, where the upper leg bone forms a ball fitting into a socket in the pelvis. This type of joint can perform movement in all directions, plus twisting of the ball.

Hinge joints, which restrict motion to a single plane, are the simplest type of joint. For example, the elbow allows the forearm to move forward and backward relative to the upper arm, but not side-to-side.

Gliding joints can be found in the skulls of a number of nonmammalian vertebrates but are also present between the lateral vertebral projections in many of them and in mammals as well. The vertebral projections are paired and extend from the front and back of each vertebra. The projections in front are a little lower, and each can slip along the undersurface of the posterior projection from the vertebra just ahead of it. This sliding joint gives stability to the vertebral column while allowing some flexibility of movement between vertebrae.

Combination joints are, as you might suppose, those that have movement characteristics of two or more joint types. The typical mammalian jaw joint is a good example. Most mammals chew food into small pieces. To chew food well, the lower jaw needs to move from side to side to get the best contact between upper and lower teeth. The lower jaw can also slip forward and backward to some extent. At the same time, the jaw joint must be shaped to allow the hingelike opening and closing of the mouth. The mammalian joint conformation thus combines features from hinge and gliding joints.

Skeletal muscles pull on bones to produce movement at joints

Skeletal muscles produce movement of the skeleton when they contract. Usually, the two ends of a skeletal muscle are attached to different bones, although some are attached to other structures, such as skin. There are two means of bone attachment: (1) Muscle fibers may connect directly to the periosteum, the bone's fibrous covering, or (2) sheets of muscle may be connected to bone by a dense cord of connective tissue called a *tendon,* which attaches to the periosteum.

One attachment of the muscle, the origin, remains relatively stationary during a contraction. The other end, the insertion, is attached to a bone that moves when the muscle contracts. For example, contraction of the quadriceps muscles of the leg causes the lower leg to rotate forward relative to the upper leg section.

Skeletal muscles produce movement of the skeleton, both in exoskeletons and endoskeletons, when they contract and cross a joint. Typically, muscles are arranged so that any movement produced by one muscle can be reversed by another. The leg **flexor** muscles, called hamstrings (figure 32.10), draw the lower leg back and upward, bending the knee. Their movement is countered by the **extensor** muscles, called quadriceps. This arrangement of mutually antagonistic muscles is important, as a muscle can only contract but cannot push.

REVIEW OF CONCEPT 32.7

Bone cells called osteocytes are embedded in the extracellular matrix in internal cavities. The outside of bones is compact and densely packed, while the interior is spongy. Intramembranous bone forms within a layer of connective tissue; endochondral bone originates with a cartilaginous model, which is then replaced with bone tissue. Bone remodeling occurs in response to repeated stresses on bones from weight or muscle use. Muscles, positioned across joints, cause movement of bones relative to each other by contracting.

- *In what ways does a bony endoskeleton overcome the limitations of an exoskeleton for terrestrial life-forms?*

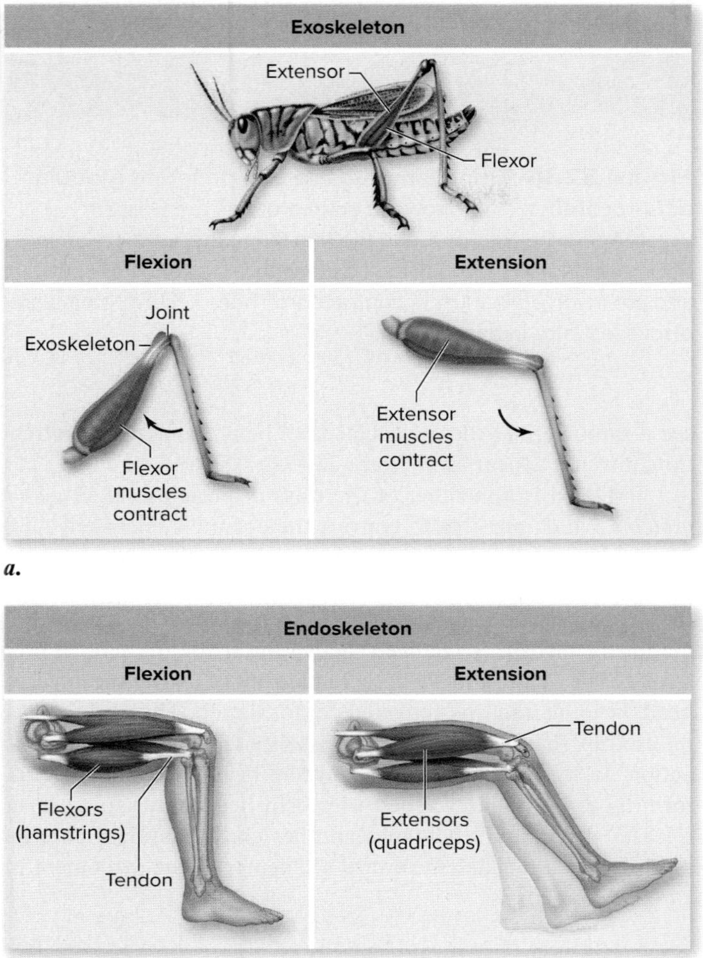
Exoskeleton
Extensor
Flexor
Flexion
Extension
Joint
Exoskeleton
Flexor muscles contract
Extensor muscles contract
a.
Endoskeleton
Flexion
Extension
Tendon
Flexors (hamstrings)
Tendon
Extensors (quadriceps)
b.

Figure 32.10 Flexor and extensor muscles of the leg. Antagonistic muscles have opposite effects in both arthropods and humans. In humans, the hamstrings, a group of three muscles, cause the lower leg to move backward relative to the upper leg, whereas the quadriceps, a group of four muscles, pull the lower leg forward.

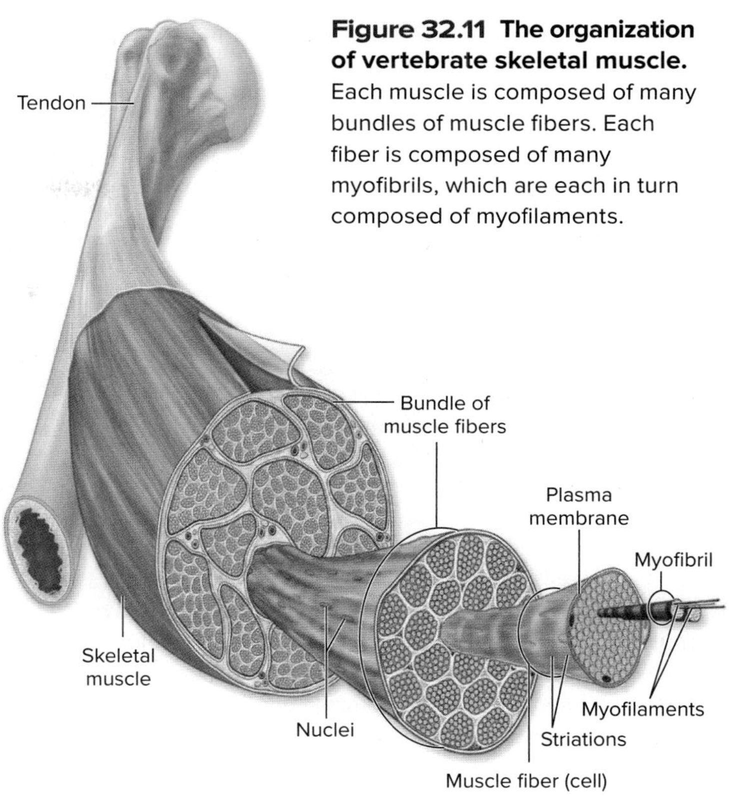
Tendon
Bundle of muscle fibers
Plasma membrane
Myofibril
Skeletal muscle
Nuclei
Myofilaments
Striations
Muscle fiber (cell)

Figure 32.11 The organization of vertebrate skeletal muscle. Each muscle is composed of many bundles of muscle fibers. Each fiber is composed of many myofibrils, which are each in turn composed of myofilaments.

32.8 Muscles Contract Because Their Myofilaments Slide

Each vertebrate skeletal muscle contains numerous muscle fibers. Each muscle fiber encloses a bundle of 4 to 20 elongated structures called **myofibrils.** Each myofibril, in turn, is composed of thick and thin **myofilaments** (figure 32.11). Muscle contraction arises by the interaction of these myofilaments. We will consider the mechanism of contraction, then the events that control when contraction occurs.

Muscle Fibers Contract as Overlapping Filaments Slide Together

LEARNING OBJECTIVE 32.8.1 Explain the sliding filament mechanism of muscle contraction.

Under a microscope, the myofibrils have alternating dark and light bands, which give skeletal muscle fiber its striped appearance. Each band in a myofibril is divided in half by a disk of protein called a *Z line* because of its appearance in electron micrographs. The thin filaments are anchored to these disks. In an electron micrograph of a myofibril, the structure of the myofibril can be seen to repeat from Z line to Z line. This repeating structure, called a **sarcomere,** is the smallest subunit of muscle contraction (figure 32.12).

A muscle contracts and shortens because its myofibrils contract and shorten. When this occurs, the myofilaments do *not* shorten; instead, the thick and thin myofilaments slide relative to each other. The thin filaments slide deeper into the A bands, making the H bands narrower until, at maximal shortening, they disappear entirely. This also makes the I bands narrower, as the Z lines are brought closer together. This is the sliding filament mechanism of contraction (figure 32.12).

The sliding filament mechanism

Electron micrographs reveal cross-bridges that extend from the thick to the thin filaments, suggesting a mechanism that might cause the filaments to slide. To understand how this is accomplished requires examining the thick and thin filaments at a molecular level. Biochemical studies show that each thick filament is composed of many subunits of the protein myosin packed together. The myosin protein consists of two subunits, each shaped like a golf club with a head region that protrudes from a long filament, with the filaments twisted together. Thick filaments are composed of many copies of myosin arranged with heads protruding from along the length of the fiber (figure 32.13). The myosin heads form the cross-bridges seen in electron micrographs.

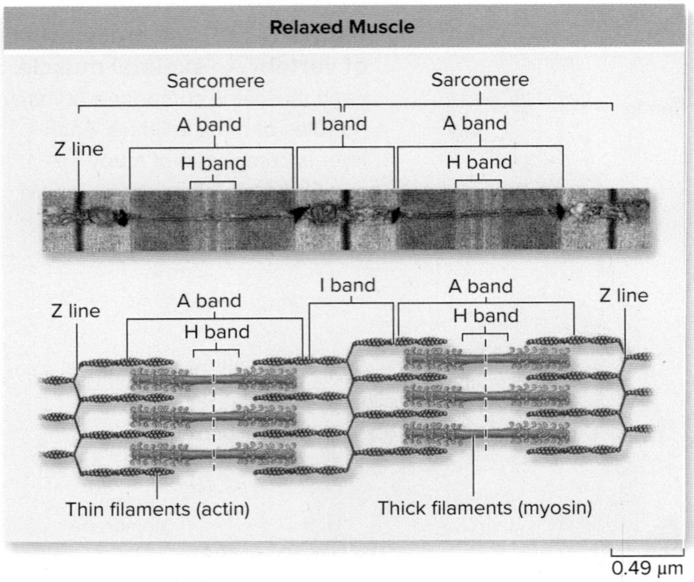
Relaxed Muscle
Sarcomere
Sarcomere
A band
I band
A band
Z line
H band
H band
Z line
A band
H band
I band
A band
H band
Z line
Thin filaments (actin)
Thick filaments (myosin)
0.49 μm

Figure 32.12 The structure of sarcomeres in relaxed muscles. The micrograph shows a segment of myofibril with two sarcomeres. The drawing below shows the arrangement of thick and thin filaments. The Z lines form the borders of each sarcomere, and the A bands represent thick filaments. The thin filaments are within the I bands and extend into the A bands interdigitated with thick filaments. The H band is the lighter-appearing, central region of the A band containing only thick filaments.

Don W. Fawcett/Science Source

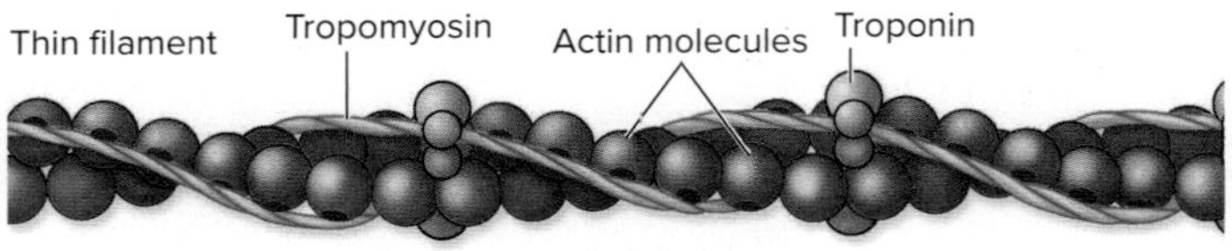
Thin filament
Tropomyosin
Actin molecules
Troponin

Figure 32.14 Thin filaments are composed of globular actin proteins. Two rows of actin proteins are twisted together in a helix to produce the thin filaments. Other proteins, tropomyosin and troponin, associate with the strands of actin and are involved in muscle contraction. These other proteins are discussed later in this section.

Each thin filament consists primarily of many globular actin proteins arranged into two fibers twisted into a double helix (figure 32.14). The actin molecules contain binding sites for the myosin head, which are critical for muscle function. If we could see a sarcomere at the molecular level before and after contraction, it would appear as in figure 32.15.

Myosin is a member of the class of proteins called *motor proteins,* which are able to convert the chemical energy in ATP into mechanical energy (refer to chapter 4). This occurs by a series of events called the cross-bridge cycle (figure 32.16). When the myosin heads hydrolyze ATP into ADP and P_i, the conformation of myosin is changed, activating it for the later power stroke. The ADP and P_i both remain attached to the myosin head, keeping it in this activated conformation. The analogy to a mousetrap, set and ready to spring, is often made to describe this action. In this set position, the myosin head can bind to actin, forming cross-bridges. When a myosin head binds to actin, it releases the P_i and undergoes another conformational change, pulling the thin filament toward the center of the sarcomere in

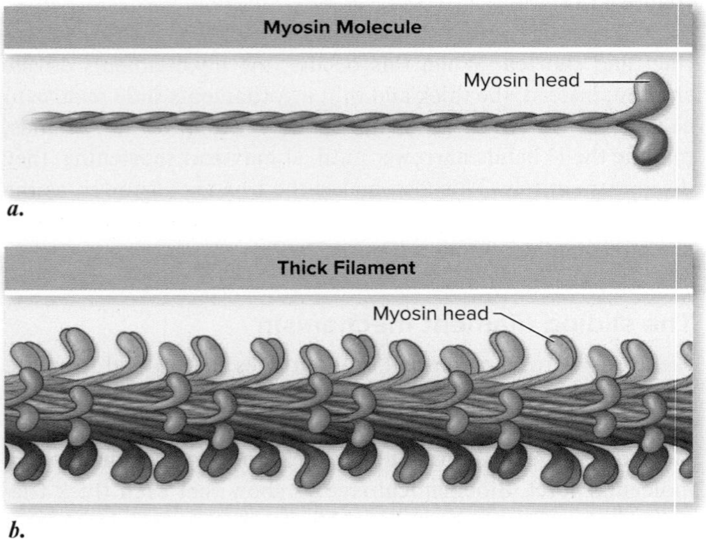
Myosin Molecule
Myosin head
a.
Thick Filament
Myosin head
b.

Figure 32.13 Thick filaments are composed of myosin. ***a.*** Each myosin molecule consists of two polypeptide chains shaped like golf clubs and wrapped around each other; at the end of each chain is a globular region referred to as the "head." ***b.*** Thick filaments consist of myosin molecules combined into bundles, from which the heads protrude at regular intervals.

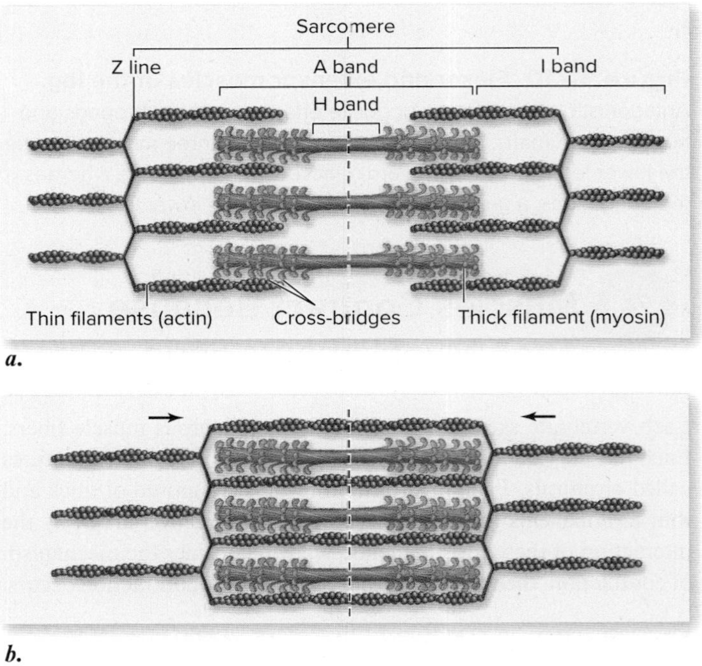
Sarcomere
Z line
A band
I band
H band
Thin filaments (actin)
Cross-bridges
Thick filament (myosin)
a.
b.

Figure 32.15 The interaction of thick and thin filaments in striated muscle sarcomeres. ***a.*** The heads on the two ends of the thick filaments are oriented in opposite directions so that the cross-bridges pull the thin filaments and the Z lines on each side of the sarcomere toward the center. ***b.*** This sliding of the filaments produces muscle contraction.

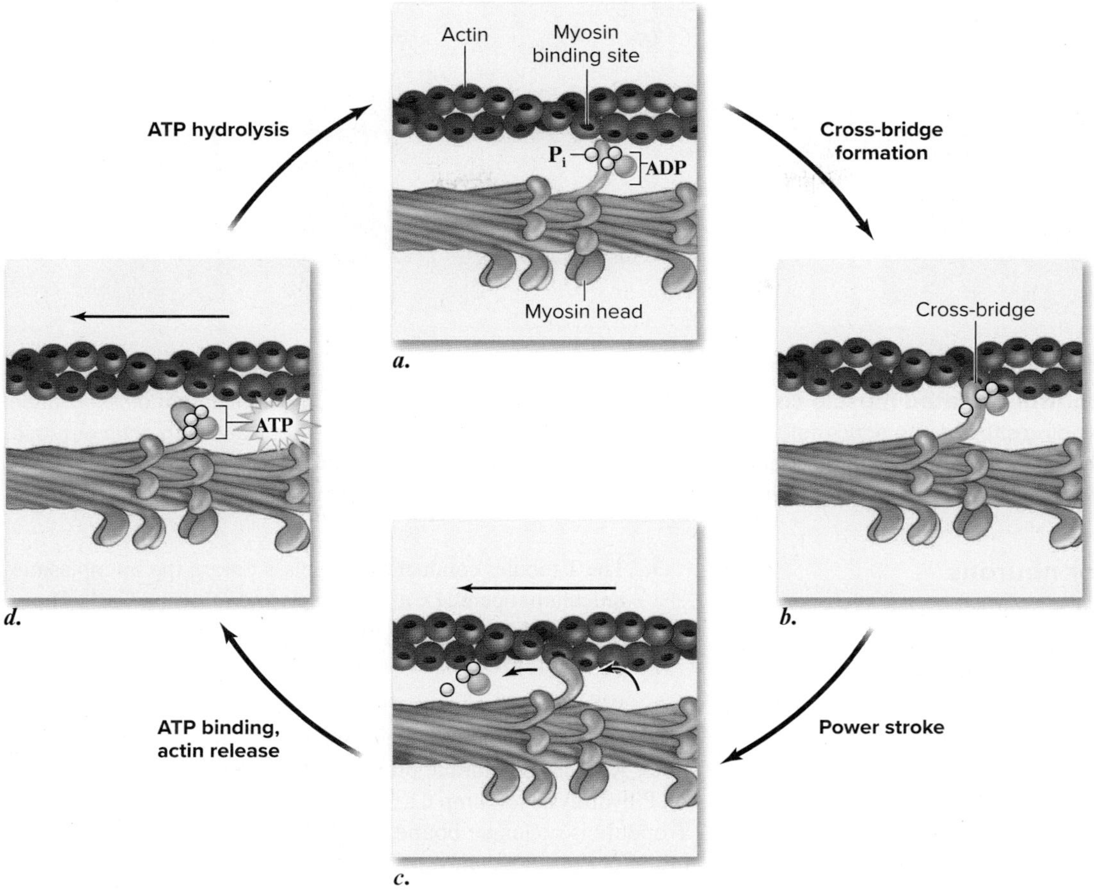

Figure 32.16 The cross-bridge cycle in muscle contraction. ***a.*** Hydrolysis of ATP by myosin causes a conformational change that moves the head into an energized state. The ADP and P_i remain bound to the myosin head, which can bind to actin. ***b.*** Myosin binds to actin, forming a cross-bridge. ***c.*** During the power stroke, myosin returns to its original conformation, releasing ADP and P_i. ***d.*** ATP binds to the myosin head, breaking the cross-bridge. ATP hydrolysis returns the myosin head to its energized conformation, allowing the cycle to begin again.

the *power stroke,* at which point it loses the ADP (figure 32.16*c*). At the end of the power stroke, the myosin head binds to a new molecule of ATP, which displaces it from actin. This cross-bridge cycle repeats as long as the muscle is stimulated to contract. This sequence of events can be thought of as pulling a rope hand-over-hand. The myosin heads are the hands and the actin fibers the rope.

In death, the cell can no longer produce ATP, and therefore the cross-bridges cannot be broken—causing the muscle stiffness of death called *rigor mortis.* A living cell, however, always has enough ATP to allow the myosin heads to detach from actin. How, then, is the cross-bridge cycle arrested so that the muscle can relax? We discuss the regulation of contraction and relaxation next.

Contraction Is Triggered by Calcium Ion Release Following a Nerve Impulse

LEARNING OBJECTIVE 32.8.2 Explain how muscle contraction is linked to a nerve impulse.

When a muscle is relaxed, its myosin heads are in the activated conformation bound to ADP and P_i, but they are unable to bind to actin. In the relaxed state, the attachment sites for the myosin heads on the actin are physically blocked by another protein, known as **tropomyosin,** in the thin filaments. Cross-bridges therefore cannot form and the filaments cannot slide.

For contraction to occur, the tropomyosin must be moved out of the way so that the myosin heads can bind to the uncovered actin-binding sites. This requires the action of **troponin,** a regulatory protein complex that holds tropomyosin and actin together. The regulatory interactions between troponin and tropomyosin are controlled by the calcium ion (Ca^{2+}) concentration of the muscle fiber cytoplasm.

When the Ca^{2+} concentration of the cytoplasm is low, tropomyosin inhibits cross-bridge formation (figure 32.17*a*). When the Ca^{2+} concentration is raised, Ca^{2+} binds to troponin, altering its conformation and shifting the troponin–tropomyosin complex. This shift in conformation exposes the myosin-binding sites on the actin. Cross-bridges can thus form, undergo power strokes, and produce muscle contraction (figure 32.17*b*).

Muscles need a reliable supply of Ca^{2+}. Muscle fibers store Ca^{2+} in a modified endoplasmic reticulum called a **sarcoplasmic reticulum (SR)** (figure 32.18). When a muscle fiber is stimulated to contract, the membrane of the muscle fiber becomes depolarized. This depolarization is transmitted deep into the muscle fiber by invaginations of the cell membrane called the **transverse tubules (T tubules).** Depolarization of the T tubules causes Ca^{2+} channels in the SR to open, releasing Ca^{2+} into the cytosol. Ca^{2+} then diffuses into the myofibrils, where it binds to troponin, altering its conformation and allowing contraction. The involvement of Ca^{2+} in muscle contraction is called **excitation–contraction coupling,** because it is the release of Ca^{2+} that links the excitation of the muscle fiber by the motor neuron to the contraction of the muscle.

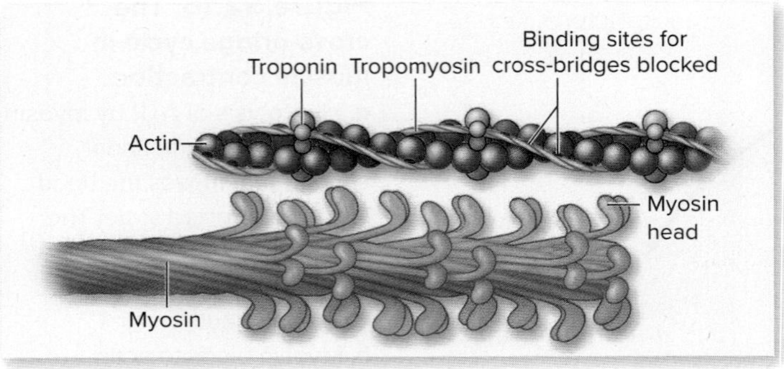

a.

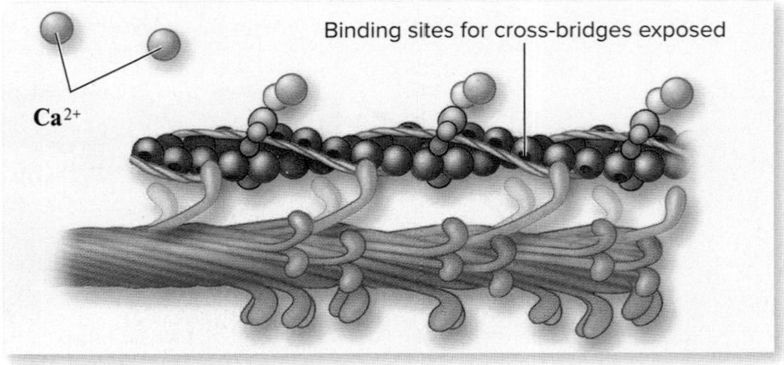

b.

Figure 32.17 How calcium controls striated muscle contraction. ***a.*** When the muscle is at rest, a long filament of the protein tropomyosin blocks the myosin-binding sites on the actin molecule. Because myosin is unable to form cross-bridges with actin at these sites, muscle contraction cannot occur. ***b.*** When Ca^{2+} binds to another protein, troponin, the Ca^{2+}–troponin complex displaces tropomyosin and exposes the myosin-binding sites on actin, permitting cross-bridges to form and contraction to occur.

Nerve impulses from motor neurons

Muscles are stimulated to contract by motor neurons. The motor neurons that stimulate skeletal muscles are called *somatic motor neurons.* The axon of a somatic motor neuron extends from the neuron cell body and branches to make synapses with a number of muscle fibers. These synapses between neurons and muscle cells are called **neuromuscular junctions** (figure 32.18). One axon can stimulate many muscle fibers, and in some animals a muscle fiber may be innervated by more than one motor neuron. However, in humans each muscle fiber has only a single synapse with a branch of one axon.

When a somatic motor neuron delivers electrochemical impulses, it stimulates contraction of the muscle fibers it innervates (makes synapses with) through the following events:

1. The motor neuron, at the neuromuscular junction, releases the neurotransmitter acetylcholine (ACh). ACh binds to receptors in the muscle cell membrane to open Na^+ channels. The influx of Na^+ ions depolarizes the muscle cell membrane.
2. The impulses spread along the membrane of the muscle fiber and are carried into the muscle fibers through the T tubules.
3. The T tubules conduct the impulses toward the sarcoplasmic reticulum, opening Ca^{2+} channels and releasing Ca^{2+}. The Ca^{2+} binds to troponin, exposing the myosin-binding sites on the actin myofilaments and stimulating muscle contraction.

When impulses from the motor neuron cease, it stops releasing ACh, in turn stopping the production of impulses in the muscle fiber. Another membrane protein in the SR then uses energy from ATP hydrolysis to pump Ca^{2+} back into the SR by active transport. Troponin is no longer bound to Ca^{2+}, so tropomyosin returns to its inhibitory position, allowing the muscle to relax.

REVIEW OF CONCEPT 32.8

Sliding of myofilaments within muscle myofibrils is responsible for contraction; it involves the motor protein myosin, which forms cross-bridges on actin fibers. The process of shortening is controlled by Ca^{2+} ions released from the sarcoplasmic reticulum. The Ca^{2+} binds to troponin, making myosin-binding sites in actin available.

- *What advantages do increased myoglobin and mitochondria confer on muscle fibers?*

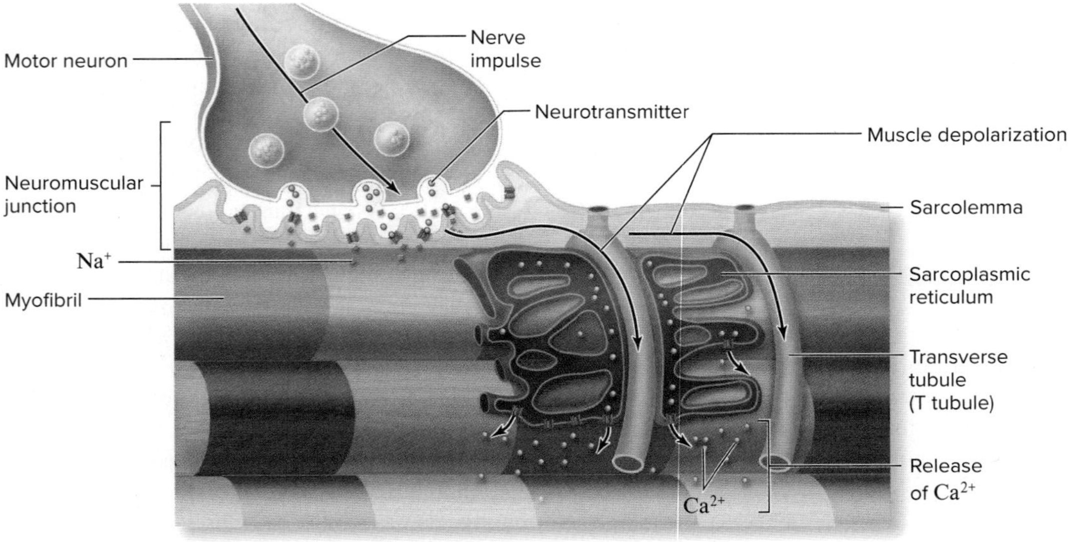

Figure 32.18 Relationship among the myofibrils, transverse tubules, and sarcoplasmic reticulum. Neurotransmitter released at a neuromuscular junction binds chemically gated Na^+ channels, causing the muscle cell membrane to depolarize. This depolarization is conducted along the muscle cell membrane and down the transverse tubules to stimulate the release of Ca^{2+} from the sarcoplasmic reticulum. Ca^{2+} diffuses through the cytoplasm to myofibrils, causing contraction.

32.9 Animal Locomotion Takes Many Forms

Animals are unique among multicellular organisms in their ability to move actively from one place to another. Locomotion requires both a propulsive mechanism and a control mechanism. There are a wide variety of propulsive mechanisms, most involving contracting muscles to generate the necessary force. In large animals, active locomotion is almost always produced by appendages that oscillate—*appendicular locomotion*—or by bodies that undulate, pulse, or undergo peristaltic waves—*axial locomotion.*

Although animal locomotion occurs in many different forms, the general principles remain much the same in all groups. The physical constraints to movement—gravity and friction—are the same in every environment, differing only in degree.

Swimmers Must Contend with Friction When Moving Through Water

LEARNING OBJECTIVE 32.9.1 Describe how swimming uses muscular force to overcome frictional drag.

For swimming animals, the buoyancy of water reduces the effect of gravity. As a result, the primary force impeding forward movement is frictional drag, so body shape is important in reducing the force needed to push through the water.

Some marine invertebrates move about using hydraulic propulsion. For example, scallops clap the two sides of their shells together forcefully, and squids and octopuses squirt water like a marine jet. In contrast, many invertebrates and all aquatic vertebrates swim.

Swimming involves pushing against the water with some part of the body. At one extreme, eels and sea snakes swim by sinuous undulations of the entire body (figure 32.19*a*). The undulating body waves are created by waves of muscle contraction alternating between the left and right axial musculature. As each body segment in turn pushes against the water, the moving wave forces the eel forward.

Other types of fish use similar mechanics but generate most of their propulsion from the posterior part of the body using the caudal (rear) fin (figure 32.19*b*). This allows considerable specialization in the front end of the body without sacrificing propulsive force. Reptiles, such as alligators, swim in the same way using movement of the tail.

Whales and other marine mammals such as sea lions returned to an aquatic lifestyle and have evolved a similar form of locomotion. Marine mammals also swim using undulating body waves; however, unlike those in fish, the waves move top-to-bottom and not side-to-side. This is a nice example of convergent evolution, but it also illustrates how history shapes subsequent evolutionary change. The mammalian vertebral column is stiffened for terrestrial life and does not allow the side-to-side movement seen in fish. When the ancestors of whales re-entered aquatic habitats, they evolved adaptation for swimming that used top-to-bottom flexing.

Many terrestrial tetrapod vertebrates are able to swim, usually through movement of their limbs. Most birds that swim, such as ducks and geese, propel themselves through the water by pushing against it with their hind legs, which typically have webbed feet. Frogs and most aquatic mammals also swim with their hind legs and have webbed feet.

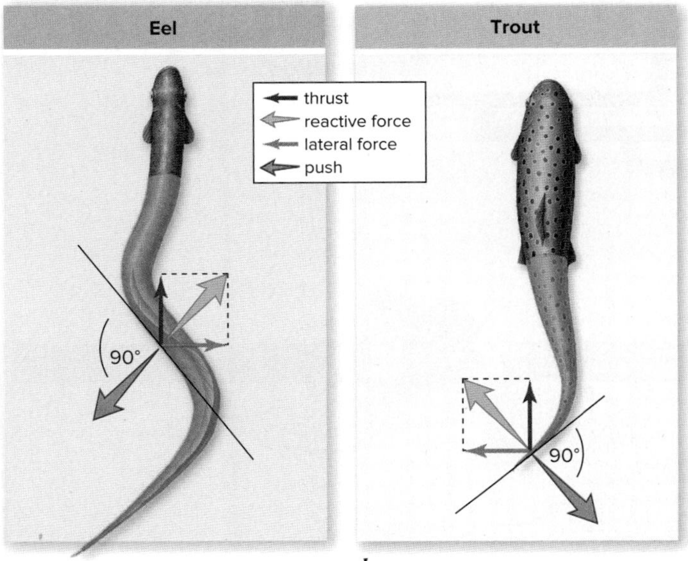

a. *b.*

Figure 32.19 Movements of swimming fishes. ***a.*** An eel pushes against the water with its whole body, whereas ***(b)*** a trout pushes only with its posterior half.

Terrestrial Locomotion Must Deal Primarily with Gravity

LEARNING OBJECTIVE 32.9.2 Describe how friction and gravity affect terrestrial locomotion.

Because air is much less dense than water, overcoming friction is less important than countering the force of gravity. This is the biggest challenge for terrestrial animals, which either move on land or fly through the air. The three great groups of terrestrial animals—mollusks, arthropods, and vertebrates—each accomplish this in different ways.

Mollusk locomotion is much slower than that of the other groups. Snails, slugs, and other terrestrial mollusks secrete a path of mucus that they glide along, pushing with a muscular foot.

Only vertebrates and arthropods are capable of rapid surface locomotion. In both groups, the body is raised above the ground and moved forward by pushing against the ground with a series of jointed appendages, otherwise known as legs.

Animals may walk on 2 legs or more than 100, but they use the same general principles. Because legs provide support as well as propulsion, it is important that their movements not shove the body's center of gravity outside the legs' zone of support for more than a short time, or the animal will fall. This need to maintain stability determines the sequence of leg movements, which are similar in vertebrates and arthropods.

The reason the walking gaits of these two groups may look different is simply the difference in numbers of legs. Vertebrates walk on two or four legs, whereas arthropods have six or more limbs. Although the many legs of arthropods increase stability during locomotion, they also appear to reduce the maximum speed.

The basic walking pattern of quadrupeds, from salamanders to most mammals, is left hind leg, right foreleg, right hind leg, left foreleg. The highest running speeds of quadruped mammals, such as the gallop of a horse, may involve the animal being

Figure 32.20 Animals that hop or leap use their rear legs to propel themselves through the air. The powerful leg muscles of this frog allow it to explode from a crouched position to a takeoff in about 100 msec.

Treat Davidson/Science Source

supported by only one leg, or even none at all. This is because the mammalian skeleton has evolved such that running can actually be a series of leaps.

Vertebrates such as kangaroos, rabbits, and frogs are effective leapers (figure 32.20). However, insects are the true Olympians of the leaping world. Many insects, such as grasshoppers, have enormous leg muscles, and some small insects can jump to heights more than 100 times the length of their body!

Flying Uses Air for Support

LEARNING OBJECTIVE 32.9.3 Describe how wings create lift.

The evolution of flight is a classic example of convergent evolution, having occurred independently four times, once in insects and three times among vertebrates (figure 32.21). All three vertebrate fliers modified the forelimb into a wing structure, but they did so in

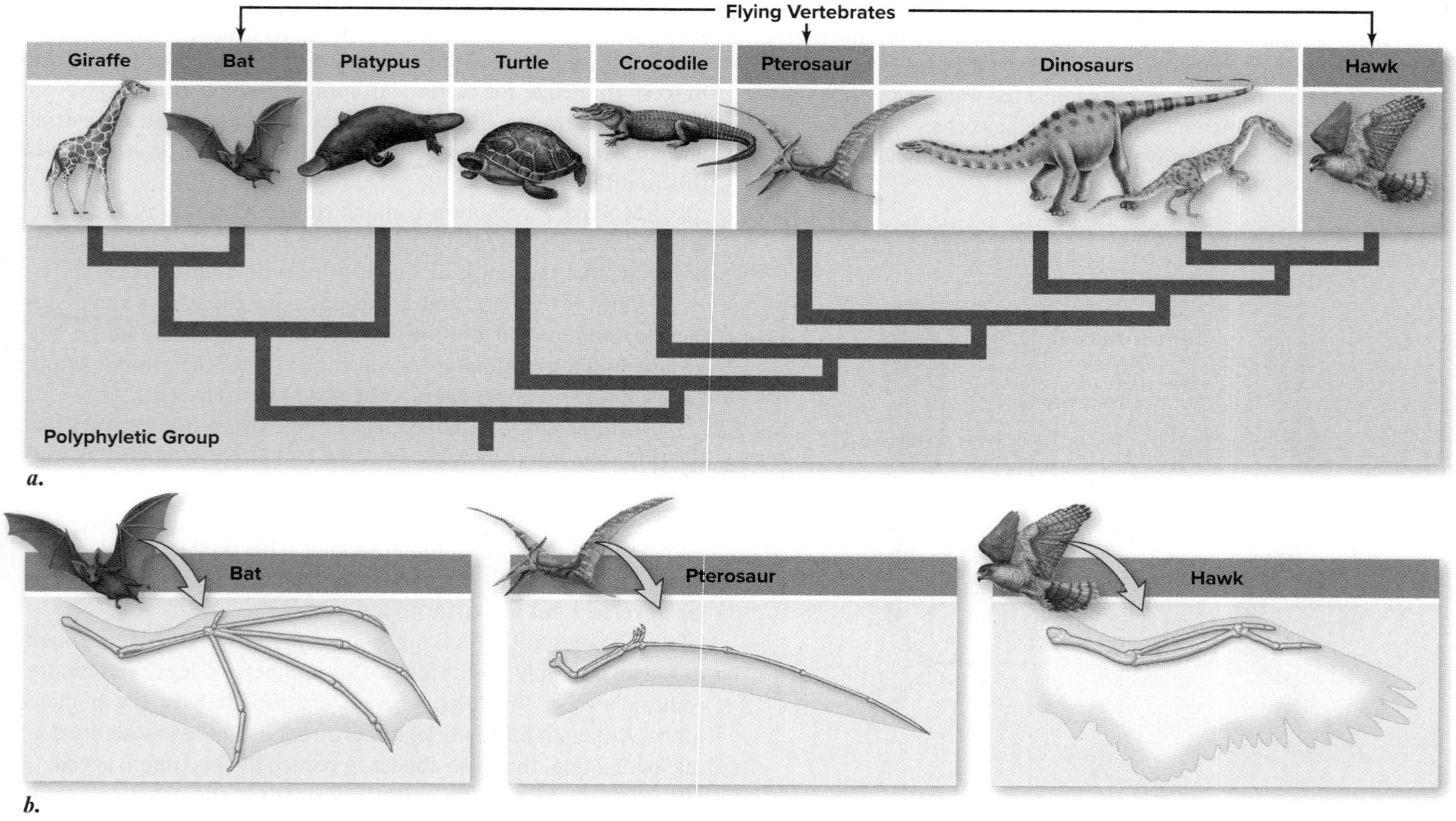

Figure 32.21 Convergent evolution of wings in vertebrates. Wings evolved independently in bats, pterosaurs, and birds, in each case by elongation of different elements of the forelimb.

different ways, illustrating how natural selection can sometimes build similar structures through different evolutionary pathways (figure 32.21*b*). In both birds and pterosaurs (an extinct group of reptiles that flourished alongside the dinosaurs), the wing is built on a single support, but in birds the wing is an elongation of the radius, ulna, and wrist bones, whereas in pterosaurs it is an elongation of the fourth finger bone. By contrast, in bats the wing is supported by multiple bones, each of which is an elongated finger bone. A second difference is that the wings of pterosaurs and bats are composed of a membrane formed from skin, whereas birds use feathers, which are modified from reptile scales.

In all groups, propulsion for flight requires pushing down against the air with wings. For insects, this provides enough lift to stay in the air. Vertebrates need greater lift and get this from the structure of their wing: the upper surface is more convex than the lower. Because air travels farther over the top surface, it moves faster. The faster a fluid moves, the lower its internal pressure. Thus, pressure is lower on top of the wing and higher on the bottom. This is the same principle used by airplane wings.

In birds and most insects, the raising and lowering of the wings is achieved by the alternate contraction of extensor muscles (elevators) and flexor muscles (depressors). Four insect orders (including those containing flies, mosquitoes, wasps, bees, and beetles) beat their wings at frequencies ranging from 100 to more than 1000 times per second, faster than nerves can carry successive impulses!

In these insects, the flight muscles are not attached to the wings at all but, instead, are attached to the stiff wall of the thorax, which is distorted in and out by their contraction. The reason these muscles can contract so fast is that the contraction of one muscle set stretches the other set, triggering its contraction in turn without waiting for the arrival of a nerve impulse.

In addition to active flight, many species have evolved adaptations—primarily flaps of skin that increase surface area and thus slow down the rate of descent—to enhance their ability to glide long distances. Gliders have done this in many ways, including flaps of skin along the body in flying squirrels, snakes, and lizards.

REVIEW OF CONCEPT 32.9

Locomotion involves friction and pressure created by body parts, often appendages, against water, air, or ground. Walking, running, and flying require supporting the body against gravity's pull. Flight is achieved when a pressure difference between air flowing over the top and bottom of a wing creates lift. Solutions to locomotion have evolved convergently many times.

- *In what ways would locomotion by a series of leaps be more advantageous than locomotion by alternation of legs?*

Which Mode of Locomotion Is the Most Efficient?

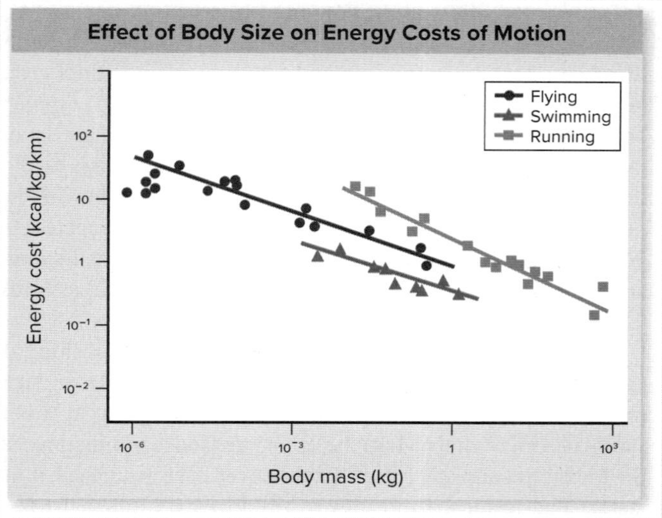

Running, flying, and swimming require more energy than sitting still, but how do they compare? The greatest differences between moving on land, in the air, and in water result from the differences in support and resistance to movement provided by water and air. The weight of swimming animals is fully supported by the surrounding water, and no effort goes into supporting the body, but running and flying animals must support the full weight of their bodies. On the other hand, water presents considerable resistance to movement, air much less, so that flying and running require less energy to push the medium out of the way.

A simple way to compare the costs of moving for different animals is to determine how much energy it takes to move. The energy cost to run, fly, or swim is in each case the energy required to move one unit of body mass over one unit of distance with that mode of locomotion. (Energy in this case is measured as a **kilocalorie (kcal)**; body mass is measured in kilograms; distance is measured in kilometers.) The graph displays three such "cost-of-motion" studies. The blue squares are running, the red circles are flying, and the green triangles are swimming. In each study, the line is drawn as the statistical "best fit" for the points. Some animals, like humans, have data in two lines, as they both run (well) and swim (poorly). Ducks have data in all three lines, as they not only fly (very well) but also run and swim (poorly).

Analysis

1. **Applying Concepts**
   a. ***Variables.*** In the graph, what is the dependent variable?
   b. ***Comparing continuous variables.*** Do the three modes of locomotion have the same or different costs?
2. **Interpreting Data**
   a. For any given mode of locomotion, what is the effect of body size on cost of moving?
   b. Is the effect of body mass the same for all three modes of locomotion? If not, which mode's cost is least affected? Why?
3. **Making Inferences**
   a. Comparing the energy costs of running versus flying for animals of the same body size, which mode of locomotion is the most expensive? Why would you expect this to be so?
   b. Comparing the energy costs of swimming to flying, which uses the least energy? Why would you expect this to be so?
4. **Drawing Conclusions** In general, which mode of locomotion is the most efficient? The least efficient? Why do you think this is so?
5. **Further Analysis** Do you think the costs of running for an athlete decrease with training? Why? How might you go about testing this?

(runner): Karl Weatherly/Getty Images; (eagle): Adam Jones/Getty Images; (fish): Stephen Frink/Digital Vision/Getty Images

Retracing the Learning Path

CONCEPT 32.1 The Vertebrate Body Has a Hierarchical Organization

32.1.1 Vertebrate Bodies Are Formed from Four Tissues Tissues are groups of cells of a single type and function. Adult vertebrate primary tissues are epithelial, connective, muscle, and nerve tissues. Organs consist of a group of different tissues that form a structural and functional unit. An organ system is a group of organs that collectively perform a function.

32.1.2 The General Body Plan of Vertebrates Is a Tube Within a Tube The tube of the digestive tract is surrounded by the skeleton and accessory organs and is enclosed in the integument. Vertebrates have both dorsal and ventral body cavities.

CONCEPT 32.2 Epithelial Tissue Covers Body Surfaces

32.2.1 Epithelium Covers Body Surfaces Epithelial cells are tightly bound together, forming a selective barrier. Epithelial cells are replaced constantly and can regenerate in wound healing.

32.2.2 Epithelial Types Reflect Their Function Epithelium is divided into two general classes: simple (one cell layer) and stratified (multiple cell layers). These are further divided into squamous, cuboidal, and columnar, based on the shape of cells.

CONCEPT 32.3 Nerve Tissue Conducts Signals Rapidly

32.3.1 Neurons Are the Primary Cells of the Nervous System Neurons have a cell body with a nucleus; dendrites, which receive impulses; and an axon, which transmits impulses away. Neuroglia help regulate the neuronal environment. Some types form the myelin sheaths that surround some axons.

CONCEPT 32.4 Connective Tissue Supports the Body

32.4.1 Connective Tissue Proper May Be Either Loose or Dense Connective tissues contain various kinds of cells in an extracellular matrix of proteins and ground substance. Connective tissue proper is divided into loose and dense connective tissue.

32.4.2 Special Connective Tissues Have Unique Characteristics All connective tissues originate from mesoderm and consist of a variety of cells within an extracellular matrix. Special connective tissues have unique cells and matrices. Cartilage is formed by chondrocytes and bone by osteocytes.

CONCEPT 32.5 Muscle Tissue Powers the Body's Movements

32.5.1 Vertebrates Possess Three Kinds of Muscle Smooth muscle is found in most organs. Involuntary smooth muscle occurs in the viscera and is composed of long, spindle-shaped cells with a single nucleus. Skeletal and cardiac muscle are multinucleate. Voluntary skeletal or striated muscle is usually attached by tendons to bones, and the cells contain contractile myofibrils. The heart is composed of cardiac muscle. This consists of striated muscle cells connected by gap junctions that allow coordination.

CONCEPT 32.6 Skeletal Systems Anchor the Body's Muscles

32.6.1 Hydrostatic Skeletons Use Water Pressure to Reinforce a Body Wall By muscular contractions, earthworms press fluid into different parts of the body, causing them to move.

32.6.2 Exoskeletons Consist of a Rigid Outer Covering The exoskeleton must be shed for the organism to grow.

32.6.3 Endoskeletons Are Composed of Hard Internal Structures Endoskeletons of vertebrates are living connective tissues that may be mineralized with calcium phosphate.

CONCEPT 32.7 Vertebrate Endoskeletons Are Made of Bone

32.7.1 Bone Tissue Is Highly Structured Bone consists of outer compact bone and inner spongy bone. The medullary cavity contains marrow, which can be red because of red blood cells or yellow because of fat cells. Bone contains bone cells, capillaries, and nerves.

32.7.2 Bones Can Be Classified by Two Modes of Development In intramembranous development, bone forms within a layer of connective tissue. In endochondral development, bone fills in a cartilaginous model.

32.7.3 Bone Remodeling Allows Bone to Respond to Use or Disuse Bone structure may thicken or thin depending on use and on forces impinging on the bone.

32.7.4 Bones Move Relative to One Another at Joints

CONCEPT 32.8 Muscles Contract Because Their Myofilaments Slide

32.8.1 Muscle Fibers Contract as Overlapping Filaments Slide Together Muscle contraction occurs when actin and myosin filaments form cross-bridges and slide relative to each other. The globular head of myosin forms a cross-bridge with actin when ATP is hydrolyzed to ADP and P_i.

32.8.2 Contraction Is Triggered by Calcium Ion Release Following a Nerve Impulse Tropomyosin, attached to actin by troponin, blocks formation of a cross-bridge. Nerve stimulation releases calcium from the sarcoplasmic reticulum and a troponin-calcium complex displaces tropomyosin.

CONCEPT 32.9 Animal Locomotion Takes Many Forms

32.9.1 Swimmers Must Contend with Friction When Moving Through Water Among vertebrates, aquatic locomotion occurs by pushing some or all of the body against the water. Many vertebrates undulate the body or tail for propulsion.

32.9.2 Terrestrial Locomotion Must Deal Primarily with Gravity Most terrestrial animals move by lifting their bodies off the ground and pushing against the ground with appendages.

32.9.3 Flying Uses Air for Support Flight involves wings pushing down against the air. Lift is created by a pressure difference as air flows above and below convex wings. In flying and gliding, convergent evolution has produced the same outcome.

Concept Overview

This Concept Overview diagrams the key concepts that were discussed in this chapter.

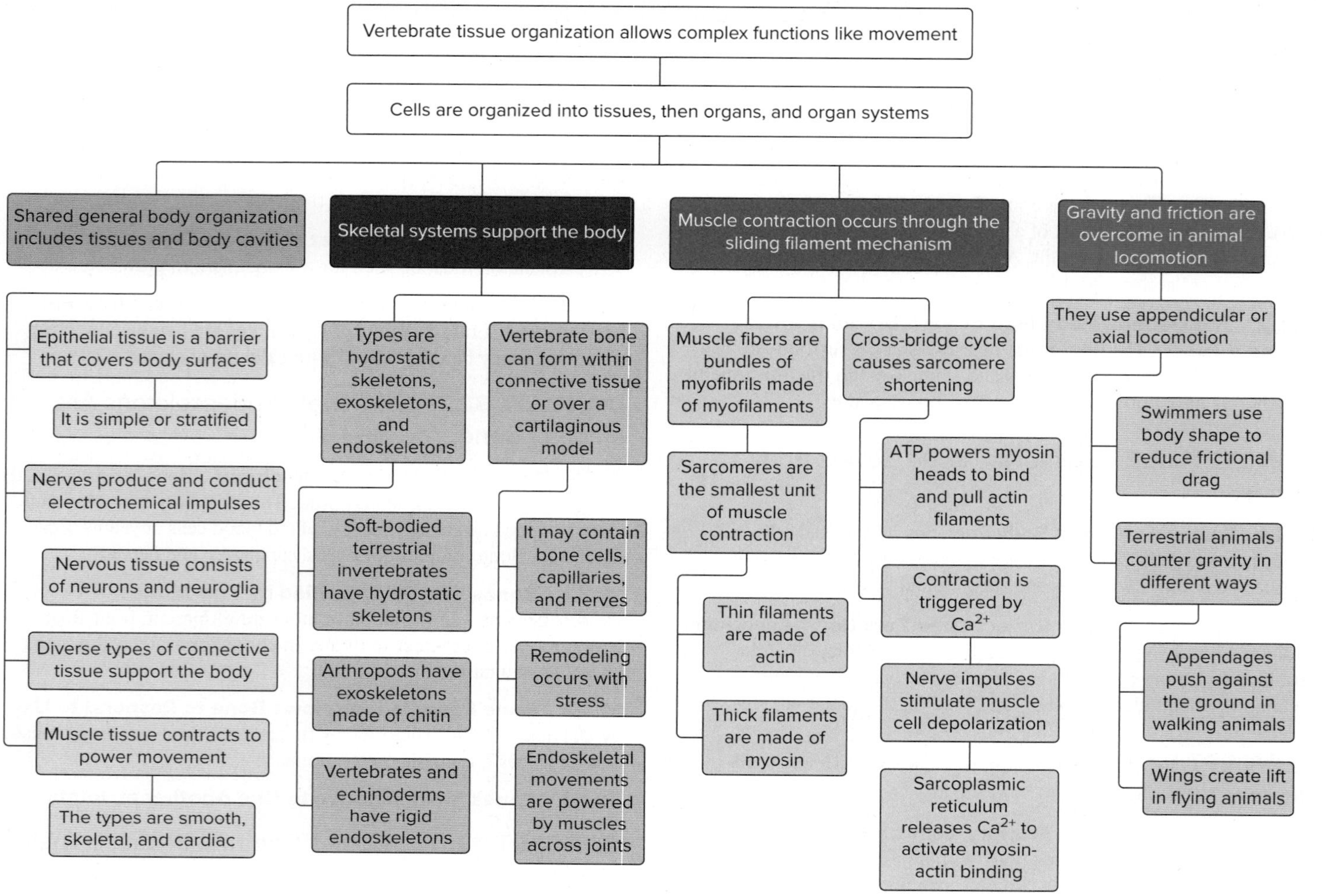

Assessing the Learning Path

Understand

1. Which of the following is NOT one of the four basic types of tissue of the adult vertebrate body?
   a. Nerve
   b. Muscle
   c. Mesoderm
   d. Connective
2. Epithelial tissues do all of the following EXCEPT
   a. form barriers or boundaries.
   b. absorb nutrients in the digestive tract.
   c. transmit information in the central nervous system.
   d. allow exchange of gases in the lung.
3. The exposed side of an epithelial tissue is the
   a. keratinized surface.
   b. apical surface.
   c. basal surface.
   d. exocrine surface.
4. The function of neuroglia is to
   a. carry messages from the PNS to the CNS.
   b. support and protect neurons.
   c. stimulate muscle contraction.
   d. store memories.
5. Nodes of Ranvier are found
   a. at the tips of dendrites.
   b. on the surface of neuroglial cells.
   c. along the axon.
   d. within the body of the neuron.
6. Connective tissues include a diverse group of cells, yet they all share
   a. a cuboidal shape.
   b. the ability to produce hormones.
   c. the ability to contract.
   d. the presence of an extracellular matrix.
7. The three types of muscle all share
   a. a structure that includes striations.
   b. a membrane that is electrically excitable.
   c. the ability to contract.
   d. the characteristic of self-excitation.

8. Endoskeletons are found in all of the following organisms EXCEPT
   a. sea urchins.
   b. sand dollars.
   c. snails.
   d. house cats.
9. Bones that form through endochondral development are typically located
   a. deep in the body.
   b. on the skull surface.
   c. in the jaw.
   d. in the dermis of the skin.
10. Haversian canals do NOT
   a. have bone laid down around them in concentric rings.
   b. run parallel to the long axis of bones.
   c. have blood vessels and nerves running through them.
   d. connect cartilage to bone.
11. The source of energy for muscle contraction is
   a. actin.
   b. myosin.
   c. sarcomeres.
   d. ATP.
12. In activating contraction, Ca^{2+} binds to
   a. actin.
   b. troponin.
   c. myosin.
   d. tropomyosin.
13. All aquatic vertebrates swim using ___________ locomotion.
   a. appendicular
   b. hydraulic
   c. axial
   d. convergent

Apply

1. What do all the organs of the body have in common?
   a. Each contains the same kinds of cells.
   b. Each is composed of several different kinds of tissue.
   c. Each is derived from ectoderm.
   d. Each can be considered part of the circulatory system.
2. Which of the following are common to all types of epithelia?
   a. An apical surface that faces outside or internally to the lumen of a space
   b. A basal surface that is connected to the basement membrane
   c. Lateral connections to form a sheet of cells
   d. All of the above
3. Suppose that an alien virus arrives on Earth. This virus causes damage to the nervous system by attacking the structures of neurons. Which of the following structures would be immune from attack?
   a. Axons
   b. Dendrites
   c. Neuroglia
   d. None of the above
4. Cartilage functions in many ways. Which of the following is NOT one of them?
   a. It makes up the hard external part of your ear.
   b. It forms the ends of bones in joints.
   c. It forms the nails on the tips of your fingers and toes.
   d. It makes spinal disks firm and flexible.
5. Skeletal muscles differ from smooth muscles in that they
   a. contain multiple nuclei.
   b. have mitochondria.
   c. have no plasma membrane.
   d. are not derived from embryonic tissue.
6. In animals with hydrostatic skeletons, the function of chaetae is to
   a. anchor the body to a hard surface.
   b. swim through the surrounding water.
   c. entrain muscle contractions.
   d. move fluid within the body cavity.
7. Growth in bone thickness occurs by adding layers of bone
   a. to replace interior cartilage.
   b. to mesenchyme.
   c. beneath the periosteum.
   d. within canaliculi.
8. Which is more likely to be found at the core of bones?
   a. Compact bone tissue
   b. Spongy bone tissue (marrow)
   c. Osteoblasts
   d. Nothing; the interior is hollow
9. The role of Ca^{2+} in the process of muscle contraction is to
   a. gather ATP for the myosin to use.
   b. cause the myosin head to shift position, contracting the myofibril.
   c. cause the myosin head to detach from the actin, causing the muscle to relax.
   d. expose myosin attachment sites on actin.
10. A bee may beat its wings 1000 times per second, faster than nerves can carry impulses. It does this by
   a. using very-large-diameter motor axons.
   b. attaching flight muscles to thorax walls rather than to wings.
   c. alternating the contraction of extensor and flexor muscles.
   d. using wings whose upper surfaces are more concave.

Synthesize

1. How much would the skin of a 60-kg individual weigh?
2. Compare the axons of motor neurons with those of interneurons. Why do you suppose this difference exists? Why not construct the axons similarly in both types of neurons?
3. Why would your heart not function well if constructed of skeletal muscle? What is the particular characteristic of cardiac muscle that is key to proper heart function?
4. Land was successfully invaded five times—by plants, fungi, mollusks, arthropods, and vertebrates. Since bodies are far less buoyant in air than in water, each of these five groups evolved a characteristic hard substance to lend mechanical support. Describe and contrast these five substances, discussing their advantages and disadvantages. Do you think plastic would have been superior to any of them?
5. Myofilaments can contract forcefully, pulling membranes attached to the two ends toward one another. Myofilaments cannot expand, however, to push membranes attached to the two ends of a myofilament apart from one another. Why is it that myofilaments can pull but not push?
6. Animals have adapted modes of locomotion to three different environments: water, land, and air. What do all three modes of locomotion have in common?

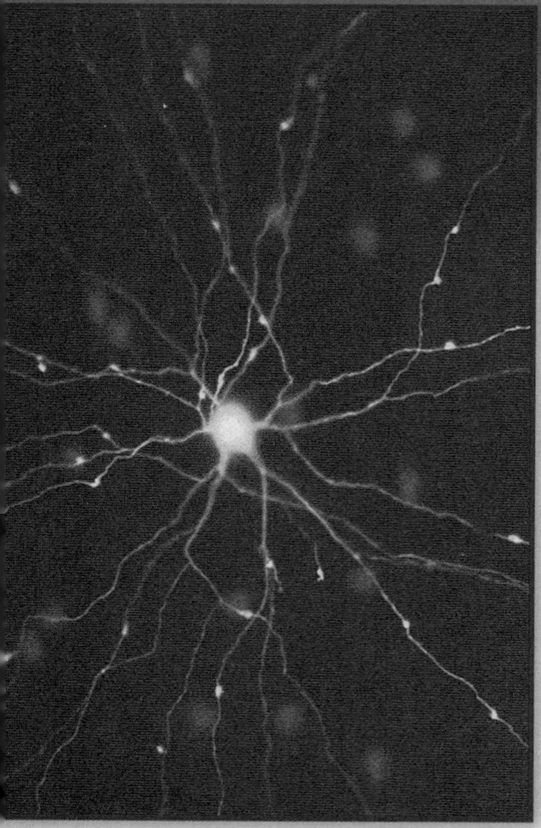

David I. Vaney, University of Queensland Australia

33 The Nervous System

Learning Path

33.1 The Nervous System Directs the Body's Actions

33.2 Neurons Maintain a Resting Potential Across the Plasma Membrane

33.3 Action Potentials Propagate Nerve Impulses

33.4 Synapses Are Where Neurons Communicate with Other Cells

33.5 The Central Nervous System Includes the Brain and Spinal Cord

33.6 The Peripheral Nervous System Consists of Both Sensory and Motor Neurons

33.7 Sensory Receptors Provide Information About the Body's Environment

33.8 Mechanoreceptors Sense Touch and Pressure

33.9 Sounds and Body Position Are Sensed by Vibration Detectors

33.10 Taste, Smell, and pH Senses Utilize Chemoreceptors

33.11 Vision Employs Photoreceptors to Perceive Objects at a Distance

Concept Overview

This Concept Overview provides a diagram of the overarching concepts that are covered in this chapter. These concepts will be expanded upon in the Concept Overview at the end of the chapter.

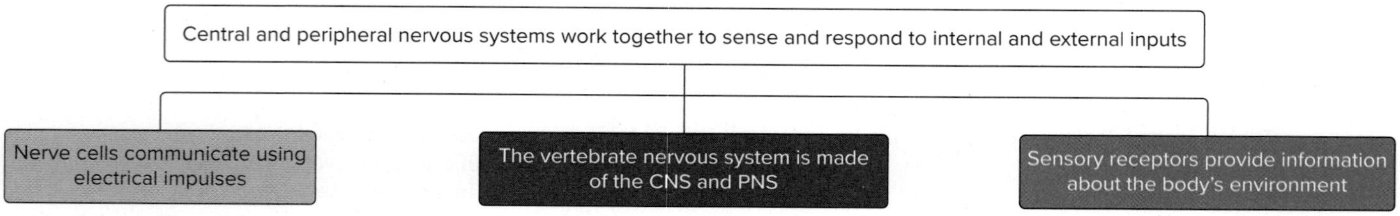
Central and peripheral nervous systems work together to sense and respond to internal and external inputs
Nerve cells communicate using electrical impulses
The vertebrate nervous system is made of the CNS and PNS
Sensory receptors provide information about the body's environment

Introduction

All animals except sponges use a network of nerve cells to gather information about the body's condition and its external environment, to process and integrate that information, and to control the body's muscles and glands. The nervous system, composed of neurons such as the one pictured on the previous page, use electrical signals as a form of rapid communication. The nervous system is a key part of the many feedback systems that maintain a constant internal environment. Sensory neurons receive input from many different kinds of receptor cells, which send information to different brain regions and, thus, are associated with the different senses. The brain distinguishes a sunset, a symphony, and searing pain only in terms of the identity of the sensory neuron carrying the action potentials and the frequency of these impulses.

33.1 The Nervous System Directs the Body's Actions

The Nervous System Is Divided into Central and Peripheral Systems

LEARNING OBJECTIVE 33.1.1 Distinguish the subdivisions of the vertebrate nervous system.

An animal must be able to respond to environmental stimuli. A fly escapes a fly swatter; the antennae of a crayfish detect food and the crayfish moves toward it. To accomplish these actions, animals must have *sensory receptors* that can detect the stimulus and *motor effectors* that can respond to it. In most invertebrate phyla and in all vertebrate classes, sensory receptors and motor effectors are organized into a nervous system.

The vertebrate nervous system is organized into the **central nervous system (CNS),** composed of the brain and spinal cord, and the **peripheral nervous system (PNS),** which consists of neurons outside the CNS. The PNS contains **sensory** (or afferent) **neurons,** which carry impulses from sensory receptors to the CNS, and **motor** (or efferent) **neurons,** which carry impulses from the CNS to effector muscles and glands. Most invertebrates and all vertebrates also have **interneurons** (or association neurons). These interneurons are found in the brain and spinal cord of vertebrates, where they help to provide more complex reflexes and higher associative functions, such as learning and memory.

The PNS can also be subdivided into the **somatic nervous system,** which is under conscious control and consists of motor neurons that stimulate skeletal muscles to contract, and the **autonomic nervous system,** which is not under conscious control and consists of neurons that regulate the activity of smooth muscles, cardiac muscle, and glands. The autonomic nervous system is divided into the sympathetic and parasympathetic divisions, which act to counterbalance each other in regulating many organ systems. Figure 33.1 illustrates the relationships among the different parts of the vertebrate nervous system.

The structure of neurons supports their function

Despite their varied appearances, most neurons have the same functional architecture (figure 33.2). The **cell body** contains the nucleus and assorted organelles. Many cytoplasmic extensions called **dendrites** act to receive signals. Motor and association neurons possess a profusion of highly branched dendrites, which allow input from many different sources simultaneously. Some neurons have extensions from the dendrites, called dendritic spines, that increase the surface area available to receive stimuli.

The surface of the cell body integrates the information arriving at its dendrites. If the resulting membrane excitation is sufficient, it triggers the conduction of impulses away from the cell body along a single **axon,** which may branch to stimulate a number of cells. These axons can be quite long, requiring

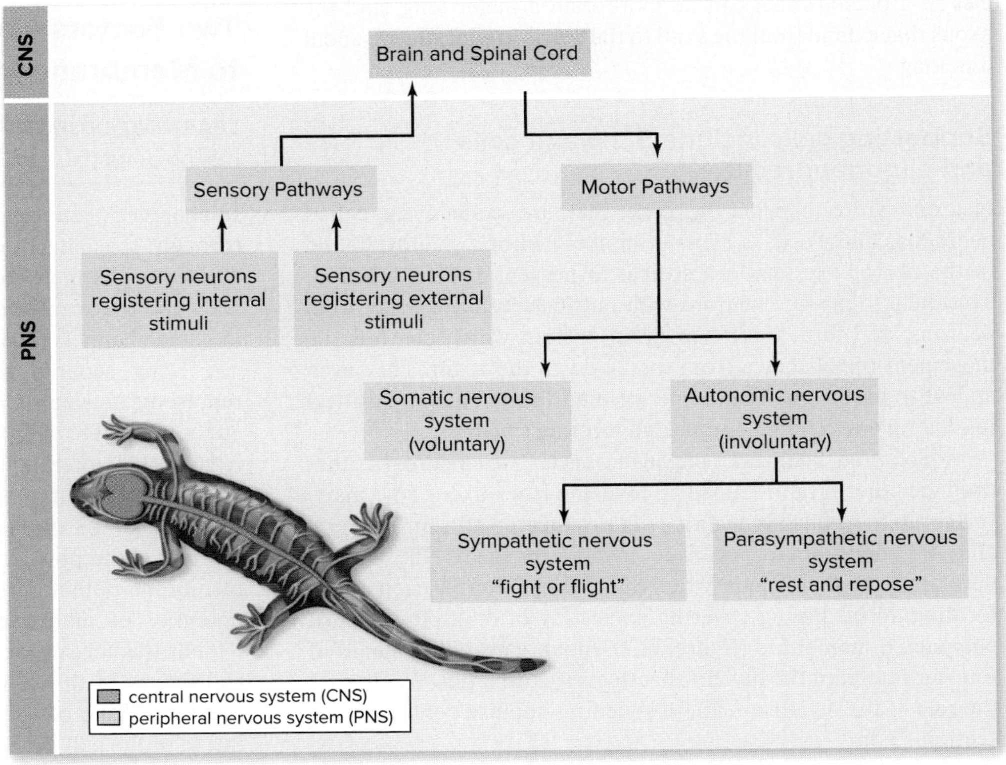

Figure 33.1 Divisions of the vertebrate nervous system. The major divisions are the central and peripheral nervous systems. The brain and spinal cord make up the central nervous system (CNS). The peripheral nervous system (PNS) includes the remainder of the nervous system outside the CNS and is divided into sensory and motor pathways. Sensory pathways can detect either external or internal stimuli. Motor pathways are divided into the somatic nervous system, which activates voluntary muscles, and the autonomic nervous system, which activates involuntary muscles. The sympathetic and parasympathetic nervous systems are subsets of the autonomic nervous system that trigger opposing actions.

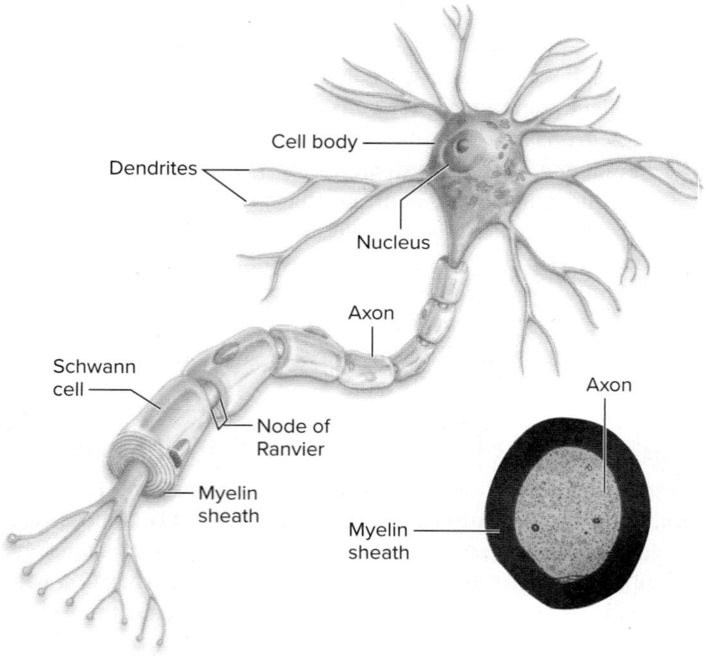

Figure 33.2 Structure of a typical vertebrate neuron. Extending from the cell body are many dendrites, which receive information and carry it to the cell body. A single axon transmits impulses away from the cell body. Many axons are encased by a myelin sheath that insulates the axon. Small gaps, called nodes of Ranvier, interrupt the sheath at regular intervals.

Don W. Fawcett/Science Source

specialized transport systems to convey material to and from the cell body via microtubule tracts. The axons controlling the muscles in a person's feet can be more than a meter long, and the axons that extend from the skull to the pelvis in a giraffe are about 3 m long.

Supporting cells include Schwann cells and oligodendrocytes

Neurons are supported by cells that are collectively called **neuroglia.** These cells, which outnumber neurons roughly 50-fold in the nervous system, are critical to nervous system function. They help to provide neurons with nutrients, remove waste from neurons, and form the blood–brain barrier, which controls the movement of substance from the blood to the brain. They even appear to participate in synaptic communication and are critical in guiding developing neurons and forming synapses.

Schwann cells and **oligodendrocytes** are neuroglia that produce the **myelin sheaths** covering the axons of many vertebrate neurons. Schwann cells produce myelin in the PNS, and oligodendrocytes produce myelin in the CNS. During development, these cells wrap themselves around each axon, forming an insulating covering consisting of multiple layers of compacted membrane (figure 33.2). Small gaps called **nodes of Ranvier** interrupt the myelin sheath every 1 to 2 μm. We discuss the role of the myelin sheath in speeding impulse conduction in section 33.3.

REVIEW OF CONCEPT 33.1

The vertebrate nervous system consists of the central nervous system (CNS) and peripheral nervous system (PNS). The PNS comprises the somatic nervous system and autonomic nervous system; the latter has sympathetic and parasympathetic divisions. A neuron consists of a cell body, dendrites that receive information, and a single axon that sends signals. Neurons carry out nervous system functions; they are supported by a variety of neuroglia.

- *Which division of the PNS is under conscious control?*

33.2 Neurons Maintain a Resting Potential Across the Plasma Membrane

Neuronal function depends on the ability to create an electric potential across the plasma membrane and then to alter this potential to propagate signals. Because cell membranes are bathed in aqueous solutions, electric charge is carried by ions, and cells create and alter electric potentials by manipulating the concentrations of a number of important ions across the plasma membrane.

When a neuron is stimulated, electrical changes in the plasma membrane spread, or propagate, from one part of the cell to another. This signaling depends on the properties of a variety of specialized membrane transport proteins. We will examine the basic electrical properties common to the membranes of animal cells that produce a membrane potential, then explore how neurons send signals as changes in membrane potential.

Two Factors Contribute to Membrane Potential

LEARNING OBJECTIVE 33.2.1 Describe the production of the resting potential.

In chapter 5, you learned that ions can cross the cell membrane through specialized membrane proteins called **ion channels.** These ion channels are specific for different ions, such as K^+ or Na^+, and they can be either leakage channels (open all the time) or gated channels (open in response to a stimulus).

Any separation of electric charges of opposite sign represents an electric potential that is capable of doing work; we encounter this when we use a flashlight, which draws current from such a potential in a battery. Like a battery, cells maintain an electric potential across the plasma membrane; in this case, the interior of the membrane is the negative pole, and the exterior is the positive pole. Because cells are very small, their membrane potential is also very small. The resting membrane potential of many vertebrate neurons ranges from −40 to −90 millivolts (mV), or 0.04 to 0.09 volts (V). For the examples and figures in this chapter, we use an average resting membrane potential value of −70 mV. The minus sign indicates that the inside of the cell is negative with respect to the outside.

Contributors to membrane potential

The inside of the cell is more negatively charged in relation to the outside because of three factors:

1. The sodium–potassium pump brings two potassium ions (K^+) into the cell for every three sodium ions (Na^+) it pumps out (figure 33.3). This helps establish and maintain high K^+ and low Na^+ concentrations inside the cell, as well as high Na^+ and low K^+ concentrations outside the cell.
2. There are more K^+ than Na^+ ion channels in the plasma membrane. This makes the membrane more permeable to K^+ than to Na^+. The higher concentration of K^+ inside the cell leads to outward diffusion, carrying positive charge outside the cell.
3. The plasma membrane is impermeable to negative ions, primarily negatively charged proteins, nucleic acids, and organic phosphates. These are responsible for a negative charge that attracts K^+ ions back into the cell.

The resting potential: Balance between two forces

The **resting potential** arises due to the action of the sodium–potassium pump and the greater permeability of the membrane to K^+. The pump moves three Na^+ outside for every two K^+ inside, which creates a small imbalance in cations outside the cell. This has only a minor effect, but the concentration gradients created by the pump are significant. The concentration of K^+ is much higher inside the cell than outside, leading to diffusion of K^+ through K^+ leakage channels that are always open. Because the membrane is not permeable to the negative ions that could counterbalance this (mainly organic phosphates, amino acids, and proteins), positive charge builds up outside the membrane and negative charge builds up inside the membrane. This electric potential is an attractive force, pulling K^+ ions back inside the cell. The balance between the diffusional force and the electrical force produces an **equilibrium potential.** By relating the work done by each type of force, we can derive a quantitative expression for this equilibrium potential (E) called the **Nernst equation.** The equation is usually formulated in terms of the concentrations of a single ion inside and outside the cell. For a positive ion with charge equal to +1, the Nernst equation for K^+ ions is

$$E_K = 58 \text{ mV} \log([K^+]\text{out}/[K^+]\text{in})$$

The calculated equilibrium potential for K^+ is −90 mV, close to the measured value of −70 mV. The calculated value for Na^+ is +60 mV, clearly not at all close to the measured value, but the leakage of a small amount of Na^+ back into the cell is responsible for lowering the equilibrium potential of K^+ to the −70 mV value observed. The resting membrane potential of a neuron can be measured and viewed or graphed using a voltmeter and a pair of electrodes, one outside and one inside the cell (figure 33.3).

REVIEW OF CONCEPT 33.2

Neurons maintain high K^+ levels inside the cell and high Na^+ levels outside the cell. Diffusion of K^+ to the outside leads to a resting potential of about −70 mV.

- *Can you imagine any events that could change the membrane's resting potential?*

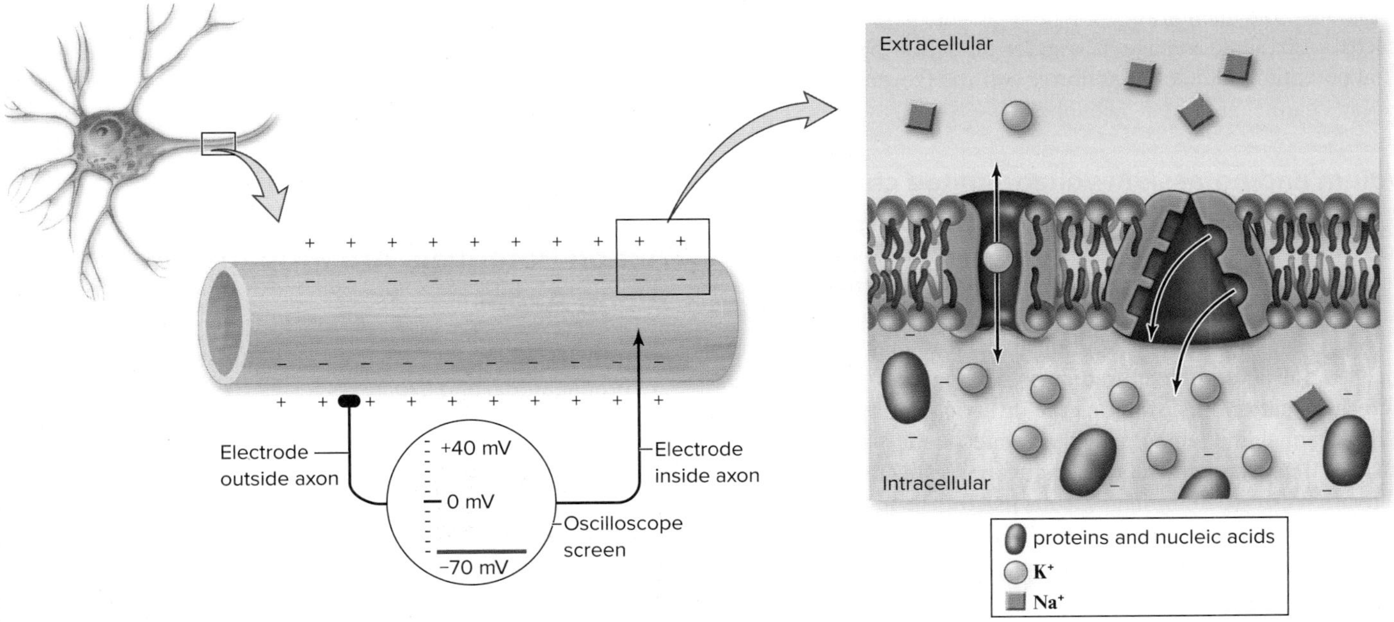

Figure 33.3 Establishment of the resting membrane potential. A voltmeter measures an electric potential of −70 mV inside relative to outside the membrane. The higher concentration of K^+ inside the cell leads to outward diffusion through ion channels, which carries positive charge outside. Negatively charged proteins and nucleic acids provide an electrical force attracting K^+ inside. This balance of electrical and diffusional forces produces the resting potential. The sodium–potassium pump maintains equilibrium by counteracting any Na^+ leakage inward and moves three Na^+ outside for every two K^+ moved inside.

33.3 Action Potentials Propagate Nerve Impulses

Neurons are unique, not because of their resting membrane potential but because of changes in membrane potential that occur in response to stimuli. Two types of changes are observed: graded potentials and action potentials. **Graded potentials** are small, *continuous* changes to the membrane potential, and action potentials are sharp, *transient* alterations of the potential. **Action potentials** are produced when the membrane potential exceeds a threshold voltage. Action potentials form the signals sent along an axon, whereas graded potentials are produced at connections between neurons and other cells.

Changes in Membrane Permeability Alter Membrane Potential

LEARNING OBJECTIVE 33.3.1 Explain how the action of voltage-gated channels produces an action potential.

The resting potential is due to the permeability of the membrane to K^+ through leakage channels. Deviations from the resting potential are due to the action of a different class of channels, called gated channels. **Gated channels** act as doors that can be either opened or closed in response to a stimulus. There are two types of gated channels: **ligand-gated channels,** which respond to a chemical signal, and **voltage-gated channels,** which respond to changes in membrane potential. Ligand-gated channels cause graded potentials that determine whether an axon will fire, and voltage-gated channels produce action potentials. We will consider the molecular basis for action potentials first.

When the membrane potential at the base of an axon exceeds a threshold voltage, it triggers an action potential. The action potential is due to the action of two voltage-gated channels: the Na^+ channel and the K^+ channel.

Sodium and potassium voltage-gated channels

The behavior of the voltage-gated Na^+ channel is complex because it has two gates: an activation gate and an inactivation gate. In its resting state, the activation gate is closed and the inactivation gate is open. When the threshold voltage is reached, the activation gate opens rapidly, leading to an influx of Na^+ ions due to both concentration and voltage gradients. After a short period, the inactivation gate closes, stopping the influx of Na^+ ions and leaving the channel in a temporarily inactivated state. The channel is returned to its resting state by the activation gate closing and the inactivation gate opening. This results in a transient influx of Na^+ that reverses the membrane's electrical polarity, or depolarizes it, in response to a threshold voltage.

The voltage-gated K^+ channel has a single activation gate that is closed in the resting state. In response to a threshold voltage, it opens slowly. With the high concentration of K^+ inside the cell, and the membrane now far from the equilibrium potential, an efflux of K^+ begins. The positive charge now leaving the cell restores the membrane to the polarity of the resting potential, or repolarizes it.

Tracing an action potential's changes

Let us now integrate these ideas to understand how the changing flux of ions produces action potentials. The action potential has three phases: a *rising phase,* a *falling phase,* and an *undershoot phase* (figure 33.4). When a threshold potential is reached, the rapid opening of the Na^+ channel causes an influx of Na^+ that shifts the membrane potential toward the equilibrium potential for Na^+ (+60 mV). This appears as the rising phase on an oscilloscope. The membrane potential never quite reaches +60 mV, because the inactivation gate of the Na^+ channel closes, terminating the rising phase. At the same time, the opening of the K^+ channel leads to K^+ diffusing out of the cell, repolarizing the membrane in the falling phase. The K^+ channels remain open longer than necessary to restore the resting potential, resulting in a slight undershoot. This entire sequence of events for a single action potential takes about a millisecond.

The nature of action potentials

Action potentials are separate, all-or-none events. An action potential occurs if the threshold voltage is reached, but not while the membrane remains below threshold. Action potentials do not add together or interfere with one another, as below-threshold potentials can. After Na^+ channels "fire," they remain in an inactivated state until the inactivation gate reopens, preventing any summing of effects. This is called the absolute refractory period, when the membrane cannot be stimulated. There is also a relative refractory period during which stimulation produces action potentials of reduced amplitude.

The production of an action potential results entirely from the passive diffusion of ions. However, at the end of each action potential the cytoplasm contains a little more Na^+ and a little less K^+ than it did at rest. Although the number of ions moved by a single action potential is tiny relative to the concentration gradients of Na^+ and K^+, eventually this would have an effect. The constant activity of the sodium–potassium pump compensates for these changes. Thus, although active transport is not required to produce action potentials, it is needed to maintain the ion gradients.

Action Potentials Are Propagated Along Axons

LEARNING OBJECTIVE 33.3.2 Describe how action potentials are propagated along axons.

The movement of an action potential through an axon is not generated by ions flowing from the base of the axon to the end. Instead, an action potential originates at the base of the axon and is re-created along the axon.

Each action potential, during its rising phase, reflects a reversal in membrane polarity. The positive charges due to influx of Na^+ can depolarize the adjacent region of membrane to threshold, so that the next region produces its own action potential. Meanwhile, the previous region of membrane repolarizes back to the resting membrane potential. The signal does not back up, because the Na^+ channels that have just "fired" are still in an inactivated state and are refractory (resistant) to stimulation.

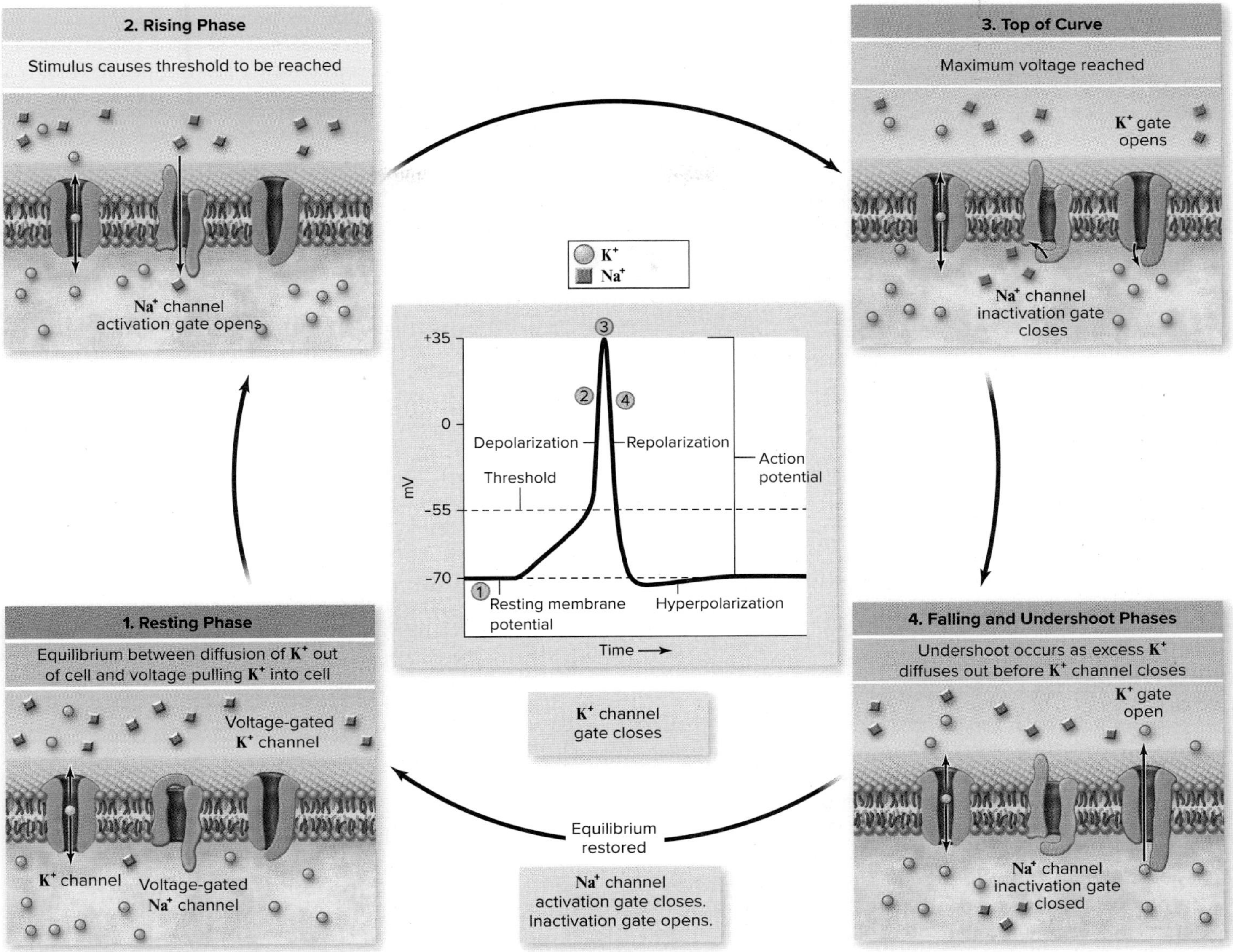

Figure 33.4 The action potential. (1) At resting membrane potential, voltage-gated ion channels are closed, but there is some diffusion of K^+. In response to a stimulus, the cell begins to depolarize, and once the threshold level is reached, an action potential is produced. (2) Rapid depolarization occurs (the rising portion of the spike), because voltage-gated sodium channel activation gates open, allowing Na^+ to diffuse into the axon. (3) At the top of the spike, Na^+ channel inactivation gates close, and voltage-gated potassium channels that were previously closed begin to open. (4) With the K^+ channels open, repolarization occurs because of the diffusion of K^+ out of the axon. An undershoot occurs before the membrane returns to its original resting potential.

The propagation of an action potential is similar to people in a stadium performing the "wave": individuals stay in place as they stand up (depolarize), raise their hands (peak of the action potential), and sit down again (repolarize). The wave travels around the stadium, but the people stay in place.

Animals have evolved two ways to increase the velocity of nerve impulses. The velocity of conduction is greater if the diameter of the axon is large or if the axon is myelinated.

Increasing the diameter of an axon increases the velocity of nerve impulses because electrical resistance is inversely proportional to cross-sectional area, which in turn is a function of diameter, so larger-diameter axons have less resistance to current flow. The axons with the largest diameter are found primarily in invertebrates.

Myelinated axons conduct impulses more rapidly than unmyelinated axons because the action potentials in myelinated axons are produced only at the nodes of Ranvier. One action potential still serves as the depolarization stimulus for the next, but the depolarization at one node spreads quickly beneath the insulating myelin to trigger the opening of voltage-gated channels at the next node. The impulses therefore appear to jump from node to node (figure 33.5) in a process called **saltatory conduction** (Latin *saltare,* "to jump").

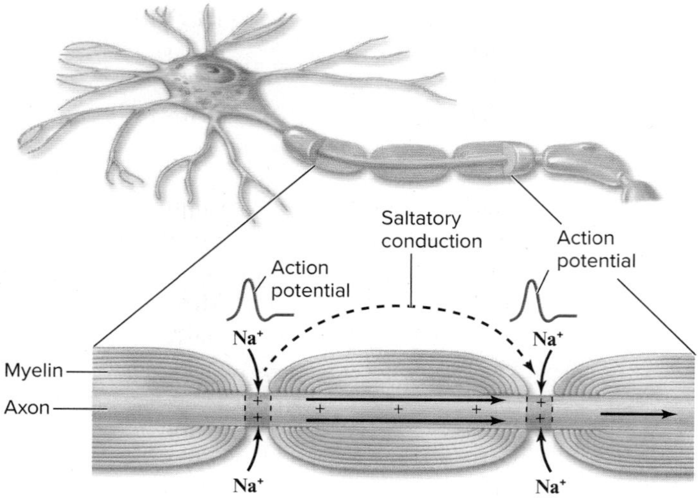

Figure 33.5 Saltatory conduction in a myelinated axon. Action potentials are produced only at the nodes of Ranvier in a myelinated axon. One node depolarizes the next node so that the action potentials can skip between nodes. As a result of fewer action potentials, saltatory conduction in a myelinated axon is more rapid than conduction in an unmyelinated axon.

To see how this speeds impulse transmission, let's return to the stadium wave analogy for propagation of action potentials. The wave moves across the seats of a crowded stadium as fans seeing the people in the adjacent section stand up are triggered to stand up. Because the wave skips empty sections, it moves around the stadium faster with more empty sections. The wave doesn't "wait" for the missing people to stand; it simply moves to the next populated section—just as the action potential jumps the insulated regions of myelin between exposed nodes.

REVIEW OF CONCEPT 33.3

Action potentials are triggered when membrane potential exceeds a threshold value. Voltage-gated Na^+ channels open, and depolarization occurs; subsequent opening of K^+ channels leads to repolarization.

- *How can only positive ions result in depolarization and repolarization of the membrane during an action potential?*

33.4 Synapses Are Where Neurons Communicate with Other Cells

A nerve signal eventually reaches the end of the axon and its branches. These then form junctions with the dendrites of other neurons, muscle cells, or gland cells. These intercellular junctions are called **synapses,** and they can be either ***electrical*** (direct connections by gap junctions) or ***chemical*** (a signal passes between the cells). These connections involve a presynaptic cell sending a signal and a postsynaptic cell receiving the signal. We will concentrate on chemical synapses.

Vertebrate Synapses Are Not Physical Connections

LEARNING OBJECTIVE 33.4.1 Describe how cells communicate across synapses.

When synapses are viewed under a light microscope, the presynaptic and postsynaptic cells appear to touch, but when viewed with an electron microscope, most have a **synaptic cleft,** a narrow space that separates these two cells (figure 33.6).

The end of the presynaptic axon contains numerous **synaptic vesicles,** each packed with chemicals called *neurotransmitters.* When action potentials arrive at the end of the axon, they stimulate the opening of voltage-gated calcium (Ca^{2+}) channels, causing a rapid inward diffusion of Ca^{2+}. This influx of Ca^{2+} leads to the fusion of synaptic vesicles with the plasma membrane and the release of neurotransmitter by exocytosis.

The higher the frequency of action potentials in the presynaptic axon, the greater the number of vesicles that release their contents of neurotransmitters. The neurotransmitters diffuse to the other side of the cleft and bind to chemical- or ligand-gated receptor proteins in the membrane of the postsynaptic cell. The action of these receptors produces graded potentials in the postsynaptic membrane.

Neurotransmitters are chemical signals in an otherwise electrical system, requiring tight control over the duration of their

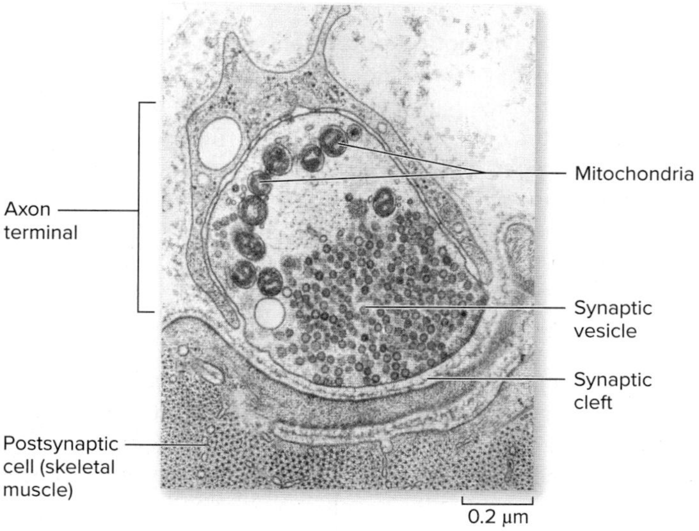

Figure 33.6 A synaptic cleft. An electron micrograph showing a neuromuscular synapse. Synaptic vesicles have been colored green for clarity.

Don W. Fawcett/Science Source

action. Neurotransmitters must be rapidly removed from the synaptic cleft to allow new signals to be transmitted. This is accomplished by a variety of mechanisms, including enzymatic digestion in the synaptic cleft, reuptake of neurotransmitter molecules by the neuron, and uptake by glial cells.

Many Different Chemicals Serve as Neurotransmitters

LEARNING OBJECTIVE 33.4.2 Contrast the effects of excitatory and inhibitory neurotransmitters.

Neurotransmitters are a chemically diverse group of molecules. They can even be gases, such as nitric oxide, or common cellular molecules. Some have wide distribution in the nervous system, and others are found only in specific types of synapses. The following are examples of representative neurotransmitters.

Acetylcholine

Acetylcholine (ACh) is the neurotransmitter used at neuromuscular junctions (figure 33.7). Acetylcholine binds to its receptor in the postsynaptic membrane, which is a ligand-gated Na^+ channel. Opening these channels in the postsynaptic membrane produces a depolarization (figure 33.8*a*) called an *excitatory postsynaptic potential (EPSP).* The EPSP, if large enough, will produce an action potential. Because the postsynaptic cell in this case is a skeletal muscle fiber, this action potential will stimulate muscle contraction, as discussed in chapter 32.

For the muscle to relax, ACh must be eliminated from the synaptic cleft. *Acetylcholinesterase (AChE),* an enzyme in the postsynaptic membrane, eliminates ACh. This enzyme cleaves ACh into inactive fragments. Nerve gas and the insecticide parathion are potent inhibitors of AChE; in humans, they can produce severe spastic paralysis and even death. Although ACh acts as a neurotransmitter between motor neurons and skeletal muscle cells, many neurons also use ACh as a neurotransmitter at their synapses with the dendrites or cell bodies of other neurons.

Amino acids

Glutamate is the major excitatory neurotransmitter in the vertebrate CNS. Excitatory neurotransmitters act to stimulate action potentials by producing EPSPs.

Glycine and γ-aminobutyric acid (GABA) are inhibitory neurotransmitters. These neurotransmitters cause the opening of ligand-gated channels for the chloride ion (Cl^-), which has a concentration gradient causing diffusion into the neuron. The negatively charged Cl^- ion makes the inside of the membrane more negative than it is at rest (figure 33.8*b*). This hyperpolarization is called an *inhibitory postsynaptic potential (IPSP).* The drug diazepam (Valium) causes its sedative and other effects by enhancing the binding of GABA to its receptors.

Biogenic amines

Dopamine is a very important neurotransmitter used in some areas of the brain controlling body movements and other functions. Degeneration of particular dopamine-releasing neurons is part of the cause of the resting muscle tremors of Parkinson disease, which is the rationale for treatment with L-dopa (L-3, 4-dihydroxyphenylalanine), a precursor for dopamine.

Serotonin is a neurotransmitter involved in the regulation of sleep, and it is implicated in various emotional states. Insufficient activity of the neurons that release serotonin may be one cause of clinical depression. Some antidepressant drugs, such as fluoxetine (Prozac), block the elimination of serotonin from the synaptic cleft; these drugs are termed *selective serotonin reuptake inhibitors,* or *SSRIs.*

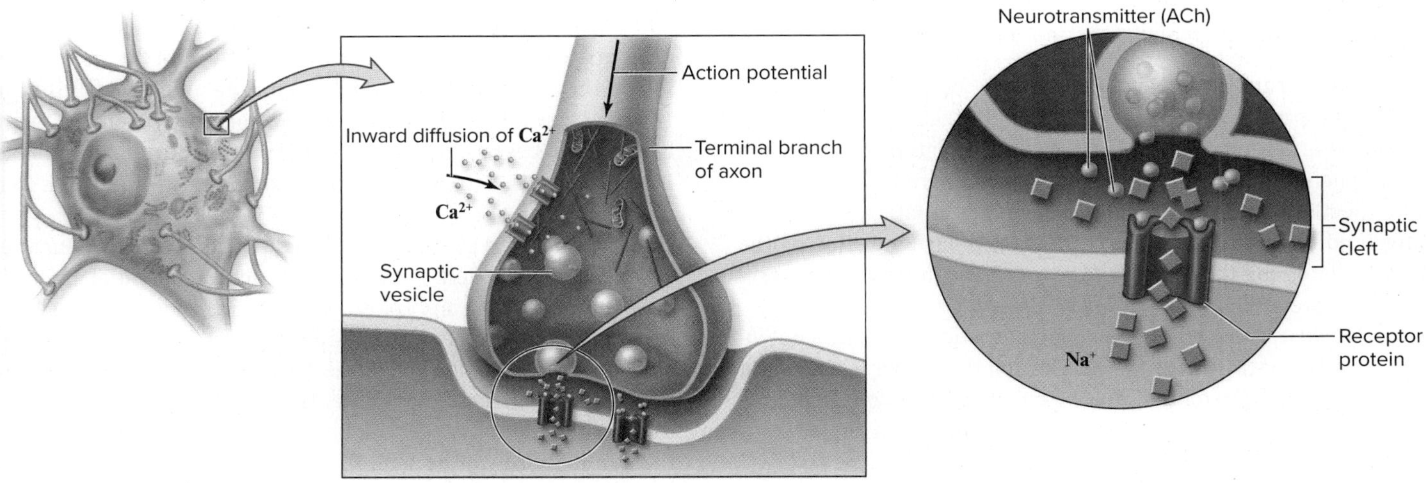

Figure 33.7 The release of neurotransmitter. Action potentials arriving at the end of an axon trigger inward diffusion of Ca^{2+}, which causes synaptic vesicles to fuse with the plasma membrane and release their neurotransmitters (acetylcholine [ACh], in this case). Neurotransmitter molecules diffuse across the synaptic gap and bind to ligand-gated receptors in the postsynaptic membrane.

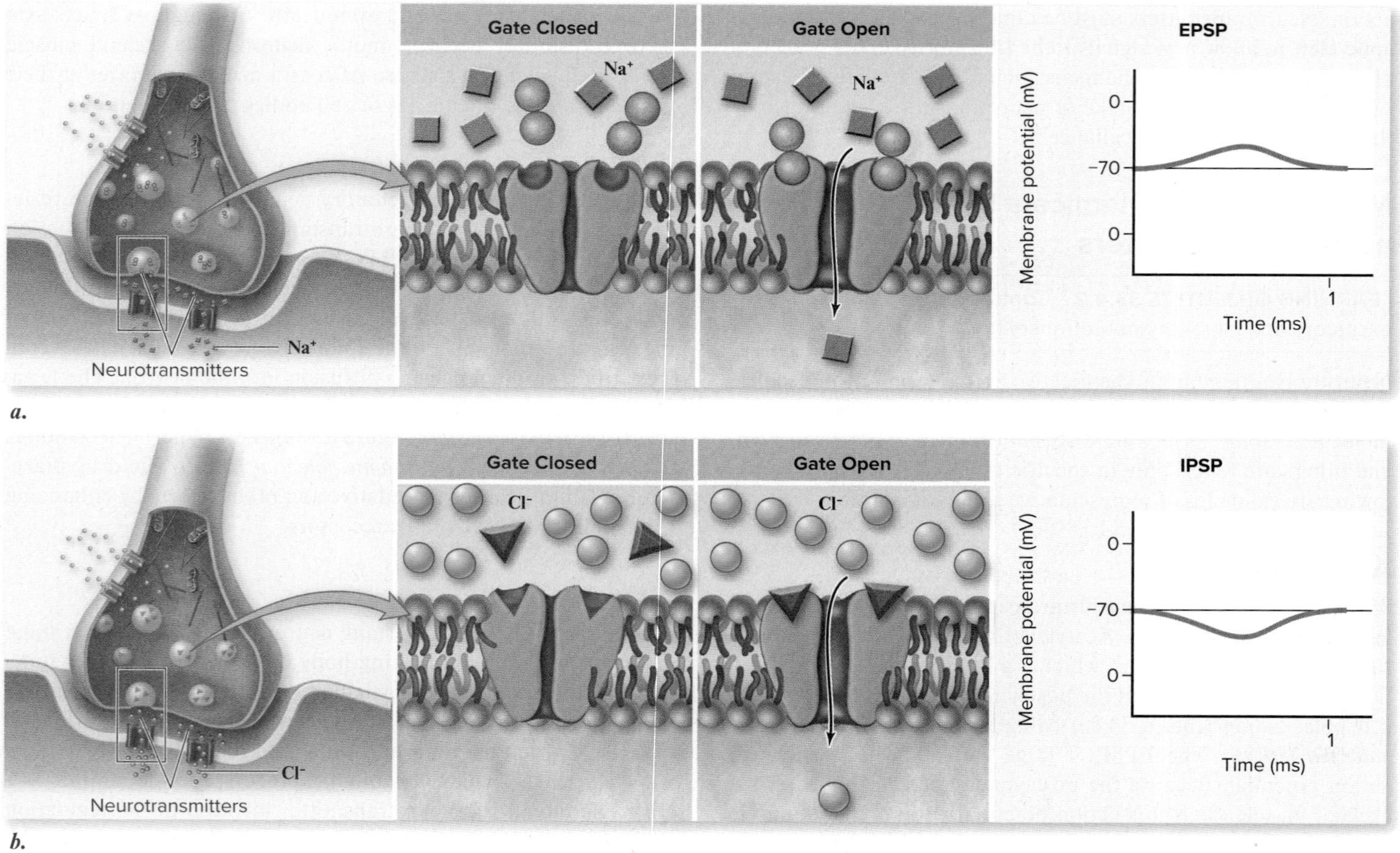

Figure 33.8 Different neurotransmitters can have different effects. ***a.*** An excitatory neurotransmitter (orange balls) promotes a depolarization, or excitatory postsynaptic potential (EPSP). ***b.*** An inhibitory neurotransmitter (purple triangles) promotes a hyperpolarization, or inhibitory postsynaptic potential (IPSP).

A Postsynaptic Neuron Must Integrate Inputs from Many Synapses

LEARNING OBJECTIVE 33.4.3 Explain how a neuron integrates the input from many other neurons.

Each postsynaptic neuron may receive both excitatory and inhibitory synapses. The EPSPs (depolarizations) and IPSPs (hyperpolarizations) from these synapses interact with each other when they reach the cell body of the neuron. Small EPSPs add together to bring the membrane potential closer to the threshold, and IPSPs subtract from the depolarizing effect of the EPSPs, deterring the membrane potential from reaching threshold. This process is called *synaptic integration.*

Because of the all-or-none characteristic of an action potential, a postsynaptic neuron is like a switch that either is turned on or remains off. Information may be encoded in the pattern of firing over time, but each neuron can only fire or not fire when it receives a signal.

The events that determine whether a neuron fires may involve many presynaptic neurons. There are two ways the membrane can reach the threshold voltage: by many different dendrites producing EPSPs that sum to the threshold voltage, or by one dendrite producing repeated EPSPs that sum to the threshold voltage. We call the first **spatial summation** and the second **temporal summation.**

In spatial summation, graded potentials due to dendrites from different presynaptic neurons that occur at the same time add together to produce an above-threshold voltage. This input does not need to be all in the form of EPSPs; all that is required is that the potential produced by summing all of the EPSPs and IPSPs is greater than the threshold voltage. When the membrane at the base of the axon is depolarized above the threshold, it produces an action potential, and a nerve impulse is sent down the axon.

In temporal summation, a single dendrite can produce sufficient depolarization to produce an action potential if it produces EPSPs that are close enough in time to sum to a depolarization that is greater than threshold. A typical EPSP can last for 15 ms, so for temporal summation to occur, the next impulse must arrive in less time. If enough EPSPs are produced to raise the membrane at the base of the axon above threshold, then an impulse will be sent.

The distinction between these two methods of summation is like building a mound on the ground with soil: you can have either many shovels that add soil to the mound until it is high enough or a single shovel that adds soil at a faster rate to build the mound. When the mound is high enough, the axon will fire.

REVIEW OF CONCEPT 33.4

Chemical synapses involve communication between neurons by neurotransmitters. These include acetylcholine, GABA, glutamate, dopamine, and serotonin. These signaling events cause graded potentials that can sum to cause an action potential.

- *How would a neurotransmitter that acts through a K^+ channel affect membrane potential?*

33.5 The Central Nervous System Includes the Brain and Spinal Cord

The central nervous system is an ancient evolutionary innovation. Essentially all bilaterian animals have a central nervous system with a ganglion at the anterior and nerve cords running the length of the body, but it is in vertebrates that this architecture reached its greatest elaboration.

Vertebrate Brains Have Three Basic Divisions

LEARNING OBJECTIVE 33.5.1 Describe the organization of the brain in vertebrates.

The complex nervous system of vertebrate animals has a long evolutionary history. Casts of the interior braincases of fossil agnathans, fishes that swam 500 MYA (refer to chapter 28), have revealed much about the early evolutionary stages of the vertebrate brain. Although small, these brains already had the three divisions that characterize the brains of all contemporary vertebrates (figure 33.9):

1. The *hindbrain,* or rhombencephalon
2. The *midbrain,* or mesencephalon
3. The *forebrain*, or prosencephalon

The hindbrain in fishes

The hindbrain was the major component of these early brains, as it still is in fishes today. Composed of the **cerebellum, pons,** and **medulla oblongata,** the hindbrain may be considered an extension of the spinal cord devoted primarily to coordinating motor reflexes. Tracts containing large numbers of axons run like cables up and down the spinal cord to the hindbrain. The hindbrain, in turn, integrates the many sensory signals coming from the muscles and coordinates the pattern of motor responses.

Much of this coordination is carried on within a small extension of the hindbrain called the cerebellum ("little cerebrum"). In more advanced vertebrates, the cerebellum plays an increasingly important role as a coordinating center for movement, and it is correspondingly larger than it is in the fishes. In all vertebrates, the cerebellum processes data on the current position and movement of each limb, the state of relaxation or contraction of the muscles involved, and the general position of the body and its relation to the outside world.

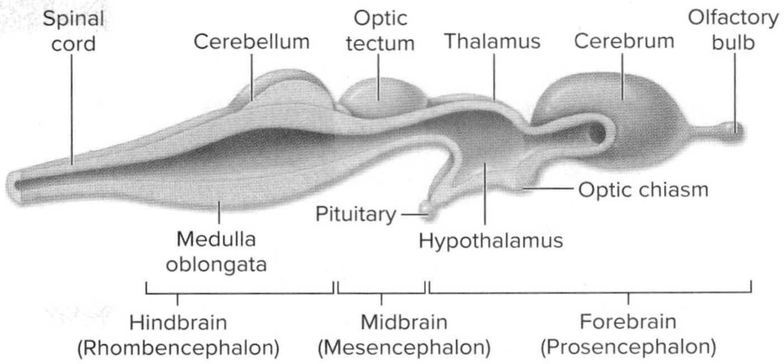

Figure 33.9 The basic organization of the vertebrate brain can be seen in the brains of primitive fishes. The brain is divided into three regions that are found in differing proportions in all vertebrates: the hindbrain, which is the largest portion of the brain in fishes; the midbrain, which in fishes is devoted primarily to processing visual information; and the forebrain, which is concerned mainly with olfaction (the sense of smell) in fishes. In terrestrial vertebrates, the forebrain plays a far more dominant role in neural processing than it does in fishes.

The dominant forebrain in more recent vertebrates

Starting with the amphibians and continuing more prominently in reptiles, birds, and mammals, the processing of sensory information is increasingly centered in the forebrain. This pattern was the dominant evolutionary trend in the further development of the vertebrate brain (figure 33.10).

The forebrain in reptiles, amphibians, birds, and mammals is composed of two elements that have distinct functions. The *diencephalon* consists of the thalamus and hypothalamus. The **thalamus** is an integration and relay center between incoming sensory information and the cerebrum. The hypothalamus participates in basic drives and emotions and controls the secretions of the pituitary gland. The **telencephalon,** or "end brain," is located at the front of the forebrain and is devoted largely to associative activity. In mammals, the telencephalon is called the cerebrum. The telencephalon also includes structures we will discuss when describing the human brain.

The expansion of the cerebrum

In examining the relationship between brain mass and body mass among the vertebrates, a remarkable difference is observed between fishes and reptiles, on the one hand, and birds and mammals, on the other. Mammals have brains that are particularly large relative to their body mass. This is especially true of porpoises and humans.

The increase in brain size in mammals largely reflects the great enlargement of the cerebrum, the dominant part of the mammalian brain. The **cerebrum** is the center for correlation, association, and learning in the mammalian brain. It receives sensory data from the thalamus and issues motor commands to the spinal cord via descending tracts of axons.

In vertebrates, the central nervous system is composed of the brain and the spinal cord. These two structures are responsible for most of the information processing within the nervous system, and they consist primarily of interneurons and neuroglia. Ascending tracts carry sensory information to the brain. Descending tracts carry impulses from the brain to the motor neurons and interneurons in the spinal cord that control the muscles of the body.

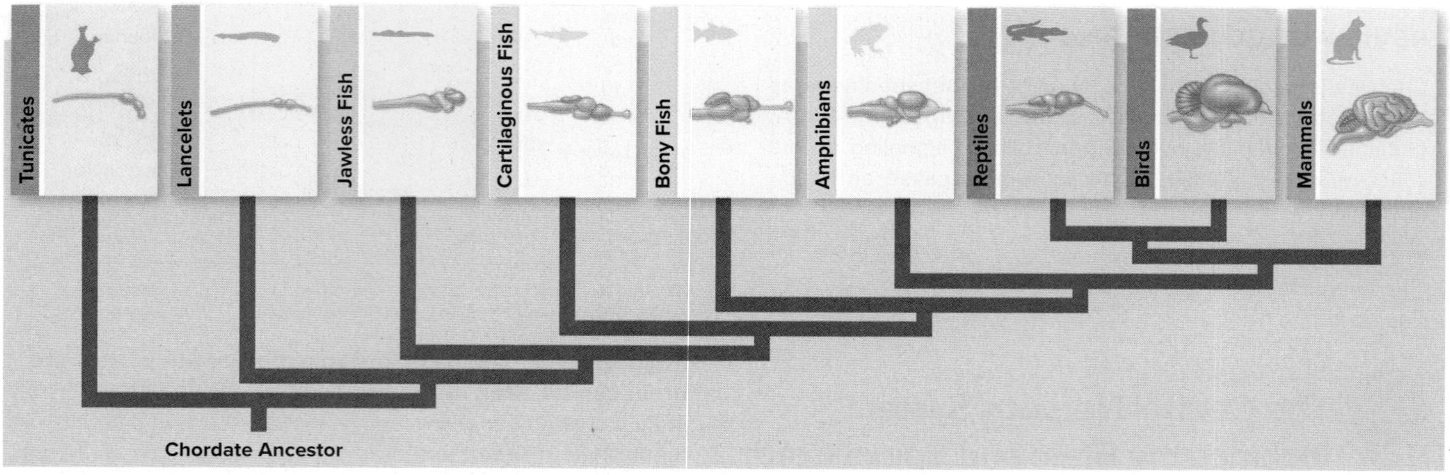

Figure 33.10 Evolution of the vertebrate brain. The relative sizes of different brain regions have changed as vertebrates have evolved. In sharks and other fishes, the hindbrain is predominant and the rest of the brain primarily processes sensory information. In amphibians and reptiles, the forebrain is far larger, and it contains a larger cerebrum devoted to associative activity. In birds, which evolved from reptiles, the cerebrum is even more pronounced. In mammals, the cerebrum covers the optic tectum and is the largest portion of the brain. The dominance of the cerebrum is greatest in humans, in whom it envelops much of the rest of the brain.

The Human Forebrain Exhibits Exceptional Information-Processing Ability

LEARNING OBJECTIVE 33.5.2 Describe the organization and functions of the various areas of the human forebrain.

The human cerebrum is so large that it appears to envelop the rest of the brain. It is split into right and left **cerebral hemispheres,** which are connected by a tract called the **corpus callosum** (figure 33.11). The hemispheres are further divided into the *frontal, parietal, temporal,* and *occipital lobes.*

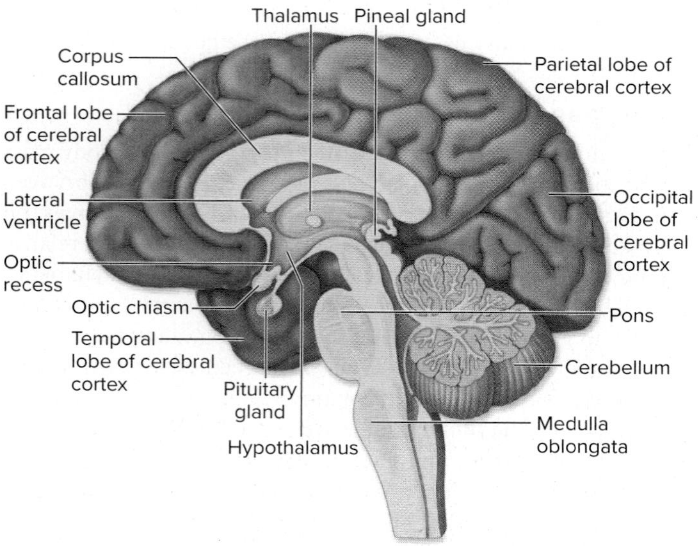

Figure 33.11 A section through the human brain. In this sagittal section showing one cerebral hemisphere, the corpus callosum, a fiber tract connecting the two cerebral hemispheres, can be clearly seen.

Each hemisphere primarily receives sensory input from the opposite, or contralateral, side of the body and exerts motor control primarily over that side. Therefore, a touch on the right hand is relayed primarily to the left hemisphere, which may then initiate movement of the right hand in response to the touch. Damage to one hemisphere due to a stroke often results in a loss of sensation and paralysis on the contralateral side of the body. Although each hemisphere is directly connected to the contralateral side, the hemispheres can influence each other through the corpus callosum and other connections between them.

The cerebral cortex

Much of the neural activity of the cerebrum occurs within a layer of gray matter only a few millimeters thick on its outer surface. This layer, called the **cerebral cortex,** is densely packed with nerve cells. In humans, it contains over 10 billion nerve cells, amounting to roughly 10% of all the neurons in the brain. The surface of the cerebral cortex is highly convoluted in some species; this is particularly true in the human brain, where the convolutions increase the surface area of the cortex threefold.

The activities of the cerebral cortex fall into one of three general categories: motor, sensory, and associative. Each of its regions correlates with a specific function. The **primary motor cortex** lies along the *gyrus* (convolution) on the posterior border of the frontal lobe, just in front of the central *sulcus* (crease). Each point on the surface of the motor cortex is associated with the movement of a different part of the body.

Just behind the central sulcus, on the anterior edge of the parietal lobe, lies the **primary somatosensory cortex.** Each point in this area receives input from sensory neurons serving skin and muscle senses in a particular part of the body. Large areas of the primary motor cortex and primary somatosensory cortex are devoted to the fingers, lips, and tongue, because of the need for manual dexterity and speech in humans. The auditory cortex lies within the temporal lobe, and different regions of this cortex deal with different sound frequencies. The visual cortex lies on the occipital lobe, with different sites processing information from different positions on the retina, equivalent to particular points in the visual fields of the eyes.

The portion of the cerebral cortex that is not occupied by these motor and sensory cortices is referred to as the **association cortex.** The site of higher mental activities, the association cortex reaches its greatest extent in primates, especially humans, where it makes up 95% of the surface of the cerebral cortex.

Thalamus and hypothalamus

The thalamus is a primary site of sensory integration in the brain. Visual, auditory, and somatosensory information is sent to the thalamus, where the sensory tracts synapse with association neurons. The sensory information is then relayed via the thalamus to the occipital, temporal, and parietal lobes of the cerebral cortex, respectively. The transfer of each of these types of sensory information is handled by specific aggregations of neuron cell bodies within the thalamus.

The hypothalamus integrates the visceral activities. It helps regulate body temperature, hunger and satiety, thirst, and—along with the limbic system—various emotional states. The hypothalamus also controls the pituitary gland, which in turn regulates many of the other endocrine glands of the body. By means of its interconnections with the cerebral cortex and with control centers in the *brainstem* (a term used to refer collectively to the midbrain, pons, and medulla oblongata), the hypothalamus helps coordinate the neural and hormonal responses to many internal stimuli and emotions.

The Spinal Cord Runs from the Brain Through the Backbone

LEARNING OBJECTIVE 33.5.3 Explain how a simple reflex works.

The spinal cord is a cable of neurons extending from the brain down through the backbone (figure 33.12). It is enclosed and protected by the vertebral column and layers of membranes called *meninges,* which also cover the brain. Inside the spinal cord are two zones.

The inner zone is gray matter and primarily consists of the cell bodies of interneurons, motor neurons, and neuroglia. The outer zone is white matter and contains cables of sensory axons in the dorsal columns and motor axons in the ventral columns. These nerve tracts may also contain the dendrites of other nerve cells. Messages from the body and the brain run up and down the spinal cord, the body's "information highway."

In addition to relaying messages, the spinal cord also functions in reflexes, the sudden, involuntary movement of muscles. Reflexes produce a rapid motor response to a stimulus because the sensory neuron signals to a motor neuron in the spinal cord, without higher-level processing. One common reflex in humans is blinking, which protects the eyes. If an object such as an insect or a cloud of dust approaches a person's eye, the eyelid blinks before they realize what has happened. The reflex occurs before the cerebrum is aware the eye is in danger.

In a few reflexes, such as the knee-jerk reflex, the sensory neuron connects directly with a motor neuron whose axon travels directly back to the muscle. Most reflexes in vertebrates, however, involve a connecting interneuron between the sensory neuron and the motor neuron. The withdrawal of a hand from a hot stove involves a relay of information from a sensory neuron through one or more interneurons to a motor neuron. The motor neuron then stimulates the appropriate muscle to contract. Notice that the sensory neuron connects to other interneurons to send signals to the brain, so although you jerked your hand away, you will still feel pain.

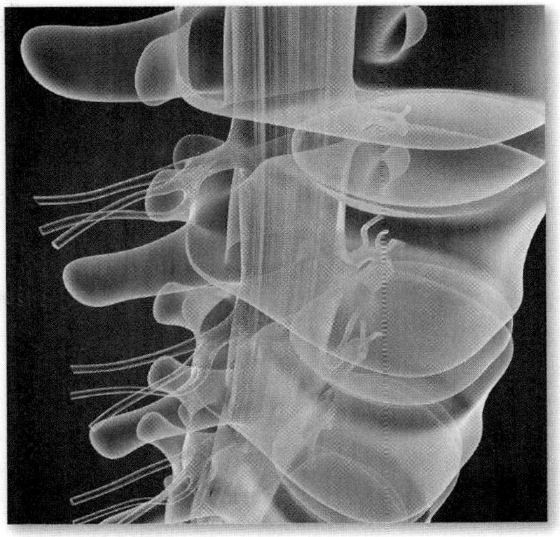

Figure 33.12 The spinal cord. A computer-generated drawing of the central portion of the spinal cord. Pairs of spinal nerves (red) can be seen extending from the main central nerve cord (green). Three vertebrae (blue) are visible, along with the discs (green) between them.

Alex Mit/Shutterstock

REVIEW OF CONCEPT 33.5

The vertebrate brain has three primary regions: the hindbrain, midbrain, and forebrain. The gray matter of the cerebral cortex on the brain's surface is where most associative activity occurs. The spinal cord relays messages to and from the brain; a reflex occurs when the spinal cord processes sensory information directly and initiates a motor response.

- *What is the advantage of having reflexes?*

33.6 The Peripheral Nervous System Consists of Both Sensory and Motor Neurons

The PNS consists of nerves, the cable-like collections of axons and **ganglia** (aggregations of neuron cell bodies; singular, *ganglion*) located outside the CNS. The function of the PNS is to receive information from the environment, to convey it to the CNS, and to carry responses to effectors such as muscle cells.

The PNS Has Somatic and Autonomic Systems

LEARNING OBJECTIVE 33.6.1 Distinguish among somatic, autonomic, sympathetic, and parasympathetic systems.

As mentioned in section 33.1, somatic motor neurons stimulate skeletal muscles to contract, and autonomic motor neurons innervate involuntary effectors—smooth muscles, cardiac muscle, and glands.

The somatic nervous system controls movements

Somatic motor neurons stimulate the skeletal muscles of the body to contract in response to conscious commands and as part of reflexes that do not require conscious control. When a particular skeletal muscle is stimulated to contract, however, its antagonist must be inhibited. In order to flex the arm, for example, the flexor muscles must be stimulated while the antagonistic extensor muscle is inhibited. Descending motor axons produce this necessary inhibition by causing hyperpolarizations (IPSPs) of the spinal motor neurons that innervate the antagonistic muscles.

The autonomic nervous system controls involuntary functions

The autonomic nervous system is composed of the *sympathetic* and *parasympathetic* divisions plus the medulla oblongata of the hindbrain, which coordinates this system. Although they differ, the sympathetic and parasympathetic divisions share several features. In both, the efferent motor pathways involve two neurons: the first has its cell body in the CNS and sends an axon to an autonomic ganglion. The second neuron has its cell body in the autonomic ganglion and sends its axon to synapse with a smooth muscle, cardiac muscle, or gland cell. This second neuron is termed the *postganglionic neuron.* Those in the parasympathetic division release ACh, and those in the sympathetic division release norepinephrine.

In general, the sympathetic nervous system is more active during physical activity, shunting blood and nutrients to active structures. This is sometimes called the fight-or-flight response. The parasympathetic system is generally involved with functions under "resting conditions," such as digestion. This active-versus-resting dichotomy is a generalization, not an absolute division of labor. For example, the sympathetic system maintains blood pressure and body temperature at rest.

The Sympathetic Division In the sympathetic division, the preganglionic neurons originate in the thoracic and lumbar regions of the spinal cord (figure 33.13, left). Most of the axons from these neurons synapse in two parallel chains of ganglia immediately outside the spinal cord. These structures are usually called the *sympathetic chain* of ganglia. The sympathetic chain contains the cell bodies of postganglionic neurons, and it is the axons from these neurons that innervate the different visceral organs.

The Parasympathetic Division The actions of the sympathetic division are antagonized by the parasympathetic division. Preganglionic parasympathetic neurons originate in the brain and the sacral regions of the spinal cord (figure 33.13, right). Because of this origin, there cannot be a chain of parasympathetic ganglia analogous to the sympathetic chain. Instead, the preganglionic axons, many of which

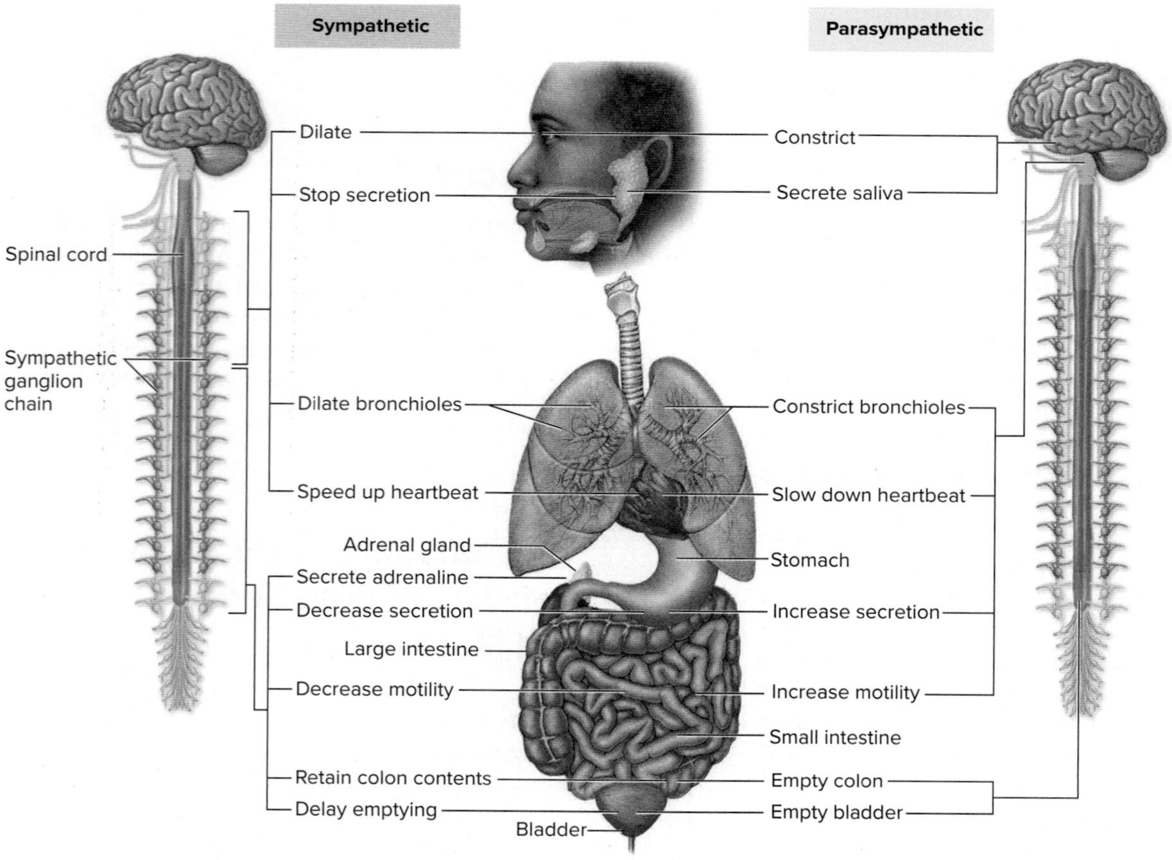

Figure 33.13 The sympathetic and parasympathetic divisions of the autonomic nervous system.

travel in the vagus (10th cranial) nerve, terminate in ganglia located near or even within the internal organs. The postganglionic neurons then regulate the internal organs by releasing ACh at their synapses. Parasympathetic nerve effects include a slowing of the heart, increased secretions and activities of digestive organs, and so on.

REVIEW OF CONCEPT 33.6

The PNS comprises the somatic and autonomic nervous systems. A spinal nerve contains sensory neurons, which carry information to the CNS, and motor neurons, which carry signals from the CNS to targets. The sympathetic system tends to affect functions involved with physical activity. The parasympathetic tends to control functions involved with rest and relaxation.

- *Why would having the sympathetic and parasympathetic divisions be more advantageous than having a single system?*

33.7 Sensory Receptors Provide Information About the Body's Environment

When we think of sensory receptors, what comes to mind are the senses of vision, hearing, taste, smell, and touch—the senses that provide information about our environment. Certainly, this external information is crucial to the survival and success of animals, but sensory receptors also provide information about internal states, such as stretching of muscles, position of the body, and blood pressure.

Sensory Receptors Detect Both External and Internal Stimuli

LEARNING OBJECTIVE 33.7.1 Explain how sensory information is conveyed from sensory receptors to the CNS.

Sensory receptors are crucial for the survival and success of all animals. These receptors fall into two broad categories, **exteroceptors** and **interoceptors.** Exteroceptors are receptors that sense stimuli outside of an animal body that are involved in the general and special senses: touch, taste, hearing, smell, and vision. Monitoring internal body state is also important. Interoceptors receive stimuli that arise from within the body, detecting changes in body position and physiology. They also transmit signals involved in actual or potential tissue damage in the form of pain.

The simplest sensory receptors are free nerve endings that respond to bending or stretching of the sensory neuron's membrane caused by changes in temperature or to chemicals such as oxygen in the extracellular fluid. Other sensory receptors are more complex, involving the association of the sensory neurons with specialized epithelial cells.

Sensory receptors can be subdivided based on the type of stimulus they detect. For example, mechanoreceptors respond to mechanical force, thermoreceptors sense temperature, nociceptors sense pain, chemoreceptors respond to chemicals, and photoreceptors respond to light. We will see how these different kinds of receptors produce the sensations we experience: hot/cold, taste, smell, touch, hearing, and vision.

Sensory information is conveyed in a four-step process

Sensory information picked up by sensory neurons is conveyed to the CNS, where the impulses are perceived in a four-step process:

1. *Stimulation*. A physical stimulus impinges on a sensory neuron or an associated, but separate, sensory receptor.
2. *Transduction*. The stimulus energy is transformed into graded potentials in the dendrites of the sensory neuron.
3. *Transmission*. Action potentials develop in the axon of the sensory neuron and are conducted to the CNS along an afferent nerve pathway.
4. *Interpretation*. The brain creates a sensory perception from the electrochemical events produced by afferent stimulation. We actually perceive the five senses with our brains, not with our sense organs.

A few vertebrate sensory systems that function well in the water, such as the electrical organs of fish, cannot function in the air and are not found among terrestrial vertebrates. In contrast, some land-dwellers have sensory systems that could not function in water, such as infrared heat detectors.

Sensory Transduction Involves Gated Ion Channels

LEARNING OBJECTIVE 33.7.2 Describe how gated ion channels work.

Sensory cells respond to stimuli because they possess **stimulus-gated ion channels** in their membranes. The sensory stimulus causes these ion channels to open or close, depending on the sensory system involved. In most cases, the sensory stimulus produces a depolarization of the receptor cell, analogous to the excitatory postsynaptic potential (EPSP) produced in a postsynaptic cell in response to a neurotransmitter. A depolarization that occurs in a sensory receptor on stimulation is referred to as a *receptor potential.*

Like an EPSP, a receptor potential is a graded potential: the larger the sensory stimulus, the greater the degree of depolarization. Receptor potentials also decrease in size with distance from their source. This prevents small, irrelevant stimuli from reaching the cell body of the sensory neuron. If the receptor potential or the summation of receptor potentials is great enough to generate a threshold level of depolarization, an action potential is produced that propagates along the sensory axon into the CNS.

The greater the sensory stimulus, the greater the depolarization of the receptor potential and the higher the frequency of action potentials. (Remember that the frequency of action potentials, not their summation, is responsible for conveying the intensity of the stimulus.)

Generally, a logarithmic relationship exists between stimulus intensity and action potential frequency. For example, if one sensory stimulus has an intensity of 100, and another of 1000, the resulting frequency of action potentials is 2 and 3, respectively. This relationship allows the CNS to interpret the strength of a sensory stimulus based on the frequency of incoming signals.

REVIEW OF CONCEPT 33.7

Exteroceptors sense external stimuli and interoceptors sense stimuli within the body. Specific sense receptors are classified by the type of stimulus they respond to. Sense information is conveyed by a process of stimulation, transduction, transmission, and interpretation.

- *How does the CNS interpret the strength of a sensory stimulus?*

33.8 Mechanoreceptors Sense Touch and Pressure

Pain Receptors Alert the Body to Potential Damage

LEARNING OBJECTIVE 33.8.1 Describe how nociceptors detect pain.

A stimulus that causes or is about to cause tissue damage is perceived as pain. The receptors that transmit impulses perceived as pain are called **nociceptors** (*noci* is derived from the Latin for hurt). They are sensitive to extreme mechanical, chemical, or thermal stimuli. Most nociceptors consist of free nerve endings located throughout the body, especially near surfaces where damage is most likely to occur. Different nociceptors may respond to extremes in temperature, very intense mechanical stimulation such as a hard impact, or specific chemicals in the extracellular fluid released by injured cells.

One kind of tissue damage can be due to extremes of temperature, and in this case the molecular details of how a noxious stimulus can result in the sensation of pain are becoming clear. A class of ion channel protein found in nociceptors, the transient receptor potential (TRP) ion channel, can be stimulated by temperature to produce an inward flow of cations, primarily Na^+ and Ca^{2+}. This depolarizing current causes the sensory neuron to fire, leading to the release of glutamate and an EPSP in neurons in the spinal cord that produce the pain response. TRP channels that respond to both hot and cold have been found. These TRP channels can also be activated by chemicals, like capsaicin. This is why eating hot peppers that contain capsaicin can induce pain.

The skin contains two populations of **thermoreceptors,** which are naked dendritic endings of sensory neurons that are sensitive to changes in temperature. These thermoreceptors contain TRP ion channels that are responsive to hot and cold.

Thermoreceptors are also found within the hypothalamus of the brain, where they monitor the temperature of the circulating blood and thus provide the CNS with information on the body's internal (core) temperature.

Several types of mechanoreceptors are present in the skin, some in the dermis and others in the underlying subcutaneous tissue (figure 33.14). These receptors contain sensory cells with ion channels that open in response to mechanical distortion of the membrane. They detect various forms of physical contact. This is the sense of touch.

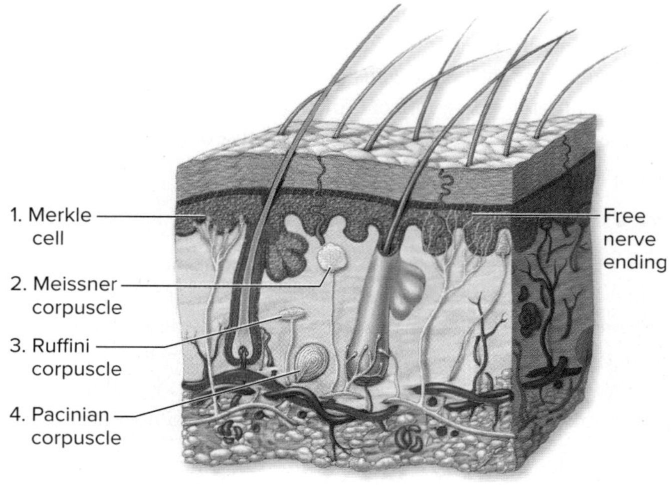

1. Merkle cell
2. Meissner corpuscle
3. Ruffini corpuscle
4. Pacinian corpuscle
Free nerve ending

Figure 33.14 Sensory receptors in human skin. Cutaneous receptors may be free nerve endings or sensory dendrites in association with other supporting structures.

Morphologically specialized receptors that respond to fine touch are most concentrated on areas such as the fingertips and face. They localize cutaneous stimuli very precisely.

Muscle Length and Tension Are Monitored by Proprioceptors

LEARNING OBJECTIVE 33.8.2 Describe how proprioceptors detect limb position and movement.

Buried within the skeletal muscles of all vertebrates except the bony fishes are **muscle spindles,** sensory stretch receptors that lie in parallel with the rest of the fibers in the muscle. Each spindle consists of several thin muscle fibers wrapped together and innervated by a sensory neuron, which becomes activated when the muscle, and therefore the spindle, is stretched.

Muscle spindles, together with other receptors in tendons and joints, are known as **proprioceptors.** These sensory receptors provide information about the relative position or movement of the animal's body parts.

Baroreceptors Detect Blood Pressure

LEARNING OBJECTIVE 33.8.3 Distinguish between proprioceptors and baroreceptors.

Blood pressure is monitored at two main sites in the body. One is the carotid sinus, an enlargement of the left and right internal carotid arteries that supply blood to the brain. The other is the aortic arch, the portion of the aorta very close to its emergence from the heart. The walls of the blood vessels at both sites contain a highly branched network of afferent neurons called baroreceptors, which detect tension or stretch in the walls.

When blood pressure decreases, the frequency of impulses produced by the baroreceptors decreases. The CNS responds to this reduced input by stimulating the sympathetic division of the autonomic nervous system, causing an increase in heart rate and vasoconstriction. Both effects help raise the blood pressure, thus maintaining homeostasis. A rise in blood pressure increases baroreceptor impulses, which conversely reduces sympathetic activity and stimulates the parasympathetic division, slowing the heart and lowering the blood pressure.

REVIEW OF CONCEPT 33.8

Nociceptors detect damage or potential damage to tissues and cause pain; thermoreceptors sense changes in heat energy; proprioceptors monitor muscle length; and baroreceptors monitor blood pressure in arteries.

- *Why is it important to detect stretching of muscles?*

33.9 Sounds and Body Position Are Sensed by Vibration Detectors

Hearing, the detection of sound waves, works better in water than in air because water transmits pressure waves more efficiently. Despite this limitation, hearing is widely used by terrestrial vertebrates to monitor their environments, communicate with other members of their species, and detect possible sources of danger.

Sound is a result of vibration, or waves, traveling through a medium, such as water or air. Detection of sound waves is possible through the action of specialized mechanoreceptors that first evolved in aquatic organisms.

The Lateral Line System of Fish Detects Low-Frequency Vibrations

LEARNING OBJECTIVE 33.9.1 Describe how the lateral line system allows fish to navigate to prey in the dark.

In addition to providing hearing, the lateral line system in fish provides a sense of "distant touch," enabling fish to sense objects that reflect pressure waves and low-frequency vibrations. This enables a fish to detect prey, as well as to swim in synchrony in a school. It also enables a blind cave fish to sense its environment by monitoring changes in the patterns of water flow past the lateral line receptors. The lateral line system is found in amphibian larvae but is lost at metamorphosis and is not present in any terrestrial vertebrate.

The lateral line system consists of hair cells within a longitudinal canal in the fish's skin that extends along each side of the body and within several canals in the head (figure 33.15*a*). The hair cells' surface processes project into a gelatinous membrane called a cupula. The hair cells are innervated by sensory neurons that transmit impulses to the brain.

Hair cells have several hairlike processes, called stereocilia, and one longer process called a **kinocilium** (figure 33.15*b*). The stereocilia are actually microvilli containing actin fibers, and the kinocilium is a true cilium that contains microtubules. Vibrations carried through the fish's environment produce movements of the cupula, which cause the processes to bend. When the stereocilia bend in the direction of the kinocilium, the associated sensory

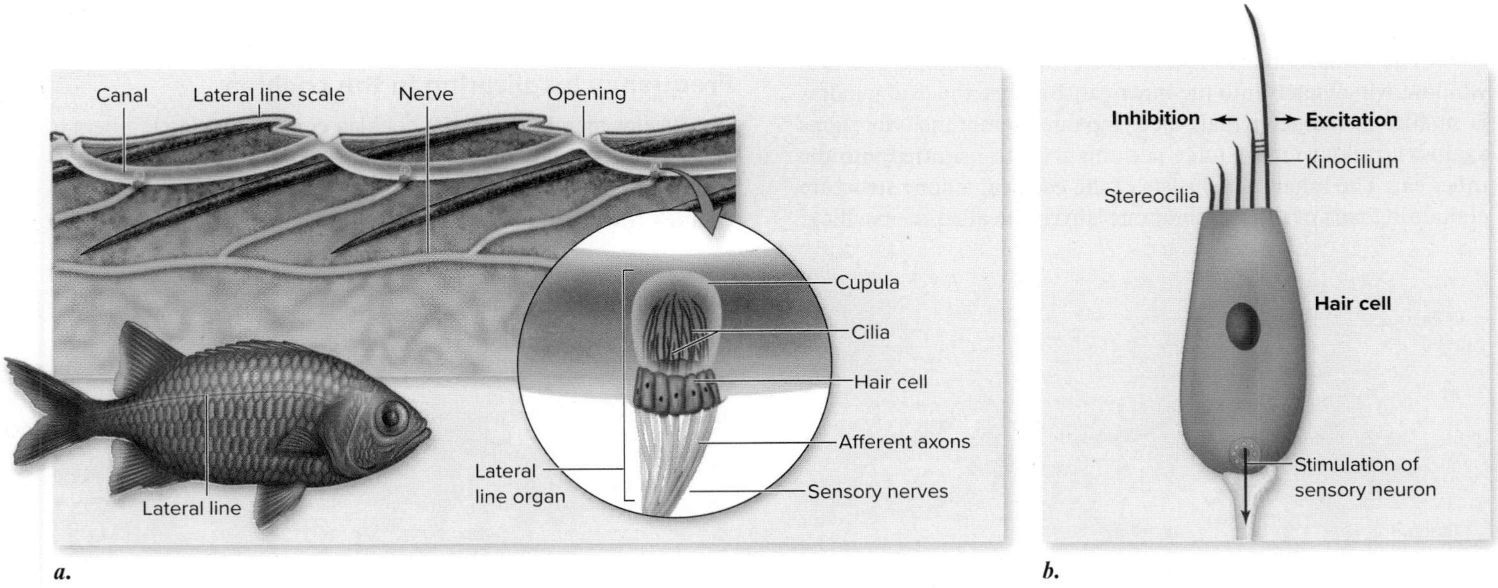

Figure 33.15 The lateral line system. ***a.*** This system consists of canals running the length of the fish's body beneath the surface of the skin. Within these canals are sensory structures containing hair cells with cilia that project into a gelatinous cupula. Pressure waves traveling through the water in the canals deflect the cilia and depolarize the sensory neurons associated with the hair cells. ***b.*** Hair cells are mechanoreceptors with hairlike cilia that project into a gelatinous membrane. The hair cells of the lateral line system (and the membranous labyrinth of the vertebrate inner ear) have a number of smaller cilia called stereocilia and one larger kinocilium. When the cilia bend in the direction of the kinocilium, the hair cell releases a chemical transmitter that depolarizes the associated sensory neuron. Bending of the cilia in the opposite direction has an inhibitory effect.

neurons are stimulated and generate a receptor potential. As a result, the frequency of action potentials produced by the sensory neuron is increased. In contrast, if the stereocilia are bent in the opposite direction, then the activity of the sensory neuron is inhibited.

Ear Structure Is Specialized to Detect Sound

LEARNING OBJECTIVE 33.9.2 Explain how sound waves in the environment lead to the production of action potentials in the inner ear.

In the ears of terrestrial vertebrates, vibrations in air may be channeled through an ear canal to the eardrum, or tympanic membrane. These structures are part of the **outer ear.** Vibrations of the tympanic membrane cause movement of one or more small bones located in a bony cavity known as the **middle ear.**

Amphibians and reptiles have a single middle-ear bone, the **stapes** (stirrup), but mammals have two others: the **malleus** (hammer) and **incus** (anvil; figure 33.16*a, b*). Where did these two additional bones come from?

The fossil record makes clear that the malleus and incus of modern mammals are derived from the two bones in the lower jaws of synapsid reptiles. Through evolutionary time, these bones became progressively smaller and came to lie closer to the stapes. Eventually, in modern mammals, they became completely disconnected from the jawbone and moved within the middle ear itself.

The middle ear is connected to the throat by the Eustachian tube, also known as the auditory tube, which equalizes the air pressure between the middle ear and the external environment. The "ear popping" you may have experienced when flying in an airplane or driving on a mountain is caused by pressure equalization between the two sides of the eardrum.

The stapes vibrates against a flexible membrane, the oval window, which leads into the **inner ear.** Because the oval window is smaller in diameter than the tympanic membrane, vibrations against it produce more force per unit area, transmitted into the inner ear. The inner ear consists of the **cochlea,** a bony structure containing part of the membranous labyrinth called the cochlear duct. The cochlear duct is located in the center of the cochlea; the area above the cochlear duct is the vestibular canal, and the area below is the tympanic canal (figure 33.16*c*). All three chambers are filled with fluid. The oval window opens to the upper vestibular canal, so that when the stapes causes it to vibrate, it produces pressure waves of fluid. These pressure waves travel down to the tympanic canal, pushing another flexible membrane, the round window, that transmits the pressure back into the middle-ear cavity.

Transduction Occurs in the Cochlea

LEARNING OBJECTIVE 33.9.3 Explain how mammals differentiate among sounds of different frequency.

As pressure waves are transmitted through the cochlea to the round window, they cause the cochlear duct to vibrate. The bottom of the cochlear duct, called the basilar membrane, is quite flexible and vibrates in response to these pressure waves. The surface of the basilar membrane contains sensory hair cells. The stereocilia from the hair cells project into an overhanging gelatinous membrane, the tectorial membrane. This sensory apparatus, consisting of the basilar membrane, hair cells with associated sensory neurons, and the tectorial membrane, is known as the organ of Corti (figure 33.16*d*).

As the basilar membrane vibrates, the cilia of the hair cells bend in response to the movement of the basilar membrane relative to the tectorial membrane. The bending of these stereocilia in one direction depolarizes the hair cells. Bending in the opposite direction repolarizes or even hyperpolarizes the membrane. The hair cells, in turn, stimulate the production of action potentials in sensory neurons that project to the brain, where they are interpreted as sound.

Frequency localization in the cochlea

The basilar membrane of the cochlea consists of elastic fibers of varying length and stiffness, like the strings of a musical instrument, embedded in a gelatinous material. At the base of the cochlea (near the oval window), the fibers of the basilar

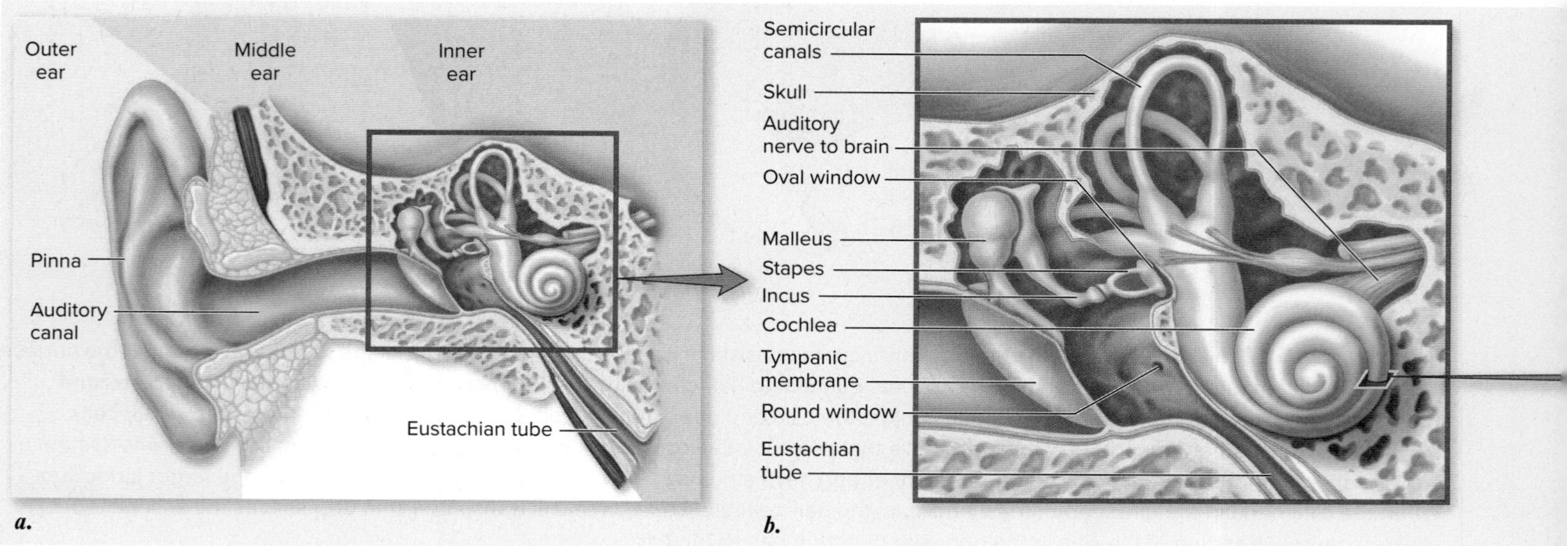
Outer ear
Middle ear
Inner ear
Pinna
Auditory canal
Eustachian tube
a.
Semicircular canals
Skull
Auditory nerve to brain
Oval window
Malleus
Stapes
Incus
Cochlea
Tympanic membrane
Round window
Eustachian tube
b.

membrane are short and stiff. At the far end of the cochlea (the apex), the fibers are 5 times longer and 100 times more flexible. Therefore, the resonant frequency of the basilar membrane is higher at the base than at the apex; the base responds to higher pitches, the apex to lower pitches.

When a wave of sound energy enters the cochlea from the oval window, it initiates an up-and-down motion that travels the length of the basilar membrane. However, this wave imparts most of its energy to that part of the basilar membrane with a resonant frequency near the frequency of the sound wave, resulting in a maximum deflection of the basilar membrane at that point. As a result, the hair cell depolarization is greatest in that region, and the afferent axons from that region are stimulated more than those of other regions. When these action potentials arrive in the brain, they are interpreted as representing a sound of a particular frequency, or pitch.

Body Position Is Detected by Gravity-Sensitive Receptors in the Inner Ear

LEARNING OBJECTIVE 33.9.4 Explain how the body's position in space is monitored by structures in the inner ear.

The evolutionary strategy of using internal calcium carbonate crystals as a way to detect vibration has also allowed the development of sensory organs that detect body position in space and movements such as acceleration.

Most invertebrates can orient themselves with respect to gravity due to a sensory structure called a **statocyst.** Statocysts generally consist of ciliated hair cells with the cilia embedded in a gelatinous membrane containing crystals of calcium carbonate. These stones, or otoliths, increase the mass of the gelatinous membrane so that it can bend the cilia when the animal's position changes. If the animal tilts to the right, for example, the otolith membrane bends the cilia on the right side and activates associated sensory neurons.

A similar structure is found in the membranous labyrinth of the inner ear of vertebrates. Though intricate, the entire structure is very small; in a human, it is about the size of a pea.

Structure of the labyrinth and semicircular canals

The receptors for gravity in most vertebrates consist of two chambers of the membranous labyrinth called the **utricle** and **saccule.** Within these structures are hair cells with stereocilia and a kinocilium, similar to those in the lateral line system of fish. The hairlike processes are embedded within a gelatinous membrane, the otolith membrane, containing calcium carbonate crystals. Because the otolith organ is oriented differently in the utricle and in the saccule, the utricle is more sensitive to horizontal acceleration (as in a moving car) and the saccule to vertical acceleration (as in an elevator). In both cases, the acceleration causes the stereocilia to bend, consequently producing action potentials in an associated sensory neuron.

The membranous labyrinth of the utricle and saccule is continuous with three **semicircular canals,** oriented in different planes so that angular acceleration in any direction can be detected. At the ends of the canals are swollen chambers called ampullae, into which protrude the cilia of another group of hair cells. The tips of the cilia are embedded within a sail-like wedge of gelatinous material called a cupula (similar to the cupula of the fish lateral line system) that protrudes into the endolymph fluid of each semicircular canal.

Action of the vestibular apparatus

When the head rotates, the fluid inside the semicircular canals pushes against the cupula and causes the cilia to bend. This bending either depolarizes or hyperpolarizes the hair cells, depending on the direction in which the cilia are bent. This is similar to the

Figure 33.16 Structure and function of the human ear. The structure of the human ear is shown in successive enlargements illustrating functional parts (***a*** to ***d***). Sound waves passing through the ear canal produce vibrations of the tympanic membrane, which causes movement of the middle-ear ossicles (the malleus, incus, and stapes) against an inner membrane (the oval window). This vibration creates pressure waves in the fluid in the vestibular and tympanic canals of the cochlea. These pressure waves cause cilia in hair cells to bend, producing signals from sensory neurons.

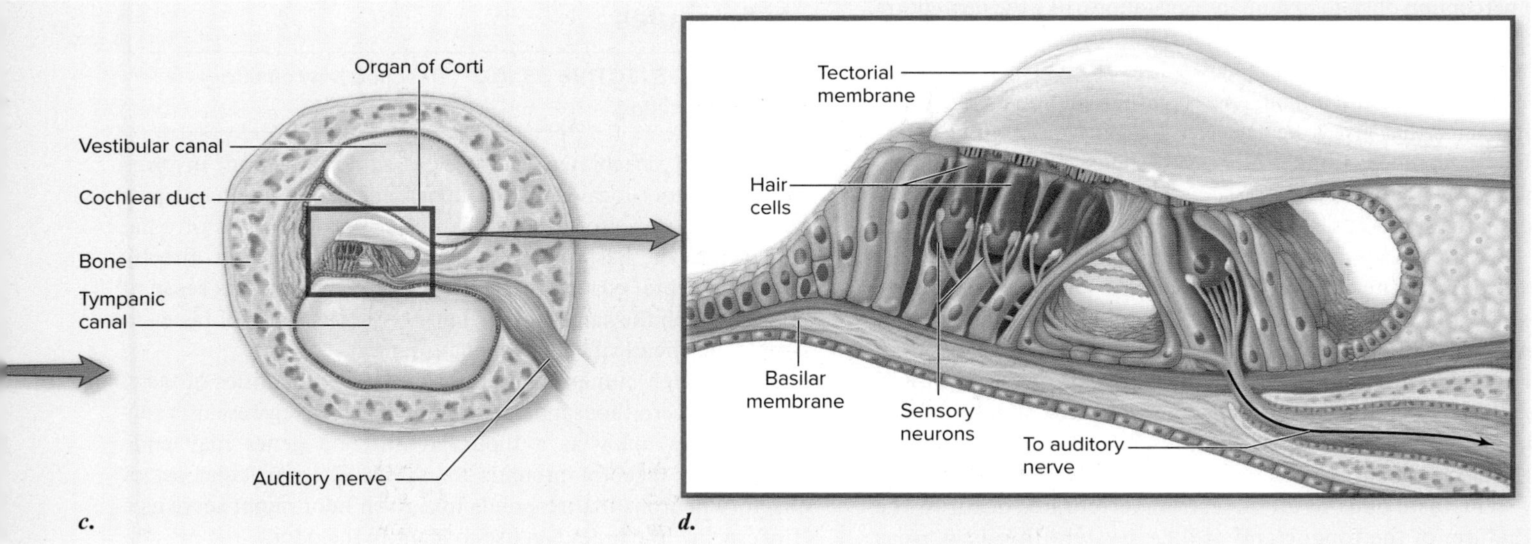

way the lateral line system works in a fish: if the stereocilia are bent in the direction of the kinocilium, a receptor potential is produced, which stimulates the production of action potentials in associated sensory neurons.

The saccule, utricle, and semicircular canals are collectively referred to as the **vestibular apparatus.** The saccule and utricle provide a sense of linear acceleration, and the semicircular canals provide a sense of angular acceleration. The brain uses information that comes from the vestibular apparatus about the body's position to maintain balance and equilibrium.

REVIEW OF CONCEPT 33.9

Sound waves cause middle ear ossicles to vibrate; fluid in the inner ear is vibrated in turn, bending hair cells and causing action potentials. In terrestrial animals, sound waves in air must transition to the fluid in the inner ear. Hair cells in the vestibular apparatus of terrestrial vertebrates provide a sense of acceleration and balance.

- *Why is a lateral line system not useful to adult amphibians?*

33.10 Taste, Smell, and pH Senses Utilize Chemoreceptors

Some sensory cells, called chemoreceptors, contain membrane proteins that can bind to particular chemicals or ligands in the extracellular fluid. In response to this chemical interaction, the membrane of the sensory neuron becomes depolarized and produces action potentials.

Taste Receptors Detect and Analyze Potential Food

LEARNING OBJECTIVE 33.10.1 List the five taste categories and describe how their receptors function.

The perception of taste in humans (gustation), like the perception of color, is a combination of physical and psychological factors. Tastes are commonly broken down into five categories: sweet, sour, salty, bitter, and umami (perception of glutamate and other amino acids that give a hearty taste to many protein-rich foods such as meat, cheese, and broths). Taste buds—collections of chemosensitive epithelial cells associated with afferent neurons—mediate the sense of taste in vertebrates. In a fish, the taste buds are scattered over the surface of the body. These are the most sensitive vertebrate chemoreceptors known. They are particularly sensitive to amino acids; a catfish, for example, can distinguish between two different amino acids at a concentration of less than 100 parts per billion (1 g in 10,000 L of water)! The ability to taste the surrounding water is very important to bottom-feeding fish, enabling them to sense the presence of food in an often murky environment.

The taste buds of all terrestrial vertebrates occur in the epithelium of the tongue and oral cavity, within raised areas called papillae. Taste buds are onion-shaped structures of between 50 and 100 taste cells; each cell has finger-like projections called microvilli, which poke through the top of the taste bud, called the taste pore (figure 33.17). Chemicals from food dissolve in saliva and contact the taste cells through the taste pore.

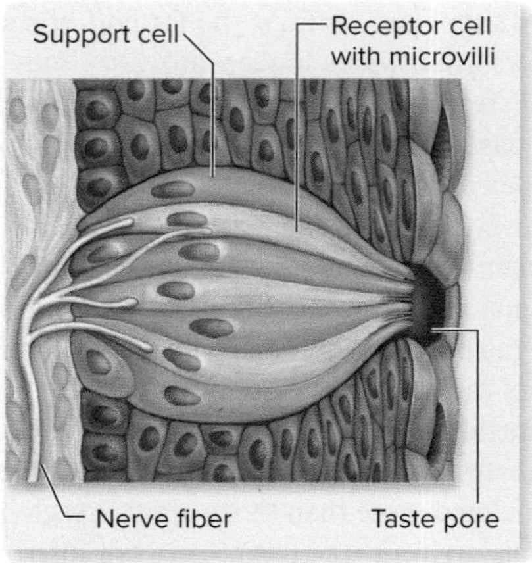

Figure 33.17 A taste bud. Individual taste buds are bulb-shaped collections of chemosensitive receptors that open out into the mouth through a pore.

Within a taste bud, the cells that detect salty tastes react directly to Na^+, and cells that detect sour taste detect H^+. The mechanisms of detection of sweet, bitter, and umami are more indirect, the substances binding to G protein–coupled receptors specific for each category.

Like vertebrates, many arthropods also have taste chemoreceptors. For example, flies, because of their mode of searching for food, have chemoreceptors able to detect sugars, salts, and other tastes in sensory hairs located on their feet. If they step on potential food, their proboscis (the tubular feeding apparatus) immediately extends to feed.

Smell Can Identify a Vast Number of Molecules

LEARNING OBJECTIVE 33.10.2 Describe how olfactory receptors function.

In terrestrial vertebrates, the sense of smell (olfaction) involves chemoreceptors located in the upper portion of the nasal passages. These receptors, whose dendrites end in tassels of cilia, project into the nasal mucosa, and their axons project directly into the cerebral cortex. A terrestrial vertebrate uses its sense of smell in much the same way that a fish uses its sense of taste—to sample the chemical environment around it.

Although humans can detect only five modalities of taste, they can discern thousands of different smells. New research suggests that as many as a thousand different genes may code for different receptor proteins for smell. The particular set of olfactory neurons that responds to a given odor might serve as a "fingerprint" the brain can use to identify the odor.

Internal Chemoreceptors Detect pH

LEARNING OBJECTIVE 33.10.3 Describe how the body monitors blood pH.

The **peripheral chemoreceptors** of the aortic and carotid bodies are sensitive to plasma pH, and the **central chemoreceptors** in the medulla oblongata of the brain are sensitive to the pH of cerebrospinal fluid. When the breathing rate is too low, the concentration of plasma CO_2 increases, producing more carbonic acid and causing a fall in the blood pH. The carbon dioxide can also enter the cerebrospinal fluid and lower the pH, thereby stimulating the central chemoreceptors. This stimulation indirectly affects the respiratory control center of the brainstem, which increases the breathing rate.

REVIEW OF CONCEPT 33.10

The five tastes humans perceive are sweet, sour, salty, bitter, and umami (amino acids). Taste and smell chemoreceptors detect chemicals from outside the body; olfactory receptors can identify thousands of different odors. Internal chemoreceptors monitor acid–base balance within the body and help regulate breathing.

- *What are the advantages of insects' having taste receptors on their feet?*

33.11 Vision Employs Photoreceptors to Perceive Objects at a Distance

The ability to perceive objects at a distance is important to most animals. Predators locate their prey, and prey avoid their predators, based on the three long-distance senses of hearing, smell, and vision. Of these, vision can act most distantly; with the naked eye, humans can see stars thousands of light years away—and a single photon is sufficient to stimulate a cell of the retina to send an action potential.

Vision Senses Light at a Distance

LEARNING OBJECTIVE 33.11.1 Compare invertebrate and vertebrate eyes.

Vision begins with the capture of light energy by **photoreceptors.** Because light travels in a straight line and arrives virtually instantaneously regardless of distance, visual information can be used to determine both the direction and the distance of an object. Other stimuli, which spread out as they travel and move more slowly, provide much less precise information.

Invertebrate eyes

Many invertebrates have simple visual systems with photoreceptors clustered in an eyespot. Simple eyespots can be made sensitive to the direction of a light source by the addition of a pigment layer that shades one side of the eye. Flatworms have a screening pigmented layer on the inner and back sides of both eyespots, allowing stimulation of the photoreceptor cells only by light from the front of the animal. The flatworm will turn and swim in the direction in which the photoreceptor cells are the least stimulated. Although an eyespot can perceive the direction of light, it cannot be used to construct a visual image.

Structure of the vertebrate eye

The human eye is typical of the vertebrate eye (figure 33.18). The "white of the eye," the **sclera,** is formed of tough connective tissue. Light enters the eye through a transparent **cornea,** which begins to focus the light. Focusing occurs because light is refracted (bent) when it travels into a medium of different density. The colored portion of the eye, the **iris,** contracts the iris muscles in bright light to decrease the size of its opening, the pupil. Light passes through the pupil to the **lens,** a transparent structure that completes the focusing of the light onto the retina at the back of the eye. The lens is attached by the suspensory ligament to the ciliary muscles.

The shape of the lens is influenced by the amount of tension in the suspensory ligament, which surrounds the lens and attaches it to the circular ciliary muscle. Because the ciliary muscle is circular, when it contracts it puts slack in the suspensory ligament, which makes the lens more rounded. The more rounded lens bends light more strongly, facilitating close vision. In distance vision, the ciliary muscles relax, making a larger circle, which tightens the suspensory ligament. The lens thus becomes

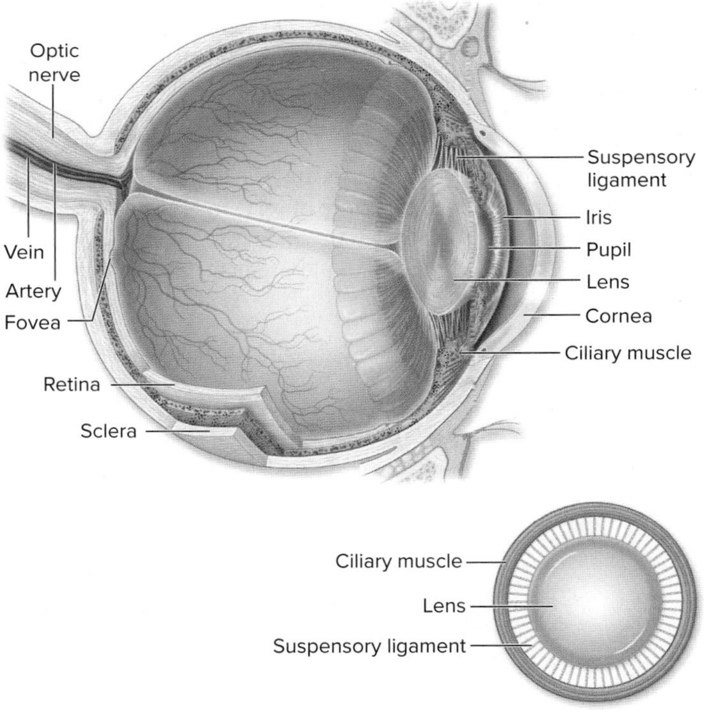

Figure 33.18 Structure of the human eye. The transparent cornea and lens focus light onto the retina at the back of the eye, which contains the photoreceptors (rods and cones). The center of each eye's visual field is focused on the fovea. Focusing is accomplished by contraction and relaxation of the ciliary muscle, which adjusts the curvature of the lens.

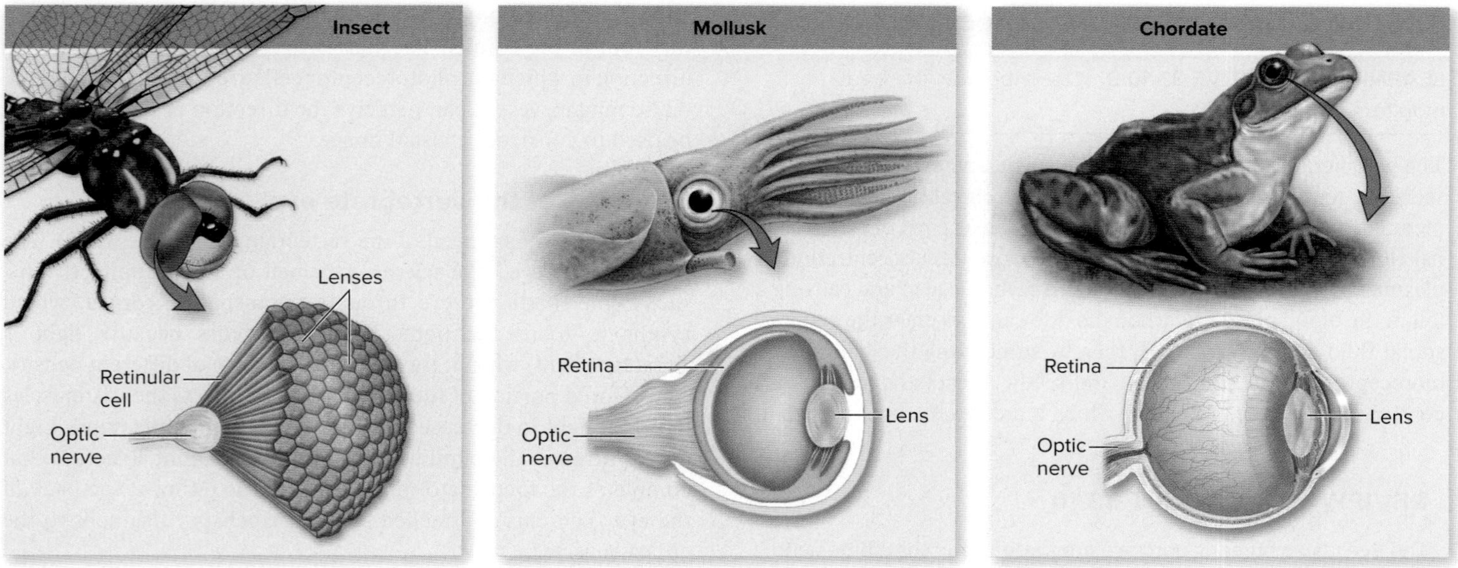
Insect
Lenses
Retinular cell
Optic nerve
Mollusk
Retina
Lens
Optic nerve
Chordate
Retina
Lens
Optic nerve

Figure 33.19 Eyes in three phyla of animals. Although they are superficially similar, these eyes differ greatly from one another in structure (refer to figure 20.16 for a detailed comparison of mollusk and chordate eye structure). Each has evolved separately and, despite the apparent structural complexity, has done so from simpler structures.

more flattened and bends light less, keeping the image focused on the retina. Interestingly, the lens of an amphibian or a fish does not change shape; these animals instead focus images by moving their lens in and out, just as you would do to focus a camera.

The evolution of eyes. The members of four phyla—annelids, mollusks, arthropods, and chordates—have evolved well-developed image-forming eyes. True image-forming eyes in these phyla are thought to have evolved independently, despite similarities in structure (figure 33.19). The photoreceptors in all of these image-forming eyes use the same light-capturing molecule (retinol), suggesting that not many alternative molecules are able to play this role.

Although the actual structure of vertebrate and invertebrate eyes appears to have evolved independently, their development may be more similar than this would predict. In the early 1990s, biologists studied eye development in both vertebrates and insects. In each case, a gene was discovered that codes for a transcription factor important in lens formation. In the mouse the gene was given the name *Pax6*, whereas the fly gene was called *eyeless* (*ey*). Loss-of-function mutations in the *ey* gene cause a complete absence of eye development; hence the name of the fly gene.

When these genes were sequenced, they proved to be highly similar. This implies that homologous *Pax6* genes are responsible for triggering lens formation in both insects and vertebrates. A striking demonstration of this homology was conducted in biologist Walter Gehring's lab where researchers created transgenic flies carrying the mouse *Pax6* gene. The mouse *Pax6* was controlled by regulatory factors that cause expression of the gene in the fly's leg. The result was the formation of an eye on the leg of a fly caused by a mouse gene (figure 33.20)!

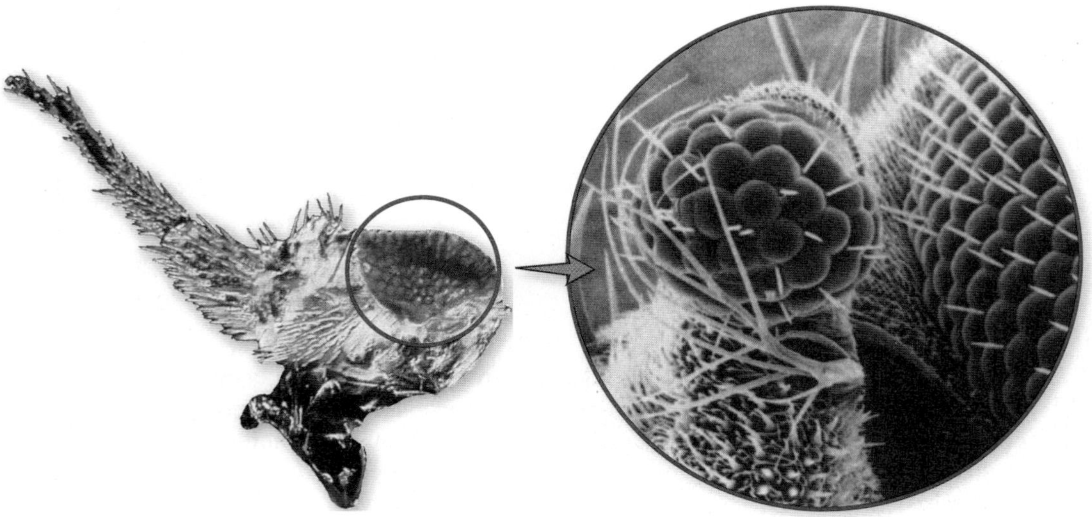

Figure 33.20 Mouse *Pax6* makes an eye on the leg of a fly. Mammalian *Pax6* and *Drosophila eyeless* are functional homologs. A fly genetically engineered to express mouse *Pax6* in the leg results in the development of an eye on this leg.

Walter J. Gehring/Biozentrum University of Basel

These results were a surprise to the evolutionary biology community, since insects and vertebrates diverged from a common ancestor more than 500 MYA. That eye development was affected by homologous genes and that these genes were similar enough that the vertebrate gene could function to initiate formation of a fly eye were completely unexpected. These findings may not be a contradiction because the complex structures of vertebrate and invertebrate eyes do appear to have evolved independently, although the initiation of eye development appears to use a similar developmental pathway. So, the actual origin of eyes may have been a single event, but the evolution of the complex structures of modern eyes occurred independently in different lineages. This also raises interesting questions about the last common ancestor to vertebrates and invertebrates, long thought to lack light-sensing capabilities.

The *Pax6* story extends to eyeless fish found in caves. Fish that live in dark caves need to rely on senses other than sight. In cavefish, *Pax6* gene expression is greatly reduced. Eyes start to develop, but then degenerate.

Vertebrate Photoreceptors Are Rod Cells and Cone Cells

LEARNING OBJECTIVE 33.11.2 Describe how photoreceptors function.

The vertebrate retina contains two kinds of photoreceptor cells, called rods and cones (figure 33.21). **Rods** are responsible for black-and-white vision in dim light. In contrast, **cones** are responsible for high visual acuity and color vision; cones have a cone-shaped outer segment. Humans have about 100 million rods and 3 million cones in each retina. Most of the cones are located in the central region of the retina known as the **fovea,** where the eye forms its sharpest image. Rods are almost completely absent from the fovea.

Structure of rods and cones

Rods and cones have the same basic cellular structure. An inner segment rich in mitochondria contains numerous vesicles filled with neurotransmitter molecules. It is connected by a narrow stalk to the outer segment, which is packed with hundreds of flattened disks stacked on top of one another. The light-capturing molecules, or photopigments, are located on the membranes of these disks (figure 33.21).

In rods, the photopigment is called **rhodopsin.** It consists of the protein opsin bound to a molecule of *cis*-retinal, which is produced from vitamin A. Vitamin A is derived from carotene, a photosynthetic pigment in plants.

The photopigments of cones, called **photopsins,** are structurally very similar to rhodopsin. Humans generally have three kinds of cones, each of which possesses a photopsin consisting of *cis*-retinal bound to a protein with a slightly different amino acid sequence. These differences shift the absorption maximum, the region of the electromagnetic spectrum that is best absorbed by the pigment (figure 33.22). The absorption maximum of the *cis*-retinal in rhodopsin is 500 nanometers (nm); in contrast, the absorption maxima of the three kinds of cone photopsins are 420 nm (blue-absorbing), 530 nm (green-absorbing), and 560 nm (red-absorbing). These differences in the light-absorbing properties of the photopsins are responsible for the different color sensitivities of the three kinds of cones, which are often referred to as simply blue, green, and red cones.

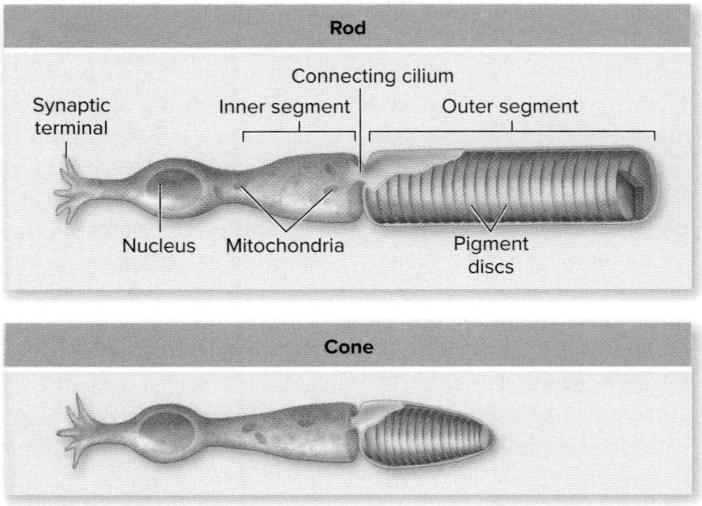

Figure 33.21 Rods and cones. The pigment-containing outer segment in each of these cells is separated from the rest of the cell by a partition through which there is only a narrow passage, the connecting cilium.

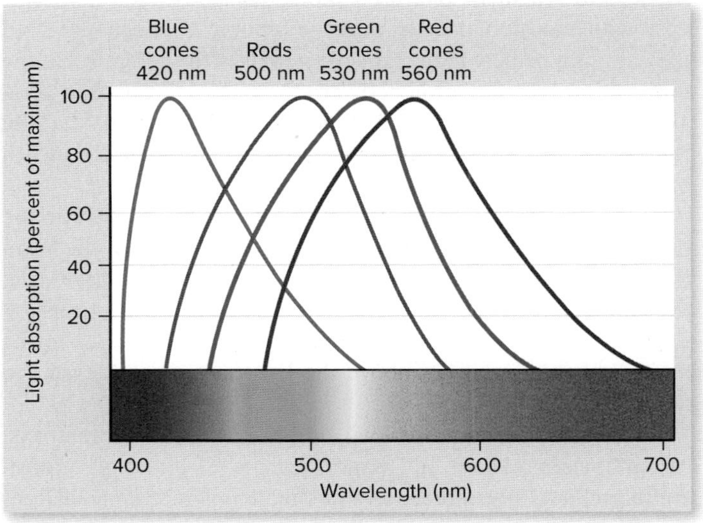

Figure 33.22 Color vision. The absorption maximum of *cis*-retinal in the rhodopsin of rods is 500 nm. However, the "blue cones" have their maximum light absorption at 420 nm, the "green cones" at 530 nm, and the "red cones" at 560 nm. The brain perceives all other colors from the combined activities of these three cones' systems.

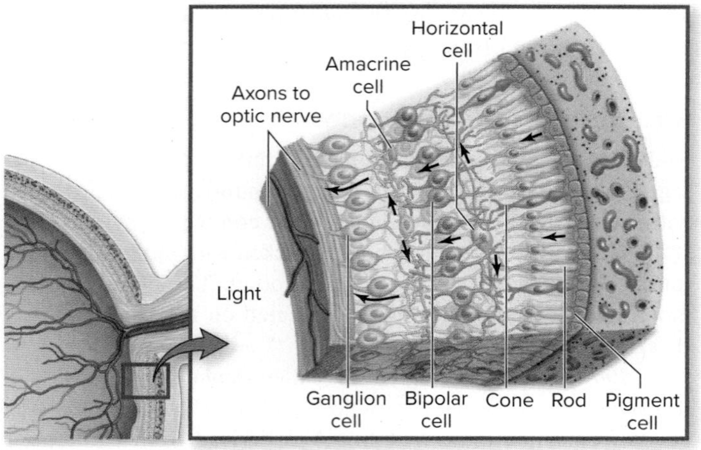

Figure 33.23 Structure of the retina. The rods and cones are at the rear of the retina, not the front. Light passes through three layers of cells in the retina to reach the rods and cones. Once the photoreceptors are activated, they stimulate bipolar cells, which stimulate ganglion cells. Thus, the flow of sensory information in the retina is opposite to the direction of light.

The **retina,** the inside surface at the back of the eye (figure 33.23), consists of a layer of pigmented epithelium covered with three layers of nervous tissue. If we start at the back of the eye and move forward towards the lens, the first layer of nervous tissue contains rod and cone cells. As we move outwards towards the lens, there is a layer of **bipolar cells,** and finally a layer of **ganglion cells.** This structure means that light coming through the lens must first pass through the ganglion and bipolar cells to reach the photoreceptors. The rods and cones synapse with the bipolar cells, and the bipolar cells synapse with the ganglion cells, which transmit impulses to the brain via the optic nerve. The ganglion cells are the only neurons that actually transmit action potentials to the brain. This means that the flow of sensory information in the retina is opposite to the path of light.

As the ganglion cells lie in the inner cavity of the eye, the optic nerve must pass back through the retina to reach the brain (figures 33.18 and 33.23), creating a blind spot. Mollusk eyes lack this blind spot because the orientation of sensing and transmitting cells is reversed, that is, the cells that send impulses to the brain lie behind and not above the light-sensing cells (refer to figure 20.16).

The retina contains two additional types of neurons, called horizontal cells and amacrine cells. Stimulation of horizontal cells by photoreceptors at the center of a spot of light on the retina can inhibit the response of photoreceptors peripheral to the center. This lateral inhibition enhances contrast and sharpens the image.

Most vertebrates, particularly those that are diurnal (active during the day), have color vision, as do many insects and some other invertebrates. Indeed, honeybees—as well as some birds, lizards, and other vertebrates—can see light in the near-ultraviolet range, which is invisible to the human eye. Color vision requires the presence of more than one photopigment in different receptor cells, but not all animals with color vision have the three-cone system characteristic of humans and other primates. Fish, turtles, and birds, for example, have four or five kinds of cones; the "extra" cones enable these animals to see near-ultraviolet light and to distinguish shades of colors that we cannot detect. On the other hand, many mammals, for example, squirrels and dogs, have only two types of cones and thus have more limited ability to distinguish different colors.

Sensory transduction in photoreceptors

The transduction of light energy into nerve impulses follows a sequence that is the opposite of the usual way that sensory stimuli are detected (figure 33.24). In the dark, the photoreceptor cells release an inhibitory neurotransmitter that hyperpolarizes the bipolar neurons. This prevents the bipolar neurons from releasing excitatory neurotransmitter to the ganglion cells that signal to the brain. In the presence of light, the photoreceptor cells stop releasing their inhibitory neurotransmitter, in effect stimulating bipolar cells. The bipolar cells in turn stimulate the ganglion cells, which transmit action potentials to the brain.

The production of inhibitory neurotransmitter by photoreceptor cells is due to ligand-gated Na^+ channels. In the dark, many of these channels are open, allowing an influx of Na^+. This flow of Na^+ in the absence of light, called the dark current, depolarizes the membrane of photoreceptor cells. In this state, the cells produce inhibitory neurotransmitter that hyperpolarizes the membrane of bipolar cells. In the light, the Na^+ channels in the photoreceptor cells rapidly close, reducing the dark current and causing the photoreceptors to hyperpolarize. In this state, they no longer produce inhibitory neurotransmitter. In the absence of inhibition, the membrane of the bipolar cells is depolarized, causing them to release an excitatory neurotransmitter to the ganglion cells.

The control of the dark current depends on the ligand for the Na^+ channels in the photoreceptor cells. These channels respond to the nucleotide cyclic guanosine monophosphate (cGMP). In the dark, the level of cGMP is high and the channels are open. The system responds to light using photopigments in the eye that are G protein–coupled receptors (refer to chapter 9). When a photopigment absorbs light, *cis*-retinal isomerizes and dissociates from the receptor protein, opsin. This alters the conformation of the opsin receptor, activating its associated G protein. The activated G protein activates the enzyme phosphodiesterase, which cleaves cGMP to GMP. The loss of cGMP causes the cGMP-gated Na^+ channels to close, reducing the dark current (figure 33.24). Each opsin is associated with over 100 regulatory G proteins, which can activate hundreds of molecules of phosphodiesterase. Each phosphodiesterase can convert thousands of cGMP to GMP, closing the Na^+ channels at a rate of about 1000 per second and inhibiting the dark current.

The absorption of a single photon of light can block the entry of more than a million Na^+, without changing K^+ permeability—the photoreceptor becomes hyperpolarized and stops releasing inhibitory neurotransmitter. Freed from

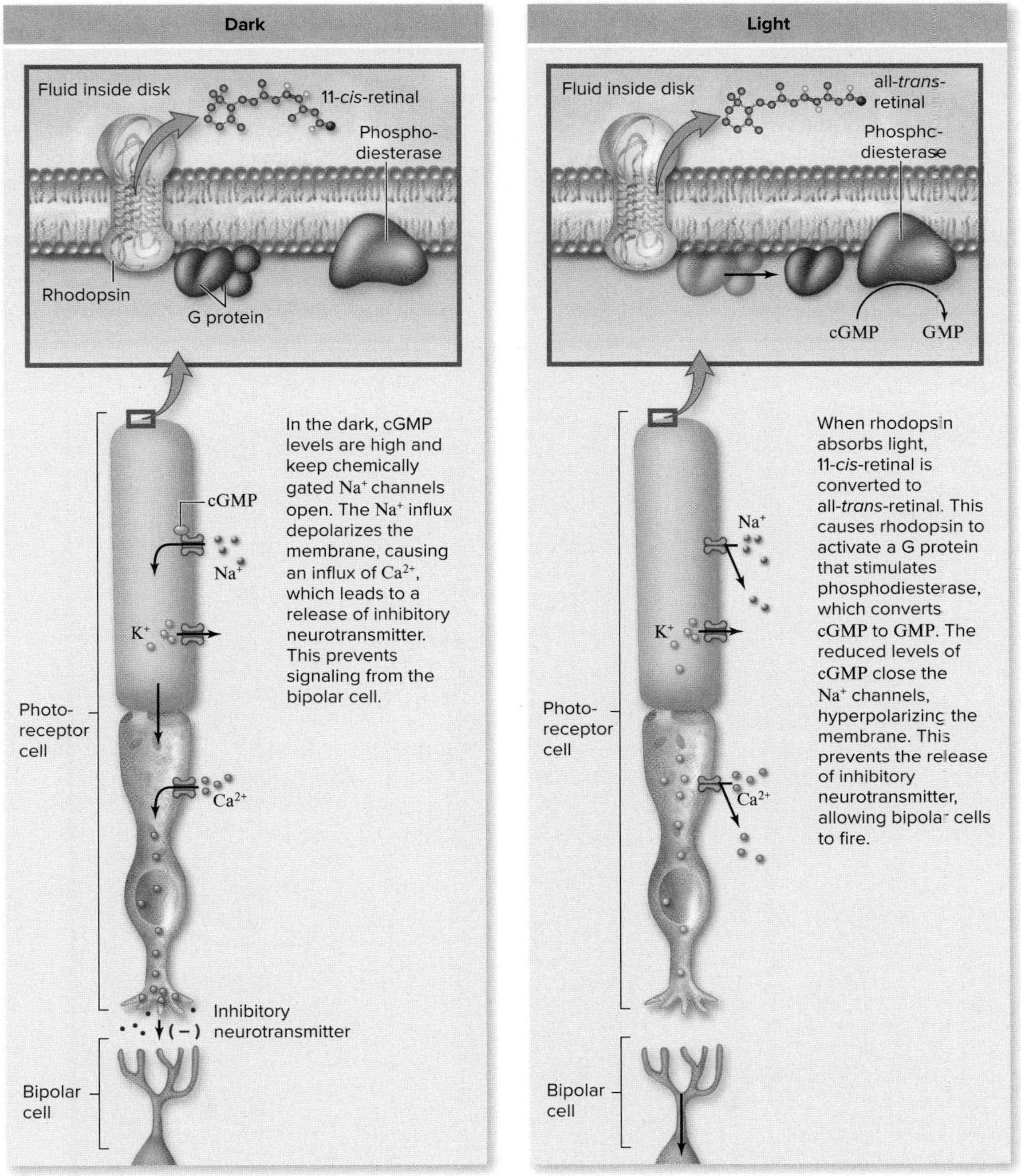

Figure 33.24 Signal transduction in the vertebrate eye. In the absence of light, cGMP keeps Na^+ channels open, causing a Na^+ influx that leads to the release of inhibitory neurotransmitter. Light is absorbed by the retinal in rhodopsin, changing its structure. This causes rhodopsin to associate with a G protein. The activated G protein stimulates phosphodiesterase, which converts cGMP to GMP. Loss of cGMP closes Na^+ channels and prevents the release of inhibitory neurotransmitter, which causes bipolar cells to stimulate ganglion cells.

inhibition, the bipolar cells activate the ganglion cells, which send impulses to the brain.

Binocular vision

Primates (including humans) and most predators have two eyes, one located on each side of the face. When both eyes are trained on the same object, the image that each eye sees is slightly different, because the views have a slightly different angle. This slight displacement of the images (an effect called parallax) permits **binocular vision,** the ability to perceive three-dimensional images and to sense depth. Having eyes facing forward maximizes the field of overlap in which this stereoscopic vision occurs.

In contrast, prey animals generally have eyes located to the sides of the head, preventing binocular vision but enlarging the overall receptive field. It seems that natural selection has favored the detection of potential predators over depth perception in many prey species. The eyes of the American woodcock *(Scolopax minor),* for example, are located at exactly opposite sides of the bird's skull so that it has a 360° field of view without turning its head.

Most birds have laterally placed eyes and, as an adaptation, have two foveas in each retina. One fovea provides sharp frontal vision, like the single fovea in the retina of mammals, and the other fovea provides sharper lateral vision.

REVIEW OF CONCEPT 33.11

Annelids, mollusks, arthropods, and chordates have independently evolved image-forming eyes. The vertebrate eye focuses light with an adjustable lens onto the retina, which contains photoreceptors. Photoreceptor rods and cones inhibit bipolar neurons in the dark. When *cis*-retinal absorbs light, it signals through a G protein–coupled receptor to remove the inhibition on ganglion cells, which then transmit a signal to the brain.

- *Can an individual with red-green color blindness learn to distinguish these two colors? Why or why not?*

Are Bigger Nerves Faster?

In this chapter, you learned that the axons of neurons can have an insulating covering called a myelin sheath, but many nerve axons do not have myelin covers. In addition, not all axons are the same. Some are thin, like fine wire, and others are much thicker. A motor axon to a human internal organ might have a diameter of 1 to 5 μm, whereas the giant motor axon to the mantle muscle of a squid is fully 500 μm in diameter. Why the great difference in size? Physics tells us that there should be a relationship between the conduction velocity of a nerve axon and its diameter. To be precise, there should be a 10-fold increase in conduction speed for a 100-fold increase in fiber diameter. The squid giant axon is 100 times thicker than the human motor axons to internal organs—is it 10 times faster? Yes. The conduction velocity of the human axon is measured at 2 m/sec, and the conduction velocity of the squid giant axon is fully 25 m/sec; the very high velocity allows squids to contract their mantles fast enough to power their jet propulsion.

Most of the motor nerve axons in a vertebrate body like yours have insulating myelin covers, allowing them to transmit signals much faster, the electrical signal jumping down the axon over the insulated segments. The photo below shows a bundle of nerve axons, many of them myelinated and looking somewhat like doughnuts. Human axons also come in a wide range of sizes, from the 5-μm-diameter axons of skin temperature receptors to the 20-μm-diameter fibers traveling to leg muscles. Would you expect fatter myelinated axons to transmit faster? Yes. Why? The transmission speed between axon node regions will depend upon how many ion channels open when an electrical impulse arrives at that node. If you think of the axon as a cylinder or pipe, then the number of ion channels exposed at a node of the axon will be proportional to the exposed surface area of the node, with the larger surface area of bigger axons exposing more ion channels and, so, transmitting the signal faster to the next node. The surface area considered at any one location of a node is simply the circumference of the axon at that point, which is the diameter times the constant pi. Thus, the velocity of a myelinated axon would be expected to be directly proportional to its diameter. Stated simply, doubling diameter should double speed. Is that the case?

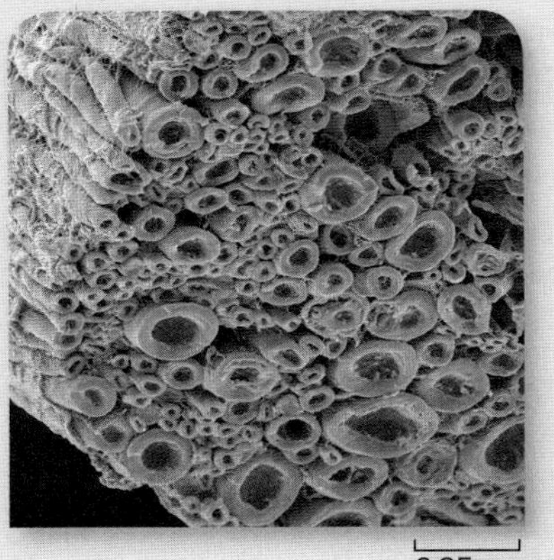

Steve Gschmeissner/Science Source

With modern electrode technology, it is possible to directly measure the conduction velocity of axons within the body of a vertebrate and, so, answer this question. The graph shows the speed of conduction of myelinated axon fibers of a cat, plotting measured speed of conductance in meters per second against axon fiber diameter measured in micrometers.

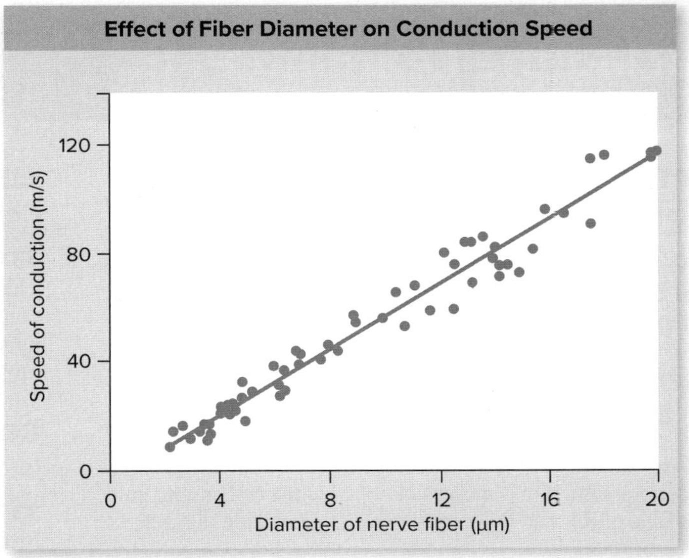

Analysis

1. **Applying Concepts**
   a. What is the dependent variable?
   b. ***Range.*** What is the range of nerve fiber diameters examined?
   c. ***Frequency.*** Which were more frequently examined, narrow diameters (less than 8 μm) or thick diameters (more than 12 μm)?
2. **Interpreting Data**
   a. For a nerve fiber diameter of 4 μm, what is the speed of conduction?
   b. For a nerve fiber twice that diameter, 8 μm, what is the speed of conduction?
   c. For a nerve fiber twice that diameter, 16 μm, what is the speed of conduction?
3. **Making Inferences**
   a. Is conduction velocity faster for larger-diameter fibers?
   b. When the nerve fiber diameter is doubled, what is the effect on speed of conductance?
4. **Drawing Conclusions** Do these data support the conclusion that conduction velocity is directly proportional to fiber diameter?
5. **Further Analysis** Do you imagine myelinated axons would transmit faster if the distance between nodes were shorter? If it were longer? How might you test this?

Retracing the Learning Path

CONCEPT 33.1 The Nervous System Directs the Body's Actions

33.1.1 The Nervous System Is Divided into Central and Peripheral Systems PNS sensory neurons carry impulses to the CNS, and motor neurons carry impulses away from the CNS.

CONCEPT 33.2 Neurons Maintain a Resting Potential Across the Plasma Membrane

33.2.1 Two Factors Contribute to Membrane Potential The sodium–potassium pump moves Na^+ outside the cell and K^+ into the cell. Leakage of K^+ also moves positive charge outside the cell.

CONCEPT 33.3 Action Potentials Propagate Nerve Impulses

33.3.1 Changes in Membrane Permeability Alter Membrane Potential Resting potential can be altered by the action of gated channels, which respond to electrical or chemical stimulus. Action potentials are the result of an influx of Na^+ followed by an efflux of K^+ through voltage-gated channels. These are separate events, each followed by a refractory period.

33.3.2 Action Potentials Are Propagated Along Axons Influx of Na^+ during an action potential causes the adjacent region to depolarize, producing its own action potential.

CONCEPT 33.4 Synapses Are Where Neurons Communicate with Other Cells

33.4.1 Vertebrate Synapses Are Not Physical Connections An action potential terminates at the end of the axon at the synapse—a gap between the axon and another cell.

33.4.2 Many Different Chemicals Serve as Neurotransmitters Neurotransmitter molecules include acetylcholine, amino acids, and biogenic amines.

33.4.3 A Postsynaptic Neuron Must Integrate Inputs from Many Synapses Excitatory postsynaptic potentials depolarize the membrane; inhibitory ones hyperpolarize it.

CONCEPT 33.5 The Central Nervous System Includes the Brain and Spinal Cord

33.5.1 Vertebrate Brains Have Three Basic Divisions The brain is divided into hindbrain, midbrain, and forebrain.

33.5.2 The Human Forebrain Exhibits Exceptional Information-Processing Ability The cerebrum is split into right and left hemispheres connected by the corpus callosum. The cerebral cortex has a motor, somatosensory, and association cortex.

33.5.3 The Spinal Cord Runs from the Brain Through the Backbone The spinal cord conveys messages and functions in reflexes. Gray matter contains cell bodies of neurons, and white matter consists of axons of sensory neurons.

CONCEPT 33.6 The Peripheral Nervous System Consists of Both Sensory and Motor Neurons

33.6.1 The PNS Has Somatic and Autonomic Systems The somatic nervous system controls movements in response to conscious commands and involuntary reflexes. The autonomic nervous system controls involuntary functions through two opposing divisions, sympathetic and parasympathetic.

CONCEPT 33.7 Sensory Receptors Provide Information About the Body's Environment

33.7.1 Sensory Receptors Detect Both External and Internal Stimuli Exteroreceptors sense external stimuli, whereas interoreceptors sense internal stimuli.

33.7.2 Sensory Transduction Involves Gated Ion Channels Sensory transduction produces a graded receptor potential. A single potential or a sum of potentials may exceed a threshold to produce an action potential.

CONCEPT 33.8 Mechanoreceptors Sense Touch and Pressure

33.8.1 Pain Receptors Alert the Body to Potential Damage Nociceptors respond to damaging stimuli perceived as pain. Thermoreceptors contain TRP ion channels.

33.8.2 Muscle Length and Tension Are Monitored by Proprioceptors Proprioceptors provide information about the relative position of body parts and degree of muscle stretching.

33.8.3 Baroreceptors Detect Blood Pressure

CONCEPT 33.9 Sounds and Body Position Are Sensed by Vibration Detectors

33.9.1 The Lateral Line System of Fish Detects Low-Frequency Vibrations Hearing, which is the detection of sound or pressure waves, works best in water and provides directional information.

33.9.2 Ear Structure Is Specialized to Detect Sound The outer ear channels sound to the eardrum. Vibrations are transferred through middle-ear bones into the cochlea.

33.9.3 Transduction Occurs in the Cochlea The basilar membrane of the cochlea consists of fibers that respond to different frequencies of sound.

33.9.4 Body Position Is Detected by Gravity-Sensitive Receptors in the Inner Ear Body position is detected by ciliated hair cells embedded in a gelatinous matrix.

CONCEPT 33.10 Taste, Smell, and pH Senses Utilize Chemoreceptors

33.10.1 Taste Receptors Detect and Analyze Potential Food

33.10.2 Smell Can Identify a Vast Number of Molecules

33.10.3 Internal Chemoreceptors Detect pH

CONCEPT 33.11 Vision Employs Photoreceptors to Perceive Objects at a Distance

33.11.1 Vision Senses Light at a Distance In the vertebrate eye, light enters through the pupil, and the lens focuses the light on the retina.

33.11.2 Vertebrate Photoreceptors Are Rod Cells and Cone Cells Rods detect black and white; cones are necessary for visual acuity and color vision. These photoreceptors inhibit bipolar cells in the dark; light-sensing receptors reverse this.

Concept Overview

This Concept Overview diagrams the key concepts that were discussed in this chapter.

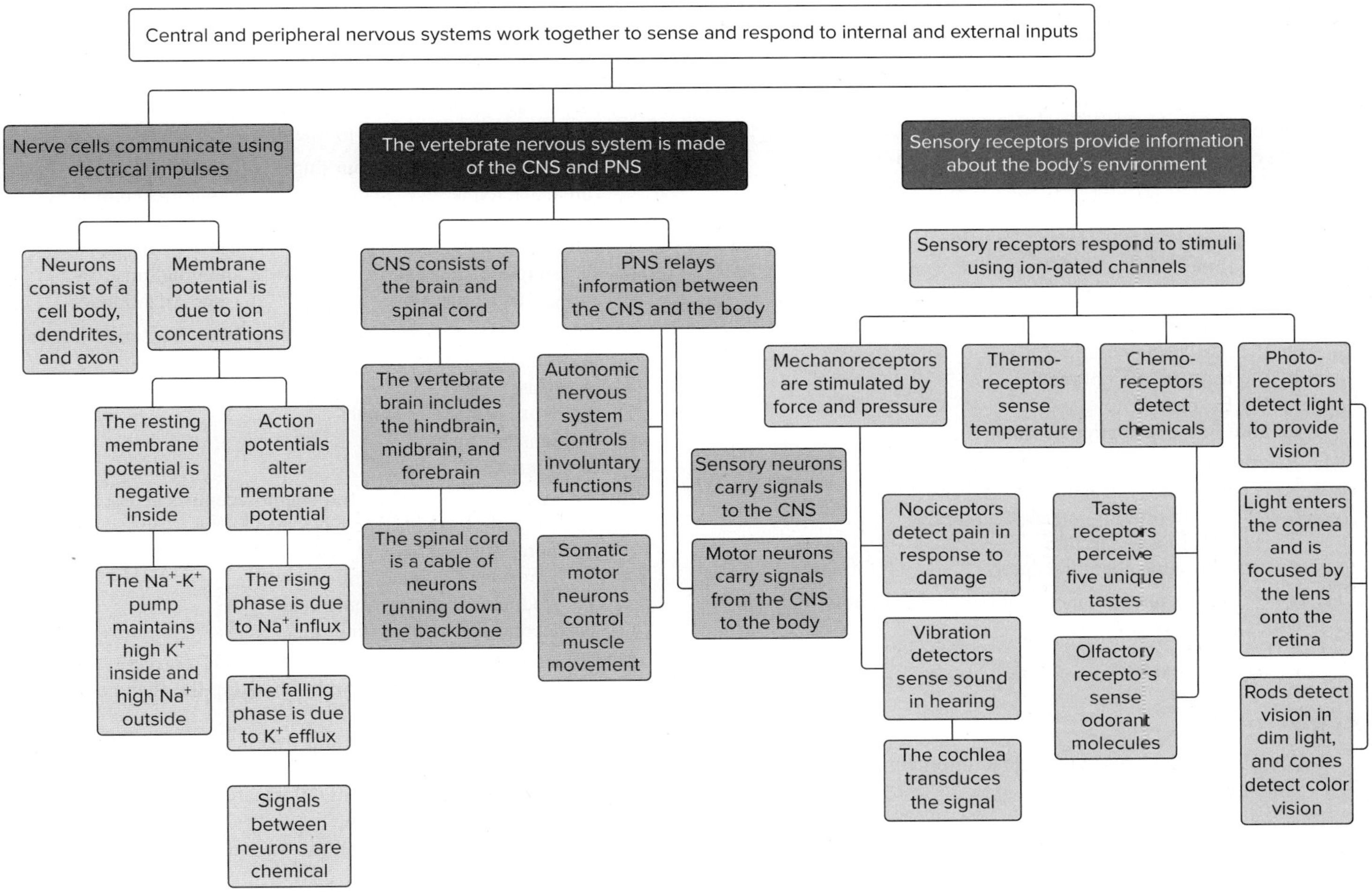

Central and peripheral nervous systems work together to sense and respond to internal and external inputs
Nerve cells communicate using electrical impulses
The vertebrate nervous system is made of the CNS and PNS
Sensory receptors provide information about the body's environment
Neurons consist of a cell body, dendrites, and axon
Membrane potential is due to ion concentrations
The resting membrane potential is negative inside
Action potentials alter membrane potential
The Na⁺-K⁺ pump maintains high K⁺ inside and high Na⁺ outside
The rising phase is due to Na⁺ influx
The falling phase is due to K⁺ efflux
Signals between neurons are chemical
CNS consists of the brain and spinal cord
PNS relays information between the CNS and the body
The vertebrate brain includes the hindbrain, midbrain, and forebrain
The spinal cord is a cable of neurons running down the backbone
Autonomic nervous system controls involuntary functions
Somatic motor neurons control muscle movement
Sensory neurons carry signals to the CNS
Motor neurons carry signals from the CNS to the body
Sensory receptors respond to stimuli using ion-gated channels
Mechanoreceptors are stimulated by force and pressure
Thermo-receptors sense temperature
Chemo-receptors detect chemicals
Photo-receptors detect light to provide vision
Nociceptors detect pain in response to damage
Vibration detectors sense sound in hearing
The cochlea transduces the signal
Taste receptors perceive five unique tastes
Olfactory receptors sense odorant molecules
Light enters the cornea and is focused by the lens onto the retina
Rods detect vision in dim light, and cones detect color vision

Assessing the Learning Path

Understand

1. Motor neurons
   a. carry impulses to the central nervous system.
   b. are used for higher functions such as memory.
   c. carry signals to muscles and glands.
   d. are only located in the central nervous system.
2. Which of the following best describes the electrical state of a neuron at rest?
   a. The inside of a neuron is more negatively charged than the outside.
   b. The outside of a neuron is more negatively charged than the inside.
   c. The inside and the outside of a neuron have the same electric charge.
   d. Potassium ions leak into a neuron at rest.
3. During an action potential,
   a. the rising phase is due to an influx of Na^+.
   b. the falling phase is due to an influx of K^+.
   c. the falling phase is due to an efflux of K^+.
   d. Both a and c
4. Inhibitory neurotransmitters
   a. hyperpolarize postsynaptic membranes.
   b. hyperpolarize presynaptic membranes.
   c. depolarize postsynaptic membranes.
   d. depolarize presynaptic membranes.

5. The corpus callosum serves as the connection between the
   a. thalamus and hypothalamus.
   b. mesencephalon and prosencephalon.
   c. left and right cerebral hemispheres.
   d. frontal lobe and occipital lobe.
6. The ____ cannot be controlled by conscious thought.
   a. motor neurons
   b. somatic nervous system
   c. autonomic nervous system
   d. skeletal muscles
7. Which of the following is NOT a category of terrestrial vertebrate sensory receptor?
   a. Mechanoreceptor
   b. Electric organ receptor
   c. Interoreceptor
   d. Chemoreceptor
8. In the fairy tale, Sleeping Beauty fell asleep after pricking her finger. What receptors respond to such painful stimulus?
   a. Mechanoreceptors
   b. Nociceptors
   c. Thermoreceptors
   d. Touch receptors
9. Baroreceptors detect
   a. blood pressure.
   b. heat.
   c. pain.
   d. tissue damage.
10. The ear detects sound by the movement of
   a. the basilar membrane.
   b. the tectorial membrane.
   c. the Eustachian tube.
   d. fluid in the semicircular canals.
11. Relative to taste, the sense of smell
   a. is less sensitive in humans.
   b. involves G protein–linked receptors.
   c. is far more acute in flies.
   d. uses the same chemoreceptors.
12. ______ is the photopigment contained within both rods and cones of the eye.
   a. Carotene
   b. *Cis*-retinal
   c. Photochrome
   d. Chlorophyll

Apply

1. A mutation that caused damage to interneurons would manifest in the
   a. entire nervous system.
   b. PNS only.
   c. CNS only.
   d. sensory systems only.
2. Imagine that you are doing an experiment on the movement of ions across neural membranes. Which of the following play a role in determining the equilibrium concentration of ions across these membranes?
   a. Ion concentration gradients
   b. Ion pH gradients
   c. Ion electrical gradients
   d. Both a and c
3. The Na^+/K^+ ATPase pump is
   a. not required for action potential firing.
   b. important for long-term maintenance of resting potential.
   c. important only at the synapse.
   d. used to stimulate graded potentials.
4. Excitatory neurotransmitters initiate an action potential in a postsynaptic neuron by opening ______ in the postsynaptic cell.
   a. Na^+ channels
   b. K^+ channels
   c. Cl^- channels
   d. Ca^{2+} channels
5. If you compare your brain with the brain of a lizard, the largest difference will be in the
   a. cerebral cortex of the forebrain.
   b. cerebral cortex of the midbrain.
   c. cerebellum of the hindbrain.
   d. thalamus of the midbrain.
6. A fight-or-flight response enables an animal to prepare for a rapid response to a dangerous or stressful situation. This is controlled by the
   a. sympathetic division of the nervous system.
   b. parasympathetic division of the nervous system.
   c. motor nervous system.
   d. somatic nervous system.
7. Sensory receptors are a diverse class of neurons. The common feature among them is that a stimulus causes
   a. the membrane to depolarize.
   b. the membrane to hyperpolarize.
   c. voltage-gated channels to open.
   d. ligand-gated channels to open.
8. Suppose that you stick your finger with a sharp pin. The area affected is very small and only one pain receptor fires. However, it fires repeatedly at a rapid rate (it hurts!). This is an example of
   a. temporal summation.
   b. spatial summation.
   c. habituation.
   d. repolarization.
9. The lateral line system in fish is most like what sensory system in terrestrial vertebrates?
   a. Vision
   b. Sense of taste or smell
   c. Sense of touch
   d. Hearing
10. When peripheral and central chemoreceptors detect a lowering of blood pH, the CNS responds by
   a. increasing plasma CO_2 levels.
   b. increasing the breathing rate.
   c. producing more carbonic acid.
   d. decreasing the breathing rate.
11. Which of the following statements is NOT correct?
   a. Vertebrates focus the eye by changing the shape of the lens.
   b. The eyes of arthropods and vertebrates use the same light-capturing molecule.
   c. Rod cells detect different colors, and cone cells detect different shades of gray.
   d. Light changes *cis*-retinal into *trans*-retinal.

Synthesize

1. Discuss how a sensory input would be relayed to the central nervous system and how it would produce reactions in the motor pathways.
2. Tetraethylammonium (TEA) blocks voltage-gated K^+ channels. What effect would TEA have on the action potentials produced by a neuron? If TEA could be applied selectively to a presynaptic neuron that releases an excitatory neurotransmitter, how would it alter the effect on the postsynaptic cell?
3. Show in a diagram the regions of the brain that have been altered in vertebrate evolution. How does this relate to human cognition?
4. Your aunt stood up suddenly at Sunday dinner and then fainted. Her fainting from standing up too quickly might involve a problem with what sensory receptor? Explain.
5. When blood pH falls too low, a potentially fatal condition known as acidosis results. In response, the body changes the breathing rate. How does the body sense this change? How does the breathing rate change?

Rene Frederic/Pixtal/age fotostock

34 Fueling the Body's Metabolism

Learning Path

34.1 Vertebrate Digestive Systems Are Tubular Tracts

34.2 Food Is Processed as It Passes Through the Digestive Tract

34.3 The Digestive Tract Is Regulated by the Nervous System and Hormones

34.4 Respiratory Systems Promote Efficient Exchange of Gases

34.5 Gills Provide for Efficient Gas Exchange in Water

34.6 Lungs Are the Respiratory Organs of Terrestrial Vertebrates

34.7 Oxygen and Carbon Dioxide Are Transported by Fundamentally Different Mechanisms

34.8 Circulating Blood Carries Metabolites and Gases to the Tissues

34.9 Vertebrate Circulatory Systems Put a Premium on Efficient Circulation

34.10 The Four Chambers of the Heart Contract in a Cycle

34.11 The Circulatory Highway Is Composed of Arteries, Capillaries, and Veins

Concept Overview

This Concept Overview provides a diagram of the overarching concepts that are covered in this chapter. These concepts will be expanded upon in the Concept Overview at the end of the chapter.

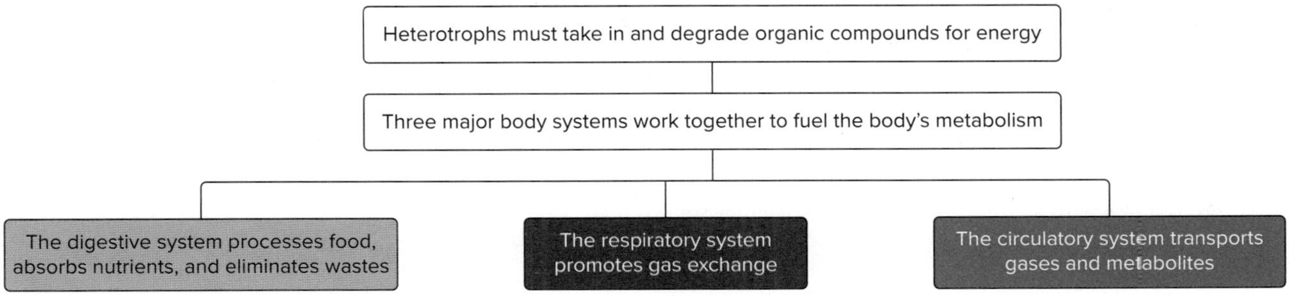

Introduction

Vertebrates like the Bengal tiger shown on the previous page are heterotrophs, fueling their bodies with organic material they consume. This material must be digested into smaller molecules, then circulated to the cells of the body. Fueling the body's metabolism thus utilizes three major body systems: digestion, circulation, and respiration. Digestion takes place in stages as food moves through the tubular digestive system. The process of respiration allows the exchange of gases from water or air. These include the oxygen needed for respiration and the carbon dioxide produced by it. Circulation moves both these gases and nutrients from where they are acquired, through a highway of vessels, to the entire body to support metabolic activity. Many structural adaptations have altered these systems as vertebrates have evolved and the environment that vertebrate animals occupy has changed.

34.1 Vertebrate Digestive Systems Are Tubular Tracts

Single-celled organisms as well as sponges digest their food intracellularly. Other multicellular animals digest their food extracellularly, within a digestive cavity. Digestive enzymes are released into a cavity that is continuous with the animal's external environment. In cnidarians and in flatworms, the digestive cavity has only one opening (refer to chapter 27). There is no specialization within this type of digestive system, called a gastrovascular cavity, because every cell is exposed to all stages of food digestion.

The Digestive Tract Is a One-Way Tube from Mouth to Anus

LEARNING OBJECTIVE 34.1.1 List the specialized zones of the vertebrate digestive tract.

Specialization occurs when the digestive tract, or alimentary canal, has a separate mouth and anus, so that transport of food is one-way. The most primitive digestive tract is seen in nematodes (phylum Nematoda), which have a simple tubular *gut* lined by an epithelial membrane. Earthworms (phylum Annelida) have a digestive tract specialized in different regions for the ingestion, storage, fragmentation, digestion, and absorption of food. All more complex animal groups, including all vertebrates, show similar specializations (figure 34.1).

The ingested food may be stored in a specialized region of the digestive tract, or it may first be subjected to physical fragmentation. This fragmentation may occur through the chewing action of teeth (in the mouth of many vertebrates) or the grinding action of pebbles (in the gizzard of earthworms and birds). Chemical digestion then occurs, breaking down the larger food molecules of polysaccharides and disaccharides, fats, and proteins into their smallest subunits.

Chemical digestion involves hydrolysis reactions that liberate the subunit molecules—primarily monosaccharides, amino acids, and fatty acids—from the food. These products of chemical digestion pass through the epithelial lining of the gut into the blood, in a process known as *absorption*. Any molecules in the food that are not absorbed cannot be used by the animal. These waste products are excreted, or defecated, from the anus.

In humans and other vertebrates, the digestive system consists of a tubular gastrointestinal tract and accessory digestive organs (figure 34.2).

Overview of the digestive tract

The initial components of the gastrointestinal tract are the mouth and the pharynx, which is the common passage of the oral and nasal cavities. The pharynx leads to the esophagus, a muscular tube that delivers food to the stomach, where some preliminary digestion occurs.

From the stomach, food passes to the small intestine, where a battery of digestive enzymes continues the digestive process. The products of digestion, together with minerals and water, are absorbed across the wall of the small intestine into the bloodstream. What remains is emptied into the large intestine, where some of the remaining water and minerals are absorbed.

In most vertebrates other than mammals, the waste products emerge from the large intestine into a cavity called the cloaca (figure 34.1), which also receives the products of the urinary and reproductive systems. In mammals, the urogenital products are kept separate, and the fecal material passes from the large intestine into the rectum and is expelled through the anus.

The accessory digestive organs include the liver, which produces *bile* (a green solution that emulsifies fat); the gallbladder, which stores and concentrates the bile; and the pancreas. The pancreas produces *pancreatic juice,* which contains digestive enzymes and bicarbonate buffer. Both bile and pancreatic juice are secreted into the first region of the small intestine, the duodenum, where they aid digestion.

REVIEW OF CONCEPT 34.1

Incomplete digestive tracts have only one opening; complete digestive tracts are flow-through, with a mouth and an anus. The digestive system of vertebrates includes mouth and pharynx, esophagus, stomach, small and large intestines, cloaca or rectum, anus, and accessory organs.

- *What might be the advantages of a one-way digestive system?*

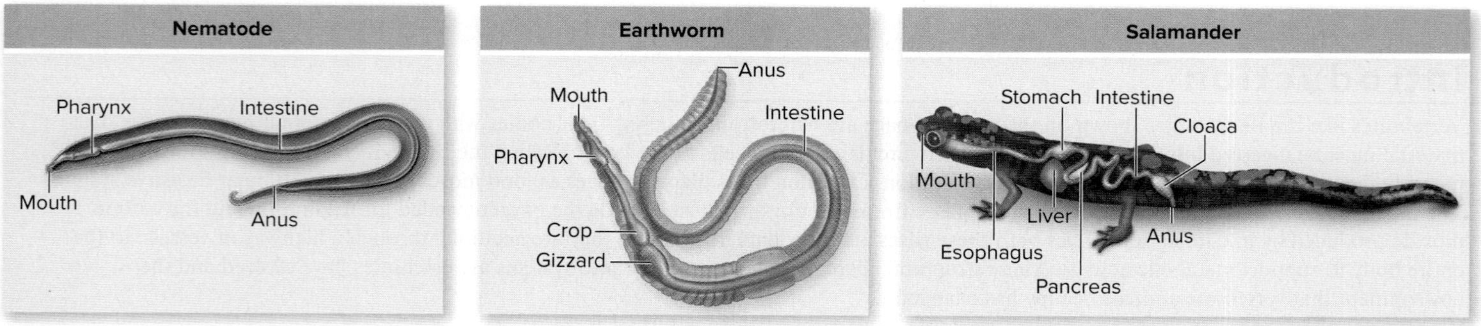

Figure 34.1 The one-way digestive tract of nematodes, earthworms, and vertebrates. One-way movement through the digestive tract allows different regions of the digestive system to become specialized for different functions.

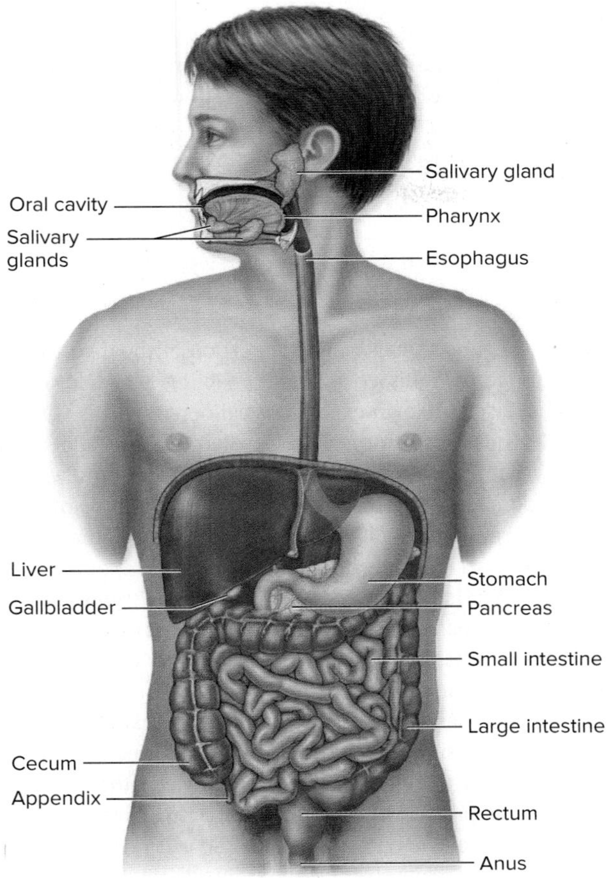
Salivary gland
Oral cavity
Pharynx
Salivary glands
Esophagus
Liver
Stomach
Gallbladder
Pancreas
Small intestine
Large intestine
Cecum
Appendix
Rectum
Anus

Figure 34.2 The human digestive system. The human digestive system consists of the oral cavity, esophagus, stomach, small intestine, large intestine, rectum, and anus and is aided by accessory organs.

34.2 Food Is Processed as It Passes Through the Digestive Tract

Specializations of the digestive systems in different kinds of vertebrates reflect the way these animals live. Birds, which lack teeth, break up food in their two-chambered stomachs. In one of these chambers, called the *gizzard,* small pebbles ingested by the bird are churned together with the food by muscular action. This churning grinds up the seeds and other hard plant material into smaller chunks that can be digested more easily.

Vertebrate Teeth Are Adapted to Different Types of Food Items

LEARNING OBJECTIVE 34.2.1 Identify adaptive variation in vertebrate tooth shape.

Many vertebrates have teeth, which are used for chewing, or mastication. This breaks food into smaller particles and mixes it with fluid secretions. The evolution of teeth is a reflection of how an animal obtains food (figure 34.3). Carnivorous mammals have pointed teeth adapted for cutting and shearing. These are used to tear off pieces of prey, not for chewing, because digestive enzymes can act directly on animal cells. Herbivores must pulverize plant tissue to release cellulose from cell walls, which can then be digested by bacteria in their rumens or cecae. These animals have large, flat teeth with ridges, which are adapted for grinding.

Human teeth are specialized for eating both plant and animal food. Viewed simply, humans are carnivores in the front of the mouth and herbivores in the back (figure 34.3). The four front teeth in the upper and lower jaws are sharp, chisel-shaped incisors used for biting and cutting. On each side of the incisors are sharp, pointed teeth called cuspids (sometimes referred to as "canine" teeth), which are used for tearing food. Behind the canines are two premolars and three molars, all with flattened, ridged surfaces for grinding and crushing food.

The mouth is a chamber for ingestion and initial processing

Inside the mouth, the tongue mixes food with a mucous solution, saliva. In humans, three pairs of salivary glands secrete saliva into the mouth through ducts in the mouth's mucosal lining. Saliva moistens and lubricates the food so that it is easier to swallow and does not abrade the tissue of the esophagus as it passes through.

Saliva also contains the hydrolytic enzyme salivary amylase, which initiates the breakdown of the polysaccharide starch into the disaccharide maltose. This digestion is usually minimal in humans, however, because most people don't chew their food very long.

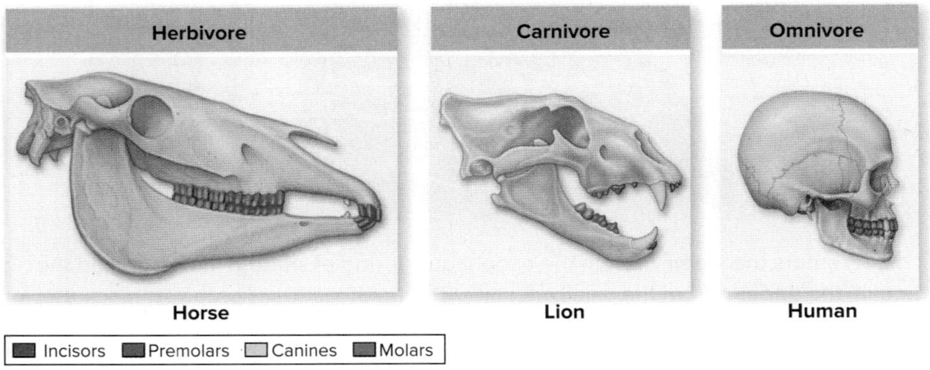
Herbivore
Carnivore
Omnivore
Horse
Lion
Human
Incisors
Premolars
Canines
Molars

Figure 34.3 Patterns of dentition depend on diet. Different vertebrates (herbivore, carnivore, or omnivore) have evolved specific variations from a generalized pattern of dentition depending on their diets.

Stimulation of salivation

The secretions of the salivary glands are controlled by the nervous system, which in humans maintains a constant flow of about half a milliliter per minute when the mouth is empty of food. This continuous secretion keeps the mouth moist.

The presence of food in the mouth triggers an increased rate of secretion. Taste buds as well as olfactory (smell) neurons send impulses to the brain, which responds by stimulating the salivary glands. The sight, sound, or smell of food can stimulate salivation markedly in many animals; in humans, thinking or talking about food can also have this effect.

Muscular Contractions of the Esophagus Move Food to the Stomach

LEARNING OBJECTIVE 34.2.2 Describe how food moves through the esophagus.

Swallowed food enters a muscular tube called the esophagus, which connects the pharynx to the stomach. The esophagus actively moves a processed lump of food, called a **bolus,** through the action of muscles. Food from a meal is stored in the stomach, where it undergoes early stages of digestion.

In adult humans, the esophagus is about 25 cm long; the upper third is enveloped in skeletal muscle for voluntary control of swallowing, whereas the lower two-thirds is surrounded by involuntary smooth muscle. The swallowing center stimulates successive one-directional waves of contraction in these muscles that move food along the esophagus to the stomach. These rhythmic waves of muscular contraction are called **peristalsis;** they enable humans and other vertebrates to swallow even if they are upside down.

In many vertebrates, the movement of food from the esophagus into the stomach is controlled by a ring of circular smooth muscle, or a *sphincter,* that opens in response to the pressure exerted by the food. Contraction of this sphincter prevents food in the stomach from moving back into the esophagus. Rodents and horses have a true sphincter at this site, and as a result they cannot regurgitate; humans lack a true sphincter. Normally, the esophagus is closed off except during swallowing.

The Stomach Is a "Holding Station" Involved in Acidic Breakdown of Food

LEARNING OBJECTIVE 34.2.3 Explain what digestive processes take place in the stomach.

The **stomach** (figure 34.4) is a saclike portion of the digestive tract. Its inner surface is highly convoluted, enabling it to fold up when empty and open out like an expanding balloon as it fills with food. For example, the human stomach has a volume of only about 50 mL when empty, but it may expand to contain 2 to 4 L of food when full.

The stomach contains a third layer of smooth muscle for churning food and mixing it with **gastric juice,** an acidic secretion of the tubular gastric glands of the mucosa. These exocrine glands contain three kinds of secretory cells: *mucus-secreting cells; parietal cells,* which secrete hydrochloric acid (HCl); and *chief cells,* which secrete **pepsinogen,** the inactive form of the protease (protein-digesting enzyme) **pepsin.**

Pepsinogen has 44 additional amino acids that block its active site. HCl causes pepsinogen to unfold, exposing the active site, which then acts to remove the 44 amino acids. This yields the active protease, pepsin. This process of secreting an inactive form that is then converted into an active enzyme outside the cell prevents the chief cells from digesting themselves. In the stomach, mucus produced by mucus-secreting cells serves the same purpose, covering the interior walls and preventing them from being digested.

Action of acid

The human stomach can produce about 2 L of HCl and other gastric secretions every day, creating a very acidic solution. The

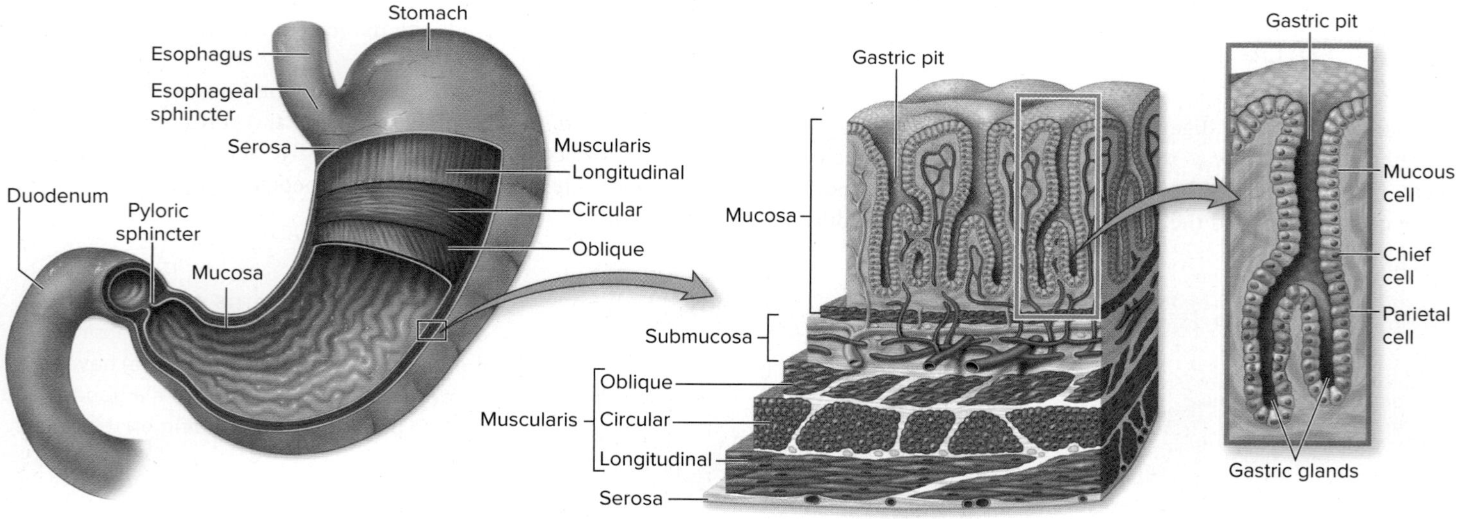

Figure 34.4 The stomach and duodenum. Food enters the stomach from the esophagus. A ring of smooth muscle called the pyloric sphincter controls the entrance to the duodenum, the upper part of the small intestine. The epithelial walls of the stomach are dotted with deep infoldings called gastric pits that contain gastric glands. The gastric glands consist of mucous cells, chief cells that secrete pepsinogen, and parietal cells that secrete HCl. Gastric pits are the openings of the gastric glands.

concentration of HCl in this solution is about 10 millimolar (m*M*), equal to a pH of 2. Thus, gastric juice is about 250,000 times more acidic than blood, the normal pH of which is 7.4.

The low pH in the stomach helps denature food proteins, making them easier to digest, and keeps pepsin maximally active. Pepsin hydrolyzes food proteins into shorter polypeptides that are not fully digested until the mixture enters the small intestine. The mixture of partially digested food and gastric juice is called **chyme.** In adult humans, only proteins are partially digested in the stomach—no significant digestion of carbohydrates or fats occurs there.

The acidic solution within the stomach also kills most of the bacteria that are ingested with the food. The few bacteria that survive the stomach and enter the intestine intact are able to grow and multiply there, particularly in the large intestine. In fact, vertebrates harbor thriving colonies of bacteria within their intestines. Bacteria that live within the digestive tracts of ruminants play a key role in the ability of these mammals to digest cellulose.

Leaving the stomach

Chyme leaves the stomach through the *pyloric sphincter* to enter the small intestine. This is where all terminal digestion of carbohydrates, lipids, and proteins occurs and where the products of digestion—amino acids, glucose, and so on—are absorbed into the blood. Only some of the water in chyme and a few substances, such as aspirin and alcohol, are absorbed through the wall of the stomach.

The Structure of the Small Intestine Is Specialized for Nutrient Uptake

LEARNING OBJECTIVE 34.2.4 Describe how the structure of the small intestine optimizes its function.

The capacity of the small intestine is limited, and its digestive processes take time. Consequently, efficient digestion requires that only relatively small amounts of chyme be introduced from the stomach into the small intestine at any one time. Coordination between gastric and intestinal activities is regulated by neural and hormonal signals.

The small intestine is approximately 4.5 m long in a living person, but 6 m long at autopsy when all the muscles have relaxed. The first 25 cm is the **duodenum;** the remainder of the small intestine is divided into the **jejunum** and the **ileum.**

The duodenum receives acidic chyme from the stomach, digestive enzymes and bicarbonate from the pancreas, and bile from the liver and gallbladder. Enzymes in the pancreatic juice digest larger food molecules into smaller fragments. This digestion occurs primarily in the duodenum and jejunum.

The epithelial wall of the small intestine is covered with tiny, finger-like projections called **villi** (singular, *villus;* figure 34.5). In turn, each epithelial cell lining the villi is covered on its apical surface (the side facing the lumen) by many foldings of the plasma membrane that form cytoplasmic extensions called **microvilli.** These are quite tiny and can be seen clearly only with an electron microscope. Under a light micrograph, the microvilli resemble the bristles of a brush, and for that reason the epithelial wall of the small intestine is also called a *brush border.*

The villi and microvilli greatly increase the surface area of the small intestine; in humans, this surface area is 300 m^2—about 3200 ft^2, larger than a tennis court! It is over this vast surface that the products of digestion are absorbed.

The microvilli also participate in digestion, because a number of digestive enzymes are embedded within the epithelial cells' plasma membranes, with their active sites exposed to the chyme. These brush border enzymes include those that hydrolyze the disaccharides lactose and sucrose, among others. Many adult humans lose the ability to produce the brush border enzyme lactase and therefore cannot digest lactose (milk sugar), a rather common condition called *lactose intolerance.* The brush border enzymes complete the digestive process that started with the action of salivary amylase in the mouth.

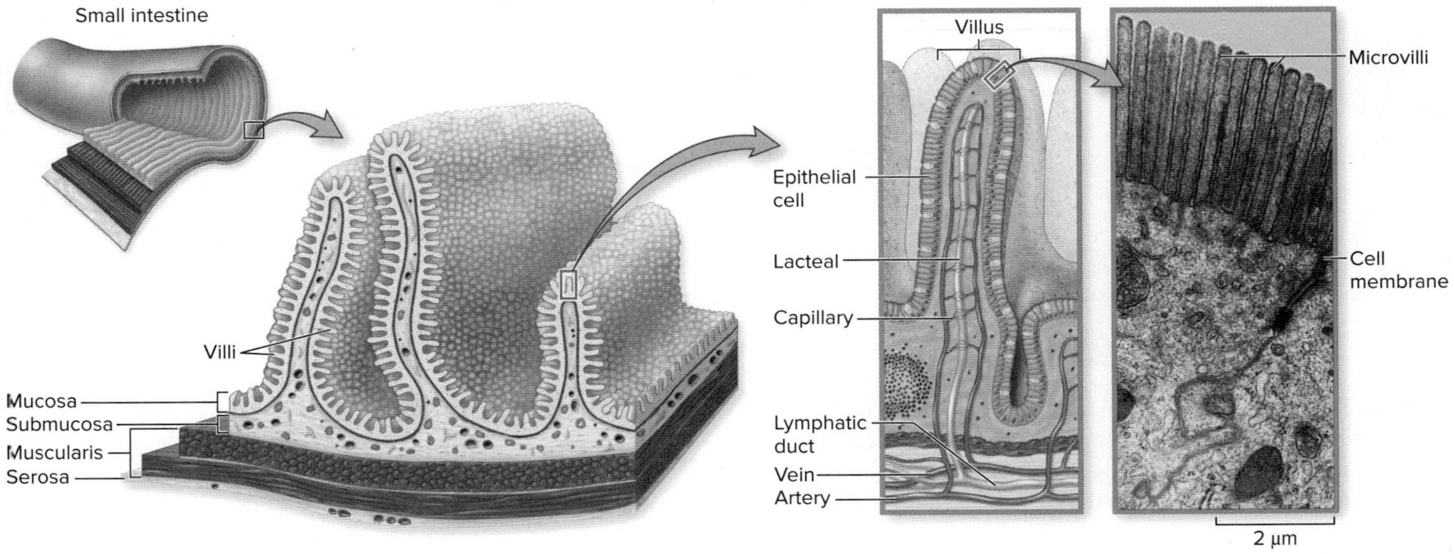

Figure 34.5 The small intestine. Successive enlargements show folded epithelium studded with villi that increase the surface area. The micrograph shows an epithelial cell with numerous microvilli.

Dennis Kunkel Microscopy/Science Source

Accessory Organs Secrete Enzymes into the Small Intestine

LEARNING OBJECTIVE 34.2.5 Name the accessory organs and describe their roles.

The main organs that aid digestion are the pancreas, liver, and gallbladder. They empty their secretions, primarily enzymes, through ducts directly into the small intestine.

Secretions of the pancreas

The pancreas (figure 34.6), a large gland situated near the junction of the stomach and the small intestine, secretes pancreatic fluid into the duodenum through the *pancreatic duct*; thus, the pancreas functions as an exocrine gland. This fluid contains a host of enzymes, including **trypsin** and **chymotrypsin,** which digest proteins; **pancreatic amylase,** which digests starch; and **lipase,** which digests fat. Like pepsin in the stomach, these enzymes are released into the duodenum primarily as inactive enzymes and are then activated by trypsin, which is first activated by a brush border enzyme of the intestine.

Pancreatic enzymes digest proteins into smaller polypeptides, polysaccharides into shorter chains of sugars, and fats into free fatty acids and monoglycerides. Digestion of proteins and carbohydrates is then completed by the brush border enzymes. Pancreatic fluid also contains bicarbonate, which neutralizes the HCl from the stomach and gives the chyme in the duodenum a slightly alkaline pH. The digestive enzymes and bicarbonate are produced by clusters of secretory cells known as **acini.**

In addition to its exocrine role in digestion, the pancreas also functions as an endocrine gland, secreting several hormones into the blood that control the blood levels of glucose and other nutrients. These hormones are produced in the **islets of Langerhans,** clusters of endocrine cells scattered throughout the pancreas. The two most important pancreatic hormones, insulin and glucagon, are described in chapter 35.

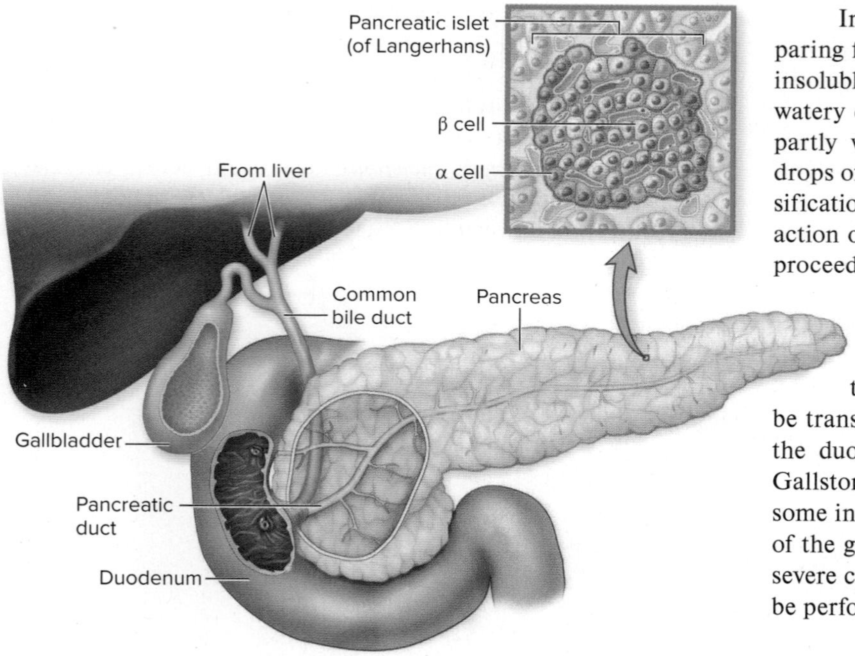

Figure 34.6 The pancreas. The pancreatic and bile ducts empty into the duodenum. The pancreas secretes pancreatic juice into the pancreatic duct. The pancreatic islets of Langerhans secrete hormones into the blood; α cells secrete glucagon, and β cells secrete insulin. The liver secretes bile, which consists of bile pigments (waste products from the liver) and bile salts. Bile salts play a role in the digestion of fats. Bile is concentrated and stored in the gallbladder until it is needed in the duodenum on the arrival of fatty food.

Liver and gallbladder

The **liver** is the largest internal organ of the body. In an adult human, the liver weighs about 1.5 kg and is the size of a football. The main exocrine secretion of the liver is bile, a fluid mixture consisting of *bile pigments* and *bile salts* that is delivered into the duodenum during the digestion of a meal.

The bile pigments do not participate in digestion; they are waste products resulting from the liver's destruction of old red blood cells and are ultimately eliminated with the feces. If the excretion of bile pigments by the liver is blocked, the pigments can accumulate in the blood and cause a yellow staining of the tissues known as *jaundice.*

In contrast, the bile salts play a very important role in preparing fats for subsequent enzymatic digestion. Because fats are insoluble in water, they enter the intestine as drops within the watery chyme. The bile salts, which are partly lipid-soluble and partly water-soluble, work as detergents, dispersing the large drops of fat into a fine suspension of smaller droplets. This emulsification action produces a greater surface area of fat for the action of lipase enzymes, and thus allows the digestion of fat to proceed more rapidly.

After bile is produced in the liver, it is stored and concentrated in the gallbladder. The arrival of fatty food in the duodenum triggers a neural and endocrine reflex that stimulates the gallbladder to contract, causing bile to be transported through the common bile duct and injected into the duodenum (these reflexes are discussed in section 34.3). Gallstones are hardened precipitates of cholesterol that form in some individuals. If these stones block the bile duct, contraction of the gallbladder causes intense pain, often felt in the back. In severe cases of blockage, surgical removal of the gallbladder may be performed.

Absorbed Nutrients Move into Blood or Lymphatic Capillaries

LEARNING OBJECTIVE 34.2.6 Explain how absorbed nutrients move into the blood or lymphatic capillaries.

After their enzymatic breakdown, proteins and carbohydrates are absorbed as amino acids and monosaccharides, respectively. They are transported across the brush border into the epithelial

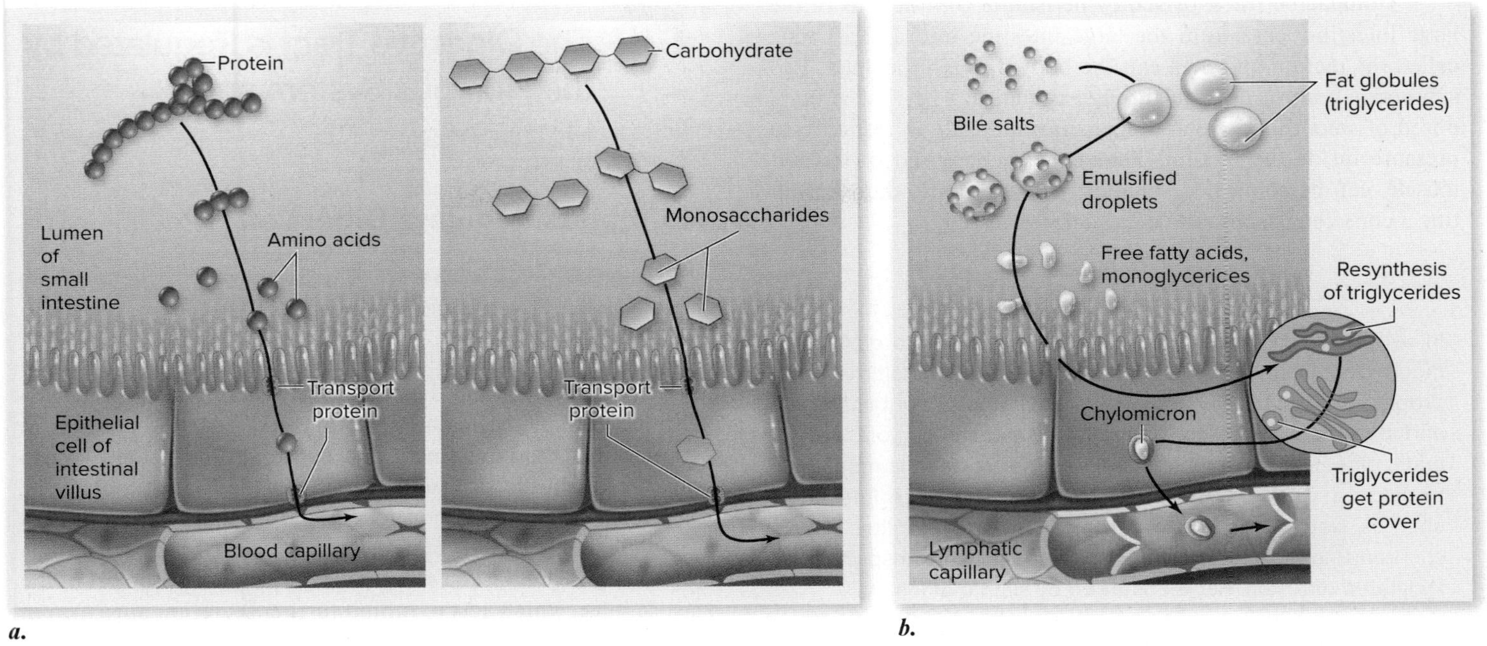

Figure 34.7 Absorption of the products of digestion. ***a.*** Monosaccharides and amino acids are transported into blood capillaries. ***b.*** Fatty acids and monoglycerides within the intestinal lumen are absorbed and converted within the intestinal epithelial cells into triglycerides. These are then coated with proteins to form structures called chylomicrons, which enter lymphatic capillaries.

cells that line the intestine by a combination of active transport and facilitated diffusion (figure 34.7*a*). Glucose is transported by coupled transport with Na^+ ions. Once they have entered epithelial cells across the apical membrane, these monosaccharides and amino acids move through the cytoplasm and are transported across the basolateral membrane and into the blood capillaries within the villi.

The blood carries these products of digestion from the intestine to the liver via the hepatic portal vein. A portal vein connects two beds of capillaries instead of returning to the heart. Because of the hepatic portal vein, the liver is the first organ to receive most of the products of digestion, except for fat. The products of fat digestion are absorbed by a different mechanism (figure 34.7*b*). Triglycerides are hydrolyzed into fatty acids and monoglycerides, which are nonpolar and can thus enter epithelial cells by simple diffusion. In the intestinal epithelial cells, they are reassembled into triglycerides and combined with proteins to form small particles called chylomicrons. These are too bulky to enter blood capillaries in the intestine, so they do not enter the hepatic portal circulation. Instead, chylomicrons are absorbed into lymphatic capillaries (discussed in section 34.11), which empty their contents into the blood in veins near the neck. Chylomicrons can make the blood plasma appear cloudy if a sample of blood is drawn after a fatty meal.

The amount of fluid passing through the small intestine in a day is startlingly large: approximately 9 L. However, almost all of this fluid is absorbed into the body rather than eliminated in the feces: about 8.5 L is absorbed in the small intestine and an additional 350 mL in the large intestine. Only about 50 g of solid and 100 mL of liquid leaves the body as feces. The normal fluid absorption efficiency of the human digestive tract can approach 99%, which is very high indeed.

The Large Intestine Eliminates Waste Material

LEARNING OBJECTIVE 34.2.7 Describe the function of the large intestine.

The large intestine, or colon, is much shorter than the small intestine but has a larger diameter. The small intestine empties directly into the large intestine at a junction where two vestigial structures, the cecum and the appendix, remain. No digestion takes place within the colon, and only about 4% of the absorption of fluids by the intestine occurs there. The inner surface has no villi; consequently, it has less than 1/30 the absorptive surface area of the small intestine. The function of the colon is to absorb water, remaining electrolytes, and products of bacterial metabolism (including vitamin K). The large intestine prepares waste material to be expelled from the body.

Bacteria live and reproduce within the large intestine, and the excess bacteria are incorporated into the refuse material, called feces. Bacterial fermentation produces gas within the colon at a rate of about 500 mL per day. This rate increases greatly after the consumption of beans or other types of vegetables, because the passage of undigested plant material (fiber) into the large intestine provides substrates for bacterial fermentation.

The human colon has evolved to process food with a relatively high fiber content. Diets that are low in fiber, which are common in the United States, result in a slower passage of food through the colon. Low dietary fiber content is thought to be associated with the high level of colon cancer in the United States.

Compacted feces, driven by peristaltic contractions of the large intestine, pass from the large intestine into a short tube called the rectum and then exit the body through the anus. Two sphincters control passage through the anus. The first is composed of smooth muscle and opens involuntarily in response to pressure inside the rectum. The second, composed of striated muscle, can be controlled voluntarily by the brain, thus permitting a conscious decision to delay defecation.

REVIEW OF CONCEPT 34.2

In vertebrates, tooth shape exhibits adaptations to diet. The mouth begins the digestion of ingested food. Peristaltic waves propel food to the stomach, where gastric juice contains acid and pepsin, a protease. In the small intestine, most enzymatic digestion takes place, and its inner surface is modified to increase surface area for absorption. The large intestine absorbs water, electrolytes, and bacterial metabolites. Glucose and amino acids are absorbed by active transport and facilitated diffusion. Fat is absorbed by simple diffusion.

- *Suppose you ate a chicken sandwich (chicken breast on bread with mayonnaise). Which of these foods would begin to be broken down in the stomach?*

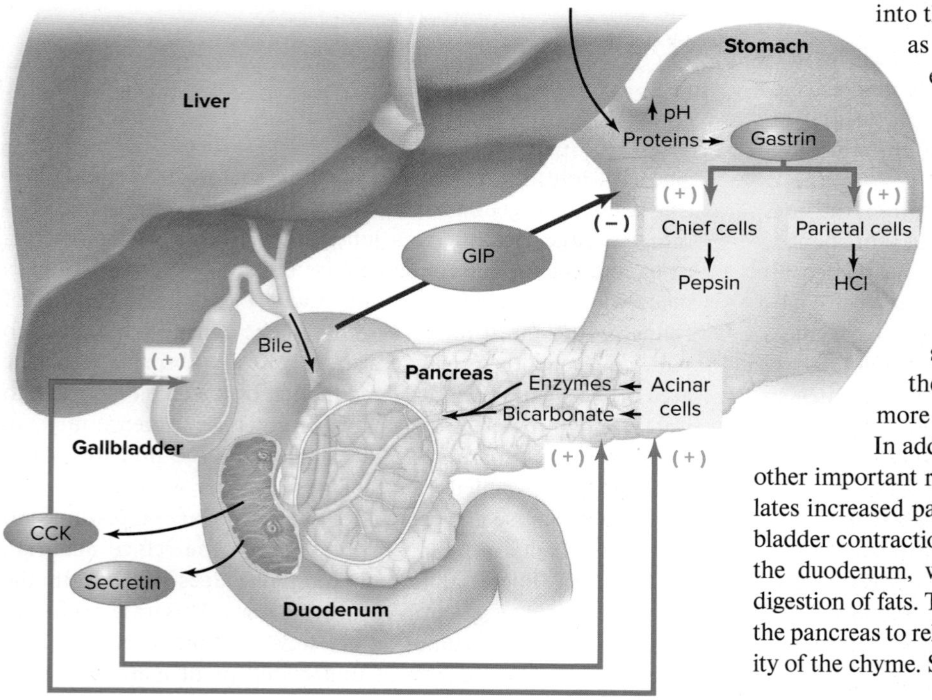

Figure 34.8 Hormonal control of the gastrointestinal tract. Gastrin, secreted by the mucosa of the stomach, stimulates the secretion of HCl and pepsinogen (which is converted into pepsin). The duodenum secretes three hormones: cholecystokinin (CCK), which stimulates contraction of the gallbladder and secretion of pancreatic enzymes; secretin, which stimulates secretion of pancreatic bicarbonate; and gastric inhibitory peptide (GIP), which inhibits stomach emptying.

34.3 The Digestive Tract Is Regulated by the Nervous System and Hormones

Hormones Regulate Digestion

LEARNING OBJECTIVE 34.3.1 Explain how the nervous system stimulates the digestive process.

The activities of the gastrointestinal tract are coordinated by the nervous system and the endocrine system. The nervous system, for example, stimulates salivary and gastric secretions in response to the sight, smell, and consumption of food. When food arrives in the stomach, proteins in the food stimulate the secretion of a stomach hormone called **gastrin,** which in turn stimulates the secretion of pepsinogen and HCl from the gastric glands (figure 34.8). The secreted HCl then lowers the pH of the gastric juice, which acts to inhibit further secretion of gastrin in a negative feedback loop. In this way, the secretion of gastric acid is kept under tight control.

The passage of chyme from the stomach into the duodenum of the small intestine inhibits the contractions of the stomach, so that no additional chyme can enter the duodenum until the previous amount can be processed. This stomach or gastric inhibition is mediated by a neural reflex and by duodenal hormones secreted into the blood. These hormones are collectively known as the **enterogastrones.** Hormonal regulation is examined in detail in chapter 35.

The major enterogastrones include cholecystokinin (CCK), **secretin,** and **gastric inhibitory peptide (GIP).** Chyme with high fat content is the strongest stimulus for CCK and GIP secretions, whereas increasing chyme acidity primarily influences the release of secretin. All three of these enterogastrones inhibit gastric motility (churning action) and gastric juice secretions; the result is that fatty meals remain in the stomach longer than nonfatty meals, allowing more time for digestion of complex fat molecules.

In addition to gastric inhibition, CCK and secretin have other important regulatory functions in digestion. CCK also stimulates increased pancreatic secretions of digestive enzymes and gallbladder contractions. Gallbladder contractions inject more bile into the duodenum, which enhances the emulsification and efficient digestion of fats. The other major function of secretin is to stimulate the pancreas to release more bicarbonate, which neutralizes the acidity of the chyme. Secretin was the first hormone ever discovered.

The Liver Removes Toxins and Functions in Homeostasis

LEARNING OBJECTIVE 34.3.2 Describe the liver's role in maintaining homeostasis.

The liver is a key organ in the breakdown of toxins. The hepatic portal vein carries blood directly from the stomach and intestine to the liver. Enzymes in the liver can modify toxic substances absorbed in the gastrointestinal tract before they reach the rest of the body. For example, ingested alcohol and some drugs are taken into liver cells and metabolized; this is one reason that the liver is often damaged as a result of long-term alcohol and drug abuse.

The liver removes a wide variety of toxins by converting them into less toxic forms. For example, the liver converts the toxic ammonia produced by intestinal bacteria into urea, a compound that can be contained safely and carried by the blood at higher concentrations.

The liver also regulates the levels of many compounds produced within the body. Steroid hormones, for instance, are converted into less active and more water-soluble forms in the liver. These molecules are then added to bile and eliminated from the body in feces or carried in blood to the kidneys and excreted in the urine.

The liver also produces most of the proteins found in blood plasma. The total concentration of plasma proteins is significant, because it must be kept within certain limits to maintain osmotic balance between blood and interstitial (tissue) fluid. If the concentration of plasma proteins drops too low, as can happen as a result of liver disease such as cirrhosis, fluid accumulates in the tissues, a condition called *edema.*

The pancreas secretes hormones that regulate the blood glucose level, in part through actions on liver cells. The neurons in the brain obtain energy primarily from the aerobic respiration of glucose obtained from the blood plasma. It is therefore vital that the blood glucose concentration not fall too low, as might happen during fasting or prolonged exercise. The hormonal control of blood glucose is discussed in detail in chapter 35.

REVIEW OF CONCEPT 34.3

Sensory input such as sight, smell, and taste stimulates salivary and gastric activity, as does the arrival of food in the stomach. The major enterogastrones are cholecystokinin (CCK), secretin, and gastric inhibitory peptide (GIP); these regulate the passage of chyme into the duodenum and the release of pancreatic enzymes and bile. The liver acts to neutralize potentially harmful toxins.

- *Would you expect anosmia, an inability to perceive scents, to affect digestion?*

34.4 Respiratory Systems Promote Efficient Exchange of Gases

Gas Exchange Involves Diffusion Across Membranes

LEARNING OBJECTIVE 34.4.1 Explain how Fick's Law of Diffusion applies to gas exchange across membranes.

One of the major physiological challenges facing all multicellular animals is obtaining sufficient oxygen and disposing of excess carbon dioxide, a physiological process called **respiration.** Oxygen is used in mitochondria for cellular respiration—the subject of chapter 7—a process that also produces CO_2 as waste. Because plasma membranes must be surrounded by water to be stable, the external environment in gas exchange is always aqueous. This is true even in terrestrial vertebrates; in these cases, oxygen from air dissolves in a thin layer of fluid that covers the respiratory surfaces.

In vertebrates, the gases O_2 and CO_2 diffuse across the aqueous layer covering the epithelial cells that line the respiratory organs. The diffusion process is passive, driven only by the difference in O_2 and CO_2 concentrations on the two sides of the membranes and their relative solubilities in the plasma membrane. For dissolved gases, concentration is usually expressed as pressure.

In general, the rate of diffusion between two regions is governed by a relationship known as **Fick's Law of Diffusion.** Fick's Law states that for a dissolved gas, the rate of diffusion *(R)* is directly proportional to the pressure difference (Δp) between the two sides of the membrane and the area *(A)* over which the diffusion occurs. Furthermore, R is inversely proportional to the distance *(d)* across which the diffusion must occur. A molecule-specific diffusion constant, *D,* accounts for the size of molecule, membrane permeability, and temperature. Shown as a formula, Fick's Law is stated as

$$R = \frac{DA\,\Delta p}{d}$$

The limits imposed by diffusion exert selective pressure to maximize the rate of diffusion. R can be increased by increasing the surface area over which diffusion occurs (A), decreasing the distance for diffusion, and increasing the effective concentration of the diffusing species (Δp). The evolution of respiratory systems has involved changes in all of these factors.

Evolution Has Acted to Maximize the Rate of Gas Diffusion

LEARNING OBJECTIVE 34.4.2 Explain how evolutionary adaptations can affect different variables of Fick's Law.

The levels of oxygen needed for cellular respiration cannot be obtained by diffusion alone over distances greater than about 0.5 mm. This restriction severely limits the size and structure of organisms that obtain oxygen entirely by diffusion from the environment. Bacteria, archaea, and protists are small enough that such diffusion can be adequate, even in some colonial forms, but most multicellular animals require structural adaptations to enhance gas exchange.

Most phyla of invertebrates lack specialized respiratory organs, but they have evolved mechanisms to improve diffusion. One mechanism is for the beating of cilia to create a water current over respiratory surfaces. Because of this continuous replenishment of water, the external oxygen concentration does not decrease along the diffusion pathway. Although oxygen molecules are passing into the organism, reducing the oxygen concentration in the water, new water continuously replaces the oxygen-depleted water. This maximizes the concentration difference, Δp, of the Fick equation.

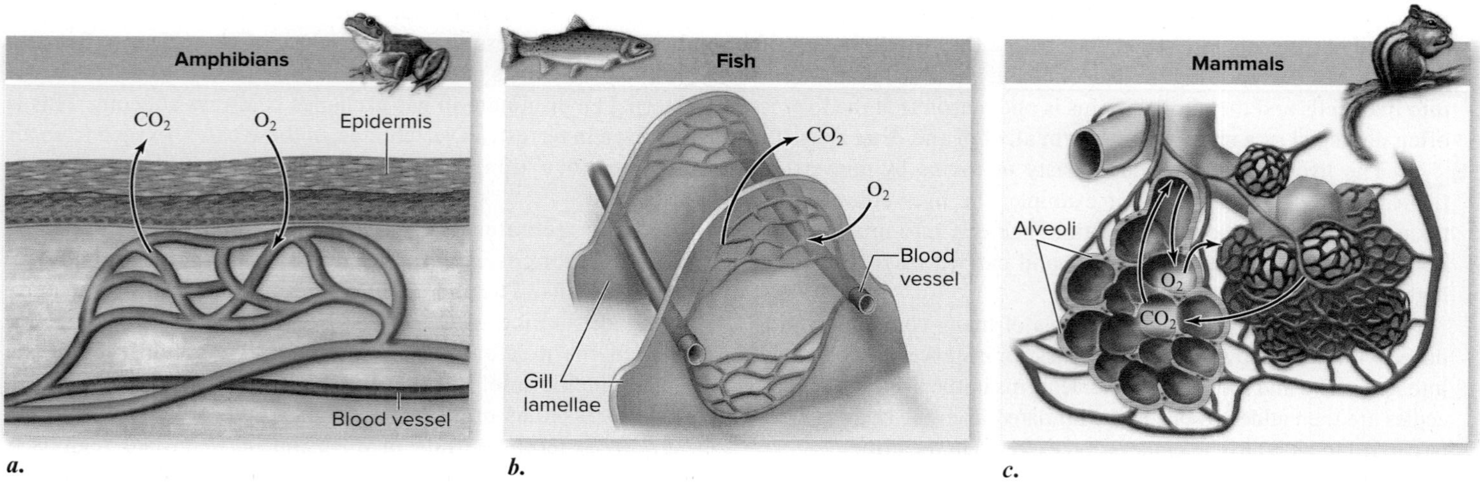

Figure 34.9 Different gas exchange systems in animals. ***a.*** Most amphibians and many other animals respire across their skin. Amphibians also exchange gases via lungs. ***b.*** The gills of fishes provide a very large respiratory surface area and countercurrent exchange. ***c.*** The alveoli in mammalian lungs provide a large respiratory surface area but do not permit countercurrent exchange.

Other invertebrates (mollusks, arthropods, echinoderms) and vertebrates possess respiratory organs—such as gills, trachea, and lungs—that increase the surface area available for diffusion (figure 34.9). These adaptations also bring the external environment (either water or air) close to the internal fluid, which is usually circulated throughout the body—such as blood or hemolymph. The respiratory organs thus increase the rate of diffusion by maximizing surface area *(A)* and decreasing the distance *(d)* the diffusing gases must travel.

REVIEW OF CONCEPT 34.4

Gases must be dissolved to diffuse across living membranes. Direction of diffusion is driven by a concentration gradient. Fick's Law states that the rate is affected by concentration difference and membrane area. Evolutionary adaptations have acted to increase concentration differences and surface area, or to reduce distance.

- *Which factor is affected by continuously beating cilia?*

34.5 Gills Provide for Efficient Gas Exchange in Water

Fish Respire with External Gills

LEARNING OBJECTIVE 34.5.1 Describe how gills take advantage of countercurrent flow.

Gills are specialized extensions of tissue that project into water. The great increase in diffusion surface area that gills provide enables aquatic organisms to extract far more oxygen from water than would be possible from their body surface alone. In this section, we concentrate on gills found in vertebrate animals.

Other moist external surfaces are also involved in gas exchange in some vertebrates and invertebrates. For example, gas exchange across the skin is a common strategy in many amphibian groups.

External gills are not enclosed within body structures. Examples of vertebrates with external gills are the larvae of many fish and amphibians, as well as amphibians such as the axolotl, which retains larval features throughout life.

One of the disadvantages of external gills is that they must constantly be moved to ensure contact with fresh water having high oxygen content. The highly branched gills, however, offer significant resistance to movement, making this form of respiration ineffective except in smaller animals. Another disadvantage is that external gills, with their thin epithelium for gas exchange, are easily damaged.

Gills of bony fishes are covered by the operculum

The gills of bony fishes are located between the oral cavity and the *opercular cavities* where the gills are housed. The two sets of cavities function as pumps that move water in one direction, first into the mouth, then across the gills, and finally out of the fish through the open **operculum**, or gill cover.

Some bony fishes that swim continuously, such as tuna, have practically immobile opercula. These fishes swim with their mouths partly open, constantly forcing water over the gills in what is known as *ram ventilation.* Most bony fishes, however, have flexible gill covers. For example, the remora, a fish that rides "piggyback" on sharks, uses ram ventilation while the shark is swimming but employs the pumping action of its opercula when the shark stops swimming.

There are between three and seven gill arches on each side of the fish's head. Each gill arch is composed of two rows of *gill filaments,* and each gill filament contains thin, membranous plates, or *lamellae,* that project out into the flow of water (figure 34.10). Water flows past the lamellae in one direction only.

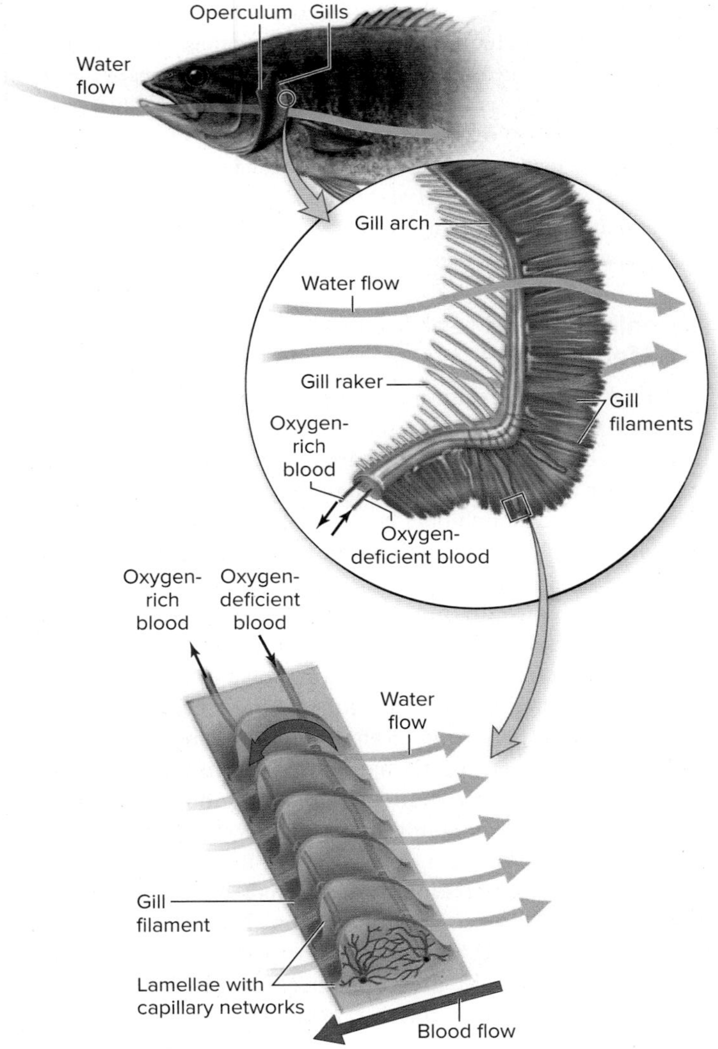

Figure 34.10 Structure of a fish gill. Water passes from the gill arch over the filaments (from left to right in the diagram). Water always passes the lamellae in a direction opposite the direction of blood flow through the lamellae. The success of the gill's operation critically depends on this countercurrent flow of water and blood.

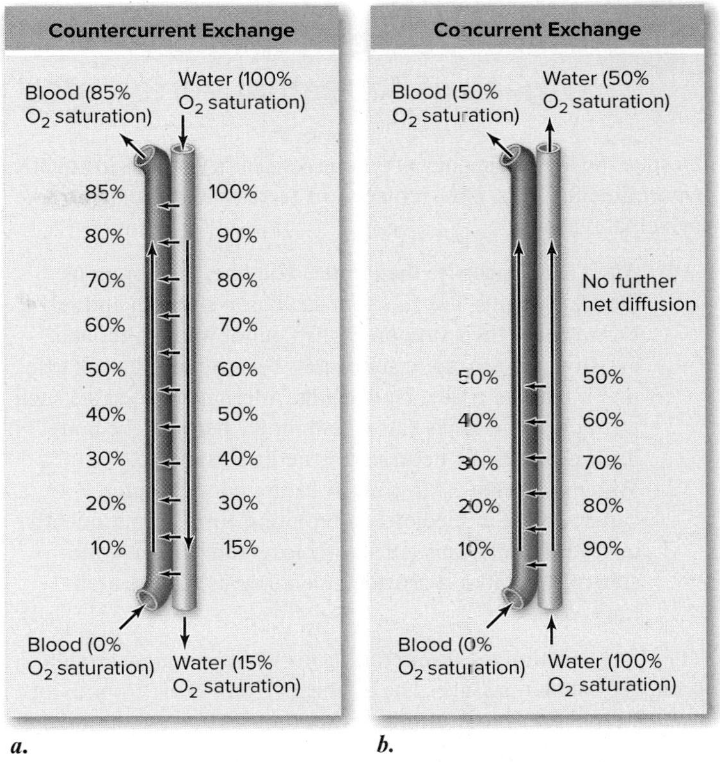

Figure 34.11 Countercurrent exchange. This process allows for the most efficient blood oxygenation. When blood and water flow in opposite directions **(*a*)**, the initial oxygen (O_2) concentration difference between water and blood is small but is sufficient for O_2 to diffuse from water to blood. As more O_2 diffuses into the blood, raising the blood's O_2 concentration, the blood encounters water with ever higher O_2 concentrations. At every point, the O_2 concentration is higher in the water, so that diffusion continues. In this example, blood attains an O_2 concentration of 85%. When blood and water flow in the same direction **(*b*)**, O_2 can diffuse from the water into the blood rapidly at first, but the diffusion rate slows as more O_2 diffuses from the water into the blood, until finally the concentrations of O_2 in water and blood are equal. In this example, blood's O_2 concentration cannot exceed 50%.

Within each lamella, blood flows opposite the direction of water movement. This kind of arrangement is called **countercurrent flow,** and it acts to maximize the oxygenation of the blood by maintaining a positive oxygen gradient along the entire pathway for diffusion, increasing Δp in Fick's Law of Diffusion. The advantages of a countercurrent flow system are illustrated in figure 34.11*a*. Countercurrent flow ensures that an oxygen concentration gradient remains between blood and water throughout the length of the gill lamellae. This permits oxygen to continue to diffuse all along the lamellae, so that the blood leaving the gills has nearly as high an oxygen concentration as the water entering the gills.

If blood and water flowed in the same direction, the flow would be *concurrent* (figure 34.11*b*). In this case, the concentration difference across the gill lamellae would fall rapidly as the water lost oxygen to the blood, and net diffusion of oxygen would cease when the levels of oxygen became the same in the water and in the blood.

Because of the countercurrent exchange of gases, fish gills are the most efficient of all respiratory organs.

REVIEW OF CONCEPT 34.5

Gills are highly subdivided structures, providing a large surface area for exchange. In countercurrent flow, blood in the gills flows opposite the direction of water to maintain a gradient difference and maximize gas exchange. Some amphibians rely on cutaneous respiration.

- *What are the anatomical requirements for a countercurrent flow system?*

34.6 Lungs Are the Respiratory Organs of Terrestrial Vertebrates

Despite the high efficiency of gills as respiratory organs in aquatic environments, gills were replaced in terrestrial animals for two principal reasons:

1. **Air is less supportive than water.** The fine, membranous lamellae of gills lack inherent structural strength and rely on water for their support. A fish out of water, although awash in oxygen, soon suffocates, because its gills collapse into a mass of tissue. Unlike gills, internal air passages such as trachea and lungs can remain open, because the body itself provides the necessary structural support.
2. **Water evaporates.** Air is rarely saturated with water vapor, except immediately after a rainstorm. Consequently, terrestrial organisms constantly lose water to the atmosphere. Gills would provide an enormous surface area for water loss.

The lung minimizes evaporation by moving air through a branched tubular passage. The tracheal system of arthropods also uses internal tubes to minimize evaporation.

The air drawn into the respiratory passages becomes saturated with water vapor before reaching the inner regions of the lung. In these areas, a thin, wet membrane permits gas exchange. Unlike the one-way flow of water that is so effective in the respiratory function of gills, gases move into and out of lungs by way of the same airway passages, a two-way flow system.

Breathing of Air Takes Advantage of Partial Pressures of Gases

LEARNING OBJECTIVE 34.6.1 Compare the breathing mechanisms of (1) amphibians and reptiles, (2) mammals, and (3) birds.

Dry air contains 78.09% nitrogen, 20.95% oxygen, 0.93% argon and other inert gases, and 0.03% carbon dioxide. Convection currents cause the atmosphere to maintain a constant composition to altitudes of at least 100 km, although the *amount* (number of molecules) of air that is present decreases as altitude increases.

Because of the force of gravity, air exerts a pressure downward on objects below it that can be measured by an apparatus called a barometer. The barometric pressure of air at sea level is 760 mm Hg, which is also defined as one atmosphere (1.0 atm) of pressure.

Each type of gas contributes to the total atmospheric pressure according to its fraction of the total molecules present. The pressure contributed by a gas is called its **partial pressure,** indicated by P_{N_2}, P_{O_2}, P_{CO_2}. These partial pressures can be thought of as the concentrations of the gases that make up air. At sea level, the partial pressures of N_2, O_2, and CO_2 are as follows:

$$P_{N_2} = 760 \times 0.7809 = 600.6\,\text{mm Hg}$$

$$P_{O_2} = 760 \times 0.2095 = 159.2\,\text{mm Hg}$$

$$P_{CO_2} = 760 \times 0.0003 = 0.2\,\text{mm Hg}$$

Humans cannot survive for long at altitudes above 6000 m. Although the air at these altitudes has the same percentage of oxygen, the atmospheric pressure is only about 380 mm Hg. This means the concentration of oxygen is only 80 mm Hg (380 × 0.2095), which is half the concentration of oxygen at sea level.

In this section, we describe the mechanisms and evolution of respiration in terrestrial vertebrates. We will begin with reptiles and amphibians, then summarize mammalian lungs and the highly specialized lungs of birds.

Amphibians and reptiles breathe in different ways

The lungs of amphibians are formed as saclike outpouchings of the gut. Although the internal surface area of these sacs is increased by folds, much less surface area is available for gas exchange in amphibian lungs than in the lungs of other terrestrial vertebrates. Each amphibian lung is connected to the rear of the oral cavity, or pharynx, and the opening to each lung is controlled by a valve, the glottis.

Amphibians do not breathe the same way as other terrestrial vertebrates. Amphibians force air into their lungs; they fill their oral cavity with air, close their mouth and nostrils, and then elevate the floor of their oral cavity. This pushes air into their lungs in the same way that a pressurized tank of air is used to fill balloons. This is called **positive pressure breathing;** in humans, it would be analogous to forcing air into a person's lungs by performing mouth-to-mouth resuscitation.

Most reptiles breathe in a different way, by expanding their rib cages by muscular contraction. This action creates a lower pressure inside the lungs than in the atmosphere, and the greater atmospheric pressure moves air into the lungs. This type of ventilation is termed **negative pressure breathing** because of the air being "pulled in" by the animal, like sucking water through a straw, rather than being "pushed in."

Mammalian lungs have greatly increased surface area

Endothermic animals, such as birds and mammals, have consistently higher metabolic rates and thus require more oxygen. Both these vertebrate groups exhibit more complex and efficient respiratory systems than ectothermic animals. The evolution of more efficient respiratory systems accommodates the increased demands on cellular respiration of endothermy.

The lungs of mammals are packed with millions of **alveoli,** tiny sacs clustered like grapes (figure 34.12). This provides each lung with an enormous surface area for gas exchange. Each alveolus is composed of an epithelium only one cell thick and is surrounded by blood capillaries with walls that are also only one cell layer thick. Thus, the distance *d* across which gas must diffuse is very small—only 0.5 to 1.5 μm.

Inhaled air is taken in through the mouth and nose past the pharynx to the larynx (voice box), where it passes through an opening in the vocal cords, the *glottis,* into a tube supported by C-shaped rings of cartilage, the trachea (windpipe). The term *trachea* is used for both the vertebrate windpipe and the respiratory tubes of arthropods, although the structures are obviously not homologous. The mammalian trachea bifurcates into right and left bronchi

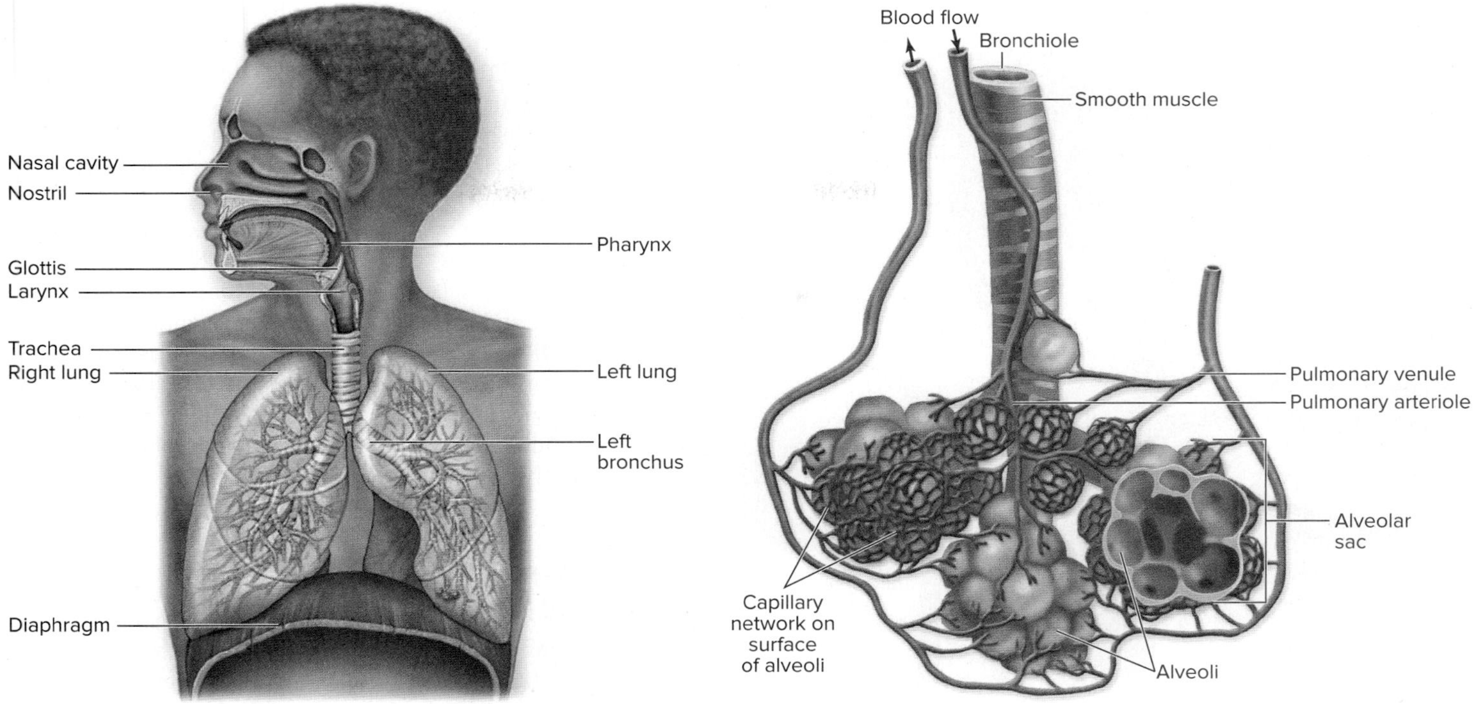

Figure 34.12 The human respiratory system and the structure of the mammalian lung. The lungs of mammals have an enormous surface area because of the millions of alveoli that cluster at the ends of the bronchioles. This provides for efficient gas exchange with the blood.

(singular, *bronchus*), which enter each lung and further subdivide into bronchioles that deliver the air into the alveoli.

The alveoli are surrounded by an extensive capillary network. All gas exchange between the air and blood takes place across the walls of the alveoli. The branching of bronchioles and the vast number of alveoli combine to increase the respiratory surface area far above that of amphibians or reptiles. In humans, each lung has about 300 million alveoli, and the total surface area available for diffusion can be as much as 80 m^2, or about 42 times the surface area of the body. Details of gas exchange at the alveolar interface with blood capillaries are described in the rest of this section.

The respiratory system of birds is a highly efficient flow-through system

The avian respiratory system is a unique structure that affords birds the most efficient respiration of all terrestrial vertebrates. Unlike the mammalian lung, which ends in blind alveoli, the bird lung channels air through tiny air vessels called parabronchi, where gas exchange occurs. Air flows through the parabronchi in one direction only, inhaled into a system of small air sacs unique to birds, then exhaled into and through the lungs. This unidirectional flow of air is similar to the unidirectional flow of water through a fish gill.

The unidirectional flow of air permits substantial respiratory efficiency: the flow of blood through the avian lung runs at a 90° angle to the air flow. This crosscurrent flow is not as efficient as the 180° countercurrent flow in fishes' gills, but it has a greater capacity to extract oxygen from the air than does a mammalian lung.

Because of these respiratory adaptations, a sparrow can be active at an altitude of 6000 m, whereas a mouse, which has a similar body mass and metabolic rate, would die at that altitude in a fairly short time from lack of oxygen.

The Mammalian Lung Has a Large Surface Area

LEARNING OBJECTIVE 34.6.2 Describe the efficiency of mammalian lungs using Fick's Law of Diffusion.

About 30 billion capillaries can be found in each lung, roughly 100 capillaries per alveolus. Thus, an alveolus can be visualized as a microscopic air bubble whose entire surface is bathed by blood. Gas exchange occurs very rapidly at this interface.

Blood returning from the systemic circulation, depleted in oxygen, has a partial oxygen pressure (Po_2) of about 40 mm Hg. By contrast, the Po_2 in the alveoli is about 105 mm Hg. The difference in pressures, namely the Δp of Fick's Law, is 65 mm Hg, leading to oxygen moving into the blood. The blood leaving the lungs, as a result of this gas exchange, normally contains a Po_2 of about 100 mm Hg. As you can see, the lungs do a very effective, but not perfect, job of oxygenating the blood. These changes in the Po_2 of the blood, as well as the changes in plasma carbon dioxide (indicated as the Pco_2), are shown in figure 34.13.

In humans and other mammals, the outside of each lung is covered by a thin membrane called the **visceral pleural membrane.** A second membrane, the **parietal pleural membrane,** lines the inner

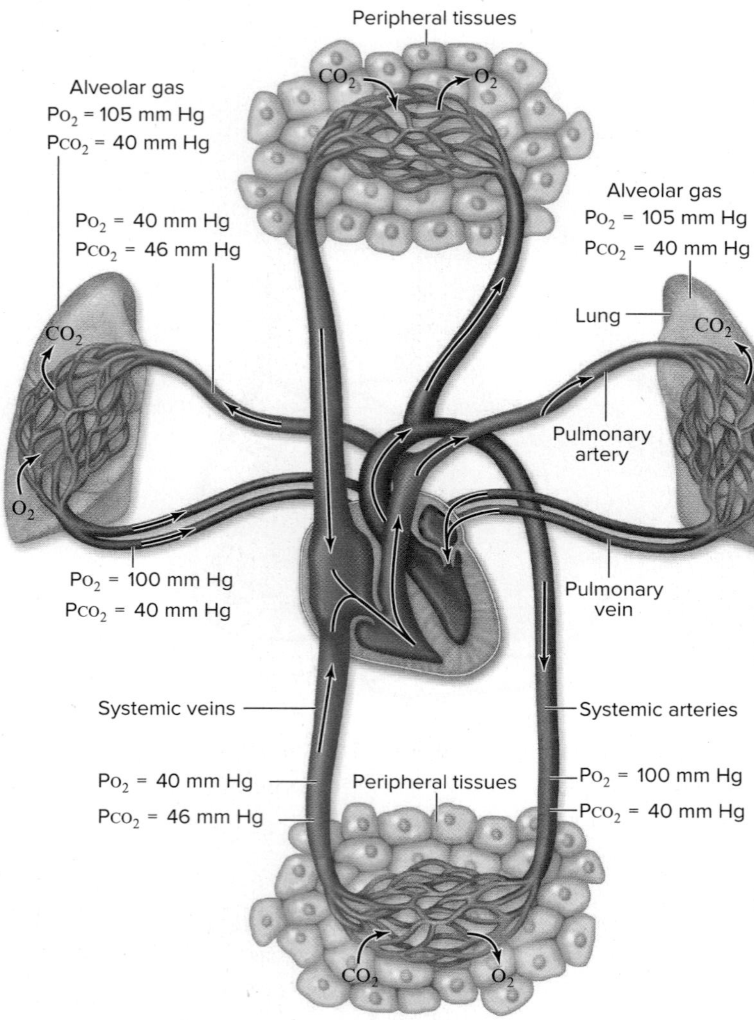
Peripheral tissues
CO_2
O_2
Alveolar gas
P_{O_2} = 105 mm Hg
P_{CO_2} = 40 mm Hg
P_{O_2} = 40 mm Hg
P_{CO_2} = 46 mm Hg
Alveolar gas
P_{O_2} = 105 mm Hg
P_{CO_2} = 40 mm Hg
Lung
CO_2
CO_2
O_2
O_2
Pulmonary artery
P_{O_2} = 100 mm Hg
P_{CO_2} = 40 mm Hg
Pulmonary vein
Systemic veins
Systemic arteries
P_{O_2} = 40 mm Hg
P_{CO_2} = 46 mm Hg
Peripheral tissues
P_{O_2} = 100 mm Hg
P_{CO_2} = 40 mm Hg
CO_2
O_2

Figure 34.13 Gas exchange in the blood capillaries of the lungs and systemic circulation. As a result of gas exchange in the lungs, the systemic arteries carry oxygenated blood with a relatively low carbon dioxide (CO_2) concentration. After the oxygen (O_2) is unloaded to the tissues, the blood in the systemic veins has a lowered O_2 content and an increased CO_2 concentration.

wall of the thoracic cavity. The space between these two membrane sheets, the **pleural cavity,** is normally very small and filled with fluid. This fluid causes the two membranes to adhere, effectively coupling the lungs to the thoracic cavity. The pleural membranes package each lung separately—if one lung collapses due to a perforation of the membranes, the other lung can still function.

The Diaphragm Expands and Contracts Lung Volume in the Respiratory Cycle

LEARNING OBJECTIVE 34.6.3 Describe how contraction of the diaphragm powers inhalation.

During inhalation, the thoracic volume is increased through contraction of two sets of muscles: the *external intercostal muscles* and the *diaphragm*. Contraction of the external intercostal muscles between the ribs raises the ribs and expands the rib cage. Contraction of the **diaphragm,** a convex sheet of striated muscle separating the thoracic cavity from the abdominal cavity, causes the diaphragm to lower and assume a more flattened shape. This expands the volume of the thorax and lungs, bringing about negative pressure ventilation, while it increases the pressure on the abdominal organs (figure 34.14*a*).

The thorax and lungs have a degree of elasticity; expansion during inhalation places these structures under elastic tension. The relaxation of the external intercostal muscles and diaphragm produces unforced exhalation because the elastic tension is released, allowing the thorax and lungs to recoil. You can produce a greater exhalation force by actively contracting your abdominal muscles—such as when blowing up a balloon (figure 34.14*b*).

Ventilation efficiency depends on lung capacity and breathing rate

A variety of terms are used to describe the volume changes of the lung during breathing. The amount of air inspired by a person at rest, about 500 ml on average, is called the **tidal volume (TV).** The additional amount of air that can be forcefully inspired, or expired, is called the **inspiratory reserve volume (IRV),** and the **expiratory reserve volume (ERV),** respectively. These measures are actually much larger than the TV, with the IRV being around 3,000 ml and the ERV about 1,200 ml. The amount of air that remains after the most vigorous expiration is called the **residual volume (RV)**. This averages around 1,300 ml and allows gas exchange to occur between breaths.

We can evaluate the health of an individual's respiratory system using a measure called the **vital capacity (VC),** which is defined as VC=(ERV+TV+IRV). Vital capacity averages 4.6 L in young males, and 3.1 L in young females, and an unusually low VC may indicate damage to the alveoli in various pulmonary disorders.

The rate and depth of breathing normally keep the blood P_{O_2} and P_{CO_2} within a normal range. If breathing is insufficient to maintain normal blood gas measurements (a rise in the blood P_{CO_2} is the best indicator), the person is hypoventilating. If breathing is excessive, so that the blood P_{CO_2} is abnormally lowered, the person is said to be **hyperventilating.**

The Central Nervous System Regulates Breathing

LEARNING OBJECTIVE 34.6.4 Describe how the central nervous system regulates breathing.

Each breath is initiated by neurons in a *respiratory control center* located in the medulla oblongata. These neurons stimulate the diaphragm and external intercostal muscles to contract, causing inhalation. When these neurons stop producing impulses, the inspiratory muscles relax and exhalation occurs. Although the muscles of breathing are skeletal muscles, they are usually controlled automatically. This control can be voluntarily overridden, however, as in hypoventilation (breath holding) or hyperventilation.

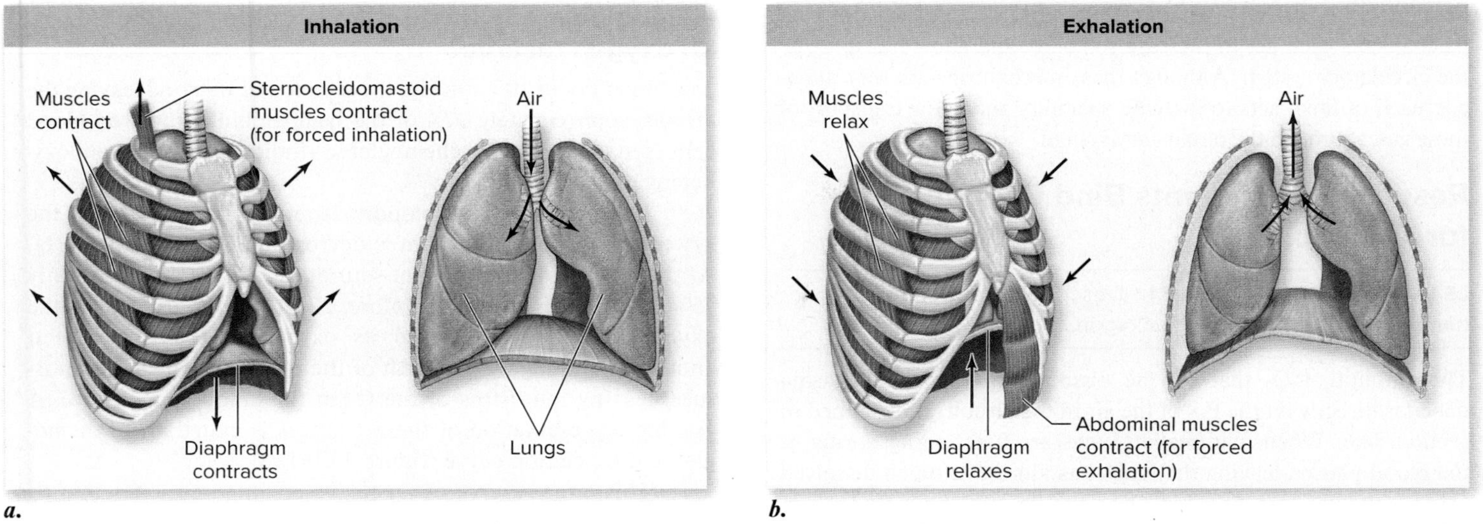

Figure 34.14 How a human breathes. ***a.*** Inhalation. The diaphragm contracts and the walls of the chest cavity expand, increasing the volume of the chest cavity and lungs. As a result of the larger volume, air is drawn into the lungs. ***b.*** Exhalation. The diaphragm and chest walls return to their normal positions as a result of elastic recoil, reducing the volume of the chest cavity and forcing air out of the lungs through the trachea. Note that inhalation can be forced by contracting accessory respiratory muscles (such as the sternocleidomastoid), and exhalation can be forced by contracting abdominal muscles.

Neurons of the medulla oblongata must be responsive to changes in blood Po_2 and Pco_2 in order to maintain homeostasis. You can demonstrate this mechanism by simply holding your breath. Your blood carbon dioxide level immediately rises, and your blood oxygen level falls. After a short time, the urge to breathe induced by the changes in blood gases becomes overpowering. The rise in blood carbon dioxide, as indicated by a rise in Pco_2, is the primary initiator, rather than the fall in oxygen levels.

A rise in Pco_2 causes an increased production of carbonic acid (H_2CO_3), which lowers the blood pH. A fall in blood pH stimulates chemosensitive neurons in the **aortic** and **carotid bodies,** in the aorta and the carotid artery. These peripheral receptors send impulses to the respiratory control center, which then stimulates increased breathing. The brain also contains central chemoreceptors that are stimulated by a drop in the pH of cerebrospinal fluid (CSF).

A person cannot voluntarily hyperventilate for too long. The decrease in plasma Pco_2 and increase in pH of plasma and CSF caused by hyperventilation suppress the reflex drive to breathe. Deliberate hyperventilation allows people to hold their breath longer—not because it increases oxygen in the blood but because the carbon dioxide level is lowered and takes longer to build back up, postponing the need to breathe.

Po_2 is usually not a significant stimulus for increased breathing rates, the exception being high altitudes where the Po_2 of the atmosphere is low. The symptoms of low oxygen at high altitude are known as mountain sickness, which may include feelings of weakness, headache, nausea, vomiting, and reduced mental function. All of these symptoms are related to the low Po_2, and breathing supplemental oxygen may remove all symptoms.

REVIEW OF CONCEPT 34.6

Lungs provide a large surface area for gas exchange while minimizing evaporation. Amphibians push air into their lungs; most reptiles and all birds and mammals pull air into their lungs by expanding the thoracic cavity. Birds have efficient, one-way air flow with crosscurrent blood flow through the lungs. Humans move an average tidal volume of 500 mL of air into and out of the lungs. The maximum amount that can be inspired and expired, the vital capacity, is indicative of respiratory health. Ventilation is regulated by CNS neurons that detect CO_2 concentration.

- *What selection pressure would bring about the evolution of birds' highly efficient lungs?*

34.7 Oxygen and Carbon Dioxide Are Transported by Fundamentally Different Mechanisms

The respiratory and circulatory systems of terrestrial organisms are coordinated to transport two gases, O_2 and CO_2, that have low solubility in water. Evolution has solved this problem in unique ways for each gas. The solubility of O_2 is increased by its binding to a protein that acts as a carrier in the blood. In many invertebrates, this protein is a hemocyanin, and in vertebrates the primary oxygen carrier is hemoglobin. The solubility of CO_2, on the other hand, is increased by a chemical reaction that converts the relatively insoluble gas into an ion. CO_2 reacts with H_2O to form H_2CO_3 (carbonic acid), which dissociates into HCO_3^- (bicarbonate) and H^+. These very different mechanisms each act to increase solubility and allow transport by the circulatory system. Although these mechanisms are very different, each of them acts to increase solubility and allow transport of these gases through the circulatory system.

Respiratory Pigments Bind Oxygen for Transport

LEARNING OBJECTIVE 34.7.1 Describe the structure of hemoglobin and how it binds oxygen.

The amount of O_2 that can be dissolved in the blood plasma depends directly on the Po_2 of the air in the alveoli, as explained in section 34.6. When mammalian lungs are functioning normally, the blood plasma leaving the lungs has almost as much dissolved O_2 as is theoretically possible, given the Po_2 of the air. Because of oxygen's low solubility, however, blood plasma can contain a maximum of only about 3 mL of O_2 per liter. But whole blood normally carries almost 200 mL of O_2 per liter. Most of the O_2 in the blood is bound to molecules of hemoglobin inside red blood cells.

Hemoglobin is a protein composed of four polypeptide chains and four organic compounds called *heme groups.* At the center of each heme group is an atom of iron, which can bind to O_2 (figure 34.15). Thus, each hemoglobin molecule can carry up to four molecules of O_2.

Hemoglobin loads up with O_2 in the alveolar capillaries of the pulmonary circulation, forming oxyhemoglobin. This molecule has a bright red color. As blood passes through capillaries in the systemic circulation, some of the oxyhemoglobin releases oxygen, becoming **deoxyhemoglobin.** Deoxyhemoglobin has a darker red color, but it imparts a bluish tinge to tissues. Illustrations of the cardiovascular system show vessels carrying oxygenated blood with a red color and vessels that carry oxygen-depleted blood with a blue color.

Hemoglobin is an ancient protein; not only is it the oxygen-carrying molecule in all vertebrates, but it also is used as an oxygen carrier by many invertebrates, including annelids, mollusks, echinoderms, flatworms, and even some protists. Many other invertebrates, however, employ different oxygen carriers, such as **hemocyanin.** In hemocyanin, the oxygen-binding atom is copper instead of iron. Hemocyanin is not found associated with blood cells but is instead one of the free proteins in the circulating fluid (hemolymph) of arthropods and some mollusks.

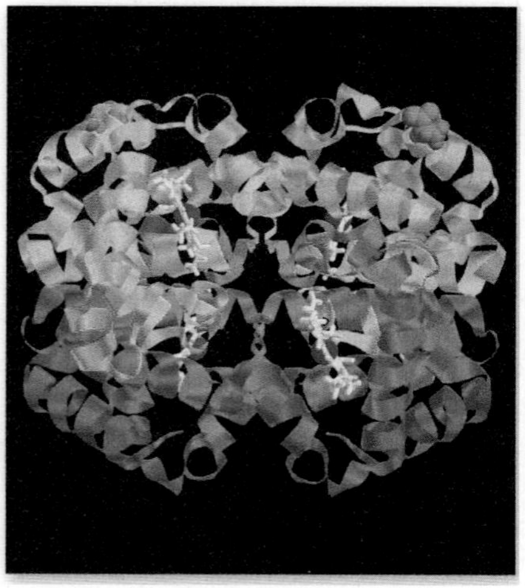

Figure 34.15 The structure of the adult hemoglobin protein. Hemoglobin consists of two α- and two β-polypeptide chains. Each of these has a heme group (in white) with a central iron atom (red ball), which can bind to a molecule of O_2.

Kenneth Eward/BioGrafx/Science Source

Hemoglobin and myoglobin provide an oxygen reserve

At a blood Po_2 of 100 mm Hg, the level found in blood leaving the alveoli, approximately 97% of the hemoglobin within red blood cells is in the form of oxyhemoglobin—indicated as a percent oxyhemoglobin saturation of 97%.

In a person at rest, blood that returns to the heart in the systemic veins has a Po_2 that is decreased to about 40 mm Hg. At this lower Po_2, the percent saturation of hemoglobin is only 75%. In a person at rest, therefore, 22% (97% minus 75%) of the oxyhemoglobin has released its oxygen to the tissues. Put another way, roughly one-fifth of the O_2 is unloaded in the tissues, leaving four-fifths of the O_2 in the blood as a reserve. A graphic representation of these changes is called an oxyhemoglobin dissociation curve (figure 34.16).

This large reserve of O_2 serves an important function. It enables the blood to supply the body's oxygen needs during exertion as well as at rest. During exercise, for example, the muscles' accelerated metabolism uses more O_2 and decreases the venous blood Po_2. The Po_2 of the venous blood could drop to 20 mm Hg; in this case, the percent saturation of hemoglobin would be only 35%. Because arterial blood would still contain 97% oxyhemoglobin, the amount of oxygen unloaded would now be 62% (97% minus 35%), instead of the 22% at rest.

In addition to this function, the O_2 reserve also ensures that the blood contains enough oxygen to maintain life for 4 to 5 min if breathing is interrupted or if the heart stops pumping.

Hemoglobin's Affinity for Oxygen Is Affected by pH and Temperature

LEARNING OBJECTIVE 34.7.2 Describe how hemoglobin's oxygen affinity changes depending on environmental conditions.

Oxygen transport in the blood is affected by other conditions, including temperature and pH. The CO_2 produced by metabolizing tissues combines with H_2O to form carbonic acid (H_2CO_3), which lowers blood pH. This reaction occurs primarily inside red blood cells, where the lowered pH reduces hemoglobin's affinity for oxygen, causing it to release oxygen more readily.

The effect of pH on hemoglobin's affinity for oxygen, known as the **Bohr effect,** or **Bohr shift,** is the result of H^+ binding to hemoglobin. It is shown graphically by a shift of the oxyhemoglobin dissociation curve to the right (figure 34.17*a*).

Increasing temperature has a similar effect on hemoglobin's affinity for oxygen (figure 34.17*b*). Because skeletal muscles produce carbon dioxide more rapidly during exercise, and because active muscles produce heat, during exercise the blood unloads a higher percentage of the oxygen it carries.

Carbon Dioxide Is Primarily Transported in Blood as Bicarbonate Ion

LEARNING OBJECTIVE 34.7.3 Explain how carbon dioxide is transported by the blood.

About 8% of the CO_2 in blood is simply dissolved in plasma; another 20% is bound to hemoglobin. Because CO_2 binds to the protein portion of hemoglobin, and not to the iron atoms

Figure 34.16 The oxyhemoglobin dissociation curve. Hemoglobin combines with O_2 in the lungs, and this oxygenated blood is carried by arteries to the body cells. After O_2 is removed from the blood to support cellular respiration, the blood entering the veins contains less O_2.

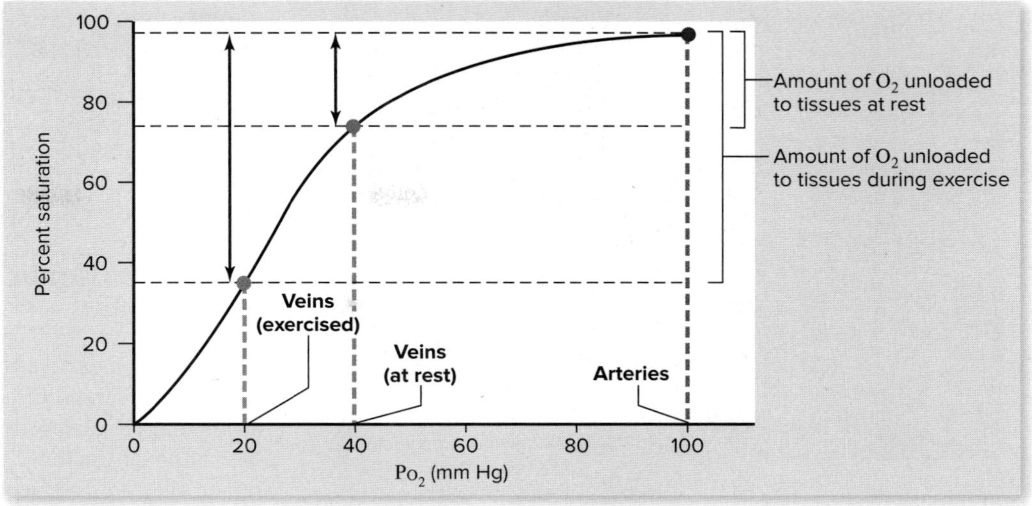

of the heme groups, it does not compete with O_2; however, it does cause hemoglobin's shape to change, lowering its affinity for O_2.

The remaining 72% of the CO_2 diffuses into the red blood cells, where the enzyme carbonic anhydrase catalyzes the reaction of CO_2 with water to form H_2CO_3. H_2CO_3 dissociates into HCO_3^- and H^+ ions. The H^+ binds to deoxyhemoglobin, reducing its affinity for oxygen and releasing oxygen into the tissues. The HCO_3^- moves out of the erythrocyte into the plasma via a transporter that exchanges one Cl^- for a HCO_3^- (this is called the "chloride shift").

This reaction removes large amounts of CO_2 from the plasma, maintaining a diffusion gradient that allows additional CO_2 to move into the plasma from the surrounding tissues (figure 34.18*a*). The formation of H_2CO_3 is also important in maintaining the acid–base balance of the blood; HCO_3^- serves as the major buffer of the blood plasma.

In the lungs, the lower P_{CO_2} of the gas mixture inside the alveoli causes the carbonic anhydrase reaction to proceed in the reverse direction, converting H_2CO_3 into H_2O and CO_2 (figure 34.18*b*). The CO_2 diffuses out of the red blood cells and into the alveoli, so that it can leave the body in the next exhalation.

Other dissolved gases are also transported by hemoglobin, most notably nitric oxide (NO), which plays an important role in vessel dilation. Carbon monoxide (CO) binds more strongly to hemoglobin than does oxygen, which is why carbon monoxide poisoning can be deadly. Victims of carbon monoxide poisoning often have bright red skin due to hemoglobin's binding with CO.

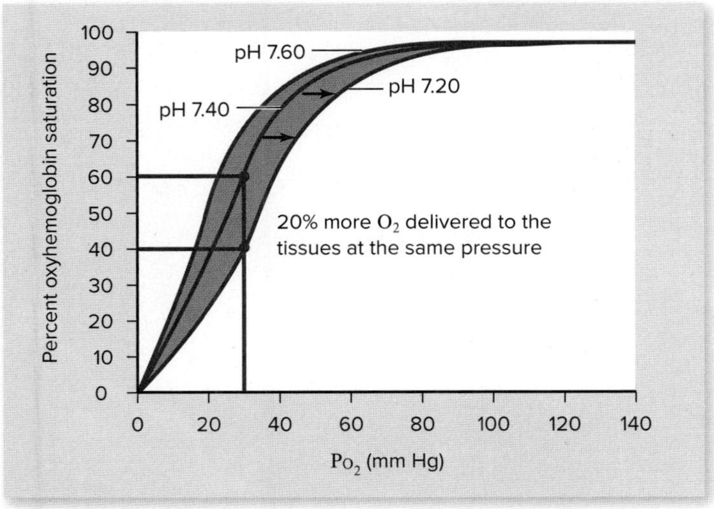

a. pH shift

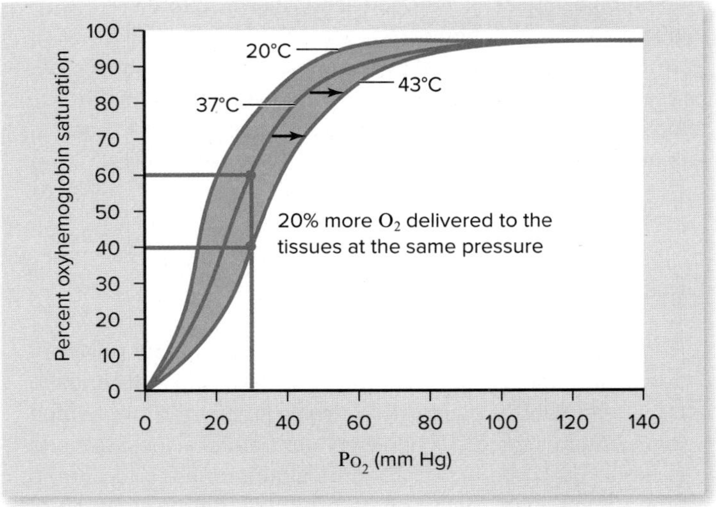

b. Temperature shift

Figure 34.17 The effect of pH and temperature on the oxyhemoglobin dissociation curve. Lower blood pH **(*a*)** and higher blood temperatures **(*b*)** shift the oxyhemoglobin dissociation curve to the right, facilitating O_2 unloading. In this example, this can be seen as a lowering of the oxyhemoglobin percent saturation from 60% to 40%, indicating that the difference of 20% more O_2 is unloaded to the tissues.

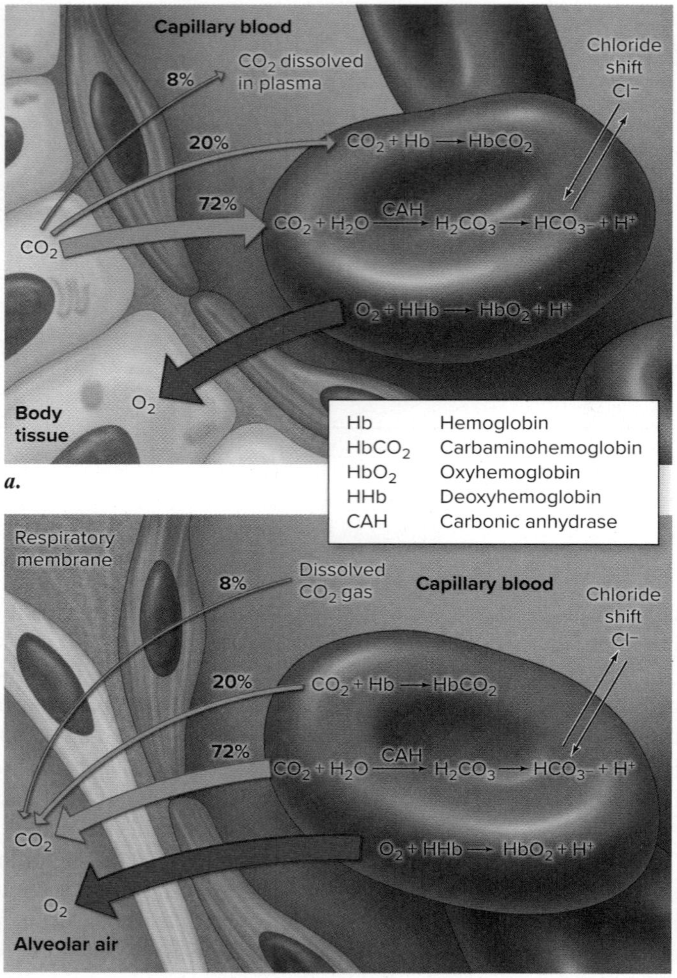

b.

Figure 34.18 The transport of carbon dioxide by the blood. ***a.*** Passage into bloodstream. CO_2 is transported in three ways: dissolved in plasma, bound to the protein portion of hemoglobin, and as bicarbonate (HCO_3^-), which forms in red blood cells. The reaction of CO_2 with H_2O to form H_2CO_3 (carbonic acid) is catalyzed by the enzyme carbonic anhydrase in red blood cells. ***b.*** Removal from bloodstream. When the blood passes through the pulmonary capillaries, these reactions are reversed so that CO_2 gas is formed, which is exhaled.

REVIEW OF CONCEPT 34.7

Hemoglobin circulating in the blood consists of four polypeptide chains, each associated with a heme group that can bind O_2. Hemoglobin's affinity for oxygen is affected by pH and temperature; more O_2 is released into tissues at lower pH and at higher temperature. Carbon dioxide is transported in the blood primarily as bicarbonate produced by the reaction of CO_2 and H_2O catalyzed by carbonic anhydrase in red blood cells.

- *What are the differences in the ways that oxygen and carbon dioxide are transported in blood?*

34.8 Circulating Blood Carries Metabolites and Gases to the Tissues

Blood Has Many Functions

LEARNING OBJECTIVE 34.8.1 List the principal functions of circulating blood.

Blood is a connective tissue composed of a fluid matrix, called **plasma,** and several different kinds of cells and other **formed elements** that circulate within that fluid (figure 34.19). Blood **platelets,** although included in figure 34.20, are not complete cells; rather, they are fragments of cells that are produced in the bone marrow. (We describe the action of platelets in blood clotting later in this section.)

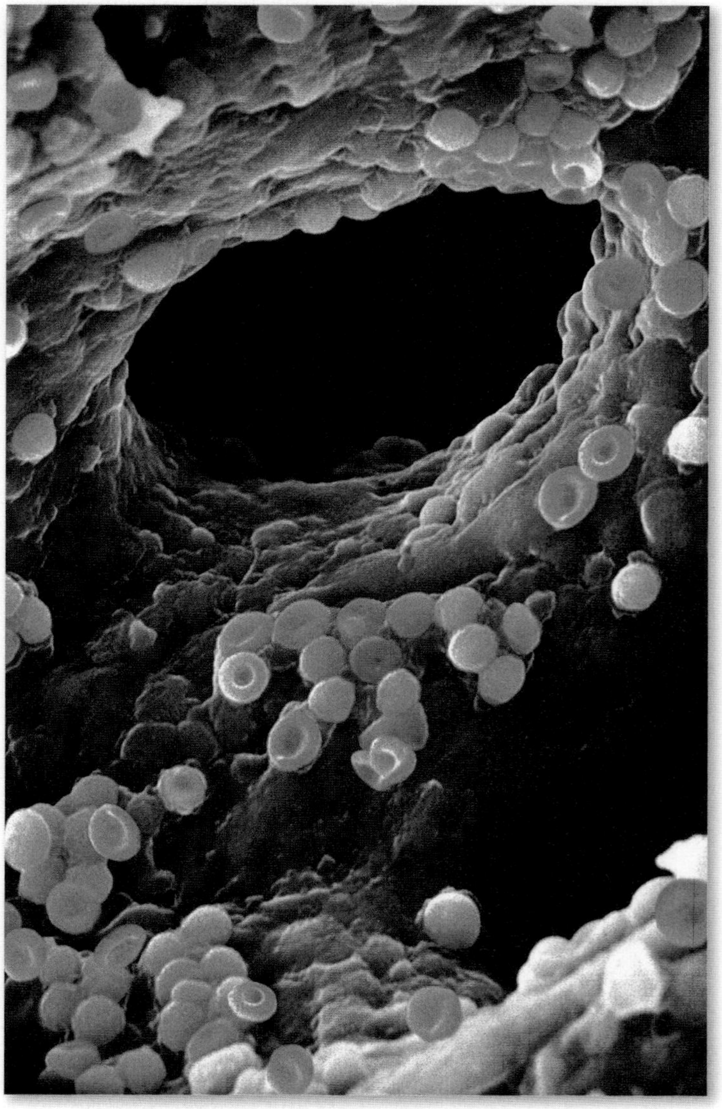

Figure 34.19 Circulating red blood cells moving through a blood vessel.

Biophoto Associates/Science Source

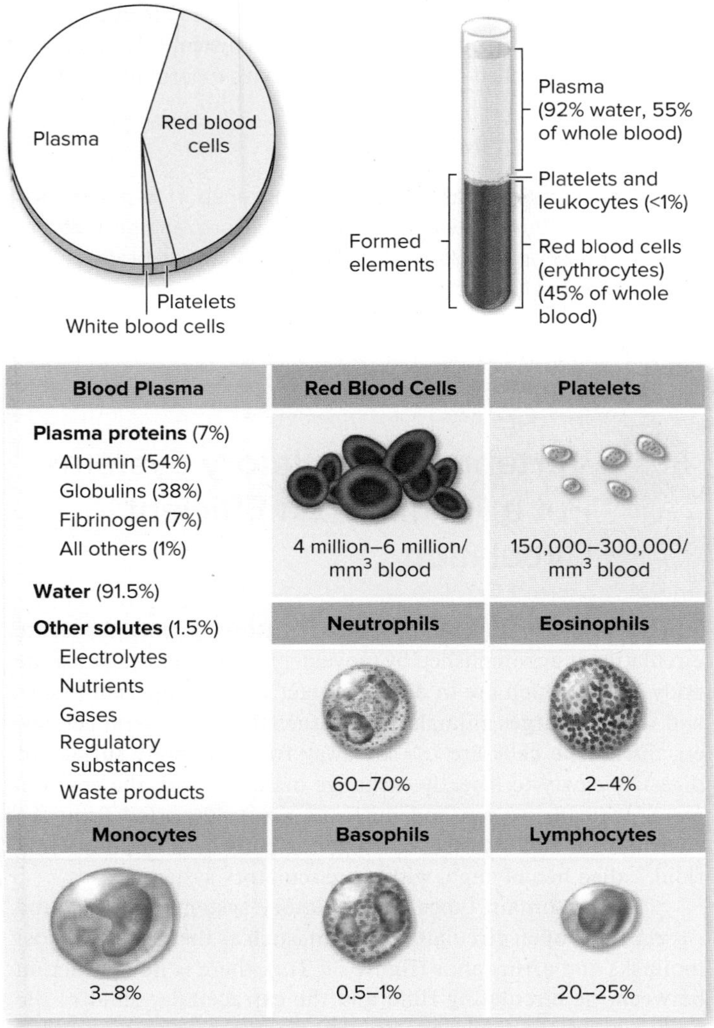
Plasma
Red blood cells
Platelets
White blood cells
Plasma (92% water, 55% of whole blood)
Platelets and leukocytes (<1%)
Formed elements
Red blood cells (erythrocytes) (45% of whole blood)
Blood Plasma
Plasma proteins (7%)
Albumin (54%)
Globulins (38%)
Fibrinogen (7%)
All others (1%)
Water (91.5%)
Other solutes (1.5%)
Electrolytes
Nutrients
Gases
Regulatory substances
Waste products
Red Blood Cells
4 million–6 million/ mm³ blood
Platelets
150,000–300,000/ mm³ blood
Neutrophils
60–70%
Eosinophils
2–4%
Monocytes
3–8%
Basophils
0.5–1%
Lymphocytes
20–25%

Figure 34.20 Composition of blood.

Circulating blood has many functions:

Transportation. All of the substances essential for cellular metabolism are transported by blood. Red blood cells transport oxygen attached to hemoglobin; nutrient molecules are carried in the plasma, sometimes bound to carriers; and metabolic wastes are eliminated as blood passes through the liver and kidneys.

Regulation. The cardiovascular system transports regulatory hormones from the endocrine glands and participates in temperature regulation. Contraction and dilation of blood vessels near the surface of the body, beneath the epidermis, help to conserve or to dissipate heat as needed.

Protection. The circulatory system protects against injury and foreign microbes or toxins introduced into the body. Blood clotting helps to prevent blood loss when vessels are damaged. White blood cells, or leukocytes, help to disarm or disable invaders such as viruses and bacteria (refer to chapter 35).

Blood Plasma Is a Fluid Matrix

LEARNING OBJECTIVE 34.8.2 Describe the solutes present in blood.

Blood plasma is the matrix in which blood cells and platelets are suspended. Interstitial (extracellular) fluids originate from the fluid present in plasma.

Although plasma is 92% water, it also contains the following solutes:

Nutrients, wastes, and hormones. Dissolved within the plasma are all of the nutrients needed by cells, including glucose, amino acids, and vitamins. Also dissolved in the plasma are wastes such as nitrogen compounds and CO_2 produced by metabolizing cells. Endocrine hormones released from glands are also carried through the blood to their target cells.

Ions. Blood plasma is a dilute salt solution. The predominant plasma ions are Na^+, Cl^-, and bicarbonate ions (HCO_3^-). In addition, plasma contains trace amounts of other ions, such as Ca^{2+}, Mg^{2+}, Cu^{2+}, K^+, and Zn^{2+}.

Proteins. As mentioned in section 34.3, the liver produces most of the plasma proteins, including albumin, which constitutes most of the plasma protein; the alpha (α) and beta (β) globulins, which serve as carriers of lipids and steroid hormones; and fibrinogen, which is required for blood clotting. Blood plasma with the fibrinogen removed is called serum.

Formed Elements Include Circulating Cells and Platelets

LEARNING OBJECTIVE 34.8.3 Distinguish among the types of formed elements present in blood.

The formed elements of blood cells and cell fragments include red blood cells, white blood cells, and platelets. Each element has a specific function in maintaining the body's health and homeostasis.

Erythrocytes

Each microliter of blood contains about 5 million ***red blood cells,*** or **erythrocytes.** The fraction of the total blood volume that is occupied by erythrocytes is called the blood's *hematocrit*; in healthy humans, the hematocrit is typically around 45%.

Each erythrocyte resembles a doughnut-shaped disk with a central depression that does not go all the way through. Mature mammalian erythrocytes lack nuclei. The erythrocytes of vertebrates contain hemoglobin, a pigment that binds and transports oxygen. In vertebrates, hemoglobin is found only in erythrocytes. In invertebrates, the oxygen-binding pigment (not always hemoglobin) is also present in plasma.

Leukocytes

Less than 1% of the cells in human blood are ***white blood cells,*** or **leukocytes**; there are only 1 or 2 leukocytes for every 1000 erythrocytes. Leukocytes are larger than erythrocytes and have nuclei. Furthermore, leukocytes are not confined to the blood, as erythrocytes are, but can migrate out of capillaries through the intercellular spaces into the surrounding interstitial (tissue) fluid.

Leukocytes come in several varieties, each of which plays a specific role in defending against invading microorganisms and other foreign substances, as described in chapter 35. **Granular leukocytes** include neutrophils, eosinophils, and basophils, which are named according to the staining properties of granules in their cytoplasm. **Nongranular leukocytes** include monocytes and lymphocytes. In healthy humans, neutrophils are the most numerous of the leukocytes, followed in order by lymphocytes, monocytes, eosinophils, and basophils.

Platelets

Platelets are cell fragments that pinch off from larger cells in the bone marrow. They are approximately 3 μm in diameter, and following an injury to a blood vessel, the liver releases *prothrombin* into the blood. In the presence of this clotting factor, fibrinogen is converted into insoluble threads of **fibrin.** Fibrin then aggregates to form the clot.

Formed Elements Arise from Stem Cells

LEARNING OBJECTIVE 34.8.4 Describe the origin of formed elements in circulating blood.

Each of the formed elements of blood has a finite life span and therefore must be constantly replaced. Many of the old cell fragments are digested by phagocytic cells of the spleen; however, many products from the old cells, such as iron and amino acids, are recycled and incorporated into new formed elements. The creation of new formed elements begins in the bone marrow.

All of the formed elements develop from **pluripotent stem cells.** The production of blood cells occurs in the bone marrow and is called **hematopoiesis.** This process generates two types of stem cells with a more restricted fate: a lymphoid stem cell that gives rise to lymphocytes and a myeloid stem cell that gives rise to the rest of the blood cells.

When the oxygen available in the blood decreases, the kidney converts a plasma protein into the hormone **erythropoietin.** Erythropoietin then stimulates the production of erythrocytes from the myeloid stem cells through a process called **erythropoiesis.**

In mammals, maturing erythrocytes lose their nuclei prior to release into circulation. In contrast, the mature erythrocytes of all other vertebrates remain nucleated. *Megakaryocytes* are examples of committed cells formed in bone marrow from stem cells. Pieces of cytoplasm are pinched off the megakaryocytes to form the platelets.

REVIEW OF CONCEPT 34.8

The circulatory system functions in the transport of materials, regulation of temperature and body processes, and protection of the body. Formed elements in blood include red blood cells, white blood cells, and platelets.

- *How does a blood platelet form?*

34.9 Vertebrate Circulatory Systems Put a Premium on Efficient Circulation

Smaller invertebrate animals lack circulatory systems. Adequate circulation is accomplished by movements of the body against the body fluids, which are in direct contact with the internal tissues and organs. Larger animals have tissues that are several cell layers thick. The cells are too far away from the body surface or digestive cavity to directly exchange materials with the environment. Instead, oxygen and nutrients are transported from the environment and digestive cavity to the body cells by an internal fluid, called hemolymph, within a circulatory system.

The two main types of circulatory systems are *open* and *closed.* In an open circulatory system, such as that found in most mollusks and arthropods (figure 34.21*a*), there is no distinction between the circulating fluid and the extracellular fluid of the body tissues. In a closed circulatory system, the circulating fluid, blood, is always enclosed within blood vessels that transport it away from and back to the heart (figure 34.21*b*). Closed circulatory systems are found in annelids, as well as chordates.

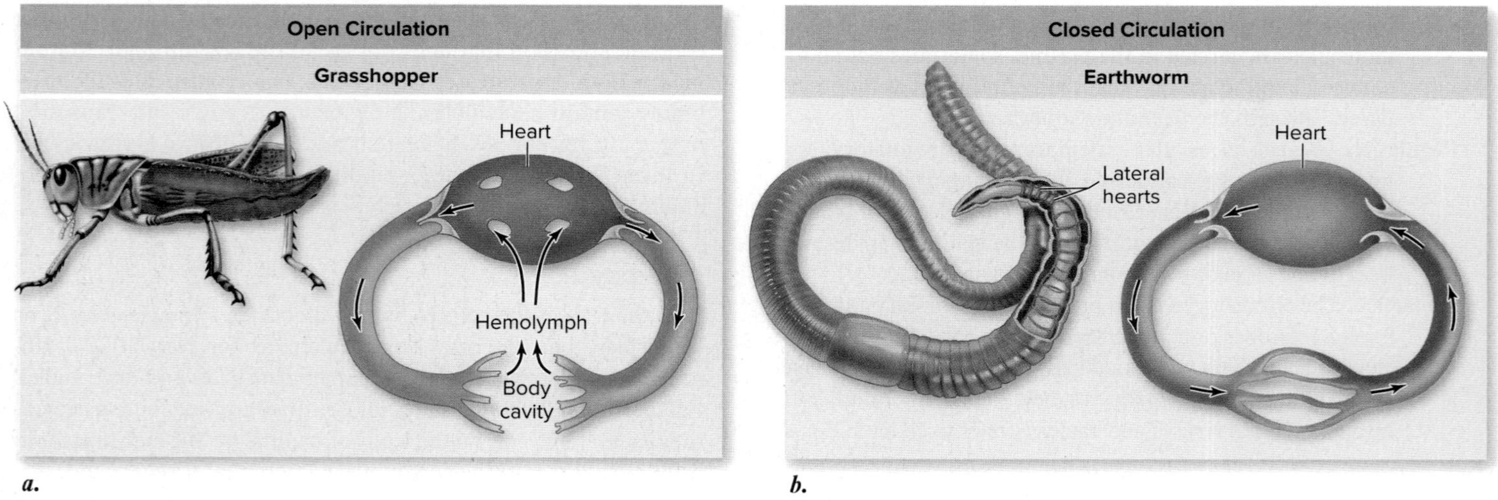

Figure 34.21 Open and closed circulatory systems ***a.*** In the insect's open circulatory system, hemolymph is pumped from a tubular heart through cavities in the body, returning to vessels for recirculation. ***b.*** In the closed circulation of the earthworm, blood from the heart remains within a system of vessels that returns it to the heart. All vertebrates also have a closed circulatory system.

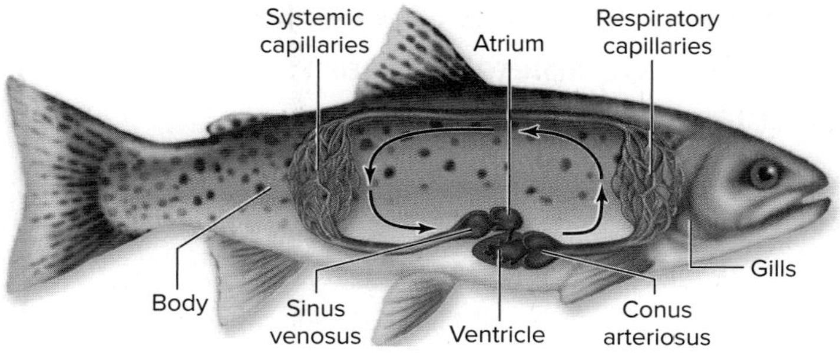

Figure 34.22 The heart and circulation of a fish. Diagram of a fish, showing the structures, in series with each other (sinus venosus, atrium, ventricles, conus arteriosus), that form two pumping chambers. Blood is pumped by the ventricle through the gills and then to the body. Blood rich in oxygen (oxygenated) is shown in red; blood low in oxygen (deoxygenated) is shown in blue.

In Fishes, a Heart Pushes Blood Through the Gills

LEARNING OBJECTIVE 34.9.1 Describe the structure of the fish heart.

Chordates ancestral to the vertebrates are thought to have had simple tubular hearts, similar to those now seen in the lancelets described in chapter 28. The heart was little more than a specialized zone of the ventral artery that was more heavily muscled than the rest of the arteries; it contracted in simple peristaltic waves.

The development of gills by fishes required a more efficient pump, and in fishes we see the evolution of a true chamber-pump heart. The fish heart is, in essence, a tube with four structures arrayed one after the other to form two pumping chambers (figure 34.22). The first two structures—the **sinus venosus** and the **atrium**—form the first chamber; the second two, the **ventricle** and the **conus arteriosus,** form the second chamber. The sinus venosus is the first to contract, followed by the atrium, the ventricle, and finally, the conus arteriosus.

Despite shifts in the relative positions of these structures, this heartbeat sequence is maintained in all vertebrates. In fish, the electrical impulse that produces the contraction is initiated in the sinus venosus; in other vertebrates, the electrical impulse is initiated by a structure homologous to the sinus venosus—the **sinoatrial (SA) node.**

After blood leaves the conus arteriosus, it moves through the gills, becoming oxygenated. Blood leaving the gills then flows through a network of arteries to the rest of the body, finally returning to the sinus venosus. This simple loop has one serious limitation: as blood passes through the capillaries in the gills, blood pressure drops significantly. This slows circulation from the gills to the rest of the body and can limit oxygen delivery to tissues.

Amphibians Evolved a Separate Pulmonary Circulation

LEARNING OBJECTIVE 34.9.2 Explain the importance of the evolution of the pulmonary vein.

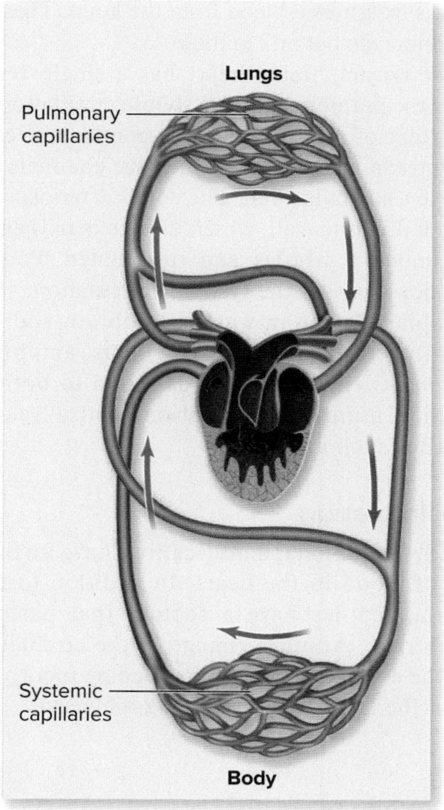

Figure 34.23 The heart and circulation of an amphibian. The frog has a three-chambered heart with two atria but only one ventricle, which pumps blood both to the lungs and to the body. Despite the potential for mixing, the oxygenated and deoxygenated bloods (red and blue lines, respectively) mix little as they are pumped to the body and lungs. Oxygenation of blood also occurs by gas exchange through the skin.

The advent of lungs in amphibians involved a major change in the pattern of circulation, a second pumping circuit (figure 34.23). After blood is pumped by the heart through the *pulmonary arteries* to the lungs, it loses much of its force as it passes through the narrow passages of the lung's capillaries. This now-sluggish circulation does not continue to the tissues of the body. Instead, in a remarkable innovation, the now-oxygenated blood is returned via the *pulmonary veins* to the heart for a second pumping before being circulated to the body's tissues. This system is termed **double circulation**: One system, the **pulmonary circulation,** moves blood between heart and lungs, and another, the **systemic circulation,** moves blood between the heart and the rest of the body.

Amphibian circulation

Optimally, oxygenated blood from lungs would go directly to tissues, rather than being mixed in the heart with deoxygenated blood returning from the body. The amphibian heart has two structural features that significantly reduce this mixing (figure 34.23). First, the atrium is divided into two chambers: the right atrium receives deoxygenated blood from the systemic circulation, and the left atrium receives oxygenated blood from the lungs. These two types of blood, therefore, do not mix in the atria.

Because an amphibian heart has a single ventricle, the separation of the pulmonary and systemic circulations is incomplete. The extent of mixing when the contents of each atrium enter the ventricle is reduced by internal channels created by recesses in the ventricular wall. The conus arteriosus is partially separated by a dividing wall, which directs deoxygenated blood into the pulmonary arteries and oxygenated blood into the *aorta,* the major artery of the systemic circulation.

Amphibians living in water can obtain additional oxygen by diffusion through their skin. Thus, amphibians have a *pulmocutaneous circuit* that sends blood to both the lungs and the skin. Cutaneous respiration is also seen in many aquatic reptiles such as turtles.

Reptilian circulation

Among reptiles, additional modifications have further reduced the mixing of blood in the heart. In addition to having two separate atria, reptiles have a septum that partially subdivides the ventricle. Another change in the circulation of reptiles is that the conus arteriosus has become incorporated into the trunks of the large arteries leaving the heart.

Mammals, Birds, and Crocodilians Have Two Completely Separated Circulatory Systems

LEARNING OBJECTIVE 34.9.3 Explain the importance of the complete septum in crocodilian, bird, and mammalian hearts.

Mammals, birds, and crocodilians have a four-chambered heart with two separate atria and two separate ventricles (figure 34.24). The hearts of birds and crocodiles exhibit some differences but overall are quite similar, which is not surprising, given their close evolutionary relationship. However, the extreme similarity between the hearts of birds and those of mammals—so alike that a single illustration can suffice for both—is a remarkable case of convergent evolution (figure 34.24).

In a four-chambered heart, the right atrium receives deoxygenated blood from the body and delivers it to the right ventricle, which pumps the blood to the lungs. The left atrium receives oxygenated blood from the lungs and delivers it to the left ventricle, which pumps the oxygenated blood to the rest of the body.

The heart in these vertebrates is a two-cycle pump. Both atria fill with blood and simultaneously contract, emptying their

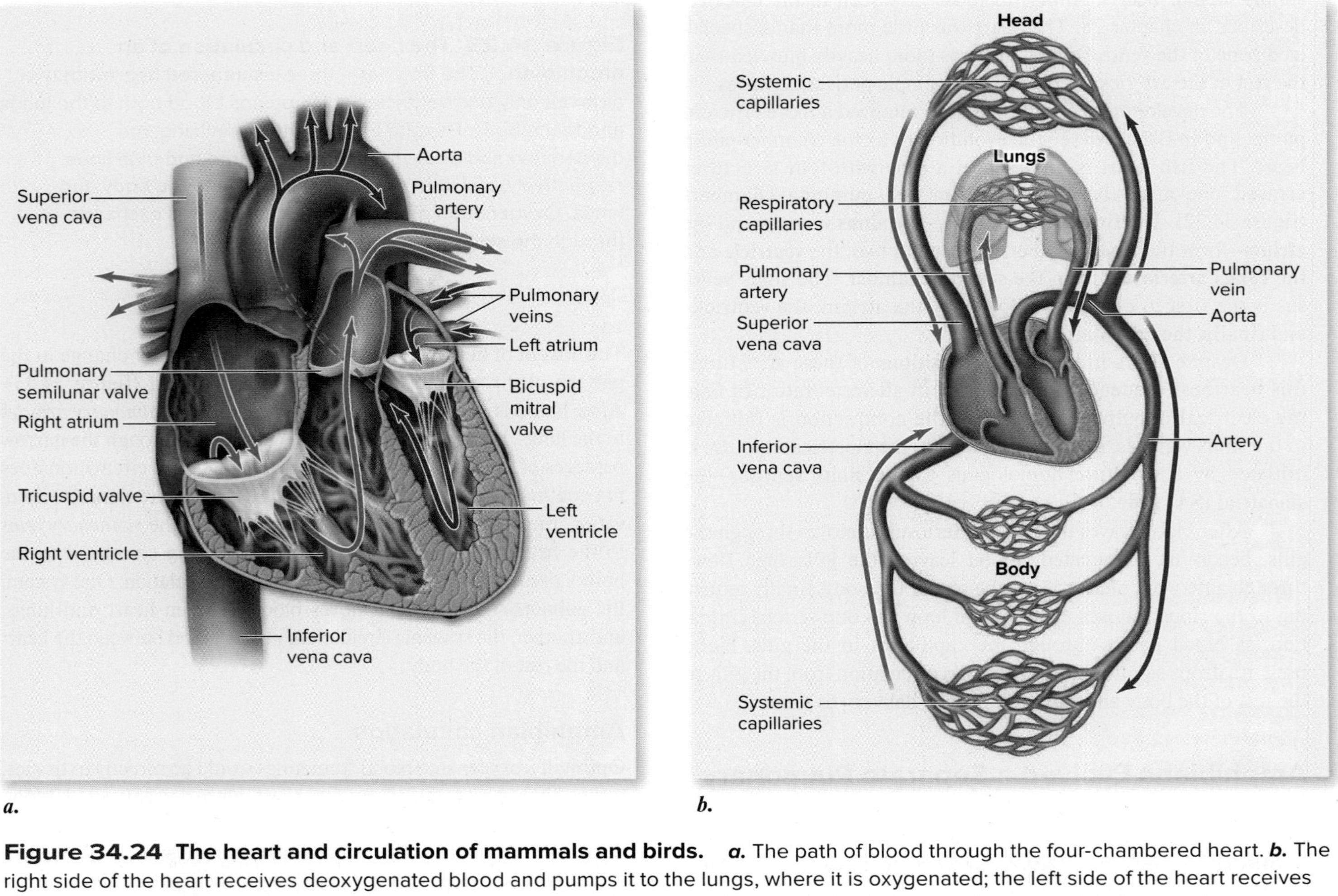

Figure 34.24 The heart and circulation of mammals and birds. ***a.*** The path of blood through the four-chambered heart. ***b.*** The right side of the heart receives deoxygenated blood and pumps it to the lungs, where it is oxygenated; the left side of the heart receives oxygenated blood and pumps it to the body. In this way, the pulmonary and systemic circulations are kept completely separate.

blood into the ventricles. Both ventricles also contract at the same time, pushing blood simultaneously into the pulmonary and systemic circulations.

The increased efficiency of the double circulatory system in mammals and birds is thought to have been important in the evolution of endothermy. More efficient circulation is necessary to support the high metabolic rate required for maintenance of internal body temperature about a set point.

Throughout the evolutionary history of the vertebrate heart, the sinus venosus has served as a **pacemaker,** the site where the impulses that initiate the heartbeat originate. Although the sinus venosus constitutes a major chamber in the fish heart, it is reduced in size in amphibians and is further reduced in reptiles. In mammals and birds, the sinus venosus is no longer present as a separate chamber, although some of its tissue remains in the wall of the right atrium. This tissue, the sinoatrial (SA) node, is still the site where each heartbeat originates, as detailed in section 34.10.

REVIEW OF CONCEPT 34.9

The chordate heart has evolved from a muscular region of a vessel to the two-chambered heart of fish, the three-chambered heart of amphibians and most reptiles, and the four-chambered heart of crocodilians, birds, and mammals. Deoxygenated blood travels in the pulmonary circuit from the heart to the lungs, then returns to the heart. Oxygenated blood travels in the systemic circuit from the heart to the body, then returns to the heart.

- *What is the physiological advantage of having separated ventricles?*

34.10 The Four Chambers of the Heart Contract in a Cycle

The Cardiac Cycle Drives the Cardiovascular System

LEARNING OBJECTIVE 34.10.1 Explain the elements of the cardiac cycle.

The heart of mammals, birds, and crocodilians goes through two contraction cycles: one of atrial contraction to send blood to the ventricles and one of ventricular contraction to send blood to the pulmonary and systemic circuits. These two contractions plus the resting period between them make up the complete **cardiac cycle**.

The cardiac cycle, commonly referred to as the heartbeat, occurs an average of 70 times per minute in humans. During the cardiac cycle, the period during muscle contraction is called **systole,** and the period of muscle relaxation phase is called **diastole**. We define the cardiac cycle as the contraction (systole) and subsequent relaxation (diastole) of the cardiac muscle in all four heart chambers.

The heart has two pairs of valves. One pair, the **atrioventricular (AV) valves,** maintains unidirectional blood flow between the atria and the ventricles. The AV valve on the right side is the **tricuspid valve,** and the AV valve on the left is the **bicuspid,** or **mitral, valve.** Another pair of valves, together called the **semilunar valves,** ensures one-way flow out of the ventricles to the arterial systems. The **pulmonary valve** is located at the exit of the right ventricle, and the **aortic valve** is located at the exit of the left ventricle. These valves open and close as the heart goes through its cycle. The closing of these valves produces the "lub-dub" sounds heard with a stethoscope.

The cardiac cycle is detailed in figure 34.25. It begins as blood returns to the resting heart through veins that empty into the right and left atria. As the atria fill, the pressure in them rises, the AV valves open, and blood flows into the ventricles. This fills the ventricles to about 80% capacity. Atrial contraction adds the remaining 20% to a final volume: about 80 mL of blood, on average, in a person at rest. These events occur while the ventricles are relaxing (ventricular diastole).

After a slight delay, the ventricles contract (ventricular systole). Contraction of each ventricle increases the pressure within each chamber, causing the AV valves to forcefully close (the "lub" sound), preventing blood from backing up into the atria. Immediately after the AV valves close, the pressure in the ventricles forces the semilunar valves open and blood flows into the arterial systems. As the ventricles relax, closing of the semilunar valves prevents backflow (the "dub" sound).

Contraction of Heart Muscle Is Initiated by Autorhythmic Cells

LEARNING OBJECTIVE 34.10.2 Describe the role of autorhythmic cells of the SA node.

As in other types of muscle, the contraction of heart muscle is stimulated by membrane depolarization (refer to chapter 33). In skeletal muscles, only nerve impulses from motor neurons usually initiate depolarization. The heart, in contrast, contains specialized "self-excitable" muscle cells called autorhythmic fibers, which can initiate periodic action potentials without neural activation.

The most important group of autorhythmic cells is the **sinoatrial (SA)** node (figure 34.26). Located in the wall of the right atrium, the SA node acts as a pacemaker for the rest of the heart by producing spontaneous action potentials at a faster rate than other autorhythmic cells. These spontaneous action potentials are due to a constant leakage of Na^+ ions into the cell that depolarize the membrane. When the threshold is reached, an action potential occurs. At the end of the action potential, the membrane is again below threshold and the process begins again. The cells of the SA node generate an action potential every 0.6 sec, equivalent to about 100 a minute. This is referred to as the **sinus rhythm,** and it can be modulated by signals from the nervous and endocrine systems. Deviation from the sinus rhythm is called arrhythmia, and can be due to a variety of causes.

Each depolarization initiated by this pacemaker is transmitted through two pathways: one to the cardiac muscle fibers of the left atrium, the other to the right atrium and the **atrioventricular (AV)** node. Once initiated, depolarization spreads quickly from one muscle fiber to another in a wave that envelops the right and left atria nearly simultaneously. The rapid spread of depolarization occurs because of connections between cells called *intercalated disks* that contain gap junctions.

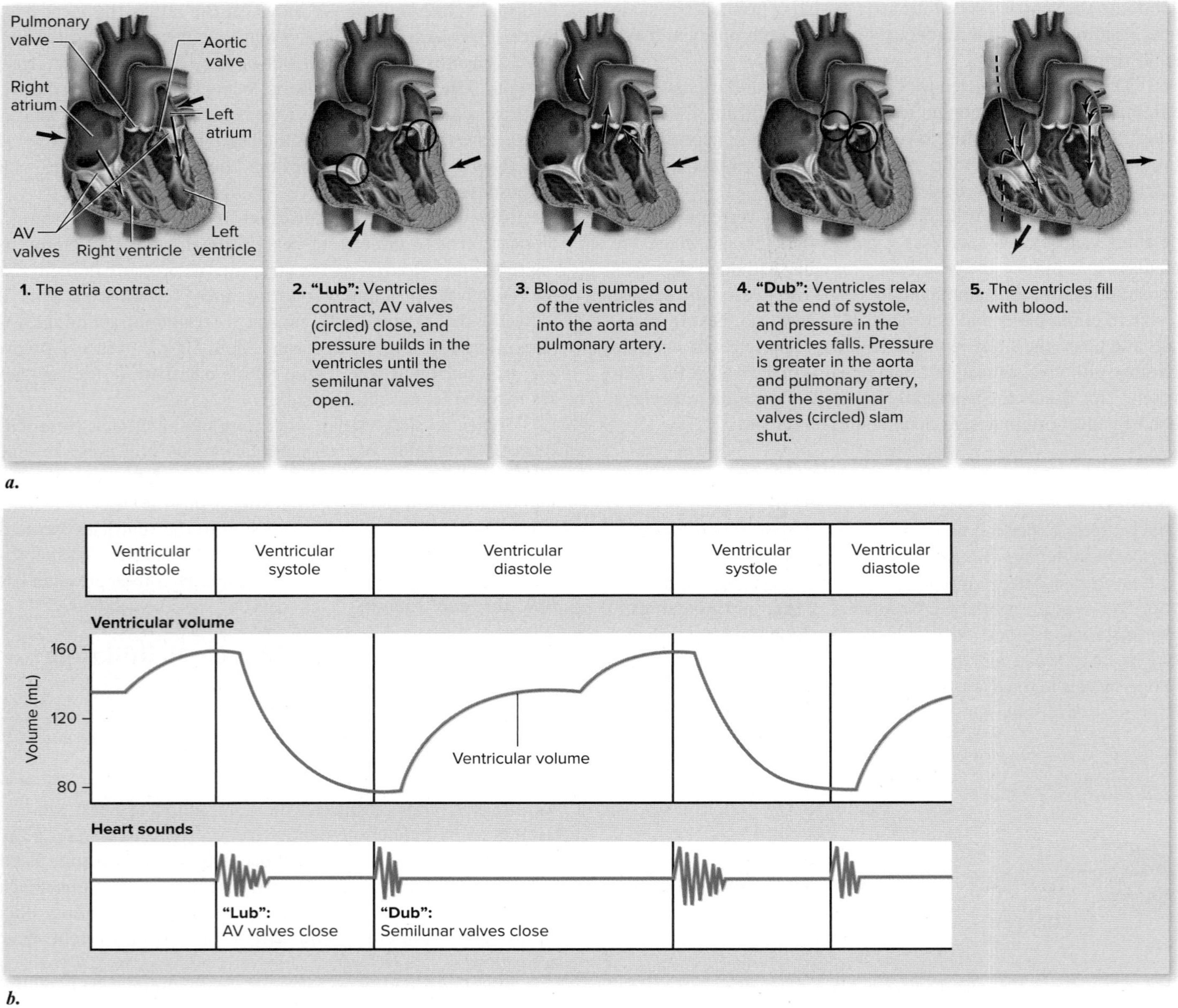

Figure 34.25 The cardiac cycle. ***a.*** Contraction and relaxation of the atria and ventricles move blood through the heart. ***b.*** Blood pressure and volume change through the cardiac cycle, shown here for the left ventricle.

A sheet of connective tissue separating the atria from the ventricles blocks the spread of excitation through muscle fibers from one chamber to the other. The AV node provides the only pathway for conduction of the depolarization from the atria to the ventricles. The fibers of the AV node are thinner and contain fewer gap junctions. This slows the depolarizing signal, delaying the contraction of the ventricles by about 0.1 sec. This allows the atria to finish contracting, filling the ventricles, before the ventricles contract.

From the AV node, the wave of depolarization is conducted rapidly over both ventricles by a network of fibers called the atrioventricular bundle, or bundle of His. These fibers relay the depolarization to Purkinje fibers, which directly stimulate the myocardial cells of the left and right ventricles, causing their almost simultaneous contraction (figure 34.26).

The stimulation of myocardial cells produces an action potential and contraction that is controlled by Ca^{2+} and the troponin–tropomyosin system, similar to that in skeletal muscle (refer to chapter 32). However, in cardiac muscle this action potential has a plateau phase absent in skeletal muscle. In cardiac muscle, the initial rising phase due to the influx of Na^+ from voltage-gated Na^+ channels is followed by an influx of Ca^{2+} through voltage-gated Ca^{2+} channels. This keeps the membrane depolarized after the Na^+ channels inactivate and leads to the opening of more voltage-gated Ca^{2+} channels in the sarcoplasmic reticulum. This additional Ca^{2+} in the cytoplasm produces a more sustained contraction. The Ca^{2+} is removed by a pump in the sarcoplasmic reticulum similar to that in skeletal muscle, and an additional carrier in the plasma membrane pumps Ca^{2+} into the interstitial space.

The electrical activity of the heart can be recorded from the surface of the body with electrodes placed on the limbs and chest. The recording, called an electrocardiogram (ECG or EKG), shows how the cells of the heart depolarize and repolarize during the

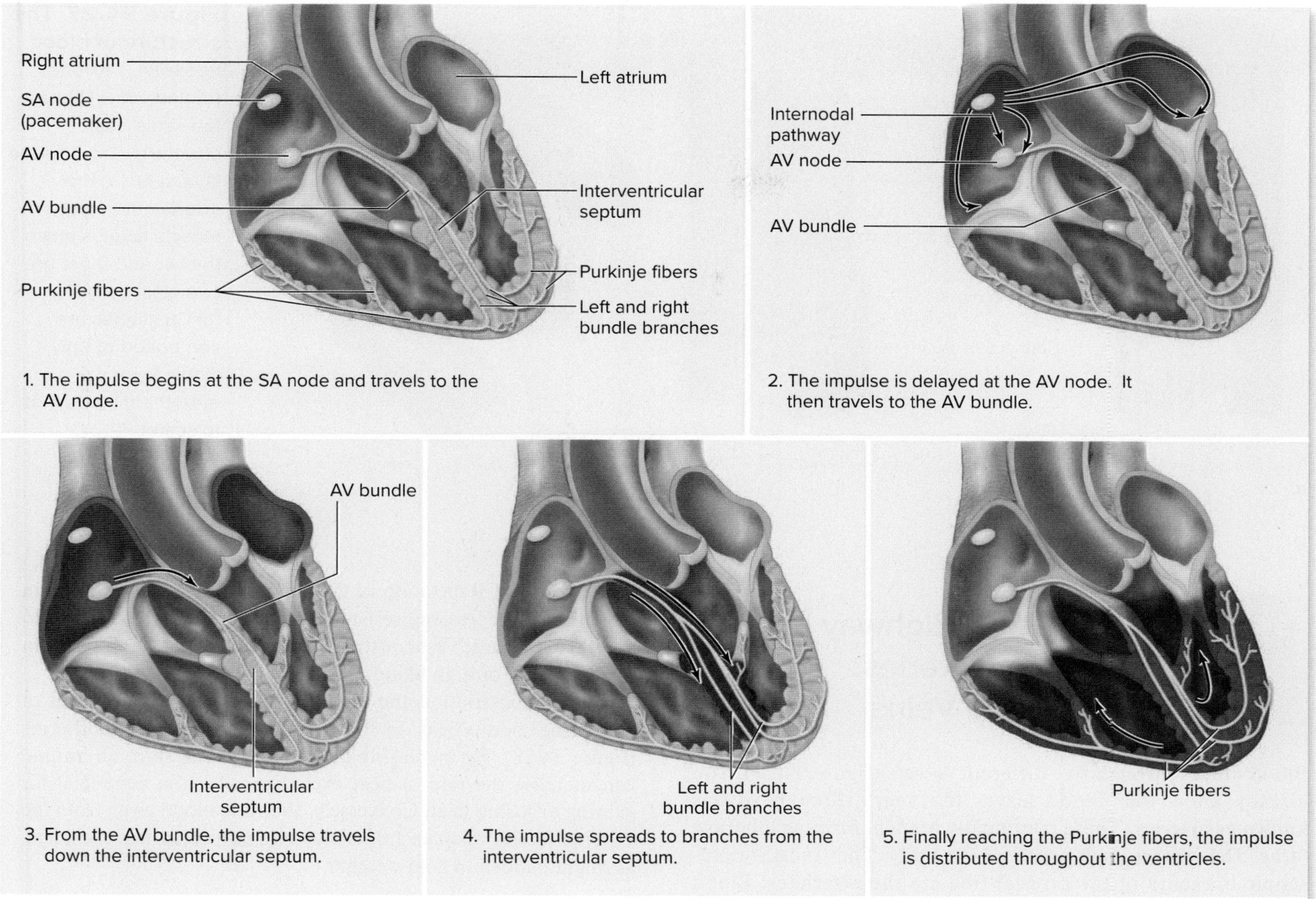
Right atrium
SA node (pacemaker)
AV node
AV bundle
Purkinje fibers
Left atrium
Interventricular septum
Purkinje fibers
Left and right bundle branches
1. The impulse begins at the SA node and travels to the AV node.
Internodal pathway
AV node
AV bundle
2. The impulse is delayed at the AV node. It then travels to the AV bundle.
AV bundle
Interventricular septum
3. From the AV bundle, the impulse travels down the interventricular septum.
Left and right bundle branches
4. The impulse spreads to branches from the interventricular septum.
Purkinje fibers
5. Finally reaching the Purkinje fibers, the impulse is distributed throughout the ventricles.

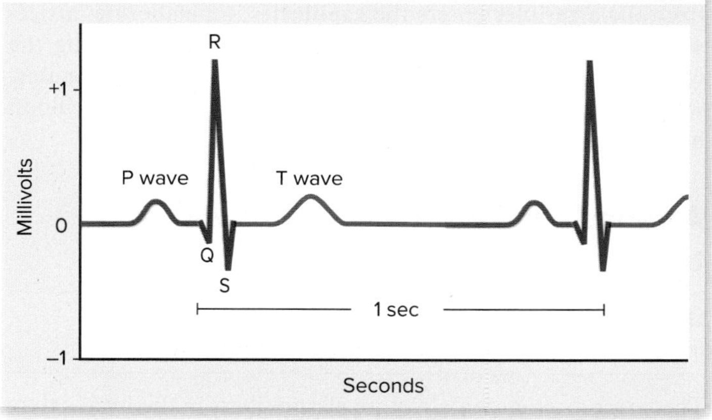
R
+1
P wave
T wave
Millivolts
0
Q
S
1 sec
−1
Seconds

Figure 34.26 The path of electrical excitation in the heart. The events occurring during contraction of the heart are correlated with the measurement of electrical activity by an electrocardiogram (ECG). The depolarization/contraction of the atrium is shown in green above and corresponds to the P wave of the ECG (also in green). The depolarization/contraction of the ventricle is shown in red above and corresponds to the QRS wave of the ECG (also in red). The T wave on the ECG corresponds to the repolarization of the ventricles. The atrial repolarization is masked by the QRS wave.

cardiac cycle (figure 34.26). Depolarization causes contraction of the heart, and repolarization causes relaxation.

Blood pressure is written as a ratio of systolic over diastolic pressure, and for a healthy person in his or her 20s, a typical blood pressure is 120/75 (measured in millimeters of mercury, or mm Hg). A systolic pressure greater than 130 mm Hg, or a diastolic pressure greater than 80 mm Hg, is considered **hypertension.** This puts an individual at risk for a variety of heart-related problems such as stroke, coronary artery disease, heart failure, and many other medical issues.

REVIEW OF CONCEPT 34.10

The cardiac cycle consists of systole and diastole; the ventricles contract at systole and relax at diastole. The SA node in the right atrium initiates waves of depolarization, which first stimulate the atria and then travel to the AV node, which stimulates the ventricles. Blood pressure is expressed as the ratio of systolic pressure over diastolic pressure.

- *What would happen without a delay between auricular and ventricular contraction?*

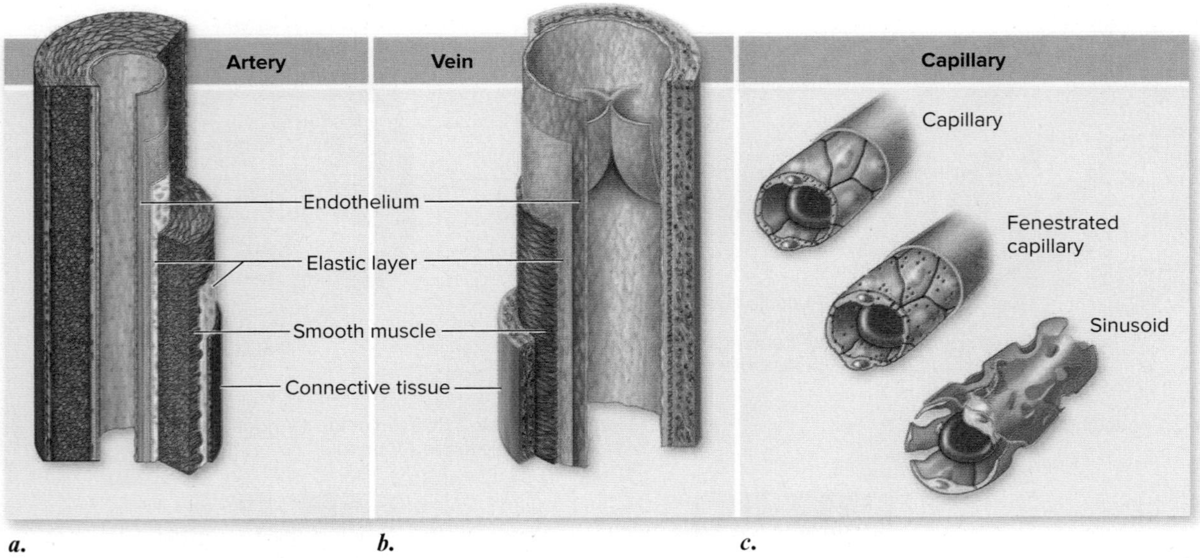
Artery
Vein
Capillary
Endothelium
Elastic layer
Smooth muscle
Connective tissue
Capillary
Fenestrated capillary
Sinusoid
a.
b.
c.

Figure 34.27 The structure of blood vessels. Arteries (*a*) and veins (*b*) have the same basic structure of four tissue layers, but in the arteries the smooth muscle layer is much thicker and there are two elastic layers. (*c*) Capillaries are composed of only a single layer of endothelial cells (not to scale).

34.11 The Circulatory Highway Is Composed of Arteries, Capillaries, and Veins

Blood moves through five different vessels (figure 34.27). You already know that blood leaves the heart through vessels known as **arteries.** These continually branch, forming a hollow "tree" that enters each organ of the body. The finest, microscopic branches of the arterial tree are the **arterioles.** Blood from the arterioles enters the **capillaries,** an elaborate latticework of very narrow, thin-walled tubes. After traversing the capillaries, the blood is collected into microscopic **venules,** which lead to larger vessels called **veins,** and these carry blood back to the heart.

Arteries and Arterioles Have Evolved to Withstand High Pressures

LEARNING OBJECTIVE 34.11.1 Explain how elasticity is important to the function of arteries and arterioles.

The larger arteries contain more elastic fibers in their walls than other blood vessels, allowing them to recoil each time they receive a volume of blood pumped by the heart. Smaller arteries and arterioles are less elastic, but their relatively thick smooth muscle layer enables them to resist bursting.

The narrower the vessel, the greater the frictional resistance to flow. In fact, a vessel that is half the diameter of another has *16 times* the frictional resistance. Resistance to blood flow is inversely proportional to the fourth power of the radius of the vessel. Therefore, within the arterial tree, the small arteries and arterioles provide the greatest resistance to blood flow.

Contraction of the smooth muscle layer of the arterioles results in **vasoconstriction,** which greatly increases resistance and decreases flow. Relaxation of the smooth muscle layer results in **vasodilation,** decreasing resistance and increasing blood flow to an organ. Chronic vasoconstriction of the arterioles can result in hypertension, or high blood pressure.

Vasoconstriction and vasodilation are important means of regulating body heat in both ectotherms and endotherms (figure 34.28). By increasing blood flow to the skin, an animal can increase the rate of heat exchange, which is beneficial for gaining or losing heat. Conversely, shunting blood away from the skin is effective when an animal needs to minimize heat exchange, as might happen in cold weather.

Capillaries Form a Vast Network for Exchange

LEARNING OBJECTIVE 34.11.2 Explain how the structure of capillaries promotes fluid and gas exchange.

The walls of capillaries, in contrast to those of arteries and veins, are composed only of endothelium, so molecules and ions can leave the blood plasma by diffusion, by filtration through pores between the cells of the capillary walls, and by transport through the endothelial cells. Therefore, exchange of gases and metabolites between the blood and the interstitial fluids and cells of the body takes place through the capillaries.

The huge number and extensive branching of the capillaries ensures that every cell in the body is within 100 μm of a capillary. On the average, capillaries are about 1 mm long and 8 μm in diameter; this diameter is only slightly larger than that of a red blood cell (5 to 7 μm). Despite the close fit, normal red blood cells are flexible enough to squeeze through capillaries without difficulty.

The rate of blood flow through vessels is governed by hydrodynamics. The smaller the cross-sectional area of a vessel, the faster fluid moves through it. Given this, flow in the capillaries would be expected to be the fastest in the system. This would not be ideal for diffusion and is actually not the case. Although each capillary is very narrow, there are so many of them that the capillaries have the greatest *total* cross-sectional area of any type of vessel. Consequently,

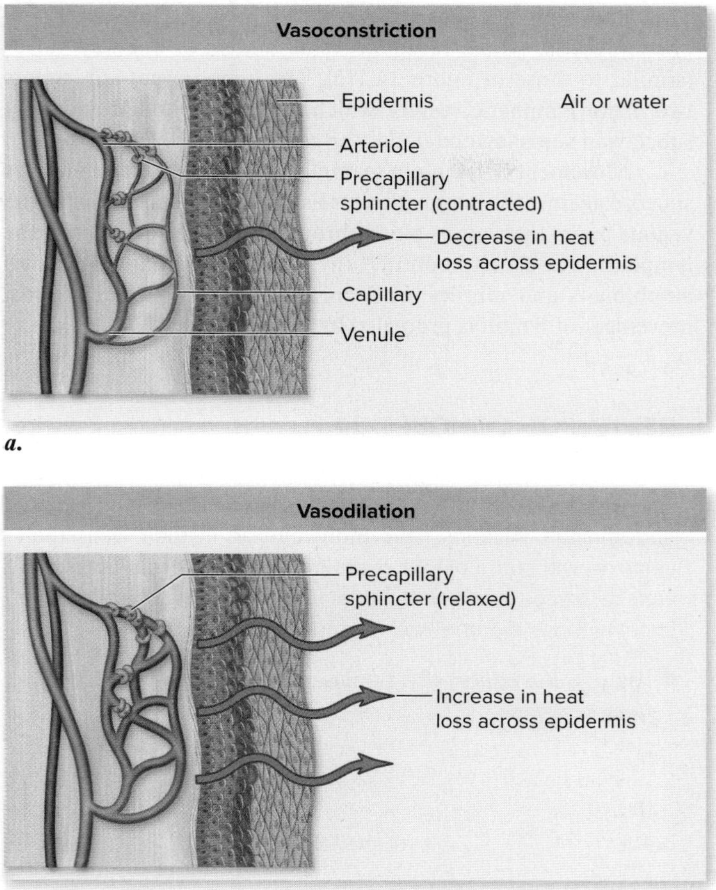

Figure 34.28 Regulation of heat exchange. The amount of heat gained or lost at the body's surface can be regulated by controlling the flow of blood to the surface. ***a.*** Constriction of surface blood vessels limits flow and heat loss when the animal is warmer than the surrounding air; when the animal is cooler than the surrounding air (not shown here), constriction minimizes heat gain. ***b.*** Dilation of these vessels increases flow and heat exchange.

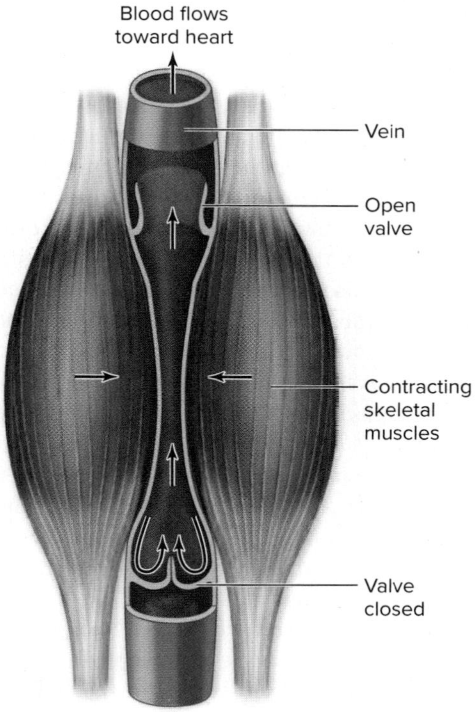

Figure 34.29 One-way flow of blood through veins. Venous valves ensure that blood moves through the veins in only one direction, back to the heart.

blood moving through capillaries goes more slowly and has more time to exchange materials with the surrounding extracellular fluid. By the time the blood reaches the end of a capillary, it has released some of its oxygen and nutrients and picked up carbon dioxide and other waste products. Blood loses pressure and velocity as it moves through the arterioles and capillaries, but as cross-sectional area decreases in the venous side, velocity increases.

Veins Have Less Muscle in Their Walls

LEARNING OBJECTIVE 34.11.3 Compare the structure of veins to that of arteries.

Venules and veins have the same tissue layers as arteries, but they have a thinner layer of smooth muscle. Less muscle is needed because the pressure in the veins is only about one-tenth that in the arteries. Most of the blood in the cardiovascular system is contained within veins, which can expand to hold additional amounts of blood. You can see the expanded veins in your feet when you stand for a long time or lower your hand to your side.

The venous pressure alone is not sufficient to return blood to the heart from the feet and legs, but several other sources of pressure provide help. Most significantly, skeletal muscles surrounding the veins can contract to move blood by squeezing the veins, a mechanism called the **venous pump.** Blood moves in one direction through the veins back to the heart with the help of **venous valves** (figure 34.29). When a person's veins expand too much with blood, the venous valves may no longer work and the blood may pool in the veins. Veins in this condition are known as varicose veins.

The Lymphatic System Reclaims Fluids That Leave the Cardiovascular System

LEARNING OBJECTIVE 34.11.4 Describe how the lymphatic system operates.

The cardiovascular system is considered a closed system because all its vessels are connected with one another—none are simply open-ended. But a significant amount of water and solutes in the blood plasma filters through the walls of the capillaries to form the interstitial (tissue) fluid. Most of the fluid leaves the capillaries near their arteriolar ends, where the blood pressure is higher; it is returned to the capillaries near their venular ends.

Fluid returns by osmosis (refer to chapter 5). Most of the plasma proteins cannot escape through the capillary pores because of their large size, and so the concentration of proteins in the plasma is greater than the protein concentration in the interstitial fluid. The difference in protein concentration produces an osmotic pressure gradient that causes water to move into the capillaries from the interstitial space.

High capillary blood pressure can cause too much interstitial fluid to accumulate. In pregnant women, for example, the enlarged uterus, carrying the fetus, compresses veins in the abdominal cavity, thereby adding to the capillary blood pressure in the woman's lower limbs. The increased interstitial fluid can cause swelling of the tissues, or **edema,** of the feet.

Edema may also result if the plasma protein concentration is too low. Fluids do not return to the capillaries but remain as interstitial fluid. Low protein concentration in the plasma may be caused either by liver disease, because the liver produces most of the plasma proteins, or by insufficient dietary protein, as occurs in starvation.

Even under normal conditions, the amount of fluid filtered out of the capillaries is greater than the amount that returns to the capillaries by osmosis. The remainder does eventually return to the cardiovascular system by way of an open circulatory system called the **lymphatic system.**

The lymphatic system consists of lymphatic capillaries, lymphatic vessels, lymph nodes, and lymphatic organs, including the spleen and thymus. Excess fluid in the tissues drains into blind-ended lymphatic capillaries with highly permeable walls. This fluid, now called **lymph,** passes into progressively larger lymphatic vessels, which resemble veins and have one-way valves (similar to those in figure 34.27*b*). The lymph eventually enters two major lymphatic vessels, which drain into the left and right subclavian veins located under the collarbones.

Movement of lymph in mammals results as skeletal muscles squeeze against the lymphatic vessels, a mechanism similar to the venous pump that moves blood through veins. In some cases, the lymphatic vessels also contract rhythmically. In many fishes, all amphibians and reptiles, bird embryos, and some adult birds, movement of lymph is propelled by **lymph hearts.**

REVIEW OF CONCEPT 34.11

Blood vessels consist of an endothelium, an elastic layer, smooth muscle, and connective tissue; capillaries have only endothelium. Arteries have more muscle in their walls than veins, to withstand higher pressure; veins have valves to prevent backflow. Excess interstitial fluid, called lymph, returns to circulation via the one-way lymphatic system.

- *What is the connection between the lymphatic and circulatory systems?*

Why Do Diabetics Excrete Glucose in Their Urine?

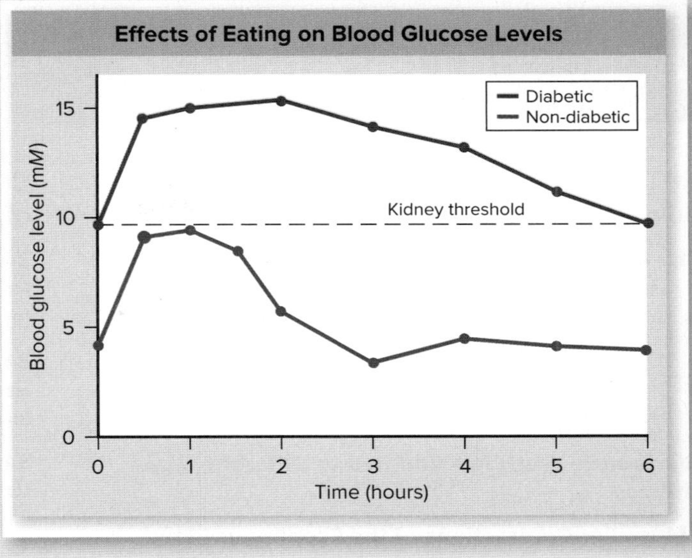

Late-onset diabetes is a serious and increasingly common disorder in which a person's cells lose their ability to respond to insulin, a hormone that is needed to trigger their uptake of glucose. As shown in the illustration below, the binding of insulin to a receptor in the plasma membrane causes the rapid insertion of glucose transporter channels into the plasma membrane, allowing the cell to take up glucose. In diabetics, however, glucose molecules accumulate in the blood while the body's cells starve for the lack of them. In mild cases, blood glucose levels rise to several times the usual value of 4 mM; in severe untreated cases, blood glucose levels may become enormously elevated, up to 25 times the usual value. A characteristic symptom of even mild diabetes is the excretion of large amounts of glucose in the urine. The name of the disorder, diabetes mellitus, means "excessive secretion of sweet urine." In non-diabetic individuals, by contrast, only trace amounts of glucose are excreted. The kidney very efficiently reabsorbs glucose molecules from the fluid passing through it. Why doesn't it do so in diabetic individuals?

The graph displays so-called glucose tolerance curves for a non-diabetic person (blue line) and a diabetic (red line). After a night without food, each individual drank a test dose of 100 g of glucose dissolved in water. Blood glucose levels were then monitored at 30-minute and 1-hour intervals. The dotted line indicates the kidney threshold, the maximum concentration of blood glucose molecules (about 10 mM) that the kidney is able to retrieve from the fluid passing through it when all of its glucose-transporting channels are being utilized full-bore.

Analysis

1. **Applying Concepts**
   a. ***Reading a curve.*** What is the immediate effect on the non-diabetic individual's blood glucose levels of consuming the test dose of glucose? How long does it take for the non-diabetic person's blood glucose level to return to the level before the test dose?
   b. ***Comparing curves.*** Is the effect any different for the diabetic person? How long does it take for the diabetic person's blood glucose level to return to the level before the test dose?
2. **Interpreting Data**
   a. Is there any point at which the non-diabetic individual's blood glucose level exceeds the kidney threshold?
   b. Is there any point at which the diabetic individual's blood glucose level does *not* exceed the kidney threshold?
3. **Making Inferences**
   a. Why do you suppose the diabetic individual took so much longer to recover from the test dose?
   b. Would you expect the non-diabetic individual to excrete glucose? Explain. The diabetic individual? Explain.
4. **Drawing Conclusions** Why do diabetic individuals secrete sweet urine?

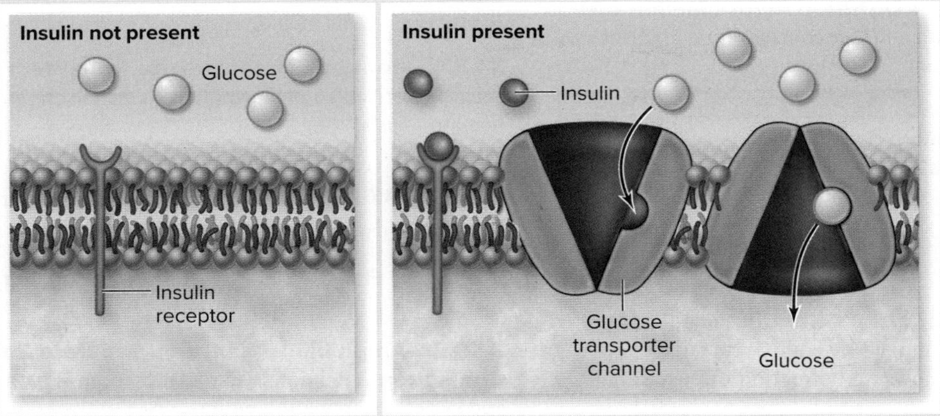

Retracing the Learning Path

CONCEPT 34.1 Vertebrate Digestive Systems Are Tubular Tracts

34.1.1 The Digestive Tract Is a One-Way Tube from Mouth to Anus The gastrointestinal tract includes the mouth and pharynx, esophagus, stomach, small and large intestines, rectum, and anus.

CONCEPT 34.2 Food Is Processed as It Passes Through the Digestive Tract

34.2.1 Vertebrate Teeth Are Adapted to Different Types of Food Items

34.2.2 Muscular Contractions of the Esophagus Move Food to the Stomach

34.2.3 The Stomach Is a "Holding Station" Involved in Acidic Breakdown of Food In the stomach, hydrochloric acid breaks down food and converts pepsinogen into pepsin.

34.2.4 The Structure of the Small Intestine Is Specialized for Nutrient Uptake The surface area of the small intestine is greatly increased by finger-like projections called villi.

34.2.5 Accessory Organs Secrete Enzymes into the Small Intestine The pancreas secretes digestive enzymes and bicarbonate. The liver secretes bile, which emulsifies fats.

34.2.6 Absorbed Nutrients Move into Blood or Lymphatic Capillaries

34.2.7 The Large Intestine Eliminates Waste Material

CONCEPT 34.3 The Digestive Tract Is Regulated by the Nervous System and Hormones

34.3.1 Hormones Regulate Digestion The activities of the GI tract are coordinated by the nervous and endocrine systems.

34.3.2 The Liver Removes Toxins and Functions in Homeostasis The liver is involved in detoxification, regulation of steroid hormone levels, and production of proteins in blood plasma.

CONCEPT 34.4 Respiratory Systems Promote Efficient Exchange of Gases

34.4.1 Gas Exchange Involves Diffusion Across Membranes The rate of diffusion increases with a higher concentration gradient and greater surface area (Fick's Law).

34.4.2 Evolution Has Acted to Maximize the Rate of Gas Diffusion Evolution has produced structures to increase diffusion, removing a limit on size of 1 mm.

CONCEPT 34.5 Gills Provide for Efficient Gas Exchange in Water

34.5.1 Fish Respire with External Gills In the gills of fishes, diffusion of gases is maximized by countercurrent exchange, where blood in gills flows in a direction opposite that of the flow of water.

CONCEPT 34.6 Lungs Are the Respiratory Organs of Terrestrial Vertebrates

34.6.1 Breathing of Air Takes Advantage of Partial Pressures of Gases The partial pressure of gases is responsible for the pressure gradient that brings about gas exchange.

34.6.2 The Mammalian Lung Has a Large Surface Area Surface area is increased by branching of bronchioles, which terminate in alveoli, where gas exchange occurs.

34.6.3 The Diaphragm Expands and Contracts Lung Volume in the Respiratory Cycle Contraction of the diaphragm and intercostal muscles creates negative pressure. Average intake of air is called tidal volume, and maximum inspiration and expiration are the vital capacity.

34.6.4 The Central Nervous System Regulates Breathing CNS neurons monitor CO_2 levels.

CONCEPT 34.7 Oxygen and Carbon Dioxide Are Transported by Fundamentally Different Mechanisms

34.7.1 Respiratory Pigments Bind Oxygen for Transport Hemoglobin allows blood to transport more oxygen than is dissolved in plasma.

34.7.2 Hemoglobin's Affinity for Oxygen Is Affected by pH and Temperature The affinity of hemoglobin for oxygen decreases as pH decreases and as temperature increases.

34.7.3 Carbon Dioxide Is Primarily Transported in Blood as Bicarbonate Ion CO_2 produced in tissues reacts with H_2O to form H_2CO_3. This dissociates into HCO_3^-, which is transported by the blood to the lungs, where the reaction is reversed.

CONCEPT 34.8 Circulating Blood Carries Metabolites and Gases to the Tissues

34.8.1 Blood Has Many Functions

34.8.2 Blood Plasma Is a Fluid Matrix Plasma is 92% water plus nutrients, hormones, ions, plasma proteins, and wastes.

34.8.3 Formed Elements Include Circulating Cells and Platelets Cells include erythrocytes, leukocytes, and platelets.

34.8.4 Formed Elements Arise from Stem Cells Blood cells are derived from pluripotent stem cells in bone marrow by hematopoiesis.

CONCEPT 34.9 Vertebrate Circulatory Systems Put a Premium on Efficient Circulation

34.9.1 In Fishes, a Heart Pushes Blood Through the Gills Fishes have a linear heart with two pumping chambers.

34.9.2 Amphibians Evolved a Separate Pulmonary Circulation Pulmonary circulation pumps blood to the lungs, and systemic circulation pumps blood to the body.

34.9.3 Mammals, Birds, and Crocodilians Have Two Completely Separated Circulatory Systems The four-chambered heart has two ventricles.

CONCEPT 34.10 The Four Chambers of the Heart Contract in a Cycle

34.10.1 The Cardiac Cycle Drives the Cardiovascular System During diastole, the atria fill until pressure opens the AV valves. During systole, the ventricles contract, which closes the AV valves and opens the semilunar valve as blood is pumped out of the heart.

34.10.2 Contraction of Heart Muscle Is Initiated by Autorhythmic Cells Contraction is initiated by the SA node, a natural pacemaker, and impulses then travel to the AV node.

CONCEPT 34.11 The Circulatory Highway Is Composed of Arteries, Capillaries, and Veins

34.11.1 Arteries and Arterioles Have Evolved to Withstand High Pressures

34.11.2 Capillaries Form a Vast Network for Exchange All diffusion occurs in the capillaries. Every cell is within 100 μm of a capillary.

34.11.3 Veins Have Less Muscle in Their Walls

34.11.4 The Lymphatic System Reclaims Fluids That Leave the Cardiovascular System

Concept Overview

This Concept Overview diagrams the key concepts that were discussed in this chapter.

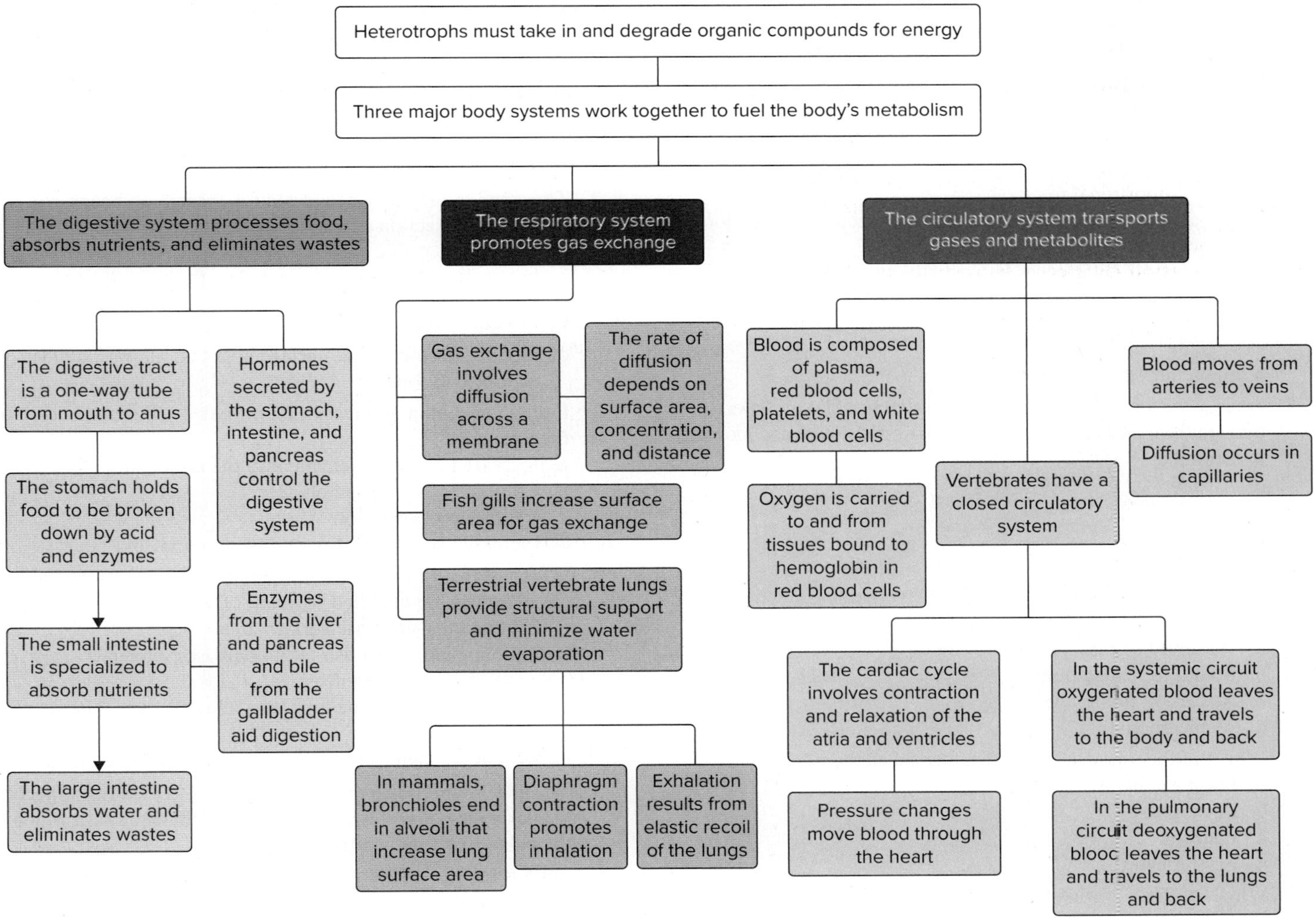

Assessing the Learning Path

Understand

1. The process of digestion occurs in which of the following structures of the vertebrate digestive system? (Select all that apply.)
   a. Esophagus
   b. Small intestine
   c. Stomach
   d. Large intestine
2. The ________ and ________ play important roles in the digestive process by producing chemicals that are required to digest proteins, lipids, and carbohydrates.
   a. liver; pancreas
   b. kidneys; appendix
   c. liver; gallbladder
   d. pancreas; gallbladder
3. The arrival of food into the stomach stimulates the release of gastrin. What is the function of gastrin?
   a. Gastrin is a protease enzyme.
   b. Gastrin activates pepsinogen.
   c. Gastrin hydrolyzes proteins.
   d. Gastrin stimulates the release of pepsinogen and HCl.
4. Fick's Law of Diffusion states that the rate of diffusion is directly proportional to
   a. the area differences between the cross section of the blood vessel and the tissue.
   b. the pressure differences between the two sides of the membrane and how large an area is available for diffusion.
   c. the pressure differences between the inside of the organism and the outside.
   d. the temperature of the gas molecule.
5. Arrange the following structures in the order in which air flows through them during inhalation, starting from the nose and mouth.
   a. Bronchus
   b. Alveoli
   c. Bronchioles
   d. Pharynx
   e. Trachea
6. Which of the following is the primary method by which carbon dioxide is transported to the lungs?
   a. Dissolved in plasma
   b. Bound to hemoglobin
   c. As carbon monoxide
   d. As bicarbonate
7. Which of the following is NOT a formed element?
   a. Plasma
   b. Leukocytes
   c. Platelets
   d. Erythrocytes
8. Which of the following is a correct description of the amphibian heart?
   a. In the amphibian heart, oxygenated and deoxygenated blood mix completely in the single ventricle.
   b. In the amphibian heart, there are two SA nodes so that contractions occur simultaneously throughout the heart.
   c. In the ventricle in the amphibian heart, internal channels reduce mixing of blood.
   d. In the amphibian heart, only the left aorta pumps oxygen obtained by diffusion through the skin.
9. The repolarization of the ventricles corresponds to which of the following on an ECG?
   a. The P wave
   b. The QRS wave
   c. The T wave
   d. It is masked by the QRS wave, so it isn't visible on the ECG.
10. Contraction of the smooth muscle layers of the arterioles
    a. increases the frictional resistance to blood flow.
    b. may be a way of increasing heat exchange through the skin.
    c. can increase blood flow to an organ.
    d. All of the above

Apply

1. The primary function of the large intestine is to concentrate wastes into solid form (feces) for release from the body. How does it accomplish this?
   a. By adding additional cells from the mucosal layer
   b. By absorbing water
   c. By releasing salt
   d. All of the above
2. The small intestine is specialized for absorption because it
   a. is the last section of the digestive tract and retains food the longest.
   b. has saclike extensions along its length that collect food.
   c. has no outlet, so food remains within it for longer periods of time.
   d. has an extremely large surface area that allows extended exposure to food.
3. Eating a meal that contains a lot of butter will trigger which of the following?
   a. Increased release of chyme into the duodenum
   b. Contraction of the gallbladder
   c. Inhibition of secretion
   d. Increased secretion of pepsinogen and HCl
4. Which of the following evolutionary adaptations to increase gas exchange are matched correctly? (Select all that apply.)
   a. Beating cilia–increase surface area
   b. Bubbling gas into a fish tank–increases oxygen concentration differences
   c. Membranes lining the lungs are one cell layer thick–decreases distance
   d. A small diameter in bacteria–decreases surface area
5. The gills found in fish are highly efficient. Which two aspects of Fick's Law of Diffusion increase the efficiency with which the gills extract oxygen from water?
6. For a given Po_2 would you expect to find a higher percent oxyhemoglobin saturation in muscles at rest or in muscles during exercise? Explain.
7. When trying to extract DNA from a blood sample at a crime scene, which of the following would not help you in isolating the DNA? (Select all that apply.)
   a. Erythrocytes
   b. Neutrophils

    c. Basophils
    d. Platelets
    e. Lymphocytes
8. Before leaving the body, a molecule of CO_2 that is generated in the cardiac muscle of the left ventricle would pass through all of the following structures EXCEPT for the
    a. right atrium.
    b. left atrium.
    c. right ventricle.
    d. left ventricle.
9. Systole is vital to heart function and begins in the heart with the
    a. activation of the AV node.
    b. activation of the SA node.
    c. opening of the voltage-gated potassium gates.
    d. opening of the semilunar valves.
10. Which of the following statements is NOT true?
    a. Only arteries carry oxygenated blood.
    b. Both arteries and veins have a layer of smooth muscle.
    c. Both arteries and veins branch out into capillaries.
    d. Blood flow through capillaries is regulated through vasoconstriction.

Synthesize

1. Chemical digestion involves hydrolysis reactions. Given what you know about hydrolysis reactions from chapter 3, write the hydrolysis reaction to break down the disaccharide lactose.
2. Starving animals often exhibit swollen bodies rather than emaciated ones in early stages of their deprivation. Why?
3. At higher elevations, the oxygen concentration is lower than at sea level. For this reason, mountain climbers carry supplemental oxygen. How does this help them in their climbs, and which variable in Fick's Law does this address?
4. What would be the effect of a carbonic anhydrase–disabling mutation, if homozygous, on human respiration?
5. Why do you think the use of erythropoietin as a drug is banned in the Olympics and in some other sports?
6. At any one time, how can the fraction of your blood present in your veins be greater than the fraction present in your arteries?

Rick & Nora Bowers/Alamy Stock Photo

35 Maintaining Homeostasis

Learning Path

35.1 Homeostasis Maintains a Constant Internal Environment

35.2 Hormones Are Chemical Messages That Direct Body Processes

35.3 The Pituitary and the Hypothalamus Are the Body's Control Centers

35.4 Peripheral Endocrine Glands Play Major Roles in Homeostasis

35.5 Animals Are Osmoconformers or Osmoregulators

35.6 The Kidney Maintains Osmotic Homeostasis in Mammals

35.7 Hormones Control Osmoregulation

35.8 The Immune System Defends the Body

35.9 Cell-Mediated Immunity Involves Helper and Killer T Cells

35.10 In Humoral Immunity, B Cells Produce Protective Antibodies

Concept Overview

This Concept Overview provides a diagram of the overarching concepts that are covered in this chapter. These concepts will be expanded upon in the Concept Overview at the end of the chapter.

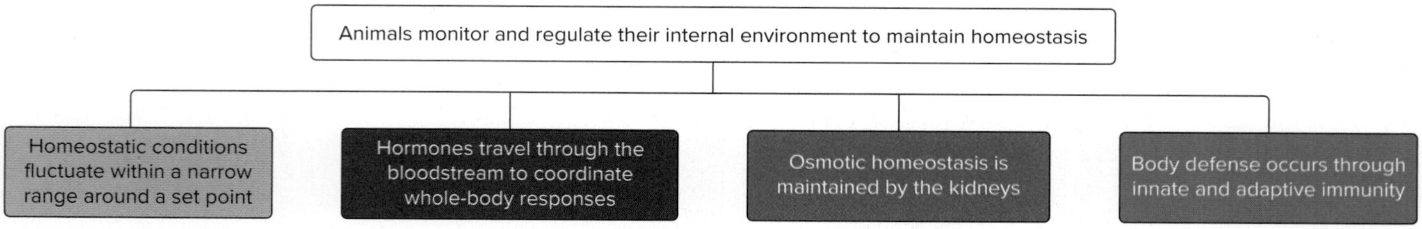

Animals monitor and regulate their internal environment to maintain homeostasis
Homeostatic conditions fluctuate within a narrow range around a set point
Hormones travel through the bloodstream to coordinate whole-body responses
Osmotic homeostasis is maintained by the kidneys
Body defense occurs through innate and adaptive immunity

Introduction

Vertebrates maintain a constant internal environment through regulatory molecules acting on multiple tissues and organ systems. In this chapter we will examine the function of the endocrine system and how it coordinates homeostatic mechanisms affecting a variety of systems. Although the majority of your body weight is water, you exist in a very dehydrating environment. The kangaroo rat pictured on the previous page lives in a desert environment, yet is so efficient with water that it does not need to drink; it generates sufficient water as a metabolic by-product. This is made possible by the osmoregulatory functions of the kidney, which control the osmotic strength of bodily fluids. Another important aspect of maintaining the body is its defenses against microorganisms and viruses. We are constantly challenged by these possible invaders, and this has led to the evolution of defensive systems like the vertebrate system of adaptive immunity.

35.1 Homeostasis Maintains a Constant Internal Environment

As animals have evolved to move about and adapt to their environments in sophisticated ways, they have become increasingly complex. Each cell is a sophisticated machine, finely tuned to carry out a precise role within the body. Such specialization of cells is possible only when extracellular conditions stay within narrow limits. Temperature, pH, glucose and oxygen concentrations, and other factors remain relatively constant. The dynamic constancy of the internal environment is called ***homeostasis.*** The term *dynamic* is used because conditions are never constant but fluctuate continuously within narrow limits.

Negative Feedback Mechanisms Keep Values Within a Narrow Range

LEARNING OBJECTIVE 35.1.1 Explain how negative feedback loops lead to homeostasis.

To maintain internal constancy, many homeostatic mechanisms use a type of control known as **negative feedback.** This is a common biochemical mechanism in which the end result of a process feeds back to limit the process. Conditions within the body, as well as outside it, are detected by specialized sensors, which may be cells or membrane receptors. If conditions deviate too far from a set point, biochemical reactions are initiated to change conditions back toward the set point. This will naturally overrun the set point, and negative feedback turns off the response.

This mechanism is analogous to the temperature setting on a space heater. A drop in room temperature is detected by a thermometer in the controls–the **sensor.** The thermostat also has a **comparator,** which closes a circuit when the temperature drops below the set point. This turns on the heater, increases room temperature until it reaches the set point, and then opens the circuit, turning off the heater.

Your body has set points for a wide variety of parameters, such as the concentration of blood glucose, the concentration of important ions (Na^+, K^+, Ca^{2+}), pH, and even the tension on some tendons. The integrating center is often a particular region of the brain or spinal cord, but in some cases it can also be cells of endocrine glands. When a deviation in a condition occurs, a message is sent to increase or decrease the activity of particular target organs, termed *effectors,* which can move the value of the condition in question back toward the set point value. When the response exceeds the set point, it is stopped by negative feedback.

Antagonistic effectors act in opposite directions

The negative feedback mechanisms that maintain homeostasis often oppose each other to produce a finer degree of control. Most factors in the internal environment are controlled by several effectors, which often have antagonistic (opposing) actions. Control by antagonistic effectors is sometimes described as "push-pull," meaning that the increasing activity of one effector is accompanied by decreasing activity of an antagonistic effector. This affords a finer degree of control than could be achieved by simply switching one effector on and off.

To return to our earlier example, room temperature can be maintained by just turning the heater on and off or turning an air conditioner on and off. A much more stable temperature is possible, however, if a thermostat controls both the air conditioner and the heater. Then the heater turns on when the air conditioner shuts off, and vice versa (figure 35.1). The regulation of the temperature of your own body uses analogous mechanisms, which we will discuss in section 35.1.3.

Positive feedback mechanisms enhance a change

In a few cases, the body uses **positive feedback** mechanisms, which push or accentuate a change farther in the same direction. In a positive feedback loop, the effector drives the value of the controlled variable even farther from the set point. As a result, systems in which there is positive feedback are highly unstable, analogous to a spark that ignites an explosion. They do not help to maintain homeostasis.

Nevertheless, such systems are important components of some physiological mechanisms. For example, positive feedback occurs in blood clotting, in which one clotting factor activates another in a cascade that leads quickly to the formation of a clot. Positive feedback also plays a role in the contractions of the uterus during childbirth. In this case, stretching of the uterus by

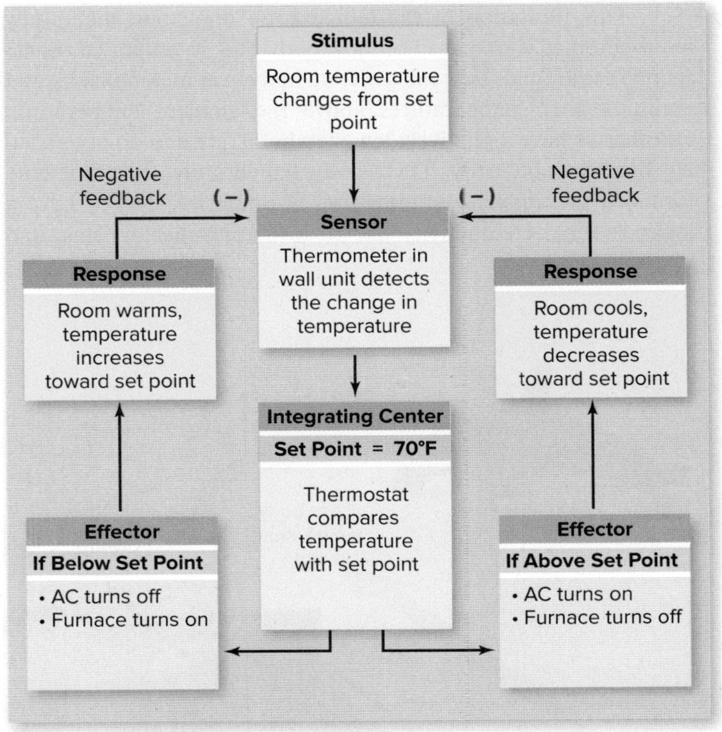

Figure 35.1 Room temperature is maintained by negative feedback and antagonistic effectors. If a thermostat senses a temperature lower than a set point, the furnace turns on and the air conditioner turns off. If the temperature is too high, the air conditioner turns on and the furnace turns off.

the fetus stimulates contraction, and contraction causes further stretching; the cycle continues until the uterus expels the fetus.

Animals Regulate Body Temperature in a Variety of Ways

LEARNING OBJECTIVE 35.1.2 Classify organisms based on temperature regulation.

Temperature is one of the most important aspects of the environment that all organisms must contend with. As we will see, some organisms have a body temperature that conforms to the environment, and others regulate their body temperature. Our classifications have changed over time, so we will consider them first.

Organisms are classified based on heat source

For many years, physiologists classified animals according to whether they maintained a constant body temperature or their body temperature fluctuated with the environment. Animals that regulated their body temperature about a set point were called *homeotherms,* and those that allowed their body temperature to conform to the environment were called *poikilotherms.*

Because homeotherms tended to maintain their body temperature above the ambient temperature, they were also colloquially called "warm-blooded"; poikilotherms were termed "cold-blooded." The problem with this terminology is that a poikilotherm in an environment with a stable temperature (such as many deep-sea fish species) has a more constant body temperature than some homeotherms.

This limitation of the temperature regulation dichotomy led to another view, based on how body heat is generated. **Endotherms** are animals that can generate body heat metabolically and maintain their temperatures above the ambient temperature. **Ectotherms** have a relatively low metabolic rate and do not generate heat metabolically. Their body temperature therefore conforms to the ambient temperature. Endotherms tend to have a lower thermal conductivity due to insulating mechanisms, and ectotherms tend to have high thermal conductivity and lack insulation.

These two terms represent ideal end points of a spectrum of physiology and adaptations. Many animals fall in between these extremes and can be considered *heterotherms.* It is a matter of judgment how a particular animal is classified if it exhibits characteristics of each group.

Ectotherms regulate temperature using behavior

Despite having low metabolic rates, ectotherms do regulate their temperature by using behavior. Most invertebrates use behavior to adjust their temperature. Many butterflies must reach a certain body temperature before they can fly. In the cool of the morning, they orient their bodies to maximize their absorption of sunlight. Other insects use a shivering reflex to warm their thoracic flight muscles.

Vertebrates other than mammals and birds are also ectothermic. This does not mean that these animals cannot maintain high and relatively constant body temperatures, but they use behavior to do this. Many ectothermic vertebrates are able to maintain temperature homeostasis, that is, are homeothermic ectotherms.

Certain large fish, marine mammals, and some sharks can maintain parts of their body at a significantly higher temperature than that of the water. They do so using countercurrent heat exchange (figure 35.2). This allows the cooler blood in the veins to be warmed through radiation of heat from the warmer blood in the arteries located close to the veins. The arteries carry warmer blood from the center of the body.

Reptiles regulate body temperature through behavior—by placing themselves in varying locations of sunlight and shade. That's why you frequently see lizards basking in the Sun. Some reptiles maximize the effect of behavioral regulation by also controlling blood flow. The marine iguana controls heart rate and the extent of dilation or contraction of blood vessels to regulate blood available for heat transfer by conduction. Increased heart rate and vasodilation allow maximal heating on land, whereas decreased

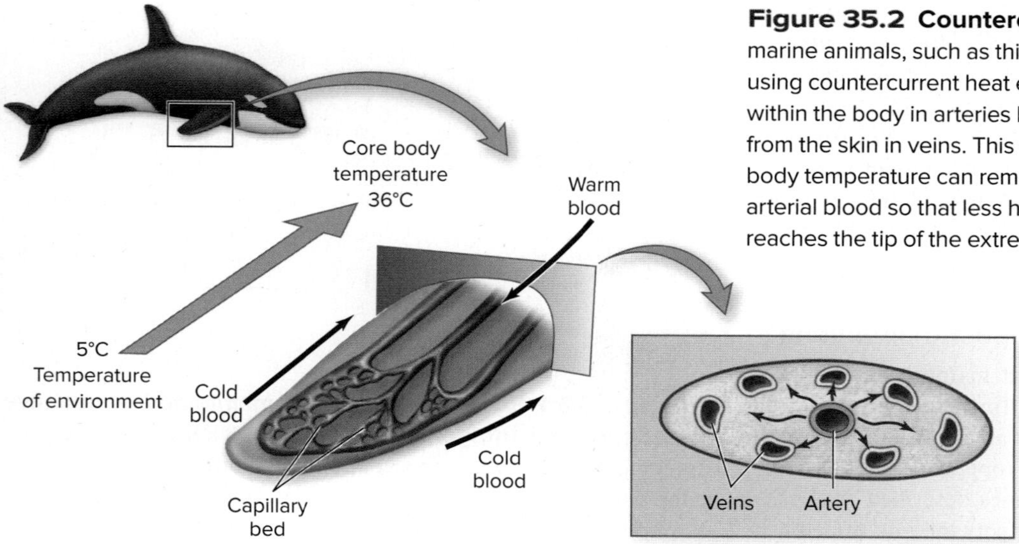

Figure 35.2 Countercurrent heat exchange. Many marine animals, such as this killer whale, limit heat loss in cold water using countercurrent heat exchange. The warm blood pumped from within the body in arteries loses heat to the cooler blood returning from the skin in veins. This warms the venous blood so that the core body temperature can remain constant in cold water, and it cools the arterial blood so that less heat is lost when the arterial blood reaches the tip of the extremity.

heart rate and vasoconstriction minimize cooling when the animal is diving for food.

In general, ectotherms have low metabolic rates, with correspondingly low intake of energy (food). It is estimated that a lizard needs only 10% of the energy intake of a mouse of comparable size. The tradeoff is that ectotherms are not capable of sustained high-energy activity.

Endotherms Create Internal Metabolic Heat for Conservation or Dissipation

LEARNING OBJECTIVE 35.1.3 Describe mechanisms for temperature homeostasis in endotherms.

When an endotherm is cold, it uses its high metabolic rate to raise its internal temperature, but that internal heat must also be dissipated when external temperatures are higher.

The simplest means of affecting heat transfer is to control the amount of blood flow to the surface of the animal. Dilating blood vessels increases the amount of blood flowing to the surface, which in turn increases thermal heat exchange and dissipation of heat. In contrast, constriction of blood vessels decreases the amount of blood flowing to the surface and decreases thermal heat exchange, limiting the amount of heat lost due to conduction.

When ambient temperatures rise, many endotherms take advantage of evaporative cooling in the form of sweating or panting. Sweating is found in some mammals, including humans, and involves the active extrusion of water from sweat glands onto the surface of the body. As the water evaporates, it cools the skin, and this cooling is transferred internally by capillaries near the surface of the skin. Panting is a similar adaptation used by some mammals and birds that takes advantage of respiratory surfaces for evaporative cooling. For evaporative cooling to be effective, the animal must be able to tolerate the loss of water.

The advantage of endothermy is that it allows sustained high-energy activity. The tradeoff for endotherms is that the high metabolic rate has a corresponding cost in requiring relatively constant and high rates of energy intake (food).

Body size and insulation

Size is one important characteristic affecting animal physiology. Changes in body mass have a large effect on metabolic rate. Smaller animals consume much more energy per unit of body mass than larger animals. This relationship is summarized in the "mouse to elephant" curve, which shows that metabolic rate versus size of mammals is not proportional (figure 35.3).

For small animals with a high metabolic rate, surface area is also large relative to their volume. In a cold environment, this can be disastrous, because they cannot produce enough internal heat to balance conductive loss through their large surface area. Thus, small endotherms in cold environments require significant insulation to maintain their body temperature. The amount of insulation can also vary seasonally and geographically, with thicker coats being needed in the north and in winter.

Large animals in hot environments have the opposite problem: although their metabolic rate is relatively low, they still produce a large amount of heat with much less relative surface area to dissipate that heat by conduction. Thus, most large endotherms in hot environments have little insulation and will use behavior to lose heat; for example, elephants flap their ears to increase

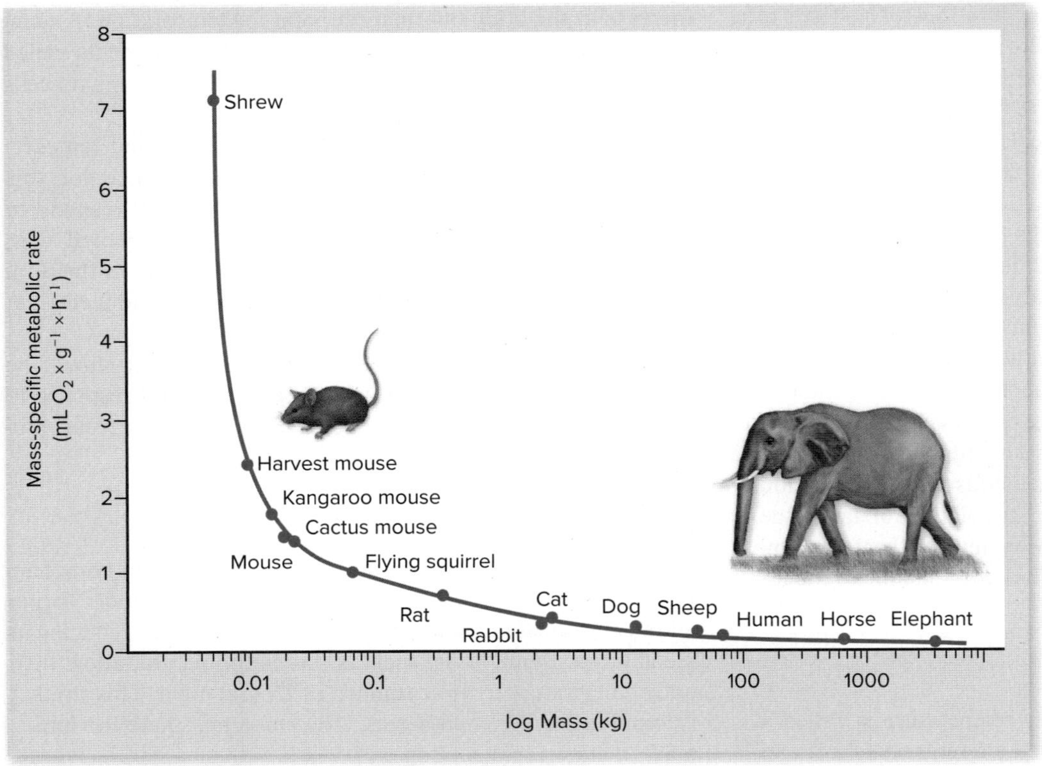

Figure 35.3 Relationship between body mass and metabolic rate in mammals. Smaller animals have a much higher metabolic rate per unit body mass than larger animals. In the figure, mass-specific metabolic rate (expressed as O_2 consumption per unit mass) is plotted against body mass. Note that the body mass axis is a logarithmic scale.

convective heat loss via the many capillaries just under the skin of the ears.

Thermogenesis

When temperatures fall below a critical lower threshold, normal endothermic responses are not sufficient to warm an animal. In this case, the animal resorts to thermogenesis, or the use of normal energy metabolism to produce heat. Thermogenesis takes two forms: shivering and nonshivering thermogenesis.

Shivering thermogenesis uses muscles to generate heat without producing useful work. It occurs in endothermic vertebrates, as well as some insects, illustrated by the earlier example of a butterfly warming its flight muscles. Shivering involves the use of antagonistic muscles to produce little net generation of movement, but hydrolysis of ATP, generating heat.

In nonshivering thermogenesis, fat metabolism is altered to produce heat instead of ATP. This involves brown fat, also called brown adipose tissue (BAT), which is distinct from white adipose tissue (WAT). Brown fat cells have a high level of the protein UCP1, which is a membrane channel that allows leakage of protons. This uncouples electron transport from oxidative phosphorylation, producing heat but no ATP. This has been studied in mice and humans. For many years it was thought that human adults lacked brown fat, but this proved wrong. In humans, brown fat is stored in small deposits in the neck and between the shoulders. This fat is highly vascularized, which allows efficient transfer of heat away from the site of production. A third kind of fat called beige fat, which can be converted from WAT to BAT, has also been identified.

Mammalian thermoregulation is controlled by the hypothalamus

A blood temperature that exceeds 37°C (98.6°F) is detected by sensory neurons in the hypothalamus (figure 35.4). This leads to stimulation of the heat-losing center in the hypothalamus. Nerves from this area cause peripheral blood vessels to dilate, bringing more blood to the surface to dissipate heat. Other nerves stimulate the production of sweat, leading to evaporative cooling. Production of hormones that stimulate metabolism is also inhibited.

When your body temperature falls below 37°C, an antagonistic set of effects is produced by the heat-promoting center in the hypothalamus. Nerves from this area cause constriction of peripheral blood vessels, thereby reducing heat transfer, and inhibiting sweating to prevent evaporative cooling. The adrenal medulla is stimulated to produce epinephrine, and the anterior pituitary to produce thyroid-stimulating hormone (TSH), both of which stimulate metabolism. TSH stimulates the thyroid to produce thyroxin, which stimulates metabolism. A combination of epinephrine and autonomic nerve stimulation of fat tissue can induce thermogenesis to produce more internal heat. As temperature rises, negative feedback to the hypothalamus reduces the heat-producing response.

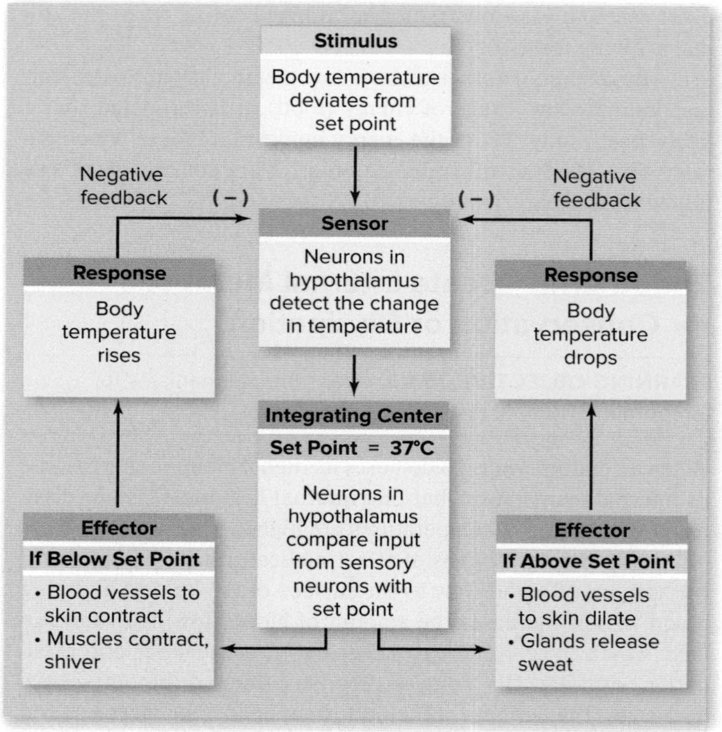

Figure 35.4 Body temperature is maintained by negative feedback and antagonistic effectors. Neurons in the hypothalamus detect changes in body temperature. The comparator (also in the hypothalamus) processes the information and activates effectors, such as surface blood vessels, sweat glands, and skeletal muscles. When a response goes beyond the set point in the other direction, negative feedback stops the response.

Fever

A substance that causes a rise in body temperature is called a **pyrogen.** This rise in temperature, called a **fever,** results from an increase in the body's normal set point for temperature. A number of gram-negative bacteria have cell wall components called endotoxins that act as pyrogens. Some circulating immune cells can produce pyrogens as well.

Fever seems to be adaptive because increased temperature can inhibit the growth of bacteria. Evidence for this comes from the observation that some ectotherms respond to pyrogens as well. When desert iguanas were injected with pyrogen-producing bacteria, they spent more time in the Sun, thereby elevating their temperature: that is, they induced fever behaviorally!

Fever is no longer viewed as something that should be treated medically, but rather as a normal and vital response to infection. However, extremely high fevers can be dangerous, producing symptoms from seizure to delirium.

Torpor

Endotherms can also reduce both metabolic rate and body temperature to produce a state of dormancy called torpor. Torpor allows an animal to reduce its need for food intake by reducing its metabolism. Some birds, such as the hummingbird, allow their body temperature to drop as much as 25°C at night. This strategy is found in smaller endotherms; larger mammals have too large a mass to allow rapid cooling.

Hibernation is an extreme state in which deep torpor lasts for several weeks to even months. In this case, the animal's temperature may drop as much as 20°C below its normal set point. The animals that practice hibernation are mainly in the mid-range of size. Smaller endotherms quickly consume more energy than they can easily store, even by reducing their metabolic rate, and very large mammals do not appear to hibernate. It was long thought that bears hibernate, but in reality their temperature is reduced by only a few degrees. They instead undergo a prolonged winter sleep. With their large thermal mass and low rate of heat loss, they do not seem to require the additional energy savings of hibernation.

REVIEW OF CONCEPT 35.1

Homeostasis is the dynamic constancy of an organism's internal environment. Negative feedback mechanisms correct deviations from a set point for internal variables such as temperature and pH. Antagonistic action of two effectors also helps maintain constancy. Ectotherms control body temperature with behavior. Mammals are endotherms that maintain a consistent body temperature by controlling metabolic rate. Thermogenesis can generate heat by uncoupling electron transport from ATP synthesis. Fever is a change in temperature set point.

- *How do antagonistic effectors and negative feedback function together?*

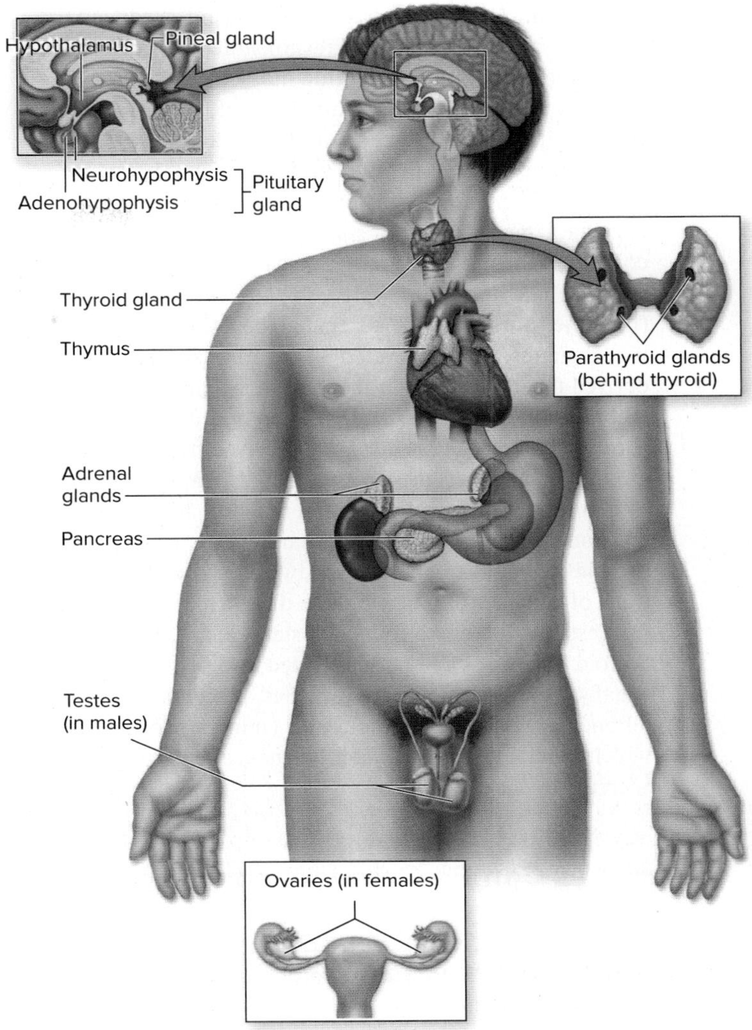

Figure 35.5 The human endocrine system. The major endocrine glands are shown, but many other organs secrete hormones in addition to their primary functions.

35.2 Hormones Are Chemical Messages That Direct Body Processes

A **hormone** is a regulatory chemical that is secreted into extracellular fluid and carried by the blood and can therefore act at a distance from its source. Organs that are specialized to secrete hormones are called *endocrine glands,* but some organs, such as the liver and the kidney, can produce hormones in addition to performing other functions. The organs and tissues that produce hormones are collectively called the **endocrine system.**

Endocrine Glands Produce Three Chemical Classes of Hormones

LEARNING OBJECTIVE 35.2.1 Describe the role of hormones in regulating body processes.

Hormones are carried to every cell in the body through blood, but only a cell with a receptor for a hormone can respond to it. The receptor proteins specifically bind the hormone and activate signal transduction pathways that produce a response to the hormone (cell communication is covered in chapter 9). The highly specific interaction between hormone and receptor enables hormones to be active at remarkably small concentrations. It is not unusual to find hormones circulating in the blood at concentrations of 10^{-8} to 10^{-10} M.

The endocrine system (figure 35.5) includes all of the organs that secrete hormones—the thyroid gland, pituitary gland, adrenal glands, and so on. Cells in these organs secrete hormones into extracellular fluid, where it diffuses into surrounding blood capillaries. For this reason, hormones are referred to as endocrine secretions. **Exocrine glands,** such as salivary glands, secrete their products into a duct leading outside the body, or into the gut.

Molecules that function as hormones must exhibit two basic characteristics. First, they must be sufficiently complex to convey regulatory information to their targets. Second, hormones must be adequately stable to resist destruction prior to reaching their target cells. Three primary chemical categories of molecules meet these requirements.

1. **Peptides and proteins** are composed of chains of amino acids. Some important examples of peptide hormones include antidiuretic hormone (9 amino acids), insulin (51 amino acids), and growth hormone (191 amino acids).

These hormones are encoded in DNA and produced by the same cellular machinery responsible for transcription and translation of other peptide molecules. The most complex are glycoproteins composed of two peptide chains with attached carbohydrates. Examples include thyroid-stimulating hormone and luteinizing hormone.

2. **Amino acid derivatives** are hormones manufactured by enzymatic modification of specific amino acids; this group comprises the biogenic amines discussed in chapter 33. They include hormones secreted by the adrenal medulla (the inner portion of the adrenal gland), thyroid, and pineal glands. Those secreted by the adrenal medulla are derived from tyrosine. Known as **catecholamines,** they include epinephrine (adrenaline) and norepinephrine (noradrenaline). Other hormones derived from tyrosine are the **thyroid hormones,** secreted by the thyroid gland. The pineal gland secretes a different amine hormone, **melatonin,** derived from tryptophan.
3. **Steroids** are lipids manufactured by enzymatic modifications of cholesterol. They include the hormones testosterone, estradiol, progesterone, aldosterone, and cortisol. Steroid hormones can be subdivided into sex steroids, usually secreted by the testes, ovaries, placenta, and adrenal cortex; and corticosteroids (mineralocorticoids, which influence salt and water balance, and cortisol), secreted by the adrenal cortex.

Hormones Can Be Categorized as Lipophilic or Hydrophilic

LEARNING OBJECTIVE 35.2.2 Differentiate between lipophilic and hydrophilic hormones.

The structure of hormones affects how they are transported and how they interact with target cells. Hormones are classified as either lipophilic (nonpolar and fat-soluble) or hydrophilic (polar and water-soluble). The lipophilic hormones include the steroid hormones and thyroid hormones. Most other hormones are hydrophilic.

This chemical difference also has a profound effect on the structure and function of the receptors for each class of hormone. Hydrophilic hormones are soluble in blood, and therefore can be easily transported, but they cannot cross the membrane. So hydrophilic hormones bind to the extracellular domain of transmembrane receptors, and the receptor itself "carries" the signal inside the cell. Lipophilic hormones are transported in the blood attached to transport proteins, but they can cross the plasma membrane to enter a target cell, where they usually act by stimulating gene expression. This then extends to the nature of the effects elicited by a hormone, where hydrophilic hormones tend to be fast-acting and of short duration (minutes to hours), whereas lipophilic hormones are slow-acting but long-lived (days to weeks).

Although this distinction is generally true, some lipophilic hormones can stimulate membrane receptors either on the plasma membrane or on internal membranes. These receptors are usually classical nuclear receptors that are linked to a lipid, anchoring them in the membrane. In rare cases, lipophilic hormones can even stimulate G protein–coupled receptors. This blurs the clear distinction between the modes of action of lipophilic and hydrophilic hormones.

Lipophilic Hormones Activate Intracellular Receptors

LEARNING OBJECTIVE 35.2.3 Explain how steroid hormone receptors activate transcription.

Lipophilic hormones can enter cells because the lipid portion of the plasma membrane does not present a barrier. Once inside the cell, the lipophilic regulatory molecules all have a similar mechanism of action (figure 35.6).

Transport and receptor binding

These hormones circulate bound to transport proteins, which make them soluble and prolong their survival in the blood. When the hormones arrive at their target cells, they dissociate from their transport proteins and pass through the plasma membrane of the cell. The hormone then binds to an intracellular receptor protein. Some steroid hormones bind to their receptors in the cytoplasm, then move as a hormone–receptor complex into the nucleus. Other steroids and the thyroid hormones travel directly into the nucleus before encountering their receptor proteins.

Activation of transcription in the nucleus

The hormone receptor, activated by binding to the hormone, is now also able to bind to specific regions of the DNA. These DNA regions, located in the promoters of specific genes, are known as **hormone response elements.** The binding of the hormone–receptor complex has a direct effect on the level of transcription at that site by activating, or in some cases deactivating, gene transcription. Receptors therefore function as *hormone-activated transcription factors* (refer to chapters 9 and 16).

The proteins that result from activation of these transcription factors have activity that changes the metabolism of the target cell in a specific fashion. When estrogen binds to its receptor in liver cells of chickens, for example, it activates the cell to produce the protein vitellogenin, which is then transported to the ovary to form the yolk of eggs.

There is now evidence that some steroid hormones can also activate membrane receptors. These can be "classical receptors" that have been anchored to the membrane, as well as G protein–coupled receptors. The action of these receptors probably explains the rapid effects observed for some steroid hormones.

Hydrophilic Hormones Activate Receptors on Target Cell Membranes

LEARNING OBJECTIVE 35.2.4 Explain how the signal carried by peptide hormones crosses the plasma membrane.

Hydrophilic hormones do not easily cross the plasma membranes of target cells. These hormones act through membrane receptors with an extracellular domain that can bind hormone, and an

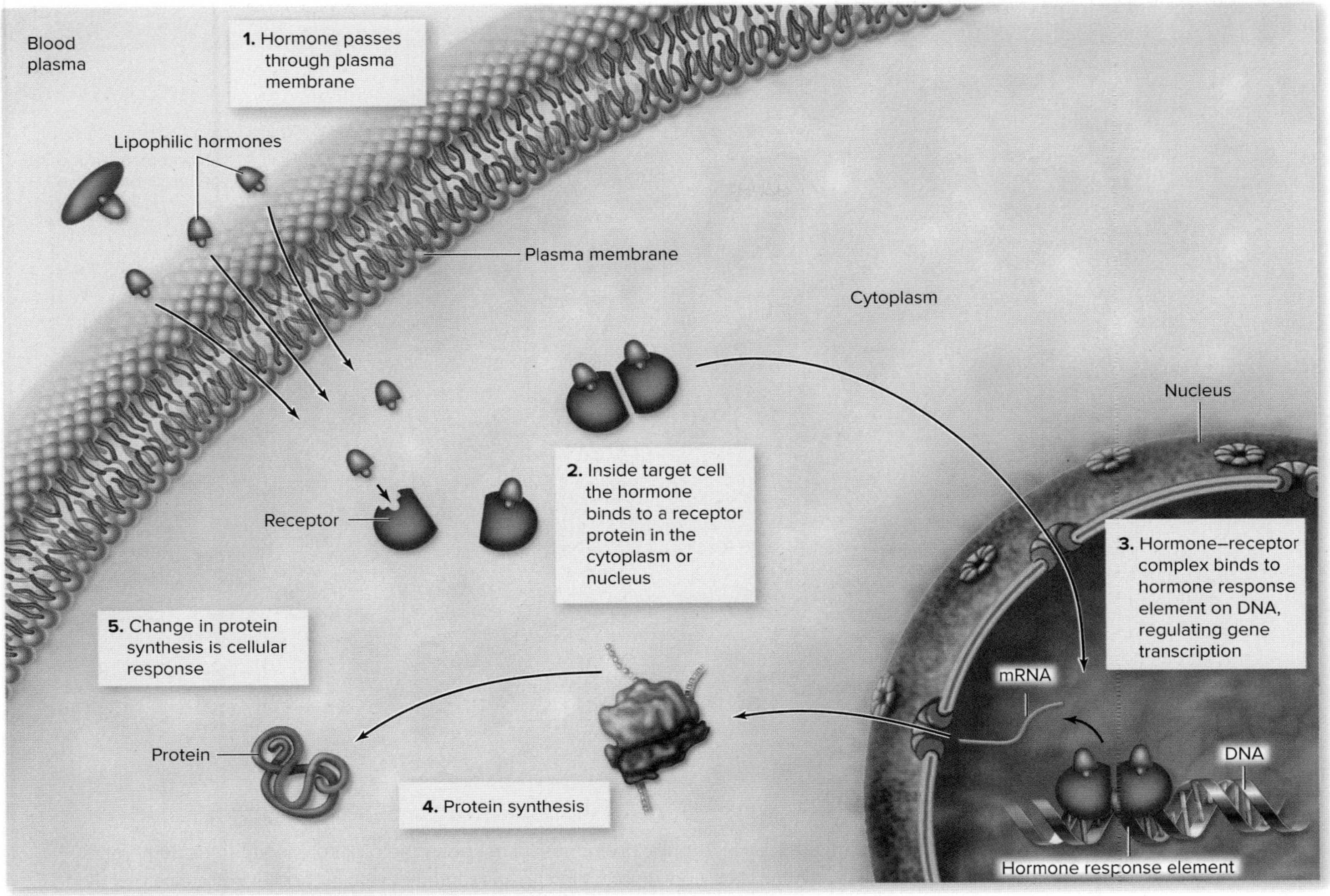
Blood plasma
1. Hormone passes through plasma membrane
Lipophilic hormones
Plasma membrane
Cytoplasm
Nucleus
Receptor
2. Inside target cell the hormone binds to a receptor protein in the cytoplasm or nucleus
3. Hormone–receptor complex binds to hormone response element on DNA, regulating gene transcription
5. Change in protein synthesis is cellular response
mRNA
Protein
DNA
4. Protein synthesis
Hormone response element

Figure 35.6 The mechanism of lipophilic hormone action. Lipophilic hormones diffuse through the plasma membrane of cells and bind to intracellular receptor proteins. The hormone–receptor complex then binds to specific regions of the DNA (hormone response elements), regulating the production of messenger RNA (mRNA). Most receptors for these hormones reside in the nucleus; if the hormone is one that binds to a receptor in the cytoplasm, the hormone–receptor complex moves together into the nucleus.

intracellular domain that can interact with cytoplasmic proteins (refer to chapter 9). These receptors fall into two categories: receptor tyrosine kinases and G protein–coupled receptors.

Receptor tyrosine kinases

The receptor tyrosine kinases (RTKs) are an important class of hormone receptor. Signaling through RTKs often results in activation of intracellular protein kinases that regulate protein function by phosphorylation. This can produce broad effects on cellular functions.

When a hormone binds to an RTK, the first response is for receptors to associate together as dimers, which stimulates their kinase activity. The associated dimers become phosphorylated, initiating signal transduction pathways through cellular proteins that bind phosphotyrosine (figure 35.7). For example, when the peptide hormone insulin binds its receptor, the response results in glucose transport proteins being placed in the plasma membrane to facilitate glucose transport. Growth hormone works through similar mechanisms, except that the phosphotyrosine-binding proteins initiate a kinase cascade (refer to chapter 10).

G protein–coupled receptors

Another large class of important hormone membrane receptors is the G protein–coupled receptors (GPCR). As you learned in chapter 9, GPCRs often act via mechanisms that increase the level of second messengers within the target cell. A number of different molecules in the cell can serve as second messengers.

In the early 1960s, Earl Sutherland showed that activation of the epinephrine receptor on liver cells increases intracellular cyclic adenosine monophosphate, or cyclic AMP (cAMP), which then serves as an intracellular second messenger. Another important hormonally regulated second-messenger system generates two lipid messengers: inositol triphosphate (IP_3) and diacylglycerol (DAG). These systems were described in chapter 9.

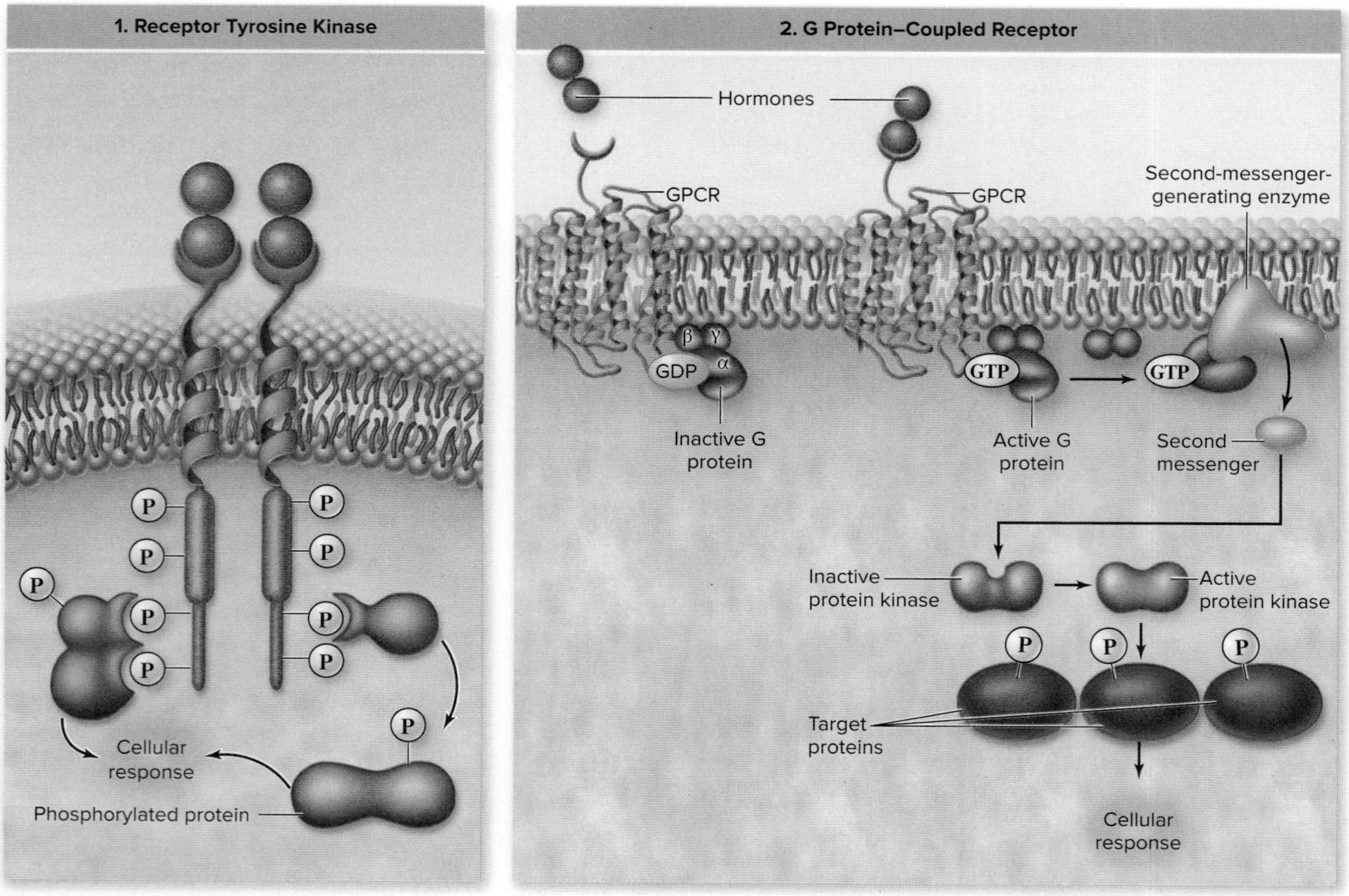

Figure 35.7 The action of hydrophilic hormones. Hydrophilic hormones do not enter the cell, but bind to transmembrane receptors. (1) Receptor tyrosine kinases (RTKs) autophosphorylate upon ligand binding, then interact with intracellular proteins that bind to phosphotyrosine to initiate signal transduction pathways. (2) G protein–coupled receptors (GPCRs) activate intermediary G proteins that activate effector proteins. These can produce second messengers, which can activate protein kinases and other signal transduction pathways.

GPCRs function by activating membrane-associated G proteins that in turn stimulate effector proteins, which often are enzymes that generate second messengers. When a hormone binds to a GPCR, the G protein is activated by exchanging GDP for GTP. The activated G protein shuttles from the receptor to the second-messenger-generating enzyme (figure 35.7). The activated enzyme then produces the second-messenger molecules, which can propagate the signal through the cell.

In the case of epinephrine, the G protein activates an enzyme called adenylyl cyclase, which catalyzes the formation of the second messenger cAMP from ATP. The second messenger formed at the inner surface of the plasma membrane then diffuses within the cytoplasm, where it binds to and activates protein kinases.

The identities of the proteins that are subsequently phosphorylated by the protein kinases vary with cell type and include enzymes, membrane transport proteins, and transcription factors. This diversity provides hormones with distinct actions in different tissues. For example, in liver cells, cAMP-dependent protein kinases activate enzymes that convert glycogen into glucose. In cardiac muscle cells in the heart, cAMP causes an increase in the rate and force of contraction. In smooth muscle, epinephrine acts through another second messenger, IP_3, to cause a release of intracellular Ca^{2+} and stimulate contraction. This enables one hormone to elicit different responses in three different cell types.

The cellular response to a hormone also depends on the type of G protein activated by the hormone's receptor. Some G proteins will activate second-messenger-producing enzymes, whereas others inhibit their second-messenger-generating enzyme. As a result, some hormones stimulate protein kinases in their target cells, and others inhibit their targets.

Duration of hydrophilic hormone effects

The binding of a hydrophilic hormone to its receptor is reversible and usually very brief; hormones soon dissociate from receptors or are rapidly deactivated by their target cells after binding. Additionally, target cells contain specific enzymes that rapidly deactivate second messengers and protein kinases. As a result, hydrophilic hormones are capable of stimulating immediate responses within cells, but they often have a brief duration of action (minutes to hours).

REVIEW OF CONCEPT 35.2

Hormones coordinate the activity of specific target cells. The three chemical classes of endocrine hormones are peptides and proteins, amino acid derivatives, and steroids. Steroids are lipophilic and cross the membrane to bind to intracellular receptor proteins. The hormone–receptor complex then acts to modulate gene expression. Hydrophilic hormones bind externally to membrane receptors that activate protein kinases directly or that operate through second-messenger systems such as cAMP or IP_3/DAG.

- *How can a single hormone, such as epinephrine, have different effects in different tissues?*

35.3 The Pituitary and the Hypothalamus Are the Body's Control Centers

The **pituitary gland,** also known as the **hypophysis,** hangs by a stalk from the hypothalamus at the base of the brain posterior to the optic chiasm. The hypothalamus is a part of the central nervous system (CNS) that has a major role in regulating body processes. Both of these structures were described in chapter 33; here we discuss in detail how they work together to bring about homeostasis and changes in body processes.

The Pituitary Is a Compound Endocrine Gland

LEARNING OBJECTIVE 35.3.1 Explain why the pituitary is considered a compound gland.

A microscopic view reveals that the gland consists of two parts, one of which, called the **anterior pituitary** or **adenohypophysis,** appears glandular. The other portion appears fibrous and is called the **posterior pituitary** or **neurohypophysis.** These two portions of the pituitary gland have different embryonic origins, secrete different hormones, and are regulated by different control systems. These two regions are conserved in all vertebrate animals, suggesting that there is an ancient and important function of each.

The posterior pituitary stores and releases two neurohormones

The posterior pituitary appears fibrous because it contains axons that originate in cell bodies within the hypothalamus and that extend along the stalk of the pituitary as a tract of fibers. This anatomical relationship results from the way the posterior pituitary is formed in embryonic development. As the floor of the third ventricle of the brain forms the hypothalamus, part of this neural tissue grows downward to produce the posterior pituitary. The hypothalamus and posterior pituitary thus remain directly interconnected by a tract of axons.

Antidiuretic hormone

The endocrine role of the posterior pituitary emerged in an unusual medical case reported in 1912. A man who had been shot in the head developed the need to urinate every 30 minutes or so, 24 hours a day. The bullet had lodged in his posterior pituitary; subsequent research demonstrated that removal of the posterior pituitary produces the same symptoms. Later a peptide hormone was isolated from the posterior pituitary, antidiuretic hormone (ADH), also called vasopressin. ADH stimulates water reabsorption by the kidneys (figure 35.8) and in doing so inhibits diuresis (urine production). Loss of the ability to produce ADH, as in the shooting victim, reduces the kidney's ability to reabsorb water, leading to excess urine production. The role of ADH in kidney function is discussed in section 35.7.

Oxytocin

The posterior pituitary also secretes **oxytocin,** a second peptide neurohormone that, like ADH, is composed of nine amino acids. In mammals, oxytocin stimulates the milk ejection reflex. During suckling, sensory receptors in the nipples send impulses to the hypothalamus, triggering the release of oxytocin. Oxytocin is also needed to stimulate uterine contractions in women during childbirth, and it is given in postterm pregnancy to induce labor.

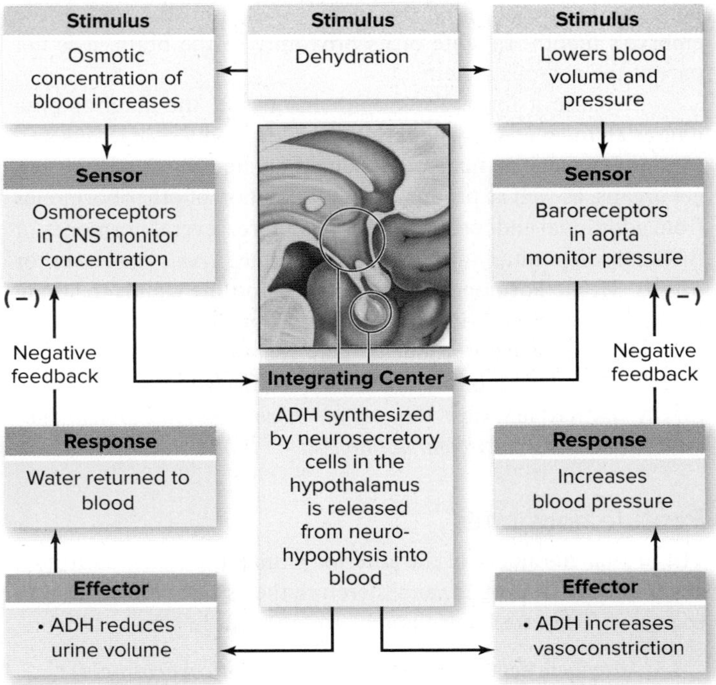

Figure 35.8 The effects of antidiuretic hormone (ADH). Dehydration increases the osmotic concentration of the blood and lowers blood pressure, stimulating the neurohypophysis to secrete ADH. ADH increases reabsorption of water by the kidneys and causes vasoconstriction, increasing blood pressure. Decreased blood osmolarity and increased blood pressure complete negative feedback loops to maintain homeostasis.

A related posterior pituitary neurohormone, *arginine vasotocin,* exerts similar effects in nonmammalian species. For example, in chickens and sea turtles, arginine vasotocin activates oviduct contraction during egg laying.

More recently, oxytocin has been identified as an important regulator of reproductive behavior. In most humans, it is thought to be involved in promoting pair bonding (leading to its being called the "cuddle hormone") as well as regulating sexual responses, including arousal and orgasm. For these effects, it most likely functions in a paracrine fashion inside the CNS, much like a neurotransmitter.

Hypothalamic production of the neurohormones

ADH and oxytocin are actually produced by neuron cell bodies located in the hypothalamus. These two neurohormones are transported along the axon tract that runs from the hypothalamus to the posterior pituitary, where they are stored. In response to the appropriate stimulation–increased blood plasma osmolarity in the case of ADH, the suckling of a baby in the case of oxytocin—the neurohormones are released by the posterior pituitary into the blood.

Because this reflex control involves both the nervous and the endocrine systems, ADH and oxytocin are said to be secreted by a **neuroendocrine reflex.**

The anterior pituitary produces seven hormones

The anterior pituitary, unlike the posterior pituitary, does not develop from growth of the brain; instead, it develops from a pouch of epithelial tissue that pinches off from the roof of the embryo's mouth. In spite of its proximity to the brain, it is not part of the nervous system.

Because it forms from epithelial tissue, the anterior pituitary is an independent endocrine gland. It produces at least seven essential hormones, many of which stimulate growth of their target organs, as well as production and secretion of other hormones from additional endocrine glands. Therefore, several hormones of the anterior pituitary are collectively termed *tropic hormones,* or *tropins.* Tropic hormones act on other endocrine glands to stimulate the secretion of hormones produced by the target gland.

The hormones produced and secreted by different cell types in the anterior pituitary can be categorized into three structurally similar families: the *peptide hormones,* the *protein hormones,* and the *glycoprotein hormones.*

Peptide hormones

The **peptide hormones** of the anterior pituitary are cleaved from a single precursor protein, and therefore they share some common sequence. They are fewer than 40 amino acids in size.

1. **Adrenocorticotropic hormone (ACTH, or *corticotropin*)** stimulates the adrenal cortex to produce corticosteroid hormones, including cortisol (in humans) and corticosterone (in many other vertebrates). These hormones regulate glucose homeostasis and are important in the response to stress.
2. **Melanocyte-stimulating hormone (MSH)** stimulates the synthesis and dispersion of melanin pigment, which darkens the epidermis of some fish, amphibians, and reptiles and can control hair pigment color in mammals.

Protein hormones

The **protein hormones** each comprise a single chain of approximately 200 amino acids, and they share significant structural similarities.

1. **Growth hormone (GH, or *somatotropin*)** stimulates the growth of muscle, bone (indirectly), and other tissues, and it is essential for proper metabolic regulation.
2. **Prolactin (PRL)** is best known for stimulating the mammary glands to produce milk in mammals; however, it has diverse effects on many other targets, including regulation of ion and water transport across epithelia, stimulation of a variety of organs that nourish young, and activation of parental behaviors.

Glycoprotein hormones

The largest and most complex hormones known, the *glycoprotein hormones* are dimers, containing alpha (α) and beta (β) subunits, each about 100 amino acids in size, with covalently linked sugar residues. The α subunit is common to all three hormones. The β subunit differs, endowing each hormone with a different target specificity.

1. **Thyroid-stimulating hormone (TSH, or *thyrotropin*)** stimulates the thyroid gland to produce the hormone thyroxine, which in turn regulates development and metabolism by acting on nuclear receptors.
2. **Luteinizing hormone (LH)** stimulates the production of estrogen and progesterone by the ovaries and is needed for ovulation in female reproductive cycles (refer to chapter 36). In males, it stimulates the testes to produce testosterone, which is needed for sperm production.
3. **Follicle-stimulating hormone (FSH)** is required for the development of ovarian follicles in females. In males, it is required for the development of sperm. FSH stimulates the conversion of testosterone into estrogen in females, and into dihydroxytestosterone in males. FSH and LH are collectively referred to as *gonadotropins.*

Hypothalamic Neurohormones Regulate the Anterior Pituitary

LEARNING OBJECTIVE 35.3.2 Describe the connections among the hypothalamus, posterior pituitary, and anterior pituitary.

The anterior pituitary, unlike the posterior pituitary, is not derived from the brain and does not receive an axon tract from the hypothalamus. Nevertheless, the hypothalamus controls the production and secretion of its hormones. This control is itself exerted hormonally rather than by means of nerve axons.

Neurons in the hypothalamus secrete two types of neurohormones, **releasing hormones** and **inhibiting hormones,** that diffuse into blood capillaries at the base of the hypothalamus (figure 35.9). These capillaries drain into small veins that run within the stalk of the pituitary to a second bed of capillaries in the anterior pituitary. This unusual system of vessels is known as the *hypothalamohypophyseal portal system.* In a portal system, two capillary beds are linked by veins. In this case, the hormone enters the first capillary bed, and the vein delivers this to the second capillary bed, where the hormone exits and enters the anterior pituitary.

Releasers

Each neurohormone released by the hypothalamus into the portal system regulates the secretion of a specific hormone in the anterior pituitary. Releasing hormones are peptide neurohormones that stimulate the release of other hormones; specifically, *thyrotropin-releasing hormone (TRH)* stimulates the release of TSH; *corticotropin-releasing hormone (CRH)* stimulates the release of ACTH; and *gonadotropin-releasing hormone (GnRH)* stimulates the release of FSH and LH. A releasing hormone for growth hormone, called *growth hormone-releasing hormone (GHRH),* has also been discovered, and TRH, oxytocin, and vasoactive intestinal peptide all appear to act as releasing hormones for prolactin.

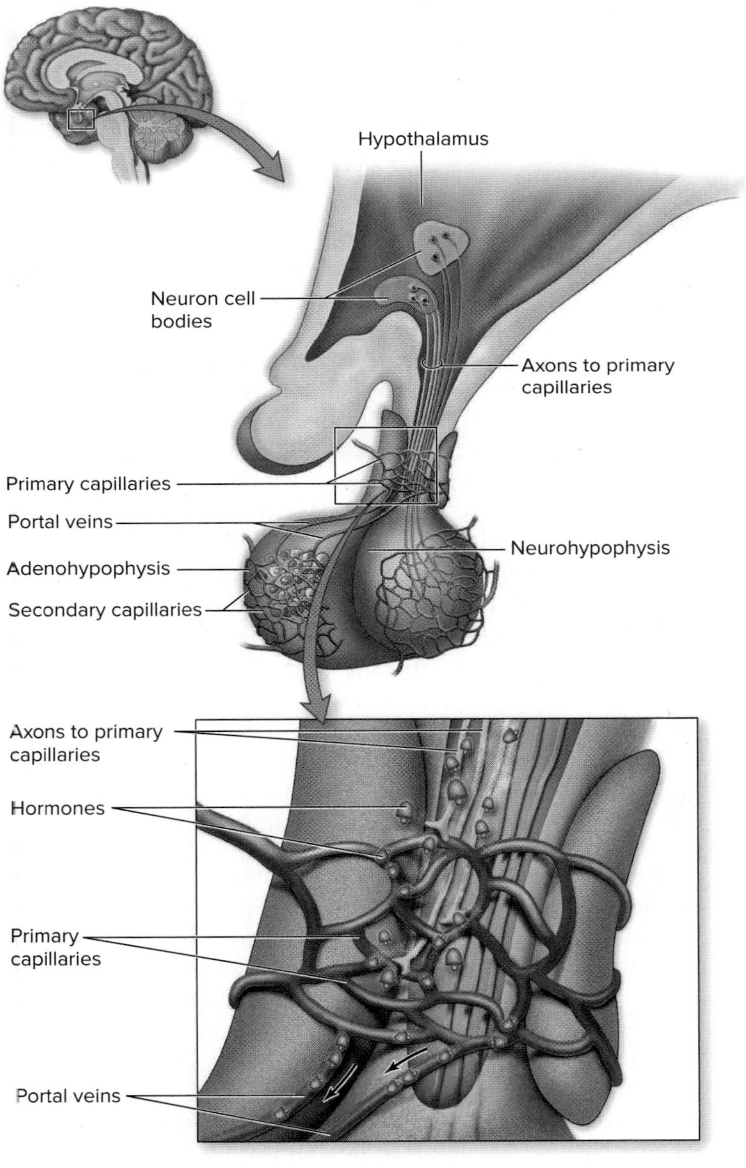

Figure 35.9 Hormonal control of the adenohypophysis by the hypothalamus. Neurons in the hypothalamus secrete hormones that are carried by portal blood vessels directly to the adenohypophysis, where they either stimulate or inhibit the secretion of hormones from the adenohypophysis.

Inhibitors

The hypothalamus also secretes neurohormones that inhibit the release of certain anterior-pituitary hormones. To date, three such neurohormones have been discovered: *somatostatin,* or *growth hormone-inhibiting hormone (GHIH),* which inhibits the secretion of GH; *prolactin-inhibiting factor (PIF),* which inhibits the secretion of prolactin and has been found to be the neurotransmitter dopamine; and *MSH-inhibiting hormone (MIH),* which inhibits the secretion of MSH.

Early experiments to explore the role of the pituitary were rather crude: they involved complete surgical removal (hypophysectomy). The hypophysectomized animals exhibited a broad spectrum of deficits, including reduced growth and development, diminished metabolism, and failure of reproduction. These diverse effects led to referring to the pituitary as the "master gland." This is due to the pituitary producing not only hormones that act directly but also tropic hormones that stimulate other endocrine glands. These tropic hormones act on the thyroid, adrenal glands, and gonads. Of anterior-pituitary hormones, growth hormone, prolactin, and MSH work primarily through direct effects, whereas the tropic hormones ACTH, TSH, LH, and FSH have endocrine glands as their exclusive targets.

Effects of growth hormone

The importance of the anterior pituitary is illustrated by a condition known as *gigantism,* characterized by excessive growth of the entire body or any of its parts. The tallest male human being ever recorded, Robert Wadlow, had gigantism (figure 35.10). Born in 1928, he stood 8 feet 11 inches tall, weighed 485 pounds, and was still growing before he died from an infection at the age of 22. The tallest female human known to medical science was Zeng Jinlian, who was 8 ft 1.75 in. Like Wadlow, she died very young (at age 17).

We now know that gigantism is caused by the excessive secretion of GH in a growing child. By contrast, a deficiency in GH secretion during childhood results in **pituitary dwarfism**—a failure to achieve normal stature.

GH stimulates protein synthesis and growth of muscles and connective tissues; it also indirectly promotes the elongation of bones by stimulating cell division in the cartilaginous epiphyseal growth plates of bones (refer to chapter 32). Researchers found that this stimulation does not occur in the absence of blood plasma, suggesting that GH must work in concert with another hormone to exert its effects on bone. We now know that GH stimulates the production of **insulin-like growth factors,** which liver and bone produce in response to stimulation by GH. The insulin-like growth factors then stimulate cell division in the epiphyseal growth plates, and thus the elongation of the bones.

Figure 35.10 The Alton giant. This photograph of Robert Wadlow of Alton, Illinois, was taken at the airport in Windsor, Ontario. Age 19 and already 8 ft 7 in, he is shown next to his 5 ft 11 in father. Born normal size, he developed a growth-hormone-secreting pituitary tumor as a young child and never stopped growing during his 22 years of life, reaching a height of 8 ft 11 in tall.

Toronto Star Archives/Getty Images

REVIEW OF CONCEPT 35.3

The posterior pituitary develops from neural tissue; the anterior pituitary develops from epithelial tissue. Axons from the hypothalamus extend into the posterior pituitary and produce neurohormones; these neurons also secrete factors that release or inhibit hormones of the anterior pituitary. Releasers stimulate secretion of hormones; TRH causes TSH release. Inhibitors suppress secretion; GHIH inhibits GH release.

- *Could someone with a pituitary tumor causing gigantism be treated with GHIH?*

35.4 Peripheral Endocrine Glands Play Major Roles in Homeostasis

Many endocrine glands are controlled by tropic hormones of the pituitary, but others, such as the adrenal medulla and the pancreas, are independent of pituitary control. Several endocrine glands develop from derivatives of the primitive pharynx, which is the most anterior segment of the digestive tract. These include the thyroid and parathyroid glands, which produce hormones that regulate processes associated with nutrient uptake, such as carbohydrate, lipid, protein, and mineral metabolism.

The Thyroid Gland Regulates Basal Metabolism and Development

LEARNING OBJECTIVE 35.4.1 Describe the actions of thyroid hormones.

The thyroid gland varies in shape in different vertebrate species, but it is always found in the neck area, anterior to the heart. In humans it is shaped like a bow tie and lies just below the Adam's apple in the front of the neck.

The thyroid gland secretes three hormones: primarily thyroxine, smaller amounts of triiodothyronine (collectively referred to as thyroid hormones), and calcitonin. Thyroid hormones are unique in being the only molecules in the body containing iodine (thyroxine contains four iodine atoms, triiodothyronine contains three).

Thyroid-related disorders

Thyroid hormones work by binding to nuclear receptors located in most cells in the body, influencing the production and activity of a large number of cellular proteins. The importance of thyroid hormones first became apparent from studies of human thyroid disorders. Adults with hypothyroidism have low metabolism due to underproduction of thyroxine, including a reduced ability to utilize carbohydrates and fats. As a result, they are often fatigued, are overweight, and often feel cold. Fortunately, because thyroid hormones are small, simple molecules, people with hypothyroidism can take thyroxine orally as a pill.

People with hyperthyroidism, by contrast, exhibit opposite symptoms: weight loss, nervousness, high metabolism, and overheating because of overproduction of thyroxine. Drugs are available that block thyroid hormone synthesis in the thyroid gland, but in some cases portions of the thyroid gland must be removed surgically or by radiation treatment.

Actions of thyroid hormones

Thyroid hormones regulate enzymes controlling carbohydrate and lipid metabolism in most cells, promoting the appropriate use of these fuels for maintaining the body's basal metabolic rate. Thyroid hormones often function cooperatively, or synergistically, with other hormones, promoting the activity of growth hormone, epinephrine, and reproductive steroids. Through these actions, thyroid hormones ensure that adequate cellular energy is available to support metabolically demanding activities.

In humans, who exhibit a relatively high metabolic rate at all times, thyroid hormones are maintained in the blood at relatively high levels. In contrast, in reptiles, amphibians, and fish, which undergo seasonal cycles of activity, thyroid hormone levels in the blood increase during periods of metabolic activation and diminish during periods of inactivity in cold months.

Some of the most dramatic effects of thyroid hormones involve their regulation of growth and development. In developing humans, thyroid hormones promote the growth of neurons and stimulate the maturation of the CNS. Children born with hypothyroidism are stunted in their growth and suffer severe intellectual impairment. Early detection allows this condition to be treated with thyroid hormone administration.

An even more dramatic role for thyroid hormones is displayed in amphibians. Thyroid hormones direct the metamorphosis of tadpoles into frogs, the transformation of an aquatic, herbivorous larval form into a terrestrial, carnivorous adult form. Removal of the thyroid gland from a tadpole prevents metamorphosis, and immature tadpoles fed pieces of a thyroid gland will undergo premature metamorphosis and become miniature frogs.

Calcium Homeostasis Is Regulated by Several Hormones

LEARNING OBJECTIVE 35.4.2 Describe the components of Ca^{2+} homeostasis.

Calcium is a vital component of the vertebrate body, both because of its being a structural component of bones and because of its role in ion-mediated processes such as muscle contraction. The thyroid and parathyroid glands act with vitamin D to regulate calcium homeostasis.

Calcitonin secretion by the thyroid

In addition to the thyroid hormones, the thyroid gland also secretes **calcitonin,** a peptide hormone that plays a role in maintaining proper levels of calcium (Ca^{2+}) in the blood. When the blood Ca^{2+} concentration rises too high, calcitonin stimulates the uptake of calcium into bones, thus lowering its level in the blood. Although calcitonin may be important in the physiology of some vertebrates, it appears less important in the day-to-day regulation of Ca^{2+} levels in adult humans. It may, however, play an important role in bone remodeling in rapidly growing children.

Parathyroid hormone (PTH)

The parathyroid glands are four small glands attached to the thyroid. Because of their size, researchers ignored them until well into the 20th century. The first suggestion that these organs have an endocrine function came from experiments on dogs: if their parathyroid glands were removed, the Ca^{2+} concentration in the dogs' blood plummeted to less than half the normal value. The Ca^{2+} concentration returned to normal when an extract of parathyroid gland was administered. However, if too much of the extract was administered, the dogs' Ca^{2+} levels rose far above normal as the calcium phosphate crystals in their bones were dissolved. It was clear that the parathyroid glands produce a hormone that stimulates the release of calcium from bone.

The hormone produced by the parathyroid glands is a peptide called **parathyroid hormone (PTH).** PTH is synthesized and released in response to falling levels of Ca^{2+} in the blood. This decline cannot be allowed to continue uncorrected, because a significant fall in the blood Ca^{2+} level can cause severe muscle spasms. A normal blood Ca^{2+} level is important for the functioning of muscles, including the heart, and for proper functioning of the nervous and endocrine systems.

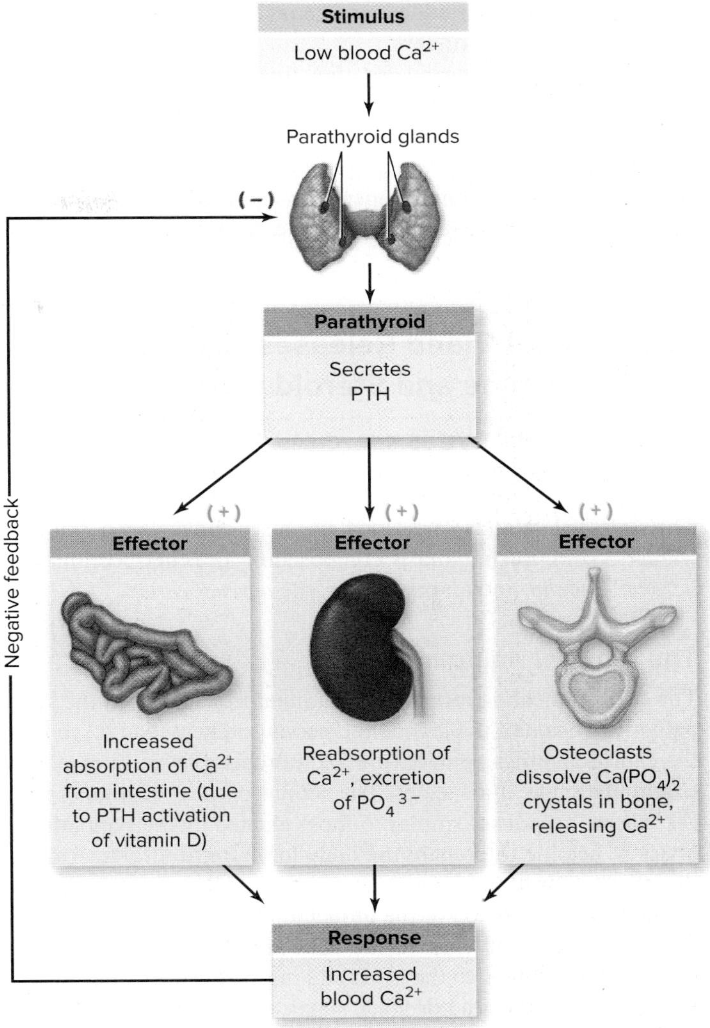

Figure 35.11 Regulation of blood Ca^{2+} levels by parathyroid hormone (PTH). When blood Ca^{2+} levels are low, PTH is released by the parathyroid glands. PTH directly stimulates the dissolution of bone and the reabsorption of Ca^{2+} by the kidneys. PTH indirectly promotes the intestinal absorption of Ca^{2+} by stimulating the production of the active form of vitamin D.

PTH stimulates the osteoclasts (bone cells) in bone to dissolve the calcium phosphate crystals of the bone matrix and release Ca^{2+} into the blood (figure 35.11). PTH also stimulates the kidneys to reabsorb Ca^{2+} from the urine and leads to the activation of vitamin D, needed for the absorption of calcium from food in the intestine.

Vitamin D

Vitamin D is produced in the skin from a cholesterol derivative in response to ultraviolet light. It is called an essential vitamin, because in temperate regions of the world a dietary source is needed to supplement the amount produced by the skin. (In the tropics, people generally receive enough exposure to sunlight to produce adequate vitamin D.) Diffusing into the blood from the skin, vitamin D is actually an inactive form of a hormone. To become activated, the molecule must gain two hydroxyl groups

(–OH); one of these is added by an enzyme in the liver, the other by an enzyme in the kidneys.

The enzyme needed for this final step is stimulated by PTH, thereby producing the active form of vitamin D known as 1,25-dihydroxyvitamin D. This hormone stimulates the intestinal absorption of Ca^{2+}, thereby helping raise blood Ca^{2+} levels so that bone can become properly mineralized. A diet deficient in vitamin D can lead to poor bone formation, a condition called rickets.

The Adrenal Gland Releases Both Catecholamine and Steroid Hormones

LEARNING OBJECTIVE 35.4.3 Compare the actions of the hormones of the adrenal medulla and the adrenal cortex.

The **adrenal glands** are located just above each kidney (figure 35.12). Each gland is composed of an inner portion, the *adrenal medulla,* and an outer layer, the *adrenal cortex.*

The adrenal medulla

The adrenal medulla receives neural input from axons of the sympathetic division of the autonomic nervous system, and it secretes the catecholamines epinephrine and norepinephrine in response to stimulation by these axons. The actions of these hormones trigger "alarm" responses similar to those elicited by the sympathetic division, helping to prepare the body for extreme efforts. Among the effects of these hormones are an increased heart rate, increased blood pressure, dilation of the bronchioles, elevation in blood glucose, reduced blood flow to the skin and digestive organs, and increased blood flow to the heart and muscles. The actions of epinephrine, released as a hormone, supplement those of neurotransmitters released by the sympathetic nervous system.

The adrenal cortex

The hormones from the adrenal cortex are all steroids and are referred to collectively as *corticosteroids. Cortisol* (also called hydrocortisone) and related steroids secreted by the adrenal cortex act on various cells in the body to maintain glucose homeostasis. In mammals, these hormones are referred to as glucocorticoids and their secretion is primarily regulated by ACTH from the anterior pituitary.

The glucocorticoids stimulate the breakdown of muscle protein into amino acids, which are carried by the blood to the liver. They also stimulate the liver to produce the enzymes needed for gluconeogenesis, which can convert amino acids into glucose. Glucose synthesis from protein is particularly important during very long periods of fasting or exercise, when blood glucose levels might otherwise become dangerously low.

Whereas glucocorticoids are important in the daily regulation of glucose and protein, they, like the adrenal medulla hormones, are also secreted in large amounts in response to stress. It has been suggested that during stress they activate the production of glucose at the expense of protein and fat synthesis.

In addition to regulating glucose metabolism, the glucocorticoids modulate some aspects of the immune response. The physiological significance of this action is still unclear, and it may be

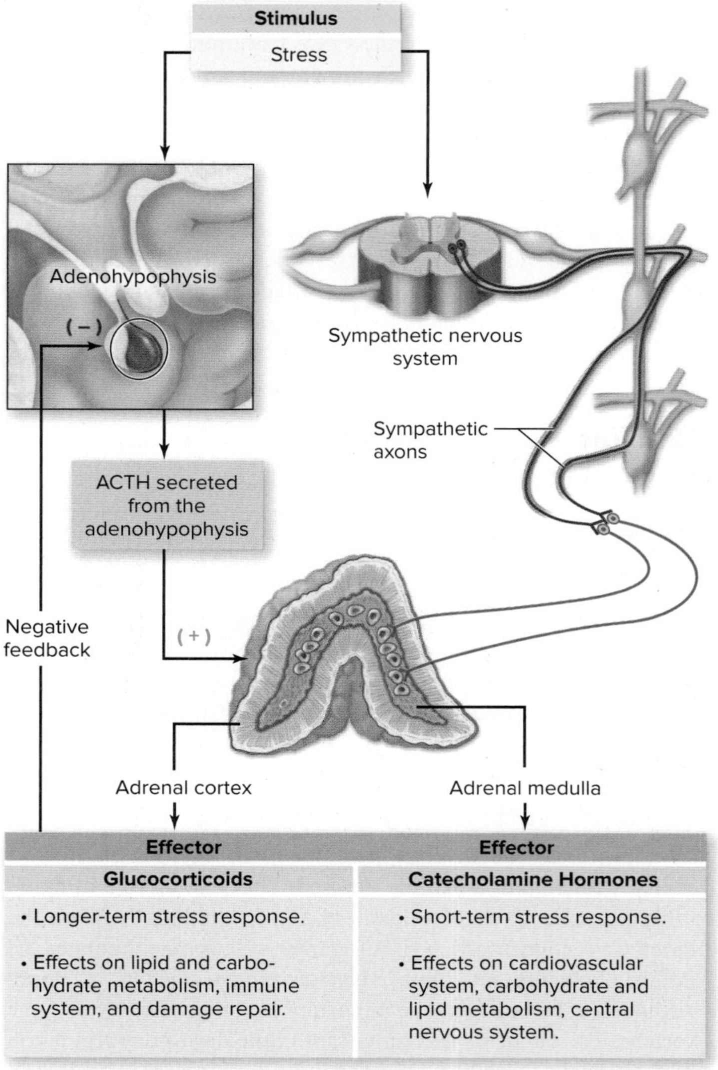

Figure 35.12 The adrenal glands. The adrenal medulla produces the catecholamines epinephrine and norepinephrine, which initiate a response to acute stress. The adrenal cortex produces steroid hormones, including the glucocorticoid cortisol. In response to stress, cortisol secretion increases glucose production and stimulates the immune response.

apparent only when glucocorticoids are maintained at elevated levels for long periods of time (such as long-term stress). Glucocorticoids are used to suppress the immune system in persons with immune disorders (such as rheumatoid arthritis) and to prevent the immune system from rejecting organ and tissue transplants. Derivatives of cortisol, such as prednisone, have widespread medical use as anti-inflammatory agents.

Aldosterone, the other major corticosteroid, is classified as a mineralocorticoid, because it helps regulate mineral balance. The secretion of aldosterone from the adrenal cortex is activated by angiotensin II, a product of the renin–angiotensin system described in section 35.7, as well as high blood K^+. Angiotensin II activates aldosterone secretion when blood pressure falls.

A primary action of aldosterone is to stimulate the kidneys to reabsorb Na^+ from the urine. (Blood levels of Na^+ decrease if Na^+ is not reabsorbed from the urine.) Sodium is the major extracellular solute; it is needed for the maintenance of normal blood volume and pressure, as well as for the generation of action potentials in neurons and muscles. Without aldosterone, the kidneys would lose excessive amounts of blood Na^+ in the urine.

Aldosterone-stimulated reabsorption of Na^+ also results in kidney excretion of K^+ in the urine. Aldosterone thus prevents K^+ from accumulating in the blood, which would lead to malfunctions in electrical signaling in nerves and muscles. Because of these essential functions performed by aldosterone, removal of the adrenal glands or diseases that prevent aldosterone secretion are invariably fatal without hormone therapy.

Pancreatic Hormones Are Primary Regulators of Carbohydrate Metabolism

LEARNING OBJECTIVE 35.4.4 Contrast the effects of insulin and glucagon on levels of blood glucose.

The pancreas is located adjacent to the stomach and is connected to the duodenum of the small intestine by the pancreatic duct. It secretes bicarbonate ions and a variety of digestive enzymes into the small intestine through this duct (refer to chapter 34), and for a long time the pancreas was thought to be solely an exocrine gland.

Insulin and glucagon are produced by different cells

In 1869, a medical student named Paul Langerhans described some unusual clusters of cells scattered throughout the pancreas, which came to be called **islets of Langerhans.** They are now more commonly called pancreatic islets. Later, it was observed that surgical removal of the pancreas caused glucose to appear in the urine, the hallmark of the disease diabetes mellitus. This led to the discovery that a hormone produced in pancreatic islets, insulin, prevents diabetes.

Insulin was not isolated until 1922 by surgeon Frederick Banting and his assistant Charles Best. They went on to use an extract purified from beef pancreas to treat a 13-year-old diabetic boy, who was not expected to survive. With that single injection, the glucose level in the boy's blood fell 25%. A more potent extract soon brought the level down to near normal. This was the first instance of successful insulin therapy.

The maintenance of a constant level of blood glucose depends on two hormones acting antagonistically: insulin, produced by beta (β) cells in the pancreatic islets, and glucagon, produced by alpha (α) cells of the islets. Insulin acts to lower blood glucose, and glucagon acts to raise it. Insulin lowers blood glucose by stimulating the removal of glucose from the blood, converting it to glycogen in liver and muscle. Glucagon raises blood glucose by stimulating the conversion of glycogen to glucose and stimulating the synthesis of glucose (figure 35.13).

After a carbohydrate-rich meal, insulin signals to the liver and skeletal muscles to remove excess glucose from the blood and store it as glycogen. When blood glucose levels decrease, as they do between meals, during periods of fasting, and during exercise, glycogen signals the liver to secrete glucose into the blood. This glucose is obtained in part from the breakdown of glycogen to glucose-6-phosphate. The phosphate group is then removed, and

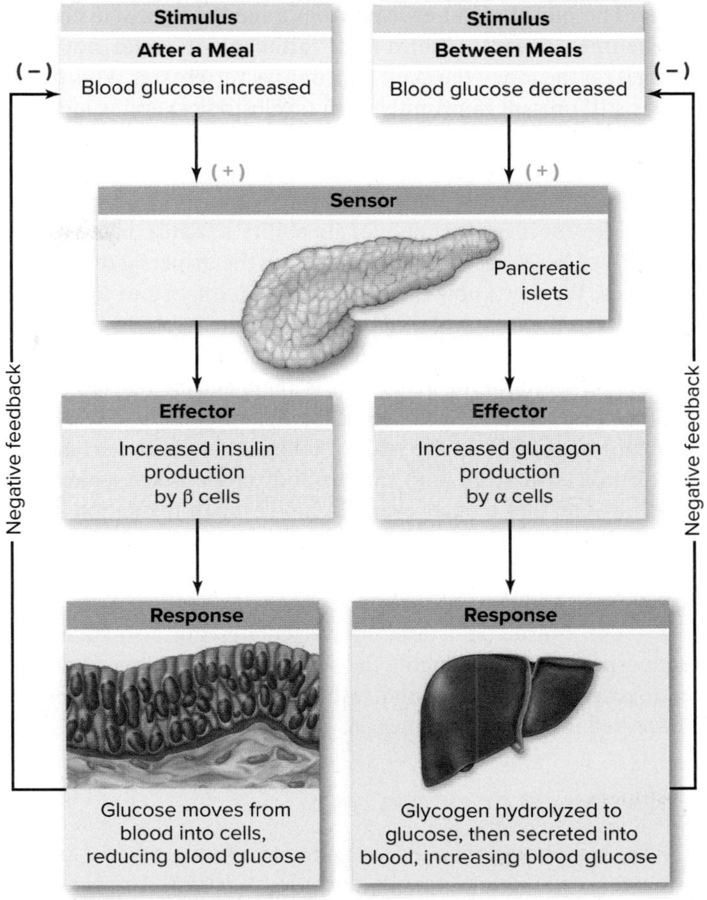

Figure 35.13 The antagonistic actions of insulin and glucagon on blood glucose. Insulin stimulates the cellular uptake of blood glucose into skeletal muscles, adipose cells, and the liver after a meal. Glucagon stimulates the hydrolysis of liver glycogen between meals, so that the liver can secrete glucose into the blood. These antagonistic effects help to maintain homeostasis of the blood glucose concentration.

free glucose is secreted into the blood. Skeletal muscles lack this enzyme; so, even though they have glycogen stores, they cannot secrete glucose into the blood. However, muscle cells use this glucose directly, since glucose-6-phosphate is a product of the first reaction in glycolysis.

If fasting or exercise continues, the liver begins to convert other molecules, such as amino acids and lactic acid, into glucose by gluconeogenesis ("new formation of glucose"). The amino acids used for gluconeogenesis are obtained from muscle protein, which explains the severe muscle wasting that occurs during starvation.

Melatonin Is Crucial to Circadian Cycles

LEARNING OBJECTIVE 35.4.5 Describe how melatonin establishes a 24-hour circadian rhythm in vertebrates.

Another major endocrine gland is the pineal gland, located in the roof of the third ventricle of the brain in most vertebrates (figure 35.5). It is about the size of a pea and is shaped like a pinecone, which gives it its name.

The pineal gland evolved from a medial light-sensitive eye (sometimes called a "third eye," although it could not form images) at the top of the skull in primitive vertebrates. This pineal eye is still present in primitive fish (cyclostomes) and some modern reptiles. In other vertebrates, however, the pineal gland is buried deep in the brain, and it functions as an endocrine gland by secreting the hormone melatonin.

Melatonin was named for its ability to cause blanching of the skin of lower vertebrates by reducing the dispersal of melanin granules. We now know that it serves as an important timing signal delivered through the blood. Melatonin levels in the blood increase in darkness and fall during the daytime.

The secretion of melatonin is regulated by activity of the *suprachiasmatic nucleus (SCN)* of the hypothalamus. The SCN is known to function as the major biological clock in vertebrates, entraining (synchronizing) various body processes to a circadian rhythm—one that repeats every 24 hr. Through regulation by the SCN, the secretion of melatonin by the pineal gland is activated in the dark. This daily cycling of melatonin release regulates sleep–wake and temperature cycles.

The ovaries and testes in vertebrates are also important endocrine glands, producing the sex steroid hormones, including estrogens, progesterone, and testosterone. These hormones are described in detail in chapter 36.

REVIEW OF CONCEPT 35.4

Major peripheral endocrine glands include the thyroid, parathyroid, and adrenal glands and the pancreas. Thyroid hormones play crucial roles in metabolism and development. Calcium homeostasis results from the action of calcitonin, parathyroid hormone, and vitamin D. The adrenal glands produce stress hormones. Insulin and glucagon, antagonists from the pancreas, help maintain blood glucose at a normal level.

- *Why does your body need two hormones to maintain blood sugar at a constant level?*

35.5 Animals Are Osmoconformers or Osmoregulators

Water in a multicellular animal's body is distributed between the intracellular and extracellular compartments. Most vertebrates maintain homeostasis for both the total solute concentration of their extracellular fluids and the concentration of specific inorganic ions, principally sodium (Na^+) and chloride (Cl^-).

Animals Control Their Osmolarity in a Variety of Ways

LEARNING OBJECTIVE 35.5.1 Differentiate between osmoconformers and osmoregulators.

You learned in chapter 5 that osmosis is the diffusion of water across a semipermeable membrane. Water moves osmotically from a more dilute solution (lower solute concentration) to a less dilute solution (higher solute concentration). The osmotic pressure of a solution is a measure of its tendency to take in water by osmosis: the higher the concentration of solute, the higher the osmotic pressure. We define the osmolarity of a solution as the number of osmotically active moles of solute per liter of solution. We also refer to the tonicity of a solution, which is the ability of a solution to change the volume of a cell by osmosis. An animal cell in a hypertonic solution loses water to the surrounding solution and shrinks, whereas an animal cell in a hypotonic solution gains water and expands. A cell in an isotonic solution shows no net water movement.

Osmoconformers live in marine environments

The osmolarity of body fluids in most marine invertebrates is the same as that of seawater. Because the extracellular fluids are isotonic to seawater, no osmotic gradient exists, and there is no tendency for water to leave or enter the body. Such organisms are termed **osmoconformers**—they are in osmotic equilibrium with their environment.

Osmoregulators in marine and freshwater environments face opposite challenges

Osmoregulators are organisms that maintain a relatively constant osmolarity of body fluids despite different concentrations in the surrounding environment. Both freshwater and marine fish are osmoregulators, but they face opposite challenges. In the case of freshwater fish, whose body fluids are hypertonic with respect to the surrounding water, these animals face two serious problems: (1) Water tends to enter the body from the environment, and (2) solutes tend to leave the body and enter the environment. Kidneys are thought to have evolved among freshwater fish. These fish solve the first problem by not drinking water and excreting a large volume of dilute urine, which is hypotonic to their body fluids. They have primitive kidneys that solve the second problem by allowing the reabsorption of ions (refer to section 35.6).

Although most groups of animals seem to have evolved first in the sea, marine bony fish (teleosts) probably evolved from freshwater ancestors. They faced significant new problems in a marine environment because their body fluids were hypotonic to seawater. Consequently, water tends to leave their bodies by osmosis across their gills, and they lose water in their urine. To compensate for this continuous water loss, marine fish drink large amounts of seawater.

Many of the divalent cations (principally, Ca^{2+} and Mg^{2+}) in the seawater that marine fish drink remain in the digestive tract and are eliminated. Some, however, are absorbed into the blood, as are the monovalent ions K^+, Na^+, and Cl^-. Most of the monovalent ions are actively transported out of the blood across the gill surfaces, whereas the divalent ions that enter the blood are secreted into the kidney tubules and excreted in the urine.

Terrestrial animals are osmoregulators

Terrestrial organisms inhabit a highly dehydrating environment, and therefore are osmoregulators that have evolved mechanisms to preserve water. Evolution has produced increasingly sophisticated systems to maintain osmotic balance, yet all depend on the simple processes of filtration of body fluids with reabsorption of

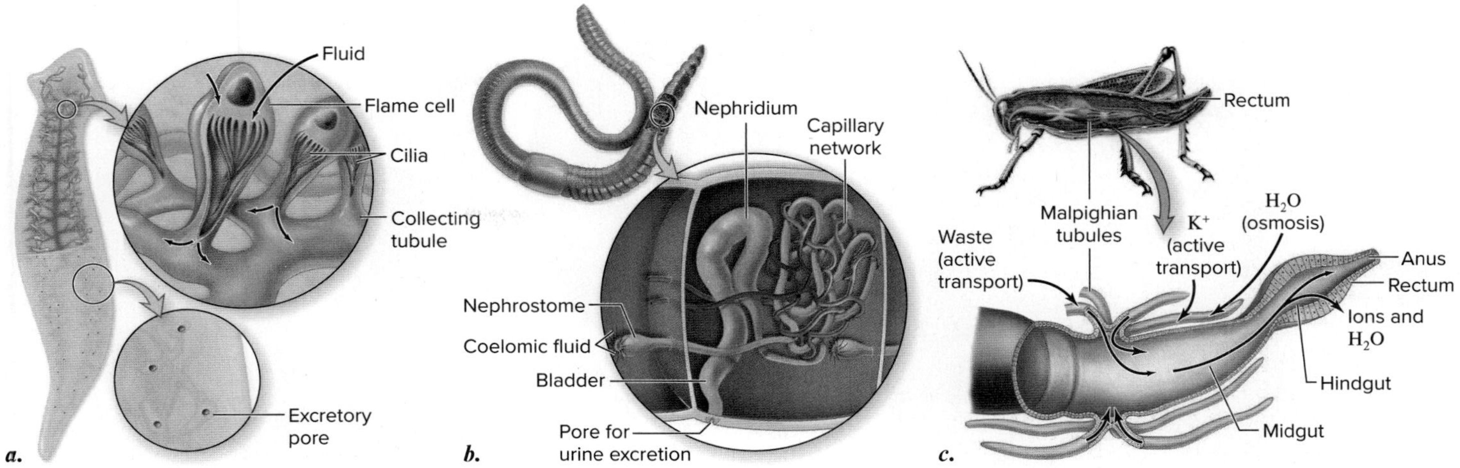

Figure 35.14 Terrestrial invertebrate osmoregulatory organs. Examples include: ***a.*** Flatworm protonephridia, ***b.*** Annelid nephridia, and ***c.*** Insect Malpighian tubules.

important solutes and water. The simplest solution among invertebrates is the **protonephridia** in flatworms that connect to excretory pores (figure 35.14*a*). Flame cells within these tubules have cilia that draw fluid in; the fluid is then expelled from pores open to the outside. As size and complexity increase, so must osmoregulatory organs. In annelids, we see the formation of **nephridia** (figure 35.14*b*). These nephridia receive a filtrate of coelomic fluid, which enters the funnel-like nephrostomes. Solutes are selectively reabsorbed from the tubules, and the remaining fluid is expelled through pores. Even more extensive reabsorption of solutes and water occurs in insects where **Malpighian tubules** extend from the digestive tract to collect water and wastes from the body's circulatory system. K^+ secreted into the tubules draws water in osmotically, and as fluid moves through the gut, water and ions are reabsorbed (figure 35.14*c*).

Among the vertebrates, sharks and their relatives are osmoconformers; all other vertebrates are osmoregulators. The maintenance of a relatively constant body fluid osmolarity has permitted vertebrates to exploit a wide variety of ecological niches. Achieving this constancy, however, requires continuous regulation. The body fluids of terrestrial vertebrates clearly have a higher concentration of water than the surrounding air, so they tend to lose water by evaporation. All reptiles, birds, and mammals, as well as amphibians on land, face this problem. The evolution of extensive urinary/osmoregulatory systems in these vertebrates has helped them retain water.

REVIEW OF CONCEPT 35.5

Physiological mechanisms help most vertebrates keep blood osmolarity and ion concentrations relatively constant. Marine invertebrates are osmoconformers, with body fluids isotonic to their environment. Most vertebrates are osmoregulators; their body fluids are either hypertonic or hypotonic compared to their environment.

- *During osmosis, does water move toward regions of higher or lower osmolarity?*

35.6 The Kidney Maintains Osmotic Homeostasis in Mammals

To maintain **fluid balance,** the amount of water and salt (NaCl) you lose each day must be the same as the amount you gain. The average human gains and loses about 2500 ml of water, and about 10.5 g of salt, a day. You lose water from exhaled breath, evaporation from the skin, and additional evaporation due to sweat, and in feces. But, by far the largest loss of water is in urine (about 1500 ml/day). This loss in urine is both highly variable, and under physiological control. In terms of salt, the greatest loss is also in urine (about 10 g). Gains of water come from metabolism, and from ingestion of food and fluids, and salt comes entirely from diet.

The maintenance of fluid balance requires the function of the kidneys, which produce urine. In humans, the kidneys are fist-sized organs located in the lower back. Each kidney receives blood from a renal artery, and from this blood, urine is produced. Urine drains from each kidney through a **ureter,** which carries the urine to a **urinary bladder**. From the bladder, urine is passed out of the body through the **urethra** (figure 35.15).

The Nephron Is the Filtering Unit of the Kidney

LEARNING OBJECTIVE 35.6.1 Name and describe the primary components of a kidney.

Within the kidney, the mouth of the ureter flares open to form a funnel-like structure, the *renal pelvis.* The renal pelvis, in turn, has cup-shaped extensions that receive urine from the renal tissue. The renal tissue is divided into an outer **renal cortex** and an inner **renal medulla.**

The kidney has three basic functions, summarized in figure 35.16:

1. *Filtration:* Fluid in the blood is filtered into the tubule system, leaving cells and large proteins in the blood and a

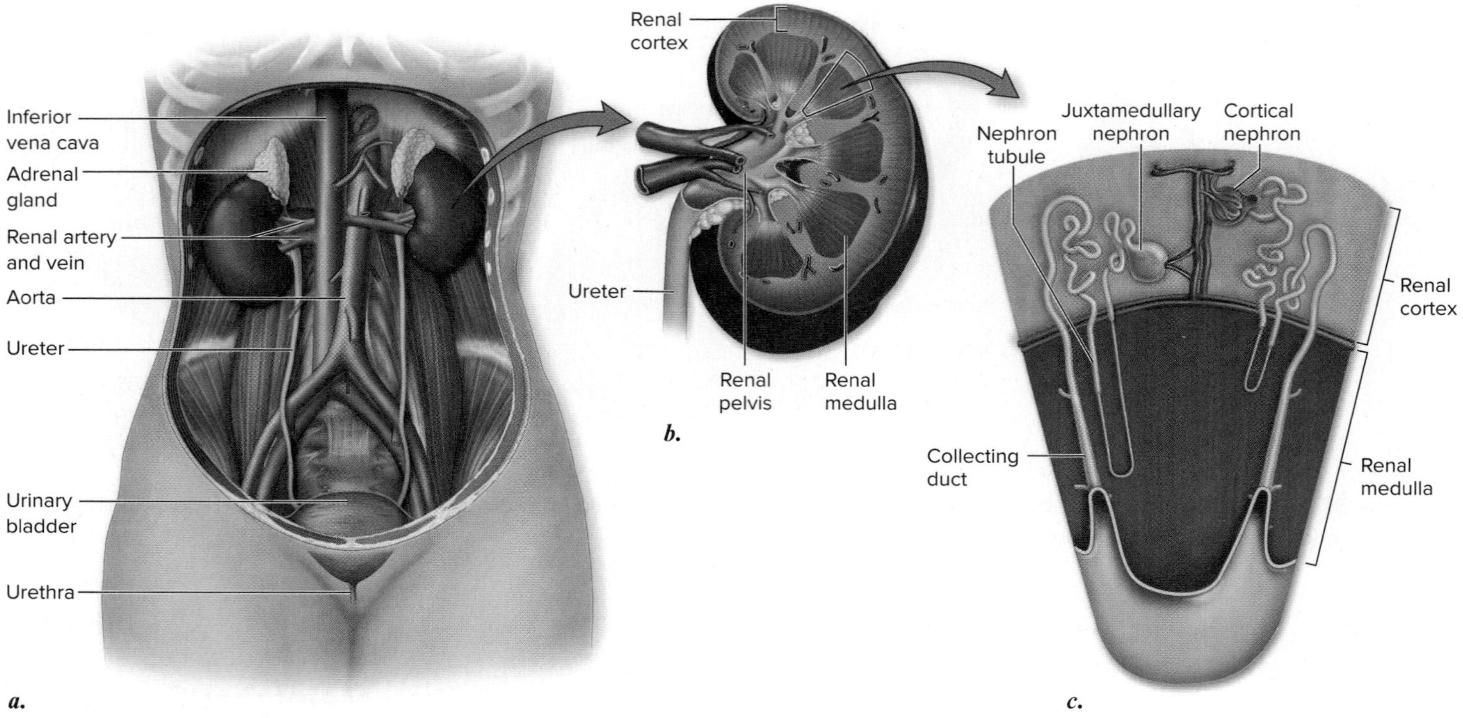

Figure 35.15 The human renal system. ***a.*** The positions of the organs of the urinary system. ***b.*** A sectioned kidney, revealing the internal structure. ***c.*** The position of nephrons in the mammalian kidney. Cortical nephrons are located predominantly in the renal cortex; juxtamedullary nephrons have long loops that extend deep into the renal medulla.

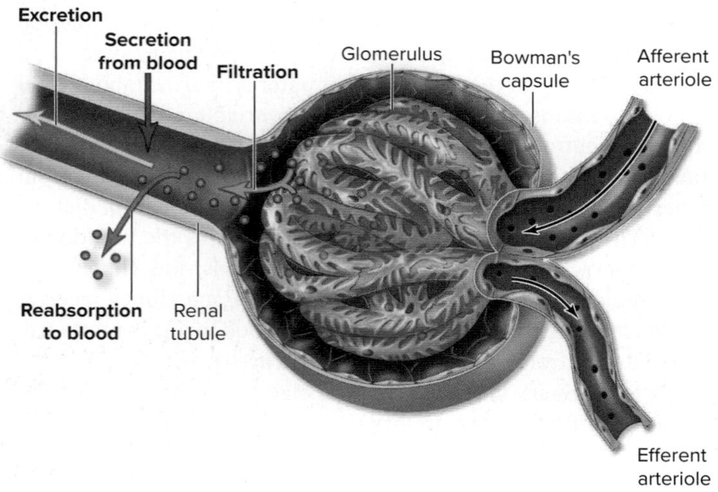

Figure 35.16 Three functions of the kidney. Molecules enter the urine by filtration out of the glomerulus and by secretion into the tubules from surrounding peritubular capillaries. Molecules that entered the filtrate can be returned to the blood by reabsorption from the tubules into surrounding peritubular capillaries. The fluid exiting the kidney is eliminated from the body by excretion through the tubule to a ureter and then to the bladder.

filtrate composed of water and all of the blood solutes. This filtrate is modified by the rest of the kidney to produce urine for excretion.

2. *Reabsorption:* Reabsorption is the selective movement of important solutes, such as glucose, amino acids, and a variety of inorganic ions, out of the filtrate in the tubule system to the extracellular fluid, then back into the bloodstream via peritubular capillaries. The process of reabsorption can utilize active or passive processes, depending on the solute. Water is also reabsorbed, and this can be controlled to regulate the amount of water loss.
3. *Secretion:* Secretion is the movement of substances from the blood into the extracellular fluid, then into the filtrate in the tubule system. Unlike reabsorption, which preserves substances in the body, this adds to what will be expelled from the body and can be used to remove toxic substances.

On a microscopic level, each kidney contains about a million functioning *nephrons.* Mammalian kidneys contain a mixture of **juxtamedullary nephrons,** which have long loops that dip deeply into the medulla, and **cortical nephrons** with shorter loops. The significance of the length of the loops will be explained later in this section.

The production of filtrate

Each nephron consists of a long tubule and associated small blood vessels (figure 35.17). First, blood is carried by an *afferent arteriole* to a tuft of capillaries in the renal cortex–the **glomerulus.** Here the blood is filtered as the blood pressure forces fluid through the porous capillary walls. Blood cells and plasma proteins are too large to enter this glomerular filtrate, but large amounts of the plasma, consisting of water and dissolved molecules, leave the vascular system at this step. The filtrate immediately enters the first region of the nephron tubules. This region, **Bowman's capsule,** envelops the glomerulus much as a large, soft

Figure 35.17 A nephron in a mammalian kidney. The nephron tubule is surrounded by peritubular capillaries in the cortex, and their vasa recta extensions surround the loop of Henle in the medulla. This capillary bed carries away molecules and ions that are reabsorbed from the filtrate.

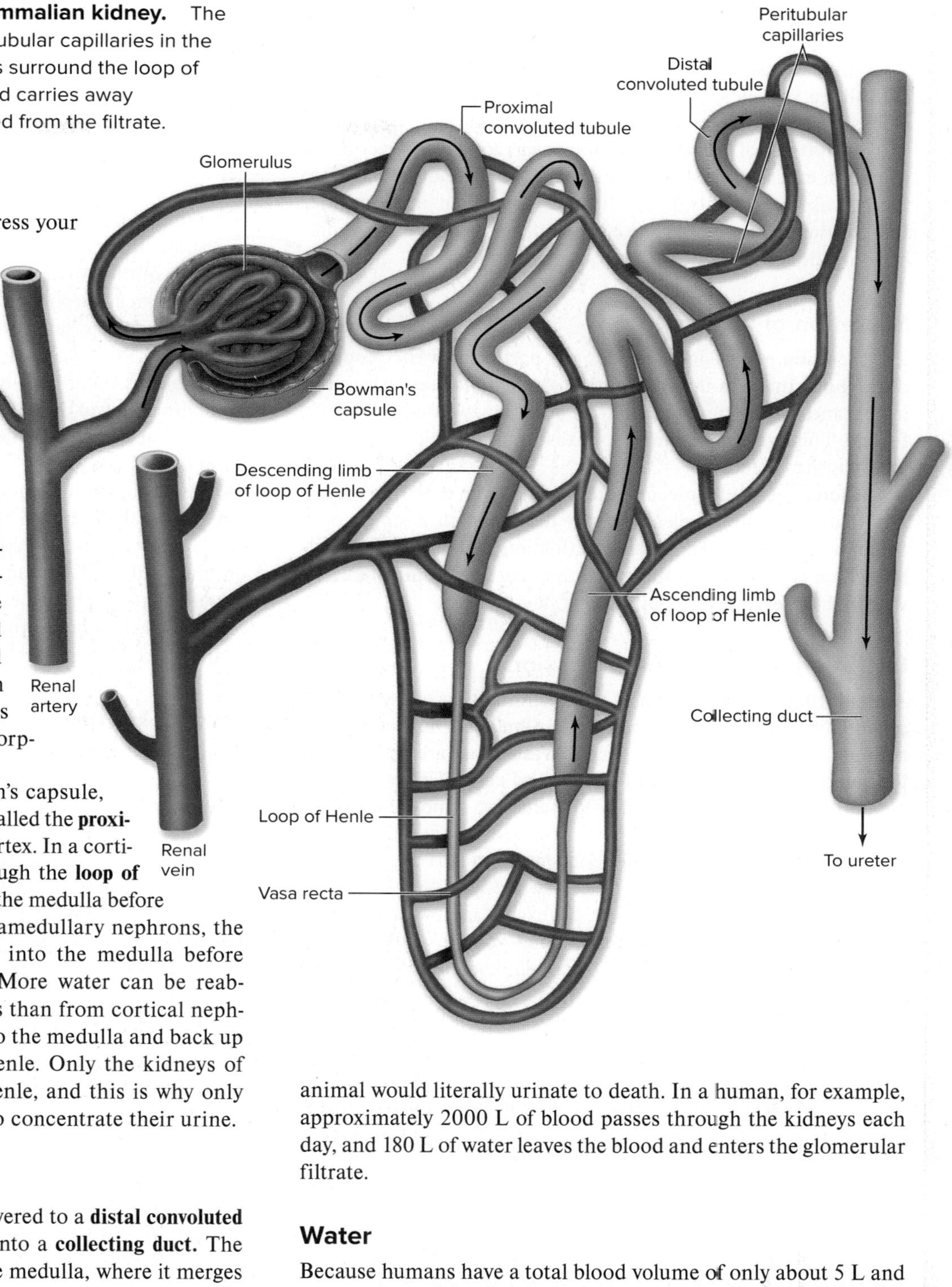

balloon surrounds your hand if you press your fist into it. The capsule has slit openings so that the glomerular filtrate can enter the system of nephron tubules.

Blood components that were not filtered out of the glomerulus drain into an *efferent arteriole,* which then empties into a second bed of capillaries, called **peritubular capillaries,** that surround the tubules. This is only one of several locations in the body where two capillary beds occur in series. In juxtamedullary nephrons, the efferent arteriole and peritubular capillaries also feed the **vasa recta** capillaries that surround the loop of Henle. As described later in this section, the peritubular capillaries are needed for the processes of reabsorption and secretion.

After the filtrate enters Bowman's capsule, it goes into a portion of the nephron called the **proximal convoluted tubule,** located in the cortex. In a cortical nephron, the fluid then flows through the **loop of Henle,** which dips only minimally into the medulla before ascending back into the cortex. In juxtamedullary nephrons, the loop of Henle extends much deeper into the medulla before ascending back up into the cortex. More water can be reabsorbed from juxtamedullary nephrons than from cortical nephrons. The fluid then moves deeper into the medulla and back up again into the cortex in a loop of Henle. Only the kidneys of mammals and birds have loops of Henle, and this is why only birds and mammals have the ability to concentrate their urine.

Collection of urine

After leaving the loop, the fluid is delivered to a **distal convoluted tubule** in the cortex that next drains into a **collecting duct.** The collecting duct again descends into the medulla, where it merges with other collecting ducts to empty its contents, now called urine, into the renal pelvis.

Water, Nutrients, and Ions Are Reabsorbed; Other Molecules Are Secreted

LEARNING OBJECTIVE 35.6.2 Describe the fate of major classes of molecules passing through the kidney.

Most of the water and dissolved solutes that enter the glomerular filtrate must be returned to the blood by reabsorption, or the animal would literally urinate to death. In a human, for example, approximately 2000 L of blood passes through the kidneys each day, and 180 L of water leaves the blood and enters the glomerular filtrate.

Water

Because humans have a total blood volume of only about 5 L and produce only 1 to 2 L of urine per day, it is obvious that each liter of blood is filtered many times per day, and most of the filtered water is reabsorbed. Water is reabsorbed from the filtrate by the proximal convoluted tubule, as it passes through the descending loop of Henle and the collecting duct. The selective reabsorption in the collecting duct is driven by an osmotic gradient produced by the loop of Henle.

Glucose and other nutrients

The reabsorption of glucose, amino acids, and many other molecules needed by the body is driven by active transport and secondary active transport (cotransport) carriers. The maximum rate of transport is reached whenever the carriers are saturated.

In the case of the renal glucose carriers in the proximal convoluted tubule, saturation occurs when the concentration of glucose in the blood (and thus in the glomerular filtrate) is about 180 mg/100 mL of blood. If a person has a blood glucose concentration in excess of this amount, as happens in diabetes mellitus, the glucose remaining in the filtrate is expelled in the urine. Indeed, the presence of glucose in the urine is diagnostic of diabetes mellitus.

Secretion of wastes

The secretion of foreign molecules and particular waste products of the body involves the transport of these molecules across the membranes of the blood capillaries and kidney tubules into the filtrate. This process is similar to reabsorption, but it proceeds in the opposite direction.

Some secreted molecules are eliminated in the urine so rapidly that they may be cleared from the blood in a single pass through the kidneys. This rapid elimination explains why penicillin, which is secreted by the nephrons, must be administered in very high doses and several times per day.

Excretion of toxins and excessions

A major function of the kidney is the elimination of a variety of potentially harmful substances that animals eat and drink. In addition, urine contains nitrogenous wastes that are products of the catabolism of amino acids and nucleic acids. Urine may also contain excess K^+, H^+, and other ions that are removed from the blood.

Urine's generally high H^+ concentration (pH 5 to 7) helps maintain the acid–base balance of the blood within a narrow range (pH 7.35 to 7.45). Moreover, the excretion of water in urine contributes to the maintenance of blood volume and pressure; the larger the volume of urine excreted, the lower the blood volume.

The purpose of kidney function is therefore homeostasis; the kidneys are critically involved in maintaining the constancy of the internal environment. When disease interferes with kidney function, it causes a rise in the blood concentration of nitrogenous waste products, disturbances in electrolyte and acid–base balance, and a failure in blood pressure regulation. Such potentially fatal changes highlight the central importance of the kidneys in normal body physiology.

Each Part of the Nephron Performs a Specific Transport Function

LEARNING OBJECTIVE 35.6.3 Relate the structure of a nephron to its function.

Approximately 180 L of isotonic glomerular filtrate enters the Bowman's capsules of human kidneys each day. After passing through the remainder of the nephron tubules, this volume of fluid would be lost as urine if it were not reabsorbed back into the blood. It is clearly impossible to produce this much urine, yet water is able to pass through a cell membrane only by osmosis, and osmosis is not possible between two isotonic solutions. Therefore, some mechanism is needed to create an osmotic gradient between the glomerular filtrate and the blood, to allow the reabsorption of water.

Proximal convoluted tubule

Virtually all the nutrient molecules in the filtrate are reabsorbed back into the systemic blood by the proximal convoluted tubule. In addition, approximately two-thirds of the NaCl and water filtered into Bowman's capsule is immediately reabsorbed across the walls of the proximal convoluted tubule.

This reabsorption is driven by the active transport of Na^+ out of the filtrate and into surrounding peritubular capillaries. Cl^- follows Na^+ passively because of electrical attraction, and water follows them both because of osmosis. Because NaCl and water are removed from the filtrate in proportionate amounts, the filtrate that remains in the tubule is still isotonic to the blood plasma.

Although only one-third of the initial volume of filtrate remains in the nephron tubule after the initial reabsorption of NaCl and water, it still represents a large volume (60 L out of the original 180 L of filtrate). Obviously, no animal can afford to excrete that much urine, so most of this water must also be reabsorbed. It is reabsorbed primarily across the wall of the collecting duct.

Loop of Henle

The function of the loop of Henle is to create a gradient of increasing osmolarity from the cortex to the medulla. This allows water to be reabsorbed by osmosis in the collecting duct as it runs down into the medulla past the loop of Henle. The descending and ascending limbs of the loop of Henle differ structurally and in their permeability to ions and water. This produces a gradient of increasing osmolarity from cortex to medulla (figure 35.18). The structure of the loop also forms another example of a countercurrent system, this time acting to increase the osmolarity of interstitial fluid. To understand the functioning of the loop of Henle, it is easiest to start in the ascending limb:

1. The entire ascending limb is impermeable to water. The thick portion of the ascending limb actively transports Na^+ out of the tubule, with Cl^- passively following. The thin ascending limb is permeable to both Na^+ and Cl^-, which move out by diffusion.
2. The descending limb is thin and permeable to water but not to NaCl. Because of the Na^+ and Cl^- lost by the ascending limb, the osmolarity of the interstitial fluid is higher than in the descending limb, and water moves out of the descending limb by osmosis. This also increases the osmolarity of the fluid in the tubule such that as it turns at the bottom, it will lose NaCl by diffusion in the thin ascending loop.
3. The loss of water from the descending limb multiplies the concentration that can be achieved at each level of the loop through the active extrusion of Na^+ (with Cl^- following passively) by the ascending limb. The longer the loop of Henle, the longer the region of interaction between the descending and ascending limbs, and the greater the total concentration that can be achieved. In a human kidney, the

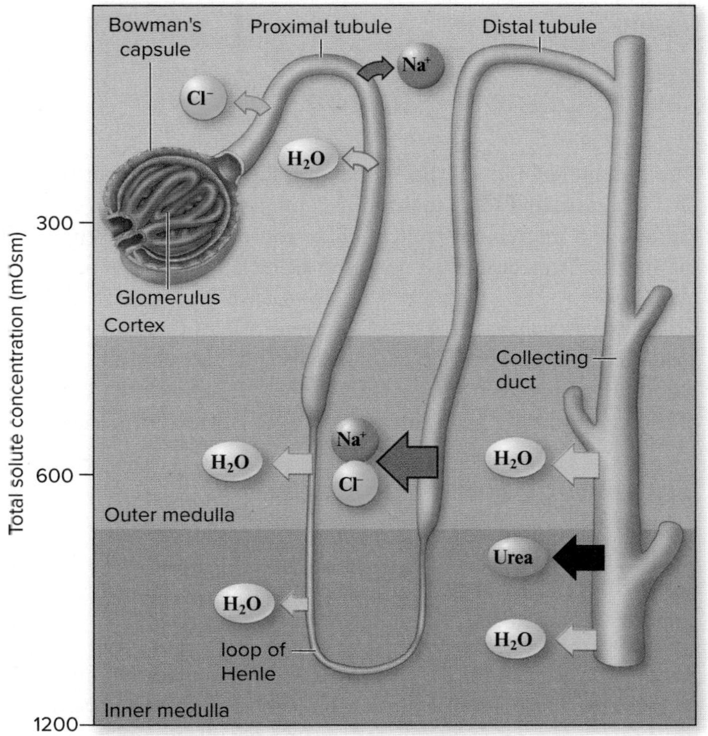

Figure 35.18 The reabsorption of salt and water in the mammalian kidney. Active transport of Na^+ out of the proximal tubules is followed by the passive movement of Cl^- and water. Active extrusion of NaCl from the ascending limb of the loop of Henle creates the osmotic gradient required for the reabsorption of water from the descending limb of the loop of Henle and the collecting duct. The two limbs of the loop form a countercurrent multiplier system that increases the osmotic gradient. The changes in osmolarity from the cortex to the medulla are indicated to the left of the figure.

concentration of filtrate entering the loop is 300 milliosmolar (mOsm), and this concentration is multiplied to more than 1200 mOsm at the bottom of the longest loops of Henle in the renal medulla.

4. The NaCl pumped out of the ascending limb of the loop is reabsorbed from the surrounding interstitial fluid into the loops of the vasa recta, so that NaCl can diffuse from the blood leaving the medulla to the blood entering the medulla. Thus, the vasa recta also functions in a countercurrent exchange, similar to the countercurrent flow of blood in the fins of large aquatic vertebrates for heat exchange (figure 35.2) and of water and blood through gills to enhance oxygen exchange (refer to chapter 34). In the case of the vasa recta, this exchange prevents the flow of blood through the capillaries from destroying the osmotic gradient established by the loop of Henle. Thus, blood can be supplied to this region of the kidney without affecting the ability of the collecting duct to selectively reabsorb water.

Because fluid flows in opposite directions in the two limbs of the loop, the action of the loop of Henle in creating a hypertonic renal medulla is known as the countercurrent multiplier system. The osmotic gradient that is established is greater than what would be produced by just active transport of salts out of the tubule system.

The high solute concentration of the renal medulla is primarily the result of NaCl accumulation by the countercurrent multiplier system, but urea also contributes to the total osmolarity of the medulla. The descending limb of the loop of Henle and the collecting duct are both permeable to urea, which leaves these regions of the nephron by diffusion.

Distal convoluted tubule and collecting duct

Because NaCl was pumped out of the ascending limb, the filtrate that arrives at the distal convoluted tubule and enters the collecting duct in the renal cortex is hypotonic (with a concentration of only 100 mOsm). The collecting duct carrying this dilute fluid now plunges into the medulla. As a result of the hypertonic interstitial fluid of the renal medulla, a strong osmotic gradient pulls water out of the collecting duct and into the surrounding blood vessels.

The osmotic gradient is normally constant, but the permeability of the distal convoluted tubule and the collecting duct to water is adjusted by a hormone, *antidiuretic hormone (ADH),* discussed in section 35.7. When an animal needs to conserve water, the posterior pituitary gland secretes more ADH, and this hormone increases the number of water channels in the plasma membranes of the collecting duct cells. This increases the permeability of the collecting ducts to water so that more water is reabsorbed and less is excreted in the urine. The animal thus excretes a hypertonic urine.

In addition to regulating water balance, the kidneys regulate the balance of electrolytes in the blood by reabsorption and secretion. For example, the kidneys reabsorb K^+ in the proximal tubule and then secrete an amount of K^+ needed to maintain homeostasis into the distal convoluted tubule (figure 35.19). The kidneys also maintain acid–base balance by excreting H^+ into the urine and reabsorbing HCO_3^-.

The reabsorption of NaCl in the distal convoluted tubule and collecting duct depends on the needs of the body and is under the control of the hormone *aldosterone.* Both ADH and aldosterone influence the distal convoluted tubule and collecting duct, although aldosterone is more significant in terms of NaCl. Hormonal control of excretion is discussed further in section 35.7.

REVIEW OF CONCEPT 35.6

Filtration removes fluid from the blood for processing. Reabsorption moves material from the renal tubules back to the blood and secretion material from the blood to the tubules. Reabsorption of glucose and amino acids occurs by active transport out of the renal tubule, whereas water moves osmotically. The loop of Henle functions to create a region of high osmolarity in the renal medulla, allowing the selective reabsorption of water.

- *The compound mannitol is filtered but cannot be reabsorbed. How would this affect the volume of urine produced?*

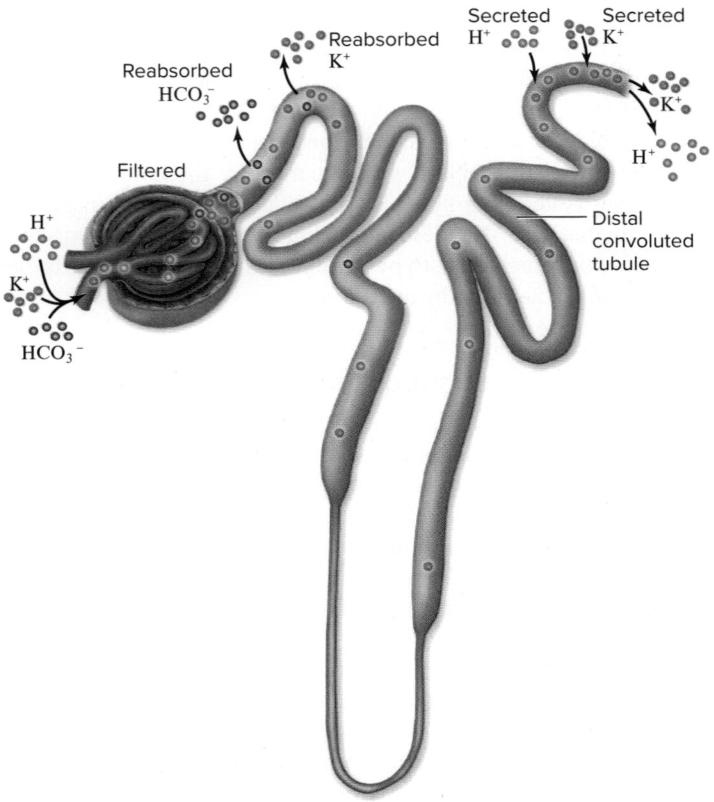

Figure 35.19 Controlling salt balance. The nephron controls the amounts of K^+, H^+, and HCO_3^- excreted in the urine. K^+ is completely reabsorbed in the proximal tubule and then secreted in hormonally regulated amounts into the distal tubule. HCO_3^- is filtered but normally completely reabsorbed. H^+ is filtered and secreted into the distal tubule, so that the final urine has an acidic pH.

35.7 Hormones Control Osmoregulation

The volume of blood, the blood pressure, and the osmolarity of blood plasma are maintained relatively constant by the kidneys, no matter how much water you drink. Acting through the mechanisms described next, the kidneys excrete a hypertonic urine when the body needs to conserve water. If you drink excess water, the kidneys excrete a hypotonic urine. The kidneys also regulate the plasma K^+ and Na^+ concentrations and blood pH within very narrow limits. These homeostatic functions of the kidneys are coordinated primarily by three hormones.

Antidiuretic Hormone Causes Water to Be Conserved

LEARNING OBJECTIVE 35.7.1 Describe how ADH maintains osmotic homeostasis.

Antidiuretic hormone (ADH) is produced by the hypothalamus and secreted by the posterior pituitary gland. The primary stimulus for ADH secretion is an increase in the osmolarity of the blood plasma. When a person is dehydrated or eats salty food, the osmolarity of plasma increases. Osmoreceptors in the hypothalamus respond to the elevated blood osmolarity by sending increasing action potentials to the integration center (also in the hypothalamus). This, in turn, triggers a sensation of thirst and an increase in the secretion of ADH (figure 35.20).

ADH causes the walls of the distal convoluted tubules and collecting ducts in the kidney to become more permeable to water. Water channels called aquaporins (refer to chapter 5) are contained within the membranes of intracellular vesicles in the epithelium of the distal convoluted tubules and collecting ducts; ADH stimulates the fusion of the vesicle membrane with the

Figure 35.20 Antidiuretic hormone stimulates the reabsorption of water by the kidneys. This action completes a negative feedback loop and helps to maintain homeostasis of blood volume and osmolarity.

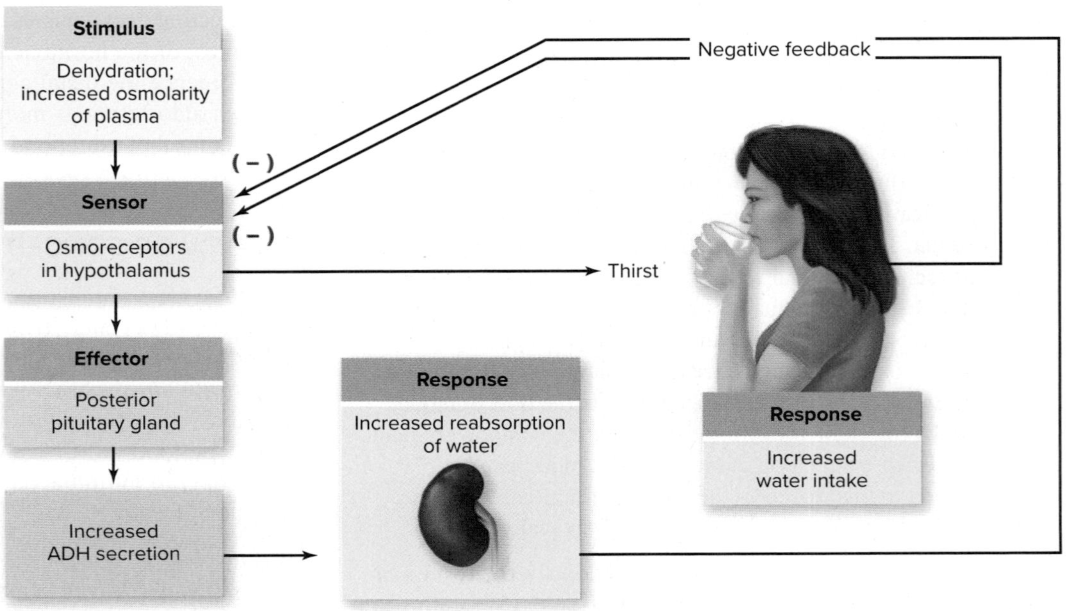

plasma membrane, similar to the process of exocytosis. The aquaporins are now in place and allow water to flow out of the tubules and ducts in response to the hypertonic condition of the renal medulla. This water is reabsorbed into the bloodstream.

When secretion of ADH is reduced, the plasma membrane pinches in to form new vesicles that contain aquaporins. This removes the aquaporins from the plasma membrane of the distal convoluted tubule and collecting duct, making them less permeable to water. Thus, more water is excreted in urine.

Under conditions of maximal ADH secretion, a person excretes only 600 mL of highly concentrated urine per day. A person who lacks ADH due to pituitary damage has the disorder known as *diabetes insipidus* and continually excretes a large volume of dilute urine. Such a person is in danger of becoming severely dehydrated and succumbing to dangerously low blood pressure.

Homeostasis via ADH action is also affected by the common drugs ethanol and caffeine, both of which inhibit secretion of ADH. This is the basis for the dehydration that is the aftereffect of drinking too much alcohol.

Aldosterone Controls Sodium Ion Concentration

LEARNING OBJECTIVE 35.7.2 Describe the relationship between control of blood osmolarity and blood pressure.

Sodium ions are the major solute in the blood plasma. When the blood concentration of Na^+ falls, therefore, the blood osmolarity also falls. This drop in osmolarity inhibits ADH secretion, causing more water to remain in the collecting duct for excretion in the urine. As a result, the blood volume and blood pressure decrease.

A decrease in extracellular Na^+ also causes more water to be drawn into cells by osmosis, partially offsetting the drop in plasma osmolarity but further decreasing blood volume and blood pressure. If Na^+ deprivation is severe, the blood volume may fall so low that blood pressure is insufficient to sustain life. For this reason, salt is necessary for life. Many animals have a "salt hunger" and actively seek salt, as when deer or cattle gather at "salt licks."

A drop in blood Na^+ concentration is normally compensated for by the kidneys under the influence of the hormone *aldosterone,* which is secreted by the adrenal cortex. Aldosterone stimulates the distal convoluted tubules and collecting ducts to reabsorb Na^+, decreasing the excretion of Na^+ in the urine. Indeed, under conditions of maximal aldosterone secretion, Na^+ may be completely absent from the urine. The reabsorption of Na^+ is followed by reabsorption of Cl^- and water, so aldosterone has the net effect of promoting the retention of both salt and water. It thereby helps to maintain blood volume, osmolarity, and pressure.

The secretion of aldosterone in response to a decreased blood level of Na^+ is indirect. Because a fall in blood Na^+ is accompanied by decreased blood volume, the flow of blood past a group of cells called the juxtaglomerular apparatus is reduced. The juxtaglomerular apparatus is located in the region of the kidney between the distal convoluted tubule and the afferent arteriole.

When blood flow is reduced, the juxtaglomerular apparatus responds by secreting the enzyme *renin* into the blood. Renin catalyzes the production of the polypeptide angiotensin I from the protein angiotensinogen. Angiotensin I is then converted by another enzyme into angiotensin II, which stimulates blood vessels to constrict and the adrenal cortex to secrete aldosterone. Thus, homeostasis of blood volume and pressure can be maintained by the activation of this renin–angiotensin–aldosterone system. In addition to stimulating Na^+ reabsorption, aldosterone also promotes the secretion of K^+ into the distal convoluted tubules and collecting ducts. Consequently, aldosterone lowers the blood K^+ concentration, helping to maintain constant blood K^+ levels in the face of changing amounts of K^+ in the diet. People who lack the ability to produce aldosterone will die if untreated because of the excessive loss of salt and water in the urine and the buildup of K^+ in the blood.

REVIEW OF CONCEPT 35.7

ADH stimulates the insertion of water channels into the cells of the distal convoluted tubule and collecting duct, increasing its reabsorption. Aldosterone promotes reabsorption of Na^+, Cl^-, and water across the distal convoluted tubule and collecting duct, as well as the secretion of K^+ into the tubules. Blood pressure is also affected by these systems, as decreasing blood volume decreases pressure.

- *What would be the effect of a compound that blocked aquaporin water channels?*

35.8 The Immune System Defends the Body

A key aspect of preserving the body's homeostasis is defending it from invasion by microbes and from takeover by cancer cells. The response of vertebrates to microbial invasion and cancerous growth occurs in two ways: innate and adaptive immunity.

Innate Immunity Recognizes Characteristic Pathogen Molecules

LEARNING OBJECTIVE 35.8.1 Describe the function of pattern recognition receptors.

Innate immunity is a response to invading pathogens involving soluble factors and a variety of different blood cells. The response is initiated by recognition of molecules found in the pathogen, which are characteristic of a group of pathogens. These are called pathogen-associated molecular patterns, or PAMPs, such as the lipopolysaccharide (LPS) found in gram-negative bacterial cell walls. Different receptor proteins exist that bind to different classes of pathogen molecules. These pattern recognition receptors can be soluble proteins or membrane proteins on the surface of blood cells.

Toll-like receptors

Some of the receptors involved in vertebrate innate immunity have wide phylogenetic distribution. One class of receptor that has been studied extensively is related to the Toll receptor, first identified in *Drosophila.* Toll-like receptors (TLRs) are found in all bilaterian animals and therefore arose prior to the divergence of protostomes and deuterostomes. In *Drosophila,* Toll is involved in dorsal-ventral patterning, but also mediates a response to fungal infection.

The number of TLRs varies in different lineages. Among deuterostomes, there has been an expansion in echinoderms, with more than 200 TLRs, whereas vertebrates have only 10. Among protostomes, molluscs and nematodes have a single TLR, but annelids have 100 TLRs. It is not clear what caused this selective expansion of TLRs in a few lineages. In vertebrates, TLRs have evolved to be part of a larger, integrated immune response.

TLRs are membrane receptors found on a variety of leukocytes, which function by binding to specific PAMPs. Known targets include gram-negative LPS, bacterial lipoproteins, bacterial peptidoglycan fragments, yeast cell-wall components, unmethylated CpG motifs in bacterial DNA, and viral RNA. This encompasses a range of pathogens that have infected vertebrates over a long period of evolutionary time.

The binding pocket of a TLR is formed by leucine-rich repeats that fold into different shapes in different receptors. This produces a variety of binding pockets that can recognize different molecules. Because these molecules are both shared and critical, characteristics of a class of pathogens, one receptor can recognize a range of pathogens. For example, the TLR that recognizes LPS will recognize most gram-negative bacteria.

Activation of innate receptors turns on signal transduction pathways that enhance the response of both innate and adaptive immune responses and provides a connection between innate and adaptive immunity. Possible responses include the inflammatory response (described later in this section), the production of antimicrobial peptides, and the production of cytokines that attract other immune cells. Together these constitute the innate response to infection.

Cytoplasmic receptors

Since the discovery of Toll and TLR proteins, a second class of receptors has been discovered localized to the cytoplasm. These internal receptors also bind to characteristic pathogen molecules and can recognize invading pathogens in the cytoplasm of cells after phagocytosis. These receptors also are part of the response to viral RNA.

Soluble receptors

Soluble receptors also circulate in serum and can respond to specific pathogen molecules, including some of the lectin family that bind to the sugar mannose found in bacterial cell walls. These lectin proteins are important in activating the complement system described later in this section.

Innate Immunity Leads to Diverse Responses to a Pathogen

LEARNING OBJECTIVE 35.8.2 Describe the inflammatory response.

Recognition of an invading pathogen by innate receptors such as the TLR family leads to a coordinated response, which includes the production of secreted signaling molecules, the production of antimicrobial peptides, and the activation of complement. This response is shown in overview in figure 35.21. The antimicrobial peptides, such as the defensins, are similar to those normally found in the integument. The cysteines in defensins interact with positively charged amino acids on the surface of a pathogen. By binding to the outer membrane of gram-negative bacteria, they can both disrupt the membrane and enhance phagocytosis. This action is also effective against enveloped viruses.

Another class of proteins induced by innate defenses that plays a key role in body defense is interferons. Interferons are important secreted signaling molecules with diverse functions. The two classes of interferons, type I and type II, have different receptors and activate different signal transduction pathways. The type I interferons are synthesized by virally infected cells and act as messengers that protect uninfected cells in the vicinity. Although viruses are still able to penetrate the neighboring cells, interferons induce the degradation of RNA and block protein

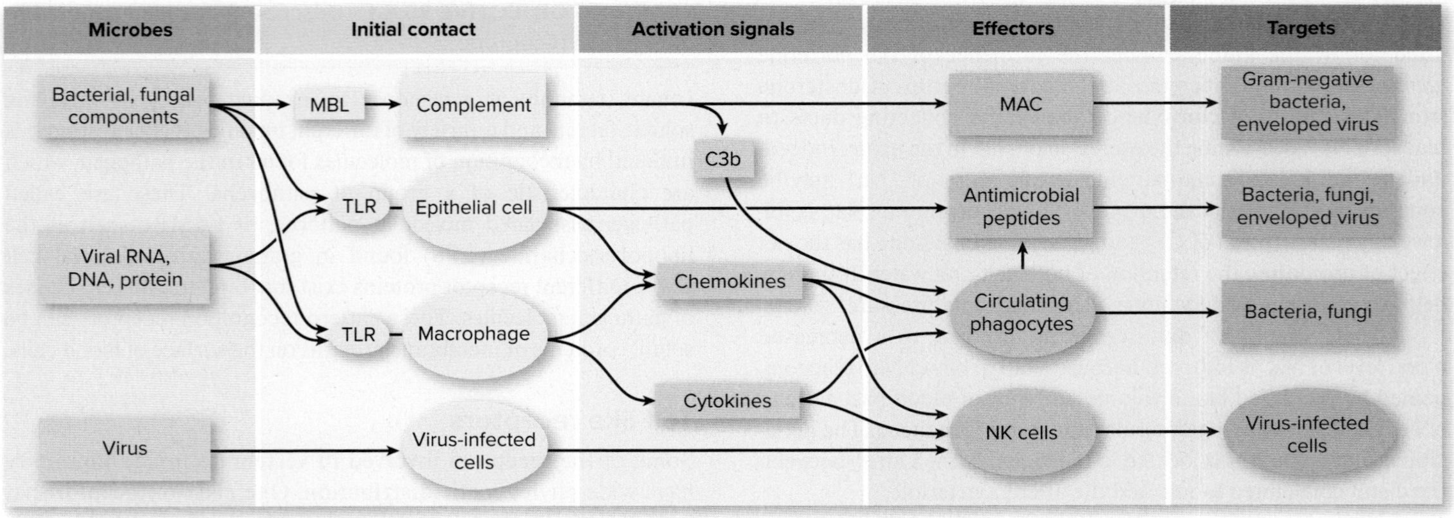

Figure 35.21 Overview of innate immunity. Pathogens have critical molecules that adhere to either membrane-bound (TLR) or soluble receptors (MBL). This results in the production of cytokines and chemokines that attract phagocytes, of antimicrobial peptides, and of the membrane attack complex (MAC) via the complement cascade. This also results in the activation of natural killer cells (NK cells).

production in these cells. Although this leads to the death of the cells, it also prevents the production and spread of the virus.

Type II interferon—in humans interferon gamma (IFN-γ)—is produced only by particular leukocytes called T lymphocytes (described later in this section) and natural killer cells. The secretion of IFN-γ by these cells is part of the immunological defense against infection and cancer.

In addition to these nonspecific defensive molecules, activation of innate immunity leads to two other responses: activation of the inflammatory response and activation of the complement pathway. Signaling from both TLRs and internal receptors can lead to the secretion of a variety of cytokines, or regulatory signaling molecules. These attract other nonspecific phagocytic cells, cause inflammation, and even signal to the adaptive immune system.

Phagocytic cells are associated with innate immunity

Among the most important innate defenses are cells that can nonspecifically kill invading pathogens. These are types of leukocytes, or white blood cells, that circulate through the body and attack pathogens within tissues. Three basic kinds of defending leukocytes—macrophages, neutrophils, and natural killer cells—have been identified, and each kills invading microorganisms differently.

Macrophages **Macrophages** ("big eaters") are large, irregularly shaped cells that kill microorganisms by ingesting them through phagocytosis. Once within the macrophage, the membrane-bound phagosome fuses with a lysosome. Fusion activates lysosomal enzymes that kill and digest the microorganism within the phagosome. Additionally, large quantities of oxygen-containing free radicals are frequently produced within the phagosome; these free radicals are very reactive and quickly degrade the pathogen.

In addition to bacteria, macrophages also engulf viruses, cellular debris, and dust particles in the lungs. Macrophages roam continuously in the extracellular fluid that bathes tissues. In response to an infection, monocytes, which are undifferentiated macrophages in the blood, squeeze through the endothelial cells of capillaries to enter the connective tissues. There, at the site of the infection, the monocytes mature into active, phagocytic macrophages.

Neutrophils **Neutrophils** are the most abundant circulating leukocytes, accounting for 50 to 70% of the peripheral blood leukocytes. They are the first type of cell to appear at the site of tissue damage or infection. Like macrophages, they squeeze between capillary endothelial cells to enter infected tissues, where they ingest a variety of pathogens by phagocytosis. Their mechanism of pathogen destruction is similar to that of macrophages, except that they produce an even greater range of reactive oxygen radicals. Neutrophils also produce defensin peptides.

Natural Killer Cells **Natural killer (NK) cells** do not attack invading microbes. Instead, they kill cells of the body that have been infected with viruses. They kill not by phagocytosis but, rather, by inducing apoptosis (programmed cell death) of the target cell. Proteins called perforins, released from the NK cells, insert into the membrane of the target cell, creating a pore in the membrane. Other NK-produced proteins, called granzymes, enter these pores and activate proteins in the target cells that induce apoptosis (figure 35.22). Macrophages ingest the resulting membrane-bounded vesicular cell debris.

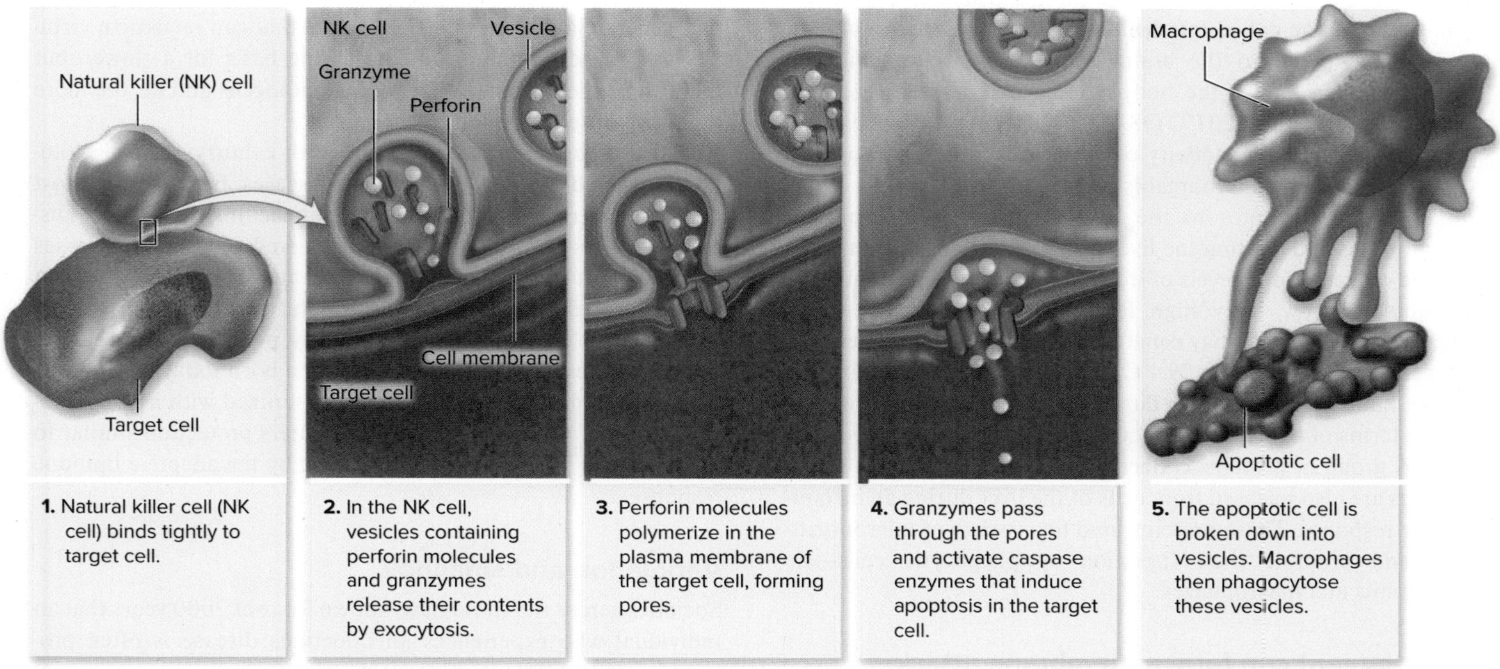

Natural killer (NK) cell
Target cell
NK cell
Vesicle
Granzyme
Perforin
Cell membrane
Target cell
Macrophage
Apoptotic cell
1. Natural killer cell (NK cell) binds tightly to target cell.
2. In the NK cell, vesicles containing perforin molecules and granzymes release their contents by exocytosis.
3. Perforin molecules polymerize in the plasma membrane of the target cell, forming pores.
4. Granzymes pass through the pores and activate caspase enzymes that induce apoptosis in the target cell.
5. The apoptotic cell is broken down into vesicles. Macrophages then phagocytose these vesicles.

Figure 35.22 How natural killer cells eliminate target cells. Natural killer cells kill virally infected cells by programmed cell death, or apoptosis. This is accomplished by secreting proteins that form pores in the cell to be killed, along with proteins that diffuse through these pores and induce apoptosis.

NK cells also attack tumor cells, often before the tumor cells have had a chance to divide sufficiently to be detectable as a tumor. The vigilant surveillance by NK cells is one of the body's most potent defenses against cancer.

The inflammatory response is a nonspecific response to infection or tissue injury

The inflammatory response involves several systems of the body, and it may be either localized or systemic. An acute response is one that generally starts rapidly but lasts for only a relatively short while.

Certain infected or injured cells release chemical alarm signals—most notably histamine, along with prostaglandins and bradykinin. These chemicals cause vasodilation of local blood vessels, increasing the flow of blood to the site and causing the area to become red and warm, two of the hallmark signs of inflammation. These chemicals also increase the permeability of capillaries in the area, producing the third hallmark sign of inflammation, the edema (tissue swelling) often associated with infection. Swelling puts pressure on nerve endings in the region, and this, in combination with the release of other mediators, leads to pain and potential loss of function, the final two hallmark signs of inflammation.

Increased capillary permeability initially promotes the migration of phagocytic neutrophils from the blood to the extracellular fluid bathing the tissues, where the neutrophils can ingest and degrade pathogens; the pus associated with some infections is a mixture of dead or dying pathogens, tissue cells, and neutrophils. The neutrophils also secrete signaling molecules that attract monocytes several hours later; as the monocytes differentiate into macrophages, they, too, engulf pathogens and the remains of the dead cells. The inflammatory response is accompanied by an acute-phase response. One manifestation of this response is an elevation of body temperature, or fever. When a macrophage with a TLR on its surface binds to an invading pathogen, the cytokine called **interleukin-1 (IL-1)** is released and carried by the blood to the brain. IL-1 causes neurons in the hypothalamus to raise the body's temperature several degrees above the usual value of 37°C (98.6°F). This increase in body temperature promotes the activity of phagocytic cells and impedes the growth of some microorganisms.

Fever contributes to the body's defense by stimulating phagocytosis and causing the liver and spleen to store iron. This storage reduces blood levels of iron, which bacteria need in large amounts to grow. Very high fevers are hazardous, however, because excessive heat may denature critical enzymes. In general, temperatures greater than 39.4°C (103°F) are considered dangerous for humans, although the duration of the fever is most important in terms of damage or mortality.

A group of proteins collectively referred to as acute-phase proteins are also released from cells of the liver during an inflammatory response. These proteins bind to a variety of microorganisms and promote their ingestion by phagocytic cells—the neutrophils and macrophages.

Complement can form a membrane attack complex

The cellular defenses of vertebrates are enhanced by a very effective chemical defense called the **complement system**—a group of approximately 30 different proteins that circulate freely in the blood plasma. Usually, they occur in an inactive form, which can enter the tissues during an inflammatory response. Complement can be activated by mannose-binding lectin protein (MBL), one of the soluble sensors of innate immunity, or can be activated by a complex series of reactions involving charged species on the surface of pathogens.

When the complement system is activated, complement proteins aggregate to form a **membrane attack complex (MAC),** which inserts itself into the pathogen's plasma membrane (or the lipid membrane around an enveloped virus), forming a pore. Extracellular fluid enters the pathogen through this pore, causing the pathogen to swell and burst. Activation of the complement proteins is also triggered in a specific fashion when antibodies (which are secreted by B lymphocytes) are bound to invading pathogens, as described in section 35.10.

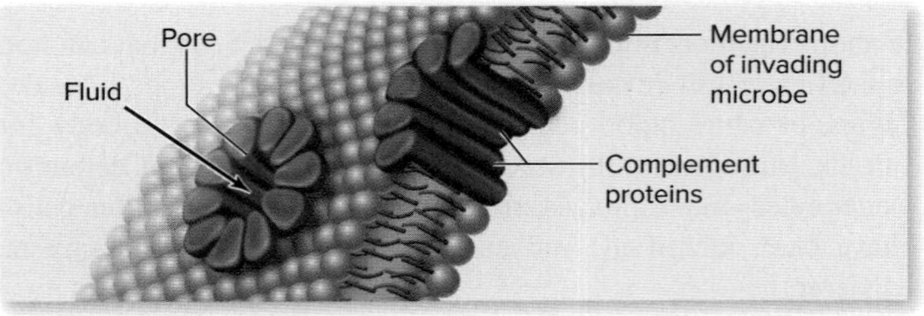

Adaptive Immunity Generates Diverse Receptors by Genetic Rearrangements

LEARNING OBJECTIVE 35.8.3 Define the characteristics of adaptive immunity.

Adaptive immunity is characterized by genetic rearrangements that generate a diverse set of molecules that can recognize virtually any invading pathogen. This is the basis for a slower, but highly specific, response to invading pathogens and for the more rapid response to a second attack.

For people born early in the 20th century, part of childhood included a variety of infectious diseases. For example, most children contracted chickenpox before reaching their teens. Diseases such as chickenpox were considered childhood diseases because most people, once recovered, never experienced them again. They developed immunity to the virus, varicella-zoster, that causes chickenpox. This immunity persists as long as the immune system remains intact. Children born today do not contract chickenpox because most are immunized with a nonpathogenic form of varicella virus, which confers protection similar to the disease. This immunity is produced by the adaptive immune response.

Variolation and smallpox

Societies across the world have known for over 2000 years that an individual who experiences an infectious disease is often protected against later infection by the same agent. Observations of this kind probably led to the practice of variolation for smallpox, which provided some protection against this dread disease. Smallpox, caused by the variola virus, was both common and deadly in the 1700s and earlier centuries. Variolation is a modern name based on the virus, but the practice was originally called inoculation. It involved taking pus from a pustule on a smallpox-infected individual, and then introducing small amounts with a needle into an uninfected individual. Variolation was common in the Ottoman Empire, and in the Middle East and north Africa, in the seventeenth century. As a result, smallpox was less serious in these regions than in western Europe. As there is not a good written record, it is not clear when or where this practice originated, but China and/or India are the most likely locations.

The practice of variolation was brought to western Europe by the writer and poet Lady Mary Wortley Montagu in 1721. As wife of the British ambassador to Turkey, she had seen the practice, and had her son inoculated in Constantinople in 1717. Upon her return to England, she had her daughter inoculated in 1721. Despite initial resistance from the medical community, Lady Montagu continued to be a proponent of variolation, and the practice was eventually accepted.

Jenner and cowpox

While the practice of variolation became common, it had an obvious danger: some people actually contracted smallpox from the procedure. The study of inoculation continued with the work of an English country doctor, Edward Jenner. Jenner observed that milkmaids infected with a milder form of pox, called cowpox, rarely contracted smallpox. Jenner set out to test the idea that cowpox could confer protection against smallpox. He inoculated a healthy child with fluid from a cowpox vesicle and later deliberately infected him with fluid from a smallpox vesicle; as he had predicted, the child did not become ill. (Jenner's experiment would be considered unethical today.)

We now know that smallpox and cowpox are caused by two different but related viruses. Jenner's patient injected with the cowpox virus mounted a defense that was also effective against a later infection of the smallpox virus. The difference between Jenner's work and variolation was that he did not use the actual infectious agent to protect against future infection. This procedure of injecting a harmless agent to confer resistance to a dangerous one is called vaccination, a topic we will return to later in this chapter.

Pasteur and avian cholera

Many years passed before anyone learned how exposure to an infectious agent could confer resistance to a disease. The next clue came more than a half century later from work on avian cholera by the famous scientist Louis Pasteur. He isolated a culture of this pathogenic bacteria that would produce the disease when injected into healthy birds. It is reported that before departing on a two-week vacation, he accidentally left this culture out on a shelf. When he returned, he injected this old culture into healthy chickens and found they became only slightly ill and then recovered. Further, those birds did not get sick when subsequently infected with fresh, disease-causing bacteria . The bacteria that had sat out had been weakened in some way, and were no longer capable of fatal infection, but they did still evoke an immune response. We now know that molecules protruding from the surfaces of the bacteria evoked active immunity in the chickens.

Antigens stimulate specific immune responses

An **antigen** is a molecule that provokes a specific immune response. The most effective antigens are large, complex molecules such as proteins. The greater their "foreignness"—their phylogenetic distance from the host—the greater the immune response they elicit.

Antigens may be components of a microorganism or a virus, but they may also be proteins or glycoproteins on the surface of transfused red blood cells or on transplanted tissue. They may also be components of foods or pollens. A large antigen is likely to have many different parts, known as *antigenic determinants,* or *epitopes* (figure 35.23), each of which can stimulate a distinct immune response.

Lymphocytes carry out the adaptive immune responses

The adaptive immune system is characterized by the following:

1. Specificity of recognition of the antigen
2. Specific recognition of a wide diversity of antigens
3. Memory, whereby the immune system responds more quickly and more intensely to an antigen it has encountered previously than to one it is meeting for the first time
4. Ability to distinguish self-antigens from nonself-antigens

The cells in the blood involved in the adaptive immune response are leukocytes derived from a stem cell line called lymphoid progenitor cells. These **lymphocytes** have receptor proteins on their surfaces that recognize specific epitopes on an antigen and direct an immune response against the antigen either in solution or on the cell surface (figure 35.24). This response is also affected by signals derived from the innate system. The innate system dominates early in infection by a new pathogen, and the adaptive response dominates in later stages of infection.

Lymphocytes and antigen recognition

Although all the receptor proteins on any one lymphocyte have the same epitope specificity, it is rare that any two lymphocytes have

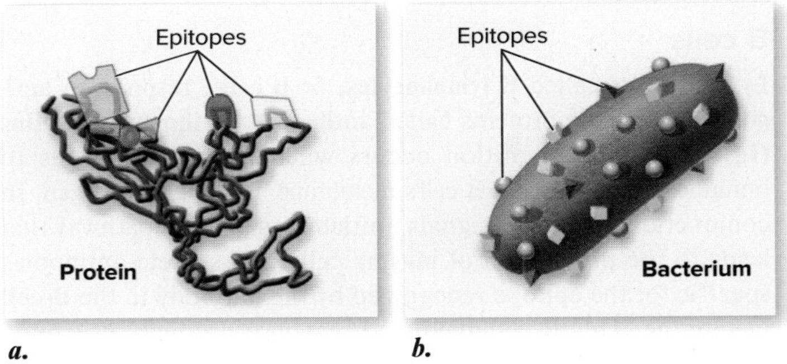

Figure 35.23 Many different epitopes are exhibited by any one antigen. ***a.*** A single protein, with associated carbohydrate, may have many different antigenic determinants called epitopes, each of which can stimulate a distinct immune response. ***b.*** A pathogen such as a bacterium has many proteins on its surface, and there are likely to be multiple copies of each. Note that the protein and the bacterium are not drawn to scale with respect to each other.

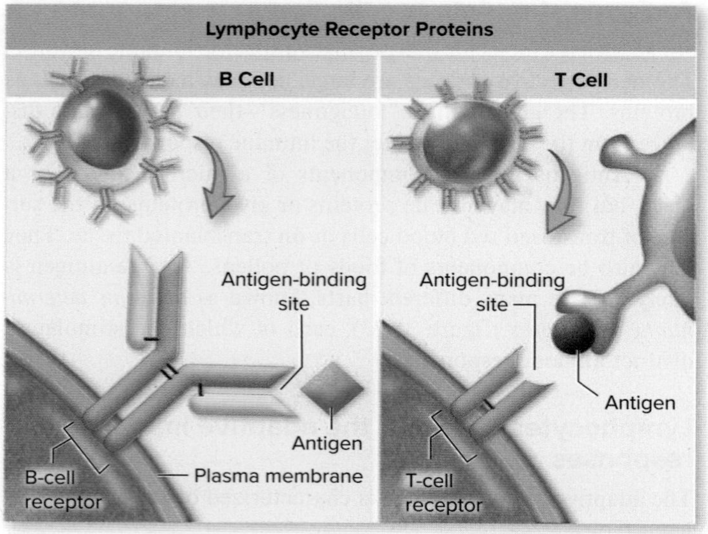
Lymphocyte Receptor Proteins
B Cell
T Cell
Antigen-binding site
Antigen-binding site
Antigen
Antigen
B-cell receptor
Plasma membrane
T-cell receptor

Figure 35.24 B- and T-cell receptors bind antigens. B-cell receptors are immunoglobulin (Ig) molecules with a characteristic Y-shaped structure. Every B cell has a single kind of Ig on its surface that binds to a single antigenic determinant. T-cell receptors are simpler than Ig molecules but also bind to specific antigenic determinants. T cells bind only to antigens bound to another cell.

identical specificities. This feature produces the diversity of immune responses that ensures that at least some epitopes of any antigen that might be encountered are recognized.

A lymphocyte that has never before encountered an antigen is referred to as a *naive lymphocyte.* When a naive lymphocyte binds to a foreign antigen, the lymphocyte is activated, causing it to divide, producing a clone of cells with identical antigen specificity, a process called **clonal selection.** Some of these cells respond immediately to the antigen, and others become memory cells, which can remain in our bodies for years and perhaps for the remainder of our lives. Memory cells are easily and rapidly activated on subsequent encounters with the same antigen.

B cells

Lymphocytes called **B lymphocytes,** or **B cells,** respond to antigens by secreting proteins called **antibodies,** or **immunoglobulins (Ig).** Antigen recognition occurs when an antigen binds to immunoglobulins on the B cell's membrane. Binding to antigen, in conjunction with other signals, initiates a signaling pathway that leads to the production of plasma cells that secrete antibodies specific for the epitope recognized by the antibody in the B-cell membrane. This B cell–mediated response producing secreted antibodies is called **humoral immunity.**

T cells

Other lymphocytes, called **T lymphocytes,** or **T cells,** do not secrete antibodies but instead regulate the immune responses of other cells or directly attack the cells that carry the specific antigens. These cells participate in the other arm of adaptive immunity, called **cell-mediated immunity.** Both cell-mediated and humoral immunity processes are described in detail in sections 35.9 and 35.10.

REVIEW OF CONCEPT 35.8

Innate immunity recognizes molecules specific to pathogens. These include bacterial lipopolysaccharide and peptidoglycan, as well as viral RNA and DNA. The inflammatory response involves signals that attract neutrophils, increase capillary permeability, activate complement, and trigger fever. Adaptive immunity recognizes individual pathogens, mounting a specific response. B cells produce circulating antibodies (humoral immunity); T cells kill pathogens or help other cells respond (cell-mediated immunity).

- *Is innate immunity nonspecific?*

35.9 Cell-Mediated Immunity Involves Helper and Killer T Cells

T cells may be characterized as either **cytotoxic T cells (T_C)** or **helper T cells (T_H).** These cells can also be identified based on cell surface markers. T_C cells have CD8 protein on their cell surface, making them $CD8^+$ cells. T_H cells have CD4 protein on their cell surface, making them $CD4^+$ cells.

Cytotoxic T Cells Eliminate Virally Infected Cells and Tumor Cells

LEARNING OBJECTIVE 35.9.1 Describe the function of cytotoxic T cells.

To be activated, both of these T-cell types must recognize peptide fragments bound to **major histocompatibility complex, or MHC,** proteins. The two cell types may be distinguished by (1) recognition of different classes of MHC proteins, which have distinct cell distributions, and (2) differing roles of the T cells after they are activated.

The MHC carries self and nonself information

The surfaces of most vertebrate cells exhibit glycoproteins encoded by the MHC. In humans, the name given to the proteins encoded by the MHC is **human leukocyte antigens (HLAs).** The genes encoding the MHC proteins are highly polymorphic (have many alleles). For example, the HLA proteins are specified by genes that are the most polymorphic known, with nearly 500 alleles detected for some of the proteins. Only rarely will two individuals have the same combination of alleles, and the HLAs are thus different for each individual.

MHC proteins on the tissue cells serve as self markers that enable an individual's immune system, specifically its T cells, to distinguish its own cells from foreign cells, an ability called *self versus nonself recognition.*

There are two classes of MHC proteins. **MHC class I proteins** are present on every nucleated cell of the body. **MHC class II proteins,** however, are found only on **antigen-presenting cells** (in addition to MHC class I); these cells include macrophages, B cells, and dendritic cells. T_C cells respond to peptides bound to MHC class I proteins, and T_H cells respond to peptides bound to MHC class II proteins.

Most of the time, the peptides bound to MHC proteins are derived from self-proteins from the individual's own cells. For this reason, it is important that T cells undergo selection in the thymus so that those that bind too strongly to peptides of self-proteins on self-MHC are eliminated. In this way, T cells are usually activated only outside the primary lymphoid organs in which they mature, when they encounter peptides of foreign proteins on self-MHC, as in the case of viral infection or cancer.

Activated cytotoxic T cells recognize "altered-self" cells, particularly those that are virally infected or tumor cells. The TCRs of cytotoxic T lymphocytes recognize peptides of endogenous antigens bound to MHC class I proteins. Peptides of endogenous antigens are generated in a cell's cytosol and then are pumped by special transport proteins into the rough endoplasmic reticulum, where they become bound to MHC class I proteins. These proteins continue on their way through the endomembrane system to the cell surface.

An endogenous antigen may be a self-protein, or it may be a viral protein produced within a virally infected cell or an unusual protein produced by a cancerous cell. T_C cells respond only to the peptides of these unusual proteins bound to self-MHC class I. T-cell activation occurs in a secondary lymphoid organ. In a lymph node, for example, T cells encounter antigen-presenting cells. Dendritic cells, in particular, often present antigens that activate T_C cells.

Because not all viruses can infect dendritic cells, the dendritic cells must ingest viruses or tumor cells and then, through a mechanism referred to as cross-presentation, place the viral or tumor peptides on MHC class I proteins. Binding of the T_C cell through its TCR and its CD8 site to the dendritic cell induces clonal expansion of the T_C cell, generating many activated T_C cells as well as memory T_C cells (figure 35.25). The activated T_C cells then circulate around the body and bind to "target" host cells that express the same combination of foreign peptide on self-MHC class I. Apoptosis of the target cell is induced by a mechanism similar to that used by NK cells.

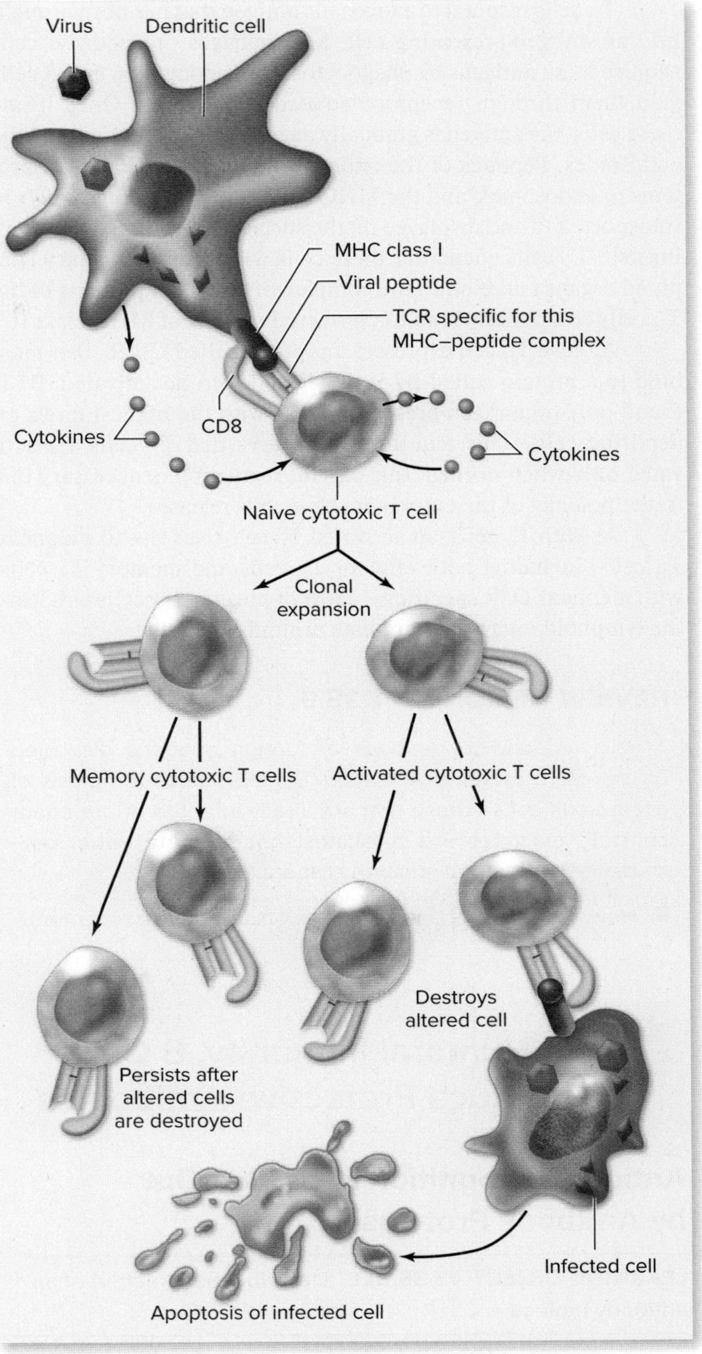

Figure 35.25 Cytotoxic T cells induce apoptosis of "altered-self" cells. Naive cytotoxic T cells are initially activated on TCR recognition of foreign peptide displayed on self-MHC class I proteins on dendritic cells in a secondary lymphoid organ. Activation results in clonal expansion and differentiation into memory cells and activated cells. Activated progeny of the T_C cell can induce apoptosis of any cell in the periphery (outside the secondary lymphoid organ) that displays the same self-MHC class I–peptide combination on its surface. This will most likely be a virally infected cell or a tumor cell.

Helper T Cells Secrete Proteins That Direct Immune Responses

LEARNING OBJECTIVE 35.9.2 Explain the role of helper T cells.

Activated helper T cells, T_H cells, secrete low-molecular-weight proteins known as **cytokines.** A vast array of cytokines is known, many but not all of which are secreted by T_H cells. These cytokines bind to specific receptors on the membranes of many other cells, particularly but not exclusively those of the immune system. On binding, they initiate signaling cascades in these cells that promote their activation or differentiation.

Because cytokines are quite potent, they are generally secreted at very low concentrations, so that they bind primarily to nearby cells. IL-1 is an exception in that it travels to the hypothalamus to induce the fever response. Different subsets of T_H cells secrete different cytokines, so it is largely the T_H cells and the cytokines they secrete that determine whether an immune response will be humoral or cell-mediated in nature.

T_H cells respond to exogenous antigen that has been brought into an antigen-presenting cell. Macrophages or dendritic cells acquire these antigens by phagocytosis or endocytosis, and B cells gain them through receptor-mediated endocytosis. Once inside these cells, the antigen is gradually degraded in increasingly acidic endosomes. Peptides of the antigen join with MHC class II proteins in endosomes, and the MHC class II–peptide complexes are transported to and displayed on the surface of the antigen-presenting cell. T_H cells encounter these cells within the secondary lymphoid organs and bind to the complexes. The CD4 proteins of the T_H cells additionally bind to conserved regions of MHC class II.

A naive T_H cell expresses a protein, called CD28, that must bind to a protein called B7 if that T cell is to be activated. B7 is found only on antigen-presenting cells with the highest levels on dendritic cells. This requirement ensures that T_H cells are activated only when needed; this careful regulation is necessary due to the potency of the cytokines these cells release.

As with T_C cells, an activated T_H cell gives rise to a clone of T_H cells, including both effector T_H cells and memory T_H cells, with identical TCR specificity. Most of the effector cells will leave the lymphoid organ and circulate around the body.

REVIEW OF CONCEPT 35.9

T cells respond to peptides of foreign antigens displayed on self-MHC proteins. Activated T_C cells induce apoptosis of altered-self cells—those that are virally infected or are tumor cells. T_H cells secrete cytokines that promote either cell-mediated or humoral immune responses.

- *How are T-cell receptors different from Toll-like receptors?*

35.10 In Humoral Immunity, B Cells Produce Protective Antibodies

Antigen Recognition Is Carried Out by Antibody Proteins

LEARNING OBJECTIVE 35.10.1 Describe the structure of an antibody molecule.

The B-cell receptors for antigen are the immunoglobulin molecules present as integral proteins in the plasma membrane. Each B cell exhibits about 10^5 immunoglobulin molecules of identical specificity for a particular epitope of an antigen. Naive B cells in secondary lymphoid organs encounter antigens. When immunoglobulin molecules on a B cell bind to a specific epitope on an antigen, and the B cell receives additional required signals, particularly cytokines secreted by T_H cells, then that B cell becomes activated, proliferating into plasma cells and memory cells.

Each plasma B cell is a miniature factory producing soluble antibodies of the same specificity as the membrane-bound antibodies of the parent B cell. These antibodies enter the lymph and blood circulation as well as the extracellular fluid, and they bind to the appropriate epitopes of antigen encountered anywhere in the body. Any one antigen may present a variety of epitopes, so that different B cells might recognize different epitopes of a single antigen.

Once immunoglobulins coat an antigen, many other cells and processes may be activated to eliminate the antigen. The immunity to avian cholera that Pasteur observed in his chickens resulted from such antibodies and from the continued presence of the progeny of the B cells that produced them.

Immunoglobulin structure reveals variable and constant regions

Each immunoglobulin molecule consists of two identical short polypeptides called **light chains** and two identical longer polypeptides called **heavy chains** (figure 35.26). The four chains in an immunoglobulin molecule are held together by disulfide bonds, forming a Y-shaped molecule (figure 35.26*a*). Each "arm" of the molecule is referred to as an Fab region, and the "stem" is the Fc region.

Antibody specificity resides in the variable region

Comparison of the amino acid sequences of many different immunoglobulin molecules has demonstrated that the specificity of immunoglobulins for antigen epitopes resides in the amino-terminal half of each Fab region. This half of the Fab has an amino acid sequence that varies from one immunoglobulin to the next and is thus designated the *variable region.* Both the light chain and the heavy chain have a variable region.

The amino acid sequence of the remainder of the immunoglobulin is relatively constant from one immunoglobulin to the next and is thus designated the *constant region* (figure 35.26*b*). Both light and heavy chains also exhibit constant regions. Careful analysis shows that light-chain constant regions of mammalian immunoglobulins consist of two different sequences, designated κ (kappa) and λ (lambda), which have apparently equivalent function. The heavy-chain constant regions consist of five different sequences: μ (mu), δ (delta), γ (gamma), α (alpha), and ε (epsilon). When each of these heavy chains is bound to either type of light chain, they give rise to a particular class of immunoglobulin: IgM, IgD, IgG, IgA, and IgE.

The variable regions of the heavy and light chains fold together to form a sort of cleft, the *antigen-binding site* (figure 35.26). The size and shape of the antigen-binding site, as well as which amino acids line its surface, determine the specificity of each immunoglobulin for an antigen epitope.

Because each immunoglobulin is composed of two identical halves, each can bind with two identical epitopes, although not generally on the same antigen, because of steric (shape) constraints. This ability to bind with two epitopes allows the formation of antigen–antibody complexes containing multiple antibody and antigen molecules.

Immunoglobulin Diversity Is Generated Through DNA Rearrangement

LEARNING OBJECTIVE 35.10.2 Explain how antibody diversity is generated.

The vertebrate immune system is capable of recognizing as foreign virtually any nonself molecule presented to it. It is estimated

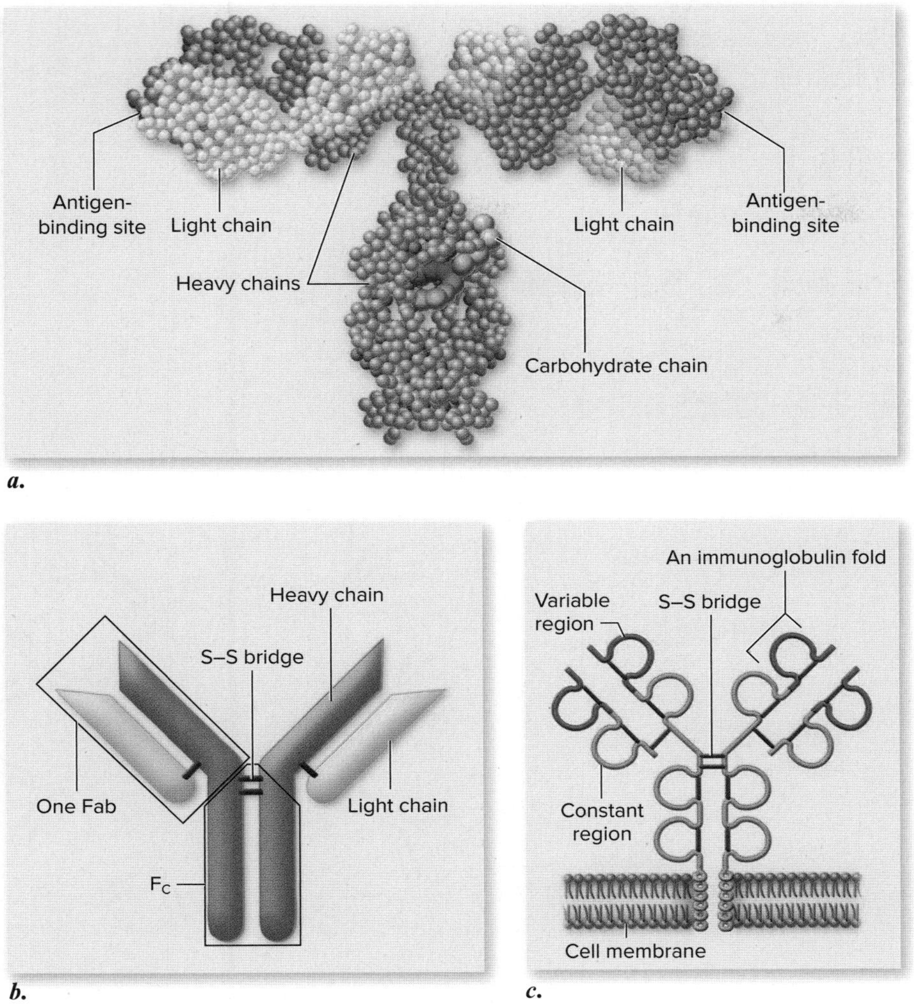

Figure 35.26 The structure of an immunoglobulin molecule. ***a.*** In this model of an immunoglobulin (Ig) molecule, amino acids in the peptide chains are represented by small spheres. The molecule consists of two heavy chains (brown) and two light chains (yellow). ***b.*** A more schematic view shows how the four chains form a Y shape, with two identical antigen-binding sites at the arms of the Y, the Fab regions, and a stem, or Fc region. The two Fab regions are joined to the Fc region by a flexible hinge. ***c.*** Ig molecule shown as a membrane protein. The variable region, shown in blue, binds to the antigen epitopes.

that human or mouse B cells can generate antibodies with over 10^{10} different antigen-binding sites. Although an individual probably does not have antibodies specific to all epitopes of an antigen, it is fairly certain that antibodies will recognize some of the epitopes, which is all that is required to generate an effective immune response. How do vertebrates generate such diversity of antigen recognition?

The answer lies in the unusual genetics of the variable region. This region in each chain of an immunoglobulin is not encoded by one single stretch of DNA but, rather, is assembled by joining two or three separate DNA segments together to produce the variable region. This process is called DNA rearrangement and is similar to the crossing over that occurs during meiosis (refer to chapter 11) but with two key differences: it occurs between loci on the same chromosome, and it is site-specific.

DNA rearrangement occurs as a progenitor B cell matures in the bone marrow. After DNA rearrangement, RNA transcription produces an mRNA that can be translated into either a heavy- or a light-chain immunoglobulin polypeptide, depending on the locus transcribed.

Cells contain homologous pairs of chromosomes, but DNA rearrangement occurs for the heavy-chain and light-chain loci on only one homolog, a process referred to as *allelic exclusion*. Thus, each B cell makes immunoglobulins of only one specificity.

Variable-region DNA rearrangements

Sequencing of human immunoglobulin heavy-chain gene loci from several different individuals shows that the locus contains a cluster of approximately 50 sequential DNA segments, termed V segments, followed by a cluster of approximately 30 smaller segments, D segments, and finally by another cluster of 6 smaller segments, J segments. Each V segment is approximately the same size as any other, but they are all of different nucleotide sequence and thus encode different amino acids; the situation is similar for the D and the J segments.

DNA rearrangement during B-cell maturation begins with a site-specific recombination event joining one of the D segments to one of the J segments (figure 35.27). Because these two sites are on the same chromosome, the intervening DNA is deleted. This is followed by another site-specific recombination joining a V segment to the rearranged DJ, again deleting the intervening DNA. Which V, which D, and which J are chosen by any cell appears to be completely random. The different combinations of V, D, and J that can be formed can produce about 9000 different heavy-chain variable-region sequences. Variable regions in the light chain are formed in a similar way, except that each is encoded by only a V and a J segment.

Other processes contribute even further to the diversity of variable-region sequence. The joining of DNA regions is somewhat imprecise, and a few nucleotides may be added to or deleted from the ends of each segment, resulting in a shift of the reading frame. B cells may end up expressing any heavy-chain variable region with any light-chain variable region during its maturation. Finally, these genes show an elevated mutation rate, termed somatic hypermutation. Taking all these processes into account has allowed the estimate of more than 10^{10} possible variable regions.

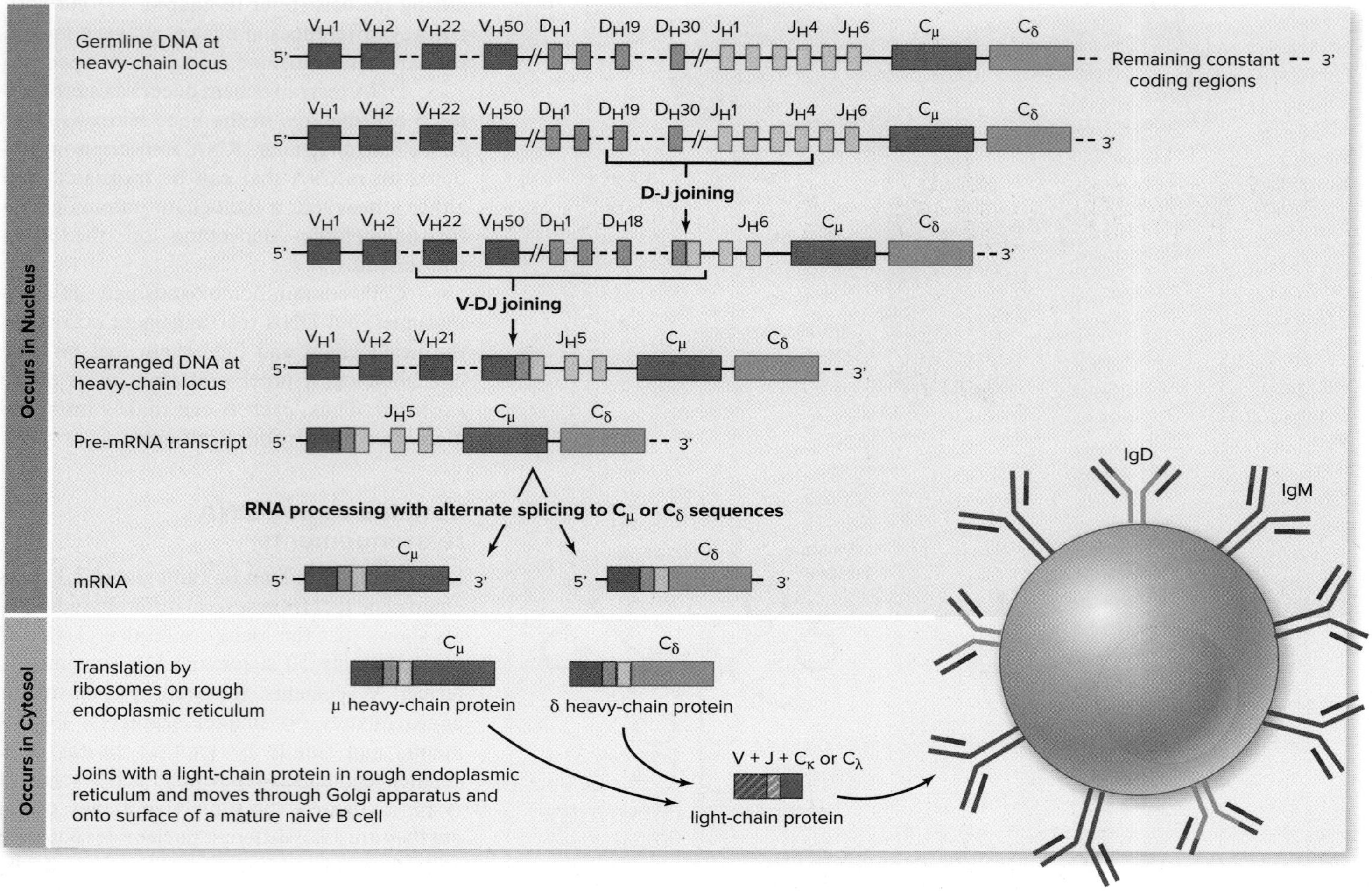

Figure 35.27 Immunoglobulin diversity is generated by rearrangement of segments of DNA. An immunoglobulin (Ig) protein is encoded by different segments of DNA: a V (variable), a D (diversity), and a J (joining) segment, plus a constant region. These are joined by a precise sequence of DNA rearrangements during maturation in the bone marrow. This first joins a D to a J segment; then this combined DJ joins to a V segment. Other cells will select other V, D, and J segments, contributing to Ig diversity. Transcription starts at the rearranged VDJ and continues through constant-region exons. Pre-mRNA splicing joins the variable region to either a μ or a δ constant region. These transcripts are translated by ribosomes on the RER to produce heavy-chain polypeptides that join with light chains (encoded by a V, a J, and a C). These proteins are transported to the cell surface, resulting in a mature naive B cell that expresses both IgM (μ constant region) and IgD (δ constant region), with the same variable region and thus the same antigen specificity.

Transcription and translation

After the DNA rearrangements that encode the variable region are complete, pre-mRNA transcripts are formed with 5′ ends that begin at the rearranged variable region–encoding segments and continue through constant-region exons encoding μ and δ constant regions.

Alternative splicing of these RNA transcripts removes any extra J segments that remain, as well as either δ or μ sequences, resulting in transcripts that all encode the same variable region but either μ or δ constant-region exons (figure 35.27). Translation results in a μ or δ heavy-chain polypeptide, which associates with a light-chain polypeptide in the rough endoplasmic reticulum. Thus, the mature naive B cell expresses both IgM and IgD on its surface, both having the same antigen-binding specificity.

The Secondary Response to an Antigen Is More Effective Than the Primary Response

LEARNING OBJECTIVE 35.10.3 Explain how vaccination prevents disease.

When a particular antigen enters the body, it must encounter naive lymphocytes with the appropriate receptors, to provoke an immune response. The first time a pathogen invades the body, only a few B or T cells may have receptors able to recognize the new pathogen's epitopes or, for infected or otherwise abnormal cells, foreign peptides bound to self-MHC. In this first encounter, a person will usually develop symptoms of illness, because only a few cells are present that can mount an immune response. The clonal expansion of the appropriate T and B cells, as well as secretion of IgM antibodies, will counteract the infection, but this takes several days (figure 35.28).

Antibodies produced by this primary response will plateau, and then decline as antibodies become bound to antigen and are cleared from circulation, and unbound antibodies are naturally metabolized. Another consequence of this response is the production of memory B and T cells that will recognize the original antigens. These memory cells are more rapidly activated than are naive lymphocytes, producing a secondary immune response that initiates and peaks much more rapidly than a primary response (figure 35.28). Further clonal expansion takes place, along with the secretion of large amounts of antibodies, generally of the IgG class, although IgA and IgE are also possible. The class of immunoglobulin produced is determined by the identity of the cytokines, produced by activated T_H memory cells, that stimulate the B cells during this secondary response.

It is advantageous for an individual to produce different classes of immunoglobulins during an immune response, because each class has a different function. During a second exposure to the same antigen, while memory cells are activated and secrete isotypes other than IgM, other naive B cells also recognize the antigen for the first time, become activated, and secrete IgM.

Memory B and T cells can survive for decades, which is why people rarely contract the same infectious disease twice. The large and rapid secondary response stops the infection before disease symptoms are even detected. The exceptions to this are pathogens that evolve quickly enough to evade the immune system, or have evolved adaptations to combat adaptive immunity.

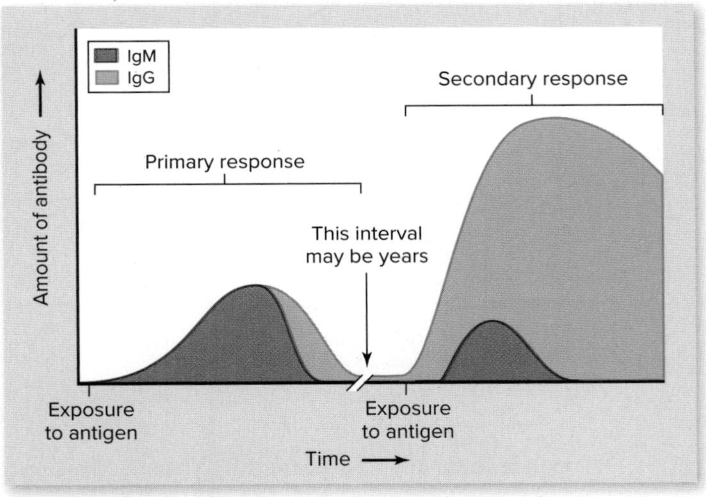

Figure 35.28 The development of active immunity. On initial exposure to an antigen, from either infection or vaccination, there will be a primary response that includes the development of lymphocyte clones, including memory cells, with receptors that bind to the antigen. This initial response will reach a plateau and decline, but if there is a later exposure to the antigen, this stimulates the memory cells to produce large amounts of antibody of the same specificity much more rapidly than during the primary immunization. The first antibodies produced during the primary response are IgM in class (red), although IgG (blue) is secreted near the end of the primary response. The majority of the antibody secreted during a secondary response is IgG, although IgA can be secreted if the antigen has activated B cells in the MALT; or, in some circumstances, such as allergies, IgE is secreted.

Vaccination can prevent infectious disease

Memory cells and the secondary immune response explain our resistance to reinfection with many diseases and the effectiveness of intentional vaccination. Vaccination with harmless antigens has been a powerful tool to reduce the toll of infectious disease. Although smallpox remains the only human pathogen we have removed entirely through vaccination, many once-common diseases have been controlled.

Effective vaccines stimulate the production of neutralizing antibodies by B cells, and also activate T cells. This will lead to the generation of memory B and T cells that are then primed to respond rapidly to the actual pathogen. Antigens for a specific pathogen can be introduced in multiple ways. Early vaccines used the actual infectious agent that had been inactivated or made less infective. Greater knowledge of both pathogens and the immune response has led to new generations of vaccines that present the immune system with a specific antigen to provoke a response (refer to chapter 17).

The aim of vaccination is to achieve herd immunity, which can protect a minority population of individuals not vaccinated, usually for health reasons. This occurs by reducing the percentage of susceptible individuals to a level where an infected individual is unlikely to encounter a susceptible one.

The COVID-19 pandemic, which spread throughout most of 2020 and in multiple waves in 2021, led to the fastest vaccine development in human history. This was possible because of a large-scale mobilization of researchers, and years of analysis of the related SARS and MERS viruses, which had identified a target antigen, the so-called spike protein. Rapid sequencing of many viral isolates, and the determination of the spike protein structure, allowed the development of both nucleic acid–and protein-based vaccines. The first two vaccines approved in the United States were also the first two mRNA vaccines ever approved for use in humans. These vaccines have performed even better than initial expectations, but have not ended the pandemic. Political resistance to being vaccinated and inefficient worldwide delivery of vaccines have combined to produce a pandemic primarily of the unvaccinated.

REVIEW OF CONCEPT 35.10

Antibodies have variable regions by which they recognize and bind to an antigen. Variable regions are encoded by joining distinct DNA segments, providing recognition diversity. Each antibody also has one of five kinds of constant region that determines its function; these five classes are IgA, IgD, IgE, IgG, and IgM. Vaccination artificially presents an antigen to elicit the primary response; when encountered later, a pathogen with this antigen is eliminated quickly by the secondary response.

- *How do Ig receptors differ from TLR innate receptors?*

How Do Sleeping Birds Stay Warm?

Adam Jones/Getty Images

Mammals and birds are endothermic—they maintain relatively constant body temperatures regardless of the temperature of their surroundings. This lets them reliably run their metabolism even when external temperatures fall—the rates of most enzyme-catalyzed reactions slow two- to threefold for every 10°C temperature drop. Your body keeps its temperature within narrow bounds around 37°C (98.6°F), and birds maintain even higher temperatures. To stay warm like this, mammals and birds continuously carry out oxidative metabolism, which generates heat. This requires a several-fold increase in metabolic rate, which is energetically expensive, particularly when the animal is not active. The logical solution is to give up the struggle to keep warm and let the body temperature drop during sleep, a condition known as *torpor*. Though humans don't adopt this approach, many other mammals and birds do. This raises an interesting question: What prevents a sleeping bird in torpor from freezing? Does its body simply adopt the temperature of its surroundings, or is there a body temperature below which metabolic heating kicks in to avoid freezing?

The graph displays an experiment examining this issue in the tropical hummingbird *Eulampis*, shown feeding in the photo. The study examines the effect on metabolic rate (measured as oxygen consumption) of decreasing air temperature. Oxygen consumption was assessed over a range of air temperatures from 3° to 37°C, for two contrasting physiological states: the blue data were collected from birds that were awake, the red data from sleeping torpid birds. The blue and red lines, called regression lines, were plotted using curve-fitting statistics that provide the best fit to the data.

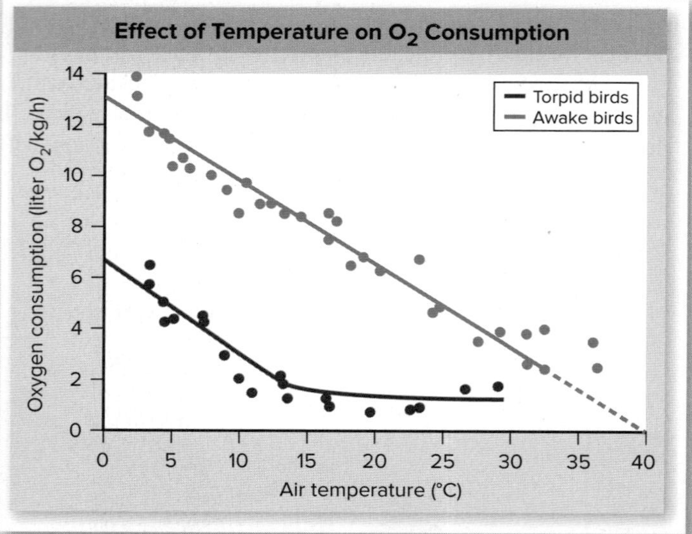

Analysis

1. **Applying Concepts**
   Comparing two data sets. Do awake hummingbirds maintain the same metabolic rate at all air temperatures? Do sleeping torpid ones? At a given temperature, which has the higher metabolic rate, an awake bird or a sleeping one?
2. **Interpreting Data**
   a. How does the oxygen consumption of awake hummingbirds change as air temperature falls? Why do you think this is so? Is the change consistent over the entire range of air temperatures examined?
   b. How does the oxygen consumption of sleeping torpid birds change as air temperature falls? Is this change consistent over the entire range of air temperatures examined? Explain any difference you detect.
   c. Are there any significant differences in the slope of the two regression lines below 15°C? What does this suggest to you?
3. **Making Inferences**
   a. For each 5-degree air temperature interval, estimate the average oxygen consumption for awake and for sleeping birds and plot the difference as a function of air temperature.
   b. Based on this curve, what would you expect to happen to a sleeping bird's body temperature as air temperatures fell from 30° to 20°C? From 15° to 5°C?
4. **Drawing Conclusions** How do *Eulampis* hummingbirds avoid becoming chilled while sleeping on cold nights?

Retracing the Learning Path

CONCEPT 35.1 Homeostasis Maintains a Constant Internal Environment

35.1.1 Negative Feedback Mechanisms Keep Values Within a Narrow Range Homeostasis uses negative feedback loops to limit responses to deviations from a set point.

35.1.2 Animals Regulate Body Temperature in a Variety of Ways Endotherms have high metabolic rates and generate heat internally; ectotherms do not.

35.1.3 Endotherms Create Internal Metabolic Heat for Conservation or Dissipation Endotherms regulate temperature by physiological changes; ectotherms use behavior.

CONCEPT 35.2 Hormones Are Chemical Messages That Direct Body Processes

35.2.1 Endocrine Glands Produce Three Chemical Classes of Hormones The three classes are peptides and proteins, amino acid derivatives, and steroids.

35.2.2 Hormones Can Be Categorized as Lipophilic or Hydrophilic Lipophilic hormones are fat-soluble and can cross the cell membrane; hydrophilic hormones are water-soluble.

35.2.3 Lipophilic Hormones Activate Intracellular Receptors Lipophilic hormones cross the plasma membrane and activate intracellular receptors to activate transcription.

35.2.4 Hydrophilic Hormones Activate Receptors on Target Cell Membranes Hydrophilic hormones bind to a membrane receptor to initiate a signal transduction pathway.

CONCEPT 35.3 The Pituitary and the Hypothalamus Are the Body's Control Centers

35.3.1 The Pituitary Is a Compound Endocrine Gland The pituitary consists of two parts: anterior and posterior. The posterior pituitary stores and releases ADH and oxytocin. The anterior pituitary produces 7 hormones.

35.3.2 Hypothalamic Neurohormones Regulate the Anterior Pituitary Hypothalamic hormones pass to the anterior pituitary and regulate its hormone production.

CONCEPT 35.4 Peripheral Endocrine Glands Play Major Roles in Homeostasis

35.4.1 The Thyroid Gland Regulates Basal Metabolism and Development The thyroid hormone thyroxine regulates basal metabolism in vertebrates.

35.4.2 Calcium Homeostasis Is Regulated by Several Hormones Blood Ca^{2+} is lowered by calcitonin and is raised by parathyroid hormone.

35.4.3 The Adrenal Gland Releases Both Catecholamine and Steroid Hormones Epinephrine and norepinephrine cause alarm responses. Corticosteroids maintain glucose levels.

35.4.4 Pancreatic Hormones Are Primary Regulators of Carbohydrate Metabolism Blood glucose is reduced by insulin and is raised by glucagon.

35.4.5 Melatonin Is Crucial to Circadian Cycles Melatonin controls daily wake–sleep cycles. Sex steroids regulate sexual development and reproduction.

CONCEPT 35.5 Animals Are Osmoconformers or Osmoregulators

35.5.1 Animals Control Their Osmolarity in a Variety of Ways Osmotic pressure is a solution's propensity to take in water.

CONCEPT 35.6 The Kidney Maintains Osmotic Homeostasis in Mammals

35.6.1 The Nephron Is the Filtering Unit of the Kidney Blood is filtered through the glomerulus into renal tubules: proximal and distal convoluted tubules, loop of Henle, and collecting duct.

35.6.2 Water, Nutrients, and Ions Are Reabsorbed; Other Molecules Are Secreted Water, glucose, amino acids, and important ions are filtered, then reabsorbed. Molecules your body wants to get rid of are secreted into the filtrate.

35.6.3 Each Part of the Nephron Performs a Specific Transport Function Reabsorption of nutrients and NaCl occurs in the proximal convoluted tubule. The loop of Henle creates a gradient of increasing osmolarity into the medulla.

CONCEPT 35.7 Hormones Control Osmoregulation

35.7.1 Antidiuretic Hormone Causes Water to Be Conserved ADH increases the permeability of the collecting duct, allowing greater reabsorption of water.

35.7.2 Aldosterone Controls Sodium Ion Concentration Aldosterone stimulates Na^+ uptake by the distal convoluted tubule.

CONCEPT 35.8 The Immune System Defends the Body

35.8.1 Innate Immunity Recognizes Characteristic Pathogen Molecules Toll-like receptors have leucine-rich regions that recognize molecules such as LPS and peptidoglycan.

35.8.2 Innate Immunity Leads to Diverse Responses to a Pathogen Binding of pathogen molecules to innate receptors leads to the inflammatory response, production of antimicrobial peptides, and cytokines that stimulate adaptive immunity.

35.8.3 Adaptive Immunity Generates Diverse Receptors by Genetic Rearrangements Surface receptors on lymphocytes recognize antigens and direct a specific immune response.

CONCEPT 35.9 Cell-Mediated Immunity Involves Helper and Killer T Cells

35.9.1 Cytotoxic T Cells Eliminate Virally Infected Cells and Tumor Cells T_C cells recognize virally infected cells and tumor cells. They destroy cells in a fashion similar to that used by NK cells.

35.9.2 Helper T Cells Secrete Proteins That Direct Immune Responses T_H cells secrete cytokines that promote both cell-mediated and humoral immunity.

CONCEPT 35.10 In Humoral Immunity, B Cells Produce Protective Antibodies

35.10.1 Antigen Recognition Is Carried Out by Antibody Proteins Immunoglobulins consist of two light-chain and two longer heavy-chain polypeptides, with variable regions.

35.10.2 Immunoglobulin Diversity Is Generated Through DNA Rearrangement Ig and TCR diversity is generated by DNA rearrangements. TCRs are similar to a single Fab region of an Ig.

35.10.3 The Secondary Response to an Antigen Is More Effective Than the Primary Response Memory cells make the secondary response more rapid on repeated exposure. This is the basis for vaccination, which can prevent infection to pathogens.

Concept Overview

This Concept Overview diagrams the key concepts that were discussed in this chapter.

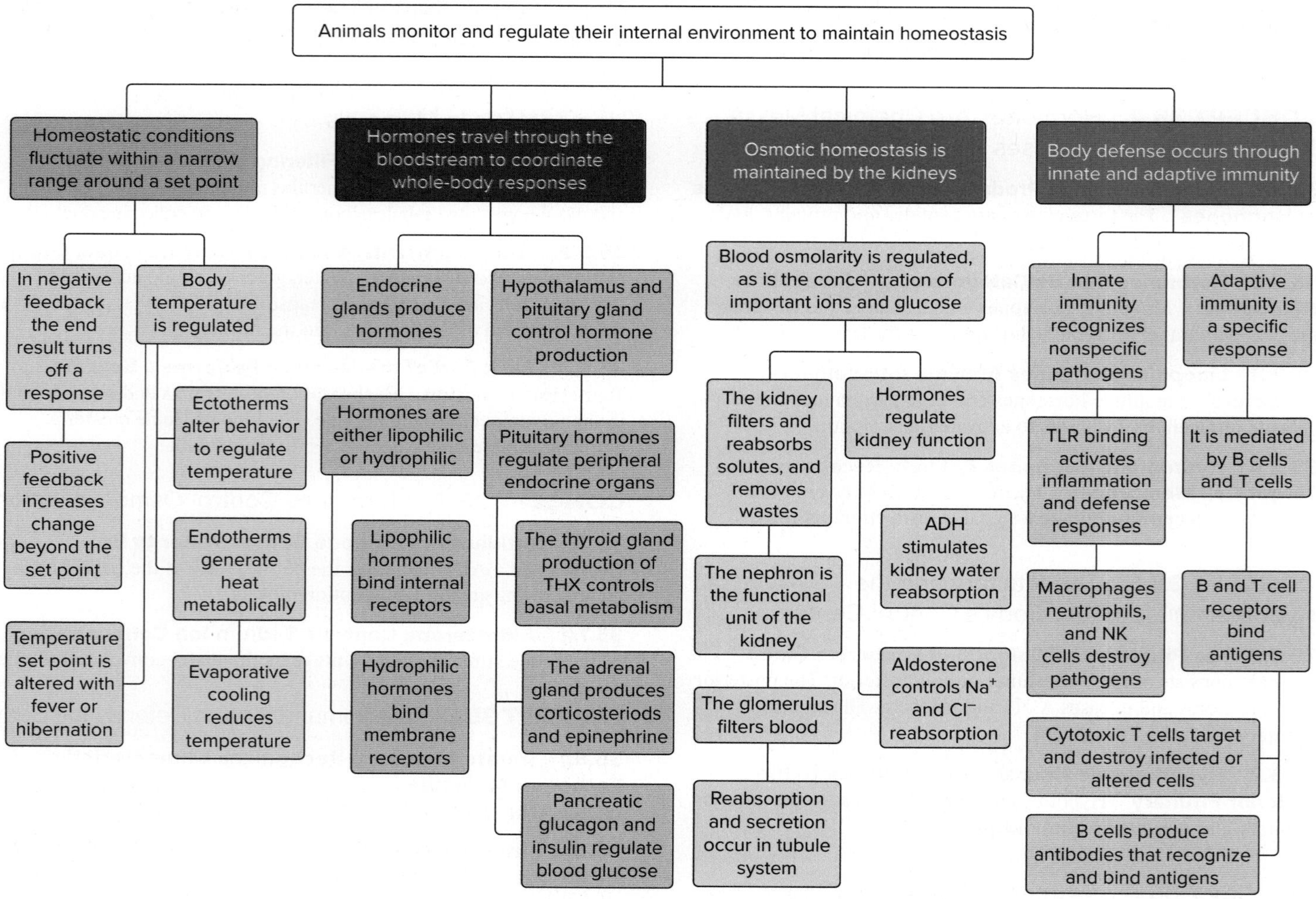

Assessing the Learning Path

Understand

1. Which of the following statements are correct? (Select all that apply.)
   a. Poikilotherms are often referred to as "warm-blooded."
   b. Endotherms use metabolic processes to generate body heat to keep warm in cold environments.
   c. Ectotherms can maintain their body temperatures because of insulation, such as blubber.
   d. Shivering is a behavior used by endotherms and ectotherms to warm their bodies.
2. Which of the following is true about lipophilic hormones?
   a. They are freely soluble in the blood.
   b. They require a transport protein in the bloodstream.
   c. They cannot enter their target cells.
   d. They are rapidly deactivated after binding to their receptors.
3. Hormones released from the pituitary gland have two different sources. Those produced by the neurons of the hypothalamus are released through the ____________, and those produced within the pituitary are released through the ____________.
   a. thalamus; hippocampus
   b. neurohypophysis; adenohypophysis
   c. right pituitary; left pituitary
   d. cortex; medulla
4. The beta cells in the islets of Langerhans secrete
   a. insulin.
   b. glucagon.
   c. calcitonin.
   d. aldosterone.
5. If a cell is placed in a hypertonic solution, it will
   a. lose water.
   b. gain water.
   c. stay the same.
   d. burst.
6. You are studying renal function in different species of mammals that are found in very different environments. You compare species from a desert environment with species from a tropical environment. The desert species would be expected to have
   a. shorter loops of Henle than the tropical species.
   b. longer loops of Henle than the tropical species.
   c. shorter proximal convoluted tubules than the tropical species.
   d. longer distal convoluted tubules than the tropical species.
7. Antidiuretic hormone (ADH) acts to
   a. increase reabsorption of NaCl in the collecting duct.
   b. increase secretion of NaCl in the collecting duct.
   c. increase reabsorption of water in the collecting duct.
   d. increase secretion of water in the collecting duct.
8. Receptors that trigger innate immune responses
   a. are antibodies recognizing specific antigens.
   b. are T-cell receptors recognizing specific antigens.
   c. recognize pathogen-associated molecular patterns.
   d. are not specific at all.
9. Which type of cell lacks MHC class II proteins?
   a. Dendritic cells
   b. Macrophages
   c. B cells
   d. T cells
10. How do vertebrates generate millions of different antibody receptor stem cells?
   a. By mutation of a few hundred genes
   b. By rearrangement of segments of DNA
   c. By antigen-elicited receptor creation
   d. None of the above

Apply

1. Which of the following scenarios correctly describes positive feedback?
   a. If the temperature increases in your room, your furnace increases its output of warm air.
   b. If you drink too much water, you produce more urine.
   c. If the price of gasoline increases, drivers decrease the length of their trips.
   d. If you feel cold, you start to shiver.
2. The classic action of steroid hormones is slower, but more long-lasting, than that of peptide hormones. In some cases steroid hormones can have rapid action. This is probably due to their
   a. entering the cell and turning on gene expression.
   b. binding to a membrane receptor to initiate a signal transduction pathway.
   c. binding to a membrane receptor to change patterns of gene expression.
   d. entering the cell and turning off gene expression.
3. Tumors that affect the pituitary can lead to decreases in some, but not all, hormones released by the pituitary. A patient with such a tumor exhibits fatigue, weight loss, and low blood sugar. This is probably due to lack of production of
   a. GH, which leads to loss of muscle mass.
   b. ACTH, which leads to loss of production of glucocorticoids.
   c. TSH, which leads to loss of production of thyroxin.
   d. ADH, which leads to excess urine production.
4. Your Uncle Sal likes to party. When he goes out drinking, he complains that he needs to urinate more often. You explain to him that this is because alcohol suppresses the release of the hormone ADH, which leads to
   a. decreased reabsorption of water from the collecting duct and concentrated urine.
   b. increased reabsorption of water from the collecting duct and concentrated urine.
   c. decreased reabsorption of water from the collecting duct and dilute urine.
   d. increased reabsorption of water from the collecting duct and dilute urine.
5. Freshwater vertebrates have a higher solute concentration in their body fluids than that of the surrounding water. Which of the following mechanisms could be used to maintain osmotic equilibrium in these animals? (Select all that apply.)
   a. They could drink excess amounts of water.
   b. They could excrete highly diluted urine.
   c. They could reabsorb NaCl in their bodies.
   d. They could excrete a highly concentrated urine.
6. A viral infection that specifically interferes with the reabsorption of ions from the glomerular filtrate would attack cells located in the
   a. Bowman's capsule.
   b. glomerulus.
   c. renal tubules.
   d. collecting duct.
7. Caffeine inhibits the secretion of ADH. Prior to an exam, you have a large coffee. During the exam, you can expect
   a. greater water reabsorption from the collecting duct.
   b. less water reabsorption from the collecting duct.

c. greater glucose reabsorption from the proximal convoluted tubule.
d. decreased potassium reabsorption from the proximal convoluted tubule.

8. A new disease is discovered that suppresses the immune system. Which of the following indicates that the disease specifically affects B cells, not helper or cytotoxic T cells?
   a. A decrease in the production of interleukin-2
   b. A decrease in interferon production
   c. A decrease in the number of plasma cells
   d. A decrease in the production of interleukin-1
9. Which of the following occurs last in an immune response?
   a. Macrophages ingest and destroy infected cells.
   b. Helper T cells are mobilized.
   c. B cells proliferate.
   d. Macrophages secrete interleukin.
10. Immunity to future invasion of a specific pathogen is accomplished by production of
   a. memory B cells.
   b. interleukin-1.
   c. plasma cells.
   d. helper T cells.

Synthesize

1. We have all experienced hunger pangs. Is hunger a positive or negative feedback stimulus? Describe the steps involved in the response to this stimulus.
2. Why do you suppose the brain goes to the trouble of synthesizing releasing hormones, rather than simply directing the production of the pituitary hormones immediately?
3. In the nephridia of annelids and the Malpighian tubules of insects, water and salts are collected from the body, only to be reabsorbed. Explain how these systems allow these terrestrial animals to maintain osmotic balance.
4. If you are lost at sea, adrift in a lifeboat with a case of liquor, and are desperately thirsty, should you drink the liquor or seawater? Explain your answer.
5. Explain how one bacterial cell can elicit many different immune responses.

Neil Bromhall/Science Source

36 Reproduction and Development

Learning Path

- **36.1** Mammals Are Viviparous
- **36.2** The Human Male Reproductive System Is Typical of Mammals
- **36.3** The Human Female Reproductive System Undergoes Cyclic Gamete Development
- **36.4** The First Step in Development Is Fertilization
- **36.5** Cells of the Early Embryo Are Totipotent
- **36.6** Cleavage Leads to the Blastula Stage
- **36.7** Gastrulation Forms the Basic Body Plan of the Embryo
- **36.8** The Body's Organs Form in Organogenesis
- **36.9** Human Development Takes Nine Months

Concept Overview

This Concept Overview provides a diagram of the overarching concepts that are covered in this chapter. These concepts will be expanded upon in the Concept Overview at the end of the chapter.

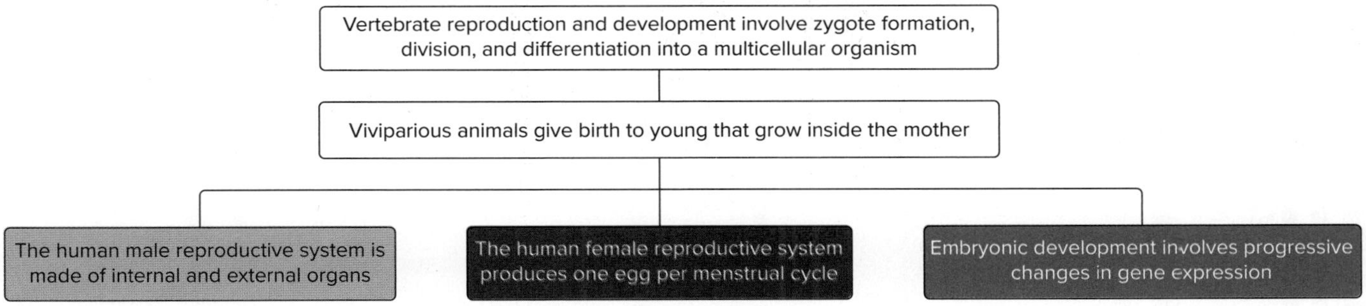
Vertebrate reproduction and development involve zygote formation, division, and differentiation into a multicellular organism
Viviparious animals give birth to young that grow inside the mother
The human male reproductive system is made of internal and external organs
The human female reproductive system produces one egg per menstrual cycle
Embryonic development involves progressive changes in gene expression

Introduction

This chapter deals with sexual reproduction and development among the vertebrates, focusing on humans. We are now in an era when long-standing questions of how organisms arise, grow, change, and mature may be answered, and new possibilities for regenerative medicine seem on the horizon. Sexual reproduction unites two haploid gametes to form a single diploid cell called a zygote. The zygote develops by a process of cell division and differentiation into a complex multicellular organism, composed of many different tissues and organs. This process involves progressive changes in gene expression that limit possible cell fates, although it is now possible to reverse this process in some cases.

36.1 Mammals Are Viviparous

Vertebrate sexual reproduction evolved in the ocean before vertebrates colonized the land. The males generally release their sperm into the water containing the eggs, where the union of the free gametes occurs. However, once vertebrates began living on land, their small and vulnerable gametes encountered a severe problem—desiccation. On land, naked gametes would soon dry up and perish. Consequently, intense selective pressure resulted in the evolution of internal fertilization in terrestrial vertebrates—that is, the introduction of male gametes directly into the female reproductive tract. By this means, fertilization still occurs in a nondesiccating environment, even when the adult animals are fully terrestrial.

Internal Fertilization Has Led to Three Strategies for Development of Offspring

LEARNING OBJECTIVE 36.1.1 Compare and contrast oviparity, ovoviviparity, and viviparity.

The vertebrates that practice internal fertilization exhibit three strategies for embryonic and fetal development, namely *oviparity, ovoviviparity,* and *viviparity.*

1. **Oviparity** is found in some bony fish, most reptiles, some cartilaginous fish, some amphibians, a few mammals, and all birds. The eggs, after being fertilized internally, are deposited outside the mother's body to complete their development.
2. **Ovoviviparity** is found in some fish and many reptiles. The fertilized eggs are retained within the mother to complete their development, but the embryos still obtain all of their nourishment from the egg yolk. The young are fully developed when they are hatched.
3. **Viviparity** is found in sharks, some amphibians, a few reptiles, and almost all mammals. The young develop within the mother and obtain nourishment directly from their mother's blood, rather than from the egg yolk.

Evolution of reproductive systems

Live birth—either viviparity or ovoviviparity—has evolved many times in vertebrates: once in mammals, but many times independently in fishes, amphibians, and reptiles (figure 36.1). The evolution of live birth appears to be a one-way evolutionary street: once it evolves, it is almost never lost. Live birth requires internal fertilization so that the eggs can develop within the body of the female; internal fertilization has evolved only once in amniotes (the clade composed of reptiles, birds, and mammals), but multiple times within fishes and amphibians.

REVIEW OF CONCEPT 36.1

Most mammals are viviparous, giving birth to young that have been nourished by the mother's body. Internal fertilization allows embryos to develop inside the female's body, leading to greater reproductive success.

- *Under what circumstances would oviparous reproduction be advantageous?*

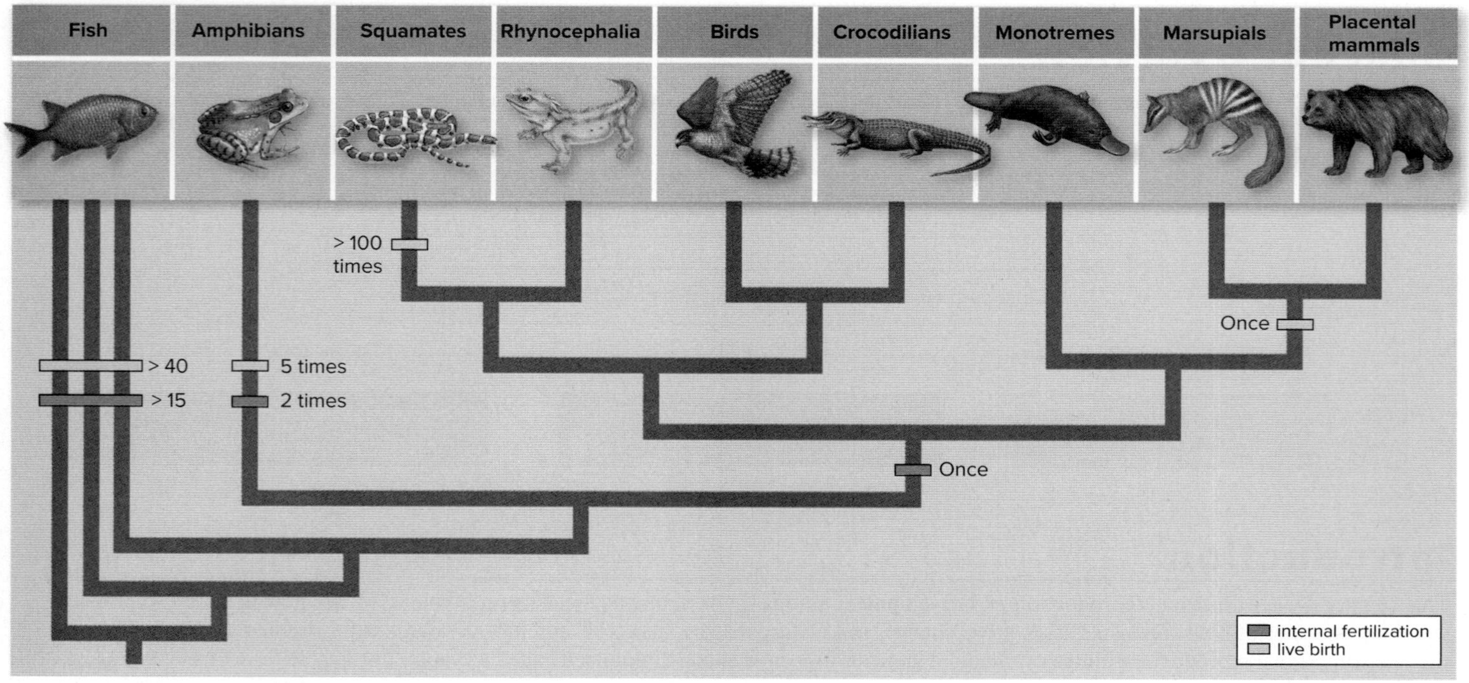

Figure 36.1 Evolution of internal fertilization and live birth in vertebrates. Although live birth has evolved many times in fishes and squamate reptiles, most species in both groups lay eggs. Evolutionary reversal from live birth to egg-laying has occurred very rarely, if at all. Estimates of the number of origins in fishes and squamates is based on detailed phylogenetic analyses within each group; uncertainty in numbers is a result of incomplete information in some groups.

36.2 The Human Male Reproductive System Is Typical of Mammals

The structures of the human male reproductive system, typical of male mammals, are illustrated in figure 36.2. When testes form in the human embryo, they develop seminiferous tubules, the sites of sperm production, beginning around 43 to 50 days after conception. At about 9 to 10 weeks, the Leydig cells, located in the interstitial tissue between the seminiferous tubules, begin to secrete testosterone (the major male sex hormone, or androgen). Testosterone secretion during embryonic development converts indifferent structures into the male external genitalia, the **penis** and the *scrotum,* the latter being a sac that contains the testes. In the absence of testosterone, these structures develop into the female external genitalia. Testosterone is also responsible at puberty for male secondary sex characteristics, such as development of the beard, a deeper voice, and thicker distribution of body hair.

Sperm Cells Are Produced by the Millions

LEARNING OBJECTIVE 36.2.1 Describe the sequence of events in spermatogenesis.

In an adult, each **testis** is composed primarily of the highly convoluted seminiferous tubules (figure 36.3, left). Although the testes are actually formed within the abdominal cavity, shortly before birth they descend through an opening called the inguinal canal into the scrotum, which suspends them outside the abdominal cavity. The scrotum maintains the testes at around 34°C, slightly lower than the core body temperature (37°C). This lower temperature is required for normal sperm development in humans.

The wall of the seminiferous tubule consists of spermatogonia, or *germ cells,* and supporting Sertoli cells. The germ cells near the outer surface of the seminiferous tubule are diploid and are the only cells that will undergo meiosis to produce gametes (refer to chapter 11). The developing gamete cells, located closer to the lumen of the tubule, are haploid.

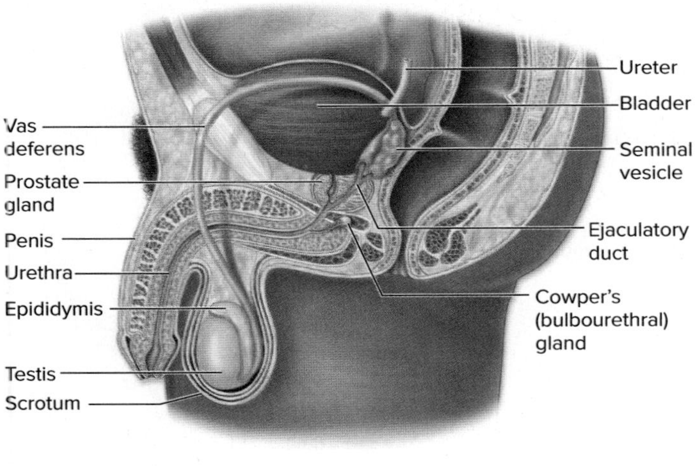

Figure 36.2 Organization of the human male reproductive system. The penis and scrotum are the external genitalia, the testes are the gonads, and the other organs are accessory sex organs, aiding the production and ejaculation of semen.

Cell divisions leading to sperm

A spermatogonium cell divides by mitosis to produce two diploid cells. One of these two cells then undergoes meiotic division to produce four haploid cells that will become sperm, while the other remains as a spermatogonium. In that way, the male never runs out of spermatogonia to produce sperm. Adult males can produce an average of 100 to 200 million sperm each day and can continue to do so throughout most of their lives.

The diploid daughter cell that begins meiosis is called a primary spermatocyte. In humans it has 23 pairs of chromosomes (46 chromosomes total), and each chromosome is duplicated, with two chromatids. The first meiotic division separates the homologous chromosome pairs, producing two haploid secondary spermatocytes. However, each chromosome still consists of two duplicate chromatids.

Each of these cells then undergoes the second meiotic division to separate the chromatids and produce two haploid cells, the **spermatids.** Therefore, a total of four haploid spermatids are produced from each primary spermatocyte (figure 36.3, right). All of these cells constitute the germinal epithelium of the seminiferous tubules, because they "germinate" the gametes.

Supporting tissues

In addition to the germinal epithelium, the walls of the seminiferous tubules contain nongerminal cells such as the Sertoli cells. These cells nurse the developing sperm and secrete products required for spermatogenesis. They also help convert the spermatids into **spermatozoa (sperm)** by engulfing their extra cytoplasm.

Sperm structure

Spermatozoa are relatively simple cells, consisting of a head, body, and flagellum (tail; figure 36.4). The head encloses a compact nucleus and is capped by a vesicle called an acrosome, which is derived from the Golgi complex. The acrosome contains enzymes that aid in the penetration of the protective layers surrounding the egg. The body and tail provide a propulsive mechanism: within the tail is a flagellum, and inside the body are a centriole, which acts as a basal body for the flagellum, and mitochondria, which generate the energy needed for flagellar movement.

Male Accessory Sex Organs Aid in Sperm Delivery

LEARNING OBJECTIVE 36.2.2 Describe semen, and explain how it is released during matings.

After the sperm are produced within the seminiferous tubules, they are delivered into a long, coiled tube called the **epididymis.** The sperm are not motile when they arrive in the epididymis, and

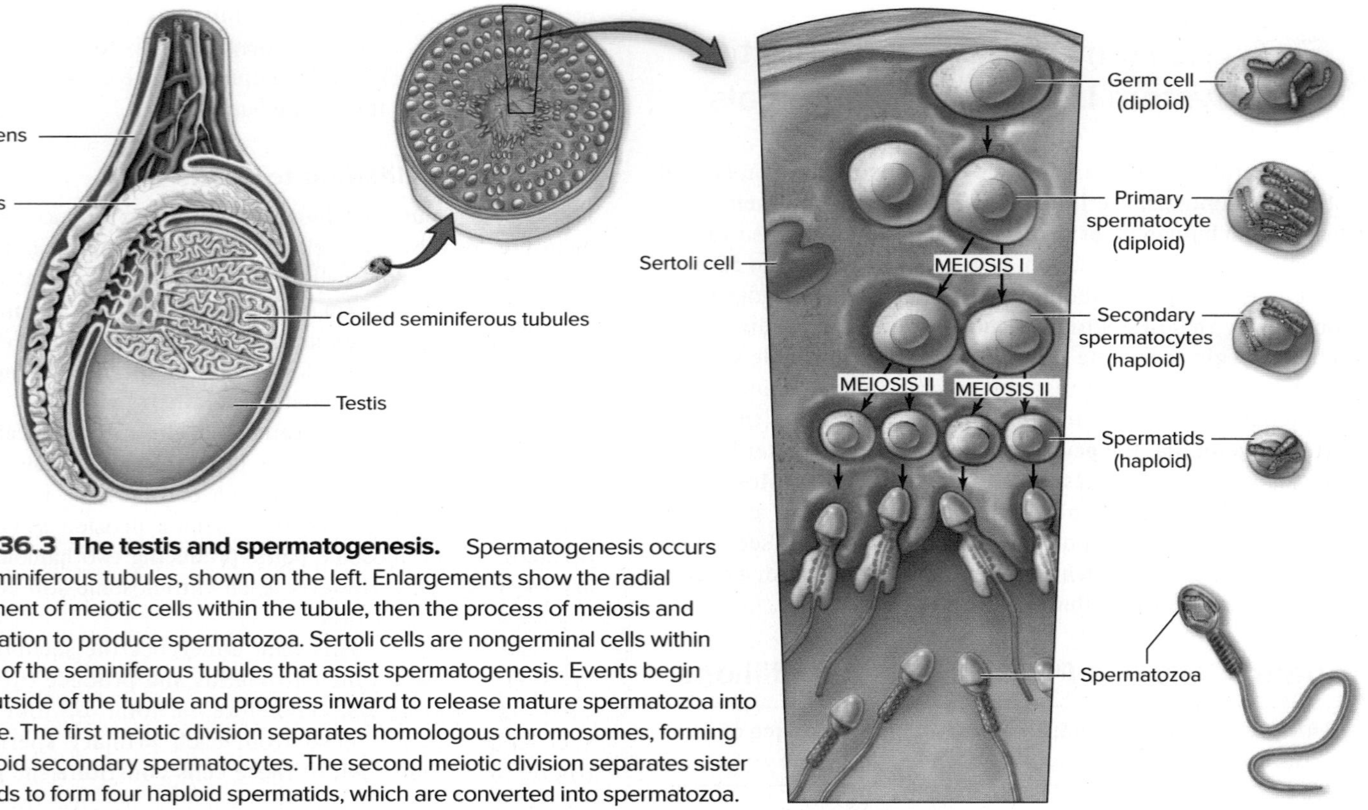

Figure 36.3 The testis and spermatogenesis. Spermatogenesis occurs in the seminiferous tubules, shown on the left. Enlargements show the radial arrangement of meiotic cells within the tubule, then the process of meiosis and differentiation to produce spermatozoa. Sertoli cells are nongerminal cells within the walls of the seminiferous tubules that assist spermatogenesis. Events begin on the outside of the tubule and progress inward to release mature spermatozoa into the tubule. The first meiotic division separates homologous chromosomes, forming two haploid secondary spermatocytes. The second meiotic division separates sister chromatids to form four haploid spermatids, which are converted into spermatozoa.

they must remain there for at least 18 hours before their motility develops. From the epididymis, the sperm enter another long tube, the **vas deferens,** which passes into the abdominal cavity via the inguinal canal.

Semen production

Semen is a complex mixture of fluids and sperm. The vas deferens from each testis joins with one of the ducts from a pair of glands called the seminal vesicles, which produce a fructose-rich fluid constituting about 60% of semen volume. From this point, the vas deferens continues as the ejaculatory duct and enters the prostate gland at the base of the urinary bladder.

In humans, the **prostate gland** is about the size of a golf ball and is spongy in texture. It contributes up to 30% of the bulk of the semen. Within the prostate gland, the ejaculatory duct merges with the urethra from the urinary bladder. The urethra

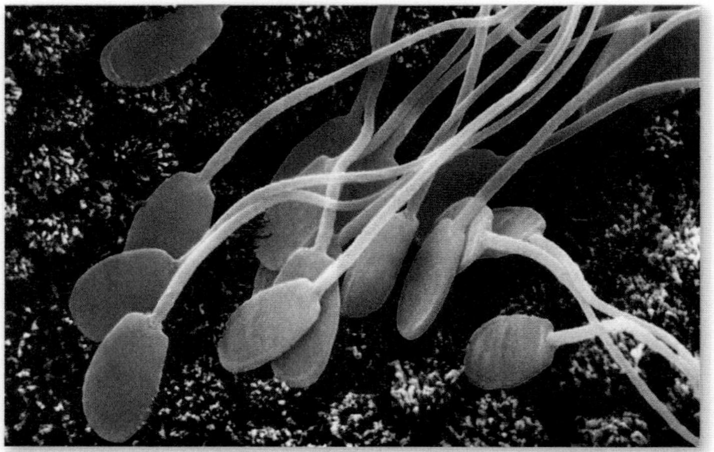
a. 20,000×

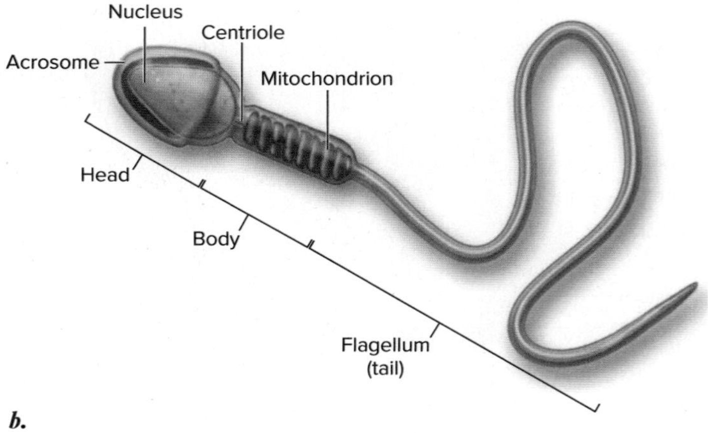

b.

Figure 36.4 Human sperm. ***a.*** A scanning electron micrograph with sperm digitally colored yellow. ***b.*** A diagram of the main components of a sperm cell.

(a): David M. Phillips/Science Source

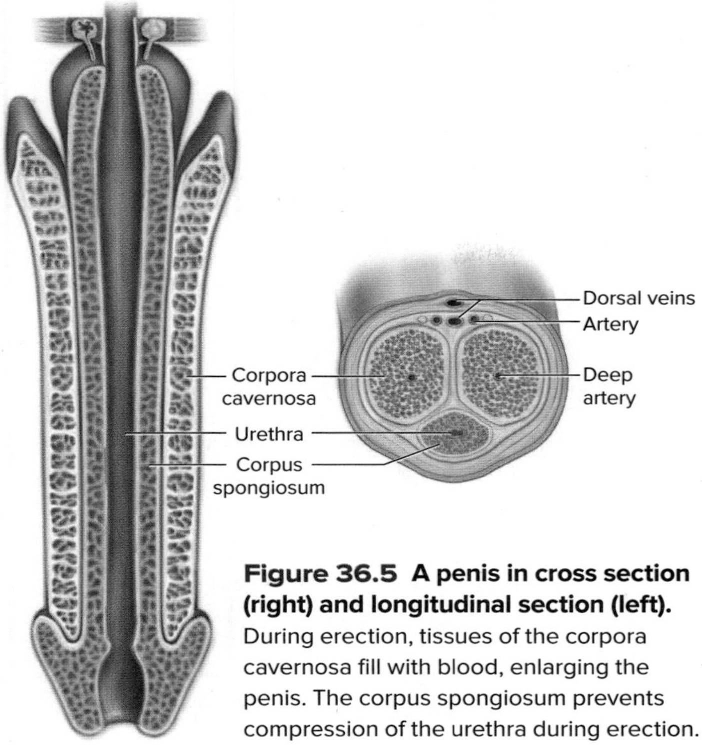

Figure 36.5 A penis in cross section (right) and longitudinal section (left). During erection, tissues of the corpora cavernosa fill with blood, enlarging the penis. The corpus spongiosum prevents compression of the urethra during erection.

carries the semen out of the body through the tip of the penis. A pair of pea-sized bulbourethral glands add secretions to make up the last 10% of semen, also secreting a fluid that lines the urethra and lubricates the tip of the penis prior to coitus (sexual intercourse).

Structure of the penis and erection

In addition to the urethra, the penis has two columns of erectile tissue, the corpora cavernosa, along its dorsal side and one column, the corpus spongiosum, along the ventral side (figure 36.5). Penile erection is produced by neurons in the parasympathetic division of the autonomic nervous system, which release nitric oxide (NO), causing arterioles in the penis to dilate. The erectile tissue becomes turgid as it engorges with blood. This increased pressure in the erectile tissue compresses the veins, so blood flows into the penis but cannot flow out.

Most mammals have a bone in the penis, called a *baculum,* which contributes to its stiffness during erection, but humans do not.

Ejaculation

The result of erection and continued sexual stimulation is ejaculation, the ejection from the penis of about 2 to 5 mL of semen containing an average of 300 million sperm. Successful fertilization requires such a high sperm count because the odds against any one sperm cell completing the journey to the egg and fertilizing it are extraordinarily high, and the acrosomes of many sperm need to interact with the egg before a single sperm can penetrate the egg (fertilization is described in section 36.4). Males with fewer than 20 million sperm per milliliter are generally considered sterile. Despite their large numbers, sperm constitute only about 1% of the volume of the semen ejaculated.

Hormones Regulate Male Reproductive Function

LEARNING OBJECTIVE 36.2.3 Explain how hormones regulate male reproductive function.

As you learned in chapter 35, the anterior pituitary gland secretes two gonadotropic hormones: follicle-stimulating hormone (FSH) and luteinizing hormone (LH). Although these hormones are named for their actions in the female, they are also involved in regulating male reproductive function (table 36.1). In males, FSH stimulates the Sertoli cells to facilitate sperm development, and LH stimulates the Leydig cells to secrete testosterone.

The principle of negative feedback inhibition applies to the control of FSH and LH secretion (figure 36.6). The hypothalamic hormone gonadotropin-releasing hormone (GnRH) stimulates the anterior pituitary gland to secrete both FSH and LH. FSH causes the Sertoli cells to release a peptide hormone called inhibin, which specifically inhibits FSH secretion. Similarly, LH

TABLE 36.1	Mammalian Reproductive Hormones
MALE	
Follicle-stimulating hormone (FSH)	Stimulates spermatogenesis via Sertoli cells
Luteinizing hormone (LH)	Stimulates secretion of testosterone by Leydig cells
Testosterone	Stimulates development and maintenance of male secondary sex characteristics, accessory sex organs, and spermatogenesis
FEMALE	
Follicle-stimulating hormone (FSH)	Stimulates growth of ovarian follicles and secretion of estradiol
Luteinizing hormone (LH)	Stimulates ovulation, conversion of ovarian follicles into corpus luteum, and secretion of estradiol and progesterone by corpus luteum
Estradiol (estrogen)	Stimulates development and maintenance of female secondary sex characteristics; prompts monthly preparation of uterus for pregnancy
Progesterone	Completes preparation of uterus for pregnancy; helps maintain female secondary sex characteristics
Oxytocin	Stimulates contraction of uterus and milk-ejection reflex
Prolactin	Stimulates milk production

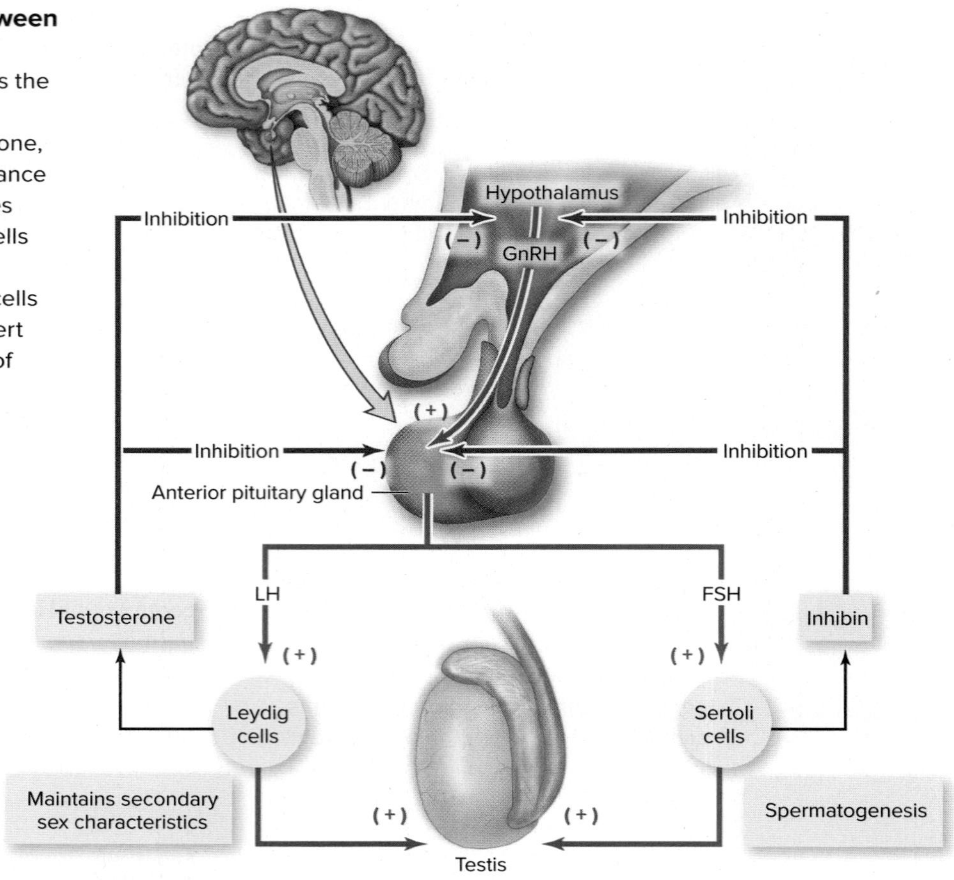

Figure 36.6 Hormonal interactions between the testes and anterior pituitary. The hypothalamus secretes GnRH, which stimulates the anterior pituitary to produce LH and FSH. LH stimulates the Leydig cells to secrete testosterone, which is involved in development and maintenance of secondary sex characteristics, and stimulates spermatogenesis. FSH stimulates the Sertoli cells of the seminiferous tubules, which facilitate spermatogenesis. FSH also stimulates Sertoli cells to secrete inhibin. Testosterone and inhibin exert negative feedback inhibition on the secretion of LH and FSH, respectively.

stimulates testosterone secretion, and testosterone feeds back to inhibit the release of LH, both directly at the anterior pituitary gland and indirectly by reducing GnRH release from the hypothalamus.

The importance of negative feedback inhibition can be demonstrated by removing the testes; in the absence of testosterone and inhibin, the secretion of FSH and LH from the anterior pituitary is greatly increased.

REVIEW OF CONCEPT 36.2

Spermatogonia divide by mitosis to produce another germ cell as well as a cell that undergoes meiosis to produce four haploid sperm cells. Semen consists of sperm from the testes and fluid from the seminal vesicles and prostate gland. Stimulation causes erection, then ejaculation of semen. Production of sperm and secretion of testosterone from the testes are controlled by FSH and LH from the anterior pituitary.

- *Would natural selection favor those males that produced more sperm? Explain your answer.*

36.3 The Human Female Reproductive System Undergoes Cyclic Gamete Development

The structures of the reproductive system in a human female are shown in figure 36.7. In contrast to the testes, the ovaries develop much more slowly. In the absence of testosterone, the female embryo develops a **clitoris** and labia majora from the same embryonic structures that produce a penis and a scrotum in males. Thus, the clitoris and penis, and the labia majora and scrotum, are said to be homologous structures. The clitoris, like the penis, contains corpora cavernosa and is therefore erectile.

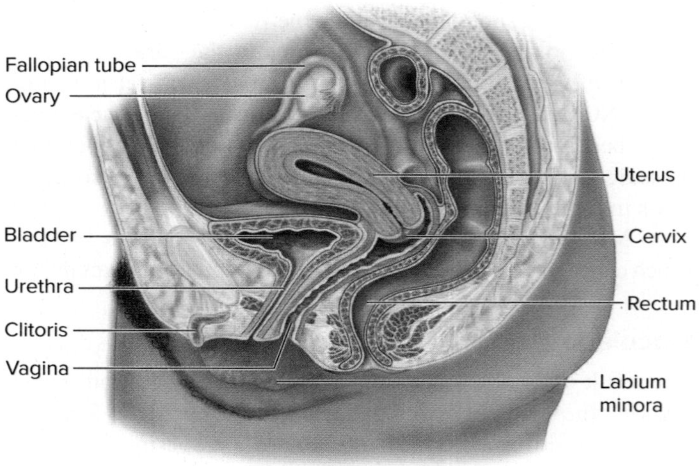

Figure 36.7 Organization of the human female reproductive system. The ovaries are the gonads, the Fallopian tubes receive the ovulated ova, and the uterus is the womb, the site of embryo development if the egg cell becomes fertilized.

Usually Only One Egg Is Ovulated per Menstrual Cycle

LEARNING OBJECTIVE 36.3.1 Describe the sequence of events in production of an oocyte.

The ovaries contain microscopic structures called ovarian follicles, each of which contains a potential egg cell called a primary oocyte and smaller **granulosa cells.**

At puberty, the granulosa cells begin to secrete the major female sex hormone, estradiol (also called estrogen), triggering *menarche,* the onset of menstrual cycling. Estradiol also stimulates the formation of the female secondary sex characteristics, including breast development and the production of pubic hair. In addition, estradiol and another steroid hormone, progesterone, help maintain the female accessory sex organs: the Fallopian tubes, uterus, and vagina.

At birth, a female's ovaries contain about 1 million follicles, each containing a **primary oocyte** that has begun meiosis but is arrested in prophase of the first meiotic division. During each menstrual cycle a group of follicles is recruited to reinitiate development, but the process of follicle maturation takes many months; at any given time, there are follicles at many different stages of maturation in the ovaries. The human menstrual cycle lasts approximately one month (28 days, on average) and can be divided in terms of ovarian activity into a follicular phase and a luteal phase, with the two phases separated by the event of **ovulation** (figure 36.8).

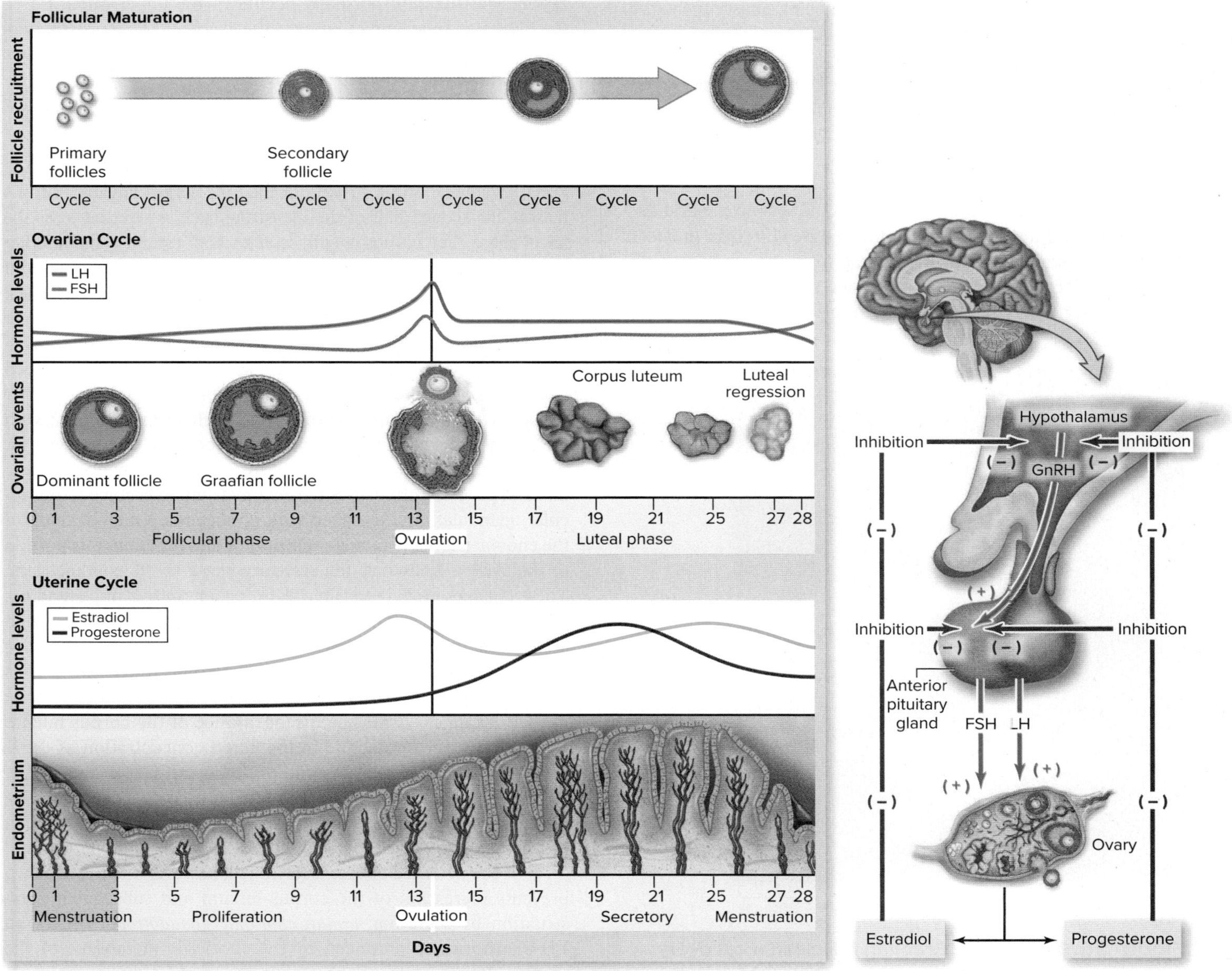

Figure 36.8 The human menstrual cycle. Left: Hormone levels during the cycle are correlated with ovulation and the growth of the endometrial lining of the uterus. Growth and thickening of the endometrium are stimulated by estradiol during the proliferative phase. Estradiol and progesterone maintain and regulate the endometrium during the secretory phase. Decline in the levels of these two hormones triggers menstruation. Right: Production of estradiol and progesterone by the anterior pituitary is controlled by negative feedback.

At the beginning of each menstrual cycle, a single dominant follicle emerges from the group that was recruited many months earlier and continues its development while the other follicles in that group enter a pathway for destruction.

Follicular phase

During each **follicular phase,** the dominant follicle achieves full maturity as a **late tertiary,** or **Graafian, follicle** under FSH stimulation. This follicle forms a thin-walled blister on the surface of the ovary. The uterus is lined with a simple columnar epithelial membrane called the endometrium; during the follicular phase, estradiol causes growth of the endometrium. This phase is therefore also known as the **proliferative phase** of the endometrium (figure 36.8).

The primary oocyte within the Graafian follicle completes the first meiotic division during the follicular phase. Instead of forming two equally large daughter cells, however, it produces one large daughter cell, the secondary oocyte (figure 36.9), and one tiny daughter cell, called a **polar body.** Thus, the secondary oocyte acquires almost all of the cytoplasm from the primary oocyte (unequal cytokinesis), increasing its chances of sustaining the early embryo, should the oocyte be fertilized. The polar body, on the other hand, disintegrates.

The secondary oocyte then begins the second meiotic division, but its progress is arrested at metaphase II. It is in this form that the potential egg cell is discharged from the ovary at ovulation, and it does not complete the second meiotic division unless it becomes fertilized in the Fallopian tube.

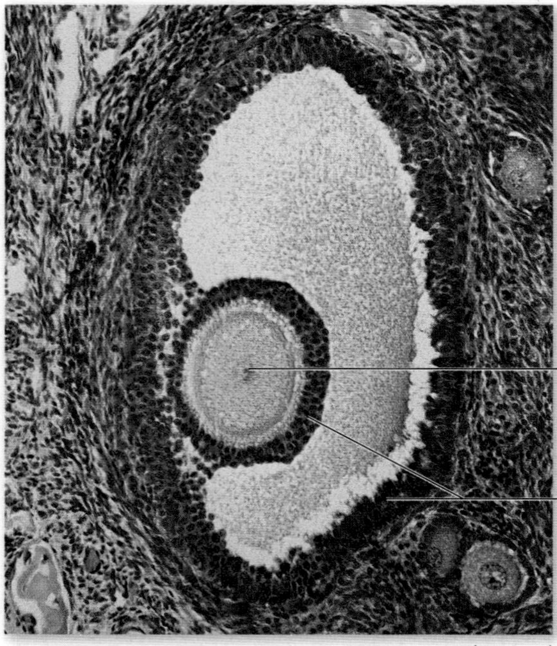

Figure 36.9 A mature Graafian follicle in a cat ovary. Note the ring of granulosa cells that surrounds the secondary oocyte. This ring will remain around the egg cell when it is ovulated, and sperm must tunnel through the ring in order to reach the plasma membrane of the secondary oocyte.
Ed Reschke

Ovulation

The increasing level of estradiol in the blood during the follicular phase stimulates the anterior pituitary gland to secrete LH at about midcycle. This sudden secretion of LH causes the fully developed Graafian follicle to burst in the process of ovulation, releasing its secondary oocyte.

The released oocyte enters the abdominal cavity near the fimbriae, the feathery projections surrounding the opening to the Fallopian tube. The ciliated epithelial cells lining the Fallopian tube draw in the oocyte and propel it through the Fallopian tube toward the uterus.

If it is not fertilized, the oocyte disintegrates within a day following ovulation. If it is fertilized, the stimulus of fertilization allows it to complete the second meiotic division, forming a fully mature ovum and a second polar body (figure 36.10). Fusion of the nuclei from the ovum and the sperm produces a diploid zygote. Fertilization normally occurs in the upper one-third of the Fallopian tube, and in humans the zygote takes approximately three days to reach the uterus and then another two to three days to implant in the endometrium (figure 36.11).

Luteal phase

After ovulation, LH stimulation completes the development of the Graafian follicle into a structure called the **corpus luteum.** For this reason, the second half of the menstrual cycle is referred to as the **luteal phase.** The corpus luteum secretes both estradiol and another steroid hormone, progesterone. The high blood levels of estradiol and progesterone during the luteal phase now exert negative feedback inhibition of FSH and LH secretion by the anterior pituitary gland. This inhibition during the luteal phase is in contrast to the stimulation exerted by estradiol on LH secretion at midcycle, which caused ovulation. The inhibitory effect of estradiol and progesterone after ovulation acts as a natural contraceptive mechanism, preventing both the development of additional follicles and continued ovulation.

During the luteal phase of the cycle, the combination of estradiol and progesterone causes the endometrium to become more vascular, glandular, and enriched with glycogen deposits. Because of the endometrium's glandular appearance and function, this portion of the cycle is known as the **secretory phase** of the endometrium. These changes prepare the uterine lining for embryo implantation.

In the absence of fertilization, the corpus luteum degenerates due to the decreasing levels of LH and FSH near the end of the luteal phase. Estradiol and progesterone, which the corpus luteum produces, inhibit the secretion of LH, the hormone needed for its survival. The disappearance of the corpus luteum results in an abrupt decline in the blood concentration of estradiol and progesterone at the end of the luteal phase, causing the built-up endometrium to be sloughed off, with accompanying bleeding. This is menstruation; the portion of the cycle in which it occurs is known as the *menstrual phase* of the endometrium.

If the ovulated oocyte is fertilized, however, the tiny embryo prevents regression of the corpus luteum and subsequent menstruation by secreting *human chorionic gonadotropin (hCG),* an LH-like hormone produced by the chorionic membrane of the embryo. By maintaining the corpus luteum, hCG keeps the levels of estradiol and progesterone high and thereby prevents menstruation, which would terminate the pregnancy. Because hCG comes from the embryonic chorion and not from the mother, it is the hormone tested for in all pregnancy tests.

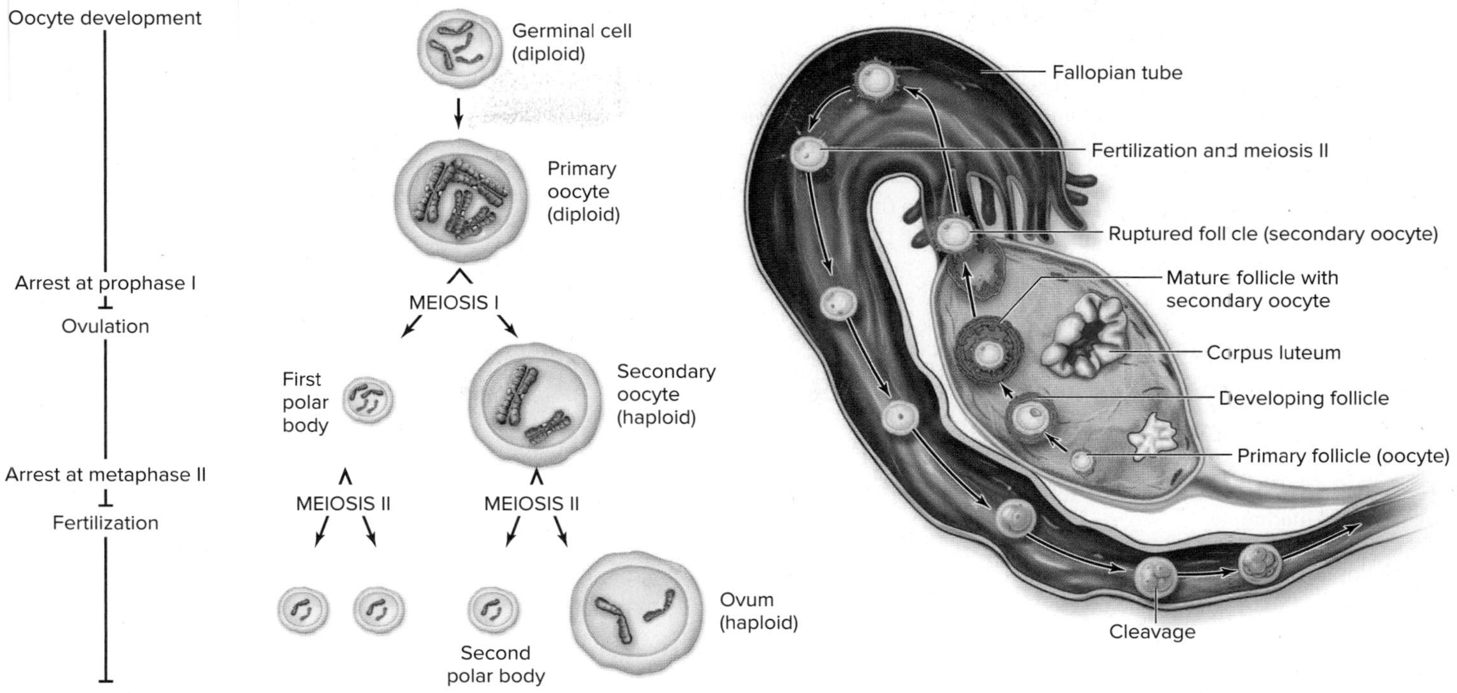

Figure 36.10 The meiotic events of oogenesis in humans. A primary oocyte is diploid. At the completion of the first meiotic division, one division product is eliminated as a polar body, and the other, the secondary oocyte, is released during ovulation. The secondary oocyte does not complete the second meiotic division until after fertilization; that division yields a second polar body and a single haploid egg, or ovum. Fusion of the haploid egg nucleus with a haploid sperm nucleus produces a diploid zygote.

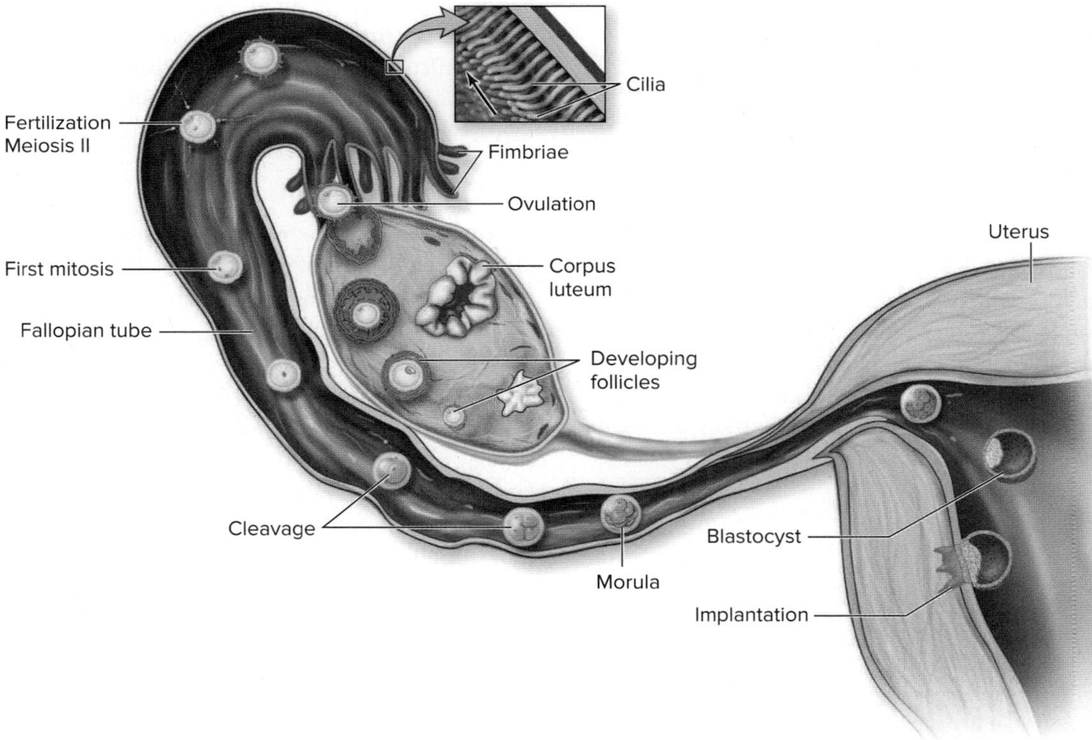

Figure 36.11 The journey of an egg. Produced within a follicle and released at ovulation, the secondary oocyte is swept into a Fallopian tube and carried along by waves of ciliary motion in the tube walls. Sperm journeying upward from the vagina penetrate the secondary oocyte; meiosis is completed and fertilization of the resulting ovum occurs within the Fallopian tube. The resulting zygote undergoes several mitotic divisions while still in the tube. By the time it enters the uterus, it is a hollow sphere of cells called a blastocyst. The blastocyst implants within the wall of the uterus, where it continues its development. (The egg and its subsequent stages have been enlarged for clarification.)

Mammals with estrous cycles

Menstruation is absent in mammals with an estrous cycle. Although such mammals (for example, dogs and cats) do cyclically shed cells from the endometrium, they don't bleed in the process. The estrous cycle is divided into four phases: proestrus, estrus, metestrus, and diestrus, which correspond to the proliferative, midcycle, secretory, and menstrual phases of the endometrium in the menstrual cycle.

Female Accessory Sex Organs Receive Sperm and Provide Nourishment and Protection to the Embryo

LEARNING OBJECTIVE 36.3.2 Compare the female reproductive tracts of mammals.

The Fallopian tubes (also called uterine tubes or oviducts) transport ova from the ovaries to the uterus. In humans, the **uterus** is a muscular, pear-shaped organ that narrows to form a neck, the cervix, which leads to the vagina.

The entrance to the vagina is initially covered by a membrane called the *hymen*. This will eventually be disrupted by vigorous activity or sexual intercourse. In the latter case, this can make the first experience painful when the hymen is ruptured.

During sexual arousal, the labia minora, clitoris, and vagina all become engorged with blood, much like the male erectile tissues. The clitoris has many sensory nerve endings and is one of the most sensitive and responsive areas for female arousal. During sexual arousal, Bartholin's glands, located near the vaginal opening, secrete a lubricating fluid that facilitates penetration by the penis. Ejaculation by the male introduces sperm cells that must then make the long swim out of the vagina and up the Fallopian tubes to encounter a secondary oocyte for fertilization to occur.

Mammals other than primates have more complex female reproductive tracts, in which part of the uterus divides to form uterine "horns," each of which leads to an oviduct. Cats, dogs, and cows, for example, have one cervix but two uterine horns separated by a septum, or wall. Marsupials, such as opossums, carry the split even further, with two unconnected uterine horns, two cervices, and two vaginas. A male marsupial has a forked penis that can enter both vaginas simultaneously.

Contraception Prevents Unwanted Pregnancy

LEARNING OBJECTIVE 36.3.3 Compare the different types of birth control.

In humans, females are sexually receptive throughout their reproductive cycle. This extended receptivity to sexual intercourse serves a second important function—it reinforces pair-bonding, the emotional relationship between two individuals. However, while this may be an important part of humans' emotional lives, not all couples desire sex to be linked to pregnancy. The prevention of pregnancy is known as birth control. Physiologically, pregnancy begins approximately a week after fertilization with successful implantation. Methods of birth control that act prior to implantation are usually termed contraception.

Most forms of contraception are temporary or reversible, the exception being sterilization. Sterilization is a surgical procedure that, in general, cannot be reversed. In females, it is called tubal ligation, which involves permanently blocking, severing, or removing the fallopian tubes. In males, vasectomy is a less invasive procedure, but also essentially irreversible. In this case the vas deferentia are severed and ligated to prevent sperm from entering the urethra.

Abstinence

The most reliable way to avoid pregnancy is to not have sexual intercourse at all, which is called *abstinence*. Of all the methods of contraception, this is the most certain. It is also the most limiting and the most difficult method to sustain for many people.

Sperm blockage

Barrier methods prevent fertilization by keeping sperm from the uterus. On the male side, this involves encasing the penis within a thin sheath, or condom. This method is easy to apply and can be very effective, but in practice it has a failure rate of 3 to 15% per year because of incorrect or inconsistent use or condom failure. Nevertheless, condom use is the most commonly employed form of contraception in the United States.

On the female side, barrier methods involve a cover over the cervix to prevent sperm entry to the uterus. This may be a relatively tight-fitting cervical cap, worn for days at a time, or a rubber dome called a diaphragm, inserted before intercourse. Because of individual anatomical differences, these must be initially fitted by a physician.

Sperm destruction

Sperm delivered to the vagina can be destroyed there with spermicidal agents, jellies, or foams. These treatments generally require application immediately before intercourse. Their failure rates vary from 10 to 25%.

Prevention of ovulation

Since about 1960, a widespread form of contraception has been the daily ingestion of birth control pills, or oral contraceptives, by women. These pills contain analogs of progesterone, sometimes in combination with estrogens. Progesterone and estradiol inhibit the secretion of FSH and LH by negative feedback during the luteal phase of the ovarian cycle. This prevents follicle development and ovulation. The hormones in birth control pills act to block ovulation, and thus prevent fertilization. Oral contraceptives have a failure rate of only 1 to 5% per year.

Prevention of embryo implantation

The insertion of an intrauterine device (IUD), a coil or other irregularly shaped object, is an effective means of contraception. These produce irritation that prevents implantation. IUDs have a failure rate of only 1 to 5%. Their high degree of effectiveness probably reflects their convenience; once they are inserted, they can be forgotten.

Another method of preventing embryo implantation is the "morning-after pill," or Plan B, which contains 50 times the dose of estrogen present in birth control pills. The pill works by temporarily stopping ovum development, by preventing fertilization, or by stopping the implantation of a fertilized ovum. This pill is intended as a method of emergency contraception.

REVIEW OF CONCEPT 36.3

Primary germ cells set aside during female development are arrested in meiosis I. After puberty, one primary oocyte in a follicle in the ovaries completes meiosis I to be released as a secondary oocyte that is arrested in meiosis II. This occurs during the follicular phase of the monthly cycle triggered by LH. If fertilized, meiosis in this oocyte will be completed. During the luteal phase of the cycle, development of other oocytes is inhibited, and if fertilization does not occur, the endometrium is sloughed off. Pregnancy can be prevented by abstinence, barrier contraceptives, and hormonal inhibition.

- *What is a possible result of multiple oocytes being ovulated?*

36.4 The First Step in Development Is Fertilization

In all sexually reproducing animals, the first step in development is the union of male and female gametes, a process called **fertilization.** Fertilization is typically external in aquatic animals. In contrast, internal fertilization is used by most terrestrial animals to provide a nondesiccating environment for the gametes.

A Sperm Must Penetrate to the Plasma Membrane of the Egg for Membrane Fusion to Occur

LEARNING OBJECTIVE 36.4.1 Describe the events necessary for fertilization to occur.

One physical challenge of sexual reproduction is for gametes to get together. Many elaborate strategies have evolved to enhance the likelihood of such encounters. For example, most marine invertebrates release hundreds of millions of eggs and sperm into the surrounding seawater on spawning; others use lunar cycles to time gamete release. Elaborate courtship behaviors are typical of many animals that utilize internal fertilization (refer to chapter 37). Fertilization itself consists of a series of coordinated events: sperm penetration and membrane fusion, egg activation, and fusion of nuclei.

Embryonic development begins with the fusion of the sperm and egg plasma membranes. But the unfertilized egg presents a challenge to this process, because it is enveloped by one or more protective coats. These protective coats include the *chorion* of insect eggs, the *jelly layer* and *vitelline envelope* of sea urchin and frog eggs, and the *zona pellucida* of mammalian eggs. Mammalian oocytes are also surrounded by a layer of supporting granulosa cells (figure 36.12). Thus, the first challenge of fertilization is that sperm have to penetrate these external layers to reach the plasma membrane of the egg.

A saclike organelle named the **acrosome** is positioned between the plasma membrane and the nucleus of the sperm head. The acrosome contains digestive enzymes, which are released by the process

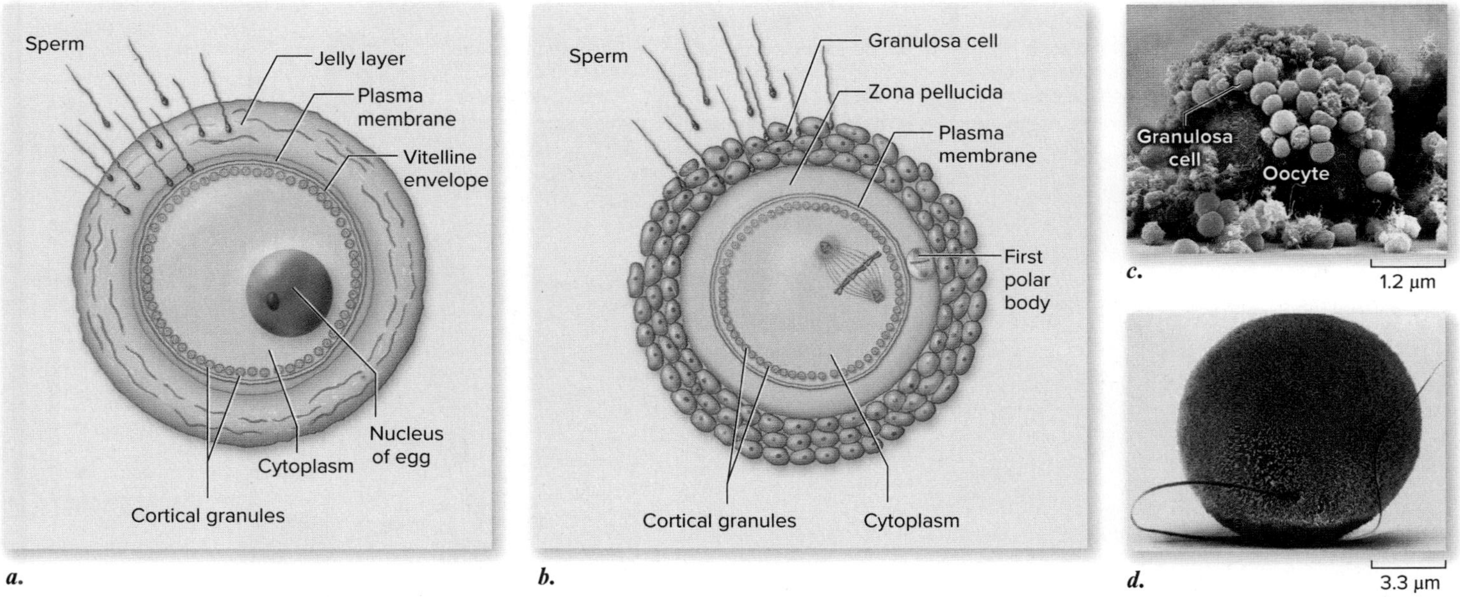

Figure 36.12 Animal reproductive cells. ***a.*** The structure of a sea urchin egg at fertilization. This diagram also shows the relative sizes of the sperm and the egg. ***b.*** A mammalian sperm must penetrate a layer of granulosa cells and then a glycoprotein layer called the zona pellucida before it reaches the oocyte membrane. The scanning electron micrographs show ***(c)*** a human oocyte surrounded by numerous granulosa cells and ***(d)*** a human sperm on an egg.

(c, d): David M. Phillips/Science Source

of exocytosis when a sperm reaches the outer layers of the egg. These enzymes create a hole in the protective layers, enabling the sperm to tunnel its way through to the egg's plasma membrane.

In sea urchin sperm, actin monomers assemble into cytoskeletal filaments just under the plasma membrane to create a long, narrow offshoot—the *acrosomal process.* The acrosomal process extends through the vitelline envelope to the egg's plasma membrane, and the sperm nucleus then passes through the acrosomal process to enter the egg.

In mice, an acrosomal process is not formed, and the entire sperm head burrows through the zona pellucida to the egg. Membrane fusion of the sperm and the egg then allows the sperm nucleus to pass directly into the egg cytoplasm. In many species, egg cytoplasm bulges out at membrane fusion to engulf the head of the sperm (figure 36.13).

Membrane Fusion Activates the Egg

LEARNING OBJECTIVE 36.4.2 List different ways that polyspermy is blocked.

After ovulation, the egg remains in a quiescent state until fusion of the sperm and egg membranes triggers reactivation of the egg's metabolism. In most species, there is a dramatic increase in the levels of free intracellular Ca^{2+} ions in the egg shortly after the sperm makes contact with the egg's plasma membrane. This increase is due to release of Ca^{2+} from internal, membrane-bounded organelles, starting at the point of sperm entry and traversing across the egg.

Scientists have been able to watch this wave of Ca^{2+} release by preloading unfertilized eggs with a dye that fluoresces when bound to free Ca^{2+}, and then fertilizing the eggs (figure 36.14). The released Ca^{2+} act as second messengers in the cytoplasm of the egg, to initiate a host of changes in protein activity. These many events initiated by membrane fusion are collectively called *egg activation.*

Blocking of additional fertilization events

Due to the large numbers of sperm released during spawning or ejaculation, many sperm are likely to reach, and try to fertilize, a single egg. Multiple fertilization, called polyspermy, produces a polyploid zygote with three or more sets of chromosomes. Polyploidy is incompatible with animal development, although it is a mechanism of speciation in plants (refer to chapter 21). In humans, about 10% of spontaneous abortions (miscarriages) are due to triploid fetuses, and about 70% of these are due to fertilization with two sperm. Not surprisingly, many animals have evolved mechanisms to prevent polyspermy.

In sea urchins, membrane contact by the first sperm results in a rapid, transient change in membrane potential of the egg, which prevents other sperm from fusing to the egg's plasma membrane. The importance of this event was shown by experiments in which sea urchin eggs were fertilized in low-sodium, artificial seawater. The change in membrane potential is mostly due to an influx of Na^+, so fertilization in low-sodium water prevents this. Under these conditions, polyspermy is much more frequent than in normal seawater.

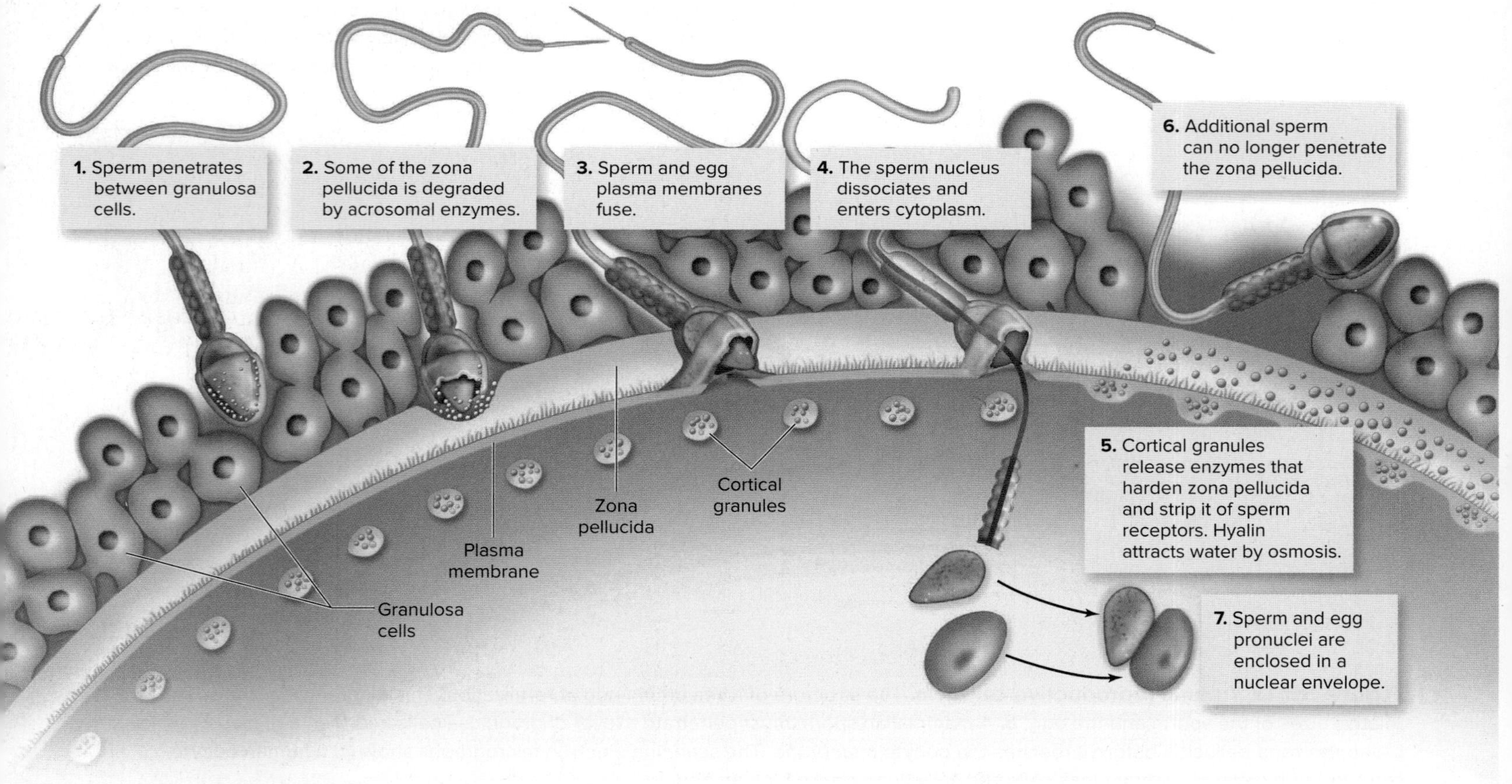

1. Sperm penetrates between granulosa cells.
2. Some of the zona pellucida is degraded by acrosomal enzymes.
3. Sperm and egg plasma membranes fuse.
4. The sperm nucleus dissociates and enters cytoplasm.
6. Additional sperm can no longer penetrate the zona pellucida.
5. Cortical granules release enzymes that harden zona pellucida and strip it of sperm receptors. Hyalin attracts water by osmosis.
7. Sperm and egg pronuclei are enclosed in a nuclear envelope.
Cortical granules
Zona pellucida
Plasma membrane
Granulosa cells

Figure 36.13 Sperm penetration and fusion. The sperm must penetrate the outer layers around the egg before fusion of sperm and egg plasma membranes can occur. Fusion activates the egg and leads to a series of events that prevent polyspermy.

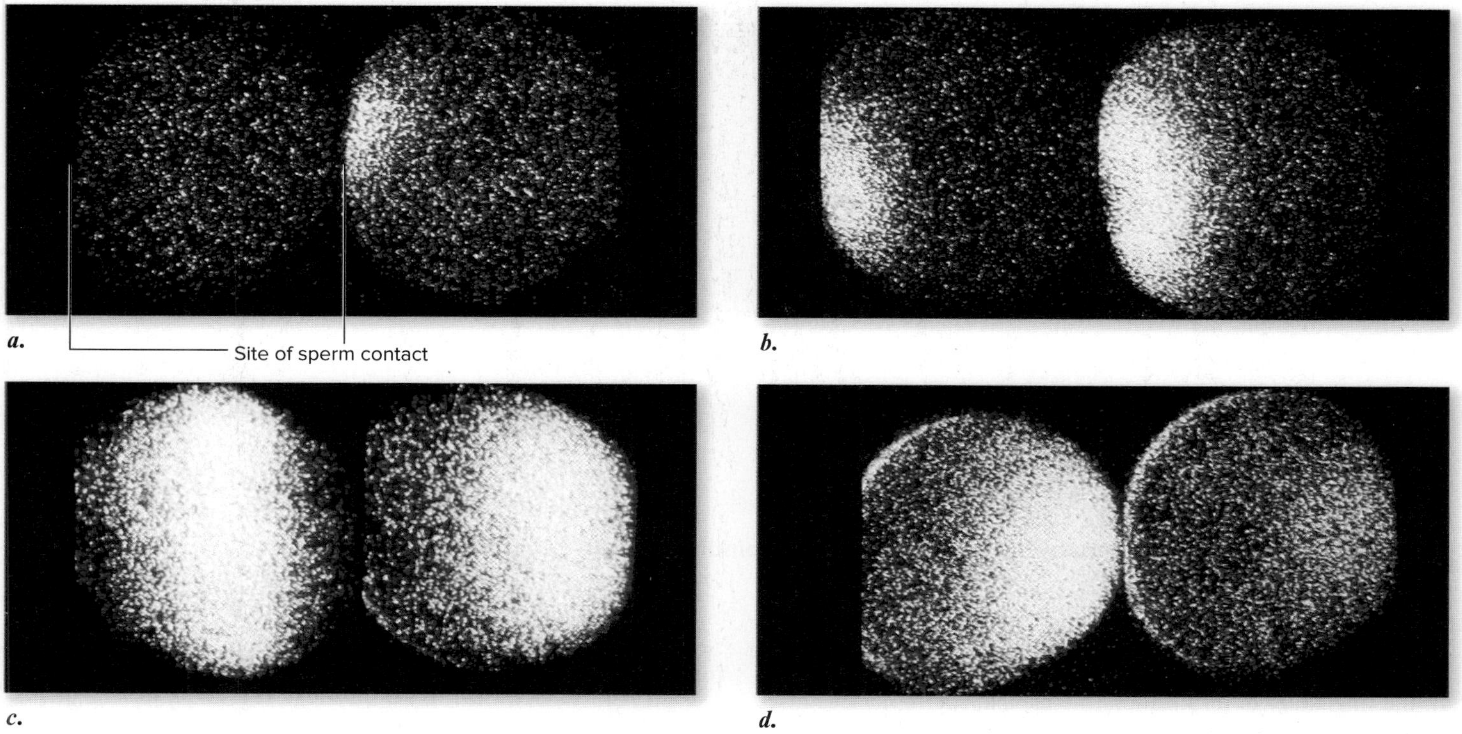
a.
Site of sperm contact
b.
c.
d.

Figure 36.14 Calcium ions are released in a wave across two sea urchin eggs following sperm contact. The bright white dots are dye molecules that fluoresce when they are bound to Ca^{2+}. The Ca^{2+} wave moves from left to right in these two eggs ***(a–d)***. The egg on the right was fertilized a few seconds before the egg on the left. The wave takes about 30 sec to cross the entire egg.

(a–d): Dr. Mathias Hafner, Mannheim University of Applied Sciences, Institute for Molecular Biology, Mannheim, Germany, and Dr. Gerald Schatten Pittsburgh Development Centre Deputy Director, Magee-Womens Research Institute, Professor and Vice-Chair of Obstetrics, Gynecology & Reproductive Sciences and Professor of Cell Biology & Physiology, Director, Division of Developmental and Regenerative Medicine, University of Pittsburgh School of Medicine Pittsburgh, PA 15213

Many animals use additional mechanisms to permanently alter the composition of the exterior egg coats, preventing any further sperm from penetrating through these layers. In sea urchins and mammals, specialized vesicles called **cortical granules,** located just beneath the plasma membrane of the egg, release their contents by exocytosis into the space between the plasma membrane and the vitelline envelope or zona pellucida, respectively. In each case, cortical granule enzymes remove critical sperm receptors from the outer coat of the egg.

Finally, the vitelline envelopes in many sea urchin species "lift off" the surfaces of the eggs when enzymes released from the cortical granules digest the connections between the vitelline envelope and the plasma membrane. *Hyalin,* a sugar-rich macromolecule, also fills this space and attracts water by osmosis, thus separating the two. Additional sperm cannot penetrate through the hardened, elevated vitelline envelope, which is now called a fertilization envelope.

Other effects of sperm penetration

In addition to the surface changes, sperm penetration can have three other effects on the egg. First, in many animals the nucleus of the unfertilized egg is not yet haploid, because it had not entered or completed meiosis prior to ovulation (figure 36.15). Fusion of the sperm plasma membrane then triggers the eggs of these animals to complete meiosis. In mammals, a single large egg with a haploid nucleus and one or more small polar bodies, which contain the other nuclei, are produced.

Second, sperm penetration in many animals triggers movements of the egg cytoplasm. In amphibian embryos, for example, the point of sperm entry is the focal point of cytoplasmic movements in the egg, and these movements ultimately establish the bilateral symmetry of the developing animal. Sperm penetration causes an outer pigmented cap of egg cytoplasm to rotate toward the point of entry, uncovering a gray crescent of interior cytoplasm opposite the point of penetration (figure 36.16). The position of this gray crescent determines the orientation of the first cell division. A line drawn between the point of sperm entry and the gray crescent would bisect the right and left halves of the future adult.

Third, activation is characterized by a sharp increase in protein synthesis and an increase in metabolic activity in general. Experiments demonstrate that the burst of protein synthesis in an activated egg uses mRNAs that were deposited into the cytoplasm of the egg during oogenesis.

In some animals, it is possible to artificially activate an egg without the entry of a sperm, simply by pricking the egg membrane. An egg that is activated in this way may go on to develop into a normal individual in certain species. In fact, a few kinds of amphibians and reptiles rely on this form of asexual reproduction (reproduction not involving the union of two genetically distinct gametes) as a primary method of reproduction. This process, called parthenogenesis, produces all female offspring.

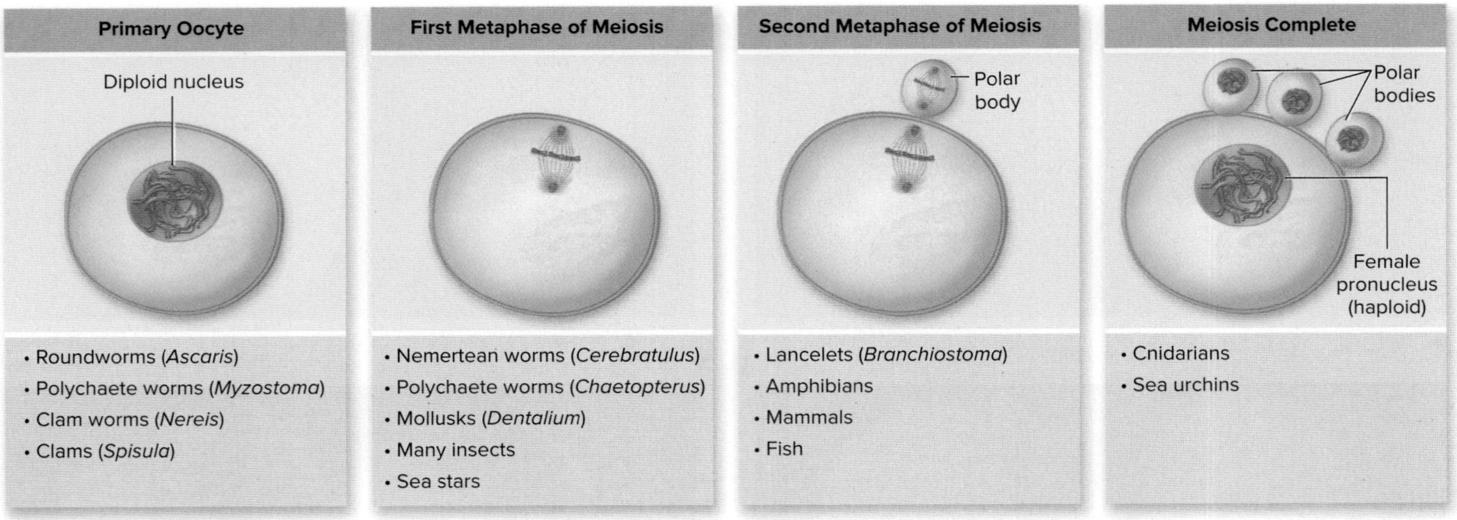
Primary Oocyte
Diploid nucleus
First Metaphase of Meiosis
Second Metaphase of Meiosis
Polar body
Meiosis Complete
Polar bodies
Female pronucleus (haploid)
• Roundworms (*Ascaris*)
• Polychaete worms (*Myzostoma*)
• Clam worms (*Nereis*)
• Clams (*Spisula*)
• Nemertean worms (*Cerebratulus*)
• Polychaete worms (*Chaetopterus*)
• Mollusks (*Dentalium*)
• Many insects
• Sea stars
• Lancelets (*Branchiostoma*)
• Amphibians
• Mammals
• Fish
• Cnidarians
• Sea urchins

Figure 36.15 Stage of egg maturation at time of sperm binding in representative animals.

Fusion of nuclei restores diploid state

In the final stage of fertilization, the haploid sperm nucleus fuses with the haploid egg nucleus to form the diploid nucleus of the zygote. The process involves migration of the two nuclei toward each other along a microtubule-based aster. A centriole that enters the egg cell with the sperm nucleus organizes the microtubule array, which is made from stored tubulin proteins in the egg's cytoplasm.

In mammals, including humans, the nuclei do not actually fuse. Instead, sperm and egg nuclear membranes each break down prior to the formation of a new diploid nucleus. A new nuclear membrane forms around the two sets of chromosomes.

REVIEW OF CONCEPT 36.4

Following penetration, fusion of sperm with the egg membrane initiates a series of events, including egg activation, blocks to polyspermy, and major rearrangements of cytoplasm. Polyspermy is blocked by changes in membrane polarity, release of enzymes that remove sperm receptors, and release of hyalin that lifts the vitelline envelope from the cell membrane. Egg and sperm nuclei then fuse to create a diploid zygote.

■ *What is the role of Ca^{2+} in egg activation?*

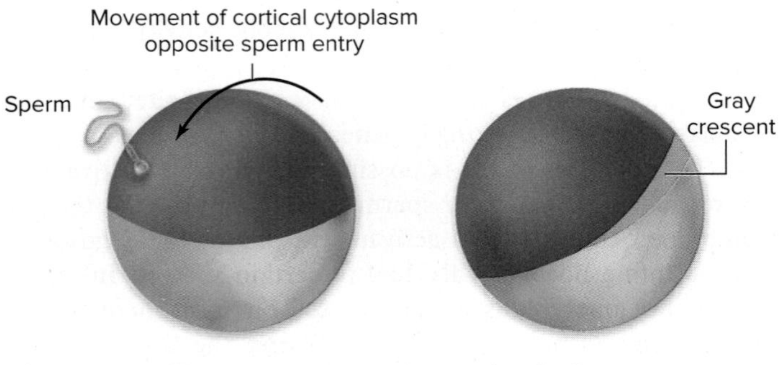
Movement of cortical cytoplasm opposite sperm entry
Sperm
Gray crescent

Figure 36.16 Gray crescent formation in frog eggs. The gray crescent forms on the side of the egg opposite the point of penetration by the sperm.

36.5 Cells of the Early Embryo Are Totipotent

At one extreme, we call a cell that can give rise to any tissue in an organism **totipotent.** In mammals, the only cells that can give rise to both the embryo and the extraembryonic membranes are the zygote and the early blastomeres from the first few cell divisions. Cells that can give rise to all of the cells in the organism's body are called **pluripotent.** A stem cell that can give rise to a limited number of cell types, such as the cells that give rise to the different blood cell types, are called **multipotent.** Then at the other extreme, **unipotent** stem cells give rise to only a single cell type, such as the cells that give rise to sperm cells in males.

Embryonic Stem Cells Are Pluripotent Cells Derived from Embryos

LEARNING OBJECTIVE 36.5.1 Describe the source and uses of embryonic stem cells.

Embryonic stem cells (ES cells) are a form of pluripotent stem cells isolated in the laboratory. These cells are made from mammalian embryos at the blastocyst stage of development. The blastocyst consists of an outer ball of cells, the trophectoderm, which will become extraembryonic structures such as the placenta, and the inner cell mass that will go on to form the embryo (see figure 36.23). Embryonic stem cells are essentially cells from the inner cell mass grown in culture (figure 36.17). In mice, these cells have been shown to be able to develop into any type of cell in the tissues of the adult. However, these cells cannot give rise to the extraembryonic tissues that arise during development, so they are pluripotent, but not totipotent.

Once these cells were found in mice, it was only a matter of time before human ES cells were derived as well. In 1998, the first

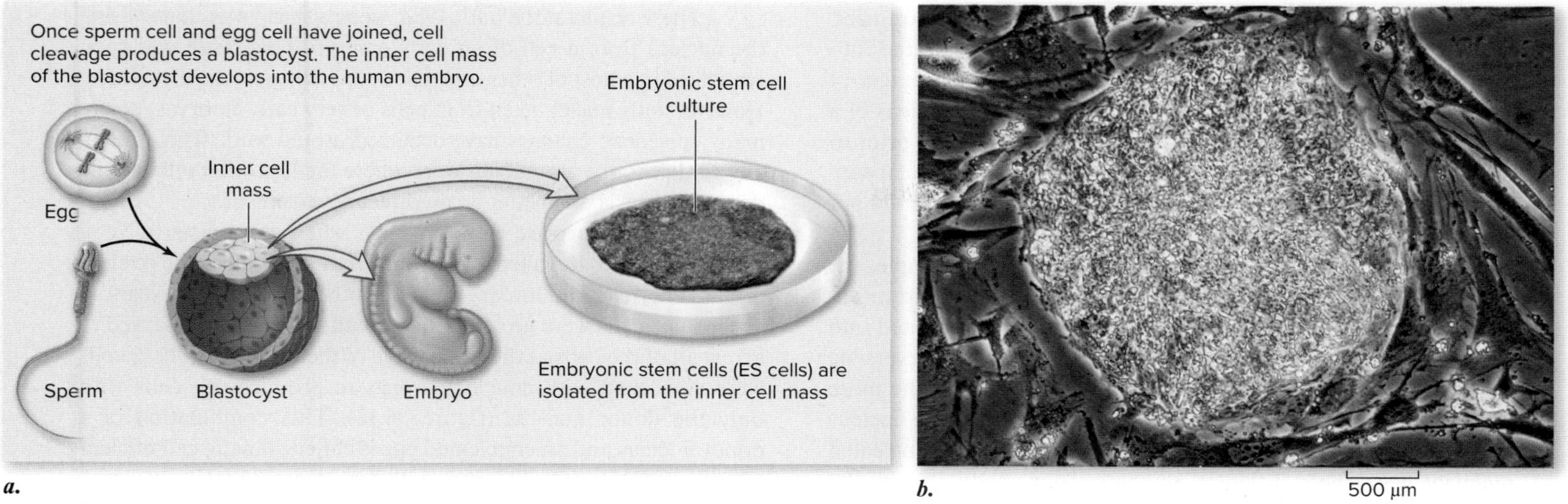

Figure 36.17 Isolation of embryonic stem cells. ***a.*** Early cell divisions lead to the blastocyst stage that consists of an outer layer and an inner cell mass, which will go on to form the embryo. Embryonic stem cells (ES cells) can be isolated from this stage by disrupting the embryo and growing the cells in culture. Stem cells removed from a six-day blastocyst can be established in culture and maintained indefinitely in an undifferentiated state. ***b.*** Human embryonic stem cells. This mass in the photograph is a colony of undifferentiated human embryonic stem cells being studied in developmental biologist James Thomson's research lab at the University of Wisconsin–Madison.

(b): Deco Images /Alamy Stock Photo

human ES cells (hES cells) were isolated and grown in culture. Although there are differences between human and mouse ES cells, there are also substantial similarities. These embryonic stem cells hold great promise for regenerative medicine based on their potential to produce any cell type. These cells have also been the source of much controversy and ethical discussion due to their embryonic origin.

Determination and differentiation in culture

In addition to their possible therapeutic uses, ES cells offer a way to study the differentiation process in culture. The manipulation of these cells by additions to the culture media will allow us to tease out the factors involved in differentiation at the level of the actual cell undergoing the process.

Using defined media, ES cells have been used to recapitulate in culture the early events in mouse development. Thus, mouse ES cells can be used to first give rise to ectoderm, endoderm, and mesoderm; then these three cell types will give rise to the different cells each germ layer is determined to become. This work is in early stages but is tremendously exciting, as it offers the promise of understanding the molecular cues that are involved in the stepwise determination of different cell types.

In humans, ES cells have been used to generate a variety of cell types in culture. For example, human ES cells have been shown to give rise to different kinds of blood cells in culture. Work is under way to produce hematopoeitic stem cells in culture, which could be used to replace such cells in patients with diseases that affect blood cells. Human ES cells have also been used to produce cardiomyocytes in culture. These cells could be used to replace damaged heart tissue after heart attacks.

Reversal of Determination Has Allowed Cloning

LEARNING OBJECTIVE 36.5.2 Explain what nuclear reprogramming is and why it is important.

The process of going from a single-celled zygote to a complex multicellular animal has fascinated biologists for as long as biology has existed as a science. One question that was of great interest, but proved challenging to answer, is whether the process requires any irreversible changes. We now know that there do not seem to be any irreversible changes to the DNA, but that the process is **epigenetic**. An epigenetic change is stable through cell division, but does not physically change the DNA. This was discussed in chapter 16 through the lens of how chromatin structure affects gene expression.

These simple conclusions are the result of many years of work by many investigators, and the exact mechanisms are still being uncovered. It is worth briefly considering the history of this work, as it is relevant to current efforts to be able to reverse the process, to create cells that can be used to replace tissue damaged by disease or injury.

Early research in amphibians

We learned in chapter 16 that during development, cells become progressively determined for particular cell fates during development. This process of determination precedes the actual process of differentiation, where cells take on their eventual fate. What was not clear was whether this process of cell differentiation involved irreversible genetic changes to cells.

This was first addressed by experiments begun in the 1950s by Robert Briggs and Thomas King, and carried on into the 1960s and 1970s by John Gurdon. These experiments used the technique of nuclear transplantation, or nuclear transfer, where the nucleus is removed from one cell and transferred into another cell with its nucleus removed (enucleated). The extent of genetic change can be assessed by transplanting the nucleus of a differentiated cell, later in development, into an egg cell prior to development. If the egg with the differentiated nucleus can support normal development, then the differentiated nucleus did not undergo any irreversible changes.

While the design of these experiments is quite simple, the results proved to be complex. On the one hand, nuclei from differentiated cells can support development when transplanted into an enucleated egg. Therefore, there have been no irreversible changes, such as the loss of genes. On the other hand, the more highly differentiated the donor cell, the less successful the nucleus in directing development when transplanted into an enucleated recipient egg. These different outcomes were reconciled by the concept of *nuclear reprogramming*. This takes into account that the process of differentiation involved epigenetic changes, which then must be reversed for the nucleus to be able to direct development.

The early work on amphibians showed that tadpoles' intestinal cell nuclei could be reprogrammed to produce viable adult frogs. These animals not only can be considered clones but also show that tadpole nuclei can be completely reprogrammed. However, nuclei from adult differentiated cells could be reprogrammed to produce only tadpoles, not viable, fertile adults.

Successful nuclear transplant in mammals

Given the success of nuclear transfer in amphibians, the process was applied to mammals, primarily mice and cattle. However, none of these experiments produced viable adults. In fact, experiments with mice where embryos were produced with only a maternal or only a paternal genetic contribution also did not produce viable adults. Further, these maternal-only and paternal-only embryos would fail to develop in different ways, with consistent abnormalities specific to maternal or paternal germ lines. This was part of the work that led to the concept of genomic imprinting (refer to chapter 13), and convinced developmental biologists that mammals could not be cloned in this way.

These results stood until 1984, when a sheep was cloned using the nucleus from a cell of an early embryo. This result was soon replicated in a host of other organisms, including pigs and monkeys. However, only nuclei taken from cells of very early embryos, before many epigenetic changes have occurred, would work. This is actually similar to the amphibian work where the less differentiated the donor nucleus is, the better it can direct development.

The real surprise came in 1996 when geneticists in Scotland were able to use a terminally differentiated cell as the nuclear donor. They used mammary cells from the udder of a six-year-old sheep. The cells were grown in tissue culture, and then starved to arrest all of the cells at the same stage of the cell cycle. These cells were combined with enucleated eggs to produce egg cells with only the donor nucleus (figure 36.18). This combination of a donor nucleus and an enucleated egg is called somatic cell nuclear transfer (SCNT). Of 277 eggs with transferred nuclei, 29 developed into embryos, and one of these ended with the live birth of a lamb named Dolly. This was the first mammal cloned from a fully differentiated cell, and the most famous lamb in human history.

Dolly proved to be healthy, and fully fertile, giving birth to six lambs by traditional methods. Thus, Dolly established beyond all dispute that determination in animals is reversible—that is, the nucleus of a fully differentiated cell can be reprogrammed to be totipotent.

Reproductive Cloning Often Fails for Lack of Proper Genomic Imprinting

LEARNING OBJECTIVE 36.5.3 Discuss the difficulties with reproductive cloning.

The term **reproductive cloning** refers to the process just described, in which scientists use SCNT to create an animal that is genetically identical to another animal. Since Dolly's birth in 1997, scientists have successfully cloned one or more cats, rabbits, rats, mice, cattle, goats, pigs, and mules. All of these procedures used some form of adult cell.

Figure 36.18 Proof that determination in animals is reversible. Scientists combined a nucleus from an adult mammary cell with an enucleated egg cell to successfully clone a sheep, named Dolly, which grew to be a normal adult and bore healthy offspring. This experiment, the first successful cloning of an adult animal, shows that a differentiated adult cell can be used to drive all of development.

(sheep): Paul Clements, File/AP Images

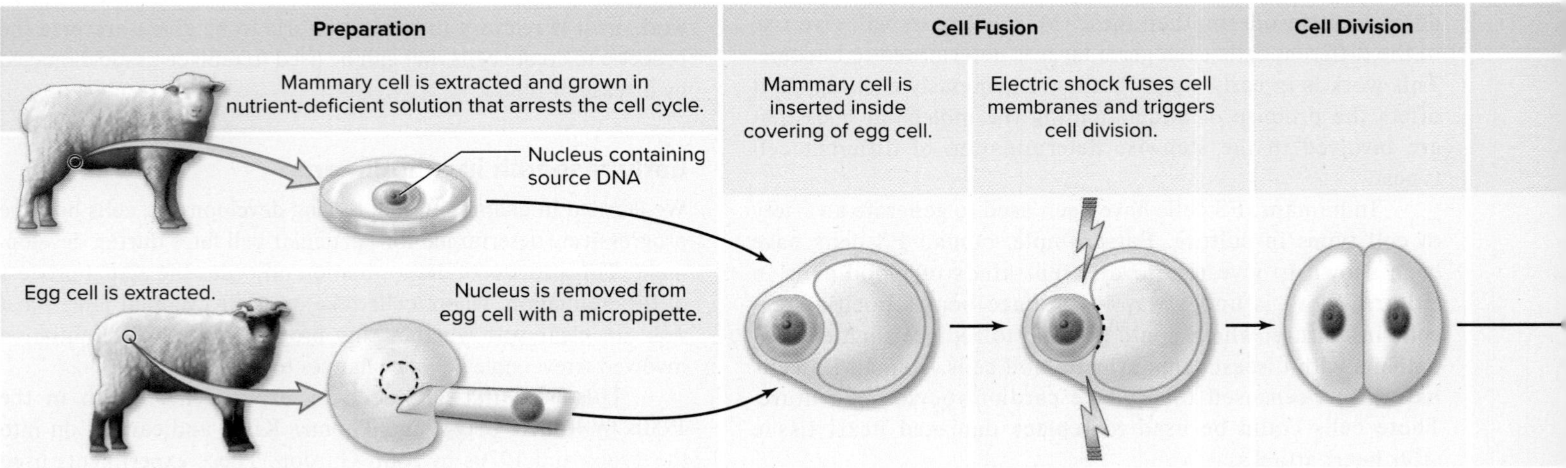
Preparation
Cell Fusion
Cell Division
Mammary cell is extracted and grown in nutrient-deficient solution that arrests the cell cycle.
Nucleus containing source DNA
Egg cell is extracted.
Nucleus is removed from egg cell with a micropipette.
Mammary cell is inserted inside covering of egg cell.
Electric shock fuses cell membranes and triggers cell division.

Low success rate and age-associated diseases

The efficiency in all reproductive cloning is quite low—only 3 to 5% of adult nuclei transferred to donor eggs result in live births. In addition, many clones that are born die soon thereafter of liver failure or infections. Many become oversized, a condition known as *large offspring syndrome (LOS).* In 2003, three of four cloned piglets developed to adulthood, but all three suddenly died of heart failure at less than six months of age.

Dolly herself was euthanized at the relatively young age of six. Although she was put down because of virally induced lung cancer, she had been diagnosed with advanced-stage arthritis a year earlier. Thus, one difficulty in using genetic engineering and cloning to improve livestock is the production of enough healthy animals.

Lack of imprinting

One reason for these problems lies in a phenomenon discussed in chapter 13: *genomic imprinting.* Imprinted genes are expressed differently depending on parental origin—that is, they are turned off in either egg or sperm, and this "setting" continues through development into the adult. Normal mammalian development depends on precise genomic imprinting.

During normal development, a variety of epigenetic programming and reprogramming events occur. The cloning protocol attempts to sidestep this and return the nucleus of a differentiated cell directly back to the state of a zygote. This involves changes in the state of the chromatin, and probably its organization as well. Until we understand all that this process involves, it is difficult to pinpoint the precise reasons for the low efficiency.

Nuclear Reprogramming Has Been Accomplished by Use of Defined Factors

LEARNING OBJECTIVE 36.5.4 Describe how nuclear reprogramming can be used to accomplish therapeutic cloning.

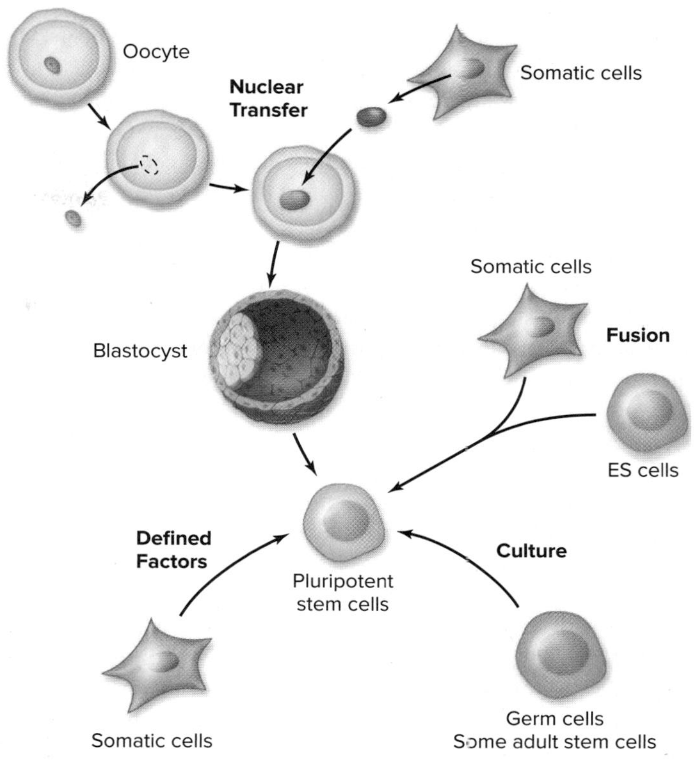
Oocyte
Nuclear Transfer
Somatic cells
Blastocyst
Somatic cells
Fusion
ES cells
Defined Factors
Pluripotent stem cells
Culture
Somatic cells
Germ cells
Some adult stem cells

Figure 36.19 Methods to reprogram adult cell nuclei. Cells taken from adult organisms can be reprogrammed to pluripotent cells in a number of different ways. Nuclei from somatic cells can be transplanted into oocytes, as during cloning. Somatic cells can be fused to ES cells created by some other means. Germ cells, and some adult stem cells, after prolonged culture appear to be reprogrammed. Recent work has shown that somatic cells in culture can be reprogrammed by the introduction of specific factors.

With the discovery of ES cells and success in the reproductive cloning of mammals, work continued on how to reprogram adult cells to pluripotency without the use of embryos (figure 36.19). One approach was to fuse an ES cell to a differentiated cell. These fusion experiments showed that the nucleus of a differentiated cell could be reprogrammed by ES cell cytoplasm. The resulting cells are tetraploid (four copies of the genome), which limits their experimental and practical utility. Another line of research showed that primordial germ cells explanted into culture can give rise to cells that act similarly to ES cells after extended time in culture.

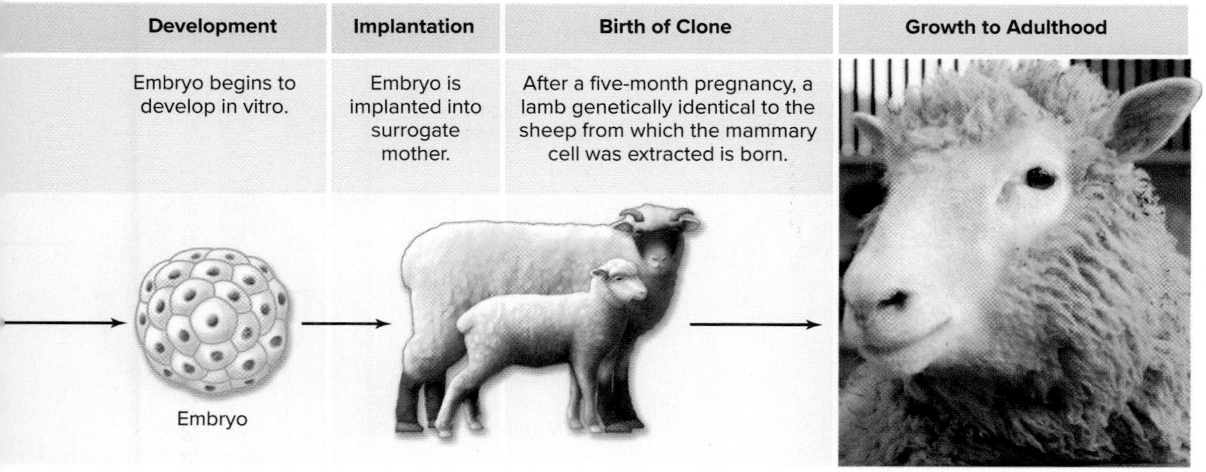
Development
Implantation
Birth of Clone
Growth to Adulthood
Embryo begins to develop in vitro.
Embryo is implanted into surrogate mother.
After a five-month pregnancy, a lamb genetically identical to the sheep from which the mammary cell was extracted is born.
Embryo

These different lines of inquiry showed that reprogramming of somatic nuclei was possible. This was complemented by the analysis of transcription factors specific to ES cells. Then in 2006, researcher Shinya Yamanaka and coworkers showed that introducing four of these transcription factors—Oct4, Sox2, c-Myc, and Klf4—could reprogram fibroblast cells in culture. These cells appeared to be pluripotent, and were named **induced pluripotent stem cells,** or **iPS** cells.

Research on nuclear reprogramming has shown that it is a multistep process. For cells reprogrammed in culture, only a subset complete each step, making the entire process inefficient. Starting from a fibroblast, cells first change shape, becoming more spherical and dividing more rapidly. They then reverse part of their developmental program to become more like epithelial cells. Finally, the stable expression of the core pluripotency regulatory factors is established. The pluripotent state is maintained by a combination of transcription factors and epigenetic changes. Whole-genome analysis of epigenetic markers shows that iPS cells do not have the same epigenetic signature as ES cells, but the significance of this is unclear.

One use of this technology is to construct ES cells from patients with genetic diseases. ES cells from patients with spinal muscular atrophy differentiate in culture into motor neurons that show the phenotype of the disease. The ability to derive disease-specific stem cells has incredible potential for research on these diseases. This allows researchers to study the specific cells affected by a genetic disease and to screen for possible therapeutics.

Pluripotent cell types themselves have potential for therapeutic applications. One way to solve the problem of graft rejection, such as in skin grafts in severe burn cases, is to produce patient-specific lines of ES cells. The first method to accomplish this has been called **therapeutic cloning** and uses the same SCNT procedure that created Dolly to assemble an embryo: the nucleus is removed from a skin cell and inserted into an egg whose nucleus has already been removed; the egg with its skin cell nucleus is allowed to form a blastocyst-stage embryo; this artificial embryo is then used to derive ES cells for transfer to injured tissue (figure 36.20).

Therapeutic cloning can address the problem of immune rejection that must be solved for stem cells to be used to repair tissues damaged by injury or disease. Because stem cells are cloned from a person's own tissues, they present as "self" to the immune system. A clinical trial in Japan using iPS cells to treat blindness due to age-related macular degeneration showed some improvement of vision. This kind of approach can be combined with new techniques that allow editing of the genome (refer to chapter 18) to combine reversing a genetic defect with tissue engineering. We are a long way from this kind of tissue engineering, but these are steps in that direction.

REVIEW OF CONCEPT 36.5

During development, cells first become determined, then differentiate into specific cell types. Embryonic stem cells are pluripotent cells that can give rise to all adult structures. Cells from early-stage embryos are pluripotent, but the nucleus of a differentiated cell must be reprogrammed. This is related to epigenetic changes such as genomic imprinting. Nuclei may be reprogrammed through the introduction of four important transcription factors. Therapeutic cloning aims to provide replacement tissue using one's own cells.

- *What changes must occur to produce a totipotent cell from a differentiated nucleus?*

36.6 Cleavage Leads to the Blastula Stage

The Blastula Is a Hollow Mass of Cells

LEARNING OBJECTIVE 36.6.1 Define the terms *cleavage* and *blastula.*

Following fertilization, the next major event in animal development is the rapid division of the zygote into a larger and larger number of smaller and smaller cells. This period of division,

Figure 36.20 How human embryos might be used for therapeutic cloning. In therapeutic cloning, after initial stages to reproductive cloning, the embryo is broken apart and its embryonic stem cells are extracted. These are grown in culture and used to replace the diseased tissue of the individual who provided the DNA. This is useful only if the disease in question is not genetic, as the stem cells are genetically identical to those of the patient.

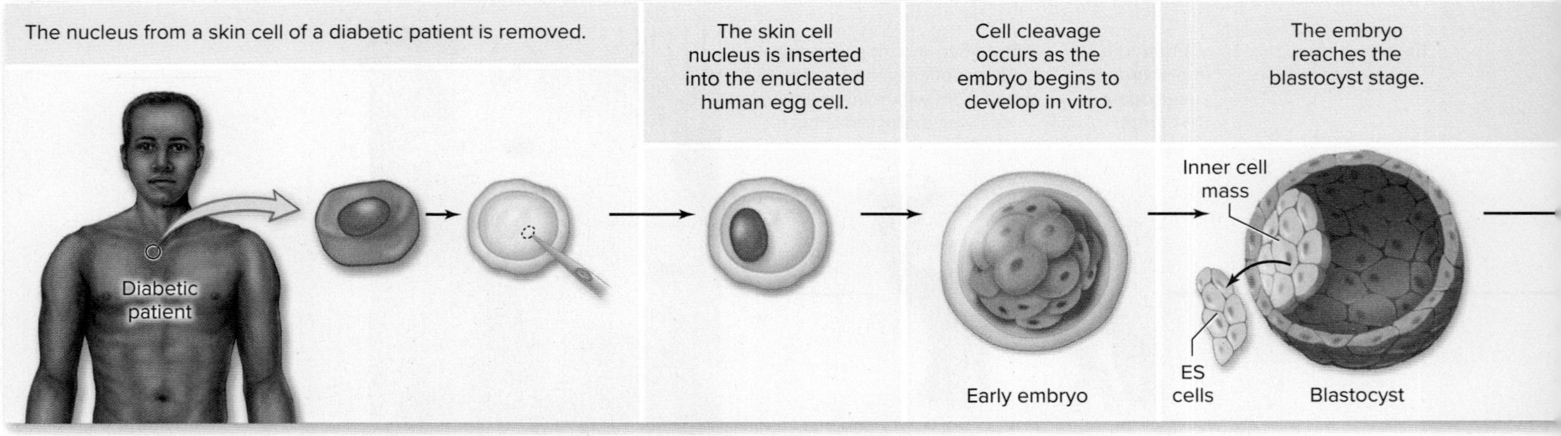

The nucleus from a skin cell of a diabetic patient is removed.
The skin cell nucleus is inserted into the enucleated human egg cell.
Cell cleavage occurs as the embryo begins to develop in vitro.
The embryo reaches the blastocyst stage.
Diabetic patient
Inner cell mass
Early embryo
ES cells
Blastocyst

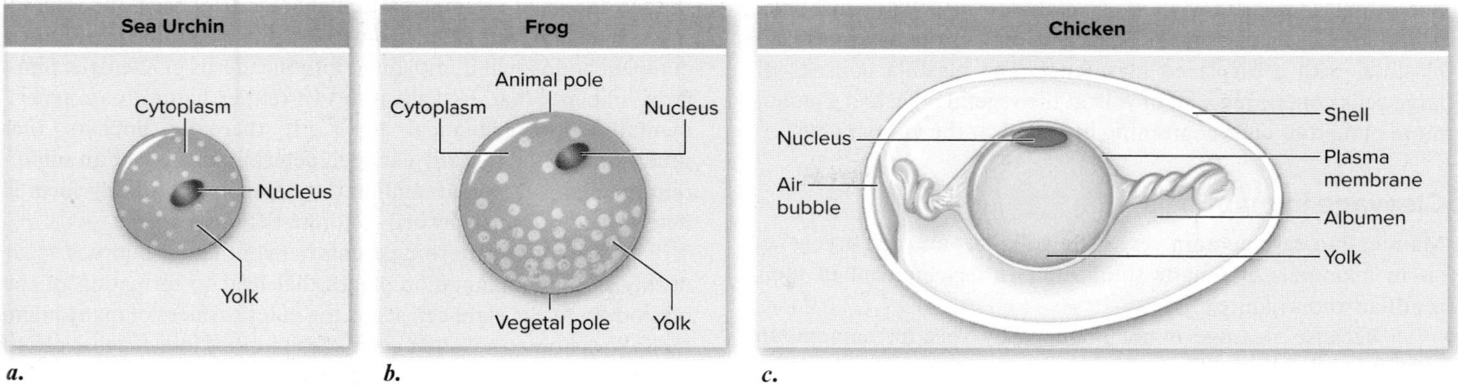
Sea Urchin
Cytoplasm
Nucleus
Yolk
Frog
Animal pole
Cytoplasm
Nucleus
Vegetal pole
Yolk
Chicken
Nucleus
Air bubble
Shell
Plasma membrane
Albumen
Yolk
a.
b.
c.

Figure 36.21 Yolk distribution in three kinds of eggs. ***a.*** In a sea urchin egg, the cytoplasm contains a small amount of evenly distributed yolk and a centrally located nucleus. ***b.*** In a frog egg, there is much more yolk, and the nucleus is displaced toward one pole. ***c.*** Bird eggs are complex, with the nucleus contained in a small disc of cytoplasm that sits on top of a large, central yolk mass.

called **cleavage,** is not accompanied by an increase in the overall size of the embryo. Each individual cell in the resulting tightly packed mass of cells is referred to as a *blastomere.* In many animals, the two ends of the egg and subsequent embryo are traditionally referred to as the **animal pole** and the **vegetal pole.** In general, the blastomeres of the animal pole go on to form the external tissues of the body, and those of the vegetal pole form the internal tissues.

In many animal embryos, the outermost blastomeres in the ball of cells produced during cleavage become joined to one another by tight junctions, belts of protein that encircle a cell and weld it to its neighbors (refer to chapter 4). These tight junctions create a seal that isolates the interior of the cell mass from the surrounding medium.

Subsequently, cells in the interior of the mass begin to pump Na^+ from their cytoplasm into the spaces between cells. The resulting osmotic gradient causes water to be drawn into the center of the embryo, enlarging the intercellular spaces. Eventually, the spaces coalesce to form a single large cavity within the embryo. The resulting hollow ball of cells is called a **blastula** (or *blastocyst* in mammals), and the fluid-filled cavity within the blastula is known as the **blastocoel.**

Cleavage Patterns Are Highly Diverse and Distinctive

LEARNING OBJECTIVE 36.6.2 Describe the different patterns of cleavage seen in animals.

Cleavage divisions are quite rapid in most species, and chapter 10 provides an overview of the conserved set of proteins that control the cell cycle in animal embryos. Cleavage patterns are quite diverse, and there are about as many ways to divide up the cytoplasm of an animal egg during cleavage as there are phyla of animals! Nonetheless, we can make some generalizations.

First, the relative amount of nutritive yolk in the egg is the characteristic that most affects the cleavage pattern of an animal embryo (figure 36.21). In eggs that contain moderate to little yolk, cleavage occurs throughout the whole egg, a pattern called **holoblastic cleavage.** This pattern of cleavage is characteristic of invertebrates such as mollusks, annelids, echinoderms, and tunicates, as well as amphibians and mammals.

Vertebrates exhibit a variety of developmental strategies involving different patterns of yolk utilization. Because yolk-rich regions divide much more slowly than areas with little yolk,

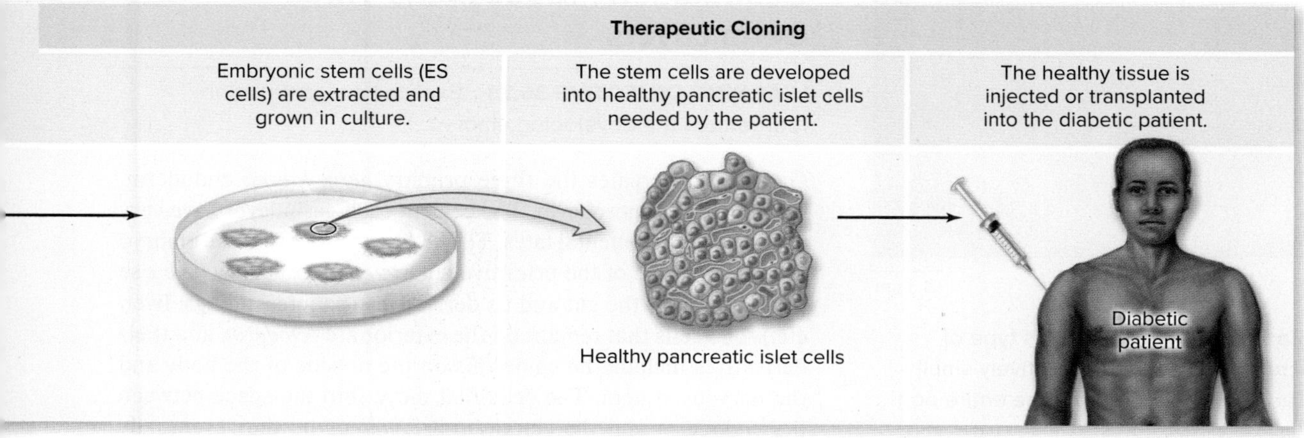
Therapeutic Cloning
Embryonic stem cells (ES cells) are extracted and grown in culture.
The stem cells are developed into healthy pancreatic islet cells needed by the patient.
The healthy tissue is injected or transplanted into the diabetic patient.
Healthy pancreatic islet cells
Diabetic patient

horizontal cleavage furrows are displaced toward the animal pole. Thus, holoblastic cleavage in frog eggs results in an asymmetrical blastula, with a displaced blastocoel. The blastula consists of large cells containing a lot of yolk at the vegetal pole, and smaller, more numerous cells containing little yolk at the animal pole.

Cleavage in mammals

Mammalian eggs contain very little yolk; however, embryogenesis in mammals has many similarities to development in their reptilian and avian relatives.

Because cleavage is not impeded by yolk in mammalian eggs, it is holoblastic (figure 36.22), forming a structure called a *blastocyst,* in which a single layer of cells surrounds a central, fluid-filled blastocoel. In addition, an **inner cell mass (ICM)** is located at one pole of the blastocoel cavity. The ICM is similar to the blastodisc of reptiles and birds, and it goes on to form the developing embryo.

The outer layer of cells, called the **trophoblast,** is similar to the cells that form the membranes underlying the tough outer shell of the reptilian egg. These cells have changed during the course of mammalian evolution to carry out a very different function: part of the trophoblast enters the maternal endometrium (the epithelial lining of the uterus) and contributes to the *placenta,* the organ that permits exchanges between the fetal and maternal blood supplies.

Blastomeres May or May Not Be Committed to Developmental Paths

LEARNING OBJECTIVE 36.6.3 Explain what regulative development is.

Viewed from the outside, cleavage-stage embryos often look like a simple ball or disc of similar cells. Mammals exhibit highly *regulative development,* in which early blastomeres do not appear to be committed to a particular fate. For example, if a blastomere is removed from an early eight-cell-stage human embryo (as is done in the process of preimplantation genetic diagnosis), the remaining seven cells of the embryo will "regulate" and develop into a complete individual if implanted into the uterus of a female. Similarly, embryos that are split into two (either naturally or experimentally) form identical twins. It therefore appears that inheritance of maternally encoded determinants is not an important mechanism in mammalian development, and body form is determined primarily by cell–cell interactions.

The earliest patterning events in mammalian embryos occur during the preimplantation stages that lead to formation of the blastocyst. At the eight-cell stage, the outer surfaces of many mammalian blastomeres flatten against each other in a process called *compaction,* which polarizes the blastomeres. The polarized blastomeres then undergo asymmetrical cell divisions. Cell lineage studies have shown that cells in the interior of the embryo most often become ICM cells of the mammalian blastocyst, whereas cells on the exterior of the embryo usually become trophoblast cells.

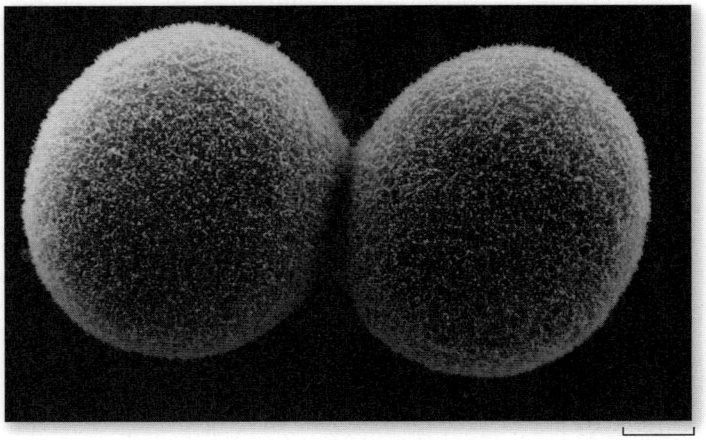

Figure 36.22 Holoblastic cleavage. In this type of cleavage, which is characteristic of eggs with relatively small amounts of yolk, cell division occurs throughout the entire egg.
David M. Phillips/Science Source

REVIEW OF CONCEPT 36.6

Rapid cell divisions during cleavage transform the zygote into a hollow ball of cells, the blastula. Yolk is the major determinant of cleavage pattern. Eggs with little yolk cleave completely; eggs with a large yolk cannot. In many animals, early cells are committed to a developmental path; in mammals, early cells are not committed.

- *If the cells of a mammalian embryo were separated at the four-cell stage, would they develop normally? What about a frog embryo at the four-cell stage?*

36.7 Gastrulation Forms the Basic Body Plan of the Embryo

In a complex series of cell shape changes and cell movements, the cells of the blastula rearrange themselves to form the basic body plan of the embryo. This process, called **gastrulation,** forms the three primary germ layers and converts the blastula into a bilaterally symmetrical embryo with a central progenitor gut and visible anterior–posterior and dorsal–ventral axes.

Gastrulation Produces the Three Germ Layers

LEARNING OBJECTIVE 36.7.1 Explain how gastrulation reorganizes the developing embryo.

Gastrulation creates the three primary *germ layers*: endoderm, ectoderm, and mesoderm. The cells in each germ layer have very different developmental fates. The cells that move into the embryo to form the tube of the primitive gut are *endoderm*; they give rise to the lining of the gut and its derivatives (pancreas, lungs, liver, etc.). The cells that remain on the exterior are *ectoderm,* and their derivatives include the epidermis on the outside of the body and the nervous system. The cells that move into the space between the endoderm and the ectoderm are *mesoderm*; they eventually form the notochord, bones, blood vessels, connective tissues, muscles, and internal organs such as the kidneys and gonads.

Cells move during gastrulation using a variety of cell shape changes. Some cells use broad, actin-filled extensions called *lamellipodia* to crawl over neighboring cells. Other cells send out narrow extensions called *filopodia,* which are used to "feel out" the surfaces of other cells or the extracellular matrix. Once a satisfactory attachment is made, the filopodia retract to pull the cell forward. Contractions of actin filament bundles are responsible for many of these cell shape changes. Cells that are tightly attached to one another via desmosomes or adherens junctions will move as cell sheets.

Gastrulation in mammals

Mammalian gastrulation begins with **delamination,** in which one sheet of cells splits into two sheets. Each migrating cell possesses particular cell-surface glycoproteins, which adhere to specific molecules on the surfaces of other cells or in the extracellular matrix. Changes in cell adhesiveness, as described in chapter 4, are key events in gastrulation. The extracellular matrix protein fibronectin and the corresponding integrin receptors of cells are essential molecules of gastrulation in many animals.

Mammalian gastrulation proceeds much the same as the process does in birds. In both types of animals, the embryo develops from a flattened collection of cells—the blastoderm in birds or the inner cell mass in mammals. Although the blastoderm of a bird is flattened because it is pressed against a mass of yolk, the inner cell mass of a mammal is flat despite the absence of a yolk mass.

In mammals, the placenta has made yolk dispensable; the embryo obtains nutrients from its mother following implantation into the uterine wall. However, the embryo still gastrulates as though it were sitting on top of a ball of yolk.

In mammals, a primitive streak forms, and cell movements through the primitive streak give rise to the three primary germ layers, much the same as in birds (figure 36.23). Similarly, mammalian embryos envelop their "missing" yolk by forming a yolk sac from extraembryonic cells that migrate away from the lower layer of the blastoderm and line the blastocoel cavity.

Extraembryonic Membranes Are an Adaptation to Life on Dry Land

LEARNING OBJECTIVE 36.7.2 Name the extraembryonic membranes in amniotes.

As an adaptation to terrestrial life, the embryos of reptiles, birds, and mammals develop within a fluid-filled *amniotic membrane,* or *amnion.* The amniotic membrane and several other membranes form from embryonic cells, but they are located outside the body of the embryo. For this reason, they are known as **extraembryonic membranes.** The extraembryonic membranes include the amnion, chorion, yolk sac, and allantois.

In mammals, the trophoblast cells of the blastocyst implant into the endometrial lining of the mother's uterus and become the chorionic membrane (figure 36.24). The part of the chorion in contact with endometrial tissue contributes to the placenta. The other part of the placenta is composed of modified endometrial tissue of the mother's uterus, as described in more detail in section 36.9. The allantois in mammals contributes blood vessels to the structure that will become the umbilical cord, so that fetal blood can be delivered to the placenta for gas exchange.

REVIEW OF CONCEPT 36.7

Gastrulation involves cell rearrangement and migration to produce ectoderm, mesoderm, and endoderm. In mammalian gastrulation, extraembryonic membranes form from embryonic cells outside the embryo's body and include the yolk sac, amnion, chorion, and allantois.

- *What kinds of cellular behaviors are necessary for gastrulation?*

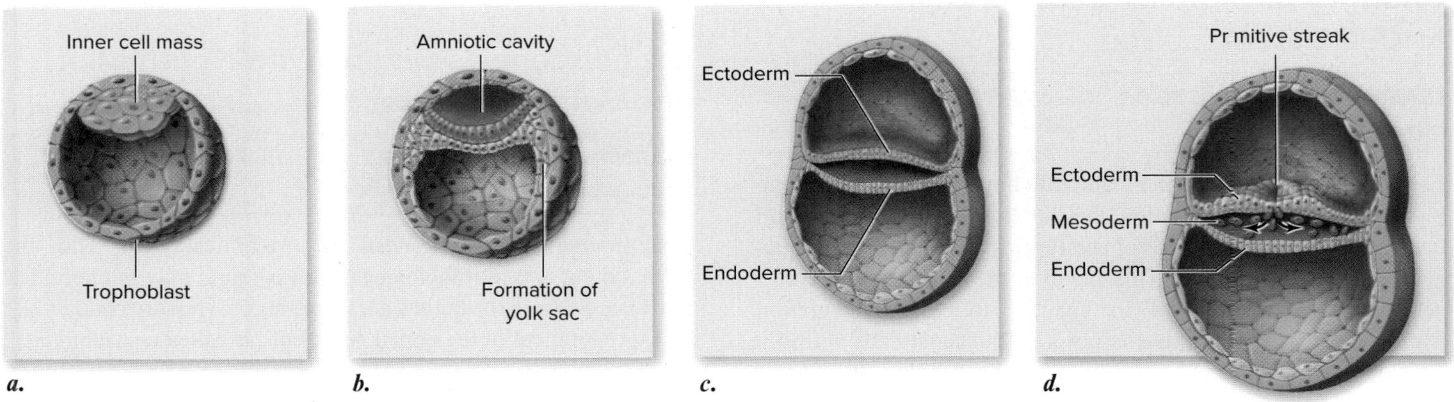

Figure 36.23 Mammalian gastrulation. ***a.*** Cross section of the mammalian blastocyst at the end of cleavage. ***b.*** The amniotic cavity forms between the inner cell mass (ICM) and the pole of the embryo. Meanwhile, the ICM flattens and delaminates into two layers that will become ectoderm and endoderm. ***b., c.*** Cells of the lower layer migrate out to line the blastocoel cavity to form the yolk sac. ***d.*** A primitive streak forms the ectoderm layer, and cells destined to become mesoderm migrate into the interior, similar to gastrulation in birds.

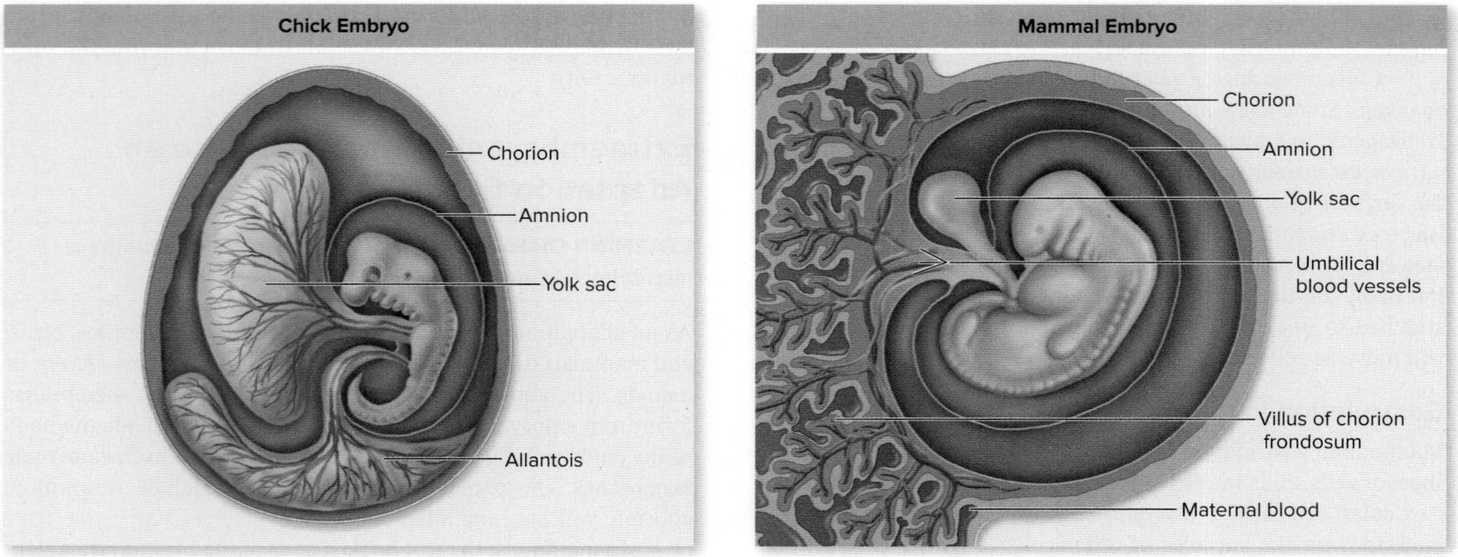
Chick Embryo
Chorion
Amnion
Yolk sac
Allantois
Mammal Embryo
Chorion
Amnion
Yolk sac
Umbilical blood vessels
Villus of chorion frondosum
Maternal blood

Figure 36.24 The extraembryonic membranes. The extraembryonic membranes in a chick embryo and a mammalian embryo share some of the same characteristics. However, in the chick, the allantois continues to grow until it eventually unites with the chorion just under the eggshell, where it is involved in gas exchange. In the mammalian embryo, the allantois contributes blood vessels to the developing umbilical cord.

36.8 The Body's Organs Form in Organogenesis

Gastrulation establishes the basic body plan and creates the three primary germ layers of animal embryos. The stage is now set for *organogenesis*–the formation of the organs in their proper locations–which occurs by interactions of cells within and between the three germ layers. Thus, organogenesis follows rapidly on the heels of gastrulation, and in many animals it begins before gastrulation is complete. Over the course of subsequent development, tissues develop into organs, and animal embryos assume their unique body form.

Changes in Gene Expression Lead to Cell Determination

LEARNING OBJECTIVE 36.8.1 Describe examples of organogenesis.

Essentially all of the cells in an animal's body have the same complement of genetic information. Despite this fact, an adult animal contains dozens to hundreds of cell types, each expressing a characteristic set of proteins. This involves mechanisms to control gene expression, explored in chapter 16. The information expressed in other cell types is not lost, but most cells within a developing organism progressively lose the capacity to express ever-larger portions of their genomes. What factors determine which genes are to be expressed in a particular cell?

To a large degree, a cell's location in the developing embryo determines its fate. By changing a cell's location, an experimenter can often alter its developmental destiny. But this is only true up to a certain point in the cell's development. At some stage, every cell's ultimate fate becomes fixed, a process referred to as *cell determination* (refer to chapter 16).

A cell's fate can be established by inheritance of cytoplasmic determinants or by interactions with neighboring cells. The process by which a cell or group of cells instructs neighboring cells to adopt a particular fate is called **induction.** If a nonporous barrier, such as a layer of cellophane, is imposed between the inducer and the target tissue, no induction takes place. In contrast, a porous filter, through which proteins can pass, does permit induction to occur.

In these experiments, researchers concluded that the inducing cells secrete a paracrine signal molecule that binds to the cells of the target tissue. Such signal molecules are capable of producing changes in the patterns of gene transcription in the target cells (refer to chapter 9).

Organogenesis Begins with Neurulation and Somitogenesis

LEARNING OBJECTIVE 36.8.2 Describe the events of neurulation and somitogenesis.

The process of organogenesis in vertebrates begins with the formation of two morphological features found only in chordates: the *notochord* and the hollow **dorsal nerve cord** (refer to chapter 28). The development of the dorsal nerve cord is called **neurulation.**

Development of the neural tube

The notochord forms from mesoderm and is first visible soon after gastrulation is complete. It is a flexible rod located along the dorsal midline in the embryos of all chordates, although its function as a supporting structure is supplanted by the subsequent development of the vertebral column in the vertebrates. After the notochord has been laid down, the region of dorsal ectodermal cells situated above the notochord begins to thicken to form the *neural plate.*

The thickening is produced by the elongation of the dorsal ectoderm cells. Those cells then assume a wedge shape because of contracting bundles of actin filaments at their apical end. This change in shape causes the neural tissue to roll up into a **neural groove** running down the long axis of the embryo. The edges of the neural groove then move toward each other and fuse, creating a long, hollow cylinder, the **neural tube** (figure 36.25). The neural tube eventually pinches off from the surface ectoderm to end up beneath the surface of the embryo's back. Regional changes, which are under control of the *Hox* gene complexes, then occur in the neural tube as it differentiates into the spinal cord and brain.

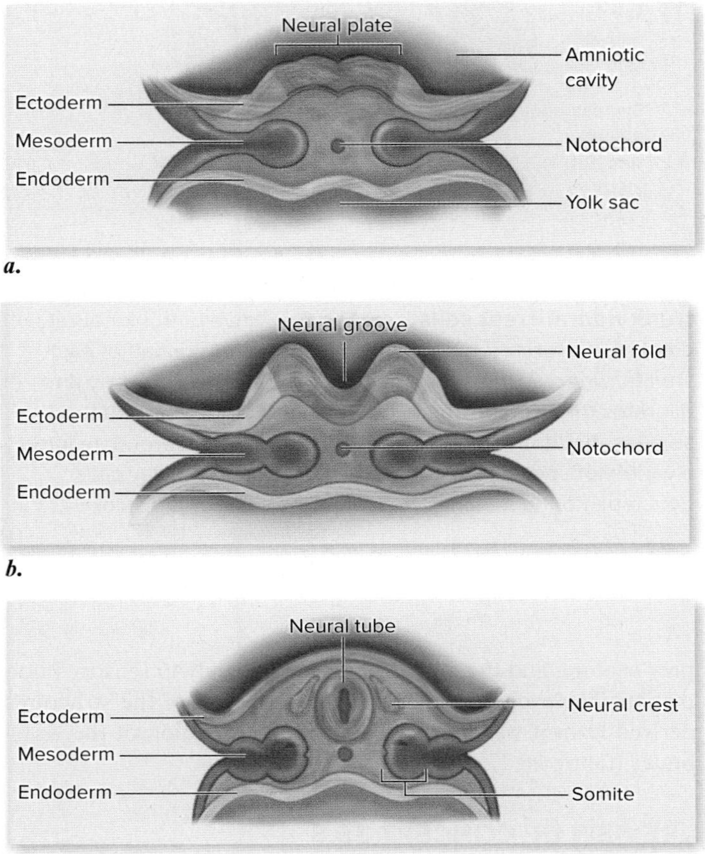

Figure 36.25 Mammalian neural tube formation. ***a.*** The neural plate forms from ectoderm above the notochord. ***b.*** The cells of the neural plate fold together to form the neural groove. ***c.*** The neural groove eventually closes to form a hollow tube called the neural tube, which will become the brain and spinal cord. As the tube closes, some of the cells from the dorsal margin of the neural tube differentiate into the neural crest, migratory cells that form a variety of structures and are characteristic of vertebrates.

Generation of somites

While the neural tube is forming from dorsal ectoderm, the rest of the basic architecture of the body is being rapidly established by changes in the mesoderm. The sheets of mesoderm on either side of the developing notochord separate into a series of rounded regions called **somitomeres.** The somitomeres then separate into segmented blocks called **somites.** The mesoderm in the head region does not separate into discrete somites but remains connected as somitomeres, which form the skeletal muscles of the face, jaws, and throat.

Somites form in an anterior–posterior wave with a regular periodicity that can be easily timed—for example, by using a vital dye, which marks cells without killing them, to mark each somite as it forms in a chick embryo. Cells at the presumptive boundary regions in the presomitic mesoderm instruct cells anterior to them to condense and separate into somites at specific times (for example, every 90 min in a chick embryo). This "clock" appears to be regulated by contact-mediated cell signaling between neighboring cells.

Somites themselves are transient embryonic structures, and soon after their formation, cells disperse and start differentiating along different pathways to ultimately form the skeleton, skeletal musculature, and associated connective tissues. The total number of somites formed is species-specific; for example, chickens form 50 somites, whereas some species of snakes form as many as 400 somites.

Some body organs, including the kidneys, adrenal glands, and gonads, develop within a strip of mesoderm that runs lateral to each row of somites. The remainder of the mesoderm, which is most ventrally located, moves out and around the endoderm and eventually surrounds it completely. As a result of this movement, the mesoderm becomes separated into two layers. The outer layer is associated with the inner body wall, and the inner layer is associated with the outer lining of the gut tube. Between these two layers of mesoderm is the *coelom* (refer to chapter 27), which becomes the body cavity of the adult.

Migratory Neural Crest Cells Differentiate into Many Cell Types

LEARNING OBJECTIVE 36.8.3 Explain the migration of neural crest cells and their role in organ development.

Neurulation occurs in all chordates; however, in vertebrates the process has an additional step. As the neural groove closes to form the neural tube, the edges pinch and form a population of cells between the neural tube and the surface ectoderm called **neural crest** cells. These neural crest cells then move away from the neural tube to colonize many different regions of the developing embryo. These cells will form many tissues, including peripheral nervous system, pigment cells, and even craniofacial structures.

The appearance of the neural crest was a key event in the evolution of vertebrates. Many of the unique vertebrate adaptations that contribute to their varied ecological roles involve structures that arise from neural crest cells. Evolutionary changes associated with the derivatives of the neural crest provided better detection of prey, a greatly improved ability to orient spatially during prey capture, and the means to respond quickly to sensory information.

The differentiation of neural crest cells depends on their migration pathway and final location. Neural crest cells migrate along one of three pathways in the embryo. Cranial neural crest cells are anterior cells that migrate into the head and neck; trunk neural crest cells migrate along one of two different pathways. Each population of neural crest cells develops into a variety of cell types.

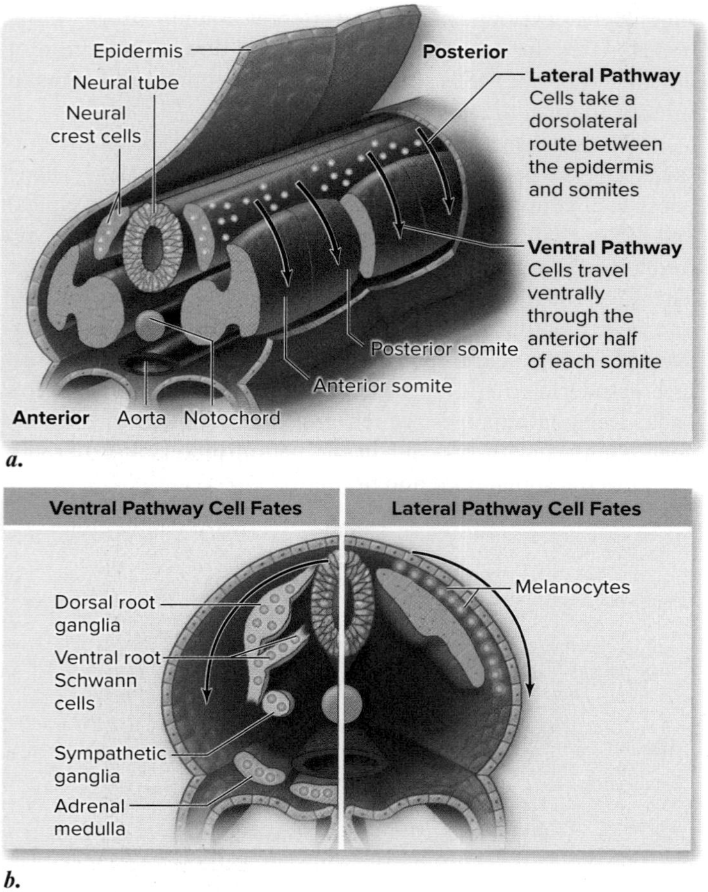

Figure 36.26 Migration pathways and cell fates of trunk neural crest cells. ***a.*** The first wave of trunk neural crest cells migrates ventrally through the anterior half of each somite, whereas the second wave of cells leaves dorsally and migrates through the space between the epidermis and the somites. ***b.*** Ventral pathway neural crest cells differentiate into a variety of specialized cell types, but lateral pathway cells develop into the melanocytes (pigment cells) of the skin.

Migration of cranial neural crest cells

Cranial neural crest cells contribute significantly to development of the skeletal and connective tissues of the face and skull, as well as differentiating into nerve and glial cells of the nervous system, and melanocyte pigment cells. Changes in the placement of cranial neural crest cells during development have led to the evolution of the great complexity and variety of vertebrate heads.

There are two waves of cranial neural crest cell migration. The first produces both dorsal and ventral structures, and the second produces only dorsal structures and makes much less cartilage and bone. Transplantation experiments indicate that the developmental potential of the cells in these two waves is identical. The differences in cell fate are due to the environment the migrating cells encounter, not to prior determination of cell fate.

Trunk neural crest cells: Ventral pathway

Neural crest cells located in more posterior positions have very different developmental fates, depending on their migration pathway. The first trunk neural crest cells that migrate away from the neural tube pass through the anterior half of each adjoining somite to ventral locations (figure 36.26*a*).

Some of these cells form the sensory neurons of the dorsal root ganglia, which send out projections to connect the periphery of the animal with the spinal cord (refer to chapter 33). Others become specialized as Schwann cells, which insulate nerve fibers to facilitate the rapid conduction of impulses along peripheral nerves. Still others form nerves of the autonomic ganglia, which regulate the activity of internal organs, and endocrine cells of the adrenal medulla (figure 36.26*b*).

Trunk neural crest cells: Lateral pathway

The second group of trunk neural crest cells migrate away from the neural tube in the space just under the surface ectoderm, to occupy this space around the entire body of the embryo. There, they will differentiate into the pigment cells of the skin (figure 36.26*a, b*). Mutations in genes that affect the survival and migration of neural crest cells lead to white spotting in the skin on ventral surfaces, as well as internal problems in other neural crest–derived tissues.

Many of the unique vertebrate adaptations that contribute to their varied ecological roles involve structures that arise from neural crest cells. The vertebrates became fast-swimming predators with much higher metabolic rates. This accelerated metabolism permitted a greater level of activity than was possible among the more primitive chordates. Other evolutionary changes associated with the derivatives of the neural crest provided better detection of prey, a greatly improved ability to orient spatially during prey capture, and the means to respond quickly to sensory information. The evolution of the neural crest and of the structures derived from it was a crucial step in the evolution of the vertebrates (figure 36.27).

REVIEW OF CONCEPT 36.8

Genetic control of organogenesis relies on conserved families of cell-signaling molecules and transcription factors. The process of neurulation forms the basic nervous system in vertebrates. Somitogenesis is the division of mesoderm into somites. Neural crest cells arise from the neural tube and migrate to many sites to form a variety of cell types. The evolution of the neural crest led to the appearance of many vertebrate-specific adaptations.

- *Are neural crest cells determined prior to migration?*

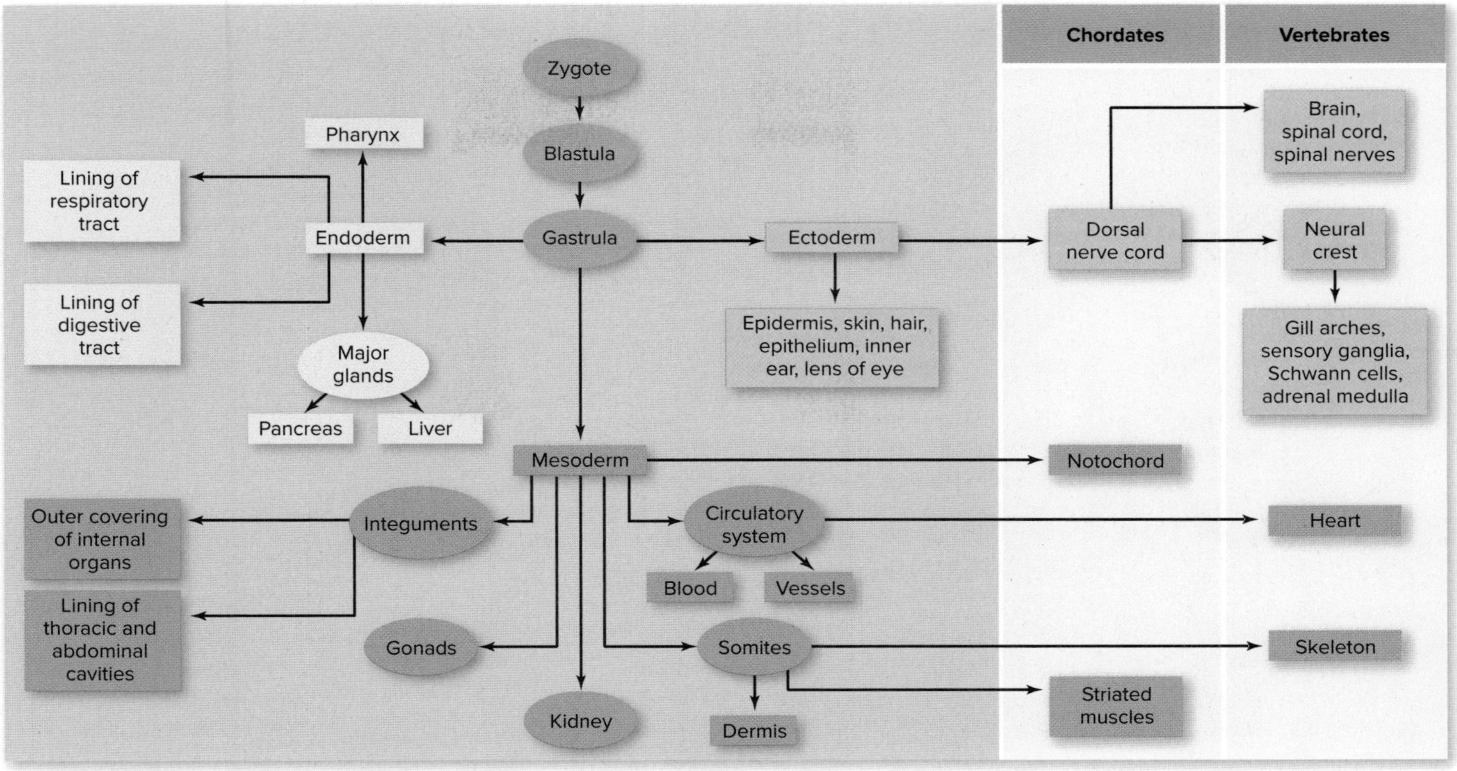

Figure 36.27 Germ-layer derivation of the major tissue types in animals. The three germ layers that form during gastrulation give rise to all the organs and tissues in the body, but the neural crest cells that form from ectodermal tissue give rise to structures that are prevalent in vertebrates, such as gill arches and bones of the face and skull.

36.9 Human Development Takes Nine Months

Human development from fertilization to birth takes an average of 266 days, or about nine months. This time is commonly divided into three periods called *trimesters.* We describe here the development of the embryo as it takes place during these trimesters. Later in this section, we summarize the process of birth, nursing of the infant, and postnatal development.

During the First Trimester, the Zygote Undergoes Rapid Development and Differentiation

LEARNING OBJECTIVE 36.9.1 Describe the major developmental events in the first trimester.

About 30 hr after fertilization, the zygote undergoes its first cleavage; the second cleavage occurs about 30 hr after that. By the time the embryo reaches the uterus, six to seven days after fertilization, it has differentiated into a blastocyst. As mentioned in section 36.6, the blastocyst consists of an inner cell mass, which will become the body of the embryo, and a surrounding layer of trophoblast cells.

The trophoblast cells of the blastocyst digest their way into the endometrial lining of the uterus in the process known as **implantation.** The blastocyst begins to grow rapidly and initiates the formation of the amnion and the chorion.

Development in the first month

During the second week after fertilization, the developing chorion and the endometrial tissues of the mother engage to form the placenta (figure 36.28). Within the placenta, the mother's blood and the blood of the embryo come into close proximity but do not mix. Gases are exchanged, however, and the placenta provides nourishment for the embryo, detoxifies certain molecules that may pass into the embryonic circulation, and secretes hormones. Certain substances, such as alcohol and drugs, including antibiotics, are not stopped by the placenta and pass from the mother's bloodstream into the embryo.

One of the hormones released by the placenta is human chorionic gonadotropin (hCG). This hormone is secreted by the trophoblast cells even before they become the chorion, and it is the hormone assayed in pregnancy tests. Human chorionic gonadotropin maintains the mother's corpus luteum. The corpus luteum, in turn, continues to secrete estradiol and progesterone, thereby preventing menstruation and further ovulations.

Gastrulation also takes place in the second week after fertilization, and the three germ layers are formed. Neurulation occurs in the third week. The first somites appear, which give rise to the muscles, vertebrae, and connective tissues. By the end of the third week, over a dozen somites are evident, and the blood vessels and gut have begun to develop. At this point, the embryo is about 2 mm long.

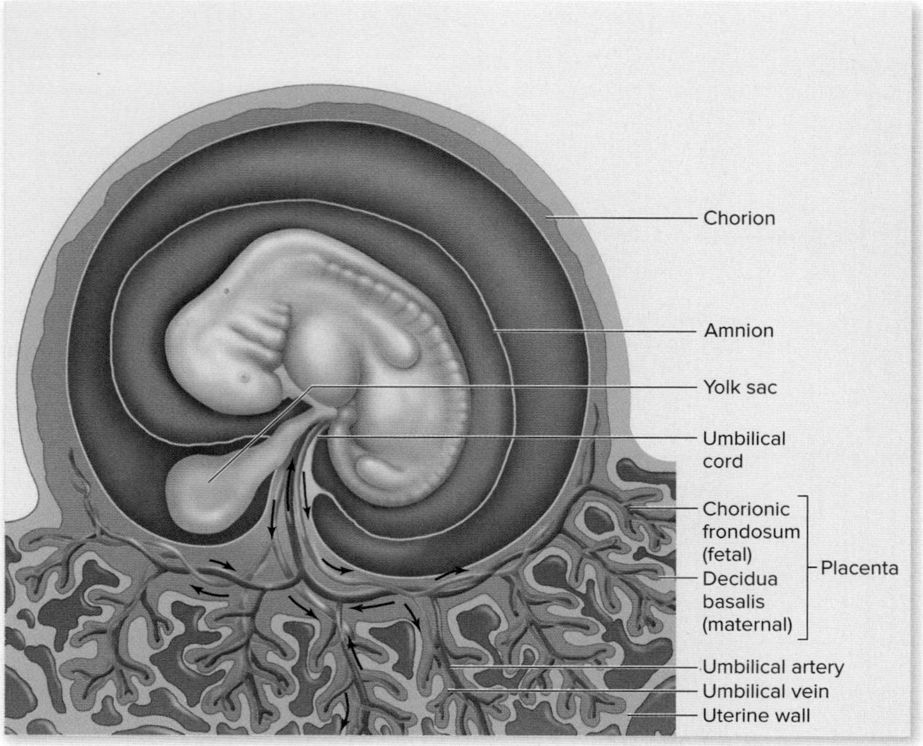

a.

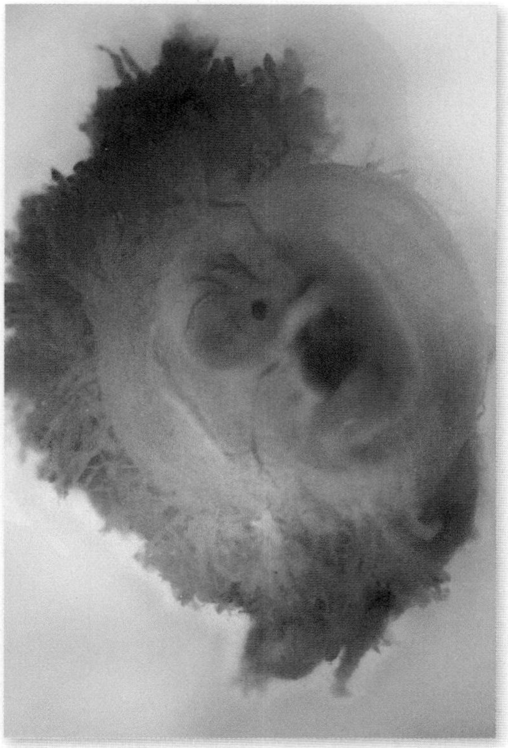

b.

Figure 36.28 Structure of the placenta. ***a.*** The placenta contains a fetal component, the chorionic frondosum, and a maternal component, the decidua basalis. Deoxygenated fetal blood from the umbilical arteries (shown in blue) enters the placenta, where it picks up oxygen and nutrients from the mother's blood. Oxygenated fetal blood returns in the umbilical vein (shown in red) to the fetus. ***b.*** Note that the seven-week embryo is surrounded by a fluid-filled amniotic sac.

(b): Scott Camazine/Medical Images/DIOMEDIA

Organogenesis begins during the fourth week (figure 36.29*a*). The eyes form. The heart forms from two fused tubes, which then bend and develop four chambers. At this point the autorythmic cells that will become the SA node (refer to chapter 34), or pacemaker, begin to produce action potentials. This is sometimes referred to as a "fetal heartbeat," but as the heart is far from completely formed, this is misleading. The SA node will pulsate rhythmically for the rest of the individual's life. At 70 beats per minute, the heart is destined to beat more than 2.5 billion times during a lifetime of 70 years.

Over 30 pairs of somites are visible by the end of the fourth week, and the arm and leg buds have begun to form. The embryo has increased in length to about 5 mm. Although the overall human body plan has formed at this point, many women are still unaware they are pregnant at this stage. Most spontaneous abortions (miscarriages), which frequently occur in the case of a defective embryo, occur by the middle of the first trimester. While data are difficult to collect, it is estimated that as few as 30–40% of fertilizations may make it to live birth.

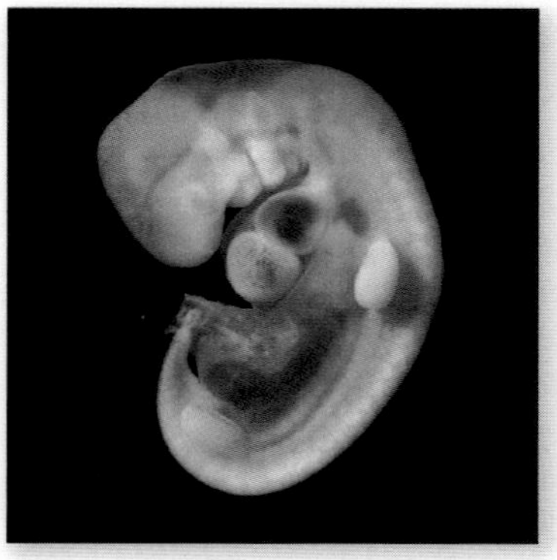

a.

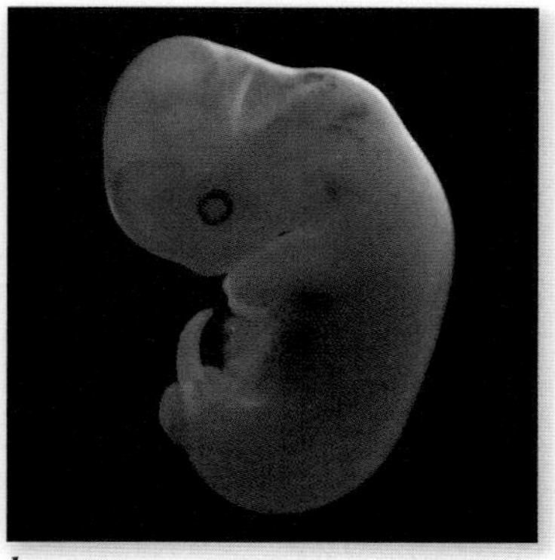

b.

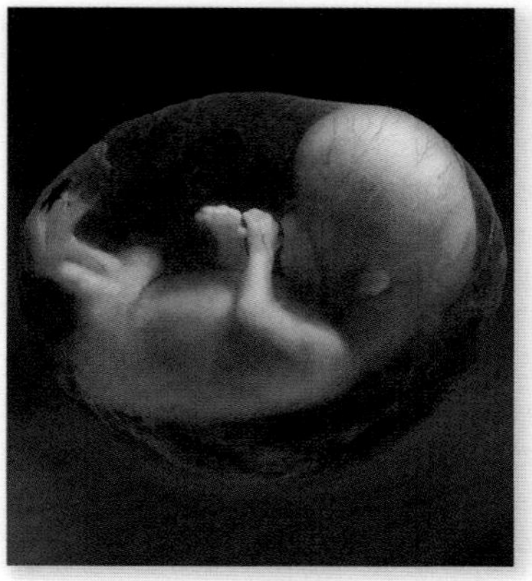

c.

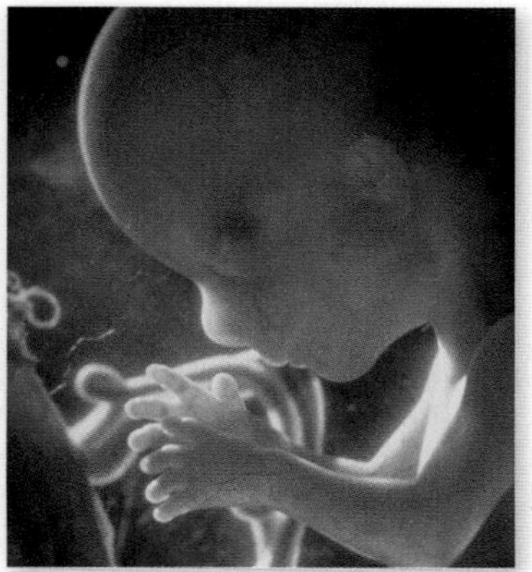

d.

Figure 36.29 The developing human. ***a.*** Four weeks. ***b.*** End of fifth week. ***c.*** Three months. ***d.*** Four months.

(a): Omikron/Science Source; (b): Kage-Mikrofotografie/age fotostock; (c): Steve Allen/Brand X Pictures/Getty Images; (d): Nestle/Petit Format/Science Source

The second month

Organogenesis continues during the second month (figure 36.29*b*). The miniature limbs of the embryo assume their adult shapes. The arms, legs, knees, elbows, fingers, and toes can all be seen—as well as a short, bony tail. The bones of the embryonic tail, an evolutionary reminder of our past, later fuse to form the coccyx.

Within the abdominal cavity, the major organs, including the liver, pancreas, and gallbladder, become evident. By the end of the second month, the embryo has grown to about 25 mm in length, weighs about 1 g, and begins to look distinctly human. The ninth week marks the transition from embryo to fetus. At this time, all of the major organs of the body have been established in their proper locations.

The third month

The nervous system develops during the third month, and the arms and legs start to move (figure 36.29*c*). The embryo begins to show facial expressions and carries out primitive reflexes such as the startle reflex and sucking.

At around 10 weeks, the secretion of hCG by the placenta declines, and the corpus luteum regresses as a result. However, menstruation does not occur, because the placenta itself secretes estradiol and progesterone (figure 36.30).

The high levels of estradiol and progesterone in the blood during pregnancy continue to inhibit the release of FSH and LH, thereby preventing ovulation. They also help maintain the uterus and eventually prepare it for labor and delivery, and they stimulate the development of the mammary glands in preparation for lactation after delivery.

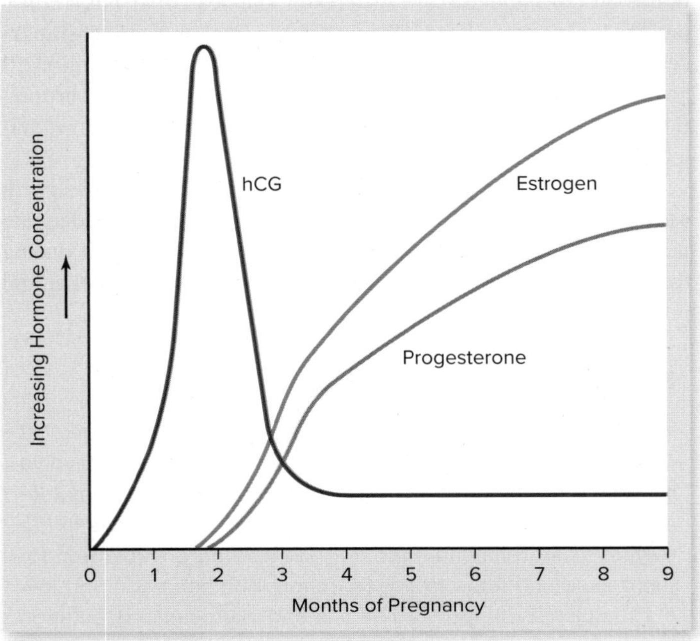

Figure 36.30 Hormonal secretion by the placenta. The placenta secretes human chorionic gonadotropin (hCG), which peaks in the second month and then declines. After five weeks, it secretes increasing amounts of estrogen and progesterone.

During the Second and Third Trimesters, Organs Mature

LEARNING OBJECTIVE 36.9.2 Describe the major developmental events in the second and third trimesters.

Bones enlarge during the fourth month (figure 36.29*d*), and by the end of the month the mother can feel the baby kicking. By the end of the fifth month, the rapid heartbeat of the fetus can be heard with a stethoscope.

Growth begins in earnest in the sixth month; by the end of that month, the fetus weighs 600 g (1.3 lb) and is over 300 mm (1 ft) long. Most of its prebirth growth is still to come, however. The fetus cannot yet survive outside the uterus without special medical intervention.

The third trimester is predominantly a period of growth and maturation of organs. The weight of the fetus doubles several times, but this increase in bulk is not the only kind of growth that occurs. Most of the major nerve tracts in the brain, as well as many new neurons (nerve cells), are formed during this period. Neurological growth is far from complete when birth takes place, however. If the fetus remained in the uterus until its neurological development were complete, it would grow too large for safe delivery through the pelvis. Instead, the infant is born as soon as the probability of its survival is high, and its brain continues to develop and produce new neurons for months after birth.

Critical Changes in Hormones Bring on Birth

LEARNING OBJECTIVE 36.9.3 Explain the hormonal control of the birth process.

In some mammals, changing hormone levels in the developing fetus initiate the process of birth. The fetuses of these mammals have an extra layer of cells in their adrenal cortex, which secrete corticosteroids that induce the uterus of the mother to manufacture prostaglandins. Prostaglandins trigger powerful contractions of the uterine smooth muscles.

In humans, fetal secretion of cortisol increases during late pregnancy, which appears to stimulate estradiol secretion by the placenta. The mother's uterus releases prostaglandins, possibly as a result of the high levels of estradiol secreted by the placenta. Estradiol also stimulates the uterus to produce more oxytocin receptors, and as a result, the uterus becomes increasingly sensitive to oxytocin.

Prostaglandins begin the uterine contractions, but then sensory feedback from the uterus stimulates the release of oxytocin from the mother's posterior pituitary gland. Working together, oxytocin and prostaglandins further stimulate uterine contractions, forcing the fetus downward (figure 36.31). This positive feedback mechanism accelerates during labor. Initially, only a few contractions occur each hour, but the rate eventually increases to one contraction every 2 to 3 min. Finally, strong contractions, aided by the mother's voluntary pushing, expel the fetus, which is now a newborn baby, or *neonate.*

After birth, continuing uterine contractions expel the placenta and associated membranes, collectively called the *afterbirth.* The umbilical cord is still attached to the baby,

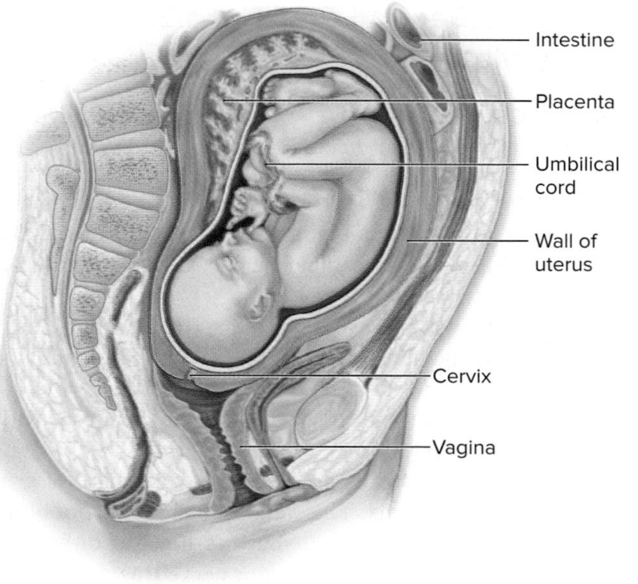
Intestine
Placenta
Umbilical cord
Wall of uterus
Cervix
Vagina

Figure 36.31 Position of the fetus just before birth. A developing fetus causes major changes in a woman's anatomy. The stomach and intestines are pushed far up, and considerable discomfort often results from pressure on the lower back. In a normal vaginal delivery, the fetus exits through the cervix, which must dilate (expand) considerably to permit passage.

and to free the newborn, a doctor or midwife clamps and cuts the cord. Blood clotting and contraction of muscles in the cord prevent excessive bleeding.

Nursing of Young Is a Distinguishing Feature of Mammals

LEARNING OBJECTIVE 36.9.4 Describe the hormonal control of lactation.

Milk production, or *lactation,* occurs in the alveoli of mammary glands when they are stimulated by the anterior-pituitary hormone prolactin. Milk from the alveoli is secreted into a series of alveolar ducts, which are surrounded by smooth muscle and lead to the nipple.

During pregnancy, high levels of progesterone stimulate the development of the mammary alveoli, and high levels of estradiol stimulate the development of the alveolar ducts. However, estradiol blocks the actions of prolactin on the mammary glands, and it inhibits prolactin secretion by promoting the release of prolactin-inhibiting hormone from the hypothalamus. During pregnancy, therefore, the mammary glands are prepared for, but prevented from, lactating. The growth of mammary glands is also stimulated by the placental hormones human chorionic somatomammotropin, a prolactin-like hormone, and human somatotropin, a growth hormone–like hormone.

When the placenta is discharged after birth, the concentrations of estradiol and progesterone in the mother's blood decline rapidly. This decline allows the anterior pituitary gland to secrete prolactin, which stimulates the mammary alveoli to produce milk. Sensory impulses associated with the baby's suckling trigger the posterior pituitary gland to release oxytocin. Oxytocin stimulates contraction of the smooth muscle surrounding the alveolar ducts, thus causing milk to be ejected by the breast. This pathway is known as the *milk let-down reflex,* and it is found in other mammals as well. The secretion of oxytocin during lactation also causes some uterine contractions, as it did during labor. These contractions help restore the tone of uterine muscles in mothers who are breast-feeding.

The first milk produced after birth is a yellowish fluid called colostrum, which is both nutritious and rich in maternal antibodies. Milk synthesis begins about three days following the birth and is referred to as the milk "coming in." Many mothers nurse for a year or longer. When nursing stops, the accumulation of milk in the breasts signals the brain to stop secreting prolactin, and milk production ceases.

Postnatal Development in Humans Continues for Years

LEARNING OBJECTIVE 36.9.5 List the events of postnatal human development.

Growth of the infant continues rapidly after birth. Babies typically double their birth weight within two months. Because different organs grow at different rates and cease growing at different times, the body proportions of infants are different from those of adults. The head, for example, is disproportionately large in newborns, but after birth it grows more slowly than the rest of the body. Such a pattern of growth, in which different components grow at different rates, is referred to as **allometric growth.**

In most mammals, brain growth is mainly a fetal phenomenon. In chimpanzees, for instance, the brain and the cerebral portion of the skull grow very little after birth, whereas the bones of the jaw continue to grow. As a result, the head of an adult chimpanzee looks very different from that of a fetal or infant chimpanzee. In human infants, by contrast, the brain and cerebral skull grow at the same rate as the jaw. Therefore, the jaw–skull proportions do not change after birth, and the head of a human adult looks very similar to that of a human fetus or infant.

The fact that the human brain continues to grow significantly for the first few years of postnatal life means that adequate nutrition and a safe environment are particularly crucial during this period for the full development of a person's intellectual potential.

REVIEW OF CONCEPT 36.9

The critical stages of human development occur in the first trimester of gestation; the subsequent six months involve growth and maturation. Growth of the brain is not complete at birth and must be completed postnatally. Hormones in the mother's blood maintain the nutritive uterine environment for the developing fetus; changes in hormone secretion and levels stimulate birth (prostaglandins and oxytocin) and lactation (oxytocin and prolactin).

- *Why are teratogens (agents that cause birth defects) most potent in the first trimester?*

Why Do STDs Vary in Frequency?

As a general rule, the incidence of a sexually transmitted disease (STD) is expected to increase with increasing frequencies of unprotected sexual contact. With the emergence of AIDS, intense publicity and education have lessened such dangerous behavior. Both the number of sexual partners and the frequency of unprotected sex have fallen significantly in the United States in the last decade. It would follow, then, that the frequencies of STDs such as syphilis, gonorrhea, and chlamydia should also be falling. Detailed yearly statistics are reported in the graph. The results are a little surprising. Although the level of gonorrhea infection has fallen since 1984, the level of chlamydia infection has been rising steadily during the same period! What are we to make of this?

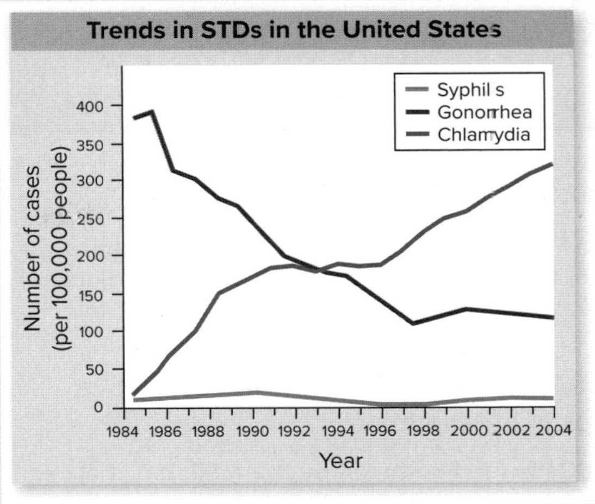

The simplest explanation of such a difference is that the two STDs are occurring in different populations, and one population has falling levels of sexual activity, while the other has rising levels. However, nationwide statistics encompass all population subgroups, and each major subgroup contains all three major STDs mentioned. Thus, this would seem an unlikely explanation for the rise in frequency of chlamydia.

A second possible explanation would be a change in the infectivity of one of the STDs. A less infective STD would tend to fall in frequency in the population, for the simple reason that fewer sexual contacts result in infection.

Syphilis is most infective in its initial stage, but this stage lasts only about a month. Most transmissions occur during the much longer second stage, marked by a pink rash and sores in the mouth. The bacteria can be transmitted at this stage by kissing or shared liquids. Any drop in infectivity of this STD would be expected to shorten this stage—but no such shortening has been observed.

Gonorrhea can be transmitted by various forms of sexual contact with an infected individual at any time during the infection. Just as with syphilis, no drop in infectivity per sexual contact has been reported.

Chlamydia offers the most interesting possibility of changes in infectivity, because of its unusual nature. *Chlamydia trachomatis* is genetically a bacterium but is an obligate intracellular parasite, much like a virus in this respect—it can reproduce only inside human cells. The red structures in the photo are chlamydia bacteria inside human cells. Like gonorrhea, chlamydia is transmitted through vaginal, anal, or oral intercourse with an infected person. As with the other two STDs, there does not appear to have been a drop in infectivity. With chlamydia, the person may show no symptoms.

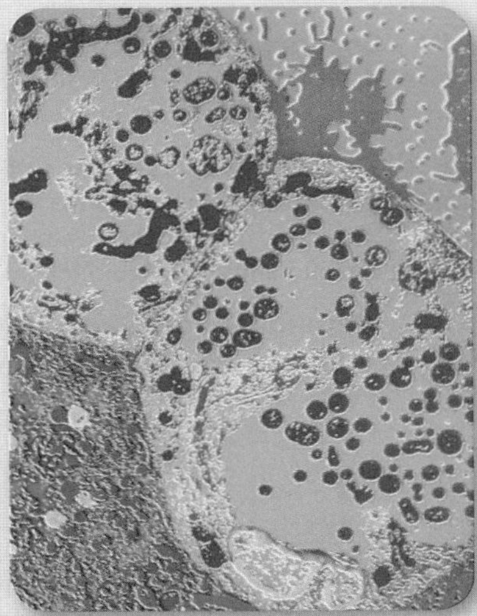
R. Dourmashkin, MD/Science Source

There is, however, a third possible explanation for why the frequency of one STD in a population might rise while the frequency of another STD in that same population falls. To grasp this third possible explanation, we need to focus on the experience of infected individuals. Unlike someone infected with syphilis and gonorrhea, who soon knows he or she is infected, a person infected with chlamydia may show no symptoms. Programs to raise public awareness of STDs might lower the sexual activity of individuals aware of their infection, but not influence the sexual behavior of individuals who are *not* aware of their infection. To assess this possibility, examine carefully the trends in the incidence in the United States of gonorrhea, chlamydia, and syphilis seen in the graph.

Analysis

1. **Applying Concepts** What is the dependent variable?
2. **Making Inferences**
   a. ***Gonorrhea.*** What is the incidence in 1984? In 1994? In 2004? Does the frequency decline or increase? In general, are individuals aware they are infected when they transmit the STD?
   b. ***Chlamydia.*** What is the incidence in 1984? In 1994? In 2004? Does the frequency decline or increase? In general, are individuals aware they are infected when they transmit the STD?
3. **Drawing Conclusions** How might heightened public awareness explain why the trend in levels of gonorrhea differs from that of chlamydia?

Retracing the Learning Path

CONCEPT 36.1 Mammals Are Viviparous

36.1.1 Internal Fertilization Has Led to Three Strategies for Development of Offspring The three strategies for development are oviparity, ovoviviparity, and viviparity.

CONCEPT 36.2 The Human Male Reproductive System Is Typical of Mammals

36.2.1 Sperm Cells Are Produced by the Millions Haploid sperm are produced by meiosis of spermatogonia. Sperm consist of a head with an acrosome, a body, and a flagellar tail.

36.2.2 Male Accessory Sex Organs Aid in Sperm Delivery The urethra of the penis transports both sperm and urine and contains two columns of erectile tissue, blood vessels, and nerves.

36.2.3 Hormones Regulate Male Reproductive Function Male reproductive function is controlled by the hormones FSH and LH and negative feedback loops.

CONCEPT 36.3 The Human Female Reproductive System Undergoes Cyclic Gamete Development

36.3.1 Usually Only One Egg Is Ovulated per Menstrual Cycle The ovarian cycle consists of a follicular phase, ovulation, and a luteal phase. At birth, primary oocytes are arrested at meiosis I. Monthly, an oocyte completes meiosis I, then is arrested in meiosis II.

36.3.2 Female Accessory Sex Organs Receive Sperm and Provide Nourishment and Protection to the Embryo The vagina receives sperm, which enter the uterus via the cervix.

36.3.3 Contraception Prevents Unwanted Pregnancy Pregnancy can be avoided by abstinence, by blocking sperm, by destroying sperm, by preventing ovulation or implantation, or by sterilization.

CONCEPT 36.4 The First Step in Development Is Fertilization

36.4.1 A Sperm Must Penetrate to the Plasma Membrane of the Egg for Membrane Fusion to Occur The sperm's acrosome releases digestive enzymes to penetrate the egg's external layer.

36.4.2 Membrane Fusion Activates the Egg Fusion of membranes triggers egg activation. Blocks to polyspermy include changes in membrane potential and alterations to the external coat of the egg. Upon egg activation, meiosis is completed.

CONCEPT 36.5 Cells of the Early Embryo Are Totipotent

36.5.1 Embryonic Stem Cells Are Pluripotent Cells Derived from Embryos Embryonic stem cells are derived from the inner cell mass. They can differentiate into any adult tissue.

36.5.2 Reversal of Determination Has Allowed Cloning Cells undergo no irreversible changes during development. Adult cells can be reprogrammed to be totipotent.

36.5.3 Reproductive Cloning Often Fails for Lack of Proper Genomic Imprinting In normal development, epigenetic marks involved in imprinting are reprogrammed. Cloning bypasses this, which can lead to problems.

36.5.4 Nuclear Reprogramming Has Been Accomplished by Use of Defined Factors Adult cells can be reprogrammed by introduction of specific genes.

CONCEPT 36.6 Cleavage Leads to the Blastula Stage

36.6.1 The Blastula Is a Hollow Mass of Cells Cleavage is a series of rapid cell divisions that produces a hollow ball of cells.

36.6.2 Cleavage Patterns Are Highly Diverse and Distinctive Cleavage in mammals, with little or no yolk, is holoblastic (involving the whole egg).

36.6.3 Blastomeres May or May Not Be Committed to Developmental Paths Mammals exhibit regulative development; the fate of early cells is not predetermined.

CONCEPT 36.7 Gastrulation Forms the Basic Body Plan of the Embryo

36.7.1 Gastrulation Produces the Three Germ Layers During gastrulation, the three germ layers differentiate: endoderm, ectoderm, and mesoderm.

36.7.2 Extraembryonic Membranes Are an Adaptation to Life on Dry Land The yolk sac, amnion, chorion, and allantois prevent desiccation and nourish and protect the embryo.

CONCEPT 36.8 The Body's Organs Form in Organogenesis

36.8.1 Changes in Gene Expression Lead to Cell Determination A cell's location in the developing embryo often determines its fate by interactions with other cells (induction).

36.8.2 Organogenesis Begins with Neurulation and Somitogenesis Neurulation is the formation of the neural tube from ectoderm near the notochord; somitogenesis is the establishment of mesoderm into units called somites.

36.8.3 Migratory Neural Crest Cells Differentiate into Many Cell Types Neural crest cells migrate to become connective tissue, nerve and glial cells, melanocytes, and other cells.

CONCEPT 36.9 Human Development Takes Nine Months

36.9.1 During the First Trimester, the Zygote Undergoes Rapid Development and Differentiation Implantation occurs at the end of the first week of pregnancy. The placenta forms and gastrulation occurs in the second week.

36.9.2 During the Second and Third Trimesters, Organs Mature

36.9.3 Critical Changes in Hormones Bring on Birth Birth is initiated by secretions of steroids from the fetal adrenal cortex that induce prostaglandins, which cause contractions.

36.9.4 Nursing of Young Is a Distinguishing Feature of Mammals Nursing involves a neuroendocrine reflex, causing the release of oxytocin and the milk let-down response.

36.9.5 Postnatal Development in Humans Continues for Years Postnatal development continues, with different organs growing at different rates—called allometric growth.

Concept Overview

This Concept Overview diagrams the key concepts that were discussed in this chapter.

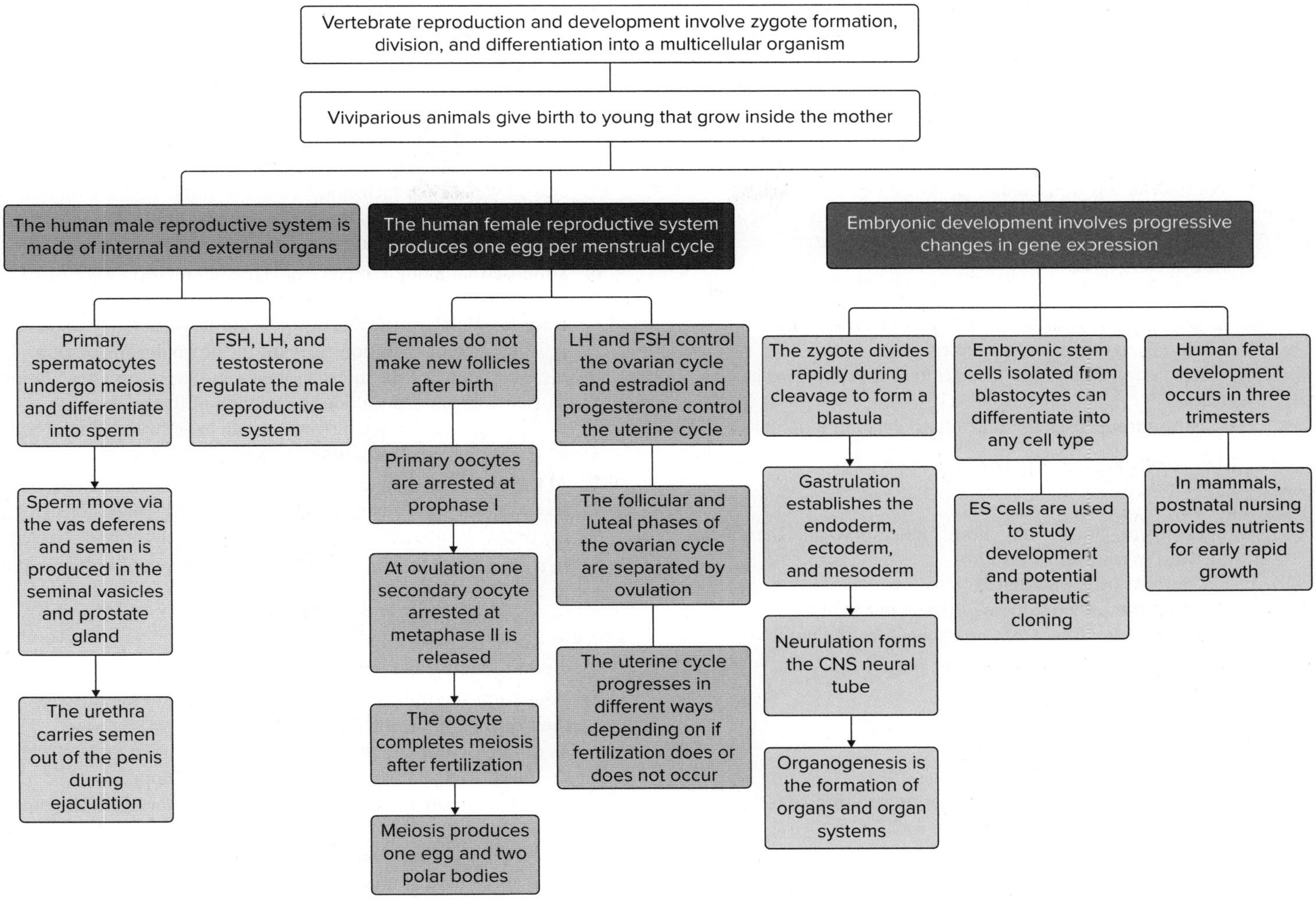

Assessing the Learning Path

Understand

1. Most mammals are viviparous. Monotremes are different because they
   a. do not exhibit live birth of their young.
   b. exhibit internal fertilization, unlike other mammals.
   c. are external fertilizers.
   d. are ovoviviparous, but fertilization is external.
2. Which of the following structures is the site of spermatogenesis?
   a. Prostate
   b. Bulbourethral gland
   c. Urethra
   d. Seminiferous tubule
3. Gametogenesis requires the conclusion of meiosis II. When does this occur in females?
   a. During fetal development
   b. At the onset of puberty
   c. After fertilization
   d. After implantation

4. Which of the following events occurs immediately after fertilization?
   a. Egg activation
   b. Polyspermy block
   c. Cytoplasm changes
   d. All of the above
5. A pluripotent cell is one that can
   a. become any cell type in an organism.
   b. produce an indefinite supply of a single cell type.
   c. produce a limited amount of a specific cell type.
   d. produce multiple cell types.
6. The fluid-filled chamber on the inside of a blastula is called the
   a. blastocyst.
   b. coelom.
   c. blastocoel.
   d. blastopore.
7. Which of the following plays the greatest role in determining how cytoplasmic division occurs during cleavage?
   a. Number of chromosomes
   b. Amount of yolk
   c. Orientation of the vegetal pole
   d. Sex of the zygote
8. Gastrulation is a critical event during development. Why?
   a. Gastrulation converts a hollow ball of cells into a bilaterally symmetrical structure.
   b. Gastrulation causes the formation of a primitive digestive tract.
   c. Gastrulation causes the blastula to develop a dorsal–ventral axis.
   d. All of the above
9. Somites
   a. begin forming at the tail end of the embryo and then move forward in a wavelike fashion.
   b. are derived from endoderm.
   c. develop into only one type of tissue per somite.
   d. may vary in number from one species to the next.
10. The transition from embryo to fetus occurs
    a. immediately after fertilization.
    b. after the first month.
    c. at the ninth week.
    d. at the seventh month.
11. Contractions of the uterus during labor are stimulated by which of the following hormones?
    a. Estrogen
    b. Prolactin
    c. Oxytocin
    d. Progesterone

Apply

1. Which of the following groups exhibits internal fertilization but never exhibits a live birth?
   a. Placental mammals
   b. Birds
   c. Amphibians
   d. Fish
2. If during development the testicles do not descend into the scrotum, a man might experience infertility, because
   a. the optimum temperature for sperm production is less than the normal core body temperature of the organism.
   b. the optimum temperature for sperm production is higher than the normal core body temperature of the organism.
   c. there is not enough room in the pelvic area for the testicles to be housed internally.
   d. it is easier for the body to expel sperm during ejaculation.
3. Which of the following is a major difference between spermatogenesis and oogenesis?
   a. Spermatogenesis involves meiosis, and oogenesis involves mitosis.
   b. Spermatogenesis is continuous, but oogenesis is variable.
   c. Spermatogenesis produces fewer gametes per precursor cell than oogenesis.
   d. All of the above
4. Mutations that affect proteins in the acrosome would impede which of the following functions?
   a. Fertilization
   b. Locomotion
   c. Meiosis
   d. Semen production
5. Assume you have the factors necessary to reprogram an adult cell, as well as the factors necessary to induce differentiation to any cell type. How could these be used to replace a specific damaged tissue in a human patient?
6. Suppose that a burst of electromagnetic radiation were to strike the blastomeres of only the animal pole of a frog embryo. Which of the following would be most likely to occur?
   a. A change or mutation relevant to the epidermis or skin
   b. A switching of the internal organs so that reverse orientation (left–right) occurs along the midline of the body
   c. The migration of the nervous system to form outside the body
   d. Failure of the reproductive system to develop
7. At the start of labor, a woman often experiences the breaking of an extraembryonic membrane that is commonly called "breaking the water." Which extraembryonic membrane is involved? Explain.
8. In a developing human, the first tissues to begin forming are the
   a. skeletal tissues.
   b. muscular tissues.
   c. nerve tissues.
   d. reproductive tissues.
9. Your Aunt Ida thinks that babies can stimulate the onset of their own labor. You tell her that
   a. among mammals the onset of labor has been most closely linked to a change in the phases of the Moon.
   b. it is the mother's circadian clock that determines the onset of labor.
   c. body weight determines the onset of labor.
   d. changes in fetal hormone levels can affect the onset of labor.

Synthesize

1. In reptiles and birds, the fetus is basically male, and fetal estrogen hormones are necessary to induce the development of female characteristics. In mammals, the reverse is true; the fetus is basically female, with fetal testosterone hormones acting to induce the development of male characteristics. Can you suggest a reason the pattern that occurs in reptiles would not work in mammals?
2. Why don't pregnant females menstruate?
3. Why are all parthenogenic parents female?
4. Given Dolly, why didn't frog nuclear transfers work?
5. Why don't mammalian eggs contain much yolk?
6. Why does gastrulation require cell movement?
7. Why is induction blocked by nonporous membranes?

Part VIII Ecology and Behavior

Mike Powles/Getty Images

37 Behavioral Biology

Learning Path

37.1 An Animal's Genome Influences Its Behavior

37.2 Learning Also Influences Behavior

37.3 Thinking Directs the Behavior of Many Animals

37.4 Migratory Behavior Is Both Innate and Learned

37.5 Animal Communication Plays a Key Role in Ecological and Social Behavior

37.6 Behavior Evolves Adaptively

37.7 Behavioral Ecology Examines the Adaptive Significance of Behaviors

37.8 Behavioral Strategies Have Evolved to Maximize Reproductive Success

37.9 Some Behaviors Decrease Fitness to Benefit Other Individuals

37.10 Group Living Has Evolved in Both Insects and Vertebrates

Concept Overview

This Concept Overview provides a diagram of the overarching concepts that are covered in this chapter. These concepts will be expanded upon in the Concept Overview at the end of the chapter.

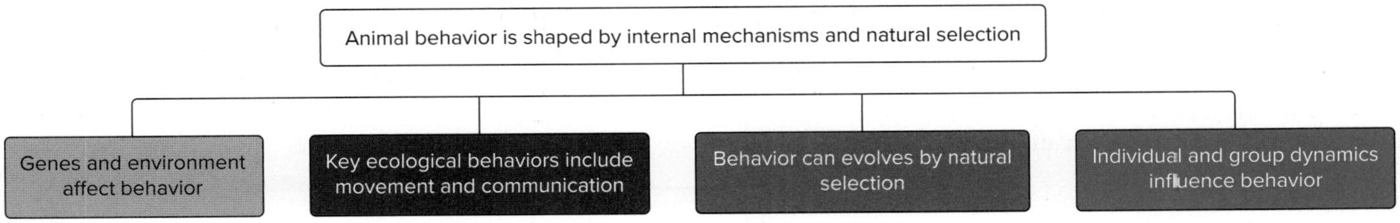
Animal behavior is shaped by internal mechanisms and natural selection
Genes and environment affect behavior
Key ecological behaviors include movement and communication
Behavior can evolves by natural selection
Individual and group dynamics influence behavior

Introduction

The study of behavior is at the center of many disciplines of biology. Observing behavior provides important insights into the workings of the brain and nervous system, the influences of genes and the environment, when and how animals reproduce, and how they adapt to their environment. Behavior is shaped by natural selection and is controlled by internal mechanisms involving genes, hormones, neurotransmitters, and neural circuits. In this chapter, we explore how behavioral biology integrates approaches from several branches of biological science to provide a detailed understanding of the mechanisms that underscore behavior and its evolution.

37.1 An Animal's Genome Influences Its Behavior

Observing animal behavior and making inferences about what one sees is at once simple and profound. Behavior is what an animal does. It is the most immediate way an animal responds adaptively to its environment. This can involve sensing environmental cues and signals, such as odors and sounds, or visual signals associated with food, predators, or mates. Behavior also concerns thinking and cognition, monitoring one's social environment, and making decisions as to whether or not to cooperate or to act altruistically. Behavior affects an animal's survival and reproductive success, making it critical to the evolutionary process.

Behavioral Differences Are Linked to Genetic Differences

LEARNING OBJECTIVE 37.1.1 Discuss the types of studies that have provided evidence to link genes and behavior.

The study of how an animal's genome influences its behavior has long been controversial: a seemingly endless debate over whether behavior is determined more by an individual's genes (nature) or by its learning and experience (nurture). The problem with this nature/nurture controversy is that the question is framed as an "either/or" proposition, which fails to consider that both instinct and experience can have significant roles, often interacting in complex ways to shape behavior.

Behavioral genetics deals with the contribution that heredity makes to behavior. It is obvious that genes, the units of heredity, are passed from one generation to the next and guide the development of the nervous system and potentially the behavioral responses it regulates. But animals may also develop in a rich social environment and have experiences that guide behavior. The relative importance of "nature" and "nurture" to behavior has been difficult to differentiate.

Artificial selection and hybrid studies link genes and behavior

Pioneering research indicated that behavioral differences among individuals result from genetic differences. Research on a variety of animals demonstrated that hybrids showed behaviors involved in nest building and courtship that were intermediate between those of parents. These early efforts to define the role of genes in behavior demonstrated that behavior can have a heritable component, but they fell short of identifying the genes involved. With the development of molecular biology, far greater precision was added to the analysis of the genetics of behavior.

Learning itself can be influenced by genes. In one classic study, rats had to find their way through a maze of blind alleys to the only exit, where a reward of food awaited them. Some rats quickly learned to zip through the maze to the food, making few mistakes, but other rats made many errors in learning the correct path. Researchers bred rats that made few errors with one another to establish a "maze-bright" group, and error-prone rats were interbred, forming a "maze-dull" group. Offspring in each group were then tested for their maze-learning ability. The offspring of maze-bright rats learned to negotiate the maze with fewer errors than their parents, but the offspring of maze-dull parents performed more poorly. Repeating this artificial selection method for several generations led to two behaviorally distinct types of rat with very different maze-learning abilities (figure 37.1). This type of study suggests the ways in which natural selection could shape behavior over time, making genes for certain abilities more prevalent.

Some Behaviors Appear to Be Controlled by a Single Gene

LEARNING OBJECTIVE 37.1.2 Explain how a single gene can influence behavior.

Artificial selection and hybrid studies only suggested a broad role for genes in behavior. Subsequent research has taken advantage of advances in molecular biology to identify the genes involved. Specific genes have been shown to influence particular behaviors in animals ranging from mice to humans.

Single genes in mice are associated with spatial memory and parenting. For example, some mice with a particular mutation have trouble remembering recently learned information about where objects are located. This forgetfulness results because

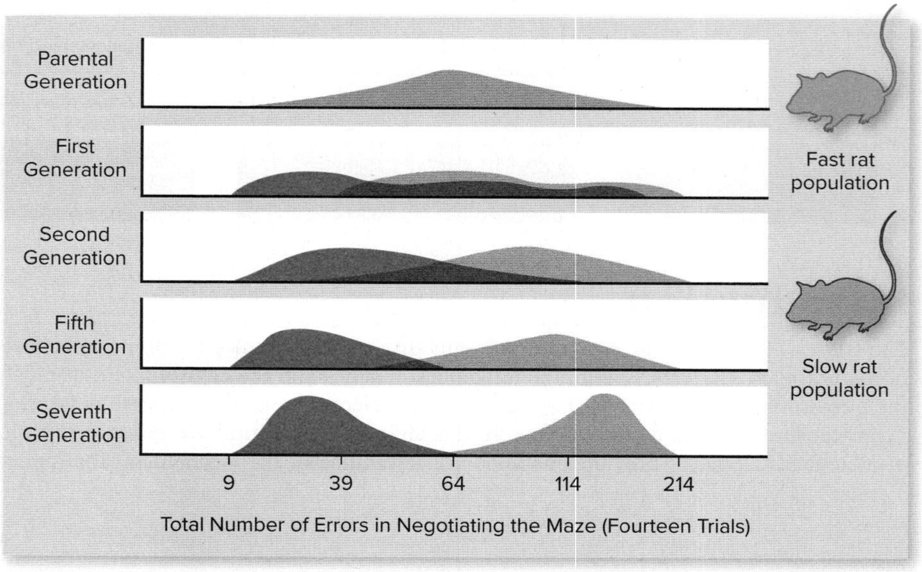

Figure 37.1 The genetics of learning. Rats that made the fewest errors in the parental population were interbred to select for rats that had improved maze-learning ability (blue), and rats that made the most errors were interbred to select for rats that were error-prone (red). The procedure was continued for seven generations.

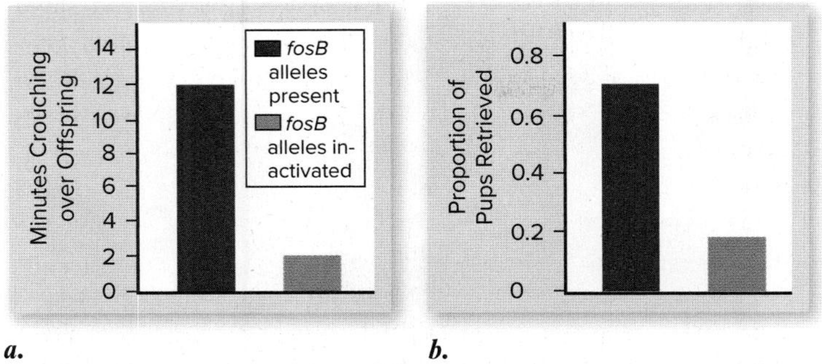

Figure 37.2 Genetically caused defect in maternal care. ***a.*** Amount of time female mice were observed crouching in a nursing posture over offspring. ***b.*** Proportion of pups retrieved when they were experimentally moved.

the mice lack the ability to produce the enzyme α-calcium-calmodulin-dependent kinase II, which plays an important role in the functioning of the hippocampus, a part of the brain important for spatial learning.

Of particular interest is the involvement of the *fosB* gene in a complex behavior: maternal care. The presence of functional *fosB* determines whether female mice nurture their young in particular ways. Females homozygous for inactive *fosB* alleles initially investigate their newborn babies but then ignore them, in contrast to the protective behavior displayed by normal females (figure 37.2). This inattentiveness appears to result from the interruption of a series of neural events. Mothers initially inspect new pups, receiving information from their auditory, olfactory, and tactile senses. These inputs to the hypothalamus cause *fosB* to be activated. The *fosB* gene encodes a protein that activates other enzymes and genes affecting neural circuitry in the hypothalamus. These neural modifications induce maternal behavior. If mothers lack *fosB* function, this process is blocked, the brain's neural circuitry is not rewired, and maternal behavior does not result. The "maternal instincts" of mice have a genetic basis!

Another fascinating example of the genetic basis of behavior concerns prairie and montane voles, two closely related species of North American rodents that differ substantially in their social behavior. Male and female prairie voles form monogamous pair bonds and share parental care, whereas montane voles are promiscuous (meaning they mate with multiple partners and go their separate ways). The act of mating leads to the release of the neuropeptides vasopressin and oxytocin, and the response to these peptides differs dramatically in each species. Injection of either peptide into prairie voles leads to pair-bonding even without mating. Conversely, injecting a chemical that blocks the action of these neuropeptides causes prairie voles not to form pair bonds after mating. By contrast, montane voles are unaffected by either of these manipulations.

These different responses have been traced to interspecific differences in brain structure (figure 37.3). The prairie vole has many receptors for these peptides in a particular part of the brain, the nucleus accumbens, which seems to be involved in the expression of pair-bonding behavior. By contrast, few such receptors occur in the same brain region in the montane vole. In laboratory experiments with prairie voles, blocking these receptors tends to prevent pair-bonding, whereas stimulating them leads to

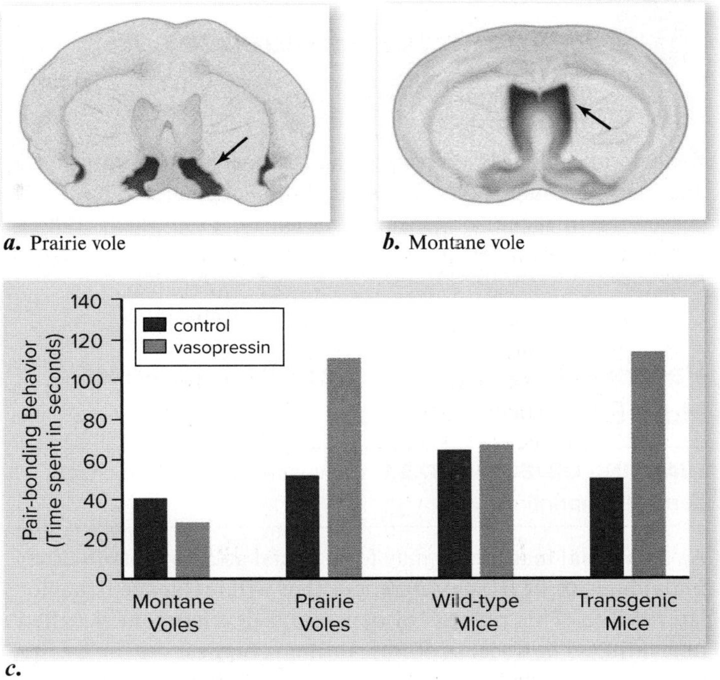

Figure 37.3 Genetic basis of differences in pair-bonding behavior in two rodent species. ***a, b.*** The prairie (*Microtus ochrogaster*) and montane (*M. montanus*) voles differ in the distribution of one type of vasopressin receptor in the brain. ***c.*** Transgenic mice created with the prairie-vole version of the receptor genes respond to injections of vasopressin by exhibiting heightened levels of pair-bonding behavior in 5-min trials compared with their response to a control injection. By contrast, normal wild-type mice (control) show no increase in such behaviors.

pair-bonding behavior. The gene that codes for the vasopressin receptor has also been identified and appears to differ between species. To test the hypothesis that this genetic difference is responsible for the differences in behavior, scientists created transgenic mice with the prairie vole vasopressin receptor and found that when injected with vasopressin, the transgenic mice exhibited pair-bonding behavior similar to that of prairie voles, whereas normal mice showed no response. The vasopressin receptor gene varies in structure among primate species that vary in degree of pair-bonding. In human males, the gene has recently been found to be associated with the strength of marital bonds and satisfaction in marriage.

REVIEW OF CONCEPT 37.1

A relationship between genes and behavior has been demonstrated in many ways, including artificial selection experiments and studies on the effects of single genes. Genes affect behavior by encoding proteins that influence the function of the nervous system; mutations altering these proteins affect behavior.

- *Suppose you wanted to create fruit flies that spent all of their time on the ground. How would you do it?*

37.2 Learning Also Influences Behavior

Behavioral biologists recognize that behavior has both genetic and learned components. Thus far in this chapter, we have discussed the influence of genes and of learning separately. But as you will see, these factors interact during development to shape behavior.

Parent–Offspring Interactions Influence How Behavior Develops

LEARNING OBJECTIVE 37.2.1 Discuss the role of the critical period in imprinting.

As an animal matures, it may form social attachments to other individuals or develop preferences that will influence behavior later in life. This process of behavioral development is called **imprinting.** The success of imprinting is highest during a critical period (roughly 13 to 16 hours after hatching, in geese). During this time, information required for normal development must be acquired. In **filial imprinting,** social attachments form between parents and offspring. For example, young birds such as ducks and geese begin to follow their mother within a few hours after hatching, and their following response results in a social bond between mother and young. The young birds' initial experience, through imprinting, can determine how social behavior develops later in life. The ethologist Konrad Lorenz showed that geese will follow the first object they see after hatching and direct their social behavior toward that object, even if it is not their mother! Lorenz raised geese from eggs, and when he offered himself as a model for imprinting, the goslings treated him as if he were their parent, following him dutifully (figure 37.4).

Figure 37.4 An unlikely parent. The eager goslings follow Konrad Lorenz as if he were their mother. He is the first object they saw when they hatched, and they have used him as a model for imprinting.

Thomas D Mcavoy/The LIFE Picture Collection/Shutterstock

Interactions between parents and offspring are key to the normal development of social behavior. The psychologist Harry Harlow gave orphaned Rhesus monkey infants the opportunity to form social attachments with two surrogate "mothers," one made of soft cloth covering a wire frame and the other made only of wire. The infants chose to spend time with the cloth mother, even if only the wire mother provided food, indicating that texture and tactile contact, rather than provision of food, may be among the key qualities in a mother that promote infant social attachment. If infant monkeys are deprived of normal social contact, their development is abnormal. Greater degrees of deprivation lead to greater abnormalities in social behavior during childhood and adulthood. Studies of orphaned human infants similarly suggest that a constant "mother figure" is important for growth and psychological development.

Recent research has revealed a biological need for the stimulation that occurs during parent–offspring interactions early in life. Female rats lick their pups after birth, and this stimulation inhibits the release of a brain peptide that can block normal growth. Pups that receive normal tactile stimulation also have more brain receptors for glucocorticoid hormones, thus a greater tolerance for stress, and longer-lived brain cells. Premature human infants who are massaged gain weight rapidly. These studies indicate that the need for social interaction is based in the brain, and that touch and other aspects of contact between parents and offspring are important for physical as well as behavioral development.

Studies of Twins Reveal a Role for Both Genes and Environment in Human Behavior

LEARNING OBJECTIVE 37.2.2 Explain the importance of identical twins to studies of behavior.

The interaction of genes and the environment can be seen in humans by comparing the behavior of identical twins (who are genetically the same), raised in the same environment or separated at birth and raised apart in different environments. Data on human twins raised together or raised apart allow researchers to determine whether similarities in behavior result from their genetic similarity or from shared environmental experiences. Twin studies indicate many similarities in a wide range of personality traits, even if twins were raised in very different environments. Other studies show that antisocial behavior in humans, for which genetic factors such as monoamine oxidase A (MAOA) deficiencies are known in individuals in the study sample, results from a combination of genes and experience during childhood. These similarities indicate that genetics plays a role in behavior even in humans, although the relative importance of genetics versus environment is still debated.

a. *b.*

Figure 37.5 Animal thinking? ***a.*** This chimpanzee is stripping the leaves from a twig, which it will then use to probe a termite nest. This behavior strongly suggests that the chimpanzee is consciously planning ahead, with full knowledge of what it intends to do. ***b.*** This sea otter is using a rock as an "anvil," bashing a clam against it to break the clam open. A sea otter will often keep a favorite rock for a long time, as though it has a clear idea of its future use of the rock. Behaviors such as these suggest that animals have cognitive abilities.

(a): Linda Koebner/Bruce Coleman Inc./Photoshot; (b): Tomalin, Norman.O/ Photoshot

REVIEW OF CONCEPT 37.2

During the critical period, offspring's engagement in certain social interactions affects behavioral development. Parent–offspring contact stimulates the release of physiological factors, such as hormones and brain receptors, crucial to growth and brain development.

■ *Why might identical twins behave differently?*

37.3 Thinking Directs the Behavior of Many Animals

For many decades, students of animal behavior flatly rejected the idea that nonhuman animals can think. Now serious attention is given to animal awareness, and the view that animals show **cognitive behavior**—that is, that they process information and respond in a manner that suggests thinking—is widely supported (figure 37.5).

Experiments Show Cognitive Behavior

LEARNING OBJECTIVE 37.3.1 Provide evidence for claims that nonhuman animals can think.

In a series of classic experiments conducted in the 1920s, a chimpanzee was left in a room with bananas hanging from the ceiling out of reach. Also in the room were several boxes lying on the floor. After some unsuccessful attempts to jump up and grab the bananas, the chimp stacked the boxes beneath the suspended bananas and climbed up to claim its prize (figure 37.6). Field researchers have observed that Japanese macaques learned to wash sand off potatoes and to float grain to separate it from sand. Chimpanzees pull leaves off a tree branch and then stick the branch into the entrance of a termite nest to "fish" for food (figure 37.5*a*). Chimps also crack open nuts using pieces of wood in a "hammer and anvil" technique. Even more remarkable is that parents appear to teach nut cracking to their offspring!

Recent studies have found that chimpanzees and other primates show amazing behaviors that provide strong evidence of cognition. Chimpanzees will eat the leaves of medicinal plants when infected with certain parasites. Chimps also cooperate with other chimps in ways that suggest an understanding of past success. Cognitive ability is not limited to primates: ravens and

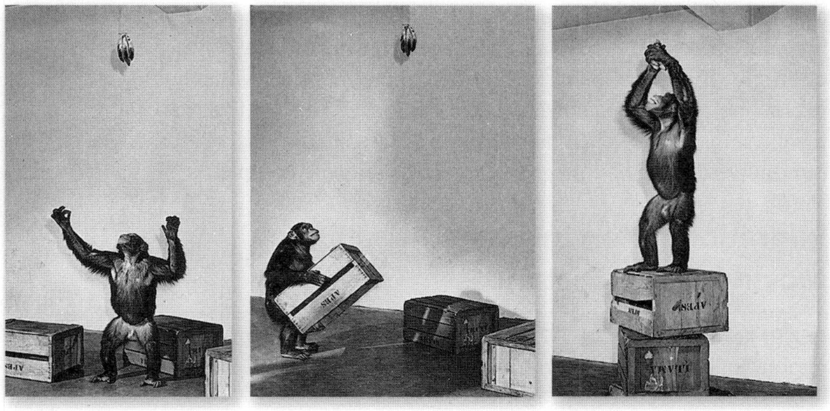

Figure 37.6 Problem solving by a chimpanzee. Unable to get the bananas by jumping, the chimpanzee devises a solution.

(all): 3LH /SuperStock

Figure 37.7 Problem solving by a raven. Confronted with a problem it has never previously faced, the raven figures out how to get the meat at the end of the string by repeatedly pulling up a bit of string and stepping on it.
Courtesy of Bernd Heinrich

related birds also show extraordinary insight and problem-solving ability (figure 37.7).

REVIEW OF CONCEPT 37.3

Research has provided compelling evidence that some nonhuman animals are able to solve problems and use reasoning, cognitive abilities once thought uniquely human.

- *Why is it important to tailor tests of cognition to the biology of the species being tested? Consider the difference between the experiments in figures 37.6 and 37.7.*

37.4 Migratory Behavior Is Both Innate and Learned

Monarch butterflies and many birds travel thousands of miles over continents to overwintering sites in the tropics. Many animals travel away from a nest during one season and then return to that nest during another season. To do so, they track cues in the environment, often showing exceptional skill at orientation. Animals with a homing instinct, such as pigeons, use recognition of complex features of the environment to return to their home. Despite decades of study, our understanding of animal orientation is far from complete.

Migration Often Involves Populations Moving Large Distances

LEARNING OBJECTIVE 37.4.1 Illustrate the process of migration, using a real-world example.

Long-range, two-way movements are known as migrations. Each fall, ducks, geese, and many other birds migrate south along flyways from Canada across the United States, heading as far as South America and then returning each spring.

Monarch butterflies also migrate each fall from central and eastern North America to their overwintering sites in several small, geographically isolated areas of coniferous forest in the mountains of central Mexico (figure 37.8). Each August, the butterflies begin a flight southward, and at the end of winter the monarchs begin the return flight to their summer breeding ranges. Two to five generations may be produced as the butterflies fly north: butterflies that migrate in the autumn to the precise locations in Mexico have never been there before!

Recent geographical range expansions by some migrating birds have revealed how migratory patterns change. When colonies of bobolinks became established in the western United

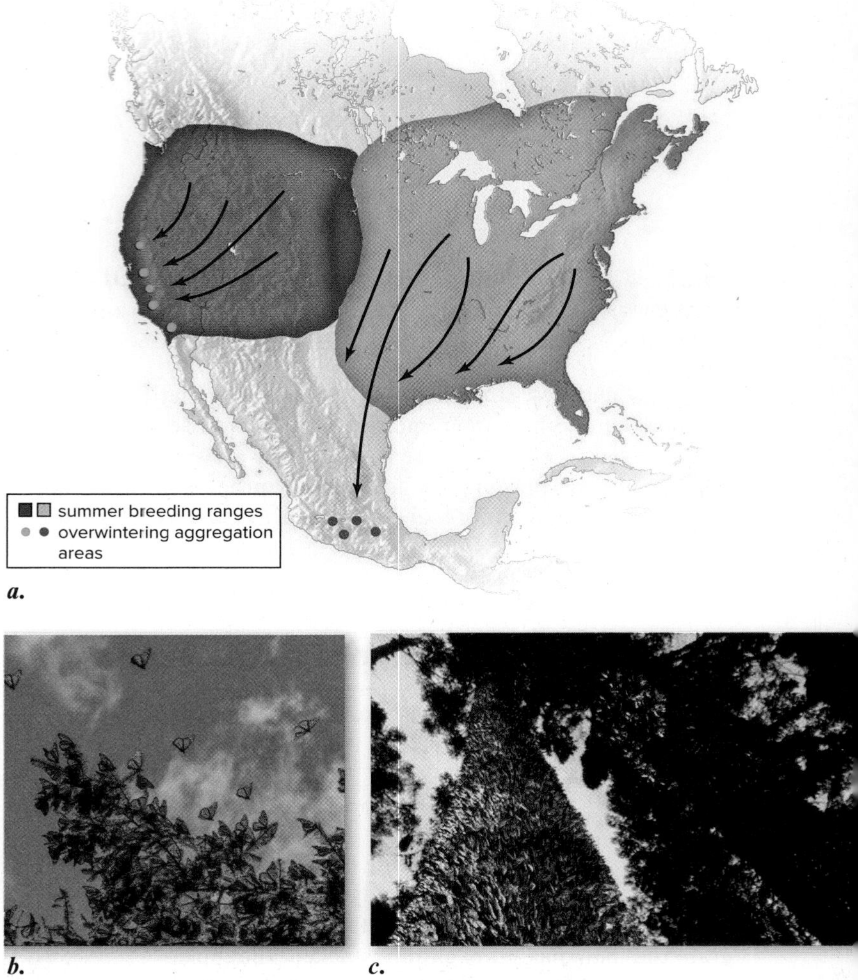
summer breeding ranges
overwintering aggregation areas
a.
b.
c.

Figure 37.8 Migration of monarch butterflies (*Danaus plexippus*). ***a.*** Monarchs from western North America overwinter in areas of mild climate along the Pacific coast. Those from eastern North America migrate over 3000 km to Mexico. ***b.*** Monarch butterflies arrive at the remote forests of the overwintering grounds in Mexico, where they **(c)** form aggregations on the tree trunks.
(b): JHVEPhoto/Shutterstock; (c): George Lepp/Getty Images

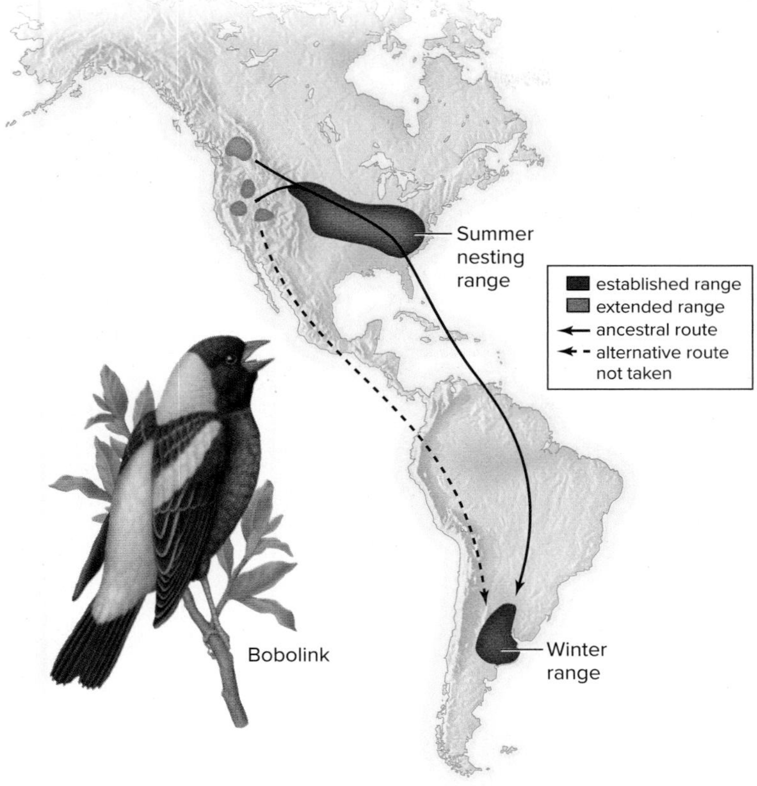

Summer
nesting
range
established range
extended range
ancestral route
alternative route
not taken
Bobolink
Winter
range

Figure 37.9 Birds on the move. The summer range of bobolinks (*Dolichonyx oryzivorus*) was recently extended to the far western United States from their established range in the Midwest. When birds in these newly established populations migrate to South America in the winter, they do not fly directly to the winter range; instead, they first fly to the Midwest and then use the ancestral flyway, going much farther than if they had flown directly to their winter range.

States, far from their normal range in the Midwest and East, they did not migrate directly to their winter range in South America. Instead, they migrated east to their ancestral range, then south along the original flyway (figure 37.9). Rather than changing the original migration pattern, they simply added a new segment. Scientists continue to study the western bobolinks to learn whether, in time, a more efficient migration path will evolve or whether the birds will always follow their ancestral course. The behavior of butterflies and birds accentuates the mysteries of the mechanism employed during migration.

Migrating Animals Must Be Capable of Orientation and Navigation

LEARNING OBJECTIVE 37.4.2 Distinguish between orientation and navigation.

To get from one place to another, animals must have a "map" (that is, know where to go) and a "compass" (use environmental cues to guide their journey). Orientation requires following a bearing such as a source of light, but navigation is the ability to set or adjust a bearing and then follow it. The former is analogous to using a compass, whereas the latter is like using a compass in conjunction with a map. The nature of the "map" that animals use is unclear.

Birds and other animals navigate by looking at the Sun during the day and the stars at night. For example, the indigo bunting uses the Sun as a guide and can compensate for movement of the Sun in the sky as the day progresses. At night, these birds use the constellations around the North Star as a compass for navigation.

Many migrating birds also have the ability to detect Earth's magnetic field and to orient themselves with respect to it when cues from the Sun or stars are not available. In an indoor cage, they will attempt to move in the correct geographical direction, even though there are no visible external cues. However, the placement of a magnet near the cage can alter the direction in which the birds attempt to move. Researchers have found magnetite, a magnetized iron ore, in the eyes and upper beaks of some birds; in addition, molecules in the eyes of some birds may also be able to detect magnetic fields. How these sensory organs function is an area of active research.

A bird's first migration appears to be innately guided by both celestial cues (birds fly mainly at night) and Earth's magnetic field. When the two cues are experimentally manipulated to give conflicting directions, the information provided by the stars seems to override the magnetic information. Recent studies, however, indicate that celestial cues indicate the general direction for migration, whereas magnetic cues indicate the specific migratory path (perhaps a turn the bird must make midroute). Experiments on starlings indicate that inexperienced birds migrate by orientation, but older birds that have migrated previously use true navigation (figure 37.10).

We know relatively little about how other migrating animals navigate. Green sea turtles migrate from Brazil halfway across the Atlantic Ocean to Ascension Island. How do they find this tiny island in the middle of the ocean, which they haven't seen in 30 years? How do the young that hatch on the island find their way to Brazil? Newly hatched turtles use wave action as a cue to head to sea, and some sea turtles use the Earth's magnetic field to maintain position in the North Atlantic, but turtle migration is still largely a mystery.

REVIEW OF CONCEPT 37.4

Migration is the long-distance movement of a population, often in a cyclic way. *Orientation* refers to following a bearing or a direction; navigation involves setting a bearing or direction based on some sort of map or memory. Many species use celestial navigation; they may also be able to detect magnetic fields when those cues are absent. The precision of animal migration remains a mystery for many species.

- *Animals as diverse as butterflies and birds migrate over long distances. Would you expect them to use different navigation systems? Why or why not?*

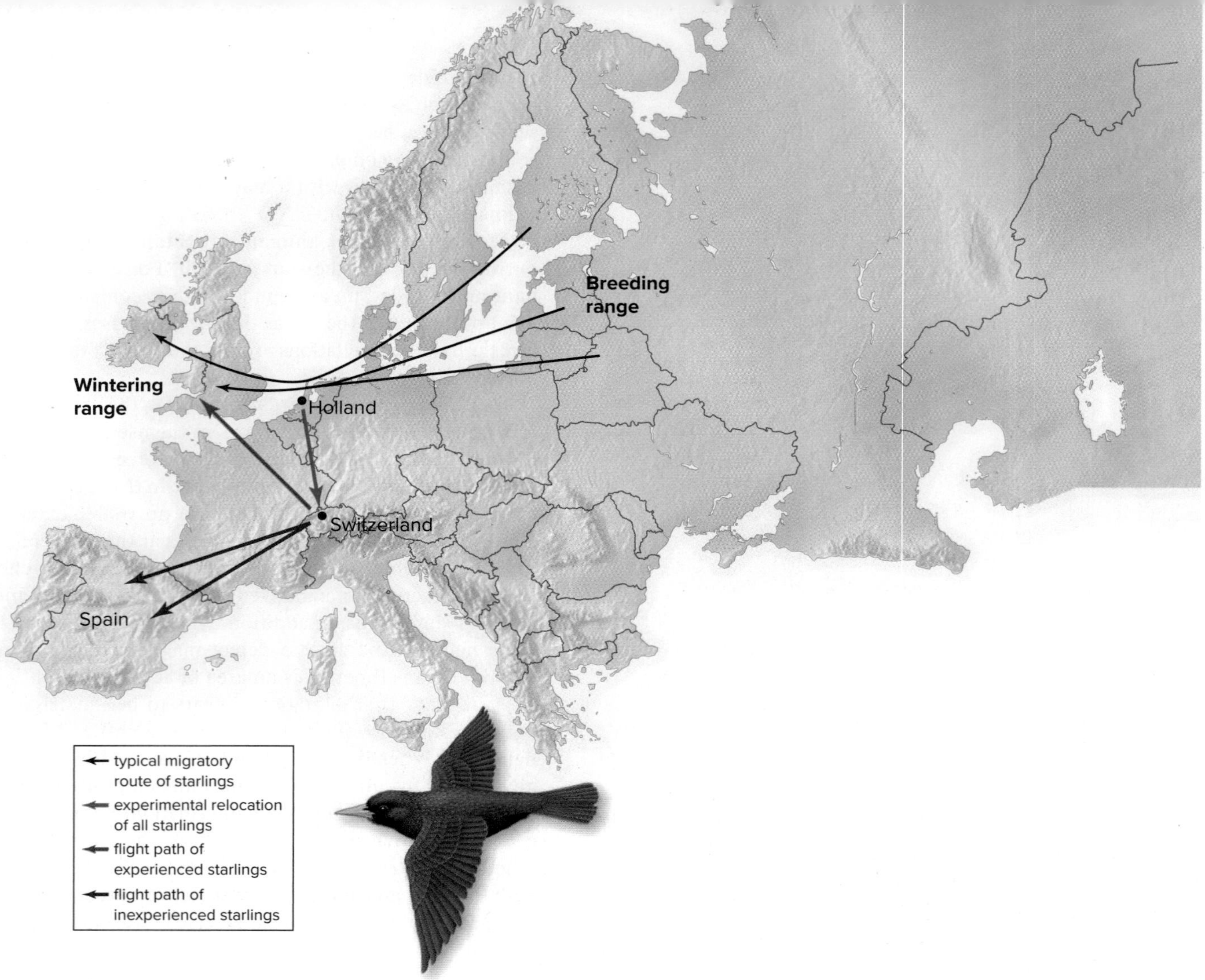

Figure 37.10 Migratory behavior of starlings (*Sturnus vulgaris*). The navigational abilities of inexperienced birds differ from those of adults that have made the migratory journey before. Starlings were captured in Holland, halfway along their full migratory route from Baltic breeding grounds to wintering grounds in the British Isles; these birds were transported to Switzerland and released. Experienced older birds compensated for the displacement and flew toward the normal wintering grounds (blue arrow). Inexperienced young birds flew in the direction they were headed prior to capture, a course that took them toward Spain (red arrows). These observations imply that inexperienced birds fly by orientation but experienced birds learn true navigation.

37.5 Animal Communication Plays a Key Role in Ecological and Social Behavior

Communication is central to species recognition and reproductive isolation and to the interactions essential to social behavior. Much research in behavior analyzes the nature of communication signals, determining how they are produced and received, and identifies their ecological roles and evolutionary origins. Communication involves several signal modalities, including visual, acoustic, chemical, electrical, and vibrational signals.

Successful Reproduction Depends on Appropriate Signals and Responses

LEARNING OBJECTIVE 37.5.1 Explain the role of courtship signals in reproductive isolation.

During courtship, animals produce signals to communicate with potential mates and with other members of their own sex. A **stimulus–response chain** sometimes occurs, in which the behavior of the female releases a behavior in the male, which in turn releases a behavior in the female, which then releases a behavior in the male, and so on, eventually, resulting in mating. These signals are usually highly species-specific.

Many studies on communication involve designing experiments to determine which key stimuli associated with an animal's visual appearance, sounds, or odors convey information about the nature of the signals produced by the sender.

Finding a mate: Communicating information about species identity

Courtship signals often restrict communication to members of the same species and, in doing so, serve a key function in reproductive isolation (refer to chapter 21). The flashes of fireflies are species-specific signals: females recognize conspecific males by their flash pattern (figure 37.11), and males recognize conspecific females by their flash response. This series of reciprocal responses provides a continuous "check" on the species identity of potential mates.

Pheromones, chemical messengers used for communication between individuals of the same species, serve as sex attractants in many animals. Female silk moths (*Bombyx mori*) produce a sex pheromone called bombykol in a gland associated with the reproductive system. The male's antennae contain numerous highly sensitive sensory receptors, which neurophysiological studies show specifically detect bombykol. In some moth species, males can detect extremely low concentrations of sex pheromone and locate females from as far as 7 km away!

Many insects, amphibians, and birds produce species-specific acoustic signals to attract mates. Bullfrog males call by inflating and discharging air from their vocal sacs, located beneath the lower jaw. Females can distinguish a conspecific male's call from those of other frogs that may be in the same habitat and calling at the same time. Male birds sing to advertise their presence and to attract females. In many species, variations in the males' songs identify individual males in a population. In these species, the song is individually specific as well as species-specific. Vibrations, like sound signals, are a form of mechanical communication used by insects, amphibians, and other animals.

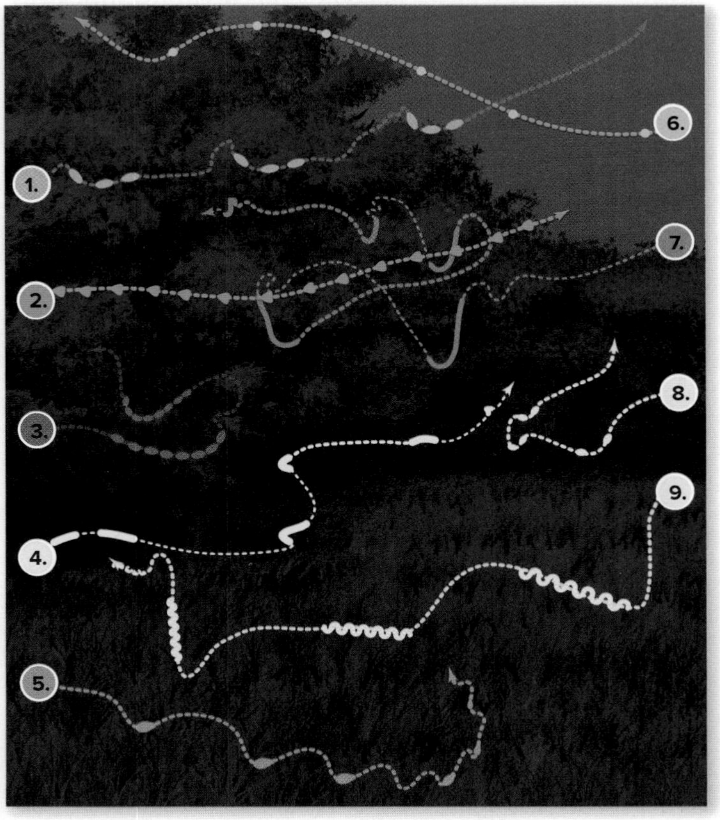

Figure 37.11 Firefly fireworks. The bioluminescent displays of these lampyrid fireflies are species-specific and serve as behavioral mechanisms of reproductive isolation. Each number represents the flash pattern of a male of a different species.

Communication Enables Information Exchange Among Group Members

LEARNING OBJECTIVE 37.5.2 Describe how honeybees communicate information on the location of food sources.

Many insects, fish, birds, and mammals live in social groups in which information is communicated between group members. For example, some individuals in mammalian societies serve as sentinels, vigilantly on the lookout for danger. When a predator appears, they give an alarm call, and group members respond by seeking shelter (figure 37.12). Social insects such as ants and honeybees produce alarm pheromones that trigger attack behavior. Ants also deposit trail pheromones between the nest and a food source to lead other colony members to food. Honeybees have an extremely complex dance language that directs hivemates to nectar sources.

The dance language of the honeybee

The European honeybee lives in colonies of tens of thousands of individuals whose behaviors are integrated into a complex, cooperative society. Worker bees may forage miles from the hive, collecting nectar and pollen from a variety of plants and switching between plant species depending on their energetic rewards. Food sources used by bees tend to occur in patches, and each patch offers much more food than a single bee can transport to the hive. A colony is able to exploit the resources of a patch because of the behavior of scout bees, which locate patches and communicate their location to hivemates through a dance language.

Figure 37.12 Alarm calling by a prairie dog (*Cynomys ludovicianus*). When a prairie dog sees a predator, it stands on its hind legs and gives an alarm call, which causes other prairie dogs to rapidly return to their burrows.

W. Perry Conway/Getty Images

a.

b.

Figure 37.13 The waggle dance of honeybees (*Apis mellifera*). ***a.*** The angle between the food source, the nest, and the Sun is represented by a dancing bee as the angle between the straight part of the dance and vertical. The food is 30° to the right of the Sun, and the straight part of the bee's dance on the hive is 30° to the right of vertical. ***b.*** A scout bee dances on a comb in the hive.

(b): Scott Camazine/Science Source

When a scout bee returns after finding a distant food source, she performs a remarkable behavior pattern called a waggle dance on a vertical comb in the darkness of the hive. The path of the bee during the dance resembles a figure-eight. On the straight part of the path (indicated with dashes in figure 37.13), the bee vibrates ("waggles") her abdomen while producing bursts of sound. The bee may stop periodically to give her hivemates a sample of the nectar carried in her crop. As she dances, she is followed closely by other bees, which soon appear at the new food source to assist in collecting food.

Experiments have demonstrated how hivemates use information in the waggle dance to locate new food sources. The scout bee indicates the direction of the food source by representing the angle between the food source, the hive, and the Sun as the deviation from vertical of the straight run of the dance performed on the hive comb. Thus, if the bee danced with the straight run pointing directly up, then the food source would be in the direction of the Sun. If the food were at a 30° angle to the right of the Sun's position, then the straight run would be oriented upward at a 30° angle to the right of vertical (figure 37.13*a*). The distance to the food source is indicated by the duration of the straight run. Computer-controlled robot bees have been used in ingenious experiments to give hivemates incorrect information: tricked bees unaware of the location of food followed the incorrect directions given by the robot scout bee's dance, demonstrating that bees use the directions coded in the dance!

Language in nonhuman primates and humans

Evolutionary biologists have sought the origins of human language in the communication systems of monkeys and apes. Some nonhuman primates have a "vocabulary" that allows individuals to signal the identity of specific predators. Different vocalizations of African vervet monkeys, for example, indicate eagles, leopards, or snakes, among other threats (figure 37.14).

a.

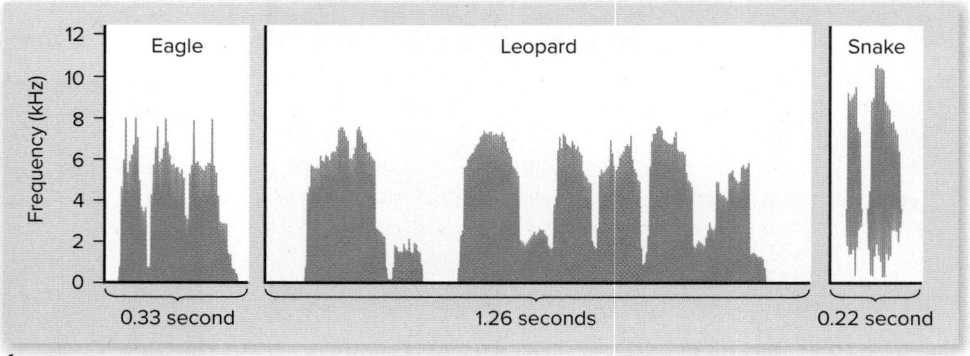

b.

Figure 37.14 Primate semantics. Vervet monkeys (*Cercopithecus aethiops*) give different alarm calls (***a***) when troop members sight an eagle, a leopard, or a snake. ***b.*** Each distinctive call elicits a different and adaptive escape behavior.

(a): The Africa Image Library/Alamy Stock Photo

The complexity of human language would at first appear to defy biological explanation, but closer examination suggests that the differences are, in fact, superficial—all languages share many basic similarities. All of the roughly 3000 languages draw from the same set of 40 consonant and vowel sounds (English uses two dozen of them), and humans of all cultures can acquire and learn them. Researchers believe these similarities reflect the way our brains handle abstract information.

REVIEW OF CONCEPT 37.5

Animal communication involves the production and reception of signals, in the form of sounds, chemicals, or movements, that primarily have an ecological function. Courtship signals are highly species-specific and serve as a mechanism of reproductive isolation. Animals living in social groups, such as honeybees, use complex systems of communication to exchange information about food and predators.

- *Two species of moth use the same sex pheromone to locate mates. Explain how these species can nevertheless be reproductively isolated.*

37.6 Behavior Evolves Adaptively

Like other phenotypes, behavioral traits can evolve by natural selection. The prerequisites for evolution by natural selection we learned in chapter 19 apply to behavior just as they do to morphological or biochemical traits. If there is genetic variation in a population that affects behavior and differences in this behavior cause differences in fitness, then natural selection can lead to evolution of the behavior.

Natural Selection Leads to Behavioral Evolution

LEARNING OBJECTIVE 37.6.1 Understand how natural selection leads to behavioral evolution.

Artificial selection experiments in the laboratory easily produce evolutionary changes in behavior. For example, when given a choice in a maze between flying up or down, fruit flies normally don't have a strong preference one way or the other. However, when researchers chose those individuals that tended more often to fly upward and allowed them to breed with each other, after 15 generations of such artificial selection, all of the flies almost always flew upward when given a choice. Similarly, selection for downward-flying flies just as quickly produced a population in which all flies almost always took the downward choice.

A particularly striking example comes from a 40-year study on captive foxes in Siberia. At the start of the experiment, all foxes reacted aggressively when a caretaker approached their cages. Nonetheless, some foxes were less aggressive than others, and these behavioral differences were consistent among trials and were genetically based: offspring inherited the aggressiveness of their parents. Every generation, researchers selected the least aggressive foxes and allowed them to breed. Forty years later, the friendly, tail-wagging, face-licking foxes are more akin to domesticated dogs than to their vulpine ancestors (refer to figure 20.9).

Researchers have also documented natural selection operating on behavioral differences among individuals in nature. For example, some bighorn sheep are bolder than others, and when mountain lion predation is common, the bolder animals have higher survival. Similarly, when sitting on a nest, Montagu's harriers (a type of hawk) differ in their response to approaching humans: some consistently remain and react aggressively, whereas others fly away. In one study, the birds that remained on the nest raised more offspring—another example of natural selection favoring one behavioral trait over another. Because variations in behavior such as these often are the result of genetic differences, such natural selection can lead to evolutionary change. Indeed, the proportion of harriers that fled from approaching humans decreased by more than 75% over the 19-year study.

Phylogenies Can Be Used to Study How Behavior Evolves Through Time

LEARNING OBJECTIVE 37.6.2 Describe how phylogenetic methods allow inference of behavior in the past.

Scientists are also using phylogenetic methods (refer to chapter 22) to study behavioral evolution. One interesting example involves poison dart frogs, small, colorful, and highly toxic frogs from Central and South America (figure 37.15). These frogs exhibit several interesting parental care behaviors: either the male or the female stays with the eggs ("attending"), keeping them moist, rotating and sometimes sitting on them. Once the eggs hatch, one of the parents—not necessarily the same one that was attending the eggs—transports the tadpoles to the water-filled plant cavities in which they live. In addition, in some cases, the female visits these cavities and lays infertile eggs, which the tadpoles then eat. In two species, all parental care is provided by females, but in other cases, some of the tasks are performed by males.

Given the distribution of these traits across species, it was not clear how they evolved. However, examination of a phylogeny makes clear that male transportation evolved first and was subsequently replaced in some clades by females taking over that duty. In addition, tadpole feeding evolved convergently twice, once in an ancestral species in which the male did the transporting and the other time in a female-transporting ancestor.

REVIEW OF CONCEPT 37.6

Natural selection can cause behavioral evolution when variation exists in a behavioral trait, that variation affects fitness, and that variation is genetically heritable. Phylogenies can allow reconstruction of the evolutionary history of a trait.

- *Some butterflies are attracted to bananas and others aren't. The ones that eat bananas survive longer and have more offspring. Explain why the trait might not evolve so that all butterflies in the population like bananas.*

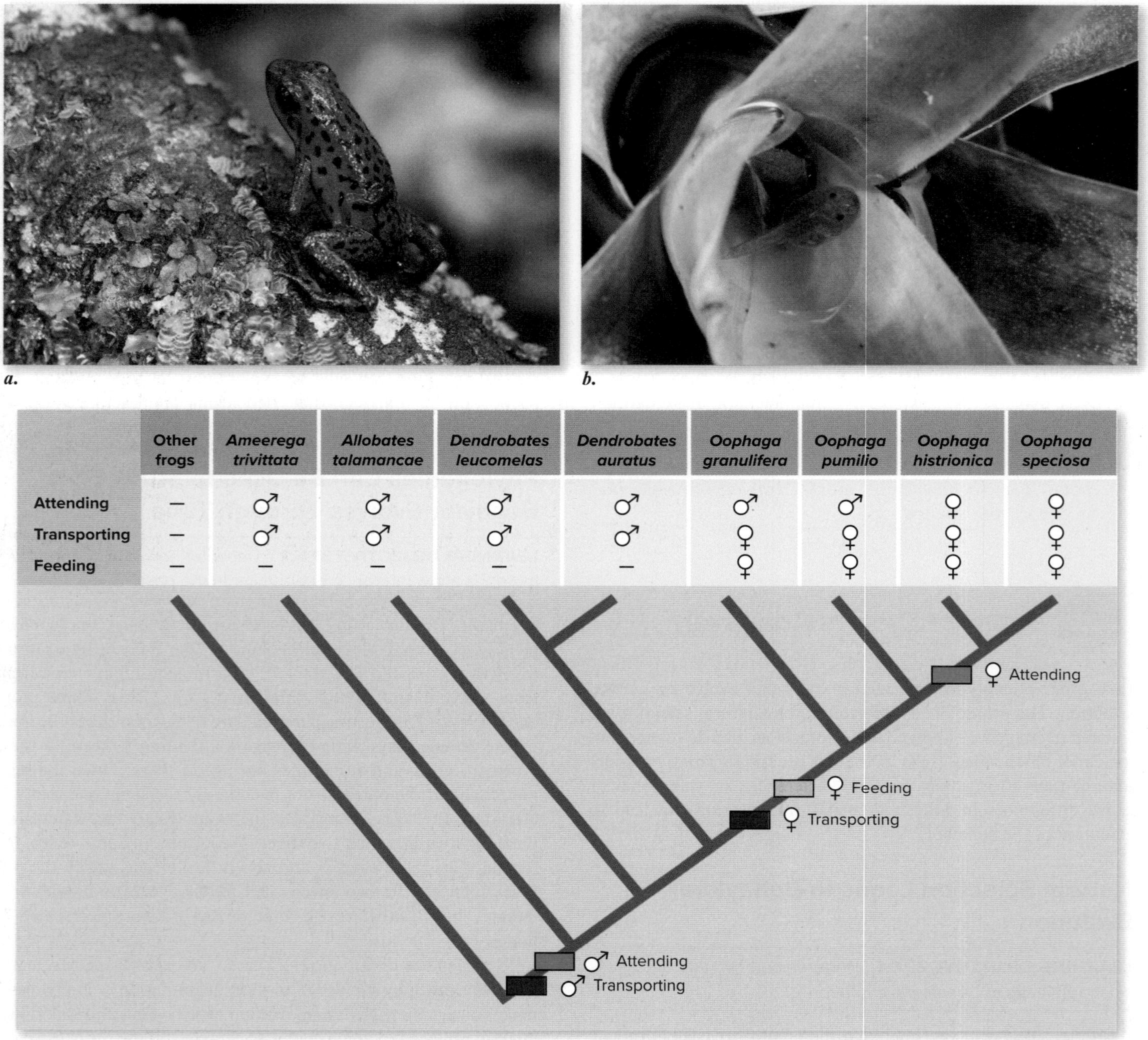
a.
b.
Other frogs
Ameerega trivittata
Allobates talamancae
Dendrobates leucomelas
Dendrobates auratus
Oophaga granulifera
Oophaga pumilio
Oophaga histrionica
Oophaga speciosa
Attending
Transporting
Feeding
♀ Attending
♀ Feeding
♀ Transporting
♂ Attending
♂ Transporting
c.

Figure 37.15 Evolution of parental care in poison dart frogs. Parents attend to the eggs, keeping them moist and rotating them, and ***(a)*** transport the hatchling tadpoles to the ***(b)*** water-filled plant cavities in which they develop. ***c.*** Phylogenetic analysis reveals the evolutionary history of parental care in these frogs.

(a) Riccardo Oggioni/Alamy Stock Photo; (b) Francesco Tomasinelli/Science Source

37.7 Behavioral Ecology Examines the Adaptive Significance of Behaviors

Behavior May Increase Fitness in Many Ways

LEARNING OBJECTIVE 37.7.1 Describe behavioral ecology.

Biologist Niko Tinbergen pioneered the study of the adaptive function of behavior. Stated simply, this is how behavior allows an animal to stay alive and reproduce successfully. For example, Tinbergen observed that after gull nestlings hatch, the parents remove the eggshells from the nest. To understand why (ultimate causation), he painted chicken eggs to resemble gull eggs (figure 37.16), which had camouflage coloration to allow them to be inconspicuous against the natural background. He distributed them throughout the area in which the gulls were nesting, placing broken eggshells with their prominent white

Figure 37.16 The adaptive value of egg coloration. Niko Tinbergen painted chicken eggs to resemble the mottled brown camouflage of gull eggs. The eggs were used to test the hypothesis that camouflaged eggs are more difficult for predators to find and thus increase the young's chances of survival.
Nina Leen/The LIFE Picture Collection/Shutterstock

interiors next to some of the eggs. As a control, he left other camouflaged eggs alone without eggshells. He then noted which eggs were found more easily by crows. Because the crows could use the white interior of a broken eggshell as a cue, they ate more of the camouflaged eggs that were near eggshells. Tinbergen concluded that eggshell removal behavior is adaptive: it reduces predation and thus increases the offspring's chances of survival.

Tinbergen is credited with being one of the founders of **behavioral ecology,** the study of how natural selection shapes behavior. This field examines the adaptive significance of behavior, or how behavior may increase survival and reproduction.

The field of behavioral ecology is concerned with two questions. First, is behavior adaptive? Although it is tempting to assume that behavior must in some way represent an adaptive response to the environment, this need not be the case. As you explored in chapter 19, traits can appear for many reasons other than natural selection, such as genetic drift, gene flow, or the correlated consequences of selection on other traits. Moreover, some traits that are no longer useful may still be present in a population because they evolved as adaptations in the past. These possibilities hold true for behavioral traits as much as for any other kind of trait.

If behavior is adaptive, the next question is, How is it adaptive? Behavioral ecologists are interested in how behavior can lead to greater reproductive success. Does a behavior enhance energy intake, thus increasing the number of offspring produced? Does it increase mating success? Does it decrease the chance of predation? Benefits and costs of behaviors, estimated in terms of energy or offspring, are often used to analyze the adaptive nature of behavior.

Foraging Behavior Can Directly Influence Energy Intake and Individual Fitness

LEARNING OBJECTIVE 37.7.2 Discuss the cost–benefit analysis of feeding behaviors.

A useful way to understand the approach of behavioral ecology is by focusing on foraging behavior. For many animals, food comes in a variety of sizes. Larger foods may contain more energy but may be harder to capture and less abundant. In addition, animals may forage for some types of food that are farther away than other types. For these animals, foraging involves a trade-off between a food's energy content and the cost of obtaining it. The net energy (estimated in calories or joules) gained by feeding on prey of each size is simply the energy content of the prey minus the energy costs of pursuing and handling it. According to **optimal foraging theory,** natural selection favors individuals whose foraging behavior is as energetically efficient as possible. In other words, animals tend to feed on prey that maximize their net energy intake per unit of foraging time.

A number of studies have demonstrated that foragers do prefer prey that maximize energy return. Shore crabs, for example, tend to feed primarily on intermediate-size mussels, which provide the greatest energy return; larger mussels yield more energy but also take considerably more energy to crack open. An experiment investigating the efficiency of this behavior is analyzed in this chapter's Inquiry & Analysis feature at the end of the chapter.

This optimal foraging approach assumes that natural selection will favor behavior that maximizes energy acquisition if the increased energy reserves lead to increases in reproductive success. In both Colombian ground squirrels and captive zebra finches, a direct relationship exists between net energy intake and the number of offspring raised; similarly, the reproductive success of orb-weaving spiders is related to how much food they can capture.

Animals have other needs besides energy, however, and sometimes these needs conflict. One obvious need is the avoidance of predators: often, the behavior that maximizes energy intake is not the one that minimizes predation risk. In this case, the behavior that maximizes fitness may reflect a trade-off between obtaining the most energy and having the least risk of being eaten. Not surprisingly, many studies have shown that a wide variety of animal species alter their foraging behavior when predators are present. Compromises, in this case a trade-off between vigilance and feeding, may thus be made during foraging.

Optimal foraging theory assumes that energy-maximizing behavior has evolved by natural selection. Therefore, it must have a genetic basis. For example, female zebra finches particularly successful in maximizing net energy intake tend to have similarly successful offspring. As young birds were removed from their mothers before they were able to leave the nest, this similarity indicates that foraging behavior probably has a genetic component. Studies on other animals show that age, experience, and learning are also important to the development of efficient foraging.

Territorial Behavior Evolves If the Benefits of Holding a Territory Exceed the Costs

LEARNING OBJECTIVE 37.7.3 Explain the evolutionary benefit of territorial behavior.

Animals often move over a large area, or home range, during their course of activity. In many species, the home range of several individuals overlaps in time or in space, but each individual defends a portion of its home range and uses it and its resources exclusively. This behavior is called **territoriality** (figure 37.17).

The defining characteristic of territorial behavior is defense against intrusion and resource use by other individuals. Territories are defended by displays advertising that territories are occupied, as well as by overt aggression. A bird sings from its perch within a territory to prevent takeover by a neighboring bird. If a potential usurper is not deterred by the song, the territory owner may attack and try to drive it away. But territorial defense has its costs. Singing is energetically expensive, and attacks can lead to injury. Using a signal (a song or visual display) to advertise occupancy can reveal a bird's position to a predator.

Why does an animal bear the costs of territorial defense? Benefits of territoriality may take the form of increased food intake due to exclusive use of resources, access to mates that live within the territory, or access to refuges from predators. Studies of nectar-feeding birds such as hummingbirds and sunbirds provide an example (figure 37.18). A bird benefits from having the exclusive use of a patch of flowers because it can efficiently harvest the nectar the flowers produce. To maintain exclusive use, the bird must actively defend the patch. The benefits of exclusive use outweigh the costs of defense only under certain conditions.

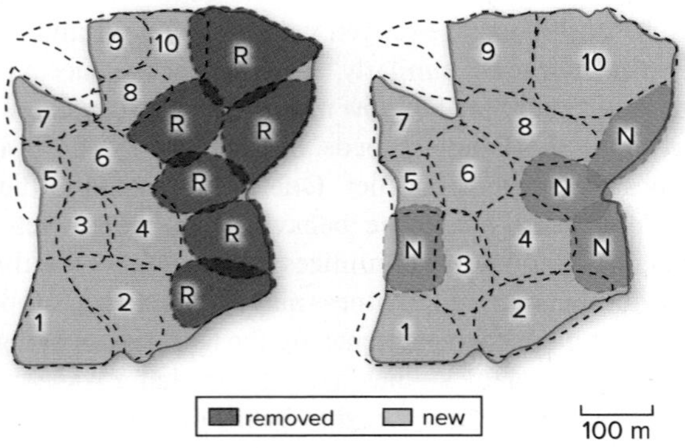

Figure 37.17 Competition for space. Territory size in birds is adjusted according to the number of competitors. When six pairs of great tits (*Parus major*) were removed from their territories (indicated by *R* in the left figure), their territories were taken over by other birds in the area and by four new pairs (indicated by *N* in the right figure). Numbers correspond to the birds present before and after.

Figure 37.18 The benefit of territoriality. Sunbirds (on the left), found in Africa and ecologically similar to New World hummingbirds (on the right), protect their food source by attacking other sunbirds that approach flowers in their territory.
(African sunbird): H Lansdown/Alamy Stock Photo;
(New World hummingbird): Gerald C. Kelley/Science Source

Sunbirds expend 3000 calories per hour chasing intruders from a territory. Whether the benefit of defending a territory exceeds this cost depends on the amount of nectar in flowers and how efficiently a bird collects it. When flowers are scarce or nectar levels are low, a nectar-feeding bird may not gain enough energy to balance the cost of defense. Under these conditions, it is not energetically advantageous to be territorial. Similarly, when flowers are abundant, a bird can efficiently meet its daily energy requirements without behaving territorially and adding the cost of defense—from an energetic standpoint, defending abundant resources isn't worth the cost, either. Territoriality therefore occurs at intermediate levels of flower availability and nectar production, when the benefits of defense outweigh the costs.

In many species, access to females is a more important determinant of territory size for males than food availability. In some lizards, males maintain enormous territories during the breeding season. These territories, which encompass the territories of several females, are much larger than required to supply enough food, and they are defended vigorously. In the nonbreeding season, by contrast, male territory size decreases dramatically, as does aggressive territorial behavior.

REVIEW OF CONCEPT 37.7

Behavioral ecology is the study of the adaptive significance of behavior—that is, how behavior affects survival and reproductive success. An economic approach estimates the energy benefits and costs of a behavior and assumes that animals gain more from a behavior than they expend, obtaining a fitness advantage. Foraging behavior and defense of a territory can be analyzed in this way. Apart from energy gains, considerations such as avoiding predators are also important to fitness.

- *The Hawaiian honeycreeper, a nectar-feeding bird, does not defend flowers that are either infrequently encountered or very abundant. Why?*

37.8 Behavioral Strategies Have Evolved to Maximize Reproductive Success

During the breeding season, animals make several important life-history "decisions" concerning their choice of mates, how many mates to have, and how much time and energy to devote to rearing offspring. These decisions are all aspects of an animal's **reproductive strategy,** a set of behaviors that presumably have evolved to maximize reproductive success.

Mating Systems Reflect the Ability of Parents to Care for Offspring

LEARNING OBJECTIVE 37.8.1 Explain why some species are generally monogamous and others are polygynous.

The number of individuals with which an animal mates during the breeding season varies among species. Mating systems include monogamy (one male mates with one female), polygyny (one male mates with more than one female; refer to figure 19.22), and polyandry (one female mates with more than one male). Only monogamous mating includes a pair bond (as seen in prairie voles). Like mate choice, mating systems have evolved to allow females and males to maximize fitness.

The option of having more than one mate may be constrained by the need for offspring care. If females and males are able to care for young, then the presence of both parents may be necessary for young to be reared successfully. Monogamy may thus be favored. Generally, this is the case for birds, in which over 90% of all species appear to be monogamous. A male may either remain with his mate and provide care for the offspring or desert that mate to search for others; both strategies may increase his fitness. The strategy that natural selection will favor depends on the requirement for male assistance in feeding or defending the offspring. In some species (such as humans), offspring are **altricial**—they require prolonged and extensive care. In these species, the need for care by two parents reduces the tendency for the male to desert his mate and seek other matings. In species in which the young are **precocial** (requiring little parental care), males may be more likely to be polygynous, because the need for their parenting is lower. In mammals, only females lactate, freeing males from feeding offspring. It follows that most mammals are polygynous.

Mating systems are strongly influenced by ecology. A male may defend a territory that holds nest sites or food sources sufficient for more than one female. If territories vary in quality or quantity of resources, a female's fitness is maximized if she mates with a male holding a high-quality territory, even if he has mated. Although a male may already have a mate, it is still more advantageous for the female to breed with a mated male holding a high-quality territory than for her to breed with an unmated male holding a low-quality territory, although she will make exceptions. This favors the evolution of polygyny.

Polyandry is relatively rare, but the evolution of multiple mating by females is becoming better understood. It is best known in birds like spotted sandpipers and jacanas living in highly productive environments such as marshes and wetlands. Here, females take advantage of the increased resources available to rear offspring by laying clutches of eggs with more than one male. Males provide all incubation and parenting, and females mate and leave eggs with two or more males.

Extra-Pair Copulations Increase Fitness in Some Monogamous Species

LEARNING OBJECTIVE 37.8.2 Explain why some individuals engage in extra-pair copulations.

The apparent "monogamy" of many bird species has been reevaluated as DNA fingerprinting (refer to chapter 17) has become commonly used to determine paternity and precisely quantify the reproductive success of individual males (figure 37.19*a*). In red-winged blackbirds (figure 37.19*b*), researchers established that half of all nests contained at least one hatchling fertilized by a male other than the territory owner; overall, 20% of the offspring were the result of such **extra-pair copulations (EPCs).**

What is the evolutionary advantage of EPCs? For males, the answer is obvious: increased reproductive success. Females, on the other hand, may mate with genetically superior individuals even if already paired with a male, thus enhancing the genes passed on to their offspring. The female produces not more offspring, but offspring of better genetic quality.

In some birds and other animals, EPCs may help females increase the amount of care they get from males to raise their offspring. This is exactly what happens in the dunnock, *Prunella modularis,* a common English bird. Females mate not only with the territory owner, but also with subordinate males that hang around the edge of the territory. If these subordinates mate a sufficient number of times with a female, they will help raise her young, presumably because they may have fathered some of these young.

Natural Selection Sometimes Favors Alternative Mating Strategies

LEARNING OBJECTIVE 37.8.3 Explain why alternative mating strategies can increase fitness.

Natural selection has led to the evolution of many ways of increasing reproductive success. For example, many species of fish have two genetic classes of males. One group is large and defends territories to obtain matings. The other group is small and adopts a completely different strategy. These males do not maintain territories, but loiter at the edge of the territories of large males. Just at the end of a male's courtship, when the female is laying her eggs and the territorial male is depositing sperm, the smaller male darts in and releases its own sperm into the water, thus fertilizing some of the eggs. If this strategy is successful, natural selection will favor the evolution of these two different male reproductive strategies.

Similar patterns are seen in other organisms. In some dung beetles, territorial males have large horns that they use to guard the chambers in which females reside, whereas genetically small males don't have horns. The smaller males can't possibly hope to outcompete the larger ones in one-on-one battles. However, they

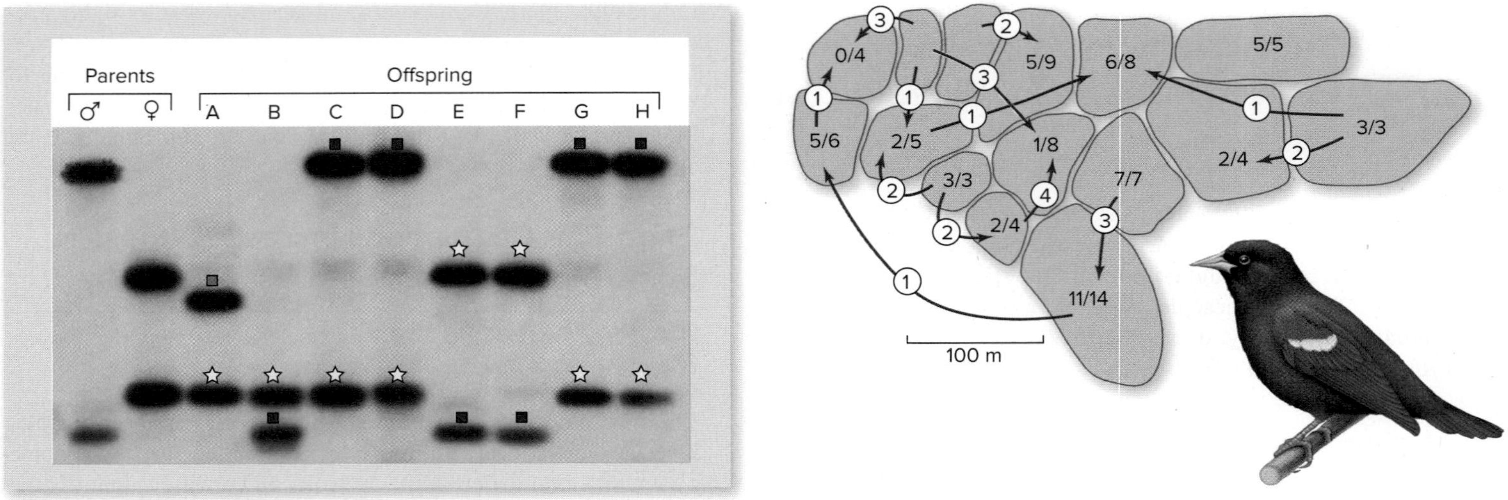

a. *b.*

Figure 37.19 The study of paternity within a territory. ***a.*** DNA fingerprinting gel from parents and offspring in a nest of pied flycatchers (*Ficedula hypoleuca*). The bands are DNA fragments of different lengths, which are alleles of a single locus. The female that lays the egg and her mate are both heterozygous and carry different alleles, which allows the parents to be identified. All offspring have one of the female's bands, indicated by a yellow star, as would be expected, given that the eggs are in her nest. Seven of the offspring have an allele from her mate, indicated by a red symbol, but the eighth, A, has an allele not carried by her mate. This indicates that this bird was the result of an extra-pair mating with another male in the population. ***b.*** Results of a DNA fingerprinting study in red-winged blackbirds (*Agelaius phoeniceus*). Fractions indicate the proportion of offspring fathered by the male in whose territory the nest occurred. Arrows indicate how many offspring were fathered by particular males outside of each territory.

(a): Dr. Douglas Ross

have a different strategy. Because of the size of their horns, the large males cannot enter the females' chambers. Taking advantage of this, the smaller males dig side tunnels and attempt to intercept and mate with the female inside her chamber. Isopods take this one step further by having three genetic size classes. The medium-sized males pass for females and enter a large male's territory in this way; males in the smallest class are so tiny, they are able to sneak in completely undetected.

A small lizard from the southwestern United States, the side-blotched lizard, illustrates another unusual mating system (figure 37.20), in this case with three different, genetically based, male strategies. One type of male is the dominant, aggressive male. These males have large territories containing many females. A second type of male is more sedentary. It has a smaller territory, usually with only one female that he guards very carefully. Finally, sneaker males have no territories. They move around a great deal and try to sneak into other males' territories to mate with the females within.

Figure 37.20 A male side-blotched lizard, *Uta stansburiana*.

Bill Gorum/Alamy Stock Photo

It turns out that none of these strategies is superior. Rather, the dominant males, being aggressive, have an advantage over the more sedentary males. However, because dominant males are so active and have such large territories, they are vulnerable to sneakers, who can slip in and mate with some of the females in their territories. Sedentary males, in turn, have an advantage over sneakers; because the sedentary males zealously guard their females, sneakers have little opportunity to mate.

The result is a situation akin to the common playground game of rock-paper-scissors. In this case, dominant beats sedentary, sedentary beats sneaker, and sneaker beats dominant. Scientists have followed populations of the side-blotched lizard through time and have found that the frequency of the three types changes in a predictable fashion. When there is a high frequency of dominants, sneakers do well and sedentaries do poorly. However, as the frequency of sneakers increases, dominants decrease and sedentaries increase. In time, as the dominants become rare and the sedentaries common, sneakers are then at a disadvantage, and dominants begin to rise in frequency, and the cycle begins anew. This is, incidentally, an example of the phenomenon of negative frequency-dependent selection discussed in chapter 19, in which each type of male has highest fitness when it is rare.

This is just a glimpse of the rich diversity in mating systems and mating tactics that have evolved. The bottom line is: if there is a way of increasing reproductive success, natural selection will favor its evolution.

REVIEW OF CONCEPT 37.8

Reproductive success influences whether males and females mate monogamously or with multiple partners. In some circumstances, both extra-pair copulations and alternative mating tactics are beneficial because they increase fitness.

- *Under what circumstances is an alternative mating tactic likely to be favored?*

37.9 Some Behaviors Decrease Fitness to Benefit Other Individuals

Understanding the evolution of altruism has been a particular challenge to evolutionary biologists, including Darwin himself. Why should an individual decrease his or her own fitness to help another? How could genes for altruism be favored by natural selection, given that the frequency of such genes should decrease in populations through time? In fact, there can be great benefits to being an altruist, even if the altruism leads an individual to forgo reproduction or even sacrifice its own life. Let's examine how this can work.

Reciprocity Theory Can Explain Altruism Between Unrelated Individuals

LEARNING OBJECTIVE 37.9.1 Explain altruism and its potential benefits.

Altruism is behavior that benefits another individual at a cost to the actor. Humans sacrificing themselves in times of war or placing themselves in jeopardy to help their children are examples, but altruism has been described in a wide variety of organisms. In many bird species, there are "helpers at the nest"—birds other than parents that assist in raising the young. In both mammals and birds, individuals that spy a predator may give an alarm call, helping others to escape, even though such an act might attract the predator's attention to the caller. In social insects like ants, workers are sterile offspring that help their mother, the colony's queen, reproduce.

Once it was thought that altruism evolved for the "good of the species." Individuals that fail to mate have even been called "altruists," with their lack of success in competition interpreted as a willingness to forgo reproduction so that the population or species does not increase in size, exhaust its resources, and go extinct. This explanation fails to explain how an allele for altruism could increase in frequency in a population if possessors of that allele have decreased fitness compared to non-altruistic individuals. Quite the contrary: we would expect natural selection to remove such alleles from the population.

Current studies of altruism note that seemingly altruistic acts can be selfish. For example, helpers at the nest are often young birds that gain valuable parenting experience by assisting established breeders; this may give them an advantage when they breed. Moreover, they may have limited opportunities to reproduce on their own, and by hanging around breeding pairs, they may inherit the territory when established breeders die.

Figure 37.21 Truth is stranger than fiction: Reciprocal altruism in vampire bats (*Desmodus rotundus*). Vampire bats do feed on the blood of large mammals, but they don't transform into people and sleep in coffins. Vampires live in groups and share blood meals. They remember which bats have provided them with blood in the past and are more likely to share with those bats that have shared with them previously. The bats here are feeding on cattle in Brazil.

Nick Gordon/ardea.com

One explanation of altruism proposes that genetically unrelated individuals may form "partnerships" in which mutual exchanges of altruistic acts occur because they benefit both participants. Partners are willing to give aid at one time and delay "repayment" for the good deed to a time in the future when they themselves are in need. In **reciprocal altruism,** the partnerships are stable because "cheaters" (nonreciprocators) are discriminated against and do not receive future aid. According to this hypothesis, if the altruistic act is relatively inexpensive, the small benefit a cheater receives by not reciprocating is far outweighed by the potential cost of not receiving future aid. Under these conditions, cheating behavior should be eliminated by selection.

Vampire bats roost in hollow trees, caves, and mines in groups of 8 to 12 individuals (figure 37.21). Because bats have a high metabolic rate, individuals that have not fed recently may die. Bats that have found a host imbibe a great deal of blood, so giving up a small amount to keep a roostmate from starvation presents no great energy cost to the donor. Vampire bats tend to share blood with past reciprocators that are not necessarily relatives. If an individual fails to give blood to a bat from which it received blood in the past, it will be excluded from future bloodsharing. Reciprocity routinely occurs in many primates, including humans.

Kin Selection Theory Proposes a Direct Genetic Advantage to Altruism

LEARNING OBJECTIVE 37.9.2 Explain kin selection and inclusive fitness.

The great population geneticist J. B. S. Haldane once passionately said in a pub that he would willingly lay down his life for two brothers or eight first cousins.

Evolutionarily speaking, this sacrifice makes sense, because for each allele Haldane received from his parents, his brothers each had a 50% chance of receiving the same allele (figure 37.22). Statistically, it is expected that two of his brothers would pass on as many of Haldane's particular combination of alleles to the next generation as Haldane himself would. Similarly, Haldane and a first cousin would share an eighth of their alleles (see figure 37.22). Their parents, who are siblings, would each share half their alleles, and each of their children would receive half of these, of which half on the average would be in common: $1/2 \times 1/2 \times 1/2 = 1/8$. Eight first cousins would therefore pass on as many of those alleles to the next generation as Haldane himself would.

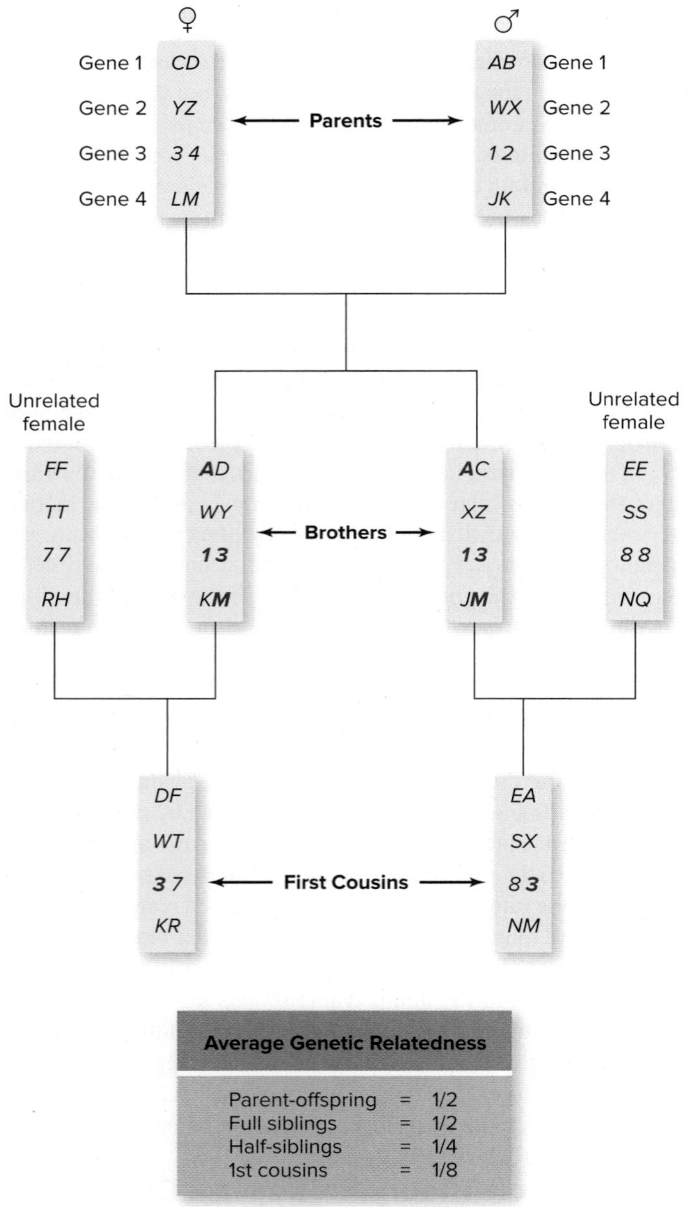

Figure 37.22 Hypothetical example of genetic relationships. On average, full siblings share half of their alleles. By contrast, cousins share only one-eighth of their alleles on average. Each letter and number represents a different allele.

The most compelling explanation for the kin-related origin of altruism was presented by one of the most influential evolutionary biologists of our time, William D. Hamilton, in 1964. Hamilton understood Haldane's point: natural selection will favor any behavior, including the sacrifice of life, that increases the propagation of an individual's alleles.

Hamilton mathematically showed that by directing aid toward close genetic relatives, an altruist may increase the reproductive success of its relatives enough to not only compensate for the reduction in its own fitness but even increase its fitness beyond what would be possible without assisting relatives. Because the altruist's behavior increases the propagation of alleles in relatives, it will be favored by natural selection. Selection that favors altruism directed toward relatives is called **kin selection.** Although the behaviors are altruistic, the genes are actually "behaving selfishly," because they encourage the organism to favor the success of copies of themselves in relatives. In other words, if an individual has a dominant allele that causes altruism, that allele is likely to be possessed by relatives as well; any action that increases the frequency of this allele in future generations will be favored, even if that action is detrimental to the actor.

Hamilton then defined reproductive success with a new concept: inclusive fitness. **Inclusive fitness** considers gene propagation through both direct reproduction (personal fitness) and indirect reproduction (the fitness of relatives). Hamilton's kin selection model predicts that altruism is likely to be directed toward close relatives. The more closely related two individuals are, the greater the potential genetic payoff, and the greater inclusive fitness. This is described by Hamilton's rule, which states that altruistic acts are favored when $rb > c$. In this expression, b and c are the benefits and costs of the altruistic act, respectively, and r is the coefficient of relatedness, the proportion of alleles shared by two individuals through common descent. For example, an individual should be willing to have one less child ($c = 1$) if such actions allow a half-sibling, which shares one-quarter of its genes ($r = 0.25$), to have five or more additional offspring ($b = 5$).

Haplodiploidy and altruism in ants, bees, and wasps

The relationship among genetic relatedness, kin selection, and altruism is most easily understood using social insects as an example. A hive of honeybees consists of a single queen, which is the sole egg-layer, and tens of thousands of her offspring, female workers with nonfunctional ovaries (figure 37.23). Honeybees are eusocial ("truly" social): their societies are defined by reproductive division of labor (only the queen reproduces), cooperative care of the brood (workers nurse, clean, and forage), and overlap of generations (the queen lives with several generations of her offspring).

Darwin was perplexed by eusociality. How could natural selection favor the evolution of sterile workers that left no offspring? Hamilton explained the origin of eusociality in hymenoptera (bees, wasps, and ants) in terms of his kin selection model. In these social species, males are haploid, produced by parthenogenetic activation of unfertilized eggs. Females are both diploid and produced by sperm from a single male (queens store sperm from a single mating). This system, called haplodiploidy, leads to unusual genetic

Figure 37.23 Reproductive division of labor in honeybees. The queen (center) is the sole egg-layer. Her daughters are sterile.

Ivan Marjanovic/Shutterstock

Figure 37.24 Kin selection in the white-fronted bee-eater (*Merops bullockoides*). Bee-eaters are small, insectivorous birds that live in Africa in large colonies. Bee-eaters often help others raise their young; helpers usually choose to help close relatives.

Imagebroker/Alamy Stock Photo

relatedness among colony members. If the queen is fertilized by a single male, then all female offspring inherit exactly the same alleles from their father. These females are the workers, as well as the queens of future colonies. They share, on average, half the alleles they get from their mother and all of the alleles they get from their father. Thus, females share 75% of their alleles with their sisters (to verify this, rework figure 37.22, but allow the father to have only one allele for each gene).

Now recall Haldane's statement of commitment to family while you read this section. If a worker should have offspring of her own, she would share only half of her alleles with her young (the other half would come from their father), in contrast to the three-quarters sharing of alleles with her sisters. Thus, because of this close genetic relatedness due to haplodiploidy, workers would propagate more of their own alleles by giving up their own reproduction to assist their mother in rearing their sisters, some of whom will be new queens, start new colonies, and reproduce.

In this way, the unusual haplodiploid system may have set the "genetic stage" for the evolution of eusociality. Indeed, eusociality has evolved at least 12 separate times in the Hymenoptera. One wrinkle in this theory, however, is that eusocial systems have evolved in other insects (thrips, weevils, and termites) and in some mammals (naked mole rats). Although thrips are also haplodiploid, termites and naked mole rats are not. Thus, although haplodiploidy may have facilitated the evolution of eusociality, other factors can influence social evolution.

Other examples of kin selection

Kin selection may explain altruism in other animals. Belding's ground squirrels give alarm calls when they spot a predator such as a coyote or a badger. Such predators may attack a calling squirrel, so giving the signal places the caller at risk. A ground squirrel colony consists of a female and her daughters, sisters, aunts, and nieces. When males mature, they disperse long distances from where they are born, so adult males in the colony are not genetically related to the females. By marking all squirrels in a colony with an individual dye pattern on their fur and by recording which individuals gave calls and the social circumstances of their calling, researchers found that females that have relatives living nearby are more likely to give alarm calls than females with no kin nearby. Males tend to call much less frequently, as would be expected, because they are not related to most colony members.

Another example is provided by the white-fronted bee-eater, a bird that lives along river banks in Africa in colonies of 100 to 200 individuals (figure 37.24). In contrast to ground squirrels, male bee-eaters usually remain in the colony in which they were born, and females disperse to join new colonies. Many bee-eaters do not raise their own offspring but instead help others. Most helpers are young, but older birds whose nesting attempts have failed may also be helpers. The presence of a single helper, on average, doubles the number of offspring that survive. Two lines of evidence support the idea that kin selection is important in determining helping behavior. First, helpers are normally males, which are usually related to other birds in the colony, and not females, which are not related. Second, when birds have the choice of helping different parents, they almost invariably choose the individuals most closely related to them.

REVIEW OF CONCEPT 37.9

Genetic and ecological factors have contributed to the evolution of altruism. Individuals may benefit directly if cooperative acts are reciprocated. Kin selection explains how altruistic acts directed toward relatives, which share alleles, increase an individual's inclusive fitness. Haplodiploidy has resulted in eusociality among some insects by increasing genetic relatedness.

- *You witness older members rescuing infants in a troupe of monkeys when a predator appears. How would you test whether this altruistic act is reciprocity or kin selection?*

37.10 Group Living Has Evolved in Both Insects and Vertebrates

Organisms from cnidarians and insects to fish, whales, chimpanzees, and humans live in social groups. To encompass the wide variety of social phenomena, we can broadly define a **society** as a group of organisms of the same species that are organized in a cooperative manner.

Insect Societies Form Efficient Colonies Containing Specialized Castes

LEARNING OBJECTIVE 37.10.1 Explain the possible advantages of group living.

Why have individuals in some species given up a solitary existence to become members of a group? One hypothesis is that individuals in groups benefit directly from social living. For example, a bird in a flock may be better protected from predators. As flock size increases, the risk of predation decreases because there are more individuals to scan the environment for predators (figure 37.25).

A member of a flock may also increase its feeding success if it can acquire information from other flock members about the location of new, rich food sources. In some predators, hunting in groups can increase success and allow the group to tackle prey too large for any one individual.

We've already discussed the origin of eusociality in the insect order Hymenoptera (ants, bees, and wasps). Additionally, all termites (order Isoptera) and a few other insect and arthropod species are eusocial. Social insect colonies are composed of different *castes,* groups of individuals that differ in reproductive ability, size, and morphology and perform different tasks. Workers nurse, maintain the nest, and forage; soldiers are large with powerful jaws specialized for defense.

Figure 37.26 Castes of ants. These leaf-cutter ants are members of different castes. The large ant is a worker carrying leaves to the nest, whereas the smaller ants protect the worker.
Amazon-Images/Alamy Stock Photo

The structure of an insect society is illustrated by leaf-cutters, which form colonies of as many as several million individuals. These ants cut leaves and use them to grow crops of fungi beneath the ground. Workers divide the tasks according to their body size: leaf cutting, defense, mulching the fungus garden, and implanting fungal hyphae (figure 37.26).

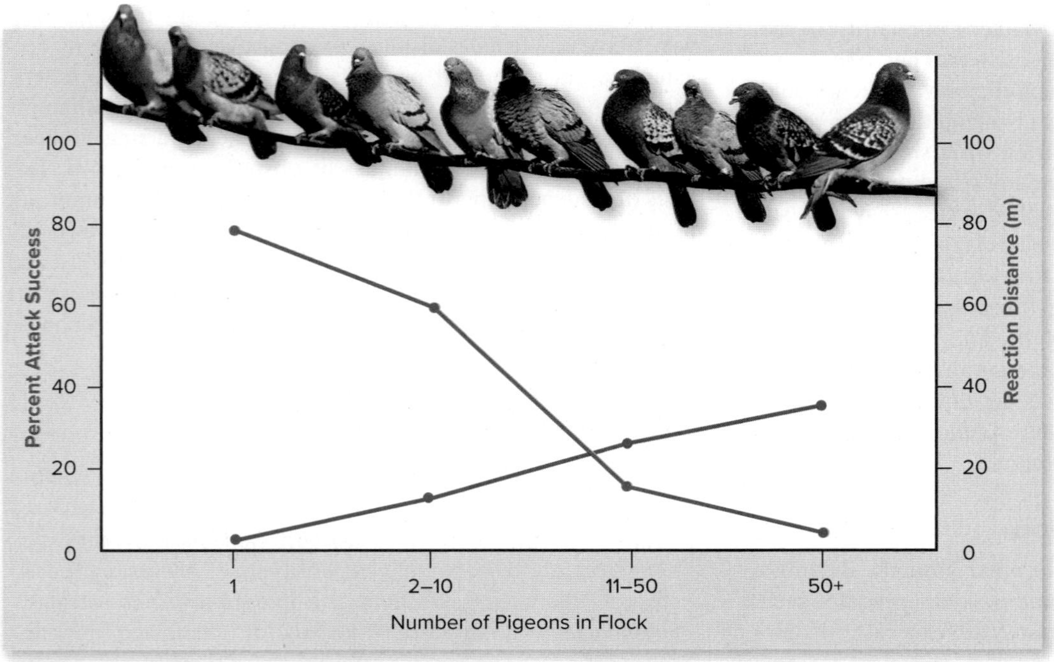

Figure 37.25 Flocking behavior decreases predation. When more pigeons are present in the flock, they can detect hawks at greater distances, allowing more time for escape. As a result, as the size of a pigeon flock increases, hawks are less successful at capturing pigeons.
Francois Gohier/Science Source

Vertebrate Societies Are Not Rigidly Organized

LEARNING OBJECTIVE 37.10.2 Contrast the natures of vertebrate and insect societies.

In contrast to the highly structured and integrated insect societies and their remarkable forms of altruism, vertebrate social groups are usually less rigidly

Figure 37.27 Foraging and predator avoidance. A meerkat sentinel on duty. Meerkats (*Suricata suricata*) are a species of highly social mongoose living in the semiarid sands of the Kalahari Desert in southern Africa. This meerkat is taking its turn to act as a lookout for predators. Under the security of its vigilance, the other group members can focus their attention on foraging.
Nigel Dennis/Science Source

organized and less cohesive. It seems paradoxical that vertebrates are generally less altruistic than insects (the exception is humans). Reciprocity and kin-selected altruism are common in vertebrate societies, although there is often more conflict among group members. Conflicts generally center on access to food and mates and occur because a vertebrate society is made up of individuals striving to improve their own fitness.

Social groups of vertebrates have a size, stability of members, number of breeding males and females, and type of mating system characteristic of a given species. Diet and predation are important factors in shaping social groups. For example, meerkats take turns watching for predators while other group members forage for food (figure 37.27).

African weaver birds, which construct nests from vegetation, provide an excellent example of the relationship between ecology and social organization. Their roughly 90 species can be divided according to the type of social group they form. One group of species lives in the forest and builds camouflaged, solitary nests. Males and females are monogamous; they forage for insects to feed their young. The second group of species nests in colonies in trees on the savanna. They are polygynous and feed in flocks on seeds.

The feeding and nesting habits of these two groups of species are correlated with their mating systems. In the forest, insects are hard to find, and both parents must cooperate in feeding the young. The camouflaged nests do not call the attention of predators to their brood. On the open savanna, building a hidden nest is not an option. Rather, savanna-dwelling weaver birds protect their young from predators by nesting in trees, which are not very abundant. This shortage of safe nest sites means that birds must nest together in colonies. Because seeds occur abundantly, a female can acquire all the food needed to rear young without a male's help. The male, free from the duties of parenting, spends his time courting many females—a polygynous mating system.

Figure 37.28 This and another mole rat species are the only known eusocial vertebrates.
Mint Images/Science Source

One exception to the general rule that vertebrate societies are not organized like those of insects is the naked mole rat, a small, hairless rodent that lives in and near East Africa (figure 37.28). Unlike other kinds of mole rats, which live alone or in small family groups, naked mole rats form large underground colonies with a far-ranging system of tunnels and a central nesting area. It is not unusual for a colony to contain 80 individuals.

Naked mole rats feed on bulbs, roots, and tubers, which they locate by constant tunneling. As in insect societies, there is a division of labor among the colony members, with some individuals working as tunnelers while others perform different tasks, depending on the size of their bodies. Large mole rats defend the colony and dig tunnels.

Naked mole rat colonies have a reproductive division of labor similar to the one normally associated with the eusocial insects. All of the breeding is done by a single female, or "queen," which has one or two male consorts. The workers, consisting of both sexes, keep the tunnels clear and forage for food.

REVIEW OF CONCEPT 37.10

Advantages of group living include protection from predators and increased feeding success. Eusocial insects form complex, highly altruistic societies that increase the fitness of the colony. The members of vertebrate societies exhibit more conflict and competition but also cooperate and behave altruistically, especially toward kin.

- *What are possible disadvantages of group living, and how do the advantages and disadvantages explain different social structures of different species of weaver birds?*

Do Crabs Eat Sensibly?

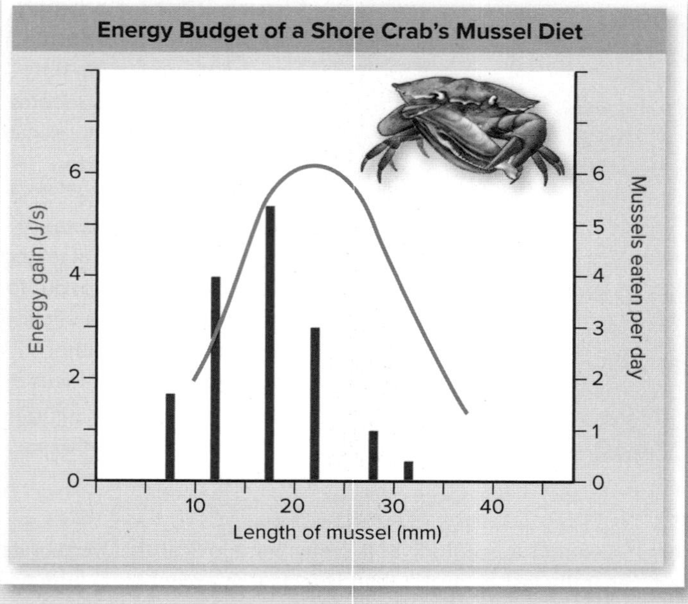
Energy Budget of a Shore Crab's Mussel Diet
Energy gain (J/s)
Mussels eaten per day
0
2
4
6
0
1
2
3
4
5
6
10
20
30
40
Length of mussel (mm)

Many behavioral ecologists claim that animals exhibit so-called optimal foraging behavior. The idea is that because an animal's choice in seeking food involves a trade-off between the food's energy content and the cost of obtaining it, evolution should favor foraging behaviors that optimize the trade-off.

Although this all makes sense, it is not at all clear that this is what animals actually do. This optimal foraging approach makes a key assumption, that maximizing the amount of energy acquired will lead to increased reproductive success. In some cases this is clearly true. As discussed in section 37.7, in ground squirrels, zebra finches, and orb-weaving spiders, researchers have found a direct relationship between net energy intake and the number of offspring raised successfully.

However, animals have other needs besides energy, and sometimes these needs conflict. One obvious "other need," important to many animals, is to avoid predators. It makes little sense for you to eat a little more food if doing so greatly increases the probability that you will yourself be eaten. Often, the behavior that maximizes energy intake increases predation risk. A shore crab foraging for mussels on a beach exposes itself to predatory gulls and other shore birds with each foray. Thus, the behavior that maximizes fitness may reflect a trade-off: obtaining the most energy with the least risk of being eaten. Not surprisingly, a wide variety of animals use a more cautious foraging behavior—becoming less active and staying nearer to cover—when predators are present.

So what does a shore crab do? To find out, an investigator studied whether shore crabs, in fact, feed on those mussels that provide the most energy, as the theory predicts. He found that the mussels on the beach he studied come in a range of sizes, from small ones less than 10 mm in length, which are easy for a crab to open but yield the least amount of energy, to large mussels over 30 mm in length, yielding the most energy but also taking considerably more energy to pry open. To obtain the most net energy, the optimal approach, described by the blue curve in the graph, would be for shore crabs to feed primarily on intermediate-size mussels about 22 mm in length. Is this, in fact, what shore crabs do? To find out, the researcher carefully monitored the size of the mussels eaten each day by the beach's population of shore crabs. The results he obtained—the numbers of mussels of each size actually eaten—are presented in the red histogram.

Analysis

1. **Applying Concepts** What is the dependent variable in the curve? In the histogram?
2. **Making Inferences**
   a. What is the most energetically optimal mussel size (in millimeters) for the crabs to eat?
   b. What size mussel (in millimeters) is most frequently eaten by crabs?
3. **Drawing Conclusions** Do shore crabs tend to feed on those mussels that provide the most energy?
4. **Further Analysis** What factors might be responsible for the slight difference in peak prey length relative to the length optimal for maximal energy gain?

Kevin Schafer/Alamy Stock Photo

Retracing the Learning Path

CONCEPT 37.1 An Animal's Genome Influences Its Behavior

37.1.1 Behavioral Differences Are Linked to Genetic Differences Breeding fast- and slow-learning rats for several generations produces two genetically distinct behavioral populations.

37.1.2 Some Behaviors Appear to Be Controlled by a Single Gene Genes can regulate behavior by producing molecular factors that influence the function of the nervous system. Mutations that alter these factors have been found to affect behavior.

CONCEPT 37.2 Learning Also Influences Behavior

37.2.1 Parent–Offspring Interactions Influence How Behavior Develops In imprinting, a young animal forms an attachment to other individuals or develops preferences that influence later behavior.

37.2.2 Studies of Twins Reveal a Role for Both Genes and Environment in Human Behavior Studies of twins raised together, or separated at birth, reveal that genetics can affect behavior.

CONCEPT 37.3 Thinking Directs the Behavior of Many Animals

37.3.1 Experiments Show Cognitive Behavior Some animals exhibit cognitive behavior and reasoning ability.

CONCEPT 37.4 Migratory Behavior Is Both Innate and Learned

37.4.1 Migration Often Involves Populations Moving Large Distances Long-range two-way movements are called migrations. Many bird species and monarch butterflies migrate yearly.

37.4.2 Migrating Animals Must Be Capable of Orientation and Navigation Orientation is the mechanism by which animals move by tracking environmental stimuli. Navigation is following a route based on orientation and some sort of "map."

CONCEPT 37.5 Animal Communication Plays a Key Role in Ecological and Social Behavior

37.5.1 Successful Reproduction Depends on Appropriate Signals and Responses Courtship signals are usually species-specific and help to ensure reproductive isolation.

37.5.2 Communication Enables Information Exchange Among Group Members Animals living in social groups, such as honeybees, use complex systems of communication to exchange information about food and predators.

CONCEPT 37.6 Behavior Evolves Adaptively

37.6.1 Natural Selection Leads to Behavioral Evolution Natural selection can lead to evolutionary change if variation exists, is genetically heritable, and leads to differences in fitness.

37.6.2 Phylogenies Can Be Used to Study How Behavior Evolves Through Time Inferring ancestral behavioral character states allows inferences about behavior of ancestral species.

CONCEPT 37.7 Behavioral Ecology Examines the Adaptive Significance of Behaviors

37.7.1 Behavior May Increase Fitness in Many Ways Behavioral ecology is the study of how behaviors affect survival and reproductive success.

37.7.2 Foraging Behavior Can Directly Influence Energy Intake and Individual Fitness Natural selection favors optimal foraging strategies in which energy acquisition (cost) is minimized and reproductive success (benefit) is maximized.

37.7.3 Territorial Behavior Evolves If the Benefits of Holding a Territory Exceed the Costs Defense of a territory can be analyzed in economic terms, comparing costs to benefits in energy gains and predator avoidance.

CONCEPT 37.8 Behavioral Strategies Have Evolved to Maximize Reproductive Success

37.8.1 Mating Systems Reflect the Ability of Parents to Care for Offspring Mating systems include monogamy, polygyny, and polyandry; they are influenced by ecology.

37.8.2 Extra-Pair Copulations Increase Fitness in Some Monogamous Species In some cases, extra-pair copulations are beneficial for both males and females.

37.8.3 Natural Selection Sometimes Favors Alternative Mating Strategies The diversity in mating strategies shows that where a strategy increases reproductive success, it will be favored.

CONCEPT 37.9 Some Behaviors Decrease Fitness to Benefit Other Individuals

37.9.1 Reciprocity Theory Can Explain Altruism Between Unrelated Individuals Mutual exchanges benefit both participants; a participant that does not reciprocate will not receive future aid.

37.9.2 Kin Selection Theory Proposes a Direct Genetic Advantage to Altruism Kin selection increases the reproductive success of relatives, increasing frequency of shared alleles, and thus increases an individual's inclusive fitness. Ants, bees, and wasps have haplodiploid reproduction, thus a high degree of gene sharing.

CONCEPT 37.10 Group Living Has Evolved in Both Insects and Vertebrates

37.10.1 Insect Societies Form Efficient Colonies Containing Specialized Castes A social system is a group organized in a cooperative manner. Social insect societies are composed of different castes that have specialized roles.

37.10.2 Vertebrate Societies Are Not Rigidly Organized Vertebrate social systems are less rigidly organized and cohesive than those of insects, with structures influenced by food availability and danger of predation.

Concept Overview

This Concept Overview diagrams the key concepts that were discussed in this chapter.

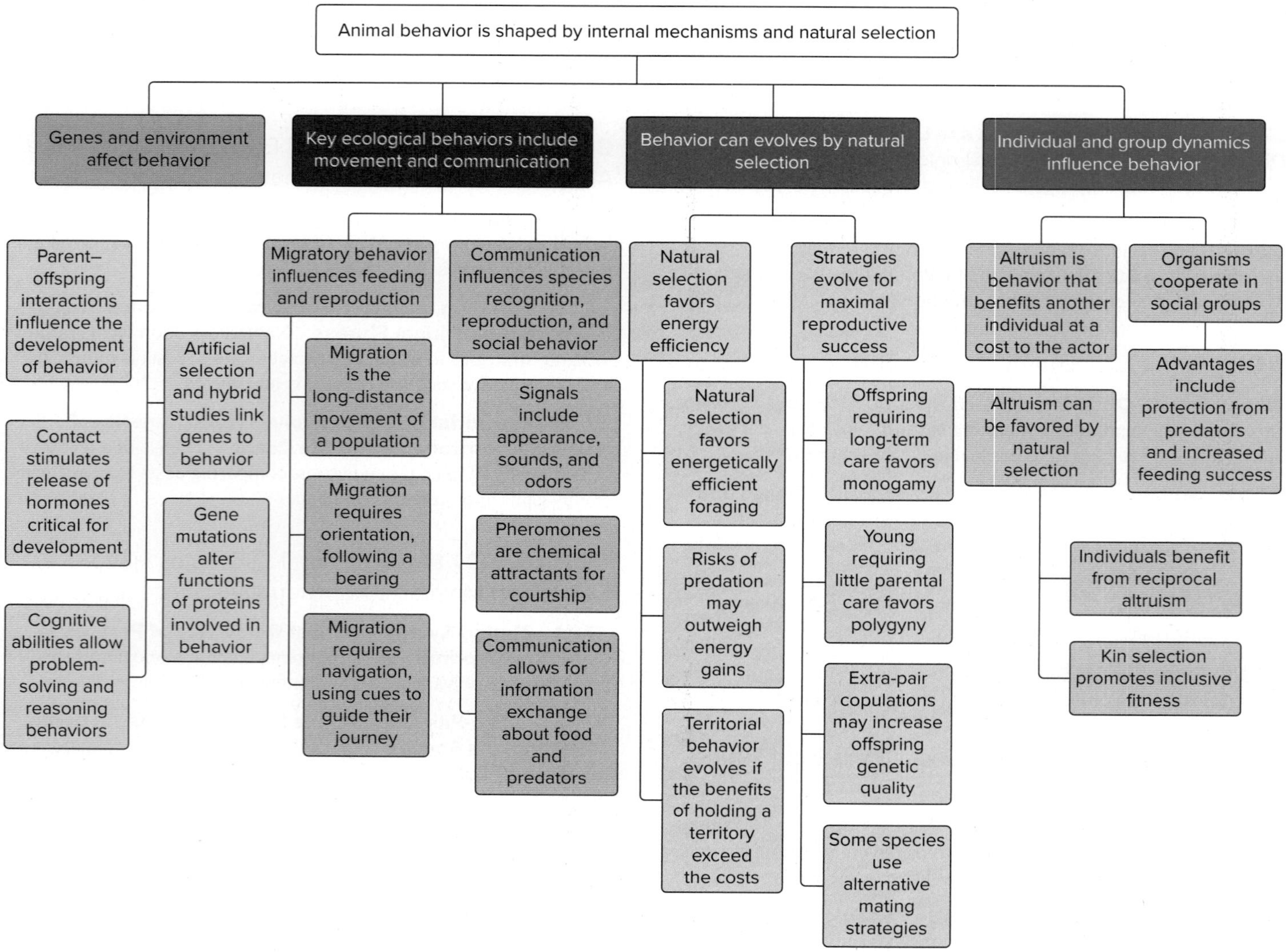

Assessing the Learning Path

Understand

1. In female mice with disabled *fosB* alleles, maternal behavior is
   a. normal.
   b. reduced.
   c. absent.
   d. increased.
2. High levels of oxytocin and vasopressin in prairie voles trigger
   a. pair-bonding.
   b. high levels of parental care.
   c. random mating.
   d. infanticide.
3. In filial imprinting,
   a. offspring form attachments to siblings.
   b. parents form attachments to their offspring.
   c. offspring form attachments to parents.
   d. None of the above
4. Which of the following is NOT an example of cognitive behavior?
   a. A sea otter keeping a favorite rock to use for cracking abalone
   b. A chimpanzee making a thin tool out of reed to fish for termites
   c. Goslings following their mother wherever she goes
   d. Macaques rinsing sand off potatoes before eating them

5. Migration among animals
   a. uses innate instructions.
   b. is learned from adults.
   c. uses celestial clues.
   d. All of the above
6. The species specificity of courtship signals implies that they play a role in
   a. reproductive isolation.
   b. altruistic alarm calling.
   c. kin selection.
   d. reciprocity.
7. Behavior will evolve if
   a. individuals vary in their behavior.
   b. variation among individuals is the result of genetic differences.
   c. individuals that vary in behavior also vary in their fitness.
   d. All of the above
8. Artificial experiments on behavior show that
   a. behaviors evolve to more active more often than to less active.
   b. most traits do not evolve quickly because variation is not genetically based.
   c. populations often evolve quickly.
   d. fruit flies can be selected to fly upward, but not downward.
9. The field of behavioral ecology asks which of the following questions?
   a. Do the sexes differ in behavior?
   b. Is behavior adaptive?
   c. Is behavior modified by experience?
   d. Is behavior developmentally determined?
10. Territorial behavior should evolve when
    a. resources are very scarce.
    b. resources are very abundant.
    c. species are monogamous.
    d. the cost of defense is less than the energy benefit gained.
11. Which of these mating strategies is always the most successful for male side-blotched lizards?
    a. Maintaining a large territory containing many females
    b. Maintaining a smaller territory containing few females, which are guarded closely
    c. Not maintaining a territory but trying to mate with females in other males' territories
    d. None of the above
12. Behaviors that are not in an individual's self-interest but are a benefit to its group are called
    a. aggressive.
    b. altruistic.
    c. cooperative.
    d. operant.
13. According to kin selection, you saving the life of your ______ would do the least for increasing your inclusive fitness.
    a. mother
    b. brother
    c. sister-in-law
    d. niece
14. Naked mole rat colonies are unusual among vertebrate societies, because
    a. workers surrender breeding rights without challenge to a few males.
    b. workers compete for breeding rights.
    c. workers share breeding rights.
    d. females select breeding males based on courtship signals.

Apply

1. In studies of maze learning in rats (figure 37.1), what would happen if, after the seventh generation, rats were randomly assigned mates regardless of their ability to learn the maze?
2. Similarities in behavior and personality between identical twins raised together versus those raised apart suggest that
   a. learning from siblings is a key factor in determining adult behavior.
   b. environment has no role in determining human behavior.
   c. human behavior has a significant genetic component.
   d. behavior is developmentally determined.
3. The difference between a chimpanzee solving a problem like reaching bananas (figure 37.6) and you solving the same problem is that
   a. any human would solve it far more quickly.
   b. you would solve it differently.
   c. you would not be able to solve it.
   d. There is no difference.
4. Bird species A has an east–west migration and navigates with true navigation. If a bird of that species is captured at the halfway point of its migration and is transported 200 km to the south of its established migration route, what should be its new direction of migration?
   a. North
   b. West
   c. Northwest
   d. Southwest
5. In courtship communication,
   a. the signal itself is always species-specific.
   b. the sign communicates species identity.
   c. a stimulus–response chain is involved.
   d. courtship signals are produced only by males.
6. Phylogenetic studies of behavior are useful when
   a. species are hard to work with in the lab, so that artificial selection studies are not possible.
   b. species vary in their behavior and a phylogeny is available for the group.
   c. natural selection has been demonstrated to act on behavior in multiple species.
   d. artificial and natural selection studies give contradictory results.
7. According to optimal foraging theory,
   a. individuals minimize energy intake per unit of time.
   b. energy content of a food item is the only determinant of a forager's food choice.
   c. time taken to capture a food item is the only determinant of a forager's food choice.
   d. a higher-energy item might be less valuable than a lower-energy item if it takes too much time to capture the larger item.
8. Females may benefit from engaging in extra-pair copulations because they
   a. produce more offspring.
   b. can maintain larger territories.
   c. may benefit from parental care provided by more than one male.
   d. spend less time choosing among possible mates.
9. Altruism
   a. is only possible through reciprocity.
   b. is only possible through kin selection.
   c. can only be explained by group selection.
   d. will only occur when the fitness benefit of a given act is greater than the fitness cost.
10. Charlie's mother has a sister who gave birth to a boy she named Richard on the same day Charlie was born. Charlie and Richard share ____________ of their alleles.
    a. half
    b. one-eighth
    c. one-sixteenth
    d. one-quarter

11. In the haplodiploidy system of sex determination, males are
   a. haploid.
   b. diploid.
   c. sterile.
   d. not present, because bees exist as single-sex populations.
12. Which factor(s) have influenced weaver bird social organization?
   a. Filial competition
   b. Predation and food type
   c. Only predation
   d. Kin selection

Synthesize

1. War is so common among human beings that it must be considered a basic behavior of our species. It appears to be absent in all other animal groups (with the possible exception of some other primates). Do you think this behavior has a genetic basis? If so, why might its evolution have been favored by natural selection?
2. Insects that sting often have black-and-yellow coloration and consequently are not eaten by predators. How can you determine if a predator has an innate avoidance of insects that are colored this way, or if the avoidance is learned?
3. In 1997, a major pigeon race from France to England across the English Channel was disrupted by the sonic boom of a *Concorde* supersonic jet airliner, a sound wave that obliterated very-low-frequency sounds (called infrasounds). Over 50,000 pigeons became disoriented and were lost. How might the birds have been using infrasounds to navigate?
4. If you were to rotate a honeybee hive 90°, would you expect the bees within the hive to be misdirected when they exited the hive to go to a food source? Explain.
5. Two populations of the same species of lizard are introduced to localities that are much colder than their native land. One population quickly evolves a new behavioral trait of basking in the sun, but the other population does not. How might you explain the different evolutionary response?
6. An altruistic act is defined as one that benefits another individual at a cost to the actor. Two theories can explain how such behavior evolves: reciprocity and kin selection. How would you distinguish between the two in a field study?

Ian Cartwright/Getty Images

38 Ecology of Individuals and Populations

Learning Path

38.1 Populations Are Groups of a Single Species in One Place

38.2 Population Growth Depends upon Members' Age and Sex

38.3 Evolution Favors Life Histories That Maximize Lifetime Reproductive Success

38.4 Environment Limits Population Growth

38.5 Resource Availability Regulates Population Growth

38.6 Earth's Human Population Is Growing Explosively

38.7 Pandemics and Human Health

Concept Overview

This Concept Overview provides a diagram of the overarching concepts that are covered in this chapter. These concepts will be expanded upon in the Concept Overview at the end of the chapter.

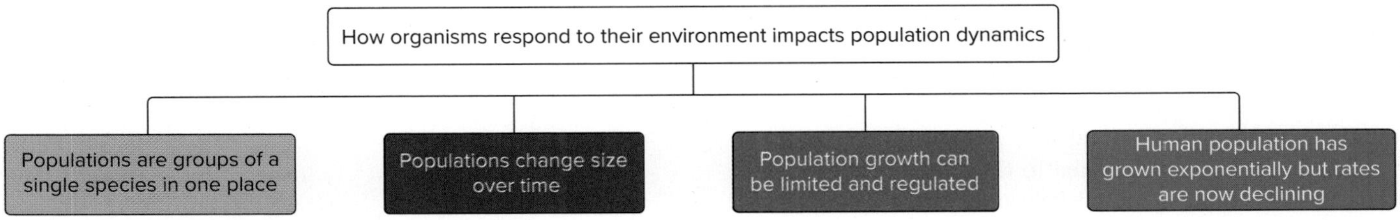

How organisms respond to their environment impacts population dynamics
Populations are groups of a single species in one place
Populations change size over time
Population growth can be limited and regulated
Human population has grown exponentially but rates are now declining

Introduction

Ecology, the study of how organisms relate to one another and to their environments, is a complex and fascinating area of biology that has important implications for each of us. In our exploration of ecological principles, we first consider how organisms respond to the abiotic environment in which they exist and how these responses affect the properties of populations, emphasizing population dynamics. In chapter 39, we discuss communities of coexisting species and the interactions that occur among them. In chapter 40, we will finish our exploration of these ideas by considering the entire biosphere and conservation biology.

38.1 Populations Are Groups of a Single Species in One Place

Organisms live as members of **populations,** groups of individuals that occur together at one place and time. In the rest of this chapter, we consider the properties of populations, focusing on factors that influence whether a population grows or shrinks, and at what rate. The explosive growth of the world's human population in the last few centuries provides a focus for our inquiry.

A Population's Geographical Distribution Is Termed Its Range

LEARNING OBJECTIVE 38.1.1 Explain how a species' geographical range can change through time.

The term *population* can be defined narrowly or broadly. This flexibility allows us to speak in similar terms of the world's human population, the population of protists in the gut of a termite, or the population of deer that inhabit a forest. Sometimes the boundaries defining a population are sharp, such as the edge of an isolated mountain lake for trout, and sometimes they are fuzzier, as when deer readily move back and forth between two forests separated by a cornfield.

Three characteristics of population ecology are particularly important: (1) population range, the area throughout which a population occurs; (2) the pattern of spacing of individuals within that range; and (3) how the population changes in size through time.

No population, not even one composed of humans, occurs in all habitats throughout the world. Most species, in fact, have relatively limited geographical ranges, and the range of some species is minuscule. For example, the Devil's Hole pupfish lives in a single spring in southern Nevada (figure 38.1), and the Socorro isopod (*Thermosphaeroma thermophilus*) is known from a single spring system in New Mexico. At the other extreme, some species are widely distributed. The common dolphin (*Delphinus delphis*), for example, is found throughout all the world's oceans.

Organisms must be adapted to the environment in which they occur. Polar bears are exquisitely adapted to survive the cold of the Arctic, but you won't find them in the tropical rainforest. Certain prokaryotes can live in the near-boiling waters of Yellowstone's geysers, but they do not occur in cooler streams nearby. Each population has its own requirements that determine where it can live and reproduce and where it cannot.

Figure 38.1 The Devil's Hole pupfish (*Cyprinodon diabolis*). This fish has the smallest range of any known vertebrate species in the world.

Stone Nature Photography/Alamy Stock Photo

Ranges undergo expansion and contraction

Population ranges are not static but change through time. These changes occur for two reasons. In some cases, the environment changes. As the glaciers retreated at the end of the last ice age, approximately 10,000 years ago, many North American plant and animal populations expanded northward. At the same time, as climates warmed, species experienced shifts in the elevations at which they could live (figure 38.2).

In addition, populations can expand their ranges when they are able to circumvent inhospitable habitat to colonize suitable, previously unoccupied areas. For example, the cattle egret is native to Africa. Sometime in the late 1800s, these birds appeared in northern South America, having made the nearly 3500-km transatlantic crossing, perhaps aided by strong winds. Since then, they

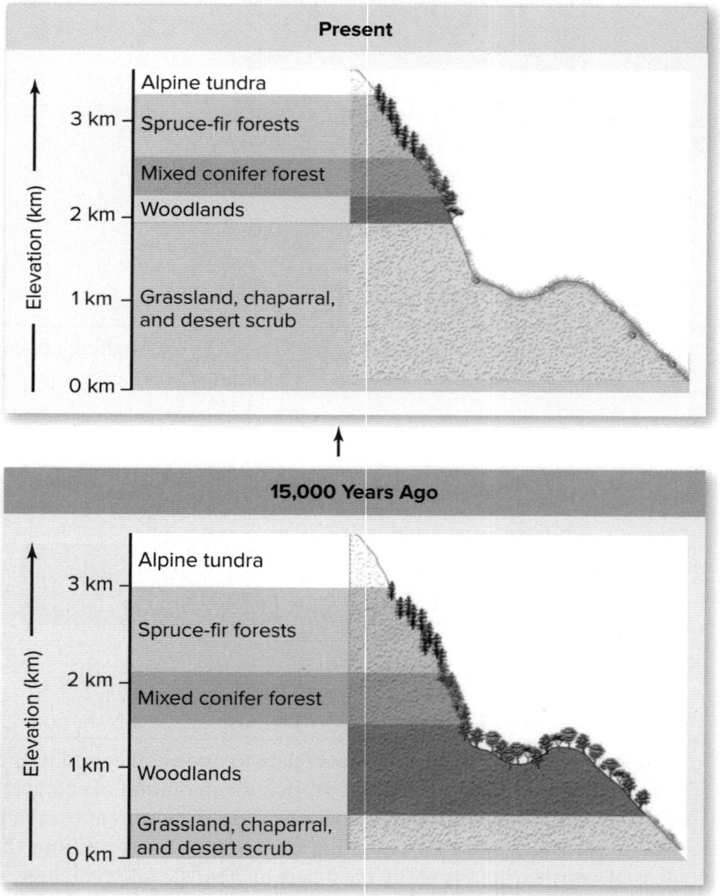

Figure 38.2 Altitude shifts in altitudinal distributions of trees in the mountains of southwestern North America. During the glacial period 15,000 years ago, conditions were cooler than they are now. As the climate warmed, tree species that require colder temperatures shifted their range upward in altitude so that they live in the climatic conditions to which they are adapted.

have steadily expanded their range and now can be found throughout most of the United States (figure 38.3).

The human effect

By altering the environment, humans have allowed some species, such as coyotes, to expand their ranges and move into areas they previously did not occupy. Moreover, humans have served as an agent of dispersal for many species. Some of these transplants have been widely successful. For example, 100 starlings were introduced into New York City in 1896 in a misguided attempt to establish every species of bird mentioned by Shakespeare. Their population steadily spread so that by 1980 they were established throughout the United States. Similar stories could be told for countless plants and animals, and the list increases every year. Unfortunately, the success of these invaders often comes at the expense of native species.

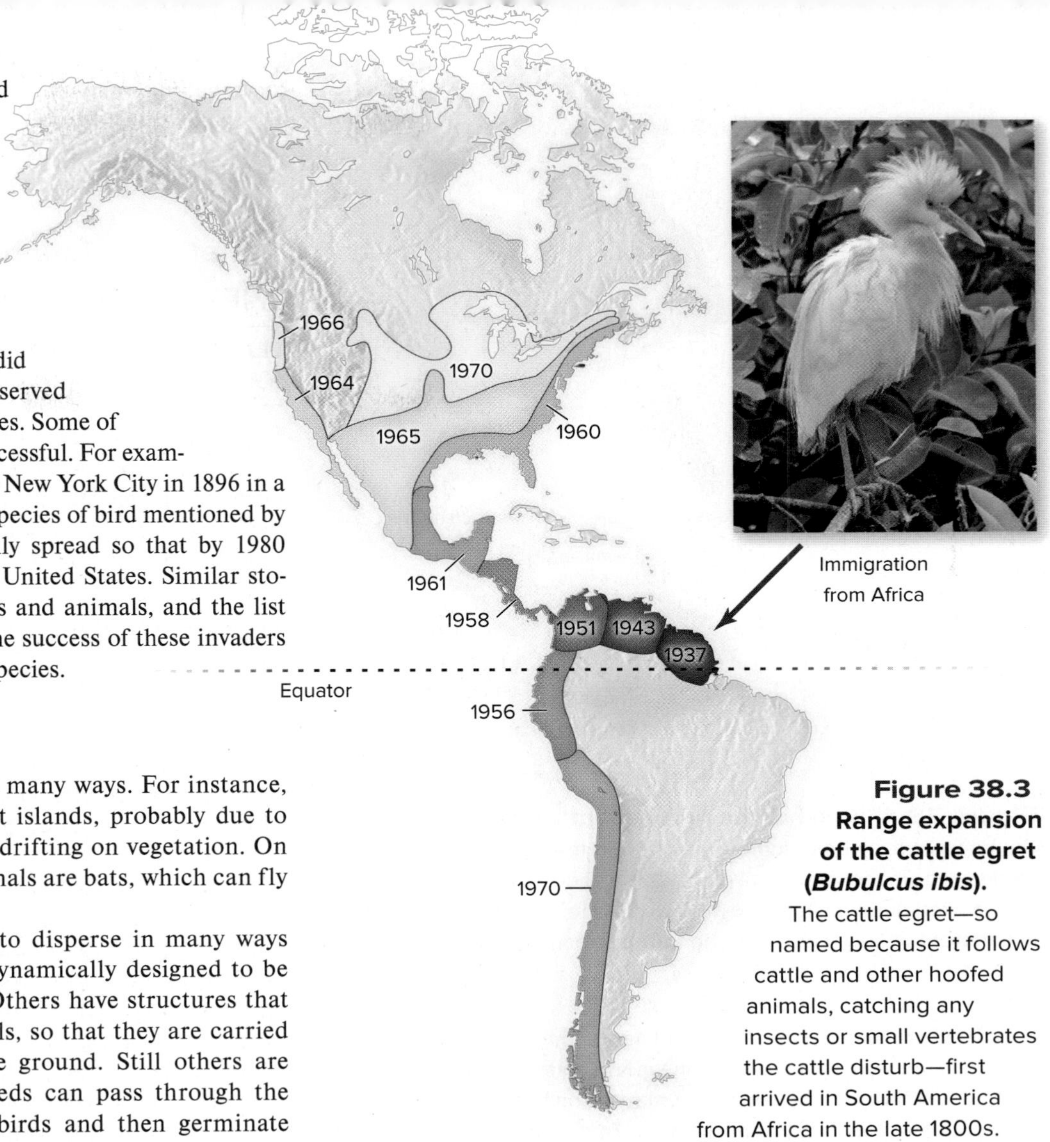

Figure 38.3 Range expansion of the cattle egret (*Bubulcus ibis*). The cattle egret—so named because it follows cattle and other hoofed animals, catching any insects or small vertebrates the cattle disturb—first arrived in South America from Africa in the late 1800s. Since the 1930s, the range expansion of this species has been well documented. The figure shows its spread over the first 100 or so years after arrival. Today, cattle egrets occur throughout almost all of South and Central America and the United States, as well as along the west coast of Canada.

(inset): Shackleford-Photography/iStock/Getty Images

Dispersal mechanisms

Dispersal to new areas can occur in many ways. For instance, lizards have colonized many distant islands, probably due to individuals or their eggs floating or drifting on vegetation. On many distant islands, the only mammals are bats, which can fly to the islands.

Seeds of plants are designed to disperse in many ways (figure 38.4). Some seeds are aerodynamically designed to be blown long distances by the wind. Others have structures that stick to the fur or feathers of animals, so that they are carried long distances before falling to the ground. Still others are enclosed in fleshy fruits. These seeds can pass through the digestive systems of mammals or birds and then germinate where they are defecated. Finally, seeds of mistletoes (*Arceuthobium*) are violently propelled from the base of the fruit in an explosive discharge. Although the probability of long-distance dispersal events leading to successful establishment of new populations is low, over millions of years many such dispersals have occurred.

Individuals in Populations Exhibit Different Spacing Patterns

LEARNING OBJECTIVE 38.1.2 Describe how individuals of a population may be distributed in space.

Another key characteristic of population structure is the way in which individuals of a population are distributed. They may be randomly spaced, uniformly spaced, or clumped.

Random spacing

Random spacing of individuals within populations occurs when they do not interact strongly with one another and when they are not affected by nonuniform aspects of their environment. Random distributions are not common in nature. Some species of trees, however, appear to be randomly distributed.

Uniform spacing

Uniform spacing within a population may often, but not always, result from competition for resources. However, this spacing is accomplished in many different ways.

In animals, uniform spacing often results from behavioral interactions, as described in chapter 37. In many species, individuals of one or both sexes defend a territory from which other individuals are excluded. These territories provide the owner with exclusive access to resources, such as food, water, hiding refuges, or mates, and can space individuals evenly across the habitat if the resources themselves are evenly distributed.

Among plants, uniform spacing is also a common result of competition for resources. Closely spaced individual plants

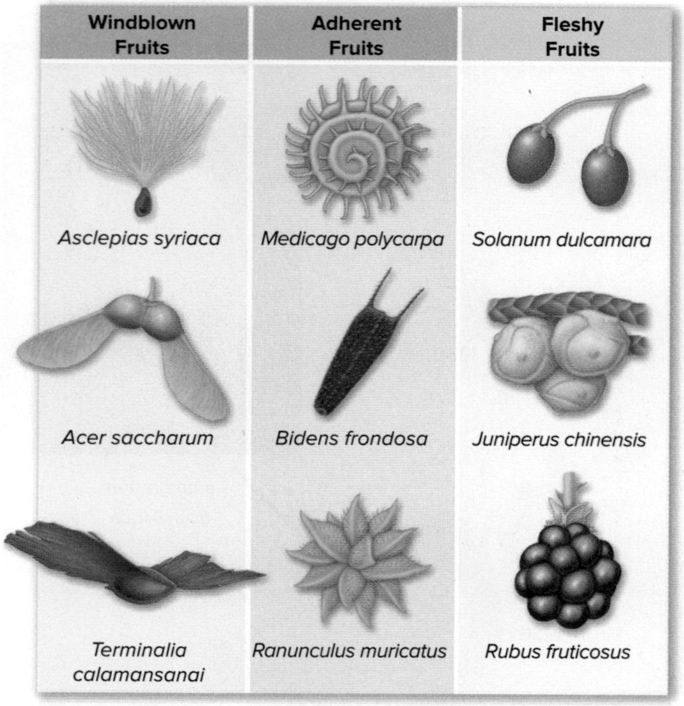
Windblown Fruits
Adherent Fruits
Fleshy Fruits
Asclepias syriaca
Medicago polycarpa
Solanum dulcamara
Acer saccharum
Bidens frondosa
Juniperus chinensis
Terminalia calamansanai
Ranunculus muricatus
Rubus fruticosus

Figure 38.4 Some of the many adaptations of seeds. Seeds have evolved a number of different means of facilitating dispersal from their maternal plant. Some seeds can be transported great distances by the wind, whereas seeds enclosed in adherent or fleshy fruits can be transported by animals.

compete for available sunlight, nutrients, or water. These contests can be direct, as when one plant casts a shadow over another, or indirect, as when two plants compete by extracting nutrients or water from a shared area. In addition, some plants, such as the creosote bush, produce chemicals in the surrounding soil that are toxic to other members of their species. In all of these cases, only plants that are spaced an adequate distance from each other will be able to coexist, leading to uniform spacing.

Clumped spacing

Individuals clump into groups or clusters in response to uneven distribution of resources in their immediate environments. Clumped distributions are common in nature because individual animals, plants, and microorganisms tend to occur in habitats defined by soil type, moisture, or other aspects of the environment to which they are best adapted.

Social interactions also can lead to clumped distributions. Many species live and move around in large groups, which go by a variety of names (for example, *flock, herd, pride*). These groupings can provide many advantages, including increased awareness of and defense against predators, decreased energy cost of moving through air and water, and access to the knowledge of all group members.

On a broader scale, populations are often most densely populated in the interior of their range and less densely distributed toward the edges. Such patterns usually result from the manner in which the environment changes in different areas, with individuals less well adapted to conditions at the extremes of their range.

A Metapopulation Comprises Distinct Populations That May Exchange Members

LEARNING OBJECTIVE 38.1.3 Distinguish between a population and a metapopulation.

Species often exist as a network of distinct populations that interact with one another by exchanging individuals. Such networks, termed **metapopulations,** usually occur in areas in which suitable habitat is patchily distributed and is separated by intervening stretches of unsuitable habitat.

Dispersal and habitat occupancy

The degree to which populations within a metapopulation interact depends on the amount of dispersal; this interaction is often not symmetrical: populations increasing in size tend to send out many dispersers, whereas populations at low levels tend to receive more immigrants than they send off. In addition, relatively isolated populations tend to receive relatively few arrivals.

Not all suitable habitats within a metapopulation's area may be occupied at any one time. For a number of reasons, some individual populations may become extinct, perhaps as a result of an epidemic disease, a catastrophic fire, or the loss of genetic variation following a population bottleneck. Dispersal from other populations, however, may eventually recolonize such areas. In some cases, the number of habitats occupied in a metapopulation may represent an equilibrium in which the rate of extinction of existing populations is balanced by the rate of colonization of empty habitats.

Source–sink metapopulations

A species may also exhibit a metapopulation structure in areas in which some habitats are suitable for long-term population maintenance but others are not. In these situations, termed **source–sink metapopulations,** the populations in the better areas (the sources) continually send out dispersers that bolster the populations in the poorer habitats (the sinks). In the absence of such continual replenishment, sink populations would have a negative growth rate and would eventually become extinct.

Metapopulations of butterflies have been studied particularly intensively. In one study, researchers sampled populations of the Glanville fritillary butterfly at 1600 meadows in southwestern Finland (figure 38.5). On average, every year, 200 populations became extinct but 114 empty meadows were colonized. A variety of factors seemed to increase the likelihood of a population's extinction, including small population size, isolation from sources of immigrants, low resource availability (as indicated by the number of flowers on a meadow), and lack of genetic variation within the population.

The researchers attribute the greater number of extinctions than colonizations to a string of very dry summers. Because none of the populations is large enough to survive on its own, continued survival of the species in southwestern Finland would appear to require the continued existence of a metapopulation network in which new populations are continually created and existing populations are supplemented by immigrants. Continued bad weather thus may doom the species, at least in this part of its range.

Metapopulations, where they occur, can have two important implications for the range of a species. First, through continuous colonization of empty patches, metapopulations prevent extinction of the species. If no such dispersal existed, then each population could eventually perish, leading to disappearance of the species from the entire area. Moreover, in source-sink metapopulations, the species occupies a larger area than it otherwise might, including marginal areas that could not support a population without a continual influx of immigrants. For these reasons, the study of metapopulations has become very important in conservation biology as natural habitats become increasingly fragmented.

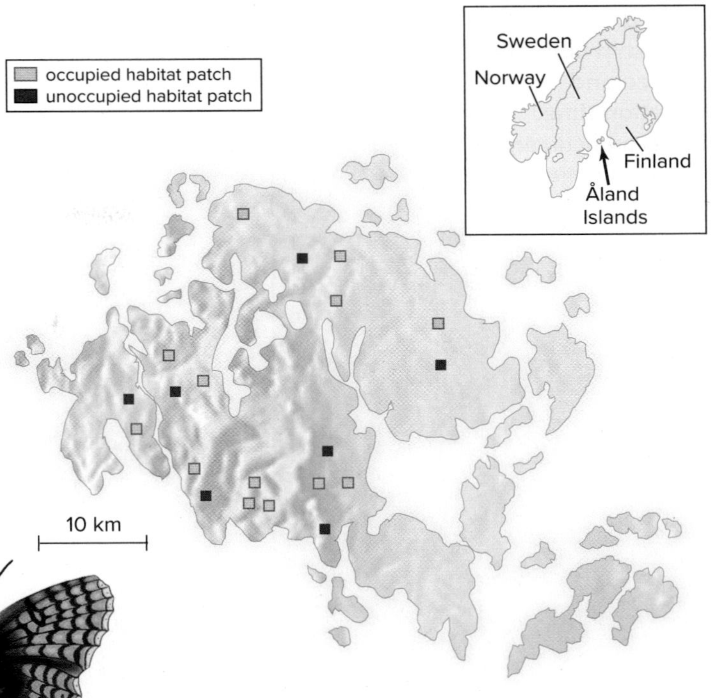

Figure 38.5 Metapopulations of butterflies. The Glanville fritillary butterfly (*Melitaea cinxia*) occurs in metapopulations in southwestern Finland on the Åland Islands. None of the populations is large enough to survive for long on its own, but continual immigration of individuals from other populations allows some populations to survive. In addition, continual establishment of new populations tends to offset extinction of established populations, although in recent years extinctions have outnumbered colonizations.

REVIEW OF CONCEPT 38.1

A population is a group of individuals of a single species existing together in an area. A population's range changes over time. Populations may form metapopulations connected by individuals that move from one group to another. The distribution of individuals in a population can be random, uniform, or clumped. The distribution is determined in part by resource availability.

- *How might the geographical range of a species change if populations cannot exchange individuals with each other?*

38.2 Population Growth Depends upon Members' Age and Sex

The dynamics of a population—how it changes through time—are affected by many factors. One important factor is the age distribution of individuals—that is, what proportion of individuals are adults, juveniles, and young.

Sex Ratio and Generation Time Affect Population Growth Rates

LEARNING OBJECTIVE 38.2.1 Define *demography.*

Demography is the quantitative study of populations. How the size of a population changes through time can be studied at two levels: as a whole or broken down into groups that share a common feature. At the most inclusive level, we can ask whether the size of the whole population is increasing, decreasing, or remaining constant. In simple terms, a population grows if births outnumber deaths, and it shrinks if deaths outnumber births. Understanding these trends is often easier if we break the population into smaller units composed of individuals of the same age, then study separately the factors affecting birth and death rates for each unit.

Population growth can be influenced by the population's sex ratio. The number of births in a population is usually directly related to the number of females; births may not be as closely related to the number of males in species in which a single male can mate with several females. In many species, males compete for the opportunity to mate with females, as you learned in chapter 37; consequently, a few males have many matings, and many males do not mate at all. In such species, the sex ratio is female-biased and does not affect population growth rates; reduction in the number of males simply changes the identities of the reproductive

Figure 38.6 **The relationship between body size and generation time.** In general, larger organisms have longer generation times, although there are exceptions.

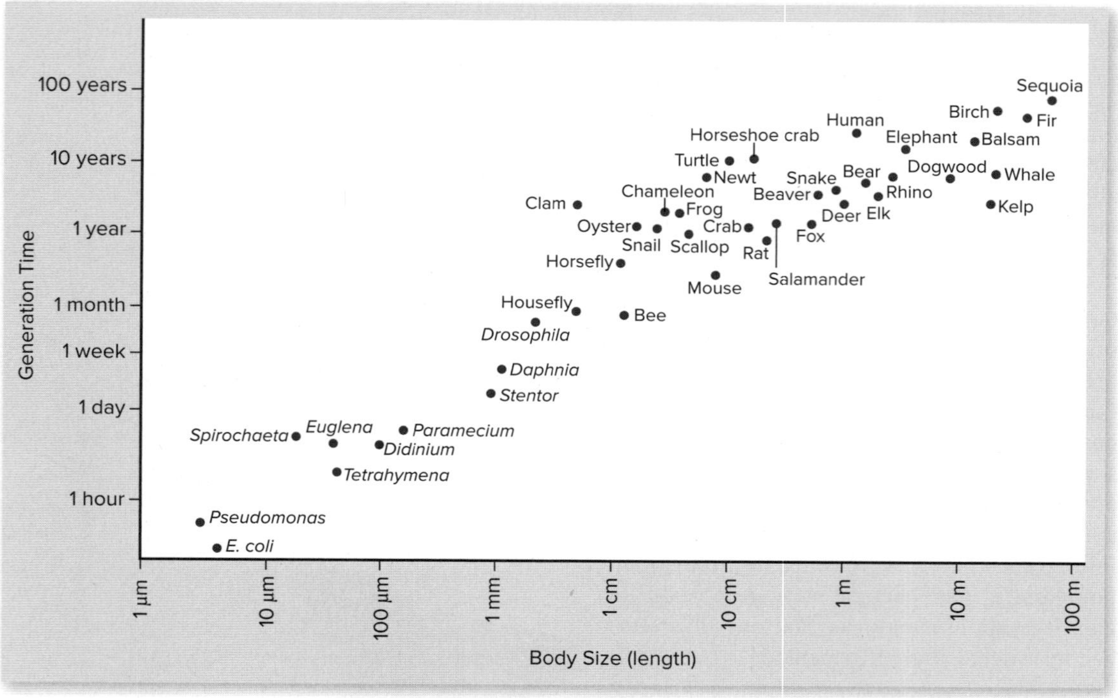

males without reducing the number of births. By contrast, among monogamous species, pairs may form long-lasting reproductive relationships, and a reduction in the number of males can then directly reduce the number of births.

Generation time is the average interval between the birth of an individual and the birth of its offspring. This factor can also affect population growth rates. Species differ greatly in generation time. Differences in body size can explain much of this variation—mice go through approximately 100 generations during the course of one elephant generation (figure 38.6). But small size does not always mean short generation time. Newts, for example, are smaller than mice but have considerably longer generation times.

In general, populations with short generations can increase in size more quickly than populations with long generations. Conversely, because generation time and life span are usually closely correlated, populations with short generation times may also diminish in size more rapidly if birth rates suddenly decrease.

Age Structure Is Determined by the Numbers of Individuals in Different Age Groups

LEARNING OBJECTIVE 38.2.2 Describe the factors that influence a species' demography.

A group of individuals of the same age is referred to as a **cohort.** In most species, the probability that an individual will reproduce or die varies through its life span. As a result, within a population every cohort has a characteristic birth rate, or **fecundity,** defined as the number of offspring produced in a standard time (for example, per year), and death rate, or **mortality,** the number of individuals that die in that period.

The relative number of individuals in each cohort defines a population's **age structure** (figure 38.7). Because different cohorts have different fecundity and death rates, age structure

Figure 38.7 Age structure in different types of populations. The grass you see in this photo is an annual plant. The individual plants are all the same age because they die every year. By contrast, the herd of bison contains individuals of varying ages because they can survive from year to year.
Glen Allison/Photodisc/Getty Images

has a critical influence on a population's growth rate. Populations with a large proportion of young individuals, for example, tend to grow rapidly, because an increasing proportion of their individuals are reproductive. Human populations in many developing countries are an example, as will be discussed in section 38.6. Conversely, if a large proportion of a population is relatively old, populations may decline. This phenomenon now characterizes Japan and some countries in Europe.

Life tables show probability of survival and reproduction through a cohort's life span

To assess how populations in nature are changing, ecologists use a **life table,** which tabulates the fate of a cohort from birth until death, showing the number of offspring produced and the number of individuals that die each year. Table 38.1 shows an example of a life table analysis from a study of the meadow grass *Poa annua.* This study follows the fate of 843 individuals through time, charting how many survive in each interval and how many offspring each survivor produces.

In table 38.1, the first column indicates the age of the cohort (that is, the number of three-month intervals from the start of the study). The second and third columns indicate the number of survivors and the proportion of the original cohort still alive at the beginning of that interval. The fifth column presents the **mortality rate,** the proportion of individuals that started that interval alive but died by the end of it. The seventh column indicates the average number of seeds produced by each surviving individual in that interval, and the last column shows the number of seeds produced relative to the size of the original cohort.

Much can be learned by examining life tables. In the case of *P. annua,* we see that both the probability of dying and the number of offspring produced per surviving individual steadily increase with age. By adding up the numbers in the last column, we get the total number of offspring produced per individual in the initial cohort. This number is almost 2, which means that for every original member of the cohort, on average 2 new individuals have been produced. A figure of 1.0 would be the break-even number, the point at which the population was neither growing nor shrinking. In this case, the population appears to be growing rapidly.

In most cases, life table analysis is more complicated than this. First, except for organisms with short life spans, it is difficult to track the fate of a cohort until the death of the last individual. An alternative approach is to construct a cross-sectional study, examining the fate of cohorts of different ages in a single period. In addition, many factors—such as offspring reproducing before all members of their parents' cohort have died—complicate the interpretation of whether populations are growing or shrinking.

TABLE 38.1 Life Table of the Meadow Grass *Poa annua* for a Cohort Containing 843 Seedlings

Age (in 3-month intervals)	Number Alive at Beginning of Time Interval	Proportion of Cohort Alive at Beginning of Time Interval (survivorship)	Deaths During Time Interval	Mortality Rate During Time Interval	Seeds Produced During Time Interval	Seeds Produced per Surviving Individual (fecundity)	Seeds Produced per Member of Cohort (fecundity × survivorship)
0	843	1.000	121	0.143	0	0.00	0.00
1	722	0.857	195	0.271	303	0.42	0.36
2	527	0.625	211	0.400	622	1.18	0.74
3	316	0.375	172	0.544	430	1.36	0.51
4	144	0.171	90	0.626	210	1.46	0.25
5	54	0.064	39	0.722	60	1.11	0.07
6	15	0.018	12	0.800	30	2.00	0.04
7	3	0.004	3	1.000	10	3.33	0.01
8	0	0.000	—		Total = 1665		Total = 1.98

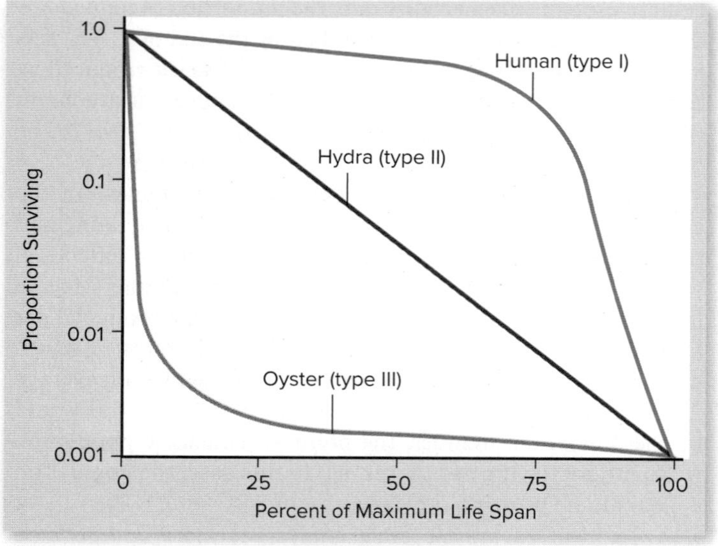

Figure 38.8 Survivorship curves. By convention, survival (the vertical axis) is plotted on a log scale. Humans have a type I life cycle, hydra (an animal related to jellyfish) type II, and oysters type III.

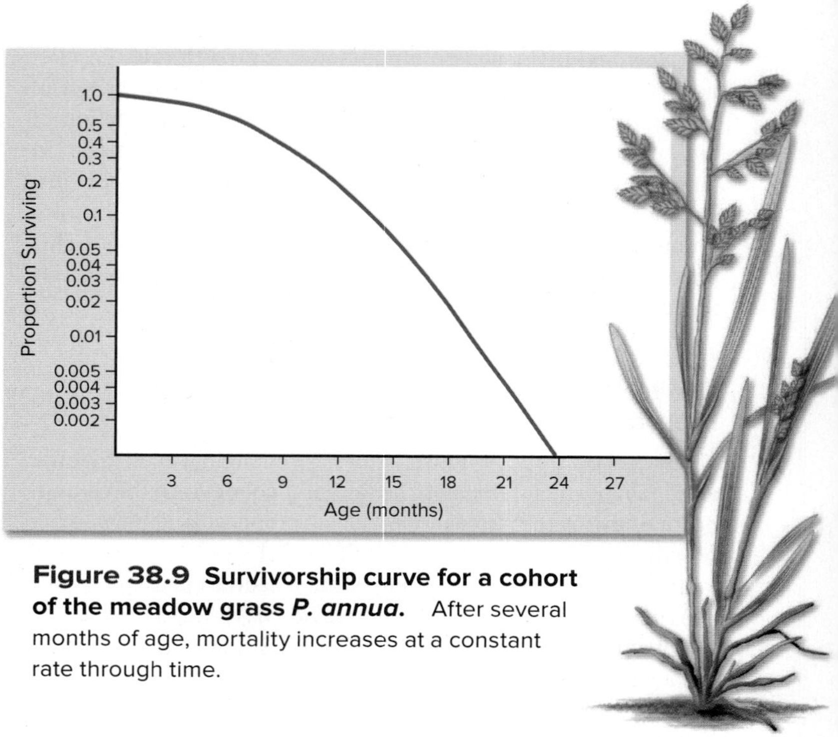

Figure 38.9 Survivorship curve for a cohort of the meadow grass *P. annua*. After several months of age, mortality increases at a constant rate through time.

Survivorship curves demonstrate how survival probability changes with age

The percentage of an original population that survives to a given age is called its **survivorship.** One way to express some aspects of the age distribution of populations is through a *survivorship curve.* Examples of different survivorship curves are shown in figure 38.8. Oysters produce vast numbers of offspring, only a few of which live to reproduce. However, once they become established and grow into reproductive individuals, their mortality rate is extremely low (type III survivorship curve). Note that in this type of curve, survival and mortality rates are inversely related. Thus, the rapid decrease in the proportion of oysters surviving produces an initially high mortality rate. In contrast, the relatively flat line at older ages indicates high survival and low mortality.

In hydra, animals related to jellyfish, individuals are equally likely to die at any age. This results in a straight survivorship curve (type II).

Finally, mortality rates in humans, as in many other animals and in protists, rise steeply later in life (type I survivorship curve).

Of course, these descriptions are just generalizations, and many organisms show more complicated patterns. Examination of the data for *P. annua,* for example, reveals that its survivorship curve is most similar to a type II curve (figure 38.9).

REVIEW OF CONCEPT 38.2

Demography is the quantitative study of populations. Demographic characteristics include age structure, life span, sex ratio, generation time, and birth and mortality rates. Population size will change with variations in mortality and birth rate in different cohorts.

- *Why might the survivorship curve of a species vary from one environment to another?*

38.3 Evolution Favors Life Histories That Maximize Lifetime Reproductive Success

Natural selection favors traits that maximize the number of surviving offspring left in the next generation by an individual organism. Two factors affect this quantity: how long an individual lives and how many young it produces each year.

Reproduction Has Significant Costs That Affect Parental Survival

LEARNING OBJECTIVE 38.3.1 Illustrate how an organism's life history is affected by reproductive trade-offs.

Why doesn't every organism reproduce immediately after its own birth, produce large families of offspring, care for them intensively, and perform these functions repeatedly throughout a long life, while outcompeting others, escaping predators, and capturing food with ease? The answer is that no one organism can do all of this, simply because not enough resources are available. Consequently, organisms allocate resources either to current reproduction or to increasing their prospects of surviving and reproducing at later life stages.

The complete life cycle of an organism constitutes its life history. All life histories involve significant trade-offs. Because resources are limited, a change that increases reproduction may decrease survival and reduce future reproduction. As one example, a Douglas fir tree that produces more cones increases its current reproductive success—but it also grows more slowly. Because the number of cones produced is a function of how large a tree is, this diminished growth will

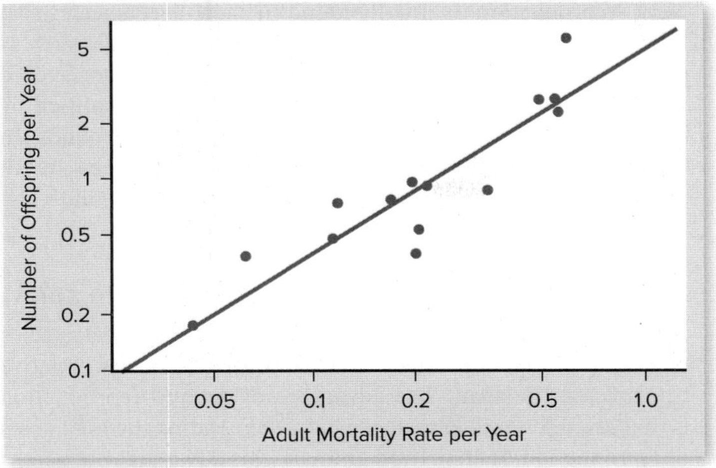

Figure 38.10 Reproduction has a price. Data from many bird species indicate that increased fecundity in birds correlates with higher mortality, ranging from the albatross (lowest) to the sparrow (highest). Birds that raise more offspring per year have a higher probability of dying during that year.

decrease the number of cones it can produce in the future. Similarly, birds that have more offspring each year have a higher probability of dying during that year or of producing smaller clutches the following year (figure 38.10). Conversely, individuals that delay reproduction may grow faster and larger, enhancing future reproduction.

In one elegant experiment, researchers changed the number of eggs in the nests of a bird, the collared flycatcher (figure 38.11). Birds whose clutch size (the number of eggs produced in one breeding event) was decreased expended less energy raising their young and thus were able to lay more eggs the next year, whereas those given more eggs worked harder and consequently produced fewer eggs the following year. Ecologists refer to the reduction in future reproductive potential resulting from current reproductive efforts as the **cost of reproduction.**

Natural selection favors the life history that maximizes lifetime reproductive success. When the cost of reproduction is low, individuals should produce as many offspring as possible, because there is little cost. Low costs of reproduction may occur when resources are abundant and may be relatively low when overall mortality rates are high. In the latter case, individuals may be unlikely to survive to the next breeding season anyway, so the incremental effect of increased reproductive efforts may have little effect on future survival.

Alternatively, when costs of reproduction are high, lifetime reproductive success may be maximized by deferring or minimizing current reproduction to enhance growth and survival rates. This situation occurs whenever costs of reproduction affect the ability of an individual to survive or decrease the number of offspring that can be produced in the future.

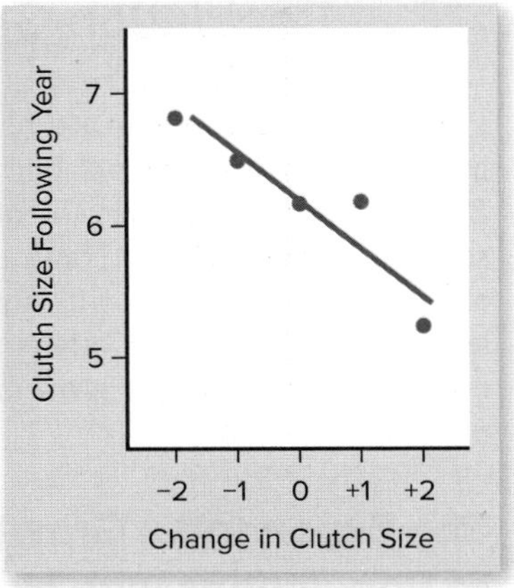

Figure 38.11 Reproductive events per lifetime. Adding eggs to nests of collared flycatchers (*Ficedula albicollis*), which increases the reproductive efforts of the female rearing the young, decreases clutch size the following year; removing eggs from the nest increases the next year's clutch size. This experiment demonstrates the trade-off between current reproductive effort and future reproductive success.

(left): Leif Ingvarson/Shutterstock

A Trade-Off Exists Between Number of Offspring and Investment per Offspring

LEARNING OBJECTIVE 38.3.2 Compare the costs and benefits of allocating resources to reproduction.

In terms of natural selection, the number of offspring produced is not as important as how many of those offspring themselves survive to reproduce. Assuming that the amount of energy to be invested in offspring is limited, a balance must be reached between the number of offspring produced and the size of each offspring (figure 38.12). This trade-off has been experimentally demonstrated in the

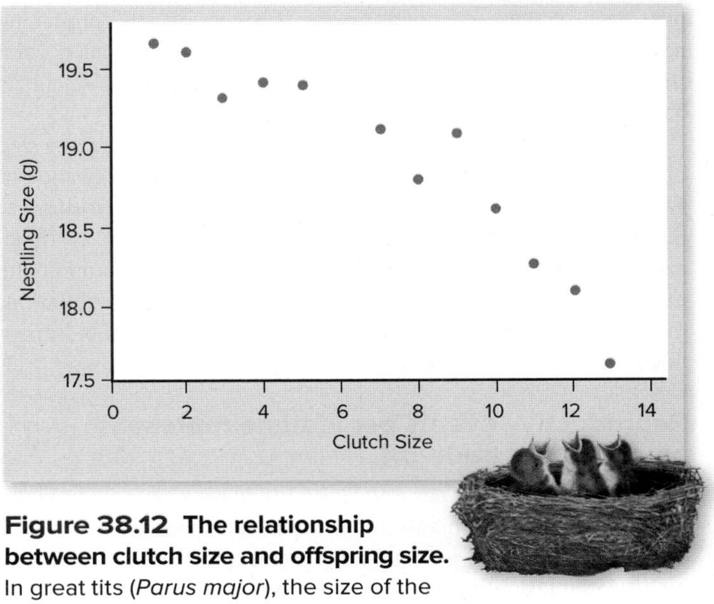

Figure 38.12 The relationship between clutch size and offspring size. In great tits (*Parus major*), the size of the nestlings is inversely related to the number of eggs laid. The more mouths they have to feed, the less the parents can provide to any one nestling. Do you think natural selection would favor producing many small young or a few large ones?

Figure 38.13 Variation in the size of baby side-blotched lizards (*Uta stansburiana*) produced by experimental manipulations. In clutches in which some developing eggs were surgically removed, the remaining offspring were larger (center) than lizards produced in control clutches in which all the eggs were allowed to develop (right). In experiments in which some of the yolk was removed from the eggs, smaller lizards hatched (left).

Barry Sinervo

side-blotched lizard, which normally lays four or five eggs at a time. When some of the eggs are removed surgically early in the reproductive cycle, the female lizard produces only one to three eggs but supplies each of these eggs with greater amounts of yolk, producing eggs and, subsequently, hatchlings that are much larger than normal (figure 38.13). Alternatively, by removing yolk from eggs, scientists have demonstrated that smaller young will be produced.

In the side-blotched lizard and many other species, the size of offspring is critical—larger offspring have a greater chance of survival. Producing many offspring with little chance of survival might not be the best strategy, but producing only a single, extraordinarily robust offspring also would not maximize the number of surviving offspring. Rather, an intermediate situation, in which several fairly large offspring are produced, should maximize the number of surviving offspring.

Reproductive events per lifetime represent an additional trade-off

The trade-off between age and fecundity plays a key role in many life histories. Annual plants and most insects focus all their reproductive resources on a single large event and then die. This life history adaptation is called **semelparity.** Organisms that produce offspring several times over many seasons exhibit a life history adaptation called **iteroparity.**

Species that reproduce yearly must avoid overtaxing themselves in any one reproductive episode so that they will be able to survive and reproduce in the future. Semelparity, or "big bang" reproduction, is usually found in short-lived species that have a low probability of staying alive between broods, such as plants growing in harsh climates. Semelparity is also favored when fecundity entails large reproductive cost, exemplified by Pacific salmon migrating upriver to their spawning grounds. In these species, rather than investing some resources in an unlikely bid to survive until the next breeding season, individuals put all their resources into one reproductive event.

Age at first reproduction correlates with life span

Among mammals and many other animals, longer-lived species put off reproduction longer than short-lived species, relative to expected life span. The advantage of delayed reproduction is that juveniles gain experience before expending the high costs of reproduction. In long-lived animals, this advantage outweighs the energy that is invested in survival and growth rather than reproduction.

In shorter-lived animals, on the other hand, time is of the essence; thus, quick reproduction is more critical than juvenile training, and reproduction tends to occur earlier.

REVIEW OF CONCEPT 38.3

Life history adaptations involve many trade-offs between reproductive cost and investment in survival. These trade-offs take a variety of forms, from laying fewer than the maximum possible number of eggs to putting all energy into a single bout of reproduction.

- *How might the life histories of two species differ if one is subject to a higher level of predation?*

38.4 Environment Limits Population Growth

Populations often remain at a relatively constant size, regardless of how many offspring are born. As you learned in chapter 1, Darwin based his theory of natural selection partly on this seeming contradiction. Natural selection occurs because of checks on reproduction, with some individuals producing fewer surviving offspring than others. To understand populations, we must consider how they grow and what factors in nature limit population growth.

The Exponential Growth Model Applies to Populations with No Growth Limits

LEARNING OBJECTIVE 38.4.1 Describe exponential growth mathematically.

The rate of population increase, r, is defined as the difference between the birth rate, b, and the death rate, d, corrected for movement of individuals into or out of the population (e, rate of movement out of the area; i, rate of movement into the area). Thus,

$$r = (b - d) + (i - e)$$

Movements of individuals can have a major influence on population growth rates. For example, the increase in human population in the United States during the closing decades of the 20th century was mostly due to immigration.

The simplest model of population growth assumes that a population grows without limits at its maximal rate and that rates of immigration and emigration are equal. This rate, called the **biotic potential,** is the rate at which a population of a given species increases when no limits are placed on its rate of growth. In mathematical terms, this is defined as

$$\frac{dN}{dt} = r_i N$$

where N is the number of individuals in the population, dN/dt is the rate of change in its numbers over time, and r_i is the intrinsic rate of natural increase for that population—its innate capacity for growth.

The biotic potential of any population is exponential (red line in figure 38.14). Even when the *rate* of increase remains constant, the actual *number* of individuals accelerates rapidly as the size of the population grows. The result of unchecked exponential growth is a population explosion.

A single pair of houseflies, laying 120 eggs per generation, could produce more than 5 trillion descendants in a year. In 10 years, their descendants would form a swarm more than 2 m thick over the entire surface of the Earth! In practice, such patterns of unrestrained growth prevail only for short periods, usually when an organism reaches a new habitat with abundant resources. Natural examples of such short periods of unrestrained growth include dandelions arriving in the fields, lawns, and meadows of North America from Europe for the first time; algae colonizing a newly formed pond; and cats introduced to an island with many birds that were previously lacking predators.

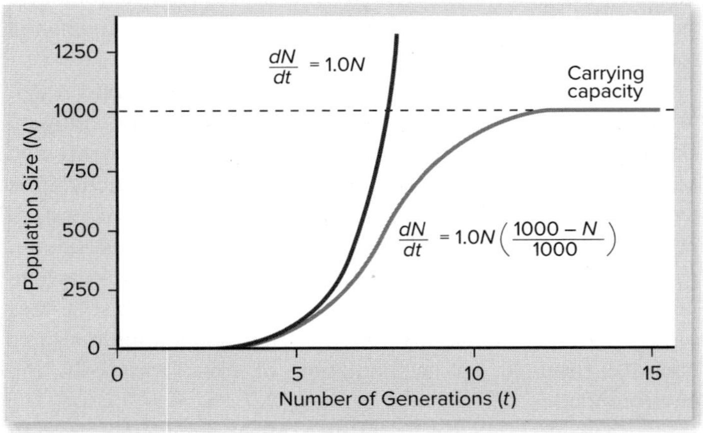

Figure 38.14 Two models of population growth. The red line illustrates the exponential growth model for a population with an r of 1.0. The blue line illustrates the logistic growth model in a population with $r = 1.0$ and $K = 1000$ individuals. At first, logistic growth accelerates exponentially; then, as resources become limited, the death rate increases and growth slows. Growth ceases when the death rate equals the birth rate. The carrying capacity (K) ultimately depends on the resources available in the environment.

The Logistic Growth Model Applies to Populations That Approach Their Carrying Capacity

LEARNING OBJECTIVE 38.4.2 Explain how carrying capacity affects the exponential growth curve.

No matter how rapidly populations grow, they eventually reach a limit imposed by shortages of important environmental factors, such as space, light, water, or nutrients. A population ultimately may stabilize at a certain size, called the carrying capacity of the particular place where it lives. The **carrying capacity,** symbolized by K, is the maximum number of individuals that the environment can support.

As a population approaches its carrying capacity, its rate of growth slows greatly, because fewer resources remain for each new individual to use. The growth curve of such a population, which is always limited by one or more factors in the environment, can be approximated by the following logistic growth equation:

$$\frac{dN}{dt} = rN\left(\frac{K - N}{K}\right)$$

In this model of population growth, the growth rate of the population (dN/dt) is equal to its intrinsic rate of natural increase (r multiplied by N, the number of individuals present at any one time), adjusted for the amount of resources available. The adjustment is made by multiplying rN by the fraction of K, the carrying capacity, still unused [$(K - N)/K$]. As N increases, the fraction of resources by which r is multiplied becomes smaller and smaller, and the rate of increase of the population declines.

Graphically, if you plot N versus t (time), you obtain a **sigmoidal growth curve** characteristic of many biological populations. The curve is called "sigmoidal" because its shape has a double curve, like the letter *S*. As the size of a population stabilizes at the carrying capacity, its rate of growth slows, eventually coming to a halt (blue line in figure 38.14).

In mathematical terms, as N approaches K, the *rate* of population growth (dN/dt) begins to slow, reaching 0 when $N = K$ (figure 38.15). Conversely, if the population size exceeds the carrying capacity, then $K - N$ will be negative, and the population will experience a negative growth rate. As the population size then declines toward the carrying capacity, the magnitude of this negative growth rate will decrease until it reaches 0 when $N = K$.

Notice that the population tends to move toward the carrying capacity regardless of whether it is initially above or below it. For this reason, logistic growth tends to return a population to the same size. In this sense, such populations are considered to be in equilibrium, because they would be expected to be at or near the carrying capacity at most times.

In many cases, real populations display trends corresponding to a logistic growth curve. This is true not only in the laboratory but also in natural populations (figure 38.16*a*). In some cases, however, the fit is not perfect (figure 38.16*b*), and as we will see in section 38.5, many populations exhibit other patterns.

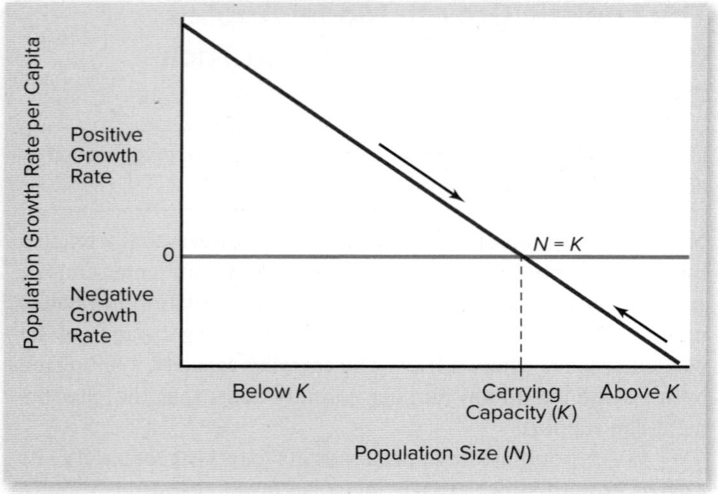

Figure 38.15 Relationship between population growth rate and population size. Populations far from the carrying capacity (*K*) have high growth rates—positive if the population is below *K*, and negative if it is above *K*. As the population approaches *K*, growth rates approach zero.

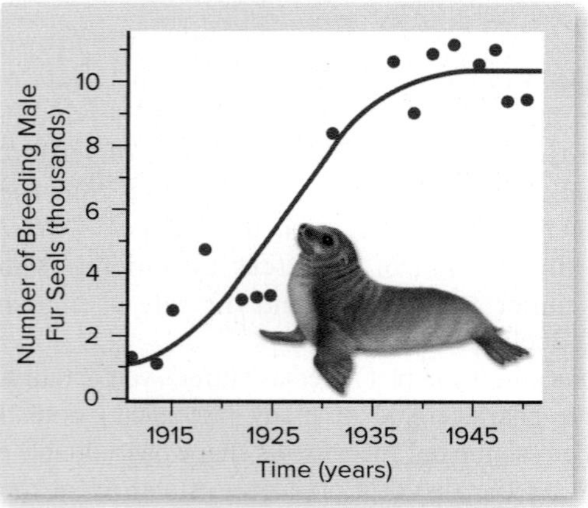

a.

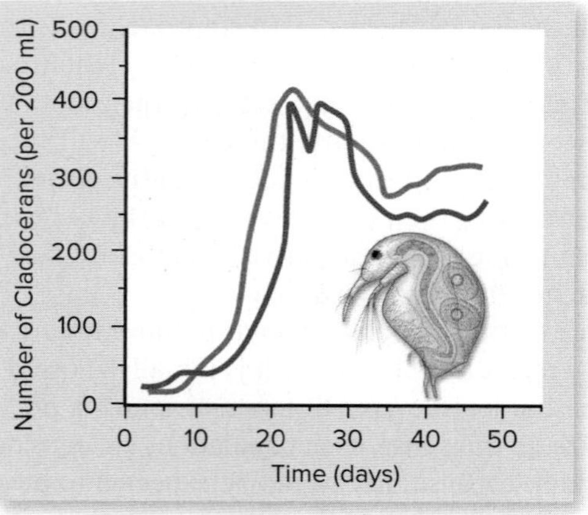

b.

Figure 38.16 Many populations exhibit logistic growth. ***a.*** A fur seal (*Callorhinus ursinus*) population on St. Paul Island, Alaska. ***b.*** Two laboratory populations of the cladoceran *Bosmina longirostris*. Note that the populations first exceeded the carrying capacity before decreasing to a size that was then maintained.

REVIEW OF CONCEPT 38.4

In exponential growth, the growth rate is equal to the intrinsic rate of increase times population size. Carrying capacity (*K*) is the largest population an environment will support. Logistic growth is limited by carrying capacity. Growth rate slows near *K*, is zero at *K*, and is negative above *K*.

- *What might cause a population's carrying capacity to change, and how would the population respond?*

38.5 Resource Availability Regulates Population Growth

A number of factors may affect population size through time. Some of these factors depend on population size and are therefore termed *density-dependent.* Other factors, such as natural disasters, affect populations regardless of size; these factors are termed *density-independent.* Many populations exhibit cyclic fluctuations in size that may result from complex interactions of factors.

Density-Dependent Effects Occur When Reproduction and Survival Are Affected by Population Size

LEARNING OBJECTIVE 38.5.1 Compare density-dependent and density-independent factors affecting population growth.

The reason population growth rates are affected by population size is that many important processes have **density-dependent effects.** That is, as population size increases, either reproductive rates decline or mortality rates increase, or both, a phenomenon termed *negative feedback* (figure 38.17).

Populations can be regulated in many different ways. When populations approach their carrying capacity, competition for resources can be severe, leading to both a decreased birth rate and an increased risk of death. This relationship is evident among the island populations of song sparrows considered in the Inquiry & Analysis feature at the end of this chapter. In addition, predators often focus their attention on a particularly common prey species, which also results in increasing rates of mortality as populations increase. High population densities can also lead to an accumulation of toxic wastes in the environment.

Behavioral changes may also affect population growth rates. Some species of rodents, for example, become antisocial, fighting more, breeding less, and generally showing signs of stress. These behavioral changes result from hormonal actions, but their ultimate cause is not yet clear; most likely, they have evolved as adaptive responses to situations in which resources are scarce. In addition, in crowded populations, the population growth rate may decrease because of an increased rate of emigration of individuals attempting to find better conditions elsewhere (figure 38.18).

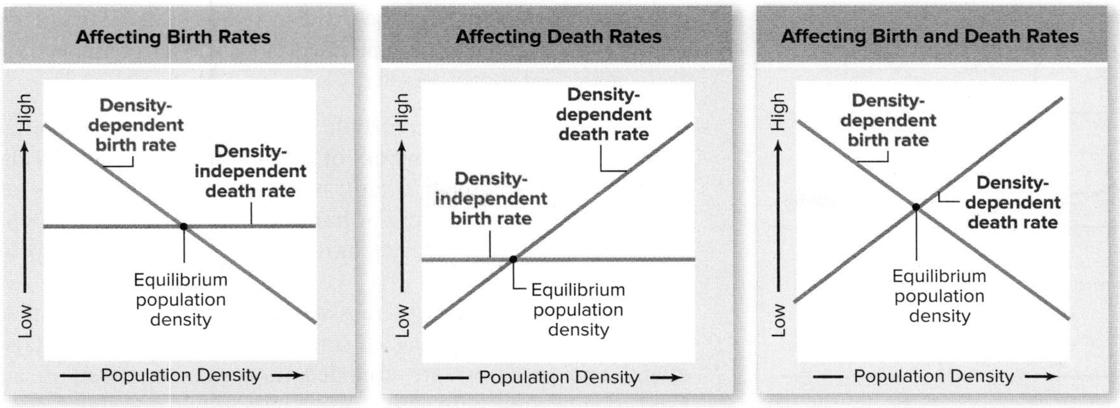
Affecting Birth Rates
High
Low
Density-dependent birth rate
Density-independent death rate
Equilibrium population density
Population Density
Affecting Death Rates
High
Low
Density-dependent death rate
Density-independent birth rate
Equilibrium population density
Population Density
Affecting Birth and Death Rates
High
Low
Density-dependent birth rate
Density-dependent death rate
Equilibrium population density
Population Density

Figure 38.17 Density-dependent population regulation. Density-dependent factors can affect birth rates, death rates, or both.

However, not all density-dependent factors are negatively related to population size. In some cases, growth rates increase with population size. This phenomenon is referred to as the **Allee effect** (after Warder Allee, who first described it) and is an example of *positive feedback.* The Allee effect can take several forms. Most obviously, in populations that are too sparsely distributed, individuals may have difficulty finding mates. Moreover, some species may rely on large groups to deter predators or to provide the necessary stimulation for breeding activities. The Allee effect is a major threat for many endangered species, which may never recover from decreased population sizes caused by habitat destruction, overexploitation, or other causes (refer to chapter 40).

Density-Independent Effects Include Environmental Disruptions and Catastrophes

LEARNING OBJECTIVE 38.5.2 Describe how density-independent factors limit population growth.

Figure 38.18 Density-dependent effects. Migratory locusts (*Locusta migratoria*) can create legendary plagues over large areas of Africa and Eurasia. At high population densities, the locusts have different hormonal and physical characteristics and take off as a swarm.

(left): Ruvan boshoff/E+/Getty Images; (right): Juniors Tierbildarchiv/Photoshot

Growth rates in populations sometimes do not correspond to the logistic growth equation. In many cases, such patterns result because growth is under the control of **density-independent effects.** In other words, the rate of growth of a population at any instant is limited by something unrelated to the size of the population.

A variety of factors may affect populations in a density-independent manner. Most of these are aspects of the external environment, such as extremely cold winters, droughts, storms, or volcanic eruptions. Individuals often are affected by these occurrences regardless of the size of the population.

Populations in areas where such events occur relatively frequently display erratic growth patterns in which the populations increase rapidly when conditions are benign but exhibit large reductions whenever the environment turns hostile (figure 38.19). Needless to say, such populations do not produce the sigmoidal growth curves characteristic of the logistic equation.

Population Cycles Are Driven by Complex Interactions

LEARNING OBJECTIVE 38.5.3 Evaluate the reasons that some populations cycle in size.

In some populations, density-dependent effects lead not to an equilibrium population size but to cyclic patterns of increase and decrease. For example, ecologists have studied cycles in hare populations since the 1820s. They have found that the North American snowshoe hare *(Lepus americanus)* follows a "10-year cycle" (in reality, the cycle varies from 8 to 11 years). Hare population numbers fall 10-fold to 30-fold in a typical cycle, and 100-fold changes can occur (figure 38.20). Two factors appear to be generating the cycle: food plants and predators.

Food plants. The preferred foods of snowshoe hares are willow and birch twigs. As hare density increases, the quantity of these twigs decreases, forcing the hares to feed on high-fiber (low-quality) food. Lower birth rates, low juvenile survivorship, and low growth rates follow. The

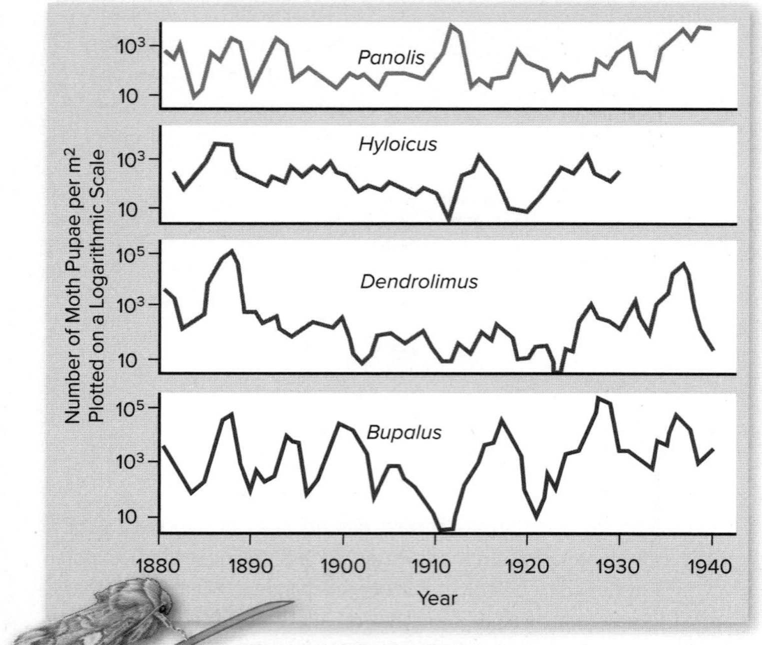

Figure 38.19 Fluctuations in the number of pupae of four moth species in Germany. The population fluctuations suggest that density-independent factors are regulating population size. The concordance in trends through time suggests that the same factors are regulating population size in all four species.

hares also spend more time searching for food, an activity that increases their exposure to predation. The result is a precipitous decline in willow and birch twig abundance and a corresponding fall in hare abundance. It takes two to three years for the quantity of mature twigs to recover.

Predators. A key predator of the snowshoe hare is the Canada lynx. The Canada lynx shows a "10-year" cycle of abundance that seems remarkably entrained to the hare abundance cycle (figure 38.20). As hare numbers increase, lynx numbers

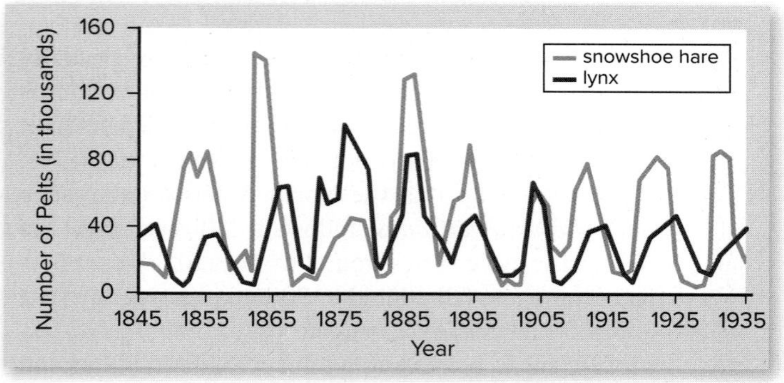

Figure 38.20 Linked population cycles of the snowshoe hare (*Lepus americanus*) and the northern lynx (*Lynx canadensis*). These data are based on records of fur returns from trappers in the Hudson Bay region of Canada. The lynx population carefully tracks that of the snowshoe hare but lags behind it slightly.

do too, rising in response to the increased availability of the lynx's food. When hare numbers fall, so do lynx numbers, their food supply depleted.

Which factor is responsible for the predator–prey oscillations? Do increasing numbers of hares lead to overharvesting of plants (a hare–plant cycle), or do increasing numbers of lynx lead to overharvesting of hares (a hare–lynx cycle)? Field experiments carried out by Charles Krebs and coworkers provided an answer.

In Canada's Yukon, Krebs set up experimental plots that contained hare populations. If food is added (no food-shortage effect) and predators are excluded (no predator effect) in an experimental area, hare numbers increase 10-fold and stay there—the cycle is lost. However, the cycle is retained if either of the factors is allowed to operate alone: exclude predators but don't add food (food-shortage effect alone), or add food in the presence of predators (predator effect alone). Thus, both factors can affect the cycle, which in practice seems to be generated by the interaction between the two.

Population cycles traditionally have been considered to occur rarely. However, a recent review of nearly 700 long-term studies (25 years or more) of trends within populations found that cycles were not uncommon; nearly 30% of the studies—including studies of birds, mammals, fish, and crustaceans—provided evidence of some cyclic pattern in population size through time, although most of these cycles are nowhere near as dramatic in amplitude as the hare–lynx cycles. In some cases, such as that of the snowshoe hare and lynx, density-dependent factors may be involved, whereas in other cases, density-independent factors, such as cyclic climatic patterns, may be responsible.

Resource Availability Affects Life History Adaptations

LEARNING OBJECTIVE 38.5.4 Distinguish between *K*-selected and *r*-selected populations.

As you have seen, some species usually maintain stable population sizes near the carrying capacity, whereas in other species population sizes fluctuate markedly and are often far below carrying capacity. The selective factors affecting such species differ markedly. Individuals in populations near their carrying capacity may face stiff competition for limited resources; by contrast, individuals in populations far below carrying capacity have access to abundant resources.

We have already described the consequences of such differences. When resources are limited, the cost of reproduction often will be very high. Consequently, selection will favor individuals that can compete effectively and utilize resources efficiently. Such adaptations often come at the cost of lowered reproductive rates. Such populations are termed ***K*-selected,** because they are adapted to thrive when the population is near its carrying capacity (*K*). Table 38.2 lists some of the typical features of *K*-selected populations. Examples of *K*-selected species include coconut palms, whooping cranes, whales, and humans.

TABLE 38.2 *r*-Selected and *K*-Selected Life History Adaptations

Adaptation	*r*-Selected Populations	*K*-Selected Populations
Age at first reproduction	Early	Late
Life span	Short	Long
Maturation time	Short	Long
Mortality rate	Often high	Usually low
Number of offspring produced per reproductive episode	Many	Few
Number of reproductions per lifetime	Few	Many
Parental care	None	Often extensive
Size of offspring or eggs	Small	Large

By contrast, in populations far below the carrying capacity, resources may be abundant. Costs of reproduction are low, and selection favors those individuals that can produce the maximum number of offspring. Selection here favors individuals with the highest reproductive rates; such populations are termed ***r*-selected.** Examples of organisms displaying *r*-selected life history adaptations include dandelions, aphids, mice, and cockroaches.

Most natural populations show life history adaptations that exist along a continuum ranging from completely *r*-selected traits to completely *K*-selected traits. Few populations show all of the traits listed in table 38.2. These attributes should be treated as generalities, with the recognition that many exceptions exist.

REVIEW OF CONCEPT 38.5

Density-dependent factors such as resource availability come into play in larger populations; density-independent factors such as natural disasters operate regardless of size. Population density may be cyclic due to complex interactions such as resource cycles and predator effects. Populations with density-dependent regulation often are near carrying capacity. In populations well below carrying capacity, natural selection favors high rates of reproduction.

- *Can a population experience both positive and negative density-dependent effects?*

38.6 Earth's Human Population Is Growing Explosively

Humans exhibit many *K*-selected life history traits, including small brood size, late reproduction, and a high degree of parental care. These life history traits evolved during the early history of hominids, when the limited resources available from the environment controlled population size.

Human Populations Have Been Growing Exponentially

LEARNING OBJECTIVE 38.6.1 Describe how the rate of human population growth has changed over time.

Throughout most of human history, our populations have been regulated by food availability, disease, and predators. Although unusual disturbances, including floods, plagues, and droughts, no doubt affected the pattern of human population growth, the overall size of the human population grew slowly during our early history.

Two thousand years ago, perhaps 130 million people populated the Earth. It took a thousand years for that number to double, and it was 1650 before it had doubled again, to about 500 million. In other words, for over 16 centuries the human population was characterized by very slow growth. In this respect, human populations resembled those of many other species with predominantly *K*-selected life history adaptations.

Starting in the early 1700s, changes in technology gave humans more control over their food supply, enabled them to develop superior weapons to ward off predators, and led to the development of cures for many diseases. At the same time, improvements in shelter and storage capabilities made humans less vulnerable to climatic uncertainties. These changes allowed humans to expand the carrying capacity of the habitats in which they lived and thus to escape the confines of logistic growth and re-enter the exponential phase of the sigmoidal growth curve.

Responding to the lack of environmental constraints, the human population has grown explosively over the last 300 years. Although the birth rate has remained unchanged at about 30 per 1000 per year over this period, the death rate has fallen dramatically, from 20 per 1000 per year to its present level of 13 per 1000 per year. The difference between birth and death rates meant that the population grew as much as 1.2% per year, although the rate has now declined to 1.1% per year.

A 1.1% annual growth rate may not seem large, but it has produced a current human population approaching 8 billion people (figure 38.21). At this growth rate, nearly 80 million people would be added to the world population in the next year, and the human population would double in 58 years. Both the current human population level and the projected growth rate have potentially grave consequences for our future.

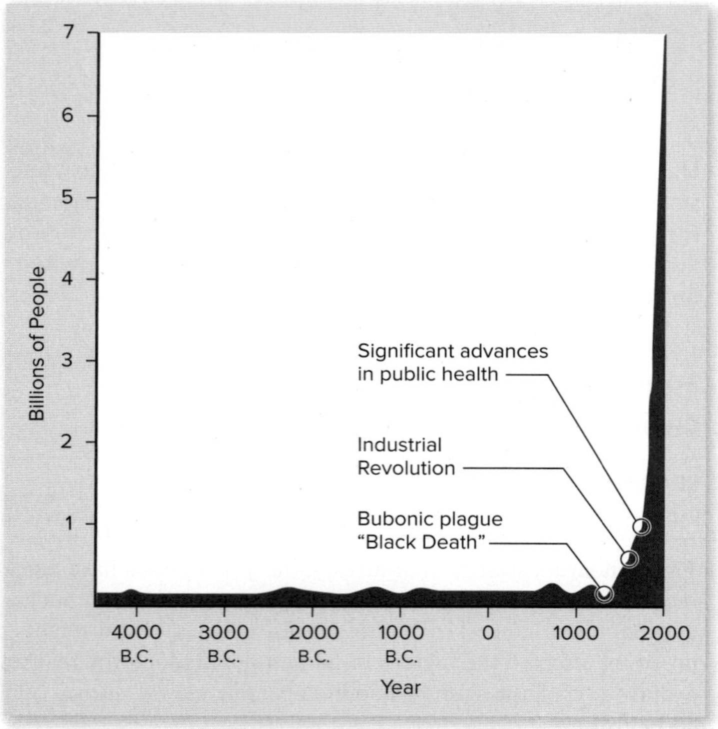

Figure 38.21 History of human population size. Temporary increases in death rate, even a severe one such as that occurring during the Black Death of the 1300s, have little lasting effect. Explosive growth began with the Industrial Revolution in the 1800s, which produced a significant, long-term lowering of the death rate. The current world population is nearing 8 billion, and at the present rate, it will double in 58 years.

Population Pyramids Reveal Birth and Death Trends

LEARNING OBJECTIVE 38.6.2 Describe the effects of age distribution on future growth.

Although the human population as a whole continues to grow rapidly in the 21st century, this growth is not occurring uniformly over the planet. Rather, most of the population growth is occurring in Africa, Asia, and Latin America (figure 38.22). By contrast, populations are decreasing in some countries in Europe.

The rate at which a population can be expected to grow in the future can be assessed graphically by means of a **population pyramid,** a bar graph displaying the numbers of people in each age category (figure 38.23). Males are conventionally shown to the left of the vertical age axis, females to the right. A human population pyramid thus displays the age composition of a population by sex. In most human population pyramids, the number of older females is disproportionately large compared with the number of older males, because females in most regions have a longer life expectancy than males.

Viewing such a pyramid, we can predict demographic trends in births and deaths. In general, a rectangular pyramid is characteristic of countries whose populations are stable, neither growing nor shrinking. A triangular pyramid is characteristic of a country that will exhibit rapid future growth because most of its population has not yet entered the childbearing years. Inverted triangles are characteristic of populations that are shrinking, usually as a result of sharply declining birth rates.

Examples of population pyramids for Sweden and Kenya in 2020 are shown in figure 38.23. The two countries exhibit very different age distributions. The nearly rectangular population pyramid for Sweden indicates that its population is not expanding, because birth rates have decreased and average life span has

Figure 38.22 Projected population growthfor the rest of this century. Developed countries are predicted to grow little; almost all of the population increase will occur in less-developed countries.

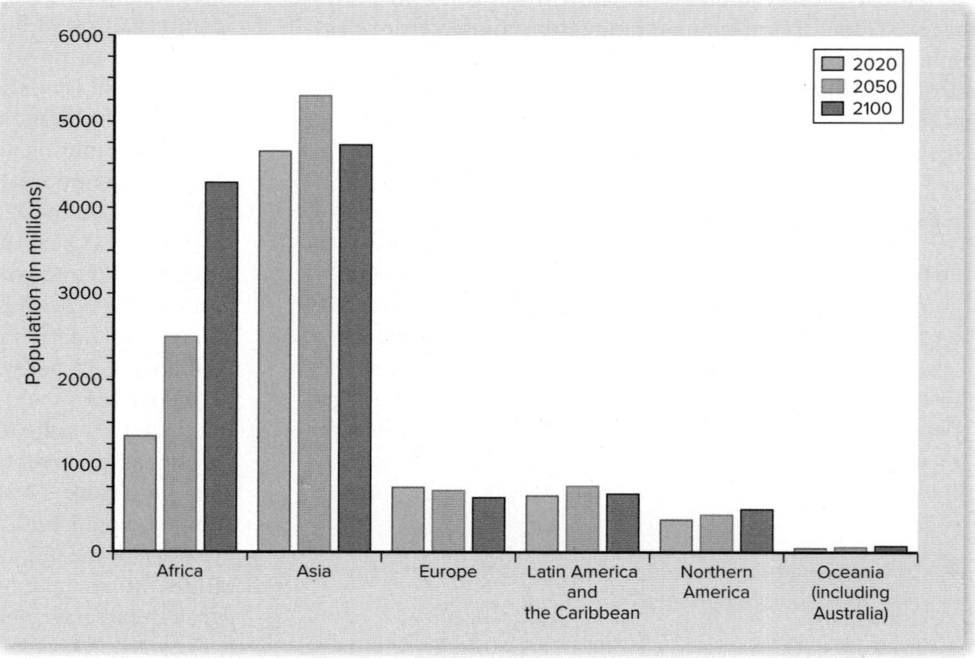

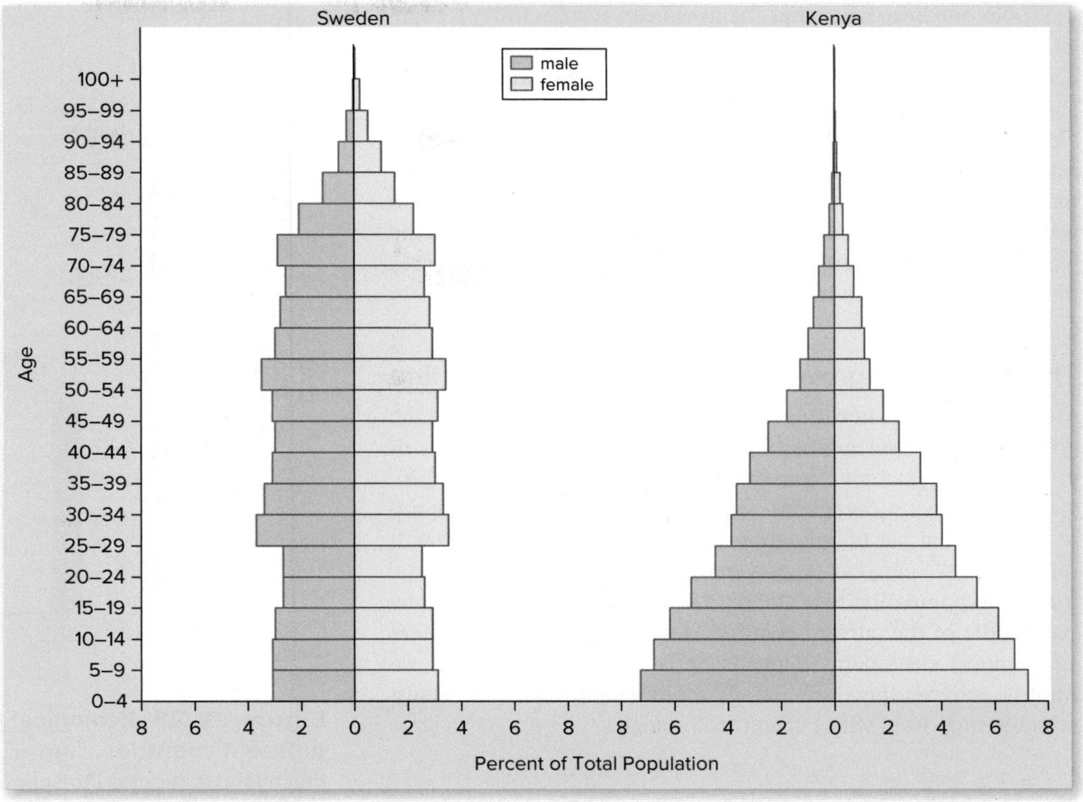

Figure 38.23 Population pyramids from 2020. Population pyramids are graphed according to a population's age distribution. Kenya's pyramid has a broad base, because of the great number of individuals below childbearing age. When the young people begin to bear children, the population will experience rapid growth. The Swedish pyramid exhibits many postreproductive individuals resulting from Sweden's long average life span.

increased. The very triangular pyramid of Kenya, by contrast, results from relatively high birth rates and shorter average life spans, which can lead to explosive future growth. The difference is most apparent when we consider that only about 18% of Sweden's population is less than 15 years old, compared with more than 40% of all Kenyans. Moreover, the fertility rate (offspring per woman) in Sweden is 1.9; in Kenya, it is 3.5. As a result, Kenya's population could double in 30 years, whereas Sweden's will remain stable.

Humanity's Future Growth Is Uncertain

LEARNING OBJECTIVE 38.6.3 Describe a typical American ecological footprint.

Earth's rapidly growing human population constitutes perhaps the greatest challenge to the future of the **biosphere,** the world's interacting community of living things. Humanity is adding 80 million people a year to its population—over a million every five days, 160 every minute! In more rapidly growing countries, the resulting population increase is staggering. India, for example, had a population of 1.05 billion in 2002; by 2050, its population likely will exceed 1.6 billion.

A key element in the world's population growth is its uneven distribution among countries. Of the billion people added to the world's population in the 1990s, 90% live in developing countries (figure 38.24). The fraction of the world's population that lives in industrialized countries is therefore diminishing. In 1950, fully one-third of the world's population lived in industrialized countries; by 1996, that proportion had fallen to one-quarter; and by 2020 the proportion had fallen to one-sixth. In the future, the world's population growth will be centered in the parts of the world least equipped to deal with the pressures of rapid growth.

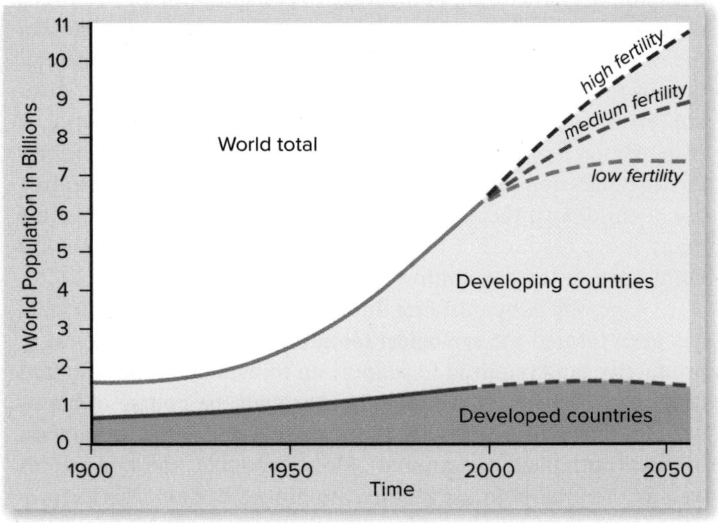

Figure 38.24 Distribution of population growth. Most of the worldwide increase in population since 1950 has occurred in developing countries. The age structures of developing countries indicate that this trend will increase in the near future. World population in 2050 likely will be between 7.3 and 10.7 billion, according to a recent United Nations study. Depending on fertility rates, the population at that time will be increasing either rapidly or slightly, or in the best case, declining slightly.

No one knows whether the world can sustain today's population of almost 8 billion people. As chapter 40 outlines, the world ecosystem is already under considerable stress. We cannot reasonably expect to expand its carrying capacity indefinitely, and indeed we already seem to be stretching the limits. Despite our using an estimated 45% of the total biological productivity of Earth's landmasses and more than one-half of all renewable sources of fresh water, between one-fourth and one-eighth of all people in the world are malnourished. If we are to avoid catastrophic increases in the death rate, birth rates must fall dramatically.

The population growth rate is beginning to decline

Most countries are devoting considerable attention to slowing the growth rate of their populations, and there are genuine signs of progress. For example, from 1984 to 2014, family planning programs in Kenya succeeded in reducing the fertility rate from 8.0 to 4.3 children per couple, thus lowering the population growth rate from 4.0% per year to 2.6% per year. Because of such efforts, the global population may stabilize at about 8.9 billion people by the middle of the current century. How many people the planet can support sustainably depends on the quality of life that we want to achieve; there are already more people than can be sustainably supported with current technologies.

Consumption in the developed world further depletes resources

Population size is not the only factor that determines resource use; per capita consumption is also important. In this respect, we in the industrialized world need to pay more attention to lessening the impact each of us makes, because even though the vast majority of the world's population is in developing countries, the overwhelming percentage of consumption of resources occurs in the industrialized countries. The wealthiest 20% of the world's population accounts for 86% of the world's consumption of resources and produces 53% of the world's carbon dioxide emissions, whereas the poorest 20% of the world is responsible for only 1.3% of consumption and 3% of carbon dioxide emissions. Looked at another way, in terms of resource use, a child born today in the industrialized world will consume many more resources over the course of his or her life than a child born in the developing world.

One way of quantifying this disparity is by calculating what has been termed the **ecological footprint,** which is the amount of productive land required to support an individual at the standard of living of a particular population through the course of his or her life. This figure estimates the acreage used for the production of food (both plant and animal), forest products, and housing, as well as the area of forest required to absorb carbon dioxide produced by the combustion of fossil fuels. As figure 38.25 illustrates, the ecological footprint of an individual in the United States is more than 8 times greater than that of someone in Nigeria.

Based on these measurements, researchers have calculated that resource use by humans is now one-third greater than the amount that nature can sustainably replace. Moreover, consumption is increasing rapidly in parts of the developing world; if all humans lived at the standard of living in the industrialized world, two additional planet Earths would be needed.

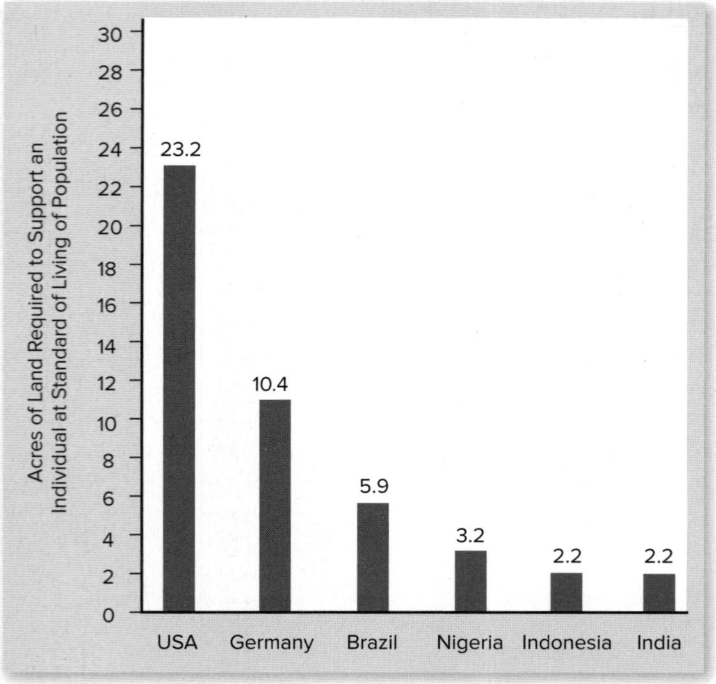

Figure 38.25 Ecological footprints of individuals in different countries. An ecological footprint calculates how much land is required to support a person through his or her life, including the acreage used for production of food, forest products, and housing, in addition to the forest required to absorb the carbon dioxide produced by the combustion of fossil fuels.

Building a sustainable world is the most important task facing humanity's future. The quality of life available to our children will depend to a large extent on our success in limiting both population growth and the amount of per capita resource consumption.

REVIEW OF CONCEPT 38.6

For most of its history, the human population increased gradually. In the last 400 years, the human population has grown exponentially; at the current rate, it would double in 58 years. Population pyramids show the number of individuals in different age categories. Pyramids with a wide base are undergoing faster growth. Growth rates overall are declining but the distribution of rates is uneven across countries. Overconsumption of resources is a challenge to sustainability.

- *Which is more important, reducing global population growth or reducing resource consumption in developed countries?*

38.7 Pandemics and Human Health

Disease outbreaks can play a major role in determining population dynamics, in some cases possibly even causing the extinction of a species.

Pandemics Occur in Plants and Animals, Including Humans

LEARNING OBJECTIVE 38.7.1 Understand the various ways pandemics have occurred in plant, animal, and human populations.

The chestnut blight, a pathogenic (disease-causing) fungus, was accidentally introduced from Asia to North America in 1904. The American chestnut, formerly a dominant tree in eastern forests, has never recovered. Similarly, the rinderpest virus, introduced with a shipment of cattle from India, wiped out almost all of the cattle and many wildlife species—including Cape buffalo, giraffe, and wildebeest—in eastern and southern Africa in the late 19th century.

In some cases, disease has even intentionally been introduced to reduce the population size of out-of-control non-native species. Rabbits, for example, were brought to Australia from Europe in the 18th century and in the absence of predators became a huge scourge, eating all available vegetation and damaging both native animal and sheep populations. When the myxoma virus was introduced by wildlife managers, more than 99% of the rabbits perished, although the population eventually rebounded.

History shows that human populations, too, have been affected by disease outbreaks. The term *pandemic* is now commonly used to refer to widespread disease outbreaks extending over multiple countries and even continents, usually in reference to human populations (*epidemic* is the term for more localized outbreaks). Perhaps the most notorious human pandemic is the bubonic plague, caused by a bacterium, *Yersinia pestis*, and spread by fleas and through contact with infected individuals. This was one of a series of plagues caused by *Y. pestis* over the last several thousand years, with the "Black Death" being the worst. This pandemic killed a third of the population of Europe in the fourteenth century (see figure 38.21). In more recent times, the Spanish flu killed more than 50 million people in 1918–1919, and the global toll of COVID-19 that erupted early in 2020 is still being calculated, with more than four million lives lost as of the end of July 2021.

Pandemics Result from Pathogens Infecting a Host Population Lacking Defenses

LEARNING OBJECTIVE 38.7.2 Describe how a disease spreads through a population.

A pandemic occurs when a pathogen becomes established in a population in which the host individuals do not have adequate immune system defenses (refer to chapter 35). This can occur when a population has no prior exposure to the pathogen, as occurs when a disease-causing agent moves from one species to another. We call these zoonotic diseases, and these emerging diseases are an increasing human health problem. In other cases, such as influenza (refer to chapter 23), the pathogen can evolve in such a way that the host individual's immune system no longer defends against it.

Degree of disease spread determines whether a pandemic occurs

An important factor determining the extent of a disease outbreak is the effectiveness of the pathogen in spreading from one host individual to another. If, on average, each infected individual infects one more individual, then the prevalence of the disease will remain constant in the population (the technical measure is termed the reproduction number, R_0, pronounced "R naught"; figure 38.26. If, on the other hand, each individual passes on the pathogen to more than one other individual ($R_0 > 1$), then the disease will increase; the greater the number of new individuals becoming infected per host, the more rapidly the disease will increase. For example, the R_0 of SARS-CoV-2, the virus that causes COVID-19, was estimated to be somewhere between 2 and 3 at the outset of the pandemic. Conversely, if each individual produces less than one new infection on average ($R_0 < 1$), then the disease will eventually disappear from the population.

The key to disease dynamics, then, is the transmission rate. Some pathogens do not spread easily (they are not very "contagious") and thus are not likely to cause an epidemic. For example, Nipah virus, transmitted to humans from pigs and bats, causes a deadly disease that kills 40% or more of humans infected, but because it has a low R_0—about 0.48—large outbreaks have not occurred.

For other diseases, reducing transmission rates is a key to stopping pandemics. That is the reason that decreasing the frequency of unprotected sexual interactions is important to combating sexually transmitted diseases such as AIDS (refer to chapter 23).

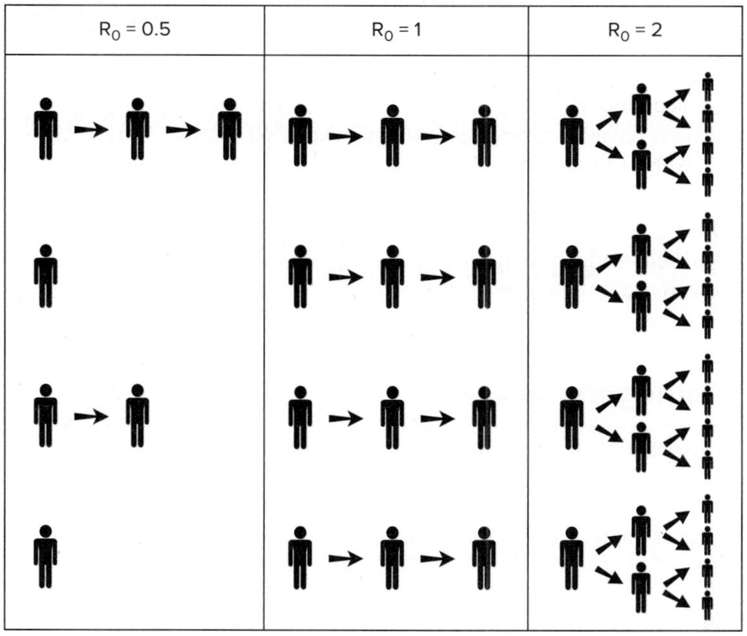

Figure 38.26 Reproduction number and pandemic outcome. The number of new individuals infected on average from each infected individual (R_0) determines whether an outbreak will occur or die out. A value of $R_0 > 1$ will lead to increasing prevalence of the disease, whereas values <1 will lead the disease to eventually disappear (as shown here, the disease is decreasing because half of the individuals do not pass it on to someone else).

Similarly, mask-wearing and maintaining social distance decreases the incidence of airborne diseases such as the flu and COVID-19.

The COVID-19 pandemic, however, has made clear that focusing on averages can be misleading, because some individuals transmit the disease more than others. According to one estimate, 10% of people infected with SARS-CoV-2 were responsible for 80% of subsequent infections. Why some individuals turn out to be "superspreaders" is not clear and probably involves both particular circumstances (for example, the individual's attendance at a crowded indoor event) and biological differences among individuals. Figuring out how to identify these individuals and prevent the clusters of disease that result is a critical area of current research.

Herd immunity provides protection

As we discussed in chapter 35, individuals can develop immunity to a disease. As the proportion of a population that is immune increases, the transmission rate of the pathogen will decline because there are fewer opportunities for it to spread from one individual to another. Eventually, R_0 will drop below 1 and, all else being equal, the disease will disappear. The phenomenon when the proportion of the population that is immune is large enough to cause the pathogen to stop spreading is called "herd immunity" (the term refers to the fact that at this point, even those who are not immune are protected by the immunity of most others). The proportion of the population that needs to be immune to produce herd immunity depends on the biology of the pathogen and particular circumstances, but often is at least 70%.

Given enough time, herd immunity will develop in a population (assuming that individuals are able to become immune), but it comes at a high cost of a large proportion of the population being exposed to the pathogen and potentially suffering adverse consequences (depending on what proportion of infected individuals develop severe disease symptoms; 40 to 50% of individuals infected with the COVID-19 virus are asymptomatic).

An alternative approach to developing herd immunity is through vaccines, which lead to a large proportion of the population becoming immune without being infected. Vaccines have been effective at completely eliminating many terrible diseases, such as smallpox and rinderpest, and bringing many others under control.

Pathogens Evolve

LEARNING OBJECTIVE 38.7.3 Explain how pathogens evolve.

It is important to remember that disease-causing agents—even viruses—can evolve. Any mutation that causes a pathogen to more effectively infect new individuals, and thus more successfully pass that mutation on to the next generation, will be favored by natural selection. Disease agents can do so in several ways. One way is to infect more individuals—that is, to become more contagious. This occurred with SARS-CoV-2 when a new variant (that is, a type of SARS-CoV-2 with several different mutations, sometimes called a "strain") appeared in late 2020 in England, and was transmitted at a rate estimated at 70% greater than the previously common strain. Subsequently, an even more contagious variant, the Delta variant, appeared in 2021 with an R_0 estimated to be between 6 and 9. This is typical of virus evolution (refer to chapter 23 on influenza virus evolution, for example).

Another way a pathogen can adapt is by evolving to overcome host individuals' acquired immunity. For example, because of its high mutation rate, the human influenza virus generates a substantial variation in surface HA and NA proteins; some of these mutations by chance render the virus unrecognizable by the human immune systems and thus are favored by natural selection (refer to chapter 23). It is because of this high rate of influenza virus evolution that flu vaccines quickly lose their effectiveness and thus new vaccines are needed every year.

Pathogens can also evolve by becoming less lethal. Although seemingly counterintuitive, the evolution of less lethality means that hosts will live longer and thus have more opportunity to pass the disease to someone else. For example, after wiping out nearly all of the rabbits in Australia, the myxoma virus evolved to become less lethal, which allowed the rabbit population to recover.

REVIEW OF CONCEPT 38.7

Disease outbreaks can play a major role in determining population dynamics, in some cases possibly even causing the extinction of a species. A pandemic occurs when a pathogen becomes established in a population in which the host individuals do not have adequate immune system defenses. An important factor determining the extent of a disease outbreak is the effectiveness of the disease-causing pathogen in spreading from one host individual to another, specifically whether R_0 is greater or less than 1. Actions can be taken to reduce pathogen spread; the development of herd immunity can do so as well. However, pathogens can evolve to counter attempts to eliminate them.

- *How would the implementation of a policy in which individuals with a disease are quarantined lead to evolution of a pathogen?*

Are Island Populations of Song Sparrows Density-Dependent?

Island populations that are isolated, receiving no visitors from other populations, provide an attractive opportunity to test the degree to which a population's growth rate is affected by its size. A population's size can influence the rate at which it grows, because an increase in the number of individuals within a population tends to deplete available resources, leading to an increased risk of death by deprivation. Also, predators tend to focus their attention on common prey, resulting in increasing rates of mortality as populations grow. However, simply knowing that a population is decreasing in numbers does not tell you that the decrease has been caused by the size of the population. Many factors, such as severe weather, volcanic eruption, and human disturbance, can influence island population sizes, too.

The graph displays data collected from 13 song sparrow populations on Mandarte Island (presented in the graph). In an attempt to gauge the effect of population size on the evolutionary success of these populations, a census was taken of each population, and for each the juvenile mortality rate was estimated. On the graph, these juvenile mortality rates have been plotted against the number of breeding adults in each population. Although the data seem scattered, the "best-fit" regression line is statistically significant (statistically significant means that there is a less than 5% chance that there is, in fact, no correlation between dependent and independent variables).

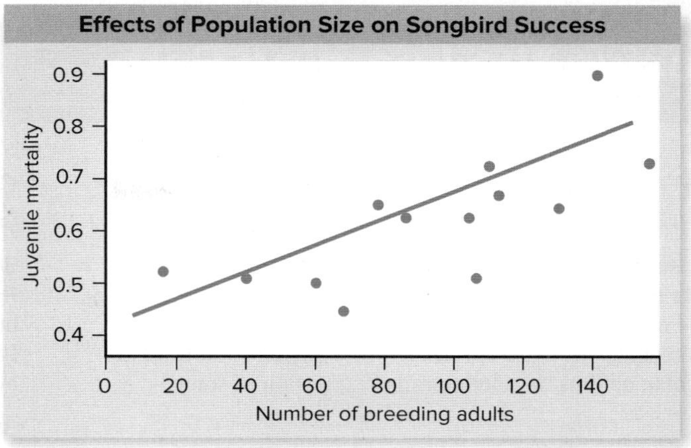

Analysis

1. **Applying Concepts**
   Analyzing scattered data. What is the size of the song sparrow population (based on breeding adults) with the lowest juvenile mortality rate? With the highest?

2. **Interpreting Data**
   a. What is the average juvenile mortality of all 13 populations, estimated from the 13 points on the graph?
   b. How many populations were observed to have juvenile mortality rates below this average value? What is the average size of these populations?
   c. How many populations were observed to have juvenile mortality rates above this average value? What is the average size of these populations?

3. **Making Inferences** Are the populations with lower juvenile mortality bigger or smaller than the populations with higher juvenile mortality?

4. **Drawing Conclusions** Do the population sizes of these song sparrows appear to exhibit density dependence?

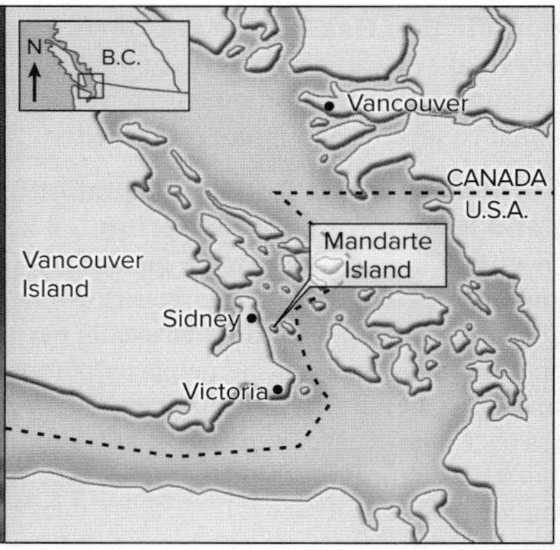

(left): Emmanuel Rondeau/Alamy Stock Photo

Retracing the Learning Path

CONCEPT 38.1 Populations Are Groups of a Single Species in One Place

38.1.1 A Population's Geographical Distribution Is Termed Its Range Populations have limited geographic ranges that expand or contract through time as the environment changes. Dispersal mechanisms allow species to cross barriers and expand ranges. Human actions have led to range expansion of some species.

38.1.2 Individuals in Populations Exhibit Different Spacing Patterns Individuals in a population are distributed randomly, distributed uniformly, or clumped. Nonrandom distributions may reflect resource distributions or competition for resources.

38.1.3 A Metapopulation Comprises Distinct Populations That May Exchange Members Exchange between populations in a metapopulation is highest when populations are large and more connected. Metapopulations may act as a buffer against extinction.

CONCEPT 38.2 Population Growth Depends upon Members' Age and Sex

38.2.1 Sex Ratio and Generation Time Affect Population Growth Rates Abundant females, a short generation time, or both can be responsible for more rapid population growth.

38.2.2 Age Structure Is Determined by the Numbers of Individuals in Different Age Groups The age structure of a population affects growth as age cohorts have characteristic fecundity and death rates. Life tables show probability of survival and reproduction through a cohort's life span. Survivorship curves show how survival probability changes with age. Some populations have high survivorship until old age, whereas in others survivorship is lowest among the youngest.

CONCEPT 38.3 Evolution Favors Life Histories That Maximize Lifetime Reproductive Success

38.3.1 Reproduction Has Significant Costs That Affect Parental Survival Because resources are limited, reproduction has a cost. Resources allocated toward current reproduction cannot be used to enhance survival and future reproduction.

38.3.2 A Trade-Off Exists Between Number of Offspring and Investment per Offspring When reproductive cost is high, fitness can be maximized by delaying reproduction, or by producing a few large-size young that have a greater chance of survival. Reproductive events per lifetime represent an additional trade-off. Semelparity is reproduction once in a single large event. Iteroparity is production of offspring several times over many seasons. Age at first reproduction correlates with life span. Longer-lived species delay first reproduction longer.

CONCEPT 38.4 Environment Limits Population Growth

38.4.1 The Exponential Growth Model Applies to Populations with No Growth Limits The rate of population increase, r, is defined as birth rate, b, minus death rate, d. Exponential growth occurs when a population is not limited by resources or by other species.

38.4.2 The Logistic Growth Model Applies to Populations That Approach Their Carrying Capacity Carrying capacity is the largest population an environment will support. Logistic growth is limited by carrying capacity. Growth rate slows near carrying capacity and is negative above it.

CONCEPT 38.5 Resource Availability Regulates Population Growth

38.5.1 Density-Dependent Effects Occur When Reproduction and Survival Are Affected by Population Size Density-dependent factors include increased competition and predation. To stabilize a population size, birth rates must decline, death rates must increase, or both.

38.5.2 Density-Independent Effects Include Environmental Disruptions and Catastrophes Density-independent factors are not related to population size and include environmental events that result in mortality.

38.5.3 Population Cycles Are Driven by Complex Interactions In some cases, population size is cyclic because of the interaction of factors such as food supply and predation.

38.5.4 Resource Availability Affects Life History Adaptations Populations at carrying capacity have adaptations to compete for limited resources; populations well below carrying capacity exhibit a high reproductive rate to use abundant resources.

CONCEPT 38.6 Earth's Human Population Is Growing Explosively

38.6.1 Human Populations Have Been Growing Exponentially Technology and other innovations have increased carrying capacity and decreased mortality.

38.6.2 Population Pyramids Reveal Birth and Death Trends Populations with many young individuals are likely to experience high growth rates as those individuals reach reproductive age.

38.6.3 Humanity's Future Growth Is Uncertain The human population is unevenly distributed. Rapid growth occurs in developing countries, whereas most resources are utilized by the industrialized world. The population growth rate has recently declined slightly. Even at this lower growth rate, the number of individuals on the planet is likely to plateau at 7 to 10 billion. A sustainable future requires limits both to population growth and to per capita resource consumption.

CONCEPT 38.7 Pandemics and Human Health

38.7.1 Pandemics Occur in Plants and Animals, Including Humans Disease outbreaks can devastate populations of plants, animals, and people. Humans have intentionally created pandemics in attempts to eradicate some pest species.

38.7.2 Pandemics Result from Pathogens Infecting a Host Population Lacking Defenses The degree to which each infected individual infects others determines whether the disease will turn into a pandemic or die out.

38.7.3 Pathogens Evolve Natural selection favors mutations that increase the ability of disease-causing organisms to survive in their hosts and infect them.

Concept Overview

This Concept Overview diagrams the key concepts that were discussed in this chapter.

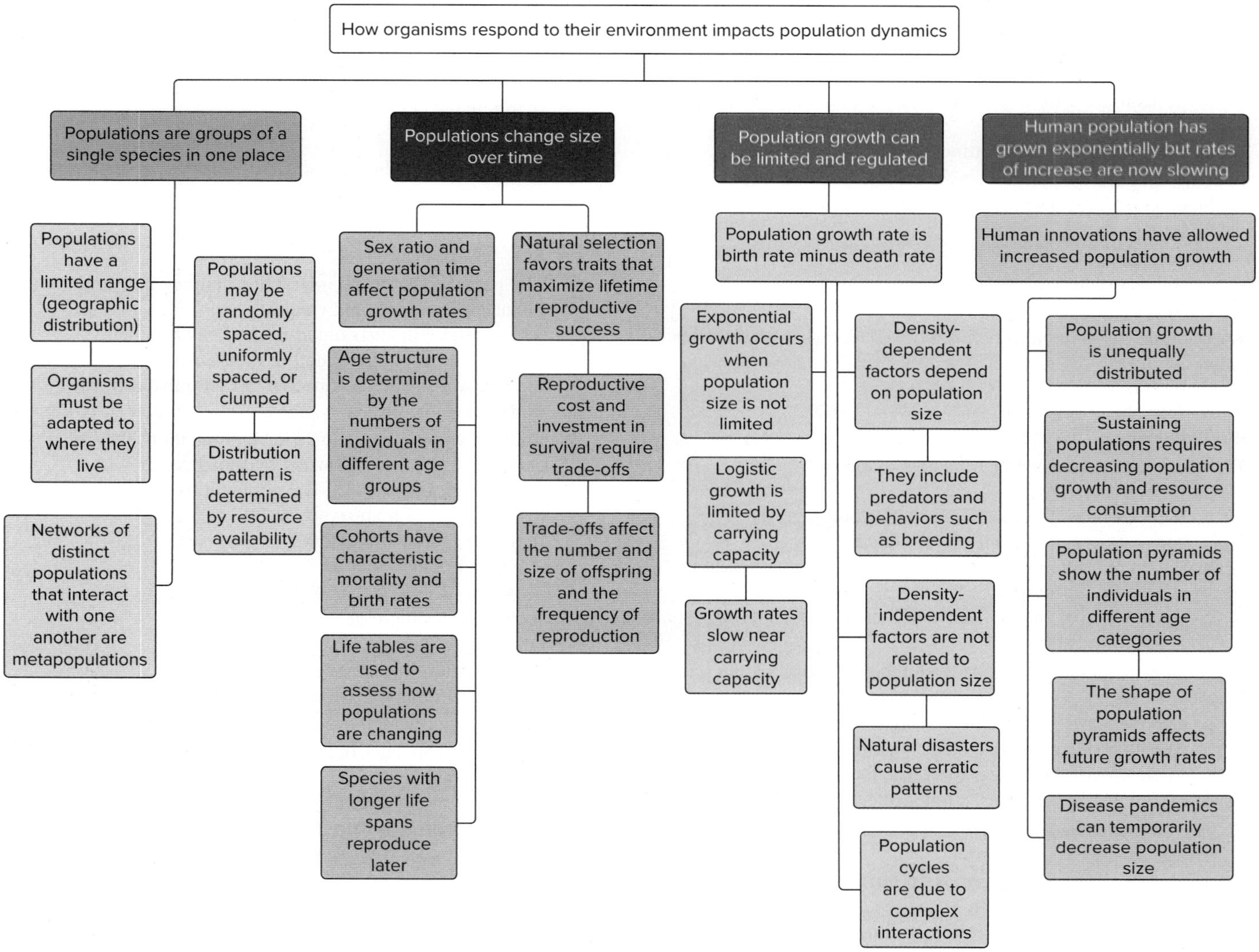

Assessing the Learning Path

Understand

1. In a population in which individuals are uniformly distributed,
   a. the population is probably well below its carrying capacity.
   b. natural selection should favor traits that maximize the ability to compete for resources.
   c. immigration from other populations is probably keeping the population from going extinct.
   d. None of the above
2. Source–sink metapopulations are distinct from other types of metapopulations, because
   a. exchange of individuals occurs only in the former.
   b. populations with negative growth rates are a part of the former.
   c. populations never go extinct in the former.
   d. all populations eventually go extinct in the former.
3. Demography is the study of changes in a population's
   a. size.
   b. ecology.
   c. social structure.
   d. range.
4. Fecundity is defined as the number of offspring produced per
   a. lifetime.
   b. brood.
   c. year.
   d. individual.
5. When ecologists talk about the cost of reproduction, they mean
   a. the reduction in future reproductive output as a consequence of current reproduction.
   b. the amount of calories it takes for all the activity used in successful reproduction.
   c. the amount of calories contained in eggs or offspring.
   d. None of the above
6. A life history trade-off between clutch size and offspring size means that
   a. as clutch size increases, offspring size increases.
   b. as clutch size increases, offspring size decreases.
   c. as clutch size increases, adult size increases.
   d. as clutch size increases, adult size decreases.
7. The difference between exponential and logistic growth rates is that
   a. exponential growth depends on birth and death rates, and logistic growth does not.
   b. in logistic growth, emigration and immigration are unimportant.
   c. both are affected by density, but logistic growth is slower.
   d. only logistic growth reflects density-dependent effects on births or deaths.
8. The maximal number of organisms that the environment can support is the
   a. biotic potential.
   b. trophic level.
   c. carrying capacity.
   d. reproductive maximum.
9. Which of the following is an example of a density-dependent effect on population growth?
   a. An extremely cold winter
   b. A tornado
   c. An extremely hot summer in which cool burrow retreats are fewer than the number of individuals in the population
   d. A drought
10. All of the following are characteristics of *K*-selected species EXCEPT
    a. they have a stable population size.
    b. they undergo stiff resource competition.
    c. they have high reproductive rates.
    d. they live in populations near their carrying capacity.
11. The current global human population is about
    a. 7 billion.
    b. 8 billion.
    c. 9 billion.
    d. 6 billion.
12. The world's population is currently growing at a rate of about ____ million people a year.

Apply

1. The potential for social interactions among individuals should be maximized when individuals
   a. are randomly distributed in their environment.
   b. are uniformly distributed in their environment.
   c. have a clumped distribution in their environment.
   d. None of the above
2. Some of a metapopulation's local populations may become extinct for all of the following reasons EXCEPT
   a. isolation from other local populations.
   b. low levels of homozygosity.
   c. low resource availability.
   d. small population size.
3. Populations grow if any of the following occur EXCEPT
   a. birth rates increase.
   b. sex ratio changes toward more males.
   c. generation time shortens.
   d. births outnumber deaths.
4. In a type I survivorship curve (figure 38.8), mortality
   a. rises steeply later in life.
   b. is the same at any age.
   c. cannot be predicted by age.
   d. is concentrated before reproduction.
5. In populations subjected to high levels of predation,
   a. individuals should invest little in reproduction so as to maximize their own survival.
   b. individuals should produce few offspring and invest little in any of them.
   c. individuals should invest greatly in reproduction, because their chance of surviving to another breeding season is low.
   d. individuals should stop reproducing altogether.
6. The logistic population growth model, $dN/dt = rN[(K - N)/K]$, describes a population's growth when an upper limit to growth is assumed. As N approaches (numerically) the value of K,
   a. dN/dt increases rapidly.
   b. dN/dt approaches 0.
   c. dN/dt increases slowly.
   d. the population becomes threatened by extinction.

7. A one-time spike in nutrients in a pond often results in a huge spike in the amount of algae in the pond. If this spike exceeded the natural carrying capacity of the pond (K), what would happen next?
   a. The population would reach a new K.
   b. The death rate would spike.
   c. The population would move to a higher equilibrium point.
   d. The population would gradually decrease to K.
8. The elimination of predators by humans
   a. will cause prey to experience exponential growth until new predators arrive or evolve.
   b. may increase the population size of a prey species if that prey's population was being regulated by predation from the predator.
   c. will lead to an increase in the carrying capacity of the environment.
   d. will lead to an Allee effect.
9. When a population pyramid has a broad base, the population
   a. will shrink.
   b. is stable.
   c. has experienced a "baby boom."
   d. will experience rapid growth.
10. The world population growth rate is
   a. stable.
   b. declining.
   c. increasing.
   d. uncertain.

Synthesize

1. What are the implications for evolutionary divergence among populations that are part of a metapopulation versus populations that are independent of other populations?
2. Suppose you wanted to keep meadow grass in your room as a houseplant. Suppose, too, that you wanted to buy an individual plant that was likely to live as long as possible. What age plant would you buy? How might the shape of the survivorship curve affect your answer?
3. You've discovered how a population growing exponentially eventually reverts to logistic growth, but can you describe a situation in which a population that exhibits logistic growth can revert to exponential growth?
4. Refer to figure 38.11. Given a trade-off between current reproductive effort and future reproductive success (the so-called cost of reproduction), would you expect old individuals to have the same "optimal" reproductive effort as young individuals?
5. Refer to figure 38.12. Because the number of offspring that a parent can produce is often a trade-off with the size of individual offspring, many circumstances lead to an intermediate number and size of offspring being favored. If the size of an offspring were completely unrelated to the quality of that offspring (its chances of surviving until it reaches reproductive age), would you expect parents to fall on the left or right side of the x-axis (clutch size)? Explain.
6. Refer to figure 38.23. Would increasing the mean generation time have the same kind of effect on population growth rate as reducing the number of children that an individual female has over her lifetime? Which effect would have a bigger influence on population growth rate? Explain.

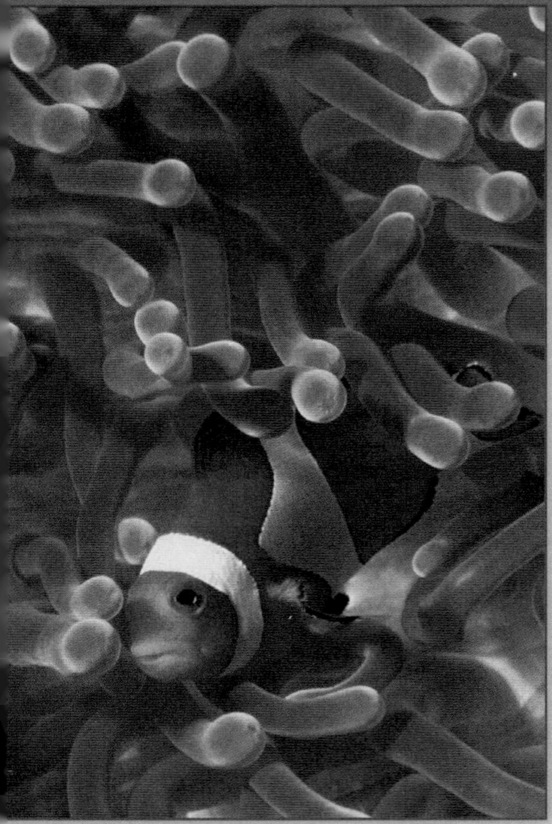

Fuse/Getty Images

39 Community Ecology and Ecosystem Dynamics

Learning Path

39.1 Competition Shapes How Species Live Together in Communities

39.2 Predator–Prey Relationships Foster Coevolution

39.3 Cooperation Among Species Can Lead to Coevolution

39.4 Ecological Succession Is a Consequence of Habitat Alteration

39.5 Chemical Elements Move Through Ecosystems in Biogeochemical Cycles

39.6 Energy Flows Through Ecosystems in One Direction

39.7 Biodiversity May Increase Ecosystem Stability

Concept Overview

This Concept Overview provides a diagram of the overarching concepts that are covered in this chapter. These concepts will be expanded upon in the Concept Overview at the end of the chapter.

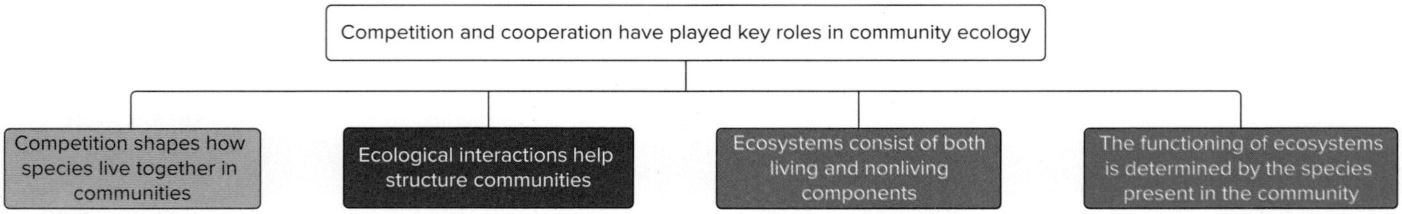

Competition and cooperation have played key roles in community ecology
Competition shapes how species live together in communities
Ecological interactions help structure communities
Ecosystems consist of both living and nonliving components
The functioning of ecosystems is determined by the species present in the community

Introduction

All the organisms that live together in a place are members of a community. The myriad of species that inhabit a tropical rainforest are a community. Indeed, every inhabited place on Earth supports its own particular array of organisms. Over time, the different species that live together have made many complex adjustments to community living, evolving together and forging relationships that give the community its character and stability. Both competition and cooperation have played key roles; in this chapter, we examine these and other factors in community ecology.

39.1 Competition Shapes How Species Live Together in Communities

A Community Is All the Species Living at One Site

LEARNING OBJECTIVE 39.1.1 Define *community.*

Almost any place on Earth is occupied by species—sometimes by many of them, as in the rainforests of the Amazon, and sometimes by only a few, as in the near-boiling waters of Yellowstone's geysers (where a number of microbial species live). The term **community** refers to the species that occur at any particular locality (figure 39.1). Communities can be characterized either by their constituent species or by their properties, such as **species richness** (the number of species present) or **primary productivity** (the amount of energy produced).

Interactions among community members govern many ecological and evolutionary processes. These interactions, such as predation and mutualism, affect the population biology of particular species—whether a population increases or decreases in abundance, for example—as well as the ways in which energy and nutrients cycle through the ecosystem. Moreover, the community context affects the patterns of natural selection faced by a species, and thus the evolutionary course it takes.

Scientists study biological communities in many ways, ranging from detailed observations to elaborate, large-scale experiments. Some studies focus on the entire community, whereas others focus only on a subset of species that are likely to interact with one another. Although scientists sometimes refer to such subsets as communities (for example, the "spider community"), the term **assemblage** is more appropriate to indicate that the species included are only a portion of those present within the entire community.

The Ecological Niche of a Species Is How It Uses Available Resources

LEARNING OBJECTIVE 39.1.2 Differentiate between fundamental and realized niches.

Each organism in a community confronts the challenge of survival in a different way. A **niche** encompasses all the ways an organism uses the resources of its environment. A niche may be described in terms of space utilization, food consumption, temperature range, appropriate conditions for mating, requirements for moisture, and other factors.

Sometimes species are not able to occupy their entire niche because of the presence or absence of other species. Species can interact with one another in a number of ways, and these interactions can have either positive or negative effects. One type of interaction, **interspecific competition,** occurs when two species attempt to use the same resource but the resource is insufficient to satisfy both. Physical interactions over access to resources—such as fighting to defend a territory or displacing an individual from a particular location—are referred to as **interference competition**; consuming the same resources is called **exploitative competition.**

Figure 39.1 An African savanna community.
A community consists of all the species—plants, animals, fungi, protists, and prokaryotes—that occur at a locality—in this case, Etosha National Park in Namibia.
Daryl & Sharna Balfour/Okopia/Science Source

The entire niche a species is capable of using, based on its physiological tolerance limits and resource needs, is called the **fundamental niche.** The actual set of environmental conditions, including the presence or absence of other species, in which the species can establish a stable population is its **realized niche.** Fundamental niches are potential; realized niches are actual. Because of interspecific interactions, the realized niche of a species may be considerably smaller than its fundamental niche.

Competition between species for niche occupancy

In a classic study, Joseph Connell of the University of California, Santa Barbara, investigated competitive interactions between two species of barnacles that grow together on rocks along the coast of Scotland. Of the two species Connell studied, *Chthamalus stellatus* lives in shallow water, where tidal action often exposes it to air, and *Semibalanus balanoides* lives lower down, where it is rarely exposed to the atmosphere (figure 39.2). Space is at a premium where the two species come into contact. In the deeper zone, *S. balanoides* could always outcompete *C. stellatus* by crowding it off the rocks, undercutting it, and replacing it even where it had begun to grow, an example of interference competition.

When Connell removed *S. balanoides* from the areas where it occurs, however, *C. stellatus* was easily able to occupy the deeper zone, indicating that no physiological or other general obstacles prevented it from becoming established there. In contrast, *S. balanoides* could not survive in the shallow-water habitats where *C. stellatus* normally occurs; it does not have the physiological adaptations to warmer temperatures that allow *C. stellatus* to occupy this zone. Thus, the fundamental niche of *C. stellatus* includes both shallow and deeper zones, but its realized niche is much narrower, because *C. stellatus* can be outcompeted by *S. balanoides* in parts of its fundamental niche. By contrast, the realized and fundamental niches of *S. balanoides* appear to be identical.

Other causes of niche restriction

Processes other than competition can also restrict the realized niche of a species. For example, the plant St. John's wort *(Hypericum perforatum)* was introduced and became widespread in open rangeland habitats in California until a specialized beetle was introduced to control it. Population size of the plant quickly decreased, and it is now found only in shady sites where the beetle cannot thrive. In this case, the presence of a predator limits the realized niche of a plant.

Competitive Exclusion Can Occur When Species Compete for Limited Resources

LEARNING OBJECTIVE 39.1.3 Describe how competitive exclusion operates.

In classic experiments carried out in 1934 and 1935, ecologist Georgii Gause studied competition among three species of *Paramecium,* a tiny protist. Each of the three species grew well in culture tubes by themselves, preying on bacteria and yeasts that fed on oatmeal suspended in the culture fluid (figure 39.3*a*). However, when Gause grew *P. aurelia* together with *P. caudatum* in the same culture tube, the numbers of *P. caudatum* always declined to extinction, leaving *P. aurelia* the only survivor (figure 39.3*b*). Why did this

Figure 39.2 Competition among two species of barnacles. The fundamental niche of *Chthamalus stellatus* includes both deep and shallow zones, but *Semibalanus balanoides* forces *C. stellatus* out of the part of its fundamental niche that overlaps the realized niche of *S. balanoides*.

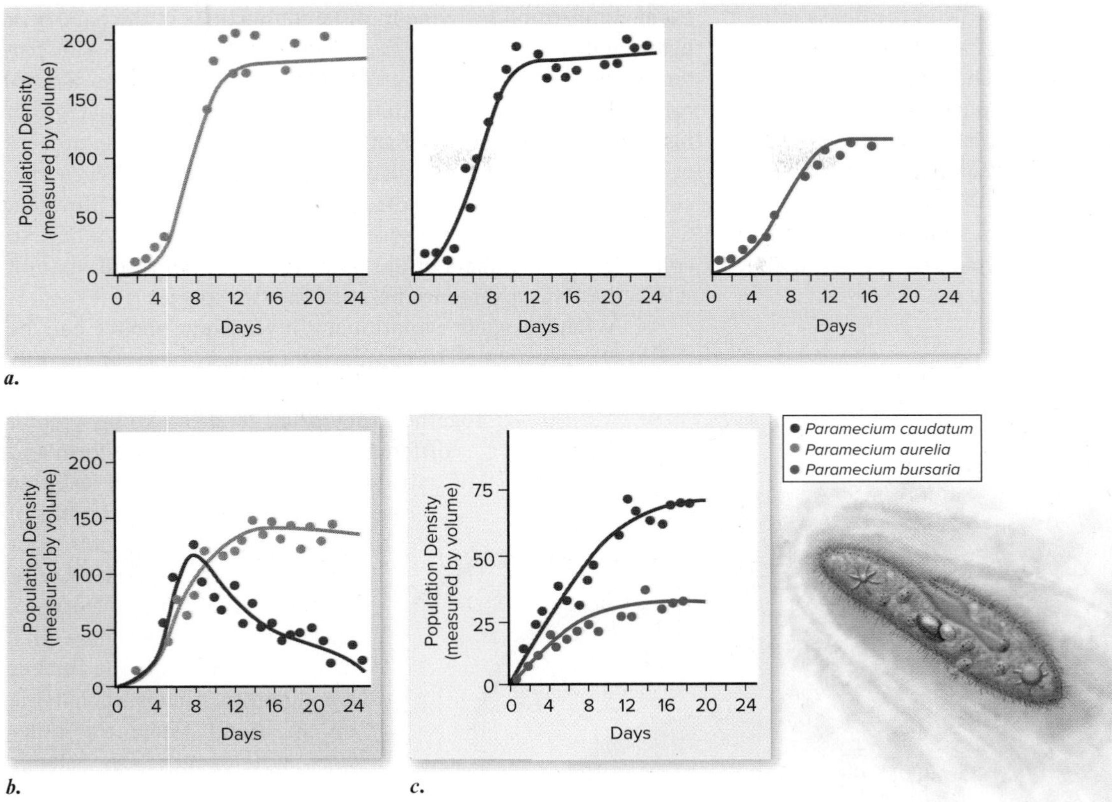

Figure 39.3 Competitive exclusion among three species of *Paramecium*. In the microscopic world, *Paramecium* is a ferocious predator that preys on smaller protists. ***a.*** In his experiments, Gause found that three species of *Paramecium* grew well alone in culture tubes. ***b.*** However, *P. caudatum* declined to extinction when grown with *P. aurelia,* because they shared the same realized niche and *P. aurelia* outcompeted *P. caudatum* for food resources. ***c.*** *P. caudatum* and *P. bursaria* were able to coexist because the two have different realized niches and thus avoided competition.

happen? Gause found that *P. aurelia* could grow six times faster than its competitor, *P. caudatum,* because it was able to better utilize the limited available resources, an example of exploitative competition.

From experiments such as this, Gause formulated what is now called the principle of **competitive exclusion.** This principle states that if two species are competing for a limited resource such as food or water, the species that uses the resource more efficiently will eventually eliminate the other locally. In other words, no two species with the same niche can coexist when resources are limiting.

Niche overlap and coexistence

In a revealing experiment, Gause challenged *Paramecium caudatum*—the defeated species in his earlier experiments—with a third species, *P. bursaria.* Because he expected these two species to also compete for the limited bacterial food supply, Gause thought one would win out, as had happened in his previous experiments. But that's not what occurred. Instead, both species survived in the culture tubes, dividing the food resources.

The explanation for the species' coexistence is simple. In the upper part of the culture tubes, where the oxygen concentration and bacterial density were high, *P. caudatum* dominated because it was better able to feed on bacteria. In the lower part of the tubes, however, the lower oxygen concentration favored the growth of a different potential food, yeast, and *P. bursaria* was better able to eat this food. The fundamental niche of each species was the whole culture tube, but the realized niche of each species was only a portion of the tube. Because the realized niches of the two species did not overlap too much, both species were able to survive. However, competition did have a negative effect on the participants (figure 39.3*c*). When grown without a competitor, both species reached much higher densities than when they were grown with a competitor.

Competitive exclusion refined

Species can and do coexist while competing for some of the same resources. Nevertheless, Gause's hypothesis predicts that when two species coexist on a long-term basis, either resources must not be limited or their niches will always differ in one or more features; otherwise, one species will outcompete the other, and the extinction of the second species will inevitably result.

Competition May Lead to Resource Partitioning

LEARNING OBJECTIVE 39.1.4 Explain how niche overlap may lead to resource partitioning.

Gause's competitive exclusion principle has a very important consequence: if competition for a limited resource is intense, then either one species will drive the other to extinction or natural selection will reduce the competition between them.

When the ecologist Robert MacArthur studied five species of warblers—small, insect-eating forest songbirds—he discovered that they appeared to be competing for the same resources. But when he studied them more carefully, he found that each species actually fed in a different part of spruce trees and, so, ate different subsets of insects. One species fed on insects near the tips of branches, a second within the dense foliage, a third on the lower branches, a fourth high on the trees, and a fifth at the very apex of the trees. Thus, each species of warbler had evolved to utilize a different portion of the spruce tree resource. They had subdivided the niche to avoid direct competition with one another. This niche subdivision is termed **resource partitioning.**

Resource partitioning often occurs in similar species that occupy the same geographical area. Such sympatric species often avoid competition by living in different portions of the habitat or by using different food or other resources (figure 39.4). This pattern of resource partitioning can result from the process of natural selection, causing initially similar species to diverge in resource use to reduce competitive pressures. This is the process of *character displacement,* discussed in chapter 21. In each species, natural selection favors those individuals most different from the other species because those individuals will not have to compete with members of the other species. As a result, natural selection drives divergence between the two species.

Whether such evolutionary divergence occurs can be investigated by comparing species whose ranges only partially overlap. Where the two species occur together, they often tend to exhibit greater differences in morphology (the form and structure of an organism) and resource use than do allopatric populations of the same species that do not occur with the other species. As an example, the two Darwin's finches in figure 39.5 have bills of similar size where the finches are allopatric (that is, each living on an island where the other does not occur). On islands where they are sympatric (that is, occur together), the two species have evolved beaks of different sizes, one adapted to larger seeds and the other to smaller ones. Character displacement such as this may play an important role in adaptive radiation, leading new species to adapt to different parts of the environment, as discussed in chapter 21.

Figure 39.4 Resource partitioning among sympatric lizard species. Species of *Anolis* lizards on Caribbean islands partition their habitats in a variety of ways. ***a.*** Some species occupy leaves and branches in the canopy of trees. ***b.*** Others use twigs on the periphery. ***c.*** Still others are found at the base of the trunk. ***d.*** Some use grassy areas in the open. When two species occupy the same part of the tree, they either utilize different-size insects as food or partition the thermal microhabitat; for example, one might be found only in the shade, whereas the other would only be found basking in sunlight.
(a–d): Jonathan Losos

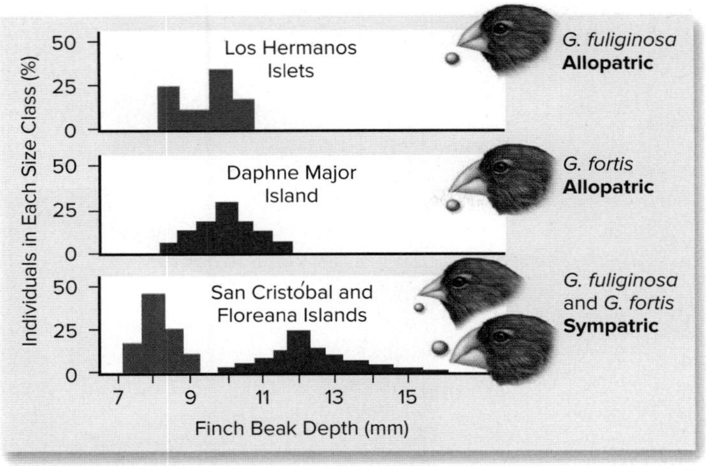

Figure 39.5 Character displacement in Darwin's finches. These two species of finches (genus *Geospiza*) have beaks of similar size when allopatric but of different sizes when sympatric.

REVIEW OF CONCEPT 39.1

A community is the sum of all the species in a particular location. A niche is how an organism utilizes available resources. A fundamental niche is the entire niche, whereas a realized niche is what a species actually utilizes. If resources are limited, two species cannot occupy the same niche indefinitely without competition driving one to local extinction. Resource partitioning allows two species to occupy a niche by reducing competition between them.

- *Under what circumstances can two species with identical niches coexist indefinitely?*

39.2 Predator–Prey Relationships Foster Coevolution

Predation is the consuming of one organism by another. In this sense, predation includes everything from a leopard capturing and eating an antelope to a deer grazing on spring grass.

Predation Strongly Influences Prey Populations

LEARNING OBJECTIVE 39.2.1 Describe the effects of predation on a population.

When experimental populations are set up under simple laboratory conditions, the predator often exterminates its prey and then becomes extinct itself, having nothing left to eat. Figure 39.6 illustrates such a situation with the predatory protist *Didinium* and its prey, *Paramecium.* If refuges are provided for the *Paramecium,* its population drops to low levels but not to extinction. Low prey population levels then provide inadequate food for the *Didinium,* causing the predator population to decrease. When this occurs, the prey population can recover.

In nature, predators often have large effects on prey populations. As the previous example indicates, however, the interaction is a two-way street: prey can also affect the dynamics of predator populations. The outcomes of such interactions are complex and depend on a variety of factors.

Prey population explosions and crashes

Some of the most dramatic examples of the interconnection between predators and their prey involve situations in which humans have either added predators to or eliminated predators from an area. For example, the elimination of large carnivores from much of the eastern United States has led to population explosions of white-tailed deer, which strip the habitat of all edible plant life within their reach. Similarly, when sea otters were hunted to near extinction on the western coast of the United States, populations of sea urchins, a principal prey item of the otters, exploded.

Conversely, the introduction of rats, dogs, and cats to many islands around the world has led to the decimation of native fauna. Populations of Galápagos tortoises on several islands are endangered by introduced rats, pigs, dogs, and cats, which eat the eggs and the young tortoises. Similarly, in New Zealand several species of birds and reptiles have been eradicated by rat predation and now occur only on a few offshore islands that the rats have not reached.

A classic example of the role predation can play in a community involves the introduction of prickly pear cactus to Australia in the 19th century. In the absence of predators, the cactus spread rapidly, so that by 1925 it had occupied 12 million hectares of rangeland in an impenetrable morass of spines that made cattle ranching difficult. To control the cactus, a predator from its natural habitat in Argentina, the moth *Cactoblastis cactorum,* was introduced, beginning in 1926. By 1940, cactus populations had been greatly reduced, and the cactus now usually occurs in small populations.

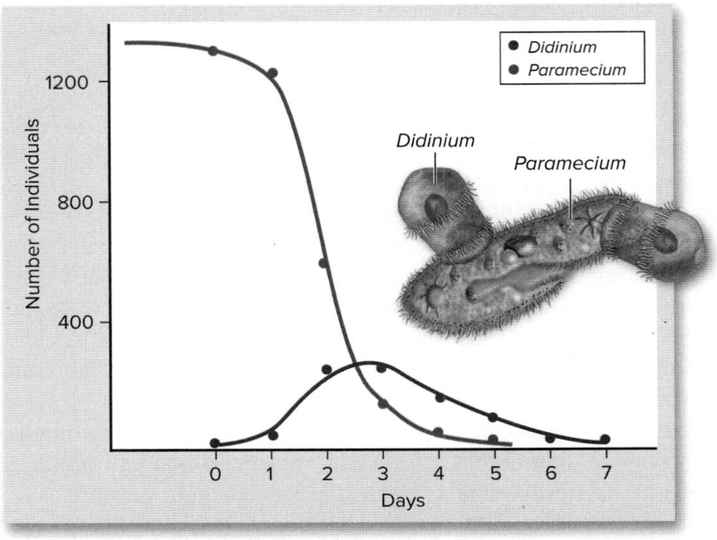

Figure 39.6 Predator–prey in the microscopic world. When the predatory *Didinium* is added to a *Paramecium* population, the numbers of *Didinium* initially rise, and the numbers of *Paramecium* steadily fall. When the *Paramecium* population is depleted, however, the *Didinium* individuals also die.

Predation and coevolution

Predation provides strong selective pressures on prey populations. Any feature that would decrease the probability of capture should be strongly favored. In turn, the evolution of such features causes natural selection to favor counteradaptations in predator populations. The process by which these adaptations are selected in lockstep fashion in two or more interacting species is termed **coevolution.** In this chapter, you'll learn more about these defenses and responses.

Plant Adaptations Defend Against Herbivores

LEARNING OBJECTIVE 39.2.2 Describe mechanisms plants use for defense against predators.

Plants have evolved many mechanisms to defend themselves from herbivores. The most obvious are morphological defenses: thorns, spines, and prickles play an important role in discouraging large plant eaters, and plant hairs, especially those that have a glandular, sticky tip, deter insect herbivores. Some plants, such as grasses, deposit silica in their leaves, both strengthening and protecting themselves. If enough silica is present, these plants are simply too tough to eat.

Chemical defenses

As significant as morphological adaptations are, the chemical defenses that occur in plants are even more widespread. Plants exhibit some amazing chemical adaptations to combat herbivores. For example, recent work demonstrates that when attacked by caterpillars, wild tobacco plants emit a chemical into the air that attracts a species of bug that feeds on that caterpillar.

The best-known of the chemical defenses of plants against herbivores are *secondary chemical compounds.* These chemicals are distinguished from primary compounds, which are the components of a major metabolic pathway. Many plants, as well as many algae, contain structurally diverse secondary compounds that either are toxic to most herbivores or disturb their metabolism greatly, preventing, for example, the normal development of larval insects. Consequently, most herbivores tend to avoid the plants that possess these compounds.

The mustard family (Brassicaceae) produces a group of chemicals known as mustard oils. These substances give the pungent aromas and tastes to plants such as mustard, cabbage, watercress, radish, and horseradish. The flavors we enjoy indicate the presence of chemicals that are toxic to many groups of insects. Similarly, plants of the milkweed family (Asclepiadaceae) and the related dogbane family (Apocynaceae) produce a milky sap that deters herbivores from eating them. In addition, these plants usually contain cardiac glycosides, molecules that can produce drastic deleterious effects on the heart function of vertebrates.

The coevolutionary response of herbivores

Certain groups of herbivores are associated with each family or group of plants protected by a particular kind of secondary compound. These herbivores are able to feed on these plants without harm, often as their exclusive food source.

a.

b.

Figure 39.7 Insect herbivores well suited to their plant hosts. ***a.*** The green caterpillars of the cabbage white butterfly (*Pieris rapae*) are camouflaged on the leaves of the cabbage and other plants on which they feed. Although mustard oils protect these plants against most herbivores, the cabbage white butterfly caterpillars are able to break down the mustard oil compounds. ***b.*** An adult cabbage white butterfly.
(a): viks74/Shutterstock; (b): tomoki1970/Shutterstock

For example, cabbage butterfly caterpillars (subfamily Pierinae) feed almost exclusively on plants of the mustard and caper families, as well as on a few other small families of plants that also contain mustard oils (figure 39.7). Similarly, caterpillars of monarch butterflies and their relatives (subfamily Danainae) feed on plants of the milkweed and dogbane families. How do these animals manage to avoid the chemical defenses of the plants, and what are the evolutionary precursors and ecological consequences of such patterns of specialization?

We can offer a potential explanation for the evolution of these particular patterns. Once the ability to manufacture mustard oils evolved in the ancestors of the caper and mustard families, the plants were protected for a time against most or all herbivores in their area. At some point, certain groups of insects—for example, the cabbage butterflies—evolved the ability to break down mustard oils and thus feed on these plants without harming themselves. Having developed this new capability, the insects were able to use a new resource without competing with other herbivores for it. Exposure to an underutilized resource often leads to evolutionary diversification and adaptive radiation.

Animal Adaptations Defend Against Predators

LEARNING OBJECTIVE 39.2.3 Describe mechanisms animals use for defense against predators.

Some animals that feed on plants rich in secondary compounds receive an extra benefit. When the caterpillars of monarch butterflies feed on plants of the milkweed family, they do not break

a. Blue jay eating monarch
b. Vomiting reaction

Figure 39.8 A blue jay learns not to eat monarch butterflies. ***a.*** This cage-reared jay had never seen a monarch butterfly before it tried eating one. ***b.*** The same jay regurgitated the butterfly a few minutes later. This bird will probably avoid trying to capture all orange-and-black insects in the future.

Figure 39.9 Vertebrate chemical defenses. Frogs of the family Dendrobatidae, abundant in the forests of Central and South America, are extremely poisonous to vertebrates; 80 different toxic alkaloids have been identified from different species in this genus. Dendrobatids advertise their toxicity with bright coloration. As a result of either instinct or learning, predators avoid such brightly colored species that might otherwise be suitable prey.

Philip Coblentz/MedioImages/SuperStock

down the cardiac glycosides that protect these plants from herbivores. Instead, the caterpillars concentrate and store the cardiac glycosides within their fat; they then pass them through the chrysalis stage to the adult and even to the eggs of the next generation.

The incorporation of cardiac glycosides protects all stages of the monarch life cycle from predators. A bird that eats a monarch butterfly quickly regurgitates it (figure 39.8) and in the future avoids the conspicuous orange-and-black pattern that characterizes the adult monarch. Some bird species have evolved the ability to tolerate the protective chemicals; these birds eat the monarchs.

Chemical defenses

Animals also manufacture and use a startling array of defensive substances. Bees, wasps, predatory bugs, scorpions, spiders, and many other arthropods use chemicals to defend themselves and to kill their own prey. In addition, various chemical defenses have evolved among many marine invertebrates, as well as a variety of vertebrates, including frogs, snakes, lizards, fishes, and some birds.

The poison-dart frogs of the family Dendrobatidae produce toxic alkaloids in the mucus that covers their brightly colored skin. These alkaloids include powerful neurotoxins that affect vertebrate predators (figure 39.9). The most poisonous species known is the golden frog, *Phyllobates terribilis.* A single individual is covered with enough alkaloid to kill 10–20 people. A number of interesting compounds have been isolated from these species, some useful in neuromuscular research. Similarly, investigations of poisonous and venomous species are under way in search of possible therapeutics.

Defensive coloration

Many insects that feed on milkweed plants are brightly colored; they advertise their poisonous nature using an ecological strategy known as warning coloration.

Showy coloration is characteristic of animals that use poisons and stings to repel predators; organisms that lack specific chemical defenses are seldom brightly colored. In fact, many have cryptic coloration—color that blends with the surroundings and thus hides the individual from predators (figure 39.10). Camouflaged animals usually do not live together in groups, because a predator that discovers one individual gains a valuable clue to the presence of others.

Figure 39.10 Cryptic coloration and form. A waved umber caterpillar (*Menophora abruptaria*) closely resembles the twig on which it is hanging.

DEA Picture Library/Getty Images

Mimicry Allows One Species to Capitalize on the Defensive Strategies of Another

LEARNING OBJECTIVE 39.2.4 Distinguish between Batesian and Müllerian mimicry.

During the course of their evolution, many species have come to resemble distasteful species that exhibit warning coloration. The mimic gains an advantage by looking like the distasteful model. Two types of mimicry have been identified: Batesian mimicry and Müllerian mimicry.

Batesian mimicry

Batesian mimicry is named for the naturalist Henry Bates, who first brought this type of mimicry to general attention in 1857. In his journeys to the Amazon region of South America, Bates discovered many instances of palatable insects that resembled brightly colored, distasteful species. He reasoned that the mimics might be avoided by predators, who would mistake the mimic for the distasteful species. This can be thought of in terms of information: the model signals to a predator via warning colors that it is unpalatable. By resembling the model, the mimic is sending the same signal, but it will now deceive a predator receiving the signal. Thus, the mimic receives the same protection as the model, despite being palatable.

Many of the best-known examples of Batesian mimicry occur among butterflies and moths. Predators of these insects must use visual cues to hunt for their prey, or mimicry would not be effective. There is also evidence that Batesian mimicry can involve nonvisual cues, such as olfactory ones, although such examples are less obvious to humans.

The butterflies that provide the models in Batesian mimicry have caterpillars that feed on one or a few closely related plant families that are strongly protected by toxic chemicals. The model butterflies incorporate the toxins from these plants into their bodies. The mimic butterflies, in contrast, have caterpillars whose feeding habits are not so restricted and feed on a number of different plant families that are not protected by toxic chemicals.

One often-studied mimic among North American butterflies is the tiger swallowtail, whose range occurs throughout the eastern United States and into Canada (figure 39.11*a*). In areas where the poisonous pipevine swallowtail occurs, female tiger swallowtails are polymorphic, and one color form is extremely similar in appearance to the pipevine swallowtail.

The caterpillars of the tiger swallowtail feed on a variety of trees, including tulip, aspen, and cherry, and neither caterpillars nor adults are distasteful to birds. Interestingly, the Batesian mimicry seen in the adult tiger swallowtail butterfly does not extend to the caterpillars: tiger swallowtail caterpillars are camouflaged on leaves, resembling bird droppings, but the pipevine swallowtail's distasteful caterpillars are very conspicuous.

Müllerian mimicry

Another kind of mimicry, **Müllerian mimicry,** was named for the biologist Fritz Müller, who first described it in 1878. In Müllerian mimicry, several unrelated but protected animal species come to resemble one another (figure 39.11*b*). If animals that resemble one another are all poisonous or dangerous, they gain an advantage, because a predator will learn more quickly to avoid them. Some predator populations even evolve an innate avoidance of species; such evolution may occur more quickly when multiple dangerous prey look alike.

In both Batesian and Müllerian mimicry, mimic and model must not only look alike but also act alike. For example, the members of several families of insects that closely resemble wasps behave surprisingly like the wasps they mimic, flying often and actively from place to place.

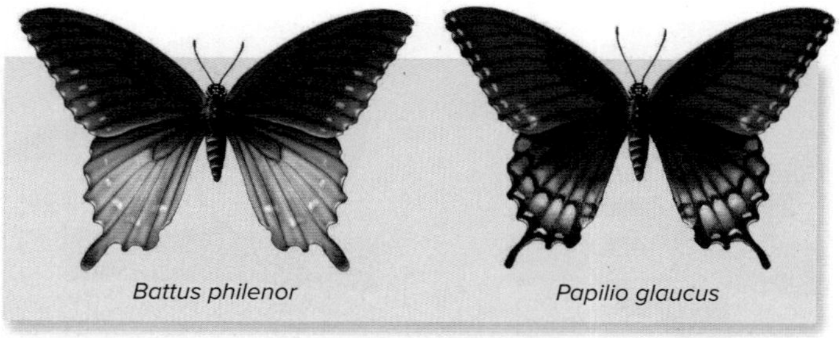

a. **Batesian mimicry:** Pipevine swallowtail butterfly (*Battus philenor*) is poisonous; Tiger swallowtail (*Papilio glaucus*) is a palatable mimic.

b. **Müllerian mimicry:** Two pairs of mimics; all are distasteful.

Figure 39.11 Mimicry. ***a.*** Batesian mimicry. Pipevine swallowtail butterflies (*Battus philenor*) are protected from birds and other predators by the poisonous compounds in their tissue derived from their diet as caterpillars. Adult pipevine swallowtails advertise their poisonous nature with warning coloration. Tiger swallowtails (*Papilio glaucus*) are Batesian mimics of the poisonous pipevine swallowtail and are not chemically protected. ***b.*** Pairs of Müllerian mimics. *Heliconius erato* and *H. melpomene* are sympatric, and *H. sapho* and *H. cydno* are sympatric. All of these butterflies are distasteful. They have evolved similar coloration patterns in sympatry to minimize predation; predators need learn only one pattern to avoid.

REVIEW OF CONCEPT 39.2

Predation is the consuming of one organism by another. High predation can drive prey populations to extinction; conversely, in the absence of predators, prey populations often explode and exhaust their resources. Defensive adaptations may evolve in prey species, such as becoming distasteful or poisonous or having defensive structures, appearance, or capabilities.

- *A nonpoisonous scarlet king snake has red, black, and yellow bands of color similar to those of the poisonous eastern coral snake. What type of mimicry is being exhibited?*

39.3 Cooperation Among Species Can Lead to Coevolution

The plants, animals, protists, fungi, and prokaryotes that live together in communities have changed and adjusted to one another continually over millions of years. We have already discussed competition and predation, but other types of ecological interactions commonly occur and lead to mutual coevolutionary adjustments.

Symbiosis Involves Long-Term Interactions

LEARNING OBJECTIVE 39.3.1 Explain the different forms of symbiosis.

Many features of flowering plants have evolved in relation to the dispersal of the plant's gametes by animals (figure 39.12). These animals, in turn, have evolved a number of special traits that enable them to obtain food or other resources efficiently from the plants they visit, often from their flowers. While doing so, the animals pick up pollen, which they may deposit on the next plant they visit, or seeds, which may be left elsewhere in the environment, sometimes a great distance from the parent plant.

In **symbiosis,** two or more kinds of organisms interact in often elaborate and more or less permanent relationships. All symbiotic relationships carry the potential for coevolution between the organisms involved, and in many instances the results of this coevolution are fascinatingly complex.

Figure 39.12 Pollination by a bat. Many flowers have coevolved with other species to facilitate pollen transfer. Insects are widely known as pollinators, but they're not the only ones: birds, bats, and even small marsupials and lizards serve as pollinators for some species. Notice the cargo of pollen on the bat's snout.
MerlinTuttle.org/Science Source

Examples of symbiosis include lichens, which are associations of certain fungi with green algae or cyanobacteria. Another important example is mycorrhizae, associations between fungi and the roots of most kinds of plants. The fungi expedite the plant's absorption of certain nutrients, and the plants in turn provide the fungi with carbohydrates (both mycorrhizae and lichens are discussed in greater detail in chapter 25). Similarly, root nodules that occur in legumes and certain other kinds of plants contain bacteria that fix atmospheric nitrogen and make it available to their host plants.

In the tropics, leaf-cutter ants are often so abundant that they can remove a quarter or more of the total leaf surface of the plants in a given area in a single year. They do not eat these leaves directly; rather, they take them to underground nests, where they chew them up and inoculate them with the spores of particular fungi. These fungi are cultivated by the ants and taken from one specially prepared bed to another, where they grow and reproduce. In turn, the fungi constitute the primary food of the ants and their larvae. The relationship between leaf-cutter ants and these fungi is an excellent example of symbiosis. Recent phylogenetic studies using DNA and assuming a molecular clock suggest that these symbioses are ancient, perhaps originating more than 50 MYA.

The major kinds of symbiotic relationships include (1) **commensalism,** in which one species benefits and the other neither benefits nor is harmed; (2) **mutualism,** in which both participating species benefit; and (3) **parasitism,** in which one species benefits but the other is harmed. Parasitism can also be viewed as a form of predation, although the organism that is preyed on does not necessarily die.

Commensalism Benefits One Species and Is Neutral to the Other

LEARNING OBJECTIVE 39.3.2 Differentiate between commensal relationships and other kinds of symbiosis.

In commensalism, one species benefits and the other is neither hurt nor helped by the interaction. In nature, individuals of one species are often physically attached to members of another. For example, epiphytes are plants that grow on the branches of other plants. In general, the host plant is unharmed and the epiphyte that grows on it benefits. An example is Spanish moss, which hangs on trees in the southern United States. This plant and other members of its genus, which is in the pineapple family, grow on trees to gain access to sunlight; they generally do not harm the trees (figure 39.13).

Similarly, various marine animals, such as barnacles, grow on other, often actively moving, sea animals, such as whales, and thus are carried passively from place to place. These "passengers" presumably gain more protection from predation than they would have if they were fixed in one place, and they reach new sources of food. The increased water circulation these animals receive as their host moves around may also be of great importance, particularly if the passengers are filter feeders. Unless the number of these passengers gets too large, the host species is usually unaffected.

Figure 39.13 An example of commensalism. Spanish moss (*Tillandsia usneoides*) benefits from using trees as a substrate, but the trees generally are not affected positively or negatively.
JLFCapture/Getty Images

When commensalism may not be commensalism

One of the best-known examples of symbiosis involves the relationships between certain small tropical fishes (clownfish) and sea anemones, shown at the beginning of this chapter. The fish have evolved the ability to live among the stinging tentacles of sea anemones, even though these tentacles would quickly paralyze other fishes that touched them. The clownfish feed on food particles left from the meals of the host anemone, remaining uninjured under remarkable circumstances.

On land, an analogous relationship exists between birds called oxpeckers and grazing animals such as cattle or antelopes (figure 39.14). The birds spend most of their time clinging to the animals, picking off parasites and other insects, carrying out their entire life cycles in close association with the host animals.

No clear-cut boundary exists between commensalism and mutualism; in each of these cases, it is difficult to be certain whether the second partner receives a benefit or not. A sea anemone may benefit by having particles of food removed from its tentacles, because it may then be better able to catch other prey. Similarly, although often thought of as commensalism, the association of grazing mammals and gleaning birds is actually an example of mutualism. The mammal benefits by having parasites and other insects removed from its body, but the birds also benefit by gaining a dependable source of food.

On the other hand, commensalism can easily transform into parasitism. Oxpeckers are also known to pick not only parasites but also scabs off their grazing hosts. Once the scab is picked, the birds drink the blood that flows from the wound. Occasionally, the cumulative effect of persistent attacks can greatly weaken the herbivore, particularly when conditions are not favorable, such as during droughts.

Mutualism Benefits Both Species

LEARNING OBJECTIVE 39.3.3 Explain how mutualism leads to coevolution.

Mutualism is a symbiotic relationship between organisms in which both species benefit. Mutualistic relationships are of fundamental importance in determining the structure of biological communities.

Figure 39.14 Commensalism, mutualism, or parasitism? In this symbiotic relationship, oxpeckers definitely receive a benefit in the form of nutrition from the ticks and other parasites they pick off their host (in this case, an impala, *Aepyceros melampus*). But the effect on the host is not always clear. If the ticks are harmful, their removal benefits the host, and the relationship is mutually beneficial. If the oxpeckers also pick at scabs, causing blood loss and possible infection, the relationship may be parasitic. If the hosts are unharmed by either the ticks or the oxpeckers, the relationship may be an example of commensalism.
P. de Graaf/Moment Open/Getty Images

Mutualism and coevolution

Some of the most spectacular examples of mutualism occur among flowering plants and their animal visitors, including insects, birds, and bats. During the course of flowering-plant evolution, the characteristics of flowers evolved in relation to the characteristics of pollinator species (refer to chapter 30). The pollinator species have also evolved, increasing their specialization for obtaining food or other substances from particular kinds of flowers.

Another example of mutualism involves ants and aphids. Aphids are small insects that suck fluids from the phloem of living plants with their piercing mouthparts. They extract a certain amount of the sucrose and other nutrients from this fluid, but they excrete much of it in an altered form through their anus. Certain ants have taken advantage of this—in effect, domesticating the aphids. Like ranchers taking cattle to fresh fields to graze, the ants carry the aphids to new plants and then consume as food the "honeydew" that the aphids excrete.

Ants and acacias: A prime example of mutualism

A particularly striking example of mutualism involves ants and certain Latin American tree species of the genus *Acacia.* In these species, certain leaf parts, called stipules, are modified as paired hollow thorns. The thorns are inhabited by stinging ants of the genus *Pseudomyrmex,* which do not nest anywhere else (figure 39.15). Like all thorns that occur on plants, the acacia thorns deter herbivores.

At the tip of the leaflets of these acacias are unique, protein-rich bodies called Beltian bodies, named after the 19th-century naturalist Thomas Belt. Beltian bodies do not occur in species of

Figure 39.15 Mutualism: Ants and acacias. Ants of the genus *Pseudomyrmex* live within the hollow thorns of certain species of acacia trees in Latin America. The nectaries at the bases of the leaves and the Beltian bodies at the ends of the leaflets provide food for the ants. The ants, in turn, supply the acacias with organic nutrients and protect the acacias from herbivores and shading from other plants.
Dreedphotography/Getty Images

Acacia that are not inhabited by ants, and their role is clear: they serve as a primary food for the ants. In addition, the plants secrete nectar from glands near the bases of their leaves. The ants consume this nectar as well, feeding it and the Beltian bodies to their larvae.

Obviously, this association is beneficial to the ants, and one can readily see why they inhabit acacias of this group. The question is, What, if anything, do the ants do for the acacia?

Whenever herbivorous insects land on an acacia inhabited by ants, the ants immediately attack and devour the herbivore. The ants that live in the acacias also help their hosts compete with other plants by cutting away any encroaching branches that touch the acacia in which they are living. They create, in effect, a tunnel of light through which the acacia can grow, even in the lush tropical rainforests of lowland Central America. In fact, when an ant colony is experimentally removed from a tree, the acacia is unable to compete successfully in this habitat. Finally, through the organic material they bring and their excretions, the ants provide the acacias with an abundant source of nitrogen.

When mutualism may not be mutualism

As with commensalism, however, things are not always as they seem. Ant–acacia associations also occur in Africa; in Kenya, several species of acacia ants occur, but only a single species is found on any one tree. One species, *Crematogaster nigriceps,* is competitively inferior to two of the other species. To prevent invasion by these other ant species, *C. nigriceps* prunes the branches of the acacia, preventing it from coming into contact with branches of other trees, which would serve as a bridge for invaders.

Although this behavior is beneficial to the ant, it is detrimental to the tree, because it destroys the tissue from which flowers are produced, essentially sterilizing the tree. In this case, what initially evolved as a mutualistic interaction has instead become a parasitic one.

Parasitism Benefits One Species at the Expense of Another

LEARNING OBJECTIVE 39.3.4 Explain how parasitism can affect community structure.

Parasitism is harmful to the prey organism and beneficial to the parasite. In many cases, the parasite kills its host, and thus the ecological effects of parasitism can be similar to those of predation. In the past, parasitism was studied mostly in terms of its effects on individuals and the populations in which they live, but in recent years researchers have realized that parasitism can be an important factor affecting community structure.

External parasites

Parasites that feed on the exterior surface of an organism are external parasites, or ectoparasites (figure 39.16). Many instances of external parasitism are known in both plants and animals. Parasitoids are insects that lay eggs in or on living hosts. This behavior is common among wasps, whose larvae feed on the body of the unfortunate host, often killing it.

Internal parasites

Parasites that live within the body of their hosts, termed **endoparasites,** occur in many different phyla of animals and protists. Internal parasitism is generally marked by much more extreme specialization than external parasitism, as shown by the many protist and invertebrate parasites that infect humans.

The more closely the life of the parasite is linked with that of its host, the more its morphology and behavior are likely to have been modified during the course of its evolution (the same is true of symbiotic relationships of all sorts). Conditions

Figure 39.16 An external parasite. The yellow vines are the flowering plant dodder (*Cuscuta*), a parasite that has lost its chlorophyll and its leaves in the course of its evolution. Because it is heterotrophic (unable to manufacture its own food), dodder obtains its food from the host plants it grows on.
Jason Patrick Ross/Shutterstock

within the body of an organism are different from those encountered outside and are apt to be much more constant. Consequently, the structure of an internal parasite is often simplified, and unnecessary armaments and structures are lost as it evolves.

Parasites and host behaviors

Many parasites have complex life cycles that require several different hosts for growth to adulthood and reproduction. Recent research has revealed the remarkable adaptations of certain parasites that alter the behavior of the host and thus facilitate transmission from one host to the next. For example, many parasites cause their hosts to behave in ways that make them more vulnerable to their predators; when the host is ingested, the parasite is able to infect the predator.

One of the most famous examples involves a parasitic flatworm, *Dicrocoelium dendriticum,* which lives in ants as an intermediate host but reaches adulthood in large herbivorous mammals such as cattle and deer. Transmission from an ant to a cow might seem difficult because cows do not normally eat insects. The flatworm, however, has evolved a remarkable adaptation. When an ant is infected, one of the flatworms migrates to the brain and causes the ant to climb to the top of vegetation and lock its mandibles onto a grass blade at the end of the day, just when herbivores are grazing (figure 39.17). The result is that the ant is eaten along with the grass, leading to infection of the grazer.

Figure 39.17 Parasitic manipulation of host behavior. Due to a parasite in its brain, an ant climbs to the top of a grass blade, where it may be eaten by a grazing herbivore, thus passing the parasite from insect to mammal.

Species Interactions Have Evolutionary Consequences

LEARNING OBJECTIVE 39.3.5 Explain how species interactions have both direct and indirect effects.

We have seen the different ways in which species can interact with one another. In nature, however, more than one type of interaction often occurs at the same time. In many cases, the outcome of one type of interaction is modified or even reversed when another type of interaction is also occurring.

Predation reduces competition

When resources are limited, a superior competitor can eliminate other species from a community through competitive exclusion. However, predators can prevent or greatly reduce exclusion by lowering the numbers of individuals of competing species.

A given predator may often feed on two, three, or more kinds of plants or animals in a given community. The predator's choice depends partly on the relative abundance of the prey options. In other words, a predator may feed on species A when it is abundant and then switch to species B when A is rare. Similarly, a given prey species may become a primary source of food for increasing numbers of species as it becomes more abundant. In this way, superior competitors may be prevented from competitively excluding other species.

Such patterns are often characteristic of communities in marine intertidal habitats. For example, in preying selectively on bivalves, sea stars prevent bivalves from monopolizing a habitat, opening up space for many other organisms (figure 39.18). When sea stars are removed from a habitat, species diversity falls precipitously, and the seafloor community comes to be dominated by a few species of bivalves.

Predation tends to reduce competition in natural communities, so it is usually a mistake to attempt to eliminate a major predator, such as wolves or mountain lions, from a community. The result may be a decrease in biological diversity.

Parasitism may counter competition

Parasites may affect sympatric species differently and thus influence the outcome of interspecific interactions. One classic experiment investigated interactions between two sympatric flour beetles, *Tribolium castaneum* and *T. confusum,* with and without a parasite, *Adelina.* In the absence of the parasite, *T. castaneum* is dominant and *T. confusum* normally becomes extinct. When the parasite is present, however, the outcome is reversed, and *T. castaneum* perishes.

Similar effects of parasites in natural systems have been observed in many species. For example, in the *Anolis* lizards of St. Maarten, the competitively inferior species is resistant to lizard malaria (a disease related to human malaria), whereas the other species is highly susceptible. In places where the parasite occurs, the competitively inferior species can hold its own and the two species coexist; elsewhere, the competitively dominant species outcompetes and eliminates it.

SCIENTIFIC THINKING

Question: *Does predation affect the outcome of interspecific competitive interactions?*
Hypothesis: *In the absence of predators, prey populations will increase until resources are limiting, and some species will be competitively excluded.*
Experiment: *Remove predatory sea stars* (Pisaster ochraceus) *from some areas of rocky intertidal shoreline and monitor populations of species the sea stars prey upon. In control areas, pick up sea stars, but replace them where they were found to control for the effects of people walking through the study area.*

a.

b.

Result: *In the absence of sea stars, the population of the mussel* Mytilus californianus *exploded, occupying all available space and eliminating many other species from the community.*
Interpretation: *What would happen if sea stars were returned to the experimental plots?*

Figure 39.18 Predation reduces competition. ***a.*** In a controlled experiment in a coastal ecosystem, Robert Paine of the University of Washington removed a key predator, sea stars (*Pisaster*). ***b.*** In response, the population of fiercely competitive mussels, a type of bivalve mollusk, exploded, effectively crowding out seven other indigenous species.
(a): NOAA National Estuarine Research Reserve Collection; (b): Ben Grib Design/Getty Images

Indirect effects

In some cases, species may not directly interact, yet the presence of one species may affect a second by way of interactions with a third. Such effects are termed indirect effects.

Many desert rodents eat seeds, and so do the ants in their community; thus, we might expect them to compete with each other. But when all rodents were removed from experimental enclosures, ant populations first increased but then declined.

The initial increase was the expected result of removing a competitor. Why did it then reverse? The answer reveals the intricacies of natural ecosystems. Rodents prefer large seeds, whereas ants prefer smaller ones. Furthermore, in this ecosystem, plants with large seeds are competitively superior to plants with small seeds. Thus, the removal of rodents led to an increase in the number of plants with large seeds, which reduced the number of small seeds available, in turn leading to a decline in ant populations. In summary, the effect of rodents on ants is complicated: a direct, negative effect of resource competition and an indirect, positive effect mediated by plant competition.

Keystone species have major effects on communities

Species whose effects on the composition of communities are greater than one might expect based on their abundance are termed **keystone species.** Predators, such as the sea star described earlier, can often serve as keystone species by preventing one species from outcompeting others, thus maintaining high levels of species richness in a community.

Some ecologists call keystone species whose activities alter the environment and create new habitats ecosystem engineers. Beavers, for example, alter the flow of water in running streams, flooding new areas. Similarly, alligators excavate deep holes at the bottoms of lakes. In times of drought, these holes are the only areas where water remains, allowing aquatic species to persist until the drought ends and the lake refills. Humans can be viewed as the ultimate ecosystem engineers, sometimes intentionally and other times unintentionally altering their environment.

REVIEW OF CONCEPT 39.3

The types of symbiosis include mutualism, in which both participants benefit; commensalism, in which one benefits and the other is neutrally affected; and parasitism, in which one benefits at the expense of the other. Mutualistic species often undergo coevolution. Ecological interactions such as predation and parasitism may lessen resource competition. These effects can be both direct and indirect.

- *How can the presence of a predator positively affect populations of a species on which it preys?*

39.4 Ecological Succession Is a Consequence of Habitat Alteration

Even when the climate of an area remains stable year after year, communities have a tendency to change from simple to complex in a process known as **succession.** This process is familiar to anyone who has seen a vacant lot or cleared woods slowly become occupied by an increasing number of species.

Succession Produces a Change in Species Composition

LEARNING OBJECTIVE 39.4.1 Distinguish between primary and secondary succession.

If a wooded area is cleared or burned and left alone, plants will slowly reclaim the area. Eventually, all traces of the clearing will disappear, and the area will again be woods. This kind of succession, which occurs in areas where an existing community has been disturbed but organisms still remain, is called **secondary succession.**

In contrast, **primary succession** occurs on bare, lifeless substrate, such as rocks, or in open water, where organisms gradually move into an area and change its nature. Primary succession occurs in lakes and on land exposed after the retreat of glaciers, as well as on volcanic islands that rise from the sea.

Glacial moraines provide an example of primary succession (figure 39.19). When glaciers recede, the exposed soil is mineral-poor, with basic pH due to carbonates in the rocks and low levels of nitrogen. Under these conditions, the first vegetation to grow is lichens, which produce acidic secretions that help break down the substrate and reduce the pH. This adds to the accumulation of soil, which is colonized by mosses, eventually building up enough nutrients in the soil for alder shrubs to take hold. Over a hundred years, the alders, which have symbiotic nitrogen-fixing bacteria, increase soil nitrogen levels, and their acidic leaves further lower soil pH. Eventually, spruce trees grow above the alders and shade them, crowding them out entirely and forming a dense spruce forest.

In a similar example, an *oligotrophic* lake—one poor in nutrients—may gradually, by the accumulation of organic matter, become *eutrophic*—rich in nutrients. As this occurs, the composition of communities will change, first increasing in species richness and then declining.

Why succession happens

Succession happens because species alter the habitat and the resources available in it in ways that favor other species. Three dynamic concepts are of critical importance in the process: establishment, facilitation, and inhibition.

1. **Establishment.** Early successional stages are characterized by weedy, *r*-selected species that are tolerant of the harsh, abiotic conditions in barren areas (chapter 38 discussed *r*-selected and *K*-selected species).
2. **Facilitation.** Weedy early successional stages introduce local changes in the habitat that favor other, less weedy species. The mosses in the Glacier Bay succession convert nitrogen to a form usable by alders. The nitrogen buildup

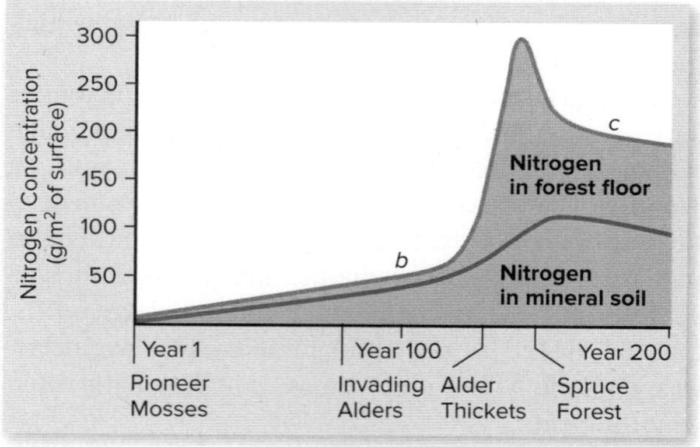

a.

b.

c.

d.

Figure 39.19 Primary succession at Alaska's Glacier Bay. ***a.*** Initially, the glacial moraine at Glacier Bay, Alaska, had little soil nitrogen. ***b.*** The first invaders of these exposed sites are pioneer moss species with nitrogen-fixing, mutualistic microbes. ***c.*** Within 20 years, young alder shrubs take hold. Rapidly fixing nitrogen, they soon form dense thickets. ***d.*** Eventually, spruce overgrow the mature alders, forming a forest.

(b): Cliff LeSergent/Alamy Stock Photo; (c): Allover images/Alamy Stock Photo; (d): Borut Trdina/Getty Images produced by the alders, though not necessary for spruce establishment, leads to more robust forests of spruce better able to resist insect attack.

3. **Inhibition.** Sometimes the changes in the habitat caused by one species, while favoring other species, also inhibit the growth of the original species that caused the changes. Alders, for example, do not grow as well in acidic soil as the spruce and hemlock that replace them.

Over the course of succession, the number of species typically increases as the environment becomes more hospitable. In some cases, however, as ecosystems mature, more *K*-selected species replace *r*-selected ones, and superior competitors force out other species, leading ultimately to a decline in species richness.

Succession in animal communities

The species of animals present in a community also change through time in a successional pattern. As the vegetation changes during succession, habitat disappears for some species and appears for others.

A particularly striking example occurred on the Krakatau islands, which were devastated by an enormous volcanic eruption in 1883. Initially composed of nothing but barren ash-fields, the three islands of the group experienced rapid successional change as vegetation became re-established. A few blades of grass appeared the next year, and within 15 years the coastal vegetation was well established and the interior was covered with dense grasslands. By 1930, the islands were almost entirely forested (figure 39.20).

The fauna of Krakatau changed in synchrony with the vegetation. Nine months after the eruption, the only animal found was a single spider, but by 1908, 200 animal species were found during a three-day exploration. For the most part, the first animals were grassland inhabitants, but as trees became established, some of these early colonists, such as the zebra dove and the long-tailed shrike (a type of predatory bird), disappeared and were replaced by forest-inhabiting species, such as fruit bats and fruit-eating birds.

Sometimes moderate levels of continual disturbance can lead to increased species richness, because species of all levels of succession may be present at any one time. Although patterns of succession of animal species have typically been caused by vegetational succession, changes in the composition of the animal community in turn have affected plant occurrences. In particular, many plant species that are dispersed or pollinated by animals could not colonize Krakatau until their dispersers or pollinators had become established. For example, fruit bats were slow to colonize Krakatau, and until they appeared, few bat-dispersed plant species were present.

REVIEW OF CONCEPT 39.4

Communities change through time by a process termed succession. Primary succession occurs on bare, lifeless substrate; secondary succession occurs where an existing community has been disturbed. Early-arriving species alter the environment in ways that allow other species to colonize, and new colonizers may have negative effects on species already present. Sometimes moderate levels of disturbance can lead to increased species richness.

- *From a community point of view, would clear-cutting a forest be better than selective harvest of individual trees?*

39.5 Chemical Elements Move Through Ecosystems in Biogeochemical Cycles

An **ecosystem** includes all the organisms that live in a particular place, plus the abiotic (nonliving) environment in which they live—and with which they interact—at that location. Ecosystems are intrinsically dynamic in a number of ways, including their processing of matter and energy. We start with matter.

a. *b.*

Figure 39.20 Succession after a volcanic eruption. A major volcanic explosion in 1883 on the island of Krakatau destroyed all life on the island. ***a.*** This photo shows a later, much less destructive eruption of the volcano. ***b.*** Krakatau, forested and populated by animals.

(a): De Agostini/C. Dani & I. Jeske/Getty Images; (b): Prisma Bildagentur AG/Alamy Stock Photo

Chemical Elements Cycle Through Ecosystems

LEARNING OBJECTIVE 39.5.1 Explain the importance of element cycling within Earth's ecosystems.

Biological systems are constantly processing matter, but the atoms that compose organisms, primarily carbon, hydrogen and oxygen, maintain their integrity even as substances are broken down and the atoms are used to assemble new compounds. The Earth has an essentially fixed number of each of the atoms of biological importance, meaning these atoms must be recycled. Each organism assembles its body from atoms that were previously in the soil, the atmosphere, other parts of the abiotic environment, or other organisms. When organisms die, their atoms are released unaltered to be used by other organisms or returned to the abiotic environment. Because of the cycling of the atomic constituents of matter, your body is likely, during your life, to contain a carbon or oxygen atom that once was part of Julius Caesar's body or Cleopatra's. Indeed, every atom in your body was once part of a star.

The atoms of the various chemical elements are said to move through ecosystems in biogeochemical cycles, a term emphasizing that the cycles of chemical elements involve not only biological organisms and processes but also geologic (abiotic) systems and processes. Biogeochemical cycles include processes that occur on many spatial scales, from cellular to planetary, and they include processes that occur on multiple timescales, from seconds (biochemical reactions) to millennia (weathering of rocks). This constant cycling of matter crosses the boundaries of ecosystems, rather than being self-contained within individual ecosystems. For example, one ecosystem might import or export carbon to others.

In this section, we consider the cycles of some major elements along with the compound water. We also present an example of biogeochemical cycles in a forest ecosystem.

Carbon, the Backbone of Organic Molecules, Cycles Through Most Ecosystems

LEARNING OBJECTIVE 39.5.2 Describe the basic carbon cycle.

Carbon is a major constituent of the bodies of organisms, because carbon atoms help form the framework of all organic compounds (refer to chapter 3); almost 20% of the weight of the human body is carbon. From the viewpoint of the day-to-day dynamics of ecosystems, carbon dioxide (CO_2) is the most significant carbon-containing compound in the abiotic environments of organisms. It makes up 0.03% of the volume of the atmosphere, meaning the atmosphere contains about 750 billion metric tons of carbon. In aquatic ecosystems, CO_2 reacts spontaneously with the water to form carbonic acid, which dissociates into H^+ and bicarbonate ion (HCO_3^-).

The basic carbon cycle

The carbon cycle is straightforward, as shown in figure 39.21. In terrestrial ecosystems, plants and other photosynthetic organisms take in CO_2 from the atmosphere and use it in photosynthesis to synthesize the carbon-containing organic compounds of which they are composed (refer to chapter 8). The process is sometimes called *carbon fixation*; *fixation* refers to metabolic reactions that make nongaseous compounds from gaseous ones.

Animals eat the photosynthetic organisms and build their own tissues by making use of the carbon atoms in the organic compounds they ingest. Both the photosynthetic organisms and the animals obtain energy during their lives by breaking down some of the organic compounds available to them, through the process of aerobic cellular respiration (refer to chapter 7). When they do this, they produce CO_2. Decaying organisms also produce CO_2. Carbon atoms returned to the form of CO_2 are available once more to be used in photosynthesis to synthesize new organic compounds.

In aquatic ecosystems, the carbon cycle is fundamentally similar, except that inorganic carbon is present in the water not only as dissolved CO_2 but also as HCO_3^-, both of which act as sources of carbon for photosynthesis by algae and aquatic plants.

Methane producers

Microbes that break down organic compounds by anaerobic cellular respiration (refer to chapter 7) provide an additional dimension to the global carbon cycle. Methanogens, for example, are microbes that produce methane (CH_4) instead of CO_2. One major source of CH_4 is wetland ecosystems, where methanogens live in the oxygen-free sediments. Methane that enters the atmosphere is oxidized abiotically to CO_2, but CH_4 that remains isolated from oxygen can persist for great lengths of time.

The rise of atmospheric carbon dioxide

Another dimension of the global carbon cycle is that over long stretches of time, some parts of the cycle may proceed more rapidly than others. These differences in rate have ordinarily been relatively minor on a year-to-year basis; in any one year, the amount of CO_2 made by the breakdown of organic compounds almost matches the amount of CO_2 used to synthesize new organic compounds.

Small mismatches, however, can have large consequences if continued for many years. The Earth's present reserves of coal were built up over geologic time. Organic compounds, such as cellulose, accumulated by being synthesized faster than they were broken down, and then they were transformed by geologic processes into the fossil fuels. Most scientists believe that the world's petroleum reserves were created in the same way.

Human burning of fossil fuels today is creating large contemporary imbalances in the carbon cycle. Carbon that took millions of years to accumulate in the reserves of fossil fuels is being rapidly returned to the atmosphere, driving the concentration of CO_2 in the atmosphere upward year by year, increasing global temperatures and lowering the pH of oceans (refer to chapter 40).

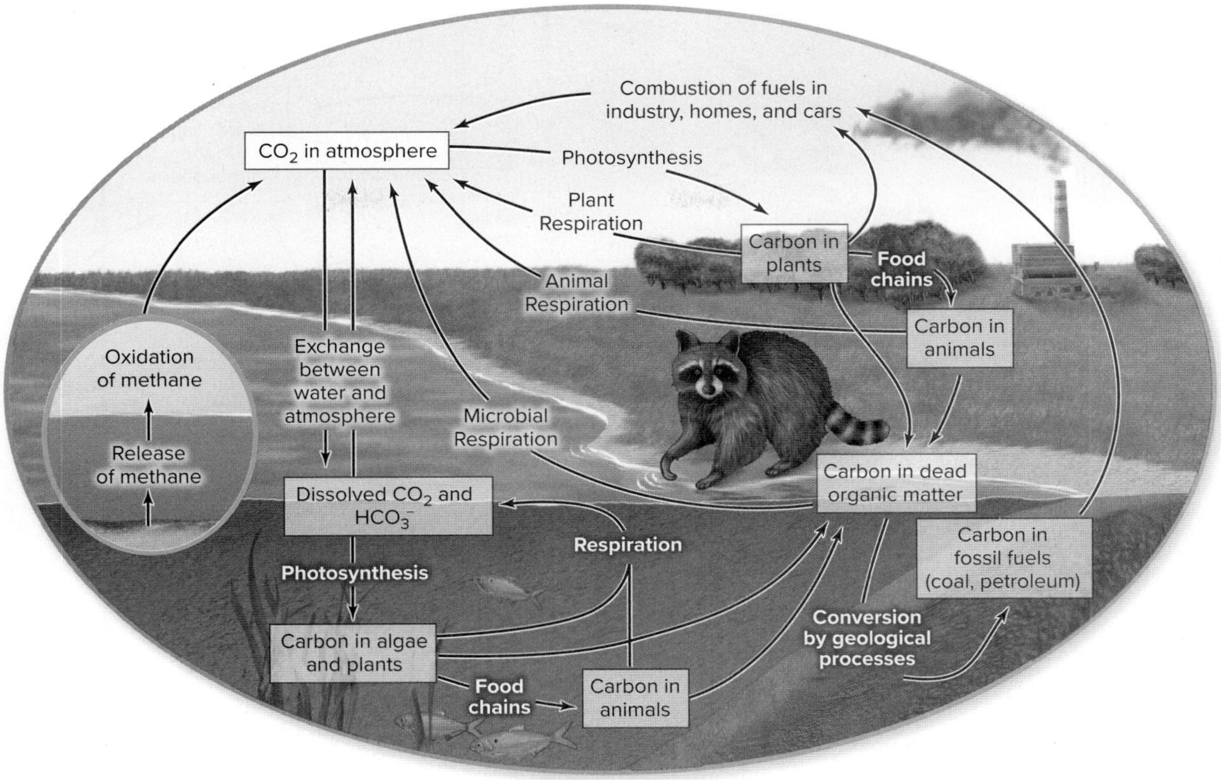

Figure 39.21 The carbon cycle. Photosynthesis by plants and algae captures carbon in the form of organic chemical compounds. Aerobic respiration by organisms and fuel combustion by humans return carbon to the form of carbon dioxide (CO_2) or bicarbonate (HCO_3^-). Microbial methanogens living in oxygen-free microhabitats, such as the mud at the bottom of a pond, might produce methane (CH_4), a gas that would enter the atmosphere and then gradually be oxidized abiotically to CO_2 (shown in green circled inset).

The Availability of Water Is Fundamental to Terrestrial Ecosystems

LEARNING OBJECTIVE 39.5.3 Describe the basic water cycle.

The water cycle, presented in figure 39.22, is probably the most familiar of all biogeochemical cycles. All life depends on the presence of water; even organisms that can survive without water in resting states require water to regain activity. The bodies of most organisms consist mainly of water. The adult human body, for example, is about 60% water by weight. The amount of water available in an ecosystem often determines the nature and abundance of the organisms present, as illustrated by the difference between forests and deserts.

Each type of biogeochemical cycle has distinctive features. A distinctive feature of the water cycle is that water is a compound, not an element, and thus it can be synthesized and broken down. It is synthesized during aerobic cellular respiration (refer to chapter 7) and chemically split during photosynthesis (refer to chapter 8). The rates of these processes are ordinarily about equal, and therefore a relatively constant amount of water cycles through the biosphere.

The basic water cycle

One key part of the water cycle is that liquid water from the Earth's surface evaporates into the atmosphere. The change of water from a liquid to a gas requires a considerable addition of thermal energy, explaining why evaporation occurs more rapidly when solar radiation beats down on a surface.

Evaporation occurs directly from the surfaces of oceans, lakes, and rivers. In terrestrial ecosystems, however, approximately 90% of the water that reaches the atmosphere passes through plants. Trees, grasses, and other plants take up water from soil via their roots, and then the water evaporates from their leaves and other surfaces by transpiration.

Evaporated water exists in the atmosphere as a gas, just like any other atmospheric gas. The water can condense back into liquid form, however, mostly because of cooling of the air. Condensation of gaseous water (water vapor) into droplets or crystals causes the formation of clouds, and if the droplets or crystals are large enough, they fall to the surface of the Earth as precipitation (rain or snow).

Groundwater

Less obvious than surface water, which we see in rivers and lakes, is water under ground—termed groundwater. Groundwater occurs in *aquifers,* which are permeable, underground layers of rock, sand, and gravel that are often saturated with water. Groundwater is the most important reservoir of water on land in many parts of the world, representing over 95% of all fresh water in the United States, for example.

Figure 39.22 The water cycle. Water circulates from the atmosphere to the surface of the Earth and back again. The Sun provides much of the energy required for evaporation.

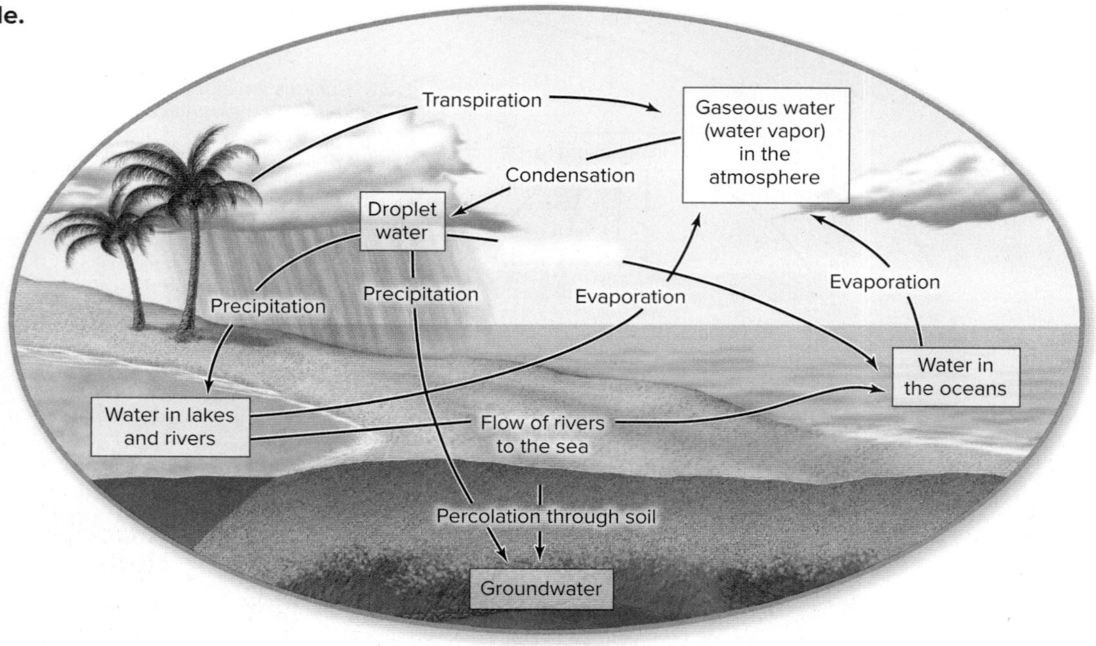

Groundwater consists of two subparts. The upper layers of the groundwater constitute the water table, which is unconfined in the sense that it flows into streams and is partly accessible to the roots of plants. The lower, confined layers of the groundwater are generally out of reach to streams and plants, but they can be tapped by wells. Groundwater is recharged by water that percolates downward from above, such as from precipitation. Water in an aquifer flows much more slowly than surface water, anywhere from a few millimeters to a meter or so per day.

In the United States, groundwater provides about 25% of the water used by humans for all purposes, and it supplies about 50% of the population with drinking water. In the Great Plains states, the deep Ogallala Aquifer is tapped extensively as a water source for agricultural and domestic needs. The aquifer is being depleted faster than it is being recharged—a local imbalance in the water cycle—posing an ominous threat to the agricultural production of the area. Similar threats exist in many of the drier portions of the globe.

Changes in ecosystems brought about by changes in the water cycle

Water is so crucial for life that changes in its supply in an ecosystem can radically alter the nature of the ecosystem. Such changes have occurred often during the Earth's geologic history.

Consider, for example, the ecosystem of the Serengeti Plain in Tanzania, famous for its seemingly endless grasslands occupied by vast herds of antelopes and other grazing animals. The semiarid grasslands of today's Serengeti were rainforests 25 MYA. Starting at about that time, mountains such as Mount Kilimanjaro rose up between the rainforests and the Indian Ocean, their source of moisture. The presence of the mountains forced winds from the Indian Ocean upward, cooling the air and causing much of its moisture to precipitate before the air reached the rainforests. The land became much drier, and the forests turned to grasslands.

Today, human activities can alter the water cycle so profoundly that major changes occur in ecosystems. Changes in rainforests caused by deforestation provide an example. In healthy tropical rainforests, more than 90% of the moisture that falls as rain is taken up by plants and returned to the air by transpiration. Plants, in a very real sense, create their own rain: the moisture returned to the atmosphere falls back on the forests.

When human populations cut down or burn the rainforests in an area, the local water cycle is broken. Water that falls as rain thereafter drains away in rivers instead of rising to form clouds and fall again on the forests. Just such a transformation is occurring today in many tropical rainforests (figure 39.23). Large areas in Brazil, for example, were transformed in the 20th century from lush tropical forest to semiarid desert, depriving many unique plant and animal species of their native habitat.

The Nitrogen Cycle Is Driven by Microbial Activity

LEARNING OBJECTIVE 39.5.4 Explain how microbial activities drive the nitrogen cycle.

Nitrogen is a component of all proteins and nucleic acids and is required in substantial amounts by all organisms; proteins are 16% nitrogen by weight. In many ecosystems, nitrogen is the chemical element in shortest supply relative to the needs of organisms. A paradox is that the atmosphere is 78% nitrogen by volume.

Nitrogen availability

How can nitrogen be in short supply if the atmosphere is so rich with it? The answer is that the nitrogen in the atmosphere is in its elemental form—molecules of nitrogen gas (N_2)—and the vast majority of organisms, including all plants and animals, have no way to use nitrogen in this chemical form.

Figure 39.23 Deforestation disrupts the local water cycle. Tropical deforestation can have severe consequences, as in the Amazon region of Brazil, where rain that used to be absorbed by the soil and eventually returned to the atmosphere by transpiration now runs off into streams and is lost from the area.
Worldwide Picture Library/Alamy Stock Photo

For animals, the ultimate source of nitrogen is nitrogen-containing organic compounds synthesized by plants or by algae or other microbes. Herbivorous animals, for example, eat plant or algal proteins and use the nitrogen-containing amino acids in them to synthesize their own proteins.

Plants and algae use a number of simple nitrogen-containing compounds as their sources of nitrogen to synthesize proteins and other nitrogen-containing organic compounds in their tissues. Two commonly used nitrogen sources are ammonia (NH_3) and nitrate ions (NO_3^-). Certain prokaryotic microbes can synthesize ammonia and nitrate from N_2 in the atmosphere, thereby constituting a part of the nitrogen cycle that makes atmospheric nitrogen accessible to plants and algae (figure 39.24). Other prokaryotes turn NH_3 and NO_3^- into N_2, making the nitrogen inaccessible. The balance of the activities of these two sets of microbes determines the accessibility of nitrogen to plants and algae.

Microbial nitrogen fixation, nitrification, and denitrification

The synthesis of nitrogen-containing compounds from N_2 is known as **nitrogen fixation.** The first step in this process is the synthesis of NH_3 from N_2; biochemists sometimes use the term *nitrogen fixation* to refer specifically to this step. After NH_3 has been synthesized, other prokaryotic microbes oxidize part of it to form NO_3^-, a process called **nitrification.**

Certain genera of prokaryotes have the ability to accomplish nitrogen fixation using a system of enzymes known as the nitrogenase complex. Most of the microbes are free-living, but on land some are found in symbiotic relationships with the roots of legumes (plants of the pea family, Fabaceae), alders, myrtles, and other plants.

Figure 39.24 The nitrogen cycle. The nitrogen cycle is complicated because it involves multiple changes in the chemical form of nitrogen. Certain prokaryotes fix atmospheric nitrogen (N_2), converting it to forms such as ammonia (NH_3) and nitrate (NO_3^-) that plants and algae can use. Other prokaryotes return nitrogen to the atmosphere as N_2 by breaking down NH_3 or other nitrogen-containing compounds. Ammonia, a gas, can enter the atmosphere directly from soils.

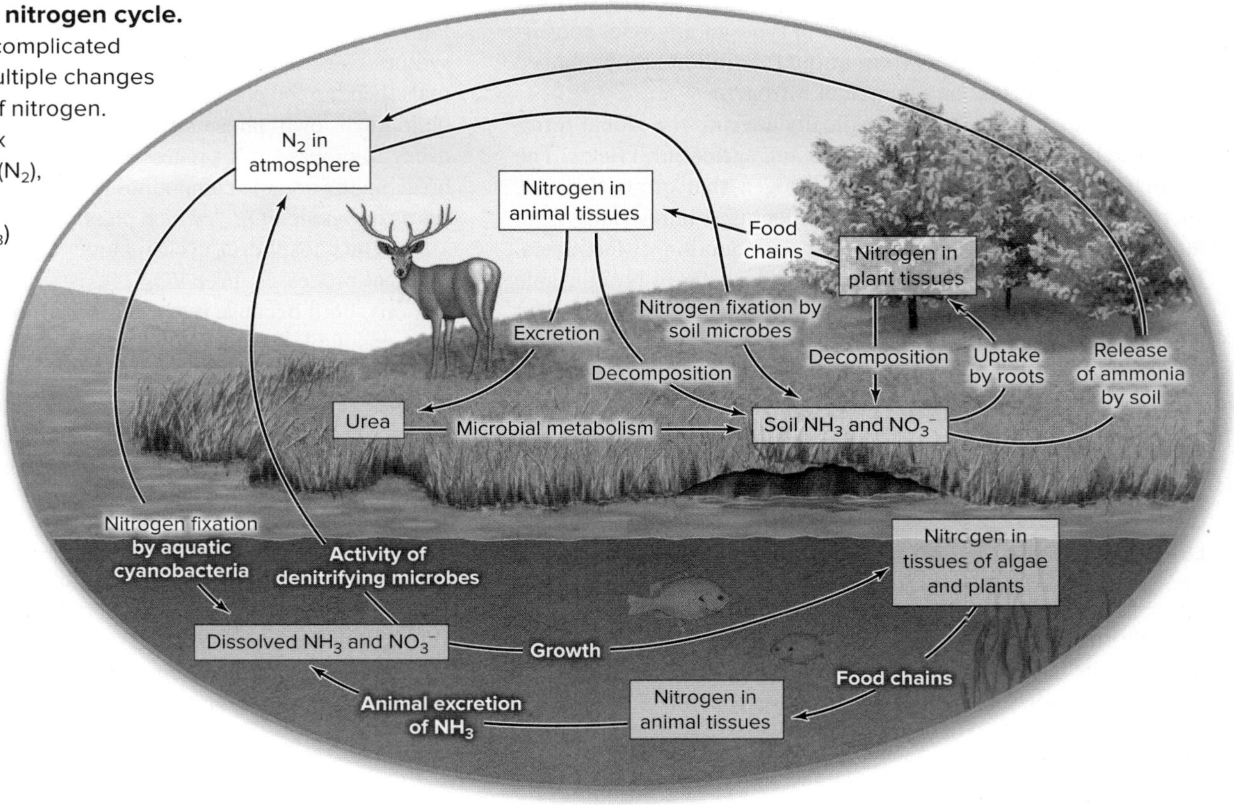

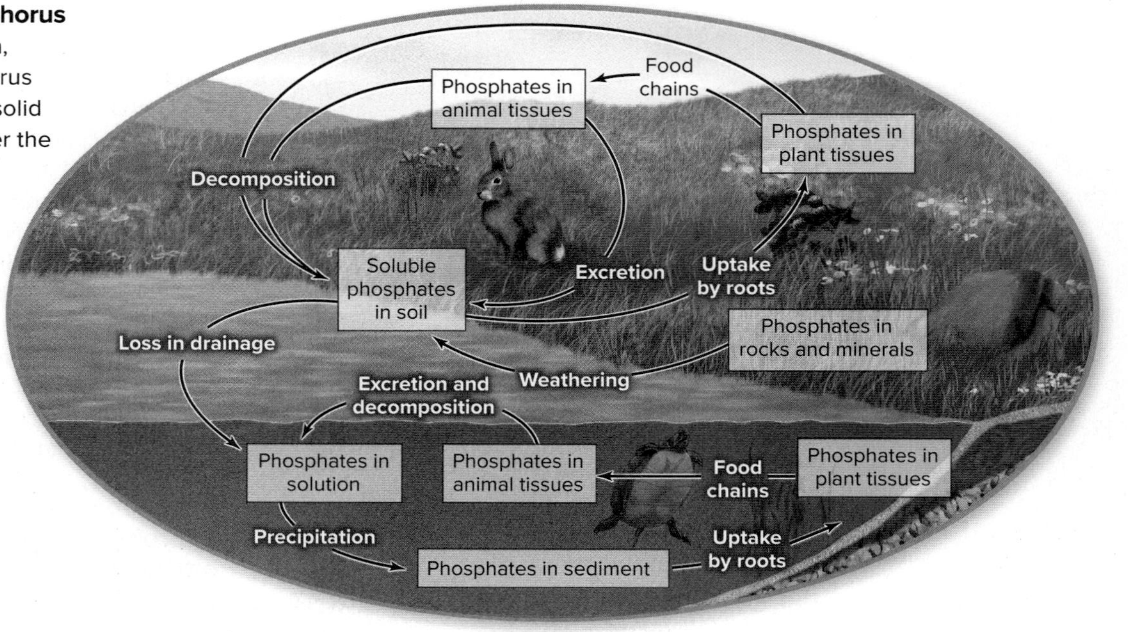

Figure 39.25 The phosphorus cycle. In contrast to carbon, water, and nitrogen, phosphorus occurs only in the liquid and solid states and thus does not enter the atmosphere.

Additional prokaryotic microbes (including both bacteria and archaea) are able to convert the nitrogen in NO_3^- into N_2 (or other nitrogen gases such as N_2O), a process termed **denitrification.** Ammonia can be subjected to denitrification indirectly by being converted first to NO_3^- and then to N_2.

Nitrogenous wastes and fertilizer use

Most animals, when they break down proteins in their metabolism, excrete the nitrogen derived from the proteins as NH_3. Humans and other mammals excrete nitrogen as urea in their urine (refer to chapter 35); a number of types of microbes convert the urea to NH_3. The NH_3 from animal excretion can be picked up by plants and algae as a source of nitrogen.

Human populations are radically altering the global nitrogen cycle by using fertilizers on lawns and agricultural fields. The fertilizers contain forms of fixed nitrogen that crops can use, such as ammonium (NH_4) salts manufactured industrially from atmospheric N_2. Partly because of the production of fertilizers, humans have already doubled the rate of transfer of N_2 in usable forms into soils and waters.

Phosphorus Cycles Through Terrestrial and Aquatic Ecosystems, but Not the Atmosphere

LEARNING OBJECTIVE 39.5.5 Compare the cycling of phosphorus to the cycling of other key elements.

Phosphorus is required in substantial quantities by all organisms; it occurs in nucleic acids, membrane phospholipids, and other essential compounds, such as adenosine triphosphate (ATP).

Unlike carbon, water, and nitrogen, phosphorus has no significant gaseous form and does not cycle through the atmosphere (figure 39.25). In this respect, the phosphorus cycle exemplifies the sorts of cycles also exhibited by calcium, silicon, and many other mineral elements. Another feature that greatly simplifies the phosphorus cycle compared with the nitrogen cycle is that phosphorus exists in ecosystems in just a single oxidation state, phosphate (PO_4^{3-}).

Phosphate availability

Plants and algae use free inorganic PO_4^{3-} in the soil or water for synthesizing their phosphorus-containing organic compounds. Animals then tap the phosphorus in plant or algal tissue compounds to build their own phosphorus compounds. When organisms die, decay microbes—in a process called phosphate remineralization—break up the organic compounds in their bodies, releasing phosphorus as inorganic PO_4^{3-}, which plants and algae can use again.

The phosphorus cycle includes critical abiotic chemical and physical processes. Free PO_4^{3-} exists in soil in only low concentrations, both because it combines with other soil constituents to form insoluble compounds and because it tends to be washed away by streams and rivers. Weathering of many sorts of rocks releases new PO_4^{3-} into terrestrial systems, but then rivers carry the PO_4^{3-} into the ocean basins. There is a large one-way flux of PO_4^{3-} from terrestrial rocks to deep-sea sediments.

Phosphates as fertilizers

Human activities have greatly modified the global phosphorus cycle since the advent of crop fertilization. Fertilizers are typically designed to provide PO_4^{3-} because crops might otherwise be short of it; the PO_4^{3-} in fertilizers is typically derived from crushed phosphate-rich rocks and bones. Detergents are another potential culprit in adding PO_4^{3-} to ecosystems, but laws now mandate low-phosphate detergents in much of the world.

REVIEW OF CONCEPT 39.5

Ecosystems consist of both living and nonliving components. Elements move between these in biogeochemical cycles. Carbon is cycled primarily by photosynthesis and respiration. Nitrogen undergoes complex cycling based on microbial activity performing nitrification, denitrification, and nitrogen fixation. Water also cycles from earth to atmosphere and is critical to ecosystems. Human activity affects these cycles.

- *Would fertilization with animal manure be less disruptive than fertilization with purified chemicals? Why or why not?*

39.6 Energy Flows Through Ecosystems in One Direction

The dynamic nature of ecosystems includes the processing of energy as well as that of matter. Energy, however, follows very different principles than does matter. Energy is never recycled. Instead, radiant energy from the Sun that reaches the Earth makes a one-way pass through our planet's ecosystems before being converted to heat and radiated back into space—that is, the Earth is an open system for energy.

Living Organisms Can Use Many Forms of Energy, but Not Heat

LEARNING OBJECTIVE 39.6.1 Relate the First and Second Laws of Thermodynamics to the biosphere.

Why is energy so different from matter? A key part of the answer is that energy exists in several different forms, such as light, chemical-bond energy, motion, and heat. Although energy is neither created nor destroyed in the biosphere (First Law of Thermodynamics), it frequently changes form (refer to chapter 6 for a review of thermodynamics).

A second key point is that organisms cannot convert heat to any of the other forms of energy. Thus, if the energy transactions occurring in living systems produce heat, the conversion is one-way; they cannot cycle that energy back into its original form.

To understand why the Earth functions as an open system with regard to energy, we need to recognize two other principles. The first is that organisms can use only certain forms of energy. For animals to live, they need energy derived from the oxidation of organic compounds from their diet. Plants can utilize energy from light to synthesize organic compounds. Neither animals nor plants (nor any other organisms) can use heat as a source of energy.

The second principle is that whenever organisms use chemical-bond or light energy, some of it is converted to heat; the Second Law of Thermodynamics states that a partial conversion to heat is inevitable. Put another way, animals and plants require chemical-bond energy and light to stay alive, but as they use these forms of energy, they convert them to heat, which they cannot use to stay alive and which they cannot cycle back into the original forms.

Fortunately for organisms, the Earth functions as an open system for energy. Light arrives every day from the Sun. Plants and other photosynthetic organisms use this sunlight to synthesize organic compounds. Animals then eat the photosynthetic organisms, making use of the chemical-bond energy in their organic molecules to stay alive. Light and chemical-bond energy are partially converted to heat at every step. In fact, the light and chemical-bond energy are ultimately converted completely to heat. The heat leaves the Earth by being radiated into outer space at invisible, infrared wavelengths of the electromagnetic spectrum. For life to continue, new light energy is always required.

The Earth's incoming and outgoing flows of radiant energy must be equal for global temperature to stay constant. One concern is that human activities are changing the composition of the atmosphere in ways that impede the outgoing flow—the so-called *greenhouse effect,* which is described in chapter 40. Heat may be accumulating on Earth, causing global climate change.

Energy Flows Through the Trophic Levels of Ecosystems

LEARNING OBJECTIVE 39.6.2 Describe the different trophic levels of ecosystems.

In chapter 7, we introduced the concepts of autotroph ("self-feeder") and heterotroph ("fed by others"). **Autotrophs** synthesize the organic compounds of their bodies from inorganic precursors such as CO_2, water, and NO_3^-, using energy from an abiotic source. Some autotrophs use light as their source of energy and therefore are **photoautotrophs;** they are the photosynthetic organisms, including plants, algae, and cyanobacteria. Other autotrophs are **chemoautotrophs,** obtaining energy by means of inorganic oxidation reactions, such as the microbes that use hydrogen sulfide available at deep-water vents. All chemoautotrophs are prokaryotic. The photoautotrophs are of greatest importance in most ecosystems, and we focus on them in the remainder of this chapter.

Heterotrophs are organisms that cannot synthesize organic compounds from inorganic precursors but, instead, live by taking in organic compounds that other organisms have made. They obtain the energy they need to live by breaking up some of the organic compounds available to them, thereby liberating chemical-bond energy for metabolic use (refer to chapter 7). Animals, fungi, and many microbes are heterotrophs.

When living in their native environments, species are often organized into chains that eat each other sequentially. For example, a species of insect might eat plants, and then a species of shrew might eat the insect, and a species of hawk might eat the shrew. Food passes through the four species in the sequence plants → insect → shrew → hawk. A sequence of species like this is termed a food chain.

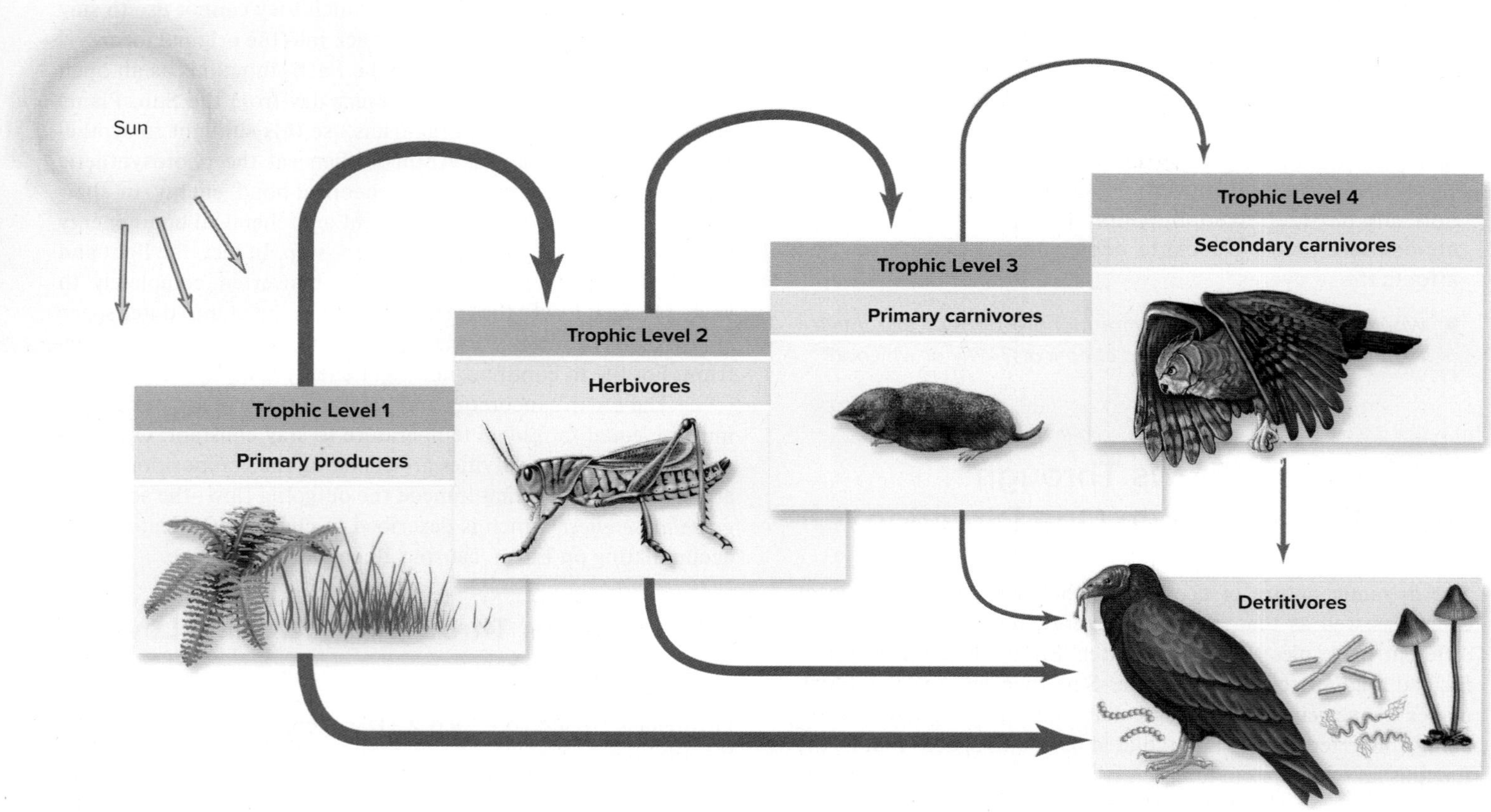

Figure 39.26 Trophic levels within an ecosystem. Primary producers such as plants obtain their energy directly from the Sun, placing them in trophic level 1. Animals that eat plants, such as plant-eating insects, are herbivores and are in trophic level 2. Animals that eat the herbivores, such as shrews, are primary carnivores and are in trophic level 3. Animals that eat the primary carnivores, such as owls, are secondary carnivores in trophic level 4. Each trophic level, although illustrated here by a particular species, consists of all the species in the ecosystem that function in a similar way in terms of what they eat. The organisms in the detritivore trophic level consume dead organic matter they obtain from all the other trophic levels.

In a whole ecosystem, many species play similar roles; there is typically not just a single species in each role. For example, the animals that eat plants might include not just a single insect species but perhaps 30 species of insects, plus perhaps 10 species of mammals. To organize this complexity, ecologists recognize a limited number of feeding, or **trophic, levels** (figure 39.26).

Definitions of trophic levels

The first trophic level in an ecosystem, called the **primary producers,** consists of all the autotrophs in the system. The other trophic levels consist of the heterotrophs—the **consumers.** All the heterotrophs that feed directly on the primary producers are placed together in a trophic level called the **herbivores.** In turn, the heterotrophs that feed on the herbivores (eating them or being parasitic on them) are collectively termed **primary carnivores,** and those that feed on the primary carnivores are called **secondary carnivores.**

Advanced studies of ecosystems need to consider that organisms often do not line up in simple linear sequences in terms of what they eat; some animals, for example, eat both primary producers and other animals. A linear sequence of trophic levels is a useful organizing principle for many purposes, however.

An additional consumer level is the **detritivore** trophic level. Detritivores differ from the organisms in the other trophic levels in that they feed on the remains of already-dead organisms; detritus in this case is dead organic matter. A subcategory of detritivores is the **decomposers,** which are mostly microbes and other minute organisms that live on and break down dead organic matter.

Concepts to describe trophic levels

Trophic levels consist of whole populations of organisms. For example, the primary-producer trophic level consists of the whole populations of all the autotrophic species in an ecosystem. Ecologists have developed a special set of terms to refer to the properties of populations and trophic levels.

The **productivity** of a trophic level is the rate at which the organisms in the trophic level collectively synthesize new organic matter (new tissue substance). **Primary productivity** is the productivity of the primary producers. An important complexity in analyzing the primary producers is that not only do they synthesize new organic matter by photosynthesis but they also break down some of the organic matter to release energy by means of aerobic cellular respiration (refer to chapter 7). The **respiration** of the primary producers, in this context, is the rate at which they break down organic compounds. **Gross primary productivity (GPP)** is simply the raw rate at which the primary producers synthesize new organic matter; **net primary productivity (NPP)** is the GPP minus the respiration of the primary producers. The NPP represents the organic matter available for herbivores to use as food.

The productivity of a heterotroph trophic level is termed **secondary productivity.** For instance, the rate at which new organic matter is made by means of individual growth and reproduction in all the herbivores in an ecosystem is the secondary productivity of the herbivore trophic level. Each heterotroph trophic level has its own secondary productivity.

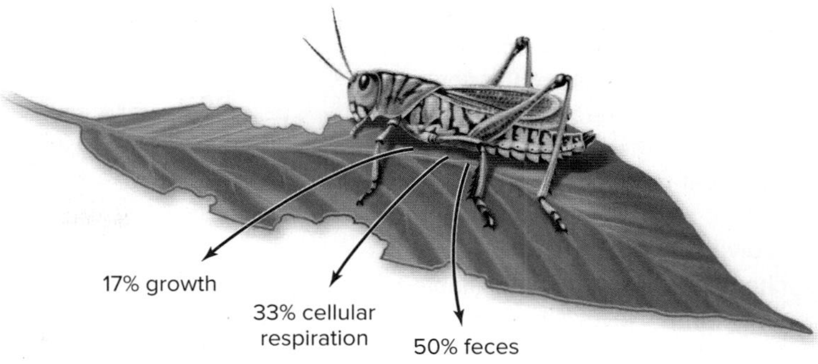

17% growth
33% cellular respiration
50% feces

Figure 39.27 The fate of ingested chemical-bond energy: Why all the energy ingested by a heterotroph is not available to the next trophic level. A heterotroph such as this herbivorous insect assimilates only a fraction of the chemical-bond energy it ingests. In this example, 50% is not assimilated and is eliminated in feces; this eliminated chemical-bond energy cannot be used by the primary carnivores. A third (33%) of the ingested energy is used to fuel cellular respiration and thus is converted to heat, which cannot be used by the primary carnivores. Only 17% of the ingested energy is converted into insect biomass through growth and can serve as food for the next trophic level, but not even that percentage is certain to be used in that way, because some of the insects die before they are eaten.

How trophic levels process energy

The fraction of incoming solar radiant energy that the primary producers capture is small. Averaged over the course of a year, something around 1% of the solar energy impinging on forests or oceans is captured. Investigators sometimes observe far lower levels but also see percentages as high as 5% under some conditions. The solar energy not captured as chemical-bond energy through photosynthesis is immediately converted to heat.

The primary producers, as noted, carry out respiration, in which they break down some of the organic compounds in their bodies to release chemical-bond energy. They use a portion of this chemical-bond energy to make ATP, which they in turn use to power various energy-requiring processes. Ultimately, the chemical-bond energy they release by respiration turns to heat.

Remember that organisms cannot use heat to stay alive. As a result, whenever energy changes form to become heat, it loses much or all of its usefulness for organisms as a fuel source. What we have seen so far is that about 99% of the solar energy impinging on an ecosystem turns to heat because it fails to be used by photosynthesis. Then some of the energy captured by photosynthesis also becomes heat because of respiration by the primary producers. All the heterotrophs in an ecosystem must live on the chemical-bond energy that is left.

An example of energy loss between trophic levels

As chemical-bond energy is passed from one heterotroph trophic level to the next, a great deal of the energy is diverted all along the way. This principle has dramatic consequences. It means that over any particular period of time, the amount of chemical-bond energy available to primary carnivores is far less than that available to herbivores, and the amount available to secondary carnivores is far less than that available to primary carnivores.

Why does the amount of chemical-bond energy decrease as energy is passed from one trophic level to the next? Consider the use of energy by the herbivore trophic level as an example (figure 39.27). After an herbivore such as a leaf-eating insect ingests some food, it produces feces. The chemical-bond energy in the compounds in the feces is not passed along to the primary carnivore trophic level. The chemical-bond energy of the food that is assimilated by the herbivore is used for a number of functions. Part of the assimilated energy is liberated by cellular respiration to be used for tissue repair, body movements, and other such functions. The energy used in these ways turns to heat and is not passed along to the carnivore trophic level. Some chemical-bond energy is built into the tissues of the herbivore and can serve as food for a carnivore. However, some herbivore individuals die of disease or accident rather than being eaten by predators.

In the end, of course, some of the initial chemical-bond energy acquired from the leaf is built into the tissues of herbivore individuals that are eaten by primary carnivores. Much of the initial chemical-bond energy, however, is diverted into heat, feces, and the bodies of herbivore individuals that carnivores do not get to eat. The same scenario is repeated at each step in a series of trophic levels (figure 39.28).

Ecologists figure, as a generality, that the amount of chemical-bond energy available to a trophic level over time is about 10% of that available to the preceding level over the same period of time. In some instances, the percentage is higher, even as high as 30%.

Heat as the final energy product

Essentially all of the chemical-bond energy captured by photosynthesis in an ecosystem eventually becomes heat as the chemical-bond energy is used by various trophic levels. To see this important point, recognize that when the detritivores in the ecosystem metabolize all the dead bodies, feces, and other materials made available to them, they produce heat just as the other trophic levels do. Of course, detritivores are sometimes eaten by carnivores, so the chemical-bond energy they consume is not all immediately lost as heat.

Figure 39.28 The flow of energy through an ecosystem. Blue arrows represent the flow of energy that enters the ecosystem as light and is then passed along as chemical-bond energy to successive trophic levels. At each step, energy is diverted, meaning that the chemical-bond energy available to each trophic level is less than that available to the preceding trophic level. Red arrows represent diversions of energy into heat. Tan arrows represent diversions of energy into feces and other organic materials useful only to the detritivores. Detritivores may be eaten by carnivores, so some of the chemical-bond energy returns to higher trophic levels.

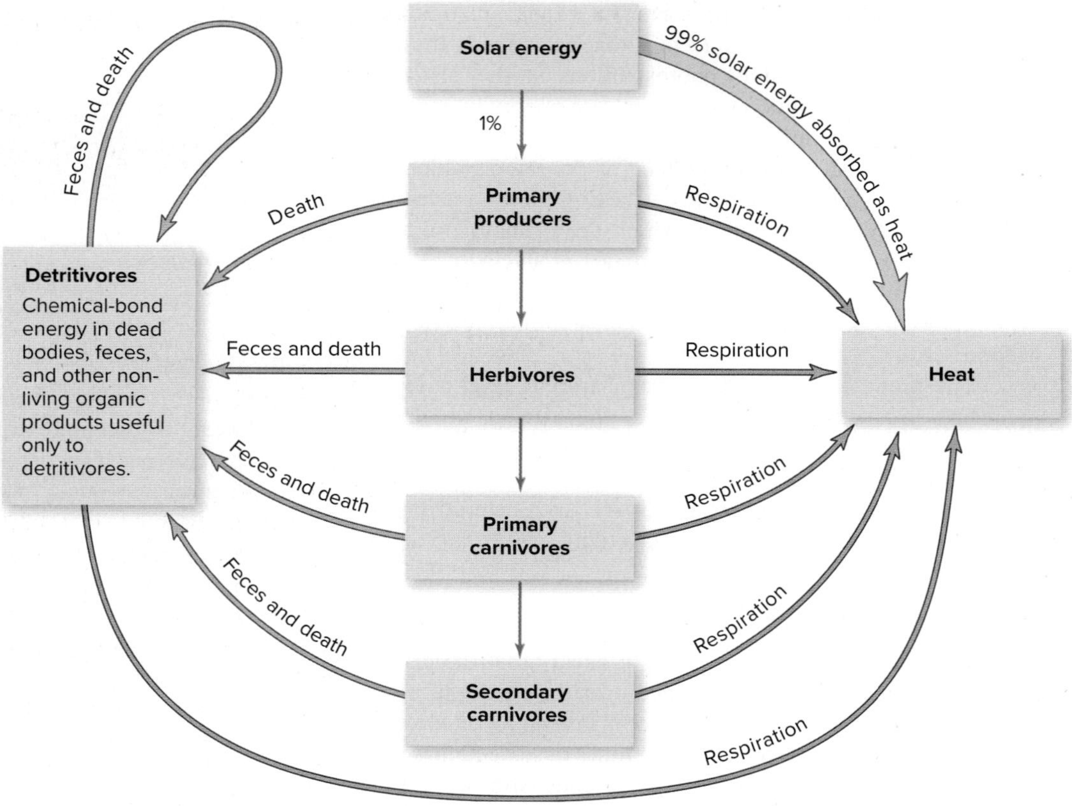

The Number of Trophic Levels Is Limited by Energy Availability

LEARNING OBJECTIVE 39.6.3 Explain how energy moves through trophic levels.

The rate at which chemical-bond energy is made available to organisms in different trophic levels decreases exponentially as energy makes its way from primary producers to herbivores and then to various levels of carnivores. To envision this critical point, assume for simplicity that the primary producers in an ecosystem gain 1000 units of chemical-bond energy over a period of time. If the energy input to each trophic level is 10% of the input to the preceding level, then the input of chemical-bond energy to the herbivore trophic level is 100 units; to the primary carnivores, 10 units; and to the secondary carnivores, 1 unit over the same period of time.

Limits on top carnivores

The exponential decline of chemical-bond energy in a trophic chain limits the lengths of trophic chains and the numbers of top carnivores an ecosystem can support. According to our model calculations, if an ecosystem includes secondary carnivores, only about one-thousandth of the energy captured by photosynthesis passes all the way through the series of trophic levels to reach these animals as usable chemical-bond energy. Tertiary carnivores would receive only one ten-thousandth. This helps explain why no predators subsist solely on eagles or lions.

The decline of available chemical-bond energy also helps explain why the numbers of individual top-level carnivores in an ecosystem tend to be low. The whole trophic level of top carnivores receives relatively little energy, yet such carnivores tend to be big: they have relatively large individual body sizes and great individual energy needs. Because of these two factors, the population numbers of top predators tend to be small.

The longest trophic chains probably occur in the oceans. Some tunas and other top-level ocean predators probably function as third- and fourth-level carnivores at times. The challenge of explaining such long trophic chains is obvious, but they are not well understood.

Humans as consumers: A case study

The flow of energy in Cayuga Lake in upstate New York (figure 39.29) helps illustrate how the energetics of trophic levels can affect the human food supply. Researchers calculated from the actual properties of this ecosystem that about 150 of each 1000 cal of chemical-bond energy captured by primary producers in the lake were transferred into the bodies of herbivores. Of these calories, about 30 were transferred into the bodies of smelt, small fish that were the principal primary carnivores in the system.

If humans ate the smelt, they gained about 6 of the 1000 cal that originally entered the system. If trout ate the smelt and humans ate the trout, the humans gained only about 1.2 cal. For human populations in general, more energy is available if plants or other primary producers are eaten than if animals are eaten—and more energy is available if herbivores rather than carnivores are consumed.

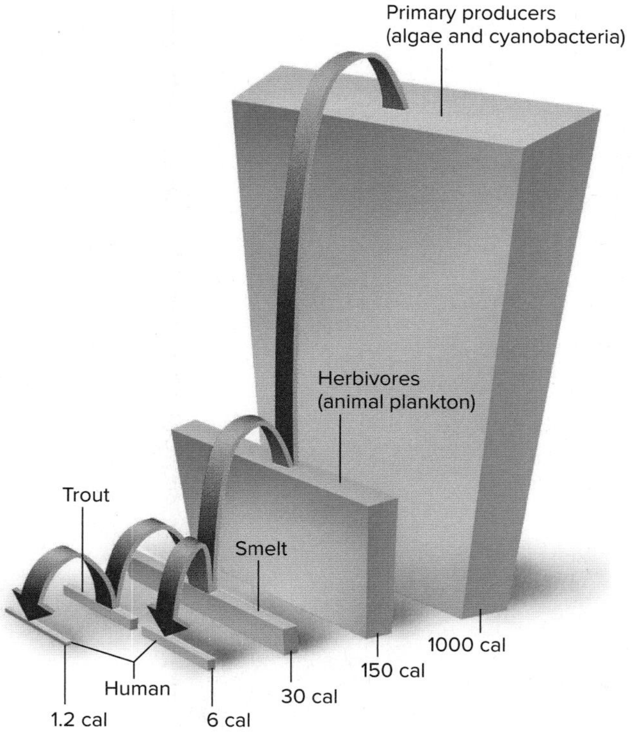
Primary producers
(algae and cyanobacteria)
Herbivores
(animal plankton)
Trout
Smelt
Human
1000 cal
150 cal
30 cal
6 cal
1.2 cal

Figure 39.29 Flow of energy through the trophic levels of Cayuga Lake. Autotrophic plankton (algae and cyanobacteria) fix the energy of the Sun, the herbivores (animal plankton) feed on them, and both are consumed by smelt. The smelt are eaten by trout. The amount of fish flesh produced per unit time for human consumption is at least five times greater if people eat smelt rather than trout, but people typically prefer to eat trout.

Ecological Pyramids Illustrate the Relationship of Trophic Levels

LEARNING OBJECTIVE 39.6.4 Compare standard and inverted biomass pyramids.

Imagine that the trophic levels of an ecosystem are represented as boxes stacked on top of each other. Imagine also that the width of each box is proportional to the productivity of the trophic level it represents. The stack of boxes will always have the shape of a pyramid; each box is narrower than the one under it because of the inviolable rules of energy flow. A diagram of this sort is called a pyramid of energy flow or pyramid of productivity (figure 39.30*a*). It is an example of an ecological pyramid.

There are several types of ecological pyramids. Pyramid diagrams can be used to represent standing crop biomass or numbers of individuals, as well as productivity.

In a **pyramid of biomass,** the widths of the boxes are drawn to be proportional to standing crop biomass. Usually, trophic levels that have relatively low productivity also have relatively little biomass present at a given time. Thus, pyramids of biomass are usually upright, meaning each box is narrower than the one below it (figure 39.30*b*). An upright pyramid of biomass is not mandated by fundamental and inviolable rules as an upright pyramid of productivity is, however. In some ecosystems, the pyramid of biomass is **inverted,** meaning that at least one trophic level has greater biomass than the one below it (figure 39.30*c*).

How is it possible for the pyramid of biomass to be inverted? Consider a common sort of aquatic system in which the primary producers are single-celled algae (phytoplankton) and the herbivores are rice-grain-size animals (such as copepods) that feed directly on the algal cells. In such a system, the turnover of the algal cells is often very rapid: the cells multiply rapidly, but the animals consume them equally rapidly. In these circumstances, the algal cells never develop a large population size or large biomass. Nonetheless, because the algal cells are very productive, the ecosystem can support a substantial biomass of the animals, a biomass larger than that ever observed in the algal population.

In a pyramid of numbers, the widths of the boxes are proportional to the numbers of individuals present in the various trophic levels (figure 39.30*d*). Such pyramids are usually, but not always, upright.

REVIEW OF CONCEPT 39.6

The flow of energy in an ecosystem is one-way. Trophic levels include primary producers, herbivores, primary carnivores, and secondary carnivores. Detritivores consume dead or waste matter from all levels. As energy passes between levels, some is lost as heat. Photosynthetic primary producers capture about 1% of solar energy. As this energy is passed through higher trophic levels, only about 10% is available to the next level. Lower trophic levels usually have greater biomass.

- *What are the different ways in which matter, such as carbon atoms, and energy move through ecosystems?*

39.7 Biodiversity May Increase Ecosystem Stability

Biodiversity is a topic that is often in the news. As we contemplate the extinction of an increasing number of species, the diversity of those that remain is an important issue for all of society.

Species Richness May Increase Stability: The Cedar Creek Studies

LEARNING OBJECTIVE 39.7.1 Define *ecosystem stability.*

Ecologists have long debated the consequences of differences in species richness—the number of species present—between communities. One theory is that species-rich communities are more stable—that is, more constant in composition and better able to resist disturbance. This hypothesis has been elegantly studied by David Tilman and colleagues at the University of Minnesota's Cedar Creek Natural History Area.

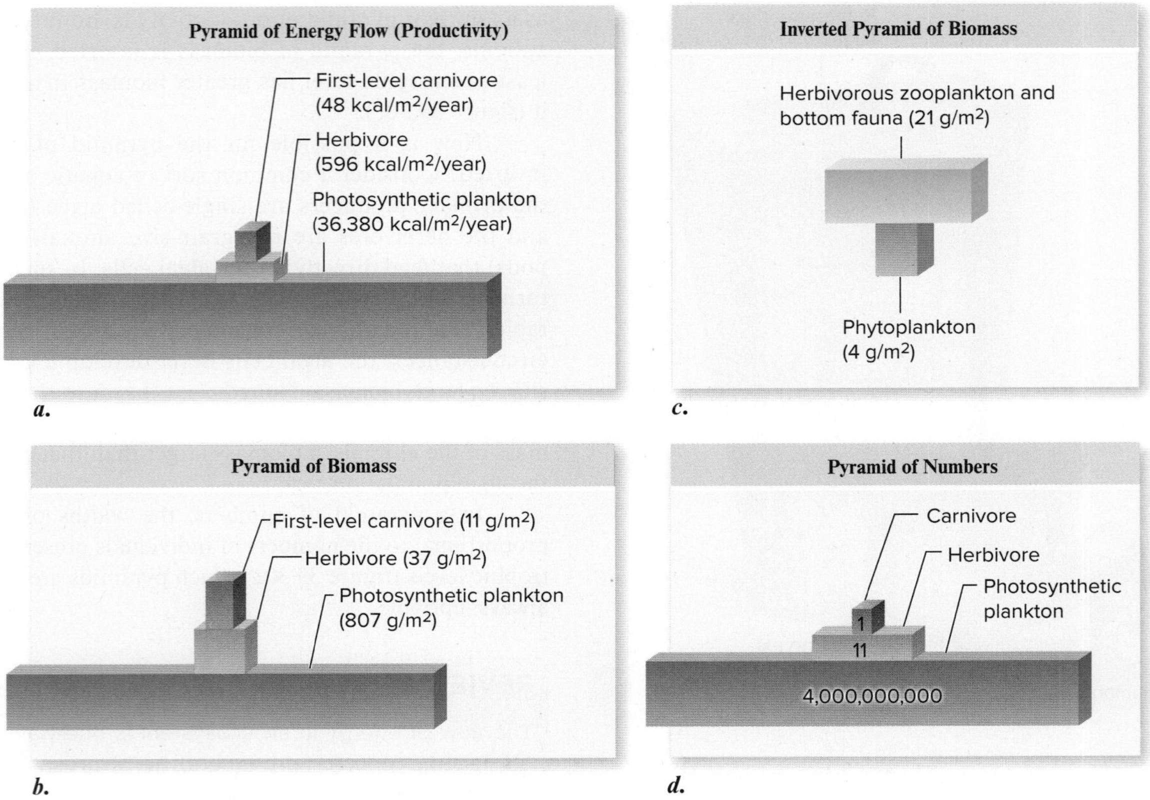

Figure 39.30 Ecological pyramids. In an ecological pyramid, successive trophic levels in an ecosystem are represented as stacked boxes, and the widths of the boxes represent the magnitude of an ecological property in the various trophic levels. Ecological pyramids can represent several different properties. ***a.*** Pyramid of energy flow (productivity). ***b.*** Pyramid of biomass of the ordinary type. ***c.*** Inverted pyramid of biomass. ***d.*** Pyramid of numbers.

Workers monitored 207 small, rectangular plots of land (8–16 m^2) for 11 years (figure 39.31*a*). In each plot, they counted the number of prairie plant species and measured the total amount of plant biomass (that is, the mass of all plants on the plot). Over the course of the study, plant species richness was related to community stability—plots with more species showed less year-to-year variation in biomass. Moreover, in two drought years, the decline in biomass was negatively related to species richness—that is, plots with more species were less affected by drought.

These findings were subsequently confirmed by an experiment in which plots were seeded with different numbers of species. Again, more species-rich plots had greater year-to-year stability in biomass, in this case over a 10-year period.

In a related experiment, when seeds of non-native plant species were added to different plots, the ability of these species to become established was negatively related to species richness (figure 39.31*b*). More diverse communities, in other words, are more resistant to invasion by new species, which is another measure of community stability.

Species richness may also affect other ecosystem processes. Tilman and colleagues monitored 147 experimental plots that varied in number of species to estimate how much growth was occurring and how much nitrogen the growing plants were taking up from the soil. They found that the more species a plot had, the greater the nitrogen uptake and total amount of biomass produced. In this study, increased biodiversity clearly appeared to lead to greater productivity.

Laboratory studies on artificial ecosystems have provided similar results. In one elaborate study, ecosystems covering 1 m^2 were constructed in growth chambers that controlled temperature, light levels, air currents, and atmospheric gas concentrations. A variety of plants, insects, and other animals were introduced to construct ecosystems composed of 9, 15, or 31 species, with the lower-diversity treatments containing a subset of the species in the higher-diversity enclosures. As with Tilman's experiments, the amount of biomass produced was related to species richness, as was the amount of carbon dioxide consumed, another measure of the productivity of the ecosystem.

Tilman's conclusion that healthy ecosystems depend on diversity is not accepted by all ecologists, however. Critics question the validity and relevance of these biodiversity studies, arguing that the more species added to a plot, the greater the probability that one species will be highly productive. To show that high productivity results from high species richness per se, rather than from the presence of particular highly productive species, experimental plots have to exhibit "overyielding"; in other words, plot productivity has to be greater than that of the single most productive species grown in isolation.

Although this point is still debated, recent work at Cedar Creek and elsewhere has provided evidence of overyielding, supporting the claim that species richness of communities enhances community productivity and stability.

SCIENTIFIC THINKING

Question: *Does species richness affect the invasibility of a community?*

Hypothesis: *The rate of successful invasion will be lower in communities with greater richness.*

Experiment: *Add seeds from the same number of non-native plants to experimental plots that differ in the number of plant species.*

a.

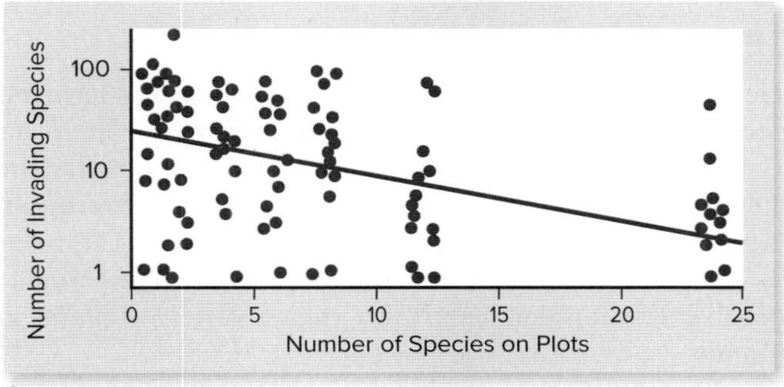

b.

Result: *Although the number of successful invasive species varies greatly, more species-rich plots on average are invaded by fewer species.*

Interpretation: *What might explain why so much variation exists in the number of successful invading species in communities with the same species richness?*

Figure 39.31 Effect of species richness on ecosystem stability. ***a.*** One of the Cedar Creek experimental plots. ***b.*** Community stability can be assessed by looking at the effect of species richness on community invasibility. Each dot represents data from one experimental plot in the Cedar Creek experimental fields. It is more difficult for non-native species to invade plots with more species.

(a): Layne Kennedy/Getty Images

Species Richness Is Influenced by Ecosystem Characteristics

LEARNING OBJECTIVE 39.7.2 Describe the effects of species richness on ecosystem function.

A number of factors are known or hypothesized to affect species richness in a community; these include loss of keystone species and moderate physical disturbance. Here we discuss three more: primary productivity, habitat heterogeneity, and climatic factors.

Primary productivity

Ecosystems differ substantially in primary productivity. Some evidence indicates that species richness is related to primary productivity, but the relationship between them is not linear: in a number of cases, ecosystems with intermediate levels of productivity tend to have the greatest number of species (figure 39.32*a*).

The underlying cause of this relationship is debated. One possibility is that levels of productivity are linked with numbers of consumers. Applying this concept to plant species richness, the argument is that at low productivity, there are few herbivores, and superior competitors among the plants are able to eliminate most other plant species. In contrast, at high productivity, so many herbivores are present that only the plant species most resistant to grazing survive, reducing species diversity. As a result, the greatest numbers of plant species coexist at intermediate levels of productivity and herbivory.

Habitat heterogeneity

Spatially heterogeneous abiotic environments are those that consist of many habitat types—different soil types, for example. These heterogeneous environments can be expected to accommodate more species of plants than spatially homogeneous environments. What's more, the species richness of animals can be expected to reflect the species richness of plants present. An example of this latter effect is in figure 39.32*b*: The number of lizard species at various sites in the American Southwest mirrors the local structural diversity of the plants.

Climatic factors

The role of climatic factors is more difficult to predict. On the one hand, more species might be expected to coexist in a seasonal environment than in a constant one, because a changing climate may favor different species at different times of the year. On the other hand, stable environments are able to support specialized species that would be unable to survive where conditions fluctuate. The number of mammal species at locations along the west coast of North America is inversely correlated with the amount of local temperature variation—the wider the variation, the fewer mammalian species—supporting the latter line of argument (figure 39.32*c*).

Tropical Regions Have the Highest Diversity, Although Reasons Are Unclear

LEARNING OBJECTIVE 39.7.3 Illustrate how multiple factors affect species richness in the tropics.

Since before Darwin, biologists have recognized that more kinds of animals and plants inhabit the tropics than the temperate regions. For many types of organisms, there is a steady increase in species richness from the Arctic to the tropics. Called a **species diversity cline,** this biogeographical gradient correlated with latitude has been reported for plants and animals, including birds (figure 39.33), mammals, and reptiles.

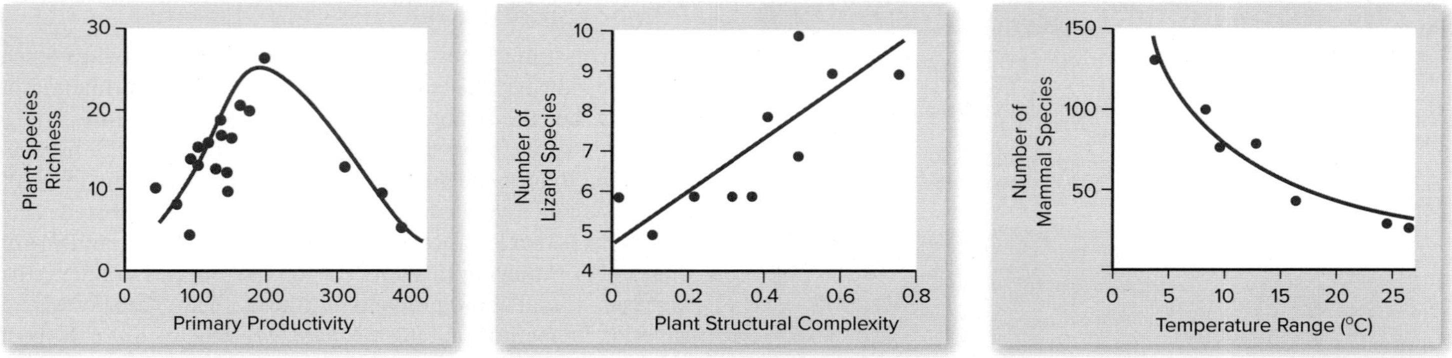

a. *b.* *c.*

Figure 39.32 Factors that affect species richness. ***a.*** *Productivity:* In plant communities of mountainous areas of South Africa, species richness of plants peaks at intermediate levels of productivity (biomass). ***b.*** *Spatial heterogeneity:* The species richness of desert lizards is positively correlated with the structural complexity of the plant cover in desert sites in the American Southwest. ***c.*** *Climate:* The species richness of mammals is inversely correlated with monthly mean temperature range along the west coast of North America.

For the better part of a century, ecologists have puzzled over this remarkable species diversity cline from the Arctic to the tropics.

The difficulty has not been in forming a reasonable hypothesis of why more species exist in the tropics but, rather, in sorting through the many reasonable hypotheses that have been proposed. Here, we consider five of the most commonly discussed suggestions.

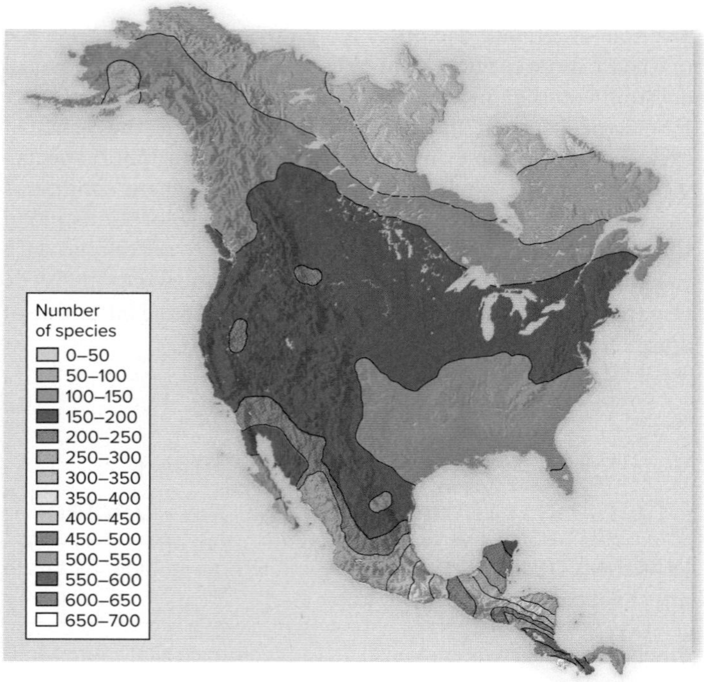

Figure 39.33 A latitudinal cline in species richness. Among North and Central American birds, a marked increase in the number of species occurs moving toward the tropics. Fewer than 100 species are found at Arctic latitudes, but more than 600 species live in southern Central America.

1. Evolutionary age

Scientists have frequently proposed that the tropics have more species than temperate regions because the tropics have existed over long, uninterrupted periods of evolutionary time, whereas temperate regions have been subject to repeated glaciations. The greater age and stability of tropical communities would have allowed complex population interactions to coevolve within them, fostering a greater variety of plants and animals.

Recent work suggests that the long-term stability of tropical communities may have been greatly exaggerated, however. An examination of pollen within undisturbed soil cores reveals that during glaciations, tropical forests contracted to a few small refuges surrounded by grassland. This suggests that the tropics have not had a continuous record of species richness over long periods of evolutionary time.

2. Increased productivity

A second often-advanced hypothesis is that the tropics contain more species because this part of the Earth receives more solar radiation than do temperate regions. The argument is that more solar energy greatly increases the overall photosynthetic activity of tropical plants growing all year long.

If we visualize the tropical forest's total resources as a pie, and its species niches as slices of the pie, we can see that a larger pie accommodates more slices. But as noted, many field studies have indicated that species richness is highest at intermediate levels of productivity. Accordingly, increasing productivity would be expected eventually to lead to lower, not higher, species richness.

3. Stability/constancy of conditions

Seasonal variation, though it does exist in the tropics, is generally substantially less than in temperate areas. This reduced seasonality might encourage specialization, with niches subdivided to partition resources and so avoid competition. The expected result would be a larger number of more specialized species in the tropics, which is what we see. Many field tests of this hypothesis have been carried out, and almost all support it, reporting larger numbers of narrower niches in tropical communities than in temperate areas.

4. Predation

Many reports indicate that predation may be more intense in the tropics. In theory, more intense predation could reduce the importance of competition, permitting greater niche overlap and thus promoting greater species richness.

5. Spatial heterogeneity

As noted, spatial heterogeneity promotes species richness. Tropical forests, by virtue of their complexity, create a variety of microhabitats and, so, may foster larger numbers of species. Perhaps the long, vertical column of vegetation through which light passes in a tropical forest produces a wide range of light frequencies and intensities, creating a greater variety of light environments and promoting species diversity.

REVIEW OF CONCEPT 39.7

An ecosystem is stable if it remains relatively constant in composition and is able to resist disturbance. Experimental field studies support the conclusion that species-rich communities are better able to resist invasion by new species, as well as having increased biomass production at the primary level, although not all ecologists agree with these conclusions. Species richness is greatest in the tropics.

- *How might primary productivity be affected if air pollution decreases the amount of sunlight reaching Earth?*

Does Clear-Cutting Forests Cause Permanent Damage?

The lumber industry practice called "clear-cutting" has been common in many states. Loggers find it more efficient to simply remove all trees from a watershed, and sort the logs out later, than to selectively cut only the most desirable mature trees. Although the open cuts seem a desolation to the casual observer, the loggers claim that new forests can become established more readily in the open cut, as sunlight now more easily reaches seedlings at ground level. Ecologists counter that clear-cutting fundamentally changes the forest in ways that cannot be easily reversed.

Who is right? The most direct way to find out is to clear-cut an area and watch it very carefully. Just this sort of massive field test was carried out in a now-classic experiment at the Hubbard Brook Experimental Forest in New Hampshire. Hubbard Brook is the central stream of a large watershed that drains a region of temperate deciduous forest in northern New Hampshire. The research team, led by professors Herbert Bormann and Gene Likens at Dartmouth College, first gathered a great deal of information about the forest watershed. Starting in 1963, they did a census of the trees, measured the flow of water through the watershed, and carefully documented the levels of minerals and other nutrients in the water leaving the ecosystem via Hubbard Brook. To keep track, they constructed concrete dams across each of the six streams that drain the forest and monitored the runoff, chemically analyzing samples. The undisturbed forest proved very efficient at retaining nitrogen and other nutrients. The small amounts of nutrients that entered the ecosystem in rain and snow were approximately equal to the amounts of nutrients that ran out of the valleys into Hubbard Brook.

Now came the test. In the winters of 1965 and 1966 the investigators felled all the trees and shrubs in 38 acres drained by one stream (as shown in the photo) and examined the water running off. The immediate effect was dramatic: the amount of water running out of the valley increased by 40%. Water that otherwise would have been taken up by vegetation and released into the atmosphere through evaporation was now simply running off. It was clear that the forest was not retaining water as well, but what about the soil nutrients, the key to future forest fertility?

The red line in the graph shows nitrogen minerals leaving the ecosystem in the runoff water of the stream draining the clear-cut area; the blue line shows the nitrogen runoff in a neighboring stream draining an adjacent uncut portion of the forest.

Northern Research Center/US Forest Service

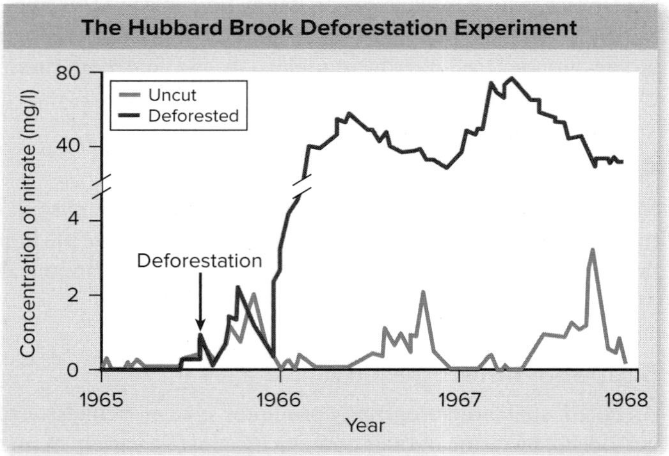

Analysis

1. **Applying Concepts**
   Scale. What is the significance of the break in the vertical axis between 4 and 40?
2. **Interpreting Data**
   a. What is the approximate concentration of nitrogen in the runoff of the uncut valley before cutting? Of the cut valley before cutting?
   b. What is the approximate concentration of nitrogen in the runoff of the uncut valley one year after cutting? Of the clear-cut valley one year after cutting?
3. **Making Inferences**
   a. Is there any yearly pattern to the nitrogen runoff in the uncut forest? Can you explain it?
   b. How does the loss of nitrogen from the ecosystem in the clear-cut forest compare with nitrogen loss from the uncut forest?
4. **Drawing Conclusions**
   a. What is the impact of this forest's trees upon its ability to retain nitrogen?
   b. Has clear-cutting harmed this ecosystem? Explain.

Retracing the Learning Path

CONCEPT 39.1 Competition Shapes How Species Live Together in Communities

39.1.1 A Community Is All the Species Living at One Site Communities are characterized by their constituent species, or properties such as species richness and productivity.

39.1.2 The Ecological Niche of a Species Is How It Uses Available Resources The fundamental niche is the entire niche a species is capable of using. The realized niche is the set of actual environmental conditions for establishment of a stable population.

39.1.3 Competitive Exclusion Can Occur When Species Compete for Limited Resources The principle of competitive exclusion states that if resources are limited, two species cannot occupy the same niche; rather, one species will be eliminated.

39.1.4 Competition May Lead to Resource Partitioning By using different resources (partitioning), sympatric species can avoid competing with each other and can coexist with reduced realized niches.

CONCEPT 39.2 Predator–Prey Relationships Foster Coevolution

39.2.1 Predation Strongly Influences Prey Populations Predation is the consuming of one organism by another. Natural selection strongly favors adaptations to prevent predation.

39.2.2 Plant Adaptations Defend Against Herbivores Plants produce secondary chemical compounds that deter herbivores. Sometimes the herbivores counter by evolving an ability to ingest the compounds and even using them for their own defense.

39.2.3 Animal Adaptations Defend Against Predators Animal adaptations include chemical defenses and defensive coloration such as warning coloration or camouflage.

39.2.4 Mimicry Allows One Species to Capitalize on the Defensive Strategies of Another In Batesian mimicry, a species that is edible evolves warning coloration similar to that of inedible species. In Müllerian mimicry, two toxic species evolve similar coloration.

CONCEPT 39.3 Cooperation Among Species Can Lead to Coevolution

39.3.1 Symbiosis Involves Long-Term Interactions Many symbiotic species have coevolved and have permanent relationships.

39.3.2 Commensalism Benefits One Species and Is Neutral to the Other Commensal relationships include epiphytes growing on large plants and barnacles growing on sea animals.

39.3.3 Mutualism Benefits Both Species Mutualism between ants and acacias involves acacias providing shelter and food for stinging ants that protect them from herbivores.

39.3.4 Parasitism Benefits One Species at the Expense of Another Many organisms have parasitic lifestyles, living on or inside host species and causing damage or disease as a result.

39.3.5 Species Interactions Have Evolutionary Consequences Species may affect one another not only through direct interactions but also through their effects on other species in the community.

CONCEPT 39.4 Ecological Succession Is a Consequence of Habitat Alteration

39.4.1 Succession Produces a Change in Species Composition Primary succession begins with a barren, lifeless substrate, whereas secondary succession occurs after an existing community is disrupted by fire, clearing, or other events.

CONCEPT 39.5 Chemical Elements Move Through Ecosystems in Biogeochemical Cycles

39.5.1 Chemical Elements Cycle Through Ecosystems The atoms of chemical elements move through ecosystems in biogeochemical cycles.

39.5.2 Carbon, the Backbone of Organic Molecules, Cycles Through Most Ecosystems The carbon cycle involves carbon dioxide, which is fixed through photosynthesis and released by respiration. Carbon is also found as bicarbonate and as methane.

39.5.3 The Availability of Water Is Fundamental to Terrestrial Ecosystems Water enters the atmosphere via evaporation and transpiration and returns as precipitation.

39.5.4 The Nitrogen Cycle Is Driven by Microbial Activity Nitrogen is usually the element in shortest supply, even though N_2 makes up 78% of the atmosphere. Nitrogen must be converted into usable forms by nitrogen-fixing microorganisms.

39.5.5 Phosphorus Cycles Through Terrestrial and Aquatic Ecosystems, but Not the Atmosphere Phosphorus is released by weathering of rocks; it flows into the oceans, where it is deposited in deep-sea sediments.

CONCEPT 39.6 Energy Flows Through Ecosystems in One Direction

39.6.1 Living Organisms Can Use Many Forms of Energy, but Not Heat The Second Law of Thermodynamics states that whenever organisms use chemical-bond or light energy, some of it is inevitably converted to heat and cannot be retrieved.

39.6.2 Energy Flows Through the Trophic Levels of Ecosystems Energy passes from organism to organism in trophic levels. The sequence through trophic levels is called a food chain.

39.6.3 The Number of Trophic Levels Is Limited by Energy Availability As energy moves through each trophic level, very little (approximately 10%) remains from the preceding trophic level. This limits the length of food chains.

39.6.4 Ecological Pyramids Illustrate the Relationship of Trophic Levels Ecological pyramids based on energy flow, biomass, or numbers of organisms are usually upright. Inverted pyramids of biomass or numbers are possible.

CONCEPT 39.7 Biodiversity May Increase Ecosystem Stability

39.7.1 Species Richness May Increase Stability: The Cedar Creek Studies The Cedar Creek studies indicate that higher species richness results in less year-to-year variation in biomass and in greater resistance to drought and invasion.

39.7.2 Species Richness Is Influenced by Ecosystem Characteristics Primary production, habitat heterogeneity, and climatic factors all affect the number of species in an ecosystem.

39.7.3 Tropical Regions Have the Highest Diversity, Although Reasons Are Unclear Higher diversity in tropical regions may reflect evolutionary time, productivity, low seasonal variation, greater predation, or spatial heterogeneity.

Concept Overview

This Concept Overview diagrams the key concepts that were discussed in this chapter.

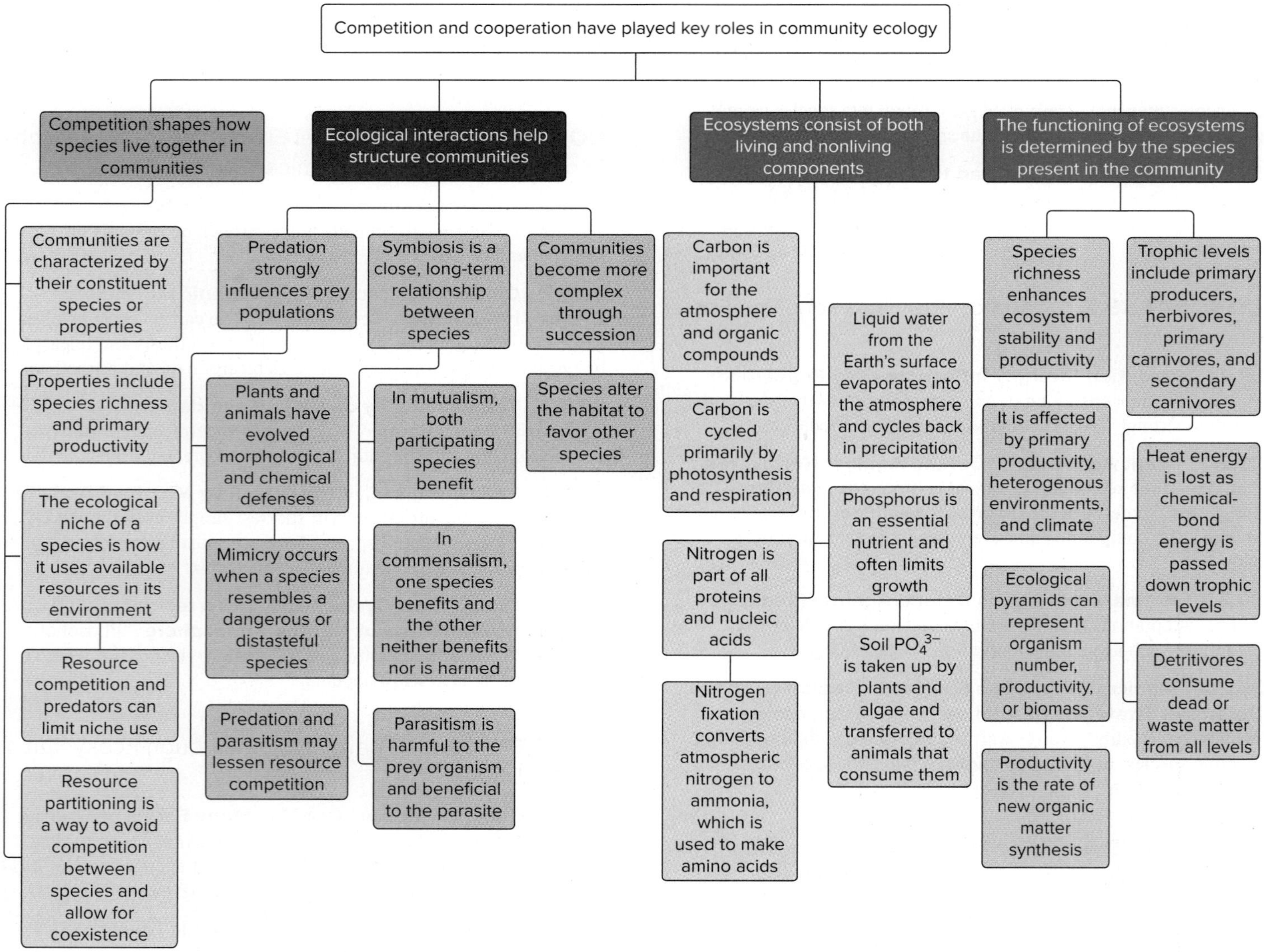

Assessing the Learning Path

Understand

1. If two species have very similar realized niches and are forced to coexist and share a limiting resource indefinitely,
   a. both species would be expected to coexist.
   b. both species would be expected to go extinct.
   c. the species that uses the limiting resource most efficiently should drive the other species extinct.
   d. both species would be expected to become more similar to each other.
2. According to the idea of coevolution between predator and prey, when a prey species evolves a novel defense against a predator,
   a. the predator is always expected to go extinct.
   b. the prey population should increase irreversibly out of control of the predator.
   c. the predator population should increase.
   d. evolution of a predator response should be favored by natural selection.
3. A relationship between two species in which one species benefits and the other is neither hurt nor helped is known as
   a. parasitism.
   b. mutualism.
   c. commensalism.
   d. competitive exclusion.
4. All of the following are locations where primary succession could occur EXCEPT
   a. land that has newly formed as a result of cooling lava.
   b. land that has been exposed by glacial melting.
   c. a tract of land that has had all of the trees removed by logging.
   d. an undersea volcano building until it breaks the surface of the water.
5. Photosynthetic organisms
   a. fix carbon dioxide.
   b. release carbon dioxide.
   c. fix oxygen.
   d. Both a and b are correct.
   e. Both a and c are correct.
6. Herbivores are
   a. primary consumers.
   b. secondary consumers.
   c. parasites.
   d. detritivores.
7. Inverted ecological pyramids of real systems usually involve
   a. energy flow.
   b. biomass.
   c. energy flow and biomass.
   d. None of the above
8. Species diversity
   a. increases with latitude as you move away from the equator to the Arctic.
   b. decreases with latitude as you move away from the equator to the Arctic.
   c. stays the same as you move away from the equator to the Arctic.
   d. increases with latitude as you move north of the equator and decreases with latitude as you move south of the equator.
9. In the Cedar Creek studies, increases in diversity
   a. led to a decrease in biomass.
   b. were linked to an increase in biomass.
   c. had no effect on biomass.
   d. led to an initial increase in biomass followed by a rapid decline in biomass.

Apply

1. Which of the following can cause the realized niche of a species to be smaller than its fundamental niche?
   a. Predation
   b. Competition
   c. Parasitism
   d. All of the above
2. Resource partitioning by sympatric species
   a. always occurs when species have identical niches.
   b. may not occur in the presence of a predator, which reduces prey population sizes.
   c. results in the fundamental and realized niches being the same.
   d. is more common in herbivores than in carnivores.
3. The presence of a predatory species
   a. always drives a prey species to extinction.
   b. can positively affect a prey species by having a detrimental effect on competing species.
   c. indicates that the climax stage of succession has been reached.
   d. None of the above
4. The yellow-and-black-striped patterns of many kinds of stinging wasps are examples of
   a. parasitism.
   b. Müllerian mimicry.
   c. commensalism.
   d. Batesian mimicry.
5. Which of the following is an example of commensalism?
   a. A tapeworm living in the gut of its host
   b. A clownfish living among the tentacles of a sea anemone
   c. An acacia tree and acacia ants
   d. Bees feeding on nectar from a flower
6. Which type of animal would probably be the last to colonize a newly formed island?
   a. Insect
   b. Bird
   c. Bat
   d. Wolf
7. Which of the statements about groundwater is NOT accurate?
   a. In the United States, groundwater provides 50% of the population with drinking water.
   b. Groundwaters are being depleted faster than they can be recharged.
   c. Groundwaters are becoming increasingly polluted.
   d. Removal of pollutants from groundwaters is easily achieved.
8. Nitrogen is often a limiting nutrient in many ecosystems, because
   a. there is much less nitrogen than carbon in the atmosphere.
   b. elemental nitrogen is very rapidly used by most organisms.
   c. nitrogen availability is being reduced by pollution due to fertilizer use.
   d. most organisms cannot use nitrogen in its elemental form.

9. The number of carnivores found at the top of an ecological pyramid is limited by the
   a. number of organisms below the top carnivores.
   b. number of trophic levels below the top carnivores.
   c. amount of biomass below the top carnivores.
   d. amount of energy transferred to the top carnivores.
10. At Cedar Creek Natural History Area, experimental plots showed reduced numbers of invaders as species diversity of plots increased,
   a. suggesting that low species diversity increases stability of ecosystems.
   b. suggesting that ecosystem stability is a function of primary productivity only.
   c. consistent with the theory that intermediate disturbance results in the highest stability.
   d. None of the above

Synthesize

1. Examine the pattern of beak size distributions of two species of finches on the Galápagos Islands illustrated in figure 39.5. One hypothesis that can be drawn from this pattern is that character displacement has taken place. Are there other hypotheses? If so, how would you test them?
2. Most chemically protected animal prey species produce toxins that make them taste bad or make their predator sick. If the prey is tasted by a predator, the prey usually dies. If it dies, how have the chemicals protected it?
3. Mimicry, common among insects, is rare among vertebrates. Why? Can you think of an example?
4. How would you experimentally test the hypothesis that the clownfish–sea anemone symbiosis is a commensal relationship?
5. At what successional stage would you characterize a field of wheat? What does this imply about its stability and productivity?
6. The occurrence of major sand storms in the Sahara Desert is well correlated with blooms of algal productivity in the Indian and Pacific Oceans. Why?
7. How can the existence of inverted pyramids of biomass be explained?
8. Many experiments by ecologists have shown that species-rich communities are more productive than species-poor ones. If this is so, how is it that American farming, based almost entirely on monocultures, is so productive?

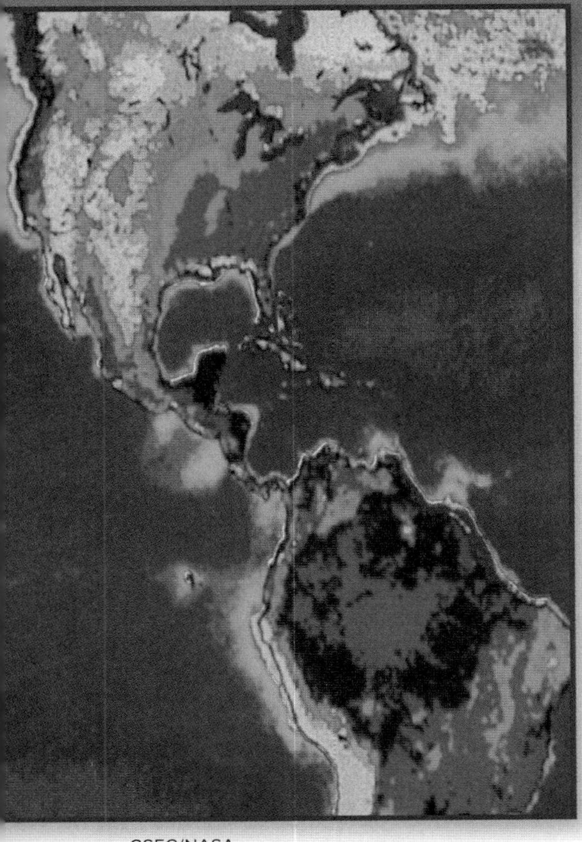

GSFC/NASA

40 The Living World

Learning Path

40.1 Ecosystems Are Shaped by Sun, Wind, and Water

40.2 Earth Has 14 Major Terrestrial Ecosystems, Called Biomes

40.3 Freshwater Habitats Occupy Less Than 2% of Earth's Surface

40.4 Marine Habitats Dominate the Earth

40.5 Humanity's Pollution Is Severely Impacting the Biosphere

40.6 Human Activity Is Altering Earth's Climate

Concept Overview

This Concept Overview provides a diagram of the overarching concepts that are covered in this chapter. These concepts will be expanded upon in the Concept Overview at the end of the chapter.

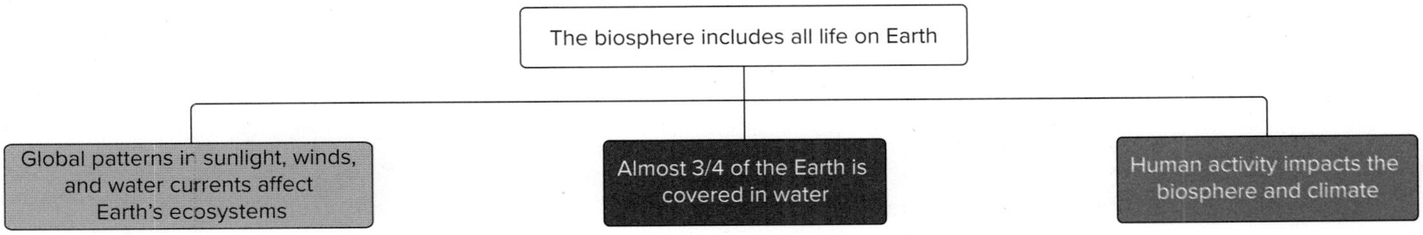

The biosphere includes all life on Earth
Global patterns in sunlight, winds, and water currents affect Earth's ecosystems
Almost 3/4 of the Earth is covered in water
Human activity impacts the biosphere and climate

Introduction

The living world, or biosphere, includes all living communities on Earth, from the profusion of life in the tropical rainforests to the planktonic communities in the world's oceans. In a very general sense, the distribution of life on Earth reflects variations in the world's abiotic environments, such as the variations in temperature and availability of water from one terrestrial environment to another. The satellite image of the Americas shown on the previous page is based on data collected over eight years. The colors are keyed to the relative abundance of chlorophyll, an indicator of the richness of biological communities. Green and dark green areas on land are areas with high primary productivity (such as thriving forests), whereas yellow areas include the deserts of the Americas and the tundra of the far north, which have lower productivity. Similarly, light blue areas along the coast are highly productive as the result of nutrient-rich upwelling currents, whereas pink areas are the least productive areas in the ocean.

40.1 Ecosystems Are Shaped by Sun, Wind, and Water

The global patterns of life on Earth are heavily influenced by (1) the amount of solar radiation that reaches different parts of the Earth and seasonal variations in that radiation, and (2) the patterns of global atmospheric circulation and the resulting patterns of oceanic circulation. Local characteristics, such as soil types and the altitude of the land, interact with the global patterns in sunlight, winds, and water currents to determine the conditions under which life exists and thus the distributions of ecosystems.

Solar Energy and the Earth's Rotation Affect Atmospheric Circulation

LEARNING OBJECTIVE 40.1.1 Explain the Coriolis effect.

The Earth receives energy from the Sun at a high rate in the form of electromagnetic radiation at visible and near-visible wavelengths. Each square meter of the upper atmosphere receives about 1400 joules per second (J/sec), which is equivalent to the output of fourteen 100-watt (W) light bulbs.

As the solar radiant energy passes through the atmosphere, its intensity and wavelength composition are modified. About half the energy is absorbed within the atmosphere, and half reaches the Earth's surface. The gases in the atmosphere absorb some wavelengths strongly while allowing other wavelengths to pass freely through. As a result, the wavelength composition of the solar energy that reaches the Earth's surface is different from that emitted by the Sun. For example, the band of ultraviolet wavelengths known as UV-B is strongly absorbed by ozone (O_3) in the atmosphere, and thus this wavelength is greatly reduced in the solar energy that reaches the Earth's surface.

How solar radiation affects climate

Some parts of the Earth's surface receive more energy from the Sun than others. These differences have a great effect on climate.

A major reason for differences in solar radiation from place to place is the fact that Earth is a sphere, or nearly so (figure 40.1*a*). The tropics are particularly warm, because the Sun's rays arrive almost perpendicular to the surface of the Earth in regions near the equator. Closer to the poles, the angle at which the Sun's rays strike (the *angle of incidence*) spreads the solar energy out over more of the Earth's surface, providing less energy per unit of surface area. As figure 40.2 shows, the highest annual mean temperatures occur near the equator (0° latitude).

The Earth's annual orbit around the Sun and its daily rotation on its own axis are also important in determining patterns of solar radiation and their effects on climate (figure 40.1*b*). The axis of rotation of the Earth is not perpendicular to the plane in which the Earth orbits the Sun. Because the axis is tilted by approximately 23.5°, a progression of seasons occurs on all parts of the Earth, especially at latitudes far from the equator. The northern hemisphere, for example, tilts toward the Sun during some months but away during others, giving rise to summer and winter.

Global circulation patterns in the atmosphere

Hot air tends to rise relative to cooler air, because the motion of molecules in the air increases as temperature increases, making it less dense. Accordingly, the intense solar heating of the Earth's surface at equatorial latitudes causes air to rise from the surface to high in the atmosphere at these latitudes. This rising air is typically rich with water vapor; one reason is that the

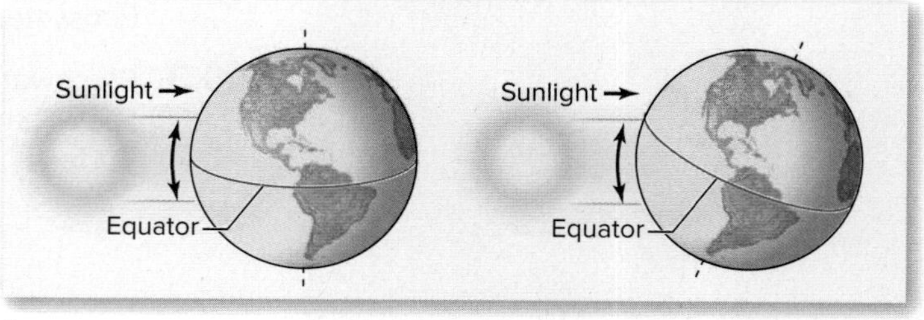

a.

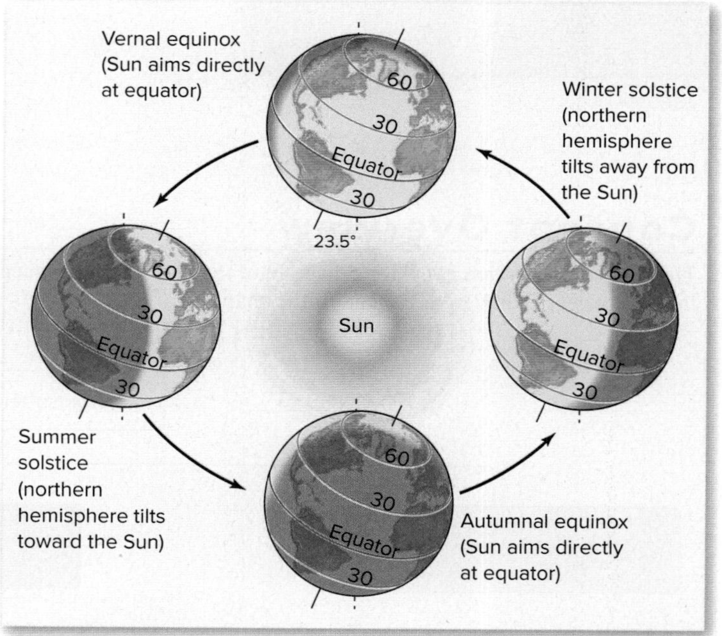

b.

Figure 40.1 Relationships between the Earth and the Sun are critical in determining the nature and distribution of life on Earth. ***a.*** A beam of solar energy striking the Earth in the middle latitudes of the northern (or the southern) hemisphere spreads over a wider area of the Earth's surface than an equivalent beam striking the Earth at the equator. ***b.*** The fact that the Earth orbits the Sun each year has a profound effect on climate. In the northern and southern hemispheres, temperature changes in an annual cycle, because the Earth's axis is not perpendicular to its orbital plane and, consequently, each hemisphere tilts toward the Sun in some months but away from the Sun in others.

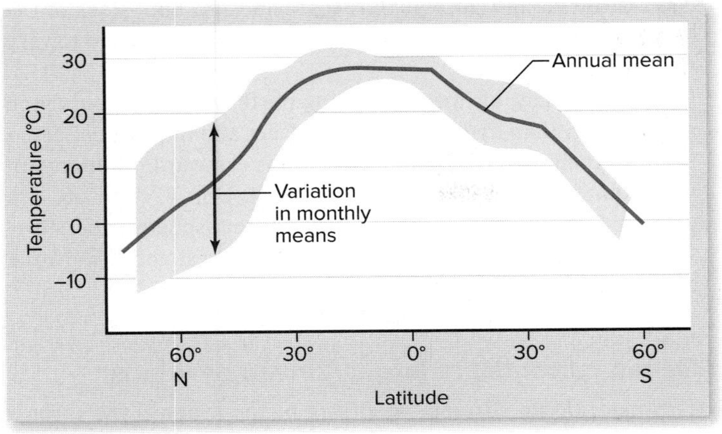

Figure 40.2 Annual mean temperature varies with latitude. The red line represents the annual mean temperature at various latitudes, ranging from near the North Pole at the left to near Antarctica at the right; the equator is at 0° latitude. At each latitude, the upper edge of the blue zone is the highest mean monthly temperature observed in all the months of the year, and the lower edge is the lowest mean monthly temperature.

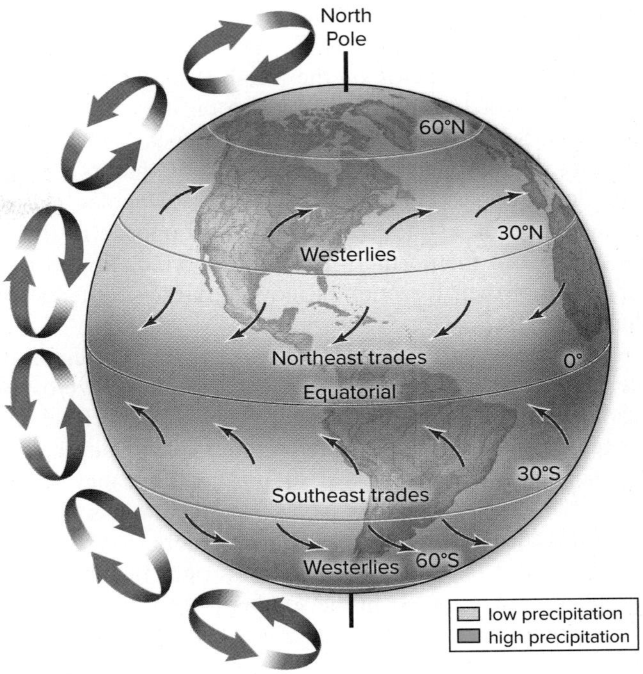

Figure 40.3 Global patterns of atmospheric circulation. The diagram shows the patterns of air circulation that prevail on average over the course of a year. Rising air that is cooled creates bands of relatively high precipitation near the equator and at latitudes near 60°N and 60°S. Air that has lost most of its moisture at high altitudes tends to descend to the surface of the Earth at latitudes near 30°N and 30°S, creating bands of relatively low precipitation. The red arrows show the winds blowing at the surface of the Earth; the blue arrows show the direction the winds blow at high altitude. The winds travel in curved paths relative to the Earth's surface, because the Earth is rotating on its axis under them (the Coriolis effect). A terminological problem to recognize is that the formal names given to winds refer to the directions from which they come, rather than the directions toward which they go; thus, the winds between 30° and 60° are called Westerlies because they come out of the west. Unfortunately, oceanographers use the opposite approach, naming water currents for the directions in which they go.

moisture-holding capacity of air increases when it is heated, and a second reason is that the intense solar radiation at the equator provides the heat needed for great quantities of water to evaporate. After the warm, moist air rises from the surface (figure 40.3), it is pushed away from the equator at high altitudes (above 10 km), to the north in the northern hemisphere and to the south in the southern hemisphere. To take the place of the rising air, cooler air flows toward the equator along the surface from both the north and the south. These air movements give rise to one of the major features of the global atmospheric circulation: air flows toward the equator in both hemispheres at the surface, rises at the equator, and flows away from the equator at high altitudes. The exact patterns of flow are affected by the spinning of the Earth on its axis, a phenomenon called the Coriolis effect.

For complex reasons, the air circulating up from the equator and away at high altitudes in both hemispheres tends to circulate back down to the surface of the Earth at about 30° of latitude, both north and south, as illustrated by the circulating red and blue arrows in figure 40.3. During the course of this movement, the moisture content of the air changes radically because of changes in temperature. Cooling dramatically decreases air's ability to hold water vapor. Consequently, much of the water vapor in the air rising from the equator condenses to form clouds and rain as the air moves upward. This rain falls in the latitudes near the equator, latitudes that experience the greatest precipitation on Earth.

By the time the air starts to descend back to the Earth's surface at latitudes near 30°, it is cold and thus has lost most of its water vapor. Although the air rewarms as it descends, it does not gain much water vapor on the way down. Many of the largest deserts occur at latitudes near 30° because of the steady descent of dry air to the surface at those latitudes. The Sahara Desert is the most dramatic example.

The air that descends at latitudes near 30° flows only partly toward the equator after reaching the surface of the Earth. Some of it flows toward the poles, helping to give rise in each hemisphere to winds that blow over the Earth's surface from 30° toward 60° latitude. At latitudes near 60°, air tends to rise from the surface toward high altitudes.

The Coriolis effect

If Earth did not rotate on its axis, global air movements would follow the simple patterns already described. Air currents—the winds—move across a rotating surface, however. Because the solid Earth rotates under the winds, the winds move in curved paths across the surface, rather than straight paths. The curvature of the paths of the winds due to Earth's rotation is termed the

Coriolis effect, after the 19th-century mathematician Gaspard-Gustave Coriolis, who described it.

If you were standing on the North Pole, the Earth would appear to be rotating counterclockwise on its axis, but if you were at the South Pole, the Earth would appear to be rotating clockwise. This property of a rotating sphere, that its direction of rotation is opposite when viewed from its two poles, explains why the direction of the Coriolis effect is opposite in the two hemispheres. In the northern hemisphere, winds always curve to the right of their direction of motion; in the southern hemisphere, they always curve to the left.

The reason for these wind patterns is that the circumference of a sphere, the Earth, changes with latitude. It is zero at the poles and 38,000 km at the equator. Thus, land surface speed changes from about 0 to 1500 km per hour going from the poles to the equator. Air descending at 30° north latitude may be going roughly the same speed as the land surface below it. As it moves toward the equator, however, it is moving more slowly than the surface below it, so it is deflected to its right in the northern hemisphere and to its left in the southern hemisphere. In other words, in both the northern and southern hemispheres, the winds blow westward as well as toward the equator. The result, as you can readily see in figure 40.3, is that winds on both sides of the equator—called the Trade Winds—blow out of the east and toward the west.

Conversely, air masses moving north from 30° are moving more rapidly than underlying land surfaces and thus are deflected again to their right, which in this case is eastward. Similarly, in the southern hemisphere, air masses between 30° and 60° are deflected eastward, to the left. In both hemispheres, therefore, winds between 30° and 60° blow out of the west and toward the east; these winds are called Westerlies.

Ocean currents are largely driven by winds

The major ocean currents are driven by the winds at the surface of the Earth, which means that indirectly the currents are driven by solar energy. The radiant input of heat from the Sun sets the atmosphere in motion as already described, and then the winds set the ocean in motion.

In the north Atlantic Ocean (figure 40.4), the global winds follow this pattern: surface winds tend to blow out of the east and toward the west near the equator, but out of the west and toward the east at midlatitudes (between 30° and 60°). Consequently,

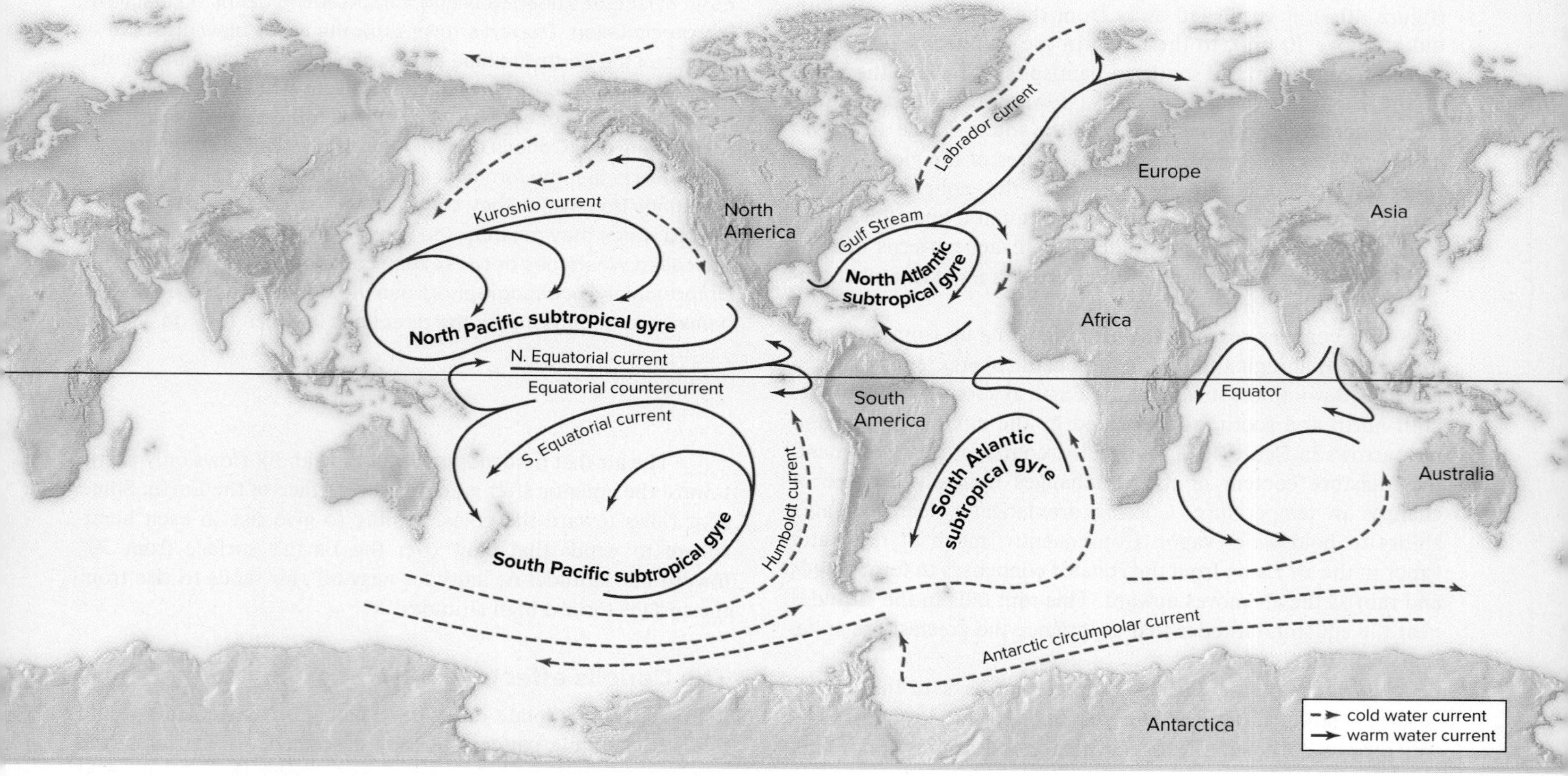

Figure 40.4 Ocean circulation. In the centers of several of the great ocean basins, surface water moves in great closed-curve patterns called gyres. These water movements affect biological productivity in the oceans and sometimes profoundly affect the climate on adjacent landmasses, as when the Gulf Stream brings warm water to the region of the British Isles.

surface waters of the north Atlantic Ocean tend to move in a giant closed curve—called a **gyre**—flowing from North America toward Europe at midlatitudes, then returning from Europe and Africa to North America at latitudes near the equator.

The Coriolis effect contributes to this clockwise closed-curve motion. Water flowing across the Atlantic toward Europe at midlatitudes tends to curve to the right and enters the flow from east to west near the equator. This latter flow also tends to curve to its right and enters the flow from west to east at midlatitudes. In the south Atlantic Ocean, the same processes occur in a sort of mirror image, and similar clockwise and counterclockwise gyres occur in the north and south Pacific Ocean.

Regional and Local Differences Affect Terrestrial Ecosystems

LEARNING OBJECTIVE 40.1.2 Explain the rain shadow effect.

The environmental conditions at a particular place are affected by regional and local effects of solar radiation, air circulation, and water circulation, not just the global patterns of these processes. In this section, we examine just a few examples of regional and local effects, focusing on terrestrial systems. These include rain shadows, monsoon winds, elevation, and the presence of microclimate factors.

Rain shadows

Deserts sometimes occur because mountain ranges intercept moisture-laden winds from the sea. When air flowing landward from the oceans encounters a mountain range (figure 40.5), the air rises and its moisture-holding capacity decreases, because it becomes cooler at higher altitude, causing precipitation to fall on the mountain slopes facing the sea.

As the air—stripped of much of its moisture—then descends on the other side of the mountain range, it is warmed and its moisture-holding capacity increases, meaning it can readily take up moisture from soils and plants.

One consequence is that the two slopes of a mountain range often differ dramatically in how moist they are; in California, for example, the eastern slopes of the Sierra Nevada mountains—facing away from the Pacific Ocean—are far drier than the western slopes. Another consequence is that a desert may develop on the dry side, the Mojave Desert being an example. The mountains are said to produce a rain shadow.

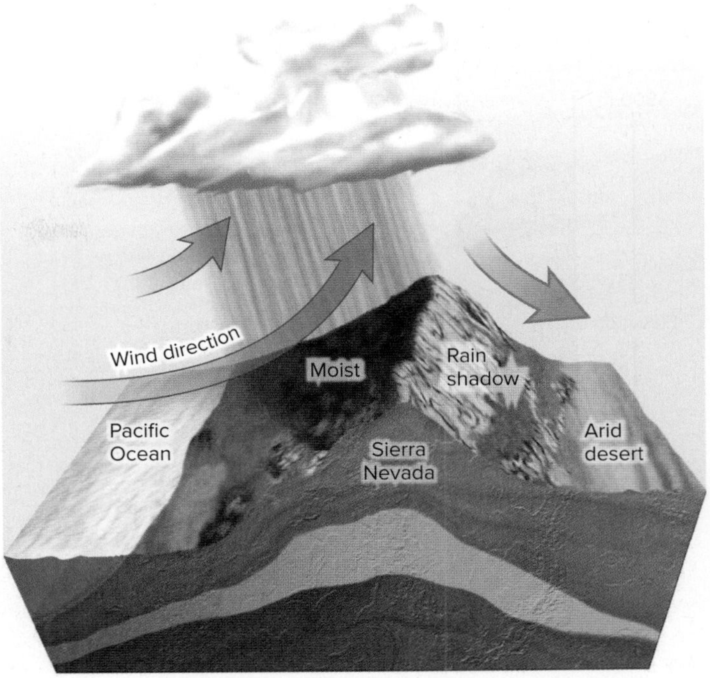

Figure 40.5 The rain shadow effect exemplified in California. Moisture-laden winds from the Pacific Ocean rise and are cooled when they encounter the Sierra Nevada mountains. As the moisture-holding capacity of the air decreases at colder, higher altitudes, precipitation occurs, making the seaward-facing slopes of the mountains moist; tall forests occur on those slopes, including forests that contain the famous giant sequoias (*Sequoiadendron giganteum*). As the air descends on the eastern side of the mountain range, its moisture-holding capacity increases again, and the air picks up moisture from its surroundings. As a result, the eastern slopes of the mountains are arid, and rain shadow deserts sometimes occur.

Monsoons

The continent of Asia is so huge that heating and cooling of its surface during the passage of the seasons cause massive regional shifts in wind patterns. During summer the surface of the Asian landmass heats up more than the surrounding oceans, but during winter the landmass cools more than the oceans. The consequence is that winds tend to blow off the water into the interior of the Asian continent in summer, particularly in the region of the Indian Ocean and western tropical Pacific Ocean. These winds reverse to flow off the continent out over the oceans in winter. These seasonally shifting winds are called the monsoons. They affect rainfall patterns, and their duration and strength can spell the difference between food sufficiency and starvation for hundreds of millions of people in the region each year.

Elevation

Another significant regional pattern is that in mountainous regions, temperature and other conditions change with elevation. At any given latitude, air temperature falls about 6°C for every 1000-m increase in elevation. The ecological consequences of the change of temperature with elevation are similar to those of the change of temperature with latitude (figure 40.6). In North America, a 1000-m increase in elevation results in a temperature drop equal to that of an 880-km increase in latitude.

Microclimates

Conditions also vary in significant ways on very small spatial scales. For example, in a forest, a bird sitting in an open patch may experience intense solar radiation, a high air temperature, and a low humidity, even while a mouse hiding under a log 5 m away may experience shade, a cool temperature, and air saturated

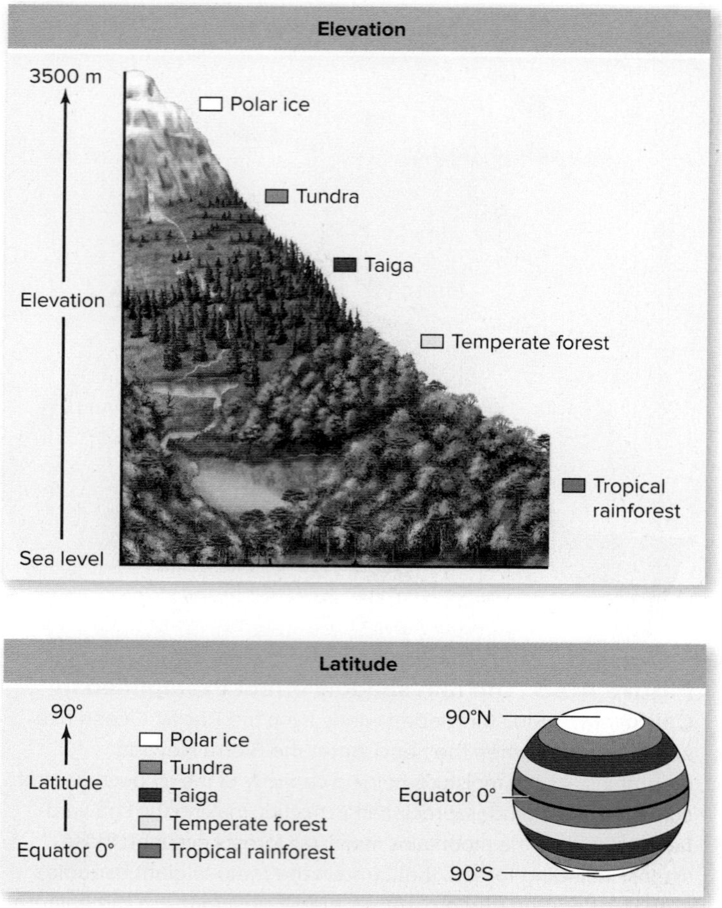

Figure 40.6 Elevation affects the distribution of biomes in much the same manner latitude does. Biomes that normally occur far north of the equator at sea level also occur in the tropics at high mountain elevations. Thus, on a tall mountain in the tropics, one might see a sequence of biomes like the one illustrated in this figure.

with water vapor. Such highly localized sets of climatic conditions are called microclimates.

In some cases, species avoid competition by adapting to use different microclimates. In previous chapters, we discussed how species diverge in resource use to minimize competitive pressures and then adapt to become more specialized to particular ecological niches. This is the process of character displacement, which underlies the process of adaptive radiation (refer to chapters 21 and 39). Most of the previous discussion of these phenomena has focused on traits such as body size, which affects the size of food an animal can eat, or other morphological features related to adaptation to different niches.

However, an alternative means of dividing niches is to evolve physiological adaptations to use different microclimates. By doing so, sympatric species encounter each other less frequently, minimizing competition for both space and food. Sympatric salamanders, for example, may be specialized for the different levels of moisture in different parts of the habitat. Similarly, in Hawaii the adaptive radiation of many types of plants is facilitated by species adapting to use different microclimates.

REVIEW OF CONCEPT 40.1

More intense solar heating of some global regions relative to others sets up global patterns of atmospheric circulation, which in turn cause global patterns of water circulation in the oceans. The Coriolis effect is caused by the Earth's spin beneath the moving air masses of the atmosphere. These patterns—plus seasonal changes—strongly affect the conditions that exist for living organisms in different parts of the world. In general, temperature declines as altitude or latitude increases.

- *How would global air movement patterns be different if the Earth turned in the opposite direction?*

40.2 Earth Has 14 Major Terrestrial Ecosystems, Called Biomes

Biomes are major types of ecosystems on land. Each biome has a characteristic appearance and is distributed over wide areas of land defined largely by sets of regional climatic conditions. Biomes are named according to their vegetational structures, but they also include the animals that are present.

Temperature and Moisture Largely Determine Biomes

LEARNING OBJECTIVE 40.2.1 Explain the factors that determine which biome is found in a region.

As you might imagine from the broad definition given for biomes, there are a number of ways to classify terrestrial ecosystems into biomes. Here we recognize eight principal biomes: (1) tropical rainforest, (2) savanna, (3) desert, (4) temperate grassland, (5) temperate deciduous forest, (6) temperate evergreen forest, (7) taiga, and (8) tundra.

Some ecologists recognize six additional biomes: polar ice; mountain zone; chaparral; warm, moist evergreen forest; tropical monsoon forest; and semidesert. Other ecologists lump these six with the eight major ones. Figure 40.7 shows the distributions of all 14 biomes.

Biomes are defined by their characteristic vegetational structures and associated climatic conditions, rather than by the presence of particular plant species. Two regions assigned to the same biome thus may differ in the species that dominate the landscape. Tropical rainforests around the world, for example, are all composed of tall, lushly vegetated trees, but the tree species that dominate a South American tropical rainforest are different from those in an Indonesian one. The similarity between such forests results from convergent evolution (refer to chapter 20).

In determining which biomes are found where, two key environmental factors are temperature and moisture. As seen in figure 40.8, if you know the mean annual temperature and mean annual precipitation in a terrestrial region, you often can predict the biome that dominates. Temperature and moisture affect ecosystems in a number of ways. One reason they are so influential is that

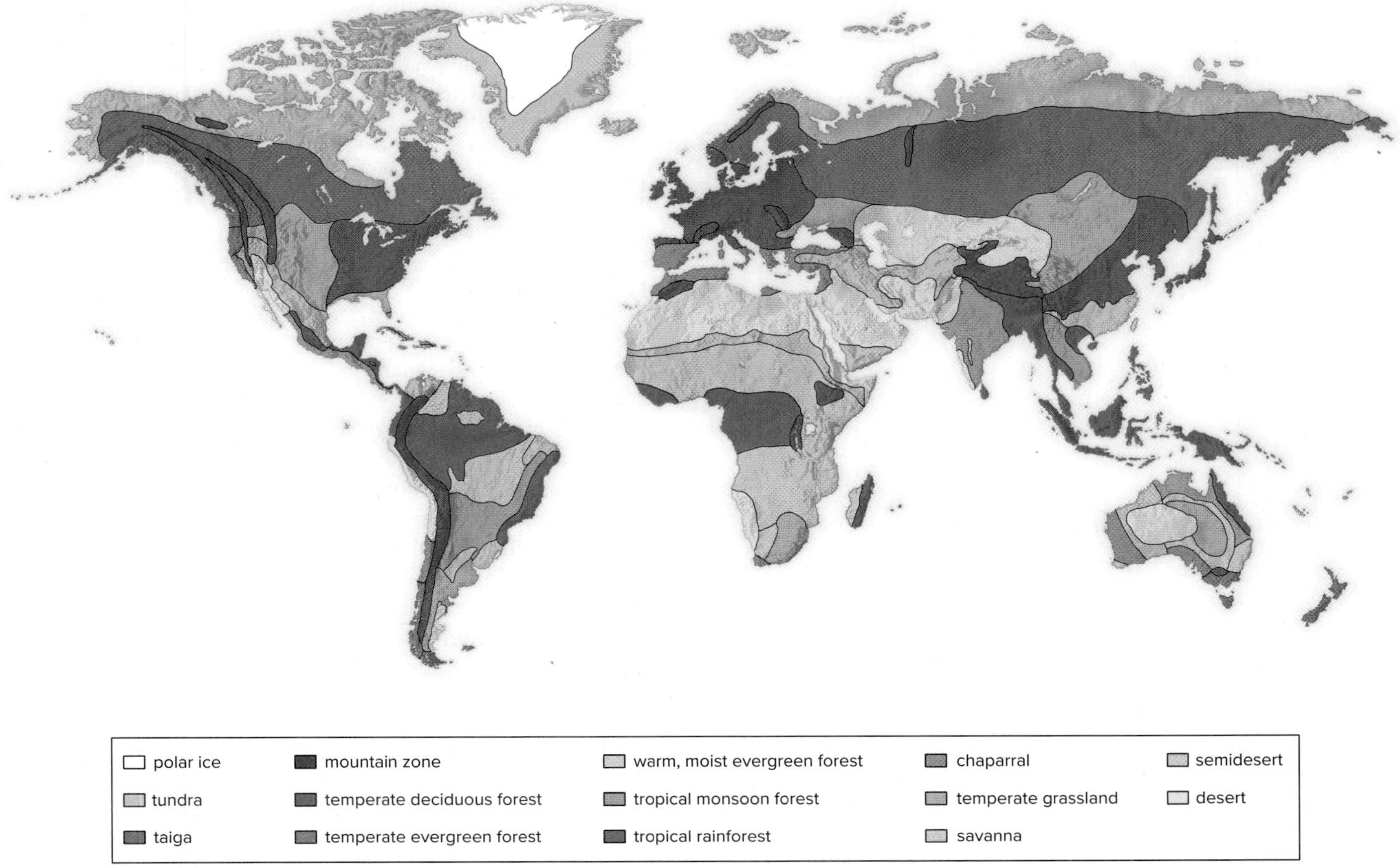

Figure 40.7 The distributions of biomes. Each biome is similar in vegetational structure and appearance wherever it occurs.

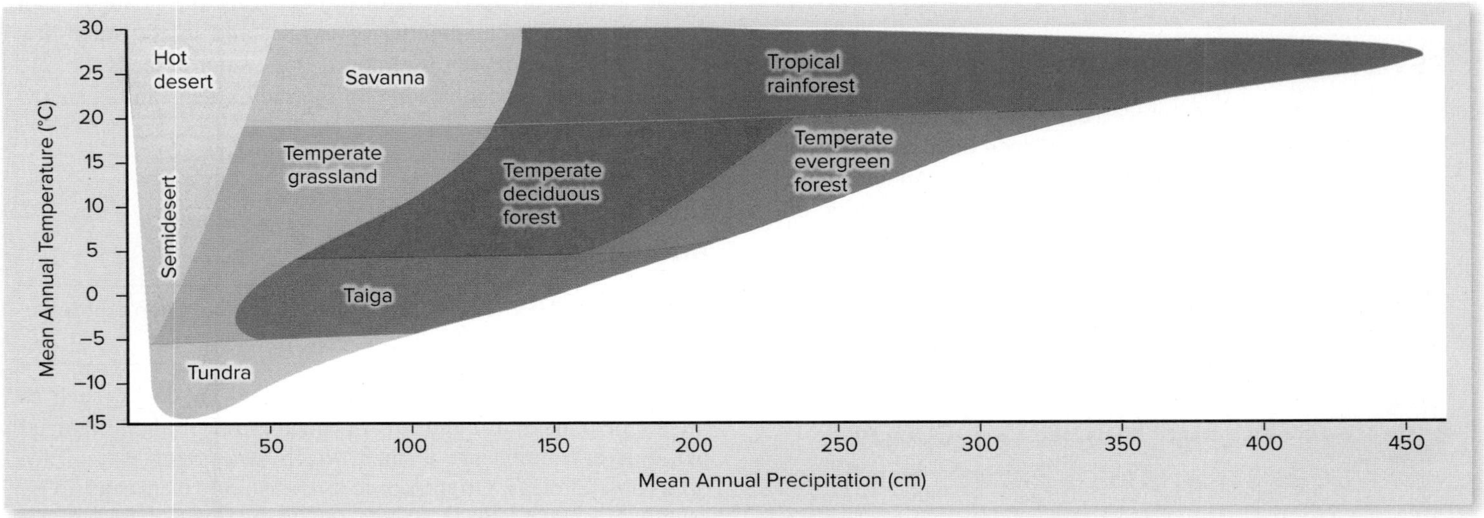

Figure 40.8 Predictors of biome distribution. Temperature and precipitation are useful predictors of biome distribution, although other factors sometimes also play critical roles.

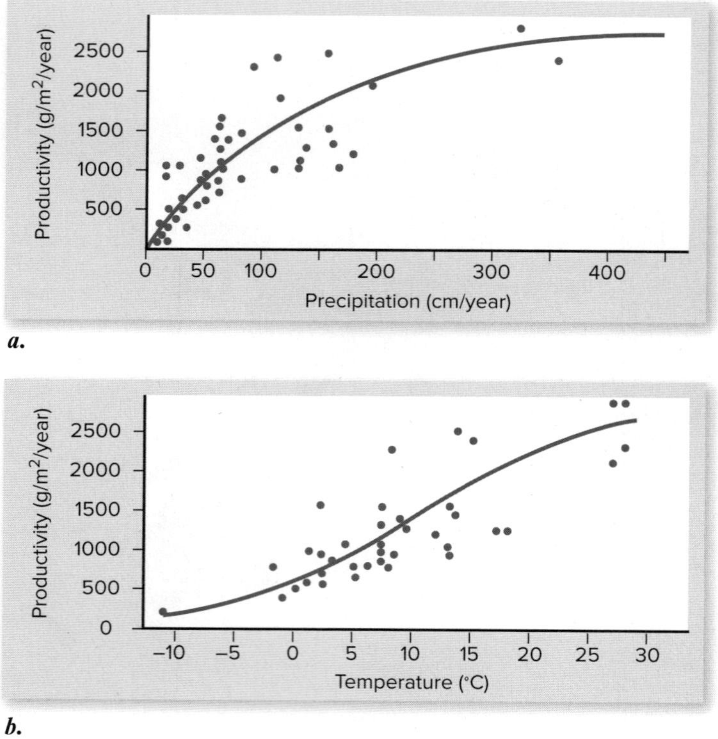

Figure 40.9 The correlations of primary productivity with precipitation and temperature. The net primary productivity of ecosystems at 52 locations around the globe correlates significantly with ***(a)*** mean annual precipitation and ***(b)*** mean annual temperature.

primary productivity is strongly correlated with them, as described in chapter 39 (figure 40.9).

Different places that are similar in mean annual temperature and precipitation sometimes support different biomes, indicating that temperature and moisture are not the only factors that can be important. Soil structure and mineral composition are among the other influential factors.

REVIEW OF CONCEPT 40.2

Major types of ecosystems called biomes can be distinguished in different climatic regions on land. These biomes are much the same wherever they are found on the Earth. Annual mean temperature and precipitation are effective predictors of biome type; however, the range of seasonal variation and the soil characteristics of a region also come into play.

- *Why do different biomes occur at different latitudes?*

40.3 Freshwater Habitats Occupy Less Than 2% of Earth's Surface

Of the major habitats, fresh water covers by far the smallest percentage of the Earth's surface: only 2%, compared with 27% for land and 71% for ocean. The formation of fresh water starts with the evaporation of water into the atmosphere, which removes most dissolved constituents, much as distillation does. When water falls back to the Earth's surface as rain or snow, it arrives in an almost pure state, although it may have picked up biologically significant dissolved or particulate matter from the atmosphere.

Life in Freshwater Habitats Depends on Oxygen Availability

LEARNING OBJECTIVE 40.3.1 Describe how the availability of oxygen influences freshwater ecosystems.

Freshwater wetlands—marshes, swamps, and bogs—are intermediate habitats between the freshwater and terrestrial realms. Wetlands are highly productive. They also play key additional roles, such as acting as water-storage basins that moderate flooding.

Primary production in freshwater bodies is carried out by single-celled algae (phytoplankton) floating in the water, by algae growing as films on the bottom, and by rooted plants such as water lilies. In addition, a considerable amount of organic matter—such as dead leaves—enters some bodies of fresh water from plant communities growing on the land nearby.

The concentration of dissolved oxygen (O_2) is a major determinant of the properties of freshwater communities. Oxygen dissolves in water just as sugar or salt does. Fish and other aquatic organisms obtain the oxygen they need by taking it up from solution. The solubility of oxygen is therefore critical.

In reality, oxygen is not very soluble in water. Consequently, even when fresh water is fully aerated and at equilibrium with the atmosphere, the amount of oxygen it contains per liter is at most 5% of that in air. This means that in terms of acquiring the oxygen they need, freshwater organisms have a far smaller margin of safety than air-breathing ones.

Oxygen is constantly added to and removed from any body of fresh water. Oxygen is added by photosynthesis and by aeration from the atmosphere, and it is removed by all organisms undergoing cellular respiration. If a lot of decaying organic matter is present in a body of water, microbes degrading this material can consume significant amounts of oxygen and affect other lifeforms. If the rate oxygen is removed from water exceeds the rate of addition, the concentration of dissolved oxygen can fall so low that many aquatic animals cannot survive.

Lakes Differ in Oxygen and Nutrient Content

LEARNING OBJECTIVE 40.3.2 Distinguish between eutrophic and oligotrophic lakes.

Bodies of fresh water that are low in algal nutrients (such as nitrate or phosphate) and low in the amount of algal material per unit of volume are termed *oligotrophic*. Such waters are often crystal clear. Oligotrophic streams and rivers tend to be high in dissolved oxygen, because the movement of the flowing water aerates them; the small amount of organic matter in the water means that oxygen is used at a relatively low rate. Similarly, oligotrophic lakes and ponds tend to be high in dissolved oxygen at all depths all year, because they also have a low rate of oxygen use. Because the water is relatively clear, light can penetrate the waters readily, allowing photosynthesis to occur through much of the water column, from top to bottom.

Eutrophic bodies of water are high in algal nutrients and often populated densely with algae. They are more likely to be low in dissolved oxygen, especially in summer. In a eutrophic body of water, decay microbes often place high demands on the oxygen available, because when thick populations of algae die, large amounts of organic matter are made available for decomposition. Moreover, light does not penetrate eutrophic waters well because of all the organic matter in the water; photosynthetic oxygen addition is therefore limited to just a relatively thin layer of water at the top.

Human activities have often transformed oligotrophic lakes into eutrophic ones. For example, when people overfertilize their lawns or fields, nitrate and phosphate from the fertilizers wash off into local water systems. Lakes that receive these nutrients become more eutrophic. A consequence is that the bottom waters are more likely to become oxygen-free during the summer. Many species of fish that are characteristic of oligotrophic lakes, such as trout, are very sensitive to oxygen deprivation. When lakes become eutrophic, these species of fish disappear and are replaced with species, like carp, that can better tolerate low oxygen concentrations. Lakes can return toward an oligotrophic state over time if steps are taken to eliminate the addition of excess nitrates, phosphates, and foreign organic matter such as sewage.

REVIEW OF CONCEPT 40.3

Freshwater habitats represent only 2% of the Earth's surface. Primary production in freshwater lakes is due to phytoplankton and is limited by oxygen levels. Eutrophic lakes are high in nutrients for algae but are low in dissolved oxygen; oligotrophic lakes are low in nutrients but high in dissolved oxygen at all depths.

- *Why can runoff containing fertilizer convert oligotrophic lakes to eutrophic lakes?*

40.4 Marine Habitats Dominate the Earth

Some of the most striking images of the Earth are those taken from space. The Earth appears as a primarily blue ball, with patches of green marking the continents. The overall appearance of blue is due to the oceans covering so much of the Earth's surface.

Open Oceans Have Low Primary Productivity

LEARNING OBJECTIVE 40.4.1 Describe the major types of ocean ecosystems.

About 71% of the Earth's surface is covered by ocean. Near the coastlines of the continents are the **continental shelves,** where the water is not especially deep (figure 40.10); the shelves, in essence, represent the submerged edges of the continents. Worldwide, the shelves average about 80 km wide, and the depth of the water over them increases from 1 m to about 130 m as one travels from the coast toward the open ocean.

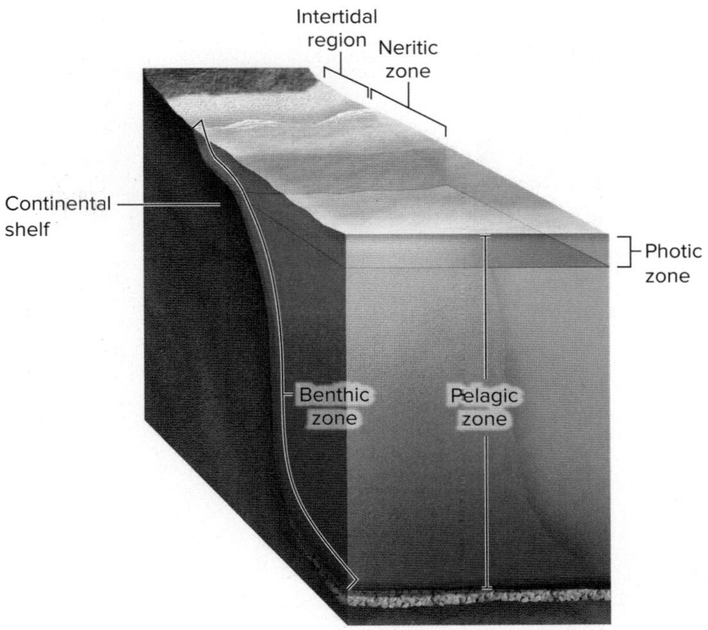

Figure 40.10 Basic concepts and terminology used in describing marine ecosystems. The continental shelf is the submerged edge of the continent. The waters over it are termed neritic and, on average, worldwide, are only 130 m deep at their deepest. The region where the tides rise and fall along the shoreline is called the intertidal region. The bottom is called the benthic zone, whereas the water column in the open ocean is called the pelagic zone. The photic zone is the part of the pelagic zone in which enough light penetrates for the phytoplankton to have a positive net primary productivity. The vertical scale of this drawing is highly compressed; whereas the outer edge of the continental shelf is 130 m deep, the open ocean, in fact, averages 35 times deeper (4000–5000 m deep).

Beyond the continental shelves, the depth suddenly becomes much greater. The average depth of the open ocean is 4000 to 5000 m, and some parts—called trenches—are far deeper, reaching 11,000 m in the Marianas Trench in the western Pacific Ocean.

In most of the ocean, the principal primary producers are phytoplankton floating in the well-lit surface waters. A revolution is currently under way in scientific understanding of the limiting nutrients for ocean phytoplankton. Primary production by the phytoplankton is presently understood to be nitrogen-limited in about two-thirds of the world's oceans, but iron-limited in about one-third. The principal known iron-limited areas are the great Southern Ocean surrounding Antarctica, parts of the equatorial Pacific Ocean, and parts of the subarctic, northeast Pacific Ocean. Where the water is shallow along coastlines, primary production is carried out not just by phytoplankton but also by rooted plants such as seagrasses and by bottom-dwelling algae, including seaweeds.

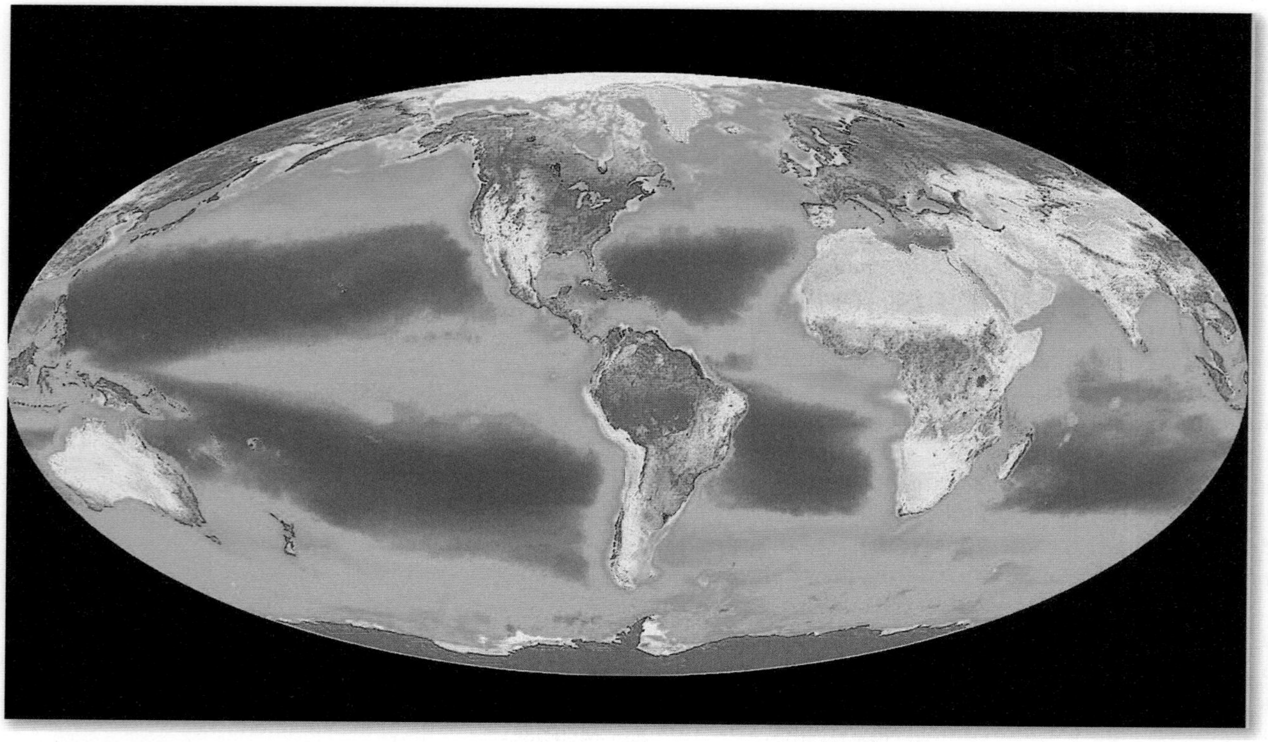

Figure 40.11 Major functional regions of the ocean. The regions classed as oligotrophic ocean (*colored dark blue*) are "biological deserts" with low productivity per unit area. Continental shelf ecosystems (green at the edge of continents) are typically medium to high in productivity. Upwelling regions (yellow at the edge of continents) are the highest in productivity per unit area and rank with the most productive of all ecosystems on Earth.

SeaWIFS, GSFC, and ORBIMAGE/NASA/Goddard Space Flight Center, and ORBIMAGE

The world's oceans are so vast that they include many different types of ecosystems (figure 40.11). Some, such as coral reefs and estuaries, are high in their net primary productivity per unit of area, but others are low in productivity per unit area. Ocean ecosystems are of four major types: open ocean waters far from land, continental shelves, upwelling regions, and deep sea.

The intensity of solar illumination in the open ocean drops from high levels at the surface to essentially zero at 200 m of depth; photosynthesis is limited to this level of the ocean, the **photic zone.** However, nutrients for phytoplankton, such as nitrate, tend to be present at low concentrations in the photic zone, because over eons of time in the past, ecological processes have exported nitrate and other nutrients from the upper waters to the deep waters, and no vigorous forces exist in the open ocean to return the nutrients to the sunlit waters.

Because of the low concentrations of nutrients in the photic zone, large parts of the open oceans are low in primary productivity per unit area and thus aptly called "biological deserts." These parts—which correspond to the centers of the great midocean gyres—are often collectively termed the *oligotrophic ocean* in reference to their low nutrient levels and low productivity.

People fish the open oceans today for only a few wide-ranging species, such as tunas and some species of squids and whales. Fishing in the open oceans is limited to relatively few species for two reasons. First, because of the low primary productivity per unit of area, animals tend to be thinly distributed in the open oceans. The only ones that are commercially profitable to catch are those that are individually large or tend to gather together in tight schools. Second, costs for traveling far from land are high. All authorities agree that as we turn to the sea to help feed the burgeoning human population, we cannot expect the open ocean regions to supply great quantities of food.

Continental Shelf Ecosystems Provide Abundant Resources

LEARNING OBJECTIVE 40.4.2 Explain why continental shelf ecosystems are more productive than open ocean ones.

Many of the ecosystems on the continental shelves are relatively high in productivity per unit area. An important reason is that the waters over the shelves—termed the **neritic waters**—tend to have relatively high concentrations of nitrate and other nutrients.

Because the waters over the shelves are shallow, they have not been subjected, over the eons of time, to the loss of nutrients into the deep sea, as the open oceans have. Over the shelves, nutrient-rich materials that sink hit the shallow bottom, and the nutrients they contain are stirred back into the water column by stormy weather. In addition, nutrients are continually replenished by runoff from nearby land.

Around 99% of the food people harvest from the ocean comes from continental shelf ecosystems or nearby upwelling regions. The shelf ecosystems are also particularly important to humankind in other ways. Mineral resources taken from the ocean, such as petroleum, come almost exclusively from the shelves. In addition, almost all recreational uses of the ocean, from sailing to scuba diving, take place on the shelves. The shelves feature prominently in these ways because they are close to coastlines and relatively shallow.

The Deep Sea Is a Cold, Dark Place

LEARNING OBJECTIVE 40.4.3 Explain how life is possible in the deep sea.

The deep sea is by far the single largest habitat on Earth, in the sense that it is a huge region characterized by relatively uniform conditions throughout the globe. The deep sea is seasonless, cold (2–5°C), totally dark, and under high pressure (400–500 atmospheres where the bottom is 4000–5000 m deep).

In most regions of the deep sea, food originates from photosynthesis in the sunlit waters far above. Such food—in the form of carcasses, fecal pellets, and mucus—can take as much as a month to drift down from the surface to the bottom, and along the way about 99% of it is eaten by animals living in the upper water column. Thus, the bottom communities receive only about 1% of the primary production and are food-poor. Nonetheless, a great many species of animals—most of them small-bodied and thinly distributed—are known to live in the deep sea. Some of the animals are bioluminescent (figure 40.12) and thereby able to communicate or attract prey by the use of light.

REVIEW OF CONCEPT 40.4

The oligotrophic ocean includes the open ocean and the deep sea, where little primary productivity occurs. Productivity is limited primarily by the nutrients nitrogen and iron. Continental shelf ecosystems tend to be moderate to high in productivity.

- *What would be the effect on productivity of adding iron to a region of ocean?*

Figure 40.12 Life in the deep sea. The luminous spot below the eye of this deep-sea fish results from the presence of a symbiotic colony of bioluminescent bacteria. Bioluminescence is a fairly common feature of mobile animals in the parts of the ocean that are so deep as to be dark. It is more common among species living partway down to the bottom than in ones living at the bottom.

Fenolio, Danté/Science Source

40.5 Humanity's Pollution Is Severely Impacting the Biosphere

We all know that human activities can cause adverse changes in ecosystems. In discussing these, it is important to recognize that creative people can often come up with rational solutions to such problems.

Chemical Pollution Is Altering the Earth

LEARNING OBJECTIVE 40.5.1 Explain how pollution alters ecosystems.

An outstanding example is provided by the history of DDT in the United States. DDT is a highly effective insecticide that was sprayed widely in the decades following World War II, often on wetlands to control mosquitoes. During the years of heavy DDT use, populations of ospreys, bald eagles, and brown pelicans—all birds that catch large fish—plummeted. Ultimately, the use of DDT was connected with the demise of these birds.

Scientists established that DDT and its metabolic products became increasingly concentrated in the tissues of animals as the compounds were passed along food chains (figure 40.13).

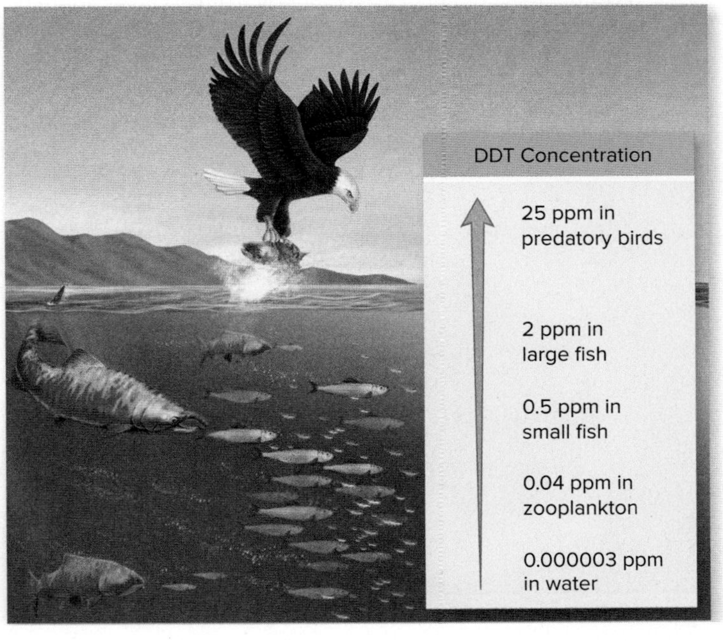
DDT Concentration
25 ppm in predatory birds
2 ppm in large fish
0.5 ppm in small fish
0.04 ppm in zooplankton
0.000003 ppm in water

Figure 40.13 Biological magnification of DDT concentration. Because all the DDT an animal eats in its food tends to accumulate in its fatty tissues, DDT becomes increasingly concentrated in animals at higher levels of the food chain. The concentrations at the right are in parts per million (ppm). Before DDT was banned in the United States, bird species that eat large fish underwent drastic population declines because metabolic products of DDT made their eggshells so thin that the shells broke during incubation.

Animals at the bottom of food chains accumulated relatively low concentrations in their fatty tissues. But the primary carnivores that preyed on them accumulated higher concentrations from eating great numbers of prey, and the secondary carnivores accumulated higher concentrations yet. Top-level carnivores, such as the birds that eat large fish, were dramatically affected by the DDT. In these birds, scientists found that metabolic products of DDT disrupted the formation of eggshells. The birds laid eggs with such thin shells that they often cracked before the young could hatch.

Researchers concluded that the demise of the fish-eating birds could be reversed by a rational plan to clean ecosystems of DDT, and laws were passed banning its use. Now, more than four decades later, populations of ospreys, eagles, and pelicans have rebounded dramatically.

Stratospheric Ozone Depletion Has Led to an Ozone "Hole"

LEARNING OBJECTIVE 40.5.2 Describe the effects of ozone depletion.

The colors of the satellite photo in figure 40.14 represent different concentrations of ozone (O_3) located 20 to 25 km above the Earth's surface in the stratosphere. Stratospheric ozone is depleted over Antarctica (purple region in the figure) to between one-half and one-third of its historically normal concentration, a phenomenon called the ozone hole.

Although depletion of stratospheric ozone is most dramatic over Antarctica, it is a worldwide phenomenon. Over the United States, the ozone concentration has been reduced by about 4%, according to the U.S. Environmental Protection Agency.

Stratospheric ozone and UV-B

Stratospheric ozone is important because it absorbs ultraviolet (UV) radiation—specifically, the wavelength called **UV-B**—from incoming solar radiation. UV-B is damaging to living organisms in a number of ways; for instance, it increases the risks of cataracts and skin cancer in people. Depletion of stratospheric ozone permits more UV-B to reach the Earth's surface and therefore increases the risks of UV-B damage. It is estimated that every 1% drop in stratospheric ozone leads to a 6% increase in the incidence of skin cancer, for example. UV exposure also may be detrimental to many types of animals, such as amphibians.

Ozone depletion and CFCs

The major cause of the depletion of stratospheric ozone is the addition of industrially produced chlorine- and bromine-containing compounds to the atmosphere. Of particular concern are chlorofluorocarbons (CFCs), used until recently as refrigerants in air conditioners and refrigerators and in manufacturing. CFCs released into the atmosphere can ultimately liberate free chlorine atoms, which in the stratosphere catalyze the breakdown of ozone molecules (O_3) to form ordinary oxygen (O_2). Ozone is continually being made and broken down, and free chlorine atoms tilt the balance toward a faster rate of breakdown.

The extreme depletion of ozone seen in the ozone hole is a consequence of the unique weather conditions that exist over

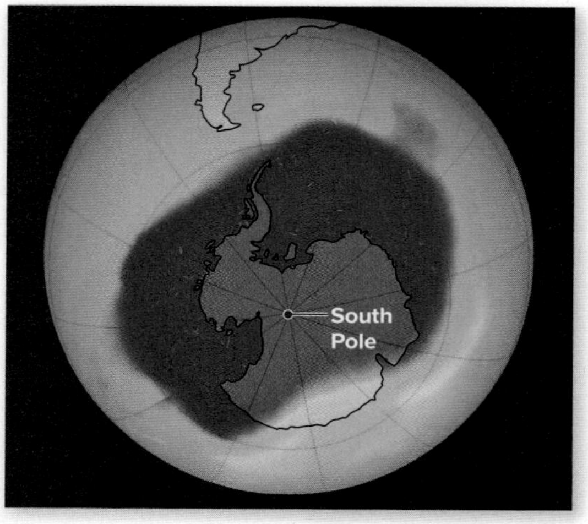

a.

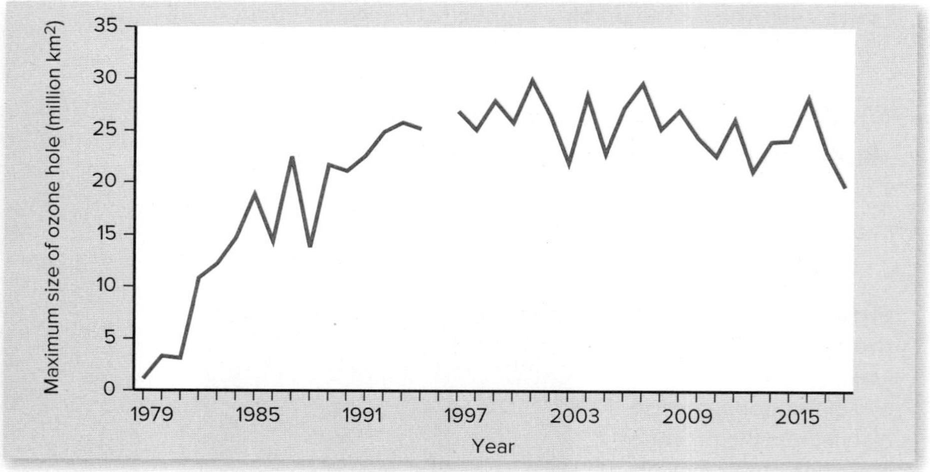

b.

Figure 40.14 The ozone hole over Antarctica. NASA satellites currently track the extent of ozone depletion in the stratosphere over Antarctica each year. Every year since about 1980, an area of profound ozone depletion, called the ozone hole, has appeared in August (early spring in the southern hemisphere) when sunlight triggers chemical reactions in cold air trapped over the South Pole during the Antarctic winter. ***a.*** In September 2006, the 11.4 million-square-mile hole (purple in the satellite image shown) covered an area larger than the United States, Canada, and Mexico combined, the largest hole ever recorded. ***b.*** The size of the ozone hole is now slowly decreasing, though year-to-year changes in temperature cause variation, such as the increase in 2015. No data are available in 1995 due to satellite instrument failure.

(a): Goddard Space Flight Center Scientific Visualization Studio/NASA

Antarctica. During the continuous dark of the Antarctic winter, a strong stratospheric wind, the polar-night jet, develops and, blowing around the full circumference of the Earth, isolates the stratosphere over Antarctica from the rest of the atmosphere.

The Antarctic stratosphere stays extremely cold (−80°C or lower) for many weeks as a consequence, permitting unique types of ice clouds to form. Reactions associated with the particles in these clouds lead to the accumulation of diatomic chlorine, Cl_2. When sunlight returns in the early Antarctic spring, the diatomic chlorine is photochemically broken up to form free chlorine atoms in great abundance, and the ozone-depleting reactions ensue.

Phase-out of CFCs

After research revealed the causes of stratospheric ozone depletion, worldwide agreements were reached to phase out the production of CFCs and other compounds that lead to ozone depletion. Manufacture of such compounds ceased in the United States in 1996, and there is now a great deal of public awareness about the importance of using "ozone-safe" alternative chemicals. The atmosphere will cleanse itself of ozone-depleting compounds only slowly, because the substances are chemically stable. Nonetheless, the problem of ozone depletion is diminishing and is expected to be substantially corrected by the second half of the 21st century.

The CFC story is an excellent example of how environmental problems arise and can be solved. Initially, CFCs were heralded as an efficient and cost-effective way to provide cooling, a clear improvement over previous technologies. At that time, their harmful consequences were unknown. Once the problems were identified, international agreements led to an effective solution, and creative technological advances led to replacements that solved the problem at little cost.

REVIEW OF CONCEPT 40.5

Pollution is the major human effects on the environment, with freshwater habitats being most threatened. Point-source pollution comes from identifiable locations, such as factories, whereas diffuse pollution comes from numerous sources, such as fertilized lawns. Ozone depletion was caused by chlorofluorocarbons. Banning CFCs has led to some recovery.

- *How do efforts to combat pollution depend on their source?*

40.6 Human Activity Is Altering Earth's Climate

By studying the Earth's history and making comparisons with other planets, scientists have determined that concentrations of certain gases in our atmosphere, particularly CO_2, maintain the average temperature on Earth about 25°C higher than it would be if these gases were absent. This fact emphasizes that the composition of our atmosphere is a keystone for life on Earth. Unfortunately, human activities are now changing the composition of the atmosphere in ways that most authorities conclude will be damaging or, in the long run, disastrous.

The Average Temperature at Earth's Surface Is Increasing

LEARNING OBJECTIVE 40.6.1 Describe global warming.

Because of changes in atmospheric composition, the average temperature of the Earth's surface is increasing, a phenomenon called global warming. As you might imagine, changes in temperature alter global wind and water-current patterns in complicated ways. This means that as the average global temperature increases, some particular regions of the world warm to a lesser extent, whereas other regions heat up to a greater extent (refer to the map in the Inquiry & Analysis feature at the end of the chapter). It also means that rainfall patterns are altered, because global precipitation patterns depend on global wind patterns. Enormous computer models are used to calculate the effects predicted in all parts of the world.

Independent computer models predict global changes

The Intergovernmental Panel on Climate Change (IPCC), which shared the 2007 Nobel Peace Prize with Al Gore for their work on global climate change, released its fifth assessment report in 2014. Based on a variety of scenarios, computer models predicted that global temperatures will increase 2.1° to 5.7°C (3.8° to 10.3°F) by the end of this century. These warnings by the IPCC were amplified in 2018 with a special report stating that although it is possible to limit temperature increase to 1.5°C by 2040, this requires immediate action to limit greenhouse gas emissions.

More ominous perhaps than temperature are some of the predictions for precipitation. For example, although northern Europe is expected to receive more precipitation than today, another recent study predicted that parts of southern Europe will receive about 20% less, disrupting natural ecosystems, agriculture, and human water supplies. Some European countries may come out ahead economically, but others will come out behind, and political relationships among countries will likely change as some shift from being food exporters to taking on the more tenuous role of requiring food imports.

Carbon Dioxide Is a Major Greenhouse Gas

LEARNING OBJECTIVE 40.6.2 Explain the link between atmospheric carbon dioxide and global warming.

Carbon dioxide is the gas usually emphasized in discussing the cause of global warming (figure 40.15), although other atmospheric gases are also involved. A monitoring station on the top

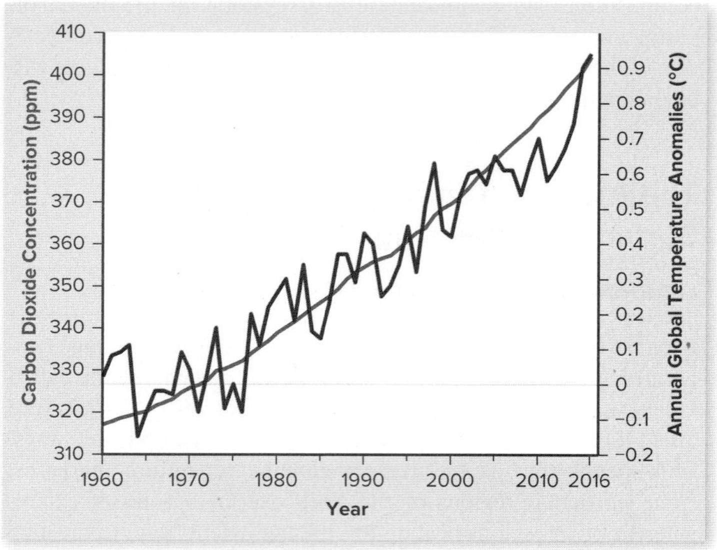

Figure 40.15 The greenhouse effect. The concentration of carbon dioxide in the atmosphere has increased steadily since the 1950s, as shown by the blue line. The red line shows the change in average global temperatures for the same period. The temperatures are recorded as anomalies relative to the mean temperature during a reference period (1951–1980). CO_2 concentration is measured as parts per million, that is, the number of CO_2 molecules compared with other molecules of gas in the atmosphere.

of the 4200-m (13,700-foot) high Mauna Loa volcano on the island of Hawaii has monitored the concentration of atmospheric CO_2 since the 1950s. This station is particularly important because it is in the middle of the Pacific Ocean, far from the great continental landmasses where most people live, and it is therefore able to monitor the state of the global atmosphere without the confounding influences of local events.

In 1958, the atmosphere was 0.031% CO_2. By 2019, the concentration had risen to 0.041%, the highest it had been in nearly a million years. All authorities agree that the cause of this steady rise in atmospheric CO_2 is the burning of coal and petroleum products by the increasing (and increasingly energy-demanding) human population.

How carbon dioxide affects temperature

The atmospheric concentration of CO_2 affects global temperature because carbon dioxide strongly absorbs electromagnetic radiant energy at some of the wavelengths that are critical for the global heat budget. Earth not only receives radiant energy from the Sun but also emits radiant energy into outer space. The Earth's temperature will be constant only if the rates of these two processes are equal.

The incoming solar energy is at relatively short wavelengths of the electromagnetic spectrum: wavelengths that are visible or near-visible. The outgoing energy from the Earth is at different, longer wavelengths. Carbon dioxide absorbs energy at certain of the important long-wave infrared wavelengths. This means that although carbon dioxide does not interfere with the arrival of radiant energy at short wavelengths, it retards the rate at which energy travels away from the Earth at long wavelengths into outer space (figure 40.16).

Carbon dioxide is often called a greenhouse gas, because its effects are analogous to those of a greenhouse. The reason that a glass greenhouse gets warm inside is that window glass is transparent to light but only slightly transparent to long-wave infrared radiation. Energy that strikes a greenhouse as light enters the greenhouse freely. Once inside, the energy is absorbed as heat and then re-radiated as long-wave infrared radiation. The infrared radiation cannot easily get out through the glass, and therefore energy accumulates inside.

Other greenhouse gases

Carbon dioxide is not the only greenhouse gas. Others include methane and nitrous oxide. The effect of any particular greenhouse gas depends on its molecular properties and concentration. For example, molecule-for-molecule, methane has about 20 times the heat-trapping effect of carbon dioxide; on the other hand, methane is less concentrated and less long-lived in the atmosphere than carbon dioxide.

Methane is produced in globally significant quantities in anaerobic soils and in the fermentation reactions of ruminant mammals, such as cows. Huge amounts of methane are presently locked up in Arctic permafrost. Melting of the permafrost could cause a sudden and large perturbation in global temperature by releasing methane rapidly.

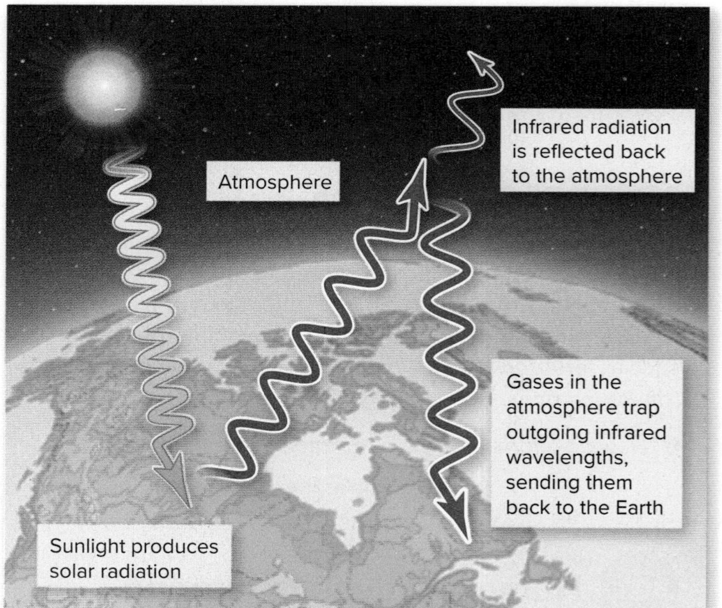

Figure 40.16 Mechanism of greenhouse warming. The solar radiation that hits the Earth is partially absorbed, heating the planet, but some is reflected back into the atmosphere. Greenhouse gases can absorb infrared radiation, preventing its escape and increasing the warming effect.

Global Temperature Change Has Affected Ecosystems in the Past and Is Doing So Now

LEARNING OBJECTIVE 40.6.3 Describe the consequences of global warming on Earth's ecosystems.

Evidence for warming occurs in many ways. For example, on a worldwide statistical basis, ice on lakes and rivers forms later and melts sooner than it used to; on average, ice-free seasons are now 2.5 weeks longer than they were a century ago. Also, the extent of ice at the North Pole has decreased substantially, and glaciers are retreating around the world (figure 40.17).

Global warming—and cooling—has occurred in the past, most recently during the ice ages and intervening warm periods. Species often responded by shifting their geographical ranges, tracking their environments. For example, a number of cold-adapted North American tree species that are now found only in the far north, or at high elevations, lived much farther south or at substantially lower elevations 10,000 to 20,000 years ago, when conditions were colder. Present-day global warming is having similar effects. For example, many butterfly and bird species have shifted northward in recent decades (figure 40.18).

Figure 40.18 Butterfly range shift. The distribution of the speckled wood butterfly, *Pararge aegeria,* in Great Britain in 1995–2009 (teal) included areas far to the north of the distribution in 1915–1939 (olive green).

Figure 40.17 Disappearing glaciers. Mount Kilimanjaro in Tanzania in 1970 (top) and 2014 (bottom). Note the decrease in glacier coverage over four decades.

(top): Dr. Bruno Messerli; (bottom): George Robertson/Alamy Stock Photo

Many migratory birds arrive earlier at their summer breeding grounds than they did decades ago. Many insects and amphibians breed earlier in the year, and many plants flower earlier (figure 40.19). Recent research shows that in Australia, wild fruit fly populations have undergone changes in gene frequencies in the past 20 years, such that populations in cool parts of the country now genetically resemble populations in warm parts.

Reef-building corals seem to have narrow margins of safety between the sea temperatures they normally experience and the maximum temperatures they can survive. Global warming is already threatening some corals by inducing mass "bleaching," a disruption of the normal and necessary symbiosis between the cnidarians and algal cells. As a result of bleaching, as well as ocean acidification and other threats, as much as 90% of the world's coral reefs could be gone by 2050.

There are reasons to think that the effects of global warming on natural ecosystems today may, overall, be more severe than those of warming events in the distant past. One concern is that the rate of warming today is rapid, and therefore evolutionary adaptations would need to occur very quickly over relatively few generations to

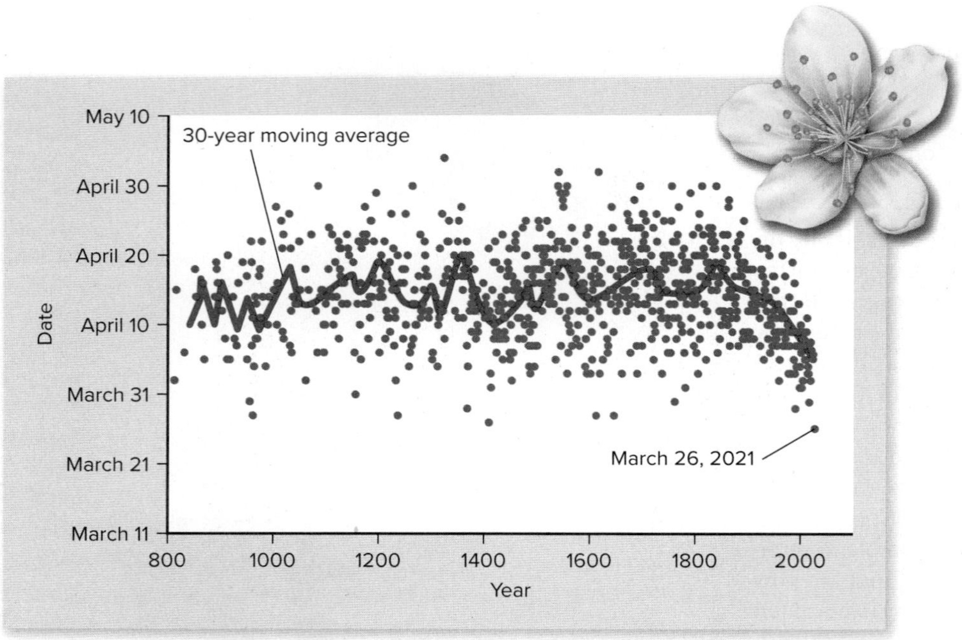

Figure 40.19 Plants are flowering earlier. Since the year 812, people in Japan have recorded when the cherry trees flower in Kyoto, Japan. Most of the increase has occurred in the last 170 years, during which time the average has moved earlier in the year by 12 days, as indicated by the red line (which is the average flowering date for the preceding 30 years). 2021 was the earliest recorded flowering date in the 1209 years since recording began. The earlier flowering results from both increased global temperatures and increased temperature in cities, "heat island effect," which results because cities have more heat-absorbing surfaces such as concrete, and fewer trees.

prevent species extinction. Another concern is that natural areas no longer cover the whole landscape but often take the form of parks completely surrounded by cities or farms. The parks are at fixed geographical locations and in general cannot be moved. If climatic conditions in a park become unsuitable for its inhabitants, the park will cease to perform its function. Moreover, the areas in which the park inhabitants might then find suitable climatic conditions are likely to be developed, rather than being protected parks.

Similarly, as temperatures increase, many montane species shift to higher altitudes to find their preferred habitat. However, eventually they can shift no higher because they reach the mountain's peak. As the temperature continues to increase, the species' habitat disappears entirely. A number of Costa Rican frog species are thought to have become extinct for this reason. The same fate may befall many Arctic species as their habitat melts away.

Global warming can affect human health and welfare in a variety of ways. Some of these changes may be beneficial, but even if they are detrimental, some countries—particularly the wealthier ones—may be able to adjust. But economically disadvantaged countries may not be able to transform as quickly, and some changes will require extremely costly countermeasures that even economically advantaged countries will be hard-pressed to afford.

Rising sea levels

During the second half of the 20th century, sea level rose at 2 to 3 cm per decade. The U.S. Environmental Protection Agency predicts that sea level is likely to rise two or three times faster in the 21st century because of two effects of global warming: (1) the melting of polar ice and glaciers adding water to the ocean and (2) higher ocean temperatures that cause an increase in volume because water expands as it warms. This rise in sea level would cause increased erosion and inundation of low-lying land and coastal marshes. As many as 200 million people would be affected by increased flooding. Should sea levels continue to rise, coastal cities and some entire islands, such as the Maldives in the Indian Ocean, would be in danger of becoming submerged.

Other climatic effects

Global warming is predicted to have a variety of other effects besides increased temperatures. In particular, the frequency or severity of extreme meteorological events—such as heat waves, droughts, severe storms, and hurricanes—is expected to increase, and El Niño events, with their attendant climatic effects, may become more common. Some evidence suggests that these effects are already evident in the increase in powerful storms, hurricanes, and the frequency of El Niño events over the past few years.

In addition, rainfall patterns are likely to shift, and those geographical areas that are already water-stressed, which are currently home to nearly 2 billion people, will likely face even graver water-shortage problems in the years to come.

Effects on agriculture

Global warming may have both positive and negative effects on agriculture. On the positive side, warmer temperatures and increased atmospheric carbon dioxide tend to increase the growth of some crops and thus may increase agricultural yields. Other crops, however, may be negatively affected. Furthermore, most crops will be affected by increased frequencies of droughts. Moreover, although crops in north temperate regions may flourish with higher temperatures, many tropical crops are already growing at their maximal temperatures, so increased temperatures may lead to reduced crop yields.

Also on the negative side, changes in rainfall patterns, temperature, pest distributions, and various other factors will require many adjustments. Such changes may come relatively easily for farmers in the developed world, but the associated costs may be devastating for those in the developing countries.

Solving the problem

Global warming is now widely recognized as an ongoing phenomenon caused in large part by human actions. Many formerly recalcitrant governments now seem poised to take action, and corporations are recognizing the opportunities provided by the need to reverse human impacts. The resulting "green" technologies and practices are becoming increasingly common. For example, investment in renewable energy sources (not including large dams) is now more than twice as great as investment in new fossil fuel plants. As a result, carbon dioxide emissions have fallen in 20 countries since 2000.

Nonetheless, we still have a long way to go. But with concerted efforts from citizens, corporations, and governments, the more serious consequences of global climate change hopefully can be averted, just as ozone depletion was reversed in the last century.

REVIEW OF CONCEPT 40.6

Carbon dioxide is a significant greenhouse gas, meaning that it prevents heat from escaping the Earth, so temperatures rise. Global warming caused by changes in atmospheric composition—most notably, CO_2 accumulation—may increase desertification and cause some habitats and species to disappear. Global warming may also melt ice caps and glaciers, altering coastlines as water levels rise. Violent weather events, disruption of water availability, and flooding of low-lying areas, as well as increased incidence of tropical diseases, may also occur.

- *In what ways does global climate change pose questions different from those posed by ozone depletion?*

How Real Is Global Warming?

The controversy over global warming has two aspects. The first contentious issue is the claim that global temperatures are rising significantly, a profound change in the Earth's atmosphere and oceans referred to as "global warming." The second contentious issue is the assertion that global warming results from elevated concentrations of carbon dioxide in the atmosphere as a consequence of the widespread burning of fossil fuels.

Resolution of the second issue requires detailed science and is only now reaching consensus acceptance. Resolution of the first issue is a simpler proposition, because it is, in essence, a data statement. The graph displays the data in question, global air temperatures for the last century and a half. Temperature data are collected from measuring stations across the globe, as shown in the image below, and averaged. The bars of the histogram represent mean yearly global air temperatures for each year since 1850. The data are presented as a histogram in which each bar presents the deviation of the value during that period from the average value determined for some standard period. In this instance, the histogram shows the deviation of each year's global mean air temperature from the mean of these values observed over a standard 30-year period between 1961 and 1990. The map indicates the distribution of these temperature deviations for the year 2019; the years 2014–2020 were the seven warmest years since record keeping began in 1880.

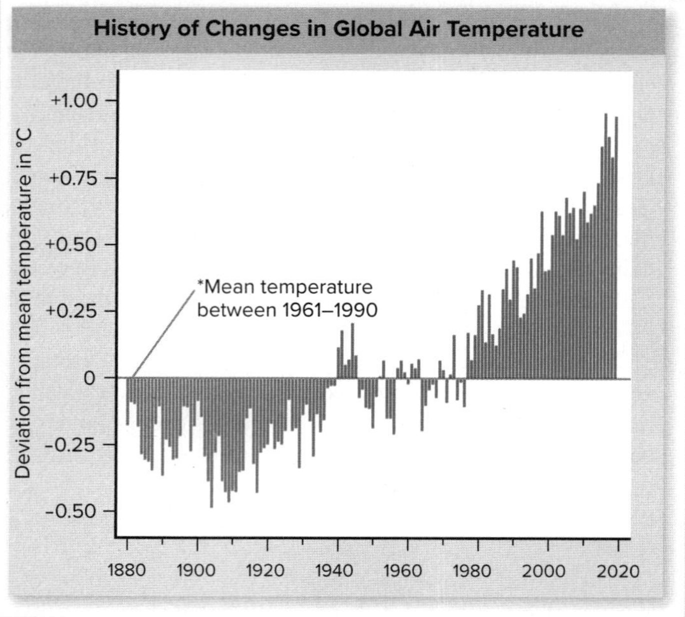

Source: GSFC/NASA

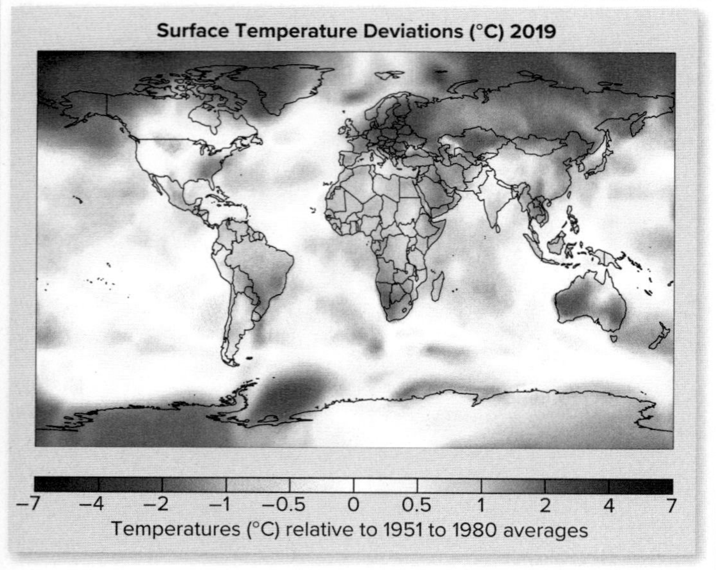

Hansen, J., et al., "Global Temperature in 2014 and 2015" http://www.columbia.edu/~jeh1/mailings/(January 16, 2015)

Analysis

1. **Applying Concepts** What fraction of the 155 years does not deviate from the 1961–1990 mean value? What fraction deviates more than +0.2°C? More than −0.2°C? More than +0.4°C? More than −0.4°C?
2. **Interpreting Data**
   a. Of the years that deviate more than +0.2°C, how many are before 1940? Between 1940 and 1980? After 1980? What fraction occurs after 1980?
   b. Of the years that deviate more than +0.4°C, how many are before 1980? After 2000? What fraction occurs after 2000?
   c. Of the years that deviate more than −0.2°C, how many are before 1940? Between 1940 and 1980? After 1980? What fraction occurs before 1940?
   d. Of the years that deviate more than −0.4°C, how many are before 1940? Before 1900? What fraction occurs before 1900?
3. **Making Inferences** If you were to pick a year at random between 1850 and 1900, would it be most likely to deviate +0.2, +0.4, 0, −0.2, or −0.4? A year between 1900 and 1940? A year after 1980?
4. **Drawing Conclusions** Has the global air temperature been warming progressively over the last century and a half?

Retracing the Learning Path

CONCEPT 40.1 Ecosystems Are Shaped by Sun, Wind, and Water

40.1.1 Solar Energy and the Earth's Rotation Affect Atmospheric Circulation The amount of solar radiation reaching the Earth's surface has a great effect on climate. The seasons result from changes in the Earth's position relative to the Sun. Hot air with its increased water content rises at the equator, then cools and loses its moisture as rain and fog, creating the equatorial rainforests. As the drier cool air of the upper atmosphere moves away from the equator and then descends to Earth, it removes moisture from the Earth's surface and creates deserts on its way back to the equator. Global currents are largely driven by winds. Four large, circular gyres in ocean currents can be found, driven by wind direction. These also are influenced by the Coriolis effect.

40.1.2 Regional and Local Differences Affect Terrestrial Ecosystems A rain shadow occurs when a range of mountains removes moisture from air moving over it from the windward side, creating a drier environment on the opposite side. For every 1000-m increase in elevation, temperature drops approximately 6°C.

CONCEPT 40.2 Earth Has 14 Major Terrestrial Ecosystems, Called Biomes

40.2.1 Temperature and Moisture Largely Determine Biomes Average annual temperature and rainfall, as well as the range of seasonal variation, determine different biomes. Eight major types of biomes are recognized: tropical rainforest, savanna, desert, temperate grassland, temperate deciduous forest, temperate evergreen forest, taiga, and tundra.

CONCEPT 40.3 Freshwater Habitats Occupy Less Than 2% of Earth's Surface

40.3.1 Life in Freshwater Habitats Depends on Oxygen Availability Freshwater habitats represent only 2% of the Earth's surface. Productivity comes from phytoplankton and is limited by oxygen. Oxygen is not very soluble in water. Oxygen is constantly added by photosynthesis of aquatic plants and removed by heterotrophs.

40.3.2 Lakes Differ in Oxygen and Nutrient Content Freshwater lakes low in algal nutrients are called oligotrophic. They are clear with high dissolved oxygen content. Eutrophic lakes are high in algal nutrients. They are cloudy with low dissolved oxygen content.

CONCEPT 40.4 Marine Habitats Dominate the Earth

40.4.1 Open Oceans Have Low Primary Productivity The ocean is divided into several zones: intertidal, neritic, photic, benthic, and pelagic zones. Phytoplankton are the primary producers in open waters, and primary production is low due to low nutrient levels. Nitrogen and iron are thought to be the two most limiting nutrients.

40.4.2 Continental Shelf Ecosystems Provide Abundant Resources Neritic waters are found over continental shelves and have higher nutrient levels. Around 99% of food harvested from the ocean comes from these ecosystems.

40.4.3 The Deep Sea Is a Cold, Dark Place The deep sea is the single largest habitat. In most regions, food in the deep sea originates from photosynthesis from above.

CONCEPT 40.5 Humanity's Pollution Is Severely Impacting the Biosphere

40.5.1 Chemical Pollution Is Altering the Earth Human activity can cause adverse changes to ecosystems. Chemical pollutants, such as DDT, may be concentrated as they move up food chains. DDT has had a severe effect on birds of prey.

40.5.2 Stratospheric Ozone Depletion Has Led to an Ozone "Hole" Increased transmission of UV-B radiation is harmful to life. Global regulation of CFCs seems to be reversing ozone depletion.

CONCEPT 40.6 Human Activity Is Altering Earth's Climate

40.6.1 The Average Temperature at Earth's Surface Is Increasing Data indicate that global temperatures are clearly increasing. Data and models predict possible temperature increases up to 5.7°C by the end of the century.

40.6.2 Carbon Dioxide Is a Major Greenhouse Gas Carbon dioxide allows solar radiation to pass through the atmosphere but prevents heat from leaving the Earth, creating warmer conditions.

40.6.3 Global Temperature Change Has Affected Ecosystems in the Past and Is Doing So Now If temperatures change rapidly, natural selection cannot occur rapidly enough to prevent many species from becoming extinct. Changing sea levels, increased frequency of extreme climatic events, direct and indirect effects on agriculture, and the expansion of tropical diseases are already beginning to affect human life.

Concept Overview

This Concept Overview diagrams the key concepts that were discussed in this chapter.

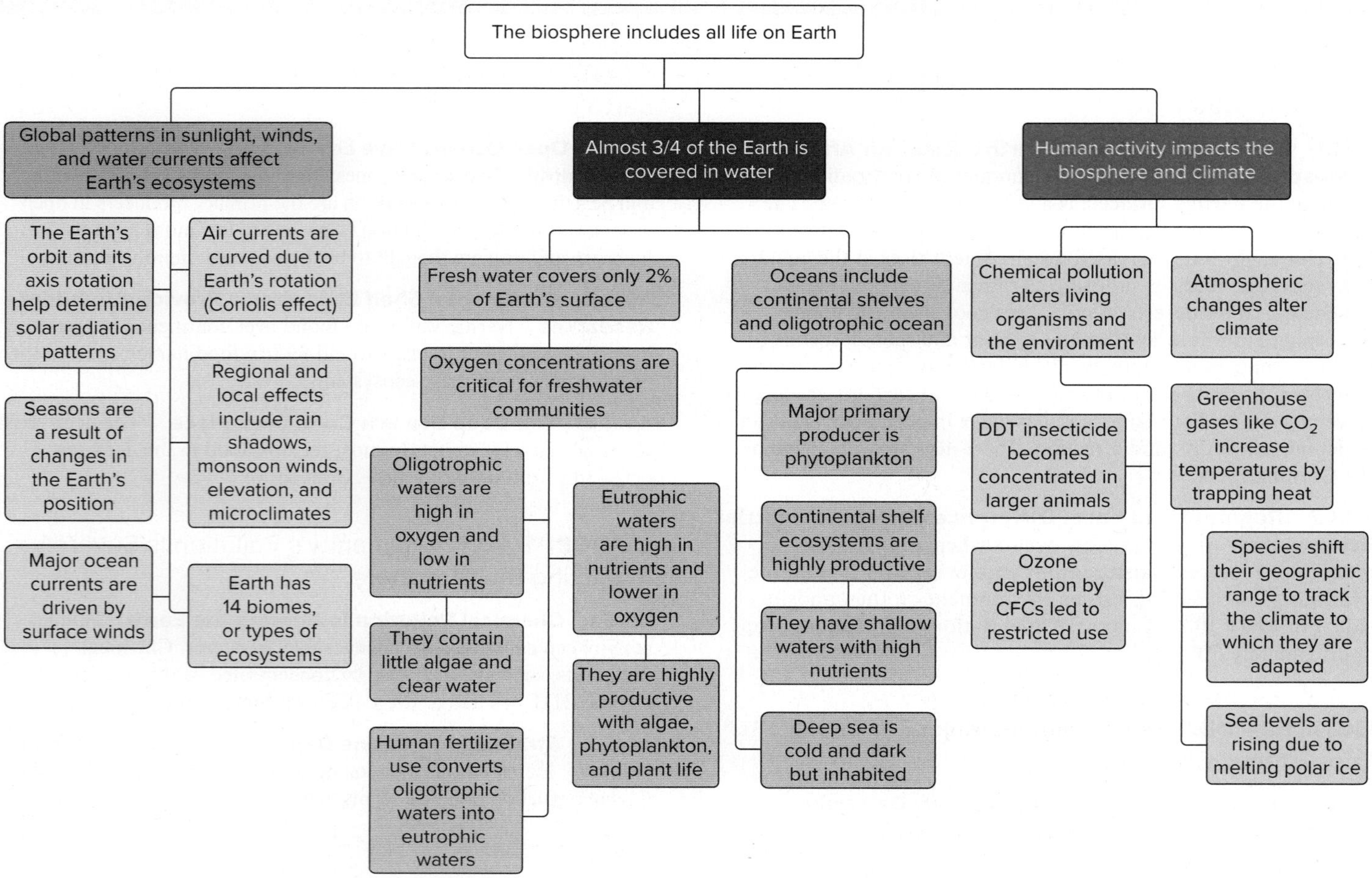
The biosphere includes all life on Earth
Global patterns in sunlight, winds, and water currents affect Earth's ecosystems
The Earth's orbit and its axis rotation help determine solar radiation patterns
Seasons are a result of changes in the Earth's position
Major ocean currents are driven by surface winds
Air currents are curved due to Earth's rotation (Coriolis effect)
Regional and local effects include rain shadows, monsoon winds, elevation, and microclimates
Earth has 14 biomes, or types of ecosystems
Almost 3/4 of the Earth is covered in water
Fresh water covers only 2% of Earth's surface
Oxygen concentrations are critical for freshwater communities
Oligotrophic waters are high in oxygen and low in nutrients
They contain little algae and clear water
Human fertilizer use converts oligotrophic waters into eutrophic waters
Eutrophic waters are high in nutrients and lower in oxygen
They are highly productive with algae, phytoplankton, and plant life
Oceans include continental shelves and oligotrophic ocean
Major primary producer is phytoplankton
Continental shelf ecosystems are highly productive
They have shallow waters with high nutrients
Deep sea is cold and dark but inhabited
Human activity impacts the biosphere and climate
Chemical pollution alters living organisms and the environment
DDT insecticide becomes concentrated in larger animals
Ozone depletion by CFCs led to restricted use
Atmospheric changes alter climate
Greenhouse gases like CO_2 increase temperatures by trapping heat
Species shift their geographic range to track the climate to which they are adapted
Sea levels are rising due to melting polar ice

Assessing the Learning Path

Understand

1. The Coriolis effect
   a. drives the rotation of the Earth.
   b. is responsible for the relative lack of seasonality at the equator.
   c. drives global wind circulation patterns.
   d. drives global wind and ocean circulation patterns.
2. In a rain shadow, air is cooled as it rises and heated as it descends, often producing wet and dry sides because the water-holding capacity of the air
   a. is directly related to air temperature.
   b. is inversely related to air temperature.
   c. is unaffected by air temperature.
   d. produces changes in air temperature.
3. What two factors are most important in biome distribution?
   a. Temperature and latitude
   b. Rainfall and temperature
   c. Latitude and rainfall
   d. Temperature and soil type
4. Oligotrophic lakes have
   a. low oxygen and high nutrient availability.
   b. high oxygen and high nutrient availability.
   c. high oxygen and low nutrient availability.
   d. low oxygen and low nutrient availability.
5. Which of the following contributes oxygen to a freshwater lake?
   a. Photosynthesis
   b. Algal growth
   c. Decomposition of organic matter
   d. Aerobic respiration of fish and invertebrates
6. Continental shelf ecosystems are high in productivity per unit area because of all of the following EXCEPT
   a. their waters are high in nitrate.
   b. they are enriched by nutrients from land runoff.
   c. being shallow, they do not lose nutrients to the deep sea.
   d. they are more biodiverse than other ocean regions.
7. The largest marine habitat on Earth is
   a. the neritic zone.
   b. the deep sea.
   c. the open ocean.
   d. the continental shelves.
8. Biological magnification occurs when
   a. pollutants increase in concentration in tissues at higher trophic levels.
   b. the effect of a pollutant is magnified by chemical interactions within organisms.
   c. an organism is placed under a dissecting scope.
   d. a pollutant has a greater-than-expected effect once ingested by an organism.
9. The Antarctic "ozone hole" is most extensive in
   a. February. c. September.
   b. March. d. August.
10. The so-called greenhouse effect is the only reason Earth's surface is not as cold as the dark side of the Moon. Greenhouse warming increases the average temperature of the Earth by __________.
    a. 70°F c. 50°C
    b. 32°F d. 25°C
11. Which of the following is a point source of pollution?
    a. Lawns
    b. Smokestacks of coal-fired power plants
    c. Factory effluent pipe draining into a river
    d. CO^2 emissions
12. Global warming affects all of the following EXCEPT
    a. rain patterns. c. ozone levels.
    b. rising sea levels. d. agriculture.

Apply

1. If the Earth were not tilted on its axis of rotation, the annual cycle of seasons in the northern and southern hemispheres
   a. would be reversed. c. would be reduced.
   b. would stay the same. d. would not exist.
2. As one travels from northern Canada south to the United States, the timberline increases in elevation. This is because as latitude
   a. increases, temperature increases.
   b. decreases, temperature increases.
   c. increases, humidity decreases.
   d. decreases, humidity decreases.
3. Which of the following biomes is NOT found south of the equator?
   a. Mountain zone
   b. Savanna
   c. Tundra
   d. Tropical monsoon forest
4. The most commonly occurring biome at the equator is
   a. deserts. c. savanna.
   b. rainforest. d. chaparral.
5. Oligotrophic lakes can be turned into eutrophic lakes as a result of human activities such as
   a. overfishing sensitive species, which disrupts fish communities.
   b. introducing nutrients into the water, which stimulates plant and algal growth.
   c. disrupting terrestrial vegetation near the shore, which causes soil to run into the lake.
   d. spraying pesticides into the water to control aquatic insect populations.
6. Amanda's backyard pond often fills with algae, and the fish she adds to the pond often die in the summer. Her friend suggests that she add a waterfall to her pond. Why would this help?
   a. The waterfall would remove the toxins from the algae.
   b. The waterfall would help to oxygenate the water.
   c. The waterfall would add nutrients to the water.
   d. The waterfall would kill the algae.
7. Open ocean waters tend to be low in primary productivity because of
   a. the low intensity of solar illumination in the photic zone.
   b. the great distances of open ocean from land.
   c. nitrate depletion in the photic zone.
   d. low levels of iron in ocean waters.

8. Life in the deep sea is fueled primarily by
   a. photosynthesis far above.
   b. deep-sea vents.
   c. deep-sea photosynthesis.
   d. chemoautotrophy.
9. If a pesticide (such as DDT) is harmless at low concentrations and used properly, how can it become a threat to nontarget organisms?
   a. After exposure to DDT, some species develop allergic reactions, even at low levels of exposure.
   b. DDT molecules can combine so that their concentration increases through time.
   c. The concentration of chemicals, such as DDT, increases at higher trophic levels.
   d. Global warming and exposure to UV-B radiation render molecules, such as DDT, increasingly potent.
10. Destruction of the ozone layer is due to
   a. global warming.
   b. the Coriolis effect.
   c. chlorofluorocarbons.
   d. chlorinated hydrocarbons.
11. If there are many greenhouse gases, why is only carbon dioxide considered a cause of global warming?
   a. The other gases do not cause global warming.
   b. Scientists are concerned about other causes; for example, release of methane from melting permafrost could have significant effects on global warming.
   c. Other gases occur in such low quantities that they have little effect on the climate.
   d. Carbon dioxide is the only gas that absorbs long-wavelength infrared radiation.
12. Biologists fear that the effects of global warming on ecosystems today may be more severe than such events in the distant past, because
   a. the world is getting hotter than it did in the past.
   b. the rate of global warming today is more rapid than in past global warming events.
   c. as a result of pollution, ecosystems are more sensitive to temperature changes than in the past.
   d. species are more specialized in today's ecosystems than in the past.

Synthesize

1. Why are most of the Earth's deserts found at approximately 30° latitude?
2. Why do increasing latitude and increasing elevation affect, in the same way, which plant species grow?
3. Why might you expect primary productivity to increase with increasing precipitation and temperature?
4. Lake Geneva in Switzerland is a large, deep freshwater lake made eutrophic by human activities. A similar lake near Seattle, Lake Washington, was also becoming eutrophic, for similar reasons. Local citizens suggested solving the problem by poisoning the algae in the lake. Why wouldn't this work?
5. In order to reduce Earth's atmospheric carbon dioxide levels, some researchers have suggested reducing oceanic carbon dioxide levels by increasing the primary productivity of the great Southern Ocean. How might this be done?
6. Why do most mineral resources taken from the ocean, such as petroleum, come almost exclusively from the continental shelves?
7. Explain why being exposed to even very tiny amounts of some chemical pollutants can, over time, be hazardous to your health (you might want to focus on endocrine disruptors such as BPA in examining the difficulties in evaluating such dangers in specific instances).
8. Why is atmospheric ozone depletion greater over the South Pole than over the North Pole?
9. Why, in the face of uniform rising atmospheric carbon dioxide levels, is the pace of global warming higher in some regions of the world than in others?
10. If the world has experienced global warming many times in the past, why should we be concerned about it happening again now?

Appendix

Answers to Review of Concepts Questions and End-of-Chapter Understand and Apply Questions

Answers to Synthesize questions and Inquiry & Analysis questions can be found in the Instructors Resources in Connect.

Chapter 1

IN-CHAPTER REVIEW OF CONCEPTS

1.1 Shared features of living systems include organization, utilization of energy, homeostasis, growth, development, reproduction, and heredity.

1.2 Evolution is an emergent property at the population level. An individual organism cannot evolve, but over time, populations do.

1.3 A scientific theory has been tested by experimentation. A hypothesis is a suggested explanation for a natural phenomenon that might be true; it serves as a starting point to explain a body of observations. When predictions generated using a hypothesis have been tested, it gains the confidence associated with a theory. A theory still cannot be "proven," however, as new data can bring about the reevaluation of an accepted theory.

1.4 No. Darwin's theory of evolution explains and describes how organisms on Earth have changed and describes a mechanism for change over time (natural selection). The theory of evolution may lend evidence in support of theories on the origin of life, but it does not explain the process.

1.5 Biology and chemistry are not completely separate fields. Biological systems function within the known laws of chemistry. There are no chemical laws unique to living systems. Biological systems evolved in a world governed by the laws of chemistry and physics.

END-OF-CHAPTER QUESTIONS

Understand

1. c 2. b 3. d 4. c
5. a 6. d 7. c 8. b
9. c 10. d

Apply

1. c 2. a 3. b 4. b
5. d 6. b 7. a, c, d

Chapter 2

IN-CHAPTER REVIEW OF CONCEPTS

2.1 If the number of protons exceeds the number of electrons, the atom has lost electrons and has a positive charge.

2.2 The noble gases are more stable than the other elements in the periodic table because they have eight (or in the case of He, two) valence electrons. A full outer energy level lends stability to an atom, rendering it inert, or nonreactive.

2.3 An ionic bond involves the loss or gain of electrons, while covalent bonds involve the sharing of electrons. An ionic bond involves the electrostatic attraction between ions with opposite charge. In a polar covalent bond, atoms share electrons unequally, because their affinity for electrons (electronegativity) differs, but the resulting molecule is not charged. Such polar molecules have an asymmetric distribution of charge.

2.4 No. While oxygen and hydrogen form polar bonds because of their difference in electronegativity, carbon and hydrogen have similar electronegativity and therefore form nonpolar bonds, and nonpolar molecules. The cohesive and adhesive properties of water are a result of its polarity. These properties would not be exhibited by nonpolar molecules.

2.5 The pH scale is logarithmic. Each unit represents a 10-fold change in hydrogen ion concentration. A change of 2 pH units indicates a 100-fold change $[H^+]$. Depending on the direction of the change, the concentration increases or decreases 100 times. For example, at pH 0, $[H^+] = 1$ and at pH 2, $[H^+] = 0.01$. Moving from pH 0 to pH 2, $[H^+]$ decreases 100 times. Moving from pH 2 to pH 0, $[H^+]$ increases 100 times.

END-OF-CHAPTER QUESTIONS

Understand

1. a 2. b 3. c 4. c
5. b 6. a 7. c 8. d
9. a 10. c 11. d 12. a
13. b 14. d 15. c

Apply

1. c 2. b 3. b 4. c
5. b 6. c 7. a 8. d
9. b

Chapter 3

IN-CHAPTER REVIEW OF CONCEPTS

3.1 Hydrolysis is the reverse reaction of dehydration (and dehydration is the reverse of hydrolysis). Hydrolysis breaks a covalent bond with the addition of water: a hydrogen atom is attached to one subunit of a macromolecule and a hydroxyl group to the other. Dehydration is a synthesis reaction during which a molecule of water is produced as a by-product (removed from the subunits) in the formation of a macromolecule.

3.2 Starch, glycogen, and cellulose are all polymers of glucose. Starch and glycogen are both energy-storage molecules composed of α-glucose subunits. These form long chains, with many C–H bonds that contribute to storing energy. These chains can be branched, and form starch granules (in plant cells) and glycogen granules (in animal cells). Cellulose is composed of β-glucose subunits, which form long, unbranched chains that create tough fibers that are ideal for a structural molecule.

3.3 Structure and function are related. If an unknown protein has a sequence similar to that of a known protein, it will form a similar structure, and we can infer its function is also similar. If an unknown protein has known functional domains or motifs, we can also use these to help predict its function.

3.4 RNA can form double strands, and sometimes it does. There can also be regions base-pairing within a single strand. There isn't really anything stopping RNA from forming double strands–it just usually doesn't.

3.5 Phospholipids have a charged group in place of one of the fatty acids in a triglyceride. This leads to an amphipathic molecule that has both hydrophobic and hydrophilic regions. Such a molecule will spontaneously form bilayer membranes in water.

END-OF-CHAPTER QUESTIONS

Understand

1. b 2. a 3. b 4. d
5. b 6. d 7. c 8. d
9. b 10. c 11. d
12. b 13. a

Apply

1. a 2. d 3. d
4. d 5. b 6. a 7. b
8. d

Chapter 4

IN-CHAPTER REVIEW OF CONCEPTS

4.1 It would really depend on what that life was like. It might change our view of cell theory if it did not have the same molecular basis as life on our planet. If it had the same molecular and cellular basis as terrestrial life on Earth, it would not change our view.

4.2 Both bacteria and archaea tend to be single cells that lack a membrane-bounded nucleus and extensive endomembrane systems. They both have cell walls, although the composition of each is different.

4.3 One would expect the cells in different organs in complex animals to have different structure based upon their specialization. In cells, like muscle cells, that require a lot of energy, you'd expect to see more mitochondria, for example.

4.4 Ribosomes on the RER do not differ from cytoplasmic ribosomes. RER ribosomes synthesize membrane proteins, proteins found in the endomembrane system, and proteins destined for export from the cell. These are targeted to the RER based on sequences in the beginning of the protein. Free ribosomes synthesize cytoplasmic proteins, nuclear proteins, mitochondrial proteins, and proteins found in organelles not derived from the endomembrane system.

4.5 The nuclear genes that encode for the proteins of organelles probably moved from the organelle to the nucleus via horizontal gene transfer.

4.6 The cytoskeleton is involved in maintenance of cell shape, movement of the cell itself, and movement of substances inside it. These things alone account for several different intricate movements.

4.7 Extracellular structures can be linked to the cytoskeleton via membrane proteins, like integrins. This allows extracellular structures to affect intracellular events.

4.8 Cell junctions help arrange cells into higher-level structures that are organized and joined in different ways. Different kinds of junctions serve different functional purposes, depending on the type of tissue. Tight junctions are found in tissue where nothing is allowed to "leak" through, as in protective coatings formed by epithelial cells. If the cells of a tissue rely heavily on cell-to-cell communication to function, gap junctions are found.

END-OF-CHAPTER QUESTIONS

Understand

1. d 2. c 3. d 4. c
5. d 6. a 7. b 8. c
9. c 10. b 11. b

Apply

1. c 2. c 3. b 4. a 5. c
6. b 7. b 8. c

Chapter 5

IN-CHAPTER REVIEW OF CONCEPTS

5.1 A cell would not be able to control its contents. Nonpolar and small polar molecules would diffuse across the membrane, and there would be no selective transport across the membrane; this requires membrane proteins. There would also be no active transport of anything into or out of the cell.

5.2 No. The nonpolar interior of the bilayer would be soluble in a nonpolar solvent. The molecules would organize with their nonpolar tails in the solvent, but the negative charge on the phosphates would cause them to repel each other, disrupting the formation of a bilayer.

5.3 Transmembrane domains are composed of hydrophobic amino acids, often arranged into α helices. The program would search for consecutive hydrophobic amino acids bounded by nonpolar or charged amino acids. Secondarily, you could search for α helical regions.

5.4 The concentration of the IV should be isotonic, relative to your blood cells. If it were hypotonic, your cells would take on water and burst; if it were hypertonic, the blood cells would lose water and shrink.

5.5 Active transport and facilitated diffusion both transport polar and charged species that cannot cross the membrane. They can also transport larger molecules, such as sugars. The difference between these forms of transport is an energy requirement. Active transport requires energy and facilitated diffusion requires a concentration gradient. This also means that facilitated diffusion can only transport a substance until the concentrations are equal across the membrane. In contrast, active transport can move a substance against a concentration gradient, or create such a gradient.

5.6 Receptor-mediated endocytosis, transport by a carrier, and catalysis by an enzyme all involve specificity: the recognition and specific binding of a molecule by a protein.

END-OF-CHAPTER QUESTIONS

Understand

1. d 2. a, b, d 3. a, b, d
4. b 5. a 6. b 7. b
8. a 9. c 10. a 11. c

Apply

1. a 2. d 3. a 4. b
5. a 6. d

Chapter 6

IN-CHAPTER REVIEW OF CONCEPTS

6.1 Since light is not an option at the bottom of the ocean, energy in the form of reduced minerals that can be oxidized, such as sulfur compounds, is used. Hydrothermal vents, which are found at the junctions of tectonic plates, provide this energy source.

6.2 The entropy of the universe is increasing, but the Earth is an open system within this universe. The planet is constantly bombarded with energy from the Sun, some of which is captured by living systems and used to create order (decrease entropy).

6.3 In the text, it is stated that the average person turns over approximately his or her body weight in ATP per day. We can calculate the amount of energy released by a 100-kg man:

Convert kg to g: $100 \text{ kg} = 1.0 \times 10^5 \text{ g} = 100{,}000$ grams

Divide the weight in grams by the molecular weight of ATP: $(100{,}000 \text{ g}) / (507.18 \text{ g/mol}) = 197.2 \text{ mol}$

Multiply by the ΔG for hydrolysis: $(197.2 \text{ mol})(7.3 \text{ kcal/mol}) = 1439 \text{ kcal}$

6.4 Enzymes can alter the rate of a reaction but not the thermodynamics of the reaction. The action of the enzyme does not change the ΔG for the reaction.

6.5 Feedback inhibition is common in anabolic pathways that synthesize metabolites. When the end product builds up, it feeds back to inhibit its own production. Catabolic pathways are involved in the degradation of compounds. Feedback inhibition makes less biochemical sense in a pathway that degrades compounds, as these are usually involved in energy metabolism, recycling, or the removal of compounds. Thus, the end product is destroyed or removed and cannot feed back. However, if a catabolic pathway breaks down larger biomolecules, yielding a compound the cell requires, the compound might feed back to slow or halt the biochemical pathway that produces it.

END-OF-CHAPTER QUESTIONS

Understand

1. b 2. a 3. b 4. b
5. a 6. d 7. b 8. a
9. d 10. c 11. c

Apply

1. c 2. a 3. c 4. a
5. b 6. a 7. a

Chapter 7

IN-CHAPTER REVIEW OF CONCEPTS

7.1 If the cell were to transfer the electrons from glucose directly to oxygen, the entire amount of free energy from the reaction would be released at once. This would be equivalent to combustion and would destroy the cell, as well as wasting this energy.

7.2 Two reactions are linked to the phosphorylation of ADP to ATP: the conversion of BPG to 3PG, and the conversion of PEP to pyruvate. As the 6-C glucose is split into 2 3-C molecules, each of these reactions occurs twice, producing a total of 4 ATP.

7.3 Electrons removed from glucose are transferred first to soluble electron carriers NAD^+ and FAD, reducing these carriers. The soluble carriers are in turn oxidized when they transfer the electrons to the electron transport chain. This allows the soluble carriers to take part in more oxidation reactions during the citric acid cycle. The electron transport chain transfers the electrons to oxygen, which is reduced to yield water.

7.4 A hole in the outer membrane would allow protons in the intermembrane space to leak out, disturbing the proton gradient across the membrane. This would reduce or stop the phosphorylation of ADP by ATP synthase.

7.5 The number of ATP synthesized by chemiosmosis depends on the number of protons translocated by electron transport and the number used by ATP synthase.

7.6 Feedback of NADH and ATP levels in the cells ensures that a cell does not invest too much of its existing ATP in the production of ATP that it does not necessarily need yet. This keeps cells from constantly producing ATP in situations that do not require the energy.

7.7 Anaerobic respiration is likely in ecosystems that have little or no free oxygen. This might include certain aquatic ecosystems and soil environments.

7.8 No, for two reasons: (1) The oxidation of fatty acids feeds acetyl units into the Krebs cycle. The primary output of the Krebs cycle is electrons that feed into the electron transport chain to eventually produce ATP by chemiosmosis. (2) The process of β oxidation that produces the acetyl units is oxygen dependent as well (β oxidation uses FAD as a cofactor for oxidation, and the $FADH_2$ is oxidized by the electron transport chain).

END-OF-CHAPTER QUESTIONS

Understand

1. c 2. b 3. d 4. d
5. b 6. d 7. b 8. c
9. d 10. c 11. d
12. b

Apply

1. b, c 2. c 3. b 4. d
5. b 6. a 7. b 8. a
9. d

Chapter 8

IN-CHAPTER REVIEW OF CONCEPTS

8.1 Both chloroplasts and mitochondria have an outer membrane and an inner membrane. The inner membrane in both forms an elaborate structure and contains an electron transport chain that moves protons across the membrane to allow for the synthesis of ATP via chemiosmosis. They also both have a soluble compartment in which a variety of enzymes carry out reactions.

8.2 All of the carbon inside the human body comes from the organic compounds we ingest. Carbon and hydrogen are the basis of all organic compounds. The carbon in these compounds was once carbon dioxide in the atmosphere, which is fixed by autotrophs.

8.3 The absorption spectrum for an individual pigment shows how much light is absorbed at different wavelengths. *Action spectrum* refers to the most effective wavelengths for a specific light-driven process—in the case of photosynthesis, wavelengths of light that promote photosynthesis are those absorbed by chlorophyll molecules.

8.4 Before photosystems were discovered, scientists had assumed that each chlorophyll molecule absorbed photons, resulting in excited electrons. This belief led scientists to predict that when photosynthetic rates were maxed out, all of a plant's pigment molecules were in use. Experimental evidence showed otherwise—that light is absorbed not by independent pigment molecules but instead by clusters of chlorophyll and accessory pigment molecules, collectively called photosystems. When photosynthetic rates are maxed out, the reaction centers of these photosystems are maxed out, not all the chlorophyll in a plant.

8.5 If the thylakoid membrane were leaky to protons, the electrochemical gradient needed to produce ATP would not exist. So, ATP would not be produced; however, NADPH could still be synthesized because electron transport would continue to occur as long as photons were still being absorbed to begin the process.

8.6 The overall product of the Calvin cycle is glucose, while it is the starting material for glycolysis. A portion of the Calvin cycle is the reverse of glycolysis: the reduction of 3-phosphoglycerate to glyceraldehyde-3-phosphate. While G3P is an important intermediate in glycolysis, it is an important product of the Calvin cycle. Both processes involve ATP: the Calvin cycle uses quite a bit of ATP, while glycolysis requires the initial investment of ATP and produces some.

8.7 To reduce photorespiration, CO_2 can be incorporated into a molecule other than RuBP, then decarboxylated to be used in the Calvin cycle. C_4 plants separate carbon fixation and decarboxylation into separate cells (the mesophyll cells and the bundle sheath cells), while CAM plants separate these processes in time (fix carbon at night and decarboxylate during the day).

END-OF-CHAPTER QUESTIONS

Understand

1. c 2. a 3. c 4. b
5. d 6. b 7. c 8. c
9. b 10. d 11. a
12. a 13. a

Apply

1. c 2. b 3. c 4. a 5. a
6. a 7. b 8. b

Chapter 9

IN-CHAPTER REVIEW OF CONCEPTS

9.1 Ligands bind to receptors based on complementary shapes. This interaction is based on molecular recognition, similar to how enzymes interact with their substrates.

9.2 A hydrophobic molecule, which can cross the membrane, would likely have an internal receptor.

9.3 Effects upon gene expression, such as those of hormones, are generally of longer duration.

9.4 Ras protein plays a central role in signaling pathways that involve growth factors. Mutations that affect Ras function can lead to cancers by disrupting the usual function of these pathways. Changes to the structure of Ras that cause it to be permanently "on" will lead to unrestrained growth. The most common such mutations affect the GTPase activity of Ras. If this activity is lost, or the ability to bind to GAP proteins to stimulate this activity, the Ras protein will stimulate MAP kinase continually.

9.5 One explanation is an evolutionary one: this is an ancient gene family with an origin early in the history of eukaryotes. Duplications to produce the main classes of this receptor type occurred prior to the origin of bilaterians. The nature of the signaling through GPCRs may also be a factor. The number of different proteins involved—receptors, G proteins, effector proteins—makes for a very flexible form of signaling.

END-OF-CHAPTER QUESTIONS

Understand

1. b 2. c 3. d 4. b
5. b 6. c 7. c 8. d
9. a 10. a 11. c

Apply

1. c 2. c 3. c 4. a
5. d 6. a

Chapter 10

IN-CHAPTER REVIEW OF CONCEPTS

10.1 The process of binary fission would likely not work as well if bacteria had many chromosomes.

10.2 No, it is not. For example, silkworms have 56 chromosomes and humans, which are more complex organisms, have 46 chromosomes.

10.3 The first irreversible step is the commitment to DNA replication.

10.4 Loss of cohesins would mean that the products of DNA replication would not be kept together. This would make normal mitosis impossible, and thus lead to aneuploid cells; it would probably be lethal.

10.5 The segregation of chromosomes that lack cohesin would be random, as they could no longer be held at metaphase and attached to opposite poles. This would likely lead to a gain or loss of chromosomes in daughter cells due to improper partitioning.

10.6 If the mutations made the Cdk unable to bind cyclin at all, no division. If the mutation blocked its ability to normally regulate the cell cycle, uncontrolled division.

10.7 Tumor-suppressor genes are genetically recessive, while proto-oncogenes are dominant. Loss of function for a tumor-suppressor gene leads to cancer, while inappropriate expression or gain of function leads to cancer with proto-oncogenes.

END-OF-CHAPTER QUESTIONS

Understand

1. d 2. b 3. b 4. a
5. d 6. d 7. b 8. a
9. d 10. b 11. b 12. b
13. d

Apply

1. b 2. b 3. a 4. c
5. b 6. d 7. b 8. c

Chapter 11

IN-CHAPTER REVIEW OF CONCEPTS

11.1 Stem cells divide by mitosis to produce one cell that can undergo meiosis and another stem cell. So, stem cells provide the body with a constant supply of germ-line cells.

11.2 No. Keeping sister chromatids together at the first division is the key to this division being reductive: reducing the chromosome number from diploid to haploid. Homologs segregate at the first division to accomplish this reduction in chromosome number.

11.3 An improper disjunction at anaphase I would result in four aneuploid gametes: two with an extra chromosome and two that were missing a chromosome. Nondisjunction at anaphase II would result in two normal gametes and two aneuploid gametes: one with an extra chromosome and one missing a chromosome.

11.4 The basic machinery of cell division is the same in mitosis and meiosis, but the behavior of the chromosomes is different. For example, DNA replication occurs prior to meiosis and mitosis; this replication is not repeated before a second division in meiosis, however.

11.5 The independent alignment of homologous pairs at metaphase I and the process of crossing over: the first shuffles the genome at the level of all chromosomes and the second shuffles the genome at the level of individual chromosomes.

END-OF-CHAPTER QUESTIONS

Understand

1. c 2. d 3. b 4. b
5. a 6. b 7. a 8. b
9. b 10. d 11. c

Apply

1. b 2. a 3. a 4. b
5. b 6. b 7. d

Chapter 12

IN-CHAPTER REVIEW OF CONCEPTS

12.1 Mendel did two important things that allowed him formulate his two principles for heredity: he was systematic in his approach, and he quantified his results. This combination of following only variants of 1 or 2 traits through successive generations, and counting the offspring in each generation, allowed him to see the mathematical regularities that underlie his model of inheritance.

12.2 In a monohybrid cross, 1/3 of tall F_2 plants are true-breeding, with 2/3 being heterozygous, and the remaining 1/4 are also true-breeding, but they are not tall plants.

12.3 The events of meiosis I are much more important in explaining Mendel's laws, specifically the segregation of homologs during anaphase I and the independent alignment of different homologous pairs at metaphase I.

12.4 The test parent would make gametes *ABC, AbC, aBC,* and *abC*. In the test cross this would give the phenotypic ratio: 1 dom (A) dom (B) dom (C):1 dom (A) rec (B) dom (C):1 rec (A) dom (B) dom (C):1 rec (A) rec (B) dom (C).

12.5 As the yellow mice must both be heterozygous, they would normally produce 3 yellow:1 normal fur, but as the homozygous yellow animal dies, the ratio would be 2 yellow:1 normal fur.

12.6 A single nucleotide change alters the sequence of the RNA transcript made from a DNA sequence. If the new base changes the sense of a codon, it will result in an amino acid substitution in the encoded protein. In the case of hemoglobin, this can affect the protein such that it aggregates and distorts the cells (sickle cell). This in turn can cause anemia, which can have a variety of effects due to reduced levels of oxygen to cells.

END-OF-CHAPTER QUESTIONS

Understand

1. b 2. c 3. b 4. b
5. b 6. c 7. d 8. a
9. a 10. d 11. d 12. b

Apply

1. b 2. c 3. a 4. c
5. c 6. b 7. b 8. c

Chapter 13

IN-CHAPTER REVIEW OF CONCEPTS

13.1 A cross of white-eyed females to red-eyed males should yield all red-eyed females and all white-eyed males. The females will be heterozygous for white and wild-type alleles, and males will carry a white allele and a Y chromosome.

13.2 Crosses between mt^- and mt^+ *Chlamydomonas* mating types would show that both parents contribute chloroplast DNA but that one parent's allele (the mt^- mating type, in this case) is inactivated when zygotes are formed between mt^- and mt^+ mating types. Another explanation is that the mt^- chloroplast DNA was from the paternal plant and because the inheritance of chloroplasts is usually maternal, the mt^- chloroplasts would not appear in the zygotes.

13.3 You can't determine this by genetic crosses because the results in the progeny would be the same. You would need some method to map the genes to the chromosome on which they reside.

13.4 90%

13.5 First division nondisjunction yields four aneuploid gametes, while second division yields only two aneuploid gametes.

13.6 A hereditary disorder is usually associated with altered protein function, but as we learn more about non-protein-coding genes, we may find genetic disorders that do not involve altered proteins.

END-OF-CHAPTER QUESTIONS

Understand

1. c 2. c 3. a 4. a
5. b 6. d 7. b 8. d
9. b 10. c 11. b

Apply

1. c 2. b 3. c 4. b
5. b 6. b 7. c

Chapter 14

IN-CHAPTER REVIEW OF CONCEPTS

14.1 The fact that proteins are made up of combinations of 20 different amino acids offers chemical complexity, which could translate to informational complexity as well.

14.2 The proper tautomeric forms are necessary for proper base-pairing, which is critical to DNA structure.

14.3 There would have been two bands after one round, one heavy and one light.

14.4 DNA Pol III is the main replication enzyme, responsible for the bulk of DNA synthesis. DNA Pol I acts on the lagging strand to remove primers and replace them with DNA. DNA Pol II does not play a role in replication but is implicated in DNA repair.

14.5 Abnormal shortening of telomeres or a lack of telomerase activity would eventually lead to the ends of chromosomes not being replicated at all; thus, the chromosomes would shorten with each replication. The DNA would no longer encode what it used to, and serious consequences would result (such as cell death).

14.6 No. Eventually, the accumulation of damage would lead to cell death.

END-OF-CHAPTER QUESTIONS

Understand

1. d 2. b 3. a 4. b
5. b 6. c 7. c 8. c
9. c 10. c 11. b 12. c
13. c

Apply

1. c 2. b 3. Tube #2 because its proportion of A:T and C:G is not 1:1
4. c 5. c 6. a 7. d 8. b

Chapter 15

IN-CHAPTER REVIEW OF CONCEPTS

15.1 There is no molecular basis for recognition between amino acids and nucleotides. tRNA is able to interact with nucleic acid by base-pairing and an enzyme can covalently attach amino acids to it.

15.2 The specificity of the genetic code would be lost.

15.3 No. Yeast is a eukaryotic organism in which the processes of transcription and translation are separated in space and time, not coupled as in the prokaryotic example, the bacteria.

15.4 No. This is a result of the evolutionary history of eukaryotes but is not necessitated by genome complexity.

15.5 Alternative splicing offers flexibility in coding information. One gene can encode multiple proteins if the product varies based upon the splicing of the gene.

15.6 This tRNA would be able to read stop codons. This could allow nonsense mutations to be viable but would cause problems, making longer-than-normal proteins. Most bacterial genes actually have more than one stop codon at the end of the gene.

15.7 Attaching amino acids to tRNAs, bringing charged tRNAs to the ribosome, and ribosome translocation.

15.8 A change in the third position of a codon is most likely to result in a silent mutation. This is due to the degeneracy of the code being primarily in the third position where changing a base does not change an amino acid.

END-OF-CHAPTER QUESTIONS

Understand

1. c 2. d 3. d 4. b
5. d 6. b 7. a 8. d
9. a 10. d 11. a 12. b
13. c

Apply

1. b 2. d 3. A/T to G/C mutations in the −10 region would affect initiation, reducing or eliminating gene expression. 4. c 5. b
6. d 7. a

Chapter 16

IN-CHAPTER REVIEW OF CONCEPTS

16.1 The control of gene expression in yeast would be more like that in a fellow eukaryote, a human, than that in a prokaryote such as *E. coli*.

16.2 It would interfere with the protein's ability to bind DNA.

16.3 If the repressor were unable to bind (to tryptophan), despite the presence of tryptophan, the operon would be turned on and transcription would occur, producing the enzymes for continued tryptophan synthesis. The mechanism of negative regulation would be disrupted.

16.4 The loss of a general transcription factor would likely be lethal, as it would affect all transcription, while the loss of a specific factor would affect only those genes controlled by the factor. The effects of the loss of a specific factor would depend on which genes were under this factor's influence.

16.5 The term *housekeeping* is used because these genes are necessary for the ordinary functioning of the cell, not specialized processes that are turned "on" or "off" depending on stimuli like internal or external changes.

16.6 RNA interference offers a mechanism to specifically target gene expression using siRNAs.

16.7 Sequence the proteins expressed during the cell's development, since some gene products are characteristic of induction (like *FGF*) and others of cytoplasmic factors (*macho*-1). If both are found, as in the case of *FGF* influencing cells with the cytoplasmic factor *macho*-1 to become notochord, the cell has become determined through a combination of both mechanisms.

END-OF-CHAPTER QUESTIONS

Understand

1. c 2. a 3. c
4. b 5. b 6. b
7. d 8. b 9. d
10. c 11. b

Apply

1. a 2. b 3. a 4. b
5. d 6. c 7. d 8. d
9. b 10. b

Chapter 17

IN-CHAPTER REVIEW OF CONCEPTS

17.1 mRNA cannot be manipulated and cloned as can DNA. Being able to convert mRNA to DNA means that it is possible to "capture" expressed genes in the form of mRNA in a DNA form that can be easily studied.

17.2 Similarities between PCR and DNA replication in cells are these: (1) both require a free 3′ OH group provided by a primer to initiate DNA synthesis, (2) both require nucleotides and a DNA polymerase to make new DNA, and (3) both require a template piece of DNA to copy.

Differences between PCR and DNA replication in cells are these: (1) the DNA polymerase used by cells is not heat stable, while the one used in PCR is heat stable; (2) the primers used in PCR are DNA, and those in cells are RNA; and (3) DNA replication in cells requires many accessory proteins that PCR does not require.

17.3 The CRISPR/Cas9 system allows direct editing of DNA in the genome of a live cell. Conventional molecular cloning allows the manipulation of DNA in vitro, which can include changes to the cloned DNA, but for this to alter the DNA in live cells, the cloned DNA must be reintroduced into the cell.

17.4 If a scientist wants to find the answer to the question "What is the in vivo function of my gene of interest?" and the gene is not essential, then a knockout is appropriate. If the question is "How does this specific genetic change to my gene of interest affect the behavior of the protein it produces, and subsequently affect the organism?" then a knockin is more appropriate.

17.5 Proteins important in medicine and research can be produced by creating recombinant DNA molecules to be expressed in bacteria. Alternatively, they can be produced in yeast or other eukaryotic cell types. For bacteria, the mRNA produced from the gene of interest is reverse transcribed to cDNA, the cDNA is cloned into a plasmid with the regulatory elements needed for expression in the bacteria, and the plasmid with the cDNA is transformed into bacteria. The bacteria make the protein, and the protein can be harvested and purified.

17.6 Grasses are wind-pollinated. Wind could easily carry transgenic pollen long distances, where it could be introduced into non-genetically modified grasses.

END-OF-CHAPTER QUESTIONS

Understand

1. c 2. b 3. d 4. b
5. b 6. b 7. b 8. d
9. a

Apply

1. c 2. c 3. a 4. c
5. a 6. d 7. c

Chapter 18

IN-CHAPTER REVIEW OF CONCEPTS

18.1 Recombination frequencies vary across the genome as a function of local chromatin structure and specific DNA sequences. cM are measured as a function of recombination frequency between two markers, so if recombination frequency varies, it follows that the apparent distance between two markers in cM may vary, given identical physical distances.

18.2 Assembling genomes using sequences from the whole-genome shotgun method is relatively quick and cheap but becomes increasingly difficult as genome size, sequence redundancy, and genomic complexity increase. Clone-contig sequencing data are easier to assemble if genomes are redundant, large, and/or complex; however, it is more labor intensive and thus expensive.

18.3 Some genomes, like that of wheat, are large, and due to genome duplication events, have many repeated regions. Those repeated sections make it hard to find overlap between clones when clone-contig sequencing is used; whole-genome shotgun sequencing is even harder to work with because the sequences are even shorter and have even less unique overlap. With genomes like that of wheat, the repetitive nature also makes genome annotation difficult; it is hard to know to which particular one of the three separate genomes a given sequence should be assigned.

18.4 It is possible that, because transposable elements can move randomly in the genome, their insertion into genes or regulator sequences will result in genetic changes that can be acted upon by selective pressures. Transposon movement can also affect chromatin structure, which might influence gene expression. Any such variation can be acted upon by natural selection.

18.5 The transcriptome provides information about the levels of gene expression to produce RNA. The absence of some RNAs and the abundance of others may give clues to the function of the cell(s) or tissue being investigated.

18.6 You would expect to share any genes in processes that are conserved when comparing plants and humans. For example, you would expect to find genes for highly conserved processes such as transcription and glycolysis. Conversely, you would not expect to find genes in humans for processes that are not performed by human cells but that are performed by plant cells. For example, you would not expect to see genes for starch biosynthesis or for proteins in photosystems, given that those are used in processes performed by plants and not humans.

END-OF-CHAPTER QUESTIONS

Understand

1. b 2. d 3. b
4. a 5. b 6. b
7. a

Apply

1. b 2. b 3. a 4. d
5. a 6. c 7. c
8. d

Chapter 19

IN-CHAPTER REVIEW OF CONCEPTS

19.1 Variation provides the raw materials for natural selection. Without variation, there are no differences among individuals, no opportunity for individuals with one trait to produce more living offspring than individuals with an alternative trait.

19.2 You could use the genotype frequencies to determine actual allele frequencies. You'd assign the variables p and q to the actual allele frequencies, and then use the Hardy–Weinberg equation ($p^2 + 2pq + q^2 = 1$) in order to determine the expected genotype frequencies. If the actual and expected genotype frequencies are the same (or not significantly different), then we consider the population to be in Hardy–Weinberg equilibrium.

19.3 These processes alter allele frequencies within a population. Natural selection, for example, causes some allele frequencies to be greater and others less than expected.

19.4 Directional selection occurs when one phenotype on either end of the phenotypic range has an adaptive advantage over other phenotypes in the population.

19.5 Directional selection, although different phenotypes had the advantage in different environments (high- or low-predation pressure).

19.6 No, none of these factors is always the most important in determining reproductive success. The cumulative effects of all these factors determine reproductive success. For example, an individual that lives longer than another, but does not mate often, will be relatively less "fit" than an individual that lives a shorter life but mates and successfully reproduces often.

19.7 How allele frequencies would change as a result of heterozygote disadvantage would depend on the relative fitness of the three genotype classes. One allele might increase or, possibly, allele frequencies might not change. In some cases, different processes can favor the same allele. For example, if an allele is favored by natural selection at the same time that gene flow brings that allele into the population, the result would be an increase in the frequency of the allele in the next generation.

19.8 The males should exhibit mate choice, as they are the sex with the greater parental investment and energy expenditure; thus, like females of most species, they should be the "choosier" sex.

END-OF-CHAPTER QUESTIONS

Understand

1. b 2. b 3. d 4. b
5. b 6. c 7. b 8. c
9. b 10. a

Apply

1. d 2. c 3. c 4. d
5. d 6. a 7. a 8. a
9. d

Chapter 20

IN-CHAPTER REVIEW OF CONCEPTS

20.1 No. If eating hard seeds causes individuals to develop bigger beaks, then the phenotype is a result of the environment, not the genotype. Changes in phenotype due to the environment are not heritable.

20.2 An experimental design that would test this hypothesis could be as simple as producing enclosures for the moths and placing equal numbers of both morphs into each enclosure and then presenting predatory birds to each enclosure. One enclosure would have a light background; the other, a dark one. After several generations, measuring the phenotype frequency of the moths should reveal clear trends—the enclosure with the dark background should house more dark-colored moths and the light-colored enclosure more light-colored moths.

20.3 If the trait that is being artificially selected for is due to environment instead of underlying genotype, then even if individuals with the desired traits are favored by selection, they would not pass the trait on to their offspring.

20.4 The rates and direction of evolutionary change vary through time because the major selective agent in most cases of natural selection is the environment. Environmental changes, like changes in climate, continental shifts, and geologic changes, have been the source of selective pressure leading to some relatively fast and major evolutionary changes across many species. Conversely, when the environment is relatively unchanged and stable, selective pressure does not change as often or as intensely, and evolutionary changes are relatively slow and minor.

20.5 Answers will vary. Mutation could provide another explanation for vestigial and homologous structures, especially in the case of vestigial structures. If a vestigial structure resulted from a mutation that had pleiotropic effects, and the other effects of the genetic anomaly were selected for, then the vestigial structure would also be selected for. In theory, homologous structures could have evolved independently in each of the species that possess them.

20.6 If the rate of evolutionary change varies through time or among genes, then the accumulation of changes may increase through time, but not at a constant rate.

20.7 Adaptive convergent evolution occurs when populations or species experience similar environmental pressures and respond, through natural selection, in similar ways. For example, penguins (birds), sharks (fish), sea lions (mammals), and even the extinct ichthyosaurs (reptiles) all exhibit the fusiform shape. All of these animals have similar environmental pressures in that they are all aquatic predators and need to be able to move swiftly and with agility in the water. Clearly, the most recent common ancestor does not have the fusiform shape; thus, the similarities are due to convergence (environment) rather than homology (ancestry). However, similar environmental pressures will not always result in convergent evolution. Sometimes, natural selection will favor different adaptations in response to the same selective environment. Penguins, for example, power their swimming with their forearms (wings), whereas sharks use their tails.

20.8 Because the idea that life has evolved through time is so strongly supported by so many different types of evidence that no single observation could disprove it. Indeed, it's hard to think of any new discovery that could show that evolution hasn't occurred.

END-OF-CHAPTER QUESTIONS

Understand

1. b 2. b 3. c 4. b
5. b 6. a 7. b 8. c
9. c 10. b 11. c

Apply

1. c 2. c 3. d 4. b
5. d 6. d 7. d

Chapter 21

IN-CHAPTER REVIEW OF CONCEPTS

21.1 The biological species concept states that members of the same species are capable of mating and producing viable, fertile offspring. If sympatric species are unable to do so, they will remain reproductively isolated and thus distinct species. Along the same lines, gene flow between populations of the same species allows for homogenization of the two populations such that they remain the same species.

21.2 In order for reinforcement to occur and complete the process of speciation, two populations must have some reproductive barriers in place prior to sympatry. In the absence of this initial reproductive isolation, we would expect rapid exchange of genes, and thus homogenization resulting from gene flow. On the other hand, if two populations are already somewhat reproductively isolated (due to hybrid infertility or a prezygotic barrier such as behavioral isolation), then we would expect natural selection to continue favoring individuals that do not mate with members of the other species and thus lead to increasing pre-mating isolation, eventually resulting in speciation.

21.3 Reinforcement occurs to the situation in which natural selection directly favors increased pre-mating isolation so that eventually two populations no longer interbreed at all and thus are different species. However, reproductive isolation can evolve for other reasons without being directly selected for. Two populations diverging in allopatry may evolve reproductive isolation incidentally as they accumulate genetic differences due to adapting to environmental differences by natural selection; alternatively, two allopatric populations may become reproductively isolated as a result of randomly evolved differences occurring through genetic drift.

21.4 Speciation by polyploidy occurs instantaneously; in a single generation, the offspring of two different parental species may be reproductively isolated because it cannot mate with either parental species and produce fertile offspring. Disruptive selection, on the other hand, requires many generations, as reproductive barriers between the two populations must evolve and be reinforced before the two would be considered separate species.

21.5 In the archipelago model, speciation occurs in isolation on different islands. Some degree of adaptive differences may also arise on separate islands. When the newly arisen species then colonize the ancestral island, character displacement then accentuates the differences among the now-sympatric species, producing a set of species adapted to use different environments. In contrast, in sympatric speciation, reproductive isolation and adaptive divergence occur simultaneously among species that are sympatric.

21.6 Certainly, evolutionary change can be punctuated; in times of changing environmental pressures we would expect adaptation to occur. The adaptations, however, do not necessarily lead to the splitting of a species—instead, one species could simply change in accordance with the environmental changes to which it is subjected. If the environment changes in a punctuated way, then evolutionary adaptation would also be punctuated, but not necessarily linked to speciation.

21.7 Unlike the previous major mass extinction events, the current mass extinction is largely attributable to human activity, including but not limited to habitat degradation, pollution, and hunting.

END-OF-CHAPTER QUESTIONS

Understand

1. a 2. d 3. d 4. b
5. c 6. a 7. b 8. d
9. a 10. a 11. b

Apply

1. a 2. b 3. b 4. a
5. a 6. d 7. d 8. c
9. Biodiversity has increased since the Cambrian period as speciation has surpassed extinction. 10. b

Chapter 22

IN-CHAPTER REVIEW OF CONCEPTS

22.1 Because of convergent evolution; two distantly related species subjected to the same environmental pressures may be more phenotypically similar than two species with different environmental pressures but a more recent common ancestor. More generally, if rates of evolution vary, then sometimes closely related species will be very different, whereas in other cases, more distantly related species will still be very similar.

22.2 In some cases in which characters diverge rapidly relative to the frequency of speciation, it can be difficult to construct a phylogeny using cladistics because the most parsimonious phylogeny may not be the most accurate. In many cases, however, cladistics is a very useful tool for inferring phylogenetic relationships among groups of organisms.

22.3 Yes, in some instances this is possible. Consider two distantly related populations within a species. Suppose they both adapt to similar environments that are different from the environment of the remaining populations. It is possible that their mating behaviors would evolve in the same way, such that they no longer mate with other populations, but do still mate with each other, if given the chance. This new species would be polyphyletic because the two populations are members of the species, but their common ancestor would not be.

22.4 The fox and the squirrel are both members of the class Mammalia. This indicates that they share the synapomorphies of mammals and that they are more closely related to each other than they are to non-mammals. Foxes are in the order Carnivora and squirrels are in the Rodentia. Taxonomy would not allow us to know whether mammals in other orders, such as whales or monkeys, are more closely related to squirrels or foxes. Examination of a phylogeny is needed for that.

22.5 The three domains are the largest groups (Archaea, Bacteria, and Eukarya), which represent the earliest branching of the tree of life. The six supergroups are contained within one of these domains, the Eukarya. These supergroups represent an attempt to distinguish the earliest branching of the Eukarya, and what large monophyletic groups resulted from these earliest branches.

END-OF-CHAPTER QUESTIONS

Understand

1. b 2. d 3. c 4. b 5. a 6. d 7. d 8. a 9. a 10. a 11. a 12. d

Apply

1. c 2. d 3. b 4. b 5. c 6. Cats have retractable claws; dogs' claws are not retractable. Cats have double eyelids and pupils that are vertical slits; dogs have single eyelids and round pupils. Dogs are pack animals, but cats are solitary. 7. c 8. d 9. b

Chapter 23

IN-CHAPTER REVIEW OF CONCEPTS

23.1 Archaea differ from both Eukarya and Bacteria in their membrane lipids. Archaeal phospholipids have different isomers of glycerol phosphate, the hydrocarbon chains are isoprenoid units, and they are joined to glycerol by ether instead of ester linkages. Archaeal membrane can also form monolayers with tetraether phospholipids.

23.2 Compare their DNA and other molecular analyses. The many metabolic tests that have been used for years have been supplanted by molecular approaches.

23.3 Transfer of genetic information in bacteria is directional, from donor to recipient, and does not involve fusion of gametes.

23.4 Prokaryotes do not have a lot of morphological features but do have diverse metabolic functions.

23.5 Pathogens tend to evolve to be less virulent. If they were too good at killing, their lifestyle would be an evolutionary dead end (there would be nothing left alive to invade).

23.6 Viruses do not contain all of the cellular machinery necessary for replication. They must use the cellular machinery of a host cell in order to be able to replicate successfully.

23.7 Any mutation that altered the protein recognized on the host cell surface by that phage. If the phage cannot bind to the host cell, it cannot infect the cell.

23.8 The most obvious problem is the high mutation rate of the HIV virus. Any vaccine targeting a specific virus will protect against that virus, but not necessarily mutated versions. The high mutation rate means that an infected person does not carry a single virus, but rather a set of viruses related by mutation. There are also safety issues with traditional killed or attenuated virus approaches. Most vaccines that have been tried use HIV genes inserted into other harmless viruses.

END-OF-CHAPTER QUESTIONS

Understand

1. c 2. c 3. c 4. b 5. a 6. c 7. a 8. d 9. d 10. a

Apply

1. d 2. a 3. b 4. c 5. Among the archaea, photosynthesis is the simplest known form of the process. It involves a single protein, bacteriorhodopsin, which uses energy from light to translocate protons across a membrane. This then provides a proton motive force for ATP synthesis. 6. d 7. c 8. c 9. b

Chapter 24

IN-CHAPTER REVIEW OF CONCEPTS

24.1 Mitochondria and chloroplasts contain their own DNA. Mitochondrial genes are transcribed within the mitochondrion, using mitochondrial ribosomes that are smaller than those of eukaryotic cells and quite similar to bacterial ribosomes. Antibiotics that inhibit protein translation in bacteria also inhibit protein translation in mitochondria. Also, both chloroplasts and mitochondria divide using binary fission, like bacteria.

24.2 Pseudopodia provide a large surface area and good traction for stable movement.

24.3 There are distinct clades in the Protista that do not share a common ancestor. Organisms commonly referred to as protists are actually a collection of a number of monophyletic clades.

24.4 Contractile vacuoles collect and expel excess water from within the Euglena.

24.5 Although gametophytes are often much smaller than the sporophytes, you could be most confident in your answer if you counted the chromosomes in the cells of each. The diploid sporophyte will have twice as many chromosomes as the haploid gametophyte.

24.6 Foraminifera feed with the podia that project through the openings in their tests.

24.7 Adaptations to drying, because fertilization relied on the aquatic environment as a medium for flagellated sperm to reach the nonmotile egg.

24.8 Comparative genomic studies of choanoflagellates and sponges would be helpful. Considering the similarities among a broader range of genes than just the conserved tyrosine kinase receptor would provide additional evidence.

END-OF-CHAPTER QUESTIONS

Understand

1. b 2. b 3. d 4. d
5. a 6. d 7. b 8. c
9. c 10. c 11. c 12. a
13. b 14. a 15. a 16. d
17. c

Apply

1. d 2. d 3. a 4. d
5. a 6. d 7. c 8. b
9. d 10. d

Chapter 25

IN-CHAPTER REVIEW OF CONCEPTS

25.1 Nuclear replication and division without cell division sometimes lead to multinucleate cells; this occurs in fungi but is unusual in animals.

25.2 The ants consume the fungus to obtain nutrients; in many respects, this has no benefit to the fungus.

25.3 Wind. It can spread spores over large distances, resulting in the spread of fungal disease.

25.4 Unlike the other fungal phyla, the Zygomycota are not monophyletic.

25.5 Microsporidians lack mitochondria, which are found in Plasmodium.

25.6 Blastocladiomycetes are free-living, have mitochondria, and show alternation of generations; microsporidia exhibit none of these characteristics.

25.7 Zygospores are more likely to be produced when environmental conditions are not favorable. Sexual reproduction increases the chances of offspring with new combinations of genes that will result in adaptation. Also, zygospores can remain dormant until growth conditions improve.

25.8 Pure cultures allow a single species' DNA to be isolated and used for phylogenetic analysis. Contamination with another species would introduce additional DNA sequences and interfere with analysis.

25.9 A dikaryotic cell has two nuclei, each with a single set of chromosomes; a diploid cell has a single nucleus with two sets of chromosomes.

25.10 Fungicides and good cultivation practices could help, as would providing spore-filtering masks to workers on contaminated fields.

END-OF-CHAPTER QUESTIONS

Understand

1. c 2. a 3. d 4. b
5. a 6. c 7. a 8. c

Apply

1. a 2. b 3. d 4. c
5. a 6. b 7. d 8. b
9. a 10. c 11. c 12. e

Chapter 26

IN-CHAPTER REVIEW OF CONCEPTS

26.1 Gametes in plants are produced by mitosis. Human gametes are produced directly by meiosis.

26.2 Mosses are extremely desiccation tolerant and can withstand a lack of water. Also, freezing temperatures at the poles are less damaging when mosses have a low water content.

26.3 The sporophyte generation has evolved to be the larger generation; therefore, an effective means of transporting water and nutrients over greater distances would be advantageous.

26.4 There was substantial climate change during that time period. Glaciers had spread, then melted and retreated. Drier climates could have contributed to the extinction of large club mosses.

26.5 The silica can increase the strength of the hollow-tube stems and would deter herbivores.

26.6 Because a pollen tube grows toward the egg, carrying the sperm within the pollen tube; this eliminates the need for a liquid medium for sperm to travel through.

26.7 The ovule is exposed to air on the scale (a modified leaf).

26.8 Increased dispersal. Animals that consume the fruit disperse the seed over longer distances than wind can disperse seed. The species can colonize a larger territory more rapidly.

END-OF-CHAPTER QUESTIONS

Understand

1. b 2. c 3. e 4. c 5. c
6. a 7. a 8. b 9. c 10. c
11. c 12. d 13. d 14. a
15. a

Apply

1. d 2. b 3. b 4. c
5. d 6. a 7. b 8. b
9. c

Chapter 27

IN-CHAPTER REVIEW OF CONCEPTS

27.1 Cephalization, the concentration of nervous tissue in a distinct head region, is intrinsically connected to the onset of bilateral symmetry. Bilateral symmetry promotes the development of a central nerve center, which in turn favors the nervous tissue concentration in the head. In addition, the onset of both cephalization and bilateral symmetry allows for the development of directional movement (bilateral symmetry) and sensory organs facing the direction in which the animal is moving (cephalization).

27.2 Phenotypic characters sometimes evolve convergently, such that taxa with the same character state are not necessarily closely related. Examination of phylogenies constructed with molecular data can often reveal such convergence when it is not previously recognized.

27.3 The cells of a truly colonial organism, such as a colonial protist, are all structurally and functionally identical; however, sponge cells are differentiated and these cells coordinate to perform functions required by the whole organism. Unlike all other animals, however, sponges do appear much like colonial organisms in that they are not composed of true tissues, and the cells are capable of differentiating from one type to another.

27.4 Tapeworms are parasitic platyhelminthes that live in the digestive system of their host. Tapeworms have a scolex, or head, with hooks for attaching to the wall of their host's digestive system. Another way in which the anatomy of a tapeworm relates to its way of life is its dorsoventrally flattened body and corresponding lack of a digestive system. Tapeworms live in their food; they absorb their nutrients directly through the body wall, and their flat bodies facilitate this form of nutrient delivery.

27.5 Bivalves filter-feed by drawing water into their shell so would have no need for a radula that involves scraping the substrate for food.

27.6 Bryozoans possess the general characteristics of animals, and not of plants (refer to section 27.1). Many animals do not move from place to place (are sessile).

27.7 One of the defining features of the arthropods is the presence of a chitinous exoskeleton. As arthropods increase in size, the exoskeleton must increase in thickness disproportionately, in order to bear the pull of the animal's muscles. This puts a limit on the size a terrestrial arthropod can reach, as the increased bulk of the exoskeleton curtails the animal's ability to move. Water is denser than air and thus provides more support; for this reason, aquatic arthropods are able to be larger than terrestrial arthropods.

27.8 Chordates have a truly internal skeleton (an endoskeleton), compared to the endoskeleton on echinoderms, which is functionally similar to the exoskeleton of arthropods. Whereas an echinoderm uses tube feet attached to an internal water vascular system for locomotion, a chordate has muscular attachments to its endoskeleton. Finally, chordates have a suite of four characteristics that are unique to the phylum–a nerve chord, a notochord, pharyngeal slits, and a postanal tail.

END-OF-CHAPTER QUESTIONS

Understand

1. b 2. e 3. a 4. a
5. b 6. d 7. b 8. b
9. a 10. a 11. b 12. d
13. c 14. c

Apply

1. c 2. c 3. c 4. b
5. c 6. d 7. a 8. c
9. c 10. a

Chapter 28

IN-CHAPTER REVIEW OF CONCEPTS

28.1 Lancelets and tunicates differ from vertebrates in that they do not have vertebrae or internal bony skeletons.

28.2 At some point, an external skeleton could become too heavy to be carried by the organism. In order to resist the pull of increasingly large muscles, the exoskeleton must dramatically increase in thickness as the animal grows larger. There is thus a limit on the size of an organism with an exoskeleton—if it gets too large, it will be unable to move due to the weight and heft of its exoskeleton.

28.3 No. As the phylogeny indicates, some fish (e.g., lungfish, class Dipnoi) are more closely related to other vertebrates (the tetrapods) than they are to all other fish.

28.4 First, tetrapods needed to be able to support their body weight and locomote on land; this challenge was overcome by the evolution of legs. Second, tetrapods needed to be able to exchange oxygen with the atmosphere; this was accomplished by the evolution of more efficient lungs than their lungfish ancestors had, as well as of cutaneous respiration. Third, since movement on land requires more energy than movement in the water, tetrapods needed a more efficient oxygen delivery system to supply their larger muscles; this was accomplished by the evolution of double-loop circulation and a partially divided heart. Finally, the first tetrapods needed to develop a way of staying hydrated in a nonaquatic environment, and these early tetrapods developed leathery skin that helped prevent desiccation.

28.5 Since amphibians remain tied to the water for their reproduction, their eggs are jelly-like and if laid on the land will quickly desiccate. Reptile eggs, on the other hand, are amniotic eggs—they are watertight and contain a yolk, which nourishes the developing embryo, and have a series of four protective and nutritive membranes.

28.6 There are two primary traits shared between birds and reptiles. First, both lay amniotic eggs. Second, they both possess scales (which cover the entire reptile body but solely the legs and feet of birds). Birds also share characteristics, such as a four-chambered heart, with only one group of reptiles: the crocodilians.

28.7 The most striking convergences between birds and mammals are endothermy, the ability to regulate body temperature internally, and possession of a four-chambered heart.

28.8 Only the hominins constitute a monophyletic group. Prosimians, monkeys, and apes are all paraphyletic—they include the common ancestor but not all descendants: the clade that prosimians share with the common prosimian ancestor excludes all anthropoids, the clade that monkeys share with the common monkey ancestor excludes hominoids, and the clade that apes share with the common ape ancestor excludes hominids.

END-OF-CHAPTER QUESTIONS

Understand

1. c 2. c 3. c 4. c 5. b
6. b 7. d 8. a 9. a
10. c 11. d 12. d 13. d
14. d 15. c 16. c 17. a

Apply

1. c 2. b 3. c 4. c
5. b 6. c 7. b 8. c
9. a 10. a 11. b

Chapter 29

IN-CHAPTER REVIEW OF CONCEPTS

29.1 Each type of growth is responsible for something different. Primary growth contributes to the increase in plant height, as well as branching. Secondary growth makes substantial contributions to the increase in girth of the plant, allowing for a much larger sporophyte generation.

29.2 Vessels transport water and are part of the xylem. The cells are dead with only the walls remaining. Cylinders of stacked vessels move water from the roots to the leaves of plants. Sieve-tube members are part of the phloem and transport nutrients. Sieve-tube members are living cells, but they lack a nucleus. They rely on neighboring companion cells to carry out some metabolic functions. Like vessels, sieve-tube members are stacked to form a cylinder.

29.3 It would be difficult for a root hair to form in the region of elongation because its base would be pulled apart by the elongation of the cell wall.

29.4 Roots are constantly growing through soil where cells are damaged and sloughed off. The tips of stems do not encounter the same abrasive physical conditions and do not require the additional protection.

29.5 Horizontally oriented leaves are exposed to sunlight from the top. Vertically oriented leaves have more limited sunlight exposure. Palisade layers are tightly packed with minimum airspace between the cells of vertical leaves to maximize photosynthetic surface area.

END-OF-CHAPTER QUESTIONS

Understand

1. c 2. a 3. d 4. c
5. d 6. d 7. b 8. b
9. b 10. b 11. c 12. a
13. d

Apply

1. a 2. c 3. b 4. b
5. c 6. b 7. d 8. a
9. a

Chapter 30

IN-CHAPTER REVIEW OF CONCEPTS

30.1 Flowers can attract pollinators, enhancing the chances of reproduction.

30.2 No. Because meiosis is involved, recombination still takes place and therefore the offspring of self-pollinating plants will not be identical to the parent. They may contain high proportions of individuals well adapted to particular habitats, however.

30.3 Only angiosperms have an endosperm that results from double fertilization. The endosperm is the nutrient source in angiosperms. Gymnosperm embryos rely on megagametophytic tissue sources for nutrients.

30.4 Trees living in cold climates with cold winters might be sensitive to temperature and may produce seed requiring a period of cold before germinating.

30.5 Features such as color and covering encourage animals to eat fruits. For example, fruits with fleshy, shiny, black, or red coverings are often eaten by birds or other vertebrates. Red fruits signal abundant food supply, and many fruits redden as they ripen. Some fruits are also aromatic.

30.6 Retaining the seed in the ground might provide greater temperature and moisture stability for the seedling until its root system is established while also keeping it shielded from the elements, predators, and UV light.

30.7 A biennial life cycle allows an organism to store up substantial reserves to be used to support reproduction during the second season. The downside to this strategy is that the plant might not survive the winter between the two growing seasons, thereby reducing its fitness to zero.

30.8 When conditions are uniform and the plant is well adapted to those constant conditions, genetic variation is not advantageous. Rather, vegetative reproduction will ensure that the genotypes that are well adapted to the current conditions are maintained.

END-OF-CHAPTER QUESTIONS

Understand

1. d 2. a 3. c 4. a
5. b 6. a 7. c 8. c
9. c 10. c 11. b 12. d
13. a 14. c 15. a 16. a

Apply

1. d 2. b 3. c 4. c
5. d 6. c 7. d 8. b
9. b 10. c

Chapter 31

IN-CHAPTER REVIEW OF CONCEPTS

31.1 Physical pressures include gravity and transpiration, as well as turgor pressure as an expanding cell presses against its cell wall. Increases in turgor pressure and other physical pressures are associated with increased water potential. Solute concentration determines whether water enters or leaves a cell via osmosis. The solute potential is the smallest amount of pressure on the side of the cell membrane with the greatest solute concentration that is necessary to stop osmosis. Water potential is the sum of the pressure from physical forces and from the solute potential.

31.2 The structure of the cell membrane allows it to be a selectively permeable barrier. Its phospholipids are oriented such that their hydrophobic tails face one another, shielded from water, and their hydrophilic heads face outward, toward water. The lipid portion of the membrane mostly serves as a barrier, blocking large or charged/polar molecules from passing; however, small, uncharged solutes can move freely. Water moves rapidly through the membrane via aquaporins, which are specific to this function. Some other molecules pass through the membrane via proteins embedded in the membrane. These proteins are specific to the types of molecules they allow to pass; some require an input of energy (to move solutes up a gradient). Thus, the membrane is a barrier that is permeable to some things but not to others (making it selective). For more information, review chapter 5.

31.3 Water could move laterally between tracheids or vessels via the pits in the walls of these structures. By moving laterally into an adjacent vessel, the flow of water can "bypass" the air bubble obstruction and continue its journey through the plant body.

31.4 It needs to be transported in the vascular tissue and cannot be moved as a gas. Once carbon dioxide is dissolved in water, it can be transported to photosynthetically active cells, where it is used in carbon fixation in the Calvin cycle.

31.5 Physical changes in the roots in response to oxygen deprivation may prevent further transport of water in the xylem. Although the leaves may be producing oxygen, it is not available to the roots. Remember, oxygen is required as the terminal electron acceptor in cellular respiration; without oxygen, cells in the root would struggle to efficiently perform cellular respiration.

31.6 Phloem liquid is rich in organic compounds, including sucrose and plant hormones dissolved in water. Fluid in the xylem consists of minerals dissolved in water.

31.7 Magnesium is an important part of the chlorophyll molecule. It is found in the center of the chlorophyll's porphyrin ring. Without sufficient magnesium, chlorophyll deficiencies will result in decreased photosynthesis and decreased yield per acre.

31.8 Abscisic acid could be isolated from root caps of several plants and applied to the stem buds of other plants of the same species. The growth—or lack of growth—of the buds could be compared with untreated controls to determine whether or not the abscisic acid had an effect.

31.9 Chlorophyll is essential for photosynthesis. Phytochromes regulate plant growth and development using light as a signal. Phytochrome-mediated responses align the plant with the light environment so that photosynthesis is maximized, which is advantageous for the plant.

31.10 The plant would not have normal gravitropic responses. Other environmental signals, including light, would determine the direction of plant growth.

END-OF-CHAPTER QUESTIONS

Understand

1. c 2. a 3. c 4. b
5. a 6. c 7. c 8. c
9. a 10. c 11. b 12. c

Apply

1. b 2. b 3. c 4. b
5. c 6. a 7. a 8. b

Chapter 32

IN-CHAPTER REVIEW OF CONCEPTS

32.1 Yes, organs can be, and are, made of multiple tissue types. For example, the heart contains muscle, connective tissue, and epithelial tissue.

32.2 The epithelium in a gland is composed of simple cuboidal and columnar epithelial cells that are specialized for secretion. A gland is formed from invaginated epithelia, and the expanded size of its constituent cell types accommodates extracellular machinery, which is required to produce glandular secretions. The epithelium that lines the inner surface of the digestive tract, which is derived from endoderm, is composed of simple columnar epithelium with microvilli at the apical surface. These projections increase the surface area of these cells for the absorption of nutrients.

32.3 Despite their impressive length, the surface-area-to-volume ratio is maintained in neurons because they are extremely thin.

32.4 Blood is considered connective tissue because it contains abundant extracellular material, the fluid plasma.

32.5 The function of heart cells requires that they be electrically connected; gap junctions allow the flow of ions between cells.

32.6 There are three limitations that terrestrial invertebrates experience due to an exoskeleton. First, animals with an exoskeleton can grow only by shedding, or molting, the exoskeleton, leaving them vulnerable to predation. Second, muscles that act upon the exoskeleton cannot strengthen and grow, as they are confined within a defined space. Finally, the exoskeleton, in concert with the respiratory system of many terrestrial invertebrates, limits the size to which these animals can grow. In order for the exoskeleton of a terrestrial animal to be strong, it has to have a sufficient surface area, and thus it has to increase in thickness as the animal gets larger. The weight of a thicker exoskeleton would impose debilitating constraints on the animal's ability to move.

32.7 First, unlike the chitinous exoskeleton, a bony endoskeleton is made of living tissue; thus, the endoskeleton can grow along with the organism. Second, because the muscles that act upon the bony endoskeleton are not confined within a rigid structure, they are able to strengthen and grow with increased use. Finally, the size limitations imposed by a heavy exoskeleton that covers the entire organism are overcome by the internal bony skeleton, which can support a greater size and weight without itself becoming too cumbersome.

32.8 Myoglobin provides oxygen to the muscles for the aerobic respiration of glucose, thus providing a higher ATP yield than anaerobic respiration. Increased mitochondria also increase the ATP productivity of the muscle by increasing the availability of cellular respiration and thus allow for sustained aerobic activity.

32.9 Locomotion via alternation of legs versus a series of leaps requires a greater degree of nervous system coordination and balance; the animal needs to constantly monitor its center of gravity in order to maintain stability. In addition, a series of leaps will cover more ground per unit time and energy expenditure than will movement by the alternation of legs.

END-OF-CHAPTER QUESTIONS

Understand

1. c 2. c 3. b 4. b
5. c 6. d 7. c 8. c
9. a 10. d 11. d 12. b
13. c

Apply

1. b 2. d 3. c 4. c
5. a 6. a 7. c 8. b
9. d 10. c

Chapter 33

IN-CHAPTER REVIEW OF CONCEPTS

33.1 The somatic division of the PNS is under conscious control.

33.2 Anything that affects the function of the sodium–potassium pump could affect the resting potential, as well as anything that causes a loss of potassium leakage through channels, or that makes the membrane permeable to other ions.

33.3 A positive current inward (influx of Na^+) depolarizes the membrane, while a positive current outward (efflux of K^+) repolarizes the membrane.

33.4 Opening a K^+ channel would hyperpolarize the membrane and represent an inhibitory postsynaptic potential.

33.5 Reflex arcs allow you to respond to a stimulus that is damaging before the information actually arises at your brain. Reflexes can protect us from harm before we can process a threat to homeostasis.

33.6 These two systems work in opposition. This may seem counterintuitive but is the basis for much of homeostasis.

33.7 The intensity of the stimulus is encoded in the frequency of action potentials. The stimulus intensity and action potential frequency have a logarithmic relationship. If a stimulus has an intensity of 100, there will be 2 action potentials; if the intensity is 1000, there will be 3 action potentials; etc.

33.8 Proprioceptors detect the stretching of muscles and subsequently relay information about the relative position and movement of different parts of the organism's body to the central nervous system. This knowledge is critical for the central nervous system; it must be able to respond to these data by signaling the appropriate muscular responses, allowing for balance, coordinated locomotion, and reflexive responses.

33.9 The lateral line system supplements the sense of hearing in fish and amphibian larvae by allowing the organism to detect minute changes in the pressure and vibrations of its environment. This is facilitated by the density of water; without an aquatic environment the adult terrestrial amphibian will no longer be able to make use of this system. On land, sound waves are more easily detectable by the sense of hearing than are vibratory or pressure waves by the similar structures of a lateral line.

33.10 Insects land on substances with their feet. Often the substances are sources of food. Having taste receptors on their feet allows them to survey the substances as potential food sources.

33.11 No, individuals who have red–green color blindness lack certain cone cells; therefore, they cannot "learn" to distinguish between the two colors. Without separate photoreceptors, the colors look the same.

END-OF-CHAPTER QUESTIONS

Understand

1. c 2. a 3. d 4. a
5. c 6. c 7. b 8. b
9. a 10. a 11. d 12. b

Apply

1. c 2. d 3. b 4. a
5. a 6. a 7. a 8. a
9. d 10. b 11. c

Chapter 34

IN-CHAPTER REVIEW OF CONCEPTS

34.1 The cells and tissues of a one-way digestive system are specialized such that ingestion, digestion, and elimination can happen concurrently, making it more efficient in terms of food processing and energy utilization. With a gastrovascular cavity, however, all of the cells are exposed to all aspects of digestion.

34.2 The sandwich represents carbohydrate (bread), protein (chicken), and fat (mayonnaise). The breakdown of carbohydrates begins with salivary amylase in the mouth. The breakdown of proteins begins in the stomach with pepsinogen, and the emulsification of fats begins in the duodenum with the introduction of bile. So it is the chicken that will begin its breakdown in the stomach.

34.3 The sight, taste, and, yes, smell of food are the triggers the digestive system needs to release digestive enzymes and hormones. The saliva and gastric secretions that are required for proper digestion and that are triggered by the sense of smell would be affected by anosmia.

34.4 Fick's law states that the rate of diffusion (R) can be increased by increasing the surface area of a respiratory surface, increasing the concentration difference between respiratory gases, and decreasing the distance the gases must diffuse:

$$R = \frac{DA\,\Delta p}{D}$$

Continually beating cilia increase the concentration difference (Δp).

34.5 Countercurrent flow systems maximize the oxygenation of the blood by increasing Δp, thus maintaining a higher oxygen concentration in the water than in the blood throughout the entire diffusion pathway. The lamellae, found within a fish's gill filaments, facilitate this process by allowing water to flow in only one direction, counter to the blood flow within the capillary network in the gill.

34.6 Birds have a more efficient respiratory system than other terrestrial vertebrates. Birds that live or fly at high altitudes are subjected to lower oxygen partial pressure and thus have evolved a respiratory system that is capable of maximizing the diffusion and retention of oxygen in the lungs. In addition, efficient oxygen exchange is crucial during flight; flying is more energetically taxing than most forms of locomotion, and without efficient oxygen exchange birds would be unable to fly even short distances safely.

34.7 Most oxygen is transported in the blood bound to hemoglobin (forming oxyhemoglobin), while only a small percentage is dissolved in the plasma. Carbon dioxide, on the other hand, is predominantly transported as bicarbonate (having first been combined with water to form carbonic acid and then dissociated into bicarbonate and hydrogen ions). A small amount of carbon dioxide is also transported bound to hemoglobin.

34.8 Platelets are cell fragments that pinch off from larger cells in the bone marrow.

34.9 The primary advantage of having two ventricles rather than one is the separation of oxygenated from deoxygenated blood. In fish and amphibians, oxygenated and deoxygenated blood mix, leading to less oxygen being delivered to the body's cells.

34.10 The delay following atrial contraction allows the atrioventricular valves to close prior to ventricular contraction. Without that delay, the contraction of the ventricles would force blood back up through the valves into the atria.

34.11 During systemic gas exchange, only about 90% of the fluid that diffuses out of the capillaries returns to the blood vessels; the rest moves into the lymphatic vessels, which then return the fluid to the circulatory system via the left and right subclavian veins.

END-OF-CHAPTER QUESTIONS

Understand

1. b, c 2. d 3. d 4. b
5. d, e, a, c, b 6. d
7. a 8. c 9. c 10. a

Apply

1. b 2. d 3. b 4. c, d
5. R is optimized by the branched nature of the gills, which increases surface area (A), and the thin epithelium of gills shortens the distance (d) for diffusion. 6. At rest. More oxygen is unloaded during exercise because of increased CO_2 production, lowering the percent saturation of oxyhemoglobin. 7. a, d
8. b 9. a 10. a

Chapter 35

IN-CHAPTER REVIEW OF CONCEPTS

35.1 Negative feedback helps to constrain the action of any effector by limiting the response. Antagonistic effectors allow a system to respond by either raising or lowering a quantity or function of interest.

35.2 The response of a particular tissue depends first on the receptors on its surface, and second on the response pathways active in a cell. There can be different receptor subtypes that bind the same hormone, and the same receptor can stimulate different response pathways.

35.3 This might lower the amount of GH in circulation; however, as a treatment, it may have unwanted side effects.

35.4 The body has two hormones to maintain blood sugar at a constant level because these hormones have antagonistic effects: one drives blood sugar up (glucagon), while another brings it back down (insulin). Insulin functions to remove glucose from the blood, converting it to glycogen in liver and muscle. Glucagon functions to raise blood glucose by stimulating the conversion of glycogen to glucose and stimulating the synthesis of glucose by the liver. In tandem, these hormones regulate blood sugar.

35.5 Water moves toward regions of higher osmolarity (from a more dilute solution to a less dilute solution).

35.6 This would increase the osmolarity within the tubule system and thus should decrease reabsorption of water. This would lead to loss of water.

35.7 Blocking aquaporin channels would prevent reabsorption of water from the collecting duct, which would result in water loss.

35.8 No, innate immunity shows some specificity for classes of molecules common to pathogens.

35.9 T-cell receptors are rearranged to generate a large number of different receptors with specific binding abilities. Toll-like receptors are not rearranged and recognize specific classes of molecules, not specific molecules.

35.10 Ig receptors are rearranged to generate many different specificities. TLR innate receptors are not rearranged and bind to specific classes of molecules.

END-OF-CHAPTER QUESTIONS

Understand

1. b, d 2. b 3. b 4. a
5. a 6. b 7. c 8. c
9. d 10. b

Apply

1. a 2. b 3. b 4. c
5. b, c 6. c 7. b 8. c
9. b 10. a

Chapter 36

IN-CHAPTER REVIEW OF CONCEPTS

36.1 Oviparity involves less parental investment, and allows the production of larger numbers of offspring. Thus oviparity is favored in environments where large numbers of offspring with little parental investment are advantageous.

36.2 In mating systems where males compete for mates, sperm competition, a form of sexual selection, is very common. In these social groups, multiple males may mate with a given female, and thus those individuals who produce the highest number of sperm have a reproductive advantage—a higher likelihood of siring the offspring.

36.3 If more than one oocyte is fertilized, then it can result in multiple births. If two oocytes are ovulated, and both are fertilized, and both survive to term, then the result would be fraternal (dizygotic) twins.

36.4 Ca^{2+} ions act as second messengers and bring about changes in protein activity that result in blocking polyspermy and increasing the rate of protein synthesis within the egg.

36.5 The nucleus must be reprogrammed. What this means exactly on the molecular level is not clear, but it probably involves changes in chromatin structure and methylation patterns.

36.6 In a mammal, the cells at the four-cell stage are still uncommitted, and thus separating them will still allow for normal development. In frogs, on the other hand, yolk distribution results in displaced cleavage; thus, at the four-cell stage the cells do not each contain a nucleus that contains the genetic information required for normal development.

36.7 The cellular behaviors necessary for gastrulation differ across organisms; however, some processes are necessary for any gastrulation to occur. Specifically, cells must rearrange and migrate throughout the developing embryo.

36.8 No—neural crest cell fate is determined by its migratory pathway.

36.9 Most of the differentiation of the embryo, in which the initial structure formation occurs, happens during the first trimester; the second and third trimesters are primarily times of growth and organ maturation, rather than the actual development and differentiation of structures. Thus, teratogens are most potent during this time of rapid organogenesis.

END-OF-CHAPTER QUESTIONS

Understand

1. a 2. d 3. c 4. d
5. a 6. c 7. b 8. d
9. d 10. c 11. c

Apply

1. b 2. a 3. b 4. a
5. These could be used to start from some other adult tissue to first reprogram this tissue, then induce the pluripotent stem cells to become the tissue of interest. The new tissue could be used to replace the damaged tissue. 6. a 7. This is the rupture of the amnion, the fluid-filled sac in which the fetus has been developing, prior to labor and delivery of the fetus. 8. c 9. d

Chapter 37

IN-CHAPTER REVIEW OF CONCEPTS

37.1 You would create a colony of fruit flies and then take the ones that spent the most time on the ground and allow them to breed with each other to form the next generation. Do that generation after generation, and most likely you'd end up with flies that very rarely actually flew.

37.2 If identical twins have different experiences growing up, those differences may shape how they behave differently as adults, despite the fact that they are genetically identical.

37.3 Species have different biological needs and capabilities. There's no way a raven could have solved the banana problem because ravens do not have arms to pick up large objects. On the other hand, a chimpanzee could have been tested with the string apparatus given to the raven.

37.4 Butterflies and birds have extremely different anatomy and physiology and thus most likely use very different navigation systems. Birds generally migrate bidirectionally, moving south during the cold months and back north during the warmer months. Usually, then, migrations are multigenerational events and it could be argued that younger birds can learn migratory routes from older generations. Butterflies, on the other hand, fly south to breed and die. Their offspring must then fly north, having never been there before.

37.5 In addition to chemical reproductive barriers, many species also employ behavioral and morphological reproductive barriers, such that even if a female moth is attracted by the pheromones of a male of another species, the two may be behaviorally or anatomically incompatible.

37.6 The trait might not be heritable. Perhaps attraction to bananas is determined entirely by where the egg is laid. In that case, natural selection on the trait would not lead to evolutionary change.

37.7 The benefits of territorial behavior must outweigh the potential costs, which may include physical danger due to conflict, energy expenditure, and the loss of foraging or mating time. In the case of a flower that is infrequently encountered, the honeycreeper would lose more energy defending the resource than it could gain by utilizing the resource. On the other hand, when resources are abundant, there is no point in wasting energy defending some flowers when there are enough for all the birds to get what they need.

37.8 Alternative mating strategies may be favored in situations where males are defending large territories and leave the female(s) unattended. This opens up opportunities for other males to have access to the females. In other situations a female may choose to mate with more males to increase her reproductive success.

37.9 Through long-term behavioral and genetic studies, you could find out whether older monkeys are rescuing closely related infants (as would be expected by kin selection) or the young of adults that had helped them (as expected by reciprocal altruism).

37.10 Disadvantages of living in groups include the amount of food needed to feed so many animals and the possibility that large groups are more likely to attract the attention of predators. Weaver birds living in savannas occur in places where predators can see them regardless and where food is abundant; thus the disadvantages do not apply. In contrast, in the forest, both food scarcity and increased likelihood of attracting predators would be major problems for groups.

END-OF-CHAPTER QUESTIONS

Understand

1. c 2. a 3. c 4. c
5. d 6. a 7. d 8. c
9. b 10. d 11. d 12. b
13. c 14. a

Apply

1. Selection for learning ability would cease, and thus there would be a change from one generation to the next in maze learning ability; it would result only from random genetic drift.
2. c 3. d 4. c 5. a
6. c 7. d 8. c 9. d
10. b 11. a 12. b

Chapter 38

IN-CHAPTER REVIEW OF CONCEPTS

38.1 If the populations in question comprised source-sink metapopulations, then the lack of immigration into the sink populations would, most likely, eventually result in the extinction of sink populations. As a result, the range of the entire species would shrink to only those populations that could sustain their population size without the input of immigrants.

38.2 The shape of the environment curve depends on the relative survival of individuals at different ages. In one environment, survival of young might be particularly low and that of adults much higher due to the predators and resources available, whereas the situation might be reversed in another habitat. Such differences would lead to differently shaped survivorship curves.

38.3 A species with high levels of predation would likely exhibit an earlier age at first reproduction and shorter inter-birth intervals in order to maximize its fitness under the selective pressure of the predation. On the other hand, species with few predators have the luxury of waiting until they are more mature before reproducing and can increase the inter-birth interval (and thus invest more in each offspring) because their risk of early mortality is decreased.

38.4 Many different factors might affect the carrying capacity of a population. For example, climate changes, even on a relatively small scale, could have large effects on carrying capacity by altering the available water and vegetation, as well as the length of the growing season and distribution of the vegetation. Regardless of the type of change in the environment, however, most populations will move toward carrying capacity; thus, if the carrying capacity is lowered, the population should decrease, and if the carrying capacity is raised, the population should increase.

38.5 Yes. Negative density-dependence occurs when factors tend to lower growth when populations are high, as might occur due to a limited amount of food available. On the other hand, positive density-dependent factors are those that increase population growth rate at high population levels. The Allee effect is one example: it may be easier to find mates when the population is large. There is no reason both effects can't be operating simultaneously. Whether a population increases or decreases would depend on the relative magnitude of the contrasting factors.

38.6 Both are extremely important if the human population is not to exceed the Earth's carrying capacity. Carrying capacity results from how many resources are available and how many resources are used by each individual—the greater the use per individual, the lower the carrying capacity. As population growth increases, the human population approaches the planet's carrying capacity; as consumption increases, the carrying capacity is lowered.

38.7 Quarantining would reduce the number of infections produced by each infected individual. This lower R_0 would lead to selective pressure for the pathogen to have weaker effects or even become asymptomatic because such people wouldn't appear to be infected and wouldn't be quarantined. If a mutation appeared with such an effect, it would be favored and would increase in frequency, diminishing the effects on the infected individual. Any variant that increased virulence in an infected individual in quarantine would also not be passed on to others. Such variants would be selected for and rapidly increase in frequency in the population.

END-OF-CHAPTER QUESTIONS

Understand

1. b 2. b 3. a 4. c
5. a 6. b 7. d 8. c
9. c 10. c 11. a 12. 84

Apply

1. c 2. d 3. b 4. a
5. c 6. b 7. d 8. c
9. d 10. d

Chapter 39

IN-CHAPTER REVIEW OF CONCEPTS

39.1 It depends upon whether we are talking about fundamental niche or realized niche. Two species can certainly have identical fundamental niches and coexist indefinitely because they can develop different realized niches within the fundamental niche. In order for two species with identical realized niches to coexist indefinitely, the resources within the niche must not be limited.

39.2 This is an example of Batesian mimicry, in which a nonpoisonous species evolves coloration similar to that of a poisonous species.

39.3 In an ecosystem with limited resources and multiple prey species, one prey species could outcompete another to extinction in the absence of a predator. In the presence of the predator, however, the prey species that would otherwise be driven to extinction by competitive exclusion is able to persist in the community if the predator preys preferentially on the competitively dominant species. The predators that lower the likelihood of competitive exclusion are known as keystone predators.

39.4 Selective harvesting of individual trees would be preferable from a community point of view. Clear-cutting would reset the community to an early stage of succession, whereas selective harvesting would retain the community at the same stage of the succession sequence.

39.5 Yes, fertilization with natural materials such as manure is less disruptive to the ecosystem than is chemical fertilization. Many chemical fertilizers, for example, contain higher levels of phosphates than does manure, and thus chemical fertilization has disrupted the natural global phosphorus cycle.

39.6 Both matter and energy flow through ecosystems by changing form, but neither can be created nor destroyed. Both matter and energy also flow through the trophic levels within an ecosystem. The flow of matter such as carbon atoms is more complex and multileveled than is energy flow, largely because it is truly a cycle. The atoms in the carbon cycle truly cycle through the ecosystem, with no clear beginning or end. The carbon is changed during the process of cycling from a solid to a gaseous state and back again. On the other hand, energy flow is unidirectional. The ultimate source of the energy in an ecosystem is the Sun. The solar energy is captured by the primary producers at the first trophic level and is changed in form from solar to chemical energy. The chemical energy is transferred from one trophic level to another, until only heat, low-quality energy, remains.

39.7 It depends on whether the amount of sunlight captured by the primary producers was affected. Currently, only approximately 1% of the solar energy in Earth's atmosphere is captured by primary producers for photosynthesis. If less sunlight reached Earth's surface, but a correlating increase in energy capture accompanied the decrease in sunlight, then the primary productivity should not be affected.

END-OF-CHAPTER QUESTIONS

Understand

1. c 2. d 3. c 4. c
5. d 6. a 7. b 8. b
9. b

Apply

1. b 2. d 3. b 4. b
5. b 6. d 7. d 8. d
9. d 10. d

Chapter 40

IN-CHAPTER REVIEW OF CONCEPTS

40.1 If the Earth rotated in the opposite direction, the Coriolis effect would be reversed. In other words, winds descending between 30° north or 30° south and the equator would still be moving more slowly than the underlying surface, so they would be deflected; however, they would be deflected to the left in the northern hemisphere and to the right in the southern hemisphere. The pattern would be reversed between 30° and 60° because the winds would be moving more rapidly than the underlying surface and would thus be deflected again in the opposite directions from normal—to the left in the northern hemisphere and to the right in the southern hemisphere. All of this would result in Trade Winds that blew from west to east and "Westerlies" that were actually "Easterlies," blowing east to west.

40.2 As with elevation, latitude is a primary determinant of climate and precipitation, which together largely determine the vegetational structure of a particular area, which in turn defines biomes.

40.3 Eutrophic lakes are high in algal nutrients and low in dissolved oxygen; when oligotrophic lakes, which are high in dissolved oxygen and low in algal nutrients, are polluted with fertilizers, they artificially receive nitrates and phosphates from the fertilizers, which transforms them into eutrophic lakes.

40.4 If iron were added to the portion of the ocean containing producers that are iron-limited (the great Southern Ocean surrounding Antarctica, parts of the equatorial Pacific Ocean, and parts of the subarctic, northeast Pacific Ocean), it might increase production in these regions. This might increase the overall productivity per unit area of the oceans.

40.5 Point-source pollutants are, in general, easier to combat because their sources are more easily identified and thus the pollutants are more easily eliminated.

40.6 While climate change and ozone depletion are both global environmental concerns due to the impact each has on human health, the environment, economics, and politics, there are some different approaches to combating and understanding each dilemma. Ozone depletion results in an increase in the ultraviolet radiation reaching the Earth's surface. Global climate change, on the other hand, results in long-term changes in sea level, ice flow, and storm activity.

END-OF-CHAPTER QUESTIONS

Understand

1. d 2. a 3. b 4. c
5. a 6. d 7. b 8. a
9. c 10. d 11. b 12. c

Apply

1. d 2. b 3. c 4. b
5. b 6. b 7. c 8. a
9. c 10. c 11. b 12. b

Index

Note: Page numbers in **bold** indicate definitions. An italic *f* indicates figures, *t* indicates tables.

A

ABA. *See* Abscisic acid (ABA)
Abdominopelvic cavity, 737, 737*f*
ABO blood groups, 247–248, **248,** 248*f*
ABP1 (auxin-binding protein), 721
Abscisic acid (ABA), **715**, 715*f*
  dormancy, 724*f*
  plant growth and, 723–724, 724*f*
  plant hormone, 714
  signaling pathway, 715
Absolute dating, 437
Absorption, in digestion, 790
Absorption spectrum, **161**
  of chlorophyll, 161–162, 161*f*
  photons efficiency, 161
Abstinence, as birth control method, 870
Acacia, mutualism of, 954, 955*f*
Acanthamoeba, 510
Acanthodians, 629
Acceptor stem, **313,** 314*f*
Accessory factors, for termination of translation, 317–318, 318*f*, 319*f*, 320*t*
Accessory pigments, **161**
  carotenoids as, 162–163, 163*f*
  phycobiliproteins, 163
Acer saccharum, seed dispersal and, 922*f*
Acetylcholine (ACh)
  muscle contraction and, 752
  as neurotransmitter, 767, 767*f*, 768*f*
Acetylcholinesterase (AChE), 767
Acetyl-CoA
  in citric acid cycle, 139, 140*f*
  β oxidation of fatty acids and, 149–150
  roles of, 151
Acetyl groups
  citric acid cycle oxidation of, 140*f*, 142
  production of, by β oxidation of fatty acids, 149–150, 151*f*
  pyruvate conversion to, 139, 139*f*
ACh. *See* Acetylcholine (ACh)
AChE (acetylcholinesterase), 767
Acid growth hypothesis, 721
Acids, **35**. *See also* Amino acids
  ABA (*See* Abscisic acid (ABA))
  hemorrhagic *E. coli* and, 111
Acini, **794**
Acoela, 600*f*
Acoelomorpha, 600*f*
Aconitase, 345
Acquired immunity, 938
Acquired immunodeficiency syndrome (AIDS), 506, 508, 509*t*, 511, 513, 516, 533, 556, 889
Acrosomal process, 872
Acrosome, **871**
ACTH (adrenocorticotropic hormone), **832**, 833
Actias luna (Luna moth), 616*f*
Actin, **81**
Actin filaments, 72*f*, 73*f*, **81**
  in cytokinesis, 209, 209*f*
  in thin filaments, 749, 750, 750*f*, 751
Actinistia, 628*f*, 631
Actinobacteria, 503*f*
Actinopoda, 535, 535*f*
Actinopterygii, 631
Action potentials, **764**
  changes in, 764, 765*f*
  nature of, 764
  propagation of, 764–766, 766*f*
Action spectrum, **162**
  determination of, 162*f*
  for photosynthesis, 162, 162*f*
Activating enzymes, **313**–314, 314*f*
Activation energy, **119**, 119*f*
  catalysis and, 121, 121*f*
  destabilization of chemical bonds by, 121, 121*f*
  enzyme lowering of, 121–122
Activators
  of enzymes, 124–125
  specific transcription factors as, 336
  of transcription, 332, 338*f*
Active sites, **122**–123, 122*f*
Active transport, **105**–108
  carrier proteins and, 106–107, 106*f*, 110*t*
  countertransport and, 107–108
  coupled, 107, 107*f*, 110*t*
Acute-phase proteins, 848
Adaptations
  life history, resource availability and, 932–933, 933*t*
  of seeds, 692–693, 694*f*
  speciation caused by, 459–460, 459*f*
  of terrestrial animals, 632–633, 632*f*
  of terrestrial plants, 570–571
Adapter proteins, **187**
Adaptive immunity, 848–850
Adaptive radiation, **432**, **462**
  character displacement and, 462–463, 463*f*
  documentation of, 463–467
  key innovations leading to, 462, 463*f*
Addition, rule of, **243**, 244
Adelina, flour beetles and, 956
Adenine, 56, 56*f*, 279, 281, 283
Adenohypophysis, **831**, 833*f*, 836*f*
Adenosine diphosphate (ADP), 57, 120
Adenosine monophosphate (AMP), 120
Adenosine triphosphate (ATP), **57**, 57*f*
  CO_2 incorporation into organic molecules and, 170–172
  energy storage by, 119–120
  hydrolysis of, 120–121, 120*f*
  muscle contraction and, 749–751, 751*f*
  protein kinase and, 182
  structure of, 119–121, 120*f*
  synthesis of (*See* ATP synthesis)
Adenylyl cyclase, **190**, 191*f*
ADH (antidiuretic hormone), 831, 831*f*, 832, 843, 844–845, 844*f*
Adherens junctions, **87**–88, 87*f*, 87*t*
Adhesive junctions, **87**–88, 87*f*, 87*t*, 88*f*
Adhesiveness, of water, 31–32, 32*f*, 32*t*
Adipose cells, **741**, 741*f*
Adipose tissue, **741**, 741*f*
ADP (adenosine diphosphate), 57, 120
Adrenal cortex, 836–837
Adrenal glands, **836**–837, 836*f*
Adrenaline, 181
Adrenal medulla, 836
Adrenocorticotropic hormone (ACTH), **832**, 833
Adventitious plantlets, **698**, 698*f*
Adventitious roots, **668**, 669*f*
Aegopodium podagraria (ground elder), 261*f*
Aepyceros melampus (impala), 954*f*
Aerial roots, **668**, 669*f*
Aerobic respiration, **133**, 135*f*, 138
  energy yield of, 146–147
  feedback inhibition and, 147, 147*f*
  in mitochondria, 144, 145*f*
Aeropyrum pernix, 396
Afferent arterioles, of nephron, 840
Afferent neurons, 774, 778
Aflatoxins, 554, 554*f*
African black-bellied seedcracker finch *(Pyrenestes ostrinus),* 414, 415*f*
African blood lily *(Haemanthus katharinae),* 206*f*
African sleeping sickness, 528, 528*f*
African weaver birds, social groups formed by, 913
Afrovenator, 635*f*
Afterbirth, 887
Agalychnis callidryas (red-eyed tree frog), 632*f*
Age
  evolutionary, tropical biodiversity and, 972
  fecundity and, 928
  at first reproduction, life span and, 928
  population pyramids and, 934, 935*f*
  telomerase and, 292
Agelaius phoeniceus (red-wing blackbird), 908*f*
Age structure, **924**–926, 924*f*
Aggregate fruits, 694*f*
Ago (Argonaute), 342–343
Agriculture
  artificial selection in, 435–436, 435*f*
  fertilizer use in, 964
  genetic engineering and, 370–373
  genome projects and, 386–387
  global temperature change and, 993–995
  no-till planting, 372
Agrobacterium tumefaciens, 366, 366*f*
Agrostis tenuis, copper tolerance in, 419, 419*f*
AIDS (acquired immunodeficiency syndrome), 506, 508, 509*t*, 511, 513, 516, 533, 556, 889
Air, composition of, 800
Alanine, 49*f*
Alarm calls, 901, 901*f*
Albinism, pedigree of, 267, 267*f*
Alders, succession and, 958
Aldosterone, 836–837, 843, 845
Aleurone, **695**, 695*f*
Alfalfa, glyphosate-resistant, 372
Algae. *See also* Brown algae; Green algae; Red algae (Rhodophyta)
  biofuel manufacture and, 371, 371*f*
  in freshwater habitats, 986–987
  plant evolution from, 569, 569*f*
Algal blooms, iron dump hypothesis and, 175
Alimentary canal. *See* Digestive system
Alkaptonuria, 301
Allantois, **634**, 634*f*, 881
Allee effect, **931**
Allee, Warder, 931
Allele frequencies, **407**
  gene flow and, 409, 409*f*
  genetic drift and, 410–411, 410*f*
  Hardy-Weinberg equation and, 407–408, 407*f*
  mutation and, 409, 409*f*
  natural selection and, 425
  nonrandom mating and, 409*f*, 410
  selection and (*See* Natural selection; Selection)
Alleles, **240**
  codominant, 247, 248*f*
  multiple, genes with, 248*f*
  pleiotropic, 246, 246*f*
Allelic exclusion, 853
Alligator, 637
Allolactose, 332–334
Allometric growth, **888**
Allomyces, 558*f*
Allomyces arbuscula, 555*f*
Allopatric species, 460, 460*f*
Allopolyploidy, **461**, 462*f*
Allosteric activators, **125**
Allosteric enzymes, 125
Allosteric inhibitor, **124**, 125*f*, 126*f*
Allosteric sites, **124**, 125*f*
Alpha (α) helix, **50**–51, 50*f*
Alpha (α)-ketoglutarate, 141*f*, 142

Alpha (α)-tubulin, 81
Alpine buttercup, adaptive radiation and, 465–467, 467f
Alternation of generations, in plants, 570–571, 570f, 682
Alternative splicing, **313**–314, 344f, 388
  of RNA, 394
Altitude, terrestrial ecosystems and, 983
Altricial offspring, **907**
Altruism, **909**
  kin selection theory and, 909–911, 910f
  reciprocal, 909
Alveolata, 487, 488f, **529**, 531–535, 531f, 532f
  bulk transport in, 108–110, 110t
  eukaryotic, 71–73, 73f
Alveoli, **800**, 801, 801f
Amacrine cells, 782, 782f
Amanita muscaria, 555f
Amanita phalloides, 561f
Ambulocetus, 438f
Ambystoma tigrinum (tiger salamander), 632f
American basswood, dormancy in, 724f
American crocodile *(Crocodylus acutus),* 637
American woodcock *(Scolopax minor),* 784
Amino acid derivatives, as hormones, 828
Amino acids, 47–**48**
  attachment to tRNA, 313, 314f
  as neurotransmitters, 766–767, 767f
  protein sequence of, 47–48, 50, 50f
  structure of, 48
Aminoacyl site (A site), **315**, 315f
Aminoacyl-tRNA synthetases, **313**–315, 314f
γ-Aminobutyric acid (GABA), 767
Amino groups, 41–42
  protein catabolism and removal of, 149, 150f
Amish people and founder effect, 411
Ammonia oxidation, 497
Amniocentesis, **270**, 271f
Amnion, **634**, 634f
Amniotes, **634**, 634f
  Synapsids and Diapsids, 634–635, 634f
Amniotic eggs, reptile, 634–635, 634f
Amniotic membrane (amnion), 881
Amoeba proteus, 539f
Amoebas
  cyst formation by, 523–524
  pseudopodia of, 523
Amoebozoa, **525**, 539–541, 539f, 540f
AMP (adenosine monophosphate), 120
Amphibians. *See also specific amphibians*
  chytridiomycosis in, 554, 554f
  circulatory systems of, 810–811, 810f
  cloning in, 875–876
  distinguishing features of, 632, 632f
  diversification of, 632–633
  egg activation in, 872
  forebrain of, 769, 770f
  lateral line systems of, 775–776, 775f
  lymphatic systems of, 815–816
  modern groups of, 633
  respiration in, 800
  species-specific acoustic signals in, 901
  sperm binding in, 874f
  sperm penetration in, 873
  thyroid hormones in, 834–835
Amplification, 359
Ampulla, echinoderm, **618**, 618f
Amylase, 127
α-Amylase, in seedlings, 695
Amylase, pancreatic, **794**
Amylopectin, 45, 46f
Amyloplasts, **80**, **665**, 729
Amylose, 45, 46f
Anabolism, **125**
Anabrus simplex (Mormon cricket), parental investment in, 422, 422f
Anaerobic respiration, **133**, 148, 148f
Analogous structures, **13**
Anaphase II, meiotic, 227f, 228, 229f
Anaphase I, meiotic, 225, 226f, 228, 228f
Anaphase, mitotic, 207f, **208**, 229f
Anaphase-promoting factor (APC), 212f, **213**
Anatomy, comparative, evolution and, 13, 13f
Ancestral characters, 476, 477f, 483
Andrias, 634
Androecium, **584**, 681, 682f
Aneuploidy, **265**, 368
Angelman syndrome (AS), 259
Angiosperm reproduction, pollination and. *See* Pollination
  asexual, 697–699, 698f, 699f
  double fertilization in, 687–688, 687f
  embryo development and, 688–689, 688f
  fruits and, 692–693, 693f, 694f
  gamete production, 682–683, 683f
  germination and, 587, 692, 695, 696f
  outcrossing and, 686–687, 686f, 687f
  regulation of flowering in, 681
  seeds and, 586f, 587, 690–693, 691f, 692f
  self-incompatibility and, 687
Angiosperms, 574f, 578t, **584**
  colonization by, 693, 694f
  distinctive characteristics of, 584, 584t
  flowers of, 584–585, 585f
  life cycle of, 585–587, 586f, 681f, 682
  life spans of, 697, 697f
  pollen formation and, 682–683, 683f
  reproduction in (*See* Angiosperm reproduction; Pollination)
  structure of, 682, 682f
Angle of incidence, 980
Animal cells, 71–73, 72f
  cytokinesis in, 209, 209f
  extracellular matrix of, 85–86, 86f
  gap junctions in, 88f, 89
  tight junctions in, 89
Animalia, 3f, 483f–485f, 485–486
Animal pole, **879**, 879f
Animals, body cavities of. *See also* specific animals and classes of animals
  adaptations defending against predators, 950–951, 951f
  artificial selection and, 435–436, 436f
  body cavities of (*See* Body cavities)
  body symmetry evolution in, 594–596, 596f
  carnivorous (*See* Carnivores)
  cell division in, 201f
  communication in, 900–901
  diversity of, 593–620, 595f–598f, 599t, 600f
  embryonic development of, 440–441, 441f, 594, 595f, 597–598, 598f
  evolutionary innovations of, 594–601, 595f, 596f, 597f, 598f, 599t, 600f
  fungal diseases of, 554–555, 554f
  fungal symbioses with, 553, 553f
  hybridization of, 456, 457
  invertebrate (*See* Invertebrates)
  knockin, 365
  knockout, 364–365, 365f
  locomotion of, 753–755
  phylogeny of, 599t, 600f–601f, 601–602
  pollination by, 683–688, 684f, 685f, 953f, 954
  segmentation in, 598–600
  sexual life cycle in, 222–223, 223f
  shared features of, 594, 595f
  spacing of individuals in populations and, 921
  succession in communities of, 958–959, 958f
  tissue evolution in, 595, 603–605, 604f, 605f
  transgenesis and, 364–366
  vertebrate (*See* Vertebrates)
Animal viruses, 511–515, 512f, 513f
Anions, **23**
Annelida, 599t, 600f, 602, **609**, 609f, 611f
Annelids, 602
  circulatory systems of, 808–809
  eyes of, 780
Annotation, genome, 388–390
Annual plants, 697, 697f
Anolis lizards
  adaptive change in, 459–460, 459f
  parasitism and competition and, 956
  resource partitioning among, 948f
Anopheles (mosquitoes). *See* Mosquitoes *(Anopheles)*
Anoxygenic photosynthesis, 157, 166, 496
Antagonistic effectors, 823, 823f
Antarctica, ozone hole over, 990
Antenna complex, **164**, 164f, 167–168
Anterior pituitary, **831**–833
Antheridia, **572**, 572f
Anthers, **584**, 585f, **682**, 682f
Anthocerotophyta, 573, 573f
Anthocyanin, 249
Anthophyta, 578t
Anthrax, 507t
Anthropoids, 644, 644f
Antibiotic resistance
  gene chips and, 368–369
  selection for, 412
  transfer of, 501–502, 504
Antibiotics, 70, 501
Antibodies, **850**. *See also* Immunoglobulins (Igs)
Anticodon loop, **313,** 314f
Antidiuretic hormone (ADH), 831, 831f, 843, 844–845, 844f
Antigen-binding sites, of antibodies, 850f, 852, 853
Antigenic determinants, 849, 849f, 850f
Antigenic drift, **513**
Antigenic shift, **514**
Antigen-presenting cells, **850**
Antigens, recognition of, **849**, 849f
  by antibody proteins, 852
  by lymphocytes, 850
Antihistamines, 179
Antioxidants, carotenoids as, 163
Antiparallel configuration, 282f, 283
Antiporters, **106**
Antisense strand, 302
Ants
  Crematogaster nigriceps, 955
  eusociality in, 910
  flatworm parasites of, 956, 956f
  fungal symbiosis with, 553, 553f
  leaf-cutter, fungal symbiosis with, 553, 553f, 953
  Pseudomyrmex, mutualism of, 954, 955f
Anura, 632f, 633
Anus, 790–791, 791f, 796
Aorta, 810
Aortic body, **803**
Aortic valve, **811**, 812f
APC (anaphase-promoting factor), **213**
APC/C (cyclosomes), 213
Apertures, 587
Apes, 644–645, 645f. *See also* Chimpanzee *(Pan troglodytes);* Gorilla *(Gorilla gorilla)*
Aphids, translocation in plants and, 717, 717f
Apical complex, 531f, 532–533, 533f
Apical meristems, **657**, 658, 658f, 659f, 665, 669, 669f
Apical surface, of epithelium, 738
Apices, **656**, 656f
Apicomplexans, 531–533, 531f, 533f
Apis mellifera (European honeybee), taxonomic classification of, 483, 902f
Aplysina longissima, 604f
Apoda, 633, 634f
Apolipoprotein B, 344
Apomixis, 697–**698**
Apoplast route, **709**, 710f
Apoprotein, 724
Apoptosis, **329**
  cytotoxic T cells and, 850–851, 851f
  p53 gene and, 215
Appendicular locomotion, 753
Aquaporins, **104**, 110t, 844–845
  osmosis and, 708–709, 709f
Aqueous solutions, 103
Aquifers, **961**
Aquificae, 502f

Arabidopsis, 573
aquaporin-like proteins in, 708
auxin transport in, 721
gravitropism in, 728, 729*f*
light response in, 724–725, 725*f*
roots of, 665
tmm mutation in, 660, 660*f*
Arabidopsis thaliana, 387*f*
Arber, Werner, 355
Arbuscular mycorrhizae, **552**, 552*f*, **553**, 553*f*, 560
Arceuthobium (mistletoe), dispersal of seeds of, 921
Archaea, 2, 2*f*, 70, 485–487, 485*f*, **486**. *See also* Prokaryotes
anaerobic respiration in, 148, 148*f*
bacteria compared with, 497, 497*f*
cell structure of, 496–497, 497*f*
characteristics of, 486–487, 487*t*
DNA replication in, 291
domain of, 485*f*, 486–487, 487*t*, 495*f*, 496–497, 497*f*
environments of, 486–487, 496
genetics of, 497
metabolism of, 497
Archaebacteria. *See* Archaea
Archaeopteryx, 437–438, 437*f*, 638, 639*f*
Archaeplastida, **525**, 526*f*, 536–539, 536*f*, 539*f*
Archegonia, **572**, 572*f*
Archenteron, **597**
Archosaurs, 481*f*, 635, 635*f*
Arcyria, 540*f*
ARFs (auxin response factors), 721
Arginine, 49*f*
Arginine vasotocin, 832
arg⁻ mutants, 301*f*, 302
Argonaute (Ago), 342–343, 343*f*
Ariolimax columbianus (banana slug), 608*f*
Aristotle, 482
Armored fish, 629–630
Arnold, William, 163–164
Aromatic amino acids, 48, 372
Arrowgrass *(Triglochin maritima)*, 588
Arrow worms, 599*t*
Artemisinin, 532
Arteries, 810, **814**, 815*f*
pulmonary, 809, 809*f*
Arterioles, **814**, 815*f*
Arthropoda, 599*t*, 601*f*, 602, 605*f*
classes of, 613*f*, 613*t*, 615–616, 617*f*
success of, 613–614, 613*t*, 614*f*
Arthropods, 599*t*, **614**
body of, 614–615, 614*f*
circulatory systems of, 808, 808*f*
classes of, 613*t*, 615–616, 615*f*, 616*f*
diversity of, 594, 615
eusociality among, 911
exoskeletons of, 745, 745*f*
eyes of, 780, 780*f*
locomotion by, 753–755
success of, 613–614, 613*t*, 614*f*
taste receptors of, 778
Artificial chromosomes, 358
Artificial selection, **12**, 411, 434–440, 436*f*, 448
behavioral evolution and, 903
behavioral genetics and, 894, 894*f*
Artificial transformation, 503
AS (Angelman syndrome), 259–260
Ascaris
chromosomes of, 222
sperm binding in, 874*f*
Asclepias syriaca, seed dispersal and, 922*f*
Ascocarp, 561, 562*f*
Ascomycetes, 551, 555
Ascomycota, 555, 555*f*, 555*t*, **560**–563, 562*f*
Ascospores, 561
Ascus, 561
Asexual reproduction, 223
in flowering plants, 697–699, 698*f*, 699*f*
in fungi, 549–550, 559–561
in protists, 525
vegetative, 672, 672*f*, 698, 698*f*
in yeasts, 562, 562*f*
Asian flu, 514
A site (aminoacyl site), **315**, 315*f*, 317, 317*f*, 318*f*
Asparagine, 49*f*
Aspartic acid, 49*f*
Aspergillus flavus, 554*f*
Asplenium rhizophyllum (walking fern), leaves of, 675
Assemblages, **945**
Assembly, **510**
in lytic cycle, 511, 512*f*
Association cortex, **771**
Association neurons, 740*f*, 761, 761*f*, 771
Assortative mating, **410**
Asteroidea, 617*f*, 618*f*
Asters, **206**, 208*f*
Atlantic green turtles *(Chelonia mydas)*, 636
Atmosphere, global circulation patterns in, 980–981, 981*f*
Atomic mass, 22–23
Atomic number, 22
Atoms, 4*f*, **5**, **22–25**
Bohr's view of structure, 22, 22*f*
chemical behavior of, electrons and, 24, 24*f*
composition of, 22–25
electronegativity of, 29, 29*t*
energy levels contained by, 24–25, 25*f*
neutral, 22
ATP. *See* Adenosine triphosphate (ATP)
ATP synthase, **134**, 144–145, 145*f*, 165, 168, 169*f*
ATP synthesis, 134–135, 135*f*
chemiosmosis and, 144–145, 144*f*, 145*f*
by glycolysis, **134**, 136, 136*f*, 137*f*, 138
mitochondrial, 80, 143–144, 145*f*
photosynthetic, 158, 166
ATP synthetase, 260
Atrioventricular (AV) node, **811**, 813*f*
Atrioventricular (AV) valves, **811**, 812*f*
Atrium, **809**, 809*f*
Attachment, 512–513
of HIV, 512–513
in lytic cycle, 510, 511*f*
SARS-CoV-2 infection, 515*f*
Auditory cortex, 770
Australia, convergent evolution and, 444, 445*f*
Australopithecines, 645–646, 646*f*
Australopithecus, 645–646, 646*f*
Australopithecus afarensis, 646*f*, 647
Australopithecus boisei, 646*f*
Australopithecus robustus, 646*f*
Autocrine signaling, 180
Autoimmune diseases, 369
Autonomic nervous system, **761**, 761*f*, 762*f*, 772–773, 772*f*
Autopolyploidy, **461**
Autosomes, **266**
Autotrophs, **132**, 504–505, 524, **965**
Auxin, 665, **720**
actions of, 721–722, 721*f*, 722*f*
discovery of, 720, 720*f*, 721*f*
effects in plants, 720–721
ethylene and, 723
plant growth and, 728
synthetic, 722
Auxin response factors (ARFs), 721
Avery, Oswald, 277
Aves, **637**. *See also* Birds
Avian cholera, 849
Avicennia germinans (black mangrove), roots of, 716, 716*f*
AV (atrioventricular) node, **811**, 813*f*
AV (atrioventricular) valves, **811**, 812*f*
Axial locomotion, 753
Axillary buds, 656, **669**, 670*f*
Axils, **672**
Axons, 740, 740*f*, 761, **762**, 762*f*
action potential propagation along, 764–766, 765*f*
myelinated, 765, 766*f*
Axopodia, of protists, 523

B

Bacillus, 497, 498*f*, 502*f*
Bacillus anthracis, 507*t*
Bacillus thuringiensis (Bt toxin), 373
Bacon, Francis, 7
Bacteria, 2, 2*f*, 485, **486**. *See also* Prokaryotes; specific bacteria
antibiotic resistance and, 412, 501–502
archaea compared with, 496–497, 497*f*
bioluminescent, 989*f*
cell division in, 199–200, 199*f*, 200*f*
cell wall of, 69, 497–498, 498*f*
characteristics of, 486, 487*t*
CRISPR/Cas9 system and, 363
in digestive tract, 793, 796
diseases caused by, 70, 505–506, 507*t*
DNA replication in, 286–290, 286*f*–290*f*, 289*t*
domain of, 486, 487*t*, 495–496, 495*f*, 496*f*
endosymbiotic, 521, 522*f*
flagella of, 496–498, 499*f*
gram-negative, 498–499, 499*f*
gram-positive, 498, 498*f*, 499, 499*f*, 502*f*
green sulfur, 166
human protein production in, 367
interior organization of, 500, 500*f*
photosynthetic, 495, 503*f*, 522–523, 522*f*
photosystem of, 166, 166*f*
purple photosynthetic, 165
purple sulfur, 160
pyrogens and, 826
shapes of, 497
symbiosis with legumes, 666
transformation in, 277, 277*f*, 358–359, 366, 366*f*
Bacteriophages, 278*f*, **279**
lysogenic cycle of, 10, 512*f*
lytic cycle of, 510, 511*f*
structure of, 508*f*
transducing, 502
Baculum, 865
Ball-and-socket joints, **748**
Banana slug *(Ariolimax columbianus)*, 608*f*
Banting, Frederick, 837
Bark, outer, 671, 671*f*
Barnacles
commensalism in, 953
interspecific competition, 945
Barometers, 800
Baroreceptors, 774–775
Barr bodies, **258**
Barrier contraceptive methods, 871
Bartholin's glands, 870
Basal body, **84**, 84*f*
Basal surface, of epithelium, 740
Basement membrane, **738**
Base-pairs, **57**, 57*f*, **281**–282, 282*f*, 283*f*
Bases, **35**
complementary, **57**
Base substitution mutations, **321**, 321*f*
Basic Local Alignment Search Tool (BLAST) algorithm, 388
Basidiomycota, 555, 555*f*, 555*t*, **560**–563
Basidium, 560, 561*f*
Basilar membrane, 776–777, 777*f*
Basophils, 807*f*
BAT (brown adipose tissue), 826
Bates, Henry, 952
Batesian mimicry, **952**, 952*f*
Batrachochytrium dendrobatidis, 554, 557, 564
Bats, 642, 642*f*
as pollinators, 685, 953*f*
vampire bats *(Desmodus rotundus)*, reciprocal altruism in, 909, 909*f*
Battus philenor (pipevine swallowtail butterfly), mimicry of, 952*f*
bc_1 complex, 143, 143*f*
B cells, **850**, 850*f*. *See also* Immunoglobulins (Igs)
Beadle, George, 7, 301–302
Beagle, H. M. S., 1
Beans, endosperm in, 691
Bears
polar, range of, 920
water, 599*t*, 600
winter sleep of, 827
Bees. *See also* Honeybees
"Bee's purple," 684, 685*f*
as pollinators, 685, 685*f*
Beetles, 616*f*
Behavior
altruistic, 909–910
cognitive, 897–898, 897*f*
communication and, 900–903
evolution of, 903, 904*f*
genomic influences on, 894–895
learning influences on, 896–897
migratory, 898–900, 898*f*–899*f*
natural selection and, 903
societies and, 912–913
Behavioral ecology, 904–905, **905**, 905*f*

Behavioral genetics, **894**
Behavioral isolation, 454–455, 454*t*, 455*f*
Belding's ground squirrels, kin selection in, 911
Beltian bodies, 954
Belt, Thomas, 954
Bergey's Manual of Systematic Bacteriology, 501, 503*f*
Berg, Paul, 355
Berries, 694*f*
Best, Charles, 837
Beta (β) carotene
 Golden Rice and, 373
 vitamin A and, 163
Beta (β)-galactosidase *(lacZ),* 332
Beta (β) oxidation, of fatty acids, 149–**150**, 151*f*
Beta prokaryotes, 503*f*
Beta (β) sheet, **50**, 50*f*, 51, 52
Beta (β) subunit, 288, 288*f*, 289, 289*f*, 291
Beta (β)-tubulin, 81
b_6f complex, 167*f*, **168**
Bicarbonate ion, carbon dioxide transport in blood as, 804–806, 806*f*
Bicuspid valves, **811**
Bidens frondosa, seed dispersal and, 922*f*
Biennial plants, **697**
Bighorn sheep, 903
Bilateral symmetry, **595**–596, 596*f*
Bilateria, 600*f*, 601, 603*f*, 605–607
Bile, 790, 793–794, 794*f*
Bile duct, 794, 794*f*
Bile pigments, 794, 794*f*
Bile salts, 794, 794*f*
Binary fission, **199**, 199*f*
Binocular vision, 644, **783**
 in primates, 644
Binomial expansion, 407
Binomials, **483**
Biochemical pathways, **125**–126, 125*f*
 feedback inhibition of, 126, 126*f*
Biocontrol services, 375
Biodiversity, 2, 6. *See also* Species richness
 of animal body plans, 594–600, 595*f*, 596*f*, 597*f*, 598*f*, 599*t*, 600*f*
 of birds, 639–640, 640*f*
 ecosystem stability and, 969–971
 evolution as source of, 16, 16*f*
 of insects, 593, 616, 616*f*
 of mollusks, 607–608
Biofilms, **496**
 bacterial, 506
Biofuels, 371, 371*f*
Biogenic amines, as neurotransmitters, 767
Biogeochemical cycles
 carbon and, 960, 961*f*
 nitrogen and, 962–963, 963*f*
 phosphorus and, 964–965, 964*f*
 water and, 961–962, 962*f*
Biogeography, **444**
 convergent evolution and, 444–445, 445*f*
 evolutionary divergence and, 445
Bioinformatics, 388, **394**
 proteomics and, 394
Biological communities, **5**
Biological models, 9
Biological species concept (BSC), **453**–457, 453*f*, 454*t*, 455*f*, 456*f*, 480, 482
Biologists, work of, 2
Biology, core concepts of, 14–16
Bioluminescent animals, in deep sea, 989, 989*f*
Biomarkers, 369, 394
Biomass, pyramid of, **969**, 970*f*
Biomes, **984**
 classification of, 984–986, 985*f*
 elevation and, 984*f*
 temperature and moisture and, 984–986, 986*f*
Bioremediation, 371
Biosphere, **5**, 5*f*, **935**
 future of, 935–936
Biotechnology. *See also* Cloning (of organisms); Genetic engineering; Recombinant DNA
 genetic engineering in industry and agriculture, 370–374
 genetic tools and medicine, 366–370
 molecular tools and genetic variation, 362–364
 polymerase chain reaction, 359–362
 recombinant DNA, 355–359
 transgenic organisms, 364–366
Biotic potential, **929**
Bipedalism, 646
Bipolar cells, **782**
Birds. *See also specific birds*
 altruism in, 909
 behavioral isolation and, 454–455, 455*f*
 body temperature of, 826, 856
 brain size of, 769
 cholera in, 849
 circulatory systems of, 810–811, 810*f*
 classification of, 480, 481*f*
 cognitive behavior of, 897, 897*f*
 DDT and, 989–990, 989*f*
 digestive system of, 790, 791
 dinosaurs and, 480, 481*f*, 638–640, 639*f*, 642
 diversity of, 640, 640*f*
 evolution of, 639–640, 640*f*
 extraembryonic membranes of, 881–882, 882*f*
 extra-pair copulations in, 907, 908*f*
 eyes of, 782
 flight of, 754*f*, 755
 flocking behavior in, 912*f*
 forebrain in, 769, 770*f*
 global temperature change and, 991, 992*f*
 key characteristics of, 638, 639*f*
 lymphatic systems of, 815–816
 migratory behavior of, 898–900, 898*f*, 899*f*
 modern, 639–640, 640*f*
 as pollinators, 684–685, 685*f*
 reproductive success of, 926–927, 927*f*
 respiration in, 801
 sex determination in, 258
 species-specific acoustic signals of, 901
 swimming by, 753
 temperature regulation in, 824
 territoriality of, 906, 906*f*
Birth
 control, 870
 hormone changes and, 887–888, 887*f*
 live, in vertebrates, 862, 862*f*
1,3-Bisphosphoglycerate (BPG), 136, 137*f*
Biston betularia (peppered moth), color of, 433, 433*f*, 434*f*
Bivalents, 223
Bivalves, 608, 609*f*
Blackburn, Elizabeth, 292
Black locust *(Robinia pseudoacacia),* leaves of, 673
Black mangrove *(Avicennia germinans),* roots of, 716, 716*f*
Black widow spider *(Latrodectus mactans),* 615*f*
Bladder, urinary, 839, 840*f*
Blades, of leaves, **673**
BLAST (Basic Local Alignment Search Tool) algorithm, 388
Blastocladiomycota, 555, 555*f*, 556, 557
Blastocoel, **879**
Blastocyst, 879, 880
Blastomeres, 879, 880
Blastopores, **597**
Blastula, 878–**879**
Blending inheritance, 407
Blood, 742. *See also* Circulatory systems
 carbon dioxide transport in, 804–806, 806*f*
 composition of, 807*f*
 formed elements in, 806, 807–808
 functions of, 806–807, 806*f*, 807*f*
 nutrient transport in, 794–795, 795*f*
 oxygen transport in, 804, 805*f*
 pH of, in swimming fish, 152
 plasma of, 806–807, 806*f*
 sickle-cell disease and, 268–269, 268*f*, 270*t*, 420–421, 421*f*
Blood cells. *See* Lymphocytes; Red blood cells (RBCs); T cells
Blood-clotting system, evolution of, 447
Blood disorders, 369
Blood groups, 247–248, 248*f*
Bloodlines, 237
Blood pressure, 812*f*, 813, 815–816, 831*f*
Blue-light receptors, in plants, 727
B lymphocytes, **850**, 850*f*
Boa constrictors, 442
Bobolink *(Dolichonyx oryzivorus),* migratory behavior of, 898–899, 899*f*
Body cavities
 evolution of, 595, 601–602
 phylogeny based on, 601–602
 of vertebrates, 737, 737*f*
Body forms
 of cnidarians, 604, 605*f*
 of mollusks, 608, 609*f*
 of Porifera, 603, 604*f*
Body plans
 of angiosperms, 688, 688*f*
 of annelids, 610, 610*f*
 diversity of, in animals, 594–600, 595*f*, 596*f*, 597*f*, 598*f*, 599*t*, 600*f*
 of vertebrates, 736–737, 736*f*
Body position detection, 775–778
Body size
 metabolic rate and, 676
 temperature regulation and, 824, 824*f*
Body temperature, regulation of. *See* Temperature regulation
Bohr effect (Bohr shift), **804**
Bohr, Niels, 22
Bollworm, cotton, 375
Bolus, **792**
Bombus (bumblebee), 684*f*
Bombykol, 901
Bombyx mori (silk moth), pheromone in, 901
Bonds, chemical. *See* Chemical bonds
Bone, 742, **745**–749. *See also* Skeletons
 compact, 748
 development of, 746*f*, 747
 intercellular, 747
 medullary, 746*f*
 movement of, 748, 749*f*
 remodeling of, 747–748, 748*f*
 spongy, 746
 structure of, 746
Bony fish, 490, 630–631, 631*f*, 838
Bormann, Herbert, 974
Borrelia burgdorferi, 507*t*
Bosmina longirostris (cladoceran), 930*f*
Bottleneck effect, 410*f*, 411
Bottle-nosed dolphin *(Tursiops truncatus),* 643*f*
Botulism, 507*t*
Bowman's capsule, **840**, 842, 843*f*
Boyer, Herbert, 355
Boysen-Jensen, Peter, 720
BPG (1,3-bisphosphoglycerate), 136, 137*f*, 171*f*
Brachiopoda, 599*t*, 600*f*, 602, **611**
Bracts, **675**
Brain
 of early australopithecines, 646
 of *Homo neanderthalensis,* 650
 of *Homo sapiens,* 648–649, 651
 human, 770–771, 770*f*
 size of, 769
 vertebrate, 769–770, 769*f*, 770*f*
Brainstem, 771
Branching diagrams, 475–476, 475*f*
Branchiostoma. See Lancelet *(Branchiostoma)*
Branch point, 312, 312*f*
Brassicaceae (mustard), chemical defense of, 950
Bread mold, 301, 558, 559*f*
Breast cancer and *HER2,* 368
Breathing. *See* Respiration; Respiratory systems
Brenner, Sydney, 303–304
Briggs, Robert, 875
Bright-field microscopes, 68*t*
Bristlecone pine *(Pinus longaeva),* 581
Bronchi, 801, 801*f*
Brown adipose tissue (BAT), 826
Brown algae, 529–530, 530*f*
 chloroplasts of, 522, 522*f*
 eukaryotic chloroplasts and, 488, 488*f*
 multicellularity of, 488–489, 524
Brownian movement, 102, 102*f*
Brown recluse spider *(Loxosceles reclusa),* 615*f*

Brown, Robert, 72, 73
Brush border, 793, 794
Bryophytes, **571**–573
adaptations for land living, 569–570
hornworts, 573, 573*f*
liverworts, 571, 571*f*
moss(es), 572–573, 572*f*, 573*f*
Bryozoa, 599*t*, 600*f*, 602, **611**, 611*f*, 620
BSC (biological species concept), **453**–457, 453*f*, 454*t*, 455*f*, 456*f*, 480, 482
Bt crops, 373
Bt toxin *(Bacillus thuringiensis),* 373
Bubulcus ibis (cattle egret), range of, 920, 921*f*
Budding
protist, 524, 525
viral, 513, 513*f*, 514
yeast, 562–563
Buffers, **35**, 35*f*
Buffon, Comte de (Georges-Louis Leclerc), 10
Bulbs, **672**, 672*f*, 698
Bulk transport, 108, 713
endocytosis and, 108–109, 108*f*, 110*t*
exocytosis and, 109, 109*f*, 110*t*
Bullfrogs, species-specific acoustic signals in, 901
Bumblebee *(Bombus),* 684*f*
Bupalus, population fluctuations in, 932*f*
Burr *(Cenchrus incertus),* 693
Butterflies, 594, 614*f*, 670, 684. *See also* Monarch butterfly
cabbage, coevolution of, 950, 950*f*
common sulfur *(Colias eurytheme),* 411
Glanville fritillary *(Melitaea cinxia),* metapopulation of, 923*f*
global temperature change and, 993, 993*f*
pipevine swallowtail *(Battus philenor),* mimicry of, 952*f*
tiger swallowtail *(Papilio glaucus),* mimicry in, 952*f*
Buttress roots, 668, 669*f*

C

Cabbage butterfly, coevolution of, 950
Cacti
leaves of, 672*f*, 675
prickly pear, population explosion of, in Australia, 949
Cactoblastis cactorum, prickly pear cactus population and, 949
Cactus finch *(Geospiza scandens),* 11*f*, 430*f*
Cadherin, 87–88, 87*f*, 87*t*
Caecilians, 632–633, 632*f*
Caenorhabditis elegans. See Roundworm *(Caenorhabditis elegans)*
Calcitonin, 343, 344*f*, **835**
Calcitonin gene-related peptide (CGRP), 343, 344*f*
Calcium
egg activation and, 872, 873*f*
hormone regulation of, 835–836, 835*f*
muscle contraction and, 751–752, 752*f*
in organisms, 26–27
as second messenger, 191, 191*f*
Calico cats, 259–260, 259*f*
Callorhinus ursinus (fur seal), 930*f*
Calories (cal), 116
Calvin cycle, 157, 158*f*, 168, 170
energy cycle of, 171–172, 172*f*
glucose production by, 170–172, 171*f*, 172*f*
outputs of, 170–171
phases of, 170
Calvin, Melvin, 170
Calyx, **682**, 682*f*
Camouflage, 950*f*, 951, 952
cAMP. *See* Cyclic adenosine monophosphate (cyclic AMP, cAMP)
CAM (crassulacean acid metabolism) plants, **174**
cAMP response protein (CRP), **334**
Campylobacter pylori, 506
Campylopterus hemileucurus (violoet sabrewing hummingbird), 685*f*
Canaliculi, 742, **746**
Cancer, **214**
breast cancer and *HER2,* 367–368, 368*f*
cell cycle control failure as cause of, 214–216, 214*f*
chromosomal mutations and, 322
genome projects and, 387
receptor tyrosine kinases and, 185–186
SIRT-7 and, 217
telomerase and, 293
whole-genome sequencing and, 215–216
Candidiasis, 554
Canola, glyphosate-resistant, 372
CAP (catabolite activator protein), 333*f*, **334**
Capillaries, **814**–815, 814*f*, 815*f*
peritubular, 840, 840*f*, **841**, 841*f*
Capsids, **507**, 508*f*
Capsule, **499**
prokaryotic, 499
Carbohydrates, **43**–47, 43*t*
absorption in animals, 794–795, 795*f*
"burning," 133
transport in plants, 716–717, 717*f*, 718*f*
Carbon. *See also* Carbohydrates; Lipids; Macromolecules; Nucleic acids; Organic molecules; Proteins
in organisms, 26–27
Carbon cycle, 960, 961*f*
Carbon dioxide
anaerobic respiration in methanogens using, 148
atmospheric, rise of, 960
energy transformation and, 15
as greenhouse gas, 991–992, 992*f*
microalgae and capture systems, 371
photorespiration and, 172–174, 173*f*
photosynthesis and, 158–163, 158*f*, 371, 960–961
respiration and, 797, 804
transpiration rate of plants and, 714–715, 715*f*
transport in blood, 804–806, 806*f*
Carbon fixation, **157**, 160, 168, 170–172, 171*f*, 172, 960, 961*f*
Carbon fixation reaction, 157
Carbonic anhydrase, 805
Carbonyl groups, 44, 44*f*
Carboxyl groups, 41*f*, 48
Carboxy terminal domain (CTD), 310, 312
Cardiac cycle, **811**, 812*f*
Cardiac muscle, 742, 743–**744**, 743*t*
Carnegiea gigantea (saguaro cactus), 685
Carnivores
limits on number of, 968
Carnivorous fungi, 550, 550*f*
Carotenoids, **162**
absorption spectra of, 161*f*
as accessory pigments, 162–163, 163*f*
Carotid body, **803**
Carpels, **584**, 584*t*, 585*f*, 682
Carrier proteins, **102**
in active transport, 106–107, 106*f*, 110*t*
in passive transport, 101–103, 103*f*, 110*t*
Carrying capacity, **929**–930, 930*f*
Cartilage, **742**, 745
Cartilaginous skeletons, of sharks, 630–631
Casparian strips, **666**, 667, 667*f*
Castes, in social insects, 912, 912*f*
Catabolism, **125**
of proteins, 149, 150*f*
Catabolite activator protein (CAP), 333*f*, **334**
Catalysis, **121**, 121*f*
Catalysts, 121
chemical reactions and, 30
Catecholamines, **828**, 836, 836*f*
Cations, **23**
Cats, coat color in, 253, 259*f*
Cattle egret *(Bubulcus ibis),* range of, 920, 921*f*
Caudata, 632*f*, 633
Caudipteryx, 638, 639*f*
Cave fish, 442, 775
Cave paintings, 648*f*
Cavitation, **713**, 713*f*
CCK (cholecystokinin), 796, 796*f*
cdc2 gene, 211
Cdks (cyclin-dependent kinases), 212–213, 212*f*, 215
cDNA (complementary DNA), **356**, 358–359, 358*f*, 388, 388*f*
CDSs (coding sequences), 388
CE (central elements), 223, 224*f*
Cedar Creek Natural History Area studies, 969
Cell body, of neurons, 740, 740*f*, **761**, 762*f*
Cell cycle
duration of, 204
growth factors and, 213
interphase of, 204–205, 205*f*
meiosis of (*See* Meiosis)
mitosis of (*See* Mitosis)
phases of, 203–204, 204*f*
regulation of, 210–214
Cell cycle control
cancer as failure of, 214–216
checkpoints and, 210, 211–212, 211*f*, 212*f*
cyclin-dependent kinases in, 212–214, 212*f*
history of investigation into, 210
in multicellular eukaryotes, 213
Cell death, programmed, 329
Cell determination, **347**
cytoplasmic determinants and, 347–348, 348*f*
gene expression and, 882
molecular basis of, 347
tracking of, 347, 347*f*
Cell division. *See also* Cell cycle
in animals, 201*f*
bacterial, 199–200, 199*f*, 200*f*
eukaryotic, 200–203, 496
in fungi, 549
importance of, 199
meiotic *vs.* mitotic, 224, 229
in plant roots, 664–665, 665*f*
in prokaryotes, 201*f*
prokaryotic, 496
in protists, 201*f*
sperm production and, 863
in yeasts, 201*f*
Cell-free DNA (cfDNA), 270
Cell junctions
adherens, 87–88, 87*f*, 87*t*
adhesive, 87–88, 87*f*, 87*t*, 88*f*
communicating, 87*t*, 88*f*, 89
septate and tight, 87*t*, 88–89, 88*f*
Cell-mediated immunity, **850**–852
Cell membranes, 66
active transport across, 105–108, 110*t*
archaeal, 497, 497*f*
bending of, 100
components of, 95–96, 97*t*
electron microscopy of, 96–97, 98*f*
endocytosis and, 108–109, 108*f*, 110*t*
exocytosis and, 109, 109*f*, 110*t*
fluid mosaic model of, 95, 96*f*
passive transport across, 101–105, 110*t*
proteins in, 99–101
selective permeability of, 102
Cell plate, **209**, 209*f*, **690**, 690*f*
Cells, 4*f*, **5**
animal (*See* Animal cells)
division of (*See* Cell division)
eukaryotic (*See* Eukaryotic cells)
gene regulation development in, 346–349
germ-line, 222–223, 223*f*
information transactions and, 15–16
muscle, 82
plant (*See* Plant cells)
prokaryotic (*See* Prokaryotic cells)
size of, 66–67, 66*f*
somatic, 222–223
structural similarities of, 68–69
visualization of, 67, 67*f*
Cell signaling, intracellular receptors and. *See* Intracellular receptors
activation of signal transduction and, 186, 186*f*
autocrine, 180
cellular protein binding to phosphotyrosine and, 187, 187*f*
direct contact, 180, 181*f*
distance from source to receptor and, 180–181, 181*f*
endocrine, 181, 181*f*

paracrine, 181, 181*f*
pathways for, 181
phosphorylation and, 181–182
synaptic, 181, 181*f*
Cell-surface identity markers, 100, 100*f*
Cell-surface markers, 96, 97*t*
Cell-surface receptors, 100, 100*f*
Cell theory, **66**
Cell-to-cell adhesion proteins, 100, 100*f*
Cell-to-cell connections
cell identity and, 86, 87*t*
cell junctions and, 87–89
Cellular level of organization, 4*f*, 5
Cellular organization, as fundamental, property of life, 3, 3*f*
Cellular respiration, **132**
ATP production by, 134–135, 135*f*
C–H bonds and, 134
citric acid cycle and, 139–142
electron transfer and, 133–134, 134*f*
eukaryotic, 495–496
prokaryotic, 495
Cellular slime molds, 540–541, 540*f*
Cellulose, 43*t*, 45–47, 46*f*
fungal breakdown of, 550–551, 557
plant cell walls and, 656, 657*f*
Cell walls, **69**, 69*f*, **72**, 73*f*
animal cell lack of, 594
archaeal, 496–497
bacterial, 70, 498, 499*f*
eukaryotic, 73, 75*f*
of fungi, 548–549
of plants, 84–85, 85*f*, 656, 657*f*
plasticity of, auxin and, 721
primary, 85, 85*f*, 89*f*
prokaryotic, 498
of protists, 524–525
secondary, 85, 85*f*
Cenchrus incertus (burr), 693, 694*f*
Centimorgan (cM), 263, 380
Centipedes, 616
Central chemoreceptors, **779**
Central dogma, **302**–303, 302*f*
Central elements (CE), 223, 224*f*
Central nervous system (CNS), **740**, **761**, 761*f*, 769–771, 769*f*, 770*f*. *See also* Brain
respiration and, 802–803
Central vacuole, **72**, 73*f*
Centrioles, **72**, 72*f*, **82**, 82*f*, 205–206, 206*f*, 208*f*
Centromere, 200, 201*f*, **205**, 205*f*, 230
Centrosomes, 82
Cephalization, **595**
Cephalochordata, 626
Cephalopods, 608, 609*f*, 610
Cephalosporins, 561
Cephalosporium, 561
Ceratium, 532*f*
Cercopithecus aethiops (vervet monkey), language in, 902*f*
Cercozoans, 536
Cerebellum, **769**, 769*f*
Cerebral cortex, **770**–771, 770*f*
Cerebral hemispheres, **770**, 770*f*
Cerebrum, **769**–770, 769*f*, 770*f*
Certhidea olivacea (warbler finch), 430*f*, 431–432, 432*f*
CF (cystic fibrosis), 54, 249*t*, 265, 269, 270*t*
CFCs (chlorofluorocarbons), stratospheric ozone and, 990–991
cfDNA (cell-free DNA), 270
cGMP (cyclic guanosine monophosphate), 185
CGRP (calcitonin gene-related peptide), 343, 344*f*
Chaetae, **610**
Chaetognatha, 599*t*, 600*f*
Chaga disease, 528
Chamerion angustifolium (fireweed), 686*f*
Channel-linked receptors, 183, 183*f*
Channel proteins, 102, 104, 105
Chaperone proteins, **53**–54, 53*f*
Chaperonins, **53**
Chara, 539, 539*f*
Character displacement, **462**–463
Characters, derived, 476–479, 477*f*
phylogenetic species concept and, 480–482, 482*f*
Character states, **476**–478
Charales, 539, 539*f*
Chargaff, Erwin, 280
Chargaff's rules, **280**, 281*f*, 282, 283
Charged amino acids, 48
Charge separation, in thylakoid reactions, 165
Charging reaction, **313**–315, 314*f*
Charophytes, **538**–539, 538*f*, 569, 569*f*
Chase, Martha, 278, 278*f*
Checkpoints, 210, **211**–212, 212*f*
Chelicerae, **615**
Chelicerata, 613*t*, 615, 615*f*
Chelonia, 636, 637*f*
Chelonia mydas (Atlantic green turtle), 636
Chemical bonds, 27–30
activation energy destabilization of, 121
chemical reaction alteration of, 30
covalent, 27, **28**–29, 28*t*
hydrogen, 27, 28*t*, 29–30
ionic, 27–28, 28*f*, 28*t*
peptide, 48, 49*f*
phosphodiester, 56, 56*f*, 279, 280*f*
van der Waals forces and, 27, 28*t*, 30
Chemical defenses
of animals, 951, 951*f*
of plants, 950
Chemical digestion, 790
Chemical elements. *See* Biogeochemical cycles; Elements
Chemical laws, living systems and, 14–15
Chemical reactions, oxidation-reduction (redox). *See* Oxidation; Oxidation-reduction reactions
anabolic, 125
biochemical pathways and, 125–126, 125*f*
bond alterations by, 30
catabolic, 125
dehydration, 42, 43*f*, 279
endergonic, **118**, 118*f*
exergonic, **118**, 118*f*
free energy changes and, 119, 121
hydrolysis, 42–43, 43*f*
reduction, 25, 117, 117*f*
spontaneous, 118–119
Chemical synapse, **181**, **766**–769
Chemiosmosis, 144–145, 144*f*
photosynthetic, 168
in thylakoid reactions, 165
Chemoautotrophs, **965**
Chemoautotrophy, 524
Chemoheterotrophs, **505**
Chemolithoautotrophs, **505**
Chemolithotrophic cells, 496
Chemoreceptors, **773**, 778–779
Chemotropic hyphae, **558**
Chestnut blight, 561
Chewing, 790, 791
Chiasmata, **223**
Chicken pox, 509*t*, 848
Chickens
influenza virus and, 513
yolk distribution in eggs of, 879*f*
Chief cells, 792, 792*f*
Chimpanzee *(Pan troglodytes)*, 645
cognitive behavior in, 897
genome of, 398
Chiral molecules, 42, 42*f*
Chitin, 43*t*, **47**, 47*f*, 745
in fungal cell walls, 548
Chitons, 609*f*
Chlamydia, 507*t*, 889
Chlamydia trachomatis, 507*t*
Chlamydomonas, 537, 573*f*
Chloramphenicol, 487*t*, 500
Chlorella, 538
photosynthesis in, 162
Chloride shift, 805
Chlorine, in organisms, 26–27
Chlorobi, 503*f*
Chloroflexi, 502*f*
Chlorofluorocarbons (CFCs), stratospheric ozone and, 990
Chlorophyll, 80, **158**
absorption spectra of, 161–162, 161*f*
as photosynthetic pigment, 161–162
structure of, 162, 162*f*
Chlorophyll a, 161
Chlorophyll b, 161
chlorophyll P_{680}, 167
chlorophyll P_{700}, 166
chlorophyll P_{840}, 166
chlorophyll P_{870}, 166
Chlorophytes, **537**–538, 537*f*
Chloroplasts, **72**, 73*f*, 79–80, 80*f*
ATP and sugar production and, 80
of dinoflagellates, 532
endosymbiosis and, 80–81, 81*f*
energy cycle of, 171–172, 172*f*
in *Euglena*, 527, 527*f*
eukaryotic, 488*f*
evolution of, 522–523, 522*f*
genetic code in, 305*t*, 306
inheritance pattern of, 272
photosynthesis and, 158–159
photosystems of, 166, 166*f*
Chloroquine, 542
Choanocytes, **603**
Choanoflagellates, 541, 541*f*, 600*f*
Cholecystokinin (CCK), 796, 796*f*
Cholera, 190, 507*t*
in birds, 849
Cholesterol, 99
Chondrichthyes, 630
Chondrocytes, 742
Chondroitin, 742
Chordata, 599*t*, 601*f*, **626**
Chordates, 599*t*, 600, 617–619, 619*f*, **626**
circulatory systems of, 808
eyes of, 780, 780*f*
neurulation in, 882–883
nonvertebrate, 626–627, 626*f*
vertebrate (*See* Vertebrates)
Chorion, **634**, 634*f*, 871
Chorionic villus sampling (CVS), **270**, 271*f*
Chromalveolata, 487, 488*f*, 525
Chromatids, **203**, 203*f*
Chromatin, **74**, 74*f*
composition of, 201, 202*f*
gene expression and structure of, 341, 341*f*
Chromatin-remodeling complexes, **340**
Chromophore, 724
Chromosomal mutations, **322**–323, 323*f*
Chromosomal rearrangements, 368
Chromosomal theory of inheritance, **257**
Chromosome maps, 381–382, 381*f*
Chromosomes, homologous. *See also entries beginning with term Genetic; Genes*
artificial, 358
assortment of, in meiosis, 257
coiling of, 202–203, 202*f*
eukaryotic, 291–293, 496
homologous (*See* Homologous chromosomes)
in meiosis, 222
in meiosis *vs.* mitosis, 229–230, 229*f*–230*f*
in metaphase, 207*f*, 208
number of, changes in, 264–265
pedigree analysis and (*See* Pedigrees)
prokaryotic, 495
in prometaphase, 207–208
recombination between F plasmid and, 503, 504*f*
replication of, 203, 203*f*, 231
sex (*See* Sex chromosomes)
species differences in, 200, 201*f*
structure of, 201, 201*f*
in telophase, 208–209
as vehicles of Mendelian inheritance, 257–260
Chthamalus stellatus, interspecies competition in, 946, 946*f*
Chylomicrons, 795
Chyme, **793**
Chymotrypsin, **794**
Chytridiomycosis, **554**, 554*f*
Chytridiomycota, 555, 555*f*, 555*t*, 556–563
Chytrids, 554, 554*f*, 555, 555*f*, 557
Cichlids, adaptive radiation in, 464–465, 466*f*
Cilia, **84**, 84*f*
eukaryotic, 73, 84, 90
of protists, 523*f*
Ciliates, **533**–535, 534*f*
Circadian rhythms, **727**
melatonin and, 837–838
Circulation
double, 809, 809*f*, 811
pulmonary, 809–810, 809*f*
systemic, 810–811, 810*f*

Circulatory systems, **596**–597. *See also* Blood
of amphibians, 633, 809–810, 809*f*
of annelids, 610
of birds, 640–641
closed, 596–597, 808, 808*f*
of endotherms, 824
of fish, 629, 635, 635*f*, 809
heart and, 811–813
of mammals and birds, 810–811, 810*f*
open, 596–597, 808, 808*f*
of reptiles, 635, 635*f*, 810
vessels of, 814–816, 814*f*
Cisternal space, **75**
Citric acid cycle
oxidation of 2-carbon acetyl groups in, 139
pyruvate conversion to 2-carbon acetyl group and, 139, 139*f*
reactions of, 140–142, 141*f*
Clades, **477**–478
monophyletic, among protists, 525–526
prokaryotic, 503*f*
Cladistics, **476**–479, 479*f*
Cladoceran *(Bosmina longirostris),* 930*f*
Cladograms, **477**, 477*f*
Cladophylls, **672**, 672*f*
Clams *(Spisula),* sperm binding in, 874*f*
Clam worms *(Nereis),* sperm binding in, 874*f*
Class, **483**, 484*f*
Classification, **480**–482
development of, 3
failure to reflect evolutionary relationships, 480, 481*f*, 482*f*
phylogenetic series concept and, 480–482, 482*f*
of prokaryotes, 500–501, 502*f*
taxonomy and, 482–485, 483*f*, 484*f*
Clathrin, 108*f*, 109
Clearance, in transcription, 307
Clear-cutting, 974
Cleavage, **879**
blastula formation, 878–879, 879*f*
in glycolysis, 136, 136*f*
holoblastic, 879–880, 880*f*
in mammals, 880, 880*f*
Cleavage furrow, **209**, 209*f*
Cleavage patterns, of animals, 597, 598*f*
Climate, change in. *See* Global climate change
changing population ranges and, 920, 920*f*
heterogeneity of, species richness and, 971, 972*f*
human activity and, 991–995
selection to match, 412
solar radiation effects on, 980–981, 981*f*
Clitoris, **866**, 866*f*, 870
Cloaca, 790, 790*f*
Clonal selection, **850**
Clone-contig assembly, **385**, 385*f*, 386
Clones. *See* Cloning (of organisms)
Cloning (molecular). *See* Molecular cloning
Cloning (of organisms)
in amphibians, 875–876
in mammals, 876, 876*f*
of plants, 698–699, 699*f*
reproductive, 876
therapeutic, 878, 878*f*
Cloning vectors, 358, 361
Closed circulatory systems, 596–**597**, 808, 808*f*
Clostridium, 502*f*
Clostridium botulinum, 507*t*
Clownfish, 954
Club mosses, 574–576, 576*f*, 578*t*
Clumped spacing, 922
Clustered regularly interspaced short palindromic repeats (CRISPR)/Cas9 system, 363, 364*f*
cM (centimorgan), 263, 380
c-Myc, 878
Cnidaria, 599*t*, 600*f*, 601, **603**–604, 605*f*
digestive system of, 790
radial symmetry in, 595, 596*f*
sperm binding in, 874*f*
CNS. *See* Central nervous system (CNS)
CNV (Copy number variations), **322**
Coactivators, **184**
transcription complex and, 338*f*
transcription factor linkage to RNA polymerase II and, 338–339
Coastal redwood *(Sequoia sempervirens),* 579, 581
Coccus, 497, 498*f*
Cochlea, **776**, 776*f*, **777**
Coconut palm *(Cocos nucifera),* 694*f*
Coding sequences (CDSs), 388–389
Coding strand, **302**
CODIS (Combined DNA Index System), 362
Codons, **303**
deciphering of, 305*t*, 306
start, 306
stop, 306
Coelacanths, 631
Coelom, **596**, 597*f*, 883
evolution of, 601–602
formation of, 598
Coenocytic hyphae, 548, 549
Coenzymes, **124**
Coevolution
commensalism and, 953–954, 954*f*
of herbivores, 950, 954
mutualism and, 954, 954*f*
parasitism and, 955–956
predation and, 950
species interactions and, 955–956, 955*f*
symbiosis and, 953, 953*f*
Cofactors, **123**–**124**
Cognitive behavior, 897–898
Cohen, Stanley, 355
Cohesins, **203**, 213, 230
Cohesiveness, of water, 31–32, 32*t*, 33*f*
Cohorts, **924**
Coleochaetales, 539, 539*f*
Coleochaete, 539, 539*f*
Coleoptile, 695, 696*f*
Coleorhiza, 695, 696*f*
Colias eurytheme (common sulfur butterfly), 411
Collagen fibers, 741, 741*f*
Collared flycatcher *(Ficedula albicollis),* reproductive success of, 927, 927*f*
Collared lizard *(Crotaphytus collaris),* 636*f*
Collecting ducts, 840*f*, **841**, 842–843
Collenchyma cells, **662**, 662*f*
Colon, 795
Colonial chlorophytes, 538, 538*f*
Colonies
of bryozoans, 611
of protists, 524
Colonization, angiosperm, 693, 694*f*
Color vision, 781, 781*f*
Colostrum, 888
Columella cells, 665, 665*f*
Columnar cells, 738, 739*t*
Columnar epithelium, simple, 738, 739*t*
Combination joints, **748**
Combined DNA Index System (CODIS), 362
Comb jellies, 603
Commensalism, **953**–954
Commensal symbiosis, **551**
Common dolphin *(Delphinus delphis),* range of, 920
Common names, 483, 483*f*
Common sulfur butterfly *(Colias eurytheme),* 411
Communicating junctions, 87*t*, 88*f*, 89
Communication
information exchange and, 901–903, 901*f*
language and (*See* Language)
reproduction and, 900–901
Communities, **945**, 945*f*
animal, succession in, 959, 959*f*
biological, **5**, 5*f*
Community ecology
biodiversity and ecosystem stability and, 969–973
biogeochemical cycles and, 959–965
competition and, 944–946
cooperation among species and, 953–957
ecological succession and, 958–959
energy flow through ecosystems and, 965–969
predator-prey relationships and, 949–952
Compact bone, **746**
Compaction, 880
Companion cells, 664, 664*f*
Comparative anatomy, evolution and, 13, 13*f*
Comparative genomics, 391–392, 392*f*
evolution and, 397–398
Comparators, **823**
Compartmentalization, in eukaryotes, **488**, 496
Competition
exploitative, 945
interference, 945
interspecific, 945
parasitism countering of, 953
predation reduction of, 953, 957*f*
between species, 945
Competitive exclusion, **947**, 947*f*
Competitive inhibitors, **124**, 125*f*
Complementary bases, **57**
pairing, 282
Complementary DNA (cDNA), **356**, 357*f*, 358–359, 388, 388*f*, 390
Complement system, **848**
Complexity, of life, 3
Compound leaves, **673**–674, 673*f*
Compound microscopes, 67
Compounds, organic, 27
Compsognathus, 638
Concurrent exchange, 799*f*
Condensation reaction, in citric acid cycle, 140, 141*f*
Condensin, **203**
Cones, **781**–782, 781*f*
Confocal microscopes, 68*t*
Confuciusornis, 640*f*
Conidium, **561**
Coniferophyta, 578*t*, 581*f*
Conifers, 578*t*, **581**–582, 581*f*
Conjugation, **501**–502, **534**
bridge, 504*f*
in ciliates, 533–534
prokaryotic, 501–502
Connective tissue, **740**
loose, 741, 741*f*
proper, 741
of vertebrates, 740
Connell, Joseph, 946
Conservative missense mutations, 321
Conservative model of DNA replication, 283, 284*f*
Constant regions, of antibodies, 852
Constitutive heterochromatin, 389
Consumers, **966**
Consumption, in developed *vs.* developing world, 936
Contig, **381**, **386**
Continental drift, 445
Continental shelves, **987**
Continuous variation, **245**
Contraception, 870–871
Contractile roots, **668**
Control group, **8**
Conus arteriosus, **809**–810, 809*f*
Convergent evolution, **444**–445, 476, 478
Cookeina tricholoma, 555*f*
Copepoda, 615*f*
Copper, in organisms, 27
Copy number variations (CNV), **322**
Corals, 603–604
global temperature change and, 991
Core polymerase, 306, 307*f*
Corepressor, 335
Coriolis effect, 981–982, 981*f*
Coriolis, Gaspard-Gustave, 982
Cork, 671, 671*f*
Cork cambium, **658**, 671, 671*f*
Corms, **672**, 698
Corn *(Zea mays)*
artificial selection and, 434–436, 435*f*, 448
endosperm of, 691, 691*f*
gene interaction in, 248–249, 249*f*
genetically modified, 373
genetic recombination in, 261–263
genome of, 392, 392*f*
glyphosate resistance of, 372
pollination of, 686
as transgenic crop, 373
Cornea, **779**, 779*f*
Corn smut, 554
Corolla, **682**, 682*f*
Corona, of rotifers, 607, 607*f*

Coronary heart disease
  single-nucleotide polymorphisms and, 265
  trans fats and, 58
Coronavirus, 514
Corpus callosum, **770**, 770f
Corpus luteum, 867f, **868**
Correns, Carl, 257, 261
Cortex
  of plant roots, **667**, 667f
  of plant stems, 670f, 671
Cortical granules, 871f, **873**
Cortical nephrons, **840**, 840f
Corticosteroids, 836
Corticotropin, 832, 833
Corticotropin-releasing hormone (CRH), 833
Cortisol, 836
  birth and, 887
Corynebacterium diphtheriae, 507t
Cost of reproduction, **927**, 932
Costus woodsonii (red button ginger flower), 685f
Cotton bollworm, 375
Cotton, genetically engineered, 372, 375
Cotyledons, 689f, **690**
Countercurrent flow, **799**, 799f
Countercurrent heat exchange, 824–825, 824f
Countertransport, 107–108
Coupled transport, 107, 107f, 110t
Courtship signals, 901, 901f
Covalent bonds, **27**, **28**–29, 28t
  nonpolar, 29
  polar, 29, 29t
  strength of, 29
COVID pandemic
  asymptomatic, 938
  causative agent, 514
  disease transmition, 938
  human viral diseases, 509t
  longer-term symptoms, 514
  new vaccine technologies, 369–370
    DNA and mRNA vaccines, 369, 370
    S protein, 370
  researchers, mobilization of, 855
  RT-PCR, 370
  in United States, 855
Cowpox, 849
Coyotes, range of, 921
C_3 photosynthesis, **170**
C_4 photosynthesis, 174, 674
C_3 plants, **173**–174
C_4 plants, **174**, 174f
Crabs, 615–616
  foraging behavior of, 905–906, 914
Cranial neural crest cells, 884
Crassulacean acid metabolism (CAM) plants, **174**
Crayfish, 615
Creighton, Harriet, 262
Crematogaster nigriceps, 955
Crenarchaeota, 502f
Crenicichla alta (Pike cichlid), 416–417, 416f
Cretaceous period, mass extinction in, 468–469, 469f
CRH (corticotropin-releasing hormone), 833
Crick, Francis, 281, 281f, 302–303
Cri du chat syndrome, 322
Crinoidea, 617f
CRISPR/Cas9 system, 363, 364f
CRISPR RNA (crRNA), 364f
Crista, **79**, 79f
Crocodile, 636f, 637
  circulatory systems of, 810–811
Crocodylia, 636f, 637
Crocodylus acutus (American crocodile), 636f, 637
Crocodylus niloticus (Nile crocodile), 637
Crocodylus porosus (estuarine crocodile), 637
Crops. *See* Agriculture
Cross-bridge cycle, 750, 751, 751f
Crossing over, **223**
  genetic variability and, 231
  matching of homologue sequences in, 232
  mechanism of, 225
Cross-pollination, 683
  transgenic crops and, 374
Crotaphytus collaris (collared lizard), 636f
CRP (cAMP response protein), **334**
crRNA (CRISPR RNA), 364f
Crustacea, 613t, 614, 614f, 615f
Crustaceans, exoskeletons of, 745
Cryptic coloration, 951, 951f
Crystals, ionic bonds in, 27–28, 28f
CTD (carboxy terminal domain), 310, 312
Ctenophora, 599t, 600f, **603**
Cuboidal cells, 738, 739t
Cuboidal epithelium, simple, 738, 739t
Cuenot, Lucien, 246
Cultural evolution, 649
Cup fungi, 561, 562f
Cuscuta (dodder), 955f
Cutaneous respiration, of amphibians, 632
Cuticle, of plants, **569**, **660**
Cutin layer, of plants, **660**
CVS (chorionic villus sampling), **270**, 271f
Cyanobacteria, 495–496, 503f, 504f, 505
  eukaryotic chloroplasts and, 488, 522–523, 522f
Cycadophyta, 578t, 582–583
Cycads, 578t, **582**–583, 583f
Cyclic adenosine monophosphate (cyclic AMP, cAMP), **184**
  as second messenger, 190, 191f
Cyclic guanosine monophosphate (cGMP), 185
Cyclic photophosphorylation, 166, 168
Cyclin-dependent kinases (Cdks), 210–214, 212f
Cyclins, 210f, **211**, 213
Cycliophora, 600f
Cyclosomes (APC/C), 213
Cynomys ludovicanus (prairie dog), alarm calling by, 901f
Cyprinodon diabolis (Devil's Hole pupfish), range of, 920f
Cysteine, 49f
Cyst formation, by protists, 523–524
Cystic fibrosis (CF), 54, 249t, 265, 269, 270t
Cytokines, **851**
Cytokinesis, **204**, 204f, 206f, 207f, **209**
  in animal cells, 209, 209f
  in fungi, 209
  in plant cells, 209, 209f
  in protists, 209
  telophase I and, 225
Cytokinins, plant growth and, **722**–723
Cytoplasm, **68**, 69f, **71**
  of egg, movement of, 873, 874f
  eukaryotic, 71, 73f, 74f, 496
  prokaryotic, 496
Cytoplasmic determinants, **347**–348, 348f
Cytoplasmic receptors, 846
Cytoproct, 534, 534f
Cytosine, 56, 56f, 279, 280f, 282–283
Cytoskeleton anchors, 100, 100f
Cytoskeletons, **72**, 72f, **81**–84
  cilia movement and, 83–84, 84f
  composition of, 81–82
  of eukaryotes, **72**
  movement of material in cells by, 82–83
Cytosol, **68**
Cytotoxic T cells (TC), **850**–852

D

2,4-D (dichlorophenoxyacetic acid), 721f, 722
Dachshunds, artificial selection and, 436, 436f
Daddy long-legs spiders, 615
DAG (diacylglycerol), **191**
Dalgarno, Lynn, 316
Daltons, 23
Danaus plexippus (monarch butterfly). *See* Monarch butterfly *(Danaus plexippus)*
Dandelions, asexual reproduction in, 697
Dark-field microscopes, 68t
Darwin, Charles Robert, 8, 10, 10f
  branching diagrams of, 475–476, 475f
  eusociality and, 910–911
  observations on evolutionary divergence, 445, 456
  On the Origin of Species by Means of Natural Selection, 10
  on reproductive strategy, 421–422
  study of evolution, 9–14
  study of finches, 10, 11f, 430–432, 430f, 464, 465f
  study of plants, 720
  theory of evolution of, 405, 411, 430–432, 436–437, 443–444
  The Voyage of the Beagle, 1, 10, 12, 430
Darwin, Francis, 720
Datura stramonium (jimsonweed), 684
Dead transposons, 390
Deamination, **149**, 150f
Death rate, 923
Decatenation, 200
Decay, radioactive, 23
Deciduous plants, 670, 715
Decomposers, 549–550, **966**
Deductive reasoning, **7**
Deer, white-tailed, population explosion of, 949
Defensins, 846
Defensive coloration, 951, 951f
Deforestation, 962, 963f
Degenerate code, **306**
Dehydration reaction, **42**, 43f
Dehydrogenations, **132**
Deinococcus-Thermus, 502f
Delamination, **881**
Delbrück, Max, 296
Deletions, 321–322, 323f
Delphinus delphis (common dolphin), range of, 920
Delta prokaryotes, 503f
Demography, **923**–924
Denaturation, **54**
  PCR and, 359, 360f
Dendrites, 740, 740f, **761**, 762f
Dendritic spines, 761
Dendrobatidae. See Poison-dart frogs *(Dendrobatidae)*
Dendrolimus, population fluctuations in, 932f
Denisovans, 648, 649
Denitrification, **963**–964
Dense connective tissue, **741**
Dense irregular connective tissue, 741
Dense regular connective tissue, 741
Density-dependent effects, **930**–931
Density-independent effects, **931**
Density, of ice, 32t, 33
Dental caries, 507t
Dental decay. *See* Tooth decay
Deoxyhemoglobin, **804**
Deoxyribonucleic acid. *See* DNA
Deoxyribose, 44f
Dephosphorylation, **182**
Derepression, **335**
Derived characters, **476**–477, 477f
  phylogenetic species concept and, 480–482, 482f
Dermal tissue. *See also* Skin
  of plants, **656**, 657t, 660–661, 689
Descriptive science, 6
Desmodus rotundus (vampire bats), reciprocal altruism in, 909, 909f
Desmosomes, 87t, **88**, 88f
Determinate development, **597**–598
Detritivores, **966**, 966f
Deuterostomes, **597**–598, 598f, 601f, 602, 617, 619, 626
Development
  of bone, 747
  determinate, 597–598
  embryonic (*See* Embryonic development)
  indeterminate, 597–598
  prenatal (*See* Embryonic development; Human prenatal development)
  regulative, 880
Devil's Hole pupfish *(Cyprinodon diabolis),* range of, 920, 920f
Dewlap, 459, 459f
Dextrorotatory forms, 42
Diabetes insipidus, 845
  nephrogenic, 104
Diabetes mellitus
  glucose transporter channel, 817f
  insulin production and Type 1, 367
Diacylglycerol (DAG), **191**, 829
Diaphragm, **802**, 803f
  in mammals, 641
Diapsids, 634–637, 635f, 636f
Diastase, 127
Diastole, **811**, 812f, 813
Diatoms, **530**, 531f

Diazepam (Valium), 767
Dicer, 342–343, 342*f*
Dichlorophenoxyacetic acid (2,4-D), 721*f*, 722
Dichogamous plants, **687**
Dicrocoelium dendriticum, parasitism of, 956
Dictyostelium discoideum, 190, 387*f*, 540*f*, 541
Didelphis virginiana (Virginia opossum), 643, 643*f*
Dideoxynucleotides, **383**
Dideoxy terminator sequencing, 383–384, 384*f*
Didinium, Paramecium population and, 949, 949*f*
Diencephalon, 769
Dietary fiber, 795
Differential-interference-contrast microscopes, 68*t*
Diffusion, 101, 102*f*
  Brownian movement and, 102
  facilitated, 102–103, 103*f*, 110*t*
Digestion
  chemical, 790
  in cnidarians, 603–604, 605*f*
  in esophagus, 792
  hormone regulation of, 796, 796*f*
  large intestine and, 795–796
  liver and, 796–797
  in mouth, 791
  nutrient transport and, 794–795, 795*f*
  in small intestine, 793–795, 793*f*
  in stomach, 792, 792*f*
Digestive enzymes, 77, 109, 557, 604, 790, 791, 793, 794, 837
Digestive system
  of flatworms, 605–606, 790
  of mammals, 641–642
  organs of, 791–796, 791*f*
  teeth and, 791–792, 791*f*
  vertebrate, 790–791, 790*f*, 791*f*
Dihybrid crosses, 243, 248, 252
Dihydroxyacetone phosphate, 137*f*
Dikaryon stage, **549**
Dikaryotic hypha, 549, 562*f*
Dinoflagellates, 531–532, 532*f*
Dinosaurs, 635–637, 636*f*
  birds as, 480, 481*f*, 639–640, 639*f*, 640*f*
  extinction of, 469
Dioecious plants, **583**
Dioxin, 722
Diphtheria, 507*t*
Diploid *(2n)*, **200**, **222**–223, 223*f*
  alternation of generations in plants and, 570, 570*f*
  asexual reproduction in plants and, 697
  partial, **501**
Diplomonads, 526, 527*f*
Dipnoi, 628*f*, 631
Direct contact, 180, 181*f*
Directional selection, 413, 414*f*
Disaccharides, 44–**45**, 45*f*
Disassortative mating, **410**
Disease. *See* Human diseases
Disease markers, 369
Disjunction, **207**
Dispersal
  of metapopulations, 922–923
  of populations, 920–921, 922*f*
  of seeds, 692–693, 693*f*, 694*f*, 921, 922*f*
Dispersal mechanisms, 921, 922*f*
Dispersive model of DNA replication, 284–285, 284*f*
Disruptive selection, **414**–415, 415*f*, 461–462
Dissociation of proteins, **54**
Distal convoluted tubule, **841**, 841*f*, 843–844, 844*f*
Diversity. *See* Biodiversity; Genetic diversity; Species richness
DNA (deoxyribonucleic acid), 4, 6, 13, 14, **15**, 16*f*, 43*t*, **55**, 55*f*
  bacterial, 199, 199*f*, 200
  base-pairing in, 57, 57*f*
  chromatin organization of, 202–203, 202*f*
  cleavage by restriction endonucleases, 355, 356*f*
  components of, 279, 280*f*
  of Denisovans, 648
  discovery as genetic material, 277–279, 277*f*, 278*f*
  double helix of, 55*f*, **57**, 281–283, 281*f*, 282*f*
  fetal, in maternal blood plasma, 270
  foreign, introduction into E. coli, 356
  functions of, 55
  genomic imprinting and, 877
  identification by hybridization, 368
  looping of, 337, 337*f*
  methylation of, 340, 343
  mutations of (*See* Mutations)
  of Neanderthals, 648, 649
  noncoding, in human genome, 388–389
  organellar inheritance and, 260–261
  packaging of, 74
  protein production and, 247–248
  reading of, 304
  rearrangement of, immunoglobulin diversity and, 852–854, 854*f*
  regulatory proteins and, 330–331
  repair of, 293–295, 294*f*
  replication of (*See* DNA replication)
  separation of fragments of, 355, 357*f*
  structural DNA, 389, 389*t*
  structure of, 56, 57*f*, 279–283
  transcription and (*See* Transcription)
  transcription factor regulatory sites and, 350
  transduction of (*See* Transduction)
  transformation of (*See* Transformation)
DNA-binding motifs, **330**–331, 331*f*
DNA fingerprinting, 362, 362*f*
DNA gyrase, **287**, 287*f*, 289*t*
DNA injection, by bacteriophages, 510, 511*f*
DNA libraries, 358–359, 358*f*
DNA ligase, **288**, 288*f*, 289*t*, 290*f*, 356, 356*f*
DNA microarrays, 368–369, **393**–394, 393*f*
DNA polymerases, **285**–286, 285*f*
  delta (Pol δ), 291
  epsilon (Pol ε), 291
  polymerase I (Pol I), **286**, 288*f*, 289*f*, 289*t*, 290*f*
  polymerase II (Pol II), **286**
  polymerase III (Pol III), **286**, 288*f*, 289, 289*f*, 289*t*, 290*f*
DNA primase, 288, 289
  eukaryotic, 291
DNA repair, 294–295, 295*f*
DNA replication, 510
  archaeal, 496
  in bacteria, 199*f*, 200
  conservative model of, 283, 284*f*
  dispersive model of, 284, 284*f*
  eukaryotic, 291–293, 291*f*, 292*f*
  meiosis and, 228, 230
  Meselson-Stahl experiment on, 284–285, 284*f*
  PCR and, **359**, 360*f*
  prokaryotic, 286–287, 286*f*–290*f*, 289*t*
  semiconservative model of, 283–286, 284*f*
  semidiscontinuous nature of, 287–289, 287*f*, 288*f*
DNA screening, 270
DNA sequences
  cladistic analysis using, 477–478, 479*f*
  PCR and, 359–362
DNA sequencing, modernization of, 383–386. *See also* Genome sequencing
DNA transfer, prokaryotic, 502–504, 505*f*
Dobzhansky, Theodosius, 470
Docking, 318, 319, 320*f*
Dodder *(Cuscuta)*, 955*f*
Dogbane (Apocynaceae), chemical defense of, 950
Dogs, artificial selection and, 435, 436*f*
Dogwood, leaves of, 675
Dolichonyx oryzivorus (bobolink), migratory behavior of, 898–899, 899*f*
Dolly (sheep), 876–877, 876*f*–877*f*
Domain Archaea, **485**–486, 496–497
Domain Bacteria, 485*f*, **486**, 486*f*
Domain Eukarya, 485*f*, **486**, 488*f*
Domains (of proteins), 52*f*, **53**
Domains (taxonomic), 2, 2*f*, 14*f*, 484*f*, 485–486. *See also* Archaea; Bacteria; Eukarya
  features of, 487*t*
  molecular data supporting, 486
  monophyletic nature of, 486
  prokaryotic, 495, 495*f*
Dominance, incomplete, 249*t*
Dominant traits, 244, 246
  crosses and, 238–239, 239*f*
  Principle of Segregation and, 239–241, 243, 252
Dopamine, **767**
Dormancy, in plants, 723–724, 724*f*
Dorsal body cavity, 737, 737*f*
Dorsal nerve cord, **882**
Dosage compensation, **258**
Double bonds, 29
Double circulation, **809**, 809*f*
Double fertilization, 585–**587**, 586*f*
  in angiosperms, 687–688, 688*f*, 691, 691*f*
Down, J. Langdon, 266
Down syndrome, 266, 266*f*, 270
Dreissena polymorpha (zebra mussel), 609
Driver mutations, 216, 387
Drosha, 342, 342*f*
Drosophila melanogaster. See Fruit flies *(Drosophila melanogaster)*
Drosophila pseudoobscura, 470
Drupes, 694*f*
Duchenne muscular dystrophy, 270*t*
Duck-billed platypus *(Ornithorhynchus anatinus)*, 483, 642, 643*f*
Ducks, 642
  influenza virus and, 513, 513*f*
  mallard, behavioral isolation and, 454–455
  pintail, behavioral isolation and, 454
Ductless glands, 738
Ducts, exocrine glands and, 738
Dugesia, 606*f*
Dunnock *(Prunella modularis)*, 907
Duodenum, 790, 792*f*, **793**, 794*f*
Duplications, **322**, 323*f*
Dutch elm disease, 561
Dwarfism
  island, 647
  pituitary, 833
Dynein, 84, 84*f*, 208

E

Ear, structure of, 776, 776*f*–777
Earthworms, 610, 610*f*
  digestive system of, 790, 790*f*
  hydrostatic skeletons of, 744–745, 744*f*
Easter lily *(Lilium candidum)*, 683*f*
Eastern gray squirrel *(Sciurus carolinensis)*, 484*f*
Eastern rat snake *(Pantherophis alleghaniensis)*, 453*f*
Ebola, 509*t*
Ecdysis, **602**, 602*f*
Ecdysozoans, 601*f*, **602**, 602*f*, 612
Echidnas, 642, 643*f*
Echinodermata, 599*t*, 601*f*, 617–618, 618*f*
Echinoderms, 599*t*, 617–618, 617*f*, 618*f*
  endoskeletons of, 745
Echinoidea, 617*f*
ECM (extracellular matrix), 85, 85*f*
Ecological footprint, **936**, 936*f*
Ecological isolation, 454, 454*t*
Ecological niches, 945–946, 946*f*
Ecological succession, **958**. *See also* Population growth; Populations
  in animal communities, 959, 959*f*
  behavioral, 904–906, 905*f*
  community (*See* Community ecology)
  population, 923, 924*f*
  primary, 958, 958*f*
  reasons for, 9538–959
  secondary, 958
  species composition and, 958–959
Ecosystem level of organization, 5, 5*f*

Ecosystems, **5**, 5*f*
atmospheric circulation and, 980–983
biomes, 985*f*, 986*f*, 9784–986
chemical element movement through (*See* Biogeochemical cycles)
energy flow through, 965–969
global temperature change and, 993–994, 993*f*
regional and local differences affecting, 983–984
stability of, biodiversity and, 969–973
water cycle changes and, 961, 962*f*
Ecotypes, 470
Ectoderm, 736, 880–881
Ectomycorrhizae, **552**–553, 552*f*
Ectoparasites, 955, 955*f*
Ectotherms, 636, **824**–825, 824*f*
Edema, 797, **816**, 848
Effectors, 823
Efferent arteriole, 841
Efferent neurons, 761, 761*f*
Efferia, 616*f*
Egg-laying mammals, 642–643, 643*f*
Eggs
activation of, 872–874, 873*f*
amniotic, of reptiles, 634, 634*f*
artificial activation of, 873
yolk distribution in, 879, 879*f*
Ejaculation, 865, 872
Eldredge, Niles, 467–468
Electrical synapses, **766**
Electric fish, behavioral isolation and, 455
Electromagnetic spectrum, 160, 161*f*, 781
Electronegativity, **29**, 29*t*
Electron microscopes, 67–68, 68*t*
cell membrane structure and, 97, 98*f*
Electrons, 22, 22*f*
in bacteria, 166
in citric acid cycle reactions, 140
in noncyclic photophosphorylation, 166–168
orbitals of, 23, 24, 24*f*
valence, 26
Electron transport, 133*f*, 134
electron carriers and, 133–134, 134*f*
in noncyclic photophosphorylation, 166–168, 167*f*, 169*f*
in photosynthetic bacteria, 165
in thylakoid reactions, 165
Electron transport chain, 134, 142
chemiosmosis and, 143–145, 144*f*, 145*f*
mutations in, 269
proton gradient produced by, 143, 143*f*
Electroreception, 455
Elements, 22–23
common, in organisms, 26–27
inert, 26, 26*f*
in living systems, atomic masses of, 25–27
movement through ecosystems (*See* Biogeochemical cycles)
periodic table of, 26, 26*f*
trace, 27
Elevation, terrestrial ecosystems and, 983, 984*f*, 993
Elimination, in humans, 795–796
Elongation
in transcription, 307–308, 308*f*
in translation, 316–318, 317*f*, 318*f*
Elongation factor, **316**–317, 317*f*
Embryo
in angiosperms, **688**, 690*f*–691*f*
chordate, 619, 619*f*
implantation of, prevention of, 870–871
seed protection of, 580, 580*f*, 691, 691*f*
Embryonic development
in angiosperms, 688–689, 688*f*, 689*f*–690*f*, 690
in animals, 440–441, 441*f*, 594, 595*f*, 597–598, 598*f*
blastula stage in, 878–880
fertilization and, 871–874
gastrulation and, 880–881
human, 441, 863
organogenesis and, 882–883
stem cells and, 369, 874–878
Embryonic stem cells (ES cells), **364**, **369**, **874**
determination and differentiation in culture, 875
fusion experiments with, 877, 877*f*
human, isolation of, 875
isolation of, 874–875, 875*f*
knockout mice and, 364
Embryo sac, **586**
of flowers, **682**, 683*f*
Emergent properties, **6**
Emerson, Robert, 163–164
Enantiomers, 42
Encephalitozoon cuniculi, 555*f*
ENCODE (Encyclopedia of DNA Elements) project, 390–391, 390*f*
Endergonic reactions, **118**, 118*f*
ATP hydrolysis and, 120, 120*f*
Endochondral bone development, 747
Endocrine diseases, 811
Endocrine glands, **738**, 827
peripheral, 834–838
Endocrine signaling, **181**, 181*f*
Endocrine system, 738, **827**–828, 827*f*
Endocytosis, **108**–109, 108*f*, 110*t*
Endoderm, 736, 880–881
Endodermis, of plant roots, **666**, 667*f*, 709–710, 710*f*
Endomembrane system, eukaryotic, **72**, 73*f*, 74*f*
Endonucleases, **286**, **355**
restriction, 355, 356*f*
Endoparasites, **955**
Endophytes, 551, 551*f*
Endoplasmic reticulum (ER), **75**, 77*f*
origins of, 521, 521*f*
protein targeting to, in eukaryotes, 318–319, 320*f*
rough, 72*f*, 73*f*, 75*f*, 76
smooth, 72*f*, 73*f*, 75*f*, 76
Endoskeletons, **745**
of echinoderms, 617, 618*f*
of vertebrates, 627, 628*f*, 745–748, 746*f*
Endosomes, 90
Endosperm, 584*t*, 688*f*, 689, 691, 691*f*
Endospores, **500**
Endosymbiosis, **80**, **488**
eukaryotes and, 487–488, 488*f*, 521–523, 521*f*, 522*f*
evidence supporting, 523
meaning of, 521
secondary, 522, 522*f*, 529
theory of, 80, 81*f*
Endosymbiotic bacteria, 521, 522*f*
Endothermy, **823**–827, 825*f*, 856
in birds, 639
in mammals, 641
Energy
activation (*See* Activation energy)
aerobic respiration and, 146–147
anaerobic respiration and, 148
ATP as (*See* Adenosine triphosphate (ATP))
cellular respiration and, 132–135
changes in (*See* Thermodynamics)
chemical reactions and, 118–119, 118*f* (*See also* Chemical reactions)
citric acid cycle and, 139–142
fatty acid oxidation and, 149–150, 151*f*
fermentation and, 148–149
flow through ecosystems, 965–969
forms used by living organisms, 965
free, 118–119
glycolysis and, 134, 136, 136*f*, 137*f*, 138
kinetic, 116, 116*f*
laws of thermodynamics and, 117–119, 965
light as, 160–161
living systems and, 15, 15*f*
locomotion and, 756
measurement of, 116
potential, 24, 116, 116*f*
protein catabolism and, 149, 150*f*
solar, atmospheric circulation and, 980–983, 981*f*, 982*f*
storage by fats, 58–59
from sunlight, 117
utilization of, as fundamental property of life, 3, 4*f*
Energy cycle, 171–172, 172*f*
Energy levels, 24–**25**, 25*f*
Energy reactions, nucleotides and, 57, 57*f*
Enhancement effect, 166, 167*f*
5-Enolpyruvylshikimate-3-phosphate (EPSP) synthase, 372, 372*f*
Enterobius vermicularis (pinworm), 613
Enterogastrones, **796**, 796*f*
Enterovirus, 509*t*
Enthalpy, **118**
Entropy, **118**, 118*f*
Envelope, viral, 507, 508*f*
Environment. *See also* Climate; Habitats; Pollution
DNA uptake from, 501, 503–504, 505*f*
extreme, 486–487, 496
gene expression and, 332
industrial melanism and, 433–434, 433*f*
phenotype effects of, 249*t*, 250, 251*f*
plant hormones and, 719–720
population growth and, 928–930
transpiration rates of plants and, 713–715, 714*f*
Enzymatic receptors, 183, 183*f*
Enzymes, 99, 100*f*, **121**
activating, 313–314, 314*f*
activation energy lowering by, 121
active sites of, 122–123, 122*f*
allosteric, 124
binding of, 127
cofactors of, 123–124
digestive, 790, 791, 794
factors influencing activity of, 124–125
forms of, 123–124
gene specification of, 301–302, 301*f*
in genetic disorders, 269
intracellular receptors acting as, 185
natural selection and, 425
nonprotein, 123
pancreatic, 794
polymorphism of, 406, 406*f*
restriction, 355, 356*f*
Enzyme-substrate complexes, **122**, 122*f*
Eosinophils, 807*f*
EPCs (extra-pair copulations), **907**, 908*f*
Ephedra, 583, 691
Ephedrine, 583
Epidermal cells, **660**
of zone of maturation, 664, 665*f*
Epidermis
of leaves, 674, 674*f*
of plants, **660**
Epididymis, **863**
Epigenetic changes, **875**
Epigenetic inheritance, **260**
Epigenetics, 339
Epinephrine, 181, 836, 836*f*
Epinephrine receptor, 192–193
Epiphytes, 953, 954*f*
Epistasis, **248,** 249*f*, 251*f*
Epithelial tissue, **736**, 738–740, 739*t*
Epithelium, **738**, 739*t*
columnar, simple, 738
cuboidal, simple, 738
keratinized, 738
regeneration of, 738
simple, 738
squamous, stratified, 738
stratified, 738
structure of, 738
Epitopes, 849, 849*f*
Epomophorus gambianus (fruit bat), 642*f*
Epsilon prokaryotes, 503*f*
EPSPs (excitatory postsynaptic potentials), 767, 768*f*, 773
EPSP (5-enolpyruvylshikimate-3-phosphate) synthase, 372, 372*f*
Epstein-Barr virus, 509*t*
Equilibrium potential, **763**, 765*f*
Equilibrium, punctuated, 620
Equilibrium systems, 15
Equisetum (horsetail), 578*t*, 579, 579*f*
Equus, 438–439
ER. *See* Endoplasmic reticulum (ER)
Eratosthenes, 7
Erection, 865, 865*f*
Erythrocytes. *See* Red blood cells (RBCs)
Erythropoiesis, **808**
Erythropoietin, **808**
Escape, in transcription, 307

ES cells. *See* Embryonic stem cells (ES cells)
Escherichia coli, 495*f*
bacteriophages of, 511
chromosome of, lac region of, 333*f*
conjugation in, 501–502, 502*f*, 503*f*
DNA of, 199, 200
DNA replication in, 286–290, 286*f*–290*f*, 289*t*
foreign DNA introduction into, 356
genes of, 311
hemorrhagic, 111
insulin production and, 367, 367*f*
interaction between genes in, selection and, 413
lac operon of, 350
transduction in, 503–504
T1 virus and, 296
E site (exit site), **315**, 315*f*, 317–318, 317*f*
Esophagus, 790, 790*f*, 791
Estradiol, 865*t*, 867–868
Estrogen
embryo implantation prevention and, 870–871
placental secretion of, 887, 887*f*
Estrous cycles, 870
ESTs (expressed sequence tags), **388**, 388*f*
Estuarine crocodile *(Crocodylus porosus),* 637
Ethanol fermentation, 148–149, 149*f*
Ethylene, effects on plants, **723**
Etiolation, 726, 726*f*
Eubacteria. *See* Bacteria
Eucalyptus, leaves of, 674
Euchromatin, **203**
Eudicots, 585
leaves of, 673–674, 674*f*
roots of, 667*f*, 668
stems of, 670–671, 670*f*
Euglena, 527–528, 527*f*
Euglenoids, 527–529, 527*f*, 528*f*
Euglenozoa, **527**–529, 527*f*, 528*f*
Eukarya, 485*f*, **486**–489, 487*t*, 488*f*
Eukaryotes, 69, 71. *See also* Protists
ATP production by aerobic respiration in, 146, 146*f*
cell division in, 199
characteristics of, 487*t*, 488–489
chromatin structure and, 340, 340*f*
chromosomal organization in, 203–204, 204*f*
chromosomes of, 291–293
circadian rhythms in, 727
earliest groups of, 488–489, 489*f*
evolution of mitosis in, 523
fossils of, 521, 521*f*
gene expression in (*See* Eukaryotic gene expression)
genomes of, 396, 397*f*
initiation in, 309, 310*f*, 316
metabolism of, 496–497
multicellular, cell cycle control in, 213
noncoding sequences on genes of, 311–313, 311*f*, 389
origins of, 487–488, 488*f*, 521–523, 521*f*, 522*f*
prokaryotes compared with, 495–496, 496*f*
protein targeting to endoplasmic reticulum in, 318–319, 320*f*
transcription in, 310–311
Eukaryotic cells, 487–489, 488*f*. *See also* Cell cycle
cytoskeleton of, 82–85
degradation and recycling in, 77–79, 77*f*
division of, 199–203
nucleus of, 71, 73–75, 73*f*–75*f*
organization of, 71–73, 73*f*, 74*f*, 76–80
origins of, 81, 81*f*
protein sorting and packaging in, 76–77, 76*f*, 77*f*
protein synthesis in, 75*f*, 76
ribosomes of, 74–75, 75*f*
Eukaryotic domain, 2, 3*f*
Eukaryotic gene expression, 319*f*, 320*t*, 329
after transcription, 341–345, 342*f*, 343*f*, 345*f*, 346*f*
chromatin structure and, 339–340, 340*f*
transcription complex and, 338, 338*f*
transcription factors and, 336–339
Eulampis (tropical hummingbird), body temperature in, 856
Eumetazoa, 600*f*–601*f*, 601, 603
Euphylls, 574*f*, 575
European honeybee *(Apis mellifera),* taxonomic classification of, 483, 902*f*
European rhinoceros beetle *(Oryctes nasicornis),* 616*f*
Euryarchaeota, 502*f*
Eusociality, 910–911, 910*f*
Eustachian tube, 776, 776*f*
Eutrophic bodies of water, 953, **987**
Evaporation, 674, 711, 712*f*, 961
Evening primrose *(Oenothera biennis),* 684
Evergreen plants, 697
Evolution. *See also* Allele frequencies; Coevolution; Systematics
allele frequencies and (*See* Allele frequencies)
of amphibians, 632, 632*f*
anatomical evidence for, 440–443
of animal body plans, 594–600, 595*f*, 596*f*, 597*f*, 598*f*, 599*t*, 600*f*
of behavior, 903, 904*f*
of birds, 639–640, 639*f*, 640*f*
birth of idea, 10
of chloroplasts, 522–523, 522*f*
of coelom, 601–602
comparative anatomy and, 13, 13*f*
comparative genomics and, 396–399
conserved characteristics and, 15, 15*f*
convergent, 444–445, 476, 478
cultural, 649
Darwin's theory of, 9–14, 405–406, 411, 431, 431*f*, 443, 443*f*, 446–447
diversity of life due to, 16, 16*f*
of eukaryotes, 488–489, 488*f*, 521–523, 521*f*, 522*f*
of eyes, 780–781, 780*f*
of fish, 490, 628–631, 629*f*, 631*f*
fitness as measure of success of, 418, 418*f*
fossil record and, 12, 13*f*, 16, 16*f*
genetic variation and, 231, 405, 405*f*
gradualism and, 467, 468, 468*f*
Hardy-Weinberg equation and, 407–408, 407*f*
of HIV, 513
in humans, 406
of humans, 643–649, 645*f*, 646*f*, 647*f*, 648*f*, 649*f*
interaction of genetic diversity and, 418–421
of jaws in fish, 629, 629*f*
of mammals, 643
mass extinctions and, 468–469, 469*f*
of mitochondria, 521–523, 522*f*
of mitosis, in eukaryotes, 523
molecular evidence for, 13–14, 14*f*, 443
natural selection and, 10, 12, 250
objections to, 446–447
pace of, 467–468, 479
phylogeny as depiction of relationships in, 475–476, 475*f*
of pollinators, 684–685, 684*f*, 685*f*
punctuated equilibrium and, 467–468, 468*f*
rates of, variation in, 397
reproductive strategies and, 421–424
reproductive success and, 926–928
of reptiles, 635–6376, 635*f*, 636*f*
of respiratory systems, 797–798, 798*f*
scope of changes produced by selection and, 436
of seed plants, 580, 580*f*
similarity of species and, 476
Simpson's hypothesis of, 490
species interactions and, 956–957, 956*f*
study of, 9–14
systematics as reconstruction of relationships in, 475–476, 475*f*
of teeth in sharks, 630
of terrestrial plants, 559–560, 569, 569*f*
as theory, 446
of tissues, 595, 603–604, 604*f*, 605*f*
transitions in, documentation by fossils, 437–438, 437*f*, 438*f*
of vascular plants, 574, 574*f*
of vertebrate brain, 769, 770*f*
of vertebrate reproductive systems, 862, 862*f*
Evolutionary age, tropical biodiversity and, 972
Evolutionary reversal, 476, 478
Excavata, 487, **525**, **526**–529, 526*f*, 527*f*, 528*f*
Excision repair, 294, 295*f*
Excitation-contraction coupling, **751**
Excitatory postsynaptic potentials (EPSPs), 767, 768*f*, 773
Excretion, 790, 790*f*, 842. *See also* Feces
of annelids, 610
of flatworms, 606
Exergonic reactions, **118**, 118*f*
Exit site (E site), **315**, 315*f*, 317–318, 317*f*
Exocrine glands, **738**, **827**
Exocytosis, **109**, 109*f*, 110*t*
Exons, **311**, 311*f*
Exon shuffling, 312
Exonucleases, **286**
Exoskeletons
in arthropods, 614–615, **745**, 745*f*
molting of, 602, 602*f*
Experiments, 7–8, 7*f*, 8*f*
Experiment treatment, **8**
Exploitative competition, **945**
Exponential population growth, 928–929, 929*f*, 933–936, 934*f*
Expressed sequence tags (ESTs), **388**, 388*f*
External intercostal muscles, 802
External parasites, 955, 955*f*
Exteroceptors, **773**
Extinctions
of Lake Victoria cichlids, 465
mass, 468–469, 469*f*
of prey population, predator extinction and, 949, 949*f*
Extracellular matrix (ECM), 85, 85*f*
Extraembryonic membranes, **881**, 882*f*
Extra-pair copulations (EPCs), **907**, 908*f*
Extremophiles, **487**, 495
Extrusion, to remove water, 105
Eye color, in fruit fly *(Drosophila melanogaster),* 257–258, 257*f*, 258*f*
eyeless gene, 780–781, 780*f*
Eyes
binocular vision and, 644, 783
of birds, 782
evolution of, 780–781, 780*f*
human, 645, 779, 779*f*, 782
invertebrate, 779, 780*f*
of vertebrates *vs.* mollusks, 441, 442*f*

F

Facilitated diffusion, **102**–103, 103*f*, 110*t*
Facultative symbiosis, **551**
FAD (flavin adenine dinucleotide), **57**–**58**
Falling phase, of action potential, 764, 765*f*
Fallopian tubes, 870
False dandelion *(Pyrrhopappus carolinianus),* 694*f*
Family, **483**, 484*f*, 485
Fats, 43*t*, 58–60. *See also* Lipids
Fatty acid desaturases, 99
Fatty acids, 58, 59*f*
β oxidation of, 149–**150**, 151*f*
Feathers, 638, 638*f*
Feces, 795, 796
energy loss between trophic levels and, 967, 967*f*
Fecundity, **924**
age and, 928
Feedback inhibition, 125, **126**, 126*f*
aerobic respiration and, 147, 147*f*
Feeding phase, of plasmodial slime molds, 540
Fermentation, **133**, 138
in colon, 795
ethanol, 148–149, 149*f*
lactic acid, 149

Ferns, 574, 574*f*, **576**-579
horsetails, 578-579, 579*f*
morphology of, 577, 577*f*
phylogeny of, 578*t*
reproduction of, 577-578, 577*f*
walking *(Asplenium rhizophyllum),* leaves of, 675
whisk, 578-579, 578*t*, 579*f*
Ferredoxin, 168
Ferritin, 345
Fertilization, **222**, 868, **871**
activation of egg and, 872-874, 873*f*
in angiosperms, 585-587, 586*f*, 687-688, 687*f*-688*f*, 691
blockage of egg activation in, 872-873
in conifers, 581-582, 582*f*
double, 687-688, 687*f*-688*f*, 691
internal, in vertebrates, 862, 862*f*
membrane fusion and, 871-872, 871*f*, 872*f*
random, genetic variability and, 231
Fertilization envelope, 873
Fertilizers
nitrogen cycle and, 962
phosphates as, 964
Fetus, position of, 887, 888*f*
Fever, **826**, 848
F_1 generation, **239**, 241*f*, 242*f*
F_2 generation, **239**, 239*f*, 240*f*, 241*f*, 242
FGF (fibroblast growth factor), 349, 349*f*
Fiber, dietary, 795
Fibers, sclerenchyma, 662
Fibrin, **808**
Fibroblast growth factor (FGF), 349, 349*f*
Fibronectin, 85, 881
Ficedula albicollis (collared flycatcher), reproductive success of, 927, 927*f*
Ficedula hypoleuca (pied flycatcher), paternity within a territory and, 908*f*
Fick's Law of Diffusion, 66, **797**, 799
Fiddleheads, 577, 577*f*
Filaments, **584**
Filariasis, 613
Filial imprinting, **896**, 896*f*
Filopodia, 881
Filoviruses, 509*t*
Filtration, by kidney, 838, 839-840, 840*f*, 841*f*
Finches
African black-bellied seedcracker finch *(Pyrenestes ostrinus),* 414, 415*f*
Darwin's study of, 10, 11*f*, 430-433, 430*f*
Galápagos (*See* Galápagos finches)
Fingers, in primates, 643-644
Fireflies, 901, 901*f*
Fire, seed release and, 692, 692*f*
Fireweed *(Chamerion angustifolium),* 686*f*
Firs, 578*t*, 581
First Law of Thermodynamics, **117**-118, 965
Fish. *See also specific fish*
alternative mating strategies of, 907
behavioral isolation and, 454
blood pH of, 152
bony, 490, 630-631, 631*f*
brain in, 769, 769*f*
circulatory systems of, 627, 635, 635*f*, 809, 809*f*
DDT and, 989-990, 989*f*
evolution of, 490, 629, 629*f*, 630*f*, 632, 632*f*
gills of, 798-799, 798*f*, 799*f*
jawed, 629, 629*f*, 630*f*
jawless, 629, 629*f*
key characteristics of, 628-629, 629*f*
lateral line system of, 775-776, 776*f*
lymphatic systems in, 816
as osmoregulators, 838
sperm binding in, 874*f*
swimming by, 753, 753*f*
taste receptors of, 778
temperature regulation in, 824-825, 824*f*
thyroid hormones in, 834
FISH (fluorescent in situ hybridization), 368, 368*f*, 381, 381*f*
Fitness, **418**
components of, 418, 418*f*
extra-pair copulations and, 907, 908*f*
of heterozygotes and homozygotes, 420-421, 421*f*
5' cap, **310**, 311*f*
Flagella, 70-**71**
eukaryotic, 73, 83-84, 84*f*
prokaryotic, 69*f*, 71, 72*f*
Flagellin, **499**, 500*f*
Flagellum
bacterial, 496, 499, 499*f*
in dinoflagellates, 531-532, 532*f*
in diplomonads, 526, 526*f*
in Euglenozoa, 527, 527*f*
in parabasalids, 527, 527*f*
prokaryotic, 496, 498, 499+f
of protists, 524
in Stramenopila, 529, 529*f*
Flame cells, **606**
Flame scallop *(Lima scabra),* 608*f*
Flatworms, 599*t*, 602, 605-607, 606*f*
Dicrocoelium dendriticum, parasitism of, 956, 956*f*
digestive system of, 790
eyes of, 779
Flavin adenine dinucleotide (FAD), **57**-58
Flavivirus, 509*t*
Flax *(Linum),* 662
Flemming, Walther, 200, 222
Flies, 616*f*
fruit (*See* Fruit flies *(Drosophila melanogaster)*)
syrphid, 700
tsetse, 528*f*
Flight
in bats, 642, 642*f*, 754, 754*f*
in birds, 638-639, 639*f*, 754-755, 754*f*
in dinosaurs, 638-639, 639*f*, 754-755, 754*f*
efficiency of, 756
in insects, 754-755
wing structure and, 754-755, 754*f*
Flocking behavior, 912, 912*f*, 922
Flooding, plant responses to, 715-716
Floral leaves, **675**
Flour beetle *(Tribolium),* parasitism and competition in, 956
Flowering plants. *See* Angiosperm reproduction; Angiosperms
Flowers, 584*t*, 669-670. *See also specific flowers*
color of, 685, 685*f*, 700
complete, 681-682, 682*f*
functions of, 586
gamete production in, 682, 683*f*
gametophyte generation in, 584-585, 585*f*
mechanical isolation and, 455
morphology of, 584, 585*f*, 682
pollinator coevolution with, 684-685, 684*f*, 685*f*
regulation of, 681
"Fluctuation test," 296
Fluid mosaic model, 95, 96*f*
Fluorescence microscopes, 68*t*
Fluorescent in situ hybridization (FISH), 368, 368*f*, 381*f*, 382
Fluoxetine (Prozac), 767
Flu viruses. *See* Influenza viruses
Flycatchers
collared *(Ficedula albicollis),* reproductive success of, 927, 927*f*
pied *(Ficedula hypoleuca),* paternity within a territory and, 908*f*
reinforcement in, 458, 458*f*
Focal adhesions, 87*t*, **88**
Follicles, Graafian (later tertiary), **868**, 868*f*
Follicle-stimulating hormone (FSH), **832**, 833, 865-866, 865*t*, 866*f*
Follicular phase, 867*f***868**, 868*f*
Food reserves, in seeds, 691, 691*f*, 693, 694*f*
Food-storage roots, **668**
Foraging behavior, 905-906
Foraminifera, 535-536, 536*f*
Forebrain
human, 770-771, 770*f*
in vertebrates, 769, 770*f*
Forensics, DNA fingerprinting and, 362, 362*f*
Formed elements, **806**, 806*f*, 807-808, 807*f*
fosB gene, 895, 895*f*
Fossil fuels
atmospheric carbon dioxide and, 371, 960
pollution caused by, 989, 989*f*
Fossils, 446
of angiosperms, 584
of birds, 639-640, 640*f*
dating of, 36, 437, 437*f*
eukaryotic, 521, 521*f*
evolution and, 12, 13*f*, 16, 16*f*
of ferns, 576
Foraminifera, 535-536, 536*f*
history of evolutionary change presented by, 436-438, 437*f*, 438*f*
of hominins, 645-646, 645*f*
of *Homo floresiensis,* 647-648
of *Homo neanderthalensis,* 648, 650
of *Homo sapiens,* 648, 649
of horses, 438-440, 438*f*-439*f*
microscopic, 495-496, 495*f*
punctuated equilibrium and, 620
systematics and, 475
Founder effect, **410**
Four o'clock *(Mirabilis jalapa),* 261
^{14}Carbon dating, 36
Fovea, **781**
Foxes, artificial selection and, 435-436, 436*f*
FOXP2 gene, 398-399, 399*f*
F plasmid, **501**
recombination between host chromosome and, 503, 503*f*, 504*f*
transfer of, 502-503, 503*f*
Frameshift mutations, **304**, 322
Franklin, Rosalind, 280-281, 281*f*
Free energy, **118**-119
Frequency-dependent selection, **419**
negative, 419-420, 420*f*
phenotypes favored by, 420
positive, 420
Freshwater habitats, 986-987
Frogs, 632-633, 632*f*
bullfrogs, species-specific acoustic signals in, 901
cell determination and, 347, 348*f*
chytridiomycosis in, 554, 554*f*
eggs of, 594, 595*f*, 871
embryonic development of, 440-441, 441*f*
global temperature change and, 991
golden *(Phyllobates terribilis),* chemical defense of, 951
homoplasy in, 477-478, 478*f*
leopard *(Rana pipiens),* postzygotic isolation in, 456, 456*f*
locomotion by, 753, 754, 754*f*
poison-dart *(Dendrobatidae),* 951, 951*f*
Rana, temporal isolation and, 455
thyroid hormones in, 834
Túngara *(Physalaemus pustulosus),* mate selection in, 424, 424*f*
yolk distribution in eggs of, 879*f*
Fronds, of ferns, 576-577, 577*f*
Fructose, 44, 44*f*, 45*f*
Fructose-1,6-bisphosphate, 137*f*
Fructose-6-phosphate, 137*f*, 170-171
Fruit bat *(Epomophorus gambianus),* 642*f*
Fruit flies *(Drosophila melanogaster)*
adaptive radiation in, 465, 467*f*
artificial selection and, 434-435, 435*f*
dosage compensation in, 258
ecotypes of, 470
enzyme polymorphism in, 406
eye color in, 257-258, 257*f*, 258*f*
genetic recombination in, 261-262, 262*f*
genome of, 386, 387*f*, 396, 397*f*, 398
homeotic mutations in, 331
miRNA genes in, 341-342
as model organism, 602
negative phototropism in, 414, 415*f*
proteins at septate junctions in, 88
reproductive isolation in, 459
sex determination in, 343
Toll receptor in, 845-846

Fruits, 584–**585**, 584*t*
ethylene and, 723
ripening of, 723
seed dispersal and, 692–693*f*, 694*f*
FSH (follicle-stimulating hormone), **832**, 833, 865–866, 865*t*, 866*f*
FtsZ protein, 200, 200*f*
Fugu rubripes (pufferfish), genome of, 387*f*
Fumarate, 141*f*, 142
Functional genomics, 392–394, 394*f*
Functional groups, **41**–42, 41*f*
Function, structure related to, 15
Fundamental niche, **946**
Fundulus heteroclitus (mummichog), 412, 425
Fungi, 2, 3*f*, 485*f*, 486
Ascomycota, 555, 555*f*, 555*t*, 560–563, 562*f*
asexual reproduction in, 549, 559, 561
Basidomycota, 555, 555*f*, 555*t*, 560–561561*f*
Blastocladiomycota, 555, 555*f*, 557, 558*f*
carnivorous, 550, 550*f*
cell wall of, 745
Chytridiomycota, 555, 555*f*, 555*t*, 556–557, 558*f*, 559*f*
cytokinesis in, 209
ecological impact of, 550–553, 550*f*, 551*f*, 552*f*
Glomeromycota, 555, 555*f*, 555*t*, 559–560
hyphae of, 548–549, 548*f*, 549*f*, 558
microsporidia, 555, 555*f*
Neocallimastigomycota, 555, 555*f*, 556
nutrient absorption by, 549–550, 550*f*
pathogenic, 553–554, 555*f*, 561
reproduction in, 548–550, 548*f*, 549*f*, 550*f*
sexual reproduction in, 549–550, 558, 559*f*, 560*f*, 562*f*
symbiotic relationships of, 550–553, 551*f*, 552*f*, 553*f*
taxonomic diversity of, 555, 555*f*, 555*t*
Zygomycota, 555, 555*f*, 556*t*, 558
Fur seal *(Callorhinus ursinus)*, 930*f*

G

GABA (γ-aminobutyric acid), 767
Galactose, 44–45, 44*f*, 45*f*
β-galactosidase *(lacZ)*, 332
Galápagos blue-footed boobies, behavioral isolation and, 455*f*
Galápagos finches
adaptive radiation in, 465, 466*f*
biological species concept and, 456
character displacement in, 948, 948*f*
Darwin's observations on evolutionary divergence and, 445, 462–463, 464
Darwin's study of, 10, 11*f*, 430–432, 430*f*
species of, 431, 432*f*
Galápagos Islands, Darwin's voyage to, 10, 11*f*
Galápagos tortoises, population endangerment of, 949
Galeocerdo cuvier (tiger shark), 630*f*
Gallbladder, 790, 791*f*, 794
Gallstones, 794
Gamete fusion, prevention of, 454*t*, 455–456
Gametes, **222**
of flowering plants, production of, 682
of plants, production of, 683*f*
Gametophytes, **570**, 570*f*
of angiosperms, 584–587, 584*t*, 585*f*, 586*f*, 682, 682*f*
of bryophytes, 571–573, 572*f*
of conifers, 581–583, 582*f*
of seed plants, 580
Gamma-aminobutyric acid (GABA), 767
Gamma prokaryotes, 502*f*
Ganglia, 740
sympathetic chain of, 772
Ganglion cells, **782**
GAP (GTPase-activating protein), 189
Gap junctions, 87*t*, 88*f*, **89**, 181
Garden pea *(Pisum sativum)*
Knight's experiments with, 237
practical considerations for use of, 237, 237*f*
Principle of Independent Assortment and, 243, 244*f*
Principle of Segregation and, 239–241, 239*f*, 240–242, 241*f*, 253
quantitative experimental design for, 237–238, 237*f*
testcross, 244, 244*f*, 244*t*
traits studied of, 237–238, 237*f*
Gar pike, 490
Garrod, Archibald, 301
Gases. *See also* Oxygen
in air, 800
greenhouse, 991–992, 992*f*
noble, 26
partial pressure of, 800–801
Gastric inhibitory peptide (GIP), **796**, 796*f*
Gastric juice, **792**
Gastrin, **796**, 796*f*
Gastropods, 607*f*, 608, 608*f*
Gastrovascular cavity, 790
Gastrulation, **880**–881
in humans, 885
in mammals, 881, 881*f*
Gated channels, **764**
Gause, Georgii, 946–947
GDP (guanosine diphosphate), 189, 189*f*
Geese, 639
filial imprinting in, 896, 896*f*
neck vertebrae of, 441
GEFs (guanine nucleotide exchange factors), 189
Gehring, Walter, 780
Gel electrophoresis, 355, 357*f*
Gene chips, 368–369
Gene cloning. *See* Cloning (of organisms)
Gene expression, eukaryotic. *See* Eukaryotic gene expression
archaeal, 497
cell development and, 346–349
chromatin structure and, 339–340, 340*f*
integrins and, 85
intracellular receptors and, 184–185, 185*f*
levels of control of, 329
organogenesis and, 882
prokaryotic, 320*t*, 329, 332–336
regulatory proteins and, 330–331
RNA and, 303
Gene flow, **409**
allele frequencies and, 409, 409*f*
evolutionary change and, 418, 420*f*
speciation and, 459
transgenic crops and, 373
Gene interactions, 248–249, 249*f*, 249*t*
selection and, 413–414
Genera, **483**, 484*f*, 485
new, 490
Generalist insect predators, 375
Generalized transduction, **502**–503, 503*f*
General transcription factors, 336, 337*f*, 338*f*
Generation time, **924**, 924*f*
Generative cells, 587
Gene regulation, prokaryotic
lac operon and, 333–334, 333*f*, 334*f*, 335*f*
transcription and, 332–333
trp operon and, 334–336, 335*f*
Genes, 15
accumulation of evolutionary differences and, 443, 443*f*
alleles of (*See* Allele frequencies; Alleles; Allelic exclusion)
Beadle-Tatum experiment on, 301–302, 301*f*
cell cycle control by, 210, 214–215, 215*f*
central dogma and, 302–303, 302*f*
cloned (*See* Cloning (of organisms); Genetic engineering)
eukaryotic, noncoding sequences on, 311–313, 311*f*
expression of (*See* Gene expression)
function determination of, 364–366, 365*f*
horizontal transfer of, 486
human genome number of, 386, 387*f*
interaction (*See* Gene interactions)
interactions between
key, conservation of, 397
with more than two alleles, 248–249, 248*f*
mutations and (*See* Mutations)
noncoding sequences on, 311–313, 311*f*, 388
novel, 396
phytochrome, 724–726, 725*f*
regulation of (*See* Gene regulation, prokaryotic)
single, influencing multiple traits, 247–248, 248*f*
trait influence of, 246
transcription and (*See* Transcription)
translation and (*See* Translation)
tumor-suppressor, 214–215, 215*f*, 387
viral, 278–279, 278*f*
Genetically modified (GM), plants, 365, 367*f*, 370–371, 372*f*
Genetic code, 303–306
Crick-Brenner experiment and, 304, 304*f*
deciphering of, 304–306, 305*t*
degenerate, 306
universality of, 306, 306*f*, 364
Genetic counseling, **269**–271
Genetic crosses
dihybrid, 242–243, 242*f*, 243
genetic mapping and, 262–263
monohybrid, 238–241, 239*f*, 240*f*, 243–245
reciprocal, 239
three-point, 264–265, 264*f*
two-point, 263–264, 264*f*
Genetic disorders, 77–78, 270*t*
chromosome nondisjunction and, 265*f*, 266–267
deletions and, 322
Garrod's observations on, 301
gene chips and, 368–369
genetic counseling and, 269–271
genomic imprinting and, 270–271, 271*f*
maternal inheritance of, 260–261
pedigrees and, 267–268
pleiotropic effects and, 246, 246*f*, 249*t*
screening for, 270–271, 271*f*
triplet repeat expansion mutations and, 322
Genetic diversity
evolution and, 405–406, 405*f*
evolutionary processes interaction with, 418–421
in human populations, 648–649, 649*f*
importance of, 231
lack of, selection and, 413
mechanisms increasing, 231
in natural populations, 406, 406*f*
Genetic drift, **410**
allele frequencies and, 410–411, 410*f*
natural selection countered by, 419
speciation and, 459, 460*f*
Genetic engineering. *See also* Cloning (of organisms); Recombinant DNA
agricultural applications of, 371–374
industrial applications of, 370–371
medical applications of, 366–370
Genetic linkage, **261**–263, 261*f*
Genetic mapping, 263–265, 264*f*, **380**
physical maps, correlation with, 382–383
Genetic markers, 270, **380**
Genetic mosaics, **258**–259, 259*f*
Genetic rearrangements, adaptive immunity and, 848–850
Genetic recombination, 261–262, 261*f*, 262*f*, 496
Genetics
archaeal, 496
behavioral, 894–895
population, 406
prokaryotic, 501–505, 503*f*, 504*f*, 505*f*
of self-pollination, 685
Genetic self-incompatibility, in plants, 687

Genetic variation
evolution and, 231
human genome and, 405–406
meiosis and, 231
molecular tools for analysis and modification of, 362–363
Genome annotation, **388**–391, 388*f*, 389*t*, 390*f*
Genomes, 14
annotation of, with functional information, 388–391
behavior and, 894–895
coding and noncoding sequences of, 388–390
comparison of, and evolutionary biology, 396–399
DNA not functionally constrained in, 478
DNA sequencing and, 383–386
eukaryotic, 397, 397*f*
of fruit flies, 386, 387*f*, 396, 397, 397*f*
human (*See* Human genome)
of mammals, 396, 397*f*
mapping of, 380–383
of mosses, 572–573, 573*f*
of Neanderthals, 648
physical maps of, 380–383, 381*f*, 382*f*
prokaryotic, 396, 396*f*
proteins encoded by, 394–395
sequencing of (*See* Genome sequencing)
size differences among, 396, 397*f*
viral, 508, 508*f*, 509*t*
in vivo editing of, 362–363, 364*f*
of wheat, 386–387, 392, 392*f*
Genome sequencing, 13, 14*f*
automated, 383, 384*f*
cancer and, 216, 387
clone-contig assembly, 385–386, 385*f*
dideoxy terminator sequencing, 383–384, 384*f*
expressed sequence tags, 388, 388*f*
humans (*See* Human genome)
massively parallel technologies, 384, 385*f*
PCR and, 361
shotgun assembly, 385, 385*f*
of whole genome, 383–384
Genome-wide association studies (GWAS), **265**
Genomic imprinting, **259**–260, 260*f*
cloning and, 876–877
Genomics. *See also* Genomes; Genome sequencing; Human genome
comparative genomics, 391–392, 392*f*
DNA sequencing modernization and, 383–386
functional genomics, 392–393
gene discoveries, 397
genome projects, revelations in agriculture and medicine, 386–387, 387*f*
mapping for, 380–383
proteomics, 394–395
Genotype frequencies, **408**
Genotypes, **240**
phenotype and, 245, 245*f*
selection favoring, 410*f*, 411–412
Geographical range. *See* Range, population
Geographic isolation, speciation and, 460–462, 461*f*
Geospiza sp., 11*f*, 430*f*, 432*f*
Geotropism, 665
Geranylgeranyl diphosphate (GGPP), 373
Germ cells, 863
Germ hypothesis, 8, 8*f*
Germination, **693**
of angiosperm seeds, 587, 693, 695, 696*f*
nutrient reserves and, 695, 696*f*
phytochrome and, 724
triggers for, 693
Germ layers, **736**
creation of, 879
derivation of major tissue types from, 884, 885*f*
Germ-line alterations, 216
Germ-line cells, **222**, 223*f*
GGPP (geranylgeranyl diphosphate), 373
GH (growth hormone), 832, 833, 834*f*
GHIH (growth hormone-inhibiting hormone), 833
GHRH (growth hormone-releasing hormone), 833
Giant viruses, 510
Giardia intestinalis, 526, 526*f*
Gibberella fujikuroi, 723
Gibberellins, plant growth and, **723**, 723*f*
Gibbon, 644–645
Gigantism, 833, 834*f*
Gila monster, 637
Gill cover, 631
Gill filaments, 798, 799*f*
Gills, 629, 798–799, 798*f*, 799*f*
Ginkgo biloba, 583, 583*f*
Ginkgophyta, 578*t*, 583, 583*f*
GIP (gastric inhibitory peptide), **796**, 796*f*
Giraffes, evolution of, 405, 405*f*, 441
Gizzard, 790, 790*f*, 791
GLABRA2 gene, 661
Glacier Bay, Alaska, primary succession at, 958, 958*f*
Glaciers, global temperature change and, 993, 993*f*
Glanville fritillary butterfly *(Melitaea cinxia),* metapopulation of, 922, 923*f*
Gliding joints, **748**
Global climate change
controversy over, 996
great iron dump hypothesis and, 175
human activity and, 991–995
Glomeromycota, 555, 555*f*, 555*t*, 559–560
Glomerulus, **840**, 841*f*
Glomus intraradices, 555*f*
Glottis, 800, 801*f*
Glucagon, 181, 837, 837*f*
Glucocorticoids, 836
Gluconeogenesis, 151, 836
Glucose, 44, 44*f*, 45*f*, 46*f*
absorption of, 794–795, 795*f*
blood, insulin and, 837, 837*f*
Calvin cycle and, 170–172, 171*f*, 172*f*
excretion of, in diabetes, 817
glucagon and, 181
glycolysis and, 136, 137*f*, 138
lac operon expression and, 332–334, 334*f*
metabolism of, glucocorticoids and, 836
in photosynthesis, 158
reabsorption by kidneys, 841
Glucose-1-phosphate, 170–171
Glucose-6-phosphate, 137*f*, 192, 192*f*, 837
Glucose transporter, 103
Glutamic acid, 49*f*
Glutamine, 49*f*
Glyceraldehyde, 44*f*
Glyceraldehyde 3-phosphate (G3P), **136**, 136*f*, 137*f*, 170–171
Glycerol, 59, 60*f*, 98
Glycine, 49*f*, 767
Glycogen, 43*t*, **45**, 46*f*, 190
Glycogen phosphorylase, 192, 192*f*
Glycogen synthase, 187, 187*f*
Glycolipids, 86, **96**
Glycolysis, **134**, 136*f*
historical background of, 138
NAD^+ regeneration and, 138, 138*f*
yield of, 136, 137*f*, 138
Glycoprotein hormones, **832**
Glycoproteins, **76**, **96**
Glyphosate, 372, 372*f*
G2/M checkpoint, **212**–213, 212*f*
GM (genetically modified), plants, 364, 365–366, 366*f*
Gnetophyta, 578*t*, 583, 583*f*
Gnetophytes, 578*t*, 582–583, 583*f*
Gnetum, 583
GnRH (gonadotropin-releasing hormone), 833, 866, 866*f*
Goblet cells, 738
Goiter, iodine deficiency, 27*f*
Golden frog *(Phyllobates terribilis),* chemical defense of, 951
Golden mean, plant structure and, 669
Golden Rice, 373, 373*f*
Golgi apparatus, 74*f*, **76**–78, 77*f*, 78*f*
Golgi bodies, **76**
root cap and, 665
Golgi, Camillo, 76
Gonadotropin-releasing hormone (GnRH), 833, 865, 865*t*, 866*f*
Gonadotropins, 832
Gondwana, 445
Gonorrhea, 507*t*, 889
Gonyaulax, 532*f*
Gorilla *(Gorilla gorilla),* 644, 644*f*
Gosling, Ray, 280
Gould, John, 430
Gould, Stephen Jay, 467–468
G3P (glyceraldehyde 3-phosphate), **136**, 136*f*, 137*f*, 170–171
GPCRs. *See* G protein-coupled receptors (GPCRs)
G_0 phase, **204**, 213–214
GPP (gross primary productivity), **967**
G protein-coupled receptors (GPCRs), 183*f*, 184, **189**
action of, 189–190, 190*f*
convergence and divergence of signals and, 192–193, 192*f*
second messengers and, 190–192, 191*f*, 829–830, 830*f*
G proteins, **184**
epinephrine receptors and, 192–193
GPCRs and, 183*f*, 184
heterotrimeric, 190
Ras, 189
receptors and effector proteins linking by, 189–190, 190*f*
Graafian follicles, **868**, 868*f*
Graded potentials, **764**
Gradualism, **467**, 468, 468*f*
Grain genomes, 392, 392*f*
Gram, Hans Christian, 498*f*
Gram-negative bacteria, **498**, 498*f*
Gram-positive bacteria, **498**, 498*f*, 502*f*
Gram stain, **498**, 498*f*
Grana, **80**, 80*f*
photosynthesis and, 157*f*, 158
Grant, Peter, 431, 457
Grant, Rosemary, 431, 457
Granular leukocytes, **807**–808
Granulosa cells, **867**
Grasp, in primates, 643
Grasses. *See also* Meadow grass *(Poa annua)*
Agrostis tenuis, copper tolerance in, 419, 419*f*
arrowgrass, 588
Kentucky bluegrass, 697
Grasshoppers, 258, 614*f*, 616, 2585
Gravitropism, **727**–729, 727*f*, 728*f*, 729*f*
Gravity
plant growth and, 727–729, 727*f*, 728*f*, 729*f*
plant perception of, 665
receptors sensitive to, in ear, 777–778
terrestrial locomotion and, 753–754, 754*f*
Gray matter, 770–771
Great tit *(Parus major)*
clutch size and offspring size of, 927*f*
territoriality in, 906*f*
Green algae, 537
Charophytes, 537–539, 537*f*, 538*f*, 569, 569*f*, 574*f*
chloroplasts of, 522
colonies of, 524, 524*f*
eukaryotic chloroplasts and, 487, 488*f*
multicellularity of, 488, 524
plant evolution from, 569, 569*f*
Greenhouse effect, 965, 991–992, 992*f*
Greenhouse gases, 982*f*, 991–992
Green sulfur bacteria, 166
Gregarines, 533, 533*f*
Greider, Carol, 292
Greyhound dogs, artificial selection and, 435, 436*f*
Griffith, Frederick, 277–278
Gross primary productivity (GPP), **967**
Ground elder *(Aegopodium podagraria),* 261*f*
Ground finch *(Geospiza fortis),* beak shape in, 431–432, 431*f*
Ground meristems, **658**, 658*t*, 666
Ground substance, **741**
Ground tissue, of plants, **656**, 657*t*, 659, 661–662, 689
Groundwater, 961–962

Growth
allometric, 888
as fundamental property of life, 3
indeterminate, 656
Growth factors, cell cycle and, 213, 215f
Growth hormone (GH), **832**, 833, 834f
Growth hormone-inhibiting hormone (GHIH), 833
Growth hormone-releasing hormone (GHRH), 833
G1/S checkpoint, **211**-212, 211f
GTP. *See* Guanosine triphosphate (GTP)
GTPase, 189
GTPase-activating protein (GAP), 189
Guanine, 56, 56f, 279, 282-283, 282f
Guanine nucleotide exchange factors (GEFs), 189
Guanosine diphosphate (GDP), 189, 189f
Guanosine triphosphate (GTP)
G protein and, 184
Ras protein and, 189, 189f
Guanylyl cyclase, 185
Guard cells, 660, 660f, 674, 674f, 714-715, 714f
Guppy *(Poecilia reticulata),* color variation in, 416-417, 417f, 418f
Gurdon, John, 875
Gustation, 778
Gut, 790, 790f
Guttation, **711**
GWAS (genome-wide association studies), 265
Gymnosperms, 574f, **580**-583, 581f, 582f, 583f
double fertilization in, 691
Gynoecium, **584**, 682, 682f
Gyres, **983**
Gyrus, 770

H

Haberlandt, Gottlieb, 722
Habitats
of animals, diversity of, 594
forest, loss of, 984-985, 984f
freshwater, 986-989
heterogeneity of, species richness and, 971, 972f
marine (*See* Oceans)
of metapopulations, 922
Haemanthus katharinae (African blood lily), 206f-207f
Haemophilus influenza, 507t
genome sequencing of, 383, 385, 386
restriction endonucleases from, 355
Hagfish, 627-629
Hair
of mammals, 641
in Stramenopila, 529, 529f
woolly, 272
Hair cells, of ear, 775f, 776
Hair-cup moss *(Polytrichum),* 572f
Hairpins, 308, 308f
Half-life, 23, 36, 437, 437f
Halocynthia aurantium (sea peach), 626f
Halophiles, 487
Halophytes, **716**
Handicap hypothesis of intersexual mate selection, **423**
Hansen disease, 507t
Haplodiploidy, 910-911
Haplodiplontic life cycles, of chlorophytes, 538, 538f
Haploid *(n),* **200**, **222**-223, 222f, 223f
alternation of generations in plants and, 570, 570f
Hardy-Weinberg equation, **407**-409, 407f
Harlow, Harry, 896
HATs (histone acetylases), **340**
Haversian canals, 746f
Haversian lamellae, 746f
Haversian system, 746f
hCG. *See* Human chorionic gonadotropin (hCG)
HDACs (histone deacetylases), 217, **340**
Head, of vertebrates, 627, 628f
Hearing, **775**-777
Heart. *See also* Cardiac muscle
of amphibians, 632
contraction of, 811-812, 812f, 813f
of fish, 809, 809f
Heart attacks, stem cells and, 369
Heart disease, trans fats and, 58
Heat, **117**-118. *See also* Temperature; Temperature regulation
as final energy product, 967, 968f
specific (*See* Specific heat)
of vaporization, 32t, **33**
Heat exchange, countercurrent, 824, 824f
Heat-losing center, 826
Heat-promoting center, 826
Heat shock proteins, 53
Heavy chains, **852**-853, 853f
Helicases, **287**, 289t
Helicobacter, 503f
Helicobacter pylori, 506, 507t
Heliconius, mimicry in, 952f
Helium, electron energy levels for, 26, 26f
Helix-turn-helix motif, **330**-331, 331f
Helper T cells (T_H), **850**, 851-852
Hematocrit, 807
Hematopoiesis, **808**
Heme groups, 804, 804f
Hemidesmosomes, 87t, **88**
Hemitrichia serpula, 540f
Hemocyanin, **804**
Hemoglobin, 50-52, 51f, **804**, 804f, 805f, 806f
function of, 246
pH and, 61
sickle-cell disease and, 268
Hemophilia, **268**-269, 270t
human mutation rates and, 322
Henri, Victor, 127
Hepadnavirus, 509t
Hepaticophyta, 571, 571f
Hepatic portal vein, 795, 796
Herbaceous plants, 697
Herbaceous stems, 670
Herbicide-resistant crops, 371-373, 372f
Herbicides, synthetic auxins as, 722
Herbivores, 961f, **962**
Herds, 920
Heredity. *See also* Chromosomes; DNA (deoxyribonucleic acid); *entries beginning with term Genetic;* Genes; Inheritance
DNA and, 15, 246
as fundamental property of life, 4
in liver, 797
Mendel's experiments on, 237-239, 239f
phenotype determination and, 246
polygenic, 248, 248f, 249t
prediction of cross results and, 244-246
Principle of Segregation and, 239-241, 239f, 240-242, 241f, 253
in ribosomes, 75f, 76
on rough endoplasmic reticulum, 76
sperm penetration and, 873
stages of solving, 237-238
in test tubes, 324
HER2 gene, 367-368, 369f
Heritability, **264**
Herpes, 509t
Herpes simplex virus, 509t
Hershey, Alfred, 278, 278f
Heterochromatin, **203**
formation of, small RNAs and, 343
Heterokaryotic hypha, **549**
Heterotherms, 824
Heterotrimeric G proteins, 190
Heterotrophs, **132**, 504, 524, 594, 966
Heterozygosity, 258
Heterozygous advantage, **420**-421, 421f
Hexapoda, 613t, 616, 616f
Hfr cells, 501, 501f, 504f
HGP (Human Genome Project), 361, 386, 387f, 389, 401
HGT. *See* Horizontal gene transfer (HGT)
HGT (horizontal gene transfer), 486, 496
Hibernation, 827
High-risk pregnancies, 269
Hill, Robin, 160
Hindbrain, 769, 769f
*Hind*III, 355
Hinge joints, **748**
Histidine, 49f
Histograms, 246
Histone acetylases (HATs), **340**
Histone deacetylases (HDACs), 217, **340**
Histone modification, 339, 340f
Histone octamer, 203
Histone proteins, **201**
HIV. *See* Human immunodeficiency virus (HIV)
HLAs (human leukocyte antigens), **850**
H1N1 virus, 514
H2N2 virus, 514
H3N2 virus, 514
Holoblastic cleavage, **879**-880, 879f
Holoenzyme, 306, 307f
Holothuroidea, 617f
Homeodomain, **331**
Homeostasis, **823**
adrenal gland and, 836-837, 836f
aldosterone and, 845
antidiuretic hormone and, 844-845, 844f
calcium, 835-836, 835f
carbohydrate metabolism and, 837, 837f
circadian cycles and, 837-838
constant internal environment and, 823-827
as fundamental property of life, 6
hormones and (*See* Hormones; *specific hormones*)
hypothalamus and, 832-833, 833f
immunity and (*See* Immunity)
liver and, 797
multicellular organisms and, 329
negative feedback mechanisms and, 823-824, 823f
osmotic, 839-844
pituitary and, 831-834
temperature and (*See* Temperature regulation)
thyroid gland and, 834-835
water balance and (*See* Kidneys; Osmoregulation)
Homeotherms, 824
Hominids, **644**-645, 645f
Hominins, 645, 645f
Homo, 645, 646f
evolution of, 645-649, 647f, 648f, 649f
Homo erectus, 647, 650
Homo ergaster, 647, 650
Homo floresiensis, 646f, 647, 647f
Homogentisic acid, 301
Homo habilis, 647, 647f
Homo heidelbergensis, 646, 648
Homokaryotic hypha, **549**
Homologous chromosomes, **200**, 203f
Creighton/McClintock experiment, 263f
crossing, over exchanges alleles, 262f
in meiosis, 222-225, 223f, 226f-228f, 229-230, 233, 261-263, 263f
Homologous recombination, between F plasmid and host chromosome, 502-503, 505f
Homologous structures, **13**, 13f, **440**, 441f
Homologues, **200**
Homo naledi, 647
Homo neanderthalensis, 648, 650
Homoplasy, **478**, 478f, 479f
rapid rates of evolutionary change and, 478-479
Homo sapiens, 645f, 648-651, 649f
Homo sapiens sapiens, 646f
Homozygosity, 241
Honeybees
dance language of, 901-902, 902f
eusociality in, 910, 911f
taxonomic classification of, 483, 485
Honeysuckle *(Lonicera hispidula),* 694f
Hong Kong flu, 514
Hooke, Robert, 66
Hookworms, 613
Hooves, 642
Horizontal cells, 782
Horizontal gene transfer (HGT), 486, 496
in eukaryotic evolution, 521
prokaryotic conjugation, 501-502, 503f, 504f

Hormone-activated transcription factors, 828
Hormone response elements, **828**
Hormones, **181**, **827**. *See also specific hormones*
adrenal, 836–837, 836*f*
calcium homeostasis and, 835–836, 835*f*
classes of, 828
digestion regulation by, 796, 796*f*
endocrine signaling and, 181, 181*f*
glycoprotein, 832
hydrophilic, 828–830, 830*f*
hypothalamic, 832–834, 833*f*
inhibiting, 832–833, 833*f*
lipophilic, 828, 829*f*
neurohormones, 831–834, 833*f*
osmoregulation and, 844–845
pancreatic, 837, 837*f*
peptide, 832
pineal, 837–838
pituitary, 831–832
in plasma, 807
protein, 828, 832
receptors for, 828–831, 830*f*
releasing, 833, 833*f*
reproductive
female, 865*t*
male, 863, 865–866, 865*t*, 866*f*
steroid, 184, 828
thyroid, 828, 834–835
tropic, 832
Horns, 642
Hornworts, **573**, 573*f*, 574*f*
Horses
evolution of, 438–440, 438*f*–439*f*
locomotion by, 753–754
thoroughbreds, 413, 413*f*
Horseshoe crabs, 615
Horsetails *(Equisetum)*, 578*t*, 579, 579*f*
Host restriction, 355
Hox genes, 397, 883
Hubbard Brook Experimental Forest, 974
Human Cell Atlas, 347
Human chorionic gonadotropin (hCG), 868
during pregnancy, 885, 887, 887*f*
Human diseases. *See also* Cancer
aquaporin dysfunction and, 104
autoimmune diseases, 369
bacterial, 70, 505–506, 507*t*
cell signaling disruptions and, 190
cystic fibrosis, 54, 246, 246*f*, 270*t*
deficiency and
vitamin A, 373
vitamin D, 836
endocrine, 827 (*See also* Diabetes mellitus)
fungal, 554
gene chips and, 368–369
genetic (*See* Genetic disorders)
hemophilia, **268–269**, 270*t*, 322
Huntington disease, **269**, 269*f*, 270*t*, 322
hypercholesterolemia, 270*t*
improper protein folding as cause of, 54
infectious, 848 (*See also* Immunity)
multiple sclerosis, 369
muscular dystrophy, 270*t*
natural selection and, 250
nematode-caused, 613
osteoporosis, 748
parasitic, 526, 526*f*, 532–533, 533*f*, 542 (*See also* Malaria)
Parkinson's disease, 767
phenylketonuria, 270, 270*t*
sexually transmitted, 527, 527*f*, 889
sickle-cell anemia, 246*f*, 270*t*
sickle-cell disease, 268, 268*f*
stem cell therapy and, 369
Tay-Sachs disease, 77–78, 270*t*
thyroid, 834
trans fats and, **58**
trypanosomes and, 527–529, 528*f*
viral, 194, 509*t*, 511–515, 513*f*, 514*f*, 848–849 (*See also* Human immunodeficiency virus (HIV))
Human genome, 250
biological function of, 391
cancer and, 215–216
costs of sequencing, 384, 386
ENCODE and functional elements in, 390–391
genetic variation, 406
noncoding DNA in, 389, 389*t*
number of genes in, 386, 387*f*
selected-effect function of, 391
similarity to genomes of animals, 397, 399*f*
Human Genome Project (HGP), 361, 386, 387*f*, 389, 396
Human Genome Sequencing Consortium, 386
Human immunodeficiency virus (HIV), 509, 509*t*
CRISPR/Cas9 system and, 363
evolution of, 513–514
infection cycle of, 511–513, 513*f*
Human leukocyte antigens (HLAs), **850**
Human mutation rates, 322
Human prenatal development
during first trimester, 885–887, 886*f*
hormonal changes and birth, 887–888, 888*f*
during second and third trimesters, 887
Human reproductive system
female, 866–871, 866*f*
fertilization and, 871–874
male, 863–866, 863*f*
Humans
ABO blood system in, 247–248, 248*f*
activities of, extinctions caused by, 469
animal population ranges and, 921
behavioral development in, 896, 897
climate change and, 991–995
as consumers, 968, 969*f*
convergent evolution in, 444–445
diseases in (*See* Human diseases)
DNA of, 291, 291*f*
ears of, 776–778, 776*f*–777*f*
eggs of, 872, 872*f*
embryonic development of, 440–441, 863
embryonic stem cells of, 874–875
endocrine system of, 827, 827*f*
evolution of, 643–649, 645*f*, 647*f*, 648*f*, 649*f*
eyes of, 644, 779, 779*f*, 782
forebrain of, 770–771, 770*f*
genetic mapping in, 263–264
genome of (*See* Human genome; Human Genome Project)
kidney of, 839–844, 840*f*
language in, 902
milk of, 888
miRNA genes in, 342
modern, 648–649
number of chromosomes in, 201, 201*f*.
oligotrophic bodies of water and, 986
pollution caused by, 989–991
population growth of, 933–936, 935*f*
postnatal development in, 888
prenatal development in (*See* Human prenatal development)
prey population explosions and crashes caused by, 949–950
races of, 649, 649*f*
reproductive system of (*See* Human reproductive system)
set points of, 823
skin of, 738, 740
sperm binding in, 873, 874*f*
survivorship curve for, 926, 926*f*
thyroid hormones in, 834–835
vermiform appendix of, 442
water cycle alterations caused by, 962, 962*f*
Hummingbirds, 639, 684, 685*f*
territoriality in, 906, 906*f*
Humoral immunity, **850**, 852–855
Huntington disease, **269**, 269*f*, 270*t*, 322
Huxley, Thomas, 8
Hyalin, 873
Hybridization, **237**
allopolyploidy and, 461, 462*f*
biological species concept limitations and, 457
early experiments on, 237–238
fluorescent in situ, 368, 368*f*, 381, 381*f*
postzygotic isolating mechanisms and hybrid fertility, 453, 454*t*, 456, 456*f*
prezygotic isolating mechanisms preventing, 454–456, 454*t*
Hybrid studies, of behavioral genetics, 894, 894*f*
Hydra, 603–604, 605*f*
survivorship curve for, 926, 926*f*
Hydration shells, 33, 33*f*
Hydrocarbons, 41
bioremediation of, 371
Hydrocortisone, 836
Hydrogen bonds, 27, 28*t*, 29–30, **30**
formation in water, 31
key properties of, 30
structure of, 32*f*
Hydrogen, in organisms, 26–27
Hydrogen ions
concentration of, pH scale and, 34, 34*f*
proton gradient, water transport in plants and, 706, 706*f*
Hydrogen sulfide, photosynthesis and, 160, 166, 166*f*
Hydrolysis reactions, 42–**43**, 43*f*
of ATP, 119–120
Hydrophilic hormones, **828**–831, 830*f*
Hydrophilic molecules, **34**
Hydrophobic exclusion, 32*t*, **34**
Hydrophobic interaction, 28*t*
Hydrophobic molecules, **34**
Hydrostatic pressure, 104
Hydrostatic skeletons, **744**–745, 744*f*
Hydroxyapatite, 747
Hydroxyl groups, 41–42, 41*f*
Hylobates, 644
Hyloicus, population fluctuations in, 932*f*
Hymen, 870
Hymenopterans, eusociality in, 910, 912*f*
Hyoseris longiloba (sunflower), 683*f*
Hypercholesterolemia, 270*t*
Hypericum perforatum (St. John's wort), niche of, 946
Hyperosmotic solutions, 104, 838
Hypertension, **813**, 814
Hyperthyroidism, 834
Hypertonic solutions, **104**, **707**, 838
Hyperventilation, **802**
Hyphae, **548**–549, 548*f*, 549*f*, 558
Hypoosmotic solutions, 104
Hypophysis, **831**. *See also* Pituitary gland
Hypothalamohypophyseal portal system, 833
Hypothalamus, 769, 769*f*, 771, 826, 826*f*, 832, 838
Hypotheses, **7**, 7*f*
testing of, 7–8, 8*f*
Hypothyroidism, 834, 835
Hypotonic solutions, **104**, **706**, 838
Hyracotherium, 438*f*, 440

I

IAA (indoleacetic acid), 721, 721*f*
IBA (indolebutyric acid), 722
Ice
density of, 32*t*, 33
entropy and, 118, 118*f*
Iceman, 36
ICGC (International Cancer Genome Consortium), 216
ICM (inner cell mass), **880**
Igs. *See* Immunoglobulins (Igs)
Iguana, temperature regulation in, 824, 826
Ileum, **793**
Illinois Long-Term Selection Experiment, 448
Illumina next-generation sequencing, 385*f*
Immune system, HIV infection and, 511–515, 513*f*
Immunity
adaptive, 848–850
cell-mediated, 850–852
glucocorticoids and, 836
humoral, 850, 852–855
innate, 845–848, 846*f*
Immunoglobulins (Igs), **850**
antigen recognition by, 852
diversity of, 852–854, 854*f*
secondary response to, 854–855, 855*f*
specificity of, 852
structure of, 852, 853*f*

Impala *(Aepyceros melampus),* 954, 954*f*
Implantation, **885**
Imprinting, **896**, 896*f*
Inclusive fitness, **910**
Incomplete dominance, **247**, 247*f*, 249*t*
Incus, **776**, 776*f*
Indels, **321**
Independent assortment, **225**
genetic variability and, 231
principle of, 243, 244*f*
Independent events, 243
Indeterminate development, 597–**598**
Indeterminate growth, **656**
Indian tent turtle *(Pangshura tentoria),* 636*f*
Indoleacetic acid (IAA), 721, 721*f*
Indolebutyric acid (IBA), 722
Induced fit, 123
Induced pluripotent stem (iPS) cells, **878**
Inducer exclusion, **334**
Induction, **332**, **882**
cell determination and, 348, 349*f*
gene regulation and, 331
Inductive reasoning, **7**
Industrial melanism, **433**–434, 433*f*
Inert elements, 26, 26*f*
Infants, human, development of, 888
Infections
inflammatory response and, 848
opportunistic, HIV infection and, 511
Inflammation, chronic, 369
Inflammatory response, 848
Influenza viruses, 509*t*
pandemics caused by, 514, 515*f*
Information, traits and, 241, 242
Information transactions, living systems and, 15–16, 16*f*
Ingenhousz, Jan, 159
Inheritance. *See also* Chromosomes; entries beginning with term Genetic; Genes; Heredity
of acquired characteristics, **10**, **405**
blending, 407
chromosomal theory, 257–259
chromosomal theory of, 257, 270
epigenetic, 273
maternal, 273
molecular basis of (*See* DNA (deoxyribonucleic acid))
organellar, 273
pedigree analysis and (*See* Pedigrees)
polygenic, 245, 246*f*, 249*t*
sex linkage, 257–259
woolly hair and, 272
Inhibiting hormones, **832**, 833*f*
Inhibitors, of enzymes, **124**–125, 125*f*
Inhibitory postsynaptic potentials (IPSPs), 767, 768, 768*f*, 772
Initiation
of eukaryotic transcription, 309, 310*f*, 316
of prokaryotic transcription, 306–307, 307*f*, 316, 316*f*
of translation, 316, 316*f*, 345
Initiation complexes, **316**, 316*f*, 336, 337*f*
Innate immunity, 845–848, 846*f*
Inner cell mass (ICM), **880**
Inner ear, **776**, 776*f*
Inner membrane, 157*f*
Innovations, evolutionary, 490
Inorganic phosphate (Pi), **120**
Inositol phosphates, as second messengers, 190–191
Inositol phospholipids, 190, 191*f*, 192*f*
Inositol triphosphate (inositol-1,4,5-triphosphate, IP_3), 829
Insecticides, 373, 412
Insectivorous leaves, **675**
Insect-resistant crops, 373
Insects, 599*t*. *See also specific insects*
altruism in, 909
diversity of, 594, 616, 616*f*
eggs of, 871
exoskeletons of, 745, 745*f*
eyes of, 780*f*
flight in, 754–755
genomes of, 398
locomotion by, 753–754
molting in, 602, 602*f*
ommatidia in, 413, 413*f*
plants trapping of, 675
as pollinators, 684–685, 684*f*, 685*f*
species-specific acoustic signals in, 901
sperm binding in, 874*f*
success of, 613–614, 613*t*, 614*f*
temperature regulation in, 824
Insect societies, 912, 912*f*
Insulin, 817, 837, 837*f*
cellular signaling by, 187, 188*f*
E. coli and production of, 367, 367*f*
Insulin-like growth factors, **833**
igf2, gene for, 259, 260*f*
Insulin receptor, 187, 187*f*
Integral membrane proteins, 100
Integument, **580**, 580*f*
Intelligent design, 446
Intercalary meristems, 658, 658*t*
Intercalated disks, 811
Intercostal muscles, external, 802
Interference competition, **945**
Interferons, 846–847
Intergovernmental Panel on Climate Change (IPCC), 991
Interkinesis, 228
Intermediate filaments, 73*f*, 74*f*, 81
Intermembrane space, **79**–80, 79*f*
Internal compartmentalization, eukaryotic, 489, 496
Internal fertilization, in vertebrates, 862, 862*f*
Internal membranes, prokaryotic, 500, 500*f*
Internal parasites, 955
International Cancer Genome Consortium (ICGC), 216
Interneurons, 740*f*, **761**, 761*f*
Internodes, **669**, 670*f*
Interoceptors, **773**
Interphase, 204–205, 206*f*
Intersexual selection, **423**, 423–424, 424*f*
Interspecific competition, **945**
Intestine
large, 791*f*, 795–796
small, 791*f*, 793–795, 793*f*
Intracellular receptors, 180*f*, 182
acting as enzymes, 185
channel-linked, 183, 183*f*
distance from source to, 180–181, 181*f*
enzymatic, 183, 183*f*
gene expression and, 184, 185*f*
G protein-coupled, 183*f*, 184, 189–193
receptor tyrosine kinases, 185–189
second messengers as, 184
steroid hormone receptors, 184, 185*f*
subtypes of, 192–193
Intramembranous bone development, 746*f*, 747
Intrasexual mate selection, 423, 423*f*
Intrasexual selection, **423**
Intrauterine devices (IUDs), 870
Introns, **311**, 311*f*, 389*t*
distribution of, 312–313
Inversions, **322**, 323*f*
Invertebrates
circulatory systems of, 808, 808*f*
diversity of, 594, 595*f*
gravity-sensitive receptors in, 777–778
hydrostatic skeletons of, 744–745, 744*f*
marine, osmolarity in, 838
as osmoregulators, 838–839, 839*f*
respiratory systems of, 798
swimming by, 753, 753*f*
temperature regulation in, 824–825
Inverted pyramid of biomass, **970**, 970*f*
In vitro mutagenesis, 362–363
Iodine, in organisms, 27
Ion channels, **102**–103, **762**
membrane potential and, 762–763, 763*f*
stimulus-gated, 773
Ionic bonds, **27**–28, 28*f*, 28*t*
Ionization, of water, 34
Ions, 23
in plasma, 807
IP_3 (Inositol triphosphate, inositol-1,4,5-triphosphate), 829
IPCC (Intergovernmental Panel on Climate Change), 991
iPS (induced pluripotent stem) cells, **878**
IPSPs (inhibitory postsynaptic potentials), 767, 768, 768*f*, 772
Iron dump hypothesis, 175
Iron, in organisms, 26–27
Irreducible complexity argument, 446
Irwin, Rebecca, 700
Island dwarfism, 647
Island trees, convergent evolution and, 445
Islets of Langerhans, **794,** 794*f*, **837**
Isocitrate, 141*f*, 142
Isoleucine, 49*f*
Isomerization reaction, in citric acid cycle, 141*f*, 142
Isomers, **42**
of sugars, 44, 45*f*
Isosmotic regulation, 105
Isosmotic solutions, 104
Isotonic solutions, **104**, 838
Isotopes, **23**
radioactive, 23–24, 23*f*, 36
Iteroparity, **928**
IUDs (intrauterine devices), 870

J

Jack pines, seed release by, 692*f*
Jacob syndrome, 267
Jagendorf acid bath experiment, 169*f*
Jaundice, 794
Jaws, in fish, 629, 629*f*
Jejunum, **793**
Jellyfish, 603–604
Jelly layer, 871, 871*f*
Jenner, Edward, 849
Jimsonweed *(Datura stramonium),* 684
Jinlian, Zeng, 833
Jointed appendages, in arthropods, 615
Joints, **748**
ball-and-socket, 748
combination, 748
gliding, 748
hinge, 748
Joules, 116
Juniperus chinensis, seed dispersal and, 922*f*
Juxtamedullary nephrons, **840**, 840*f*

K

Kalanchoë daigremontiana, 698, 698*f*
leaves of, 675
Kangaroos, 643*f*
locomotion by, 754
Karyogamy, 560–561
Karyotype, **200**
analysis of, 270
kcals (kilocalories), 116, **756**
Kentucky bluegrass, 697
Keratin, 81, 738
Keratinized epithelium, 738
Kettlewell, Bernard, 433
Key innovations, **462**, 463*f*
Keystone species, **957**
Khorana, H. Gobind, 304–306
Kidneys
functions of, 839–840
nephrons of, 839–841, 841*f*, 842–844, 842*f*, 843*f*
nutrient reabsorption and, 841–842
waste secretion and, 842
"Killer" *Paramecium* strains, 534
Killifish *(Rivulus hartii),* 416–417, 416*f*
Kilocalories (kcals), 116, **756**
Kinase cascades, **187**–188, 188*f*
Kinesin, 83
Kinetic energy, **116**, 116*f*
Kinetochores, **205**, 205*f*, 208, 225–226, 230
Kinetoplastids, 527–529, 528*f*
Kingdoms, 2, 3*f*, **483**, 484*f*, 485–486, 485*f*
King, Thomas, 875
Kinocilium, **775**, 775*f*, 777
Kinorhyncha, 601*f*
Kin selection, 909, **910**, 911, 911*f*
Kipukas, 464
Kleiber, Max, 676
Kleiber's law, 676
Klinefelter syndrome, 267
Knee-jerk reflex, 771
Knockin mice, **365**
Knockout mice, **364**, 365*f*
Koehn, Richard, 425

Kölreuter, Josef, 237
Komodo dragon, 637
Krakatau islands, succession in, 959, 959*f*
Krebs, Charles, 932
Krebs cycle. *See* Citric acid cycle
Krill, 615
K-selected populations, **932**, 933*t*
Kuhn, Thomas, 9
Kurosawa, Eiichi, 723

L

Labia minora, 870
Labrador retriever coat color, due to epistasis, 249*f*
Labyrinth, 777
lacA (lactose transacetylase), 333, 333*f*
Lacewings, behavioral isolation and, 455*f*
lac operon, **332**
  glucose levels and, 333–334, 335*f*
  lac repressor and, 333, 333*f*, 334*f*
  transcription factor binding site and, 350
Lactate dehydrogenase, 412
Lactation, 888
Lactic acid fermentation, 149
Lactose, 45, 332–333, 333*f*, 334*f*
Lactose intolerance, 793
Lactose permease *(lacY)*, 332, 333*f*
Lactose transacetylase *(lacA)*, 332, 333*f*
Lactuca (wild lettuce), temporal isolation and, 455
Lacunae, **742**
lacY (lactose permease), 332, 333*f*
lacZ (β-galactosidase), 332
Lagging strand, **287**, 287*f*
  synthesis of, 288, 288*f*
Lakes
  oligotrophic, 986–987
  pollution of, 989
  succession and, 958
Lake Victoria cichlid fish, adaptive radiation in, 464–465, 466*f*
Lamarck, Jean-Baptiste, 10, 405
Lambda phage, 502
Lamellae, 798, 799*f*
Lamellipodia, 881
Laminaria, 530*f*
Lampreys, 627–629
Lamp shells, 599*t*
Lancelet *(Branchiostoma),* 626, 626*f*
  sperm binding in, 874*f*
Langerhans, Paul, 837
Language
  genomics and, 398, 399*f*
  in humans, 902
  in nonhuman primates, 902–903, 902*f*
Lanugo, 441
Large ground finch *(Geospiza magnirostris),* 11*f*, 430*f*
Large intestine, 791*f*, 795–796
Large offspring syndrome (LOS), 877
Lariat, 312
Lateral elements (LE), 223, 224*f*
Lateral line system, **630**, 775, 776*f*
Lateral meristems, **657**, 659*f*
Later tertiary follicles, **868**, 868*f*
Latrodectus mactans (black widow spider), 615*f*
LE (lateral elements), 223, 224*f*
Leading strand, **287**, 287*f*
  synthesis of, 288, 288*f*
Leaf-cutter ants, fungal symbiosis with, 553, 553*f*, 953
Leaflets, 673, 674*f*
Leafy sea-dragon *(Phycodurus eques),* 629*f*
Leaping, 754, 754*f*
Learning
  behavior and, 896–897
  genetics and, 894, 894*f*
Leaves
  arrangement of, 669–670, 670*f*
  chloroplasts in, 157*f*
  compound, 673–674, 673*f*
  evolution of, 575, 575*f*
  external structure of, 673–674, 673*f*
  internal structure of, 674, 674*f*
  modified, 675
  simple, 673, 674*f*
Leber hereditary optic neuropathy (LHON), 260
Leclerc, Georges-Louis, 10
Legs, of amphibians, 633
Legumes, 666
  fruits of, 694*f*
  nitrogen fixation by, 963
Leishmaniasis, 528
Lemurs, 643, 645*f*
Lens, **779**
  fly and mouse development of, 780–781, 780*f*
Lenticels, 671, 671*f*
Leopard frog *(Rana pipiens),* postzygotic isolation in, 456, 456*f*
Lepidoptera, as pollinators, 684
Leprosy, 507*t*
Lepus americanus (snowshoe hare), population cycle of, 931–932, 932*f*
let-7 gene, 341
Leucine, 49*f*
Leucine zipper, **331**, 331*f*
Leucoplasts, 80
Leukemia
  chronic myelogenous leukemia, 381–382, 381*f*
  stem cell therapy and, 369
Leukocytes, 742, **807**
Levorotatory forms, 42
Leydig cells, 863
LH (luteinizing hormone), **832**, 833, 865–866, 865*t*, 866*f*
LHON (Leber hereditary optic neuropathy), 260
Lichens, **551**–552, 552*f*
  succession and, 958
  symbiosis of, 953
Life
  biologists' study of, 2–3
  characteristics of, 4–5
  chemical and physical laws of, 14–15
  definition of, 3–4
  diversity of (*See* Biodiversity; Species richness)
  emergent properties of, 6
  hierarchical organization and, 4*f*–5*f*, 5
Life history adaptations, resource availability and, 932–933, 933*t*
Life span, age at first reproduction and, 928
Life tables, **925**, 925*t*
Ligand-gated channels, **764**, 782
Ligands, **180**, 180*f*
Light. *See also* Solar energy
  directional growth of plants and, 726–727, 726*f*
  as energy form, 160–161
  intensity of, rate of photosynthesis and, 163–164, 164*f*
  plant growth and, 724–727
  transpiration rates and, 715
  ultraviolet (*See* Ultraviolet light)
Light chains, **852**–853, 853*f*
Light-dependent reactions, **158**, 165–166
Light-harvesting complex, 164, 164*f*
Light-independent reactions, **157**, 157–158, 159*f*
Light microscopes, 67, 67*f*, 68*t*
Lignin, **662**
  fungal breakdown of, 550–551, 557
Likens, Gene, 974
Lilium candidum (Easter lily), 683*f*
Lima scabra (flame scallop), 608*f*
Limestone, Foraminifera in, 535–536, 536*f*
LINEs (long interspersed elements), **390**
lin-4 gene, 341
lin-14 gene expression, 341, 341*f*
Linkage, genetic, **261**–263, 261*f*
Linnaean taxonomy, 2, 483, 483*f*, 484*f*, 485
Linnaeus, Carolus, 10, 483
Linum (flax), 662
Liochlorophis vernalis, 636*f*
Lions *(Panthera leo),* 643*f*
  ecological isolation of, 454
Lipase, **794**
Lipid droplets, **76**
Lipid raft, 96
Lipids, 43*t*, **58**. *See also* Phospholipids
  energy storage by, 58–59
  fluid mosaic model and, 95, 95*f*, 96*f*
  hydrophobicity of, 58, 59*f*
  membrane formation from, 59–60, 60*f*
Lipophilic hormones, **828**, 829*f*
Lipopolysaccharide (LPS), **499**, 845
Lips, of humans, 738
Little paradise kingfisher *(Tanysiptera hydrocharis),* phenotypic differentiation in, 460, 461*f*
Live birth, in vertebrates, 862–863, 862*f*
Liver, 790, 791*f*, **794**, 797
  insulin and, 837, 837*f*
Liverworts, **571**, 571*f*, 574*f*
Living systems
  atomic masses of elements in, 25–27
  chemical and physical laws, 14–15
  energy and matter transformation, 15, 15*f*
  information transactions and, 15–16, 16*f*
Lizards, 636*f*, 637–638
  Anolis (*See Anolis* lizards)
  side-blotched *(Uta stansburiana),* 908, 908*f*, 928*f*
  temperature regulation in, 825
  territoriality in, 906
Lobe-finned fish, 630*t*, 631
Lobsters, 615, 616*f*
Locomotion
  in air (*See* Flight)
  appendicular, 753
  axial, 753
  efficient modes of, 756
  flying, 754–755, 754*f*
  terrestrial, 753–754, 754*f*
  in water, 753, 753*f*
Locomotor organelles, of protists, 523
Locusta migratoria (migratory locust), population density of, 931*f*
Logarithmic scales, 35
Logistic growth, 929–930, 929*f*, 930*f*
Lolium perenne, 551, 551*f*
Long interspersed elements (LINEs), **390**
Longleaf pines *(Pinus palustris),* 581*f*
Long terminal repeats (LTRs), **390**
Lonicera hispidula (honeysuckle), 694*f*
Loop of Henle, **841**, 841*f*, 842–843, 842*f*
Loose connective tissue, **741**, 741*f*
Lophophorates, 611, 611*f*
Lophotrochozoans, 600*f*, **602**, 602*f*, 607
Lord Howe Island, 461
Lorenz, Konrad, 896, 896*f*
Loricifera, 600*f*
Lorises, 644, 644*f*
LOS (large offspring syndrome), 877
Loxosceles reclusa (brown recluse spider), 615*f*
LPS (lipopolysaccharide), **499**, 845
LTRs (long terminal repeats), **390**
Ludwigia peruviana (Peruvian primrose), 685*f*
Lumen, **75**
Luna moth *(Actias luna),* 616*f*
Lungfish, 631
Lungs
  amphibian, 632
  central nervous system and, 802
  diaphragm and, 802, 803*f*
  mammalian, 800–802, 801*f*, 802*f*
Luria, Salvadore, 296
Luteal phase, 867, 867*f*, **868**
Luteinizing hormone (LH), **832**, 833, 865–866, 865*t*, 866*f*
Lycophytes, 574–576, 574*f*, 575*f*, 576*f*, 578*t*
Lyme disease, 507*t*
Lymph, **816**
Lymphatic system, 815–**816**
Lymph hearts, **816**
Lymphocytes, 807*f*, **849**, 849*f*
  antigen recognition and, 849–850
  B cells, 850
  naive, 850
Lynx canadensis (northern lynx), population cycles of, 932, 932*f*
Lyon, Mary, 339
Lysine, 49*f*
Lysogen, **510**
Lysogenic cycle, bacteriophage, 510, 512*f*
Lysogenic phages, **510**
Lysosomal storage disorders, 77–78
Lysosomes, 72, 72*f*, **78**–79, 78*f*
Lysozyme, 122*f*
Lytic cycle, bacteriophage, 510, 511*f*

M

MAC (membrane attack complex), **848**
MacArthur, Robert, 947
macho-1 gene, 348
MacLeod, Colin, 277
Macrocystis pyrifera, 530*f*
Macromolecules, 4*f*, **5**, 43*t*. *See also* Carbohydrates; Lipids; Nucleic acids; Proteins
  dehydration reaction and, 42, 43*f*
  hydrolysis reaction and, 42–44, 43*f*
Macronucleus, **534**, 534*f*
Macronutrients, **718**
Macrophages, **847**
Macropus rufus (red kangaroo), 643*f*
Madreporite, **618**, 618*f*
Magnesium, in organisms, 26–27
Magnetosomes, electron micrograph of, 71*f*
Maize. *See* Corn *(Zea mays)*
Majerus, Michael, 433–434
Major histocompatibility complex (MHC) proteins, 87, **850**–851, 851*f*
Malaria
  fitness and, 421, 421*f*
  Plasmodium and, 532–533, 533*f*, 542, 556
  sickle-cell allele and, 250, 251*f*
  treatment window for, 542
Malate
  in citric acid cycle, 140, 140*f*
  in C_4 plants, 173*f*, 174, 174*f*
Mallard ducks, behavioral isolation and, 454, 454*f*
Malleus, **776**, 777*f*
Malpighian tubules, **839**, 839*f*
Malthus, Thomas, *An Essay on the Principle of Population,* 11–12
Maltose, 45, 45*f*
Mammals. *See also specific animals*
  blockage of egg activation in, 872–873
  body temperature of, 826
  brain in, 769, 769*f*
  chorionic membrane in, 881, 882*f*
  circulatory systems of, 810–811, 810*f*
  cleavage in, 880, 880*f*
  cloning in, 876–877, 876*f*–877*f*
  common characteristics of, 640–643, 642*f*, 643*f*
  egg-laying, 642, 643*f*
  eggs of, 871, 871*f*
  estrous cycles in, 870
  evolution of, 642
  excretion in, 794
  flying, 642, 642*f*
  gastrulation in, 881, 881*f*
  genomes of, 396, 397*f*
  kidney of, 839–844, 840*f*
  locomotion by, 753–754
  lungs of, 800–802, 801*f*, 802*f*
  lymphatic systems in, 816
  modern, 642–643, 643*f*
  neck vertebrae of, 441
  neural tube in, 882–883, 883*f*
  placental, 444, 444*f*, 643, 643*f*
  pouched, 643, 643*f*
  regulative development in, 880
  reproduction in (*See* Human reproductive system)
  respiration in, 800–801, 801*f*
  ruminant, cellulose digestion in, 557
  sperm binding in, 873, 874*f*
  temperature regulation in, 826–827
  uterus of, 870
  viviparity in, 862–863
  X chromosomes in, 257–259
Mammary glands, 642
  lactation and, 888
Manatees, 442, 442*f*
Mandibles, **613**
Mandrill *(Mandrillus sphinx),* 644*f*
Mandrillus sphinx (mandrill), 644*f*
Mannose-binding lectin protein (MBL), 848
Mantle, **608**
MAP (mitogen-activated protein) kinases, **187**, 213, 215*f*
Mapping, genetic, 262*f*, 263–265
Maps/mapping, genomic
  chromosome, 381–382, 381*f*
  genetic, 382, 383
  physical, 380–383, 381*f*, 382*f*
  restriction, 380–381, 381*f*
  sequence-tagged site, 382, 382*f*
Map units, **263**
Marchantia, 571*f*
Margulis, Lynn, 521
Marine habitats. *See* Oceans
Marine mammals, swimming by, 753
Marine predators, convergent evolution in, 444, 445*f*
Markers, proteins as, 86
Marrow cavity, **746**
Marsilea, 577
Marsupials, **643**, 643*f*
  convergent evolution and, 444–445, 445*f*
  uterus of, 870
Martin, John, 175
Mass, 22
Mass extinctions, **468**–469, 469*f*
Massively parallel sequencing, 384, 384*f*
Mass spectroscopy (mass spec), 394, 395*f*
Mastication, 791
Maternal age, Down syndrome and, 266, 266*f*
Maternal inheritance, **260**–261
Mate selection
  intersexual, 423, 424*f*
  intrasexual, 423, 423*f*
  parental investment and, 421–422
Mating. *See also entries beginning with term Reproductive*
  alternative strategies of, 907–909, 908*f*
  nonrandom, allele frequencies and, 409*f*, 410
Matrix
  of connective tissue, **740**–742
  of mitochondria, **80**, 81*f*
Matter
  atoms and, 22
  energy transactions and, 15
Maturation-promoting factor (MPF), 210–211, 210*f*
Mature mRNA, **310**
  editing of, 344
Mayr, Ernst, 453, 460
MBL (mannose-binding lectin protein), 848
McCarty, Maclyn, 278
McClintock, Barbara, 262
McKusick, Victor, 269
Meadow grass *(Poa annua)*
  life table of, 925, 925*t*
  survivorship curve for, 926, 926*f*
Measles, 194, 509*t*
Mechanical isolation, 454*t*, 455
Mechanoreceptors, **774**, 774–775
Mediators, transcription factor linkage to RNA polymerase II and, 337–338
Medicago polycarpa, seed dispersal and, 922*f*
Medicine. *See also* Genetic disorders; Human diseases
  cancer genome projects, 387
  genetic engineering applications in, 366–370
  glucocorticoids in, 836
  insulin therapy and, 837
Medulla oblongata, **769**, 769*f*
Medullary bone, **746**
Medullary cavity, 746
Medusae, 604, 605*f*
Meerkats, social groups formed by, 913, 913*f*
Megakaryocytes, 808
Megapascals (MPa), **706**
Megaphylls, **673**
Meiosis, **222**, 224*f*
  chromosome assortment in, 261–263
  chromosomes in, 222, 222*f*
  crossing over in, 223, 231–233
  distinct features of, 229–230
  genetic variation and, 231
  mitosis contrasted with, 223–224
  nondisjunction and, 265–267, 266*f*
  in plants, 570, 570*f*
  process of, 224–228
Melanocyte-stimulating hormone (MSH), **832**
Melatonin, **828**, 837–838
Melitaea cinxia (Glanville fritillary butterfly), metapopulation of, 923, 923*f*
Membrane attack complex (MAC), **848**
Membrane potential, 102, 762–763, 763*f*
Membranes
  amniotic, 881
  basilar, 776, 777*f*
  cell (*See* Cell membranes)
  endoplasmic reticulum (*See* Endoplasmic reticulum (ER))
  epithelial, 738
  extraembryonic, 881, 882*f*
  internal, 500, 500*f*
  phospholipids in, 59–60, 60*f*
  plasma (*See* Plasma membranes)
  pleural, 802
  thylakoid (*See* Thylakoid membrane)
  undulating, 527, 527*f*
Memory cells, secondary response to antigens and, 855
Menarche, 867
Mendeleev, Dmitri, 25
Mendel, Gregor, 236, **237**, 237*f*, 407
  experimental design by, 237–238, 238*f*
  experiments conducted by, 237, 238*f*
  extending Mendel's model, 245–249
  Principle of Independent Assortment, 242–243, 242*f*
  Principle of Segregation, 238–241, 239*f*
  testcross, 244, 244*f*
  traits studied by, 238–239, 239*f*
Mendelian ratio, **240**
Meninges, 771
Menophra abruptaria (waved umber caterpillar), cryptic coloration of, 951*f*
Menstrual cycle, human, 867–870, 867*f*
Menstrual phase, 868
Mereschkowsky, Konstantin, 521
Meristems, 657*f*, 658*t*
  apical, 658, 658*f*, 659*f*, 665, 669–670, 669*f*
  ground, **658**, 666*F*
  intercalary, 658
  lateral, 658, 659*f*
  primary, 658, 689
Merops bullockoides (white-fronted bee-eater), kin selection in, 911, 911*f*
Meselson, Matthew, 284–285
Meselson-Stahl experiment, 284–285, 284*f*
Mesenchymal stem cells, **369**
Mesenchyme, **747**
Mesoderm, 736, 880–881
Mesohyl, **603**
Mesophyll, 157*f*, **674**, 674*f*
Messenger RNA (mRNA), **75**, 250, **303**
  complementary DNA from, 356, 358*f*, 359
  degradation of, 345
  mature, 310, 344
  reverse transcription PCR and, 360–361, 361*f*
  splicing of, 312–313, 312*f*
  transcription of, 309
Metabolic pathways, intermediates connecting, 150–151
Metabolic rate
  of ectotherms, 824–825
  of endotherms, 825–827, 825*f*, 856
  size related to, 676
  thyroid hormones and, 834–835
  torpor and, 826–827
Metabolism, **125**–126
  biochemical pathways and, 125–126, 125*f*, 126*f*
  as emergent property of life, 6
  eukaryotic, 496
  feedback inhibition and, 126, 126*f*
  prokaryotic, 496–497, 504
  staged energy release in, 133–134
Metamorphosis, **616**
Metaphase II, in meiosis, 227*f*, 228, 229*f*
Metaphase I, meiotic, 225, 226*f*, 228*f*, 230*f*
Metaphase, mitotic, 207*f*, **208**, 208*f*, 229*f*
Metaphase plate, 208, 208*f*

Metapopulations, **922**-923
Metastases, **214**
Methane
as greenhouse gas, 992
production of, 960
Methanococcoides burtonii, 495*f*
Methanogenesis, 497
Methanogens, **148**, **487**, 960
Methionine, 49*f*
Methyl-G cap, 310
Methyl groups, 41*f*
Metrarabdotos, 620
MHC class II proteins, **850**
MHC class I proteins, **850**
MHC (major histocompatibility complex) proteins, 87, **850**-851, 851*f*
Mice
behavioral genetics in, 894, 894*f*
genome of, 386, 387*f*, 400
knockin, 365
knockout, 364-365, 365*f*
pocket, 412*f*
Micelles, 60, 60*f*
Microalgae, biofuels and, 371, 371*f*
Microbial degradation, 371
Microbodies, **78**
Microclimates, 983-984
Microfossils, **495**, 495*f*
Micrognathozoa, 600*f*
Micronucleus, **534**, 534*f*
Micronutrients, **718**, 719*f*
Microphylls, **673**
Micropyle, **581**
Micro-RNA (miRNA), 303, **342**
biogenesis and function of, 342, 342*f*
in gene expression, 342
noncoding DNA and, 389
siRNAs distinguished from, 343
Microscopes
electron (*See* Electron microscopes)
light, 67, 67*f*, 68*t*
Microsporidia, 555, 555*f*, 556, 556*f*
Microtubule-organizing centers, 82
Microtubules, 72*f*, 73*f*, **81**
in metaphase, 207*f*
in prometaphase, 207-208
Microtus (voles), behavioral genetics in, 895, 895*f*
Microvilli, 67, **793**
eukaryotic, 73
in small intestine, 738, 793, 793*f*
Middle ear, **776**-777, 776*f*
Middle lamella, **85**, 85*f*, 209
Miescher, Friedrich, 279
Migratory behavior, 898-899, 898*f*-900*f*
Migratory locusts *(Locusta migratoria),* population density of, 931*f*
MIH (MSH-inhibiting hormone), 833
Milk, human, 888
Milk let-down reflex, 888
Milkweed (Asclepiadaceae), chemical defense of, 950
Millipedes, 594, 595*f*, 616
Mimicry, Batesian, 952, 952*f*
Mimivirus, 510
Minerals, transport in plants, 707*f*, 711*f*, 713
Minimal medium, 301-302
Mirabilis jalapa (four o'clock), 261
mi-RNA. *See* Micro-RNA (miRNA)
Mirounga angustirostris. See Northern elephant seal *(Mirounga angustirostris)*
Mismatch repair (MMR), 294
Missense mutations, **321**, 321*f*
Mistletoe *(Arceuthobium),* dispersal of seeds of, 921
Mites, 615
Mitochondria, **71**, **80**, 81*f*
ATP synthesis in, 80, 144, 145*f*
endosymbiosis and, 81, 81*f*
energy cycle of, 171-172, 172*f*
eukaryotic, 71, 73*f*, 74*f*, 488, 488*f*
evolution of, 521-522, 522*f*
genetic code in, 306
Mitochondrion-related organelles (MROs), 526
Mitogen-activated protein (MAP) kinases, **187**, 213, 215*f*
Mitogens, 187
Mitosis, **200**, **204**, 206*f*-207*f*
anaphase of, 207*f*, 208, 229*f*
chromosome segregation in, 204
cytokinesis and, 204, 204*f*, 207*f*, 209
evolution in eukaryotes, 523
in fungi, 549
meiosis contrasted with, 223-224
metaphase of, 207*f*, 208, 208*f*, 229*f*
in plants, 570, 570*f*
prometaphase of, 206*f*, 207-208
prophase of, 206, 206*f*, 229*f*
telophase of, 207*f*, 208-209, 229*f*
Mitral valves, **810**
Mixotrophs, 524
MMR (mismatch repair), 294
Mobile genetic elements, 390
Models, 9
Moisture, biomes and, 981, 981*f*, 983*f*
Molar concentration, **34**
Molds
bread, 558, 559*f*
spores of, 549
Molecular clocks, 443, 443*f*, **479**
Molecular cloning, 356, 357*f*. *See also* Cloning (of organisms)
Molecular evidence
animal phylogenetic tree and, 599*t*, 600*f*-601*f*, 601-602, 602*f*
evolution and, 13-14, 14*f*
Molecular markers, 265, 380
Molecular motors, 82-83, 83*f*
Molecules, 4*f*, **5**, **27**
bonds in, 27-30, 28*t*
Brownian movement of, 102, 102*f*
chiral, 42, 42*f*
hydrophilic, 34
hydrophobic, 34
nonpolar, in water, 34
organic (*See* Carbohydrates; Lipids; Macromolecules; Nucleic acids; Organic molecules; Proteins)
polar, water as solvent for, 33, 33*f*
Moles, **34**
Mollicutes, 70
Mollusca, 599*t*, 600*f*, 602, **608**
Mollusks, 599*t*, 602, 607-609, 608*f*, 609*f*. *See also specific mollusks*
circulatory systems of, 808, 808*f*
eyes of, 442, 442*f*, 780, 780*f*
locomotion by, 753
sperm binding in, 874*f*
Molting, of exoskeleton, 602, 602*f*, 745
Molybdenum, in organisms, 27
Monarch butterfly *(Danaus plexippus)*
adaptation defending against predators, 951, 951*f*
coevolution of, 953
migration of, 898, 900*f*
Monkeys, 645, 645*f*
as pollinators, 685
vervet *(Cercopithecus aethiops),* language in, 902, 902*f*
Monocots, 585
leaves of, 675
roots of, 667*f*, 668
stems of, 669-671, 670*f*
Monocytes, 807*f*, 848
Monoecious plants, **686**, 687*f*
Monohybrid cross, 239-240, 239*f*, 242*f*, 243-244
Monokaryotic hypha, **549**
Monomers, **42**
Mononucleosis, 509*t*
Monophyletic groups, **480**, 481*f*, 482*f*, **483**
domains as, 486
Monopolar attachment, 225
Monosaccharides, **44**, 44*f*, 45*f*
Monosomics, **266**
Monotremes, **642**, 643*f*
Monounsaturated fats, 58
Monsoons, 983
Montagu's harriers, 903
Morels *(Morchella esculenta),* 561, 562*f*
Morgan, Thomas Hunt, 257, 263
Mormon crickets *(Anabrus simplex),* parental investment in, 422, 422*f*
Morning-after pill, 871
Morphogenesis, in angiosperms, 689-690, 690*f*
Mortality, **924**
Mortality rate, **925**
Mosaics, genetic, **258**-259, 259*f*
Mosquitoes *(Anopheles),* 532-533, 533*f*, 542
genome of, 387*f*, 398
Moss animals, 599*t*, 611, 611*f*
Mosses, 571, 572*f*, 574*f*
club, 574, 576, 576*f*, 578*t*
distribution of, 572
genomes of, 572-573, 573*f*
succession and, 958
Moths, 616*f*
behavioral isolation and, 454-455
peppered *(Biston betularia),* color of, 433-434, 433*f*, 434*f*
as pollinators, 684
population fluctuations in, 932*f*
silk *(Bombyx mori),* pheromone in, 901
Motifs of proteins, **52**, 52*f*
Motor effectors, 761
Motor neurons, 740*f*, 752, 752*f*, **761**, 761*f*, 786
Motor proteins, 207-208, 750
Mouse. *See* Mice
Mouth, 790, 790*f*
Movement, flagella and. *See* Flagella; Locomotion
of animals, 594, 595*f*
of annelids, 610
as characteristic of life, 3
skeletal muscle and, 748, 749*f*
MPa (megapascals), **706**
MPF (M phase-promoting factor), **210**-211
MPF (maturation-promoting factor), 210-211, 210*f*
M-phase cyclin, 213
M phase-promoting factor (MPF), **210**-211
mRNA. *See* Messenger RNA (mRNA)
MROs (mitochondrion-related organelles), 526
MSH (melanocyte-stimulating hormone), **832**
MSH-inhibiting hormone (MIH), 833
Mucus-secreting cells, 792, 792*f*
Müller, Fritz, 952
Müllerian mimicry, **952**, 952*f*
Multicellularity
in animals, 594
cell communication in, 180-182
in chlorophytes, 538, 538*f*
in eukaryotes, 488-489
in fungi, 548-549, 549*f*
gas exchange and, 798
in protists, 524
Multienzyme complexes, **123**, 123*f*
pyruvate oxidation and, 139
Multiple fruits, 694*f*
Multiple sclerosis, 369
Multiplication, rule of, **243**
Multipotent stem cells, **874**
Mummichog *(Fundulus heteroclitus),* 412, 425
Muscle cells, 82
Muscle contraction
calcium ions and, 751-752, 752*f*
of heart muscle, 811-812, 813*f*
myofilaments and, 749, 749*f*
sliding of filaments and, 749-751, 750*f*, 751*f*
Muscle fibers, **742**-743
contraction of, 749-750, 749*f*
Muscles
cardiac, 743-744, 743*t*
proprioceptors and, 774
skeletal, 742-744, 743*t*, 748, 749*f*
smooth (visceral), 742, 743*t*
striated, 742
Muscle spindles, **774**
Muscle tissue, **736**, 742-744
Muscular dystrophy, 270*t*
Mussels, predation and, 957, 957*f*
Mustard *(Brassicaceae),* chemical defense of, 950
Mutagens, **293**
Mutations
allele frequencies and, 409, 409*f*
base substitution, 321, 321*f*
cancer and, 214-216, 387
chromosomal, 322-323, 323*f*
DNA repair and, 293-295, 294*f*
driver, 216
expansion, triplet repeat, 322
frameshift, 304, 322
guard cell stomata position and, 660, 660*f*
human rates of, 322
missense, 321, 321*f*
natural selection countered by, 419
nonsense, 321, 321*f*
nutritional, 301

Mutations (*Continued*)
oncogenic, 216
phenotype and, 247, 247*f*
in plants, 689, 689*f*
point, 321–322, 321*f*
random *vs.* directed, 296
silent, 321, 321*f*
triplet repeat expansion, 322
in vitro mutagenesis and, 362–363
Mutualism, **953**, 954, 954*f*
Mutualistic symbiosis, **551**
Mutually exclusive events, 243
Mycelium, **548**, 549*f*
primary, 561, 561*f*
secondary, 560, 561*f*
Mycobacterium leprae, 507*t*
Mycobacterium tuberculosis, 396, 506, 507*t*
Mycoplasma, 70, 507*t*
Mycorrhizae, 552–553, 552*f*, 560, 953
Myelinated axons, 765–766, 766*f*
Myelin sheaths, 740, **762**, 762*f*, 785
Myofibrils, **743**, **749**, 750*f*
Myofilaments, **749**, 749*f*
Myoglobin, 804
Myosin, **83**, 749, 750*f*
Myriapoda, 613*t*, 616
Myxini, 629, 630

N

n. See Haploid *(n)*
2*n*. *See* Diploid (2*n*)
NAA (naphthalene acetic acid), 722
NAD+. *See* Nicotinamide adenine dinucleotide
NADH
electron transport and, 133–134, 133*f*
oxidation of, in glycolysis, 136, 137*f*
oxidation-reduction reactions and, 132, 132*f*
NADH dehydrogenase, **143**, 143*f*, 260
NADPH. *See* Nicotinamide adenine dinucleotide phosphate (NADPH)
NADP reductase, 168, 169*f*
Naive lymphocytes, 850
Naked mole rat, social groups formed by, 913
Naphthalene acetic acid (NAA), 722
Nathans, Daniel, 355
Natural killer (NK) cells, **847–848**, 847*f*
Natural selection, 10, **12**, 250
alternative mating strategies and, 907–909, 908*f*
behavior and, 903
biological species concept and, 456–457
conditions necessary for, 411
convergent evolution and, 444–445
criticism of, 446–447
enzyme polymorphism and, 425
experimental study of, 415–417, 417*f*, 418*f*
forces countering, 419
reinforcement of reproductive isolation by, 458, 458*f*
reproductive success and, 926
speciation and, 460, 460*f*
Natural transformation, 503, 505*f*
Nauplius, **616**
Nautilus pompilius, 608*f*
Navigation, by birds, 899, 900*f*
NDI (nephrogenic diabetes insipidus), 104
Neanderthals, 648, 650
Necator, 613
Nectaries, 585
nectin-4 gene, 194
Negative control, of gene expression, **332**
Negative feedback, **823**–824, 823*f*, 930, 931*f*
Negative gravitropic responses, 728
Negative pressure breathing, **800**
Negative-strand viruses, 509
Neisseria gonorrhoeae, 507*t*
Nematocysts, **604**, 605*f*
Nematocytes, **604**, 605*f*
Nematoda, 599*t*, 601*f*
Nematodes, 601–602, **612**, 612*f*
carnivorous fungi and, 550, 550*f*
digestive system of, 790, 790*f*
genome of, 387*f*
human diseases caused by, 613
Nemertea, 599*t*, 600*f*
Nemertean (Cerebratulus), sperm binding in, 874*f*
Nemertean worms *(Cerebratulus),* sperm binding in, 874*f*
Neocallimastigomycota, 555, 555*f*, **557**
Neocallimastix, 557
Neonates, 887
Neotyphodium, 551, 551*f*
Nephridia, **608**, **839**, 839*f*
Nephrogenic diabetes insipidus (NDI), 104
Nephrons, 839, 842–844, 842*f*, 843*f*
cortical, 840, 840*f*
filtrate production by, 840–841, 841*f*
juxtamedullary, 840, 840*f*
urine collection by, 841
Nereis virens, 609*f*
Neritic waters, **988**, 988*f*
Nernst equation, **763**
Nerves, 740, 740*f*
Nerve tissue, **736**, 740
Nervous system
action potentials and, 764–766, 765*f*
autonomic, 761, 761*f*, 772–773, 772*f*
central, 740, 761, 761*f*, 769–771, 769*f*, 770*f*
chemoreceptors of, 778–779
mechanoreceptors of, 774–775
neurons of (*See* Neurons)
parasympathetic, 761, 761*f*, 772–773, 772*f*
peripheral, 740, 761, 761*f*, 772–773, 772*f*
photoreceptors of, 781–784, 781*f*, 783*f*
sensory receptors of, 761, 773
somatic, 761, 772
supporting cells of, 762, 762*f*
sympathetic, 761, 761*f*, 772, 772*f*
synapses of, 766–768, 766*f*
vertebrate, 769, 769*f*
vibration detectors of, 775–778
Net primary productivity (NPP), **967**
Neural crest, 627, **883**
cell migration from, 883–884, 884*f*, 885*f*
Neural groove, **883**
Neural plate, 883
Neural tube, **883**, 883*f*
Neuroendocrine reflex, **832**
Neuroglia, **740**, **762**
Neurohormones
hypothalamic, 832–833, 833*f*
pituitary, 831–832
Neurohypophysis, **831**–832
Neuromuscular junctions, **752**, 752*f*
Neurons, **740**
interneurons (association), 740*f*, 761, 761*f*
membrane potential and, 762–763, 763*f*
motor (efferent), 740*f*, 752, 752*f*, 761, 761*f*
postganglionic, 772
postsynaptic, 768
preganglionic, 772–773
sensory (afferent), 740*f*, 761, 761*f*
size of, 785
structure of, 761–762, 762*f*
Neurospora crassa, 301, 301*f*
Neurotransmitters, **181**, 181*f*, 766–767
photoreceptors and, 781–784, 781*f*, 783*f*
Neurulation, **882**–883
in humans, 885–886
Neutral atoms, 23
Neutrons, 22, 22*f*
Neutrophils, 807*f*, **847**
Newton, Isaac, 7
New World monkeys, 644, 644*f*
New Zealand alpine buttercups, adaptive radiation and, 465–467, 467*f*
Next-generation sequencing (NGS), 361, 384, 385*f*
Niches, **945**–948
fundamental, 945
interspecies competition for occupancy of, 945, 945*f*
overlap of, coexistence and, 947, 947*f*
realized, 946
restriction of, 946
Nicotiana tabacum, 699*f*
Nicotinamide adenine dinucleotide (NAD+), **57**, **132**
as electron carrier, 133, 134*f*
oxidation-reduction reactions and, 132, 132*f*
regeneration of, in glycolysis, 138, 138*f*
Nicotinamide adenine dinucleotide phosphate (NADPH)
CO_2 incorporation into organic molecules and, 170–172
in photosynthesis, 158*f*, 160, 166–167
Nile crocodile *(Crocodylus niloticus),* 637
9 + 2 structure, 83, 84*f*
Nirenberg, Marshall, 305–306
Nitric oxide (NO), 180
receptor, 185
Nitrification, 505, **963**
Nitrogen
electron energy levels for, 26*f*
in organisms, 26–27
Nitrogen cycle, 962–964, 963*f*
Nitrogen fixation, **963**
Nitroglycerin, 185
Nitrous oxide, as greenhouse gas, 992
NK (natural killer) cells, **847–848**, 847*f*
30-nm fiber, **202**, 202*f*
NO (nitric oxide). *See* Nitric oxide (NO)
Noble gases, 26
Nociceptors, **774**
Noctiluca, 532*f*
Nodes, of plant stems, **670**, 670*f*
Nodes of Ranvier, 740, **762**, 762*f*
Noncompetitive inhibitors, **124**, 125*f*
Nonconservative missense mutations, 321
Noncyclic photophosphorylation, **166–168**, 167*f*
Nondisjunction, 207, **265**
of autosomes, 266, 266*f*
of sex chromosomes, 266–267
Nonequilibrium systems, 15
Nonextreme archaea, **487**
Nongranular leukocytes, **808**
Noninvasive prenatal testing, 270–271
Nonpolar amino acids, 48, 49*f*
Nonpolar covalent bonds, **29**
Nonprotein enzymes, 123
Nonrandom mating, allele frequencies and, 409*f*, 410
Nonsense mutations, **321**, 321*f*
Nonshivering thermogenesis, 826
Nonvertebrate chordates, 626, 626*f*
Norepinephrine, 836, 836*f*
Normal distribution, 248
Northern elephant seal *(Mirounga angustirostris)*
bottleneck effect and, 411
intrasexual selection in, 423, 423*f*
Northern lynx *(Lynx canadensis),* population cycles of, 932, 932*f*
Notochord, 882
NPP (net primary productivity), **967**
Nucellus, **581**
Nuclear envelope, 73*f*, **74**, 75*f*
breakdown of, 206
in telophase, 208
Nuclear lamina, 74, 74*f*
Nuclear lamins, 74
Nuclear pores, 73*f*, **74**, 74*f*, 75*f*
Nuclear receptor superfamily, 184
Nuclear reprogramming, 877–878, 877*f*–878*f*
Nuclear transplant, 876, 876*f*–877*f*
Nucleic acids, 43*t*, **55**–56, 56*f*, **279**. *See also* DNA (deoxyribonucleic acid); RNA (ribonucleic acid)
energy reactions and, 57, 57*f*
as nucleotide polymers, 55–56, 56*f*
viral, 508–509, 508*f*, 509*t*
Nucleoid region, **496**
Nucleoids, **68**, 69*f*
bacterial, 199
prokaryotic, 495
Nucleolus, 73*f*, **74**
eukaryotic, 73*f*, 74*f*
Nucleoplasm, 74*f*
Nucleosomes, **201–202**, 202*f*
Nucleotides, 55–56, 56*f*, 279–280, 280*f*
Nucleus, 68, **71**, **73**
in diplomonads, 526
of eukaryotic cells, 71, 73–75, 73*f*–75*f*

in fungal cells, 549
origins of, 521, 521*f*
somatic, reprogramming of, 876–878, 877*f*–878*f*
in telophase, 208–209
Nutrients
absorption by plants, 709–711, 709*f*, 712*f*
absorption of, 794–795, 795*f*
in freshwater habitats, 986–987
fungi absorption of, 550, 550*f*
macronutrients, 718
micronutrients, 718, 719*f*
plant requirements for, 718–719, 719*f*
in plasma, 807
reabsorption by kidneys, 841–842
in seeds, 691–692, 691*f*, 695, 696*f*
Nutrition
of fish, 628
of fungi, 550, 550*f*
of protists, 524
of vascular plants, 574, 574*f*
Nutritional mutants, **301**–302

O

Oak trees *(Quercus)*
ecological isolation of, 454
fibers in, 663
Obligate symbiosis, **551**
Ocean currents, 982–983, 982*f*
Oceans
continental shelf ecosystems in, 988–989
deep, 989, 989*f*
global temperature change and, 991
marine mammals and, swimming by, 753
oligotrophic, 987
open, productivity of, 987–988, 987*f*, 988*f*
predators in, convergent evolution in, 444, 445*f*
Octet rule, **26**, 26*f*, 28
Octopus *(Octopus vulgaris)*, 608, 608*f*
Oenothera biennis (evening primrose), 684
Offspring
number of, reproductive success and, 926–927, 926*f*, 927*f*
parental ability to care for, mating systems and, 907, 908*f*
parent-offspring interactions and behavior and, 896, 896*f*
Oil content, of corn, 448
Okazaki fragments, **287**–288, 291
Okazaki, Reiji, 287
Okazaki, Tsuneko, 287
Old World monkeys, 645, 645*f*
Olfaction, 778
Oligodendrocytes, **762**
Oligotrophic bodies of water, 958, 986–987
OMIM (Online Mendelian Inheritance in Man), 269
Ommatidia, 413, 413*f*
Oncogenes, 186, **214**, 387
Oncogenic mutations, 216
One-gene/one-polypeptide hypothesis, **302**
Online Mendelian Inheritance in Man (OMIM), 269
On the Origin of Species by Means of Natural Selection (Darwin), 10, 12, 405, 453
Onychophora, 599*t*, 601*f*
Oocytes
primary, 867, 868, 868*f*
secondary, 868, 868*f*, 869*f*, 870
Oogenesis, 869*f*
Oomycetes, 530–531
Open circulatory systems, 596–**597**, 808, 808*f*
Open reading frames (ORFs), **388**
Operators, **332**
Opercular cavities, 798–799, 799*f*
Operculum, **631**, **798**
Operons, **309**, **332**
Ophiuroidea, 617*f*
Opisthokonta, **526**, 526*f*, 539
Opportunistic infections, HIV infection and, 511
Optimal foraging theory, **905**–906
Optimum temperature, 124, 124*f*
Orangutan, 645
Orbitals of electrons, **23**, 24, 24*f*
Orders, **483**, 484*f*, 485
new, 490
ORFs (open reading frames), **388**
Organellar inheritance, 260–261
Organelles, 4*f*, **5**, **69**, **71**–73, 73*f*. *See also* Chloroplasts; Endosomes; Mitochondria
breakdown of, 78–79, 78*f*
DNA replication, bacterial, 289–290, 289*f*, 289*t*, 290*f*
locomotor, of protists, 523
prokaryotic cells and, 69*f*, 70
Organic compounds, 27
oxidation of, 132–135
Organic molecules, **41**–43. *See also* Carbohydrates; Lipids; Macromolecules; Nucleic acids; Proteins
building of, by Calvin cycle, 170–172, 171*f*, 172*f*
functional groups in, 41–42, 41*f*
transport in plants, 717, 718*f*
Organismal level of organization, 4*f*–5*f*, 5
Organisms, **5**, 5*f*
elements common in, 26–27
Organ of Corti, 776, 777*f*
Organogenesis, 882–883
in humans, 885–887, 886*f*
Organs, 4*f*, **5**, 736, 736*f*
of vertebrates, 627–628, 628*f*
Organ systems, **5**, 5*f*, **736**, 736*f*
animal, 596–597, 597*f*
oriC, 286, 291
Orientation, for migration, 899, 899*f*
Origin of replication, 199, 199*f*
Ornithorhynchus anatinus (duck-billed platypus), 642, 643*f*
Oryctes nasicornis (European rhinoceros beetle), 616*f*
Oryza sativa. *See* Rice *(Oryza sativa)*
Oscillating selection, **420**, 476
Osmoconformers, **838**
Osmolarity, 838–839
Osmoregulation, 838, 839*f*
hormone control of, 844–845
kidney and, 839–844
Osmoregulators, **838**–839, 839*f*
Osmosis, 103, **104**–105, 104*f*, 110*t*, **708**
aquaporins and, 709, 709*f*
Osmotic balance, maintenance of, 105
Osmotic concentration, **104**
Osmotic pressure, **104**, 105*f*
Ossicles, **618**
Osteoblasts, **747**
Osteocytes, **742**
Osteoporosis, 748
Ostracoderms, 629
Ostrich, 639
Ötzi, 36
Outcrossing, 683, 685–686, 686*f*, 687*f*
transgenic crops and, 373
Outer bark, 671, 671*f*
Outer ear, **776**, 776*f*
Outer membrane, 157*f*
Outgroup comparison, 477
Outgroups, **477**
Ovaries of flowers, **585**, 585*f*, **682**, 682*f*
Oviducts, 870
Oviparity, **862**
Ovoviviparity, **862**
Ovulation, **867**–870, 867*f*–869*f*
prevention of, 870
Ovule, **580**, 580*f*, 584, 585*f*
Owls, 639
Oxaloacetate
in C_4 plants, 174, 174*f*
regeneration, in citric acid cycle, 141*f*, 142
Oxidation, **25**, **117**, 117*f*
in citric acid cycle, 141*f*, 142
in glycolysis, 136, 136*f*
of organic compounds, 132–135
in photosynthesis, 158
β Oxidation, of fatty acids, 149–**151**, 151*f*
Oxidation-reduction reactions, 25, **117**, 117*f*
energy harvest by, 132, 133*f*
in photosynthesis, 158, 163, 165, 165*f*
Oxidative phosphorylation, **134**–135
Oxpeckers, 954, 954*f*
Oxygen
life in freshwater habitats and, 986–987
in organisms, 26–27
photorespiration and, 172
in photosynthesis, 158, 160
respiration and, 797, 804–806
transport in blood, 804–806, 806*f*
Oxygenic photosynthesis, 157, 157*f*, 166, 496
Oxytocin, **831**–832, 833, 865*t*
birth and, 887
lactation and, 888
OY genotype, 267
Oyster mushroom *(Pleurotus ostreatus)*, 550, 550*f*
Oysters
shell evolution of, 438
survivorship curve for, 926, 926*f*
Ozone hole, 293, 990–991, 990*f*

P

Paal, Arpad, 720
Pacemaker, **811**
Pacific salmon, semelparity in, 928
Paine, Robert, 957*f*
Pain receptors, 774, 774*f*
Paired appendages, of fish, 627
Pakicetus, 438*f*
PAM (protospacer adjacent motif), 364*f*
PAMPs (pathogen-associated molecular patterns), 845
Pan, 644
Pancreas, 790, 790*f*, 791*f*
enzymes secreted by, 794
hormones secreted by, 837, 837*f*
Pancreatic amylase, **794**
Pancreatic duct, 794, 794*f*
Pancreatic islets, 837
Pancreatic juice, 790, 793
Pandemics, influenza, 514, 514*f*
Pangshura tentoria (Indian tent turtle), 636*f*
Panolis, population fluctuations in, 932*f*
Panthera leo (lion). *See* Lions *(Panthera leo)*
Pantherophis alleghaniensis (eastern rat snake), 453*f*
Panting, 825
Pan troglodytes. *See* Chimpanzee *(Pan troglodytes)*
Papermaking, 663
Papilio glaucus (tiger swallowtail butterfly), mimicry in, 952, 952*f*
Parabasalids, 526–527, 527*f*
Paracrine signaling, **181**, 181*f*
Parallax, 783
Paramecium, 534–535, 534*f*
competitive exclusion among species of, 946, 947*f*
Didinium predation on, 949, 949*f*
extrusion in, 105
sexual reproduction in, 525*f*
Paramyxoviruses, 509*t*
Paraphyletic groups, **480**, 481*f*, 482, 482*f*
Parasites, **551**
external (ectoparasites), 955, 955*f*
flatworms, 606
host behaviors and, 956, 956*f*
internal (endoparasites), 955
microsporidia, 556, 556*f*
Parasitic roots, **668**
Parasitism, **953**, 955–956, 955*f*
competition countering by, 956
of kinetoplastids, 528–529, 528*f*
Plasmodium, 532–533, 533*f*, **540**, 542, 556
Parasympathetic division, of autonomic nervous system, 761, 761*f*, 772–773, 772*f*
Parathyroid hormone (PTH), **835**, 835*f*
Parazoa, 600*f*, 603
Parenchyma cells, **661**–662, 662*f*, 667
Parental investment, **421**–422
Parent-offspring interactions, behavior and, 896, 896*f*
Parietal cells, 792, 792*f*
Parietal pleural membrane, **801**
Parkinson's disease, 767
Parrots, 639
Parsimony, principle of, 478–479, 478*f*
Parthenogenesis, 873
Partial diploids, **501**
Partial pressure, **800**
Particle bombardment, 366
Parus major. *See* Great tit *(Parus major)*
Passenger mutations, 387
Passeriformes, 639, 640*f*

Passive transport
by diffusion, 102–103, 103*f*, 110*t*
by osmosis, 103–105, 104*f*, 110*t*
Pasteur, Louis, 8, 849
Pathogen-associated molecular patterns (PAMPs), 845
Pathogens, **551**
fungal, 553–554, 554*f*, 561
recognition by innate immunity, 845–846
responses to, 846–848, 846*f*
Pavo cristatus (peacock), sexual selection in, 422, 422*f*
Pax6 gene, 780–781, 780*f*
Payen, Anselme, 127
PCNA (proliferating cell nuclear antigen), 291
PCR. *See* Polymerase chain reaction
PDGF (platelet-derived growth factor), **213**
Peacock *(Pavo cristatus),* sexual selection in, 422, 422*f*
Peat moss *(Sphagnum),* 572
Pedicels, **584**
Pedigrees, **267**
albinism and, 267–268, 267*f*
hemophilia and, 268–269
Huntington disease and, 269, 269*f*
mutant alleles and human diseases, 269, 270*t*
sickle-cell disease and, 268–269, 268*f*
Pellicle, in *Euglena,* 527, 527*f*
Pelycosaurs, 634
Penetration, **510**
in lytic cycle, 510, 511*f*
Penicillin, 499, 561
Penicillium chrysogenum, 561
Penis, **863**, 863*f*, 865, 865*f*
PEP (phosphoenolpyruvate), 135*f*, 173–174, 174*f*
PEP carboxylase, 173–174
Peppered moth *(Biston betularia),* color of, 433–434, 433*f*
Pepsin, **792**
Pepsinogen, **792**
Peptic ulcers, 506, 507*t*
Peptide bonds, 48, 48*f*
formation of, 317, 318*f*
Peptide hormones, **832**
Peptides, 828
Peptidoglycan, 70, 497–498, 499*f*
Peptidyl site (P site), **315**, 315*f*, 316–317, 318*f*
Peptidyl transferase, **315**, 316–317, 318*f*
Perch, 490
Perennial plants, **696**–697, 697*f*
Pericardial cavity, 737, 737*f*
Pericarp, 692
Pericentriolar material, **82**, 205
Pericycle, **667**, 667*f*
Periderm, 671, 671*f*
Periodic table of elements, **26**, 26*f*
Peripheral chemoreceptors, **779**
Peripheral nervous system (PNS), **740**, **761**, 761*f*, 771
autonomic, 772–773, 772*f*
somatic, 772
Peristalsis, **792**
Peritoneal cavity, 737, 737*f*
Peritubular capillaries, **840**, 841*f*
Permian period, mass extinction in, 468, 469*f*
Peroxisomes, **79**
of eukaryotes, 71–73, 72*f*, 74*f*
Peruvian primrose *(Ludwigia peruviana),* 685*f*
Pesticides, resistance to, selection for, 412, 412*f*
Petals, **584**
Petioles, **673**
Petromyzontida, 628*f*
Petromyzontida , 629
2PG (2-phosphoglycerate), 137*f*
3PG (3-phosphoglycerate), 137*f*
PGA (3-phosphoglycerate), 170
p53 gene, 215
pH. *See also* Acids; *Bases*
of blood of swimming fish, 152
chemoreceptors and, 778
enzyme activity and, 124, 124*f*
oxygen transport in blood and, 804–805, 805*f*
protein function and, 61
stabilization by buffers, 35, 35*f*
in stomach, 793
of urine, 842
Phage. *See* Bacteriophages
Phage lambda, 502
Phagocytic cells, 847
Phagocytosis, 77–78, 108*f*, **109**, 110*t*
Phagotrophs, 524
Pharynx, 790, 790*f*
Phase change, **681**
Phase-contrast microscopes, 68*t*
Phelloderm, 671, 671*f*
Phenotypes, **240**
alleles affecting multiple aspects of, 413
changing, oscillating selection and, 420
environment effects on, 245, 245*f*, 249*t*
frequency-dependent selection favored of, 419–420, 420*f*
genotype and, 247, 247*f*
with greater fitness, frequency of, 418
intermediate, selection and, 414–415, 415*f*
multiple genes impacts on, 245, 246*f*
Phenylalanine, 49*f*
Phenylketonuria, 270, 270*t*
Pheromones, **454**, **901**
Phloem, **574**, 664, 664*f*, 667, 730
organic molecule transport in, 716–717, 717*f*, 718*f*
Phloem loading, **717**
Phlox, 684
Phosphatases, **182**, 182*f*
Phosphate groups, 41*f*, 43*t*, 60, 60*f*
Phosphates, 964
Phosphate-to-oxygen (P/O) ratio, calculation of, 146–147
Phosphodiester backbone, 281, 282*f*
Phosphodiester bonds, 56, 56*f*, **279**, 280*f*
Phosphoenolpyruvate (PEP), 137*f*, 173–174, 174*f*
2-Phosphoglycerate (2PG), 137*f*
3-Phosphoglycerate (3PG), 137*f*
3-Phosphoglycerate (PGA), 170
Phospholipid bilayer, 95, 97*t*
anchoring of proteins in, 101
fluid nature of, 98–99, 99*f*
formation of, 98–99, 98*f*
Phospholipids, 43*t*, **59**–60, 60*f*
archaeal, 497, 497*f*
Phospholipid sheets, **95**, 95*f*
Phosphorus, 964, 964*f*
in organisms, 26–27
Phosphorylase kinase, 192, 192*f*
Phosphorylation, **182**. *See also* Photophosphorylation
in glycolysis, 136, 136*f*, 137*f*
oxidative, 134–135
protein function and, 181–182
of proteins, 182, 182*f*
receptor tyrosine kinases and, 186, 186*f*
substrate-level, 134, 135*f*
Phosphorylation cascades, 187–188, 188*f*
Phosphotyrosine domains, cellular protein binding to, 187, 187*f*
Photic zone, 987*f*, **988**
Photoautotrophs, **505**, **965**
Photoefficiency, 162
Photoelectric effect, **161**, 162*f*
Photoheterotrophs, **505**
Photolyase, 294, 294*f*
Photomorphogenesis, **724**
Photons, **160**, 161*f*
Photophosphorylation
chemiosmotic mechanism for, 168, 169*f*
cyclic, 166, 168
noncyclic, 166–168, 167*f*
Photopsins, **781**
Photoreceptors, **779**
invertebrate, 779
sensory transduction in, 782–783, 783*f*
vertebrate, 781–784, 781*f*, 783*f*
Photorepair, **294**, 294*f*
Photorespiration, **172**–174, 173*f*
Photosynthesis, **157**
anoxygenic, 157, 166, 496
archaeal, 505
bacterial, 496, 506
beginning of, 695, 696*f*
C_3, **170**
C_4, 174, 674
Calvin cycle and, 170–172, 171*f*, 172*f*
carbon dioxide and, 371, 960–961
carotenoids in, 162–163, 163*f*
as chemical process, 158–160, 159*f*
chemiosmosis and, 168–169
chlorophyll in, 161–162, 164–165
chloroplasts and, 158, 166, 167*f*
in dinoflagellates, 532
in *Euglena,* 528, 528*f*
leaf structure and, 674
light and, 160–161
light capture for, 164–165, 165*f*
noncyclic photophosphorylation and, 166–168, 167*f*
oxygenic, 157, 157*f*, 166, 496
oxygen source in, 160
photorespiration and, 172–174, 173*f*
in protists, 524
rate of, light intensity and, 163–164, 164*f*
in red algae (Rhodophyta), 537, 537*f*
respiration and, 171–172
stages of, 157–158, 158*f*
stored nutrients for growth before, 691, 691*f*
thylakoid membrane and, 165–166
water as electron donor in, 160
Photosynthetic bacteria, 496, 503*f*, 505
chloroplast evolution from, 522–523, 522*f*
photosystem of, 166
plasma membrane in, 165
purple, 166, 166*f*
Photosynthetic pigments. *See* Accessory pigments; Pigments
Photosystems, **158**, 163–164
bacterial, 166, 166*f*
of chloroplasts, 166, 166*f*
of plants, 166–168, 167*f*
Phototrophs, 524
Phototropism, **724**, 726–727, 726*f*
pH scale, 34–35, 34*f*
Phycobiliproteins, 163
Phycobilisomes, 537
Phycodurus eques (leafy sea-dragon), 629*f*
Phycoerythrin, 537
Phyllobates terribilis (golden frog), chemical defense of, 951
Phyllotaxy, **669**
Phylogenetic species concept (PSC), **480**–482, 482*f*
Phylogenetic trees, **13**–14, 14*f*, 475–476, 475*f*
Phylogeny, **475**
of animals, 599*t*, 600*f*–601*f*, 601–602, 602*f*
behavioral evolution and, 903, 904*f*
as depiction of evolutionary relationships, 475–476, 475*f*
of fungi, 555, 555*f*, 555*t*
homoplasy and, 479
molecular clock and, 479
plant, 569, 569*f*, 578*t*
protist, 525, 526*f*
statistical approaches to, 479
of vertebrates, 628*f*–629*f*
Phylum, **483**, 484*f*, 485
Physalaemus pustulosus (Túngara frog), mate selection in, 424, 424*f*
Physcomitrella patens, 573
Physical laws, living systems and, 14–15
Physical maps of genome, **380**–383, 381*f*, 382*f*
Phytochrome, **724**
growth responses and, 725–726, 726*f*
plant growth and, 725–726, 725*f*
Phytophthora infestans, 531
Phytoplankton, in open oceans, 987
P_i (inorganic phosphate), **120**
Pied flycatchers *(Ficedula hypoleuca),* paternity within a territory and, 908*f*
PIF (prolactin-inhibiting factor), 833
Pigeons, 448
Pigments, **160**. *See also* Chlorophyll
accessory (*See* Accessory pigments)
bile, 794
in flowers, 684–685, 685*f*
photosynthetic, in Euglena, 527, 528*f*
respiratory, 804, 804*f*

Pigs
influenza virus and, 514
insulin from, 367
transgenic, 306*f*
Pike cichlid *(Crenicichla alta)*, 416–417, 417*f*
Pillbugs, 615
Pilobolus, 555*f*
Pilus, **499**
Pineal gland, 837–838
Pines, 578*t*, 581, 581*f*, 582*f*
Pinocytosis, 108*f*, **109**, 110*t*
Pintail ducks, behavioral isolation and, 454, 455*f*
Pinus longaeva (bristlecone pine), 581
Pinus palustris (longleaf pine), 581*f*
Pinworms *(Enterobius vermicularis)*, 613
PIP_2 (phosphatidylinositol-4,5-bisphosphate), 191, 191*f*, 192*f*
Pipevine swallowtail butterfly *(Battus philenor)*, mimicry of, 952, 9452*f*
Piromyces communis, 555*f*
Pisum sativum. *See* Garden pea *(Pisum sativum)*
Pith, 667, 670, 670*f*
Pithovirus, 510
Pits, in tracheids, 663, 663*f*
Pituitary dwarfism, **833**
Pituitary gland, **831**
anterior, 832–834
posterior, 832–833
PKA (protein kinase A), 190, 191*f*, 192
PKC (protein kinase C), 191
Placenta, 641, 641*f*, 643, 880, 881
hormone secretion by, 887, 888*f*
structure of, 885, 886*f*
Placental mammals, 643, 643*f*
convergent evolution and, 444, 445*f*
Placodermi, 629
Placozoa, 600*f*
Plague, 507*t*
Plantae, 3, 3*f*, 485–486, 485*f*
Plant body
diagram of, 656*f*
primary, 658
secondary, 658
Plant cells, 71–73, 73*f*
cell walls of, 85, 85*f*, 657, 657*f*
cytokinesis in, 209, 209*f*
osmotic pressure and, 105*f*
plasmodesmata in, 89, 89*f*
Plant growth
directional, 726–727, 726*f*
gravity and, 727–729, 727*f*, 728*f*, 729*f*
hormones guiding, 719–724
indeterminate, 656
light and, 724–727
Plant hormones
growth and, 719–724, 728
transpiration rates and, 715, 715*f*
Plant receptor kinases, **186**
Plants. *See also specific plants and types of plants*
adaptations defending against herbivores, 950, 950*f*
adaptations to water stress, 715–716
alternation of generations in, 570, 570*f*, 682
annual, 697, 697*f*
artificial selection and, 435–436, 436*f*
biennial, 697
bryophytes, 571–573, 571*f*, 572*f*, 573*f*
C_3, 173–174
C_4, 173*f*, 174, 174*f*
CAM, 174
classification of, 480, 482*f*
cloning of, 698–699, 699*f*
crop (*See* Agriculture)
detection of spacing by, 726
evolution from algae, 569*f*, 570
flowering (*See* Angiosperms; Flowers)
fungal pathogens of, 554–555, 554*f*, 561
fungal symbioses with, 553, 553*f*
growth of (*See* Plant growth)
heterochromatin formation in, 343
hormones and (*See* Plant hormones)
hybridization of, 456, 457
life spans of, 696–697, 697*f*
limitation of water loss by, 715
mineral transport in, 707*f*, 710*f*, 713
nutrient absorption by, 709–710, 709*f*, 712*f*
nutrient requirements of, 718, 719*f*
organic molecule transport in, 716–717, 717*f*, 718*f*
perennial, 696–697, 697*f*
photorespiration in, 172–174, 173*f*
photosynthesis in (*See* Photosynthesis; Photosystems)
photosystems of, 166–168, 167*f*
phylogeny of, 569, 569*f*, 578*t*
reproduction by (*See* Angiosperms; Flowers; Plants)
reproduction by seed (*See* Angiosperm reproduction; Angiosperms; Gymnosperms; Seeds)
RNA-silencing pathways in, 343
spacing of individual in populations and, 921–922
terrestrial, evolution of, 559–560, 569, 569*f*
terrestrial, land adaptations of, 571
transgenic, 365, 366*f*, 371–373, 372*f*, 373*f*
transpiration by, 711–715, 712*f*
vacuoles of, 79, 79*f*
vascular (*See* Vascular plants)
water transport in, 706–711, 712*f*, 730
Plaques, dental, 506
Plasma, **806**–807, 807*f*
Plasma membranes, **69**, **71**, **95**
archaeal, 497, 497*f*
of eukaryotes, 70, 72*f*, 73*f*
Plasmids, 358, 358*f*
conjugative, 501, 503*f*, 504*f*
Plasmodesmata, **72**, 73*f*, 87*t*, **89**, 89*f*
in charophytes, 539
Plasmodial slime molds, 540, 540*f*
Plasmodium, 532–533, 533*f*, **540**, 542, 556
Plasmodium falciparum, 421, 421*f*, 542
Plasmolysis, **707**
Plastids, **80**
Plastocyanin, 168
Plastoquinone, 168
Platelet-derived growth factor (PDGF), **213**
Platelets, 742, **806**, 807, 807*f*
Platyhelminthes, 599*t*, 600*f*, **602**, **605**–606, 606*f*
Platyspiza crassirostris (vegetarian tree finch), 430–431, 430*f*, 432*f*
Pleiotropic alleles, **246**, 246*f*, 249*t*
Pleiotropy, 413
Plesiomorphies, **477**
Plesiosaurs, neck vertebrae of, 441
Pleural cavity, 737, 737*f*, **802**
Pleural membrane
parietal, 801
visceral, 801
Pleurotus ostreatus (oyster mushroom), 550, 550*f*
Plumatella, 611*f*
Pluripotent stem cells, **808**, **874**–875, 875*f*, **878**
induced, 878
Pneumatophores, **668**, 669*f*, 716, 716*f*
Pneumonia, 507*t*
PNS. *See* Peripheral nervous system (PNS)
Poa annua. *See* Meadow grass *(Poa annua)*
Pocket mice, 412*f*
Pocket protein, 215
Podia, of Foraminifera, 536, 536*f*
Poecilia reticulata (guppy), color variation in, 416–417, 416*f*, 417*f*
Poikilotherms, 824
Poinsettia, leaves of, 675
Point mutations, **321**–322, 321*f*
Poison-dart frogs *(Dendrobatidae)*
behavioral evolution of, 903, 904*f*
chemical defense of, 951, 951*f*
Polar bears, range of, 920
Polar bodies, 873, 874*f*
Polar covalent bonds, **29**, 29*t*
Polarity, **477**
of characters, 477
Polar microtubule, 207*f*
Polar molecules, water as solvent for, 33, 33*f*
Polar tubes, 556, 556*f*
Polar uncharged amino acids, 48, 48*f*
Polio, 509*t*
Pollen, 580
of angiosperms, 586–587
of conifers, 581–582, 582*f*
formation of, 682, 683*f*
Pollen grains, **682**, 683*f*
Pollen tube, **580**
in angiosperms, 687, 687*f*–688*f*
Pollination, **587**, 683
by animals, 684–685, 684*f*, 685*f*, 953*f*, 954
cross-pollination, 373, 686–687, 686*f*, 687*f*
double fertilization and, 687–688, 687*f*–688*f*
flower color and, 685, 685*f*, 700
self-pollination, 685–686
by wind, 684
Pollinators, 585, **684**–685, 684*f*, 685*f*, 943*f*, 954
flower color and, 685, 685*f*, 700
Pollution
chemical, 989–990, 989*f*
climate change and, 959, 991–995
color of moths and, 433–434, 433*f*
ozone hole and, 990–991, 990*f*
Polyandry, 907
Poly-A polymerase, 310, 311*f*
Polychaete worms
Chaetopterus, sperm binding in, 874*f*
Myzostoma, sperm binding in, 874*f*
Polydactylism, 411
Polygenic inheritance, **245**, 246*f*, 249*t*
Polygyny, 907
Polymerase chain reaction (PCR), **359**
DNA replication and, 359, 360*f*
DNA sequencing and, 361
reverse transcription PCR, 360–361, 361*f*
in vitro mutagenesis and, 362–363
Polymers, **42**
Polymorphisms, 265, **406**, 406*f*, 425
Polynucleotides, 56
Polypeptides, **48**
Polyphyletic groups, **480**, 481*f*
Polyploidy, **461**, 462*f*, 872
Polyps, 604, 605*f*
Polysaccharides, **45**–47
Polyspermy, 872
Polytrichum (hair-cup moss), 572*f*
Polyunsaturated fats, **58**
Ponds, oligotrophic, 986–987
Pongo, 644
Pons, **769**, 770*f*
Populational level of organization, 5–6, 5*f*
Population cycles, 931–932, 932*f*
Population genetics, **406**
Population growth
environmental limits of, 931, 932*f*
human, exponential, 933–936
Malthus' views on, 11–12, 12*f*
projected growth rate, 933, 934*f*
reproductive success and, 926–927
resource availability regulation of, 930–933
sex and age of members and, 923–926
Population pyramids, **934**, 935*f*
Populations, **5**, 5*f*, **920**
biotic potential of, 929
exchange of members between, 922–923
growth of (*See* Population growth)
range of, 920–921, 920*f*
size of, interspecies competition and, 17
spacing patterns of individuals in, 921–922
P/O (phosphate-to-oxygen) ratio, calculation of, 146–147
p-orbitals, 24, 24*f*
Pore proteins, 101, 101*f*
Porifera, 599*t*, 600*f*, 601, **603**, 604*f*
Porin proteins, 51
Porphyra, 537
Porphyrin rings, 162, 162*f*
Positive control, of gene expression, **331**
Positive feedback, **823**, 931
Positive gravitropic responses, 728
Positive pressure breathing, **800**
Positive-strand viruses, 509
Posterior pituitary, **831**–832
Postganglionic neuron, 772

Postsynaptic neurons, 768
Postzygotic isolating mechanisms, **453**, 454*t*, 456, 456*f*
Potassium, in organisms, 26–27
Potassium voltage-gated channels, 764
Potatoes, 672, 672*f*
Potential energy, 24, **116**, 116*f*
  C–H bonds and, 132–133
Pouched mammals, 643, 644*f*
Power stroke, 751
Prader-Willi syndrome (PWS), 259
Prairie dog (*Cynomys ludovicanus*), alarm calling by, 901*f*
PRC2 (polycomb repressive complex), 339
Precipitation
  acid, 984, 985*f*
  biomes and, 984*f*, 985*f*
  climate change and, 991
  global temperature change and, 993
Precocial offspring, **907**
Predation, **949**
  animal defenses and, 950–951, 951*f*
  coevolution and, 950
  competition reduction by, 956, 957*f*
  influence on prey populations, 949–950, 949*f*
  mimicry and, 952, 952*f*
  plant defenses against herbivores and, 950, 9450*f*
  tropical biodiversity and, 969
Predators, selection to avoid, 411–412, 412*f*
Prednisone, 836
Preganglionic neurons, 772
Pregnancy, 868. *See also* Human prenatal development
  high-risk, 269
  prevention of, 870–871
Preimplantation genetic screening, **271**, 271*f*
Pre-mRNA splicing, **312**
Prenatal development. *See* Embryonic development; Human prenatal development
Prenatal testing, noninvasive, 270–271
Pressure-flow model, **717**
Pressure potential, **708**, 708*f*
Pressure tolerant archaea, 487
Prezygotic isolating mechanisms, **453**, 454–456, 454*t*
Prickly pear cactus, population explosion of, in Australia, 949
Pride, 922
Priestly, Joseph, 159
Primaquine, 542
Primary carnivores, **966**, 966*f*
Primary meristems, 658, 658*t*, 689
Primary motor cortex, **770**–771, 770*f*
Primary mycelium, 560–561, 561*f*
Primary oocytes, **867**, 868, 869*f*
Primary photoevent, in thylakoid reactions, 165
Primary plant body, **658**
Primary producers, **966**, 966*f*
Primary productivity, **945**, **966**
  of open oceans, 987–988, 987*f*, 988*f*
  species richness and, 969–970, 971*f*
Primary somatosensory cortex, **770**, 770*f*
Primary spermatocytes, 863
Primary structure of proteins, **47**–48, **50**, 50*f*
Primary succession, **958**, 958*f*
Primary tissues, **658**, 736
Primary transcripts, **310**
  splicing of, 311
Primary walls, **85**, 85*f*
Primates, **643**. *See also* Apes; Humans; Monkeys
  distinct features of, 643
  evolution of humans, 643–649, 645*f*, 647*f*, 648*f*, 649*f*
  nonhuman, language in, 902–903, 902*f*
Primer, for DNA polymerases, 286
Primers, 359
Priming reactions, in glycolysis, 136, 136*f*
Primitive streak, 881
Primordium, **584**
Primosomes, 289
Principle of Independent Assortment, 241–243, 242*f*, **243**, 244*f*
Principle of parsimony, **478**–479, 478*f*
Principle of Segregation, 238, **240**, 240*f*, 241, 253
PRL. *See* Prolactin (PRL)
Probability, prediction of genetic cross results using, 243–244
Problem solving, in animals, 897–898, 897*f*, 898*f*
Procambium, **658**, 658*t*
Processivity, 288
Prochloron, 70, 70*f*
Productivity, **966**
  primary, 945, 966, 969, 970*f*
  secondary, 967
  of a trophic level, 966
  tropical biodiversity and, 971
Product rule, **243**–244
Products, of chemical reactions, 30
Progesterone, 865*t*, 867, 868
  lactation and, 888
  placental secretion of, 887*f*, 888
Programmed cell death, 329
Progymnosperms, **580**
Prokaryotes, 2, 2*f*, **68**, 503*f*. *See also* Archaea; Bacteria
  anaerobic respiration in, 148, 148*f*
  animal *vs.* plant cells, 86*t*
  cell division in, 199, 201*f*
  cell walls of, 498, 498*f*
  classification of, 500–501, 502*f*–503*f*
  diseases caused by, 505–506, 507*t*
  DNA replication in, 286–290, 286*f*–290*f*, 289*t*
  DNA transfer in, 501–504, 503*f*, 504*f*, 505*f*
  domains of, 495–496, 495*f*, 496*f*
  eukaryotes compared with, 496, 496*f*
  external structures of, 498–500, 499*f*
  gene expression in, 320*t*, 329, 331–336
  genomes of, 397, 397*f*
  initiation in, 306–307, 307*f*, 316, 316*f*
  interior organization of, 500, 500*f*
  metabolism in, 496–497, 506
  microfossils of, 495, 495*f*
  nitrogen fixation, nitrification, and denitrification and, 963–964
  numbers of, 494
  range of, 920
  shapes of, 497
  transcription in, 306–309, 332
Prokaryotic cells, 69
  architecture of, 496–497, 497*f*
  flagella of, 69*f*, 71, 71*f*, 496, 498, 499*f*
  organelles of, 68, 69*f*
  size of, 496
Prolactin (PRL), **832**, 865*t*
  lactation and, 888
Prolactin-inhibiting factor (PIF), 833
Proliferating cell nuclear antigen (PCNA), 291
Proliferative phase, **868**
Proline, 49*f*
Prometaphase, 206*f*, **207**–208
Promoters, **306**
  eukaryotic, 309
  prokaryotic, 306–307, 307*f*
  transcription factors and, 337, 337*f*
Prophage, **510**
Prophase II, meiotic, 227*f*, 228, 229*f*
Prophase I, meiotic, 224–225, 226*f*, 228*f*
Prophase, mitotic, **206**, 206*f*, 228*f*
Proprioceptors, **774**
Prop roots, **668**, 669*f*
Prosimians, **644**, 644*f*
Prostaglandins, 43*t*, 58
  birth and, 887
Prostate gland, **864**
Proteasomes, **78**–79, 78*f*, **724**
Protein-encoding genes, 389*t*
Protein hormones, 828, **832**
Protein kinase A (PKA), 190, 191*f*, 192
Protein kinase C (PKC), 191
Protein kinases, 182, 182*f*, **183**
  cascade of, signal amplification by, 187–188, 188*f*
Protein-protein interactions, mapping of, 392
Proteins, 43*t*
  absorption of, 795, 795*f*
  acute-phase, 848
  adapter, 187
  amino acids in, 47–48
  auxin-binding, 721
  cAMP response, 334
  carrier (*See* Carrier proteins)
  catabolism of, 149, 150*f*
  catabolite activator, 333*f*, 334
  in cell membranes, 99–101
  as cell-surface identity markers, 100, 100*f*
  as cell-surface receptors, 100, 100*f*
  as cell-to-cell adhesion proteins, 100, 100*f*
  in centromeres, 205, 205*f*
  channel, 102–103, 110*t*
  chaperone, 53, 53*f*
  of cytoskeleton, 82–83
  as cytoskeleton anchors, 100, 100*f*
  denaturation of, 54, 54*f*
  dissociation of, 54
  domains of, 52*f*, **53**
  as enzymes, 99, 100*f*
  evolution and, 13
  functions of, 47
  G (*See* G protein-coupled receptors (GPCRs); G proteins)
  heat shock, 53
  histone, 203
  as hormones, 828, 832
  improbability of, 446
  integral membrane, 96
  interior cell membrane network of, 97*t*
  major histocompatibility complex, 87, 850–851, 851*f*
  mannose-binding lectin, 848
  motifs of, 52, 52*f*
  motor, 207–208, 750
  peptide bonds in, 48, 48*f*, 49*f*
  pH and function of, 61
  phenotype determination by, 245
  in phospholipid bilayer, anchoring of, 101
  phosphorylation of, 182, 182*f*
  in plasma, 807
  pocket, 215
  pore, 101, 101*f*
  porin, 51
  predicting structure and function of, 394–396, 396*f*
  primary structure of, 47–48, 50, 50*f*
  quaternary structure of, 50*f*, 52
  Rec8, 230–231
  receptor, 180, 180*f*, 828
  regulatory, 329, 331–333
  renaturation of, 54
  scaffold, 187–188, 188*f*
  secondary structure of (*See* Secondary structure of proteins)
  shugoshin, 230
  single-strand binding, 287, 289*t*
  structural maintenance of chromosome, 199
  structure of, 48, 50–53, 50*f*
  surface, 86–87, 87*t*
  synthesis of (*See* Protein synthesis)
  tertiary structure of, 50*f*, 51–52, 51*f*
  translation repressor, 345
  transmembrane, 96, 97*t*, 828
  transporter, 99, 100*f*
  transport of, 69, 77–78, 77*f*, 828
Protein synthesis. *See also* Translation
Proteobacteria, 503*f*
Proteome, **312**, 394
Proteomics, **394**
  bioinformatics and, 394
  protein structure and function prediction, 394–395, 396*f*
Prothrombin, 808
Protista, 2, 3*f*, 483–485, 485*f*
Protists, 600*f*. *See also specific protists*
  Amoebozoa, 525, 526*f*, 539–541, 539*f*, 540*f*, 541*f*
  Archaeplastida, 525, 526*f*, 536–539, 537*f*, 538*f*, 539*f*
  cell division in, 201*f*
  cell wall of, 745
  Chromalveolata, 487, 488*f*, 525, 526*f*
  cytokinesis in, 209
  diversity of, 523–525, 523*f*, 524*f*, 525*f*

Excavata, 526–529, 526*f*, 527*f*, 528*f*
as first eukaryotes, 521–523, 521*f*, 522*f*
genetic code in, 305
Opisthokonta, 525, 526*f*, 539
phylogeny of, 525–526, 526*f*
SAR group, 487, 488*f*, 525, 526*f*, 529–536, 530*f*, 531*f*, 532*f*, 533*f*, 534*f*, 535*f*, 536*f*
unicellular, 523, 523*f*
Protoderm, **658**, 658*t*
Protonephridia, **839**, 839*f*
Proton gradient, in light-dependent reactions, 165, 168
Proton pumps, 143, 144*f*, 706, 706*f*
Protons, 22, 22*f*
Proto-oncogenes, **214**
Protoplasts, regeneration of, 698–699, 699*f*
Protospacer adjacent motif (PAM), 364*f*
Protostomes, **597**–598, 598*f*, 600*f*–601*f*, 601–602, 603*f*
Prototheria, **642**
Proximal convoluted tubule, **841**, 841*f*, 842
Prozac (fluoxetine), 767
Prunella modularis (dunnock), 907
Prunus cerasifera (purple-leaf plum tree), 673
PSC (phylogenetic species concept), **480**–482, 482*f*
Pseudocoelom, **596**, 597*f*
Pseudogenes, 389, 389*t*
Pseudomyrmex ants, mutualism of, 954, 955*f*
Pseudopodia, of amoeba, 523, 540, 540*f*
Pseudopods, of Actinopoda, 535, 535*f*
Pseudostratified columnar epithelium, 739*t*
P site (peptidyl site), **315**, 315*f*, 316–317, 318*f*
Pterophyta, 576*f*
Pterophytes, 574, 574*f*, 576–579, 577*f*, 578*t*, 579*f*
Pterosaurs, 754*f*, 755
PTH (parathyroid hormone), **835**, 835*f*
Ptychodiscus, 532*f*
Pufferfish *(Fugu rubripes)*, genome of, 387*f*, 396, 397*f*
Pufferfish *(Tetraodon nigroviridis)*, genome of, 386
Pulmocutaneous circuit, 810
Pulmonary arteries, 809, 809*f*
Pulmonary circulation, **809**–810, 809*f*
Pulmonary valve, **811**, 812*f*
Pulmonary veins, 809, 810*f*
of amphibians, 632
Punctuated equilibrium, 467–**468**, 468*f*, 620
Punnett square, **241**, 241*f*, 244*f*
Purines, 56, 56*f*, **279**, 280*f*
Purple-leaf plum tree (Prunus cerasifera), 673
Purple photosynthetic bacteria, 165–166, 166*f*
Purple sulfur bacteria, 165
PWS (Prader-Willi syndrome), 259
Pyloric sphincter, 793
Pyramid of biomass, **969**, 970*f*
Pyrenestes ostrinus (African black-bellied seedcracker finch), 414, 415*f*
Pyrimidines, 56, 56*f*, **279**, 280*f*
Pyrogen, **826**
Pyrrhopappus carolinianus (false dandelion), 693, 694*f*
Pyruvate
in C_4 plants, 174, 174*f*
in ethanol fermentation, 148–149, 149*f*
in glycolysis, 136, 136*f*, 137*f*
oxidation of, 139, 139*f*
Pyruvate dehydrogenase, 123, 123*f*, 139

Q

Quadruped locomotion, 753–754
Quantitative traits, **245**
Quaternary structure of proteins, 50*f*, **52**
Quercus. See Oak trees
Quiescent center, in root apical meristem, 665, 665*f*
Quinine, 542
Quinone, 165

R

Rabbits, locomotion by, 754
Rabies, 509*t*
Races, 649, 649*f*
Radial cleavage, **597**, 598*f*
Radial symmetry, **595**, 596*f*
Radicle, **693**
Radioactive dating, 36, 437, 437*f*
Radioactive decay, 23
^{14}carbon dating and, 23
Radioactive isotopes, **23**, 23*f*
half life of, 23, 36
Radiolarians, **535**
Radula, **608**
Rain shadows, 983, 983*f*
Ram ventilation, 798
Rana pipiens (leopard frog), postzygotic isolation in, 456, 456*f*
Random mutagenesis, 363
Range, population
dispersal mechanisms and, 921, 922*f*
expansion and contraction of, 920–921, 920*f*
human effect on, 921
Ranunculus muricatus, seed dispersal and, 922*f*
Raphes, 530, 531*f*
Ras protein, 187, 189
Rats, learning and genetics in, 894, 894*f*
Ravens, cognitive behavior in, 897, 898*f*
Ray-finned fish, 628*f*
Ray initials, of vascular cambium, 663
Rays, 630
of parenchyma cells, 663
Rb (retinoblastoma susceptibility gene), **214**–215, 215*f*
RBCs. *See* Red blood cells (RBCs)
RBS (ribosome-binding sequence), **316**
Reabsorption, by kidneys, 839, 840*f*, 841–844
Reactants, 30
Reaction centers, 164, 164*f*–167*f*, **165**–168
Reactions, chemical. *See* Chemical reactions
Reading frame, **304**, 304*f*
Read length, 384
Realized niche, **946**
Reasoning
deductive, **7**
inductive, **7**
Receptacles, **584**
Receptor kinases
plant, 186
tyrosine (*See* Receptor tyrosine kinases (RTKs))
Receptor-mediated endocytosis, 108*f*, **109**, 110*t*
Receptor potentials, 773
Receptor proteins, **180**, 181*f*, 828
Receptor tyrosine kinases (RTKs), **185**
binding to, 829, 830*f*
inactivation of, 187
phosphorylation of, 186, 186*f*
protein binding to, 187, 187*f*
signal amplification and, 187–188, 188*f*
Recessive traits, **239**
crosses and, 238–239, 239*f*, 240*f*
Principle of Segregation and, 238–241, 240*f*, 252
Reciprocal altruism, **909**, 909*f*
Reciprocal crosses, **239**
Recognition helix, 330
Recombinant DNA, **355**. *See also* Genomics
construction of, 355–359
DNA fragments and, 355–356, 356*f*
DNA libraries and, 358–359, 358*f*
gel electrophoresis and, 355, 357*f*
knockout mice and, 364–365
protein production and, 367, 368*f*
Recombination
directed away from functional mouse genome, 400
between F plasmid and host chromosome, 501, 505*f*
Recombination frequency, **263**
Rec8 protein, 230–231
Rectum, 791*f*, 796
Red algae (Rhodophyta), 537, 537*f*
chloroplasts of, 522, 522*f*
eukaryotic chloroplasts and, 488, 488*f*
multicellularity in, 489, 524
Red blood cells (RBCs), 742, 806*f*, **807**
facilitated diffusion in, 103
osmotic pressure and, 105*f*
Red Button Ginger flower *(Costus woodsonii)*, 685*f*
Red-eyed tree frog *(Agalychnis callidryas)*, 632*f*
Red kangaroo *(Macropus rufus)*, 644*f*
Redox reactions. *See* Oxidation-reduction reactions
Red tides, 532, 532*f*
Reduction, **25**, **117**, 117*f*
in photosynthesis, 158
Reduction division, 223–224
Reductionism, **9**
Red-wing blackbirds *(Agelaius phoeniceus)*, 908*f*
Redwood trees, 578*t*, 581
Reflexes, 771
Regeneration
of epithelial tissue, 738
of protoplasts, 698, 699*f*
Regenerative medicine, 369
Regulative development, 880
Regulatory proteins, **329**
gene expression and, 330–331
Reich, Peter, 676
Reinforcement, **458**
speciation and, 458*f*, 459–460
Relative dating, 437
Release phase, **510**
in lytic cycle, 510, 511*f*
Releasing hormones, **833**, 833*f*
Remodeling, of bone, 747–748, 748*f*
Remora, ram ventilation in, 798
Renal cortex, **839**, 840*f*
Renal medulla, **839**, 840*f*
Renaturation of proteins, 54
Renin, 845
Renner, Otto, 730
Replication
of DNA (*See* DNA replication)
of HIV, 512, 512*f*
viral, 508–510
Replication fork, **288**
eukaryotic, 291
prokaryotic, 286
Replicons, **286**, 291
Replisome, 287*f*, 289–290, 290*f*
Repression, **332**
Repressors, **332**, 333, 333*f*
Reproduction
angiosperm (*See* Angiosperm reproduction)
asexual (*See* Asexual reproduction)
bacteriophage, 510 511*f*
communication and, 900–901
cost of, 926–927, 927*f*
fungal, 548–550, 548*f*, 549*f*, 550*f*
human (*See* Human reproductive system)
sexual (*See* Sexual reproduction)
viral, 508–510
Reproductive cloning, **876**–877
Reproductive isolating mechanisms, **453**
postzygotic, 453, 454*t*, 456, 456*f*
prezygotic, 454–456, 454*t*
reinforcement of natural selection by, 458
Reproductive isolation, 453–457, 454*t*
random changes causing, 459
Reproductive leaves, **675**
Reproductive strategies
alternative, 907–909, 908*f*
offspring care and, 907, 908*f*
Reproductive success
mate competition and mate choice and, 423–424
number of offspring produced and, 925, 926*f*
parental survival and, 926–927, 927*f*
sex differences in, 421–422, 422*f*, 423*f*
Reproductive systems
of flatworms, 606
human (*See* Human reproductive system)
oxytocin and, 832
vertebrate evolution of, 862, 862*f*

Reptiles. *See also specific reptiles*
bird classification as, 480, 481*f*
brain in, 769, 769*f*
circulation in, 810
egg activation in, 872, 872*f*
evolution of, 636*f*, 637
key characteristics of, 636*f*, 637
modern, 635–637, 636*f*
respiration in, 800
temperature regulation in, 825
thyroid hormones in, 834
RER (rough endoplasmic reticulum), 72*f*, 73*f*, 75*f*, **76**
Reservoir, in *Euglena,* 527, 527*f*
Resolution, of microscopes, 67
Resource partitioning, **947**–948, 948*f*, 949*f*
Resources
competitive exclusion and, 956, 957*f*
consumption in developed *vs.* developing world, 936, 936*f*
of continental shelves, 987–988
population growth and availability of, 930–933
Respiration, **797**, **967**
aerobic, 133, 133*f*, 138
in amphibians and reptiles, 800
anaerobic, 133, 148, 148*f*
in birds, 801
cellular (*See* Cellular respiration)
cutaneous, of amphibians, 632
in mammals, 800–801, 801*f*
neural control of, 803
partial pressure of gases and, 800–801
photorespiration, 172–174
photosynthesis and, 170–172
of primary producers, 966
Respiratory control center, 802
Respiratory pigments, 804, 804*f*
Respiratory systems
of birds, 6438
evolution of, 797–798, 798*f*
of fish, 798–799, 798*f*, 799*f*
of invertebrates, 798
oxygen and carbon dioxide transport and, 803–805
of reptiles, 635
of terrestrial vertebrates, 800–803
Resting potential, 762–**763**, 763*f*
Restriction endonucleases, **355**, 356*f*
Restriction maps, **380**, 381*f*
Restriction point, 212
Reticulons, 100
Retina, 781–**782**
Retinal, 161–162
Retinoblastoma, 214
Retinoblastoma susceptibility gene (Rb), **214**–215, 215*f*
Retinol, 780
Retroviruses, **302**, 356, 509
Reversal, evolutionary, 476, 478
Reverse transcriptase, **302**, 509
making cDNA libraries, 358–359
making DNA copies from RNA using, 356–358, 358*f*
Reverse transcription PCR (RT-PCR), 360–361, 360*f*
Reverse transcription quantitative PCR (RT-qPCR), 360–361, 361*f*
Rhabdovirus, 509*t*
Rhinomuraena quaesita (ribbon eel), 629*f*
Rhizaria, 487, 488*f*, 525, 526*f*, 535–536, 535*f*, 536*f*
Rhizomes, 577, **672**, 672*f*, **698**
Rhizophydium, 555*f*
Rhizopus (bread mold), 558, 559*f*
Rhodophyta, **537**, 537*f*. *See also* Red algae (Rhodophyta)
Rhodopsin, **781**
Rhombencephalon, 769, 769*f*
Rhynchocephalia, 636–637, 636*f*
Ribbon eel *(Rhinomuraena quaesita),* 629*f*
Ribbon worms, 599*t*
Ribonucleic acid. *See* RNA (ribonucleic acid)
Ribose, 44*f*
Ribosomal RNA (rRNA), **75**, 123, **303**, 309
Ribosome-binding sequence (RBS), **316**
Ribosomes, **69**, 69*f*, 73*f*, 74*f*, **75**, 75*f*
eukaryotic, 496
functions of, 315, 315*f*
prokaryotic, 496, 500
translocation of, 317
tRNA-binding sites on, 314–315, 315*f*
Ribozymes, 123
Ribulose 1,5-bisphosphate (RuBP), **170**
Ribulose bisphosphate carboxylase/ oxygenase (rubisco), **170**
Rice *(Oryza sativa)*
genome of, 386, 387*f*, 392, 392*f*, 397, 397*f*
Golden Rice, 373, 373*f*
Rickets, 836
Rickettsia, 503*f*
eukaryotic mitochondria and, 488
Rickettsia typhi, 507*t*
Rigor mortis, 751
RISC (RNA-induced silencing complex), 343*f*, 3342
Rising phase, of action potential, 764, 765*f*
Rivulus hartii (killifish), 416–417, 416*f*
RNA (ribonucleic acid), 43*t*, **55**, 55*f*
alternative splicing of, 394, 394*f*
as enzyme, 123
functions of, 57
gene expression and, 302–303
messenger (*See* Messenger RNA (mRNA))
micro-(*See* Micro-RNA (miRNA))
noncoding, 389, 389*t*
pre-miRNA, 342–343
pri-miRNA, 342
in protein synthesis, 245
ribosomal, 75, 123, 303
small (*See* Small RNAs)
small interfering (*See* Small interfering RNAs (siRNAs))
splicing of, 312*f*, 313*f*
structure of, 57
transfer (*See* Transfer RNA (tRNA))
RNAi (RNA interference), 342, 343*f*, 363
RNA-induced silencing complex (RISC), 342, 343*f*
RNA interference (RNAi), 342, 343*f*, 363
RNA polymerases, 302, 303*f*, **306**
eukaryotic, 309
general transcription factors and, 337, 337*f*
polymerase I, **309**
polymerase II, **309**, 310, 338, 339*f*
polymerase III, **309**
prokaryotic, 306–309, 307*f*
regulatory proteins and, 329
RNA polymerases and, 329
RNA sequencing (RNA-seq), 394
RNA viruses, 509, 509*t*
Robber fly, 616*f*
Robinia pseudoacacia (black locust), leaves of, 673
Rodhocetus, 438*f*
Rods, 498, **781**–782, 781*f*, 782*f*
Rolling-circle replication, 501, 504*f*
Root cap, **665**, 665*f*
Root cap cells, 665, 665*f*
Root hairs, **661**, 661*f*, 666–667
nutrient absorption through, 709, 709*f*
Root pressure, **711**
Roots, 664
adventitious, 668, 669*f*
evolution of, 574–575
first, 695
gravitropism of, 728–729, 728*f*
modified for specialized functions, 668
nutrient absorption through, 709–711, 709*f*, 712*f*
regions of, 664–668, 665*f*
tissues of, 656–657, 657*f*
Root system, **656**–657, 656*f*
Ross, Ronald, 542
Rotifera, 599*t*, 600*f*, **606**–607, 607*f*
Rough endoplasmic reticulum (RER),72*f*, 73*f*, 75*f*, **76**
translation in, 318–319, 320*f*
Roundworm *(Caenorhabditis elegans),* 599*t*, 602, 612–613, 612*f*
chromosomes of, 222
let-7 gene in, 341
lin-14 gene in, 341, 341*f*
sperm binding in, 874*f*
Rous sarcoma virus, 214
Royal hemophilia, 268
R plasmids, **501**
R point, 212
rRNA (ribosomal RNA), **75**, 123, **303**, 309
r-selected populations, **933**, 933*t*
RTKs. *See* Receptor tyrosine kinases (RTKs)
RT-PCR (reverse transcription PCR), 360–361, 360*f*
RT-qPCR (reverse transcription quantitative PCR), 360–361, 361*f*
Rubisco (ribulose bisphosphate carboxylase/oxygenase), **170**
RuBP (ribulose 1,5-bisphosphate), **170**
Rubus fruticosus, seed dispersal and, 922*f*
Rule of addition, **243**
Rule of eight, 26, 26*f*
Rule of multiplication, **243**–244
Ruminants, 557
Runners, **672**, 672*f*, **698**
Running, 756

S

Saccharomyces cerevisiae, 387*f*, 562*f*, 563
Saccule, **777**
Sager, Ruth, 260
Saguaro cactus *(Carnegiea gigantea),* 685
Saimiri oerstedii (squirrel monkey), 644*f*
Salamanders, 632–633, 632*f*, 753
Saliva, 791–792
Salivary amylase, 791–792
Salmonella typhi, 507*t*
Salt
plant adaptations to high concentrations of, 588, 716
reabsorption by nephron, 842–843, 843*f*, 844*f*
Saltatory conduction, **765**, 766*f*
Samaras, 694*f*
Sand fleas, 615
Sanger, Frederick, 48, 383
SA (sinoatrial) node, **809**, **811**, 813*f*
Sap, **716**
Saprolegnia, 531
Sarcomeres, **749**, 750*f*, 751*f*
Sarcoplasmic reticulum (SR), **751**, 752*f*
SAR group, **525**
Alveolata, 487, 488*f*, 525, 526*f*, 529, 531–535, 531*f*, 532*f*, 533*f*, 534*f*
Rhizaria, 487, 488*f*, 525, 526*f*, 535–536, 535*f*, 536*f*
Stramenopila, 487, 488*f*, 526, 526*f*, **529**–531, 530*f*, 531*f*
SARS (severe acute respiratory syndrome), 509*t*
SARS-CoV-2 virus, 369
Saturated fats, **58**, 59*f*
Saturation, 103
Scaffold proteins, 187–188, 188*f*
Scanning electron microscopes, 67, 68*t*
SCARECROW (SCR) gene, 665–666, 666*f*
scarecrow (scr) mutation, in *Arabidopsis,* 728, 729*f*
Scars, on plant stems, 670, 670*f*
Schizogony, 524
Schizymenia borealis, 537
Schleiden, Matthias, 66
Schwann cells, **762**, 762*f*
Schwann, Theodor, 66
Science
descriptive, 6
as social construction, 9
Scientific method, 6
Scientific process, 6–9
Sciurus carolinensis (Eastern gray squirrel), 484*f*
Sclera, **779**, 779*f*
Sclereids, 662, 662*f*
Sclerenchyma cells, **662**, 662*f*
SCN (suprachiasmatic nucleus), 838
SCNT (somatic cell nuclear transfer), 876–877, 876*f*–877*f*
SCR (SCARECROW) gene, 665, 666*f*
scr (scarecrow) mutation, in *Arabidopsis,* 728, 729*f*
Scrotum, 863, 863*f*
Scrub oak *(Quercus dumosa),* ecological isolation of, 454
Scutellum, **695**
Sea anemones, 603–604, 954
Sea lions, swimming by, 753

Sea peach *(Halocynthia aurantium),* 626*f*
Seas. *See* Oceans
Sea spiders, 615
Sea stars, 617, 617*f*, 618*f*
  Pisaster, predation and, 957, 957*f*
  sperm binding in, 874*f*
Sea urchins
  egg activation in, 872–873, 873*f*
  eggs of, 871, 871*f*
  sperm binding in, 874*f*
  yolk distribution in eggs of, 879*f*
Sea walnuts, 599*t*
Seaweeds, 529–530, 530*f*
Secondary carnivores, **966**, 966*f*
Secondary chemical compounds, as plant defenses, 950
Secondary endosymbiosis, **522**, 522*f*, 529
Secondary growth, **659**, 670, 671*f*
Secondary mycelium, 560*f*, 561
Secondary oocytes, 868, 868*f*, 869*f*, 870
Secondary plant body, **658**
Secondary productivity, **967**
Secondary sexual characteristics, **423**
Secondary spermatocytes, 863
Secondary structure of proteins, **50**–51, 50*f*
  domains and, 52*f*, 53
  motifs and, 52, 52*f*
Secondary succession, **958**
Secondary tissues, **658**
Secondary walls, 85, 85*f*
Second Law of Thermodynamics, **117**, 117*f*, 965
  violation by evolution, 446
Second messengers, **184**
  calcium as, 191, 192*f*
  cyclic AMP as, 190, 191*f*
  inositol phosphates as, 190–191
Second-messenger systems, 829–830
Secretin, **794,** 796, 796*f*
Secretion, by kidney, 840, 840*f*
Secretory phase, **868**
Secretory vesicles, 72*f*, **77**
Securin, 213
Seed bank, **695**
Seed coat, 580, 580*f*, 586, **691**, 691*f*
Seed plants, 574. *See also* Angiosperm reproduction; Angiosperms; Gymnosperms
  angiosperms, 574*f*, 578*t*, 584–587, 584*t*, 585*f*, 586*f*
  evolution of, 580, 580*f*
  gymnosperms, 574*f*, 580–583, 581*f*, 582*f*, 583*f*
  phylogeny of, 578*t*
Seeds
  adaptations of, 692, 692*f*
  in angiosperms, 585, 586*f*, 690–692, 691*f*, 692*f*
  in conifers, 581–582, 581*f*
  development of, 691, 691*f*
  dispersal of, 692–693, 693*f*, 694*f*, 921, 922*f*
  embryo protection by, 580, 580*f*, 691–692, 691*f*
  evolution of, 573–575
  germination of, 587, 692, 695, 696*f*, 726
  stratification of, 693
Segmental duplications, 389, 389*t*
Segmentation
  in animals, 598, 600, 609
  in Annelida, 609–610, 610*f*
  in arthropods, 614, 614*f*
  in Myriapoda, 616
Segmented worms, 599*t*, 609–610, 609*f*, 610*f*
Segregation, **237**
  Principle of, 239–241, 241*f*, 253
  of traits, 237–238, 252
Selaginella moellendorffii, 576*f*
Selection, **411**
  artificial (*See* Artificial selection)
  climatic conditions and, 412
  clonal, 850
  conditions for, 411
  determining agent of, 434
  directional, 414, 415*f*
  disruptive, 414, 415*f*, 461
  frequency-dependent, 419–420, 420*f*
  industrial melanism and, 433–434, 433*f*
  kin, 909–911, 910*f*
  limits to, 413–414
  mate (*See* Mate selection)
  natural (*See* Natural selection)
  oscillating, 420, 476
  for pesticide and microbial resistance, 412
  predator avoidance and, 411–412, 412*f*
  scope of changes produced by, 436
  sexual, 418, 422–424, 422*f*, 423*f*
  stabilizing, 414, 415, 415*f*, 416*f*
Selective permeability, **102**
Selective serotonin reuptake inhibitors (SSRIs), 767
Self-incompatibility, genetic, in plants, 687
Self-organization, 15, 15*f*
Self-pollination, 683, 685
Self *versus* nonself recognition, 850–851
Semelparity, **928**
Semen, **864**
Semibalanus balanoides, interspecies competition in, 946, 946*f*
Semicircular canals, **777**
Semiconservative model of DNA replication, 283–286, 284*f*
Semilunar valves, **811**
Senescence, in plants, **697**
Sense strand, 302
Sensors, **823**
Sensory exploitation theory of mate selection, **424**, 424*f*
Sensory neurons, 740*f*, **761**, 761*f*
Sensory receptors, 761, 773
Sensory transduction, in photoreceptors, 782, 783*f*
Sepals, **584**
Separase, 213
Septate junctions, 87*t*, **88**
Septation, **200**, 200*f*
Septum, **200**
  bacterial formation of, 199–200, 199*f*, 200*f*
Sequence-tagged sites (STSs), 382, 382*f*
Sequoia sempervirens (coastal redwood), 581
SER (smooth endoplasmic reticulum), 72*f*, 73*f*, 75*f*, **78**
Serengeti Plain, 962
Serine, 49*f*
Serine-threonine kinases, 182*f*, 186
Serotonin, **767**
Sertoli cells, 863
Set point, 823
Severe acute respiratory syndrome (SARS), 509*t*
Sex chromosomes, **257**, 257–259, 258*t*
  adult body cells of mammals and, 258–259
  dosage compensation and, 258
  nondisjunction of, 266
  white eye trait on, 257–258
  X chromosome inactivation and, 258–259, 259*f*, 339–340
  Y chromosome in humans, 257
Sex determination, 258
  in *Drosophila,* 343
Sex-linked traits, **268**, 268*f*
Sex ratio, population growth and, 923
Sexual dimorphism, **423**
Sexually transmitted diseases (STDs), 527, 527*f*, 889
Sexual reproduction, **222**–223, 223*f*, **524**. *See also* Meiosis
  in angiosperms, 585–587, 586*f*
  in animals, 594
  in conifers, 581–582, 582*f*
  in eukaryotes, 488
  in fungi, 549–550, 558–563, 559*f*, 560*f*, 562*f*
  in Ginkgo trees, 583, 583*f*
  human (*See* Human reproductive system)
  in mosses, 572, 572*f*
  in protists, 525, 526*f*
Sexual selection, 418, **423**
  sex differences in, 422–424, 422*f*, 423*f*
Shade leaves, **675**
Shared derived characters, **476**–479, 477*f*
  phylogenetic species concept and, 480–482, 482*f*
Sharks, 630–631, 631*f*
  temperature regulation in, 824–825, 824*f*
Sharp-nosed torrent frog *(Taudactylus acutirostris),* 564
Sheep, cloning of, 876–877, 876*f*–877*f*
Shells
  of diatoms, 530, 531*f*
  of Foraminifera, 535–536, 535*f*, 536*f*
Shine-Dalgarno sequence, 316
Shine, John, 316
Shipworms, 608–609
Shivering thermogenesis, 826
Shoots
  elongation of, phytochrome and, 726
  tissues of, 656–657, 657*f*
Shoot system, **656**–657, 656*f*
Short interspersed elements (SINEs), **390**
Short-nosed echidna *(Tachyglossus aculeatus),* 643*f*
short root (shr) mutation, in *Arabidopsis,* 728, 729*f*
Short tandem repeats (STRs), **265**, 362, 362*f*
Shotgun assembly, **385**–386, 385*f*
Shrimps, 615
shr (short root) mutation, in *Arabidopsis,* 728, 729*f*
Shugoshin proteins, 233
Siamese cats, coat color in, 245, 245*f*
Sickle-cell allele, 250, 251*f*, 321
Sickle-cell anemia, 246*f*, 270*t*, 421, 421*f*
Sickle-cell disease, 268, 268*f*
Side-blotched lizard *(Uta stansburiana)*
  alternative mating strategies of, 908, 908*f*
  clutch size and offspring size of, 927*f*
Sieve cells, 664
Sieve-tube members, 664, 664*f*
Sigmoidal growth curve, **929**, 929*f*
Signaling, cellular. *See* Cell signaling
Signal recognition particle (SRP), **303**, 318
Signal sequence, **318**
Signal transduction, 180–**181**. *See also* Cell signaling
  cellular receptors and (*See* Intracellular receptors)
  pathways of, 181
Sildenafil (Viagra), 185
Silent mutations, **321**, 321*f*
Silk moths *(Bombyx mori),* pheromone in, 901
Silver fox, 435, 436*f*
Simple columnar epithelium, 738, 739*t*
Simple cuboidal epithelium, 738, 739*t*
Simple epithelial membranes, 738
Simple leaves, 673, 673*f*
Simple sequence repeats (SSRs), **389**, 389*t*
Simple sugars, 44, 44*f*
Simpson, George, 490
SINEs (short interspersed elements), **390**
Singer, S. Jonathan, 95
Single bonds, 29
Single-nucleotide polymorphisms (SNPs), **265**, 394
Single nucleotide variation (SNV), **321**–322, 321*f*
Single-strand binding (SSB) protein, 287, 289*t*
Sinoatrial (SA) node, **809**, **811**, 813*f*
Sinosauropteryx, 639*f*
Sinus venosus, **809**, 809*f*, 811
SIR2, 217
siRNAs. *See* Small interfering RNAs (siRNAs)
SIRT-7, 217
Sirtuins, 217
Sister chromatids, **203**
  cohesion, 225, 230
  in meiosis, 223, 224–225, 224*f*
  in mitosis, 203, 203*f*, 204, 205*f*, 206, 207, 207*f*
6-PGD gene, 414
Skates, 630
Skeletal muscles, **742**–743, 743*t*
  movement and, 748, 749*f*
Skeletons. *See* Endoskeletons
  of birds, 638
  exoskeletons, 614–615, 745, 745*f*
  hydrostatic, 744–745, 744*f*
  of sharks, 630–631, 631*f*
  vertebrate, 627, 628*f*, 744–745, 745*f*

Skin
human, 738, 741
of reptiles, 634
Skulls, of amniotes, 634, 634*f*
S-layer, **497**
"Sliding clamp," 288, 288*f*
Sliding filament mechanism, 749–751, 750*f*, 751*f*
Slime molds (Amoebozoans), 525, 526*f*, 539, 539*f*
cellular, 540–541, 540*f*
plasmodial, 540, 540*f*
Small interfering RNAs (siRNAs), 303, **341**
biogenesis and function of, 342, 342*f*
miRNAs distinguished from, 342
Small intestine, 791*f*, 793–794, 793*f*
Small nuclear ribonucleoprotein particles (snRNPs), **312**, 312*f*
Small nuclear RNA (snRNA), **303**
Smallpox, 509*t*
Smallpox virus, 849
Small RNAs, 303
gene expression after transcription by, 341–343, 341*f*
heterochromatin formation and, 343
protective role of, 343
SMC (structural maintenance of chromosome) proteins, 199
Smell receptors, 778
Smooth endoplasmic reticulum (SER), 72*f*, 73*f*, 75*f*, **78**
Smooth muscle, **742**, 743*t*
Snakes, 636*f*, 637
eastern rat snake *(Pantherophis alleghaniensis)*, 453*f*
Snake venom, evolution of, 446
Snowshoe hare *(Lepus americanus)*, population cycle of, 931–932, 932*f*
SNPs (single-nucleotide polymorphisms), **265**, 394
snRNA (small nuclear RNA), **303**
snRNPs (small nuclear ribonucleoprotein particles), **312**, 312*f*
SNV (single nucleotide variation), **321**–322, 321*f*
Social construction of science, 9
Social groups, animal, information exchange in, 901, 901*f*
Societies, **912**
insect, 912, 912*f*
vertebrate, 912–913, 912*f*
Socorro isopod *(Thermosphaeroma thermophilus)*, range of, 920
Sodium. *See also* Salt
aldosterone and, 836–837, 845
in organisms, 26–27
Sodium channels, ligand-gated, 782
Sodium-potassium pump, **106**–107, 106*f*, 110*t*
Sodium voltage-gated channels, 764
Soil, plants and, 159
Solanum dulcamara, seed dispersal and, 922*f*
Solar energy. *See also* Ultraviolet light
atmospheric circulation and, 980–983, 980*f*, 981*f*
greenhouse gases and, 991–992, 992*f*
oceans and, 987
Solenoid, **202**, 202*f*
Solubility, in water, 32*t*, 33, 33*f*
Soluble receptors, 846
Solute potential, **708**, 708*f*
Solutes, 33, 33*f*, **103**–104
Solutions
aqueous, 103
hypertonic (hyperosmotic), 104, 708, 838
hypotonic (hypoosmotic), 104, 708, 838
isotonic (isosmotic), 104, 838
osmolarity of, 838
tonicity of, 708, 838
Solvents, water as, 33, 33*f*, **103**
Somatic cell nuclear transfer (SCNT), 876–877, 876*f*–877*f*
Somatic cells, **222**–223
Somatic motor neurons, 752, 752*f*
Somatic nervous system, **761**, 761*f*, 772
Somatostatin, 833
Somatotropin, 832
Somites, **883**
Somitogenesis, 882
Somitomeres, **883**
Song sparrow, population size of, 939
s-orbitals, 24, 24*f*
Sori, 576–**577**
"SOS response," 295
Sounds, detection of, 775–778
Source-sink metapopulations, **922**–923, 923*f*
Southern beech tree, 445
Soybeans
glyphosate resistance of, 372
as transgenic crop, 373
Spacing patterns, of individuals in populations, 921–923
Spallanzani, Lazzaro, 642
Spanish flu, 514
Spanish moss *(Tillandsia usneoides)*, 953, 954*f*
Spatial heterogeneity, tropical biodiversity and, 973
Spatial summation, **768**
Special connective tissues, **741**–742
Specialization, in vertebrate digestive tract, 790, 790*f*, 791*f*
Specialized transduction, **502**–503
Speciation
adaptation and, 459, 459*f*
adaptive radiation and, 462–467, 463*f*
geographic separation and, 460–462
random changes and, 459
reinforcement of, 458, 458*f*
Species, **5**, 5*f*
allopatric, 458–459, 458*f*, 460, 461*f*
competition among, 17
composition of, ecological succession, 958–959
ecotypes as, 470
identity of, communication about, 901, 901*f*
of insects, 616, 616*f*
interactions of (*See* Commensalism; Mutualism; Parasitism; Symbiosis)
keystone, 957
naming of, 483, 483*f*, 484*f*
similarity of, 476
sympatric, 453, 455, 455*f*, 461
taxonomy and, 482–485, 483*f*, 484*f*
Species concept
biological, 453–457, 480, 482
phylogenetic, 480–482, 482*f*
Species diversity cline, **971**, 972*f*
Species richness, **945**
ecosystem characteristics influencing, 971
ecosystem stability and, 969–973, 972*f*
in tropics, 972*f*, 9671–973
Specific heat, **32**
of water, 32–33, 32*t*
Specific transcription factors, **336**
Speech. *See* Language
Spermatids, **863**
Spermatocytes, 863
Spermatozoa (sperm), **863**
blockage of, as birth control method, 870
delivery of, 863–865
destruction of, as birth control method, 870
fertilization and (*See* Fertilization)
fusion with egg, 871–872, 871*f*, 872*f*
penetration of, effects other than fertilization, 873, 874*f*
production of, 863, 864*f*
structure of, 863, 864*f*
Sperm competition, **423**
Sphagnum (peat moss), 57
S (synthesis) phase, **204**–205, 204*f*
Sphenodon punctatus (tuatara), 636, 636*f*
Sphincters
anal, 796
esophageal, 792
pyloric, 793
Spiders, 615, 615*f*
Spinal cord, 769, 769*f*
Spindle, 15, 15*f*, **206**
chromosome attachment to, 207–208
Spindle checkpoint, 211*f*, **212**, 212*f*, 213
Spindle pole bodies, 549
Spines (leaves), **675**
Spiny fish, 629, 629*f*
Spiral cleavage, **597**, 598*f*
Spirillum, 497
Spirochetes, 497, 502*f*
Spliceosomes, **310**, 311–**312**, 312*f*
Sponges, 599*t*, 601, 601*f*, 603, 604*f*
Spongy bone, **746**
Spontaneous chemical reactions, 119
Spontaneous generation, 8, 8*f*
Sporangia, **540**
of ferns, 577
of mosses, 572, 572*f*
of slime molds, 540, 540*f*
Sporangiophores, **559**
Spore mother cells, **570**
Spores
of chytrids, 557, 557*f*
of conifers, 581, 582*f*
fungal, 549–550, 550*f*
in plants, 570, 570*f*
Sporocytes, **570**
Sporophytes, **570**, 570*f*
of angiosperms, 587
of bryophytes, 571–573, 572*f*, 573*f*
of whisk ferns and horsetails, 578–579, 579*f*
Spruce trees, 578*t*, 581
succession and, 958
Squamata, 636*f*, 637
Squamous cells, 738
Squamous epithelium
simple, 739*t*
stratified, 738, 739*t*
Squirrel monkey *(Saimiri oerstedii)*, 644*f*
SR (sarcoplasmic reticulum), **751**, 752*f*
Src gene, 214
SRP (signal recognition particle), **303**, 318
SRY gene, 273
SSB (single-strand binding) protein, 290*f*
SSRIs (selective serotonin reuptake inhibitors), 767
SSRs (simple sequence repeats), **389**, 389*t*
Stabilizing selection, 414*f*, **415**, 416*f*
Stacked GM crops, 373
Stahl, Franklin, 284–285
Stamens, **584**, 681, 682*f*
Stapes, **776**–777, 777*f*
Starches, 43*t*, **45**, 46*f*, 171
Starlings *(Sturnus vulgaris)*
migratory behavior of, 899, 900*f*
range of, 921
START checkpoint, 212
Start codons, **306**
Start site, **306**–307, 307*f*
Stasis, **467**
Statocysts, **777**
STDs (sexually transmitted diseases), 527, 527*f*, 889
Stele, **667**
Stem cells
embryonic, 364, 369, 874–875, 875*f*, 877, 877*f*
formed elements arising from, 808
induced pluripotent, 878
mesenchymal, 369
multipotent, 874
pluripotent, 808, 874–878
unipotent, 874
Stem cell therapy, 369–370
Stems
evolution of, 574–575
external structure of, 669–670, 669*f*, 670*f*
gravity and, 728, 728*f*
internal structure of, 670–671, 670*f*, 671*f*
modified, 672, 672*f*
phototropism in, 726*f*, 727
Ste5 protein, 188
Stereocilia, 775, 776*f*, 777
Stereoisomers, 42
Stern, Curt, 262
Steroid hormone receptors
action of, 184
coactivators and, 184
gene expression and, 184–185, 185*f*
hormone binding by, 828
Steroid hormones, 184, **828**
Steroids, 43*t*, 58
Stigma, **585**
Stimulation, response to, 3
Stimulus-gated ion channels, **773**

Stimulus-response chain, **900**
Stipules, 670, **673**
St. John's wort *(Hypericum perforatum),* niche of. 946
Stolons, **672**, 672*f*, **698**
Stomach, 790, 790*f*, 791*f*, **792**-793, 792*f*
hemorrhagic *E. coli* in, 111
Stomata, **172**, **570**
of hornworts, 573
of plant guard cells, **660**, 660*f*, 674, 674*f*, 714-715, 714*f*
in plant leaves, 172, 173*f*, 174, 575
Stop codons, **306**
Stramenopila, 487, 488*f*, 526, 526*f*, **529**-531, 530*f*, 531*f*
Stratification, of seeds, **693**
Stratified epithelial membranes, **738**
Stratified squamous epithelium, 738, 739*t*
Strauss, Sharon, 700
Strawberry plants, 672, 672*f*
Streptococcus, 507*t*
Streptococcus mutans, 507*t*
Streptococcus pneumoniae, transformation in, 277, 277*f*
Streptococcus sabrinus, 507*t*
Streptophytes, 537-539, 569, 569*f*
Striated muscle, 742
Stroma, 81, 81*f*, 157*f*, **158**
STRs (Short tandem repeats), **265**, 362, 362*f*
Structural isomers, 42
Structural maintenance of chromosome (SMC) proteins, 199
Structure, function related to, 15
STSs (sequence-tagged sites), 382, 382*f*
Sturgeon, 490
Sturnus vulgaris (starlings). *See* Starlings *(Sturnus vulgaris)*
Sturtevant, Alfred, 262
Styles, **585**
Suberin, 666, 671, **710**
Subspecies, **453**, 453*f*
Substrate-level phosphorylation
ATP synthesis by, 133, 133*f*
in citric acid cycle, 139, 139*f*
Substrates, **121**
active sites of enzymes and, 122-123, 122*f*
enzyme binding to, 127
β Subunit, 288, 288*f*
Succession. *See* Ecological succession
Succinate, 141*f*, 142
Suckers, **698**
Sucrase, 122*f*
Sucrose, 45, 45*f*, 170
Sugarbeets, glyphosate-resistant, 372
Sugarcane, genome of, 392, 392*f*
Sugars. *See also* Glucose
isomers of, 44, 45*f*
simple, 44, 44*f*, 45*f*
transport in plants, 716-717, 717*f*, 718*f*
Sulcus, 770
Sulfate, anaerobic respiration in sulfur bacteria using, 148, 148*f*
Sulfhydryl groups, 41*f*
Sulfur bacteria, 505
anaerobic respiration in, 148, 148*f*
green, 166
purple, 160
Sulfur, in organisms, 26-27
Sunflower *(Hyoseris longiloba),* 683*f*
Sunflowers, transgenes and, 374
Sunlight. *See* Light; Solar energy; Ultraviolet light
Supercoiling, **287**, 287*f*
Suprachiasmatic nucleus (SCN), 838
Surface-area-to-volume ratio, 66, 66*f*
Surface proteins, 86-87, 87*t*
Surface tension, **31**, 32*f*
Survivorship, **926**
Survivorship curves, 926, 926*f*
Suspensors, **689**
Sutherland, Earl, 829-830
Sutton, Walter, 257
Sweating, 825, 826
Swifts, 639
Swim bladder, **630**, 631*f*
Swimming, 753, 753*f*, 756
Swine flu, 514
Symbiosis, **953**, 953*f*
of bacteria and legumes, 666
commensal, 551
facultative, 551
fungal, 550-555, 551*f*, 552*f*, 553*f*
mutualistic, 551
obligate, 551
Symmetry
bilateral, 595-596, 596*f*
radial, 595, 596*f*
Sympathetic chain, 772
Sympathetic division, of autonomic nervous system, 761, 761*f*, 772, 772*f*
Sympatric species, **453**, 454, 455*f*, 462
Symplast route, **709**, 710*f*
Symplesiomorphies, **477**
Symporters, **106**
Synapomorphy, **477**-478
Synapses, **766**
neurotransmitters and, 766
postsynaptic neurons and, 768
vertebrate, 766-767, 766*f*
Synapsids, **634**-635, 634*f*, 635*f*
Synapsis, **223**, 224*f*, 225
Synaptic cleft, **766**, 766*f*
Synaptic integration, 768
Synaptic signaling, **181**, 181*f*
Synaptic vesicles, **766**
Synaptonemal complex, **223**, 224*f*, 232
Syngamy, **222**
Synteny, 392, 392*f*
Synthesis, in DNA replication, 359, 360*f*
Synthesis phase, **510**
in lytic cycle, 510, 511*f*
Synthesis (S) phase, **204**-205, 204*f*
Synthetic auxins, 721
Syphilis, 889
Syrphid fly, 700
Systematics, **475**
cladistics and, 476-479, 477*f*, 478*f*, 479*f*
classification and, 480-482, 481*f*, 482*f*
domains and, 485-489, 486*f*, 487*t*, 488*f*
molecular clock and, 479-480
as reconstruction of evolutionary relationships, 475-476, 475*f*
statistical approaches to, 479
taxonomy and, 483-485, 483*f*, 484*f*
Systemic circulation, **809**-810, 809*f*
Systole, **811**, 812*f*
Szostak, Jack, 292-293

T

2,4,5-T (2,4,5-trichlorophenoxyacetic acid), 722
Tachyglossus aculeatus (short-nosed echidna), 643*f*
TAFs (transcription-associated factors), 336, 337*f*
Tagmata, **614**
Tandem duplication, 322
Tanysiptera hydrocharis (little paradise kingfisher), phenotypic differentiation in, 460, 461*f*
Taproots, 668
Taq polymerase, **359**
Tardigrada, 599*t*, 601*f*
Tarsiers, 644, 644*f*, 645*f*
Taste buds, 778, 778*f*
Taste receptors, 778, 778*f*
TATA boxes, **309**, 336, 337*f*
Tatum, Edward, 7, 301-302
Taudactylus acutirostris (sharp-nosed torrent frog), 564
Tautomers, 281
Taxol, 581
Taxon, 483
Taxonomy, **483**
of fungi, 554-555, 555*f*, 555*t*
Linnaean, 483, 483*f*, 484*f*, 485
Tay-Sachs disease, 78, 270*t*
TB (tuberculosis), 506, 507*t*
TC (cytotoxic T cells), **850**-851
TCA (tricarboxylic acid) cycle. *See* Citric acid cycle
T cells, **850**. *See* T cells
cytotoxic, 850-851
helper, 850, 851-852
HIV infection of, 511-514, 513*f*
TCGA (The Cancer Genome Atlas), 216
Teeth. *See also* Tooth decay
evolution of, in sharks, 630
of mammals, 641
vertebrate, 791-792, 792*f*
Telencephalon, **769**
Telomerase, 293, 293*f*
Telomeres, **292**
Telophase II, meiotic, 226, 228*f*, 229*f*
Telophase I, meiotic, 226, 227*f*, 228*f*
Telophase, mitotic, 207*f*, **208**-209, 228*f*
Temperate phages, **510**
Temperature. *See also* Heat; Thermodynamics
biomes and, 984-986, 985*f*, 986*f*
chemical reactions and, 30
enzyme activity and, 124, 124*f*
global change in, 991, 992*f*
greenhouse gases and, 992
increase in, human activity and, 991
oxygen transport in blood and, 804
photorespiration and, 172-174, 173*f*
photosynthesis and, 158-160, 159*f*
transpiration rates and, 713
Temperature regulation. *See also* Endothermy
ectothermy and, 636, 824-825, 824*f*
fever and, 826
heterothermy and, 824
homeothermy and, 824
hypothalamus and, 826, 826*f*
poikilothermy and, 824
thermogenesis and, 826
types of, 824
by vasodilation and vasoconstriction, 814, 815*f*
Template strand, **302**-**303**
Temporal isolation, 454*t*, 455
Temporal summation, **768**
Tendrils, **672**, 672*f*
Tensile strength, of water, **711**
Terminal buds, 670, 670*f*
Terminalia calamansanai, seed dispersal and, 922*f*
Termination, 289
accessory factors required for, 317, 318*f*, 319*f*, 320*t*
of eukaryotic transcription, 309-310, 310*f*
of prokaryotic transcription, 306, 307*f*
of translation, 318, 318*f*
Terminators, **306**
Terminus of replication, 199*f*, 286
Terpenes, 43*t*, 58
Terrestrial life
animal adaptations to, 633, 634*f*
plant adaptations to, 569-570
Territoriality, **906**, 906*f*
Tertiary structure of proteins, 50*f*, **51**-52, 51*f*
Testcross, **244**, 244*f*, 244*t*
genetic mapping and, 263, 263*f*
Testes, **863**, 863*f*, 864*f*
Testosterone, 863, 865*t*
Tetracycline, 500
Tetrads, 223
Tetrahedrons, 31
Tetraodon nigroviridis (pufferfish), genome of, 386
Tetraploids, 461
Tetrapods, **632**
diversification of, 632-633
evolution of, 632, 633
TF (transverse filaments), 224, 224*f*
T_H (helper T cells), **850**, 851-852
Thalamus, **769**, 770*f*, 771
The Cancer Genome Atlas (TCGA), 216
Theories, scientific, 8-**9**
Therapeutic cloning, 877-**878**, 878*f*-879*f*
Therapsids, 633
Theria, **642**
Thermodynamics, 116, **117**-119
activation energy and, 119-120, 120*f*
First Law of, 117, 965
free energy and, 118-119
Second Law of, 117-118, 118*f*, 965
Thermogenesis, 826
Thermophiles, 487, 502*f*
Thermoproteus tenax, anaerobic respiration in, 148*f*
Thermoreceptors, **774**, 774*f*
Thermoregulation. *See* Temperature regulation
Thermosphaeroma thermophilus (Socorro isopod), range of, 920
Thermotogae, 502*f*
Thermus aquaticus, 359

Theropods, 635
Thick filaments, 749–750, 750*f*
Thin filaments, 749–750, 750*f*
Thiomargarita namibia, 496
Thoracic breathing, of reptiles, 634
Thoracic cavity, 737, 737*f*
Thorns, 675
Thoroughbred horses, 413, 413*f*
Three-point crosses, **264**–265, 264*f*
3' poly-A tail, **310**, 311*f*
Threonine, 49*f*
Thrombocytes, 742
Thunnus albacares (yellowfin tuna), 629*f*
Thylakoid membrane
  photosynthesis and, 158
  reactions taking place in, 165–166
Thylakoids, **80**, 80*f*
Thymine, 56, 56*f*, 279, 282, 282*f*, 303
Thymine dimer, 294, 294*f*
Thyroid gland
  calcitonin secretion by, 835
  disorders of, 834
Thyroid hormones, **828**, 834, 835
  actions of, 834–835
Thyroid, iodine deficiency, 27*f*
Thyroid-stimulating hormone (TSH), 826, **832**
Thyrotropin, **832**, 833
Thyrotropin-releasing hormone (TRH), 833
Ticks, 615
Tiger salamander *(Ambystoma tigrinum)*, 632*f*
Tigers, ecological isolation of, 454
Tiger shark *(Galeocerdo cuvier)*, 630*f*
Tiger swallowtail butterfly *(Papilio glaucus)*, mimicry in, 952, 952*f*
Tight junctions, 87*t*, 88–**89**, 88*f*
Tiktaalik, 438, 631*f*
Tillandsia usneoides (Spanish moss), 953, 954*f*
Tilman, David, 969–971
Tinbergen, Niko, 904–905, 905*f*
Ti (tumor-inducing) plasmid, **366**, 366*f*
Tissue culture, of plants, 698, 699*f*
Tissue injury, inflammatory response and, 848
Tissues, 4*f*, **5**, 736*f*. *See also* Connective tissue
  animal, 594–595
  cell organization in, 69–70
  connective, 736
  dermal, of plants, 656, 657*t*, 689, 6560–661
  epithelial, 736, 738, 739*t*, 740
  evolution of, 595, 603–604, 604*f*, 605*f*
  ground, of plants, 656, 657*t*, 658, 661–662, 689
  muscle, 736, 742–744, 749–752, 749*f*
  nerve, 736, 740
  skeletal, of vertebrates, 744–749, 745*f*
  vascular, of plants, 656, 657*t*, 659, 662–664, 663*f*, 689
Tissue systems, **656**
TLRs (toll-like receptors), 845–846
T lymphocytes. *See* T cells
TMM gene, 660
Toads, 633
  Bufo, ecological isolation of, 454
  Xenopus laevis, recombinant DNA and, 355
Toes, in primates, 643
Toll-like receptors (TLRs), 845–846
Tonicity, **708**, 838
Tonoplasts, 73*f*, **79**
Too many mouths (tmm) mutation, 660, 660*f*
Tooth decay, 506, 507*t*
Topoisomerases, **287**
Topological state, 287
Topology, 287
Torpor, 826–827, 856
Torsional strain, 287, 287*f*
Tortoises, 636, 636*f*
Totipotent cells, **874**
Toxins, excretion of, 842
Toxoplasma gondii, 533, 533*f*
TP53 gene, 215
Trace elements, 27
Trachea, 800, 801*f*
Tracheids, 571, 663, 663*f*
  cavitation and, 713, 713*f*
  tensile strength of water in, 713
Tracheophytes, **574**. *See also* Vascular plants
Transcription, **302**–303, 303*f*
  coupling of, to translation, 308, 308*f*
  in eukaryotes, 309–310
  hormone-activated transcription factors and, 828
  immunoglobulins and, 853, 854*f*
  prokaryotic, 306–309, 332–333
  regulation after, 341–346
Transcription-associated factors (TAFs), 336, 337*f*
Transcription bubble, **307**, 307*f*
Transcription complex, 338, 338*f*
Transcription factors, **309**, 310*f*
  active in embryonic stem cells, 878
  binding site location of, 350
  enhancers and, 337, 337*f*
  general, 336, 337*f*
  linkage to RNA polymerase II, 337–338
  specific, 336
Transcription units, **306**
Transcriptome, **312**, 394
Transcripts, primary, **310**
  splicing of, 310
Transduction
  generalized, 502–503, 505*f*
  in prokaryotes, 501, 504–505, 505*f*
  specialized, 502–503
  viral, 505*f*, 502–503
Trans fats, **58**
Transfer RNA (tRNA), **75**, **303**
  amino acid attachment to, 313–315, 314*f*
  bifunctionality of, 313, 314*f*
  charged, 313–314, 314*f*
  initiator, 316, 316*f*
  ribosome binding sites for, 314–315, 315*f*
  transcription of, 308
Transformation, **277**
  in bacteria, 277–278, 277*f*
  DNA libraries and, 358–359, 358*f*
  in plants, 366, 366*f*
  in prokaryotes, 501, 504–505, 505*f*
Transgenic crops, 371–373, 373*f*
Transgenic organisms
  gene function and, 364–366
  pigs, 306*f*
  plants and, 366, 366*f*
  universality of genetic code and, 364
Translation, **302**, 303
  control of, 344
  coupling of, to transcription, 308, 308*f*
  elongation in, 316–318, 317*f*, 318*f*
  immunoglobulins and, 853
  initiation of, 316, 316*f*
  termination of, 317, 318*f*
  wobble pairing in, 317
Translation factors, 344
Translation repressor proteins, **344–345**
Translocation Down syndrome, 266
Translocations, **322**, 323*f*, **716**
Transmembrane domains, **100**-101, 101*f*
Transmembrane proteins, 96, 97*t*
  hormone binding to, 829
Transmembrane route, **709**, 710*f*
Transmission electron microscopes, 67, 68*t*
Transpiration, **663**
  environment and, 713–715, 714*f*
  by plants, 711–715, 712*f*
Transporter proteins, 99, 100*f*
Transport proteins, 89
  hormone binding to, 828
Transposable elements (transposons), 389, 389*t*, 390
Transverse filaments (TF), 224*f*
Transverse tubules (T tubules), **751**, 752*f*
Tree fern, 576*f*
Tree rings, 670, 671*f*
Trees
  chestnut blight and, 561
  clear-cutting of, 974
  deforestation and, 962, 963*f*
  Dutch elm disease and, 561
  fir, 578*t*, 581
  island, convergent evolution and, 445
  oak (*See* Oak trees *(Quercus)*)
  pine, 578*t*, 581–582, 581*f*, 582*f*
  purple-leaf plum *(Prunus cerasifera)*, 673
  redwood, 578*t*, 581
  southern beech, 445
  spruce, 578*t*, 581, 958
Trees (phylogenetic), **13**–14, 14*f*, 475–476, 475*f*
TRH (thyrotropin-releasing hormone), 833
Triacylglycerols, 58, 59*f*
Tribolium (flour beetle), parasitism and competition in, 956
Tricarboxylic acid (TCA) cycle. *See* Citric acid cycle
Trichechus manatus (West Indian manatee), 442, 442*f*
Trichinella, 613
Trichinosis, 613
2,4,5-Trichlorophenoxyacetic acid (2,4,5-T), 722
Trichomes, **661**, 661*f*
Trichomonas vaginalis, 527, 527*f*
Tricuspid valve, **811**
Tridacna, 608
Triglochin maritima (arrowgrass), 588
Triglycerides, **58**, 795
Trimesters, 885
Triple-binding assay, 305
Triple bonds, 29
Triplet repeat expansion mutations, 322
Triticum aestivum (wheat), 386–387
tRNA. *See* Transfer RNA (tRNA)
Trophic levels, **966**, 966*f*
  concepts to describe, 966
  definitions of, 966
  ecological pyramids and, 969, 970*f*
  energy flow through, 965–966
  energy loss between, 967, 967*f*, 968*f*
  energy processing by, 967
  heat as final energy product and, 967
  limits on number of, 968
Trophoblast, **880**
Tropical hummingbird *(Eulampis)*, body temperature in, 856
Tropic hormones (tropins), 832
Tropics, biodiversity in, 971–973, 972*f*
Tropins (tropic hormones), 832
Tropomyosin, **751**
Troponin, **751**
trp operon, **332**
  regulation of, 334–335, 335*f*
True-breeding plants, **237**, 238
Trunk neural crest cells, 884, 884*f*
Trypanosoma cruzi, 529
Trypanosomes, 527–529, 528*f*
Trypanosomiasis, 528, 528*f*
Trypsin, **794**
Tryptophan, 49*f*, 332, 334–335, 335*f*, 721, 721*f*
Tsetse fly, 528*f*
TSH (thyroid-stimulating hormone), 826, **832**
T tubules (transverse tubules), **751**, 752*f*
Tuatara *(Sphenodon punctatus)*, 636–637, 636*f*
Tuberculosis (TB), 506, 507*t*
Tubers, **672**, 672*f*, 698
Tubulins, 82, **205**
Tumor-inducing (Ti) plasmid, **366**, 366*f*
Tumors, **214**
  breast cancer and HER2, 368, 368*f*
  plasmid induced, 366, 366*f*
Tumor-suppressor genes, **214**–215, 215*f*, 387
Túngara frog *(Physalaemus pustulosus)*, mate selection in, 424, 424*f*
Tunicates, 347–348, 348*f*, 626, 626*f*
Turgor pressure, **105**
  carbohydrate movement in phloem and, 717, 718*f*
  transpiration and, 714–715, 715*f*
Turner syndrome, 267
Tursiops truncatus (bottle-nosed dolphin), 643*f*
Turtles, 636–637, 637*f*
T1 virus, 296
Twin studies, of behavior, 897
2PG (2-phosphoglycerate), 137*f*
Two-point crosses, 263, 264*f*
Typhoid fever, 507*t*

Typhus, 507*t*
Tyrosinase, 245, 245*f*
Tyrosine, 49*f*
Tyrosine kinases, 182*f*

U

Ubiquinone, 143, 143*f*
Ubiquitin, 213
Ultraviolet light, 293
  flower color and, 684, 685*f*
Ulva, 538, 538*f*
Undershoot phase, of action potential, 764, 765*f*
Undulating membranes, in parabasalids, 527, 527*f*
Undulation, of flagella, 83
Unicellularity, of prokaryotes, 495
Uniform spacing, 921
Uniporters, **106**
Unipotent stem cells, **874**
Unsaturated fats, **58**, 59*f*
Uracil, 56, 56*f*, 279, 302
Ureters, **839**, 840*f*
Urethra, **839**, 840*f*
Urinary bladder, **839**, 840*f*
Urine, 839
  collection of, 841, 841*f*
  glucose excretion in, in diabetes, 817
Urochordata, 626, 626*f*
Ustilago maydis, 554, 554*f*
Uta stansburiana (side-blotched lizard). *See* Side-blotched lizard *(Uta stansburiana)*
Uterine horns, 870
Uterine tubes, 870
Uterus, **870**
Utricle, **777**
UV-B, **990**
  uvrABC complex, 294–295, 295*f*
UV radiation, 293

V

Vaccines, secondary response to antigens and, 855
Vacuoles, **79**, 79*f*
  of ciliates, 534, 534*f*
Vagina, 870
Valence electrons, **26**
Valine, 49*f*
Valium (diazepam), 767
Valley oak *(Quercus lobata),* ecological isolation of, 454
Vampire bats *(Desmodus rotundus),* reciprocal altruism in, 909, 909*f*
van Beneden, Edouard, 222
van der Waals forces, 27, 28*t*, **30**
van Helmont, Jan Baptista, 159
van Leeuwenhoek, Anton, 66
van Niel, C. B., 160
Variable regions, of antibodies, 853, 853*f*, 854*f*
Variables, **8**
Varicella-zoster virus, 509*t*, 848
Variola virus, 509*t*
Vasa recta, **841**, 841*f*, 843
Vascular bones, **747**
Vascular cambium, **658**, 670, 671*f*
Vascular plants
  dermal tissue of, 660–661
  evolution of, 574–575, 574*f*, 575*f*
  ground tissue of, 661–662
  leaves of, 673–675
  lycophytes, 574–576, 574*f*, 575*f*, 576*f*, 578*t*
  meristems of, 656–659, 657*f*–659*f*, 658*t*
  nutrient distribution in, 574, 574*f*
  phloem of, 664, 664*f*
  phylogeny of, 578*t*
  pterophytes, 576–579, 577*f*, 578*f*, 578*t*, 579*f*
  root system of, 656, 656*f* (*See also* Roots)
  shoot system of, 656–657, 656*f*
  stems of, 669–673
  xylem of, 662–663, 663*f*
Vascular tissues, of plants, **656**, 657*t*, 659, 662–664, 663*f*, 689–690
Vas deferens, **864**
Vasoactive intestinal peptide, 833
Vasoconstriction, **814**, 815*f*
Vasodilation, **814**, 815*f*
Vasopressin, 831, 831*f*895, 895*f*
Vectors, cloning, 358, 358*f*
Vegetal pole, **879**, 879*f*
Vegetarian tree finch (Platyspiza crassirostris), 430*f*, 431–432, 432*f*
Vegetative reproduction, 672–673, 672*f*, 698, 698*f*
Veins
  of human circulatory system, **814**–816, 814*f*, 815*f*
  of leaves, **673**, 673*f*
  pulmonary, of vertebrate circulatory system, 809, 809*f*
Velociraptor, 638, 639*f*
Velvet worms, 599*t*
Venomous lizards, 637
Venomous snakes, 637
Venous pump, **815**
Venous valves, **815**, 815*f*
Venter, Craig, 386
Ventral body cavity, 737, 737*f*
Ventricle, **809**, 809*f*
Venules, **814**
Vernalization, **681**
Vertebrae, neck, number of, 441
Vertebral column, 627, 628*f*
  of fish, 628–629
Vertebrates, 618–619, 619*f*, **627**. *See also* Mammals; *specific vertebrates*
  body cavities of, 737, 737*f*
  body plan of, 736–737, 736*f*
  brain in, 769, 769*f*, 770*f*
  circulatory systems of, 808–811, 808*f*
  common features of, 627, 628*f*
  digestive systems of, 790–791, 790*f*, 791*f*
  diversity of, 594, 595*f*
  ears of, 775–776, 775*f*–776*f*
  evolution of terrestrial, 633, 633*f*
  eyes of, 779–784, 779*f*, 780*f*
  flight in, 754–755, 754*f*
  gravity-sensitive receptors in, 777–778
  internal fertilization in, 862, 862*f*
  live birth in, 862–863, 862*f*
  locomotion by, 753–754, 754*f*
  nervous system of, 761–762, 761*f*, 766–768, 766*f*, 769–771, 769*f*, 770*f* (*See also* Nervous system)
  photoreceptors of, 781–784, 781*f*, 783*f*
  phylogeny of, 627*f*–628*f*
  reproductive system evolution in, 862, 862*f*
  respiratory systems of, 800–803
  sex determination in, 258
  societies of, 912–913, 913*f*
  swimming by, 753, 753*f*
  taste buds of, 778, 778*f*
  teeth of, 791–792, 792*f*
  temperature regulation in, 824–825
Vervet monkeys *(Cercopithecus aethiops),* language in, 902, 902*f*
Vesicles, **72**
Vessels, **583**
  angiosperm, 584*t*
  gymnosperm, 583
Vessels, of plants, 663, 663*f*
  cavitation and, 713, 713*f*
  tensile strength of water in, 711
Vestibular apparatus, **777**
Vestigial structure, **442**, 442*f*
Viagra (Sildenafil), 185
Vibration detectors, 775–778
Vibrio cholerae, 190, 507*t*
  chromosomes of, 496
Villi, small intestinal, 793, 793*f*
Vimentin, 81
Violet Sabrewing hummingbird *(Campylopterus hemileucurus),* 685*f*
Virginia opossum *(Didelphis virginiana),* 643, 644*f*
Viridiplantae, 489
Virions, 507
Viruses, 495
  animal, 511–514, 513*f*, 515*f*
  bacterial (*See* Bacteriophages)
  genes of, 278, 278*f*
  genomes of, 508–509, 508*f*, 509*t*
  giant, 510
  human diseases caused by, 194, 509*t*, 511–515, 513*f*, 515*f*, 848–849 (*See also* Human immunodeficiency virus (HIV))
  negative-strand, 509
  positive-strand, 509
  replication of, 509
  reverse transcriptase and, 302
  structure of, 508–510, 508*f*
  transduction by, 503–504, 504*f*
Viscera, 742
Visceral muscle, 742, 743*t*
Visceral pleural membrane, **801**
Vision
  binocular, 644, 783
  invertebrate eyes and, 779–780, 780*f*
  in primates, 643
  vertebrate eyes and, 779–782, 779*f*, 780*f*
Visual cortex, 771
Vitamin A deficiency, 163, 373
Vitamin D, parathyroid hormone and, 835–836
Vitelline envelope, 871, 871*f*, 873
Viviparity, **862**, 862*f*
Vivipary, 724
Voles *(Microtus),* behavioral genetics in, 895, 895*f*
Voltage-gated channels, **764**
Volvox, 538, 538*f*
  colonies of, 524, 524*f*
Vorticella, 523, 523*f*
The Voyage of the Beagle (Darwin), 430

W

Wadlow, Robert, 833, 834*f*
Walking fern *(Asplenium rhizophyllum),* leaves of, 675
Wallace, Alfred Russel, 12
Warbler finch *(Certhidea olivacea),* 430, 430*f*, 432*f*
Wasps, eusociality in, 910
Wastes
  excretion of (*See* Excretion; Feces; Urine)
  nitrogenous, 964
  in plasma, 807
  secretion by kidneys, 842
Water, 31–35
  antidiuretic hormone and, 844–845, 844*f*
  cohesion and adhesion of, 31–32, 31*f*, 32*f*, 32*t*
  conduction in plants, 662–664, 663*f*
  density of, 32
  diffusion across lipid bilayer, 104
  as electron donor for photosynthesis, 160
  entropy and, 118, 118*f*
  excess, plant responses to, 715–716, 716*f*
  heat of vaporization of, 32*t*, 33
  hydrogen bond formation by, 31, 31*f*
  ionization of, 34
  loss of, plant limitation of, 715
  nonpolar molecules and, 34
  in photosynthesis, 165
  polarity of, 31
  properties of, 31–34, 32*t*
  reabsorption by kidneys, 841–843, 842*f*, 843*f*
  as solvent for polar molecules, 33, 33*f*
  specific heat of, 32–33, 32*t*
  tensile strength of, 713
  transpiration rate of plants and, 713–715, 715*f*
  transport in plants, 706–711, 706*f*, 707*f*, 712*f*, 730
Water bears, 599*t*
Water cycle, 961–962, 962*f*, 963*f*
Waterfowl, 639
Water homeostasis. *See* Osmoregulation
Water mold, 530–531
Water potential, **706**–707, 707*f*, 708–709, 708*f*, 711, 712*f*, 714
Water-storage roots, **668**, 669*f*
Water-vascular system, of echinoderms, 618, 618*f*
Watson, James, 281, 281*f*, 283
Waved umber caterpillar *(Menophra abruptaria),* cryptic coloration of, 951*f*
Wavelength, energy and, 161, 161*f*
Weather. *See also* Climate
  extreme, global temperature change and, 991
Weight, 22
Weinberg, Wilhelm, 407
Welwitschia, 583, 583*f*

Went, Frits, 720, 720*f*
WEREWOLF (WER) gene, 665, 666, 666*f*
West Indian manatee *(Trichechus manatus),* 442, 442*f*
Whales
evolution of, 438, 438*f*
swimming by, 753
Wheat Genome Project, 386-387
Wheel animals, 599*t*, 607, 607*f*
Whisk ferns, 578-579, 578*t*, 579*f*
White blood cells. *See* Leukocytes; Lymphocytes; T cells
White Cliffs of Dover, 536, 536*f*
White-eyed females, 258*f*
White-fronted bee-eater *(Merops bullockoides),* kin selection in, 911, 911*f*
White matter, 771
White-tailed deer, population explosion of, 949
Whorls, **584**, 682, 682*f*
Wild lettuce *(Lactuca),* temporal isolation and, 455
Wilkins, Maurice, 280-281
Wind
Coriolis effect and, 981-982, 981*f*
ocean currents and, 982-983, 982*f*
pollination by, 684
Window leaves, **675**
Wings, evolution of, 754-755, 754*f*
Wobble pairing, **317**
Woese, Carl, 485
WOL (WOODEN LEG) gene, 690, 690*f*
Woodpecker finch *(Cactospiza pallida),* 11*f*, 430*f*
Woodpeckers, 639
Woody plants, 696
Woody stems, 670, 670*f*
Woolly hair, 272
Worms. *See also* Earthworms; Flatworms; Polychaete worms; Roundworm; *specific worms*
clam (Nereis), sperm binding in, 874*f*
phylogeny of, 599*t*

X

X chromosome, **257**-258
inactivation of, 258-259, 259*f*, 339
nondisjunction of, 265
Xenopus laevis (toad), 355
X-inactivation specific transcript (Xist), 339
X-linked traits, **258**, 258*f*
XO genotype, 267
X-ray diffraction, 281, 281*f*
XXX genotype, 266
XXY genotype, 267
Xylem, **574**, 583, 662-663, 663*f*, 667-668, 730
mineral transport in, 707*f*, 710*f*, 713
transpiration and, 711, 712*f*, 713
water transport in, 706, 706*f*, 707*f*, 711-713
XYY genotype, 267

Y

Yamanaka, Shinya, 878
Yarrow plant, 470
Y chromosome, **257**-259
nondisjunction of, 265
Yeasts, 548, 550
asexual reproduction in, 561, 561*f*
cell cycle checkpoints and, 212, 212*f*
cell division in, 201*f*, 223
ethanol fermentation in, 148-149, 149*f*
genome of, 387*f*
genomics and, 367
heterochromatin formation in, 343
Yellow fever, 509*t*
Yellowfin tuna *(Thunnus albacares),* 629*f*
Yersinia pestis, 507*t*
Yolk sac, **634**, 634*f*, 881

Z

Zamecnik, Paul, 324
Z diagrams, 167, 167*f*
Zea mays. See Corn *(Zea mays)*
Zebra finches, 905
Zebra mussel *(Dreissena polymorpha),* 609
Zeng Jinlian, 833
Zinc fingers, **331**
Zinc, in organisms, 27
Z lines, 749, 750*f*
Zoecium, 611
Zona pellucida, 871, 871*f*, 872
Zone of cell division, **665**, 666*f*
Zone of elongation, **665**-666
Zone of maturation, 665*f*, **666**-668, 666*f*, 667*f*, 668*f*
Zoospores, 531
of chytrids, 557, 557*f*
Zygomycota, 555, 555*f*, 556*t*, **558**-559
Zygosporangium, **558**-559
Zygospores, 558
Zygote, **222**, 885-886